Bacillus subtilis and Other Gram-Positive Bacteria

BIOCHEMISTRY, PHYSIOLOGY, AND MOLECULAR GENETICS

Bacillus subtilis
and Other Gram-Positive Bacteria

BIOCHEMISTRY, PHYSIOLOGY, AND MOLECULAR GENETICS

EDITOR IN CHIEF

Abraham L. Sonenshein
Department of Molecular Biology and Microbiology
Tufts University, Boston, Massachusetts

EDITORS

James A. Hoch
Division of Cellular Biology
Department of Molecular and Experimental Medicine
The Scripps Research Institute, La Jolla, California

Richard Losick
Department of Cellular and Developmental Biology
Harvard University, Cambridge, Massachusetts

AMERICAN SOCIETY FOR MICROBIOLOGY, WASHINGTON, D.C.

Cover photograph courtesy of M. L. Higgins, J. Bylund, and P. J. Piggot

American Society for Microbiology
1325 Massachusetts Ave., N.W.
Washington, DC 20005

Library of Congress Cataloging-in-Publication Data

Bacillus subtilis and other gram-positive bacteria : biochemistry, physiology, and molecular genetics / editor in chief, Abraham L. Sonenshein; editors, James A. Hoch, Richard Losick.
p. cm.
Includes index.
ISBN 1-55581-053-5
1. Bacillus subtilis. 2. Gram-positive bacteria. 3. Molecular microbiology. 4. Bacterial genetics. I. Sonenshein, A. L., 1944-. II. Hoch, James A. III. Losick, Richard.
[DNLM; 1. Bacillus subtilis. 2. Gram-Positive Bacteria. QW 127.5.B2 B1252]
QR82.B3B333 1993
589.9'5--dc20
DNLM/DLC
for Library of Congress 92-48294
CIP

CONTENTS

CONTRIBUTORS

C. Anagnostopoulos • Laboratoire de Génétique Microbienne, Institut National de la Recherche Agronomique, 78352 Jouy en Josas Cedex, France

Dwight Anderson • University of Minnesota, 18-246 Moos Tower, 515 Delaware Street, S.E., Minneapolis, Minnesota 55455

M. V. Arbige • Genencor International Inc., 180 Kimball Way, South San Francisco, California 94080

A. R. Archibald • Department of Microbiology, The Medical School, The University of Newcastle upon Tyne, Framlington Place, Newcastle upon Tyne NE2 4HH, United Kingdom

Arthur I. Aronson • Department of Biological Sciences, Purdue University, West Lafayette, Indiana 47907

Vasco Azevedo • Laboratoire de Génétique Microbienne, Institut National de la Recherche Agronomique, Domaine de Vilvert, 78352 Jouy en Josas Cedex, France

Simon Baumberg • Department of Genetics, University of Leeds, Leeds LS2 9JT, United Kingdom

David Bol • Department of Biological Sciences and The Program in Molecular and Cell Biology, University of Maryland Baltimore County, Baltimore, Maryland 21045

B. A. Bulthuis • Genencor International Inc., 180 Kimball Way, South San Francisco, California 94080

Michael G. Caparon • Department of Molecular Microbiology, Washington University School of Medicine, St. Louis, Missouri 63110

Michael J. Chamberlin • Division of Biochemistry and Molecular Biology, University of California, Berkeley, California 94720

Glenn H. Chambliss • Department of Bacteriology, University of Wisconsin-Madison, Madison, Wisconsin 53706

Bruce M. Chassy • Department of Food Science, University of Illinois, ABL 103, 1302 W. Pennsylvania Avenue, Urbana, Illinois 61801

K. F. Chater • John Innes Institute, John Innes Centre, Norwich NR4 7UH, United Kingdom

David Cheo • Department of Biological Sciences and The Program in Molecular and Cell Biology, University of Maryland Baltimore County, Baltimore, Maryland 21045

Stewart T. Cole • Unité de Génétique Moléculaire Bactérienne, Institut Pasteur, 28 Rue du Docteur Roux, 75724 Paris Cedex 15, France

D. Crabb • Genencor International Inc., 180 Kimball Way, South San Francisco, California 94080

John E. Cronan, Jr. • Department of Microbiology and Department of Biochemistry, University of Illinois, Urbana, Illinois 61801

Diego de Mendoza • Departamento de Microbiologia, Facultad de Ciencias Bioquimicas y Farmaceuticas, Universidad Nacional de Rosario, Suipacha 531, 2000 Rosario, Argentina

David Dubnau • Department of Microbiology, Public Health Research Institute, 455 First Avenue, New York, New York 10016

S. Dusko Ehrlich • Laboratoire de Génétique Microbienne, Institut National de la Recherche Agronomique, Domaine de Vilvert, 78352 Jouy en Josas Cedex, France

J. Errington • Sir William Dunn School of Pathology, South Parks Road, Oxford OX1 3RE, United Kingdom

Matthew J. Fagan • Department of Biology, University of California at San Diego, La Jolla, California 92093-0116

Eugenio Ferrari • Genencor International Inc., 180 Kimball Way, South San Francisco, California 94080

Pamela S. Fink • Department of Microbiology and Immunology, Wright State University, Dayton, Ohio 45435

Susan H. Fisher • Department of Microbiology, Boston University School of Medicine, 80 East Concord Street, Boston, Massachusetts 02118

Peter Fortnagel • Abteilung für Mikrobiologie, Institut für Allgemeine Botanik, Universität Hamburg, Ohnhorststrasse 18, D-2000 Hamburg, Germany

Roberto Grau • Departamento de Microbiologia, Facultad de Ciencias Bioquimicas y Farmaceuticas, Universidad Nacional de Rosario, Suipacha 531, 2000 Rosario, Argentina

Christopher J. Green • SRI International, 333 Ravenswood Avenue, Menlo Park, California 94025

Alexandra Gruss • Laboratoire de Génétique Microbienne, Institut National de la Recherche Agronomique, 78352 Jouy en Josas Cedex, France

I. C. Hancock • Department of Microbiology, The Medical School, The University of Newcastle upon Tyne, Framlington Place, Newcastle upon Tyne NE2 4HH, United Kingdom

C. R. Harwood • Department of Microbiology, The Medical School, The University of Newcastle upon Tyne, Framlington Place, Newcastle upon Tyne NE2 4HH, United Kingdom

Lars Hederstedt • Department of Microbiology, University of Lund, Sölvegatan 21, S-223 62 Lund, Sweden

Tina M. Henkin • Department of Biochemistry and Molecular Biology, Albany Medical College, Albany, New York 12208

Dennis Henner • Department of Cell Genetics, Genentech Inc., 460 Point San Bruno Boulevard, South San Francisco, California 94080

James A. Hoch • Division of Cellular Biology, Department of Molecular and Experimental Medicine, The Scripps Research Institute, 10666 North Torrey Pines Road, La Jolla, California 92037-1093

Christian Hoischen • Department of Biology, University of California at San Diego, La Jolla, California 92093-0116

David A. Hopwood • John Innes Institute, John Innes Centre, Norwich NR4 7UH, United Kingdom

F. Marion Hulett • Department of Biological Sciences, Laboratory for Molecular Biology, University of Illinois at Chicago, Chicago, Illinois 60680

Mitsuhiro Itaya • Mitsubishi Kasei Institute of Life Sciences, 11, Minamiooya, Machida-shi, Tokyo 194, Japan

Laurent Jannière • Laboratoire de Génétique Microbienne, Institut National de la Recherche Agronomique, 78352 Jouy en Josas Cedex, France

Alisha S. Jarnagin • Genencor International Inc., 180 Kimball Way, South San Francisco, California 94080

Helen M. Kieser • John Innes Institute, John Innes Centre, Norwich NR4 7UH, United Kingdom

Tobias Kieser • John Innes Institute, John Innes Centre, Norwich NR4 7UH, United Kingdom

Ursula Klingel • Department of Genetics, University of Leeds, Leeds LS2 9JT, United Kingdom

Frank Kunst • Unité de Biochimie Microbienne, Centre National de la Recherche Scientifique, URA 1300, Institut Pasteur, 25 rue du Docteur Roux, 75724 Paris Cedex 15, France

Mohamed A. Marahiel • FB Chemie, Philips-Universität Marburg, Marburg, Germany

Leticia Màrquez-Magaña • Division of Biochemistry and Molecular Biology, University of California, Berkeley, California 94720

Charles P. Moran, Jr. • Department of Microbiology and Immunology, Emory University School of Medicine, Atlanta, Georgia 30322

Tarek Msadek • Unité de Biochimie Microbienne, Centre National de la Recherche Scientifique, URA 1300, Institut Pasteur, 25 rue du Docteur Roux, 75724 Paris Cedex 15, France

Cynthia M. Murphy • Department of Food Science, University of Illinois, ABL 103, 1302 W. Pennsylvania Avenue, Urbana, Illinois 61801

Vasantha Nagarajan • Central Research and Development Division, E. I. du Pont de Nemours & Company, Wilmington, Delaware 19880-0228

Michiko M. Nakano • Department of Biochemistry and Molecular Biology, Louisiana State University Medical Center, 1501 Kings Highway, Shreveport, Louisiana 71130-3932

Richard Novick • Public Health Research Institute, 455 First Avenue, New York, New York 10016

Mario Noyer-Weidner • Max-Planck-Institut für Molekulare Genetik, Ihnestrasse 73, DW-1000 Berlin 33, Germany

Per Nygaard • Institute of Biological Chemistry B, University of Copenhagen, DK-1307 Copenhagen K, Denmark

George W. Ordal • Department of Biochemistry, College of Medicine, University of Illinois, Urbana, Illinois 61801

Peter A. Pattee • Department of Microbiology, Immunology and Preventive Medicine, 205 Science 1, Iowa State University, Ames, Iowa 50011

Henry Paulus • Department of Metabolic Regulation, Boston Biomedical Research Institute, 20 Staniford Street, Boston, Massachusetts 02114, and Department of Biological Chemistry and Molecular Pharmacology, Harvard Medical School, Boston, Massachusetts 02115

Marta Perego • Istituto di Tecnica Farmaceutica, Università degli Studi di Parma, Via M. D'Azeglio 85, 43100 Parma, Italy, and Dipartimento di Genetica e Microbiologia, Università degli Studi di Pavia, Via Abbiategrasso 207, 27100 Pavia, Italy

John B. Perkins • OmniGene, Inc., 763D Concord Avenue, P.O. Box 9002, Cambridge, Massachusetts 02139-9002

Janice G. Pero • OmniGene, Inc., 763D Concord Avenue, P.O. Box 9002, Cambridge, Massachusetts 02139-9002

Patrick J. Piggot • Department of Microbiology and Immunology, Temple University School of Medicine, 3400 North Broad Street, Philadelphia, Pennsylvania 19140-5196

Fergus G. Priest • Department of Biological Sciences, Heriot Watt University, Edinburgh EH14 4AS, Scotland

Cheryl L. Quinn • Molecular Immunology Group, Institute of Molecular Medicine, University of Oxford, John Radcliffe Hospital, Headington, Oxford OX3 9DU, England

Georges Rapoport • Unité de Biochimie Microbienne, Centre National de la Recherche Scientifique, URA 1300, Institut Pasteur, 25 rue du Docteur Roux, 75724 Paris Cedex 15, France

Bernard Reilly • University of Minnesota, 18-246 Moos Tower, 515 Delaware Street, S.E., Minneapolis, Minnesota 55455

Jonathan Reizer • Department of Biology, University of California at San Diego, La Jolla, California 92093-0116

Fernando Rojo • Centro de Biología Molecular (CSIC-UAM), Universidad Autónoma, Cantoblanco, 28049 Madrid, Spain

Milton H. Saier, Jr. • Department of Biology, University of California at San Diego, La Jolla, California 92093-0116

Margarita Salas • Centro de Biología Molecular (CSIC-UAM), Universidad Autónoma, Cantoblanco, 28049 Madrid, Spain

Richard M. Sayre • Nematology Laboratory, Agricultural Research Service, U.S. Department of Agriculture, 10300 Baltimore Avenue, Beltsville, Maryland 20705

Brian F. Schmidt • Genencor International Inc., 180 Kimball Way, South San Francisco, California 94080

Harold H. Schreier • Center of Marine Biotechnology, University of Maryland, 600 East Lombard Street, Baltimore, Maryland 21202, and Department of Biological Sciences, University of Maryland, Baltimore County, Baltimore, Maryland 21228

J. Schultz • Genencor International Inc., 180 Kimball Way, South San Francisco, California 94080

June R. Scott • Department of Microbiology and Immunology, Emory University Health Sciences Center, Atlanta, Georgia 30322

Pascale Serror • Laboratoire de Génétique Microbienne, Institut National de la Recherche Agronomique, Domaine de Vilvert, 78352 Jouy en Josas Cedex, France

Peter Setlow • Department of Biochemistry, University of Connecticut Health Center, Farmington, Connecticut 06030-3305

Alan Sloma • Novo-Nordisk Biotech, Davis, California 95616

Issar Smith • Department of Microbiology, The Public Health Research Institute, New York, New York 10016

Abraham L. Sonenshein • Department of Molecular Biology and Microbiology, Tufts University School of Medicine, 136 Harrison Avenue, Boston, Massachusetts 02111-1800

Michel Steinmetz • Laboratoire de Génétique des Micro-organismes, 78850 Thiverval-Grignon, France

Charles R. Stewart • Department of Biochemistry and Cell Biology, Rice University, Houston, Texas 77251

Mark A. Strauch • Division of Cellular Biology, Department of Molecular and Experimental Medicine, The Scripps Research Institute, 10666 North Torrey Pines Road, La Jolla, California 92037

Robert L. Switzer • Department of Biochemistry, University of Illinois, 1209 West California, Urbana, Illinois 61801

Harry W. Taber • Wadsworth Center for Laboratories and Research, New York State Department of Health, Albany, New York 12201

Curtis B. Thorne • Department of Microbiology, University of Massachusetts, Amherst, Massachusetts 01003

Thomas A. Trautner • Max-Planck-Institut für Molekulare Genetik, Ihnestrasse 73, DW-1000 Berlin 33, Germany

Patricia S. Vary • Department of Biological Sciences, Northern Illinois University, DeKalb, Illinois 60115

Robert Luis Vellanoweth • Department of Cellular and Structural Biology, University of Texas Health Science Center, 7703 Floyd Curl Drive, San Antonio, Texas 78284-7762

Barbara S. Vold • Syva Company, P. O. Box 10058, Palo Alto, California 94303

R. G. Wake • Department of Biochemistry, University of Sydney, Sydney, New South Wales 2006, Australia

Neil E. Welker • Department of Biochemistry, Molecular Biology, and Cell Biology, Northwestern University, Evanston, Illinois 60208

Charles Yanofsky • Department of Biological Sciences, Stanford University, Stanford, California 94305-5020

Ronald E. Yasbin • Department of Biological Sciences and The Program in Molecular and Cell Biology, University of Maryland Baltimore County, Baltimore, Maryland 21045

H. Yoshikawa • Department of Genetics, Osaka University Medical School, Osaka 565, Japan

Michael Young • Department of Biological Sciences, University of Wales, Aberystwyth, Dyfed SY23 3DA, United Kingdom

Philip Youngman • Department of Genetics, University of Georgia, Athens, Georgia 30602

Stanley A. Zahler • Section of Genetics and Development, Division of Biological Sciences, Cornell University, Ithaca, New York 14853-2703

Howard Zalkin • Department of Biochemistry, Purdue University, West Lafayette, Indiana 47907

Peter Zuber • Department of Biochemistry and Molecular Biology, Louisiana State University Medical Center, 1501 Kings Highway, Shreveport, Louisiana 71130-3932

PREFACE

In the history of prokaryotic biology, only a few species have received highly intensive study. The rise to prominence of *Bacillus subtilis* as a subject of modern microbiological investigation can be traced to the success of John Spizizen, in the late 1950s, in demonstrating genetic transformation of a particular isolate of *B. subtilis* by purified DNA. Previously, transformation had been successful only with the pneumococcus. Using a mutant in tryptophan biosynthesis, isolated by Burkholder and Giles after heavy mutagenesis of the standard Yale University strain, Spizizen and Costa Anagnostopoulos determined the growth conditions that induce genetic competence, paving the way for all future studies in *B. subtilis* of gene structure, organization, and regulation and of the mechanisms of transformation and recombination. This highly transformable *trp* mutant, called strain 168, became the standard parental strain for most subsequent studies using either genetic or biochemical approaches. With a few years after Spizizen's discovery, Pierre Schaeffer used the new technology to break open the study of microbial differentiation by integrating genetics, biochemistry, and morphology. The subsequent development of *B. subtilis* as a subject of intensive study was therefore dependent on and supported by the twin pillars of an all-important technological advance and the insight that bacterial differentiation was of enormous intrinsic interest and had become amenable to detailed analysis. At present, *Escherichia coli* and its closest relatives are the only prokaryotes as well understood as is *B. subtilis*.

The genus *Bacillus* incorporates many species of gram-positive, rod-shaped, aerobic, endospore-forming bacteria (see chapter 1). Most species normally inhabit the soil or rotting plant materials. Their usually strict aerobiosis differentiates *Bacillus* species from *Clostridium* species, which are generally strict anaerobes. Except for *B. anthracis*, the *Bacillus* spp. are considered to be nonpathogenic or, at most, opportunistic pathogens for humans and animals. At least some species (e.g., *B. subtilis* var. *natto*) are sufficiently innocuous to be eaten in large quantities on a regular basis.

The division of the bacterial world according to the results of an early cytological reaction (the Gram stain) has proved to have a solid foundation in bacterial structure and evolution. It is the multilayered structure of the cell wall that allows gram-positive bacteria to retain a crystal violet-iodine precipitate when exposed to organic solvents. The outer membrane and thin cell wall of gram-negative bacteria are unable to protect the crystals. The significance of this division has been demonstrated by evolutionary studies of rRNA sequences, protein sequences, and DNA homologies. Even the molecular mass of the major RNA polymerase sigma factor seems to correlate strongly with Gram-staining properties. Moreover, the ability of plasmids to replicate and genes to be expressed in various host cells seems to be closely related to the Gram reaction. On the other hand, there are important exceptions to this rule, e.g., plasmids that replicate in both gram-positive and gram-negative bacteria and genes from a gram-positive bacterium that are more readily expressed in a gram-negative bacterium than they are in other gram-positive bacteria. Moreover, the special features associated with growth under particular environmental conditions, such as anaerobiosis, thermophily, and alkalophily, impose demands that cut across the gram-positive/gram-negative divide.

This book is meant to provide fundamental information about gram-positive bacteria; it is modeled on the two-volume set entitled *Escherichia coli and Salmonella typhimurium: Cellular and Molecular Biology* (American Society for Microbiology). These earlier volumes have proven to be an outstanding resource for the facts about bacterial physiology, genetics, biochemistry, and regulation. One could be led to believe that their regular updating would be a sufficient resource for understanding the entire bacterial world. We now know, however, that the *E. coli* paradigm is not always useful when applied to distantly related microbes. Thus, the interesting physiology of many gram-positive species, their role in pathogenesis, and their long and successful utilization in the fermentation industry make an up-to-date account of their special properties long overdue. It should be noted that several important phenomena, such as lysogeny, conjugative transposition, and alternative sigma factors for RNA polymerase, were discovered first in gram-positive bacteria. Some of these phenomena have been studied in these organisms at an unmatched level of detail. Moreover, the developmental biology and secondary metabolism of *Bacillus*, *Clostridium*, and *Streptomyces* have drawn researchers to study stationary-phase events in these bacteria for many decades, providing theoretical and experimental paradigms for all bacteria.

Diversity within a common framework is the hallmark of the bacterial world. Recent interest in understanding the detailed physiology of diverse bacteria has revealed similar but inexact counterparts to the genetic organization, metabolic pathways, and mechanisms of gene regulation found in *E. coli*. For instance, *E. coli* regulates expression of genes for degradation of carbon and nitrogen sources by controlling the activities of positive regulatory proteins that stimulate transcription of large groups of genes. The evidence to date for gram-positive bacteria is that most carbon and nitrogen metabolism genes are controlled by negative regulatory proteins, and no evidence for very large regulons has been found. Moreover, in *E. coli*, the signal compound for carbon metabolism is cyclic AMP. In gram-positive bacteria, cyclic AMP is either absent or irrelevant to the regulation of carbon metabolism genes.

There are interesting nuances to this pattern of diversity. Some of the regulatory proteins active in gram-positive bacteria are clearly homologs of proteins found in all bacteria. They fall into various families of related proteins (e.g., the two-component systems, LysR family members, *lac-gal* repressor homologs, sigma factors, antitermination proteins, etc.). The *E. coli* glutamine synthetase and nitrogen metabolism regulon is controlled by a two-component system. In *B. subtilis*, two-component systems regulate stationary-phase gene expression, degradative enzyme synthesis, phosphorus metabolism, and genetic competence, but not nitrogen metabolism. Instead, the *B. subtilis* glutamine synthetase gene is regulated by a repressor homologous to the *E. coli* MerR repressor. (In *Clostridium acetobutylicum*, a third mechanism that may involve antisense RNA seems to be responsible for regulation of glutamine synthetase synthesis.) Similarly, in *E. coli*, a specific sigma factor, σ^{54}, recognizes promoter sites in the nitrogen metabolism regulon; the homologous protein in *B. subtilis* directs transcription of a carbon metabolism operon.

The predominant mechanisms of transcription attenuation in gram-positive and gram-negative bacteria also appear to be different. For many amino acid biosynthesis operons in *E. coli* and its relatives, the translatability of the RNA sequence upstream of the first structural gene of the operon determines whether transcription of the operon will continue or be aborted. That is, movement of ribosomes along the nascent mRNA is the determining factor. In the *trp* operon of *B. subtilis*, regulation by transcriptional attenuation also occurs, but in this case a specific protein acts as a termination factor when tryptophan is in sufficient supply. Terminator or antiterminator proteins are thought to regulate the *B. subtilis* purine biosynthesis, pyrimidine biosynthesis, levansucrase, and tyrosine tRNA synthetase transcription units as well.

The detailed mechanism of regulation of chemotaxis also differs in *E. coli* and *B. subtilis*. Both organisms use similar proteins to encode flagella, chemotactic receptors, switch functions, and the flagellar motor. In *E. coli*, the activity of the receptors (or methyl-accepting chemotaxis proteins) is modified by cycles of methylation and demethylation. The *B. subtilis* receptors, by contrast, maintain the same degree of methylation under all conditions, but exchange their methyl groups with an as yet unidentified regulator protein. This regulator seems to link the receptors to the flagellar motor. In addition, phosphorylation of a major regulatory protein, CheY, has opposite effects on the flagellar motor in the two organisms.

The ability of bacteria to find multiple ways to achieve the same end is also seen in the organization of biochemical pathways and the genes that encode the relevant enzymes. For example, the reactions catalyzed by anthranilate synthetase (the first step in tryptophan biosynthesis) and *p*-aminobenzoate synthetase are homologous. In *E. coli* and *Serratia marcescens*, the glutamine amidotransferase components of these enzyme complexes are encoded in separate genes in the *trp* and *pab* operons. In *B. subtilis*, however, a single gene, located in the *pab* operon, is used; the glutamine amidotransferase functions with both enzymes. For synthesis of threonine and methionine, *B. subtilis* and all other gram-positive bacteria use a single homoserine dehydrogenase. *E. coli* K-12 has two pathway-specific enzymes, each of which is a bifunctional aspartokinase-homoserine dehydrogenase. The gram-positives have, in these cases, opted to maximize efficient use of DNA; the gram-negatives achieve efficiency by maximizing the specificity of regulation. Gram-negative bacteria do not always choose pathway specificity over multifunctionality. *B. subtilis* has two carbamoylphosphate synthetases, whose synthesis or activity is regulated separately by arginine or uridine nucleotides, while *E. coli* has a single enzyme.

Such differences in metabolism or its regulation do not necessarily reflect overall evolutionary distances among species. In various bacteria, lysine is synthesized by the seven-step epimerase pathway, by the four-step dehydrogenase pathway, or by both pathways. The determining factor is whether *meso*-diaminopimelate is incorporated in the cell wall. This diversity does not correlate with the gram-positive/gram-negative dichotomy; even *Bacillus* species differ in this regard. A second point of divergence is the nature of the acyl blocking group used in lysine and methionine biosynthesis. *E. coli* uses succinylated intermediates, while *B. subtilis* uses acetylated intermediates. Most other gram-positive bacteria use acetyl blocking groups for methionine synthesis, but many use succinyl groups for the lysine pathway. This issuc influences overall metabolism in a fundamental way, since the need to make succinyl coenzyme A (the succinyl donor) for amino acid biosynthesis forces some cells to express part of the Krebs cycle for that purpose only.

In summary, the contrasting themes of conservation and divergence, of homology and diversity, will be found throughout this book whether stated explicitly or not. Realizing that *B. subtilis* is no more perfect a paradigm for all gram-positives than is *E. coli* for all microorganisms, we sought to include as much material as possible about other representatives of the gram-positive world. In addition, we introduce the book with a series of monographs designed to acquaint the reader with some of the special properties of various gram-positive bacteria that make them particularly interesting subjects for investigation. We feel that this aspect of our endeavor has not been fully realized, however. Even though it is true that the breadth and depth of knowledge about *B. subtilis* cannot be matched for any other gram-positive bacterium, our ability to provide a comprehensive treatment has been hampered by the constraints of time, page allotment, and the sheer diversity of the gram-positive world. Thus, we present the known facts about cell structure, intermediary metabolism, synthesis of useful products, gene organization, and gene regulation for *B. subtilis* along with substantial, but not exhaustive, information about other gram-positive bacteria. In some cases, such as morphogenesis and transcriptional control of phage ϕ29, a particular system has been presented in detail as a paradigm. In many cases, useful comparisons with gram-negative bacteria, particularly *E. coli*, are drawn. We have purposely avoided detailed discussions of important aspects of developmental biology, since this topic has been and will be explored in considerable depth in other books.

Given the biological interest of gram-positive bacteria and the recent increase in available information, it is not surprising that the American Society for Microbiology raised the possibility of compiling this book as a companion to the *E. coli-S. typhimurium* "Bible." That they did so is in great measure attributable to the efforts of our late colleague, Helen R. Whiteley. Dr. Whiteley was a leader in molecular genetic analysis of *B. subtilis* transcription and *B. thuringiensis* insect toxin gene structure and regulation. As the long-time head of the ASM Publications Board, she was instrumental in assuring the quality of the Society's journals and books and instigated a number of new projects, including the suggestion for a book of this type.

It should be obvious that it would have been impossible to assemble this book without the contributions of the many authors. We are extraordinarily grateful to them for their cheerful participation, timely responses to deadlines, and assiduous incorporation of reviewers' comments. We also offer special thanks to Patrick J. Piggot, who organized and edited the series of chromosome maps. Many anonymous and unpaid reviewers helped us to assure the factual accuracy and logical flow of the presentation. We hereby absolve them of any blame for inaccuracies that remain.

Abraham L. Sonenshein
James A. Hoch
Richard M. Losick

I. GRAM-POSITIVE BACTERIA

1. Systematics and Ecology of *Bacillus*

FERGUS G. PRIEST

The biology of the genus *Bacillus* is dominated by the endospore, which is the unifying feature of an otherwise diverse collection of saprophytic bacteria. The process of spore formation offers a superb model system for the molecular biology of differentiation, is associated with the production of a range of biotechnologically important products such as insect toxins and peptide antibiotics, and yields a dormant structure that is unique in its resistance to adverse physical and chemical agents and its longevity in the environment.

Microbiologists quickly realized the importance of the spore as a taxonomic feature. It is readily seen microscopically and provides a simple defining characteristic for the family *Bacillaceae*, the spore-forming bacteria. Within this family, the genus *Bacillus* was established to include the rod-shaped bacteria that grew in the presence of air, thus distinguishing them from the strictly anaerobic *Clostridium* spp. During the early part of this century, microbiologists named many new species of *Bacillus*, often with scant, largely morphological descriptions that were synonymous with those of established species. This tendency became self-generating, since it was virtually impossible to identify new isolates when a comparison had to be made with 50 or 100 poorly described existing species. Much simpler to invent a new name for the new isolate! By the 1930s, some 200 *Bacillus* species had been named. At this point, a major contribution to the systematics of the genus *Bacillus* was initiated by Nathan R. Smith at what is now the Northern Regional Research Laboratory at Peoria, Ill. Together with Francis E. Clark and Ruth E. Gordon, he examined 1,134 *Bacillus* strains that had, on receipt, 158 species names and assigned 1,114 of these strains to just 19 species (76). Revised and supplemented descriptions of these *Bacillus* species were subsequently published by Gordon and her colleagues (21). These studies form the framework of the current classification of the genus that appears in *Bergey's Manual of Systematic Bacteriology* (9).

Nevertheless, it became apparent that Gordon and her colleagues had been too severe in their reduction of the genus to so few species, and during the past 2 decades, several old species names have been reintroduced and new species have been recognized. Today, there are 65 validly described species in the genus (Table 1; 10). This undoubtedly still represents an underestimate of the genetic and physiological diversity in the genus, and strains of new species are continually being isolated and described. In this chapter, I shall review current ideas concerning the systematics of the aerobic endospore-forming bacteria and show how an appreciation of their classification and identification makes possible a better understanding of their ecology.

BACILLUS AND RELATED GENERA

Aerobic endospore-forming bacteria are currently assigned to four genera in the family *Bacillaceae*. The limits to the genus *Bacillus* are set principally on morphological grounds. Branching mycelial bacteria that differentiate into aerial and substrate mycelia similar to those of *Streptomyces* spp. but that form true endospores are placed in the genus *Thermoactinomyces*. Since the turn of the century, this genus has been included in the order *Actinomycetales*, but recent chemotaxonomic analyses are consistent with its assignment to the family *Bacillaceae*. In particular, the endospores are typical of *Bacillus* spores and contain dipicolinic acid. These spores are borne on both aerial and substrate mycelia and may be located either adjacent to the mycelium (sessile) or at the tips of short sporophores. The cell walls of *Thermoactinomyces* spp. contain *meso*-diaminopimelic acid (*meso*-A_2pm), as do those of most *Bacillus* spp., and the major menaquinones in the membrane are unsaturated, with seven or nine isoprene units, again like those of most *Bacillus* spp. (42). The DNA base composition of thermoactinomycete species is around 50 to 54% G+C, which is much lower than the 60% or more found in the true actinomycetes (30). Finally, 16S rRNA analyses have placed *Thermoactinomyces* species firmly alongside other thermophilic *Bacillus* spp., and the sequences show very low homology with actinomycete rRNA sequences (80). Thermoactinomycetes therefore seem to be bacilli that have adopted a mycelial morphology, presumably through selection in their habitat, which is largely associated with plant materials and composts and is similar to that of streptomycetes (42).

At the other morphological extreme, cocci that differentiate into true endospores are placed in the genus *Sporosarcina*. Currently, two species are recognized: *Sporosarcina halophila* and *Sporosarcina ureae*. These bacteria share some characteristics with the micrococci; in particular, the cell wall lacks *meso*-A_2pm as the diamino acid in the peptidoglycan and contains lysine. This is, however, consistent with *Bacillus pasteurii* and *Bacillus sphaericus*, which also have cell walls based on lysine rather than *meso*-A_2pm. This relationship is further strengthened by rRNA sequence determinations which show that *S. ureae* diverged from *B. pasteurii* relatively recently (79). The DNA base composition is also low (around 40% G+C), which differentiates *Sporosarcina* spp.

Fergus G. Priest • Department of Biological Sciences, Heriot Watt University, Edinburgh EH14 4AS, Scotland.

Table 1. Allocation of some *Bacillus* species to groups on the basis of phenotypic similarities

Species[a]	Mol % G+C[b]	RNA group[c]	Characteristics of group
Group 1			All species are facultative anaerobes and grow strongly in absence of oxygen. Acid is produced from variety of sugars. Endospores are ellipsoidal and swell the mother cell.
B. alvei	46	3	
B. amylolyticus	53	3	
"*B. apiarius*"	ND	ND	
B. azotofixans	52	3	
B. circulans	39	1	
B. glucanolyticus	48	ND	
B. larvae	38	3	
B. lautus	51	1	
B. lentimorbus	38	1	
B. macerans	52	3	
B. macquariensis	40	3	
B. pabuli	49	3	
B. polymyxa	44	3	
B. popilliae	41	1	
B. psychrosaccharolyticus	44	1	
B. pulvifaciens	44	3	
B. thiaminolyticus	53	ND	
B. validus	54	3	
Group II			All species produce acid from variety of sugars including glucose. Most are able to grow at least weakly in absence of oxygen, particularly if nitrate is present. Spores are ellipsoidal and do not swell the mother cell.
B. alcalophilus	37	UG	
B. amyloliquefaciens	43	1	
B. anthracis	33	1	
B. atrophaeus	42	1	
"*B. carotarum*"	ND	ND	
B. firmus	41	1	
B. flexus	38	ND	
B. laterosporus	40	5	
B. lentus	36	1	
B. licheniformis	45	1	
B. megaterium	37	1	
B. mycoides	34	1	
B. niacini	38	ND	
B. pantothenticus	37	1	
B. pumilus	41	1	
B. simplex	41	1	
B. subtilis	43	1	
B. thuringiensis	34	1	
Group III			These strict aerobes do not produce acid from sugars; names in brackets are exceptions. They produce ellipsoidal spores that swell the mother cell.
(*B. alginolyticus*)	48	ND	
"*B. aneurinolyticus*"	42	UG	
B. azotoformans	39	1	
B. badius	44	1	
B. brevis	47	4	
(*B. chondroitinus*)	47	ND	
"*B. freudenreichii*"	44	ND	
B. gordonae	55	3	
Group IV			All species produce spherical spores that may swell the mother cell and contain L-lysine or ornithine in cell wall. All species are strictly aerobic, but some have limited ability to produce acid from sugars.
("*B. aminovorans*")	40	ND	
B. fusiformis	36	2	
B. globisporus	40	2	
B. insolitus	36	2	
B. marinus	39	ND	
B. pasteurii	38	2	
(*B. psychrophilus*)	42	2	
B. sphaericus[d]	37	2	
Group V			These thermophilic species all grow optimally at >50°C. Physiologically and morphologically, they are heterogeneous, but most produce oval spores that swell the mother cell.
B. coagulans	44	1	
"*B. flavothermus*"	61	ND	
B. kaustophilus	53	5	
B. pallidus	40	ND	
B. schlegelii	64	ND	
B. smithii	39	1	
B. stearothermophilus	52	5	

Continued

Table 1—*Continued*

Species[a]	Mol % G+C[b]	RNA group[c]	Characteristics of group
B. thermocatenulatus	69	5	
B. thermocloacae	42	ND	
B. thermodenitrificans	52	ND	
B. thermoglucosidasius	45	5	
B. thermoleovorans	55	5	
B. thermoruber	57	ND	
B. tusciae	58	ND	
Group VI			Thermophilic, acidophilic species with membraneous ω-alicyclic fatty acids.
A. acidocaldarius	60	6	
A. acidoterrestris	52	6	
A. cycloheptanicus	56	6	
Unassigned species			
B. benzoevorans	41	1	
B. fastidiosus	35	1	
B. naganoensis	45	ND	

[a] Names in quotation marks refer to taxa that do not appear in the *Approved Lists of Bacterial Names* or its supplements (46a, 75a) and therefore have not been validly published. Names in parentheses refer to species that are atypical of the general description.
[b] Given either as the value for the type strain or as the mean of a range for several strains.
[c] RNA groups are based on the work of Ash et al. (4), Collins (10a), and Wisotzkey et al. (91), as shown in Fig. 1A. UG, ungrouped; ND, no data available.
[d] *B. sphaericus* includes at least five "species" of round-spored strains (see text).

from *Micrococcus* spp., which have a high G+C content.

Finally, bacteria isolated initially from chicken feeds and subsequently from the rhizospheres of plants have been placed in the genus *Sporolactobacillus* because they share characteristics with both *Bacillus* and *Lactobacillus* spp. (53). In particular, they differentiate into true endospores, but like the lactobacilli, they lack catalase, grow optimally under microaerophilic conditions, and conduct lactic acid fermentation. It seems unlikely that these bacteria warrant a separate genus, since physiologically they resemble *Bacillus coagulans* (although the latter is catalase positive, it conducts a homolactic fermentation) and rRNA analyses show significant homology between *Sporolactobacillus inulinus* and *B. coagulans* (81).

Aerobic endospore-forming bacteria excluded from the genera *Sporolactobacillus*, *Sporosarcina*, and *Thermoactinomyces* are by definition placed in the genus *Bacillus*. As mentioned above, this genus encompasses more than 60 species. Compared with many other genera, the genus *Bacillus* is large and represents too great a genetic diversity for a single genus. Although there is no generally accepted definition of a procaryotic genus, it has been suggested that the maximum genetic diversity allowable in the species of a genus should be that represented by a spread of chromosomal base composition of 10 to 12% G+C. That is, if species in a genus differ by more than this figure, they must be so distantly related as to be assigned to separate genera (59). If we accept this definition, the range of base composition in the genus *Bacillus* (36% G+C for *Bacillus cereus* to more than 60% G+C for various thermophiles) must signify great phylogenetic divergence and therefore more than one genus. Similarly, rRNA sequences have revealed as much phylogenetic divergence in the genus *Bacillus* as in the combined families of *Enterobacteriaceae* and *Vibrionaceae* (81).

From a purely practical viewpoint, the size of the genus *Bacillus* complicates identification of new isolates, since a comparison with many species descriptions must be made before identification of an isolate can be effected. Even with computer-assisted systems, at least 30 tests must be done to achieve a reasonably accurate identification (61). These tests entail time and expense that few laboratories are prepared to countenance. Ecological studies are therefore restricted to the easily recognized species or, more commonly, make no attempt at species identification and simply list isolates as *Bacillus* spp. There are therefore both theoretical and practical grounds for splitting *Bacillus* into several genera, but we should first consider briefly some aspects of the evolution of the endospore-forming bacteria.

Evolution of the Endospore-Forming Bacteria

Current views about the evolutionary branching patterns (cladistics) of *Bacillus* and related genera are derived entirely from 16S rRNA sequence comparisons derived initially as catalogs (81) and later as full sequences (4, 68). These studies indicate that the gram-positive bacteria form two independent lines of descent, i.e., the high-G+C actinomycete lineage and the low-G+C clostridial lineage (92). The genus *Bacillus* is included in the latter; it diverged from its closest clostridial relative (perhaps *Clostridium thermosaccharolyticum* or *Clostridium thermoaceticum*) at a measure of sequence similarity of about 0.4. This divergence would be consistent with the development of several other aerobic lineages, both gram positive and gram negative, and is thought to correspond to the occurrence of oxygen on the planet at high concentrations about 700 to 800 million years ago.

A problem introduced by phylogenetic analyses of *Bacillus* spp. and relatives is the relationship of non-

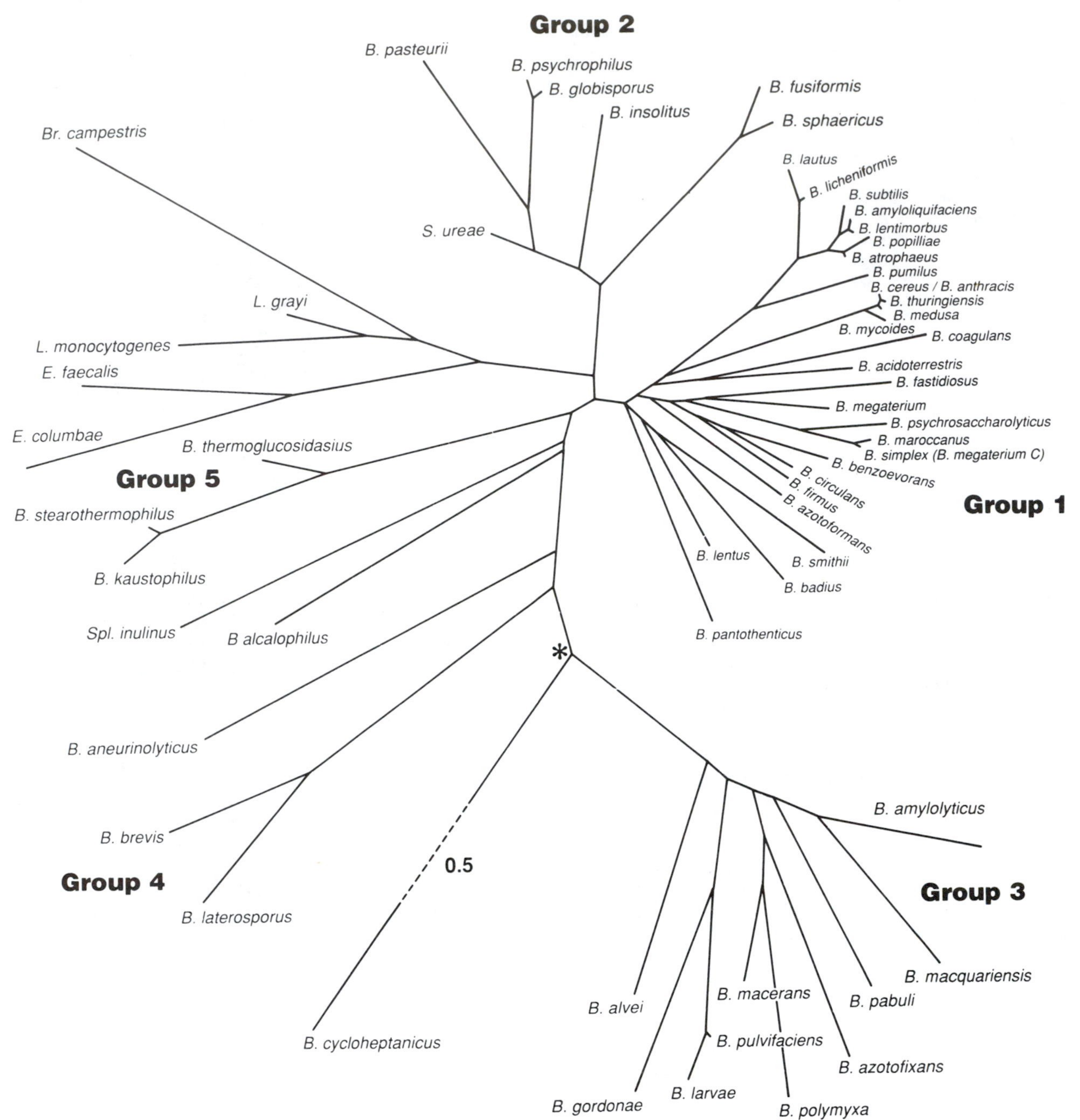

Figure 1. Phylogenetic trees of some members of the genus *Bacillus* based on 16S rRNA sequence analysis. (A) Reproduced with permission from reference 4. Note that *B. acidoterrestris* is misplaced and should be assigned to the *B. cycloheptanicus* branch (10a). (B) Reproduced with permission from reference 68.

sporulating taxa to endosporeformers. Early rRNA cataloging studies revealed a closer relationship of some nonsporulating species such as *Planococcus citreus*, *Filibacter limicola*, and *Caryophanon latum* to "*Bacillus aminovorans*," *Bacillus globisporus*, and *B. sphaericus*, respectively, than these *Bacillus* spp. show to *Bacillus subtilis* (79). Unfortunately, there are few full RNA sequence comparisons of *Bacillus* spp. with nonsporulating related species, so these observations may be influenced by artifacts introduced by the cataloging process. Nevertheless it is unlikely that these errors would affect the pattern significantly. We are thus confronted with the inclusion of nonsporulating bacteria into the genus *Bacillus* if we are to adhere strictly to cladistics, a point which has not been addressed seriously.

Moreover, assuming the low-G+C-content gram-positive bacteria to be an evolutionary branch of significant depth, it is apparent that endospore-forming bacteria encompass all the lactic acid bacteria, the staphylococci, and various other genera (81, 92). This genealogy raises some fascinating evolutionary questions. Aspects of endospore formation and structure in *Bacillus* and *Clostridium* spp. are highly conserved, making it likely that sporulation evolved once only. Given the greater phylogenetic depth of the genus *Clostridium* than of the genus *Bacillus* (92), the progenitor of the low-G+C-content gram-positive line must have been an anaerobic sporeformer. It follows that genera *Kurthia*, *Pediococcus*, and *Lactobacillus* and other members of the group must have lost the ability to differentiate into spores. Alternatively, but on current evidence more unlikely, sporulation may not have been a single evolutionary event but may have arisen in two phylogenetic lines (*Bacillus* and *Clostridium* spp.).

Bacillus: One Genus or Several Genera?

Reclassifying *Bacillus* spp. into several genera is not a new idea. In the past, the genus "*Aerobacillus*" has been suggested for *Bacillus polymyxa*, "*Zymobacillus*" has been suggested for *Bacillus macerans*, and "*Thermobacillus*" has been suggested for the thermophiles, to mention just a few suggestions (reviewed in reference 19). Moreover, the morphological division of *Bacillus* spp. into (i) species that produce oval endospores that do not distend the mother cell, (ii) species that produce oval endospores that distend the mother cell, and (iii) species that produce spherical endospores was a useful, informal division of the genus that simplified identification of new isolates by providing three taxa of relatively few species each.

It is perhaps time to consider splitting the genus in a more natural way by using phenetic and chemotaxonomic data in addition to phylogenetic information. An essential feature must be that the new genera must be readily distinguished by phenotypic features (85). Initial indications of how *Bacillus* spp. might be subdivided came from numerical analysis of phenotypic features. Three comprehensive and independent studies in this area have provided essentially congruous results (43, 63, 64), and the main findings are summarized in Table 1. Although comprehensive phylogenetic studies of *Bacillus* spp. have been published recently (4, 68), I have emphasized the phenetic classification for the reasons mentioned above and because the two published phylogenies are preliminary, are based only on 16S rRNA sequences, and in several instances are at variance with each other. This might be explained by the cladistic classifications being based on single reference strains (29). Thus, I have indicated the phylogenetic assignments wherever possible but present a traditional or natural classification that is generally consistent with phylogenetic observations. There are relatively few differences between the cladistic and phenetic arrangements.

It is possible to allocate many *Bacillus* species to one of six taxa that have distinguishable physiologies and, perhaps surprisingly, are generally consistent with the division of the genus based on spore morphologies.

B. polymyxa group (group I)

Group I strains (*B. polymyxa* group) are phenetically coherent but not entirely homogeneous phylogenetically (Fig. 1A). Interestingly, all of the group I species listed in Table 1 differentiate into oval endospores that distend the sporangium. They have fairly complex nutritional requirements, and none are prototrophs. The taxon is typified by *B. polymyxa* and includes species that are true facultative anaerobes. Some, such as *B. macerans*, prefer to grow in the absence of oxygen. Major fermentation pathways differ. *B. macerans* operates an acidogenic fermentation that switches to an ethanolic fermentation at acid pH; *B. polymyxa* produces 2,3-butanediol in large amounts. Other species, such as *Bacillus circulans* and *Bacillus glucanolyticus*, use mixed acid fermentations. Some reference strains, notably those of *B. circulans* and *Bacillus lautus* (Fig. 1A), show a closer evolutionary relationship with the *B. subtilis* group than with the *B. polymyxa* group, and this relationship warrants further study.

A second conflict between the cladistic and phenetic classifications concerns the obligate insect pathogenic species *Bacillus larvae*, *Bacillus lentimorbus*, and *Bacillus popilliae*. These catalase-negative, facultatively anaerobic bacteria have fastidious growth requirements, fail to grow in nutrient broth, and sporulate heavily only in the hemolymph of their insect hosts (37). They are phenetically most similar to members of the *B. polymyxa* group, but *B. popilliae* and *B. lentimorbus* have been assigned to the *B. subtilis* group by rRNA sequencing (4).

B. subtilis group (group II)

Group II species (*B. subtilis* group) are phylogenetically and phenetically consistent. *B. subtilis* is an appropriate representative of the taxon that includes many common names (Table 1). These bacteria all produce acids from a range of sugars and some, *B. cereus* and *Bacillus licheniformis* in particular, are facultative anaerobes. However, the latter grows poorly anaerobically and can use glucose only under anaerobic conditions. Although *B. subtilis* is generally regarded as an aerobe, it can grow and sporulate slowly under strict anaerobic conditions. Given glucose, with nitrate as a terminal electron acceptor, it grows strongly anaerobically. These bacteria are therefore an intermediate stage between the true facultative anaerobes of the group I strains and the strict aerobes in groups III and IV. This is reflected in their production of acids from several sugars. All these bacteria produce oval endospores that do not swell the mother cell and are generally located centrally or subterminally.

Several species assigned to the *B. subtilis* group by rRNA sequence comparisons are halotolerant and associated with marine and estuarine habitats. These include *Bacillus lentus*, *Bacillus firmus*, and several unnamed intermediate taxa (23). The alkaliphilic species have been associated with *B. firmus*, and indeed, alkaliphilic variants of this bacterium are common (22). However, alkaliphilic *Bacillus* spp. form a heterogeneous group that encompasses strict aerobes and facultative anaerobes and includes morphologies representing all three sporulation patterns (17, 31). It is therefore an oversimplification to assign alkaliphilic *Bacillus* spp. to the *B. firmus* group, and it appears that alkaliphily is a polyphyletic trait that has evolved several times within the genus *Bacillus*.

Bacillus brevis group (group III)

Group III strains (*B. brevis* group) are poorly defined. Indeed, the representative species, *B. brevis*, is taxonomically heterogeneous (64). These bacteria are strict aerobes that do not produce acid from sugars. Most species ("*Bacillus aneurinolyticus*," *Bacillus badius*, *B. brevis*, and "*Bacillus freudenreichii*") have an oxidative metabolism and produce an alkaline reaction in peptone media. Others, for example, *Bacillus azotoformans*, are noted for anaerobic respiration with nitrate as electron acceptor. *Bacillus alginolyticus* and *Bacillus chondroitinus* are exceptions that produce acid from sugars (48), so their assignment to this group may be incorrect. Indeed, both *B. azotoformans* and *B. badius* are included in the *B. subtilis* group on the basis of rRNA sequences, and *Bacillus gordonae* is included in the *B. polymyxa* group (4). To further complicate the situation, the facultative anaerobe *Bacillus laterosporus* has considerable sequence homology with *B. brevis*, and the two presumably diverged relatively recently (4, 68).

B. sphaericus group (group IV)

Bacillus spp. that differentiate into spherical endospores have been allocated to group IV (*B. sphaericus* group). Interestingly, with the exception of those of "*B. aminovorans*," the cell walls of these bacteria do not contain *meso*-A_2pm (as do all other members of the *Bacillus* genus) but contain lysine or, as in the case of *Bacillus insolitus*, ornithine as the diamino acid. These bacteria are strictly oxidative and in most cases will not use sugars as a source of carbon or energy, preferring acetate or amino acids such as glutamate as carbon sources (2). From the phylogenetic viewpoint, they form a homogeneous group with close homology to *S. ureae* (79).

Thermophiles (groups V and VI)

The thermophilic bacilli represent a heterogeneous collection that I have collated simply on the basis of thermophily. Physiologies range from facultative anaerobes with a lactic fermentation, such as *B. coagulans*, to strict aerobes with no action on sugars (*Bacillus thermocloacae*). Hydrocarbon-utilizing strains, such as *Bacillus thermoleovorans*, have also been described.

At least three independent lines of descent are represented by the thermophiles. Some acidophilic species contain unique fatty acids such as ω-cyclohexane (in "*Bacillus acidocaldarius*" and "*Bacillus acidoterrestris*" [13]) and ω-cycloheptane (in "*Bacillus cycloheptanicus*" [14]). Recent 16S rRNA sequence determinations have supported the allocation of these three species to a new genus, *Alicyclobacillus* (91), and their atypical biochemistry warrants their allocation to a separate phenetic group (Table 1).

A second phylogenetic nucleus is represented by *Bacillus kaustophilus*, *Bacillus stearothermophilus*, and *Bacillus thermoglucosidasius*, which form a distinct lineage within the genus *Bacillus* (but compare Fig. 1A with B, which groups *B. stearothermophilus* with *B. subtilis*). A third origin for thermophily lies with *B. coagulans* and *Bacillus smithii* in the *B. subtilis* group (Fig. 1A). The range of G+C content among the thermophiles is 39 to 69%, and the varied phylogenetic allocations indicate that the group is polyphyletic, i.e., that thermophily has arisen several times in the genus. This is perfectly reasonable, since it is believed that thermophilic growth has a restricted genetic basis (75) and probably makes the allocation of all thermophiles to a single genus untenable.

It is not clear how all *Bacillus* spp. can be allocated with conviction to one of these five groups. For example, *Bacillus psychrophilus* produces a round spore but unlike the oxidative bacteria in group IV is a facultative anaerobe using sugars as carbon sources. Similarly, the round-spore-forming, thermophilic, hydrogen-oxidizing, chemolithoautotrophs *Bacillus schlegelii* and *Bacillus tusciae* present problems and have been included in group V simply on the basis of thermophily. *Bacillus benzoeformans* (58) and a methanol-utilizing thermophile (1) comprise endospore-forming cells in filamentous sheathed trichomes more similar morphologically and physiologically to *Caryophanon latum* than to most bacilli. It is to be hoped that comparative taxonomic studies will indicate more appropriate places for these species.

BACILLUS SPECIES

The basic taxonomic unit in bacterial systematics has long been recognized as the species, and yet there remains controversy as to the most appropriate way to define this concept. The traditional view is that a species comprises a group of strains that show a high degree of overall phenotypic similarity and that differ considerably from strains in related groups (82). This definition is similar to that adopted by Gordon and her collaborators in their comprehensive study of *Bacillus* taxonomy (21), although they extended the definition by suggesting that the description of a species should include both freshly isolated strains and laboratory cultures, since strain phenotypes can become modified on repeated subculture on laboratory media. These phenotypic definitions have several limitations, not the least being that in a genus such as *Bacillus*, the sugar fermentation tests used for *B. circulans* and other facultative anaerobes are of no use whatsoever for bacteria with an oxidative metabolism that cannot metabolize carbohydrates. Speciation therefore lacked uniformity; sugar-utilizing species were generally divided into tight homogeneous groups, while the oxidative taxa were classified by negative criteria (failure to metabolize carbohydrates) into apparently homogeneous groups subsequently found to be genetically diverse.

It has become established that DNA-DNA hybridization is the "gold standard" for the designation of bacterial species. Extensive studies with the *Enterobacteriaceae* and other procaryotic groups have led to the recommendation that strains within a species should share at least 70% DNA sequence homology when measured at nonstringent temperatures and that the hybrids should generally show a thermal instability (ΔT_m) close to zero (85). A further refinement of the species concept is the definition of a subspecies within a species. Different subspecies should show 60 to 70% sequence homology and ΔT_ms between 2 and 5°C. This definition of the subspecies is relevant to some areas of *Bacillus* classification.

DNA sequence homology is a valuable criterion for speciation in the genus *Bacillus*. Many studies have now shown that strains within well-established species show levels of sequence homology greater than 70% and often greater than 80%. Interspecies homology levels are usually minimal and well below 40%, even when strains of highly related species such as *B. subtilis* and *B. licheniformis*, which share 10 to 15% homology, are compared (72). Thus, there is a discontinuous spread of homology values, with high intraspecies homology and low interspecies homology.

B. subtilis

The designation in 1936 of the Marburg strain as the neotype strain of *B. subtilis* clarified early confusion among *B. subtilis*, *B. cereus*, and *B. licheniformis*. Following the studies of Gordon et al. (21), *B. subtilis* became established as a taxon distinguishable from the closely related *B. licheniformis* by its inability to grow well anaerobically and from *Bacillus pumilus* by its secretion of amylase and use of nitrate as a terminal electron acceptor under anaerobic conditions. These species are considered the "subtilis group" because of their otherwise similar physiological properties.

Bacteria used for the commercial production of α-amylase were originally isolated and named *Bacillus amyloliquefaciens* by Fukumoto (17a). Although Gordon et al. (21) failed to distinguish these strains from *B. subtilis* strains, it had been shown previously that strains from amylase fermentations had low levels of DNA sequence homology with authentic *B. subtilis* strains (86) and produced an amylase with properties different from those of the *B. subtilis* amylase (87). These DNA sequence homologies were confirmed by others (56, 59, 72). Average values are 17 to 36% homology between strains of *B. amyloliquefaciens* and *B. subtilis* and 9 to 17% homology between strains of *B. amyloliquefaciens* and *B. licheniformis*. Thus, *B. amyloliquefaciens* was established as a genomic species, and subsequent studies revealed that several phenotypic characteristics such as lactose fermentation, gluconate utilization, and secretion of DNase and carboxymethyl cellulase readily distinguished it from *B. subtilis* (49, 62).

Other high-amylase-producing strains related to *B. subtilis* were described mainly from the Japanese fermented soybean product natto. "*Bacillus amylosacchariticus*" and "*Bacillus natto*" strains, however, shared considerable DNA sequence homology with *B. subtilis* Marburg (85 and 70%, respectively) and could be considered strains of *B. subtilis* (72). Interestingly, sequence analysis of the α-amylase genes from these bacteria shows that the liquefying amylase of *B. amyloliquefaciens* shares little homology with the saccharifying amylase of *B. subtilis* and yet closely resembles the liquefying amylases of *B. licheniformis* and *B. stearothermophilus*. The saccharifying amylases of "*B. amylosacchariticus*," "*B. natto*," and *B. subtilis* are, however, virtually identical (see chapter 62).

Strains of *B. subtilis* that produce a black soluble pigment when grown on media containing a source of carbohydrate such as glucose were originally labeled variety "*aterrimus*," and those that produced a black pigment following growth on tyrosine were labeled variety "*niger*." Recently, Nakamura has demonstrated that strains labeled variety "*niger*" are genetically unrelated to *B. subtilis* (they share about 20 to 30% DNA sequence homology with *B. subtilis*) and form a distinct genomic species for which he suggested the name *Bacillus atrophaeus* (50). However, variety "*aterrimus*" strains share high (84 to 100%) DNA homology with *B. subtilis*. In this instance, phenotypic properties for the distinction of *B. atrophaeus* from *B. subtilis* were few, but the two taxa could be distinguished by numerical taxonomy (64).

It is therefore important to be aware of the evolving delineation of *B. subtilis* as a phylogenetically and phenetically homogeneous species. Many early studies of amylase biosynthesis, for example, were misleading because *B. amyloliquefaciens* and *B. subtilis* were confused. Similarly, *B. atrophaeus* could be readily confused with *B. subtilis*, but it is a different species and will behave differently from *B. subtilis* in molecular and genetic studies.

Bacillus anthracis, *B. cereus*, *Bacillus mycoides*, and *Bacillus thuringiensis*

The *B. cereus* group is important medically and industrially and yet is one of the most taxonomically confusing areas of the genus. Gordon et al. (21) confirmed the earlier opinions of Smith et al. (76) that *B. cereus* should be considered the parent species and that the other taxa should be regarded as subspecies of this species. This opinion generated criticism from medical microbiologists and insect pathologists who were involved in recognizing anthrax in humans and animals or diseases in insects caused by *B. thuringiensis*. Consequently, the subspecies recommendations were not adopted, and the four taxa in the group were each awarded species status in recognition of their unique properties.

B. anthracis is the causative agent of anthrax in humans and animals (see chapter 8). The three toxin genes of *B. anthracis* are located on a large (114-MDa) plasmid, pXO1, and the capsule gene is located on a smaller (60-MDa) plasmid, pXO2 (67). Avirulent pXO1- and/or pXO2-deficient strains are virtually indistinguishable from *B. cereus*, and the question arises whether the two organisms should be classified as the same species. Published DNA hybridization data are unhelpful, since the two strains of *B. anthracis* studied share 94 and 56% homology with a *B. cereus* reference strain (35), but recent studies have demonstrated consistently low levels of DNA sequence homology between strains of *B. anthracis* and *B. cereus*, which are appropriate to their classification as separate species (10a). Certainly, the two taxa can be distinguished phenetically by numerical taxonomy (43), and there are some phenotypic tests such as sensitivity to penicillin (*B. cereus* possesses a chromosomal β-lactamase, whereas *B. anthracis* is virtually always penicillin sensitive), hemolysin activity (*B. anthracis* is negative, and *B. cereus* is positive), and tyrosine decomposition (*B. anthracis* is negative, and *B. cereus* is positive) that can be used to distinguish the species. It is important to establish the classification of these taxa, since it could have a bearing on the ability of strains to accommodate pXO1 and pXO2.

B. cereus strains may also be pathogenic to animals through the synthesis of various toxins, including phospholipase C, hemolysins, diarrheogenic enterotoxin, and emetic toxin. Of these, it seems likely that the emetic strains, which are phenotypically distinguishable from *B. cereus* sensu stricto (44), represent a distinct genomic species (10a). Although *B. cereus* strains may contain numerous plasmids, no correlation between any plasmid and any of the agresins has yet been established (83).

B. thuringiensis differs from *B. cereus* in producing a crystal protein toxin that is pathogenic to various insects, notably lepidoptera, diptera, and coleoptera (see chapter 64). In all other respects, *B. thuringiensis* and *B. cereus* are virtually identical, and there are no differential features as there are for distinguishing *B. anthracis* from *B. cereus* (63, 64). DNA sequence homology data are again unhelpful, since the few studies to date (35, 78) reveal intermediate levels of homology (45 to 80%) between strains of *B. cereus* and *B. thuringiensis*. Moreover, a preliminary study of multilocus enzyme electrophoresis has failed to reveal distinct boundaries (93). However, specialized techniques such as pyrolysis gas chromatography or pyrolysis mass spectrometry, in which whole cells are pyrolized in an inert atmosphere and the products are analyzed to provide a pattern of peaks, can distinguish *B. cereus* from *B. thuringiensis* (55), presumably from the toxin components, which form a major component of the *B. thuringiensis* cell (chapter 64).

In most cases, the toxin genes of *B. thuringiensis* are located on large conjugative plasmids, although chromosomal genes have been characterized in some strains. These plasmids are transmissible by conjugation between *B. thuringiensis* strains in vivo in infected insects (32), and laboratory transfer to strains of *B. cereus* has been demonstrated (20). Such *B. cereus* transconjugants synthesize crystal proteins and since the dividing line between *B. cereus* and *B. thuringiensis* is so dubious, these organisms could be considered to have changed species to *B. thuringiensis*. *B. thuringiensis* has been divided into at least 34 serotypes or serovars based on flagellar (H) antigens. These serovars form the basis of the subspecies names given to *B. thuringiensis* and provide a useful classification for ordering the ever-increasing number of *B. thuringiensis* isolates (12).

Little is known about *B. mycoides*. Gordon et al. (21) decided that its lack of motility and distinctive rhizoidal phenotype were insufficient to warrant species status and reduced the species to a subspecies of *B. cereus*, but at present, *B. mycoides* has been afforded species status.

Thus, the *B. cereus* group provides an interesting taxonomic problem. rRNA sequences of representatives of the four subspecies are virtually identical and within the variation expected of a single species (3). The relationships will be unravelled only by sensitive typing schemes capable of distinguishing taxa at the subspecies level (25).

B. sphaericus

The bacteria of group IV in Table 1 are interesting because of their distinctive cell wall composition (lack of *meso*-A_2pm) and strictly oxidative metabolism. Their inability to use sugars as sources of carbon and energy negates many of the traditional taxonomic tests and has resulted in mesophilic, round-spored bacilli all being allocated to *B. sphaericus*. The basis of *B. sphaericus* taxonomy was established by Krych et al. (40), who used DNA hybridization to establish five DNA homology groups. Subsequently, strains representing these homology groups were studied by using carbon source utilization tests (rather than sugar fermentations) and other features appropriate to oxidative bacteria, and the five homology groups were shown to be phenotypically distinct (2).

One group of particular importance includes strains pathogenic for certain mosquito larvae. These bacteria produce crystal proteins similar to those of *B. thuringiensis*, although the toxin genes are located on the chromosome rather than on plasmids. All pathogens were allocated by DNA reassociation to DNA homology group II, which was shown to be a species separate from homology group I (*B. sphaericus* sensu stricto), with which there was only 15 to 29% homology. Thus, the mosquito pathogens are a different species from *B. sphaericus* despite the same name being used for both pathogenic and nonpathogenic types. In contrast, *B. cereus* and *B. thuringiensis* are allocated different names although they are virtually identical bacteria!

Within homology group II, all pathogens were allocated to subgroup IIA. Bacteria in this subgroup share 60 to 70% sequence homology with subgroup IIA strains, and the melting temperatures of hybrid duplexes were 2 to 6°C below that of the homologous duplex DNA (40). This is consistent with these two groups being designated subspecies of a single species (25), but for registration for use as an insecticide in the United States, the name *B. sphaericus* has been retained for the mosquito pathogens of group IIA.

Gene Exchange within and between Species

That *Bacillus* species constitute discrete genetic and phenetic entities raises interesting questions with regard to gene exchange and the evolution of the species. If lateral gene exchange were common, it might be expected that species would form a continuum of hybrid forms, each containing shared genetic information. Even where such continua were suspected, such as the *B. circulans* "complex" (21) or the *B. firmus-B lentus* "series" (23), DNA reassociation studies have now shown that each group consists of several discrete species (51, 59), and yet, chromosomal gene exchange in the environment seems to be reasonably common in some species. Strains of *B. subtilis* growing together in soil have been shown to exchange blocks of linked genes by transformation. Such gene exchange was shown to lead to extensive reorganization of the genotypic structure of the population (24). In comparison, population genetic studies of *Escherichia coli* (54) and *Salmonella typhimurium* (73) indicate for these species a clonal structure in which chromosomal gene exchange is uncommon. These enterobacterial species therefore comprise populations of largely independent clones of strains, between which lateral chromosomal gene transfer is rare but plasmid transfer is commonplace. The full evolutionary significance of the exchange of chromosomal DNA within populations of *B. subtilis* is, as yet, unclear, but analyses of other naturally transformable species, such as *Haemophilus influenzae* (47), reveals greater genetic heterogeneity and less clonality than in *E. coli*, suggesting that *B. subtilis* will also be genetically diverse.

Extension of the gene transfer experiments to interspecies DNA exchange between *B. licheniformis* and *B. subtilis* revealed that bidirectional chromosomal gene exchange was common in broth and soil cultures, particularly following outgrowth from spores. The resulting hybrid types were presumably merodiploids of some kind and were phenotypically mixed. However, the hybrids were subsequently resolved into one of the parental forms (16). Thus, species are naturally maintained, at least in the *B. subtilis* group, not by placing a barrier to gene exchange but by subsequent "correction" of exchanged genes.

Although the genetic isolation of species is a major feature of bacterial evolution, this does not mean that lateral transfer of genes cannot take place. Conjugal transfer of plasmid genes between *B. cereus* and *B. subtilis* in soils has been demonstrated (84). Particular evidence for lateral gene transfer between "species" of aerobic spore-forming bacteria comes from the study of amylase genes from two strains of *B. stearothermophilus*. The amylase genes of these strains were shown to be virtually identical despite the fact that differences in their 5S rRNA sequences revealed that the two strains were phylogenetically dissimilar and probably diverged over 400 million years ago. It was suggested that the amylase genes were introduced from an external source relatively recently (probably less than 20 million years ago) and perhaps were initially plasmid borne (69).

ECOLOGY OF *BACILLUS* SPP.

The endospore is the most important aspect of *Bacillus* ecology for several reasons. First, because heat treatment is the most common selective isolation procedure for the recovery of *Bacillus* spp. from the environment, most studies concentrate on the endospore to the exclusion of vegetative cells. Second, because the spore is a dormant structure of great longevity, ecological studies are often simply estimates of the accumulation of spores in an environment rather than an assessment of the contribution of the bacterium to the environment. Nevertheless, a large number of spores of a particular species in a habitat is strongly indicative of previous or continuing growth and metabolism in that niche. For example, the first *B. thuringiensis* subsp. *israelensis* strain was isolated as a spore from the mud (containing dead mosquito larvae) from a stagnant pool in the Negev desert of Israel (45). These pools are prime mosquito-breeding habitats during the winter rains and into early spring. During this period, the bacteria grow until the pools dry out in summer, leaving behind many spores and few vegetative cells.

It is generally believed that endospore-forming bacteria remain dormant for long periods and germinate in conditions conducive to growth to spend short periods in the vegetative state. However, in certain conditions when the environment remains consistent, the vegetative state may be the norm. These habitats would include the acid hot springs in which *A. acidocaldarius* lives (11); the rumens of cattle, with which several facultatively anaerobic strains are associated (89); or aerobic sewage sludge treatment plants (15, 71). In other instances, such as the isolation of strict alkaliphilic spores from acid soils or thermophilic spores from sea or cold lake sediments, it is clear that the bacteria are essentially irrelevant to that environment and have simply accumulated as spores (reviewed in reference 65).

The endospore is important for the dispersion of *Bacillus* spp. Spores are readily blown about in dust and air currents and are prevalent in animal feces. This is well illustrated by the distribution of the thermophilic *Bacillus* spp. Nonsporulating strictly thermophilic bacteria are often geographically isolated and restricted to particular thermal areas. For example, certain phenotypic groupings of *Thermus* strains tend to be associated with specific geographical locations, suggesting that they are evolving independently (90). However, a recent extensive study of thermophilic *Bacillus* spp. showed that species were globally distributed and that no specific types were associated with geographic origins (88).

B. polymyxa Group (Group I)

The facultative anaerobes in group I tend to have complex growth requirements, including needs for various amino acids and vitamins. These organisms are uncommon in soils of poor nutrient status and frequent in rotting plant materials and composts. Several species, such as *Bacillus azotofixans*, *B. polymyxa*, and the sporolactobacilli, are associated with the rhizospheres and rhizoplanes of plants. Some species fix atmospheric nitrogen; in one study, *B. polymyxa* was found to be the predominant nitrogen-fixing bacterium in temperate-forest and tundra soils (34). It has long been suspected that such bacteria receive nutrients from the roots of plants and in turn provide fixed nitrogen (66, 74). It is thought that this symbiotic relationship of *B. polymyxa*- and *B. macerans*-like organisms may enable Marram grass to populate nutrient-poor sand dunes in Wales (66). *B. polymyxa* and related organisms are also important in the development of Canadian wheat. Spring wheat plants inoculated with *B. polymyxa* may receive up to 10% of their total nitrogen from the bacteria (41). Interestingly, the enhancement of wheat growth may not be entirely due to fixed nitrogen, and there is evidence for the secretion of plant growth-enhancing substances such as gibberellins by root-associated bacteria. Nitrogen fixation may be associated with other ecosystems; *B. macerans* has been implicated in ruminal nitrogen fixation in sheep (33).

Several insect commensal and pathogenic species are included in group I. Facultative anaerobes such as *B. circulans* may be isolated from the feces of many healthy insects, but some of the species have a more intimate relationship with insects. "*Bacillus apiarius*," *Bacillus pulvifaciens*, and *Bacillus thiaminolyticus* have been isolated from living and dead honeybee larvae but are not thought to be pathogenic (7). Similarly, *Bacillus alvei* is common in bees suffering from European foulbrood but is not pathogenic to honeybees.

The obligate insect pathogens that sporulate (heavily) only in the hemolymph of living insects include *B. popilliae*, a pathogen of scarabaeid beetles. After the spore is ingested by the insect larva, the spore germinates, grows, sporulates in the larva, resulting in the deposition of spores into the environment. Studies in New Jersey and Delaware showed the presence of *B. popilliae* 25 to 30 years after their original application (37). *B. larvae*, on the other hand, is to be avoided, since it is the causative agent of American foulbrood in honeybees.

B. subtilis Group (Group II)

Bacteria related to *B. subtilis* are commonly encountered and easily identified. The soil is the reservoir of these bacteria. From the soil, they are transferred to various associated environments including plants and plant materials, foods, animals, and marine and freshwater habitats.

B. cereus, *B. licheniformis*, *B. pumilus*, and *B. subtilis* are prevalent in soils, particularly low-nutrient soils. They are also common on straw and cereals, including rice and pulses, which they presumably colonize from wind-blown soil particles and dust. The association with plants may be less superficial than at first realized and may be used to agricultural benefit. Bacterial inoculants are used to improve crop performance by increasing nutrient availability to the plant (see above for nitrogen fixation by *B. polymyxa*), stimulating plant growth (38), and suppressing plant diseases. It is in this last category that bacteria related to *B. subtilis* are being studied. Indeed, *B. subtilis* itself has been used to control several types of fungal disease associated with fruits, vegetables, field crops, and flower crops (26), largely through the secretion of antifungal antibiotics. Moreover, a *B. cereus* strain has been shown to have biocontrol activity in preventing damping off of alfalfa seedlings by *Phytophthera* spp. (28). Interestingly, this bacterium also enhances root nodulation of soybean plants both in the field and in a growth chamber (27). Take-all disease of wheat (caused by *Gaumannomyces graminis*) is effectively suppressed by *B. mycoides* and *B. pumilus*, but no mechanism has been suggested (8).

These bacteria contaminate various foodstuffs, but because the bacteria are considered nonpathogenic, they are largely ignored. An exception is *B. cereus*, which may be responsible for two distinct forms of food poisoning characterized by diarrhea or vomiting. Both forms arise from the ingestion of a toxin remaining after excessive growth of pathogenic types of *B. cereus*. The most common source of the emetic type of food poisoning is reheated cooked rice, while the diarrheal syndrome is associated with a wider range of foods, including meats, vegetables, puddings, and sauces. Both are intoxications that arise when *B. cereus* spores survive cooking temperatures, germinate, and grow in nonrefrigerated cooked foods. Since the emetic toxin is thermostable (survives at 126°C for 1.5 h), it remains active after the food is reheated (reviewed in reference 83).

Foodstuffs colonized predominantly by *Bacillus* spp. of group II include cocoa and spices. Cocoa is prepared by fermenting cacao beans in their pods and then drying, roasting, and grinding the beans. During the fermentation, *B. subtilis*, *B. pumilus*, *B. licheniformis*, *Bacillus megaterium*, and some thermophilic *Bacillus* spp. multiply to about 10^6 bacteria per g and constitute 90% of the bacterial flora. Subsequent drying and roasting reduces this population, but even when handled hygienically, cocoa will contain about 10^3 *Bacillus* cells per g. Of other dried foods, spices are particularly highly contaminated with *Bacillus* spores. Again, *B. subtilis* and relatives, including *B. cereus*, predominate and are thought to originate from the plants and from soil (reviewed in reference 60).

B. subtilis and relatives are also associated with animals both large and small. Although these bacilli can be isolated from the feces of most animals, this probably represents transient habitation of the gut following ingestion of spores with food. However, *Bacillus* spp. may be important members of the rumen ecosystem, and *B. circulans*, *B. laterosporus*, and *B. licheniformis* have been implicated in hemicellulose conversion in the rumens of cows, sheep, and goats (89). *B. thuringiensis* is widely distributed in soils (46) and on plant leaves (77) and is responsible for intoxications in the larvae of various insects (see chapter 64). It is not clear how important the insect is in the distribution of these bacteria, since, unlike the insect pathogens of group I, *B. thuringiensis* is not dependent on growth in the larvae for sporulation. Nevertheless, growth in the guts of both target and nontarget insects probably contribute to the dissemination of these bacteria (57).

It is thought that a large proportion of the *Bacillus* spp. in seawater represents runoff of soil bacteria into the sea. This would explain the higher numbers of *Bacillus* in coastal waters (8 to 20% of total bacterial population) compared with oceanic sites (0 to 7%) and the occurrence of thermophilic *Bacillus* and *Thermoactinomyces* spp. in coastal and estuarine sediments. However, species such as *B. licheniformis* and *B. subtilis* are universally distributed and dominate the marine flora to such an extent that they could be considered primary inhabitants of the oceans. Other species, such as *B. cereus*, have been found in nonpolluted sites only. *Bacillus* spp. generally occur more frequently in sediments than in the water column; at lower levels, *Bacillus* spores may account for up to 80% of the total heterotrophic flora (reviewed in reference 5).

B. firmus and related halotolerant bacteria are particularly common in marine and estuarine habitats and salt marshes (22, 23), but *Bacillus* spp. are rarer in more concentrated saline habitats such as salterns. Many of these bacteria are pigmented red, orange, or yellow and are of uncertain taxonomic positions (23, 43, 64).

Alkaliphilic bacilli are widespread in soils. They can be found in acidic soils but are more common in high-pH soils, where numbers approach 10^6/g (31).

B. brevis Group (Group III)

Group III bacteria are widely distributed in the environment, but the soil seems to be their primary habitat, with marine and freshwater habitats secondary. Since they are of little interest other than for the production of the antibiotics gramicidin and tyrocidin, little is known of the ecology of these bacteria or their roles in the environment.

B. sphaericus Group (Group IV)

B. sphaericus strains, both mosquito pathogens and other types, are widely distributed in soil, sediments from pools and lakes, drainage ditches, and other sites suitable for larval growth (18). One of the advantages of *B. sphaericus* as a mosquito control agent is that it persists in the environment longer than does *B. thuringiensis*. When infected larvae die, the bacteria may grow and sporulate in the cadaver, thus returning spores to the environment. This recycling may be responsible for the reisolation of applied strains several months or even years after initial application (52).

The oxidative metabolisms of many of these species are associated with growth at high pH, and several round-spored bacteria similar to *B. sphaericus* are alkaliphilic. Similarly, *B. pasteurii* and *S. ureae* are strongly ureaclastic and are frequent in urban soils associated with humans and dogs. They are also commonly isolated from urinals (9).

Several psychrophilic species that form round spores and lack *meso*-A_2pm in their cell walls were originally isolated from soil and mud (*Bacillus marinus* was isolated from marine sediments). Two studies have shown that such species are prevalent in refrigerated and frozen foods, where they could be responsible for spoilage (reviewed in reference 60).

Thermophiles (Groups V and VI)

The thermophiles are heterotrophic bacteria, aerobes or facultative anaerobes, and are associated with a variety of thermal and nonthermal sites. *B. kaustophilus*, *B. stearothermophilus*, *Bacillus thermodenitrificans*, *B. thermoglucosidasius*, and *B. thermoleovorans* can be isolated from soils and muds in temperate areas, although several are associated with heated materials such as the hot infusion waters from beet sugar refineries (36). Industrial effluents such as those from thermophilic aerobic sewage treatment plants are heavily populated with thermophilic *Bacillus* spp., and several recent isolates such as *Bacillus pallidus* (71) and *B. thermocloacae* (15) were derived from these systems, as were some unusual methanol-utilizing strains (1). "*Bacillus flavothermus*" and *Bacillus thermocatenulatus* were isolated from hot springs and other geothermal areas. Self-heating composts also enrich for thermophilic *Bacillus* spp. such as *Bacillus thermoruber* and thermoactinomycetes (reviewed in reference 75).

B. coagulans and *B. smithii* are slightly acidophilic and thermotolerant rather than thermophilic. Their microaerophilic metabolism is especially suitable for growth in canned milk, vegetables, and fruits, where they are responsible for "flat sours," a problem that has largely been circumvented by improved hygiene. Other thermophilic *Bacillus* spp. are notably acidophilic. *Alicyclobacillus acidocaldarius* has been isolated from acid hot springs worldwide and from volcanically heated soils (11, 75). It grows optimally at pH 2 to 6 and up to 70°C. The other acidophiles (*Alicyclobacillus acidoterrestris* and *Alicyclobacillus cycloheptanicus*) are less extreme and can be recovered from most soils.

Two hydrogen-oxidizing facultative chemolithoautotrophs, *B. schlegelii* and *B. tusciae*, have been described. Originally isolated from the sediment of a eutrophic lake in Switzerland, *B. schlegelii* has also been isolated from a settling pond of a sugar factory in Germany and a volcanic soil in Antartica (6, 70). *B. schlegelli* is capable of growing with carbon dioxide or carbon monoxide as a sole carbon and energy source (39).

CONCLUDING REMARKS

The systematics of the genus *Bacillus* has entered an interesting phase. Demands from biotechnological industries for new or improved products have led to the isolation of many new endospore-forming bacteria. This has stimulated interest in the ecology and habitats of these organisms (see reference 60). Many of these new species have interesting properties not usually associated with *Bacillus* spp. Methanol utilization, aerobic sewage and waste treatment, and biological control of fungal pathogens of plants are some of the varied applications of *Bacillus* spp. that complement the traditional products such as extracellular enzymes, insect toxins, and peptide antibiotics. These new strains must be classified, if only for the prosaic requirements of patent applications, and this necessity has stimulated an appreciation of the difficulties surrounding the systematics of these bacteria.

Interest in the systematics and ecology of *Bacillus* spp. has also been generated by the use of recombinant strains in industry and the potential for genetically engineered strains for biological control. Release of these strains into the environment, either deliberate or accidental, will require detection and monitoring of indigenous and introduced populations.

Fortunately, these requirements have occurred at a time when microbial taxonomy and ecology are undergoing major developments through the application of phylogenetic analyses based largely on rRNA sequences and the development of nucleic acid-based probes (4, 25). The increased number of well-defined *Bacillus* species that will be described together with the physiological diversity that they will undoubtedly represent will present a formidable challenge to the taxonomist, but the outcome should be a fascinating insight into the evolutionary history of these bacteria and perhaps of endospore formation itself.

Acknowledgments. This work was supported by grants from the Science and Engineering Research Council and the German Culture Collection. I am grateful to Michael Goodfellow for many useful comments during the preparation of the manuscript.

REFERENCES

1. **Al-Awadhi, N., T. Egli, G. Hamer, and E. Wehrl.** 1989. Thermotolerant and thermophilic solvent-utilizing methylotrophic aerobic bacteria. *Syst. Appl. Microbiol.* **11:**207–216.
2. **Alexander, B., and F. G. Priest.** 1990. Numerical classification and identification of *Bacillus sphaericus* including some strains pathogenic for mosquito larvae. *J. Gen. Microbiol.* **136:**362–370.
3. **Ash, C., J. A. E. Farrow, M. Forsch, E. Stackebrandt, and M. D. Collins.** 1991. Comparative analysis of *Bacillus anthracis* and *Bacillus cereus* and related species on the basis of reverse transcriptase sequencing of 16S rRNA. *Int. J. Syst. Bacteriol.* **41:**343–346.
4. **Ash, C., J. A. E. Farrow, S. Wallbanks, and M. D. Collins.** 1991. Phylogenetic heterogeneity of the genus *Bacillus* revealed by comparative analysis of small-subunit-ribosomal RNA sequences. *Lett. Appl. Microbiol.* **13:**202–206.
5. **Bonde, G. J.** 1981. *Bacillus* from marine habitats, allocation to phena established by numerical techniques, p. 181–215. *In* R. C. W. Berkeley and M. Goodfellow (ed.), *The Aerobic Endospore-Forming Bacteria, Classification and Identification.* Academic Press, London.
6. **Bonjour, F., and M. Aragno.** 1984. *Bacillus tusciae*, a new species of thermoacidophilic, facultatively chemolithoautotrophic, hydrogen oxidizing sporeformer from a geothermal area. Arch. Microbiol. **139:**397–401.
7. **Bucher, G. E.** 1981. Identification of bacteria found in insects, p. 7–33. *In* H. D. Burges (ed.), *Microbial Control of Pests and Plant Diseases.* 1970–1980. Academic Press, London.
8. **Capper, A. L., and R. Campbell.** 1986. The effect of artificially inoculated antagonistic bacteria on the prevalence of take-all disease of wheat in field experiments. *J. Appl. Bacteriol.* **60:**159–160.
9. **Claus, D., and R. C. W. Berkeley.** 1986. Genus *Bacillus* Cohn 1982, p. 1105–1139. *In* P. H. A. Sneath (ed.), *Bergey's Manual of Systematic Bacteriology*, vol. 2. The Williams & Wilkins Co., Baltimore.
10. **Claus, D., and D. Fritze.** 1989. Taxonomy of *Bacillus*, p. 5–26. *In* C. R. Harwood (ed.), *Biotechnology Handbooks, 2. Bacillus.* Plenum Press, New York.

10a. **Collins, M. D.** Unpublished data.

11. **Darland, G., and T. D. Brock.** 1971. An acidophilic spore-forming bacterium. *J. Gen. Microbiol.* **67:**9–15.
12. **de Barjac, H., and F. Frachon.** 1990. Classification of *Bacillus thuringiensis*. *Entomophaga* **35:**233–240.
13. **Deinhard, G., P. Blanz, K. Poralla, and E. Alton.** 1987. *Bacillus acidoterrestris* sp. nov., a new thermotolerant acidophile isolated from different soils. *Syst. Appl. Microbiol.* **10:**47–53.
14. **Deinhard, G., J. Saar, W. Krischte, and K. Poralla.** 1987. *Bacillus cycloheptanicus* sp. nov., a new thermophile containing ω-cycloheptane fatty acids. *Syst. Appl. Microbiol.* **10:**68–73.
15. **Demharter, W., and R. Hensel.** 1989. *Bacillus thermocloacae* sp. nov., a new thermophilic species from sewage sludge. *Syst. Appl. Microbiol.* **11:**272–276.
16. **Duncan, K. E., C. A. Istock, J. B. Graham, and N. Ferguson.** 1989. Genetic exchange between *Bacillus subtilis* and *Bacillus licheniformis*: variable hybrid stability and the nature of bacterial species. *Evolution* **43:**1585–1609.
17. **Fritze, D., J. Flossdorf, and D. Claus.** 1990. Taxonomy of alkaliphilic *Bacillus* strains. *Int. J. Syst. Bacteriol.* **40:**92–97.

17a. **Fukumoto, J.** 1943. Studies on the production of bacterial amylase. 1. Isolation of bacteria producing potent amylases and their distribution. *J. Agric. Chem. Soc. Japan* **19:**487–503. (In Japanese.)

18. **Geurineau, M., B. Alexander, and F. G. Priest.** 1991. Isolation and identification of *Bacillus sphaericus* strains pathogenic for mosquito larvae. *J. Invertebr. Pathol.* **57:** 325–333.
19. **Gibson, T., and R. E. Gordon.** 1974. *Bacillus*, p. 529–550. *In* R. E. Buchanan and N. E. Gibbons (ed.), *Bergey's Manual of Determinative Bacteriology.* The Williams & Wilkins Co., Baltimore.
20. **Gonzalez, J. M., B. S. Brown, and B. C. Carlton.** 1982. Transfer of *Bacillus thuringiensis* plasmids coding for δ-endotoxin among strains of *B. thuringiensis* and *B. cereus*. *Proc. Natl. Acad. Sci. USA* **79:**6951–6955.
21. **Gordon, R. E., W. C. Haynes, and C. H.-N. Pang.** 1973. *The Genus Bacillus.* United States Department of Agriculture, Washington, D.C.
22. **Gordon, R. E., and J. L. Hyde.** 1982. The *Bacillus firmus-Bacillus lentus* complex and pH 7.0 variants of some alkalophilic strains. *J. Gen. Microbiol.* **128:**1109–1116.
23. **Gordon, R. E., J. L. Hyde, and J. A. Moore, Jr.** 1977. *Bacillus firmus-Bacillus lentus*: a series or one species. *Int. J. Syst. Bacteriol.* **27:**256–262.
24. **Graham, J. B., and C. A. Istock.** 1979. Gene exchange and natural selection cause *Bacillus subtilis* to evolve in soil culture. *Science* **204:**637–639.
25. **Grimont, F., and P. A. D. Grimont.** 1990. DNA fingerprinting, p. 249–279. *In* E. Stackebrandt and M. Goodfellow (ed.), *Nucleic Acid Techniques in Bacterial Systematics.* Wiley Interscience, Chichester, United Kingdom.
26. **Hall, T. J., and W. E. E. Davis.** 1990. Survival of *Bacillus subtilis* in silver and sugar maple seedlings over a two-year period. *Plant Disease* **74:**608–609.
27. **Halversen, L. J., and J. Handelsman.** 1991. Enhancement of soybean nodulation by *Bacillus cereus* UW85 in the field and in a growth chamber. *Appl. Environ. Microbiol.* **57:**2767–2770.
28. **Handelsman, J., S. Raffel, E. H. Mester, and S. Raffel.** 1990. Biological control of damping off of alfalfa seedlings with *Bacillus cereus* UW85. *Appl. Environ. Microbiol.* **56:**713–718.
29. **Hartford, T., and P. H. A. Sneath.** 1988. Distortion of taxonomic structure from DNA relationships due to different choice of reference strains. *Syst. Appl. Microbiol.* **10:**241–250.
30. **Hirst, J., C. R. Bailey, and F. G. Priest.** 1991. Deoxyribonucleic acid sequence homology among some strains of *Thermoactinomyces*. *Lett. Appl. Microbiol.* **13:**35–38.
31. **Horikoshi, K., and T. Akiba.** 1982. *Alkalophilic Microorganisms, a New Microbial World.* Springer Verlag, New York.
32. **Jarrett, P., and M. Stephenson.** Plasmid transfer between strains of *Bacillus thuringiensis* infecting *Galleria mellonella* and *Spodoptera littoralis*. *Appl. Environ. Microbiol.* **56:**1608–1614.
33. **Jones, K., and J. G. Thomas.** 1974. Nitrogen fixation by the rumen contents of sheep. *J. Gen. Microbiol.* **85:**97–101.
34. **Jurgensen, M. F., and C. B. Davey.** 1971. Nonsymbiotic nitrogen-fixing micro-organisms in forest and tundra soils. *Plant Soil* **34:**341–356.
35. **Kaneko, T., R. Nozaki, and K. Aizawa.** 1978. Deoxyribonucleic acid relatedness between *Bacillus anthracis*, *Bacillus cereus* and *Bacillus thuringiensis*. *Microbiol. Immunol.* **22:**639–641.
36. **Klaushofer, H., F. Hollaus, and G. Pollach.** 1971. Microbiology of beet sugar manufacture. *Process Biochem.* **6:**39–41.
37. **Klein, M. G.** 1981. Advances in the use of *Bacillus popilliae* for pest control, p. 183–192. *In* H. D. Burges (ed.), *Microbial Control of Pests and Plant Diseases*, 1970–1980. Academic Press, London.
38. **Kloepper, J. W., R. Lifshitz, and R. M. Zablotowicz.** 1989. Free-living bacterial inocula for enhancing crop productivity. *Trends Biotechnol.* **7:**39–44.
39. **Krüger, B., and O. Meyer.** 1984. Thermophilic bacilli growing with carbon monoxide. *Arch. Microbiol.* **139:** 402–406.
40. **Krych, V. K., J. L. Johnson, and A. A. Yousten.** 1980.

Deoxyribonucleic acid homologies among strains of *Bacillus sphaericus*. *Int. J. Syst. Bacteriol.* **30:**476–484.

41. **Kucey, R. H. N.** 1988. Alteration of wheat root systems and nitrogen fixation by associative nitrogen-fixing bacteria measured under field conditions. *Can. J. Microbiol.* **34:**735–739.
42. **Lacey, J., and T. Cross.** 1989. Genus *Thermoactinomyces* Tsiklinsky 1899, 501[AL], p. 2574–2585. *In* S. T. Williams, M. E. Sharpe, and J. G. Holt (ed.), *Bergey's Manual of Systematic Bacteriology*, vol. 4. The Williams & Wilkins Co., Baltimore.
43. **Logan, N., and R. C. W. Berkeley.** 1981. Classification and identification of the genus *Bacillus* using API tests, p. 106–140. *In* R. C. W. Berkeley and M. Goodfellow (ed.), *The Aerobic Endospore-Forming Bacteria: Classification and Identification*. Academic Press, London.
44. **Logan, N. N., B. J. Capel, J. Melling, and R. C. W. Berkeley.** 1987. Distinction between emetic and other strains of *Bacillus cereus* using the API system and numerical methods. FEMS Microbiol. Lett. **5:**373–375.
45. **Margalit, J., and D. Dean.** 1985. The story of *Bacillus thuringiensis* var. *israelensis* (B.t.i.). *J. Am. Mosq. Control Assoc.* **1:**1–7.
46. **Martin, P. A. W., and R. S. Travers.** 1989. Worldwide abundance and distribution of *Bacillus thuringiensis* isolates. *Appl. Environ. Microbiol.* **55:**2437–2442.

46a. **Moore, W. E. C., and L. V. H. Moore.** 1989. *Index of the Bacterial and Yeast Nomenclatural Changes*. American Society for Microbiology, Washington, D.C.

47. **Musser, J. M., S. J. Barenkamp, D. M. Granoff, and R. K. Selander.** 1986. Genetic relationships of serologically nontypable and serotype strains of *Haemophilus influenzae*. *Infect. Immun.* **52:**183–191.
48. **Nakamura, L. K.** 1987. *Bacillus alginolyticus* sp. nov. and *Bacillus chondroitinus* sp. nov., two alginate-degrading species. *Int. J. Syst. Bacteriol.* **37:**284–286.
49. **Nakamura, L. K.** 1987. Deoxyribonucleic acid relatedness of lactose-positive *Bacillus subtilis* and *Bacillus amyloliquefaciens*. *Int. J. Syst. Bacteriol.* **37:**444–445.
50. **Nakamura, L. K.** 1989. Taxonomic relationship of black-pigmented *Bacillus subtilis* strains and a proposal for *Bacillus atrophaeus* sp. nov. *Int. J. Syst. Bacteriol.* **39:**295–300.
51. **Nakamura, L. K., and J. Swezey.** 1983. Taxonomy of *Bacillus circulans* Jordan 1980: base composition and reassociation of deoxyribonucleic acid. *Int. J. Syst. Bacteriol.* **33:**46–52.
52. **Nicholas, L., J. Dossou-Yovo, and J.-M. Hougard.** 1987. Persistence and recycling of *Bacillus sphaericus* 2362 spores in *Culex quinquefasciatus* breeding sites in West Africa. *Appl. Microbiol. Biotechnol.* **25:**341–345.
53. **Norris, J. R.** 1981. *Sporosarcina* and *Sporolactobacillus*, p. 322–357. *In* R. C. W. Berkeley and M. Goodfellow (ed.), *The Aerobic Endospore-Forming Bacteria: Classification and Identification*. Academic Press, London.
54. **Ochman, H., and R. K. Selander.** 1984. Evidence for clonal population structure in *Escherichia coli*. *Proc. Natl. Acad. Sci. USA* **81:**198–201.
55. **O'Donnel, A. G., and J. R. Norris.** 1981. Pyrolysis gas liquid chromatographic studies, p. 141–179. *In* R. C. W. Berkeley and M. Goodfellow (ed.), *The Aerobic Endospore-Forming Bacteria, Classification and Identification*. Academic Press, London.
56. **O'Donnel, A. G., J. R. Norris, R. C. W. Berkeley, D. Claus, T. Kaneko, N. A. Logan, and R. Nozaki.** 1981. Characterization of *Bacillus subtilis*, *Bacillus pumilus*, *Bacillus licheniformis* and *Bacillus amyloliquefaciens* by pyrolysis gas liquid chromatography, deoxyribonucleic acid-deoxyribonucleic acid hybridization, biochemical tests, and API systems. *Int. J. Syst. Bacteriol.* **30:**448–459.
57. **Pantuwatana, S., and J. Sattabongkot.** 1990. Comparison of development of *Bacillus thuringiensis* subsp. *israelensis* and *Bacillus sphaericus* in mosquito larvae. *J. Invertebr. Pathol.* **55:**189–201.
58. **Pichinoty, F., and J. Asselineau.** 1984. Morphologie et cytologie de *Bacillus benzoevorans*, une nouvelle espèce filamenteuse, engainée et mesophile, dégradant divers acides aromatiques et phénols. *Ann. Microbiol. (Inst. Pasteur)* **135B:**199–207.
59. **Priest, F. G.** 1981. DNA homology in the genus *Bacillus*, p. 33–57. *In* R. C. W. Berkeley and M. Goodfellow (ed.), *The Aerobic Endospore-Forming Bacteria: Classification and Identification*. Academic Press, London.
60. **Priest, F. G.** 1989. Isolation and identification of aerobic endospore-forming bacteria, p. 27–56. *In* C. R. Harwood (ed.), *Biotechnology Handbooks*, vol. 2. *Bacillus*. Plenum Press, New York.
61. **Priest, F. G., and B. Alexander.** 1988. A frequency matrix for the probabilistic identification of some bacilli. *J. Gen. Microbiol.* **134:**3011–3018.
62. **Priest, F. G., M. Goodfellow, L. A. Shute, and R. C. W. Berkeley.** 1987. *Bacillus amyloliquefaciens* sp. nov., nom. rev. *Int. J. Syst. Bacteriol.* **37:**69–71.
63. **Priest, F. G., M. Goodfellow, and C. Todd.** 1981. The genus *Bacillus*: a numerical analysis, p. 91–103. *In* R. C. W. Berkeley and M. Goodfellow (ed.), *The Aerobic Endospore-Forming Bacteria: Classification and Identification*. Academic Press, London.
64. **Priest, F. G., M. Goodfellow, and C. Todd.** 1988. A numerical classification of the genus *Bacillus*. *J. Gen. Microbiol.* **134:**1847–1882.
65. **Priest, F. G., and R. Grigorova.** 1991. Methods for studying the ecology of endospore-forming bacteria, p. 565–591. *In* R. Grigorova and J. R. Norris (ed.), *Methods in Microbiology*, vol. 22. Academic Press, London.
66. **Rhodes-Roberts, M. E.** 1981. The taxonomy of some nitrogen-fixing *Bacillus* species with special reference to nitrogen fixation, p. 315–335. *In* R. C. W. Berkeley and M. Goodfellow (ed.), *The Aerobic Endospore-Forming Bacteria: Classification and Identification*. Academic Press, London.
67. **Robertson, D. L., T. S. Bragg, S. Simpson, R. Kaspar, W. Xie, and M. T. Tippets.** 1990. Mapping and characterization of the *Bacillus anthracis* plasmids pXO1 and pXO2, p. 55–58. *In* P. C. B. Turnbull (ed.), *Proceedings of the International Workshop on Anthrax*. Salisbury medical bulletin no. 68, special supplement. Salisbury Medical Society, Salisbury, England.
68. **Rössler, D., W. Ludwig, K.-H. Schleifer, C. Lin, T. J. McGill, J. D. Wisotzkey, P. Jurtshuk, Jr., and G. E. Fox.** 1991. Phylogenetic diversity in the genus *Bacillus* as seen by 16S rRNA sequencing studies. *Syst. Appl. Microbiol.* **14:**266–269.
69. **Satoh, H., H. Nishida, and K. Isono.** 1988. Evidence for movement of the α-amylase gene into two phylogenetically distant *Bacillus stearothermophilus* strains. *J. Bacteriol.* **170:**1034–1040.
70. **Schenk, A., and M. Aragno.** 1979. *Bacillus schlegelii*, a new species of thermophilic facultatively chemolithoautotrophic bacterium oxidizing molecular hydrogen. *J. Gen. Microbiol.* **115:**333–341.
71. **Scholz, T., W. Demharter, R. Hensel, and O. Kandler.** 1987. *Bacillus pallidus* sp. nov., a new thermophilic species from sewage. *Syst. Appl. Microbiol.* **9:**91–96.
72. **Seki, T., and Y. Oshima.** 1989. Taxonomic position of *B. subtilis*, p. 7–25. *In* B. Maruo and M. Yoshikawa (ed.), *Bacillus subtilis: Molecular Biology and Industrial Applications*. Elsevier, Amsterdam.
73. **Selander, R. K., P. Beltran, and N. H. Smith.** 1991. Evolutionary genetics of *Salmonella*, p. 25–57. *In* R. K. Selander, A. G. Clark, and T. S. Whittam (ed.), *Evolution at the Molecular Level*. Sinauer Associates, Sunderland, Mass.
74. **Seldin, L., J. D. Van Elsas, and E. G. C. Penido.** 1984. *Bacillus azotofixans* sp. nov., a nitrogen-fixing species

from Brazilian soils and grass roots. *Int. J. Syst. Bacteriol.* **34:**451–456.

75. **Sharp, R. J., D. White, and P. W. Riley.** 1992. Heterotrophic thermophilic bacilli, p. 19–50. *In* E. J. Kristjansson (ed.), *Thermophilic Eubacteria*. CRC Press Inc., Boca Raton, Fla.

75a. **Skerman, V. B. D., V. McGowan, and P. H. A. Sneath.** 1989. *Approved Lists of Bacterial Names*. American Society for Microbiology, Washington, D.C.

76. **Smith, N. R., R. E. Gordon, and F. E. Clark.** 1952. *Aerobic Spore-Forming Bacteria*. U.S. Department of Agriculture, Washington, D.C.

77. **Smith, R. A., and G. A. Couche.** 1991. The phylloplane as a source of *Bacillus thuringiensis* isolates. *Appl. Environ. Microbiol.* **57:**311–315.

78. **Somerville, H. J., and M. L. Jones.** 1972. DNA competition experiments within the *Bacillus cereus* group of bacilli. *J. Gen. Microbiol.* **73:**257–265.

79. **Stackebrandt, E., W. Ludwig, M. Weizenegger, S. Dorn, T. J. McGill, G. E. Fox, C. R. Woese, W. Schubert, and K.-H. Schleiffer.** 1987. Comparative 16S rRNA oligonucleotide analyses and murein types of round-spore-forming bacilli and non-spore-forming relatives. *J. Gen. Microbiol.* **133:**2523–2529.

80. **Stackebrandt, E., and C. R. Woese.** 1981. Towards a phylogeny of the actinomycetes and related organisms. *Curr. Microbiol.* **5:**197–202.

81. **Stackebrandt, E., and C. R. Woese.** 1981. The evolution of prokaryotes. *Symp. Soc. Gen. Microbiol.* **32:**1–31.

82. **Staley, J. T., and N. R. Krieg.** 1986. Classification of procaryotic organisms: an overview, p. 965–968. *In* P. H. A. Sneath (ed.), *Bergey's Manual of Systematic Bacteriology*, vol. 2. The Williams & Wilkins Co., Baltimore.

83. **Turnbull, P., J. Kramer, and J. Melling.** 1990. *Bacillus*, p. 187–210. *In* M. T. Parker and D. I. Duerden (ed.), *Topley & Wilson's Principles of Bacteriology, Virology and Immunity*, vol. 2, 8th ed. Edward Arnold, London.

84. **Van Elsas, J. D., J. M. Govaert, and J. A. Van Veen.** 1987. Transfer of plasmid pFT30 between bacilli in soil as influenced by bacterial population dynamics and soil conditions. *Soil Biol. Biochem.* **19:**639–647.

85. **Wayne, L. G., D. J. Brenner, R. R. Colwell, P. A. D. Grimont, O. Kandler, M. I. Krichesvky, L. H. Moore, W. E. C. Moore, R. G. E. Murray, E. Stackebrandt, M. P. Starr, and H. G. Truper.** 1987. Report of the ad hoc committee on reconciliation of approaches to bacterial systematics. *Int. J. Syst. Bacteriol.* **37:**463–464.

86. **Welker, N. E., and L. L. Campbell.** 1967. Unrelatedness of *Bacillus amyloliquefaciens* and *Bacillis subtilis*. *J. Bacteriol.* **94:**1124–1130.

87. **Welker, N. E., and L. L. Campbell.** 1967. Comparison of the α-amylase of *Bacillus subtilis* and *Bacillus amyloliquefaciens*. *J. Bacteriol.* **94:**1131–1135.

88. **White, D., F. G. Priest, and R. J. Sharp.** 1989. Preliminary taxonomic studies on 1000 isolates of thermophilic bacilli, p. 387. *In* M. S. da Costa, J. C. Duarte, and R. A. D. Williams (ed.), *Microbiology of Extreme Environments and Its Potential for Biotechnology*. Elsevier Applied Science, New York.

89. **Williams, A. G., and S. E. Withers.** 1983. *Bacillus* spp. in the rumen ecosystem. Hemicellulose depolymerases and glycoside hydrolases of *Bacillus* spp. and rumen isolates grown under anaerobic conditions. *J. Appl. Bacteriol.* **55:**283–292.

90. **Williams, R. A. D.** 1989. Biochemical taxonomy of the genus *Thermus*, p. 82–97. *In* M. S. da Costa, J. C. Duarte, and R. A. D. Williams (ed.), *Microbiology of Extreme Environments and Its Potential for Biotechnology*. Elsevier Applied Science, New York.

91. **Wisotzkey, J. D., P. Jurtshuk, Jr., G. E. Fox, G. Deinhard, and K. Poralla.** 1992. Comparative sequence analyses on the 16S rRNA (rDNA) of *Bacillus acidocaldarius*, *Bacillus acidoterrestris*, and *Bacillus cycloheptanicus* and proposal for creation of a new genus, *Alicyclobacillus* gen. nov. *Int. J. Syst. Bacteriol.* **42:**263–269.

92. **Woese, C. R.** 1987. Bacterial evolution. *Microbiol. Rev.* **51:**221–271.

93. **Zahner, V., H. Momen, C. A. Salles, and L. Rabinowitch.** 1989. A comparative study of enzyme variation in *Bacillus cereus* and *Bacillus thuringiensis*. *J. Appl. Bacteriol.* **67:**275–282.

2. *Staphylococcus*

RICHARD NOVICK

As Koch recognized in 1878, distinct diseases are produced by gram-positive cocci that have different patterns of growth: in pairs, chains, or clusters. The last group, the staphylococci, are nonmotile, facultatively aerobic, glucose-fermenting, gram-positive cocci distinguished by growth as irregular clusters (Gr. *staphyle*, bunch of grapes; Fig. 1 and 2) and by pentaglycine cross-bridges in their peptidoglycans. Most species are natural inhabitants of the mammalian skin and mucous membranes and have no other important habitat except when involved in infection. Infections caused by staphylococcal species include deep and superficial abscesses, endocarditis, mastitis, osteomyelitis, pneumonia, meningitis, wound infections, and sepsis. In addition, they cause several toxinoses, including food poisoning, toxic epidermal necrolysis, and toxic shock syndrome (TSS).

MORPHOLOGY

The diameter of an individual coccus is 0.7 to 1.2 μm. Cells in old cultures or those ingested by phagocytes may be gram-negative. The characteristic cell clustering (Fig. 2) is most striking on solid media; it arises because staphylococci divide in three successive perpendicular planes and the daughter cells do not separate completely (94). The formation of irregular aggregates is attributable to attachments eccentric to the plane of division as well as to actual movement of cells from the centric position.

GROWTH AND METABOLISM

Staphylococci are facultative aerobes, and growth of some strains is enhanced by increased CO_2 tension. Colonies are opaque, sharply defined, round, regular, and convex. The classic golden yellow of *Staphylococcus aureus* colonies is caused by carotenoids and develops slowly over several days. Pigment is not produced in the presence of glucose (65a), during anaerobic growth or in liquid culture, and nonpigmented variants occur frequently (10^{-2} to 10^{-4} per colony) in many strains.

Most *S. aureus* strains are hemolytic; however, hemolytic patterns are highly variable. Four distinct hemolysins (α, β, γ, and δ) with interstrain variabilities and with different properties and erythrocyte (RBC) species specificities are known.

Staphylococci are among the hardiest of all non-spore-forming bacteria. Some strains are relatively resistant to heat (withstanding 60°C for 30 min) and to most disinfectants.

Most strains will grow readily in a chemically defined medium containing glucose, salts, 14 amino acids, thiamine, and nicotinic acid. The amino acid requirements, however, are complex and poorly understood. Amino acid requirements vary from strain to strain and are often partial. If one of these partially required amino acids is removed, the organism may adapt after a prolonged lag period, suggesting that at least one step in the biosynthetic pathway for that amino acid is only weakly functional.

On complex media, *S. aureus* grows well over a wide range of pH (4.8 to 9.4) and temperature (25 to 43°C), showing a minimum doubling time of 30 to 40 min with vigorous aeration. Under aerobic conditions, catalase is produced and acid is formed from glucose, mannitol, xylose, lactose, sucrose, maltose, and glycerol. *S. aureus* is the only *Staphylococcus* species that ferments mannitol anaerobically. *S. aureus* has a high salt tolerance, and media containing 1.3 to 1.7 M NaCl are used for its selective enrichment.

CLASSIFICATION

The genus *Staphylococcus* contains some 20 distinct species. On the basis of DNA reassociation kinetics, strains of the same species have 80 to 100% sequence identity, whereas different species never have more than 20% identity. Nevertheless, the various species have much in common, can participate broadly in genetic exchange, and probably possess a single common pool of plasmids and transposons. Most of the available molecular and genetic data are for *S. aureus*; other species are increasingly being investigated at the molecular genetic level (66a). Figure 3 is a dendrogram showing the relation of staphylococci to other eubacteria based on 16S rRNA sequence similarities (48). Note especially that micrococci are very distant from staphylococci and that the family *Micrococcaceae*, which formerly included both, is therefore not a valid taxon.

Because pathogenicity has been found to correlate better with the ability to produce coagulase than with pigmentation, all coagulase-positive staphylococci of human origin are now grouped as *S. aureus*. Two groups of animal-specific, coagulase-variable strains have been granted species status: *Staphylococcus intermedius* and *Staphylococcus hyicus*.

Coagulase-negative staphylococci show much greater heterogeneity than do coagulase-positive staphylococci, and as the tools of modern taxonomy have been applied, the number of distinct biotypes that can arguably be regarded as separate species has increased from two in the 1974 edition of *Bergey's Manual* (9a) to 20 in a 1984 listing (24a). Differentia-

Richard Novick • Public Health Research Institute, 455 First Avenue, New York, New York 10016.

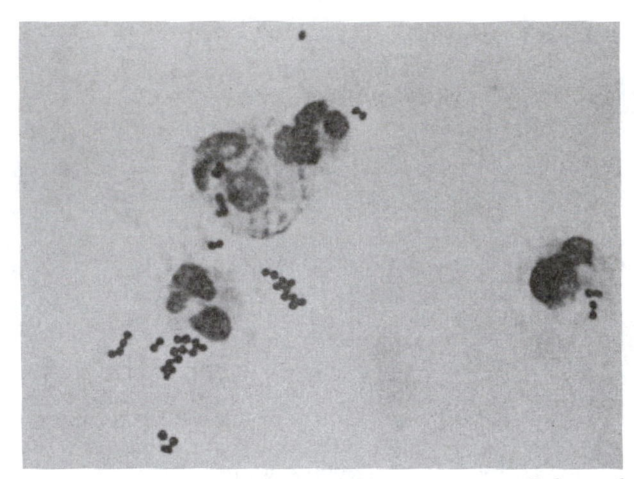

Figure 1. Gram stain of exudate containing intracellular and extracellular staphylococci. Magnification, ×650. (From reference 99a with permission.)

tion of these species on the basis of biochemical characteristics is consistent with the oligonucleotide patterns of the 16S RNA. Nevertheless, there is far from general agreement on the speciation of staphylococci.

EPIDEMIOLOGY AND BIOTYPING OF *S. AUREUS*

As is generally true of bacteria, staphylococci possess a wide spectrum of variable traits that represent variable genes or variably expressed genes. These variations are readily apparent, as they affect many of the organism's most visible phenotypic features, including patterns of exoprotein production, antibiotic resistance, surface antigens, phage sensitivity, etc. Table 1 lists the remarkable assemblage of nonessential, variable traits that are carried by accessory genetic elements and collectively determine the uniqueness of the genus and the individuality of particular strains of any one staphylococcal species. These traits include resistance to antibiotics and other substances and formation of extracellular enzymes, toxins, and other proteins that are involved in pathogenicity. Study of the genotypic basis of these variable traits has revealed variable genes, such as those carried by plasmids, transposons, and other heterologous genetic elements, and variably expressed genes, such as those belonging to the *agr* global regulon. These studies are reviewed below. Independently, variable traits have been used for the development of empirical typing schemes intended for the tracking of strains in a clinical-epidemiological setting. These include the classic phage-typing and serotyping systems and more recently developed systems based on exoprotein patterns and lysotypes. These are all based on unknown genotypic variables, are subject to poorly characterized variations, and are not really satisfactory for their intended purpose. Newer typing systems based on known genotypic variables such as plasmid profiles, chromosomal transposon profiles, restriction fragment length polymorphisms, electrophoretic-mobility variants, ribosomal-DNA polymorphisms, etc., are much more reliable and are gradually replacing the older methods. A particularly promising system is that based on chromosomal transposons, which generate a great deal of diversity by their movement but once established are remarkably stable (Fig. 4) (38).

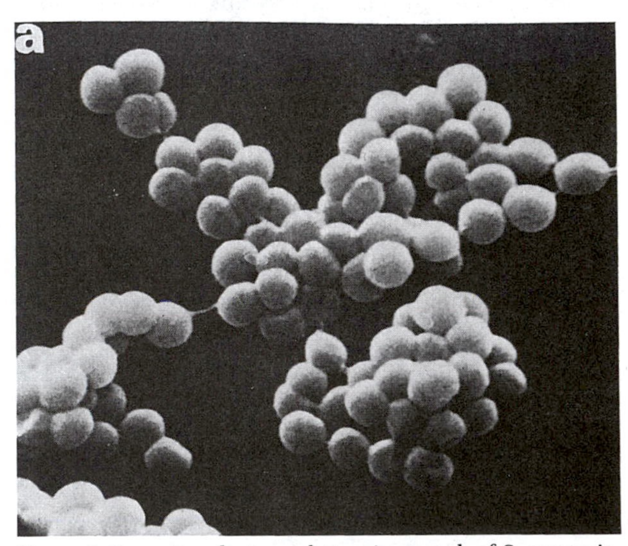

Figure 2. Scanning electron photomicrograph of *S. aureus* in serum-salts broth. (From reference 97a with permission.)

KNOWN OR SUSPECTED DETERMINANTS OF PATHOGENICITY

Staphylococci, particularly *S. aureus*, elaborate a large number of surface factors and secreted proteins, many of which have been implicated in pathogenicity. These are described in the following sections, and their known or suspected roles in infection are outlined thereafter.

Cellular Antigens

Capsules

A few laboratory strains of *S. aureus* have capsules, often composed of glucosaminuronic acid. The prototype encapsulated strain (the Smith strain) gives rise reversibly to nonencapsulated variants (91), implying the existence of a genotypic switch.

Polysaccharide A

Species-specific surface carbohydrate antigens of *S. aureus* and *Staphylococcus epidermidis* (polysaccharides A and B, respectively) have been identified as teichoic acids. Polysaccharide A is a linear ribitol teichoic acid with *N*-acetylglucosamine attached at C-4 of ribitol and with D-alanine attached at approximately 50% of the C-2 atoms. The antigenic determinant is the glucosamine residue, which may be in either α or β glycosidic linkage. Most strains have teichoic acids with both anomers, but some have only one; hence, tests for species identification and antibodies require antisera and teichoic acids, respectively, with both specificities. A glycerol lipoteichoic acid is also present.

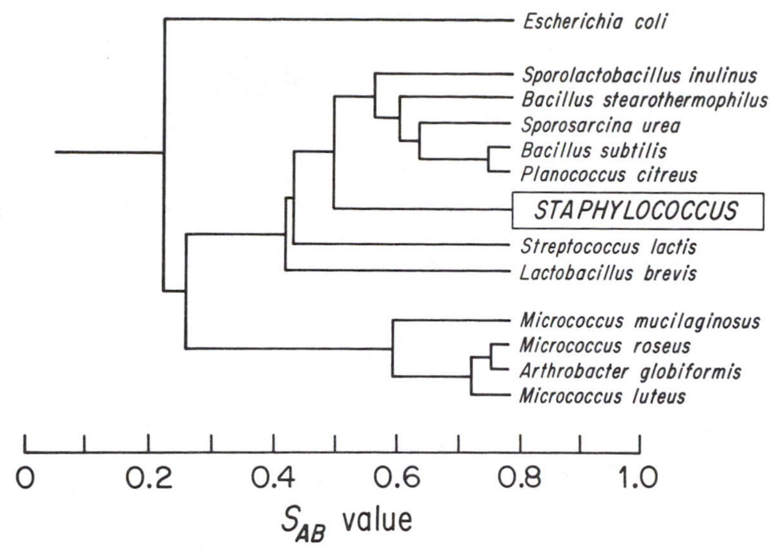

Figure 3. Dendrogram of phylogenetic relationships based on homologies of 16S rRNA. The scale represents the similarity coefficient (S_{AB} value; the percentage of shared oligonucleotides in staphylococcal nuclease digests. (From reference 48 with permission.)

Protein A

Virtually all *S. aureus* strains possess as a surface component the 42-kDa protein A anchored C terminally in the cell membrane by a stop-transfer signal and having the N-terminal region protruding through the wall; some is also released extracellularly. Protein A can induce specific antibodies and will react with their Fab portions. In addition, protein A on the bacterial surface or in solution interacts nonspecifically with the Fc portion of immunoglobulins of virtually all mammalian species. All subclasses of human immunoglobulin G (IgG), except IgG3, and some IgM and IgA2 samples are adsorbed by staphylococci, and the immunoglobulin-protein A interaction has been used for a wide variety of immunological techniques.

Protein A-Fc interactions in vivo have a variety of immunological effects, many of which are manifested as allergic reactions, including anaphylaxis, complement activation, histamine release, and mitogenicity

Table 1. Accessory genetic traits in *S. aureus*

Trait or protein	Location of encoding gene			
	Plasmid	Phage	Chromosomal transposon or HI[a]	Chromosome
Resistance				
Penicillin (β-lactamase)	×		×	
Methicillin			×	
Tetracycline	×		×	
MLS[b]	×		×	
Streptomycin	×			
Spectinomycin			×	
Kanamycin-neomycin	×			
Gentamicin	×		×	
Chloramphenicol	×			
Ethidium bromide	×			
Mercury	×		×	
Arsenate	×		×	
Cadmium	×		×	
Exoproteins and toxins				
α-Hemolysin				×
Staphylokinase		×		
Exfoliatin A				×
Exfoliatin B	×			
TSST-1			×	
Enterotoxin A		×		
Enterotoxin B			×	

[a] HI, inserted DNA segment of heterologous origin.
[b] MLS, macrolide, lincosamide, streptogramin B.

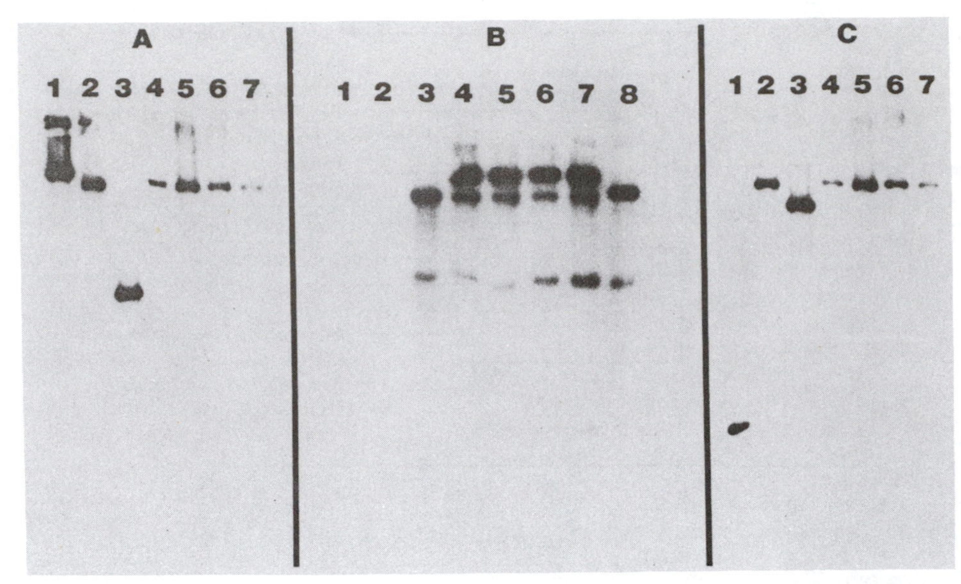

Figure 4. Southern blotting with probes for chromosomal transposons. Lanes 4 through 7 in each panel contain *Cla*1 restriction digests of chromosomal DNA from TSS-causing *S. aureus* strains that are epidemiologically related. Lanes 1 through 3 contain digests of chromosomal DNA from epidemiologically unrelated TSS strains. (A) Probe specific for the β-lactamase gene; (B) probe specific for Tn*554*; (C) probe specific for the TSST-1 gene. (From reference 38 with permission.)

for human B but not T lymphocytes; additionally, while chemotactic factors are generated, phagocytosis is inhibited, presumably because of interference with opsonic (phagocytosis-promoting) antibodies through binding to the Fc receptors of the phagocytes. Recent studies by Foster and coworkers have shown that protein A is an important virulence factor in a mouse mammary infection model (18).

Clumping factor (bound coagulase)

Most nonencapsulated strains of *S. aureus* clump when suspended in plasma or in fibrinogen solutions. It is believed that the clumping factor is a surface coagulase.

Adhesins

Like other bacteria, staphylococci have specific surface proteins (adhesins) that enable them to bind to matrix proteins such as laminin, fibronectin, and collagen and to host cellular surfaces (96). These binding activities are thought to be involved in colonization of the intracellular matrix, invasion of tissue cells, and resistance to phagocytosis (44). A fibronectin binding protein of M_r 197,000 has been purified and is distinct from protein A and clumping factor.

Extracellular Proteins

Staphylococci, especially *S. aureus*, produce a wide variety of immunogenic exoproteins. Some are toxins, some contribute to pathogenicity by attacking the intercellular matrix, and others attack host cells directly. These are accessory proteins, of which most, during optimal growth in rich media, are synthesized at the end of exponential growth or in early stationary phase. Under suboptimal conditions such as Mg^{2+} deficiency, which may occur in infected tissue, they may be synthesized throughout the exponential phase (53).

Coagulase

Culture filtrates of *S. aureus* clot the plasma of many animal species as a result of production of the clotting factor coagulase, the standard marker for *S. aureus*. However, a few wild-type strains (identifiable by DNA hybridization) as well as mutants are coagulase negative. Antigenically distinct coagulases occur, and a metalloproteinase with coagulaselike activity is produced by some coagulase-negative strains.

Clotting requires interaction in plasma with a coagulase-reacting factor, which is probably a derivative of prothrombin: a coagulase–coagulase-reacting-factor complex converts fibrinogen to fibrin. Although the fibrinopeptides released are the same as those released with thrombin, the process differs from normal clotting in that the multiple accessory factors, including Ca^{2+}, are not required and the clot is more friable and does not retract.

Hydrolases

Staphylokinase, like streptokinase and urokinase, causes clot dissolution by activating conversion of the proenzyme plasminogen to the fibrinolytic enzyme plasmin. Its gene resides on a phage (85). The nuclease of *S. aureus* has both endonuclease and exonuclease activities on both DNA and RNA, producing 3′ nucleotides. The lipases are assayed by testing for the ability to produce opacity on egg yolk agar or to split Tween detergents. Lipase production appears to contribute to survival of the organism on skin. Most *S. aureus* strains produce hyaluronidase and one or more proteases; three have been identified. A recently discovered enzyme detoxifies staphylocidal fatty acids by linking them to cholesterol. Lysostaphin, a lytic en-

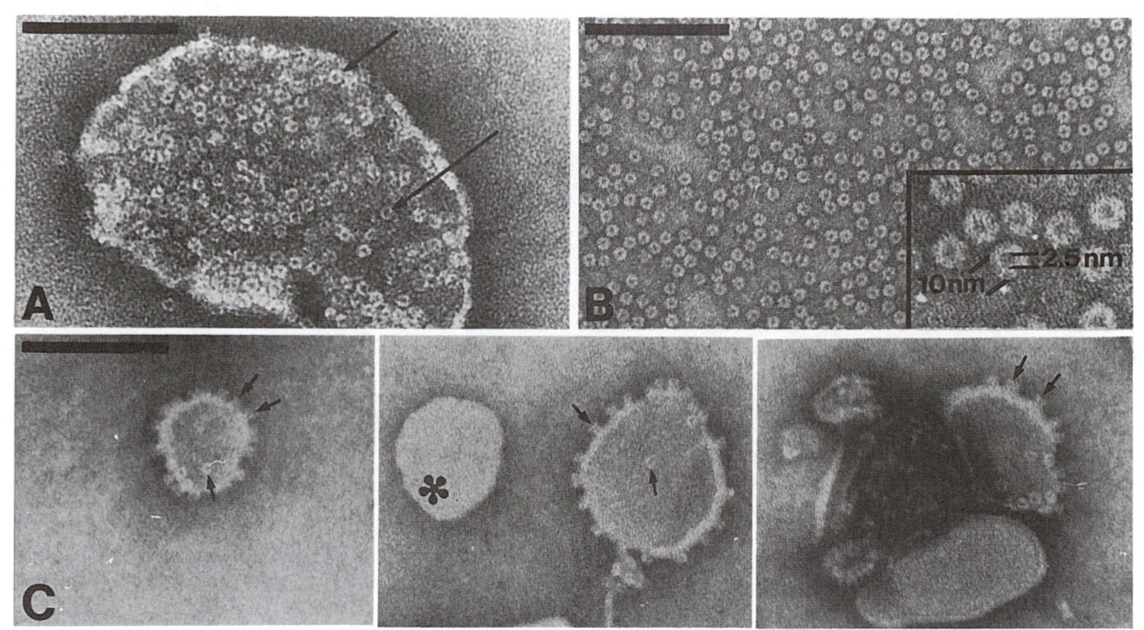

Figure 5. (A) Negatively stained fragment of rabbit RBC lysed with *S. aureus* α-toxin. Numerous 10-nm ring-shaped structures are seen over the membrane (arrows). (B) Isolated α-toxin hexamers in detergent solution. (C) Lecithin liposomes carrying reincorporated α-toxin hexamers. The hexamers are seen as stubs along the edge of the liposomal membrane and as rings over the membrane (arrows). Characteristically, liposomes that escape incorporation of the toxin are impermeable to the stain (asterisk). (From reference 9 with permission.)

zyme produced by a strain of *Staphylococcus simulans*, attacks the pentaglycine bridge between peptidoglycan chains and is thus specific for staphylococci. It is used for diagnostic purposes and is also the mainstay of virtually all research requiring lysis of staphylococcal cells.

Hemolysins

Four different protein hemolysins of *S. aureus* are now recognized; all produce clear β-hemolysis, but they differ in RBC species specificity and mechanism of action. Tissue cells may also be damaged: some of the hemolysins produce local necrosis and are lethal for experimental animals. Individual strains may produce any combination of the four hemolysins.

α-Hemolysin (α-toxin) is a potent cytotoxin and is the principal hemolysin of human strains of *S. aureus*. It is most active against rabbit RBCs; human RBCs are not susceptible, but human platelets and tissue culture cells are affected. In experimental animals, α-hemolysin causes dermal necrosis after local injection and is lethal at the level of a few nanograms per kilogram of body weight when given systemically; the main effect appears to be spasms of vascular smooth muscle. The specific receptor on the RBC membrane is a sialoglycoprotein. α-Hemolysin is secreted as a water-soluble monomer of 34,000 Da that on contact with a membrane receptor rearranges to form a cylindrical hexamer that penetrates the membrane (9). As Fig. 5 shows, this cylinder not only traverses the membrane but also projects above the surface. Negative staining reveals a pore 2 to 3 nm wide on the surface, but its diameter within the membrane is not known. This structure resembles the lytic C5b-9(m) complex of complement.

β-Hemolysin is produced commonly by animal strains but by only 10 to 20% of human isolates. It is a "hot-cold" hemolysin: its lytic effects are not fully developed unless mixtures with blood (or blood agar cultures) are placed at low temperature following incubation at 37°C. β-Hemolysin is a sphingomyelinase C, of M_r 30,000, that is activated by Mg^{2+} but not by Ca^{2+}; it splits sphingomyelin into *N*-acylsphingosine and phosphorylcholine. Sheep, human, and guinea pig RBCs, in that order, contain decreasing amounts of sphingomyelin and are decreasingly sensitive. β-Hemolysin is cytotoxic for a variety of tissue culture cells, and large doses are toxic for experimental animals.

γ-Hemolysin consists of two basic proteins acting in concert. Rabbit, human, and sheep RBCs are susceptible, whereas horse and fowl RBCs are not. Agar and other sulfated polymers inhibit γ-hemolysin, and so it is not active on blood agar plates. Cholesterol and many other lipids are also inhibitory.

δ-Hemolysin is an amphipathic 26-residue peptide that is secreted as a primary gene product (17). It occurs as heterogeneous aggregates with subunits of M_r 5,000. It acts, possibly as a direct surfactant, on various cell types, including RBCs, leukocytes, cultured mammalian cells, and bacterial protoplasts, and is not species specific. δ-Hemolysin is produced by most human strains of *S. aureus* and is frequently produced by pathogenic *S. epidermidis* strains (87). β-Hemolysin and δ-hemolysin are strikingly synergistic, and this synergism is the basis for the classic Elek-Levy test (16), which identifies both. Other hemolysins also exhibit synergism with β-hemolysin, and one such is produced by *Staphylococcus lugdunensis*. This however, is a 46-kDa peptide rather than the standard δ-hemolysin (87a).

Pyrogenic exotoxins

S. aureus strains elaborate a series of toxins that have in common several pathogenic activities: they are pyrogenic, at least partly as a consequence of interleukin-1 induction; they are immunosuppressive, owing to potent mitogenicity for suppressor T-lymphocytes; they dramatically enhance the toxicity of gram-negative endotoxin by blockading the clearance function of the reticuloendothelial systems; and they cause erythroderma by evoking delayed hypersensitivity. They are generally not serologically cross-reactive, but an individual sensitized by any one of them will show delayed hypersensitivity to each of the others. Included in this group are staphylococcal pyrogenic exotoxins A and B, staphylococcal enterotoxins (of which there are five distinct serotypes), and TSST-1 (toxic shock syndrome toxin 1), the cause of toxic shock syndrome. Some 50% of *S. aureus* isolates produce one or more enterotoxins, 15% produce TSST-1, and a smaller fraction produce pyrogenic exotoxin A or B. Other toxins in this group are produced by *Streptococcus pyogenes*, including the classic erythrogenic toxin and the closely related (or identical) toxin SPEA, which is the cause of streptococcal toxic shock. A large majority of the TSST-1-producing *S. aureus* strains have a characteristic biotype that includes sensitivity to typing phage 29 or 52 or both and the presence of chromosomally located resistances to Cd^{2+}, AsO_4^{2-}, and penicillin.

Other exotoxins

Panton-Valentine leukocidin, produced by most *S. aureus* strains, acts only on human and rabbit polymorphonuclear cells and macrophages. It has two components (F and S). Component S first binds to ganglioside GM_1 (the cholera toxin receptor) and activates an endogenous membrane-bound phospholipase, A_2. The products then bind component F, inducing a K^+-specific ion channel in the membrane and hence cytolysis.

Exfoliatin (epidermolytic toxin) causes a variety of dermatologic lesions. This relatively heat-stable and acid-labile protein of M_r 24,000 is produced by approximately 5% of *S. aureus* strains, mostly of phage group II. It occurs as two antigenic variants: ETB, occurring primarily in strains of phage group II, is plasmid coded; ETA, produced by strains of various phage types, is chromosomal (100). Many strains produce both types. The toxin acts by cleaving the stratum granulosum of the epidermis, probably by splitting desmosomes that link the cells of this layer.

Antibiotic resistance

Although antibiotic resistance is not, strictly speaking, a determinant of pathogenicity, it has played such a prominent role in clinical staphylococcal disease and is such an important feature of the organism in its clinical existence that it seems appropriate to include a brief description here.

Soon after antibiotics were introduced into clinical medicine, resistant strains of *S. aureus* appeared among clinical isolates, and the organism has continued to respond in this manner to the introduction of new drugs. Penicillin resistance mediated by a powerful β-lactamase appeared in the early 1950s and was rapidly followed by resistance to macrolide antibiotics, aminoglycosides, and tetracyclines. During succeeding years, the frequency of resistant strains increased rapidly, and in many clinical settings, multiple antibiotic resistance is now the rule. The determinants of these resistances have probably always existed in natural populations of staphylococci and other bacteria, a fact that could account for the rapidity with which the resistant organisms became prevalent. For example, some 5 to 10% of stored pre-antibiotic-era *S. aureus* isolates show β-lactamase-mediated penicillin resistance. The introduction of "penicillinase-resistant" penicillin derivatives such as methicillin around 1960 was followed shortly by the emergence of resistant strains, and these are now one of the principal causes of therapeutic failure in hospital-acquired infections. Methicillin resistance (*mec*) is caused by a new, possibly multifunctional penicillin-binding protein, PBP-2a (24), that has much lower affinity for β-lactam compounds than the other penicillin-binding proteins and has been added to the standard set, presumably by acquisition from some other species. Molecular fingerprinting analysis suggests that *mec* was acquired just once by *S. aureus*, probably around 1960, and that all extant strains are descendants of this initial clone (36b). The *mec* determinant resides on a complex inserted element of 30 to 40 kb that is associated with determinants of resistance to mercury and to antibiotics other than methicillin, including tetracycline and erythromycin (51). As with other bacteria, individual staphylococcal strains tend to accumulate multiple determinants of resistance even under conditions in which they are exposed to only a single antibiotic. Although plasmids and transposons are certainly involved, the actual evolutionary mechanism underlying this phenomenon has yet to be explained; one of its consequences has been the emergence of epidemic hospital strains of *S. aureus* resistant to virtually all useful antibiotics, including methicillin but thus far excepting vancomycin. These strains are currently a significant cause of nosocomial infections in many parts of the world.

ROLE IN PATHOGENICITY

Although the precise roles of individual bacterial factors in staphylococcal disease have not been clearly defined, virtually all of the toxic and enzymatic products have been implicated. α-Toxin is lethal for animals and seems likely to play a role in overwhelming septicemic disease, both δ-hemolysin and Panton-Valentine leukocidin destroy human polymorphonuclear leukocytes (which may be why *S. aureus* is killed less efficiently than coagulase-negative organisms by phagocytosis), lipase may be important in the development of boils, other enzymes detoxify bactericidal fatty acids released from lipids in response to the infection, and coagulase probably contributes to the localization and persistence of the lesions by walling them off from phagocytic cells (although a fibrin coating on individual bacteria does not impede their phagocytosis). Other factors seem to enhance the spread of lesions: hyaluronidase breaks down interstitial hyaluronic acid, proteases degrade collagen and elastin, and staphylokinase causes clot lysis, antagonizing the action of coagulase.

Consistent with such differential effects are observations that strains isolated from boils are high in lipase and low in hyaluronidase, whereas the reverse is true of strains isolated from spreading lesions such as bullous impetigo. On the other hand, strains that lack one or more of these several substances have been isolated from lesions, and similar mutants appear to be no less virulent in experimental infections.

The role of bacterial surface antigens is also unclear. The rare constitutively encapsulated strains are more virulent for animals (no data are available for humans), probably because encapsulated staphylococci are relatively resistant to phagocytosis. There is suggestive evidence that other strains form capsules in vivo and lose them on cultivation in vitro, so that capsule formation may be more important in pathogenesis than is currently appreciated. Protein A, displaying a wide variety of biological effects consequent to its interactions with immunoglobulins (particularly inhibition of phagocytosis), has recently been shown to have a clear role in virulence. Surface proteins (adhesins), which enable the organisms to bind to various tissue components and cell surfaces, may impede phagocytosis, are likely to be involved in colonization, and may also be involved in invasiveness. Data linking them specifically to virulence are not yet available.

Because no single factor is decisive, virulent and avirulent strains of *S. aureus* cannot be defined sharply, as can rough and smooth pneumococci or Tox^+ and Tox^- strains of *Corynebacterium diphtheriae*. Even epidemic strains demonstrate no qualitative or quantitative difference in presumed virulence factors and indeed may be less virulent for animals than conventional strains.

Even more problematic is the mechanism of pathogenicity of *S. epidermidis* and other opportunistic coagulase-negative species, which have recently equalled and sometimes surpassed *S. aureus* as nosocomial pathogens. These strains, which are especially adept at colonizing indwelling catheters and vascular prostheses, generally do not elaborate any of the factors associated with *S. aureus* virulence. They possess adhesins and often produce an extracellular "slime" that may impede phagocytosis. Whether these factors are sufficient to account for pathogenesis is presently unknown.

With *S. aureus*, certain conditions (toxinoses) are caused by preformed toxins and do not require the presence of viable organisms. In these cases, pathogenicity is much clearer than in infections. Important examples include staphylococcal food poisoning, caused by preformed enterotoxin in the food; exfoliative skin disease (e.g., impetigo, scalded skin syndrome), caused by the exfoliative toxins; and TSS, caused by TSST-1 and probably also by several of the enterotoxins. Additionally, evidence that δ-hemolysin may be responsible for neonatal necrotizing enterocolitis, a life-threatening condition primarily affecting premature infants and often caused by δ-hemolysin-producing coagulase-negative staphylococci, has recently been obtained (87).

GENETICS

The staphylococcal genome consists of a chromosome of 2.8×10^6 nucleotide pairs plus various accessory genetic elements including plasmids, transposons, prophages, and uncharacterized chromosomal insertions of heterologous DNA. The chromosome of strain NTCC8325 has been mapped, single-handedly, by Pattee and coworkers, who used both genetic and physical methods (77). Some 100 loci have been mapped, including phenotypic markers, silent transposon insertions, and most of the 16 *Sma*I restriction sites. Several biosynthetic operons have been defined, and these resemble their counterparts in other bacteria. The genotypic basis for absolute or partial mutational requirements has not been determined. A recent version of this map is presented in chapter 34 of this volume.

PLASMIDS

S. aureus

Most naturally occurring *S. aureus* strains contain plasmids ranging in size from approximately 1 to 60 kb and falling into four general classes (recently reviewed by Novick [66] and listed in Table 2). Class I consists of small (1- to 5-kb) multicopy (15 to 60 copies per cell) plasmids that either are cryptic or carry a single resistance determinant. Rarely, a plasmid carries two markers. Markers include those for tetracycline, erythromycin, chloramphenicol, streptomycin, neomycin, bleomycin, quaternary amine, and cadmium. Twelve of these plasmids have been sequenced and assigned to four families on the basis of homology of replicon functions. Plasmids of this class replicate by an asymmetric rolling-circle mechanism similar to that of the filamentous single-stranded coliphage. This mechanism is used by all known small plasmids from gram-positive bacteria (23, 66). All encode initiator (Rep) proteins that act by introducing a site-specific nick in the leading-strand replication origin (34), and all regulate copy numbers at the level of initiator synthesis. In the best-studied of these plasmids, the pT181 family, this regulation is accomplished by antisense RNAs, or countertranscripts, that cause attenuation of the Rep protein mRNA (70). Each plasmid contains a large dyad element that serves as the initiation signal for lagging-strand replication (22) and includes a site (RS_B) for sequence-specific interplasmid cointegrate formation mediated by an unknown host function (72). Some members of this class also contain a site-specific recombination function, *pre* (19), that acts at a different recombination site, RS_A. Most of the currently available cloning vectors for gram-positive bacteria are derived from these plasmids (see below). A functional map of the prototype of this class, pT181, is presented in Fig. 6.

Class II plasmids are larger (15 to 30 kb); have lower copy numbers (four to six per cell); and carry some combination of resistance to β-lactam antibiotics (β-lactamase), macrolides, and a variety of heavy metal

Table 2. Staphylococcal plasmids[a]

Class(es)	Family	Plasmid	Copy no.	Size (kb)	Incompatibility group	Phenotype[b]
I[c]	pT181	pT181	22	4.4	3(C)	Tc^r
		pT127	50	4.4	3	Tc^r
		pC221	22	4.6	4(D)	Cm^r
		pC223		4.6	10(J)	Cm^r
		pUB112		4.1	9(I)	Cm^r
		pS194	22	4.4	5(E)	Sm^r
		pCW7		4.2	14(N)	Cm^r
	pC194	pC194	15	2.9	8(H)	Cm^r
		pUB110	10	4.5	13(M)	Km^r Bl^r
		pOX6		3.2		Cd^r
		pRBH1[d]				Km^r
		pBC16[d]		4.5	13(M)	Tc^r
	pSN2	pSN2	50	1.3		Cryptic
		pTCS1		1.3		Cryptic
		pE12	10	2.2	12(L)	Em^r
		pIM13	10	2.1	12(L)	Em^r
		pE5		2.1		Em^r
		pT48		2.1		Em^r
		pNE131		2.1		Em^r
	pE194	pE194	55	3.7	11(K)	Em^r
II and III	IIα	pI524	5	31.8	1(A)	Pc^r Cd^r Pb^r Hg^r Om^r Asa^r Asi^r Sb^r Bin$^+$
	IIα	pI258	5	28.2	1(A)	Asa^r Em^r Bin$^-$
	IIβ	pII147	5	32.6	2(B)	Pc^r Cd^r Pb^r Hg^r Om^r Bin$^-$ Asa^r Bi^{hs}
	IIα	pI9789		19.7		Cd^r Pb^r Hg^r Om^r Asa^r Asi^r Sb^r
	III	pG01		52		Gm^r Tp^r Tra$^+$ Eb^r Qa^r

[a] Only those plasmids mentioned in the text are listed here (25, 26, 29, 30, 32, 33, 39, 40, 52, 55, 56, 78, 79, 84, 92). An extensive list has been presented by Lyon and Skurray (50).

[b] Abbreviations: Pc, penicillin; Cd, cadmium ion; Pb, lead ion; Hg, mercuric ion; Om, organomercurial compounds; Asa, arsenate ion; Asi, arsenite ion; Sb, antimonyl ion; Bin, site-specific inversion; Em, erythromycin; Bi, bismuth ion; hs, hypersensitivity; Gm, gentamicin; Tp, trimethoprim; Eb, ethidium bromide; Qa, quaternary amines; Tra, conjugative proficiency; Tc, tetracycline; Cm, chloramphenicol; Sm, streptomycin; Km, kanamycin; Bl, bleomycin.

[c] Class I plasmids are listed here only if their sequences differ significantly or if they are discussed in the text.

[d] *B. subtilis* plasmids listed because of their close relation to pUB110 (see text).

ions (arsenic, cadmium, lead, and mercury) (73), some of which are known or predicted to be transposable. Most of the resistance genes are inducible. A vector system based on the pI258 *bla* determinant has been developed (97). These plasmids also encode initiator proteins (103) but use the theta replication mechanism rather than the rolling circle (88).

Class III consists of considerably larger (30- to 60-kb) plasmids that carry a determinant of conjugative transfer (*tra*) plus some combination of resistance markers, including those for gentamicin, penicillin, quaternary amines, and trimethoprim, some of which are transposable, and a number of insertion element (IS)-like sequences. Plasmids that appear to be composites or recombinants of members of these three classes and plasmids that appear to have resulted from interplasmid transposon movement have been isolated (21, 50). The map of a typical conjugative class III plasmid, pG01 (92), is presented in Fig. 7.

A few plasmids that do not seem to belong to any of these three well-defined classes have been identified. These have not been studied in any detail and are provisionally placed in a fourth class (66).

Other Species

Plasmids similar or identical to *S. aureus* plasmids of classes I through III are common in a variety of coagulase-negative staphylococcal species (20, 81, 99). Additionally, plasmids of classes I and III can be readily transferred by protoplast transformation (class I) or conjugation-mobilization (classes I and III; see below) among various staphylococcal species. Plasmids of class I have been found throughout the gram-positive bacteria, and these can readily be transferred among all gram-positive species tested.

Many of the cloning vectors that have been developed for gram-positive bacteria are based on the small multicopy plasmids from *S. aureus*. Several of these are listed in Table 3 and described in Fig. 8.

BACTERIOPHAGES

Most *S. aureus* strains are multiply lysogenic, and the temperate phages are usually UV inducible and typically integrate at unique chromosomal sites by the Campbell mechanism (43). Phages have morphologies typical of temperate phages from other species and fall into three main serological groups, A, B, and F (2), of which group B contains most of the known transducing phages, including ϕ11, ϕ147 (unpublished data), and typing phages 53, 79, 80, and 83. This grouping is well correlated with DNA sequence similarities (27). ϕ11, a prototypical group B transducing phage (65), has a latency of $\approx$60 min and a burst size

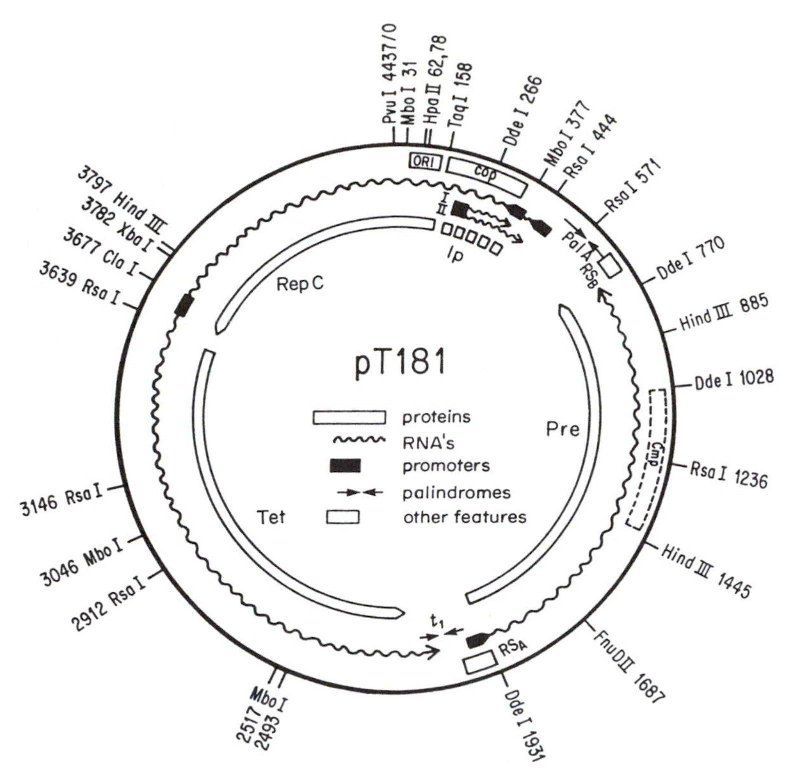

Figure 6. Functional genetic map of pT181. Important restriction sites are given with nucleotide positions. Wavy lines represent known transcripts; solid blocks represent known promoters; heavy lines represent reading frames known to encode proteins; lp, putative *repC* leader peptide. Functional elements (in counterclockwise order: cop, copy control; ORI, replication origin; RepC, initiator protein coding sequence; Tet, tetracycline resistance determinant (whether there are one or two *tet* genes is still uncertain); t_1, probable termination signal for *tet*; RS_A, recombination site A, *pre* promoter; Pre, *pre* coding sequence; Cmp, competition determinant; RS_B, recombination site B; Pal A, palindrome A; *repC* and countertranscript promoters.

of ≈250, and it requires Ca^+ for growth as well as adsorption. Host protein synthesis is shut down 30 to 40 min after prophage induction (10). Temperature- and suppressor-sensitive as well as clear-plaque and virulent mutants have been isolated (39, 65; unpublished data). Lysogenization frequency is low (between 1 and 10%) and can be increased to >90% by the addition of a growth-inhibitory concentration of chloramphenicol (e.g., 5 to 10 μg/ml) (unpublished observations). Its 45-kb genome is circularly permuted, terminally redundant, and flush ended (46). It has been restriction mapped (5, 46), and early and late regions plus a number of individual genes, including a late switch gene that is required for late protein synthesis and shutdown of host protein synthesis, have been identified (10, 39). Its *pac* site has been mapped (5), and it is assumed that packaging is by sequential headfuls. Available physical and genetic data are summarized in Fig. 9. Virulent T-like phages (82) but no lambdalike *cos*-containing phage have been described for staphylococci.

Several examples of lysogenic conversion have been documented; prophages have been found to carry the genes for enterotoxin A (8) and staphylokinase (35) and also transposons (63) and class I plasmids (28). Additionally, prophages have been observed to cause negative lysogenic conversion, owing to the presence of phage attachment sites within certain structural genes. Two well-studied examples are the lipase (41) and β-hemolysin (13) structural genes.

TRANSPOSONS

A variety of transposons have been described, each specifying resistance to one or two antibiotics. These transposons and their important features are listed in Table 4 (see Murphy [58] for a review). Tn*551* from *S. aureus*, its close relative Tn*917* from *Streptococcus faecalis*, and Tn*552* and its relatives are class II transposons in the Tn*3* family and transpose to more or less random sites, creating 5- or 6-nucleotide target duplications. Tn*4001* (and presumably Tn*3581*) are class I transposons, similar in organization to Tn*5* or Tn*10*. A directly repeated 1.35-kb DNA element flanking the *mer* operon in plasmid pI258 has been designated IS*431* (6). Homologous insertion sequences are widespread in *S. aureus*, and the flanking repeats of Tn*4001*, designated IS*256* (49), are very similar or identical to IS*431* (6). Both IS*431* and IS*256* have been assumed to be independently transposable, justifying their designation as IS. Both Tn*551* and Tn*917* carry the classic macrolide-lincosamide-streptogramin B determinant, which is constitutive in Tn*551* and inducible in Tn*917*. Tn*551* and Tn*917* have strong hot spots, fortunately different, thus tending to favor dif-

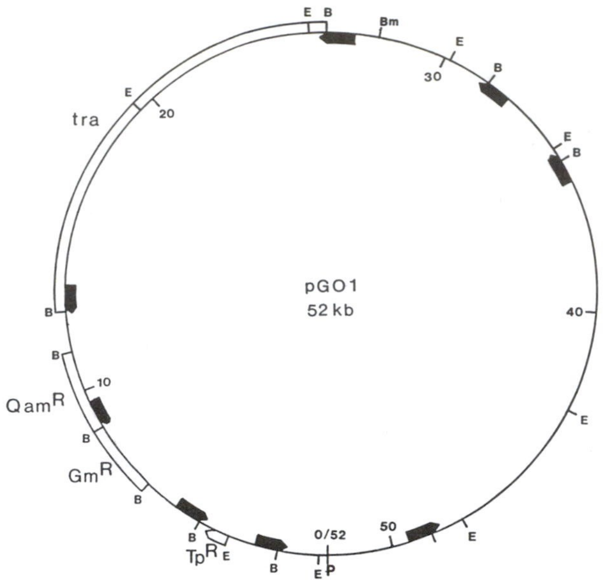

Figure 7. Map of staphylococcal conjugative plasmid pG01. Kilobase coordinates begin at 6 o'clock at the single *Pst*I site (P). Dark arrows are IS-like elements; direction of each arrow indicates the direction of transcription of the single open reading frame on IS*431*, to which these IS-like elements are homologous. Open boxes are antimicrobial resistance genes or the conjugative transfer (*tra*) region. The arrow at the end of the box designating the trimethoprim resistance gene (Tp^R) indicates the direction of transcription. The arrow at 5 o'clock indicates the specific *Eco*RI fragment in which a mobile element encoding penicillinase (*bla*) inserts. While the element is not present on pG01, its insertion site has been mapped on other homologous conjugative plasmids. Other abbreviations: Gm^R, gentamicin resistance; Qam^R, quaternary ammonium-ethidium bromide resistance. Restriction endonuclease cleavage sites are *Eco*RI (E), *Bgl*II (B), and *Bam*HI (Bm).

ferent target sites (77). Tn*916* uses a different set of sites (31). Tn*4001* and Tn*3581*, carrying Gm^r Km^r Nm^r markers, use target sites not highly preferred by Tn*551* and Tn*917* (77). Transposon delivery is usually via thermosensitive replication-defective (Tsr) vectors such as pRN3032 (69) and pRN5101 (67). Transposition to the chromosome occurs at a frequency of 10^{-3} to 10^{-5} per cell; class II and class III plasmids are good targets for many of these transposons. Tn*551* seems unable to transpose to class I plasmids (21a).

Tn*552* and Related Transposons

Tn*552* is a 6.7-kb β-lactamase transposon of the Tn*3* type that is found intact or as remnants at a variety of plasmid and chromosomal locations (83). Transposition of intact transposons of this type has been observed in only a very small number of cases (98). As with Tn*3*, Tn*552* resolvase (*bin*) is encoded in a region flanking the *res* site at which it acts (83). This *bin-res* complex, present in certain plasmids and on the chromosome of certain *S. aureus* strains without the rest of the transposon, serves as a hot spot for Tn*552* transposition, first described by Asheshov (4). The insertion of Tn*552* in this region may create an inverted repeat of the *bin-res* element, and in such cases, one observes site-specific *rec*-independent inversion at high frequency (62). Additionally, the *bin-res* complex can mediate high-frequency reversible cointegrate formation between elements that carry it (61).

Tn*554* and Tn*3582*

Tn*554* is an *S. aureus* transposon that carries Em^r and Sp^r markers and has several remarkable features. It transposes with a frequency approaching 100% to a single chromosomal site (63) and neither contains terminal repeats nor generates target duplications (60). It blocks transposition of a second copy by occupation of its *att* site and also by a specific, *trans*-acting interference determinant located at one end of the transposon (57). In naturally occurring strains, it has been found at several additional (2°) sites, and it can transpose to a 2° site on class II plasmids at a frequency of $<10^{-8}$ (63). This 2° site is related to one of the chromosomal secondary sites. The closely related Tn*3582* occurs naturally on a conjugative plasmid and behaves as a hitchhiking transposon (93), by which is meant that following entry of the carrier plasmid by conjugation, it immediately transposes to its chromosomal *att* site. Tn*554* has been sequenced, and three genes involved in transposition have been identified (7, 59). Two of these genes are related to the lambda Int family of recombinases (3).

OTHER VARIABLE GENETIC ELEMENTS

Variable genetic elements occur in some but not other strains of a given species. Aside from plasmids, prophages, IS, and transposons, a number of variable

Table 3. Cloning vectors

Plasmid	Size (kb)	Copy no.	Useful features	Reference
pRN5543	2.9	90–100	pUC19 polylinker in unique *Hin*dIII site	80
pRN6441	3.7	50–60	pUC18 polylinker in unique *Pst*I site[a], Tsr[b]	19a
pRN6725	4.9	90–100	pSK265 with β-lactamase promoter + 2/3 of *blaZ* gene cloned to *Hin*dIII-*Xba*I sites of polylinker	65a
pWN1818 and pWN1819	10.5		Derivatives of pWN101 with *bla* promoter and Shine-Dalgarno sequence replaced by pUC18 or pUC19 polylinker	97
pHV33	7.3		pC194-pBR322 cointegrate-shuttle vector	78a
pEM9940	12	2–5	Tn*554* carried by pT181 (Tsr)	56a
pLS1	4.4		Replicates in gram-positive and -negative species	39a

[a] The cloned polylinker is flanked by *Pst*I sites.
[b] Temperature-sensitive replication.

Table 4. Transposon delivery vectors

Plasmid	Replicon	Features	Transposon	Frequency	Selection for jumps	Reference
pRN3206	pI258	Tsr, Cd^r	Tn*551*	10^{-4}–10^{-5}	Em 43°	65b
pLTV1	pE194	Tsr, Tc^r	Tn*917*	10^{-4}–10^{-5}	Em 43°	104
pEM9940	pT181	Tsr, Tc^r	Tn*554*	1.0	Em	56a
pPQ61	pI258	Tsr, Cd^r	Tn*4001*, Tn*551*	10^{-4}–10^{-5}	Em 43° Gm 43°	76a

chromosomal determinants have been characterized. Typically, these contain one or more identifiable markers flanked by additional unique sequences. Three examples in *S. aureus* are the determinants of TSST-1, staphylococcal enterotoxin B, and Mc^r. Additionally, a large set of probably linked chromosomal antibiotic resistance genes has been identified in a series of widely dispersed Mc^r hospital isolates sensitive to typing phage 88 (86). Among the variable elements, the TSST-1 element may occupy at least two different chromosomal locations (11). These elements may turn out to be transposons, prophages, or integrated plasmids; however, presently available data are insufficient to permit any such specific identification, and the elements have provisionally been designated heterologous insertions (11). Variable genetic elements that lack phenotypic markers are provisionally designated IS sequences. Two examples have been identified in staphylococci (see, for example, Fig. 7). Elements with sequences very similar or identical to those of the flanking repeats of Tn*4001* have been found in various plasmid and chromosomal locations without the rest of the transposon and are referred to collectively as IS*256* (92). A second IS-like element (IS*257*/IS*431*) has also been identified at various locations (6, 92). To date, neither of these has been observed either to transpose or to promote plasmid cointegrate formation.

GENETIC EXCHANGE

Staphylococci participate in several different types of genetic exchange, including generalized transduction, DNA-mediated transformation of either intact cells or protoplasts, plasmid-mediated conjugation, and protoplast fusion. Conjugative transfer has been demonstrated for plasmid but not chromosomal markers. Staphylococci also exchange plasmids with *Bacillus* species and streptococci. Electroporation is being used with increasing frequency, especially for plasmid transfer. Staphylococcal strains often show strong restriction barriers to interstrain and interspecific transfer; various restriction enzyme determinants have been identified, and a number of staphylococcal restriction enzymes have been developed for research. Restrictionless mutants have been isolated and found to be good recipients for genetic transfer. Of these, one in particular, RN4220 (37), is able to accept DNA from *Escherichia coli*.

General and Site-Specific Recombination

The *S. aureus recA* gene, identified by mutation and characterized in the early 1970s (102), has recently been cloned and found to be homologous with *recA* determinants of other eubacteria (15a). RecA$^-$ mutants show residual generalized recombination frequencies of about 10^{-4}-fold, implying the existence of other *rec* systems. Several examples of site-specific recombination have been described, including phage integration (which closely resembles that seen in other bacteria) (42, 43), reversible inversion (62), and site-specific plasmid cointegrate formation (72) mediated by plasmid-coded recombinases.

REGULATION OF GENE EXPRESSION

β-Lactamase

The first inducible gene to be described in *S. aureus*, namely, that for β-lactamase, has a regulatory system that is parallel to that of *Bacillus lichenformis* and similar to that of *Bacillus cereus*. This mechanism appears to involve a signal transduction pathway in which one regulatory element, the putative transmembrane penicillin-binding protein BlaR1, binds the inducer externally and evidently activates an internal regulatory protein, BlaR2, encoded by an unlinked gene. There is also a classic repressor encoded by *bla*I, which presumably binds to the *bla* operator, a strong hairpin including part of the promoter. It is assumed that the active form of BlaR2 directly interacts with the repressor, activating transcription of the gene. Induction of β-lactamase in certain gram-negative bacteria has recently been demonstrated, appears to be even more complex, and may also involve signal transduction (45).

lac

The only classic regulatory system that has been studied in *S. aureus* is the *lac* system (recently reviewed by Oskouian et al. [75]). The lactose pathway is substrate inducible and catabolite repressible, though as with other gram-positive bacteria, catabolite repression does not involve cyclic AMP. The system consists of two transcription units, of which one contains the repressor (*lacR*) gene and the other contains the seven metabolic genes, each preceded by a strong Shine-Dalgarno site, as is typical for genes of gram-positive bacteria. Lactose is phosphorylated at the galactose-6 position during transport by the phosphotransferase system and then split by phospho-β-galactosidase to glucose and galactose-6-phosphate, the internal inducer. *E. coli* β-galactosidase does not cleave lactose-phosphate and consequently does not function in *S. aureus*.

The repressor, LacR, contains a putative helix-turn-helix motif and possesses significant homology with DeoR, the repressor of the *deo* operon, which also uses

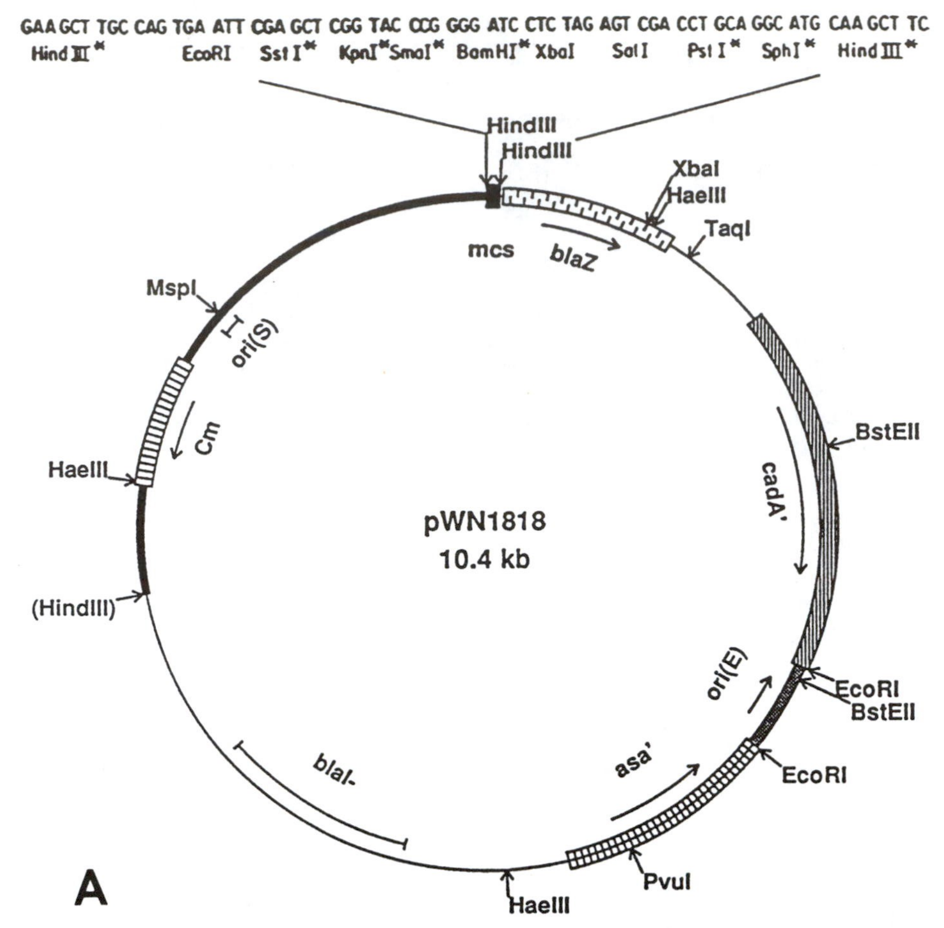

Figure 8. Maps of cloning vectors useful for *S. aureus*. (A) pWN1818 (97). pWN1818 is a promoter-probe vector derived from pA07 (74), a fusion of *cadA* (*blaZ*) containing the 6.9-kb *Eco*RI-B fragment of p1258 (71) to a 0.5-kb segment containing the ColE1 origin, ori(E) (12 to 9 o'clock). pC194 (9 to 12 o'clock) was inserted into the unique *Hin*dIII site of pA07, the *bla* promoter was eliminated by *Bal* 31 digestion, and finally, the pUC18 polylinker (Mc^S) was inserted as shown (97). Versions of this vector with stop codons in all three reading frames, with the polylinker in the opposite orientation, or with the *bla* Shine-Dalgarno sequence deleted are available. (HindIII), a *Hin*dIII site that has been filled in and religated; cadA' and asa', truncated *cadA* and *asa* genes; blaI, *bla* control region with a constitutive mutation. (B) pRN6725 (36). pRN6725 is a promoter vector derived by inserting the *Hin*dIII-*Xba*I fragment containing the *bla* promoter plus two-thirds of the *blaZ* structural gene into the polylinker region of pRN5543, a pC194 derivative. DNA inserted into the polylinker is transcribed from the *bla* promoter, which can be made inducible by providing the *bla* repressor in *trans*. P-bla, β-lactamase promoter; blaZ', truncated *blaZ*; palB, dyad region of unknown function; rep, *rep* gene of pC194; ori, leading-strand replication origin; cat, Cm^r gene; palA, lagging-strand replication origin; ori, origin. (C) Protein A fusion vectors pRIT16 and pRIT21-23. These vectors were constructed by cloning to pEMBL9 a 1.1-kb fragment containing a 3' truncated *spa* gene lacking the coding region for the C-terminal membrane-spanning domain of protein A and containing the pUC18 polylinker sites just 3' to the *spa* fragment. Staphylococcal plasmid pC194 was then inserted between ColE1 and *spa* regions to give pRIT16, and a transcription termination signal (T) was inserted just past the polylinker region to give pRIT21-23. Cloning to the polylinker in this vector will, if in frame, produce fusion proteins that may be periplasmic in *E. coli* and secreted in *S. aureus* (depending on the fused protein). The native *spa* promoter is present and is supplemented by the *E. coli lacUV5* promoter (arrow) (1). (D) pPL703. This plasmid consists of a 1,250-bp *Pst*I-*Bgl*II fragment of *Bacillus pumilus* NCIB8600 DNA inserted between the *Eco*RI and *Bam*HI sites of pUB110 by use of a 21-bp *Eco*RI-*Pst*I fragment from m13mp7 (54, 101). The promoterless gene *cat-86* resides within the 1,250-bp fragment, and the gene is followed by an efficient transcription termination signal, designated *ter* (64). The pUB110 portion of pPL703 provides an origin of replication and a neomycin resistance gene (Neo^r). *cat-86* specifies chloramphenicol acetyltransferase when the gene is transcriptionally activated by inserting a promoter into any of four unique restriction sites 5' to *cat-86*: *Eco*RI, *Bam*HI, *Sal*I, and *Pst*I. RBS-1, RBS-2, and RBS-3 designate the approximate locations of ribosome-binding sites identified by their complementarity to *B. subtilis* 26S rRNA. Since the *cat-86* regulatory sequences (→←) are intact, chloramphenicol inducibility is retained. Accordingly, with a promoter-containing derivative of pPL703, cloning of any gene in frame into the *cat-86* coding sequence will result in a chloramphenicol-inducible fusion protein. Reprinted from Lovett et al. (47) by kind permission from the publishers. Ori, origin.

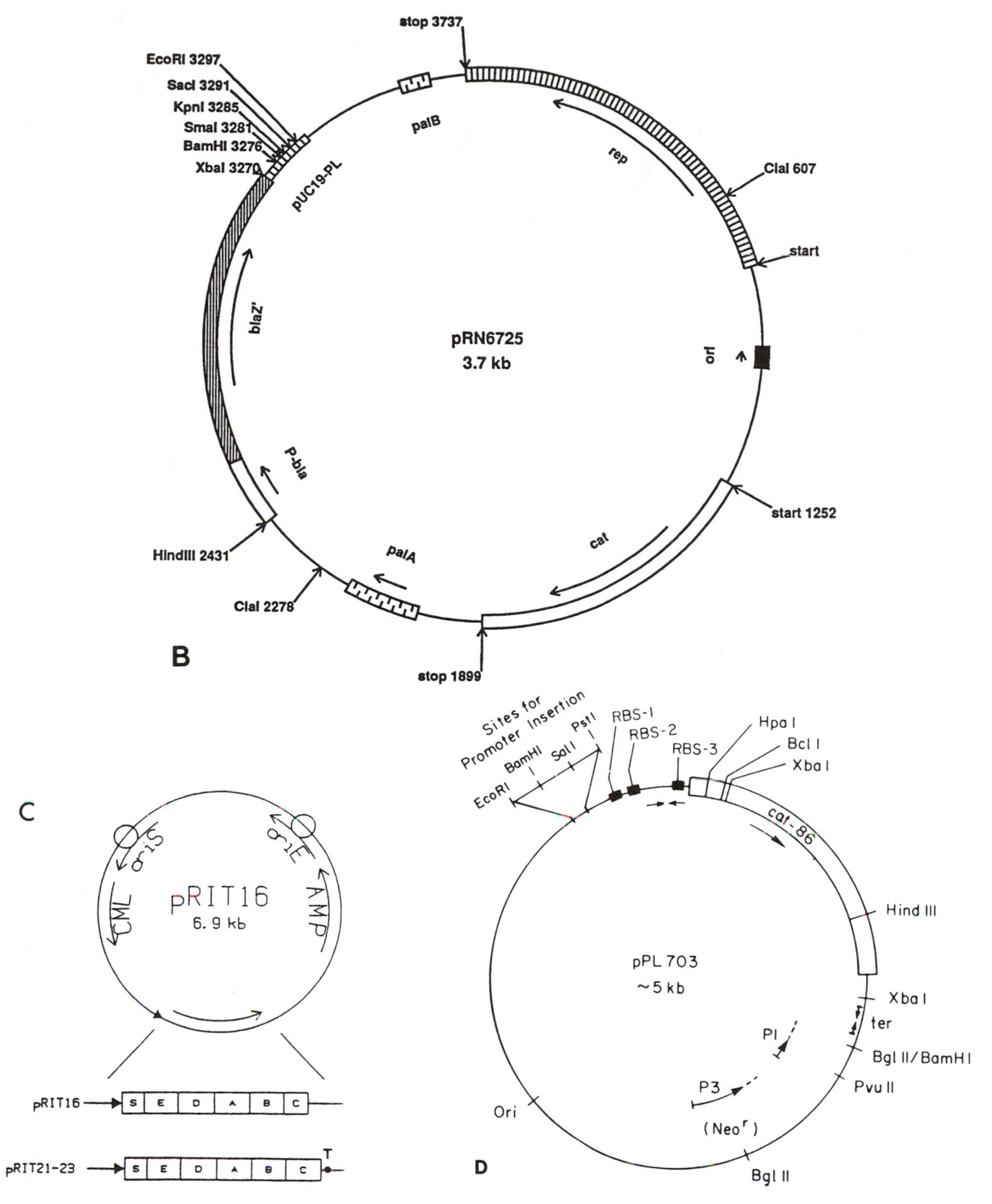

Figure 8. *Continued.*

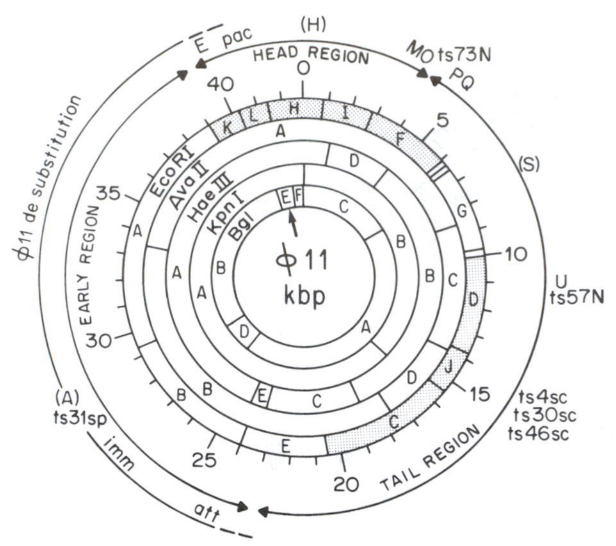

Figure 9. Phage ϕ11 map. Restriction sites indicated were mapped by Bachi (5), Lofdahl et al. (46), and Novick et al. (68). The arrow pointing to *Bgl*II-E indicates the *pac* site (5); kbp refers to the circular scale outside the *Eco*RI map. Early, head, and tail regions were mapped by suppressor-sensitive mutations (39), and these regions were correlated with the physical map by cloning (46) (the shaded fragments cause high-frequency transduction when cloned to a plasmid [see text]) and by analysis of the ϕ11::pI258 plasmid-phage recombinant ϕ11*de* (46, 47). Temperature-sensitive mutations (ts) isolated by this laboratory (N) (unpublished data), by Cohen et al. (sc) (12), or by Sjostrom and Philipson (sp) (90) have been localized by cloning. Capital letters represent genes mapped by complementation of *sus* mutants (39); *att* has been mapped by sequencing of the cloned prophage junctions (43), and *imm* has been mapped by cloning (46).

a phosphorylated carbohydrate, deoxyribose-5-phosphate, as an inducer.

Other Resistance Genes

Plasmid-coded staphylococcal resistance to antibiotics that inhibit protein synthesis (erythromycin and chloramphenicol) is regulated by a novel mechanism that has not been seen in gram-negative bacteria, namely, translational attenuation. In each case, the resistance gene mRNA is preceded by a long leader that encodes one or more short peptides. Inducing concentrations of the respective antibiotics causes stalling of ribosomes translating the leader peptide at a position that modifies the downstream folding of the leader so as to promote translation of the resistance gene. The translational attenuation mechanism for Cm^r and Em^r has also been described in bacilli (14, 15). In the case of *tetM*, the mechanism is slightly different, involving transcriptional attenuation. In the absence of tetracycline, only a short leader transcript is produced. In the presence of tetracycline, the short leader disappears and is replaced by the full-length transcript. There is a strong potential hairpin in the leader peptide coding sequence, and it is presumed that translation of the leader peptide promotes the formation of this hairpin, which terminates transcription, and that inhibition by tetracycline of leader peptide synthesis blocks formation of the stem, permitting readthrough transcription (36a).

Plasmid-coded heavy-metal resistance in *S. aureus* is also inducible, and the details have been determined in the case of Hg^r. The pI258 *mer* operon, consisting of a regulatory gene (*merR*), three genes involved in transport (*merT*), and two inactivating genes (*merA* and *merB*) (71), is similar to that of other bacteria, is flanked by IS sequences, and is likely to be (or to have been) transposable. The regulator, MerR, is a helix-turn-helix protein that binds to the operator-promoter region of the operon and appears to be a transcriptional activator (reviewed by Silver and Laddaga [89]). A unique feature of mercury resistance is the hypersensitivity that is seen with *merA* and *merB* mutants and is due to transport of the toxic ions.

agr

The production of most *S. aureus* exoproteins and at least some surface proteins is regulated by a complex biphasic global regulatory system whose details are just beginning to emerge. One component of the biphasic system is the global regulator *agr* (for accessory gene regulation), which consists of a pair of divergent operons. One of these, transcribed from promoter P2, contains four open reading frames (*agrA*, *agrB*, *agrC*, and *agrD*), all four of which are required for activity of the system and two of which (*agrA* and *agrB*) correspond to the two components of the classic sensory transduction systems in bacteria. *agrA*, the most distal, is presumed to be the response regulator, and *agrB* is presumed to be the sensory transducer. This two-component system, which responds to unknown signals that occur early in the exponential phase in flask cultures, serves only to activate its own promoter (P2) plus that of the divergent operon (P3), and it is a product of the P3 operon that activates the target genes. Remarkably, this product turns out to be the primary 517-nucleotide transcript itself, known as RNAIII, rather than any translation product (36). The other component of the biphasic system is a temporal signal that occurs at the end of exponential growth (95). Both *agr* and the temporal signal are necessary for transcriptional activation of the up-regulated proteins; neither alone has any significant effect. Surface proteins such as protein A, however, are repressed by *agr*, and in this case, production of RNAIII itself is sufficient; the temporal signal is not involved (95). Preliminary evidence for an *agr* system in other staphylococci exists, and it is predicted that streptococci will also be found to have such a system.

REFERENCES

1. **Abraham, L. J., and J. I. Rood.** 1988. The *Clostridium perfringens* chloramphenicol resistance transposon Tn4451 excises precisely in *Escherichia coli. Plasmid* **19:**164–168.
2. **Anderson, E. S., and R. E. O. Williams.** 1956. Bacteriophage typing of enteric pathogens and staphylococcus and its use in epidemiology. *J. Clin. Pathol.* **9:**94–127.
3. **Argos, P., A. Landy, K. Abremski, J. B. Egan, E. Haggard-Ljungquist, R. H. Hoess, M. L. Kahn, B. Kalionis, S. B. L. Narayana, L. S. Pierson III, N. Sternberg, and J. M. Leong.** 1986. The integrase family of site-specific

recombinases: regional similarities and global diversity. *EMBO J.* **5:**433–440.

4. **Asheshov, E. H.** 1969. The genetics of penicillinase production in *Staphylococcus aureus* strain PS80. *J. Gen. Microbiol.* **59:**289–301.
5. **Bachi, B.** 1980. Physical mapping of the *Bgl* I, *Bgl* II, *Pst* I and *Eco* R1 restriction fragments of staphylococcal phage ϕ11 DNA. *Mol. Gen. Genet.* **180:**391–398.
6. **Barberis-Maino, L., B. Berger-Bachi, H. Weber, W. D. Beck, and F. H. Kayser.** 1987. IS431, a staphylococcal insertion sequence-like element related to IS26 from *Proteus vulgaris*. *Gene* **59:**107–113.
7. **Bastos, M. C. F., and E. Murphy.** 1988. Transposon Tn554 encodes three products required for transposition. *EMBO J.* **7:**2935–2941.
8. **Betley, M. J., and J. J. Mekalanos.** 1985. Staphylococcal enterotoxin A is encoded by phage. *Science* **229:**185–187.
9. **Bhakdi, S., and J. Tranum-Jensen.** 1984. Mechanism of complement cytolysis and the concept of channel-forming proteins. *Philos. Trans. R. Soc. London Ser. B* **306:** 311–324.

9a. **Buchanan, R. E., and N. E. Gibbons (ed.).** 1974. *Bergey's Manual of Determinative Bacteriology.* Williams & Wilkins, Baltimore.

10. **Chapple, R., and P. R. Stewart.** 1987. Polypeptide synthesis during lytic induction of phage 11 of *Staphylococcus aureus. Virology* **68:**1401–1409.
11. **Chu, M. C., B. N. Kreiswirth, P. A. Pattee, R. P. Novick, M. E. Melish, and J. J. James.** 1988. Association of toxic shock toxin-1 determinant with a heterologous insertion at multiple loci in the *Staphylococcus aureus* chromosome. *Infect. Immun.* **56:**2702–2708.
12. **Cohen, S., H. M. Sweeney, and S. K. Basu.** 1977. Mutations in prophage ϕ11 that impair the transducibility of their *Staphylococcus aureus* lysogens for methicillin resistance. *J. Bacteriol.* **129:**237–245.
13. **Coleman, D. C., J. P. Arbuthnott, H. M. Pomeroy, and T. H. Birkbeck.** 1986. Cloning and expression in *Escherichia coli* and *Staphylococcus aureus* of the beta-lysin determinant from *Staphylococcus aureus*: evidence that bacteriophage conversion of beta-lysin activity is caused by insertional inactivation of the beta-lysin determinant. *Microb. Pathog.* **1:**549–564.
14. **Dubnau, D.** 1984. Translational attenuation: the regulation of bacterial resistance to the macrolide-lincosamide-streptogramin B antibiotics. *Crit. Rev. Biochem.* **16:**103–132.
15. **Duvall, E. J., N. P. Ambulos, Jr., and P. S. Lovett.** 1987. Drug-free induction of a chloramphenicol acetyltransferase gene in *Bacillus subtilis* by stalling ribosomes in a regulatory leader. *J. Bacteriol.* **169:**4235–4241.

15a. **Ehrlich, S. D.** Personal communication.

16. **Elek, S. O., and E. Levy.** 1954. The nature of discrepancies between hemolysins in culture filtrates and plate hemolysin patterns of staphylococci. *J. Pathol. Bacteriol.* **68:**31–40.
17. **Fitton, J. E., A. Dell, and W. V. Shaw.** 1980. The amino acid sequence of the delta haemolysin of *Staphylococcus aureus. FEBS Lett.* **115:**209–212.
18. **Foster, T. J., M. O'Reilly, P. Phonimdaeng, J. Cooney, A. H. Patel, and A. J. Bramley.** 1990. Genetic studies of virulence factors of *Staphylococcus aureus*. Properties of coagulase and γ-toxin, α-toxin, β-toxin and protein A in the pathogenesis of *S. aureus* infections, p. 403–420. *In* R. P. Novick (ed.), *Molecular Biology of the Staphylococci*. VCH Publishers, New York.
19. **Gennaro, M. L., J. Kornblum, and R. P. Novick.** 1987. A site-specific recombination function in *Staphylococcus aureus* plasmids. *J. Bacteriol.* **169:**2601–2610.

19a. **Gennaro, M. L., and R. P. Novick.** Unpublished data.

20. **Gotz, F., J. Zabielski, L. Philipson, and M. Lindberg.** 1983. DNA homology between the arsenate resistance plasmid pSX267 from *Staphylococcus xylosus* and the penicillinase plasmid pI258 from *Staphylococcus aureus. Plasmid* **9:**126–137.
21. **Gray, G. S.** 1983. Characterization of plasmids in aminocyclitol-resistant *Staphylococcus aureus*: electron microscopic and restriction endonuclease analysis. *Plasmid* **9:**159–181.

21a. **Gruss, A., and R. Novick.** Unpublished data.

22. **Gruss, A., H. F. Ross, and R. P. Novick.** 1987. Functional analysis of a palindromic sequence required for normal replication of several staphylococcal plasmids. *Proc. Natl. Acad. Sci. USA* **84:**2165–2169.
23. **Gruss, A. D., and S. D. Ehrlich.** 1989. The family of highly interrelated single-stranded deoxyribonucleic acid plasmids. *Microbiol. Rev.* **53:**231–241.
24. **Hartman, B. J., and A. Tomasz.** 1984. Low-affinity penicillin binding protein associated with β-lactam resistance in *Staphylococcus aureus. J. Bacteriol.* **158:** 513–516.

24a. **Holt, J. G. (ed.).** 1984. *Bergey's Manual of Systematic Bacteriology.* William & Wilkins, Baltimore.

25. **Horinouchi, S., and B. Weisblum.** 1982. Nucleotide sequence and functional map of pC194, a plasmid that specifies inducible chloramphenicol resistance. *J. Bacteriol.* **150:**815–825.
26. **Horinouchi, S., and B. Weisblum.** 1982. Nucleotide sequence and functional map of pE194, a plasmid that specifies inducible resistance to macrolide, lincosamide, and streptogramin type B antibiotics. *J. Bacteriol.* **150:**804–814.
27. **Inglis, B., H. Waldron, and P. R. Stewart.** 1987. Molecular relatedness of *Staphylococcus aureus* typing phages measured by DNA hybridization and by high resolution thermal denaturation analysis. *Arch. Virol.* **93:**69–80.
28. **Inoue, M., and S. Mitsuhashi.** 1976. Recombination between phage S1 and the Tc resistant gene on *Staphylococcus aureus* plasmid. *Virology* **72:**322–329.
29. **Iordanescu, S., and M. Surdeanu.** 1980. New incompatibility groups for *S. aureus* plasmids. *Plasmid* **4:**256–260.
30. **Iordanescu, S., M. Surdeanu, P. Della Latta, and R. Novick.** 1978. Incompatibility and molecular relationships between small staphylococcal plasmids carrying the same resistance marker. *Plasmid* **1:**468–479.
31. **Jones, J., S. Yost, and P. Pattee.** 1987. Transfer of the conjugal tetracycline resistance transposon Tn*916* from *Streptococcus faecalis* to *Staphylococcus aureus* and identification of some insertion sites in the staphylococcal chromosome. *J. Bacteriol.* **169:**2121–2131.
32. **Khan, S. A., and R. P. Novick.** 1982. Structural analysis of plasmid pSN2 in *Staphylococcus aureus*: no involvement in enterotoxin B production. *J. Bacteriol.* **149:**642–649.
33. **Khan, S. A., and R. P. Novick.** 1983. Complete nucleotide sequence of pT181, a tetracycline resistance plasmid from *Staphylococcus aureus. Plasmid* **10:**151–159.
34. **Koepsel, R. R., R. W. Murray, W. D. Rosenblum, and S. A. Khan.** 1985. The replication initiator protein of plasmid pT181 has sequence-specific endonuclease and topoisomerase-like activities. *Proc. Natl. Acad. Sci. USA* **82:**6845–6849.
35. **Kondo, I., S. Itoh, and Y. Yoshizawa.** 1981. Staphylococcal phages mediating the lysogenic conversion of staphylokinase. *Zentralbl. Bakteriol. Suppl.* **10:**357–362.
36. **Kornblum, J., B. Kreiswirth, S. J. Projan, H. Ross, and R. P. Novick.** 1990. *agr*: a polycistronic locus regulating exoprotein synthesis in *Staphylococcus aureus*, p. 373–402. *In* R. P. Novick (ed.), *Molecular Biology of the Staphylococci*. VCH Publishers, New York.

36a. **Kornblum, J., and R. P. Novick.** Unpublished data.

36b. **Kreiswirth, B., J. Kornblum, R. D. Arbeit, W. Eisner, J. N. Maslow, A. McGeer, D. E. Low, and R. P. Novick.**

1993. Evidence for a clonal origin of methicillin resistance in *S. aureus*. *Science* **259:**227–230.
37. **Kreiswirth, B., S. Lofdahl, M. Betley, M. O'Reilly, P. Schlievert, M. Bergdoll, and R. P. Novick.** 1983. The toxic shock syndrome exotoxin structural gene is not detectably transmitted by a prophage. *Nature* (London) **305:**709–712.
38. **Kreiswirth, B. N., G. R. Kravitz, P. M. Schlievert, and R. P. Novick.** 1986. Nosocomial transmission of a strain of *Staphylococcus aureus* causing toxic shock syndrome. *Ann. Intern. Med.* **105:**704–707.
39. **Kretschmer, P. J., and J. B. Egan.** 1975. Genetic map of the staphylococcal bacteriophage ϕ11. *J. Virol.* **16:**642–651.
39a. **Lacks, S. A., P. Lopez, B. Greenberg, and M. Espinosa.** 1986. Identification and analysis of genes for tetracycline resistance and replication functions in the broad-host-range plasmid pLS1. *J. Mol. Biol.* **4:**753–765.
40. **Lampson, B. C., and J. T. Parisi.** 1986. Naturally occurring *Staphylococcus epidermidis* plasmid expressing constitutive macrolide-lincosamide-streptogramin B resistance contains a deleted attenuator. *J. Bacteriol.* **166:**479–483.
41. **Lee, C. Y., and J. J. Iandolo.** 1986. Lysogenic conversion of staphylococcal lipase caused by insertion of the bacteriophage phage L54a genome into the lipase structural gene. *J. Bacteriol.* **166:**385–391.
42. **Lee, C. Y., and J. J. Iandolo.** 1986. Integration of staphylococcal phage L54a occurs by site-specific recombination: structural analysis of the attachment sites. *Proc. Natl. Acad. Sci. USA* **83:**5474–5478.
43. **Lee, C. Y., and J. J. Iandolo.** 1988. Structural analysis of staphylococcal bacteriophage phi 11 attachment sites. *J. Bacteriol.* **170:**2409–2411.
44. **Lindberg, M., K. Jonsson, H. Muller, H. Jonsson, C. Signas, M. Hook, R. Raja, G. Raucci, and G. M. Anantharamaiah.** 1990. Fibronectin-binding proteins in *S. aureus*, p. 327–356. *In* R. P. Novick (ed.), *The Molecular Biology of the Staphylococci*. VCH Publishers, New York.
45. **Lindquist, S., M. Galleni, F. Lindberg, and S. Normark.** 1989. Signalling proteins in enterobacterial AmpC β-lactamase regulation. *Mol. Microbiol.* **3:**1091–1102.
46. **Lofdahl, S., J. Zabielski, and L. Philipson.** 1981. Structure and restriction enzyme maps of the circularly permuted DNA of staphylococcal bacteriophage ϕ11. *J. Virol.* **37:**784–794.
47. **Lovett, P. S., D. M. Williams, E. J. Duvall, and S. Mongkilsuk.** 1984. Chloramphenicol inducibility of foreign gene expression in *B. subtilis*, p. 275–283. *In* A. T. Ganesan and J. A. Hoch (ed.), *Genetics and Biotechnology of Bacilli*, vol. 1. Academic Press, Inc., New York.
48. **Ludwig, W., K. Schleifer, G. E. Fox, E. Seewaldt, and E. Stackebrandt.** 1981. A phylogenetic analysis of staphylococci, *Peptococcus saccharolyticus* and *Micrococcus mucilaginosus*. *J. Gen. Microbiol.* **125:**357–366.
49. **Lyon, B. R., M. T. Gillespie, and R. A. Skurray.** 1987. Detection and characterization of IS256, an insertion sequence in *Staphylococcus aureus*. *J. Gen. Microbiol.* **133:**3031–3038.
50. **Lyon, B. R., and R. Skurray.** 1987. Antimicrobial resistance of *Staphylococcus aureus*: genetic basis. *Microbiol. Rev.* **51:**88–134.
51. **Matthews, P. R., B. Inglis, and P. R. Stewart.** 1990. Clustering of resistance genes in the *mec* region of the chromosome of *S. aureus*, p. 69–83. *In* R. P. Novick (ed.), *Molecular Biology of the Staphylococci*. VCH Publishers, New York.
52. **McKenzie, T., T. Hoshino, T. Tanaka, and N. Sueoka.** 1986. The nucleotide sequence of pUB110: some salient features in relation to replication and its regulation. *Plasmid* **15:**93–103.
53. **Mills, J. T., Y. Tsai, M. Kendrick, R. K. Hickman, and E. H. Kass.** 1985. Control of production of toxic-shock-syndrome toxin-1 (TSST-1) by magnesium ion. *J. Infect. Dis.* **151:**1158–1161.
54. **Mongkolsuk, S., Y. Chiang, R. B. Reynolds, and P. S. Lovett.** 1983. Restriction fragments that exert promoter activity during post-exponential phase growth of *B. subtilis*. *J. Bacteriol.* **155:**1399–1406.
55. **Monod, M., C. Denoya, and D. Dubnau.** 1986. Sequence and properties of pIM13, a macrolide-lincosamide-streptogramin B resistance plasmid from *Bacillus subtilis*. *J. Bacteriol.* **167:**138–147.
56. **Muller, R. E., T. Ano, T. Imanaka, and S. Aiba.** 1986. Complete nucleotide sequences of *Bacillus* plasmids pUB110dB, pRBH1 and its copy mutants. *Mol. Gen. Genet.* **202:**169–171.
56a. **Murphy, E.** Personal communication.
57. **Murphy, E.** 1983. Inhibition of Tn554 transposition: deletion analysis. *Plasmid* **10:**260–269.
58. **Murphy, E.** 1989. Transposable elements in gram-positive bacteria, p. 269–288. *In* D. E. Berg, and M. M. Howe (ed.), *Mobile DNA*. American Society for Microbiology, Washington, D.C.
59. **Murphy, E., L. Huwyler, and M. Bastos.** 1985. Transposon Tn554: complete nucleotide sequence and isolation of transposition-defective and antibiotic-sensitive mutants. *EMBO J.* **4:**3357–3365.
60. **Murphy, E., and S. Lofdahl.** 1984. Transposition of Tn554 does not generate a target duplication. *Nature* (London) **307:**292–294.
61. **Murphy, E., and R. Novick.** 1980. Site-specific recombination between plasmids of *Staphylococcus aureus*. *J. Bacteriol.* **141:**316–326.
62. **Murphy, E., and R. P. Novick.** 1979. Physical mapping of *S. aureus* penicillinase plasmid pI524: characterization of an invertible region. *Mol. Gen. Genet.* **175:**19–30.
63. **Murphy, E., S. Phillips, I. Edelman, and R. P. Novick.** 1981. Tn554: isolation and characterization of plasmid insertions. *Plasmid* **5:**292–305.
64. **Nilsson, B., L. Abrahmsen, and M. Uhlen.** 1985. Immobilization and purification of enzymes with staphylococcal protein A gene fusion vectors. *EMBO J.* **4:**1075–1080.
65. **Novick, R.** 1967. Properties of a cryptic high-frequency transducing phage in *Staphylococcus aureus*. *Virology* **33:**155–166.
65a. **Novick, R. P.** Unpublished data.
65b. **Novick, R. P.** 1978. The mechanism of translocation in bacteria. *Brookhaven Symp. Biol.* **29:**272–276.
66. **Novick, R. P.** 1989. Staphylococcal plasmids and their replication. *Annu. Rev. Microbiol.* **43:**537–565.
66a. **Novick, R. P. (ed.).** 1990. *Molecular Biology of the Staphylococci*. VCH Publishers, New York.
67. **Novick, R. P., G. K. Adler, S. J. Projan, S. Carleton, S. Highlander, A. Gruss, S. A. Khan, and S. Iordanescu.** 1984. Control of pT181 replication. I. The pT181 copy control function acts by inhibiting the synthesis of a replication protein. *EMBO J.* **3:**2399–2405.
68. **Novick, R. P., I. Edelman, and S. Lofdahl.** 1986. Small *Staphylococcus aureus* plasmids are transduced as linear multimers which are formed and resolved by replicative processes. *J. Mol. Biol.* **192:**209–220.
69. **Novick, R. P., I. Edelman, M. D. Schwesinger, D. Gruss, E. C. Swanson, and P. A. Pattee.** 1979. Genetic translocation in Staphylococcus aureus. *Proc. Natl. Acad. Sci. USA* **76:**400–404.
70. **Novick, R. P., S. Iordanescu, S. J. Projan, J. Kornblum, and I. Edelman.** 1989. pT181 plasmid replication is regulated by a countertranscript-driven transcriptional attenuator. *Cell* **59:**395–404.
71. **Novick, R. P., E. Murphy, T. J. Gryczan, E. Baron, and I. Edelman.** 1979. Penicillinase plasmids of *Staphylococcus aureus*: restriction-deletion maps. *Plasmid* **2:**109–129.

72. **Novick, R. P., S. J. Projan, W. Rosenblum, and I. Edelman.** 1984. Staphylococcal plasmid cointegrates are formed by host- and phage-mediated general *rec* systems that act on short regions of homology. *Mol. Gen. Genet.* **195:**374–377.
73. **Novick, R. P., and C. Roth.** 1968. Plasmid-linked resistance to inorganic salts in *Staphylococcus aureus. J. Bacteriol.* **95:**1335–1342.
74. **Oka, A., N. Nomura, M. Hiroyuki, and K. Sugimoto.** 1979. Nucleotide sequence of small ColE1 derivatives: structure of the regions essential for autonomous replication and colicin E1 immunity. *Mol. Gen. Genet.* **172:**151–159.
75. **Oskouian, B., E. L. Rosey, F. B. Breidt, Jr., and G. C. Stewart.** 1991. The lactose operon of *S. aureus*, p. 99–112. *In* R. P. Novick (ed.), *The Molecular Biology of the Staphylococci*. VCH Publishers, New York.
76. **Patel, A. H., J. Kornblum, B. Kreiswirth, R. P. Novick, and T. J. Foster.** 1992. Regulation of the protein A-encoding gene in *Staphylococcus aureus. Gene* **114:**25–34.

76a. **Pattee, P.** Personal communication.

77. **Pattee, P. A., H. Lee, and J. P. Bannantine.** 1990. Genetic and physical mapping of the chromosome of *Staphylococcus aureus*, p. 41–58. *In* R. P. Novick (ed.), *Molecular Biology of the Staphylococci*. VCH Publishers, New York.
78. **Peyru, G., L. F. Wexler, and R. P. Novick.** 1969. Naturally-occurring penicillinase plasmids in *Staphylococcus aureus. J. Bacteriol.* **98:**215–221.

78a. **Primrose, S. B., and S. D. Ehrlich.** 1981. Isolation of plasmid deletion mutants and study of their instability. *Plasmid* **6:**193–201.

79. **Projan, S. J., and R. P. Novick.** 1988. Comparative analysis of five related staphylococcal plasmids. *Plasmid* **19:**203–221.
80. **Ranellin, D. M., C. L. Jones, M. B. Johns, G. J. Mussey, and Khan.** 1985. Molecular cloning of staphylococcal enterotoxin B gene in *Escherichia coli* and *Staphylococcus aureus. Proc. Natl. Acad. Sci. USA* **82:**5850–5854.
81. **Recsei, P., A. Gruss, and R. Novick.** 1987. Cloning, sequence, and expression of the lysostaphin gene from *Staphylococcus simulans. Proc. Natl. Acad. Sci. USA* **84:**1127–1131.
82. **Rees, P. J., and B. A. Fry.** 1981. The morphology of staphylococcal bacteriophage K and DNA metabolism in infected *Staphylococcus aureus. J. Gen. Virol.* **53:**293–307.
83. **Rowland, S. J., and K. G. Dyke.** 1989. Characterization of the staphylococcal β-lactamase transposon Tn552. *EMBO J.* **8:**2761–2773.
84. **Ruby, C., and R. P. Novick.** 1975. Plasmid interactions in *Staphylococcus aureus*. Nonadditivity of compatible plasmid DNA pools. *Proc. Natl. Acad. Sci. USA* **72:**5031–5035.
85. **Sako, T., S. Sawaki, T. Sakurai, S. Ito, Y. Yoshizawa, and I. Kondo.** 1983. Cloning and expression of the staphylokinase gene of *Staphylococcus aureus* in *Escherichia coli. Mol. Gen. Genet.* **190:**271–277.
86. **Schaefler, S.** 1982. Bacteriophage-mediated acquisition of antibiotic resistance by *Staphylococcus aureus* type 88. *Antimicrob. Agents Chemother.* **21:**460–467.
87. **Scheifele, D. W., G. L. Bjornson, R. A. Dyer, and J. E. Dimmick.** 1987. Delta-like toxin produced by coagulase-negative staphylococci is associated with neonatal necrotizing enterocolitis. *Infect. Immun.* **55:**2268–2273.

87a. **Sheehan, D.** Personal communication.

88. **Sheehy, R. J., and R. P. Novick.** 1975. Studies on plasmid replication. IV. Replicative intermediates. *J. Mol. Biol.* **93:**237–253.
89. **Silver, S., and R. A. Laddaga.** 1990. Molecular genetics of heavy metal resistance systems of *Staphylococcus* plasmids, p. 531–549. *In* R. P. Novick (ed.), *Molecular Biology of the Staphylococci*. VCH Publishers, New York.
90. **Sjostrom, J., and L. Philipson.** 1974. Role of the ϕ11 phage genome in competence of *Staphylococcus aureus. J. Bacteriol.* **119:**19–32.
91. **Smith, R. M., J. T. Parisi, L. Vical, and J. N. Baldwin.** 1977. Nature of the genetic determinant controlling encapsulation in *Staphylococcus aureus* Smith. *Infect. Immun.* **17:**231–234.
92. **Thomas, W. D., Jr., and G. L. Archer.** 1989. Identification and cloning of the conjugative transfer region of the *Staphylococcus aureus* plasmid pG01. *J. Bacteriol.* **171:**684–691.
93. **Townsend, D. E., S. Bolton, N. Ashdown, D. I. Annear, and W. B. Grubb.** 1986. Conjugative, staphylococcal plasmids carrying hitch-hiking transposons similar to Tn554: intra- and interspecies dissemination of erythromycin resistance. *Aust. J. Expression Biol. Med. Sci.* **64:**367–379.
94. **Tzagoloff, H., and R. P. Novick.** 1977. Geometry of cell division in *Staphylococcus aureus. J. Bacteriol.* **129:**343–350.
95. **Vandenesch, F., J. Kornblum, and R. P. Novick.** 1991. A temporal signal, independent of *agr*, is required for *hla* but not *spa* transcription in *Staphylococcus aureus. J. Bacteriol.* **173:**6313–6320.
96. **Wadstrom, T., J. Erdei, M. Paulsson, and A. S. Naidu.** 1990. Binding of collagen and vitronectin to *S. aureus* and coagulase-negative staphylococci (CNS), p. 357–371. *In* R. P. Novick (ed.), *Molecular Biology of the Staphylococci*. VCH Publishers, New York.
97. **Wang, P., S. J. Projan, K. Leason, and R. P. Novick.** 1987. Translational fusion with a secretory enzyme as an indicator. *J. Bacteriol.* **169:**3082–3087.

97a. **Watanakunakorn, C., R. J. Fass, A. S. Klainer, and M. Hamburger.** 1971. Light and scanning-beam electron microscopy of wall-defective *Staphylococcus aureus* induced by lysostaphin. *Infect. Immun.* **4:**73–78.

98. **Weber, D. A., and R. V. Goering.** 1988. Tn4201, a β-lactamase transposon in *Staphylococcus aureus. Antimicrob. Agents Chemother.* **32:**1164–1169.
99. **Weinstein, R. A., S. A. Kabins, C. Nathans, H. M. Sweeney, H. W. Jaffe, and S. Cohen.** 1982. Gentamicin-resistant staphylococci as hospital flora: epidemiology and resistance plasmids. *J. Infect. Dis.* **145:**374–382.

99a. **White, A., and G. F. Grooks.** 1977. Furunculosis, pyoderma, and impetigo, p. 785–793. *In* P. D. Hoeprich (ed.), *Infectious Diseases*, 2nd ed. Harper & Row, Hagerstown, Md.

100. **Wiley, B. B., and M. Rogolsky.** 1977. Molecular and serological differentiation of staphylococcal exfoliative toxin synthesized under chromosomal and plasmid control. *Infect. Immun.* **18:**487–494.
101. **Williams, D. M., J. Duvall, and P. S. Lovett.** 1981. Cloning restriction fragments that promote expression of a gene in *Bacillus subtilis. J. Bacteriol.* **146:**1162–1165.
102. **Wyman, L., R. V. Goering, and R. P. Novick.** 1974. Genetic control of chromosomal and plasmid recombination in *Staphylococcus aureus. Genetics* **76:**681–702.
103. **Wyman, L., and R. P. Novick.** 1974. Studies on plasmid replication. IV. Complementation of replication-defective mutants by an incompatibility-deficient plasmid. *Mol. Gen. Genet.* **135:**149–161.
104. **Youngman, P.** Personal communication.

3. *Clostridium*

MICHAEL YOUNG and STEWART T. COLE

INTRODUCTION

The genus *Clostridium* is not a natural taxonomic grouping but a heterogeneous assemblage of obligately anaerobic, gram-positive, endospore-forming, rod-shaped bacteria. Some 80 different species, most of which conform to these four criteria, are currently recognized. Many species are only poorly characterized. The heterogeneity of the group is indicated by the remarkable range of deoxy (G+C) [d(G+C)] contents of their DNAs, from a minimum of 24 mol% for *Clostridium pasteurianum* to a maximum of 55 mol% for *Clostridium barkeri* (45, 118).

The complete 16S rRNA sequences of eight different species have been determined. Comparative analysis suggests that six of these species form a coherent taxonomic grouping distinct from the genus *Bacillus*, whereas two others are more closely allied to the mycoplasmas (40). This chapter focuses on three species, *Clostridium perfringens*, *Clostridium acetobutylicum*, and *Clostridium thermocellum*, only the first of which features in the analysis just given of 16S rRNA sequences.

There have been several recent compilations of research on the clostridia (120, 121, 198, 261), and three books largely or entirely devoted to these bacteria have recently been published (159, 211, 248). This chapter briefly summarizes three important aspects of their biology: the molecular bases of pathogenesis, solventogenesis, and polysaccharide hydrolysis. The three organisms mentioned above continue to play a cardinal role in investigations of these topics. For more extensive coverage and for information about other organisms and topics not dealt with here, such as taxonomy, Fe-S proteins, bioenergetics, and general metabolism, consult the reviews cited above.

Gene transfer technology has recently been developed for *C. perfringens* and *C. acetobutylicum*, and as a result, these two organisms have assumed particular importance as model pathogenic and solventogenic species, respectively. The current status of their genetics is summarized in the next section. The potential for undertaking genetic analysis has rekindled interest in the biotechnological exploitation of the clostridia (57). Moreover, there is growing awareness that these organisms can now be employed for analysis of a variety of fundamental problems of general biological interest. Some of these uses are outlined in the final section of this chapter.

GENERAL GENETICS

Genome Organization

The application of pulsed-field gel electrophoresis to large DNA fragments has allowed the genome size to be estimated for a number of clostridia (38, 42, 245). There is considerable variation from one species to another and even, in the case of *C. acetobutylicum*, from strain to strain (Table 1). This variation serves to underline the remarkable diversity of the genus noted above.

C. perfringens is the best-studied species. Detailed maps of the single circular chromosome of about 3,600 kbp have been obtained for a number of strains, and likely positions of the origins and termini of replication have been identified (38, 42). The 10 rRNA operons are organized divergently around the putative origin of replication and are situated in a region containing many tRNA genes, constituting about 30% of the chromosome (73). This arrangement is analogous to that seen in *Bacillus subtilis* (114). Twenty-seven genetic loci, including those for both virulence genes and housekeeping functions, have been located on the current version of the *C. perfringens* genetic map (Fig. 1). Comparative mapping of several different strains showed the *C. perfringens* chromosome to display a reasonably conserved organization in terms of both known genes and rare restriction sites, although a number of polymorphic regions, often associated with genes for virulence factors, were detected (42). A heat-stable enterotoxigenic strain showed the greatest deviation from the general organization; its genome was some 400 kb smaller because of a large deletion near the putative replication terminus.

The physical map of the 6,500-kbp chromosome of *C. acetobutylicum* NCIMB 8052 is nearing completion (246). A probable 14 rRNA operons and 12 genes, 6 of which encode enzymes concerned with fermentative metabolism, have been situated on the map. Approximate estimations of genome size are also available for *Clostridium difficile*, *C. pasteurianum*, *Clostridium stercorarium*, *C. thermocellum*, and *C. tyrobutyricum* (Table 1).

Plasmids

Plasmids are commonly encountered among the pathogenic clostridia (30, 60, 196). Although these plasmids are often cryptic, they sometimes bear the genes for virulence functions. This is best exemplified

Michael Young • Department of Biological Sciences, University of Wales, Aberystwyth, Dyfed SY23 3DA, United Kingdom. **Stewart T. Cole** • Unité de Génétique Moléculaire Bactérienne, Institut Pasteur, 28 Rue du Docteur Roux, 75724 Paris Cedex 15, France.

Table 1. Genetic features of some species of *Clostridium*

Organism	Chromosome size[a] (Mbp)	Mapped	% d(G+C)	*rrn* operons	Indigenous plasmids	Transposon mutagenesis	Electrotrans-formation
C. perfringens	3.2–3.6	+	25	10	+	+	+
C. tetani	?	–	26	?	+	+	?
C. botulinum	?	–	27	?	+	+	?
C. difficile	3.2	–	?	?	+	+	?
C. acetobutylicum							
ATCC 824	4.4	–	28	?	–[b]	+	+
NCP 262	2.9	–	?	?	–[b]	+	–
NCIMB 8052	6.5	+	?	14	–[b]	+	+
C. thermocellum ATCC 27401	3.5	–	46	?	–	?	–
C. pasteurianum ATCC 6013	3.9	–	26	?	–	–	–
C. tyrobutyricum DSM 1460	2.5	–	28	?	–	–	–
C. stercorarium NCIMB 11754	3.0	–	?	?	–	–	–

[a] Data are from references 38, 206, 207, 245, and 246.
[b] Cryptic plasmids have been found in other strains of *C. acetobutylicum* (230).

by *Clostridium botulinum*, the causative agent of botulism, in which the gene encoding the type G botulism toxin is plasmid encoded (66). This could facilitate dissemination of the toxin gene by conjugative transfer to normally nonpathogenic organisms such as *Clostridium butyricum*. Similarly, the tetanus toxin gene is encoded by a plasmid in *Clostridium tetani* (71). In *C. perfringens*, the λ, β, and ε toxins are plasmid encoded (28, 42, 60), although this species more commonly harbors large R factors conferring resistance to antibiotics such as tetracycline, chloramphenicol, and erythromycin or clindamycin (2, 3, 30, 101, 200, 213). Drug resistance has also been encountered in other clinically important species such as *C. difficile* (possibly a result of conjugation with *C. perfringens*) but has not been described thus far amongst solventogenic or cellulolytic organisms.

Small plasmids encoding bacteriocins are common among the pathogenic clostridia (30, 142, 143), and bacteriocin production in *C. butyricum* has also been reported (160). The most extensively characterized bacteriocinogenic plasmid is *C. perfringens* plasmid pIP404, which carries 10 genes, including those for production of, immunity to, and release of bacteriocin BCN5 (30, 49, 74–76). To date, no plasmids have been found in the cellulolytic clostridia.

Bacteriophages

Clostridia, like most other bacteria, harbor and play host to a variety of bacteriophages (172, 173, 218), both lytic and lysogenic, that can be involved in lysogenic conversion. The genes for type C and D botulism toxins are phage borne, while those of types A, B, E, and F are chromosomal and might therefore be associated with a prophage (67). The chromosomal attachment sites of phages ϕ29 and ϕ59 have been located on the *C. perfringens* chromosome (39), but no functions have been attributed to these phages. Bacteriophages have been used for strain typing in *C. difficile* (215) and as a tool for transfection of *C. acetobutylicum* (188, 190, 192).

A single-stranded phage in *C. acetobutylicum* has recently been described (124), but generally speaking, very little is known at the molecular level about clostridial phages. Surprisingly, bacteriophages able to infect *C. thermocellum* have not yet been found.

Transposons

Two transposons, Tn*4451* and Tn*4452*, which are associated with chloramphenicol resistance, are carried by conjugative R factors in *C. perfringens* (4, 30, 212). These two closely related elements are 6.2 kb in size, excise readily on conjugation or on cloning in *Escherichia coli*, and share extensive sequence homology, but they differ in 0.4-kb segments at their right ends. Tn*4451* has been studied quite extensively (4, 5). It appears to duplicate a 2-bp sequence on insertion and to contain a −35 promoter element capable of expressing adjacent genes. In *C. difficile*, a tetracycline resistance determinant that shares substantial homology with Tn*916* (*tetM*) from *Enterococcus faecalis* has been found, thus suggesting that transposon exchange may occur among unrelated clinically important species (94, 168). No indigenous transposons have been described among other pathogenic clostridia, nor have such transposons been reported for either solvent-forming or cellulolytic species.

Nevertheless, transposon mutagenesis is a useful tool for genetic analysis of various clostridia (reviewed in references 62 and 260). The conjugative enterococcal transposon Tn*916* and its close relative, Tn*1545*, have been used as insertional mutagens, and both *Enterococcus faecalis* and *E. coli* have served successfully as donors in filter matings. Tn*916* inserts into the chromosomes of *C. tetani*, *C. botulinum*, *C. perfringens*, *C. difficile*, and several strains of *C. acetobutylicum* (25, 26, 137, 147, 169, 198, 236, 250), and Tn*1545* has also been employed in the last-named species (245, 250).

Gene Transfer and Vector Development

A dominant selective marker was an essential prerequisite for the development of gene transfer systems, and streptococcal antibiotic resistance genes represented an obvious choice. In early studies with *C. acetobutylicum*, prototype vectors (159) employed the erythromycin resistance gene of pAMβ1 (32). Vectors developed for gene transfer experiments with *C. per-*

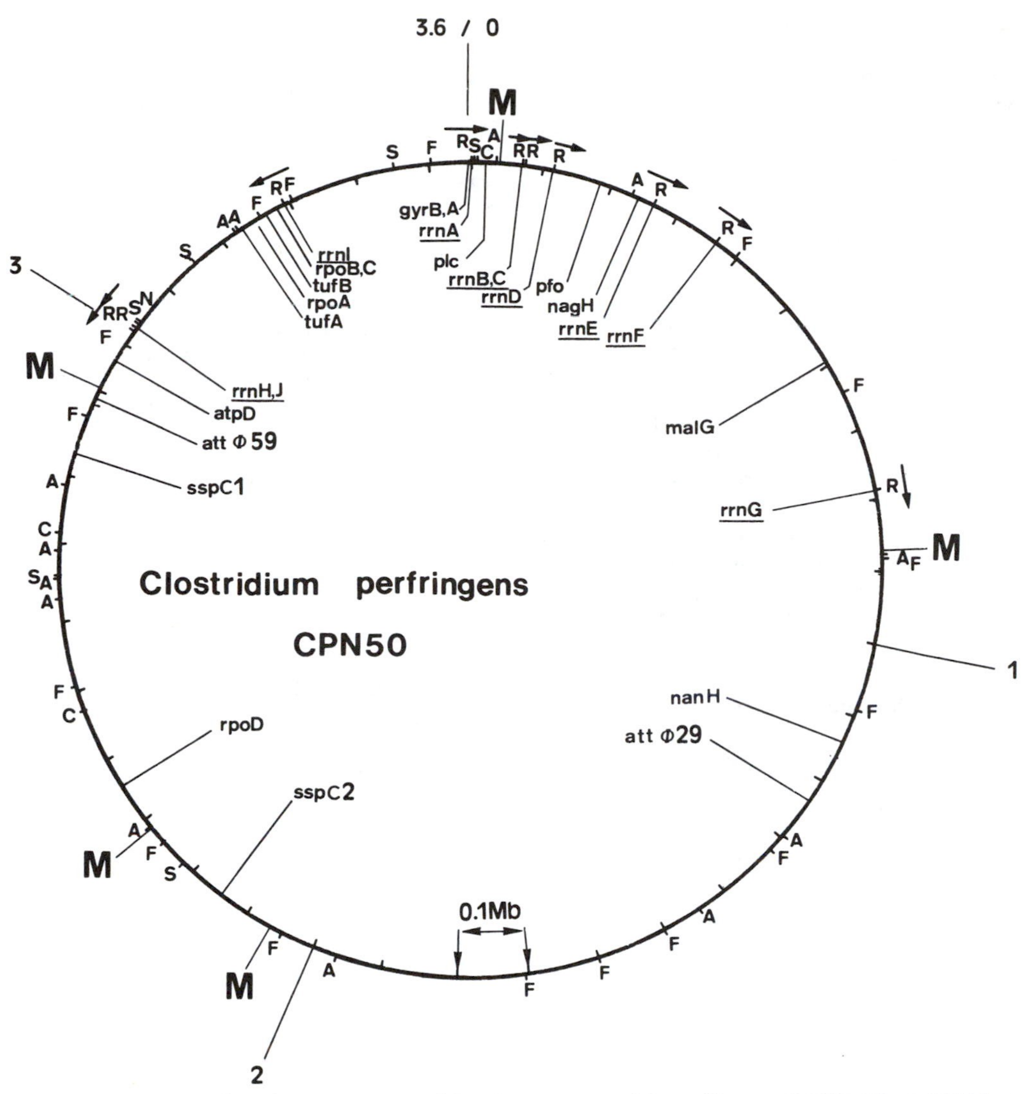

Figure 1. Combined physical and genetic maps of the chromosome of *C. perfringens* CPN50. The 3,600-kbp genome is divided into 100-kbp segments. Restriction sites for *Apa*I (A), *Fsp*I (F), *Mlu*I (M), *Sac*II (C), and *Sma*I (S) are indicated on the outside of the circle, as are the positions of clusters of rare restriction sites (*Sac*II-*Sac*II-*Sma*I-*Sma*I-*Nru*I-*Sma*I-*Nru*I) (R), which correspond to the locations of the 10 *rrn* operons. Genetic markers that have been assigned to the map are indicated on the inside of the circle (see references 38, 39, 42, 73, and 198 for further details).

fringens carried homologous tetracycline or chloramphenicol resistance markers derived from the indigenous R plasmids found in this organism (219, 220). The introduction of electroporation in the late 1980s greatly facilitated the transformation of clostridial strains and has largely superseded the protoplast transformation procedures used previously, one of which permits recovery of up to 10^6 transformants per μg of DNA in *C. acetobutylicum* (11, 188, 192). At present, strains of both *C. perfringens* (7, 8, 123, 210) and *C. acetobutylicum* (151, 176) can reproducibly be transformed by electroporation. Successful electrotransformation of other clostridial species, such as *C. thermocellum*, has not yet been described. Conjugative gene transfer from *E. coli* donors by using vectors containing the origin of transfer (*oriT*) of the IncP plasmid RK2 may be a feasible alternative, especially for strains recalcitrant to electroporation (229, 247).

A variety of cloning vectors are now available (see reference 158 for a recent review). A series of shuttle plasmids based on the broad-host-range replicon pAMβ1 has been constructed (175, 178, 224) and successfully used to introduce and express cloned genes in *C. acetobutylicum* (157, 176, 177). Likewise, a versatile shuttle vector (217) based on the well-characterized bacteriocinogenic plasmid pIP404 (75, 76) is available for work with *C. perfringens*. Unlike the majority of plasmids from gram-positive bacteria (92), neither of these plasmids employs a single-stranded replicative intermediate, and recombinant derivatives do not generally suffer structural instability (112). Another family of vectors based on the single-stranded DNA plasmid pCB101 (11, 31, 50, 140, 161, 247) is also available for use in *C. acetobutylicum*.

GENE EXPRESSION

A characteristic feature of the genes from many but not all clostridial species is their extremely high d(A+T) content, which is reflected in their unusual codon usages and also in promoter recognition by RNA polymerase.

RNA Polymerase

Transcriptionally active preparations of RNA polymerase holoenzyme have been purified from both *C. perfringens* and *C. acetobutylicum* and characterized with respect to their subunit composition and cofactor requirements (77, 185). In both cases, subunits analogous to the α, β, and β' components of other eubacterial RNA polymerases were identified, and for the *C. acetobutylicum* enzyme, the σ subunit was shown to cross-react with antibodies raised against the σ^{70} protein of *E. coli*. To date, no direct evidence for the existence of additional σ factors required for activation of global responses is available, although circumstantial evidence (105) supports the view that these factors will be found.

Transcriptional Signals

Promoters recognized by the major form of RNA polymerase have been identified and characterized in vivo in *C. tetani*, *C. botulinum*, and *C. thermocellum* (21, 27) and both in vivo and in vitro in *C. acetobutylicum*, *C. pasteurianum*, and *C. perfringens* (73, 77, 81, 89, 113, 170, 171). These contain −35 and −10 elements strongly resembling those of *E. coli* and *B. subtilis* vegetative promoters.

The largest collection of promoters has been established for *C. perfringens* (198), and one striking feature observed was the significant paucity of cytosine residues throughout the region likely to make initial contact with RNA polymerase. Among the other biochemically studied promoters are a class in *C. thermocellum*, resembling those recognized in *B. subtilis* by σ^{D}-containing-RNA polymerase (162), and a group of UV-inducible promoters with unusual sequences from *C. perfringens* (77, 78).

Codon Usage

Catalogs of the codons used in *C. perfringens* and *C. acetobutylicum* mRNAs reveal an extraordinary bias towards codons in which A and U predominate (198, 261). This is particularly striking for amino acids with six synonymous codons, such as arginine and leucine, which are encoded preferentially by AGA and UUA, respectively. These two codons are rarely used in *E. coli*, and a significant body of evidence indicates that the availability of the cognate tRNAs can be a rate-limiting factor for expression in this host (47, 145). Although reasonable amounts of clostridial proteins of average size (30 to 50 kDa) can be produced in *E. coli*, the levels of expression of the larger clostridial genes (>900 codons), such as those encoding the tetanus toxin or *C. difficile* cytotoxin A, are extremely low (65, 239). Further discussion of the effects of these translation barriers can be found in reference 49. By contrast, the codon usage in *C. thermocellum* is less biased and is closer to that found in *B. subtilis*. These facts are well illustrated by the finding that the wobble position of <60% of codons is occupied by A or U in *C. thermocellum*, whereas the corresponding values for *C. perfringens* and *C. acetobutylicum* are >80% (261).

MOLECULAR GENETICS OF PATHOGENIC CLOSTRIDIA

Of the many pathogenic clostridia, only *C. perfringens* has been subjected to a classical genetic analysis. In the early days, a defined chemical medium for growth was established (29) and a number of auxotrophic mutants were isolated (214). Although other workers characterized additional mutants, including several with defects in sporulation (61), progress was slow until the power of recombinant DNA technology was brought to bear in the 1980s.

C. perfringens

C. perfringens has much more advanced molecular genetics than all the other pathogenic clostridia. Shuttle vectors, transformation systems, chromosomal maps, and several gene libraries are available (198). Initial emphasis was on the cloning and characterization of known or potential virulence genes likely to be of importance in the various *C. perfringens*-mediated pathologies such as gas gangrene, necrotic enteritis, and food poisoning (86, 143). Several research groups reported the cloning and molecular analysis of the *plc* (phospholipase C) and *pfoA* (perfringolysin O) genes, which encode, respectively, the α and θ toxins, believed to be the major cytolytic factors implicated in gas gangrene (136, 174, 202, 216, 226, 227, 231–233). An interesting feature of *pfoA* is its close linkage to *pfoR*, a regulatory locus encoding a putative transcriptional activator and the first such locus identified in any clostridial species (216).

Other candidate virulence genes that have been analyzed are *nagH* (41) and *nanH* (195), which encode, respectively, hyaluronidase and sialidase, hydrolytic enzymes capable of destroying connective tissue, and *etx* (106), the gene coding for the ε toxin, a potent virulence factor important in veterinary medicine. Among the other genes characterized to date are those encoding the rRNAs (73), histidine decarboxylase (235), and small acid-soluble proteins found in spores (105). The *cpe* gene has also been studied (97, 110, 234). Its expression is linked to sporulation, and its product, enterotoxin, is the sole factor required for food poisoning (87, 142). The various toxin genes that have been cloned from *C. perfringens* and other toxigenic clostridia are listed in Table 2.

In addition to analysis of genes concerned with toxin production, a major subject of genetic research in *C. perfringens* has been the molecular basis for antibiotic resistance. This was recently reviewed in reference 197. The intensive and often indiscriminate use of antibiotics in veterinary and human medicine has resulted in the emergence of resistant strains of bacteria that display a number of interesting properties. Tetracycline resistance in *C. perfringens* is due to

Table 2. Cloned clostridial toxin genes

Organism	Toxin	Gene	Location	Activity	References
C. perfringens	α	*plc*	Chromosome	Phospholipase-sphingomyelinase	38, 42, 136, 174, 202, 226, 231
C. perfringens	Enterotoxin	*cpe*	Variable	Ion channel	87, 97, 110, 148, 234
C. perfringens	ε	*etx*	Plasmid	?	42, 106
C. perfringens	μ	*nagH*	Chromosome	Hyaluronidase	41, 42
C. perfringens	Sialidase	*nanH*	Chromosome	Neuraminidase	195
C. perfringens	θ (perfringolysin O)	*pfoA*	Chromosome	Pore-forming cytotoxin	216, 232, 233
C. tetani	Tetanus	*tet*	Plasmid	Neurotoxin (central nervous system)	65, 70, 71
C. botulinum	Botulism	*bot*	Variable	Neurotoxin (peripheral nervous system)	27, 85, 99, 126, 189, 225, 243
C. butyricum	Botulism	*bot*	?	Neurotoxin (peripheral nervous system)	72, 85, 189
C. difficile	A	*toxA*	Chromosome	Enterotoxin	59, 117, 184, 205, 238, 251
C. difficile	B	*toxB*	Chromosome	Cytotoxin	237, 251

a single, almost universally distributed tetracycline resistance determinant, *tetP*, located on the ubiquitous conjugative plasmid pCW3 (2, 3, 6). Intriguingly, this resistance gene has not been found in other genera or in *C. difficile* (1). Chloramphenicol resistance is associated with two distinct determinants, *catP* (220) and *catQ* (19). CatP is identical to CatD from *C. difficile* (254), which suggests that direct genetic exchange, possibly involving conjugative transposons, may occur between these two clostridial species (199). In contrast, the *catQ* gene appears to be confined to *C. perfringens*, although its product is more closely related to chloramphenicol acetyltransferases from *Staphylococcus aureus* than to CatP. *C. perfringens* plasmids commonly confer resistance to macrolide-lincosamide antibiotics (erythromycin, clindamycin, and lincomycin), owing to the presence of the *ermP* determinant (24) located between two direct repeats of about 1,200 bp (197). This gene is identical in sequence to the *erm* determinant of pAMβ1 from *Enterococcus faecalis* (32), which suggests that gene transfer can also occur between *C. perfringens* and other gram-positive bacteria in nature.

C. tetani

Since tetanus is one of the major causes of human morbidity and since there is a large market for an antitetanus vaccine, it is not surprising that genetic studies of *C. tetani* have been devoted almost exclusively to the tetanus toxin gene. This gene was cloned by two groups interested in understanding the molecular biology and mode of action of tetanus toxin (65, 70, 71), one of the most potent neurotoxins known, and improving the existing vaccine. Among the many interesting findings was the observation that the structural gene, composed of 1,315 codons rich in d(A+T), was expressed very poorly in *E. coli*. However, reasonable expression levels of a truncated gene encoding the C fragment could be obtained following removal of the "rare" codons, and recombinant fragment C offers great potential as a subunit vaccine (144, 145).

C. botulinum

Botulism is a generally fatal condition resulting from ingestion of food containing a powerful neurotoxin released by *C. botulinum*. Like the tetanus toxin, the botulism toxin also consists of heavy and light polypeptide chains, but it differs from tetanus toxin in its tropism, as it is active on the peripheral nervous system, where it blocks the release of acetylcholine. Seven toxin variants are known, and the genes for four of them (A, C1, D, and E) have been cloned and sequenced (27, 85, 99, 126, 189, 225, 243), as have the genes for three toxins from *C. butyricum* that are very closely related to type E (72, 85, 189). Current studies are aimed at the development of diagnostic tests for use in human medicine and the food industry.

C. difficile

Antibiotic-associated colitis and, more rarely, pseudomembranous colitis both result from colonization of the lower intestine of humans by *C. difficile* (79). The two toxins, A and B, produced by this organism probably represent the major virulence determinants.

Both toxin A, an enterotoxin, and toxin B, a potent cytotoxin, are very large polypeptides of 2,710 and 2,366 amino acid residues, respectively (59, 117, 184, 237, 238, 251), each containing 38 copies of a 20- or 21-residue repeat at the COOH terminus. In the case of toxin A, the repeats are involved in carbohydrate binding, and like the toxin B repeats, they share conserved sequences with a family of streptococcal ligand-binding polypeptides (251). The precise role of toxins A and B in the pathogenesis of *C. difficile* is currently being assessed by reverse genetics, and a series of toxin-deficient mutants has been constructed by gene replacement (252).

The other aspect of *C. difficile* genetics that has been studied is the problem of antibiotic resistance and its significance in the context of antibiotic-associated colitis. Conjugative transposons conferring resistance to tetracycline (TetM) and macrolides have been found in the chromosome of *C. difficile* (93, 94, 168),

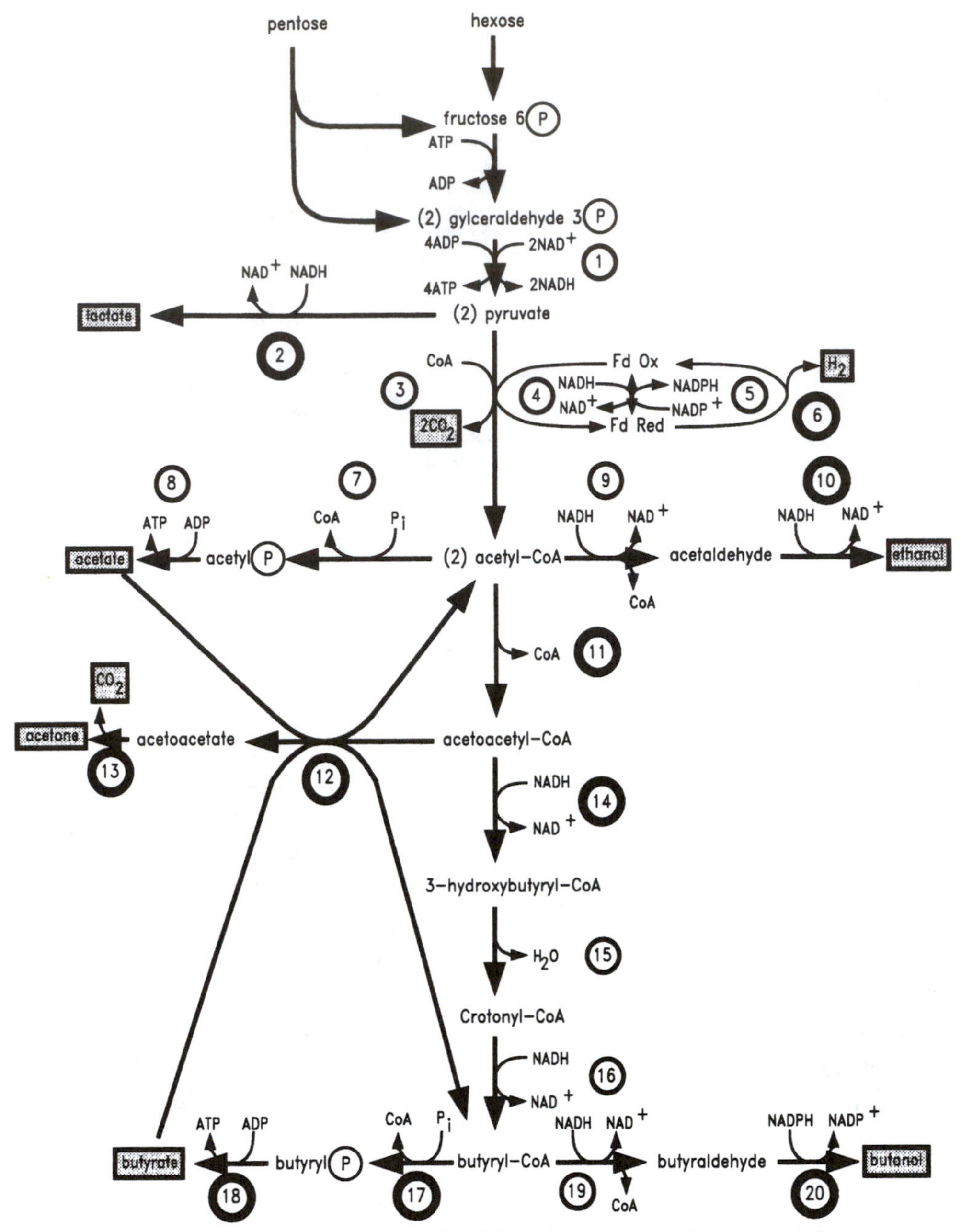

Figure 2. ABE fermentation in *C. acetobutylicum* (modified from that given in reference 261). Enzymes are indicated by numbers as follows (numbers in bold circles denote enzymes whose genes have been cloned in *E. coli*): 1, glyceraldehyde-3-phosphate–NAD oxidoreductase; 2, lactate dehydrogenase (51, 156); 3, pyruvate-ferredoxin oxidoreductase; 4, NADH-ferredoxin oxidoreductase; 5, NADPH-ferredoxin oxidoreductase; 6, hydrogenase (cloned from *C. pasteurianum* [155]); 7, phosphate acetyltransferase; 8, acetate kinase; 9, acetaldehyde dehydrogenase; 10, ethanol dehydrogenase (263, 264); 11, acetyl-CoA acetyltransferase (thiolase) (156, 181); 12, acetoacetyl-CoA-acetate–butyrate-CoA transferase (44, 81); 13, acetoacetate decarboxylase (80, 180); 14, 3-hydroxybutyryl-CoA dehydrogenase (262); 15, crotonase; 16, butyryl-CoA dehydrogenase; 17, phosphate butyryltransferase (43, 156); 18, butyrate kinase (43, 156); 19, butyraldehyde dehydrogenase; 20, butanol dehydrogenase (156, 183). P, phosphate.

and a chloramphenicol resistance gene has been cloned and sequenced (253, 254).

THE SOLVENTOGENIC CLOSTRIDIA

A number of clostridia ferment sugary substrates to produce solvents and alcohols. The most extensively characterized organism is *C. acetobutylicum*, well known for the acetone-butanol-ethanol (ABE) fermentation that it undertakes (16, 121, 193). This organism played an important role in industrial-scale production of bulk chemical feedstocks and is associated with the establishment of the State of Israel in 1949 as a result of the Balfour declaration. Chaim Weizmann,

Table 3. Cloned genes concerned with fermentative metabolism

Gene(s)	Enzyme	Strain and reference(s)	Comments
adc	Acetoacetate decarboxylase	ATCC 824 (180), DSM 792 (80)	Convergently transcribed with *atoDA*
adh	Alcohol dehydrogenase	NCP 262 (263, 264)	Adjacent to *fadB*
atoB	Thiolase	ATCC 824 (181), NCIMB 8052 (156)	
atoDA	CoA transferase	ATCC 824 (44), DSM 792 (81)	Convergently transcribed with *adc*
bdh	Butanol dehydrogenase	ATCC 824 (183), *Clostridium tetanomorphum* (156)	Two genes
butK	Butyrate kinase	ATCC 824 (43), NCIMB 8052 (156)	
fadB	β-Hydroxybutyryl CoA dehydrogenase	NCP 262 (262)	Adjacent to *adh*
fdh	Ferredoxin	*C. pasteurianum* ATCC 6013 (88)	
hyd	Hydrogenase I	*C. pasteurianum* ATCC 6013 (155)	
ldh	Lactate dehydrogenase	ATCC 824 (51), NCIMB 8052 (156)	
ptb	Phosphate transbutyrylase	ATCC 824 (43) NCIMB 8052 (156)	

who discovered ABE fermentation while working in Manchester, England, became the first president of the State of Israel.

Physiology

ABE fermentation (Fig. 2) is biphasic. Initially, electron flow is directed towards the production of volatile fatty acids (mostly acetate and butyrate) that accumulate in the growth medium, with a concomitant lowering of the external pH. In response to an unknown stimulus, a number of additional enzymes are induced. Their concerted action leads to reassimilation and reduction of volatile fatty acids to neutral solvents (butanol and acetone). The enzymes induced include acetoacetate decarboxylase, acetoacetyl-coenzyme A (CoA)-acetate–butyrate-CoA transferase, butanol dehydrogenase, and butyraldehyde dehydrogenase (9, 17, 18, 64, 98, 107, 108, 151, 179, 258). After the switch to solventogenesis, further carbon and electron flow is then directed towards production of acetone and butanol instead of acetate and butyrate.

The physiology and biochemistry of ABE fermentation have been extensively studied (see reference 22 for a recent review), and several factors that regulate carbon and electron flux during the solventogenic fermentation are known (15). Carbon-limited growth prevents solvent production (and, incidentally, sporulation; cf. *B. subtilis* [163, 164]), suggesting that high ATP concentrations favor solventogenesis (153). Solvent formation is also favored by phosphate limitation (12), perhaps because this limits the activity of phosphate transbutyrylase, which is involved in butyrate production. Butanol formation is enhanced by factors that reduce the effectiveness with which hydrogenase can dispose of excess reducing power, such as a high partial pressure of hydrogen (58, 259) or treatment with carbon monoxide (56, 125, 152–154).

The onset of solventogenesis usually coincides with the onset of spore formation (119, 138, 139). Solventogenesis is a kind of stress response, permitting the bacteria effectively to detoxify their environment. Production of *C. acetobutylicum* equivalents of the DnaK and GroEL heat shock proteins is stimulated during the switch from acid to solvent formation (186). As is the case for *B. subtilis* (242), the *grpE*, *dnaK*, and *dnaJ* genes of *C. acetobutylicum* (strain DSM 1731) form a polycistronic operon together with an additional heat shock gene, designated *orfA* (corresponding to *orf-39* in *B. subtilis*), of unknown function, that lies upstream from *grpE* (171). Similarly, the *groES* and *groEL* heat shock genes of the same strain are organized to form a bicistronic operon (170). Primer extension analysis revealed that both operons are apparently transcribed from vegetative promoters, providing no evidence for the presence in *C. acetobutylicum* of a specialized heat shock σ factor analogous to the *rpoH* gene product, σ^{32}, of *E. coli* (91, 132). In both operons, however, a 10- to 11-bp inverted repeat motif, which may play an important regulatory role in the heat shock response, is located between the transcription and translation start sites. A very similar motif is found at an equivalent position in a number of other bacterial genes that are mostly concerned with the heat shock response (170, 171, 242).

Genetics

A genetic approach is now being employed to analyze the solventogenic fermentation of *C. acetobutylicum*. Mutants deficient in some of the relevant enzymes, such as butanol dehydrogenase and CoA transferase, are available (48, 63, 68, 111, 122, 194). Possible regulatory mutants simultaneously lacking several enzyme activities have also been obtained (25, 48). Asolventogenic mutants can show pleiotropic defects in other processes such as spore formation, granulose production, capsule synthesis, and motility (138, 139, 182, 191, 249). Some strains of *C. acetobutylicum* (e.g., NCIMB 8052) show an interesting phenomenon known as "degeneration" and lose the ability to form solvents, spores, and granulose after long periods in culture (221). Sporogenic revertants selected from such strains simultaneously reacquire the ability to make solvents and granulose (249).

Recently, streptococcal conjugative transposons have been used to generate mutants unable to make solvents; some of these mutants are apparently regulatory (25). One of them appears to contain an insertion in the gene encoding the threonyl tRNA corresponding to the rare ACG codon, which has been found so far only in genes concerned with formation of acetone and uptake and metabolism of minor C and N substrates in *C. acetobutylicum* (204). Apart from suggesting that expression of these genes is under some kind of translational control, this finding has implications for the expression of heterologous genes in *C. acetobutylicum* (see section on codon usage above). There is an interesting parallel here with the *bldA* locus of *Streptomyces coelicolor*, which encodes a leucyl tRNA corresponding to the rare TTA codon. This codon appears to be confined to genes principally expressed during the transition from vegetative growth to differentiation (aerial mycelium formation and antibiotic production) in streptomycetes (135).

Several of the genes encoding enzymes concerned with fermentative metabolism have been cloned in *E. coli* (Table 3). Some (e.g., *ptb* [phosphate transbutyrylase], *butK* [butyrate kinase], *adh* [alcohol dehydrogenase], and *ldh* [lactate dehydrogenase]) were isolated by functional complementation of suitable auxotrophic strains. Others (e.g., *atoB* [thiolase], *bdh* [butanol dehydrogenase], *atoDA* [acetoacetyl-CoA-acetate–butyrate-CoA transferase], *adc* [acetoacetate decarboxylase], and *hyd* [hydrogenase]) were isolated by screening DNA libraries with oligonucleotide probes or by using immunochemical methods.

Some of these functionally related genes are clustered. The *atoDA* genes encoding the two subunits of the CoA transferase are adjacent (44) and cotranscribed in the ATCC 824 and DSM 1731 strains (81). They are the distal genes in an operon whose 4.1-kb transcript probably includes other as-yet-uncharacterized genes lying upstream. This is reminiscent of the organization of the *ato* operon in *E. coli* (116), which contains, in addition, a thiolase (*atoB*) gene. However, in *C. acetobutylicum* ATCC 824, the *atoB* gene is not adjacent to *atoDA* (22).

The *adc* gene lies immediately downstream from *atoDA* (80, 180) and is convergently transcribed from a promoter closely resembling the consensus vegetative promoters of clostridia and other gram-positive bacteria (81). These latter results provide no evidence for the involvement of a novel σ factor in the induction of this solventogenic enzyme. Production of both *adc* and *atoDA* mRNAs is enhanced during the shift from acidogenesis to solventogenesis, but it precedes the appearance of solvents in the culture medium by several hours (81). Recent experiments have indicated that the production of acetoacetate decarboxylase (and phosphate transbutyrylase) can be enhanced between 20- and 40-fold by cloning the corresponding structural genes on multicopy plasmids and reintroducing them into the ATCC 824 strain via electroporation (151).

The *ptb* and *butK* genes of strain ATCC 824 may also form part of an operon; they are transcribed from the same DNA strand and separated by no more than 250 bp (43). Three alcohol dehydrogenase genes have been isolated from *C. acetobutylicum*. One, from the NCP 262 strain (263, 264), encodes an NADPH-specific enzyme. It has been denoted *adh*, although it shows substantial activity with either ethanol or butanol as substrate. Two genes, denoted *bdh*, have been isolated on a single restriction fragment of strain ATCC 824 DNA (22). Their products differ in substrate (acetaldehyde versus butyraldehyde) and cofactor (NADH versus NADPH) specificities and may correspond to the two separable enzyme activities previously characterized (64). The roles played by these gene products in butanol and ethanol production are far from clear at present. Sequence inspection revealed that an open reading frame immediately upstream from the *adh* gene in the NCP 262 strain encodes β-hydroxybutyryl-CoA dehydrogenase (*fadB*) (262).

Several of the strain NCIMB 8052 counterparts of these genes have been located at three widely separated positions on the physical map of the bacterial chromosome, which currently contains four gaps (246). The *atoDA* and *atoB* genes lie on adjacent *Apa*I fragments. This arrangement could be fortuitous, but these genes might be organized in an operon (as in *E. coli*) containing a rare restriction site. A second thiolase gene has also been detected in the NCIMB 8052 strain (156). The *ptb* and *butK* genes lie within the same 30-kbp genome segment in the NCIMB 8052 strain and appear to be organized in an operon (156), as may also be the case in the ATCC 824 strain (43). The *ldh* gene lies elsewhere on the NCIMB 8052 chromosome.

Further analysis of these and other genes concerned with solventogenesis should soon provide a coherent picture of the regulation of biphasic ABE fermentation.

POLYSACCHARIDE UTILIZATION

Cellulose

There is an extensive literature on polysaccharide-degrading enzymes from clostridia (reviewed in references 10, 20, and 36). Most work has been done on cellulases from thermophilic species. Annual world production of cellulose has been estimated at some 4×10^{10} metric tons (52), and work on clostridial cellulases has been impelled by the prospect of microbial production of ethanol from inexpensive cellulosic materials (141). The availability of cloned genes whose products can degrade cellulose (or any other relevant polysaccharide; see below) could have other important implications for industrial fermentation processes. "Improved" organisms endowed with the ability to use a range of different polysaccharides would provide, at one and the same time, additional flexibility and access to inexpensive, alternative fermentation substrates (e.g., wastes from other processes).

Cellulolysis, however, is an interesting phenomenon in its own right. The enzymatic degradation of a highly recalcitrant, insoluble, crystalline substrate is not a trivial problem. Our understanding of cellulolysis is far from complete.

Broadly speaking, three classes of enzymes are directly involved in cellulose degradation: cellobiohydrolases (exoglucanases) attack the nonreducing end of the glycan chain, splitting off cellobiose units; endoglucanases attack the chain randomly at internal positions, yielding a heterogeneous mixture of cellodextrin products; and β-glucosidases hydrolyze low-molecular-weight cellodextrins to their monomer glucose units. A single enzyme showing both exo- and endoglucanase activities, depending on the substrate, has been reported from *C. stercorarium* (34).

Although cellobiohydrolase genes have only rarely been isolated from clostridia (34, 166), many endoglucanase genes have been isolated and characterized. The predicted products of these cloned genes are similar to cellulases from other sources (bacterial and fungal) and comprise characteristic domains, the order of which can vary from one gene product to another (20, 36, 84). Many of the gene products fall into group A, as defined by hydrophobic cluster analysis (103).

The cellulases from *C. thermocellum* are the most studied, but another thermophilic species, *C. stercorarium*, and mesophilic species such as *Clostridium cellulolyticum* and *Clostridium cellulovorans* are now attracting attention (36). *C. thermocellum*, in common with many other cellulolytic bacteria, produces from the cell surface characteristic excrescences with groups of cellulosome complexes at their extremities (129–131). The cellulosome appears to be a more or less organized assemblage of about 18 different polypeptides ranging in size from 48 to >200 kDa (53, 128) whose concerted action enables *C. thermocellum* to adhere to crystalline cellulose fibrils and degrade them.

It has proved difficult to fractionate the cellulosome into its biologically active constituent polypeptides. Two components, S_8 (82 kDa; also denoted S_S) and S_L (250 kDa; also denoted S_1), account for >50% of the total cellulosome protein (256). The S_L polypeptide is heavily glycosylated and apparently lacks glucanase activity (82). It seems to act as an anchor or scaffold for many (all?) of the remaining cellulosome components (83). S_L is required for the attachment of S_S (S_8) to crystalline cellulose (255); in the absence of S_L, S_S has no activity on crystalline cellulose (Avicel) but retains the ability to degrade a substituted cellulose derivative, carboxymethyl cellulose.

Fifteen different endoglucanase genes have been isolated to date from banks of the ATCC 27405 (=NCIMB 10682) strain of *C. thermocellum* in *E. coli* (100). They are all located on the bacterial chromosome and tend to lie in monocistronic operons. All eight that have been sequenced to date encode an N-terminal signal sequence characteristic of secreted proteins and a major catalytic domain. Most of them also have a duplicated segment of 22 amino acids with a variable spacer (10 to 17 residues). Deletion analysis has shown that this duplicated motif is not essential for catalytic activity with soluble substrates (46, 90, 95, 257). It seems to be a diagnostic feature of cellulosome-associated glucanases. The *celC* gene product, which lacks it (209), is not a cellulosome component, whereas the *xynZ* gene product, a xylanase containing this motif (90), presumably is. Engineered derivatives of CelD and XynZ lacking this motif fail to bind to S_L (228), which suggests that the function of the motif is to promote the association of cellulosome enzymes with the scaffolding protein S_L.

The expression of *C. thermocellum* endoglucanases has been investigated by using organisms growing on cellobiose. Under these conditions, *celA*, *celD*, and *celF* are most actively transcribed at the end of exponential growth, suggesting that a form of catabolite repression controls their transcription (162). Both *celA* and *celD* have dual promoters. The major *celA* promoter used in vivo has distinct similarities with σ^D promoters of *B. subtilis*, and a minor σ^A promoter is located upstream (21). Strong promoters have been used to overexpress several endoglucanase genes in *E. coli*, giving appreciable quantities of enzymatically active, thermostable proteins.

Unlike most other cellulolytic clostridia, *C. stercorarium* produces a relatively simple cellulase system comprising only an endoglucanase and an exoglucanase, both of which act on crystalline cellulose (35, 54, 55). Both genes have been cloned and sequenced (36, 208), and their predicted products contain one or two copies of a pair of distinct substrate-binding domains, each comprising about 150 residues (115). The two genes are adjacent on the bacterial chromosome and lie in the same transcriptional orientation, suggesting that they are cotranscribed and coordinately regulated (36).

At present, it is not clear why most cellulolytic clostridia, including *C. thermocellum*, require a substantial battery of cellulolytic enzymes to accomplish what *C. stercorarium* can apparently achieve with only two.

Hemicellulose

The cellulose fibrils in plant cell walls are normally embedded in an amorphous hemicellulose matrix containing branched heteropolysaccharides of variable compositions, depending on their origins. This hemicellulose material may contain xylose, mannose, and a number of other sugars as major monosaccharide components. According to the predominant monomer, hemicelluloses are termed xylans, mannans, etc. Organisms that hydrolyze cellulose in nature also elaborate hemicellulases, which act to expose cellulose fibrils in the substrate. Xylans are a major component of most hemicelluloses. They are degraded by xylanases and xylosidases.

C. thermocellum grows only poorly on hemicellulose and not at all on xylose (244). Nevertheless, it produces several xylanolytic enzymes, some of which are cellulosome components (90, 127, 165). Presumably, these enzymes mainly serve to expose cellulose fibrils in naturally occurring substrates. Other cellulolytic clostridia, such as *C. stercorarium*, probably make better use of this hemicellulose material and elaborate an impressive battery of enzymes concerned with xylan utilization (23, 33). At least nine distinct genes encoding xylanolytic enzymes have been isolated from this organism (203).

The ability to grow on hemicellulose materials is not the sole prerogative of cellulolytic clostridia. Xylanolytic enzymes and *xyn* genes have been isolated from both the ATCC 824 and NCP 262 strains of *C. acetobutylicum* (133, 134, 265, 266).

Starch

Many clostridia produce α-amylases, β-amylases, and pullulanases (reviewed in reference 36). Some enzymes, such as the product of the *Clostridium thermohydrosulfuricum apu* gene, have been denoted amylopullulanases, since the same active site is involved in hydrolysis of both α-1,4 and α-1,6 linkages (146, 187).

Six genes concerned with starch hydrolysis have been cloned from thermophilic organisms, and sequence information is available in most cases. Clostridial amylases fall into two distinct families, of which one comprises the α-amylases and pullulanases and the other comprises the β-amylases. The deduced products of all sequenced genes contain secretory signal peptides and have composite structures with both catalytic and substrate-binding regions (36).

The clostridial *amy* genes are inducible, and glucose repression of the *C. thermohydrosulfuricum apu* gene has also been demonstrated (109). The canonical σ^A promoter of the *apu* gene is flanked by two palindromes (150). The downstream palindrome is similar to the *amyO* operator of the *B. subtilis amyE* gene, which suggests that the mechanism of catabolite repression of *apu* expression in *C. thermohydrosulfuricum* may be similar to that of *amyE* in *B. subtilis* (102, 240, 241).

Lying upstream from the *amyA* gene of *C. thermosulfurogenes* (13) are two open reading frames that encode proteins that may be involved in the uptake of starch degradation products. These are extremely hydrophobic and are similar to binding-protein-dependent transport systems of *E. coli* (14).

PROSPECTS FOR FUTURE RESEARCH

The mesophilic clostridia might usefully be employed for experiments aimed at elucidating the biological consequences of having DNA with a low d(G+C) content. In some species, the low content approaches the theoretical minimum of the conventional genetic code (69). Intergenic regions are often substantially richer in d(A+T) than are coding regions, which may explain why streptococcal conjugative transposons, such as Tn*1545*, that preferentially insert into d(A+T)-rich target sequences tend to produce silent mutations in *C. acetobutylicum*.

Why the DNA of mesophilic clostridia is so d(A+T) rich is not known, and to our knowledge, no systematic investigation of the fate of d(G+C)-rich sequences artificially introduced into mesophilic clostridia has yet been conducted. In addition to the remarkable bias in codon usage previously mentioned, an exceptionally low d(G+C) content has many other consequences, since it imposes constraints on a variety of metabolic transactions. For example, it has important implications for DNA-protein interactions, and as already noted elsewhere (77), in experiments with promoters of equivalent strength, *C. perfringens* RNA polymerase has a distinct preference for d(A+T)-rich templates. Moreover, DNA substantially enriched in two of its four component building blocks must necessarily have a reduced informational content. Regulatory proteins most often interact with target sequences lying in intergenic regions, which are even poorer in d(G+C) than are coding sequences (261). Perhaps the targets for regulatory proteins tend to be those parts of intergenic regions with a higher-than-average d(G+C) content. Alternatively, clostridial regulatory proteins, having coevolved with the genome, may recognize very d(A+T)-rich sites (like the integrase-excisionase system of streptococcal conjugative transposons). To confer an adequate degree of specificity, longer target sequences may be required than are necessary with DNA of average d(G+C) content. As the sequences of more genes and their associated regulatory elements become available, these hypotheses can be evaluated.

It has been proposed that negative supercoiling plays an important role in the regulation of gene expression (104). Such coiling might be of comparatively lesser importance in organisms whose DNAs have low d(G+C) contents, since the intergenic regions tend to be particularly d(A+T) rich, and localized melting associated with RNA polymerase binding may be less dependent on torsional stresses introduced by helicases. In this context, it will be of interest to compare the sequences and activities of DNA gyrases from mesophilic clostridia with those from other organisms, such as *E. coli* and *B. subtilis*, whose DNAs are much more d(G+C) rich.

Base-pairing interactions between nucleic acids become progressively weaker and there is an increasing chance of random mispairing as the d(G+C) content decreases. It is noteworthy that the rRNA genes of mesophilic clostridia have a d(G+C) content that is almost twice the genomic average (73). This high d(G+C) content presumably reflects unknown functional constraints required to ensure the adoption of a stable and specific secondary structure in the rRNA. It is also striking that the thermophilic clostridia have DNAs with much higher d(G+C) contents than those of the mesophilic clostridia (Table 1). Many of these organisms grow at temperatures that would presumably lead to appreciable melting of DNA with a lower d(G+C) content.

The molecular basis of the metabolic switch from acid to solvent production, which is unique to the solventogenic clostridia, is likely to be a major preoccupation in the next few years. Like many other stress responses, it may involve a sensor-regulator couple (222) that relays information (concerning the external pH perhaps) to the transcriptional apparatus. Stress responses may also involve specific σ factors (223) for the transcription of particular stress-associated genes, but the currently available evidence (81, 170, 171) does not favor this hypothesis. A number of groups are actively investigating this topic, and it will be interesting to investigate whether similar regulatory devices have been adopted by the various apparently disparate strains of *C. acetobutylicum*.

It is surprising, perhaps, that almost no attention has yet been paid to a comparative analysis of *spo* genes in *Bacillus* and *Clostridium* spp. It has long been known (201) that the sporulation processes in these two groups of organisms are morphologically similar, yet by molecular taxonomic criteria, organisms in the two genera appear to have diverged in the distant evolutionary past. Two *ssp* genes have been isolated from *C. perfringens* (105), and internal segments of what are believed to be *spo0A* homologs have been obtained from several species of *Clostridium* (37). A gene whose product is similar to that of the *spoIID* gene of *B. subtilis* has been found in the NCP 262 strain of *C. acetobutylicum* (96). These findings suggest that this common developmental process in bacilli and clostridia does indeed reflect a fundamental underlying similarity at the genetic level. Is sporulation an ancient evolutionary trait? To what extent has there been horizontal gene transfer between *Bacillus* and *Clostridium* species? Answers to these questions may be forthcoming once more clostridial *spo* genes have been isolated and analyzed.

Little is known about the molecular basis of the various degrees of O_2 sensitivity displayed by different clostridia. It is generally accepted that most aerobic organisms contain superoxide dismutase and catalase and that most anaerobic organisms do not. However, there are exceptions, including both aerobes that lack these enzymes and anaerobes that contain them (167). Other factors appear to be of importance in helping the organism overcome the potentially damaging effects of free radicals generated in the presence of oxygen. Now that genetic analysis in the model species is possible, it would be timely to examine whether aerobic organisms contain genes able to enhance the aerotolerance of either *C. perfringens* or *C. acetobutylicum*.

During the 1990s, considerable progress toward unraveling the molecular mechanisms employed by various pathogenic clostridia in their respective pathologies is likely to be made. The application of reverse genetics will lead to a better understanding of the events involved in gas gangrene and pseudomembranous colitis. The attenuated strains that will result from this research may well find application as live vaccines in human and especially veterinary medicine. Pulpy kidney, tetanus, and necrotic enteritis of farm animals are all diseases that could be tackled by this immunoprophylactic approach.

Finally, it would be misleading to leave the impression that the many genetic tools currently available for use in *C. perfringens* and *C. acetobutylicum* have reached the same degree of sophistication as those currently employed in *B. subtilis*. Nevertheless, vectors for detecting promoter activity and regulating the expression of cloned genes have been developed (158). A major lacuna in genetic analysis of both organisms is a lack of genetic exchange involving homologous recombination with the bacterial chromosome. This process is a necessary prerequisite for undertaking the gene disruption and allelic replacement experiments that are of cardinal importance for analyzing the roles of specific genes in biological processes in vivo. The encouraging results very recently obtained concerning gene replacement in *C. difficile* (252) and gene disruption with integrational plasmids in the NCIMB 8052 strain of *C. acetobutylicum* (246) bode well.

Acknowledgments. We thank all of our colleagues, who kindly supplied advice and unpublished information. Work in Paris was supported by grants from the Fondation pour la Recherche Medicale and the Institut Pasteur. Work in Aberystwyth was supported by the EEC BAP programme and the SERC Biotechnology Directorate.

REFERENCES

1. **Abraham, L. J., D. I. Berryman, and J. I. Rood.** 1988. Hybridization analysis of the class P tetracycline resistance determinant from the *Clostridium perfringens* R-plasmid, pCW3. *Plasmid* **19:**113–120.
2. **Abraham, L. J., and J. I. Rood.** 1985. Cloning and analysis of the *Clostridium perfringens* tetracycline resistance plasmid, pCW3. *Plasmid* **13:**155–162.
3. **Abraham, L. J., and J. I. Rood.** 1985. Molecular analysis of transferable tetracycline resistance plasmids from *Clostridium perfringens*. *J. Bacteriol.* **161:**636–640.
4. **Abraham, L. J., and J. I. Rood.** 1987. Identification of Tn*4451* and Tn*4452*, chloramphenicol resistance transposons from *Clostridium perfringens*. *J. Bacteriol.* **169:**1579–1584.
5. **Abraham, L. J., and J. I. Rood.** 1988. The *Clostridium perfringens* chloramphenicol resistance transposon Tn*4451* excises precisely in *Escherichia coli*. *Plasmid* **19:**164–168.
6. **Abraham, L. J., A. J. Wales, and J. I. Rood.** 1985. Worldwide distribution of the conjugative *Clostridium perfringens* tetracycline resistance plasmid, pCW3. *Plasmid* **14:**37–46.
7. **Allen, S. P., and H. P. Blaschek.** 1988. Electroporation-induced transformation of intact cells of *Clostridium perfringens*. *Appl. Environ. Microbiol.* **54:**2322–2324.
8. **Allen, S. P., and H. P. Blaschek.** 1990. Factors involved in the electroporation-induced transformation of *Clostridium perfringens*. *FEMS Microbiol. Lett.* **70:**217–220.
9. **Andersch, W., H. Bahl, and G. Gottschalk.** 1983. Level of enzymes involved in acetate, butyrate, acetone and butanol formation by *Clostridium acetobutylicum*. *Eur. J. Appl. Microbiol. Biotechnol.* **18:**327–332.
10. **Aubert, J. P., P. Béguin, and J. Millet.** 1993. Genes and proteins involved in cellulose and xylan degradation by *Clostridium thermocellum*, p. 412–422. *In* M. Sebald (ed.), *Genetics and Molecular Biology of Anaerobic Bacteria*. Springer-Verlag, New York.
11. **Azeddoug, H., J. Hubert, and G. Reysset.** Stable inheritance of shuttle vectors based on plasmid pIM13 in a mutant strain of *Clostridium acetobutylicum*. *J. Gen. Microbiol.* **138:**1371–1378.
12. **Bahl, H., W. Andersch, and G. Gottschalk.** 1982. Continuous production of acetone and butanol by *Clostridium acetobutylicum* in a two-stage phosphate limited chemostat. *Eur. J. Appl. Microbiol. Biotechnol.* **15:**201–205.
13. **Bahl, H., G. Burchhardt, A. Spreinat, K. Haeckel, A. Wienecke, B. Schmidt, and G. Antranikian.** 1991. α-Amylase of *Clostridium thermosulfurogenes* EM1: nucleotide sequence of the gene, processing of the enzyme, and comparison to other α-amylases. *Appl. Environ. Microbiol.* **57:**1554–1559.
14. **Bahl, H., G. Burchhardt, and A. Wienecke.** 1991. Nucleotide sequence of two *Clostridium thermosulfurogenes* EM1 genes homologous to *E. coli* genes encoding integral membrane components of binding protein-dependent transport systems. *FEMS Microbiol. Lett.* **81:**83–88.
15. **Bahl, H., and G. Gottschalk.** 1984. Parameters affecting solvent production by *Clostridium acetobutylicum* in continuous culture. *Biotechnol. Bioeng. Symp.* **14:**215–223.
16. **Bahl, H., and G. Gottschalk.** 1988. Microbial production of butanol/acetone, p. 1–30. *In* H. J. Rehm and G. Reed (ed.), *Biotechnology*, vol. 6b. VCH Verlagsgesellschaft, Weinheim, Germany.
17. **Ballongue, J., J. Amine, E. Masion, H. Petitdemange, and R. Gay.** 1985. Induction of acetoacetate decarboxylase in *Clostridium acetobutylicum*. *FEMS Microbiol. Lett.* **29:**273–277.
18. **Ballongue, J., R. Janati-Idrissi, H. Petitdemange, and R. Gay.** 1989. Correlation between solvent production and level of solventogenic enzymes in *Clostridium acetobutylicum*. *J. Appl. Bacteriol.* **67:**611–617.
19. **Bannam, T. L., and J. I. Rood.** 1991. The relationship between the *Clostridium perfringens catQ* gene product and chloramphenicol acetyltransferases from other bacteria. *Antimicrob. Agents Chemother.* **35:**471–476.
20. **Béguin, P.** 1990. Molecular biology of cellulose degradation. *Annu. Rev. Microbiol.* **44:**219–248.
21. **Béguin, P., M. Rocancourt, M. C. Chebrou, and J.-P. Aubert.** 1986. Mapping of mRNA encoding endogluca-

nase A from *Clostridium thermocellum*. *Mol. Gen. Genet.* **202:**251–254.
22. **Bennett, G. N., and D. J. Petersen.** 1993. Cloning and expression of *Clostridium acetobutylicum* genes involved in solvent production, p. 317–343. *In* M. Sebald (ed.), *Genetics and Molecular Biology of Anaerobic Bacteria*. Springer-Verlag, New York.
23. **Berenger, J. F., C. Frixon, J. Bigliardi, and N. Creuzet.** 1985. Production, purification, and properties of thermostable xylanase from *Clostridium stercorarium*. *Can. J. Microbiol.* **31:**635.
24. **Berryman, D. I., and J. I. Rood.** 1989. Cloning and hybridization analysis of *ermP*, a macrolide-lincosamide-streptogramin B resistance determinant from *Clostridium perfringens*. *Antimicrob. Agents Chemother.* **33:**1346–1353.
25. **Bertram, J., A. Kuhn, and P. Dürre.** 1990. Tn*916*-induced mutants of *Clostridium acetobutylicum* defective in regulation of solvent formation. *Arch. Microbiol.* **153:**373–377.
26. **Bertram, J., M. Strätz, and P. Dürre.** 1991. Natural transfer of conjugative transposon Tn*916* between gram-positive and gram-negative bacteria. *J. Bacteriol.* **173:**443–448.
27. **Binz, T., H. Kurazono, M. Wille, J. Frevert, K. Wernars, and H. Niemann.** 1990. The complete sequence of botulinum neurotoxin type A and comparison with other clostridial neurotoxins. *J. Biol. Chem.* **265:**9153–9158.
28. **Blaschek, H. P., and M. Solberg.** 1981. Isolation of a plasmid responsible for caseinase activity in *Clostridium perfringens* ATCC 3626B. *J. Bacteriol.* **147:**262–266.
29. **Boyd, M. J., M. A. Logan, and A. A. Tytell.** 1948. The growth requirements of *Clostridium perfringens* BP6K. *J. Biol. Chem.* **174:**1013–1025.
30. **Bréfort, G., M. Magot, H. Ionesco, and M. Sebald.** 1977. Characterization and transferability of *Clostridium perfringens* plasmids. *Plasmid* **1:**52–66.
31. **Brehm, J. K., A. Pennock, H. M. S. Bullman, M. Young, J. D. Oultram, and N. P. Minton.** 1992. Physical characterization of the replication origin of the cryptic plasmid pCB101 isolated from *Clostridium butyricum* NCIB 7423. *Plasmid* **28:**1–13.
32. **Brehm, J., G. Salmond, and N. Minton.** 1987. Sequence of the adenine methylase gene of the *Streptococcus faecalis* plasmid pAMβ1. *Nucleic Acids Res.* **15:**3177.
33. **Bronnenmeier, K., C. Ebenbichler, and W. L. Staudenbauer.** 1990. Separation of the cellulolytic and xylanolytic enzymes of *Clostridium stercorarium*. *J. Chromatogr.* **521:**301–310.
34. **Bronnenmeier, K., K. P. Rücknagel, and W. L. Staudenbauer.** 1991. Purification and properties of a novel type of exo-1,4-β-glucanase (avicelase-II) from the cellulolytic thermophile *Clostridium stercorarium*. *Eur. J. Biochem.* **200:**379–385.
35. **Bronnenmeier, K., and W. L. Staudenbauer.** 1988. Resolution of *Clostridium stercorarium* cellulase by fast protein liquid chromatography (FPLC). *Appl. Microbiol. Biotechnol.* **27:**432–436.
36. **Bronnenmeier, K., and W. L. Staudenbauer.** The molecular biology and genetics of substrate utilization in clostridia. *In* D. R. Woods (ed.), *The Clostridia and Biotechnology*, in press. Butterworths, Boston.
37. **Brown, D. P., L. Ganova-Raeva, and P. Youngman.** 1992. Personal communication.
38. **Canard, B., and S. T. Cole.** 1989. Genome organization of the anaerobic pathogen *Clostridium perfringens*. *Proc. Natl. Acad. Sci. USA* **86:**6676–6680.
39. **Canard, B., and S. T. Cole.** 1990. Lysogenic phages of *Clostridium perfringens*: mapping of the chromosomal attachment sites. *FEMS Microbiol. Lett.* **66:**323–326.
40. **Canard, B., T. Garnier, B. Lafay, R. Christen, and S. T. Cole.** 1992. Phylogenetic analysis of the pathogenic anaerobe *Clostridium perfringens* using the 16S rRNA sequence. *Int. J. Syst. Bacteriol.* **42:**312–314.
41. **Canard, B., T. Garnier, B. Saint-Joanis, and S. T. Cole.** Molecular genetic analysis of the *nagH* gene encoding the hyaluronidase (Mu toxin) of *Clostridium perfringens*. Submitted for publication.
42. **Canard, B., B. Saint-Joanis, and S. T. Cole.** 1992. Genomic diversity and organisation of virulence genes in the pathogenic anaerobe *Clostridium perfringens*. *Mol. Microbiol.* **6:**1421–1429.
43. **Cary, J. W., D. J. Petersen, E. T. Papoutsakis, and G. N. Bennett.** 1988. Cloning and expression of *Clostridium acetobutylicum* phosphotransbutyrylase and butyrate kinase genes in *Escherichia coli*. *J. Bacteriol* **170:**4613–4618.
44. **Cary, J. W., D. J. Petersen, E. T. Papoutsakis, and G. N. Bennett.** 1990. Cloning and expression of *Clostridium acetobutylicum* ATCC 824 acetoacetyl-coenzyme A:acetate/butyrate:coenzyme A-transferase in *Escherichia coli*. *Appl. Environ. Microbiol.* **56:**1576–1583.
45. **Cato, E. P., W. L. George, and S. M. Finegold.** 1986. Genus *Clostridium* Prazmowski 1880, 23^{AL}, p. 1141–1200. *In* S. P. H. A. Sneath, S. N. Mair, M. E. Sharpe, and J. G. Holt (ed.), *Bergey's Manual of Systematic Bacteriology*, vol. 2. The Williams & Wilkins Co, Baltimore.
46. **Chauvaux, S., P. Béguin, J.-P. Aubert, K. M. Bhat, L. A. Gow, T. M. Wood, and A. Bairoch.** 1990. Calcium-binding affinity and calcium-enhanced activity of *Clostridium thermocellum* endoglucanase D. *Biochem. J.* **265:**261–265.
47. **Chen, G.-F. T., and M. Inouye.** 1990. Suppression of the negative effect of minor arginine codons on gene expression; preferential usage of minor codons within the first 25 codons of the *Escherichia coli* genes. *Nucleic Acids Res.* **18:**1465–1473.
48. **Clark, S. W., G. N. Bennett, and F. B. Rudolph.** 1989. Isolation and characterization of mutants of *Clostridium acetobutylicum* ATCC 824 deficient in acetoacetyl-coenzyme A:acetate/butyrate:coenzyme A-transferase (EC 2.8.3.9) and in other solvent pathway enzymes. *Appl. Environ. Microbiol.* **55:**970–976.
49. **Cole, S. T., and T. Garnier.** 1993. Molecular genetic studies of UV-inducible bacteriocin production in *Clostridium perfringens*, p. 248–254. *In* M. Sebald (ed.), *Genetics and Molecular Biology of Anaerobic Bacteria*. Springer-Verlag, New York.
50. **Collins, M. E., J. D. Oultram, and M. Young.** 1985. Identification of restriction fragments from two cryptic *Clostridium butyricum* plasmids that promote the establishment of a replication-defective plasmid in *Bacillus subtilis*. *J. Gen. Microbiol.* **131:**2097–2105.
51. **Contag, P. R., M. G. Williams, and P. Rogers.** 1990. Cloning of a lactate dehydrogenase gene from *Clostridium acetobutylicum* B643 and expression in *Escherichia coli*. *Appl. Environ. Microbiol.* **56:**3760–3765.
52. **Coughlan, M. P.** 1985. The properties of fungal and bacterial cellulases with comment on their production and application. *Biotechnol. Genet. Eng. Rev.* **3:**39–109.
53. **Coughlan, M. P., K. Hon-nami, H. Hon-nami, G. Ljungdahl, J. J. Paulin, and W. E. Rigsby.** 1985. The cellulolytic enzyme complex of *Clostridium thermocellum* is very large. *Biochem. Biophys. Res. Commun.* **130:**904–909.
54. **Creuzet, N., J. F. Berenger, and C. Frixon.** 1983. Characterization of exoglucanase and synergistic hydrolysis of cellulose in *Clostridium stercorarium*. *FEMS Microbiol. Lett.* **20:**347–350.
55. **Creuzet, N., and C. Frixon.** 1983. Purification and characterization of an endoglucanase from a newly isolated thermophilic anaerobic bacterium. *Biochimie* **65:**149–156.
56. **Datta, R., and J. G. Zeikus.** 1985. Modulation of ace-

tone-butanol-ethanol fermentation by carbon monoxide and organic acids. *Appl. Environ. Microbiol.* **49:**522–529.

57. **Dixon, B.** 1990. Anaerobes at work. *Bio/Technology* **8:**591.
58. **Doremus, M. G., J. C. Linden, and A. R. Moreira.** 1985. Agitation and pressure effects on acetone-butanol fermentation. *Biotechnol. Bioeng.* **27:**852–860.
59. **Dove, C. H., S. Z. Wang, S. B. Price, C. J. Phelps, D. M. Lyerly, T. D. Wilkins, and J. L. Johnson.** 1990. Molecular characterization of the *Clostridium difficile* toxin A gene. *Infect. Immun.* **58:**480–488.
60. **Duncan, C. L., E. A. Rokos, C. M. Christenson, and J. I. Rood.** 1978. Multiple plasmids in different toxigenic types of *Clostridium perfringens*: possible control of beta toxin production, p. 246–248. *In* D. Schlessinger (ed.), *Microbiology—1978*. American Society for Microbiology, Washington, D.C.
61. **Duncan, C. L., D. H. Strong, and M. Sebald.** 1972. Sporulation and enterotoxin production by mutants of *Clostridium perfringens*. *J. Bacteriol.* **110:**378–391.
62. **Dürre, P.** Transposons in clostridia. *In* D. R. Woods (ed.), *The Clostridia and Biotechnology*, in press. Butterworths, Boston.
63. **Dürre, P., A. Kuhn, and G. Gottschalk.** 1986. Treatment with allyl alcohol selects specifically for mutants of *Clostridium acetobutylicum* deficient in butanol synthesis. *FEMS Microbiol. Lett.* **36:**77–81.
64. **Dürre, P., A. Kuhn, M. Gottwald, and G. Gottschalk.** 1987. Enzymatic investigations on butanol dehydrogenase and butyraldehyde dehydrogenase in extracts of *Clostridium acetobutylicum*. *Appl. Microbiol. Biotechnol.* **26:**268–272.
65. **Eisel, U., W. Jarausch, K. Goretzki, A. Henschen, J. Engels, U. Weller, M. Hudel, E. Habermann, and H. Niemann.** 1986. Tetanus toxin: primary structure, expression in *E. coli*. *EMBO J.* **5:**2495–2502.
66. **Eklund, M. W., F. T. Poysky, L. M. Mseitif, and M. S. Strom.** 1988. Evidence for plasmid-mediated toxin and bacteriocin production in *Clostridium botulinum* type G. *Appl. Environ. Microbiol.* **54:**1405–1408.
67. **Eklund, M. W., F. T. Poysky, S. M. Reed, and C. A. Smith.** 1971. Bacteriophage and the toxigenicity of *Clostridium botulinum* type C. *Science* **172:**480–482.
68. **El Kanouni, A., A. M. Junelles, R. Janati-Idrissi, H. Petitdemange, and R. Gay.** 1989. *Clostridium acetobutylicum* mutants isolated for resistance to the pyruvate halogen analogs. *Curr. Microbiol.* **18:**139–144.
69. **Elton, R. A.** 1973. The relationship of DNA base composition and individual protein composition in microorganisms. *J. Mol. Evol.* **2:**263–276.
70. **Fairweather, N. F., and V. A. Lyness.** 1986. The complete nucleotide sequence of tetanus toxin. *Nucleic Acids Res.* **14:**7809–7812.
71. **Fairweather, N. F., V. A. Lyness, D. J. Pickard, G. Allen, and R. O. Thomson.** 1986. Cloning, nucleotide sequencing, and expression of tetanus toxin fragment C in *Escherichia coli*. *J. Bacteriol.* **165:**21–27.
72. **Fujii, N., K. Kimura, T. Yashiki, T. Indoh, T. Murakami, K. Tsuzuki, N. Yokosawa, and K. Oguma.** 1991. Cloning of a DNA fragment encoding the 5′-terminus of the botulinum type E toxin gene from *Clostridium butyricum* strain BL6340. *J. Gen. Microbiol.* **137:**519–525.
73. **Garnier, T., B. Canard, and S. T. Cole.** 1991. Cloning, mapping and molecular characterization of the rRNA operons of *Clostridium perfringens*. *J. Bacteriol.* **173:**5431–5438.
74. **Garnier, T., and S. T. Cole.** 1986. Characterization of a bacteriocinogenic plasmid from *Clostridium perfringens* and molecular genetic analysis of the bacteriocin-encoding gene. *J. Bacteriol.* **168:**1189–1196.
75. **Garnier, T., and S. T. Cole.** 1988. Complete nucleotide sequence and genetic organization of the bacteriocinogenic plasmid, pIP404, from *Clostridium perfringens*. *Plasmid* **19:**134–150.
76. **Garnier, T., and S. T. Cole.** 1988. Identification and molecular genetic analysis of replication functions of the bacteriocinogenic plasmid pIP404 from *Clostridium perfringens*. *Plasmid* **19:**151–160.
77. **Garnier, T., and S. T. Cole.** 1988. Studies of UV-inducible promoters from *Clostridium perfringens in vivo* and *in vitro*. *Mol. Microbiol.* **2:**607–614.
78. **Garnier, T., S. F. J. Le Grice, and S. T. Cole.** 1988. Characterization of the promoters for two UV-inducible transcriptional units carried by plasmid pIP404 from *Clostridium perfringens*, p. 211–214. *In* A. T. Ganesan, and J. A. Hoch (ed.), *Genetics and Biotechnology of Bacilli*, vol. 2, Academic Press, Inc., San Diego.
79. **George, W. L.** 1989. Antimicrobial agent-associated diarrhea and colitis, p. 661–678. *In* S. M. Finegold and W. L. George (ed.), *Anaerobic Infections in Humans*. Academic Press, Inc., San Diego.
80. **Gerischer, U., and P. Dürre.** 1990. Cloning, sequencing, and molecular analysis of the acetoacetate decarboxylase gene region from *Clostridium acetobutylicum*. *J. Bacteriol.* **172:**6907–6918.
81. **Gerischer, U., and P. Dürre.** 1992. mRNA analysis of the *adc* gene region of *Clostridium acetobutylicum* during the shift to solventogenesis. *J. Bacteriol.* **174:**426–433.
82. **Gerwig, G. J., P. de Waard, J. P. Kamerling, J. F. G. Vliegenthart, E. Morgenstern, R. Lamed, and E. A. Bayer.** 1989. Novel *O*-linked carbohydrate chains in the cellulase complex (cellulosome) of *Clostridium thermocellum*. *J. Biol. Chem.* **264:**1027–1035.
83. **Gerwig, G. J., J. P. Kamerling, J. F. G. Vliegenthart, E. Morag, R. Lamed, and E. A. Bayer.** 1991. Primary structure of *O*-linked carbohydrate chains in the cellulosome of different *Clostridium thermocellum* strains. *Eur. J. Biochem.* **196:**115–122.
84. **Gilkes, N. R., B. Henrissat, D. G. Kilburn, R. C. Miller, and R. A. J. Warren.** 1991. Domains in microbial β-1,4-glycanases: sequence conservation, function, and enzyme families. *Microbiol. Rev.* **55:**303–315.
85. **Gimenez, J., J. Foley, and B. R. DasGupta.** 1988. Neurotoxin type E from *Clostridium botulinum* and *Clostridium butyricum*; partial sequence and comparison. *FASEB J.* **2:**A1750.
86. **Granum, P.** 1990. *Clostridium perfringens* toxins involved in food poisoning. *Int. J. Food Microbiol.* **10:**101–112.
87. **Granum, P. E., and G. S. A. B. Stewart.** 1993. Molecular biology of *Clostridium perfringens* enterotoxin, p. 235–247. *In* M. Sebald (ed.), *Genetics and Molecular Biology of Anaerobic Bacteria*. Springer-Verlag, New York.
88. **Graves, M. C., G. T. Mullenbach, and J. C. Rabinowitz.** 1985. Cloning and nucleotide sequence of the *Clostridium pasteurianum* ferredoxin gene. *Proc. Natl. Acad. Sci. USA* **82:**1653–1657.
89. **Graves, M. C., and J. C. Rabinowitz.** 1986. *In vivo* and *in vitro* transcription of the *Clostridium pasteurianum* ferredoxin gene. Evidence for "extended" promoter elements in Gram-positive organisms. *J. Biol. Chem.* **261:**11409–11415.
90. **Grépinet, O., M. C. Chebrou, and P. Béguin.** 1988. Nucleotide sequence and deletion analysis of the xylanase gene (*xynZ*) of *Clostridium thermocellum*. *J. Bacteriol.* **170:**4582–4588.
91. **Grossman, A. D., J. W. Erickson, and C. A. Gross.** 1984. The *htpR* gene product of *E. coli* is a sigma factor for heat-shock promoters. *Cell* **38:**383–390.
92. **Gruss, A., and S. D. Ehrlich.** 1989. The family of highly interrelated single-stranded deoxyribonucleic acid plasmids. *Microbiol. Rev.* **53:**231–241.
93. **Hächler, H., B. Berger-Bächi, and F. H. Kayser.** 1987. Genetic characterization of a *Clostridium difficile* eryth-

romycin-clindamycin resistance determinant that is transferable to *Staphylococcus aureus*. *Antimicrob. Agents Chemother.* **31:**1039–1045.

94. **Hächler, H., F. H. Kayser, and B. Berger-Bächi.** 1987. Homology of a transferable tetracycline resistance determinant of *Clostridium difficile* with *Streptococcus* (*Enterococcus*) *faecalis* transposon Tn*916*. *Antimicrob. Agents Chemother.* **31:**1033–1038.
95. **Hall, J., G. P. Hazlewood, P. J. Barker, and H. J. Gilbert.** 1988. Conserved reiterated domains in *Clostridium thermocellum* endoglucanases are not essential for catalytic activity. *Gene* **69:**29–38.
96. **Hancock, K. R., S. J. Reid, J. D. Santangelo, and D. R. Woods.** Cloning and sequencing of a gene from *Clostridium acetobutylicum* that shows homology to the *Bacillus* sporulation gene, *spoIID*. Submitted for publication.
97. **Hanna, P. C., A. P. Wnek, and B. A. McClane.** 1989. Molecular cloning of the 3′ half of the *Clostridium perfringens* enterotoxin gene and demonstration that this region encodes receptor-binding activity. *J. Bacteriol.* **171:**6815–6820.
98. **Hartmanis, M. G. N., and S. Gatenbeck.** 1984. Intermediary metabolism in *Clostridium acetobutylicum*: levels of enzymes involved in the formation of acetate and butyrate. *Appl. Environ. Microbiol.* **47:**1277–1283.
99. **Hauser, D., M. W. Eklund, H. Kurazono, T. Binz, H. Niemann, D. M. Gill, P. Boquet, and M. R. Popoff.** 1990. Nucleotide sequence of *Clostridium botulinum* C1 neurotoxin. *Nucleic Acids Res.* **18:**4924.
100. **Hazlewood, G. P., M. P. M. Romaniec, K. Davidson, O. Grépinet, P. Béguin, J. Millet, O. Raynaud, and J.-P. Aubert.** 1988. A catalogue of *Clostridium thermocellum* endoglucanase, β-glucosidase and xylanase genes cloned in *Escherichia coli*. *FEMS Microbiol. Lett.* **51:** 231–236.
101. **Heefner, D. L., C. H. Squires, R. J. Evans, B. J. Kopp, and M. J. Yarus.** 1984. Transformation of *Clostridium perfringens*. *J. Bacteriol.* **159:**460–464.
102. **Henkin, T. M., F. J. Grundy, W. L. Nicholson, and G. H. Chambliss.** 1991. Catabolite repression of α-amylase gene expression in *Bacillus subtilis* involves a trans-acting gene product homologous to the *Escherichia coli lacI* and *galR* repressors. *Mol. Microbiol.* **5:**575–584.
103. **Henrissat, B., M. Claeyssens, P. Tomme, L. Lemesle, and J.-P. Mornon.** 1989. Cellulase families revealed by hydrophobic cluster analysis. *Gene* **81:**83–95.
104. **Higgins, C. F., C. J. Dorman, and N. Ni Bhriain.** 1989. Environmental influences on DNA supercoiling: a novel mechanism for the regulation of gene expression, p. 421–432. *In* K. Drlica and M. Riley (ed.), *The Bacterial Chromosome*. American Society for Microbiology, Washington, D.C.
105. **Holck, A., H. Blom, and P. E. Granum.** 1990. Cloning and sequencing of the genes encoding the acid soluble spore proteins from *Clostridium perfringens*. *Gene* **91:** 107–111.
106. **Hunter, S. E. C., I. N. Clarke, D. C. Kelly, and R. W. Titball.** 1992. Cloning and nucleotide sequence of the *Clostridium perfringens* epsilon-toxin gene and its expression in *Escherichia coli*. *Infect. Immun.* **60:**102–110.
107. **Husemann, M. H. W., and E. T. Papoutsakis.** 1989. Enzymes limiting butanol and acetone formation in continuous and batch cultures of *Clostridium acetobutylicum*. *Appl. Environ. Microbiol.* **31:**435–444.
108. **Husemann, M. H. W., and E. T. Papoutsakis.** 1989. Comparison between *in vivo* and *in vitro* enzyme activities in continuous and batch fermentations of *Clostridium acetobutylicum*. *Appl. Microbiol. Biotechnol.* **30:** 585–595.
109. **Hyun, H. H., and J. G. Zeikus.** 1985. Regulation and genetic enhancement of glucoamylase and pullulanase production in *Clostridium thermohydrosulfuricum*. *J. Bacteriol.* **164:**1146–1152.
110. **Iwanejko, L. A., M. N. Routledge, and G. S. A. B. Stewart.** 1989. Cloning in *Escherichia coli* of the enterotoxin gene from *Clostridium perfringens* type A. *J. Gen. Microbiol.* **135:**903–909.
111. **Janati-Idrissi, R., A. M. Junelles, A. El Kanouni, H. Petitdemange, and R. Gay.** 1987. Selection de mutants de *Clostridium acetobutylicum* defectifs dans la production d'acetone. *Ann. Microbiol. (Inst. Pasteur)* **138:**313–323.
112. **Jannière, L., C. Bruand, and S. D. Ehrlich.** 1989. Structurally stable *Bacillus subtilis* cloning vectors. *Gene* **87:**53–61.
113. **Janssen, P. J., W. A. Jones, D. T. Jones, and D. R. Woods.** 1988. Molecular analysis and regulation of the *glnA* gene of the gram-positive anaerobe *Clostridium acetobutylicum*. *J. Bacteriol.* **170:**400–408.
114. **Jarvis, E. D., R. L. Widom, G. La Fauci, Y. Setoguchi, I. R. Richter, and R. Rudner.** 1988. Chromosomal organization of rRNA operons in *Bacillus subtilis*. *Genetics* **120:**625–635.
115. **Jauris, S., K. P. Rücknagel, W. H. Schwarz, P. Kratzsch, K. Bronnenmeier, and W. L. Staudenbauer.** 1990. Sequence analysis of the *Clostridium stercorarium celZ* gene encoding a thermoactive cellulase (avicelase-I): identification of catalytic and cellulose-binding domains. *Mol. Gen. Genet.* **223:**258–267.
116. **Jenkins, L. S., and W. D. Nunn.** 1987. Genetic and molecular characterization of the genes involved in short-chain fatty acid degradation in *Escherichia coli*. *J. Bacteriol.* **169:**42–52.
117. **Johnson, J. L., C. H. Dove, S. B. Price, T. W. Sickles, C. J. Phelps, and T. D. Wilkins.** 1988. The toxin A gene of *Clostridium difficile*, p. 115–123. *In* J. M. Hardie and S. P. Borriello (ed.), *Anaerobes Today*. John Wiley & Sons, Ltd., Chichester, United Kingdom.
118. **Johnson, J. L., and B. S. Francis.** 1975. Taxonomy of the clostridia: ribosomal ribonucleic acid homologies among species. *J. Gen. Microbiol.* **88:**229–244.
119. **Jones, D. T., A. van der Westhuizen, S. Long, E. R. Allcock, S. J. Reid, and D. R. Woods.** 1982. Solvent production and morphological changes in *Clostridium acetobutylicum*. *Appl. Environ. Microbiol.* **43:**1434–1439.
120. **Jones, D. T., and D. R. Woods.** 1986. Gene transfer, recombination, and gene cloning in *Clostridium acetobutylicum*. *Microbiol. Sci.* **3:**19–22.
121. **Jones, D. T., and D. R. Woods.** 1986. Acetone-butanol fermentation revisited. *Microbiol. Rev.* **50:**484–524.
122. **Junelles A.-M., R. Janati-Idrissi, A. El Kanouni, H. Petitdemange, and R. Gay.** 1987. Acetone-butanol fermentation by mutants selected for resistance to acetate and butyrate halogen analogues. *Biotechnol. Lett.* **9:**175–178.
123. **Kim, A. Y., and H. P. Blaschek.** 1989. Construction of an *Escherichia coli-Clostridium perfringens* shuttle vector and plasmid transformation of *Clostridium perfringens*. *Appl. Environ. Microbiol.* **55:**360–365.
124. **Kim, A. Y., and H. P. Blaschek.** 1991. Isolation of a filamentous viruslike particle from *Clostridium acetobutylicum* NCIB 6444. *J. Bacteriol.* **173:**530–535.
125. **Kim, B. H., P. Bellows, R. Datta, and J. G. Zeikus.** 1984. Control of carbon and electron flow in *Clostridium acetobutylicum* fermentations: utilization of carbon monoxide to inhibit hydrogen production and to enhance butanol yields. *Appl. Environ. Microbiol.* **48:**764–770.
126. **Kimura, K., N. Fujii, K. Tsuzuki, T. Murakami, T. Indoh, N. Yokosawa, and K. Oguma.** 1991. Cloning of the structural gene for *Clostridium botulinum* type C1 toxin and whole nucleotide sequence of its light-chain component. *Appl. Environ. Microbiol.* **57:**1168–1172.
127. **Kohring, S., J. Wiegel, and F. Mayer.** 1990. Subunit

composition and glycosidic activities of the cellulase complex from *Clostridium thermocellum* JW20. *Appl. Environ. Microbiol.* **56:**3798–3804.

128. **Lamed, R., and E. A. Bayer.** 1988. The cellulosome concept: exocellular/extracellular enzyme reactor centers for efficient binding and cellulolysis, p. 101–116. *In* J. P. Aubert, P. Béguin, and J. Millet (ed.), *FEMS Symposium No. 43: Biochemistry and Genetics of Cellulose Degradation.* Academic Press, London.
129. **Lamed, R., J. Naimark, E. Morgenstern, and E. A. Bayer.** 1987. Specialized cell surface structures in cellulolytic bacteria. *J. Bacteriol.* **169:**3792–3800.
130. **Lamed, R., E. Setter, and E. A. Bayer.** 1983. Characterization of a cellulose-binding, cellulase-containing complex in *Clostridium thermocellum. J. Bacteriol.* **156:** 828–836.
131. **Lamed, R., E. Setter, R. Kenig, and E. A. Bayer.** 1983. The cellulosome: a discrete cell surface organelle of *Clostridium thermocellum* which exhibits separate antigenic, cellulose-binding and various cellulolytic activities. *Biotechnol. Bioeng. Symp.* **13:**163–181.
132. **Landick, R., V. Vaughn, E. T. Lau, R. A. VanBogelen, J. W. Erickson, and F. C. Neidhardt.** 1984. Nucleotide sequence of the heat shock regulatory gene of *E. coli* suggests its protein product may be a transcription factor. *Cell* **38:**175–182.
133. **Lee, S. F., and C. W. Forsberg.** 1987. Isolation and some properties of a β-D-xylosidase from *Clostridium acetobutylicum* ATCC 824. *Appl. Environ. Microbiol.* **53:**651–654.
134. **Lee, S. F., C. W. Forsberg, and J. B. Rattray.** 1987. Purification and characterization of two endoxylanases from *Clostridium acetobutylicum* ATCC 824. *Appl. Environ. Microbiol.* **53:**644–650.
135. **Leskiw, B. K., M. J. Bibb, and K. F. Chater.** 1991. The use of a rare codon specifically during development? *Mol. Microbiol.* **5:**2861–2867.
136. **Leslie, D., N. Fairweather, D. Pickard, G. Dougan, and M. Kehoe.** 1989. Phospholipase C and haemolytic activities of *Clostridium perfringens* alpha-toxin cloned in *Escherichia coli*: sequence and homology with *Bacillus cereus* phospholipase C. *Mol. Microbiol.* **3:**383–392.
137. **Lin, W. J., and E. A. Johnson.** 1991. Transposon Tn*916* mutagenesis in *Clostridium botulinum. Appl. Environ. Microbiol.* **57:**2946–2950.
138. **Long, S., D. T. Jones, and D. R. Woods.** 1984. The relationship between sporulation and solvent production in *Clostridium acetobutylicum* P262. *Biotechnol. Lett.* **6:**529–534.
139. **Long, S., D. T. Jones, and D. R. Woods.** 1984. Initiation of solvent production, clostridial stage, and endospore formation in *Clostridium acetobutylicum. Appl. Environ. Microbiol.* **20:**256–261.
140. **Luczak, H., H. Schwarzmoser, and W. L. Staudenbauer.** 1985. Construction of *Clostridium butyricum* plasmids and transfer to *Bacillus subtilis. Appl. Microbiol. Biotechnol.* **23:**114–122.
141. **Lynd, L. R., J. H. Cushman, R. J. Nichols, and C. E. Wyman.** 1991. Fuel ethanol from cellulosic biomass. *Science* **251:**1318–1323.
142. **Mahony, D. E.** 1974. Bacteriocin susceptibility of *Clostridium perfringens*: a provisional typing schema. *Appl. Microbiol.* **28:**172–176.
143. **Mahony, D. E.** 1979. Bacteriocin, bacteriophage and other epidemiological typing methods for the genus *Clostridium*, p. 1–30. *In* T. Bergan and J. R. Norris (ed.), *Methods in Microbiology*, vol. 13. Academic Press, Inc., New York.
144. **Makoff, A. J., S. P. Ballantine, A. E. Smallwood, and N. F. Fairweather.** 1989. Expression of tetanus toxin fragment C in *E. coli*: its purification and potential use as a vaccine. *Bio/Technology* **7:**1043–1046.
145. **Makoff, A. J., M. D. Oxer, M. A. Romanos, N. F. Fairweather, and S. Ballantine.** 1989. Expression of tetanus toxin fragment C in *E. coli*: high level expression by removing rare codons. *Nucleic Acids Res.* **17:** 10191–10202.
146. **Mathupala, S., B. C. Saha, and J. G. Zeikus.** 1990. Substrate competition and specificity at the active site of amylopullulanase from *Clostridium thermohydrosulfuricum. Biochem. Biophys. Res. Commun.* **166:**126–132.
147. **Mattsson, P. M., and P. Rogers.** 1989. Identification of some transposon insertion sites in the chromosome of *Clostridium acetobutylicum* after transfer of the tetracycline resistance transposon Tn*916* from *Streptococcus faecalis*, abstr. O-39, p. 311. Abstr. 89th Annu. Meet. Am. Soc. Microbiol.
148. **McClane, B. A., P. C. Hanna, and A. P. Wnek.** 1988. *Clostridium perfringens* enterotoxin. *Microb. Pathog.* **4:**317–323.
149. **McDonel, J. L.** 1986. Toxins of *Clostridium perfringens* types A, B, C, D and E, p. 477–517. *In* F. Dorner, and J. Drews (ed.), *Pharmacology of Bacterial Toxins.* Permagon Press, Oxford.
150. **Melasniemi, H., M. Paloheimo, and L. Hemiö.** 1990. Nucleotide sequence of the α-amylase-pullulanase gene from *Clostridium thermohydrosulfuricum. J. Gen. Microbiol.* **136:**447–454.
151. **Mermelstein, L. D., N. E. Welker, G. N. Bennett, and E. T. Papoutsakis.** 1992. Expression of cloned homologous fermentative genes in *Clostridium acetobutylicum* ATCC 824. *Bio/Technology* **10:**190–195.
152. **Meyer, C. L., J. K. McLaughlin, and E. T. Papoutsakis.** 1985. The effect of CO on growth and product formation in batch cultures of *Clostridium acetobutylicum. Biotechnol. Lett.* **7:**37–42.
153. **Meyer, C. L., and E. T. Papoutsakis.** 1989. Increased levels of ATP and NADH are associated with increased solvent production in continuous cultures of *C. acetobutylicum. Appl. Microbiol. Biotechnol.* **30:**450–459.
154. **Meyer, C. L., J. W. Roos, and E. T. Papoutsakis.** 1986. Carbon monoxide gassing leads to alcohol production and butyrate uptake without acetone formation by *Clostridium acetobutylicum. Appl. Microbiol. Biotechnol.* **24:**159–167.
155. **Meyer, J., and J. Gagnon.** 1991. Primary structure of hydrogenase I from *Clostridium pasteurianum. Biochemistry* **30:**9697–9704.
156. **Minton, N. P.** 1992. Personal communication.
157. **Minton, N. P., J. K. Brehm, J. D. Oultram, T.-J. Swinfield, and D. E. Thompson.** 1988. Construction of plasmid vector systems for gene transfer in *Clostridium acetobutylicum*, p. 125–134. *In* J. M. Hardie and S. P. Borriello (ed.), *Anaerobes Today.* John Wiley & Sons, Chichester, United Kingdom.
158. **Minton, N. P., J. K. Brehm, T.-J. Swinfield, S. M. Whelan, M. L. Rodger, N. Bodsworth, and J. D. Oultram.** Clostridial cloning vectors. *In* D. R. Woods (ed.), *The Clostridia and Biotechnology*, in press. Butterworths, Boston.
159. **Minton, N. P., and D. J. Clarke (ed.).** 1989. *Clostridia. Biotechnology Handbooks*, vol. 3. Plenum Publishing Co., New York.
160. **Minton, N. P., and J. G. Morris.** 1981. Isolation and partial characterization of three cryptic plasmids from strains of *Clostridium butyricum. J. Gen. Microbiol.* **127:**325–331.
161. **Minton, N. P., and J. D. Oultram.** 1988. Host-vector systems for gene cloning in *Clostridium. Microbiol. Sci.* **5:**310–315.
162. **Mishra, S., P. Béguin, and J.-P. Aubert.** 1991. Transcription of *Clostridium thermocellum* endoglucanase genes *celF* and *celD. J. Bacteriol.* **173:**80–85.
163. **Monot, F., and J. M. Engasser.** 1983. Production of acetone and butanol by batch and continuous culture of

Clostridium acetobutylicum under nitrogen limitation. *Biotechnol. Lett.* **5**:213–218.
164. **Monot, F., J. M. Engasser, and H. Petitdemange.** 1983. Regulation of acetone butanol production in batch and continuous cultures of *Clostridium acetobutylicum. Biotechnol. Bioeng. Symp.* **13**:207–216.
165. **Morag, E., E. A. Bayer, and R. Lamed.** 1990. Relationship of cellulosomal and noncellulosomal xylanases of *Clostridium thermocellum* to cellulose-degrading enzymes. *J. Bacteriol.* **172**:6098–6105.
166. **Morag, E., I. Halevy, E. A. Bayer, and R. Lamed.** 1991. Isolation and properties of a major cellobiohydrolase from the cellulosome of *Clostridium thermocellum. J. Bacteriol.* **173**:4155–4162.
167. **Morris, J. G.** 1988. Oxygen toxicity; protective strategies. *FEMS Symp.* **44**:84–98.
168. **Mullany, P., M. Wilks, I. Lamb, C. Clayton, B. Wren, and S. Tabaqchali.** 1990. Genetic analysis of a tetracycline resistance element from *Clostridium difficile* and its conjugal transfer to and from *Bacillus subtilis. J. Gen. Microbiol.* **136**:1343–1349.
169. **Mullany, P., M. Wilks, and S. Tabaqchali.** 1991. Transfer of Tn*916* and Tn*916ΔE* into *Clostridium difficile*: demonstration of a hot-spot for these elements in the *C. difficile* genome. *FEMS Microbiol. Lett.* **79**:191–194.
170. **Narberhaus, F., and H. Bahl.** 1992. Cloning, sequencing, and molecular analysis of the *groESL* operon of *Clostridium acetobutylicum. J. Bacteriol.* **174**:3282–3289.
171. **Narberhaus, F., K. Giebeler, and H. Bahl.** 1992. Molecular characterization of the *dnaK* gene region of *Clostridium acetobutylicum*, including *grpE*, *dnaJ*, and a new heat shock gene. *J. Bacteriol.* **174**:3290–3299.
172. **Nieves, B. M., F. Gil, and F. J. Castillo.** 1981. Growth inhibition activity and bacteriophage and bacteriocin-like particles associated with different species of *Clostridium. Can. J. Microbiol.* **27**:216–225.
173. **Ogata, S., and M. Hongo.** 1979. Bacteriophages of the genus *Clostridium. Adv. Appl. Microbiol.* **25**:241–273.
174. **Okabe, A., T. Shimizu, and H. Hayashi.** 1989. Cloning and sequencing of a phospholipase C gene of *Clostridium perfringens. Biochem. Biophys. Res. Commun.* **160**: 33–39.
175. **Oultram, J. D., A. Davies, and M. Young.** 1987. Conjugal transfer of a small plasmid from *Bacillus subtilis* to *Clostridium acetobutylicum* by cointegrate formation with plasmid pAMβ1. *FEMS Microbiol. Lett.* **42**:113–119.
176. **Oultram, J. D., M. Loughlin, T.-J. Swinfield, J. K. Brehm, D. E. Thompson, and N. P. Minton.** 1988. Introduction of plasmids into whole cells of *Clostridium acetobutylicum* by electroporation. *FEMS Microbiol. Lett.* **56**:83–88.
177. **Oultram, J. D., H. Peck, J. K. Brehm, D. Thompson, T.-J. Swinfield, and N. P. Minton.** 1988. Introduction of genes for leucine biosynthesis from *Clostridium pasteurianum* into *Clostridium acetobutylicum. Mol. Gen. Genet.* **214**:177–179.
178. **Oultram, J. D., and M. Young.** 1985. Conjugal transfer of plasmid pAMβ1 from *Streptococcus lactis* and *Bacillus subtilis* to *Clostridium acetobutylicum. FEMS Microbiol. Lett.* **27**:129–134.
179. **Palosaari, N., and P. Rogers.** 1988. Purification and properties of the inducible coenzymeA-linked butyraldehyde dehydrogenase from *Clostridium acetobutylicum. J. Bacteriol.* **170**:2971–2976.
180. **Petersen, D. J., and G. N. Bennett.** 1990. Purification of acetoacetate decarboxylase from *Clostridium acetobutylicum* ATCC 824 and cloning of the acetoacetate decarboxylase gene in *Escherichia coli. Appl. Environ. Microbiol.* **56**:3491–3498.
181. **Petersen, D. J., and G. N. Bennett.** 1991. Cloning of the *Clostridium acetobutylicum* ATCC 824 acetyl coenzyme A acetyltransferase (thiolase; EC 2.3.1.9) gene. *Appl. Environ. Microbiol.* **57**:2735–2741.
182. **Petersen, D. J., and G. N. Bennett.** 1991. Enzymatic characterization of a nonmotile, nonsolventogenic *Clostridium acetobutylicum* ATCC 824 mutant. *Curr. Microbiol.* **23**:253–258.
183. **Petersen, D. J., R. W. Welch, F. B. Rudolph, and G. N. Bennett.** 1991. Molecular cloning of an alcohol (butanol) dehydrogenase gene cluster from *Clostridium acetobutylicum* ATCC 824. *J. Bacteriol.* **173**:1831–1834.
184. **Phelps, C. J., D. L. Lyerly, J. L. Johnson, and T. D. Wilkins.** 1991. Construction and expression of the complete *Clostridium difficile* toxin A gene in *Escherichia coli. Infect. Immun.* **59**:150–153.
185. **Pich, A., and H. Bahl.** 1991. Purification and characterization of the DNA-dependent RNA polymerase from *Clostridium acetobutylicum. J. Bacteriol.* **173**:2120–2124.
186. **Pich, A., F. Narberhaus, and H. Bahl.** 1990. Induction of heat shock proteins during initiation of solvent formation in *Clostridium acetobutylicum. Appl. Microbiol. Biotechnol.* **33**:697–704.
187. **Plant, A. R., R. M. Clemens, H. W. Morgan, and R. M. Daniel.** 1987. Active-site- and substrate-specificity of *Thermoanaerobium* Tok6-B1 pullulanase. *Biochem. J.* **246**:537–541.
188. **Podvin, L., G. Reysset, J. Hubert, and M. Sebald.** 1988. Recent developments in the genetics of *Clostridium acetobutylicum*, p. 135–140. *In* J. M. Hardie, and S. P. Boriello (ed.), *Anaerobes Today*. John Wiley & Sons, Chichester, United Kingdom.
189. **Poulet, S., D. Hauser, M. Quanz, H. Niemann, and M. Popoff.** 1992. Sequences of the botulinal neurotoxin E derived from *Clostridium botulinum* type E (strain Beluga) and *Clostridium butyricum* (strains ATCC 43181 and ATCC 43755). *Biochem. Biophys. Res. Commun.* **183**:107–113.
190. **Reid, S. J., E. R. Allcock, D. T. Jones, and D. R. Woods.** 1983. Transformation of *Clostridium acetobutylicum* protoplasts with bacteriophage DNA. *Appl. Environ. Microbiol.* **45**:305–307.
191. **Reysenbach, A. L., N. Ravenscroft, S. Long, D. T. Jones, and D. R. Woods.** 1986. Characterization, biosynthesis, and regulation of granulose in *Clostridium acetobutylicum. Appl. Environ. Microbiol.* **52**:275–281.
192. **Reysset, G., J. Hubert, L. Podvin, and M. Sebald.** 1988. Transfection and transformation of *Clostridium acetobutylicum* strain N1-4081 protoplasts. *Biotechnol. Technol.* **2**:199–204.
193. **Rogers, P.** 1986. Genetics and biochemistry of *Clostridium* relevant to development of fermentation processes. *Adv. Appl. Microbiol.* **31**:1–60.
194. **Rogers, P., and N. Palosaari.** 1987. *Clostridium acetobutylicum* mutants that produce butyraldehyde and altered quantities of solvents. *Appl. Environ. Microbiol.* **53**:2761–2766.
195. **Roggentin, P., B. Rothe, F. Lottspeich, and R. Schauer.** 1988. Cloning and sequencing of a *Clostridium perfringens* sialidase gene. *FEBS Lett.* **238**:31–34.
196. **Rokos, E. A., J. I. Rood, and C. L. Duncan.** 1978. Multiple plasmids in different toxigenic types of *Clostridium perfringens. FEMS Microbiol. Lett.* **4**:323–326.
197. **Rood, J. I.** 1993. Antibiotic resistance determinants of *Clostridium perfringens*, p. 141–155. *In* M. Sebald (ed.), *Genetics and Molecular Biology of Anaerobic Bacteria*. Springer-Verlag, New York.
198. **Rood, J. I., and S. T. Cole.** 1991. Molecular genetics and pathogenesis of *Clostridium perfringens. Microbiol. Rev.* **55**:621–648.
199. **Rood, J. I., S. Jefferson, T. L. Bannam, J. M. Wilkie, P. Mullany, and B. W. Wren.** 1989. Hybridization analysis of three chloramphenicol resistance determinants from *Clostridium perfringens* and *Clostridium difficile. Antimicrob. Agents Chemother.* **33**:1569–1574.

200. **Rood, J. I., V. N. Scott, and C. L. Duncan.** 1978. Identification of a transferable resistance plasmid (pCW3) from *Clostridium perfringens*. *Plasmid* **1:**563–570.
201. **Roper, G., J. A. Short, and P. D. Walker.** 1976. The ultrastructure of *Clostridium perfringens* spores, p. 279–296. *In* A. M. Baker, J. Wold, D. J. Ellar, G. H. Dring, and G. W. Gould (ed.), *Spore Research*. Academic Press, London.
202. **Saint-Joanis, B., T. Garnier, and S. T. Cole.** 1989. Gene cloning shows the alpha toxin of *Clostridium perfringens* to contain both sphingomyelinase and lecithinase activities. *Mol. Gen. Genet.* **219:**453–460.
203. **Sakka, K., Y. Kojima, K. Yoshikawa, and K. Shimada.** 1990. Cloning and expression in *Escherichia coli* of *Clostridium stercorarium* strain F-9 genes related to xylan hydrolysis. *Agric. Biol. Chem.* **54:**337–342.
204. **Sauer, U., and P. Dürre.** 1993. Possible function of $tRNA^{Thr}_{ACG}$ in regulation of solvent formation in *Clostridium acetobutylicum*. *FEMS Microbiol. Lett.* **100:**147–154.
205. **Sauerborn, M., and C. von Eichel-Streiber.** 1990. Nucleotide sequence of *Clostridium difficile* toxin A. *Nucleic Acids Res.* **18:**1629–1630.
206. **Saulnier, C.** 1992. Personal communication.
207. **Schwarz, W. H.** 1992. Personal communication.
208. **Schwarz, W. H., S. Jauris, M. Kouba, K. Bronnenmeier, and W. L. Staudenbauer.** 1989. Cloning and expression of *Clostridium stercorarium* cellulase genes in *Escherichia coli*. *Biotechnol. Lett.* **11:**461–466.
209. **Schwarz, W. H., S. Schimming, K. P. Rücknagel, S. Burgschwaiger, G. Kreil, and W. L. Staudenbauer.** 1988. Nucleotide sequence of the *celC* gene encoding endoglucanase C of *Clostridium thermocellum*. Gene **63:**23–30.
210. **Scott, P. T., and J. I. Rood.** 1989. Electroporation-mediated transformation of lysostaphin-treated *Clostridium perfringens*. *Gene* **82:**327–333.
211. **Sebald, M. (ed.).** 1993. *Genetics and Molecular Biology of Anaerobic Bacteria*. Springer-Verlag, New York.
212. **Sebald, M., D. Bouanchaud, and G. Bieth.** 1975. Nature plasmidique de la resistance à plusieurs antibiotiques chez *C. perfringens* type A, souche 659. *C.R. Acad. Sci. Ser. D* **280:**2401–2404.
213. **Sebald, M., and G. Bréfort.** 1975. Transfert du plasmide tétracycline-chloramphenicol chez *Clostridium perfringens*. *C.R. Acad. Sci. Ser. D* **281:**317–319.
214. **Sebald, M., and R. N. Costilow.** 1975. Minimal growth requirements for *Clostridium perfringens* and isolation of auxotrophic mutants. *Appl. Microbiol.* **29:**1–6.
215. **Sell, T. L., D. R. Schaburg, and D. R. Fekety.** 1983. Bacteriophage and bacteriocin typing scheme for *Clostridium difficile*. *J. Clin. Microbiol.* **17:**1148–1152.
216. **Shimizu, T., A. Okabe, J. Minami, and H. Hayashi.** 1991. An upstream regulatory sequence stimulates expression of the perfringolysin O gene of *Clostridium perfringens*. *Infect. Immun.* **59:**137–142.
217. **Sloan, J., T. A. Warner, P. T. Scott, T. L. Bannam, D. I. Berryman, and J. I. Rood.** 1992. Construction of a sequenced *Clostridium perfringens-Escherichia coli* shuttle plasmid. *Plasmid* **27:**207–219.
218. **Smith, H. W.** 1959. The bacteriophages of *Clostridium perfringens*. *J. Gen. Microbiol.* **21:**622–630.
219. **Squires, C. H., D. L. Heefner, R. J. Evans, B. J. Kopp, and M. J. Yarus.** 1984. Shuttle plasmids for *Escherichia coli* and *Clostridium perfringens*. *J. Bacteriol.* **159:**465–471.
220. **Steffen, C., and H. Matzura.** 1989. Nucleotide sequence analysis of a chloramphenicol-acetyltransferase coding gene from *Clostridium perfringens*. *Gene* **75:**349–354.
221. **Stephens, G. M., R. A. Holt, J. C. Gottschal, and J. G. Morris.** 1985. Studies on the stability of solvent production by *Clostridium acetobutylicum* in continuous culture. *J. Appl. Bacteriol.* **59:**597–605.
222. **Stock, J. B., A. J. Ninfa, and A. M. Stock.** 1989. Protein phosphorylation and regulation of adaptive responses in bacteria. *Microbiol. Rev.* **50:**450–490.
223. **Stragier, P., and R. Losick.** 1990. Cascades of sigma factors revisited. *Mol. Microbiol.* **4:**1801–1806.
224. **Swinfield, T.-J., J. D. Oultram, D. E. Thompson, J. K. Brehm, and N. P. Minton.** 1990. Physical characterisation of the replication region of the *Streptococcus faecalis* plasmid pAMβ1. *Gene* **87:**79–90.
225. **Thompson, D. E., J. K. Brehm, J. D. Oultram, T. J. Swinfield, C. S. Shone, T. Atkinson, J. Melling, and N. P. Minton.** 1990. The complete amino acid sequence of the *Clostridium botulinum* type A neurotoxin, deduced by nucleotide sequence analysis of the encoding gene. *Eur. J. Biochem.* **189:**73–81.
226. **Titball, R. W., S. E. C. Hunter, K. L. Martin, B. C. Morris, A. D. Shuttleworth, T. Rubidge, D. W. Anderson, and D. C. Kelly.** 1989. Molecular cloning and nucleotide sequence of the alpha-toxin (phospholipase C) of *Clostridium perfringens*. *Infect. Immun.* **57:**367–376.
227. **Titball, R. W., H. Yeoman, and S. E. C. Hunter.** 1993. Gene cloning and organisation of the alpha-toxin of *Clostridium perfringens*, p. 211–226. *In* M. Sebald (ed.), *Genetics and Molecular Biology of Anaerobic Bacteria*. Springer-Verlag, New York.
228. **Tokatlidis, K., S. Salamitou, P. Béguin, P. Dhurjati, and J.-P. Aubert.** 1991. Interaction of the duplicated segment carried by *Clostridium thermocellum* cellulases with cellulosome components. *FEBS Lett.* **291:**185–188.
229. **Trieu-Cuot, P., C. Carlier, P. Martin, and P. Courvalin.** 1987. Plasmid transfer by conjugation from *Escherichia coli* to Gram-positive bacteria. *FEMS Microbiol. Lett.* **48:**289–294.
230. **Truffaut, N., and M. Sebald.** 1983. Plasmid detection and isolation in strains of *Clostridium acetobutylicum* and related species. *Mol. Gen. Genet.* **189:**178–180.
231. **Tso, J. Y., and C. Siebel.** 1989. Cloning and expression of the phospholipase C gene from *Clostridium perfringens* and *Clostridium bifermentans*. *Infect. Immun.* **57:**468–476.
232. **Tweten, R. K.** 1988. Cloning and expression in *Escherichia coli* of the perfringolysin O (theta toxin) gene from *Clostridium perfringens* and characterization of the gene product. *Infect. Immun.* **56:**3228–3234.
233. **Tweten, R. K.** 1988. Nucleotide sequence of the gene for perfringolysin O (theta toxin) from *Clostridium perfringens*: significant homology with the genes for streptolysin O and pneumolysin. *Infect. Immun.* **56:**3235–3240.
234. **van Damme-Jongsten, M., K. Wernars, and S. Notermans.** 1989. Cloning and sequencing of the *Clostridium perfringens* enterotoxin gene. *Antonie van Leeuwenhoek* **56:**181–190.
235. **van Poelje, P. D., and E. E. Snell.** 1990. Cloning, sequencing, expression and site-directed mutagenesis of the gene from *Clostridium perfringens* encoding pyruvoyl-dependent histidine decarboxylase. *Biochemistry* **29:**132–139.
236. **Volk, W. A., B. Bizzini, K. R. Jones, and F. L. Macrina.** 1988. Inter- and intrageneric transfer of Tn*916* between *Streptococcus faecalis* and *Clostridium tetani*. *Plasmid* **19:**255–259.
237. **von Eichel-Streiber, C., R. Laufenberg-Feldmann, S. Sartingen, J. Schulze, and M. Sauerborn.** 1990. Cloning of *Clostridium difficile* toxin B gene and demonstration of high N terminal homology between toxin A and toxin B. *Med. Microbiol. Immunol.* **179:**271–279.
238. **von Eichel-Streiber, C., and M. Sauerborn.** 1990. *Clostridium difficile* toxin A carries a C-terminal repetitive structure homologous to the carbohydrate binding re-

gion of streptococcal glycosyltransferases. *Gene* **96:** 107–113.

239. **von Eichel-Streiber, C., D. Suckau, M. Wachter, and U. Hadding.** 1989. Cloning and characterisation of overlapping DNA fragments of the toxin A gene of *Clostridium difficile. J. Gen. Microbiol.* **135:**55–64.
240. **Weickert, M. J., and G. H. Chambliss.** 1989. Genetic analysis of the promoter region of the *Bacillus subtilis* α-amylase gene. *J. Bacteriol.* **171:**3656–3666.
241. **Weickert, M. J., and G. H. Chambliss.** 1990. Site-directed mutagenesis of a catabolite repression operator sequence in *Bacillus subtilis. Proc. Natl. Acad. Sci. USA* **87:**6238–6242.
242. **Wetzstein, M., U. Völker, J. Dedio, S. Löbau, U. Zuber, M. Scheisswohl, C. Herget, M. Hecker, and W. Schumann.** 1992. Cloning, sequencing, and molecular analysis of the *dnaK* locus from *Bacillus subtilis. J. Bacteriol.* **174:**3300–3310.
243. **Whelan, S. M., M. J. Elmore, N. J. Bodsworth, T. Atkinson, and N. P. Minton.** 1992. The complete amino acid sequence of the *Clostridium botulinum* type E neurotoxin, derived by nucleotide sequence analysis of the encoding gene. *Eur. J. Biochem.* **204:**657–667.
244. **Wiegel, J., C. P. Mothershed, and J. Puls.** 1985. Differences in xylan degradation by various non-cellulolytic thermophilic anaerobes and *Clostridium thermocellum. Appl. Environ. Microbiol.* **49:**656–659.
245. **Wilkinson, S. R., and M. Young.** Wide diversity of genome size among different strains of *Clostridium acetobutylicum. J. Gen. Microbiol.*, in press.
246. **Wilkinson, S. R., and M. Young.** Unpublished data.
247. **Williams, D. R., D. I. Young, and M. Young.** 1990. Conjugative plasmid transfer from *Escherichia coli* to *Clostridium acetobutylicum. J. Gen. Microbiol.* **136:**819–826.
248. **Woods, D. R. (ed.).** *The Clostridia and Biotechnology*, in press. Butterworths, Boston.
249. **Woolley, R. C., and J. G. Morris.** 1990. Stability of solvent production by *Clostridium acetobutylicum* in continuous culture: strain differences. *J. Appl. Bacteriol.* **69:**718–728.
250. **Woolley, R. C., A. Pennock, R. J. Ashton, A. Davies, and M. Young.** 1989. Transfer of Tn*1545* and Tn*916* to *Clostridium acetobutylicum. Plasmid* **22:**169–174.
251. **Wren, B. W.** 1991. A family of clostridial and streptococcal ligand-binding proteins with conserved C-terminal repeat sequences. *Mol. Microbiol.* **5:**797–803.
252. **Wren, B. W.** 1992. Personal communication.
253. **Wren, B. W., P. Mullany, C. Clayton, and S. Tabaqchali.** 1988. Molecular cloning and genetic analysis of a chloramphenicol acetyltransferase determinant from *Clostridium difficile. Antimicrob. Agents Chemother.* **32:** 1213–1217.
254. **Wren, B. W., P. Mullany, C. Clayton, and S. Tabaqchali.** 1989. Nucleotide sequence of a chloramphenicol acetyltransferase gene from *Clostridium difficile. Nucleic Acids Res.* **17:**4877.
255. **Wu, J. D. H., and A. L. Demain.** 1988. Proteins of the *Clostridium thermocellum* cellulase complex responsible for degradation of crystalline cellulose, p. 117–131. *In* J.-P. Aubert, P. Béguin, and J. Millet (ed.), *FEMS Symposium No. 43: Biochemistry and Genetics of Cellulose Degradation.* Academic Press, London.
256. **Wu, J. D. H., W. H. Orme-Johnson, and A. L. Demain.** 1988. Two components of an extracellular protein aggregate of *Clostridium thermocellum* together degrade crystalline cellulose. *Biochemistry* **27:**1703–1709.
257. **Yagüe, E., P. Béguin, and J.-P. Aubert.** 1990. Nucleotide sequence and deletion analysis of the cellulase-encoding gene *celH* of *Clostridium thermocellum. Gene* **89:**61–67.
258. **Yan, R. T., C. X. Zhu, C. Golemboski, and J. S. Chen.** 1988. Expression of solvent-forming enzymes and onset of solvent production in batch cultures of *Clostridium beijerinckii. Appl. Environ. Microbiol.* **54:**642–648.
259. **Yerushalmi, L., B. Volesky, and T. Szezesny.** 1985. Effect of increased hydrogen partial pressure on the acetone-butanol fermentation by *Clostridium acetobutylicum. Appl. Microbiol. Biotechnol.* **22:**103–107.
260. **Young, M.** Development and exploitation of conjugative gene transfer in clostridia. *In* D. R. Woods (ed.), *The Clostridia and Biotechnology*, in press. Butterworths, Boston.
261. **Young, M., N. P. Minton, and W. L. Staudenbauer.** 1989. Recent advances in the genetics of the clostridia. *FEMS Microbiol. Rev.* **63:**301–326.
262. **Youngleson, J. S., D. T. Jones, and D. R. Woods.** 1989. Homology between hydroxybutyryl and hydroxyacyl coenzyme A dehydrogenase enzymes from *Clostridium acetobutylicum* fermentation and vertebrate fatty acid β-oxidation pathways. *J. Bacteriol.* **171:**6800–6807.
263. **Youngleson, J. S., W. A. Jones, D. T. Jones, and D. R. Woods.** 1989. Molecular analysis and nucleotide sequence of the *adh1* gene encoding a NADPH-dependent butanol dehydrogenase in the Gram-positive anaerobe *Clostridium acetobutylicum. Gene* **78:**355–364.
264. **Youngleson, J. S., J. D. Santangelo, D. T. Jones, and D. R. Woods.** 1988. Cloning and expression of *Clostridium acetobutylicum* alcohol dehydrogenase gene in *Escherichia coli. Appl. Environ. Microbiol.* **54:**676–682.
265. **Zappe, H., D. T. Jones, and D. R. Woods.** 1987. Cloning and expression of a xylanase gene from *Clostridium acetobutylicum* P262 in *Escherichia coli. Appl. Microbiol. Biotechnol.* **27:**57–63.
266. **Zappe, H., W. A. Jones, and D. R. Woods.** 1990. Nucleotide sequence of a *Clostridium acetobutylicum* P262 xylanase gene (*xynB*). *Nucleic Acids Res.* **18:**2179.

4. *Streptococcus*

JUNE R. SCOTT and MICHAEL G. CAPARON

The streptococci are a heterogenous group of gram-positive cocci that includes organisms commonly found among the normal flora of humans as well as organisms that can cause both mild and life-threatening diseases. Also in this group are organisms that inhabit the environment outside the human body, organisms that can cause disease in animals, and organisms that are essential for many industrial processes (especially in the dairy industry). Among the characteristics shared by these bacteria is a fermentative metabolism that produces lactic acid as an end product. Another trait shared by many streptococci is the arrangement of the individual bacteria into chains, which occurs because the organisms remain attached after division and the cells all tend to divide along the same plane.

No single scheme has been entirely successful in classifying this diverse collection of bacteria. In a medical setting, the appearance of streptococcal colonies on media containing sheep blood has traditionally been used for identification. The streptococci are classified as beta-hemolytic if complete hemolysis of the blood cells in the agar surrounding the colonies is observed, alpha-hemolytic if a characteristic greenish color is seen, or nonhemolytic if the blood cells are not affected.

Streptococci can also be grouped on the basis of serological reactivity of a cell wall-associated carbohydrate, according to a scheme that was developed by Rebecca Lancefield (59). However, some species are not included in the Lancefield scheme, most notably *Streptococcus pneumoniae* and the viridans group (i.e., *Streptococcus sanguis*, *Streptococcus mutans*, and *Streptococcus salivarius*). Because the streptococci are so heterogeneous, reclassification is common. Some species have recently been removed from the genus *Streptococcus* to form the genera *Enterococcus* (formerly Lancefield group D streptococci) and *Lactococcus* (formerly Lancefield group N). While traditional schemes are still in common use, a modern taxonomy based on nucleic acid hybridizations and comparison of rRNA sequences is being developed (87).

Several important discoveries have resulted from the modern molecular genetic study of the streptococci. The ability of *S. pneumoniae* to be transformed by exogenous DNA was used by Avery et al. in their classic experiments that showed that DNA is the genetic material (3). In general outline, the processing of DNA for transformation of the pneumococcus appears to be similar to that of the other well-studied gram-positive organism, *Bacillus subtilis*. However, in *S. pneumoniae*, the ability to take up DNA (competence for transformation) is regulated by production of an extracellular protein called competence factor (103), and no such factor has been described for *B. subtilis*. It is possible, therefore, that regulation of DNA uptake differs in these two organisms. Further work on transformation in *S. pneumoniae* has yielded information on the enzymology of DNA uptake and DNA repair (4, 20, 65, 81, 82, 103) and on the biology of restriction enzymes (58). However, other groups of streptococci have remained relatively intractable genetically. For example, only some isolates of *S. mutans* and the Challis strain of *Streptococcus gordonii* (formerly *S. sanguis*) are naturally competent for transformation. Genetic techniques available for manipulation of *Streptococcus pyogenes* are described by Caparon and Scott (18).

Two new types of genetic elements transferred by conjugation have been identified in streptococci. One type, the conjugative transposons, is the subject of chapter 41 in this volume. These elements transpose from a DNA molecule in one cell to a DNA molecule in another cell by a conjugation mechanism that involves cell-cell contact. No accessory conjugative DNA element is required for this transfer. Transposition of these elements proceeds through a covalently closed intermediate form of the transposon by a novel mechanism of recombination.

The other new type of transferrable element, characterized in *Enterococcus faecalis* by Clewell's group, is a class of conjugative plasmids that are transferred by a pheromone-inducible transfer system (21). This system, which is the best-understood conjugation system in gram-positive bacteria, is very efficient, even in matings conducted in liquid. In this system, a plasmid-free enterococcus can secrete a small peptide pheromone that induces a plasmid-containing donor cell to express a surface protein called aggregation substance. This allows the donor to bind to a receptor on the surface of the recipient and promotes the cell-cell contact that leads to formation of a mating aggregate. The plasmid is then transferred to the recipient cell and represses further pheromone synthesis. The new plasmid-containing host can act as a donor of the plasmid in additional pheromone-induced mating events.

Other conjugative plasmids generally promote DNA transfer at a much lower efficiency and do so only when the cells are forced into close contact by being placed on a solid matrix like a filter or an agar surface. Presumably, the induction of aggregates accounts for the high efficiency of the pheromone-induced system in *E. faecalis*. Currently, little is known about the

June R. Scott • Department of Microbiology and Immunology, Emory University Health Sciences Center, Atlanta, Georgia 30322. **Michael G. Caparon** • Department of Molecular Microbiology, Washington University School of Medicine, St. Louis, Missouri 63110.

mechanics and biochemistry of DNA transfer in conjugation in gram-positive organisms. To investigate these, the pheromone-induced *E. faecalis* mating system should serve as an excellent model.

STREPTOCOCCI AND DISEASE

Although the streptococci have been studied for their genetic systems and their roles in some industrial processes, much of the interest in these organisms comes from the fact that they cause disease in both humans and animals. For example, the enterococci, which are normal flora of the gut, are sometimes involved in urinary tract infections and also frequently cause infective endocarditis. Endocarditis can also result from infection by members of the viridans group. Group B streptococci (*Streptococcus agalactiae*) are a common cause in infants and adults of neonatal meningitis and sepsis, diseases which appear to be on the rise (41). *S. pneumoniae* is a frequent cause of ear infections in children and remains one of the leading causes of morbidity and mortality in compromised individuals. The production of acid in dental plaque by *S. mutans* plays a central role in the development of caries, which is considered the most widespread of all human diseases. Finally, the group A streptococci (*S. pyogenes*) cause a large number of different disease syndromes. This group will be the main focus of this chapter.

The dissemination of different antibiotic resistances among the pathogenic streptococci is limiting the use of many of these drugs in treatment. Traditionally, most streptococcal species have been very sensitive to the penicillins. However, a penicillin-binding protein that confers resistance to this antibiotic has been identified in some *S. pneumoniae* isolates (112). The same binding protein has also been identified in isolates of streptococcal species of the normal oral flora (28), suggesting the possible future spread of penicillin resistance among these organisms. In addition, a plasmid-borne β-lactamase was found in a strain of *E. faecalis* (69) which appears to have received this plasmid from a *Staphylococcus* species (68). This finding highlights the need for further investigations of streptococcal pathogenesis to provide alternative therapy regimens and possibly to develop vaccines to protect against streptococcal diseases. This is particularly true for the group A streptococci, which are currently sensitive to penicillin but cause the most varied types of infections.

The group A streptococci are the causative agents of several serious diseases, such as necrotizing fasciitis, scarlet fever, sepsis, and a recently recognized toxic shock-like syndrome (5, 99), as well as suppurative infections of the skin and throat, such as impetigo, erysipelas, and pharyngitis ("strep throat"). While the suppurative infections are generally self-limiting and not life threatening, they are treated immediately with penicillin to prevent the possible serious sequelae, such as rheumatic fever and acute glomerulonephritis. At this time, no group A streptococcal isolate resistant to penicillin has been reported.

MAJOR VIRULENCE FACTORS OF GROUP A STREPTOCOCCI

The group A streptococci can secrete a large number of different proteins, including some with activities that suggest a role in pathogenesis. These proteins include the plasminogen-activating streptokinases, which may aid in dissemination of the bacteria in tissue (62); a proteinase that may be involved in invasive infections (9, 14); and the pyrogenic exotoxins types A, B, and C. The pyrogenic exotoxins (reviewed in reference 2) are responsible for the systemic symptoms of scarlet fever and have been implicated in the toxic streptococcal syndrome (5, 43, 99, 105). These exotoxins and/or their genes have been sequenced. It appears that exotoxin B (SPE B) is closely related to (and may be identical to) the precursor form of the proteinase (42). The gene encoding one or both of these proteins is present in all group A streptococcal isolates that have been examined. Exotoxin A (SPE A) is both structurally and functionally related to enterotoxin B, an exotoxin of *Staphylococcus aureus* (50). Exotoxin C (SPE C) appears to be unrelated to the other exotoxins, but its poor immunogenicity has severely hampered further investigations (88).

In addition to the exoproteins, a number of surface structures have been implicated as virulence factors of the group A streptococci. A hyaluronic acid capsule surrounds the bacterium and may contribute to virulence (106, 108, 111). However, because the capsule is loosely associated with the cell and because it is easily degraded by both human and streptococcal hyaluronidases, its role in pathogenesis has been difficult to evaluate. Surface proteins may also contribute to virulence. These include the C5a peptidase, which can prevent polymorphonuclear leukocytes from recognizing chemotactic signals (107), and the immunoglobulin receptor proteins, which can bind the Fc components of immunoglobulin G (101) and immunoglobulin A (61). Recently, a surface molecule called protein F, which mediates binding of fibronectin by the streptococci, has been shown to be of critical importance for adherence of these organisms to respiratory epithelial cells (40). It seems likely, therefore, that this molecule plays an essential role in the virulence of group A streptococci.

Unfortunately, a rigorous examination of the role of any of these different factors in the pathogenesis of group A streptococcal infections has not been possible, because there is no representative animal model for any of the diseases caused by this organism. Instead, streptococcal virulence has traditionally been evaluated in the phagocytosis assay developed by Lancefield (60), in which a small number of streptococci are mixed with an aliquot of freshly drawn human blood. This mixture is rotated end over end for 3 h, and the number of streptococci in the mixture at the end of the incubation period is determined. Virulent streptococci are defined as organisms that not only survive but also multiply during incubation, while avirulent streptococci fail to multiply and are killed. Lancefield's experiments indicated that the abilities of virulent streptococci to survive in this assay correlates with their abilities to escape phagocytosis and killing by polymorphonuclear leukocytes in the blood. Furthermore, this antiphagocytic property depends on the expression of a streptococcal surface protein designated the M protein. This protein will be the major focus of the rest of this chapter.

THE MAJOR VIRULENCE FACTOR OF GROUP A STREPTOCOCCI IS M PROTEIN

Introduction to M Protein

M protein is considered the major virulence factor of group A streptococci (*S. pyogenes*) because it protects the bacteria from phagocytosis by polymorphonuclear leukocytes. For recent reviews of the M protein, see Manjula (63), Fischetti (30), and Scott (90). In electron micrographs of *S. pyogenes*, the M protein appears as fibrils on the bacterial surface (102). The M molecule is a highly alpha-helical coiled-coil dimer with a structure resembling that of tropomyosin (66, 79) and other intermediate filament proteins in this family. The coiled-coil structure results from a seven-residue periodicity in which residues a and d are hydrophobic; b, c, and f tend to be polar; and e and g are frequently charged. The hydrophobic amino acids are meshed within the two coils, and the polar residues are on the outside of the dimer. The charged amino acids allow the formation of salt bridges between residues at similar positions of the repeat on both monomers in the coiled-coil. The dimeric M molecule is attached to *S. pyogenes* through its carboxy terminus (79).

Infection by *S. pyogenes* causes production of antibodies that include those that react with M protein. Among the anti-M antibodies are some that are opsonic; i.e., they allow the bacteria to be phagocytized. These antibodies are protective against later infection by group A streptococci. However, as is the case for many immunogenic surface proteins of pathogens, the M protein is antigenically variable among different strains of streptococci. Currently, more than 80 different antigenic types of M protein are recognized, and the scheme for classification of group A streptococci is based on these types. Because opsonic antibodies tend to be type specific, antigenic variation among M proteins allows repeated infection of one host by different strains of streptococci. The existence of the many M-protein variants provides the molecular biologist with an important tool that should help in analysis of the relation of structure to function in the M molecule.

The M protein or a derivative of it is being considered as a possible antistreptococcus vaccine. There are two serious difficulties with this approach, however. First, the plethora of antigenic types of M protein poses a serious problem, and second, many anti-M-protein antibodies cross-react with human proteins, including some in human cardiac tissue (22, 24, 25). This reaction may result from the structural similarities of some heart proteins and the M molecule and may be involved in causing the rheumatic heart disease sometimes seen as a sequela to infection by *S. pyogenes*. Pursuit of the M protein for vaccine purposes is thus based on the hope of discovering one or more protective epitopes shared among many M-protein types that differ from epitopes of human proteins.

Because of its importance in virulence and epidemiology and its possible use as a vaccine, the M protein has served as a major focus of research on the group A streptococci. From these analyses, we have learned about several basic biological mechanisms that appear to have general applicability. These include the mechanism of antigenic and size variation of the protein and the mechanism of attachment of proteins to the surface of gram-positive cocci.

Structural Domains

The first complete delineation of the structure of an M protein was obtained by translation of the complete sequence of the structural gene for M6, *emm*6.1 (46). Therefore, the M6.1 molecule serves as the prototype with which other M proteins have been compared. These comparisons led to the idea that M proteins have three major regions (34), which are described in this section (Fig. 1). Further molecular analysis using deletions and rearrangements is required to validate the current view of the M molecule.

Conserved C terminus

The carboxy-terminal region of the M molecule (Fig. 1) is similar to the C-terminal regions of other surface proteins of gram-positive cocci (e.g., protein A of *Staphylococcus aureus* and the streptococcal proteins G, H, etc.). This C-terminal region is composed of 19 hydrophobic amino acids, which presumably constitute a membrane-spanning domain, followed by six charged amino acids at the C terminus. Immediately N terminal to the hydrophobic region is a proline-glycine-rich segment expected to be involved in spanning the streptococcal cell wall. It has been proposed that these residues provide the flexibility needed for the protein to snake through the peptidoglycan layer (34). Together, these three segments of the molecule are referred to as the "anchor domain" of the M protein.

The sequence LPSTGE ends four residues N terminal to the hydrophobic amino acids, within the "wall domain." The conservation of this sequence among surface proteins of gram-positive cocci (including protein A of *Staphylococcus aureus* [37], immunoglobulins A- and G-binding proteins of *S. pyogenes* [35, 36], a cell wall protease of *Streptococcus cremoris* [56], and surface proteins of *S. pyogenes* [44, 89] and *S. mutans* [29, 55, 73]) at the amino acid and DNA levels suggests that the sequence may be functionally important for attaching the protein to the bacterial surface (33, 75).

A similar motif is found among eukaryotic proteins attached by a glycosyl phosphatidylinositol (GPI) link to the membrane. It is proposed that such proteins are attached to GPI in a reaction that involves cleavage of the transmembrane segment from the protein. The conserved motif is thought to contain the sequence at which the cleavage and transfer to GPI occurs. By analogy, it was suggested that a cleavage occurs after the serine in LPSTGE of M6 and that the remaining M6 molecule is then transferred to a similar anchoring moiety in the streptococcal membrane (34, 75). Although this membrane component has not yet been identified, the M protein appears to bind to membrane and not to a cell wall component, since streptococcal protoplasts retain the M protein under some conditions.

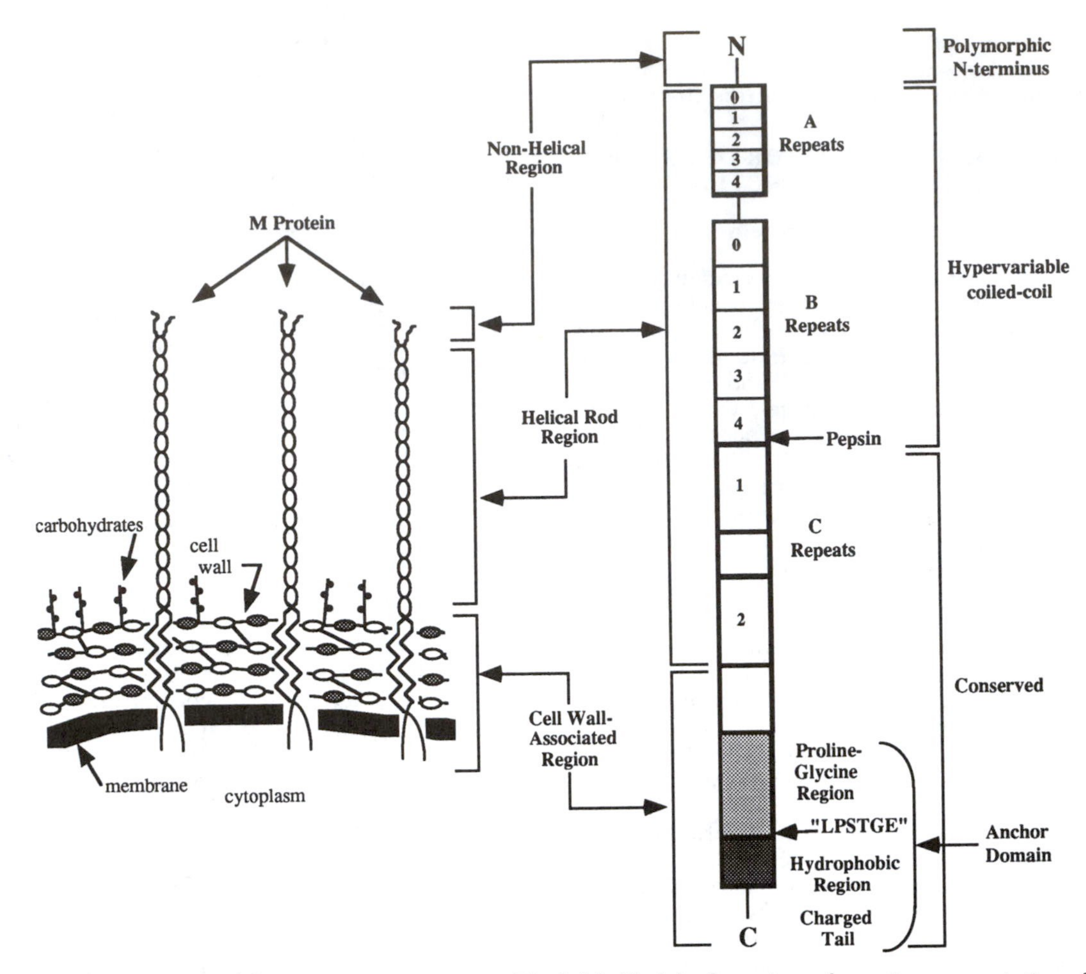

Figure 1. The M protein of the group A streptococcus. The left half of the figure is a schematic representation of the dimeric coiled-coil M-protein molecules and the streptococcal cell wall. M-protein dimers, cell membrane, cytoplasm, peptidoglycan (cell wall), and group A-specific carbohydrate(s) are shown. The right half of the figure shows the domains of the M6.1 protein, which has served as a prototype for the structures of all M proteins. The brackets between the two halves of the figure indicate the locations of the main structural features of the M molecule. Shown is the cell wall-associated region, which remains after trypsin digestion of streptococcal cells; the helical rod region, which forms the coiled-coil structure; and the nonhelical amino terminus (N), which is the region most distal to the streptococcal cell wall. Specific domains of the M6 molecule are shown, including the five A repeats of 14 amino acids each, the five B repeats of 25 amino acids each, and the two C repeats of 27 amino acids each. Also shown is the anchor domain, which consists of the carboxy-terminal (C) cytoplasmic charged tail, a membrane-spanning hydrophobic region, and a flexible region rich in proline and glycine residues that traverses the cell wall. LPSTGE indicates the location of an amino acid motif conserved among many gram-positive surface proteins that is also found in eukaryotic proteins attached to membranes by a GPI linkage. The arrow at Pepsin indicates the site at which the M molecule is cleaved preferentially by pepsin. Brackets on the far right of the figure show locations of the three major domains of all M-protein molecules: the polymorphic N terminus, the hypervariable coiled-coil domain, and the conserved domain, which is highly similar among the different M proteins that have been studied.

One activity identified in streptococcal membranes (even from an M^- strain) releases most of the M protein from protoplasts. The protein released lacks the membrane anchor, and the segment missing may begin within the LPSTGE sequence (75). The missing segment may have been removed at the time the M protein was attached by its serine to the streptococcal membrane, which would support the analogy to eukaryotic GPI-linked proteins given above. Since the M protein appears to be an excellent model for bacterial surface proteins, the functions of the regions within its anchor domain are currently under active investigation.

On the surface of the streptococcus, about one-third of the M6.1 molecule is resistant to trypsin digestion, suggesting that this region of the protein is buried within the cell wall (Fig. 1) (74). The protected region corresponds approximately to the C-terminal third of the M6.1 molecule (Fig. 1).

The C-terminal region is highly conserved among the 10 different types of M proteins whose sequences have been studied, i.e., types 5, 6, 19, 24, 30, and 55 (45); type 1 (39; GenBank accession number X07933); type 2 (11); type 12 (84); type 49 (38); and the "M-like proteins" (an immunoglobulin A receptor encoded by Arp [35] and an immunoglobulin G receptor called protein H [36]). Hybridization and polymerase chain reaction studies suggest sequence conservation of this region among additional M types as well (80, 83, 92, 93).

There are two apparent reasons for this evolutionary conservation: first, because it is not exposed, the C-terminal region is less sensitive to selective pressure by the immune system, and second, the structure of this region is important for attachment of the M molecule to the streptococcal surface and so cannot vary greatly. However, just N terminal to the region buried in the streptococcal cell wall in the M6 protein lie two direct nontandem repeated regions (Fig. 1) (46). These are encoded by the 81-bp C repeats. This part of the molecule is also highly conserved among proteins of different M types (11, 45), although its location external to the streptococcal cell is indicated by its susceptibility to trypsin digestion (74). It is not subject to the variation seen further N terminal in the other repeats of the protein (see below), suggesting that the sequence of the C repeat region may be functionally important.

Group A streptococci have been divided into two classes based on the serological types of their M-associated proteins (MAP), antigens that copurify with the M protein (109, 110). The MAP has not been purified and characterized and may actually represent a fragment of the M protein (30). The organisms that fall into class I by this scheme tend not to produce opacity factor, an apoprotease that is responsible for causing horse serum agar to turn opaque (85). Although the basis of this classification has not been elucidated, it seems to correlate with the reactivity of the M protein C repeat region to monoclonal antibody 10F5, which was made against the M6.1 protein (10). Most class I streptococci react with this monoclonal antibody, and most class II strains do not. This classification is not clear-cut and absolute, probably because there is a continuum of M-protein sequences and evolution is proceeding rapidly (see below). However, this scheme is intriguing because it has been suggested to correlate with the disease potential of the streptococcus. Of the strains examined, most that had been associated with outbreaks of rheumatic fever appear to fall into class I.

Molecular analyses indicate that the class which an M-protein-producing strain of group A streptococcus falls into is determined by a few amino acids within the highly conserved C repeat region of the molecule (11). Whether the correlation of properties on which this classification is based results from the evolutionary history of the streptococci (perhaps the strains of class I derive from a different ancestor than those of class II) or has functional significance remains to be determined.

Hypervariable coiled-coil structure

All the M proteins whose sequences are currently available in whole or in part display a repeated structure, as was first recognized from the partial amino acid sequences of the M5 and M24 molecules (8, 49, 64). That long repeats are also present at the DNA level was shown by the complete DNA sequence of *emm*6.1 (46), which contains two extensive series of tandem direct repeats. There are five A repeats of 42 bp (or 10 repeats of 21 bp), of which the central three repeats are essentially identical and the external two repeats are more divergent. This is followed by five B repeats of 75 bp with the same type of conservation of the internal elements. The existence of such an extensive series of DNA repeat regions suggested the possibility of homologous recombination and/or slipped-strand mispairing during DNA replication, which would result in deletion or addition of repeat units (46). This situation has been shown to result in the observed hypervariability of this region of M-protein molecules (47), which leads to rapid alteration in size and antigenicity of the main part of the protein (see below).

Polymorphic N terminus

In the M6 molecule and most of the other M molecules whose structures are known, there is a short non-alpha-helical region N terminal to the repeats. This region shows no homology among M proteins of different serotypes, tends to be negatively charged, and has been suggested as being responsible for type specificity of the M protein (30). This suggestion is based on studies of antibodies made to sequential synthetic peptides corresponding to the sequence of M6.1. The only peptide that generated opsonic antibodies was that corresponding to the N-terminal 21 amino acids, the first 12 of which precede the first A repeat (51). In addition, synthetic peptides derived from the first 20 residues of M5 stimulated production of type-specific opsonic antibodies that did not cross-react with other types of M proteins (26, 27). This occurred also with M6 (7, 51), M19 (15), and M1 (57). In addition, an opsonic monoclonal anti-M6 antibody (3B8) appears to recognize a peptide in the A repeat region, although there is significant nonlinear conformational character to the recognition site of this antibody (52, 53). However, for M5, some opsonic antibodies that are cross-reactive and opsonize several M types appear to recognize epitopes at other locations (23, 86). The critical experiment to test the function of the non-alpha-helical N terminus is to replace the N-terminal region of M6 by the N-terminal region of a different M type. This experiment has not yet been reported.

Role of Repeats in Sequence Variation

Antigenic variation of surface proteins that induce protective antibodies is a common theme among pathogens. This variation is often induced by homologous recombination of the expressed gene copy with DNA from a nonexpressed pseudogene elsewhere in the chromosome, as is the case for the variable surface proteins of trypanosomes (14a) or the major pilin protein of the gonococcus (94). The M protein, however, uses a different mechanism.

When the first M-protein structural gene was cloned (from an M6 strain) and hybridized with chromosomal DNA from its strain of origin, only a single copy of the gene (called *emm*6.1) was detected (93). Thus, there are no partially homologous sequences elsewhere in the chromosome of that *Streptococcus* sp. to participate in gene conversion. More recently, several strains of group A streptococci with two copies of an *emm*-like gene have been described (12, 38, 54). However, in no case have more than two copies been seen.

In addition to the antigenic variation, there is extensive size variation among M proteins of different strains of group A streptococci (32). The apparent molecular weights of these molecules as determined by sodium dodecyl sulfate-polyacrylamide gel electrophoresis varied by about 40,000, and even among M6 proteins from different strains, the variation was about 19,000. In the laboratory, mutants of M6 strain D471 with M proteins of different sizes were detected at a frequency of 2×10^{-4}/CFU, which is much higher than expected for a random mutation. Such extensive variation would be expected if discrete numbers of repeat segments (see above) were lost and gained within the gene (46).

To test this hypothesis, the sequences of the *emm6* genes of spontaneous mutants arising in the laboratory were determined (47). As predicted, spontaneous mutants of strain D471 that produced smaller M proteins had *emm6* genes with a discrete number of repeat units deleted. Furthermore, from analysis of a single outbreak in a child care center, several M6 strains were distinguished, all of which were presumably derived from the same ancestral *Streptococcus* sp. The *emm* genes in these strains differed from each other by a discrete number of A and/or B repeat units. Thus, size decrease among M proteins occurs by loss of repeat elements, which presumably results either from homologous recombination or from slipped-strand mispairing (46).

What about antigenic variation? Since in both the A and B regions of M6.1 the external repeats in each group of five repeats differ significantly from the canonical sequence of the repeat element, events that involve these external repeats will generate new DNA sequence and, possibly, new amino acid sequences (46). To test whether this affects important antigenic epitopes, an opsonic monoclonal antibody (3B8) was used (52). One of the spontaneous deletion mutants was altered sufficiently that it was no longer efficiently opsonized by this antibody, thus demonstrating that the repeats play a pivotal role in increasing the antigenic diversity of the M protein. This experiment also strongly suggests that the division into M types is artificial, since there appears to be a continuum of sequences of M proteins and possibly of potential M types (91).

In summary, unlike the case for variation of surface proteins of other pathogens, the events that generate M-protein variants involve intragenic homologous elements. This variation is frequent and probably results from slipped-strand mispairing during DNA replication and/or from homologous recombination. The isolation of a mutant of a group A streptococcus with a mutation in the *recA* gene may make it possible to determine which mechanism is responsible for the frequent variation in *emm* gene structure.

Function of M Protein

The mechanism by which M protein protects the streptococcus from phagocytosis is not yet clear. Phagocytosis of M$^-$ streptococci appears to require the alternate complement pathway (13, 78). It has been suggested that M protein provides resistance to this process because it binds complement factor H, thus interfering with the complement cascade (48). Competition for factor H binding by peptides corresponding to fragments of M6.1 suggests that factor H binds to the spacer region between the C repeats (31). It seemed possible that the specific sequence recognized by factor H was necessary for the function of the M protein and that this requirement might provide the explanation for the lack of variation of this region of the molecule. (Recombination between homologous sequences in the two C repeats would delete the spacer.) However, in vitro deletion of the DNA encoding the factor H-binding peptide does not interfere with binding by factor H (77), indicating that recognition and interaction of the M protein and factor H are complex.

It has also been suggested that the M protein is involved in attachment of the streptococcus to its host either directly (104) or by acting as an anchor for the streptococcal lipoteichoic acid (7, 71, 72), and it has been suggested that fibronectin, a glycoprotein found in saliva, may serve as the receptor on host cells for streptococcal binding (95, 96). To test this directly, the *emm* gene was deleted from an M6 strain by using in vitro DNA manipulation (70), and this strain was compared with its isogenic M$^+$ parent for attachment to human buccal and tonsillar cells and to tonsillar tissue (19). No difference in the number of streptococcal units that adhered was detected in these strains (19). Furthermore, both strains bind fibronectin equally well (19). Thus, the M protein is not the primary streptococcal ligand and is not required for attachment of the primary streptococcal ligand responsible for adherence of group A streptococci to their host cells. However, when attached to human tonsillar cells, the M$^+$ strain formed clumps, while its M$^-$ derivative did not (19). (Similar clumping was not observed on human buccal cells.) This finding suggests that one bacterium attaches to host tissue by some specific non-M-protein ligand and that additional bacteria then bind to the first one via the M protein (Fig. 2). If this is correct, M protein would still be important for promoting colonization of the human host, even though it is not the primary ligand.

Recently, a candidate for the primary streptococcal ligand has been identified, and its structural gene has been cloned (40). The protein, named F, binds fibronectin; an insertion in the structural gene for protein F prevents adherence of the streptococcal strain to pharyngeal cells in culture. Further work on this gene and its product should lead to an understanding of the process of colonization of the human throat by group A streptococci.

REGULATION OF M PROTEIN SYNTHESIS

mry Gene

A positive regulator for transcription of *emm6* was identified by insertion mutagenesis with the conjugative transposon Tn*916* (17). The regulator was named *mry* for M-protein RNA yield, and the insertion mutation was called *mry-1*. The *mry-1* mutant shows a reduction in the amount of M protein of about 50-fold compared with the amount in the wild-type parent, as

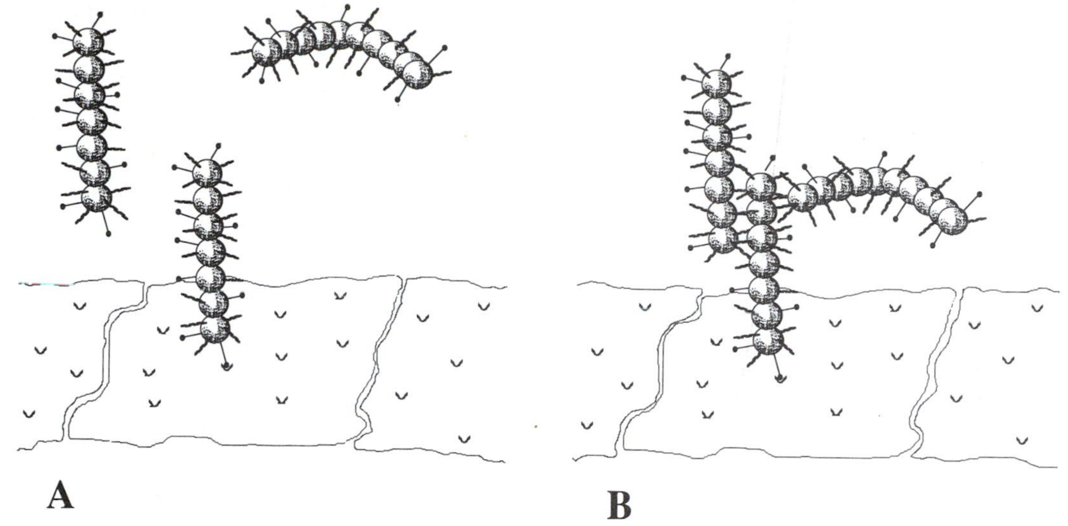

Figure 2. Model for the roles of different proteins in the attachment of *S. pyogenes* to epithelial cells. (A) Chains composed of individual streptococcal cells are shown in the process of attaching to an epithelial cell. Both M protein ( ) and protein F ( ), a fibronectin-binding protein, are shown on the surface of the streptococcal cells. Molecules of fibronectin ( ) are shown on the surface of the epithelial cell. The streptococcal chain in the center of the figure has adhered to the epithelial cell by a mechanism that involves the binding of protein F to a molecule of fibronectin. (B) Once a chain of streptococci has attached by its protein F to a tonsillar epithelial cell, the M protein of the attached organisms interacts with other streptococci to cause bacterial aggregation. This type of M-protein-mediated interstreptococcal interaction may contribute to colonization of the host.

determined by Western blot (immunoblot) analysis. The insertion in *mry-1* was determined by sequence analysis to lie 244 bases upstream of the open reading frame for the Mry protein (76). The roles of these bases and the location(s) of the promoter(s) for *mry* have not yet been determined.

The sequence of the Mry protein suggests that this protein is a member of the response regulator class of signal transducing systems (often called two-component regulatory systems) (76). In the simplest of these systems, a surface-exposed histidine kinase responds to environmental changes by autophosphorylation of a histidine residue, and this phosphate group is transferred to the aspartate residue of a cytoplasmic protein that activates transcription of one or more operons (1, 67, 100). However, Mry differs from other members of this class by being about twice as big (molecular weight of 62,000) and having two response regulator domains and two potential helix-turn-helix DNA-binding domains. It further differs in that the predicted phosphorylatable response regulator domains are at the C terminus of the Mry molecule.

There are two potential in-frame start sites for the Mry protein, and proteins of both predicted sizes have been seen on a gel following in vitro transcription and translation of this gene in an *Escherichia coli* extract (76). The mechanism of action of *mry* and its regulation remain to be determined.

Effect of Mry on Other Virulence Determinants

Expression of several bacterial virulence determinants in one organism is often controlled by a single regulatory protein, and often these regulators are members of the two-component family (67). Since Mry is probably also a member of this family, it seemed possible that Mry might regulate other potential virulence genes of group A streptococci. In an M12 strain, deletion mutations that reduce expression of both M12 and the C5a peptidase (a potential virulence factor in this strain) have been designated *virR* (97, 98). These deletions lie within the coding sequence of *mry*. However, with cultures grown under normal laboratory conditions, *mry* mutants produce the same amounts of streptokinase and capsule as their parental isogenic mry^+ control strains (76).

Environmental Regulation of M Protein

Another property of signal transducing regulatory systems is the response to changes in environmental conditions. We have recently determined that transcription of the M protein is sensitive to such environmental changes. Specifically, by using a chloramphenicol acetyltransferase gene (*cat*) fused to the *emm* gene promoter, we have found that transcription is increased when the streptococci are grown in a high-CO_2 atmosphere (16). Mry is required for expression of *emm* under all conditions tested, and the mechanism of regulation by environmental conditions is currently under study.

PROSPECTS FOR THE FUTURE

A greater understanding of the relation of the structure of the M protein to its function should help in the rational design of an antistreptococcal vaccine and should also increase our understanding of the way in which coiled-coil fibrous proteins can act. Because the regulatory system(s) of the M protein appears to have some unique features, it is hoped that elucidation of this system will reveal mechanisms that can serve as prototypes for regulation of other genes, especially in the more poorly studied gram-positive cocci.

Acknowledgments. Work in the laboratory of J.R.S. was supported by Public Health Service grant AI20723 from the National Institutes of Health and by a grant from the American Heart Association (Georgia affiliate). Work in the laboratory of M.G.C. was supported by Public Health Service grant AI30643 from the National Institutes of Health and by a grant from the American Heart Association.

REFERENCES

1. **Albright, L. M., E. Huala, and F. M. Ausubel.** 1989. Prokaryotic signal transduction mediated by sensor and regulator protein pairs. *Annu. Rev. Genet.* **23:**311–336.
2. **Alouf, J. E.** 1980. Streptococcal toxins (streptolysin O, streptolysin S, erythrogenic toxin). *Pharmacol. Ther.* **11:**661–717.
3. **Avery, O. T., C. M. MacLeod, and M. McCarty.** 1944. Studies on the chemical nature of the substance inducing transformation of pneumococcal types. Induction of transformation by a deoxyribonucleic acid fraction isolated from pneumococcus type III. *J. Exp. Med.* **79:**137–159.
4. **Balganesh, T. S., and S. A. Lacks.** 1985. Heteroduplex DNA mismatch repair system of *Streptococcus pneumoniae*: cloning and expression of the *hexA* gene. *J. Bacteriol.* **162:**979–984.
5. **Bartter, T., A. Dascal, K. Carroll, and F. J. Curley.** 1987. "Toxic strep syndrome": a manifestation of group A streptococcal infection. *Arch. Intern. Med.* **148:**1421–1424.
6. **Beachey, E. H., and I. Ofek.** 1976. Epithelial cell binding of group A streptococci by lipoteichoic acid on fimbriae denuded of M protein. *J. Exp. Med.* **143:**759–771.
7. **Beachey, E. H., and J. M. Seyer.** 1986. Protective and non-protective epitopes of chemically synthesized peptides of the amino-terminal region of type 6 streptococcal M protein. *J. Immunol.* **136:**2287–2291.
8. **Beachey, E. H., J. M. Seyer, and A. H. Kang.** 1978. Repeating covalent structure of M protein. *Proc. Natl. Acad. Sci. USA* **75:**3163–3167.
9. **Beletskaya, L. V., and E. V. Gnezditskaya.** 1982. The effect of streptococcal proteinase on connective tissue, p. 141–142. *In* S. E. Holm and P. Christensen (ed.), *Basic Concepts of Streptococci and Streptococcal Diseases: Proceedings of the VIIIth International Symposium on Streptococci and Streptococcal Diseases*, June 1981. Reedbooks, Ltd., Chertsey, Surrey, England.
10. **Bessen, D., K. F. Jones, and V. A. Fischetti.** 1989. Evidence for two distinct classes of streptococcal M protein and their relationship to rheumatic fever. *J. Exp. Med.* **169:**269–283.
11. **Bessen, D. E., and V. A. Fischetti.** 1990. Differentiation between two biologically distinct classes of group A streptococci by limited substitutions of amino acids within the shared region of M protein-like molecules. *J. Exp. Med.* **172:**1757–1764.
12. **Bessen, D. E., and V. A. Fischetti.** 1992. Nucleotide sequences of two adjacent M or M-like protein genes of group A streptococci: different RNA transcript levels and identification of a unique immunoglobulin A-binding protein. *Infect. Immun.* **60:**124–135.
13. **Bisno, A. L.** 1979. Alternate complement pathway activation by group A streptococci: role of M-protein. *Infect. Immun.* **26:**1172–1176.
14. **Bjorck, L., P. Akesson, M. Bohus, J. Trojnar, M. Abrahamson, I. Olafsson, and A. Grubb.** 1989. Bacterial growth blocked by a synthetic peptide based on the structure of a human proteinase inhibitor. *Nature* (London) **337:**385–386.

14a. **Borst, P., and G. A. M. Cross.** 1982. Molecular basis for trypanosome antigenic variation. *Cell* **29:**291–303.

15. **Bronze, M. S., E. H. Beachey, and J. B. Dale.** 1988. Protective and heart-crossreactive epitopes located within the N-terminus of type 19 streptococcal M protein. *J. Exp. Med.* **167:**1849–1859.
16. **Caparon, M. G., J. Perez-Casal, and J. R. Scott.** Unpublished data.
17. **Caparon, M. G., and J. R. Scott.** 1987. Identification of a gene that regulates expression of M protein, the major virulence determinant of group A streptococci. *Proc. Natl. Acad. Sci. USA* **84:**8677–8681.
18. **Caparon, M. G., and J. R. Scott.** 1990. Genetic manipulation of pathogenic streptococci. *Methods Enzymol.* **204:**556–586.
19. **Caparon, M. G., D. S. Stephens, A. Olsen, and J. R. Scott.** 1991. Role of M protein in adherence of group A streptococci. *Infect. Immun.* **59:**1811–1817.
20. **Claverys, J. P., and S. A. Lacks.** 1986. Heteroduplex deoxyribonucleic acid base mismatch repair in bacteria. *Microbiol. Rev.* **50:**133–165.
21. **Clewell, D.** 1981. Plasmids, drug resistance, and gene transfer in the genus *Streptococcus*. *Microbiol. Rev.* **45:**409–436.
22. **Dale, J. B., and E. H. Beachey.** 1982. Protective antigenic determinant of streptococcal M protein shared with sarcolemmal membrane protein of human heart. *J. Exp. Med.* **156:**1165–1176.
23. **Dale, J. B., and E. H. Beachey.** 1984. Unique and common protective epitopes among different serotypes of group A streptococcal M proteins defined with hybridoma antibodies. *Infect. Immun.* **46:**267–269.
24. **Dale, J. B., and E. H. Beachey.** 1985. Multiple heart-cross-reactive epitopes of streptococcal M proteins. *J. Exp. Med.* **161:**113–122.
25. **Dale, J. B., and E. H. Beachey.** 1985. Epitopes of streptococcal M protein shared with cardiac myosin. *J. Exp. Med.* **162:**583–591.
26. **Dale, J. B., and E. H. Beachey.** 1986. Localization of protective epitopes of the amino terminus of type 5 streptococcal M protein. *J. Exp. Med.* **163:**1191.
27. **Dale, J. B., J. M. Seyer, and E. H. Beachey.** 1983. Type specific immunogenicity of a chemically synthesized peptide fragment of type 5 streptococcal M protein. *J. Exp. Med.* **158:**1727–1732.
28. **Dowson, C. G., A. Hutchison, N. Woodford, A. P. Johnson, R. C. George, and B. G. Spratt.** 1990. Penicillin-resistant viridans streptococci have obtained altered penicillin-binding protein genes from penicillin-resistant strains of *Streptococcus pneumoniae*. *Proc. Natl. Acad. Sci. USA* **87:**5858–5862.
29. **Ferretti, J. J., R. R. B. Russell, and M. L. Dao.** 1989. Sequence analysis of the wall-associated protein precursor of *Streptococcus mutans* antigen A. *Mol. Microbiol.* **3:**469–478.
30. **Fischetti, V. A.** 1989. Streptococcal M protein: molecular design and biological behavior. *Clinical Microbiol. Rev.* **2:**285–314.
31. **Fischetti, V. A.** 1991. Streptococcal M protein. *Sci. Am.* **264:**58–65.
32. **Fischetti, V. A., K. F. Jones, and J. R. Scott.** 1985. Size variation of the M protein in group A streptococci. *J. Exp. Med.* **161:**1384–1401.
33. **Fischetti, V. A., V. Pancholi, and O. Schneewind.** 1990. Conservation of a hexapeptide sequence in the anchor region of surface proteins from Gram-positive cocci. *Mol. Microbiol.* **4:**1603–1605.
34. **Fischetti, V. A., D. A. D. Parry, B. L. Trus, S. K. Hollingshead, J. R. Scott, and B. N. Manjula.** 1988. Conformational characteristics of the complete se-

quence of group A streptococcal M6 protein. *Proteins Struct. Funct. Genet.* **3:**60–69.

35. **Frithz, E., L.-O. Heden, and G. Lindahl.** 1989. Extensive sequence homology between IgA receptor and M protein in *Streptococcus pyogenes. Mol. Microbiol.* **3:**1111.
36. **Gomi, H., T. Hozumi, S. Hattori, C. Tagawa, F. Kishimoto, and L. Bjorck.** 1990. The gene sequence and some properties of protein H. *J. Immunol.* **144:**4046.
37. **Guss, B., M. Uhlen, B. Nilsson, M. Lindberg, J. Sjoquist, and J. Sjodaht.** 1984. Region X, the cell-wall-attachment part of staphylococcal protein A. *Eur. J. Biochem.* **138:**413–420.
38. **Haanes, E. J., and P. P. Cleary.** 1989. Identification of a divergent M protein gene and an M protein-related gene family in *Streptococcus pyogenes* serotype 49. *J. Bacteriol.* **171:**6397–6408.
39. **Haanes-Fritz, E., W. Kraus, V. Burdett, J. B. Dale, E. H. Beachey, and P. P. Cleary.** 1988. Comparison of the leader sequences of four group A streptococcal M protein genes. *Nucleic Acids Res.* **16:**4667–4677.
40. **Hanski, E., and M. G. Caparon.** 1992. Protein F, a fibronectin-binding protein, is an adhesin of the group A streptococcus. *Proc. Natl. Acad. Sci. USA* **89:**6172–6176.
41. **Harvey, R. C., M. M. Farley, T. Stull, J. D. Smith, A. Shuchat, and D. S. Stephens.** 1991. Population-based assessment of *Streptococcus agalactiae* (group B streptococcus) meningitis and bacteremia in metropolitan adults, abstr. 1059. Program Abstr. 31st Intersci. Conf. Antimicrob. Agents Chemother. 1991.
42. **Hauser, A. R., and P. M. Schlievert.** 1990. Nucleotide sequence of the streptococcal pyrogenic exotoxin type B gene and relationship between the toxin and the streptococcal proteinase precursor. *J. Bacteriol.* **172:**4536–4542.
43. **Hauser, A. R., D. L. Stevens, E. L. Kaplan, and P. M. Schlievert.** 1991. Molecular analysis of pyrogenic exotoxins from *Streptococcus pyogenes* isolates associated with toxic shock-like syndrome. *J. Clin. Microbiol.* **29:**1562–1567.
44. **Heath, D. G., and P. P. Cleary.** 1989. Fc-receptor and M-protein genes of group A streptococci are products of gene duplication. *Proc. Natl. Acad. Sci. USA* **86:**4741–4745.
45. **Hollingshead, S., V. Fischetti, and J. R. Scott.** 1987. A highly conserved region present in transcripts encoding heterologous M proteins of group A streptococcus. *Infect. Immun.* **55:**3237–3239.
46. **Hollingshead, S. K., V. A. Fischetti, and J. R. Scott.** 1986. Complete nucleotide sequence of type 6 M protein of the group A streptococcus: repetitive structure and membrane anchor. *J. Biol. Chem.* **261:**1677–1686.
47. **Hollingshead, S. K., V. A. Fischetti, and J. R. Scott.** 1987. Size variation in group A streptococcal protein is generated by homologous recombination between intragenic repeats. *Mol. Gen. Genet.* **207:**196–203.
48. **Horstmann, R. D., H. J. Sievertsen, J. Knobloch, and V. A. Fischetti.** 1987. Antiphagocytic activity of streptococcal M protein: selective binding of complement control protein factor H. *Proc. Natl. Acad. Sci. USA* **85:**1657–1661.
49. **Hosein, B., M. McCarty, and V. A. Fischetti.** 1979. Amino acid sequence and physicochemical similarities between streptococcal M protein and mammalian tropomyosin. *Proc. Natl. Acad. Sci. USA* **76:**3765–3768.
50. **Johnson, L. P., J. J. L'Italien, and P. M. Schlievert.** 1986. Streptococcal pyrogenic exotoxin type A (scarlet fever toxin) is related to *Staphylococcus aureus* enterotoxin B. *Mol. Gen. Genet.* **203:**354–356.
51. **Jones, K. F., and V. A. Fischetti.** 1988. The importance of the location of antibody binding on the M6 protein for opsonization and phagocytosis of group A M6 streptococci. *J. Exp. Med.* **167:**1114–1123.
52. **Jones, K. F., S. K. Hollingshead, J. R. Scott, and V. A. Fischetti.** 1988. Spontaneous M6 protein size mutants of group A streptococci display variation in antigenic and opsonogenic epitopes. *Proc. Natl. Acad. Sci. USA* **85:**8271–8275.
53. **Jones, K. F., S. K. Hollingshead, J. R. Scott, and V. A. Fischetti.** Unpublished data.
54. **Kehoe, M. A., T. P. Poirier, E. H. Beachey, and K. N. Timmis.** 1985. Cloning and genetic analysis of serotype 5 M protein determinant of group A streptococci: evidence for multiple copies of the M5 determinant in the *Streptococcus pyogenes* genome. *Infect. Immun.* **48:**190–197.
55. **Kelly, C., P. Evans, L. Bergmeier, S. F. Lee-Fox, A. Progulske, A. C. Harris, A. Aitken, A. S. Bleiweis, and T. Lerner.** 1990. Sequence analysis of a cloned streptococcal surface antigen I/II. *FEBS Lett.* **258:**127–132.
56. **Kok, J., K. J. Leenhouts, A. J. Haandrikman, A. M. Ledeboer, and G. Venema.** 1988. Nucleotide sequence of the cell wall proteinase gene of *Streptococcus cremoris* Wg2. *Appl. Environ. Microbiol.* **54:**231–238.
57. **Kraus, W., E. Hannes-Fritz, P. P. Cleary, J. Seyer, J. Dale, and E. Beachey.** 1987. Sequence and type-specific immunogenicity of the amino-terminal region of type 1 streptococcal M protein. *J. Immunol.* **139:**3084–3090.
58. **Lacks, S. A., B. M. Mannarelli, S. S. Springhorn, and B. Greenberg.** 1986. Genetic basis of the complementary *Dpn*I and *Dpn*II restriction systems of *S. pneumoniae*: an intercellular cassette mechanism. *Cell* **46:**993–1000.
59. **Lancefield, R. C.** 1933. A serological differentiation of human and other groups of hemolytic streptococci. *J. Exp. Med.* **57:**571–595.
60. **Lancefield, R. C.** 1957. Differentiation of group A streptococci with a common R antigen into three serological types, with special reference to the bactericidal test. *J. Exp. Med.* **106:**525–544.
61. **Lindahl, G., and B. Akerstrom.** 1989. Receptor for IgA in group A streptococci: cloning of the gene and characterization of the protein expressed in *Escherichia coli. Mol. Microbiol.* **3:**239–247.
62. **Malke, H., and J. Ferretti.** 1984. Streptokinase: cloning, expression, and excretion by *Escherichia coli. Proc. Natl. Acad. Sci. USA* **81:**3557–3561.
63. **Manjula, B. N.** 1988. Molecular aspects of the phagocytosis resistance of group A streptococci. *Eur. J. Epidemiol.* **4:**289–300.
64. **Manjula, B. N., and V. A. Fischetti.** 1980. Tropomyosin-like seven residue periodicity in three immunologically distinct streptococcal M proteins and its implications for the antiphagocytic property of the molecule. *J. Exp. Med.* **151:**695–708.
65. **Martin, B., H. Prats, and J. P. Claverys.** 1985. Cloning of the *hexA* mismatch-repair gene of *Streptococcus pneumoniae* and identification of the product. *Gene* **34:**293–303.
66. **McLachlan, A. D., and M. Stewart.** 1975. Tropomyosin coiled-coil interactions: evidence for an unstaggered structure. *J. Mol. Biol.* **98:**293–304.
67. **Miller, J. F., J. J. Mekalanos, and S. Falkow.** 1989. Coordinate regulation and sensory transduction in the control of bacterial virulence. *Science* **243:**916–922.
68. **Murray, B. E.** 1987. Plasmid-mediated beta-lactamase in *Enterococcus faecalis*, p. 83–86. *In* J. J. Ferretti and R. Curtiss (ed.), *Streptococcal Genetics*. American Society for Microbiology, Washington, D.C.
69. **Murray, B. E., and B. Mederski-Samoraj.** 1983. Transferable beta-lactamase: a new mechanism for *in vitro* penicillin resistance in *Streptococcus faecalis. J. Clin. Invest.* **72:**1168–1171.
70. **Norgren, M., M. G. Caparon, and J. R. Scott.** 1989. A method for allelic replacement that uses the conjugative transposon Tn*916*: deletion of the *emm*6.1 allele in

Streptococcus pyogenes JRS4. *Infect. Immun.* **57:**3846–3850.
71. **Ofek, I., E. H. Beachey, W. Jefferson, and G. L. Campbell.** 1975. Cell membrane-binding properties of group A streptococcal lipoteichoic acid. *J. Exp. Med.* **141:**990–1003.
72. **Ofek, I., W. A. Simpson, and E. H. Beachey.** 1982. Formation of molecular complexes between a structurally defined M protein and acylated or deacylated lipoteichoic acid of *Streptococcus pyogenes*. *J. Bacteriol.* **149:**426–433.
73. **Olsson, A., M. Eliasson, B. Guss, B. Nilsson, U. Hellman, M. Lindberg, and M. Uhlen.** 1987. Structure and evolution of the repetitive gene encoding streptococcal protein G. *Eur. J. Biochem.* **168:**319–324.
74. **Pancholi, V., and V. Fischetti.** 1988. Isolation and characterization of the cell-associated region of group A streptococcal M6 protein. *J. Bacteriol.* **170:**2618–2624.
75. **Pancholi, V., and V. A. Fischetti.** 1989. Identification of an endogenous membrane anchor-cleaving enzyme for group A streptococcal M protein. *J. Exp. Med.* **170:**2119–2133.
76. **Perez-Casal, J., M. G. Caparon, and J. R. Scott.** 1991. Mry, a *trans*-acting positive regulator of the M protein gene of *Streptococcus pyogenes* with similarity to the receptor proteins of two-component regulatory systems. *J. Bacteriol.* **173:**2617–2624.
77. **Perez-Casal, J., V. A. Fischetti, and J. R. Scott.** Unpublished data.
78. **Peterson, P. K., D. Schmeling, P. P. Cleary, B. J. Wilkinson, Y. Kim, and P. G. Quie.** 1979. Inhibition of alternative complement pathway opsonization by group A streptococcal M protein. *J. Infect. Dis.* **139:**575–585.
79. **Phillips, G. N., Jr., P. F. Flicker, C. Cohen, B. N. Manjula, and V. A. Fischetti.** 1981. Streptococcal M protein: alpha-helical coiled-coil structure and arrangement on the cell surface. *Proc. Natl. Acad. Sci. USA* **78:**4689–4693.
80. **Podbielski, A., B. Melzer, and R. Lutticken.** 1991. Application of the polymerase chain reaction to study the M protein(-like) gene family in beta-hemolytic streptococci. *Med. Microbiol. Immunol.* **180:**213–227.
81. **Prats, H., B. Martin, and J. P. Claverys.** 1985. The *hexB* mismatch repair gene of *Streptococcus pneumoniae*: characterization, cloning, and identification of the product. *Mol. Gen. Genet.* **200:**482–489.
82. **Radnis, B. A., D. K. Rhee, and D. A. Morrison.** 1990. Genetic transformation in *Streptococcus pneumoniae*: nucleotide sequence and predicted amino acid sequence of *recP*. *J. Bacteriol.* **172:**3669–3674.
83. **Relf, W. A., and K. S. Sriprakash.** 1990. Limited repertoire of the D-terminal region of the M protein in *Streptococcus pyogenes*. *FEMS Microbiol. Lett.* **71:**345–350.
84. **Robbins, J. C., J. G. Spanier, S. J. Jones, W. J. Simpson, and P. P. Cleary.** 1987. *Streptococcus pyogenes* type 12 M protein gene regulation by upstream sequences. *J. Bacteriol.* **169:**5633–5640.
85. **Saravani, G. A., and D. R. Martin.** 1990. Opacity factor from group A streptococci is an apoproteinase. *FEMS Microbiol. Lett.* **68:**35–40.
86. **Sargent, S. J., E. H. Beachey, C. E. Corbett, and J. B. Dale.** 1987. Sequence of protective epitopes of streptococcal M proteins shared with cardiac sarcolemmal membranes. *J. Immunol.* **139:**1285–1290.
87. **Schleifer, K. H., and R. Kilpper-Balz.** 1987. Molecular and chemotaxonomic approaches to the classification of streptococci, enterococci and lactococci: a review. *Syst. Appl. Microbiol.* **10:**1–19.
88. **Schlievert, P. M., K. M. Bettin, and D. W. Watson.** 1979. Production of pyrogenic exotoxin by groups of streptococci: association with group A. *J. Infect. Dis.* **140:**676.
89. **Schneewind, O., K. F. Jones, and V. A. Fischetti.** 1990. Sequence and structural characterization of the trypsin-resistant T6 surface protein of group A streptococci. *J. Bacteriol.* **172:**3310–3317.
90. **Scott, J. R.** 1990. The M protein of group A *Streptococcus*: evolution and regulation, p. 177–203. *In* B. M. Iglewski and V. L. Clark (ed.), *The Bacteria*, vol. XI. *Molecular Basis of Bacterial Pathogenesis*. Academic Press, Inc., San Diego, Calif.
91. **Scott, J. R., S. K. Hollingshead, and V. A. Fischetti.** 1988. The evolution of M proteins of group A streptococci, p. 63–75. *In* F. C. Cabello and C. Pruzzo (ed.), *Bacteria, Complement and the Phagocytic Cell*. Springer-Verlag, Berlin.
92. **Scott, J. R., S. K. Hollingshead, K. F. Jones, and V. A. Fischetti.** 1987. Structural and genetic relationships of the family of M protein molecules of group A streptococci, p. 93–97. *In* J. Ferretti and R. Curtiss (ed.), *Streptococcal Genetics*. American Society for Microbiology, Washington, D.C.
93. **Scott, J. R., W. M. Pulliam, S. K. Hollingshead, and V. A. Fischetti.** 1985. Relationship of M protein genes in group A streptococci. *Proc. Natl. Acad. Sci. USA* **82:**1822–1826.
94. **Seifert, H. S., and M. So.** 1988. Genetic mechanisms of bacterial antigenic variation. *Microbiol. Rev.* **52:**327–336.
95. **Simpson, W. A., and E. H. Beachey.** 1983. Adherence of group A streptococci to fibronectin on oral epithelial cells. *Infect. Immun.* **39:**275–279.
96. **Simpson, W. A., H. S. Courtney, and I. Ofek.** 1987. Interactions of fibronectin with streptococci: the role of fibronectin as a receptor for *Streptococcus pyogenes*. *Rev. Infect. Dis.* **9:**S351–S359.
97. **Simpson, W. J., D. LaPenta, C. Chen, and P. P. Cleary.** 1990. Coregulation of type 12 M protein and streptococcal C5a peptidase genes in group A streptococci: evidence for a virulence regulon controlled by the *virR* locus. *J. Bacteriol.* **172:**696–700.
98. **Spanier, J. G., S. J. C. Jones, and P. Cleary.** 1984. Small DNA deletions creating avirulence in *Streptococcus pyogenes*. *Science* **225:**935–938.
99. **Stevens, D. L., M. H. Tanner, J. Winship, R. Swarts, K. M. Ries, P. M. Schlievert, and E. Kaplan.** 1989. Severe group A streptococcal infections associated with a toxic shock-like syndrome and scarlet fever toxin A. *N. Engl. J. Med.* **321:**1–7.
100. **Stock, J. B., A. J. Ninfa, and A. M. Stock.** 1989. Protein phosphorylation and regulation of adaptive responses in bacteria. *Microbiol. Rev.* **53:**450–490.
101. **Svensson, M., P. Christensen, and D. Schalen.** 1986. Monoclonal opsonic mouse antibodies specific for streptococcal IgG Fc-receptor. *J. Med. Microbiol.* **22:**251–256.
102. **Swanson, J., K. C. Hsu, and E. C. Gotschlich.** 1969. Electron microscope studies on streptococci. I. M. Antigen. *J. Exp. Med.* **130:**1063–1075.
103. **Tomasz, A., and J. L. Mosser.** 1966. On the nature of the pneumococcal activator substance. *Proc. Natl. Acad. Sci. USA* **55:**58–66.
104. **Tyewska, S. K., V. A. Fischetti, and R. J. Gibbons.** 1988. Binding selectivity of *Streptococcus pyogenes* and M-protein to epithelial cells differs from that of lipoteichoic acid. *Curr. Microbiol.* **16:**209–216.
105. **Watson, D. W.** 1979. Characterisation and pathobiological properties of group A streptococcal pyrogenic exotoxins, p. 62–63. *In* M. T. Parker (ed.), *Pathogenic Streptococci: Proceedings of the VIIth International Symposium on Streptococci and Streptococcal Diseases*, September 1978. Reedbooks, Ltd., Chertsey, Surrey, England.
106. **Wessels, M. R., A. E. Moses, J. B. Goldberg, and T. J. DiCesare.** 1991. Hyaluronic acid capsule is a virulence

factor for mucoid group A streptococci. *Proc. Natl. Acad. Sci. USA* **88:**8317–8321.

107. **Wexler, D. E., R. D. Nelson, and P. P. Cleary.** 1983. Human neutrophil chemotactic response to group A streptococci: bacteria-mediated interference with complement-derived factors. *Infect. Immun.* **39:**239–246.
108. **Whitnack, E., A. L. Bisno, and E. H. Beachey.** 1981. Hyaluronate capsule prevents attachment of group A streptococci to mouse peritoneal macrophages. *Infect. Immun.* **31:**985–991.
109. **Widdowson, J. P.** 1980. The M-associated protein antigens of group A streptococci, p. 125–147. *In* S. E. Read and J. B. Zabriskie (ed.), *Streptococcal Diseases and the Immune Response*. Academic Press, Inc., New York.
110. **Widdowson, J. P., W. R. Maxted, and A. M. Pinney.** 1976. Immunological heterogeneity among the M-associated protein antigens of group A streptococci. *J. Med. Microbiol.* **9:**73.
111. **Wilson, A.** 1959. The relative importance of the capsule to the M-antigen in determining colony form of group A streptococci. *J. Exp. Med.* **109:**257–269.
112. **Zighelboim, S., and A. Tomasz.** 1980. Penicillin-binding proteins of multiply antibiotic resistant South African strains of *Streptococcus pneumoniae*. *Antimicrob. Agents Chemother.* **17:**433–442.

5. *Lactococcus* and *Lactobacillus*

BRUCE M. CHASSY and CYNTHIA M. MURPHY

The lactic acid bacteria are a diverse group of organisms that are ubiquitous in the environment, occupying niches ranging from plant surfaces to the gastrointestinal tracts of many animals. These bacteria are somewhat arbitrarily grouped together because of their association with the preparation of fermented food and dairy products. They are used in the production of a variety of products such as yogurt, fermented milks, cheeses, sourdough bread, sour mash whiskeys, wine, cured meats, sausages, pickled vegetables, and silage (fermented animal feed). The organisms can also be used on an industrial scale for the production of lactic acid. The basis for the commercial application of lactic acid bacteria derives from a common biochemical pathway, i.e., conversion of various carbohydrates to lactic acid. The lactic acid bacteria are gram-positive, catalase-negative, asporogenous, facultative anaerobes. On this basis, members of the genera *Streptococcus* and *Enterococcus* could also be regarded as lactic acid bacteria, but they are not considered here since they are not associated with food products (see chapter 4). The scope of this chapter will be limited to discussions of *Lactobacillus*, *Lactococcus*, *Leuconostoc*, *Pediococcus*, and *Streptococcus thermophilus* (168).

Deliberate fermentation by one or more lactic acid bacteria is one of the oldest known techniques for food preservation. The sour aromatic flavors imparted by lactic fermentation are also desirable traits in fermented products. Food, dairy, and industrial fermentations by lactic acid bacteria constitute a worldwide industry of great economic importance. Lactic acid has found widespread use as an industrial chemical intermediate, an acidulant, and a food ingredient. Perhaps for these reasons, lactic acid bacteria have been the focus of an intensive burst of research activity over the last 10 years. This chapter will summarize and abstract the major areas of research. The historical association of the lactic acid bacteria with milk fermentation has encouraged a substantial amount of research on the biochemistry and genetics of lactose and milk protein catabolism, to which major portions of this chapter are devoted (for reviews see references 25, 43, and 118).

LACTIC ACID BACTERIA: *LACTOBACILLUS, LACTOCOCCUS, LEUCONOSTOC, PEDIOCOCCUS,* AND *STREPTOCOCCUS THERMOPHILUS*

Physical Traits and Appearance

Lactobacillus spp. are rod-shaped bacteria that are usually straight; however, under certain conditions, spiral or coccobacillary forms may appear. Lactobacilli are often found in pairs or chains of various lengths. The optimal growth temperatures for most strains range from 30 to 40°C but may be as high as 45°C in thermophilic strains. In contrast, *Lactococcus* spp. are ovoid cocci usually found in pairs or short chains. The optimal growth temperature for lactococci is 30°C. *Leuconostoc* spp. tend to be spherical to lenticular cocci found in pairs or chains. The optimal growth temperatures for *Leuconostoc* spp. range from 20 to 30°C. *Pediococcus* spp. are spherical cocci that divide in multiple planes to form tetrads. The optimal growth temperature for the pediococci is usually around 30°C. *S. thermophilus* often appear as ovoid pairs or chains with an optimum growth temperature of 45°C (168).

Taxonomy and Nomenclature

The taxonomy of the lactic acid bacteria is undergoing major change. Classifications were based originally on easily determined phenotypes such as those involving biochemical reactions. Newer classification schemes use rRNA sequence analysis or other phylogenetic techniques. The current nomenclature for each of the genera considered here is given below. The lactobacilli were grouped by the Orla-Jensen system in three groups based largely on biochemical characteristics. The three major clusters of the lactococci were originally assigned to different species; however, on the bases of DNA and protein homologies, these organisms have been regrouped into a single species as *Lactococcus lactis* subsp. *lactis*, *Lactococcus lactis* subsp. *diacetylactis*, and *Lactococcus lactis* subsp. *cremoris*. The genus *Leuconostoc* consists of four significant species. *Leuconostoc mesenteroides* has three subspecies: *mesenteroides*, *dextranicum*, and *cremoris*. *Leuconostoc lactis*, *Leuconostoc paramesenteroides*, and *Leuconostoc oenos* complete the genus. The genus *Pediococcus* contains a number of divergent species. They are more likely to be confused with micrococci than classified as lactic acid bacteria. Finally, on the bases of nucleic acid hybridization data and long-chain-fatty-acid studies, it has been proposed that *S. thermophilus* be renamed *S. salivarius* subsp. *thermophilus* (168).

Phylogeny

Oligonucleotide cataloging of 16S rRNA indicates that all of the lactic acid bacteria together with the bacilli and the streptococci constitute a supercluster within the clostridial subbranch of the gram-positive

Bruce M. Chassy and Cynthia M. Murphy • Department of Food Science, University of Illinois, Urbana, Illinois 61801.

eubacteria that have low guanine-plus-cytosine contents in their DNAs. The *Lactobacillus* line of descent diverges at about the same similarity coefficient as *Bacillus* and *Streptococcus* lines. The divergence from a primordial clostridiumlike common ancestor may have occurred as long as 2 billion years ago. The low similarity coefficients found between species within the genus *Lactobacillus* indicate that the divisions in the genus are phylogenetically deep and are, therefore, ancient (170). DNA-RNA relatedness studies of the genus *Leuconostoc* indicate that there are three groups within the genus: *Leuconostoc mesenteroides* subsp. *cremoris*, *Leuconostoc oenos*, and all others, which can be relegated to a single group. The complex relationships within the *Pediococcus-Lactobacillus-Leuconostoc* cluster are typified by the observation that *Leuconostoc paramesenteroides* is more closely related to *Lactobacillus viridescens* and *Lactobacillus confusus* than to *Leuconostoc oenos* (120, 168). Members of the genus *Pediococcus* show 0 to 50% homology by DNA-DNA hybridization analysis (168). 16S rRNA sequence analysis also demonstrates considerable phylogenetic heterogeneity within the genus *Pediococcus* (34). The pediococci, like the leuconostocs, fall within the *Lactobacillus* supercluster. *S. thermophilus* is phylogenetically located within the *Streptococcus* line. rRNA data indicate that this strain is closely related to *Streptococcus agalactiae*, *Streptococcus dysgalactiae*, and *Streptococcus acidominimus* (168). The lactococci also reside on the *Streptococcus* line and appear to be most closely related to the Lancefield group N streptococci (33). It is thus clear that the term lactic acid bacteria represents a useful functional definition for a diverse group of food and dairy microorganisms.

Metabolism and Fermentation Pathways

The lactic acid bacteria are nutritionally fastidious organisms; however, defined media capable of supporting the growth of many species have been reported. Lactic acid bacteria may require several amino acids, vitamins, or other metabolic intermediates, end products, or cofactors (168). The organisms exhibit generation times of 25 to several hundred minutes under optimal conditions; they are frequently thought of as slow growing relative to organisms such as *Bacillus* spp., which are capable of rapid aerobic proliferation (168). With a few exceptions, the lactic acid bacteria derive all of their energy from conversion of glucose to lactate or to lactate and a mixture of other products via the homofermentative or heterofermentative pathway. During homolactic fermentation, ~85 to 95% of the sugar utilized is converted to lactic acid; in contrast, during heterolactic fermentation, only ~50% of the sugar utilized is converted to lactic acid. The remaining products of heterolactic fermentation are carbon dioxide and ethanol, which is sometimes further converted to acetate. These pathways generate energy and ATP by nonoxidative substrate-level phosphorylation. The net energy yield of the homofermentative pathway is 2 molecules of ATP per molecule of glucose. Under strictly anaerobic conditions, the net energy yield of the heterofermentative pathway is 1 molecule of ATP per molecule of glucose; however, 2 molecules of ATP are generated in an aerobic or microaerophilic environment because of the presence of an alternative electron sink for cofactor regeneration. Some strains have also been shown to derive energy from a pH gradient established by carrier-mediated lactate efflux (55).

Role in Human Nutrition and Health

The lactic acid bacteria may have beneficial or detrimental effects on human health (for a review, see reference 56). The nutritional advantages of the consumption of lactic acid bacteria have long been debated. The nutritional value of fermented dairy products that can be consumed by lactose-intolerant individuals is a noncontroversial benefit derived from fermentation of milk by lactic acid bacteria. As much as 90% of the world's adult population may suffer from lactase insufficiency or deficiency. Certain other health benefits of yogurt and other fermented foods are also becoming well established. Medical benefits of lactic acid bacteria have also been studied. For example, heat-killed cells of *S. thermophilus* elicit an antitumor activity against fibrosarcoma in mice that appears to be T lymphocyte mediated (83). Also, hydrogen peroxide produced by *Lactobacillus acidophilus*, the predominant member of the normal human vaginal microflora, in combination with mammalian peroxidase and the appropriate halide cofactor has a viricidal effect on human immunodeficiency virus type 1 in vitro (93).

Although the lactic acid bacteria are generally regarded as nonpathogenic microorganisms, they can function as opportunistic pathogens under exceptional conditions. *Leuconostoc* spp., *Pediococcus* spp., and *Lactobacillis* spp. have been reported to cause clinical infections, including bacteremia, meningitis, and peritonitis. Since vancomycin is often the antibiotic of choice for therapy, treatment difficulties arise, because *Leuconostoc* spp., *Pediococcus* spp., and *Lactobacillis* spp. are often misidentified in clinical laboratories and are frequently vancomycin resistant. Increased use of vancomycin in broad-spectrum-antibiotic regimens is cited as the probable cause for the increase in infections caused by vancomycin-resistant lactic acid bacteria (54, 64, 172).

SIGNIFICANT BIOCHEMICAL TRAITS

Lactose Metabolism

The Lac-PTS. Many of the lactic acid bacteria are found in milk or are used in the preparation of dairy products. Rapid metabolism of lactose is a common biochemical characteristic of such strains. *Lactococcus lactis* (116), *Lactobacillus casei* (26), and *Staphylococcus aureus* (70, 124) possess an unusual pathway for lactose transport and hydrolysis that appears to be very limited in its distribution in nature (see Fig. 5 in chapter 11, this volume). The system consists of two biochemical steps unique to lactose metabolism: (i) transport of lactose into the cell with concomitant phosphorylation by the lactose phosphoenolpyruvate-dependent phosphotransferase system (Lac-PTS) and

Table 1. Sequence similarities of LacE, LacF, and LacG

Sources	% Similarity in sequences of:		
	Enzyme II	P-β-Gal	Factor III
Lactobacillus casei and *S. aureus*	45.3	53.4	46.2
Lactobacillus casei and *Lactococcus lactis*	48.3	54.1	49.1
S. aureus and *Lactococcus lactis*	71.6	82.1	74.8

(ii) hydrolysis of lactose 6-phosphate to glucose and galactose 6-phosphate by β-D-phosphogalactoside galactohydrolase (P-β-Gal). The Lac-PTS is composed of enzyme II^{Lac} (LacE), an integral 55-kDa membrane protein responsible for the vectorial phosphorylation and translocation of lactose into the cell as lactose 6-phosphate, and factor III^{Lac} (LacF), a 39-kDa trimeric peripheral membrane protein that phosphorylates enzyme II^{Lac} (26). LacG denotes the soluble cytoplasmic 54-kDa lactose 6-phosphate-hydrolyzing enzyme P-β-Gal (143). The genes encoding each of these proteins in staphylococci, streptococci, and lactobacilli have been isolated by molecular cloning and sequenced (2, 3, 20, 44, 45, 135, 143). We are not aware of any reports of the Lac-PTS pathway occurring in bacteria other than streptococci, enterococci, lactococci, lactobacilli, and staphylococci.

Glucose formed by the hydrolysis of lactose 6-phosphate is converted to glucose 6-phosphate and further metabolized by the glycolytic pathway. Galactose 6-phosphate can also be formed by the action of a separate Gal-PTS, which has been reported to occur in both *Lactococcus lactis* and *Lactobacillus casei* (30). The metabolism of galactose 6-phosphate requires the additional gene products of the tagatose 6-phosphate pathway gene cluster (*lacABCD*) encoding galactose-6-phosphate isomerase, tagatose-6-phosphate kinase, and tagatose-1,6-diphosphate aldolase (Fig. 5, in chapter 11, this volume). These three enzymes are analogous to the first three enzymes of glycolysis and result in the formation of glyceraldehyde 3-phosphate and dihydroxyacetone phosphate. *lacA* and *lacB* apparently encode nonidentical subunits of galactose-6-phosphate isomerase. The *lacC* gene codes for tagatose-6-phosphate kinase. The tagatose-1,6-diphosphate aldolase is encoded by the *lacD* gene. The genes encoding the enzymes of the tagatose-phosphate pathway of *Lactococcus lactis* (189) and *S. aureus* (135) have been isolated by molecular cloning and sequenced. In *Lactococcus lactis* (188, 189), the tagatose-phosphate pathway and Lac-PTS genes are found in a single large transcriptional unit (regulon) containing eight genes. The gene order is *lacABCDFEGX*; the function of *lacX* remains to be elucidated. *lacX* does not occur in staphylococci, in which the gene order observed is *lacABCDFEG* (135). The genes of the tagatose-phosphate pathway of *Lactobacillus casei* have not yet been characterized; the Lac-PTS gene order, *lacEGF*, is not the same as that observed for lactococci and staphylococci. The deduced sequences (based on DNA sequence analysis) of proteins LacG, LacF, and LacE are more similar than would be expected for such phylogenetically diverse organisms. The *Lactococcus lactis* and *S. aureus* proteins are more than 70% similar; moreover, the protein relatedness remains greater than 50% when the *Lactobacillus casei* sequences are compared with those of *Lactococcus lactis* and *S. aureus* (Table 1). The proteins encoded by the gram-positive lactose regulon appear to be evolutionarily distinct from those of other gene families thus far documented. Only distant relationships between the *lac* gene products and those of other gene families have been observed (2, 3, 20, 45, 143). For example, it has been suggested on the basis of protein sequence homology that LacG is encoded by a member of the same gene family, *bglB*, that encodes the β-D-phosphoglucoside glucohydrolase of *Escherichia coli* (143). On the basis of several limited regions of significant similarity, *lacE* has been postulated to have an ancestral relationship with the cellobiose permease of *E. coli* (149). The deduced amino sequence of tagatose-6-phosphate kinase of *Lactococcus lactis* is similar to those of *E. coli* Pfk-2 phosphofructokinase (26%) and *S. aureus* LacC (61%) (189).

The limited distribution and unique composition of the Lac-PTS pathway make its evolutionary origin difficult to assess. Several pertinent obervations suggest a possible origin. In *Lactococcus lactis* and *Lactobacillus casei*, the genes encoding the Lac-PTS are almost always plasmid encoded (26, 117). In staphylococci, the Lac-PTS and tagatose 6-phosphate pathway genes are located on the chromosome (19, 135). The Lac-PTS–P-β-Gal genes were found to be chromosomally encoded in one strain of *Lactobacillus casei* (28). Moreover, a cryptic copy of the P-β-Gal gene has been found on the chromosome of one *Lactobacillus casei* strain (159). The lactose plasmids are frequently conjugal and allow the transmission of Lac-PTS genes from lactose-positive to non-lactose-fermenting lactic streptococci. It is interesting to note that the majority of *Lactococcus lactis* strains isolated from plants, thought to be their natural ecological reservoir, ferment lactose poorly if at all. The lactose regulon of *Lactococcus lactis* is immediately followed by an *iso*-IS*1* sequence (135). All of these observations lead to the speculation that the Lac-PTS and possibly the tagatose 6-phosphate pathway evolved recently and were disseminated horizontally across genus lines by plasmid-mediated transfer of transposable elements. The recent emergence of mammals, and with them the sugar lactose, may have served as the selective pressure.

The *lac* operon. Strains of *S. thermophilus* and the majority of *Lactobacillus* spp. contain a pathway of lactose metabolism that is phenotypically analogous to that encoded by the classic *lac* operon genes, *lacY* and *lacZ*. Biochemical characterization reveals that lactose is transported into the cell without phosphorylation in these strains. β-Galactosidase activity can be detected in the cytoplasm of cells that utilize this pathway; on agar plates containing 5-bromo-4-chloro-indolyl-β-D-galactopyranoside, blue colonies are formed. Pediococci also appear to metabolize lactose through the action of a β-galactosidase (11). The β-galactosidase of *Lactobacillus delbrueckii* subsp. *bulgaricus* is composed of two identical subunits of about 114 kDa each (155). The β-galactosidase- and lactose permease-encoding genes of *S. thermophilus* (141, 156) and *Lactobacillus delbrueckii* subsp. *bulgaricus* (104, 155) have been isolated by molecular cloning in *E. coli* and sequenced. The deduced protein sequences of these β-galactosidases were found to be related (at the 30 to 50% level) with a number of other LacZ-like proteins. The similarities found in the LacZ-like gene products of *S. thermophilus* and *Lactobacillus delbrueckii* subsp. *bulgaricus* point to a common ancestral relationship with the β-galactosidases of *Clostridium acetobutylicum* (63), *E. coli* (*lacZ* and *ebgA*), and *Klebsiella pneumoniae* but not with *Bacillus stearothermophilus* (*bgaB*) (155). The *lacZ* genes of some strains of *Lactobacillus delbrueckii* subsp. *bulgaricus* have been observed to undergo spontaneous deletions that give rise to a Lac^- phenotype (122). In one instance, an unusual 72-bp direct-duplication event reversed the loss of β-galactosidase activity caused by a 30-bp deletion in *lacZ* (123).

Analysis of the gene encoding the lactose permease activity of *S. thermophilus* (141) revealed an open reading frame (ORF) upstream of the *lacZ* gene that is capable of encoding a 69.5-kDa protein. Although it could be demonstrated that the gene product catalyzed both lactose exchange and uptake reactions in intact cells and membrane vesicles, its sequence bore little resemblance to that of the *E. coli* lactose permease, LacY. The N-terminal region shows distant similarity to the melibiose carrier of *E. coli*, while the C-terminal domain appears to be distantly related to enzyme III^{Glc} of *E. coli* (141). It was suggested that the N-terminal region functions as the transmembrane portion of the permease that serves in lactose transport and that the C-terminal, factor III-like domain is regulatory region. The gene has been designated *lacS*. An analogous gene has been isolated from *Lactobacillus delbrueckii* subsp. *bulgaricus* (104). The LacS proteins of *Lactobacillus delbrueckii* subsp. *bulgaricus* and *S. thermophilus* exhibit about 80% conservation of sequence. Many strains of *S. thermophilus* do not utilize galactose and expel the galactose portion of lactose into the medium during growth on lactose-containing media. Recently, it was demonstrated that *S. thermophilus* has a lactose/galactose antiporter activity that allows galactose efflux to drive lactose uptake (75). Thus, it appears that this gram-positive *lac* operon is only partly related to the one found in *E. coli*. The *lac* genes of *S. thermophilus* and *Lactobacillus* spp. are organized differently than those of *E. coli*. The genes encoding two enzymes of the Leloir pathway for the metabolism of galactose 1-phosphate, mutarotase, and UDP-glucose 4-epimerase are located immediately before (5′ of) the genes for *lacS* and *lacZ* in *S. thermophilus* (142).

Two exceptional *Lactobacillus casei* strains have β-galactosidase activity in addition to the Lac-PTS (78). These atypical strains have a plasmid-encoded β-galactosidase gene that has been cloned into *E. coli* from the 28-kbp plasmid pLZ15 and sequenced (28). The plasmid pLZ15 also carries multiple copies of a DNA sequence that hybridizes to the *Lactobacillus casei* insertion element ISL*1* (160, 161). A plasmid-encoded β-galactosidase gene was isolated from a strain of *Leuconostoc lactis* and sequenced (39). Somewhat surprisingly, the β-galactosidase-encoding genes isolated from *Leuconostoc lactis* and *Lactobacillus casei* are nearly identical, with >98% sequence identity (28, 39). It is interesting to note that both genes are plasmid encoded, the sequence similarity terminates abruptly on either side of the coding sequences, and the *Lactobacillus casei*-derived plasmid, pLZ15, contains insertion sequence-like elements near the coding region. Both plasmids also encode a noncontiguous lactose permease. The coding sequences contain two overlapping out-of-frame ORFs, *lacL* and *lacM*, whose expression appears to be translationally coupled. Analysis of the *lacL*-derived polypeptide shows 30 to 50% similarity to the N-terminal two-thirds of *lacZ* gene products isolated from gram-positive and gram-negative bacteria. *lacL*-encoded sequences correspond to the C-terminal regions of LacZ proteins. The *lacL* and *lacM* genes are assumed to have arisen from a deletion event in a parental *lacZ*-like gene. In the process, about 8% of the sequence of *lacZ* was deleted. The region it encodes is apparently unnecessary for enzymatic activity. Active β-galactosidase isolated from *Lactobacillus casei* has a molecular weight of around 220,000. The enzyme, an $\alpha_2\beta_2$ type, is composed of two identical 38-kDa subunits and two identical 70-kDa subunits.

Proteinases

Although the specific amino acids required may be species and strain dependent, all lactic acid bacteria have an absolute growth requirement for amino acids. Milk contains a high concentration of amino acids bound in the form of protein; however, milk is not a rich source of free amino acids. The free amino acids present in milk are sufficient to support approximately 15 to 25% of the growth observed for *Lactococcus lactis* in milk (121). The remaining required amino acids must be obtained from proteolysis of milk proteins. Strains of lactic acid bacteria that develop well in milk are usually proteolytic. In addition to one or more protease activities, strains must possess a variety of peptidases and peptide transport proteins to obtain individual amino acids (for recent reviews, see references 58, 96, and 99).

A variety of proteolytic activities have been identified in lactococci. These proteinases have been shown to vary in substrate specificity, size, isoelectric point, pH optima, dependence on Ca^{2+}, and antigenic determinants (96). Extracellular protease production in *Lactococcus lactis* is an unstable and variable trait. The molecular basis of instability and variability can now be explained as loss, deletion, or rearrangement of proteinase-encoding plasmids (117). Instability may be mediated by insertion sequence-like elements that have been identified on proteinase plasmids (61). It has been demonstrated that insertion of the proteinase genes into the host chromosome results in proteinase-stable strains that develop rapidly in milk (102). The plasmid-encoded serine proteinase genes of *Lac-*

tococcus lactis subsp. *cremoris* SK11 (192) and Wg2 (97) and *Lactococcus lactis* subsp. *lactis* NCDO763 (92) have been isolated and studied in detail. In the nucleotide sequence of the DNA fragments isolated from strains Wg2 and NCDO763, two divergent ORFs were identified and called *prtM* and *prtP*. *prtP* encodes the structural gene for a 1,902-residue-containing pre-pro-proteinase; *prtM*, known to be required for converting pre-pro-proteinase to mature proteinase, could potentially encode a 32-kDa protein (58, 60). The two *prtP* genes are very similar, differing in coding potential by only 18 amino acids. A comparison of the PrtPs of strains Wg2 and SK11 reveals only 44 amino acid changes; a 60-amino-acid duplication in the SK11 PrtP contributes to its larger size (1,962 amino acid residues). The PrtPs contain both a signal sequence of 33 amino acids and a proregion of about 154 amino acids. The mature proteinases have molecular weights of 187,000. Despite size differences, the *Lactococcus lactis* PrtPs somewhat resemble the serine proteases isolated from *Bacillus* spp. Most of the 1,200-amino-acid C-terminal region, which is involved in specificity and cell anchoring, is not required for catalytic activity. Proline-rich domains in the C-terminal region serve as anchoring sequences that retain the mature proteinase in the cell membrane. In a self-digestion process that is stimulated by low Ca^{2+} concentrations and inhibited by high Ca^{2+} concentrations, a C-terminal-truncated active proteinase can be released from the cell. Three conserved regions with high degrees of similarity to those found in the much smaller subtilisin are found in the N-terminal domains of PrtP. In these regions of the Wg2 protease, subtilisin active-center residues Asp-32, His-64, and Ser-221 are found at positions Asp-217, His-281, and Ser-620, respectively (58, 96). Although none has been studied in detail, extracellular proteases have been reported in lactobacilli and *S. thermophilus* (99). A strain of *Lactobacillus casei* was found to possess a cell-bound proteinase activity with properties similar to those of the lactococcal serine proteinases; chromosomal DNA from this strain hybridized to fragments of the *prtP* and *prtM* genes of *Lactococcus lactis*. Preliminary mapping analysis ponits to some conservation of restriction endonuclease cutting sites (95).

Peptidases and Amino Acid Transport

The proteinases of lactic acid bacteria produce a complicated mixture of polypeptides that the bacteria must break down into individual amino acids. The literature contains reports describing aminopeptidases, di- and tripeptidases, aryl-peptidyl amidases, aminopeptidase P, proline iminopeptidase, prolinase, prolidase, X-prolyl-dipeptidyl aminopeptidases, endopeptidases, and carboxypeptidases in various species of lactic acid bacteria. Considerable attention has been devoted to the biochemical isolation and characterization of peptidases; recently, genetic characterization has begun (for reviews, see references 58, 96, 99, 113, and 182). Since the peptidases characterized seem to be primarily intracellular, it appears that lactic acid bacteria possess transport systems for polypeptides as well as for free amino acids.

Broad-specificity aminopeptidases capable of hydrolyzing di-, tri-, and tetrapeptides have been reported to occur in *Lactococcus lactis* subsp. *cremoris* (86), *Lactobacillus acidophilus* (112), and *Lactobacillus lactis* (50). Aminopeptidases have also been noted in the pediococci (12). These enzymes preferentially hydrolyze peptides with more than two amino acids. All of the enzymes characterized to date are metalloenzymes that lack carboxypeptidase or endopeptidase activity. Recently, a fragment of DNA that complements the *pepN* mutation in *E. coli* has been isolated from *Lactococcus lactis* by molecular cloning, and the DNA sequence has been determined (178). The lactococcal *pepN* gene encodes an intracellular lysyl-aminopeptidase of 95 kDa.

Proteolysis of casein produces proline-rich polypeptides that are not substrates for the aminopeptidases or the di- and tripeptidases. Aminopeptidase P, proline iminopeptidase, prolinase, prolidase, and X-prolyl-dipeptidyl aminopeptidases serve the lactic acid bacteria for the degradation of proline-containing peptides. A number of these "proline peptidases" have been purified and characterized biochemically. Prolidase has been isolated from *Lactococcus lactis* subsp. *cremoris* (85); X-prolyl-dipeptidyl-aminopeptidase has been isolated from *Lactobacillus delbrueckii* subsp. *bulgaricus*, *Lactobacillus lactis*, and *S. thermophilus* (6, 119); and prolyl-dipeptidyl aminopeptidase has been isolated from *Lactobacillus helveticus* (88). The distinct physical and biochemical identities of each of these activities underscore the complexity of proline-peptide metabolism in the lactic acid bacteria. This complexity is further demonstrated by the observation that mutants of *Lactobacillus delbrueckii* subsp. *bulgaricus* that lack X-prolyl-dipeptidyl-aminopeptidase grow at a slightly reduced growth rate and produce a lower biomass (6).

The genes encoding the X-prolyl-dipeptidyl aminopeptidases of two strains of *Lactococcus lactis* subsp. *lactis* have been cloned, characterized at the nucleotide level, and found to have 99% identity (114, 131). The genes, designated *pepXP*, contain an ORF capable of encoding a 763-residue protein (88 kDa), which was expressed in an enzymatically active form in *Bacillus subtilis*. No signal sequences indicative of secreted proteins were present. No analog could be found in the protein or nucleic acid data bases. In both strains, a divergent ORF of unknown function was present; its relation to *pepXP* remains undetermined.

Di- and tripeptidases but not carboxypeptidases have been isolated from a number of species and strains of lactic acid bacteria (96). There have been only two reports of endopeptidases in the lactic acid bacteria. Two unusual endopeptidases with substrate specificities limited to specific casein fragments were purified from extracts of *Lactococcus lactis* subsp. *cremoris* (195, 196).

The finding that some lactic acid bacteria require peptides for growth (167) correlates well with the intracellular location determined for the peptidases that have been studied in detail. However, free amino acids per se can be transported into lactic acid bacteria. In lactococci, three kinds of amino acid transport systems have been characterized (100): (i) an ATP-dependent uptake of glutamine, glutamate, and asparagine; (ii) a specialized ornithine/arginine antiporter;

and (iii) generalized systems for the pmf-driven uptake of basic, neutral, branched-chain, aromatic, and individual amino acids (96, 100). Free L-proline cannot be transported into *Lactococcus lactis* (167). It seems clear that lactic acid bacteria possess a number of systems for the transport of peptides and polypeptides. The intracellular localization of peptidases, the growth requirement of some strains for peptides, and the observed import of peptides experimentally support this conclusion. None of these peptide transport systems has been explored at the molecular level.

Secreted Proteins

The rational modification of food, dairy, or industrial bacterial starter cultures requires an understanding of their protein secretory systems. While much attention has been paid to the secretion of proteins by *E. coli* and *B. subtilis*, only a few secreted proteins of the lactic acid bacteria have been investigated in detail. The serine proteases described above are translated with 33-residue N-terminal secretion signal peptides (92, 97, 192) that conform to the major consensus features articulated for signal peptides (190). On the other hand, the polypeptide antibiotic nisin, which is secreted from many strains of lactococci (23), and lacticin F (127), the bacteriocin of *Lactobacillus acidophilus*, have atypical N-terminal sequences that do not conform to the consensus rules. The signal peptide sequence of a lactococcal bacteriocin, lactococcin A, is 21 residues long according to known N-terminal sequence data and the DNA sequence of the gene; however, the predicted cleavage site at residue 20 would require further N-terminal processing for removal of an additional glycine residue (74, 180). A major obstacle to further understanding in this area has been the paucity of good descriptions of extracellular proteins in the lactic acid bacteria. Judged by the types and quantities of protein they produce, the lactic acid bacteria do not appear to be prodigious protein secretors. Two species of lactobacilli, *Lactobacillus amylophilus* (129) and *Lactobacillus amylovorus* (128), are known to secrete α-amylases and degrade insoluble starch (76). Perhaps analysis of these genes will yield additional valuable information. Many strains of lactic acid bacteria synthesize complex extracellular homo- or heteropolysaccharides through the actions of secreted and cell-bound enzymatic activities (for a review, see reference 24). The biochemistry of these polysaccharides and the enzymes that produce them is complex and largely unexplored at the molecular level; however, the diversity of exopolysaccharides suggests a plentiful and diverse source of secreted enzymes responsible for their biosynthesis. Genes encoding multiple extracellular glucosyl- and fructosyl-transferases have been found in oral streptococci but have not been characterized in the lactic acid bacteria (24). The "ropey" character imparted by certain starter cultures is thought to be due to exopolysaccharide synthesis. In this regard, it may prove fortuitous that the mucoid or ropey character of some *Lactococcus lactis* strains appears to be plasmid associated.

All strains of *Lactococcus lactis* subsp. *lactis* have a common 45-kDa surface protein antigen called Usp45 (179). The gene, *usp-45*, was cloned and expressed in *E. coli*. Usp45 was secreted efficiently into the periplasmic space of the *E. coli* host. Translation of a 1,383-bp ORF in *usp-45* predicts a protein of 44.6 kDa. A 27-amino-acid signal peptide precedes the mature protein. The composition and structural features of the signal peptide correlate well with those observed in other bacteria. Usp45 is related to P54, an extracellular protein of *Enterococcus faecium*. The two proteins have similar sequences and common structural characteristics but are of unknown function (179). Since Usp45 is efficiently secreted in both *E. coli* and *Lactococcus lactis*, its characterization opens the way for biochemical and genetic analyses of protein secretion in lactic acid bacteria.

Recently, secretion signal peptide-encoding sequences present in the *Lactococcus lactis* chromosome were selected in *E. coli* and subcloned into *Lactococcus lactis* and *B. subtilis* (163). Sequences that supported the secretion of TEM β-lactamase could be detected by the appearance of ampicillin-resistant colonies. The leader peptide sequence identified in each case included a strong positive charge, a hydrophobic core, and a putative cutting site. Furthermore, the signal peptides isolated by this method were functional in *Lactococcus lactis*, *B. subtilis*, and *Lactobacillus plantarum*.

GENETIC CHARACTERIZATION

Genes Isolated from Lactic Acid Bacteria and Characterized

To date, the nucleotide sequences of more than 50 genes isolated from lactic acid bacteria, primarily the lactococci and lactobacilli, are published and available in nucleic acid data banks. Many of these genes are discussed elsewhere in this chapter. Examination of the sequences has revealed homology with sequences from other gram-positive bacteria; however, some significant differences are also apparent. The heterogeneity of species noted earlier and the relative scarcity of sequence from a single genus and species (as is available for *E. coli* or *B. subtilis*) makes sweeping generalizations about regulatory sequences in lactic acid bacteria difficult. In only a few cases are in vitro or in vivo data available to directly confirm the function of putative regulatory elements present in the DNA sequences. The following paragraphs tabulate sequence features observed for genes isolated from lactococci and lactobacilli. Table 2 presents sequences 5′ to transcription initiation sites of lactococcal genes thought to contain promoter elements; Table 3 presents similar sequences from lactobacilli. Sequences resembling ribosome-binding sites (RBS) in transcripts of lactococcal genes are given in Table 4; RBS-like sequences of lactobacilli can be found in Table 5. The name of each gene or gene product and the literature citation(s) are also listed in each table. The 16S rRNA genes of *Lactobacillus catenaformis* (193) and *Lactobacillus confusus* (197) and the sequence of an insertion element, *Lactobacillus casei* ISL*1* (161), are not included in the tables.

Control of Transcription

Promoters and promoterlike sequences. The sites of initiation of transcription are known for at least 12 genes of lactococcal origin (Table 2). For brevity, additional similar promoters that were isolated by using a secretion signal probe vector (163) or a promoter probe vector are not shown (94). A series of six promoters isolated by use of a promoter probe vector in *B. subtilis* (187) have sequences showing similarity to the gram-positive consensus promoter sequences. Consensus promoterlike sequences associated with intact genes cloned from lactococci are less certain; as can be seen in Table 2, canonical −35 or −10 sequences may be absent from the regions 5′ to well-expressed genes such as *lacR* and *usp-45*. The spacing between features seems variable as well. Transcription start sites have been determined for six genes isolated from *Lactobacillus* species (Table 3). Sequences resembling −35 and −10 sites were identified 5′ to each gene. Agreement between these sites and the consensus sequence varies from 50 to 100% at −10 and is less clearly defined at −35. It is sometimes difficult to identify promoterlike elements 5′ to *Lactobacillus* genes. This may reflect the researcher's natural bias to search out an *E. coli* σ70- or *B. subtilis* σ43-like consensus promoterlike element. Table 3 lists six genes for which reasonable −10 and −35 sequences can be proposed; however, the exact transcription start sites are unknown. It is unclear whether the difficulty represents a real difference in the consensus sequences used by lactobacilli or whether too few examples have been examined. A promoter probe vector has been used to isolate from *S. thermophilus* chromosomal fragments that have promoterlike activity in *S. thermophilus* or *Lactococcus lactis* (166). DNA sequencing revealed the −10 and −35 sequences of consensus *E. coli* σ70- or *B. subtilis* σ43-associated promoters. Thus, promoter probes seem to be useful for isolation of canonical promoters when applied to the lactic acid bacteria. This conclusion is based largely on −35 and −10 sequences; however, it is not yet known whether there are additional sequences outside of the traditional boxes that control promoter strength. While some promoters found in lactic acid bacteria appear to deviate from the gram-positive or gram-negative consensus sequences, many will drive expression in heterologous cloning hosts, yet the converse is not always true: foreign genes preceded by strong consensus promoters may not necessarily be transcribed well in the lactic acid bacteria. For example, in vitro studies of transcription by *Lactobacillus acidophilus* RNA polymerase using five *E. coli*-derived promoters (Table 3) revealed that only three were functional (132). Two *Lactobacillus acidophilus*-derived DNA fragments with sequences not markedly different from those of the *E. coli* promoters also served as promoters.

Repressors. The first example of a repressor in lactic acid bacteria has recently been described. LacR, which is associated with the lactose-inducible Lac-PTS, has been cloned and the nucleotide sequence has been determined (188). *lacR* contains an ORF of 861 bp, and the deduced amino acid sequence of LacR shows that it is related to DeoR, FucR, and GutR of *E. coli*. An N-terminal helix-turn-helix motif, frequently found in DNA-binding proteins, is present in LacR. In *Lactococcus lactis*, the expression of *lacR* is repressed by growth on lactose. Expression of *lacR*, which is under the control of the constitutive *prtP* promoter, repressed the activity of the Lac-PTS in *Lactococcus lactis*.

The xylose genes of *Lactobacillus pentosus* require induction to be expressed. A gene that could encode a protein bearing resemblance (58%, allowing conservative amino acid exchanges) to the xylose repressor (XylR) of *B. subtilis* has been identified in the xylose operon of *Lactobacillus pentosus* (108). Although its function has not been demonstrated in vivo, it contains the characteristic helix-turn-helix motif of XylR. An operatorlike sequence was also identified 5′ to the *xylA* gene.

Positive regulators. A gene required for the expression of malolactic fermentation in *Lactococcus lactis*

Table 2. *Lactococcus lactis*-derived transcription initiation sites[a]

Promoter	Sequence	Spacing (no. of residues)		Reference(s)
		−35 to −10	−10 to start	
P21	TTTTTTCTTGACAGAAGAAGGCGAAAAATGGTATTATATTTAGGTACTGTT	18	8	187
P23	CCTAAGACTGATGACAAAAAGAGCAAATTTTGATAAAATAGTATTAGAATT	17	6	187
P32	GTGAGCTTGGACTAGAAAAAAACTTCACAAAATGCTATACTAGGTAGGTAA	22/17	4	187
P44	AGCTAAACTCTTGTTTTACTTGATTTTATGTTAAAATAATTAATGAGTGTA	15/18	8/5	187
P59	ACATTAAATTCTTGACAGGGAGAGATAGGTTTGATAGAATATAATAGTTGT	17	5	187
prtP	ATTTTCGTTGAATTTGTTCTTCAATAGTATATAATATAATAGTATATAATA	21/15	5	60, 97, 187
prtM	ACCCTACGCTTGATGTAGTTAAGATTATATTATATAATATTATATACTATT	17	7	60, 97, 187
Prenisin	TAAACGGCTCTGATTAAATTCTGAAGTTTGTTAGATACAATGATTTCGTTC	20	4	23
p9B4-6 ORF A1	TAACATTTGTTAACGAGTTTTATTTTTATATAATCTATAATAGATTTATAA	23/20	4	180
p9B4-6 ORF B1	ACATTTGTTAACGAGTTTTATTTTTATATAATCTATAATAGATTTATAAAA	23/20	6	180
usp-45	TCATAAAGAAATATTAAGGTGGGGTAGGAATAGTATAATATGTTTATTCAA	22/14	6	180
lacR	TAATTTTTTGTTTTTTTTTATTTGTTTTTTTAAAAAATAGATAACACCGTT	19	7	188
Consensus[b]	---C-T---TTG--- ------------T---- T-A--AAAA----TA			183
E. coli	TTGACA -------17-------- TATAAT---7---Pu[c]			66
Gram positive[d]	TA-AAAAA--GTTGACA ---A--A---A-T- TG-TATAATAATAT			57

[a] Adapted from Van de Guchte (183).
[b] Consensus *Lactococcus lactis* transcription initiation sequence derived from the bases present at a given position in at least 70% of the sequences listed in this table.
[c] Pu, purine.
[d] Consensus sequence for transcription initiation sites over a wide range of gram-positive bacteria.

Table 3. *Lactobacillus* promoters: sequences near initiation of transcription[a]

Promoter[b]	Sequence					Spacing		Reference
		−35		−10		−35 to −10	−10 to start	
Based on mapped transcription site[b]								
Lactobacillus confusus L-2-HicDH	TGAGTTGTTCAG	TGGTGA	ACAACGTCTCTTTTTTG	TATAAT	GAACA T	17	5	106
Lactobacillus bulgaricus lacP	TAGTAAAACATC	TTGGTT	TATTTAGTAAACAAGTC	TATACT	GTAATTAT	17	7	104
Lactobacillus casei L-LDH[c]	AAGTCTGTCAATT	TTGTTT	CGGCGAATTGATAAT GTGT	TATACT	CACA A	18, 19	4, 14	90
Lactobacillus casei D-2-HicDH	TAAAGATGCTTT	TTCACA	GAGGCATCTTGTA	TACGGT	GAGT G	13	4	105
Lactobacillus casei fgs	CTGGCTAAGC	TCATAA	GTTAGACCCATTTTTCTGG	TATGAT	GAAGCT A	19	6	176
Lactobacillus spp. *hdcA*	AAGTTCAATAAAACGC	TTTACT	TTTTTATTTTTGAGGT	TATAAT	GAAAATCG	16	7	186
Recognized in vitro by *Lactobacillus acidophilus* RNA polymerase[d]								
pRNL1 P1	AAAGAAGCC	TTAACT	GGCTTCGATCTGGG	TATAAT	TTTTTT A	15	6	132
pRNL1 P2	AATTTTTTT	TTGTAT	TTATTTTTACTAGTTTTATGC	TATAAT	AAATTCGA	21	7	132
pBr322 *bla* P3	CTAAATACA	TTCAAA	TATGTATCCGCTCATGA	GACAAT	AACCCCTG	17	7	132
E. coli trp	AATGAGCTG	TTGACA	ATTAATCATCGAACTAG	TTAACT	AGTACGCA	17	7	132
tacUC	AATGAGCTG	TTGACA	ATTAATCATCGCGGCTCG	TATAAT	GTGTGG A	18	6	132
pBR322 *bla* P1[e]	*ACACGGTGC*	*CTGACT*	*GCGTTAGCAATTTAACTGTGA*	*TAAACT*	*ACCGCATT*	21	7	132
pBR322 *tet* P2[e]	*TTCTCATGT*	*TTGACA*	*GCTTATCATCGTAAGC*	*TTTAAT*	*GCGGGTAG*	16	7	132
Putative promoters associated with expressed genes[f]								
Lactobacillus acidophilus laf	TAAAAGTAATT	TTAGAC	ACAAATAGAACAATATTGG	TCAATT	TTATATCTTAAGAT	18		127
Lactobacillus casei DHFR	GATATTTTCGACT	TTGACA	TGAAGGATATTAAGTTGC	TTAATT	ACGATCCTTATCCG	18		4
Lactobacillus casei dae	GATACTGCTGGTC	TTGGTT	TATCTATACGGTTACAGCGG	TATTAA	CAACAGTAAATTCA	20		69
Lactobacillus plantarum D-LDH	TTTTCACAATTT	ATGATT	TTATTTCTGGCATTTACTGT	TATAAT	GTTTAATGTATCAAT	20		173
Lactobacillus plantarum L-LDH	AGAAGCATGTTCAAT	TCCACA	CTTGCCTGGCGCAAATATGT	TACAAT	ACAACCGTAAAACTA	20		173
Lactobacillus plantarum pyrE	CAATTACGCAGGC	TGCAGA	ACCAGTTCAAGCTTACCGAGA	TATTAT	GACAGAATGGAGTA	21		17
Gram-positive consensus sequence[g]	Ta-AAAAA--G	TTGAcA	---a--A---a-T-TG-	TATAAT	AAtAt	17	7	57
E. coli[g]	a------t--	TTGACa	t--t--------t-tg-	TAtAAT	-------Pu	17	7	115

[a] Regions resembling consensus −35 and −10 sequences according to the original reference are underlined; double underlining indicates alternative sequences. Transcription starts are denoted by boldface letters.

[b] LDH, lactate dehydrogenase; HicDH, hydroxyisocaproate dehydrogenase; *fgs*, folylpoly-γ-glutamate synthetase; *hdcA*, histidine decarboxylase.

[c] −35 and −10 sites were determined in *E. coli*, not *Lactobacillus* spp.

[d] Promoters recognized by *Lactobacillus acidophilus* RNA polymerase were described by Natori et al. (132).

[e] Not used by *Lactobacillus acidophilus* RNA polymerase, as indicated by italics.

[f] The following sequences with features resembling consensus −10 and −35 signals were observed 5′ to the coding regions of the following genes: *Lactobacillus acidophilus* lacteriocin F (Laf; 127), *Lactobacillus casei* dihydrofolate reductase (4), *Lactobacillus casei* D-alanine activating enzyme (69), *Lactobacillus plantarum* D-lactic dehydrogenase (173), *Lactobacillus plantarum* L-lactic dehydrogenase (173), and *Lactobacillus plantarum pyrE* (17).

[g] Capital letters indicate >70% frequency, and lowercase letters indicate >40% frequency.

Table 4. RBS used by lactococci in initiation of translation[a]

Source	Sequence of putative ribosome-binding site	Spacing (no. of residues) from RBS to ATG	Reference(s)
mleR	UGUUAAAAUAGCUUCUAGGAGUAUAGCCAUG	8	150
thyA	-----AAUUAAAAAAGAUAGGAAAAUUUCAUG	7	153
X-PDAP[b]	----------UAUUACGGAGGAUUUAAAAUG	9	114, 131
lacG	-CAAAUGCUUUUUUUGAAAGGACUUACACUUAUG	9	16, 46, 47
pTR2030hsp	---GAUUUUUAGAACAGGGAGUAGGUAAAUG	8	73
pWV101*repA*	-GCUACUUGUUUUUGAUAAGGUAAUUAUAUCAUG	9	103
pSH71*repC*	-----------UUGAUUGGAGUUUUUUAAAUG	9	46
p9B4-6 ORF A1	-----UUAUAAAAAUAAGGAGAUUAUUAUG	7	180
p9B4-6 ORF A2	UAUAAAAAUUGAAAGGAUUCAGGUACUAAAAUG	13	180
p9B4-6 ORF A3	--CCAGUACACUAAAGAAAGGCUUACAAAUUAAUG	10	180
p9B4-6 ORF B2	--CGGGUUGCACCAUUGAGGAUUAGUUAAGAUAUG	14, 11	180
p9B4-6 ORF C2	-CUAAUAAAAAAGAACUGAGGUUUAGAGUUAAUG	12	180
Φ_vML3 lysin	-----------ACCGUGGGAACAAAUG	4	158
ISS*1* transposase gene	-AAUAAAAAUGACAGCGAGGAUAUAUCAAUG	10, 7	140, 152
IS*904* transposase gene	---AAGGUUUCUCGCUCAGGUUUCUAUGAAUACAUG	15	48, 147
prtP	---ACUUUUGGAAAGUGGAGGAUAUUGGAUG	9	60, 97
prtM	ACUGUAAGCAUUUCAGAGGAGACCGAAUCGAUG	10	60, 97
usp-45	--GGAGGAAAAAUUAAAAAAGAACAGUUAUG	5	180
Prenisin gene	--AAAAUAAAUUAUAAGGAGGCACUCAAAAUG	10	23, 48, 84
ORF 32	-GGUAAAAAAAUAUUCGGAGGAAUUUUGAAAUG	11	187
ORF 44	GUGAUGUGUGAGGGAAAGGAGUCGCUUUUAUG	9	187
citP	AUAAGACAGAUAUAAAUGGAGAUAGAAUUAUG	9	40

[a] Table adapted from reference 183, p. 16–17.
[b] X-PDAP, X-prolyl-dipeptidyl-aminopeptidase.

subsp. *lactis* was cloned and characterized at the nucleotide level (150). Designated *mleR*, it encodes a 34-kDa gene product whose sequence bears some resemblance to LysR and other positive regulators of gram-negative bacteria.

Initiation of Translation: RBS

The 16S rRNA sequences of a number of *Bacillus* spp., lactococci, lactobacilli, and other lactic acid bacteria are known (62, 111, 193). Sequences with complementarity to the 3′ terminus of 16S rRNA can be found in the regions preceding the translation start codons of the genes of lactic acid bacteria thus far characterized (Tables 4 and 5). As can be seen in Table 4, lactococcal genes are preceded by Shine-Dalgarno sequences that are somewhat weaker (average $\Delta G = -11$ kcal [1 cal = 4.184 J] per mol) than those of *Bacillus* spp. (average $\Delta G = -16$ kcal/mol); in this regard, they resemble those of *E. coli* (average $\Delta G = -11$ kcal/mol [183]). The apparently weak RBS of *Lactococcus lactis* does not seem to interfere with expression. Usp45 is expressed well in spite of a weak Shine-Dalgarno sequence (AAAG; Table 4). The putative *Lactobacillus* RBS sequences are significantly different from those found in the lactococci (Table 5). The sequence AGGAGG is strongly conserved, and the sequences (average $\Delta G = -17$ kcal/mol) more closely resemble those of *Bacillus* spp. than those of lactococci. Additional possible effects of sequences 5′ to the start codon or within the coding sequence have not been analyzed in detail (183). However, translational coupling, optimal reading frame overlap, and spacing in *Lactococcus lactis* have been carefully studied (184). Partially overlapping reading frames at the stop-start codons can be used to enhance expression. It is interesting to note that the *lacL* and *lacM* genes of *Lactobacillus casei* and *Leuconostoc lactis* are overlapped, thus presumably ensuring a 1:1 synthesis of β-galactosidase subunits (28, 39).

Codon Usage

Statistical analysis reveals a biased codon usage by lactococci and lactobacilli that resembles that found in *E. coli*. While more selectivity is found than is observed with *B. subtilis*, preference is given to codons other than those preferred by *E. coli*. AUG is strongly favored as the start codon in lactococci, while GUG and UUG are rare start codons in lactobacilli. *Lactococcus lactis* infrequently (<5%) employs AGA and AGG as arginine codons; similarly, it rarely employs CUG and CUC as leucine codons. Lactobacilli, like *E. coli*, seldom use AUA as an isoleucine codon, CUA as a leucine codon, and AGG and, particularly, AGA (<1%) as arginine codons. It is not clear that every required tRNA is available in the cell. Genes with a heavy concentration of the low-use codons noted above probably encounter difficulty in expression; nevertheless, an expression problem stemming from codon distribution has yet to be demonstrated in the lactic acid bacteria.

GENETIC MANIPULATION

Transformation by Electroporation

Reports that whole washed cells of the gram-negative bacterium *E. coli* and the gram-positive bacteria *Lactococcus lactis*, *Lactobacillus casei*, and *S. thermophilus* could be transformed by electroporation in the presence of plasmid DNA sparked renewed interest in electrical methods of transformation (for reviews, see references 29 and 177). To date, more than 100 species

Table 5. Putative ribosome binding sites of *Lactobacillus* spp.

Source[a]	Sequence[b]	No. of bp from Shine-Dalgarno sequence to AUG	Reference
Lactobacillus casei derived			
lacG (P-β-Gal)	UUAACAGGAGGUUAUUAAGCAAUG	10	143
lacE (enzyme II)	CGCUCAGGAGGAAAAGACUCAUG	9	3
lacF (factor III)	UGAAUCGGAGGGAAAAUGAUG	7	2
lacL (β-galactosidase ORF 1)	UUGAAAGGAGCUUCCUCAUG	6	28
Dihydrofolate reductase	CUCAAAGGAGGGGUCUCAGAUG	7	90
Thymidylate synthase	GUAAAAGGAGCAGAGCAGACAAUG	11	138
D-Alanine activating enzyme	UUUAAAGGGGGAAACCGCGAUG	9	69
trp operon (ORF Y)	AUCGAAGGAGGUCUUGGGUAUG	9	133
trpD	CAAUUUGGAGGCAUUCAUCAAUG	9	133
trpC	GCGUAAGGACGUGGUGGCAUG	9	133
trpF	GGUUAAGGAGGUUCCCAAAAUG	8	133
trpB	AGCAAAAGAGGACAUCAUCUAUG	8	133
trpA	UACAAAGGAGUCGAUG	3?	133
Folylpolyglutamate synthase	AAAUAAGAAAGUUAACAAUG	6	176
L-LDH	AAGAAAGGAUGAUAUCACCAUG	10	90
D-HicDH	UGGAAAGGAAGUUUAACACAUG	7	105
From other lactobacilli			
lacZ (*L. bulgaricus*)	UUAGAAGGGAAGAAUUAGAAAAUG	9	155
lacY (*L. bulgaricus*)	CCUAAAGGAGAAUUUCAUG	6	104
lysA (*L. bulgaricus*)	ACCAGUGGAGGUUAAGAAGAACAAUG	?	15
hdcA	UUAUUAGGAGGUCUAAUUAUG	8	186
L-HicDH (*L. confusus*)	AAUUUGGGGGAUAUUAUG	5–6	106
xylR (*L. pentosus*)	AUUACGAGAGGGUGAUGUCGUG	8–9	108
xylA (*L. pentosus*)	UAACUUGGAGGCAAGAAUUAUG	7	108
xylB (*L. pentosus*)	AUGUGAGGAGGAUGCAAAAUG	8	108
pyrE (*L. plantarum*)	CAGAAUGGAGUAACUAAUAUG	7	17
laf (*Lactobacillus* spp.)	AUUUUAGGAGGUUUCUAUCAUG	8	127
Unknown ORF 2 (*L. helveticus*)	AGGUGAAGAAGAAAUCAUG	5	79
Helveticin J (*L. helveticus*)	UUUUUCGGAGGUUUUAUUAUG	7	79
Unknown ORF 3 (*L. helveticus*)	CUGUCAGAAAGACAGUUAUG	7	79
D-LDH (*L. plantarum*)	UAUAUAGGAGGAAUUUUGUAAUG	9	173
L-LDH (*L. plantarum*)	AGGAGAGGAUGACUAUUUUUUG	10	173
16S RNA 3′→5′ complements			
B. subtilis	AGAAAGGAGGUGAUC...		62
Lactococcus lactis	AAAAGGAGGU...		111
Lactobacillus casei	AGAAAGGAGGUGAU...		193
Observed consensus sequences			
Lactobacillus casei RBS consensus	--aAAGGAGG--auc		
Lactobacillus RBS consensus	u-aaAGGAGG--au-		

[a] LDH, lactate dehydrogenase; HicDH, hydroxyisocaproate dehydrogenase.
[b] Capital letters indicate bases that occur at >70% frequency, and lowercase letters indicate frequencies of >40%. Boldface letters denote the rare start codons UUG and GUG.

of bacteria have been successfully transformed by electroporation. The method replaces tedious, time-consuming, and frequently unreliable protoplast transformation techniques that had taken years to develop for the lactic acid bacteria (25). The frequency of transformation observed with gram-positive bacteria is generally lower than that observed with *E. coli*, but yields of 10^7 transformants per μg of DNA are possible with some strains. In these strains, library construction is possile. Most strains give frequencies suitable for exchange of shuttle vectors between strains. It is not uncommon to observe large differences in transformation frequency between two closely related strains. Unfortunately, some strains are refractory to electrotransformation. The reasons for these difficulties are usually not fully understood, but a number of possible causes can be postulated. The dense composition of the gram-positive cell wall may present a barrier to entry of DNA in some strains. The efficiency of transformation of *Listeria monocytogenes* can be raised from 300 to 8×10^5 transformants per μg of DNA by incorporation of low concentrations of the cell wall synthesis inhibitor penicillin G in the growth medium (136). Gentle treatment of cells with muralytic enzymes prior to electrotransformation has been used to "soften" cells of lactococci in order to raise the transformation efficiency (146). Incompatibility between transforming vectors and resident plasmids may also lower frequencies of transformation. Incompatibility can interfere with electroporation of *Lactococcus lactis* (185) and lactobacilli (145). Other possible barriers include restriction systems, nonspecific nucleases, nonexpressed marker genes, nonex-

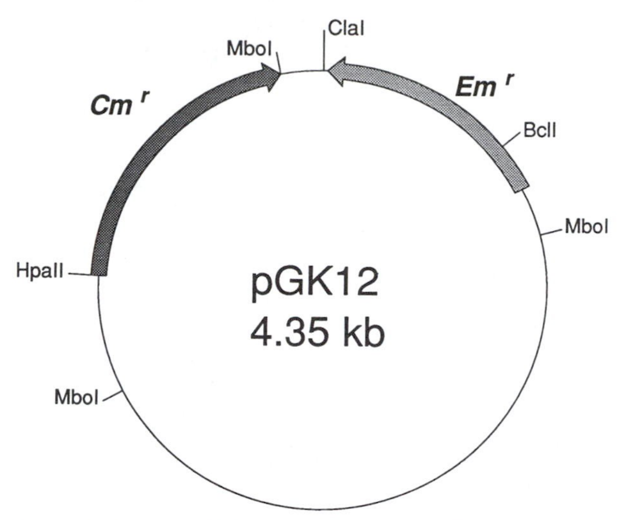

Figure 1. The broad-host-range vector pGK12 is derived from a group N streptococcus cryptic plasmid replicon and the chloramphenicol and erythromycin resistance determinants of pC194 and pE194, respectively, by a strategy similar to that described in the legend to Fig. 2 (98).

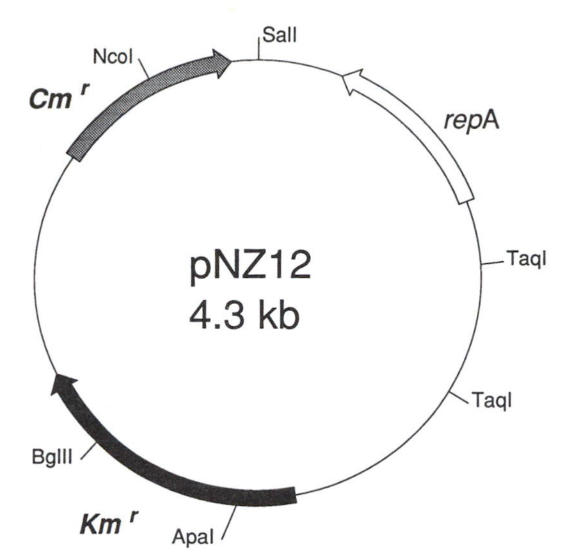

Figure 2. The broad-host-range cloning vector pNZ12 originates from the ligation of a *Taq*I restriction fragment of a high-copy-number *Lactococcus lactis* subsp. *cremoris* cryptic plasmid with a fragment of DNA encoding antibiotic resistance markers but not an origin of replication followed by transformation into *B. subtilis* (46). Km^r and Cm^r derive from *S. aureus* plasmids pUB110 and pC194, respectively.

pressed replication-essential functions, and interference with host-essential functions.

Plasmids and Vectors

A wide assortment of potential vectors is available for use with the lactic acid bacteria. Plasmids with a variety of functional properties (i.e., lactose metabolism, proteinase activity, citrate metabolism, phage resistance, and nisin production and resistance) were known for some time to be present in lactococcal strains (117). Most strains also contain one or more small cryptic plasmids suitable for vector construction. pGK12 (Fig. 1 [98]) and pNZ12 (Fig. 2 [46]) are two examples of vectors that function in many strains of lactic acid bacteria. These plasmids can also be used to transform lactobacilli (27, 109, 110), leuconostocs (38), and a wide variety of bacteria, since the broad-host-range lactococcal replicon functions in gram-positive and gram-negative hosts. Vectors based on indigenous plasmid replicons have also been used successfully in *S. thermophilus* (118). It has been reported that pAMβ1 and pIP501 can enter lactic acid bacteria by conjugation (for a review, see reference 52); in addition, pAMβ1 can be introduced via electroporation (110). pIL251 (164), a conjugation-deficient mini-pAMβ1 containing multiple cloning sites in a synthetic polylinker, can be used as a cloning vector in lactococci and lactobacilli. Numerous vectors that function in *Bacillus* spp. will function in at least some of the lactic acid bacteria. Vectors based on replicons having a single-stranded rolling-circle mode of replication (e.g., pC194, pE194, pUB110, or derivatives such as pBD64) are frequently of value. It has recently been demonstrated in *Lactobacillus casei* that the stability of such vectors is enhanced by the presence of a functionally active (−) origin (162).

Small multicopy plasmids of lactococci and lactobacilli have been analyzed in detail, and the complete nucleotide sequences of several of the plasmids have been determined (18, 21, 31, 42, 51, 68, 77, 81, 82, 103, 174, 194). Many of these plasmids have been used in the construction of cloning vectors for lactic acid bacteria.

Heterologous Gene Expression

The first example of heterologous gene expression in lactic acid bacteria was associated with the transfer of conjugal plasmids such as pAMβ1 into lactococci. Both replication and antibiotic (macrolide-lincosamide-streptogramin B) resistance-determining genes needed to be expressed in the heterologous host for conjugation to be observed. The markers that are most frequently used in the lactic acid bacteria, chloramphenicol resistance from pC194 and macrolide-lincosamide-streptogramin B resistance from pE194 or pAMβ1, are expressed from their own regulatory signals in a wide variety of hosts. Setting aside the plasmid-associated replication and resistance functions, *lacZ* of *E. coli* was one of the first heterologous genes to be expressed in *Lactococcus lactis* (46). The α-amylase of *B. subtilis* has been coupled with the α-galactosidase of *Cyamosis tetragonoloba* in the construction of a vector for the analysis of divergent promoters in *Lactococcus lactis* (59). Bovine prochymosin (47), *B. subtilis* neutral protease (183), *E. coli* bacteriophage T4 lysozyme (183), the α-amylase of *B. stearothermophilus* (165), a clostridial β-galactosidase gene (137), and TEM β-lactamase (163) have also been expressed in *Lactococcus lactis*.

Heterologous genes have also been expressed in *S. thermophilus* (118) and a few strains of lactobacilli. Examples are the expression of α-amylase and endoglucanase genes in *Lactobacillus plantarum* (154), the cloning and expression of α-amylase from *Bacillus amyloliquefaciens* into *Lactobacillus plantarum* (80), and the transfer of the *B. stearothermophilus* α-amylase into five *Lactobacillus* spp. (32). The bacterial luciferase (*lux*) genes have also been reported to function in *Lactococcus lactis* and *Lactobacillus casei* (1). Finally, the *xyl* genes of *Lactobacillus pentosus* can be used to complement xylose metabolism mutations in *Lactobacillus plantarum* and *Lactobacillus casei* (144). The limited experience gained thus far with heterologous gene expression in the lactic acid bacteria indicates that heterologous genes can function efficiently in these hosts. Expression signals derived from the same or a closely related host function particularly well.

Integration of DNA

Homologous recombination has been used to introduce DNA into the chromosome of *Lactococcus lactis* via a chromosomal target copy of the transposon Tn*919* (67). A plasmid-encoded proteinase gene of *Lactococcus lactis* has been integrated into the chromosome by means of a Campbell-type insertion mechanism (102). Gene replacement at the *pepXP* locus by double cross-over recombination has also been observed (101). An account of integration and expression of α-amylase and endoglucanase genes in the *Lactobacillus plantarum* chromosome has been published (154). Under selective pressure, unstable plasmids integrate into the chromosome of *Lactobacillus plantarum* (151). These reports indicate that homologous recombination systems that can be used for the replacement of genes are functional in some strains of lactic acid bacteria.

OTHER TOPICS

We should not close without mentioning the many areas of research it has not been possible to discuss in this chapter. To facilitate further reading, recent citations follow each of the topics mentioned below.

Significant amounts of information on the metabolism, enzymology, transport physiology, and energetics of these organisms have been omitted (87, 100, 175). Unique traits such as the malolactic fermentation of wine (8, 130) and the contribution of citrate fermentation to the production of diacetyl and butter flavor (40) have been carefully examined.

Phages are of critical interest, since phage attacks often interfere with food and dairy fermentations (41). Phage molecular biology and phage resistance genes have received particular attention. Much recent information about the phages of lactic acid bacteria is available (5, 9, 15, 32, 53, 71–73, 89, 91, 148, 157, 158).

Transposons and insertion sequence elements in the lactic acid bacteria have been identified and examined in detail (7, 36, 52, 61, 139, 140, 147, 160, 161).

The lactic acid bacteria are used to preserve food for reasons that go beyond their abilities to lower pH by the production of lactic acid. Numerous antagonistic activities of lactic acid bacteria have been observed in food and feed fermentations mediated by lactococci, lactobacilli, streptococci, pediococci, and leuconostocs (for a review, see reference 108). The production and modes of action of natural antibiotics and bacteriocins produced by many strains of lactic acid bacteria are being examined at the molecular level (10, 13, 14, 22, 23, 37, 48, 65, 79, 84, 125–127, 134, 169, 171, 180, 181, 191).

CONCLUSIONS

During the past decade, the attention focused on lactic acid bacteria has greatly increased our understanding of the biochemistry, physiology, and genetics of these microorganisms. Within a few years, we should have the knowledge to explain many of the distinctive features of the lactic acid bacteria at the detailed molecular level. It is already possible to design, construct, and evaluate strains that might have increased utility in food and industrial applications, even though significant gaps in our knowledge remain to be filled. We hope that this chapter dissuades the reader from the commonly held belief that little is known about the biology, biochemistry, and genetics of lactic acid bacteria.

NOTE ADDED IN PROOF

Since the research for this chapter was completed, the amount of information available on the genetics and biochemistry of the lactic acid bacteria has increased substantially. A number of additional genes from *Lactococcus*, *Lactobacillus*, and *Pediococcus* have been cloned and sequenced, and lactic acid bacteria have served as hosts for expression of heterologous genes from a variety of prokaryotes and eukaryotes (see, for example, references 76a, 107a, 113a, and 168a). New studies on the integration of DNA sequences into the chromosomes of lactic acid bacteria (e.g., references 23a and 152a) move us ever closer to creating stable, genetically altered commercial starter cultures. The remarkable pace of recent advances in genetic manipulation and in our understanding of metabolism, enzymology, transport physiology, energetics, and phage biology should ensure that a detailed picture of the biology, biochemistry, and genetics of the lactic acid bacteria will be available in the near future.

Acknowledgments. We thank Joseph DeParasis and José-Luis Montelongo for library research and bibliographic assistance.

REFERENCES

1. **Ahmad, K. A., and G. S. A. B. Stewart.** 1988. Cloning of the *lux* genes into *Lactobacillus casei* and *Streptococcus lactis*: phosphate-dependent light production. *Biochem. Soc. Trans.* **1988:**1068.
2. **Alpert, C. A., and B. M. Chassy.** 1988. Molecular cloning and nucleotide sequence of the factor III(lac) gene of *Lactobacillus casei*. Gene **62:**277–288.
3. **Alpert, C. A., and B. M. Chassy.** 1990. Molecular cloning and DNA sequence of *lacE*, the gene encoding the lactose-specific enzyme II of the phosphotransferase system of *Lactobacillus casei*: evidence that a cysteine residue is essential for sugar phosphorylation. *J. Biol. Chem.* **265:**22561–22570.
4. **Andrews, J., G. M. Clore, R. W. Davies, A. M. Gronenborn, B. Gronenborn, D. Kalderone, P. C. Papadopoulos, S. Schafer, P. F. G. Simms, and R. Stancombe.** 1985. Nucleotide sequence of the dihydrofolate reductase gene of methotrexate resistant *Lactobacillus casei*. Gene **35:**217–222.
5. **Arendt, E. K., H. Neve, and W. P. Hammes.** 1990. Characterization of phage isolates from a phage-carry-

ing culture of *Leuconostoc oenos* 58N. *Appl. Microbiol. Biotechnol.* **34:**220–224.

6. **Atlan, D., P. Laloi, and R. Portalier.** 1990. *X*-Prolyl-dipeptidyl aminopeptidase of *Lactobacillus delbrueckii* subsp. *bulgaricus*: characterization of the enzyme and isolation of deficient mutants. *Appl. Environ. Microbiol.* **56:**2174–2179.
7. **Aukrust, T., and I. F. Nes.** 1988. Transformation of *Lactobacillus plantarum* with the plasmid pTV1 by electroporation. *FEMS Microbiol. Lett.* **52:**127–132.
8. **Battermann, G., and F. Radler.** 1991. A comparative study of malolactic enzyme and malic enzyme of different lactic acid bacteria. *Can. J. Microbiol.* **37:**211–217.
9. **Benbadis, L., M. Faelen, P. Slos, A. Fazel, and A. Mercenier.** 1990. Characterization and comparison of virulent bacteriophages of *Streptococcus thermophilus* isolated from yogurt. *Biochimie* **72:**855–862.
10. **Berry, E. D., M. B. Liewen, R. W. Mandigo, and R. W. Hutkins.** 1990. Inhibition of *Listeria monocytogenes* by bacteriocin-producing *Pediococcus* during the manufacture of fermented semidry sausage. *J. Food Prot.* **53:**194–197.
11. **Bhowmik, T., and E. H. Marth.** 1990. β-Galactosidase of *Pediococcus* species—induction, purification and partial characterization. *Appl. Microbiol. Biotechnol.* **33:**317–323.
12. **Bhowmik, T., and E. H. Marth.** 1990. Peptide-hydrolysing enzymes of *Pediococcus* species. *Microbios* **62:** 197–211.
13. **Bhunia, A. K., M. C. Johnson, B. Ray, and N. Kalchayanand.** 1991. Mode of action of pediocin AcH from *Pediococcus acidilactici* H on sensitive bacterial strains. *J. Appl. Bacteriol.* **70:**25–33.
14. **Biswas, S. R., P. Ray, M. C. Johnson, and B. Ray.** 1991. Influence of growth conditions on the production of a bacteriocin, pediocin AcH, by *Pediococcus acidilactici* H. *Appl. Environ. Microbiol.* **57:**1265–1267.
15. **Boizet, B., Y. Lahbib-Mansais, L. Dupont, P. Ritzenthaler, and M. Mata.** 1990. Cloning, expression and sequence analysis of endolysin-encoding gene of *Lactobacillus bulgaricus* bacteriophage MV1. *Gene* **94:**61–68.
16. **Boizet, B., D. Villeval, P. Slos, M. Novel, G. Novel, and A. Mercinier.** 1988. Isolation and structural analysis of the phospho-β-galactosidase gene from *Streptococcus lactis* Z268. *Gene* **62:**249–261.
17. **Bouia, A., F. Bringel, L. Frey, A. Belarbi, A. Guyonvarch, B. Kammerer, and J. C. Hubert.** 1990. Cloning and structure of the *pyrE* gene of *Lactobacillus plantarum* CCM1904. *FEMS Microbiol. Lett.* **69:**233–238.
18. **Bouia, A., F. Bringel, L. Frey, B. Kammerer, A. Belarbi, A. Goyonvarch, and J. C. Hubert.** 1989. Structural organization of pLP-1, a cryptic plasmid from *Lactobacillus plantarum* CCM1904. *Plasmid* **22:**185–192.
19. **Breidt, F. J., and G. C. Stewart.** 1986. Cloning and expression of the gene phospho-β-galactosidase of *Staphylococcus aureus* in *Escherichia coli*. *J. Bacteriol.* **166:**1061–1066.
20. **Breidt, F. J., and G. C. Stewart.** 1987. Nucleotide and deduced amino acid sequences of the *Staphylococcus aureus* phospho-β-galactosidase gene. *Appl. Environ. Microbiol.* **53:**969–973.
21. **Bringel, F., L. Frey, and J. C. Hubert.** 1989. Characterization, cloning, curing, and distribution in lactic acid bacteria of pLP1, a plasmid from *Lactobacillus plantarum* CCM 1904 and its use in shuttle vector construction. *Plasmid* **22:**193–202.
22. **Broadbent, J. R., and J. K. Kondo.** 1991. Genetic construction of nisin-producing *Lactococcus lactis* subsp. *cremoris* and analysis of a rapid method for conjugation. *Appl. Environ. Microbiol.* **57:**517–524.
23. **Buchman, G. W., S. Banerjee, and J. N. Hansen.** 1988. Structure,expression and evolution of a gene encoding the precursor of nisin, a small protein antibiotic. *J. Biol. Chem.* **263:**16260–16266.

23a. **Casey, J., C. Daly, and G. F. Fitzgerald.** 1992. Controlled integration into the *Lactococcus* chromosome of the PCI829-encoded abortive infection gene from *Lactococcus lactis* subsp. *lactis* UC811. *Appl. Environ. Microbiol.* **58:**3283–3291.

24. **Cerning, J.** 1990. Exocellular polysaccharides produced by lactic acid bacteria. *FEMS Microbiol. Lett.* **87:**113–130.
25. **Chassy, B. M.** 1987. Prospects for the genetic manipulation of lactobacilli. *FEMS Microbiol. Rev.* **46:**297–312.
26. **Chassy, B. M., and C. A. Alpert.** 1989. Molecular characterization of the plasmid-encoded lactose-PTS of *Lactobacillus casei*. *FEMS Microbiol. Lett.* **63:**157–166.
27. **Chassy, B. M., and J. L. Flickinger.** 1987. Transformation of *Lactobacillus casei* by electroporation. *FEMS Microbiol. Lett.* **44:**173–177.
28. **Chassy, B. M., J. L. Flickinger, and J. Thompson.** Unpublished data.
29. **Chassy, B. M., A. Mercenier, and J. Flickinger.** 1988. Transformation of bacteria by electroporation. *Trends Biotechnol.* **6:**303–309.
30. **Chassy, B. M., and J. Thompson.** 1983. Regulation and characterization of the galactose:phosphoenolpyruvate-dependent phosphotransferase system in *Lactobacillus casei*. *J. Bacteriol.* **154:**1204–1214.
31. **Cocconcelli, P. S., M. J. Gasson, L. Morelli, and V. Bottazzi.** 1991. Single-stranded DNA plasmid, vector construction and cloning of *Bacillus stearothermophilus* α-amylase in *Lactobacillus*. *Res. Microbiol.* **142:**643–652.
32. **Coffey, A. G., G. F. Fitzgerald, and C. Daly.** 1991. Cloning and characterization of the determinant for abortive infection of bacteriophage from lactococcal plasmid pC1829. *J. Gen. Microbiol.* **137:**1355–1362.
33. **Collins, M. D., U. Rodrigues, C. Ash, M. Aguirre, J. A. E. Farrow, A. Martinezmurcia, B. A. Phillips, A. M. Williams, and S. Wallbanks.** 1991. Phylogenetic analysis of the genus Lactobacillus and related lactic acid bacteria as determined by reverse transcriptase sequencing of 16S rRNA. *FEMS Microbiol. Lett.* **77:**5–12.
34. **Collins, M. D., A. M. Williams, and S. Wallbanks.** 1990. The phylogeny of *Aerococcus* and *Pediococcus* as determined by 16S rRNA sequence analysis. *FEMS Microbiol. Lett.* **70:**296–301.
35. **Copeland, W. C., J. D. Domena, and J. D. Roberts.** 1989. The molecular cloning, sequence and expression of the *hdc* B gene from *Lactobacillus* 30A. *Gene* **85:**259–266.
36. **Cosby, W. M., L. T. Axelsson, and W. J. Dobrogosz.** 1989. Tn*917* transposition in *Lactobacillus plantarum* using the highly temperature-sensitive plasmid pTV1ts as a vector. *Plasmid* **22:**236–243.
37. **Daeschel, M. A., D. S. Jung, and B. T. Watson.** 1991. Controlling wine malolactic fermentation with nisin and nisin-resistant strains of *Leuconostoc oenos*. *Appl. Environ. Microbiol.* **57:**601–603.
38. **David, S., G. Simons, and W. M. De Vos.** 1991. Plasmid transformation by electroporation of *Leuconostoc paramesenteroides* and its use in molecular cloning. *Appl. Environ. Microbiol.* **55:**1483–1489.
39. **David, S., H. Stevens, M. Van Rield, G. Simons, and W. M. De Vos.** 1992. *Leuconostoc lactis* β-galactosidase is encoded by two overlapping genes. *J. Bacteriol.* **174:** 4475–4481.
40. **David, S., M. E. van der Rest, A. J. M. Driessen, G. Simons, and W. M. DeVos.** 1990. Nucleotide sequence and expression in *Escherichia coli* of the *Lactococcus lactis* citrate permease gene. *J. Bacteriol.* **172:**5789–5794.
41. **Davidson, B. E., I. B. Powell, and A. J. Hillier.** 1990. Temperate bacteriophages and lysogeny in lactic acid bacteria. *FEMS Microbiol. Rev.* **87:**79–90.

42. **De Rossi, E., P. Brigidi, M. Rossi, D. Matteuzzi, and G. Riccardi.** 1991. Characterization of Gram-positive broad host-range plasmids carrying a thermophilic replicon. *Res. Microbiol.* **142:**389–396.
43. **DeVos, W. M.** 1987. Gene cloning and expression in lactic streptococci. *FEMS Microbiol. Rev.* **46:**281–295.
44. **De Vos, W. M., I. Boerrigter, R. J. Van Rooyen, B. Reiche, and W. Hengstenberg.** 1990. Characterization of the lactose-specific enzymes of the phosphotransferase system in *Lactococcus lactis*. *J. Biol. Chem.* **265:**22554–22560.
45. **DeVos, W. M., and M. J. Gasson.** 1989. Structure and expression of the *Lactococcus lactis* gene for phospho-β-galactosidase (*lacG*) in *Escherichia coli* and *L. lactis*. *J. Gen. Microbiol.* **135:**1833–1846.
46. **De Vos, W. M., and G. Simons.** 1988. Molecular cloning of lactose genes in the dairy streptococci: the phospho-β-galactosidase genes and their expression products. *Biochimie* **70:**461–473.
47. **De Vos, W. M., G. Simons, and S. David.** 1989. Gene organization and expression in the mesophilic lactic acid bacteria. *J. Dairy Sci.* **72:**3398–3405.
48. **Dodd, H. M., N. Horn, and M. J. Gasson.** 1990. Analysis of the genetic determinant for production of the peptide antibiotic nisin. *J. Gen. Microbiol.* **136:**555–566.
49. **Dower, W. J., B. M. Chassy, J. T. Trevors, and H. P. Blaschek.** 1992. Protocols for the transformation of bacteria by electroporation, p. 485–500. *In* D. C. Chang et al. (ed.), *Guide to Electroporation and Electrofusion*. Academic Press, Inc., San Diego, Calif.
50. **Eggimann, B., and H. Bachmann.** 1980. Purification and partial characterization of an aminopeptidase from *Lactobacillus lactis*. *Appl. Environ. Microbiol.* **40:**876–882.
51. **Feirtag, J. M., J. P. Petzel, E. Pasalodos, K. A. Baldwin, and L. L. McKay.** 1991. Thermosensitive plasmid replication, temperature-sensitive host growth, and chromosomal plasmid integration conferred by *Lactococcus lactis* subsp. *cremoris* lactose plasmids in *Lactococcus lactis* subsp. *lactis*. *Appl. Environ. Microbiol.* **57:**539–548.
52. **Fitzgerald, G. F., and M. J. Gasson.** 1988. *In vivo* gene transfer systems and transposons. *Biochimie* **70:**489–502.
53. **Forsman, P., and T. Alatossava.** 1991. Genetic variation of *Lactobacillus delbrueckii* subsp. *lactis* bacteriophages isolated from cheese processing plants in Finland. *Appl. Environ. Microbiol.* **57:**1805–1812.
54. **Friedland, I. R., M. Snipelisky, and M. Khoosal.** 1990. Meningitis in a neonate caused by *Leuconostoc* sp. *J. Clin. Microbiol.* **28:**2125–2126.
55. **Gaetje, G., V. Mueller, and G. Gorrschalk.** 1991. Lactic acid excretion via carrier-mediated facilitated diffusion in *Lactobacillus helveticus*. *Appl. Microbiol. Biotechnol.* **34:**778–782.
56. **Gilliland, S. E.** 1990. Health and nutritional benefits from lactic acid bacteria. *FEMS Microbiol. Lett.* **87:**175–188.
57. **Graves, M. C., and J. C. Rabinowitz.** 1986. *In vivo* and *in vitro* transcription of the *Clostridium pasteurianum* ferredoxin gene. *J. Biol. Chem.* **261:**11409–11415.
58. **Haandrikman, A. J.** 1990. Maturation of cell envelope-associated proteinase of *Lactococcus lactis*. Ph.D. dissertation. Rijksuniversiteit, Groningen, The Netherlands.
59. **Haandrikman, A. J., L. Hamoen, K. J. Leenhouts, J. Kok, and G. Venema.** Development and use of a broad host-range vector for the simultaneous analysis of divergent promoters. Submitted for publication.
60. **Haandrikman, A. J., J. Kok, H. Laan, S. Soemitro, A. M. Ledeboer, W. N. Konings, and G. Venema.** 1989. Identification of a gene required for maturation of an extracellular lactococcal serine proteinase. *J. Bacteriol.* **171:**2789–2794.
61. **Haandrikman, A. J., C. Van Leeuwen, J. Kok, P. Vos, W. M. de Vos, and G. Venema.** 1990. Insertion elements on lactococcal protease plasmids. *Appl. Environ. Microbiol.* **56:**1890–1896.
62. **Hager, P. W., and J. C. Rabinowitz.** 1985. Translational specificity in *Bacillus subtilis*, p. 1–31. *In* D. A. Dubnau (ed.), *The Molecular Biology of the Bacilli*. Academic Press, Inc., Orlando, Fla.
63. **Hancock, K. R., E. Rockman, C. A. Young, L. Pearce, I. S. Maddox, and D. B. Scott.** 1991. Expression and nucleotide sequence of the *Clostridium acetobutylicum* β-galactosidase gene cloned in *Escherichia coli*. *J. Bacteriol.* **173:**3084–3095.
64. **Handwerger, S., H. Horowitz, K. Coburn, A. Kolokathis, and G. P. Wormser.** 1990. Infection due to *Leuconostoc* sp.—6 cases and review. *Rev. Infect. Dis.* **12:**602–610.
65. **Hardings, C. D., and B. G. Shaw.** 1990. Antimicrobial activity of *Leuconostoc gelidum* against closely related species and *Listeria monocytogenes*. *J. Appl. Bacteriol.* **69:**648–654.
66. **Harley, C. B., and R. P. Reynolds.** 1987. Analysis of *E. coli* promoter sequences. *Nucleic Acids Res.* **15:**2343–2361.
67. **Hayes, F., J. Law, C. Daly, and G. F. Fitzgerald.** 1990. Integration and excision of plasmid DNA in *Lactococcus lactis* ssp. *lactis*. *Plasmid* **24:**81–89.
68. **Hayes, F., P. Vos, G. F. Fitzgerald, W. M. De Vos, and C. Daly.** 1991. Molecular organization of the minimal replicon of novel, narrow-host-range, lactococcal plasmid pCI305. *Plasmid* **25:**16–26.
69. **Heaton, M. P., and F. C. Neuhaus.** 1992. Biosynthesis of D-alanyl lipooteichoic acid: cloning, nucleotide sequence and expression of the *Lactobacillus casei* gene for the D-alanine activating enzyme. *J. Bacteriol.* **174:**4707–4717.
70. **Hengstenberg, W., W. K. Penberthy, K. L. Hill, and M. L. Morse.** 1968. Metabolism of lactose by *Staphylococcus aureus*. *J. Bacteriol.* **96:**2187–2188.
71. **Hill, C., I. J. Massey, and T. R. Klaenhammer.** 1991. Rapid method to characterize lactococcal bacteriophage genomes. *Appl. Environ. Microbiol.* **57:**283–288.
72. **Hill, C., L. A. Miller, and T. R. Klaenhammer.** 1990. Cloning, expression, and sequence determination of a bacteriophage fragment encoding bacteriophage resistance in *Lactococcus lactis*. *J. Bacteriol.* **172:**6419–6426.
73. **Hill, C., L. A. Miller, and T. R. Klaenhammer.** 1990. Nucleotide sequence and distribution of the pTR2030 resistance determinant (*hsp*) which aborts bacteriophage infection in lactococci. *Appl. Environ. Microbiol.* **56:**2255–2258.
74. **Holo, H., O. Nilssen, and I. F. Nes.** 1991. Lactococcin A, a new bacteriocin from *Lactococcus lactis* subsp. *cremoris*: isolation and characterization of the protein and its gene. *J. Bacteriol.* **173:**3879–3887.
75. **Hutkins, R. W., and C. Ponne.** 1991. Lactose uptake driven by galactose efflux in *Streptococcus thermophilus*—evidence for a galactose-lactose antiporter. *Appl. Environ. Microbiol.* **57:**941–944.
76. **Imam, S. H., A. Burgess-Cassler, G. L. Cote, S. H. Gordon, and F. L. Baker.** 1991. A study of cornstarch granule digestion by an unusually high molecular weight α-amylase secreted by *Lactobacillus amylovorus*. *Curr. Microbiol.* **22:**365–370.
76a. **Ishino, Y., P. Morgenthaler, H. Hottinger, and D. Soll.** 1992. Organization and nucleotide sequence of the glutamine synthetase *glnA* gene from *Lactobacillus delbreuckii* subsp. *bulgaricus*. *Appl. Environ. Microbiol.* **58:**3165–3169.
77. **Jahns, A., A. Schaefer, A. Geis, and M. Terber.** 1991. Identification, cloning and sequencing of the replica-

tion region of *Lactococcus lactis* subsp. *lactis* biovar *diacetylactis* Bu2 citrate plasmid pSL2. *FEMS Microbiol. Lett.* **80**:253–258.

78. **Jimeno, J., M. Casey, and F. Hofer.** 1984. The occurrence of β-galactosidase and P-β-galactosidase in *Lactobacillus casei* strains. *FEMS Microbiol. Lett.* **25**:275–278.
79. **Joerger, M. C., and T. R. Klaenhammer.** 1990. Cloning, expression and nucleotide sequence of the *Lactobacillus helveticus* 481 gene encoding the bacteriocin helveticin. *J. Bacteriol.* **172**:6339–6347.
80. **Jones, S., and P. J. Warner.** 1990. Cloning and expression of α-amylase from *Bacillus amyloliquefaciens* in a stable plasmid vector in *Lactobacillus plantarum*. *Lett. Appl. Microbiol.* **11**:214–219.
81. **Josson, K., T. Scheirlinck, F. Michiels, C. Platteeuw, P. Stanssens, H. Joos, P. Dhaese, M. Zabeau, and J. Mahillon.** 1989. Characterization of a Gram-positive broad-host-range plasmid isolated from *Lactobacillus hilgardii*. *Plasmid* **21**:9–20.
82. **Josson, K., P. Soetaert, F. Michiels, H. Joos, and J. Mahillon.** 1990. *Lactobacillus hilgardii* plasmid pLAB 1000 consists of two functional cassettes commonly found in other gram-positive organisms. *J. Bacteriol.* **172**:3089–3099.
83. **Kaklij, G. S., S. M. Kelkar, M. A. Shenoy, and K. B. Sainis.** 1991. Antitumor activity of *Streptococcus thermophilus* against fibrosarcoma: role of T-cells. *Cancer Lett.* **56**:37–44.
84. **Kaletta, C., and K. Entian.** 1989. Nisin, a peptide antibiotic: cloning and sequencing of the *nisA* gene and posttranslational processing of its peptide product. *J. Bacteriol.* **171**:1597–1601.
85. **Kaminogawa, S., N. Azuma, I. K. Hwang, Y. Suzuki, and K. Yamauchi.** 1984. Isolation and characterization of a prolidase from *Streptococcus cremoris* H61. *Agric. Biol. Chem.* **48**:3035–3040.
86. **Kaminogawa, S., T. Ninomiya, and K. Yamauchi.** 1984. Aminopeptidase profiles of lactic streptococci. *J. Dairy Sci.* **67**:2483–2492.
87. **Kashket, E. R.** 1987. Bioenergetics of lactic acid bacteria: cytoplasmic pH and osmotolerance. *FEMS Microbiol. Rev.* **46**:233–244.
88. **Khalid, N. M., and E. M. Marth.** 1990. Purification and partial characterization of a prolyl-dipeptidyl aminopeptidase from *Lactobacillus helveticus* CNRZ 32. *Appl. Environ. Microbiol.* **56**:381–388.
89. **Kim, J. H., and C. A. Batt.** 1991. Nucleotide sequence and deletion analysis of a gene coding for a structural protein of *Lactococcus lactis* bacteriophage F4-1. *Food Microbiol.* **8**:27–36.
90. **Kim, S. F., S. F. Baek, and M. Y. Pack.** 1991. Cloning and nucleotide sequence of the *Lactobacillus casei* lactate dehydrogenase gene. *Appl. Environ. Microbiol.* **57**: 2413–2417.
91. **Kim, S. G., and C. A. Batt.** 1991. Identification of a nucleotide sequence conserved in *Lactococcus lactis* bacteriophages. *Gene* **98**:95–100.
92. **Kiwaki, M., H. Ikemura, M. Shimizu-Kadota, and A. Hirashima.** 1989. Molecular characterization of a cell wall associated proteinase gene of *Streptococcus lactis* NCDO763. *Mol. Microbiol.* **3**:359–369.
93. **Klebanoff, S. J., and R. W. Coombs.** 1991. Viricidal effect of *Lactobacillus acidophilus* on human immunodeficiency virus type 1: possible role in heterosexual transmission. *J. Exp. Med.* **174**:289–292.
94. **Koivula, T., M. Sibakov, and I. Palva.** 1991. Isolation and characterization of *Lactococcus lactis* subsp. *lactis* promoters. *Appl. Environ. Microbiol.* **57**:333–340.
95. **Kojic, M., D. Fira, A. Banina, and L. Topisirovic.** 1991. Characterization of the cell wall-bound proteinase of *Lactobacillus casei* HN14. *Appl. Environ. Microbiol.* **57**: 1753–1757.
96. **Kok, J.** 1990. Genetics of the proteolytic system of the lactic acid bacteria. *FEMS Microbiol. Lett.* **87**:15–42.
97. **Kok, J., K. J. Leenhouts, A. J. Haandrikman, A. M. Ledeboer, and G. Venema.** 1988. Nucleotide sequence of the cell wall proteinase gene of *Streptococcus lactis* Wg2. *Appl. Environ. Microbiol.* **54**:231–238.
98. **Kok, J., J. M. B. M. van der Vossen, and G. Venema.** 1984. Construction of plasmid cloning vectors for lactic streptococci which also replicate in *Bacillus subtilis* and *Escherichia coli*. *Appl. Environ. Microbiol.* **48**:726–731.
99. **Kok, J., and G. Venema.** 1988. Genetics of proteinases of lactic acid bacteria. *Biochimie* **70**:475–488.
100. **Konings, W. N., B. Poolman, and A. J. M. Driessen.** 1989. Bioenergetics and solute transport in lactococci. *Crit. Rev. Microbiol.* **16**:419–476.
101. **Leenhouts, K.** 1990. Development and use of plasmid integration systems for lactococci. Ph.D. dissertation. Rijksuniversiteit, Groningen, The Netherlands.
102. **Leenhouts, K. J., J. Gietema, J. Kok, and G. Venema.** 1991. Chromosomal stabilization of proteinase genes in *Lactococcus lactis*. *Appl. Environ. Microbiol.* **57**:2568–2575.
103. **Leenhouts, K. J., B. Tolner, S. Bron, J. Kok, G. Venema, and J. Seegers.** 1991. Nucleotide sequence and characterization of the broad-host-range lactococcal plasmid pWV01. *Plasmid* **26**:55–66.
104. **Leong-Morgenthaler, P., M. C. Zwahlen, and H. Hottinger.** 1991. Lactose metabolism in *Lactobacillus bulgaricus*: analysis of the primary structure and expression of the genes involved. *J. Bacteriol.* **173**:1951–1957.
105. **Lerch, H. P., H. Blocker, H. Kallwas, J. Hoppe, H. Tsai, and J. Collins.** 1989. Cloning, sequencing and expression in *Escherichia coli* of the D-2-hydroxyisocaproate dehydrogenase gene of *Lactobacillus casei*. *Gene* **78**:47–57.
106. **Lerch, H.-P., R. Frank, and J. Collins.** 1989. Cloning, sequencing and expression of the L-2-hydroxyisocaproate dehydrogenase-encoding gene of *Lactobacillus confusus* in *Escherichia coli*. *Gene* **83**:263–270.
107. **Lindgren, S. E., and W. J. Dobrogosz.** 1990. Antagonistic activities of lactic acid bacteria in food and feed fermentations. *FEMS Microbiol. Lett.* **87**:149–164.

107a. **Llanos, R. M., A. J. Hillier, and B. E. Davidson.** 1992. Cloning, nucleotide sequence, expression, and chromosomal location of *idh*, the gene encoding L-dextro lactate dehydrogenase from *Lactococcus lactis*. *J. Bacteriol.* **174**:6956–6964.

108. **Lokman, B. C., P. van Santen, J. C. Verdoes, R. J. Leer, M. Posno, and P. H. Pouwels.** 1991. Organization and characterization of three genes involved in D-xylose catabolism in *Lactobacillus pentosus*. *Mol. Gen. Genet.* **230**:161–169.
109. **Luchansky, J. B., E. G. Kleeman, R. R. Raya, and T. R. Klaenhammer.** 1989. Genetic transfer sysems for delivery of plasmid DNA, deoxyribonucleic acid, to *Lactobacillus acidophilus* ADH: conjugation, electroporation, and transduction. *J. Dairy Sci.* **72**:1408–1417.
110. **Luchansky, J. B., M. C. Tennant, R. R. Raya, and T. R. Klaenhammer.** 1989. Identification of temperature-dependent replication of pGK12 and direct cloning via electroporation in *Lactobacillus acidophilus* ADH. *J. Dairy Sci.* **72**:113.
111. **Ludwig, W. E., E. Seewaldt, R. Kipper-Balz, K. H. Schliefer, L. Magrum, C. R. Woese, G. E. Fox, and E. Stackebrandt.** 1985. The phylogenetic position of *Streptococcus* and *Enterococcus*. *J. Gen. Microbiol.* **131**:543–551.
112. **Machuga, E. J., and D. H. Ives.** 1984. Isolation and characterization of an aminopeptidase from *Lactobacillus acidophilus* R-26. *Biochim. Biophys. Acta* **789**:26–36.
113. **Marshall, V. M. E., and B. A. Law.** 1984. The physiology and growth of dairy lactic acid bacteria, p. 67–98. *In* F. L. Davies and B. A. Law (ed.), *Advances in the*

Microbiology and Biochemistry of Cheese and Fermented Milk. Elsevier Applied Science Publishers, London.

113a. **Marugg, J. D., C. F. Gonzalez, B. S. Kunka, A. M. Ledeboer, M. J. Pucci, M. Y. Toonen, S. A. Walker, L. C. Zoetmulder, and P. A. Vanden Bergh.** 1992. Cloning, expression, and nucleotide sequence of genes involved in production of pediocin PA-1, a bacteriocin from *Pediococcus acidilactici* pAC1.0. *Appl. Environ. Microbiol.* **58:**2360–2367.

114. **Mayo, B., J. Kok, K. Venema, W. Bockelmann, M. Teuber, H. Reinke, and G. Venema.** 1991. Molecular cloning and sequence analysis of the *X*-prolyldipeptidyl aminopeptidase gene from *Lactococcus lactis* subsp. *cremoris*. *Appl. Environ. Microbiol.* **57:**38–44.

115. **McClure, W. R.** 1985. Mechanism and control of transcription initiation in prokaryotes. *Annu. Rev. Biochem.* **54:**171–204.

116. **McKay, L. L.** 1982. Regulation of lactose metabolism in dairy streptococci, p. 153–182. *In* R. Davies (ed.), *Developments in Food Microbiology*. Applied Science Publishers Ltd., Essex, England.

117. **McKay, L. L.** 1983. Functional properties of plasmids in lactic streptococci. *Antonie van Leeuwenhoek J. Microbiol.* **49:**259–274.

118. **Mercenier, A.** 1990. Molecular genetics of *Streptococcus thermophilus*. *FEMS Microbiol. Lett.* **87:**61–78.

119. **Meyer, J., and R. Jordi.** 1987. Purification and characterization of *X*-prolyldipeptidyl-aminopeptidase from *Lactobacillus lactis* and from *Streptococcus thermophilus*. *J. Dairy Sci.* **70:**738–745.

120. **Milliere, J. B., A. G. Mathot, P. Schmitt, and C. Divies.** 1989. Phenotypic characterization of *Leuconostoc* species. *J. Appl. Bacteriol.* **67:**529–542.

121. **Mills, O. E., and T. D. Thomas.** 1981. Nitrogen sources for growth of lactic streptococci in milk. *N.Z. J. Dairy Sci. Technol.* **15:**43–55.

122. **Mollet, B., and M. Delley.** 1990. Spontaneous deletion formation within the β-galactosidase gene of *Lactobacillus bulgaricus*. *J. Bacteriol.* **172:**5670–5676.

123. **Mollet, B., and M. Delley.** 1991. A β-galactosidase deletion mutant of *Lactobacillus bulgaricus* reverts to generate an active enzyme by internal DNA sequence duplication. *Mol. Gen. Genet.* **227:**17–21.

124. **Morse, M. L., K. L. Hill, J. B. Egan, and W. Hengstenberg.** 1968. Metabolism of lactose by *Staphylococcus aureus* and its genetic basis. *J. Bacteriol.* **95:**2270–2274.

125. **Mortvedt, C. I., and I. F. Nes.** 1990. Plasmid-associated bacteriocin production by a *Lactobacillus sake* strain. *J. Gen. Microbiol.* 136:1601–1608.

126. **Mortvedt, C. I., J. Nissen-Meyer, K. Sletten, and I. F. Nes.** 1991. Purification and aminio acid sequence of lactocin S, a bacteriocin produced by *Lactobacillus sake* L45. *Appl. Environ. Microbiol.* **57:**1829–1834.

127. **Muriana, P. M., and T. R. Klaenhammer.** 1991. Cloning, phenotypic expression, and DNA sequence of the gene for lactocin F, an antimicrobial peptide produced by *Lactobacillus* spp. *J. Bacteriol.* **173:**1779–1788.

128. **Nakamura, L. K.** 1981. *Lactobacillus amylovorus*, a new starch-hydrolyzing species from cattle waste-corn fermentations. *Int. J. Syst. Bacteriol.* **31:**56–63.

129. **Nakamura, L. K., and C. D. Crowell.** 1979. *Lactobacillus amylophilus*, a new starch-hydrolyzing species from swine waste-corn fermentation. *Dev. Ind. Microbiol.* **20:**531–540.

130. **Naouri, P., P. Chagnaud, A. Arnaud, and P. Galzy.** 1990. Purification and properties of a malolactic enzyme from *Leuconostoc oenos* ATCC 23278. *J. Basic Microbiol.* **30:**577–585.

131. **Nardi, M., M. C. Chopin, A. Chopin, M. M. Cals, and J. C. Gripon.** 1991. Cloning and DNA sequence analysis of an *X*-prolyldipeptidyl aminopeptidase gene from *Lactococcus lactis* subsp. *lactis* NCDO 763. *Appl. Environ. Microbiol.* **57:**45–50.

132. **Natori, Y., Y. Kano, and F. Imamoto.** 1988. Characterization and promoter selectivity of *Lactobacillus acidophilus* RNA polymerase. *Biochimie* **70:**1765–1774.

133. **Natori, Y., Y. Kano, and F. Imamoto.** 1990. Nucleotide sequences and genomic constitution of five tryptophan genes of *Lactobacillus casei*. *J. Biochem.* (Tokyo) **107:**248–255.

134. **Okereke, A., and T. J. Montville.** 1991. Bacteriocin inhibition of *Clostridium botulinum* spores by lactic acid bacteria. *J. Food Prot.* **54:**349–353.

135. **Oskouian, B., and G. C. Stewart.** 1990. Repression and catabolite repression of the lactose operon of *Staphylococcus aureus*. *J. Bacteriol.* **172:**3804–3812.

136. **Park, S. F., and G. S. A. B. Stewart.** 1990. High efficiency transformation of *Listeria monocytogenes* by electroporation of penicillin-treated cells. *Gene* **94:**129–132.

137. **Pilledge, C. J., and L. E. Pearce.** 1991. Expression of a β-galactosidase gene from *Clostridium acetobutylicum* in *Lactococcus lactis* subsp. *lactis*. *J. Appl. Microbiol.* **71:**78–85.

138. **Pinter, K., V. J. Davisson, and D. V. Santi.** 1988. Cloning, sequencing, and expression of the *Lactobacillus casei* thymidylate synthase gene. *DNA* **7:**235–241.

139. **Polzin, K. M., and L. L. McKay.** 1991. Identification, DNA sequence, and distribution of IS*981*, a new, high-copy-number insertion sequence in lactococci. *Appl. Environ. Microbiol.* **57:**734–743.

140. **Polzin, K. M., and M. Shimizu-Kadota.** 1987. Identification of a new insertion element, similar to gram-negative IS*26*, on the lactose plasmid of *Streptococcus lactis* ML3. *J. Bacteriol.* **169:**5481–5488.

141. **Poolman, B., T. J. Royer, S. E. Mainzer, and B. F. Schmidt.** 1989. Lactose transport system of *Streptococcus thermophilus*: a hybrid protein with homology to the melibiose carrier and enzyme III of phosphoenolpyruvate-dependent phosphotransferase systems. *J. Bacteriol.* **171:**244–253.

142. **Poolman, B., T. J. Royer, S. E. Mainzer, and B. F. Schmidt.** 1990. Carbohydrate utilization in *Streptococcus thermophilus*: characterization of the genes for aldolase-1-epimerase (mutarotase) and UDP-glucose 4-epimerase. *J. Bacteriol.* **172:**4037–4047.

143. **Porter, E. V., and B. M. Chassy.** 1988. Nucleotide sequence of the β-D-phosphogalactoside galactohydrolase gene of *Lactobacillus casei*: comparison to analogous *pbg* genes of other gram-positive organisms. *Gene* **62:**263–276.

144. **Posno, M., P. T. H. M. Heuvelmans, M. J. F. Van Giezen, B. C. Lokman, R. J. Leer, and P. H. Pouwels.** 1991. Complementation of the inability of *Lactobacillus* strains to utilize D-xylose with D-xylose catabolism-encoding genes of *Lactobacillus pentosus*. *Appl. Environ. Microbiol.* **57:**2764–2766.

145. **Posno, M., R. J. Leer, N. Vanluijk, M. J. F. Vangeizen, and P. T. H. M. Heuvelmans.** 1991. Incompatibility of *Lactobacillus* vectors with replicons derived from small cryptic *Lactobacillus* plasmids and segregational instability of the introduced vectors. *Appl. Environ. Microbiol.* **57:**1822–1828.

146. **Powell, I. B., M. G. Achen, A. J. Hillier, and B. E. Davidson.** 1988. A simple and rapid method for genetic transformation of lactic streptococci by electroporation. *Appl. Environ. Microbiol.* **54:**655–660.

147. **Rauch, P. J. G., M. M. Beerthuyzen, and W. M. De Vos.** 1990. Nucleotide sequence of IS904 from *Lactococcus lactis* subsp. *lactis* strain NIZO R5. *Nucleic Acids Res.* **18:**4253–4254.

148. **Raya, R. R., E. G. Kleeman, J. B. Luchansky, and T. R. Klaenhammer.** 1989. Characterization of the temperate bacteriophage PHI-ADH and plasmid transduction on *Lactobacillus acidophilus* ADH. *Appl. Environ. Microbiol.* **55:**2206–2213.

149. **Reizer, J., A. Reizer, and M. H. Saier.** 1990. The cellobiose permease of *Escherichia coli* consists of three proteins and is homologous to the lactose permease of *Staphylococcus aureus*. *Res. Microbiol.* **141:**1061–1067.
150. **Renault, P. C., C. Gaillardin, and H. Heslot.** 1989. Product of the *Lactococcus lactis* gene required for malolactic fermentation is homologous to a family of positive regulators. *J. Bacteriol.* **171:**3108–3114.
151. **Rixon, J. E., G. P. Hazlewood, and H. J. Gilbert.** 1990. Integration of an unstable plasmid into the chromosome of *Lactobacillus plantarum*. *FEMS Microbiol. Lett.* **71:**105–110.
152. **Romero, D. A., and T. R. Klaenhammer.** 1990. Characterization of insertion sequence IS*946*, and iso-ISS*1* element, isolated from the conjugative lactococcal plasmid pTR2030. *J. Bacteriol.* **172:**4151–4160.
152a.**Romero, D. A., and T. R. Klaenhammer.** 1992. IS*946*-mediated integration of heterologous DNA into the genome of *Lactococcus lactis* subsp. *lactis*. *Appl. Environ. Microbiol.* **58:**699–702.
153. **Ross, P., F. O'Gara, and S. Condon.** 1990. Cloning and characterization of the thymidylate synthase gene from *Lactococcus lactis* subsp. *lactis*. *Appl. Environ. Microbiol.* **56:**2156–2163.
154. **Scheirlinck, T., J. Mahillon, H. Joos, P. Dhaese, and F. Michiels.** 1989. Integration and expression of α-amylase and endoglucanase genes in the *Lactobacillus plantarum* chromosome. *Appl. Environ. Microbiol.* **55:**2130–2137.
155. **Schmidt, B. F., R. M. Adams, C. Requadt, S. Power, and S. E. Mainzer.** 1989. Expression and nucleotide sequence of the *Lactobacillus bulgaricus* β-galactosidase gene cloned in *Escherichia coli*. *J. Bacteriol.* **171:**625–635.
156. **Schroeder, C. J., C. Robert, G. Lenzen, L. L. McKay, and A. Mercenier.** 1991. Analysis of the *lacZ* sequences from 2 *Streptococcus thermophilus* strains—comparison with the *Escherichia coli* and *Lactobacillus bulgaricus* β-galactosidase sequences. *J. Gen. Microbiol.* **137:**369–380.
157. **Shearman, C., H. Underwood, K. Jury, and M. Gasson.** 1989. Cloning and DNA sequence analysis of a *Lactococcus* bacteriophage lysin gene. *Mol. Gen. Genet.* **218:**214–221.
158. **Shearman, C. A., S. Hertwig, M. Teuber, and M. J. Gsson.** 1991. Characterization of the prolate-headed lactococcal bacteriophage phi-vML3: location of the lysin gene and its DNA homology with other prolate-headed phages. *J. Gen. Microbiol.* **137:**1285–1292.
159. **Shimizu-Kadota, M.** 1987. Properties of lactose plasmid pYL101 in *Lactobacillus casei*. *Appl. Environ. Microbiol.* **53:**2987–2991.
160. **Shimizu-Kadota, M., J. L. Flickinger, and B. M. Chassy.** 1988. Evidence that *Lactobacillus casei* insertion element ISL*1* has a narrow host range. *J. Bacteriol.* **170:**4976–4978.
161. **Shimizu-Kadota, M., M. Kiwaki, H. Hirokawa, and N. Tsuchida.** 1985. ISL*1*: a new transposable element in *Lactobacillus casei*. *Mol. Gen. Genet.* **200:**193–198.
162. **Shimizu-Kadota, M., H. Shibahara-Sone, and H. Ishiwa.** 1991. Shuttle plasmid vectors for *Lactobacillus casei* and *Escherichia coli* with a minus origin. *Appl. Environ. Microbiol.* **57:**3292–3300.
163. **Sibakov, M., T. Koivula, A. Von Wright, and I. Palva.** 1991. Secretion of TEM β-lactamase with signal sequences isolated from the chromosome of *Lactococcus lactis* subsp. *lactis*. *Appl. Environ. Microbiol.* **57:**341–348.
164. **Simon, D., and A. Chopin.** 1988. Construction of a vector plasmid family and its use for molecular cloning in *Streptococcus lactis*. *Biochimie* **70:**559–566.
165. **Simons, G. H., M. van Asseldonk, G. Rutten, A. Braks, M. Nijhuis, M. Hornes, and W. M. de Vos.** 1990. Analysis of secretion signals of lactococci. *FEMS Microbiol. Rev.* **87:**24.
166. **Slos, P., J. C. Bourquin, Y. Lemoine, and A. Mercenier.** 1991. Isolation and characterization of chromosomal promoters of *Streptococcus salivarius* subsp. *thermophilus*. *Appl. Environ. Microbiol.* **57:**1333–1339.
167. **Smid, E. J., R. Plapp, and W. N. Konings.** 1989. Peptide uptake is essential for growth of *Lactococcus lactis* on the milk protein casein. *J. Bacteriol.* **171:**6135–6140.
168. **Sneath, P. H. A., N. S. Mair, M. E. Sharpe, and J. G. Holt (ed.).** 1986. *Bergey's Manual of Systematic Bacteriology*, vol. 2. The Williams & Wilkins Co., Baltimore.
168a.**Somkuti, G. A., D. K. Y. Solaiman, and D. H. Steinberg.** 1992. Expression of *Streptomyces* sp. cholesterol oxidase in *Lactobacillus casei*. *Appl. Microbiol. Biotechnol.* **37:**330–334.
169. **Spelhaug, S. R., and S. K. Harlander.** 1989. Inhibition of foodborne bacterial pathogens by bacteriocins from *Lactococcus lactis* and *Pediococcus pentosaceus*. *J. Food Prot.* **52:**856–862.
170. **Stackebrandt, E., V. J. Fowler, and C. R. Woese.** 1983. A phylogenetic analysis of lactobacilli, *Pediococcus pentosaceus* and *Leuconostoc mesenteroides*. *Syst. Appl. Microbiol.* **4:**326–337.
171. **Steen, M. T., Y. J. Chung, and J. N. Hansen.** 1991. Characterization of the nisin gene as part of a polycistronic operon in the chromosome of *Lactococcus lactis* ATCC 11454. *Appl. Environ. Microbiol.* **57:**1181–1188.
172. **Swenson, J. M., R. R. Facklam, and C. Thornsberry.** 1990. Antimicrobial susceptibility of vancomycin-resistant *Leuconostoc*, *Pediococcus*, and *Lactobacillus* species. *Antimicrob. Agents Chemother.* **34:**543–549.
173. **Taguchi, H., and T. Ohta.** 1991. D-Lactate dehydrogenase is a member of the D-isomer-specific 2-hydroxyacid dehydrogenase family: cloning, sequencing and expression in *Escherichia coli* of the L-lactate dehydrogenase gene of *Lactobacillus plantarum*. *J. Biol. Chem.* **266:**12588–12594.
174. **Takiguchi, R., H. Hashiba, K. Aoyama, and S. Ishii.** 1989. Complete nucleotide sequence and characterization of a cryptic plasmid from *Lactobacillus helveticus* subsp. *jugurti*. *Appl. Environ. Microbiol.* **55:**1653–1655.
175. **Thompson, J.** 1987. Regulation of sugar transport and metabolism in lactic acid bacteria. *FEMS Microbiol. Rev.* **46:**221–231.
176. **Toy, J., and A. L. Bognar.** 1990. Cloning and expression of the gene encoding *Lactobacillus casei* folylpoly-gamma-glutamate synthetase in *Escherichia coli* and determination of its primary structure. *J. Biol. Chem.* **265:**2492–2499.
177. **Trevors, J. T., B. M. Chassy, W. J. Dower, and H. P. Blaschek.** 1992. Electrotransformation of bacteria by plasmid DNA, p. 265–290. *In* D. C. Chang et al. (ed.), *Guide to Electroporation and Electrofusion*. Academic Press, Inc., San Diego, Calif.
178. **Van Alen-Boerrighter, I. J., R. Baankreis, and W. M. De Vos.** 1991. Characterization and overexpression of the *Lactococcus lactis pepN* gene and localization of its product, aminopeptidase N. *Appl. Environ. Microbiol.* **57:**2555–2561.
179. **Van Asseldonk, M., G. Rutten, M. Oteman, R. J. Seizen, W. J. De Vos, and G. Simons.** 1990. Cloning of *usp45*, a gene encoding a secreted protein from *Lactococcus lactis* ssp. *lactis* MG1363. *Gene* **95:**155–160.
180. **Van Belkum, M. J., B. J. Hayema, R. E. Jeeninga, J. Kok, and G. Venema.** 1991. Organization and nucleotide sequences of two lactococcal bacteriocin operons. *Appl. Environ. Microbiol.* **57:**492–498.
181. **Van Belkum, M. J., J. Kok, G. Venema, H. Holo, I. F. Nes, W. N. Konings, and T. Abee.** 1991. The bacteriocin lactococcin A specifically increases permeability of lactococcal cytoplasmic membranes in a voltage-indepen-

dent, protein-mediated manner. *J. Bacteriol.* **173:**7934–7941.

182. **Van Boven, A., and W. N. Konings.** 1986. Uptake of peptides by microorganisms. *Neth. Milk Dairy J.* **40:** 117–127.
183. **Van de Guchte, M.** 1991. Heterologous gene expression in *Lactococcus lactis*. Ph.D. dissertation. Rijksuniversiteit, Groningen, The Netherlands.
184. **Van de Guchte, M., J. Kok, and G. Venema.** 1991. Distance-dependent translational coupling and interference in *Lactococcus lactis*. *Mol. Gen. Genet.* **227:**65–71.
185. **van der Lelie, D., J. M. B. M. van der Vossen, and G. Venema.** 1988. Effect of plasmid incompatibility on DNA transfer to *Streptococcus cremoris*. *Appl. Environ. Microbiol.* **54:**865–871.
186. **Vanderslice, P., W. Copeland, and J. Robertus.** 1986. Cloning and nucleotide sequence of wild type and a mutant histidine decarboxylase from *Lactobacillus* 30a. *J. Biol. Chem.* **261:**15186–15191.
187. **Van der Vosen, J. M. B. M., D. van der Lelie, and G. Venema.** 1987. Isolation and characterization of *Streptococcus cremoris* Wg2-specific promoters. *Appl. Environ. Microbiol.* **53:**2452–2457.
188. **Van Rooijen, R. J., and W. M. De Vos.** 1991. Molecular cloning, transcriptional analysis, and nucleotide sequence of *lacR*, a gene encoding the repressor of the lactose phosphotransferase system of *Lactococcus lactis*. *J. Biol. Chem.* **265:**18499–18503.
189. **Van Rooijen, R. J., S. Van Schalkwijk, and W. M. De Vos.** 1991. Molecular cloning, characterization, and nucleotide sequence of the tagatose-6-phosphate pathway gene cluster of the lactose operon of *Lactococcus lactis*. *J. Biol. Chem.* **266:**7176–7181.
190. **Von Heijne, G., and L. Abrahamsén.** 1989. Species specific variation in signal peptide design. Implication for protein secretion in foreign hosts. *FEBS Lett.* **244:** 439–446.
191. **Von Wright, A., S. Wessels, S. Tynkkynen, and M. Saarela.** 1990. Isolation of a replication region of a large lactococcal plasmid and use in cloning of a nisin resistance determinant. *Appl. Environ. Microbiol.* **56:**2029–2035.
192. **Vos, P., G. Simons, R. J. Siezen, and W. M. de Vos.** 1989. Primary structure and organization of the gene for a prokaryotic cell envelope-located serine protease. *J. Biol. Chem.* **264:**13579–13585.
193. **Weisburg, W. G., J. G. Tully, D. L. Rose, J. P. Petzel, H. Oyaizu, D. Yang, L. Mandelco, J. Sechrest, T. G. Lawrence, J. Van Etten, J. Maniloff, and C. R. Woese.** 1989. A phylogenetic classification of the mycoplasmas: basis 'for their classification. *J. Bacteriol.* **171:**6455–6467.
194. **Xu, F., L. E. Pearce, and P. L. Yu.** 1991. Genetic analysis of a lactococcal plasmid replicon. *Mol. Gen. Genet.* **227:**33–39.
195. **Yan, T.-R., K. N. S. Azuma, and K. Yamauchi.** 1987. Purification and characterization of a novel metalloendopeptidase from *Streptococcus cremoris* H61. *Eur. J. Biochem.* **163:**259–265.
196. **Yan, T.-R., N. Azuma, S. Kaminogawa, and K. Yamauchi.** 1987. Purification and characterization of a substrate-size-recognizing metalloendopeptidase from *Streptococcus cremoris* H61. *Appl. Environ. Microbiol.* **53:**2296–2302.
197. **Yang, D., and C. R. Woese.** 1989. Phylogenetic structure of the *Leuconostocs*—an interesting case of a rapidly evolving organism. *Syst. Appl. Microbiol.* **12:**145–149.

6. *Streptomyces*

K. F. CHATER and D. A. HOPWOOD

Phylogenetically, gram-positive bacteria can be divided into two major groups, with either a low or a high proportion of G+C in their DNAs. The low-G+C grouping includes, in addition to *Bacillus* spp., such genera as *Clostridium*, *Staphylococcus*, *Streptococcus*, *Lactococcus*, *Lactobacillus*, *Thermoactinomyces*, and *Mycoplasma*. The high-G+C group, which may loosely be called "actinomycetes," includes genera with diverse morphological characteristics (68). The simplest, such as members of the genera *Micrococcus*, *Arthrobacter*, and *Corynebacterium*, can reproduce by binary fission. Other organisms, such as some strains of *Mycobacterium* and *Nocardia*, exhibit a degree of filamentous growth followed by fragmentation. At the other extreme, seen in, for example, *Actinoplanes* and *Streptomyces* spp., filamentous growth is so extensive as to form a coherent branching network of hyphae (a mycelium), with dispersion taking place by the production of spores in morphologically specialized structures, the sporangia or aerial hyphae.

One actinomycete genus, *Streptomyces*, has become preeminent for genetic research (96). This preeminence reflects several attributes of the organisms: the ease of isolating them from soil (where streptomycetes are important in the breakdown of intractable organic debris through the action of hydrolytic exoenzymes [164]) and the convenience of cultivating them on simple defined media in the laboratory; the importance of streptomycetes to medicine, agriculture, and industry as the source of an astonishing diversity of secondary metabolites with antibiotic, growth-promoting, herbicidal, antihelminthic, and pharmacological activities; the fascinating morphological complexity of the colonies, which can be truly regarded as multicellular and as consisting of distinct "tissues" (32); and the availability of natural systems of genetic exchange that are readily harnessed for genetic analysis and more recently have been supplemented by versatile systems for recombinant DNA work (96).

Because important secondary metabolites are produced by different species and strains, there has always been a tendency for research to be done on diverse examples of the genus, so that in sharp contrast to the situation for *Bacillus* genetics, we know quite a lot about many different species but not nearly so much about a paradigm strain equivalent to *Bacillus subtilis* 168. Nevertheless, one strain, *Streptomyces coelicolor* A3(2), which makes at least four secondary metabolites, none known to be of direct applied value (thus allowing the organism to acquire "model" status), has by far the most extensively studied and manipulatable genetics of any streptomycete. A second species, *Streptomyces griseus*, owes its importance to three attributes: the most famous *Streptomyces* antibiotic, streptomycin, though no longer extensively used in medicine, was discovered in *S. griseus*; some isolates of *S. griseus* can sporulate abundantly in liquid culture (an ability not shown by most streptomycetes); and *S. griseus* produces and responds to a low-molecular-weight pheromone, A-factor, that is necessary for streptomycin production and sporulation. Even *S. griseus* research has, however, been spread over several different strains.

GENETIC MANIPULATION OF *STREPTOMYCES* SPP.

The inherent immobility of mycelial growth means that when hyphae converge, they may make permanent contact, a situation that should favor fusion and genetic exchange. Indeed, spores harvested from agar-grown mixed cultures of genetically marked derivatives of a common parental strain have almost always been found to contain recombinants. Particularly in *S. coelicolor* A3(2), this trait has been harnessed for genetic analysis, leading to a detailed chromosomal linkage map (99) that in turn has led to rational construction of strains with desired genotypes, the genetic classification of phenotypically similar mutants, and the elucidation of a combined genetic and physical map (126; see also chapter 35, this volume).

Although crosses within strain lineages are usually successful, those between strains originating from different wild isolates seldom yield recombinants. In principle, there might be several reasons for this failure: inhibition of one strain by another because of antibiotics or bacteriocins may prevent effective hyphal contact; differences in patterns of DNA modification and restriction may prevent interaction at the DNA level (restriction modification systems are widespread in the genus and include many enzymes familiar to molecular biologists, notably *Sal*I, *Sca*I, *Sst*I, *Sph*I, *Sst*II, *Stu*I, and *Sno*I); nucleotide sequence divergence interferes with recombination at the molecular level (such limited information as there is indicates that homologous genes of different *Streptomyces* species may diverge by up to 10 to 20% at the nucleotide sequence level [13]); and differences in gross genome organization, such as inversions, between isolates might lead to disadvantageous genome arrangements in recombinants (no examples of this are known).

The exchange of DNA by conjugation, when it has been studied adequately, depends entirely on the sex factor activity of plasmids (101). Most species contain plasmids in some form, although standard plasmid isolation techniques have given positive results in only about 20% of strains (101). By far the majority of

K. F. Chater and D. A. Hopwood • John Innes Institute, John Innes Centre, Norwich NR4 7UH, United Kingdom.

Streptomyces plasmids, including linear and integrating examples (see below), are self-transmissible and can act as fertility factors even when they are smaller than 10 kb (101). Streptomycetes can also act as plasmid recipients in matings with *Escherichia coli* carrying certain mobilizable plasmids (162) and will act as hosts for the replication of some promiscuous plasmids from gram-negative bacteria (69). Natural plasmid transfer can augment in vitro genetic manipulation since it can often bypass the need for artificial polyethylene glycol-stimulated introduction of in vitro constructs into protoplasts, which is the route usually followed in *Streptomyces* molecular genetics (96). Protoplast transformation suffers from the disadvantage that conditions for protoplast regeneration (and, to a much lesser extent, protoplast formation) tend to be somewhat strain specific, and in some strains [but not *S. coelicolor* A3(2)], protoplasting and regeneration may cause or select mutations (143). No well-accredited system of natural, competence-mediated transformation has been described for *Streptomyces* spp.

Transformation and transfection of *Streptomyces* protoplasts are often highly efficient, even with large constructs up to at least 40 kb: more than 10^7 transformants (or plaques, in the case of phage DNA) per μg of DNA may be obtained, making it fairly straightforward to construct libraries (96). The ease of introducing DNA constructs directly from *E. coli* varies, but some strains (including *S. coelicolor*) are much more transformable by DNA passaged through *E. coli* hosts lacking the DNA methylases encoded by the *dam*, *dcm*, and *hsp* genes (152). (In an interesting reciprocal situation, DNA from *Streptomyces cyaneus* could be cloned efficiently in *E. coli* only when an *mcrA mcrB* mutant [i.e., one lacking the ability to restrict DNA containing methylcytosine] was used as recipient [223].) Alternatively, for some purposes, it may be appropriate to use *Streptomyces* hosts that naturally (e.g., *Streptomyces lividans* [96], *Streptomyces ambofaciens* [184], *Streptomyces griseofuscus* [184]) or as a result of mutagenesis (e.g., *Streptomyces fradiae* [160] and *Streptomyces rimosus* [113]) lack restriction systems. With such hosts, bifunctional cosmid vectors able to replicate in *E. coli* and *Streptomyces* spp. (184) may be particularly useful, although these do not always give stable libraries.

Protoplasts also serve another purpose. Polyethylene glycol at a higher concentration than is usually used for transformation (i.e., 50% rather than 20% [wt/vol]) induces very efficient protoplast fusion. The resulting zygotes seem to contain complete genomes from both parents (94), in contrast to the partial zygotes that typify natural matings (93). Moreover, the process of regeneration probably involves multiple genome replications in an expanding protoplast before hyphae emerge, and during this process, there may be many successive recombination events (105). This tends to loosen linkage, hindering genetic mapping of distant markers but aiding that of close markers (102, 111). The tendency to randomize the genetic differences between two parents after protoplast fusion is a valuable attribute in industrial strain improvement, since it allows extreme diversity to be generated by recombination between divergent lines.

Naturally occurring generalized transduction has been unambiguously demonstrated for only two species: *Streptomyces venezuelae* and *S. fradiae* (31, 211). Extensive searches for transducing phages have been unsuccessful for *S. coelicolor* A3(2) (212). This situation is particularly unfortunate now that transposon mutagenesis is developing (see below), since such phages would provide a ready means of proving, by 100% cotransduction, that a mutant phenotype is caused by a transposon insertion and would make it easy to move transposon-induced mutations into different strains.

Derivatives of transposons and integrating phages that permit fusions of chromosomal transcription units to the reporter genes *xylE* (2, 20) or *luxAB* (201) now allow gene regulation to be studied in situ in the genome, thus complementing the use of a variety of plasmid and phage promoter-probe vectors to study the activities of promoters removed from their normal contexts (20, 116). Increasingly, DNA clones are being selected by means other than mutant complementation, such as hybridization to heterologous or synthetic probes (e.g., see reference 63). Such approaches often necessitate the use of cloned DNA to generate mutations so that the role of the DNA can be assessed. These gene disruptions and replacements can be achieved by a variety of methods (127) with plasmid vectors innately or conditionally unable to replicate in *Streptomyces* spp. or temperate phage vectors lacking their *attP* sites.

It is perhaps appropriate to draw attention to progress in developing systems for genetic manipulation in other actinomycetes. Cloning vectors and effective protoplast transformation are available for *Saccharopolyspora erythraea* (224), *Amycolatopsis mediterranea* (154), *Micromonospora olivasterospora* (80), and *Corynebacterium glutamicum*, and a plethora of advances have also been made recently in *Mycobacterium* spp. (207). A significant fraction of this progress has depended on approaches first used in *Streptomyces* spp.

THE GENOME

The Chromosome

All the essential genes of *S. coelicolor* A3(2) lie on a circular chromosome that is larger (~8 Mb) than those of *E. coli* and *B. subtilis* (~4.5 Mb); *S. ambofaciens* has a similar genome size (145). Details of the *S. coelicolor* genetic and physical maps are given in chapter 35 of this volume. From one species to another, the linkage maps appear to be broadly similar (40), though there is obviously heterogeneity in specialized genetic information such as that for antibiotic production. The chromosomes of various species include very long segments of nonrepetitive DNA that contain no genes essential for growth in laboratory conditions. This fact is shown by the occurrence of viable strains with very large deletions; the record appears to be 2 Mb for *S. ambofaciens* (144). So far, only relatively small deletions have been found in the very long "3 o'clock silent region" of the *S. coelicolor* A3(2) chromosome (64; see chapter 35). The deletion mutants of various species may show phenotypic changes, especially affecting aerial mycelium formation, pigment and antibiotic production, and resistance to antibiotics (115). One view of these deletable regions is that they have arisen through successive insertion events following the natural acquisition of "foreign" DNA. Once such accessory DNA occupied a part of the chromosome, it could have provided a nonessential region into which further insertions could be made without interfering with normal growth.

Deletions are often accompanied by the amplification of DNA close to the deletion. The amplified DNA sequences vary in unit length from a few to several tens of kilobases, and it is not unusual for the amplified sequence to make up at least 20 to 30% of the total DNA. The mechanism(s) of amplification has so far been difficult to analyze, although models have been proposed (57, 81, 115, 196, 236).

Chromosome replication and segregation in streptomycetes present some unusual features. During hyphal growth, individual cellular compartments contain many copies of the chromosome. Close to the growing hyphal tip, replication and segregation give rise to separate single nucleoids, but in older hyphal regions, the chromosomes appear to replicate without observable segregation (136). During branch formation, which is always from such older compartments, segregation of one compound nucleoid seems to occur, and the progeny of this nucleoid populate the new branch (135). A third pattern of segregation is seen during sporulation septation of aerial hyphae (see below). The analysis of all these situations will be helped by recent successes in isolating origins of replication of the *S. lividans* and *S. coelicolor* chromosomes (29, 237). Many of the features of these regions closely resemble those of other bacterial *oriC* regions.

Plasmids

Representative plasmids of several kinds have been analyzed (97, 100, 101, 115). Studies of small, high-copy-number, covalently closed circular plasmids (up to about 10 kb) have shown a typical organization that includes a presumptive replicase gene next to a replication origin, a gene for primary plasmid transfer (in remarkable contrast to the 20 to 30 kb needed for transfer of plasmids among gram-negative bacteria), and two or more genes that appear to mediate migration ("spread") of plasmid copies into a hyphal network following primary transfer (121–123, 206). The expression of these last genes (*spd*) causes partial inhibition of the hyphae receiving the plasmid and can be lethal if uncontrolled by *kor* (kill-override) genes. The transient inhibition associated with plasmid spread gives rise to visible "pocks" surrounding donor colonies growing in a lawn of a recipient strain. Where studied, replication of small, high-copy-number plasmids seems to involve the production of a single-stranded DNA intermediate that accumulates to a significant extent in many of the vectors derived from such plasmids because of the elimination of a site (Sti) needed for efficient second-strand synthesis (52). Derivatives of small plasmids of this kind have found many applications as vectors (96, 97). One naturally temperature-sensitive replicon (pGM5 [173]) and one induced temperature-sensitive mutant plasmid (pMT660 [12]), but especially the former, have extra benefits as "suicide" vectors in gene replacement, gene disruption, and transposon delivery. Most of the plasmids can be introduced into and maintained in a wide variety of *Streptomyces* hosts. Interestingly, pIJ101-based plasmids rarely appear to be inherited through spores in *Saccharopolyspora erythraea*, thereby providing a useful suicide system (225).

The larger, low-copy-number plasmid SCP2 (30 kb) has been widely and successfully used to generate cloning vectors with a high capacity (96, 97). All of these are based on a mutant form, SCP2*, with enhanced fertility and pock-forming properties. Studies of SCP2 molecular biology are only now beginning.

Plasmids of an unusual class, apparently widespread among actinomycetes, are those that typically occupy a chromosomal location but are capable of excision, transfer, and inheritance when their host encounters a strain lacking the plasmid. Nearly all known examples first became apparent after contact of the donor strain with *S. lividans*, leading to the formation of pocks in which the autonomous form of the plasmid is present in the recipient. In general, these plasmids are intermediate in size (10 to 15 kb) and copy number (ca. 5) and exhibit transfer, spread, and fertility properties like those of other *Streptomyces* plasmids. In all examples sufficiently analyzed, attachment sites of the plasmid and chromosome have about 50 to 112 nucleotides of common sequence, which overlaps one end of a tRNA gene (3, 17, 163, 185, 204). Different tRNA genes are used by different plasmids. This situation has been found for some *att* sites of temperate phages of gram-negative bacteria (185).

More unusual still are the linear plasmids. The first to be described, pSLA2, is 17 kb long and has terminal inverted repeats 614 bp long with protein molecules bound to the 5′ ends (87). A 12-kb linear plasmid of *Streptomyces clavuligerus* has been completely sequenced; it has terminal inverted repeats 969 bp long that are homologous to those of pSLA2 (233). Much larger linear plasmids are very widespread (128). In *S. coelicolor*, SCP1, one of the few *Streptomyces* plasmids encoding a recognizable phenotype (the production of methylenomycin), is a "giant" linear plasmid of some 350 kb with proteins bound to the 5′ ends of its 80-kb terminal inverted repeats (129). SCP1 can integrate into the *S. coelicolor* chromosome (103, 130), causing changes in the fertility pattern (as, for example, in so-called NF strains, which donate chromosomal markers on either side of the SCP1 integration site at high frequency to SCP1$^-$ recipients).

Phages

Streptomyces phages are readily isolated from soil, and several (mostly temperate) have been characterized in some detail and exploited for genetic manipulation (31). They have an overall resemblance to many phages of other bacteria. Thus, most have similar morphology, including a more or less simple tail and a polyhedral head. Some package their DNA by a head-filling mechanism, and others package it by site-specific cleavage of DNA concatemers. Some lysogenize by *att* site-mediated chromosomal integration, and others form plasmid prophages. The most-studied example, ϕC31, superficially resembles coliphage λ: it has a repressor gene, *att* site, integration and excision genes, and cohesive DNA ends. However, there are also extensive differences from λ. For example, the repressor gene is transcribed and translated from three different start points to give three differently sized proteins, none of which closely resembles known phage repressors (200); early transcription appears to be from structurally unusual promoters distributed throughout the early region (200); *attB* and *attP* sites have a common core of only 3 nucleotides (138); the integrase is different from known integrases (138); and the *cos* ends contain 10-nt 3′ (rather than 5′, as with λ) single-stranded extensions (132).

Although ϕC31 has been the phage most used for vector development (20, 97), other phages (TG1 [65], SAt1 [178], and R4 [215]) have also been manipulated as vectors. An interesting example of a specialized use of a phage is provided by the very wide host range virus FP43 (76), which packages DNA in headfuls rather than by site-specific cleavage and can be made to transduce DNA of plasmids containing a specific segment of its DNA (168). Plasmids carrying the *cos* sites of phages R4 or ϕC31 can also be transduced by the respective phages (132, 172). In addition, the *att-int* region of ϕC31 has been exploited in the construction of integrating plasmid vectors (139, 150).

In addition to a variety of interactions of *Streptomyces* phages with classical host restriction modification systems, interaction of ϕC31 with *S. coelicolor* has revealed a host resistance system that seems particularly adapted to a multicellular organism. The wild-type host is Pgl$^+$ (phage growth limitation), and when infected by ϕC31, hyphal compartments undergo a normal infection cycle and are lysed, but the phages released are dramatically impaired in their ability to lyse neighboring cells (43). Pgl is an unstable phenotype: mutations between Pgl$^+$ and Pgl$^-$ occur at a frequency of about 10^{-3} per spore (43).

Movable Genetic Elements

In addition to various integrating plasmids and phages, elements more closely resembling "classical" transposons and insertion sequences abound in *Streptomyces* spp. (39). One representative of the ubiquitous Tn*3*-like class of transposons, Tn*4556*, occurs in an *S. fradiae* strain (44); it has been completely sequenced (195) and extensively manipulated (201). Derivatives of Tn*4556* have found use in the mutagenesis of cloned DNA (45, 48), and more recent experiments have begun to justify the hope that these derivatives could be effective in generalized transposon mutagenesis (189, 201, 234). Another transposon for such purposes was engineered from an insertion sequence, IS*493* from *S. lividans*, by the introduction of selectable resistance markers between the two open reading frames to give Tn*5096* (202). Derivatives of Tn*4556* and IS*493* that allow reporter genes such as *luxAB* and *xylE* to be expressed from adjacent host promoters have been constructed (2, 201).

IS*493* of *S. lividans* is related to IS*112* of *Streptomyces albus* G. Other IS*112*-related sequences are also present in *S. lividans* and in many other *Streptomyces* spp. (187). Another family of insertion sequence elements (including IS*110* of *S. coelicolor*, IS*116* of *S. clavuligerus*, and IS*900* and IS*901* of *Mycobacterium* spp.) is also very widespread among actinomycetes, and related elements have been found in other organisms as distant as *Pseudomonas atlantica* (IS*492*) and *Thermus thermophilus* (IS*1000*) (72, 141, 149). Members of the latter class of elements differ from classical insertion sequence elements in lacking terminal inverted repeats and in failing to generate target site duplications on integration. They usually show some preference for insertion at sites with particular sequence features (72).

S. coelicolor contains two chromosomally integrated copies of an apparently different and somewhat larger (2.55-kb) mobile element, IS*117*, which also occurs as a low-abundance free circular form (the "minicircle" [151]), believed to be a transposition intermediate. When propagated on a convenient vector (usually an *E. coli* plasmid) and introduced into IS*117*-free species (such as *S. lividans*), IS*117* integrates efficiently into a preferred attachment site. This site shows only slight homology with the *att* site of the element itself beyond a 3-base common core sequence at which crossing over takes place without the generation of a target site duplication (85). In an *S. lividans* host that lacks a primary chromosomal *att* site, lower-frequency IS*117* integration at secondary sites occurs (84). IS*117* has been manipulated as a vector for the stable integration of DNA into the chromosomes of a wide variety of actinomycetes (127). Interestingly, the largest open reading frame in IS*117* is related to that of members of the IS*110* family described above (85).

GENE EXPRESSION

Transcription

The RNA polymerase of *Streptomyces* spp. conforms to the prokaryotic stereotype. It is generally sensitive to the same antibiotics and contains β, β', and two α subunits supplemented by various σ factors (22, 119, 227): the RNA polymerase complex may indeed contain as many as 25 different polypeptides (58). It can utilize promoters of more than one class from various gram-negative and gram-positive bacteria (8, 118, 227), providing genetic evidence that some of these *Streptomyces* σ factors have functional homologs in other bacteria.

Many *Streptomyces* genes are transcribed from two or more promoters (210). A particularly striking case is provided by the four promoters of the *S. coelicolor* *dagA* (agarase) gene, which are distributed over a region of about 250 bp (24). Each promoter is transcribed in vitro by an RNA polymerase containing a different σ factor (26). Combined biochemical and genetic analyses have so far revealed at least eight σ factors in *S. coelicolor* (22). The essential major vegetative σ factor, σ^{66}, specified by *hrdB* (19), is very similar to σ^{70} of *E. coli* and σ^{A} of *B. subtilis* (194). Remarkably, *S. coelicolor* contains three other genes, *hrdA*, *hrdC*, and *hrdD*, that are predicted from sequencing to specify homologs of the principle σ factor (217, 218). A similar situation is found in many *Streptomyces* spp. Disruption of *hrdA*, *hrdC*, and *hrdD*, either separately or in a single genome, led to no observable phenotypic effects in normal laboratory conditions, whereas disruption of *hrdB* appeared to be lethal (23, 25). Thus, *hrdA*, *hrdC*, and *hrdD* cannot adequately substitute for *hrdB*. In *Streptomyces aureofaciens*, *hrdC* is replaced by a fifth *hrd* gene, *hrdE*, which is more similar to *hrdB* (134).

A representative target for σ^{66} is the p4 promoter of *dagA*, which, predictably, resembles the consensus sequence for the major class of bacterial promoters at −10 and −35 (19). Compilations of *Streptomyces* promoters reveal that while some conform to this consensus, the rest display great diversity (210). This suggests that several more σ factors remain to be discovered. Some of this complexity may well turn out to be associated with the relatively complex differentiation of *Streptomyces* colonies.

Many *Streptomyces* genes are followed by inverted repeat sequences presumed to act as terminators, though only a few of these have been shown to cause termination (51, 175). There is no apparent obligatory requirement for such hairpin loop terminators to be followed by a run of U residues, and no termination factors have been described.

Translation of mRNA

Although translation initiation in *Streptomyces* spp. follows the same general rules as in other eubacteria, there are several exceptions and qualifications (210). Untranslated leader sequences are often comparatively long. Conventional ribosome-binding sites, though usually present, are often less extensively complementary to sequences near the 3′ end of 16S rRNA than is the case for gram-positive bacteria with low-G+C DNA (169). GUG is often used instead of AUG as an initiation codon, perhaps correlating with the high average G+C content of genomic DNA (nevertheless, a few examples of putative UUG start codons have been described [60, 149]). It is not uncommon (especially in transcripts encoding antibiotic resistance) to find no leader at all, translation being initiated at an AUG or GUG codon at the 5′ end of the mRNA (120, 210). There is some evidence that the efficiency of translation initiation may be affected by the nature of the first few codons; changes at these positions from rare codons to synonymous codons common in *Streptomyces* genes significantly increased expression of the *Vibrio harveyi luxAB* genes in *S. coelicolor* (18). In a detailed analysis of an opposite situation, *Streptomyces* genes were more strongly expressed in *E. coli* when those of their first few codons ending in G or C were changed to synonyms with A or T (71). There is little information on the translation initiation factors of *Streptomyces* spp.

Over long windows, codons in *Streptomyces* genes often have ca. 70% G+C in position 1, ca. 50% in position 2, and ca. 90% in position 3, a distribution consistent with the high G+C content of the DNA and permitting the relatively easy recognition of open reading frames in DNA sequences (9). Some codons (notably CTG, TTT, and TTA) are very rare (232), and there is evidence that TTA is absent from all genes needed for vegetative growth (see below). There is little information about the synthesis of functional charged tRNA species or their delivery to the ribosomes beyond the observation in hybridization and DNA-sequencing procedures that tRNA genes are not part of operons encoding rRNA and that many of them do not determine the 3′ end (CCA) presumed to be present in the mature tRNA (182, 190). Translation termination is usually at a UGA or, less often, a UAG triplet, with UAA being very rare as a stop codon (232).

IMPORTANCE OF EXTRACELLULAR ENZYMES AND OTHER PROTEINS IN THE LIFE AND BIOTECHNOLOGICAL EXPLOITATION OF *STREPTOMYCES* SPP.

Streptomycetes are particularly well adapted to growth on such recalcitrant organic macromolecules as cellulose or lignin (164), an ability that is potentially highly relevant to the exploitation of tree and other plant residues by humans. Hence, a substantial interest in the molecular biology of streptomycete cellulases, xylanases, and ligninases has developed (e.g., see reference 220). The mycelial growth habits of streptomycetes coupled with their production of suitable exoenzymes enables them to colonize and penetrate plant residues in the soil.

Most of the proteins secreted by streptomycetes have relatively long N-terminal signal peptides that are removed during secretion (15, 115) (tyrosinase is an exception, having no clear signal sequence). The mechanisms of secretion in streptomycetes have not been investigated, a glaring omission, since the process is so important to their life styles. A novel aspect of the extracellular activities of streptomycetes is the occurrence of proteins that specifically inhibit certain hydrolytic enzymes. For example, several species produce proteinaceous inhibitors of proteases (209, 214) and α-amylases (133, 174). The roles of these proteins in the life of the organisms are not obvious. It is also unclear why so many streptomycetes secrete β-lactamases. The genes for these enzymes and regulation of the genes have been popular objects of *Streptomyces* genetic analysis (66, 219).

This ability to produce a plethora of extracellular enzymes has led to the investigation of streptomycetes as expression-secretion hosts for foreign proteins (15). An important aspect of this technology is the avoidance of proteolysis, so extracellular proteases (which are widespread among streptomycetes) have also become a focus of attention (e.g., see reference 86).

PRIMARY METABOLISM

Carbon Metabolism

In addition to the highly substrate-inducible extracellular hydrolytic enzymes (in most cases, the actual inducer has not been identified but is likely to be a breakdown product of the macromolecule that can enter the cells), streptomycetes are also well able to use more readily available material, such as simple sugars. Some such carbon sources, notably glucose, repress transcription of many exoenzymes (such as α-amylase [221]) and chitinase [50]), implying that these soluble carbon sources are preferred. Glucose is also often preferred to other soluble carbon sources, as shown by glucose repression of the operons for glycerol and galactose utilization (1, 197). The *gyl* and *gal* operons are highly inducible by the relevant substrates, with different transcription levels resulting from different combinations of inducer and repressing glucose. Interestingly, enzymes specified by the *gyl* operon are repressed to intermediate levels by various other alternative carbon sources (192).

Uptake of carbon sources is not easy to study in filamentous organisms because of the difficulty of preparing physiologically uniform suspensions of hyphae, but several general conclusions have been drawn (90). Constitutive transport systems often have comparatively low substrate affinity, whereas inducible systems have high affinity (in the micromolar range). Phosphoenolpyruvate-dependent phosphotransferase systems (188) have not been found (176). Instead, it seems likely that transport is coupled to phosphorylation of the transported molecule by cytoplasmic kinases. This immediately implies that the mechanisms by which available glucose is "perceived," leading to glucose repression, must differ from that in *E. coli*, in which the phosphotransferase system fulfills this role (188). In *Streptomyces* spp., it seems that the activity of glucose kinase may bring about glucose repression of several operons, since glucose kinase deficiency leads to the release of the operons from such repression (88). The role of glucose kinase in this repression may be indirect (through influences on glucose transport or on the production of glucose 6-phosphate or its further metabolites), but it may play a more direct role; there is extensive homology between it and the regulatory proteins specified by *nagC* in *E. coli* and *xylR* in *B. subtilis* (1a). In *Streptomyces* spp., it also seems that cyclic AMP plays no role in glucose repression, because there is no systematic relationship between intracellular cyclic-AMP levels and glucose repression, and exogenous cyclic-AMP derivatives do not relieve catabolite repression (88).

Glucose is metabolized in *S. coelicolor* through glycolysis, the hexose monophosphate pathway, and the Krebs cycle; many of the relevant enzymes have been purified, and their genetic determinants have been cloned (53, 112). Alternative carbon sources, such as glycerol or galactose, enter these central pathways by the normal enzymatic steps, which are often encoded by polycistronic operons. The *gyl* operon consists of an upstream regulatory gene (*gylR*), a dual promoter region, and a transcription unit in which the promoter-distal genes show some evidence of regulation by control of termination within the operon (198). In the *gal* operon, no regulatory gene has been identified. There is differential regulation of the second two genes in the *gal* operon, because they are preceded by an apparently constitutive promoter utilized by a form of RNA polymerase holoenzyme different from that active on the promoter of the whole *gal* operon (226).

S. coelicolor does not utilize lactose very efficiently, and its β-galactosidase is a very large extracellular enzyme (21). This, together with the apparent failure to take up 5-bromo-4-chloro-3-indolyl-β-D-galactopyranoside (X-Gal) and poor structural stability and expression of the *E. coli lacZ* gene in *Streptomyces* spp. (131), has interfered with the use of *lacZ* as a reporter gene for *Streptomyces* spp., an unfortunate circumstance for *Streptomyces* genetics!

Amino Acid Metabolism and Biosynthetic Pathways

Few combined physiological and genetic approaches to amino acid biosynthesis have been reported for *Streptomyces* spp. A notable exception involves glutamine synthetase (GS), which incorporates ammonia into glutamate to produce glutamine and is a well-known regulatory focal point in bacteria (62). Streptomycetes have a typical prokaryotic GS (GSI) whose activity is regulated at the transcription level by an unlinked positively acting gene (*glnR*) (62) and posttranslationally by adenylylation (62) (a property shared with *E. coli* GS but not with *B. subtilis* GS). DNA sequence and Southern analyses (5, 140) also suggest that many streptomycetes have a second GS (GSII), of a type found in nitrogen-fixing prokaryotes as well as in eukaryotes.

In many bacteria, genes for amino acid biosynthesis and degradation are highly regulated at the transcription level by the relevant amino acid. This appears to be true only seldom for streptomycetes (92). It has been suggested that the kind of regulation of amino acid pools depends on whether the amino acid in question is a precursor of a secondary metabolite (92). Thus, in *S. coelicolor*, in which proline is a precursor of the red-pigmented antibiotic undecylprodigiosin, examples of proline anabolic and catabolic genes are not regulated by externally added proline, which is incorporated into the antibiotic; in contrast, the genes for tryptophan biosynthesis are more clearly regulated, though apparently by growth rate rather than by a tryptophan-sensitive regulatory system (92). Histidase levels are strongly inducible by histidine (137), whereas the histidine biosynthetic enzymes are relatively slightly repressible by histidine (92); thus, the histidine pool may be principally regulated by control of catabolism. The histidase of *S. griseus* has unusual growth phase-dependent posttranslational regulation, being reversibly inactivated at the initiation of sporulation (137).

SECONDARY METABOLISM

The Switch between Primary and Secondary Metabolism

Most *Streptomyces* antibiotics are not produced during the period of most active vegetative growth of the colony. Instead, they typically appear as the growth rate slows. In colonies growing on a solid surface, this slowdown occurs as the aerial hyphae begin to develop from the substrate mycelium, whereas in most liquid-grown cultures, it takes place at a "transition stage" as biomass production changes from the quasiexponential toward the stationary phase. It has been argued (42) that such timing of antibiotic production is adaptive in helping to defeat marauding microorganisms that could otherwise steal the nutrients released by lysis of the substrate mycelium, which are properly destined to fuel development of the reproductive phase of the colony. The genetic and physiological factors responsible for the switch between primary and secondary metabolism are still largely obscure, but recent developments in genetic knowledge and techniques at least offer good prospects for elucidating some of the key regulatory circuits in the not-too-distant future. Two kinds of approaches, only now converging, are being used to throw light on the switch mechanism.

On the one hand, global physiological controls, some also implicated in the regulation of primary metabolism (see above), have been considered as potential switching devices. Certainly there is much evidence that either specific carbon or nitrogen sources or P_i at concentrations above particular threshold values can prevent the onset of antibiotic production (157). More recently, the level of ppGpp has been proposed as a key element (177), but work with both *S. coelicolor* (208) and *S. clavuligerus* (4) suggests that increased ppGpp per se may be insufficient, and a reduced growth rate may also be necessary.

The second approach has been to seek mutants pleiotropically defective in the production of more than one antibiotic in the same organism, or DNA segments that exert pleiotropic effects, stimulatory or depressive, on antibiotic production when they are present as extra, cloned copies. In *S. coelicolor* A3(2), this search has led to the identification of the *abs*, *aba*, and *afs* genes and the *mia* sequence, which affect only secondary metabolism, and the various *bld* genes, which are needed for production of both aerial mycelium and one or more antibiotics (see below). The very existence of genes of these two classes strongly indicates that regulatory elements act at levels "above" that of an individual biosynthetic pathway, both at a level of secondary metabolism separate from aerial mycelial development and at a level including both processes (95).

The sequences and the likely natures of the gene products for several of the genes that affect only secondary metabolism have been determined. However, many questions have yet to be addressed, such as where and when their products are made, whether they are regulated posttranscriptionally, what signal transduction pathways they take part in, whether they interact with each other, and how they interface with particular sets of genes for secondary metabolism and associated pathway-specific regulatory genes. Although several researchers have presented hypothetical regulatory hierarchies involving such genes (30, 34, 95, 106), we will confine our account to a brief review of information about the genes themselves.

Two genes, *absA* and *absB*, were discovered by analyzing morphologically normal mutants defective in production of both pigmented antibiotics of *S. coelicolor* (30). The *afsB* locus was identified by studies of *S. coelicolor* mutants that fail to make functional homologs of A-factor (see below); the mutants were subsequently divided into two classes, with *afsA* being defective only in A-factor production and *afsB* mutants proving to be generally defective in secondary metabolism (78). The *afsA* gene of *S. griseus* has been characterized (110), but there is no molecular information about *afsB*, *absA*, or *absB*. The *afsR* gene was identified by its ability at high copy number to restore antibiotic production to *afsB* mutants (it was originally believed to be the wild-type allele of *afsB* itself [108, 205]). This is a property surprisingly retained by each of two subdomains of the large AfsR protein, one suggested to include a DNA-binding region (109). AfsR is also a target for phosphorylation by a protein kinase specified by *afsK*, a gene closely linked to *afsR* (107). The *afsR-afsK* gene products are not overtly similar to typical prokaryotic protein kinase/response regulator pairs. However, the $afsQ_1$ and $afsQ_2$ genes, a contiguous gene pair cloned by an ability at multiple copy number to provoke actinorhodin overproduction in *S. lividans*, encode putative products of this sort (117). Their significance in *S. coelicolor* is unclear, because disruption of neither $afsQ_1$ nor $afsQ_2$ affected antibiotic production under the conditions studied (117). The *abaA* locus was cloned in a similar way and found to contain several open reading frames (61), whose sequences have not given clues to their function. At least one of the open reading frames is essential for production of three of the *S. coelicolor* antibiotics. The *mia* sequence, in contrast to the *afs* and *aba* sequences, depresses antibiotic production when cloned at high copy number in *S. coelicolor* (30).

The existence of morphologically normal mutants that make no antibiotics shows that (at least in the conditions tested) antibiotics are generally not instrumental in morphological differentiation. Nevertheless, the fact that most *bld* mutants are also deficient in antibiotic synthesis reveals genetically determined components common to both processes. The extent to which these are truly regulatory components is the focus of much current research. Among the perhaps 8 to 10 known *bld* genes of *S. coelicolor*, only 2 have been sequenced. The *bldB* gene product is a small protein with no sequence features giving strong clues to its function (238). On the other hand, the primary function of the *bldA* gene product is clear: it is the tRNA for the rare leucine codon UUA (142, 148). Since *bldA* mutants (including a constructed deletion mutant [146]) grow apparently normally vegetatively, it seems that TTA codons are probably absent from all genes needed during growth, a deduction supported by published DNA sequences. On the other hand, regulatory and resistance genes associated with secondary metabolism in *S. coelicolor* and other *Streptomyces* species often contain TTA codons, giving rise to the plausible notion that the TTA codon provides a means of regulating secondary metabolism and, through presumed TTA-containing *bld* genes, growth of aerial mycelium (59, 147). This attractive idea has so far proved difficult to substantiate because the tRNA level seems to be strongly growth-phase dependent in some experimental conditions (142, 146) but not in others (70). Perhaps *bldA* exerts a significant regulatory role only in certain conditions and other regulatory elements are more important in different circumstances.

Mutations in the *bldA* homolog of *S. griseus* generate a phenotype similar to that of *S. coelicolor bldA* mutants (166), strengthening the view that the observed distribution of TTA codons is more than a statistical freak. In addition, the regulation of secondary metabolism and sporulation in some *S. griseus* strains exhibits an aspect apparently absent from *S. coelicolor*: the requirement for an extracellular autoregulatory molecule, A-factor (124). (*S. coelicolor* makes an A-factor-like substance, but no role for it in *S. coelicolor* has yet been found [78].) A-factor, a γ-butyrolactone (6), accumulates in *S. griseus* cultures and probably diffuses freely into and out of the cells. At concentrations in the nanomolar range, it associates with a cytoplasmic A-factor-binding protein, causing this protein to release its repression of secondary metabolism and differentiation (107). The mechanism of repression is uncertain, although at least one further regulatory step seems to be necessary before genes specific for streptomycin biosynthesis can be activated (222). Other streptomycetes use similar autoregulators. Thus, an A-factor-like *Streptomyces virginiae* butyrolactone binds to a protein homologous to the *E. coli* NusG protein, directly or indirectly causing virginiamycin biosynthesis to be switched on (180).

Clusters of Antibiotic Biosynthetic Genes

Elucidation of the organization of the sets of genes for biosynthesis of several diverse antibiotics has been one of the successes of *Streptomyces* genetics (158, 191). Strikingly, the genes for each individual antibiotic are, without well-established exceptions, clustered together in a series of contiguous operons typically occupying from ca. 15 kb to as much as 100 kb. The clusters usually also include pathway-specific regulatory genes and one or more genes for resistance to the organism's own antibiotic (35). Where critical evidence exists, all these gene sets are chromosomally located except for the cluster of methylenomycin biosynthetic genes, which are on the SCP1 plasmid in *S. coelicolor* and the pSV1 plasmid of *S. violaceoruber*.

At least two of the clusters, those for production of actinorhodin by *S. coelicolor* (28, 59, 60, 77, 155, 159) and tetracenomycin by *Streptomyces glaucescens* (7, 75, 114, 213), have been completely sequenced. This information, combined with transcriptional analysis, has revealed that both clusters contain a series of convergent and divergent operons. Similar conclusions can be drawn from studies of other sets of biosynthetic genes for which molecular genetic analysis is also far advanced, such as those for oxytetracycline in *S. rimosus* (167); tylosin in *S. fradiae* (86a); erythromycin in *Saccharopolyspora erythraea* (82); bialaphos in *Streptomyces hygroscopicus* (91); streptomycin and hydroxystreptomycin in *S. griseus* and *S. glaucescens*, respectively (55); avermectin in *S. avermitilis* (153); and undecylprodigiosin (156) and methylenomycin (37, 175) in *S. coelicolor*.

Clustering could have obvious significance in an evolutionary context: a "package" of structural, regulatory, and resistance genes would be an effective entity for horizontal gene transfer, whereas a partial set of biosynthetic genes or a complete set without a self-protective gene would usually provide little selective advantage or actually be deleterious (although recruitment of extra genes to extend a pathway or cause it to branch could be beneficial in some cases). Alternatively, or in addition, the arrangement of genes in a cluster probably allows significant *in cis* regulatory mechanisms to operate. Several such possibilities have been discussed in the literature, although none is yet established. One concerns the possible coordinate or sequential expression of resistance and biosynthetic genes, a possibility based on the argument that obligate expression of a resistance gene before antibiotic production itself could occur would provide a fail-safe protection against self-poisoning (41, 54). In several cases, a resistance gene is transcribed from a region containing divergent promoters (10, 11, 27, 54, 175); this might facilitate regulatory interplay between the resistance gene and the divergent gene. Another *cis*-acting regulatory mechanism was suggested by the discovery of a resemblance between the StrR regulatory gene product (which regulates streptomycin biosynthesis) and the λ anti-terminator protein Q; only after *strR* is expressed could transcription perhaps proceed into downstream genes (54).

Clustering of functionally related antibiotic biosynthetic genes could also bear on the efficiency of their cotranslation. Sequencing studies have revealed many examples of "translational coupling" between pairs of biosynthetic genes, with the stop codon of the upstream gene overlapping the start codon of the downstream gene, a device believed to favor stoichiometric production of the corresponding proteins (67, 235). In at least one case, involving two genes in the actinorhodin polyketide synthase gene cluster of *S. coelicolor*, experimental uncoupling of two coupled genes did in fact result in inefficient expression of one or more downstream genes (125).

Biochemical Genetics of Antibiotic Biosynthesis

Major problems in the analysis of antibiotic biosynthetic pathways by traditional biochemical and chemical procedures have been caused by the low concentrations of the pathway enzymes, their lability in cell-free preparations, and the chemical instability of many highly reactive intermediates of the pathways. As a result, the pathways even of important antibiotics that have been industrial products for decades, such as β-lactams, aminoglycosides, and macrolides, have until recently remained poorly characterized. A wealth of new information is now coming from molecular genetic analysis. Notable examples are provided by the chemical class of natural products called polyketides, which includes macrolides, tetracyclines, anthracyclines, avermectins, and many others. A mechanistic correspondence between their biosynthesis and that of fatty acids has long been postulated (179) and has recently been shown to reflect a clear evolutionary relationship between the genes for the synthases of the two classes of compounds (98, 104, 186). These enzymes catalyze the sequential assembly of carbon chains of various lengths from simple carboxylic acids (typically, residues of acetate, propionate, and butyrate), at the same time introducing an enormous variety in the patterns of side groups on the chain by the reduction or otherwise of its keto groups and by the order of addition of different kinds of building units. The ways in which these variations are genetically "programmed" into the synthase are still obscure but are being analyzed by a combination of genetic and chemical approaches. Already, an unexpected finding is that some actinomycete synthases consist of a series of separate protein subunits for each class of enzymic function, like the *E. coli* fatty acid synthase (7, 193), while other synthases are built from large multifunctional proteins containing "modules" of active sites, each resembling the fatty acid synthase of vertebrates (47, 56).

The subject of biochemical genetics of antibiotic production is too wide for adequate treatment here, but a few observations on the important topic of the biosynthesis of β-lactam (penicillin-related) antibiotics should be made. Just as large, multifunctional enzymes control the orderly extension and modification of the carbon skeletons of complex polyketides, so large enzymes control the assembly of the δ-(L-α-aminoadipyl)-L-cysteinyl-D-valine (ACV) oligopeptide skeleton of β-lactams (46). These ACV synthase enzymes are related to the synthases of peptide antibiotics such as gramicidin produced by *Bacillus* spp. (46). The genetic determinants for ACV synthase and cyclase enzymes involved in β-lactam biosynthesis in *Streptomyces* spp. are notably similar to those in fungi and simple bacteria, which suggests gene transfer between these groups relatively recently in their evolutionary history (199).

MORPHOLOGICAL DIFFERENTIATION

Surface-grown *Streptomyces* colonies may be considered multicellular organisms with several distinct cell types. Sometimes, aspects of this differentiation can be made to occur in liquid medium and thereby become accessible to physiological and biochemical analyses. For example, large-scale, fairly synchronous sporulation of liquid-grown *S. griseus* can readily be achieved in suitable conditions of nutritional shift-down (see reference 89 for further examples). Nevertheless, most of our understanding of *Streptomyces* morphological differentiation comes from genetical work on *S. coelicolor* and *S. griseus*. Information from these two systems is rather different, though this difference probably mainly reflects varied experimental approaches rather than fundamental biological differences.

Considerably fewer genes are specific to and essential for aerial mycelium formation and sporulation, as judged by the classification of about 100 *S. coelicolor* morphological mutants, than is the case for endospore formation in *B. subtilis* (33). This fact, the morphological details of the process, and the results of the limited molecular analysis so far achieved all indicate that differentiation in *Streptomyces* spp. has an origin independent from the origin of differentiation in *B. subtilis*. Its detailed elucidation is therefore likely to reveal substantial novelty. Here, we deal with aerial mycelium formation separately from sporulation. This separation is not only convenient but also scientifically appropriate, since molecular connections between the two morphologically associated processes have not yet been found.

Aerial Mycelium Formation

It is difficult to be sure how many *Streptomyces bld* genes are known at present, because not all have been subjected to extensive genetic analysis. Ten would probably be a reasonable approximation (33, 230). Surprisingly, only three of the known mutant classes (*bldB* and *bldI* [79] and *bldF* [183]) are unconditional, and none of the known mutants is defective only in aerial mycelium formation: they all show various degrees of impairment in secondary metabolism (see above). One would expect there to be some genes that specify structural components peculiar to aerial hyphae and in which mutations eliminate aerial mycelium but not secondary metabolism. The absence of such mutants could be explained in numerous ways, ranging from the anecdotal (not enough *bld* mutants have yet been isolated) through the frustrating (there may be functional overlap between such genes) to the scientifically exciting (the construction of aerial hyphae may be necessary for secondary metabolism to take place in surface-grown cultures through some kind of "morphological coupling").

Most classes of *bld* mutants lack one or more of the Saps (spore-associated proteins) normally present on the surfaces of aerial hyphae and spores (74, 230). This has been particularly closely studied for one of the Saps (the tiny, apparently glycosylated SapB) because of the availability of anti-SapB antibodies (230). Several *bld* mutants can be stimulated to develop almost normally by growth near a wild-type strain, apparently because of the secretion of SapB: the application of purified SapB alone can induce a brief period of normal development (230). Mutants isolated by virtue of nonproduction of SapB antigen are deficient in aerial mycelium (230). This gives rise to the notion that SapB and possibly other Saps can assemble on the outsides of hyphae and contribute structurally to aerial growth. Some Saps (A, C, D, and E) are certainly synthesized by the normal protein synthesis route, but it is still an open question whether SapB might be nonribosomally synthesized (229). In any case, it is intriguing that Saps C, D, and E are encoded by the SCP1 plasmid (165) and are therefore dispensable for aerial mycelium development, since SCP1$^-$ cultures are morphologically normal.

The intracellular events controlling aerial growth of *S. coelicolor* are also poorly understood. Thus, although the reason *bldA* mutants fail to make antibiotics is clear for those cases where known pathway-specific regulatory genes contain TTA codons (see above), there is no such clear explanation of their morphological defect. It has been suggested that this defect may have a metabolic foundation, involving the metabolism of storage compounds (32, 33). According to this hypothesis, the degradation of storage compounds provides an increase in osmotic pressure and hence turgor pressure that drives extension of aerial hyphae. This notion has centered chiefly on glycogen deposition, which occurs in hyphae at the interfaces between substrate and aerial mycelium in *bldA*$^+$ but not *bldA* mutant strains (181). This implies that a TTA-containing gene might influence glycogen metabolism. Clear pointers to other modes of action of *bld* genes are lacking, although it is interesting to note that a *bld* gene from *S. griseus* and its homolog in *S. coelicolor* both contain a TTA codon, providing a putative *bldA*-dependent target (166).

The role of A-factor in stimulating streptomycin production in *S. griseus* has been described above. Its effects on aerial morphological differentiation are achieved at least partially through the A-factor-binding repressor protein, since a mutant lacking this protein shows premature sporulation as well as streptomycin production (171).

Metamorphosis of Aerial Hyphae into Spores

Metamorphosis of aerial hyphae into spores has been studied by using mutants that produce a white aerial mycelium that fails to take on the gray color typical of mature spores (the *whi* mutants). These mutants have defined at least eight *whi* loci scattered round the chromosome, the majority of which are blocked at early stages in sporulation (33). Three of these loci (*whiB*, *whiE*, and *whiG*) have been cloned and sequenced (38, 48, 49), leading to the first study of regulatory interactions in what is probably a central controlling pathway. In what follows, we try to reduce the results to a simplified account of the events during sporulation.

Once initiated, aerial hyphae continue to extend (supplied, perhaps, by nutrients partly derived from storage compounds such as glycogen as well as by the lysing substrate mycelium), until at some point, a sporulation-specific RNA polymerase form containing a σ factor specified by *whiG* accumulates to a concentration sufficient to cause a switch from this indeterminate growth into sporulation septation (38). During sporulation septation, the tips of the coenocytic aerial hyphae become regularly and synchronously subdivided over lengths of many tens of microns into cylindrical compartments destined to form spores. The synchronous formation of the multiple septa is accompanied by segregation of single genomes into each compartment (228) and by a second round of glycogen accumulation only in the sporulating hyphal tips (14). For this septation to take place at all also requires the products of the *whiA*, *whiB*, and *whiH* genes (170). The *whiC* and *whiI* genes also contribute to the proper coordination of septation (33). The mode of action of these five genes is unclear, though the ~10-kDa WhiB protein somewhat resembles transcription factors in having predicted α helices near its acidic N terminus, in its cysteine-rich central region, and near its basic C terminus (49). Moreover, promoters of the *whiE* cluster, which encodes a spore pigment that is deduced to be a polyketide, show dramatically reduced abilities to activate a reporter gene when introduced into *whiA*, *whiB*, and *whiH* mutants on a promoter-probe plasmid (16). Surprisingly, *whiA*, *whiB*, and *whiH* are also needed for detectable activation of the same reporter gene when the promoter of the *whiG* (σ factor) gene is used (18). σ^{WhiG} is not required for transcription of *whiB* or *whiE* (16, 203). Why, then, do *whiG* mutants fail to make spore pigment? Possibly, they are deficient in the provision of metabolic precursors for biosynthesis of the pigment.

It seems that a significant but physically undetectable level of *whiG* transcription may go on independently of the *whiA* and *whiB* genes, since two promoters that depend in vivo on *whiG* and *whiH* but not on *whiA* and *whiB* have been cloned (216). An interesting aspect of these promoters is their resemblance, at appropriate positions, to conserved features of promoters for motility or chemotaxis genes recognized in *B. subtilis* by the σ^D form of RNA polymerase and in *E. coli* and *Salmonella typhimurium* by the σ^F form (83). These σ factors closely resemble σ^{WhiG} in the regions believed to make contact with the −10 and −35 regions of promoters. This again emphasizes the distinct origins of sporulation in *Streptomyces* spp. and *B. subtilis*, since σ^D plays no detectable role in endospore formation.

DUPLICATION OF FUNCTIONALLY OR STRUCTURALLY RELATED GENES IN *STREPTOMYCES* SPP.

In the course of this chapter, we have mentioned the occurrence of multiple copies of genes encoding principal σ-factor homologs and of two quite different genes for glutamine synthase. Such an apparent lack of parsimony extends further: thus, many species also contain more than one gene for GroEL (a heat shock protein and molecular chaperonin [73, 161]), and three genes encoding proteins resembling elongation factor Tu are also often present (231). Some of this redundancy may be related to the need to express the relevant functions in physiologically distinct cell types. The same rationalization may also account for the frequent occurrence of individual genes and transcription units with more than one promoter.

CHALLENGES FOR THE FUTURE

It should be clear from the large number of open questions raised in this chapter that *Streptomyces* geneticists are not short of challenging scientific problems. A recent essay (36) enumerated some of these problems, and here we merely leave the reader with a select list of a dozen. (i) How do hyphal tips elongate, how do lateral and aerial branches form, and how are these activities coordinated with DNA replication? (ii) What are the crucial signals that cause switching to secondary metabolism? (iii) How can the roles and interactions of the multiplicity of apparent regulatory genes involved in antibiotic synthesis be rationalized? (iv) How do the *whi* genes that control sporulation interact with each other, and how do their actions have the observed morphological consequences? (v) What roles, if any, do storage compounds play in morphological development, and how are they used in the mobilization of nutrients to growing tips of aerial hyphae? (vi) What is the importance of Saps? (vii) Can the growing knowledge of secondary metabolism be harnessed for widespread industrial use? (It can certainly illuminate fundamental problems of bioorganic chemistry.) (viii) What are the mechanisms of protein secretion in *Streptomyces* spp., and how can this secretion be made use of for biotechnology? (ix) How many different forms of RNA polymerase holoenzyme are present in *Streptomyces* spp., and why is there such diversity? (x) How can a more complete "gene tool kit"—including facile transposon mutagenesis, the availability of generalized transduction, the construction of ordered genomic libraries, and the further development of reporter systems such as the *luxAB* luminescence genes that permit spatial analysis of gene expression in colonies—be developed? (xi) What properties are encoded by the large, often unstable, "silent" regions of *Streptomyces* chromosomes? Is there much in common between these regions in different *Streptomyces* spp. isolated from soil, and if not, how great is the "gene pool" represented by them? (xii) What aspects of *Streptomyces* biology are relevant to other actinomycetes? For example, what is the phenotype of a *bldA* mutant in simpler genera such as *Micrococcus* and *Corynebacterium*? Are the genes involved in cell fragmentation of simple actinomycetes the same as those involved in sporulation septation in *Streptomyces*? And, on the other hand, what genes account for the genus-defining distinctive morphological properties of other actinomycetes, such as sporangia, motile spores, and different cell wall types?

REFERENCES

1. **Adams, C. W., J. A. Fornwald, F. J. Schmidt, M. Rosenberg, and M. E. Brawner.** 1988. Gene organization and structure of the *Streptomyces lividans gal* operon. *J. Bacteriol.* **170:**203–212.

1a. **Angell, S., E. Schwarz, and M. J. Bibb.** 1992. The glucose kinase gene of *Streptomyces coelicolor* A3(2): its nucleotide sequence, transcriptional analysis and role in glucose repression. *Mol. Microbiol.* **6:**2833–2844.

2. **Baltz, R. H., D. R. Hahn, M. A. McHenney, and P. J. Solenberg.** 1992. Transposition of Tn*5096* and related transposons in *Streptomyces* species. *Gene* **115:**61–65.

3. **Bar-Nir, D., A. Cohen, and M. E. Goedeke.** 1992. tDNA^{ser} sequences are involved in the excision of *Streptomyces griseus* plasmid pSG1. *Gene* **122:**71–76.

4. **Bascarom, V., L. Sanchez, C. Hardisson, and A. F. Braña.** 1991. Stringent response and initiation of secondary metabolism in *Streptomyces clavuligerus*. *J. Gen. Microbiol.* **137:**1625–1634.

5. **Behrmann, I., D. Hillemann, A. Pühler, E. Strauch, and W. Wohlleben.** 1990. Overexpression of *Streptomyces viridochromogenes* gene (*glnII*) encoding a glutamine synthetase similar to those of eucaryotes confers resistance against the antibiotic phosphinothricyl-alanyl-alanine. *J. Bacteriol.* **172:**5326–5334.

6. **Beppu, T.** 1992. Secondary metabolites as chemical signals for differentiation. *Gene* **115:**159–165.

7. **Bibb, M. J., S. Biro, H. Motamedi, J. F. Collins, and C. R. Hutchinson.** 1989. Analysis of the nucleotide sequence of the *Streptomyces glaucescens tcmI* genes provides key information about the enzymology of polyketide antibiotic biosynthesis. *EMBO J.* **8:**2727–2736.

8. **Bibb, M. J., and S. N. Cohen.** 1982. Gene expression in *Streptomyces*: construction and application of promoter-probe plasmid vectors in *Streptomyces lividans*. *Mol. Gen. Genet.* **187:**265–277.

9. **Bibb, M. J., P. R. Findlay, and M. W. Johnson.** 1984. The relationship between base composition and codon usage in bacterial genes and its use for the simple and reliable identification of protein-coding sequences. *Gene* **30:**157–166.

10. **Bibb, M. J., and G. R. Janssen.** 1987. Unusual features of transcription and translation of antibiotic resistance genes in two antibiotic-producing *Streptomyces* species, p. 309–318. *In* M. Alačević, D. Hranueli, and Z. Toman (ed.), *Genetics of Industrial Microorganisms*, part B. Pliva, Zagreb, Yugoslavia.

11. **Bibb, M. J., J. M. Ward, and S. N. Cohen.** 1985. Nucleotide sequences encoding and promoting expression of three antibiotic resistance genes indigenous to *Streptomyces*. *Mol. Gen. Genet.* **199:**26–36.

12. **Birch, A. W., and J. Cullum.** 1985. Temperature-sensitive mutants of the *Streptomyces* plasmid pIJ702. *J. Gen. Microbiol.* **131:**1299–1303.

13. **Bolotin, A., and S. Biro.** 1989. Nucleotide sequence of the putative regulatory gene and major promoter region of the *Streptomyces griseus* glycerol operon. *Gene* **87:**151–152.

14. **Braña, A. F., M. B. Manzanal, and C. Hardisson.** 1980. Occurrence of polysaccharide granules in sporulating hyphae of *Streptomyces viridochromogenes*. *J. Bacteriol.* **144:**1139–1142.

15. **Brawner, M., G. Poste, M. Rosenberg, and J. Westpheling.** 1991. *Streptomyces*: a host for heterologous gene expression. *Curr. Opin. Biotechnol.* **2:**674–681.

16. **Brian, P.** Personal communication.

17. **Brown, D. P., K. B. Idler, and L. Katz.** 1990. Characterization of the genetic elements required for site-specific integration of plasmid pSE211 in *Saccharopolyspora erythraea*. *J. Bacteriol.* **172:**1877–1888.

18. **Brown, G. L.** Personal communication.

19. **Brown, K. L., S. Wood, and M. J. Buttner.** 1992. Isolation and characterization of the major vegetative RNA polymerase of *Streptomyces coelicolor* A3(2): renaturation of a sigma subunit using GroEL. *Mol. Microbiol.* **6:**1133–1139.

20. **Bruton, C. J., E. P. Guthrie, and K. F. Chater.** 1991. Phage vectors that allow monitoring of secondary metabolism genes in *Streptomyces*. *Bio/Technology* **9:**652–656.

21. **Burnett, W. W., M. Brawner, D. P. Taylor, L. R. Fare, J. Henner, and T. Eckhardt.** 1985. Cloning and analysis of an exported β-galactosidase and other proteins from *Streptomyces lividans*, p. 441–444. *In* L. Lieve (ed.), *Microbiology—1985*. American Society for Microbiology, Washington, D.C.

22. **Buttner, M. J.** 1989. RNA polymerase heterogeneity in *Streptomyces coelicolor* A3(2). *Mol. Microbiol.* **3:**1653–1659.

23. **Buttner, M. J., K. F. Chater, and M. J. Bibb.** 1990. Cloning, disruption and transcriptional analysis of three RNA polymerase sigma factor genes of *Streptomyces coelicolor* A3(2). *J. Bacteriol.* **172:**3367–3378.

24. **Buttner, M. J., I. M. Fearnley, and M. J. Bibb.** 1987. The agarase gene (*dagA*) of *Streptomyces coelicolor* A3(2): nucleotide sequence and transcriptional analysis. *Mol. Gen. Genet.* **209:**101–109.

25. **Buttner, M. J., and C. G. Lewis.** 1992. Construction and characterisation of *Streptomyces coelicolor* A3(2) mutants that are multiply deficient in the non-essential *hrd*-encoded RNA polymerase sigma factors. *J. Bacteriol.* **174:**5165–5167.

26. **Buttner, M. J., A. M. Smith, and M. J. Bibb.** 1988. At least three RNA polymerase holoenzymes direct transcription of the agarase gene (*dagA*) of *Streptomyces coelicolor* A3(2). *Cell* **52:**599–607.

27. **Caballero, J. L., F. Malpartida, and D. A. Hopwood.** 1991. Transcriptional organization and regulation of an antibiotic export complex in the producing *Streptomyces* culture. *Mol. Gen. Genet.* **228:**372–380.

28. **Caballero, J. L., E. Martinez, F. Malpartida, and D. A. Hopwood.** 1991. Organisation and functions of the *act*VA region of the actinorhodin biosynthetic gene cluster of *Streptomyces coelicolor*. *Mol. Gen. Genet.* **230:**401–412.

29. **Calcutt, M. J., and F. J. Schmidt.** 1992. Conserved gene arrangement in the origin region of the *Streptomyces coelicolor* chromosome. *J. Bacteriol.* **174:**3220–3226.

30. **Champness, W., P. Riggle, T. Adamidis, and P. Vandervere.** 1992. Identification of *Streptomyces coelicolor* genes involved in regulation of antibiotic synthesis. *Gene* **115:**55–60.

31. **Chater, K. F.** 1986. *Streptomyces* phages and their application to *Streptomyces* genetics, p. 119–158. *In* S. W. Queener and L. E. Day (ed.), *The Bacteria. A Treatise on Structure and Function*, vol. 9. *Antibiotic-Producing Streptomyces*. Academic Press, Inc., Orlando, Fla.

32. **Chater, K. F.** 1989. Multilevel regulation of *Streptomyces* differentiation. *Trends Genet.* **5:**372–376.
33. **Chater, K. F.** 1989. Sporulation in *Streptomyces*, p. 277–299. *In* I. Smith, R. A. Slepecky, and P. Setlow (ed.), *Regulation of Procaryotic Development*. American Society for Microbiology, Washington, D.C.
34. **Chater, K. F.** 1990. The improving prospects for yield increase by genetic engineering in antibiotic-producing streptomycetes. *Bio/Technology* **8:**115–121.
35. **Chater, K. F.** 1992. Genetic regulation of secondary metabolic pathways in *Streptomyces*. *CIBA Found. Symp.* **171:**144–162.
36. **Chater, K. F.** 1992. *Streptomyces* genetics in the 1990s. *World J. Microbiol. Biotechnol.* **8**(Suppl. 1):18–21.
37. **Chater, K. F., and C. J. Bruton.** 1985. Resistance, regulatory and production genes for the antibiotic methylenomycin are clustered. *EMBO J.* **4:**1893–1897.
38. **Chater, K. F., C. J. Bruton, K. A. Plaskitt, M. J. Buttner, C. Mendez, and J. Helmann.** 1989. The developmental fate of *Streptomyces coelicolor* hyphae depends crucially on a gene product homologous with the motility sigma factor of *Bacillus subtilis*. *Cell* **59:**133–143.
39. **Chater, K. F., D. J. Henderson, M. J. Bibb, and D. A. Hopwood.** 1988. Genome flux in *Streptomyces coelicolor* and other streptomycetes and its possible relevance to the evolution of mobile antibiotic resistance determinants, p. 7–42. *In* A. J. Kingsman, K. F. Chater, and S. M. Kingsman (ed.), *Transposition*. Cambridge University Press, Cambridge.
40. **Chater, K. F., and D. A. Hopwood.** 1984. *Streptomyces* genetics, p. 229–286. *In* M. Goodfellow, M. Mordarski, and S. T. Williams (ed.), *Biology of the Actinomycetes*. Academic Press, London.
41. **Chater, K. F., and D. A. Hopwood.** 1989. Antibiotic biosynthesis in *Streptomyces*, p. 129–150. *In* D. A. Hopwood and K. F. Chater (ed.), *Genetics of Bacterial Diversity*. Academic Press, London.
42. **Chater, K. F., and M. J. Merrick.** 1979. Streptomycetes, p. 93–114. *In* J. H. Parish (ed.), *Developmental Biology of Prokaryotes*. Blackwell Scientific Publications, Oxford.
43. **Chinenova, T. A., N. M. Mkrtumian, and N. D. Lomovskaya.** 1982. Genetic study of a novel phage resistance character in *Streptomyces coelicolor* A3(2). *Genetika* **18:**1945–1952.
44. **Chung, S. T.** 1987. Tn*4556*, a 6.8 kb transposable element of *Streptomyces fradiae*. *J. Bacteriol.* **169:**4436–4441.
45. **Chung, S. T.** 1989. Transposition of Tn*4556* in *Streptomyces*. *Dev. Ind. Microbiol.* **29:**81–88.
46. **Coque, J. J. R., J. F. Martin, J. G. Calzada, and P. Liras.** 1991. The cephamycin biosynthetic genes *pcbAB*, encoding a large multidomain peptide synthetase, and *pcbC* of *Nocardia lactamdurans* are clustered together in an organization different from the same genes in *Acremonium chrysogenum* and *Penicillium chrysogenum*. *Mol. Microbiol.* **5:**1125–1133.
47. **Cortes, J., S. F. Haydock, G. A. Roberts, D. J. Bevitt, and P. F. Leadlay.** 1990. An unusually large multifunctional polypeptide in the erythromycin-producing polyketide synthase of *Saccharopolyspora erythraea*. *Nature* (London) **348:**176–178.
48. **Davis, N. K., and K. F. Chater.** 1990. Spore colour in *Streptomyces coelicolor* A3(2) involves the developmentally regulated synthesis of a compound biosynthetically related to polyketide antibiotics. *Mol. Microbiol.* **4:**1679–1691.
49. **Davis, N. K., and K. F. Chater.** 1992. The *Streptomyces coelicolor whiB* gene encodes a small transcription factor-like protein dispensable for growth but essential for sporulation. *Mol. Gen. Genet.* **232:**351–358.
50. **Delić, I., P. Robbins, and J. Westpheling.** 1992. Direct repeat sequences are implicated in the regulation of two *Streptomyces* chitinase promoters that are subject to carbon catabolite control. *Proc. Natl. Acad. Sci. USA* **89:**1885–1889.
51. **Deng, Z., T. Kieser, and D. A. Hopwood.** 1987. Activity of a *Streptomyces* transcriptional terminator in *Escherichia coli*. *Nucleic Acids Res.* **15:**2665–2675.
52. **Deng, Z., T. Kieser, and D. A. Hopwood.** 1988. "Strong incompatibility" between derivatives of the *Streptomyces* multi-copy plasmid pIJ101. *Mol. Gen. Genet.* **214:** 286–294.
53. **Dijkhuizen, L.** Personal communication.
54. **Distler, J., C. Braun, A. Ebert, and W. Piepersberg.** 1987. Gene cluster for streptomycin biosynthesis in *Streptomyces griseus*: analysis of a central region including the major resistance gene. *Mol. Gen. Genet.* **208:** 204–210.
55. **Distler, J., K. Mansouri, G. Mayer, M. Stockmann, and W. Piepersberg.** 1992. Streptomycin biosynthesis and its regulation in streptomycetes. *Gene* **115:**105–111.
56. **Donadio, S., M. J. Staver, J. B. McAlpine, S. J. Swanson, and L. Katz.** 1991. Modular organization of genes required for complex polyketide biosynthesis. *Science* **252:**675–679.
57. **Dyson, P., and H. Schrempf.** 1987. Genetic instability and DNA amplification in *Streptomyces lividans* 66. *J. Bacteriol.* **169:**4796–4803.
58. **Farkašovský, M., J. Kormanec, and M. Kollárová.** 1991. RNA polymerase heterogeneity in *Streptomyces aureofaciens*; characterization by antibody-linked polymerase assay. *FEMS Microbiol. Lett.* **90:**57–62.
59. **Fernández-Moreno, M. A., J. L. Caballero, D. A. Hopwood, and F. Malpartida.** 1991. The *act* cluster contains regulatory and antibiotic export genes, direct targets for translational control by the *bldA* tRNA gene of *Streptomyces*. *Cell* **66:**769–780.
60. **Fernández-Moreno, M. A., E. Martínez, L. Boto, D. A. Hopwood, and F. Malpartida.** 1992. Nucleotide sequence and deduced functions of a set of co-transcribed genes of *Streptomyces coelicolor* A3(2) including the polyketide synthase for the antibiotic actinorhodin. *J. Biol. Chem.* **267:**19278–19290.
61. **Fernández-Moreno, M. A., A. J. Martín-Triana, E. Martínez, J. Niemi, H. M. Kieser, D. A. Hopwood, and F. Malpartida.** 1992. *abaA*, a new pleiotropic regulatory locus for antibiotic production in *Streptomyces coelicolor*. *J. Bacteriol.* **174:**2958–2967.
62. **Fisher, S. H.** 1992. Glutamine synthesis in *Streptomyces*—a review. *Gene* **115:**13–17.
63. **Fishman, S. E., K. Cox, J. L. Larson, P. A. Reynolds, E. T. Seno, W.-K. Yeh, R. van Frank, and C. L. Hershberger.** 1987. Cloning genes for the biosynthesis of a macrolide antibiotic. *Proc. Natl. Acad. Sci. USA* **84:** 8248–8252.
64. **Flett, F., and J. Cullum.** 1987. DNA deletions in spontaneous chloramphenicol-sensitive mutants of *Streptomyces coelicolor* A3(2) and *Streptomyces lividans* 66. *Mol. Gen. Genet.* **207:**499–502.
65. **Foor, F., and N. Morin.** 1990. Construction of a shuttle vector consisting of the *Escherichia coli* plasmid pACYC177 inserted into the *Streptomyces cattleya* phage TG1. *Gene* **94:**109–113.
66. **Forsman, M., B. Haggstrom, L. Lindgren, and B. Jaurin.** 1990. Molecular analysis of β-lactamases from four species of *Streptomyces*: comparison of amino acid sequences with those of other β-lactamases. *J. Gen. Microbiol.* **136:**589–598.
67. **Gold, L., and G. Stormo.** 1987. Translational initiation, p. 1302–1307. *In* F. C. Neidhardt, J. L. Ingraham, K. B. Low, B. Magasanik, M. Schaechter, and H. E. Umbarger (ed.), *Escherichia coli and Salmonella typhimurium: Cellular and Molecular Biology*, vol. 2. American Society for Microbiology, Washington, D.C.
68. **Goodfellow, M., and T. Cross.** 1984. Classification, p. 7–164. *In* M. Goodfellow, M. Mordarski and S. T.

Williams (ed.), *The Biology of the Actinomycetes*. Academic Press, London.

69. **Gormley, E. P., and J. Davies.** 1991. Transfer of plasmid RSF1010 by conjugation from *Escherichia coli* to *Streptomyces lividans* and *Mycobacterium smegmatis*. *J. Bacteriol.* **173:**6705–6708.
70. **Gramajo, H., E. Takano, and M. J. Bibb.** Stationary phase production of the antibiotic actinorhodin is transcriptionally regulated. *Mol. Microbiol.*, in press.
71. **Gramajo, H. C., J. White, C. R. Hutchinson, and M. J. Bibb.** 1991. Overproduction and localization of components of the polyketide synthase of *Streptomyces glaucescens* involved in the production of the antibiotic tetracenomycin C. *J. Bacteriol.* **173:**6475–6483.
72. **Green, E. P., M. L. V. Tizard, M. T. Moss, J. Thompson, D. J. Winterbourne, J. J. McFadden, and J. Hermon-Taylor.** 1989. Sequence and characteristics of IS*900*, an insertion element identified in a human Crohn's disease isolate of *Mycobacterium paratuberculosis*. *Nucleic Acids Res.* **17:**9063–9073.
73. **Guglielmi, G., P. Mazodier, C. J. Thompson, and J. Davies.** 1991. A survey of the heat shock response in four *Streptomyces* species reveals two *groEL*-like genes and three GroEL-like proteins in *Streptomyces albus*. *J. Bacteriol.* **173:**7374–7381.
74. **Guijarro, J., R. Santamaria, A. Schauer, and R. Losick.** 1988. Promoter determining the timing and spatial localization of transcription of a cloned *Streptomyces coelicolor* gene encoding a spore-associated polypeptide. *J. Bacteriol.* **170:**1895–1901.
75. **Guilfoile, P. G., and C. R. Hutchinson.** 1992. Sequence and transcriptional analysis of the *Streptomyces glaucescens tcmAR* tetracenomycin C resistance and repressor gene loci. *J. Bacteriol.* **174:**3651–3658.
76. **Hahn, D. R., M. A. McHenney, and R. H. Baltz.** 1991. Properties of the streptomycete temperate bacteriophage FP43. *J. Bacteriol.* **173:**3770–3775.
77. **Hallam, S. E., F. Malpartida, and D. A. Hopwood.** 1988. DNA sequence, transcription and deduced function of a gene involved in polyketide antibiotic synthesis in *Streptomyces coelicolor*. *Gene* **74:**305–320.
78. **Hara, O., S. Horinouchi, T. Uozumi, and T. Beppu.** 1983. Genetic analysis of A-factor synthesis in *Streptomyces coelicolor* A3(2) and *Streptomyces griseus*. *J. Gen. Microbiol.* **129:**2939–2944.
79. **Harasym, M., L.-H. Zhang, K. Chater, and J. Piret.** 1990. The *Streptomyces coelicolor* A3(2) *bldB* region contains at least two genes involved in morphological development. *J. Gen. Microbiol.* **136:**1543–1550.
80. **Hasegawa, M.** 1992. A novel, highly efficient gene-cloning system in *Micromonospora* applied to the genetic analysis of fortimicin biosynthesis. *Gene* **115:**85–91.
81. **Häusler, A., A. Birch, W. Krek, J. Piret, and R. Hütter.** 1989. Heterogeneous genomic amplification in *Streptomyces glaucescens*: structure, location and junction sequence analysis. *Mol. Gen. Genet.* **217:**437–446.
82. **Haydock, S. F., J. A. Dowson, N. Dhillon, G. A. Roberts, J. Cortes, and P. F. Leadlay.** 1991. Cloning and sequence analysis of genes involved in erythromycin biosynthesis in *Saccharopolyspora erythraea*: sequence similarities between EryG and a family of S-adenosylmethionine-dependent methyltransferases. *Mol. Gen. Genet.* **230:**120–128.
83. **Helmann, J. D.** 1991. Alternative sigma factors and the control of flagellar gene expression. *Mol. Microbiol.* **5:**2875–2882.
84. **Henderson, D. J., D.-F. Brolle, T. Kieser, R. E. Melton, and D. A. Hopwood.** 1990. Transposition of IS*117* (the *Streptomyces coelicolor* A3(2) mini-circle) to and from a cloned target site and into secondary chromosomal sites. *Mol. Gen. Genet.* **224:**65–71.
85. **Henderson, D. J., D. J. Lydiate, and D. A. Hopwood.** 1989. Structural and functional analysis of the mini-circle, a transposable element of *Streptomyces coelicolor* A3(2). *Mol. Microbiol.* **10:**1307–1318.
86. **Henderson, G., P. Krygsman, C. J. Lu, C. C. Davey, and L. T. Malek.** 1987. Characterization and structure of genes for proteases A and B from *Streptomyces griseus*. *J. Bacteriol.* **169:**3778–3784.

86a. **Hershberger, C. L., B. Arnold, J. Larson, P. Skatrud, P. Reynolds, P. Szoke, P. R. Rosteck, Jr., J. Swartling, and G. McGilvray.** 1989. Role of giant linear plasmids in the biosynthesis of macrolide and polyketide antibiotics, p. 147–155. *In* C. L. Hershberger, S. W. Queener, and G. Hegeman (ed.), *Genetics and Molecular Biology of Industrial Microorganisms*. American Society for Microbiology, Washington, D.C.

87. **Hirochika, H., K. Nakamura, and K. Sakaguchi.** 1984. A linear DNA plasmid from *Streptomyces rochei* with an inverted terminal repetition of 614 base pairs. *EMBO J.* **3:**761–766.
88. **Hodgson, D. A.** 1980. Carbohydrate utilization in *Streptomyces coelicolor* A3(2). Ph.D. thesis. University of East Anglia, Norwich, United Kingdom.
89. **Hodgson, D. A.** 1992. Differentiation in actinomycetes, p. 407–440. *In* S. Mohan, C. Dow, and J. A. Cole (ed.), *Prokaryotic Structure and Function: a New Perspective*. Cambridge University Press, Cambridge.
90. **Hodgson, D. A.** Personal communication.
91. **Holt, T. G., C. Chang, C. Laurent-Winter, T. Murakami, J. I. Garrells, J. E. Davies, and C. J. Thompson.** 1992. Global changes in gene expression related to antibiotic biosynthesis in *Streptomyces hygroscopicus*. *Mol. Microbiol.* **6:**969–980.
92. **Hood, D. W., R. Heidstra, U. K. Swoboda, and D. A. Hodgson.** 1992. Molecular genetic analysis of proline and tryptophan biosynthesis in *Streptomyces coelicolor* A3(2): interaction between primary and secondary metabolism—a review. *Gene* **115:**5–12.
93. **Hopwood, D. A.** 1967. Genetic analysis and genome structure in *Streptomyces coelicolor*. *Bacteriol. Rev.* **31:** 373–403.
94. **Hopwood, D. A.** 1981. Genetic studies with bacterial protoplasts. *Annu. Rev. Microbiol.* **35:**237–272.
95. **Hopwood, D. A.** 1988. Towards an understanding of gene switching in *Streptomyces*, the basis of sporulation and antibiotic production. The Leeuwenhoek Lecture, 1987. *Proc. R. Soc. B* **235:**121–138.
96. **Hopwood, D. A., M. J. Bibb, K. F. Chater, T. Kieser, C. J. Bruton, H. M. Kieser, D. J. Lydiate, C. P. Smith, J. M. Ward, and H. Schrempf.** 1985. *Genetic Manipulation of Streptomyces. A Laboratory Manual*. John Innes Foundation, Norwich, United Kingdom.
97. **Hopwood, D. A., M. J. Bibb, T. Kieser, and K. F. Chater.** 1987. Plasmid and phage vectors for gene cloning and analysis in *Streptomyces*. *Methods Enzymol.* **153:**116–166.
98. **Hopwood, D. A., and C. Khosla.** 1992. Genes for polyketide secondary metabolic pathways in microorganisms and plants. *CIBA Found. Symp.* **171:**88–112.
99. **Hopwood, D. A., and T. Kieser.** 1990. The *Streptomyces* genome, p. 147–162. *In* K. Drlica and M. Riley (ed.), *The Bacterial Chromosome*. American Society for Microbiology, Washington, D.C.
100. **Hopwood, D. A., and T. Kieser.** Conjugative plasmids of *Streptomyces*. *In* D. B. Clewell (ed.), *Bacterial Conjugation*, in press. Plenum Press, New York.
101. **Hopwood, D. A., T. Kieser, D. J. Lydiate, and M. J. Bibb.** 1986. *Streptomyces* plasmids: their biology and use as cloning vectors, p. 159–229. *In* S. W. Queener and L. E. Day (ed.), *The Bacteria. A Treatise on Structure and Function*. Academic Press, Inc., Orlando, Fla.
102. **Hopwood, D. A., T. Kieser, H. M. Wright, and M. J. Bibb.** 1983. Plasmids, recombination and chromosome mapping in *Streptomyces lividans* 66. *J. Gen. Microbiol.* **129:**2257–2269.

103. **Hopwood, D. A., D. J. Lydiate, F. Malpartida, and H. M. Wright.** 1984. Conjugative plasmids in *Streptomyces*, p. 615–634. *In* D. Helinski, S. N. Cohen, D. B. Clewell, D. A. Jackson, and A. Hollaender (ed.), *Plasmids in Bacteria*. Plenum Press, New York.
104. **Hopwood, D. A., and D. H. Sherman.** 1990. Molecular genetics of polyketides and its comparison to fatty acid biosynthesis. *Annu. Rev. Genet.* **24:**37–66.
105. **Hopwood, D. A., and H. M. Wright.** 1978. Bacterial protoplast fusion: recombination in fused protoplasts of *Streptomyces coelicolor*. *Mol. Gen. Genet.* **162:**307–317.
106. **Horinouchi, S., and T. Beppu.** 1987. A-factor and regulatory network that links secondary metabolism with cell differentiation in *Streptomyces*, p. 41–48. *In* M. Alačević, D. Hranueli, and Z. Toman (ed.), *Proceedings of Fifth International Symposium on the Genetics of Industrial Microorganisms*. Pliva, Zagreb, Yugoslavia.
107. **Horinouchi, S., and T. Beppu.** 1992. Regulation of secondary metabolism and cell differentiation in *Streptomyces*: A-factor as a microbial hormone and the AfsR protein as a component of a two-component regulatory system. *Gene* **115:**167–172.
108. **Horinouchi, S., O. Hara, and T. Beppu.** 1983. Cloning of a pleiotropic gene that positively controls biosynthesis of A-factor, actinorhodin and prodigiosin in *Streptomyces coelicolor* A3(2) and *Streptomyces lividans*. *J. Bacteriol.* **155:**1238–1248.
109. **Horinouchi, S., M. Kito, M. Nishiyama, K. Furuya, S.-K. Hong, K. Miyake, and T. Beppu.** 1990. Primary structure of AsfR, a global regulatory protein for secondary metabolite formation in *Streptomyces coelicolor* A3(2). *Gene* **95:**49–56.
110. **Horinouchi, S., H. Suzuki, M. Nishiyama, and T. Beppu.** 1989. Nucleotide sequence and transcriptional analysis of the *Streptomyces griseus* gene (*afsA*) responsible for A-factor biosynthesis. *J. Bacteriol.* **171:**1206–1210.
111. **Hranueli, D., J. Pigac, T. Smokvina, and M. Alačević.** 1983. Genetic interactions in *Streptomyces rimosus* mediated by conjugation and by protoplast fusion. *J. Gen. Microbiol.* **129:**1415–1422.
112. **Hunter, I. S.** Personal communication.
113. **Hunter, I. S., and E. J. Friend.** 1984. 'Restriction-deficient' mutants of industrial *Streptomyces*. *Biochem. Soc. Trans.* **12:**643–644.
114. **Hutchinson, C. R.** Personal communication.
115. **Hütter, R., and T. Eckhardt.** 1988. Genetic manipulation, p. 89–184. *In* M. Goodfellow, S. T. Williams, and M. Mordarski (ed.), *Actinomycetes in Biotechnology*. Academic Press, London.
116. **Ingram, C., M. Brawner, P. Youngman, and J. Westpheling.** 1989. *xylE* functions as an efficient reporter gene in *Streptomyces* spp: use for the study of *galP1*, a catabolite-controlled promoter. *J. Bacteriol.* **171:**6617–6624.
117. **Ishizuka, H., S. Horinouchi, H. M. Kieser, D. A. Hopwood, and T. Beppu.** 1992. A putative two-component regulatory system involved in secondary metabolism in *Streptomyces* spp. *J. Bacteriol.* **174:**7585–7594.
118. **Jaurin, B., and S. N. Cohen.** 1984. *Streptomyces lividans* RNA polymerase recognizes and uses *Escherichia coli* transcription signals. *Gene* **28:**83–91.
119. **Jones, G. H.** 1979. Purification of RNA polymerase from actinomycin producing and nonproducing cells of *Streptomyces antibioticus*. *Arch. Biochem. Biophys.* **198:** 195–204.
120. **Jones, R. L., J. C. Jaskula, and G. R. Janssen.** 1992. In vivo translational start site selection on leaderless mRNA transcribed from the *Streptomyces fradiae aph* gene. *J. Bacteriol.* **174:**4753–4760.
121. **Kataoka, M., T. Seki, and T. Yoshida.** 1991. Five genes involved in self-transmission of pSN22, a *Streptomyces* plasmid. *J. Bacteriol.* **173:**4220–4228.
122. **Kataoka, M., T. Seki, and T. Yoshida.** 1991. Regulation and function of the *Streptomyces* plasmid pSN22 genes involved in pock formation and inviability. *J. Bacteriol.* **173:**7975–7981.
123. **Kendall, K. J., and S. N. Cohen.** 1988. Complete nucleotide sequence of the *Streptomyces lividans* plasmid pIJ101 and correlation of the sequence with genetic properties. *J. Bacteriol.* **170:**4634–4651.
124. **Khokhlov, A. S., L. N. Anisova, I. I. Tovarova, E. Y. Kleiner, I. V. Kovalenko, O. I. Krasilnikova, E. Y. Kornitskaya, and S. A. Pliner.** 1973. Effect of A-factor on the growth of asporogenous mutants of *Streptomyces griseus*, not producing this factor. *Z. Allg. Mikrobiol.* **13:**647–655.
125. **Khosla, C., S. Ebert-Khosla, and D. A. Hopwood.** 1992. Targeted gene replacements in a *Streptomyces* polyketide synthase gene cluster: role for the acyl carrier protein. *Mol. Microbiol.* **6:**3237–3249.
126. **Kieser, H. M., T. Kieser, and D. A. Hopwood.** 1992. A combined genetic and physical map of the *Streptomyces coelicolor* A3(2) chromosome. *J. Bacteriol.* **174:**5496–5507.
127. **Kieser, T., and D. A. Hopwood.** 1991. Genetic manipulation of *Streptomyces*: integrating vectors and gene replacement. *Methods Enzymol.* **204:**430–458.
128. **Kinashi, H., and M. Shimaji.** 1987. Detection of giant linear plasmids in antibiotic producing strains of *Streptomyces* by the OFAGE technique. *J. Antibiot.* **40:**913–916.
129. **Kinashi, H., and M. Shimaji-Murayama.** 1991. Physical characterization of SCP1, a giant linear plasmid from *Streptomyces coelicolor*. *J. Bacteriol.* **173:**1523–1529.
130. **Kinashi, H., M. Shimaji-Murayama, and T. Hanafusa.** 1992. Integration of SCP1, a giant linear plasmid, into the *Streptomyces coelicolor* chromosome. *Gene* **115:**35–41.
131. **King, A. A., and K. F. Chater.** 1986. The expression of the *Escherichia coli lacZ* gene in *Streptomyces*. *J. Gen. Microbiol.* **132:**1739–1752.
132. **Kobler, L., G. Schwertfirm, H. Schmieger, A. Bolotin, and I. Sladkova.** 1991. Construction and transduction of a shuttle vector bearing the *cos* site of *Streptomyces* phage ϕC31 and determination of its cohesive ends. *FEMS Microbiol. Lett.* **78:**347–354.
133. **Koller, K.-P., and G. Reiss.** 1989. Heterologous expression of the α-amylase inhibitor gene cloned from an amplified genomic sequence of *Streptomyces tendae*. *J. Bacteriol.* **171:**4953–4957.
134. **Kormanec, J., M. Farkašovský, and L. Potúčková.** 1992. Four genes in *Streptomyces aureofaciens* containing a domain characteristic of principal sigma factors. *Gene* **122:**63–70.
135. **Kretschmer, S.** 1987. Nucleotide segregation pattern during branching in *Streptomyces granaticolor* mycelia. *J. Basic Microbiol.* **27:**203–206.
136. **Kretschmer, S., and C. Kummer.** 1987. Increase of nucleoid size with increasing age of hyphal region in vegetative mycelia of *Streptomyces granaticolor*. *J. Basic Microbiol.* **27:**23–27.
137. **Kroening, T. A., and K. E. Kendrick.** 1987. In vivo regulation of histidine ammonia-lyase activity from *Streptomyces griseus*. *J. Bacteriol.* **169:**823–829.
138. **Kuhstoss, S., and R. N. Rao.** 1991. Analysis of the integration function of the streptomycete bacteriophage ϕC31. *J. Mol. Biol.* **222:**897–908.
139. **Kuhstoss, S., M. A. Richardson, and R. N. Rao.** 1991. Plasmid vectors that integrate site-specifically in *Streptomyces* spp. *Gene* **97:**143–146.
140. **Kumada, Y., E. Takano, K. Nagaoka, and C. J. Thompson.** 1990. *Streptomyces hygroscopicus* has two glutamine synthetase genes. *J. Bacteriol.* **172:**5343–5351.

141. **Kunze, Z. M., S. Wall, R. Appelberg, M. T. Silva, F. Portaels, and J. J. McFadden.** 1991. IS*901*, a new member of a widespread class of atypical insertion sequences, is associated with pathogenicity in *Mycobacterium avium*. *Mol. Microbiol.* **5:**2265–2272.
142. **Lawlor, E. J., H. A. Baylis, and K. F. Chater.** 1987. Pleiotropic morphological and antibiotic deficiencies result from mutations in a gene encoding a tRNA-like product in *Streptomyces coelicolor* A3(2). *Genes Dev.* **1:**1305–1310.
143. **Leblond, P., P. Demuyter, J. M. Simonet, and B. Decaris.** 1990. Genetic instability and hypervariability in *Streptomyces ambofaciens*: towards an understanding of a mechanism of genome plasticity. *Mol. Microbiol.* **4:**707–714.
144. **Leblond, P., P. Demuyter, J.-M. Simonet, and B. Decaris.** 1991. Genetic instability and associated genome plasticity in *Streptomyces ambofaciens*: pulsed-field gel electrophoresis evidence for large DNA alterations in a limited genomic region. *J. Bacteriol.* **173:**4229–4233.
145. **Leblond, P., F. X. Francou, J.-M. Simonet, and B. Decaris.** 1990. Pulsed-field gel electrophoresis analysis of the genome of *Streptomyces ambofaciens* strains. *FEMS Microbiol. Lett.* **72:**79–88.
146. **Leskiw, B. K.** Personal communication.
147. **Leskiw, B. K., M. J. Bibb, and K. F. Chater.** 1991. The use of a rare codon specifically during development? *Mol. Microbiol.* **5:**2861–2867.
148. **Leskiw, B. K., E. J. Lawlor, J. M. Fernandez-Abalos, and K. F. Chater.** 1991. TTA codons in some genes prevent their expression in a class of developmental, antibiotic-negative *Streptomyces* mutants. *Proc. Natl. Acad. Sci. USA* **88:**2461–2465.
149. **Leskiw, B. K., M. Mevarech, L. S. Barritt, S. E. Jensen, D. J. Henderson, D. A. Hopwood, C. J. Bruton, and K. F. Chater.** 1990. Discovery of an insertion sequence, IS*116*, from *Streptomyces clavuligerus* and its relatedness to other transposable elements from actinomycetes. *J. Gen. Microbiol.* **136:**1251–1258.
150. **Lomovskaya, N. D.** Personal communication.
151. **Lydiate, D. J., H. Ikeda, and D. A. Hopwood.** 1986. A 2.6 kb DNA sequence of *Streptomyces coelicolor* A3(2) which functions as a transposable element. *Mol. Gen. Genet.* **203:**79–88.
152. **MacNeil, D. J.** 1988. Characterization of a unique methyl-specific restriction system in *Streptomyces avermitilis*. *J. Bacteriol.* **170:**5607–5612.
153. **MacNeil, D. J., J. L. Occi, K. M. Gewain, T. MacNeil, P. H. Gibbons, C. L. Ruby, and S. J. Danis.** 1992. Complex organization of the *Streptomyces avermitilis* genes encoding the avermectin polyketide synthase. *Gene* **115:**119–125.
154. **Madoń, J., and R. Hütter.** 1991. Transformation system for *Amycolatopsis* (*Nocardia*) *mediterranei*: direct transformation of mycelium with plasmid DNA. *J. Bacteriol.* **173:**6325–6331.
155. **Malpartida, F.** Personal communication.
156. **Malpartida, F., J. Niemi, R. Navarrete, and D. A. Hopwood.** 1990. Cloning and expression in a heterologous host of the complete set of genes for biosynthesis of the *Streptomyces coelicolor* antibiotic undecylprodigiosin. *Gene* **93:**91–99.
157. **Martin, J. F., and A. L. Demain.** 1980. Control of antibiotic synthesis. *Microbiol. Rev.* **44:**230–251.
158. **Martin, J. F., and P. Liras.** 1989. Organization and expression of genes involved in the biosynthesis of antibiotics and other secondary metabolites. *Annu. Rev. Microbiol.* **43:**173–206.
159. **Martínez, E., M. A. Fernández-Moreno, J. L. Caballero, D. A. Hopwood, and F. Malpartida.** Unpublished data.
160. **Matsushima, P., K. L. Cox, and R. H. Baltz.** 1987. Highly transformable mutants of *Streptomyces fradiae* defective in several restriction systems. *Mol. Gen. Genet.* **206:**393–400.
161. **Mazodier, P., G. Guglielmi, J. Davies, and C. J. Thompson.** 1992. Characterization of the *groEL*-like genes in *Streptomyces albus*. *J. Bacteriol.* **173:**7382–7386.
162. **Mazodier, P., R. Petter, and C. Thompson.** 1989. Intergeneric conjugation between *Escherichia coli* and *Streptomyces* species. *J. Bacteriol.* **171:**3583–3585.
163. **Mazodier, P., C. Thompson, and F. Boccard.** 1990. The chromosomal integration site of the *Streptomyces* element pSAM2 overlaps a putative tRNA gene conserved among actinomycetes. *Mol. Gen. Genet.* **222:**431–434.
164. **McCarthy, A. J., and S. T. Williams.** 1992. Actinomycetes as agents of biodegradation in the environment—a review. *Gene* **115:**189–192.
165. **McCormick, J.** Personal communication.
166. **McCue, L. A., J. Kwak, M. J. Babcock, and K. E. Kendrick.** 1992. Molecular analysis of sporulation in *Streptomyces griseus*. *Gene* **115:**173–179.
167. **McDowall, K. J., D. Doyle, M. J. Butler, C. Binnie, M. Warren, and I. S. Hunter.** 1991. Molecular genetics of oxytetracycline production by *Streptomyces rimosus*, p. 105–116. *In* D. Noack, H. Krügel, and S. Baumberg (ed.), *Genetics and Product Formation in Streptomyces*. Plenum Press, New York.
168. **McHenney, M. A., and R. H. Baltz.** 1988. Transduction of plasmid DNA in *Streptomyces* spp. and related genera by bacteriophage FP43. *J. Bacteriol.* **170:**2276–2282.
169. **McLaughlin, J. R., C. L. Murray, and J. C. Rabinowitz.** 1981. Unique features of the ribosome binding site of the Gram-positive *Staphylococcus aureus* β-lactamase gene. *J. Biol. Chem.* **256:**11283–11291.
170. **McVittie, A. M.** 1974. Ultrastructural studies on sporulation in wild-type and white colony mutants of *Streptomyces coelicolor*. *J. Gen. Microbiol.* **81:**291–302.
171. **Miyake, K., T. Kuzuyama, S. Horinouchi, and T. Beppu.** 1990. The A-factor-binding protein of *Streptomyces griseus* negatively controls streptomycin production and sporulation. *J. Bacteriol.* **172:**3003–3008.
172. **Morino, T., K. Takagi, T. Nakamura, T. Takita, H. Saito, and H. Takahashi.** 1986. Studies of cosmid transduction in *Streptomyces lividans* and *Streptomyces parvulus*. *Agric. Biol. Chem.* **50:**2493–2497.
173. **Muth, G., B. Nussbaumer, W. Wohlleben, and A. Pühler.** 1989. A vector system with temperature-sensitive replication for gene disruption and mutational cloning in streptomycetes. *Mol. Gen. Genet.* **219:**341–348.
174. **Nagaso, H., S. Saito, H. Saito, and H. Takashashi.** 1988. Nucleotide sequence and expression of a *Streptomyces griseosporus* proteinaceous alpha-amylase inhibitor (HaimII) gene. *J. Bacteriol.* **170:**4451–4457.
175. **Neal, R. J., and K. F. Chater.** 1991. Bidirectional promoter and terminator regions bracket *mmr*, a resistance gene embedded in the *Streptomyces coelicolor* A3(2) gene cluster encoding methylenomycin production. *Gene* **100:**75–83.
176. **Novotna, J., and Z. Hošťálek.** 1985. Phosphorylation of hexoses in *Streptomyces aureofaciens*: evidence that the phosphoenolpyruvate:sugar phosphotransferase system is not operative. *FEMS Microbiol. Lett.* **28:**347–350.
177. **Ochi, K.** 1987. Metabolic initiation of differentiation and secondary metabolism by *Streptomyces griseus*: significance of the stringent response (ppGpp) and GTP content in relation to A-factor. *J. Bacteriol.* **169:**3608–3616.
178. **Ogata, S., H. Suenaga, Y. Koyama-Miyoshi, and S. Hayashida.** 1984. Cloning of the *his* gene of *Streptomyces azureus* in temperate phage SAt1. *J. Gen. Appl. Microbiol.* **30:**405–409.
179. **O'Hagan, D.** 1991. *The Polyketide Metabolites*. Ellis Horwood, Chichester, United Kingdom.
180. **Okamoto, S., T. Nihira, H. Kataoka, A. Suzuki, and Y. Yamada.** 1992. Purification and molecular cloning of a

butyrolactone autoregulator receptor from *Streptomyces virginiae*. *J. Biol. Chem.* **267:**1093–1098.

181. **Plaskitt, K. A.** Personal communication.
182. **Plohl, M., and V. Gamulin.** 1990. Five transfer RNA genes lacking CCA termini are clustered in the chromosome of *Streptomyces rimosus*. *Mol. Gen. Genet.* **222:** 129–134.
183. **Puglia, A. M.** Personal communication.
184. **Rao, R. N., M. A. Richardson, and S. Kuhstoss.** 1987. Cosmid shuttle vectors for cloning and analysis of *Streptomyces* DNA. *Methods Enzymol.* **153:**166–198.
185. **Reiter, W. D., P. Palm, and S. Yeats.** 1989. Transfer RNA genes frequently serve as integration sites for prokaryotic genetic elements. *Nucleic Acids Res.* **17:** 1907–1914.
186. **Robinson, J.** 1991. Polyketide synthase complexes: their structure and function in antibiotic biosynthesis. *Proc. R. Soc. B* **332:**107–114.
187. **Rodicio, M. R., M. A. Alvarez, and K. F. Chater.** 1991. Isolation and genetic analysis of IS*112*, an insertion sequence responsible for the inactivation of the *Sal*I restriction-modification system of *Streptomyces albus* G. *Mol. Gen. Genet.* **225:**142–147.
188. **Saier, M. H.** 1985. *Mechanism and Regulation of Carbohydrate Transport in Bacteria*. Academic Press, Inc., Orlando, Fla.
189. **Schauer, A. T., A. D. Nelson, and J. B. Daniel.** 1991. Tn*4563* transposition in *Streptomyces coelicolor* and its application to isolation of new morphological mutants. *J. Bacteriol.* **173:**5060–5067.
190. **Sedlmeier, R., and H. Schmieger.** 1991. tRNA genes in *Streptomyces lividans*: structure, organisation and construction of suppressor tRNA, p. 65–73. *In* S. Baumberg, H. Krügel, and D. Noack (ed.), *Genetics and Product Formation in Streptomyces*. Plenum Press, New York.
191. **Seno, E. T., and R. H. Baltz.** 1989. Structural organization and regulation of antibiotic biosynthesis and resistance genes in actinomycetes, p. 1–48. *In* S. Shapiro (ed.), *Regulation of Secondary Metabolism in Actinomycetes*. CRC Press, Inc., Boca Raton, Fla.
192. **Seno, E. T., and K. F. Chater.** 1983. Glycerol catabolism enzymes and their regulation in wild-type and mutant strains of *Streptomyces coelicolor* A3(2). *J. Gen. Microbiol.* **129:**1403–1413.
193. **Sherman, D. H., F. Malpartida, M. J. Bibb, H. M. Kieser, M. J. Bibb, and D. A. Hopwood.** 1989. Structure and deduced function of the granaticin-producing polyketide synthase gene cluster of *Streptomyces violaceoruber* Tü22. *EMBO J.* **8:**2717–2725.
194. **Shiina, T., K. Tanaka, and H. Takahashi.** 1991. Sequence of *hrdB*, an essential gene encoding sigma-like transcription factor of *Streptomyces coelicolor* A3(2): homology to principal sigma factors. *Gene* **107:**145–148.
195. **Siemieniak, D. R., J. L. Slightom, and S. T. Chung.** 1990. Nucleotide sequence of *Streptomyces fradiae* transposon Tn*4556*: a class II transposon related to Tn*3*. *Gene* **86:**1–9.
196. **Simonet, J.-M., D. Schneider, J.-N. Volff, A. Darym, and B. Decaris.** 1992. Genetic instability in *Streptomyces ambofaciens*: inducibility and associated genome plasticity. *Gene* **115:**49–54.
197. **Smith, C. P., and K. F. Chater.** 1988. Structure and regulation of controlling sequences for the *Streptomyces coelicolor* glycerol operon. *J. Mol. Biol.* **204:**569–580.
198. **Smith, C. P., and K. F. Chater.** 1988. Cloning and transcription analysis of the entire glycerol utilization (*gylABX*) operon of *Streptomyces coelicolor* A3(2) and identification of a closely associated transcription unit. *Mol. Gen. Genet.* **211:**129–137.
199. **Smith, D. J., M. K. R. Burnham, J. H. Bull, J. E. Hodgson, J. M. Ward, P. Browne, J. Brown, B. Barton, A. J. Earl, and G. W. Turner.** 1992. β-Lactam antibiotic biosynthetic genes have been conserved in clusters in prokaryotes and eukaryotes. *EMBO J.* **9:**741–747.
200. **Smith, M. C. M., C. J. Ingham, C. E. Owen, and N. T. Wood.** 1992. Gene expression in the temperate *Streptomyces* phage φC31. *Gene* **115:**43–48.
201. **Sohaskey, C. D., H. Im, A. D. Nelson, and A. T. Schauer.** 1992. Tn*4556* and luciferase: synergistic tools for visualizing transcription in *Streptomyces*. *Gene* **115:**67–71.
202. **Solenberg, P. J., and R. H. Baltz.** 1991. Transposition of Tn*5096* and other IS*493* derivatives in *Streptomyces griseofuscus*. *J. Bacteriol.* **173:**1096–1104.
203. **Soliveri, J., K. L. Brown, M. J. Buttner, and K. F. Chater.** 1992. Two promoters for the *whiB* sporulation gene of *Streptomyces coelicolor* A3(2) and their activities in relation to development. *J. Bacteriol.* **174:**6215–6220.
204. **Sosio, M., J. Madon, and R. Hütter.** 1989. Excision of pIJ408 from the chromosome of *Streptomyces glaucescens* and its transfer to *Streptomyces lividans*. *Mol. Gen. Genet.* **218:**169–176.
205. **Stein, D., and S. N. Cohen.** 1989. A cloned regulatory gene of *Streptomyces lividans* can suppress the pigment deficiency phenotype of different developmental mutants. *J. Bacteriol.* **171:**2258–2261.
206. **Stein, D. S., K. J. Kendall, and S. N. Cohen.** 1989. Identification and analysis of transcriptional regulatory signals for the *kil* and *kor* loci of *Streptomyces* plasmid pIJ101. *J. Bacteriol.* **171:**5768–5775.
207. **Stover, C. K., V. F. de la Cruz, T. R. Fuerst, J. E. Burlein, L. A. Benson, L. T. Bennett, G. P. Bansal, J. F. Young, M. H. Lee, G. F. Hatfull, S. B. Snapper, R. G. Garletta, W. R. Jacobs, Jr., and B. R. Bloom.** 1991. New use of BCG for recombinant vaccines. *Nature* (London) **351:**456–460.
208. **Strauch, E., E. Takano, H. Baylis, and M. J. Bibb.** 1991. The stringent response in *Streptomyces coelicolor* A3(2). *Mol. Microbiol.* **5:**289–298.
209. **Strickler, J. E., T. M. Berka, J. Gorniak, J. Fornwald, R. Keys, J. J. Rowland, M. Rosenberg, and D. P. Taylor.** 1992. Two novel *Streptomyces* protein protease inhibitors: purification, activity, cloning and expression. *J. Biol. Chem.* **267:**3236–3241.
210. **Strohl, W. R.** 1992. Compilation and analysis of DNA sequences associated with apparent streptomycete promoters. *Nucleic Acids Res.* **5:**961–974.
211. **Stuttard, C.** 1989. Generalized transduction in *Streptomyces* species, p. 157–162. *In* C. L. Hershberger, S. W. Queener, and G. Hegeman (ed.), *Genetics and Molecular Biology of Industrial Microorganisms*. American Society for Microbiology, Washington, D.C.
212. **Stuttard, C.** Personal communication.
213. **Summers, R. G., E. Wendt-Pienkowski, H. Motamedi, and C. R. Hutchinson.** 1992. Nucleotide sequence of the *tcmII-tcmIV* region of the tetracenomycin C biosynthetic gene cluster of *Streptomyces glaucescens* and evidence that the *tcmN* gene encodes a multifunctional cyclase-dehydratase-*O*-methyl transferase. *J. Bacteriol.* **174:**1810–1820.
214. **Taguchi, S.** 1992. *Streptomyces* subtilisin inhibitor: genetical characterization and its application. *Actinomycetologica* **6:**9–20.
215. **Takahashi, H., T. Isogai, T. Morino, H. Kojima, and H. Saito.** 1982. Development of phage vector systems in *Streptomyces*, p. 61–65. *In* Y. Ikeda and T. Beppu (ed.), *Proceedings of the IVth International Symposium on Genetics of Industrial Microorganisms, 1982*. Kodansha, Tokyo.
216. **Tan, H.** 1991. Molecular genetics of developmentally regulated promoters in *Streptomyces coelicolor* A3(2). Ph.D. thesis, University of East Anglia, Norwich, United Kingdom.
217. **Tanaka, K., T. Shiina, and H. Takahashi.** 1988. Multiple principal sigma factor homologs in eubacteria:

identification of the "rpoD box." *Science* **242:**1040–1042.

218. **Tanaka, K., T. Shiina, and H. Takahashi.** 1991. Nucleotide sequence of genes *hrdA, hrdC* and *hrdD* from *Streptomyces coelicolor* A3(2) having similarity to *rpoD* genes. *Mol. Gen. Genet.* **229:**234–240.
219. **Urabe, H., and H. Ogawara.** 1992. Nucleotide sequence and transcriptional analysis of activator-regulator proteins for β-lactamase in *Streptomyces cacaoi. J. Bacteriol.* **174:**2834–2842.
220. **Vats-Mehta, S., P. Bouvrette, F. Shareck, R. Morosoli, and D. Kluepfel.** 1990. Cloning of a second xylanase-encoding gene of *Streptomyces lividans* 66. *Gene* **86:**119–122.
221. **Virolle, M. J., and M. J. Bibb.** 1988. Cloning, characterization and regulation of an α-amylase gene from *Streptomyces limosus. Mol. Microbiol.* **2:**197–208.
222. **Vujaklija, D., K. Ueda, S.-K. Hong, T. Beppu, and S. Horinouchi.** 1991. Identification of an A-factor-dependent promoter in the streptomycin biosynthetic cluster of *Streptomyces griseus. Mol. Gen. Genet.* **229:**119–128.
223. **Wang, P., S. S. Harvey, P. F. G. Sims, and P. Broda.** The construction of *Streptomyces cyaneus* genomic libraries in *Escherichia coli* is dependent upon the use of Mcr-deficient strains. *Gene* **119:**127–129.
224. **Weber, J. M., J. O. Leung, S. J. Swanson, K. B. Idler, and J. B. McAlpine.** 1991. An erythromycin derivative produced by targeted gene disruption in *Saccharopolyspora erythraea. Science* **252:**114–117.
225. **Weber, J. M., and R. Losick.** 1988. The use of a chromosome integration vector to map erythromycin resistance and production genes in *Saccharopolyspora erythraea* (*Streptomyces erythreus*). *Gene* **68:**173–180.
226. **Westpheling, J., and M. Brawner.** 1989. Two transcribing activities are involved in expression of the *Streptomyces* galactose operon. *J. Bacteriol.* **171:**1355–1361.
227. **Westpheling, J., M. Ranes, and R. Losick.** 1985. RNA polymerase heterogeneity in *Streptomyces coelicolor. Nature* (London) **313:**22–27.
228. **Wildermuth, H., and D. A. Hopwood.** 1970. Septation during sporulation in *Streptomyces coelicolor. J. Gen. Microbiol.* **60:**51–59.
229. **Willey, J.** Personal communication.
230. **Willey, J., R. Santamaria, J. Guijarro, M. Geistlich, and R. Losick.** 1991. Extracellular complementation of a developmental mutation implicates a small sporulation protein in aerial mycelium formation by *Streptomyces coelicolor. Cell* **65:**641–650.
231. **Woudt, L. P., K. Rietveld, M. Verdurmen, J. van Haarlem, G. P. van Wezel, E. Vijgenboom, and L. Bosch.** Three *tuf*(-like) genes in *Streptomyces ramocissimus.* Submitted for publication.
232. **Wright, F., and M. J. Bibb.** 1992. Codon usage in the G+C-rich *Streptomyces* genome. *Gene* **113:**55–65.
233. **Wu, X., and K. Roy.** 1993. Complete nucleotide sequence of a linear plasmid from *Streptomyces clavuligerus* and characterization of its RNA transcripts. *J. Bacteriol.* **175:**37–52.
234. **Yagi, Y.** 1990. Transposition of Tn*4560* in *Streptomyces avermitilis. J. Antibiot.* **43:**1204–1206.
235. **Yanofsky, C., and I. P. Crawford.** 1987. The tryptophan operon, p. 1453–1472. *In* F. C. Neidhardt, J. L. Ingraham, K. B. Low, B. Magasanik, M. Schaechter, and H. E. Umbarger (ed.), *Escherichia coli and Salmonella typhimurium: Cellular and Molecular Biology,* vol. 2. American Society for Microbiology, Washington, D.C.
236. **Young, M., and J. Cullum.** 1987. A plausible mechanism for large-scale chromosomal DNA amplification in streptomycetes. *FEMS Microbiol. Lett.* **212:**10–14.
237. **Zakrzewska-Czerwinska, J., and H. Schrempf.** 1992. Characterization of an autonomously replicating region from the *Streptomyces lividans* chromosome. *J. Bacteriol.* **174:**2688–2693.
238. **Zhang, L., G. Hintermann, and J. Piret.** 1991. Detection and expression of the *bldB* gene product during morphological differentiation in *Streptomyces coelicolor*, abstr. no. P1-086. *In* Abstracts, *International Symposium on Biology of Actinomycetes*. Madison, Wis.

7. *Pasteuria*, Metchnikoff, 1888

RICHARD M. SAYRE

HISTORICAL INTRODUCTION

In fleeing the Ukrainian city of Odessa during late summer of 1887, Elie Metchnikoff abandoned his directorship of a newly established research institute. But in doing this, he escaped public humiliation and possible civil charges that might have arisen from the botched field trial of a vaccine to control anthrax in sheep. Apparently, the vaccine contained virulent cells of *Bacillus anthracis* Cohn, 1872, because many thousands of sheep were killed in the test. A few weeks later, undaunted by the failure, he arrived at the Paris Laboratory of Pasteur and received from Pasteur not only a research position but also a laboratory in the newly formed Pasteur Institute (3).

Seemingly to thank his benefactor, Metchnikoff hurriedly prepared for publication a manuscript in which he named a new bacterial genus in honor of Pasteur. Specimens of the bacterium had been brought from Odessa to Paris in his luggage. Observations of the bacteria that he had begun in Odessa were finalized in Paris. Parasitized adult water fleas, *Daphnia magna* Straus, were individually mounted and crushed on microscope slides to release the bacterium's life stages for viewing. Perhaps he worked with populations of parasitized water fleas that were also brought from Odessa in glass vials. With the help of M. Roux, photographer for the Pasteur Institute, an illustrative plate with text on the bacterial parasite of water fleas appeared a few months later under the title "*Pasteuria ramosa*, un représentant des bactéries à division longitudinale" (18).

Having emphasized in the title the bacterium's mode of division, Metchnikoff's (18) thinking roughly translates as follows: "*Pasteuria* sp. was able to undergo as many as five longitudinal divisions at the same time, giving it a characteristic fan shape. The peculiarity of *Pasteuria* is therefore its ability to divide in just one plane, longitudinal. Supposing that the most primitive bacteria possess the ability to divide in three different directions, one can easily agree that in most forms, it is transverse division which is conserved or maintained, while in the minority of cases, longitudinal division becomes primary." His observations and speculative statements may be some of the earliest made in the literature regarding bacterial division and its evolutionary history (18).

Over the years, this paper with its concept of longitudinal division and the accompanying drawings showing "stalked" spores (Fig. 1) has enticed investigators to seek out the bacterial parasite of water fleas (6). Conversely, Migula (19) dismissed the bacterium straightaway as being a myxomycete or myxobacterium. Still others, having failed to find a bacterium as illustrated by Metchnikoff (18), concluded that he was mistaken regarding the bacterium's morphology. From their reevaluation of *P. ramosa*, they concluded that a budding bacterial species of the *Blastobacter* group was probably the organism that Metchnikoff had observed (7, 34). Still others, not aware of the bacterium's published taxonomic description, placed their observed parasites of water fleas in different taxa (8, 46).

Sayre and Wergin (29) (Fig. 1) noted the original description and drawings of the life cycle of *P. ramosa* and compared them with those of the bacterial parasite of root-knot nematode, *Meloidogyne incognita* (Kofoid and White, 1919) Chitwood, 1949. They searched for *P. ramosa* and "rediscovered" it infecting *Moina rectirostris* (Leydig, 1860), a member of the Daphnidae (30). No evidence of longitudinal division was found, but their morphological studies of the bacterium supported much of Metchnikoff's original description for *P. ramosa* (23, 24). The primary colonies were cauliflowerlike. Daughter colonies were formed by fragmentation of mother colonies, and the daughter colonies in turn produced quartets and doublets of sporangia and finally the single sporangium, consisting of a conical stem, swollen middle cell, and endogenous spore (Fig. 2). The genus *Pasteuria* as described by Metchnikoff was conserved (9, 36) and later more fully documented (26, 27; Table 1).

There has been little additional information on the bacterial parasites of water fleas beyond that in publications mentioned above. Consequently, emphasis from here on shifts to species of *Pasteuria* that parasitize plant-parasitic nematodes. Aspects of this association may seem atypical because of the bacterium's unusual morphology and life style. However, it is the best-documented parasitic association between plant-parasitic nematodes and bacteria.

Thorne (43), adhering to Cobb's (2) 1906 placement of the organism among the protozoans, described the nematode parasite as a microsporidium and named it *Duboscqia penetrans*. He could not have realized its true bacterial nature, mainly because electron microscopic techniques were not available to him and the concept of the prokaryotic cell had not yet been introduced. Through electron microscopic studies, Mankau and Imbriani (14, 15) established the prokaryotic nature of the organism. Given the information available, its taxonomic placement in the bacterial species "*Bacillus penetrans*" (Thorne, 1940) Mankau, 1975, was warranted (13). The rediscovery by Sayre et al. (30) of *P. ramosa*, Metchnikoff, 1888, and its morphological similarities to "*B. penetrans*"

Richard M. Sayre • Nematology Laboratory, Agricultural Research Service, U.S. Department of Agriculture, 10300 Baltimore Avenue, Beltsville, Maryland 20705.

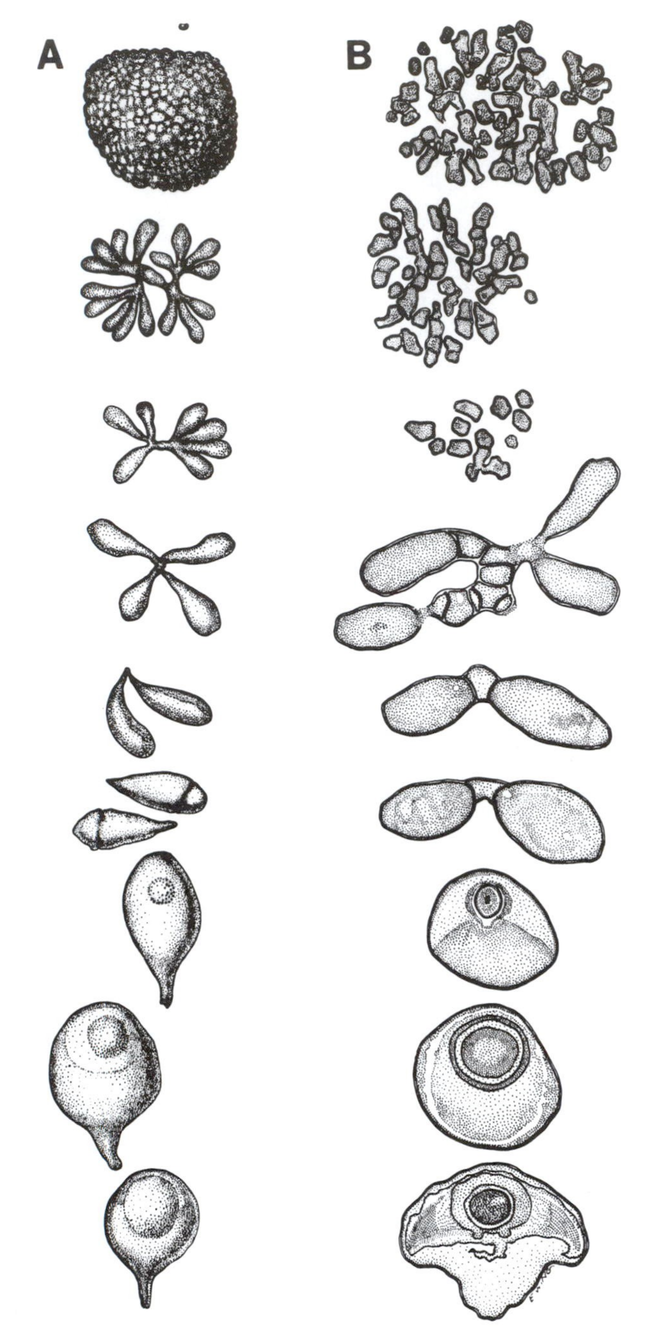

Figure 1. Line drawings of life stages of *P. ramosa* and *P. penetrans*. (A) Life stages of *P. ramosa*, beginning with the cauliflower stage followed by the daughter colony stage. Next are the quartets of sporangia, then doublets, and finally, single sporangia in which endospores are formed. (Drawings are from Metchnikoff [18].) (B) Comparable life stages of *P. penetrans*, the bacterial parasite of the root-knot nematode *M. incognita* (from reference 30).

plus attention to certain nomenclatural issues by Starr et al. (9, 24, 36) led to the designation of *Pasteuria penetrans* (ex Thorne, 1940) Sayre and Starr, 1985, for the bacterial parasite of the root-knot nematodes (25).

LIFE CYCLE OF *PASTEURIA* SPP.

Attachment of Endospores to Nematode Juveniles

When vermiform life stages (i.e., including four molts plus adults) of susceptible nematodes move through soil, their cuticles become encumbered by the endospores of their respective *Pasteuria* spp. (Fig. 3). In cross attachment studies with endospores of *Pasteuria* spp. from different nematode hosts, there has been no reciprocal attachment between those endospores coming from different genera of plant nematodes (17, 26). Generally, endospores appear to recognize and attach only to the hosts in which they were originally found and are able to complete their life cycle, but cross attachment and parasitism between closely related species within the same genus (i.e., *Meloidogyne* and *Heterodera* spp.) do occur. Also, endospores of a nematode species may attach to another life stage of its host (e.g., adult males), as in the case of *Pasteuria nishizawae*, but these attached endospores are unable to germinate and then penetrate and colonize the male nematode (32) (Fig. 4).

Soil environmental factors influence the attachment process. For example, for adhesion of a mature endospore to occur, its basal attachment surfaces must be exposed and physical contact with the nematode's cuticle must be made. When mature parasitized females are manually crushed for laboratory examination, most of the endospores revealed are surrounded by sporangial walls and thin exosporia. These two coverings are removed or degraded in the soil by some unknown process(es) that is a necessary prelude for attachment of endospores to the cuticle of juveniles (Fig. 5).

The endospore-to-endospore cycle was found to depend on soil temperature (37). Temperature determines not only the duration of the cycle but also the number of endospores produced per female root-knot nematode. A temperature of 25°C was found favorable for development of 2.2×10^6 endospores per female in 38 days. At 20°C, the duration of the nematode's life cycle increased from 85 to 100 days, with 2.4×10^6 endospores per female. At 30°C, the bacterium rapidly colonized the female, resulting in a lower number of endospores, at 1.4×10^6 endospores per female.

Endospore Germination

Germination of *P. penetrans* endospores occurs about 6 to 8 days after the endospore-encumbered nematode enters the plant root and begins to feed (29). The germ tube emerges through a central opening in the basal attachment layer and penetrates the cuticle of the nematode (Fig. 6). After entering the hypodermal tissue of the nematode, the germ tube develops into a filamentous microcolony consisting of a dichotomously branched septate mycelium. Close scrutiny of the vegetative stage reveals its bacterial nature (30).

Vegetative Stage

The vegetative stages of *P. penetrans* within its host have been described elsewhere in the literature (13,

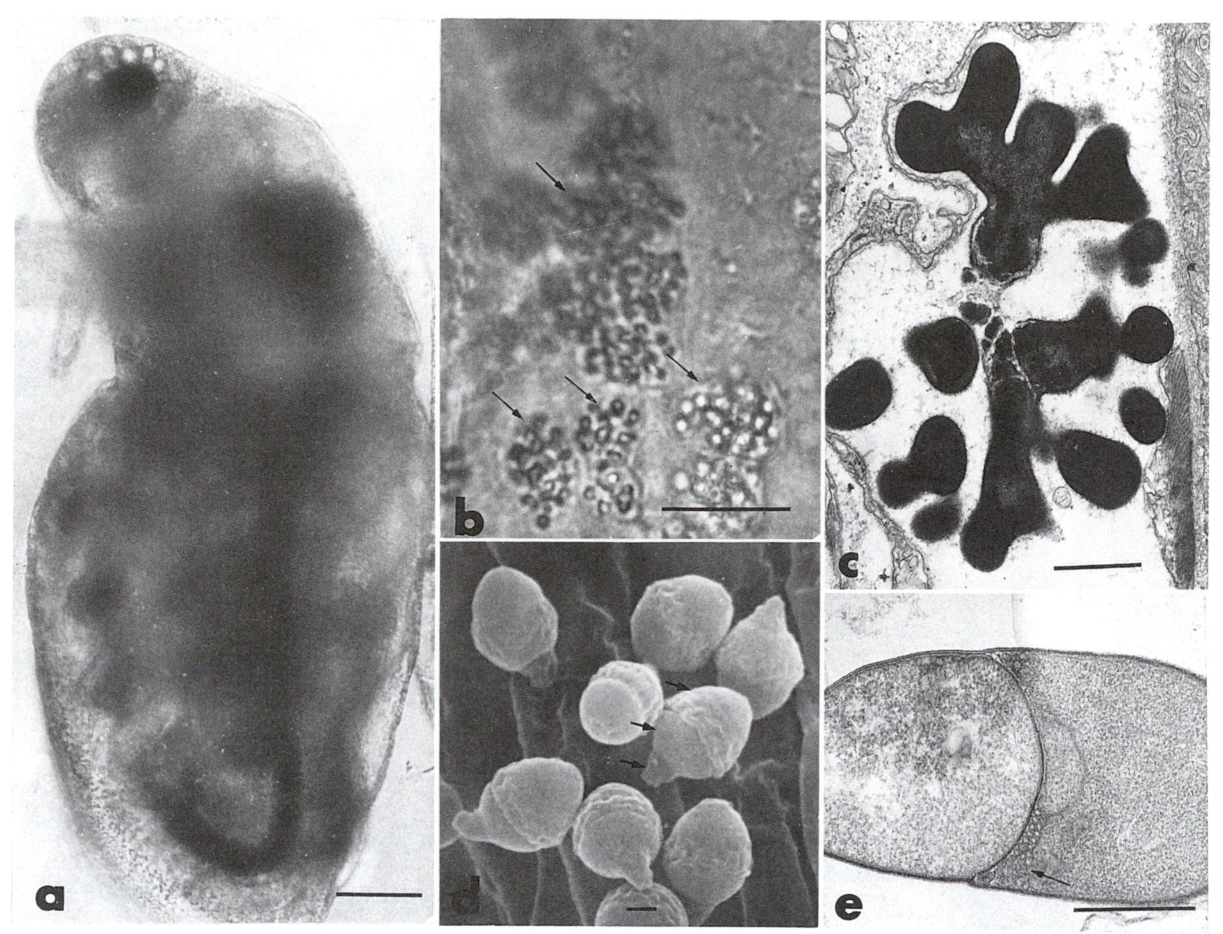

Figure 2. Life stages of *P. ramosa* infecting the water flea *Moina rectirostris*. (a) Adult female that has been rendered partially opaque by the myriad of bacteria in its coelom. Bar = 100 μm (from reference 30). (b) Photomicrograph of cauliflowerlike colonies of *P. ramosa* (arrows) under the carapace of *M. rectirostris*. Bar = 10 μm (from reference 23). (c) Ultrathin section showing fine structure of a filamentous colony of *P. ramosa* developing within tissues of the water flea. Separation of the filamentous strands results from lysis of intercalary cells. The concept of longitudinal division was probably based on observing and monitoring the fragmentation of a microcolony, with the subsequent expansion of its individual filaments during formation of endospores. Bar = 2.5 μm (from reference 24). (d) Scanning electron micrograph of several sporangia, some clearly showing the three external divisions noted by Metchnikoff (18): a conical stem, a swollen middle cell, and an apical endogenous spore (arrows). Bar = 1 μm (from reference 24). (e) Mesosome (arrow) associated with the developing septum found in an early sporangial stage of *P. ramosa*. Bar = 1 μm (from reference 23).

15, 29). Only organelles characteristic of the prokaryotic cell have been found in the few described bacterial parasites of plant-parasitic nematodes that have been examined to date. When intercalary cells in the microcolony lyse, daughter colonies form. The process continues, resulting in a large number of daughter colonies containing fewer but larger vegetative cells. Eventually, quartets of developing sporangia predominate in the nematode's body cavity. These quartets separate into doublets of sporangia and finally separate into a single sporangium that will eventually contain an endospore (Fig. 7).

Endospore Formation

In each species of *Pasteuria* so far examined, the fine structures of its mature endospore are unique (5, 23–25, 27, 28, 31, 32, 35; Table 1). However, in their immature developmental stages during endogenous spore formation, all species share the typical sequence of the gram-positive endospore-forming bacterium (Fig. 8). Briefly, the familiar sequence includes the formation of a transverse septum within the endospore mother cell, condensation of a forespore from the anterior protoplast, formation of multilayered walls about the forespore, and finally, lysis of the sporangial wall, releasing an endospore that is tolerant of heat and desiccation and survives long storage periods.

Soil Phase

Endospores of *Pasteuria* sp. are released into the soil when the plant root with its complement of parasi-

Table 1. Comparison of *P. ramosa*, *P. penetrans*, *Pasteuria thornei*, and *Pasteuria nishizawae*

Trait[a]	*P. ramosa*	*P. penetrans*	*P. thornei*	*P. nishizawae*
Colony shape	Like cauliflower floret	Spherical to cluster of elongated grapes	Small elongated clusters	Spherical to cluster of elongated grapes
Sporangium				
Shape	Teardrop	Cup	Rhomboidal	Cup
Diam (μm)				
LM	3.3–4.1[b]	4.5 ± 0.3	3.5 ± 0.2	5.3 ± 0.3
TEM	2.12–2.77[c]	3.4 ± 0.2	2.4 ± 0.2	4.4 ± 0.3
Ht (μm)				
LM	4.8–5.7[b]	3.6 ± 0.3	3.1 ± 0.2	4.3 ± 0.3
TEM	3.4–4.35[c]	2.5 ± 0.2	2.2 ± 0.2	3.1 ± 0.3
Sporangial wall's fate at maturity of endospore	Remains rigidly in place; external markings divide sporangium into three parts	Basal portion collapses inward on developed endospore; no clear external markings	Remains rigid, sometimes collapsing at bases; no clear external markings	Basal portion collapses on developed endospore; no clear external markings
Exosporium	Not observed	Present; relatively smooth surface	Present; smooth	Present; velutinous to hairy surface
Stem cell	Remains attached to most sporangia	Rarely seen; attachment of second sporangium sometimes observed	Neither stem cell nor second sporangium seen	Occasionally seen
Endospores				
Shape	Oblate spheroid, ellipsoid, narrowly elliptic in section	Oblate spheroid, ellipsoid, broadly elliptic in section	Oblate spheroid, ellipsoid, sometimes almost spherical, narrowly elliptic in section	Oblate spheroid, ellipsoid, narrowly elliptic in section
Orientation of major axis to sporangium base	Vertical	Horizontal	Horizontal	Horizontal
Diam (μm)				
LM	2.1–2.4[d]	2.1 ± 0.2	1.6 ± 0.1	2.1 ± 0.2
TEM	1.2–1.5[c]	1.4 ± 0.1	1.3 ± 0.2	1.6 ± 0.2
Ht (μm)				
LM		1.7 ± 0.2	1.5 ± 0.1	1.7 ± 0.1
TEM	1.37–1.61[c]	1.1 ± 0.1	1.1 ± 0.1	1.3 ± 0.1
Partical epicortical wall	Not observed	Surrounds endospore laterally, not basal or polar areas	Surrounds endospore somewhat sublaterally	Entirely surrounds endospore
Pore				
Presence	Absent	Present	Present	Present
Characteristics		Basal annular opening formed from thickened outer wall	Basal cortical wall thins to expose inner endospore	Thickness of basal wall and depth of pore constant
Diam (μm)		0.3 ± 0.1	0.1 ± 0.0	0.2 + 0.0
Parasporal structures				
Fibers, origin and orientation	Long primary fibers arise laterally from cortical wall, bending sharply downward to yield numerous secondary fibers arrayed internally toward granular matrix	Fibers arise directly from cortical wall, gradually arching downward to form attachment layer of numerous shorter fibers	Long fibers arise directly from cortical wall, bending sharply downward to form attachment layer of numerous shorter fibers	Same as *P. penetrans* but additional layer is formed on obverse surface of endospore
Matrix, at maturity	Persists as fine granular material	Becomes coarsely granular; lysis occurs; sporangial wall collapses; base is vacuolate	Persists, but more granular; some strands are formed and partial collapse may occur	Persist; numerous strands are formed and partial collapse may occur
Host	Cladocerans: *Daphnia* and *Moina* spp.	Root-knot nematodes: *Meloidogyne incognita*	Root-lesion nematodes: *Pratylenchus brachyurus*	Cyst nematodes: *Heterodera* spp.
Completes life cycle in nematode juveniles		No; only in adult	Yes; in all larval stages and adult	No; only in adults
Location in host	Hemocoel and musculature; sometimes found attached to coelom walls	Pseudocoelom and musculature; no attachment to coelom walls	Pseudocoelom and musculature; no attachment to coelom walls	Pseudocoelom and musculature; no attachment to coelom walls

(Continued)

Table 1—*Continued*

Trait[a]	*P. ramosa*	*P. penetrans*	*P. thornei*	*P. nishizawae*
Attachment of spores	Spores not observed to attach to or accumulate on surface of cladoceran	Spores accumulate in large numbers on cuticular surface	Spores accumulate in large numbers on cuticular surface	Spores accumulate on juveniles and adult males
Mode of penetration	Not known; suspected to occur through gut wall by hyphal strand	Direct penetration of nematode	Direct penetration suspected but not seen	Direct penetration suspected but not seen
Source of host	Pond mud, fresh water	Soil, plants	Soil, plants	Soil, plants

[a] LM, light microscopy; TEM, transmission electron microscopy.
[b] Data are from reference 30.
[c] Data are from reference 27.
[d] Data are from reference 25.

tized root-knot nematode females decomposes during the fall and winter months. Because the endospores are not actively motile in soil, contact with a susceptible nematode must be mediated by the motility of the nematode juvenile and by passive movements of bacterial endospores. Such passive movements depend on many factors. These probably include rate of soil water percolation, sizes of soil pore openings, surface charges of soil particles, tillage practices, and, to a lesser extent, activities of other soil invertebrates.

PATHOGENESIS

No reports in the literature involve the application of Koch's postulates (rules of proof in disease etiology) to any bacterial disease of plant-parasitic nematodes.

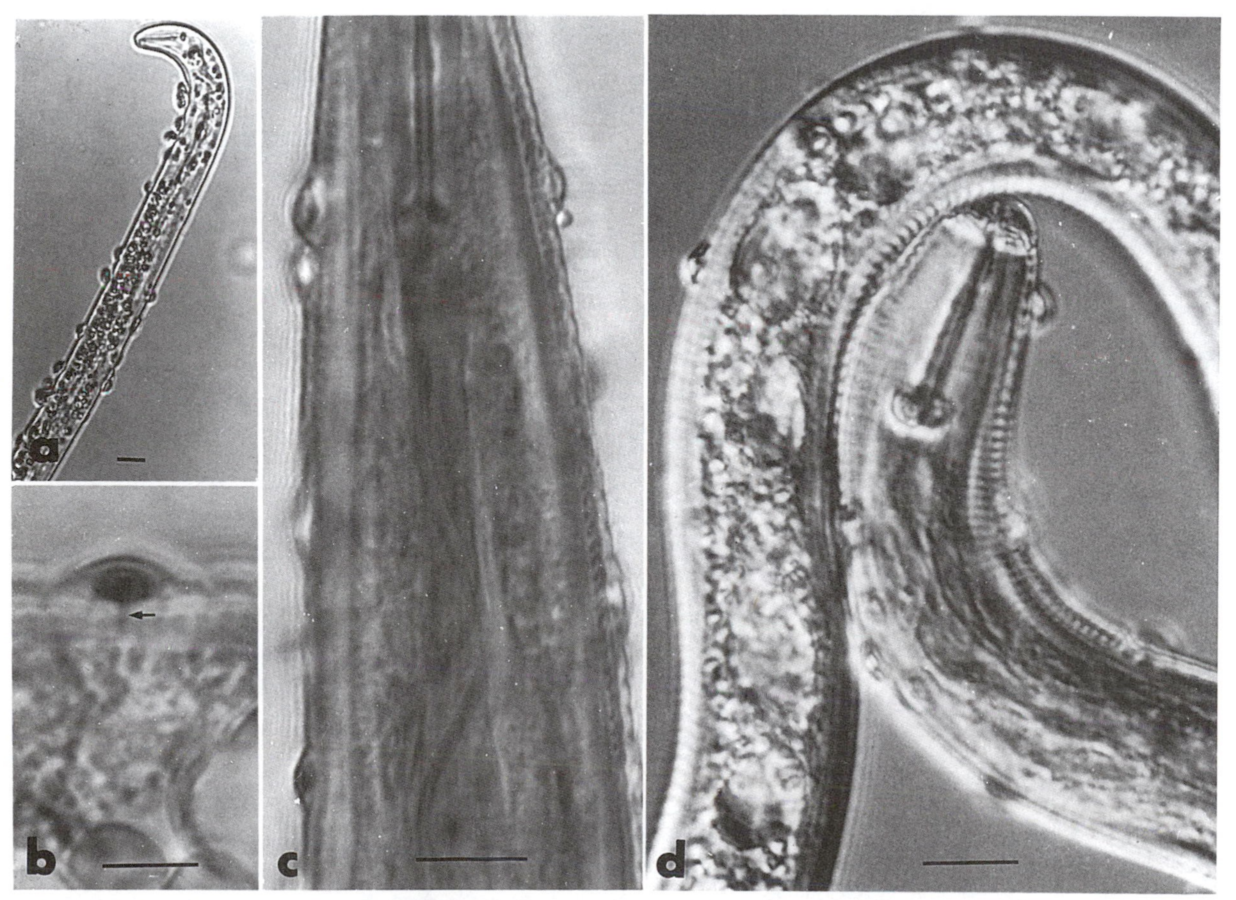

Figure 3. Attachment of endospores of *Pasteuria* spp. to vermiform stages of plant-parasitic nematodes. (a) Endospores of *P. penetrans* on the cuticle of the root-knot nematode *M. incognita*. Bar = 10 μm (from reference 32). (b) Endospore on the cuticle of a *Tylenchorhynchus* sp., showing an infection germ tube (arrow) entering the nematode's body. Bar = 5 μm. (c) Numerous endospores on the cuticle of a *Tylenchorhynchus* sp. Bar = 10 μm. (d) Endospores on the cuticle of the soybean cyst nematode *Heterodera glycines*. Bar = 10 μm (from reference 31).

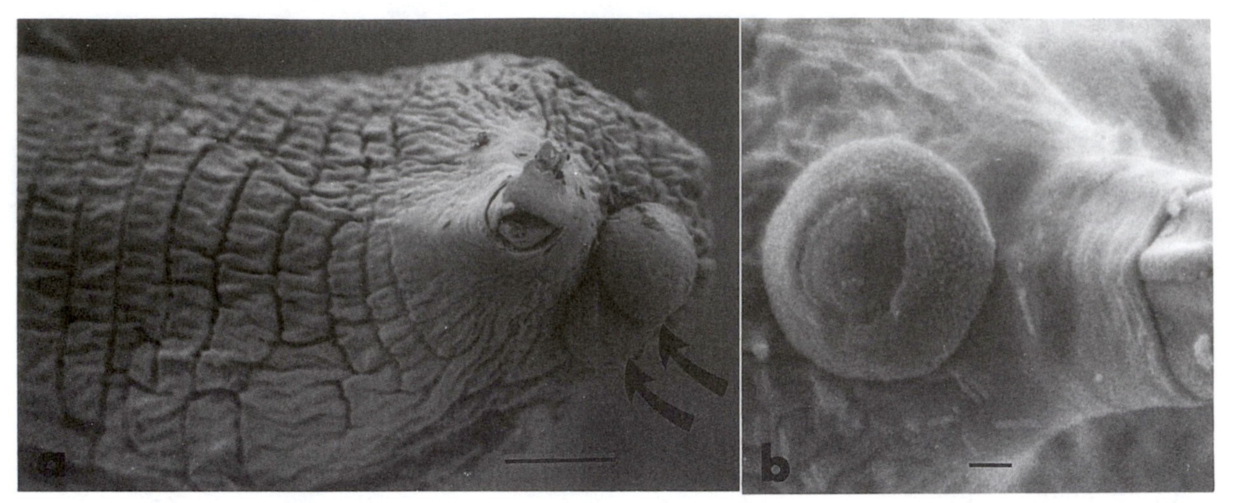

Figure 4. Scanning electron micrographs of endospores on the cuticle of the male soybean cyst nematode *H. glycines*. (a) Endospores (arrows) near the slightly extended spicules of the male. These spores are covered with the velutinous layer of the exosporium. Bar = 5 μm (from reference 32). (b) Lacking the exosporium, the central dome of the endospore is visible. This spore is attached near the base of the spicule. Bar = 1 μm (from reference 31).

Consequently, pathogenicity in the strict sense has not been firmly established for a species of *Pasteuria* that causes a disease in a nematode species. This deficiency stems from lack of a method for axenic isolation of the bacterium. Cultivation of *P. penetrans* in *M. incognita* females on oligoxenic excised tomato root culture has been reported in at least two instances (44, 45). This three-member cultivation may bring the bacterium a bit closer to axenic culture but not close enough for use in establishing Koch's postulates.

The current inability to axenically cultivate these bacteria is not from any lack of trying. Even Metchnikoff (18) mentioned his attempts at cultivation with gelatin, glycerin gels, and other nutrient media. It is generally accepted among investigators that eventual utilization of these bacteria as pest control agents depends on their cultivation in a relatively inexpensive medium. An inexpensive nematicidal product is needed so there can be widespread applications of it to soils infested with pest nematodes. With this goal in mind, most laboratories investigating *Pasteuria* species have attempted the organism's cultivation, but at this time, none has been fully successful (22, 47).

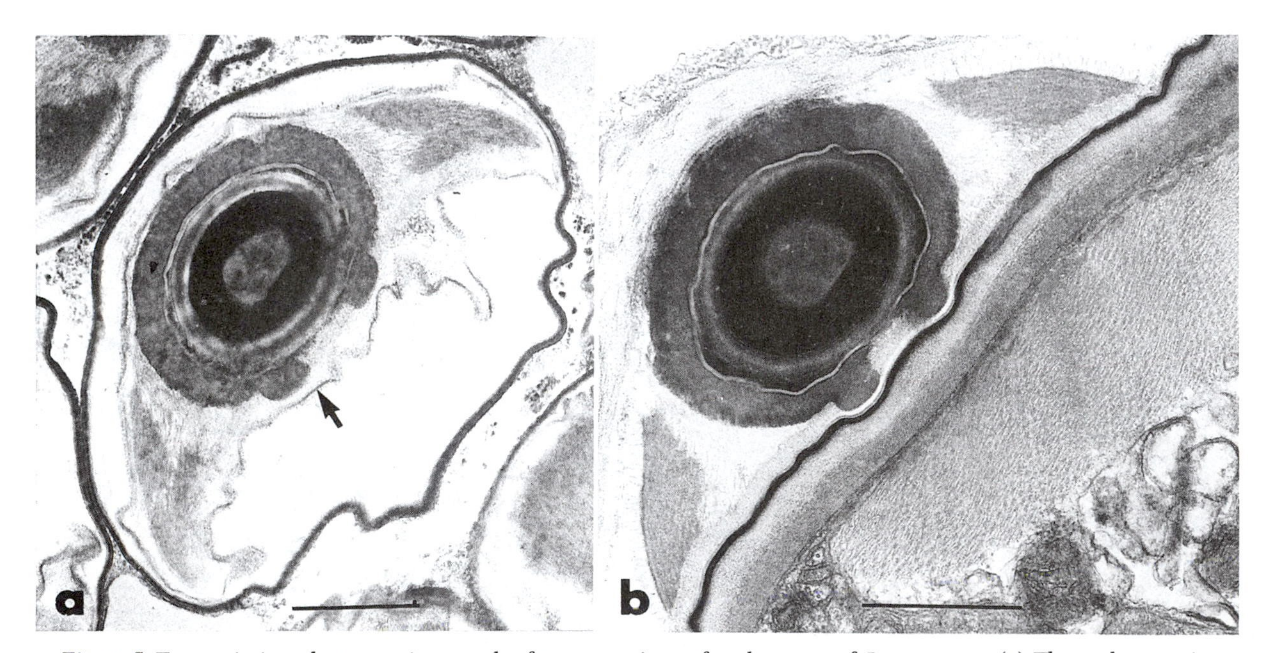

Figure 5. Transmission electron micrograph of cross-sections of endospores of *P. penetrans*. (a) The endospore is enclosed within its sporangial wall and thinner inner exosporium (arrows). Bar = 1 μm. (b) Cross-section of an endospore attached to the cuticle of a juvenile root-knot nematode. Bar = 1 μm.

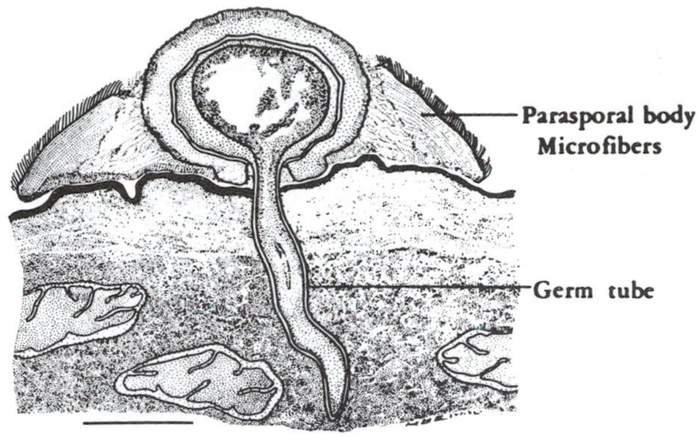

Figure 6. Line drawing showing the germ tube passing through the cuticle into the hypodermis of a juvenile of the root-knot nematode. (Drawn from an electron micrograph [29].) Bar = 1 μm (from reference 25).

ULTRASTRUCTURAL DIVERSITY IN SPORANGIAL TYPES

Investigators have examined several different plant nematode species parasitized by *Pasteuria* spp. and found various sporangial sizes in these different hosts (5, 23, 26, 27, 35). The diversity in fine structures of sporangia has been exploited as a means of separating the several species of *Pasteuria* found in various nematode hosts (27, 32, 35; Fig. 9, Table 1).

While these measurable morphological differences were significant, there are probably even more striking differences that cannot yet be measured, namely, the physiological differences involved in associations between a bacterium and its distinct nematode host. The widely differing physiologies of diverse nematodes parasitized by these bacteria certainly must require for successful parasitism by the bacterium an equally varied physiological response from each *Pasteuria* sp. One component of the wide differences is in feeding mode and habitat class of each parasitized nematode, i.e., free-living, predacious, motile endoparasitic, sedentary endoparasitic, and ectoparasitic. This listing mirrors more or less the durations of their life cycles, which vary from a few days for the free-living forms to 2 years or more for some of the ectoparasitic forms. For a *Pasteuria* sp. to be a successful parasite, it must have the necessary physiological compatibility with the various life cycles of different nematodes.

HOST RANGE AND GEOGRAPHICAL DISTRIBUTION

Pasteuria spp., including those designated under their earlier synonyms of "*D. penetrans*" and "*B. penetrans*," have been reported from about 204 different nematode species belonging to some 95 nematode genera in 10 orders. The bacterium has been reported in at least 12 states of the United States, in 51 countries, and on five continents (26, 42).

EFFECTIVENESS AS A BIOCONTROL AGENT

In the future, the introduction of *Pasteuria* spp. into agricultural soils may provide effective methods for biological control of some plant-parasitic nematode diseases. These control practices would lessen our current reliance on the use of chemicals to control nematodes and might help improve groundwater quality. Mankau and Prasad (12, 16) presented some of the first data on biocontrol of root-knot nematodes by *P. penetrans*. Their field trials suggested that small amounts of endospore-infested soil were an effective means of introducing the bacterium into soil to obtain control of root-knot nematodes.

Another method of introducing the bacterium into soil was developed by Stirling and Wachtel (40). In their method, large quantities of roots infested with root-knot nematodes parasitized by the bacterium were air dried and ground into fine powder. This powder was then bioassayed for its potential for controlling nematodes. Juvenile nematodes were added to aerated water extracts of the powder that contained suspended endospores. After 24 h in the suspension, juveniles were recovered and examined for endospores carried on their cuticles. The percentage of the population carrying endospores and the number of endospores on each juvenile served as indicators of the ability of the product to later parasitize root-knot nematodes in field soils. Stirling (38, 39) found that galling of tomato roots and number of nematodes in the soil harvest were reduced significantly when the powdered material was incorporated at 212 to 600 mg of powder per kg of soil. Effective biocontrol was achieved when at least 80% of the bioassayed juveniles were encumbered with 10 or more endospores per nematode. A few examples of natural suppression of nematode populations in agricultural soils have been associated with the occurrence of a *Pasteuria* sp. In South Australia, Stirling and White (41) conducted a survey of vineyards near the towns of Loxton and Cooltong and found *P. penetrans* occurring more frequently in the vineyard soils that had been in production for 25 years than in soils that had been planted to grapes for fewer than 10 years. Thirty-three to 40% of female nematodes in the older vineyards were parasitized, suggesting that the bacterium might be acting as a biological control agent of the nematode in the older vineyards.

From these same vineyards, Bird and Brisbane (1) examined seven soils for their suppressiveness of the reproduction of the root-knot nematode *Meloidogyne javanica* (Treub, 1885) Chitwood, 1949. Soils from the Cooltong district significantly reduced the numbers of egg masses produced by the root-knot nematodes. Suppression of nematodes in these soils was found to be biological in origin, as suppression stopped when the soil was autoclaved. Where inhibition of nematode reproduction occurred, endospores of *P. penetrans* were present in the adult females, replacing egg production. Unrelated soil bacterial isolates taken from these same soils showed no inhibitory effect on *M. javanica*. The presence of the *Pasteuria* sp. appeared to be the predominant soil factor that could be associated with the decline in nematode population.

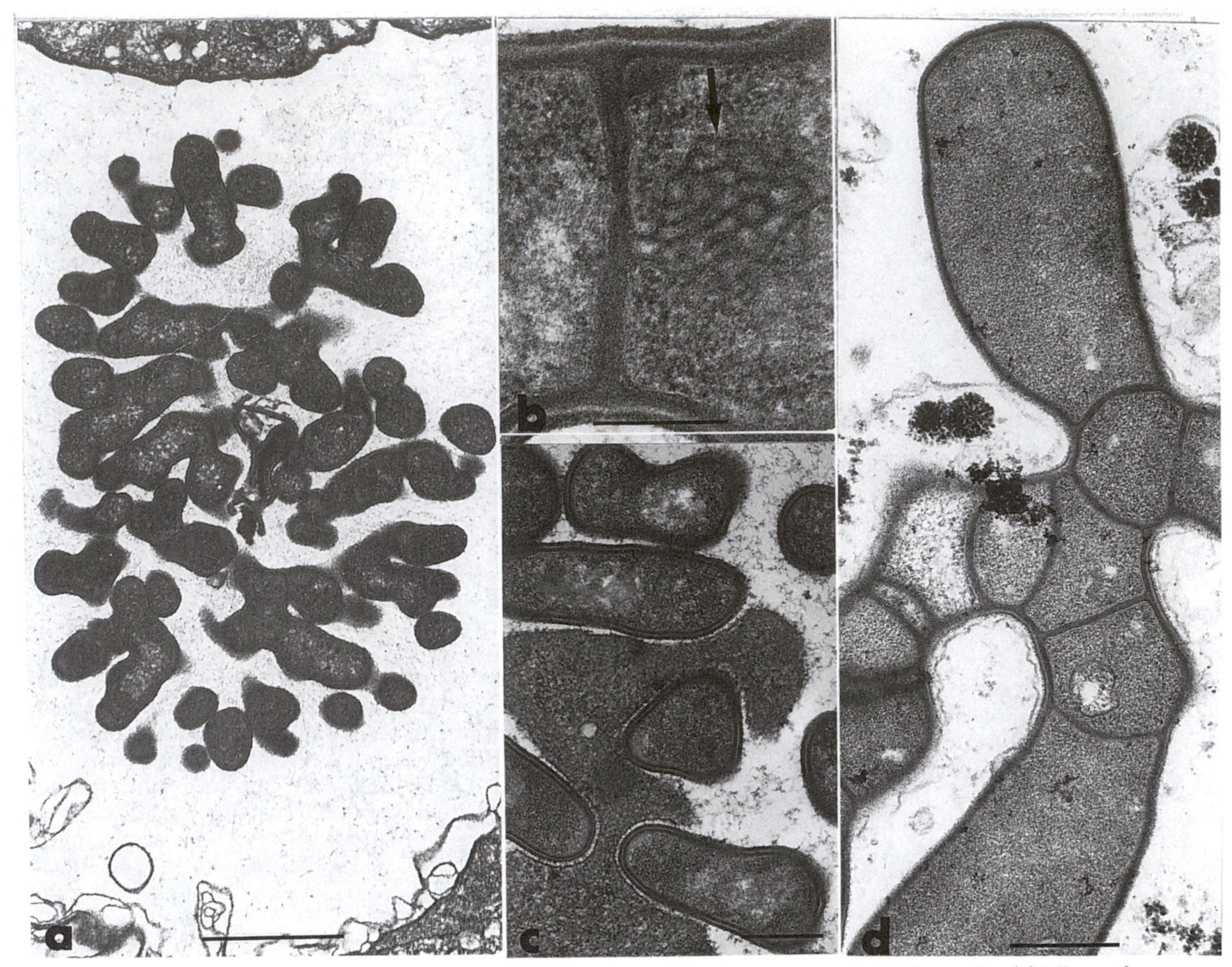

Figure 7. Vegetative life stages of *P. penetrans* parasitizing the root-knot nematode *M. incognita*. (a) Microcolony in the nematode's pseudocoelom. Bar = 0.2 μm (from reference 24). (b) Formation of a septum within a cell of a microcolony. A mesosome (arrow) is associated with the developing septum. Bar = 0.5 μm. (c) Cross-section of bacterial strands within tissues of a nematode. Bar = 0.5 μm (from reference 24). (d) Late stage in a microcolony, showing ovate tips that become mother cells for endospores. Bar = 0.5 μm.

Similarly, in newly established experimental plots, Nishizawa (21) monitored populations of rice cyst nematodes that were in monocultures of upland rice. Periodically, he sampled plots for their populations of the rice cyst nematode *Heterodera elachista* (Oshima) 1974. During a 6-year test period, populations of the nematode rose to about 380 cysts per 100 g of wet soil in the fourth year and then, after the appearance of a *Pasteuria* sp. in the plots, fell to a low of 40 cysts per 100 g of wet soil in the final year.

Another association between a declining population of root-knot nematodes, *Meloidogyne arenaria* (Neal, 1889) Chitwood, 1949, and the occurrence of *P. penetrans* was observed in Georgia soils (20). Field plots of 2 hectares located at Tifton, Ga., were used for 2 decades in research on plant-parasitic nematodes. Despite the continuous cropping of the plots to host plants of the nematode, nematode numbers decreased to the extent that yields of peanuts were no longer economically affected. These nematode-suppressive soils were bioassayed by using migratory second-stage larvae of *M. arenaria*. Larvae extracted from the suppressive soil carried the endospores. Some individuals had as many as 25 endospores on their cuticles. In greenhouse experiments using this nematode-suppressive soil, the incidence of root knot galling was found to be inversely related to the numbers of endospores found in the soil. When these soils were autoclaved, the suppression of root-knot nematode disease was lost. This suggests that the control agent was biological in origin. Minton and Sayre concluded that the naturally occurring population of *P. penetrans* acted as a biological control agent of *M. incognita* (20).

From the few observations of field decline of pest nematode populations, a parallel can be drawn with the first documented study of a nematode-suppressive soil, done by Gair et al. (4). Briefly, the continuous cropping of oats did not, as would be expected, result in high populations of the cereal cyst nematode *Heterodera avenae* Wollenweber, 1924, in the oat and barley fields of southern England. Three species of fungi that parasitized females and cysts of the nematode were found (10). Apparently, these fungal popu-

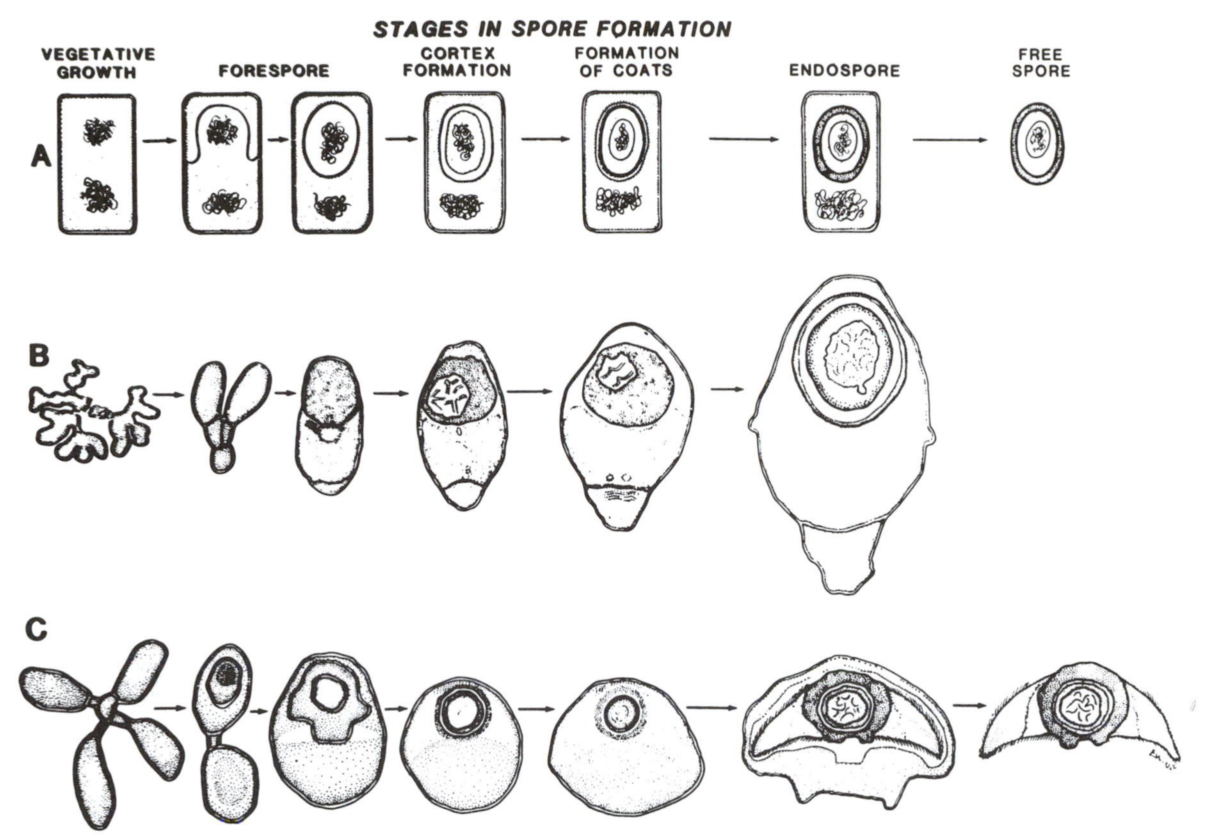

Figure 8. Line drawings of cross-sections showing generalized development of bacterial sporangia leading to endospore formation (A). These are compared with sporogenous stages of *P. ramosa* (B) and *P. penetrans* (C). Aside from parasporal structures of the two *Pasteuria* species, their life stages are very similar to those of other endospore-forming bacteria (from reference 25).

lations kept the number of cereal cyst nematodes below threshold densities that caused economic loss. From the field observations and data, a typical pattern of biological control emerges, whether the control agent is fungal or bacterial. Nematode-suppressive soils do occur naturally and have been discussed by Schneider (33) and Kerry (11).

Kerry (11) suggested a theoretical nematode population curve that depicts the onset of biological control but extends over several years in soils that are in crop monocultures. During the initial induction period, a nematode antagonist(s) increases in numbers but does not exert any appreciable control over nematodes. This period may be as long as 5 years, as is the case for *H. avenae* on oats. After the antagonists have increased in numbers, the nematode population falls below the threshold for economic damage. The long periods of continuous association between bacterial antagonists and nematodes appear to be necessary for the induction of a suppressive soil. The emergence and increase in population of a natural control agent of nematodes are most probable under annual monocropping or in perennial crops (11). The examples of nematode-suppressive soils that have been discussed appear to support this theory. Each example has involved a monocultured crop or a perennial host resulting in the apparent control of a nematode.

CONCLUSIONS

Pasteuria spp. appear to have characteristics that make them ideal control agents for protecting crops against several plant-parasitic nematodes. The bacteria pose no threat to the environment or to other soil invertebrates, because each species has a narrow host range and appears to be self-perpetuating only in soils that contain the targeted host nematode. Generally, endospores appear to withstand heat and desiccation, survive long periods of storage, and resist some soil pesticides (16). Besides these characteristics, the ideal nematicidal product would also have a reasonable shelf life, a necessary requirement for commercialization of the bacteria.

If the bacterium is eventually cultured in vitro and proves an effective biological control agent, the laborious procedures for the determination of species, which depend largely on ultrastructural differences, could be put aside. More-modern taxonomic methods would be able to keep pace with the numerous isolates needed to control the many distinctly different nematode problems. The classification scheme currently employed in cataloging the numerous isolates of *Bacillus thuringiensis* might be adapted to track the various effective biological control agents of nematodes.

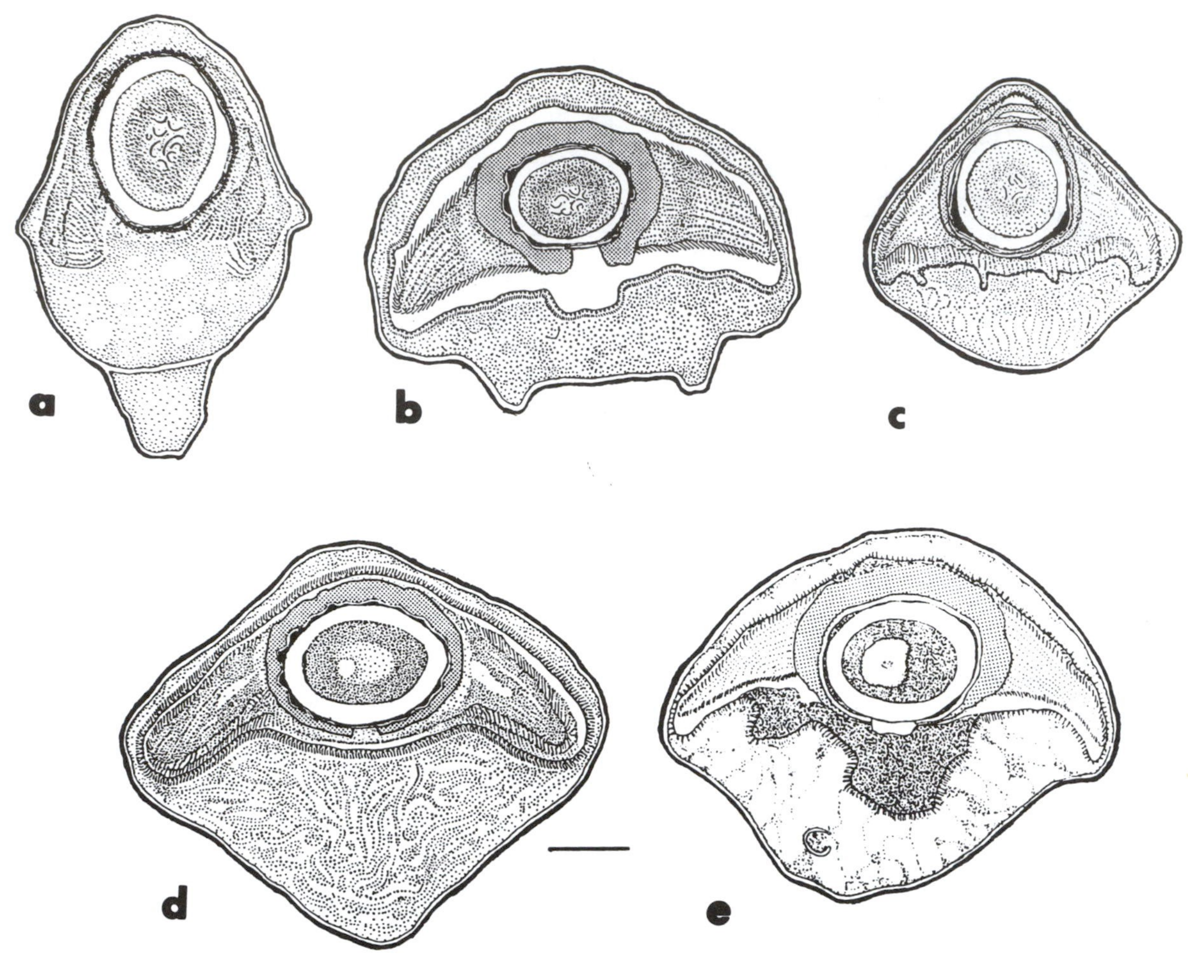

Figure 9. Line drawings of cross-sections of sporangia showing the diversity of fine structures among several species of *Pasteuria*. (a) *P. ramosa*; (b) *P. penetrans*; (c) *Pasteuria thornei*; (d) *P. nishizawae*; (e) *Pasteuria* sp. found in the pea cyst nematode *Heterodera goettingiana* Liebscher, 1892. Bar = 1 μm (modified and extended from reference 32).

Acknowledgments and Dedication. In preparing figures, the proficient technical help offered by Naeema Latif and Robert Reiss is acknowledged. Comments and suggestions of W. Friedman, Jean M. Schmidt, and Robert Gherna were helpful in drafting the final form of the manuscript. Conservation of the genus *Pasteuria* in the sense of Metchnikoff was largely due to the sustained efforts of the late M. P. Starr, whose expertise I sorely missed during the manuscript's preparation. The nine collaborative studies with Dr. Starr were drawn on heavily for this review and are gratefully acknowledged. I dedicate this presentation to his memory.

REFERENCES

1. **Bird, A. F., and P. G. Brisbane.** 1988. The influence of *Pasteuria penetrans* in field soils on the reproduction of root-knot nematodes. *Rev. Nematol.* **11:**75–81.
2. **Cobb, N. A.** 1906. *Fungus Maladies of the Sugar Cane, with Notes on Associated Insects and Nematodes*, 2nd ed. Hawaiian Sugar Planters Association bulletin no. 5. Hawaiian Sugar Planters Association, Honolulu.
3. **de Kruif, P.** 1926. *The Microbe Hunters*, p. 207–233. Cornwall Press, Inc., Cornwall, N.Y.
4. **Gair, R., P. L. Mathias, and P. N. Harvey.** 1969. Studies on cereal nematode populations and cereal yields under continuous or intensive culture. *Ann. Appl. Biol.* **63:**503–512.
5. **Giblin-Davis, R. M., L. L. McDaniel, and F. G. Bilz.** 1990. Isolates of the *Pasteuria penetrans* group from phytoparasitic nematodes in Bermuda grass turf. *J. Nematol.* **22**(Suppl. 4)**:**750–762.
6. **Henrici, A. T., and D. E. Johnson.** 1935. Studies of freshwater bacteria. II. Stalked bacteria, a new order of schizomycetes. *J. Bacteriol.* **30:**61–93.
7. **Hirsch, P.** 1972. Re-evaluation of *Pasteuria ramosa*, Metchnikoff 1888, a bacterium pathogenic for *Daphnia* species. *Int. J. Syst. Bacteriol.* **22:**112–116.
8. **Jirovec, O.** 1939. *Dermocystidium vejdovskyi* n. sp., ein neuer Parasit des Hechtes, nebst einer Bemerkung über *Dermocystidium daphniae* (Rühberg). *Arch. Protistenkd.* **92:**137–146.
9. **Judicial Commission.** 1986. Opinion 61. Rejection of the type strain of *Pasteuria ramosa* (ATCC 27377) and conservation of the species *Pasteuria ramosa* Metchnikoff, 1888 on the basis of the type descriptive material. *Int. J. Syst. Bacteriol.* **36:**119.
10. **Kerry, B. R.** 1981. Progress in the use of biological agents for control of nematodes, p. 79–90. *In* G. C. Papavizas (ed.), *Biological Control in Crop Production* (BARC Symposium no. 5). Allenheld and Osmum, Totowa, N.J.
11. **Kerry, B. R.** 1987. Biological control, p. 233–263. *In* R. H. Brown and B. R. Kerry (ed.), *Principles and Practices of Nematode Control in Crops*. Academic Press, Inc., New York.

12. **Mankau, R.** 1973. Utilization of parasites and predators in nematode pest management ecology, p. 129–143. *In Proceedings: Tall Timbers Conference on Ecological Animal Control by Habitat Management*, 24–25 February 1972. Tall Timbers Research Station, Tallahassee, Fla.
13. **Mankau, R.** 1975. *Bacillus penetrans* n. comb. causing a virulent disease of plant-parasitic nematodes. *J. Invertebr. Pathol.* **26:**333–339.
14. **Mankau, R.** 1975. Prokaryote affinities of *Duboscqia penetrans* Thorne. *J. Protozool.* **21:**31–34.
15. **Mankau, R., and J. L. Imbriani.** 1975. The life cycle of an endoparasite in some Tylenchid nematodes. *Nematologica* **21:**89–94.
16. **Mankau, R., and N. Prasad.** 1972. Possibilities and problems in the use of a sporozoan endoparasite for biological control of plant parasitic nematodes. *Nematropica* **2:**7–8.
17. **Mankau, R., and N. Prasad.** 1977. Infectivity of *Bacillus penetrans* in plant-parasitic nematodes. *J. Nematol.* **9:**40–45.
18. **Metchnikoff, E.** 1888. *Pasteuria ramosa*, un représentant des bactéries à division longitudinale. *Ann. Inst. Pasteur* (Paris) **2:**165–170.
19. **Migula, W.** 1900. *System der Bakerien*. Gustav Fisher, Jena, Germany.
20. **Minton, N. A., and R. M. Sayre.** 1989. The suppressive influence of *Pasteuria penetrans* discovered in Georgia on the reproduction of the peanut root-knot nematode, *Meloidogyne incognita*. *J. Nematol.* **24:**574–575.
21. **Nishizawa, T.** 1987. A decline phenomenon in a population of upland cyst nematode, *Heterodera elachista*, caused by the bacterial parasite, *Pasteuria penetrans*. *J. Nematol.* **19:**546.
22. **Reise, R. W., K. Hackett, and R. N. Heuttel.** 1991. Limited *in vitro* cultivation of *Pasteuria nishizawae*. *J. Nematol.* **23:**547–548.
23. **Sayre, R. M., J. R. Adams, and W. P. Wergin.** 1979. Bacterial parasite of a cladoceran: morphology, development in vivo, and taxonomic relationships with *Pasteuria ramosa*. *Int. J. Syst. Bacteriol.* **29:**252–262.
24. **Sayre, R. M., R. L. Gherna, and W. P. Wergin.** 1983. Morphological and taxonomic reevaluation of *Pasteuria ramosa* Metchnikoff, 1888 and *"Bacillus penetrans"* Mankau, 1975. *Int. J. Syst. Bacteriol.* **33:**636–649.
25. **Sayre, R. M., and M. P. Starr.** 1985. *Pasteuria penetrans* (ex Thorne 1940) nom. rev., comb. n., sp. n., a mycelial and endospore-forming bacterium parasitic in plant-parasitic nematodes. *Proc. Helminthol. Soc. Washington* **52:**149–165.
26. **Sayre, R. M., and M. P. Starr.** 1988. Bacterial diseases and antagonisms of nematodes, vol. 1, p. 69–101. *In* G. O. Poinar, Jr., and H.-B. Jansson (ed.), *Diseases of Nematodes*. CRC Press, Inc., Boca Raton, Fla.
27. **Sayre, R. M., and M. P. Starr.** 1989. Genus *Pasteuria* Metchnikoff, 1888, p. 2601–2615. *In* S. T. Williams (ed.), *Bergey's Manual of Systematic Bacteriology*, vol. 4. The Williams & Wilkins Co., Baltimore.
28. **Sayre, R. M., M. P. Starr, A. M. Golden, W. P. Wergin, and B. Y. Endo.** 1988. Comparison of *Pasteuria penetrans* from *Meloidogyne incognita* with a related mycelial and endospore-forming bacterial parasite from *Pratylenchus brachyurus*. *Proc. Helminthol. Soc. Washington* **55:**28–49.
29. **Sayre, R. M., and W. P. Wergin.** 1977. Bacterial parasite of a plant nematode: morphology and ultrastructure. *J. Bacteriol.* **129:**1091–1101.
30. **Sayre, R. M., W. P. Wergin, and R. E. Davis.** 1977. Occurrence in *Monia* [sic] *rectirostris* (Cladocera: Daphnidae) of a parasite morphologically similar to *Pasteuria ramosa* Metchnikoff, 1888. *Can. J. Microbiol.* **23:**1573–1579.
31. **Sayre, R. M., W. P. Wergin, T. Nishizawa, and M. P. Starr.** 1991. Light and electron microscopy of a bacterial parasite from the cyst *Heterodera glycines*. *J. Helminthol. Soc. Washington* **58:**69–81.
32. **Sayre, R. M., W. P. Wergin, J. Schmidt, and M. P. Starr.** 1991. *Pasteuria nishizawae* sp. nov., a mycelial and endospore-forming bacterium parasitic on cyst nematodes of the genera *Heterodera* and *Globodera*. *Res. Microbiol.* **142:**551–564.
33. **Schneider, R. W.** 1982. *Suppressive Soils and Plant Disease*. The American Phytopathological Society, St. Paul, Minn.
34. **Staley, J. T.** 1973. Budding bacteria of the *Pasteuria-Blastobacter* group. *Can. J. Microbiol.* **19:**609–614.
35. **Starr, M. P., and R. M. Sayre.** 1988. *Pasteuria thornei* sp. nov. and *Pasteuria penetrans sensu stricto* emend., mycelial and endospore-forming bacteria parasitic, respectively, on plant parasitic nematodes of the genera *Pratylenchus* and *Meloidogyne*. *Ann. Inst. Pasteur/Microbiol.* (Paris) **139:**11–31.
36. **Starr, M. P., R. M. Sayre, and J. M. Schmidt.** 1983. Assignment of ATCC 27377 to *Planctomyces staleyi* sp. nov. and conservation of *Pasteuria ramosa* Metchnikoff, 1888 on the basis of type descriptive material. Request for an opinion. *Int. J. Syst. Bacteriol.* **33:**666–671.
37. **Stirling, G. R.** 1981. Effect of temperature on infection of *Meloidogyne javanica* by *Bacillus penetrans*. *Nematologica* **27:**458–462.
38. **Stirling, G. R.** 1984. Biological control of *Meloidogyne javanica* with *Bacillus penetrans*. *Phytopathology* **74:**55–60.
39. **Stirling, G. R.** 1988. Biological control of plant-parasitic nematodes, p. 93–139. *In* G. O. J. Poinar and H.-B. Jansson (ed.), *Diseases of Nematodes*. CRC Press, Inc., Boca Raton, Fla.
40. **Stirling, G. R., and M. F. Wachtel.** 1980. Mass production of *Bacillus penetrans* for the biological control of root-knot nematodes. *Nematologica* **26:**308–312.
41. **Stirling, G. R., and A. M. White.** 1982. Distribution of a parasite of root-knot nematodes in south Australian vineyards. *Plant Dis.* **66:**52–53.
42. **Sturhan, D.** 1988. New host and geographical records of nematode-parasitic bacteria of the *Pasteuria penetrans* group. *Nematologica* **34:**350–356.
43. **Thorne, G.** 1940. *Duboscqia penetrans* n. sp. (Sporozoa: Microsporidia, Nosematidae), a parasite of the nematode *Pratylenchus pratensis* (de Man) Filipjev. *Proc. Helminthol. Soc. Washington* **7:**51–53.
44. **Verdejo, S., and B. A. Jaffee.** 1989. Reproduction of *Pasteuria penetrans* in a tissue-culture system containing *Meloidogyne javanica* and *Agrobacterium rhizogenes*-transformed roots. *Phytopathology* **78:**1284–1286.
45. **Verdejo, S., and R. Mankau.** 1986. Culture of *Pasteuria penetrans* in *Meloidogyne incognita* on oligoxenic excised tomato root culture. *J. Nematol.* **18:**635.
46. **Weiser, J.** 1943. Beiträge zur Entwicklungsgeschichte von *Dermocystidium daphniae* Jirovec. *Zool. Anz.* **142:** 200–205.
47. **Williams, A. B., G. R. Stirling, A. C. Hayward, and J. Perry.** 1989. Properties and attempted culture of *Pasteuria penetrans*, a bacterial parasite of root-knot nematode (*Meloidogyne javanica*). *J. Appl. Bacteriol.* **67:**145–156.

8. *Bacillus anthracis*

CURTIS B. THORNE

Bacillus anthracis, the causative agent of anthrax, is the only member of the genus *Bacillus* that is capable of causing epidemic disease in humans and other mammals. Before a vaccine for domestic animals became available, anthrax was a serious disease among farm animals in many parts of the world. Instances of anthrax among humans are usually traceable, either directly or indirectly, to contact with infected animals. For example, cases of anthrax were once fairly common among employees in processing plants who worked with contaminated wool, hair, and hides. Although the use of an effective veterinary vaccine and the availability of antibiotic treatment have reduced the frequency of animal and human cases of anthrax, the disease remains endemic in various regions of the world. The epidemiology and clinical characteristics of anthrax, as well as methods of isolation and identification of *B. anthracis*, have recently been summarized (126), and these aspects are not discussed in any detail in this chapter.

THE ORGANISM

B. anthracis is similar in many respects to *Bacillus cereus* and *Bacillus thuringiensis*. Characteristically *B. anthracis* grows in longer chains than does *B. cereus* or *B. thuringiensis*, and unlike most strains of the last two species, it is nonmotile. Reported G+C contents of *B. anthracis* DNA range from 32 to 37 mol%, which is very similar to those reported for DNAs of *B cereus* and *B. thuringiensis* (9, 45, 74). Virulent strains of *B. anthracis* are easily distinguished from the two other species by virtue of the facts that all virulent strains produce capsules and that colonies are thus very mucoid under conditions favorable for capsule synthesis (see below). However, the distinction between *B. cereus* or *B. thuringiensis* and noncapsulated strains of *B. anthracis* is more difficult. Much has been written concerning ways to distinguish between *B. anthracis* and other *Bacillus* species (e.g., see references 1, 15, 17, 50, 51, 94, and 99). However, some of the most distinguishing characteristics of *B. anthracis* are attributable to its plasmids, pXO1 and pXO2, and strains cured of one or both of the plasmids are not easily identified.

THE *B. ANTHRACIS* CAPSULE

The capsule is one of the virulence factors of *B. anthracis*. Synthesis of the capsule is encoded by plasmid pXO2 (29, 128). The capsule is composed largely, if not entirely, of glutamyl polypeptide. There is some uncertainty whether glutamyl polypeptide is the only constituent of the capsule (2, 118). The fact that in *B. anthracis* the glutamyl polypeptide adheres to the cell as a capsule while in other organisms, such as *B. licheniformis*, a similar polypeptide is released into the medium suggests that in *B. anthracis* there may be other structural constituents of the capsule. If there are other constituents in the *B. anthracis* capsule, they are most likely not linked chemically to the polypeptide. When capsulated cells of *B. anthracis* are autoclaved in water, the polypeptide is released, and it can be isolated in pure form by precipitation with ethanol.

The glutamyl polypeptide of *B. anthracis* is composed entirely of the D isomer of glutamic acid (35, 41, 107). Chemical evidence (13, 14, 120, 131) and results of experiments on specific enzymatic hydrolysis (107) indicate that the glutamyl residues are connected largely, if not entirely, through the gamma carboxyl groups (see reference 12 for a review).

Some other species of *Bacillus* also produce glutamyl polypeptide. Among those most studied are *Bacillus megaterium* (34, 119, 121), *Bacillus subtilis* (*natto*) (36), and *Bacillus licheniformis* (53, 114). (The organism referred to as *B. subtilis* 9945A in these last two references has since been classified as *B. licheniformis* 9945A.) Preparations of polypeptides from the various species have been reported to contain various proportions of the D and L isomers of glutamic acid. However, it was shown (114) that *B. licheniformis* actually produces two glutamyl polypeptides, one composed of the D isomer and one composed of the L isomer. The isolated D and L peptides possess the interesting characteristic of coprecipitating stoichiometrically from acid solution; separately, however, each is soluble (108, 114). This explains earlier observations that preparations of peptide from *B. anthracis* were soluble in acid, while various preparations from *B. subtilis* were insoluble in acid. The report (130) that an enzyme quantitatively released L-glutamic acid from glutamyl polypeptide preparations containing both isomers of glutamic acid is further evidence that the polypeptide produced by *B. licheniformis* is a mixture of two separate peptides. In *B. licheniformis*, the proportion of total glutamyl polypeptide that was composed of the D or L isomer was determined by the concentrations of Mn^{2+}, Co^{2+}, or Zn^{2+} ions in the growth medium, with higher concentrations favoring synthesis of peptide containing the D isomer (53). In contrast to these results, determinations did not show any significant amount of L-glutamic acid in polypeptide preparations from *B. anthracis* (107).

Leonard and Housewright (52) demonstrated polyglutamic acid synthetase activity in disrupted cells of

Curtis B. Thorne • Department of Microbiology, University of Massachusetts, Amherst, Massachusetts 01003.

B. licheniformis. Labelled L-glutamic acid was the precursor of both D- and L-glutamic acids incorporated into polypeptide, and D-glutamic acid was not an intermediate. Purified *B. licheniformis* γ-glutamyl peptidase, which does not hydrolyze D-α-polyglutamic acid or L-α-polyglutamic acid but does hydrolyze γ-polyglutamic acid containing either L- or D-glutamic acid, was used to show that the glutamic acid in the polypeptide synthesized in the cell-free reaction was γ linked. These studies on the mechanism of glutamyl polypeptide synthesis by *B. licheniformis* were later expanded by Troy and Gardner (26, 123, 124). A membranous polyglutamic acid synthetase complex was shown to catalyze a series of reactions in which L-glutamic acid was activated, racemized, and polymerized. In those studies, the polymer synthesized in vitro apparently contained only the D isomer of glutamic acid. Thus, it is not clear how the biosynthetic reactions for synthesis of the L- and D-polypeptides differ. Nor is it clear how the concentrations of Mn^{2+}, Co^{2+}, or Zn^{2+}, referred to above, influence the proportions of the two peptides synthesized by growing cultures of *B. licheniformis*.

Genes involved in glutamyl polypeptide synthesis by *B. licheniformis* are located on the chromosome, and some of them have been mapped by transduction with phage SP-15 (66). In *B. anthracis*, capsule synthesis is encoded by plasmid pXO2 (29, 128), and three genes have been identified (64) (see below). Synthesis of glutamyl polypeptide by cell-free preparations of *B. anthracis* has not been reported. However, it has been shown that in *B. anthracis*, as in *B. licheniformis*, the precursor of the glutamic acid in the D-polypeptide is L-glutamic acid rather than D-glutamic acid (107).

An outstanding difference between *B. anthracis* and *B. licheniformis* with respect to glutamyl polypeptide synthesis is the requirement of bicarbonate by *B. anthracis*. Virulent strains of *B. anthracis* are encapsulated when they grow in an infected host, but when grown in vitro, they produce capsules only in the presence of bicarbonate and an atmosphere rich in CO_2. On agar medium containing bicarbonate and incubated in 20% CO_2, colonies of encapsulated cells of *B. anthracis* are very mucoid (S type), but in the absence of added bicarbonate and CO_2, the cells do not form capsules and the colonies are rough (R). Upon prolonged incubation, occasional mucoid colonies of encapsulated cells produce R-type outgrowth. When such outgrowth is subcultured in the presence or absence of bicarbonate and CO_2, it produces colonies of the R type. Culturing *B. anthracis* in a bicarbonate medium in a CO_2-rich atmosphere offers a selective advantage for the R type (108), and broth cultures originating from the mucoid encapsulated type but grown under these conditions may contain from <1 to $>90\%$ of the R type depending on the age of the culture and the strain used. This change from the virulent S type to the nonvirulent R type has been known for many years, and Sterne (103, 104) demonstrated that the R type could be used as an effective vaccine to immunize animals against anthrax. An important question regarding the use of the R type as a vaccine dealt with whether such strains could revert to the virulent S type. McCloy (65) made the observation that noncapsulated cells of *B. anthracis* are lysed by virulent phage, while encapsulated cells are resistant. A method (108) based on this observation that could detect 1 to 10 S cells in the presence of 10^8 or more R cells was developed. When the method was applied to a series of R-type mutants, some of them were found to revert to the S type, but in others, no reversion could be detected. Furthermore, the use of phage as a selective agent permitted the isolation of mutants that were able to synthesize capsules in the absence of added bicarbonate or CO_2, i.e., CO_2-independent mutants. Meynell (68) made similar observations. It is of interest that the R-type strains isolated by Sterne and used as live vaccines were shown not to revert.

Once it was discovered that genes encoding capsule synthesis by *B. anthracis* are located on plasmid pXO2 (see below), it became clear why there are reverting and nonreverting R mutants. The Sterne vaccine strains have been cured of the capsule plasmid and thus cannot revert to the S type. R mutants that can revert have been shown to retain the capsule plasmid. By moving pXO2 from such strains by conjugation or by transduction to cells that had been cured of pXO2, we showed that mutations conferring the R phenotype were located on the plasmid (46, 47). In addition, mutations that conferred CO_2 independence for capsule synthesis were also shown to be located on the plasmid.

The exact role of bicarbonate in capsule synthesis has not been ascertained. Experiments with labeled CO_2 (21) led to the conclusion that the essential role of CO_2 or bicarbonate is not that of a carbon source. Other studies (67, 69) supported the idea that capsule synthesis depends not on the atmospheric concentration of CO_2 but on the concentration of bicarbonate ion in the medium and suggested that the role of bicarbonate is that of a regulatory agent. Further evidence that bicarbonate is involved in the regulation of capsule synthesis is presented below in the discussion of the capsule plasmid pXO2.

THE *B. ANTHRACIS* TOXIN

In 1937, Sterne (103) proposed that the virulence of *B. anthracis* is determined by the capsule and by another factor that he referred to as "factor A." He proposed that (i) immunity to anthrax produced by vaccine strains was due to factor A, (ii) a strain may become nonvirulent by loss of the ability to produce factor A while retaining the ability to produce capsules, and (iii) a strain may become nonvirulent by loss of the ability to produce capsules while retaining the ability to produce factor A. These proposals represent a simplified version of the current status of our knowledge of virulence of *B. anthracis*.

After studying anthrax in laboratory animals, Cromartie et al. (18, 19) concluded that the histopathological changes observed could be explained by the production of a diffusible poison, and they showed that crude extracts of anthrax lesions produced characteristic edema when injected into the skin of rabbits. The edema fluid was capable of immunizing rabbits against anthrax. Gladstone (28) demonstrated the presence in *B. anthracis* culture filtrates of protective antigen that would immunize animals against challenge doses of virulent strains. The presence of a specific lethal toxin that was related immunologically to the protective antigen and was presumably the same as the toxic material originally found in anthrax lesions was demonstrated by Smith et al. (97). For several years, it was believed that the anthrax toxin was made up of only two components (98, 116), but it was later established that the toxin actually consists of three proteins (5, 100). The early literature on the toxin components is confusing, because British workers referred to the three components as factors I, II, and III, and American workers referred to them as edema factor (EF), protective antigen (PA), and lethal factor (LF), respectively.

All three components of anthrax toxin have been purified. Their molecular weights are approximately 85,000 for PA, 83,000 for LF, and 89,000 for EF (55). None of the three components is toxic by itself. However, a combination of PA and EF injected intradermally into rabbits or guinea pigs produces edema (23, 116), and intravenous injection of PA and LF causes death (5). Genes for synthesis of all three components are carried on plasmid pXO1, and all three genes have been cloned and sequenced (see below).

Pezard et al. (75) constructed *B. anthracis* strains that were each deficient in production of one of the three toxin components and tested them for virulence in mice. Mutants deficient in LF or PA were not virulent, but those deficient in EF production were lethal. Edema formation was observed with the LF^- mutant. EF^- and LF^- mutants were less efficient at inducing lethality and edema, respectively, than was the parental Sterne strain, which produced all three toxin components. The results suggested that the three toxin components might act synergistically in the infected animal to cause lethality and edema.

Initial attempts to produce toxin in vitro were unsuccessful, but Harris-Smith et al. (37) succeeded in obtaining toxin production in a medium containing a large amount of serum. Thorne et al. (116) found that culture filtrates were toxic when *B. anthracis* was grown under conditions for PA production (112) in either a hydrolyzed casein medium or a completely synthetic medium but only when the media were supplemented with serum or gelatin. Culture filtrates obtained after growth of the organism in either medium without added protein were completely nontoxic, although the PA titers were equivalent to those of toxic culture filtrates from medium containing added protein. Further experimentation showed that *B. anthracis* produced all three toxin components in media without serum or gelatin but that in the absence of added proteins, EF and LF were adsorbed to the sintered-glass filters commonly used at that time for removing cells from culture fluids. The adsorbed toxin components could be eluted from the glass filter with alkaline buffer, and the EF and LF could be separated by further fractionation (5). Current methods for separation and purification of the three toxin components have been described elsewhere (49, 56, 79).

The EF component of anthrax toxin was shown by Leppla (54, 55) to be a calmodulin-dependent adenylate cyclase catalyzing intracellular cyclic AMP production from host ATP. Purified EF had no detectable adenylate cyclase activity when assayed in the absence of calmodulin. When EF in combination with PA was added to monolayer cultures of Chinese hamster ovary (CHO) cells, the cells elongated, which is characteristic of the change brought about by elevated cyclic AMP concentrations. Direct assay of cyclic AMP in the intoxicated cells confirmed the presence of elevated concentrations. Binding of EF to the CHO cells was dependent on the concentration of PA as well as the time of PA addition. PA could be added before EF or at the same time; if EF was added before PA, binding of EF did not occur. Binding of EF was blocked by the addition of LF, and this was consistent with the earlier observation that LF blocked the action of EF in animal tests (100).

These observations led to a proposed model (54, 55) for the mechanism of action of anthrax toxin. The model is an extension of a suggestion made earlier (72) that toxic mixtures and PA are fixed at the same tissue site and that toxic mixtures are fixed through their PA component. According to the model, PA first binds to specific cell receptors, creating new sites that can be used by both LF and EF to get into the cell. Once EF is inside the cell, it interacts with calmodulin to form an adenylate cyclase. It is assumed that LF also reacts within the cytoplasm, although no enzymatic activity has as yet been assigned to LF, and its actual target has not been identified.

Additional experiments have supported the model described above (57, 95, 96). Following binding of the 83-kDa PA to specific cell receptors, it is cleaved by a cellular protease, liberating a 20-kDa fragment and leaving the 63-kDa carboxyl-terminal end (PA63) bound to the cell receptor. PA63 has a site apparently not exposed in intact PA to which LF or EF binds. The PA63-LF or PA63-EF complex is then internalized, and its toxic activity is expressed in the cytosol. Low concentrations of trypsin can cleave PA to produce PA63, and the cleavage occurs in the sequence Arg-164–Lys-165–Lys-166–Arg-167. The PA gene was mutagenized in vitro to delete amino acid residues 163 to 168, and the mutant PA protein carrying the deletion was produced and purified from *B. subtilis*. The mutant protein was not cleaved by trypsin or by target cell protease, and it did not form a toxic complex with LF. The facts that the mutant PA would competitively block binding of wild-type PA to susceptible cells and would compete with wild-type PA in lethality tests with LF-PA mixtures in rats confirmed that the mutant PA retained the ability to bind to cell surface receptors.

Recent work (22, 81) has demonstrated that PA does in fact interact with a specific receptor on the surfaces of cells. Pretreatment of cells with a variety of proteases strongly inhibited PA binding, suggesting that the receptor may be a protein or at least partially protein. Using monoclonal antibodies and proteins encoded by deletant derivatives of the PA gene, Little and Lowe (61) determined that the domain of PA that binds to the cell receptor is located in the COOH-terminal region.

Linker insertion mutagenesis was employed by Quinn et al. (80) to create structural disruptions of LF to facilitate mapping functional domains. Insertions in the NH_2-terminal one-third of the protein eliminated toxicity and binding to PA, and insertions in the COOH-terminal one-third of the protein eliminated toxicity but did not affect binding to PA. These results support the hypotheses that structures required for binding to PA are located in the NH_2-terminal domain of LF and structures responsible for toxic activity are located in the COOH-terminal domain.

As stated above, the mechanism of action of LF has not been elucidated. Friedlander (25) demonstrated that LF in combination with PA lysed mouse peritoneal macrophages, as evidenced by loss of lactic dehydrogenase activity. This in vitro system for studying anthrax lethal toxin activity should expedite our understanding of the toxic reaction. For example, with this system, it has been demonstrated that calcium is required for expression of lethal toxin activity (8).

Role of Bicarbonate in Toxin Synthesis

Gladstone (28) demonstrated that bicarbonate was a necessary ingredient of the medium for production of PA in vitro by *B. anthracis*, and later work has shown that bicarbonate is required for the synthesis of all three components of the anthrax toxin. It was proposed several years ago (76) that bicarbonate increases the permeability of *B. anthracis*, resulting in a release of toxin that otherwise would accumulate inside the cell. However, more recent experiments (3) failed to demonstrate accumulation of toxin in *B. anthracis* cells grown in the presence or absence of bicarbonate. Strange and Thorne (106) showed that at pHs below 7.0, secreted PA was rapidly degraded by proteases present in the culture supernatant fluid. They suggested that the role of bicarbonate was that of a buffer to maintain the pH of the medium above 7.0. However, other buffers that maintained the medium at an elevated pH were not as effective as bicarbonate with respect to supporting maximum yields of PA. This suggested that bicarbonate had another specific effect in addition to its role as a buffer.

Recent work (3) has shown that bicarbonate functions in transcriptional regulation of the protective antigen gene (*pag*; see below). Data from hybridization experiments in which RNA prepared from *B. anthracis* was probed for PA mRNA show that bicarbonate is required for increased transcription of the PA gene in cells grown in a synthetic medium (R medium [83]) commonly used for PA production; transcription of *pag* was low in cells grown in rich medium and was not stimulated by bicarbonate. In further experiments, the promoter region of *pag* was fused to the chloramphenicol acetyltransferase gene (*cat-86*) of pPL703 (73) and introduced by electroporation into $pXO1^+$ (Tox^+; see below) and $pXO1^-$ (Tox^-) cells of *B. anthracis*. Analysis of chloramphenicol acetyltransferase in the transformants confirmed the results obtained by hybridization. Data obtained with the *pag–cat-86* fusion also revealed that plasmid pXO1 was required for stimulation of *pag* transcription by bicarbonate, suggesting that a *trans*-acting factor encoded by pXO1 was involved in activation of *pag* transcription.

Recent experiments (40) provide further evidence that bicarbonate is involved in regulation of the toxin genes. Insertion mutants obtained by transposon Tn*917* mutagenesis of pXO1 have a variety of phenotypes. Of particular interest are some of those in which the insertion maps outside the toxin structural genes. One class, in which synthesis of all three toxin components is abolished, maps between the PA and EF structural genes. A second class of mutants, which overproduces PA and LF in the presence as well as in the absence of added bicarbonate and CO_2, has Tn*917* inserted upstream of the LF structural gene. These results suggest that at least two regions of pXO1 are involved in regulation of toxin synthesis.

B. ANTHRACIS PLASMIDS

Virulent strains of *B. anthracis* carry two plasmids, pXO1 (70, 111) and pXO2 (29, 128). Genes for toxin synthesis are located on pXO1, and genes for capsule synthesis are located on pXO2. The toxin plasmid was originally referred to as pBA1 (90, 129). The capsule plasmid is also known as pTE702 (63, 64, 127, 128).

pXO1

Robertson et al. (86) prepared a restriction map of pXO1 and estimated the size of the plasmid to be about 184 kb. The three structural genes, *pag*, *cya*, and *lef*, for synthesis of PA, EF, and LF, respectively, are located on the plasmid. All three genes have been cloned and sequenced (see below).

Restriction analyses have shown differences in the restriction map of a 77-kb toxin-encoding region of pXO1 from Sterne (U.S. Army Medical Research Institute of Infectious Diseases [USAMRIID]) strain and those of pXO1 plasmids from several other strains, including Weybridge, Weybridge A, Anvax, Vollum, Ames, and New Hampshire (38). These differences are shown in Fig. 1. The restriction map generated in my laboratory of the toxin-encoding region of pXO1 from the Sterne (USAMRIID) strain is very similar to that reported by Robertson et al. (86) for a Sterne strain that is probably the same as the Sterne (USAMRIID) strain. The differences in restriction maps could best be explained by an inversion of approximately 40 kb of DNA, as depicted in Fig. 1. Further evidence for such an inversion was provided by a series of DNA-DNA hybridizations that showed that (i) the 34.8- and 14.6-kb *Bam*HI fragments of pXO1 from strain Weybridge A shared homology with the 29-kb *Bam*HI fragment of pXO1 from Sterne (USAMRIID), (ii) the 19- and 8.6-kb *Pst*I fragments of pXO1 from Sterne (USAMRIID) shared homology with the 18.7- and 9.3-kb *Pst*I fragments of pXO1 from Weybridge A, and (iii) the LF gene cloned from Sterne (USAMRIID) and shown to be on the 29-kb *Bam*HI fragment of pXO1 from that strain (87) hybridized to the 34.8-kb *Bam*HI fragment of pXO1 from strain Weybridge A. Among toxin plasmids carried by the seven *B. anthracis* strains mentioned above, pXO1 from Sterne (USAMRIID) appears to be unique with respect to having an inverted arrangement of toxin genes. Thus far, no phenotypic characteristics have been attributed to the inversion. I have not been able to trace the history of this particular Sterne strain, and its relationship to other Sterne-type strains is not known. Whether the inverted segment of DNA constitutes a transposon, defective or otherwise, or some other type of invert-

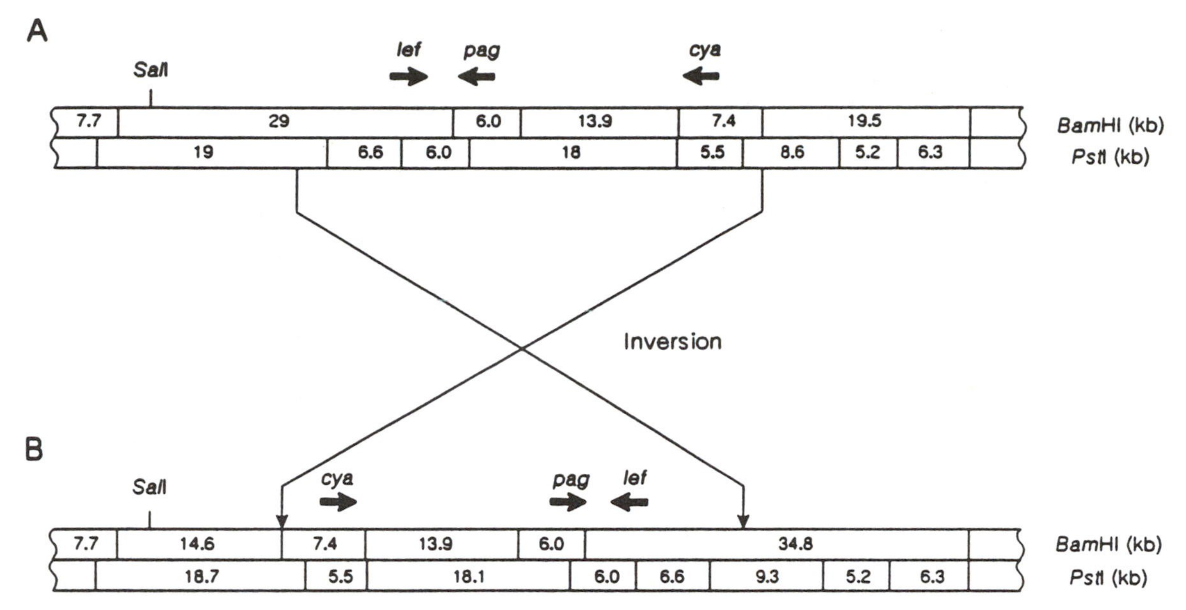

Figure 1. *Bam*HI and *Pst*I restriction maps of toxin-encoding region of pXO1. A 77-kb region of pXO1 encompassing the toxin structural genes is shown. (A) pXO1 from *B. anthracis* Sterne (USAMRIID). (B) pXO1 from *B. anthracis* Weybridge A. The maps suggest that an inversion has occurred, as depicted in the diagram. The inverted region represents approximately 40 kb. Arrows indicate the approximate locations and directions of transcription of the *pag*, *lef*, and *cya* genes as shown on the restriction map of pXO1 reported by Robertson et al. (86).

ible element remains to be seen. Knowledge of the nucleotide sequence of the presumed terminal areas of the segment might help answer this question.

No structural genes other than those involved in toxin synthesis have been identified on pXO1. However, in addition to losing toxin production, derivatives of the Weybridge A strain cured of pXO1 differed from the parent strain in colonial morphology, sporulated earlier and at a higher frequency than did the uncured parent, grew considerably more poorly than did the parent strain on minimal medium, and were more sensitive than was the parent strain to certain *B. cereus* bacteriophages (89, 90, 111). When cured derivatives were reinfected with pXO1 by conjugation (see below), the original phenotypes of the uncured Weybridge A strain were restored.

The Weybridge A strain (4), which is a mutant derived from the Weybridge (Sterne) strain (112), is oligosporogenous (Osp$^+$) at 37°C but sporulates normally at 30°C. Unlike the parent Weybridge strain, it grows well on a minimal medium composed of salts, glucose, thiamine, and six amino acids: glutamic acid, glycine, methionine, proline, serine, and threonine (89). However, Weybridge A derivatives cured of pXO1 grow very poorly on this medium. Addition of other individual amino acids, purines, pyrimidines, or vitamins did not improve growth of cured strains, although the addition of several amino acids together did result in improved growth. Thus, it appears very unlikely that loss of the plasmid results in acquisition of specific growth requirements. It seems more likely that loss of the plasmid affects regulatory activities.

The difference in growth of cured and uncured strains on minimal medium appears to be related to the change in sporulation characteristics of cured cells. Cured cells sporulate very early on minimal medium, and cell yields may be limited by the sporulation process. Asporogenous mutants occur at high frequencies in populations of cured cells, and such mutants grow as well as the uncured parent strain on minimal medium.

Cured cells are more sensitive to three bacteriophages, CP-51, CP-2, and CP-20 (89, 90). Although the phages infect both cured and uncured strains, the phage yields produced on cured cells are considerably greater than those produced on the parent strain. Furthermore, under normal conditions used for assay of these phages at 30°C, they produce distinct plaques in lawns of cured cells but fail to produce visible plaques on lawns of cells carrying pXO1. A pXO1-encoded restriction system was ruled out as the cause of this phenomenon, and the fact that the phages formed plaques on Spo$^-$ mutants carrying pXO1 suggested that there was no direct interference by the plasmid itself (89).

The PA gene

The PA gene was first cloned in *Escherichia coli* (129) by ligating into pBR322 a 6-kb *Bam*HI fragment containing the PA gene from pXO1 isolated from *B. anthracis* Sterne (USAMRIID). Two clones containing the recombinant plasmids pSE24 and pSE36 were isolated and shown to produce very low quantities (5 to 10 ng/ml) of PA that was biologically and serologically active. The PA was present in cell lysates but not in culture supernatant fluids.

The PA gene from the recombinant plasmid pSE36 of *E. coli* was cloned into *B. subtilis* (43) by ligating the 6-kb *Bam*HI fragment into pUB110. Two transformants, designated PA1 and PA2 and carrying the recombinant plasmids pPA101 and pPA102, respectively, were isolated. In both recombinant plasmids, part of the original 6-kb fragment was deleted, but the complete PA gene remained intact. Both strains produced more PA than did the *B. anthracis* Sterne strain from which the PA gene was derived, and PA was present in the culture supernatant fluid. In contrast to *B. anthracis*, the *B. subtilis* recombinant strains did not require addition of bicarbonate to the growth medium for PA production. Immunization with cells of each of the recombinant strains protected guinea pigs against an otherwise lethal challenge with *B. anthracis* spores and protected rats against intravenous challenge with anthrax lethal toxin.

The nucleotide sequence of the PA gene cloned in *E. coli* has been determined (134). The open reading frame has 2,292 bp; 2,205 of these encode the 735 amino acids found in the secreted PA. Preceding this region are 29 codons that encode the signal peptide.

The EF gene

The EF gene from pXO1 carried by *B. anthracis* Sterne has been cloned in two laboratories. Mock et al. (71) cloned the gene in plasmid pACYC184 carried by *E. coli* TP610, which was adenylate cyclase deficient. The strain was first transformed with plasmid pVUC-1, which encodes synthesis of a synthetic model of calmodulin (84), and used as a recipient for cloning the pXO1 fragments. Recombinant clones were screened for adenylate cyclase production on MacConkey agar plates containing maltose. Clones unable to synthesize cyclic AMP could not ferment maltose and produced white colonies; clones that produced cyclic AMP could ferment maltose and were red. Several Cya^+ clones were isolated; the sizes of the inserts ranged from 6 to 9 kb. From results of subcloning experiments, the DNA region encoding the adenylate cyclase was narrowed to a 3- to 4-kb fragment.

Tippetts and Robertson (117) also cloned the *B. anthracis cya* gene in *E. coli*. They screened a recombinant DNA library of pXO1 from *B. anthracis* Sterne for hybridization with a synthetic oligonucleotide that was specific for the first five amino acids of EF.

The nucleotide sequence of the EF gene cloned in *E. coli* has been determined (88). The open reading frame has 2,400 bp; 2,301 of these encode the mature EF protein, which contains 767 amino acids. The remaining 99 bp encode a signal peptide containing 33 amino acids.

The LF gene

The LF gene has also been cloned. A recombinant DNA library of pXO1 from *B. anthracis* Sterne was created by ligating fragments to pUC8 and transforming *E. coli* with the recombinant plasmids (87). Colonies carrying recombinant plasmids were screened by an immunological assay for LF production. Transcription of the LF gene in *E. coli* was apparently under the control of its *B. anthracis* promoter. The LF protein produced in *E. coli* was not secreted, but it could be demonstrated in sonic extracts. It was biochemically active; in the presence of PA, it was toxic to mouse macrophages.

The nucleotide sequence of the cloned LF DNA has also been determined (10, 85). The open reading frame has 2,427 bp; 2,328 of these encode the 776 amino acids of the mature secreted protein. The remaining 99 encode a signal peptide composed of 33 amino acids. The N termini of the sequenced LF and EF genes share extensive homology, and Bragg and Robertson (10, 85) suggested that the homologous regions represent the PA-binding domains.

pXO2

Genes for capsule synthesis in *B. anthracis* are found on the approximately 95-kb plasmid designated pXO2 or pTE702. Restriction maps of pTE702 (127) and pXO2 (32, 86) are very similar, if not identical, and no differences in properties have thus far been reported. Thus, it appears likely that all encapsulated strains of *B. anthracis* carry the same plasmid.

Proof that pXO2 is involved in capsule synthesis came from experiments in which the plasmid was transferred by phage CP-51-mediated transduction and by a mating system in which plasmid transfer was mediated by a *B. thuringiensis* fertility plasmid, pXO12 (29). Cells of *B. cereus* as well as cells of *B. anthracis* previously cured of pXO2 became Cap^+ after they were infected with pXO2.

Genes required for capsule synthesis were cloned from the capsule plasmid into *E. coli* via the shuttle vector pHY300PLK (63). They were then subcloned into pUB110 and introduced into *B. anthracis* by protoplast transformation. The *cap* genes were expressed in *E. coli* and *B. anthracis*; both organisms were observed microscopically to produce capsules, and CO_2 was required for capsule formation. However, the amount of capsular material produced by the recombinant strains was much less than that produced by *B. anthracis* carrying the wild-type capsule plasmid. When the cloned genes were complemented by a derivative of the wild-type plasmid in which one or more *cap* genes was insertionally inactivated, wild-type amounts of capsular material were synthesized. Makino et al. (63) concluded from these results that the cloned plasmid fragment contains all the genes necessary for capsule synthesis but that additional DNA sequences are required to achieve wild-type amounts of capsular material.

Three genes (*capB*, *capC*, and *capA*, in that order of arrangement) were reported to be involved in capsule synthesis, and from minicell analysis, it was concluded that the three genes encode proteins of 44, 16, and 46 kDa, respectively (64). The nucleotide sequence of the complete *cap* region was determined (64). There were three open reading frames: *capB* (397 amino acids; molecular weight, 44,872), *capC* (149 amino acids; molecular weight, 16,522), and *capA* (411 amino acids; molecular weight, 46,420). The three genes were transcribed in the same direction, and each had its own promoter. In *E. coli* minicells, the three proteins were associated with the membrane. Makino et al. (64) concluded that the three proteins are membrane-associated enzymes mediating the synthesis of poly-D-glutamate in *B. anthracis*.

Transposon mutagenesis with Tn*917* in *B. anthracis* has produced several different classes of mutants with respect to the requirements of bicarbonate and CO_2 for capsule synthesis (33, 133). Insertion of Tn*917* at various sites on pXO2 resulted in the following phenotypes. One class produces greater amounts of capsular material than wild type in the presence of CO_2

and bicarbonate; a second class is Cap$^+$ when grown with or without CO_2 and bicarbonate; a third class is Cap$^+$ when grown in air or CO_2, and bicarbonate inhibits their growth; a fourth class requires CO_2 for growth and is Cap$^+$ only in the presence of CO_2 and bicarbonate; a fifth class is Cap$^+$ when grown in air without bicarbonate, and growth is inhibited by CO_2; a sixth class retains pXO2 and is Cap$^-$ under all conditions of growth. In all of these mutants, the insertion has been shown to be in pXO2. However, another mutant in which the insertion is in the chromosome has the same phenotype, i.e., overproduction of capsular material, as the first class described above (32). These mutants, which are currently being analyzed, suggest that bicarbonate regulation of capsule synthesis in *B. anthracis* is a complex process.

GENETICS AND PHYSIOLOGY

Nutritional Requirements

The nutritional requirements of *B. anthracis* are complicated. For most strains, the only absolute requirements for growth are methionine and thiamine; however, no strains have been reported to grow in a glucose-salts medium supplemented only with methionine and thiamine. Gladstone (27) grew *B. anthracis* in a synthetic medium containing 18 amino acids, glucose, and salts. He reported that *B. anthracis* could be "trained" to utilize ammonia as the sole source of nitrogen by gradually reducing the number of amino acids in the medium during serial subculture. However, with ammonia as the sole nitrogen source, growth was very poor and delayed. Brewer et al. (11) devised a synthetic medium that contained 19 amino acids, uracil, adenine, guanine, ribose, glucose, thiamine, and salts. Cell yields approached those obtained in a hydrolyzed casein medium.

In my own laboratory, where we have worked principally with the Sterne (Weybridge) strain of *B. anthracis*, simpler media have allowed satisfactory growth for many purposes. Although the Sterne (Weybridge) strain grows reasonably well only when several amino acids are provided, it does not require specific amino acids except methionine. Therefore, one can use media prepared with various groups of amino acids and thus isolate auxotrophic mutants requiring particular amino acids. One such medium contains nine amino acids, glucose, thiamine, and salts. An even simpler medium contains only glutamic acid, glycine, methionine, proline, serine, and threonine in addition to glucose, thiamine, and salts. This medium has allowed the isolation of several auxotrophic mutants; among them are mutants requiring anthranilic acid, tryptophan, valine, leucine, phenylalanine, purine, uracil, nicotinic acid, riboflavin, and arginine plus uracil (probably analogous to *pyrA* mutants in *B. subtilis*).

D-Amino Acid Transaminases

Cell extracts of *B. anthracis* were shown to be active in transamination reactions involving D amino acids (115), and some of the D transaminations catalyzed by *B. anthracis* were different from those demonstrated in *B. licheniformis* (referred to as *B. subtilis* in the original publication) (113). Several amino acids were tested, and with the exceptions of glutamic acid and alanine, only the L isomers transaminated with α-ketoglutaric acid, and only the D isomers transaminated with pyruvate. As with preparations from *B. licheniformis*, stability data and results of ammonium sulfate fractionation indicated that the *B. anthracis* enzymes catalyzing the D transaminations were distinct from those active with L amino acids. One of the most active reactions was that between D-phenylalanine and pyruvate, forming alanine and phenylpyruvate. Cell extracts also contained an active alanine racemase, and it appears that alanine may be a connecting link between the metabolism of L and D amino acids. Thus, these organisms should be able to convert one isomer of an amino acid to a mixture of the two isomers by way of transamination and racemization of alanine.

Gene Exchange Systems

Transduction

Two generalized transducing phages, CP-51 and CP-54, were reported for *B. anthracis* several years ago (109, 110, 135). Both phages were originally isolated from soil in experiments in which *B. cereus* was used as the host organism, and both were found to carry out transduction in *B. anthracis* and *B. thuringiensis* as well as in *B. cereus*. The two phages are serologically related but not identical, and both are cold labile. Both are quite lytic for all three species and thus create some problems by lysing potential transductants. Inactivation of PFU with UV light and plating transduction mixtures in the presence of phage antiserum increase the number of transductants obtained. More recently, the use of a temperature-sensitive mutant of CP-51 reduced the number of potential transductants lost by lysis. Transducing lysates are prepared at a permissive temperature, but adsorption and selection of transductants are carried out at a nonpermissive temperature.

In addition to transduction of chromosomal markers (109), CP-51 is also useful for transduction of plasmids (92) among the species *B. anthracis*, *B. cereus*, and *B. thuringiensis*. The molecular mass of CP-51 DNA is about 60 MDa (136); thus, the phage can package and transduce reasonably large plasmids. There was no problem in obtaining Cap$^+$ transductants of *B. cereus* or strains of *B. anthracis* previously cured of pXO2 (approximately 63 MDa) when phage was grown on a donor carrying pXO2 (29, 30). Stepanov et al. (101) reported that CP-54 is also capable of transducing plasmids between strains of *B. cereus* and *B. anthracis*. They used a mutant of CP-54, designated CP-54ant, that had improved adsorption and a shortened latent period and was apparently more efficient than the wild-type phage in transducing plasmids.

A more recently isolated phage, designated TP-21, also carries out generalized transduction, including transduction of plasmids, in *B. anthracis*, *B. cereus*, and *B. thuringiensis* (91, 93). TP-21 is an interesting phage in that its prophage exists as a plasmid, the first plasmid prophage to be reported in *Bacillus* species. It was isolated from *B. thuringiensis* subsp. *kurstaki*, in which it exists as a plasmid, and it was found to have a fairly wide host range among *B. thuringiensis*, *B. cereus*, and *B. anthracis* strains.

Transformation

Several procedures for plasmid transformation (62, 78, 102) or electroporation (20, 78) of *B. anthracis* have been described. Reported frequencies are quite low, but the methods appear to be adequate for introducing relatively small plasmids into *B. anthracis*. In my laboratory, electroporation is the method of choice. Transformation of *B. anthracis* with chromosomal DNA has not been reported.

Transfer of plasmids by mating

The *B. thuringiensis* fertility plasmids, pXO11, pXO12, pXO13, pXO14, pXO15, and pXO16, isolated from various strains, function efficiently in *B. anthracis* (4, 82). *B. anthracis* recipients inheriting one of the fertility plasmids from *B. thuringiensis* are, in turn, effective donors to other strains of *B. anthracis* as well as to *B. cereus* and *B. thuringiensis*. Plasmid transfer was demonstrated to be unaffected by the addition of DNase to mating mixtures, and cell-free filtrates of donor cultures were ineffective, eliminating transformation and transduction as mechanisms of plasmid transfer. Cell-to-cell contact was demonstrated to be necessary, indicating a conjugationlike process (4). The number of transcipients derived from a mating with a *B. anthracis* donor increased 164-fold between 2 and 4 h of mixed incubation in broth and increased slowly after that. No transcipients could be detected before 2 h of mixed incubation, suggesting that a period of growth of donors and recipients together was required before plasmid transfer could occur. Transcipients were undetectable in mating mixtures prepared from donors and recipients that had been grown separately for increasing periods (4 to 16 h) before they were mixed.

The frequency of pXO12-mediated transfer of the 4.2-kb tetracycline resistance plasmid, pBC16 (7), was as high as 8×10^{-1} transcipients per donor. Larger plasmids were transferred at considerably lower frequencies. For example, pXO2, which had been tagged with Tn*917* to permit direct selection of the transferred plasmid, was transferred at a frequency 200- to 1,000-fold lower than the frequency of pBC16 transfer. Physical analysis of transferred plasmids suggested that pBC16 was transferred by donation and that the large *B. anthracis* plasmids, pXO1 and pXO2, were transferred by conduction (30). No alterations in pBC16 could be detected following its transfer. However, in the majority of transcipients that acquired pXO1 or pXO2 in pXO12-mediated matings, the transferred plasmid contained a copy of the *B. thuringiensis* transposon Tn*4430* (58) inherited from pXO12. Furthermore, several transcipients from independent matings contained cointegrates of pXO12 and pXO1 or pXO2, suggesting that Tn*4430* functions as a mediator of cointegrate formation between the conjugative plasmid pXO12 and the plasmid to be transferred by conduction. Other experiments (30) suggested that Tn*4430* is not unique in this function; in matings in which the donor carried pXO12::Tn*917* and pXO2, many of the Cap$^+$ transcipients inherited a pXO2 plasmid in which a copy of Tn*917* but not of Tn*4430* was inserted. Whether the five other fertility plasmids mentioned above contain Tn*4430* has not been determined; however, it was shown by DNA-DNA hybridization that homology exists among pXO11, pXO12, pXO13, and pXO14 (82), and it was suggested that Tn*4430* may account for part of the homology. pXO15 and pXO16 were not included in the homology tests.

Among the six *B. thuringiensis* conjugative plasmids mentioned above, only pXO12 carried genes for synthesis of parasporal crystals (4, 82). *B. anthracis* transcipients that inherited pXO12 produced parasporal crystals that in appearance and toxicity resembled those produced by the parent *B. thuringiensis* strain (111).

The *B. subtilis* (*natto*) plasmid pLS20 mediates interspecies plasmid transfer among a variety of *Bacillus* species, including *B. anthracis* (48). In general, the frequencies of pBC16 transfer mediated by pLS20 were not as high as those mediated by the *B. thuringiensis* fertility plasmids. In addition to pBC16, pLS20 also mobilized the *Staphylococcus aureus* plasmid pUB110 and the *B. subtilis* (*natto*) plasmid pLS19. Mobilization of several other plasmids, including pTV1 and pTV24 (138), pC194 (31), and pE194 (132), could not be demonstrated.

The mobilizable shuttle vector pAT187 (122) was shown to be transferred by conjugation from *E. coli* (pRK212.1) to *B. anthracis* (16). The *pag* gene of pXO1, as well as an insertionally inactivated derivative thereof, was cloned into pAT187. The clones were introduced into *E. coli*(pRK212.1) by transformation and transferred to *B. anthracis* by conjugation. Transfer frequencies were reported to be low, on the order of 10^{-8}, but reproducible. Thus, this shuttle system appears to be a useful and convenient method for transferring cloned genes from *E. coli* to *B. anthracis*.

Transposon Mutagenesis

In addition to the transposon mutagenesis of pXO1 and pXO2 described above, the temperature-sensitive transposition selection vector pTV1 (137) containing the 5.6-kb macrolide-lincosamide-streptogramin B resistance transposon Tn*917* has also been used for chromosomal insertions in *B. anthracis* (39). A variety of auxotrophic mutants were isolated and confirmed by transduction to result from insertion of Tn*917*.

The streptococcal tetracycline resistance transposon Tn*916* (24) was introduced into *B. anthracis* from *Streptococcus faecalis* by conjugation and used successfully to create Tn*916* insertion mutations in the chromosome (44). Ivins et al. (44) selected specifically for mutants requiring aromatic amino acids.

ANTHRAX VACCINES

The vaccine commonly used to immunize livestock against anthrax consists of live spores of a toxigenic, noncapsulated strain, i.e., a strain that has been cured of the capsule plasmid, pXO2, but still carries the toxin plasmid, pXO1. This type of strain was demonstrated by Sterne (104, 105) to be very effective in immunizing animals. The original strain used by

Sterne, strain 34F2, or derivatives thereof are still being used today. In Russia, live spores of a Sterne-type strain, designated STI, are used for livestock and humans (59). Capsulated nontoxigenic (pXO1$^-$ pXO2$^+$) strains are completely ineffective in providing protection against anthrax (42).

In the United States, the current anthrax vaccine for humans consists of antigens from the culture supernatant fluid of a Sterne-type strain, V770-NP1-R, adsorbed to aluminum hydroxide (77). The principal antigen in the vaccine is PA. Cultures of the producing strain are grown anaerobically on a large scale in a chemically defined medium. A similar preparation is also produced and used in England for immunizing humans. The *B. anthracis* strain used in England is a derivative of the original Sterne 34F2 strain. Following growth of the strain under static conditions in a hydrolyzed casein medium (112), the supernatant fluid is filtered through a mixed cellulose ester membrane. Most of the EF and LF are adsorbed by the filter. The PA is then precipitated from the filtrate with alum (6).

There are some serious problems associated with the PA vaccines described above. Several doses have to be given initially, and frequent booster doses are necessary to sustain antibody titers. A more serious problem is that these vaccines do not protect experimental animals against all strains of *B. anthracis*. For example, immunization with live spores of a Sterne-type strain protected guinea pigs against challenge doses of 27 strains of *B. anthracis*, but in parallel tests, 9 of these strains were found to be vaccine resistant in guinea pigs immunized with the PA vaccine (60). Guinea pigs immunized with individual toxin components and having antibody titers comparable to those of guinea pigs immunized with live spores were not protected when challenged with the vaccine-resistant strains. Little and Knudson (60) concluded that antibodies to toxin may not be sufficient to protect against all strains of *B. anthracis* and that other antigens may be important in anthrax immunity. This is in agreement with the finding that anti-PA and anti-LF titers may not necessarily correlate with protective immunity against anthrax infection (125).

Acknowledgments. The experimental work from this laboratory was supported by contracts DAMD17-80-C-0099, DAMD17-85-C-5212, and DAMD17-91-C-1100 from the U.S. Army Medical Research Acquisition Activity.

REFERENCES

1. **Ash, C., J. A. E. Farrow, M. Dorsch, E. Stackebrandt, and M. D. Collins.** 1991. Comparative analysis of *Bacillus anthracis*, *Bacillus cereus*, and related species on the basis of reverse transcriptase sequencing of 16S rRNA. *Int. J. Syst. Bacteriol.* **41:**343–346.
2. **Avakyan, A. A., L. N. Katz, K. N. Levina, and I. B. Pavlova.** 1965. Structure and composition of the *Bacillus anthracis* capsule. *J. Bacteriol.* **90:**1082–1095.
3. **Bartkus, J. M., and S. H. Leppla.** 1989. Transcriptional regulation of the protective antigen gene of *Bacillus anthracis*. *Infect. Immun.* **57:**2295–2300.
4. **Battisti, L., B. D. Green, and C. B. Thorne.** 1985. Mating system for transfer of plasmids among *Bacillus anthracis*, *Bacillus cereus*, and *Bacillus thuringiensis*. *J. Bacteriol.* **162:**543–550.
5. **Beall, F. A., M. J. Taylor, and C. B. Thorne.** 1962. Rapid lethal effect in rats of a third component found upon fractionating the toxin of *Bacillus anthracis*. *J. Bacteriol.* **83:**1274–1280.
6. **Belton, F. C., and R. E. Strange.** 1954. Studies on a protective antigen produced *in vitro* from *Bacillus anthracis*: medium and methods of production. *Br. J. Exp. Pathol.* **35:**144–152.
7. **Bernhard, K., H. Schrempf, and W. Goebel.** 1978. Bacteriocin and antibiotic resistance plasmids in *Bacillus cereus* and *Bacillus subtilis*. *J. Bacteriol.* **133:**897–903.
8. **Bhatnagar, R., Y. Singh, S. H. Leppla, and A. M. Friedlander.** 1989. Calcium is required for the expression of anthrax lethal toxin activity in the macrophage-like cell line J774A.1. *Infect. Immun.* **57:**2107–2114.
9. **Bohm, R., and G. Spath.** 1990. The taxonomy of *Bacillus anthracis* according to DNA-DNA hybridization and G+C content, p. 29–31. *In* P. C. Turnbull (ed.), *Proceedings of the International Workshop on Anthrax, Winchester, England.* Salisbury medical bulletin, no. 68, special supplement. Salisbury Medical Society, Salisbury, England.
10. **Bragg, T. S., and D. L. Robertson.** 1989. Nucleotide sequence and analysis of the lethal factor gene (*lef*) from *Bacillus anthracis*. *Gene* **81:**45–54.
11. **Brewer, C. R., W. G. McCullough, R. C. Mills, W. G. Roessler, E. J. Herbst, and A. F. Howe.** 1946. Studies on the nutritional requirements of *Bacillus anthracis*. *Arch. Biochem.* **10:**65–75.
12. **Bricas, E., and C. Fromageot.** 1953. Naturally occurring peptides. *Adv. Protein Chem.* **8:**1–125.
13. **Bruckner, V., M. Kajtar, J. Kovacs, H. Nagy, and J. Wein.** 1958. Synthese des immunspezifischen, polypeptid-artigen Haptens der anthrax-subtilis Bacillen-Gruppe. Ein synthetischer Beweis der Konstitution der naturlichen Poly-glutaminsauren. *Tetrahedron* **2:**211–240.
14. **Bruckner, V., J. Kovacs, and G. Denes.** 1953. Structure of poly-D-glutamic acid isolated from capsulated strains of *B. anthracis*. *Nature (London)* **172:**508.
15. **Burdon, K. L., and R. D. Wende.** 1960. On the differentiation of anthrax bacilli from *Bacillus cereus*. *J. Infect. Dis.* **107:**224–234.
16. **Cataldi, A., E. Labruyere, and M. Mock.** 1990. Construction and characterization of a protective antigen-deficient *Bacillus anthracis* strain. *Mol. Microbiol.* **4:**1111–1117.
17. **Cole, H. B., J. W. Ezzell, K. F. Keller, and R. J. Doyle.** 1984. Differentiation of *Bacillus anthracis* and other *Bacillus* species by lectins. *J. Clin. Microbiol.* **19:**48–53.
18. **Cromartie, W. J., W. L. Bloom, and D. W. Watson.** 1947. Studies on infection with *Bacillus anthracis*. I. A histopathological study of skin lesions produced by *B. anthracis* in susceptible and resistant animal species. *J. Infect. Dis.* **80:**1–13.
19. **Cromartie, W. J., D. W. Watson, W. L. Bloom, and R. J. Heckly.** 1947. Studies on infection with *Bacillus anthracis*. II. The immunological and tissue damaging properties of extracts prepared from lesions of *B. anthracis* infections. *J. Infect. Dis.* **80:**14–27.
20. **Dunny, G. M., L. N. Lee, and D. J. LeBlanc.** 1991. Improved electroporation and cloning vector system for gram-positive bacteria. *Appl. Environ. Microbiol.* **57:**1194–1201.
21. **Eastin, J. D., and C. B. Thorne.** 1963. Carbon dioxide fixation in *Bacillus anthracis*. *J. Bacteriol.* **85:**410–417.
22. **Escuyer, V., and R. J. Collier.** 1991. Anthrax protective antigen interacts with a specific receptor on the surface of CHO-K1 cells. *Infect. Immun.* **59:**3381–3386.
23. **Fish, D. C., B. G. Mahlandt, J. P. Dobbs, and R. E. Lincoln.** 1968. Purification and properties of in vitro-

produced anthrax toxin components. *J. Bacteriol.* **95:** 907–918.

24. **Franke, A. E., and D. B. Clewell.** 1981. Evidence for a chromosome-borne resistance transposon (Tn*916*) in *Streptococcus faecalis* that is capable of "conjugal" transfer in the absence of a conjugative plasmid. *J. Bacteriol.* **145:**494–502.
25. **Friedlander, A. M.** 1986. Macrophages are sensitive to anthrax lethal toxin through an acid-dependent process. *J. Biol. Chem.* **261:**7123–7126.
26. **Gardner, J. M., and F. A. Troy.** 1979. Chemistry and biosynthesis of the poly(γ-D-glutamyl) capsule in *Bacillus licheniformis*. Activation, racemization, and polymerization of glutamic acid by a membranous polyglutamyl synthetase complex. *J. Biol. Chem.* **254:**6262–6269.
27. **Gladstone, G. P.** 1939. Inter-relationships between amino acids in the nutrition of *B. anthracis*. *Br. J. Exp. Pathol.* **20:**189–200.
28. **Gladstone, G. P.** 1946. Immunity to anthrax: protective antigen present in cell-free culture filtrates. *Br. J. Exp. Pathol.* **27:**394–418.
29. **Green, B. D., L. Battisti, T. M. Koehler, C. B. Thorne, and B. E. Ivins.** 1985. Demonstration of a capsule plasmid in *Bacillus anthracis*. *Infect. Immun.* **49:**291–297.
30. **Green, B. D., L. Battisti, and C. B. Thorne.** 1989. Involvement of Tn*4430* in transfer of *Bacillus anthracis* plasmids mediated by *Bacillus thuringiensis* plasmid pXO12. *J. Bacteriol.* **171:**104–113.
31. **Gryczan, T. J., S. Contente, and D. Dubnau.** 1978. Characterization of *Staphylococcus aureus* plasmids introduced by transformation into *Bacillus subtilis*. *J. Bacteriol.* **134:**318–329.
32. **Guaracao-Ayala, A. I., and C. B. Thorne.** Unpublished data.
33. **Guaracao-Ayala, A. I., and C. B. Thorne.** 1991. Transposon mutagenesis of *Bacillus anthracis* capsule plasmid pXO2, abstr. D-120, p. 98. Abstr. 91st Gen. Meet. Am. Soc. Microbiol. 1991.
34. **Guex-Holzer, S., and J. Tomcsik.** 1956. The isolation and chemical nature of capsular and cell-wall haptens in a *Bacillus* species. *J. Gen. Microbiol.* **14:**14–25.
35. **Hanby, W. E., and H. N. Rydon.** 1946. The capsular substance of *Bacillus anthracis*. *Biochem. J.* **40:**297–309.
36. **Hara, T., Y. Fujio, and S. Ueda.** 1982. Polyglutamate production by *Bacillus subtilis* (*natto*). *J. Appl. Biochem.* **4:**112–120.
37. **Harris-Smith, P. W., H. Smith, and J. Keppie.** 1958. Production *in vitro* of the toxin of *Bacillus anthracis* previously recognized *in vivo*. *J. Gen. Microbiol.* **19:**91–103.
38. **Hornung, J., and C. B. Thorne.** Unpublished data.
39. **Hornung, J. M., A. I. Guaracao-Ayala, and C. B. Thorne.** 1989. Transposon mutagenesis in *Bacillus anthracis*, abstr. H-256, p. 212. Abstr. 89th Annu. Meet. Am. Soc. Microbiol. 1989.
40. **Hornung, J. M., and C. B. Thorne.** 1991. Insertion mutations affecting pXO1-associated toxin production in *Bacillus anthracis*, abstr. D-121, p. 98. Abstr. 91st Gen. Meet. Am. Soc. Microbiol. 1991.
41. **Ivanovics, G., and V. Bruckner.** 1937. Die chemische Struktur der Kapsel substanz des Milzbrandbazillus und der serologisch identischen spezifischen Substanz des *Bacillus mesentericus*. *Z. Immunitatsforsch.* **90:**304–318.
42. **Ivins, B. E., J. W. Ezzell, Jr., J. Jemski, K. W. Hedlund, J. D. Ristroph, and S. H. Leppla.** 1986. Immunization studies with attenuated strains of *Bacillus anthracis*. *Infect. Immun.* **52:**454–458.
43. **Ivins, B. E., and S. L. Welkos.** 1986. Cloning and expression of the *Bacillus anthracis* protective antigen gene in *Bacillus subtilis*. *Infect. Immun.* **54:**537–542.
44. **Ivins, B. E., S. L. Welkos, G. B. Knudson, and D. J. LeBlanc.** 1988. Transposon Tn*916* mutagenesis in *Bacillus anthracis*. *Infect. Immun.* **56:**176–181.
45. **Kaneko, T., R. Nozaki, and K. Aizawa.** 1978. Deoxyribonucleic acid relatedness between *Bacillus anthracis*, *Bacillus cereus* and *Bacillus thuringiensis*. *Microbiol. Immunol.* **22:**639–641.
46. **Koehler, T. M.** 1987. Plasmid-related differences in capsule production by *Bacillus anthracis* and characterization of a fertility plasmid from *Bacillus subtilis* (*natto*). Ph.D. dissertation. University of Massachusetts, Amherst.
47. **Koehler, T. M., R. E. Ruhfel, B. D. Green, and C. B. Thorne.** 1986. Plasmid-related differences in capsule production by *Bacillus anthracis*, abstr. H-178, p. 157. Abstr. 86th Annu. Meet. Am. Soc. Microbiol. 1986.
48. **Koehler, T. M., and C. B. Thorne.** 1987. *Bacillus subtilis* (*natto*) plasmid pLS20 mediates interspecies plasmid transfer. *J. Bacteriol.* **169:**5271–5278.
49. **Larson, D. K., G. J. Calton, S. F. Little, S. H. Leppla, and J. W. Burnett.** 1988. Separation of three exotoxic factors of *Bacillus anthracis* by sequential immunosorbent chromatography. *Toxicon* **26:**913–921.
50. **Lawrence, D., S. Heitefuss, and H. S. H. Seifert.** 1991. Differentiation of *Bacillus anthracis* from *Bacillus cereus* by gas chromatography whole-cell fatty acid analysis. *J. Clin. Microbiol.* **29:**1508–1512.
51. **Leise, J. M., C. H. Carter, H. Friedlander, and S. W. Fried.** 1959. Criteria for identification of *Bacillus anthracis*. *J. Bacteriol.* **77:**655–660.
52. **Leonard, C. G., and R. D. Housewright.** 1963. Polyglutamic acid synthesis by cell-free extracts of *Bacillus licheniformis*. *Biochim. Biophys. Acta* **73:**530–532.
53. **Leonard, C. G., R. D. Housewright, and C. B. Thorne.** 1958. Effects of some metallic ions on glutamyl polypeptide synthesis by *Bacillus subtilis*. *J. Bacteriol.* **76:** 499–503.
54. **Leppla, S. H.** 1982. Anthrax toxin edema factor: a bacterial adenylate cyclase that increases cyclic AMP concentrations in eucaryotic cells. *Proc. Natl. Acad. Sci. USA* **79:**3162–3166.
55. **Leppla, S. H.** 1984. *Bacillus anthracis* calmodulin-dependent adenylate cyclase: chemical and enzymatic properties and interactions with eucaryotic cells. *Adv. Cyclic Nucleotide Protein Phosphorylation Res.* **17:**189–198.
56. **Leppla, S. H.** 1988. Production and purification of anthrax toxin. *Methods Enzymol.* **165:**103–116.
57. **Leppla, S. H., A. M. Friedlander, and E. Cora.** 1987. Proteolytic activation of anthrax toxin bound to cellular receptors, p. 111–112. *In* F. Fehrenback, J. E. Alouf, P. Falmagne, W. Goebel, J. Jeljaszewicz, D. Jurgens, and R. Rappuoli (ed.), *Bacterial Toxins*. Gustav Fischer Verlag, Stuttgart, Germany.
58. **Lereclus, D., J. Mahillon, G. Menou, and M.-M. Lecadet.** 1986. Identification of Tn*4430*, a transposon of *Bacillus thuringiensis* functional in *Escherichia coli*. *Mol. Gen. Genet.* **204:**52–57.
59. **Lesnyak, O. T., and R. A. Saltykov.** 1970. Comparative assessment of the immunogenicity of anthrax vaccine strains. *Zh. Mikrobiol. Epidemiol. Immunobiol.* **1970:**32–35.
60. **Little, S. F., and G. B. Knudson.** 1986. Comparative efficacy of *Bacillus anthracis* live-spore vaccine and protective-antigen vaccine against anthrax in the guinea pig. *Infect. Immun.* **52:**509–512.
61. **Little, S. F., and J. R. Lowe.** 1991. Location of receptor-binding region of protective antigen from *Bacillus anthracis*. *Biochem. Biophys. Res. Commun.* **180:**531–537.
62. **Makino, S., C. Sasakawa, I. Uchida, N. Terakado, and M. Yoshikawa.** 1987. Transformation of a cloning vector, pUB110, into *Bacillus anthracis*. *FEMS Microbiol. Lett.* **44:**45–48.

63. **Makino, S., C. Sasakawa, I. Uchida, N. Terakado, and M. Yoshikawa.** 1988. Cloning and CO_2-dependent expression of the genetic region for encapsulation from *Bacillus anthracis*. *Mol. Microbiol.* **2**:371–376.
64. **Makino, S.-I., I. Uchida, N. Terakado, C. Sasakawa, and M. Yoshikawa.** 1989. Molecular characterization and protein analysis of the *cap* region, which is essential for encapsulation in *Bacillus anthracis*. *J. Bacteriol.* **171**:722–730.
65. **McCloy, E. W.** 1951. Studies on a lysogenic *Bacillus* strain. I. A bacteriophage specific for *Bacillus anthracis*. *J. Hyg.* **49**:114–125.
66. **McCuen, R. W., and C. B. Thorne.** 1971. Genetic mapping of genes concerned with glutamyl polypeptide production by *Bacillus licheniformis* and a study of their relationship to the development of competence for transformation. *J. Bacteriol.* **107**:636–645.
67. **Meynell, E., and G. G. Meynell.** 1964. The roles of serum and carbon dioxide in capsule formation by *Bacillus anthracis*. *J. Gen. Microbiol.* **34**:153–164.
68. **Meynell, E. W.** 1963. Reverting and non-reverting rough variants of *Bacillus anthracis*. *J. Gen. Microbiol.* **32**:55–60.
69. **Meynell, G. G., and E. Meynell.** 1966. The biosynthesis of poly D-glutamic acid, the capsular material of *Bacillus anthracis*. *J. Gen. Microbiol.* **43**:119–138.
70. **Mikesell, P., B. E. Ivins, J. D. Ristroph, and T. M. Dreier.** 1983. Evidence for plasmid-mediated toxin production in *Bacillus anthracis*. *Infect. Immun.* **39**:371–376.
71. **Mock, M., E. Labruyere, P. Glaser, A. Danchin, and A. Ullmann.** 1988. Cloning and expression of the calmodulin-sensitive *Bacillus anthracis* adenylate cyclase in *Escherichia coli*. *Gene* **64**:277–284.
72. **Molnar, D. M., and R. A. Altenbern.** 1963. Alterations in the biological activity of protective antigen of *Bacillus anthracis* toxin. *Proc. Soc. Exp. Biol. Med.* **114**:294–297.
73. **Mongkolsuk, S., Y.-W. Chiang, R. B. Reynolds, and P. S. Lovett.** 1983. Restriction fragments that exert promoter activity during postexponential growth of *Bacillus subtilis*. *J. Bacteriol.* **155**:1399–1406.
74. **Normore, W. M.** 1973. Guanine-plus-cytosine composition of the DNA of bacteria, fungi, algae, and protozoa, p. 585–740. *In* A. L. Laskin and H. A. Lechavalier (ed.), *Handbook of Microbiology*, vol. II. *Microbial Composition*. The Chemical Rubber Co., Washington, D.C.
75. **Pezard, C., P. Berche, and M. Mock.** 1991. Contribution of individual toxin components to virulence of *Bacillus anthracis*. *Infect. Immun.* **59**:3472–3477.
76. **Puziss, M., and M. B. Howard.** 1963. Studies on immunity in anthrax. XI. Control of cellular permeability by bicarbonate ion in relation to protective antigen elaboration. *J. Bacteriol.* **85**:237–243.
77. **Puziss, M., L. C. Manning, J. W. Lynch, E. Barclay, I. Abelow, and G. G. Wright.** 1963. Large-scale production of protective antigen of *Bacillus anthracis* in anaerobic cultures. *Appl. Microbiol.* **11**:330–334.
78. **Quinn, C. P., and B. N. Dancer.** 1990. Transformation of vegetative cells of *Bacillus anthracis* with plasmid DNA. *J. Gen. Microbiol.* **136**:1211–1215.
79. **Quinn, C. P., C. C. Shone, P. C. B. Turnbull, and J. Melling.** 1988. Purification of anthrax-toxin components by high-performance anion-exchange, gel-filtration and hydrophobic-interaction chromatography. *Biochem. J.* **252**:753–758.
80. **Quinn, C. P., Y. Singh, K. R. Klimpel, and S. H. Leppla.** 1991. Functional mapping of anthrax toxin lethal factor by in-frame insertion mutagenesis. *J. Biol. Chem.* **266**: 20124–20130.
81. **Raziuddin, A., and A. Friedlander.** 1991. Anthrax protective antigen receptor: Identification of a protective antigen binding protein by chemical cross-linking, abstr. B-301, p. 75. Abstr. 91st Gen. Meet. Am. Soc. Microbiol. 1991.
82. **Reddy, A., L. Battisti, and C. B. Thorne.** 1987. Identification of self-transmissible plasmids in four *Bacillus thuringiensis* subspecies. *J. Bacteriol.* **169**:5263–5270.
83. **Ristroph, J. D., and B. E. Ivins.** 1983. Elaboration of *Bacillus anthracis* antigens in a new defined culture medium. *Infect. Immun.* **39**:483–486.
84. **Roberts, D. M., R. Crea, M. Malecha, G. Alvarado-Urbina, R. H. Chiarello, and D. M. Watterson.** 1985. Chemical synthesis and expression of a calmodulin gene designated for site-specific mutagenesis. *Biochemistry* **24**:5090–5098.
85. **Robertson, D. L., and T. S. Bragg.** 1990. Nucleotide sequence of the lethal factor (*lef*) and edema factor (*cya*) genes from *Bacillus anthracis*: elucidation of the EF and LF functional domains, p. 59. *In* P. C. B. Turnbull (ed.), *Proceedings of the International Workshop on Anthrax, Winchester, England.* Salisbury medical bulletin, no. 68, special supplement. Salisbury Medical Society, Salisbury, England.
86. **Robertson, D. L., T. S. Bragg, S. Simpson, R. Kaspar, W. Xie, and M. T. Tippetts.** 1990. Mapping and characterization of the *Bacillus anthracis* plasmids pXO1 and pXO2, p. 55–58. *In* P. C. B. Turnbull (ed.), *Proceedings of the International Workshop on Anthrax, Winchester, England.* Salisbury medical bulletin, no. 68, special supplement. Salisbury Medical Society, Salisbury, England.
87. **Robertson, D. L., and S. H. Leppla.** 1986. Molecular cloning and expression in *Escherichia coli* of the lethal factor gene of *Bacillus anthracis*. *Gene* **44**:71–78.
88. **Robertson, D. L., M. T. Tippetts, and S. H. Leppla.** 1988. Nucleotide sequence of the *Bacillus anthracis* edema factor gene (*cya*): a calmodulin-dependent adenylate cyclase. *Gene* **73**:363–371.
89. **Robillard, N. J.** 1984. Changes associated with plasmid loss in *Bacillus anthracis*. Ph.D. dissertation. University of Massachusetts, Amherst.
90. **Robillard, N. J., T. M. Koehler, R. Murray, and C. B. Thorne.** 1983. Effects of plasmid loss on the physiology of *Bacillus anthracis*, abstr. H-54, p. 115. Abstr. 83rd Annu. Meet. Am. Soc. Microbiol. 1983.
91. **Ruhfel, R. E.** 1989. Physical and genetic characterization of the *Bacillus thuringiensis* subsp. *kurstaki* HD-1 extrachromosomal temperate phage TP-21. Ph.D. dissertation. University of Massachusetts, Amherst.
92. **Ruhfel, R. E., N. J. Robillard, and C. B. Thorne.** 1984. Interspecies transduction of plasmids among *Bacillus anthracis*, *B. cereus*, and *B. thuringiensis*. *J. Bacteriol.* **157**:708–711.
93. **Ruhfel, R. E., and C. B. Thorne.** 1988. Physical and genetic characterization of the *Bacillus thuringiensis* subsp. *kurstaki* HD-1 extrachromosomal temperate phage TP-21, abstr. H-4, p. 145. Abstr. 88th Annu. Meet. Am. Soc. Microbiol. 1988.
94. **Sadler, D. F., J. W. Ezzell, Jr., K. F. Keller, and R. J. Doyle.** 1984. Glycosidase activities of *Bacillus anthracis*. *J. Clin. Microbiol.* **19**:594–598.
95. **Singh, Y., V. K. Chaudhary, and S. H. Leppla.** 1989. A deleted variant of *Bacillus anthracis* protective antigen is non-toxic and blocks anthrax toxin action *in vivo*. *J. Biol. Chem.* **264**:19103–19107.
96. **Singh, Y., K. R. Klimpel, C. P. Quinn, V. K. Chaudhary, and S. H. Leppla.** 1991. The carboxyl-terminal end of protective antigen is required for receptor binding and anthrax toxin activity. *J. Biol. Chem.* **266**:15493–15497.
97. **Smith, H., J. Keppie, and J. L. Stanley.** 1955. The chemical basis of the virulence of *Bacillus anthracis*. V. The specific toxin produced by *B. anthracis in vivo*. *Br. J. Exp. Pathol.* **36**:460–472.
98. **Smith, H., D. W. Tempest, J. L. Stanley, P. W. Harris-Smith, and R. C. Gallop.** 1956. The chemical basis of

the virulence of *Bacillus anthracis*. VII. Two components of the anthrax toxin: their relationship to known immunizing aggressins. *Br. J. Exp. Pathol.* **37:**263–271.

99. **Somerville, H. J., and M. L. Jones.** 1972. DNA competition studies within the *Bacillus cereus* group of bacilli. *J. Gen. Microbiol.* **73:**257–265.
100. **Stanley, J. L., and H. Smith.** 1961. Purification of factor I and recognition of the third factor of anthrax toxin. *J. Gen. Microbiol.* **26:**49–66.
101. **Stepanov, A. S., S. V. Gavrilov, O. B. Puzanova, T. M. Grigoryeva, and R. R. Azizbekyan.** 1989. Transduction of plasmids in *Bacillus anthracis* by bacteriophage CP-54. *Mol. Genet. Microbiol. Virol.* **1:**14–19.
102. **Stepanov, A. S., O. B. Puzanova, S. Y. Dityatkin, O. G. Loginova, and B. N. Ilyashenko.** 1990. Glycine-induced cryotransformation of plasmids into *Bacillus anthracis*. *J. Gen. Microbiol.* **136:**1217–1221.
103. **Sterne, M.** 1937. Variation in *Bacillus anthracis*. *Onderstepoort J. Vet. Sci. Anim. Ind.* **8:**271–349.
104. **Sterne, M.** 1939. The immunization of laboratory animals against anthrax. *Onderstepoort J. Vet. Sci. Anim. Ind.* **13:**313–317.
105. **Sterne, M.** 1939. The use of anthrax vaccines prepared from avirulent (uncapsulated) variants of *Bacillus anthracis*. *Onderstepoort J. Vet. Sci. Anim. Ind.* **13:**307–312.
106. **Strange, R. E., and C. B. Thorne.** 1958. Further purification studies on the protective antigen of *Bacillus anthracis* produced *in vitro*. *J. Bacteriol.* **76:**192–202.
107. **Thorne, C. B.** 1956. Capsule formation and glutamyl polypeptide synthesis by *Bacillus anthracis* and *Bacillus subtilis*, p. 68–80. *In* E. T. C. Spooner and B. A. D. Stocker (ed.), *Bacterial Anatomy*. Cambridge University Press, Cambridge.
108. **Thorne, C. B.** 1960. Biochemical properties of virulent and avirulent strains of *Bacillus anthracis*. *Ann. N.Y. Acad. Sci.* **88:**1024–1033.
109. **Thorne, C. B.** 1968. Transduction in *Bacillus cereus* and *Bacillus anthracis*. *Bacteriol. Rev.* **32:**358–361.
110. **Thorne, C. B.** 1978. Transduction in *Bacillus thuringiensis*. *Appl. Environ. Microbiol.* **35:**1109–1115.
111. **Thorne, C. B.** 1985. Genetics of *Bacillus anthracis*, p. 56–62. *In* L. Leive, P. F. Bonventre, J. A. Morello, S. Schlesinger, S. D. Silver, and H. C. Wu (ed.), *Microbiology—1985*. American Society for Microbiology, Washington, D.C.
112. **Thorne, C. B., and F. C. Belton.** 1957. An agar diffusion method for titrating *Bacillus anthracis* immunizing antigen and its application to a study of antigen production. *J. Gen. Microbiol.* **17:**505–516.
113. **Thorne, C. B., C. G. Gomez, and R. D. Housewright.** 1955. Transamination of D-amino acids by *Bacillus subtilis*. *J. Bacteriol.* **69:**357–362.
114. **Thorne, C. B., and C. G. Leonard.** 1958. Isolation of D- and L-glutamyl polypeptides from culture filtrates of *Bacillus subtilis*. *J. Biol. Chem.* **233:**1109–1112.
115. **Thorne, C. B., and D. M. Molnar.** 1955. D-Amino acid transamination in *Bacillus anthracis*. *J. Bacteriol.* **70:** 420–426.
116. **Thorne, C. B., D. M. Molnar, and R. E. Strange.** 1960. Production of toxin *in vitro* by *Bacillus anthracis* and its separation into two components. *J. Bacteriol.* **79:**450–455.
117. **Tippetts, M. T., and D. L. Robertson.** 1988. Molecular cloning and expression of the *Bacillus anthracis* edema factor toxin gene: a calmodulin-dependent adenylate cyclase. *J. Bacteriol.* **170:**2263–2266.
118. **Tomcsik, J.** 1956. Bacterial capsules and their relation to the cell wall, p. 41–67. *In* E. T. C. Spooner and B. A. D. Stocker (ed.), *Bacterial Anatomy*. Cambridge University Press, Cambridge.
119. **Torii, M.** 1956. Optical isomers of glutamic acid composing bacterial glutamyl polypeptides. *Med. J. Osaka Univ.* **6:**1043–1046.
120. **Torii, M.** 1959. Studies on the chemical structure of bacterial glutamyl polypeptides by hydrazinolysis. *J. Biochem.* **46:**189–200.
121. **Torii, M., O. Kurimura, S. Utsumi, H. Nozu, and T. Amano.** 1959. Decapsulation of *Bacillus megaterium*. *Biken J.* **2:**265–276.
122. **Trieu-Cuot, P., C. Carlier, P. Martin, and P. Courvalin.** 1987. Plasmid transfer by conjugation from *Escherichia coli* to Gram-positive bacteria. *FEMS Microbiol. Lett.* **48:**289–294.
123. **Troy, F. A.** 1973. Chemistry and biosynthesis of the poly(γ-D-glutamyl) capsule in *Bacillus licheniformis*. I. Properties of the membrane-mediated biosynthetic reaction. *J. Biol. Chem.* **248:**305–315.
124. **Troy, F. A.** 1973. Chemistry and biosynthesis of the poly(γ-D-glutamyl) capsule in *Bacillus licheniformis*. II. Characterization and structural properties of the enzymatically synthesized polymer. *J. Biol. Chem.* **248:**316–324.
125. **Turnbull, P. C. B., M. G. Broster, J. A. Carman, R. J. Manchee, and J. Melling.** 1986. Development of antibodies to protective antigen and lethal factor components of anthrax toxin in humans and guinea pigs and their relevance to protective immunity. *Infect. Immun.* **52:**356–363.
126. **Turnbull, P. C. B., and J. M. Kramer.** 1991. *Bacillus*, p. 296–303. *In* A. Balows, W. J. Hausler, Jr., K. L. Herrmann, H. D. Isenberg, and H. J. Shadomy (ed.), *Manual of Clinical Microbiology*, 5th ed. American Society for Microbiology, Washington, D.C.
127. **Uchida, I., K. Hashimoto, S.-I. Makino, C. Sasakawa, M. Yoshikawa, and N. Terakado.** 1987. Restriction map of a capsule plasmid of *Bacillus anthracis*. *Plasmid* **18:**178–181.
128. **Uchida, I., T. Sekizaki, K. Hashimoto, and N. Terakado.** 1985. Association of the encapsulation of *Bacillus anthracis* with a 60-megadalton plasmid. *J. Gen. Microbiol.* **131:**363–367.
129. **Vodkin, M. H., and S. H. Leppla.** 1983. Cloning of the protective antigen gene of *Bacillus anthracis*. *Cell* **34:** 693–697.
130. **Volcani, B. E., and P. Margalith.** 1957. A new species (*Flavobacterium polyglutamicum*) which hydrolyzes the γ-L-glutamyl bond in polypeptides. *J. Bacteriol.* **74:**646–653.
131. **Waley, S. G.** 1955. The structure of bacterial polyglutamic acid. *J. Chem. Soc.* **1955:**517–522.
132. **Weisblum, B., M. Y. Graham, T. Gryczan, and D. Dubnau.** 1979. Plasmid copy number control: isolation and characterization of high-copy-number mutants of plasmid pE194. *J. Bacteriol.* **137:**635–643.
133. **Welkos, S. L.** 1991. Plasmid-associated virulence factors of non-toxigenic (pXO1$^-$) *Bacillus anthracis*. *Microb. Pathog.* **10:**183–198.
134. **Welkos, S. L., J. R. Lowe, F. Eden-McCutchan, M. Vodkin, S. H. Leppla, and J. J. Schmidt.** 1988. Sequence and analysis of the DNA encoding protective antigen of *Bacillus anthracis*. *Gene* **69:**287–300.
135. **Yelton, D. B., and C. B. Thorne.** 1970. Transduction in *Bacillus cereus* by each of two bacteriophages. *J. Bacteriol.* **102:**573–579.
136. **Yelton, D. B., and C. B. Thorne.** 1971. Comparison of *Bacillus cereus* bacteriophages CP-51 and CP-53. *J. Virol.* **8:**242–253.
137. **Youngman, P. J., J. B. Perkins, and R. Losick.** 1983. Genetic transposition and insertional mutagenesis in *Bacillus subtilis* with *Streptococcus faecalis* transposon Tn*917*. *Proc. Natl. Acad. Sci. USA* **80:**2305–2309.
138. **Youngman, P. J., J. B. Perkins, and R. Losick.** 1983. Construction of a cloning site near one end of Tn*917* into which foreign DNA may be inserted without affecting transposition in *Bacillus subtilis* or expression of the transposon-borne *erm* gene. *Plasmid* **12:**1–9.

II. METABOLISM AND ITS REGULATION

9. Introduction to Metabolic Pathways

ABRAHAM L. SONENSHEIN

OVERVIEW

Carbon Utilization

The central pathways of carbon dissimilation (the glycolytic pathway and the Krebs cycle, shown in Fig. 1) are conserved in virtually all organisms, but bacteria differ from plants and animals and from each other in the details of specific biosynthetic and degradative pathways. Figure 1 also indicates the branch points at which the central pathways produce precursors for important cellular constituents. Not all branches exist in all organisms (e.g., some are natural auxotrophs). This scheme takes into account the characteristics of metabolism that differentiate gram-positive and gram-negative bacteria and the differing pathways used by various gram-positive species. It should be noted, however, that not all pathways are fully described for any one gram-positive species or for gram-positive bacteria as a group. The detailed biosynthetic pathways for amino acids, purines, pyrimidines, and vitamins are the subject of the chapters that follow.

The central pathways of carbon metabolism are fed not only by glucose but also by other hexoses and disaccharides (see chapter 11 [Steinmetz]), pentoses (metabolized through ribose 5-phosphate), fatty acids (through acetyl coenzyme A), various amino acids, and a variety of two-, three- and four-carbon compounds (which enter the main pathway through glyceraldehyde 3-phosphate, pyruvate, acetyl coenzyme A, or the Krebs cycle) (Fig. 2). Among the amino acids that serve as carbon sources, glutamate, glutamine, histidine, arginine, and proline are converted to 2-ketoglutarate, alanine is converted to pyruvate, and aspartate and asparagine are converted to fumarate. In some gram-positive bacteria, arginine gives rise to succinate instead of 2-ketoglutarate.

Not all of the molecules of substrate that enter the glycolytic pathway are fully catabolized to CO_2 and H_2O. Depending on availability of nutrients, a significant fraction of the pyruvate or acetyl coenzyme A generated by catabolism may be converted to by-products that are excreted into the medium (Fig. 1). The specific products are characteristic of the species, growth conditions, and growth phase. In the case of *Clostridium acetobutylicum*, these products, primarily acetone and butanol (Fig. 1), have important industrial applications. More typically, the excreted products are acids (acetate or lactate; Fig. 1) whose production during growth lowers the pH of the medium. As cells use up available supplies of favored substrates, such as glucose, the excreted acids are transported back into the cell (raising the pH of the medium) and further metabolized through the Krebs cycle. Production and utilization of these by-products are tightly regulated. Note that utilization of acetate as sole carbon source requires a functional glyoxylate cycle (Fig. 2), a pathway absent in some gram-positive bacteria. Moreover, use of two-, three-, or four-carbon compounds as sole carbon sources requires a mechanism of gluconeogenesis to provide the backbones for cell wall and teichoic acid polysaccharides. Many of the glycolytic enzymes catalyze reversible reactions but some do not. For the latter reactions, additional enzymes must be available (see chapter 12 [Fortnagel] for details).

Nitrogen Utilization

The major pathway for assimilation of nitrogen is through glutamine (Fig. 1 and chapter 20 [Schreier]). That is, exogenously supplied or endogenously produced ammonium ion is fixed to glutamate by glutamine synthetase. Glutamine serves as the donor of virtually all amino and amide groups in cellular components, either by the direct participation of glutamine in biosynthetic reactions or through the action of glutamate as a substrate for transamination reactions. Some gram-positive bacteria have a second enzyme, glutamate dehydrogenase, that assimilates ammonium ion if supplied in high concentration. In commonly used strains of *Bacillus subtilis*, however, glutamate dehydrogenase is strictly a catabolic enzyme. Some biosynthetic enzymes that normally use glutamine as amino donor, e.g., CTP synthetase, can function, albeit less inefficiently, when supplied with ammonium ion at high concentration.

The step that links metabolism of carbon and nitrogen is conversion of 2-ketoglutarate to glutamate and glutamine. It is not surprising that in gram-negative bacteria the ultimate determinant of expression of nitrogen metabolism genes (e.g., genes for glutamine synthetase, nitrogen fixation, amino acid transport, and degradation) is the ratio of the intracellular concentrations of 2-ketoglutarate (which reflects carbon availability) and glutamine (a reflection of nitrogen availability). It is reasonable to suppose that gram-positive bacteria monitor the same ratio.

Sources of ammonium ion include amino acids (e.g., glutamate, histidine, arginine, alanine, aspartate, asparagine, proline), amino sugars (e.g., glucosamine), γ-aminobutyric acid, nitrate, and urea (Fig. 2). Pathways for utilization of these compounds are described in chapter 16 (Fisher). Most gram-positive bacteria are unable to fix molecular nitrogen. Species that do

Abraham L. Sonenshein • Department of Molecular Biology and Microbiology, Tufts University School of Medicine, Boston, Massachusetts 02111-1800.

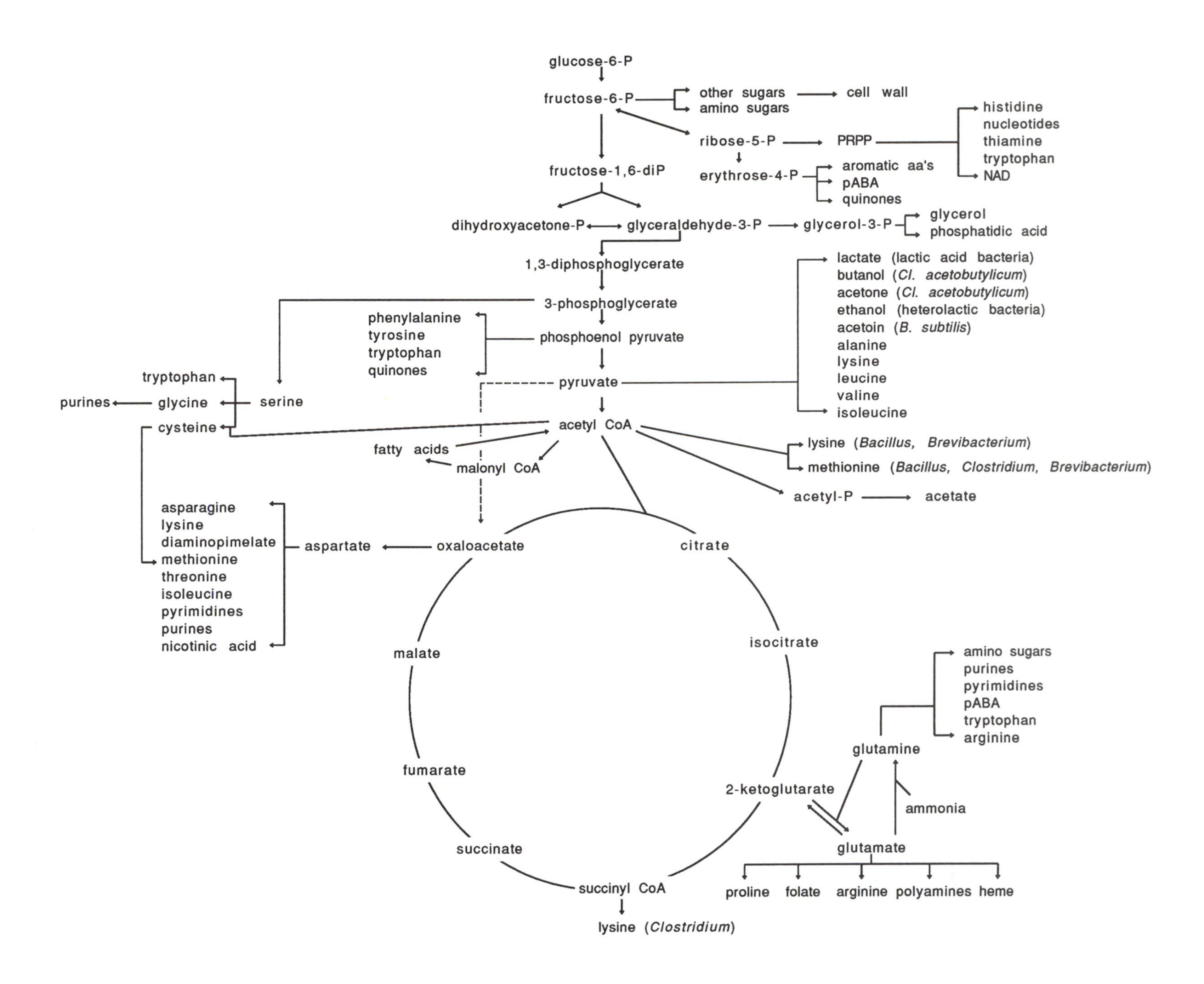

glucose-6-P
fructose-6-P
other sugars
amino sugars
cell wall
ribose-5-P
PRPP
histidine
nucleotides
thiamine
tryptophan
NAD
erythrose-4-P
aromatic aa's
pABA
quinones
fructose-1,6-diP
dihydroxyacetone-P
glyceraldehyde-3-P
glycerol-3-P
glycerol
phosphatidic acid
1,3-diphosphoglycerate
3-phosphoglycerate
phenylalanine
tyrosine
tryptophan
quinones
phosphoenol pyruvate
pyruvate
lactate (lactic acid bacteria)
butanol (Cl. acetobutylicum)
acetone (Cl. acetobutylicum)
ethanol (heterolactic bacteria)
acetoin (B. subtilis)
alanine
lysine
leucine
valine
isoleucine
tryptophan
purines
glycine
serine
cysteine
acetyl CoA
fatty acids
malonyl CoA
lysine (Bacillus, Brevibacterium)
methionine (Bacillus, Clostridium, Brevibacterium)
acetyl-P
acetate
asparagine
lysine
diaminopimelate
methionine
threonine
isoleucine
pyrimidines
purines
nicotinic acid
aspartate
oxaloacetate
citrate
isocitrate
malate
amino sugars
purines
pyrimidines
pABA
tryptophan
arginine
glutamine
fumarate
2-ketoglutarate
ammonia
succinate
glutamate
succinyl CoA
proline
folate
arginine
polyamines
heme
lysine (Clostridium)

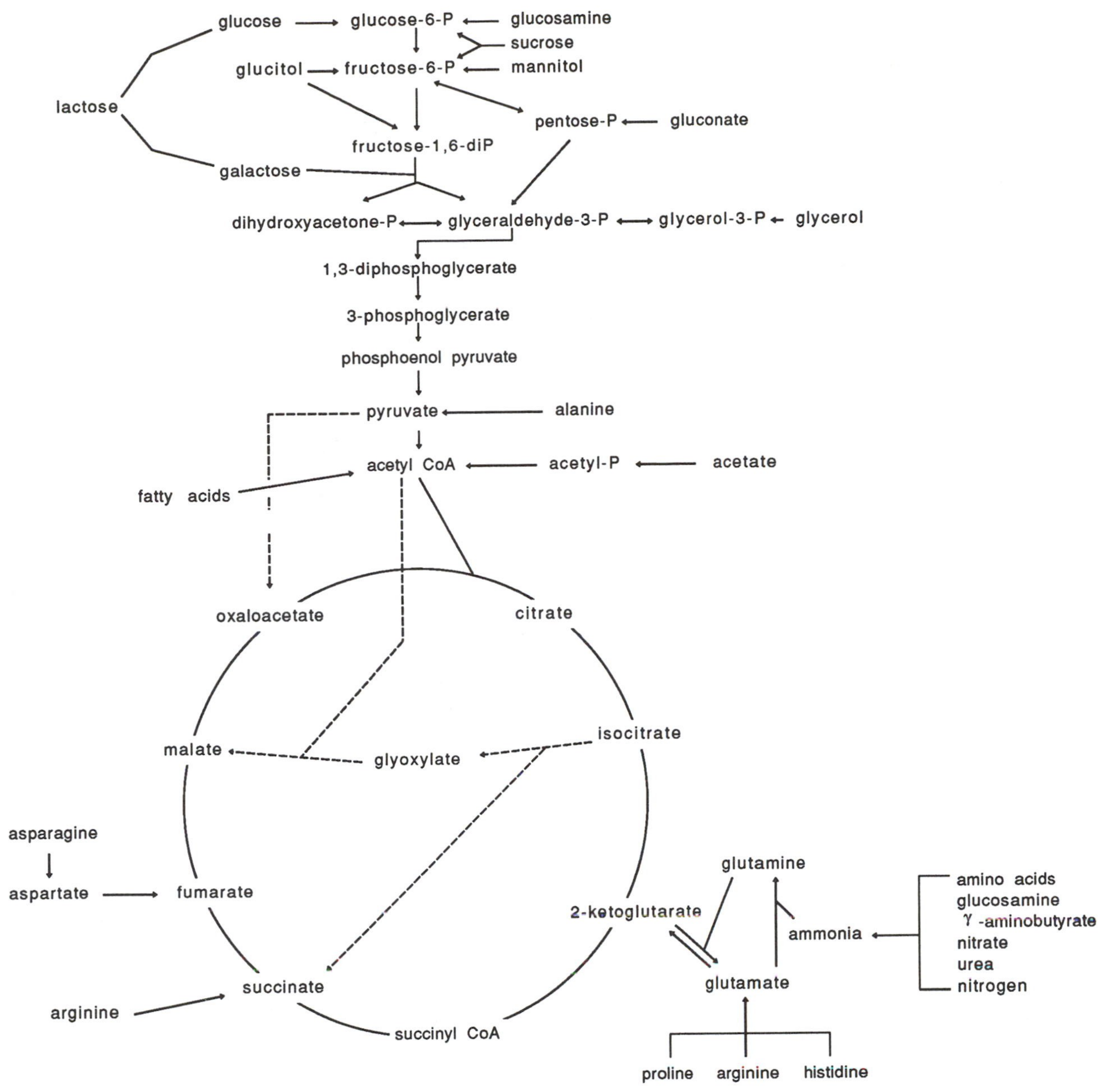

Figure 2. Pathways for utilization of carbon and nitrogen sources other than glucose and ammonia. Abbreviation: P, phosphate.

so include *Bacillus polymyxa*, *Bacillus azotofixans*, *Bacillus macerans*, and *Clostridium pasteurianum*.

ORPHAN PATHWAYS

Some biosynthetic pathways in gram-positive bacteria have received so little attention that they could not be the subject of separate chapters. These include those for synthesis of L- and D-alanine and histidine and the intersecting pathways for glycine, serine, and cysteine. Current knowledge about these systems is therefore summarized here.

Histidine Biosynthesis

The pathway for biosynthesis of histidine is thought to be essentially the same in gram-negative and gram-positive bacteria (see Fig. 3 and reference 38). Few of

Figure 1. The central pathways of glucose dissimilation, ammonia assimilation, and production of biosynthetic precursors. The linkage between carbon and nitrogen metabolism through conversion of 2-ketoglutarate to glutamate and glutamine is also indicated. Abbreviations: P, phosphate; PRPP, phosphoribosylpyrophosphate; pABA, *p*-aminobenzoate.

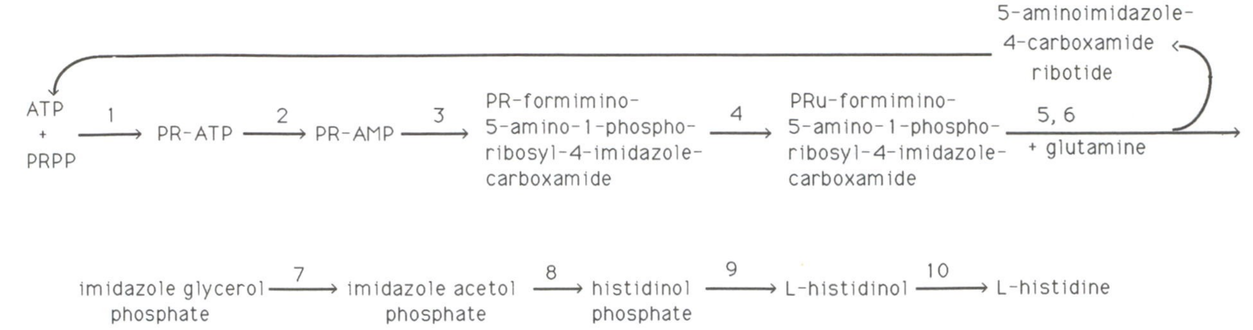

Figure 3. Pathway for histidine biosynthesis. The precursors, ATP and phosphoribosylpyrophosphate (PRPP), are converted to histidine in 10 steps. The enzymes that carry out each step and the genes that encode them are listed in Table 1. Abbreviations: PR, phosphoribosyl; PRu, phosphoribulosyl.

the enzymes from gram-positive bacteria have been characterized, but genes whose products seem to have functional or sequence homology with genes from *Escherichia coli* and *Salmonella typhimurium* have been identified for several organisms. Detailed information is available for *Streptomyces coelicolor* and *Lactococcus lactis* (see Table 1). The *his* genes of *S. coelicolor* are organized in three unlinked clusters (3, 31), one of which has been sequenced nearly in its entirety (21, 22). Judging by comparisons of DNA sequences and by missing enzyme activities in mutant strains, it appears that the *S. coelicolor* enzymes imidazoleglycerol-phosphate dehydratase and histidinol-phosphate phosphatase are encoded by unlinked genes that are not coordinately regulated (3, 21, 31), whereas in *E. coli* and *S. typhimurium* the two enzyme activities are carried out by a single bifunctional protein (38).

Most of the histidine biosynthetic genes from *L. lactis* have been cloned and sequenced (6). By analysis of sequence similarities and ability of subclones to complement mutants of *E. coli* and *B. subtilis*, it was possible to assign most of the deduced open reading frames to specific enzymes (Table 1). As in *S. coelicolor*, imidazoleglycerol-phosphate dehydratase and histidinol-phosphate phosphatase activities are apparently encoded by separate genes (6). In fact, no candidate for the *L. lactis* histidinol-phosphate phosphatase gene could be found in the cloned cluster. The *hisIE* gene from enterics encodes a bifunctional enzyme (38); the same situation seems to pertain for *L. lactis* (6).

In *Staphylococcus aureus*, a cluster of eight genes has been identified within which mutations lead to accumulation of characteristic intermediates, allowing tentative assignments of gene identities by comparison with *S. typhimurium* (1, 17, 18; Table 1). Four linkage groups of *his* genes have been discovered in *Micrococcus luteus* (16). One of these loci is linked to the *trp* operon.

Mutants blocked in 9 of the 10 steps of histidine biosynthesis have been isolated from *B. subtilis* (4). The mutations in all but one of these mutant classes map to the *hisA* locus. This major cluster has been cloned in a λ phage (36), but has not yet been sequenced. The *hisH* gene, unlinked to *hisA*, lies just downstream of the *trp* operon in both *B. subtilis* and *Bacillus megaterium* (2, 11). The product of *hisH* is a histidinol-phosphate amidotransferase/tyrosine-phenylalanine transaminase (11, 27). *B. subtilis hisH* lies just upstream of *tyrA*, which encodes a second tyrosine-phenylalanine transaminase (11, 27). As a result, *hisH* mutants require histidine, but not tyrosine or phenylalanine; single mutants in the *tyrA* gene have

Table 1. Genes and enzymes for histidine biosynthesis

Enzymatic step	Enzyme name[a]	Gene name					
		E. coli-S. typhimurium	*S. coelicolor* (old)	*S. coelicolor* (new)	*L. lactis*	*S. aureus*	*M. luteus*
1	ATP phosphoribosyltransferase	*hisG*	*hisH*?	*hisG* or *E*	*hisG*	*hisG*	
2	PR-ATP pyrophosphohydrolase	*hisIE*	*hisH*?	*hisG* or *E*	*hisIE*		
3	PR-AMP cyclohydrolase	*hisIE*	*hisE*	*hisI*	*hisIE*	*hisE*	
4	PR-formimino-amino-phosphoribosyl-imidazole-carboxamide isomerase	*hisA*	*hisI*, *C*, or *F*	*hisA* or *H*	*hisA*		
5	Glutamine amidotransferase	*hisH*	*hisI*, *C*, or *F*	*hisA* or *H*	*hisH*		
6	Cyclase	*hisF*	*hisI*, *C*, or *F*	*hisF*	*hisF*		
7	Imidazoleglycerol-phosphate dehydratase	*hisB*	*hisB*	*hisBd*	*hisB*	*hisB*	*hisB*
8	Histidinol-phosphate aminotransferase	*hisC*	*hisG*	*hisC*	*hisC*	*hisC*	*hisC*
9	Histidinol phosphatase	*hisB*	*hisD*	*hisBpx*			
10	Histidinol dehydrogenase	*hisD*	*hisA*	*hisD*	*hisD*	*hisD*	*hisD*

[a] PR, phosphoribosyl.

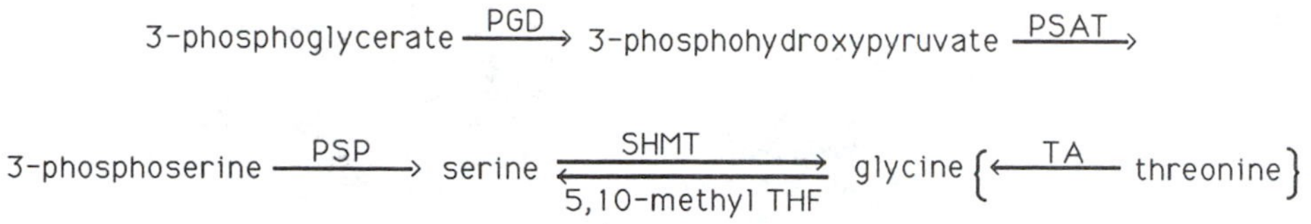

3-phosphoglycerate —PGD→ 3-phosphohydroxypyruvate —PSAT→

3-phosphoserine —PSP→ serine ⇄(SHMT / 5,10-methyl THF) glycine {←TA— threonine}

Figure 4. Presumed pathways for biosynthesis of serine and glycine in gram-positive bacteria. The 3-phosphoglycerate pathway seems to be the major route to serine and glycine in most gram-positive bacteria. In certain *Clostridium* species, however, glycine is made from threonine (shown in brackets in figure). Abbreviations: PGD, 3-phosphoglycerate dehydrogenase; PSAT, 3-phosphoserine aminotransferase; PSP, 3-phosphoserine phosphatase; SHMT, serine hydroxymethyltransferase; THF, tetrahydrofolate; TA, threonine aldolase.

no phenotype. Transcription of *hisH* appears to come from two promoters (11). Some transcription is due to read-through from the *trp* operon promoter; the remainder begins at a promoter located within the *trpB* or *trpA* genes (11). Neither promoter is regulated by histidine.

Biosynthesis of Serine and Glycine

The major route of serine biosynthesis in *B. subtilis* (30) and *M. luteus* (26) is the 3-phosphoglycerate pathway characteristic of most bacteria (Fig. 4). The first enzyme of this pathway, 3-phosphoglycerate dehydrogenase, is feedback inhibited by serine (25, 30). It is likely that most gram-positive bacteria interconvert serine and glycine through the enzyme L-serine hydroxymethyltransferase (Fig. 4), since this activity has been detected in *Sarcina albida* (7) and *Arthrobacter globiformis* (19) and in some strains of *B. subtilis*, *Micrococcus*, and *Diplococcus glycinophilus* (7). In *Clostridium acidi-urici*, however, a different pathway for interconversion seems to function. In this case, glycine and formaldehyde condense to form serine (13). Whether this represents the normal pathway for serine synthesis in this organism or reflects intermediates in purine degradation is unknown.

Mutations at two *B. subtilis* loci cause serine auxotrophy (23, 35). The *serA* gene has been cloned and sequenced; it encodes a protein with strong similarity to *E. coli* 3-phosphoglycerate dehydrogenase (33). SerA$^-$ mutants require either serine or glycine for growth (23, 35). The *serC* locus is defined by a transposon insertion mutation that causes serine auxotrophy. Growth is improved if threonine and serine are both provided (35). There are three unlinked loci at which mutations cause glycine auxotrophy (see *B. subtilis* genetic map, chapter 29 [Anagnostopoulos et al.]). No detailed information is available about the likely gene products.

An entirely different pathway to glycine is found in *Clostridium kluyveri* and *C. pasteurianum*. In these organisms, glycine is made primarily from threonine through the activity of threonine aldolase (14; Fig. 4). This is consistent with the minor contribution of serine to the C_1 pool in these organisms. Instead, most of the C_1 comes from pyruvate as a result of the activity of pyruvate-formate lyase (14).

Cysteine Biosynthesis

There is little detailed information available about the biochemical steps used by gram-positive bacteria to transport and assimilate sulfur into organic compounds. For most bacteria, the primary inorganic sulfur source is sulfate ion. Its transport, reduction, and incorporation into cysteine in *E. coli* require at least 14 gene products and significant expenditure of energy (20). Some of the features of the *E. coli* system are known to be conserved in *B. subtilis*. The first two enzymes of sulfate utilization, ATP sulfurylase and adenosine-5′-phosphosulfate kinase, and enzyme activities that convert activated sulfate to sulfite and sulfide and catalyze incorporation into cysteine are present in both *B. subtilis* and *E. coli* (28, 29). Both activation of sulfate and reduction to sulfide are repressed during growth in cysteine-containing medium (28, 29). Glutathione represses the part of the pathway leading to sulfite, but only in *B. subtilis* (28). Five *S. coelicolor* cistrons (12) and four *B. subtilis* cistrons (see *B. subtilis* genetic map, chapter 29) have been identified within which mutations cause various blocks in cysteine biosynthesis. On the basis of the absence of enzyme activity in a mutant strain, it is likely that the *B. subtilis cysA* gene encodes serine transacetylase (15).

The mechanisms for assimilation of sulfate and synthesis of cysteine are likely to be similar in most gram-positive bacteria. The detailed study of this pathway should prove to be of great interest and importance.

Synthesis of L-Alanine and D-Alanine

L-Alanine is probably synthesized by transamination of pyruvate. An alanine-requiring mutant of *B. subtilis* has been isolated and its mutation has been mapped (24). It is likely that this mutation inactivates the transaminase, but there is no direct evidence for this conclusion. Alanine dehydrogenase activity is induced during sporulation (32, 37); this enzyme seems to be required only for catabolism, since an *ald* mutant is unable to utilize alanine as sole carbon source (34) and does not require alanine for growth (9). The *ald* gene has been cloned and sequenced (32).

The only pathway to D-alanine is by racemization of L-alanine. The *B. subtilis* gene for alanine racemase (*dal*) has been cloned and sequenced (8). This gene probably encodes the only such enzyme since a Dal$^-$ mutant has an absolute requirement for D-alanine (5, 10).

REFERENCES

1. **Burke, M. E., and P. A. Pattee.** 1972. Histidine biosynthetic pathway in *Staphylococcus aureus*. *Can. J. Microbiol.* **18:**569–576.

2. **Callahan, J. P., I. P. Crawford, G. F. Hess, and P. S. Vary.** 1983. Cotransductional mapping of the *trp-his* region of *Bacillus megaterium*. *J. Bacteriol.* **154:**1455–1458.
3. **Carere, A., S. Russi, M. Bignami, and G. Sermonti.** 1973. An operon for histidine biosynthesis in *Streptomyces coelicolor*. I. Genetic evidence. *Mol. Gen. Genet.* **123:**219–224.
4. **Chapman, L. F., and E. W. Nester.** 1969. Gene-enzyme relationships in histidine biosynthesis in *Bacillus subtilis*. *J. Bacteriol.* **97:**1444–1448.
5. **Dedonder, R., J.-A. Lepesant, J. Lepesant-Kejzlarova, A. Billault, M. Steinmetz, and F. Kunst.** 1977. Construction of a kit of reference strains for rapid genetic mapping of *Bacillus subtilis* 168. *Appl. Environ. Microbiol.* **33:**989–993.
6. **Delorme, C., S. D. Ehrlich, and P. Renault.** 1992. Histidine biosynthesis genes in *Lactococcus lactis* subsp. *lactis*. *J. Bacteriol.* **174:**6571–6579.
7. **Ema, M., T. Kakimoto, and I. Chibata.** 1979. Production of L-serine by *Sarcina albida*. *Appl. Environ. Microbiol.* **37:**1053–1058.
8. **Ferrari, E., D. J. Henner, and M. Y. Yang.** 1985. Isolation of an alanine racemase gene from *Bacillus subtilis* and its use for plasmid maintenance in *B. subtilis*. *Bio/Technology* **3:**1003–1007.
9. **Freese, E., S. W. Park, and M. Cashel.** 1964. The developmental significance of alanine dehydrogenase in *Bacillus subtilis*. *Proc. Natl. Acad. Sci. USA* **51:**1164–1172.
10. **Heaton, M. P., R. B. Johnston, and T. L. Thompson.** 1988. Controlled lysis of bacterial cells utilizing mutants with defective synthesis of D-alanine. *Can. J. Microbiol.* **34:**256–261.
11. **Henner, D. J., L. Band, G. Flaggs, and E. Chen.** 1986. The organization and nucleotide sequence of the *Bacillus subtilis hisH*, *tyrA* and *aroE* genes. *Gene* **49:**147–152.
12. **Hopwood, D. A., K. F. Chater, J. E. Dowding, and A. Vivian.** 1973. Advances in *Streptomyces coelicolor* genetics. *Bacteriol. Rev.* **37:**371–405.
13. **Hougland, A. E., and J. V. Beck.** 1979. The formation of serine from glycine and formaldehyde by cell free extracts of Clostridium acidi-urici. *Microbios* **24:**151–157.
14. **Jungermann, K. A., W. Schmidt, F. H. Kirchiawy, E. H. Rupprecht, and R. K. Thauer.** 1970. Glycine formation via threonine and serine aldolase. Its interrelation with the pyruvate formate lyase pathway of one-carbon unit synthesis in *Clostridium kluyveri*. *Eur. J. Biochem.* **16:**424–429.
15. **Kane, J. F., R. L. Goode, and J. Wainscott.** 1975. Multiple mutations in *cysA14* mutants of *Bacillus subtilis*. *J. Bacteriol.* **121:**204–211.
16. **Kane-Falce, C., and W. E. Kloos.** 1975. A genetic and biochemical study of histidine biosynthesis in *Micrococcus luteus*. *Genetics* **79:**361–376.
17. **Kloos, W. E., and P. A. Pattee.** 1965. A biochemical characterization of histidine-dependent mutants of *Staphylococcus aureus*. *J. Gen. Microbiol.* **39:**185–194.
18. **Kloos, W. E., and P. A. Pattee.** 1965. Transduction analysis of the histidine region in *Staphylococcus aureus*. *J. Gen. Microbiol.* **39:**195–207.
19. **Kochi, H., and G. Kikuchi.** 1969. Reactions of glycine synthesis and glycine cleavage catalyzed by extracts of *Arthrobacter globiformis* grown on glycine. *Arch. Biochem. Biophys.* **132:**359–369.
20. **Kredich, N. M.** 1987. Biosynthesis of cysteine, p. 419–428. *In* F. C. Neidhardt, J. L. Ingraham, K. B. Low, B. Magasanik, M. Schaechter, and H. E. Umbarger (ed.), *Escherichia coli and Salmonella typhimurium: Cellular and Molecular Biology*. American Society for Microbiology, Washington, D.C.
21. **Limauro, D., A. Avitabile, C. Capellano, M. A. Puglia, and C. B. Bruni.** 1990. Cloning and characterization of the histidine biosynthetic gene cluster of *Streptomyces coelicolor* A3(2). *Gene* **90:**31–41.
22. **Limauro, D., A. Avitabile, C. Capellano, M. A. Puglia, and C. B. Bruni.** 1991. Cloning and characterization of the histidine biosynthetic gene cluster of *Streptomyces coelicolor* A3(2) (Correction). *Gene* **101:**161–162.
23. **Mahler, I., R. Warburg, D. J. Tipper, and H. O. Halvorson.** 1984. Cloning of an unstable *spoIIA-tyrA* fragment from *Bacillus subtilis*. *J. Gen. Microbiol.* **130:**411–421.
24. **Mattioli, R., M. Bazzicalupo, G. Federici, E. Gallori, and M. Polsinelli.** 1979. Characterization of mutants of *Bacillus subtilis* resistant to S-(2-aminoethyl) cysteine. *J. Gen. Microbiol.* **114:**223–225.
25. **Nelson, J. D., Jr., and H. B. Naylor.** 1971. Control of serine biosynthesis in *Micrococcus lysodeikticus*: inhibition of phosphoglyceric acid dehydrogenase. *Can. J. Microbiol.* **17:**25–30.
26. **Nelson, J. D., Jr., and H. B. Naylor.** 1971. The synthesis of L-serine by *Micrococcus lysodeikticus*. *Can. J. Microbiol.* **17:**73–77.
27. **Nester, E. W., and A. L. Montoya.** 1976. An enzyme common to histidine and aromatic amino acid biosynthesis in *Bacillus subtilis*. *J. Bacteriol.* **126:**699–705.
28. **Pasternak, C. A.** 1962. Sulphate activation and its control in *Escherichia coli* and *Bacillus subtilis*. *Biochem. J.* **85:**44–49.
29. **Pasternak, C. A., R. J. Ellis, M. C. Jones-Mortimer, and C. E. Crichton.** 1965. The control of sulphate reduction in bacteria. *Biochem. J.* **96:**270–275.
30. **Ponce-de-Leon, M. M., and L. I. Pizer.** 1972. Serine biosynthesis and its regulation in *Bacillus subtilis*. *J. Bacteriol.* **110:**895–904.
31. **Russi, S., A. Carere, A. Siracusano, and A. Ballio.** 1973. An operon for histidine biosynthesis in *Streptomyces coelicolor*. II. Biochemical evidence. *Mol. Gen. Genet.* **123:**225–232.
32. **Siranosian, K. J., K. Ireton, and A. D. Grossman. (Massachusetts Institute of Technology).** 1992. Personal communication.
33. **Sirokin, A., and S. D. Ehrlich. (Institut National de la Recherche Agronomique).** 1992. Personal communication.
34. **Trowsdale, J., and D. A. Smith.** 1975. Isolation, characterization, and mapping of *Bacillus subtilis* 168 germination mutants. *J. Bacteriol.* **123:**85–95.
35. **Vandeyar, M. A., and S. A. Zahler.** 1986. Chromosomal insertions of Tn*917* in *Bacillus subtilis*. *J. Bacteriol.* **167:**530–534.
36. **Walton, D. A., A. Moir, R. Morse, I. Roberts, and D. A. Smith.** 1984. The isolation of λ phage carrying DNA from the histidine and isoleucine-valine regions of the *Bacillus subtilis* chromosome. *J. Gen. Microbiol.* **130:**1577–1586.
37. **Warren, S. C.** 1968. Sporulation in *Bacillus subtilis*. Biochemical changes. *Biochem. J.* **109:**811–818.
38. **Winkler, M. E.** 1987. Biosynthesis of histidine, p. 395–411. *In* F. C. Neidhardt, J. L. Ingraham, K. B. Low, B. Magasanik, M. Schaechter, and H. E. Umbarger (ed.), *Escherichia coli and Salmonella typhimurium: Cellular and Molecular Biology*. American Society for Microbiology, Washington, D.C.

10. Transport Mechanisms

MILTON H. SAIER, JR., MATTHEW J. FAGAN, CHRISTIAN HOISCHEN, and JONATHAN REIZER

Gram-positive bacteria represent just one small branch of the tree of living organisms, all of which are believed to have descended from a common ancestor (37, 38). It is therefore not surprising that the transport systems found in gram-positive bacteria are structurally and evolutionarily related to those found in other eubacteria, archaebacteria, and eukaryotes (170). In fact, we currently know of no single type of transport system that is unique to gram-positive bacteria. Since representatives of most classes of permeases have been characterized more extensively in gram-negative bacteria or eukaryotes than in gram-positive bacteria, this chapter provides comparative information with a focus on the unique features of specific well-characterized transport systems in representative gram-positive bacteria. For a summary of these unique features, see Conclusions and Perspectives below.

Gram-positive bacteria have been divided into four subcategories, and only two of these, the high-G+C species and the low-G+C species, possess typical gram-positive cell walls (204). For these organisms, most of the detailed analyses of transport mechanisms have been performed with members of the low-G+C group. Consequently, this review focuses primarily on studies involving these organisms. The genera within this group that are considered here primarily include *Bacillus*, *Clostridium*, *Staphylococcus*, *Streptococcus*, *Enterococcus*, and *Lactobacillus*. *Mycoplasma* and *Acholeplasma*, also to be included in the discussion, represent wall-less variants of this grouping. Several of the classes of transport systems to be discussed have been reviewed elsewhere. References to such reviews are provided.

ATPASES: ATP-DRIVEN ACTIVE TRANSPORT SYSTEMS

The solute-transporting ATPases of living organisms fall into four principal classes that are distinguished by their subunit compositions, their sensitivities to characteristic inhibitors, their dependencies on various anions and cations, and the presence or absence of a phosphoenzyme intermediate.

P-Type ATPases

The P-type (or E_1E_2-type) ATPases of eukaryotes include the Na^+,K^+ ATPases found in the plasma membranes of higher animals, the gastric H^+,K^+ ATPases of mammals, the plasma membrane and sarcoplasmic reticular Ca^{2+} ATPases, and the H^+ ATPases of plants, fungi, and lower eukaryotes (46, 47). They share the following properties with the sequenced P-type ATPases of bacteria: (i) they possess a large polypeptide chain (>70 kDa) with regions of striking sequence similarity; (ii) they function by a mechanism that involves a high-energy aspartyl phosphate intermediate; (iii) they exhibit sensitivity to vanadate, usually in the micromolar concentration range; and (iv) they exist in either of two noncovalent intermediate states, designated E_1 and E_2. One, two, or three dissimilar protein subunits may be present. For a schematic depiction of these proteins, see Fig. 1 (46).

The first of the bacterial P-type ATPases to be characterized was the K^+-transporting Kdp system of *Escherichia coli*. It consists of three subunits: KdpA (59 kDa), KdpB (72 kDa), and KdpC (20.5 kDa). All three are essential, integral membrane proteins. Homology of KdpB with the eukaryotic ATPases has been demonstrated. KdpB contains the aspartyl phosphorylation site and possibly the ATP-binding site characteristic of the eukaryotic enzymes. On the basis of mutant analyses, it has been suggested that KdpA bears the cation recognition site and serves as the K^+ channel. It is largely membrane embedded, and of 20 mutations analyzed that altered cation affinity and/or specificity, 16 were in *kdpA*. Of the remaining four mutations of this class, three were in *kdpB* and one was in *kdpC* (46). The function of KdpC is not well defined.

A Kdp-like system from *Bacillus acidocaldarius* has been partially characterized (9). Like the *E. coli* Kdp system, it has high (micromolar) affinity for K^+ and low (millimolar) affinity for Rb^+, is inhibited by 1 mM vanadate, and is induced in cells grown in the presence of low K^+ concentrations. Antibodies against the *E. coli* KdpB protein cross-react with a 70-kDa *Bacillus* protein, and purification revealed the presence of three subunits of apparent molecular sizes by sodium dodecyl sulfate-polyacrylamide gel electrophoresis (SDS-PAGE) of 70, 44, and 23 kDa (68). It is therefore clear that gram-positive bacteria possess a Kdp-like P-type ATPase.

Physiological and biochemical evidence suggests that *Enterococcus faecalis* possesses a Na^+ extrusion system that is driven by ATP (71, 72, 98). First, accumulation of $^{22}Na^+$ by everted membrane vesicles of *E. faecalis* is resistant to the action of agents that dissipate the proton electrochemical gradient but is dependent on ATP. Second, the vesicles contain a Na^+-stimulated ATPase that is distinct from the known F-type ATPase but correlates with Na^+ trans-

Milton H. Saier, Jr., Matthew J. Fagan, Christian Hoischen, and Jonathan Reizer • Department of Biology, University of California at San Diego, La Jolla, California 92093-0116.

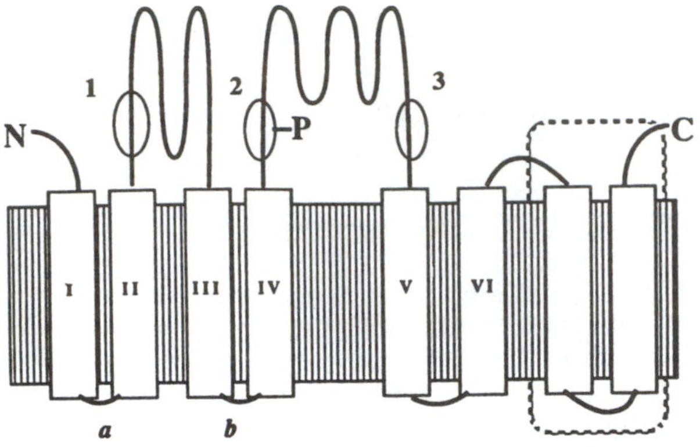

Figure 1. Generalized structure of the large subunit of P-type (E_1E_2) ATPases of bacteria and eukaryotes. The enzymes are embedded in the phospholipid bilayer of the membrane (shaded area), with six to eight putative membrane-spanning helices (I through VI, which are observed in most P-type ATPases, and two or more additional transmembrane helices that are present in most eukaryotic enzymes but are absent from the bacterial enzymes [boxed area at the C terminus]). Both the N terminus (N) and the C terminus (C) are on the cytoplasmic side of the membrane for those enzymes localized to the cytoplasmic membrane of the cell. Highly conserved regions, circled in the figure, include the following: 1, a region possibly involved in cation binding; 2, the aspartyl phosphorylation site (DKTGTI/LT); and 3, the ATP-binding site. Transmembrane helices I and II make up hairpin structure a, while transmembrane helices III and IV form hairpin structure b. Segment 1, used for phylogenetic tree construction (see text and reference 47), includes hairpin structures a and b as well as the included cytoplasmic loop between transmembrane helices II and III. Segment 2 includes the large cytoplasmic loop between transmembrane helices IV and V. The C-terminal regions of the proteins, from transmembrane helix V on, are poorly conserved in or absent from the bacterial enzymes. The sequenced P-type ATPases found in gram-positive bacteria include the Cd^{2+} ATPase of *S. aureus* and the K^+ ATPase of *E. faecalis*. (Reproduced from an article by W. Epstein [*Philos. Trans. R. Soc. London Ser. B* **326**:479–486, 1990] with permission of the Royal Society.)

port activity in whole cells. Finally, sodium extrusion activity in whole cells that is modified by physiological conditions or by mutation correlates with Na^+-dependent ATPase activity measured in vitro. These results suggest that Na^+ extrusion is mediated by an ATPase, very likely a P-type ATPase.

E. faecalis and *Streptococcus sanguis* appear to possess Ca^{2+}-dependent ATPases that normally function in Ca^{2+} extrusion (87, 99). Everted membrane vesicles derived from these organisms accumulated $^{45}Ca^{2+}$ when specifically incubated with ATP and Mg^{2+}, and the uptake process was resistant to inhibition by dicyclohexyl carbodiimide (DCCD), an inhibitor of F-type ATPases. Uptake did not require a proton electrochemical gradient.

Ambudkar et al. (4) reconstituted ATP-dependent Ca^{2+} transport in phospholipid vesicles after extraction of proteins from *E. faecalis* or *S. sanguis* membranes with the detergent octyl-β-D-glucoside. ATP-dependent Ca^{2+} transport activity was comparable to that found in the original membrane vesicles that had been extracted with detergent. Transport in the reconstituted system was resistant to ionophores and DCCD but sensitive to vanadate (in micromolar concentrations), suggesting the involvement of a P-type ATPase.

A Cd^{2+}-resistance ATPase (CadA) from *Staphylococcus aureus* confers heavy-metal resistance by pumping Cd^{2+} out of the cell. CadA is a 727-residue, 78-kDa, plasmid-encoded P-type ATPase with the basic structural elements and regions of homology observed for the eukaryotic P-type ATPases (129, 186, 187). Some evidence suggests that it catalyzes $2H^+$-Cd^{2+} exchange (199). It is similar in some respects to KdpB of *E. coli* and closely similar to a recently sequenced Cu^{2+}-transporting ATPase of humans. A Cd^{2+} efflux system in *Bacillus* spp. that resembles the corresponding enzyme from *S. aureus* has recently been identified (103a).

Interestingly, the N terminus of the *S. aureus* CadA protein possesses a short region that exhibits striking sequence similarity to a segment of mercuric reductase (39% sequence identity; 11 S.D. over a 31-amino-acid segment) and a segment of the periplasmic mercury-binding protein designated MerP (47). This observation has led to the suggestion that the homologous region of CadA is involved in Cd^{2+} binding. This conserved region contains cysteine residues (positions 23 and 26 in CadA) in all three enzymes. The initial binding domain that encompasses these cysteyl residues is about 100 residues longer than, for example, the comparable cytoplasmic region of the Kdp K^+ transport ATPase of *E. coli*, and the presence of this extra structure has been postulated as playing a role in the direction of Cd^{2+} pumping (129, 186, 187). An N-terminal segment homologous to that in the CadA protein is also found in the FixI ATPase of *Rhizobium meliloti*, and the cysteyl residues are conserved (47). Although the cation specificity of this ATPase is not known, these observations suggest that FixI may transport a divalent cation.

In other respects, the CadA protein resembles other P-type ATPases with respect to its domain structure (Fig. 1). Following the extended N-terminal cytoplasmic domain comes the first of three transmembrane hairpin structures. Each of these hairpins consists of two closely spaced putative transmembrane helical segments. Only four residues may compose the extracellular loop between the two transmembrane segments.

A large, highly conserved, hydrophilic domain of about 190 residues follows the first transmembrane hairpin structure (hairpin structure a in Fig. 1). It is possible that in the Cd^{2+} ATPase, as in the eukaryotic enzymes, this segment serves two functions: (i) to move the transported cation from its presumed initial binding site in the N-terminal domain of CadA to a site nearer the membrane surface (186, 187) and (ii) to hydrolyze the phosphorylated aspartyl residue (D-415) in the protein (179).

Next comes the second transmembrane hairpin structure (b in Fig. 1), where the two presumptive transmembrane helical segments in CadA extend from residues 336 to 356 and from residues 364 to 384. This conserved region may be part of the cation channel. An invariant prolyl residue in the latter loop is characteristic of virtually all P-type ATPases. In CadA, this prolyl residue is surrounded by two cysteyl residues, C-371 and C-373. These two residues and those at the N terminus are the only four cysteyl residues in the protein. They may compose part of the cation-binding site, as cysteyl residues can function in Cd^{2+} complexation. Importantly, this CPC sequence is conserved in the FixI ATPase of *R. meliloti* (47). In the eukaryotic, sarcoplasmic reticular Ca^{2+} ATPase, mutation of the conserved prolyl residue decreases the affinity of the enzyme for Ca^{2+} by 10-fold (see reference 47 for a review).

Sandwiched between the second and third transmembrane helical hairpin structures is a large (240-residue) cytoplasmic domain that includes the ATP-binding site (circled region 3 in Fig. 1; residues 489 to 492 in CadA) as well as the site of phosphorylation (circled region 2 in Fig. 1; D-415 in CadA). The region adjacent to the phosphorylation site may function in catalysis of autophosphorylation. A presumptive α-helical loop (residues 618 to 626) is predicted to make a Mg^{2+}-mediated salt bridge with the terminal phosphate of ATP (179). This region is strongly conserved in P-type ATPases. Unlike the eukaryotic enzymes, CadA ends shortly after the third transmembrane hairpin structure (Fig. 1).

In addition to CadA, a second protein, CadC, appears to be essential for unidirectional Cd^{2+} efflux, as defects in this protein have been reported to alter the relative rates of uptake versus efflux without altering induced expression of the *cad* operon (211). CadC is a soluble protein with a predominance of charged residues. It shows weak sequence identity with the DNA-binding protein ArsR of the arsenate resistance system (see below). It also exhibits sequence similarity to CadX, an undefined open reading frame in the sequence of an otherwise unrelated Cd^{2+} resistance system designated CadB (42a).

Two distinct P-type ATPases have been shown to be present in *E. faecalis*. The first of these is activated by divalent cations such as Mg^{2+} and Mn^{2+} and may transport protons and/or potassium ions. This enterococcal enzyme is phosphorylated on an aspartyl residue and is inhibited by 3 mM vanadate (55). It has the mobility in SDS-PAGE of a 78-kDa protein.

The second ATPase of *E. faecalis* has been characterized genetically and is homologous to other P-type ATPases. This 63-kDa protein exhibits 30% sequence identity with CadA of *S. aureus* and 24% identity with KdpB of *E. coli* (191). It is clearly distinct from the biochemically characterized ATPase discussed above, as disruption of its chromosomal gene does not result in loss of the former ATPase activity. It is probably a K^+-transporting enzyme.

Phylogenetic trees defining the relatedness of 47 dissimilar sequenced P-type ATPases have recently been constructed (47). The eukaryotic enzymes fall into three principal clusters: (i) those specific for Ca^{2+}, (ii) those specific for Na^+ and K^+ as well as the gastric enzymes specific for H^+ and K^+, and (iii) the H^+-specific ATPases of plants, fungi, and lower eukaryotes. While the Mg^{2+}-specific ATPase of *Salmonella typhimurium* clusters with the eukaryotic ATPases, all other sequenced bacterial ATPases cluster in a distinct but diverse subfamily. The last four proteins include the *E. coli* Kdp K^+ transport system, the *S. aureus* Cd^{2+} transport system, the K^+-transporting ATPase of *E. faecalis*, and the cation-translocating ATPase of *R. meliloti*. The bacterial ATPases were in general more distant from each other than were members of a specific group of eukaryotic enzymes, which is consistent with the fact that at least some of them exhibit different cation specificities (47).

F_0F_1 ATPases

The H^+-translocating ATP synthase-ATPases (F_0F_1 or F-type ATPases) in the membranes of mitochondria, chloroplasts, and bacteria utilize an electrochemical gradient of H^+ or Na^+ to synthesize ATP in a reversible process that in the reverse direction can result in the generation of an electrochemical gradient as a consequence of ATP hydrolysis. The bacterial F_0F_1 ATPases generally consist of a peripheral membrane five-subunit (α through ε) catalytic complex designated F_1 and an integral membrane three-subunit (a through c) proton translocation complex designated F_0. In the absence of F_0, F_1 can serve as an ATPase, and in the absence of F_1, F_0 can serve as a passive proton-specific channel (56, 178). The enzyme has been identified and partially characterized in a number of gram-positive bacterial genera, including *Bacillus*, *Clostridium*, *Micrococcus*, *Streptococcus*, *Lactobacillus*, *Mycoplasma*, and *Mycobacterium*. The β subunit from the gram-positive, spore-forming, thermophilic bacterium PS3 is strikingly similar to the β subunits from mitochondria, chloroplasts, and other bacteria. Construction of a phylogenetic tree of these β subunits revealed that the β subunit from PS3, the only gram-positive bacterial protein included in this study, is on a branch of its own (3). It is, however, clear that β subunits from *Bacillus megaterium*, *Streptococcus mutans*, *S. sanguis*, and *Streptococcus sobrinus* are strikingly similar to the PS3 β subunit (16, 145). Additionally, the catalytic subunits of the V-type ATPases and components of the flagellar apparati of *S. typhimurium* and *Bacillus subtilis* are homologous to the β subunits of F_0F_1 ATPases. The V-type ATPases will be discussed in the next section. The homologous flagellar protein of *S. typhimurium* is 48% identical to the corresponding flagellar protein from *B. subtilis*. They are both about 29% identical to the β subunit of the *E. coli* ATP synthase and somewhat less similar to the catalytic subunits of the V-type ATPases of the bread mold *Neurospora crassa* and the archaebacterium *Sulfolobus acidocaldarius* (1, 201). Some of the biochemically better characterized F-type ATPases from the gram-positive genuses will be discussed below.

Dissociation and reconstitution of an F_0F_1 ATPase complex was first accomplished by Kagawa and his coworkers in the late 1970s by using a purified enzyme from the thermophilic gram-positive bacterium PS3 (for reviews, see references 106 and 176). Three functions were defined for the various subunits: ATPase function was attributed to the F_1 subunits α, β, and γ; transmembrane H^+ channel function was attributed to the F_0 subunits a, b, and c; and "gate" function, i.e., connecting the ATPase with channel activity, was attributed to the γ, δ, and ε subunits of F_1 and the a and b subunits of F_0.

Alkalophilic bacilli including *Bacillus firmus* and *Bacillus alcalophilus* possess F_0F_1 ATPases that have recently been purified, characterized, and reconstituted (78, 82, 83, 104). These ATPases resemble the *E.*

coli and PS3 enzymes in terms of subunit compositions and physical properties. Of primary interest is the fact that both alkalophilic organisms grow optimally in the pH range 10 to 12. Since the cytoplasmic pH is normally maintained at about pH 8.5, it seemed possible that the ATPases translocate Na^+ instead of H^+, as had been demonstrated for *Propionigenium modestum* (see, for example, reference 83). This possibility was rendered more likely by the fact that flagellar motion and solute uptake in these alkalophic bacteria are apparently energized by the Na^+ electrochemical gradient. In spite of these considerations, the reconstituted ATPases were not stimulated by Na^+, and ATP synthesis was blocked by H^+-transporting uncouplers (protonophores), arguing against a Na^+-coupled ATP synthesis mechanism. Further, purified enzyme reconstituted in proteoliposomes catalyzed H^+ and not Na^+ translocation. These results convincingly show that H^+ and not Na^+ is the coupling ion translocated via the F_0F_1 ATPase of alkalophilic bacilli.

Early reports had suggested that *Lactobacillus casei* possesses an unusual F_0F_1 ATPase with subunit composition and stoichiometry different from those of ATPases from other bacteria (121, 122). However, in a more recent report, Muntyan et al. (124) have come to the opposite conclusion. The analyses reported revealed that the purified enzyme resembles other F_0F_1 ATPases in all essential structural and functional aspects examined.

Clostridium pasteurianum was reported to possess an ATPase of subunit composition and properties that might be suggestive of a V-type ATPase (see below). On the other hand, a closely related clostridial species, *Clostridium thermoaceticum*, apparently has a classic H^+-translocating F_0F_1 ATPase (89). The purified F_1 portion has a molecular size of 370 kDa, and four subunits (60, 55, 37, and 17 kDa) with apparent molar ratios of 3:3:1:1 were revealed by SDS-PAGE. This solubilized DCCD-insensitive enzyme could be reassociated with stripped membranes to reconstitute DCCD-sensitive ATPase activity as expected for an F-type ATPase. The intact enzyme and the F_1 moiety were both inhibited by azide and 7-chloro-4-nitrobenz-2-oxa-1,3-diazole (89). No evidence for an ε subunit was obtained, but such a subunit might well have remained associated with the membrane fraction. It remains to be ascertained whether the apparent absence of an ε subunit in the F_1 domain of the *C. thermoaceticum* enzyme represents a fundamental structural difference or is merely the consequence of an experimental technicality.

Purification and properties of an F_0F_1 ATPase from the strictly aerobic gram-positive bacterium *Micrococcus lysodeikticus* have been reported (23). The enzyme exhibited the expected subunit composition except that part of the α subunit may have been proteolytically clipped. Like the ATPases of chloroplasts, *Mycobacterium phlei*, and *Azotobacter vinelandii*, this enzyme was latent and exhibited ATPase activity only after mild trypsin treatment. No clear explanation for latency, a characteristic of the enzymes from strictly aerobic organisms, is at hand. However, in several bacteria, the smallest of the F_1 moiety subunits, ε, appears to be a natural inhibitor of the ATPase. Treatments (mild trypsin or heat) that cause dissociation of the ε subunit from F_1 unmask the ATPase activities of such latent enzymes, although the same treatments of active F-type ATPases do not give rise to stimulation (89). It has yet to be established that the ε subunit bears responsibility for latency of the *Micrococcus* enzyme.

Evidence for the presence of a Na^+-translocating F_0F_1 ATPase in *Mycoplasma gallisepticum* has been reported (181). This enzyme may function in volume regulation in this wall-less organism. The data presented by Shirvan et al. (181) clearly indicate that Na^+ rather than H^+ is the transported (extruded) species in response to ATP hydrolysis. The enzyme was inhibited by DCCD (50 μM) but not by vanadate (100 μM), and although some of the bacterial P-type ATPases are inhibited only when higher concentrations of vanadate are used, the results suggest the involvement of an F-type rather than a P-type ATPase. ATPase activity in membrane preparations was dependent on Mg^+ and stimulated by Na^+. The membranes contained a 52-kDa protein that cross-reacted with antibodies raised against the β subunit of the *E. coli* F_0F_1 ATPase. Thus, the Na^+-stimulated ATPase may be an electrogenic Na^+ pump of the F type. Further work will be required to establish this fact, however, particularly because of the presence of a distinct Na^+-independent ATPase in this organism (181).

The species of gram-positive bacteria discussed above with respect to their F-type ATPases all fall within the low-G+C classification. Recently, *Streptomyces lividans*, a member of the high-G+C group, has been examined with respect to the properties of its membrane-bound ATPase (76), and the enzyme appears to resemble other F-type ATPases. The F_1 moiety could be released from the membrane with a buffer of low ionic strength in the presence of EDTA, and the expected five subunits were identified (apparent molecular sizes of 58, 50, 30, 28, and 13 kDa). Immunological cross-reactivity with the F_1 portion of the *E. coli* enzyme was demonstrated, but the purified F_1 moiety of the *S. lividans* ATPase could not be functionally reconstituted with stripped membranes of *E. coli*. The enzyme was maximally active in the presence of Ca^{2+}, and Mg^{2+} exerted only an inhibitory effect on the Ca^{2+}-dependent activity. These results serve to demonstrate that the ATPase of *S. lividans* is a classic F-type enzyme that differs substantially from the *E. coli* enzyme.

V-Type ATPases

V-type ATPases of eukaryotic vacuoles, mammalian kidney plasma membranes, and cytoplasmic membranes of archaebacteria resemble F-type ATPases in several respects (8, 32, 62, 126, 192). (i) Both types are large multisubunit enzymes (molecular sizes, ca. 500 kDa). (ii) Both consist of a peripheral-membrane catalytic sector, which can be released as a water-soluble ATPase, and an integral-membrane protein channel complex. (iii) Each hydrophilic sector contains three copies of the catalytic subunit (the β subunits of the F-type ATPases and the A or 70-kDa subunits of the V-type ATPases) as well as three copies of a presumed regulatory subunit (the α subunits of the F-type

ATPases and the B or 60-kDa subunits of the V-type ATPases). (iv) At least some hydrophilic sectors of the V-type ATPases also contain single copies of several minor subunits, probably corresponding to the γ, δ, and ε subunits of the F-type ATPases. (v) Sequence data establish that the V-type catalytic and regulatory subunits are both homologous to the catalytic and regulatory subunits of the F-type ATPases. (vi) The hydrophobic c-proteolipid constituents of the V-type ATPases are homologous to the c subunits of the F_0 sectors of the F-type ATPases, and each contains a conserved intramembrane glutamate residue that probably reacts with DCCD. (vii) The three aforementioned subunits, A, B, and C, are found in all V-type ATPases, although the presence of some of the other subunits in some of these enzymes (i.e., the bacterial, fungal, and plant enzymes) is questionable. (viii) The complex of catalytic and regulatory subunits of the mammalian and archaebacterial (*Sulfolobus, Methanosarcina,* and *Halobacterium* species) V-type ATPases are similar to those of the F-type ATPases: subunits A and B of these enzymes alternate in the same ways as do the β and α subunits of the F-type ATPases.

In spite of these similarities, it is clear that V-type ATPases diverged from F-type ATPases relatively early, before individual F-type ATPases diverged from each other. The degree of sequence similarity among homologous V-type subunits or among homologous F-type subunits is much greater than between the corresponding V- and F-type ATPase subunits. Homologous V- and F-type ATPase subunits differ from each other with respect to the presence of extensive deletions and insertions of up to 100 amino acyl residues, and these deletions and insertions are conserved within each class of ATPase. Moreover, inhibitor sensitivities of most V-type ATPases of mammalian, plant, fungal, and archaebacterial origin are similar to each other but different from those of F-type ATPases. These enzymes are inhibited by DCCD as are the F-type ATPases, but they are also inhibited by 7-chloro-4-nitrobenzo-2-oxa-1,3-diazole and nitrate, and the eukaryotic enzymes are also inhibited by bafilomycin and *N*-ethylmaleimide, which are not inhibitors of the F-type ATPases. The V-type ATPases are not inhibited by classic F-type ATPase inhibitors such as oligomycin or azide. In addition, the arrangement of genes within the operon encoding the three known subunits of the *S. acidocaldarius* ATPase is totally different from that within the *E. coli unc* operon and the gram-positive PS3 ATPase operon. Finally, the c subunit of eukaryotic V-type ATPases (17 kDa) corresponds to an internal duplication of the c subunit of the F-type ATPases and consequently is twice as large. Surprisingly, the archaebacterial c subunit is the size of the F-type ATPase c subunit.

Recent work has provided evidence for V-type ATPases in eubacteria. An ATPase (molecular weight, 360,000) purified from plasma membranes of the thermophilic gram-negative eubacterium *Thermus thermophilus* (210) consisted of four recognizable subunits with molecular sizes of 66, 55, 30, and 12 kDa. It was not inhibited by azide but was inhibited by nitrate and *N*-ethylmaleimide. N-terminal sequence analyses revealed that both the 66- and the 55-kDa subunits were more similar to the catalytic and regulatory subunits of V-type ATPases than to the catalytic and regulatory subunits of F-type ATPases. This evidence suggested that V-type ATPases are not restricted to eukaryotes and archaebacteria.

Biochemical evidence suggests that gram-positive bacteria also possess V-type ATPases. *E. faecalis* possesses a Na^+-translocating ATPase that is stimulated by Na^+ and Li^+ but not by other monovalent cations. The activity of this ATPase is enhanced by growth in the presence of Na^+ under alkaline conditions, and it is believed to play a role in Na^+ circulation, particularly when the organism is grown under conditions of alkaline pH (91, 92). The enzyme was reported to be resistant to vanadate and DCCD and possibly to exchange Na^+ for K^+. Washing the *E. faecalis* membranes with EDTA resulted in loss of Na^+-stimulated ATPase activity, regardless of whether the soluble fraction or the stripped membranes were assayed. However, when the two fractions were reunited, reconstitution of Na^+-dependent ATPase activity was observed (93). These experiments argue against the possibility that the enzyme is a P-type ATPase.

Further analyses revealed that the EDTA-extracted soluble fraction contained a 330-kDa protein that was inducible under the conditions normally used for induction of the Na^+-dependent ATPase (94). It was lacking from a mutant *E. faecalis* strain defective for the ATPase. Three protein bands with molecular sizes of 73, 52, and 38 kDa were identified in the 330-kDa protein. The Na^+-stimulated ATPase of native membranes was shown to be inhibited by both nitrate and *N*-ethylmaleimide, inhibitors of vacuolar H^+ ATPases of eukaryotes and archaebacteria. On the basis of these observations, it was suggested that *E. faecalis* possesses a V-type ATPase that exhibits cation specificity for Na^+, but sequence data will be required to establish this possibility.

An early report noted that the membrane-bound ATPase of vegetative *C. pasteurianum* is apparently atypical of F_0F_1 ATPases with respect to its subunit composition and regulatory properties (24). This fact led to the possibility that this enzyme is a V-type ATPase. Only four protein components were separated by SDS-PAGE (molecular sizes, 65, 57, 43, and 15 kDa). The soluble form of the enzyme was reported to contain the three larger subunits in a ratio of 2:1:2. The 15-kDa subunit, present in the membrane fraction, was not quantitated. While the 65-kDa subunit was shown to bind 4-chloro-7-nitrobenzofurazan, the 15-kDa subunit bound DCCD. Both agents were inhibitory to the ATPase activity.

Effector molecules (phosphoenolpyruvate [PEP] and fructose 1,6-diphosphate) changed the kinetic parameters of the ATPase drastically, decreasing the K_m to below 2 mM and lowering the V_{max}. They also altered the pH profile of the enzyme. Only in the presence of these agents were hyperbolic kinetics observed.

When the ATPase was reconstituted into proteoliposomes that also contained bacteriorhodopsin, which pumps protons in response to illumination, ATP synthesis could be demonstrated (25). Inhibitors that blocked ATPase activity also blocked ATP synthesis. The properties of the enzyme generally suggest a V-type ATPase, but further work will be required to establish this possibility. Concern regarding the conclusion that *C. pasteurianum* possesses a V-type

ATPase is augmented in view of the report of Ivey and Ljungdahl (89) showing that the closely related *C. thermoaceticum* possesses a classic F-type ATPase. It is worthy of note, however, that several of the properties reported for the *C. thermoaceticum* enzyme differed from those reported for the *C. pasteurianum* enzyme.

ABC-Type ATPases

In gram-negative bacteria, numerous solutes are transported across the cytoplasmic membrane by multicomponent transport systems with a common organization (54). Among the best characterized of these ABC-type (or ATP-binding cassette-type) systems are transporters in enteric bacteria that exhibit specificity for (i) sugars such as maltose and maltodextrins, (ii) amino acids such as histidine and leucine, (iii) peptides (including dipeptides and oligopeptides), (iv) anions such as sulfate and phosphate, (v) organo-iron complexes, (vi) metals such as iron and nickel, (vii) opines, (viii) vitamins such as vitamin B_{12}, (ix) di- and tricarboxylates, and (x) proteins such as hemolysins. The SecD and SecF proteins, which are constituents of the protein secretion and membrane protein insertion apparatuses in *E. coli*, show limited sequence similarity to certain integral membrane constituents of the ABC-type permeases (58), but it has not yet been possible to establish that these secretory (Sec) proteins are homologous to the ABC-type permeases (155). The oxyanion transporter encoded by the plasmid-born arsenical resistance genes of both gram-positive and gram-negative bacteria may also fall within this group, but sequence comparisons have been insufficient to establish this possibility (see below). Of the bacterial transporters that clearly fall into the ABC class, eight clusters of evolutionarily related permeases have been suggested on the basis of sequence comparisons of the periplasmic protein constituents (194a). The eight clusters are specific for (i and ii) carbohydrates and iron, (iii and iv) amino acids and opines, (v) peptides and nickel, (vi and vii) inorganic and organic anionic species, and (viii) iron complexes and vitamin B_{12}.

The common protein components of these systems include two transmembrane proteins that usually span the membrane about six times each, one or two peripheral-membrane ATP-binding protein(s) localized on the cytoplasmic side of the membrane, and a high-affinity solute-binding protein. In gram-negative bacteria, this ligand-specific binding protein is periplasmic; in gram-positive bacteria, it is extracellular but bound to the membrane. The transmembrane protein components serve as the solute-specific channel, the peripheral-membrane ATP-binding protein energizes the system and sometimes serves a regulatory role, and the ligand-binding protein confers specificity and high affinity for the substrates. This last protein constituent is lacking if the system catalyzes solute efflux.

The presence of ABC-type transport systems in gram-positive bacteria has only recently been documented. Gilson et al. (61) and Dudler et al. (41) provided evidence for the presence of such systems in *Streptococcus pneumoniae* and a mycoplasma species, *Mycoplasma hyorhinis*. Since these organisms lack an outer membrane and consequently have no periplasm, their solute-specific binding proteins are lipoproteins with an N-terminal glyceride-cysteine, which allows them to be tethered to the external surface of the cell membrane. This hydrophobic anchor maintains the binding protein in the proximity of the external face of the integral cell membrane components of the transport system.

The genes for two such solute-binding lipoproteins in *S. pneumoniae* were sequenced (61). The first, designated MalX, is a maltose-inducible maltodextrin-binding lipoprotein that is 27% identical to MalE, the periplasmic binding protein of *E. coli*. The second, designated AmiA, exhibits 24% identical residues with OppA, the oligopeptide-specific periplasmic binding protein of *S. typhimurium* (61, 79, 194a). The periplasmic binding proteins of the gram-negative transport systems are essential constituents of maltodextrin and oligopeptide transporters, respectively.

The amino-terminal parts of the *E. coli malE* and *S. pneumoniae malX* gene products represent "signal peptides," which are characteristic of proteins that are exported through the cytoplasmic membrane. In the case of MalE, it was directly shown that the first 26 residues are cleaved from the remainder of the protein following export (42). In the case of MalX, the region of the potential cleavage site belongs to a well-defined category; it carries a sequence L-V-A–C-G-S, which corresponds to the consensus sequence for the precursors of lipoproteins (L-*Y*-*Z*-cleavage site-C-*y*-*z*, where *Y* is A, S, V, Q, or T; *Z* is G or A; *y* is S, G, A, N, Q, or D; and *z* is S, A, N, or Q) (reviewed in references 206 and 209). It thus appears likely that the mature MalX protein is a lipoprotein. Lipoproteins are exported through the cytoplasmic membrane in the same manner as other secreted proteins, but following cleavage, the amino-terminal cysteine is thioacylated to give a lipo-amino acid. This lipophilic modification is thought to be responsible for the membrane anchorage of a number of exported proteins (128). Based on the observed sequence comparisons and the biochemical requirements of the system, membrane attachment of MalX is likely to occur through the same mechanism. MalX would thus be expected to be exposed to the outer face of the membrane as an anchored but otherwise water-soluble protein. It is interesting to note that a mutant *E. coli* MalE protein that is anchored by its uncleaved amino-terminal signal peptide to the external face of the cytoplasmic membrane can still operate in transport (52).

The situation discussed above is comparable to that of the *amiA* and *oppA* gene products. In the case of OppA of *S. typhimurium*, an N-terminal peptide of 23 residues is cleaved from the remainder of the protein following export (80). For AmiA of *S. pneumoniae*, the region of the potential cleavage site carries the sequence L-A-A–C-S-S, which corresponds exactly to the consensus of the lipoprotein precursors.

A gene cluster with the same organization as ABC-type systems from gram-negative bacteria has been described in *M. hyorhinis* (41). The mature extracytoplasmic component, the so-called p37 protein, presents the characteristics of a lipoprotein: the amino-terminal sequence of the mature protein starts with C-S-N-, corresponding to the lipoprotein consensus sequence, and the p37 protein is bound to the membrane. The sequence before the potential cleavage site is less typical (A-I-S-cleavage site), but the putative signal sequence is unusual, as it contains four phenylalanine residues (41). Since very little is known about signal sequences in mycoplasma, it is possible that they differ significantly from those in other microorganisms.

The considerations mentioned above lead to the proposal that MalX and AmiA in *Streptococcus* spp. and p37 in *Mycoplasma* spp. are the functional equivalents of periplasmic solute-binding proteins in gram-negative bacteria. In the case of AmiA, the other components of the transport system are encoded by genes at the *ami* locus (2), and the substrates are likely to be oligopeptides. For MalX, the other components of the transport system have yet to be identified, but the transport substrates are likely to be maltodextrins. For the *Mycoplasma* transporter, which includes the p37 protein, the other components are known, but the substrates transported have yet to be identified.

A systematic search for lipoproteins in *Bacillus licheniformis* and *Bacillus cereus* revealed sets of lipoproteins that were released from protoplasts by mild trypsin treatment, suggesting an orientation to the outside of the membrane (128). In addition, the β-lactamases involved in resistance to penicillins are periplasmic proteins in gram-negative bacteria but are present in gram-positive bacteria such as *B. licheniformis* and *S. aureus* in substantial amounts in lipoprotein, membrane-bound forms (127). This fact suggests that lipoproteins in gram-positive bacteria play roles equivalent to those of many of the free periplasmic proteins from *E. coli* (128).

Examination of the gene products of the six open reading frames of the *ami* locus of *S. pneumoniae*, *amiABCDEF*, revealed that all but the AmiB protein are homologous to components of the oligopeptide permeases of *S. typhimurium* and *E. coli*. Intriguingly, AmiB was found to be homologous to ArsC, a cytosolic modifier subunit of the oxyanion pump encoded by the arsenical resistance operon of the R factor R773 from *E. coli* (see Note Added in Proof, below, and reference 22). Thus, AmiA is the solute-binding lipoprotein analog of the periplasmic binding proteins of gram-negative bacteria as noted above; AmiC and AmiD are the hydrophobic, transmembrane-channel-forming constituents of the system, and AmiE and AmiF are two homologous ATP-binding proteins localized to the cytoplasmic side of the cell membrane that presumably energize peptide uptake.

Mutations at the *ami* locus of *S. pneumoniae* have pleiotropic effects. Initially isolated on the basis of increased resistance to aminopterin (183), these mutations were shown to confer sensitivity to an imbalance in the extracellular concentrations of the branched-chain amino acids leucine, isoleucine, and valine (182). They also conferred increased resistance to methotrexate (198) and Celiptium (2-*N*-methylhydroxyellipticinium) (174). Selective alteration of the transport kinetics for several amino acids was reported for a mutation in the *ami* locus and correlated with a decrease in the transmembrane electric potential as deduced from measurement of tetramethylphosphonium accumulation (198). This complex phenotype is not entirely explicable in terms of the assumed function of the *ami* gene products as components of a peptide transport system. Nevertheless, the abilities of wild-type strains of *S. pneumoniae* (but not *ami* mutants) to utilize leucine- and arginine-containing peptides to satisfy their needs for these two auxotrophic requirements clearly argue that the Ami transporter recognizes oligopeptides as substrates.

Recently, the genes encoding the protein components of an oligopeptide transport system in *B. subtilis* were sequenced, and the encoded system was shown to play a role in the initiation of sporulation and genetic competence (136, 164). A mutation, originally designated *spo0K*, that gives rise to a stage zero phenotype (81) mapped within the five-cistron operon that encodes the transport system. All five encoded proteins were highly homologous to the corresponding components of the oligopeptide transporters of *S. typhimurium* and *E. coli*.

Studies with toxic peptide analogs provided evidence that this operon does indeed encode a peptide transport system. The deduced amino acid sequences of the five open reading frames were very similar to those of the OppA, -B, -C, -D, and -F proteins of *S. typhimurium*. The OppA protein was the least conserved, especially at the amino terminus, with 182 (33%) identical residues compared with the equivalent *S. typhimurium* protein. The ATP-binding proteins OppD and OppF were more highly conserved, with 165 (54%) identical residues. The hydrophobic OppB and OppC proteins fell in between these two extremes. The degree of conservation in every case was high and allowed unequivocal assignment of these sequences as the *B. subtilis opp* operon. Interestingly, unlike the Opp system of *S. typhimurium*, one of the two *B. subtilis* ATP-binding proteins, OppF, was not required for peptide transport or sporulation, although all of the other protein constituents of the transport system were required. The fact that OppF was dispensable means that OppD may possibly form a functional homodimer that eliminates the need for OppF altogether. Surprisingly, OppF is required for genetic competence (164).

As expected, the OppA peptide-binding protein had a signal sequence characteristic of lipoproteins with an amino-terminal lipo-amino acid anchor. Cellular location studies revealed that OppA of *B. subtilis* was associated with the cell during exponential growth but was released into the medium in stationary phase. The OppA protein was shown to be required for transport of peptides via this system. Whether or not release into the medium serves a biological function or whether the protein can be recaptured by the same or other cells following its release has not been clarified. It is possible that modification of the lipoprotein derivative provides a mechanism for regulating retention and release of the protein from the cell, a process that might be directly or indirectly important to the regulation of sporulation.

In this connection, it has been postulated that the accumulation of peptides derived from the degraded cell wall (the peptidoglycan) plays a signaling role in the initiation of sporulation and that the sporulation defect in *opp* mutants results from an inability to transport these peptides (136). The rationale for this notion is as follows. Peptidelike molecules probably play a general role in signaling differentiation in all sporulating organisms (65). It is possible that small peptides composed of protein amino acyl residues linked exclusively by α-peptide bonds cannot serve a signaling role because they are rapidly degraded by intracellular peptidases. However, Opp proteins in *S. typhimurium* and *E. coli* can transport peptidase-resistant cell wall peptides released from the peptidoglycan during growth (64), and the same would be expected of the gram-positive homolog. It is therefore possible that cell wall peptides serve to signal the onset of stationary phase and to initiate sporulation. Since *opp* mutants cannot recycle cell wall peptides, the peptides do not accumulate internally, and this presumed sporulation signal is not generated. An association of murein components with sporulation in *Myxococcus xanthus* has also been reported (180).

Enteric bacteria possess three genetically distinct peptide uptake systems with overlapping substrate specificities. The first system, the oligopeptide permease (Opp), mediates the uptake of peptides containing up to five amino acid residues with practically no specificity for the nature of the amino acid chains (67, 79, 133). As noted above, this system plays an essential role in the recycling of peptides released from the cell wall during growth (64). The second system, a dipeptide permease (Dpp), like Opp, is a periplasmic binding-protein-dependent transport system, but its specificity is limited to dipeptides. The third system, a tripeptide permease (Tpp), is less well characterized but has the greatest affinity for hydrophobic tripeptides. Expression of the *tpp* genes is induced during anaerobic growth, but the physiological role of the Tpp system is still unclear (60, 90).

Recently, a dipeptide transport system expressed early during sporulation in *B. subtilis* and dependent on the *spo0A* gene product for expression has been characterized (117, 189). The genes encoding the five protein constituents of the *dciA* operon (*dciAA* through *-E*) were found to include homologs of the genes of the Opp system of both *B. subtilis* and enteric bacteria as well as of genes of the Dpp system of *E. coli* (see below; 131). Interestingly, when the DciA system of *B. subtilis*, which clearly catalyzes dipeptide uptake, is defective, it does not interfere with sporulation. It may facilitate adaptation from nutrient-rich conditions to nutrient-poor conditions (117). Thus, it may actually antagonize sporulation by providing an alternative mode of growth. That is, it and the products of many other starvation-induced genes may function to improve the nutritional state of the cell both by generating new nutrients (i.e., through the generation of antibiotics and extracellular macromolecular degradative enzymes) and by improving the cell's ability to utilize meager supplies of limiting nutrients (i.e., through the induction of high-affinity transport systems).

The five proteins encoded within the *dciA* operon, DciAA through -E, were screened against the data base for similarity of their sequences with those of other sequenced proteins. The protein encoded by the first open reading frame, DciAA, did not exhibit significant sequence similarity to any other sequenced protein. DciAB was a hydrophobic protein (six putative transmembrane segments) that exhibited 53% identity with OppB of *B. subtilis* and 42% identity with OppB of *S. typhimurium*. DciAC was similar to DciAB in being hydrophobic, and it most resembled the OppC proteins of *B. subtilis* and *S. typhimurium*. DciAD proved to be an ATP-binding protein homologous to OppD and OppF. Finally, DciAE resembled the external di- and oligopeptide-binding proteins. This last protein was shown to be essential for dipeptide utilization, although tripeptides could still be utilized by strains lacking the protein. The tripeptides were apparently utilized via the *Bacillus* Opp system (117). Evidence for a tripeptide-specific transporter in *B. subtilis* analogous to that in the corresponding enteric systems has not been reported.

Guilfoile and Hutchinson (66) have presented the sequences of two genes (*drrA* and *drrB*) that apparently encode a two-component drug resistance transport system from *Streptomyces peucetius*. One of the two proteins (DrrA) was found to be similar to the family of ATP-binding proteins that energize solute transport via the ABC family of permeases (54). The other protein, DrrB, was hydrophobic, with six putative transmembrane α-helical segments, and was reported to show no significant sequence similarity to other known transport proteins. These two proteins were proposed to function together in the export of daunorubicin and doxorubicin, both of which are produced by *S. peucetius*, and induction of the syntheses of these transport proteins was coordinate with induction of the syntheses of the drug biosynthetic enzymes. A parallel between bacterial drug resistance and mammalian multidrug resistance was proposed (66).

Sequence comparison studies revealed that the drug resistance transporter of *S. peucetius* (DrrAB) and two nodulation gene products (NodIJ) of *Rhizobium leguminosarum* are homologous to proteins encoded by three sets of genes that constitute the capsular polysaccharide export systems in gram-negative bacteria: KpsTM of *E. coli*, BexABC of *Haemophilus influenzae*, and CtrDCB of *Neisseria meningitidis* (156). These five systems apparently form a subfamily within the family of ABC-type transporters. Three of the systems making up this subfamily (Drr, Nod, and Kps) may function with the participation of a single integral membrane constituent, while the other two systems (Bex and Ctr) probably depend on the simultaneous presence of two dissimilar integral membrane constituents. This observation and other published evidence suggest that the transmembrane channels of ABC-type transporters can be formed of homo- or heterooligomers, as is true of several other classes of transport systems (156).

The common organization of high-affinity ABC-type transport systems, many of which are homologous, for the transport of very different substrates, regardless of the bacterial or eukaryotic species from which they are derived, clearly suggests a common evolutionary origin. The ancestral systems probably existed before extensive divergence of the species occurred ($>1.5 \times 10^9$ years ago) (130). It is interesting to consider whether the ancestral system included a solute-binding lipoprotein that evolved into a periplasmic soluble protein or whether the reverse occurred. If, as has been proposed (204), the gram-negative cell type was ancestral, then free, periplasmic binding proteins would be expected to have preceded the lipoproteins.

A-Type ATPases

A plasmid-encoded arsenical resistance ATPase responsible for the active extrusion of arsenite, antimonite, and arsenate has been extensively characterized in *E. coli*, and sequence data for the corresponding genes from two *Staphylococcus* species have been presented (185).

The structure of the *E. coli* system has recently been reviewed (162). It consists of three subunits, ArsA, ArsB, and ArsC. The ArsA polypeptide chain is a "fusion dimer" with two potential ATP-binding sites (two ATP-binding motifs), the first (but not the second) of which has been shown to be essential for the active extrusion of arsenicals (95, 163). It is present in the membrane-bound complex as a dimer. Since each polypeptide chain contains two ATP-binding motifs, the complex may thus be a functional tetramer (120). It is interesting to note that the ATP-binding constituent of the ribose permease (an ABC-type ATPase) exhibits ATP-binding repeats like those of many of the eukaryotic homologs of this transporter. ArsA is homologous to NifH, an iron-containing nitrogenase subunit from *Anabaena* spp. that hydrolyzes ATP to energize molecular nitrogen reduction (162).

ArsB is the hydrophobic constituent of the oxyanion transport system that presumably functions as the transmembrane channel. It exhibits limited sequence similarity with the transmembrane constituent of the ribose permease, RbsC, and a protein implicated in protein secretion in *E. coli*, SecF (58). However, the degree of sequence similarity observed is insufficient to establish homology (147a).

The third constituent of the oxyanion permease of *E. coli*, the ArsC protein, is considered a "specificity" protein. In its absence, cells bearing the ArsA and ArsB proteins are resistant to arsenite and antimonite but not arsenate. ArsC in conjunction with ArsA and ArsB apparently confers resistance to arsenate. It is homologous to the AmiB protein involved in peptide transport in *Streptococcus* spp. (see the previous section) and also to an unidentified open reading frame found within a *nif* (nitrogen-fixing) region of the chromosome of *A. vinelandii* (162; see Note, below).

Physiological studies and sequence analyses of the plasmid-encoded *ars* genes of *Staphylococcus xylosus* and *S. aureus* have revealed that the mechanism of resistance to arsenite, antimonite, and arsenate is the same in these gram-positive bacteria as in *E. coli*. The operon is similar to that from *E. coli* except that the *arsA* gene is lacking. In fact, no *arsA* gene has been found on either of the *Staphylococcus* plasmids. It seems likely that a chromosomally encoded protein serves the function of ArsA.

In addition to arsenate, arsenite, and antimonite, many other ions, both cations and anions, are toxic to bacterial cells (184–187). Some of these are essential nutrients required in very low amounts. Such toxic ions include mercury, cadmium, zinc, cobalt, nickel, copper, iron, and chromate among others. For virtually all of these ions, transport systems exist either for their active uptake or for their active extrusion, and in some cases (i.e., those ions that are both essential and toxic), systems for both processes may be present in a single cell. These transport systems may be encoded by chromosomal and/or plasmid-borne genes. Thus, the cytoplasmic concentrations of the ions can be tightly regulated. In some cases, the resistance mechanisms are different in the examined gram-positive and gram-negative bacteria, but in most cases, structural and functional overlaps are observed. The genetic regulatory mechanisms and mechanistic aspects of the transport processes have recently been reviewed (188), and the reader is referred to that comprehensive review for a summary of the available information.

FACILITATORS: UNIPORTERS AND SECONDARY ACTIVE TRANSPORT SYSTEMS

Glycerol Uniporters

In all prokaryotes (109) and eukaryotes (110) studied to date, exogenous glycerol has been reported to enter the cytoplasm by passive or facilitated diffusion. Two different mechanisms of facilitation appear to operate in eukaryotes and prokaryotes, however. Eukaryotic cells, such as human erythrocytes, transport glycerol by a saturable process, possibly by a carrier-type mechanism in which the substrate-binding site of the permease is capable of movement so that it can alternate between one state accessible to the external side of the membrane and another state accessible to the cytoplasmic side of the membrane (110). By contrast, glycerol appears to cross the *E. coli* cytoplasmic membrane by a nonsaturable proteinaceous pore-type mechanism (73, 109). The glycerol facilitator is encoded by the *glpF* gene (11). This gene and the *glpK* structural gene, encoding glycerol kinase, make up an operon that is under the control of GlpR, the glycerol repressor (27, 69).

The *glpF* gene product allows the entry of straight-chain polyols including tetritols, pentitols, and hexitols in addition to glycerol (73). The rates of transport are related primarily to the size and shape of the substrate and not to the stereo arrangement of the hydroxyl groups. Ring sugars and phosphorylated compounds do not cross the membrane, but urea, glycine, and glyceraldehyde, all straight-chain compounds with different properties, are transported.

The glycerol facilitators of three bacterial species, *E. coli*, *B. subtilis*, and *Streptomyces coelicolor* have been fully or partially sequenced (86, 125, 190), and an open reading frame encoding a protein homologous to the three sequenced glycerol facilitators has been identified in *Lactococcus lactis* (118). The *Lactococcus* gene is the only one of these homologous genes that is not within an operon that also includes the gene encoding glycerol kinase. This fact may be suggestive of an alternative function.

These four transport proteins are strikingly similar. For example, the *S. coelicolor* and *E. coli* proteins are 46% identical throughout the regions of the two proteins for which sequence data are available. Similarly, the *B. subtilis* and *E. coli* proteins are 33% identical, while the gram-positive proteins are 34 to 44% identical with each other. These proteins are members of a large family of channel proteins of animal, plant, fungal, and bacterial origins that all presumably function in the transport of water or small molecules (132, 155). Although the biochemical functions of the eukaryotic proteins are not known, they generally function in developmental or cell-cell communication processes, as revealed by the physiological defects that result from genetic lesions in the structural genes for these proteins. Resolution of their biochemical mechanisms of action and their specificities will be of great interest.

Solute:Cation Symporters

Gram-positive bacteria of the genera *Bacillus*, *Enterococcus*, *Streptococcus*, *Lactobacillus*, *Staphylococcus*, and *Clostridium* have been extensively characterized with respect to the presence of solute:cation (H^+ or Na^+) symport systems that allow accumulation or extrusion of solutes against a concentration gradient in the presence of a membrane potential that is negative inside. Among the solutes transported by such a mechanism in these bacteria are amino acids, peptides, simple sugars, oligosaccharides, organic phosphate esters, organic acids (including mono, di-, and tricarboxylic acids), inorganic cations and anions, and vitamins. Since this topic has recently been reviewed elsewhere in considerable detail (101, 115a), we will treat the subject selectively rather than exhaustively.

Glucitol is taken up by *B. subtilis* cells by an inducible active-transport system that is under the control of the glucitol repressor GutR. Mutants that lacked both GutR and glucitol dehydrogenase, the product of the *gutB* gene, accumulated [^{14}C]glucitol in an apparently unaltered form against a large concentration gradient (>75-fold when the external concentration was 1 mM [21]). Addition of excess nonradioactive glucitol to the cell suspension resulted in rapid expulsion of the accumulated radioactive sugar. The product was characterized as D-glucitol by chromatographic and enzymatic techniques. The K_m for uptake was 0.8 mM, and the V_{max} was 145 nmol of glucitol taken up per min per mg of protein. Inhibitors of electron transfer, i.e., sodium azide (10 mM), 2,4-dinitrophenol (2 mM), and potassium cyanide (10 mM), all inhibited uptake and lowered the accumulation level of the solute. These results are in accordance with expectation for a proton symport system.

Gluconate appears to be accumulated in *E. faecalis* via the PEP-dependent phosphotransferase system (PTS), but in most other gram-positive bacteria examined (*B. subtilis* [10, 119], *C. pasteurianum* [12], and *Arthrobacter pyridinolis* [115]), a proton motive force (PMF)-driven proton symport mechanism appears to be operative. In all cases, the systems are inducible by growth in the presence of gluconate, and this induction is prevented by the presence of glucose, suggesting that expression of the gluconate catabolic system is subject to catabolite repression. The system in *B. subtilis* has a K_m for gluconate of 30 μM and a V_{max} of 90 μmol/min/g (dry weight) of cells (10, 119). The transport system appears to be highly specific for gluconate, as no other compounds tested except mannonate (K_i = 2.6 mM) inhibited uptake appreciably. Electron transport inhibitors (cyanide [1 mM] and 2-heptyl-4-hydroxyquinoline-*N*-oxide [40 μM]) and uncouplers [tetrachlorosalicylanilide [40 μM] and 2,4-dinitrophenol [1 mM]) each inhibited [^{14}C]gluconate uptake, suggesting that an active-transport mechanism, most likely involving proton symport, is operative.

In 1986, Fujita and coworkers reported the nucleotide sequence of the gluconate (*gnt*) operon of *B. subtilis* (53). Four genes, designated *gntRKPZ*, were found. The gluconate permease (GntP) consisted of 448 amino acids. Its function was defined by insertional inactivation and deletion analysis of the *gntP* gene. Its size and degree of hydrophobicity were characteristic of numerous facilitators found in bacteria and eukaryotes (74, 112). GntP, although not strikingly similar in sequence to other solute:cation facilitators (symporters [146]), undoubtedly uses such a symport facilitation mechanism for gluconate uptake in *B. subtilis* and several other gram-positive bacteria. It will be interesting to see whether the gluconate permease of *E. faecalis*, which is a phosphotransferase-type system, is evolutionarily related to the gluconate permease of *B. subtilis*. PTS permeases have been shown to exhibit weak sequence similarity with several facilitators, but in no case yet examined has this degree of similarity been sufficient to establish a common evolutionary origin (147, 207).

Recent evidence suggests that the glucitol and gluconate permeases may be regulated by a unique mechanism involving the protein kinase-generated serine phosphorylated derivative of the phosphocarrier protein HPr of the *B. subtilis* PTS. This novel regulatory mechanism is discussed below.

Solute:Solute Antiporters

Obligatory antiporters are well known in virtually all living organisms that have been examined for these transport systems. Among the best characterized are the HCO_3^-/Cl^- anion exchanger of the human erythrocyte, the mitochondrial ATP/ADP exchanger, and the sugar phosphate:phosphate exchanger of *E. coli* (113, 170). The last-mentioned system is also found in gram-positive bacteria such as *L. lactis* (4, 5, 113, 114). Because information concerning the sugar phosphate:phosphate exchange system has recently been reviewed in detail (114), we focus here on two other gram-positive bacterial exchange transporters, the arginine:ornithine exchanger and the lactose:galactose exchanger of *Lactococcus* spp. While the former system appears to function largely as an exchange transporter, the latter system can probably also function efficiently by lactose:H^+ symport. This possibility would allow active uptake of lactose either in exchange for cytoplasmic galactose or in response to a membrane potential that is negative inside. These antiport systems are shown schematically in Fig. 2.

The arginine deiminase pathway results in the conversion of arginine to ornithine in a process that is coupled to substrate-level phosphorylation of ADP (Fig. 2). The driving force for arginine uptake in *L. lactis* is provided by the accumulation of cytoplasmic ornithine, which is excreted in a one-for-one exchange with extracellular arginine. The inducibly synthesized antiporter has been characterized in whole cells and membrane vesicles (39, 138, 196). Sequence analyses, described below, provide evidence that antiporters are homologous with symporters. Rapid uncoupler (protonophore)-insensitive heterologous exchange of

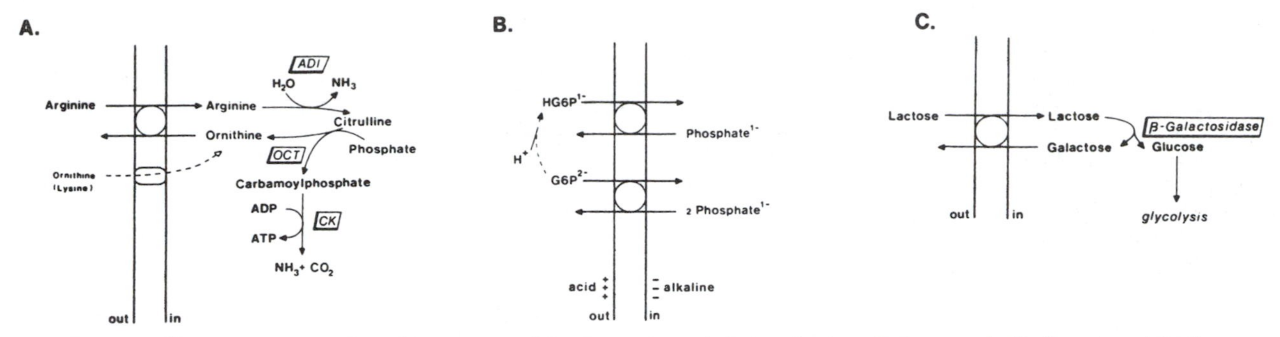

Figure 2. Three antiporters found in gram-positive bacteria and their roles in cellular metabolic function. (A) The arginine/ornithine antiporter and the arginine deiminase pathway of gram-positive bacteria such as *L. lactis*. (B) The sugar-6-phosphate/phosphate antiporter, illustrating its two transport modes. (C) The lactose/galactose antiporter and its involvement in sugar metabolism (modified from reference 101 with permission). Abbreviations: ADI, arginine deiminase; OCT, ornithine carbamoyltransferase; CK, carbamoyl kinase; $HG6P^{1-}$, protonated glucose 6-phosphate; $G6P^{2-}$, nonprotonated glucose 6-phosphate.

external arginine for internal ornithine occurs at rates that are more than 50-fold greater than those of arginine:H^+ symport or of unidirectional ornithine efflux. The rate of [^{14}C]ornithine efflux is enhanced at least 500-fold by addition of external arginine or ornithine. The V_{max} of arginine uptake was shown to increase with increasing internal ornithine concentration, with a half-maximal rate observed at 42 μM cytoplasmic ornithine. A ping-pong mechanism was proposed (39), suggesting that a single binding site is alternately exposed on the two sides of the membrane.

The Arg/Orn exchanger can be reversibly inactivated by sulfhydryl reagents such as *p*-chloromercuribenzene sulfonate. The reactive sulfhydryl group is presumably localized to the external surface of the permease. It may be near the substrate-binding site, since either arginine or ornithine, when present in the external fluid, protects against inactivation.

The exchanger is synthesized in response to arginine, which serves as an inducer. Ornithine and citrulline (Fig. 2) are not inducers. Glucose can repress synthesis of the antiporter as well as of the three cytoplasmic enzymes of the arginine deiminase pathway. Thus, the pathway seems to be sensitive to a form of catabolite repression, but the mechanism of this repressive effect is unknown (137).

Although the sequence of the Arg/Orn exchanger of gram-positive bacteria is not yet available, the corresponding sequence of a protein from *Pseudomonas aeruginosa* which catalyzes Arg/Orn exchange has been determined (200). This latter protein was found to be homologous to a large number of amino acid transporters of bacteria and eukaryotes, and it clustered on the phylogenetic tree together with two other potential antiporters, both from *E. coli*, one specific for putrescine (and maybe agmatine) and the other specific for cadaverine (and maybe lysine) (10a, 155). The former system may resemble a putrescine:agmatine antiport system in *E. faecalis* (40). These observations emphasize the structural and functional similarities between symporters and antiporters and suggest a unified transport mechanism. This suggestion is substantiated by the characteristics of the lactose/galactose exchanger discussed below.

Hutkins and Ponne (88) have recently characterized the transport mechanism of the lactose permease (LacS) of *Streptococcus thermophilus* T52. When this organism is grown in medium containing excess lactose, galactose is released into the extracellular medium. When starved and deenergized lactose-grown cells were preloaded with high concentrations (≤50 mM) of galactose and suspended in medium containing lactose, rapid galactose efflux occurred. These galactose-loaded cells accumulated [^{14}C]lactose more rapidly and to a greater extent than did unloaded cells. Further, when galactose-loaded cells were suspended in carbohydrate-free medium, a PMF approaching 90 mV was formed. Nonloaded cells maintained a PMF of about 50 mV. This observation suggested that unidirectional galactose efflux could contribute to the PMF, presumably by a galactose:H^+ symport mechanism (138). When the PMF was abolished by addition of an uncoupler (a protonophore) or an inhibitor of the F_0F_1 ATPase, uptake of lactose in exchange for internal galactose still occurred. These results support the conclusion that the lactose permease of *S. thermophilus* can catalyze both lactose-galactose antiport and lactose:H^+ or galactose:H^+ symport. Both of these processes, antiport and symport, may be of physiological significance. Under conditions in which the glucosyl but not the galactosyl moiety of lactose is rapidly metabolized following hydrolysis of the disaccharide, the intracellularly accumulated galactose can be utilized to drive lactose uptake in exchange for the intracellular galactose. On the other hand, if galactose is metabolized as rapidly as glucose, no cytoplasmic galactose will accumulate, and lactose can be accumulated via PMF-driven lactose:H^+ symport. The availability of these alternative mechanisms allows the streptococcal cell to utilize either of two different forms of energy, a PMF or a sugar concentration gradient, for the accumulation of lactose, depending on which source is available.

The gene encoding the lactose permease of *S. thermophilus* has been shown to encode a protein of 634 amino acids (molecular size, 69 kDa). The protein possesses an amino-terminal hydrophobic 49-kDa moiety and a carboxy-terminal hydrophilic 20-kDa domain that respectively are homologous to the melibiose permease of *E. coli* (23% identity throughout its length) and enzyme IIA^{Glc} of the *E. coli* PTS (40% identity) (139). Since lactose and galactose are not phosphorylated in transit by a PTS-mediated mechanism, it is most likely that the enzyme IIA^{Glc}-like moiety of the permease serves a regulatory role. Possibly the energy-coupling proteins of the PTS phosphorylate the C-terminal domain in a process that regulates the activity of the permease (101, 168). Indeed, PTS-dependent phosphorylation of the enzyme IIA^{Glc}-like domain of the *S. thermophilus* lactose permease was recently demonstrated (138).

Although LacS of *S. thermophilus* and LacY of *E. coli* do not exhibit overall amino acid sequence similarity, a region similar to the proposed catalytic region of LacY was identified in the former permease. Site-directed substitutions of the histidyl residues in these regions, i.e., histidine 376 of LacS and histidine 322 of LacY, provided evidence that these histidines and the surrounding conserved regions play a role in energy transduction and sugar recognition (138). Interestingly, a glutamate residue in this region (corresponding to glutamate 325 of the *E. coli* LacY protein) was shown to be conserved in several permeases, including the LacS proteins of *S. thermophilus* and *Lactobacillus bulgaricus*; the LacY proteins of *E. coli* and *Klebsiella pneumoniae*; and the raffinose (RafB), melibiose (MelB), and glucuronate (GusB) permeases of *E. coli* (138). Substitution of this residue (glutamate 325) in LacY of *E. coli* led to a transport protein that failed to catalyze lactose/H^+ symport or lactose efflux. The mutant protein did catalyze efficient lactose:lactose exchange and counterflow (19, 20). Similar substitution of the corresponding glutamate residue in MelB (glutamate 361) also affected the translocation of the sugar and the cation but not the recognition of these substrates (141).

Porters for Solute Efflux

Coryneform bacteria are known to actively excrete amino acids under certain conditions, and they are widely used for the commercial production of amino acids and other metabolites (96, 97, 195). Considerable information concerning the physiology of these organisms and the metabolic events leading to amino acid overproduction is available. However, information concerning the transport processes involved in these efflux processes is only now emerging.

Several models for amino acid excretion by *Corynebacterium glutamicum* have been proposed. On the basis of studies showing that efficient glutamate secretion correlates with alterations in the lipid composition of the cytoplasmic membrane, it was suggested that amino acid excretion is due to a leaky membrane with altered physical properties (17, 31, 123). For the excretion of lysine, Luntz et al. (111) proposed a process mediated by pores that open and close in response to osmotic pressure. Clement et al. (26) postulated that the export of glutamate is catalyzed by functional inversion of the normal uptake system. Recently, evidence suggesting that secretion of glutamate, isoleucine, and lysine occurs via specific excretion carriers has been presented (14, 44, 84). It has also been reported that efflux of 5′-IMP (195) and NAD (175) in *Brevibacterium ammoniagenes* is mediated by efflux-specific systems.

Biotin limitation is one of the conditions that induces glutamate secretion. Biotin-limited producer cells were analyzed with respect to basic energetic and kinetic parameters and compared with biotin-supplemented nonproducers (84, 85). Secreting cells showed high internal glutamate concentrations, a large membrane potential, substantial K^+ gradients, and secretory specificity for glutamate. The efflux process was observed to be regulated and to depend on the metabolic and energetic state of the cells. The leaky membrane model was thus ruled out, and the inversion-of-uptake model was shown to be improbable. The data presented could best be accounted for by assuming the presence of a secretion-specific carrier. High rates of efflux via this carrier are apparently dependent on specific alterations in the cytoplasmic membrane as well as on metabolic conditions.

The uptake system for glutamate in *C. glutamicum* is inducible by glutamate. It shows an apparent K_m of 0.5 to 1.3 mM under both induced and uninduced conditions (103). It is regulated by the internal K^+ concentration as well as by the internal pH. High uptake rates can be observed only when the internal K^+ concentration is above 200 mM and the internal pH is higher than 6.6. The possibility of a secondary transport mechanism for the uptake of glutamate via H^+, Na^+, or K^+ cotransport or via anion antiport was excluded (102). Accumulation ratios of more than 2×10^5 were observed. The unidirectional uptake rate correlated with the cytosolic-ATP content and the ATP/ADP ratio. These facts are evidence for a primary active transport system exhibiting similarities to ABC-type transport systems.

Isoleucine excretion in *C. glutamicum* can be induced by addition of its precursor, 2-ketobutyrate. By studying the secretory process with respect to various energetic and kinetic parameters, (i.e., membrane potential, rates of uptake and efflux, concentration gradients of isoleucine and ions), Ebbighausen et al. (44) obtained evidence for a specific efflux carrier for isoleucine. Secretion by passive diffusion or by a functionally inverted uptake system could again be ruled out. Isoleucine efflux is specific and independent of the concentration gradient. Excreting cells exhibit a large membrane potential and substantial ion gradients, and they are able to excrete isoleucine against a chemical gradient. It was demonstrated that the secretory system depends on the energetic status of the cells. A decrease in the membrane potential by addition of valinomycin inhibited efflux completely.

Uptake of isoleucine occurs in *C. glutamicum* via a common energy-dependent carrier for branched-chain amino acids (isoleucine, leucine, and valine), with similar affinities for all substrates (43). The uptake system depends on the membrane potential and is stimulated by Na^+ ions. It was postulated that isoleucine is taken up by a secondary, active, Na^+-coupled, symport mechanism involving a binary transport complex.

Recently, Bröer and Krämer (14) demonstrated the existence of a specific efflux carrier for lysine in *C. glutamicum*. Using a mutant strain with deregulated lysine biosynthesis, they showed that lysine excretion is not due to passive diffusion, since the excretion system exhibited substrate specificity, saturation kinetics, a pH optimum, and a high activation energy. The possibility of pore-mediated lysine excretion was excluded, since the positively charged lysine molecules were shown to be excreted against their chemical gradient and the membrane potential. The secretory system was also shown to differ from the uptake system. The secretory carrier was shown to be a secondary transport system (15) driven by the lysine chemical gradient, the proton chemical potential, and the membrane potential. Bröer and Krämer presented a kinetic model for lysine export that postulates that lysine is excreted in symport with two OH^- ions in a single step. This proposed mechanism cannot be distinguished from an antiport mechanism in which two protons are taken up. Assuming the former possibility, the substrate-loaded carrier was thought to be uncharged, while the unloaded carrier was considered positively charged.

Bröer and Krämer (13) demonstrated that resting cells of *C. glutamicum* accumulate radiolabeled lysine by an electroneutral lysine-lysine antiport process that is not affected by uncouplers or ionophores. They also provided evidence for net lysine uptake in lysine-depleted cells of a lysine auxotroph. Net uptake could be detected in wild-type cells and was shown to be due to heterologous electrogenic antiport of lysine for alanine. It was postulated that the cells switch between homologous and heterologous antiport depending on their metabolism.

THE PTS: A GROUP TRANSLOCATION SYSTEM

General Features of the System

The phosphoenolpyruvate (PEP):sugar phosphotransferase system (PTS) is a complex enzyme system that is responsible for the detection, transmembrane transport, and phosphorylation of its numerous sugar substrates in both gram-negative and gram-positive prokaryotes. Normally, PEP phosphorylates enzyme I (EI) of the PTS, initiating the phosphoryl transfer chain (Fig. 3). Phosphohistidyl-enzyme I then phosphorylates a histidyl residue in the small, heat-stable phosphoryl carrier protein of the PTS, HPr. This phosphoprotein then transfers its phosphoryl group to a histidyl residue in one of the sugar-specific proteins of the PTS designated enzyme (or domain) IIA (169, 171). Phosphoprotein IIA then phosphorylates enzyme (or domain) IIB either on a cysteyl residue or on a histidyl residue, depending on the particular enzymes under study. Enzyme (or domain) IIC then transfers the phosphoryl group to the incoming sugar, which is transported via the IIC protein. It is the complex of the three enzyme II domains or proteins, IIA, IIB, and IIC, that functions coordinately in the transport and phosphorylation of sugar substrates (Fig. 3).

The PTS functions in the regulation of various bacterial physiological processes and is itself subject to elaborate regulatory control. Recent work has revealed some interesting structural features of the systems present in a number of diverse gram-positive bacterial species. Carbohydrate uptake by bacteria can be regulated by any of several distinct mechanisms. Seven such mechanisms were discussed by Saier (166), Reizer et al. (149), and Reizer and Peterkofsky (152). These mechanisms include (i) competition for a common permease, (ii) inhibition by intracellular sugar-phosphate, (iii) inhibition by the

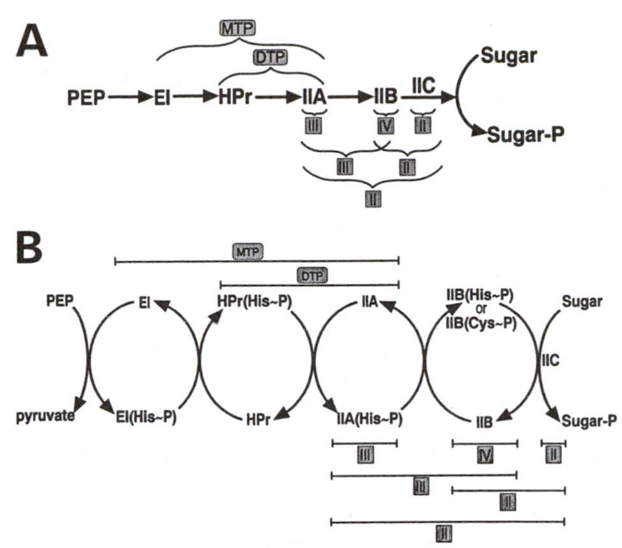

Figure 3. Generalized phosphoryl transfer reactions catalyzed by the proteins of the PTS. (A) Linear scheme indicating the proteins but not the nature of their derivatives. (B) Reaction scheme illustrating the individual phosphoryl transfer reactions and the residues phosphorylated within each protein. For definitions of the proteins involved, see reference 171, where alternative protein (domain) designations for these proteins have been used. Thus, IIA has been called enzyme III, IIB has been called enzyme IV, and IIC has been called enzyme II. For some systems, fused IIA-IIB domains have been referred to as enzymes III, fused IIB-IIC domains have been referred to as enzymes II, and fused IIA-IIB-IIC domains have been referred to as enzymes II. Regardless of the state of fusion of the IIA, IIB, and IIC proteins-domains, they are referred to collectively as the enzyme II complexes. MTP and DTP are the multiphosphoryl transfer protein and diphosphoryl transfer protein, respectively, of the fructose-specific phosphotransferases of *R. capsulatus* and *S. typhimurium*, respectively.

proton-electrochemical gradient, (iv) PTS-mediated regulation involving the glucose-specific IIA protein, (v) competition for phospho-HPr, (vi) regulation by ATP-dependent phosphorylation of a regulatory seryl residue in HPr, and (vii) expulsion of intracellular sugars.

A transcriptional regulatory mechanism involving PTS-mediated phosphorylation of transcriptional antiterminators has been unveiled in recent years (see chapter 48). The PTS proteins involved in antiterminator phosphorylation in the sucrose system of *B. subtilis* may differ from those utilized in the β-glucoside system of *E. coli* (6, 7, 28, 30, 145a, 155, 177). One can anticipate that PTS-mediated regulation of transcription will be a general feature of catabolic systems and that a variety of transcriptional factors (including classical repressors and activators) will be targets of this type of regulation (167, 172).

Two early reviews focused on the PTS in *S. aureus* (70, 75). More recent reviews have discussed the PTS in gram-positive bacteria but have emphasized work with gram-negative organisms (140, 166). A volume of broad scope (153) and a comprehensive review of the PTS in gram-positive bacteria (149) were published in 1987 and 1988, respectively. The material concerning the PTS in this review focuses on aspects of the PTS specific to gram-positive bacteria that have been elucidated since 1987. The reader is referred to the above-mentioned references for consideration of the earlier literature.

Enzymes I

Genes encoding enzymes I of the PTS have been sequenced from three gram-positive bacteria: *B. subtilis* (154), *Staphylococcus carnosus* (100), and *Streptococcus salivarius* (57). The deduced protein products were shown to be homologous not only to the enzymes I of gram-negative bacteria but also to PEP synthase of *E. coli* and to bacterial and plant pyruvate:phosphate dikinases (154). Multiple alignments and computer analyses of these 11 proteins have been published, and the phylogenetic tree of this protein family was constructed (154). The analyses revealed regions of sequence similarity and divergence. There are four regions (A through D) of striking conservation for the 11 proteins, and three additional regions (E through G) of striking conservation specific to the enzymes I. Only region A is functionally defined, however. This region contains the active-site histidyl residue that is known to be phosphorylated in all protein members of this family, and it is preceded by a threonyl residue that was shown to be phosphorylated in the maize pyruvate, phosphate dikinase (160). Regions B and C may be involved in PEP binding and/or phosphoryl transfer. Region D contains the conserved cysteyl residue that presumably corresponds to the essential, catalytically active cysteyl residue present in each of the 11 enzymes. The regions that are strongly conserved in the enzymes I (E through G) but not in the other enzymes might function in HPr recognition or in phosphoryl transfer to this protein. A putative DNA-binding (helix-turn-helix) region has also been identified in the enzyme I and pyruvate, phosphate dikinase members of this family (108, 154). An improved procedure for the overproduction and purification of the *B. subtilis* enzyme I that allows quick preparation of gram quantities of the protein from *B. subtilis* has recently been devised (155).

HPr Proteins

HPr-encoding genes or their protein products from the following gram-positive bacteria have been sequenced: *B. subtilis*, *S. aureus*, *S. carnosus*, and *E. faecalis*. Multiple alignments of these proteins with the sequenced HPr proteins and FPr (fructose-inducible HPr-like) domains of gram-negative bacteria have been reported. The phylogenetic tree of these proteins (154) is shown in Fig. 4. The pattern of the tree shown in Fig. 4 was found to be qualitatively similar to that for the enzymes I, with some interesting differences. The gram-positive HPrs cluster together, and these proteins are more than 60% identical. The multiple alignment revealed three regions of sequence similarity in all HPr proteins, A through C. The A region includes the enzyme I-dependent phosphorylatable histidine. The B region includes the gram-positive HPr kinase-dependent phosphorylatable seryl residue (Ser-46) that is conserved in all HPrs. The study of site-specific mutations of this region of the *B. subtilis*

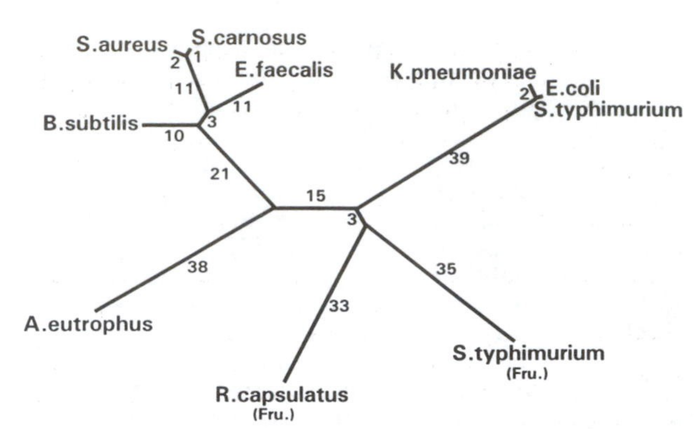

Figure 4. Phylogenetic tree of 10 sequenced proteins that make up the family of HPr proteins and HPr protein domains. Relative evolutionary distances are given adjacent to the branches. The programs of Doolittle and Feng (38) and Feng and Doolittle (51) were used to calculate relative distances. The HPr protein domains of the diphosphoryl transfer protein (DTP) and the multiphosphoryl transfer protein (MTP) encoded within the fructose operons of *S. typhimurium* and *R. capsulatus*, respectively, are denoted (Fru.). Four main clusters are apparent: (i) HPr proteins of gram-positive bacteria, i.e., *B. subtilis* (63, 149), *S. aureus* (149), *S. carnosus* (45), and *E. faecalis* (36); (ii) HPr proteins of enteric bacteria, i.e., *S. typhimurium* (18, 142), *E. coli* (33, 165), and *K. pneumoniae* (197); (iii) HPr protein domains of the DTP and MTP proteins of *S. typhimurium* (59) and *R. capsulatus* (208), respectively; and (iv) HPr of *Alcaligenes eutrophus* (144).

HPr has provided evidence for the involvement of Ser-46 in regulatory interactions (150, 158, 159). The C region contains two adjacent acidic residues (aspartate 69 and glutamate 70) that are believed to be important for catalytic function as revealed by site-specific mutagenesis studies of the *E. coli* protein (99a). Interestingly, the C region is not well conserved in the HPr protein domains of the diphosphoryl transfer protein and the multiphosphoryl transfer protein of *S. typhimurium* and *Rhodobacter capsulatus*, respectively.

HPr of gram-positive bacteria can be phosphorylated on a histidyl residue (histidine 15) by PEP and enzyme I and on a seryl residue (serine 46) by ATP and an HPr(Ser) kinase. The latter enzyme is allosterically activated by cytoplasmic intermediates of carbohydrate metabolism such as fructose 1,6-diphosphate and gluconate 6-phosphate (34, 149–151). Phosphorylation of the seryl residue inhibits phosphorylation of the histidyl residue about 100-fold. Thus, phosphorylation of serine 46 potentially reduces phosphoryl transfer via the PTS to less than 1% of the normal rate.

In vivo studies (150, 158, 159, 194) did not provide evidence for the suggestion (35) that this mechanism serves as a device for regulating the activity of the PTS in whole cells. This fact, as well as the recent demonstration of HPr, HPr(Ser) kinase, and HPr(Ser-P) phosphatase in heterofermentative lactobacilli (148, 149, 161) and *Acholeplasma laidlawii*, all of which lack a functional PTS (155), prompted consideration of an alternative physiological function for the HPr(Ser) kinase-HPr(Ser-P) phosphatase system (33a). Synthesis of gluconate kinase in a wild-type *B. subtilis* strain was inducible by growth in the presence of gluconate

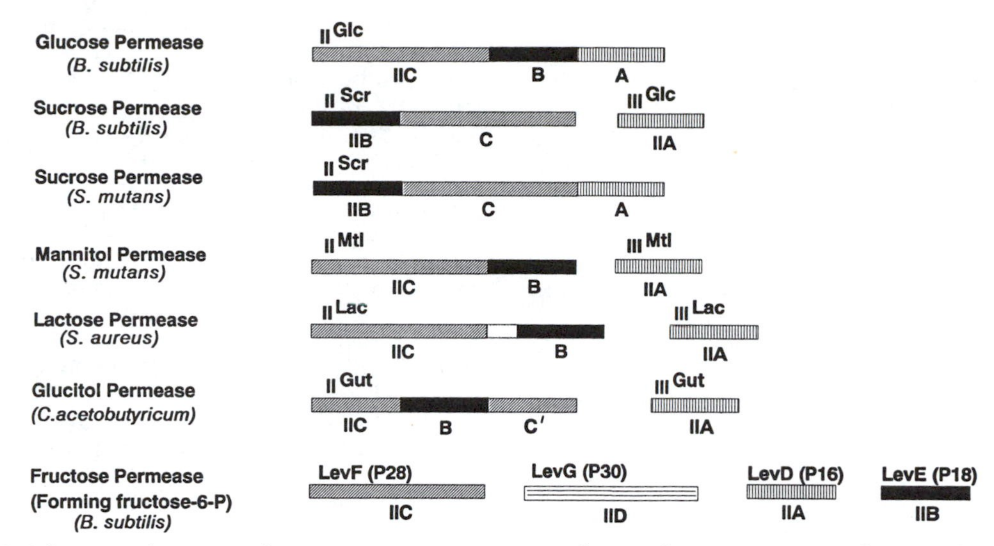

Figure 5. Schematic depiction of representative gram-positive bacterial PTS permeases showing the different known permutations of the constituent proteins and domains. Portions or domains of the various permeases are indicated as follows: transmembrane hydrophobic domain (IIC) ( ), domain bearing the first phosphorylation site (IIA) ( ), domain bearing the second phosphorylation site (IIB) ( ), transmembrane partially hydrophobic domain of unknown function (IID) ( ), and nonhomologous domain or region of unknown function ( ). All permeases shown except the last two exhibit convincing regions of homology. The uniform domain nomenclature is provided below the bars, which represent the various domains-proteins (see reference 171 for discussion of PTS protein nomenclature). Alternative designations for the proteins are given above the bars. References for most of the permeases shown can be found in Saier et al. (173) and Lengeler et al. (105) with the following exceptions: the glucitol permease of *C. acetobutyricum* (123a), the mannitol permease of *S. mutans* (86a), and the fructose permease (forming fructose 6-phosphate) of *B. subtilis* (116). The C′ domain in the glucitol enzyme II refers to the second half of the hydrophobic (IIC) domain.

and strongly repressed by inclusion of glucose or mannitol, both PTS sugars in *B. subtilis*, in the growth medium. An isogenic *B. subtilis* strain with a single mutation (S46A) in the chromosomal structural gene for HPr that changed the regulatory seryl residue to alanine so that ATP-dependent phosphorylation of this protein could not occur, abolished the repressive effect of the PTS sugars on gluconate kinase synthesis. Very similar behavior was observed for another catabolic enzyme, glucitol dehydrogenase, involved in the catabolism of the non-PTS sugar glucitol.

The mechanism of the repressive effect was further investigated. The same strains were examined with respect to inhibition of the uptake of [^{14}C]gluconate in cells induced by growth in the presence of both gluconate and the PTS sugar glucose or mannitol. The PTS sugar inhibited gluconate uptake into intact cells if synthesis of the uptake system for that PTS sugar had been induced prior to initiation of the uptake experiment. This result provided evidence that the apparent catabolite repression was in part due to exclusion of the inducer, gluconate, from the cytoplasm of the cell.

In contrast to the situation for the PTS-mediated regulatory system in gram-negative enteric bacteria (167), HPr(Ser-P) (rather than enzyme IIA^{Glc}) may be the species that binds to the allosteric regulatory site on the cytoplasmic side of the gluconate permease in *B. subtilis* to inhibit its activity. However, since phosphorylation of the active histidyl residue in HPr inhibits HPr(Ser) phosphorylation, as noted previously (148, 150, 158), the binding of this seryl phosphorylated PTS protein to the non-PTS permease may be functionally equivalent to binding of the free form of the PTS protein IIA^{Glc} in *E. coli* and other enteric bacteria. These preliminary results therefore suggest that gram-positive and gram-negative bacteria have evolved two functionally equivalent but mechanistically distinct processes in order to allow the PTS to regulate cytoplasmic levels of non-PTS sugar inducers. Such mechanisms may provide bacteria with the capacity to sense the availability of various carbohydrates in their environment and respond by the selection of the preferred carbon and energy source. It is important to emphasize, however, that the suggested mechanism in gram-positive bacteria is hypothetical and needs to be rigorously established or refuted.

The three-dimensional structure of the *B. subtilis* HPr protein has recently been determined (77, 202, 203). This small, heat-stable protein consists of an open-face β sandwich formed of four antiparallel β strands serving as the underlying bread packed against three α helices serving as the overlying spread. As revealed by the X-ray structure, the β sheet curls back on itself so that the regulatory seryl residue (serine 46) is close to the active-site histidyl residue (histidine 15). Thus, the presence of the negatively charged phosphoryl group at position 46 may inhibit introduction of the phosphoryl group at the active site by electrostatic repulsion or by inhibition of enzyme I binding. The same explanations apply to inhibition of seryl residue 46 phosphorylation by prior phosphorylation of histidyl residue 15. These three-dimensional structural studies therefore provide the first information leading to an understanding of the regulatory interactions in molecular detail.

Enzyme II Complexes

Genes encoding PTS permeases specific for six sugars have been fully or largely sequenced from gram-positive bacteria as of January 1992 (Fig. 5; 147). These include genes for (i) the glucose permease of *B. subtilis* (partial sequence available); (ii) the sucrose permeases of *B. subtilis* and *S. mutans*; (iii) the mannitol permease of *S. mutans* (86a); (iv) the lactose permeases of *S. aureus*, *L. casei*, and *L. lactis*; (v) the glucitol permease of *Clostridium acetobutyricum* (123a); and (vi) the fructose 6-phosphate-forming fructose permease of *B. subtilis*. The domain structures of these permeases are illustrated in Fig. 5. Although most of these proteins are homologous (147), their domain structures differ. Thus, the glucose permease of *B. subtilis* consists of a single polypeptide chain with the domain order IICBA. The two sequenced sucrose permeases of *B. subtilis* (SacP and SacX) consist of IIBC polypeptide chains, and the IIA domain utilized to phosphorylate the enzyme IIBC is that of the glucose permease (194). In *S. mutans*, the IIA domain is part of the sucrose permease, localized to the C terminus of the protein, and consequently, this protein has the domain structure IIBCA.

The sequenced mannitol permease of *S. mutans* consists of two proteins, the IICB protein and the IIA protein. The lactose permeases of gram-positive bacteria are of the same domain structure as the mannitol permease, but a short nonhomologous segment separates the IIC and IIB domains. The glucitol permease from *Clostridium* spp. also consists of two polypeptide chains, but in this case, as in that of the *E. coli* protein, the IIB domain may be sandwiched between the two homologous halves of the IIC domain (see references 147 and 171 for discussions of the protein from *E. coli*). The glucitol permease is not demonstrably homologous to the above-mentioned PTS proteins. Finally, the fructose 6-phosphate-forming fructose permease of *B. subtilis* consists of four polypeptide chains, designated IIA, IIB, IIC, and IID. This permease is homologous to the mannose permease of *E. coli* and the sorbose permease of *K. pneumoniae* but not to the other permeases shown in Fig. 5. The structural and evolutionary relationships of these systems have been discussed previously (147, 171).

The *E. coli* and *B. subtilis* IIA^{Glc} proteins are functionally interchangeable (29, 159, 194). Furthermore, the three-dimensional structures of the *E. coli* and *B. subtilis* IIA^{Glc} proteins have been determined (48–50, 107, 134, 135, 193, 205), and site-specific mutagenesis studies of both proteins revealed that two histidyl residues are required for activity (143, 159). One of these residues (His-90 in the *E. coli* IIA^{Glc} and His-83 in the *B. subtilis* protein) is the site of phosphorylation. The other histidyl residue (His-75 in the *E. coli* protein and His-68 in the *B. subtilis* IIA^{Glc}) is required for transfer of the phosphoryl group to the IIB^{Glc} domain of the glucose permease but not for transfer from phospho-HPr to IIA^{Glc} (143, 159). In the three-dimensional structure, these two histidyl residues are juxtaposed in perpendicular array and are probably hydrogen bonded to each other (107). Homologous IIA^{Glc}-like domains were found in the lactose:H^+ symport permeases of *S. thermophilus* and *L. bulgaricus*. The former protein was shown to be phosphorylated by HPr(His~P), and this phosphorylation event presumably plays a regulatory role (138).

Two conserved histidyl residues are also found in the IIA^{Fru} and IIA^{Mtl} proteins, which are homologous to each other but are not demonstrably homologous to the family of IIA^{Glc} proteins. The relative positions of these two histidines in the linear sequences of the former proteins are entirely different from those found in the latter proteins (157). Interestingly, the PTS members of this protein (domain) family, are homologous to a class of proteins of gram-negative bacteria, one of which presumably affects transcription of the nitrogen-regulatory σ^{54}-dependent operons in *K. pneumoniae* (157). These observations suggest that PTS-catalyzed phosphorylation may provide a regulatory link between carbon and nitrogen assimilation in bacteria. Specific regulatory properties of PTS proteins have been described in earlier reports (148, 150, 152, 158, 166, 167) and in chapters 11 and 50 of this book.

CONCLUSIONS AND PERSPECTIVES

In this review, we have considered several major classes of transport proteins that form permease systems found in gram-positive bacteria. These transport systems are energized by ATP hydrolysis, consumption of chemiosmotic energy in the form of ion gradients and membrane potentials, or phosphoryl transfer from PEP to the sugar substrate in the phosphotransferase-catalyzed group translocation process. All three types of systems are found in gram-negative bacteria as well as in gram-positive bacteria, and representatives within all permease classes except the group-translocating PTS permeases are also found in eukaryotes (169). Although for the purpose of this review we treat these classes of proteins as though they were distinct, there is evidence that at least some members of different classes of transport systems share structural and functional characteristics as well as evolutionary origins. This possibility is discussed elsewhere (169).

Although the transport proteins derived from gram-positive bacteria are, in general, related to transport proteins of other bacteria and eukaryotes, the gram-positive bacterial systems exhibit some unique properties. In some cases, these systems are characterized well enough that they provide information that clearly complements or contrasts with that obtained from the study of related transport systems of other organisms.

(i) Analyses of the Cd^{2+} ATPase of *S. aureus* have provided information about the cation substrate-binding site(s) that has not been forthcoming from studies with other P-type ATPases.

(ii) Both heavy-metal resistance mechanisms and genetic regulatory mechanisms controlling the syntheses of the responsible transporters appear to differ in well-defined gram-positive and gram-negative bacteria.

(iii) Some F-type ATPases of gram-positive bacteria (specifically, some of those derived from strictly aerobic organisms) may lack an ε subunit, and this fact may be of importance with respect to the regulation of their activities.

(iv) The first Na^+-translocating V-type ATPase to be discovered may have been detected and characterized in a gram-positive bacterium.

(v) The unique membrane-embedded lipoprotein nature of the gram-positive solute-binding protein constituents of ABC-type solute-translocating ATPases, which accounts for their membrane attachment, appears to reflect the unique structure of the gram-positive bacterial envelope. The release of some of these proteins into the medium during stationary phase and their potential recapture upon restoration of growth represent further unique features.

(vi) The proposed role of the oligopeptide transport system and cell wall-derived peptides in providing signals for initiation of the sporulation cycle in *B. subtilis* may prove applicable to other developmental systems.

(vii) The proposed role of the dipeptide transport system in maintenance of the vegetative state in *B. subtilis* may represent a developmental process allowing adaptation from nutrient-rich to nutrient-poor conditions without initiation of the sporulation cycle. The effect of the dipeptide transport system on the state of differentiation may thus antagonize that of the oligopeptide transporter and thereby provide an alternative route to sporulation under near-starvation conditions. The use of transport systems for the provision of antagonistic developmental signals may prove to be applicable to other organisms that undergo differentiation.

(viii) The oligopeptide transporter (Ami) of *S. pneumoniae* was found to have a substrate-modifier subunit (AmiB) homologous to the ArsC subunit of the Ars transporter. No other ABC-type transporter has been reported to possess such a subunit. Its presence in both the Ami and Ars systems provides evidence that these two transporters are related and that the Ars system is therefore a structurally divergent ABC-type ATPase, even though homology cannot be established on the basis of sequence comparisons (see Note, below).

(ix) The operon encoding the Ars transport system of *S. aureus* is organized differently from that of the *E. coli* system. Since the *arsA* gene is not present within the plasmid-encoded *S. aureus ars* operon, the possibility of a hybrid system, derived in part from chromosomally encoded genes, must be considered.

(x) The gene encoding the GlpF-like (glycerol facilitator-like) protein of *L. lactis* was not linked to a *glpK* (glycerol kinase-encoding) gene as other sequenced *glpF* genes are. This fact suggests a distinct function and/or genetic regulatory system in *L. lactis* unlike that found in the other well-characterized bacteria.

(xi) The GerAII alanine receptor for germination in *B. subtilis* is likely to prove to be an alanine transporter according to the sequence comparison analyses reported. This fact suggests a developmental role for transport proteins in addition to those noted above for the oligopeptide and dipeptide transporters in the *Bacillus* sporulation cycle.

(xii) The efflux studies with corynebacteria described above represent the most detailed analyses of their type for any organism. Analogous information concerning the distinct nature of uptake versus efflux pathways is not yet available for other prokaryotic or eukaryotic species.

(xiii) The PTS-mediated phosphorylation of sucrose regulon-specific antiterminator proteins of *B. subtilis* has been demonstrated, and the functional significance of this process has been established (see above and chapter 11).

(xiv) The HPr kinases that phosphorylate serine 46 in gram-positive HPr proteins provide a novel mechanism of metabolic (and consequent genetic) regulation.

(xv) The demonstration of HPr, HPr(Ser) kinase, and HPr(Ser-P) phosphatase in heterofermentative lactobacilli and *A. laidlawii*, which lack a functional PTS, suggests that the ATP-dependent kinase-catalyzed phosphorylation of HPr plays a regulatory role in a metabolic or transport pathway that is distinct from the PTS.

(xvi) The covalent attachment of the IIA^{Glc} domain of the *B. subtilis* glucose permease to the other domains of the enzyme II complex (IICB) and the corresponding covalent attachment of the IIA^{Scr} domain of the *S. mutans* sucrose permease to the IIBC domains of the sucrose enzyme II complex represent unique features of these gram-positive PTS permeases. The *S. mutans* enzyme is the only sucrose permease thus far characterized that possesses its own sugar-specific IIA domain or protein.

(xvii) The three-dimensional structures of a PTS protein and a PTS protein domain of *B. subtilis*, HPr and IIA^{Glc}, respectively, have been solved by both two- and three-dimensional nuclear magnetic resonance and by X-ray crystallography. These are the first PTS proteins for which high-resolution three-dimensional structural data are available.

(xviii) The demonstration of a IIA^{Glc}-like domain in the sequenced lactose:H^+ symport permease of *S. thermophilus* and the demonstration that this domain can be phosphorylated by the PTS suggests that a unique regulatory mechanism for control of the activity of this permease may be at hand.

(xix) Finally, definition of the phylogenetic relationships for various classes of homologous transport proteins leads to clear confirmation of the conclusion that gram-positive bacteria represent a distinct and coherent branch of the eubacteria.

NOTE ADDED IN PROOF

After completion of this manuscript, the ArsC protein of the arsenic resistance operon of *S. aureus* plasmid pI258 was shown to be an arsenate reductase, converting intracellular arsenate [As(V)] to arsenite [As(III)]. Arsenite is then exported from the cells via the arsenite-antimonite efflux system. In vitro, thioredoxin plus dithiothreitol could serve as reducing agents (90a). Thus, the AmiB protein of *S. pneumoniae* and the protein encoded by the unidentified open reading frame found within the *nif* (nitrogen fixing) region of the chromosome of *A. vinelandii* (162), which are homologous to ArsC, may similarly be reductases rather than specificity components of transport systems.

The *glp* (glycerol) operon of *E. coli* which encodes the facilitator (GlpF) and the kinase (GlpK) also encodes a novel protein (GlpX) of unknown function. The gene order is *glpFKX* (198a).

The glycerol facilitator of *E. coli* has been incorporated into frog oocytes, and the activity has been shown to be that of a nonspecific channel for straight-chain carbon compounds as reported previously using different methodology (73, 155).

The complete sequence of the glucose enzyme II of the *B. subtilis* PTS has recently been published (212).

REFERENCES

1. **Albertini, A. M., T. Caramori, W. D. Crabb, F. Scoffone, and A. Galizzi.** 1991. The *flaA* locus of *Bacillus subtilis* is part of a large operon coding for flagellar structures, motility functions, and an ATPase-like polypeptide. *J. Bacteriol.* **173:**3573–3579.
2. **Alloing, G., M.-C. Trombe, and J.-P. Claverys.** 1990. The *ami* locus of the Gram-positive bacterium *Streptococcus pneumoniae* is similar to binding protein-dependent transport operons of Gram-negative bacteria. *Mol. Microbiol.* **4:**633–644.
3. **Amann, R., W. Ludwig, and K. H. Schleifer.** 1988. β subunit of ATP synthase: a useful marker for studying the phylogenetic relationship of eubacteria. *J. Gen. Microbiol.* **134:**2815–2821.
4. **Ambudkar, S. V., A. R. Lynn, P. C. Maloney, and B. P. Rosen.** 1986. Reconstitution of ATP-dependent calcium transport from streptococci. *J. Biol. Chem.* **261:**15596–15600.
5. **Ambudkar, S. V., L. A. Sonna, and P. C. Maloney.** 1986. Variable stoichiometry of phosphate-linked anion exchange in *Streptococcus lactis*: implications for the mechanism of sugar phosphate transport by bacteria. *Proc. Natl. Acad. Sci. USA* **83:**280–284.
6. **Amster-Choder, O., F. Houman, and A. Wright.** 1989. Protein phosphorylation regulates transcription of the beta-glucoside utilization operon in *E. coli*. *Cell* **58:**847–855.
7. **Amster-Choder, O., and A. Wright.** 1990. Regulation of activity of a transcriptional anti-terminator in *E. coli* by phosphorylation *in vivo*. *Science.* **249:**540–542.
8. **Anraku, Y., N. Umemoto, R. Hirata, and Y. Wada.** 1989. Structure and function of the yeast vacuolar membrane proton ATPase. *J. Bioenerg. Biomembr.* **21:** 589–603.
9. **Bakker, E. P., A. Borchard, M. Michels, K. Altendorf, and A. Siebers.** 1987. High-affinity potassium uptake system in *Bacillus acidocaldarius* showing immunological cross-reactivity with the Kdp system from *Escherichia coli*. *J. Bacteriol.* **169:**4342–4348.
10. **Baxter, L., S. Torrie, and M. McKillen.** 1974. D-Gluconate transport in *Bacillus subtilis*. *Biochem. Soc. Trans.* **2:**1370–1372.

10a. **Bennett, G.** Personal communication.

11. **Berman-Kurtz, M., E. C. C. Lin, and D. P. Richey.** 1971. Promoter-like mutant with increased expression of the glycerol kinase operon of *Escherichia coli*. *J. Bacteriol.* **106:**724–731.
12. **Booth, I. R., and J. G. Morris.** 1975. Proton-motive force in the obligately anaerobic bacterium *Clostridium pasteurianum*: a role in galactose and gluconate uptake. *FEBS* Lett. **59:**153–157.
13. **Bröer, S., and R. Krämer.** 1990. Lysine uptake and exchange in *Corynebacterium glutamicum*. *J. Bacteriol.* **172:**7241–7248.
14. **Bröer, S., and R. Krämer.** 1991. Lysine excretion by *Corynebacterium glutamicum*. I. Identification of a specific secretion carrier system. *Eur. J. Biochem.* **202:**131–135.
15. **Bröer, S., and R. Krämer.** 1991. Lysine excretion by *Corynebacterium glutamicum*. II. Energetics and mechanism of the transport system. *Eur. J. Biochem.* **202:** 137–143.
16. **Brusilow, W. S., M. A. Scarpetta, C. A. Hawthorne, and W. P. Clark.** 1989. Organization and sequence of the genes coding for the proton-translocating ATPase of *Bacillus megaterium*. *J. Biol. Chem.* **264:**1528–1533.
17. **Bunch, A. W., and R. E. Harris.** 1986. The manipulation of microorganisms for the production of secondary metabolites. *Biotechnol. Genet. Eng. Rev.* **4:**117–144.
18. **Byrne, C. R., R. S. Monroe, K. A. Ward, and N. M. Kredich.** 1988. DNA sequences of the *cysK* regions of *Salmonella typhimurium* and *Escherichia coli* and linkage of the *cysK* regions to *ptsH*. *J. Bacteriol.* **170:**3150–3157.
19. **Carrasco, N., L. M. Antes, M. S. Poonian, and H. R. Kaback.** 1986. Lac permease of *Escherichia coli* histidine-322 and glutamic acid-325 may be components of a charge-relay system. *Biochemistry* **25:**4486–4488.
20. **Carrasco, N., I. B. Püttner, L. M. Antes, J. A. Lee, J. D. Larigan, J. S. Lolkema, P. D. Roepe, and H. R. Kaback.** 1989. Characterization of site-directed mutants in the *lac* permease of *Escherichia coli*. 2. Glutamate-325 replacements. *Biochemistry* **28:**2533–2539.
21. **Chalumeau, H., A. Delobbe, and P. Gay.** 1978. Biochemical and genetic study of D-glucitol transport and catabolism in *Bacillus subtilis*. *J. Bacteriol.* **134:**920–928.
22. **Chen, C.-M., T. K. Misra, S. Silver, and B. P. Rosen.** 1986. Nucleotide sequence of the structural genes for an anion pump. The plasmid-encoded arsenical resistance operon. *J. Biol. Chem.* **261:**15030–15038.
23. **Chung, Y., and M. R. Salton.** 1988. Purification and properties of the latent F_0F_1-ATPase from *Micrococcus lysodeikticus*. *Mikrobios* **54:**187–205.
24. **Clarke, D. J., F. M. Fuller, and J. G. Morris.** 1979. The proton-translocating adenosine triphosphatase of the obligately anaerobic bacterium *Clostridium pasteurianum*. 1. ATPase phosphohydrolase activity. *Eur. J. Biochem.* **98:**597–612.
25. **Clarke, D. J., and J. G. Morris.** 1979. The proton-translocating adenosine triphosphatase of the obligately anaerobic bacterium *Clostridium pasteurianum*. 2. ATP synthetase activity. *Eur. J. Biochem.* **98:**613–620.
26. **Clement, Y., B. Escoffier, M. C. Trombe, and G. Lanéelle.** 1984. Is glutamate excreted by its uptake system in *Corynebacterium glutamicum*? A working hypothesis. *J. Gen. Microbiol.* **130:**2589–2594.
27. **Cozzarelli, N. R., W. B. Freedberg, and E. C. C. Lin.** 1968. Genetic control of the L-α-glycerophosphate system in *Escherichia coli*. *J. Mol. Biol.* **31:**371–387.
28. **Crutz, A. M., M. Steinmetz, S. Aymerich, R. Richter, and D. Le Coq.** 1990. Induction of levansucrase in *Bacillus subtilis*: an antitermination mechanism negatively controlled by the phosphotransferase system. *J. Bacteriol.* **172:**1043–1050.
29. **Dean, D. A., J. Reizer, H. Nikaido, and M. H. Saier, Jr.** 1990. Regulation of the maltose transport system of *Escherichia coli* by the glucose-specific enzyme III of the PTS: characterization of inducer exclusion-resistant mutants and reconstitution of inducer exclusion in proteoliposomes. *J. Biol. Chem.* **265:**21005–21010.
30. **Debarbouille, M., M. Arnaud, A. Fouet, A. Klier, and G. Rapoport.** 1990. The *sacT* gene regulating the *sacPA* operon in *Bacillus subtilis* shares strong homology with transcriptional antiterminators. *J. Bacteriol.* **172:**3966–3973.
31. **Demain, A. L., and J. Birnbaum.** 1968. Alteration of permeability for the release of metabolites from the microbiol cell. *Curr. Top. Microbiol. Immunol.* **46:**1–25.
32. **Denda, K., J. Konishi, K. Hajiro, T. Oshima, T. Date, and M. Yoshida.** 1990. Structure of an ATPase operon of an acidothermophilic archaebacterium, *Sulfolobus acidocaldarius*. *J. Biol. Chem.* **265:**21509–21513.
33. **De Reuse, H., A. Roy, and A. Danchin.** 1985. Analysis of the *ptsH-ptsI-crr* region in *Escherichia coli* K-12: nucleotide sequence of the *ptsH* gene. *Gene* **35:**199–207.

33a. **Deutscher, J., et al.** Unpublished data.

34. **Deutscher, J., and R. Engelmann.** 1984. Purification and characterization of an ATP-dependent protein kinase from *Streptococcus faecalis*. *FEMS Microbiol. Lett.* **23:**157–162.

35. **Deutscher, J., U. Kessler, A. Alpert, and W. Hengstenberg.** 1984. Bacterial phosphoenolpyruvate-dependent phosphotransferase system: P-Ser-HPr and its possible regulatory function. *Biochemistry* **23:**4455–4460.
36. **Deutscher, J., B. Pevec, K. Beyreuther, H.-H. Kiltz, and W. Hengstenberg.** 1986. Streptococcal phosphoenolpyruvate-sugar phosphotransferase system: amino acid sequence and site of ATP-dependent phosphorylation of HPr. *Biochemistry* **25:**6543–6551.
37. **Doolittle, R. F., K. L. Anderson, and D.-F. Feng.** 1989. Estimating the prokaryote-eukaryote divergence time from protein sequences, p. 73–85. *In* B. Fernholm, K. Bremer, and H. Jörnvall (ed.), *The Hierarchy of Life.* Elsevier Science Publishing, Inc., New York.
38. **Doolittle, R. F., and D.-F. Feng.** 1990. Nearest neighbor procedure for relating progressively aligned amino acid sequences. *Methods Enzymol.* **183:**659–669.
39. **Driessen, A. J. M., D. Molenaar, and W. N. Konings.** 1989. Kinetic mechanism and specificity of the arginine-ornithine antiporter of *Lactococcus lactis. J. Biol. Chem.* **264:**10361–10370.
40. **Driessen, A. J. M., E. J. Smid, and W. N. Konings.** 1988. Transport of diamines by *Enterococcus faecalis* is mediated by an agmatine-putrescine antiporter. *J. Bacteriol.* **170:**4522–4527.
41. **Dudler, T., C. Schmidhauser, R. W. Parish, R. E. H. Wettenhall, and T. Schmidt.** 1988. A mycoplasma high-affinity transport system and the *in vivo* invasiveness of mouse sarcoma cells. *EMBO J.* **7:**3963–3970.
42. **Duplay, P., H. Bedouelle, A. Fowler, I. Zabin, W. Saurin, and M. Hofnung.** 1984. Sequences of the *malE* gene and of its product, the maltose-binding protein of *Escherichia coli* K12. *J. Biol. Chem.* **259:**10606–10613.
42a. **Dyke, K. G. H.** Unpublished data.
43. **Ebbighausen, H., B. Weil, and R. Krämer.** 1989. Transport of branched-chain amino acids in *Corynebacterium glutamicum. Arch. Microbiol.* **151:**238–244.
44. **Ebbighausen, H., B. Weil, and R. Krämer.** 1989. Isoleucine excretion in *Corynebacterium glutamicum*: evidence for a specific efflux carrier system. *Appl. Microbiol. Biotechnol.* **31:**184–190.
45. **Eiserman, R., R. Fischer, U. Kessler, A. Neubauer, and W. Henstenberg.** 1991. Staphylococcal phosphoenolpyruvate-dependent phosphotransferase system. Purification and protein sequencing of the *Staphylococcus carnosus* histidine-containing protein, and cloning and DNA sequencing of the *ptsH* gene. *Eur. J. Biochem.* **197:**9–14.
46. **Epstein, W.** 1990. Bacterial transport ATPases, p. 87–110. *In* T. A. Krulwich (ed.), *The Bacteria,* vol. 12. Academic Press, Inc., San Diego, Calif.
47. **Fagan, M. J., and M. H. Saier, Jr.** P-type ATPases of eukaryotes and bacteria: sequence comparisons and construction of phylogenetic trees. Submitted for publication.
48. **Fairbrother, W. J., J. Cavanagh, H. J. Dyson, A. G. Palmer III, S. L. Sutrina, J. Reizer, M. H. Saier, Jr., and P. E. Wright.** 1991. Polypeptide backbone resonance assignments and secondary structure of *Bacillus subtilis* enzyme III^{glc} determined by two-dimensional and three-dimensional heteronuclear NMR spectroscopy. *Biochemistry* **30:**6896–6907.
49. **Fairbrother, W. J., G. P. Gippert, J. Reizer, M. H. Saier, Jr., and P. E. Wright.** 1992. Low resolution solution structure of the *Bacillus subtilis* glucose permease IIA domain derived from heteronuclear three-dimensional NMR spectroscopy. *FEBS Lett.* **296:**148–152.
50. **Fairbrother, W. J., A. G. Palmer III, M. Rance, J. Reizer, M. H. Saier, Jr., and P. E. Wright.** 1992. Assignment of the aliphatic ^1H and ^{13}C resonances of the *Bacillus subtilis* glucose permease IIA domain using double- and triple-resonance heteronuclear three-dimensional NMR spectroscopy. *Biochemistry* **31:**4413–4425.
51. **Feng, D.-F., and R. F. Doolittle.** 1990. Progressive alignment and phylogenetic tree construction of protein sequences. *Methods Enzymol.* **183:**375–387.
52. **Fikes, J. D., and P. J. Bassford, Jr.** 1987. Export of unprocessed precursor maltose-binding proteins to the periplasm of *Escherichia coli* cells. *J. Bacteriol.* **169:**2352–2359.
53. **Fujita, Y., T. Fujita, Y. Miwa, J. Nihashi, and Y. Aratani.** 1986. Organization and transcription of the gluconate operon, *gnt*, of *Bacillus subtilis. J. Biol. Chem.* **261:**13744–13753.
54. **Furlong, C. E.** 1987. Osmotic-shock-sensitive transport systems, p. 768–796. *In* F. C. Neidhardt, J. L. Ingraham, K. B. Low, B. Magasanik, M. Schaechter, and H. E. Umbarger (ed.), *Escherichia coli and Salmonella typhimurium*: cellular and molecular biology, vol. 1. American Society for Microbiology, Washington, D.C.
55. **Fürst, P., and M. Solioz.** 1987. Formation of a β-aspartyl phosphate intermediate by the vanadate-sensitive ATPase of *Streptococcus faecalis. J. Biol. Chem.* **260:**50–52.
56. **Futai, M., T. Noumi, and M. Maeda.** 1989. ATP synthase (H^+-ATPase): results by combined biochemical and molecular biological approaches. *Annu. Rev. Biochem.* **58:**111–136.
57. **Gagnon, G., C. Vadeboncoeur, R. C. Lévesque, and M. Frenette.** 1992. Cloning, sequencing and expression in *Escherichia coli* of the *ptsI* gene encoding for the enzyme I phosphoenolpyruvate:sugar phosphotransferase transport system from *Streptococcus salivarius. Gene* **121:**71–78.
58. **Gardell, C., K. Johnson, A. Jacq, and J. Beckwith.** 1990. The *secD* locus of *E. coli* codes for two membrane proteins required for protein export. *EMBO J.* **9:**3209–3216.
59. **Geerse, R. H., F. Izzo, and P. W. Postma.** 1989. The PEP:fructose phosphotransferase system in *Salmonella typhimurium*: FPr combines enzyme III^{Fru} and pseudo-HPr activities. *Mol. Gen. Genet.* **216:**517–525.
60. **Gibson, M. M., M. Price, and C. F. Higgins.** 1984. Genetic characterization and molecular cloning of the tripeptide permease (*tpp*) genes of *Salmonella typhimurium. J. Bacteriol.* **160:**122–130.
61. **Gilson, E., G. Alloing, T. Schmidt, J.-P. Claverys, R. Dudler, and M. Hofnung.** 1988. Evidence for high affinity binding-protein dependent transport systems in Gram-positive bacteria and in mycoplasma. *EMBO J.* **7:**3971–3974.
62. **Gogarten, J. P., H. Kibak, P. Dittrich, L. Taiz, E. J. Bowman, B. J. Bowman, M. F. Manolson, R. J. Poole, T. Date, T. Oshima, J. Konishi, K. Denda, and M. Yoshida.** 1989. Evolution of the vacuolar H^+-ATPase: implications for the origin of eukaryotes. *Proc. Natl. Acad. Sci. USA* **86:**6661–6665.
63. **Gonzy-Tréboul, G., M. Zagorec, M.-C. Rain-Gion, and M. Steinmetz.** 1989. Phosphoenolpyruvate:sugar phosphotransferase system of *Bacillus subtilis*: nucleotide sequence of *ptsX, ptsH,* and the 5′-end of *ptsI* and evidence for a *ptsHI* operon. *Mol. Microbiol.* **3:**103–112.
64. **Goodell, E. W., and C. F. Higgins.** 1987. Uptake of cell wall peptides by *Salmonella typhimurium* and *Escherichia coli. J. Bacteriol.* **169:**3861–3865.
65. **Grossman, A. D., and R. Losick.** 1988. Extracellular control of spore formation in *Bacillus subtilis. Proc. Natl. Acad. Sci. USA* **85:**4369–4373.
66. **Guilfoile, P. G., and C. R. Hutchinson.** 1991. A bacterial analog of the *mdr* gene of mammalian tumor cells is present in *Streptomyces peucetius,* the producer of daunorubicin and doxorubicin. *Proc. Natl. Acad. Sci. USA* **88:**8553–8557.
67. **Guyer, C. A., D. G. Morgan, N. Osheroff, and J. V.**

Staros. 1985. Purification and characterization of a periplasmic oligopeptide binding protein from *Escherichia coli*. *J. Biol. Chem.* **260:**10812–10816.
68. **Hafer, J., A. Siebers, and E. P. Bakker.** 1989. The high-affinity K^+-translocating ATPase complex from *Bacillus acidocaldarius* consists of three subunits. *Mol. Microbiol.* **3:**487–495.
69. **Hayashi, S., and E. C. C. Lin.** 1965. Product induction of glycerol kinase in *Escherichia coli*. *J. Mol. Biol.* **14:**515–521.
70. **Hays, J. B.** 1978. Group translocation transport systems, p. 43. *In* B. P. E. Rosen (ed.), *Microbiology Series*, Vol. 4. *Bacterial Transport*. Marcel Dekker, New York.
71. **Heefner, D. L., and F. M. Harold.** 1980. ATP-linked sodium transport in *Streptococcus faecalis*. *J. Biol. Chem.* **255:**11396–11402.
72. **Heefner, D. L., and F. M. Harold.** 1982. ATP-driven sodium pump in *Streptococcus faecalis*. *Proc. Natl. Acad. Sci. USA* **79:**2798–2802.
73. **Heller, K. B., E. C. C. Lin, and T. H. Wilson.** 1980. Substrate specificity and transport properties of the glycerol facilitator of *Escherichia coli*. *J. Bacteriol.* **144:** 274–278.
74. **Henderson, P. J. F.** 1990. The homologous glucose transport proteins of prokaryotes and eukaryotes. *Res. Microbiol.* **141:**316–328.
75. **Hengstenberg, W.** 1977. Enzymology of carbohydrate transport in bacteria. *Curr. Top. Microbiol. Immunol.* **77:**97.
76. **Hensel, M., G. Deckers-Hebestreit, and K. Altendorf.** 1991. Purification and characterization of the F_1 portion of the ATP synthase (F_1F_0) of *Streptomyces lividans*. *Eur. J. Biochem.* **202:**1313–1319.
77. **Herzberg, O., P. Reddy, J. Reizer, and G. Kapadia.** 1992. Structure of the histidine-containing phosphocarrier protein HPr from *Bacillus subtilis* at 2.0 Å resolution. *Proc. Natl. Acad. Sci. USA* **89:**2499–2503.
78. **Hicks, D. B., and T. A. Krulwich.** 1990. Purification and reconstitution of the F_1F_0 ATP synthase from alkalophilic *Bacillus firmis* 0F4. Evidence that the enzyme translocates H^+ but not Na^+. *J. Biol. Chem.* **265:**20547–20554.
79. **Hiles, I. D., M. P. Gallagher, D. J. Jamieson, and C. F. Higgins.** 1987. Molecular characterization of the oligopeptide permease of *Salmonella typhimurium*. *J. Mol. Biol.* **195:**125–142.
80. **Hiles, I. D., and C. F. Higgins.** 1986. Peptide transport by *Salmonella typhimurium*: the periplasmic oligopeptide-binding protein. *Eur. J. Biochem.* **158:**561–567.
81. **Hoch, J. A.** 1976. Genetics of bacterial sporulation. *Adv. Genet.* **18:**69–99.
82. **Hoffmann, A., and P. Dimroth.** 1990. The ATPase of *Bacillus alcalophilus*. Purification and properties of the enzyme. *Eur. J. Biochem.* **194:**423–430.
83. **Hoffmann, A., and P. Dimroth.** 1991. The ATPase of *Bacillus alcalophilus*. Reconstitution of energy-transducing functions. *Eur. J. Biochem.* **196:**493–497.
84. **Hoischen, C., and R. Krämer.** 1989. Evidence for an efflux carrier system involved in the secretion of glutamate by *Corynebacterium glutamicum*. *Arch. Microbiol.* **151:**342–347.
85. **Hoischen, C., and R. Krämer.** 1990. Membrane alteration is necessary but not sufficient for effective glutamate secretion in *Corynebacterium glutamicum*. *J. Bacteriol.* **172:**3409–3416.
86. **Holmberg, C., L. Beijer, B. Rutberg, and L. Rutberg.** 1990. Glycerol catabolism in *Bacillus subtilis*: nucleotide sequence of the genes encoding glycerol kinase (*glpK*) and glycerol-3-phosphate dehydrogenase (*glpD*). *J. Gen. Microbiol.* **136:**2367–2375.
86a. **Honeyman, A.** Personal communication.
87. **Houng, H., A. R. Lynn, and B. P. Rosen.** 1986. ATP-driven calcium transport in membrane vesicles of *Streptococcus sanguis*. *J. Bacteriol.* **168:**1040–1044.
88. **Hutkins, R. W., and C. Ponne.** 1991. Lactose uptake driven by galactose efflux in *Streptococcus thermophilus*: evidence for a galactose-lactose antiporter. *Appl. Environ. Microbiol.* **57:**941–944.
89. **Ivey, D. M., and L. G. Ljungdahl.** 1986. Purification and characterization of the F_1-ATPase from *Clostridium thermoaceticum*. *J. Bacteriol.* **165:**252–257.
90. **Jamieson, D. J., and C. F. Higgins.** 1984. Anaerobic and leucine-dependent expression of a peptide transport gene in *Salmonella typhimurium*. *J. Bacteriol.* **160:**131–136.
90a. **Ji, G., and S. Silver.** 1992. Reduction of arsenate to arsenite by the ArsC protein of the arsenic resistance operon of *Staphylococcus aureus* plasmid pI258. *Proc. Natl. Acad. Sci. USA* **89:**9474–9478.
91. **Kakinuma, Y., and K. Igarashi.** 1990. Amplification of the Na^+-ATPase of *Streptococcus faecalis* at alkaline pH. *FEBS Lett.* **261:**135–138.
92. **Kakinuma, Y., and K. Igarashi.** 1990. Mutants of *Streptococcus faecalis* sensitive to alkaline pH lack Na^+-ATPase. *J. Bacteriol.* **172:**1732–1735.
93. **Kakinuma, Y., and K. Igarashi.** 1990. Release of the component of *Streptococcus faecalis* Na^+-ATPase from the membranes. *FEBS Lett.* **271:**102–105.
94. **Kakinuma, Y., and K. Igarashi.** 1990. Some features of the *Streptococcus faecalis* Na^+-ATPase resemble those of the vacuolar type ATPases. *FEBS Lett.* **271:**102–105.
95. **Karkaria, C. E., and B. P. Rosen.** 1990. Mutagenesis of a nucleotide binding site of an anion-translocating ATPase. *J. Biol. Chem.* **265:**7832–7836.
96. **Kinoshita, S.** 1985. Glutamic acid bacteria, p. 115–142. *In* A. Demain and N. Solomon (ed.), *Biology of Industrial Microorganisms*. The Benjamin/Cummings Publishing Co., Redwood City, Calif.
97. **Kinoshita, S., S. Udaka, and M. Shimino.** 1957. Studies on amino acid fermentation. I. Production of L-glutamic acid by various microorganisms. *J. Gen. Appl. Microbiol.* **3:**193–205.
98. **Kinoshita, N., T. Unemoto, and H. Kobayashi.** 1984. Sodium-stimulated ATPase in *Streptococcus faecalis*. *J. Bacteriol.* **158:**844–848.
99. **Kobayashi, H., J. V. Brunt, and F. M. Harold.** 1978. ATP-linked calcium transport in vesicles of *Streptococcus faecalis*. *J. Biol. Chem.* **253:**2085–2092.
99a. **Koch, S., et al.** Unpublished data.
100. **Kohlbrecher, D., R. Eisermann, and W. Hengstenberg.** 1992. Staphylococcal phosphoenolpyruvate-dependent phosphotransferase system: molecular cloning and nucleotide sequence of the *Staphylococcus carnosus ptsI* gene and expression and complementation studies of the gene product. *J. Bacteriol.* **174:**2208–2214.
101. **Konings, W. N., B. Poolman, and A. J. Driessen.** 1989. Bioenergetics and solute transport in lactococci. *Crit. Rev. Microbiol.* **16:**419–476.
102. **Krämer, R., and C. Lambert.** 1990. Uptake of glutamate in *Corynebacterium glutamicum*. II. Evidence for a primary active transport system. *Eur. J. Biochem.* **194:** 937–944.
103. **Krämer, R., C. Lambert, C. Hoischen, and H. Ebbighausen.** 1990. Uptake of glutamate in *Corynebacterium glutamicum*. I. Kinetic properties and regulation by internal pH and potassium. *Eur. J. Biochem.* **194:**929–935.
103a. **Krulwich, T.** Personal communication.
104. **Krulwich, T. A., D. B. Hicks, D. Seto-Young, and A. A. Guffanti.** 1988. The bioenergetics of alkalophilic bacilli. *Crit. Rev. Microbiol.* **16:**15–36.
105. **Lengeler, J. W., F. Titgemeyer, A. P. Vogler, and B. M. Wöhrl.** 1990. Structures and homologies of carbohydrate:phosphotransferase system (PTS) proteins. *Phil. Trans. R. Soc. Lond. B.* **326:**489–504.

106. **Leonard, J. E., C. Lee, A. Apperson, S. S. Dills, and M. H. Saier, Jr.** 1981. The role of membranes in the transport of small molecules, p. 1–52. *In* B. K. Gosh (ed.), *Organization of Prokaryotic Cell Membranes*, vol. I. CRC Press, Boca Raton, Fla.
107. **Liao, D.-I., G. Kapadia, P. Reddy, M. H. Saier, Jr., J. Reizer, and O. Herzberg.** 1991. Structure of the IIA domain of the glucose permease of *Bacillus subtilis* at 2.2 Å resolution. *Biochemistry* **30:**9583–9594.
108. **LiCalsi, C., T. S. Crocenzi, E. Freire, and S. Roseman.** 1991. Sugar transport by the bacterial phosphotransferase system: structural and thermodynamic domains of enzyme I of *Salmonella typhimurium*. *J. Biol. Chem.* **266:**19519–19527.
109. **Lin, E. C. C.** 1976. Glycerol dissimilation and its regulation in bacteria. *Annu. Rev. Microbiol.* **30:**535–578.
110. **Lin, E. C. C.** 1977. Glycerol utilization and its regulation in mammals. *Annu. Rev. Microbiol.* **46:**765–795.
111. **Luntz, M. G., H. I. Zhdanova, and G. I. Bourd.** 1986. Transport and excretion of L-lysine in *Corynebacterium glutamicum*. *J. Gen. Microbiol.* **132:**2137–2146.
112. **Maloney, P. C.** 1990. A consensus structure for membrane transport. *Res. Microbiol.* **141:**374–383.
113. **Maloney, P. C., S. V. Ambudkar, V. Anantharam, L. A. Sonna, and A. Varadhachary.** 1990. Anion-exchange mechanisms in bacteria. *Microbiol. Rev.* **54:**1–17.
114. **Maloney, P. C., S. V. Ambudkar, J. Thomas, and L. Schiller.** 1984. Phosphate/hexose 6-phosphate antiport in *Streptococcus lactis*. *J. Bacteriol.* **158:**238–245.
115. **Mandel, K. G., and T. A. Krulwich.** 1979. D-Gluconate transport in *Arthrobacter pyridinolis*: metabolic trapping of a protonated solute. *Biochim. Biophys. Acta* **552:**478–491.
115a. **Marger, M. D., and M. J. Saier, Jr.** 1993. A major superfamily of transmembrane facilitators that catalyze uniport, symport and antiport. *Trends Biol. Sci.* **18:**13–20.
116. **Martin-Verstraete, I., M. Debarbouille, A. Klier, and G. Rapoport.** 1990. Levanase operon of *Bacillus subtilis* includes a fructose-specific phosphotransferase system regulating the expression of the operon. *J. Mol. Biol.* **214:**657–671.
117. **Mathiopoulos, C., J. P. Mueller, F. J. Slack, C. G. Murphy, S. Patankar, G. Bukusoglu, and A. L. Sonenshein.** 1991. A *Bacillus subtilis* dipeptide transport system expressed early during sporulation. *Mol. Microbiol.* **5:**1903–1913.
118. **Mayo, B., J. Kok, K. Venema, W. Bockelmann, M. Teuber, H. Reinke, and G. Venema.** 1991. Molecular cloning and sequence analysis of the X-prolyl dipeptidyl aminopeptidase gene from *Lactococcus lactis* subsp. *cremoris*. *Appl. Environ. Microbiol.* **57:**38–44.
119. **McKillen, M. N., and J. H. Rountree.** 1973. D-Gluconate transport in *Bacillus subtilis*. *Biochem. Soc. Trans.* **1:**442–445.
120. **Mei-Hsu, C., P. Kaur, C. E. Karkaria, R. F. Steiner, and B. P. Rosen.** 1991. Substrate-induced dimerization of the ArsA protein, the catalytic component of an anion-translocating ATPase. *J. Biol. Chem.* **266:**2327–2332.
121. **Mileykovskaya, E. I., A. N. Abuladze, S. S. Kormer, and D. N. Ostrovsky.** 1987. Some peculiarities of functioning of H^+-ATPase from the membranes of the anaerobic bacterium *Lactobacillus casei*. *Eur. J. Biochem.* **167:**367–370.
122. **Mileykovskaya, E. I., A. N. Abuladze, and D. N. Ostrovsky.** 1987. Subunit composition of the H^+-ATPase complex from anaerobic bacterium *Lactobacillus casei*. *Eur. J. Biochem.* **168:**703–708.
123. **Milner, J. L., L. B. Vink, and J. M. Wood.** 1987. Transmembrane amino acid flux in bacterial cells. *Crit. Rev. Biotechnol.* **5:**1–47.
123a. **Minton, N.** Personal communication.
124. **Muntyan, M. A., I. V. Mesyanzhinova, Y. M. Milgrom, and V. P. Skulachev.** 1990. The F_1-type ATPase in anaerobic *Lactobacillus casei*. *Biochim. Biophys. Acta* **1016:**371–377.
125. **Muramatsu, S., and T. Mizuno.** 1989. Nucleotide sequence of the region encompassing the *glpKF* operon and its upstream region containing a bent DNA sequence of *Escherichia coli*. *Nucleic Acids Res.* **17:**4378.
126. **Nelson, N.** 1989. Structure, molecular genetics, and evolution of vacuolar H^+-ATPases. *J. Bioenerg. Biomembr.* **21:**553–571.
127. **Nielsen, J. B. K., M. P. Caulfield, and J. O. Lampen.** 1981. Lipoprotein nature of *Bacillus licheniformis* membrane penicillinase. *Proc. Natl. Acad. Sci. USA* **78:**3511–3515.
128. **Nielsen, J. B. K., and J. O. Lampen.** 1982. Glyceride-cysteine lipoproteins and secretion by gram-positive bacteria. *J. Bacteriol.* **152:**315–322.
129. **Nucifora, G., L. Chu, T. K. Misra, and S. Silver.** 1989. Cadmium resistance from *Staphylococcus aureus* plasmid pI258 *cadA* gene results from cadmium-efflux ATPase. *Proc. Natl. Acad. Sci. USA* **86:**3544–3548.
130. **Ochman, H., and A. C. Wilson.** 1987. Evolution in bacteria: evidence for a universal substitution rate in cellular genomes. *J. Mol. Evol.* **26:**74–86.
131. **Olson, E. R., D. S. Dunyak, L. M. Jurss, and R. A. Poorman.** 1991. Identification and characterization of *dppA*, and *Escherichia coli* gene encoding a periplasmic dipeptide transport protein. *J. Bacteriol.* **173:**234–244.
132. **Pao, G. M., L.-F. Wu, K. D. Johnson, H. Höfte, M. J. Chrispeels, G. Sweet, N. N. Sandal, and M. H. Saier, Jr.** 1991. Evolution of the MIP family of integral membrane transport proteins. *Mol. Microbiol.* **5:**33–37.
133. **Payne, J. W., and C. Gilvarg.** 1968. Size restriction on peptide utilization in *Escherichia coli*. *J. Biol. Chem.* **243:**6391–6399.
134. **Pelton, J. G., D. A. Torchia, N. D. Meadow, C. Y. Wong, and S. Roseman.** 1991. ^1H, ^{15}N, and ^{13}C NMR signal assignments of III^{Glc}, a signal-transducing protein of *Escherichia coli*, using three-dimensional triple-resonance techniques. *Biochemistry* **30:**10043–10057.
135. **Pelton, J. G., D. A. Torchia, N. D. Meadow, C. Y. Wong, and S. Roseman.** 1991. Secondary structure of the phosphocarrier protein III^{Glc}, a signal-transducing protein of *Escherichia coli*, determined by heteronuclear three-dimensional NMR spectroscopy. *Proc. Natl. Acad. Sci. USA* **88:**3479–3483.
136. **Perego, M., C. F. Higgins, S. R. Pearce, M. P. Gallagher, and J. A. Hoch.** 1991. The oligopeptide transport system of *Bacillus subtilis* plays a role in the initiation of sporulation. *Mol. Microbiol.* **5:**173–185.
137. **Poolman, B., A. J. M. Driessen, and W. N. Konings.** 1987. Regulation of arginine-ornithine exchange and the arginine deiminase pathway in *Streptococcus lactis*. *J. Bacteriol.* **169:**5597–5604.
138. **Poolman, B., R. Modderman, and J. Reizer.** 1992. Lactose transport system of *Streptococcus thermophilus*: the role of histidine residues. *J. Biol. Chem.* **267:**9150–9157.
139. **Poolman, B., T. J. Royer, S. E. Mainzer, and B. F. Schmidt.** 1989. Lactose transport system of *Streptococcus thermophilus*: a hybrid protein with homology to the melibiose carrier and enzyme III of phosphoenolpyruvate-dependent phosphotransferase systems. *J. Bacteriol.* **171:**244–253.
140. **Postma, P. W., and J. W. Lengeler.** 1985. Phosphoenolpyruvate:carbohydrate phosphotransferase system of bacteria. *Microbiol. Rev.* **49:**232–269.
141. **Pourcher, T., H. K. Sarkar, M. Bassilana, H. R. Kaback, and G. Leblanc.** 1990. Histidine-94 is the only important histidine residue in the melibose permease of *Escherichia coli*. *Proc. Natl. Acad. Sci. USA* **87:**468–472.
142. **Powers, D. A., and S. Roseman.** 1984. The primary

structure of *Salmonella typhimurium* HPr, a phosphocarrier protein of the phosphoenolpyruvate:glycose phosphotransferase system. *J. Biol. Chem.* **259:**15212–15214.

143. **Presper, K. A., C. Y. Wong, L. Liu, N. D. Meadow, and S. Roseman.** 1989. Site-directed mutagenesis of the phosphocarrier protein III^{Glc}, a major signal-transducing protein in *Escherichia coli. Proc. Natl. Acad. Sci. USA* **86:**4052–4055.
144. **Pries, A., H. Priefert, N. Kruger, and A. Steinbuchel.** 1991. Identification and characterization of two *Alcaligenes eutrophus* gene loci relevant to the poly(β-hydroxybutyric acid)-leaky phenotype which exhibit homology to *ptsH* and *ptsI* of *Escherichia coli. J. Bacteriol.* **173:**5843–5853.
145. **Quivey, R. G., Jr., R. C. Faustoferri, W. A. Belli, and J. S. Flores.** 1991. Polymerase chain reaction amplification, cloning, sequence determination and homologies of streptococcal ATPase-encoding DNAs. *Gene* **97:** 63–68.
145a. **Rapaport, G.** Personal communication.
146. **Reizer, A., J. Deutscher, M. H. Saier, Jr., and J. Reizer.** 1991. Analysis of the gluconate (*gnt*) operon of *Bacillus subtilis. Mol. Microbiol.* **5:**1081–1089.
147. **Reizer, A., G. M. Pao, and M. H. Saier, Jr.** 1991. Evolutionary relationships among the permease proteins of the bacterial phosphoenolpyruvate:sugar phosphotransferase system. Construction of phylogenetic trees and possible relatedness to proteins of eukaryotic mitochondria. *J. Mol. Evol.* **33:**179–193.
147a. **Reizer, A., J. Reizer, and M. H. Saier, Jr.** Unpublished data.
148. **Reizer, J.** 1989. Regulation of sugar uptake and efflux in Gram-positive bacteria. *FEMS Microbiol. Rev.* **63:**149–156.
149. **Reizer, J., J. Deutscher, F. Grenier, J. Thompson, W. Hengstenberg, and M. H. Saier, Jr.** 1988. The phosphoenolpyruvate:sugar phosphotransferase system in Gram-positive bacteria: properties, mechanism and regulation. *Crit. Rev. Microbiol.* **15:**297–338.
150. **Reizer, J., J. Deutscher, and M. H. Saier, Jr.** 1989. Metabolite-sensitive, ATP-dependent, protein kinase-catalyzed phosphorylation of HPr, a phosphocarrier protein of the phosphotransferase system in Gram-positive bacteria. *Biochimie* **71:**989–991.
151. **Reizer, J., M. J. Novotny, W. Hengstenberg, and M. H. Saier, Jr.** 1984. Properties of ATP-dependent protein kinase from *Streptococcus pyogenes* that phosphorylates a seryl residue in HPr, a phosphocarrier protein of the phosphotransferase system. *J. Bacteriol.* **160:**333–340.
152. **Reizer, J., and K. Peterkofsky.** 1987. Regulatory mechanism for sugar transport in Gram-positive bacteria, p. 333–364. *In* J. Reizer and A. Peterkofsky (ed.), *Sugar Transport and Metabolism in Gram-Positive Bacteria.* Ellis Horwood, Chichester, England.
153. **Reizer, J., and A. Peterkofsky (ed.).** 1987. *Sugar Transport and Metabolism in Gram-Positive Bacteria.* Ellis Horwood, Chichester, England.
154. **Reizer, J., A. Reizer, C. Hoischen, and M. H. Saier, Jr.** Nucleotide sequence of the *ptsI* gene of *Bacillus subtilis* and phylogenetic relationship with other phosphoenolpyruvate utilizing and synthesizing enzymes. *Protein Sci.*, in press.
155. **Reizer, J., A. Reizer, and M. H. Saier, Jr.** Unpublished data.
156. **Reizer, J., A. Reizer, and M. H. Saier, Jr.** 1992. A new subfamily of bacterial ABC-type transport systems catalyzing export of drugs and carbohydrates. *Protein Sci.* **1:**1326–1332.
157. **Reizer, J., A. Reizer, M. H. Saier, Jr., and G. R. Jacobson.** 1992. A proposed link between nitrogen and carbon metabolism involving protein phosphorylation in bacteria. *Protein Sci.* **1:**722–726.
158. **Reizer, J., S. L. Sutrina, M. H. Saier, Jr., G. C. Stewart, A. Peterkofsky, and P. Reddy.** 1989. Mechanistic and physiological consequences of HPr(Ser) phosphorylation on the activities of the phosphoenolpyruvate:sugar phosphotransferase system in Gram-positive bacteria: studies with site-specific mutants of HPr. *EMBO J.* **8:**2111–2120.
159. **Reizer, J., S. L. Sutrina, L.-F. Wu, J. Deutscher, and M. H. Saier, Jr.** 1992. Functional interactions between proteins of the phosphoenolpyruvate:sugar phosphotransferase systems of *Bacillus subtilis* and *Escherichia coli. J. Biol. Chem.* **267:**9158–9169.
160. **Roeske, C. A., R. M. Kutny, R. J. Budde, and R. Chollet.** 1988. Sequence of the phosphothreonyl regulatory site peptide from inactive maize leaf pyruvate, orthophosphate dikinase. *J. Biol. Chem.* **263:**6683–6687.
161. **Romano, A. H., G. Brino, A. Peterkofsky, and J. Reizer.** 1987. Regulation of β-galactoside transport and accumulation in heterofermentative lactic acid bacteria. *J. Bacteriol.* **169:**5589–5596.
162. **Rosen, B. P.** 1990. The plasmid-encoded arsenical resistance pump: an anion-translocating ATPase. *Res. Microbiol.* **141:**336–341.
163. **Rosen, B. P., C. Mei-Hsu, P. Kaur, C. E. Karkaria, and R. F. Steiner.** 1991. Substrate-induced dimerization of the ArsA protein, the catalytic component of an anion-translocating ATPase. *J. Biol. Chem.* **266:**2327–2332.
164. **Rudner, D. Z., J. R. LeDeaux, K. Ireton, and A. D. Grossman.** 1991. The *spo0K* locus of *Bacillus subtilis* is homologous to oligopeptide permease and is required for sporulation and competence. *J. Bacteriol.* **173:**1388–1398.
165. **Saffen, D. W., K. A. Presper, T. L. Doering, and S. Roseman.** 1987. Sugar transport by the bacterial phosphotransferase system. Molecular cloning and structural analysis of the *Escherichia coli ptsH, ptsI*, and *crr* genes. *J. Biol. Chem.* **262:**16241–16253.
166. **Saier, M. H., Jr.** 1985. *Mechanisms and Regulation of Carbohydrate Transport in Bacteria.* Academic Press, Inc., Orlando, Fla.
167. **Saier, M. H., Jr.** 1989. Involvement of the bacterial phosphotransferase system in diverse mechanisms of transcriptional regulation. *Res. Microbiol.* **140:**349–354.
168. **Saier, M. H., Jr.** 1989. Protein phosphorylation and allosteric control of inducer exclusion and catabolite repression by the bacterial phosphoenolpyruvate:sugar phosphotransferase system. *Microbiol. Rev.* **53:**109–120.
169. **Saier, M. H., Jr., and J. Reizer.** 1990. Domain shuffling during evolution of the proteins of the bacterial phosphotransferase system. *Res. Microbiol.* **141:**1033–1038.
170. **Saier, M. H., Jr., and J. Reizer.** 1991. Families and superfamilies of transport proteins common to prokaryotes and eukaryotes. Curr. Opin. Struct. Biol. **1:**362–368.
171. **Saier, M. H., Jr., and J. Reizer.** 1992. Proposed uniform nomenclature for the proteins and protein domains of the bacterial phosphoenolpyruvate:sugar phosphotransferase system based on structural, evolutionary and functional considerations. *J. Bacteriol.* **174:**1422–1438.
172. **Saier, M. H., Jr., L.-F. Wu, and J. Reizer.** 1990. Regulation of bacterial processes by three types of protein phosphorylating systems. *Trends Biochem. Sci.* **15:**391–395.
173. **Saier, M. H., Jr., M. Yamada, B. Erni, K. Suda, J. Lengeler, R. Ebner, P. Argos, B. Rak, K. Schnetz, C. A. Lee, G. C. Stewart, F. Breidt, Jr., E. B. Waygood, K. G. Peri, and R. F. Doolittle.** 1988. Sugar permeases of the bacterial phosphoenolpyruvate-dependent phosphotransferase system: sequence comparisons. *FASEB J.* **2:**199–208.
174. **Sautereau, A. M., and M. C. Trombe.** 1986. Electric transmembrane potential mutation and resistance to

the cationic and amphiphilic antitumoral drugs derived from pyridocarbazole, 2-*N*-methylellipticinium and 2-*N*-methyl-9-hydroxyellipticinium, in *Streptococcus pneumoniae. J. Gen. Microbiol.* **132:**2637–2641.
175. **Schmid, G., and G. Auling.** 1989. Alterations of the membrane composition induced by manganese depletion are late events in the nucleotide fermentation with *Brevibacterium ammoniagenes* ATCC 6872. *Agric. Biol. Chem.* **53:**1783–1788.
176. **Schneider, E., and K. Altendorf.** 1987. Bacterial adenosine 5′-triphosphate synthase (F_1F_0): purification and reconstitution of F_0 complexes and biochemical and functional characterization of their subunits. *Microbiol. Rev.* **51:**477–497.
177. **Schnetz, K., and B. Rak.** 1990. Beta-glucoside permease represses the *bgl* operon of *Escherichia coli* by phosphorylation of the antiterminator protein and also interacts with glucose-specific enzyme-III. The key element in catabolite control. *Proc. Natl. Acad. Sci. USA* **87:**5074–5078.
178. **Senior, A. E.** 1990. The proton-translocating ATPase of *Escherichia coli. Annu. Rev. Biophys. Biophys. Chem.* **19:**7–41.
179. **Serrano, R., and F. Portillo.** 1990. Catalytic and regulatory sites of yeast plasma membrane H^+-ATPase studied by directed mutagenesis. *Biochim. Biophys. Acta* **1018:**195–199.
180. **Shimkets, L. J., and D. Kaiser.** 1982. Murein components rescue developmental sporulation of *Myxococcus xanthus. J. Bacteriol.* **152:**462–470.
181. **Shirvan, M. H., S. Schuldiner, and S. Rottem.** 1989. Volume regulation in *Mycoplasma gallisepticum*: evidence that Na^+ is extruded via a primary Na^+ pump. *J. Bacteriol.* **171:**4417–4424.
182. **Sicard, A. M.** 1964. A new synthetic medium for *Diplococcus pneumoniae* and its use for the study of reciprocal transformation at the *amiA* locus. *Genetics* **50:**31–44.
183. **Sicard, A. M., and H. Ephrussi-Taylor.** 1965. Genetic recombination in DNA-induced transformation of pneumococcus. II. Mapping the *amiA* region. *Genetics* **52:**1207–1227.
184. **Silver, S.** 1991. Bacterial heavy metal resistance systems and possibility of bioremediation, p. 265–287. *In* D. Kamely et al. (ed.), *Biotechnology, Bridging Research and Applications*. Kluwer Academic Publishers, Norwell, Mass.
185. **Silver, S., and R. A. Laddaga.** 1990. Molecular genetics of heavy metal resistances in *Staphylococcus* plasmids, p. 531–549. *In* R. P. Novik (ed.), *Molecular Biology of Staphylococci*. VCH Publishers, New York.
186. **Silver, S., T. K. Misra, and R. A. Laddaga.** 1989. DNA sequence analysis of bacterial toxic heavy metal resistances. *Biol. Trace Element Res.* **21:**145–163.
187. **Silver, S., G. Nucifora, L. Chu, and T. K. Misra.** 1989. Bacterial resistance ATPases: primary pumps for exporting toxic cations and anions. *Trends Biochem. Sci.* **14:**76–80.
188. **Silver, S., and M. Walderhaug.** 1992. Gene regulation of plasmid- and chromosome-determined inorganic ion transport in bacteria. *Microbiol. Rev.* **56:**195–228.
189. **Slack, F. J., J. M. Mueller, M. A. Strauch, C. Mathiopoulos, and A. L. Sonenshein.** 1991. Transcriptional regulation of a *Bacillus subtilis* dipeptide transport operon. *Mol. Microbiol.* **5:**1903–1913.
190. **Smith, C. P., and K. F. Chater.** 1988. Structure and regulation of controlling sequences for the *Streptomyces coelicolor* glycerol operon. *J. Mol. Biol.* **204:**569–580.
191. **Solioz, M., S. Mathews, and P. Fürst.** 1987. Cloning of the K^+-ATPase of *Streptococcus faecalis*. Structural and evolutionary implications of its homology to the KdpB protein of *Escherichia coli. J. Biol. Chem.* **262:**7358–7362.
192. **Stone, D. K., B. P. Crider, T. C. Südhof, and X. Xie.** 1989. Vacuolar proton pumps. *J. Bioenerg. Biomembr.* **1:**605–620.
193. **Stone, M. J., W. J. Fairbrother, A. G. Palmer III, J. Reizer, M. H. Saier, Jr., and P. E. Wright.** 1992. The backbone dynamics of the *Bacillus subtilis* glucose permease IIA domain determined from ^{15}N NMR relaxation measurements. *Biochemistry* **31:**4394–4406.
194. **Sutrina, S. L., P. Reddy, M. H. Saier, Jr., and J. Reizer.** 1990. The glucose permease of *Bacillus subtilis* is a single polypeptide chain that functions to energize the sucrose permease. *J. Biol. Chem.* **265:**18581–18589.
194a. **Tam, R., and M. H. Saier, Jr.** 1993. Structural, functional, and evolutionary relationships among the extracellular solute-binding receptors of bacteria. *Microbiol. Rev.*, in press.
195. **Teshiba, S., and A. Furuya.** 1984. Mechanisms of 5′-inosinic acid accumulation by permeability mutants of *Brevibacterium ammoniagenes*. IV. Excretion mechanisms of 5′-IMP. *Agric. Biol. Chem.* **48:**1311–1317.
196. **Thompson, J.** 1987. Ornithine transport and exchange in *Streptococcus lactis. J. Bacteriol.* **169:**4147–4153.
197. **Titgemeyer, F., R. Eisermann, W. Hengstenberg, and J. W. Lengeler.** 1990. The nucleotide sequence of *ptsH* gene from *Klebsiella pneumoniae. Nucleic Acids Res.* **18:**1898.
198. **Trombe, M. C., G. Laneelle, and A. M. Sicard.** 1984. Characterization of *Streptococcus pneumoniae* mutant with altered electric transmembrane potential. *J. Bacteriol.* **158:**1109–1114.
198a. **Truniger, V., W. Boos, and G. Sweet.** 1992. Molecular analysis of the *glpFKX* regions of *Escherichia coli* and *Shigella flexneri. J. Bacteriol.* **174:**6981–6991.
199. **Tynecka, Z., Z. Gos, and J. Zajac.** 1981. Energy-dependent efflux of cadmium coded by a plasmid resistance determinant in *Staphylococcus aureus. J. Bacteriol.* **147:**313–319.
200. **Verhoogt, H. J. C., H. Smit, T. Abee, M. Gamper, A. J. M. Driessen, D. Haas, and W. N. Konings.** 1992. *arcD*, the first gene of the *arc* operon for anaerobic arginine catabolism in *Pseudomonas aeruginosa*, encodes an arginine-ornithine exchanger. *J. Bacteriol.* **174:**1568–1573.
201. **Vogler, A. P., M. Homma, V. M. Irikura, and R. M. Macnab.** 1991. *Salmonella typhimurium* mutants defective in flagellar filament regrowth and sequence similarity of FliI to F_0F_1, vacuolar, and archaebacterial ATPase subunits. *J. Bacteriol.* **173:**3564–3572.
202. **Wittekind, M., J. Reizer, J. Deutscher, M. H. Saier, Jr., and R. E. Klevit.** 1989. Common structural changes accompany the functional inactivation of HPr by seryl phosphorylation or by serine to aspartate substitution. *Biochemistry* **28:**9908–9912.
203. **Wittekind, M., J. Reizer, and R. E. Klevit.** 1990. Sequence-specific ^1H NMR resonance assignments of *Bacillus subtilis* HPr: use of spectra obtained from mutants to resolve spectral overlap. *Biochemistry* **29:**7191–7200.
204. **Woese, C. R.** 1987. Bacterial evolution. *Microbiol. Rev.* **51:**221–271.
205. **Worthylake, D., N. D. Meadow, S. Roseman, D. I. Liao, O. Herzberg, and S. J. Remington.** 1991. 3-Dimensional structure of the *Escherichia coli* phosphocarrier protein III^{Glc}. *Proc. Natl. Acad. Sci. USA* **88:**10382–10386.
206. **Wu, H. C.** 1987. Posttranslational modification and processing of membrane proteins in bacteria, p. 37–71. *In* M. Inouye (ed.), *Bacterial Outer Membranes as Model Systems*. Wiley-Interscience, New York.
207. **Wu, L.-F., and M. H. Saier, Jr.** 1990. On the evolutionary origins of the bacterial phosphoenolpyruvate:sugar phosphotransferase system. *Mol. Microbiol.* **4:**1219–1222.
208. **Wu, L.-F., J. M. Tomich, and M. H. Saier, Jr.** 1990.

Structure and evolution of a multidomain multiphosphoryl transfer protein. Nucleotide sequence of the *fruB(HI)* gene in *Rhodobacter capsulatus* and comparisons with homologous genes from other organisms. *J. Mol. Biol.* **213:**687–703.

209. **Yamaguchi, K., F. Yu, and M. Inouye.** 1988. A single amino acid determinant of the membrane localization of lipoproteins in *E. coli. Cell* **53:**423–432.
210. **Yokoyama, K., T. Oshima, and M. Yoshida.** 1990. *Thermus thermophilus* membrane-associated ATPase. *J. Biol. Chem.* **265:**21946–21950.
211. **Yoon, K. P., and S. Silver.** 1991. A second gene in the *Staphylococcus aureus cadA* cadmium resistance determinant of plasmid pI258. *J. Bacteriol.* **173:**7636–7642.
212. **Zagorec, M., and P. W. Postma.** 1992. Cloning and nucleotide sequence of the *ptsG* gene of *Bacillus subtilis. Mol. Gen. Genet.* **234:**325–328.

11. Carbohydrate Catabolism: Pathways, Enzymes, Genetic Regulation, and Evolution

MICHEL STEINMETZ

The study of sugar utilization systems such as the *Escherichia coli lac* operon has played a major role in the development of modern biology. There are still those who love such systems, and this review discusses what they have found in *Bacillus subtilis* and other gram-positive bacteria. Their work is in some cases driven by biotechnological or medical questions (How do lactic acid bacteria involved in the dairy industry ferment lactose? Why do sucrose-rich diets increase the incidence of dental caries caused by *Streptococcus mutans*?) but has generated a body of data that can be compared with what is known about other bacteria, especially enteric bacteria. Carbohydrate degradation pathways in gram-positive bacteria are generally not unique; furthermore, the units constituting regulons (genes encoding enzymes and permeases) are generally related, often closely, to units playing the same roles in other bacteria. On the other hand, the organization and regulation of each regulon are unique. Such regulation involves a great diversity of regulatory proteins (repressors, activators, antiterminators, etc.) that interact with DNA or RNA targets, and these regulators are themselves often controlled by protein cascades. These regulatory systems are generally (elegant) variations on themes previously observed in enteric bacteria, but novel mechanisms have also been uncovered. This body of regulatory and structural data is of more than parochial or esthetic interest. One of the major current aims of biologists is to understand the working of biological macromolecules, with the practical aim of creating new activities or specificities. Bacterial systems involved in peripheral metabolism (e.g., catabolism of sugars, amino acids, or nucleosides; biosynthesis of secondary compounds and antibiotics; antibiotic resistance; or other detoxification processes) offer a rich and accessible source of natural genes and catalytic activities. The study of these systems will help us define protein domains involved in the binding of DNA, nucleotides, coenzymes, etc., and understand (and then modify) protein properties.

Even when the gene products involved in a pathway common to many different bacteria are related, the gene organization is often species specific. This variability, a general feature of all the peripheral metabolism regulons, strongly contrasts with the highly conserved organization of operons mediating essential functions (92, 97). Genes involved in peripheral metabolism thus constitute a dynamic subset in the bacterial genome and may therefore be particularly informative for studies of recent evolution.

CATABOLIC CAPACITIES OF *B. SUBTILIS* AND GRAM-POSITIVE BACTERIA: AN OVERVIEW

Enteric bacteria such as *E. coli* and *Salmonella typhimurium* can use any one of about a hundred different carbohydrates (sugars, polyols, and carboxylates) as the sole source of carbon and energy (58). The range of the best-studied gram-positive species appears to be lower. As shown in this review, data exist for the metabolism of fewer than 20 carbohydrates by *B. subtilis* 168. This apparent inferiority may be due at least in part to lack of experimentation. The capacities of enteric bacteria, extensively documented for a long time because of their taxonomic interest, are the collective properties of a large number of strains. A quite different image appears when individual strains are compared: while *B. subtilis* 168 is unable to use galactose or lactose, it can grow on sucrose, xylose, or cellobiose, which wild-type *E. coli* K-12 cannot do (58). In addition, *B. subtilis* has a highly developed ability to produce extracellular degradative enzymes. At least six extracellular proteases (see chapter 63 of this volume) provide the opportunity to use amino acids and short peptides as carbon sources. *B. subtilis* also secretes polysaccharide-degrading enzymes whose enzymatic properties and biotechnological value are described elsewhere (see chapter 62, this volume). Regulation of one of these enzymes (levanase) and of an extracellular fructosyltransferase (levansucrase [Lvs]) are reviewed in this chapter. Regulation of α-amylase is discussed elsewhere (see chapter 15 of this volume). Carbohydrate catabolism in other gram-positive bacteria is in most cases less well documented than that in *B. subtilis*. Recent reviews have focused on sugar catabolism in *Streptomyces* spp. (47) and lactic acid-producing bacteria (12, 19, 111).

PATHWAYS IN *B. SUBTILIS*

Transport

In *B. subtilis*, as in most bacteria, carbohydrates are transported into the cell by a great diversity of systems (see chapter 10). Some systems result in intracellular accumulation of unmodified sugar, and others, belonging to the phosphoenolpyruvate-dependent sugar-phosphotransferase system (PTS) and referred to here as PTS-permeases, couple transport and phosphorylation of sugar and polyol species (referred to as PTS-sugars). The phosphoryl group is transferred

Michel Steinmetz • Laboratoire de Génétique des Micro-organismes, 78850 Thiverval-Grignon, France.

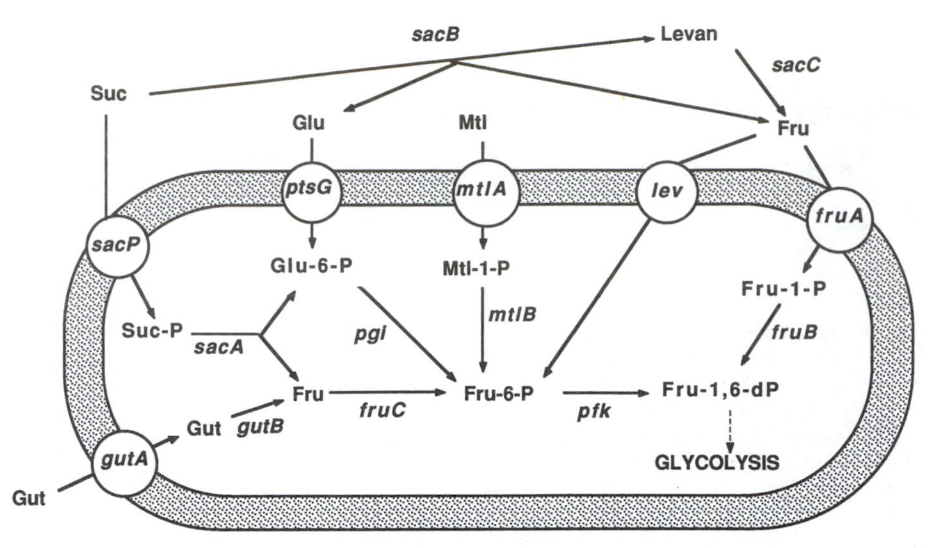

Figure 1. Catabolism of hexoses, linear hexitols, and sucrose in *B. subtilis*. Transport and enzymatic steps are symbolized by arrows. The names of the structural genes (in some cases putative) of the relevant proteins are indicated in italics (*lev* = *levD*, *levE*, *levF*, and *levG*). All these genes are both mapped (except *pgi*, for phosphoglucoisomerase) and inducible (except *fruC*, *pfk*, and *pgi*). Various controls prevent (or reduce) simultaneous expression of alternative pathways for sucrose and fructose (see text). Transfructosylation from sucrose by Lvs (*sacB*) results mainly in a mixture of glucose, fructose, and levan. *sacC* encodes levanase. Fru, Glu, Gut, Mtl, and Suc: fructose, glucose, glucitol, mannitol, and sucrose, respectively; Suc-P, sucrose-phosphate (phosphorylated on position 6 of glucose moiety). Other sugar-phosphates are designated according to standard conventions; for example, Glu-6-P and Fru-1,6-dP designate glucose 6-phosphate and fructose 1,6-diphosphate.

from phosphoenolpyruvate to sugar-specific PTS-permeases by a common phosphorylation cascade beginning with the *ptsI* gene product. *B. subtilis ptsI* mutants are, as expected, unable to use PTS-sugars including glucose, fructose, and sucrose. In addition, these mutants are unable to use carbon sources, such as glycerol, that are not transported by PTS-permeases (36). These catalytic and regulatory roles of the PTS are reviewed in detail elsewhere (see chapter 10).

From Hexoses and Hexitols to Fructose 1,6-Diphosphate

In *B. subtilis*, hexoses and linear hexitols enter the glycolytic pathway via fructose 1,6-diphosphate. *B. subtilis* has two unique features (Fig. 1). In enteric bacteria, both mannitol and glucitol are converted to fructose 1,6-diphosphate via parallel pathways beginning with a specific PTS-permease (58). In *B. subtilis*, this type of pathway exists only for mannitol. Glucitol is transported via a nonphosphorylating permease, and the resulting intracellular hexitol is then dehydrogenated to fructose. The permease and dehydrogenase are inducible by glucitol and are inactive in *gutA* and *gutB* mutants, respectively (35). Intracellular fructose resulting from glucitol dehydrogenation is converted to fructose 6-phosphate and then into fructose 1,6-diphosphate by two constitutive enzymes, fructokinase and phosphofructokinase, that are defective in the *fruC* and *pfk* mutants, respectively (35, 37). However, this pathway cannot be considered the prototype for the glucitol pathway in gram-positive bacteria, because the spore-forming *Clostridium acetobutylicum*, a close relative of *B. subtilis*, possesses a PTS-dependent glucitol pathway and the genes involved are homologous to those of *E. coli* (68). Both glucitol permease and dehydrogenase are constitutive in *B. subtilis gutR* mutants selected for their capacities to use D-xylitol as a carbon source (35). This suggests that this glucitol pathway could have (recently?) arisen from a pentitol pathway (59). The *gutR* and *fruC* mutations are linked to the *gutAB* locus (35). The *B. subtilis* PTS-dependent mannitol pathway is induced by the presence of mannitol and is affected by mutations in the *mtlAB* locus (34). This pathway is mediated in *Staphylococcus carnosus* by proteins homologous to their equivalents in *E. coli* (21).

The second unique feature in *B. subtilis* is the presence of two fructose pathways, each initiated by a PTS-permease (Fig. 1). The first, similar to that of enteric bacteria, was identified both biochemically and genetically by the characterization of mutants affected at the *fruAB* locus. This pathway is inducible and seems to be the major route for the entry of fructose, since the *fruA* and *fruB* mutants, respectively deficient in the first and second steps of the pathway (Fig. 1), grow very poorly on fructose (37). The four-gene cluster (*levDEFG*) responsible for the second route was identified (67) as part of the fructose-inducible levanase operon (see below). These four genes encode a fructose-specific PTS-permease (Lev) that accumulates fructose 6-phosphate intracellularly in contrast to the major *fruA*-dependent route, which leads to fructose 1-phosphate. The Lev permease is related to enterobacterial permeases specific for hexoses other than fructose (see chapter 10). Lev might have easily evolved from such permeases, since its intracellular vectorial product, fructose 6-phosphate, can be metabolized via constitutive enzymes involved in several pathways (Fig. 1).

B. subtilis 168 is also able to use D-glucose, D-mannose (36), and hexosamines including glucosamine and *N*-acetylglucosamine but not D-galactose or D-sorbose (103). *ptsG* mutants are affected in the gene for the glucose PTS-permease (43, 109); they are impaired in their growth on glucose and many hexoses and disaccharides, including cellobiose, glucosamine, maltose, mannose, *N*-acetylglucosamine, sucrose, and trehalose (103). These pleiotropic effects could be at least partly due to regulatory functions of *ptsG*.

Galactosidases and Glucosidases

Several gram-positive bacteria are able to catabolize lactose by means of diverse pathways. *B. subtilis* 168, which is both galactose and lactose minus, contains a galactokinase gene (see below) and possesses at least two β-galactosidase (βGal) genes (20, 117). The βGal activities pose problems for genetic and physiological studies that use fusions to the *E. coli lacZ* reporter gene. βGal is weakly expressed during stationary phase in the absence of specific inducer and can therefore hamper studies of (weakly) expressed *lacZ* fusions. A *bgaA* (formerly *lacA*) mutation abolishing this activity has been isolated and mapped (20). βGal genes can be decryptified by mutations. This can cause problems when mutations that derepress fusions to *lacZ* are screened for, a common approach to the analysis of gene regulation. Such a cryptic gene (*bgaX*), unlinked to *bgaA*, has been cloned and expressed in *E. coli* (117) and mapped between *purB* and *tre* (see chapter 31, this volume). A Δ(*bgaX*) mutant has been constructed (103); however, another βGal activity can still be decryptified in the mutant (62a).

The physiological substrates of these βGal (presumably unidentified β-galactosides) are unknown. The derepression of the βGal encoded by *bgaA* in stationary phase in the absence of any specific inducer is a phenomenon that is both paradoxical and relatively frequently observed in *B. subtilis*: extracellular proteases and amylase are produced under the same conditions (see chapters 15 and 63). Possibly these derepressions are last attempts by the bacterium to exploit the medium; if so, this attempt is probably more often profitable in natural niches than under laboratory conditions!

B. subtilis slowly metabolizes α-methyl glucoside through an apparently inducible pathway (25).

Pentoses, Disaccharides, and Other Carbohydrates

Catabolism of L-arabinose, D-xylose, sucrose, glycerol, and gluconate and the corresponding regulons are described below. *myo*-Inositol, a cyclic hexitol, is catabolized through an incompletely defined pathway involving an inducible *myo*-inositol dehydrogenase whose gene was recently cloned (26, 32a). In addition to sucrose, *B. subtilis* 168 is able to use at least three disaccharides: cellobiose, maltose, and trehalose (103). A *tre* mutation abolishing the capacity to utilize trehalose has been mapped (see chapter 29).

Gdh

Genomes contain many genes whose function is unknown. This is the case for the *B. subtilis* glucose dehydrogenase (Gdh) gene, which is expressed in the forespore compartment during sporulation (32). The *gdh* gene encoding this activity belongs to an operon whose expression is controlled at the level of transcription (76). Several (noncontiguous) genes homologous to *gdh* were recently identified by sequencing chromosomal fragments of the *sacA-sacY* region (40). The role(s) of these *gdh*-like genes is unknown. A *gdhA* (null) mutant was constructed. This alteration, which abolishes Gdh activity in sporulating cells, has no effect on sporulation efficiency or on germination in rich medium (76). This shows that Gdh is not required for sporulation or for germination under these conditions. Of course, Gdh might be required under some specific germination conditions, as previously proposed (108). This also shows that the *gdh*-like genes do not encode Gdh activity or are not expressed in sporulating cells under the conditions tested.

Cryptic Genes

Cryptic genes are apparently bona fide genes (devoid of deletions, nonsense codons, etc.) whose expression is insignificant in the wild-type strain. At least one mutation is required to "decryptify" their expression, conferring on the mutant a novel activity. For instance *E. coli* K-12 carries four gene systems that when decryptified code for the assimilation of β-glucosides (73, 79). The K-12 strain also carries a lactase (*ebgA*) gene that allows lactose assimilation by Δ(*lacZ*) mutants when the gene is decryptified by at least two mutations, one affecting the regulation of *ebgA* and the other affecting the activity of the product of *ebgA* (44). Several hypotheses have been offered to explain the paradoxical existence of such genes. The cryptic and decryptified states could be alternatively useful to the strain, allowing, respectively, protection against poisonous substrates and utilization of profitable carbon sources (79). Alternatively, these genes could be involved in unknown functions, for example, hydrolysis of unidentified β-galactosides in the case of EbgA. In such a case, the decryptification mutations would render the genes able to promote another function. A third possibility is discussed below in the case of the *B. subtilis* Lvs (*sacB*) gene. In *B. subtilis*, the *bgaX* gene mentioned above and a *galK*-like gene that complements an *E. coli galK* (galactokinase) mutant (40) can be considered cryptic genes.

B. SUBTILIS OPERONS AND REGULONS STUDIED AT THE MOLECULAR LEVEL

L-Arabinose Operon

The *araA*, *-B*, and *-D* genes, which encode L-arabinose isomerase, L-ribulose kinase, and L-ribulose-5-phosphate isomerase, respectively, are induced by L-arabinose and are structurally homologous to their *E. coli* counterparts. They probably constitute an operon with the order A-B-D, which is different from the B-A-D order in *E. coli* (84). An *araC* locus unlinked to *araABD* was defined by mutations conferring both resistance to D-fucose and constitutive expression of arabinose enzymes (85).

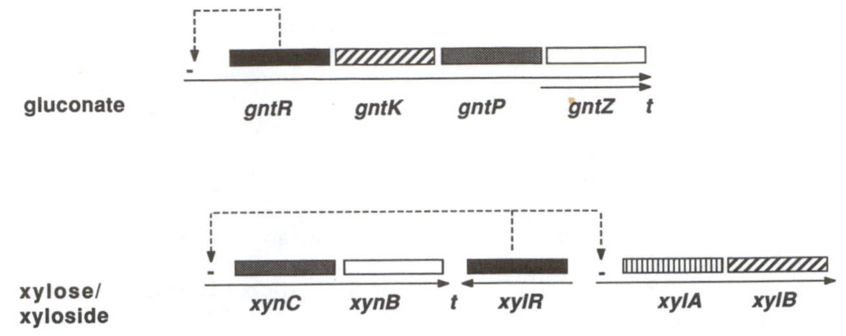

Figure 2. Organization of the gluconate (*gnt*) operon and the xyloside-xylose (*xyn-xyl*) regulon in *B. subtilis*. The repressor of the *gnt* operon is encoded by the first gene (*gntR*) of the operon; *gntK*, *-P*, and *-Z* very likely encode gluconate kinase, gluconate permease, and 6-phosphogluconate dehydrogenase, respectively; *gntZ* is transcribed from both the GntR-repressed promoter and a constitutive promoter within *gntP*. The repressor of the xylose-xyloside regulon (XylR) controls transcription of the *xynCB* and *xylAB* operons; XynB is a xylosidase; XynC appears to be a membrane protein that might be a permease of xyloside and/or xylose; and *xylA* and *xylB* encode xylose isomerase and xylulose kinase, respectively. t, terminator.

Gluconate (*gnt*) Operon

B. subtilis is able to grow on D-gluconate by means of an incompletely defined pathway. The first three steps are transport, phosphorylation, and dehydrogenation (26). No information on subsequent enzymatic steps is available. Since *B. subtilis* appears to lack the Entner-Doudouroff pathway (41), it was suggested that gluconate enters central metabolism through the pentose-phosphate cycle (26, 77). The *B. subtilis* four-gene *gntRKPZ* operon (Fig. 2) maps between genes involved in *myo*-inositol assimilation and the *fdp* gene encoding the key enzyme of gluconeogenesis, fructose diphosphatase (26). The operon is transcribed from a gluconate-inducible promoter upstream from *gntR*. *gntK* and *gntP* encode gluconate kinase and permease, respectively (31). GntK is homologous to bacterial kinases specific for fucose, glycerol, and xylulose (77, 100). The *gntZ* gene product is very likely 6-phosphogluconate dehydrogenase, judging by its homology with bacterial and mammalian 6-phosphogluconate dehydrogenases (77). Since a *B. subtilis gntZ* null mutant is able to grow on gluconate (29), it was suggested that *B. subtilis* contains a second gene for 6-phosphogluconate dehydrogenase (77). However, an alternative route cannot be excluded. The terminator of the operon was mapped downstream from the *gntZ* gene, which is transcribed from three promoters, i.e., the inducible promoter of the operon and two overlapping promoters mapped to the *gntP-gntZ* intergenic region (31). The reason for this unique regulation of *gntZ* is unknown.

A multicopy plasmid harboring the *gnt* leader region renders the operon partly constitutive, indicating that an operator present on the fragment titrated a repressor (27). This repressor, encoded by the first gene (*gntR*) of the operon (28), is homologous to several *E. coli* putative regulatory proteins (77). The induction of the *gntRKP*(*Z*) operon does not result in a large increase in the amount of *gntR* gene product. Although the *gntK* and *-P* gene products are induced 50- to 100-fold by gluconate, the measured amount of GntR increases only 4-fold after addition of inducer. This phenomenon, which results from an unknown mechanism, was interpreted as being a way to avoid the synthesis of wasteful amounts of repressor in the presence of gluconate (30).

Glycerol Catabolic Regulon

Like enteric bacteria and *Streptomyces coelicolor* (see below), *B. subtilis* catabolizes glycerol via an inducible pathway that involves transport, phosphorylation, and then dehydrogenation leading to the glycolytic intermediate dihydroxyacetone phosphate. This pathway is performed by the products of genes constituting the glycerol regulon, whose intracellular inducer is glycerol 3-phosphate. Mutants affected in this route are severely impaired for glycerol utilization (61); therefore, the role of a hypothetical second (constitutive) pathway involving dehydrogenation to dihydroxyacetone and subsequent phosphorylation (83) is unknown.

The *B. subtilis* catabolic glycerol regulon comprises genes mapped in at least two loci (60, 61). Mutations conferring resistance to fosfomycin and abolishing transport of glycerol 6-phosphate but not of glycerol have been mapped at the *glpT* locus (60). Genes essential for glycerol catabolism and unlinked to *glpT* are clustered in the order *glpP-glpF-glpK-glpD* (see Fig. 4). The last two encode glycerol kinase and glycerol 3-phosphate dehydrogenase, respectively, which are structurally related to their *E. coli* equivalents (48). *glpP* encodes a positive regulator of the regulon (49). GlpP might be involved in a glycerol-dependent antitermination process at a palindromic sequence (terminator) located between *glpK* and *glpD*, which are transcribed in the same orientation. The *glpK-glpD* intergenic region also contains a constitutive promoter upstream from the terminator. Therefore, *glpD* is an inducible and at least partially autonomous transcription unit (48, 49). Sequence data suggest a similar control of the *glpFK* operon via a similar conditional terminator (50). The *glpP* gene region contains open reading frames whose products do not share sequence similarity with antiterminators of the BglG-SacY family (see below) or with known *E. coli* regulators of glycerol metabolism (50).

The glycerol transport system in *B. subtilis* is a

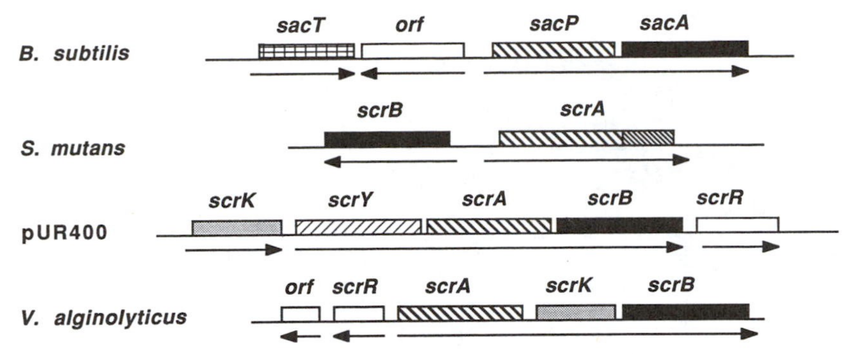

Figure 3. Sucrose gene clusters in four bacteria. The comparison includes two gram-positive systems, from *B. subtilis* and *S. mutans*, and two gram-negative systems (sucrose operon in *V. alginolyticus*, a close relative of enteric bacteria, and sucrose genes present on the *S. typhimurium* pUR400 plasmid; the pUR400 sucrose regulon appears to be closely related to the chromosomal *K. pneumoniae* sucrose system). The four enzymes $II^{sucrose}$, encoded by *sacP* (in *B. subtilis*) or *scrA* genes, are structurally homologous. The *S. mutans* ScrA protein contains a C-terminal extension containing a second phosphorylation site (enzyme $III^{sucrose}$ domain). The gene designated *sacA* in *B. subtilis* or *scrB* in *S. mutans* and *V. alginolyticus* encodes related phosphosucrases (the pUR400 *scrB* gene sequence is unknown). Expression of the *S. mutans* sucrose genes is stimulated by the presence of different sugars in the medium (see text). The three other systems are specifically induced by sucrose via activation of the SacT antiterminator in *B. subtilis* or inactivation of ScrR repressors in the case of the pUR400 or *V. alginolyticus* systems. The two ScrR repressors are not strongly similar, but both belong to the LacI-GalR superfamily; the pUR400-encoded ScrR repressor binds fructose; the two gram-negative regulons are dependent on the cyclic-AMP receptor protein–cyclic-AMP complex. The *scrK* and *scrY* genes encode fructokinase and a sucrose porin (related to the LamB maltoporin), respectively. *orf* encodes unknown functions (probably not involved in sucrose metabolism) (see references 7, 15, 87, and 89 and references therein). The *S. xylosus* sucrose genes were recently cloned and sequenced; the gene products are homologous to those mentioned above. *scrA* (encoding a short enzyme II) and *scrB* are unlinked. Both are sucrose inducible via a negative control. Similar operatorlike sequences are found in both leader regions (114a).

"facilitator" protein encoded by *glpF* (50). However, the PTS controls the glycerol regulon in *B. subtilis*: *ptsI* mutants are unable to grow on glycerol (36). The mechanism leading to this effect is unknown. Polymers of glycerol phosphate are major constituents of cell wall techoic acid in *B. subtilis*, implying either an active anabolic pathway or transport of glycerol from the medium. The dual glycerol metabolisms, both anabolic and catabolic, could have selected (partially) autonomous controls for *glpD* on the one hand and for *glpFK* on the other.

Twin Sucrose (*sac*) Regulons

There are two sucrose pathways in *B. subtilis*. The first involves transport and phosphorylation of sucrose by a PTS-permease and hydrolysis of intracellular sucrose 6-phosphate by a sucrose-6-phosphate hydrolase, referred to hereafter as phosphosucrase (Fig. 1). An identical pathway is observed in lactic acid-producing bacteria such as *S. mutans*, in *Vibrio alginolyticus*, and in enteric bacteria including some *S. typhimurium* strains and most *Klebsiella pneumoniae* strains that harbor plasmid and chromosomal sucrose genes, respectively (see legend to Fig. 3). The *B. subtilis* phosphosucrase, which is encoded by the *sacA* gene (24), and the *sacP* gene product (23), which is a membrane subunit of the sucrose-specific PTS-permease (enzyme $II^{sucrose}$), are structurally related to their equivalents in the bacteria cited above. However, the gene organization and regulation of these sucrose regulons are very variable (Fig. 3). The phosphosucrases of the SacA family share sequence similarities with the *B. subtilis* levanase and two sucrases, the yeast "invertase" encoded by SUC2, and the enteric *rafD* gene product (4, 67). *sacP* and *sacA* constitute an operon whose induction by sucrose is discussed below.

There is a second mode of sucrose metabolism in *B. subtilis* that is performed by the exoenzyme Lvs (encoded by *sacB*). Lvs transfers the fructosyl group from sucrose to acceptors, including water and the fructose polymer levan; this results in two main physiological reactions performed by Lvs in the presence of sucrose, i.e., hydrolysis into glucose plus fructose and levan elongation (10, 106). Therefore, Lvs is both anabolic and catabolic (Fig. 1). Initiation of levan synthesis is presumably also performed by Lvs by means of successive additions of fructosyl groups to an acceptor sucrose molecule. Lvs, which does not belong to the SacA-invertase-levanase family, shares significant similarities with the *S. mutans* fructosyl-transferase (99), which catalyzes, from sucrose, the synthesis of a fructan containing mainly a 1—2 bond; the major bonds in levan are 6—2 (10).

E. coli and many bacterial strains expressing the *sacB* gene are sensitive to sucrose (107). This sensitivity, whose cause is unknown, has been used as a positive selection procedure in a variety of gram-negative bacteria, in the gram-positive *Corynebacterium glutamicum*, and in the cyanobacterium *Anabaena* spp., to entrap insertion sequence elements and/or to select genetic rearrangements (see references 9, 11, 38, and 80 and references therein).

Both the *sacB* gene and the *sacPA* operon are sucrose inducible. However, the external sucrose concentrations required for maximal expression of these transcriptional units are different: *sacPA* is fully in-

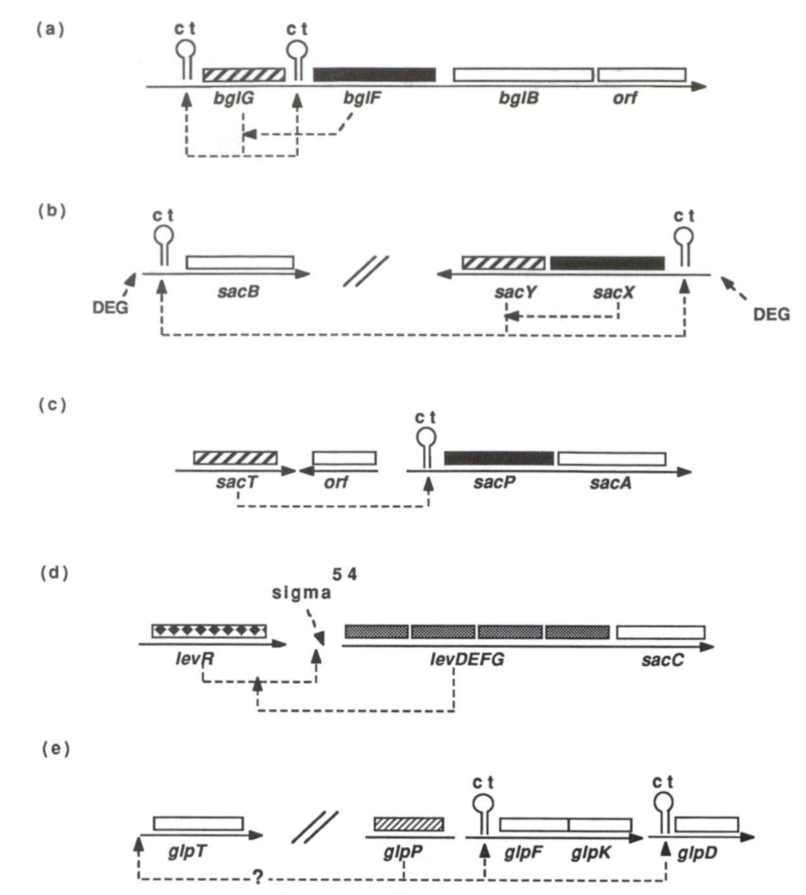

Figure 4. Regulons dependent on antiterminators and/or PTS control: comparison of regulation and genetic organization. The first three systems (a, b, and c) involve elements both structurally and functionally homologous: antiterminators, conditional terminators (ct), and enzymes II. (a) *E. coli* β-glucoside (*bgl*) operon model. In the absence of β-glucoside inducer, transcription of the operon from the promoter is prevented by two conditional terminators bracketing the antiterminator *bglG* gene. In these conditions, BglG function is inhibited by phosphorylation by the *bglF* gene product (enzyme $II^{\beta\text{-glucoside}}$). The presence of β-glucoside in the medium results in its transport and phosphorylation by enzyme $II^{\beta\text{-glucoside}}$. This results in phosphate group rerouting. Dephosphorylated antiterminator then allows full transcription of the operon. BglB is the phospho-β-glucosidase (3, 91). (b) *B. subtilis sacB* (Lvs) regulon. This regulon comprises the *sacB* gene and the unlinked *sacXY* regulator operon. Both appear to be controlled by means of a regulatory cascade similar to that of the *bgl* operon, i.e., positively by both sucrose and SacY antiterminator and negatively by SacX, a putative enzyme $II^{sucrose}$. Several regulatory cross-talks with the *sacPA-sacT* regulon (see below) are not shown. The *deg* genes (DEG; see chapter 50) activate transcription from both *sacB* and *sacXY* promoters (13, 14, 98, 104). (c) *B. subtilis sacPA* operon. Transcription of this operon is positively controlled by sucrose and the antiterminator encoded by the linked *sacT* gene; *orf-2* appears not to play a role in this regulation (15). The mechanism of SacT activation by the presence of sucrose is unknown (see text). (d) *B. subtilis* levanase operon. The first four genes encode an enzyme $II^{fructose}$ (very poorly related to those mentioned above) controlling the LevR positive regulator. The fructose-dependent induction cascade appears similar to that of the *bgl* operon but appears not to involve antitermination. LevR is a transcriptional activator binding upstream from the operon promoter and requiring σ^{54} as a cofactor (16, 17, 67). (e) *B. subtilis* glycerol regulon. This regulon comprises at least two unlinked gene clusters and three promoters likely controlled by the *glpP* gene product. This regulator (unrelated to BglG) appears to function as an antiterminator at conditional terminators just upstream from *glpD* and *glpFK* (48–50).

duced by 1 mM sucrose and repressed by high sucrose concentrations, while full induction of *sacB* is obtained at around 30 mM, which is close to the K_m of Lvs for sucrose. *sacPA* but not *sacB* is subject to repression by glucose (57, 105). Sucrose induction of the *sac* genes is controlled by two sets of regulatory elements very similar to each other and to those involved in induction of the *E. coli bgl* operon by β-glucosides (Fig. 4). The promoters of *sacB* and *sacPA* are, like that of the *bgl* operon, immediately followed by conditional, rho-independent terminators that prevent transcription of the structural genes in the absence of inducer (15, 98, 106). The proteins responsible for the induction of *sacB* and *sacPA* are encoded by the *sacXY* operon and the *sacT* gene, respectively. SacY, SacT, and the *E. coli* BglG regulator of the *bgl* operon are very similar positive regulators required for full transcription via antitermination of the genes they control (14, 15, 64, 91). The BglG and SacY antiterminators do not appear to be activated by binding of the inducer or one of its metabolic derivatives (3, 14). In the case of the *bgl* operon, the signal of inducer presence is transmitted to BglG through the *bglF* gene product, which is the β-glucoside-specific PTS-permease. Biochemical and genetic evidence has shown that BglF can phosphorylate (inactivate) and dephosphorylate (activate) BglG in the absence and the presence, respectively, of inducer (3). Genetic arguments suggest a similar control of SacY by the *sacX* gene product, which is homologous to SacP, the *B. subtilis* sucrose PTS-permease (14).

Binding of the BglG antiterminator to its RNA target has been demonstrated in vitro (51). The BglG-binding site comprises a 32-nucleotide imperfect palindromic sequence whose 3′ region overlaps the 5′ end of the conditional terminator. It was therefore suggested that binding of BglG to this sequence stabilizes the formation of a stem-loop structure that prevents formation of the alternative stem-loop structure that causes termination (51). Similar sequences overlap the *sac* terminators. The *sacPA* and *sacB* 32-nucleotide sequences differ from each other at 3 positions. In vitro mutagenesis has been used to show that these differences are responsible for the specific induction of *sacB* and *sacPA* by SacY and SacT, respectively (5). Three problems that have been solved in the case of the *bgl* operon still remain in the case of the *sac* genes. (i) The role of SacX remains to be demonstrated biochemically. (ii) It is unknown whether SacT activity is modulated by phosphorylation-dephosphorylation, and if it is, whether SacP is its kinase. (iii) Although BglF is a complete PTS-permease containing two phosphorylation sites, both SacP and SacX belong to the family of "short" membrane PTS-proteins that contain only one such site (see chapter 10). Therefore, what is the PTS element (enzyme $III^{sucrose}$) that transfers the phosphoryl group from cytoplasmic PTS-proteins to these SacP and SacX membrane proteins? According to the model proposed for SacY activation-inactivation (14), enzyme $III^{sucrose}$ must play a regulatory role. Apparently, the glucose PTS-permease encoded by *ptsG* can serve as an enzyme $III^{sucrose}$, at least in vitro (109). However, a *ptsG* deletion does not significantly affect the induction of *sacB* or *sacPA* by sucrose (55).

The duality of sucrose metabolism in *B. subtilis* and the anabolic role of Lvs have parallels with sucrose metabolism in *S. mutans* (see below). It must also be noted that *sacB* is virtually cryptic in the wild-type strain, at least under laboratory conditions (57, 104). Mutations affecting the *deg* genes, which belong to a complex regulatory network, are required to decryptify *sacB* (see chapter 50). It has been suggested that *sacB* crypticity reflects a requirement for two inducers, i.e., sucrose at high concentration and an unidentified inducer that signals via the *deg* gene products a specific ecological niche in which synthesis of levan would be useful for the cell (104).

Xylose Regulons in *B. subtilis* and Elsewhere

The presence of D-xylose in the medium induces the expression in *B. subtilis* of at least two operons, *xynC-xynB* and *xylA-xylB*, involved in xylose and xyloside catabolism. These operons and the repressor-encoding *xylR* gene are clustered on the chromosome of strain 168 (46; Fig. 2). This gene cluster is located near *thyA* (see chapter 31), far from the *xynA* gene which is involved in xylan degradation (81). Parallel studies with xylose genes from the *B. subtilis* 168 and W23 strains showed that the presence on a multicopy plasmid of sequences in the vicinity of the *xynC* and *xylA* promoters partially derepress the regulon by titrating the XylR repressor. The operators of XylR are partially palindromic 25-bp sequences overlapping the transcription start sites (33, 46, 52).

The xylose regulons of enteric bacteria also comprise several transcription units clustered on the chromosome (54, 96). The pathway appears to be identical to and to involve enzymes structurally related to those of several other bacteria, including *B. subtilis* and *Streptomyces* species. Three characteristics should be noted: (i) genetic experiments show convincingly that induction of the regulon by xylose in *S. typhimurium* LT2 is mediated by a positive regulator (96); (ii) no xylosidase activity in *E. coli* or *S. typhimurium* has been described; and (iii) the xylose regulon is cryptic in the wild-type *E. coli* K-12 strain (58). The organization of the *xylR-xylAB* gene cluster (Fig. 2) is identical in *B. subtilis*, *Bacillus licheniformis*, and *Bacillus megaterium*, although in *Staphylococcus xylosus*, *xylR* is oriented as the *xylAB* operon is; the promoters of *xylA* in *B. licheniformis*, *B. megaterium*, and *S. xylosus* are followed by putative operators similar to that in *B. subtilis* (88, 100).

Bacterial D-xylose isomerases, especially thermostable enzymes from *Streptomyces* species, are widely used in the starch industry because they catalyze the conversion of glucose into a mixture of fructose and glucose. Determination of the *Streptomyces* xylose isomerase structure showed a typical "β-barrel" organization. In *Streptomyces rubiginosus*, the *xylA* and *xylB* genes are divergently transcribed from overlapping promoters (see reference 116 and references therein). The gene of a still more thermostable enzyme was recently cloned from a thermophile bacterium (18).

Levanase (*lev-sacC*) Operon

The unravelling of the levanase operon revealed an interesting and complex system of regulation. The exoenzyme levanase was first identified as a third saccharolytic enzyme in *B. subtilis*. Its synthesis is rendered constitutive by mutations mapping at the *sacL* locus. The enzyme, which is not inducible by sucrose in the wild-type strain, was able to hydrolyze sucrose and at least two fructans, inulin and levan, which are fructose polymers with β-1,2 and β-2,6 bonds, respectively (53). The N-terminal part of the levanase protein is very similar to the product of the *Bacillus polymyxa lel* gene, an inulin-hydrolyzing, probably cytoplasmic enzyme (6); less similar to the *Saccharomyces cerevisiae* invertase; and still more divergent from bacterial phosphosucrases such as the *sacA* product (65). A deletion of a significant part of the C-terminal end of the protein does not abolish its saccharolytic activity (93).

The *sacL* locus turned out to contain six genes involved in fructan utilization: a pentacistronic *lev* operon, inducible by low amounts of fructose in the medium, and the *levR* regulatory gene lying immediately upstream. The levanase structural gene (*sacC*) is the distal gene of the operon. The first four genes, *levD*, *-E*, *-F*, and *-G*, encode the subunits of a PTS-permease for fructose (67). The levanase operon is repressed by fructose concentrations of more than 15 mM and by glucose (66). The mechanism(s) of these repressions is still unknown. *ptsI* (null) mutants are levanase constitutive (34). Furthermore, levanase-constitutive mutations previously mapped in the *sacL* locus turned out to lie in *levD* and *levE* (67). Taken together, these observations suggest that fructose induction of the operon involves a mechanism related to those of the *E. coli bgl* and *B. subtilis sacB* systems discussed above. However, the operon promoter is not followed by a transcription terminator of the *bgl-sac* family.

LevR is a positive regulator that is homologous to two classes of regulators. Its N-terminal domain shares sequence similarities with transcriptional activators of the NifA family, and its C-terminal domain is similar to antiterminators of the BglG family. A *levR* null mutation abolishes induction of the operon; on the other hand, a deletion of the C-terminal domain (homologous to antiterminators) renders the operon constitutive (16). These observations strongly suggest that LevR is a transcription activator that in the absence of inducer is inactivated by the Lev permease by phosphorylation in the region similar to BglG. In the presence of external fructose, the sugar is used as a phosphoryl acceptor and is transported into the cell by the Lev permease. As in the *bgl-sac* systems, this rerouting of phosphoryl groups results in LevR dephosphorylation and activation of LevR ability to enhance transcription (16; Fig. 4). LevR is dependent on the minor σ^{54} factor to activate the levanase operon, as are other activators of the NifA family. This was first demonstrated in *E. coli* (16) and subsequently in *B. subtilis* after cloning of the relevant gene; inactivation of the *B. subtilis* σ^{54} gene (*sigL*) showed that this factor is required both for expression of the levanase operon and for utilization of diverse amino acids as nitrogen sources (17; see also chapter 16).

WELL-DOCUMENTED SYSTEMS IN OTHER GRAM-POSITIVE BACTERIA

Streptomyces spp.: Galactose, Glycerol, and Xylose

The *S. coelicolor* glycerol (*gyl*) operon is glycerol inducible and glucose repressible; it comprises at least three genes cotranscribed in the following order: *gylA*, *gylB*, and *glyX*. *gylB* and *gylX* encode glycerol 3-phosphate dehydrogenase and a protein of unknown function, respectively (102). The *gylA* gene, whose 5′ part has been sequenced (102), was presumed to encode glycerol kinase; however, the predicted N-terminal part of the GylA protein shows no similarities to either the *E. coli* or the *B. subtilis* glycerol kinase (48). In fact, the 3-kb *gylA* region encodes two functions. The 5′ and 3′ parts of *gylA* encode, respectively, a membrane protein homologous to the glycerol facilitator of both *E. coli* and *B. subtilis* and the glycerol kinase activity (101). Immediately upstream of the *gylABX* operon and transcribed in the same orientation lies the *gylR* gene, which probably encodes the positive regulator of the operon. GylR contains a convincing helix-turn-helix domain (102) that looks like those of several *E. coli* DNA-binding regulatory proteins, especially AsnC, Fnr, TrpR, and the recently identified lclR regulator (71). Sequence similarity extending largely outside the helix-turn-helix domain is observed with AsnC (probably an activator of the asparagine synthetase A gene) and still more with lclR (probably a repressor of the acetate *aceBAK* operon). Transcription of *gylR* itself is inducible by glycerol and appears weakly repressed by glucose (102).

Transcription of the three-gene galactose operon of *Streptomyces lividans* is subject to induction by galactose and repression by glucose (22). The regulator(s) involved in this control has not been characterized but may bind galactose 6-phosphate (Gal-6-P) (1) rather than galactose as in *E. coli*. The *S. lividans gal* operon can complement an *E. coli* strain with the entire *gal* operon deleted. The gene products present regions similar to regions in their equivalents in *E. coli* and the yeast *Saccharomyces carlsbergensis*. The gene order in *S. lividans* (*galTEK*) (1) differs from that in *E. coli* (*galETK*) (or in *Lactobacillus helveticus*; see below). A weak constitutive promoter was found within the *galT-galE* intergenic region (22). The regulated and the constitutive promoters appear to be recognized by different RNA polymerases (115).

As discussed above, D-xylose isomerases from *Streptomyces* species were extensively studied because of their biotechnological interest.

Sucrose Metabolism by Cariogenic Streptococci

Streptococci such as *S. mutans* are natural inhabitants of the human oral cavity and play an important role in the etiology of dental caries (62). Because a sucrose-rich diet is an aggravating cofactor for this pathology, sucrose metabolism in these organisms has been extensively studied (12). Carbohydrate assimilation by these bacteria results in excretion of lactic acid, which alters the surface of the tooth. However, sucrose catabolism has a major additional effect by stimulating colonization of (adhesion to) the surface of the tooth. As in *B. subtilis*, two sucrose metabolic pathways coexist in *S. mutans*: an extracellular anabolic and catabolic metabolism and an intracellular PTS-dependent catabolism. The latter is very similar to that observed in *B. subtilis* and in gram-negative bacteria and involves structurally related gene products (86). However, the organization of the *S. mutans* sucrose gene cluster (*scrA-scrB*) is unique (Fig. 3). Regulation of *scrA* by the carbon source appears complex; transcription of *scrA*, which was monitored by using a *lacZ* transcriptional fusion, is maximal when sucrose is the carbon source, two times lower with glucose or glucitol, and seven times lower with fructose or maltose (87).

Extracellular sucrose metabolism in oral streptococci is mediated by several transferases and leads to synthesis of fructan and glucan; glucan appears to be involved in adhesion to the tooth surface. The *S. mutans* fructosyltransferase is structurally related to the *B. subtilis* Lvs (99). In some strains, glucosyltransferase activity is encoded by several tandemly organized genes (39, 112). The fructan and glucan polymers may also be subsequently metabolized (95). This delayed catabolism also parallels that observed in *B. subtilis* with levanase. The complexity of sucrose metabolism by streptococci can be explained. Their ecological niche, the oral cavity, is the place of rapid food transit. In the presence of sucrose, the intracellular catabolism of sucrose (and of hexoses liberated by transferase activities) brings an immediate supply of carbon and energy, allowing cell growth and division. The synthesis of extracellular polymers by transferases both allows adhesion of daughter cells and constitutes a food stock that can subsequently be used when the food supply has been swallowed.

The role of sucrose metabolism in *S. mutans* virulence has been analyzed by reverse genetics in several laboratories. For example, it was recently shown that a mutant defective in the intracellular sucrose pathway causes caries in rats at levels similar to those of the wild-type strain (63). As noted by those authors, this result must be viewed in light of the multiple means that *S. mutans* has for assimilating sucrose and its hexose products. Furthermore, these experiments involved monoinfection of gnotobiotic rats with mutant or wild-type strains (63). It would be interesting to analyze the competitiveness of the mutant in double-infection experiments.

Staphylococcus aureus Lactose Operon and Related Systems in Lactic Acid Bacteria: the Tagatose-Phosphate Pathway

The staphylococcal *lac* (chromosomal) genes are arranged in a heptacistronic *lacABCDFEG* operon (Fig. 5) whose transcription is inducible by addition of either lactose or galactose to the culture medium and is subject to catabolite repression (70, 72). A similar (often plasmid-borne) *lac* operon is found in several *Lactococcus lactis* strains involved in industrial dairy fermentation (19). The *lacE* and *lacF* gene products constitute the lactose PTS-permease (Fig. 5). The product of the *lacG* gene is a phospho-β-galactosidase that hydrolyzes lactose-phosphate to glucose and Gal-6-P. The *lacE*, *-F*, and *-G* plasmid genes of another industrial lactic bacterium, *Lactobacillus casei* 64H, have also been characterized (2); unexpectedly, the *L. lactis* LacE, LacF, and LacG proteins are more closely related to their *S. aureus* homologs than to those of *L. casei* 64H (2, 19); furthermore, *lacF* is the last gene of the operon in *L. casei* 64H (Fig. 5).

Gal-6-P is metabolized via a pathway different from the Leloir pathway found in *E. coli*, eukaryotes, and *Streptomyces* spp. (58). The so-called tagatose 6-phosphate pathway, whose enzymes are encoded by the first four genes of the operon (82, 114), converts Gal-6-P into two triose-phosphate molecules (Fig. 5). The tagatose 6-phosphate pathway genes could have evolved, at least in part, by duplication of glycolysis genes. The sequence similarity between LacC and an *E. coli* phosphofructokinase supports this suggestion (114). As noted below, a completely different lactose system exists in other lactic acid bacteria, including an *L. casei* strain.

Regulation of the *S. aureus lac* operon has been intensively studied. Gal-6-P was found to be the intracellular inducer of the system (70), presumably through binding with the regulator encoded by the *lacR* gene, which is located immediately upstream from the operon promoter. The LacR repressor shares sequence similarity with three *E. coli* repressors, DeoR, GutR, and FucR, which bind phosphorylated sugars; catabolite repression appears to involve an unidentified repressor whose DNA target overlaps that of LacR (72). Several similar cases (catabolite repression mediated by a negative regulator) have been described for *B. subtilis* (see chapter 15). The *L. lactis lac* operon, which is controlled by a regulator very similar to the *S. aureus* LacR repressor, presents several unique features (Fig. 5). Transcription of the *L. lactis lacR* gene is stimulated by glucose (113).

Non-PTS-Dependent Lactose-Galactose Systems in Gram-Positive Bacteria

The lactic bacteria *Streptococcus thermophilus* and *Lactobacillus bulgaricus* possess inducible lactose operons encoding a unique (nonphosphorylating) permease (*lacS* gene) and a βGal (*lacZ* gene) related to the *E. coli lacZ* gene product (56, 74, 90, 94). Similar βGal genes, and probably similar pathways, are present in other lactic acid bacteria, *Leuconostoc paramesenteroides*, a strain of *L. casei* (19), and two *Clostridium* species (8, 45).

In *S. thermophilus*, the *lacSZ* operon is immediately preceded by genes encoding two enzymes of the Leloir galactose pathway. Transcription of these genes is high in lactose medium and low in glucose medium (75). However, like most *S. thermophilus* strains, the strain used for this study is galactose minus and excretes the galactose moiety when it is grown on lactose medium (75). Gal$^+$ mutants can be selected from a *S. thermophilus* strain; this phenotype is rapidly lost when the mutant is grown in nonlimiting lactose concentrations (110). It was suggested that galactose excretion (lactose-galactose exchange) energizes lactose transport (75; see also chapter 10). The Gal$^-$ phenotype of these strains could be the result of recent cryptification of a pathway.

On the other hand, *L. helveticus* possesses a functional Leloir galactose pathway; the *gal* genes responsible for this pathway appear to form an operon with a unique gene organization (69); four different organizations are found in *E. coli*, *L. helveticus*, *S. lividans*, and *S. thermophilus*.

CONCLUDING REMARKS

Pathways

It has been known for decades that catabolic pathways are conserved elements in prokaryotes and more generally throughout all groups of living organisms. There are, of course, exceptions. Three unique carbohydrate pathways in gram-positive bacteria have been highlighted here: pathways for glucitol and fructose utilization in *B. subtilis* and the tagatose-phosphate pathway for lactose-galactose degradation in *S. aureus* and some lactic acid bacteria. The tagatose-phosphate pathway (Fig. 5) is a radical alternative to the otherwise ubiquitous Leloir galactose pathway. It is not known which of these pathways is the more recent innovation. *Streptomyces* species and *L. helveti-*

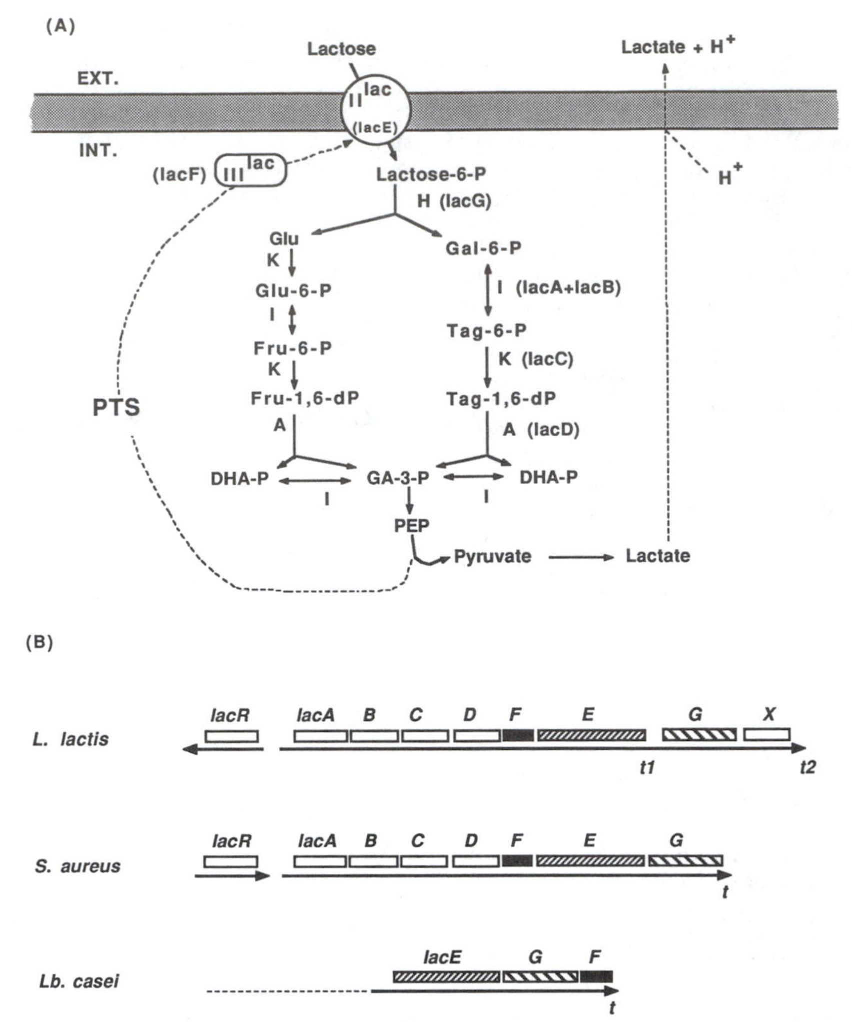

Figure 5. PTS-dependent lactose systems of *S. aureus* and some lactic acid-producing bacteria: pathways and gene organization. (A) Transport and catabolism of lactose via the tagatose 6-phosphate pathway. Lactose is transported and phosphorylated by enzyme II^{Lac} (II^{lac}) and then hydrolyzed into glucose (catabolized by the classical glycolysis pathway) and Gal-6-P, which is isomerized to tagatose 6-phosphate (Tag-6-P). Tag-6-P is converted, by means of a pathway similar to that of its isomer, fructose 6-phosphate (Fru-6-P), into two molecules of triose-phosphate (dihydroxyacetone-phosphate [DHA-P] and glyceraldehyde 3-phosphate [GA-3-P]) and then into phosphoenolpyruvate (PEP). Dephosphorylation of PEP to pyruvate results partly in ATP production (not shown) and partly in phosphorylation of proteins of the cytoplasmic PTS cascade. This cascade ends with phosphorylation of the membrane enzyme II^{Lac} protein by the cytoplasmic enzyme III^{Lac} (III^{lac}) protein. In lactic acid bacteria, lactate produced from pyruvate is excreted with protons; this coexcretion contributes to energizing the proton motive force. A, H, I, and K, aldolases, hydrolases, isomerases, and kinases, respectively. The genes encoding the enzymes of the pathway are indicated in parentheses. (B) Organization of the *lac* operons in *L. lactis*, *S. aureus*, and *L. casei*. Arrows indicate the transcripts; t, t1, and t2 are transcriptional terminators. Three major features distinguish the *L. lactis lac* operon from that of *S. aureus*: the orientation of the *lacR* gene, the presence of a weak terminator (t1) downstream from *lacE*, and the presence of an eighth gene, *lacX* of unknown function, at the 3′ end of the operon. The 5′ end of the *L. casei* operon has not yet been sequenced, but its transcript is similar in size to that of *S. aureus*. (Adapted from references 2, 19, 72, and 114.)

cus have functional Leloir pathways, and other gram-positive bacteria, including *S. thermophilus* and *B. subtilis*, appear to possess elements of this pathway.

The diversity is such that it is impossible to define a typically gram-positive pattern of carbohydrate catabolism.

Horizontal Transfer

The sequence homologies in bacterial carbohydrate enzymes that mediate identical functions but are of different origins suggest that ancestral enzymes were present in a common ancestral prokaryote or that the relevant genes spread after species divergence by horizontal transfer. Since the major, if not the sole, sources of some sugars (for instance, vascular plants for sucrose and mammals for lactose) are more recent than the divergence between the major bacterial taxonomic groups, it is likely that horizontal transfer has played an important role, as is generally accepted for antibiotic resistance. Since interspecific homologies are often observed for several proteins involved in a common pathway, these gene transfers presumably involved operons (or gene clusters) rather than single genes.

Enzyme Evolution

Enzymes with different but related specificities sometimes constitute superfamilies, which may have derived from a common ancestor (42, 78). In vivo experimental approaches show that a protein with a new activity can sometimes be produced by a very small number of mutations in the gene for a preexisting protein (see references 44 and 59 and references therein). The appearance in the course of evolution of a new sugar would be expected to provide a selective advantage to bacteria with mutations that result in the capacity to assimilate the new carbon source. One such superfamily includes several bacterial phosphosucrases, the *B. subtilis* levanase, and both yeast and bacterial sucrases. Another superfamily is made up of the phospho-β-glucosidases, for example, the *E. coli bglB* gene product, and the phospho-β-galactosidases of gram-positive origin (42). Analysis of such homologies can help us understand and thereby modify enzyme specificity constructively.

Genetic Organization and Regulation

The gene organization of regulons for the catabolism of five sugars (arabinose, galactose, glycerol, sucrose, and xylose) was elucidated for both enteric and gram-positive bacteria (*B. subtilis* or *Streptomyces* spp.); there is no conservation of organization between any of these pairs of regulons. The same is true for the sucrose gene clusters (Fig. 3); in this case, there is no conservation of gene organization, even when one compares two gram-positive or two gram-negative systems. It is also true for the xylose regulons or the galactose (Leloir pathway) genes. On the other hand, the glycerol gene clusters encoding the membrane "facilitator," the kinase, and the glycerol-phosphate dehydrogenase appear similarly organized in *B. subtilis* and *S. coelicolor*, even though their control elements appear to be different.

A nearly complete picture of the specific control of regulons for the catabolism of four sugars (galactose, glycerol, sucrose, and xylose) both in enteric bacteria and in *B. subtilis* or *Streptomyces* spp. is available. There is no conservation of control systems between any of these pairs of regulons. Furthermore, a virtually systematic inversion of the control mode (positive versus negative regulation) is observed. However, most regulators of the gram-positive regulons involved in sugar catabolism belong to families that also contain members in enteric bacteria: SacY and SacT (for sucrose metabolism in *B. subtilis*) are homologous to BglG (β-glucoside utilization in *E. coli*); LacR and GylR from *S. aureus* and *S. coelicolor* are homologous to several enteric DNA-binding regulators; a close homolog of LacI exists in *B. subtilis* (see chapter 15). The ancestors of these regulators therefore either were present before divergence of these bacterial phyla or spread more recently by horizontal transfer.

Invention and/or Change of Control Systems

The ideal genetic control system prevents wasteful synthesis of degradative proteins in the absence of their substrate (and therefore demands inducibility) or when a better alternative pathway can function (requiring catabolite repression). Such a system also prevents accumulation of poisons in the cell (due to transport from the medium or de novo synthesis by, respectively, permeases or enzymes, which are never completely specific) and futile metabolic cycles, which may exist when a metabolite can enter both catabolic and anabolic pathways (e.g., glycerol in *B. subtilis* and galactose in *E. coli* and *Streptomyces* spp.). Thus, a bacterium that by mutation or gene transfer acquires constitutive or imperfectly controlled genes is subject to various selective constraints that depend on both intrinsic and ecological factors. Comparing similar systems, for example, the *B. subtilis sacPA* and enteric (pUR400) sucrose operons (Fig. 3), can be illuminating. The intracellular inducer of the latter is fructose. This appears to be a good regulatory solution in enteric bacteria, in which intracellular (unphosphorylated) fructose is produced only by hydrolysis of sucrose (or raffinose in rare strains). This regulation would be unsatisfactory in *B. subtilis*, in which intracellular fructose is produced during glucitol catabolism. This example may help explain why a particular mode of control for a given pathway can be used by one bacterium but not another. Similarly, a DNA-binding regulator acquired by genetic transfer may recognize and bind sequences within essential genes of the recipient and disturb their expression. In brief, when genes and their control elements are acquired by genetic transfer, specific constraints in the recipient cell may result in adoption, adaptation, or replacement of the original control.

Acknowledgments. I express deep appreciation to my colleagues for useful information and/or comments: S. Aymerich, S. Bezzate, R. Brückner, A. M. Crutz, P. Glaser, W. Hillen, D. Hodgson, C. Holmberg, H. Kuramitsu, D. Le Coq, R. Losick, A. Mercenier, R. Richter, B. Rutberg, P. Serror, C. Smith, A. L. Sonenshein, and G. C. Stewart. This work was supported by the Centre National de la Recherche Scientifique and the Institut National de la Recherche Agronomique.

REFERENCES

1. **Adams, C. V., J. A. Fornwald, F. J. Schmidt, M. Rosenberg, and M. E. Brawner.** 1988. Gene organization and structure of the *Streptomyces lividans gal* operon. *J. Bacteriol.* **170:**203–212.
2. **Alpert, C. A., and B. M. Chassy.** 1990. Molecular cloning and DNA sequence of *lacE*, the gene encoding the lactose-specific enzyme II of the phosphotransferase system of *Lactobacillus casei*. *J. Biol. Chem.* **265:**22561–22568.
3. **Amster-Choder, O., F. Houman, and A. Wright.** 1989. Protein phosphorylation regulates transcription of the β-glucoside utilization operon in *E. coli*. *Cell* **58:**847–855.
4. **Asladinis, C., K. Schmid, and K. Schmitt.** 1989. Nucleotide sequences and operon structure of plasmid-borne genes mediating uptake and utilization of raffinose in *Escherichia coli*. *J. Bacteriol.* **171:**6753–6763.
5. **Aymerich, S., and M. Steinmetz.** 1992. Specificity determinants and structural features in the RNA target of

the bacterial antiterminator proteins of the BglG/SacY family. *Proc. Natl. Acad. Sci. USA* **89:**10410–10414.

6. **Bezzate, S., M. Steinmetz, and S. Aymerich.** Unpublished data.
7. **Blatch, G. L., and D. R. Woods.** 1991. Nucleotide sequence and analysis of the *Vibrio alginolyticus scrR* repressor-encoding gene (*scrR*). *Gene* **101:**17–23.
8. **Burchhardt, G., and H. Bahl.** 1991. Cloning and analysis of the β-galactosidase-encoding gene from *Clostridium thermosulfurogenes* EM1. *Gene* **106:**13–19.
9. **Cai, Y.** 1991. Characterization of insertion sequences IS*892* and related elements from the cyanobacterium *Anabaena* sp. strain PCC 7120. *J. Bacteriol.* **173:**5771–5786.
10. **Chambert, R., and G. Gonzy-Tréboul.** 1976. Levansucrase of *B. subtilis*: kinetic and thermodynamic aspects of the transfructosylation process. *Eur. J. Biochem.* **62:**55–64.
11. **Charles, T. C., and T. M. Finan.** 1991. Analysis of a 1600-kilobase *Rhizobium meliloti* megaplasmid using defined deletions generated *in vivo*. *Genetics* **127:**5–20.
12. **Chassy, B.** 1983. Sucrose metabolism and glycosyltransferase activity in oral streptococci, p. 3–10. *In* R. J. Doyle and J. E. Ciardi (ed.), *Glucosyltransferases, Glucans, Sucrose, and Dental Caries.* IRL Press, Washington, D.C.
13. **Crutz, A. M., and M. Steinmetz.** 1992. Transcription of the *Bacillus subtilis sacX* and *sacY* genes, encoding regulators of sucrose metabolism, is both inducible by sucrose and controlled by the DegS-DegU signalling system. *J. Bacteriol.* **174:**6087–6095.
14. **Crutz, A. M., M. Steinmetz, S. Aymerich, R. Richter, and D. Le Coq.** 1990. Induction of levansucrase in *Bacillus subtilis*: an antitermination mechanism negatively controlled by the phosphotransferase system. *J. Bacteriol.* **172:**1043–1050.
15. **Débarbouillé, M., A. Fouet, M. Arnaud, A. Klier, and G. Rapoport.** 1990. The *sacT* gene regulating the *sacPA* operon in *Bacillus subtilis* shares strong homology with transcriptional antiterminators. *J. Bacteriol.* **172:**3966–3973.
16. **Débarbouillé, M., I. Martin-Verstraete, A. Klier, and G. Rapoport.** 1991. The transcriptional regulator LevR of *Bacillus subtilis* has domains homologous to both σ^{54}- and phosphotransferase system-dependent regulators. *Proc. Natl. Acad. Sci. USA* **88:**2212–2216.
17. **Débarbouillé, M., I. Martin-Verstraete, A. Klier, and G. Rapoport.** 1991. The *Bacillus subtilis sigL* gene encodes an equivalent of σ^{54} from gram-negative bacteria. *Proc. Natl. Acad. Sci. USA* **88:**9092–9096.
18. **Dekker, K., H. Yamagata, K. Sakaguchi, and S. Ukada.** 1991. Xylose (glucose) isomerase gene from the thermophile *Thermus thermophilus*: cloning, sequencing, and comparison with other thermostable xylose isomerases. *J. Bacteriol.* **173:**3078–3083.
19. **De Vos, W. M.** 1991. Disaccharide utilization in lactic acid bacteria, p. 447–457. *In* H. Heslot, J. Davies, J. Florent, L. Bobichon, G. Durand, and L. Penasse (ed.), *Proceedings of the 6th International Symposium on the Genetics of Industrial Microorganisms.* Société Française de Microbiologie, Paris.
20. **Errington, J., and C. Vogt.** 1990. Isolation and characterization of mutations in the gene encoding an endogenous *Bacillus subtilis* β-galactosidase and its regulator. *J. Bacteriol.* **172:**488–490.
21. **Fischer, R., R. Eisermann, B. Reiche, and W. Hengstenberg.** 1989. Cloning, sequencing, and overexpression of the mannitol-specific enzyme-III-encoding gene of *Staphylococcus carnosus*. *Gene* **82:**249–257.
22. **Fornwald, J. A., F. J. Schmidt, C. W. Adams, M. Rosenberg, and M. E. Brawner.** 1987. Two promoters, one inducible and one constitutive, control transcription of the *Streptomyces lividans* galactose operon. *Proc. Natl. Acad. Sci. USA* **84:**2130–2134.
23. **Fouet, A., M. Arnaud, A. Klier, and G. Rapoport.** 1987. *Bacillus subtilis* sucrose specific enzyme II of the phosphotransferase system. Expression in *Escherichia coli* and homology to enzymes II from enteric bacteria. *Proc. Natl. Acad. Sci. USA* **84:**8773–8777.
24. **Fouet, A., A. Klier, and G. Rapoport.** 1986. Nucleotide sequence of the sucrase gene of *Bacillus subtilis*. *Gene* **45:**221–225.
25. **Freeze, E., W. Klofat, and E. Galliers.** 1970. Commitment to sporulation and induction of glucose-phosphoenolpyruvate-transferase. *Biochim. Biophys. Acta* **22:**265–289.
26. **Fujita, Y., and T. Fujita.** 1983. Genetic analysis of a pleiotropic deletion mutation (Δ*igf*) in *Bacillus subtilis*. *J. Bacteriol.* **154:**864–869.
27. **Fujita, Y., and T. Fujita.** 1986. Identification and nucleotide sequence of the promoter region of the *Bacillus subtilis* gluconate operon. *Nucleic Acids Res.* **14:**1237–1252.
28. **Fujita, Y., and T. Fujita.** 1987. The gluconate operon *gnt* of *Bacillus subtilis* encodes its own transcriptional negative regulator. *Proc. Natl. Acad. Sci. USA* **84:**4524–4528.
29. **Fujita, Y., and T. Fujita.** 1989. Effect of mutations causing gluconate kinase or gluconate permease deficiency on expression of the *Bacillus subtilis gnt* operon. *J. Bacteriol.* **171:**1751–1754.
30. **Fujita, Y., T. Fujita, and Y. Miwa.** 1990. Evidence for posttranscriptional regulation of synthesis of the *Bacillus subtilis* Gnt repressor. *FEBS Lett.* **267:**71–74.
31. **Fujita, Y., T. Fujita, Y. Miwa, J. Nihashi, and Y. Aratani.** 1986. Organization and transcription of the gluconate operon, *gnt*, of *Bacillus subtilis*. *J. Biol. Chem.* **261:**13744–13753.
32. **Fujita, Y., A. Ramaley, and E. Freeze.** 1977. Location and properties of glucose dehydrogenase in sporulating cells and spores of *Bacillus subtilis*. *J. Bacteriol.* **132:** 282–293.

32a. **Fujita, Y., K. Shindo, Y. Miwa, and K. Yoshida.** 1991. *Bacillus subtilis* inositol dehydrogenase-encoding gene (*idh*): sequence and expression in *Escherichia coli*. *Gene* **108:**121–125.

33. **Gärtner, D., M. Geissendörfer, and W. Hillen.** 1988. Expression of the *Bacillus subtilis xyl* operon is repressed at the level of transcription and is induced by xylose. *J. Bacteriol.* **170:**3102–3109.
34. **Gay, P.** 1979. Ph.D. thesis. Université Paris VI, Paris.
35. **Gay, P., H. Chalumeau, and M. Steinmetz.** 1983. Chromosomal localization of *gut*, *fruC*, and *pfk* mutations affecting glucitol catabolism in *Bacillus subtilis*. *J. Bacteriol.* **153:**1133–1137.
36. **Gay, P., P. Cordier, M. Marquet, and A. Delobbe.** 1973. Carbohydrate metabolism and transport in *Bacillus subtilis*. A study of *ctr* mutations. *Mol. Gen. Genet.* **121:**355–368.
37. **Gay, P., and A. Delobbe.** 1973. Fructose transport in *Bacillus subtilis*. *Eur. J. Biochem.* **79:**363–373.
38. **Gay, P., D. Le Coq, M. Steinmetz, T. Berkelman, and C. I. Kado.** 1985. Positive selection procedure for entrapment of insertion sequence elements in gram-negative bacteria. *J. Bacteriol.* **164:**918–921.
39. **Giffard, P. M., C. L. Simpson, C. P. Milward, and N. A. Jacques.** 1991. Molecular characterization of a cluster of at least two glucosyltransferase genes in *Streptococcus salivarius* ATCC 25975. *J. Gen. Microbiol.* **137:**2577–2593.
40. **Glaser, P. (Institut Pasteur, Paris).** 1991. Personal communication.
41. **Goldman, M., and H. J. Blumenthal.** 1963. Pathways of glucose in *Bacillus subtilis*. *J. Bacteriol.* **86:**303–311.
42. **Gonzales-Candelas, L., D. Ramon, and J. Polaina.** 1990. Sequences and homology analysis of two genes encoding β-glucosidases from *Bacillus polymyxa*. *Gene* **95:**31–38.
43. **Gonzy-Tréboul, G., J. H. De Vaard, M. Zagorec, and P. W. Postma.** 1991. The glucose permease of the phosphotransferase system of *Bacillus subtilis*: evidence for II^{Glc} and III^{Glc} domains. *Mol. Microbiol.* **5:**1241–1249.
44. **Hall, B. G., P. W. Betts, and J. C. Wootton.** 1989. DNA

sequence analysis of artificially evolved *ebg* enzyme and *ebg* repressor genes. *Genetics* **123:**635–648.

45. **Hancock, K. R., E. Rockman, C. A. Young, L. Pearce, I. S. Maddox, and D. B. Scott.** 1991. Expression and nucleotide sequence of the *Clostridium acetobutylicum* β-galactosidase gene cloned in *Escherichia coli. J. Bacteriol.* **173:**3084–3095.
46. **Hastrup, S.** 1988. Analysis of the *Bacillus subtilis* xylose regulon, p. 79–83. *In* A. T. Ganesan and J. A. Hoch (ed.), *Genetics and Biotechnology of Bacilli*, vol. 2 Academic Press, Inc., New York.
47. **Hodgson, D.** Primary metabolism-carbon catabolism. *In* E. M. Wellington (ed.), *Streptomyces*, in press. Plenum Biotechnology Handbooks, New York.
48. **Holmberg, C., L. Beijer, B. Rutberg, and L. Rutberg.** 1990. Glycerol catabolism in *Bacillus subtilis*: nucleotide sequence of the genes encoding glycerol kinase (*glpK*) and glycerol-3-phosphate dehydrogenase (*glpD*). *J. Gen. Microbiol.* **136:**2367–2375.
49. **Holmberg, C., and B. Rutberg.** 1991. Expression of the gene encoding glycerol-3-phosphate dehydrogenase (*glpD*) in *Bacillus subtilis* is controlled by antitermination. *Mol. Microbiol.* **5:**2891–2900.
50. **Holmberg, C., L. Rutberg, and B. Rutberg (University of Lund).** 1991. Personal communication.
51. **Houman, F., M. R. Diaz-Torres, and A. Wright.** 1990. Transcriptional antitermination in the *bgl* operon of *E. coli* is modulated by a specific RNA binding protein. *Cell* **62:**1153–1163.

51a. **Jäger, W., A. Schäfer, A. Pühler, G. Labes, and W. Wohlleben.** 1992. Expression of the *Bacillus subtilis sacB* gene leads to sucrose sensitivity in the gram-positive bacterium *Corynebacterium glutamicum* but not in *Streptomyces lividans*. *J. Bacteriol.* **174:**5462–5465.

52. **Kreuzer, P., D. Gärtner, R. Allmansberger, and W. Hillen.** 1989. Identification and sequence analysis of the *Bacillus subtilis* W23 *xylR* gene and *xyl* operator. *J. Bacteriol.* **171:**3840–3845.
53. **Kunst, F., M. Steinmetz, J. A. Lepesant, and R. Dedonder.** 1977. Presence of a third sucrose hydrolysing enzyme in *Bacillus subtilis*: constitutive levanase synthesis by mutants of *Bacillus subtilis* Marburg 168. *Biochimie* **59:**287–292.
54. **Lawlis, V. B., M. S. Dennis, E. Y. Chen, D. H. Smith, and D. J. Henner.** 1984. Cloning and sequencing of the xylose isomerase and xylulose kinase genes of *Escherichia coli. Appl. Environ. Microbiol.* **47:**15–21.
55. **Le Coq, D., A. M. Crutz, R. Richter, and M. Steinmetz.** Unpublished data.
56. **Leong-Morgenthaler, P., M. C. Zwahlen, and H. Hottinger.** 1991. Lactose metabolism in *Lactobacillus bulgaricus*: analysis of the primary structure and expression of the genes involved. *J. Bacteriol.* **173:**1951–1957.
57. **Lepesant, J. A., F. Kunst, M. Pascal, J. Lepesant-Kejzlarova, M. Steinmetz, and R. Dedonder.** 1976. Specific and pleiotropic regulatory mechanisms in the sucrose system of *Bacillus subtilis* 168, p. 58–69. *In* D. Schlessinger (ed.), *Microbiology—1976*. American Society for Microbiology, Washington, D.C.
58. **Lin, E. C. C.** 1987. Dissimilatory pathways for sugars, polyols, and carboxylates, p. 244–284. *In* F. C. Neidhart, J. L. Ingraham, K. B. Low, B. Magasanik, M. Schaechter, and H. E. Umbarger (ed.), *Escherichia coli and Salmonella typhimurium: Cellular and Molecular Biology*, vol. 1. American Society for Microbiology, Washington, D.C.
59. **Lin, E. C. C., A. J. Hacking, and J. Aguilar.** 1976. Experimental models of acquisitive evolution. *BioScience* **26:**548–555.
60. **Lindgren, V.** 1978. Mapping of a genetic locus that affects glycerol 3-phosphate in *Bacillus subtilis. J. Bacteriol.* **133:**667–670.
61. **Lindgren, V., and L. Rutberg.** 1976. Genetic control of the *glp* system in *Bacillus subtilis. J. Bacteriol.* **127:**1047–1057.
62. **Loesche, W. L.** 1986. Role of *Streptococcus mutans* in human dental decay. *Microbiol. Rev.* **50:**353–380.

62a. **Losick, R. (Harvard University).** 1992. Personal communication.

63. **Macrina, F. L., K. R. Jones, C. A. Alpert, B. M. Chassy, and S. M. Michalek.** 1991. Repeated DNA sequence involved in mutations affecting transport of sucrose in *Streptococcus mutans* V403 via the phosphoenolpyruvate phosphotransferase system. *Infect. Immun.* **59:**1535–1543.
64. **Mahadevan, S., and A. Wright.** 1987. A bacterial gene involved in transcription antitermination: regulation at a Rho-independent terminator in the *bgl* operon of *E. coli. Cell* **50:**485–494.
65. **Martin, I., M. Debarbouillé, E. Ferrari, A. Klier, and G. Rapoport.** 1987. Characterization of the levanase gene of *Bacillus subtilis* which shows homology to yeast invertase. *Mol. Gen. Genet.* **208:**177–184.
66. **Martin, I., M. Debarbouillé, A. Klier, and G. Rapoport.** 1989. Induction and metabolite regulation of levanase synthesis in *Bacillus subtilis. J. Bacteriol.* **171:**1885–1892.
67. **Martin-Verstraete, I., M. Debarbouillé, A. Klier, and G. Rapoport.** 1990. Levanase operon of *Bacillus subtilis* includes a fructose-specific phosphotransferase system regulating the expression of the operon. *J. Mol. Biol.* **214:**657–671.
68. **Minton, N. P., S. P. Chambers, W. J. Mitchell, and J. K. Brehm.** 1991. Program Abstr. 6th Int. Conf. Bacilli, abstr. M1.
69. **Mollet, B., and N. Pilloud.** 1991. Galactose utilization in *Lactobacillus helveticus*: isolation and characterization of the galactokinase (*galK*) and galactose-1-phosphate uridyl transferase (*galT*) genes. *J. Bacteriol.* **173:**4464–4473.
70. **Morse, M. L., K. L. Hill, J. B. Egan, and W. Hengstenberg.** 1968. Metabolism of lactose by *Staphylococcus aureus* and its genetic basis. *J. Bacteriol.* **95:**2270–2274.
71. **Nègre, D., J.-C. Cortay, I. G. Old, A. Galinier, C. Richaud, I. Saint-Girons, and A. J. Cozzone.** 1991. Overproduction and characterization of the *iclR* gene product of *Escherichia coli* K12 and comparison with that of *Salmonella typhimurium* LT2. *Gene* **97:**29–37.
72. **Oskouian, B., and G. C. Stewart.** 1990. Repression and catabolite repression of the lactose operon of *Staphylococcus aureus. J. Bacteriol.* **172:**3804–3812.
73. **Parker, L. L., and B. G. Hall.** 1990. Characterization and nucleotide sequence of the cryptic *cel* operon of *Escherichia coli* K12. *Genetics* **124:**455–471.
74. **Poolman, B., T. J. Royer, S. E. Mainzer, and B. F. Schmidt.** 1989. Lactose transport system of *Streptococcus thermophilus*: a hybrid protein with homology to the melibiose carrier and enzyme III of phosphoenolpyruvate-dependent phosphotransferase systems. *J. Bacteriol.* **171:**244–253.
75. **Poolman, B., T. J. Royer, S. E. Mainzer, and B. F. Schmidt.** 1990. Carbohydrate utilization in *Streptococcus thermophilus*: characterization of the genes for aldose 1-epimerase (mutarotase) and UDPglucose 4-epimerase. *J. Bacteriol.* **172:**4037–4047.
76. **Rather, P. N., and C. P. Moran, Jr.** 1988. Compartment-specific transcription in *Bacillus subtilis*: identification of the promoter for *gdh. J. Bacteriol.* **170:**5086–5092.
77. **Reizer, A., J. Deutscher, M. H. Saier, Jr., and J. Reizer.** 1991. Analysis of the gluconate (*gnt*) operon *Bacillus subtilis. Mol. Microbiol.* **5:**1081–1089.
78. **Reizer, J., A. Reizer, and M. H. Saier, Jr.** 1990. The cellobiose permease of *Escherichia coli* consists of three proteins and is homologous to the lactose permease of *Staphylococcus aureus. Res. Microbiol.* **141:**1061–1067.
79. **Reynolds, A. E., J. Felton, and A. Wright.** 1981. Inser-

tion of DNA activates the cryptic *bgl* operon in *Escherichia coli* K12. *Nature (London)* **293:**625–629.

80. **Romantschuk, M., G. Y. Richter, P. Mukhopadhyay, and D. Mills.** 1991. IS*801*, an insertion sequence element isolated from *Pseudomonas syringae* pathovar. *Mol. Microbiol.* **5:**617–622.
81. **Roncero, M. I. G.** 1983. Genes controlling xylan utilization by *Bacillus subtilis*. *J. Bacteriol.* **156:**257–263.
82. **Rosey, E. L., B. Oskouian, and G. C. Stewart.** 1991. Lactose metabolism by *Staphylococcus aureus*: characterization of *lacABCD*, the structural genes of the tagatose 6-phosphate pathway. *J. Bacteriol.* **173:**5992–5998.
83. **Saheb, S. A.** 1972. Etude de deux mutants du métabolisme du glycérol chez *Bacillus subtilis*. *Can. J. Microbiol.* **18:**1315–1325.
84. **Sa-Nogueira, I., and H. de Lancastre.** 1991. Program Abstr. 6th Int. Conf. Bacilli., abstr. T9.
85. **Sa-Nogueira, I., H. Paveia, and H. de Lancastre.** 1988. Isolation of constitutive mutants for L-arabinose utilization in *Bacillus subtilis*. *J. Bacteriol.* **170:**2855–2857.
86. **Sato, Y., F. Poy, G. R. Jacobson, and H. K. Kuramitsu.** 1989. Characterization and sequence analysis of the *scrA* gene encoding enzyme II^{Scr} of the *Streptococcus mutans* phosphoenolpyruvate-dependent sucrose phosphotransferase system. *J. Bacteriol.* **171:**263–271.
87. **Sato, Y., Y. Yamamoto, R. Suzuki, H. Kizaki, and H. K. Kuramitsu.** 1991. Construction of *scrA*::*lacZ* gene fusion to investigate regulation of the sucrose PTS of *Streptococcus mutans*. *FEMS Microbiol. Lett.* **79:**339–346.
88. **Scheler, A., T. Rygus, R. Allmansberger, and W. Hillen.** 1991. Molecular cloning, structure, promoters and regulatory elements for transcription of the *Bacillus licheniformis* encoded regulon for xylose utilization. *Arch. Microbiol.* **155:**526–534.
89. **Schmid, K., R. Ebner, K. Jahreis, J. W. Lengeler, and F. Tigemeyer.** 1991. The sugar-specific porin, ScrY, is involved in sucrose uptake in enteric bacteria. *Mol. Microbiol.* **5:**941–950.
90. **Schmidt, B. F., R. M. Adams, C. Requadt, S. Power, and S. E. Mainzer.** 1989. Expression and nucleotide sequence of the *Lactobacillus bulgaricus* β-galactosidase gene cloned in *Escherichia coli*. *J. Bacteriol.* **171:**625–635.
91. **Schnetz, K., and B. Rak.** 1988. Regulation of the bgl operon of *Escherichia coli* by transcriptional antitermination. *EMBO J.* **7:**3271–3278.
92. **Scholzen, T., and E. Arndt.** 1991. Organization and nucleotide sequence of ten ribosomal protein genes from the region equivalent to the spectinomycin operon in the archaebacterium *Halobacterium marismortui*. *Mol. Gen. Genet.* **228:**70–80.
93. **Schörgendorfer, K., H. Scharb, and R. M. Lafferty.** 1988. Molecular characterization of *Bacillus subtilis* levanase and a C-terminal deleted derivative. *J. Biotechnol.* **7:**247–258.
94. **Schroeder, C. J., C. Robert, G. Lenzen, L. L. McKay, and A. Mercenier.** 1991. Analysis of the *lacZ* sequence from two *Streptococcus thermophilus* strains: comparison with the *Escherichia coli* and *Lactococcus bulgaricus* β-galactosidase sequences. *J. Gen. Microbiol.* **137:**369–380.
95. **Schroeder, V., S. M. Michalek, and F. L. Macrina.** 1989. Biochemical characterization and evaluation of virulence of a fructosyltransferase-deficient mutant of *Streptococcus mutans* V403. *Infect. Immun.* **57:**3560–3569.
96. **Shamanna, D. K., and K. E. Sanderson.** 1979. Genetics and regulation of the D-xylose utilization in *Salmonella typhimurium* LT2. *J. Bacteriol.* **139:**71–79.
97. **Shazand, K., P. Hwang, J. Tucker, J. C. Rabinowitz, T. Leighton, and M. Grunberg-Manago.** 1990. Program. Abstr. Conf. *Bacillus subtilis* Genome, Paris, abstr. P.66.
98. **Shimotsu, H., and D. Henner.** 1986. Modulation of *Bacillus subtilis* levansucrase gene expression by sucrose and regulation of the steady-state mRNA level by *sacU* and *sacQ* genes. *J. Bacteriol.* **168:**380–388.
99. **Shiroza, T., and H. K. Kuramitsu.** 1988. Sequence analysis of the *Streptococcus mutans* fructosyltransferase gene and flanking regions. *J. Bacteriol.* **170:**810–816.
100. **Sizemore, C., E. Buchner, T. Rygus, C. Witke, F. Götz, and W. Hillen.** 1991. Organization, promoter analysis and transcriptional regulation of the *Staphylococcus xylosus* xylose utilization operon. *Mol. Gen. Genet.* **227:**377–384.
101. **Smith, C. P. (University of Manchester).** 1991. Personal communication.
102. **Smith, C. P., and K. F. Chater.** 1988. Structure and regulation of controlling sequences for the *Streptomyces coelicolor* glycerol operon. *J. Mol. Biol.* **204:**569–580.
103. **Steinmetz, M.** Unpublished data.
104. **Steinmetz, M., and S. Aymerich.** 1990. The *Bacillus subtilis sac-deg* system: how and why?, p. 303–311. *In* M. Zukowski, A. T. Ganesan, and J. A. Hoch (ed.), *Genetics and Biotechnology of Bacilli*, vol. 3. Academic Press, Inc., New York.
105. **Steinmetz, M., D. Le Coq, and S. Aymerich.** 1989. Induction by sucrose of saccharolytic enzymes in *Bacillus subtilis*: evidence for two partially interchangeable regulatory pathways. *J. Bacteriol.* **171:**1519–1523.
106. **Steinmetz, M., D. Le Coq, S. Aymerich, G. Gonzy-Tréboul, and P. Gay.** 1985. The DNA sequence of the gene for the secreted *Bacillus subtilis* enzyme levansucrase. *Mol. Gen. Genet.* **200:**220–228.
107. **Steinmetz, M., D. Le Coq, H. Ben Djemia, and P. Gay.** 1983. Analyse génétique de *sacB*, gène de structure d'une enzyme sécrétée, la lévane-saccharase de *Bacillus subtilis*. *Mol. Gen. Genet.* **191:**138–144.
108. **Strauss, N.** 1983. Role of glucose dehydrogenase in germination of *Bacillus subtilis* spores. *FEMS Microbiol. Lett.* **20:**379–384.
109. **Sutrina, S. L., P. Reddy, M. H. Saier, Jr., and J. Reizer.** 1990. The glucose permease of *Bacillus subtilis* is a single polypeptide chain that functions to energize the sucrose permease. *J. Biol. Chem.* **265:**18581–18589.
110. **Thomas, T. D., and V. L. Crow.** 1984. Selection of galactose-fermenting *Streptococcus thermophilus* in lactose-limited chemostat cultures. *Appl. Environ. Microbiol.* **48:**186–191.
111. **Thompson, J.** 1987. Regulation of sugar transport and metabolism in lactic acid bacteria. *FEMS Microbiol. Rev.* **46:**221–232.
112. **Ueda, S., T. Shiroza, and H. K. Kuramitsu.** 1988. Sequence analysis of the *gtfC* gene from *Streptococcus mutans* GC5. *Gene* **69:**101–109.
113. **Van Rooijen, R. J., and W. M. De Vos.** 1990. Molecular cloning, transcription analysis, and nucleotide sequence of *lacR*, a gene encoding the repressor of the lactose phosphotransferase system of *Lactococcus lactis*. *J. Biol. Chem.* **265:**18499–18503.
114. **Van Rooijen, R. J., S. Vanschalkwijk, and W. M. De Vos.** 1991. Molecular cloning, characterization, and nucleotide sequence of the tagatose 6-phosphate pathway gene cluster of the lactose operon of *Lactococcus lactis*. *J. Biol. Chem.* **266:**7176–7182.

114a. **Wagner, E., and R. Brückner (University of Tübingen).** 1992. Personal communication.

115. **Westpheling, J., and M. Brawner.** 1989. Two transcribing activities are involved in expression of the *Streptomyces* galactose operon. *J. Bacteriol.* **171:**1355–1361.
116. **Wong, H. C., Y. Ting, H.-C. Lin, F. Reichert, K. Myambo, K. W. K. Watt, P. T. Toy, and R. J. Drummond.** 1991. Genetic organization and regulation of the xylose degradation genes in *Streptomyces rubiginosus*. *J. Bacteriol.* **173:**6849–6858.
117. **Zagorec, M., and M. Steinmetz.** 1991. Construction of a derivative of Tn*917* containing an outward directed promoter and its use in *Bacillus subtilis*. *J. Gen. Microbiol.* **137:**107–112.

12. Glycolysis

PETER FORTNAGEL

In eubacteria, the glycolytic pathway is the central and constitutive route of carbohydrate metabolism. The reactions of glycolysis have been extensively summarized for *Escherichia coli* and *Salmonella typhimurium* (14). Although our knowledge of individual enzymes, their properties, and their catalytic activities is sufficient, data on metabolic fluxes under different nutritional conditions and the relation of these fluxes to enzymic properties are surprisingly scarce (14, 88).

In *Bacillus* spp., the situation is even more complex than in the above-mentioned two organisms, since growth and sporulation often have different metabolic requirements (29, 40, 82). There are many open questions concerning carbon flow, especially in complex media. Not only are glycolytic reactions necessary for cellular growth, but endospore formation during the post-exponential phase requires nutrients and low-molecular-weight building blocks that are often supplied by glycolytic reactions. To maintain pools of certain metabolites during growth or sporulation and simultaneously to prevent the accumulation of interfering metabolites requires delicate regulation (for a review, see reference 40).

To study the consequences of limitations and accumulations of different metabolites, *Bacillus subtilis* mutants blocked at various steps in the glycolytic pathway or at side reactions have been isolated. Their enzymic defects have been well characterized, and the influences of the metabolic blocks on growth and cellular development have been investigated. The most complete set of such mutants was collected by Ernst Freese. They were either isolated in his laboratory or donated by others. The data base of his strain collection is available from me; strains will be available in the future from the *Bacillus* Genetic Stock Center (The Ohio State University). No comparable collection exists for any other gram-positive bacterium.

The enzymic reactions of glycolysis and associated pathways are shown in Fig. 1. In this figure, the nomenclature of the Freese mutant collection is used, although some of the strains originated in other laboratories. In some cases, two mutants have been listed if it was obvious from the data base that they were in fact individual and independently isolated mutants or if they originated from a different parental strain. In other cases, a typical strain was selected from the data base. The glycolytic mutants described here are summarized in Table 1. The names of the corresponding genes and their map positions were taken from references 61 and 62. For more detailed information about pathways for utilization of sugars other than glucose, see chapter 11.

UPTAKE OF SUGARS: PEP-SUGAR PHOSPHOTRANSFERASE

The phosphoenolpyruvate (PEP)-sugar phosphotransferase system (PTS) has been previously described for a number of bacteria, both gram positive and gram negative (for reviews, see references 45, 65, 73, and 74 and chapter 10 of this volume). This enzyme system is present in *B. subtilis* (for a review, see reference 40) and is responsible for the transport of glucose (49), methyl-α-D-glucopyranoside (9, 23, 48), glucosamine (49), fructose (30, 50, 60), mannose, mannitol, and sucrose (44). Enzyme I (30, 54), Hpr (33, 34, 48), and several enzymes II (23, 30) have been characterized genetically or biochemically (or both). The structural genes for enzyme I, Hpr, and some enzymes II have been cloned (13, 33, 34, 54).

In *B. subtilis*, the inhibition by glucose of the uptake of glycerol (70) and the existence of a *crr*-like gene (33) gave the first evidence of an enzyme III^{Glc}-like function that in *E. coli* is responsible both for transporting glucose and for establishing catabolite repression of the utilization of other carbohydrates. Recent results (32) give evidence for II^{Glc} and III^{Glc} domains in the putative product of this gene, but the mechanism of their function is not clear. The transport of glucose by the PTS system yields glucose 6-phosphate (and glucose 1-phosphate) as the intracellular precursor(s) for all subsequent steps in glycolysis.

Glucose induces a specific permease (enzyme II^{Glc}) (18). The capacity for this induction declines after exponential growth ceases in nutrient sporulation medium (NSM). Therefore, glucose has to be present during growth to repress sporulation; if added at the onset of stationary phase or later, no inhibitory action of glucose is observed (18). As expected, enzyme I-deficient mutants (e.g., strain 61309) do not grow on the sugars transported by the PTS system, and their sporulation in complex media is not inhibited by these sugars.

GLYCOLYTIC REACTIONS

Phosphoglucomutase

Phosphoglucomutase converts glucose 6-phosphate to glucose 1-phosphate. This is the first step in biosynthesis of glucose-teichoic acid. *gtaC*, the gene for phosphoglucomutase, was mapped by density transfer experiments (46) close to the glycerol operon (*glpPKD*), but *gtaC* is not a component of this operon.

In mutants ($GtaC^-$) blocked in this enzyme (e.g., strains 61155 and 61370) (67), the glycosylation of

Peter Fortnagel • Abteilung für Mikrobiologie, Institut für Allgemeine Botanik, Universität Hamburg, Ohnhorststrasse 18, D-2000 Hamburg, Germany.

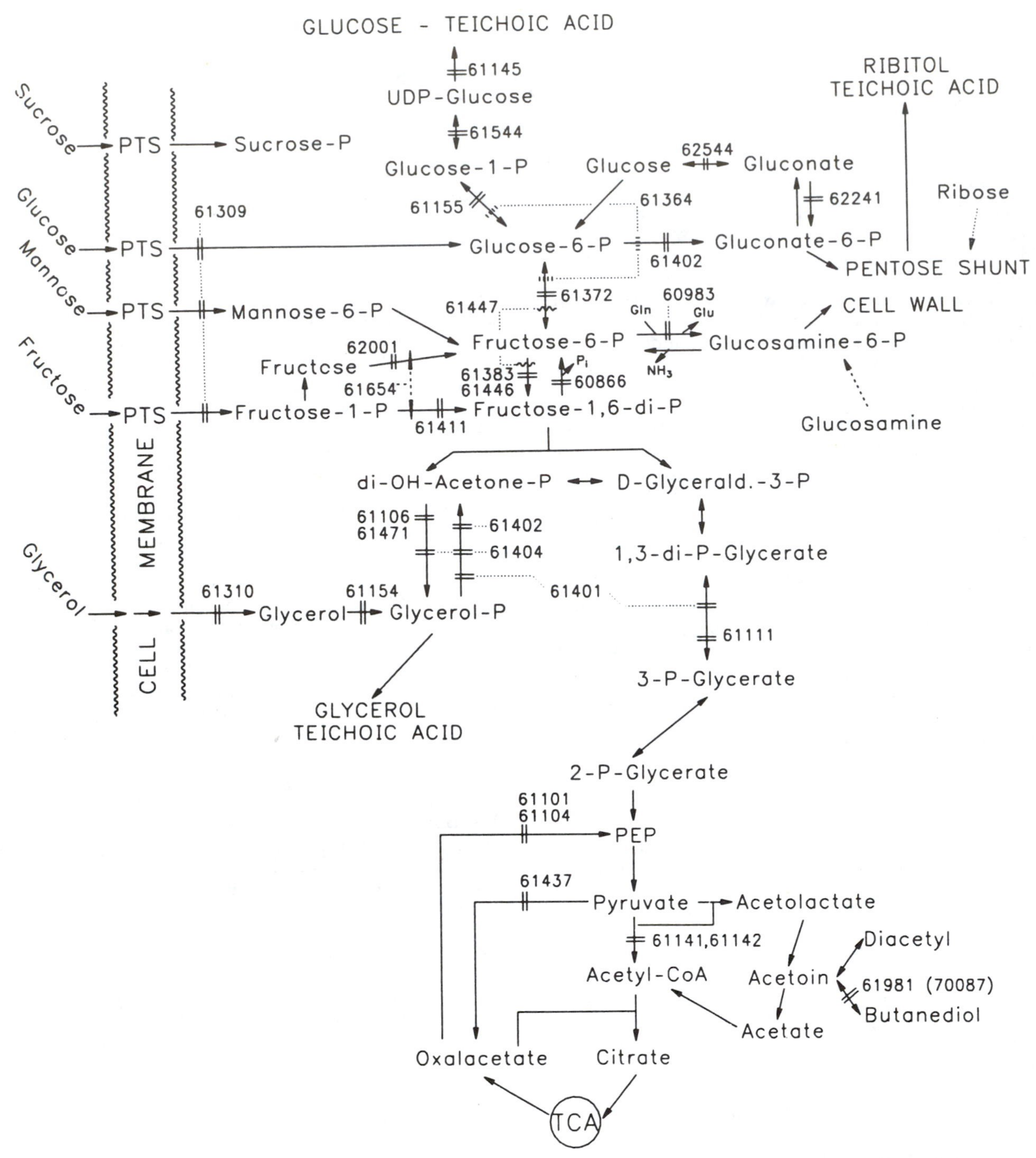

Figure 1. Glycolytic pathway and side reactions in *B. subtilis*. Numbers represent the mutants available in the Ernst Freese strain collection. The Freese mutant nomenclature is used throughout, although some of the mutants originated in different laboratories. Acetyl-CoA, acetyl coenzyme A.

teichoic acid is no longer possible. The mutants show normal growth rates in NSM, but their final cell yields are slightly reduced. These yields can be restored to normal by addition of glucose to the medium. Sporulation of this mutant is normal, as are the inhibitory influences of different sugars such as glucose, fructose, glucosamine, and glycerol.

Mutant strains accumulate glucose 1-phosphate if they are grown in the presence of galactose (67). This accumulation is highly toxic to the cells. It causes spheroplast formation and subsequent cell lysis. Using a set of double and triple mutants, Prasad and Freese demonstrated that in fact, glucose 1-phosphate is the metabolite responsible for this reaction (67). Spheroplast formation occurs because peptidoglycan synthesis is blocked. The inhibition of muramic acid

Table 1. Glycolytic enzymes and genes that encode them

Enzyme	E.C. no.	Gene	Map position (°)	Mutant(s)[a]	Reference(s)
PEP-hexose-phosphotransferase enzyme I	2.7.3.9	*ptsI*		61309	18, 29, 33, 49, 54
Phospho-carrier protein		*ptsH*	118		29, 33, 48
Enzyme II (glucose)	2.1.1.69	*ptsG*			
Enzyme II (fructose)		*fruA*	123		28, 30, 60
Enzyme II (mannose)					40
Enzyme II (sucrose)		*sacP*	330		13, 44, 50
Phosphoglucomutase	2.7.5.1	*gtaC*		61155, 61364*	17, 19, 41, 63, 67
UDP-glucose-pyrophosphorylase	2.7.7.9	*gtaB*	308	61544	63, 67
UDP-glucose-teichoic acid glycosyltransferase (teichoic acid glycosyltransferase)		*gtaA*	309	61145	63
Glucokinase	2.7.1.2				
Glucose dehydrogenase	1.1.1.47	*gdh*	34	70127	1, 6, 27, 43, 86
Gluconate kinase	2.7.1.12			62241	
Glucose-6-phosphate dehydrogenase	1.1.1.43	*gpd*		61402, 61364*	17, 19, 67
Phosphoglucoseisomerase	5.3.1.9	*pgi*		61372, 61364*, 61447%	17, 19, 41, 66, 67
Glutamine-fructose-6-phosphate aminotransferase	2.6.1.16	*gca*		60983	20
Fructokinase (mannokinase)	2.7.1.4	*fruC*	51	62001, 61654+	28
Mannose-6-phosphate isomerase	5.3.1.8				
Phosphofructokinase	2.7.1.11	*pfk*	255	61383, 61446, 61447%	17, 19, 28, 41, 66
Fructose-1,6-bisphosphatase	3.1.3.11	*fdpA*	344	60866	25
Fructose-1-phosphate kinase	2.7.1.	*fruB*	123	61411, 61654+	28, 31, 66
Fructose-1,6-bisphosphate aldolase	4.1.2.13	*tsr*			84
Triose-phosphate isomerase	5.3.1.1				41
Glycerol uptake				61310	57, 70
Glycerol kinase	2.7.1.30	*glpK*		61154	39
Glycerol-3-phosphate dehydrogenase (NAD)	1.1.1.8	*glpD*	75	61402, 61404^	46, 52
Glycerol-3-phosphate dehydrogenase (NAD independent)				61106, 61471, 61401", 61404^	22, 57
Glyceraldehyde-3-phosphate dehydrogenase	1.2.1.12	*gap*			21, 87
3-Phosphoglycerate kinase	2.7.2.3	*pgk*		61111, 61401"	19, 22, 57, 66
Phosphoglycerate mutase	2.7.5.3				56, 78, 79, 80, 88
Enolase	4.2.1.11	*eno*			10, 61, 77
Pyruvate kinase	2.7.1.40				
PEP carboxykinase	4.1.1.32			61101, 61104	11, 22
Pyruvate dehydrogenase	1.2.4.1	*pdhA*	126	61141, 61142	2, 15, 35, 37, 39a
Pyruvate carboxylase	6.4.1.1	*pycA*	149	61437	5, 11, 53
Acetoin (diacetyl) reductase				61981 (70087)	47

[a] From the E. Freese collection. %, *, +, ", and ^ indicate mutants with multiple glycolytic blocks.

production by the accumulation of glucose 1-phosphate is caused by direct inhibition of phosphoglucosamine mutase or glucosamine-1-phosphate acetyltransferase.

UDP-Glucose-Pyrophosphorylase and Teichoic Acid Glycosyltransferase

Glucose-teichoic acid synthesis from glucose 1-phosphate proceeds via UDP-glucose-pyrophosphorylase and teichoic acid glycosyltransferase (UDP-glucose-polyglycerol teichoic acid glycosyltransferase). Defects in the genes for these two enzymes (*gtaA* and *gtaB*, respectively) cause resistance to bacteriophage ϕ29 (63). Correct adsorption of phage ϕ29 requires a glycosylated poly(glycerolphosphate) (63). Mutants of the *gtaA* linkage group are resistant to ϕ29 but remain sensitive to phage ϕ25. Mutants of the *gtaA* linkage group have normal colony morphology, but *gtaB* and *gtaC* mutants show smooth colonies. Except for resistance to ϕ29, no changes in growth properties or ability to sporulate have been reported.

The *gtaA* and *gtaB* markers are closely linked (map positions 309° and 308°, respectively). *gtaC*, the structural gene for phosphoglucomutase, is unlinked to the other *gta* genes.

Glucokinase

B. subtilis has a soluble ATP-dependent glucokinase (or hexokinase). This enzyme seems to be constitutive; its specific activity does not change significantly during growth or sporulation (18), and glucose has no influence on its synthesis. The glucokinase phosphorylates glucose but not α-methylglycoside to glucose 6-phosphate.

This enzymatic activity appears to be necessary for the phosphorylation of glucose that might enter the cells at a slow rate when the PTS system is not functioning. The major function of the enzyme, however, seems to be in disaccharide metabolism.

Glucose Dehydrogenase

Glucose dehydrogenase converts glucose to gluconate, a precursor of pentoses. The glucose dehydrogenase operon has been mapped, cloned, and sequenced (43, 85, 86). The operon consists of two structural genes. The promoter-proximal open reading frame codes for a 31.5-kDa protein of unknown function. Its structure as predicted from the amino acid sequence suggests that it might be a membrane protein. The promoter-distal gene codes for the 31.5-kDa subunit of glucose dehydrogenase. As in other dehydrogenases, the active form of the enzyme is a tetramer of this subunit.

Expression of both genes occurs only after forespores have been formed and exclusively in the forespore compartment (3). Expression of the operon under the control of different promoters during growth and early during sporulation before forespores are made does not interfere with growth or sporulation.

Glucose dehydrogenase has been purified from *Bacillus megaterium* (59) and *B. subtilis* (36, 68). Although the sequence homologies are quite high (12), dissociation of the tetrameric active form into subunits and reassociation of the subunits differ significantly (36).

After nitrosoguanidine treatment, germination-negative mutants (*gerB*) devoid of glucose dehydrogenase were isolated. They do not germinate when provided with a mixture of glucose, fructose, asparagine, and KCl, but they do germinate on L-alanine (83). Inactivation of the glucose dehydrogenase gene or the preceding open reading frame by insertional mutagenesis, however, does not influence growth or sporulation. The true function of either protein is therefore not clear.

Gluconate Kinase

Gluconate is transported in *B. subtilis*, as in *E. coli*, via an H^+ symporter and is subsequently phosphorylated to gluconate 6-phosphate, which can be utilized via the pentose shunt.

A mutant devoid of gluconate kinase activity (strain 62241) exists, but no experimental data describing its growth and sporulation properties are available. The entire gluconate operon *gntRKPZ*, which encodes the transcriptional repressor of the operon (*gntR*), gluconate kinase (*gntK*), the permease (*gntP*), and the 6-phosphogluconate dehydrogenase (*gntZ*), has been cloned and sequenced (26, 69). The gluconate repressor has strikingly high homology to several putative regulatory proteins in *E. coli*. The gluconate kinase has high homology to xylulose kinase, glycerol kinase, and fucose kinase (69). The function of the *gntZ* gene product as 6-phosphogluconate dehydrogenase was postulated from the homology of the predicted gene product with other 6-phosphate-dehydrogenases from bacteria and animals (69).

Glucose-6-Phosphate Dehydrogenase

Another way into the pentose shunt is from glucose 6-phosphate to gluconate 6-phosphate. This reaction is catalyzed by glucose-6-phosphate dehydrogenase. No mutants that are blocked in the reaction have been isolated, and no data on the enzyme itself are available.

Phosphoglucoseisomerase

The interconversion of glucose 6-phosphate and fructose 6-phosphate is catalyzed by phosphoglucoseisomerase. *pgi* mutants lacking this enzyme (e.g., strain 61372) grow very slowly on glucose. Apparently, the pentose shunt, which is required under these conditions, is quite inefficient compared with glycolysis. A second function of the enzyme is to produce glucose 6-phosphate during growth on glycerol, malate, gluconate, or other compounds not otherwise interconvertible with glucose.

Glutamine-Fructose-6-Phosphate Aminotransferase and 2-Amino-2-Deoxy-D-Glucose-6-Phosphate Ketol-Isomerase

De novo synthesis of glucosamine 6-phosphate in *B. subtilis* proceeds via glutamine-fructose-6-phosphate aminotransferase with fructose 6-phosphate and glutamine as substrates. Mutants that lack glutamine-fructose-6-phosphate aminotransferase activity (i.e., the biosynthetic enzyme) (20) strictly require glucosamine for growth in minimal and complex sporulation media. Sporulation is severely blocked in these mutants. Single or multiple doses of glucosamine given at various times fail to cure the sporulation defect. When glucosamine is continuously provided in small quantities, sporelike particles that have normal spore coats but no cortex are produced. They are resistant to octanol but not to heat.

Mutants blocked in glutamine-fructose-6-phosphate aminotransferase will grow on glucosamine as sole source of carbon, since the biosynthetic but not the degradative pathway is affected. Glucosamine 6-phosphate can be generated efficiently from external glucosamine via the PTS system. Glucosamine 6-phosphate is metabolized by the deaminating 2-amino-2-deoxy-D-glucose-6-phosphate ketol-isomerase. No mutants blocked in the glucosamine-degradative pathway have been reported except for those blocked in the PTS system enzyme I, which are defective in glucosamine transport.

Fructose-1-Phosphate Kinase and Fructokinase (Mannokinase)

Genes coding for the enzymes that catalyze the three initial steps of fructose metabolism have been mapped (28). Fructose 1-phosphate is created by the PTS-driven fructose uptake system.

In the presence of ATP, fructose 1-phosphate can be phosphorylated to fructose 1,6-bisphosphate by fructose-1-phosphate kinase, the gene product of the *fruB* gene. A defect in this enzyme causes the accumulation of fructose 1-phosphate (31).

In a second reaction, free fructose can be formed in wild-type cells of *B. subtilis* by the dephosphorylation of fructose 1-phosphate. Fructokinase, the product of the *fruC* gene, phosphorylates this free intracellular fructose to fructose 6-phosphate. Since *fruB* mutants accumulate significant amounts of fructose 1-phosphate, this pathway from free fructose to fructose 6-phosphate does not appear to be very efficient.

A fructose-1-phosphate kinase-negative mutant (strain 61411) and a double mutant lacking both fructose-1-phosphate kinase and fructokinase accumulate fructose 1-phosphate. The mutants grow on glucose and sporulate with high efficiency on NSM. While these spores do not initiate germination in a medium containing fructose, glucose, and asparagine, they respond normally to alanine or to mannose-glucose-asparagine. This response shows that fructose has to be converted to fructose 1,6-diphosphate to allow germination (66), a conversion that is not possible in the mutant.

Mannose-6-Phosphate Isomerase

The mannose 6-phosphate originating from the PTS-mannose transport system is converted to fructose 6-phosphate by mannose 6-phosphate isomerase. This reaction can be used to accumulate fructose 6-phosphate in a phosphofructokinase-negative mutant (see below). No mutant or enzyme data concerning mannose-6-phosphate isomerase have been reported.

Phosphofructokinase (Fructose 6-Phosphate-1-Kinase)

Phosphofructokinase catalyzes the ATP-dependent phosphorylation of fructose 6-phosphate to fructose 1,6-bisphosphate. Mutants defective in this enzyme are not able to catabolize fructose 6-phosphate. The mutants grow on fructose, which can enter the down pathway via fructose 1-phosphate and fructose 1,6-bisphosphate and the up pathway via fructose 1-phosphate, fructose, and fructose 6-phosphate. Growth on glycerol or malate is possible but with reduced doubling times. No growth is possible with glucose or mannose as sole source of carbon. The mutants sporulate with a time delay in NSM. In the presence of glucose, sporulation is delayed for an extended time. Other sugars, such as fructose, mannose, glucosamine, or glycerol, inhibit sporulation significantly. Apparently, all conditions that cause the accumulation of fructose 6-phosphate cause this inhibition (16, 17).

Fructose-1,6-Bisphosphatase

The manganese-requiring enzyme fructose-1,6-bisphosphatase catalyzes the dephosphorylation of fructose 1,6-bisphosphate to yield fructose 6-phosphate. The enzyme has been purified, and the reaction in *B. subtilis* (24) and *Bacillus licheniformis* has been characterized biochemically (58). Whereas in enterobacteria this enzyme is required for gluconeogenic growth, this is not the case for *Bacillus* spp. A *B. subtilis* mutant with a defect in this enzyme grew on any carbon source that allowed growth of the wild-type strain (25). During hexose catabolism or during gluconeogenesis, this reaction can be bypassed. The enzyme is also not essential for sporulation. The only existing mutant (strain 60866), however, has the pleiotropic phenotype: it does not produce inositol dehydrogenase and gluconate kinase. The reason for this failure is not clear (25).

Fructose-1,6-Bisphosphate Aldolase

The enzyme in bacteria that splits fructose 1,6-bisphosphate to dihydroxyacetone phosphate and glyceraldehyde 3-phosphate is a metal-dependent class II aldolase (14). Enzyme synthesis appears to be constitutive in *B. subtilis*. Until recently, there were no known mutants of this enzyme. The gene that undoubtedly encodes aldolase was found downstream of the *spo0F* gene in a locus (*orfY-tsr*) previously identified as a site of mutations that cause rRNA synthesis to be temperature sensitive (84). Interestingly, mutations causing a similar phenotype were recently described for *E. coli* (75, 76). In neither case has the phenotype caused by the mutations been explained.

Triose-Phosphate Isomerase

Triose-phosphate isomerase interconverts dihydroxyacetone phosphate and glyceraldehyde 3-phosphate. No mutants defective for this enzyme have been reported in *Bacillus* spp.

Glycerol Kinase (Glycerol Uptake)

The metabolism of glycerol in *E. coli* (7, 8, 42), *B. subtilis* (38, 46, 51, 52, 57, 71, 72), and *Streptomyces coelicolor* (81) has been the subject of extensive studies. The uptake of glycerol occurs via a specific transport system that is induced by glycerol and glycerol 3-phosphate and repressed by PTS-driven glucose uptake. Intracellular glycerol is subsequently phosphorylated by an ATP-dependent kinase and converted to dihydroxyacetone phosphate by an NAD-independent glycerol phosphate dehydrogenase. This sequence is essential for growth on glycerol and for the establishment of the glycerol-dependent suppression of sporulation (57).

Mutants defective in the uptake of glycerol (e.g., strain 61310) or that do not exhibit glycerol kinase activity (e.g., strain 61154) fail to grow with glycerol as sole carbon source but do grow on glycerol phosphate plus glucose (46, 52, 57). In complex sporulation media, sporulation of the glycerol kinase mutant is not significantly reduced (57). The sporulation frequency of the mutant is not influenced by the addition of glycerol. In the uptake mutant (e.g., strain 61310), sporulation is reduced to about 5%, but the nature of the block is obscure, since in addition glycerol uptake being blocked, PTS-dependent transport is significantly hampered.

Addition of glycerol phosphate to complex media has a stimulatory influence on growth of both mutants. Apparently, glycerol phosphate per se can enter the cells, but no quantitative data are available (57).

Glycerol-3-Phosphate Dehydrogenase

B. subtilis has two glycerol-3-phosphate dehydrogenases. One is an NAD-dependent enzyme (E.C. 1.1.1.94) that is necessary for synthesis of glycerol phosphate during growth of the organism on glucose. Mutants blocked in this enzyme (e.g., strains 61106 and 61471) strictly require glycerol or glycerol phosphate for growth (57). The mutants sporulate normally if glycerol phosphate is supplied from external sources (glycerol phosphate or multiple small doses of glycerol to prevent glycerol repression of sporulation). When glycerol-requiring mutants reach stationary phase in nutrient sporulation medium, dramatic morphological changes occur. The cell membrane collapses and retracts from the cell wall. Simultaneously, ATP is released into the medium. This occurs in a mutant (strain 61106) that is defective in the NAD-dependent glycerol phosphate dehydrogenase and contains glycerol-teichoic acid as well as in a strain (61471) with the same block but containing ribitol teichoic acid (21). Obviously, the lack of glycerol phosphate itself and not the lack of teichoic acid is the cause for the morphological changes.

The second glycerol phosphate dehydrogenase (E.C. 1.1.99.5) is NAD independent (46). Mutants blocked in this enzyme grow on glucose as sole carbon source but will not grow on glycerol (57). In NSM containing 2.5 mM glycerol, L-α-glycerol-phosphate is accumulated intracellularly and subsequently excreted into the culture fluid. The maximal level exceeds 0.12 mM. The accumulation of L-α-glycerophosphate from external glycerol in this mutant suppresses both growth and sporulation. After the end of growth in sporulation media containing glycerol, cells develop several asymmetric septa at both cell poles. Cellular development is halted at this stage, and no spores are made.

The genes for glycerol metabolism in *B. subtilis* were initially mapped by density transfer experiments, transformation, and PBS1 transduction (46). The gene order was determined as *glpP* (a positive regulator protein), *glpK* (the structural gene for glycerol kinase), and *glpD* (the structural gene for glycerol-3-phosphate dehydrogenase). The *glpPKD* region has now been cloned as a 7-kb fragment, and the genes *glpK* and *glpD* have been sequenced (38). They code for a predicted 54-kDa glycerol kinase and a 63-kDa glycerol-3-phosphate dehydrogenase. Both proteins are in fact formed in a coupled in vitro transcription-translation system from the cloned 7-kb fragment (38).

As in *E. coli* and *B. subtilis*, the glycerol operon in *S. coelicolor* consists of three genes (here termed *gylABX*) (81). *gylA* codes for the glycerol kinase, *gylB* codes for the glycerol-3-phosphate dehydrogenase, and *gylX* codes for a protein with an unknown function that is not essential for glycerol metabolism. The transcription of this operon is glycerol induced and glucose repressed and occurs from a tandem promoter structure. A positively acting regulatory protein is postulated as the product of the *gylR* gene, which was identified directly upstream of the glycerol operon (81).

Glyceraldehyde-3-Phosphate Dehydrogenase

Glyceraldehyde-3-phosphate dehydrogenase is a key enzyme in glycolysis and gluconeogenesis. It reversibly catalyzes the oxidative phosphorylation of D-glyceraldehyde 3-phosphate to 1,3-diphosphoglycerate with NAD as cofactor.

No mutants with defects in this enzyme are available in the strain collection.

Recently, the glyceraldehyde-3-phosphate dehydrogenase genes from *B. subtilis* (87) and *Bacillus stearothermophilus* (4) have been cloned and sequenced. Similarities with eukaryotic glyceraldehyde-3-phosphate dehydrogenase subunits average about 50% for the 334-amino-acid polypeptide from *B. subtilis*. The two *Bacillus* enzymes have 80% amino acid similarity.

The key role of glyceraldehyde-3-phosphate dehydrogenase in the overall activity of the glycolytic pathway obtains not only in *Bacillus* spp. but also in *Streptococcus* spp. In streptococci, this enzyme is rate limiting if starved cells recover growth on carbohydrates such as lactose (64).

3-Phosphoglycerate Kinase

3-Phosphoglycerate kinase plays a central role in glycolysis. It represents the only connection between the hexose part of glycolysis and the PEP-pyruvate-tricarboxylic acid cycle section. In a mutant lacking 3-phosphoglycerase kinase activity (strain 61111 of the Freese collection; 19), the glycolytic pathway is separated into two disconnected sections, which Freese named the upper and lower subdivisions (19). The mutant will not grow on single carbon sources from either the upper or the lower subdivision. It requires one from each subdivision, e.g., glucose and acetate or glycerol and malate.

The properties of the mutant help reveal the direction of carbon flow in NSM during growth and sporulation. In NSM, growth of the mutant is slow, and low cell numbers are obtained (Fig. 2). Addition of carbon sources from the upper subdivision of the glycolytic pathway do not improve growth; high amounts of glucose (50 mM) even reduce the growth rate by a factor of 2. Addition of carbon sources from the lower subdivision of the glycolytic pathway improve growth considerably. This means that in NSM, the flow of carbon is from the upper to the lower subdivision. In order to grow at a high rate and to elevate cell densities, metabolites of the lower section must be produced by this flux. Since this is no longer possible in the mutant, the metabolites must be provided externally. The normal drain of the metabolites of the upper part of the pathway is blocked in any case.

Although growth in NSM without any supplements is limited, the mutant eventually sporulates well (frequency of >70%). This means that for normal sporulation in NSM, the nutrients in the medium are sufficient for both metabolic subdivisions at the low cell yields. The metabolites of the upper part that remain after the end of growth do not have inhibitory influences on sporulation. If glucose is added to the mutant in NSM, sporulation is blocked as in the standard strain. If malate is added, growth improves, but at high concentrations of malate, cell lysis is observed. Still, the cells that survive lysis sporulate.

A second mutant (strain 61401) is additionally deficient in the NAD-independent glycerol phosphate dehydrogenase. The properties of this mutant with re-

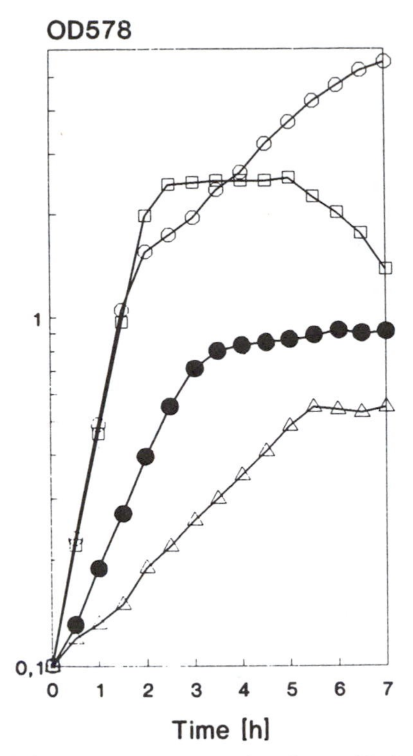

Figure 2. Influence of glucose and malate added to NSM on growth of 3-phosphoglycerate mutant 61111. The growth rate increased and growth reached a higher optical density in NSM plus 10 mM malate (□) than in NSM alone (●), but lysis occurred after the end of growth. In NSM supplemented with 10 mM malate and 50 mM glucose (○), this lysis did not occur. Addition to NSM of 50 mM glucose alone (△) caused a twofold reduction in the growth rate. OD578, optical density at 578 nm.

spect to the 3-phosphoglycerate kinase are similar to those of the strain lacking only this activity.

Phosphoglycerate Mutase

Phosphoglycerate mutase has been purified from both *B. subtilis* (88) and *B. megaterium* (80), and its biochemical properties have been investigated. The enzyme is a single polypeptide of 74 kDa (61 kDa in *B. megaterium*). It has an absolute and specific requirement for Mn^{2+}.

In the absence of added manganese, cells accumulate 3-phosphoglycerate from any carbohydrate in the medium. Growth stops early, and cells do not sporulate (56). If sufficient Mn^{2+} is added, the accumulation occurs only late in forespores. Apparently, Mn^{2+} is not present in this cell compartment at a concentration high enough to permit activity of phosphoglycerate mutase (78). This late accumulation of 3-phosphoglycerate does not prevent sporulation. In *B. megaterium*, the 3-phosphoglycerate formed in the forespore is stored in the spore. It is utilized rapidly during spore germination.

Enolase

Enolase has been purified from cells and germinated spores of *B. megaterium* in a two-step procedure. The mass of the native enzyme is 335 kDa, with the subunit having a molecular mass of 42 kDa. Mg^{2+} is an essential cofactor (77).

The *eno* gene has apparently been defined by mutations and mapped, but no published data are available (10, 61).

Pyruvate Kinase

Pyruvate kinase is constitutively found in *Bacillus* cells. It is responsible for the conversion of PEP to pyruvate. No mutants of the enzyme have been reported.

Under physiological conditions, this enzyme dephosphorylates PEP in an irreversible reaction. The reverse reaction in vitro requires high ATP concentrations that exceed the normal cellular levels by far.

Pyruvate Dehydrogenase Complex

The pyruvate dehydrogenase complexes of *B. subtilis* and *B. stearothermophilus* have been purified. Their structures are similar to those of the complexes in other bacteria and mammals (14, 35, 37). Whereas the *E. coli* enzyme complex contains three proteins (55), all other complexes consist of four proteins: E1α (42 kDa; *B. subtilis*) and E1β (36 kDa), the two components of the dehydrogenase; E2 (48 kDa), the acyl transferase component; and E3 (50 kDa), the lipoamide dehydrogenase component.

The *B. subtilis* pyruvate dehydrogenase operon has been cloned (as secretory S complex) and sequenced (35). The gene order is *pdhABCD*. *pdhA* (E1α) and *pdhB* (E1β) are transcribed from a common promoter, but *pdhC* (E2) has its own predicted promoter structure in front, though it is not clear whether this promoter or the *pdhAB* promoter is utilized for the transcription of all three genes in vivo. The occurrence in the operon of a third promoter, which lies upstream of the *pdhD* (E3) gene, is consistent with cell physiology. The pyruvate dehydrogenase complex and the oxoglutarate dehydrogenase complex are closely related enzyme complexes. Both have E3 as a constituent enzyme. Since both enzyme complexes are substrate induced, the *pdhD*-promoter makes it possible to express the last gene independently of the other genes of the *pdh* operon. The *pdhD* gene of *Staphylococcus aureus* encodes a polypeptide whose sequence is 73% identical to those of *pdhD* gene products from *B. subtilis* and *B. stearothermophilus* (34a).

Some mutants (strains 61141 and 61142) are deficient in pyruvate dehydrogenase activity. The mutation of strain 61142 has been mapped within the *pdhA* gene (i.e., within the E1α subunit) (39a). The mutants require acetate for growth in minimal media. The mutants excrete pyruvate and subsequently acetoin. Their growth rate in complex media (NSM or NSM with 0.1 M potassium phosphate, pH 6.5) is only slightly lowered (15), but the final cell yield is significantly reduced. Addition of 70 mM acetate restores normal growth rates and improves the sporulation frequency from 1×10^{-5} to 1.3×10^{-1}.

In NSM, a flux of carbon from pyruvate to acetyl coenzyme A is necessary for good growth and sporulation. This is not possible in the *pdh* mutants. As a result, sporulation is blocked unless acetate is supplied from the outside. Growth but not sporulation can be restored to normal levels with 2,3-butanediol, citrate, or glutamate but not with acetoin, malate, oxaloacetate, or pyruvate.

Pyruvate Carboxylase and PEP Carboxykinase

Constitutive pyruvate carboxylase is the major CO_2-fixing enzyme in *B. subtilis*. It is essential for growth on glucose, but it is not necessary for sporulation in complex sporulation media (11). A mutant blocked in this enzyme is unable to make oxaloacetate and therefore grows only if minimal media are supplied with citric acid cycle compounds. Growth in NSM is retarded, but the spore frequency is about normal. The wild-type enzyme is strongly activated by acetyl coenzyme A. Apparently, the flow of carbon into the tricarboxylic acid cycle is maintained in *B. subtilis* by pyruvate kinase and pyruvate carboxylase. This process is different from that in enteric bacteria, where PEP carboxylase is responsible for the formation of oxaloacetate.

PEP carboxykinase is the major enzyme controlling the reverse path from oxaloacetate to PEP. A PEP carboxykinase-negative mutant cannot grow on citric acid cycle compounds but grows normally on glucose. Such a mutant grows well in sporulation media but lyses rapidly after the end of growth. Even among surviving cells, very few heat-resistant spores are produced. Lysis can be prevented by addition of carbon sources such as gluconate. Addition of gluconate partially restores sporulation but only if multiple doses of the additional carbon source are applied whenever the cells appear to be on the verge of lysis (11). PEP carboxykinase is apparently involved in gluconeogenesis when the flow of carbon shifts from down to up, for instance, during sporulation, when substrates like acetate, acetoin, or pyruvate, which accumulate during growth, are utilized.

PEP carboxykinase activity is strongly repressed by glucose. This repression makes sense. It helps prevent a loss of ATP by the cycle of the three enzymes that convert PEP to pyruvate (pyruvate kinase), pyruvate to oxaloacetate (pyruvate carboxylase), and oxaloacetate to PEP (PEP carboxykinase).

CONCLUSIONS

By using the set of mutants given in Fig. 1 and Table 1, it is possible to interrupt glycolytic and side reactions at various positions. This causes deficiencies in various glycolytic intermediates and/or accumulation of others. Feeding of carbon sources that enter glycolysis at various steps can cure the deficiencies or enhance the accumulation. With this experimental approach, the flow of carbon during growth and sporulation and the influence of individual glycolytic metabolites have been evaluated. This took considerable effort. The data reviewed here were collected over more than 20 years in different laboratories. It is therefore understandable that the experimental approaches differed from one investigator to the other and that points of interest changed considerably with time. One would like to have information on all possible growth and sporulation conditions for all the mutants in order to get the complete picture, but in most cases, one or two carbon sources were not tested or the results were not published. Therefore, this chapter can summarize current knowledge and raise a number of outstanding issues but cannot state the definite roles of all the glycolytic reactions during growth and sporulation. Too many questions remain open for a final analysis of the interdependencies and interrelatedness of the individual steps of this central metabolic route.

Acknowledgments. I thank Ramon L. Tate from the Computer Systems Laboratory of the National Institutes of Health for his effort and his help in recovering the Ernst Freese strain collection data base from an old computer system and for transforming it to a personal-computer-compatible format.

REFERENCES

1. **Bach, J. A., and H. L. Sadoff.** 1962. Aerobic sporulating bacteria. I. Glucose dehydrogenase of *Bacillus cereus*. *J. Bacteriol.* **83:**699–707.
2. **Boudreaux, D. P., E. Eisenstadt, T. Ijima, and E. Freese.** 1981. Biochemical and genetic characterization of an auxotroph of *Bacillus subtilis* altered in the acetyl-CoA: acyl-carrier-protein transacetylase. *Eur. J. Biochem.* **115:** 175–181.
3. **Branlant, G., G. Flesch, and C. Branlant.** 1983. Molecular cloning of the glyceraldehyde-3-phosphate dehydrogenase genes of *Bacillus stearothermophilus* and *Escherichia coli* and their expression in *Escherichia coli*. *Gene* **25:**1–7.
4. **Branlant, C., T. Oster, and G. Branlant.** 1989. Nucleotide sequence determination of the DNA region coding for *Bacillus stearothermophilus* glyceraldehyde-3-phosphate dehydrogenase and of the flanking DNA regions required for its expression in *Escherichia coli*. *Gene* **75:**145–155.
5. **Buxton, R. S.** 1978. A heat-sensitive lysis mutant of *Bacillus subtilis* 168 with a low activity of pyruvate carboxylase. *J. Gen. Microbiol.* **105:**175–185.
6. **Chaudry, G. R., Y. S. Halpern, C. Saunders, N. Vasantha, J. B. Schmidt, and E. Freese.** 1984. Mapping of the glucose dehydrogenase gene in *Bacillus subtilis*. *J. Bacteriol.* **160:**607–611.
7. **Cozzarelli, N. R., W. B. Freedberg, and E. C. C. Lin.** 1968. Genetic control of the L-α-glycerolphosphate system in *Escherichia coli*. *J. Mol. Biol.* **31:**371–387.
8. **Cozzarelli, N. R., and E. C. C. Lin.** 1966. Chromosomal location of the structural gene for glycerol kinase in *Escherichia coli*. *J. Bacteriol.* **91:**1763–1766.
9. **Delobbe, A., R. Haguenauer, and G. Rapoport.** 1971. Studies on the transport of α-methyl-D-glucopyranoside in *Bacillus subtilis* 168. *Biochimie* **53:**1015–1021.
10. **Dhaese, P.** Unpublished data.
11. **Diesterhaft, M. D., and E. Freese.** 1973. Role of pyruvate carboxylase, phosphoenol-pyruvate carboxykinase, and malic enzyme during growth and sporulation of *Bacillus subtilis*. *J. Biol. Chem.* **246:**6062–6070.
12. **Fortnagel, P., K. A. Lampel, K.-D. Neitzke, and E. Freese.** 1986. Sequence homologies of glucose-dehydrogenase of *Bacillus megaterium* and *Bacillus subtilis*. *J. Theor. Biol.* **120:**489–497.
13. **Fouet, A., A. Klier, and G. Rapoport.** 1986. Nucleotide sequence of the sucrase gene of *Bacillus subtilis*. *Gene* **45:**221–225.
14. **Fraenkel, D. G.** 1987. Glycolysis, pentose phosphate pathway, and Entner-Douderoff pathway, p. 142. *In* F. C. Neidhardt, J. L. Ingraham, K. B. Low, B. Magasanik, M.

Schaechter, and H. E. Umbarger (ed.), *Escherichia coli and Salmonella typhimurium: Cellular and Molecular Biology*, vol. 1. American Society for Microbiology, Washington, D.C.

15. **Freese, E., and U. Fortnagel.** 1969. Growth and sporulation of *Bacillus subtilis* mutants blocked in the pyruvate dehydrogenase complex. *J. Bacteriol.* **99:**745–756.
16. **Freese, E., and J. E. Heinze.** 1984. Metabolic and genetic control of bacterial sporulation, p. 101–173. *In* A. Hurst, G. Gould, and J. Dring (ed.), *The Bacterial Spore*, vol. 2. Academic Press, Inc., New York.
17. **Freese, E., T. Ichikawa, K. Y. Oh, E. B. Freese, and C. Prasad.** 1974. Deficiencies or excess of metabolites interfering with differentiation. *Proc. Natl. Acad. Sci. USA* **71:**4188–4193.
18. **Freese, E., W. Klofat, and E. Galliers.** 1970. Commitment to sporulation and induction of glucose-phosphoenolpyruvate transferase. *Biochim. Biophys. Acta* **222:**265–289.
19. **Freese, E., Y. K. Oh, E. B. Freese, M. D. Diesterhaft, and C. Prasad.** 1972. Suppression of sporulation of *Bacillus subtilis*, p. 212. *In* H. O. Halvorson, R. S. Hanson, and L. L. Campbell (ed.), *Spores V*. American Society for Microbiology, Washington, D.C.
20. **Freese, E. B., R. M. Cole, W. Klofat, and E. Freese.** 1970. Growth, sporulation, and enzyme defects of glucosamine mutants of *Bacillus subtilis*. *J. Bacteriol.* **101:**1046–1062.
21. **Freese, E. B., and Y. K. Oh.** 1974. Adenosine 5′-triphosphate release and membrane collapse in glycerol-requiring mutants of *Bacillus subtilis*. *J. Bacteriol.* **120:**507–515.
22. **Freese, E. B., N. Vasantha, and E. Freese.** 1979. Induction of sporulation in developmental mutants of *Bacillus subtilis*. *Mol. Gen. Genet.* **170:**67–74.
23. **French, A., and S. Chang.** 1978. The phosphoenolpyruvate:methyl-α-D-glucoside phosphotransferase system in *Bacillus subtilis* Marburg: kinetic studies of enzyme II and evidence for a phosphoryl enzyme II intermediate. *Biochimie* **60:**1283–1287.
24. **Fujita, Y., and E. Freese.** 1979. Purification and properties of fructose-1,6-bisphosphatase of *Bacillus subtilis*. *J. Biol. Chem.* **254:**5340–5349.
25. **Fujita, Y., and E. Freese.** 1981. Isolation of a *Bacillus subtilis* mutant unable to produce fructose-bisphosphatase. *J. Bacteriol.* **145:**760–767.
26. **Fujita, Y., T. Fujita, Y. Miwa, J. Nihashi, and Y. Aratani.** 1986. Organization and transcription of the gluconate operon, gnt, of *Bacillus subtilis*. *J. Biol. Chem.* **261:**13744–13753.
27. **Fujita, Y., R. Ramaley, and E. Freese.** 1977. Location and properties of glucose dehydrogenase in sporulating cells and spores of *Bacillus subtilis*. *J. Bacteriol.* **132:**282–293.
28. **Gay, P., H. Chalumeau, and M. Steinmetz.** 1983. Chromosomal localization of *gut*, *fruC*, and *pfk* mutations affecting genes involved in *Bacillus subtilis* D-glucitol catabolism. *J. Bacteriol.* **153:**1133–1137.
29. **Gay, P., P. Cordier, M. Marquet, and A. Delobbe.** 1973. Carbohydrate metabolism and transport in *Bacillus subtilis*. A study of ctr mutations. *Mol. Gen. Genet.* **121:**355–368.
30. **Gay, P., and A. Delobbe.** 1977. Fructose transport in *Bacillus subtilis*. *Eur. J. Biochem.* **79:**363–373.
31. **Gay, P., and G. Rapoport.** 1970. Etude des mutants depourvus de fructose-1-phosphate-kinase chez *B. subtilis*. *C.R. Acad. Sci.* **271:**374–377.
32. **Gonzy-Tréboul, G., J. H. Dewaard, M. Zagorec, and P. W. Postma.** 1991. The glucose permease of the phosphotransferase system of *Bacillus subtilis*—evidence for IIglc and IIIglc domains. *Mol. Microbiol.* **5:**1241–1249.
33. **Gonzy-Tréboul, G., and M. Steinmetz.** 1987. Phosphoenolpyruvate:sugar phosphotransferase system of *Bacillus subtilis*: cloning of the region containing the *ptsH* and *ptsI* genes and evidence for a *crr*-like gene. *J. Bacteriol.* **169:**2287–2290.
34. **Gonzy-Tréboul, G., M. Zagorec, M.-C. Rain-Guion, and M. Steinmetz.** 1989. Phosphoenolpyruvate:sugar phosphotransferase system of *Bacillus subtilis*: nucleotide sequence of *ptsX*, *ptsH* and 5′-end of *ptsI* and evidence for a *ptsHI* operon. *Mol. Microbiol.* **3:**103–112.

34a. **Hemilä, H.** 1991. Lipoamide dehydrogenase of *Staphylococcus aureus*: nucleotide sequence and sequence analysis. *Biochim. Biophys. Acta* **1129:**119–123.

35. **Hemilä, H., A. Plava, L. Paulin, S. Arvidson, and I. Plava.** 1990. Secretory S complex of *Bacillus subtilis*: sequence analysis and identity to pyruvate dehydrogenase. *J. Bacteriol.* **172:**5052–5063.
36. **Hilt, W., G. Pfleiderer, and P. Fortnagel.** 1991. Glucose dehydrogenase from *Bacillus subtilis* expressed in *Escherichia coli*. I. Purification, characterization and comparison with glucose dehydrogenase from *Bacillus megaterium*. *Biochim. Biophys. Acta* **1076:**298–304.
37. **Hodgson, J. A., P. N. Lowe, and R. N. Perham.** 1983. Wild-type and mutant forms of the pyruvate dehydrogenase multienzyme complex from *Bacillus subtilis*. *Biochem. J.* **211:**463–472.
38. **Holmberg, C., L. Beijer, B. Rutberg, and L. Rutberg.** 1990. Glycerol catabolism in *Bacillus subtilis*: nucleotide sequence of the genes encoding glycerol kinase (*glpK*) and glycerol-3-phosphate dehydrogenase (*glpD*). *J. Gen. Microbiol.* **136:**2367–2375.
39. **Holmberg, C., and B. Rutberg.** 1989. Cloning of the glycerol kinase gene of *Bacillus subtilis*. *FEMS Microbiol. Lett.* **58:**1151–1156.

39a. **Jin, S., and A. L. Sonenshein.** Personal communication.

40. **Klier, A. F., and G. Rapoport.** 1988. Genetics and regulation of carbohydrate catabolism in Bacillus. *Annu. Rev. Microbiol.* **42:**65–95.
41. **Klofat, W., G. Picciolo, E. Chappelle, and E. Freese.** 1969. Production of adenosine triphosphate in normal cells and sporulation mutants of *Bacillus subtilis*. *J. Biol. Chem.* **244:**3270–3276.
42. **Koch, J. P., S. Hayashi, and E. C. C. Lin.** 1964. The control of dissimilation of glycerol and L-α-glycerolphosphate in *Escherichia coli*. *J. Biol. Chem.* **239:**3106–3108.
43. **Lampel, K. A., B. Uratani, G. R. Chaudry, R. F. Ramaley, and S. Rudikoff.** 1986. Characterization of the developmentally regulated *Bacillus subtilis* glucose dehydrogenase gene. *J. Bacteriol.* **166:**238–243.
44. **Lepesant, J.-A., and R. Dedonder.** 1968. Transport du saccharose chez *B. subtilis*. *C.R. Acad. Sci.* **267:**1109–1112.
45. **Lin, E. C. C.** 1970. The genetics of bacterial transport. *Annu. Rev. Genet.* **4:**225–262.
46. **Lindgren, V., and L. Rutberg.** 1974. Glycerol metabolism in *Bacillus subtilis*: gene enzyme relationships. *J. Bacteriol.* **119:**431–442.
47. **Lopez, J., B. Thoms, and P. Fortnagel.** 1973. Mutants of *Bacillus subtilis* blocked in acetoin reductase. *Eur. J. Biochem.* **40:**479–483.
48. **Marquet, M., M. C. Creignou, and R. Dedonder.** 1976. The phosphoenolpyruvate methyl-α-D-glycoside phosphotransferase system in *Bacillus subtilis* Marburg: purification and identification of the phosphocarrier protein (Hpr). *Biochimie* **58:**435–441.
49. **Marquet, M., M. C. Wagner, and R. Dedonder.** 1971. Separation of components of the phosphoenolpyruvate-glucose-phosphotransferase system from *Bacillus subtilis* Marburg. *Biochimie* **53:**1131–1134.
50. **Marquet, M., M.-C. Wagner, A. Delobbe, P. Gay, and G. Rapoport.** 1970. Mise en evidence de systeme de phosphotransferases dans le transport du glucose, du fructose et du saccharose chez *B. subtilis*. *C.R. Acad. Sci.* **271:**449–452.
51. **Mindich, L.** 1968. Pathway for oxidative dissimilation of glycerol in *Bacillus subtilis*. *J. Bacteriol.* **96:**565–566.

52. **Mindich, L.** 1970. Membrane synthesis in *Bacillus subtilis*: isolation and properties of strains bearing mutations in glycerol metabolism. *J. Mol. Biol.* **49:**415–432.
53. **Mueller, J. P., and H. W. Taber.** 1988. Genetic regulation of cytochrome aa3 in *Bacillus subtilis*, p. 91. *In* A. T. Ganesan and J. A. Hoch (ed.), *The Genetics and Biotechnology of Bacilli*. Academic Press, Inc., San Diego, Calif.
54. **Niaudet, B., P. Gay, and R. Dedonder.** 1975. Identification of the structural gene of PEP-phosphotransferase enzyme I in *Bacillus subtilis* Marburg. *Mol. Gen. Genet.* **136:**337–349.
55. **Nimmo, H. G.** 1987. The tricarboxylic acid cycle and anapleurotic reactions, p. 156. *In* F. C. Neidhardt, J. L. Ingraham, K. B. Low, B. Magasanik, M. Schaechter, and H. E. Umbarger (ed.), *Escherichia coli and Salmonella typhimurium: Cellular and Molecular Biology*, vol. 1. American Society for Microbiology, Washington, D.C.
56. **Oh, Y., and E. Freese.** 1976. Manganese requirement of phosphoglycerate mutase and its consequence for growth and sporulation of *Bacillus subtilis*. *J. Bacteriol.* **127:**739–746.
57. **Oh, Y., E. B. Freese, and E. Freese.** 1973. Abnormal septation and inhibition of sporulation by accumulation of L-α-glycerolphosphate in *Bacillus subtilis* mutants. *J. Bacteriol.* **113:**1034–1045.
58. **Oppenheim, D. J., and R. W. Bernlohr.** 1975. Purification and regulation of fructose-1,6-bisphosphatase from *Bacillus licheniformis*. *J. Biol. Chem.* **250:**3024–3033.
59. **Pauly, H. E., and G. Pfleiderer.** 1975. D-Glucose dehydrogenase from *Bacillus megaterium* M1286. Purification, properties, and structure. *Hoppe Seyler's Physiol. Chem.* **365:**1613–1623.
60. **Perret, J., and P. Gay.** 1979. Kinetic study of a phosphoryl exchange reaction between fructose and fructose-1-phosphate catalysed by the membrane bound enzyme II of the phosphoenolpyruvate:fructose-1 phosphotransferase system of Bacillus subtilis. *Eur. J. Biochem.* **102:**237–246.
61. **Piggot, P. J.** 1989. Revised genetic map of *Bacillus subtilis* 168, p. 1. *In* I. Smith, R. A. Slepecky, and P. Setlow (ed.), *Regulation of Procaryotic Development*. American Society for Microbiology, Washington, D.C.
62. **Piggot, P. J., M. Amjad, J.-J. Wu, H. Sandoval, and J. Castro.** 1990. Genetic and physical maps of *Bacillus subtilis*, p. 493. *In* C. R. Harwood and S. M. Cutting (ed.), *Molecular Biological Methods for Bacillus*. Wiley, Chichester, England.
63. **Pooley, H. M., D. Paschoud, and D. Karamata.** 1987. The gtaB marker in *Bacillus subtilis* 168 is associated with a deficiency in UDP-glucose pyrophosphorylase. *J. Gen. Microbiol.* **133:**3481–3493.
64. **Poolman, B., B. Bosman, J. Kiers, and W. N. Konings.** 1987. Control of glycolysis by glyceraldehyde-3-phosphate dehydrogenase in *Streptococcus cremoris* and *Streptococcus lactis*. *J. Bacteriol.* **169:**5887–5890.
65. **Postma, P. W.** 1987. Phosphotransferase system for glucose and other sugars, p. 127. *In* F. C. Neidhardt, L. J. Ingraham, K. B. Low, B. Magasanik, M. Schaechter, and H. E. Umbarger (ed.), *Escherichia coli and Salmonella typhimurium*: *Cellular and Molecular Biology*, vol. 1. American Society for Microbiology, Washington, D.C.
66. **Prasad, C., M. Diesterhaft, and E. Freese.** 1972. Initiation of spore germination in glycolytic mutants of *Bacillus subtilis*. *J. Bacteriol.* **110:**321–328.
67. **Prasad, C., and E. Freese.** 1974. Cell lysis of *Bacillus subtilis* caused by intracellular accumulation of glucose-1-phosphate. *J. Bacteriol.* **118:**1111–1122.
68. **Ramaley, R. F., and N. Vasantha.** 1983. Glycerol protection and purification of *Bacillus subtilis* glucose dehydrogenase. *J. Biol. Chem.* **258:**12558–12565.
69. **Reizer, A., J. Deutsche, M. H. Saier, and J. Reizer.** 1991. Analysis of the gluconate (*gnt*) operon of *Bacillus subtilis*. *Mol. Microbiol.* **5:**1081–1089.
70. **Reizer, J., M. J. Novotny, I. Stuiver, and M. H. Saier, Jr.** 1984. Regulation of glycerol uptake by the phosphoenolpyruvate sugar phosphotransferase system in *Bacillus subtilis*. *J. Bacteriol.* **159:**243–250.
71. **Saheb, S. A.** 1972. Etude de deux mutants du metabolisme du glycerol chez *Bacillus subtilis*. *Can. J. Microbiol.* **18:**1315–1325.
72. **Saheb, S. A.** 1972. Permeation du glycerol et sporulation chez *Bacillus subtilis*. *Can. J. Microbiol.* **18:**1307–1313.
73. **Saier, M. H.** 1985. Mechanisms and regulation of carbohydrate transport in bacteria. Academic Press, Inc., New York.
74. **Saier, M. H., R. D. Simoni, and H. Roseman.** 1970. The physiological behaviour of enzyme I and heat stable protein mutant of bacterial phosphotransferase system. *J. Biol. Chem.* **245:**5870–5875.
75. **Singer, M., P. Rossmiessl, B. M. Cali, H. Liebke, and C. A. Gross.** 1991. The *Escherichia coli ts*8 mutation is an allele of *fda*, the gene encoding fructose-1,6-diphosphate aldolase. *J. Bacteriol.* **173:**6242–6248.
76. **Singer, M., W. A. Walter, B. M. Cali, P. Rouviere, H. H. Liebke, R. L. Gourse, and C. A. Gross.** 1991. Physiological effects of the fructose-1,6-diphosphate aldolase *ts*8 mutation on stable RNA synthesis in *Escherichia coli*. *J. Bacteriol.* **173:**6249–6257.
77. **Singh, R. P., and P. Setlow.** 1978. Enolase from spores and cells of *Bacillus megaterium*: two-step purification of the enzyme and some of its properties. *J. Bacteriol.* **134:**353–355.
78. **Singh, R. P., and P. Setlow.** 1978. Phosphoglycerate mutase in developing forespores of *Bacillus megaterium* may be regulated by the intrasporal level of free manganous ion. *Biochem. Biophys. Res. Commun.* **82:**1–5.
79. **Singh, R. P., and P. Setlow.** 1979. Regulation of phosphoglycerate phosphomutase in developing forespores and dormant and germinated spores of *Bacillus megaterium* by the level of free manganous ions. *J. Bacteriol.* **139:**889–898.
80. **Singh, R. P., and P. Setlow.** 1979. Purification and properties of phosphoglycerate phosphomutase from spores and cells of *Bacillus megaterium*. *J. Bacteriol.* **137:**1024–1027.
81. **Smith, C. P., and K. F. Chater.** 1988. Structure and regulation of controlling sequences for the *Streptomyces coelicolor* glycerol operon. *J. Mol. Biol.* **204:**569–580.
82. **Sonenshein, A. L.** 1989. Metabolic regulation of sporulation and other stationary phase phenomena, p. 109. *In* I. Smith, R. A. Slepecky, and P. Setlow (ed.), *Regulation of Procaryotic Development*. American Society for Microbiology, Washington, D.C.
83. **Strauss, N.** 1983. Role of glucose dehydrogenase in germination of *Bacillus subtilis*. *FEMS Microbiol. Lett.* **20:**379–384.
84. **Trach, K., J. W. Chapman, P. Piggot, D. LeCoq, and J. A. Hoch.** 1988. Complete sequence and transcriptional analysis of the *spo0F* region of the *Bacillus subtilis* chromosome. *J. Bacteriol.* **170:**4194–4208.
85. **Uratani, B., K. A. Lampel, R. H. Lipsky, and E. Freese.** 1984. Characterization of the gene for glucose dehydrogenase and flanking genes of *Bacillus subtilis*, p. 71. *In* J. A. Hoch and P. Setlow (ed.), *Biology of Microbial Differentiation*. American Society for Microbiology, Washington, D.C.
86. **Vasantha, N., B. Uretani, R. F. Ramaley, and E. Freese.** 1983. Isolation of a developmental gene of *Bacillus subtilis* and its expression in *Escherichia coli*. *Proc. Natl. Acad. Sci. USA* **80:**785–789.
87. **Viaene, A., and P. Dhaese.** 1989. Sequence of the glyceraldehyde-3-phosphate dehydrogenase gene from *Bacillus subtilis*. *Nucleic Acids Res.* **17:**1251.
88. **Watabe, K., and E. Freese.** 1979. Purification and properties of the manganese-dependent phosphoglycerate mutase of *Bacillus subtilis*. *J. Bacteriol.* **137:**773–778.

13. The Krebs Citric Acid Cycle

LARS HEDERSTEDT

The citric acid cycle (CAC) has several functions in aerobic bacteria. Together with the pyruvate dehydrogenase multienzyme complex (PDHC), it completely oxidizes pyruvate and provides membrane-bound respiratory systems with reducing equivalents. Another important function is to supply intermediates, e.g., 2-oxoglutarate, succinyl coenzyme A (CoA), and oxaloacetate, for anabolism. Several intermediates of the cycle are also generated by peripheral catabolic processes such as amino acid degradation. During each complete turn of the CAC, one acetyl group in the form of acetyl-CoA enters the cycle by being condensed to oxaloacetate, yielding citrate and CoA (Fig. 1). In eight steps, involving the actions of 10 enzymes, the two carbon atoms of the entering acetyl group are released as CO_2 and oxaloacetate is regenerated. Reducing equivalents in the form of NAD(P)H are generated in the steps with pyruvate, isocitrate, 2-oxoglutarate, and malate as substrates. In the succinate-to-fumarate step, two electrons are transferred directly to the respiratory chain, i.e., to menaquinone.

All enzymes of the CAC, except the succinate:menaquinone oxidoreductase, which is tightly bound to the cytoplasmic membrane and also functions as an integral part of the respiratory chain, can be isolated as water-soluble proteins. In vivo, different CAC enzymes may be arranged in organized aggregates, or "metabolons," which favor transport of intermediates between the active sites of successive enzymes (154). Such multienzyme clusters made up of fumarase, malate dehydrogenase, citrate synthase, aconitase, and isocitrate dehydrogenase have been found in lysates of several bacterial species, including *Bacillus subtilis* (8). Whether individual "soluble" CAC enzymes have nonrandom localizations within the bacterial cell could be studied by immunoelectron microscopy.

Several so-called anaplerotic reactions serve to balance the CAC when intermediates are drained off or supplied by connecting anabolic and catabolic pathways. Two examples involve pyruvate carboxylase and malic enzyme. Pyruvate carboxylase (28, 159), which is strongly activated by acetyl-CoA, helps replenish oxaloacetate by catalyzing the reaction pyruvate + CO_2 + ATP $\rightarrow$ oxaloacetate + ADP + P_i. Malic enzyme converts malate into pyruvate by catalyzing the reaction malate + $NAD(P)^+ \leftrightarrow$ pyruvate + CO_2 + NAD(P)H + H^+. Both NAD and NADP can be used by the purified *B. subtilis* malic enzyme, but NAD is a more-efficient substrate (28). The major direction of the reaction catalyzed by malic enzyme is in vivo from malate to pyruvate. Oxidation of malate to oxaloacetate can be carried out by malic enzyme together with pyruvate carboxylase, thus bypassing malate dehydrogenase. However, this bypass is not efficient, as is demonstrated by the poor growth of malate dehydrogenase mutants on malate (28).

An overview of the biochemistry and genetics of CAC enzymes in *B. subtilis* is presented in this chapter. Data on other *Bacillus* species and other genera of gram-positive bacteria are provided for comparisons and in cases where information on *B. subtilis* is lacking. As will be evident from this chapter, our knowledge about different CAC enzymes in *B. subtilis* is very unbalanced. For reviews on various aspects of the CAC cycle, see references 91, 153, 156, and 162.

CAC IN *B. SUBTILIS*

Physiology

B. subtilis, being a strict aerobe, runs a complete CAC, as is demonstrated by enzyme activity measurements with cell extracts or purified enzymes and by the ability of this bacterium to grow on most of the intermediates of the CAC as sole carbon source. *B. subtilis* transformable Marburg strains grow in minimal salts medium N (43) supplemented with one of the following carboxylic acids: pyruvate, citrate, glutamate (converted to 2-oxoglutarate by the action of glutamate dehydrogenase), fumarate, malate, or oxaloacetate (39). Acetate does not sustain growth of *B. subtilis*, and some strains seem unable to grow on succinate, although both of these compounds can be transported into the cell and metabolized (15, 39, 43a). Oxidation of [^{14}C]acetate to $^{14}CO_2$ is efficient only in *Bacillus* cells of early stationary growth phase, as shown for *Bacillus cereus* (59). Acetyl-CoA formation from acetate can be catalyzed by acetokinase and phosphotransacetylase (both of these enzyme activities are found in early-stationary-phase cells [39]) or by acetyl-CoA synthase. The presence of the latter enzyme in *B. subtilis* is indicated by cloning and characterization of the *acsA* gene located at 263° on the chromosome (52). Inability of *B. subtilis* to grow on acetate as sole carbon source is probably due to the absence of a glyoxylate shunt like that in *Escherichia coli* (116) or of a similar pathway that salvages carbon atoms of acetate from being completely oxidized to CO_2 and thereby provides carbon for anabolism (34, 43a).

The levels of enzymes of the CAC, as determined by the specific activities of cell extracts of *B. subtilis* and *B. cereus*, peak when batch cultures enter stationary growth phase (39, 58, 119, 120). This peak is pronounced in cells cultured on media that allow sporulation to occur and is correlated with the onset of sporulation. Peaks in citrate synthase, aconitase, and

Lars Hederstedt • Department of Microbiology, University of Lund, Sölvegatan 21, S-223 62 Lund, Sweden.

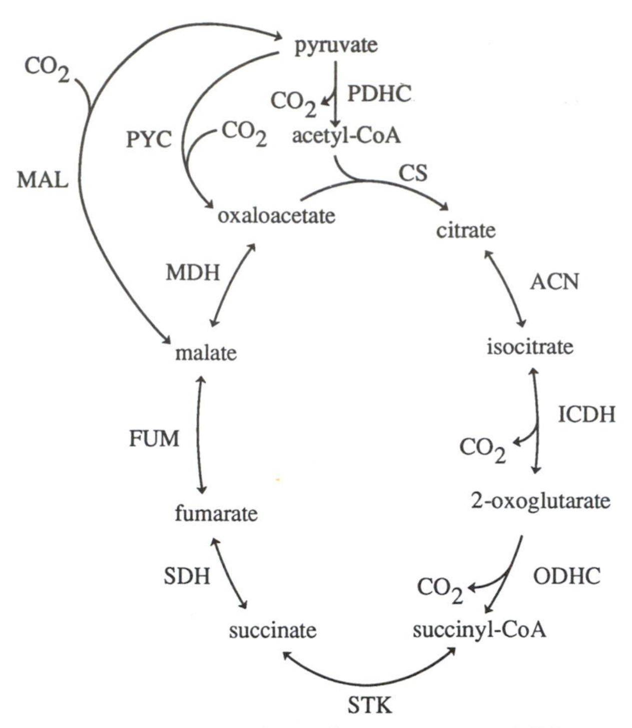

Figure 1. Krebs CAC with anaplerotic reactions. ACN, aconitase; CS, citrate synthase; FUM, fumarase; ICDH, isocitrate dehydrogenase; MAL, malic enzyme; MDH, malate dehydrogenase; ODHC, 2-oxoglutarate dehydrogenase complex; PDHC, pyruvate dehydrogenase complex; PYC, pyruvate carboxylase; SDH, succinate dehydrogenase (succinate:menaquinone reductase); STK, succinate thiokinase.

2-oxoglutarate dehydrogenase multienzyme complex (ODHC) activities also occur in cells growing in minimal glucose medium if the drug decoyinine is added to the culture (158). Whether this surge of CAC enzymes is absolutely required for sporulation to proceed is uncertain (152, 158). Mutants with CAC defects, however, generally do not sporulate normally (119).

Genetics

An intact CAC is not required for vegetative growth of *B. subtilis*. Mutants defective in each step of the CAC, except for that catalyzed by succinate thiokinase, have been isolated (15, 39, 43, 43a, 71, 82, 97, 121, 147). After radiation, chemical, or heat mutagenesis of vegetative cells or spores, the first selection in most screenings for CAC mutants has been defective sporulation on plates (colony morphology and pigment formation) (43, 147) or deficient metabolism of amino acids in the presence of a low concentration of glucose (acidification of the medium around the mutant colony assessed by using a pH indicator in the plate or by seeing a halo on plates with calcium carbonate) (15, 71). Potential CAC-defective mutants can then be found among those unable to grow on minimal medium with lactate. To identify the specific CAC defect in mutant clones, it is convenient to use zymogram staining of colonies (71). Dissimilation analysis with, for instance, [^{14}C]glutamate or [^{14}C]succinate is often informative (15, 39). The CAC intermediate preceding the block in the mutant will be accumulated and is resolved by, e.g., thin-layer chromatography followed by autoradiography. Finally, the enzymatic defect(s) is confirmed by in vitro activity measurements of cell extracts. It is not known why mutants defective in succinate thiokinase have escaped isolation. It is not likely that such mutants would be unconditionally lethal, since *E. coli* succinate thiokinase-deficient mutants can grow aerobically on succinate, fumarate, and malate by generating succinyl-CoA from 2-oxoglutarate (100).

As in other systems, different mutants have proved to be extremely valuable tools for molecular genetic studies of the genes encoding CAC enzymes (gene organization and control of expression), for investigations of the structures and functions of individual enzymes, and for physiological in vivo studies.

Gene-enzyme relationships have been elucidated largely through the use of mutants, and the assignments have in the cases of PDHC (*pdhABCD*), citrate synthase (*citA*), aconitase (*citB*), ODHC (*odhAB*), succinate:menaquinone reductase (*sdhCAB*), and fumarase (*citG*) been confirmed by DNA sequence analysis of cloned genes. Table 1 summarizes the present state of knowledge concerning structural genes for CAC enzymes in *B. subtilis*. The genes or genetic loci for all the enzymes of the CAC except *citE* (succinate thiokinase) have been mapped on the chromosome as indicated in Fig. 2. The *citB* gene has been located on the physical map of *B. subtilis* 168 within the 730-kb *Sfi*I fragment SF1 (7), which is identical to fragment AS in the map constructed by Itaya and Tanaka (87). Clearly, the structural genes are not linked on the chromosome, with the exception of *citC* and *citH*, which are close but apparently not adjacent (49). The organization of the loci on the chromosome is different from that in *E. coli*, where the genes encoding citrate synthase (*gltA*), succinate:quinone reductase (*sdhCDAB*), the unique subunits of ODHC (*sucAB*), and succinate thiokinase (*sucCD*) are clustered (for a review, see reference 108). However, the structural genes for the subunits of CAC enzyme complexes in *B. subtilis* are, as in *E. coli*, organized in operons, i.e., *pdhABCD*, *odhAB*, and *sdhCAB* (Fig. 2).

Mutants defective in the anaplerotic enzyme pyruvate carboxylase have also been isolated; they require a CAC intermediate or aspartate for growth (14a, 28, 83). Pyruvate carboxylase is encoded by the *pycA* (*aspA*) gene located at 149° (128). The terminal part of *pycA* has been cloned on a fragment containing the *ctaA* gene (114).

Control: General Aspects

Because of the amphibolic character of the CAC, its control is rather complex and is far from understood at the molecular level in *Bacillus* species. Enzyme activities of individual CAC enzymes found in cell extracts vary depending on growth conditions (e.g., medium composition), growth stage at which the cells are harvested, and particular strain or mutant analyzed. Whether the differences observed in enzyme activities reflect allosteric or other regulation directly on the enzyme or variations in cellular concentrations

Table 1. Genes for CAC enzymes of *B. subtilis*

Enzyme	Gene symbol	Map position (°)[a]	Sequence data available	Accession no.	Reference(s)
PDHC					
Pyruvate decarboxylase (E1p)	*pdhAB*	126	+	M57435	80
Dihydrolipoamide acetyltransferase (E2p)	*pdhC*	126	+	M57435	80
Lipoamide dehydrogenase (E3)	*pdhD*	127	+	M54735	80
Citrate synthase	*citA1*	90	+	?	88, 89
	citA2	260	+	?	88, 89
Aconitase	*citB*	173	+ (partial)	M16776	30
Isocitrate dehydrogenase	*citC*	259	– (cloned)		126
ODHC					
2-Oxoglutarate decarboxylase (E1o)	*odhA*	181	+	X54805	136
Dihydrolipoamide succinyltransferase (E2o)	*odhB*	181	+	M27141	18
Lipoamide dehydrogenase (E3)	*pdhD*	127	+	M57435	80
Succinate thiokinase	*citE*	?	–		
Succinate:menaquinone reductase					
Fp subunit	*sdhA*	252	+	M16753	127
Ip subunit	*sdhB*	252	+	M16753	127
Cytochrome b_{558} subunit	*sdhC*	252	+	M16753	96
Fumarase	*citG*	289	+	X01701	107
Malate dehydrogenase	*citH*	259	–		

[a] Map positions are according to those given in references 88 and 128.

of enzyme as the result of variations in the rate of enzyme synthesis or degradation is often ambiguous in the literature.

By way of illustration, let us consider the effect on CAC enzymes of growing *B. subtilis* cells in three different media: (i) NB, a broth medium containing amino acids and peptides but poor in rapidly metabolizable carbohydrate; (ii) NB plus a high concentration of glucose; and (iii) minimal salts medium with glucose. Growth in NB results in general derepression of CAC enzymes. Under these conditions, the cycle mainly has the role of energy production, since most intermediates for anabolic reactions are provided in the medium. At the end of exponential growth, when the more rapidly metabolizable carbon sources have been exhausted, there is a further peak in CAC enzyme activities as metabolites such as acetate and acetoin are converted to acetyl-CoA and oxidized by the CAC to generate energy for postexponential-phase processes like sporulation. Growth in NB-glucose medium, in contrast, results in strong repression of CAC enzymes; i.e., they are catabolite repressed. Under these conditions, the anabolic functions of the cycle are minimal, and there is also a relatively small demand for CAC enzymes for energy production. The situation in cells growing in MG is in between that in NB and NB-glucose, since the cycle is in this case required for anabolic functions. Catabolite repression is a general phenomenon in microorganisms and has recently been reviewed for *B. subtilis* (36). It is relevant to observe that cyclic AMP and cyclic-AMP acceptor protein do not seem to be effector molecules in carbon catabolite repression in *B. subtilis*, unlike the well-documented case in *E. coli* (13, 36). As mentioned above, derepression of certain CAC enzymes in cells growing in minimal glucose medium occurs if decoyinine is added to the medium. This drug induces sporulation in *B. subtilis* by blocking the synthesis of guanine nucleotides.

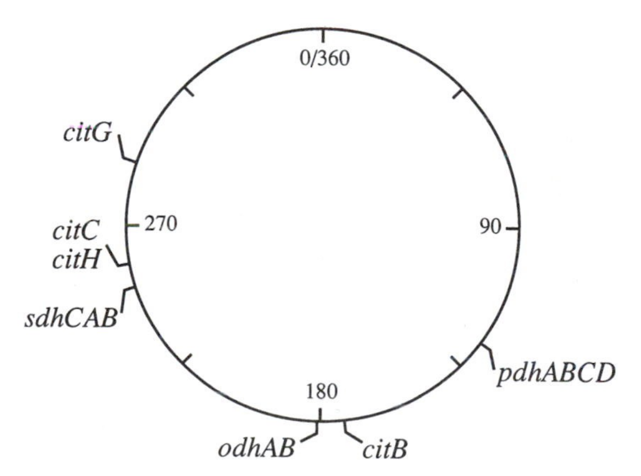

Figure 2. Chromosomal localization of CAC genes in *B. subtilis*. See Table 1 for details and references.

In addition to general regulation, as sketched above, the tricarboxylic (citrate-to-isocitrate) and dicarboxylic (succinate-to-oxaloacetate) parts of the cycle are differently regulated by various metabolites (119, 120, 162). It is beyond the space limits of this chapter to discuss all of the collected observations and findings on regulation of the CAC in gram-positive bacteria. The literature and present state of knowledge on the control of individual enzymes at the molecular level are, however, discussed in the following sections.

PDHC

The 2-oxoacid dehydrogenase multienzyme complexes constitute an enzyme family that includes PDHC, ODHC, and the branched-chain oxo acid dehydrogenase complex (BCODH) (for recent reviews, see references 124, 141, and 167). These complexes catalyze the following generalized reaction: RCOCOOH + NAD^+ + CoASH → RCO-SCoA + NADH + H^+ + CO_2, where CoASH is free, reduced CoA, –SCoA is oxidized, covalently bound CoA, and R is an acyl group. Each complex consists of multiple copies of three different enzymes: a thiamine PP_i-dependent enzyme, E1; a

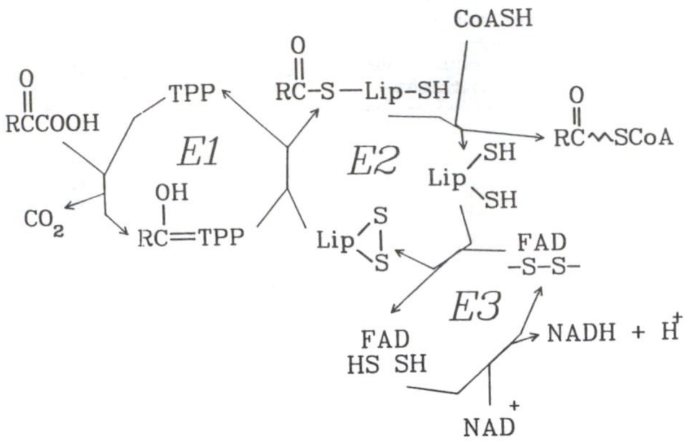

CoASH
O ‖ RCCOOH
TPP
O ‖ RC–S—Lip-SH
O ‖ RC∿SCoA
E1
E2
SH Lip SH
CO_2
OH | RC=TPP
Lip S | S
FAD –S–S–
E3
FAD HS SH
NADH + H^+
NAD^+

Figure 3. Reaction sequences of the 2-oxo acid dehydrogenase multienzyme complexes. Lip, lipoic acid; TPP, thiamine PP_i; CoASH, free, reduced CoA; –SCoA, oxidized, covalently bound CoA; R, acyl group.

lipoic acid-containing enzyme, E2; and a flavoenzyme, E3. E1 and E2 are unique for each multienzyme complex, whereas the same E3 enzyme functions in the different complexes in most cells. The suffixes p, o, and b are used to denote subunits of PDHC, ODHC, and BCODH, respectively. The three enzymes catalyze consecutive partial reactions (Fig. 3), giving the essentially irreversible net reaction shown above. Twenty-four or 60 copies of E2, depending on the particular enzyme and species, form a core of octahedral or icosahedral symmetry. E1 dimers and E3 dimers are bound to the edges and the faces, respectively, of the core, making the fully assembled multienzyme complex larger than a ribosome: a diameter of about 270 Å (27.0 nm) and a molecular weight of about 5×10^6 for the octahedral complexes (54, 124).

Oxidative decarboxylation of pyruvate to yield acetyl-CoA and reduced NAD is catalyzed by the PDHC; i.e., R in the generalized reaction shown above is in this case a methyl group. The enzymes of PDHC are pyruvate decarboxylase (E1p, pyruvate dehydrogenase [lipoamide]; EC 1.2.4.1), dihydrolipoamide acetyltransferase (E2p; EC 2.3.1.12), and dihydrolipoamide dehydrogenase (E3; EC 1.8.1.4). In many organisms, E1p is split into two polypeptides, E1pα and E1pβ, which form a heterotetramer, $\alpha_2\beta_2$. The α subunit is thought to catalyze decarboxylation of pyruvate, and the β subunit is thought to catalyze the reductive acylation of lipoamide. E2 of PDHC from different species has a rather well-defined domain structure (57). Starting from the N terminus, it has one to three lipoyl domains, each consisting of about 80 residues, and one lipoyl moiety covalently bound to the side chain of a lysine residue in a somewhat conserved sequence. The lipoyl domain(s) is followed by a 25- to 30-residue sequence rich in proline, alanine, and charged residues. This segment is very flexible and connects the lipoyl domain with a domain of about 30 residues which is required for E3 binding. A large "catalytic" domain, which harbors the transacetylase active site and is required for assembly of the core, is constituted by the C-terminal part of the E2 polypeptide. The three-dimensional structure of the E1p catalytic domain of PDHC from *Azotobacter vinelandii* (without the lipoyl and E3-binding domains) has recently been determined to 2.6-Å (0.26-nm) resolution by X-ray crystallography (101). This structure shows that the cubic core is built up by eight trimers and is hollow, with the CoA acylation active site inside the cube. The lipoyl domain of E2p from *Bacillus stearothermophilus* (comprising residues 1 through 85) has by genetic engineering been produced in *E. coli*, and its structure has been analyzed by two-dimensional ^{1}H nuclear magnetic resonance spectroscopy (26, 27). The data obtained show that the lipoyl-lysine (residue 42) is located in a sharp turn in a β-sheet secondary structure. The E3 dimer crystal structure of *A. vinelandii* PDHC has also been determined at a high resolution (102). This structure belongs to the family of disulfide oxidoreductases that have noncovalently bound flavin adenine dinucleotides (FAD) and active-site intrachain disulfide bridges (102). The structure of E3 (466 amino acid residues) is very similar to that of glutathione reductase, although only 26% of the residues in the primary structures of the two proteins are identical. The surface residues of the E3 dimer that are responsible for binding of the protein to the E2 core assembly have not yet been identified.

PDHC has been isolated from *B. subtilis* (84), *B. stearothermophilus* (81), and *Streptococcus faecalis* (*Enterococcus faecalis*) (5, 151). Table 2 contains a comparison of PDHC from these gram-positive bacteria with those from *E. coli*, *Acholeplasma laidlawii*, and mammalian mitochondria. PDHC of the gram-positive bacteria and *A. laidlawii* are similar to those

Table 2. Properties of PDHC from different organisms

Source of PDHC	Symmetry of core	No. of: E1p subunits	No. of: Lipoyl groups in E2	Mol wt (10^{-3})[a]: E1p	Mol wt (10^{-3})[a]: E2p	Mol wt (10^{-3})[a]: E3	Reference(s)
B. subtilis	Icosahedral	2	1	α, 42 (41–43); β, 35 (36–38)	48 (59–66)	50 (54–63)	80, 84, 94
B. stearothermophilus	Icosahedral	2	1	α, 41 (42); β, 35 (36)	46 (57)	49 (54)	11, 61, 81, 123
S. aureus	Not determined	2	1	α, ? (46); β, ? (41)	46 (71)	50 (60)	3, 77, 79
S. faecalis	Icosahedral	2	2	α, ? (42); β, ? (30–36)	56 (97–100)	? (55)	5, 124, 151
E. coli	Octahedral	1	3	99	66	51	6, 108
A. laidlawii	Not determined	2	2	α, >39; β, 36	57	>36	160
Mammals	Icosahedral	2	2	α, 40; β, 35	59	50	92, 157

[a] Values within parentheses for the gram-positive bacteria are those estimated from SDS-PAGE. Other values were deduced from the DNA sequences of cloned genes, and those for mammalian proteins are for the processed forms.

of mitochondria in having a split E1p. The gram-negative bacteria, e.g., *E. coli* (108) and *A. vinelandii* (12), have one E1p polypeptide with a size roughly corresponding to the sum of E1pα and E1pβ. This difference in E1p organization is correlated with the quatenary structure of the core; complexes with a single large E1p are of octahedral symmetry, whereas those with a split E1p have icosahedral symmetry (Table 2). E2p of gram-positive bacteria seems generally to have one or two lipoyl domains compared with three in the gram-negative bacteria. That of *Saccharomyces cerevisiae* has one lipoyl domain (117). The significance of the number of lipoyl domains is not obvious, since engineered *E. coli* E2p with only one lipoyl domain is functional (53).

The structural genes for E1pα, E1pβ, E2p, and E3 in gram-positive bacteria are designated *pdhA*, *pdhB*, *pdhC*, and *pdhD*, respectively. Such genes have been cloned and sequenced from *B. subtilis* (80), *B. stearothermophilus* (11, 61), and *Staphylococcus aureus* (all four genes have been cloned, but the full sequences of only *pdhC* and *pdhD* have been reported) (3, 77, 79). The *pdhB* and *pdhC* genes of *S. faecalis* have also been cloned, but the sequence of only the latter gene has been published (5). It is peculiar that the *B. subtilis*, *S. aureus*, and *A. laidlawii pdh* genes were cloned by serendipity. In the former two cases, phage lambda clones with *pdh* DNA were found by immunoscreening with antiserum against proteins that were thought to be involved in protein export (3, 79, 80). The *A. laidlawii* genes were obtained by immunoscreening of a plasmid library in *E. coli* with antibodies at that time believed to be directed against an acylated membrane protein (160). It is interesting that a putative *Neurospora crassa* mitochondrial membrane protein, MRP3, has from sequence comparisons been identified as E2p (145). Still it is not clear whether these experimental connections are artifacts due to the physicochemical properties of PDHC subenzymes or whether they indicate that these proteins have a physiological function in membranes.

Gram-positive bacteria seem to have *pdh* genes clustered and in a conserved organization as judged from the cases of *B. subtilis*, *B. stearothermophilus*, *S. aureus*, and *S. faecalis*. The gene order is *pdhA-pdhB-pdhC-pdhD*, and all genes are transcribed from A to D. In *B. subtilis*, *pdhD* is flanked by two identified genes, *pdhD-slp-ldc/cad*. They encode a 124-residue lipoprotein and a lysine decarboxylase, respectively (78) (the accession number for the *cad* sequence is X58433). Both *slp* and *ldc/cad* are transcribed from the DNA strand opposite to that from which *pdhD* is transcribed. An open reading frame (ORF5) flanks the *pdhA* gene and has the same polarity as the *pdh* genes (80). The gene flanking *pdhD* in the *B. stearothermophilus* chromosome is not the same as that in *B. subtilis*; i.e., it is an open reading frame corresponding to 80 residues and transcribed in the same direction as the *pdh* genes (11).

Several transcripts from the *pdh* gene cluster have been found in *B. subtilis* and *S. aureus*; a 5- to 6-kb *pdhABCD* mRNA, an ~3-kb *pdhCD* mRNA, and a 1.6- to 1.7-kb *pdhD* mRNA (3, 80). Thus, *pdhABCD* is arranged as an operon, but the *pdhD* gene can also be separately expressed. A similar transcriptional organization has been found for the corresponding *E. coli* genes: *aceE* (E1p), *aceF* (E2p), and *lpd* (E3) (108). The reason for this organization is probably that the *pdhD-lpd* gene product, E3, is also a component of ODHC and the gene expression must be coordinated with that of the *odhAB-sucAB* genes. Putative σ^A promoter regions are present in the *B. subtilis pdh* cluster at positions that correlate with the three detected transcripts; these promoter regions are located 40 to 100 bp upstream of the respective translation initiation codons of *pdhA*, *pdhC*, and *pdhD* (80). Metabolic regulation of *pdh* gene expression in *Bacillus* species remains to be investigated.

Attempts to inactivate *pdh* by insertional mutagenesis in both *B. subtilis* and *S. aureus* have failed, suggesting that PDHC is essential in these bacteria (3, 80). *B. subtilis* mutants defective in PDHC have, however, been isolated after chemical mutagenesis and characterized (43a, 82). Mutants specifically defective in PDHC require acetate for growth on glucose minimal medium and were originally designated AceA$^-$ (43a). CitL mutants, which are defective in E3, are deficient in both PDHC and ODHC (82). The *aceA* locus lies within the *pdhA* gene at 126° on the chromosome (13a, 80, 88). The *citL* locus may be identical to *pdhD*, which encodes E3, but no *citL* mutations have to my knowledge been sequenced to confirm this possibility. A mutant with the *citL22* mutation has very low or no PDHC, ODHC, and BCODH activities but contains 1 to 5% of the lipoamide dehydrogenase activity of wild-type cells (82, 94). It has not been established whether this residual activity is due to a low expression of *pdhD*, an only partially defective E3, or the presence in *B. subtilis* of more than one lipoamide dehydrogenase, as is the case in *E. coli* (138). Two PDHC-defective *B. subtilis* mutants (*aceA*) isolated by Freese and Fortnagel (43a) have been biochemically characterized by Hodgson et al. (84). Both mutants contain assembled enzymatically defective PDHC. One mutant has an inactive E1 component that is only loosely bound to the E2p-E3 complex. The other mutant (originally it was reported to absolutely require acetate for growth, but in the hands of Hodgson et al. [84], it grew slowly in the absence of acetate) has a defective E1 component with an ~10-fold-increased K_m for pyruvate. This latter mutant enzyme is also cold labile; E1 is lost from E2p-E3 at low temperatures, e.g., 0 to 4°C. It is not known whether the mutation(s) of the respective mutant is in *pdhA* or *pdhB*. BCODH plays an important role in *B. subtilis* by supplying branched carboxylic acids for branched-chain fatty acid synthesis. The *B. subtilis* PDHC is considered to also display branched-chain 2-oxo acid dehydrogenase activity, i.e., to have dual catalytic functions. This conclusion is based on the finding that PDHC and BCODH could not be separated by physical or genetic means (13a, 94, 164a). For example, both the E1p mutants described above lack PDHC and BCODH activities. Contradictory data suggesting the presence of both a branched chain 2-oxo acid decarboxylase (E1b component) and pyruvate decarboxylase (E1p component) in *B. subtilis* have been presented; the two enzymes have very similar α and β subunits (122).

The E1pα and E1pβ polypeptides are conserved in different organisms (163). For example, *B. subtilis* E1pβ has 36 and 45% sequence identity to the corresponding proteins in human mitochondria and *A. laidlawii*, respectively (80, 160). Some of the residues conserved in all E1pα have been assigned to functions such as thiamine PP_i binding and subunit interaction (163). PDHC activity in mammals is regulated by phosphorylation-dephosphorylation of serine residues in the E1pα subunit (135, 167). The inactivating phosphorylation site in the human E1pα is the underlined serine in the sequence Gly-His-His-<u>Ser</u>-Thr-Ser-Asp-Asp. The corresponding sequence in E1pα from *B. subtilis* and *B. stearothermophilus* is **Gly**-Pro-**His**-<u>Thr</u>-Met-**Ser**/Ala-Gly-**Asp**. Thus, the Ser in the mammalian protein is replaced by Thr, evidence that prokaryotic PDHC is not regulated by phosphorylation. Russell and Guest have made an extensive sequence comparison of E2 sequences from PDHC, ODHC, and BCODH of different organisms (145). The comparison shows homology between E2 polypeptides of different multienzyme complexes and reveals that all E2 variants are likely to have evolved from a common ancestral protein. It should be noted that the E2p, E2o, and E2b sequences form subfamilies of more closely related proteins. Residues that are conserved in all E2 polypeptides are a Lys residue in the lipoyl domain(s), to which lipoate is covalently bound, and an active-site His residue close to the C terminus in the catalytic domain. The relatively large number of E2 sequences now available (145) together with the crystal structure of *A. vinelandii* E2p catalytic domain (101) and nuclear magnetic resonance structures of the lipoyl domain of *B. stearothermophilus* E1p (26) and the E3-binding domain of *E. coli* E2o (139a) make it possible to predict rather detailed structures from sequence information (54). The same is true for E3, where one crystal structure is known (102) and many homologous primary structures are available (77).

The lipoyl group is posttranslationally attached to E2 polypeptides by an enzyme(s) not present in the final multienzyme complex. Lipoylation activity was first studied by Reed et al., who used *S. faecalis* PDHC as an experimental system (134). It is not known whether there is one specific enzyme for each type of E2 polypeptide. Two E2p lipoylation activities have been found in *E. coli* (14), and these enzymes apparently exhibit sequence specificity, since the lipoyl domain of *B. stearothermophilus* E2p is lipoylated in *E. coli* (27) but that of bovine E2b is not (51).

CITRATE SYNTHASE

Citrate synthase (EC 4.1.3.7) catalyzes a condensation reaction: acetyl-CoA + oxaloacetate + $H_2O \rightarrow$ citrate + CoA. This reaction is essentially irreversible in vivo. For reviews on the enzymology and comparative biochemistry of this enzyme, see references 91, 162, and 164. Citrate synthases of gram-positive bacteria and eukaryotic cells are homodimers of about 40- to 50-kDa subunits. The structure of the dimeric porcine heart enzyme with different ligands is known from studies by X-ray crystallography (164). Citrate synthases of *E. coli* and many other gram-negative bacteria are, in contrast, homohexameric and allosterically inhibited by micromolar concentrations of NADH (25, 38, 162).

Purified *Bacillus megaterium* citrate synthase is a protein of 84 ± 5 kDa that consists of two subunits of 40 ± 4 kDa and that has a synthase activity that is not sensitive to NADH (K_i = 7.5 mM) (25, 140). Molecular and enzyme kinetic properties of *B. subtilis* (90) and *Bacillus* (*coagulans*) sp. strain C4 (149) citrate synthase are very similar to those of *B. megaterium*. The *B. subtilis* enzyme has been reported to be unstable and therefore was not purified to homogeneity; its activity is not sensitive to NADH (90). The *ctsA* gene encoding the *Bacillus* (*coagulans*) citrate synthase was recently cloned by complementation of an *E. coli* mutant (149). The deduced amino acid sequence, confirmed by N-terminal sequence analysis, has 373 residues (41.9 kDa) with 30% sequence identity to that of the *E. coli* citrate synthase encoded by the *gltA* gene. The $citA_1$ gene encoding *B. subtilis* citrate synthase has been cloned, also by complementation of an *E. coli* mutant strain (88, 89). The deduced primary structure is homologous to that of the *E. coli* enzyme, and there is complete conservation of residues thought to be part of the active site. When $citA_1$ was inactivated, *B. subtilis* cells were unexpectedly found to still sporulate normally and to contain citrate synthase activity, indicating the presence of multiple genes for this enzyme. Indeed, a second citrate synthase has recently been purified, and the corresponding gene, $citA_2$, which encodes a 41-kDa polypeptide, has been cloned (88, 89). It is not clear whether the previously (90) partially purified and kinetically analyzed enzyme preparation is encoded by $citA_1$ or $citA_2$ or both. Transcription of the $citA_1$ gene increases when cells enter stationary growth phase in a broth medium (88, 89), which suggests that the citrate synthase encoded by this isogene has a function in stationary growth phase or sporulation. Evidence for two citrate synthase isoenzymes, a 250-kDa NADH-sensitive form and a 100-kDa NADH-insensitive form, has been obtained from studies with different *Pseudomonas* species (109, 110). The amino acid sequence of the NADH-sensitive citrate synthase of *Pseudomonas aeruginosa*, as deduced from the DNA sequence of the cloned gene (31), shows 28% identity to the polypeptide encoded by *ctsA* of the *Bacillus* (*coagulans*) sp. (149). The two *citA* gene variants found in *B. subtilis* may encode enzymes of different oligomeric states and regulatory properties.

Few *B. subtilis* mutants specifically defective in citrate synthase have been reported. Such mutants are expected to require citrate, isocitrate, or glutamate (2-oxoglutarate) for growth (Fig. 1). Carls and Hanson isolated a mutant with temperature-sensitive growth on minimal glucose plates (15). This mutant required citrate or glutamate for growth at 50°C but grew without supplementation at 37°C. Analysis of partially purified citrate synthase from the mutant confirmed that the activity was temperature sensitive and showed that the enzyme had normal kinetic properties except that it was not inhibited by ATP to the same degree as the wild-type enzyme. A mutation in the structural gene for the synthase is strongly suggested from these results. The chromosomal location of this mutation and its relation to the $citA_1$ and $citA_2$ genes are not known. Citrate synthase is coordinately regulated with aconitase (37, 119; see next section).

ACONITASE

The isomerization of citrate to isocitrate with *cis*-aconitate as an enzyme-bound intermediate is catalyzed by aconitase (EC 4.2.1.3; citrate [isocitrate] hydrolyase): citrate ↔ H_2O + aconitate ↔ isocitrate. The structure of porcine aconitase is known (139). It is an iron-sulfur protein that in its enzymatically active form contains a [4Fe-4s] center. One iron (Fe_a) is readily lost from this center under aerobic conditions, resulting in a [3Fe-4S] cluster and an inactive enzyme. Incubation with ferrous iron and a reducing agent, preferably a thiol, restores the tetranuclear center and enzyme activity (9). Of particular interest is the fact that aconitase contains an iron-sulfur center, although the enzyme does not catalyze a net oxidation-reduction reaction.

Purified *B. subtilis* aconitase, which is encoded by the *citB* gene (Table 1 and Fig. 2), is a 120-kDa polypeptide as determined by sodium dodecyl sulfate-polyacrylamide gel electrophoresis (SDS-PAGE) and is probably monomeric (30). It is larger than aconitases from *S. cerevisiae* mitochondria (779 residues; 85.6 kDa) (48), porcine heart mitochondria (754 residues; 82.7 kDa) (148, 171), and *E. coli* (890 residues; 97.5 kDa) (130, 131). Mammalian and *S. cerevisiae* mitochondrial aconitase are monomers. Only the first 141 residues of *B. subtilis* aconitase are known, as deduced from the *citB* sequence and N-terminal analysis of the purified enzyme (30). This short sequence is homologous to those of porcine and *E. coli* (*acn* gene product) aconitases and constitutes part of structural domain I; residues Asp-135 and His-136 in the *B. subtilis* enzyme most likely form an active-site Asp-His pair. The extra residues of the relatively larger *B. subtilis* enzyme must be internal or located at the C-terminal end of the polypeptide.

B. subtilis aconitase is also deactivated in the presence of air and can be reactivated. This suggests the presence of a labile iron-sulfur center, although such a presence has not been directly demonstrated (30, 40). The iron-sulfur component in *E. coli* aconitase is ligated to three conserved cysteine residues: Cys-435, Cys-500, and Cys-503. Interestingly, the primary structure of *E. coli* aconitase is very similar (53% identity) to that of the human iron-responsive-element-binding protein (IRE-BP), whereas similarity to the sequence of mitochondrial aconitase is only about 28% (130, 144). From this similarity between aconitase and IRE-BP, it has been suggested that IRE-BP is an iron-sulfur protein and that bacterial aconitase may also have an iron-responsive regulatory function in the cell (130). In the presence of a low (<1 mM) concentration of transition metal chelator compounds, e.g., *o*-phenanthroline, the growth of *B. subtilis* cells is slower, sporulation is blocked, and the aconitase-catalyzed step of the CAC is inhibited (40).

The *citB* gene is located at 173° between the *gltA* and *glnA* markers (Fig. 2; 120). Mutants deficient in aconitase activity have been isolated, but in only a few cases has the responsible mutation been mapped on the chromosome (15, 39, 147). The *citB* mutants require isocitrate or 2-oxoglutarate for growth and are blocked in sporulation at stage O or I. A 7-kb DNA fragment containing the promoter-proximal end of *B. subtilis citB* has been cloned, and its sequence has been analyzed (30, 143). The translational start codon of *citB* has been confirmed by N-terminal sequence analysis of aconitase purified from *B. subtilis* SMY; the initiator formylmethionine is not present in the final protein (30). A transcriptional start site has been identified 68 bp upstream from the start of the *citB* reading frame. Transcription from this site is by RNA polymerase containing σ^A (29, 143). Just upstream of the *citB* promoter region is a gene of unknown function that is transcribed from the strand opposite that from which *citB* is transcribed (143). Transcription from this divergent gene seems to be induced in stationary growth phase (29).

Intracellular citrate is required for *citB* expression (29, 41). Aconitase is catabolite repressed in minimal glucose medium, and this repression is more extensive if glutamine (2-oxoglutarate) is also present in the medium, although glutamine in the absence of a rapidly metabolizable carbon source does not cause repression (118, 143). Derepression of aconitase occurs in late exponential growth phase in a nutrient broth medium or when decoyinine is added to cells growing in a minimal glucose medium. The molecular basis for this regulation has been studied in detail by Sonenshein and coworkers (29, 30, 41, 42, 143) by exploiting the cloned *citB* promoter region. From these experimental data, it is concluded that *citB* expression is regulated mainly at the mRNA level and by negative regulator proteins. Evidence for such a regulator has been obtained from gel shift experiments and other types of data (41). Using a mutagenesis approach in combination with a *citB-lacZ* transcriptional fusion, the sequence −73 to −54 in the *citB* promoter region has been found critical for functional catabolite repression. Single base mutations at many positions (but not all positions) in this region result in transcription of *citB* that is insensitive to glucose plus glutamine. According to one proposed model, repressor molecules bind to the sequence of dyad symmetry in the −65 region and to a sequence closer to the promoter site, maybe the repeat at position −32 to −27 (41). The bound putative dimeric repressor could block the binding of RNA polymerase and may be activated by pyruvate or 2-oxoglutarate (35); alternatively, it may be inactivated by citrate (36).

Catabolite repression-resistant *citB* mutants are derepressed in minimal glucose plus glutamine medium but still show, like the wild type, growth stage-dependent expression of aconitase in a nutrient broth medium; i.e., *citB* transcription in cells is relatively repressed in early exponential growth stage and is derepressed at the end of exponential growth. This demonstrates that catabolite repression and the temporal control of *citB* gene expression are distinct and act by different mechanisms (41). The steady-state cellular concentration of *citB* mRNA in *B. subtilis* growing in a broth medium varies with growth stage; it remains fairly constant between 1 h before the end of exponential growth phase (t_{-1}) and t_1 but is dramatically reduced at t_2 and is essentially absent at T_4 as assessed by S1 nuclease protection analysis (29).

ISOCITRATE DEHYDROGENASE

Isocitrate dehydrogenase (EC 1.1.1.42) catalyzes the following reaction: isocitrate + $NADP^+$ ↔ 2-oxoglutarate + CO_2 + NADPH + H^+. This enzyme has been isolated from an alkalophilic *Bacillus* sp. (150) and from *B. subtilis* 168 (purified 500-fold but not to homogeneity) (132). Isocitrate dehydrogenase from both these species is probably homodimeric and has a reported mass of 80 to 90 kDa, which is similar to those of *B. stearothermophilus* (86) and other bacteria, including *E. coli*. The mass of the *B. subtilis* monomer is 44 kDa as determined by the expression of the cloned *citC* gene in an *E. coli* system (126). The alkalophilic *Bacillus* enzyme has a high substrate specificity for β-$NADP^+$; no activity was observed with α-$NADP^+$ or β-NAD^+ (150). *Bacillus* isocitrate dehydrogenase can be inhibited in vitro by certain CAC intermediates such as 2-oxoglutarate and oxaloacetate and also by various nucleotides (38, 150). However, it is uncertain whether these compounds have a significant regulatory effect on the enzyme in vivo. The *E. coli* $NADP^+$-linked isocitrate dehydrogenase is not under allosteric control but is converted between active and inactive forms by phosphorylation (116). The specific kinase and dephosphorylase are under allosteric control. The amount of phosphorylated (inactive) isocitrate dehydrogenase is high in cells in which the glyoxylate shunt is operating, for instance, during growth on acetate. It is not known whether isocitrate dehydrogenase is phosphorylated in *Bacillus* species. Isocitrate lyase (EC 4.1.3.1), which catalyzes the first enzyme in the glyoxylate pathway, seems to be present in alkalophilic *Bacillus* spp. (150). Both the *B. subtilis* and the alkalophilic *Bacillus* isocitrate dehydrogenases are inhibited in a concerted fashion by glyoxylate and oxaloacetate. This inhibition, also seen with the *E. coli* enzyme, is not considered physiologically relevant (116).

In contrast to the case with aconitase, the addition of glutamate to minimal glucose medium does not cause further repression of isocitrate dehydrogenase in *B. subtilis*. *B. subtilis* mutants lacking isocitrate dehydrogenase require glutamate or another source of 2-oxoglutarate for growth (15, 147). Isocitrate dehydrogenase-defective mutants show derepressed levels of aconitase that are probably due to citrate accumulation (120, 147). The structural gene *citC*, located at 259° on the chromosome (Fig. 2), has been cloned by complementation of an *E. coli icd* mutant (126). The cloned gene carried on a plasmid resulted in increased isocitrate dehydrogenase activity in *B. subtilis* and *E. coli*.

ODHC

ODHC belongs to the same family of enzymes as PDHC (see the section on pyruvate dehydrogenase above). ODHC is composed of multiple copies of 2-oxoglutarate decarboxylase (E1o; 2-oxoglutarate dehydrogenase [lipoamide]; EC 1.2.4.2), dihydrolipoamide succinyltransferase (E2o; EC 2.3.1.61), and lipoamide dehydrogenase (E3; EC 1.8.1.4). E1o and E2o are exclusively found in ODHC. The multienzyme complex catalyzes the following net reaction: 2-oxoglutarate + NAD^+ + CoA → succinyl-CoA + NADH + H^+ + CO_2. Compared with PDHC, ODHC is more conserved in organization; i.e., the E1o component is one polypeptide, E2o has one lipoyl domain, and the symmetry of the core is octahedral in all cases analyzed so far (124).

Reports on the isolation of ODHC from gram-positive bacteria are lacking, but substantial biochemical information is available indirectly from the cloned *B. subtilis* structural genes for E1o (*odhA*, previously called *citK*) and E2o (*odhB*, previously called *citM*) (Table 1) and from mutant studies (16, 56, 82). *odhA* and *odhB* are organized in an operon located close to the terminus of the chromosome (Fig. 2): *ilvA*-(SPβ prophage)-*kauA-odhAB-terC-gltA* (16, 34, 50, 82). Weiss and Wake have established a restriction map over this region of the *B. subtilis* chromosome (161). That map, combined with sequence data (18, 136), shows that *odhAB* and *terC* are about 96 kbp apart, which corresponds to a distance of about 8° on the circular chromosomal map of strain 168. The *odhAB* region, with several kilobase pairs of flanking DNA from both sides, has been cloned on plasmids in *E. coli* (17–19), but complete sequence data are available only for the *odh* genes (18, 136).

The promoter region, defined to a 375-bp fragment and with a putative σ^A-type promoter, directs the synthesis of a 4.5-kb *odhAB* dicistronic mRNA (136). The 5′ end of this mRNA corresponds to a site located about 112 bp upstream from the translational start codon of *odhA*. Indirect data indicate a possible σ^A promoter in the 3′ end of the *odhA* sequence, possibly allowing *odhB* to also be expressed separately (18). Such an organization would be similar to that of the *B. subtilis pdh* operon (80). A separate *odhB* transcript has, however, not been detected by Northern (RNA) blot analysis (18, 136). The structural genes for *E. coli* E1o and E2o (*sucA* and *sucB*) are located in a cluster of CAC genes; those encoding citrate synthase (*gltA*) and succinate:quinone reductase (*sdhCDAB*) are on the 5′ flank, and those for succinate thiokinase (*sucCD*) are on the 3′ flank (108). It will be interesting to learn whether the genes for *B. subtilis* succinate thiokinase are linked to the *odh* region.

The physical half-life of the *B. subtilis odhAB* full-length mRNA, about 1.5 min in cells growing in a broth medium, is not affected by growth stage or the presence of glucose in the medium (136). However, the steady-state cellular concentration of this mRNA is very growth stage dependent; the concentration increases somewhat during exponential growth phase, peaks at the end of exponential phase, and rapidly decreases to undetectable levels as the culture continues into stationary phase (136). Catabolite repression of ODHC is executed primarily on the transcriptional level and can be explained by down-regulation of the *odhAB* promoter (136). These findings are similar to those made for *citB* (see above). An apparent difference between the promoter regions of *citB*, *odhAB*, and *sdhCAB* is that the presence of the promoter region on a high-copy-number plasmid partially relieves glucose repression of *citB* but apparently not of *odh* or *sdh* (29, 105, 106, 136). This difference merits further investigation in light of recent advances in the understanding of the *citB* promoter region.

B. subtilis E1o consists of 937 residues (105.3 kDa) as deduced from the DNA sequence and confirmed by the size of the polypeptide and N-terminal analysis (136). The amino acid sequence is homologous to that of E1o of gram-negative bacteria and *S. cerevisiae* (37 to 41% sequence identity) but only distantly related to that of the unsplit E1p of gram-negative bacteria. *B. subtilis* E2o, predicted to contain 417 residues (46.0 kDa; mass estimated from SDS-PAGE is about 60 kDa) (18), is very similar in sequence to other E2o polypeptides. It shows homology also to E2p polypeptides (145); e.g., *B. subtilis* E2o and E2p are 36% identical (80). The single lipoyl group in *B. subtilis* E2o is most likely bound to Lys-42. *B. subtilis* E2o polypeptides seem to be lipoylated also in *E. coli*, since expression of *odhB* in an *E. coli* E2o-deficient mutant complements the ODHC defect (17). Similarly, an *E. coli* E1o-defective mutant has been shown to be complemented by *B. subtilis* E1o expressed from *odhA* on a plasmid (18). These in vivo complementation data and results from corresponding in vitro experiments (16, 17) show that the subenzymes of the *E. coli* and *B. subtilis* ODHC are similar in three-dimensional structure and function.

ODHC-defective *B. subtilis* mutants are available (15, 34, 47, 170), and several of these have been analyzed in more detail (16, 35, 82, 56). Such mutants can grow on minimal glucose medium. This is in contrast to *E. coli* ODHC-defective mutants, which require succinate in minimal glucose medium for aerobic growth. The step catalyzed by ODHC in the CAC is irreversible, explaining why a block in this step does not allow the cell to grow on tricarboxylic acids, lactate, or glutamate and why a dicarboxylic acid (succinate, fumarate, malate, or oxaloacetate [aspartate]) is required for growth at wild-type rates on a poor carbon source like maltose or lactate (34, 147). Fisher and Magasanik (34, 35) have analyzed intracellular metabolite pools in an *odhA* mutant and shown that 2-oxoglutarate is accumulated in a medium where growth is limited by the concentration of glucose. Accumulation of radioactive 2-oxoglutarate has been demonstrated in ODHC mutants when they are allowed to assimilate [^{14}C]glutamate or [^{14}C]succinate (15, 39, 147).

Rutberg and Hoch (147) isolated 11 ODHC mutants after chemical mutagenesis of *B. subtilis* and classified them as CitK$^-$ or CitD$^-$ depending on the specific activity of a second enzyme, isocitrate dehydrogenase, in mutant cell extracts. Those with a low activity were designated CitD$^-$; they also grew more poorly on tryptose blood agar base plates and were generally nonmotile. The mutations that cause the CitD$^-$ phenotype are usually large deletions encompassing *odhA* and often also *odhB* (18, 56, 82, 147). Most *B. subtilis* 168 strains contain the prophage SPβ integrated into the chromosome close to the *odhAB* operon (128). SPβ is a specialized transducing phage for *odh* (*citK*) markers (170). When the phage is excised, e.g., after induction with a mutagen, flanking sequences sometimes accompany the phage DNA, resulting in a deletion in the bacterial chromosome. ODHC mutants (the CitK$^-$ phenotype) fall into two complementation groups corresponding to OdhA$^-$ and OdhB$^-$ mutants. Whether a mutant is defective in E1o or E2o can be assessed by in vitro complementation assays (16, 82) or by direct subenzyme activity measurements. Many *odhA* (*citK*) mutations but only two *odhB* (*citM*) mutations have been reported (16, 34). These two *odhB* mutations have been mapped (17) and recently identified by DNA sequence analysis (56). The *odhB1* mutation is an A-to-G transition predicted to result in the replacement of Gly-355 in *B. subtilis* E2o with Asp. The *odhB17* mutation is a T-to-G transversion resulting in the replacement of Leu-257 with Trp. Gly-355 corresponds to Ala-578 in the homologous *A. vinelandii* E2p (101) and is followed by five residues that are conserved in *E. coli*, *A. vinelandii*, and *B. subtilis* E2o (145). In the crystal structure of the E2p catalytic domain, Ala-578 is located in the N-terminal end of β-strand I in the central β sheet of the protein. Leu-257 of *B. subtilis* E2o is a conserved residue corresponding to Leu-478 in the *A. vinelandii* E2p sequence. This Leu is in the *Azotobacter* E2p positioned at the end of α helix 3 (see, e.g., Fig. 3 in reference 101 for structural details). Wild-type E1o and mutant E2o polypeptides are found in close to normal amounts in mutants carrying the respective *odhB* mutation (56). It has not yet been established whether the two *B. subtilis* mutant E2o assemble into a core or if the core (if assembled) is functional in binding the peripheral subunits.

SUCCINATE THIOKINASE

Succinate thiokinase (succinyl-CoA synthase; EC 6.2.1.5) catalyzes the following reaction: succinyl-CoA + ADP + P_i ↔ succinate + ATP + CoA. This enzyme from various organisms contains two subunits, α and β. Weitzman (162) has shown two classes of succinate thiokinases. One class consists of "large" (140- to 185-kDa) enzymes with $\alpha_2\beta_2$ structures. Enzymes of this class are found in gram-negative bacteria (108) and archaebacteria (25). The second class contains "small," $\alpha\beta$ enzymes (about 75 kDa). These enzymes are found in mitochondria, e.g., from porcine heart and liver, and in gram-positive bacteria, e.g., *B. megaterium* and *B. stearothermophilus*. Few data are available on succinate thiokinase in gram-positive bacteria, but the native size and certain enzyme kinetic properties have been studied (162). The *Bacillus* enzymes appear very specific for ADP; i.e., they show no activity with GDP.

SUCCINATE:MENAQUINONE REDUCTASE

Succinate in the CAC is oxidized to fumarate by the tightly membrane bound enzyme succinate:quinone oxidoreductase (SQR; EC 1.3.5.1): succinate + Q ↔ fumarate + QH_2, where Q is a quinone. The reducing equivalents (electrons) released in the oxidation are transferred directly from the enzyme to the quinone (93), which functions in the membrane as an electron carrier between membrane-bound redox proteins, e.g., dehydrogenases and oxidases (see chapter 14 in this volume). Gram-positive aerobic bacteria contain menaquinone as the only type of quinone in the membrane; *Bacillus* species usually contain menaquinone-7. Succinate dehydrogenase activity (succinate: [acceptor] oxidoreductase; EC 1.3.99.1) is measured with water-soluble redox dyes instead of quinone as electron acceptor. The assay with dyes is simple, but the activity of the complete enzyme system is not measured, and enzyme preparations can show succinate dehydrogenase activity but no quinone reductase activity.

Comprehensive reviews of the biochemistry and

Table 3. Molecular data on succinate:quinone reductase from gram-positive bacteria compared with data for *E. coli* and mammalian cells[a]

Organism	Molecular mass (kDa)			FAD_{cov}[b] in Fp
	Fp	Ip	Cytochrome *b*	
Bacillus subtilis	65[c]	28[c]	+ (23)[c]	+
Bacillus sp. strain PS3	63	26	+	+
Micrococcus luteus	72	30	+ (17 + 15)	+
Staphylococcus aureus	62	30	+	+
Mycobacterium phlei	62	26	?	–
Escherichia coli	64[c]	27[c]	+ (14 + 13)[c]	+
Mammal	70	27	+ (15 + 13)	+

[a] Data are from references 1, 22, 23, 64, 73, 75, 96, 108, 127, 133, 151a, and my own unpublished data.
[b] FAD_{COV}, covalently linked FAD.
[c] Calculated from the DNA sequence.

Table 4. Localization and characteristics of prosthetic groups of *B. subtilis* succinate:menaquinone reductase[a]

Prosthetic group	Subunit localization	E_m (mV)	EPR signal	Reference(s)
FAD_{cov}	Fp	?	?	63, 99
$[2Fe\text{-}2S]^{(2+,1+)}$	Ip	−80	g = 2.035, 1.945, 1.889	72, 99
$[3Fe\text{-}4S]^{(1+,0)}$	Ip	−25	g_{max} = 2.023	72, 99
$[4Fe\text{-}4S]^{(2+,1+)}$	Ip	−240	?	72, 99
$cytb_H$	Cytochrome	+65	g_{max} = 3.68	55
$cytb_L$	Cytochrome	−95	g_{max} = 3.42	55

[a] EPR, electron paramagnetic resonance; FAD_{cov}, covalently linked FAD.

genetics of SQR from prokaryotic and eukaryotic cells have very recently been published (1, 73). SQR consists of a membrane-extrinsic part that in bacteria faces the cytoplasm and in several cases can be released from the membrane as a water-soluble protein. This water-soluble part can reduce certain redox dyes but not quinone. The extrinsic part is comprised of two protein subunits, a flavoprotein (Fp) and an iron-sulfur protein (Ip). These two SQR subunits from different organisms are highly conserved with respect to both the contents of prosthetic groups and the primary structure. Fp (about 65 kDa) contains, with a few exceptions, an FAD molecule covalently attached in isoalloxazine 8(α)-His linkage to the polypeptide. Ip (about 30 kDa) contains three iron-sulfur centers of different types; one [2Fe-2S] center, one [3Fe-4S] center, and one [4Fe-4S] center. The Fp and Ip subunits are tightly anchored to the membrane by an integral membrane protein. This anchor protein spans the membrane and is generally a cytochrome *b* composed of one large polypeptide (about 25 kDa) or two smaller polypeptides (11 to 17 kDa each). In SQR from different organisms, the amino acid sequences of the anchor proteins are far less conserved than are the Fp and Ip subunits. The biochemistry, genetics, and biogenesis of SQR have been studied in detail, with *B. subtilis* as a model system. Present knowledge of this enzyme, summarized in this chapter, has been acquired from studies involving various *B. subtilis* mutants (4, 44–47, 69, 71, 72, 74, 85, 97–99, 104, 105, 146). The few data available for SQR from other gram-positive bacteria are summarized in Table 3 or mentioned elsewhere in this chapter.

B. subtilis SQR has been purified to homogeneity in the presence of detergent (55). It consists of Fp, Ip, and a cytochrome b_{558} subunit in the stoichiometry 1:1:1 (Table 3). Cytochrome b_{558} anchors the Fp and Ip subunits to the membrane and contains two protoheme IX molecules (55, 62, 70). The two heme components show different low-temperature light absorption spectra, electron paramagnetic resonance signals, and midpoint redox potentials and are designated $cytb_H$ (high potential) and $cytb_L$ (low potential) (55, 66). $Cytb_H$ under steady-state conditions is reducible by succinate (55). The prosthetic groups of *B. subtilis* SQR and some of their properties are presented in Table 4. FAD, iron-sulfur centers, and probably $cytb_H$ function in electron transfer from the dicarboxylate (succinate) catalytic site on the Fp subunit to the menaquinone active site associated with the cytochrome b_{558} polypeptide.

The *B. subtilis* SQR polypeptides are encoded by *sdhC* (cytochrome b_{558}), *sdhA* (Fp), and *sdhB* (Ip) (Table 1). These genes are organized in an *sdhCAB* operon at 252° on the chromosome (Fig. 2). (The *sdh* locus was originally designated *citF* [146], and the genes for cytochrome b_{558}, Fp, and Ip were first called *sdhA*, *sdhB*, and *sdhC*, respectively [71], according to the Demerec guidelines but were then renamed [127] *sdhC*, *sdhA*, and *sdhB*, respectively, to conform with the genetic nomenclature used for *E. coli*.) The *sdh* genes were cloned from *B. subtilis* by the complementation of *sdh* mutations (60, 95). An almost-9-kb DNA sequence is known for the *sdh* region of the *B. subtilis* 168 chromosome (20, 21, 24, 96, 125, 127). The *ask* (aspartokinase II) and *gerE* (germination locus) genes flank the *sdhCAB* operon together with two unidentified open reading frames, ORFX (148 codons) and ORFY (147 codons): *ask*-ORFX-*sdhCAB*-ORFY-*gerE*. *ask*, *sdh*, putative ORFY, and *gerE* are transcribed in the direction shown, whereas ORFX would be transcribed in the opposite direction. Deletions reaching from the terminal end of *ask* to various positions in *sdhA* and *sdhB* have been introduced into the *B. subtilis* chromosome (45, 46). Mutants with such deletions are viable, and no particular phenotype, except Ask^- and Sdh^-, has been detected. Presently available data do not exclude the possibility that the ORFX or ORFY gene products have some role in expression, regulation, or assembly of functional SQR.

A promoter, probably of the σ^A type, directs transcription of a 3.45-kb *sdh* mRNA (105, 106, 137). The −35 and −10 regions of the *sdh* promoter have been confirmed by promoter-down single-base-pair substitutions within them (45, 73, 105). The transcriptional start site is located approximately 90 bp upstream of the start of the *sdhC* coding region (105). This start site and the length of the mRNA suggests that ORFY is not present on the *sdhCAB* mRNA. The steady-state concentration of *sdh* mRNA is four times higher in cells grown in a broth medium than in cells grown in the same medium containing 1% glucose (105, 106). This difference is mainly due to transcriptional down-regulation of the *sdh* operon, similar to what has been found for *citB*, *odhAB*, and *citG*. The cellular concentration of *sdh* mRNA seems to be strongly dependent on the growth stage of the cell, and this dependence has been suggested to be primarily the result of a variable stability of the transcript. The half-life of *sdh* mRNA is 2 to 3 min in exponential growth phase but only about 0.4 min in early stationary growth phase (106, 137). The mechanism(s) for degradation of *sdh* mRNA is not known, but the 5′ end of the transcript is important for stability (104). The half-life of the transcript is shorter than normal if the *sdhC* gene is less frequently translated; i.e., a mutation in the ribosome-binding sequence, which decreases the translation of *sdhC* about 10-fold and has only a minor polar effect on the translation of the *sdhA* gene, destabilizes the *sdh* transcript to a half-life of 0.5 to 1.3 min. By contrast, introduction of a translational stop codon in the middle of the *sdhC* gene does not cause a drastic decrease in stability of the *sdh* mRNA.

B. subtilis mutants with a defective SQR can be found among CAC-defective mutants isolated after mutagenesis of cells or spores (71, 121). Mutants with less than 1% of normal SQR activity grow poorly on minimal plates with citrate and glutamate, and those with less than 10% of wild-type activity cannot efficiently metabolize amino acids; e.g., they produce acidic colonies on agar plates containing nutrient broth with a low concentration of glucose (47). A rather large number of mutations affecting the expression, assembly, or activity of SQR have been collected. Many of these mutations have been mapped on the chromosome and within the *sdh* region by using three-factor transformation crosses (71, 97, 127). About a dozen different mutations on the nucleotide level have been identified (47, 65, 73, 99, 105). No mutant with a mutation outside the *sdh* region that drastically affects SQR has been reported.

During biogenesis of SQR in *B. subtilis* and after translation is completed, Fp and Ip appear first in the cytoplasm as water-soluble polypeptides (74, 75). The prosthetic groups of Fp and Ip are incorporated into the cytoplasm. It is not known by what mechanism FAD becomes covalently bound to the side chain of a specific His residue (His-40); i.e., is it by an autocatalytic process or by the action of factors not present in the final enzyme (65). *B. subtilis* Fp synthesized from the cloned *sdhCAB* operon on a plasmid in *E. coli* is not flavinylated despite normal processing of the N terminus (67). This could suggest a requirement for specific (cell specific) factors, because the Fp of the *E. coli* SQR is flavinylated in the same cells. Cytochrome b_{558} is synthesized without a cleaved signal sequence in the N-terminal part of the protein (67) and seems to be incorporated into the membrane as apocytochrome in concert with synthesis (45, 74). In the absence of the cytochrome b_{558} anchor polypeptide or under heme starvation, the Fp and Ip subunits remain in the cytoplasm; i.e., the membrane-extrinsic subunits cannot be bound to apocytochrome. Either the apocytochrome and the holocytochrome in the membrane have different conformations or the heme in cytochrome b_{558} directly contributes to the binding of Fp and Ip (74, 75, 85). It has been demonstrated that the heme environment in cytochrome b_{558} differs depending on whether Fp and Ip is bound (55). Neither the Fp nor the Ip subunit can individually bind to cytochrome b_{558} in the membrane (74); i.e., they can be bound only as heterodimers. SQR is also assembled in mutants lacking the covalently bound FAD, but these enzymes cannot oxidize succinate even though all three iron-sulfur centers with normal properties are present (65, 72, 99).

The Fp consists of 296 residues (67, 127). Information from sequence comparisons in conjunction with mutant data have helped identify functionally important residues and structural domains in Fp (reviewed in reference 73). Segments in the N-terminal region close to His-40 and a segment around residue 370 probably make contacts with the AMP part of the FAD molecule in the folded protein (65, 127). A mutant Fp subunit with Asp instead of Gly at position 47 lacks FAD, possibly because of steric hindrance in the FAD-binding "pocket" (99). Residue Ala-252 is located in or close to the dicarboxylate-binding site according to site-specific mutagenesis experiments that demonstrate (i) that exchange of this residue with Cys does not inactivate the enzyme but renders it very sensitive to thiol-modifying reagents and (ii) that substrates protect the enzyme against inactivation (69). The adjacent Arg-253 possibly makes a bidentate ionic pair with one carboxyl of succinate-fumarate and thereby orients the substrate at the active site. Maybe the conserved Arg-113 makes a similar pair with the other carboxyl of the substrate, since exchange of the close and conserved Gly-116 with Glu inactivates *B. subtilis* SQR (65). The unprotonated form of His-236 in the triad His-Pro-Thr may have a proton receptor function at succinate oxidation (73).

Ip has 252 residues and contains 11 conserved Cys residues arranged in three clusters (127). These Cys residues, or at least 10 of them, are ligands to the iron-sulfur centers (1, 73). The first cluster of Cys around residue 70 contains ligands to the [2Fe-2S] center, and residues 1 to 147 constitute a stable domain that apparently can interact with the Fp subunit (4, 72, 98). This suggests that Ip consist of two domains. The N-terminal domain has a ferredoxinlike structure and contains the [2Fe-2S] center. The second domain would be composed of the C-terminal half of Ip and contains both the [3Fe-4S] and the [4Fe-4S] centers. The midpoint redox potentials of the iron-sulfur centers in *B. subtilis* SQR (Table 4) are very similar to those of *Micrococcus luteus* SQR (23) but different from those in mammalian and *E. coli* SQR in that the potential of the [3Fe-4S] center is roughly 50 mV lower in gram-positive bacteria. The relatively low potential of the [3Fe-4S] center has been explained by the presence of menaquinone (E_m = −74 mV) as electron acceptor in the membrane of the gram-positive bacteria, whereas aerobic *E. coli* and mammalian mitochondria contain ubiquinone (E_m = +112 mV). The low potential of the center adjusts the enzyme to the potential of the electron acceptor in the membrane (1, 73). The *B. subtilis* SQR is apparently a better fumarate reductase (menaquinol oxidase) than SQR (93).

The 203-residue cytochrome b_{558} has five transmembrane, probably α-helical segments (45, 67, 76, 96). The N terminus is in the cytoplasm, and the C terminus is on the outside of the cytoplasmic membrane. Studies using biophysical techniques (44, 55, 66), mutant data (45, 47, 68), and results from sequence comparisons (73) strongly suggest that His-28, His-70, His-113, and His-155 function as axial ligands to the irons of the two hemes. Tentatively, His-28 and

His-113 are axial ligands to one heme, and His-70 and His-155 are axial ligands to the other heme. These four His residues are located in or at the borders of four different predicted transmembrane segments. It has not been established which His pair ligates cytb_H and cytb_L (55). Why the cytochrome, similar to *Wolinella succinogenes* fumarate reductase but unlike other SQRs studied so far, contains two heme components is not understood.

FUMARASE

Fumarate and malate are readily interconverted by the action of fumarase (fumarate hydratase; EC 4.2.1.2): fumarate + H_2O ↔ L-malate. Mammalian, *S. cerevisiae*, and *B. subtilis* fumarases are homotetramers of 200 kDa. Eukaryotic organisms contain fumarase both in the mitochondrion and in the cytoplasm, and these two isoenzymes differ only in their N termini. In *S. cerevisiae*, the two isoenzymes are encoded by the same nuclear gene; two transcripts with different 5′ ends result in one product that is retained in the cytoplasm and one that is translocated into the mitochondrion (166). *E. coli* has three fumarase genes, *fumA*, *fumB*, and *fumC* (54, 165). FumA and FumB are class I fumarases, which are 120-kDa homodimeric, labile iron-sulfur proteins. FumC is similar to the yeast, mammalian, and *B. subtilis* fumarases, i.e., tetrameric and rather stable. *Bradyrhizobium japonicum* seems to contain two different fumarases, one each of class I and class II (2).

The structural gene for *B. subtilis* fumarase, *citG* (Table 1), was cloned by complementation of an *E. coli fumA* mutant with a recombinant lambda clone (111, 112). A 49-kDa polypeptide is encoded by the cloned *citG* gene, confirming the 50.4-kDa mass deduced from the nucleotide sequence (107). *B. subtilis* fumarase (CitG) shows 56% amino acid sequence identity with the fumarase of *S. cerevisiae* (166) and 65% identity with *E. coli* FumC. The *gerA* gene precedes *citG* but is transcribed in the opposite direction (32, 172). In the 369-bp region separating the open reading frames of *gerA* and *citG*, there is an open reading frame (*orfA*) encoding a putative 61-residue polypeptide. This putative protein does not seem involved in regulation of *citG* expression (129). Two promoters, *citGp1* and *citGp2*, direct transcription of *citG* (33). The former promoter is of the σ^A type and is located upstream of *orfA*. This promoter is functional in *E. coli* and seems to have two transcriptional start sites separated by 13 bp. In *B. subtilis*, transcription from this promoter is relatively low and is independent of growth medium and growth phase (113, 129). Transcription from *citGp2* occurs with RNA polymerase containing σ^H (33, 113, 129, 155) and is strongly dependent on the growth medium; transcription is repressed in minimal glucose medium containing Casamino Acids and derepressed in minimal lactate medium (129). This pattern reflects that of fumarase activities found in *B. subtilis* grown in different media, suggesting that fumarase is largely regulated at the transcriptional level. *B. subtilis* fumarase mutants (15) growing in a rich (broth) medium have, like ODHC and SQR mutants, repressed levels of malate dehydrogenase activity (120). Addition of malate to the medium, as expected, enhances the growth of a fumarase mutant (120).

MALATE DEHYDROGENASE

Oxidation of malate in the CAC is catalyzed by malate dehydrogenase (L-malate:NAD oxidoreductase; EC 1.1.1.37): L-malate + NAD^+ ↔ oxaloacetate + NADH + H^+. Eukaryotic cells contain two isoenzymes, a cytoplasmic and a mitochondrial malate dehydrogenase. Both of these are encoded by nuclear genes and are homodimers with about 35-kDa subunits. The crystal structure is known for porcine heart mitochondrial (142) and cytoplasmic (10) malate dehydrogenases. These two isozymes have similar structures, although the primary structures show only about 20% sequence identity (103). The *E. coli* enzyme is a 70-kDa homodimer very similar to the mitochondrial isozyme; the deduced amino sequence of the *E. coli* enzyme shows about 50% identity to that of the mitochondrial enzymes of *S. cerevisiae* and porcine heart (103).

Malate dehydrogenase of *Bacillus* species, e.g., *B. subtilis* and *B. stearothermophilus*, has a quaternary structure different from those of the enzymes mentioned above. Purified *Bacillus* enzymes are homotetramers with 30- to 35-kDa subunits (115, 168, 169). The occurrence of larger and smaller types of malate dehydrogenase seems not to be correlated with gram-positive or gram-negative bacteria as citrate synthase and succinate thiokinase are (162). The *B. subtilis* and *B. stearothermophilus* malate dehydrogenases must be very similar, since they cross-react immunologically. However, antisera against the *Bacillus* enzymes do not recognize the *E. coli* malate dehydrogenase and vice versa (115).

Little information on the genetics of malate dehydrogenase in gram-positive bacteria is available, but *B. subtilis* mutants defective in this CAC enzyme have been isolated (15, 120). Such $CitH^-$ mutants can grow on the tricarboxylic acid glutamate or malate (15, 28). Oxaloacetate is obtained in these mutants by the concerted action of malic enzyme and pyruvate carboxylase. The *citH* locus is located in the same region of the chromosome as *citC* (Table 1 and Fig. 2), but *citH* and *citC* are separated by the *polA* locus (49). It has not been established that *citH* is the structural gene for malate dehydrogenase.

NOTE ADDED IN PROOF

The citrate synthase encoded by the *B. subtilis* $citA_2$ gene is required for synthesis of glutamate and for sporulation (88). Deletion of the $citA_1$ gene has no detectable phenotype, but this deletion increases the severity of the sporulation defect of a $citA_2$ mutant.

Acknowledgments. I am grateful to T. Clementz, M. Hansson, and I. Schröder for generous help with computers. This work was supported by grants from the Swedish Natural Science Research Council, the Swedish Medical Research Council, and the Emil och Wera Cornells Stiftelse.

REFERENCES

1. **Ackrell, B. A. C., M. K. Johnson, R. P. Gunsalus, and G. Cecchini.** 1992. Structure and function of succinate

dehydrogenase and fumarate reductase, p. 229–297. *In* F. Müller (ed.), *Chemistry and Biochemistry of Flavoenzymes*, vol III. CRC Press, Boca Raton, Fla.

2. **Acuna, G., S. Ebeling, and H. Hennecke.** 1991. Cloning, sequencing, and mutational analysis of the *Bradyrhizobium japonicum fumC*-like gene: evidence for the existence of two different fumarases. *J. Gen. Microbiol.* **137:**991–1000.
3. **Adler, L.-Å, and S. Arvidson.** 1988. Cloning and expression in *Escherichia coli* of genes encoding a multiprotein complex involved in secretion of proteins from *Staphylococcus aureus*. *J. Bacteriol.* **170:**5337–5343.
4. **Aevarsson, A., and L. Hederstedt.** 1988. Ligands to the 2Fe iron-sulfur center in succinate dehydrogenase. *FEBS Lett.* **232:**298–302.
5. **Allen, A. G., and R. N. Perham.** 1991. Two lipoyl domains in the dihydrolipoamide acetyltransferase chain of the pyruvate dehydrogenase multienzyme complex of *Streptococcus faecalis*. *FEBS Lett.* **287:**206–210.
6. **Allison, N., C. H. Williams, Jr., and J. R. Guest.** 1988. Overexpression and mutagenesis of the lipoamide dehydrogenase of *Escherichia coli*. *Biochem. J.* **256:**741–749.
7. **Amjad, M., J. M. Castro, H. Sandoval, J.-J. Wu, M. Yang, D. J. Henner, and P. J. Piggot.** 1990. An *Sfi*I restriction map of the *Bacillus subtilis* 168 genome. *Gene* **101:**15–21.
8. **Barnes, S. J., and P. D. J. Weitzman.** 1986. Organization of citric acid cycle enzymes into a multienzyme cluster. *FEBS Lett.* **201:**267–270.
9. **Beinert, H., and A. J. Thomson.** 1983. Three-iron clusters in iron-sulfur proteins. *Arch. Biochem. Biophys.* **222:**333–361.
10. **Birktoft, J. J., R. A. Bradshaw, and L. J. Banaszak.** 1987. Structure of porcine heart cytoplasmic malate dehydrogenase: combining X-ray diffraction and chemical sequence data in structural studies. *Biochemistry* **26:**2722–2734.
11. **Borges, A., C. F. Hawkins, L. C. Packman, and R. N. Perham.** 1990. Cloning and sequence analysis of the genes encoding the dihydrolipoamide acetyltransferase and dihydrolipoamide dehydrogenase components of the pyruvate dehydrogenase multienzyme complex of *Bacillus stearothermophilus*. *Eur. J. Biochem.* **194:**95–102.
12. **Bosma, H. J., A. de Kok, A. H. Westphal, and C. Veeger.** 1984. The composition of the pyruvate dehydrogenase complex from *Azotobacter vinelandii*. *Eur. J. Biochem.* **142:**541–549.
13. **Botsford, J. L., and J. G. Harman.** 1992. Cyclic AMP in procaryotes. *Microbiol. Rev.* **56:**100–122.

13a. **Boudreaux, D. P., E. Eisenstadt, T. Iijima, and E. Freese.** 1981. Biochemical and genetic characterization of an auxotroph of *Bacillus subtilis* altered in the acyl-CoA:acyl-carrier-protein transacylase. *Eur. J. Biochem.* **115:**175–181.

14. **Brookfield, D. E., J. Green, S. T. Ali, R. S. Machado, and J. R. Guest.** 1991. Evidence for two protein-lipoyl activities in *Escherichia coli*. *FEBS Lett.* **295:**13–16.

14a. **Buxton, R. S.** 1978. A heat-sensitive lysis mutant of *Bacillus subtilis* 168 with low activity of pyruvate carboxylase. *J. Gen. Microbiol.* **105:**175–185.

15. **Carls, R. A., and R. S. Hanson.** 1971. Isolation and characterization of tricarboxylic acid cycle mutants of *Bacillus subtilis*. *J. Bacteriol.* **106:**848–855.
16. **Carlsson, P., and L. Hederstedt.** 1986. In vitro complementation of *Bacillus subtilis* and *Escherichia coli* 2-oxoglutarate dehydrogenase complex mutants and genetic mapping of *B. subtilis citK* and *citM* mutations. *FEMS Microbiol. Lett.* **37:**373–378.
17. **Carlsson, P., and L. Hederstedt.** 1987. *Bacillus subtilis citM*, the structural gene for dihydrolipoamide transsuccinylase: cloning and expression in *Escherichia coli*. *Gene* **61:**217–224.
18. **Carlsson, P., and L. Hederstedt.** 1989. Genetic characterization of *Bacillus subtilis odhA* and *odhB*, encoding 2-oxoglutarate dehydrogenase and dihydrolipoamide transsuccinylase, respectively. *J. Bacteriol.* **171:**3667–3672.
19. **Carlsson, P., and L. Hederstedt.** Unpublished data.
20. **Chen, N.-Y., F.-M. Hu, and H. Paulus.** 1987. Nucleotide sequence of the overlapping genes for the subunits of *Bacillus subtilis* aspartokinase II and their control genes. *J. Biol. Chem.* **262:**8787–8798.
21. **Chen, N.-Y., J.-J. Zhang, and H. Paulus.** 1989. Chromosomal location of the *Bacillus subtilis* aspartokinase II gene and nucleotide sequence of the adjacent genes homologous to *uvrC* and *trx* of *Escherichia coli*. *J. Gen. Microbiol.* **135:**2931–2940.
22. **Crowe, B. A., and P. Owen.** 1983. Molecular properties of succinate dehydrogenase isolated from *Micrococcus luteus* (*lysodeikticus*). *J. Bacteriol.* **153:**1493–1501.
23. **Crowe, B. A., P. Owen, D. S. Patil, and R. Cammack.** 1983. Characterization of succinate dehydrogenase from *Micrococcus luteus* (*lysodeikticus*) by electron-spin-resonance spectroscopy. *Eur. J. Biochem.* **137:**191–196.
24. **Cutting, S., and J. Mandelstam.** 1986. The nucleotide sequence and transcription during sporulation of the *gerE* gene of *Bacillus subtilis*. *J. Gen. Microbiol.* **132:** 3013–3024.
25. **Danson, M. J., S. C. Black, D. L. Woodland, and P. A. Wood.** 1985. Citric acid cycle enzymes of the archaebacteria: citrate synthase and succinate thiokinase. *FEBS Lett.* **179:**120–124.
26. **Dardel, F., E. D. Laue, and R. N. Perham.** 1991. Sequence-specific ^{1}H-NMR assignments and secondary structure of the lipoyl domain of the *Bacillus stearothermophilus* pyruvate dehydrogenase multienzyme complex. *Eur. J. Biochem.* **201:**203–209.
27. **Dardel, F., L. C. Packman, and R. N. Perham.** 1990. Expression in *Escherichia coli* of a sub-gene encoding the lipoyl domain of the pyruvate dehydrogenase complex of *Bacillus stearothermophilus*. *FEBS Lett.* **264:**206–210.
28. **Diesterhaft, M. D., and E. Freese.** 1973. Role of pyruvate carboxylase, phosphoenolpyruvate carboxykinase, and malic enzyme during growth and sporulation of *Bacillus subtilis*. *J. Biol. Chem.* **248:**6062–6070.
29. **Dingman, D. W., M. Rosenkrantz, and A. L. Sonenshein.** 1987. Relationship between aconitase gene expression and sporulation in *Bacillus subtilis*. *J. Bacteriol.* **169:**3068–3075.
30. **Dingman, D. W., and A. L. Sonenshein.** 1987. Purification of aconitase from *Bacillus subtilis* and correlation of its N-terminal amino acid sequence with the sequence of the *citB* gene. *J. Bacteriol.* **169:**3062–3067.
31. **Donald, L. J., G. F. Molgat, and H. W. Duckworth.** 1989. Cloning, sequencing and expression of the gene for NADH-sensitive citrate synthase of *Pseudomonas aeruginosa*. *J. Bacteriol.* **171:**5542–5550.
32. **Feavers, I. M., J. S. Miles, and A. Moir.** 1985. The nucleotide sequence of a spore germination gene (*gerA*) of *Bacillus subtilis* 168. *Gene* **38:**95–102.
33. **Feavers, I. M., V. Price, and A. Moir.** 1988. The regulation of the fumarase (*citG*) gene of *Bacillus subtilis* 168. *Mol. Gen. Genet.* **211:**465–471.
34. **Fisher, S. H., and B. Magasanik.** 1984. Synthesis of oxaloacetate in *Bacillus subtilis* mutants lacking the 2-ketoglutarate dehydrogenase enzymatic complex. *J. Bacteriol.* **158:**55–62.
35. **Fisher, S. H., and B. Magasanik.** 1984. 2-Ketoglutarate and the regulation of aconitase and histidase formation in *Bacillus subtilis*. *J. Bacteriol.* **158:**379–382.
36. **Fisher, S. H., and A. L. Sonenshein.** 1991. Control of

carbon and nitrogen metabolism in *Bacillus subtilis*. *Annu. Rev. Microbiol.* **45:**107–135.

37. **Flechtner, V. R., and R. S. Hanson.** 1969. Coarse and fine control of citrate synthase from *Bacillus subtilis*. *Biochem. Biophys. Acta* **184:**252–262.
38. **Flechtner, V. R., and R. S. Hanson.** 1970. Regulation of the tricarboxylic acid cycle in bacteria. A comparison of citrate synthases from different bacteria. *Biochem. Biophys. Acta* **222:**253–264.
39. **Fortnagel, P., and E. Freese.** 1968. Analysis of sporulation mutants. II. Mutants blocked in the citric acid cycle. *J. Bacteriol.* **95:**1431–1438.
40. **Fortnagel, P., and E. Freese.** 1968. Inhibition of aconitase by chelation of transition metals causing inhibition of sporulation in *Bacillus subtilis*. *J. Biol. Chem.* **243:**5289–5295.
41. **Fouet, A., S.-F. Jin, G. Raffel, and A. L. Sonenshein.** 1990. Multiple regulatory sites in the *Bacillus subtilis citB* promoter region. *J. Bacteriol.* **172:**5408–5415.
42. **Fouet, A., and A. L. Sonenshein.** 1990. A target for carbon source-dependent negative regulation of the *citB* promoter of *Bacillus subtilis*. *J. Bacteriol.* **172:**835–844.
43. **Freese, E., and P. Fortnagel.** 1967. Analysis of sporulation mutants. I. Response of uracil incorporation to carbon sources, and other mutant properties. *J. Bacteriol.* **94:**1957–1969.

43a. **Freese, E., and U. Fortnagel.** 1969. Growth and sporulation of *Bacillus subtilis* mutants blocked in the pyruvate dehydrogenase complex. *J. Bacteriol.* **99:**745–756.

44. **Fridén, H., M. R. Cheesman, L. Hederstedt, K. K. Andersson, and A. J. Thomson.** 1990. Low temperature EPR and MCD studies on cytochrome *b*-558 of the *Bacillus subtilis* succinate:quinone oxidoreductase indicate bis-histidine coordination of the heme iron. *Biochem. Biophys. Acta* **1041:**207–215.
45. **Fridén, H., and L. Hederstedt.** 1990. Role of His residues in *Bacillus subtilis* cytochrome *b*558 for haem binding and assembly of succinate:quinone oxidoreductase (complex II). *Mol. Microbiol.* **4:**1045–1056.
46. **Fridén, H., L. Hederstedt, and L. Rutberg.** 1987. Deletion of the *Bacillus subtilis sdh* operon. *FEMS Microbiol. Lett.* **41:**203–206.
47. **Fridén, H., L. Rutberg, K. Magnusson, and L. Hederstedt.** 1987. Genetic and biochemical characterization of *Bacillus subtilis* mutants defective in expression and function of cytochrome *b*-558. *Eur. J. Biochem.* **168:**695–701.
48. **Gangloff, S. P., D. Marguet, and G. J.-M. Lauquin.** 1990. Molecular cloning of the yeast mitochondrial aconitase gene (*ACO1*) and evidence of a synergistic regulation of expression by glucose plus glutamate. *Mol. Cell. Biol.* **10:**3551–3561.
49. **Gass, K. B., and N. R. Cozzarelli.** 1973. Further genetic and enzymological characterization of the three *Bacillus subtilis* deoxyribonucleic acid polymerases. *J. Biol. Chem.* **248:**7688–7700.
50. **Goldstein, B. J., and S. A. Zahler.** 1976. Uptake of branched-chain α-keto acids in *Bacillus subtilis*. *J. Bacteriol.* **127:**667–670.
51. **Griffin, T. A., R. M. Wynn, and D. T. Chuang.** 1990. Expression and assembly of mature apotransacylase (E_{2b}) of bovine branched-chain α-keto acid dehydrogenase complex in *Escherichia coli*. *J. Biol. Chem.* **265:**12104–12110.
52. **Grundy, F. J., and T. H. Henkin.** 1992. Characterization of the *Bacillus subtilis acsA* gene, encoding acetyl-CoA synthetase. Poster abstr. no. 80. Eleventh International Spores Conference. Woods Hole, Mass.
53. **Guest, J. R., and H. M. Lewis.** 1985. Genetic reconstruction and functional analysis of repeating lipoyl domains in the pyruvate dehydrogenase multienzyme complex of *Escherichia coli*. *J. Mol. Biol.* **185:**743–754.
54. **Guest, J. R., and G. C. Russell.** 1992. Complexes and complexities of the citric acid cycle in *Escherichia coli*. *Curr. Top. Cell. Regul.* **33:**231–247.
55. **Hägerhäll, C., R. Aasa, C. von Wachenfeldt, and L. Hederstedt.** 1992. Two hemes in *Bacillus subtilis* succinate:menaquinone oxidoreductase (complex II). *Biochemistry* **31:**7411–7421.
56. **Hägerhäll, C., M. Jönsson, and L. Hederstedt.** Unpublished data.
57. **Hanamaaijer, R., A. Janssen, A. de Kok, and C. Veeger.** 1988. The dihydrolipoyltransacetylase component of the pyruvate dehydrogenase complex from *Azotobacter vinelandii*. *Eur. J. Biochem.* **174:**593–599.
58. **Hanson, R. S., V. R. Srinivasan, and H. O. Halvorson.** 1963. Biochemistry of sporulation. II. Enzymatic changes during sporulation of *Bacillus cereus*. *J. Bacteriol.* **86:**45–50.
59. **Hanson, R. S., V. R. Srinivasan, and H. O. Halvorson.** 1963. Biochemistry of sporulation. I. Metabolism of acetate by vegetative and sporulating cells. *J. Bacteriol.* **85:**451–460.
60. **Hasnain, S., R. Sammons, I. Roberts, and C. M. Thomas.** 1985. Cloning and deletion analysis of a genomic segment of *Bacillus subtilis* coding for the *sdhA, B, C* (succinate dehydrogenase) and *gerE* (spore germination) loci. *J. Gen. Microbiol.* **131:**2269–2279.
61. **Hawkins, C. F., A. Borges, and R. N. Perham.** 1990. Cloning and sequence analysis of the gene encoding the α and β subunits of the E1 component of the pyruvate dehydrogenase multienzyme complex of *Bacillus stearothermophilus*. *Eur. J. Biochem.* **191:**337–346.
62. **Hederstedt, L.** 1980. Cytochrome *b* reducible by succinate in an isolated succinate dehydrogenase-cytochrome *b* complex from *Bacillus subtilis*. *J. Bacteriol.* **144:**933–940.
63. **Hederstedt, L.** 1983. Succinate dehydrogenase mutants of *Bacillus subtilis* lacking covalently bound flavin in the flavoprotein subunit. *Eur. J. Biochem.* **132:**589–593.
64. **Hederstedt, L.** 1986. Molecular properties, genetics, and biosynthesis of *Bacillus subtilis* succinate dehydrogenase complex. *Methods Enzymol.* **126:**399–414.
65. **Hederstedt, L.** 1987. Covalent binding of FAD to *Bacillus subtilis* succinate dehydrogenase, p. 729–735. *In* D. E. Edmondson and D. M. McCormick (ed.), *Flavins and Flavoproteins*. Walter de Gruyter & Co., Berlin.
66. **Hederstedt, L., and K. K. Andersson.** 1986. Electron-paramagnetic-resonance spectroscopy of *Bacillus subtilis* cytochrome b_{558} in *Escherichia coli* membranes and in succinate dehydrogenase complex from *Bacillus subtilis* membranes. *J. Bacteriol.* **167:**735–739.
67. **Hederstedt, L., T. Bergman, and H. Jörnvall.** 1987. Processing of *Bacillus subtilis* succinate dehydrogenase and cytochrome *b*-558 polypeptides. *FEBS Lett.* **213:**385–390.
68. **Hederstedt, L., C. Hägerhäll, H. Fridén, and R. Aasa.** Unpublished data.
69. **Hederstedt, L., and L.-O. Hedén.** 1989. New properties of *Bacillus subtilis* succinate dehydrogenase altered at the active site. *Biochem. J.* **260:**491–497.
70. **Hederstedt, L., E. Holmgren, and L. Rutberg.** 1979. Characterization of a succinate dehydrogenase complex solubilized from the cytoplasmic membrane of *Bacillus subtilis* with the nonionic detergent Triton X-100. *J. Bacteriol.* **138:**370–376.
71. **Hederstedt, L., K. Magnusson, and L. Rutberg.** 1982. Reconstitution of succinate dehydrogenase in *Bacillus subtilis* by protoplast fusion. *J. Bacteriol.* **152:**157–165.
72. **Hederstedt, L., J. J. Maguire, A. J. Waring, and T. Ohnishi.** 1985. Characterization by electron paramagnetic resonance and studies on subunit location and assembly of the iron-sulfur clusters of *Bacillus subtilis* succinate dehydrogenase. *J. Biol. Chem.* **260:**5554–5562.
73. **Hederstedt, L., and T. Ohnishi.** 1992. Progress in succi-

nate:quinone oxidoreductase research, p. 163–198. *In* L. Ernster (ed.), *Molecular Mechanisms in Bioenergetics*, in press. Elsevier, Amsterdam.

74. **Hederstedt, L., and L. Rutberg.** 1980. Biosynthesis and membrane binding of succinate dehydrogenase in *Bacillus subtilis*. *J. Bacteriol.* **144:**941–951.
75. **Hederstedt, L., and L. Rutberg.** 1981. Succinate dehydrogenase: a comparative review. *Microbiol. Rev.* **45:** 542–555.
76. **Hederstedt, L., and L. Rutberg.** 1983. Orientation of succinate dehydrogenase and cytochrome b_{558} in the *Bacillus subtilis* cytoplasmic membrane. *J. Bacteriol.* **153:**57–65.
77. **Hemilä, H.** 1991. Lipoamide dehydrogenase of *Staphylococcus aureus*: nucleotide sequence and sequence analysis. *Biochem. Biophys. Acta* **1129:**119–123.
78. **Hemilä, H.** 1991. Sequence of a PAL-related lipoprotein from *Bacillus subtilis*. *FEMS Microbiol. Lett.* **82:**37–42.
79. **Hemilä, H., A. Palva, L. Paulin, L. Adler, S. Arvidson, and I. Palva.** 1991. Secretory S complex identified as pyruvate dehydrogenase. *Res. Microbiol.* **142:**779–785.
80. **Hemilä, H., A. Palva, L. Paulin, S. Arvidson, and I. Palva.** 1990. Secretory S complex of *Bacillus subtilis*: sequence analysis and identity to pyruvate dehydrogenase. *J. Bacteriol.* **172:**5052–5063.
81. **Henderson, C. E., and R. N. Perham.** 1980. Purification of the pyruvate dehydrogenase multienzyme complex of *Bacillus stearothermophilus* and resolution of its four component polypeptides. *Biochem. J.* **189:**161–172.
82. **Hoch, J. A., and H. J. Coukoulis.** 1978. Genetics of the α-ketoglutarate dehydrogenase complex of *Bacillus subtilis*. *J. Bacteriol.* **133:**265–269.
83. **Hoch, J. A., and J. Mathews.** 1972. Genetic studies in *Bacillus subtilis*, p. 113–116. *In* H. O. Halvorson, R. Hanson, and L. L. Campbell (ed.), *Spores V*. American Society for Microbiology, Washington, D.C.
84. **Hodgson, J. A., P. N. Lowe, and R. N. Perham.** 1983. Wild-type and mutant forms of the pyruvate dehydrogenase multienzyme complex from *Bacillus subtilis*. *Biochem. J.* **211:**463–472.
85. **Holmgren, E., L. Hederstedt, and L. Rutberg.** 1979. Role of heme in synthesis and membrane binding of succinic dehydrogenase in *Bacillus subtilis*. *J. Bacteriol.* **138:**377–382.
86. **Howard, R. L., and R. R. Becker.** 1970. Isolation and some properties of the triphosphopyridine nucleotide isocitrate dehydrogenase from *Bacillus stearothermophilus*. *J. Biol. Chem.* **245:**3186–3194.
87. **Itaya, M., and T. Tanaka.** 1991. Complete physical map of the *Bacillus subtilis* 168 chromosome constructed by a gene-directed mutagenesis method. *J. Mol. Biol.* **220:** 631–648.
88. **Jin, S., and A. L. Sonenshein.** Unpublished data.
89. **Jin, S., and A. L. Sonenshein.** 1992. Citrate synthase genes in *Bacillus subtilis*. Abstr. 84. Eleventh International Spores Conference. Woods Hole, Mass.
90. **Johnson, D. E., and R. S. Hanson.** 1974. Bacterial citrate synthases: purification, molecular weight and kinetic mechanism. *Biochim. Biophys. Acta* **350:**336–353.
91. **Kay, J., and P. D. J. Weitzman (ed.).** 1987. Krebs' citric acid cycle—half a century and still turning. *Biochem. Soc. Symp.* **54:**1–198.
92. **Koike, K., S. Ohta, Y. Urata, Y. Kagawa, and M. Koike.** 1988. Cloning and sequencing of cDNA encoding α and β subunits of human pyruvate dehydrogenase. *Proc. Natl. Acad. Sci. USA* **85:**41–45.
93. **Lemma, E., C. Hägerhäll, V. Geisler, U. Brandt, G. von Jagow, and A. Kröger.** 1991. Reactivity of the *Bacillus subtilis* succinate dehydrogenase complex with quinones. *Biochim. Biophys. Acta* **1059:**281–285.
94. **Lowe, P. N., J. A. Hodgson, and R. N. Perham.** 1983. Dual role of a single multienzyme complex in the oxidative decarboxylation of pyruvate and branched-chain 2-oxo acids in *Bacillus subtilis*. *Biochem. J.* **215:** 133–140.
95. **Magnusson, K., L. Hederstedt, and L. Rutberg.** 1985. Cloning and expression in *Escherichia coli* of *sdhA*, the structural gene for cytochrome b_{558} of the *Bacillus subtilis* succinate dehydrogenase complex. *J. Bacteriol.* **162:**1180–1185.
96. **Magnusson, K., M. K. Phillips, J. R. Guest, and L. Rutberg.** 1986. Nucleotide sequence of the gene for cytochrome b_{558} of the *Bacillus subtilis* succinate dehydrogenase complex. *J. Bacteriol.* **166:**1067–1071.
97. **Magnusson, K., B. Rutberg, L. Hederstedt, and L. Rutberg.** 1983. Characterization of a pleiotropic succinate dehydrogenase-negative mutant of *Bacillus subtilis*. *J. Gen. Microbiol.* **129:**917–922.
98. **Maguire, J. J., and L. Hederstedt.** 1989. EPR characterization of soluble fragments of succinate dehydrogenase from mutant strains of *Bacillus subtilis*. *FEBS Lett.* **256:**195–199.
99. **Maguire, J. J., K. Magnusson, and L. Hederstedt.** 1986. *Bacillus subtilis* mutant succinate dehydrogenase lacking covalently bound flavin: identification of the primary defect and studies on the iron-sulfur clusters in mutated and wild-type enzyme. *Biochemistry* **25:**5202–5208.
100. **Mat-Jan, F., C. R. Williams, and D. P. Clark.** 1989. Anaerobic growth defects resulting from gene fusions affecting succinyl-CoA synthetase in *Escherichia coli* K12. *Mol. Gen. Genet.* **215:**276–280.
101. **Mattevi, A., G. Obmolova, E. Schulze, K. H. Kalk, A. H. Westphal, A. de Kok, and W. G. J. Hol.** 1992. Atomic structure of the cubic core of the pyruvate dehydrogenase multienzyme complex. *Science* **255:**1544–1550.
102. **Mattevi, A., A. J. Schierbeek, and W. G. J. Hol.** 1991. Refined crystal structure of lipoamide dehydrogenase from *Azotobacter vinelandii* at 2.2 Å resolution. *J. Mol. Biol.* **220:**975–994.
103. **McAlister-Henn, A.** 1988. Evolutionary relationships among the malate dehydrogenases. *Trends Biochem. Soc.* **13:**178–181.
104. **Melin, L., H. Fridén, E. Dehlin, L. Rutberg, and A. von Gabain.** 1990. The importance of the 5′-region in regulating the stability of *sdh* mRNA in *Bacillus subtilis*. *Mol. Microbiol.* **4:**1881–1889.
105. **Melin, L., K. Magnusson, and L. Rutberg.** 1987. Identification of the promoter of the *Bacillus subtilis sdh* operon. *J. Bacteriol.* **169:**3232–3236.
106. **Melin, L., L. Rutberg, and A. von Gabain.** 1989. Transcriptional and posttranscriptional control of the *Bacillus subtilis* succinate dehydrogenase operon. *J. Bacteriol.* **171:**2110–2115.
107. **Miles, J. S., and J. R. Guest.** 1985. Complete nucleotide sequence of the fumarase gene (*citG*) of *Bacillus subtilis* 168. *Nucleic Acids Res.* **13:**131–140.
108. **Miles, J. S., and J. R. Guest.** 1987. Molecular genetic aspects of the citric acid cycle of *Escherichia coli*. *Biochem. Soc. Symp.* **54:**45–65.
109. **Mitchell, C. G., and P. D. J. Weitzman.** 1986. Molecular size diversity of citrate synthases from *Pseudomonas* species. *J. Gen. Microbiol.* **132:**737–742.
110. **Mitchell, C. G., S. O'Neil, H. C. Reeves, and P. D. J. Weitzman.** 1986. Separation of isoenzymes of citrate synthase and isocitrate dehydrogenase by fast protein liquid chromatography. *FEBS Lett.* **196:**211–214.
111. **Moir, A.** 1983. The isolation of λ transducing phages carrying the *citG* and *gerA* genes of *Bacillus subtilis*. *J. Gen. Microbiol.* **129:**303–310.
112. **Moir, A., I. M. Feavers, and J. R. Guest.** 1984. Characterization of the fumarase gene of *Bacillus subtilis* 168 cloned and expressed in *Escherichia coli* K12. *J. Gen. Microbiol.* **130:**3009–3017.
113. **Moir, A., and V. A. Price.** 1990. Sigma H-directed

transcription from a *citG* promoter is metabolically regulated, p. 277–286. *In* M. M. Zukowski, A. T. Ganesan, and J. A. Hoch (ed.), *Genetics and Biotechnology of Bacilli*. Academic Press, Inc., New York.

114. **Mueller, J. P., and H. W. Taber.** 1989. Isolation and sequence of *ctaA*, a gene required for cytochrome aa_3 biosynthesis and sporulation in *Bacillus subtilis*. *J. Bacteriol.* **171:**4967–4978.

115. **Murphey, W. H., C. Barnaby, F. J. Lin, and N. O. Kaplan.** 1967. Malate dehydrogenases. *J. Biol. Chem.* **242:**1548–1559.

116. **Nimmo, H. G.** 1987. The tricarboxylic acid cycle and anaplerotic reactions, p. 156–169. *In* J. L. Ingraham, K. B. Low, B. Magasanik, M. Schaechter, and H. E. Umbarger (ed.), *Escherichia coli and Salmonella typhimurium: Cellular and Molecular Biology*, vol. 1. American Society for Microbiology, Washington, D.C.

117. **Niu, X.-D., K. S. Browning, R. H. Behal, and L. J. Reed.** 1988. Cloning and nucleotide sequence of the gene for dihydrolipoamide acetyltransferase from *Saccharomyces cerevisiae*. *Proc. Natl. Acad. Sci. USA* **85:**7546–7550.

118. **Ohné, M.** 1974. Regulation of aconitase synthesis in *Bacillus subtilis*: induction, feedback repression, and catabolite repression. *J. Bacteriol.* **117:**1295–1305.

119. **Ohné, M.** 1975. The citric acid cycle of *Bacillus subtilis*. Ph.D. thesis. Karolinska Institutet, Stockholm.

120. **Ohné, M.** 1975. Regulation of the dicarboxylic acid part of the citric acid cycle in *Bacillus subtilis*. *J. Bacteriol.* **122:**224–234.

121. **Ohné, M., B. Rutberg, and J. A. Hoch.** 1973. Genetic and biochemical characterization of mutants of *Bacillus subtilis* defective in succinate dehydrogenase. *J. Bacteriol.* **115:**738–745.

122. **Oku, H., and T. Kaneda.** 1988. Biosynthesis of branched-chain fatty acids in *Bacillus subtilis*. *J. Biol. Chem.* **263:**18386–18396.

123. **Packman, L. C., A. Borges, and R. N. Perham.** 1988. Amino acid sequence analysis of the lipoyl and peripheral subunit-binding domains in the lipoate acetyltransferase component of the pyruvate dehydrogenase complex from *Bacillus stearothermophilus*. *Biochem. J.* **252:**79–86.

124. **Perham, R. N.** 1991. Domains, motifs and linkers in 2-oxo acid dehydrogenase multienzyme complexes: a paradigm in the design of a multifunctional protein. *Biochemistry* **30:**8501–8512.

125. **Petricek, M., L. Rutberg, and L. Hederstedt.** 1989. The structural gene for aspartokinase II in *Bacillus subtilis* is closely linked to the *sdh* operon. *FEMS Microbiol. Lett.* **61:**85–88.

126. **Phang, C.-H., and K. Jeyaseelan.** 1988. Isolation and characterization of *citC* gene of *Bacillus subtilis*, p. 97–100. *In* A. T. Ganesan and J. A. Hoch (ed.), *Genetics and Biotechnology Bacilli*, vol. 2. Academic Press, Inc., New York.

127. **Phillips, M. K., L. Hederstedt, S. Hasnain, L. Rutberg, and J. R. Guest.** 1987. Nucleotide sequence encoding the flavoprotein and the iron-sulfur protein subunits of the *Bacillus subtilis* PY79 succinate dehydrogenase complex. *J. Bacteriol.* **169:**864–873.

128. **Piggot, P. J.** 1989. Revised genetic map of *Bacillus subtilis* 168, p. 1–41. *In* I. Smith, R. A. Slepecky, and P. Setlow (ed.), *Regulation of Procaryotic Development*. American Society for Microbiology, Washington, D.C.

129. **Price, V. A., I. M. Feavers, and A. Moir.** 1989. Role of sigma H in expression of the fumarase gene (*citG*) in vegetative cells of *Bacillus subtilis* 168. *J. Bacteriol.* **171:**5933–5939.

130. **Prodromou, C., P. J. Artymiuk, and J. R. Guest.** 1992. The aconitase of *Escherichia coli*. *Eur. J. Biochem.* **204:**599–609.

131. **Prodromou, C., M. J. Haynes, and J. R. Guest.** 1991. The aconitase of *Escherichia coli*: purification of the enzyme and molecular cloning and map location of the gene (*acn*). *J. Gen. Microbiol.* **137:**2505–2515.

132. **Ramaley, R. F., and M. O. Hudock.** 1973. Purification and properties of isocitrate dehydrogenase (NADP) from *Thermus aquaticus* YT-1, *Bacillus subtilis* 168 and *Chlamydomonas reinhardti* Y-2. *Biochim. Biophys. Acta* **315:**22–36.

133. **Reddy, T. L. P., and M. M. Weber.** 1986. Solubilization, purification, and characterization of succinate dehydrogenase from membranes of *Mycobacterium phlei*. *J. Bacteriol.* **167:**1–6.

134. **Reed, L. J., F. R. Leach, and M. Koike.** 1958. Studies on a lipoic acid-activating system. *J. Biol. Chem.* **232:**123–142.

135. **Reed, L. J., and S. J. Yeaman.** 1987. Pyruvate dehydrogenase. *Enzymes* **18:**77–95.

136. **Resnekov, O., L. Melin, P. Carlsson, M. Mannerlöv, A. von Gabain, and L. Hederstedt.** 1992. Organization and regulation of the *Bacillus subtilis odhAB* operon, which encodes two of the subenzymes of the 2-oxoglutarate dehydrogenase complex. *Mol. Gen. Genet.* **234:**285–296.

137. **Resnekov, O., L. Rutberg, and A. von Gabain.** 1990. Changes in the stability of specific mRNA species in response to growth stage in *Bacillus subtilis*. *Proc. Natl. Acad. Sci. USA* **87:**8355–8359.

138. **Richarme, G.** 1989. Purification of a new dihydrolipoamide dehydrogenase from *Escherichia coli*. *J. Bacteriol.* **171:**6580–6585.

139. **Robbins, A. H., and C. D. Stout.** 1989. Structure of activated aconitase: formation of the [4Fe-4S] cluster in the crystal. *Proc. Natl. Acad. Sci. USA* **86:**3639–3643.

139a. **Robien, M. A., G. M. Clore, J. G. Omichinski, R. N. Perham, E. Appella, K. Sakaguchi, and A. M. Gronenborn.** 1992. Three-dimensional solution structure of the E3-binding domain of the dihydrolipoamide succinyltransferase core from the 2-oxoglutarate dehydrogenase multienzyme complex of *Escherichia coli*. *Biochemistry* **31:**3463–3471.

140. **Robinson, M. S., M. J. Danson, and P. D. J. Weitzman.** 1983. Citrate synthase from a Gram-positive bacterium. *Biochem. J.* **213:**53–59.

141. **Roche, T. E., and M. S. Patel (ed.).** 1989. Alpha-keto acid dehydrogenase complexes: organization, regulation and biomedical ramifications. *Ann. N.Y. Acad. Sci.* **573:**1–474.

142. **Roderick, S. L., and L. J. Banaszak.** 1986. The three-dimensional structure of porcine heart mitochondrial malate dehydrogenase at 3.0-Å resolution. *J. Biol. Chem.* **261:**9461–9464.

143. **Rosenkranz, M. S., D. W. Dingman, and A. L. Sonenshein.** 1985. *Bacillus subtilis citB* gene is regulated synergistically by glucose and glutamine. *J. Bacteriol.* **164:**155–164.

144. **Rouault, T. A., C. D. Stout, S. Kaptain, J. B. Harford, and R. D. Klausner.** 1991. Structural relationship between an iron-regulated RNA-binding protein (IRE-BP) and aconitase: functional implications. *Cell* **64:**881–883.

145. **Russell, G. C., and J. R. Guest.** 1991. Sequence similarities within the family of dihydrolipoamide acyltransferases and discovery of a previously unidentified fungal gene. *Biochem. Biophys. Acta* **1076:**225–232.

146. **Rutberg, B., L. Hederstedt, E. Holmgren, and L. Rutberg.** 1978. Characterization of succinic dehydrogenase mutants of *Bacillus subtilis* by crossed immunoelectrophoresis. *J. Bacteriol.* **136:**304–311.

147. **Rutberg, B., and J. A. Hoch.** 1970. Citric acid cycle: gene-enzyme relationships in *Bacillus subtilis*. *J. Bacteriol.* **104:**826–833.

148. **Rydén, L., L.-G. Öfverstedt, H. Beinert, M. H. Emptage, and M. C. Kennedy.** 1984. Molecular weight of beef heart aconitase and stoichiometry of the components of its iron-sulfur cluster. *J. Biol. Chem.* **259:**3141–3144.

149. **Schendel, F. J., P. R. August, R. Anderson, R. S. Han-**

son, and M. C. Flickinger. 1992. Cloning and nucleotide sequence of the gene coding for citrate synthase from a thermotolerant *Bacillus* sp. *Appl. Environ. Microbiol.* **58:**335–345.

150. **Shikata, S., K. Ozaki, S. Kawai, S. Ito, and K. Okamoto.** 1988. Purification and characterization of $NADP^+$-linked isocitrate dehydrogenase from an alkalophilic *Bacillus*. *Biochim. Biophys. Acta* **952:**282–289.
151. **Snoep, J. L., A. H. Westphal, J. A. E. Benen, M. J. Teixeira de Mattos, O. Neijssel, and A. de Kok.** 1992. Isolation and characterization of the pyruvate dehydrogenase complex of anaerobically grown *Enterococcus faecalis* NCTC 775. *Eur. J. Biochem.* **203:**245–250.

151a. **Sone, N.** Personal communication.

152. **Sonenshein, A. L.** 1989. Metabolic regulation of sporulation and other stationary-phase phenomena, p. 109–128. *In* I. Smith, R. A. Slepecky, and P. Setlow (ed.), *Regulation of Procaryotic Development*. American Society for Microbiology, Washington, D.C.
153. **Spencer, M. E., and J. R. Guest.** 1987. Regulation of citric acid cycle genes in facultative bacteria. *Microbiol. Sci.* **4:**164–168.
154. **Srere, P. A.** 1990. Citric acid cycle redux. *Trends Biochem. Sci.* **15:**411–412.
155. **Tatti, K. M., H. L. Carter III, A. Moir, and C. P. Moran, Jr.** 1989. Sigma H-directed transcription of *citG* in *Bacillus subtilis*. *J. Bacteriol.* **171:**5928–5932.
156. **Thauer, R. K.** 1988. Citric-acid cycle, 50 years on. Modifications and an alternative pathway in anaerobic bacteria. *Eur. J. Biochem.* **176:**497–508.
157. **Thekkumkara, T. J., L. Ho, I. D. Wexler, G. Pons, T.-C. Liu, and M. S. Patel.** 1988. Nucleotide sequence of a cDNA for the dihydrolipoamide acetyltransferase component of human pyruvate dehydrogenase complex. *FEBS Lett.* **240:**45–48.
158. **Uratani-Wong, B., J. M. Lopez, and E. Freese.** 1981. Induction of citric acid cycle enzymes during initiation of sporulation by guanine nucleotide deprivation. *J. Bacteriol.* **146:**337–344.
159. **Utter, M. F., R. E. Barden, and B. L. Taylor.** 1975. Pyruvate carboxylase: an evaluation of the relationships between structure and mechanism and between structure and catalytic activity. *Adv. Enzymol.* **42:**1–72.
160. **Wallbrandt, P., V. Tegman, B.-H. Jonsson, and Å. Wieslander.** 1992. Identification and analysis of the genes coding for the putative pyruvate dehydrogenase enzyme complex of *Acholeplasma laidlawii*. *J. Bacteriol.* **174:**1388–1396.
161. **Weiss, A. S., and R. G. Wake.** 1983. Restriction map of DNA spanning the replication terminus of the *Bacillus subtilis* chromosome. *J. Mol. Biol.* **171:**119–137.
162. **Weitzman, P. D. J.** 1981. Unity and diversity in some bacterial citric acid-cycle enzymes. *Adv. Microb. Physiol.* **22:**185–244.
163. **Wexler, I. D., S. G. Hemalatha, and M. S. Patel.** 1991. Sequence conservation in the α and β subunits of pyruvate dehydrogenase and its similarity to branched-chain α-keto acid dehydrogenase. *FEBS Lett.* **282:**209–213.
164. **Wiegand, G., and S. J. Remington.** 1986. Citrate synthase: structure, control and mechanism. *Annu. Rev. Biophys. Biophys. Chem.* **15:**97–117.

164a. **Willecke, K., and A. B. Pardee.** 1971. Fatty acid-requiring mutant of *Bacillus subtilis* defective in branched chain α-keto acid dehydrogenase. *J. Biol. Chem.* **246:**5264–5272.

165. **Woods, S. A., J. S. Miles, and J. R. Guest.** 1988. Sequence homologies between argininosuccinase, aspartase and fumarase: a family of structurally-related enzymes. *FEMS Microbiol. Lett.* **51:**181–186.
166. **Wu, M., and A. Tzagoloff.** 1987. Mitochondrial and cytoplasmic fumarases in *Saccharomyces cerevisiae* are encoded by a single nuclear gene *FUM1*. *J. Biol. Chem.* **262:**12275–12282.
167. **Yeaman, S. J.** 1986. The mammalian 2-oxoacid dehydrogenases: a complex family. *Trends Biochem.* **11:**293–296.
168. **Yoshida, A.** 1965. Enzymic properties of malate dehydrogenase of *Bacillus subtilis*. *J. Biol. Chem.* **240:**1118–1124.
169. **Yoshida, A.** 1965. Purification and chemical characterization of malate dehydrogenase of *Bacillus subtilis*. *J. Biol. Chem.* **240:**1113–1117.
170. **Zahler, S. A., R. Z. Korman, R. Rosenthal, and H. E. Hemphill.** 1977. *Bacillus subtilis* bacteriophage SPβ: localization of the prophage attachment site and specialized transduction. *J. Bacteriol.* **129:**556–558.
171. **Zheng, L., P. C. Andrews, M. A. Hermodson, J. E. Dixon, and H. Zalkin.** 1990. Cloning and structural characterization of porcine heart aconitase. *J. Biol. Chem.* **265:**2814–2821.
172. **Zuberi, A. R., A. Moir, and I. M. Feavers.** 1987. The nucleotide sequence and gene organization of the *gerA* spore germination operon of *Bacillus subtilis* 168. *Gene* **51:**1–11.

14. Respiratory Chains

HARRY W. TABER

INTRODUCTION

The goal of this chapter is to review briefly some major aspects of the physiology, biochemistry, and molecular genetics of *Bacillus* respiratory chains currently under active study and to suggest some neglected areas to which effort could be directed. It will necessarily reflect the author's biases, but the references will lead the diligent reader to a fuller or more balanced treatment of certain issues. Certain omissions are deliberate. Discussion of the best-studied dehydrogenase complex (succinate dehydrogenase) is to be found in chapter 13 (this volume) because the enzyme can be considered part of the tricarboxylic acid cycle as much as part of the respiratory chain. Similarly, although the flavins are prosthetic groups for the respiratory chain dehydrogenases, their biosynthesis is treated in a separate chapter (see chapter 23, this volume) that has a wider purview. Recent bibliographic references frequently are given precedence over earlier references when the latter can be readily traced from the recent literature. This has been done to contain what would otherwise be an ungovernable reference list and is not meant necessarily to confer more importance on the recent literature.

It is now well over 60 years since the first of David Keilin's classical papers on cytochrome spectra and the nature of cellular respiration were published (58, 59). Microbiologists and molecular geneticists (not to mention biochemists) of the modern era would do well to recall that *Bacillus subtilis* was the first bacterial species in which the characteristic absorption band spectrum of the cytochromes was observed (see reference 60, p. 144–145). However, because of the early and rapid development of genetic technologies for members of the family *Enterobacteriaceae* and the tremendous advances in cell physiological and metabolic studies that ensued (84), respiratory chains in organisms such as *Escherichia coli* and *Salmonella typhimurium* received much closer scrutiny than those in other groups (2, 6, 70, 72, 92). This imbalance is only beginning to be redressed. *Bacillus* spp. present an impressive array of bioenergetic solutions to the problems of maximizing energy production in changing and often hostile environments. Among *Bacillus* spp., there are thermophilic, alkaliphilic, and facultative species as well as obligate aerobes perhaps loosely linked by evolution but with rather similar genetic structures encoding their bioenergetic components.

Membrane-Associated Energy Conservation: a Brief Summary

Membrane-bound respiratory chains catalyze the transfer of reducing equivalents from a reduced substrate to an oxidant (55). This transfer is spontaneous and is based on differences in oxidation-reduction potential among the members of the chain. A release of free energy is associated with transfer of reducing equivalents, and energy is conserved as a transmembrane electrochemical gradient of protons. In *Bacillus* spp. and other bacteria, this gradient is used to drive energy-requiring reactions, including solute transport, ATP synthesis via oxidative phosphorylation, and motility. According to the chemiosmotic theory of Mitchell (see discussion in references 4 and 45), this driving force is generated as a Gibbs free energy change associated with the transfer of protons from the outside of the cell to the inside. This free energy change is most usefully expressed as a proton electrochemical gradient, $\Delta\mu H^+$, and is given by the following equation.

$$\Delta\mu H^+ = -RT \ln \frac{[H^+{}_o]}{[H^+{}_i]} + F\Delta\Psi \text{ (in J} \cdot \text{mol}^{-1}) \quad (1)$$

where $[H^+{}_o]$ and $[H^+{}_i]$ are proton concentrations in the bulk extracellular and intracellular aqueous phases, respectively; $\Delta\Psi$ is the membrane potential (the difference between the electric potentials of the bulk aqueous phases inside and outside the cell); and R, T, and F are, respectively, the gas constant (in Joules per Kelvin), the temperature (in degrees Kelvin), and the Faraday constant (in Joules per volt). Equation 1 is commonly expressed as a voltage, which requires rearrangement of the terms to define the proton motive force, or Δp.

$$\Delta p = \frac{\Delta\mu H^+}{F}$$

$$= \Delta\Psi - \left\{ \left(\frac{RT}{F}\right) \ln \left(\frac{[H^+{}_o]}{[H^+{}_i]}\right) \right\} \text{ (in volts)} \quad (2)$$

Equation 2 can then be simplified to yield the following expression (for 25°C).

$$\Delta p = \Delta\Psi - 59\Delta pH \text{ (in millivolts)} \quad (3)$$

where $\Delta pH = pH_{in} - pH_{out}$. Thus, it is possible to calculate Δp by experimental measurement of $\Delta\Psi$ and ΔpH. Such measurements have yielded Δp values for many neutrophilic bacteria of approximately -200 mV, and this value remains relatively constant in the face of modest downward excursions in extracellular pH away from neutrality. Because of the homeostatic

Harry W. Taber • Wadsworth Center for Laboratories and Research, New York State Department of Health, Albany, New York 12201.

control of intracellular pH by bacteria, this constancy requires that $\Delta\Psi$ be adjusted to compensate for changes in ΔpH.

Because of space limitations, no attempt will be made here to discuss further the principles underlying establishment and maintenance of the proton motive force except to stress that the components of the respiratory chain provide the mechanisms that allow the outward (transmembrane) movement of protons and lead to formation of the proton motive force. Thus, control of the concentration of respiratory chain components (dehydrogenases, iron-sulfur proteins, cytochromes, quinones, and terminal oxidases or reductases) is a prerequisite to cell functionality, particularly in response to changing extracellular availability of reducing equivalents or oxidants.

An unusually lucid introduction to membrane-associated energy conservation in bacteria can be found in an article by Jones (56). The volumes edited by Haddock and Hamilton (45), Anthony (4), and Krulwich (64) can be consulted for comprehensive discussions of bacterial bioenergetics. It should be pointed out that the chemiosmotic theory is not universally accepted in the form that Mitchell originally proposed it, that is, that protons appear in the bulk aqueous phase exterior to the bacterial cell membrane and are drawn from that phase to perform work. Cogent arguments that call into question whether the proton motive force is an energetically significant intermediate in electron transport-driven phosphorylation have been marshaled (e.g., see reference 61). The alternative is, of course, to posit respiratory chain components directly coupled (at least under certain conditions) to membrane-bound energy-consuming processes such as ATP synthesis and solute transport. This topic will be taken up again in the section on the alkaliphilic *Bacillus* spp.

PROSTHETIC GROUPS OF RESPIRATORY COMPONENTS: BIOSYNTHESIS AND MOLECULAR GENETICS OF HEME FORMATION

The cytochromes, a large and diverse set of respiratory chain components, all contain hemes as essential parts of their reaction centers. The chemistry of these hemes contributes to the specific properties of the different cytochromes, and the formation of appropriate amounts of each of the hemes and timely insertion of hemes into newly synthesized apocytochromes are necessary not only to maintain correct stoichiometry within the respiratory chain but also to ensure that free hemes (which are highly reactive) do not accumulate within the cell's cytoplasm or membrane systems.

Structures of the several heme types are shown in Fig. 1. The basic structure is that of an iron tetrapyrrole, with modifications to the periphery of the tetrapyrrole ring. These modifications help establish midpoint potentials for the reaction centers in which the hemes are involved (and thereby position the various cytochromes within the sequence of electron carriers). They also play a role in binding the hemes to their respective apoproteins; for example, heme C is joined covalently to apocytochrome *c* by thioether linkages between the C-2–C-4 groups and cysteines within the polypeptide chain; other hemes are noncovalently bound.

Biosynthesis

Formation of hemes in *B. subtilis* (and presumably in the other *Bacillus* spp.) originates with glutamate and follows the C_5 pathway for formation of 5-aminolevulinic acid (86) and then uroporphyrinogen III, as shown in Fig. 2 (see references 47, 87, and 104 for reviews of gene-enzyme relationships and relevant enzymology). The genes encoding the steps shown in Fig. 2 form a cluster located at 245° on the *B. subtilis* genetic map (7, 8, 77, 88). This cluster has been cloned and sequenced and appears to be organized as an operon (Fig. 3) (47, 87). All of the genes in the *hemAX CDBL* operon except *hemL* (encoding glutamate-1-semialdehyde 2,1-aminotransferase), in which a deletion did not give rise to 5-aminolevulinic acid or heme auxotrophy (47), are essential for heme formation. This latter finding suggests either that the conversion of glutamate-1-semialdehyde to 5-aminolevulinic acid can occur spontaneously or that a second copy of the gene for the aminotransferase is present elsewhere in the chromosome. The *hemX* gene is predicted to encode a hydrophobic protein of unknown function. Cloning of other *hem* loci (8, 76) from *B. subtilis* should provide interesting material for studies on regulation of heme formation. The isolation of mutants specifically defective in synthesis of heme A, D, or O and the cloning of the genes encoding these heme-specific enzymes is an important issue for the future. Presumably, these genes will be found to be regulated in concert with the apocytochromes that have the cognate hemes as their prosthetic groups.

ELECTRON TRANSPORT: COMPONENTS, PATHWAYS, AND GENE-PROTEIN RELATIONSHIPS

Terminal Oxidases

Biochemistry

The terminal oxidases are defined as heme proteins or copper-heme proteins that oxidize lower-potential reductants and use the electrons to reduce oxygen to water. Some are known to function as proton pumps, but even those that do not cause transmembrane proton movement contribute to the proton motive force by consuming protons on the inner side of the cell membrane during the reduction of oxygen (108, 109). In laboratories worldwide, there is intense interest in the structure, function, and regulation of bacterial terminal oxidases. This interest has two principal sources: (i) a recognition that these oxidases are structurally and functionally homologous to their eukaryotic mitochondrial counterparts and (ii) a growing concern with the mechanisms used by bacteria to modify their bioenergetic capacities in response to environmental changes. There is now ample evidence that the bacterial oxidases constitute a family of proteins with highly conserved functional domains and quinol oxidase or cytochrome oxidase activity (17, 42, 100, 102, 103).

The oxidases from *Bacillus* spp. have two, three, or

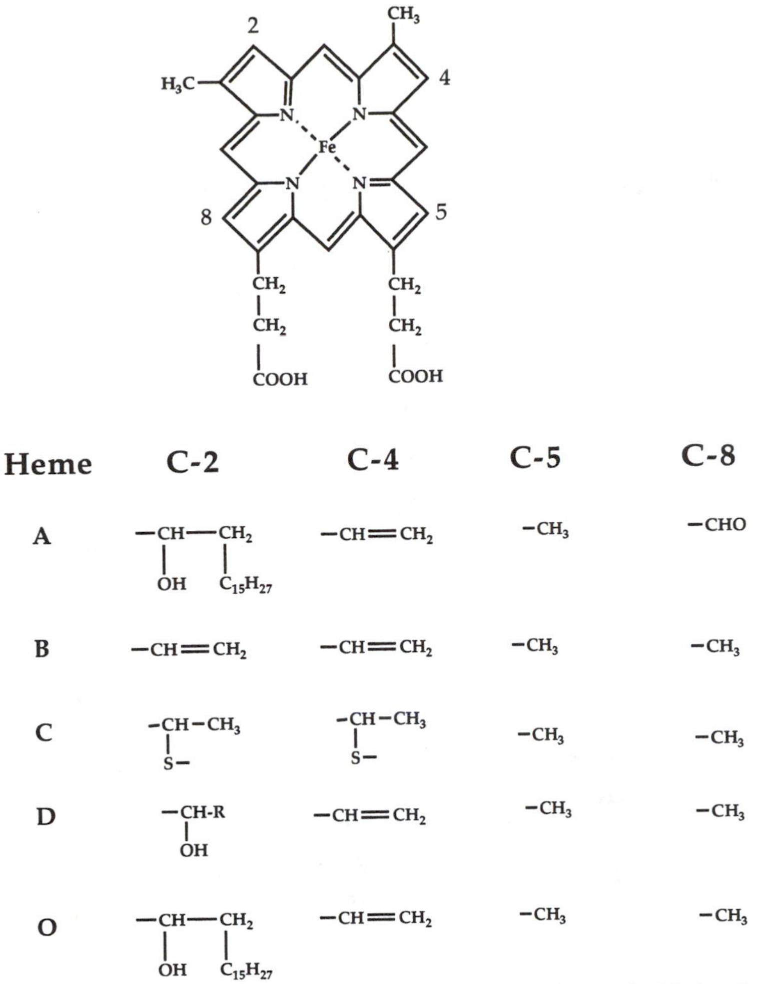

Heme	C-2	C-4	C-5	C-8
A	$-CH(OH)-CH_2(C_{15}H_{27})$	$-CH=CH_2$	$-CH_3$	$-CHO$
B	$-CH=CH_2$	$-CH=CH_2$	$-CH_3$	$-CH_3$
C	$-CH(S-)-CH_3$	$-CH(S-)-CH_3$	$-CH_3$	$-CH_3$
D	$-CH(OH)-R$	$-CH=CH_2$	$-CH_3$	$-CH_3$
O	$-CH(OH)-CH_2(C_{15}H_{27})$	$-CH=CH_2$	$-CH_3$	$-CH_3$

Figure 1. Structures of hemes found in bacterial cytochromes. In heme D, the 8,9 double bond is saturated to create a dihydroporphyrin (chlorin) ring structure. The group at C-2 of heme A is hydroxyethylfarnesyl. Heme O has been proposed (94) to have a grouping at C-2 similar to that of heme A but with a methyl rather than a formyl group at C-1.

four subunits (Table 1), but the enzymes from some other species, such as *Thermus thermophilus* (140), appear to have the necessary reaction centers and functionality confined to a single polypeptide chain (Table 1). In several instances, there is ambiguity about the precise subunit structure, because lower-molecular-weight subunits appear not to be bound tightly to the larger subunits in which the reaction centers are contained. In general, though, subunits I, II, and III in bacterial oxidases have sequences similar to those of the corresponding subunits encoded by mitochondrial DNA in eukaryotes, with two hemes and a copper localized to subunit I and a second copper (if present) in subunit II. In the case of *Paracoccus denitrificans* cytochrome aa_3, subunit III is necessary for functional assembly of subunits I and II (46). A fourth subunit, referred to as IVB (to stress its lack of homology to the eukaryotic subunit IV), has been identified as part of the subunit gene operons for A-type oxidases in *B. subtilis* (101) and *Bacillus* strain PS3 (52) and subsequently has been isolated with the PS3 oxidase complex (114). The method of oxidase preparation has marked effects on the subunit composition, and it is likely that further understanding of the molecular genetics of these enzymes will result in more accurate reconstruction of their subunit compositions. Attention will then need to be directed to the functions of these subunits.

Among *Bacillus* spp., terminal oxidases that have heme A, B, or O forming part of the heme-copper binuclear center for interaction with oxygen in subunit I have been isolated (Table 1). In addition, heme C covalently bound to subunit II may form part of the site of interaction with reductants in the caa_3 oxidases (Table 1; 100, 102). By comparing it with cytochrome *d* (or *bd*) of *E. coli*, heme D in *Bacillus* cytochrome *d* has been inferred to react with oxygen. This inference is supported by spectroscopic evidence that this center binds carbon monoxide (49).

The in vivo reductants for the terminal oxidases are in general not yet defined. For example, although the PS3 caa_3 oxidase will oxidize cytochrome *c*, caa_3 can be combined with a bc_1 complex from the same organism to form a super complex that has quinol oxidase activity that is not enhanced by addition of cytochrome *c* (113). This ambiguity as to proximal

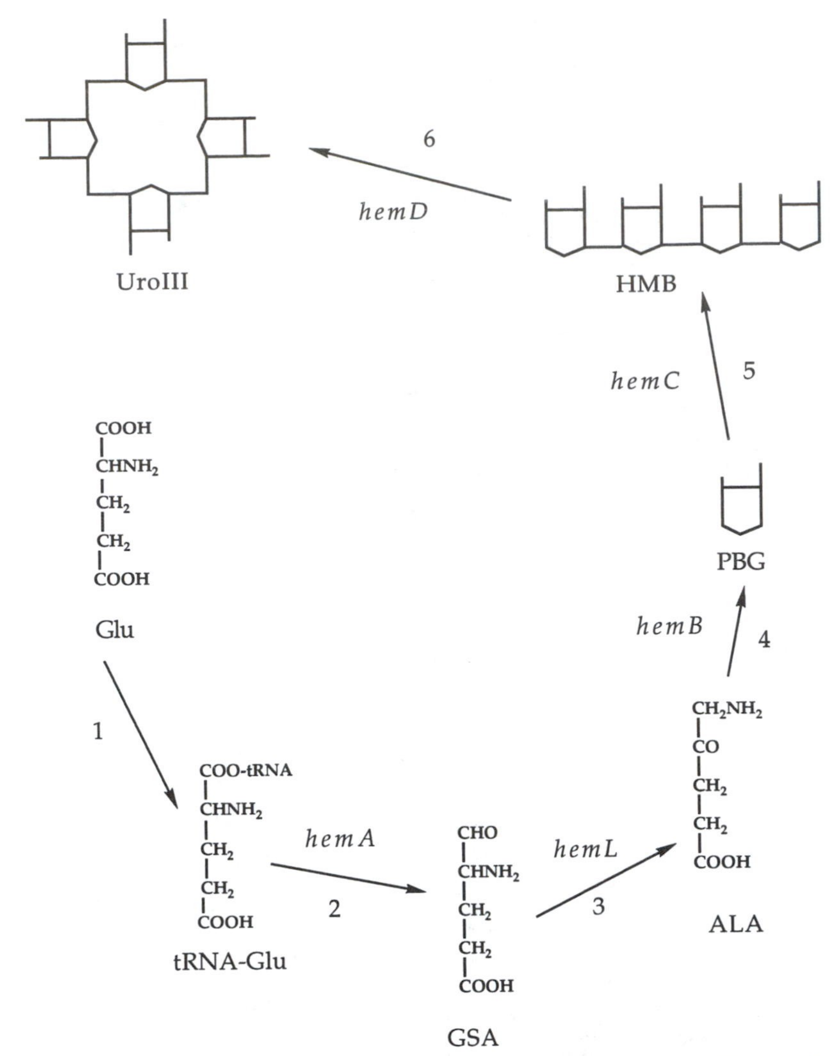

Figure 2. Gene-enzyme relationships in the biosynthesis of uroporphyrinogen III (UroIII) from glutamate (Glu) in *B. subtilis* (47). Protoporphyrin is formed from UroIII by sequential decarboxylation and dehydrogenation reactions, and protoheme is formed from protoporphyrin by incorporation of Fe^{2+}. Enzyme 1, glutamyl-tRNA synthase; enzyme 2, NAD(P)H:glutamyl-tRNA reductase; enzyme 3, glutamate-1-semialdehyde (GSA) 2,1-aminotransferase; enzyme 4, porphobilinogen (PBG) synthase; enzyme 5, hydroxymethylbilane (HMB) synthase; enzyme 6, UroIII synthase. ALA, 5-aminolevulinic acid.

reductant is exemplified by the cytochrome aa_3 of *B. subtilis*, which because of its biochemical similarity to the mitochondrial oxidase was thought to be a cytochrome *c* oxidase. The recent work of Glaser and his colleagues (99), however, shows that the (translated) gene for subunit II does not contain the copper-binding motif characteristic of cytochrome *c*-oxidizing complexes. This result, taken together with the findings of Lauraeus et al. (68) showing that the purified caa_3 oxidase will oxidize cytochrome *c* but that the aa_3 complex oxidizes only quinols at reasonable rates, strongly suggests that *B. subtilis* cytochrome aa_3 is a menaquinol oxidase comparable in function to the cytochrome *o* ubiquinol oxidase of *E. coli*.

Cytochrome *o* was originally discovered in bacteria

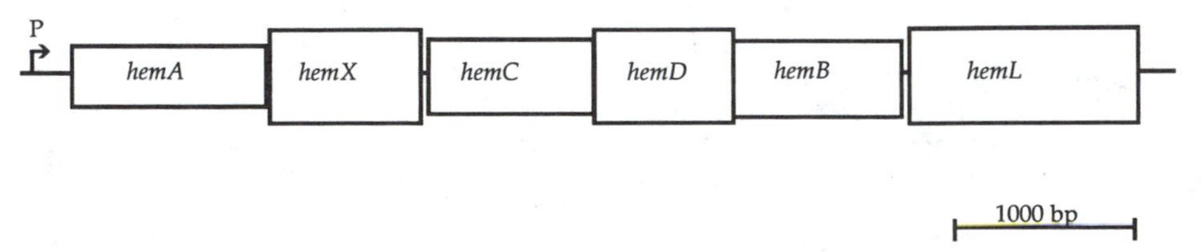

Figure 3. Organization of the *hemAXCDBL* gene cluster (based on data in reference 47). Gene functions are as given in Fig. 2. P, promoter.

Table 1. Terminal oxidases identified in *Bacillus* spp. and several other bacterial species[a]

Organism	Oxidase[b]	Heme[c]	Subunit size(s) (kDa)	Reductant[d]	Reference(s)
Bacillus sp. strain PS3	*o* (B-558)	B	60, 30, 16	Cytochrome *c*	89, 90, 110, 112
	caa_3	A	56, 38, 22, 12	Cytochrome *c*	85, 89, 90, 116, 117
	cao	O	56, 38, 22, 12	Cytochrome *c*	110
B. firmus	*o*	ND	ND	ND	49
	caa_3	A	56, 40, 14	Cytochrome *c*	49, 63
	d	ND	65, 60	ND	49
B. cereus	aa_3	A	51, 31	Cytochrome *c*	40, 41
	caa_3	A	51, 37	Cytochrome *c*	40
	o	ND	ND	ND	
	d	ND	ND	ND	30, 32
Bacillus sp. strain YN-2000	*aco*	B (?)	52, 42	Cytochrome *c*	97
B. subtilis	*o*	ND	ND	ND	26, 129
	caa_3	A	69, 40, 23, 13	Cytochrome *c*	68
	aa_3	A	57, 37, 21	Cytochrome *c*, quinols[e]	21, 22, 24–26, 68, 99
	d	ND	ND	ND	128
B. megaterium	*o*	ND	ND	ND	13
	aa_3	ND	ND	ND	13
	d	ND	ND	ND	51
B. stearothermophilus	*o* (B-558)	ND	ND	ND	111
	caa_3	A	57, 37, 22	Cytochrome *c*	23, 111
	cao	O	ND	ND	111
E. coli	*o*	O	66, 36, 18, 13	Ubiquinol	3
	d	D	58, 43	Ubiquinol	3
Sulfolobus acidocaldarius	aa_3	A	63, 58, 19[f]	Caldariella quinol	1, 70a
T. thermophilus	caa_3	A	55, 33	Cytochrome *c*	73, 139
	ba_3	A	35	Cytochrome *c*	140

[a] This summary combines data from measurements on purified enzymes, membrane vesicle preparations, and whole cells as well as inferences from sequences of oxidase subunit genes. Where possible, references are given for studies in which purified enzymes have been isolated, even though earlier reports may have identified the oxidases spectroscopically in whole cells or membrane preparations. The reference for *E. coli* is a convenient summary of oxidase properties in that organism. General reviews and discussions can be found in Poole (90, 91), Ludwig (71), Saraste (100, 102), and Gennis (42). Cloning of oxidase genes from many organisms is proceeding rapidly (see, e.g., reference 105).

[b] Nomenclature of the oxidases is based on spectroscopic properties of the enzymes either in whole cells or purified preparations. If the latter are available, careful heme analyses are necessary in order to establish stoichiometry as well as identity. See the text for discussion of some difficulties in establishing the biochemical properties of *o*-type oxidases. ND, not determined.

[c] Heme type present in the binuclear center at which reduction of oxygen occurs.

[d] In most studies, the cytochrome *c* used as reductant was not a cytochrome *c* indigenous to the organism from which the oxidase was isolated. For exceptions, consult the references.

[e] The *B. subtilis* aa_3 oxidase probably uses menaquinol in vivo as reductant.

[f] The *S. acidocaldarius* oxidase appears to be a complex of three proteins homologous to cytochrome *b* and cytochrome oxidase subunits I and II but with only heme A bound to the complex (70a).

by Castor and Chance (14), who, on the basis of its carbon monoxide-binding properties and photochemical action spectrum for relief of carbon monoxide inhibition, suggested that it is a terminal oxidase. This appears to be its function in *Bacillus* spp. Cytochrome *o* has been identified spectroscopically or spectrophotometrically in *Bacillus megaterium*, *B. subtilis*, *Bacillus firmus*, *Bacillus stearothermophilus*, and PS3 (Table 1) but has been purified only from PS3. For concise reviews of cytochrome *o*-type oxidases in *Bacillus* spp., consult references 110, 111, and 112. In *E. coli*, cytochrome *o* functions as a ubiquinol oxidase. Because of the visible absorption spectra of the reduced and reduced-CO forms, its prosthetic group had been thought to be protoheme; however, the *E. coli* enzyme has recently been shown to possess a new heme, for which the name heme O has been proposed (93, 94). Indeed, recent usage has moved toward consistency with the "aa_3" nomenclature by referring to cytochrome *o* as oo_3. This nomenclature (which also originated with Keilin) considers the two hemes in a terminal oxidase to have the same chemistry but differing reactivities; i.e., the o_3 (or a_3) center interacts directly with oxygen, while the *o* and *a* centers in the respective oxidases are positioned in subunit I to receive electrons directly from reductant or from the copper center in subunit II and convey them to the o_3 and a_3 centers.

Difficulties with nomenclature for the *o*-type cytochromes are likely to continue, since Sone's laboratory has found a novel cytochrome *o* (B-558) terminal oxidase in PS3 and *B. stearothermophilus* that has heme B (protoheme) prosthetic groups, while the *cao* oxidase (Table 1) has heme O as its oxygen-reactive center (110, 112). The necessity for carefully studying the chemistry of the heme groups in *o*-type oxidases is suggested by the conclusions of Qureshi et al. (97) that, according to spectral evidence alone, the *aco* oxidase of *Bacillus* YN-2000 contains heme B. Similarities in the spectra of hemes B and O can readily lead to confusion between the two hemes.

Molecular genetics and regulation

The subunits for the terminal oxidases are encoded by genes that appear to be organized in operon structures. This organization has been established for the caa_3 (101, 131) and aa_3 (99) oxidases of *B. subtilis*,

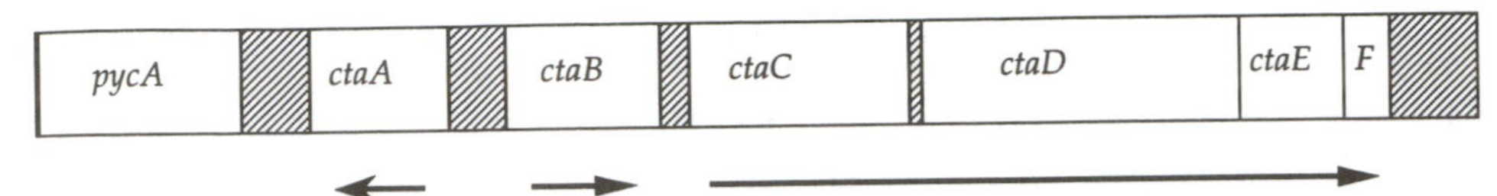

Figure 4. Organization of the *ctaA*, *ctaB*, and *ctaCDEF* genes in *B. subtilis* (81–83, 101, 131). The *ctaA* gene constitutes a verified monocistronic transcriptional unit (82, 83), while the *ctaB* and *ctaCDEF* transcripts are inferred from localization of putative promoter sequences. Arrangement of the *cta* genes appears to be similar in the thermophilic *Bacillus* sp. strain PS3 (52) and the alkaliphile *B. firmus* (95).

cytochrome caa_3 of PS3 (52), and cytochrome caa_3 of *B. firmus* (95). The structure of the caa_3 operon (*ctaC DEF*) of *B. subtilis* is shown in Fig. 4 with the structures of the regulatory genes *ctaA* and *ctaB* located adjacent to *ctaCDEF* (81, 82, 131; see below). A similar organization has been observed for the cytochrome *o* (*cyo*) subunit genes of *E. coli* (3, 103, 131). It seems likely that other *Bacillus* spp. will prove to have similar genetic organizations for their oxidase subunit genes. Investigators should keep in mind the findings with *P. denitrificans*, in which additional, unlinked genes that appear to encode subunit isoforms have been found (98).

It has been known since the mid-1950s (15, 16, 107) that the terminal oxidases of *Bacillus* spp. are regulated. A number of later studies confirmed and expanded on the earlier findings concerning effects of carbon sources, phase of growth, and oxygen tension (5, 32, 74, 120, 122, 123, 129). The studies of Frade and Chaix (36, 37) on the influence of extracellular pH on synthesis of cytochrome aa_3 in *Bacillus coagulans* need to be reconsidered in the light of recent findings (50) that suggest that promoters for respiratory chain components respond to environmental parameters such as pH. With the availability of cloned oxidase genes, there is now the opportunity to supplement these physiological approaches with studies of gene expression.

Regulatory factors that specifically affect expression of oxidase subunit genes have not been identified, although several types of mutations are known to affect terminal oxidase formation in *B. subtilis* (33, 34, 54, 81, 82, 119, 120, 122, 131). Menaquinone deficiency mutations (33, 34) cause *a*-type cytochromes (but not cytochrome *o*) to be down-regulated by an unknown mechanism. Similarly, the mechanisms by which mutations in *strC* (119), *ctaA* (54, 81–83), and *ctaB* (131) lead to loss of *a*-type cytochromes from *B. subtilis* are not yet understood. It is known that the kinetics of *ctaA* transcription are abnormal in a *strC* background (83). The facility with which regulatory mutations affecting *a*-type cytochromes can be isolated by their resistance to streptomycin should make it possible to reconstruct the regulatory network controlling these terminal oxidases (5, 74, 81, 82, 119, 127).

Cytochromes Other Than Terminal Oxidases

Cytochrome c

Cytochrome *c* is the only heme protein that has a covalently bound heme (Fig. 1). Cytochromes *c* have been isolated from several *Bacillus* species, often as multiple forms (e.g., see references 20, 26, 78, 133, 134, 138), but the exact functions of these cytochromes are not yet clear (see reference 133 for a review). By analogy with cytochromes *c* in other bacteria and in mitochondria, it is likely that under aerobic conditions, these *Bacillus* cytochromes transfer electrons from lower-potential components to terminal oxidases. However, the mechanism of transfer probably differs in detail because of the strong membrane-binding characteristics of the cytochromes *c* from *Bacillus* spp. (26, 133, 134). This membrane-binding property presumably is necessary because of the lack of an outer membrane in gram-positive bacteria. The outer membranes in gram-negative bacteria and mitochondria are thought to protect against loss of cytochrome *c* from the outer surfaces of the membranes, where it binds and serves as reductant for terminal oxidases.

von Wachenfeldt and Hederstedt (133) labeled *B. subtilis* with radiolabelled 5-aminolevulinic acid and studied the biochemistry and genetics of a prominent 13-kDa component. This cytochrome *c* is encoded by the *cccA* gene, which is adjacent to *sigA* at 223° on the *B. subtilis* chromosome. Although the two genes are arranged in tandem, they appear to be transcribed independently. At least one other cytochrome *c* was revealed by deletion of *cccA*; whether this is the heme C component of cytochrome caa_3 was not reported.

Cytochrome *b*-558

Cytochrome *b*-558 forms part of complex II of the respiratory chain (succinate:quinone oxidoreductase [EC 1.3.5.1]) and is discussed in chapter 13. The succinate dehydrogenase complex has been the subject of intense study by Hederstedt and his collaborators (39, 48) and will not be dealt with here. As further illustration of problems with the nomenclature of bacterial heme proteins, one must be careful not to confuse this cytochrome *b*-558 with the cytochrome *b*-558 described by Sone et al. (112), which functions as a cytochrome *c* oxidase in *Bacillus* sp. strain PS3.

Cytochrome *b*-cytochrome c_1 complex

The cytochrome *b*-cytochrome c_1 complex (the bc_1 complex) mediates quinol oxidation and commonly has cytochrome *c* reductase activity. Such complexes have been partially purified from *Bacillus* sp. strain PS3 (113, 115) and *B. subtilis* (26). The complex from *B. subtilis* has cytochrome *b* and c_1 components with midpoint potentials more positive than that of menaquinone but also more positive than that of the cytochrome *c* isolated in the same study (26). The bc_1 complex from PS3 was isolated as a super complex with the caa_3 terminal oxidase from this organism (113) and appeared capable of transferring electrons from quinol directly to the oxidase. The PS3 bc_1 complex had four subunits, containing two equivalents of heme B and two of heme C. Unpublished observations cited in reference 113 also indicate that a high-potential iron-sulfur center is present. Thus, it appears likely that bacilli have bc_1 complexes comparable to those of other bacteria in subunit organization and functionality.

Dehydrogenases

In terms of total electron flow through the respiratory chain, the two most important dehydrogenases are those responsible for oxidizing succinate and NADH. A discussion of succinate dehydrogenase can be found in chapter 13. NADH dehydrogenase has been purified from *B. subtilis* (9, 11); it contains flavin adenine dinucleotide as a prosthetic group, is specific for NADH, and reduces menaquinone. Whether the dehydrogenase functions in transmembrane proton pumping is unknown; iron-sulfur centers have been detected neither in the isolated enzyme nor (by electron paramagnetic resonance measurements) in intact wild-type membrane preparations. Although catalytic activity is found within a single subunit of 63 kDa, there may be other subunits that interact weakly in situ with the enzyme. Molecular genetic studies could, as with the caa_3 terminal oxidase, shed considerable light on the detailed biochemistry of NADH dehydrogenase.

Menaquinone

Biochemistry

Gram-positive bacteria possess only menaquinone as their redox quinone component of the respiratory chain, although many gram-negative species have both menaquinone and ubiquinone (18). Archaebacteria also have menaquinone (and lack ubiquinone) but sometimes have unique quinone structures replacing menaquinone (see, e.g., reference 19). The midpoint potential of the menaquinone-menaquinol couple places this component in the lower portion of the respiratory chain, between dehydrogenases and the bc_1 complex. Terminal oxidases such as *B. subtilis* cytochrome aa_3 also appear to oxidize menaquinone (68, 99). There has been some theoretical difficulty in placing menaquinone in electron transfer chains catalyzing succinate oxidation, because the midpoint potential of the succinate-fumarate couple is more positive (by about 0.1 V) than that of menaquinone. However, Lemma et al. (69) have demonstrated by liposome reconstitution experiments that menaquinone can serve to mediate electron flow from succinate via purified *B. subtilis* succinate dehydrogenase to menaquinone and thence to the bc_1 complex. Previous studies had shown that menaquinone was obligatory in NADH oxidase activity (10, 12).

Relatively few studies of the electron transfer functions of menaquinone in facultative *Bacillus* spp. have been done, despite extensive results obtained with *E. coli* (70, 130, 135, 137). This reflects the tendency to utilize *B. subtilis* (a strict aerobe) because of its advantages for molecular genetic studies. Nitrate respiration is a commonly used means for electron transport-based energy generation anaerobically, and nitrate reductases have been isolated from several *Bacillus* spp. (27, 62, 132). Ketchum et al. (62) recently described a menaquinol-nitrate oxidoreductase from *Bacillus halodenitrificans* that was formed when the organism was grown anaerobically in nitrate-containing medium. This enzyme complex consisted of three subunits, one of which is a cytochrome *b*. One equivalent of molybdenum and two (4Fe-4S) iron-sulfur centers (as determined from the electron paramagnetic resonance spectra) are present per heme. Perhaps extension of these technologies to a more genetically amenable species such as *Bacillus licheniformis* would provide insight into anaerobic regulatory mechanisms in *Bacillus* spp.

Biosynthesis and regulation

Menaquinone (MK) is composed of a naphthoquinone ring system to which is joined a variable number of isoprenyl units; MK-7, which has a seven-unit isoprenoid side chain, is shown in Fig. 5. MK-7 is the predominant isoprenoid homolog found in *Bacillus* spp. Figure 5 also outlines the biosynthetic steps from chorismate, in the shikimate pathway, to MK-7. In *B. subtilis*, mutations have been obtained in each of the genes shown except *menA* (75, 124). These *men* mutants were isolated by a positive selection procedure that relies on an intact respiratory chain for transport of aminoglycoside antibiotics (33, 121–125, 127); thus, *men* mutants are resistant to multiple aminoglycoside selection. The screening procedure would not have identified mutations either in *menA* or in the genes encoding formation of the polyprenyl side chain (see reference 126 for a review of this synthetic sequence). The *men* genes were found to be clustered (124) and recently have been cloned and sequenced (28, 79, 80, 98b). Two major operons within the gene cluster, the *menCD* and *menBE* operons, have been identified by transcript mapping. The *men* gene cluster of *E. coli* appears to have a somewhat similar organization (75a, 106). The presence in *B. subtilis* of several additional promoters (Fig. 6) was inferred from genetic disruption experiments, and readthrough transcription appears to be extensive. The reasons for this complex organization are not obvious, although it is likely to have regulatory significance (28). Several open reading frames in the cluster (*icsM*, ORF5, ORF8) were not defined by *men* mutations. *icsM* appears to encode a menaquinone-specific isochorismate synthase isoform; its sequence is similar to that of the *entC* gene of *E. coli*. Organisms with deletions of *icsM* are Men$^+$; furthermore, the *icsM* promoter is not iron regulated as is the *entC* promoter in *E. coli* (98a). Consideration of Fig. 7 shows that isochorismate is the branch point for biosynthesis of dihydroxybenzoic acid-based iron chelators. It therefore seems likely that a second, iron starvation-responsive "*icsF*" gene exists elsewhere in the *B. subtilis* chromosome. Even in nonstarvation conditions, the *icsF* gene product can supply sufficient isochorismate for the rather modest needs of menaquinone biosynthesis. This situation is in contrast to the findings of Kaiser and Leistner (57) with *E. coli*, in which an *entC* mutant did not grow under anaerobic conditions known to require menaquinone formation (suggesting that only a single isochorismate synthase is present in *E. coli*).

The *B. subtilis men* ORF1 gene has been mutagenized, and the mutants have been analyzed (94a). When

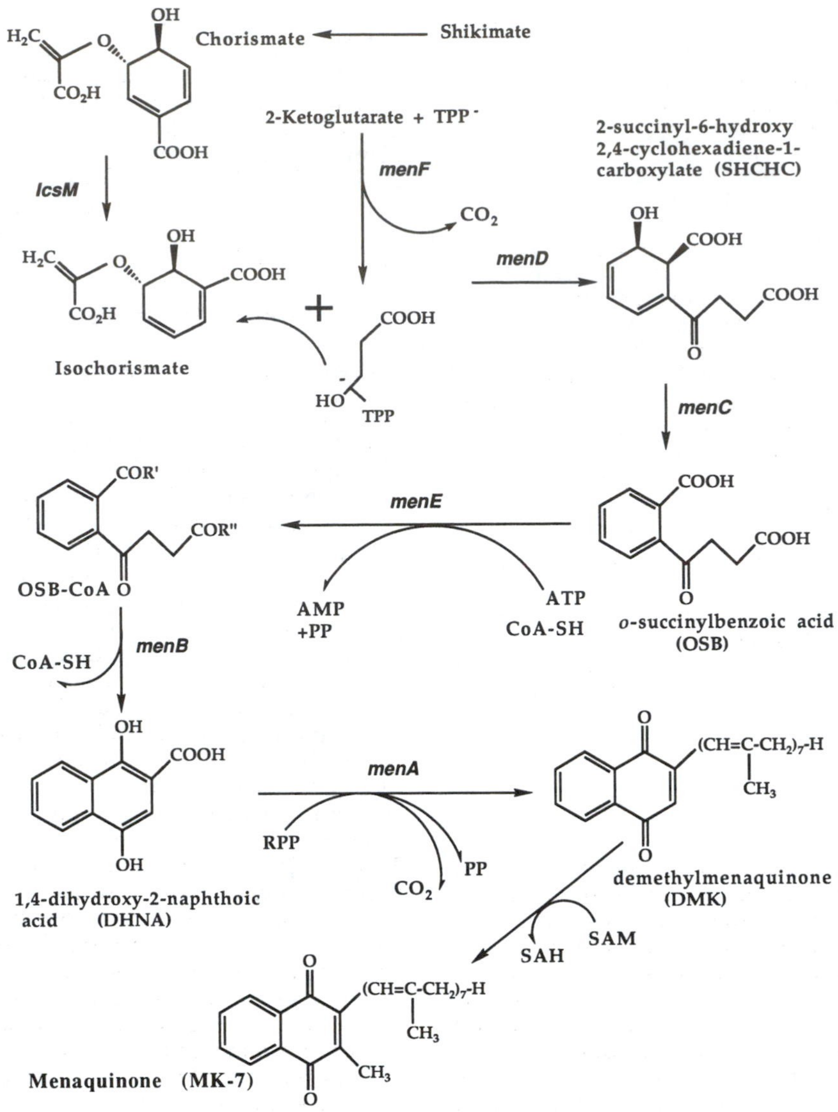

Figure 5. Gene-enzyme relationships in the biosynthesis of menaquinone from chorismate in *B. subtilis*. This scheme is in part based on results from *E. coli*, but the relationships appear to be identical in both species. *icsM*, menaquinone-specific isochorismate synthase; *menF*, menaquinone-specific 2-ketoglutarate dehydrogenase; *menD*, 2-succinyl-6-hydroxy-2,4-cyclohexadiene-1-carboxylate synthase; *menC*, *o*-succinylbenzoic acid synthase; *menE*, *o*-succinylbenzoic acid coenzyme A synthetase; *menB*, 1,4-dihydroxy-2-naphthoic acid synthase; *menA*, polyisoprenyl transferase; SAM, *S*-adenosylmethionine; TPP^-, thiamine pyrophosphate anion; CoA-SH, coenzyme A-sulfhydryl; RPP, polyisoprenyl pyrophosphate; and SAH, *S*-adenosylhomocysteine. R′ and R″ in OSB-CoA denote the positional indeterminacy of CoA-SH.

B. subtilis is growing on glucose as a carbon source, the ORF1 gene product appears to be necessary for proper regulation of the accumulation and subsequent utilization of glycolytic end products such as acetoin, butanediol, and acetate. Its presence in the *men* gene cluster may provide a way for the cell to coordinate expression of a central respiratory chain component with other enzymes required for oxidative

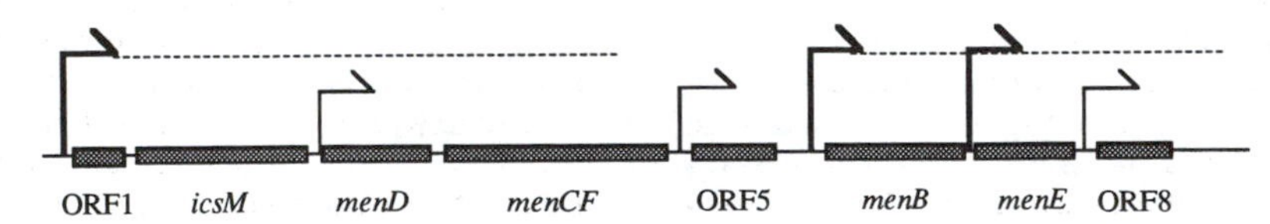

Figure 6. Sequence and transcriptional organization of the *men* gene cluster in *B. subtilis* (28, 79, 80, 98b). The *icsM* and *men* gene functions are shown in Fig. 4 and 5; open reading frames 1, 5, and 8 have not been identified with specific enzymatic activities. Bold arrows denote experimentally established sites of transcription initiation; light arrows indicate sites inferred from integrational disruption experiments. See text for details.

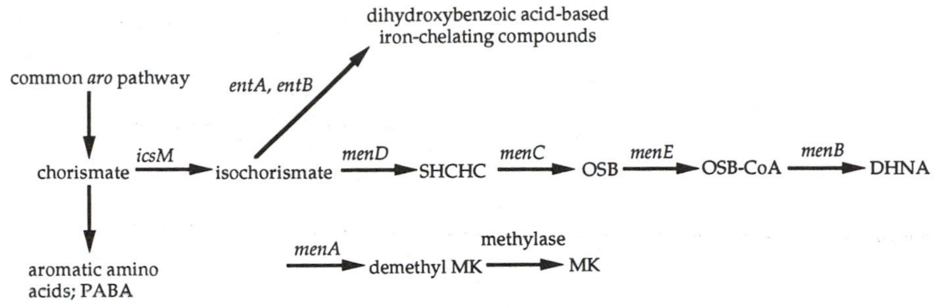

Figure 7. Metabolic relationship of menaquinone biosynthesis to formation of dihydroxybenzoic acid-based iron-chelating compounds in *B. subtilis*. A similar scheme can be drawn for *E. coli*, with *entC* replacing *icsM*. For abbreviations, see the legend to Fig. 5. PABA, *p*-aminobenzoic acid.

metabolism. In this connection, it may be relevant that the ORF1 promoter is activated by acidic extracellular pH in the presence of glycolytic end products (50). Within this promoter, as in several other promoters for genes encoding oxidative functions such as tricarboxylic acid cycle enzymes (118) and *ctaA* (83), is found a sequence motif that may be important for the regulation of these promoters by glycolytic end products.

RESPIRATORY CHAIN ADAPTATIONS IN THE ALKALIPHILIC BACILLI

Alkaliphilic *Bacillus* species such as *B. firmus* OF4 maintain a cytoplasmic pH of 8.2 to 8.5, which, although higher than that of the typical neutrophile (7.5), is very much less than the extracellular pH of 10.5 that is typical of conditions in which alkaliphiles grow and respire (53, 65, 67, 96). Acidification of the cytoplasm relative to the external milieu is maintained by activity both of the respiratory chain and of an electrogenic Na^+/H^+ antiporter (66). According to the chemiosmotic theory, Δp across the cytoplasmic membrane must be acid outside and positive. However, in alkaliphiles, the ΔpH component of Δp is reversed (i.e., acid inside relative to outside) and quite large at the alkaline extracellular pHs optimal for growth. The net (bulk) Δp under these conditions becomes energetically insufficient to drive oxidative phosphorylation via the ATP synthase reaction, yet the alkaliphile grows perfectly well.

Studies of the pH dependence of ATP synthesis in response to a diffusion potential (Δp formed artificially across the membranes of starved cells by ionic manipulations) showed that this bulk-phase Δp promotes ATP synthesis at pH 7.5 but not at pH 10.5. From this type of result and having excluded a number of other possibilities, Guffanti and Krulwich have proposed a "parallel coupling model" for oxidative phosphorylation (44) that might have applicability not only to alkaliphiles but also to other species. At nearly neutral pH, facultative alkaliphiles would utilize a standard chemiosmotic mechanism to generate a sufficient Δp for oxidative phosphorylation via bulk-phase proton movement inward through the transmembrane ATP synthase. At alkaline pH, however, such movement could not occur because of a pH-regulated gate in the synthase; instead, a respiratory chain proton pump would directly transfer protons to the synthase, resulting in ATP formation in the absence of a substantial transmembrane Δp.

This model has much to recommend it, although intramembranal proton transfer is difficult to conceptualize and test. An intriguing finding is the markedly enhanced synthesis of respiratory chain components at alkaline pH in the alkaliphiles, possibly a mechanism for increasing the frequency of collision between proton transfer complexes and the ATP synthase (44).

REGULATION OF RESPIRATORY CHAIN FORMATION DURING SPORULATION

The sporulating cell has a basic metabolic problem: to sustain respiration and energy production long enough to allow completion of the sporulation sequence, to block respiration in the spore, and to reinitiate respiration at the onset of germination. A number of studies have addressed regulation of the respiratory chain and/or requirements for individual components (29–31, 34, 35, 40, 43, 51, 83, 121, 122, 129, 136). Much of the attention has focused on menaquinone and the terminal oxidases. *B. subtilis* mutants lacking *a*-type oxidases did not sporulate (122) and exhibited unusual membrane morphologies (38). Dormant spores of *B. subtilis* had higher concentrations of *o*-type than of *a*-type oxidases (129), perhaps a consequence of differential synthesis of the two oxidase types in the forespore membrane, as in *Bacillus cereus* (31). Of the two *a*-type oxidases in *B. cereus*, cytochrome aa_3 was found to be synthesized both in vegetative and in sporulating cells, but cytochrome caa_3 was formed only in sporulating cells (40); whether there is mother cell- or forespore-specific localization of either oxidase is not known.

Menaquinone formation is increased at the beginning of sporulation of *B. subtilis* (35), and menaquinone deficiency blocks sporulation (34). If menaquinone in *B. cereus* is inactivated by near-UV light, respiration during sporulation is completely inhibited (30); furthermore, menaquinone contents of forespore membranes declined sharply during sporulation of *B. cereus* (29). NADH oxidation was very low in membrane preparations derived from *B. cereus* spores but was stimulated by addition of menadione (29). The presence of substantial levels of NADH dehydrogenase and cytochromes in spores of *B. cereus* suggested that the block in spore respiration was due to a programmed menaquinone deficiency introduced during

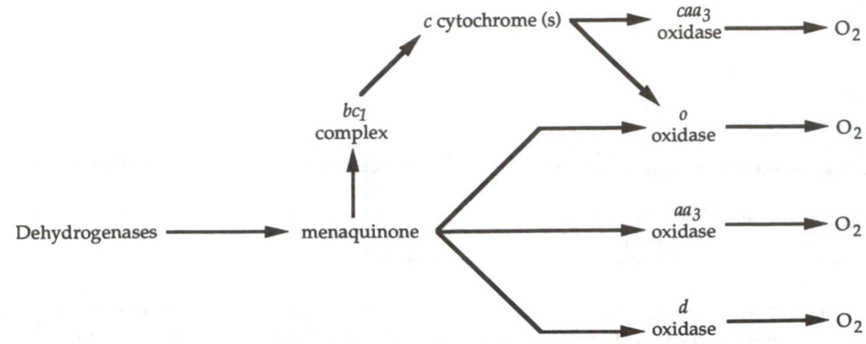

Figure 8. Possible paths for aerobic electron transfer in the bacilli. See the text for discussion.

the development of the spore. However, in *B. subtilis*, Weber and Broadbent (136) found that menaquinone was present in mature spores at levels comparable to those in vegetative cells but that dehydrogenase levels were low. It thus appears that different *Bacillus* species have developed a variety of strategies for interrupting the respiratory chain in the dormant spore. Experimental materials, in the form of respiratory chain mutants, cloned genes, and purified enzyme complexes, for pursuing issues of gene expression, enzyme localization, and metabolic signaling that may be subject to the extracellular and intracellular environment unique to the sporulating cell are available.

PROSPECTS

The strong similarities in the biochemical natures of respiratory chain components among *Bacillus* spp. are probably not a testament to the evolutionary relatedness of this group but to the conservatism of evolution in a much broader field. The usage of menaquinone (to the exclusion of ubiquinone), for example, not only in all gram-positive bacteria but also in the majority of archaebacteria suggests that formation of the unique features of bacterial respiratory chains constituted a very early event in the evolutionary development of cellular biochemistry. The same might be said for the terminal oxidases, which have been preserved with remarkable similarity throughout the eubacterial and archaebacterial kingdoms. As work with *Bacillus* respiratory chains progresses to studies of the molecular nature of the bc_1 complex, NADH dehydrogenase, cytochrome *o*, and cytochrome *d*, we will undoubtedly find further structural and functional similarities to well-studied bacteria such as *E. coli*. Indeed, as shown in Fig. 8, the paths for electron flow through *Bacillus* respiratory chains to oxygen have remarkable similarities to those in the *Enterobacteriaceae* (92). The surprises are likely to come from studies on regulation of the genes for these components and the meaning of this regulation to oxidative metabolism. Anaerobiosis among *Bacillus* spp. clearly deserves more effort. Also, one hopes that as part of the growing recognition that sporulation is a particularly complicated response to stationary-phase growth conditions, studies of respiratory chain regulation will be reintegrated into studies of membrane biogenesis, which is, after all, one of the most striking morphological events during spore formation.

Acknowledgments. A number of investigators have communicated results in advance of publication, and their kindness is appreciated. I thank John Mueller, Belinda Rowland, and Xuan Qin for helpful discussions and Jeff Driscoll both for discussions and for assistance with graphics and word processing. Work in my laboratory has been supported in part by grant GM44547 from the National Institutes of Health.

REFERENCES

1. **Anemüller, S., and G. Schäfer.** 1989. Cytochrome aa_3 from the thermoacidophilic archaebacterium *Sulfolobus acidocaldarius*. *FEBS Lett.* **244:**451–455.
2. **Anraku, Y.** 1988. Bacterial electron transfer chains. *Annu. Rev. Biochem.* **57:**101–132.
3. **Anraku, Y., and R. B. Gennis.** 1987. The aerobic respiratory chain of *Escherichia coli*. *Trends Biochem. Sci.* **12:**262–266.
4. **Anthony, C. (ed.).** 1988. *Bacterial Energy Transduction*. Academic Press, Inc., New York.
5. **Arrow, A. S., and H. W. Taber.** 1986. Streptomycin accumulation by *Bacillus subtilis* requires both a membrane potential and cytochrome aa_3. *Antimicrob. Agents Chemother.* **29:**141–146.
6. **Bentley, R., and R. Meganathan.** 1987. Biosynthesis of the isoprenoid quinones ubiquinone and menaquinone, p. 512–520. *In* F. C. Neidhardt, J. L. Ingraham, K. B. Low, B. Magasanik, M. Schaechter, and H. E. Umbarger (ed.), *Escherichia coli and Salmonella typhimurium: Cellular and Molecular Biology*, vol. 1. American Society for Microbiology, Washington, D.C.
7. **Berek, I., A. Miczak, and G. Ivanovics.** 1974. Mapping of the delta-aminolevulinic acid dehydrase and porphobilinogen deaminase loci in *Bacillus subtilis*. *Mol. Gen. Genet.* **132:**233–239.
8. **Berek, I., A. Miczak, I. Kiss, G. Ivanovics, and I. Durko.** 1975. Genetic and biochemical analysis of hemin dependent mutants of *Bacillus subtilis*. *Acta Microbiol. Acad. Sci. Hung.* **22:**157–167.
9. **Bergsma, J., M. B. M. van Dongen, and W. N. Konings.** 1982. Purification and characterization of NADH dehydrogenase from Bacillus subtilis. *Eur. J. Biochem.* **128:** 151–157.
10. **Bergsma, J., K. E. Meihuizen, W. van Oeveren, and W. N. Konings.** 1982. Restoration of NADH oxidation with menaquinones and menaquinone analogues in membrane vesicles from the menaquinone-deficient *Bacillus subtilis* aroD. *Eur. J. Biochem.* **125:**651–657.
11. **Bergsma, J., R. Strijker, J. Y. E. Alkema, H. G. Seijen,**

and W. N. Konings. 1981. NADH dehydrogenase and NADH oxidation in membrane vesicles from *Bacillus subtilis*. *Eur. J. Biochem.* **120:**599–606.
12. **Bisschop, A., and W. N. Konings.** 1976. Reconstitution of reduced nicotinamide adenine dinucleotide oxidase activity with menadione in membrane vesicles from the menaquinone-deficient *Bacillus subtilis* aroD. Relation between electron transfer and active transport. *Eur. J. Biochem.* **67:**357–365.
13. **Broberg, P. L., and L. Smith.** 1967. The cytochrome system of *Bacillus megaterium* KM. The presence and properties of two CO-binding cytochromes. *Biochim. Biophys. Acta* **131:**479–489.
14. **Castor, L. N., and B. Chance.** 1959. Photochemical determinations of the oxidases of bacteria. *J. Biol. Chem.* **234:**1587–1592.
15. **Chaix, P., and J. F. Petit.** 1956. Etude des differents spectres cytochromique de *Bacillus subtilis*. *Biochim. Biophys. Acta* **22:**66–71.
16. **Chaix, P., and J. F. Petit.** 1957. Influence du taux de croissance sur la constitution du spectra hématinique de *Bacillus subtilis*. *Biochim. Biophys. Acta* **25:**481–486.
17. **Chepuri, V., L. J. Lemieux, D. C.-T. Au, and R. B. Gennis.** 1990. The sequence of the *cyo* operon indicates substantial structural similarities between the cytochrome *o* ubiquinol oxidase of *Escherichia coli* and the aa_3-type family of the cytochrome *c* oxidases. *J. Biol. Chem.* **265:**11185–11192.
18. **Collins, M. D., and D. Jones.** 1981. Distribution of isoprenoid quinone structure types in bacteria and their taxonomic implications. *Microbiol. Rev.* **45:**316–354.
19. **Collins, M. D., and T. A. Langworthy.** 1983. Respiratory quinone composition of some acidophilic bacteria. *Syst. Appl. Microbiol.* **4:**295–304.
20. **Davidson, M. W., K. A. Gray, D. B. Knaff, and T. A. Krulwich.** 1988. Purification and characterization of two soluble cytochromes from the alkalophile *Bacillus firmus* RAB. *Biochim. Biophys. Acta* **933:**470–477.
21. **De Vrij, W., A. Azzi, and W. N. Konings.** 1983. Structural and functional properties of cytochrome *c* oxidase from *Bacillus subtilis* W23. *Eur. J. Biochem.* **131:**97–103.
22. **De Vrij, W., A. J. M. Driessen, K. J. Hellingwerf, and W. N. Konings.** 1986. Measurements of the protonmotive force generated by cytochrome *c* oxidase from *Bacillus subtilis* in proteoliposomes and membrane vesicles. *Eur. J. Biochem.* **156:**431–440.
23. **De Vrij, W., R. I. R. Heyne, and W. N. Konings.** 1989. Characterization and application of a thermostable primary transport system: cytochrome *c* oxidase from *Bacillus stearothermophilus*. *Eur. J. Biochem.* **178:**763–770.
24. **De Vrij, W., and W. N. Konings.** 1987. Kinetic characterization of cytochrome *c* oxidase from *Bacillus subtilis*. *Eur. J. Biochem.* **166:**581–587.
25. **De Vrij, W., B. Poolman, W. N. Konings, and A. Azzi.** 1986. Purification, enzymatic properties and reconstitution of cytochrome *c* oxidase from *Bacillus subtilis*. *Methods Enzymol.* **126:**159–173.
26. **De Vrij, W., B. van den Burg, and W. N. Konings.** 1987. Spectral and potentiometric analysis of cytochromes from *Bacillus subtilis*. *Eur. J. Biochem.* **166:**589–595.
27. **Downey, R. J.** 1966. Nitrate reductase and respiratory adaptation in *Bacillus stearothermophilus*. *J. Bacteriol.* **91:**634–641.
28. **Driscoll, J. R., and H. W. Taber.** 1992. Sequence organization and regulation of the *Bacillus subtilis menBE* operon. *J. Bacteriol.* **174:**5063–5071.
29. **Escamilla, J. E., B. Barquera, R. Ramirez, A. Garcia-Horsman, and P. Del Arenal.** 1988. Role of menaquinone in inactivation and activation of the *Bacillus cereus* forespore respiratory system. *J. Bacteriol.* **170:**5908–5912.
30. **Escamilla, J. E., and M. C. Benito.** 1984. Respiratory system of vegetative and sporulating *Bacillus cereus*. *J. Bacteriol.* **160:**473–477.
31. **Escamilla, J. E., R. Ramírez, P. Del Arenal, and A. Aranda.** 1986. Respiratory systems of the *Bacillus cereus* mother cell and forespore. *J. Bacteriol.* **167:**544–550.
32. **Escamilla, J. E., R. Ramírez, I. P. Del Arenal, G. Zarzoza, and V. Linares.** 1987. Expression of cytochrome oxidases in *Bacillus cereus*: effects of oxygen tension and carbon source. *J. Gen. Microbiol.* **133:**3549–3555.
33. **Farrand, S. K., and H. W. Taber.** 1973. Pleiotropic menaquinone-deficient mutant of *Bacillus subtilis*. *J. Bacteriol.* **115:**1021–1034.
34. **Farrand, S. K., and H. W. Taber.** 1973. Physiological effects of menaquinone deficiency in *Bacillus subtilis*. *J. Bacteriol.* **115:**1035–1044.
35. **Farrand, S. K., and H. W. Taber.** 1974. Changes in menaquinone concentration during growth and early sporulation in *Bacillus subtilis*. *J. Bacteriol.* **117:**324–326.
36. **Frade, R., and P. Chaix.** 1973. Influence du pH sur la synthèse de l'oxidase (cytochrome a_3) chez *Bacillus coagulans*, microorganisme thermophile, capable d' "adaptation réspiratoire." *Biochim. Biophys. Acta* **325:**424–432.
37. **Frade, R., and P. Chaix.** 1976. Etude du mécanisme de l'influence du pH extracellulaire sur la synthèse du complexe oxydasique (cytochromes *a* + *a*3) chez *Bacillus coagulans*: relation avec "l'effet glucose" et rôle de la coproporphyrine III excretée. *Biochim. Biophys. Acta* **423:**573–585.
38. **Freese, E. B.** 1973. Unusual membranous structures in cytochrome *a*-deficient mutants of *Bacillus subtilis*. *J. Gen. Microbiol.* **75:**187–190.
39. **Friden, H., and L. Hederstedt.** 1990. Role of His residues in *Bacillus subtilis* cytochrome *b*-558 for haem binding and assembly of succinate:quinone oxidoreductase (complex II). *Mol. Microbiol.* **4:**1045–1056.
40. **Garcia-Horsman, J. A., B. Barquera, and J. E. Escamilla.** 1991. Two different aa_3-type cytochromes can be purified from the bacterium *Bacillus cereus*. *Eur. J. Biochem.* **199:**761–768.
41. **Garcia-Horsman, J. A., B. Barquera, D. Gonzalez-Halphen, and J. E. Escamilla.** 1991. Purification and characterization of two-subunit cytochrome aa_3 from *Bacillus cereus*. *Mol. Microbiol.* **5:**197–205.
42. **Gennis, R. B.** 1991. Some recent advances relating to prokaryotic cytochrome *c* reductases and cytochrome *c* oxidases. *Biochim. Biophys. Acta* **1058:**21–24.
43. **Goodman, S. R., B. L. Marrs, R. J. Narconis, and R. E. Olson.** 1976. Isolation and description of a menaquinone mutant from *Bacillus licheniformis*. *J. Bacteriol.* **125:**282–289.
44. **Guffanti, A. A., and T. A. Krulwich.** 1992. A model for a non-chemiosmotic mode of energization of oxidative phosphorylation by alkaliphilic *Bacillus firmus* OF4. Personal communication.
45. **Haddock, B. A., and W. A. Hamilton (ed.).** 1977. *Microbial Energetics*. Society for General Microbiology Symposium, vol. 27. Cambridge University Press, Cambridge.
46. **Haltia, T., M. Finel, N. Harms, T. Nakari, M. Raitio, M. Wikstrom, and M. Saraste.** 1989. Deletion of the gene for subunit III leads to defective assembly of bacterial cytochrome oxidase. *EMBO J.* **8:**3571–3579.
47. **Hansson, M., L. Rutberg, I. Schröder, and L. Hederstedt.** 1991. The *Bacillus subtilis hemAXCDBL* gene cluster, which encodes enzymes of the biosynthetic pathway from glutamate to uroporphyrinogen III. *J. Bacteriol.* **173:**2590–2599.
48. **Hederstedt, L.** 1986. Molecular properties, genetics and

biosynthesis of *Bacillus subtilis* succinate dehydrogenase complex. *Methods Enzymol.* **126:**399–414.

49. **Hicks, D. B., R. J. Plass, and P. G. Quirk.** 1991. Evidence for multiple terminal oxidases, including cytochrome *d*, in facultatively alkaliphilic *Bacillus firmus* OF4. *J. Bacteriol.* **173:**5010–5016.
50. **Hill, K. F., J. P. Mueller, and H. W. Taber.** 1990. The *Bacillus subtilis menCD* promoter is responsive to extracellular pH. *Arch. Microbiol.* **153:**355–359.
51. **Hogarth, C., J. J. Wilkinson, and D. J. Ellar.** 1977. Cyanide resistant electron transport in sporulating *Bacillus megaterium* KM. *Biochim. Biophys. Acta* **461:**109–123.
52. **Ishizuka, M., K. Machida, S. Shimada, A. Magi, T. Tsuchiya, T. Ohmori, Y. Suoma, M. Gonda, and N. Sone.** 1990. Nucleotide sequence of the genes coding for four subunits of cytochrome *c* oxidase from the thermophilic bacterium PS3. *J. Biochem.* **108:**866–873.
53. **Ivey, D. M., D. B. Hicks, A. A. Guffanti, G. Sobel, and T. A. Krulwich.** 1990. The problem of the electrochemical proton potential in alkaliphilic bacteria. *Mosbach Colloq.* **41:**105–113.
54. **James, W. S., F. Gibson, P. Taroni, and R. K. Poole.** 1989. The cytochrome oxidases of *Bacillus subtilis*: mapping of a gene affecting cytochrome aa_3 and its replacement by cytochrome *o* in a mutant strain. *FEMS Microbiol. Lett.* **58:**277–282.
55. **Jones, C. W.** 1977. Aerobic respiratory systems in bacteria, p. 23–59. *In* B. A. Haddock and W. A. Hamilton (ed.), *Microbial Energetics*. Society for General Microbiology Symposium, vol. 27. Cambridge University Press, Cambridge.
56. **Jones, C. W.** 1987. Membrane-associated energy conservation in bacteria: a general introduction, p. 1–82. *In* C. Anthony (ed.), *Bacterial Energy Transduction*. Academic Press, Inc., New York.
57. **Kaiser, A., and E. Leistner.** 1990. Role of the *ent*C gene in enterobactin and menaquinone biosynthesis in *Escherichia coli*. *Arch. Biochem. Biophys.* **276:**331–335.
58. **Keilin, D.** 1925. On cytochrome, a respiratory pigment, common to animals, yeast, and higher plants. *Proc. R. Soc. Ser. B* **98:**312–339.
59. **Keilin, D.** 1929. Cytochrome and respiratory enzymes. *Proc. R. Soc. Ser. B* **104:**206–252.
60. **Keilin, D.** 1966. *The History of Cell Respiration and Cytochrome.* Cambridge University Press, Cambridge.
61. **Kell, D. B.** 1987. Protonmotive energy-transducing systems: some physical principles and experimental approaches, p. 429–490. *In* C. Anthony (ed.), *Bacterial Energy Transduction*. Academic Press, Inc., New York.
62. **Ketchum, P. A., G. Denaniaz, J. LeGall, and W. J. Payne.** 1991. Menaquinol-nitrate oxidoreductase of *Bacillus halodenitrificans*. *J. Bacteriol.* **173:**2498–2505.
63. **Kitada, M., and T. A. Krulwich.** 1984. Purification and characterization of the cytochrome oxidase from alkaliphilic *Bacillus firmus* RAB. *J. Bacteriol.* **158:**963–966.
64. **Krulwich, T. A. (ed.).** 1990. *The Bacteria*, vol. XII. *Bacterial Energetics*. Academic Press, Inc., New York.
65. **Krulwich, T. A., and A. A. Guffanti.** 1989. Alkalophilic bacteria. *Annu. Rev. Microbiol.* **43:**435–463.
66. **Krulwich, T. A., and A. A. Guffanti.** 1989. The Na^+ cycle of extreme alkalophiles: a secondary Na^+/H^+ antiporter and Na^+/solute symporters. *J. Bioenerg. Biomembr.* **21:**663–677.
67. **Krulwich, T. A., D. B. Hicks, D. Seto-Young, and A. A. Guffanti.** 1988. The bioenergetics of alkalophilic bacilli. *Crit. Rev. Microbiol.* **16:**15–36.
68. **Lauraeus, M., T. Haltia, M. Saraste, and M. Wikstrom.** 1991. *Bacillus subtilis* expresses two kinds of haem A-containing terminal oxidases. *Eur. J. Biochem.* **197:** 699–705.
69. **Lemma, E., C. Hagerhall, V. Geisler, U. Brandt, G. von Jagow, and A. Kröger.** 1991. Reactivity of the *Bacillus subtilis* succinate dehydrogenase complex with quinones. *Biochim. Biophys. Acta* **1059:**281–285.
70. **Lin, E. C. C., and D. R. Kuritzkas.** 1987. Pathways for anaerobic electron transport, p. 201–221. *In* F. C. Neidhardt, J. L. Ingraham, K. B. Low, B. Magasanik, M. Schaechter, and H. E. Umbarger (ed.), *Escherichia coli and Salmonella typhimurium: Cellular and Molecular Biology*, vol. 1. American Society for Microbiology, Washington, D.C.

70a. **Lübben, M., B. Kolmerer, and M. Saraste.** 1992. An archaebacterial terminal oxidase combines core structures of two mitochondrial respiratory complexes. *EMBO J.* **11:**805–812.

71. **Ludwig, B.** 1987. Cytochrome c oxidase in prokaryotes. *FEMS Microbiol. Rev.* **46:**41–56.
72. **Maloney, P. C.** 1987. Coupling to an energized membrane: role of ion-motive gradients in the transduction of metabolic energy, p. 222–243. *In* F. C. Neidhardt, J. L. Ingraham, K. B. Low, B. Magasanik, M. Schaechter, and H. E. Umbarger (ed.), *Escherichia coli and Salmonella typhimurium: Cellular and Molecular Biology*, vol. 1. American Society for Microbiology, Washington, D.C.
73. **Mather, M. W., P. Springer, and J. A. Fee.** 1991. Cytochrome oxidase genes from *Thermus thermophilus*. Nucleotide sequence and analysis of the deduced primary structure of subunit IIc of cytochrome caa_3. *J. Biol. Chem.* **266:**5025–5035.
74. **McEnroe, A. S., and H. W. Taber.** 1984. Correlation between cytochrome aa_3 concentrations and streptomycin accumulation in *Bacillus subtilis*. *Antimicrob. Agents Chemother.* **26:**507–512.
75. **Meganathan, R., R. Bentley, and H. Taber.** 1981. Identification of *Bacillus subtilis men* mutants which lack *o*-succinylbenzoyl-coenzyme A synthetase and dihydroxynaphthoate synthase. *J. Bacteriol.* **145:**328–332.

75a. **Meganathan, R., and M. Hudspeth.** Personal communication.

76. **Miczák, A., I. Berek, and G. Ivánovics.** 1976. Mapping the uroporphyrinogen decarboxylase, coproporphyrinogen oxidase and ferrochetalase loci in Bacillus subtilis. *Mol. Gen. Genet.* **146:**85–87.
77. **Miczák, A., B. Prágai, and I. Berek.** 1979. Mapping the uroporphyrinogen III cosynthase locus in *Bacillus subtilis*. *Mol. Gen. Genet.* **174:**293–295.
78. **Miki, K., and K. Okunuki.** 1969. Cytochromes of *Bacillus subtilis*. Purification and spectral properties of cytochromes c_{-550} and c_{-554}. *J. Biochem.* **66:**831–854.
79. **Miller, P., J. Mueller, K. Hill, and H. Taber.** 1988. Transcriptional regulation of a promoter in the *men* gene cluster of *Bacillus subtilis*. *J. Bacteriol.* **170:**2742–2748.
80. **Miller, P., A. Rabinowitz, and H. Taber.** 1988. Molecular cloning and preliminary genetic analysis of the *men* gene cluster of *Bacillus subtilis*. *J. Bacteriol.* **170:**2735–2741.
81. **Mueller, J. P., and H. W. Taber.** 1988. Genetic regulation of cytochrome aa_3 in *Bacillus subtilis*, p. 91–95. *In* A. T. Ganesan and J. A. Hoch (ed.), *Genetics and Biotechnology of Bacilli*, vol. 2. Academic Press, Inc., New York.
82. **Mueller, J. P., and H. W. Taber.** 1989. Isolation and sequence of *ctaA*, a gene required for cytochrome aa_3 biosynthesis and sporulation in *Bacillus subtilis*. *J. Bacteriol.* **171:**4967–4978.
83. **Mueller, J. P., and H. W. Taber.** 1989. Structure and expression of the cytochrome aa_3 controlling gene *ctaA* of *Bacillus subtilis*. *J. Bacteriol.* **171:**4979–4986.
84. **Neidhardt, F. C., J. L. Ingraham, K. B. Low, B. Magasanik, M. Schaechter, and H. E. Umbarger (ed.).** 1987. *Escherichia coli and Salmonella typhimurium: Cellular and Molecular Biology*. American Society for Microbiology, Washington, D.C.

85. **Nicholls, P., and N. Sone.** 1984. Kinetics of cytochrome *c* and TMPD oxidation by cytochrome *c* oxidase from the thermophilic bacterium PS3. *Biochim. Biophys. Acta* **767:**240–247.
86. **O'Neill, G. P., M.-W. Chen, and D. Söll.** 1989. Delta-aminolevulinic acid biosynthesis in *Escherichia coli* and *Bacillus subtilis* involves formation of glutamyl-tRNA. *FEMS Microbiol. Lett.* **60:**255–260.
87. **Petricek, M., L. Rutberg, I. Schröder, and L. Hederstedt.** 1990. Cloning and characterization of the *hemA* region of the *Bacillus subtilis* chromosome. *J. Bacteriol.* **172:**2250–2258.
88. **Piggot, P. J.** 1989. Revised genetic map of *Bacillus subtilis* 168, p. 1–41. *In* I. Smith, R. A. Slepecky, and P. Setlow (ed.), *Regulation of Procaryotic Development*. American Society for Microbiology, Washington, D.C.
89. **Poole, R. K.** 1981. Ligand-binding cytochromes a_3, *c* and *o* in membranes from the thermophilic bacterium PS3. *FEBS Lett.* **133:**255–259.
90. **Poole, R. K.** 1983. Bacterial cytochrome oxidases. A structurally and functionally diverse group of electron-transfer proteins. *Biochim. Biophys. Acta* **726:**205–243.
91. **Poole, R. K.** 1988. Bacterial cytochrome oxidase, p. 231–291. *In* C. Anthony (ed.), *Bacterial Energy Transduction*. Academic Press, Inc., New York.
92. **Poole, R. K., and W. J. Ingledew.** 1987. Pathways of electrons to oxygen, p. 170–200. *In* F. C. Neidhardt, J. L. Ingraham, K. B. Low, B. Magasanik, M. Schaechter, and H. E. Umbarger (ed.), *Escherichia coli and Salmonella typhimurium: Cellular and Molecular Biology*, vol. 1. American Society for Microbiology, Washington, D.C.
93. **Puustinen, A., M. Finel, T. Haltia, R. P. Gennis, and M. Wikström.** 1991. Properties of the two terminal oxidases of *Escherichia coli*. *Biochemistry* **30:**3936–3942.
94. **Puustinen, A., and M. Wikström.** 1991. The heme groups of cytochrome *o* from *Escherichia coli*. *Proc. Natl. Acad. Sci. USA* **88:**6122–6126.
94a. **Qin, X., and H. Taber.** Unpublished data.
95. **Quirk, P. G., A. A. Guffanti, and T. A. Krulwich.** 1992. Cloning and characterization of genes encoding the caa_3-type terminal oxidase from alkaliphilic *Bacillus firmus* OF4. Abstr. Annu. Meet. Biophys. Soc., p. A285.
96. **Quirk, P. G., A. A. Guffanti, R. J. Plass, S. Clejan, and T. A. Krulwich.** 1991. Protonophore-resistance and cytochrome expression in mutant strains of the facultative alkaliphile *Bacillus firmus* OF4. *Biochim. Biophys. Acta* **1058:**131–140.
97. **Qureshi, M. H., I. Yomoto, T. Fujiwara, Y. Fukumori, and T. Yamana.** 1990. A novel *aco*-type cytochrome-*c* oxidase from a facultative alkalophilic *Bacillus*: purification, and some molecular and enzymatic features. *J. Biochem.* **107:**480–485.
98. **Raitio, M., J. M. Pispa, T. Metso, and M. Saraste.** 1990. Are there isoenzymes of cytochrome *c* oxidase in *Paracoccus denitrificans*? *FEBS Lett.* **261:**431–435.
98a. **Rowland, B.** Unpublished data.
98b. **Rowland, B., et al.** Unpublished data.
99. **Santana, M., F. Kunst, M. F. Hullo, G. Rapoport, A. Danchin, and P. Glaser.** 1992. Molecular cloning, sequencing and physiological characterization of the *qox* operon from *Bacillus subtilis*, encoding the aa_3-600 quinol oxidase. *J. Biol. Chem.* **267:**10225–10231.
100. **Saraste, M.** 1990. Structural features of cytochrome oxidase. *Q. Rev. Biophys.* **23:**331–366.
101. **Saraste, M., L. Holm, L. Lemieux, M. Lubben, and J. van der Oost.** 1991. The happy family of cytochrome oxidases. *Biochem. Soc. Trans.* **19:**608–612.
102. **Saraste, M., T. Metso, T. Nakari, T. Jalli, M. Lauraeus, and J. van der Oost.** 1991. The *Bacillus subtilis* cytochrome *c* oxidase. Variations on a conserved theme. *Eur. J. Biochem.* **195:**517–525.
103. **Saraste, M., M. Raitio, T. Jalli, V. Chepuri, L. Lemieux, and R. B. Gennis.** 1988. Cytochrome *o* from *Escherichia coli* is structurally related to cytochrome aa_3. *Ann. N.Y. Acad. Sci.* **550:**314–324.
104. **Schröder, I., L. Hederstedt, C. G. Kannangara, and S. P. Gough.** 1992. Glutamyl-tRNA reductase activity in *Bacillus subtilis* is dependent on the *hemA* gene product. *Biochem. J.* **281:**843–850.
105. **Shapleigh, J. P., and R. B. Gennis.** 1992. Cloning, sequencing and deletion from the chromosome of the gene encoding subunit I of the aa_3-type cytochrome *c* oxidase of *Rhodobacter sphaeroides*. *Mol. Microbiol.* **6:**635–642.
106. **Sharma, V., K. Suvarna, R. Meganathan, and M. E. S. Hudspeth.** 1992. Menaquinone (vitamin K_2) biosynthesis: nucleotide sequence and expression of the *menB* gene from *Escherichia coli*. *J. Bacteriol.* **174:**5057–5062.
107. **Smith, L.** 1954. Bacterial cytochromes. Difference spectra. *Arch. Biochem. Biophys.* **50:**299–314.
108. **Sone, N.** 1989. Energy transducing complexes in bacterial respiratory chains. *Subcell. Biochem.* **14:**279–337.
109. **Sone, N.** 1990. Respiration-driven proton pumps, p. 1–32. *In* T. A. Krulwich (ed.), *The Bacteria*, vol. XII. *Bacterial Energetics*. Academic Press, Inc., New York.
110. **Sone, N., and Y. Fujiwara.** 1991. Haem O can replace haem A in the active site of cytochrome *c* oxidase from thermophilic bacterium PS3. *FEBS Lett.* **288:**154–158.
111. **Sone, N., and Y. Fujiwara.** 1991. Effects of aeration during growth of *Bacillus stearothermophilus* on proton pumping activity and change of terminal oxidases. *J. Biochem.* **110:**111–116.
112. **Sone, N., E. Kutoh, and K. Sato.** 1990. A cytochrome *o*-type oxidase of the thermophilic bacterium PS3 grown under air-limited conditions. *J. Biochem.* **107:** 597–602.
113. **Sone, N., M. Sekimachi, and E. Kutoh.** 1987. Identification and properties of a quinol oxidase super-complex composed of a bc_1 complex and cytochrome oxidase in the thermophilic bacterium PS3. *J. Biol. Chem.* **262:**15386–15391.
114. **Sone, N., S. Shimada, T. Ohmori, Y. Souma, M. Gonda, and M. Ishizuka.** 1990. A fourth subunit is present in cytochrome *c* oxidase from the thermophilic bacterium PS3. *FEBS Lett.* **262:**249–252.
115. **Sone, N., and T. Takagi.** 1990. Monomer-dimer structure of cytochrome *c* oxidase and cytochrome bc_1 complex from the thermophilic bacterium PS3. *Biochim. Biophys. Acta* **1020:**207–212.
116. **Sone, N., and Y. Yanagita.** 1982. A cytochrome aa_3-type terminal oxidase of a thermophilic bacterium. Purification, properties and proton pumping. *Biochim. Biophys. Acta* **682:**216–226.
117. **Sone, N., F. Yokoi, T. Fu, S. Ohta, T. Metso, M. Raitio, and M. Saraste.** 1988. Nucleotide sequence of the gene coding for cytochrome oxidase subunit I from the thermophilic bacterium PS3. *J. Biochem.* **103:**606–610.
118. **Sonenshein, A. L.** 1989. Metabolic regulation of sporulation and other stationary-phase phenomena, p. 109–129. *In* I. Smith, R. A. Slepecky, and P. Setlow (ed.), *Regulation of Procaryotic Development*. American Society for Microbiology, Washington, D.C.
119. **Staal, S. P., and J. A. Hoch.** 1972. Conditional dihydrostreptomycin resistance in *Bacillus subtilis*. *J. Bacteriol.* **110:**202–207.
120. **Taber, H.** 1974. Isolation and properties of cytochrome *a* deficient mutants of *Bacillus subtilis*. *J. Gen. Microbiol.* **81:**435–444.
121. **Taber, H.** 1980. Functions of vitamin K_2 in microorganisms, p. 177–187. *In* J. W. Suttie (ed.), *Vitamin K Metabolism and Vitamin K-Dependent Proteins*. University Park Press, Baltimore, Md.
122. **Taber, H., S. K. Farrand, and G. M. Halfenger.** 1972. Genetic regulation of membrane components in *Bacillus subtilis*, p. 140–147. *In* H. O. Halvorson, R. S.

Hanson, and L. L. Campbell (ed.), *Spores V*. American Society for Microbiology, Washington, D.C.

123. **Taber, H., and E. Freese.** 1974. Sporulation properties of cytochrome *a*-deficient mutants of *Bacillus subtilis*. *J. Bacteriol.* **120:**1004–1011.
124. **Taber, H., and G. M. Halfenger.** 1976. Multiple aminoglycoside-resistant mutants of *Bacillus subtilis* deficient in accumulation of kanamycin. *Antimicrob. Agents Chemother.* **9:**251–259.
125. **Taber, H. W., E. A. Dellers, and L. R. Lombardo.** 1981. Menaquinone biosynthesis in *Bacillus subtilis*: isolation of *men* mutants and evidence for clustering of *men* genes. *J. Bacteriol.* **145:**321–327.
126. **Taber, H. W., J. P. Mueller, P. F. Miller, and A. S. Arrow.** 1987. Bacterial uptake of aminoglycoside antibiotics. *Microbiol. Rev.* **51:**439–457.
127. **Taber, H. W., B. J. Sugarman, and G. M. Halfenger.** 1981. Involvement of menaquinone in active accumulation of gentamicin by *Bacillus subtilis*. *J. Gen. Microbiol.* **123:**143–149.
128. **Takahashi, I., and K. Ogura.** 1982. Prenyltransferases of *Bacillus subtilis*: undecaprenyl pyrophosphate synthetase and geranylgeranyl pyrophosphate synthetase. *J. Biochem.* **92:**1527–1537.
129. **Tochikubo, K.** 1971. Changes in terminal respiratory pathways of *Bacillus subtilis* during germination, outgrowth, and vegetative growth. *J. Bacteriol.* **108:**652–661.
130. **Unden, G.** 1988. Differential roles for menaquinone and demethylmenaquinone in anaerobic electron transport of *E. coli* and their *fnr*-independent expression. *Arch. Microbiol.* **150:**499–503.
131. **van der Oost, J., C. von Wachenfeld, L. Hederstedt, and M. Saraste.** 1991. *Bacillus subtilis* cytochrome oxidase mutants: biochemical analysis and genetic evidence for two aa_3-type oxidases. *Mol. Microbiol.* **5:**2063–2072.
132. **van't Riet, J., F. B. Wientjes, J. Van Doorn, and R. J. Planta.** 1979. Purification and characterization of the respiratory nitrate reductase of *Bacillus licheniformis*. *Biochim. Biophys. Acta* **576:**347–360.
133. **von Wachenfeldt, C., and L. Hederstedt.** 1990. *Bacillus subtilis* 13-kilodalton cytochrome *c*-550 encoded by *cccA* consists of a membrane-anchor and a heme domain. *J. Biol. Chem.* **265:**13939–13948.
134. **von Wachenfeldt, C., and L. Hederstedt.** 1990. *Bacillus subtilis* holo-cytochrome *c*-550 can be synthesized in aerobic *Escherichia coli*. *FEBS Lett.* **270:**147–151.
135. **Wallace, B. J., and I. G. Young.** 1977. Role of quinones in electron transport to oxygen and nitrate in *Escherichia coli*. Studies with a $ubiA^-$ $menA^-$ double quinone mutant. *Biochim. Biophys. Acta* **461:**84–100.
136. **Weber, M. M., and D. A. Broadbent.** 1975. Electron transport in membranes from spores and from vegetative and mother cells of *Bacillus subtilis*, p. 411–417. *In* P. Gerhardt, R. N. Costilow, and H. L. Sadoff (ed.), *Spores VI*. American Society for Microbiology, Washington, D.C.
137. **Wissenbach, U., A. Kröger, and G. Unden.** 1990. The specific functions of menaquinone and demethylmenaquinone in anaerobic respiration with fumarate, dimethyl-sulfoxide, trimethylamine N-oxide and nitrate by *Escherichia coli*. *Arch. Microbiol.* **154:**60–66.
138. **Wooley, K. J.** 1987. The *c*-type cytochromes of the Gram-positive bacterium *Bacillus licheniformis*. *Arch. Biochem. Biophys.* **254:**376–379.
139. **Yoshida, T., and J. A. Fee.** 1984. Studies on cytochrome c oxidase activity of the c_1aa_3 complex from *Thermus thermophilus*. *J. Biol. Chem.* **259:**1031–1036.
140. **Zimmermann, B. H., C. I. Nitsche, J. A. Fee, F. Rusnak, and E. Munck.** 1988. Properties of a copper-containing cytochrome ba_3: a second terminal oxidase from the extreme thermophile *Thermus thermophilus*. *Proc. Natl. Acad. Sci. USA* **85:**5779–5783.

15. Carbon Source-Mediated Catabolite Repression

GLENN H. CHAMBLISS

Catabolite repression is a regulatory mechanism by which the cell (i) coordinates metabolism of carbon and energy sources to maximize efficiency and (ii) regulates other metabolic processes as well. The objective of this review is to examine what is known about the molecular mechanisms by which catabolite repression operates in gram-positive bacteria, especially *Bacillus subtilis*. Herein, I use the term catabolite repression to refer specifically to carbon source-mediated regulation and not to regulation by nitrogen or other nutrient sources.

Originally termed glucose repression, the phenomenon of catabolite repression has been known for 50 years (11, 38). The presence of glucose in the culture medium was observed to repress the synthesis of enzymes required for the metabolism of other less rapidly utilized carbohydrates. The repression was found to be a general phenomenon in which rapidly metabolized carbohydrates suppress the utilization of less readily metabolized sugars. As a result, selective repression of enzyme synthesis establishes priorities in the utilization of various carbon and energy sources. Catabolite repression of protein synthesis is not restricted to carbohydrate catabolic enzymes. The synthesis of secondary metabolites, including antibiotics, by both prokaryotic and eukaryotic microorganisms (35) is subject to glucose repression either directly or indirectly. Developmental pathways such as spore formation and the synthesis of certain extracellular enzymes and toxins are repressed by readily metabolized carbohydrates (4, 5, 12, 22, 28, 44, 51, 52).

The phenomenon of catabolite repression has been best characterized in *Escherichia coli* (30–32, 57). The addition of glucose to a culture of *E. coli* causes a transient, very strong repression of genes subject to catabolite repression. This is followed by an extended repression that lasts as long as sufficient glucose is in the medium. This latter repression is less severe than the initial transient repression. In addition to the effect it has on repression, glucose blocks the uptake of a number of carbohydrates, thereby preventing the intracellular concentrations of these carbohydrates from reaching levels adequate to fully induce synthesis of the enzymes involved in their metabolism. This phenomenon is called inducer exclusion. Both inducer exclusion and long-term repression have been observed in gram-positive bacteria, but transient repression has not been well defined.

CATABOLITE REPRESSION IN *E. COLI*

For comparative purposes, I shall briefly review the basic mechanism by which *E. coli* accomplishes this type of global regulation. Catabolite repression in *E. coli* is interference with a positive regulatory mechanism in which the catabolite repressor protein (CRP or CAP) in complex with the cyclic nucleotide cyclic AMP (cAMP) binds to a specific site in the promoter region of catabolite repression-sensitive genes or operons (30–32, 57). This binding activates transcription of the catabolite repression-sensitive operon (if it has also been induced). The consensus CRP-binding site, A A N T G T G A N_2 T N_4 C A, is generally located upstream of the −35 regions of promoters subject to catabolite repression. Whether CRP bound at this site physically contacts RNA polymerase in facilitating promoter utilization is an open question.

The binding of CRP at CRP-binding sites is determined by the intracellular level of cAMP, which in turn is regulated, at least in part, by the rate of cAMP synthesis by adenylate cyclase. The activity of adenylate cyclase, which converts ATP into cAMP with the liberation of pyrophosphate, is stimulated by the phosphorylated form of enzyme III^{Glu} of the phosphotransferase system (PTS) for carbohydrate uptake (43 and chapter 10, this volume). Thus, when glucose (or another PTS sugar) is present, it will be taken up rapidly by the cell and will act as a sink for transfer of phosphate through the PTS, and consequently, the pool of enzyme III^{Glu} will consist predominantly of the nonphosphorylated form. In the absence of enzyme III^{Glu}-PO_4, adenylate cyclase activity will drop to a low basal level, thus causing a decrease in cAMP concentration. Ultimately, a level of cAMP that is insufficient for complex formation with CRP is reached. This will prevent CRP binding to catabolite repression-regulated promoters. Without CRP being bound, these promoters, which in general are weak, are not utilized well, even though they may be induced.

In the absence of glucose (or other PTS sugars), enzyme III^{Glu} accumulates in the phosphorylated form, which is stimulatory to adenylate cyclase. In this situation, cAMP accumulates sufficiently to allow cAMP-CRP complex formation and binding to CRP sites, thus activating transcription from catabolite-regulated promoters that have been induced. Because the concentration of cAMP within a cell is not sufficient to saturate CRP, relatively small changes in cAMP concentration can have rather pronounced effects on the binding of CRP at CRP-binding sites (57).

Not only is enzyme III^{Glu} important in regulating cAMP synthesis, it is also directly involved in inducer exclusion. Nonphosphorylated enzyme III^{Glu} binds to sugar permeases in the presence of the cognate sugar and inhibits the activities of the permeases (43).

Glenn H. Chambliss • Department of Bacteriology, University of Wisconsin-Madison, Madison, Wisconsin 53706.

CATABOLITE REPRESSION IN GRAM-POSITIVE BACTERIA

In the three genera of gram-positive bacteria that have been studied best, the molecular mechanism by which catabolite repression functions appears to be fundamentally different from that in *E. coli*. In none of the gram-positive organisms does cAMP play a central role. Also, catabolite repression seems to be mediated by a negative regulatory mechanism.

Staphylococcus aureus

In *S. aureus*, glucose represses the synthesis of catabolic enzymes required for the metabolism of alternative carbon sources, such as those of the *lac* operon (42), as well as the synthesis of several exoproteins, including the enterotoxins SEA, SEB, SEC, α-hemolysin, β-hemolysin, and lipase (2, 9, 15, 28, 44). Although the mechanism by which glucose represses gene expression is not known, cAMP is not involved (22). Glucose suppresses the steady-state level of enterotoxin type C mRNA, presumably by repressing transcription of *sec* (44). In their studies of the *S. aureus lac* operon, Oskouian and Stewart (42) found that deletion of a repeat sequence located upstream of *lacP* reduced binding of a hypothetical negatively acting CRP. Because the repressing activity could be titrated out by a plasmid with a relatively low copy number (about 20 copies per cell) carrying the *lacP* region, it was proposed that the catabolite repression activity observed might be specific to the *lac* operon (42). This also raises the possibility that glucose repression of this *lac* operon is mediated primarily by inducer exclusion.

Streptomyces spp.

Glucose repression is a well-established phenomenon in *Streptomyces* spp. It affects the utilization of a number of sugars (13, 21, 25, 26, 50), synthesis of extracellular enzymes such as agarase (3) and α-amylase (58), production of certain antibiotics (35), and spore formation (4). Preliminary observations indicate that glucose causes inducer exclusion and represses gene expression at the transcription level (21).

Very little is known about the molecular mechanism by which glucose repression operates in *Streptomyces* spp., but it does not appear to involve cAMP (6, 20). Glucose repression-resistant mutants of *Streptomyces coelicolor* A3(2) isolated on the basis of their abilities to grow in the presence of 2-deoxyglucose were defective in glucose kinase activity (21, 26). This was true even for mutants that grew well on glucose as the sole carbon source (26). A fragment of *S. coelicolor* DNA that restored glucose kinase activity and glucose repression was cloned (26).

Putative promoters of chitinase genes of *Streptomyces plicatus*, which are induced by chitin or its hydrolysis products and repressed by glucose, have been examined (7). A *cis*-acting base pair substitution mutation in the promoter region of the *chi-63* gene rendered transcription from this promoter constitutive and glucose resistant. This mutation, located at position −17 with regard to the transcription start site, is situated in a 12-bp direct repeat sequence. This same or a very similar direct repeat is found in all chitinase promoters thus far analyzed (37). It is hypothesized that the mutation affects the operator site for a repressor that mediates induction by chitin and that glucose repression acts primarily by inducer exclusion (7).

B. subtilis

The synthesis of a number of catabolic enzymes in *B. subtilis* is subject to catabolite repression, as are the synthesis of extracellular enzymes and the process of endospore formation (12). To date, few studies have been directed at elucidating the molecular mechanism by which carbon source-mediated catabolite repression operates. In their now-classic study, Schaeffer et al. (47) concluded that sporulation is repressed by a carbon-, nitrogen-, and phosphorus-containing metabolite during rapid growth. Limitation for carbon, nitrogen, or phosphorus would deplete the concentration of the critical metabolite and derepress sporulation. Unfortunately, the identity of such a metabolite has not been established unambiguously.

Mutants capable of sporulating in the presence of sugars have been isolated (56). A number of different mutants were isolated on the basis of the genetic locus affected and the sugar used in selection (54). One of the mutations, *crsA*, renders both sporulation and synthesis of α-amylase resistant to repression by glucose (unpublished results). The molecular basis for the glucose resistance phenotype is not evident, but the *crsA* mutation affects the gene for the σ^A subunit of RNA polymerase (8, 55). That the *crsA* mutation affects both amylase synthesis and sporulation suggests that both processes may be subject to the same catabolite repression mechanism or to two overlapping mechanisms.

Primarily on the basis of studies of α-amylase synthesis, a molecular model for the mechanism of catabolite repression in *B. subtilis* can be proposed. In this model, a repressor protein is activated by the presence of glucose or other readily metabolized carbohydrates including glycerol. Activation of the repressor could occur by either the binding or the detachment of an effector molecule whose intracellular concentration is dictated by the availability of the repressing carbohydrate. Alternatively, the repressor could be activated by modification such as phosphorylation or dephosphorylation. Once activated, the repressor would bind to an operator sequence located, in the case of α-amylase, near the transcription start site, thus preventing transcription of the gene or operon subject to catabolite repression. This model differs from the catabolite repression mechanism in *E. coli* in that regulation is negatively controlled in the *B. subtilis* model. Also, if an effector molecule is involved, it is certainly not cAMP (18, 23, 48).

The operator region alluded to in the model was first identified genetically (39–41). A mutant generated by *N*-methyl-*N*′-nitro-*N*-nitrosoguanidine (NTG) mutagenesis that synthesized α-amylase in the presence of excess glucose (2%) was isolated. Through complementation analyses, the mutation was shown to affect a *cis*-acting regulatory element. Genetic mapping revealed that the mutation was within or near the *amyR* region (40), which was previously identified as a region important in regulating amylase production (62–64). After cloning and sequencing, the original mutation, *gra-10*, was found to be a G-C–to–A-T transition at position +5 from the start site of transcription of the amylase gene (41). Examination of the DNA sequences flanking the mutation revealed that the region was homologous (70% identity) to the *lac* operator (45, 49) and, with about 90% identity, to the consensus *gal* operator of *E. coli* (10). The region of greatest similarity to the *E. coli* operators was 14 bp long and extended from positions −3 to +11 immediately downstream of the amylase promoter. The *lacO630*(Con) (33, 53) and *galO34*(Con) (24) mutations are apparently analogous to *gra-10* in that they affect the corresponding G-C base pair in the *lac* and *gal* operators, rendering them ineffective. Because catabolite repression of α-amylase synthesis occurs at the level of transcription, the similarity of the operatorlike region at the transcription start site to known operators suggested that the operatorlike region might be the site at which a repressor protein would bind when the culture medium contained glucose or other readily metabolized carbohydrates.

The first two independently isolated NTG-generated mutants able to produce α-amylase in the presence of glucose contained identical mutations (unpublished data). This introduced the possibility that the G-C base pair at position +5 was the only position vital for catabolite repression or was at least the most critical position. Alternatively, there was concern that the mutagenesis procedure was biased. To evaluate the contributions of various base pairs within the region to catabolite repression of α-amylase synthesis, oligonucleotide-directed mutagenesis was undertaken (61). The *Hpa*I site, located on the 3′ side of the original mutation, was used to select for most of the mutants, though not all. This bias resulted in most of the lesions being located in the right half of the operator region. Even so, the specific changes allowed the deduction of the critical sequence as T G T/A A A N C ↓ G N T N A/T C A, where underlined letters represent the most critical bases, N is any base, and the vertical arrow denotes an axis of symmetry. On the basis of two mutations that increased the twofold symmetry of the relevant region by making the right half more complementary to the left half and simultaneously increasing the strength of the region as an operator, the optimal sequence T G T A A G C G C T T A C A was deduced.

If the operator region downstream of the α-amylase promoter is indeed the site at which a repressor of the catabolite repression system binds and if catabolite repression of α-amylase is part of a global regulatory mechanism, then one might expect to find sequences homologous to the amylase operator (*amyO*) in or near the promoters of other genes subject to catabolite repression. Screening the promoter regions of a number of *Bacillus* genes either known to be or suspected of being catabolite repressed led to the identification of sequences with greater or lesser similarity to the sequence of the amylase operator. From these sequences, a consensus sequence was deduced, and it was identical to that derived from mutagenesis studies (61). This identity could be taken as evidence of the global nature of the mechanism responsible for catabolism repression of α-amylase synthesis.

Recently, two independent groups have found that sites with strong homology to *amyO* are important for catabolite repression of the *hut* (11a) and *xyl* (27) operons of *B. subtilis*. Surprisingly, the operator sequences are not located near the promoters of these operons but are within the first structural gene in each operon. This location suggests that the simplistic model presented above, in which the binding of a repressor at *amyO* blocks initiation of transcription from the amylase promoter, might be inadequate to fully explain the mechanism operating in the *hut* and *xyl* operons.

The existence of an operator sequence downstream of the amylase promoter and required for catabolite repression suggests the existence of a repressor molecule that would bind to the operator. Transposon mutagenesis was employed to identify genes potentially encoding such a repressor (19, 39). A scheme was devised to select for mutants capable of transcribing the amylase promoter in the presence of glucose and to screen such mutants for *trans*-acting mutations conferring catabolite repression resistance. The scheme involved placing the promoterless *cat-86* gene of plasmid pPL603 under the control of the α-amylase promoter (40). Cells carrying this plasmid construct were resistant to >50 μg of chloramphenicol per ml when grown in nutrient broth medium lacking added sugar. Addition of 2% glucose to the medium rendered the cells sensitive to 50 μg of chloramphenicol per ml; thus, glucose-resistant mutants could be selected directly. By incorporating starch into the selection plates, mutant colonies could also be screened for their abilities to produce amylase from the wild-type copy of the amylase gene in the chromosome.

This selection and screening procedure was used to isolate 63 independent mutants that could make α-amylase and chloramphenicol acetyltransferase in the presence of excess glucose. Mutations fell into four categories based on their locations on the *B. subtilis* genetic map. One group of mutations was located between *argA* and *aroG*, a second group was mapped to a position near *metC*, and a third group lay to the 5′ side of *amyE* but separated from it. The chromosomal location of the fourth group has not yet been determined. Since this last group contains more than 12 members, more than one genetic locus may be represented.

All of the mutants are similar in phenotype in that they produce α-amylase in the presence of glucose, fructose, maltose, mannitol, and sucrose. In nine of the mutants, α-amylase synthesis remained sensitive to repression by glycerol, while in the rest, it was resistant. These nine mutants fell into two of the above-mentioned groups, i.e., groups I and III. Interestingly, only 8 of the 13 group III mutants were glycerol sensitive, which could mean that two or more closely linked loci are affected in this group of mutants. One way to explain the observation that not all mutants resistant to repression by glucose are resistant to glycerol, even though many are, is to propose that catabolite repression by these two carbohydrates is exerted through a branched pathway in which one branch is involved in monitoring glucose and other sugars and the other is involved in monitoring glycerol. Information from each branch would be fed into a central pathway that would in turn lead to repression of genes sensitive to catabolite repression. Thus, mutations affecting components of the central pathway would impair repression by both glucose and glycerol, whereas mutations affecting one of the branches would affect repression only by the corresponding carbohydrate.

As a start in trying to sort out relationships among the mutated genes and to unravel the network by which catabolite repression operates, the gene *ccpA*, which is mutated in the group I mutant, was cloned (19). Its base sequence revealed that it encodes a protein, CcpA, of 38 kDa. The deduced amino acid sequence of CcpA is homologous to sequences of the *gal* and *lac* repressor proteins of *E. coli*. CcpA has 31% amino acid identity with GalR and 25% identity with LacI. Amino acids in GalR and LacI are 25% identical. The region of greatest similarity resides in the amino-terminal half of the protein, containing the DNA-binding domain of LacI (1). The amino-terminal portion of CcpA is predicted to form an α-helix–turn–α-helix structure characteristic of LacI and related repressor proteins (46, 59). Judging from the conservation of amino acid sequence and mutagenesis studies of the LacI protein, key amino acid residues are critical for repressor function (17, 29). Nearly all of the key residues are conserved in CcpA (19). This strongly suggests that CcpA is a DNA-binding protein and that it is likely to bind an operator sequence resembling *galO* or *lacO*. Because mutation of the gene encoding CcpA leads to resistance to glucose repression of α-amylase synthesis, an obvious preliminary conclusion is that CcpA is a repressor protein involved in mediating catabolite repression.

According to sequence homology, CcpA also belongs to the GalR family of repressor proteins (60). Members of this family, which includes proteins from both gram-positive and gram-negative bacteria, have conserved features in their DNA-binding, dimerization, and effector-binding domains. Most of the proteins in this family are sugar-binding repressors (GalR, GalS, LacI, MalI, FruR, RbtR, and RafR), but two of them bind nucleosides (CytR and PurR). CcpA is homologous to both the DNA-binding and effector-binding domains of the GalR family of repressors (60). This fact suggests that the effector for CcpA is most likely a sugar or a nucleoside, possibly glucose itself.

As an alternative approach to identifying possible repressor proteins, extracts of glucose-grown *B. subtilis* cells were examined for proteins that would bind specifically to the amylase operator (*amyO*). One such activity has been partially purified (unpublished results). After passage over DNA-agarose, Sephacryl S-300, and heparin agarose columns, the active cell extract fraction still contains five protein bands in sodium dodecyl sulfate-polyacrylamide gel electrophoresis. The major band, which contains about one-half the total protein in the sample, has a molecular mass of approximately 18 kDa. None of the proteins in the sample is larger than 25 kDa. This fraction binds DNA fragments containing *amyO* as measured by gel retardation (16). In DNase I footprinting experiments, the binding activity protected the region from −10 to +27, within which *amyO* is centered. The active fraction selectively blocked transcription by purified Eσ^A RNA polymerase from the amylase promoter without affecting transcription from a control promoter. These results are consistent with the activity being a repressor molecule that binds specifically to *amyO*.

An important question to be answered is whether the CcpA protein is responsible for the *amyO*-binding activity observed in cell extract. The answer to this question is not yet known completely, but there is evidence that it is not. Crude cell extract preparations (after elution from DNA-agarose) from *ccpA* mutants contain an activity that gives a footprint on *amyO*-containing DNA that is identical to that obtained for the binding activity from wild-type cells. CcpA produced in *E. coli* has DNA-binding activity, but in DNase I footprinting analysis, it protects the region of the amylase promoter from −13 to +20, which is slightly different from the binding activity found in *B. subtilis* extracts. Rabbit antiserum prepared against CcpA does not cross-react with the partially purified *B. subtilis* DNA-binding proteins. Also, catabolite repression of α-amylase synthesis is not completely relieved in *ccpA* mutants. These mutants make about one-half as much amylase in the presence of glucose as in its absence.

Is catabolite repression of α-amylase part of a global regulatory system? It is possible that *B. subtilis* contains multiple catabolite repression systems and that the regulation of α-amylase is an example of one isolated system. Evidence from a number of sources indicates that this is not the case. The *ccpA* mutation has pleiotropic effects. Not only does it make α-amylase synthesis resistant to glucose repression, but it does the same for sporulation (unpublished observations). In addition, the production of acetoin by wild-type cells in response to glucose in the medium is eliminated by the *ccpA* mutation. (The *ccpA* gene has been shown to be allelic to *alsA* [19], which is involved in the regulation of acetolactate synthase activity [65, 66].) Finally, the *ccpA* mutation relieves glucose repression of levanase synthesis (34), the *hut* operon (11a), and β-glucanase synthesis (53a). These observations suggest that the *ccpA* gene plays an important role in coordinating global gene expression in response to sugars in the medium.

In spite of the above evidence, other observations suggest the presence of multiple catabolite repression systems in *B. subtilis*. The existence of additional *cis*-acting regulatory elements associated with catabolite repression but having no apparent homology to *amyO* supports the notion of multiple systems. A region upstream of the *citB* promoter is associated with glucose regulation of *citB* expression (14). Deletion of the material between positions −84 and −67 deregulates the *citB* promoter. Within this region, there is no obvious sequence homology to *amyO*.

A region reported to be involved in glucose repression of the *B. subtilis gnt* (gluconate) operon was localized to the area of positions +136 through +148, which is within the coding portion of the *gntR* gene (36). A consensus sequence, A T T G A A A G, was deduced on the basis of similar sequences downstream of the promoters of *Bacillus* genes known to be subject to catabolite repression (36). At first glance, this fact suggests that catabolite repression of the *gnt* operon is subject to different regulatory mechanisms than is amylase. However, examination of the material flanking the *gnt* consensus sequence reveals a stretch of bases with great similarity to *amyO*, as is shown in the following comparison, where N is any base.

```
gnt sequence:    A T T G A   A A G C G G T A C   C A
                     | | |   | |   | |   |       | |
amyO consensus:      T G T/A A A N C G N T N A/T C A
```

The C located at the third position from the 3′ end of the *gnt* sequence is the only clear mismatch. Although an exactly corresponding variant of *amyO* has not been isolated, a variant with a G at this position has been (61). In this variant, glucose repressed amylase synthesis by about 75%, which is very similar to the extent of glucose repression observed in the construct used in identifying the *gnt* catabolite repression sequence (31). This striking homology to *amyO* suggests that the *gnt* operon may be subject to the same catabolite repression system as α-amylase and that the reported consensus sequence is actually a portion of the above described *amyO* operator.

SUMMARY

In contrast to catabolite repression in *E. coli*, that in *B. subtilis* and the other gram-positive bacteria examined is accomplished by a negative regulatory mechanism in which a hypothetical repressor protein binds to an operatorlike site, exemplified by *amyO*, and represses transcription of catabolite-sensitive genes and operons. The evidence upon which the negative regulatory model is based comes from several sources. *cis*-acting mutations that relieve catabolite repression of specific genes or operons have been isolated in *B. subtilis*, *S. aureus*, and *S. plicatus*. In *S. aureus*, cloning of a *cis*-acting region upstream of *lacP* on a multicopy plasmid relieved catabolite repression of the chromosomal copy of the *lac* operon, suggesting that a repressor had been titrated by the *cis*-acting site. A *trans*-acting mutation in *B. subtilis* that relieves glucose repression of α-amylase synthesis and the *hut* operon affects the gene encoding a protein, CcpA, that specifically binds to the *cis*-acting site *amyO*.

The location of the operator site at the start site of transcription of the α-amylase gene suggests that the repressor protein may block RNA synthesis from the amylase promoter in much the same way that LacI blocks transcription of the *E. coli lac* operon. This explanation may not be sufficient to explain the situation with regard to the *xyl*, *hut*, and *gnt* operons, where the operator sites are located well downstream of the transcription start site.

Catabolite repression in *B. subtilis* appears to be a global system that regulates the expression of genes (*amyE*) and operons (*hut*, *xyl*, and possibly *gnt*) involved in carbohydrate metabolism. Because the *ccpA* mutation renders sporulation resistant to repression by glucose, this process may be subject either directly or indirectly to this global regulatory system. Because *citB* does not appear to be subject to the same regulatory controls, there may be more than one catabolite repression system in *B. subtilis*.

The mechanism by which *B. subtilis* monitors its environment for the presence of readily metabolized carbohydrates and translates that information into regulatory signals is unknown. The homology of CcpA to known repressors of the GalR family suggests that the signal transduction system may involve an effector molecule such as sugar or nucleotide. In none of the gram-positive bacteria examined is cAMP the putative effector molecule.

REFERENCES

1. **Adler, K., K. Beyreuther, E. Fanning, N. Geisler, B. Gronenborn, A. Klemm, B. Muller-Hill, M. Pfahl, and A. Schmitz.** 1972. How *lac* repressor binds to DNA. *Nature* (London) **237:**322–327.
2. **Bjorklind, A., and S. Arvidson.** 1980. Mutants of *Staphylococcus aureus* affected in the regulation of exoprotein synthesis. *FEMS Microbiol. Lett.* **7:**203–206.
3. **Buttner, M. J., A. M. Smith, and M. J. Bibb.** 1988. At least three different RNA polymerase holoenzymes direct transcription of the agarase gene (*dagA*) of *Streptomyces coelicolor* A3(2). *Cell* **52:**599–609.
4. **Champness, W.** 1988. New loci required for *Streptomyces coelicolor* morphological and physiological differentiation. *J. Bacteriol.* **170:**1168–1174.
5. **Chater, K. F.** 1984. Morphological and physiological differentiation in *Streptomyces*, p. 89–115. *In* R. Losick and L. Shapiro (ed.), *Microbial Development*. Cold Spring Harbor Laboratory, Cold Spring Harbor, N.Y.
6. **Chatterjee, S., and L. C. Vining.** 1982. Catabolite repression in *Streptomyces venezuelae*. Induction of β-galactosidase, chloramphenicol production, and intracellular cyclic adenosine 3′,5′ monophosphate concentrations. *Can. J. Microbiol.* **28:**311–317.
7. **Delic, I., P. Robbins, and J. Westpheling.** 1992. Direct repeat sequences are implicated in the regulation of two chitinase promoters that are subject to carbon catabolite control in *Streptomyces*. *Proc. Natl. Acad. Sci. USA* **89:** 1885–1889.
8. **Doi, R. H., M. Gitt, L.-F. Wang, C. W. Price, and F. Kawamura.** 1985. Major sigma factor, sigma-43, of *Bacillus subtilis* RNA polymerase and interacting *spo0* products are implicated in catabolite control of sporulation, p. 157–161. *In* J. Hoch and P. Setlow (ed.), *Molecular Biology of Microbial Differentiation*. American Society for Microbiology, Washington, D.C.
9. **Duncan, J. L., and G. J. Cho.** 1972. Production of staphylococcal alpha toxin. II. Glucose repression of toxin formation. *Infect. Immun.* **6:**689–694.
10. **Ebright, R. H.** 1986. Proposed amino acid-base pair contacts for 13 sequence-specific DNA binding proteins, p. 207–219. *In* D. L. Oxender (ed.), *Protein Structure, Folding, and Design*. Alan R. Liss, Inc., New York.
11. **Epps, H. M. R., and E. F. Gale.** 1942. The influence of the presence of glucose during growth on the enzymic activities of *Escherichia coli*: comparison of the effect with that produced by fermentation acids. *Biochem. J.* **36:** 619–623.

11a. **Fisher, S.** Personal communication.

12. **Fisher, S. H., and A. L. Sonenshein.** 1991. Control of carbon and nitrogen metabolism in *Bacillus subtilis*. *Annu. Rev. Microbiol.* **45:**107–135.
13. **Fornwald, J. A., F. J. Schmidt, C. W. Adams, M. Rosenberg, and M. E. Brawner.** 1987. Two promoters, one inducible and one constitutive, control transcription of the *Streptomyces lividans* galactose operon. *Proc. Natl. Acad. Sci. USA* **84:**2130–2134.
14. **Fouet, A., and A. L. Sonenshein.** 1990. A target for carbon source-dependent negative regulation of the *citB* promoter of *Bacillus subtilis*. *J. Bacteriol.* **172:**835–844.
15. **Gaskill, M. E., and S. A. Khan.** 1988. Regulation of the enterotoxin B gene in *Staphylococcus aureus*. *J. Biol. Chem.* **263:**6276–6280.

16. **Gilman, M. Z., R. N. Wilson, and R. A. Weinberg.** 1986. Multiple protein binding sites in the 5′-flanking region regulate c-*fos* expression. *Mol. Cell. Biol.* **6:**4305–4316.
17. **Gordon, A. J. E., P. A. Burns, D. F. Fy, F. Yatagi, F. L. Allen, M. J. Horsfall, J. A. Halliday, J. Gray, C. Bernelot-Moens, and B. W. Glickman.** 1988. Missense mutation in the *lacI* gene of *Escherichia coli*. Inferences on the structure of the repressor protein. *J. Mol. Biol.* **200:**239–251.
18. **Hanson, R. S., J. A. Peterson, and A. A. Yousten.** 1970. Unique biochemical events in bacterial sporulation. *Annu. Rev. Microbiol.* **24:**53–90.
19. **Henkin, T. M., F. J. Grundy, W. L. Nicholson, and G. H. Chambliss.** 1991. Catabolite repression of an amylase gene expression in *Bacillus subtilis* involves a *trans*-acting gene product homologous to *Escherichia coli lacI* and *galR* repressors. *Mol. Microbiol.* **5:**575–584.
20. **Hodgson, D. A.** 1980. Carbohydrate utilization in *Streptomyces coelicolor* A3(2). Ph.D. thesis, University of East Anglia, Norwich, United Kingdom.
21. **Hodgson, D. A.** 1982. Glucose repression of carbon source uptake in *Streptomyces coelicolor* A3(2) and its perturbation in mutants resistant to 2-deoxyglucose. *J. Gen. Microbiol.* **128:**2417–2430.
22. **Iandolo, J. J., and W. J. Shafer.** 1977. Regulation of staphylococcal enterotoxin B. *Infect. Immun.* **16:**610–616.
23. **Ide, M.** 1971. Adenylcyclase of bacteria. *Arch. Biochem. Biophys.* **144:**262–268.
24. **Ikeda, H., E. T. Seno, C. J. Bruton, and K. F. Chater.** 1984. Genetic mapping, cloning and physiological aspects of the glucose kinase gene of *Streptomyces coelicolor. Mol. Gen. Genet.* **196:**501–507.
25. **Ingram, C., M. Brawner, P. Youngman, and J. Westpheling.** 1989. *xylE* functions as an efficient reporter gene in *Streptomyces* spp.: use for the study of *galP1*, a catabolite-controlled promoter. *J. Bacteriol.* **171:**6617–6624.
26. **Irani, M., L. Orosz, S. Busby, T. Taniguchi, and S. Adhya.** 1983. Cyclic AMP-dependent constitutive expression of *gal* operon: use of repressor titration to isolate operator mutations. *Proc. Natl. Acad. Sci. USA* **80:**4775–4779.
27. **Jacob, S., R. Allmansberger, D. Gartner, and W. Hillen.** 1991. Catabolite repression of the operon for xylose utilization from *Bacillus subtilis* W23 is mediated at the level of transcription and depends on a *cis* site in the *xylA* reading frame. *Mol. Gen. Genet.* **229:**189–196.
28. **Jarvis, A. W., R. C. Lawrence, and G. G. Pritchard.** 1975. Glucose repression of enterotoxins A, B and C and other extracellular proteins in staphylococci in batch and continuous culture. *J. Gen. Microbiol.* **86:**75–87.
29. **Kleina, L. G., and J. H. Miller.** 1990. Genetic studies of the *lac* repressor. XIII. Extensive amino acid replacements generated by the use of natural and synthetic suppressors. *J. Mol. Biol.* **212:**295–318.
30. **Magasanik, B.** 1961. Catabolite repression. Cold Spring Harbor Symp. Quant. Biol. **26:**249–256.
31. **Magasanik, B.** 1970. Glucose effects: inducer exclusion and repression, p. 189–219. *In* J. R. Beckwith and D. Zipser (ed.), *The Lactose Operon*. Cold Spring Harbor Laboratory, Cold Spring Harbor, N.Y.
32. **Magasanik, B., and F. C. Neidhardt.** 1987. Regulation of carbon and nitrogen utilization, p. 1318–1325. *In* F. C. Neidhardt, J. L. Ingraham, K. B. Low, B. Magasanik, M. Schaechter, and H. E. Umbarger (ed.), *Escherichia coli and Salmonella typhimurium: Cellular and Molecular Biology*, vol. 2. American Society for Microbiology, Washington, D.C.
33. **Maquat, L. E., K. Thornton, and W. S. Reznikoff.** 1980. *lac* promoter mutations located downstream from the transcription start site. *J. Mol. Biol.* **139:**537–549.
34. **Martin, I., M. Debarbouille, A. Klier, and G. Rapoport.** 1989. Induction and metabolite regulation of levanase synthesis in *Bacillus subtilis*. *J. Bacteriol.* **171:**1885–1892.
35. **Martin, J. F., and A. L. Demain.** 1980. Control of antibiotic biosynthesis. *Microbiol. Rev.* **44:**230–251.
36. **Miwa, Y., and Y. Fujita.** 1990. Determination of the *cis* sequence involved in catabolite repression of the *Bacillus subtilis gnt* operon; implication of a consensus sequence in catabolite repression in the genus *Bacillus*. *Nucleic Acids Res.* **18:**7049–7053.
37. **Miyashita, K., T. Fujii, and Y. Sawada.** 1991. Molecular cloning and characterization of chitinase genes from *Streptomyces lividans* 66. *J. Gen. Microbiol.* **137:**2065–2072.
38. **Monod, J.** 1947. The phenomenon of enzymatic adaptation. *Growth* **11:**223–289.
39. **Nicholson, W. L.** 1987. Regulation of α-amylase synthesis in *Bacillus subtilis*. Ph.D. thesis, University of Wisconsin, Madison.
40. **Nicholson, W. L., and G. H. Chambliss.** 1985. Isolation and characterization of a *cis*-acting mutation conferring catabolite resistance to α-amylase synthesis in *Bacillus subtilis*. *J. Bacteriol.* **161:**875–881.
41. **Nicholson, W. L., Y.-K. Park, T. M. Henkin, M. Won, M. J. Weickert, J. A. Gaskell, and G. H. Chambliss.** 1987. Catabolite repression-resistant mutations of the *Bacillus subtilis* alpha-amylase promoter affect transcription levels and are in an operator-like sequence. *J. Mol. Biol.* **198:**609–618.
42. **Oskouian, B., and G. C. Stewart.** 1990. Repression and catabolite repression of the lactose operon of *Staphylococcus aureus*. *J. Bacteriol.* **172:**3804–3812.
43. **Postma, P. W.** 1987. Phosphotransferase system for glucose and other sugars, p. 127–141. *In* F. C. Neidhardt, J. L. Ingraham, K. B. Low, B. Magasanik, M. Schaechter, and H. E. Umbarger (ed.), *Escherichia coli and Salmonella typhimurium: Cellular and Molecular Biology*, vol. 2. American Society for Microbiology, Washington, D.C.
44. **Regassa, L. B., J. L. Couch, and M. J. Betley.** 1991. Steady-state staphylococcal enterotoxin type C mRNA is affected by a product of the accessory gene regulator (*agr*) and by glucose. *Infect. Immun.* **59:**955–962.
45. **Sadler, J. R., H. Sasmor, and J. L. Betz.** 1983. A perfectly symmetric *lac* operator binds the *lac* repressor very tightly. *Proc. Natl. Acad. Sci. USA* **80:**6785–6789.
46. **Sauer, R. T., R. R. Yocum, R. F. Doolittle, M. Lewis, and C. O. Pabo.** 1982. Homology among DNA-binding proteins suggests use of a conserved super-secondary structure. *Nature* (London) **298:**447–451.
47. **Schaeffer, P., J. Millet, and J. D. Aubert.** 1965. Catabolite repression of bacterial sporulation. *Proc. Natl. Acad. Sci. USA* **54:**704–711.
48. **Setlow, P.** 1973. Inability to detect cyclic AMP in vegetative or sporulating cells or dormant spores of *Bacillus megaterium*. *Biochem. Biophys. Res. Commun.* **52:**365–372.
49. **Simons, A., D. Tils, B. von Wilcken-Bergmann, and B. Muller-Hill.** 1984. Possible ideal *lac* operator: *Escherichia coli lac* operator-like sequences from eukaryotic genomes lack the central G-C pair. *Proc. Natl. Acad. Sci. USA* **81:**1624–1628.
50. **Smith, C. P., and K. F. Chater.** 1988. Cloning and transcriptional analysis of the entire glycerol utilization (*glyABX*) operon of *Streptomyces coelicolor* A3(2) and identification of a closely associated transcriptional unit. *Mol. Gen. Genet.* **211:**129–137.
51. **Smith, J. L., M. M. Bencivengo, R. L. Buchanan, and C. A. Kunsch.** 1987. Effect of glucose analogs on the synthesis of staphylococcal enterotoxin A. *J. Food Safety* **8:**139–146.
52. **Smith, J. L., M. M. Bencivengo, and C. A. Kunsch.** 1986. Enterotoxin A synthesis in *Staphylococcus aureus*: inhibition by glycerol and maltose. *J. Gen. Microbiol.* **132:** 3375–3380.
53. **Smith, T. F., and J. R. Sadler.** 1971. The nature of lactose

operator constitutive mutations. *J. Mol. Biol.* **59:**273–305.

53a. **Stuhlkes, J.** Personal communication.

54. **Sun, D., and I. Takahashi.** 1982. Genetic mapping of catabolite-resistant mutants of *Bacillus subtilis. Can. J. Microbiol.* **28:**1242–1251.
55. **Sun, D., and I. Takahashi.** 1984. A catabolite-resistance mutation is localized in the *rpo* operon of *Bacillus subtilis. Can. J. Microbiol.* **30:**423–429.
56. **Takahashi, I.** 1979. Catabolite repression-resistant mutants of *Bacillus subtilis. Can. J. Microbiol.* **25:**1283–1287.
57. **Ullmann, A., and A. Danchin.** 1983. Role of cyclic AMP in bacteria. *Adv. Cyclic Nucleotide Res.* **15:**1–53.
58. **Virolle, M. J., and M. J. Bibb.** 1988. Cloning, characterization and regulation of an alpha-amylase gene from *Streptomyces limosus. Mol. Microbiol.* **2:**197–208.
59. **von Wilcken-Bergmann, B., and B. Muller-Hill.** 1982. Sequence of the *galR* gene indicates a common evolutionary origin of *lac* and *gal* repressor in *Escherichia coli. Proc. Natl. Acad. Sci. USA* **79:**2427–2431.
60. **Weickert, M., and S. Adhya.** 1992. A family of bacterial regulators homologous to Gal and Lac repressors. *J. Biol. Chem.* **267:**15869–15874.
61. **Weickert, M., and G. Chambliss.** 1990. Site-directed mutagenesis of a catabolite repression operator sequence in *Bacillus subtilis. Proc. Natl. Acad. Sci. USA* **87:**6238–6242.
62. **Yamaguchi, K., H. Matsuzaki, and B. Maruo.** 1969. Participation of a regulator gene in the α-amylase production of *Bacillus subtilis. J. Gen. Appl. Microbiol.* **15:** 97–107.
63. **Yamaguchi, K., Y. Nagata, and B. Maruo.** 1974. Genetic control of the rate of α-amylase synthesis in *Bacillus subtilis. J. Bacteriol.* **119:**410–415.
64. **Yuki, S.** 1968. On the gene controlling the rate of α-amylase production in *Bacillus subtilis. Biochem. Biophys. Res. Commun.* **31:**182–187.
65. **Zahler, S. A., L. G. Benjamin, B. S. Glatz, P. F. Winter, and B. J. Goldstein.** 1976. Genetic mapping of the *alsA, alsR, thyA, kauA,* and *citD* markers in *Bacillus subtilis*, p. 35–43. *In* D. Schlessinger (ed.), *Microbiology—1976.* American Society for Microbiology, Washington, D.C.
66. **Zahler, S. A., N. Najimudin, D. S. Kessler, and M. A. Vandeyar.** 1990. α-Acetolactate synthesis in *Bacillus subtilis*, p. 25–42. *In* Z. Barak, D. M. Chipman, and J. V. Schloss (ed.), *Biosynthesis of Branched Chain Amino Acids.* VCH Verlagsgesellschaft, Berlin.

16. Utilization of Amino Acids and Other Nitrogen-Containing Compounds

SUSAN H. FISHER

Gram-positive bacteria can use ammonium (NH_4^+) and a number of other compounds as sole sources of nitrogen (for a listing of nitrogen sources utilizable by *Bacillus subtilis*, see Table 1). For their constituent nitrogen atoms to be incorporated into cellular macromolecules, these compounds must be transported into the cell and enzymatically degraded to NH_4^+ or glutamate. NH_4^+ is subsequently assimilated into glutamine by glutamine synthetase (GS) and glutamate dehydrogenase (for a discussion of gram-positive bacteria that contain glutamate dehydrogenase enzymes, see chapter 20). Glutamate and glutamine then serve as nitrogen donors for the synthesis of all nitrogen-containing compounds in the cell. Although NH_4^+ derived from the degradation of nitrogen-containing compounds could be incorporated into asparagine by an NH_4^+-dependent asparagine synthetase, this enzyme has not been reported in any gram-positive bacterium (see chapter 20).

The synthesis of enzymes required for the utilization of nitrogen-containing compounds is generally induced by their substrates or related compounds. In addition, the expression of many of these catabolic enzymes and transport systems is regulated in response to the nitrogen (or carbon) composition of the growth medium. This regulation allows a cell growing in the presence of multiple nitrogen (or carbon) sources to preferentially use those compounds that result in the fastest growth rates.

In enteric bacteria such as *Escherichia coli*, *Salmonella typhimurium*, and *Klebsiella aerogenes*, GS and a number of amino acid permeases and degradative enzymes are synthesized at reduced rates during growth in the presence of high levels of NH_4^+ (48, 63). During NH_4^+-limited growth, the expression of these enzymes is elevated by the global nitrogen regulatory (Ntr) system. The key components of the Ntr system are the NtrB and NtrC proteins, which are encoded by the *glnA-ntrB-ntrC* operon. The NtrB protein, a protein kinase, phosphorylates and dephosphorylates the NtrC protein in response to the intracellular ratio of glutamine to 2-ketoglutarate. The phosphorylated form of NtrC, a DNA-binding protein, activates transcription of nitrogen-regulated genes when the glutamine/2-ketoglutarate ratio is low. Not all genes are directly regulated by NtrC. During NH_4^+ limitation in *K. aerogenes*, the *nac* gene product is required for the activation of expression of several genes (urease, histidine, and proline utilization) (10, 11, 46). The Nac protein belongs to the LysR family of transcriptional regulatory proteins (10). Transcription of the *nac* gene is activated by the Ntr system in response to NH_4^+ restriction, and the Nac protein activates expression of these nitrogen-regulated genes.

The regulation of nitrogen metabolism has been most extensively studied in *B. subtilis*, where this regulation differs considerably from that seen in enteric bacteria. The fastest growth of *B. subtilis* occurs in medium containing glutamine as the sole nitrogen source, while the optimal nitrogen source for enteric bacteria is NH_4^+ (26). A global Ntr system does not appear to be present in *B. subtilis* insofar as the GS regulatory protein, GlnR, is not known to regulate expression of any other genes. Only a limited number of NH_4^+-generating degradative enzymes, whose expression is repressed by NH_4^+, have been identified in *B. subtilis*. This may reflect, in part, the small number of catabolic enzymes and transport systems whose expression has been carefully examined. However, NH_4^+ is not the preferred nitrogen source for *B. subtilis*, and several degradative pathways are regulated in response to nitrogen-containing compounds other than NH_4^+. Expression of proline oxidase and the histidine degradation enzymes is not significantly elevated during NH_4^+-restricted growth, but their synthesis is severely inhibited by growth in the presence of amino acids (5, 25). Glutamine, not NH_4^+, represses the expression of the first two enzymes in the arginine degradative pathway (9).

B. subtilis can also circumvent growth inhibition resulting from restricted nitrogen availability by initiating sporulation. Amino acid limitation is thought to induce sporulation in nutrient broth sporulation medium. However, sporulation may not be the favored response of *B. subtilis* for surviving nutrient limitation. Immediately following the cessation of growth in sporulation medium, *B. subtilis* attempts a number of alternative responses (increased motility, synthesis of degradative enzymes, antibiotic synthesis, competence) that, if successful, could overcome the deficiency in nutrient availability and allow growth to resume (73; see also chapter 54). The elevated expression of the histidine-, proline-, and arginine-degradative enzymes at the onset of stationary growth in nutrient broth sporulation medium suggests that these amino acids are utilized by the cell during its adaptation to the nutrient-limited growth environment.

This chapter discusses the catabolism of amino acids and other nitrogen-containing compounds for which information is available. The degradative pathways for many nitrogen-containing compounds utilized by *B. subtilis* and other gram-positive bacteria have not been studied.

Susan H. Fisher • Department of Microbiology, Boston University School of Medicine, Boston, Massachusetts 02118.

Table 1. Compounds used as nitrogen sources by *B. subtilis* 168[a]

Compound	Growth[b]
None	±
NH_4Cl	++
L-Alanine	++
Allantoin	+
L-Arginine	++
L-Aspartate	++
L-Asparagine	+
γ-Amino *n*-butyric acid	+
Glucosamine	+[c]
N-Acetylglucosamine	±[c]
L-Glutamate	+
L-Glutamine	+++
L-Histidine	+
L-Isoleucine	+
KNO_3	+
L-Ornithine	+
L-Proline	++
L-Threonine	+
Urea	+
L-Valine	+

[a] *B. subtilis* 168 cells pregrown on minimal plates containing glucose and NH_4^+ as carbon and nitrogen sources were streaked onto plates containing 0.5% glucose and 0.2% of the indicated nitrogen sources, and the plates were incubated for 2 days at 37°C (25). Plates contained Noble agar (Difco). No growth was observed on the other nine L-amino acids (25).

[b] +++, opaque 3- to 4-mm colonies; ++, opaque 2-mm colonies; +, opaque 1-mm colonies; ±, translucent 0.5- to 1-mm colonies.

[c] Syntheses of the enzymes required for the degradation of *N*-acetylglucosamine and glucosamine are subject to regulation by catabolite repression in *B. subtilis* (7). When citrate replaced glucose as the carbon source in plates containing either *N*-acetylglucosamine or glucosamine as the nitrogen source, larger colonies were seen.

AMINO ACID UTILIZATION

Aspartate and Asparagine

L-Asparaginase (coded for by *asnA*) catalyzes the formation of NH_4^+ and L-aspartate from L-asparagine. L-Aspartate can be degraded to NH_4^+ and fumarate by L-aspartase (coded for by *asnB*). These two enzymes are encoded by a dicistronic operon (*asnAB*) in *B. subtilis* (74). Since disruption of the *asnA* gene results in the inability to grow with asparagine as a carbon or nitrogen source, only one asparaginase enzyme is present in *B. subtilis* (74). In contrast, *E. coli* has two asparaginase isozymes, i.e., a constitutive cytoplasmic enzyme and a periplasmic enzyme that is induced together with aspartase by anaerobic growth (37, 38). These enzymes are encoded by separate cistrons in *E. coli* (37, 38, 78). *K. aerogenes* synthesizes a constitutive periplasmic asparaginase whose expression is activated by the Ntr system during NH_4^+-restricted growth (46, 47).

When *B. subtilis* grows vegetatively in minimal medium containing a good source of NH_4^+, the Asn enzymes are synthesized only if aspartate or asparagine is present in the growth medium (74). It is not known whether aspartate, asparagine, or a metabolite derived from these compounds is the actual *asn* inducer. The Asn enzymes are synthesized constitutively in the *aspH* mutant during growth in medium containing high levels of NH_4^+ (35, 74). The *aspH* mutation is 80% cotransformable with *asnB* and lies in a gene whose product regulates *asnAB* transcription in *trans*.

During NH_4^+-limited growth in both *B. subtilis* and *Bacillus licheniformis*, asparaginase is synthesized in the absence of exogenous inducer (4, 30). It is possible that the *asn* inducer is generated intracellularly under these growth conditions. Alternatively, the *asn* operon may be regulated by a second system that is AspH-independent and responds to NH_4^+ availability. This second regulatory system may involve the GS protein. The GS structural gene (*glnA*) and the GS repressor gene (*glnR*) are required for the nitrogen source-dependent regulation of GS expression in *B. subtilis* (see chapter 20). Mutations in either gene result in the inability to repress GS transcription in the presence of good nitrogen sources (65, 67). Asparaginase is synthesized constitutively in *glnA* mutants but not in *glnR* mutants (4). Since the GS protein is thought to regulate *glnA* expression by transmitting the availability of good nitrogen sources to the GlnR protein, GS may play a similar role in signal transduction in *asn* regulation. (GS may also play a role in regulation of other nitrogen metabolism genes in *B. subtilis* [see below].)

During sporulation, the Asn enzymes are preferentially expressed in the forespore compartment in *B. subtilis* and *Bacillus cereus*, and both enzymes are found in the *B. subtilis* mature spore (2, 74). The *B. subtilis asn* operon is transcribed by a σ^A-dependent promoter during vegetative growth in rich medium, and expression of this promoter is elevated by the *aspH* mutation (74). However, during both exponential and stationary phases in sporulation medium, *asn* transcription occurs from a second promoter whose initiation site is 7 bp upstream of the start site for the σ^A promoter (74). The nucleotide sequence upstream of this second promoter shows no similarity to the consensus sequences for any known *B. subtilis* sigma factor. Since *asn* transcription occurs from different promoters in cells growing in rich medium or in sporulation medium, its expression may respond to distinct regulatory signals under these growth conditions.

Aspartate is transported into *B. subtilis* by two systems, a high-affinity system energized by the proton motive force and a low-affinity system (77). Mutants in the high-affinity system (*aspT*) have been isolated by their resistance to DL-*threo*-β-hydroxyaspartate (77). During growth in nutrient broth sporulation medium, the rates of uptake of aspartate and other amino acids increase toward the end of the exponential growth phase and decline during sporulation in *B. subtilis* and *Bacillus megaterium* (19, 20). At least nine amino acid transport carriers with K_ms in the micromolar range are present in *B. subtilis* membrane vesicles (42).

Arginine

Multiple pathways for L-arginine degradation are present in microorganisms (see references 1 and 21 for descriptions of pathways in gram-negative bacteria). The enzymes of the arginase degradative pathway are found in *B. subtilis* and *B. licheniformis* (Fig. 1) (9, 44). The molecular masses of the arginase enzymes from

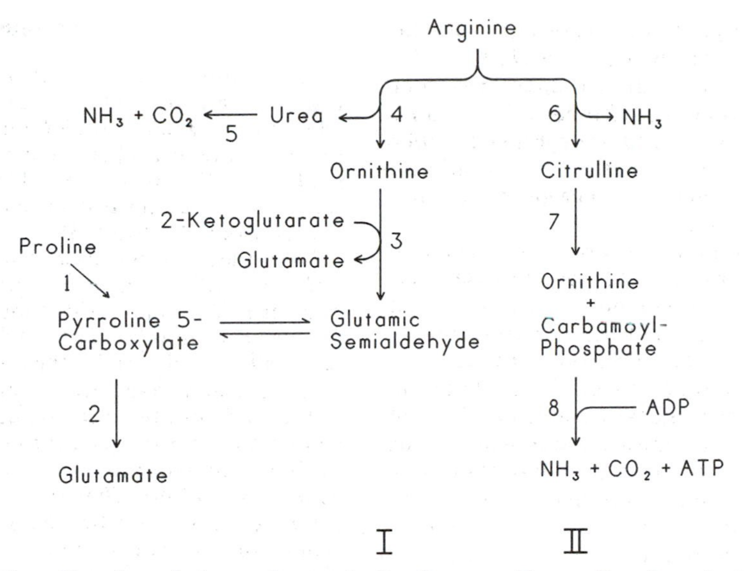

Figure 1. Arginine and proline degradative pathways in *Bacillus* spp. The proline degradative enzymes are as follows: 1, proline oxidase; 2, pyrroline-5-carboxylate dehydrogenase. The enzymes of the arginase degradative pathway (I) are as follows: 3, ornithine transaminase; 4, arginase; 5, urease. The enzymes of the deiminase pathway (II) are as follows: 6, arginine deiminase; 7, ornithine carbamoyltransferase; 8, carbamate kinase. Glutamate semialdehyde is spontaneously converted to pyrroline 5-carboxylate, the more-stable cyclic form of glutamate semialdehyde. The last step in the arginase degradative pathway and proline degradation was reported to be catalyzed by different pyrroline-5-carboxylate dehydrogenase isozymes in *B. subtilis* (23). However, the bacterial strain used in these studies (23) was subsequently found to be *B. licheniformis* (45).

B. subtilis (296,000 kDa) and *B. licheniformis* (260,000 kDa) are similar, and the activities of both enzymes are competitively inhibited by ornithine (36, 70). Since urease is induced by arginine in *B. subtilis*, urea derived from arginine is likely to be metabolized to NH_4^+ by this enzyme (4). In *B. licheniformis*, lactic bacteria, and clostridia, arginine is degraded by a second route, which is similar to the deiminase pathway found in pseudomonads (Fig. 1) (15, 16). This second catabolic system is likely to be present in *B. subtilis* and other *Bacillus* strains (9, 57).

In *Streptomyces griseus*, arginine is converted to succinate through a sequence of intermediates including 4-guanidinobutyramide, 4-guanidinobutyrate, 4-aminobutyrate, and succinate semialdehyde by an arginine-inducible catabolic system (75). The first enzyme of the pathway, arginine decarboxyoxidase, has been purified and shown to contain a flavoprotein (55). Arginine is likely to be degraded by this route in other *Streptomyces* strains. Guanidinobutyrate ureohydrolase, the third enzyme of this pathway, is arginine inducible in *Streptomyces coelicolor*, *Streptomyces lividans*, and *Streptomyces clavuligerus* (58). While arginase activity is also present in these *Streptomyces* strains, arginase levels do not increase during growth in arginine-containing medium (6, 58).

In *B. subtilis* and *B. licheniformis*, the first two enzymes in the arginase-degradative pathway, arginase and ornithine transaminase, are induced by arginine, ornithine, citrulline, and proline (9, 44). The ability of proline to induce these enzymes is likely to reflect the convergence of the arginine- and proline-degradative pathways at their final step, where pyrroline 5-carboxylate is converted to glutamate (Fig. 1). Pyrroline 5-carboxylate or glutamic semialdehyde may be the true inducer of the arginase catabolic system. In *B. subtilis*, the proline-dependent induction of these enzymes is severely inhibited in the presence of either NH_4^+ or glutamine, while only glutamine represses their induction by arginine (9). This makes physiological sense because, as judged by growth rates, *B. subtilis* prefers either NH_4^+ or glutamine to proline as sole source of nitrogen, but only glutamine is a better nitrogen source than arginine (9). Synthesis of the arginase-degradative enzymes is subject to glucose catabolite repression in *B. licheniformis* (44). Arginase activity cannot be detected in cells growing rapidly in sporulation medium but appears at the onset of stationary phase in both *B. subtilis* and *B. licheniformis* (24, 62). Mature *B. subtilis* spores contain arginase activity (24).

Resistance to the toxic arginine analog arginine hydroxamate has been used to isolate *B. subtilis* mutants (*ahr*) defective in the utilization of arginine, citrulline, and/or ornithine as sole nitrogen source (34). These mutations map at four loci, *ahrA*, *ahrB*, *ahrC*, and *ahrD* (52). Since these *ahr* mutations alter the expression of both arginase and ornithine transaminase, they are likely to be regulatory mutations (34, 52). In *ahrC* mutants, the arginine biosynthetic enzymes are expressed constitutively and the enzymes of the arginase catabolic system cannot be induced (34). *ahrC* mutations lie in an open reading frame that has 27% identity with the *E. coli* arginine repressor protein (72). The AhrC protein binds to the promoter region of the *B. subtilis* arginine biosynthetic gene, *argC*, in vitro (72; see chapter 21). The inability of *ahrC* mutants to induce the arginase degradative enzymes indicates that the AhrC protein is also involved in their regulation.

No *B. subtilis* or *B. licheniformis* mutants containing lesions in structural genes for the degradative enzymes of the arginase pathway have been isolated. β-Galactosidase expression in one *B. subtilis* Tn*917-lacZ* transposon insertion mutant is AhrC dependent and inducible by arginine, ornithine, and citrulline (8). However, this insertion mutant synthesizes wild-type levels of arginase and ornithine transaminase and has no growth defect on medium containing arginine as a nitrogen source.

Arginine and ornithine utilization is also defective in *B. subtilis sigL* mutants (22). In addition, *sigL* mutants cannot utilize isoleucine or valine as sole sources of nitrogen and are defective in transcription of the levanase operon. The derived amino acid sequence of the *sigL* polypeptide is similar to that of the σ^{54} protein found in gram-negative bacteria. The role of the *sigL* protein in arginine catabolism is not understood. An open reading frame with significant homology to other arginine permeases has been shown to be located in an operon whose expression is dependent on the σ^{54} protein (29). It is possible that this open reading frame encodes an arginine permease in *B. subtilis*. However, because insertional mutagenesis of the putative arginine permease does not result in the inability to utilize arginine as a nitrogen source, the arginine phenotype of the *sigL* mutant cannot be attributed solely to the inability to express this permease. This result does suggest that *B. subtilis* is likely to have multiple arginine transport systems, as has been observed in *E. coli* and yeast cells.

The arginine deiminase degradative pathway directly provides the cell with both energy and nitrogen compounds, since 1 mol of ATP is produced per mol of arginine consumed (Fig. 1). During anaerobic growth, the deiminase degradative pathway is preferentially expressed in *B. licheniformis*, while the arginase catabolic system is the main route for arginine degradation in aerobically grown *B. licheniformis* cells (15). *B. licheniformis* mutants deficient in carbamate kinase, the third enzyme of the deiminase pathway, are unable to use arginine during anaerobiosis but can still degrade arginine by the arginase pathway during aerobic growth (21). In contrast, a *B. licheniformis* mutant unable to induce the first two enzymes of the arginase degradative pathway can use arginine as a nitrogen source only during anaerobiosis (15). The deiminase pathway is repressed by both aerobic (O_2) and anaerobic respiration. Nitrate, a substrate for anaerobic respiration, represses the deiminase catabolic system and induces the arginase degradative enzymes during anaerobic growth in *B. licheniformis* (15).

In *Clostridium sporogenes*, the highest level of arginine deiminase, the first enzyme of the deiminase pathway, is seen at the onset of stationary phase in sporulation medium (76). The other enzymes of this pathway, ornithine carbamoyltransferase and carbamate kinase, are expressed constitutively throughout exponential and stationary phases. Expression of arginine deiminase but not of ornithine carbamoyltransferase and carbamate kinase appears to be regulated by catabolite repression, because arginine deiminase levels are reduced by growth in medium containing either glucose or glycerol.

Proline

In *B. subtilis* and *B. licheniformis*, the proline-degradative enzymes (Fig. 1) are induced by proline (5, 44). In *B. subtilis*, this induction is inhibited if the growth medium also contains glucose and amino acids (5). Proline oxidase levels in *B. subtilis* also increase at the onset of stationary growth in nutrient sporulation medium (25).

Both the proline oxidase and the pyrroline-5-carboxylate dehydrogenase enzymatic activities are present in the same multifunctional membrane-associated polypeptide (*putA*) in *E. coli* and *S. typhimurium* (49). Interestingly, the *S. typhimurium* PutA protein also acts as a negative regulatory protein, binding to the *putA* control region and repressing *putA* expression in the absence of proline (56). *put* expression is proline inducible in *E. coli*, *S. typhimurium*, and *K. aerogenes* (47, 49). The Ntr system and the Nac gene product regulate proline oxidase synthesis in *K. aerogenes*, while its expression is subject to glucose catabolite repression in *S. typhimurium* and *E. coli* (46, 47, 49).

Alanine

L-Alanine dehydrogenase catalyzes the degradation of L-alanine to NH_4^+ and pyruvate, with NAD^+ as a cofactor. Alanine dehydrogenase from *B. cereus* is composed of six identical subunits (42,000 Da each) and can form a hybrid enzyme with *B. subtilis* alanine dehydrogenase subunits in vitro (60).

B. subtilis mutants deficient in alanine dehydrogenase (*ald*) are unable to utilize alanine as a sole source of nitrogen or carbon (28). In addition, *ald* mutants sporulate poorly, producing many nonrefractile spores during sporulation in medium containing alanine (27). Sporulation-defective mutants generated by Tn*917* mutagenesis that contain a transposon insertion in the *ald* structural gene have been identified (31). The reason(s) for the poor sporulation of *ald* mutants has not been determined. One possibility is that these mutants accumulate alanine (and other amino acids [see below]) and as a consequence initiate sporulation inefficiently. Alternatively, reduced intracellular pools of pyruvate could cause the sporulation defect in these mutants. Alanine can still initiate germination of spores produced by an *ald* mutant, although *ald* spores lose their refractility slowly under these germination conditions and show delayed outgrowth in minimal medium containing alanine (27). *ald* spores germinate as rapidly as wild-type spores in rich medium.

Alanine dehydrogenase is maximally induced in *B. subtilis* and *B. licheniformis* by the addition of L-alanine to minimal medium (12, 66). In addition, 10 other L-amino acids and 12 other D-amino acids can partially induce alanine dehydrogenase synthesis in *B. subtilis* (12). Berberich et al. have proposed that *ald* induction by these other amino acids results from their conversion to either L-alanine or D-alanine by transamination (12). Indeed, the majority of the L-amino acid (and D-amino acid) inducers can be converted to L-alanine (and D-alanine) by transamination in vitro. D-Alanine appears to be the true inducer of *ald* expression, because only D-alanine and other

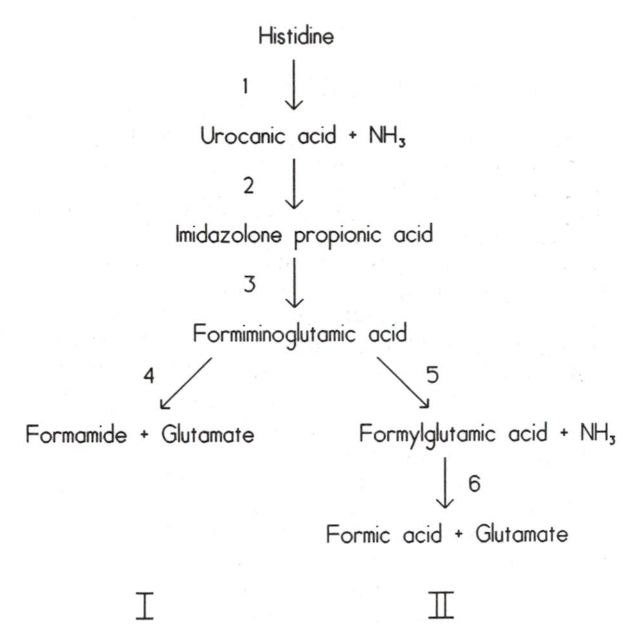

Figure 2. Histidine degradative pathways. Pathway I is found in enteric bacteria and *B. subtilis,* while pathway II is present in pseudomonads and *S. coelicolor.* The enzymes are as follows: 1, histidase (*hutH*); 2, urocanase (*hutU*); 3, imidazolonepropionate hydrolase (*hutI*); 4, formiminoglutamic acid formiminohydrolase (*hutG*); 5, formiminoglutamic acid iminohydrolase (*hutF*); 6, formylglutamic acid amidohydrolase (*hutG*).

D-amino acids can induce *ald* expression in alanine racemase mutants. (Alanine racemase interconverts L-alanine and D-alanine.)

A second mechanism may regulate *ald* expression in *B. subtilis*. Transcription of the *ald* structural gene can be induced by the addition of decoyinine to cells growing in glucose minimal medium in the absence of alanine (31). Decoyinine, an inhibitor of purine nucleotide synthesis, is thought to induce sporulation in exponentially growing cells by lowering the levels of intracellular pools of guanine nucleotides (73).

Alanine dehydrogenase appears to be required for alanine degradation in other gram-positive bacteria. An *S. clavuligerus* mutant that synthesizes significantly reduced levels of alanine dehydrogenase activity is unable to utilize alanine as a sole source of carbon and nitrogen (14). Expression of Ald activity is induced by growth in the presence of alanine (or NH_4^+; see chapter 20) in *S. clavuligerus*.

Histidine

L-Histidine is degraded by two partially overlapping pathways in bacteria (Fig. 2). In *B. subtilis,* the four enzymes that catalyze the degradation of histidine to NH_4^+, glutamate, and formamide are encoded in a multicistronic operon (*hutPHUIG*) (Fig. 3) (17, 39, 54). Upstream of the *hutH* gene is an open reading frame (*hutP*) whose protein product positively regulates *hut* expression in *trans* (54). Transcriptional antitermination may be involved in *hut* regulation, because the noncoding region between the *hutP* and

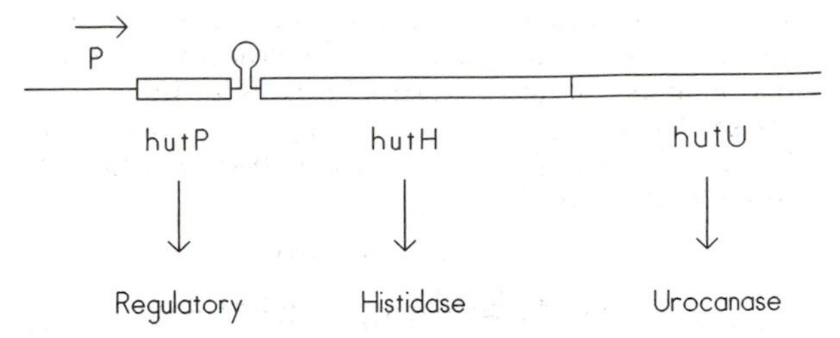

Figure 3. Genetic organization of the *B. subtilis hut* genes. The physical map of the *hutPHU* genes is taken from reference 54. Abbreviations: P, *hut* promoter region; *hutP, hut* regulatory gene; *hutH,* histidase structural gene; *hutU,* urocanase structural gene.

hutH genes contains a potential stem-loop structure (Fig. 3) (54).

Hut expression in *B. subtilis* is induced by histidine and repressed by rapidly metabolized carbon sources such as glucose (17). Growth in the presence of amino acids severely inhibits synthesis of the Hut enzymes (5). Since amino acid repression of Hut and proline oxidase synthesis is not altered in *B. subtilis* mutants defective in either glucose catabolite repression or nitrogen regulation, a distinctive regulatory system that responds to amino acid availability appears to be present in *B. subtilis*. Hut expression is also activated during the onset of stationary phase and sporulation in nutrient broth sporulation medium (25).

During growth in minimal medium containing poor sources of carbon, *spo0* mutants synthesize higher levels of histidase, the first enzyme of histidine degradation, than do wild-type cells (13). *spo0A abrB* mutants express histidase at wild-type levels (25). This fact suggests that the overproduction of histidase in *spo0* mutants results from elevated levels of the AbrB protein and that the AbrB protein positively regulates Hut expression under these growth conditions (see chapter 52).

Although histidine is degraded by the same pathway in *B. subtilis* and enteric bacteria, the genetic organization and regulation of the Hut enzymes differ significantly between these bacteria. The *K. aerogenes hut* genes are organized as two closely linked operons (47). A repressor, the product of the HutC gene, regulates the expression of both operons. Hut expression is induced by urocanate, the product of histidase enzymatic activity. In both *K. aerogenes* and *S. typhimurium,* the *hut* genes are regulated in response to carbon availability by the catabolite gene activator protein–cyclic-AMP complex (47, 48). Hut expression can also be activated during nitrogen starvation by the Ntr system and the Nac gene product in *K. aerogenes* (10, 46, 47).

S. coelicolor and the pseudomonads utilize the same pathway for histidine degradation (Fig. 2) (18, 41). Although the synthesis of the Hut enzymes is inducible by urocanate in *S. coelicolor, S. griseus,* and *S. clavuligerus,* there is no evidence for regulation by carbon or nitrogen availability in any *Streptomyces* strain (6, 41, 43). Examination of the mechanism of histidine induction in *S. griseus* showed that histidase (*hutH*), the first enzyme of histidine degradation, is synthesized constitutively and maintained in an inactive form until inducer is added to the growth medium (43). In contrast, induction of urocanase (*hutU*), the second enzyme of histidine degradation, requires concurrent protein synthesis (40). Unlike the *B. subtilis* and *K. aerogenes hutH* genes, the *S. griseus hutH* structural gene appears to be encoded as a monocistronic transcript (79).

Leucine, Phenylalanine, and Valine

Dehydrogenase enzymes may play a role in the degradation of phenylalanine in *Bacillus badius* (3), of valine in *Streptomyces* spp. (53, 61), and of leucine in *B. cereus* (68).

UTILIZATION OF OTHER NITROGEN SOURCES

Urea

Urease is expressed constitutively in *B. subtilis*, *S. coelicolor*, and *S. clavuligerus* cells grown in medium containing urea or nitrogen compounds that are poor sources of NH_4^+ (4, 6, 25).

Nitrogen regulation of urease (and asparaginase) expression in *B. subtilis* partially overlaps with the system that mediates GS regulation in response to nitrogen availability (4). Since neither urease nor GS expression is repressed during growth in the presence of good sources of nitrogen in *glnA* mutants, the GS protein appears to be involved in signaling nitrogen availability to the systems regulating GS and urease (and asparaginase [see above]). Nitrogen regulation of β-galactosidase synthesis in two *B. subtilis* strains containing the Tn*917-lacZ* insertions *nrg-21*::Tn*917-lacZ* and *nrg-29*::Tn*917-lacZ* was also found to require the wild-type GS protein (4). The observation that the wild-type GS protein is required for the regulation of all genes or gene products whose expression is elevated during NH_4^+ limitation in *B. subtilis* (GS, urease, asparaginase, *nrg-21*, and *nrg-29*) suggests that a common signal may mediate this regulation.

Nitrate

Nitrate reductase activity is found in *B. subtilis* cells growing in the presence of nitrate under semianaerobic conditions (32). The AbrB protein appears to positively regulate synthesis of nitrate reductase, because enzyme levels are elevated in *spo0A* and *spo0B* mutants but not in *spo0A abrB* mutants (32, 33, 51).

When *Streptomyces venezuelae* is grown in minimal medium containing a mixture of NH_4^+ and nitrate, NH_4^+ is preferentially utilized and nitrate consumption does not begin until the NH_4^+ present in the growth medium has been exhausted (69). This nitrogen regulation of nitrate utilization appears to occur at the level of nitrate uptake. The level of nitrate reductase expression is not altered by the presence of NH_4^+ in the growth medium, and the enzymatic activity of nitrate reductase is not inhibited in vitro by NH_4^+ or other nitrogen-containing metabolic intermediates.

Peptides

Amino acids and small peptides produced by the degradation of extracellular polypeptides can also supply *B. subtilis* with nutrients during growth and sporulation. Oligopeptide and dipeptide transport systems have been genetically identified in *B. subtilis*. The *spo0K* locus contains an operon (*opp*) of five open reading frames that are homologous to the oligopeptide transport system of gram-negative bacteria (59, 64). Two ATP-binding proteins are present in the *opp* operon. One of these proteins (*oppD*) is required for peptide transport and sporulation, while the second (*oppF*) is necessary for the development of genetic competence. The Opp transport system is present in vegetatively growing cells and is thought to be required for uptake of one or more signal peptides involved in sporulation regulation.

A dipeptide transport system is encoded by the multicistronic *dciA* operon (50, 71). Unlike the Opp transport system, the *dciA* transport system is not required for successful sporulation. During exponential growth, *dciA* transcription is independently repressed by glucose or amino acids. *dciA* expression also increases at the onset of sporulation. Growth-phase regulation of *dciA* expression is due in part to negative regulation by the AbrB protein, a repressor of early sporulation genes. *dciA* expression is not activated in *spo0A* mutants during stationary growth phase, but this activation does occur, albeit slowly, in *spo0A abrB* mutants. The purified AbrB protein binds to the *dciA* promoter region in vitro, partially overlapping the RNA polymerase-binding region. However, since AbrB mutants do not transcribe *dciA* constitutively during vegetative growth, other factors must regulate the expression of this operon in response to the growth state of the culture.

Acknowledgments. I thank S. Baumberg, M. Debarbouille, P. Glaser, A. Grossman, K. Kendrick, and P. Setlow for communicating results before publication and L. Wray for helpful comments on the manuscript. Unpublished work from my laboratory was supported by research grant AI23168 from the U.S. Public Health Service.

REFERENCES

1. **Abdelal, A. T.** 1979. Arginine catabolism by microorganisms. *Annu. Rev. Microbiol.* **33:**139–168.
2. **Andreoli, A. J., J. Saranto, N. Caliri, E. Escamilla, and E. Pina.** 1978. Comparative studies of proteins from forespore and mother cell compartments of *Bacillus cereus*, p. 260–264. *In* G. Chambliss and J. C. Vary (ed.), *Spores VII*. American Society for Microbiology, Washington, D.C.
3. **Asano, Y., A. Nakazawa, K. Endo, Y. Hibino, M. Ohmori, N. Numano, and K. Kondo.** 1987. Phenylalanine dehydrogenase of *Bacillus badius*. *Eur. J. Biochem.* **168:**153–159.
4. **Atkinson, M. R., and S. H. Fisher.** 1991. Identification of genes and gene products whose expression is activated during nitrogen-limited growth in *Bacillus subtilis*. *J. Bacteriol.* **173:**23–27.
5. **Atkinson, M. R., L. V. Wray, and S. H. Fisher.** 1990. Regulation of the histidine and proline degradative enzymes by amino acid availability in *Bacillus subtilis*. *J. Bacteriol.* **172:**4758–4765.

6. **Bascarán, V., C. Hardisson, and A. F. Braña.** 1989. Regulation of nitrogen catabolic enzymes in *Streptomyces clavuligerus*. *J. Gen. Microbiol.* **135:**2465–2474.
7. **Bates, C. J., and C. A. Pasternak.** 1965. Further studies on the regulation of amino acid sugar metabolism in *Bacillus subtilis*. *Biochem. J.* **96:**147–154.
8. **Baumberg, S. (Leeds University).** 1992. Personal communication.
9. **Baumberg, S., and C. R. Harwood.** 1979. Carbon and nitrogen repression of arginine catabolic enzymes in *Bacillus subtilis*. *J. Bacteriol.* **137:**189–196.
10. **Bender, R. A.** 1991. The role of the NAC protein in the nitrogen regulation of *Klebsiella aerogenes*. *Mol. Microbiol.* **5:**2575–2580.
11. **Bender, R. A., P. M. Snyder, R. Bueno, M. Quinto, and B. Magasanik.** 1983. Nitrogen regulation system of *Klebsiella aerogenes*: the *nac* gene. *J. Bacteriol.* **156:**444–446.
12. **Berberich, R., M. Kaback, and E. Freese.** 1968. D-Amino acids as inducers of L-alanine dehydrogenase in *Bacillus subtilis*. *J. Biol. Chem.* **243:**1006–1011.
13. **Boylan, S. A., K. T. Chun, B. A. Edson, and C. W. Price.** 1988. Early-blocked sporulation mutants alter expression of enzymes under carbon control in *Bacillus subtilis*. *Mol. Gen. Genet.* **212:**271–280.
14. **Braña, A. F., N. Paiva, and A. L. Demain.** 1986. Pathways and regulation of ammonium assimilation in *Streptomyces clavuligerus*. *J. Gen. Microbiol.* **132:**1305–1317.
15. **Broman, K., N. Lauwers, V. Stalon, and J.-M. Wiame.** 1978. Oxygen and nitrate in utilization by *Bacillus licheniformis* of the arginase and arginine deiminase routes of arginine catabolism and other factors affecting their synthesis. *J. Bacteriol.* **135:**920–927.
16. **Broman, K., V. Stalon, and J.-M. Wiame.** 1975. The duplication of arginine catabolism and the meaning of the two ornithine carbamoyltransferases in *Bacillus licheniformis*. *Biochem. Biophys. Res. Commun.* **66:**821–827.
17. **Chasin, L. A., and B. Magasanik.** 1968. Induction and repression of the histidine-degrading enzymes of *Bacillus subtilis*. *J. Biol. Chem.* **243:**5165–5178.
18. **Consevage, M. W., R. D. Porter, and A. T. Porter.** 1985. Cloning and expression in *Escherichia coli* of histidine utilization genes for *Pseudomonas putida*. *J. Bacteriol.* **162:**132–146.
19. **Cooney, P. H., P. Fawcett Whiteman, and E. Freese.** 1977. Media dependence of commitment in *Bacillus subtilis*. *J. Bacteriol.* **129:**901–907.
20. **Cooney, P. H., and E. Freese.** 1976. Commitment to sporulation in *Bacillus megaterium* and uptake of specific compounds. *J. Gen. Microbiol.* **95:**381–390.
21. **Cunin, R., N. Glansdorff, A. Piérard, and V. Stalon.** 1986. Biosynthesis and metabolism of arginine in bacteria. *Microbiol. Rev.* **50:**314–352.
22. **Debarbouille, M., I. Martin-Verstraete, F. Kunst, and G. Rapoport.** 1991. The *Bacillus subtilis sigL* gene encodes an equivalent of sigma 54 from Gram-negative bacteria. *Proc. Natl. Acad. Sci. USA* **88:**9092–9096.
23. **De Hauwer, G., R. Lavalle, and J. M. Wiame.** 1964. Étude de la pyrroline déshydrogénase et de la régulation du catabolisme de l'arginine et de la proline chez *Bacillus subtilis*. *Biochim. Biophys. Acta* **81:**257–269.
24. **Deutscher, M. P., and A. Kornberg.** 1968. Biochemical studies of bacterial sporulation and germination. VIII. Patterns of enzyme development during growth and sporulation of *Bacillus subtilis*. *J. Biol. Chem.* **243:**4653–4660.
25. **Fisher, S. H., and M. R. Atkinson.** Unpublished data.
26. **Fisher, S. H., and A. L. Sonenshein.** 1991. Control of carbon and nitrogen metabolism in *Bacillus subtilis*. *Annu. Rev. Microbiol.* **45:**107–135.
27. **Freese, E., and M. Cashel.** 1965. Initial stages of germination, p. 144–151. *In* L. L. Campbell and H. O. Halvorson (ed.), *Spores III*. American Society for Microbiology, Ann Arbor, Mich.
28. **Freese, E., S. W. Park, and M. Cashel.** 1964. The developmental significance of alanine dehydrogenase in *Bacillus subtilis*. *Proc. Natl. Acad. Sci. USA* **51:**1164–1172.
29. **Glaser, P. (Pasteur Institute).** 1992. Personal communication.
30. **Golden, K. J., and R. W. Bernhohr.** 1985. Nitrogen catabolite repression of the L-asparaginase of *Bacillus licheniformis*. *J. Bacteriol.* **164:**938–940.
31. **Grossman, A. (Massachusetts Institute of Technology).** 1991. Personal communication.
32. **Guespin-Michel, J. F.** 1971. Phenotypic reversion in some early blocked sporulation mutants of *Bacillus subtilis*. Genetic study of polymyxin resistant partial revertants. *Mol. Gen. Genet.* **112:**243–254.
33. **Guespin-Michel, J. F., M. Piechaud, and P. Schaeffer.** 1970. Constitutivité vis-à-vis du nitrate de la nitrate-réductase chez les mutants asporogènes précoces de *Bacillus subtilis*. *Ann. Inst. Pasteur* **119:**711–718.
34. **Harwood, C. R., and S. Baumberg.** 1977. Arginine hydroxamate-resistant mutants of *Bacillus subtilis* with altered control of arginine metabolism. *J. Gen. Microbiol.* **100:**177–188.
35. **Iijama, T., M. D. Diesterhaft, and E. Freese.** 1977. Sodium effect of growth on aspartate and genetic analysis of a *Bacillus subtilis* mutant with high aspartate activity. *J. Bacteriol.* **129:**1440–1447.
36. **Issaly, I. M., and A. S. Issaly.** 1974. Control of ornithine carbamoyltransferase activity by arginase in *Bacillus subtilis*. *Eur. J. Biochem.* **49:**485–495.
37. **Jennings, M. P., and I. R. Beacham.** 1990. Analysis of the *Escherichia coli* gene encoding L-asparaginase II, *ansB* and its regulation by cyclic AMP receptor and FNR proteins. *J. Bacteriol.* **172:**1491–1498.
38. **Jerlstrom, P. G., D. A. Bezjak, M. P. Jennings, and I. R. Beacham.** 1989. Structure and expression in *Escherichia coli* K-12 of the L-asparaginase I-encoding *ansA* gene and its flanking region. *Gene* **78:**37–46.
39. **Kaminskas, E., and B. Magasanik.** 1970. Sequential synthesis of histidine-degrading enzymes in *Bacillus subtilis*. *J. Biol. Chem.* **245:**3549–3555.
40. **Kendrick, K. (Ohio State University).** 1991. Personal communication.
41. **Kendrick, K. E., and M. L. Wheelis.** 1982. Histidine dissimilation in *Streptomyces coelicolor*. *J. Gen. Microbiol.* **128:**2029–2040.
42. **Konings, W. N., and E. Freese.** 1972. Amino acid transport in membrane vesicles of *Bacillus subtilis*. *J. Biol. Chem.* **247:**2408–2418.
43. **Kroening, T. A., and K. E. Kendrick.** 1989. Cascading regulation of histidase activity in *Streptomyces griseus*. *J. Bacteriol.* **171:**1100–1105.
44. **Laishley, E. J., and R. W. Bernlohr.** 1968. Regulation of arginine and proline catabolism in *Bacillus licheniformis*. *J. Bacteriol.* **96:**322–329.
45. **Legrain, C., V. Stalon, J.-P. Noullez, A. Mercenier, J.-P. Simon, K. Boman, and J.-M. Wiame.** 1977. Structure and function of ornithine carbamoyltransferases. *Eur. J. Biochem.* **80:**401–409.
46. **Macaluso, A., E. A. Best, and R. A. Bender.** 1990. Role of the *nac* gene product in the nitrogen regulation of some NTR-regulated operons of *Klebsiella aerogenes*. *J. Bacteriol.* **172:**7249–7255.
47. **Magasanik, B.** 1982. Genetic control of nitrogen assimilation in bacteria. *Annu. Rev. Genet.* **16:**135–168.
48. **Magasanik, B., and F. C. Neidhardt.** 1987. Regulation of carbon and nitrogen utilization, p. 1318–1325. *In* F. C. Neidhardt, J. L. Ingraham, K. B. Low, B. Magasanik, M. Schaechter, and H. Umbarger (ed.), *Escherichia coli and Salmonella typhimurium: Cellular and Molecular Biology*, vol. 2. American Society for Microbiology, Washington, D.C.

49. **Maloy, S. R.** 1987. The proline utilization operon, p. 1513–1519. *In* F. C. Neidhardt, J. L. Ingraham, K. B. Low, B. Magasanik, M. Schaechter, and H. Umbarger (ed.), *Escherichia coli and Salmonella typhimurium: Cellular and Molecular Biology*, vol. 2. American Society for Microbiology, Washington, D.C.
50. **Mathiopoulos, C., J. P. Mueller, F. J. Slack, C. G. Murphy, S. Patanker, G. Bukusoglu, and A. L. Sonenshein.** 1991. A *B. subtilis* dipeptide transport system expressed early during sporulation. *Mol. Microbiol.* **5:**1903–1913.
51. **Michel, J. F., B. Cami, and P. Schaeffer.** 1968. Sélection de mutants de *Bacillus subtilis* bloqués au début de la sporulation. I. Mutants aporogènes pléotropes sélectionnés par croissance en milieu au nitrate. *Annu. Inst. Pasteur* **114:**11–20.
52. **Mountain, A., and S. Baumberg.** 1980. Map locations of some mutations conferring resistance to arginine hydroxamate in *Bacillus subtilis* 168. *Mol. Gen. Genet.* **178:**691–701.
53. **Navarrete, R. M., J. A. Vara, and C. R. Hutchinson.** 1990. Purification of an inducible L-valine dehydrogenase of *Streptomyces coelicolor* A3(2). *J. Gen. Microbiol.* **136:**273–281.
54. **Oda, M., A. Sugishita, and K. Furukawa.** 1988. Cloning and nucleotide sequences of histidase and regulatory genes in the *Bacillus subtilis hut* operon and positive regulation of the operon. *J. Bacteriol.* **170:**3199–3205.
55. **Olomucki, A., D. B. Pho, R. Lebar, L. Delcambe, and N. V. Thoai.** 1968. Arginine oxygenase decarboxylante V. Purification et nature flavinique. *Biochem. Biophys. Acta* **151:**353–366.
56. **Ostrovsky de Spicer, P., K. O'Brien, and S. Maloy.** 1991. Regulation of proline utilization in *Salmonella typhimurium*: a membrane-associated dehydrogenase binds DNA in vitro. *J. Bacteriol.* **173:**211–219.
57. **Ottow, J. C. G.** 1974. Arginine dihydrolase activity in species of the genus *Bacillus* revealed by thin-layer chromatography. *J. Gen. Microbiol.* **84:**209–213.
58. **Padilla, G., Z. Hindle, R. Callis, A. Corner, M. Ludovice, P. Liras, and S. Baumberg.** 1991. The relationship between primary and secondary metabolism in *Streptomyces*, p. 35–45. *In* S. Baumberg, H. Krügel, and D. Noack (ed.), *Genetics and Product Formation in Streptomyces*. Plenum Press, New York.
59. **Perego, M., C. F. Higgins, S. R. Pearce, M. P. Gallagher, and J. A. Hoch.** 1991. The oligopeptide transport system of *Bacillus subtilis* plays a role in the initiation of sporulation. *Mol. Microbiol.* **5:**173–185.
60. **Porumb, H., D. Vancea, L. Mureşan, E. Presecan, I. Lascu, I. Petrescu, T. Porumb, R. Pop, and O. Bârzu.** 1987. Structural and catalytic properties of L-alanine dehydrogenase from *Bacillus cereus*. *J. Biol. Chem.* **262:**4610–4615.
61. **Priestley, N. D., and J. A. Robinson.** 1989. Purification and catalytic properties of L-valine dehydrogenase from *Streptomyces cinnamonensis*. *Biochem. J.* **261:**853–861.
62. **Ramaley, R. F., and R. E. Bernlohr.** 1966. Postlogarithmic phase metabolism of sporulating microorganisms. *J. Biol. Chem.* **241:**620–623.
63. **Reitzer, L. J., and B. Magasanik.** 1987. Ammonium assimilation and the biosynthesis of glutamine, glutamate, aspartate, asparagine, L-alanine, and D-alanine, p. 302–320. *In* F. Neidhardt, J. L. Ingraham, K. B. Low, B. Magasanik, M. Schaechter, and H. Umbarger (ed.), *Escherichia coli and Salmonella typhimurium: Cellular and Molecular Biology*, vol. 1. American Society for Microbiology, Washington, D.C.
64. **Rudner, D. Z., J. R. LeDeaux, K. Ireton, and A. D. Grossman.** 1991. The *spo0K* locus of *Bacillus subtilis* is homologous to the oligopeptide permease locus and is required for sporulation and competence. *J. Bacteriol.* **173:**1388–1398.
65. **Schreier, H. J., S. W. Brown, K. D. Hirschi, J. F. Nomellini, and A. L. Sonenshein.** 1989. Regulation of the *Bacillus subtilis* glutamine synthetase gene expression by the product of the *glnR* gene. *J. Mol. Biol.* **210:**51–63.
66. **Schreier, H. J., T. M. Smith, and R. W. Bernlohr.** 1982. Regulation of nitrogen catabolic enzymes in *Bacillus* spp. *J. Bacteriol.* **151:**971–975.
67. **Schreier, H. J., and A. L. Sonenshein.** 1986. Altered regulation of the *glnA* gene in glutamine synthetase mutants of *Bacillus subtilis*. *J. Bacteriol.* **167:**35–43.
68. **Schutte, H., W. Hummerl, H. Tsai, and M.-R. Kula.** 1985. L-Leucine dehydrogenase from *Bacillus cereus*. Appl. Microbiol. Biotechnol. **22:**306–317.
69. **Shapiro, S., and L. C. Vining.** 1984. Suppression of nitrate utilization by ammonium and its relationship to chloramphenicol production in *Streptomyces venezuelae*. *Can. J. Microbiol.* **30:**798–804.
70. **Simon, J.-P., and V. Stalon.** 1976. Purification and structure of arginase of *Bacillus licheniformis*. *Biochimie* **58:**1419–1421.
71. **Slack, F. J., J. P. Muellar, M. A. Strauch, C. Mathiopoulos, and A. L. Sonenshein.** 1991. Transcriptional regulation of a *Bacillus subtilis* dipeptide transport operon. *Mol. Microbiol.* **5:**1915–1925.
72. **Smith, M. C. M., L. Czaplewski, A. K. North, S. Baumberg, and P. G. Stockley.** 1989. Sequences required for regulation of arginine biosynthesis promoters are conserved between *Bacillus subtilis* and *Escherichia coli*. *Mol. Microbiol.* **3:**23–28.
73. **Sonenshein, A. L.** 1989. Metabolic regulation of sporulation and other stationary-phase phenomena, p. 109–130. *In* I. Smith, R. A. Slepecky, and P. Setlow (ed.), *Regulation of Procaryotic Development*. American Society for Microbiology, Washington, D.C.
74. **Sun, D., and P. Setlow.** 1991. Cloning, nucleotide sequence and expression of the *Bacillus subtilis ans* operon, which codes for L-asparaginase and L-aspartase. *J. Bacteriol.* **173:**3831–3845.
75. **Thoai, N. V., F. Thome-Beau, and A. Olomucki.** 1966. Induction and spécificité des enzymes de la nouvelle voie catabolique de l'arginine. *Biochim. Biophys. Acta* **115:**73–80.
76. **Venugopal, V., and G. B. Nadkarni.** 1977. Regulation of the arginine dihydrolase pathway in *Clostridium sporogenes*. *J. Bacteriol.* **131:**683–695.
77. **Whiteman, P. A., T. Iijima, M. D. Diesterhaft, and E. Freese.** 1978. Evidence for a low affinity but high velocity aspartate transport system needed for rapid growth of *Bacillus subtilis* on aspartate as sole carbon source. *J. Gen. Microbiol.* **107:**297–307.
78. **Willis, R. C., and C. A. Woolfolk.** 1974. Asparagine utilization in *Escherichia coli*. *J. Bacteriol.* **118:**231–241.
79. **Wu, P.-C., T. A. Kroening, P. J. White, and K. E. Kendrick.** 1992. Purification of histidase from *Streptomyces griseus* and nucleotide sequence of the *hutH* structural gene. *J. Bacteriol.* **174:**1647–1655.

17. Regulation of Phosphorus Metabolism

F. MARION HULETT

Phosphorus is an essential element for all cells as a component of essential biomolecules and in energy metabolism. P_i is a component of membrane phospholipids, nucleic acids, and many posttranslationally modified proteins. In aerobic organisms, intracellular P_i enters the organophosphate pool by substrate-level phosphorylation, as in glycolysis, or via oxidative phosphorylation, as in ATP formation. In anaerobic organisms, P_i can additionally be incorporated into the organophosphate pool via mixed acid fermentation, as in acetylphosphate.

In soil, the natural environment of *Bacillus subtilis* and many other gram-positive bacteria, P_i is the major limiting nutrient for biological growth and is often present at levels 2 to 3 orders of magnitude lower than those of other required ions (45). One major reason for the limited availability of P_i is the relative abundance of Al, Fe, and Ca, which form insoluble compounds with phosphate. Soil bacteria, including *B. subtilis*, have evolved complex regulatory systems for utilizing this limiting nutrient efficiently.

Usable phosphate compounds in the soil include P_i esters, organic phosphate esters, and phosphonates. *B. subtilis* uses organic phosphate esters and P_i esters. At least one *Bacillus* species, *Bacillus cereus*, has also been reported to use phosphonates as a P_i source (31, 47, 48). The *phn* locus for phosphonate utilization in *Escherichia coli* has been cloned and characterized (6, 59, 61). Phosphonate utilization is cryptic in *E. coli* K-12 and related strains, but most *E. coli* strains, including *E. coli* B strains, are Phn^+.

The focus of this chapter is the genes that are expressed in response to phosphate starvation, i.e., genes of the phosphate stimulon. Special emphasis is given to genes of the PHO regulon, which are dependent on the PhoP-PhoR two-component regulators.

APases OF *B. SUBTILIS* AND OTHER GRAM-POSITIVE BACTERIA

Alkaline phosphatase (APase) enzymes are ubiquitous, being found in nearly all organisms studied, and have been the enzymes of choice as reporters for phosphate regulation. APase catalyzes the hydrolysis of phosphate monoesters, forming a phosphorylated enzyme intermediate, E-PO_4. For *E. coli* APase, serine 102 becomes phosphorylated (9). Dephosphorylation of the serine is limiting at low pH, but at pH 9 to 10, the enzyme is maximally active, with product dissociation being the rate-limiting step. The more likely physiological reaction of APases is a phosphoryl transfer reaction to alcohols, which occurs maximally at pH 7.5 to 8.5 but at a lower specific activity. The APases of *B. subtilis* are useful as models of enzymes synthesized in response to decreased phosphate concentration, and their genes can be used to examine the roles of various regulators involved in phosphate metabolism.

The study of the regulation of phosphate metabolism in *Bacillus* species in general and in *B. subtilis* specifically has been complicated by a feature that increases the potential importance of this process: APases are encoded by multiple structural genes. Genetic analysis was hampered because mutations in one APase gene were masked by other functioning members of the multigene family. *B. subtilis* has an APase family consisting of at least four structural genes. Multiple structural genes for APases have not been found in other organisms except for mammals, which have four APase genes (20, 21, 30, 44). *E. coli* contains a single structural gene for APase that is induced when the organism is starved for phosphate, resulting in the synthesis of APase that is secreted into the periplasmic space (2). *B. subtilis* APases are induced by either of two conditions or in some cases by both: phosphate starvation or sporulation.

Two unlinked APase structural genes, *phoA* (formerly *phoAIV*) and *phoB* (formerly *phoAIII*), from *B. subtilis* have been cloned (22, 24). *phoA* was mapped at approximately 73° on the *B. subtilis* chromosome (4), and *phoB* was mapped at approximately 50° (28). Characterization of the genes showed that sequences encoding the mature APase proteins are 64% identical. The deduced protein sequences for the mature proteins are 63% identical. The *phoA* and *phoB* genes code for predicted mature proteins of 47,149 and 45,935 Da, respectively (24).

Models of the *Bacillus* APase structures were constructed using the coordinates of the refined crystal structure of *E. coli* APase (24, 29). The *E. coli* protein is the only APase for which the X-ray structure is known, making it a valuable prototype for protein structure-function studies. Extensive physical-chemical characterizations of the *E. coli* APase protein range from the mechanism of catalysis involving three metal atoms per active site (9) to the crystal structure (29). The mature *E. coli* enzyme is a dimer with two zinc ions and one magnesium ion in each active-site region. Analysis of the *Bacillus* APases by using the *E. coli* structure indicated that the 10-stranded sheet structure, which forms the core of the *E. coli* APase enzyme and functions to position amino acid residues important to catalytic activity, metal binding, and phosphate binding, was conserved among the APases. Where exact residues are not retained, conservative replacements often occur. However, when other re-

F. Marion Hulett • Department of Biological Sciences, Laboratory for Molecular Biology, University of Illinois at Chicago, Chicago, Illinois 60680.

gions of the protein are compared with the *E. coli* enzyme, significant differences appear. Three sizable surface loops of *E. coli* APase are deleted in the *Bacillus* APases, and a small domain involved in subunit interactions in *E. coli* APase is replaced by a larger domain in the *Bacillus* enzymes. The subunit interface residues are more variable than the average for the whole structure and much more variable than residues in the core structure. The amino-terminal segment of the *E. coli* structure, which has more extensive contact with the companion subunit than with the subunit of which it is a part, is completely absent in both *Bacillus* APases.

Insertional mutations in *phoA* and *phoB* showed that both enzymes contribute to total APase produced during phosphate starvation induction, but only *phoB* contributes to total sporulation APase. A mutation in APaseA (the product of *phoA*) reduced total APase specific activity by 60 to 75% during phosphate starvation induction (28), while a mutation in APase B (the product of *phoB*) resulted in 25 to 40% reduction (4). In *phoA phoB* double mutants, total APase was reduced to less than 1% of wild-type levels during phosphate starvation (27). Mutations in *phoB* but not *phoA* affected total sporulation APase, reducing total APase activity to approximately 55% of wild-type levels. A second sporulation APase, APaseC, has been isolated from sporulating cells. The amino-terminal sequence of APaseC is different from that of APaseA or APaseB (25).

Promoter-*lacZ* fusion studies and enzyme isolation confirmed that the *phoA* gene is induced during phosphate starvation (27) but not during sporulation. The *phoB* promoter-*lacZ* fusion is induced during either phosphate starvation or sporulation (4, 7).

A fourth APase, APaseD, with a low specific activity, was isolated from phosphate-starved cells (25). The subunit molecular weight of APaseD is 49,000, and the amino-terminal sequence matches the amino-terminal sequence of a phosphodiesterase-APase characterized earlier by Yamane et al. (62–64).

APases FROM OTHER GRAM-POSITIVE BACTERIA

It is likely that *Bacillus licheniformis*, like *B. subtilis*, has several APase genes. Secreted APases and APases from different cell fractions have similar physical and chemical properties (15, 19, 26, 32, 40, 50, 54–56). These proteins are basic proteins (pI 9.5 to 10), as are the *B. subtilis* APases, but are larger than most prokaryotic APases, having a molecular weight of 60,000 instead of 45,000. Approximately 80% of one of the *B. licheniformis* APase genes has been cloned and sequenced (32, 33). The partial sequence shows similarity to sequences of the other APases in that the first 8 strands of the 10-membered beta-sheet structure, the ligands to metal-binding sites, and the phosphorylation site are conserved. The similarities of this sequence to those of the *B. subtilis* APases (36 and 38%) are surprisingly not much greater than similarities to *E. coli* APase or human APases (28 to 30%) (33). Atomic absorption spectroscopy indicated that *B. licheniformis* APases contain two atoms of cobalt per monomer and only trace amounts of zinc, which suggests that they are cobalt metalloenzymes (55).

An APase gene from *Streptococcus faecalis* has recently been cloned and sequenced (49). Its putative product is more similar to the *B. subtilis* enzymes (63 and 55%) than to the *E. coli* or human APases. In fact, the *Streptococcus* APase is more similar to the *B. subtilis* enzymes than is the *B. licheniformis* APase (63).

The gene for an APase that is about 46% identical to the *E. coli* APase has been cloned from *Streptomyces griseus* and sequenced (36). This gene, *strK*, is located in a cluster of streptomycin production genes and specifically cleaves streptomycin-6-phosphate. It also cleaves streptomycin-3″-phosphate, albeit more slowly.

B. cereus secretes APase and two lipases capable of phospholipase C activity when P_i concentrations become growth limiting (18). High levels of these enzymes are secreted in response to low P_i concentrations, and the APase secretion is correlated with a transient P_i increase in the medium. This transient increase in P_i is accompanied by a recommencement of cell growth. Guddal et al. (18) suggested that the phospholipases may degrade phospholipids to compounds such as phosphorylethanolamine or phosphorylserine, which might serve as substrates for APase. A phosphonatase (31) and a phosphonate transport system (47, 48) have been identified in *B. cereus*. Taken together, these data suggest that *B. cereus* has an efficient P_i-repressed phosphorus retrieval and scavenging system.

An extracellular APase was isolated from the culture fluid of *Micrococcus sodonensis* (13, 14). The molecular weight of this metalloenzyme was approximately 80,000. Interestingly, the activity could be restored to the apoenzyme with calcium but not with zinc, suggesting that this APase may be a Ca metalloenzyme.

REGULATION OF *B. SUBTILIS* APase SYNTHESIS DURING PHOSPHATE STARVATION

Genetic analysis has shown that at least three *trans*-acting regulators are involved in controlling synthesis of the members of the APase family during phosphate starvation. Two genes, *phoP* and *phoR*, that show similarity to prokaryotic two-component signal transduction regulators (51, 52) are required for phosphate starvation induction of APases (41). PhoP is 40% identical to the PhoB protein of *E. coli*, the transcription activator protein for the PHO regulon of *E. coli*. The kinase for PhoB of *E. coli*, PhoR, and the carboxyl-terminal three-fourths of PhoR from *B. subtilis* show significant similarity (52). The *B. subtilis* PhoR has an amino-terminal extension of 137 amino acids not found in the *E. coli* PhoR. An additional gene, *spo0A*, which is involved in regulation of gene expression during the transition from exponential to stationary growth and is essential for sporulation, also influenced phosphate starvation-induced APase production. In contrast to original reports (23), a mutation in the *spo0A* gene results in hyperinduction of total APase (25). (A number of commonly used *spo0* strains have been shown to contain secondary mutations affecting the *phoP-phoR* operon [25].) Spo0A, a member of the regulator class of two-component systems, is active when phosphorylated by either of two kinases, KinA or KinB, and possibly other, as-yet-unidentified kinases via a phosphorelay involving several other *spo0* gene products (5).

Regulation studies of *phoB*, the structural gene for APaseB, showed that separate promoters are responsible for transcription during phosphate starvation and sporulation induction (7). It was determined by using *phoB-lacZ* fusions and primer extension experiments that mutations in *phoP* or *phoR* reduce expression from the promoter used during phosphate starvation induction to less than 5% of that observed in wild-type cultures (3, 7). Increased levels of APaseB were observed in *spo0A* strains that hyperinduced total phosphate starvation-induced APases (25). Thus, expression from the *phoB* promoter during phosphate starvation is reduced in *phoP* and *phoR* mutants, just as total APase synthesis is reduced (11, 12, 41) and is hyperinduced in *spo0A* strains. Current evidence suggests that Spo0A does not act directly at the *phoB* (APaseB) promoter to negatively regulate or indirectly through AbrB, since no binding of either protein to the promoter could be detected in gel retardation or footprinting studies (57). Furthermore, a second mutation at the *abrB* locus does not suppress an APase hyperinduction phenotype caused by a *spo0A* mutation (25).

Similar studies under sporulation induction conditions showed that a second promoter is responsible for *phoB* transcription during sporulation and that it is dependent on *spo0* and *spoII* genes (7, 8).

The *phoA* gene, which encodes APaseA, is expressed during phosphate starvation but not during sporulation under the induction conditions reported. Expression of *phoA* is regulated by *phoP*, *phoR*, and *spo0A* (25, 27).

The characterization of these two APase genes and the PHO regulon regulatory genes *phoP* and *phoR* confirms the earlier biochemical data of LeHegarat and Anagnostopoulos (34), Glenn (11), and Glenn and Mandelstam (12), which showed that mutations in *phoP* result in reduced amounts of APase and phosphodiesterase enzymes unaltered in their physical-chemical characteristics. Each group correctly concluded that the *phoP* locus was a regulatory locus.

CONTROL OF THE PHO REGULON

Sequencing data (51, 52) suggested that the genes for the PHO regulon regulators, *phoP* and *phoR*, lie in an operon. One possibility for the role of *spo0A* in PHO regulon regulation is to regulate transcription of the *phoP-phoR* operon. This hypothesis was tested by using *phoP-lacZ* transcriptional fusions (27). In a wild-type strain, JH642, induction of the *phoP-phoR* operon correlates with the induction of total APase during phosphate starvation of the culture. Transcription from the *phoP-phoR* promoter is hyperinduced in spo0A mutant strains, in which expression of APases is hyperinduced, suggesting that the role of Spo0A in PHO regulon regulation may be at the level of *phoP-phoR* transcription (25). Current evidence suggests that Spo0A does not act directly at the *phoP-phoR* operon promoter, since binding could not be detected in gel retardation or footprinting studies. Transcription from the *phoP-phoR* promoter in the *phoP* mutant strain examined is similar to transcription in JH642 (27), suggesting that PhoP does not regulate transcription of the operon encoding PhoP and PhoR.

There may be significant differences in the sensory transduction mechanism by which the PHO regulons of *E. coli* and *B. subtilis* control gene expression during phosphate starvation. Although the structural portions of the kinases and response regulators, PhoR and PhoP of *B. subtilis* and PhoR and PhoB of *E. coli*, are highly conserved, certain differences in the regulation of transcription of the *E. coli phoB-phoR* operon and the *B. subtilis phoP-phoR* operon are apparent. First, PhoB is an activator of transcription of the *E. coli phoB-phoR* operon (17), but current evidence suggests that PhoP is not necessary for transcription of the *B. subtilis phoP-phoR* operon. Second, null mutations in *phoR* of *B. subtilis* cause loss of APase activity, while *phoR* amber mutants have an intermediate constitutive phenotype in *E. coli* K-12 strains or a clonal variation phenotype in other *E. coli* strains because of cross regulation or networking (60), the process by which other protein kinases, such as CreC (PhoM) (1), act as surrogate phosphorylators of PhoB. Third, Spo0A and other members of the phosphorelay (Spo0B and Spo0F) are required for wild-type expression of PHO regulon genes in *B. subtilis* (25), perhaps acting indirectly in the negative transcriptional regulation of the response regulator, PhoP, and the histidine kinase, PhoR, for that regulon. There are no known homologs or functional equivalents to these *spo* genes in *E. coli*.

OTHER PHOSPHATE-REGULATED GENES

Promoters in *B. subtilis* that are induced by phosphate starvation have been identified by Tn*917-lacZ* fusions (27). Sixteen of the promoter fusions that showed the strongest induction under phosphate starvation conditions and no induction under phosphate repletion conditions were characterized. These promoters are not clustered but show a broad distribution on the chromosome (Fig. 1).

Analysis of the impact of *phoP* or *phoR* mutations on the function of each promoter fusion resulted in the identification of three classes of phosphate starvation-inducible promoters based on their dependencies on the PHO regulon regulators PhoP and PhoR, as shown in Table 1.

The first class, consisting of five promoters, is dependent on PhoP and PhoR. One member of this class, the *phoB* promoter, responds to sporulation induction and carbon starvation in addition to phosphate starvation. This gene, the structural gene for APaseB, has two promoters, of which one functions during phosphate starvation and the other functions during sporulation (7). It has not yet been determined which of these two promoters, if either, responds to carbon starvation.

The second class, consisting of six promoters, functions independently of PhoP but still requires PhoR. This class of promoter fusions suggests the interesting possibility that PhoR may function with a second

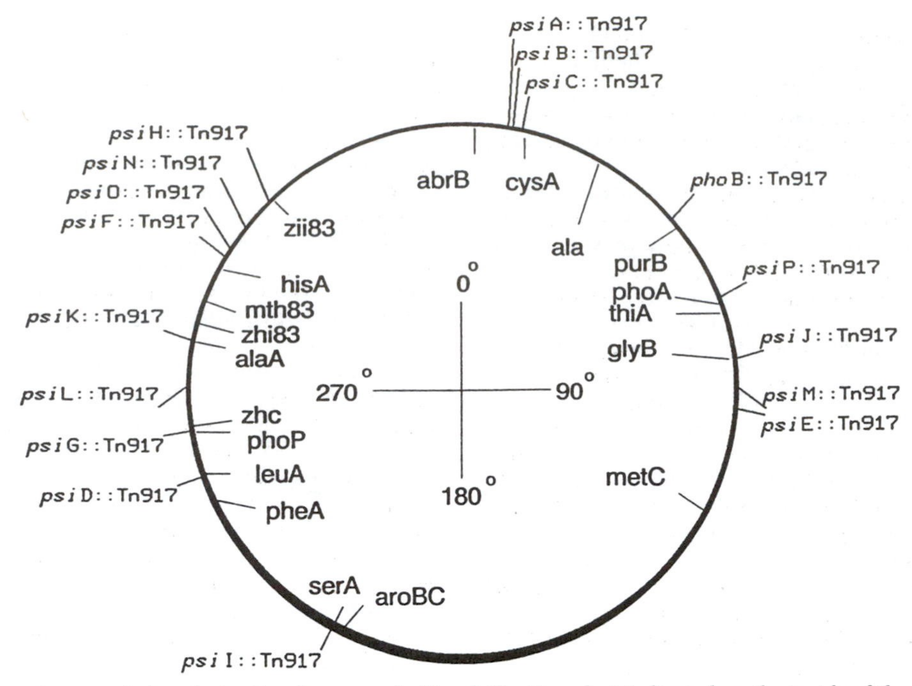

Figure 1. Genetic map of phosphate stimulon genes in *B. subtilis*. Gene loci indicated on the inside of the circle are for reference and include all the loci used in mapping Tn*917* insertions (46, 58). The *psi* promoters (on the outside of the circle) show phosphate starvation-inducible expression; they were identified by screening a Tn*917-lacZ* fusion library. *phoB* was given a *pho* designation because the Tn*917* interruption was in the structural gene for APaseB. *phoP* identifies the *phoP-phoR* operon regulatory locus. *phoA* identifies the mapping location of the gene encoding APaseA.

response regulator, in addition to PhoP, to regulate a different set of genes during phosphate starvation. All of the promoters in the second class were also induced by carbon starvation or sporulation induction or both.

The third class of promoters requires neither PhoP nor PhoR for phosphate starvation induction. The presence of this class of promoters suggests that part of the phosphate stimulon in *B. subtilis* (possibly entire regulons) is regulated independently of the PHO regulon regulators. Two promoters in this class responded only to phosphate starvation induction, while the remaining four can also be induced by carbon starvation or sporulation. In two cases, the Tn*917* insertions that created the promoter-*lacZ* fusions caused mutations that reduced sporulation.

Taken together, the data obtained from the three classes of Tn*917*-promoter fusions strongly suggest that the phosphate depletion stimulus induces more than one regulon in *B. subtilis*, with the PhoP-PhoR controlled promoters constituting one regulon in that stimulon. A number of genes of known function are controlled by phosphate depletion, but their dependence on PhoP or PhoR has not been established.

degQ

degQ encodes a small polypeptide that is one of two small polypeptides involved in the regulation of synthesis of a number of degradative enzymes (42) in *B. subtilis*. An increase in degradative-enzyme production as the result of overproduction of DegQ is absolutely dependent on a functional DegS-DegU two-component system (42). A second two-component system, ComP-ComA, which is involved in competence control, also controls expression of *degQ*. However, the induction of *degQ* during phosphate limitation is not dependent on either of these two-component regulatory systems. Whether the PhoR-PhoP two-component system is involved in *degQ* expression during phosphate limitation is not yet known.

Teichoic Acid/Teichuronic Acid Regulation

Teichoic acid is the cell wall anionic polymer synthesized by *B. subtilis* under phosphate repletion conditions. Teichoic acid synthesis ceases, and synthesis of teichuronic acid, an anionic polymer that contains no phosphate, is initiated under phosphate-limiting growth conditions (16, 37). Sequences upstream of genes involved in teichoic acid synthesis are similar (12 of 18 positions matching) to the *pho* box consensus characteristic of *E. coli* PhoB-PhoR-regulated genes (35). The significance of this sequence in *B. subtilis*, if any, is unknown. We have shown that sequences in the *phoB* and *phoP* promoter regions that are similar to those in the *E. coli* pho box are not involved in phosphate regulation in *B. subtilis* (7, 27).

dciA

The *dciA* operon of *B. subtilis*, which encodes a dipeptide transport system, was identified during a screen for genes that are turned on early during sporulation (38, 39). Transcription of this operon is induced when sporulation is induced by decoyinine or when cultures are starved for phosphate. The *dciA*

Table 1. Classification of phosphate starvation-inducible promoters based on their PhoP-PhoR dependencies (27)

Promoter type	Gene	Induction conditions[a]	Phenotype[b]
phoP-phoR dependent	*phoB*	P, C, S	APaseIII
	psiJ	P	APase^{v,hy}, motility^{hy}
	psiK	P	ND
	psiL	P	APase^{s-}
	psiO	P	ND
phoP independent, *phoR* dependent	*psiA*	P, C, S	APase^{v,hy}
	psiD	P, S	APase^{v,hy}
	psiI	P, S	APase^{v,low}
	psiF	P, C, S	ND
	psiP	P, C	ND
	psiG	P, S	ND
phoP-phoR independent	*psiB*	P, C, S	ND
	psiE	P, C, S	ND
	psiH	P, S	HP Spo$^+$ LP Spo$^-$
	psiM	P	APase^{v,low}
	psiN	P	ND
	psiC	P, C, S	Spo$^-$

[a] P, phosphate starvation; C, carbon starvation; S, sporulation induction.

[b] Phenotypic changes in levels of APase production (phosphate starvation inducible or sporulation inducible), motility, and ability to sporulate due to Tn*917* insertion mutations were determined. APase^v, vegetative APase (phosphate starvation-inducible APase); APase^s, sporulation APase; ^{hy}, hyperinduction of APase activity; ^{low}, reduced APase activity; Spo$^-$, sporulation deficient; Spo$^+$, wild-type sporulation level; ND, no phenotype observed as the result of Tn*917* insertion; HP, high-phosphate growth conditions; LP, low-phosphate growth conditions.

transcription induced by decoyinine is dependent on *spo0A* and *spo0H* (39, 53).

gsiA

A second phosphate starvation-inducible promoter was identified among clones expressed early during sporulation. Induction of the *gsiA* operon results from glucose deprivation, phosphate deprivation, decoyinine induction of sporulation, or exhaustion of nutrient broth. Both glucose and phosphate deprivation induction of *gsiA* are dependent on the ComP-ComA two-component regulatory system (43).

SUMMARY

Every gene whose expression is regulated by phosphate concentration is a part of the phosphate stimulon, the PHO regulon being part of that stimulon. Genes regulated by the PHO regulon show overlapping regulation with other global regulatory systems outside of the phosphate stimulon. All genes that are known to require PhoP and PhoR for expression are also controlled by Spo0A. This is consistent with the thought that phosphorylated Spo0A controls genes needed for various stationary-phase phenomena and not only those needed for sporulation, since phosphate-starved cultures enter postexponential growth because of limited phosphate availability. Further analysis of promoters controlled by phosphate starvation and a two-component system other than PhoP-PhoR, such as *gsiA*, which requires ComP-ComA, may reveal additional network circuitry between the two-component systems.

Further studies are required for understanding the mechanism of PHO regulon regulation involving PhoP and PhoR. It will be interesting to determine whether PhoR does communicate with a second response regulator in addition to PhoP. If so, what metabolic function is served by the PhoR-X branch of phosphate regulation? How are phosphate starvation-inducible genes that do not require PhoP or PhoR regulated?

The mechanisms of P_i transport and phosphate assimilation in *B. subtilis* are unknown. No systems similar to the high-K_m or low-K_m specific transport systems in *E. coli* (10, 60) have been identified.

A role for phosphate depletion in induction of differentiation leading to sporulation has been suggested. It is not known whether there is a sensor kinase activity dedicated to monitoring phosphate availability or whether PhoR is important in sporulation.

REFERENCES

1. **Amemura, M., K. Makino, H. Shinagawa, and A. Nakata.** 1990. Cross talk to the phosphate regulon of *Escherichia coli* by PhoM protein: PhoM is a histidine protein kinase and catalyzes phosphorylation of PhoB and PhoM-open reading frame 2. *J. Bacteriol.* **172:**6300–6307.
2. **Berg, P.** 1981. Cloning and characterization of the *Escherichia coli* gene coding for alkaline phosphatase. *J. Bacteriol.* **146:**660–667.
3. **Bookstein, C.** 1990. Cloning and characterization of the *Bacillus subtilis* genes *phoAIII* and *XPAC*. Ph.D. thesis, University of Illinois at Chicago, Chicago.
4. **Bookstein, C., C. W. Edwards, N. V. Kapp, and F. M. Hulett.** 1990. The *Bacillus subtilis* 168 alkaline phosphatase III gene: impact of a *phoAIII* mutation on total alkaline phosphatase synthesis. *J. Bacteriol.* **172:**3730–3737.
5. **Burbulys, D., K. A. Trach, and J. A. Hoch.** 1991. Initiation of sporulation in *B. subtilis* is controlled by a multicomponent phosphorelay. *Cell* **64:**545–552.
6. **Chen, C.-M., Q. Ye, Z. Zhu, B. L. Wanner, and C. T. Walsh.** 1990. Molecular biology of carbon-phosphorus bond cleavage: cloning and sequencing of the *phn* (*psiD*) genes involved in alkylphosphonate uptake and C-P lyase activity in *Escherichia coli* B. *J. Biol. Chem.* **265:**4461–4471.
7. **Chesnut, R. S., C. Bookstein, and F. M. Hulett.** 1991. Separate promoters direct expression of *phoAIII*, a member of the *Bacillus subtilis* alkaline phosphatase multigene family, during phosphate starvation and sporulation. *Mol. Microbiol.* **5:**2181–2190.
8. **Chesnut, R. S., and F. M. Hulett.** Unpublished data.
9. **Coleman, J. E., and P. Gettins.** 1983. Alkaline phosphatase, solution structure, and mechanism. *Adv. Enzymol.* **55:**381–452.
10. **Elvin, C. M., C. M. Hardy, and H. Rosenberg.** 1987. Molecular studies on the phosphate (inorganic) transport (*pit*) gene of *Escherichia coli*: identification of the *pit** gene product and physical mapping of the *pit-gar* region of the chromosome. *Mol. Gen. Genet.* **204:**477–484.
11. **Glenn, A. R.** 1975. Alkaline phosphatase mutants of *Bacillus subtilis*. *Aust. J. Biol. Sci.* **28:**323–330.
12. **Glenn, A. R., and J. Mandelstam.** 1971. Sporulation in *Bacillus subtilis* 168: comparison of alkaline phosphatase from sporulating and vegetative cells. *Biochem. J.* **123:**129–138.
13. **Glew, R. H., and E. C. Heath.** 1971. Studies on the extracellular alkaline phosphatase of *Micrococcus sod-*

onensis isolation and characterization. *J. Biol. Chem.* **246:**1556–1565.
14. **Glew, R. H., and E. C. Heath.** 1971. Studies on the extracellular alkaline phosphatase of *Micrococcus sodonensis* factors effecting secretion. *J. Biol. Chem.* **246:** 1566–1574.
15. **Glynn, J. A., S. D. Schaffel, J. M. McNicholas, and F. M. Hulett.** 1977. Biochemical localization of the alkaline phosphatase of *Bacillus licheniformis* MC14 as a function of culture age. *J. Bacteriol.* **129:**1010–1019.
16. **Grant, W. D.** 1979. Cell wall teichoic acid as a reserve phosphate source in *Bacillus subtilis*. *J. Bacteriol.* **137:**35–43.
17. **Guan, C.-D., B. Wanner, and H. Inouye.** 1983. Analysis of regulation of *phoB* expression using *phoB* cat fusion. *J. Bacteriol.* **156:**710–717.
18. **Guddal, P. H., T. Johansen, K. Schulstad, and C. Little.** 1989. Apparent phosphate retrieval system in *Bacillus cereus*. *J. Bacteriol.* **171:**5702–5706.
19. **Hansa, J. G., M. La Porta, M. Kuna, R. Reimschussel, and F. M. Hulett.** 1981. A soluble alkaline phosphatase from *Bacillus licheniformis* MC14: histochemical localization, purification, characterization and comparison with the membrane-associated alkaline phosphatase. *Biochim. Biophys. Acta* **657:**390–401.
20. **Henthorn, P. S., B. J. Knoll, M. Raducha, K. N. Rothblum, C. Slaughter, M. Weiss, M. A. Lafferty, T. Fischer, and H. Harris.** 1986. Products of two common alelles at the locus for human placental alkaline phosphatase differ by seven amino acids. *Proc. Natl. Acad. Sci. USA* **83:**5597–5601.
21. **Hirano, K., K. Kusano, Y. Matsumoto, T. Stigbrand, S. Iino, and K. Hayahsi.** 1989. Intestinal-like alkaline phosphatase expressed in normal human adult kidney. *Eur. J. Biochem.* **183:**419–423.
22. **Hulett, F. M., C. Bookstein, and K. Jensen.** 1990. Evidence for two structural genes for alkaline phosphatase in *Bacillus subtilis*. *J. Bacteriol.* **172:**735–740.
23. **Hulett, F. M., and K. Jensen.** 1988. Critical roles of *spo0A* and *spo0H* in vegetative alkaline phosphatase production in *Bacillus subtilis*. *J. Bacteriol.* **170:**3765–3768.
24. **Hulett, F. M., E. E. Kim, C. Bookstein, N. V. Kapp, C. W. Edwards, and H. W. Wyckoff.** 1991. *Bacillus subtilis* alkaline phosphatases III and IV. *J. Biol. Chem.* **266:** 1077–1084.
25. **Hulett, F. M., W. Liu, S. Birkey, and L. Shi.** Unpublished data.
26. **Hulett, F. M., K. Stuckmann, D. B. Spencer, and T. Sanopoulou.** 1986. Purification and characterization of the secreted alkaline phosphatase of *Bacillus licheniformis* MC14: identification of a possible precursor. *J. Gen. Microbiol.* **132:**2387–2395.
27. **Kapp, N. V.** 1992. Analysis of the phosphate stimulon of *Bacillus subtilis*. Ph.D. thesis. University of Illinois at Chicago, Chicago.
28. **Kapp, N. V., C. W. Edwards, R. S. Chesnut, and F. M. Hulett.** 1990. The *Bacillus subtilis phoAIV* gene: effects of *in vitro* inactivation on total alkaline phosphatase production. *Gene* **96:**95–100.
29. **Kim, E. E., and H. W. Wycoff.** 1989. Structure of alkaline phosphatases. *Clin. Chim. Acta* **186:**175–188.
30. **Knoll, B. J., K. N. Rothblum, and M. Longley.** 1988. Nucleotide sequence of the human placental alkaline phosphatase gene. *J. Biol. Chem.* **263:**12020–12027.
31. **La Nauze, J. M., H. Rosenberg, and D. C. Shaw.** 1970. The enzyme cleavage of the carbon-phosphorus bond: purification and properties of phosphonatase. *Biochim. Biophys. Acta* **212:**332–350.
32. **Lee, J. K., C. W. Edwards, and F. M. Hulett.** 1991. *Bacillus licheniformis* APase I gene promoter: a strong well-regulated promoter in *B. subtilis*. *J. Gen. Microbiol.* **137:**1127–1133.
33. **Lee, J. K., H. W. Wyckoff, and F. M. Hulett.** Unpublished data.
34. **Le Hegarat, J.-C., and C. Anagnostopoulos.** 1973. Purification, subunit structure and properties of two repressible phosphohydrolases of *Bacillus subtilis*. *Eur. J. Biochem.* **39:**525–539.
35. **Makino, K., H. Shinagawa, M. Amemura, K. Kimura, and A. Nakata.** 1988. Regulation of the phosphate regulon of *Escherichia coli*: activation of *pstS* transcription by PhoB protein *in vitro*. *J. Mol. Biol.* **203:**85–95.
36. **Mansouri, K., and W. Piepersberg.** 1991. Genetics of streptomycin production in *Streptomyces griseus*: nucleotide sequence of five genes, *strFGHIK*, including a phosphatase gene. *Mol. Gen. Genet.* **228:**459–469.
37. **Manuël, C., M. Young, and D. Karamata.** 1991. Genes concerned with synthesis of poly(glycerol phosphate), the essential teichoic acid in *Bacillus subtilis* strain 168, are organized in two divergent transcription units. *J. Gen. Microbiol.* **137:**929–941.
38. **Mathiopoulos, C., J. P. Mueller, F. J. Slack, C. G. Murphy, S. Patankar, G. Bukusoglu, and A. L. Sonenshein.** 1991. A *Bacillus subtilis* dipeptide transport system expressed early during sporulation. *Mol. Microbiol.* **5:**1903–1913.
39. **Mathiopoulos, C., and L. Sonenshein.** 1989. Identification of *Bacillus subtilis* genes expressed early during sporulation. *Mol. Microbiol.* **3:**1071–1081.
40. **McNicholas, J. M., and F. M. Hulett.** 1977. Electron microscopic histochemical localization of alkaline phosphatase(s) in *Bacillus licheniformis* MC14. *J. Bacteriol.* **129:**501–515.
41. **Miki, T., A. Minami, and Y. Ikeda.** 1965. The genetics of alkaline phosphatase formation in *Bacillus subtilis*. *Genetics* **52:**1093–1100.
42. **Msadek, T., F. Kunst, A. Klier, and G. Rapoport.** 1991. DegS-DegU and ComP-ComA modulator-effector pairs control expression of the *Bacillus subtilis* plieotropic regulatory gene *degQ*. *J. Bacteriol.* **173:**2366–2377.
43. **Mueller, P., G. Bukusoglu, and A. L. Sonenshein.** Personal communication.
44. **Ogata, S., Y. Hayashi, N. Takami, and Y. Ikehara.** 1988. Chemical characterization of the membrane-anchoring domain of human placental alkaline phosphatase. *J. Biol. Chem.* **263:**10489–10494.
45. **Ozanne, P. G.** 1980. Phosphate nutrition of plants—a general treatise, p. 559–585. *In* E. Khasawneh (ed.), *The Role of Phosphorus in Agriculture*. American Society of Agronomy, Madison, Wis.
46. **Piggot, P. J., and J. A. Hoch.** 1985. Revised genetic linkage map of *Bacillus subtilis*. *Microbiol. Rev.* **49:**158–179.
47. **Rosenberg, H., and J. M. La Nauze.** 1967. The metabolism of phosphonates by microorganisms: the transport of aminoethylphosphonic acid in *Bacillus cereus*. *Biochim. Biophys. Acta* **141:**79–90.
48. **Rosenberg, H., N. Medveczky, and J. M. La Nauze.** 1969. Phosphate transport in *Bacillus cereus*. *Biochim. Biophys. Acta* **193:**159–167.
49. **Rothschild, C. B., R. P. Ross, and A. Claiborne.** 1991. Molecular analysis of the gene encoding alkaline phosphatase in *Streptococcus faecalis* 10C1, p. 45–48. *In* G. M. Dunny, P. P. Cleavy, and L. L. McKay (ed.), *Genetics and Molecular Biology of Streptococci, Lactococci, and Enterococci*. American Society for Microbiology, Washington, D.C.
50. **Schaffel, S. D., and F. M. Hulett.** 1978. Alkaline phosphatase from *Bacillus licheniformis*, solubility dependent on magnesium, purification and characterization. *Biochim. Biophys. Acta* **526:**457–467.
51. **Seki, T., H. Yoshikawa, H. Takahashi, and H. Saito.** 1987. Cloning and nucleotide sequence of *phoP*, the regulatory gene for alkaline phosphatase and phospho-

diesterase in *Bacillus subtilis*. *J. Bacteriol.* **169:**2913–2916.

52. **Seki, T., H. Yoshikawa, H. Takahashi, and H. Saito.** 1988. Nucleotide sequence of the *Bacillus subtilis phoR* gene. *J. Bacteriol.* **170:**5935–5938.
53. **Slack, F. J., J. P. Mueller, M. A. Strauch, C. Mathiopoulos, and A. L. Sonenshein.** 1991. Transcriptional regulation of a *Bacillus subtilis* dipeptide transport operon. *Mol. Microbiol.* **5:**1915–1925.
54. **Spencer, D. B., J. G. Hansa, K. V. Stuckmann, and F. M. Hulett.** 1982. Membrane-associated alkaline phosphatase from *Bacillus licheniformis* that requires detergent for solubilization: lactoperoxidase ^{125}I localization and molecular weight determination. *J. Bacteriol.* **150:**826–834.
55. **Spencer, D. B., and F. M. Hulett.** 1981. Effect of cobalt on synthesis and activation of *Bacillus licheniformis* MC14 alkaline phosphatase. *J. Bacteriol.* **145:**926–933.
56. **Spencer, D. B., and F. M. Hulett.** 1981. Lactoperoxidase ^{125}I localization of salt extractable alkaline phosphatase on the cytoplasmic membrane of *Bacillus licheniformis*. *J. Bacteriol.* **145:**934–945.
57. **Strauch, M. A., J. A. Hoch, and F. M. Hulett.** Unpublished data.
58. **Vandeyar, M. A., and S. A. Zahler.** 1986. Chromosomal insertions of Tn*917* in *Bacillus subtilis*. *J. Bacteriol.* **167:**530–534.
59. **Wackett, L. P., B. L. Wanner, C. P. Venditti, and C. T. Walsh.** 1987. Involvement of the phosphate regulon and the *psiD* locus in the carbon-phosphorus lyase activity of *Escherichia coli* K-12. *J. Bacteriol.* **169:**1753–1756.
60. **Wanner, B. L.** 1990. Phosphorus assimilation and its control of gene expression in *Escherichia coli*, p. 152–163. *In* G. Hauska, and R. Thauer (ed.), *41st Mosbach Colloquium: The Molecular Basis of Bacterial Metabolism*. Springer-Verlag, Heidelberg.
61. **Wanner, B. L., and J. A. Boline.** 1990. Mapping and molecular cloning of the *phn* (*psiD*) locus for phosphonate utilization in *Escherichia coli*. *J. Bacteriol.* **172:**1186–1196.
62. **Yamane, K., and F. M. Hulett.** Unpublished data.
63. **Yamane, K., and B. Maruo.** 1978. Purification and characterization of extracellular soluble and membrane-bound insoluble alkaline phosphatases possessing phosphodiesterase activities in *Bacillus subtilis*. *J. Bacteriol.* **134:**100–107.
64. **Yamane, K., T. Miki, H. Saito, Y. Ikeda, and B. Maruo.** 1976. Isolation of a mutant secreting extracellular soluble alkaline phosphatase in *Bacillus subtilis*. *J. Bacteriol. Chem.* **40:**2181–2185.

18. Biosynthesis of the Aspartate Family of Amino Acids

HENRY PAULUS

BIOSYNTHESIS OF DIAMINOPIMELATE, LYSINE, METHIONINE, AND THREONINE

General Features of the Pathway

Diaminopimelate, lysine, methionine, and threonine derive most of their carbon atoms from L-aspartate, and these amino acids are therefore often referred to as the aspartate family. Their biosynthesis is effected by a complex pathway involving common intermediates from which multiple branches lead to the end products (Fig. 1). The so-called aspartate pathway has several features that distinguish it from other pathways of amino acid biosynthesis and lend its study particular interest in the contexts of bacterial physiology and biochemical evolution. One interesting aspect is that at least a portion of the pathway must function even when all the amino acid constituents of proteins are readily available from other sources. Another is the great evolutionary diversity of the pathway, with fundamental differences between eukaryotes and prokaryotes, among prokaryotes, and even within a single genus.

Function of the pathway in growth and sporulation

A major function of the aspartate pathway in growing bacteria is provision of amino acids for protein synthesis. However, the pathway also provides another important product, *meso*-diaminopimelate, which plays a dual role as an intermediate in the biosynthesis of lysine and as a constituent of the cell wall peptidoglycan of many bacteria, including most *Bacillus* species (120). Accordingly, diaminopimelate biosynthesis from aspartate must continue in *Bacillus subtilis* even when the protein amino acids are available from exogenous sources, for example, during growth in rich media, which are usually deficient in diaminopimelate. Under such conditions, the other pathways of amino acid biosynthesis are largely nonfunctional, owing to repression and feedback inhibition by the amino acid end products. The control of the reactions leading from aspartate to diaminopimelate in *B. subtilis* is therefore fundamentally different from the control of other steps in amino acid biosynthesis.

Another unique feature of the aspartate pathway in *B. subtilis* is that it plays a critical role not only in growth but also in sporulation. During sporulation of *B. subtilis*, there is little net synthesis of protein, amino acids for protein synthesis being made available by protein turnover (e.g., see reference 73). On the other hand, two products of the aspartate pathway that cannot be derived from protein turnover are required by sporulating cells of *B. subtilis*. These two are dipicolinate, which constitutes up to 10% of the dry weight of the spore, and *meso*-diaminopimelate, a component of the spore cortex peptidoglycan. The reaction sequence from aspartate to *meso*-diaminopimelate must remain operational during sporulation to allow synthesis of these substances.

Phylogenetic differences in the reaction sequence

In the course of evolution, the aspartate pathway has diverged to a much greater extent than most other metabolic pathways. In animals, the aspartate pathway is entirely absent: lysine, methionine, and threonine are essential amino acids that are obtained from the diet, and diaminopimelate has no known metabolic function except as a constituent of a sleep-promoting factor (74), which is probably of bacterial origin. In fungi, lysine, methionine, and threonine are synthesized, but the pathway of lysine synthesis differs from that used by bacteria. Instead of being synthesized from aspartate via diaminopimelate, lysine is derived from glutamate by an entirely different reaction sequence with L-α-aminoadipate as a key intermediate. This difference between bacteria and fungi can be rationalized by the fact that diaminopimelate is a component of bacterial but not fungal cell walls, whereas L-α-aminoadipate is required for the biosynthesis of the β-lactam antibiotics, which are characteristic fungal products. Bacteria such as *Streptomyces lipmanii*, which has a typical bacterial cell wall containing LL-diaminopimelate but behaves like fungi by producing L-α-aminoadipate-containing antibiotics, may thus represent an evolutionary link between bacterial and fungal lysine metabolisms. In *S. lipmanii* and related species, lysine is synthesized by the aspartate pathway with diaminopimelate as an intermediate, and L-α-aminoadipate is a product of lysine catabolism (65, 72). On the other hand, plants synthesize lysine by the aspartate pathway, even though diaminopimelate has no other known function. In this case, the determining factor in the choice of pathway may be the evolutionary relationship between prokaryotes and chloroplasts, a major site of amino acid biosynthesis in plants (e.g., see reference 28).

Another type of evolutionary divergence occurs in

Henry Paulus • Department of Metabolic Regulation, Boston Biomedical Research Institute, Boston, Massachusetts 02114, and Department of Biological Chemistry and Molecular Pharmacology, Harvard Medical School, Boston, Massachusetts 02115.

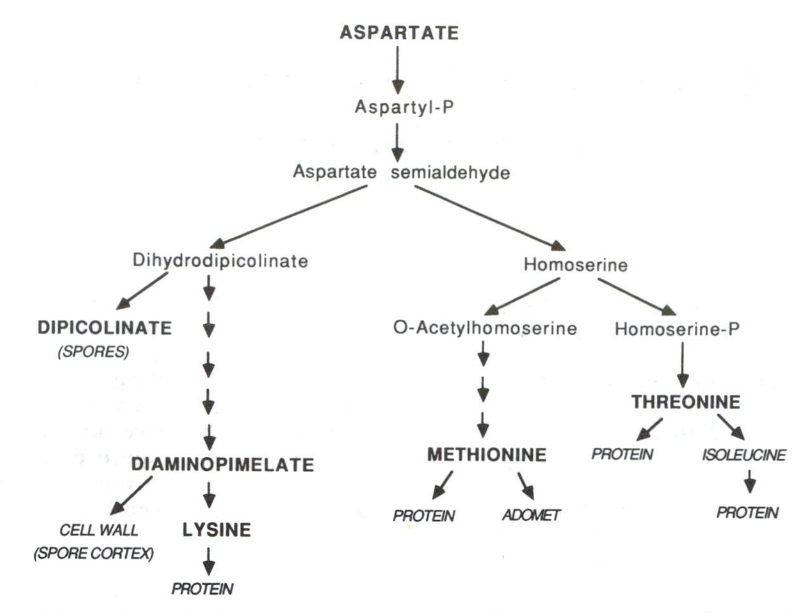

Figure 1. Overview of the pathway for biosynthesis of the aspartate family of amino acids. *ADOMET*, *S*-adenosylmethionine.

the branch of the aspartate pathway that converts aspartate semialdehyde to lysine. In *B. subtilis* and all other *Bacillus* species in which the peptidoglycan of vegetative cells contains *meso*-diaminopimelate, the lysine branch involves seven enzymatic steps, which are collectively termed the "epimerase pathway" because they include the epimerization of LL-diaminopimelate to *meso*-diaminopimelate. On the other hand, *Bacillus sphaericus* and a few other *Bacillus* species with a vegetative cell peptidoglycan that contains lysine instead of *meso*-diaminopimelate carry out the conversion of aspartic semialdehyde to lysine by the four-step "dehydrogenase pathway" (see Fig. 5), in which four of the steps of the epimerase pathway are replaced by a single reaction catalyzed by *meso*-diaminopimelate dehydrogenase (5, 88). The metabolic rationale for this dichotomy is not clear, and some species such as *Bacillus macerans* (5) and *Corynebacterium glutamicum* (121) can synthesize lysine by either route.

A third kind of phylogenetic difference occurs also in the lysine branch of the pathway and concerns the acyl blocking group that is used to prevent the spontaneous cyclization of L-2-amino-6-ketopimelate. The acyl group is introduced during the conversion of tetrahydrodipicolinate to L-2-amino-6-ketopimelate and then removed after transamination to LL-α,ε-diaminopimelate when the possibility of cyclization has passed (see Fig. 5). All *Bacillus* species examined use an acetyl blocking group, whereas many other bacteria, including *Escherichia coli* and a number of other gram-negative and gram-positive species, use succinylated intermediates (146).

An acylation-deacylation cycle also occurs in the methionine branch of the aspartate pathway but with a different purpose than in the lysine branch, namely, to convert the primary hydroxyl group of homoserine into a more effective leaving group for subsequent nucleophilic displacement by cysteine (see Fig. 6). Many kinds of acyl groups, ranging from succinyl in *E. coli* to acetyl in fungi and phosphoryl in plants, are used in methionine biosynthesis. Bacteria use either acetylated or succinylated intermediates in methionine biosynthesis, with *O*-acetylhomoserine being used by most gram-positive genera, including *Bacillus* and *Corynebacterium* spp., and *O*-succinylhomoserine being used by gram-negative bacteria, including *E. coli* (62). It is interesting that many species (e.g., *E. coli* and *B. subtilis*) use the same acyl donor in methionine and lysine synthesis in spite of the different functions of acylation in the two processes; however, this correlation does not always hold (e.g., *C. glutamicum*).

A different type of phylogenetic variation in the aspartate pathway is its extension in some bacterial species to produce another product, dipicolinate. This product is derived in a single step from dihydrodipicolinate, an intermediate in lysine biosynthesis (Fig. 1). Dipicolinate is a major constituent of the bacterial spore, and its synthesis is confined to those bacteria (*Bacillus*, *Clostridium*, and *Sporosarcina* spp. and members of the actinomycete group) that form heat-resistant endospores. Consistent with its singular function as a component of the spore, dipicolinate is synthesized only during sporulation and not during growth phase.

In conclusion, the aspartate pathway shows an exceptionally high degree of phylogenetic variation. A possible explanation for this evolutionary instability is that the efficient regulation of complex pathways with multiple end products poses serious problems and that the selective pressure to solve these problems encourages evolutionary divergence in response to different metabolic needs. This notion is consistent with the large number of different mechanisms for the control of the aspartate pathway that have evolved in the eubacteria and even within the genus *Bacillus*.

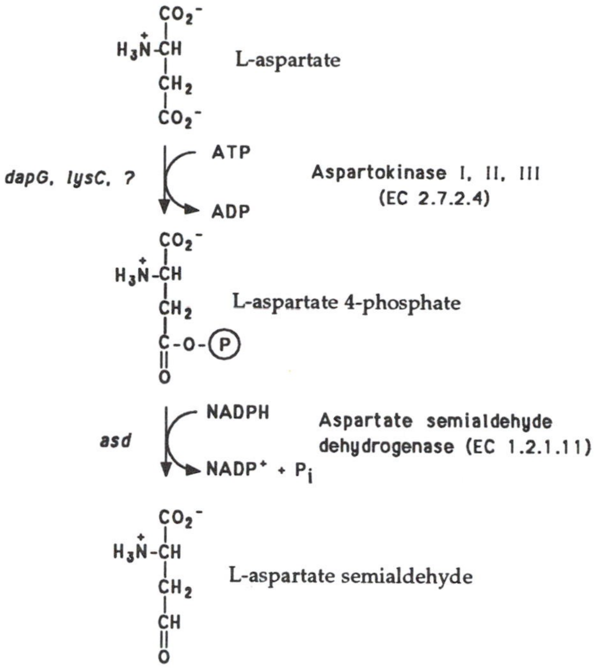

CO_2^-
$H_3\overset{+}{N}$-CH
CH_2
CO_2^-
L-aspartate

dapG, lysC, ?
ATP
ADP
Aspartokinase I, II, III
(EC 2.7.2.4)

CO_2^-
$H_3\overset{+}{N}$-CH
CH_2
C-O-(P)
O
L-aspartate 4-phosphate

asd
NADPH
$NADP^+ + P_i$
Aspartate semialdehyde
dehydrogenase (EC 1.2.1.11)

CO_2^-
$H_3\overset{+}{N}$-CH
CH_2
CH
O
L-aspartate semialdehyde

Figure 2. Common reactions in the biosynthesis of diaminopimelate, lysine, threonine, and methionine.

Some of these mechanisms will be discussed in subsequent sections of this chapter.

The primary focus of this review will be on *B. subtilis* and closely related species. Studies of other gram-positive bacteria will not be reviewed as comprehensively but will be introduced by way of complementing the discussion of the corresponding reactions in *B. subtilis*. This review will therefore slight the very extensive literature on the aspartate pathway in coryneform bacteria. These bacteria have been important sources for the industrial production of lysine and threonine among other useful products (see chapter 6).

The Common Pathway

Two reactions are common to the biosynthesis of all end products of the aspartate pathway: the transfer of phosphate from ATP to the β-carboxyl group of L-aspartate, which is catalyzed by the enzyme aspartokinase, and the subsequent NADPH-dependent reduction of the acyl-phosphate moiety of β-aspartyl phosphate by aspartate semialdehyde dehydrogenase to yield aspartate semialdehyde (Fig. 2). These reactions are chemically analogous to the reactions in the glycolytic sequence catalyzed by 3-phosphoglycerate kinase and glyceraldehyde-3-phosphate dehydrogenase. The reaction catalyzed by aspartokinase is the committing step in the utilization of L-aspartate for amino acid biosynthesis and is the major target of feedback control in the aspartate pathway.

Aspartokinase

The existence of multiple aspartokinase isozymes in *B. subtilis* was recognized as early as 1971 (46, 114), but it was not until recently that three aspartokinase isozymes were clearly defined as separate entities by the examination of mutants that lacked one or more of these enzymes (41, 162, 163). The three *B. subtilis* aspartokinases, like the three *E. coli* aspartokinase isozymes, differ in their feedback control mechanisms, suggesting functional specialization. However, the feedback control patterns of the aspartokinase isozymes of *B. subtilis* are quite different from those of the *E. coli* isozymes, and the designation of the isozymes in both organisms by the roman numerals I, II, and III does not imply analogous functions. The three aspartokinase isozymes of *B. subtilis*, whose properties are summarized in Table 1, are numbered in the order in which they elute from the column when extracts of *B. subtilis* are subjected to gel filtration,

Table 1. Aspartokinase isozymes of *B. subtilis*[a]

Property	Aspartokinase I	Aspartokinase II	Aspartokinase III
Structural gene	*dapG*	*lysC*	?
Mol wt	≥250,000	122,900	≈120,000
Subunit mol wt	42,900	43,700 (α), 17,700 (β)	?
Feedback inhibitor(s)	*meso*-Diaminopimelate	L-Lysine	L-Threonine and L-lysine[b]
Inhibition kinetics[c]	Noncompetitive	Noncompetitive	Competitive
K_i (mM) for feedback inhibitor	0.025[d]	0.1	0.15 each[e]
Shape of inhibition curve	Hyperbolic	Hyperbolic	Sigmoid
Monovalent-cation requirement	Not specific	K^+ or NH_4^+	K^+ or NH_4^+
Apparent K_m (mM) for L-aspartate[f]	3	1	17
pH range[g]	6.0–9.5	?	6.6–8.2
Corepressor of enzyme synthesis	None known	L-Lysine	L-Threonine
Relative enzyme level in rich and minimal media	Rich ≈ minimal	Rich << minimal	Rich << minimal
Change in enzyme level in stationary phase	None	Decline	Decline

[a] Data were compiled from references 24, 41, 93–95, 114, 162, and 163.
[b] The two inhibitors act synergistically, with little inhibition at physiological concentrations of each alone.
[c] With respect to the substrate L-aspartate.
[d] Measured at pH 8.0 (160).
[e] With L-threonine and L-lysine varied together at equimolar concentrations.
[f] Measured at 1 mM ATP.
[g] Range over which activity is ≥50% of maximum.

with aspartokinase I emerging with the void volume in most gel filtration media and being followed by the partially overlapping peaks of aspartokinases II and III (41, 114, 162).

Diaminopimelate-sensitive aspartokinase (aspartokinase I)

B. subtilis. Aspartokinase I differs from the other aspartokinase isozymes not only by its large apparent size but also by its activity, which is controlled by the cell wall constituent *meso*-diaminopimelate rather than by a protein amino acid. Moreover, its level is not significantly affected by the presence of exogenous amino acids and remains high even after the cells have entered stationary phase (41, 114). These properties suggest that the primary function of aspartokinase I is to provide precursors for the biosynthesis of *meso*-diaminopimelate during both growth and sporulation and for the synthesis of dipicolinate during sporulation, when the other aspartokinase isozymes are not present at significant levels (see below). The functional specialization of the aspartokinase isozymes will be discussed in a subsequent section.

Inhibition of aspartokinase I by *meso*-diaminopimelate is noncompetitive with respect to the substrate L-aspartate and is characterized by hyperbolic inhibitor saturation curves (114). The substrate saturation curves are also hyperbolic, even in the presence of *meso*-diaminopimelate, and the K_m for L-aspartate is not affected by the inhibitor. These properties suggest that aspartokinase I represents a V-type allosteric system (96). The enzyme is active over a broad pH range and is stimulated about twofold by monovalent cations.

Aspartokinase I has not been purified to any significant extent, and virtually nothing is known about its structure as a protein. The relatively high molecular weight observed in crude enzyme preparations ($\geq$250,000) is not necessarily a measure of the intrinsic size of aspartokinase I but could reflect the molecular weight of a complex with other proteins. Indeed, the aspartokinase I gene, which has recently been cloned and sequenced, encodes a polypeptide of nearly the same size (404 amino acids; M_r = 42,900) as the major subunit of aspartokinase II of *B. subtilis* (24). The deduced amino acid sequence of aspartokinase I is similar to the sequences of *B. subtilis* aspartokinase II (36% identity) and the aspartokinases of other organisms (24).

Two types of mutants with an altered aspartokinase I have been characterized. A temperature-sensitive mutant, originally isolated on the basis of its conditional lysis phenotype (13), was recently found to have a thermolabile aspartokinase I (115). The locus of the mutation, originally termed *lssD*, has been renamed *dapG* to reflect its status as the structural gene for aspartokinase I. Another type of mutant, which has an aspartokinase I refractory to feedback inhibition by *meso*-diaminopimelate, has been isolated (163), but the mutation has not yet been genetically mapped.

Other *Bacillus* species. Enzymes resembling aspartokinase I have also been identified in other *Bacillus* species. A *meso*-diaminopimelate-sensitive aspartokinase that has been partially purified from *Bacillus stearothermophilus* resembles aspartokinase I of *B. subtilis* in many respects, including noncompetitive inhibition by *meso*-diaminopimelate at 55°C, the magnitudes of the K_i for *meso*-diaminopimelate and the K_m for L-aspartate, a broad pH optimum, and mild stimulation by monovalent cations (76). During growth of *B. stearothermophilus* in defined media, the level of the enzyme is not affected by the addition of lysine and threonine and is relatively constant throughout exponential and early stationary phases (76). On the other hand, in a complex sporulation medium, the level of the diaminopimelate-sensitive aspartokinase increases about 10-fold in the course of sporulation, whereas no such increase is observed in an asporogenous mutant (77). Unlike aspartokinase I of *B. subtilis*, the diaminopimelate-sensitive aspartokinase of *B. stearothermophilus* has a molecular weight of only 110,000 (76). A *meso*-diaminopimelate-sensitive aspartokinase activity has also been described in *Bacillus cereus*, where it represents the major aspartokinase activity during sporulation, its level increasing two- to threefold during the transition from exponential growth to sporulation (54). The "feedback-insensitive" aspartokinase activities described earlier for *B. cereus* (2, 35) and *Bacillus licheniformis* (43), which were originally thought to be produced by the modification of a lysine-sensitive enzyme but were not further characterized, are probably diaminopimelate-sensitive aspartokinases.

Lysine-sensitive aspartokinase (aspartokinase II)

B. subtilis. The first aspartokinase to be purified to homogeneity from extracts of *B. subtilis* was the lysine-sensitive isozyme aspartokinase II (93). Both the activity and the synthesis of aspartokinase II are regulated by lysine, and the activity of the enzyme declines rapidly at the onset of stationary phase or during glucose starvation (42), suggesting that aspartokinase II functions primarily to provide precursors for the synthesis of lysine as a building block for proteins.

The inhibition of aspartokinase II by L-lysine is noncompetitive with respect to L-aspartate, and the L-aspartate saturation curves are hyperbolic both in the absence and the presence of L-lysine (93). Like aspartokinase I, lysine-sensitive aspartokinase thus appears to be an allosteric V system. The lysine analog, *S*-(2-aminoethyl)-L-cysteine, is a much weaker inhibitor than L-lysine (K_i = 1 mM), and other amino acids related to lysine, including L-ornithine, D-lysine, and *meso*-diaminopimelate, here no effect on activity. Threonine, either alone or in combination with lysine, has no inhibitory effect. The activity of aspartokinase II, unlike that of aspartokinase I, depends specifically on the presence of K^+ or NH_4^+ (95).

Aspartokinase II has a molecular weight of about 120,000 and is composed of equimolar amounts of two dissimilar subunits, termed α and β, with molecular weights of about 43,000 and 17,000, respectively, indicative of an $\alpha_2\beta_2$ subunit structure (93). The threonine-plus-lysine-sensitive aspartokinases from *Bacillus polymyxa* (7), *C. glutamicum* (61), and *Corynebacterium flavum* (33) have similar molecular weights and the same kind of subunit structure, even though the feedback inhibition patterns of these enzymes suggest that the enzymes are functionally more closely related to *B. subtilis* aspartokinase III (see below). Indeed, structural investigations of *B. subtilis* aspartokinase II were in large part modeled on earlier studies of the *B. polymyxa* enzyme, and the results obtained with the two enzymes show a close correspondence (8, 93). Studies of the products of renaturation of the separated aspartokinase subunits and their mixtures as well as studies of cross-linking suggest that the aspartokinase subunits are arranged in a "sandwich" structure of the type $\beta\alpha\alpha\beta$, with no direct contacts between the β subunits but with an α-β domain of bonding that has self-associative properties which allow the formation of α tetramers and β dimers when the separated subunits are renatured. An explanation for the tendency of the α-β domain to self-associate was provided by the discovery that the β subunit corresponds to the carboxyl-terminal portion of the larger α subunit. This conclusion was based on chemical and immunological comparisons of the subunits (94) and on the observation that a single deletion in the cloned aspartokinase II genes causes the truncation of both aspartokinase subunits, each by the same amount, which can be the case only if the aspartokinase α and β subunits share a coding region (11). The quaternary structure of aspartokinase II can thus be understood in terms of two types of symmetrical bonding domains, one between the N-terminal portions of the α subunits and the other between the C-terminal part of an α subunit and the corresponding portion of the β subunit. In the context of this model, it is interesting that prolonged treatment of the purified α subunit with trypsin under nondenaturing conditions leads to the production of a single polypeptide with a molecular weight slightly higher than that of the β subunit, suggesting that the C-terminal portion of the α subunit corresponding to the β subunit consists of a compact globular domain connected to the N-terminal portion containing the catalytic site (see below) by a protease-sensitive linker (10, 101). The α subunit of aspartokinase II thus probably consists of two globular domains, one involved in α-α and the other involved in α-β interactions, as illustrated in Fig. 3A.

The function of the β subunit is still problematic. The subunits of aspartokinase II can be separated by gel filtration in the presence of 4.2 M urea and renatured by dialysis or dilution (93). A large fraction (80 to 90%) of aspartokinase activity is recovered on renaturation of an equimolar mixture of subunits; when the purified subunits are renatured separately, β subunit yields no aspartokinase activity, whereas α subunit yields substantial aspartokinase activity that is virtually indistinguishable from the intact native enzyme activity with respect to both catalytic and allosteric properties. It appears that the β subunit plays no essential role in the structure or function of the purified enzyme or, as shown by subsequent studies of mutant forms of *B. subtilis* aspartokinase II (25), in vivo when the enzyme is expressed in a heterologous *E. coli* system. In light of the fact that the aspartokinases from several other *Bacillus* species, including *B. polymyxa* (7), *Bacillus brevis* (49), and the thermophilic methylotrophic *Bacillus* sp. strain MGA3 (117), and from *C. glutamicum* (61) and *C. flavum* (33) have an analogous subunit structure, it is unlikely that the β subunit lacks a specific function. However, it is clear that the function of the β subunit is exercised neither in vitro nor in a heterologous system, suggesting that it is related to a specific aspect of amino acid metabolism in the organisms in which these enzymes occur, perhaps in the interaction with other enzymes to form multienzyme complexes.

The genes encoding the two subunits of aspartokinase II have been cloned (11), and sequence analysis has shown that the subunits are encoded by in-phase overlapping genes, with the β subunit being translated from a translation initiation codon preceded by a strong ribosome-binding site located within the coding region for the α subunit (23). The molecular weights of the α and β subunits deduced from the nucleotide sequence are 43,700 and 17,700, respectively, giving a molecular weight of 122,900 for intact aspartokinase II. Studies involving specific deletions and site-directed mutagenesis showed that the translation of the α and β subunits occurs independently (24). Nucleotide sequence analysis of the genes for the threonine-plus-lysine-sensitive aspartokinases from *C. glutamicum* (60) and *C. flavum* (33) suggests that their subunits, which show considerable sequence similarity to the α and β subunits of *B. subtilis* aspartokinase II, are encoded by in-phase overlapping genes.

Other *Bacillus* species. Lysine-sensitive aspartokinase isozymes have also been studied in several other *Bacillus* species, including *B. cereus* (35, 54), *B. licheniformis* (43, 133), and *B. brevis* (50), but these enzymes have not been sufficiently characterized to determine whether they are indeed homologs of aspartokinase II of *B. subtilis*. The *B. brevis* aspartokinase appears to have an $\alpha_2\beta_2$ subunit structure resembling those of the *B. subtilis* aspartokinase II and the *B. polymyxa* aspartokinase (49). The aspartokinase of *B. licheniformis* has been reported to be inhibited by aspartate semialdehyde (133), but in view of the very high concentrations (20 mM) of DL-aspartate semialdehyde necessary to effect 50% inhibition and the known interference of this metabolite with the assay procedure used in these studies (31), the physiological significance of this effect needs to be reexamined. The gene for the lysine-sensitive aspartokinase from the methylotrophic *Bacillus* sp. strain MGA3 has recently been cloned and sequenced and found to consist of in-phase overlapping genes for subunits with a high degree of similarity to the α and β subunits of aspartokinase II (117).

Threonine-plus-lysine-sensitive aspartokinase (aspartokinase III)

B. subtilis. The first survey of aspartokinase isozymes in *B. subtilis* was done with the Marburg strain (ATCC 6051) and revealed the presence of a threonine-plus-lysine-sensitive aspartokinase (114). Subsequent studies of *B. subtilis* aspartokinases were done on derivatives of strain 168 and showed a lysine-sensitive aspartokinase to be the predominant isozyme (46, 93, 94), the gene for which was subsequently cloned and sequenced (11, 23). Originally, it was assumed that the threonine-plus-lysine-sensitive aspartokinase and the lysine-sensitive enzymes represented allelic variants of the same isozyme in different *B. subtilis* strains, and both enzymes were referred to as aspartokinase II (101). It was not until Graves and Switzer (41) carefully analyzed the aspartokinase activities of a deriv-

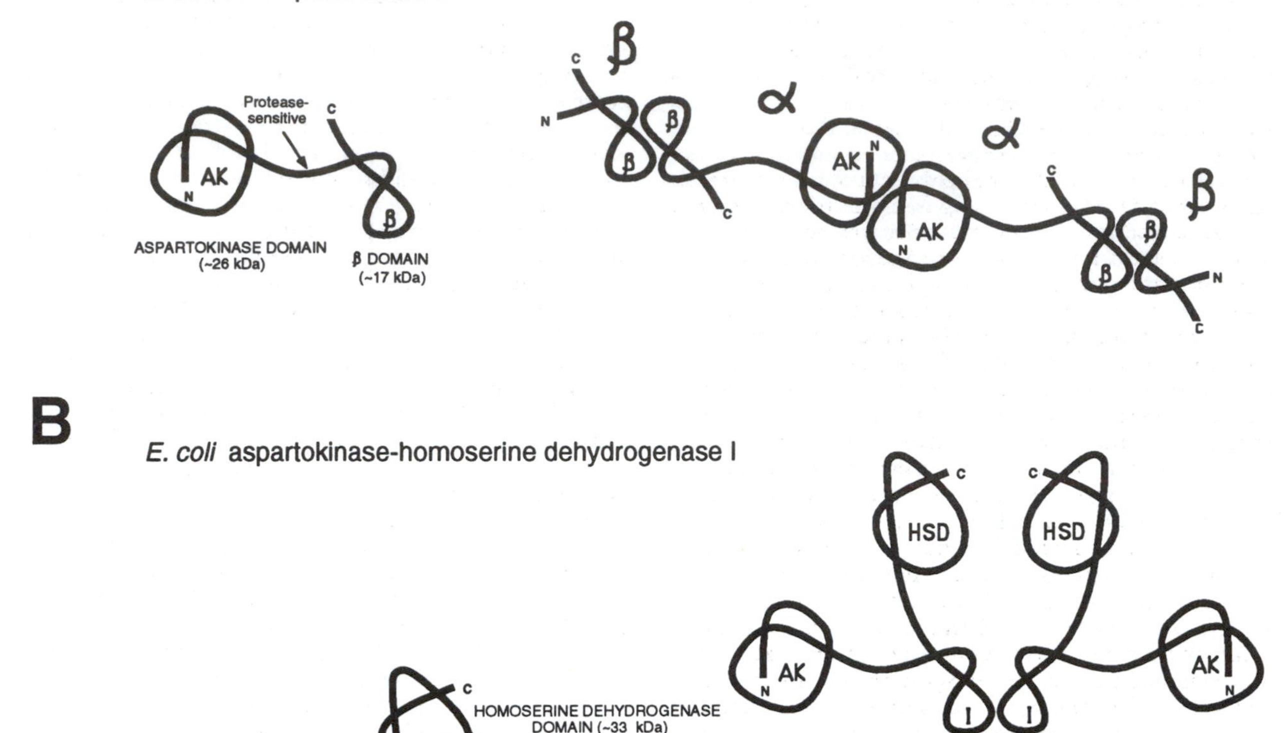

Figure 3. Postulated folding domains of aspartokinases. (A) Proposed biglobular structure of the α subunit of *B. subtilis* aspartokinase II (left) and arrangements of the α and β subunits in the native enzyme (right). (B) Proposed triglobular structure of a subunit of *E. coli* aspartokinase-homoserine dehydrogenase I (left) and arrangements of subunits in the native tetrameric enzyme (right) (modified from reference 27). AK, aspartokinase domain; HSD, homoserine dehydrogenase domain; I, interdomain.

ative of *B. subtilis* 168 in which the gene for the lysine-sensitive isozyme had been disrupted (25) that the lysine-plus-threonine-sensitive aspartokinase was recognized as a distinct aspartokinase isozyme, aspartokinase III. The levels of aspartokinases II and III vary reciprocally in different strains of *B. subtilis*, aspartokinase II being the predominant isozyme in derivatives of *B. subtilis* 168 and aspartokinase III predominating in *B. subtilis* Marburg (162) and W23 (164). Aspartokinase III is not expressed during growth in rich media, during stationary phase, or during sporulation, and its activity is repressed by threonine, suggesting that the primary role of aspartokinase III, like that of aspartokinase II, is in the biosynthesis of amino acids as precursors for protein synthesis.

Aspartokinase III is inhibited in a synergistic manner by L-threonine and L-lysine, a type of control often referred to as concerted or multivalent feedback inhibition. L-Threonine and L-lysine alone at physiological concentrations ($\leq$1 mM) have little effect on aspartokinase activity, but in combination, they effectively inhibit the enzyme at much lower concentrations (41, 114). Unlike feedback inhibition of the other aspartokinase isozymes, the inhibition of aspartokinase III by L-threonine plus L-lysine is competitive with respect to the substrate L-aspartate, and the high apparent K_m for L-aspartate is increased even further in the presence of the inhibitors (114). The activity of aspartokinase III is also modulated in a complex manner by high concentrations of certain nonpolar amino acids (L-methionine, L-leucine, and L-phenylalanine); this observation suggests the existence of a specific binding site for nonpolar amino acids, but its physiological significance is uncertain. Aspartokinase III resembles aspartokinase II in that its activity is completely dependent on K^+ or NH_4^+, with Na^+ being ineffective.

Since aspartokinase III of *B. subtilis* has been only partially purified, little is known about its structure except that the enzyme elutes from gel filtration columns just slightly after aspartokinase II, suggesting a molecular weight close to 120,000 (41, 162). The gene for aspartokinase III has been neither cloned nor mapped on the *B. subtilis* chromosome. Mutant strains of *B. subtilis* that lack aspartokinase III have been isolated (162, 163), but it is not clear whether the mutations affect the structural gene of aspartokinase II or a regulatory locus.

Other *Bacillus* species. Threonine-plus-lysine-sensitive aspartokinases also occur in other *Bacillus* species. The aspartokinase from *B. polymyxa* has been purified to homogeneity and studied in considerable detail, especially with respect to its subunit structure (7, 8, 102, 103). As mentioned above, the structure of the *B. polymyxa* aspartokinase resembles that of *B. subtilis* aspartokinase II; whether it also resembles the structure of *B. subtilis* aspartokinase III, perhaps even more closely, cannot be addressed until the latter has been elucidated. The catalytic and regulatory properties of the *B. polymyxa* aspartokinase are in many respects similar to those of aspartokinase III except for a complex concentration dependence of the catalytic parameters at higher temperatures (102). Like aspartokinase III, the *B. polymyxa* aspartokinase has a binding site for nonpolar L-amino acids, even though the two enzymes differ in their relative specificities and patterns of response to nonpolar amino acids (103). Other threonine-plus-lysine-sensitive aspartokinases that have been partially purified and characterized include the threonine-repressible enzymes from *B. licheniformis* (43) and *B. brevis* (50, 51), and one of the aspartokinase isozymes of *B. stearothermophilus* (75); these enzymes resemble aspartokinase III of *B. subtilis* in many respects. Threonine-plus-lysine-sensitive aspartokinases that have been described but not studied in detail include a threonine-repressible enzyme of *B. sphaericus* (6) and a threonine-plus-lysine-repressible aspartokinase of *Bacillus megaterium* 7581 (21).

Coryneform bacteria. The coryneform bacteria appear to have only a single aspartokinase, which is controlled by concerted feedback inhibition by L-threonine and L-lysine. The aspartokinase from *Brevibacterium* (*Corynebacterium*) *flavum* has been partially purified and studied in detail (125, 127). Its properties are strikingly similar to those of aspartokinase III of *B. subtilis*, including relatively high apparent K_ms for L-aspartate and the presence of a low-affinity binding site for nonpolar amino acids. Aminoethylcysteine-resistant mutants with altered aspartokinases in which concerted inhibition by L-threonine and L-lysine (or L-aminoethylcysteine) is abolished have been isolated from *B. flavum* (127) and *C. glutamicum* (63); in the former, inhibition by a high concentration of L-lysine or L-aminoethylcysteine alone is not affected (127). The coding regions for the threonine-plus-lysine-sensitive aspartokinases from *C. glutamicum* (60, 61, 138) and *C. flavum* (33) have been cloned, sequenced, and found to consist of in-phase overlapping genes whose deduced products have considerable sequence similarity to the α and β subunits of aspartokinase II from *B. subtilis*. These studies also provide insights into the amino acid residues essential for feedback inhibition by showing that the genes for wild-type aspartokinase and for the feedback-resistant enzyme from an aminoethylcysteine-resistant mutant differ by a single nucleotide substitution, resulting in the replacement of serine 301 by a tyrosine residue. The mutation, which leads to insensitivity to feedback inhibition, occurs in the coding region common to the α and β subunits; consequently, the feedback-resistant aspartokinase has amino acid substitutions in both types of subunits. Strains that carry the genes for the normal α and β subunits on the chromosome and the gene for a mutant β subunit on a multicopy plasmid have a partially feedback-resistant aspartokinase; however, it remains to be determined whether full feedback resistance can be achieved if only one of the subunits is altered (60, 138).

Functional specialization of the aspartokinase isozymes

Both the activities and the syntheses of the three aspartokinase isozymes are controlled by different end products, and the primary function of each isozyme can thus be defined by its pattern of feedback control. However, the question arises whether there is a certain degree of redundancy that may allow the three aspartokinases to substitute for each other's functions. With respect to aspartokinases II and III, the answer is that the function of either isozyme can be fulfilled by another aspartokinase. The gene for aspartokinase II can be disrupted without deleterious effect on growth under ordinary laboratory conditions (25, 162). Conversely, mutants with chromosomal lesions affecting both aspartokinase II and III grow normally when aspartokinase II is restored by transformation with a plasmid carrying its gene (162, 163). Indeed, the fact that the levels of aspartokinases II and III vary reciprocally in different strains of *B. subtilis* suggests that the functions of these isozymes are to some extent redundant.

On the other hand, aspartokinase I does not seem to be able to substitute for the other isozymes, and strains that lack both aspartokinases II and III are unable to grow unless supplemented with the amino acid end products L-lysine, L-methionine, and L-threonine, in spite of normal levels of aspartokinase I (162, 163). Pseudorevertants of such strains, which can grow in the absence of the amino acid end products but still lack aspartokinases II and III, have altered aspartokinases I that have normal catalytic activities but are completely refractory to feedback inhibition by *meso*-diaminopimelate (163). It appears, therefore, that the ability of aspartokinase I to function in the synthesis of protein amino acids during growth is constrained by its feedback inhibition by *meso*-diaminopimelate. The biochemical basis of this limitation becomes clear when we consider that wild-type aspartokinase I is 50% inhibited by 25 μM *meso*-diaminopimelate, a concentration of *meso*-diaminopimelate much lower than that needed for the conversion of *meso*-diaminopimelate to lysine by diaminopimelate decarboxylase, which has a K_m for *meso*-diaminopimelate of 1,000 μM (113). As a result, inhibition of

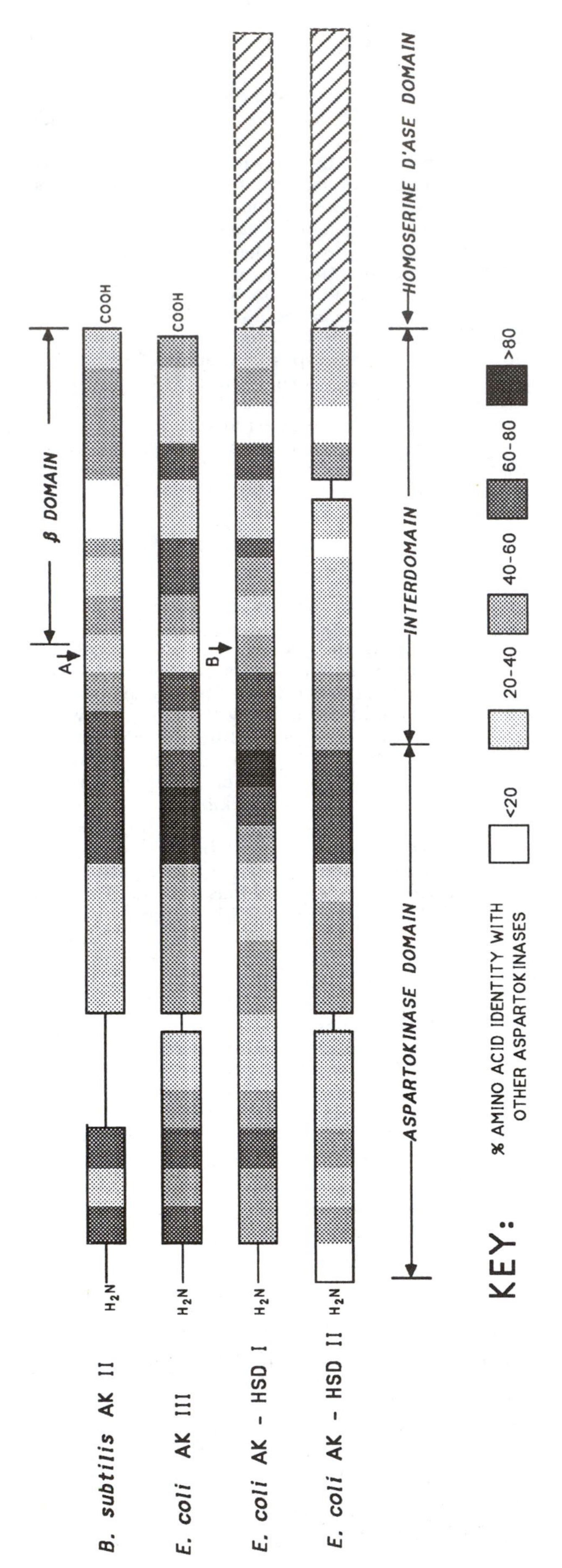

aspartokinase I by *meso*-diaminopimelate will throttle lysine synthesis; however, it will not interfere with the utilization of *meso*-diaminopimelate for cell wall synthesis by UDP-*N*-acetylmuramoyl-L-alanyl-D-glutamate:*meso*-diaminopimelate ligase, because the K_m of this enzyme for *meso*-diaminopimelate is much lower (36 μM for the enzyme from *E. coli* [85]; the corresponding enzyme from *B. subtilis* has not been studied).

The opposite situation, i.e., possible substitution of aspartokinases II and III for aspartokinase I, has been examined in *dapG* mutants with a thermosensitive aspartokinase I (115). In rich media, these mutants undergo lysis at 45°C, presumably because aspartokinases II and III are repressed, and thermal inactivation of aspartokinase I thus precludes the synthesis of precursors for *meso*-diaminopimelate. In minimal medium, the *dapG* mutants continue to grow at a substantial rate after shift to 45°C, suggesting that aspartokinases II and III can substitute for inactivated aspartokinase I. Nevertheless, the growth rate under these conditions is significantly less than that of the parent strain, an indication that the other isozymes cannot fully compensate for the loss of aspartokinase I in spite of their considerably higher total catalytic activities (41, 162). On the other hand, in minimal medium supplemented with L-lysine, the *dapG* mutants rapidly cease growth after temperature shift and eventually undergo lysis (115). Evidently, aspartokinases II and III cannot compensate for the loss of aspartokinase I under these conditions, even though their total activities measured in cell extracts are similar in unsupplemented and lysine-supplemented minimal media (41, 162). One possible interpretation of these results is that aspartokinase II, the predominant aspartokinase activity in unsupplemented medium, can partially substitute for aspartokinase I but that aspartokinase III, which predominates during growth in lysine-supplemented medium, is unable to do so. Since the failure of aspartokinase III to provide adequate levels of precursors for diaminopimelate synthesis cannot be ascribed to feedback inhibition, which requires the presence of both lysine and threonine, this explanation suggests that the three aspartokinase isozymes contribute to different precursor pools that are differentially utilized for synthesis of the various pathway end products.

Comparison with the *E. coli* aspartokinases

A comparison of the amino acid sequences of *B. subtilis* aspartokinase II and the three *E. coli* aspartokinases (Fig. 4) reveals considerable similarity, not only in certain highly conserved regions but also to some extent throughout the molecules (23). Such extensive sequence similarity suggests similar patterns of polypeptide folding, a conclusion that is supported by the observation that, under mild conditions, trypsin cleaves only a single bond at similar

Figure 4. Sequence homology of *B. subtilis* aspartokinase II and the *E. coli* aspartokinases. A and B are the sites at which trypsin cleaves *B. subtilis* aspartokinase II and *E. coli* aspartokinase-homoserine dehydrogenase I, respectively. AK, aspartokinase; HSD, homoserine dehydrogenase.

positions in *B. subtilis* aspartokinase II and *E. coli* aspartokinase-homoserine dehydrogenase I. The postulated biglobular structure of *B. subtilis* aspartokinase II, which consists of an aspartokinase domain and a β-subunit domain (Fig. 3A), may thus correspond to the two amino-terminal domains, i.e., the aspartokinase domain and the interdomain, of the triglobular *E. coli* aspartokinase-homoserine dehydrogenase I (Fig. 3B) (27). Yet in spite of such structural conservatism, which suggests a common evolutionary origin, there are a number of basic differences between the aspartokinase isozymes and their modes of regulation in *E. coli* and *B. subtilis*.

The first difference concerns the isozyme pattern. In *E. coli*, the aspartokinase isozymes are neatly geared to respond to the requirements of protein synthesis, with each isozyme being controlled by one of the protein amino acid end products, lysine, methionine, or threonine (27). On the other hand, *E. coli* seems to make no provision for ensuring the synthesis of the cell wall precursor diaminopimelate when there is a surfeit of protein amino acids; perhaps growth of *E. coli* in rich media is possible only because of leakiness in the control of the aspartokinase isozymes by their end products. In contrast, the aspartokinase pattern of *B. subtilis* is not as well suited for ensuring the balanced synthesis of protein amino acids. The lack of participation of methionine in the control of one of the isozymes is a puzzling omission, while the participation of L-lysine in the feedback inhibition of two aspartokinases (II and III) seems unnecessarily redundant. On the other hand, *B. subtilis* appears to be especially well adjusted to cope with growth in rich media and stationary phase by the presence of an aspartokinase isozyme controlled by diaminopimelate and unaffected by the levels of protein amino acids.

A second major difference between the *E. coli* and *B. subtilis* aspartokinases concerns the structures of the enzymes. Two of the *E. coli* aspartokinases are bifunctional proteins that also have homoserine dehydrogenase activity (27). None of the *B. subtilis* aspartokinases is associated with homoserine dehydrogenase, which in *B. subtilis* is a monofunctional protein encoded by a separate gene (100). Rather, a structural hallmark of *B. subtilis* aspartokinase II, which also applies to all aspartokinases from related species studied to date, is its heteropolymeric nature, which involves two dissimilar subunits encoded by in-phase overlapping genes. This type of protein structure is relatively rare in nature, and its function is still unknown. Thus, in spite of the similarities in primary and probably tertiary structures of the *E. coli* and *B. subtilis* aspartokinases (Fig. 3 and 4), there are basic differences in the manner in which the polypeptide chains are organized, the physiological significance of which remains to be elucidated.

Aspartate semialdehyde dehydrogenase

The enzyme that catalyzes the second step in the aspartate biosynthetic pathway has not yet been extensively studied in *B. subtilis*. Stable L-phase variants of *B. subtilis* and *B. licheniformis* that specifically lack aspartate semialdehyde dehydrogenase have been isolated (144, 152). The observations that *asd* mutant strains can be obtained without the use of mutagenic agents and that the resulting L-form phenotype can be transferred to normal strains by transformation at frequencies consistent with a single genetic marker (151) suggest that *B. subtilis* has only a single aspartate semialdehyde dehydrogenase. The gene for aspartate semialdehyde dehydrogenase from *B. subtilis* has recently been cloned by complementation of the auxotrophic defect in an *E. coli* strain with an *asd* deletion (24). The putative *asd* gene has been sequenced, and the deduced translation product, a 346-residue polypeptide with an M_r of 37,700, shows extensive similarity to the aspartate semialdehyde dehydrogenases from *Streptococcus mutans* (17), *C. glutamicum* (60), and *C. flavum* (33) but very little similarity to the corresponding enzyme from *E. coli* (48). The low degree of similarity with the *E. coli asd* gene suggests that aspartate semialdehyde dehydrogenases from gram-positive species are not very closely related to the enzyme from *E. coli*, even though they can functionally substitute for the latter.

The modulation of aspartate semialdehyde dehydrogenase activity by growth conditions and metabolites has been examined in several *Bacillus* species. In *B. cereus*, the level of the enzyme increases during sporulation in parallel with levels of aspartokinase I, dihydrodipicolinate synthase, and dihydrodipicolinate reductase (35). In *B. brevis*, a biphasic change occurs in the level of aspartate semialdehyde dehydrogenase, with a threefold decline in specific activity at the end of exponential growth followed by an increase to the original level in the course of sporulation (108). The enzyme from *B. brevis* differs from all other aspartate semialdehyde dehydrogenases described in its sensitivity to inhibition by amino acids, with 5 mM L-threonine or L-leucine reducing its activity by about 50% (108). Aspartate semialdehyde dehydrogenase of *B. sphaericus* is neither repressed nor inhibited by amino acids (6), but the level of the *B. megaterium* enzyme is repressed 40% by the presence of methionine (22).

The genes for aspartate semialdehyde dehydrogenase from *S. mutans* and coryneform bacteria have been cloned and sequenced. The *asd* gene from *S. mutans* encodes a 357-residue polypeptide (M_r = 38,900) that complements an *asd* lesion in *E. coli* in spite of the low sequence similarity between the enzymes from the two organisms (17). The *asd* genes from aminoethylcysteine-resistant strains of *C. glutamicum* (61) and *C. flavum* (33), also cloned by complementation of an *asd* mutant strain of *E. coli*, encode 344-residue polypeptides (M_r = 36,200) with a high degree of sequence similarity to the aspartate semialdehyde dehydrogenase from *S. mutans*.

The Branch Leading to Diaminopimelate and Lysine

The aspartate pathway splits after the synthesis of aspartate semialdehyde, one branch leading to biosynthesis of diaminopimelate and lysine and the other leading to biosynthesis of threonine and methionine. In *B. subtilis* and all other *Bacillus* species except *B. sphaericus*, *Bacillus globisporus*, and *Bacillus pasteurii*, the diaminopimelate-lysine branch involves seven en-

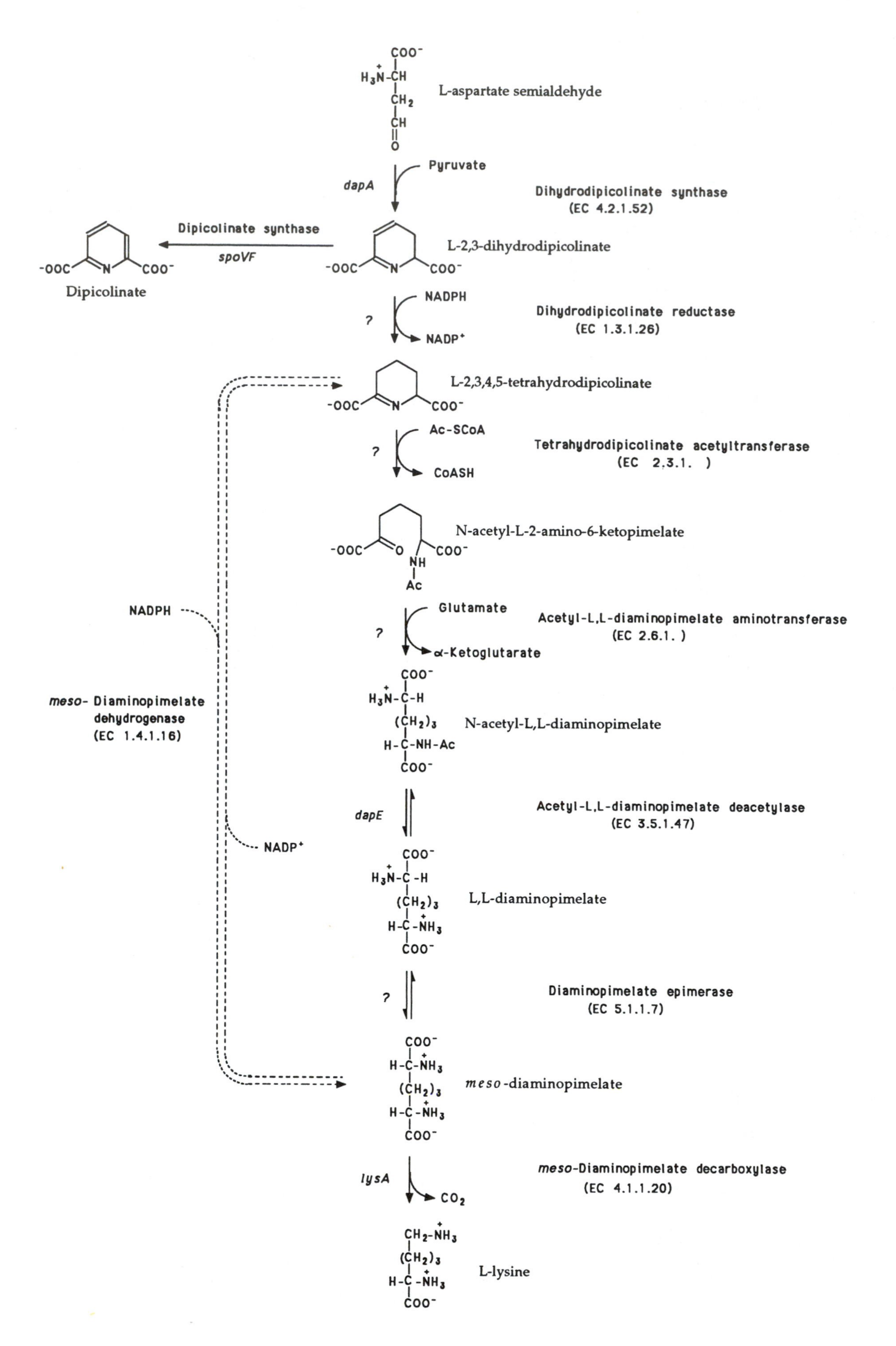
L-aspartate semialdehyde
Pyruvate
dapA
Dihydrodipicolinate synthase
(EC 4.2.1.52)
Dipicolinate synthase
spoVF
L-2,3-dihydrodipicolinate
Dipicolinate
NADPH
?
NADP+
Dihydrodipicolinate reductase
(EC 1.3.1.26)
L-2,3,4,5-tetrahydrodipicolinate
Ac-SCoA
?
CoASH
Tetrahydrodipicolinate acetyltransferase
(EC 2.3.1.)
N-acetyl-L-2-amino-6-ketopimelate
Glutamate
?
α-Ketoglutarate
Acetyl-L,L-diaminopimelate aminotransferase
(EC 2.6.1.)
NADPH
meso- Diaminopimelate
dehydrogenase
(EC 1.4.1.16)
N-acetyl-L,L-diaminopimelate
dapE
Acetyl-L,L-diaminopimelate deacetylase
(EC 3.5.1.47)
NADP+
L,L-diaminopimelate
?
Diaminopimelate epimerase
(EC 5.1.1.7)
meso-diaminopimelate
lysA
CO2
meso-Diaminopimelate decarboxylase
(EC 4.1.1.20)
L-lysine

zymatic steps, with the first six, which lead to *meso*-diaminopimelate, functioning in both protein and cell wall synthesis and only the final step, the decarboxylation of *meso*-diaminopimelate to lysine, exclusively contributing to protein synthesis (Fig. 5). During sporulation, the diaminopimelate-lysine branch also serves for biosynthesis of 2,6-dipicolinate, which is produced in a single step from the first intermediate in the pathway, dihydrodipicolinate. For lysine biosynthesis, *B. sphaericus*, *B. globisporus*, and *B. pasteurii* use the four-step dehydrogenase pathway, in which four of the reactions of the seven-step epimerase pathway are replaced by a reaction catalyzed by *meso*-diaminopimelate dehydrogenase (Fig. 5).

Only a few of the steps in the diaminopimelate-lysine branch of the aspartate pathway in *B. subtilis* have been studied to any significant extent. One reason for the lack of progress in genetic analysis is that the first six steps in the pathway constitute the only source of *meso*-diaminopimelate. Because *B. subtilis*, unlike *E. coli*, cannot transport *meso*-diaminopimelate, mutations causing the complete loss of any of the enzymes required for diaminopimelate synthesis are lethal. Attempts to circumvent this problem by searching for temperature-sensitive mutants (13, 15, 16, 115) or for mutations that give rise to stable L forms, which can be maintained under conditions in which cell wall biosynthesis is not essential, have been made (144). Although these approaches have led to the isolation of mutants with defects in aspartokinase I (115) and aspartate semialdehyde dehydrogenase (144), the only mutants obtained in the diaminopimelate-lysine branch of the aspartate pathway are ones with a defective *N*-acetyl-LL-diaminopimelate deacylase (115). A plausible explanation for the difficulty in obtaining mutants affected in the other five steps in the biosynthesis of *meso*-diaminopimelate is that because of their dual function in cell wall and protein synthesis, these reactions may be catalyzed by two sets of isozymes, so that two mutations would be required for the complete abolition of enzyme activity (115).

Dihydrodipicolinate synthase

B. subtilis. The committing step in the conversion of aspartate semialdehyde to diaminopimelate and lysine involves the condensation of aspartate semialdehyde with pyruvate to yield the cyclic intermediate, L-2,3-dihydrodipicolinate, which is catalyzed by dihydrodipicolinate synthase. In *B. subtilis*, the level of dihydrodipicolinate synthase is not affected by the presence of lysine in the culture medium but increases two- to threefold during the late stages of sporulation just prior to the production of dipicolinate, suggesting that dipicolinate synthesis imposes an increased demand for this enzyme (20, 53, 143). The increase in dihydrodipicolinate synthase levels during sporulation requires new RNA synthesis (20, 53) but is thought not to involve the synthesis of a new isozyme, for no significant differences in the physical and catalytic properties of dihydrodipicolinate synthase isolated from vegetative and sporulating cells have been detected.

Some of the properties of dihydrodipicolinate synthase from *B. subtilis* are summarized in Table 2. A partially purified enzyme from sporulating cells assayed at pH 7.4 was found to have sigmoid substrate saturation curves and a low apparent K_m for aspartate semialdehyde (20), but these properties were probably artifacts due to stabilization of the enzyme by pyruvate at this suboptimal pH, a phenomenon also observed with dihydrodipicolinate synthase from *B. licheniformis* (45). A more highly purified enzyme preparation, assayed by a slightly different procedure at pH 8.5, had a higher apparent K_m for aspartate semialdehyde and exhibited hyperbolic pyruvate saturation curves (153). Optimal enzyme activity is observed at pH 9.5, and the greatest stability is observed above pH 10. The end products L-lysine, *meso*-diaminopimelate, and dipicolinate are not inhibitory (20, 153). In this respect, dihydrodipicolinate synthase from *B. subtilis* differs from the corresponding enzyme in *E. coli*, which is subject to feedback inhibition by L-lysine (160). The lack of feedback control allows the *B. subtilis* enzyme to function in the biosynthesis of diaminopimelate and dipicolinate during sporulation even at high intracellular lysine concentrations. The molecular weight of highly purified preparations of dihydrodipicolinate synthase from sporulating cells has been estimated at 124,000 by gel filtration under several different conditions, and a similar value was obtained for a enzyme preparation from vegetative cells (153). The subunit molecular weight has been tentatively estimated at 30,000, suggesting that the native enzyme is a tetramer. The deduced translation product of a gene adjacent to the gene for aspartokinase I has a high degree of sequence similarity with dihydrodipicolinate synthase of *E. coli* and probably is the corresponding *B. subtilis* enzyme (24). It consists of 292 amino acid residues with a predicted molecular weight of 31,100.

Other *Bacillus* species. Dihydrodipicolinate synthase has also been studied in other *Bacillus* species. The enzyme has been purified to homogeneity from *B. licheniformis* (45), and some of its properties are summarized in Table 2. The sigmoid pyruvate saturation curves and low apparent K_m for aspartate semialdehyde found in earlier studies (132) are artifacts caused by the effects of the substrates on enzyme stability (45). L-Lysine, *meso*-diaminopimelate, and dipicolinate have no effect on the activity of dihydrodipicolinate synthase from *B. licheniformis* (132). The molecular weight of the purified enzyme was estimated to be between 108,000 and 118,000, with a subunit molecular weight of 28,000, which is consistent with a tetrameric structure (45). The level of dihydrodipicolinate synthase in *B. cereus* undergoes a fourfold increase during sporulation, but the molecular weights and catalytic properties of the enzyme from vegetative and sporulating cells are indistinguishable (55).

Figure 5. The branch of the aspartate pathway leading to biosynthesis of diaminopimelate and lysine. The reaction catalyzed by diaminopimelate dehydrogenase, indicated by the dotted arrows, occurs in a few *Bacillus* species and in coryneform bacteria but not in *B. subtilis*. Ac-SCoA, acetyl-CoA; CoASH, coenzyme A; Ac, acetyl.

Table 2. Enzymes of the diaminopimelate-lysine pathway

Property	*B. subtilis*	*B. licheniformis*	*B. subtilis*	*B. cereus*	*B. sphaericus*	*B. subtilis*
Enzyme	Dihydrodipicolinate synthase	Dihydrodipicolinate synthase	Dihydrodipicolinate reductase	Dihydrodipicolinate reductase	*meso*-Diaminopimelate dehydrogenase	*meso*-Diaminopimelate decarboxylase
Structural gene	*dapA*	?[a]	?	?	?	*lysA*
Mol wt	124,000	≈110,000	74,000	155,000	80,000	≈105,000
Subunit mol wt	32,000	28,000	18,500	?	40,000	48,900
K_m for substrate(s)	3 mM (aspartate semialdehyde), 1 mM (pyruvate)	5.5 mM (aspartate semialdehyde), 2.5 mM (pyruvate)	0.77 mM (dihydrodipicolinate), 0.07 mM (NADPH) [FMN prosthetic group]	0.062 mM (dihydrodipicolinate), 0.008 mM (NADPH) [not a flavoprotein]	0.24 mM (α-amino-ε-ketopimelate), 0.2 mM (NADPH), 12.5 mM (NH_4^+)	1 mM (*meso*-diaminopimelate), 2 μM (pyridoxal phosphate)
Kinetic mechanism	?	?	Ping pong	Ordered sequential	Ordered sequential	?
pH range[b]	8.0–10.5	Optimum at 8.6	5.0–7.5	6.0–9.0	6.5–9.0	7.5–9.0
Feedback inhibitor	None known	None known	Dipicolinate	Dipicolinate	*meso*-Diaminopimelate	L-lysine
Inhibition kinetics[c]			Noncompetitive	Noncompetitive	Uncompetitive	Competitive
K_i (mM) feedback inhibitor			0.3	0.09	1.67 (90% inhibition)	1.5
Other inhibitors or activators	None known	None known	1,10-phenanthroline (K_i = 0.05 mM)	No inhibition by 1,10-phenanthroline	*p*-Chloromercuribenzoate, $HgCl_2$	Requires high ionic strength
Corepressor of enzyme synthesis	None known	None known	?	?	None known	L-Lysine
Change in enzyme level in stationary phase	Increase	Increase	?	?	?	Decline
References	20, 24, 53, 143, 153	45, 132	67, 68, 70	69, 70	6, 86	39, 113, 154

[a] ?, not studied.
[b] Range over which activity is ≥50% of maximum.
[c] With respect to the substrate that is a pathway intermediate.

Neither the addition of amino acids to the culture medium nor starvation for lysine affects the levels of dihydrodipicolinate synthase in vegetative or sporulating cells of *B. cereus*. Dihydrodipicolinate synthase partially purified from sporulating cells of *B. megaterium* is also insensitive to feedback inhibition by L-lysine but has significantly lower apparent K_ms for aspartate semialdehyde and pyruvate than the enzymes of *B. subtilis* and *B. licheniformis* (145). Mutants of *B. megaterium* lacking dihydrodipicolinate synthase are defective in the synthesis of both diaminopimelate and lysine during growth and of dipicolinate during sporulation (37). The fact that both the growth and sporulation defects are corrected in revertants indicates that a single dihydrodipicolinate synthase can fulfill both functions in *B. megaterium*. It should be noted that the isolation of this type of mutant was possible because *B. megaterium*, unlike *B. subtilis*, has a transport system for diaminopimelate, so that the absence of dihydrodipicolinate synthase is not unconditionally lethal (38).

The vegetative cell wall of *B. sphaericus* does not contain diaminopimelate (5), and the sole function of the diaminopimelate-lysine branch of the aspartate pathway in this organism is therefore the synthesis of lysine. Dihydrodipicolinate synthase of *B. sphaericus* differs from the corresponding enzyme of other *Bacillus* species by being subject to feedback inhibition by lysine, inhibition being noncompetitive, with a K_i of 0.6 mM (6). The interesting question of whether a lysine-insensitive isozyme of dihydrodipicolinate synthase is produced in sporulating cells of *B. sphaericus* to meet the dipicolinate and diaminopimelate requirements of spores has not yet been addressed.

Coryneform bacteria. The gene for dihydrodipicolinate synthase of *C. glutamicum*, *dapA*, has been cloned by complementation of *E. coli dapA* strains and found to be closely linked to *dapB*, the gene for dihydrodipicolinate reductase (29, 158). The *dapA* gene encodes a 301-residue polypeptide (9). Dihydrodipicolinate synthase of *C. glutamicum* resembles the *B. subtilis* enzyme in that neither its activity nor its level is affected by lysine or any other amino acid examined (30, 139, 158). Mutants with defective dihydrodipicolinate synthase require high levels of *meso*-diaminopimelate (50 mM) for optimal growth and overproduce threonine (129). The overproduction of threonine by mutants with reduced dihydrodipicolinate synthase activity is consistent with the notion that the competition of dihydrodipicolinate synthase and homoserine dehydrogenase for aspartate semialdehyde regulates the flow of intermediates into the two main branches of the aspartate pathway.

Dihydrodipicolinate reductase

Dihydrodipicolinate reductase has been purified to homogeneity from sporulating cells of *B. subtilis* (67), and its properties are summarized in Table 2. The enzyme is a flavoprotein with flavin mononucleotide (FMN) as prosthetic group and carries out the NADPH-dependent reduction of dihydrodipicolinate to tetrahydrodipicolinate (67, 68). The reaction proceeds by a ping-pong mechanism, with enzyme-bound FMN presumably serving as an intermediate hydrogen carrier, which is consistent with the observed diaphorase activity of the enzyme. Dihydrodipicolinate reductase is most active at pH 6 and is inhibited by low concentrations of the chelators 1,10-phenanthroline and dipicolinate (67, 68). Inhibition by the chelators seems not to involve metal ion binding, for removal of 1,10-phenanthroline by dialysis restores activity. Inhibition occurs at physiological concentrations of dipicolinate and is noncompetitive with respect to dihydrodipicolinate but competitive with respect to NADPH, the saturation curves for dipicolinate as well as those for NADPH in the presence of dipicolinate being slightly sigmoid (68). It has been suggested that the inhibition of dihydrodipicolinate reductase by dipicolinate functions as a regulator in sporulating cells by diverting the common intermediate dihydrodipicolinate from the synthesis of diaminopimelate to dipicolinate as the latter begins to accumulate in the spore (68). Diaminopimelate and lysine do not inhibit the enzyme (67). The molecular weight of dihydropipicolinate reductase from *B. subtilis* is 74,000 in the native state and 18,500 when denatured, which is consistent with a tetrameric structure; on the other hand, each mole of native flavoprotein contains only 2 mol of bound FMN (67).

Dihydrodipicolinate reductase of *B. licheniformis* is quite similar to the *B. subtilis* enzyme (70). However, both these enzymes differ strikingly from the dihydrodipicolinate reductases of *B. cereus* (Table 2) and *B. megaterium*, which are not flavoproteins and catalyze the reduction of dihydrodipicolinate by an ordered sequential mechanism (70). Their pH optima are 7.4, their activities are not affected by 1,10-phenanthroline, and their molecular weights are about 150,000, which is twice that of the enzymes of *B. subtilis* and *B. licheniformis* (69). In spite of these differences, the dihydrodipicolinate reductases of *B. cereus*, *B. megaterium*, *B. subtilis*, and *B. licheniformis* are all inhibited noncompetitively by low concentrations of dipicolinate, which is consistent with the notion that inhibition by dipicolinate serves a regulatory function (69). The relationship between the enzyme purified from *B. cereus* by Kimura and Goto (69) and the dipicolinate-insensitive dihydrodipicolinate reductase described by Forman and Aronson (35), whose level increases 10-fold during sporulation, is not clear. Dihydrodipicolinate reductase of *B. sphaericus* is similar to the enzymes of *B. cereus* and *B. megaterium* in having a pH optimum above 7 and being more sensitive to inhibition by dipicolinate than by 1,10-phenanthroline; its level is modestly repressed in culture media supplemented with L-lysine (6). In *B. brevis*, the level of dihydrodipicolinate reductase is not affected by amino acid supplementation but declines by 90% toward the end of exponential growth and then increases fourfold during sporulation; the activity of the enzyme is not affected by lysine or diaminopimelate, but 1 mM dipicolinate inhibits it by 65% (108).

The gene for dihydrodipicolinate reductase of *C. glutamicum* has been cloned by complementation of *E. coli dapB* strains (29, 158). Neither the activity nor the level of dihydrodipicolinate reductase from *C. glutamicum* is affected by lysine or any other amino acid examined, but the effect of dipicolinate has not been investigated (30).

Tetrahydrodipicolinate acetyltransferase, *N*-acetyl-LL-diaminopimelate aminotransferase, *N*-acetyl-LL-diaminopimelate deacylase, and diaminopimelate epimerase

The enzymes that catalyze the four terminal reactions in the biosynthesis of *meso*-diaminopimelate by the epimerase pathway have not been studied in *B. subtilis*. As mentioned in the introduction to this chapter, these reactions differ from the corresponding ones in many other bacterial species by involving acetylated rather than succinylated intermediates; moreover, in some gram-positive bacteria, the four terminal steps of the epimerase pathway are bypassed by a single reaction catalyzed by diaminopimelate dehydrogenase.

The only known mutations affecting the diaminopimelate branch of the aspartate pathway of *B. subtilis* are in *dapE*, the gene encoding *N*-acetyl-LL-diaminopimelate deacylase. The mutants, which were isolated as temperature-sensitive lysis strains (13, 16), accumulate *N*-acetyl-LL-diaminopimelate and have much reduced levels of the deacylase when extracts of cells grown at a permissive temperature are assayed at 37°C (16, 115). When shifted to the nonpermissive temperature, *dapE* mutants are unable to grow in lysine-free media and undergo rapid lysis in the presence of lysine, an indication that the product of the *dapE* locus is essential for the biosynthesis of both *meso*-diaminopimelate and L-lysine (115).

Unlike *B. subtilis*, *B. megaterium* has a diaminopimelate transport system, and mutants with a defect in *meso*-diaminopimelate synthesis can therefore be isolated as diaminopimelate auxotrophs. Nevertheless, no *B. megaterium* mutants affected in diaminopimelate synthesis other than strains lacking *N*-acetyl-LL-diaminopimelate deacylase have been obtained (116, 135), which is consistent with the notion that many steps in the diaminopimelate pathway are catalyzed by multiple enzymes.

Diaminopimelate epimerase, the enzyme catalyzing the final step in the biosynthesis of *meso*-diaminopimelate, has been partially purified from extracts of *B. megaterium* (148). Its activity depends on the addition of a thiol compound, but there is no evidence for a pyridoxal phosphate requirement. The equilibrium constant for the epimerization of LL-diaminopimelate to *meso*-diaminopimelate is 2, and the much higher K_m of diaminopimelate epimerase for the *meso*-isomer than for LL-diaminopimelate (100 versus 6.7 mM) may serve to minimize product inhibition.

The enzymes catalyzing the epimerase pathway have been assayed in extracts of *C. glutamicum* and found to prefer succinylated to acetylated substrates; moreover, *N*-succinyldiaminopimelate was detected in the cytoplasm of a lysine-excreting strain of *C. glutamicum* (121). These observations suggest that, like most bacteria other than *Bacillus* spp., the coryneform bacteria use succinylated intermediates for lysine biosynthesis. It should be noted, however, that a crude enzyme preparation from *Brevibacterium lactofermentum* has been reported to be about twice as active in the synthesis of *N*-acyl-ε-keto-α-aminopimelate with acetyl coenzyme A (acetyl-CoA) as substrate as it is with succinyl-CoA (139).

meso-Diaminopimelate dehydrogenase

The enzyme *meso*-diaminopimelate dehydrogenase, which was first described in *B. sphaericus* (87) and subsequently in other bacteria (88), catalyzes the reductive amination of tetrahydrodipicolinate or its noncyclic isomer L-2-amino-6-ketopimelate to *meso*-diaminopimelate, thereby providing a one-step alternative to the four-step reaction sequence for the biosynthesis of *meso*-diaminopimelate discussed in the preceding section. The study of lysine auxotrophs of *B. sphaericus* that lack diaminopimelate dehydrogenase and of their revertants established that this enzyme plays an essential role in diaminopimelate synthesis in this organism (147). The enzyme has been purified to homogeneity from extracts of *B. sphaericus* (86), and its properties are summarized in Table 2. The activity of *meso*-diaminopimelate dehydrogenase is not affected by L-lysine, but *meso*-diaminopimelate inhibits the reductive amination of L-2-amino-6-ketopimelate in a manner consistent with product inhibition of a reversible, ordered ter-bi mechanism (86). Even if not allosteric, the inhibition by *meso*-diaminopimelate (90% inhibition at 1.67 mM) is in a physiologically significant range (6). The native enzyme has a molecular weight of 80,000 and is composed of 40,000-Da subunits that bind 1 mol of NADPH each (86).

In spite of the fact that the vegetative peptidoglycan of *B. sphaericus* lacks diaminopimelate (5), so that the only function of *meso*-diaminopimelate dehydrogenase in *B. sphaericus* is the synthesis of lysine, the level of the enzyme is not repressed by growth in the presence of lysine (6). Other *Bacillus* species with lysine-containing peptidoglycans, including *B. globisporus* and *B. pasteurii*, also use the dehydrogenase pathway for lysine biosynthesis (5). A curious anomaly is *B. macerans*, whose peptidoglycan contains *meso*-diaminopimelate and which has all the enzymes for the synthesis of diaminopimelate by the epimerase pathway but also significant levels of *meso*-diaminopimelate dehydrogenase (5).

Although *C. glutamicum* has a full complement of the enzymes for diaminopimelate synthesis by the epimerase pathway using succinylated intermediates, it also has a high level of diaminopimelate dehydrogenase (121). Studies involving strains with a disrupted diaminopimelate dehydrogenase gene show that both the epimerase and the dehydrogenase pathways contribute to lysine biosynthesis (121). The level of diaminopimelate dehydrogenase is not affected by the presence of amino acids in the growth medium, and lysine has no effect on the activity of the enzyme (30). The gene for diaminopimelate dehydrogenase (*ddh* or *dapY*) has been cloned by its ability to restore lysine synthesis to an *E. coli* mutant (*dapD*) with a defective epimerase pathway (58, 158). The effective complementation of an *E. coli dapD* defect by *C. glutamicum ddh* shows that the corynebacterial gene can be functionally expressed in an organism that ordinarily uses a different pathway for lysine biosynthesis. The nucleotide sequence of the *ddh* (*dapY*) gene has been determined and found to encode a 320-residue polypeptide (57).

meso-Diaminopimelate decarboxylase

The final step in the biosynthesis of lysine is the pyridoxal phosphate-dependent decarboxylation of *meso*-diaminopimelate, which is catalyzed by diaminopimelate decarboxylase. It is the only step in the diaminopimelate-lysine branch of the aspartate pathway of *B. subtilis* that does not have a dual function, and diaminopimelate decarboxylase is the only enzyme in this reaction sequence that is subject to feedback regulation in *B. subtilis*. Its level is reduced about 50% by growth in the presence of lysine (39, 113) and more than 90% in a rich medium (113). The effect of growth conditions on the level of diaminopimelate decarboxylase has also been studied in other *Bacillus* species. In *B. cereus*, the enzyme is repressed by growth in the presence of lysine, its level declines near the end of the exponential growth phase, and no activity can be detected in sporulating cells at the time of dipicolinate synthesis (40). In *B. brevis*, the level of diaminopimelate decarboxylase is not affected by the addition of lysine to growth media and is reduced only 50% by growth in rich media but declines during sporulation (108). Diaminopimelate decarboxylase from *B. sphaericus*, which uses the dehydrogenase pathway for lysine biosynthesis and has a lysine-sensitive dihydrodipicolinate synthase, is neither repressed nor inhibited by L-lysine (6).

The properties of *meso*-diaminopimelate decarboxylase have been studied in crude extracts from *B. subtilis* (113) and are summarized in Table 2. The enzyme is most active at high pH and high ionic strength and is inhibited by L-lysine in a manner competitive with respect to *meso*-diaminopimelate and noncompetitive with respect to pyridoxal phosphate. The inhibition of diaminopimelate decarboxylase by L-lysine occurs at physiological concentrations of this amino acid and may well play a significant role in regulating the conversion of *meso*-diaminopimelate to lysine. An aminoethylcysteine-resistant mutant of *B. subtilis* with a lysine-insensitive diaminopimelate decarboxylase has been described (123). The molecular weight of diaminopimelate decarboxylase has been estimated at 105,000 by gel filtration (113), but the enzyme has not been further purified and characterized. The gene for diaminopimelate decarboxylase, *lysA*, has been cloned (59, 155) and sequenced (154). It encodes a 440-residue polypeptide (M_r = 48,900) whose residues are 30% identical to those of the deduced product of the *lysA* gene of *E. coli* (154).

The gene for diaminopimelate decarboxylase from *C. glutamicum* encodes a 445-residue polypeptide that has considerable sequence similarity (38% identity) with the deduced product of the *B. subtilis lysA* gene and also with diaminopimelate decarboxylase of *E. coli* (158, 159).

Dipicolinate synthase

2,6-Dipicolinate is produced by only a few bacterial genera, and its production is confined to a short period during bacterial sporulation. It is not essential for the structure or metabolism of growing cells, but its absence results in heat-sensitive spores. The synthesis of dipicolinate from dihydrodipicolinate, an intermediate in lysine biosynthesis, occurs in a single step and is catalyzed by the enzyme dipicolinate synthase. The nonenzymatic conversion of dihydrodipicolinate to dipicolinate has also been described (19, 66, 71), but its physiological significance is uncertain.

The synthesis of dipicolinic acid was first demonstrated in extracts of sporulating cells of *B. megaterium* (3) and subsequently of *B. subtilis* (19, 20). The properties of dipicolinate synthase have not been studied in detail because this enzyme is extremely unstable, and attempts to stabilize it sufficiently for purification have been unsuccessful (19, 20). When detergent-treated cells were used to minimize instability, the levels of dipicolinate synthase were negligible in vegetative cells of *B. subtilis* but increased rapidly about 4 h after the onset of sporulation in parallel with a twofold rise in dihydrodipicolinate synthase levels and just prior to the appearance of dipicolinate (19, 20). Mutants with a defect in a locus originally termed *dpa* (4) but now often referred to as *spoVF* (107) produce heat-sensitive spores in the absence of dipicolinate but have normal growth rates in synthetic media, a phenotype consistent with a defective dipicolinate synthase (4).

The Branch Leading to Threonine

The other major branch of the aspartate pathway leads to the conversion of aspartate semialdehyde to threonine and methionine (Fig. 6). A key intermediate in this branch is L-homoserine, which serves as common precursor for the biosynthesis of threonine and methionine. Whereas in *E. coli* both threonine and methionine participate in the control of homoserine formation from aspartate semialdehyde, in *B. subtilis*, this reaction is controlled by threonine alone, and methionine participates only in the regulation of the reactions that convert homoserine to methionine and *S*-adenosylmethionine.

Homoserine dehydrogenase

B. subtilis. The branch point enzyme homoserine dehydrogenase is the counterpart of dihydrodipicolinate synthase in controlling the utilization of aspartate semialdehyde for the biosynthesis of threonine and methionine by catalyzing the NADPH-dependent reduction of L-aspartate semialdehyde to L-homoserine. In contrast to *E. coli* K-12, which has two isozymes of a bifunctional aspartokinase-homoserine dehydrogenase (27), *B. subtilis* and other gram-positive bacteria each have only a single homoserine dehydrogenase, which is devoid of aspartokinase activity. The homoserine dehydrogenase of *B. subtilis* has not been purified, and the few studies of its properties, summarized in Table 3, must be viewed with reservations on account of the unreliability of NADPH-linked enzyme assays in crude bacterial extracts. The enzyme has been described as sensitive (142) or resistant (156) to threonine inhibition; the most reliable results are probably those obtained with homoserine dehydrogenase expressed from the cloned *B. subtilis hom* gene in *E. coli*, which shows 50% inhibition by 0.5 mM L-threonine (100). The level of homoserine dehydrogenase is relatively constant throughout exponential growth but is reduced to about one-half by supplementation of minimal media with L-methionine, L-threonine, L-isoleucine, or com-

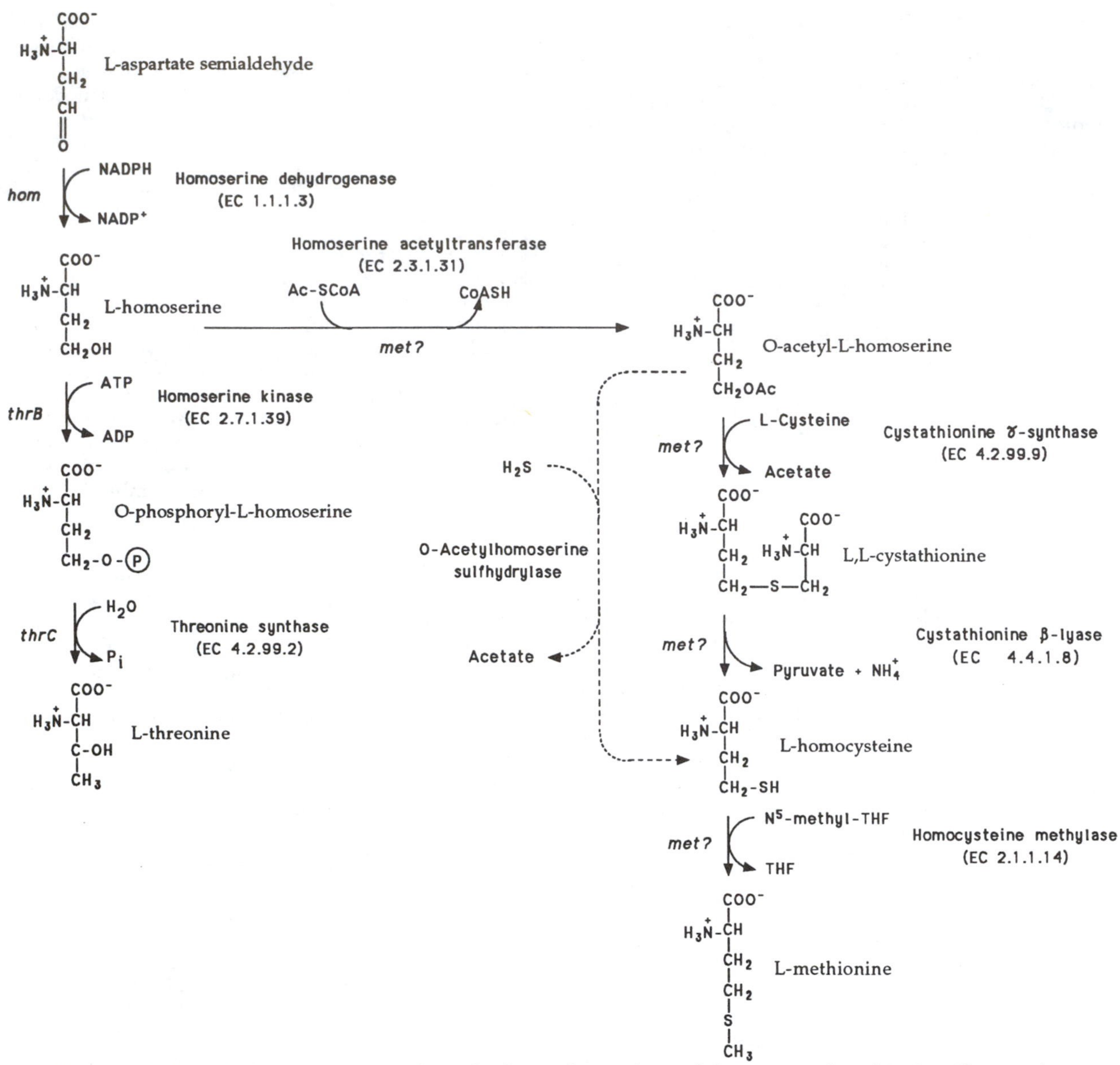

Figure 6. The branch of the aspartate pathway leading to biosynthesis of threonine and methionine. The reaction catalyzed by *O*-acetylhomoserine sulfhydrylase, indicated by dotted arrows, occurs in coryneform bacteria. CoASH, coenzyme A; Ac-SCoA, acetyl-CoA; Ac, acetyl; THF, tetrahydrofolate.

binations thereof and declines at the onset of sporulation (156). The *hom* gene encodes a 433-residue polypeptide of M_r 47,400 (100). The deduced amino acid sequence of the *B. subtilis* homoserine dehydrogenase has significant similarity (25% identical residues) to that of the C-terminal homoserine domain of the bifunctional aspartokinase-homoserine dehydrogenases I and II of *E. coli* except for a 108-residue C-terminal extension, which has no counterpart in the *E. coli* enzymes and may play a role in feedback inhibition.

Coryneform bacteria. The homoserine dehydrogenase of coryneform bacteria has been studied in some detail (Table 3). The enzyme has been partially purified from *B. flavum*, and its properties have been characterized (89). The reduction of aspartate semialdehyde occurs by an ordered bi-bi mechanism, the binding of substrates being in the order NADPH and aspartate semialdehyde. The pathway end product, threonine, is a potent inhibitor. Inhibition is of the mixed type, with sigmoid inhibitor saturation curves, and declines significantly with increasing pH, with little inhibition above pH 8.0. Isoleucine is a less-effective inhibitor, and its action is competitive with respect to aspartate semialdehyde and less sensitive to pH than is inhibition by L-threonine. The threonine analog α-amino-β-hydroxyvalerate is about 20 times less effective than threonine as an inhibitor (126). Homoserine dehydrogenases of α-amino-β-hydroxyvalerate-resistant mutants of *B. flavum* are much less sensitive to inhibition by threonine and isoleucine (126). Threonine stabilizes homoserine dehydrogenase at low ionic strength, where the enzyme is ordinarily

Table 3. Enzymes of the threonine-methionine pathway

Property	*B. subtilis*	*B. flavum*	*B. subtilis*	*B. flavum*	*B. subtilis*	*C. glutamicum*	*B. polymyxa*	*B. flavum*
Enzyme	Homoserine dehydrogenase	Homoserine dehydrogenase	Homoserine kinase	Homoserine kinase	Threonine synthase	Threonine synthase	Homoserine acetyltransferase	Homoserine acetyltransferase
Structural gene	*hom*	*hom*	*thrB*	*thrB*	*thrC*	*thrC*	?[a]	?
Mol wt	?	≈250,00	?	55,000	90,000 and 45,000	?	40,000	?
Subunit mol wt	47,400	46,100[b]	33,500	32,600[b]	47,100	54,500	40,000	?
K_m (mM) for substrates	?	0.2 (aspartate semialdehyde), 0.04 (NADPH)	?	0.8 (L-homoserine), 1.2 (ATP)	0.025 (homoserine-phosphate)	?	0.8 (L-homoserine), 0.8 (acetyl-CoA)	2.8 (L-homoserine), 0.05 (acetyl-CoA)
pH optimum	?	6.0	?	7.4	?	?	7.8	8.0
Feedback inhibitors	L-Threonine	L-Threonine, L-isoleucine	L-Threonine	L-Threonine	None known	?	L-Methionine and *S*-adenosyl-L-methionine[c]	*S*-Adenosyl-L-methionine, L-methionine
Inhibition kinetics[d]	Competitive	Mixed (threonine), competitive (isoleucine)	?	Competitive		?	Competitive	Noncompetitive
K_i for feedback inhibitor	0.5 mM	0.06 mM (threonine), 5 mM (isoleucine)	≈10 mM	8 mM		?	0.35 mM (methionine at 2 μM *S*-adenosylmethionine), 6 μM (*S*-adenosylmethionine at 0.1 mM methionine)	0.25 mM (*S*-adenosylmethionine),[e] 5 mM (methionine)[e]
Other inhibitors or activators	None known	None known	None known	Twofold activation by K^+ or NH_4^+	P_i (K_i = 0.5 mM), SO_4^{2-} (K_i = 1 mM), cysteine (K_i = 10 mM)	?	Zn^{2+} (K_i = 3 μM)	*p*-Hydroxymercuribenzoate, cysteine, and other thiols
Corepressor(s) of enzyme synthesis	Methionine, threonine, isoleucine	Methionine	None known	None known	None known	?	Methionine	Methionine
References	99, 131, 142, 156	89, 104, 126	98, 131, 142	91, 104	98, 118, 131, 142	47	149, 150	92, 128

[a] ?, Not studied.
[b] Homoserine kinase of *C. glutamicum*.
[c] The two inhibitors act synergistically.
[d] With respect to the substrate that is a pathway intermediate.
[e] Most-sensitive enzyme preparations.

quite unstable (89). Homoserine dehydrogenase from threonine-insensitive mutants is much more stable under these conditions, and its stability is not affected by threonine (126). Gel filtration measurements suggest molecular weights of about 250,000 for both the parent and the mutant enzymes (126).

Whereas the activity of homoserine dehydrogenase of *B. flavum* is controlled primarily by threonine, its level is controlled by methionine, being reduced about fourfold when minimal medium is supplemented with L-methionine (90, 126). It is interesting that homoserine dehydrogenase and dihydrodipicolinate synthase of *B. flavum* have similar K_ms for aspartate semialdehyde (0.2 mM) but that the level of the former is about 15 times higher during growth in minimal medium, suggesting that in the absence of feedback inhibition and repression, considerably more aspartate semialdehyde will be utilized for the synthesis of threonine and methionine than for the synthesis of diaminopimelate and lysine (89).

The genes for homoserine dehydrogenase and the other threonine biosynthetic enzymes of the related species *C. glutamicum* have been cloned and sequenced (34, 47, 104). The *hom* gene encodes a 445-residue polypeptide (M_r = 46,100) that has a high degree of similarity (31% identity) to the deduced amino acid sequence of *B. subtilis* homoserine dehydrogenase, including the 100-residue C-terminal segment that is absent from the multifunctional *E. coli* aspartokinase-homoserine dehydrogenases (104). The gene for a threonine-insensitive homoserine dehydrogenase from an α-amino-β-hydroxyvalerate-resistant mutant of *C. glutamicum* differs in its C-terminal portion from the parent enzyme by a change in the codon for glycine 378 to encode glutamate (109). The role of glycine 378 in feedback inhibition was confirmed by site-directed mutagenesis (109). Since the homoserine dehydrogenase domains of the bifunctional *E. coli* aspartokinase-homoserine dehydrogenases do not contain feedback control sites (27), these observations suggest that the 100-residue C-terminal domain common to the homoserine dehydrogenases of *B. subtilis* and *C. glutamicum* but absent from the *E. coli* enzymes may function primarily in feedback inhibition. It has been suggested that a region of homoserine dehydrogenase not far from glycine 378, which shows significant sequence similarity to the corresponding region of *B. subtilis* homoserine dehydrogenase as well as to segments of several different threonine dehydratases and serine/threonine-specific protein kinases, may represent the binding site for the feedback inhibitor threonine (109).

Homoserine kinase and threonine synthase

The enzymes catalyzing the two steps leading from homoserine to threonine in *B. subtilis* have not been studied in a systematic manner. Homoserine kinase, which catalyzes the first of these reactions, the ATP-dependent phosphorylation of homoserine to homoserine phosphate, is weakly inhibited by threonine when assayed at saturating concentrations of L-homoserine (Table 3) (142). Although feedback control by threonine can be rationalized because homoserine kinase catalyzes the first reaction specifically devoted to threonine biosynthesis and is also observed with the corresponding enzyme from *E. coli* (27), this effect needs to be reexamined at physiological inhibitor and substrate concentrations before its regulatory significance can be evaluated.

Threonine synthase of *B. subtilis* (Table 3), which catalyzes the pyridoxal phosphate-dependent conversion of homoserine phosphate to L-threonine, is competitively inhibited by low concentrations of sulfate and phosphate and also by cysteine (118, 142). Gel filtration resolves the enzyme into components with molecular weights of 90,000 and 45,000, with the larger species predominating in the presence of 0.2 M KCl (118). An interesting aspect of threonine synthase is that it can also catalyze, albeit at a much lower rate, the mechanistically similar pyridoxal phosphate-dependent deamination of homoserine phosphate or threonine to α-ketobutyrate, the first intermediate in the conversion of threonine to isoleucine, whose formation is normally catalyzed by the enzyme threonine dehydratase (118, 119, 131). The conversion of homoserine phosphate to threonine and to α-ketobutyrate proceeds with the same K_m, which is 1,000 times less than the K_m for the threonine synthase-catalyzed conversion of threonine to α-ketobutyrate (119, 131). Pseudorevertants of isoleucine-requiring strains with defective threonine dehydratases, in which either threonine or homoserine can suppress the isoleucine requirement, have been isolated, suggesting that threonine synthase can also catalyze the formation of α-ketobutyrate in vivo (131, 141, 142). One class of revertants involves an unlinked mutation, *sprA*, that causes derepression of the threonine operon, with the elevated level of threonine synthase compensating for its low threonine dehydratase activity; another class, *sprB*, is due to a mutation in the gene for threonine synthase which enhances its threonine dehydratase activity. The threonine dehydratase activity of threonine synthase differs from that of threonine dehydratase itself in being absolutely dependent on the presence of P_i, being sensitive to inhibition by low concentrations of sulfate, having a much higher K_m for L-threonine, and lacking feedback inhibition by isoleucine (131). It has been suggested that the ability of threonine synthase to convert homoserine phosphate or threonine to α-ketobutyrate may have a selective advantage by preventing the accumulation of high concentrations of threonine that would otherwise throttle methionine biosynthesis by completely inhibiting homoserine dehydrogenase (131).

The genes for homoserine kinase and threonine synthase, *thrB* and *thrC*, respectively, have been cloned and sequenced (99). The *thrB* gene encodes a 298-residue polypeptide (M_r = 33,500) with extensive sequence similarity (26% identity) to *E. coli* homoserine kinase. The deduced product of *thrC* is 351 residues long (M_r = 47,100), and its alignment with *E. coli* threonine synthase shows that the *B. subtilis thrC* gene product lacks the 48 N-terminal residues of the larger *E. coli* polypeptide, with 34% of the remaining residues being identical. There is also significant sequence similarity between the threonine synthases of *B. subtilis* and *E. coli* and the threonine dehydratase of *Saccharomyces cerevisiae* (99), which is of interest with respect to the ability of the enzyme from *B.*

subtilis to catalyze the threonine dehydratase reaction (131).

Homoserine kinase from the coryneform bacterium *B. flavum* has been partially purified and characterized (91). The substrates L-homoserine and ATP bind to the enzyme in random order, and L-threonine is a weak competitive inhibitor with respect to L-homoserine (Table 3). The molecular weight of homoserine kinase of *B. flavum*, as estimated by gel filtration, is 55,000. In *C. glutamicum*, the gene for homoserine kinase (*thrB*) encodes a 319-residue polypeptide with an M_r of 32,600 that has high degrees of sequence similarity (31 and 25% identity, respectively) to the corresponding gene products of *B. subtilis* and *E. coli* (104). The gene for threonine synthase (*thrC*) contains the coding region for a 489-residue polypeptide with an M_r of 54,500 (47). The size of the putative translation product is closer to that of threonine synthase of *E. coli* (428 residues) than to that of the *B. subtilis* enzyme (351 residues); indeed, threonine synthase of *C. flavum* has a greater degree of sequence similarity to the *E. coli* than to the *B. subtilis* enzyme (22 versus 15% identity) and has in common with the *E. coli* enzyme the approximately 50-residue N-terminal domain that is absent in *B. subtilis* homoserine kinase.

The Branch Leading to Methionine

The reactions leading to the biosynthesis of methionine from homoserine are shown in Fig. 6. The acetylation of homoserine by homoserine *O*-acetyltransferase sets the stage for subsequent nucleophilic displacement of the acetyl group by cysteine catalyzed by cystathionine γ-synthase. LL-Cystathionine, the first sulfur-containing intermediate, is converted to L-homocysteine by cystathionine β-lyase-catalyzed elimination of pyruvate and ammonia followed by methyl transfer from N^5-methyltetrahydrofolate, which is catalyzed by homocysteine methylase, to yield L-methionine. In coryneform bacteria, the reactions catalyzed by cystathionine γ-synthase and cystathionine β-lyase are bypassed by the *O*-acetylhomoserine sulfhydrylase-catalyzed direct conversion of *O*-acetylhomoserine to homocysteine (Fig. 6). Methionine has two roles: as a precursor for protein synthesis and as the source of the methyl donor *S*-adenosylmethionine; its dual function is reflected in the participation of *S*-adenosylmethionine in the control of methionine biosynthesis.

Homoserine acetyltransferase

Whereas methionine biosynthesis in *E. coli* involves *O*-succinylhomoserine (27), the acylation of homoserine in *B. subtilis* specifically requires acetyl-CoA (14). Nevertheless, homoserine *O*-acetyltransferase of *B. subtilis* resembles the corresponding *O*-succinyltransferase of *E. coli* (79) in its feedback control pattern, being synergistically inhibited by methionine and *S*-adenosylmethionine (14). Homoserine acetyltransferase of *B. subtilis* has not been studied in detail, but an enzyme from *B. polymyxa* with similar catalytic and regulatory properties (summarized in Table 3) has been purified to homogeneity and extensively characterized (149). Since homoserine acetyltransferase is repressed by growth in the presence of methionine (150), enzyme purification made use of a methionine auxotroph of *B. polymyxa* grown with limiting methionine. The reaction catalyzed by homoserine acetyltransferase proceeds by a ping-pong mechanism, as indicated by the kinetic pattern and the ability to catalyze a homoserine-acetylhomoserine exchange reaction in the absence of acetyl-CoA. L-Methionine and *S*-adenosyl-L-methionine inhibit competitively with respect to both substrates and act in a highly synergistic manner. Ethionine inhibits the enzyme about one-third as effectively as methionine. Substrate and inhibitor saturation curves are strictly hyperbolic. Homoserine acetyltransferase is very sensitive to inhibition by Zn^{2+}. No other divalent metal ions tested affect the activity of the enzyme, and inhibition by Zn^{2+} can be completely reversed by dilution. The molecular weight of purified homoserine acetyltransferase of *B. polymyxa*, both in the native state and after denaturation with sodium dodecyl sulfate, is 40,000, indicating that the enzyme is a monomer. The monomeric nature of the enzyme explains the absence of cooperativity in the substrate and the inhibitor kinetics noted above. The question of whether homoserine acetyltransferase also functions as a monomer within the cell was addressed by assaying the enzyme in situ, using cells permeabilized to small molecules by treatment with toluene. Under such conditions, the enzyme exhibits a sigmoid response to inhibitor concentration, with Hill coefficients as high as 2.2, suggesting that at intracellular protein concentrations, homoserine acetyltransferase functions as a dimer or higher oligomer (150).

A striking property of homoserine acetyltransferase of *B. polymyxa* is its extreme instability, the purified enzyme having a half-life of 15 min at 0°C (149). Although the in vitro stability of the enzyme is considerably enhanced by the presence of 20% ethylene glycol and 1 mM EDTA, a circumstance that allowed purification to homogeneity, albeit in low yield (149), the enzyme is also quite unstable in vivo, with half-lives of 40 and 10 min at 37°C in intact and permeabilized cells, respectively (150). The instability of homoserine acetyltransferase is highly temperature dependent both in vitro and in vivo and seems to be the major factor limiting the growth of *B. polymyxa* in minimal media at high temperatures. This limitation manifests itself particularly upon nutritional shift-down at 39°C, which is followed by extraordinarily long lag periods when methionine is absent, probably because derepression of homoserine acetyltransferase is thwarted by rapid thermal inactivation of the enzyme (150). It is interesting that the thermal stability of homoserine transsuccinylase, the enzyme that catalyzes the first step of methionine biosynthesis in *E. coli*, also limits the growth of that organism in minimal media at elevated temperatures (111, 112).

Coryneform bacteria also use acetylated intermediates in methionine biosynthesis, and homoserine acetyltransferase of *B. flavum* has been partially purified and characterized (92, 128). Homoserine acetyltransferase of *B. flavum* resembles the corresponding enzyme of *Bacillus* species in catalyzing the formation of *O*-acetylhomoserine by a ping-pong mechanism and being repressed by methionine (92) but differs in its pattern of feedback inhibition (Table 3). Initial studies (92) failed to show any kind of feedback inhibition, but it was subsequently found that the enzyme had irreversibly lost sensitivity to allosteric control during extraction and storage (128). Homoserine acetyltransferase activities in extracts prepared in the presence of thiols and stored in buffers containing glycerol are highly sensitive to inhibition by *S*-adenosylmethionine. Methionine is a much weaker inhibitor and shows no synergistic effect with *S*-adenosylmethionine; inhibition by methionine is probably not of physiological significance. Inhibition by *S*-adenosylmethionine is noncompetitive with respect to both substrates, and *S*-adenosylhomocysteine is nearly as effective an inhibitor. The enzyme is also inhibited by high concentrations of thiols, including cysteine and homocysteine, and by thiol reagents, such as *p*-hydroxymercuribenzoate. At subinhibitory concentrations, both types of compounds modulate feedback inhibition, with thiols preventing the loss of sensitivity to feedback inhibition and *p*-hydroxymercuribenzoate promoting such loss. The properties of homoserine acetyltransferase of *C. glutamicum* resemble those reported for the enzyme from *B. flavum*, but the question of whether the enzyme preparations studied had been desensitized to inhibition by *S*-adenosylmethionine has not been addressed (64). Ethionine- and selenomethionine-resistant mutants of *C. glutamicum* with elevated levels of homoserine acetyltransferase have been isolated and found to excrete *O*-acetylhomoserine into the culture medium (64). There is no evidence for the formation of *O*-succinylhomoserine in *B. flavum* or *C. glutamicum* (64, 92).

Later steps in methionine biosynthesis

The steps leading from *O*-acetylhomoserine to methionine in *B. subtilis* have not been studied. Nevertheless, the fact that certain methionine-requiring mutants of *B. subtilis* can also grow when supplemented with homocysteine or cystathionine (116) suggests that the biosynthetic reaction sequence involves cystathionine as an intermediate. On the other hand, no methionine auxotrophs of *B. flavum* capable of growth on cystathionine have been found, although methionine-requiring mutants that can grow on *O*-acetylhomoserine, homocysteine, or methionine have been isolated (98). The absence of *O*-acetylhomoserine sulfhydrylase in extracts of methionine auxotrophs that are able to grow on homocysteine but not on *O*-acetylhomoserine and the presence of wild-type levels of this enzyme in prototrophic revertants suggest that the biosynthesis of methionine in *B. flavum* involves the direct formation of homocysteine from acetylhomoserine and H_2S without cystathionine as intermediate (98). Some of the properties of partially purified preparations of acetylhomoserine sulfhydrylase of *B. flavum* have been characterized (98). The pyridoxal phosphate-dependent enzyme has a molecular weight of about 360,000, is most active at high pH, and is highly specific for *O*-acetylhomoserine (K_m = 2 mM), being completely inactive with *O*-succinylhomoserine, and its activity is not significantly affected by physiological concentrations of any of the intermediates or end products of the methionine biosynthetic pathway. Low levels of β-cystathionase as well as of cystathionine-δ-synthase, which is active with either *O*-succinylhomoserine or *O*-acetylhomoserine (62), are found in extracts of coryneform bacteria, but these enzymes are not affected in mutants blocked between O-acetylhomoserine and homocysteine, suggesting that they do not play a critical role in methionine biosynthesis (98).

Gene Organization and Regulation

meso-Diaminopimelate biosynthesis

The structural gene for the diaminopimelate-sensitive aspartokinase I, *dapG*, was found by Roten et al. (115) to correspond to *lssD*, the locus for a class of mutations leading to a temperature-sensitive lysis phenotype (13). *dapG* maps on the *B. subtilis* chromosome at 144°, which is close to *spoVF*, the putative structural gene for dipicolinate synthase (4, 107). The genes for aspartokinase I (*dapG*) and dihydrodipicolinate synthase (*dapA*) are adjacent to the gene for aspartate semialdehyde dehydrogenase (*asd*) and were cloned by complementation of an *asd* deletion strain of *E. coli* (24). The genes were tentatively identified by comparison of the deduced amino acid sequences of the gene products with those of known proteins (24). The putative *asd* gene product (346 amino acids) has 57% identity with aspartate semialdehyde dehydrogenase of *S. mutans* (357 amino acids) (17), the putative *dapG* gene product (404 amino acids) has 36% of its amino acids identical to those of aspartokinase II of *B. subtilis* (408 amino acids) (23), and the putative *dapA* gene product (292 amino acids) and dihydrodipicolinate synthase of *E. coli* (293 amino acids) (110) have 44% of their amino acids identical. The *dapG* gene was also identified by disruption, which causes the loss of aspartokinase I and a phenotype identical to that reported (115) for temperature-sensitive *dapG* mutants at nonpermissive temperatures (24). As shown in Fig. 7, the genes are arranged in the order *asd-dapG-dapA* and probably constitute an operon. It is not certain at this time whether *orfY* and *orfX*, unidentified open reading frames upstream of *asd* that encode 297- and 200-residue polypeptides, respectively, are functionally related to the *asd-dapG-dapA* cluster. However, it is of interest that disruption of *orfX* leads to an oligosporogenous phenotype with heat-labile spores that can be partially corrected by growth in the presence of dipicolinate (24) and is similar to the phenotype of the *dpa* (*spoVF*) mutants described by Balassa et al. (4), which may have defective dipicolinate synthases. Whether *orfY* and *orfX* correspond to *spoVF*, which is known to be closely linked to *dapG* (115), and constitute the structural genes for subunits of dipicolinate synthase remains to be determined.

There are no obvious transcription initiation or termination sites between the coding regions of the *asd-dapG-dapA* cluster. The 91-bp *asd-dapG* intercistronic region is exceptionally long for *B. subtilis*, where adjacent cistrons usually overlap, and contains an extensive inverted repeat, which on transcription could fold into either of two stable conformations (illustrated in Fig. 8). In one of these structures, the ribosome-binding site for *dapG* is sequestered in a hairpin structure, suggesting that the intercistronic spacer may function in translational attenuation of *dapG* expression. Such down-regulation of *dapG* may

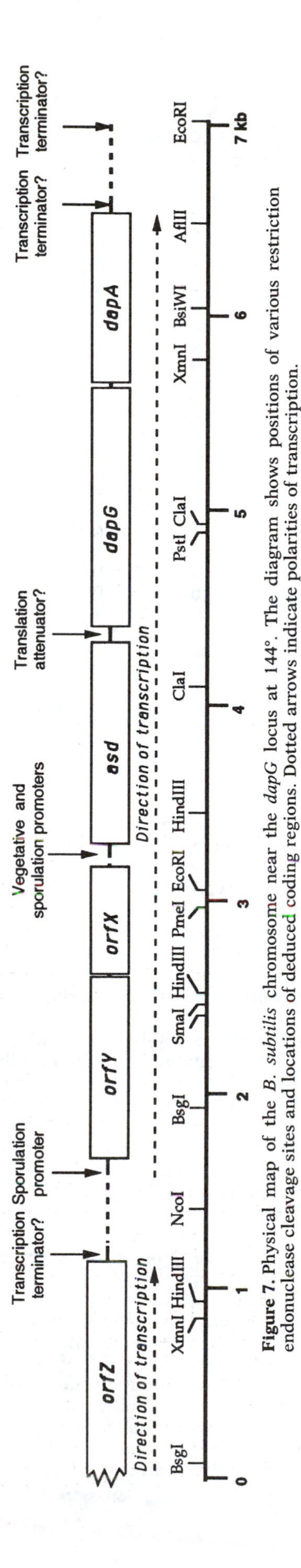

Figure 7. Physical map of the *B. subtilis* chromosome near the *dapG* locus at 144°. The diagram shows positions of various restriction endonuclease cleavage sites and locations of deduced coding regions. Dotted arrows indicate polarities of transcription.

be advantageous when aspartate semialdehyde dehydrogenase is required in larger amounts than aspartokinase I, for example, during growth in minimal medium when aspartate semialdehyde dehydrogenase needs to service all three aspartokinase isozymes, providing precursors not only for diaminopimelate but also for the biosynthesis of lysine, methionine, threonine, and isoleucine. The observation that plasmids containing the entire *B. subtilis asd* gene but only portions of the open reading frame upstream of *asd* can complement an *asd* deletion in *E. coli* is consistent with the presence of a promoter recognizable by *E. coli* RNA polymerase in the untranslated 125-bp region just upstream of *asd* (24). Evidence that the same promoter functions in growing cells of *B. subtilis* was provided by transcript mapping through primer extension (24). At the onset of sporulation, the use of this promoter diminishes progressively, and 5 h after initiation of sporulation, all transcripts start at two new sites, of which one is upstream of *orfY* and the other is just downstream from the vegetative transcription start site between *orfX* and *asd* (24). It is not yet clear to what extent the transcription of *orfY* and *orfX*, which probably corresponds to *spoVF*, is coupled to that of the *asd-dapG-dapA* cluster.

The gene for aspartate semialdehyde dehydrogenase adjacent to *dapG* at 144°, which was identified in the studies described above (24), is probably identical to the locus Ward described (144), whose mutation gives rise to stable L forms and which seems to define the only gene for aspartate semialdehyde dehydrogenase (149). The only other gene involved in diaminopimelate synthesis that has been mapped is *dapE*, which encodes *N*-acetyl-LL-diaminopimelate deacylase and is located at 125° on the *B. subtilis* chromosome (13, 16). On the other hand, no mutants of *B. subtilis* with defects in dihydrodipicolinate synthase have been isolated, perhaps owing to the presence of two isofunctional genes. Although biochemical evidence suggests that there is only a single form of dihydrodipicolinate synthase, the possible existence of isozymes has not been rigorously excluded. It is thus possible that the *dapA* gene adjacent to *dapG* described above (24) is not the only locus encoding dihydrodipicolinate synthase. Recombinant plasmids with 37- and 7.5-kb inserts of *B. subtilis* chromosomal DNA have been reported to complement *dapA*, *dapB*, *dapC*, *dapD*, *dapE*, and *lysA* mutants of *E. coli*, suggesting that genes for dihydrodipicolinate synthase and other enzymes of diaminopimelate and lysine synthesis may be clustered on the *B. subtilis* chromosome (1, 122). In view of the identification of *dapA*, *dapE*, and *lysA* loci in disparate regions of the *B. subtilis* chromosome, the existence of such a cluster implies the existence of isofunctional genes.

Lysine biosynthesis

Both the activity (93) and the synthesis (41, 46, 80) of aspartokinase II are regulated by lysine. The structural gene for aspartokinase II, *lysC*, has been mapped at 253° on the *B. subtilis* chromosome by a combination of nucleotide sequencing and genetic methods (26, 105). *lysC* is near the locus of a class of mutations, previously referred to as *aecA*, that confer resistance to the lysine analog aminoethylcysteine, cause over-

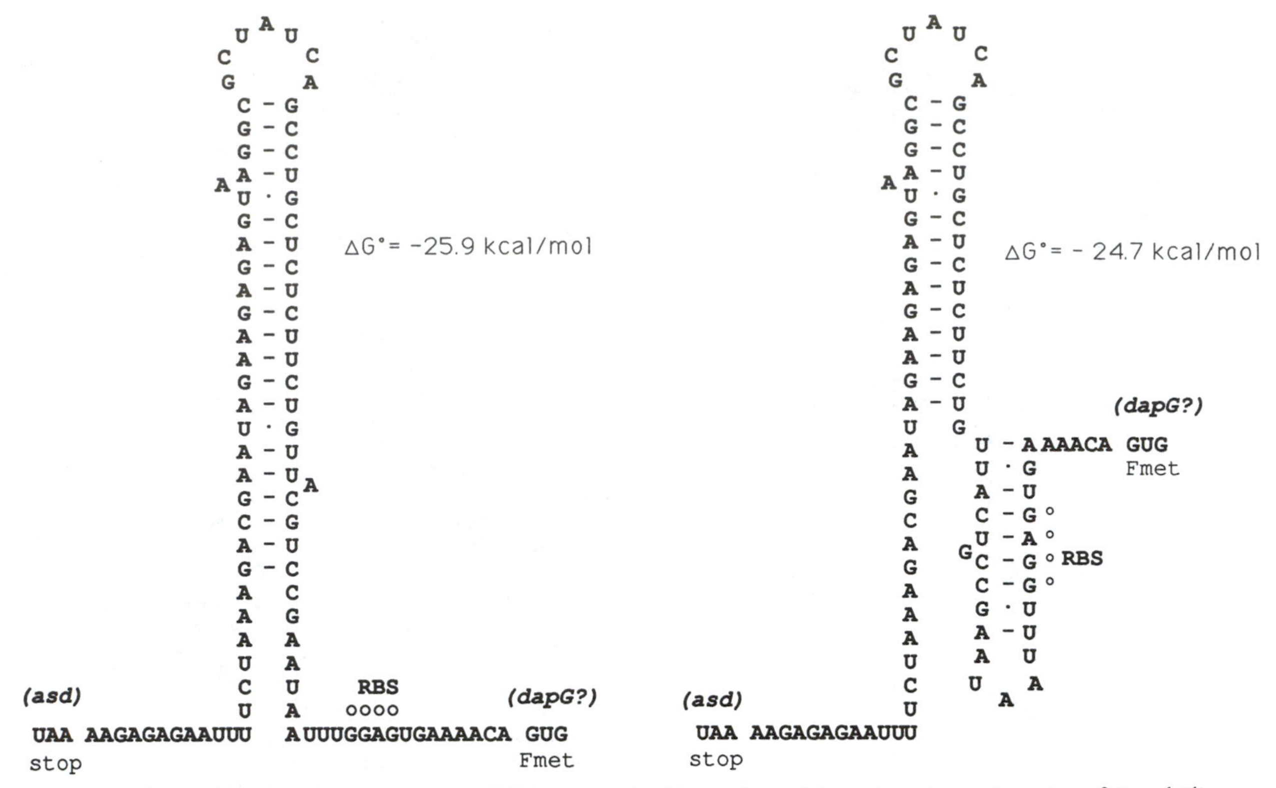

Figure 8. Possible secondary structures of RNA transcribed from the *asd-dapG* intercistronic region of *B. subtilis* (24). Stabilization free energies were estimated by using the parameters of Freier et al. (36). RBS, ribosome-binding site; Fmet, formylmethionyl.

production of aspartokinase II (82, 84, 123, 143, 157), and are now known to be point mutations in the *lysC* control region (80). The in-phase overlapping coding regions for the aspartokinase II subunits are the only structural genes encoded by the *lysC* operon, being immediately followed by a putative rho-independent transcription terminator and preceded by an unusually extensive control region (Fig. 9). Transcript mapping (23) and site-directed mutagenesis (80) have shown that the expression of aspartokinase II involves a single transcription initiation site, about 330 bp upstream of the aspartokinase coding region, and that the −35 and −10 promoter elements overlap the reading frame of the adjacent *uvrC* (formerly *uvrB*) gene (Fig. 9). The overlapping genes for the α and β subunits of aspartokinase II are independently translated from the *lysC* transcript, with synthesis of the β subunit utilizing a ribosome-binding site and translation start site within the coding region of the larger α subunit (25). In exponentially growing cultures of *B. subtilis*, the efficiencies of translation initiation at the separate α- and β-subunit start sites are exactly balanced so as to yield equimolar amounts of the subunits; on the other hand, there is twofold overproduction of α subunit in outgrowing spores of *B. subtilis* and threefold overproduction of β subunit when the cloned *lysC* operon is expressed in *E. coli* (12). The *lysC* coding region is immediately followed by an inverted repeat that resembles a rho-independent transcription terminator except that the consecutive thymidylates are within the inverted repeat, potentially allowing the terminator to function bidirectionally (23). A bidirectional rho-independent transcription terminator in *E. coli* has been described (44), and the *lysC* operon and a converging unidentified transcription unit may likewise share a single transcription terminator. Functional overlaps seem to be a hallmark of the *lysC* operon of *B. subtilis*, which encodes in-phase overlapping genes and has transcription initiation and termination elements that overlap adjacent operons.

The function of the 330-bp *lysC* leader region (23) remains to be elucidated. It undoubtedly plays an important role in the regulation of aspartokinase II synthesis by lysine, for all *aecA* mutations analyzed to date, which cause between 20- and 100-fold derepression of aspartokinase II in lysine-containing media, can be linked to nucleotide substitutions or small deletions in this region (80, 81). The *lysC* leader region, shown in Fig. 10, contains a number of potential control sites: (i) an inverted repeat centered on residue 40, which is the site of six independent *aecA* mutations; (ii) an open reading frame starting at residue 82 that encodes a potential 24-residue polypeptide with tandem lysine codons near the N terminus, which is the site of five *aecA* mutations; and (iii) a set of four inverted repeats, of which the first overlaps the coding region of the putative leader peptide and the last resembles a rho-independent transcription terminator and of which three can assume alternate hairpin loops reminiscent of the terminator and antiterminator structures of the tran-

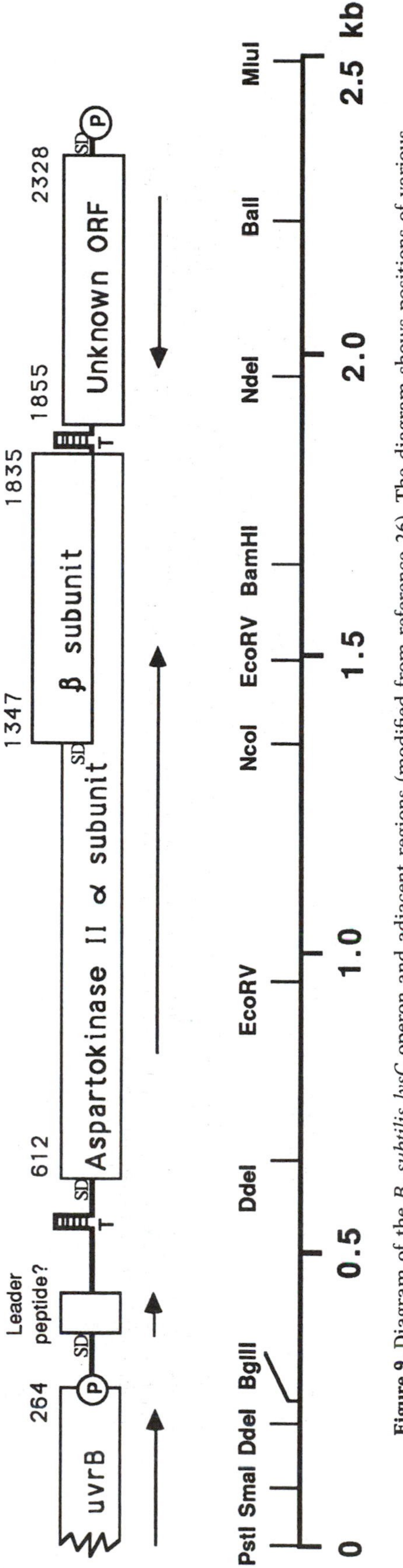

Figure 9. Diagram of the *B. subtilis lysC* operon and adjacent regions (modified from reference 26). The diagram shows positions of various restriction endonuclease cleavage sites and locations of deduced coding regions. Arrows indicate the polarities of transcription. P, promoter; SD, ribosome-binding site; T, transcription terminator; ORF, open reading frame.

scription attenuators commonly encountered in biosynthetic operons of members of the family *Enterobacteriaceae* (78). It is interesting that the lysine-sensitive aspartokinases from *E. coli* and the methylotrophic *Bacillus* sp. strain MGA3 also have extensive untranslated leader transcripts. The untranslated *lysC* leader region of *E. coli* consists of 308 nucleotides but differs from that of *B. subtilis lysC* by the absence of elements resembling a transcription attenuator (18). The *Bacillus* sp. strain MGA3 leader transcript is 366 nucleotides long and can form two alternate hairpin loops, one of them a potential rho-independent transcription terminator which may function similarly to the transcription attenuator of the *B. subtilis trp* operon (130), but which differs from the *B. subtilis lysC* leader by the absence of a region coding for a lysine-rich leader peptide (117).

The nucleotide sequence of the *lysC* leader region is identical in derivatives of *B. subtilis* 168 and Marburg (ATCC 6051), suggesting that the much reduced levels of aspartokinase II (and the correspondingly higher levels of aspartokinase III) observed in Marburg strains are due to differences in amino acid pool sizes and not to different control mechanisms (162). The apparent induction of aspartokinase II observed on supplementation of minimal media by methionine (41) may be due to an indirect effect of methionine on intracellular lysine concentrations. On the other hand, the involvement of possible *trans*-acting elements in the control of aspartokinase II synthesis, even though no such element has been identified, cannot be ruled out.

The low levels of aspartokinase II (41, 94) observed in sporulating cells of *B. subtilis* result both from a reduced rate of synthesis and from the specific degradation of aspartokinase II under conditions of nutrient limitation (42). The degradation of aspartokinase II is similar to that of other biosynthetic enzymes in *B. subtilis*, such as aspartate transcarbamoylase and glutamine-phosphoribosylpyrophosphate amidotransferase, in being induced by starvation for glucose, ammonium, or a required amino acid and in being blocked by inhibitors of protein synthesis (e.g., chloramphenicol) and energy production (e.g., fluoroacetate). On the other hand, the starvation-induced degradation of aspartokinase II differs from that of other biosynthetic enzymes by its independence from the *relA* gene, suggesting the existence of multiple degradative pathways (42).

Another step in the aspartate pathway that is specifically controlled by lysine is the last reaction in lysine biosynthesis, which is catalyzed by diaminopimelate decarboxylase. The gene for diaminopimelate decarboxylase, *lysA*, maps at 210° on the *B. subtilis* chromosome (56). It is preceded by −35 and −10 consensus sequences recognized by the major vegetative RNA polymerase and immediately followed by an inverted repeat resembling a rho-independent terminator (154). In *E. coli*, a positive regulatory element, *lysR*, whose product is essential for *lysA* expression, is transcribed divergently from *lysA* (134), but none of the deduced products of the open reading frames surrounding *B. subtilis lysA* have any similarity to the *lysR* product (154). However, a mutation that leads to the loss of diaminopimelate decarboxylase but maps at 250°, remote from the *lysA* gene,

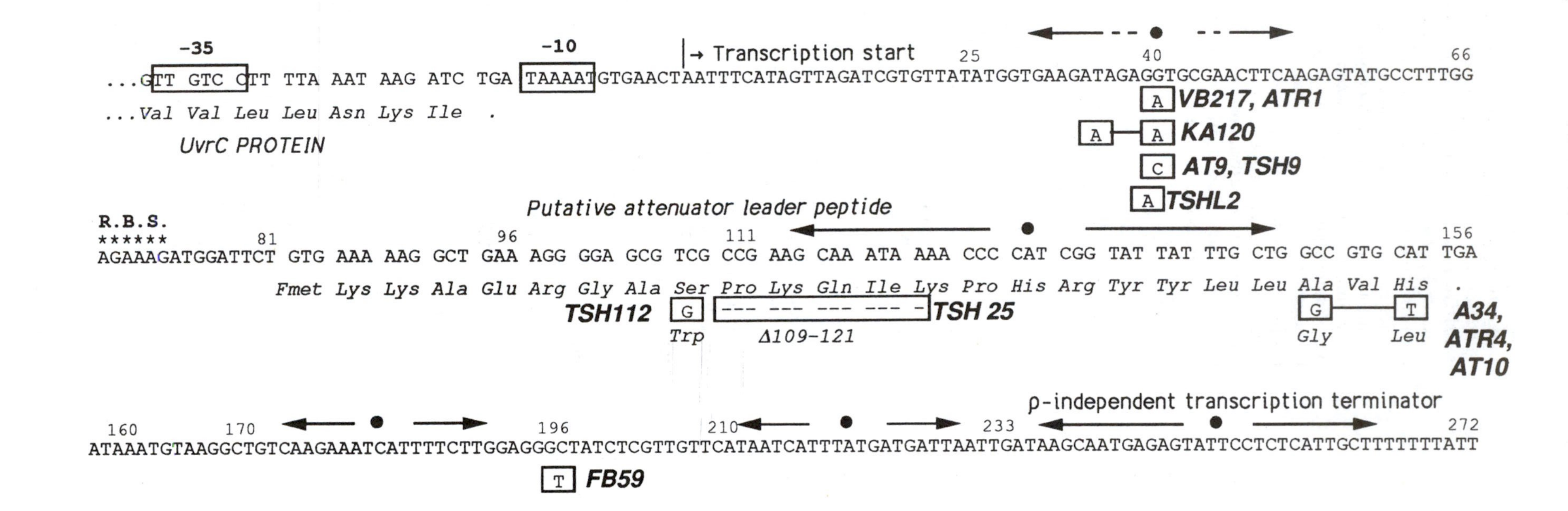

Figure 10. Nucleotide sequence of the untranslated leader region of the *B. subtilis lysC* operon (23, 80, 81). The sequence from *B. subtilis* 168 is shown in its entirety. Sequences from *aecA* strains VB217, FB59, and KA120 (80) and from A34, ATR1, ATR4, AT9, AT10, TSH9, TSH25, TSH112, and TSHL2 (81) are shown only where they differ from that of strain 168. The sequence is annotated to show elements of potential regulatory significance, with inverted repeats indicated by arrows and ribosome-binding sites (R.B.S.) indicated by asterisks. Nucleotide residues are numbered from the transcription start site.

may represent a positive control element analogous to *lysR* (124).

Expression of the *lysA* gene of *C. glutamicum* depends on the presence upstream of the *lysA* coding region of a 2.35-kb region containing an open reading frame (*orfX*) with a potential 550-residue polypeptide product (83). *orfX* appears to be cotranscribed with the *lysA* gene, and a frameshift mutation in *orfX* has a polar effect on *lysA* expression. On the other hand, deletion of most of *orfX* has no effect on *lysA* expression, indicating that *orfX* and its product do not play an essential role and are thus not analogous to *E. coli lysR*. An earlier report that *lysA* in *C. glutamicum* is repressed by lysine (30) has not been substantiated (83).

Threonine biosynthesis

Synthesis of the threonine-plus-lysine-sensitive aspartokinase III of *B. subtilis* is repressed by threonine and induced by lysine (41). The induction by lysine could be either a direct effect of lysine or, more likely, the effect of decreased intracellular threonine concentrations caused by the repression of aspartokinase II, the major aspartokinase activity in minimal media. Mutants lacking aspartokinase III have been described (162), but it is not clear whether the mutations affect the structural gene or a regulatory locus, and the chromosomal location of the gene for aspartokinase III is still unknown. On the other hand, a regulatory mutation, *spr*, leading to about 20-fold derepression of aspartokinase III has recently been mapped to the interval between *pheB* and *spo0B* at 240° on the *B. subtilis* chromosome (161). The *spr* mutants were isolated as suppressors of *ilvA*, similar to the *sprA* mutants (141, 142), and had a derepressed threonine operon as well as derepressed levels of aspartokinase III (164). The *pheB-spo0B* interval has been entirely sequenced (140) and contains no candidate gene for aspartokinase III, suggesting that *spr* may encode a pleiotropic *trans*-acting element controlling the expression of aspartokinase III and the threonine operon.

The genes for the enzymes that catalyze the conversion of homoserine to threonine, *hom* (homoserine dehydrogenase), *thrB* (homoserine kinase), and *thrC* (threonine synthase), form a cluster at 284° on the *B. subtilis* chromosome (131). This gene cluster has been cloned and sequenced and seems to constitute a transcription unit (99, 100), which is consistent with the earlier observation that an unlinked mutation, *sprA*, leads to parallel derepression of all three genes (131, 141, 142). The genes are arranged in the order *hom-thrC-thrB*, with the translation termination codon of the *hom* gene overlapping the translation start codon of the adjacent *thrC* locus, which similarly overlaps the *thrB* gene (99). It is interesting that the order of *thrC* and *thrB* is the reverse of that found in *E. coli* (27). Although there is a consensus promoter sequence for the major vegetative RNA polymerase 75 residues upstream of the translation start of the *hom* gene (100) and a potential rho-independent transcription terminator just downstream of the *thrB* translation termination site (99), the roles of these potential transcription elements remain to be defined by direct transcript analysis. The possibility that the extensive unidentified open reading frames both upstream and directly downstream of the *hom-thrC-thrB* cluster (99, 100) are part of the same transcription unit cannot be excluded.

The level of homoserine dehydrogenase is relatively constant throughout exponential growth but is reduced to about one-half by supplementation of minimal media with L-methionine, L-threonine, or L-isoleucine (156). On the other hand, the more-than-20-fold derepression produced by the *sprA* mutations (141, 142) suggests that the threonine operon is significantly repressed even during growth in minimal medium.

In coryneform bacteria, which have only a single aspartokinase, the genes for threonine biosynthesis are organized quite differently from those in *B. subtilis*, which appears to place a premium on the supply of precursors for the diaminopimelate-lysine branch of the aspartate pathway, with two of the aspartokinase isozymes being specialized for this purpose and the gene for aspartate semialdehyde dehydrogenase being associated with the gene for the constitutive diaminopimelate-sensitive aspartokinase. In contrast, metabolite flux in coryneform bacteria favors the threonine-methionine branch of the aspartate pathway, the catalytic capacity of homoserine dehydrogenase exceeding that of dihydrodipicolinate synthase by more than 1 order of magnitude (89) unless countermanded by threonine, which participates in the control of the single aspartokinase as well as of homoserine dehydrogenase. The genes for the threonine-plus-lysine-sensitive aspartokinases of *C. glutamicum* and *C. flavum* differ in only a few nucleotide residues and have been designated *lysC* (80) and *ask* (33), respectively, the functional implications of the designation *lysC* being somewhat misleading. They are composed of in-phase overlapping reading frames for the α and β subunits, which are capable of independent translation in a manner similar to that of the genes for the subunits of *B. subtilis* aspartokinase II (33, 60, 61). The *ask* (*lysC*) gene is separated by a 23-bp spacer from *asd*, the gene encoding aspartate semialdehyde dehydrogenase, with which it forms an *ask-asd* transcription unit (33, 60, 61). An interesting feature of the *ask-asd* operon derived from aminoethylcysteine-resistant strains of *Corynebacterium* spp. is the occurrence of two transcription start sites, i.e., a major promoter upstream of *ask* and a minor one within *ask* (33, 61). Transcription initiation at the upstream promoter leads to coordinate transcription of *ask* and *asd*, whereas initiation at the internal promoter presumably produces a transcript encoding only the aminoethylcysteine-resistant aspartokinase β subunit (138) and aspartate semialdehyde dehydrogenase (33, 61). RNA mapping studies suggest that transcription starts 35 and 46 residues upstream of the aspartokinase α- and β-subunit translation initiation sites, respectively (33). However, it has also been proposed (61) that the internal promoter is located well within the β-subunit coding region near the site of nucleotide substitution responsible for aminoethylcysteine resistance. Expression of the *ask-asd* operon is not repressed by any of the end products of the aspartate pathway (90). It is interesting that in *B. subtilis*, the *asd* gene is also associated with the gene for a constitutive aspartokinase isozyme (*dapG*), whereas in *E. coli*, in which all three aspartokinase isozymes are subject to end product repression, *asd* is transcribed as an independent unit (48).

The organization of the genes for homoserine dehydrogenase and the other threonine biosynthetic enzymes of *C. glutamicum* differs from that in both *B. subtilis* and *E. coli* in that the gene for homoserine dehydrogenase (*hom*) directly precedes *thrB*, and *thrC* is not linked to the *hom-thrB* cluster (34, 47, 104). The *hom* and *thrB* coding regions are separated by 10 bp, and *thrB* is followed by a potential rho-independent transcription terminator (104). Transcript mapping has shown *hom-thrB* to be a transcription unit, with transcription beginning 88 residues upstream of the homoserine dehydrogenase-coding region (104). Transcription of *hom* and *thrB* in coryneform bacteria (*B. flavum*) is specifically repressed by methionine (89, 90), thus ensuring multivalent control of the threonine-sensitive homoserine dehydrogenase by the two end products in whose synthesis it participates. On the other hand, the repression by methionine of homoserine kinase, which is cotranscribed with homoserine dehydrogenase, is more difficult to rationalize unless homoserine kinase also plays a role in methionine synthesis, as it does in plants. Deletion analysis and transcript mapping of *thrC* place the promoter and the transcription start site within 38 and 6 bp, respectively, of the site of translation initiation, with no clearly discernible ribosome-binding site (47). The expression of *thrC* in *B. flavum* is not significantly repressed by any of the end products of the aspartate pathway (90).

Methionine biosynthesis

Four classes of mutants with defects in methionine biosynthesis have been identified: *metA*, *metB*, *metC*, and *metD*. The mutations map at different sites on the *B. subtilis* chromosome (110°, 197°, 118°, and 100°, respectively; e.g., see reference 106), which probably are the loci of the genes for the four enzymes of methionine biosynthesis. No studies on gene-enzyme relationships or the control of expression of these genes have been published. In *B. polymyxa*, homoserine acetyltransferase, which catalyzes the first specific step in methionine biosynthesis, is repressed by growth in media containing methionine (150).

Methionine-requiring strains of *B. flavum* have been classified into three groups on the basis of their abilities to grow on *O*-acetylhomoserine and/or homocysteine; each group presumably has a defect in one of the steps of the alternate pathway for methionine biosynthesis (Fig. 6) that operates in coryneform bacteria (98). The first two enzymes of methionine biosynthesis in *B. flavum*, homoserine transacetylase and acetylhomoserine sulfhydrylase, are repressed by methionine (98).

BIOSYNTHESIS OF ASPARTATE

The major route for the biosynthesis of aspartate from glycolytic intermediates involves the carboxylation of pyruvate to oxaloacetate and subsequent transamination (Fig. 11). Pyruvate carboxylase, a biotin enzyme, catalyzes the ATP-dependent carboxylation of pyruvate, with acetyl-CoA as an essential activator

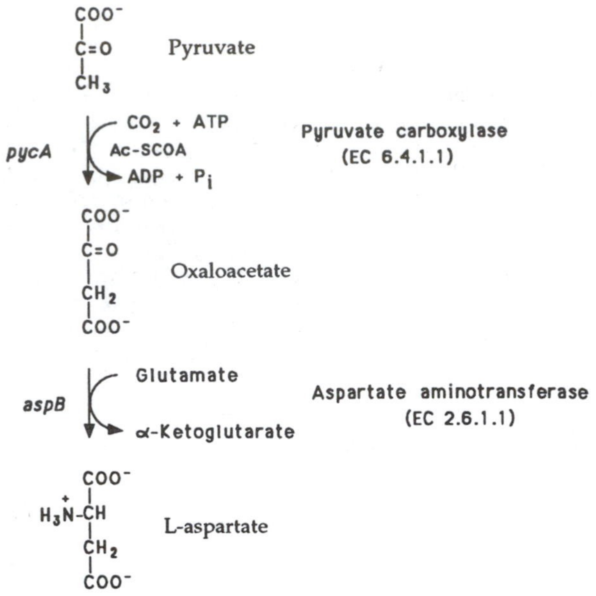

Figure 11. Biosynthesis of L-aspartate. Ac-SCoA, acetyl-CoA.

(32). *B. subtilis*, unlike the *Enterobacteriaceae*, cannot synthesize oxaloacetate by the carboxylation of phosphoenolpyruvate (32). Mutants deficient in pyruvate carboxylase are unable to grow on glucose but grow slowly on malate as a carbon source (32) or when supplemented with aspartate or glutamate (52). The genetic locus for pyruvate carboxylase (*pycA*; formerly designated *aspA*) maps at 127° on the *B. subtilis* chromosome (97, 115). A temperature-sensitive lysis mutant with an aspartate requirement at nonpermissive temperatures that was earlier thought to be defective in pyruvate carboxylase (15) was recently found to have a temperature-sensitive aspartokinase I (115).

aspB mutants lack glutamate-oxaloacetate transaminase activity and require aspartate for growth, suggesting that the transaminase encoded by *aspB* is the only enzyme capable of synthesizing aspartate from oxaloacetate (52, 56). In spite of its essential role in aspartate biosynthesis, the level of glutamate-oxaloacetate transaminase is not reduced during growth on aspartate as carbon source (56). The *aspB* locus maps at 198° on the *B. subtilis* chromosome, but neither the gene nor its product has been studied further, despite their central importance in amino acid metabolism. Glutamate-oxaloacetate transaminase has recently been purified and cloned from the thermophilic *Bacillus* sp. strain YM-2. The enzyme is highly specific for amino transfer between L-aspartate and oxaloacaetate, with K_ms of 3 and 2 mM, respectively, and catalyzes the transamination reaction by a ping-pong bi-bi mechanism, with highest activity at 70°C (136). The native enzyme has a molecular weight of about 97,000 and is composed of 42-kDa subunits, each of which contains one molecule of pyridoxal phosphate (136). The cloned gene encodes a 392-residue polypeptide (M_r = 42,700) whose amino acid sequence has much lower similarity (13% identity) to that of *E. coli* aspartate transaminase than to that of the corresponding enzyme from the archaebacterium *Sulfolobus solfataricus* (25% identity) (137).

Acknowledgments. I thank M. T. Follettie, M. C. Flickinger, and F. J. Schendel for providing me with manuscripts before publication. The work on the aspartate pathway conducted in my laboratory has been supported by grants from the National Science Foundation.

REFERENCES

1. **Aleksieva, Z. M., T. N. Shevtchenko, and S. S. Malyuta.** 1985. A study of lysine operon organization in *Bacillus subtilis*. *Biopolim. Kletka* **1**:156–158.
2. **Aronson, A. I., E. Henderson, and A. Tincher.** 1967. Participation of the lysine pathway in dipicolinic acid synthesis of *Bacillus cereus* T. *Biochem. Biophys. Res. Commun.* **26**:454–460.
3. **Bach, M. L., and C. Gilvarg.** 1966. Biosynthesis of dipicolinic acid in sporulating *Bacillus megaterium*. *J. Biol. Chem.* **241**:4563–4566.
4. **Balassa, G., P. Milhaud, E. Raulet, M. T. Silva, and J. C. F. Sousa.** 1979. A *Bacillus subtilis* mutant requiring dipicolinic acid for the development of heat-resistant spores. *J. Gen. Microbiol.* **110**:365–379.
5. **Bartlett, A. T. M., and P. J. White.** 1985. Species of *Bacillus* that make a vegetative peptidoglycan containing lysine lack diaminopimelate epimerase but have diaminopimelate dehydrogenase. *J. Gen. Microbiol.* **131**:2145–2152.
6. **Bartlett, A. T. M., and P. J. White.** 1986. Regulation of the enzymes of lysine biosynthesis in *Bacillus sphaericus* NCTC 9602 during vegetative growth. *J. Gen. Microbiol.* **132**:3169–3177.
7. **Biswas, C., E. Gray, and H. Paulus.** 1970. Multivalent feedback inhibition of aspartokinase in *Bacillus polymyxa*. III. Purification and subunit structure of the enzyme. *J. Biol. Chem.* **245**:4900–4906.
8. **Biswas, C., and H. Paulus.** 1973. Multivalent feedback inhibition of aspartokinase in *Bacillus polymyxa*. IV. Arrangement and function of the subunits. *J. Biol. Chem.* **248**:2894–2900.
9. **Bonassie, S., J. Oreglia, and A. M. Sicard.** 1990. Nucleotide sequence of the *dapA* gene from *Corynebacterium glutamicum*. *Nucleic Acids Res.* **18**:6421.
10. **Bondaryk, R.** 1984. Ph.D. thesis. Harvard University, Cambridge, Mass.
11. **Bondaryk, R. P., and H. Paulus.** 1985. Cloning and structure of the gene for the subunits of aspartokinase II from *Bacillus subtilis*. *J. Biol. Chem.* **260**:585–591.
12. **Bondaryk, R. P., and H. Paulus.** 1985. Expression of the gene for *Bacillus subtilis* aspartokinase II in *Escherichia coli*. *J. Biol. Chem.* **260**:592–597.
13. **Brandt, C., and D. Karamata.** 1987. Thermosensitive *Bacillus subtilis* mutants which lyse at the nonpermissive temperature. *J. Gen. Microbiol.* **133**:1159–1170.
14. **Brush, A., and H. Paulus.** 1971. The enzymatic formation of O-acetylhomoserine in *Bacillus subtilis* and its regulation by methionine and S-adenosylmethionine. *Biochem. Biophys. Res. Commun.* **45**:735–741.
15. **Buxton, R. S.** 1978. A heat-sensitive lysis mutant of *Bacillus subtilis* 168 with a low activity of pyruvate decarboxylase. *J. Gen. Microbiol.* **105**:175–185.
16. **Buxton, R. S., and J. B. Ward.** 1980. Heat-sensitive lysis mutants of *Bacillus subtilis* 168 blocked at three different stages of peptidoglycan synthesis. *J. Gen. Microbiol.* **120**:283–293.
17. **Cardineau, G. A., and R. Curtiss III.** 1987. Nucleotide sequence of the *asd* gene of *Streptococcus mutans*. Identification of the promoter region and evidence for attenuator-like sequences preceding the structural gene. *J. Biol. Chem.* **262**:3344–3353.
18. **Cassan, M., J. Ronceray, and J. C. Patte.** 1983. Nucleotide sequence of the promoter region of the *E. coli lysC* gene. *Nucleic Acids Res.* **11**:6157–6166.
19. **Chasin, L. A., and J. Szulmajster.** 1967. Biosynthesis of dipicolinic acid in *Bacillus subtilis*. *Biochem. Biophys. Res. Commun.* **29**:648–654.
20. **Chasin, L. A., and J. Szulmajster.** 1969. Enzymes of dipicolinic acid biosynthesis in *Bacillus subtilis*, p. 133–147. *In* L. L. Campbell (ed.), *Spores IV*. American Society for Microbiology, Washington, D.C.
21. **Chatterjee, M.** 1986. Aspartokinase of lysine excreting and non-excreting strain of *Bacillus megaterium*. *Curr. Sci.* **55**:1176–1179.
22. **Chatterjee, M., and P. J. White.** 1982. Activities and regulation of the enzymes of lysine biosynthesis in a lysine-excreting strain of *Bacillus megaterium*. *J. Gen. Microbiol.* **128**:1073–1081.
23. **Chen, N. Y., F. M. Hu, and H. Paulus.** 1987. Nucleotide sequence of the overlapping genes for the subunits of *Bacillus subtilis* aspartokinase II and their control region. *J. Biol. Chem.* **262**:8787–8798.
24. **Chen, N. Y., S. Q. Jiang, D. A. Klein, and H. Paulus.** Organization and nucleotide sequence of the *Bacillus subtilis* diaminopimelate operon, a cluster of genes encoding the first three enzymes of diaminopimelate synthesis and dipicolinate synthase. *J. Biol. Chem.* **268**, in press.
25. **Chen, N. Y., and H. Paulus.** 1988. Mechanisms of expression of the overlapping genes of *Bacillus subtilis* aspartokinase II. *J. Biol. Chem.* **263**:9526–9532.
26. **Chen, N. Y., J. J. Zhang, and H. Paulus.** 1989. Chromosomal location of the *Bacillus subtilis* aspartokinase II gene and nucleotide sequence of the adjacent genes homologous to *uvrC* and *trx* of *Escherichia coli*. *J. Gen. Microbiol.* **135**:2931–2940.
27. **Cohen, G. N., and I. Saint-Girons.** 1987. Biosynthesis of threonine, lysine and methionine, p. 429–444. *In* F. C. Neidhardt, J. L. Ingraham, K. B. Low, B. Magasanik, M. Schaechter, and H. E. Umbarger (ed.), *Escherichia coli and Salmonella typhimurium: Cellular and Molecular Biology*, vol. 1. American Society for Microbiology, Washington, D.C.
28. **Coruzzi, G. M.** 1991. Molecular approaches to the study of amino acid biosynthesis in plants. *Plant Sci.* **74**:145–155.
29. **Cremer, J., L. Eggeling, and H. Sahm.** 1990. Cloning the *dapA-dapB* cluster of the lysine-secreting bacterium *Corynebacterium glutamicum*. *Mol. Gen. Genet.* **224**:317–324.
30. **Cremer, J., C. Treptow, L. Eggeling, and H. Sahm.** 1988. Regulation of the enzymes of lysine biosynthesis in *Corynebacterium glutamicum*. *J. Gen. Microbiol.* **134**:3221–3229.
31. **Datta, P., and L. Prakash.** 1966. Aspartokinase of *Rhodopseudomonas spheroides*. Regulation of enzyme activity by aspartate β-semialdehyde. *J. Biol. Chem.* **241**:5827–5835.
32. **Diesterhaft, M. D., and E. Freese.** 1973. Role of pyruvate carboxylase, phosphoenolpyruvate carboxykinase, and malic enzyme during growth and sporulation of *Bacillus subtilis*. *J. Biol. Chem.* **248**:6062–6070.
33. **Follettie, M. T., O. P. Peoples, and A. J. Sinskey.** Structure and expression analysis of the *Corynebacterium flavum* N13 *ask-asd* operon. Submitted for publication.
34. **Follettie, M. T., H. K. Shin, and A. J. Sinskey.** 1988. Organization and regulation of the *Corynebacterium glutamicum hom-thrB* and *thrC* loci. *Mol. Microbiol.* **2**:53–62.
35. **Forman, M., and A. Aronson.** 1972. Regulation of dipicolinic acid biosynthesis in sporulating *Bacillus cereus*. Characterization of enzymatic changes and analysis of mutants. *Biochem. J.* **126**:503–513.
36. **Freier, S. M., R. Kierzek, J. A. Jaeger, N. Sugimoto, M. H. Caruthers, T. Neilson, and D. H. Turner.** 1986. Improved free energy parameters for predictions of

RNA duplex stability. *Proc. Natl. Acad. Sci. USA* **83:** 9373–9377.
37. **Fukuda, A., and C. Gilvarg.** 1968. The relationship of dipicolinate and lysine biosynthesis in *Bacillus megaterium. J. Biol. Chem.* **243:**3871–3876.
38. **Gally, D., C. R. Harwood, and A. R. Archibald.** 1991. Diaminopimelate uptake by *Bacillus megaterium*: influence of growth conditions and other amino acids. *Lett. Appl. Microbiol.* **12:**54–58.
39. **Grandgenett, D. P., and D. P. Stahly.** 1971. Repression of diaminopimelate decarboxylase by L-lysine in different *Bacillus* species. *J. Bacteriol.* **105:**1211–1212.
40. **Grandgenett, D. P., and D. P. Stahly.** 1971. Control of diaminopimelate decarboxylase by L-lysine during growth and sporulation of *Bacillus cereus. J. Bacteriol.* **106:**551–560.
41. **Graves, L. M., and R. L. Switzer.** 1990. Aspartokinase III, a new isozyme in *Bacillus subtilis* 168. J. Bacteriol. **172:**218–233.
42. **Graves, L. M., and R. L. Switzer.** 1990. Aspartokinase II from *Bacillus subtilis* is degraded in response to nutrient limitation. *J. Biol. Chem.* **265:**14947–14955.
43. **Gray, B. H., and R. W. Bernlohr.** 1969. The regulation of aspartokinase in *Bacillus licheniformis. Biochim. Biophys. Acta* **178:**248–261.
44. **Grundström, T., and B. Jaurin.** 1982. Overlap between *ampC* and *frd* operons of the *Escherichia coli* chromosomes. *Proc. Natl. Acad. Sci. USA* **79:**1111–1115.
45. **Halling, S. M., and D. P. Stahly.** 1976. Dihydrodipicolinic acid synthase from *Bacillus licheniformis*. Quaternary structure, kinetics, and stability in the presence of sodium chloride and substrates. *Biochim. Biophys. Acta* **452:**580–596.
46. **Hampton, M. L., N. G. McCormick, N. C. Behforouz, and E. Freese.** 1971. Regulation of two aspartokinases in *Bacillus subtilis. J. Bacteriol.* **108:**1129–1134.
47. **Han, K. S., J. A. Archer, and A. J. Sinskey.** 1990. The molecular structure of the *Corynebacterium glutamicum* threonine synthase gene. *Mol. Microbiol.* **4:**1693–1702.
48. **Haziza, C., P. Stragier, and J. C. Patte.** 1982. Nucleotide sequence of the *asd* gene of *Escherichia coli*: absence of a typical attenuation signal. *EMBO J.* **1:**379–384.
49. **Hitchcock, M. J. M.** 1976. Ph.D. thesis. University of Melbourne, Melbourne, Australia.
50. **Hitchcock, M. J. M., and B. Hodgson.** 1976. Lysine- and lysine-plus-threonine-inhibitable aspartokinases in *Bacillus brevis. Biochim. Biophys. Acta* **445:**350–363.
51. **Hitchcock, M. J. M., B. Hodgson, and J. L. Linforth.** 1980. Regulation of lysine- and lysine-plus-threonine-inhibitable aspartokinases in *Bacillus brevis. J. Bacteriol.* **142:**424–432.
52. **Hoch, J. A., and J. Mathews.** 1972. Genetic studies in *Bacillus subtilis*, p. 113–116. *In* H. O. Halvorson, R. Hanson, and L. L. Campbell (ed.), *Spores V*. American Society for Microbiology, Washington, D.C.
53. **Hoganson, D. A., R. L. Irgens, R. H. Doi, and D. P. Stahly.** 1975. Bacterial sporulation and regulation of dihydrodipicolinate synthase in ribonucleic acid polymerase mutants of *Bacillus subtilis. J. Bacteriol.* **124:** 1628–1629.
54. **Hoganson, D. A., C. D. Smith, and D. P. Stahly.** 1978. Regulation of aspartokinase activity in *Bacillus cereus*, p. 304–307. *In* G. Chambliss and J. C. Vary (ed.), *Spores VII*. American Society for Microbiology, Washington, D.C.
55. **Hoganson, D. A., and D. P. Stahly.** 1975. Regulation of dihydrodipicolinate synthase during growth and sporulation of *Bacillus cereus. J. Bacteriol.* **124:**1344–1350.
56. **Iijima, T., M. D. Diesterhaft, and E. Freese.** 1977. Sodium effect of growth on aspartate and genetic analysis of a *Bacillus subtilis* mutant with high aspartase activity. *J. Bacteriol.* **129:**1441–1447.
57. **Ishino, I., T. Mizukami, K. Yamaguchi, R. Katsumata, and K. Araki.** 1987. Nucleotide sequence of the *meso*-diaminopimelate D-dehydrogenase gene from *Corynebacterium glutamicum. Nucleic Acids Res.* **15:**3917.
58. **Ishino, I., T. Mizukami, K. Yamaguchi, R. Katsumata, and K. Araki.** 1988. Cloning and sequencing of the *meso*-diaminopimelate D-dehydrogenase gene (*ddh*) of *Corynebacterium glutamicum. Agric. Biol. Chem.* **52:** 2903–2909.
59. **Jenkinson, H. F., and J. Mandelstam.** 1983. Cloning of the *Bacillus subtilis lys* and *spoIIIB* genes in phage ϕ105. *J. Gen. Microbiol.* **129:**2229–2240.
60. **Kalinowski, J., B. Bachmann, G. Thierbach, and A. Pühler.** 1990. Aspartokinase genes *lysC*α and *lysC*β overlap and are adjacent to the aspartate semialdehyde dehydrogenase gene *asd* in *Corynebacterium glutamicum. Mol. Gen. Genet.* **224:**317–324.
61. **Kalinowski, J., J. Cremer, B. Bachmann, L. Eggeling, H. Sahm, and A. Pühler.** 1991. Genetic and biochemical analysis of the aspartokinase from *Corynebacterium glutamicum. Mol. Microbiol.* **5:**1197–1204.
62. **Kanzaki, H., M. Kobayashi, T. Nagasawa, and H. Yamada.** 1986. Distribution of two kinds of cystathionine γ-synthase in various bacteria. *FEMS Microbiol. Lett.* **33:**65–68.
63. **Kase, H., and K. Nakayama.** 1974. Mechanism of L-threonine and L-lysine production by analog-resistant mutants of *Corynebacterium glutamicum. Agric. Biol. Chem.* **38:**993–1000.
64. **Kase, H., and K. Nakayama.** 1974. Production of O-acetyl-*L*-homoserine by methionine analog-resistant mutants and regulation of homoserine-O-transacetylase in *Corynebacterium glutamicum. Agric. Biol. Chem.* **38:**2021–2030.
65. **Kern, B. A., D. Hendlin, and E. Inamine.** 1980. L-Lysine-ε-aminotransferase involved in cephamycin C synthesis in *Streptomyces lactamdurans. Antimicrob. Agents Chemother.* **17:**679–685.
66. **Kimura, K.** 1975. Pyridine-2,6-dicarboxylic acid (dipicolinic acid) formation in *Bacillus subtilis*. I. Nonenzymatic formation of dipicolinic acid from pyruvate and aspartic semialdehyde. *J. Biochem.* **75:**961–967.
67. **Kimura, K.** 1975. A new flavin enzyme catalyzing the reduction of dihydrodipicolinate in sporulating *Bacillus subtilis*. I. Purification and properties. *J. Biochem.* **77:**405–413.
68. **Kimura, K., and T. Goto.** 1975. A new flavin enzyme catalyzing the reduction of dihydrodipicolinate in sporulating *Bacillus subtilis*. II. Kinetics and regulatory function. *J. Biochem.* **77:**415–420.
69. **Kimura, K., and T. Goto.** 1977. Dihydrodipicolinate reductases from *Bacillus cereus* and *Bacillus megaterium*. I. Purification and properties. *J. Biochem.* **81:** 1367–1373.
70. **Kimura, K., T. Goto, and S. Ujita.** 1978. Two differentiatable types of dihydrodipicolinate reductases from sporeforming bacilli, p. 308–311. *In* G. Chambliss and J. C. Vary (ed.), *Spores VII*. American Society for Microbiology, Washington, D.C.
71. **Kimura, K., and T. Sasakawa.** 1975. Pyridine-2,6-dicarboxylic acid (dipicolinic acid) formation in *Bacillus subtilis*. I. Non-enzymatic and enzymatic formations of dipicolinic acid from α,ε-diketopimelic acid and ammonia. *J. Biochem.* **78:**381–390.
72. **Kirkpatrick, J. R., L. E. Doolin, and O. W. Godfrey.** 1973. Lysine biosynthesis in *Streptomyces lipmanii*: implications in antibiotic biosynthesis. *Antimicrob. Agents Chemother.* **4:**542–550.
73. **Kornberg, A., J. A. Spudich, D. L. Nelson, and M. P. Deutscher.** 1968. Origins of proteins in sporulation. *Annu. Rev. Biochem.* **37:**51–78.
74. **Krueger, J. M., J. R. Pappenheimer, and M. L. Karnovsky.** 1982. The composition of sleep-promoting fac-

tor isolated from human urine. *J. Biol. Chem.* **257:**1664–1669.

75. **Kuramitsu, H. K.** 1970. Concerted feedback inhibition of aspartokinase from *Bacillus stearothermophilus*. I. Catalytic and regulatory properties. *J. Biol. Chem.* **245:** 2991–2997.
76. **Kuramitsu, H. K., and S. Yoshimura.** 1971. Catalytic and regulatory properties of *meso*-diaminopimelate-sensitive aspartokinase from *Bacillus stearothermophilus*. *Arch. Biochem. Biophys.* **147:**683–691.
77. **Kuramitsu, H. K., and S. Yoshimura.** 1972. Elevated diaminopimelate-sensitive aspartokinase activity during sporulation of *Bacillus stearothermophilus*. *Biochim. Biophys. Acta* **264:**152–164.
78. **Landick, R., and C. Yanofsky.** 1987. Transcription attenuation, p. 1276–1301. *In* F. C. Neidhardt, J. L. Ingraham, K. B. Low, B. Magasanik, M. Schaechter, and H. E. Umbarger (ed.), *Escherichia coli and Salmonella typhimurium: Cellular and Molecular Biology*, vol. 2. American Society for Microbiology, Washington, D.C.
79. **Lee, C. W., J. M. Ravel, and W. Shive.** 1966. Multimetabolite control of a biosynthetic pathway by sequential metabolites. *J. Biol. Chem.* **241:**5479–5480.
80. **Lu, Y., N. Y. Chen, and H. Paulus.** 1991. Identification of *aecA* mutations in *Bacillus subtilis* as nucleotide substitutions in the untranslated leader region of the aspartokinase II operon. *J. Gen. Microbiol.* **137:**1135–1143.
81. **Lu, Y., T. N. Shevtchenko, and H. Paulus.** Unpublished observations.
82. **Magnusson, K., B. Rutberg, L. Hederstedt, and L. Rutberg.** 1983. Characterization of a pleiotropic succinate dehydrogenase-negative mutant of *Bacillus subtilis*. *J. Gen. Microbiol.* **129:**917–922.
83. **Marcel, T., J. A. C. Archer, D. Mengin-Lecreulx, and A. J. Sinskey.** 1990. Nucleotide sequence and organization of the upstream region of the *Corynebacterium lysA* gene. *Mol. Microbiol.* **4:**1819–1830.
84. **Mattioli, R., M. Bazzicolupo, G. Federici, E. Gallori, and M. Polsinelli.** 1979. Characterization of mutants of *Bacillus subtilis* resistant to S-(2-aminoethyl)cysteine. *J. Gen. Microbiol.* **114:**223–225.
85. **Michaud, C., D. Megnin-Lecreulx, J. van Heijenoort, and D. Blanot.** 1990. Over-production, purification and properties of the uridine-diphosphate-N-acetylmuramoyl-L-alanyl-D-glutamate:*meso*-2,6-diaminopimelate ligase from *Escherichia coli*. *Eur. J. Biochem.* **194:**853–861.
86. **Misono, H., and K. Soda.** 1980. Properties of *meso*-α,ε-diaminopimelate dehydrogenase from *Bacillus sphaericus*. *J. Biol. Chem.* **255:**10599–10605.
87. **Misono, H., H. Togawa, T. Yamomoto, and K. Soda.** 1976. Occurrence of *meso*-α,ε-diaminopimelate dehydrogenase in *Bacillus sphaericus*. *Biochem. Biophys. Res. Commun.* **72:**89–93.
88. **Misono, H., H. Togawa, T. Yamomoto, and K. Soda.** 1979. *meso*-α,ε-Diaminopimelate dehydrogenase: distribution and the reaction product. *J. Bacteriol.* **137:**22–27.
89. **Miyajima, R., and I. Shiio.** 1970. Regulation of aspartate family amino acid biosynthesis in *Brevibacterium flavum*. III. Properties of homoserine dehydrogenase. *J. Biochem.* **68:**311–319.
90. **Miyajima, R., and I. Shiio.** 1971. Regulation of aspartate family amino acid biosynthesis in *Brevibacterium flavum*. IV. Repression of the enzymes in threonine biosynthesis. *Agric. Biol. Chem.* **35:**424–430.
91. **Miyajima, R., and I. Shiio.** 1972. Regulation of aspartate family amino acid biosynthesis in *Brevibacterium flavum*. V. Properties of homoserine kinase. *J. Biochem.* **71:**219–226.
92. **Miyajima, R., and I. Shiio.** 1973. Regulation of aspartate family amino acid biosynthesis in *Brevibacterium flavum*. VII. Properties of homoserine O-transacetylase. *J. Biochem.* **73:**1061–1068.
93. **Moir, D., and H. Paulus.** 1977. Properties and subunit structure of aspartokinase II from *Bacillus subtilis*. *J. Biol. Chem.* **252:**4648–4654.
94. **Moir, D., and H. Paulus.** 1977. Immunological and chemical comparison of the nonidentical subunits of aspartokinase II from *Bacillus subtilis*. *J. Biol. Chem.* **252:**4655–4661.
95. **Moir, D. T.** 1977. Ph.D. thesis. Harvard University, Cambridge, Mass.
96. **Monod, J., J. Wyman, and J. P. Changeux.** 1965. On the nature of allosteric transitions: a plausible model. *J. Mol. Biol.* **12:**88–105.
97. **Mueller, J. P., and H. W. Taber.** 1989. Isolation and sequence of *ctaA*, a gene required for cytochrome aa_3 biosynthesis and sporulation in *Bacillus subtilis*. *J. Bacteriol.* **171:**4967–4978.
98. **Ozaki, H., and I. Shiio.** 1982. Methionine biosynthesis in *Brevibacterium flavum*: properties and essential role of O-acetylhomoserine sulfhydrylase. *J. Biochem.* **91:** 1163–1171.
99. **Parsot, C.** 1986. Evolution of biosynthetic pathways: a common ancestor for threonine synthase, threonine dehydratase and D-serine dehydratase. *EMBO J.* **5:**3013–3019.
100. **Parsot, C., and G. N. Cohen.** 1988. Cloning and nucleotide sequence of the *Bacillus subtilis hom* gene coding for homoserine dehydrogenase. Structural and evolutionary relationships with *Escherichia coli* aspartokinases-homoserine dehydrogenases I and II. *J. Biol. Chem.* **263:**14654–14660.
101. **Paulus, H.** 1984. Regulation and structure of aspartokinase in the genus *Bacillus*. *J. Biosci.* **6:**403–418.
102. **Paulus, H., and E. Gray.** 1967. Multivalent feedback inhibition of aspartokinase in *Bacillus polymyxa*. I. Kinetic studies. *J. Biol. Chem.* **242:**4980–4986.
103. **Paulus, H., and E. Gray.** 1968. Multivalent feedback inhibition of aspartokinase in *Bacillus polymyxa*. II. Effect of nonpolar L-amino acids. *J. Biol. Chem.* **243:** 1349–1355.
104. **Peoples, O. P., W. Liebl, M. Bodis, P. J. Maeng, M. T. Follettie, J. A. Archer, and A. J. Sinskey.** 1988. Nucleotide sequence and fine structure analysis of the *Corynebacterium glutamicum hom-thrB* operon. *Mol. Microbiol.* **2:**63–72.
105. **Petricek, M., L. Rutberg, and L. Hederstedt.** 1989. The structural gene for aspartokinase II in *Bacillus subtilis* is closely linked to the *sdh* operon. *FEMS Microbiol. Lett.* **61:**85–88.
106. **Piggot, P. J., and J. A. Hoch.** 1985. Revised genetic linkage map of *Bacillus subtilis*. *Microbiol. Rev.* **49:**158–179.
107. **Piggot, P. J., A. Moir, and D. A. Smith.** 1981. Advances in the genetics of *Bacillus subtilis* differentiation, p. 29–39. *In* H. S. Levinson, A. L. Sonenshein, and D. J. Tipper (ed.), *Sporulation and Germination*. American Society for Microbiology, Washington, D.C.
108. **Rao, A. S.** 1985. Regulation of lysine and dipicolinic acid biosynthesis in *Bacillus brevis* ATCC 10068: significance of derepression of the enzymes during the change from vegetative growth to sporulation. *Arch. Microbiol.* **141:**143–150.
109. **Reinscheid, D. J., B. J. Eikmanns, and H. Sahm.** 1991. Analysis of a *Corynebacterium glutamicum hom* gene coding for a feedback-resistant homoserine dehydrogenase. *J. Bacteriol.* **173:**3228–3230.
110. **Richaud, F., C. Richaud, P. Ratet, and J. C. Patte.** 1986. Chromosomal location and nucleotide sequence of the *Escherichia coli dapA* gene. *J. Bacteriol.* **166:**297–300.
111. **Ron, E. Z., and B. D. Davis.** 1971. Growth rate of

Escherichia coli at elevated temperatures: limitation by methionine. *J. Bacteriol.* **107:**391–396.

112. **Ron, E. Z., and M. Shani.** 1971. Growth rate of *Escherichia coli* at elevated temperatures: reversible inhibition of homoserine *trans*-succinylase. *J. Bacteriol.* **107:** 397–400.
113. **Rosner, A.** 1975. Control of lysine biosynthesis in *Bacillus subtilis*: inhibition of diaminopimelate decarboxylase by lysine. *J. Bacteriol.* **121:**20–28.
114. **Rosner, A., and H. Paulus.** 1971. Regulation of aspartokinase in *Bacillus subtilis*. The separation and properties of two isofunctional enzymes. *J. Biol. Chem.* **246:**2965–2971.
115. **Roten, C. A. H., C. Brandt, and D. Karamata.** 1991. Genes involved in *meso*-diaminopimelate synthesis in *Bacillus subtilis*: identification of the gene encoding aspartokinase I. *J. Gen. Microbiol.* **137:**951–962.
116. **Saleh, F., and P. J. White.** 1979. Metabolism of DD-2,6-diaminopimelic acid by a diaminopimelate-requiring mutant of *Bacillus megaterium*. *J. Gen. Microbiol.* **115:** 95–100.
117. **Schendel, F. J., and M. C. Flickinger.** 1992. Cloning and nucleotide sequence of the gene coding for aspartokinase II from a thermophilic methylotrophic *Bacillus* sp. *Appl. Environ. Microbiol.* **58:**2806–2814.
118. **Schildkraut, I.** 1974. Ph.D. Thesis. University of Miami, Coral Gables, Fla.
119. **Schildkraut, I., and S. Greer.** 1973. Threonine synthetase-catalyzed conversion of phosphohomoserine to α-ketobutyrate in *Bacillus subtilis*. *J. Bacteriol.* **115:**777–785.
120. **Schleiffer, K. H., and O. Kandler.** 1972. Peptidoglycan types of bacterial cell walls and their taxonomic implications. *Bacteriol. Rev.* **36:**407–477.
121. **Schrumpf, B., A. Schwarzer, J. Kalinowski, A. Pühler, L. Eggeling, and H. Sahm.** 1991. A functionally split pathway for lysine synthesis in *Corynebacterium glutamicum*. *J. Bacteriol.* **173:**4510–4516.
122. **Shevtchenko, T. N., O. V. Okunev, Z. M. Aleksieva, and S. S. Malyuta.** 1984. Expression of genes for the biosynthesis of lysine from *Bacillus subtilis* in cells of *Escherichia coli*. *Tsitol. Genet.* **1:**58–60.
123. **Shevtchenko, T. N., H. O. Timashova, and Z. M. Aleksieva.** 1989. Mutants of *Bacillus subtilis* resistant to S-2-aminoethyl-L-cysteine. *Genetika* **25:**1937–1945.
124. **Shevtchenko, T. N., H. O. Timashova, Z. M. Aleksieva, N. V. Rotnin, and S. S. Malyuta.** 1988. *Bacillus subtilis* mutants auxotrophic for lysine. *Mol. Genet. Mikrobiol. Virusol.* **6:**33–37.
125. **Shiio, I., and R. Miyajima.** 1969. Concerted inhibition and its reversal by end products of aspartate kinase in *Brevibacterium flavum*. *J. Biochem.* **65:**849–859.
126. **Shiio, I., R. Miyajima, and S. Hakamori.** 1970. Homoserine dehydrogenase genetically desensitized to the feedback inhibition in *Brevibacterium flavum*. *J. Biochem.* **68:**859–866.
127. **Shiio, I., R. Miyajima, and K. Sano.** 1970. Genetically desensitized aspartokinase to the concerted feedback inhibition in *Brevibacterium flavum*. *J. Biochem.* **68:** 701–710.
128. **Shiio, I., and H. Ozaki.** 1981. Feedback inhibition by methionine and S-adenosylmethionine, and desensitization of homoserine acetyltransferase in *Brevibacterium flavum*. *J. Biochem.* **89:**1493–1500.
129. **Shiio, I., A. Yokota, Y. Toride, and S. Sugimoto.** 1989. Threonine production by dihydrodipicolinate synthase-defective mutants of *Brevibacterium flavum*. *Agric. Biol. Chem.* **53:**41–48.
130. **Shimotsu, H., M. I. Kuroda, C. Yanofsky, and D. J. Henner.** 1986. Novel form of transcription attenuation regulates expression of the *Bacillus subtilis* tryptophan operon. *J. Bacteriol.* **166:**461–471.
131. **Skarstedt, M. T., and S. B. Greer.** 1973. Threonine synthetase of *Bacillus subtilis*. The nature of an associated dehydratase activity. *J. Biol. Chem.* **248:**1032–1044.
132. **Stahly, D. P.** 1969. Dihydrodipicolinate synthase of *Bacillus licheniformis*. *Biochim. Biophys. Acta* **191:**439–451.
133. **Stahly, D. P., and R. W. Bernlohr.** 1967. Control of aspartokinase during development of *Bacillus licheniformis*. *Biochim. Biophys. Acta* **146:**467–476.
134. **Stragier, P., F. Richaud, F. Borne, and J. C. Patte.** 1983. Regulation of diaminopimelate decarboxylase synthesis in *Escherichia coli*. I. Identification of a *lysR* gene encoding an activator of the *lysA* gene. *J. Mol. Biol.* **168:**307–320.
135. **Sundharadas, G., and C. Gilvarg.** 1967. Biosynthesis of α,ε-diaminopimelic acid in *Bacillus megaterium*. *J. Biol. Chem.* **242:**3983–3988.
136. **Sung, M. H., K. Tanizawa, H. Tanaka, S. Kuramitsu, H. Kagamiyama, K. Hirotsu, A. Okamoto, T. Higuchi, and K. Soda.** 1991. Thermostable aspartate aminotransferase from a thermophilic *Bacillus* species. Gene cloning, sequence determination, and preliminary X-ray characterization. *J. Biol. Chem.* **266:**2567–2572.
137. **Sung, M. H., K. Tanizawa, H. Tanaka, S. Kuramitsu, H. Kagamiyama, and K. Soda.** 1990. Purification and characterization of thermostable aspartate aminotransferase from a thermophilic *Bacillus* species. *J. Bacteriol.* **172:**1345–1351.
138. **Thierbach, G., J. Kalinowski, B. Bachmann, and A. Pühler.** 1990. Cloning of a DNA fragment from *Corynebacterium glutamicum* conferring aminoethyl cysteine resistance and feedback resistance to aspartokinase. *Appl. Microbiol. Biotechnol.* **32:**443–448.
139. **Tosaka, O., and K. Takinami.** 1978. Pathway and regulation of lysine biosynthesis in *Brevibacterium lactofermentum*. *Agric. Biol. Chem.* **42:**95–100.
140. **Trach, K., and J. A. Hoch.** 1989. The *Bacillus subtilis spo0B* stage 0 sporulation operon encodes an essential GTP-binding protein. *J. Bacteriol.* **171:**1362–1371.
141. **Vapnek, D., and S. Greer.** 1971. Suppression by derepression in threonine dehydratase-deficient mutants of *Bacillus subtilis*. *J. Bacteriol.* **106:**615–625.
142. **Vapnek, D., and S. Greer.** 1971. Minor threonine dehydratase encoded within the threonine synthetic region of *Bacillus subtilis*. *J. Bacteriol.* **106:**983–993.
143. **Vold, B., J. Szulmajster, and A. Carbone.** 1975. Regulation of dihydrodipicolinate synthase and aspartate kinase in *Bacillus subtilis*. *J. Bacteriol.* **121:**970–974.
144. **Ward, J. B.** 1975. Peptidoglycan synthesis in L-phase variants of *Bacillus subtilis* and *Bacillus licheniformis*. *J. Bacteriol.* **124:**668–678.
145. **Webster, F. H., and R. V. Lechowich.** 1970. Partial purification and characterization of dihydrodipicolinic acid synthetase from sporulating *Bacillus megaterium*. *J. Bacteriol.* **101:**118–126.
146. **Weinberger, S., and C. Gilvarg.** 1970. Bacterial distribution of the use of succinyl and acetyl blocking groups in diaminopimelic acid biosynthesis. *J. Bacteriol.* **101:** 323–324.
147. **White, P. J.** 1983. The essential role of diaminopimelate dehydrogenase in the biosynthesis of lysine by *Bacillus sphaericus*. *J. Gen. Microbiol.* **129:**739–749.
148. **White, P. J., B. Lejeune, and E. Work.** 1969. Assay and properties of diaminopimelate epimerase from *Bacillus megaterium*. *Biochem. J.* **113:**589–601.
149. **Wyman, A., and H. Paulus.** 1975. Purification and properties of homoserine transacetylase from *Bacillus polymyxa*. *J. Biol. Chem.* **250:**3897–3903.
150. **Wyman, A., E. Shelton, and H. Paulus.** 1975. Regulation of homoserine transacetylase in whole cells of *Bacillus polymyxa*. *J. Biol. Chem.* **250:**3904–3908.
151. **Wyrick, P. B., M. McConnell, and H. J. Rogers.** 1973.

Genetic transfer of the stable L form state to intact bacterial cells. *Nature* (London) **244:**505–507.

152. **Wyrick, P. B., and H. J. Rogers.** 1973. Isolation and characterization of cell-wall defective variants of *Bacillus subtilis* and *Bacillus licheniformis*. *J. Bacteriol.* **116:** 456–465.
153. **Yamakura, F., Y. Ikeda, K. Kimura, and T. Sasakawa.** 1974. Partial purification and some properties of pyruvate-aspartic semialdehyde condensing enzyme from sporulating *Bacillus subtilis*. *J. Biochem.* **76:**611–621.
154. **Yamamoto, J., M. Shimizu, and K. Yamane.** 1991. Molecular cloning and analysis of nucleotide sequence of the *Bacillus subtilis lysA* gene region using *B. subtilis* phage vectors and a multi-copy plasmid, pUB110. *Agric. Biol. Chem.* **55:**1615–1626.
155. **Yamane, K., Y. Takeichi, T. Masuda, F. Kawamura, and H. Saito.** 1982. Construction and physical map of a *Bacillus subtilis* specialized transducing phage ρ11 containing *Bacillus subtilis* lys^+ gene. *J. Gen. Appl. Microbiol.* **28:**417–428.
156. **Yeggy, J. P., and D. P. Stahly.** 1980. Sporulation and regulation of homoserine dehydrogenase in *Bacillus subtilis*. *Can. J. Microbiol.* **26:**1386–1391.
157. **Yeh, E. C., and W. Steinberg.** 1978. The effect of gene position, gene dosage and a regulatory mutation on the temporal sequence of enzyme synthesis accompanying outgrowth of *Bacillus subtilis* spores. *Mol. Gen. Genet.* **158:**287–296.
158. **Yeh, P., A. M. Sicard, and A. J. Sinskey.** 1988. General organization of the genes specifically involved in the diaminopimelate-lysine biosynthetic pathway of *Corynebacterium glutamicum*. *Mol. Gen. Genet.* **212:**105–111.
159. **Yeh, P., A. M. Sicard, and A. J. Sinskey.** 1988. Nucleotide sequence of the *lysS* gene of *Corynebacterium glutamicum* and possible mechanisms for modulation of its expression. *Mol. Gen. Genet.* **212:**112–119.
160. **Yugari, Y., and C. Gilvarg.** 1965. The condensation step in diaminopimelate biosynthesis. *J. Biol. Chem.* **240:** 4710–4716.
161. **Zeigler, D. R., and H. Paulus.** Unpublished data.
162. **Zhang, J. J., F. M. Hu, N. Y. Chen, and H. Paulus.** 1990. Comparison of the three aspartokinase isozymes in *Bacillus subtilis* Marburg and 168. *J. Bacteriol.* **172:**701–708.
163. **Zhang, J. J., and H. Paulus.** 1990. Desensitization of *Bacillus subtilis* aspartokinase I to allosteric inhibition by *meso*-diaminopimelate allows aspartokinase I to function in amino acid biosynthesis during exponential growth. *J. Bacteriol.* **172:**4690–4693.
164. **Zhang, J. J., and H. Paulus.** Unpublished data.

19. Biosynthesis of Aromatic Amino Acids

DENNIS HENNER and CHARLES YANOFSKY

In all organisms capable of synthesizing the three aromatic amino acids (phenylalanine, tyrosine, and tryptophan), biosynthesis proceeds from the common precursor, chorismate, via what is termed the common aromatic, or shikimate, pathway. Chorismate also serves as precursor of folic acid, ubiquinone, enterochelin, and other minor aromatic metabolites. The intermediates and enzymatic reactions involved in the formation of chorismate and its conversion to tryptophan are believed to be identical in all organisms. However, there are several known pathways of phenylalanine and tyrosine biosynthesis (3, 10). The genes and operons responsible for aromatic amino acid biosynthesis are organized differently in different bacterial species, reflecting different evolutionary histories and possibly the functional and regulatory constraints experienced by each species. We shall see that gene arrangements in *Bacillus subtilis* may have both metabolic and regulatory explanations.

B. subtilis and *Escherichia coli* differ significantly in the regulatory strategies they use to control expression of genes of aromatic amino acid metabolism. In *B. subtilis*, many of these genes are organized into common, overlapping, or interdependent transcriptional units, making cross-pathway regulation rather than specific regulation a preferred regulatory strategy. These cross-pathway interactions apparently have led to the adoption of more-complex regulatory relationships than are observed for the same genes in enteric bacteria. In addition, biosynthetic intermediates as well as end products serve as feedback inhibitors in *B. subtilis*, whereas in *E. coli*, only the aromatic amino acids fill this role. In *E. coli* and other enteric bacteria, regulatory strategies tend to be somewhat more operon specific and regulatory interactions influencing several pathways do not appear to be as common.

In this chapter, we describe the genes, enzymes, and reactions of aromatic amino acid biosynthesis in *B. subtilis*. The chromosomal locations of the various operons are summarized in Table 1. We also present our current understanding of regulation of aromatic amino acid metabolism in this bacillus. We compare the structural, functional, and regulatory features of aromatic amino acid metabolism in *B. subtilis* with that of *E. coli*.

ENZYMATIC PATHWAYS AND REGULATION

Common Aromatic Pathway

The pathway for synthesis of the common precursors of tryptophan, phenylalanine, and tyrosine is shown in Fig. 1. All of the enzymatic steps in *B. subtilis* have been characterized and shown to be identical with those in *E. coli* (16, 26a, 45, 50, 55). The first enzyme in the pathway, 3-deoxy-D-*arabino*-heptulosonate 7-phosphate (DAHP) synthase, catalyzes the conversion of erythrose 4-phosphate and phosphoenol pyruvate to DAHP. Some strains of *B. subtilis* have DAHP synthases with different properties (22, 40). The original Marburg strain of *B. subtilis* has a monofunctional DAHP synthase. This monofunctional enzyme is inhibited by prephenate but not by chorismate (40). Strain 168, which was derived from the Marburg strain by X-ray irradiation, has a bifunctional DAHP synthase that also exhibits chorismate mutase activity (22). This DAHP synthase activity is inhibited by prephenate and also by chorismate, although at a 10-fold-higher concentration (22). The two enzymes show identical chromatographic behaviors, both appear to be tetramers containing four identical subunits, and they exhibit almost identical kinetics and K_ms for their substrates (40). The bifunctional enzyme has a single binding site for prephenate and chorismate that is noncompetitive for the substrates of DAHP synthase (22). Partial proteolysis of the enzyme releases a fragment with DAHP synthase activity that has lost chorismate mutase activity and inhibition by prephenate and chorismate (22). It has been postulated that the bifunctional enzyme arose from the monofunctional enzyme by conversion of the allosteric site used for binding prephenate into an active site for chorismate mutase activity (40).

In the next three steps in the aromatic amino acid pathway, DAHP is converted first to 3-dehydroquinate (DHQ), then to 3-dehydroshikimate, and then to shikimic acid; these reactions are catalyzed by DHQ synthase, DHQ dehydratase, and shikimate dehydrogenase, respectively. The enzymes have not been characterized other than to show that their activities can be measured in crude extracts of *B. subtilis* (26a, 45). The gene encoding DHQ synthase, *aroB*, has been cloned and sequenced (15). The predicted amino acid sequence is 40% identical to that of the *E. coli* DHQ synthase (15).

Shikimate kinase catalyzes the conversion of shikimate and ATP to shikimate-3-phosphate. Shikimate kinase can be isolated in a complex with the bifunctional DAHP synthase-chorismate mutase, and its presence in this complex appears to be essential for its activity (21, 44). Shikimate kinase activity is inhibited by prephenate at concentrations similar to those that inhibit DAHP synthase activity (44). Proteolysis studies with DAHP synthase-chorismate mutase suggest

Dennis Henner • Department of Cell Genetics, Genentech Inc., 460 Point San Bruno Boulevard, South San Francisco, California 94080. **Charles Yanofsky** • Department of Biological Sciences, Stanford University, Stanford, California 94305-5020.

Table 1. Locations of operons involved in aromatic amino acid biosynthesis

Gene(s)	Map position (°)[a]
aroA(G)[b]	265
aroC	210
aroD	230
aroFBH	210
aroI	25
aroJ	Unmapped
pheA	245
hisH, tyrA, aroE	210
trpEDCFBA	210
trpG[c]	10
trpS	100
mtrAB	215

[a] Map positions are based on the 360° genetic linkage map compiled by Piggot and Hoch (54).

[b] The chorismate mutase encoded by a gene given the designation *aroG* has been shown to be part of a bifunctional DAHP synthase enzyme, whose specifying gene is designated *aroA*. Only certain strains have this bifunctional enzyme, as described in detail in the text.

[c] Formerly designated *trpX*.

that its activation site for shikimate kinase is close to but distinct from the chorismate mutase active site (21).

5-Enolpyruvoylshikimate-3-phosphate (EPSP) synthase converts phosphoenolpyruvate and shikimate 3-phosphate into EPSP, which is then converted by chorismate synthase into chorismate, the terminal metabolite of the common aromatic amino acid pathway. From this point, the pathway diverges into the terminal amino acid pathways and several vitamin synthesis pathways. EPSP synthase and chorismate synthase activities have been demonstrated in extracts of *B. subtilis*, but the enzymes have not been otherwise characterized (45). The gene encoding chorismate synthase has been isolated and sequenced (15); the deduced amino acid sequence shows about 38% identity with the equivalent polypeptides from *E. coli* and *Salmonella typhimurium* (15).

Regulation of Activities of Enzymes of the Common Pathway

B. subtilis uses a strategy different from that of *E. coli* to regulate the flow of metabolites into the common aromatic amino acid pathway. In *E. coli*, there are three DAHP synthase isoenzymes, each of which is specifically inhibited by one of the three aromatic amino acids (55). Thus, each amino acid influences the flow of carbon into the pathway. In *B. subtilis*, however, the aromatic amino acids exhibit no significant inhibition of DAHP synthase (26a). Rather, prephenate inhibits DAHP synthase and regulates the entry of metabolites into the common pathway (26a).

An excess of phenylalanine and tyrosine would

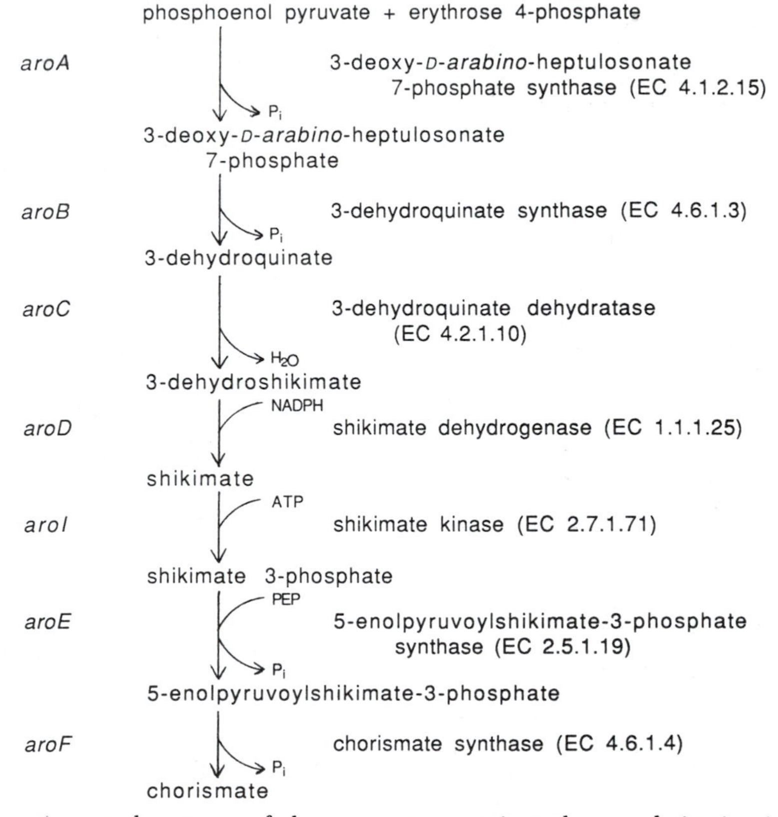

Figure 1. Genes, reactions, and enzymes of the common aromatic pathway culminating in the synthesis of chorismate, the precursor of phenylalanine, tyrosine, and tryptophan. Structures of the compounds depicted in Fig. 1 through 4 can be found in Pittard (55).

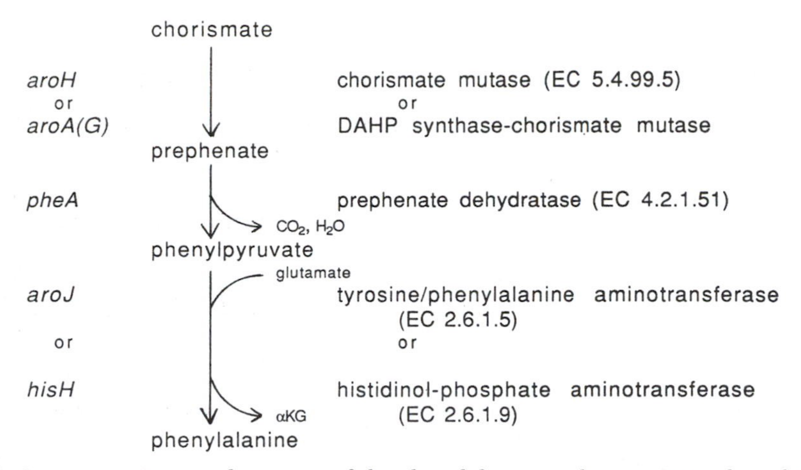

Figure 2. Genes, reactions, and enzymes of the phenylalanine pathway. αKG, α-ketoglutarate.

feedback-inhibit the further conversion of prephenate (discussed below). The buildup of prephenate would then inhibit DHAP synthase and reduce the flow of metabolites into the common pathway. This strategy could potentially starve the cell for tryptophan, whose synthesis pathway branches from the common pathway at chorismate. Chorismate mutase, which catalyzes the conversion of chorismate to prephenate, is product inhibited by prephenate (48). This product inhibition presumably is sufficient to channel the chorismate into the tryptophan pathway. Perhaps the low content of tryptophan in proteins allows a reduced metabolite flow through the common pathway to provide sufficient tryptophan and also the vitamins for which chorismate is a precursor. A more thorough discussion of these regulatory events can be found in Kane et al. (32).

One intermediate enzyme in the common pathway, shikimate kinase, is also inhibited by chorismate and prephenate (44). The fact that shikimate kinase can be found in a complex with the bifunctional DAHP synthase-chorismate mutase (44) might explain the mechanism of inhibition but does not explain the necessity for regulation of this intermediate step. In *E. coli*, there are two shikimate kinases. That fact has been used to postulate that shikimate might have once been the starting point for aromatic amino acid synthesis or that an unknown pathway branches from shikimate (55). The regulation of the activity of the shikimate kinase in *B. subtilis* supports this speculation. Shikimate and quinate can be used as carbon sources by some organisms.

Phenylalanine Pathway

Both the phenylalanine and tyrosine pathways begin with the conversion of chorismate to prephenate by chorismate mutase (Fig. 2). Two chorismate mutase activities can be found in various strains of *B. subtilis*. One is the bifunctional DAHP synthase-chorismate mutase discussed above. The other is a monofunctional chorismate mutase (12). Strains with the active monofunctional chorismate mutase exhibit about 10-fold-higher levels of activity than those that contain only the bifunctional enzyme (32). The monofunctional enzyme is presumably the "original" chorismate mutase, but strains with the lower levels of activity are prototrophic and exhibit only subtle growth differences in the presence of certain amino acid analogs such as D-tyrosine (32, 40). The monofunctional chorismate mutase has been purified to apparent homogeneity and exists as a homodimer with a subunit molecular weight of 14,600. It exhibits classical Michaelis-Menton kinetics, with a K_m of 100 μM and a k_{cat} of 50/s per subunit (12). The gene encoding the enzyme has been cloned; the deduced amino acid sequence bears only minimal similarity to those of other chorismate mutases (12). Neither enzyme exhibits any apparent regulation by amino acids, but both exhibit product inhibition by prephenate (48).

Prephenate is the last common precursor of phenylalanine and tyrosine. Prephenate dehydratase catalyzes the conversion of prephenate to phenylpyruvate. Phenylalanine strongly inhibits the activity of this enzyme; tryptophan and tyrosine inhibit it to a lesser extent (6, 57).

Phenylpyruvate is converted to phenylalanine by an aminotransferase. Two such aminotransferases have been identified in *B. subtilis* (51). The first is a histidinol phosphate aminotransferase, which catalyzes both the conversion of imidazole acetol phosphate to histidinol phosphate and the terminal aminotransferase reactions in the tyrosine and phenylalanine pathways. The gene for this enzyme has been cloned, and the deduced amino acid sequence shows about 32% identity with histidinol phosphate aminotransferase sequences from other sources (13). Deletion of the gene for this enzyme causes histidine auxotrophy (13). The second aminotransferase is specific for tyrosine and phenylalanine; it is not capable of carrying out the imidazole acetol phosphate aminotransferase reaction (51). The presence of either enzyme is sufficient to satisfy the phenylalanine and tyrosine requirements for growth. The aminotransferases have not been characterized in any detail.

Tyrosine Pathway

The chorismate mutase and aminotransferase reactions in the tyrosine pathway are carried out by the same enzymes discussed above for the phenylalanine

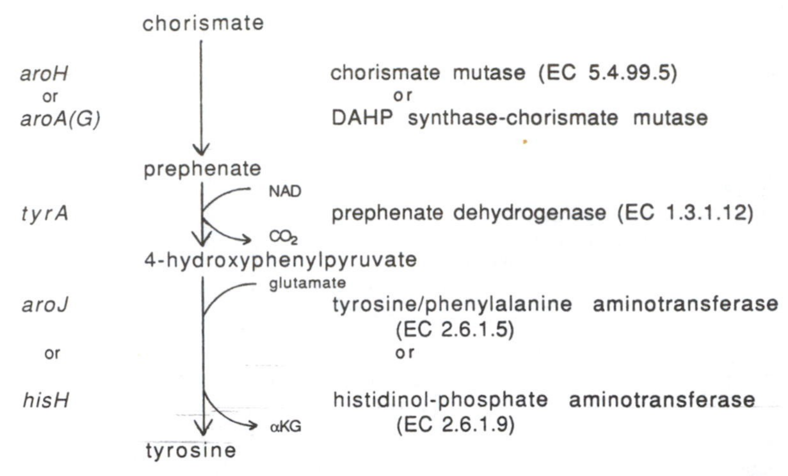

Figure 3. Genes, reactions, and enzymes of the tyrosine pathway. αKG, α-ketoglutarate.

pathway. The only specific step in the tyrosine pathway is catalyzed by prephenate dehydrogenase (Fig. 3). Prephenate dehydrogenase activity is inhibited more than 90% by 0.1 mM tyrosine (48). Tryptophan and phenylalanine can also inhibit the enzyme's activity, albeit at concentrations 2 orders of magnitude higher than that of tyrosine (48). The gene for prephenate dehydrogenase has been cloned and sequenced (13). Its deduced amino acid sequence exhibits only moderate similarity to those of other prephenate dehydrogenases, with about 15 to 17% identity over part of its sequence.

Gene Regulation for the Common Pathway

With the exception of the genes of the tryptophan pathway, regulation of expression of genes concerned with aromatic amino acid biosynthesis has not been examined in great detail. Studies in the late 1960s showed that synthesis of DAHP synthase (*aroA*) and shikimate kinase (*aroI*) was repressed moderately (three- to fivefold) by the presence of tyrosine in the culture medium (46, 49). Neither tryptophan nor phenylalanine had a repressive effect, although the presence of both phenylalanine and tyrosine moderately increased repression over that caused by tyrosine alone (49). Synthesis of the other enzymes of the common aromatic amino acid pathway apparently was not repressed by the presence of aromatic acids (49). Why tyrosine alone regulates the synthesis of two enzymes of the common pathway is not clear.

Regulation of Genes of Tyrosine and Phenylalanine Biosynthesis

Synthesis of the bifunctional DAHP synthase-chorismate mutase [*aroA*(*G*)] is repressed by tyrosine, as discussed above. Synthesis of the monofunctional chorismate mutase (*aroH*) was shown to be moderately repressed by the presence of aromatic amino acids, but no attempt was made to differentiate which amino acid was responsible for this effect (41). Synthesis of prephenate dehydratase (*pheA*), the only gene specific to phenylalanine synthesis, is not repressed by phenylalanine or either of the other aromatic amino acids (49). Tyrosine was shown to modestly repress the synthesis of prephenate dehydrogenase (*tyrA*), i.e., about three- to fivefold (49, 58). However, the sequence of the *hisH-tyrA-aroE* operon (13) sheds no light on possible mechanisms for this repression. There are no reports on regulation of synthesis of the phenylalanine- and tyrosine-specific aminotransferase (*aroJ*).

Cross-Pathway Regulation of Aromatic Amino Acid Enzyme Synthesis by Histidine

As early as 1963 (52), it was recognized that histidine had some influence on regulation of expression of genes of the aromatic amino acid pathway. It was shown that histidine could repress the synthesis of both DAHP synthase and prephenate dehydrogenase (46). Later, single-step mutants that were derepressed for enzymes of both aromatic amino acid biosynthesis and histidine biosynthesis were isolated (5). The genes for the derepressed histidine enzymes were not linked to any of the genes for aromatic amino acid biosynthesis (5). A locus responsible for this phenotype was mapped (47), but the biochemical basis of this cross-pathway regulation was not determined.

There are further interconnections between histidine and aromatic amino acid biosynthesis. The gene encoding histidinol phosphate aminotransferase (which can also catalyze the terminal steps in the phenylalanine and tyrosine pathways) is located immediately downstream of the cluster of *trp* genes and is cotranscribed with the *tyrA* and *aroE* genes. Histidine has also been shown to stimulate the activity of anthranilate synthase, the first enzyme of the tryptophan pathway (31). Finally, a common metabolite, 5-phosphoribosyl-1-pyrophosphate, is used in the synthesis of both tryptophan and histidine. These links might be relics of a more primitive regulatory pathway that once interconnected histidine and aromatic amino acid biosyntheses.

Tryptophan Pathway

The pathway of synthesis of tryptophan from the common aromatic precursor chorismate is shown in

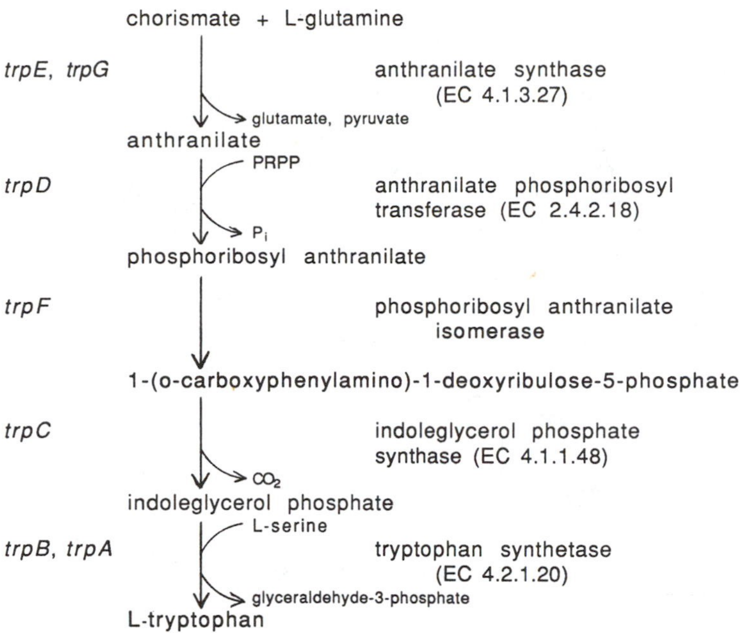

Figure 4. Genes, reactions, and enzymes of the tryptophan pathway. PRPP, 5′-phosphoryl-1-pyrophosphate.

Fig. 4. This sequence of reactions requires seven enzymatic functions, designated A through G (7, 68). The polypeptide domains responsible for performing these functions have been assigned letters corresponding to the enzymatic functions. The enzymes of tryptophan biosynthesis are readily measured in extracts of most bacterial species, including *B. subtilis* (18). However it has been observed that phosphorylribosyl transferase (TrpD) and indoleglycerol phosphate synthase (TrpC) of *B. subtilis* are particularly labile (18). Stabilizing conditions for these enzymes have been developed (18).

The *B. subtilis* enzymes that catalyze the reactions of tryptophan formation have been only partially characterized. Of those examined, the observed properties generally resemble those of the corresponding proteins of *E. coli* (66). However, there are several significant differences in gene organization that are reflected in protein structure. In *E. coli*, the seven enzymatic functions necessary for tryptophan synthesis are encoded in five genes that constitute a single operon. These functions are performed by two enzyme complexes, $E_2(G \cdot D)_2$ and B_2A_2 (a dot separating letters representing individual domains indicates that the two domains are joined covalently), and a single bifunctional enzyme, C · F. In *B. subtilis*, the seven enzymatic functions are encoded in seven genes (Fig. 4); these specify two enzyme complexes, EG and B_2A_2, that resemble *E. coli* complexes, with each of the remaining three polypeptides functioning independently. In *B. subtilis*, six of the seven *trp* genes (Fig. 5) are contiguous (4) and are organized as a region of clustered genes of aromatic amino acid biosynthesis that has been called a supraoperon (58). The seventh gene, *trpG*, resides in an operon primarily concerned with folic acid biosynthesis (27, 62) (Fig. 5). In *B. subtilis*, *trpD* immediately follows *trpE*, but the TrpD polypeptide does not appear to associate with the EG enzyme complex. TrpC and TrpF of *B. subtilis* are encoded by adjacent genes (Fig. 5), whereas as mentioned above, in *E. coli*, these genes are fused in the order *trpC-trpF*. Detailed biochemical and structural analyses with the *E. coli* C · F enzyme have shown that although the C and F domains are joined covalently, they have independent active sites (56). The genes *trpB* and *trpA* follow *trpF* in the genomes of both organisms; their products form the tetrameric tryptophan synthetase enzyme complex (Fig. 4). Several *B. subtilis* mutants in which more than one Trp pathway enzymatic activity appears to have been lost as a consequence of a single mutational event have been isolated (65). Although these findings suggest polypeptide interactions, this possibility is not supported by most of the existing data.

Products of other biosynthetic pathways are required for tryptophan formation from chorismate; these products include L-glutamine, 5-phosphoribosy-1-pyrophosphate, and L-serine.

Anthranilate synthase

Anthranilate synthase of *B. subtilis*, the enzyme complex catalyzing the initial step specific to tryptophan synthesis, appears to be a 1:1 complex of TrpE and TrpG polypeptides (30). The TrpE polypeptide contains a tryptophan binding site that provides feedback inhibition of anthranilate synthase activity and hence regulation of entry of chorismate into the tryptophan pathway (48). The 1:1 subunit composition of the *B. subtilis* anthranilate synthase is not certain but seems likely on the basis of sizing column analyses (30). In most (but not all) organisms for which the composition of the anthranilate synthase complex has been accurately determined, the enzyme is a tetramer (69). In many organisms, the *trpG* domain is fused to a second Trp domain to form a multifunctional polypeptide. In *E. coli*, the TrpG and TrpD domains are fused (68), whereas in *Neurospora crassa*, the TrpG

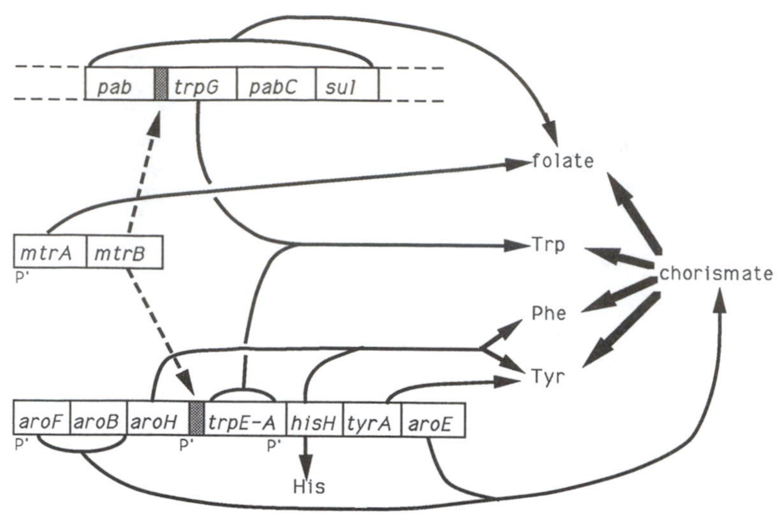

Figure 5. Metabolic and regulatory interactions involving the folate operon, *mtr* operon, and aromatic tryptophan supraoperon. Heavy arrows indicate metabolic pathways converting chorismate to each of its end products. Lighter arrows designate genes that encode enzymes involved in synthesis of the various metabolites. The *trpG* polypeptide participates in the biosynthesis of both tryptophan and *p*-aminobenzoic acid, a component of folic acid. Both *aroH* and *hisH* encode enzymes involved in phenylalanine and tyrosine syntheses. Dotted arrows indicate sites of action of the MtrB polypeptide in the respective transcripts. The transcription regulation site preceding *trpE* and the translation regulation site preceding *trpG* are designated by shaded boxes. The locations of known promoters for these operons are indicated by P'. Transcription is from left to right within each operon shown. Additional information is provided in the text.

domain resides in a trifunctional polypeptide, G · C · F (59). As mentioned above, in *B. subtilis*, *trpE* immediately precedes *trpD* in the *trp* gene cluster, whereas *trpG* is located in the folic acid operon (Fig. 5). The TrpG polypeptide functions as a glutamine amidotransferase component of two enzyme complexes, TrpE-TrpG, which is responsible for *o*-aminobenzoate (anthranilate) formation from chorismate in the tryptophan pathway, and PabB-TrpG, which is responsible for *p*-aminobenzoate formation from chorismate in folate biosynthesis (27, 28). Dual-function polypeptides of the TrpG type have been called amphibolic polypeptides (34). The TrpE polypeptide alone can catalyze anthranilate formation from chorismate in the presence of high ammonia concentrations at high pHs, but when TrpE is complexed with TrpG, glutamine is used efficiently as an amino donor at physiological pHs (24). Since in *E. coli* the TrpG domain is fused to the TrpD domain, the *E. coli* complex, $E_2(G \cdot D)_2$, catalyzes both anthranilate formation and the subsequent step in tryptophan biosynthesis. In *E. coli*, a separate glutamine amidotransferase, the product of *pabA*, participates in *p*-aminobenzoic acid formation (35). However, an additional enzyme is required to complete synthesis of *p*-aminobenzoate in both organisms (53). PabA and TrpG of *E. coli* and TrpG of *B. subtilis* are homologous polypeptides (7). Interestingly, TrpG of *B. subtilis* is more closely related to PabA than to TrpG of *E. coli* (7).

There are species of enteric bacteria in which *trpG* is not fused to *trpD*. However, in these species, there is a separate PabA polypeptide, as in *E. coli*. For example, *trpG* of *Serratia marcescens* is located between *trpE* and *trpD* in its *trp* operon (43). In *B. subtilis*, there is no indication that a *trpG*-like sequence ever was located between *trpE* and *trpD* (2). Unlike the *trp* gene organization in *B. subtilis*, the gene for the amphibolic TrpG polypeptide of *Acinetobacter calcoaceticus* is associated with other *trp* genes (33).

Other Trp enzymes

The TrpD, TrpC, and TrpF polypeptides of *B. subtilis* appear to be discrete functional proteins; their migrations on sizing columns suggest that each exists as a monomer. Tryptophan synthetase of *B. subtilis*, like the enzyme of *E. coli*, is a tetrameric complex containing two TrpA and two TrpB polypeptides. The three-dimensional structures of the phosphoribosylanthranilate isomerase-indoleglycerol phosphate synthase (TrpF · TrpC) of *E. coli* (56) and the tryptophan synthetase $TrpB_2TrpA_2$ complex of *S. typhimurium* have been determined (23). Comparison of the sequences of the TrpF, TrpC, TrpB, and TrpA domains of *B. subtilis* with those of *E. coli* and *S. typhimurium* suggests that these three-dimensional models will prove applicable to the *B. subtilis* proteins.

Comparison of *trp* coding regions and polypeptides of *B. subtilis* and *E. coli*

The *trp* polypeptides of *B. subtilis* are generally homologous to their corresponding polypeptides of other species and have shown greatest similarity to those of *Pseudomonas aeruginosa* (7). The sequence data that are available suggest that all present-day functional *trp* polypeptide domains evolved from common ancestral domains. The extent of sequence divergence varies among the *trp* polypeptide domains (7), with the *trpB* polypeptide sequence varying the least. The extent of sequence conservation presumably reflects the fraction of the amino acid residues of each domain that has an important function. The properties of the Trp enzymes of *B. subtilis*, *Bacillus pumilus*, and *Bacillus alvei* have been compared, and some differences have been noted (19). Gene-polypeptide relationships in these bacilli appear to be the same.

Regulation of Genes of Tryptophan Biosynthesis in *B. subtilis*

Early studies of regulation of tryptophan biosynthesis in *B. subtilis* were based on in vivo physiological studies, mutant analyses, and measurements of enzyme levels and activities. The findings in these investigations established that the six proteins encoded by the *trp* gene cluster were generally synthesized coordinately upon tryptophan starvation (18) and that the first enzyme of the pathway, anthranilate synthase, was feedback inhibited by tryptophan (K_i = 0.07 mM) (48). It was also shown that one of the *trp* genes, *trpG* (27), which encodes the glutamine amidotransferase or small subunit of anthranilate synthase), was not in the *trp* gene cluster but was associated with genes concerned with folic acid biosynthesis; nevertheless, *trpG* expression was observed to be regulated by tryptophan (27). Complex regulatory responses were observed in experiments that explored relationships between synthesis of tryptophan and other amino acids, notably phenylalanine, tyrosine, and histidine (58). Mutants altered at a single locus, designated *mtr*, conferred resistance to tryptophan analogs (20, 52); resistance was shown to be due to synthesis of high levels of the Trp enzymes, suggesting that the *mtr* locus was regulatory (20). The *mtr* locus was mapped to a location preceding the genes *aroF* and *aroB*, which lie just upstream of the *trp* gene cluster (17). Some *mtr* mutants exhibited a partial phenylalanine requirement (20), while in others, there were increased levels of tryptophan, tyrosine, and histidine pathway enzymes (58, 64). A temperature-sensitive tryptophanyl-tRNA synthetase mutant that produced elevated levels of the tryptophan biosynthetic enzymes was isolated (63). These and other observations suggested that regulation of the genes of tryptophan metabolism in *B. subtilis* was complex and that synthesis of the tryptophan pathway enzymes was responsive to products of other metabolic pathways.

In 1984, the nucleotide sequence of the *trp* gene cluster and its flanking regions was reported (14). This sequence information served as the basis for experiments designed to determine explanations for the regulatory observations made in earlier studies.

The *trp* promoter

The transcription start site of the *trp* promoter was located approximately 200 bp upstream of the start codon for the TrpE polypeptide (60). The DNA segment responsible for the organism's sensitivity to tryptophan analogs and response to mutations at the *mtr* locus was localized to the *trp* leader segment, the region between the *trp* promoter and the *trpE* start codon (61). An important finding was that when introduced into *B. subtilis*, a multicopy plasmid containing the *trp* promoter and leader region conferred resistance to the tryptophan analog 5-methyltryptophan (60). Resistance was shown to be due to elevated expression of the resident *trp* genes of the bacterium (37). It was proposed that multiple copies of the promoter-leader region increased *trp* gene expression by providing excess *trp* transcripts that titrated a limiting amount of some regulatory molecule (61). The normal function of this regulatory molecule, then, would be to limit *trp* gene expression whenever the intracellular tryptophan concentration was high (60). Transcription of the plasmid-borne *trp* leader region was essential for this effect, but other promoters could substitute for the *trp* promoter (60).

The *trp* promoter contains elements recognizable as −35 and −10 regions comparable to the consensus promoter sequences recognized by the major forms of *E. coli* and *B. subtilis* RNA polymerases. There is no palindromic sequence suggestive of an operator in the vicinity of the *trp* promoter. Rather, the sequence of the leader region has features typical of the leader regions of operons regulated by transcription attenuation (37, 39).

Analysis of regulatory features of the leader region

The predicted sequence of the leader segment of the transcript produced following transcription initiation at the *trp* promoter would allow the transcript to fold and form alternative hairpin structures (61) of the type that have been characterized in the leader transcript of the *trp* operon of *E. coli* (39) (Fig. 6). The comparable though unrelated sequence of the *B. pumilus trp* leader transcript could conceivably fold to form similar alternative hairpin structures (38). In each organism, the alternative *trp* leader RNA hairpins have sequences in common; hence, formation of the earlier structure, designated the antiterminator, would prevent formation of the latter structure, the terminator. Mapping of the 3′ end of the in vivo transcript established that in the presence of tryptophan, transcription generally terminates following synthesis of the presumed terminator structure (61). It was also shown that whenever tryptophan was growth limiting, termination was considerably reduced and transcription continued into the *trp* structural genes. These observations established that transcription attenuation is used to regulate expression of the clustered *trp* genes of *B. subtilis*. In vitro transcription analyses with templates bearing overlapping deletions established that formation of the antiterminator prevented transcription termination (61). Termination was prevented because the alternative, terminator structure could not form when the antiterminator structure existed.

Ribosome stalling is responsible for regulation of attenuation in the *trp* and other operons of *E. coli* and other bacterial species (39). The *trp* leader transcripts of *B. subtilis* and *B. pumilus* do not code for leader peptides; hence, transcription attenuation control in these organisms must involve some mechanism other than ribosome stalling at Trp codons. Deletion analyses and in vivo titration experiments identified a repeated sequence (AGAATGAGTT, beginning at nucleotide positions +36 and +69 of the *B. subtilis* transcript) that was responsible for the regulatory titration effect (37, 61). These RNA segments presumably serve as binding sites for the Mtr regulatory molecule (Fig. 6). The repeat sequence beginning at position +69 is within the 5′ strand of the antitermi-

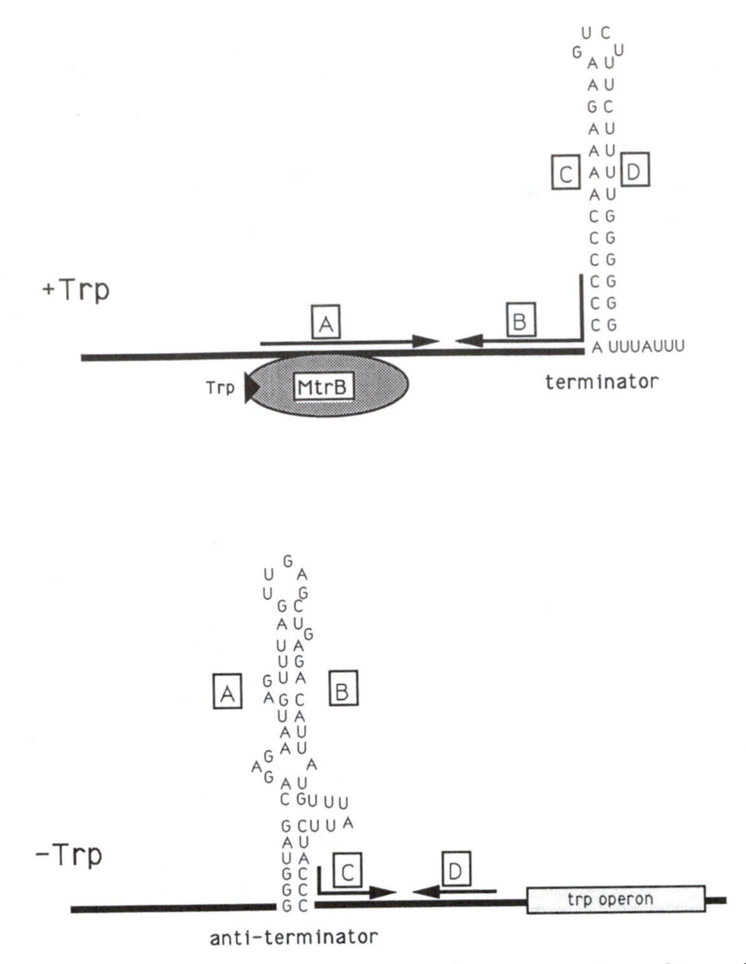

Figure 6. Model of regulation of transcription termination at the attenuator located immediately distal to the *trp* promoter in the supraoperon of *B. subtilis*. The alternative terminator and antiterminator RNA structures and the proposed mechanism of action of the tryptophan-activated MtrB protein are indicated. A and B denote RNA segments that form the antiterminator hairpin, and C and D denote RNA segments that form the terminator structure. Since stems B and C overlap by 4 nucleotides (arrows B and C), formation of antiterminator and terminator structures is mutually exclusive.

nator RNA hairpin. The Mtr protein is believed to act by preventing formation of the antiterminator hairpin structure. It may bind to the +69 sequence alone or to both this sequence and the +36 sequence, forming an RNA loop that disrupts the antiterminator. These hypotheses have yet to be tested. In vivo studies suggest that MtrB, a polypeptide product of the *mtr* locus, is necessary for regulation of *trp* gene expression in *B. subtilis* (11, 37).

Regulation of synthesis of TrpG, the amphibolic glutamine amidotransferase

The TrpG polypeptide, an essential component of the anthranilate synthase complex that catalyzes the initial reaction unique to tryptophan biosynthesis, is encoded by a gene located in an operon specifying polypeptides concerned with the synthesis of *p*-aminobenzoic acid and the pterin precursor of folic acid (Fig. 5) (62). Insertional inactivation of *trpG* in fact results in a requirement for tryptophan and a partial requirement for *p*-aminobenzoic acid (62). However, in high-pH media containing high concentrations of ammonia, TrpG function is not essential for either tryptophan or *p*-aminobenzoic acid synthesis. *trpG* is preceded by *pabB*, the structural gene for the large subunit of *p*-aminobenzoate synthase, with which TrpG presumably associates (62). The Shine-Dalgarno sequence for *trpG* overlaps the stop codon of *pabB*. The sequence surrounding and including the *trpG* Shine-Dalgarno sequence is homologous to the Mtr-binding-site sequence in the leader segment of the *trp* transcript (62). It has been suggested that translation of the *trpG* coding region is regulated by Mtr binding at the segment containing the Shine-Dalgarno sequence (62). This would explain coordinate regulation of *trpG* and *trpEDCFBA* expression by tryptophan. Preliminary in vivo studies indicate that the Mtr protein does regulate translation of the *trpG* coding region (36).

Characteristics of the *mtr* operon

The *mtr* operon of *B. subtilis* was cloned, sequenced, and found to contain two genes, *mtrA* and *mtrB*, encoding 22- and 8-kDa polypeptides, respectively (11). MtrA is homologous to GTP cyclohydrolase I of rats (1), and MtrB has sequence similarities to RegA, the RNA-binding protein of phage T4 (11). The mutational changes in several *mtr* mutants were determined; these changes were either in the presumed promoter for the operon or within *mtrB* (11). Attempts to inactivate MtrA were unsuccessful, suggesting that this protein has an essential function. These studies preceded the discovery that MtrA catalyzes the first (essential) step in pterin biosynthesis in the folic acid pathway (1). Several *mtrB* mutations were *trans* dominant, suggesting that the MtrB protein functions as a multimeric protein (11). A multimeric state might facilitate RNA looping between MtrB molecules bound at the two binding site sequences in the *trp* leader transcript. However, as mentioned above, RNA binding and RNA looping have yet to be demonstrated. Regulatory studies performed with *E. coli* containing plasmids encoding MtrA and/or MtrB have established that only MtrB is required for regulation by attenuation in the *B. subtilis trp* leader region (1).

In vivo studies suggest that transcription attenuation in the *trp* leader region of *B. subtilis* is directly regulated by tryptophan rather than by charged $tRNA^{Trp}$ (37). It is assumed that tryptophan activates the MtrB protein, which then binds to its specific RNA target sequences in the transcripts of the *trp* leader region and the folate operon.

SUPRAOPERON ORGANIZATION

Upstream *aroFBH* Operon

The *aroFBH* operon overlaps the *trp* promoter (Fig. 5); the *aroH* stop codon is within the *trp* promoter, and the *trp* attenuator is the (initial) transcription terminator for the *aroFBH* operon (12). In vivo studies have shown that an upstream transcript enters the *trp* leader region (61); this transcript probably initiates at the *aroFBH* promoter. It is not known which metabolite(s) regulates transcription initiation in the *aroFBH* operon. However, since deletion of the *trp* promoter does not eliminate expression of the clustered *trp* genes (61), it is likely that the *aroFBH* promoter provides a functional *trp* transcript. However, introduction of a *trp* promoter deletion into an *mtr* mutant does not confer 5-methyltryptophan resistance, which implies that expression from the upstream promoter alone is insufficient to provide levels of *trp* biosynthetic enzymes sufficient to confer drug resistance (61).

Punctuation and Overlaps Between Trp Polypeptide Coding Regions of *B. subtilis*

Each of the six Trp polypeptide coding regions within the *trp* gene cluster (*trpE*, *trpD*, *trpC*, *trpF*, *trpB*, and *trpA*) is preceded by an appropriately spaced Shine-Dalgarno sequence that is complementary to the 3′ end of *B. subtilis* 16S rRNA (14). This organization suggests that ribosomes can initiate at each of the *trp* polypeptide coding regions. For several pairs of *trp* genes, the coding regions overlap, placing the stop codon of the preceding gene distal to the start codon of the subsequent gene (14). This arrangement exists for *trpE* and *trpD*, *trpD* and *trpC*, *trpF* and *trpB*, and *trpB* and *trpA*. Overlapping coding regions are common in other operons of *B. subtilis* (9). It is conceivable that in this organism, gene overlaps are often used to facilitate coordinate translation of functionally related coding regions. Coordinating TrpG synthesis with TrpE synthesis is a special problem, since, as mentioned above, *trpG* is not within the *trp* gene cluster and its product is a component of two enzyme complexes.

Downstream Overlapping *hisH-tyrA-aroE* Operon

In early in vivo studies of aromatic amino acid biosynthesis in *B. subtilis*, it was observed that histidine, tyrosine, or tryptophan synthesis influenced the cellular levels of enzymes involved in the biosynthesis of these three amino acids (5, 29, 46, 58). It was also found that histidine reversed the growth inhibitory effects of 5-methyltryptophan and could stimulate synthesis and secretion of tryptophan (25, 31). These suggestive relationships were more firmly established when it was shown that three genes encoding enzymes involved in histidine, tyrosine, and aromatic biosynthesis are located immediately downstream from the *trp* genes on the *B. subtilis* chromosome (Fig. 5) (58). The enzymes specified are histidinol phosphate aminotransferase (*hisH*), prephenate dehydrogenase (*tyrA*), and enolpyruvylshikimate-5-phosphate synthetase (*aroE*). Furthermore, in regulatory studies, it was demonstrated that expression of the *hisH*, *tyrA*, and *aroE* genes was subject to tryptophan control, indicating that there might be some readthrough transcription beyond *trpA* (64). When the nucleotide sequence of the region immediately distal to the *trp* genes was determined (13), it became evident that *hisH*, *tyrA*, and *aroE* were in the same transcription unit as the *trp* genes, lending support to the notion that all of these genes reside in a single supraoperon.

Only 10 nucleotides separate *trpA* from *hisH*, 48 nucleotides separate *hisH* and *tyrA*, and 10 nucleotides separate *tyrA* from *aroE*. The stop codon of *aroE* is within a region which, as RNA, could fold to form a typical Rho-factor-independent transcription terminator, suggesting that this structure is the distal terminator for the supraoperon. Deletion and integrative plasmid studies verified that the genes concerned were required for histidine, tyrosine, and aromatic biosynthesis (13). It was also shown that expression of *hisH-tyrA-aroE* could be initiated at the *trp* promoter and also at a promoter located within *trpA* (13). A *hisH-lacZ* translational fusion was constructed, integrated into the chromosome, and analyzed in regulatory studies (13). The modest regulation by tryptophan was consistent with the interpretation that both the tryptophan-regulated *trp* promoter and the promoter located in *trpA* can direct *hisH* transcription. There was no evidence of regulation by histidine, in distinction to previous regulatory studies employing direct enzyme assays with mutants and wild type (58, 64). The explanation for these differences is not known.

OTHER CONSIDERATIONS

Other Genes of Aromatic Amino Acid Biosynthesis

With the exception of *aroJ*, which encodes the phenylalanine and tyrosine aminotransferases, all of the other genes involved in aromatic amino acid biosynthesis have been located on the *B. subtilis* genetic map (Table 1). None of these genes is clustered with other genes involved in aromatic amino acid biosynthesis. Since these genes have not been cloned and sequenced, it is not known whether they and other genes are organized in operons.

Other Features

The aromatic amino acid permeases of *B. subtilis* have not been studied extensively. *B. subtilis* forms a permease that transports phenylalanine and tyrosine but not tryptophan or any other amino acid (8). Transport of phenylalanine and tyrosine by this system has been shown to be competitive (8).

All of the *trp* biosynthetic enzymes of *B. subtilis* except TrpC and TrpA contain tryptophan. In *E. coli*, TrpE and TrpA lack tryptophan, and this deficiency allows preferential synthesis of these two polypeptides when cells are severely starved of tryptophan (67). Tryptophan deficiency in the *trp* enzymes reaches the extreme in *Methanobacterium thermoautotrophicum* Marburg. In this organism, all of the Trp polypeptides except TrpB lack tryptophan (42). The seven *trp* genes of this species are clustered in the order *trpEGCFBAD* (42), and all their polypeptide products are homologous to those of other bacterial species. The significance of tryptophan distribution in the Trp proteins of *B. subtilis* has not been examined.

Comparisons with Other Gram-Positive Microorganisms

The common pathway leading to chorismate appears to proceed via the same enzymatic steps in every organism that has been studied to date. However, a number of regulatory strategies that control the flow of metabolites through the pathway have evolved. These regulatory strategies appear to be common to broad groupings of microorganisms and correlate well with phylogenetic trees constructed by sequencing 16S rRNA (2b). As discussed above, *B. subtilis* DAHP synthase is feedback inhibited by chorismate and prephenate; this strategy is termed sequential feedback inhibition. The closely related staphylococci exhibit a similar regulatory strategy. Nocardiae and mycobacteria have a DAHP synthase that is feedback inhibited by tyrosine, while the enzyme from clostridia is inhibited by phenylalanine. Still other gram-positive organisms, such as streptomycetes, have unregulated DHAP synthases.

All gram-positive organisms studied to date synthesize phenylalanine exclusively via the prephenate dehydratase pathway. The organisms studied include *Bacillus*, *Staphylococcus*, *Streptococcus*, and *Brevibacterium* species and *Acholeplasma laidlawii*. The prephenate dehydratase activity from each of these organisms exhibits a pattern of regulation termed metabolic interlock, whereby metabolites outside the pathway exert allosteric effects. For example, the *B. subtilis* enzyme is inhibited by tryptophan and activated by either methionine or leucine (25a). There is variation in the details of regulation in different gram-positive organisms; e.g., the *Brevibacterium glutamicus* enzyme absolutely requires tyrosine for activity, and leucine fails to activate at all in either *A. laidlawii* or *B. glutamicus* (26).

Two enzymatic pathways from prephenate to tyrosine exist in gram-positive organisms. The first, outlined in Fig. 3, has a dehydrogenase followed by an aminotransferase. This is the pathway found in *Bacillus* spp. The other pathway transaminates prephenate to arogenate (originally called pretyrosine) and then utilizes an arogenate dehydrogenase to form tyrosine. This pathway is found in such organisms as *Brevibacterium*, *Corynebacterium*, and *Nocardia* spp. (9a, 24a).

As mentioned previously, the same seven enzymatic functions are used in the biosynthesis of tryptophan in all organisms that have this synthetic capability. The genes encoding the polypeptide domains responsible for these seven functions are organized differently in the chromosomes of many species. There are numerous examples of gene fusions that provide bi- or multifunctional proteins that perform two or more of the tryptophan pathway reactions. In *B. pumilus*, the *trp* gene arrangement and regulatory features appear to be identical to those of *B. subtilis*. In other gram-positive species in which *trp* genes have been studied, i.e., *Lactobacillus casei* and *Brevibacterium lactofermentum*, many of the *trp* genes appear to be clustered (7). In *B. lactofermentum* in particular (41a), the seven enzymatic functions are encoded in six genes arranged in an operon organized identically to that of the enterobacterium *Serratia marcescens*. The gene order in these two species is *trpE*, *-G*, *-D*, *-C·F*, *-B*, and *-A*. A remarkable feature of the *B. lactofermentum trp* operon is that its regulatory region closely resembles those of enteric bacteria rather than that of *B. subtilis* (58a). In addition, the *trp* polypeptides of *B. lactofermentum* show considerably greater sequence similarity to those of enteric species than to those of *B. subtilis* or *L. casei* (7). The authenticity of the *Brevibacterum* species that was the source of the *trp* operon that was sequenced has been questioned, but codon usage in the *trp* genes does reflect the high G+C content expected of the members of this species. The possible evolutionary implications of these findings with regard to lateral gene transfer have been discussed elsewhere (7, 25b).

The literature on the biosynthesis of aromatic amino acids is particularly rich. The enzymology, reaction mechanisms, feedback controls, and genetic organization have been studied in great detail. Comparative studies have been used to confirm and enrich evolutionary conclusions. Interested readers should consult reviews by Bentley (2a), Byng et al. (2b), Crawford (7), and Jensen (25b) for further details.

Acknowledgments. We thank Roy Jensen for helpful discussions and for providing unpublished material.

REFERENCES

1. **Babitzke, P., P. Gollnick, and C. Yanofsky.** 1992. The *mtrAB* operon of *Bacillus subtilis* encodes GTP cyclohydrolase I (MtrA), an enzyme involved in folic acid biosynthesis, and MtrB, regulator of tryptophan biosynthesis. *J. Bacteriol.* **174:**2059–2064.
2. **Band, L., H. Shimotsu, and D. J. Henner.** 1984. Nucleotide sequence of the *Bacillus subtilis trpE* and *trpD* genes. *Gene* **27:**55–65.

2a. **Bentley, R.** 1990. The shikimate pathway—a metabolic tree with many branches. *Crit. Rev. Biochem. Mol. Biol.* **25**:307–384.

2b. **Byng, G. S., J. F. Kane, and R. A. Jensen.** 1982. Diversity in the routing and regulation of complex biochemical pathways as indicators of microbial relatedness. *Crit. Rev. Microbiol.* **9**:227–252.

3. **Camakaris, H., and J. Pittard.** 1983. Tyrosine biosynthesis, p. 339–350. *In* K. M. Herrmann and R. L. Somerville (ed.), *Amino Acids: Biosynthesis and Genetic Regulation*. Addison Wesley Publishing Co., Reading, Mass.

4. **Carlton, B. C., and D. D. Whitt.** 1969. The isolation and genetic characterization of mutants of the tryptophan system of *Bacillus subtilis*. *Genetics* **62**:445–460.

5. **Chapman, L. F., and E. W. Nester.** 1968. Common element in the repression control of enzymes of histidine and aromatic amino acid biosynthesis in *Bacillus subtilis*. *J. Bacteriol.* **96**:1658–1663.

6. **Coats, J. H., and E. W. Nester.** 1967. Regulation reversal mutation: characterization of end product-activated mutants of *Bacillus subtilis*. *J. Biol. Chem.* **242**:4948–4955.

7. **Crawford, I. P.** 1989. Evolution of a biosynthetic pathway: the tryptophan paradigm. *Annu. Rev. Microbiol.* **43**:567–600.

8. **D'Ambrosio, S. M., G. I. Glover, S. O. Nelson, and R. A. Jensen.** 1973. Specificity of the tyrosine-phenylalanine transport system in *Bacillus subtilis*. *J. Bacteriol.* **115**: 673–681.

9. **Ebbole, D. J., and H. Zalkin.** 1987. Cloning and characterization of a 12-gene cluster from *Bacillus subtilis* encoding nine enzymes for *de novo* purine nucleotide synthesis. *J. Biol. Chem.* **262**:8274–8287.

9a. **Fazel, A. M., J. R. Bowen, and R. A. Jensen.** 1980. Arogenate (pretyrosine) is an obligatory intermediate of L-tyrosine biosynthesis: confirmation in a microbial mutant. *Proc. Natl. Acad. Sci. USA* **77**:1270–1273.

10. **Garner, C., and K. M. Herrmann.** 1983. Biosynthesis of phenylalanine, p. 339–350. *In* K. M. Herrmann and R. L. Somerville (ed.), *Amino Acids: Biosynthesis and Genetic Regulation*. Addison Wesley Publishing Co., Reading, Mass.

11. **Gollnick, P., S. Ishino, M. I. Kuroda, D. J. Henner, and C. Yanofsky.** 1990. The *mtr* locus is a two-gene operon required for transcription attenuation in the *trp* operon of *Bacillus subtilis*. *Proc. Natl. Acad. Sci. USA* **87**:8726–8730.

12. **Gray, J. V., B. Golinelli-Pimpaneau, and J. R. Knowles.** 1990. Monofunctional chorismate mutase from *Bacillus subtilis*: purification of the protein, molecular cloning of the gene, and overexpression of the gene product in *Escherichia coli*. *Biochemistry* **29**:376–383.

13. **Henner, D. J., L. Band, G. Flaggs, and E. Chen.** 1986. The organization and nucleotide sequence of the *Bacillus subtilis hisH*, *tyrA* and *aroE* genes. *Gene* **49**:147–152.

14. **Henner, D. J., L. Band, and H. Shimotsu.** 1984. Nucleotide sequence of the *Bacillus subtilis* tryptophan operon. *Gene* **34**:169–177.

15. **Henner, D. J., P. Gollnick, and A. Moir.** 1990. Analysis of an 18 kilobase pair region of the *Bacillus subtilis* chromosome containing the *mtr* and *gerC* operons and the *aro-trp-aro* supraoperon, p. 657–666. *In* H. Heslot, J. Davies, J. Florent, L. Bobichon, G. Durand, and L. Penasse (ed.), *6th International Symposium on Genetics of Industrial Microorganisms*. Societe Francaise de Microbiologie, Strasbourg, France.

16. **Hoch, J. A., and E. W. Nester.** 1973. Gene-enzyme relationships of aromatic acid biosynthesis in *Bacillus subtilis*. *J. Bacteriol.* **116**:59–66.

17. **Hoch, S. O.** 1974. Mapping of the 5-methyltryptophan resistance locus in *Bacillus subtilis*. *J. Bacteriol.* **117**:315–317.

18. **Hoch, S. O., C. Anagnostopoulos, and I. P. Crawford.** 1969. Enzymes of the tryptophan operon of *Bacillus subtilis*. *Biochem. Biophys. Res. Commun.* **35**:838–844.

19. **Hoch, S. O., and I. P. Crawford.** 1973. Enzymes of the tryptophan pathway in three *Bacillus* species. *J. Bacteriol.* **116**:685–693.

20. **Hoch, S. O., C. W. Roth, I. P. Crawford, and E. W. Nester.** 1971. Control of tryptophan biosynthesis by the methyltryptophan resistance gene in *Bacillus subtilis*. *J. Bacteriol.* **105**:38–45.

21. **Huang, L., A. L. Montoya, and E. W. Nester.** 1974. Characterization of the functional activities of the subunits of 3-deoxy-D-arabinoheptulosonate 7-phosphate synthetase-chorismate mutase from *Bacillus subtilis* 168. *J. Biol. Chem.* **249**:4473–4479.

22. **Huang, L., W. M. Nakatsukasa, and E. Nester.** 1974. Regulation of aromatic amino acid biosynthesis in *Bacillus subtilis* 168: purification, characterization and subunit structure of the bifunctional enzyme 3-deoxy-D-arabinoheptulosonate 7-phosphate synthetase-chorismate mutase. *J. Biol. Chem.* **249**:4467–4472.

23. **Hyde, C. C., S. A. Ahmed, E. A. Padlan, E. W. Miles, and D. R. Davies.** 1988. Three-dimensional structure of the tryptophan synthase $\alpha_2\beta_2$ multienzyme complex from *Salmonella typhimurium*. *J. Biol. Chem.* **263**:17857–17871.

24. **Ito, J., E. C. Cox, and C. Yanofsky.** 1969. Anthranilate synthase, an enzyme specified by the tryptophan operon of *Escherichia coli*: purification and characterization of component I. *J. Bacteriol.* **97**:725–733.

24a. **Jensen, R. A.** Personal communication.

25. **Jensen, R. A.** 1969. Antimetabolite action of 5-methyltryptophan in *Bacillus subtilis*. *J. Bacteriol.* **97**:1500–1501.

25a. **Jensen, R. A.** 1969. Metabolic interlock: regulatory interactions exerted between biochemical pathways. *J. Biol. Chem.* **244**:2816–2823.

25b. **Jensen, R. A.** 1992. An emerging outline of the evolutionary history of aromatic amino acid biosynthesis, p. 205–236. *In* R. P. Mortlock (ed.), *The Evolution of Metabolic Function*, in press. Telford Press, West Caldwell, N.J.

26. **Jensen, R. A., T. A. d'Amato, and L. I. Hochstein.** 1988. An extreme-halophile archaebacterium possesses the interlock type of prephenate dehydratase characteristic of the gram-positive eubacteria. *Arch. Microbiol.* **149**:365–371.

26a. **Jensen, R. A., and E. W. Nester.** 1965. The regulatory significance of intermediary metabolites: control of aromatic acid biosynthesis by feedback inhibition in *Bacillus subtilis*. *J. Mol. Biol.* **12**:468–481.

27. **Kane, J. F.** 1977. Regulation of a common amidotransferase subunit. *J. Bacteriol.* **132**:419–425.

28. **Kane, J. F., W. M. Holmes, and R. A. Jensen.** 1972. Metabolic interlock: the dual function of a folate pathway gene as an extra-operonic gene of tryptophan biosynthesis. *J. Biol. Chem.* **247**:1587–1596.

29. **Kane, J. F., and R. A. Jensen.** 1970. Enzyme induction in the tryptophan biosynthetic pathway in *Bacillus subtilis*. *Biochem. Biophys. Res. Commun.* **38**:1161–1167.

30. **Kane, J. F., and R. A. Jensen.** 1970. The molecular aggregation of anthranilate synthase in *Bacillus subtilis*. *Biochem. Biophys. Res. Commun.* **41**:328–333.

31. **Kane, J. F., and R. A. Jensen.** 1970. Metabolic interlock: the influence of histidine on tryptophan biosynthesis in *Bacillus subtilis*. *J. Biol. Chem.* **245**:2384–2390.

32. **Kane, J. F., S. L. Stenmark, D. H. Calhoun, and R. A. Jensen.** 1971. Metabolic interlock: the role of the subordinate type of enzyme in the regulation of a complex pathway. *J. Biol. Chem.* **246**:4308–4318.

33. **Kaplan, J. B., P. Goncharoff, A. Seibold, and B. P. Nichols.** 1984. The nucleotide sequence of the *Acinetobacter calcoaceticus trpGDC* gene cluster. *Mol. Biol. Evol.* **1**:456–472.

34. **Kaplan, J. B., W. K. Merkel, and B. P. Nichols.** 1985. Evolution of glutamine amidotransferase genes: nucleotide sequences of the *pabA* genes from *Salmonella typhimurium*, *Klebsiella aerogenes* and *Serratia marcescens*. *J. Mol. Biol.* **183**:327–340.

35. **Kaplan, J. B., and B. P. Nichols.** 1983. Nucleotide sequence of *Escherichia coli pabA* and its evolutionary relationship to *trp(G)D*. *J. Mol. Biol.* **168:**451–468.
36. **Kothe, G., P. Gollnick, and C. Yanofsky.** Unpublished data.
37. **Kuroda, M. I., D. Henner, and C. Yanofsky.** 1988. *cis*-acting sites in the transcript of the *Bacillus subtilis trp* operon regulate expression of the operon. *J. Bacteriol.* **170:**3080–3088.
38. **Kuroda, M. I., H. Shimotsu, D. J. Henner, and C. Yanofsky.** 1986. Regulatory elements common to the *Bacillus pumilus* and *Bacillus subtilis trp* operons. *J. Bacteriol.* **167:**792–798.
39. **Landick, R., and C. Yanofsky.** 1987. Transcription attenuation, p. 1276–1301. *In* F. C. Neidhardt, J. L. Ingraham, K. B. Low, B. Magasanik, M. Schaechter, and H. E. Umbarger (ed.), *Escherichia coli and Salmonella typhimurium: Cellular and Molecular Biology*, vol. 2. American Society for Microbiology, Washington, D.C.
40. **Llewellyn, D. J., A. Daday, and G. D. Smith.** 1980. Evidence for an artificially evolved bifunctional 3-deoxy-D-arabinoheptulosonate-7-phosphate synthase-chorismate mutase in *Bacillus subtilis*. *J. Biol. Chem.* **255:**2077–2084.
41. **Lorence, J. H., and E. W. Nester.** 1967. Multiple molecular forms of chorismate mutase in *Bacillus subtilis*. *Biochemistry* **6:**1541–1552.
41a. **Matsui, K., K. Sano, and E. Ohtsubo.** 1986. Complete nucleotide and deduced amino acid sequences of the *Brevibacterium lactofermentum* tryptophan operon. *Nucleic Acids Res.* **14:**10113–10114.
42. **Miele, L., R. Stettler, R. Banholzer, M. Kotik, and T. Leisinger.** 1991. Tryptophan gene cluster of *Methanobacterium thermoautotrophicum* Marburg: molecular cloning and nucleotide sequence of a putative *trpEGCFBAD* operon. *J. Bacteriol.* **173:**5017–5023.
43. **Miozzari, G., and C. Yanofsky.** 1979. Gene fusion during the evolution of the *trp* operon in Enterobacteriaceae. *Nature* (London) **277:**486–489.
44. **Nakatsukasa, W. M., and E. W. Nester.** 1972. Regulation of aromatic amino acid biosynthesis in *Bacillus subtilis* 168. 1. Evidence for and characterization of a trifunctional enzyme complex. *J. Biol. Chem.* **247:**5972–5979.
45. **Nasser, D., and E. W. Nester.** 1967. Aromatic amino acid biosynthesis: gene-enzyme relationships in *Bacillus subtilis*. *J. Bacteriol.* **94:**1706–1714.
46. **Nester, E. W.** 1968. Cross pathway regulation: effect of histidine on the synthesis and activity of enzymes of aromatic acid biosynthesis in *Bacillus subtilis*. *J. Bacteriol.* **96:**1649–1657.
47. **Nester, E. W., B. Dale, A. Montoya, and B. Vold.** 1974. Cross pathway regulation of tyrosine and histidine synthesis in *Bacillus subtilis*. *Biochim. Biophys. Acta* **361:**59–72.
48. **Nester, E. W., and R. A. Jensen.** 1966. Control of aromatic acid biosynthesis in *Bacillus subtilis*: sequential feedback inhibition. *J. Bacteriol.* **91:**1594–1598.
49. **Nester, E. W., R. A. Jensen, and D. S. Nasser.** 1969. Regulation of enzyme synthesis in the aromatic amino acid pathway of *Bacillus subtilis*. *J. Bacteriol.* **97:**83–90.
50. **Nester, E. W., J. H. Lorence, and D. S. Nasser.** 1967. An enzyme aggregate involved in the biosynthesis of aromatic amino acids in *Bacillus subtilis*. Its possible function in feedback regulation. *Biochemistry* **6:**1553–1562.
51. **Nester, E. W., and A. L. Montoya.** 1976. An enzyme common to histidine and aromatic amino acid biosynthesis in *Bacillus subtilis*. *J. Bacteriol.* **126:**699–705.
52. **Nester, E. W., M. Schafer, and J. Lederberg.** 1963. Gene linkage in DNA transfer: a cluster of genes concerned with aromatic biosynthesis in *Bacillus subtilis*. *Genetics* **48:**529–551.
53. **Nichols, B. P., A. M. Seibold, and S. Z. Doktor.** 1989. *para*-aminobenzoate synthesis from chorismate occurs in two steps. *J. Biol. Chem.* **264:**8597–8601.
54. **Piggot, P. J., and J. A. Hoch.** 1985. Revised genetic linkage map of *Bacillus subtilis*. *Microbiol. Rev.* **49:**158–179.
55. **Pittard, A. J.** 1987. Biosynthesis of the aromatic amino acids, p. 368–394. *In* F. C. Neidhardt, J. L. Ingraham, K. B. Low, B. Magasanik, M. Schaechter, and H. E. Umbarger (ed.), *Escherichia coli and Salmonella typhimurium: Cellular and Molecular Biology*, vol. 1. American Society for Microbiology, Washington, D.C.
56. **Priestle, J. P., M. G. Grutter, J. L. White, M. G. Vincent, M. Kania, E. Wilson, T. S. Jardetsky, K. Kirschner, and J. N. Jansonius.** 1987. Three-dimensional structure of the bifunctional enzyme N-(5′-phosphoribosyl)anthranilate isomerase-indole-3-glycerol-phosphate synthase from *Escherichia coli*. *Proc. Natl. Acad. Sci. USA* **84:**5690–5694.
57. **Rebello, J. L., and R. A. Jensen.** 1970. Metabolic interlock: the multi-metabolite control of prephenate dehydratase activity in *Bacillus subtilis*. *J. Biol. Chem.* **245:** 3738–3744.
58. **Roth, C. W., and E. W. Nester.** 1971. Co-ordinate control of tryptophan, histidine and tyrosine enzyme synthesis in *Bacillus subtilis*. *J. Mol. Biol.* **62:**577–589.
58a. **Sano, K., and K. Matsui.** 1987. Structure and function of the *trp* operon control regions of *Brevibacterium lactofermentum*, a glutamic-acid-producing bacterium. *Gene* **53:** 191–200.
59. **Schechtman, M. G., and C. Yanofsky.** 1983. Structure of the trifunctional *trp*-1 gene from *Neurospora crassa* and its aberrant expression in *Escherichia coli*. *J. Mol. Appl. Genet.* **2:**89–93.
60. **Shimotsu, H., and D. J. Henner.** 1984. Characterization of the *Bacillus subtilis* tryptophan promoter region. *Proc. Natl. Acad. Sci. USA* **81:**6315–6319.
61. **Shimotsu, H., M. I. Kuroda, C. Yanofsky, and D. J. Henner.** 1986. Novel form of transcription attenuation regulates expression of the *Bacillus subtilis* tryptophan operon. *J. Bacteriol.* **166:**461–471.
62. **Slock, J., D. P. Stahly, C.-Y. Han, E. W. Six, and I. P. Crawford.** 1990. An apparent *Bacillus subtilis* folic acid biosynthetic operon containing *pab*, an amphibolic *trpG* gene, a third gene required for synthesis of *para*-aminobenzoic acid, and the dihydropteroate synthase gene. *J. Bacteriol.* **172:**7211–7226.
63. **Steinberg, W.** 1974. Temperature-induced derepression of tryptophan biosynthesis in a tryptophanyl-transfer ribonucleic acid synthetase mutant of *Bacillus subtilis*. *J. Bacteriol.* **117:**1023–1034.
64. **Weigent, D. A., and E. W. Nester.** 1976. Regulation of histidinol phosphate aminotransferase synthesis by tryptophan in *Bacillus subtilis*. *J. Bacteriol.* **128:**202–211.
65. **Whitt, D. D., and B. C. Carlton.** 1968. Characterization of mutants with single and multiple defects in the tryptophan biosynthetic pathway in *Bacillus subtilis*. *J. Bacteriol.* **96:**1273–1280.
66. **Yanofsky, C., and I. P. Crawford.** 1987. The tryptophan operon, p. 1453–1472. *In* F. C. Neidhardt, J. L. Ingraham, K. B. Low, B. Magasanik, M. Schaechter, and H. E. Umbarger (ed.), *Escherichia coli and Salmonella typhimurium: Cellular and Molecular Biology*, vol. 2. American Society for Microbiology, Washington, D.C.
67. **Yanofsky, C., and J. Ito.** 1966. Nonsense codons and polarity in the tryptophan operon. *J. Mol. Biol.* **21:**313–334.
68. **Yanofsky, C., T. Platt, I. P. Crawford, B. P. Nichols, G. E. Christie, H. Horowitz, M. vanCleemput, and A. M. Wu.** 1981. The complete nucleotide sequence of the tryptophan operon of *Escherichia coli*. *Nucleic Acids Res.* **9:**6647–6668.
69. **Zalkin, H.** 1980. Anthranilate synthase: relationships between bifunctional and monofunctional enzymes, p. 123–149. *In* H. Bisswanger and E. Schmincke-Ott (ed.), *Multifunctional Enzymes*. John Wiley & Sons, Inc., New York.

20. Biosynthesis of Glutamine and Glutamate and the Assimilation of Ammonia

HAROLD J. SCHREIER

INTRODUCTION

Growth of bacteria in a simple salts medium containing glucose and ammonia requires the biosynthesis of amino acids and other nitrogen-containing compounds from ammonia and intermediary metabolites. While de novo synthesis may be accomplished by direct incorporation of ammonia, the vast majority of nitrogen atoms found in macromolecules are derived from the amino acids glutamate and glutamine. The amino groups of both compounds are utilized for the production of other amino acids, and the amide group of glutamine is used directly for the synthesis of purines, pyrimidines, glucosamine 6-phosphate, tryptophan, histidine, *p*-aminobenzoate, and NAD. Thus, glutamate and glutamine are principal precursors for all nitrogenous compounds, and their synthesis is a potential control point. In the enteric bacteria, for instance, the mechanisms that regulate glutamine synthesis are intricately coupled to the regulation of nitrogen metabolism pathways. These mechanisms have been studied extensively and have been reviewed elsewhere (73–75, 96, 120).

For members of the genus *Bacillus* and other gram-positive bacteria, elucidation of pathways for assimilation of ammonia is critical for understanding how these organisms regulate nitrogen flow through the mainstream of metabolism. Such knowledge may provide clues regarding signals used to sense nitrogen deprivation, thereby triggering the processes leading to the differentiated state. One can imagine, for instance, that regulation of early sporulation genes in *Bacillus subtilis* may be affected by the signals utilized for controlling ammonia assimilation, which may be metabolites that are end products or by-products of the pathways themselves. Such signals may also be relevant to mechanisms regulating growth rate and secondary metabolite production.

Regulation of the enzymes involved in glutamine and glutamate biosynthesis in *Bacillus* spp. and their roles in ammonia assimilation will be the focus of this chapter. Assimilatory processes in *Clostridium* and *Streptomyces* spp. have only recently been studied and will be addressed in turn. Control of nitrogen catabolic pathways in *B. subtilis* is described in a recent review (47) as well as in chapter 16 in this book.

Enzymes of Ammonia Assimilation

In bacteria, the biosynthesis of glutamine occurs solely via the action of glutamine synthetase (GS; EC 6.3.1.2), where Me^{2+} is either Mg^{2+} or Mn^{2+}: NH_3 + glutamate + Me^{2+}ATP $\rightarrow$ glutamine + Me^{2+}ADP + P_i. Glutamate synthesis is accomplished by at least four different routes: (i) by the reductive amination of α-ketoglutarate via glutamate synthase (GOGAT; EC 1.4.1.13): glutamine + α-ketoglutarate + NAD(P)H + H^+ $\rightarrow$ 2 glutamate + NAD(P)$^+$; (ii) by the reductive amination of α-ketoglutarate via glutamate dehydrogenase (GDH; EC 1.4.1.4): NH_3 + α-ketoglutarate + NAD(P)H + H^+ $\rightarrow$ glutamate + NAD(P)$^+$; (iii) by a transamination reaction involving the transfer of the α-amino group from one amino acid to the keto group of α-ketoglutarate, e.g., via glutamate:α-ketoglutarate aminotransferase; and (iv) by the degradation of other amino acids, for instance, glutamine by glutaminase, histidine by the *hut* pathway, and proline by the *put* pathway.

For most bacteria, assimilation of ammonia is accomplished through the synthesis of glutamine and glutamate. Alanine dehydrogenase (ADH; EC 1.4.1.1) and asparagine synthetase (EC 6.3.1.1) have also been implicated in assimilation, and their roles will be discussed below.

Physiology of Ammonia Assimilation

In the enteric bacteria, GDH is responsible for ammonia assimilation when cells are growing in ammonia-rich media. Under these conditions, GS is maintained at low steady-state levels solely to satisfy the glutamine pool requirement (96, 120). In ammonia-limiting media or in the presence of nonpreferred nitrogen sources, GS activity is increased and the coupling of GS and GOGAT reactions affect ammonia assimilation. Most bacterial GDHs have a lower affinity for ammonia [K_m > 5 mM] than does GS [$K_m \approx$ 0.5 mM], which precludes the use of GDH for glutamate production under limiting nitrogen conditions. Utilization of GS and GOGAT under these conditions allows the organism to scavenge ammonia at the expense of energy.

GS, GOGAT, and GDH are found in almost all members of the genera *Bacillus*, *Clostridium*, and *Streptomyces* that have been examined. Differences occur in the pathway used for assimilation and in regulation of the pathway in response to the nitrogen source in the medium.

Assimilation in *Bacillus* spp.

The *Bacillus* spp. can be separated into three groups based on the pathway used for assimilation. One

Harold J. Schreier • Center of Marine Biotechnology, University of Maryland, 600 East Lombard Street, Baltimore, Maryland 21202, and Department of Biological Sciences, University of Maryland Baltimore County, Baltimore, Maryland 21228.

group, represented by *B. subtilis*, employs GS and GOGAT for assimilation. The second, which contains most members of the genus, utilizes all three enzymes, depending on nutritional environment. The third group uses only GDH for assimilation, a characteristic of some N_2-fixing *Bacillus* spp.

Assimilation of ammonia in derivatives of *B. subtilis* 168 and SMY is solely accomplished through the coupled action of GS and GOGAT. Mutations in either enzyme result in an inability to grow with ammonia as nitrogen source (14, 28, 45), and mutant strains impaired in ammonia assimilation are defective in GS or GOGAT (25). As discussed below, GDH activity is detectable only in cells grown under special conditions, and the enzyme is utilized exclusively for glutamate catabolism (reverse reaction shown in route ii above). An exception is *B. subtilis* PCI 219, which expresses GDH in a nitrogen-dependent manner (67). High GDH levels are found in an excess-ammonia culture (76), a feature similar to that described for the enteric bacteria and other *Bacillus* spp. (see below). The origin of this strain has not been described, raising the possibility that its classification as *B. subtilis* is incorrect.

Expression of GS in *B. subtilis* is related to the presence of readily available nitrogen sources (29, 89, 95). Enzyme levels are lowest when the medium contains glutamine, a preferred nitrogen source. Moderate or high expression occurs when cells are grown in the presence of limiting concentrations of ammonia or nonpreferred nitrogen sources such as histidine or arginine. Differences in GS expression probably reflect the demand for glutamine production during growth; the intracellular glutamine pool is maintained in the 2 to 5 mM range (29).

For GOGAT, levels vary approximately 10-fold and depend not only on nitrogen availability but also on the nature of the nitrogen source (28, 89). The presence in the medium of glutamate or amino acids that are degraded directly to glutamate, such as histidine and arginine, results in low GOGAT levels. Under these conditions, GOGAT activity is not necessary, since the intracellular glutamate requirement is met by amino acid catabolism. An exception is glutamine, which results in intermediate GOGAT levels. While glutamine may be catabolized to glutamate by glutaminase, the low GS levels found in glutamine cultures suggest that glutaminase action is not significant and that increased GOGAT expression is required to provide glutamate. GOGAT levels are highest in the presence of ammonia or a nonpreferred nitrogen source that cannot be catabolized to form glutamate, such as nitrate (28, 89).

Levels of GS and GOGAT in other *Bacillus* spp., such as *Bacillus licheniformis* and *Bacillus megaterium*, vary in approximately the same manner as that described for *B. subtilis* (38, 40, 54, 78, 79, 111). GDH is also found in these bacteria and plays a role in assimilation under certain physiological conditions. In *B. megaterium*, for instance, GDH levels are highest in excess-ammonia cultures, suggesting its use for assimilation under these conditions (54). The low affinity that GDH has for ammonia (see below) indicates that GS and GOGAT are responsible for assimilation under nitrogen-limited conditions. This is consistent with the observation that uptake of [^{13}N] ammonia in a nitrogen-starved culture occurs primarily by incorporation of label into glutamine (66).

In *B. licheniformis*, GDH levels are lowest when glutamate is the sole source of nitrogen (91, 111). On the other hand, enzyme levels are 35-fold higher when ammonia is the sole nitrogen source, a condition that also results in very high GOGAT activities (111). Under these conditions, GDH and GOGAT may have different roles. Chemostat studies suggest that the capacity to catalyze both catabolic and anabolic reactions may enable GDH to be used as a valve for maintaining static glutamate pools (78). Such a role is consistent with the kinetic properties of the enzyme from *B. licheniformis* (see below).

^{15}N nuclear magnetic resonance studies have shown that nitrogen supplied to nitrogen-fixing *Bacillus polymyxa* as ammonia, nitrate, or N_2 is incorporated primarily into the amino group of glutamate, suggesting that assimilation occurs via GDH (61). The unusually high affinity of GDH for ammonia (K_m = 2.9 mM) as well as the small variations in GS and GOGAT levels observed under limited- and excess-nitrogen conditions support a major role for GDH in assimilation (61). A similar role was described for *Bacillus macerans* (62).

Assimilation during sporulation

At the onset of stationary phase, GS, GOGAT, and GDH levels are adjusted to match alterations associated with changing environmental conditions. In a glucose and glutamine medium, for instance, GS and GOGAT activities increase upon glutamine deprivation (89, 103). For GS, intermediate to high levels remain constant throughout stationary phase, satisfying a requirement for glutamine during development rather than assuming a role in assimilating extracellular ammonia (36, 89, 103). It is not known whether GS expression is compartment specific or whether the enzyme is packaged into the endospore. On the other hand, brief transient increases in the levels of GOGAT and GDH at t_0 are followed by a rapid drop in specific activity (29, 104). The rapid decline observed for GOGAT in *B. subtilis* (29) and *B. licheniformis* (104) and for GDH in *B. licheniformis* and *Bacillus cereus* (91) is similar to that observed for enzymes undergoing proteolysis at this stage (118) and suggests that GOGAT and GDH are not required during sporulation.

Assimilation in other gram-positive bacteria

Ammonia assimilation has been studied in *Clostridium kluyveri* (63), *Clostridium butyricum* (63), and *Clostridium pasteurianum* (68). For *C. kluyveri* GS, GOGAT, and GDH levels are lower (four- to sevenfold) in N_2-fixing (nitrogen-limited) cultures than when ammonia is used as the nitrogen source. The properties of the *C. kluyveri* GDH (K_m of 12 mM) and the intracellular concentrations of ammonia (13.8 and 0.99 mM in ammonia and N_2 cultures, respectively) suggest that GDH is the major route of utilization in ammonia-grown cells and that assimilation in N_2-grown cells is accomplished by GS and GOGAT (63). On the other hand, GDH activity is either not detectable or is present at very low levels in ammonia-grown or N_2-fixing cultures of *C. butyricum* and *C. pasteurianum*, which utilize GS and GOGAT for assimilation (63, 68). Since high GDH levels are obtained when *C. butyricum* is grown in complex medium, GDH may be used only for catabolic purposes in both organisms, even though the low K_m (2.8 mM) for ammonia might be consistent with an assimilatory role (63).

The physiology of GS, GDH, and GOGAT expression has been examined in only a few members of the *Streptomyces* spp. As in the *Bacillus* spp., enzyme levels are influenced by ammonia availability and the nitrogen source in the medium. The level of GS varies as much as 10-fold in *Streptomyces coelicolor* (48) and *Streptomyces clavuligerus* (16), with the highest activity found in cells grown in the presence of nonpreferred nitrogen sources or limiting ammonia. Addition of ammonia to nitrogen-limited cultures of *Streptomyces cattleya* or *S. coelicolor* results in a rapid decline in GS activity (16, 48, 117), a phenomenon observed for GS from the enteric bacteria. The effect is due to posttranslational control and is discussed below.

In *S. coelicolor*, GOGAT levels vary in approximately the same manner as those of GDH, displaying as much as 10-fold regulation (43). The highest activities are observed in the presence of ammonia; a nitrogen source that can be converted to glutamate results in low or intermediate levels. Genetic studies (see below) have demonstrated that GDH is used for assimilation when nitrogen is in excess and that the GS-GOGAT pathway is responsible for assimilation under nitrogen-limited conditions. On the other hand, *S. clavuligerus* assimilates ammonia solely by the GS-GOGAT pathway (16). In this organism, GDH is not detectable under any physiological condition, and GOGAT activity is not appreciably regulated by the nitrogen source (16).

Regulation of Nitrogen Catabolic Pathways

The utilization of a source of nitrogen other than ammonia involves induction of specific enzyme systems necessary for degrading or converting the nitrogen source to ammonia and other components. In the enteric bacteria, many nitrogen catabolic pathways are induced as a consequence of nitrogen source limitation, and their regulation is coupled to GS control (reviewed in reference 96). Enzyme systems involved in the degradation of nitrogen-containing compounds are regulated by nitrogen catabolite repression control. This control is accomplished via a global regulatory network, Ntr, whose components sense carbon and nitrogen availability, modify regulatory proteins, and modulate expression of target genes. Ntr includes the *ntrB* and *ntrC* gene products, NR_{II} and NR_I, respectively, which are members of the two-component family of bacterial regulatory proteins. The NR_{II} protein is a kinase/phosphatase; the NR_I protein is the phosphorylatable response regulator. Nitrogen-carbon balance is reflected in the relative pool sizes of glutamine and α-ketoglutarate, respectively. In concert with two other regulatory proteins (a uridylyltransferase and the P_{II} protein), these metabolites indirectly control NR_{II} activity. Nitrogen-limited conditions stimulate the phosphorylation of NR_I by NR_{II}, which results in the conversion of NR_I into an activator protein. This protein stimulates transcription at promoters of several nitrogen metabolism genes by utilizing the σ^{54} form of RNA polymerase.

Presently, there is no evidence of any system analogous to Ntr that regulates nitrogen catabolic pathways in *Bacillus* spp. (47). Regulation of some stationary-phase and sporulation gene expression in *B. subtilis* has been shown to involve proteins that are members of two-component regulatory systems, but none of these proteins has been shown to play any role in nitrogen metabolism (115). While a σ^{54} homolog has also been found in *B. subtilis*, there is no evidence that this factor is involved in control of nitrogen metabolism, although the ability to utilize some amino acids as nitrogen sources was impaired in a strain with this protein deleted (27).

In *Bacillus* spp., expression of some nitrogen utilization pathways, such as those involved in utilization of histidine and proline, is not affected by the nitrogen source (4, 47, 111). However, nitrogen limitation stimulates expression of other nitrogen-related enzyme systems, such as asparaginase, urease, and amidase in *B. subtilis* (4, 93). Expression of these systems appears to be subject to a global system of control involving GS (see below and chapter 16 in this book).

BIOSYNTHESIS OF GLUTAMINE

Three different classes of bacterial GS have been described. GSI is composed of 12 identical subunits arranged in two superimposed hexagonal rings and is found in most bacteria, including members of the family *Enterobacteriaceae* and *Bacillus*, *Clostridium*, and *Streptomyces* spp. GSII is the form prevalent among eukaryotes; it is octameric. The GSII subunit is approximately 100 amino acids smaller than that of GSI, lacking some of the carboxy-terminal portion of the GSI subunit. Members of the family *Rhizobiaceae* and some *Streptomyces* spp. (see below) produce both GSI and GSII. GSIII is found in the genus *Bacteroides* and has a subunit molecular weight of 75,000; the subunits appear to be arranged as a hexamer (55).

In *E. coli* and related bacteria, GS activity is regulated by posttranslational control and feedback inhibition (see reference 96 for a review). The reversible modification of 1 to 12 GS subunits occurs via adenylylation, which is accomplished by a regulatory cascade involving at least three other proteins. Modification results in decreased activity and is a function of the availability of ammonia in the medium; GS with an adenylylation state of 10 to 12 is found in media containing excess nitrogen. The components of the cascade include adenylyltransferase, uridylyltransferase, and the P_{II} accessory protein. The uridylyltransferase and P_{II} proteins are also involved in regulating Ntr components, and regulation of the modification cascade, like that of the regulatory cascade, is dependent on the relative intracellular levels of glutamine and α-ketoglutarate. Under nitrogen-limited conditions, the demand for glutamine exceeds that for α-ketoglutarate and adenylylation is inhibited, resulting in active GS.

Once modified, adenylylated GS is subject to feedback inhibition by a variety of end products of glutamine metabolism, including alanine, glycine, histidine, tryptophan, AMP, CTP, carbamyl phosphate, and glucosamine-6-phosphate.

Properties of GS Enzymes from *Bacillus* spp.

In *Bacillus* spp., the GS subunit has a molecular weight of ≈50,000 (32, 36, 56, 58). The enzyme from *B. subtilis* differs significantly from the *E. coli* enzyme in amino acid composition, susceptibility to digestion with carboxypeptidase A, and immunochemical properties (30). The *B. subtilis* subunit is 444 amino acids long, as deduced from the nucleotide sequence of the cloned gene (84, 116), and the derived amino acid sequence is most similar to that of the *Clostridium acetobutylicum* enzyme. GS enzymes from *B. subtilis* and *C. acetobutylicum* differ from those of other prokaryotes in a region known to contain several regulatory and catalytic mutations (see below) (116).

Modification of the *Bacillus* GS via an adenylylation mechanism has been ruled out by both in vitro and in vivo studies. The *B. subtilis* enzyme cannot be modified by the *E. coli* adenylylation system in vitro (30, 32), and regulation of GS activity expressed from the cloned *B. subtilis* gene in *E. coli* is not dependent on adenylylation proteins (106). While residues around the *E. coli* GS adenylylation site are found in the *B. subtilis* enzyme, the inability to observe adenylylation by *E. coli* components may be due to differences in the three-dimensional structure surrounding the site (116). Posttranslational control by other mechanisms has also been ruled out. For both *B. subtilis* and *B. licheniformis*, no differences in the kinetic parameters of the biosynthetic activities of enzymes purified from cells grown under excess-nitrogen and nitrogen-limited conditions can be detected (30, 36). Furthermore, alterations in GS activity suggestive of regulation by modification are not detected upon addition of excess nitrogen to an ammonia-limited culture of *B. subtilis* (46). On the other hand, treatment of the *B. subtilis* enzyme with iodoacetamide results in changes in catalytic parameters that are qualitatively similar to changes seen upon adenylylation of *E. coli* GS (31). Alkylation is influenced by substrates and feedback inhibitors and influences a cysteine residue located in an active-site region (31, 81). Regulation by sulfhydryl group modification in vivo is difficult to establish, and the physiological significance of these results has not been determined.

The activity of GS in *Bacillus* spp. appears to be influenced by feedback inhibition. Although some end products may be involved in its regulation, the enzyme is not as sensitive to the vast array of end products of glutamine metabolism as is the enzyme from enteric bacteria. The primary effectors of GS activity in *B. licheniformis* appear to be alanine and glutamine (36, 37). Alanine displays a fractional inhibition value, $I_{0.5}$, of 5.1 mM, which is within the range of the pool size for this amino acid (37). Regulation by alanine is significant, since the pool of this amino acid increases dramatically during late exponential phase and sporulation (22). Inhibition by glutamine is moderated by glutamate. At high concentrations of glutamate (50 mM), glutamine binds to one allosteric site per enzyme unit with an $I_{0.5}$ of 2.2 mM; these concentrations are within the normal range of the pool sizes of these amino acids (37). At intermediate glutamate concentrations (20 mM), there seem to be two glutamine binding sites, with the allosteric site having an $I_{0.5}$ of 0.4 mM (37). In *B. licheniformis*, the glutamate pool drops from 100 mM during growth to approximately 15 mM during stationary phase. Differences in inhibition by glutamine and alanine, therefore, may reflect a need for regulating activity during stationary phase and sporulation. The stage of growth may also be important in the inhibition of enzyme activity by AMP, which has an $I_{0.5}$ of 0.4 mM (37). Intracellular AMP concentrations rise from approximately 0.1 to 0.9 mM just at the onset of stationary phase in glucose-ammonia minimal medium, making AMP another potential inhibitor once the cell enters stationary phase (34, 35).

For *B. subtilis* GS activity, the effects of various inhibitors are a function of the divalent cation used for enzyme activation. When Mg^{2+} is present, glutamine and AMP (each at 5 mM) inhibit enzyme activity 98 and 95%, respectively, with inhibition being partially competitive with respect to all three substrates; the inhibition observed by glutamine is greatly potentiated by AMP (33). On the other hand, in the presence of Mn^{2+}, glutamine and AMP are not inhibitory, while glycine, CTP, and alanine are, inhibiting from 25 to 50% when they are present at 5 mM (33).

Apparently, many properties of the *B. subtilis* enzyme are dependent on the divalent cation used for activation. Both Mn^{2+} and Mg^{2+} activate glutamine synthesis, and the Mg^{2+}-dependent activity is intrinsically less stable than the Mn^{2+}-dependent activity (30, 32). Complex kinetics are observed; they depend on the divalent cation used in the assay. For Mn^{2+}, full activity is attained at a Mn^{2+}-to-ATP ratio of 1:1, and activity decreases when Mn^{2+} is present in excess relative to ATP. Linear kinetics occur for ATP (K_{mapp} of 0.2 mM) and ammonia (K_{mapp} of 0.4 mM), while glutamate shows patterns indicative of product inhibition (K_{mapp} of 0.8 mM at low concentrations) (30). For Mg^{2+}, 4.5 times more Mg^{2+} than ATP is required for full activity, and saturation for all three substrates cannot be obtained, indicating that negative cooperativity may be involved in binding ATP, glutamine, and ammonia (30). Similar studies examining the effects of manganese and magnesium on activity and the kinetic and feedback inhibition properties of GS from *B. cereus* (57, 82) and *B. licheniformis* (36, 57, 58, 71) have also been reported.

While the physiological significance of the effects observed for Mn^{2+} and Mg^{2+} has not been determined, it may be significant that the sensitivity of enzyme activity to fluctuations in the levels of nucleotides and glutamine is dependent on the relative concentrations of Mn^{2+}, Mg^{2+}, and ATP. In *E. coli*, divalent-ion specificity is determined by the adenylylation-deadenylylation state. Adenylylation leads to a dependence on manganese and greater sensitivity to feedback inhibition; deadenylylation leads to a requirement for magnesium and a loss in sensitivity.

It has been suggested that fluctuations in the relative concentrations of Mn^{2+} and Mg^{2+} during growth and sporulation in *Bacillus* are relevant to the regulation of several key enzymes in metabolism (21, 88, 123). Thus, under certain physiological conditions, the ratios of these two cations may be important in the regulation of the *B. subtilis* GS.

GS from thermophilic *Bacillus stearothermophilus* and *Bacillus caldolyticus* have been studied in some detail. The *B. stearothermophilus* GS has kinetic and biochemical characteristics that are similar to those of other *Bacillus* spp. (124). On the other hand, *B. caldolyticus*, an organism capable of growth at 85 to 90°C, has been found to produce two kinetically distinct GS enzymes, E_I and E_{II} (125, 127). Both enzymes are dodecamers composed of identical subunits, and neither cross-reacts with antibodies reactive with *E. coli* or *B. subtilis* GS (127). The two enzymes differ significantly from each other in their amino acid composition, isoelectric points, and inhibitory properties (127). Interestingly, each responds to metal ions in a different manner; E_I is more active with Mg^{2+} than with Mn^{2+}, and E_{II} is more active with Mn^{2+} than with Mg^{2+}. In addition, both enzymes exhibit substrate K_ms that are a function of the divalent metal ion present (127). The Mg^{2+}-dependent E_I enzyme is more strongly inhibited than E_{II} by glycine, alanine, and serine, whereas the Mn^{2+}-dependent E_{II} is more strongly inhibited by glutamine and AMP (126). Regulation of the two enzymes in a complementary fashion results in an overall net control of GS activity in a manner similar to the control observed for the *B. subtilis* GS. Although the relationships of the two enzymes to divalent metal ions and corresponding effector metabolites are opposite that of the *B. subtilis* GS, the evolution of such a system in *B. caldolyticus* may indicate that divalent cations are important for regulation in this organism.

Properties of GS Enzymes from Other Gram-Positive Bacteria

Clostridium spp.

There is no evidence for regulation of *Clostridium* GS activity by adenylylation (68). Kinetic parameters of purified *C. pasteurianum* GS are not affected by treatment with snake venom phosphodiesterase, suggesting the absence of adenylylation (69), and regulation of the cloned *C. acetobutylicum* GS gene in *E. coli* is not dependent on the adenylylation system in vivo (121).

The *C. acetobutylicum* GS is a dodecamer with an apparent subunit molecular weight of ≈59,000 (121), and as in *B. subtilis*, the subunit is 444 amino acids long (60). As described above for *B. subtilis* GS, the enzyme shares polypeptide sequences with other GSI proteins in regions implicated in substrate binding, but it lacks the tyrosine that is the target of *E. coli* GS adenylylation (60).

Krishnan et al. (69) purified and characterized GS from *C. pasteurianum*. In the presence of Mg^{2+}, biosynthetic activity is approximately 23-fold higher than in the presence of Mn^{2+}. The Mg^{2+}-dependent activity has a high affinity for ammonia ($K_{m_{app}}$ of 0.34 mM for the ammonia analog hydroxylamine). Glutamate displays biphasic kinetics with apparent K_ms of 22.2 and 3.8 mM for high (10 to 150 mM) and low (<10 mM) glutamate concentrations, respectively. Negative cooperativity is observed for glutamate binding, indicating that sequential binding of glutamate to subunits produces new low-affinity binding sites, thus rendering the enzyme less sensitive to changes in glutamate concentration.

The *C. pasteurianum* GS displays weak feedback inhibition properties limited to only a few amino acids: alanine, serine, and glycine, when added singly at a concentration of 5 mM, inhibit the Mg^{2+}-dependent biosynthetic activity at least 50% (69). Unlike the case with the *Bacillus* enzyme, glutamine does not have any effect on activity, although its effect on the Mn^{2+}-dependent activity and possible effects of purine or pyrimidine metabolites were not reported (69).

Streptomyces spp.

Although little is known about how GS activity is regulated in *Streptomyces* spp., it is clear that these organisms have evolved GS systems that are very different from those of other gram-positive bacteria. There is clear evidence for two distinct GS types in some species. GSI is composed of 12 identical subunits and is similar to other prokaryotic GS proteins (90). Unlike the enzyme from *B. subtilis* or *C. acetobutylicum*, GSIs from *S. cattleya* and *S. coelicolor* are regulated by adenylylation (48, 117). The low GSI activity observed in crude extracts of *S. cattleya* grown under excess-nitrogen conditions can be enhanced when extracts are subjected to treatment with snake venom phosphodiesterase (117). Radiolabeling studies indicated that modification includes incorporation of an AMP moiety into the GSI protein, a reaction consistent with adenylylation (117). Similar results were obtained with the enzyme from *S. coelicolor* (48). The *S. coelicolor* GSI open reading frame is more similar to those of GS proteins from gram-negative bacteria than to that of GS from *B. subtilis* or *C. acetobutylicum* (132). Furthermore, the tyrosine residue utilized for adenylylation in *E. coli* is conserved in the *S. coelicolor* sequence (132).

The mechanisms utilized for controlling GSI activity by adenylylation have not been determined. Genetic studies have suggested that glutamine or a metabolite derived from glutamine is involved in regulating adenylylation (48), although the nature of this control is unknown. Interestingly, crude extracts prepared from *E. coli* are not capable of adenylylating purified *S. cattleya* GS and vice versa (90). Furthermore, *S. coelicolor* GSI is not adenylylated when expressed in *E. coli* (132). Thus, either the *Streptomyces* GSI protein cannot be used as a substrate for the *E. coli* modification system, or certain regulatory proteins are not present in the heterologous system.

The GSII enzyme is a heat-labile protein composed of eight identical subunits, and nothing is known about how enzymatic activity is regulated. Genes coding for this enzyme have been found in several *Streptomyces* spp. (5, 70). The primary sequence shows that the enzyme subunit is homologous to that of eukaryotic GS as well as to that of GSII from nitrogen-fixing bacteria (5, 70). While the role of GSII in assimilation is not understood, the enzyme may be important for glutamine biosynthesis during the production of bialophos, an antibiotic containing the GS inhibitor phosphinothricin (5, 70).

Control of GS Expression in *Bacillus* spp.

Regulation of *glnA* in *B. subtilis*

In enteric bacteria, the GS structural gene, *glnA*, is found in the *glnA ntrB ntrC* operon, and transcription is primarily dependent on the Ntr system (reviewed in reference 96). The factors involved in activating Ntr genes (discussed above) also act to regulate *glnA* expression. Thus, under excess-nitrogen conditions, the level of *glnA* mRNA is low because of the absence of activating NR_I protein. Upon nitrogen deprivation, *glnA* transcription is stimulated by the NR_I-$E\sigma^{54}$ apparatus. In this manner, *glnA* transcription is intimately coupled to Ntr control, and *glnA* expression responds to the nitrogen-carbon balance in the cell. Glutamine synthesis is, therefore, a reflection of both the rate of *glnA* transcription and the activity of existing molecules. Both are adjusted as the organism encounters new growth conditions.

Since GS in *Bacillus* spp. is not regulated by covalent modification, differences in enzyme levels are solely due to control at the level of transcription (44, 106, 113, 116). Measurements of *glnA* transcription using either the cloned gene as a probe or expression from an operon fusion have shown that *glnA* mRNA levels are regulated 9- to 10-fold, varying with the nitrogen source in the same manner as does GS (44, 105, 106, 113).

Autoregulation of GS

The characteristics of a number of *B. subtilis glnA* mutants led to the conclusion that regulation of GS expression in this organism is unlike that found for the enteric bacteria. In addition to synthesizing GS altered in both catalytic and feedback inhibition properties, these mutants produce elevated levels of GS antigen (25, 97). The *glnA100* mutation, for instance, causes an 18-fold increase in Mn^{2+}-dependent GS activity and a 2-fold increase in GS antigen (26). These studies implicated GS in regulating its own synthesis (26).

Additional evidence for an autoregulatory role was obtained from studying the cloned gene in *E. coli*. When contained in a lambda vector or when present on a plasmid, the *B. subtilis glnA* locus complements the glutamine requirement of an *E. coli glnA* mutant, and GS expression from the cloned DNA exhibits nutritional regulation (44, 50). Mutations in *E. coli* Ntr and adenylylation systems did not effect regulation (106). Deletions removing most or all of *glnA* resulted in constitutive expression; those removing the carboxy-terminal third of the gene resulted in partial control (106).

The influences of various *glnA* mutations on *glnA* expression were studied in *B. subtilis* strains containing integrated *glnA-lacZ* transcriptional fusions (113). In the wild-type background, β-galactosidase expression from the *glnA* promoter was regulated in the same manner as GS, being repressed under excess-nitrogen conditions. Mutations that altered GS activity or a *glnA* deletion, which eliminated detectable GS activity or antigen, resulted in derepressed *lacZ* expression under excess-nitrogen conditions. A *glnA* mutation (*glnA33*) that confers partial resistance to methionine sulfoximine, a glutamine analog, also showed altered regulation (109). In addition to altered feedback characteristics, GS enzymes from these mutants display decreased affinity for either glutamate or ammonia (26, 109), and the mutations in several strains that overproduce GS have been found to reside in regions contributing to the enzyme's active site (109, 134). Thus, the participation of GS in its own regulation may involve some aspect of its catalytic activity.

The absence of *glnA* mutants defective in transcriptional regulation but displaying wild-type catalytic properties does not suggest that these functions are inseparable. Although mutations are found in at least three different active-site regions (109, 134), these regions may define overlapping domains that participate in regulation as well as catalysis. Regardless of regulatory domain location, the protein does not affect transcription by directly binding to *glnA* regulatory sequences (see below). On the other hand, GS may be utilized for sensing and/or relaying a nitrogen-dependent signal to other proteins involved in regulation; the nature of such a control will be discussed below.

glnRA operon

In *B. subtilis*, the *glnA* gene maps at 152° between *dnaA* and *thyA*. While there has been disagreement as to its exact location (26, 45, 97), cloning and fine-structure mapping verified the *dnaA-glnA-thyA* gene order (44, 116). In *B. subtilis*, *glnA* is followed by two small open reading frames having the potential to code for hydrophobic polypeptides with unknown functions (84, 116). There is no evidence that either is cotranscribed with *glnA*. An apparent rho-independent termination site is found immediately after the *glnA* sequences (84, 116), and the distal open reading frame is preceded by a sequence having homology with the $E\sigma^A$ consensus (116).

In both *B. subtilis* and *B. cereus*, *glnA* is the second gene of a dicistronic operon shown schematically in Fig. 1a (80, 84, 116). The first gene, *glnR*, codes for GlnR, a protein that acts with GS to negatively regulate transcription (105). Mutations in *glnR* result in derepressed GS levels under excess-nitrogen conditions; restoration of wild-type regulation occurs when *glnR* is provided in *trans* (105).

glnR gene product

GlnR is 135 amino acids long and has a deduced molecular weight of 15,834 (116). The *B. subtilis* protein has been purified to homogeneity from an *E. coli* overproducing strain (105); GlnR from *B. cereus* has also been purified (83). In both cases, the protein is a dimer in solution, and the subunit molecular weight is in agreement with that predicted from the DNA sequence. The amino acid sequence between residues 59 and 83 has the potential to form an α-helix–turn–α-helix (HTH) structure similar to the motif found in one class of bacterial regulatory proteins (105), and mutations within this region, between residues 70 and 83, abolish repressor activity (108). Evidence that GlnR directly affects transcription was obtained from several studies. In the absence of added factors, purified GlnR binds to DNA containing the *glnRA* promoter region with high affinity ($K_{dapp} = 10^{-11}$ M [17]) and protects this region from DNAse and restriction

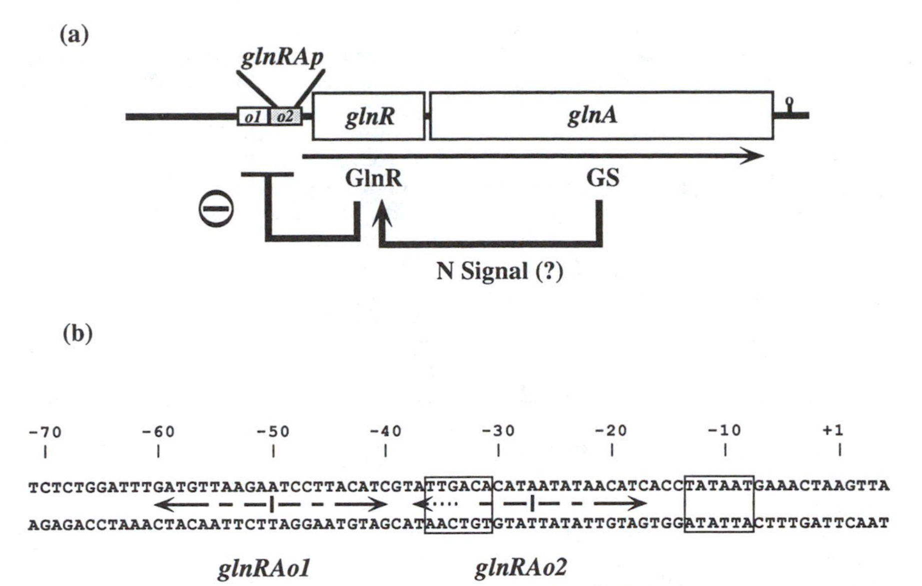

Figure 1. *glnRA* operon in *B. subtilis*. (a) Structure of the operon and model for control. Autogenous control of *glnRA* expression from the *glnRA* promoter, *glnRAp*, involves *glnR* and *glnA* gene products GlnR and GS, respectively. GlnR represses transcription under excess-nitrogen conditions by binding to *glnRA* operators *glnRAo*$_1$ and *glnRAo*$_2$, thereby inhibiting transcription initiation. GlnR activity is thought to be mediated in part by GS, which may be used to relay information about the nitrogen state of the cell. (b) Sequence of the *glnRA* promoter region from −71 to +5 relative to the start point of transcription (116). Boxed sequences indicate −10 and −35 promoter regions. Diverging arrows indicate symmetrical *glnRAo*$_1$ (−40 to −60) and quasisymmetrical *glnRAo*$_2$ (−17 to −37) sites shown to be involved in GlnR binding.

endonuclease digestion (17, 105). Similarly, binding of GlnR to specific promoter region sequences has been demonstrated by dimethylsulfate modification studies in vitro (17, 53, 110) and in vivo (53). Finally, GlnR inhibits transcription initiation from the *glnRA* promoter in vitro (17).

glnRA promoter and operator

Transcription from the *glnRA* promoter in *B. subtilis* requires the Eσ^A form of RNA polymerase; *E. coli* Eσ^{70} polymerase will initiate from the same start point (17, 103). The −35 and −10 regions of the promoter are identical to the σ^A consensus and are separated by 17 bp (116). There is no evidence for transcription initiation by other polymerases during exponential growth or stationary phase. Upstream from the promoter lies an A-T-rich region that is characteristic of many *B. subtilis* promoters. Elimination of this region does not affect promoter efficiency (107, 110).

Regulation of transcription involves the binding of a GlnR dimer to each of two operators, *glnRAo*$_1$ and *glnRAo*$_2$ (Fig. 1b) (53). The upstream site, *glnRAo*$_1$, is a 21-bp symmetrical sequence lying between positions −40 and −60 and is homologous to operator sites utilized by the HTH class of DNA-binding proteins (110). This operator is necessary for nitrogen-dependent control, since deletions or mutations that disrupt dyad symmetry result in constitutive GS expression (110).

The *glnRAo*$_2$ operator is a quasisymmetrical sequence extending from approximately positions −17 to −37 (Fig. 1b). The downstream side of the element is homologous to the *glnRAo*$_1$ half-site, and GlnR binding over the entire *glnRAo*$_2$ region occurs in a manner similar to that of the upstream site (17, 53). When separated, *glnRAo*$_1$ is a better substrate for GlnR binding than is *glnRAo*$_2$ (85). However, when the sites are joined, GlnR interacts with both sites equally in a noncooperative manner (17). In vivo, both *glnRAo*$_1$ and *glnRAo*$_2$ are occupied under nitrogen excess conditions and are vacant in a nitrogen-limited culture (53).

The roles of the two operators in *glnRA* control are not understood. *glnRAo*$_1$ is essential for nitrogen-dependent control, but *glnRAo*$_2$ may be redundant. Sequential binding of GlnR to *glnRAo*$_1$ and then to *glnRAo*$_2$ might be a mechanism for obtaining different levels of transcription, with occupancy at both sites resulting in full repression. However, such a model is not consistent with available data, and certain base substitutions within the downstream side of *glnRAo*$_2$ do not affect regulation (53). Furthermore, although a *glnRAo*$_1$ site is present upstream of the *B. cereus glnRA* promoter region, *glnRAo*$_2$-like sequences are not found in this organism (80).

Evidence that *glnRAo*$_2$ sequences may bind one or more factors other than GlnR has come from in vivo studies (53). Protection from modification at positions within *glnRAo*$_2$ was observed in both wild-type and *glnR* mutant strains grown under derepressing conditions. GS is not directly responsible for the effect, since this protein does not bind to DNA (53). Thus, GlnR-independent regulation may be involved in *glnRA* control under certain physiological conditions. Loci affecting GS expression that are unlinked to *glnRA* have been identified by genetic studies (2, 25, 116). A mutation in *outB*, for instance, inhibits derepression of a *glnR-lacZ* fusion, causing 10- to 44-fold lower β-galactosidase expression under nitrogen-limited conditions (2). The relationship of *outB* to *glnRA* control is not known, but it may be significant that extracts prepared from *B. subtilis* contain more than one protein with binding activity for the *glnRA* regulatory region (85).

Model for *glnRA* control

A model for *glnRA* control in *B. subtilis* is shown in Fig. 1a. Inhibition of transcription by GlnR alone is not sufficient to explain repression under excess-nitrogen conditions, since GS is also required for such control. Direct interaction of GS with GlnR may occur through subunit contacts, or alternatively, GlnR may receive from GS some signal that reflects the nitrogen state of the cell. Evidence for the former has been obtained from experiments examining GlnR binding in vitro. DNase protection and electrophoretic gel mobility shift assays showed that GS stimulates the binding and repressor activities of GlnR from *B. subtilis* approximately fourfold (17). A similar effect was reported for GlnR from *B. cereus* (83). The significance of these interactions in vivo as well as the factors regulating them require clarification. GS catalytic activity may also be critical in determining GS and GlnR-DNA interactions.

Since purified GlnR binds operator DNA with high affinity in the absence of other factors, it follows that some mechanism must be available to prevent repression under nitrogen-limited conditions. There is no evidence that covalent modification is involved in regulating GlnR activity. GlnR from extracts of *B. subtilis* grown under limited- and excess-nitrogen conditions is not altered with respect to size or charge (85). On the other hand, regulation of GlnR activity by some metabolite or protein factor has been suggested from genetic studies. Mutations in *glnR* resulting in a protein unable to recognize nitrogen-limited conditions are clustered within the carboxy portion of the protein (105, 108). Deletion analysis demonstrated that the last seven residues, from 129 to 135, are part of a domain essential for recognizing nitrogen conditions (108), since removal of these sequences results in the inability to derepress under nitrogen-limiting conditions (108). However, full repression is not observed under excess-nitrogen conditions, and the deletion may have an altered subunit conformation, thereby not allowing proper DNA binding. Alternatively, the deletion might have removed a domain involved in subunit-subunit interaction, which may inhibit effective dimerization.

Although several models can explain the information at hand, it is clear that *glnRA* control in *B. subtilis* is not simple. While the regulation of GS activity and synthesis in the enteric bacteria can be correlated to the intracellular concentrations of glutamine and α-ketoglutarate, no such relationship can be applied in *Bacillus* spp. Changes in the levels of glutamine, ammonia, or glutamate are not consistent with their utilization in regulating GS expression in *B. subtilis* or *B. licheniformis* (29, 78, 105, 111, 112). Thus, the signals used to monitor nitrogen availability are presently unknown.

Pleiotropic effects of some *glnA* mutants and role of GS in sporulation

Early studies by Schaeffer et al. (101) suggested that sporulation in *B. subtilis* is repressed by a catabolite containing carbon and nitrogen, which could be glutamine or some product derived from glutamine. In fact, a GS mutant of *B. megaterium* was able to sporulate in a medium containing excess glucose and ammonia, as if synthesis of glutamine were necessary for repression (39). Later experiments showed that a nucleotide (possibly GTP) is the more direct precursor of the repressing metabolite and that the role of glutamine is to serve as a precursor in purine biosynthesis (41).

No studies utilizing Gln$^-$ mutants have shown any direct role for GS in the control of sporulation. However, the phenotypes of certain *glnRA* regulatory mutations have provided some evidence for the participation of GS in some facet of sporulation control. Certain *glnRA*o_1 mutations that result in constitutive GS expression also affect sporulation initiation (108). Expression of *spoIID* and glucose dehydrogenase during stationary phase is delayed in these mutants, which suggests that unregulated synthesis of glutamine and/or GS itself may be inhibiting some aspect of initiation (108).

In addition to influencing sporulation initiation, GS has been implicated in the regulation of a variety of other metabolic processes. A number of *B. subtilis glnA* mutants are pleiotropically altered in the regulation of carbohydrate metabolism and expression of histidine degradation enzymes (45, 46). One mutant, containing the *glnA22* allele, has altered diauxic growth patterns in the presence of various glucose-repressed carbon sources, possibly implicating GS or glutamine in glucose metabolism or catabolite repression (45). In the presence of glucose, histidine, and glutamine, histidase levels in the mutant are four- to sixfold higher than they are in wild-type cells grown under the same conditions (45, 46). Reversion of this mutation to prototrophy resulted in the restoration of wild-type levels. Similarly, expression of urease, asparaginase, and the *nrg* genes also appears to be subject to a control that involves GS (4). These genes and enzymes are found at high levels only when ammonia is limiting and are constitutively expressed in *glnA* mutants (4) (see chapter 16 in this book). Thus, GS may transmit information about the availability of nitrogen to a number of metabolic regulatory factors in a manner similar to the role it plays in its own regulation.

Regulation of *glnA* Expression in Other Gram-Positive Bacteria

Clostridium glnA gene

The cloned *glnA* gene from *C. acetobutylicum* complements the glutamine auxotrophy of an *E. coli glnA* mutant, and its expression in *E. coli* is regulated by the nitrogen source (121). Control by an Ntr-like system was not observed, since deletion of host *ntrBC* genes did not affect regulation and the 2-kb fragment of DNA downstream of the *C. acetobutylicum glnA* gene did not complement an *ntrBC* deletion in the host (121).

Sequence analysis of the *C. acetobutylicum glnA* gene region revealed several interesting features that may relate to control in this organism (Fig. 2) (60). Two putative promoter sites, separated by 30 bp, with

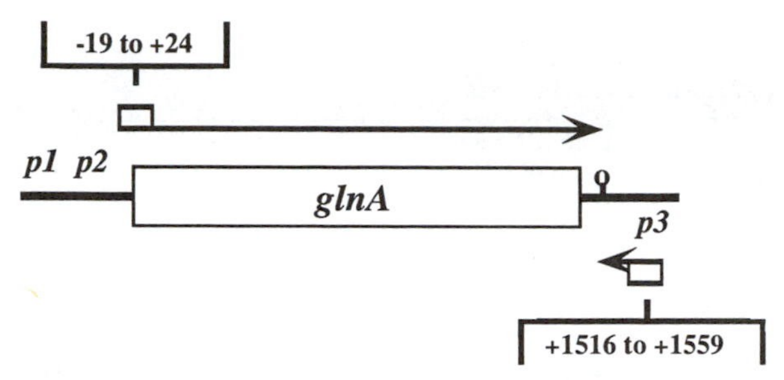

Figure 2. *glnA* gene of *C. acetobutylicum*. The schematic diagram shows orientation of the *glnA* transcript as directed from either of two promoters, *p1* and *p2*, and a putative anti-*glnA* transcript expressed from promoter *p3* from the opposite strand. Boxes at the beginning of each message represent regions whose sequences, shown relative to the start point of the *glnA* open reading frame, are complementary to each other (see text).

sequences similar to the *E. coli* σ^{70} or *B. subtilis* σ^{A} RNA polymerase consensus recognition sites, were identified upstream of the GS coding region. Three regions of dyad symmetry are located upstream of the distal promoter; one such region overlaps the promoter. The roles of these promoters and upstream elements have not been addressed; however, a 43-nucleotide sequence, including the presumed *glnA* ribosome-binding site and first 8 codons, is very similar to a sequence located beyond the 3′ end of the coding region and a terminatorlike element (60). These downstream sequences are followed by another putative promoter sequence, *p3*, which is situated on the opposite DNA strand and oriented towards the 3′ end of the structural gene (Fig. 2). Thus, transcription from this promoter could result in production of an RNA complementary to the 5′ end of *glnA* mRNA. This antisense message may be involved in regulation, since deletion of these sequences results in altered regulation of the cloned DNA in *E. coli* (60). An RNA corresponding to the antisense message has been detected in *C. acetobutylicum*, and its production is regulated by the nitrogen source (130). The mechanism by which downstream sequences are involved in regulating GS production in *C. acetobutylicum* is presently unknown.

Control of *glnA* expression in *Streptomyces* spp.

S. coelicolor glnA, a monocistronic gene encoding the GSI subunit, was cloned by complementation of an *E. coli* glutamine auxotroph (132). Unlike expression of the cloned genes from other gram-positive bacteria, expression of the cloned *S. coelicolor* gene in *E. coli* is not subject to nitrogen-dependent regulation (132). However, transcription in *E. coli* was shown to initiate from vector sequences instead of from the *S. coelicolor glnA* promoter (132). Since *glnA* mRNA is regulated in response to nitrogen availability in *S. coelicolor*, the lack of control in *E. coli* suggested the absence of transcription and/or other regulatory factors. The promoter region sequences are homologous to the *Streptomyces* consensus vegetative promoter but only in the −10 region; no enteric Ntr-like sequences are present (132). Thus, transcription from the *glnA* promoter in *S. coelicolor* probably requires the presence of activator proteins, an alternative form of RNA polymerase, or both. One such protein may be coded for by the *glnR* gene, which is required for *glnA* expression in *S. coelicolor* (131). Strains containing *glnR* mutations do not produce detectable amounts of *glnA* message, and transcription can be restored by introduction of a *glnR*-containing plasmid.

Regulation of *glnB* (or *glnII*), which encodes the GSII subunit in *Streptomyces* spp., has not been examined, and its relationship to *glnA* control is unknown. Enteric-like regulatory factors may be involved in *glnB* control in the symbiotic actinomycete *Frankia* sp. and other plant symbionts (see reference 98 and references therein), since the gene appears to be expressed from an Ntr-like promoter under nitrogen-limited conditions. It will be interesting to determine whether *glnB* control in *Streptomyces* occurs via similar mechanisms.

GLUTAMATE BIOSYNTHESIS IN *BACILLUS* SPP.

In both gram-positive and gram-negative bacteria, the intracellular glutamate pool is by far the largest of the amino acid pools, accounting for between 52 and 89% of the total amino acid content, depending on the organism (79, 119). In enteric bacteria, the glutamate pool concentration ranges from 5 to 10 mM (79, 119). On the other hand, in *Bacillus* spp., glutamate is maintained from approximately 40 to >100 mM and may represent 5 to 10% of the dry weight of the cell (3, 8, 9, 22, 29, 79, 111, 112, 119). The large glutamate pool in *Bacillus* spp. demonstrates that the amino acid not only is necessary for the synthesis of glutamine and other amino acids but also provides the cell with carbon and nitrogen precursors for spore biogenesis and may play a role in supplying energy for the starving cell during spore formation. Glutamate may also be used for equalizing the external osmotic strength of the growth medium (22). In *B. cereus*, glutamate appears to influence the UV resistance, dipicolinic acid content, and heat resistance of spores (18). Thus, given the demand for glutamate during growth and sporulation, it is not surprising that a number of *Bacillus* spp. produce both GDH and GOGAT under similar conditions of growth. However, in *B. subtilis*, GOGAT alone is sufficient to provide the cell with its glutamate requirement, since a mutation in either of the enzyme's subunits causes glutamate auxotrophy. The auxotrophy can be relieved by adding glutamate or a metabolite that can be converted to glutamate, such as glutamine, histidine, or aspartate, to the medium.

GOGAT

Properties of GOGAT in *Bacillus* spp.

GOGAT from several *Bacillus* spp. have been purified to homogeneity, and the characteristics and properties of enzymes from various sources are summarized in Table 1. The enzyme is an iron-sulfur flavoprotein composed of two nonidentical subunits. The large subunit binds glutamine and catalyzes the following glutaminase-like reaction: glutamine + $H_2O \rightarrow$ glutamate + NH_3.

The small subunit transfers ammonia to α-ketoglut-

Table 1. GOGAT and GDH in gram-positive bacteria

Enzyme and organism	Preferred cofactor	K_{mapp} for substrate	Subunit structure (mol wt)	Probable role	Reference(s)
GOGAT					
B. licheniformis	NADPH	NADPH, 13 μM; glutamine, 8 and 100 μM[a]; α-ketoglutarate, 6 and 50 μM[a]	$\alpha\beta$ (158,000 and 54,000)		104
B. macerans	NADPH				62
B. megaterium	NADPH	NADPH, 7.1 μM; glutamine, 200 mM; α-ketoglutarate, 9 μM	$\alpha_4\beta_4$ (142,000 and 54,000)		54
B. polymyxa	NADPH				61
B. stearothermophilus	NADPH	NADPH, 22 μM; glutamine, 29 μM; α-ketoglutarate, 15 μM	? (native partially purified enzyme = 160,000)		102
B. subtilis PCI 219	NADPH	NADPH, 6μM; glutamine, 100 μM[b]; α-ketoglutarate, 7 μM[b]	$\alpha\beta$ (160,000 and 56,000)		76
C. butyricum	NADH				63
C. kluyveri	NADPH				63
C. pasteurianum	NADH	Glutamine, 200 μM; α-ketoglutarate, 200 μM			24
S. clavuligerus	NADH				16
S. coelicolor	NADH				43
Streptomyces noursei	NADH				51
Streptomyces venezuelae	NADH				114
GDH					
B. licheniformis	NADPH	NADPH, 120 μM; ammonia, 5.5 mM; α-ketoglutarate, 6.7 mM; glutamate, 39 mM	α_6 (50,000)	Catabolism and assimilation[c]	91
B. macerans	NADPH	Ammonia, 2.9 mM; α-ketoglutarate, 0.38 mM		Assimilation	62
B. megaterium	NADPH	NADPH, 8.7 μM; ammonia, 22 mM; α-ketoglutarate, 0.36 mM; $NADP^+$, 50 μM; glutamate, 29 mM	α_6 (47,000)	Catabolism and assimilation[c]	54
B. polymyxa	NADPH	Ammonia, 2.9 mM; α-ketoglutarate, 1.4 mM		Assimilation	61
B. subtilis 168	NADH			Catabolism	3, 65
B. subtilis PCI 219	NADH	NAD^+, $\approx$0.2 mM; glutamate, $\approx$50 mM	α_6 (57,000)	Biosynthesis and catabolism	67
B. thuringiensis	NADH			Biosynthesis and catabolism	3
Thermophilic *Bacillus* sp.	NADPH	NADPH, 53 μM; ammonia, 21 mM; α-ketoglutarate, 1.3 mM; $NADP^+$, 300 μM; glutamate, 11 mM	? (native $\approx 2 \times 10^6$)	Biosynthesis and catabolism	42
Clostridium sp. strain SB_4	NADH	NADH, 10 μM; ammonia, 0.32 mM; α-ketoglutarate, 0.65 mM; NAD^+, 10 μM; glutamate, 1.8 mM	? (native = 275,000)	Catabolism and assimilation (?)	129
C. butyricum	NADPH	Ammonia, 2.8 and 25 mM[a]; glutamate, 8.9 mM		Catabolism and assimilation[c]	63
C. kluyveri	NADPH	Ammonia, 12 mM		Assimilation[c]	63
C. pasteurianum	NADH			Catabolism	68
S. coelicolor	NADH			Assimilation[c]	43
S. fradiae	GDH_I: NADH	Ammonia, 90 mM; glutamate, 1.5 mM		Catabolism	122
	GDH_{II}: NADPH	NADPH, 70 μM; ammonia, 31 mM; α-ketoglutarate, 1.5 mM; $NADP^+$, 120 μM; glutamate, 29 mM	α_4 (49,000)	Biosynthesis	

[a] Biphasic kinetics with respect to indicated substrate.
[b] Enzyme exhibited biphasic kinetics with respect to glutamine or α-ketoglutarate; values were obtained from initial velocity experiments performed in the presence of a fixed concentration of one substrate while the other substrates were varied.
[c] GDH may be used for ammonia assimilation under excess-nitrogen conditions.

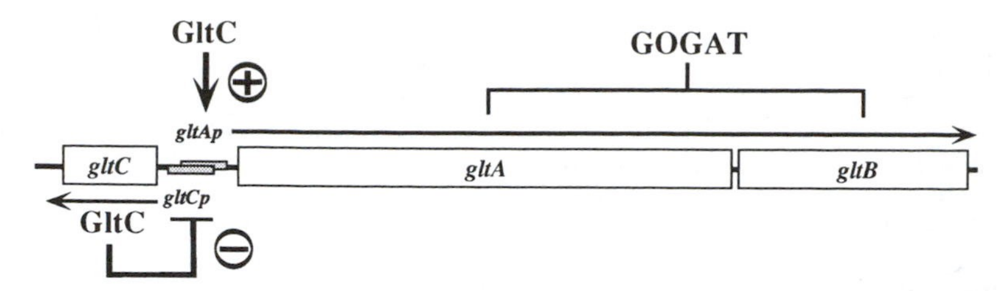

Figure 3. *gltAB* and *gltC* genes of *B. subtilis*. Synthesis of GOGAT large and small subunits, coded for by *gltA* and *gltB*, respectively, are shown to occur from a monocistronic message, although such has not been established. The GltC protein, produced by *gltC*, represses its own synthesis and is also required for activation of transcription from the *gltAp* promoter under glutamate-limited growth conditions. The overlap of the two promoters and the presence of several repeated sequences within the *gltAp* and *gltCp* regions are thought to play roles in regulating *gltA*(*B*) transcription (see text).

arate in a GDH-like manner as follows: NH_3 + α-ketoglutarate + NADPH + H^+ → glutamate + $NADP^+$.

GOGAT from *Bacillus* spp. has a nearly absolute requirement for NADPH; NADH will support only a small fraction of the activities of the enzymes from *B. megaterium* and *B. subtilis* (54, 76, 104). While GOGAT is capable of catalyzing a GDH-like reaction, the ammonia-dependent activity is normally only 2 to 4% of the glutamine-dependent activity and is not physiologically significant (104). In *B. megaterium*, hyperbolic kinetics were obtained for glutamine, α-ketoglutarate, and NADPH, and these substrates were found to bind with high affinity (Table 1) (54). On the other hand, the *B. licheniformis* (104) and *B. subtilis* (76) enzymes displayed cooperative kinetics for glutamine and α-ketoglutarate, a characteristic reported only for α-ketoglutarate in the ammonia-dependent reaction of the *E. coli* GOGAT (15).

For *Bacillus* spp. and most other bacteria, GOGAT has been shown to be relatively insensitive to regulation by feedback inhibition. A variety of metabolites has been examined for inhibitory capability, and very few displayed significant inhibition at concentrations that were physiologically relevant. In *B. subtilis*, L-malate was shown to be a potent effector, yielding 70% inhibition at a concentration of 10 mM (76).

$NADP^+$ was found to be an effective inhibitor of GOGAT activity in *B. licheniformis*. The cofactor was competitive with respect to NADPH, displaying a K_i of 20 μM, and 50% inhibition was obtained when the ratio of $NADP^+$ to NADPH was 2:1 (104). However, it was concluded that regulation by the $NADP^+$ pool is probably not significant, since $NADP^+$/NADPH ratios do not approach 2:1 when measured during exponential growth, at the onset of stationary phase, or 2 h into sporulation (104).

Regulation of GOGAT expression

In *B. subtilis*, changes in enzyme levels have been shown to be due to differences in steady-state levels of GOGAT mRNA, and regulation appears to be primarily exerted at the level of transcription (10). In the enteric bacteria, control of GOGAT expression has not been characterized well, and it is unclear whether the genes for GOGAT are subject to regulation by the Ntr system.

Genetic studies

Early genetic studies of the enzymes of the tricarboxylic acid cycle established the presence of a locus called *gltA*, which was defined by mutations causing auxotrophy for glutamate as well as a deficiency in GOGAT activity. The *gltA* gene was mapped at 160°, near *citK*, in the vicinity of the chromosomal terminus. Isolation of a *B. subtilis* GOGAT mutant genetically linked to but distinct from *gltA*, called *gltB*, indicated the presence of two cistrons whose products were the two GOGAT subunits; an extract of the *gltB1* allele was capable of complementing, in vitro, an extract of the *gltA292* allele (28). Bohannon et al. (10) identified the wild-type alleles of the *gltA292* and *gltB1* mutations from banks of *B. subtilis* DNA cloned in phage λ (10). The locations of *gltA* and *gltB* on the cloned DNA were determined by analyzing Tn*917* insertions and examining transcription by pulse-chase studies (10). The relative sizes of the apparent transcription units and analyses of *E. coli* minicells containing various subclones of *gltA* and *gltB* indicated that *gltA* and *gltB* corresponded to the large and small subunits of GOGAT, respectively (10, 11). While DNA fragments containing *gltA* and *gltB* are contiguous, it is not known whether the two genes form an operon (10). In *E. coli* and *Salmonella typhimurium*, the large and small subunits of GOGAT are encoded by *gltB* and *gltD*, respectively, which are cotranscribed (72, 87).

The *gltC* gene

Among the glutamate auxotrophs isolated via Tn*917* mutagenesis was a group mapping outside of *gltA* and *gltB* (10). These Tn*917* insertions were found to lie within a gene, called *gltC*, whose transcription was found to be divergent from that of *gltA* and whose transcriptional control sequences overlap those of *gltA* (Fig. 3) (10). Strains containing *gltC* mutations produce minimal quantities of *gltA* mRNA and aberrantly regulated levels of GOGAT activity. Since *gltC* could act in *trans* to correct these defects, it was suggested that the *gltC* gene product is a positive regulator of GOGAT synthesis (12).

Transcription of *gltC* does not appear to be appreciably regulated by the nitrogen source used for growth (10). Disruption of *gltC* by Tn*917* insertion results in increased steady-state levels of *gltC* mRNA, an effect opposite to that observed for *gltA* message (12). On the basis of these experiments, it was proposed that the *gltC* gene product acts in a negative manner to control its own transcription (12).

Sequence analysis of the *gltC* gene showed an open

reading frame that could code for a protein having a molecular weight of 34,795 (12). A protein of the appropriate size was synthesized in *E. coli* when the *gltC* gene was placed under the control of a coliphage promoter (13). The sequence of the predicted GltC protein showed significant homology to sequences of several proteins containing the HTH motif of a DNA-binding domain. Mutations in this region result in glutamate auxotrophy, indicating that the HTH domain is necessary for protein function (59). The protein appears to be a member of the *E. coli* LysR family of regulatory proteins, which are a class of positive regulatory factors containing highly conserved residues throughout the length of the protein (12).

gltA-gltC promoter region

Transcription start points for *gltA* and *gltC* are separated by approximately 50 bp; the *gltC* −35 sequence is positioned between the *gltA* −35 and −10 regions and on the opposite strand (12). Both promoter regions are homologous to the −10 and −35 *B. subtilis* Eσ^A consensus sequences, although the *gltA* −35 (TTGTTT) fit is poor, a fact that may be correlated to its dependence on a positive regulator. Nine copies of a direct repeat with the consensus sequence 5′-ATATTGTTT-3′ appear in the region of DNA between the 5′ ends of *gltA* and *gltC* (12). At least two copies of the repeat appear within *gltC* promoter sequences, overlapping the *gltA* −35 region. These sequences are also similar to those found in the target DNAs of several other LysR family proteins. Thus, the binding of GltC to these sequences may be relevant to *gltC-gltA* control.

Model for *gltC-gltA* control in *B. subtilis*

Although it is not known how *gltC* is involved in GOGAT control, a model involving the binding of GltC to the repeated segments found within the *gltC-gltA* promoter region has been proposed (Fig. 3) (12). Under all growth conditions, GltC limits initiation frequency from its own promoter, thereby sustaining low concentrations of *gltC* message. In the presence of limiting glutamate, GltC becomes an activator, perhaps via some conformational change, allowing for *gltA* transcription. The stimulation of *gltA* transcription by GltC probably involves the interaction of the protein with RNA polymerase. Such a role would explain why certain rifampin-resistant RNA polymerase mutants display little or no detectable GOGAT activity (99, 100).

Activation of *gltA* transcription requires that GltC recognize the glutamate status of the cell by interacting directly or indirectly with some metabolic factor. While the identity of the metabolite used to effect GltC activity is not known, physiological studies have suggested that glutamine may play a role in controlling GOGAT synthesis in *B. subtilis*. Measurements of GOGAT activity and the intracellular pool of glutamine in *B. subtilis* grown in the presence of various nitrogen sources indicated an inverse relationship between enzymatic activity and the glutamine pool size (28, 29). Similarly, when cultures growing under conditions that yielded high GOGAT levels were shifted to conditions that resulted in low GOGAT activity, the glutamine pool increased approximately 8-fold within the first 30 min and then GOGAT specific activity decreased by 2.5-fold 15 min later (29). This relationship was maintained in the *glnA22* strain of *B. subtilis*, which had glutamine pool concentrations that were increased sixfold and GOGAT levels that were decreased fivefold compared with those of wild type (46). Determining whether glutamine is directly involved in GOGAT expression and GltC activity in *B. subtilis* will require the analysis of factors necessary for control in vitro. For *B. licheniformis*, no correlation between glutamine and GOGAT levels was detected when cultures were grown in the presence of a variety of nitrogen sources (111, 112). However, the occurrence of both GOGAT and GDH in *B. licheniformis* suggests that this organism has evolved a mechanism for GOGAT control that may be different from that in *B. subtilis*. The GOGAT gene structure in *B. licheniformis* has not been determined.

GDH

In the enteric bacteria, GDH is a completely dispensable enzyme. Strains deficient in GDH are capable of utilizing the GOGAT pathway for glutamate production and have no discernible phenotype (see reference 96 for a review). Regulation of enzyme synthesis differs from organism to organism, and little is known about specific mechanisms involved in its control. In *Klebsiella aerogenes*, the *nac* gene, which is involved in the responses of certain genes to nitrogen limitation, plays a role in repressing GDH synthesis under excess-nitrogen conditions (6).

While GDH has been studied in several *Bacillus* spp., the role the enzyme plays in ammonia assimilation is not clear. Strains lacking GDH have not been isolated, and the involvement of GDH in metabolism has been extrapolated either from kinetic or physiological studies. Table 1 summarizes the properties of GDH from several *Bacillus* spp.

Properties of GDH from *B. subtilis*

The search for GDH activity in crude extracts of *B. subtilis* was initially unsuccessful because of the nature of the assay used and the growth conditions examined. While GDH isolated from most bacteria prefers NADPH as a substrate for the anabolic reaction (glutamate synthesis), an NADH-dependent GDH has been purified from *B. subtilis* (64, 67). The enzyme is considerably more active with NADH as the cofactor than with NADPH, a property also observed for the enzymes from *B. stearothermophilus* (102) and *Bacillus thuringiensis* (3). The enzyme is similar to most prokaryotic GDHs, being a hexamer of identical subunits and having a native molecular weight of approximately 300,000 (65, 67).

In *B. subtilis*, GDH levels are influenced by the presence of glutamate in the medium (64). Maximum GDH levels occur during mid-exponential phase in cells grown in rich media or in minimal media containing glutamate as the sole source of carbon and nitrogen (65). The inclusion of glucose in the minimal medium results in as much as a 100-fold decrease in GDH specific activity (65). Thus, the enzyme is probably utilized in a catabolic capacity and may be regulated by a carbon catabolite control mechanism

(65). A catabolic role for GDH is consistent with the notion that GOGAT is solely responsible for glutamate biosynthesis in *B. subtilis*. Furthermore, GDH catabolic activity is inhibited allosterically by oxaloacetate and citrate and competitively by α-ketoglutarate. The levels of all of these metabolites are directly dependent on the organism's carbon balance (67). Because nonhyperbolic kinetics were observed for the catabolic reaction, K_ms have not been reported. However, extrapolation of data presented by Kimura et al. (67) indicates a K_m for glutamate in the range of 25 to 50 mM, a physiologically significant level. On the other hand, purified enzyme displays 20 to 50 times more anabolic activity than glutamate-deaminating activity. A complete understanding of the role GDH plays in glutamate catabolism in *B. subtilis* will require studies using GDH mutants.

GDHs from other *Bacillus* spp.

As can be seen in Table 1, GDHs from most *Bacillus* spp. require NADPH; activity in the presence of NADH is either low or undetectable. The difference in cofactor requirement compared with that of *B. subtilis* may reflect utilization of the enzyme in an assimilatory capacity. As discussed above, such a role has been suggested for GDH in *B. macerans* (62) and *B. polymyxa* (61), two organisms having enzymes with moderate affinities for ammonia (Table 1). The *B. licheniformis* GDH is similar to the *B. polymyxa* and *B. macerans* enzymes with respect to its affinity for ammonia (Table 1). However, the saturation curve for this substrate is not linear, and the K_m shown in Table 1 is an approximation. The low affinity for α-ketoglutarate and the observation that high concentrations of ammonia enhance enzymatic activity (91) suggest that the enzyme may be important only when *B. licheniformis* is growing in the presence of limiting ammonia. Such is also the case for GDH in *B. megaterium* and the thermophilic *Bacillus* spp., which have very low affinities for ammonia (Table 1).

In addition to an assimilatory role, GDH from *B. licheniformis*, *B. megaterium*, and the thermophilic *Bacillus* spp. may function in a catabolic capacity under certain physiological conditions. While these enzymes have low affinities for glutamate, their K_ms for glutamate are well within or below the glutamate pool concentration ranges observed in *Bacillus* spp. (Table 1). However, since none of the NADPH-dependent enzymes have been shown to be influenced by allosteric control or feedback inhibition, it is not clear how these organisms regulate both anabolic and catabolic activities.

Glutaminase

The production of glutamate from glutamine is accomplished by the enzyme L-glutamine aminohydrolase (EC 3.5.1.2), also known as glutaminase. In *E. coli*, there are two glutaminase activities: glutaminase A is an inducible enzyme that responds to nitrogen metabolites, cyclic AMP, and the stage of culture growth, and glutaminase B is a constitutive activity that is probably the sum of all glutaminase activities associated with amidotransferases (92). Control of glutaminase activity is important to ammonia assimilation, since significant levels of both activities are present in ammonia-limited cells, a condition that results in highly active GS. How *E. coli* regulates glutaminase in order to avoid futile cycling is not understood.

Glutaminase activity has been detected in extracts of *B. subtilis* (46) and *B. licheniformis* (23). In *B. licheniformis*, two distinct glutaminase activities could be measured based on the assay of enzymatic activity at pHs 7 and 9 (23). Activity at pH 7 was similar to the glutaminase B activity of *E. coli* and was not significantly affected by growth conditions. On the other hand, activity at pH 9 was approximately 10-fold higher in glucose-glutamine cultures than in cultures grown in the presence of glucose and ammonia, a response similar to that shown by glutaminase A of *E. coli*. This high activity did not change when ammonia was added to the glucose-glutamine cultures, but a further threefold increase was observed when glutamine was the sole source of carbon and nitrogen. Thus, the glutaminase A-like activity appears to be used for glutamine catabolism and may be regulated by a carbon catabolite repression mechanism.

The influence of glutaminase on glutamine production is difficult to assess without an understanding of factors involved in regulating enzyme synthesis and activity. While the extracellular concentration of glutamine may be relevant in determining glutaminase expression, the intracellular glutamine pool may not be important in enzyme control. In *B. licheniformis*, intracellular glutamine concentrations determined in cultures grown in the presence of different nitrogen sources (46) did not show any correlation with the glutaminase activity found under similar conditions (23). Similarly, the increased glutamine pool found in the *glnA22* mutant of *B. subtilis* did not effect glutaminase expression, which was at wild-type levels (47). Finally, the moderate expression of GOGAT in both *B. subtilis* and *B. licheniformis* in glutamine-grown cultures suggests that catabolism by glutaminase may not be significant.

Glutamate-α-Ketoglutarate Transaminase and Glutamate-Pyruvate Transaminase

Both glutamate-α-ketoglutarate transaminase and glutamate-pyruvate transaminase activities have been detected in a number of *Bacillus* spp., including *B. cereus* (18, 20), *B. licheniformis* (78), *B. megaterium* (66), *B. polymyxa* (61), and *B. thuringiensis* (3) and have been assayed throughout growth and sporulation (20). The reversible transfer of an amino group between amino acids and α-keto acids and the constitutive nature of these activities suggest that these pathways establish and maintain an equilibrium between the amino acid and α-keto acid pools. However, since nothing is known about the enzymes that catalyze these reactions in *Bacillus* spp., such a role can only be speculative. In the enteric bacteria, three classes of transaminases have been characterized; these enzymes favor the utilization of keto acids to form the cognate amino acids, with glutamate as the donor of the amino group (96).

GLUTAMATE BIOSYNTHESIS IN OTHER GRAM-POSITIVE BACTERIA

Clostridium spp.

In *Clostridium* spp., neither the properties nor the regulation of GDH or GOGAT has been studied in any detail. An NADH-dependent GDH purified from *Clostridium* sp. strain SB_4 displayed an uncharacteristically high affinity for ammonia (Table 1) (129). The enzyme was not found to be regulated allosterically, and high turnover rates for NADH oxidation and NAD^+ reduction indicated that this enzyme participated in the energy-yielding processes of this organism (129). On the other hand, extracts of *C. kluyveri* were found to contain an NADPH-dependent GDH displaying a K_m for ammonia of 12 mM (63). The pattern of expression under limiting- and excess-nitrogen conditions indicated that GDH is used for assimilation only when ammonia is in excess. Nitrogen-limited cultures possess little GDH activity, and an NADPH-dependent GOGAT is utilized for glutamate production under these conditions.

Unlike the case for the *Bacillus* spp., an NADH-dependent GOGAT was found in extracts of *C. butyricum* (63) and *C. pasteurianum* (24). GDH activity could not be found in these organisms when they were grown in minimal media under either excess-nitrogen or nitrogen-limited conditions. In complex medium, however, an NADPH-dependent GDH activity was detected in *C. butyricum* (63). The enzyme displayed biphasic kinetics with respect to ammonia, having K_ms for ammonia of 2.8 and 25 mM in the concentration ranges of 1 to 25 and 25-100 mM, respectively (63). While the moderate affinity for ammonia in the 1 to 25 mM range hints at an assimilatory role, the enzyme's presence in complex media suggests that it may be synthesized solely for the production of α-ketoglutarate.

Streptomyces spp.

GOGAT and GDH have been detected in extracts of several different *Streptomyces* spp., and their properties are shown in Table 1. For *S. clavuligerus*, GOGAT appears to be solely responsible for glutamate biosynthesis, since GOGAT mutants require glutamate for growth in all media (16). In contrast to the GOGAT from *Bacillus* spp., the enzyme from several *Streptomyces* spp. has been shown to require NADH for activity (Table 1).

Both NADH- and NADPH-dependent GDH activities are found in extracts of *Streptomyces fradiae* (122). While NADH-dependent GDH levels respond to the nitrogen source, NADPH-dependent activity does not. Purified NADPH-dependent GDH is composed of four identical subunits, a characteristic not shared by other bacterial GDHs, and has an absolute requirement for NADPH. Neither enzyme appears to be important in ammonia assimilation, as suggested by the high K_m for ammonia (Table 1). In fungi and yeast cells, NADPH- and NAD^+-dependent GDHs are used for glutamate biosynthesis and catabolism, respectively (128). It will be interesting to determine whether the occurrence of two enzymes is more widespread among *Streptomyces* spp. and what their roles are.

In *S. coelicolor*, both physiological and genetic studies revealed that this organism probably utilizes GDH for assimilation when ammonia is in excess (43). GDH levels are at their highest in media containing ammonia as the nitrogen source, and growth under these conditions was observed for a GOGAT mutant (Glt^-), indicating that glutamate production was accomplished via GDH. The Glt^- strain also did not require glutamate when glutamine or aspartate was present in the medium. These amino acids provide glutamate via glutaminase (for glutamine) and the glutamate:α-ketoglutarate transaminase (for aspartate) reactions. The ineffectiveness of GDH for glutamate production under nitrogen-limited conditions was illustrated by the requirement for glutamate in the Glt^- strain when the nitrogen source was alanine or asparagine. Arginine or histidine, however, could not satisfy the glutamate requirement, perhaps a consequence of another mutation effecting *hut* or *aut* expression (43). In either case, the finding that a Glt^- GDH^- strain required glutamate for growth under all conditions indicates that GOGAT is necessary for glutamate production in *S. coelicolor* (43).

OTHER ROUTES OF AMMONIA ASSIMILATION

ADH

In *Bacillus* spp., ADH, which catalyzes the reversible amination of pyruvate to form alanine, has been studied for its role in the catabolism of L-alanine during germination (49) and sporulation (49, 77). Several lines of evidence indicate that the enzyme is not involved in ammonia assimilation in *Bacillus* spp. First, ADH is dispensable in *B. subtilis* and *B. megaterium*, since mutants lacking ADH grow normally in minimal medium (38, 49). Second, in *B. megaterium*, the kinetics of incorporation of $[^{13}N]$ammonia into the amino acid pool was not consistent with utilization of alanine as a primary assimilation route (66). Similar results were obtained for the uptake of $[^{15}N]$ammonia by *B. polymyxa* (61). Third, the pattern of ADH expression in *B. subtilis* and other *Bacillus* spp. suggests that it is utilized solely for alanine deamination, since levels are at their highest when alanine is present in the medium (3, 7, 77, 78, 111). Finally, the high K_m for ammonia makes it unlikely that the enzyme functions effectively for alanine synthesis under nitrogen-limited conditions (52, 86, 133), and alanine biosynthesis may be accomplished by glutamate:pyruvate transaminase (18, 38, 61, 78).

Ammonia assimilation via the ADH pathway has been suggested as an alternative route to the GS-GOGAT pathway in *S. clavuligerus*, an organism having no detectable GDH activity (1). The enzyme exhibits a K_m for ammonia of 20 mM, and the pattern of ADH expression is consistent with its use for ammonia assimilation when *S. clavuligerus* is grown in excess-ammonia medium (1). That ADH is not utilized directly for ammonia assimilation was demonstrated by using mutants devoid of ADH, which showed the same pattern of nitrogen source utilization as the wild type (16).

Asparagine Synthetase

In *E. coli* and *K. aerogenes*, two asparagine synthetases have been identified. One, encoded by *asnA*, catalyzes the ammonia-dependent amidation of aspartate to form asparagine; the other, the *asnB* product, uses glutamine as substrate in place of ammonia. Utilization of either enzyme for asparagine biosynthesis depends on the nitrogen source found in the medium, and control of enzyme synthesis by the Ntr system has been documented (96). Asparagine biosynthesis has not been studied in *Bacillus* spp. or most other gram-positive organisms; purification of an ammonia-dependent enzyme from *Lactobacillus arabinosus* (94) and *Streptococcus bovis* (19) has been reported.

Acknowledgments. The collaboration of Stuart Brown, Tim Donohue, Sue Fisher, Mark Rosenkrantz, and members of my laboratory, Jeanine Gutowski, Kendal Hirschi, John Nomellini, and Chris Rostkowski, as well as the inspiration and collaboration of Bob Bernlohr and Linc Sonenshein, are acknowledged. Work from my laboratory was supported in part by Public Health Service grant GM39541 from the National Institute of General Medical Sciences.

REFERENCES

1. **Aharonowitz, Y., and C. G. Friedrich.** 1980. Alanine dehydrogenase of the β-lactam producer *Streptomyces clavuligerus*. *Arch. Microbiol.* **125:**137–142.
2. **Albertini, A. M., and A. Galizzi.** 1990. The *Bacillus subtilis outB* gene is highly homologous to an *Escherichia coli ntr*-like gene. *J. Bacteriol.* **172:**5482–5485.
3. **Aronson, J. N.** 1975. Ammonia assimilation and glutamate catabolism by *Bacillus thuringiensis*. *Microbiology* **1975:**444–449.
4. **Atkinson, M. R., and S. H. Fisher.** 1991. Identification of genes and gene products whose expression is activated during nitrogen-limited growth in *Bacillus subtilis*. *J. Bacteriol.* **173:**23–27.
5. **Behrmann, I., D. Hillemann, A. Puhler, E. Strauch, and W. Wohlleben.** 1990. Overexpression of a *Streptomyces viridochromogenes* gene (*glnII*) encoding glutamine synthetase similar to those of eucaryotes confers resistance against the antibiotic phosphoinothricyl-alanyl-alanine. *J. Bacteriol.* **172:**5326–5334.
6. **Bender, R. A.** 1991. The role of the Nac protein in the nitrogen regulation of *Klebsiella aerogenes*. *Mol. Microbiol.* **5:**2575–2580.
7. **Berberich, R., M. Kaback, and E. Freese.** 1968. D-Amino acids as inducers of L-alanine dehydrogenase in *Bacillus subtilis*. *J. Biol. Chem.* **243:**1006–1011.
8. **Bernlohr, R. W.** 1967. Changes in amino acid permeation during sporulation. *J. Bacteriol.* **93:**1031–1044.
9. **Bernlohr, R. W., H. J. Schreier, and T. J. Donohue.** 1984. Enzymes of glutamate and glutamine biosynthesis in *Bacillus licheniformis*. *Curr. Top. Cell. Regul.* **24:** 145–152.
10. **Bohannon, D. E., M. S. Rosenkrantz, and A. L. Sonenshein.** 1985. Regulation of *Bacillus subtilis* glutamate synthase genes by the nitrogen source. *J. Bacteriol.* **163:**957–964.
11. **Bohannon, D. E., H. J. Schreier, and A. L. Sonenshein.** Unpublished data.
12. **Bohannon, D. E., and A. L. Sonenshein.** 1989. Positive regulation of glutamate biosynthesis in *Bacillus subtilis*. *J. Bacteriol.* **171:**4718–4727.
13. **Bohannon, D. E., and A. L. Sonenshein.** 1990. GltC, the positive regulator of glutamate synthase gene expression, p. 141–145. *In* J. A. Hoch and A. T. Ganesan (ed.), *Genetics and Biotechnology of Bacilli*. Academic Press, Inc., New York.
14. **Bott, K. F., G. Reysset, J. Gregoire, D. Islert, and J.-P. Aubert.** 1977. Characterization of glutamine requiring mutants of *Bacillus subtilis*. *Biochem. Biophys. Res. Commun.* **79:**996–1003.
15. **Bower, S., and H. Zalkin.** 1983. Chemical modification and ligand binding studies with *Escherichia coli* glutamate synthase. *Biochemistry* **22:**1613–1620.
16. **Brana, A. F., N. Paiva, and A. L. Demain.** 1986. Pathways and regulation of ammonium assimilation in *Streptomyces clavuligerus*. *J. Gen. Microbiol.* **132:**1305–1317.
17. **Brown, S. B., and A. L. Sonenshein.** Submitted for publication.
18. **Buono, F., R. Testa, and D. G. Lundgren.** 1966. Physiology of growth and sporulation in *Bacillus cereus*. I. Effect of glutamic and other amino acids. *J. Bacteriol.* **91:**2291–2299.
19. **Burchall, J. J., E. C. Reichelt, and M. J. Wolin.** 1964. Purification and properties of the asparagine synthetase of *Streptococcus bovis*. *J. Biol. Chem.* **239:**1794–1798.
20. **Charba, J. F., and H. M. Nakata.** 1977. Role of glutamate in the sporogenesis of *Bacillus cereus*. *J. Bacteriol.* **130:**242–248.
21. **Charney, J., W. P. Fisher, and C. P. Hegarty.** 1951. Manganese as an essential element for sporulation in the genus *Bacillus*. *J. Bacteriol.* **62:**145–148.
22. **Clark, V. L., D. E. Peterson, and R. W. Bernlohr.** 1972. Changes in free amino acid production and intracellular amino acid pools of *Bacillus licheniformis* as a function of culture age and growth media. *J. Bacteriol.* **112:**715–725.
23. **Cook, W. H., J. H. Hoffman, and R. W. Bernlohr.** 1981. Occurrence of an inducible glutaminase in *Bacillus licheniformis*. *J. Bacteriol.* **148:**365–367.
24. **Dainty, R. H.** 1972. Glutamate biosynthesis in *Clostridium pasteurianum* and its significance in nitrogen metabolism. *Biochem. J.* **126:**1055–1056.
25. **Dean, D. R., and A. I. Aronson.** 1980. Selection of *Bacillus subtilis* mutants impaired in ammonia assimilation. *J. Bacteriol.* **141:**985–988.
26. **Dean, D. R., J. A. Hoch, and A. I. Aronson.** 1977. Alteration of the *Bacillus subtilis* glutamine synthetase results in overproduction of the enzyme. *J. Bacteriol.* **131:**981–987.
27. **Debarbouille, M., I. Martin-Verstraete, F. Kunst, and G. Rapoport.** 1991. The *Bacillus subtilis sigL* gene encodes an equivalent of σ^{54} from gram-negative bacteria. *Proc. Natl. Acad. Sci. USA* **88:**9092–9096.
28. **Deshpande, K. L., J. R. Katze, and J. F. Kane.** 1980. Regulation of glutamate synthase from *Bacillus subtilis* by glutamine. *Biochem. Biophys. Res. Commun.* **95:**55–60.
29. **Deshpande, K. L., J. R. Katze, and J. F. Kane.** 1981. Effect of glutamine on enzymes of nitrogen metabolism in *Bacillus subtilis*. *J. Bacteriol.* **145:**768–774.
30. **Deuel, T., and E. R. Stadtman.** 1970. Some kinetic properties of *Bacillus subtilis* glutamine synthetase. *J. Biol. Chem.* **245:**5206–5213.
31. **Deuel, T. F.** 1971. *Bacillus subtilis* glutamine synthetase. Specific catalytic changes associated with limited sulfhydryl modification. *J. Biol. Chem.* **246:**599–605.
32. **Deuel, T. F., A. Ginsberg, J. Yeh, E. Shelton, and E. R. Stadtman.** 1970. *Bacillus subtilis* glutamine synthetase. *J. Biol. Chem.* **245:**5195–5205.
33. **Deuel, T. F., and S. Prusiner.** 1974. Regulation of glutamine synthetase from *Bacillus subtilis* by divalent cations, feedback inhibitors and L-glutamine. *J. Biol. Chem.* **249:**257–264.

34. **Donohue, T. J., and R. W. Bernlohr.** 1978. Carbon and nitrogen catabolite repression, metabolite pools, and the regulation of sporulation in *Bacillus licheniformis*, p. 293–298. *In* G. Chambliss and J. C. Vary (ed.), *Spores VII*. American Society for Microbiology, Washington, D.C.
35. **Donohue, T. J., and R. W. Bernlohr.** 1978. Effect of cultural conditions on the concentrations of metabolic intermediates during growth and sporulation of *Bacillus licheniformis*. *J. Bacteriol.* **135:**363–372.
36. **Donohue, T. J., and R. W. Bernlohr.** 1981. Properties of the *Bacillus licheniformis* A5 glutamine synthetase purified from cells grown in the presence of ammonia or nitrate. *J. Bacteriol.* **147:**589–601.
37. **Donohue, T. J., and R. W. Bernlohr.** 1981. Regulation of the activity of the *Bacillus licheniformis* A5 glutamine synthetase. *J. Bacteriol.* **148:**174–182.
38. **Elmerich, C.** 1972. Le cycle du glutamate, point de depart du metabolisme de l'azote, chez *Bacillus megaterium*. *Eur. J. Biochem.* **27:**216–224.
39. **Elmerich, C., and J. Aubert.** 1972. Role of glutamine synthetase in the repression of bacterial sporulation. *Biochem. Biophys. Res. Commun.* **46:**892–897.
40. **Elmerich, C., and J.-P. Aubert.** 1971. Synthesis of glutamate by a glutamine: 2-oxo-glutarate amidotransferase (NADP oxidoreductase) in *Bacillus megaterium*. *Biochem. Biophys. Res. Commun.* **42:**371–376.
41. **Elmerich, C., and J.-P. Aubert.** 1975. Involvement of glutamine synthetase and the purine nucleotide pathway in repression of bacterial sporulation, p. 385–390. *In* P. Gerhardt, R. N. Costilow, and H. L. Sadoff (ed.), *Spores VI*. American Society for Microbiology, Washington, D.C.
42. **Epstein, I., and N. Grossowicz.** 1975. Purification and properties of glutamate dehydrogenase from a thermophilic *Bacillus*. *J. Bacteriol.* **122:**1257–1264.
43. **Fisher, S. H.** 1989. Glutamate synthesis in *Streptomyces coelicolor*. *J. Bacteriol.* **171:**2372–2377.
44. **Fisher, S. H., M. S. Rosenkrantz, and A. L. Sonenshein.** 1984. Glutamine synthetase gene of *Bacillus subtilis*. *Gene* **32:**427–438.
45. **Fisher, S. H., and A. L. Sonenshein.** 1977. Glutamine requiring mutants of *Bacillus subtilis*. *Biochem. Biophys. Res. Commun.* **79:**987–996.
46. **Fisher, S. H., and A. L. Sonenshein.** 1984. *Bacillus subtilis* glutamine synthetase mutants pleiotropically altered in glucose catabolite repression. *J. Bacteriol.* **157:**612–621.
47. **Fisher, S. H., and A. L. Sonenshein.** 1991. Control of carbon and nitrogen metabolism in *Bacillus subtilis*. *Annu. Rev. Microbiol.* **45:**107–135.
48. **Fisher, S. H., and J. L. V. Wray.** 1989. Regulation of glutamine synthetase in *Streptomyces coelicolor*. *J. Bacteriol.* **171:**2378–2383.
49. **Freese, E., S. W. Park, and M. Cashel.** 1964. The developmental significance of alanine dehydrogenase in *Bacillus subtilis*. *Proc. Natl. Acad. Sci. USA* **51:**1164–1172.
50. **Gardner, A. L., and A. I. Aronson.** 1984. Expression of the *Bacillus subtilis* gene for glutamine synthetase in *Escherichia coli*. *J. Bacteriol.* **158:**967–971.
51. **Grafe, U., H. Bocker, and H. Thrum.** 1977. Regulative influence of o-aminobenzoic acid on the biosynthesis of nourseothricin in cultures of *Streptomyces noursei* JA3890b. II. Regulation of glutamine synthetase and the role of the glutamine synthetase/glutamate synthase pathway. *Z. Allg. Mikrobiol.* **17:**201–209.
52. **Grimshaw, C. E., and W. W. Cleland.** 1981. Kinetic mechanism of *Bacillus subtilis* L*-alanine dehydrogenase*. *Biochemistry* **20:**5650–5655.
53. **Gutowski, J. C., and H. J. Schreier.** 1992. Interaction of the *Bacillus subtilis glnRA* repressor with operator and promoter regions in vivo. *J. Bacteriol.* **174:**671–681.
54. **Hemmila, I. A., and P. I. Mantsala.** 1978. Purification and properties of glutamate synthase and glutamate dehydrogenase from *Bacillus megaterium*. *Biochem. J.* **173:**45–52.
55. **Hill, R. T., J. R. Parker, H. J. K. Goodman, D. T. Jones, and D. R. Woods.** 1988. Molecular analysis of a novel glutamine synthetase of the anaerobe *Bacteroides fragilis*. *J. Gen. Microbiol.* **135:**3271–3279.
56. **Hsu, R., S. J. Singer, P. Keim, T. F. Deuel, and R. L. Heinrikson.** 1977. Structural studies of *Bacillus subtilis* glutamine synthetase. Further purification, sulfhydryl groups, and the NH_2-terminal amino acid sequence. *Arch. Biochem. Biophys.* **178:**644–651.
57. **Hubbard, J. S., and E. R. Stadtman.** 1967. Regulation of glutamine synthetase. II. Patterns of feedback inhibition in microorganisms. *J. Biol. Chem.* **93:**1045–1055.
58. **Hubbard, J. S., and E. R. Stadtman.** 1967. Regulation of glutamine synthetase. V. Partial purification and properties of glutamine synthetase from *Bacillus licheniformis*. *J. Bacteriol.* **94:**1007–1015.
59. **Janssen, P., and A. L. Sonenshein.** Personal communication.
60. **Janssen, P. J., W. A. Jones, D. T. Jones, and D. R. Woods.** 1988. Molecular analysis and regulation of the *glnA* gene of the gram-positive anaerobe *Clostridium acetobutylicum*. *J. Bacteriol.* **170:**400–408.
61. **Kanamori, K., R. L. Weiss, and J. D. Roberts.** 1987. Ammonia assimilation in *Bacillus polymyxa*. ^{15}N NMR and enzymatic studies. *J. Biol. Chem.* **262:**11038–11045.
62. **Kanamori, K., R. L. Weiss, and J. D. Roberts.** 1987. Role of glutamate dehydrogenase in ammonia assimilation in nitrogen-fixing *Bacillus macerans*. *J. Bacteriol.* **169:** 4692–4695.
63. **Kanamori, K., R. L. Weiss, and J. D. Roberts.** 1989. Ammonia assimilation pathways in nitrogen-fixing *Clostridium kluyverii* and *Clostridium butyricum*. *J. Bacteriol.* **171:**2148–2154.
64. **Kane, J. F., and K. L. Deshpande.** 1979. Properties of glutamate dehydrogenase from *Bacillus subtilis*. *Biochem. Biophys. Res. Commun.* **88:**761–767.
65. **Kane, J. F., J. Wakim, and R. S. Fischer.** 1981. Regulation of glutamate dehydrogenase in *Bacillus subtilis*. *J. Bacteriol.* **148:**1002–1005.
66. **Kim, C.-H., and T. C. Hollocher.** 1982. ^{13}N isotope studies on the pathway of ammonia assimilation in *Bacillus megaterium* and *Escherichia coli*. *J. Bacteriol.* **151:**358–366.
67. **Kimura, K., A. Miyakawa, T. Imai, and T. Sadakawa.** 1977. Glutamate dehydrogenase from *Bacillus subtilis* PCI 219. *J. Biochem.* **81:**467–476.
68. **Kleiner, D.** 1979. Regulation of ammonium uptake and metabolism by nitrogen fixing bacteria. III. *Clostridium pasteurianum*. *Arch. Microbiol.* **120:**263–270.
69. **Krishnan, I. S., R. K. Singhal, and R. D. Dua.** 1986. Purification and characterization of glutamine synthetase from *Clostridium pasteurianum*. *Biochemistry* **25:**1589–1599.
70. **Kumada, Y., E. Takano, K. Nagaoka, and C. J. Thompson.** 1990. *Streptomyces hygroscopicus* has two glutamine synthetase genes. *J. Bacteriol.* **172:**5343–5351.
71. **Leonard, C. G., R. D. Housewright, and C. B. Thorne.** 1962. Effects of metal ions on the optical specificity of glutamine synthetase and glutamyl transferase of *Bacillus licheniformis*. *Biochim. Biophys. Acta* **62:**432–434.
72. **Madonna, M. J., R. L. Fuchs, and J. E. Brenchley.** 1985. Fine-structure analysis of *Salmonella typhimurium* glutamate synthase genes. *J. Bacteriol.* **161:**353–360.
73. **Magasanik, B.** 1982. Genetic control of nitrogen assimilation in bacteria. *Annu. Rev. Genet.* **16:**135–168.
74. **Magasanik, B.** 1988. Reversible phosphorylation of an enhancer binding protein regulates the transcription of bacterial nitrogen utilization genes. *Trends Biochem. Sci.* **13:**475–479.

75. **Magasanik, B., and F. C. Neidhardt.** 1987. Regulation of carbon and nitrogen utilization, p. 1321–1325. *In* F. C. Neidhardt, J. L. Ingraham, K. B. Low, B. Magasanik, M. Schaechter, and H. E. Umbarger (ed.), *Escherichia coli and Salmonella tyhimurium: Cellular and Molecular Biology*, vol. 2. American Society for Microbiology, Washington, D.C.
76. **Matsuoka, K., and K. Kimura.** 1986. Glutamate synthase from *Bacillus subtilis* PCI 219. *J. Biochem.* **99:** 1087–1100.
77. **McCowen, S. M., and P. V. Phibbs, Jr.** 1974. Regulation of alanine dehydrogenase in *Bacillus licheniformis*. *J. Bacteriol.* **118:**590–597.
78. **Meers, J. L., and L. K. Pedersen.** 1972. Nitrogen assimilation by *Bacillus licheniformis* organisms growing in chemostat cultures. *J. Gen. Microbiol.* **70:**277–286.
79. **Meers, J. L., D. W. Tempest, and C. M. Brown.** 1970. 'Glutamine(amide):2-oxoglutarate amino transferase oxido-reductase (NADP)', an enzyme involved in the synthesis of glutamate by some bacteria. *J. Gen. Microbiol.* **64:**187–194.
80. **Nakano, Y., K. Chiaki, E. Tanaka, K. Kimura, and K. Horikoshi.** 1989. Nucleotide sequence of the glutamine synthetase gene (*glnA*) and its upstream region from *Bacillus cereus*. *J. Biochem.* **106:**209–215.
81. **Nakano, Y., M. Itoh, E. Tanaka, and K. Kimura.** 1990. Identification of the reactive cysteinyl residue and ATP binding site in *Bacillus cereus* glutamine synthetase by chemical modification. *J. Biochem.* **107:**180–183.
82. **Nakano, Y., and K. Kimura.** 1987. Independent bindings of Mn^{2+} and Mg^{2+} to the active site of *B. cereus* glutamine synthetase. *Biochem. Biophys. Res. Commun.* **142:**475–482.
83. **Nakano, Y., and K. Kimura.** 1991. Purification and characterization of a repressor for the *Bacillus cereus glnRA* operon. *J. Biochem.* **109:**223–228.
84. **Nakano, Y., E. Tanaka, C. Kato, K. Kimura, and K. Horikoshi.** 1989. The complete nucleotide sequence of the glutamine synthetase gene (*glnA*) of *Bacillus subtilis*. *FEMS Microbiol. Lett.* **57:**81–86.
85. **Nomellini, J. F., and H. J. Schreier.** Unpublished data.
86. **Ohshima, T., and K. Soda.** 1979. Purification and properties of alanine dehydrogenase from *Bacillus sphaericus*. *Eur. J. Biochem.* **100:**29–39.
87. **Oliver, G., G. Gosset, R. Sanchez-Pescador, E. Loyoza, L. M. Ku, N. Flores, B. Becerril, F. Valle, and F. Bolivar.** 1987. Determination of the nucleotide sequence for the glutamate synthase structural genes of *Escherichia coli* K-12. *Gene* **60:**1–11.
88. **Opheim, D. J., and R. W. Bernlohr.** 1975. Purification and regulation of fructose-1,6-bisphosphatase from *Bacillus licheniformis*. *J. Biol. Chem.* **250:**3024–3033.
89. **Pan, F. L., and J. G. Coote.** 1979. Glutamine synthetase and glutamate synthase activities during growth and sporulation of *Bacillus subtilis*. *J. Gen. Microbiol.* **112:** 373–377.
90. **Paress, P. S., and S. L. Streicher.** 1985. Glutamine synthetase of *Streptomyces cattleya*: purification and regulation of synthesis. *J. Gen. Microbiol.* **131:**1903–1910.
91. **Phibbs, P. V., Jr., and R. W. Bernlohr.** 1971. Purification, properties and regulation of glutamic dehydrogenase of *Bacillus licheniformis*. *J. Bacteriol.* **106:**375–385.
92. **Prusiner, S.** 1973. Glutaminases of *Escherichia coli*: properties, regulation and evolution, p. 293–316. *In* S. Prusiner and E. R. Stadtman (ed.), *The Enzymes of Glutamine Metabolism*. Academic Press, Inc., New York.
93. **Ramaley, R. F., N. Fernald, and T. DeVries.** 1972. Dicarboxylate ω-amidase of *Bacillus subtilis*-168: evidence for a membrane-associated form. *Arch. Biochem. Biophys.* **153:**88–94.
94. **Ravel, J. M.** 1970. Asparagine synthetase (*Lactobacillus arabinosus*). *Methods Enzymol.* **17(A):**722–726.
95. **Rebello, J. L., and N. Strauss.** 1969. Regulation of glutamine synthase in *Bacillus subtilis*. *J. Bacteriol.* **148:**653–658.
96. **Reitzer, L. J., and B. Magasanik.** 1987. Ammonia assimilation and the biosynthesis of glutamine, glutamate, aspartate, asparagine, L-alanine, and D-alanine, p. 302–320. *In* F. C. Neidhardt, J. L. Ingraham, K. B. Low, B. Magasanik, M. Schaechter and H. E. Umbarger (ed.), *Escherichia coli and Salmonella tyhimurium: Cellular and Molecular Biology*, vol. 1. American Society for Microbiology, Washington, D.C.
97. **Reysset, G.** 1981. New class of *Bacillus subtilis* glutamine-requiring mutants. *J. Bacteriol.* **148:**653–658.
98. **Rochefort, D. A., and D. R. Benson.** 1990. Molecular cloning, sequencing, and expression of the glutamine synthetase II (*glnII*) gene from the actinomycete root nodule symbiont *Frankia* sp. strain CpI1. *J. Bacteriol.* **172:**5335–5342.
99. **Ryu, J.-I.** 1978. Pleiotropic effect of a rifampicin-resistant mutation in *Bacillus subtilis*. *J. Bacteriol.* **135:**408–414.
100. **Ryu, J.-I.** 1979. Ribonucleic acid polymerase mutation affecting glutamate synthase activity in and sporulation of *Bacillus subtilis*. *J. Bacteriol.* **139:**652–656.
101. **Schaeffer, P., J. Millet, and J.-P. Aubert.** 1965. Catabolic repression of bacterial sporulation. *Proc. Natl. Acad. Sci. USA* **54:**704–711.
102. **Schmidt, C. N. G., and L. Jervis.** 1982. Partial purification and characterization of glutamate synthase from a thermophilic bacillus. *J. Gen. Microbiol.* **128:**1713–1718.
103. **Schreier, H. J.** Unpublished data.
104. **Schreier, H. J., and R. W. Bernlohr.** 1984. Purification and properties of glutamate synthase from *Bacillus licheniformis*. *J. Bacteriol.* **160:**591–599.
105. **Schreier, H. J., S. W. Brown, K. D. Hirschi, J. F. Nomellini, and A. L. Sonenshein.** 1989. Regulation of *Bacillus subtilis* glutamine synthetase gene expression by the product of the *glnR* gene. *J. Mol. Biol.* **210:**51–63.
106. **Schreier, H. J., S. H. Fisher, and A. L. Sonenshein.** 1985. Regulation of expression from the *glnA* promoter of *Bacillus subtilis* requires the *glnA* gene product. *Proc. Natl. Acad. Sci. USA* **82:**3375–3379.
107. **Schreier, H. J., K. D. Hirschi, and C. A. Rostkowski.** 1990. *cis*-acting sequences regulating *glnRA* expression in *Bacillus subtilis*, p. 81–87. *In* J. A. Hoch and A. T. Ganesan (ed.), *Genetics and Biotechnology of Bacilli*. Academic Press, Inc., New York.
108. **Schreier, H. J., and C. A. Rostkowski.** Unpublished data.
109. **Schreier, H. J., C. A. Rostkowski, and E. M. Kellner.** 1993. Altered regulation of the *glnRA* operon in a *Bacillus subtilis* mutant that produces methionine sulfoximine-tolerant glutamine synthetase. *J. Bacteriol.* **175:**892–897.
110. **Schreier, H. J., C. A. Rostkowski, J. F. Nomellini, and K. D. Hirschi.** 1991. Identification of DNA sequences involved in regulating *Bacillus subtilis glnRA* expression by the nitrogen source. *J. Mol. Biol.* **220:**241–253.
111. **Schreier, H. J., T. M. Smith, and R. W. Bernlohr.** 1982. Regulation of nitrogen catabolic enzymes in *Bacillus* spp. *J. Bacteriol.* **151:**971–975.
112. **Schreier, H. J., T. M. Smith, T. J. Donohue, and R. W. Bernlohr.** 1981. Regulation of nitrogen metabolism and sporulation in *Bacillus licheniformis*, p. 138–141. *In* H. Levinson, A. L. Sonenshein, and D. J. Tipper (ed.), *Sporulation and Germination*. American Society for Microbiology, Washington, D.C.
113. **Schreier, H. J., and A. L. Sonenshein.** 1986. Altered regulation of the *glnA* gene in glutamine synthetase mutants of *Bacillus subtilis*. *J. Bacteriol.* **167:**35–43.
114. **Shapiro, S., and L. C. Vining.** 1983. Nitrogen metabo-

lism and chloramphenicol production in *Streptomyces venezuelae*. *Can. J. Microbiol.* **29:**1706–1714.

115. **Smith, I.** 1989. Initiation of sporulation, p. 185–210. *In* I. Smith, R. Slepecky, and P. Setlow (ed.), *Regulation of Prokaryotic Development*. American Society for Microbiology, Washington, D.C.
116. **Strauch, M. A., A. I. Aronson, S. W. Brown, H. J. Schreier, and A. L. Sonenshein.** 1988. Sequence of the *Bacillus subtilis* glutamine synthetase gene region. *Gene* **71:**257–265.
117. **Streicher, S., and B. Tyler.** 1981. Regulation of glutamine synthetase activity by adenylylation in the gram-positive bacterium *Streptomyces cattleya*. *Proc. Natl. Acad. Sci. USA* **78:**229–233.
118. **Switzer, R. L., M. R. Maurizi, J. Y. Wong, and K. L. Flom.** 1979. Selective inactivation and degradation of enzymes in sporulating bacteria, p. 65–79. *In* D. E. Atkinson and C. F. Fox (ed.), *Modulation of Protein Function*. Academic Press, Inc., New York.
119. **Tempest, D. W., J. L. Meers, and C. M. Brown.** 1970. Influence of environment on the content and composition of microbial free amino acid pools. *J. Gen. Microbiol.* **64:**174–185.
120. **Tyler, B.** 1978. Regulation of the assimilation of nitrogen compounds. *Annu. Rev. Biochem.* **47:**1127–1162.
121. **Usdin, K. P., H. Zappe, D. T. Jones, and D. R. Woods.** 1986. Cloning, expression and purification of glutamine synthetase from *Clostridium acetobutylicum*. *Appl. Environ. Microbiol.* **52:**413–419.
122. **Vancurova, I., A. Vancura, J. Volc, J. Kopecky, J. Neuzil, G. Basarova, and V. Behal.** 1989. Purification and properties of NADP-dependent glutamate dehydrogenase from *Streptomyces fradiae*. *J. Gen. Microbiol.* **135:**3311–3318.
123. **Vasantha, N., and E. Freese.** 1979. The role of manganese in growth and sporulation of *Bacillus subtilis*. *J. Gen. Microbiol.* **112:**329–336.
124. **Wedler, F. C., J. Carfi, and A. E. Ashour.** 1976. Glutamine synthetase of *Bacillus stearothermophilus*. Regulation, site interactions, and functional information. *Biochemistry* **15:**1749–1755.
125. **Wedler, F. C., R. M. Kenney, A. E. Ashour, and J. Carfi.** 1978. Two regulatory isozymes of glutamine synthetase from *Bacillus caldolyticus*, an extreme thermophile. *Biochem. Biophys. Res. Commun.* **81:**122–126.
126. **Wedler, F. C., D. S. Shreve, K. E. Fisher, and D. J. Merkler.** 1981. Complementarity of regulation for the two glutamine synthetases from *Bacillus caldolyticus*, an extreme thermophile. *Arch. Biochem. Biophys.* **211:** 276–287.
127. **Wedler, F. C., D. S. Shreve, R. M. Kenney, A. E. Ashour, J. Carfi, and S. G. Rhee.** 1980. Two glutamine synthetases from *Bacillus caldolyticus*, an extreme thermophile. *J. Biol. Chem.* **255:**9507–9516.
128. **Wiame, J.-M., M. Grenson, and H. N. Arst, Jr.** 1985. Nitrogen catabolite repression in yeasts and filamentous fungi. *Adv. Microb. Physiol.* **26:**1–88.
129. **Winnacker, E. L., and H. A. Barker.** 1970. Purification and properties of a NAD-dependent glutamate dehydrogenase from *Clostridium* SB_4. *Biochim. Biophys. Acta* **212:**225–242.
130. **Woods, D. R.** Personal communication.
131. **Wray, L. V., Jr., M. R. Atkinson, and S. H. Fisher.** 1991. Identification and cloning of the *glnR* locus, which is required for transcription of the *glnA* gene in *Streptomyces coelicolor* A3(2). *J. Bacteriol.* **173:**7351–7360.
132. **Wray, L. V., Jr., and S. H. Fisher.** 1988. Cloning and nucleotide sequence of the *Streptomyces coelicolor* gene encoding glutamine synthetase. *Gene* **71:**247–256.
133. **Yoshida, A., and E. Freese.** 1965. Enzymatic properties of alanine dehydrogenase of *Bacillus subtilis*. *Biochim. Biophys. Acta* **96:**248–262.
134. **Zhang, J., M. Strauch, and A. I. Aronson.** 1989. Glutamine auxotrophs of *Bacillus subtilis* that overproduce glutamine synthetase antigen have altered conserved amino acids in or near the active site. *J. Bacteriol.* **171:**3572–3574.

21. Biosynthesis of Arginine, Proline, and Related Compounds

SIMON BAUMBERG and URSULA KLINGEL

PATHWAYS, ENZYMES, AND CONTROL OF ENZYME ACTIVITY

General aspects of arginine and proline metabolism in prokaryotes have been covered in various reviews (those since 1980 include references 8, 10, 16, 46, and 57). As might be expected, most work in this area has been done with *Escherichia coli* and *Salmonella typhimurium*. By comparison, studies with gram-positive bacteria are regrettably fragmentary.

Pathways

Figure 1 depicts schematically the routes of arginine, proline, and polyamine metabolism in *Bacillus subtilis* and (for comparison) *E. coli*; enzymes and reactions are listed in Table 1. In different organisms, glutamate is converted into arginine in seven or eight steps according to whether a transacetylase (enzyme ArgJ; the product of *argJ*) simultaneously removes the acetyl group from *N*-acetylornithine and transfers it to glutamate to form *N*-acetylglutamate or whether deacetylation of *N*-acetylornithine and synthesis of *N*-acetylglutamate are carried out by separate enzymes, i.e., ArgE and ArgA, respectively. *E. coli* has only the ArgA-ArgE combination; other bacteria such as pseudomonads have ArgJ and ArgA, which fulfills an anaplerotic role (10). *Bacillus* species seem to use the ArgJ enzyme system (see below), as reported also for other gram-positive bacteria of the genera *Corynebacterium*, *Micrococcus*, and *Streptomyces* (62; see also, for *Streptomyces coelicolor*, reference 56a).

Arginine catabolism occurs by a variety of routes in prokaryotes; one strain of *Pseudomonas putida* alone may possess at least four pathways (10, 61). The major route in *Bacillus* spp. appears to be via arginase and ornithine aminotransferase (OAT) to glutamic γ-semialdehyde and glutamate (see chapter 16). Another route, the arginine deiminase (ADI) pathway, has been described for *Bacillus licheniformis* (6) and at least some strains of *B. subtilis* (43) as well as for other gram-positive bacteria such as *Streptococcus faecalis* and species of *Lactobacillus* and *Clostridium* (10). In this pathway, arginine is deiminated and the resulting citrulline is converted to ornithine and carbamoyl phosphate by ornithine carbamoyltransferase (OCT) in the reverse of the biosynthetic reaction; the ornithine may then be metabolized further, while the carbamoyl phosphate may be converted to ATP, bicarbonate, and ammonia by the agency of carbamate kinase. There seem to be three possible roles for the ADI pathway in different gram-positive bacteria. (i) In organisms such as *S. faecalis* (10), the ADI pathway generates ATP by anaerobic catabolism of arginine. (ii) With *Lactobacillus* species such as *Lactobacillus leichmannii*, which apparently do not possess a carbamoyl phosphate synthetase (CPS), the prime role of the pathway is to generate carbamoyl phosphate (23). (iii) In species of *Clostridium*, the ADI pathway yields energy from the ornithine provided; the latter is then fermented on its own or participates in the Strickland reaction, whereby the oxidation of one amino acid is coupled to the reduction of another (for further details, see chapter 16 or reference 10). *Streptomyces griseus* and probably other species of *Streptomyces* catabolize arginine via 4-guanidinobutyramide and 4-guanidinobutyrate (10, 60). Routes of arginine or ornithine catabolism utilizing the polyamine pathways are mentioned below.

The pathways of proline metabolism in *Bacillus* and *Streptomyces* spp. seem to be much as in members of the family *Enterobacteriaceae*. The same intermediates, glutamic γ-semialdehyde and Δ^1-pyrroline 5-carboxylate (P5C), which are in equilibrium with each other, are involved in both biosynthesis and catabolism. The final formation and initial breakdown of proline are catalyzed by different enzymes (P5C reductase and proline oxidase, respectively), as are the formation and catabolism of the intermediates (γ-glutamokinase and glutamic γ-semialdehyde dehydrogenase for the former and P5C dehydrogenase for the latter). It has been reported that *B. licheniformis* possesses two P5C dehydrogenases, whose synthesis is controlled by arginine or proline (14). Other gram-positive bacteria seem to show similar routes of proline biosynthesis and catabolism. The cyclodeaminase reaction, which yields proline from ornithine in *Clostridium* species (7, 61), seems to function primarily for ornithine catabolism rather than proline biosynthesis.

Putrescine and other polyamines are required by bacteria, as by eukaryotes, for optimal growth, though exactly why seems still obscure (for a general review, see reference 59). In addition, agmatine and putrescine, the products of arginine and ornithine decarboxylases, respectively, can represent intermediates in the breakdown of these amino acids. It appears that in gram-positive bacteria, putrescine is generally synthesized from ornithine that has been produced either during arginine biosynthesis or via arginine catabolism (10). Arginine decarboxylase occurs infrequently, though it has been reported in *Mycobacterium* spp. (64).

Simon Baumberg and Ursula Klingel • Department of Genetics, University of Leeds, Leeds LS2 9JT, United Kingdom.

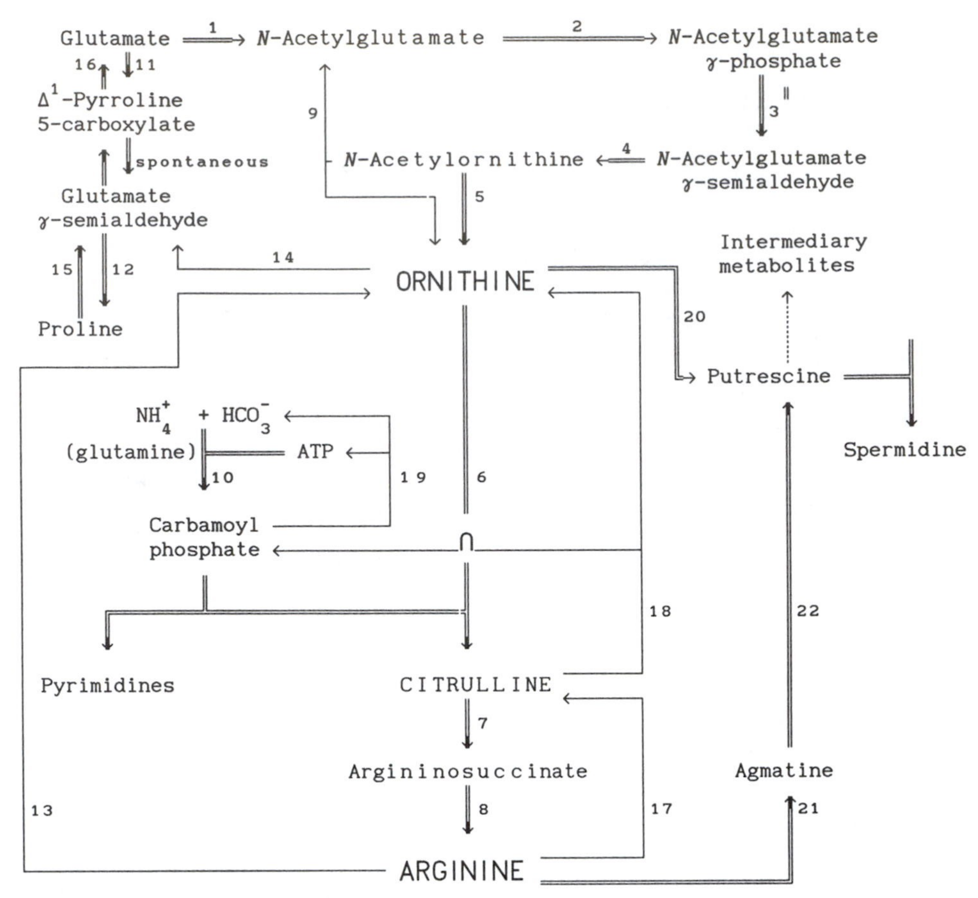

Figure 1. Pathways of arginine, proline, and polyamine metabolism in *E. coli* and *B. subtilis*. ═, pathways found in both organisms; —, pathways found only in *B. subtilis*. Numbered enzymes mediating the various steps are listed in Table 1. The following pathways are referred to in this chapter but not indicated in this figure: the ornithine cyclase pathway of proline biosynthesis and catabolism (*Clostridium* spp.) and the guanidinobutyrate pathway of arginine catabolism (*Streptomyces* spp.).

Enzymes of Arginine Biosynthesis

In vitro activities of enzymes of arginine biosynthesis

ArgD, ArgF, and ArgH enzyme activities can readily be demonstrated in *B. subtilis* cell extracts (18, 29, 34). In our experience (62a), no ArgB, ArgC, or ArgG enzyme activity can be demonstrated in *B. subtilis* sonicates or cells permeabilized by various means. Similar difficulties, if perhaps less extreme, seem to have been noted for the enzymes of tryptophan biosynthesis (20). The reason for these observations is unknown. A reported activity for ArgB (62) is very low, and its significance is uncertain. However, Sakanyan et al. (50) have recently reported measurable ArgB activities in extracts of *Bacillus stearothermophilus*.

Which pathway for ornithine synthesis is used in *Bacillus* spp.?

It has been suggested that *B. subtilis*, like *E. coli*, uses the ArgA-ArgE pathway to synthesize ornithine (62, 63). This now appears not to be the case. Sakanyan et al. (50) reported an arginine-repressible ArgJ activity in *B. stearothermophilus*, *B. subtilis*, and *B. licheniformis*. In addition, Devine and O'Reilly (15a) have found within the "early" *arg* operon of *B. subtilis* (see below) a DNA sequence whose predicted product was identified as ArgJ by homology with the enzyme from *Neisseria gonorrhoeae* (31, 32). The utilization of the ArgJ pathway would bring *B. subtilis* into line with other gram-positive bacteria such as *Streptomyces* spp. (56a, 62). It would also be expected in this case that ArgB rather than ArgA would be subject to feedback inhibition by arginine (62), but this has not been reported. Sakanyan et al. (50) report for *B. stearothermophilus* slight inhibition of both ArgA and ArgJ by arginine.

Characteristics of arginine biosynthetic enzymes

Little work has been done on the properties of most of these enzymes. Sequences of *B. subtilis argC* and *argF* yield predicted proteins similar in size and with 30 to 40% amino acid identity to their *E. coli* homologs (38, 56). In a detailed report on the *B. subtilis* ArgF enzyme (OCT) (39), Neway and Switzer showed that it is found as a mixture of dimers, tetramers, and hexamers of the ca. 44-kDa subunit. Other biosyn-

Table 1. Enzyme reactions of arginine, proline, and polyamine metabolism as numbered in Fig. 1

Enzyme No.	Enzyme Name	Reaction catalyzed	Gene designation[a]	Abbreviation[b]
1	*N*-Acetylglutamate synthase	Glutamate + AcSCoA → *N*-acetylglutamate + HSCoA[c]	*argA*	
2	*N*-Acetylglutamokinase	*N*-Acetylglutamate + ATP → *N*-acetylglutamyl γ-phosphate + ADP	*argB*	
3	*N*-Acetylglutamylphosphate reductase	*N*-Acetylglutamyl γ-phosphate + NADPH + H^+ → *N*-acetylglutamic γ-semialdehyde + NADP	*argC*	
4	*N*-Acetylornithine δ-transaminase	*N*-Acetylglutamic γ-semialdehyde + glutamate → *N*-acetylornithine + α-oxoglutarate	*argD*	
5	*N*-Acetylornithinase	*N*-Acetylornithine + H_2O → ornithine + acetate	*argE*	
6	Ornithine carbamoyltransferase (anabolic)	Ornithine + carbamoyl phosphate → citrulline + P_i	*argF*	OCT
7	Argininosuccinate synthetase	Citrulline + aspartate + ATP → argininosuccinate + AMP + P_i	*argG*	
8	Argininosuccinase	Argininosuccinate → arginine + fumarate	*argH*	
9	Ornithine acetyltransferase	*N*-Acetylornithine + glutamate → ornithine + *N*-acetylglutamate	*argJ*	
10	Carbamoyl phosphate synthetase	Glutamine + 2ATP + HCO_3^- → carbamoyl phosphate + 2ADP + P_i	*carAB*, *cpa-pyrA*	CPS
11				
a	γ-Glutamyl kinase	Glutamate + ATP → glutamyl γ-phosphate + ADP	*proA*	
b	γ-Glutamyl phosphate reductase	Glutamyl γ-phosphate + NADPH + H^+ → glutamic γ-semialdehyde (↔ Δ^1-pyrroline 5-carboxylate) + $NADP^+$		
12	Δ^1-Pyrroline 5-carboxylate reductase	Δ^1-Pyrroline 5-carboxylate + NAD(P)H + H^+ → proline + NAD(P)	*proC*	P5C reductase
13	Arginase	Arginine + H_2O → ornithine + urea		
14	Ornithine aminotransferase	Ornithine + α-oxoglutarate → glutamic γ-semialdehyde (→ Δ^1-pyrroline 5-carboxylate) + glutamate		OAT
15	Proline oxidase	Proline + O_2 → Δ^1-pyrroline 5-carboxylate + H_2O		
16	Δ^1-Pyrroline 5-carboxylate dehydrogenase	Δ^1-Pyrroline 5-carboxylate + $NAD(P)^+$ → glutamate + NAD(P)H + H^+		P5C dehydrogenase
17	Arginine deiminase	Arginine + H_2O → citrulline + NH_4^+		ADI
18	Ornithine carbamoyltransferase (catabolic)	Citrulline + P_i → ornithine + carbamoyl phosphate		OCT
19	Carbamate kinase	Carbamoyl phosphate + ADP → ATP + NH_4^+ + HCO_3^-		
20	Ornithine decarboxylase	Ornithine → putrescine + CO_2	*speC*	
21	Arginine decarboxylase	Arginine → agmatine + CO_2	*speA*	
22	Agmatine ureohydrolase	Agmatine + H_2O → putrescine + urea	*speB*	

[a] Where possible, gene designations follow those for *E. coli* (2) except that *cpa* and *pyrA* refer to the arginine- and pyrimidine-specific CPSs, respectively.
[b] Except as indicated here, enzymes are abbreviated according to the gene designation, e.g., *N*-acetylglutamate synthase may also be referred to as ArgA.
[c] AcSCoA, acetyl coenzyme A; HSCoA, coenzyme A.

thetic OCTs are purely hexameric (10). This enzyme is proteolyzed and thereby inactivated during sporulation (40). The control of its activity is of some interest. If a cell were to contain appreciable activities of the ArgF, ArgG, and ArgH enzymes of arginine synthesis together with arginase, a "futile cycle" of arginine synthesis and breakdown would be set up, with wasteful consumption of ATP (in the production of carbamoyl phosphate for OCT-catalyzed condensation with ornithine to give citrulline). Organisms with genes for all these enzymes have various ways of avoiding the problem (10, 12). In *B. subtilis* (24), as in *Saccharomyces cerevisiae* (33), arginase binds to OCT in the presence of ornithine and arginine, with consequent inhibition of the OCT activity, a process termed "epiarginasic" (33) or "epienzymic" (24) inhibition.

B. subtilis possesses two CPSs (the products of the *cpa* and *pyrA* genes) (45). It thereby differs from gram-negative bacteria, which have only one CPS, and rather resembles microbial eukaryotes such as *S. cerevisiae* (12). The two CPSs, both of which utilize glutamine with high efficiency and ammonia with low efficiency, show distinct physical and biochemical properties, and the controls of their synthesis and activity differ (48). Synthesis of CPS A, of ca. 200 kDa, is repressed in the presence of arginine, but its activity seems not to be subjected to modulation by low-molecular-weight compounds. Synthesis of CPS P, of 90 to 100 kDa, is repressed in the presence of uracil, inhibited by uridine nucleotides, and activated by phosphoribosylpyrophosphate and GMP. A result of these regulatory patterns is that in minimal medium, growth of a mutant lacking CPS A is inhibited by uracil, while growth of a mutant lacking CPS P is inhibited by arginine. In contrast, *Streptomyces* spp. appear to possess, like *E. coli*, a single CPS, as indicated by the isolation of mutants simultaneously requiring arginine and uracil in *S. coelicolor* and *Streptomyces lividans* and also *Streptomyces clavuligerus* (49a).

GENETIC ORGANIZATION AND REGULATION OF ENZYME SYNTHESIS

We will first discuss the situation in *Bacillus* spp. and then report on present knowledge of *Streptomyces* spp., the only other group for which there is significant information.

Genes of Arginine, Proline, and Polyamine Metabolism

In *B. subtilis*, the eight genes of arginine biosynthesis (including *cpa*, which encodes the arginine-controlled CPS; see above) are located in two clusters, one comprising the genes involved in steps up to and including citrulline (position 102°) and the other comprising the two steps between citrulline and arginine (position 260°). The gene order in the first cluster is *argCJBD-cpa-argF* (37); a promoter-operator region lies just upstream of the *argC* gene (11, 30, 55; see also below). This may be the only promoter for the operon, since insertions in *argC* are strongly polar on expression of *argF* (52a). A similar sequence has been described for *B. stearothermophilus* (51; reported in reference 50). In the *argGH* cluster, the promoter is thought to be upstream of *argG* (52a). It is worth noting that the *Bacillus* gene arrangement differs from that found in *E. coli*, in which the *argECBH* cluster (divergently transcribed from a complex promoter-operator region between *argC* and *argE*) is physically separated from the other genes, each of which has a unique locus (16). In *Proteus mirabilis*, the corresponding cluster comprises *argECBGH* (49).

The *argCJBD-cpa-argF* clusters from *B. subtilis* and *B. stearothermophilus* were first cloned in *E. coli* by selecting for complementation of arginine auxotrophs (36, 51). The 12-kbp *Eco*RI fragment containing the *B. subtilis* cluster proved highly unstable when present on its own in *E. coli* on high- or medium-copy-number vectors; relative stability required the additional presence of a 6-kbp *B. subtilis* chromosomal fragment (36). It was later shown (54) that the instability resulted from the deleterious effect of overexpression of one or more of the *arg* cluster genes (though why this should be so is still unclear); the 6-kbp fragment carries the *ahrC* gene that encodes the biosynthesis repressor (see below).

Mutations leading to proline auxotrophy in *B. subtilis* have been mapped to the *pro* locus at position 115°. The growth requirements of all such mutants so far described can be satisfied by proline as well as by arginine, ornithine, or citrulline (17). This phenotype is consistent with lesions in *proA* or *proB*, i.e., a blockage before glutamate γ-semialdehyde, with which Δ^1-pyrroline 5-carboxylate (P5C) is spontaneously interconvertible, since this intermediate is derivable from ornithine through the agency of OAT. These *pro* mutants can still utilize proline as sole nitrogen source (17), confirming the independence of steps in glutamate γ-semialdehyde synthesis and catabolism. It is also possible that *B. subtilis* can make proline from ornithine by means of an ornithine cyclodeaminase activity, which occurs in clostridia (7) and many other bacteria (61).

No mutants with abnormalities in polyamine metabolism seem to have been obtained in *Bacillus* spp., and the genes involved remain to be discovered.

Control of Expression of Genes Encoding Enzymes of Arginine, Proline, or Polyamine Metabolism

Wild-type regulatory phenotype

In studies of the control of synthesis of arginine biosynthetic enzymes, it is most often levels of OCT that are measured; other enzymes of the pathway, such as ArgA-ArgJ, ArgD, and ArgH, give broadly similar results when investigated. In the following, it will be assumed, unless specified otherwise, that OCT assays reflect differential rates of synthesis of the biosynthetic enzyme encoded by *argF*. In general in *Bacillus* spp., OCT activities are repressed in the presence of adequate levels of arginine and at least in early stages of the growth cycle in minimal media. Some early suggestions, usually relating to *B. subtilis*, that repression was lifted toward the end of exponential growth (29, 63) may be accounted for by the presence of levels of arginine in the medium so low as to allow the medium to become exhausted (17). However, there remain accounts of derepression at this stage that probably cannot be explained away (15). Recently, a possible basis for such observations in *B. subtilis* has been revealed by Devine and colleague (15a), who have found that the *argCJBD-cpa-argF* operon is negatively regulated by Spo0A. It will be very interesting to explore the significance and implications of this apparent gearing of a typical intermediary metabolism system to overall physiological state.

Although the control of arginine catabolism is dealt with in chapter 16 of this volume, we will mention here the control of synthesis of the catabolic enzymes arginase and OAT, since this relates to the properties of arginine regulatory mutants to be discussed in the next section. Generally, in *Bacillus* spp., these enzymes are induced in the presence of arginine (14, 18, 27), proline, or the precursor ornithine or citrulline, though in these last two cases, in *B. subtilis* at least, the biosynthetic enzymes are not repressed (34). This somewhat unexpected relationship between the biosynthetic and catabolic systems has not been explained.

As far as we know, there are no reports on the regulation of enzyme levels in proline biosynthesis (for proline catabolism, see chapter 16) or polyamine metabolism in *Bacillus* spp. whether in the wild type or in mutants.

Regulatory mutants in arginine metabolism

Kisumi et al. (26) first showed that the analog arginine hydroxamate could be used to select *B. subtilis* mutants that overproduce arginine. A variety of arginine hydroxamate-resistant (Ahr) mutants were subsequently characterized (18, 34, 35). Mutations at three loci, *ahrA*, *ahrB*, and *ahrC* (positions 342°, 328°, and 219°, respectively), confer the Ahr phenotype in minimal media (34); mutations at other loci, including *ahrD* (99°), confer this phenotype only in the presence of the precursor ornithine or citrulline (35). Mutations at *ahrA*, *ahrB*, or *ahrC* always affect the levels of both arginase and OAT and are therefore unlikely to lie in the structural gene for either enzyme. Effects on these enzymes vary appreciably with different mutations at the same locus; generally, *ahrA* mutations lead to complete loss of arginase and considerable loss of OAT activities, whereas an *ahrB* or *ahrC* mutation leaves at least basal levels of the enzymes. A significant finding (18, 34) was that whereas mutations in *ahrA*, *ahrB*, or *ahrD* have no effect on levels of the biosynthetic enzymes, *ahrC* mutations lead to simultaneous loss of repressibility of the biosynthetic enzymes and of inducibility of the catabolic enzymes, with the implication that these two controls share at least one common component.

Further work has shown (see below) that the AhrC product is a repressor of the biosynthetic enzymes and probably an activator of the catabolic enzymes. AhrB may be involved in arginine uptake, the evidence being as follows. It has recently been shown (13) that an insertion mutant of the *B. subtilis sigL* gene, whose product is an alternative σ factor equivalent to the σ^{54} of gram-negative bacteria, cannot grow with arginine or ornithine as sole nitrogen source. P. Glaser and coworkers (unpublished results) have discovered a gene, located ca. 20 kbp downstream of the *sacPA* cluster, whose product shares homology with a number of arginine permeases. This gene lies in an operon whose expression has been shown to be σ^{54} dependent. It is thus tempting to speculate that this gene encodes a *B. subtilis* arginine permease. Although insertions within this gene or others of the σ^{54}-dependent operon did not lead to an arginine utilization deficiency, this failure could reflect the existence of multiple arginine uptake systems as found in *E. coli* and yeast cells. The location of the putative arginine permease gene is consistent with its being *ahrB*, although the results given above caution against a ready identification of the two. The roles of AhrA and AhrD remain unknown.

ahrC gene, its product AhrC, and specific DNA binding of AhrC

The serendipitous cloning of *ahrC* on a DNA fragment that stabilized plasmids carrying the *argC* operon has been described above. Starting with an *ahrC*::Tn*917* insertion in the *B. subtilis* chromosome, North et al. (42) sequenced the gene. The predicted AhrC polypeptide of 149 amino acids shows 27% identity (concentrated particularly towards the C terminus) to the *E. coli* ArgR polypeptide (28); the latter has a role as a repressor analogous to that of AhrC but is not known to function as an activator. The repressor activity of ArgR is conventional in that ArgR binds to sites overlapping or close to promoters for the genes whose expression it controls; these genes (termed the arginine regulon) make up the *E. coli argECBH* cluster and other genes specific to arginine biosynthesis as well as the *argR* gene itself and the *carAB* genes. These last encode the subunits of CPS and are under joint arginine and pyrimidine nucleotide control in *E. coli* (16). It is expected that the repressor functions of ArgR and AhrC will prove to have similar mechanisms. AhrC and ArgR are unusual among DNA-binding regulatory proteins in that both are hexameric (11, 28).

DNA-binding sites for AhrC would be expected at several locations corresponding to control sites for the *B. subtilis* arginine regulon, the minimum being sites associated with the *argCJBD-cpa-argF* and *argGH* clusters, and presumably with the catabolism loci (see below). Arginine-dependent AhrC binding protects two regions of the *argC* operon against DNAse I and hydroxyl radical attack (11). The first of these, located between positions −60 and −9 (taking the transcription start point as +1), is termed $argC_{O1}$. The second lies between positions +100 and +133 and is termed $argC_{O2}$; it lies wholly within the *argC* structural gene. This topography of binding sites is similar to that seen in the *E. coli lac* and *gal* operons. Protection at $argC_{O2}$ occurs only at appreciably higher AhrC concentrations than are required for protection at $argC_{O1}$; nevertheless, binding at $argC_{O2}$ can be shown to be necessary for maximal repression in vivo. The results of gel retardation experiments are in accordance with the protection data. A model has been put forward (11) whereby two AhrC dimers bind along $argC_{O1}$ and a third binds along $argC_{O2}$; it is suggested that the DNA between the two forms a loop, but this possibility has not been proven.

The question of cross-recognition of operators by AhrC and ArgR has been examined elsewhere (53). AhrC represses OCT effectively in vivo in *E. coli* and binds in vitro to a DNA fragment containing the *argR* promoter. On the other hand, ArgR does not repress the *B. subtilis argC* promoter in *E. coli* and binds only weakly to this promoter in vitro. Interestingly, there is no evidence that ArgR has binding sites within any of the *E. coli arg* or *car* genes. This implies that AhrC may manage to bind *E. coli arg* promoters effectively without benefit of a downstream binding site. AhrC can also substitute for ArgR in plasmid ColE1 monomerization via site-specific recombination (58); it seems that ArgR (and AhrC) binding to a site upstream of the *cer* site at which crossing over takes place causes DNA bending, which in turn facilitates the recombination event.

Less information is available on the activator role of AhrC, in part because genes of arginine catabolism have hitherto not been mapped or cloned (for physiological studies, see chapter 16 and references 1, 3, 27, and 52). However, a Tn*917-lacZ* insertion (41) in which the regulatory phenotype of *lacZ* expression is the same as that for the arginine catabolic enzymes arginase and OAT, namely, AhrC-dependent inducibility in the presence of arginine, ornithine, or citrulline, has been isolated. Despite this, the insertion mutant utilized arginine or its precursors as nitrogen sources and showed wild-type levels of arginase and OAT. A 300-bp DNA fragment lying about 4 kbp upstream of the Tn*917-lacZ* insertion has been shown by gel retardation to bind AhrC; this fragment also contains a 12-bp sequence identical to one within $argC_{O1}$ (26a). Deletions removing the AhrC-binding region do not express *lacZ*; the deletion approaching closest to the Tn*917-lacZ* insertion ends ca. 2 kbp upstream of the insertion. These data are consistent with a model in which AhrC acts as an activator of transcription initiation at a promoter adjacent to the binding site. An alternative model is that AhrC represses an operon containing the gene for a hypothetical repressor, which in turn represses the promoter directing transcription of *lacZ*. However, the deletion data noted above require that these deletions, with end points from ca. 4 to ca. 2 kbp upstream of the Tn*917-lacZ* insertion, all permit adequate transcription of the hypothetical repressor gene while rendering this gene insensitive to AhrC, which seems improbable.

Although as noted above, the *E. coli* ArgR protein acts as far as is known only as a repressor and not as an activator, at least one *E. coli* regulatory protein of amino acid metabolism, TyrR, can act in either capacity (47), though in both cases with biosynthesis genes. It has recently been shown (25) that in *Pseudomonas aeruginosa*, at least some genes of arginine biosynthesis and catabolism are subject to a common regulatory element(s). Finally, the existence of homologous regulatory proteins of arginine metabolism in the evolutionarily widely separated *B. subtilis* and *E. coli* suggests (barring the rather implausible intervention of horizontal gene transfer) that the organism in which the common ancestral protein evolved may have lived a quite different kind of life from either of its descendants; attempts to explain the mode of regulation by reference to present-day ecological niches therefore seem problematic.

It is noteworthy that in *B. subtilis* as in *E. coli*, arginine biosynthesis involves no element of transcriptional attenuation. Reasons that this may be so in *E. coli* have been discussed by Cunin et al. (9). They point out several possible reasons: (i) tRNAs for the preferred codons CGU and CGC have inosine in the "wobble" position of the anticodon, which could weaken codon-anticodon interaction such that a stretch of CGU-CGC-CGA in a hypothetical leader transcript could impede ribosome movement even if arginine were in excess, leading to permanent "deattenuation"; (ii) the presence of many CG pairs in the leader RNA could make the RNA polymerase pause, either through GC richness or via formation of hairpins; (iii) use of less-common arginine codons such as AGA or AGG would still give a weaker interaction; (iv) AGA and AGG might create inadvertent Shine-Dalgarno sequences, thus disturbing translation.

Streptomyces spp.: arginine genes and regulation

Standard linkage maps of *S. coelicolor* (e.g., see reference 22) show loci *argA*, *argB*, *argC*, and *argD*, the first three being closely linked. These designations do not, however, correspond to specific genes. *argA*, *argC*, and *argD* mutants respond to ornithine or citrulline as well as arginine, suggesting blocks in the early part of the pathway; *argB* mutants respond only to arginine, implying blocks in *argG* or *argH* (22). The status of the unique *argD* mutant as specifically affected in arginine biosynthesis is apparently questionable (21a). Mutations to proline auxotrophy map at a further locus, *proA*; these mutants do not respond to arginine, possibly because in *Streptomyces* spp., levels of arginase are low and show only weak inducibility by arginine (44). Mutations to simultaneous auxotrophy for arginine and uracil are known in *S. coelicolor* and *S. lividans* and have been reported also for *S. clavuligerus* (49a). Their existence suggests the presence of a single CPS as in *E. coli* rather than two as in *Bacillus* spp.

Work on the genes of arginine biosynthesis presumably corresponding to the *argAC* locus indicates the existence in *S. coelicolor* and *S. clavuligerus* of a cluster (probably an operon), *argCJB*, with an organization that may be similar to that in *B. subtilis* (19, 28a, 44).

In many *Streptomyces* spp., *argG* deletions are associated with remarkable genetic events involving massive deletions (sometimes of many hundreds of base pairs [5]) and amplifications (see, e.g., chapters by Schrempf, Altenbuchner and Eichenseer, Cullum et al., and Piendl et al. in reference 4). Mapping of *argG* has not been described in detail, but the gene may lie in the so-called "silent arcs" of the genetic map (22), where genes are only sparsely distributed; manifestation of large deletions preferentially in these regions may merely reflect a relative absence of genes encoding indispensable functions.

Repression of enzymes of arginine biosynthesis by exogenous arginine has previously been believed to be slight, i.e., not more than two- or threefold for OCT (44). This was regarded as being in line with other amino acid biosynthesis systems in this group, where strong end product repression effects have been generally accepted as being absent (21). Recently, however, it has been shown by P. Liras and her colleagues for *S. clavuligerus* (personal communication) and confirmed in our laboratory for *S. coelicolor* (56a) that high external concentrations of arginine (ca. 15 to 20 mM) yield repression of at least some biosynthetic enzymes by factors on the order of 10 to 20. Preliminary evidence from the laboratories of P. Liras and us (56a) suggests the existence in these organisms of an arginine repression system possibly mediated by a protein homologous with AhrC-ArgR. As for enzymes of arginine catabolism, the low levels and inducibility of arginase have been commented on above; the enzyme guanidinobutyrate ureohydrolase of the 4-guanidinobutyramide–4-guanidinobutyrate catabolic pathway is, however, induced 5- to 10-fold by arginine (44).

Arginine, like many amino acids, is a precursor of many of the secondary metabolites (including antibiotics) elaborated by *Streptomyces* spp. For instance, it donates two guanidino groups in the synthesis of the streptidine moiety of streptomycin in producers of the latter such as *S. griseus* and provides the hydroxyornithine residue in the formation of clavulanic acid by *S. clavuligerus* (it is understood in this case that industrial high-level production of clavulanic acid has been greatly aided by manipulation of arginine biosynthesis in production strains). It would be interesting to know whether regulation of arginine metabolism differs between organisms in which it serves as a secondary metabolite precursor and others in which it does not.

Streptomyces spp.: proline genes and regulation

An important summary of work from D. A. Hodgson's laboratory has recently appeared (21). A cloned *S. coelicolor* fragment contained three open reading frames, the first homologous to *E. coli proB* and the third homologous to *E. coli proA*; the second showed no homology to any sequences in the data base. These open reading frames lay downstream of a region of unusual structure including extensive repeats. Use of a promoter probe vector with *xylE* as reporter gene showed that each of the open reading frames was transcribed from a separate promoter. Exogenous proline caused a twofold increase in expression from the *proA* and *proB* promoters; P5C reductase, the third enzyme of proline biosynthesis, showed a twofold increase in activity in extracts of mycelium grown with exogenous proline or glutamate. The enzymes of proline degradation (proline oxidase and P5C dehydrogenase) were not induced by proline in *S. coelicolor*, though they have been reported to be ca. 12-fold inducible in *S. clavuligerus*. Growth in glutamate, glutamine, or arginine led to a 10- to 20-fold drop in levels of P5C dehydrogenase in *S. coelicolor*.

An intriguing connection between proline catabolism and transport and the production of a secondary metabolite, undecylprodigiosin, of which proline is a precursor, was found in *S. coelicolor*. Mutants that were resistant to proline analogs and unable to use proline as sole nitrogen or carbon source were isolated. These isolates were variably deficient in the enzymes of proline catabolism (proline oxidase and P5C dehydrogenase) and could not transport proline. They also overproduced undecylprodigiosin. These results are consistent with a role for secondary metabolism as "overspill" for excess precursor that in this case, because of mutation, can neither be catabolized nor exported from the cell.

Acknowledgments. We and our present and past collaborators thank the many people with whom we have over the years discussed the genetics and physiology of arginine metabolism. Many of these people, especially N. Glansdorff, V. Stalon, M. Débarbouillé, P. Glaser, S. H. Fisher, D. A. Hodgson, W. Maas, K. Devine, V. Sakanyan, D. J. Sherratt, R. L. Switzer, and their respective colleagues, have provided preprints and unpublished data.

REFERENCES

1. **Atkinson, M. R., and S. H. Fisher.** 1990. Identification of genes and gene products whose expression is activated during nitrogen-limited growth in *Bacillus subtilis*. *J. Bacteriol.* **173:**23–27.
2. **Bachmann, B. J.** 1990. Linkage map of *Escherichia coli* K-12, edition 8. *Microbiol. Rev.* **54:**130–197.
3. **Baumberg, S., and C. R. Harwood.** 1979. Carbon and nitrogen repression of arginine catabolic enzymes in *Bacillus subtilis*. *J. Bacteriol.* **137:**189–196.
4. **Baumberg, S., H. Krügel, and D. Noack (ed.).** 1991. *Genetics and Product Formation in Streptomyces*. Plenum Press, New York.
5. **Birch, A., A. Häusler, M. Vögtli, W. Krek, and R. Hütter.** 1989. Extremely large chromosomal deletions are intimately involved in genetic instability and genomic rearrangements in *Streptomyces glaucescens*. *Mol. Gen. Genet.* **217:**447–458.
6. **Broman, K., N. Lauwers, V. Stalon, and J. M. Wiame.** 1978. Oxygen and nitrate in utilization by *Bacillus licheniformis* of the arginase and arginine deiminase routes of arginine catabolism and other factors affecting their synthesis. *J. Bacteriol.* **135:**920–927.
7. **Costilow, R. N., and L. Laycock.** 1969. Reactions involved in the conversion of ornithine to proline in clostridia. *J. Bacteriol.* **100:**622–667.
8. **Cunin, R.** 1983. Regulation of arginine biosynthesis in prokaryotes, p. 53–79. *In* K. M. Hermann and R. L. Sommerville (ed.), *Biotechnology Series 3: Amino Acid Biosynthesis and Genetic Regulation*. Addison-Wesley, New York.
9. **Cunin, R., T. Eckhardt, J. Piette, A. Boyen, A. Piérard, and N. Glansdorff.** 1983. Molecular basis for modulated regulation of gene expression in the arginine regulon of *Escherichia coli* K12. *Nucleic Acids Res.* **11:**5007–5019.
10. **Cunin, R., N. Glansdorff, A. Piérard, and V. Stalon.** 1986. Biosynthesis and metabolism of arginine in bacteria. *Microbiol. Rev.* **50:**314–352.
11. **Czaplewski, L. G., A. K. North, M. C. M. Smith, S. Baumberg, and P. G. Stockley.** 1992. Purification and initial characterization of AhrC: the regulator of arginine metabolism genes in *Bacillus subtilis*. *Mol. Microbiol.* **6:**267–275.
12. **Davis, R. H.** 1986. Compartmental and regulatory mechanisms in the arginine pathways of *Neurospora crassa* and *Saccharomyces cerevisiae*. *Microbiol. Rev.* **50:**280–313.
13. **Débarbouillé, M., I. Martin-Verstraete, F. Kunst, and G. Rapoport.** 1991. The *Bacillus subtilis sigL* gene encodes an equivalent of σ^{54} from gram negative bacteria. *Proc. Natl. Acad. Sci. USA* **88:**9092–9096.
14. **De Hauwer, G., R. Lavallé, and J. M. Wiame.** 1964. Etude de la pyrroline dehydrogenase et de la régulation du catabolisme de l'arginine et de la proline chez *Bacillus subtilis*. *Biochim. Biophys. Acta* **81:**257–269.
15. **Deutscher, M. P., and A. Kornberg.** 1968. Biochemical studies of bacterial sporulation and germination. VIII. Patterns of enzyme development during growth and sporulation of *Bacillus subtilis*. *J. Biol. Chem.* **243:**4653–4660.

15a. **Devine, K., and M. O'Reilly.** Personal communication.

16. **Glansdorff, N.** 1987. Biosynthesis of arginine and polyamines, p. 321–344. *In* F. C. Neidhart, J. L. Ingraham, K. B. Low, B. Magasanik, H. Schaechter, and H. E. Umbarger (ed.), *Escherichia coli and Salmonella typhimurium: Cellular and Molecular Biology*, vol. 1. American Society for Microbiology, Washington, D.C.
17. **Harwood, C. R.** 1974. Genetic control of arginine enzymes in the bacterium *Bacillus subtilis*. Ph.D. thesis, University of Leeds, Leeds, United Kingdom.
18. **Harwood, C. R., and S. Baumberg.** 1977. Arginine hydroxamate-resistant mutants of *Bacillus subtilis* with altered control of arginine metabolism. *J. Gen. Microbiol.* **100:**177–188.
19. **Hindle, Z.** 1990. A study of genes of arginine biosynthesis from Streptomyces. Ph.D. thesis, University of Leeds, Leeds, United Kingdom.
20. **Hoch, S. O., C. Anagnostopoulos, and I. P. Crawford.** 1969. Enzymes of the tryptophan operon of *Bacillus subtilis*. *Biochem. Biophys. Res. Commun.* **35:**838–844.
21. **Hood, D. W., R. Heidstra, U. K. Swoboda, and D. A. Hodgson.** 1992. Molecular genetic analysis of proline and tryptophan biosynthesis in *Streptomyces coelicolor* A3(2): interaction between primary and secondary metabolism—a review. *Gene* **115:**5–12.

21a. **Hopwood, D. A.** Personal communication.

22. **Hopwood, D. A., M. J. Bibb, K. F. Chater, T. Kieser, C. J. Bruton, H. M. Kieser, D. J. Lydiate, C. P. Smith, J. M. Ward, and H. Schrempf.** 1985. *Genetic Manipulation of Streptomyces: a Laboratory Manual*. The John Innes Foundation, Norwich, United Kingdom.
23. **Hutson, J. Y., and M. Downing.** 1968. Pyrimidine biosynthesis in *Lactobacillus leichmannii*. *J. Bacteriol.* **96:**1249–1254.
24. **Issaly, I. M., and A. S. Issaly.** 1974. Control of ornithine carbamoyltransferase by arginase in *Bacillus subtilis*. *Eur. J. Biochem.* **49:**485–495.
25. **Itoh, Y., and H. Matsumoto.** 1992. Mutations affecting regulation of the anabolic *argF* and the catabolic *aru* genes in *Pseudomonas aeruginosa* PAO. *Mol. Gen. Genet.* **231:**417–425.
26. **Kisumi, M., J. Kato, M. Suguira, and I. Chibata.** 1971. Production of L-arginine by arginine hydroxamate-resistant mutants of *Bacillus subtilis*. *Appl. Microbiol.* **22:**987–991.

26a. **Klingel, U.** Unpublished data.

27. **Laishley, E. J., and R. W. Bernlohr.** 1968. Regulation of

arginine and proline catabolism in *Bacillus licheniformis*. *J. Bacteriol.* **96:**322–329.
28. **Lim, D., J. D. Oppenheim, T. Eckhardt, and W. K. Maas.** 1987. Nucleotide sequence of the *argR* gene of *Escherichia coli* K12 and isolation of its product, the arginine repressor. *Proc. Natl. Acad. Sci. USA* **84:**6697–6701.
28a. **Ludovice, M., P. Carrachas, and P. Liras.** Personal communication.
29. **Mahler, I., J. Neumann, and J. Marmur.** 1963. Studies of genetic units controlling arginine biosynthesis in *Bacillus subtilis*. *Biochim. Biophys. Acta* **72:**69–79.
30. **Mann, N. H., A. Mountain, R. N. Munton, M. C. M. Smith, and S. Baumberg.** 1984. Transcription analysis of a *Bacillus subtilis arg* gene following cloning in *Escherichia coli* in an initially unstable hybrid plasmid. *Mol. Gen. Genet.* **197:**75–81.
31. **Martin, P. R., and M. H. Mulks.** 1992. Molecular characterization of the *argJ* mutation in *Neisseria gonorrhoeae* strains with requirements for arginine, hypoxanthine, and uracil. *Infect. Immun.* **60:**970–975.
32. **Martin, P. R., and M. H. Mulks.** 1992. Sequence analysis and complementation studies of the *argJ* gene encoding ornithine acetyltransferase from *Neisseria gonorrhoeae*. *J. Bacteriol.* **174:**2694–2701.
33. **Messenguy, F., and J. M. Wiame.** 1969. The control of ornithine transcarbamylase activity by arginase in *Saccharomyces cerevisiae*. *FEBS Lett.* **3:**47–49.
34. **Mountain, A., and S. Baumberg.** 1980. Map locations of some mutations conferring resistance to arginine hydroxamate in *Bacillus subtilis* 168. *Mol. Gen. Genet.* **178:**691–701.
35. **Mountain, A., and S. Baumberg.** 1984. *Bacillus subtilis* 168 mutants resistant to arginine hydroxamate in the presence of ornithine or citrulline. *J. Gen. Microbiol.* **130:**1247–1252.
36. **Mountain, A., N. H. Mann, R. N. Munton, and S. Baumberg.** 1984. Cloning a *Bacillus subtilis* restriction fragment complementing auxotrophic mutants of eight *Escherichia coli* genes of arginine biosynthesis. *Mol. Gen. Genet.* **197:**82–89.
37. **Mountain, A., J. McChesney, M. C. M. Smith, and S. Baumberg.** 1986. Gene sequence within a cluster in *Bacillus subtilis* encoding early enzymes of arginine synthesis as revealed by cloning in *Escherichia coli*. *J. Bacteriol.* **165:**1026–1028.
38. **Mountain, A., M. C. M. Smith, and S. Baumberg.** 1990. Nucleotide sequence of the *Bacillus subtilis argF* gene encoding ornithine carbamoyltransferase. *Nucleic Acids Res.* **18:**4594.
39. **Neway, J. O., and R. L. Switzer.** 1983. Purification, characterization, and physiological function of *Bacillus subtilis* ornithine carbamoyltransferase. *J. Bacteriol.* **155:**512–521.
40. **Neway, J. O., and R. L. Switzer.** 1983. Degradation of ornithine transcarbamylase in sporulating *Bacillus subtilis* cells. *J. Bacteriol.* **155:**522–530.
41. **North, A. K.** 1989. Analysis of a putative cloned *arg* repressor gene from *Bacillus subtilis*. Ph.D. thesis, University of Leeds, Leeds, United Kingdom.
42. **North, A. K., M. C. M. Smith, and S. Baumberg.** 1989. Nucleotide sequence of a *Bacillus subtilis* arginine regulatory gene and homology of its product to the *Escherichia coli* arginine repressor. *Gene* **80:**29–38.
43. **Ottow, J. C. G.** 1974. Arginine dihydrolase activity in species of the genus *Bacillus* revealed by thin-layer chromatography. *J. Gen. Microbiol.* **84:**209–213.
44. **Padilla, G., Z. Hindle, R. Callis, A. Corner, M. Ludovice, P. Liras, and S. Baumberg.** 1991. The relationship between primary and secondary metabolism in streptomycetes, p. 35–45. *In* S. Baumberg, H. Krügel, and D. Noack (ed.), *Genetics and Product Formation in Streptomyces*. Plenum Press, New York.
45. **Paulus, T. J., and R. L. Switzer.** 1979. Characterization of pyrimidine-repressible and arginine-repressible carbamoylphosphate synthetases from *Bacillus subtilis*. *J. Bacteriol.* **137:**82–91.
46. **Piérard, A.** 1983. Evolution des systèmes de synthèse et d'utilisation du carbamoylphosphate, p. 55–61. *In* G. Hervé (ed.), *L'Évolution des Protéines*. Masson, Paris.
47. **Pittard, A. J., and B. E. Davidson.** 1991. TyrR protein of *Escherichia coli* and its role as repressor and activator. *Mol. Microbiol.* **5:**1585–1592.
48. **Potvin, B., and H. Gooder.** 1975. Carbamylphosphate synthesis in *Bacillus subtilis*. *Biochem. Genet.* **13:**125–143.
49. **Prozesky, O. W., W. O. K. Grabnow, S. van der Merwe, and J. N. Coetzee.** 1973. Arginine cluster in *Proteus-Providence* group. *J. Gen. Microbiol.* **77:**237–240.
49a. **Rudd, B. A. M.** Personal communication.
50. **Sakanyan, V., A. Kochikyan, I. Mett, C. Legrain, D. Charlier, A. Piérard, and N. Glansdorff.** 1992. A re-examination of the pathway for ornithine biosynthesis in a thermophilic and two mesophilic *Bacillus* species. *J. Gen. Microbiol.* **138:**125–130.
51. **Sakanyan, V. A., A. S. Hoysepyan, I. L. Mett, A. V. Kochikyan, and P. K. Petrosan.** 1990. Molecular cloning and structural-functional analysis of the arginine biosynthesis genes of the thermophilic bacterium *Bacillus stearothermophilus*. *Genetika* **26:**1915–1925.
52. **Schreier, H. J., T. M. Smith, and R. M. Bernlohr.** 1982. Regulation of nitrogen catabolism in *Bacillus* sp. *J. Bacteriol.* **151:**971–975.
52a. **Smith, M. C. M.** Unpublished data.
53. **Smith, M. C. M., L. Czaplewski, A. K. North, S. Baumberg, and P. G. Stockley.** 1989. Sequences required for regulation of arginine biosynthesis promoters are conserved between *Bacillus subtilis* and *Escherichia coli*. *Mol. Microbiol.* **3:**23–28.
54. **Smith, M. C. M., A. Mountain, and S. Baumberg.** 1986. Cloning in *Escherichia coli* of a *Bacillus subtilis* arginine repressor gene through its ability to confer structural stability on a fragment carrying genes of arginine biosynthesis. *Mol. Gen. Genet.* **205:**176–182.
55. **Smith, M. C. M., A. Mountain, and S. Baumberg.** 1986. Sequence analysis of the *Bacillus subtilis argC* promoter region. *Gene* **49:**53–60.
56. **Smith, M. C. M., A. Mountain, and S. Baumberg.** 1990. Nucleotide sequence of the *Bacillus subtilis argC* gene encoding N-acetylglutamate-gamma-semialdehyde dehydrogenase. *Nucleic Acids Res.* **18:**4595.
56a. **Soutar, A.** Unpublished results.
57. **Stalon, V.** 1985. Evolution of arginine metabolism, p. 227–308. *In* H. K. Schleifer and E. Stackebrandt (ed.), *Evolution of Procaryotes*. Academic Press, Inc., New York.
58. **Stirling, C. J., G. Szatmari, G. Stewart, M. C. M. Smith, and D. J. Sherrat.** 1988. The arginine repressor is essential for plasmid-stabilising site-specific recombination of the ColE1 *cer* locus. *EMBO J.* **7:**4389–4395.
59. **Tabor, C. W., and H. Tabor.** 1985. Polyamines in microorganisms. *Microbiol. Rev.* **49:**81–99.
60. **Thoai, N. V., F. Thome-Beau, and A. Olomucki.** 1966. Induction et specificité des enzymes de la nouvelle voie catabolique de l'arginine. *Biochim. Biophys. Acta* **115:**73–80.
61. **Tricot, C., V. Stalon, and C. Legrain.** 1991. Isolation and characterization of *Pseudomonas putida* mutants affected in arginine, ornithine and citrulline catabolism: function of the arginine oxidase and arginine succinyltransferase pathways. *J. Gen. Microbiol.* **137:**2911–2918.
62. **Udaka, S.** 1966. Pathway-specific pattern of control of arginine biosynthesis in bacteria. *J. Bacteriol.* **91:**617–621.
62a. **Vernon, D. I., and S. Baumberg.** Unpublished data.
63. **Vogel, R. H., and H. J. Vogel.** 1963. Acetylated intermediates of arginine synthesis in *Bacillus subtilis*. *Biochim. Biophys. Acta* **69:**174–176.
64. **Zeller, A., L. S. Van Orden, and A. Vogtli.** 1954. Enzymology of mycobacteria. VII. Degradation of guanidine derivatives. *J. Biol. Chem.* **209:**429–455.

22. Biosynthesis of the Branched-Chain Amino Acids

PAMELA S. FINK

For most of the gram-positive bacteria, biosynthesis of the branched-chain amino acids (isoleucine, valine, and leucine) occurs via the same pathway used by gram-negative bacteria, fungi, and plants (9, 60). Beginning with either threonine or pyruvate, the cells synthesize isoleucine or valine, respectively. An intermediate in valine biosynthesis, α-ketoisovalerate, is the starting material for leucine synthesis.

Branched-chain amino acid biosynthesis has been most thoroughly characterized in *Bacillus subtilis*; hence, most of the information in this chapter will come from studies of *B. subtilis*. However, novel mechanisms for biosynthesis in other gram-positive genera will also be described. In addition, the organization of the genes responsible for isoleucine, valine, and leucine biosynthesis and proposed regulatory mechanisms for expression of those genes will be discussed and compared with those for the corresponding genes in the gram-negative bacterium *Escherichia coli*.

BIOSYNTHESIS

The Pathways

The major pathways for the synthesis of the branched-chain amino acids in vegetatively growing *B. subtilis* cells are shown in Fig. 1. The first step in the path leading specifically to isoleucine production is the deamination of threonine to form α-ketobutyrate, a step mediated by the enzyme threonine deaminase. In *B. subtilis* cells that have a defective threonine deaminase, this step may also be mediated by a secondary activity of the enzyme threonine synthetase (55).

For the next four reaction steps, the same enzymes catalyze the reactions leading to the formation of either isoleucine or valine; i.e., the reactions occur in parallel. First, an active acetyladehyde derived from pyruvate is transferred to α-ketobutyrate (isoleucine biosynthesis) or to another pyruvate (valine biosynthesis) by the enzyme acetohydroxy acid synthase (AHAS). *B. subtilis* cells defective in this enzyme activity can still synthesize valine but not isoleucine by producing another enzyme, acetolactate synthase, which forms α-acetolactate (27, 72). This same type of enzymatic activity has also been seen by Lessie and Thorne in *Bacillus licheniformis* (37).

The acetohydroxy acids are then converted to dihydroxy acids by acetohydroxy acid isomeroreductase and to branched-chain keto acids by dihydroxy acid dehydratase. The final step involves a transamination of the branched-chain keto acids to isoleucine or valine and is presumably mediated by a branched-chain amino acid aminotransferase. Holtzclaw and Chapman (28) reported that there was no detectable transaminase C activity (i.e., an alanine-valine transaminase) in *B. subtilis*. In contrast, Temeyer and Chapman (59) reported finding transaminase B (i.e., the branched-chain amino acid transaminase) in *B. subtilis*. Despite efforts by Zahler and coworkers to generate strains of *B. subtilis* that are defective for this aminotransferase, no such mutants have been recovered (71). This suggests that multiple transaminases capable of catalyzing the final step of branched-chain amino acid biosynthesis exist in *B. subtilis* (6).

The biosynthetic pathway leading to leucine production is called the isopropylmalate pathway. The enzyme α-isopropylmalate synthase mediates the condensation of α-ketoisovalerate, an intermediate in valine biosynthesis, with acetyl coenzyme A to form α-isopropylmalate. The next two steps are catalyzed by isopropylmalate dehydratase and β-isopropylmalate dehydrogenase, and the final transamination step that converts α-ketoisocaproate to leucine is catalyzed by a branched-chain amino acid aminotransferase.

The branched-chain amino acid biosynthetic pathway of *B. subtilis* very closely resembles the pathways found in *E. coli* and *Salmonella typhimurium*, with only a few differences. In these gram-negative bacteria, there appear to be multiple AHASs with at least four isozymes in *E. coli* (4) and two in *S. typhimurium* (15). In addition, distinct aminotransferases that convert branched-chain keto acids to amino acids have been identified: the branched-chain amino acid transaminase, which catalyzes synthesis of all three amino acids (52); the alanine-valine transaminase, which synthesizes valine and leucine (7); and the tyrosine aminotransferase, which synthesizes only leucine (19).

In *Bacillus* spp., some of the intermediates in branched-chain amino acid biosynthesis, specifically the branched-chain keto acids, are the immediate precursors of synthesis of the branched-chain fatty acids that are an integral part of the cell membrane (32). Branched-chain keto acids are converted to the corresponding fatty acids by branched-chain ketoacid dehydrogenase, an enzyme closely related and possibly identical to pyruvate dehydrogenase (40, 69). This additional use of these biochemical intermediates does not occur in gram-negative bacteria, since they do not have branched-chain fatty acids in their cell membranes (50).

Enzymes of the Pathway

For many biosynthetic pathways, the first enzyme in the path leading to a specific product is subject to

Pamela S. Fink • Department of Microbiology and Immunology, Wright State University, Dayton, Ohio 45435.

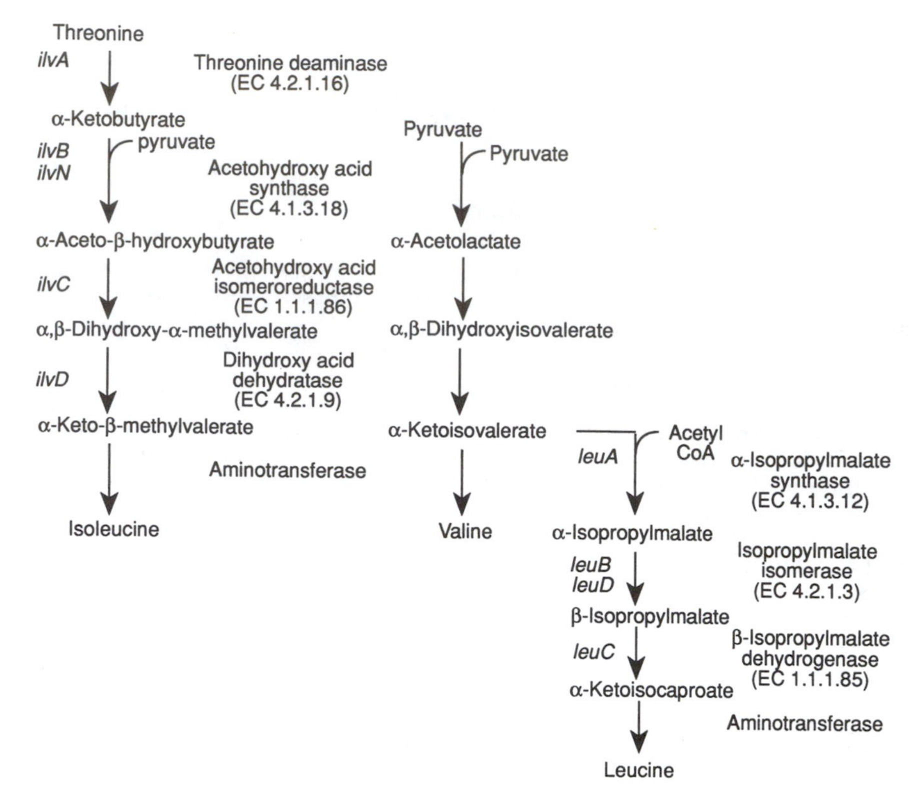

Figure 1. Biosynthetic pathways for isoleucine, valine, and leucine formation in *B. subtilis*. Enzymes catalyzing the reactions and genes encoding the enzymes are indicated.

feedback inhibition by that product; the branched-chain amino acid pathway is no exception. Hatfield and Umbarger (25) demonstrated that in *B. subtilis*, threonine deaminase, which catalyzes the first step in isoleucine synthesis, is inhibited by isoleucine. Similarly, Ward and Zahler (64) reported that the *B. subtilis* α-isopropylmalate synthetase, which mediates the first step in leucine production, is inhibited by leucine.

Only a few of the enzymes in the pathway have been isolated from gram-positive bacteria and thoroughly characterized. Threonine deaminase was isolated from *B. subtilis* cells by Hatfield and Umbarger, who showed that the enzyme is a multisubunit structure, has a molecular weight of 200,000, and requires pyridoxal phosphate to function (24). They also found that activity of the enzyme is competitively inhibited by isoleucine and noncompetitively inhibited by valine (25).

The *B. subtilis* gene encoding threonine deaminase has been cloned and sequenced, and the deduced amino acid sequence corresponds to that of a protein with a molecular weight of approximately 47,000 (2, 3), which is somewhat smaller than its counterpart from either *E. coli* or *S. typhimurium* (molecular weight of 56,200 [11, 36, 58]). In Fig. 2, the predicted amino acid sequence of the *B. subtilis* enzyme is compared with the sequences of the *E. coli* biosynthetic and biodegradative threonine deaminases. The *B. subtilis* enzyme, consisting of 422 amino acid residues, is smaller than the biosynthetic *E. coli* enzyme (515 residues) but larger than the biodegradative one (329 residues [13]). All three enzymes have a lysine residue at approximately position 61, which is believed to be the site of pyridoxal phosphate binding (13, 48). In fact, most of the catalytic domains (C) are well conserved in all three proteins, but the regulatory domains (R) show some striking differences. Several of the regulatory domains in the biosynthetic *E. coli* threonine deaminase are absent in the biodegradative enzyme and in the *B. subtilis* enzyme; however, the sequence thought to be the isoleucine binding site in domain R6 (58) appeared both in the biosynthetic enzyme of *E. coli* and in the *B. subtilis* enzyme. This finding correlates well with the fact that the activities of these two enzymes but not that of the biodegradative *E. coli* enzyme are inhibited by isoleucine (13, 25, 61).

The *B. subtilis* genes that encode AHAS (*ilvB* and *ilvN*) and acetohydroxy acid isomeroreductase (*ilvC*) have been isolated. The amino acid sequences predicted from the nucleotide sequence reveal that AHAS consists of a large catalytic subunit (the *ilvB* product) with a molecular weight of 61,390 and a smaller regulatory subunit (the *ilvN* product) with a molecular weight of 18,000 (62). The larger subunit is the same size as the corresponding AHAS subunit of *E. coli* (67); however, the smaller subunit is closest to the size of the regulatory subunit of a minor, valine-sensitive form of the enzyme in *E. coli*, AHAS III (56).

The predicted amino acid sequence of the *B. subtilis* acetohydroxy acid isomeroreductase reveals a protein with a molecular weight of 36,500 (62). This is considerably smaller than the corresponding enzyme from *E. coli*, which has a molecular weight of 54,000 (65). In addition, the predicted amino acid sequences of the

```
B.s. ilvA  MKPLLKENSLIQVKHILKAHQNVKDVVIH..TPLQRNDRLSERYECNIYL   48
E.c. ilvA  MadsqplsgapegaeyLRAvlrapvyeaaqvTPLQkmekLSsRlDnvIlV
E.c. tdc   ....mhitydLpVaiddiieakqrlagriykygmpRsnyfSERckgeIfL
                                                           --

           KREDLQVVRSFKLRGAYHKMKQLSSEQTENGVVCASAGNHAQGVAFSCKH   98
           KREDrQpVHSFKLRGAYamMagLteEQkahGVItASAGNHAQGVAFSsaR
           KfEnmQrtgSFKIRGAFnklssLtdaekrkGVVacSAGNHAQGVslSCam
           -----C1----------             --------C2--------

           LGIHGKIFMPSTTPRQKVSQVELFGKGFIDIILTGDTFDDVYKSAAECCE  148
           LGVkalIvMPtaTadiKVdaVrgFGgevIlhganfDeakakaielsqqqf
           LGIdGKVvMPkgaPKsKVaatcdYsaevVlhgdnfndtiakvseivEmeg

           AESRTFIHPFDDPDVMAGQGTLAVEILNDIDTEPHFLFASVGGGGLLSGV  198
           f...TwVpPFDhPmViAGQGTLALELLqqdahld.rVFvpVGGGGLaaGV
           r...iFIpPYDDPkViAGQGTIgLEImeDLydvd.nVivpIGGGGLIaGI
                ------------C3----------            -----------C4-

           GTYLKNVSPDTKVIAVEPAGAASYFESNKAGHVVTLDKIDKFVDGAAVKK  248
           avlIKQLmPqiKVIAVEaedsAclkaaldAGHpVdLpRVglFaEGvAVKR
           avaIKsInPtiRVIgVqsenvhgmaaSfhsGeitThrttgtlaDGcdVsR
           ---------------

           IGEETFRTLETVVDDILLVPEGKVCTSILELYNECAVVAEPAGALSVAAL  298
           IGDETFRlcqeyLDDIItVdsdaICaamkDLFeDvraVAEPsGALaLAgm
           pGnlTYeiVreLVDDIVLVsedeIrnSmIaLiQrnkVVtEgAGALacAAL
                        --C5-- ----------------R1------------

           DLYKDQIKGKNVVC..VVSGGNNSIGRMQEMKERSLIFEGLQH.......  339
           kkYialhniRgerlahILSGaNvn.............FhGLrysercelg
           lsgKldqyiqNrktvsIISGGNiDLsRvsqitgfvda*............ (329)
           ----            --------.............R2-----------

           ..................................................
           eqreallavtipeekgsflkfcqllggrsvtefnyrfadaknacifvgvr
           ..................................................
                                    ----R3----

           ..................................................
           lsrgleerkeilqmlndggysvvdlsddemaklhvrymvggrpshplqer
           ..................................................
                                  --------R4--------  -------

           YFIVNFPQRAGALREFLDEVLGPNDDITRFEYTKKNNKSNGPALVGIELQ  389
           lysfeFPespGALlrFLntLqtywn.IslFhY..............hyrs
           ..................................................
           --R5--------------       --------.............----

           NKADYGPLIERMNKKPFHYVEVNKDEDLFHLLI*..............    422
           hgtDYGrVLaafelgdhepdfetrlnELgydchdetnnpafrfflag*   (515)
           ................................................
           -R6--------             ---------R7-------
```

Figure 2. Comparison of deduced amino acid sequences of the *B. subtilis* (*B.s. ilvA*), *E. coli* biosynthetic (*E.c. ilvA*), and *E. coli* biodegradative (*E.c. tdc*) threonine deaminases. C1 to C5 and R1 to R7 are the catalytic and regulatory domains, respectively, described by Taillon et al. (58).

two proteins show no homology, indicating that the enzyme may have evolved from another reductase in *B. subtilis* (62).

The activities of the enzymes in the leucine biosynthetic pathway have been examined in *B. subtilis* by Ward and Zahler (64). Those investigators found that the activity of α-isopropylmalate synthetase, the first enzyme involved in the pathway, is inhibited by leucine and stimulated by the ion Mn^{2+}. The optimum pH for activity of this enzyme is 8.1. The α-isopropylmalate isomerase is very unstable in extracts but is partly stabilized by the presence of its substrate, α-isopropylmalate. In crude extracts, the β-isopropylmalate dehydrogenase is the most stable of the leucine-specific enzymes, and the optimum pH for assaying enzyme activity is 8.5.

Biosynthetic Pathways in Other Gram-Positive Bacteria

Although most of the gram-positive bacteria synthesize branched-chain amino acids by the reactions outlined in Fig. 1, some *Clostridium* species have alternative pathways. Both *Clostridium pasteurianum* and *Clostridium sporogenes* synthesize α-ketobutyrate by the reductive carboxylation of propionate (8, 44). In addition, *C. sporogenes* can synthesize isoleucine from α-methylbutyrate (44) and valine from isobutyrate (43). These alternate pathways are not observed in aerobic bacteria, because only anaerobes have the highly reduced environments required for reductive carboxylation to occur (44).

The isopropylmalate pathway for leucine formation seems to be the most commonly used pathway in gram-positive bacteria other than *B. subtilis* (57). Enzymes specific for the isopropylmalate pathway have been identified in a number of bacteria, including *Bacillus polymyxa*, *Bacillus megaterium*, and *Micrococcus* and *Streptomyces* spp. In addition, several genes have been isolated from gram-positive bacteria and used to restore prototrophy to leucine-requiring *E. coli* strains, indicating that the corresponding enzyme also exists in the other bacteria. For example, the gene encoding β-isopropylmalate dehydrogenase has been cloned from two different *Clostridium* species, *Clostridium acetobutylicum* and *Clostridium butyricum* (17, 29). In addition, the gene specific for α-isopropylmalate synthase has been cloned from *Streptomyces rochei* (26).

GENETICS OF BRANCHED-CHAIN AMINO ACID BIOSYNTHESIS

Overview of *B. subtilis* and *E. coli* Genes

Many of the genes whose products are involved in branched-chain amino acid biosynthesis in *B. subtilis* have been identified (14, 51), and their places in the pathways are shown in Fig. 1. *ilv* genes encode proteins that function in the valine and/or isoleucine biosynthetic pathways, and *leu* gene products catalyze reactions in the leucine pathway.

Threonine deaminase is encoded by a gene designated *ilvA* for both *B. subtilis* and *E. coli*. For *B. subtilis*, the large catalytic subunit of AHAS is encoded by the *ilvB* gene, and the smaller regulatory subunit is encoded by the *ilvN* gene. In contrast, *E. coli* (and *S. typhimurium* [15]) contains multiple copies of genes that encode different forms of this enzyme: *ilvB* and *ilvN* are specific for the large and small subunits, respectively, of AHAS I, a valine-sensitive form of the enzyme; *ilvG* and *ilvM* are specific for the large and small subunits of AHAS II, a valine-insensitive isozyme; *ilvI* and *ilvH* are specific for the large and small subunits of AHAS III, another valine-sensitive isozyme; and *ilvJ* is specific for AHAS IV, another valine-insensitive isozyme (4). In addition, another locus, *ilvF*, is a cryptic gene that encodes a valine-resistant form of AHAS that is stable at elevated temperatures (1). The purpose of all of these different isozymes of AHAS in *E. coli* presumably is to allow synthesis of isoleucine even in the presence of valine. It is not certain how *B. subtilis* and other gram-positive bacteria handle this dilemma. The next enzyme in the pathway, dihydroxy acid dehydratase, is encoded by the *ilvD* gene in both *E. coli* and *B. subtilis*.

The *leuA* gene encodes the first enzyme of the leucine biosynthetic pathway, α-isopropylmalate synthase, in both kinds of bacteria. In *E. coli*, the subunits of the next enzyme in the pathway, isopropylmalate isomerase, are encoded by the *leuC* and *leuD* genes, while in *B. subtilis*, the corresponding genes are called *leuB* and *leuD*. Using specialized transducing particles of the bacteriophage SPβ carrying the *ilv-leu* genes in complementation tests, Mackey and Zahler (42) demonstrated that the *leuD* gene exists in *B. subtilis*. Finally, β-isopropylmalate dehydrogenase is the product of the *leuC* gene in *B. subtilis* and the *leuB* gene in *E. coli*.

In *E. coli*, several genes play a role in the last step in the biosynthesis of the branched-chain amino acids, the transamination reaction. The *ilvE* gene encodes the branched-chain amino acid aminotransferase, the *avtA* gene encodes the alanine-valine transaminase used in valine synthesis, and the *tyrB* gene encodes the tyrosine-repressible transaminase used in leucine synthesis (6). Despite efforts to locate transaminase genes in *B. subtilis*, no such loci have been identified (71). The only type of *B. subtilis* mutant that cannot use the branched chain α-keto acids to satisfy isoleucine and valine requirements has a defect in the α-keto acid uptake mechanism, which is encoded by the *kauA* gene (20).

Several other genes ancillary to branched-chain amino acid biosynthesis have been identified in both *E. coli* and *B. subtilis*. Many of these genes play a regulatory role, controlling the expression of the genes involved in biosynthesis. For example, in *E. coli*, *ilvY* and *ilvR* encode positive regulators for the *ilvC* (54, 65) and *ilvGMEDA* (31) transcription units, respectively; *ileR* and *leuJ* encode repressors that negatively regulate expression of the *ilv* and *leu* operons, respectively (30).

In *B. subtilis*, one locus that apparently controls expression of the *ilv* and *leu* genes is *azlA*, which is tightly linked to the *ilv-leu* cluster. Cells carrying a mutation in this gene are resistant to the leucine analog 4-azaleucine because of overexpression of the *ilv* and *leu* genes (64). The mechanism of action of an *azlA* mutation will be discussed below. Another gene, *ilvX*, may play a role in regulating the level of expression of the *ilv-leu* cluster (62). A mutation in *ilvX* leads to overexpression of the *ilv-leu* genes, but the gene lies near *thrA*, which is not linked to the *ilv-leu* cluster.

Enzymes that primarily function in other pathways can also play a role in branched-chain amino acid biosynthesis. For example, threonine synthetase, encoded by the *thrC* gene (previously called *thrB* [51]), has a secondary activity that converts phosphohomoserine to α-ketobutyrate (53). This activity enables cells that have defects in the *ilvA* gene to produce isoleucine.

The main function of the *alsRSD* genes in *B. subtilis* is a degradative one; the gene products catalyze reactions involved in acetoin production from pyruvate (14, 21). The *alsS* gene product is α-acetolactate synthase (14), which can function in place of AHAS in the valine biosynthetic pathway. This activity is usually observed in an *ilvB* auxotrophic strain; these cells become Val^+ in the presence of isoleucine and acetate, which induces expression of the *alsRSD* genes (72). Zahler et al. found that a defect in the *alsR* gene led to constitutive production of α-acetolactate synthase regardless of the level of acetate in the medium, indicating that this locus controls expression of the

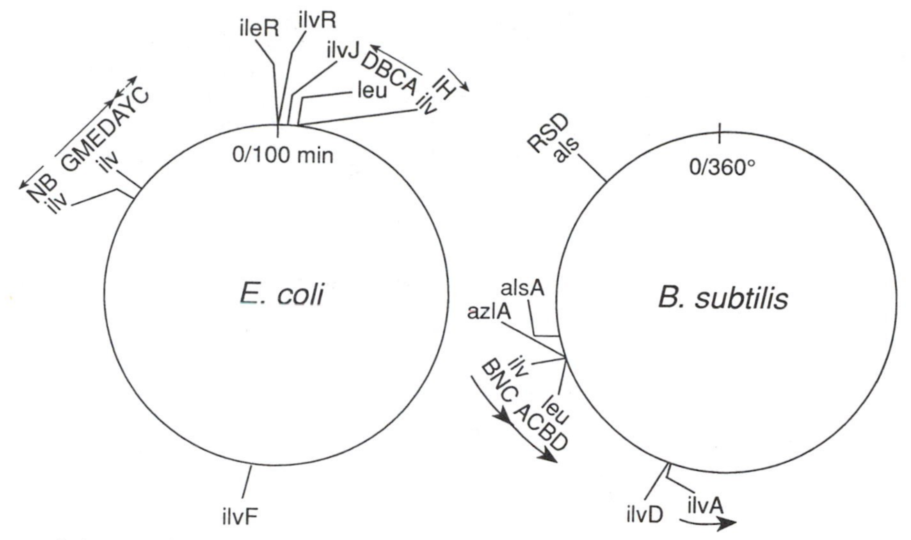

Figure 3. Maps of the *E. coli* and *B. subtilis* chromosomes showing the locations of genes that function during branched-chain amino acid biosynthesis.

alsRSD cluster (72). The other gene in the cluster, *alsD*, encodes α-acetolactate decarboxylase, but this enzyme apparently plays no part in valine biosynthesis. The *alsA* gene is believed to encode a positive regulator of the *alsRSD* cluster; mutations in *alsA* (also known as *ccpA*) result in no expression of α-acetolactate synthase either in the presence of acetate or in an *alsR* mutant (14, 25a, 72).

Organization of *ilv* and *leu* Genes on the Chromosome

The genes involved in branched-chain amino acid biosynthesis have been located on the chromosomal maps of *E. coli* and *B. subtilis* (4, 14, 51), abbreviated forms of which appear in Fig. 3. Many of the *ilv* and *leu* genes in both types of bacteria appear to be clustered and organized in operons. An obvious difference between the two bacteria appears to be exactly which genes are assembled to form the operons.

A more compact organization of genes occurs in *B. subtilis* (5), reflecting the absence of multiple genes for AHAS. A majority of the *ilv* and *leu* genes lie in one operon, the *ilvBNC-leuACBD* cluster, which appears at about 250° on the chromosomal map. The *azlA* locus, which functions in genetic regulation of the *ilv-leu* cluster, may actually lie within the *ilvB* gene (see below).

The other two *ilv* genes, *ilvA* and *ilvD*, lie almost a quarter of the chromosome away, at about 194° and 195°, respectively. Although these genes appear to be adjacent, they are actually physically separated by an operon consisting of two genes that are involved in thymidine biosynthesis, *thyB* and *dfrA*. *thyB* encodes a minor form of thymidylate synthetase, and *dfrA* encodes dihydrofolate reductase; there is no obvious relationship between isoleucine-valine biosynthesis and thymidine biosynthesis. It appears, then, that the two *ilv* genes probably do not constitute an operon.

The *alsS* gene, which encodes α-acetolactate synthase, lies in an operon with the *alsR* and *alsD* genes in the order *alsRSD*; the *alsR* gene is probably responsible for regulating the expression of the operon, since it is very closely linked to the other two genes and since a defect in *alsR* leads to constitutive expression of the α-acetolactate synthase. Another gene that probably encodes a positive regulator of the *alsRSD* cluster, *alsA*, lies near the *ilv-leu* cluster.

Clustering of the *ilvBNC-leu* genes is a common theme in *Bacillus* species. Lessie and Thorne (37) reported that in *B. licheniformis*, the *ilvB*, *ilvC*, and *leu* genes lie in the same gene order and the cluster is flanked by the same genes that surround the *B. subtilis ilv-leu* cluster, namely, *argA* and *pheA*. In *B. megaterium*, Garbe et al. found that the *ilv* and *leu* genes apparently have a slightly different order, i.e., *ilv-leuABC* (18). This is the order of the *leu* genes in *E. coli*, given that *leuC* in *E. coli* and *leuB* in *Bacillus* species encode the enzyme β-isopropylmalate dehydrogenase. In addition, the *ilv-leu* genes of *B. megaterium* appear to be linked to the *pheA* gene; however, contrary to mapping in *B. subtilis*, in which the *ilv* genes lie adjacent to *argA*, the *B. megaterium ilv* genes lie adjacent to *pheA*.

Organization of *ilv* and *leu* Genes in Other Gram-Positive Bacteria

Some information on *ilv* and *leu* gene organization in other gram-positive bacteria is available. Pattee et al. (49) demonstrated that the genes lie in the following order on the *Staphylococcus aureus* chromosome (41): *ilvA*, *ilvB*, *ilvC*, *ilvD*, *leuA* (*leuBCD*). The precise arrangement of the *leu* genes could not be determined. Pattee et al. (49) also identified a locus, called *ilvR*, that is very tightly linked to the *ilv-leu* cluster. A defect in this gene resulted in overexpression of all three branched-chain amino acids; perhaps the *ilvR* gene is analogous to the *azlA* gene of *B. subtilis*.

A chromosomal map for *Clostridium perfringens* has recently been reported (10), but the *ilv* and *leu* genes have not been located on this map. At least some of the *leu* genes must exist in *Clostridium* species, because the genes have been cloned in *E. coli* (17, 29). In fact, the *leuB* and *leuC* genes are adjacent to one another in *C. pasteurianum*, because both genes lie on a single restriction fragment (47). Whether the organization of *ilv* and *leu* genes in *Clostridium* spp. and other gram-positive bacteria is similar to that in *Bacillus* spp. or *S. aureus* remains to be seen.

GENETIC REGULATION

Regulation of Biosynthetic Genes in General

The goal in regulating expression of biosynthetic genes is to ensure that gene products are synthesized only when the cells need the end products of the specific biosynthetic pathway. To this end, many bacteria have devised a mechanism whereby the RNA transcripts that would lead to the production of the gene products in question are prematurely ended or attenuated (70).

The attenuation model of genetic control is well documented for a number of gene clusters in *E. coli* (34) and *B. subtilis* (73). A simplified version of attenuation is as follows. In *E. coli*, transcription of an amino acid biosynthetic operon depends on how efficiently a small peptide encoded upstream of the structural gene is translated. This peptide is encoded by a region of the transcript that contains several overlapping regions of dyad symmetry, the terminator-antiterminator region. In addition, the peptide contains numerous codons specific for the amino acid in question. If the peptide is translated efficiently, i.e., if there is an ample supply of that amino acid in the cell, a stem-and-loop structure resembling a rho-independent transcription terminator forms in the transcript just prior to the structural genes, and transcription ends. If, on the other hand, the cell is limited for the amino acid, translation of the leader peptide stalls, and another hairpin loop forms and prevents the terminator from forming. Transcription thus proceeds through the rest of the operon.

In *B. subtilis*, there is a similar attenuation mechanism, but it is more generally applicable to any biosynthetic operon because its function is not coupled to peptide synthesis. Within the transcript upstream of the structural genes lie several overlapping regions of dyad symmetry. One of these hairpin loops resembles a rho-independent transcription terminator. Another stem and loop that precludes formation of the terminator can form, but this happens only when the amino acid (or nucleotide) product is limited. When the product is in excess, it binds to a protein, and the whole complex binds to a specific site in the RNA such that formation of the terminator is the favored response. Transcription then terminates before entering the structural genes.

If the biosynthetic genes are part of a polycistronic operon, the expression of all of the genes can be coordinately controlled. This will ensure that the proper balance of gene products is generated so that the biosynthetic pathway is not stalled at any one step. In addition, if the genes are clustered and transcribed as a single mRNA with the coding regions for adjacent polypeptides in close proximity, translation of the genes may proceed more efficiently. The concept of this "translational coupling" has been well documented for several operons (23, 73).

The regulation of *ilv* and *leu* gene expression will be discussed below. A brief overview of what occurs in *E. coli* will be presented first, and then what occurs in *B. subtilis* will be described.

Regulation of *ilv* and *leu* Gene Expression

Most of the *ilv* and *leu* gene clusters that have been examined in *E. coli* appear to be regulated by the leader peptide-transcription attenuation mechanism. For example, the *leuACBD*, *ilvBN*, and *ilvGMEDA* operons all appear to contain the leader peptide and terminator-antiterminator structures in the transcribed region just upstream of the first structural gene of the operon (35, 45, 67, 68).

Other mechanisms of regulation are also important in branched-chain amino acid biosynthesis in *E. coli*. In addition to the major promoter upstream of *ilvG* in the *ilvGMEDA* operon, there are at least two other internal promoters. These regulate expression of *ilvE* (38, 66) and *ilvA* (39). Independent expression of the *ilvA* gene is required by *E. coli* during the shift from anaerobic to aerobic growth. Threonine deaminase activity leads to an increase of α-ketobutyrate, which serves as an "alarmone" in *E. coli* to regulate the levels of intracellular metabolic intermediates during the switch to aerobic growth conditions (12).

The *ilvC* gene is not at all regulated by attenuation. When levels of the substrates for the isomeroreductase reaction build up in the cells, IlvY, a member of the LysR family of bacterial regulatory proteins, binds to the acetohydroxy acid substrates and then to the *ilvC* promoter region, stimulating transcription (65). The *ilvY* gene thus acts as a positive regulator of *ilvC* gene expression.

Several regulatory proteins also play a role in controlling coordinate expression of the other widely scattered *ilv* and *leu* genes. The *ileR* gene product encodes a protein that interacts with the branched-chain amino acids and binds near the *ilvGMEDA* promoter to down-regulate expression of the operon. Mutations in *ileR* lead to constitutive expression of the *ilvGMEDA* genes, indicating that the *ileR* product acts as a repressor (30). The *leuJ* gene product acts in a similar manner on the *leu* gene cluster (30). Finally, the *ilvR* gene has been identified by deletion analysis; the gene product serves as an activator of the *ilvGMEDA* genes to increase expression of this cluster (31).

Genetic regulation of the *ilv* and *leu* genes in gram-positive bacteria has not been nearly so well-characterized as regulation of these genes in gram-negative bacteria. In the early 1970s, Ward and Zahler found that the *ilvB*, *ilvC*, and *leu* genes are tightly linked on the chromosome of *B. subtilis* (63) and that limiting levels of leucine cause a coordinate derepression of the enzymes encoded by these genes. They also found a mutation that maps just upstream of the *ilvB* gene, called *azlA*, that causes an increased production of all of the gene products in the operon. From this information, they proposed that the genes encoding these enzymes may be under a single control mechanism (64).

```
5'...ATCTTCCATG TTTATATAAA ATTAAATAAT TCTGGTGTAT TCTTGATAGA -150
     TTAATAAAAA AATTATAAAA ACTTTTAACA TTTGCAATTC CTTTTGTACC -100
     AATAATGAAA GCGTATACAA TATAGATTGA TTATTCAAAA TTGTCTAATA  -50
     ATTTTAAAAA ATGCTGTTGA CACTGCGTCC AAAGCGGCGT AATATGAGTT
    +1
     CAACAAAAGA TAAATGCAAG CTTCACAAGC GAAAATCATC GCAGTATGAT  +50
     TCTAAAAAAA TGAAAAACAA ACGACCTTCT TGAACAGCTG GAAAGCCGTT +100
     CCGAAGGACT TGAATATGAA AAGCGCAGAT GAGGATAAGT AGCCTTGATA +150
     AAGTTTTCCA CAGAGAACCG GGTTAGCTGA GAACCGGCGA AGCTTTACAA +200
     GGTGAACTCG CCTCAGAGTG CCAGCCTGAA ATGACAGTAG GACTTGGCCG +250
     GTGAACTTGA TTCACTGCTT ACTAAAGCGG ATAGAAATAT CCATGAGACG +300

                                                A  >>>>>>>
     GCCGATTAAC AGGCCGTAAA CAAGGGTGGT ACCGCGGAAA GAAAAGCCTT +350

     >>>>>>>>>> >>>>>>>>>> >>>    <<<< <<<<<<<<<<  <<<<<<<<<
                                  C  >>>>>>>> >
     TTCGCCCCTT TTAGCTATCG CAGTTACTGC GGGCTGATTG TGGGCGGAAG +400

     <<<<<   B
                     <<<< <<<<   D
     GGCTTTTTTT ATTGAATAAT CAGCTATCTA GCTAATGAAA AGATGATCTT +450
     TAAAGGATGA AAATCCAAAA GGAGGAACTA AA ATG GGG ACT AAT GTA
                       ... ......            Met Gly Thr Asn Val
     CAG GTG GAT TCA GCA...3'
     Gln Val Asp Ser Ala...
```

Figure 4. Nucleotide sequence of the upstream regulatory region of the *B. subtilis ilvB* gene. Underlined sequences represent the −35 and −10 segments of the promoter, +1 indicates the transcription start site, and dots signify the Shine-Dalgarno site before the ATG start of the *ilvB* open reading frame. Arrowheads indicate regions of inverted repeats occurring in transcripts that originate at the promoter. These sequences can form potential stem-and-loop structures A:B and C:D (shown in Fig. 5).

This control mechanism has recently been characterized; it appears to involve an attenuator somewhat similar to others that have been described in *B. subtilis* (71, 73). That is, its function does not appear to be dependent on translation of a leader peptide. The region upstream of the *ilvB* gene has been cloned, and its nucleotide sequence has been determined (62); the sequence is shown in Fig. 4. Transcription of the *ilvBC-leu* cluster begins some 480 bp upstream of the start codon for the *ilvB* gene (Fig. 4); the sequences TTGACA and TAATAT, which are separated by 17 bp, represent the −35 and −10 regions of the promoter, respectively (22, 62, 71). On the basis of insertion studies with the transposon Tn*917*, this promoter is believed to function in the expression of all genes in the cluster. Tn*917*, which contains transcription terminator signals, can generate auxotrophic strains that require all three branched-chain amino acids, the so-called *liv* mutants, if it inserts downstream of this putative promoter region (62). Several years ago, Lessie and Thorne reported finding *liv* mutants following UV irradiation of *B. licheniformis* (37). Since the gene order appears to be the same in the two species, the species may share a common regulatory mechanism for *ilv-leu* gene expression, i.e., transcription of the entire cluster from a promoter that lies upstream of the *ilvB* gene.

Using *lacZ* fusions to monitor gene expression in the *B. subtilis ilv-leu* cluster, Vandeyar (62) found that there was a low basal level of expression even in the presence of all three branched-chain amino acids. Perhaps this expression is needed to maintain a constant supply of branched-chain intermediates for isoleucine and valine biosynthesis in the presence of leucine or for membrane biosynthesis (32). The presence of leucine alone represses synthesis of the whole operon (22, 62), which correlates with the previous work of Ward and Zahler on enzyme activity levels (64). When leucine is limiting in the cells, there is a fourfold increase in the level of transcription of the genes (62), indicating that intracellular leucine levels control expression of the operon.

In the long untranslated leader transcript, there are several regions of potential dyad symmetry (indicated by arrowheads in Fig. 4). These regions lie upstream of the presumed Shine-Dalgarno sequence that precedes the ATG start codon of the *ilvB* gene. The structure of one potential stem and loop, A:B, is depicted in Fig. 5. This molecular configuration has a calculated free energy of −35 kcal/mol (1 cal = 4.184 J), which means that it is a fairly stable structure (62, 71). Vandeyar has proposed that the formation of this stem and loop, which resembles a rho-independent transcription terminator, actually causes termination of the *ilv-leu* transcript in the presence of leucine (62).

In in vitro transcription experiments using *B. subtilis* RNA polymerase, Grandoni et al. (22) found that 90% of the transcripts that originated at the promoter upstream of *ilvB* terminated at the structure described by Vandeyar (62). When leucine is limiting, a *trans*-acting protein factor might interfere with the formation of the putative terminator, resulting in antitermination (22, 71). In addition, there are other potential stem-and-loop regions in this area, e.g., sequences C and D in Fig. 4 and 5, that overlap the putative terminator and could interfere with the formation of the A:B hairpin loop. It appears, then, that the default state for regulation of the *ilv-leu* cluster involves ending the transcript at the terminator in the presence of leucine and allowing readthrough only when leucine is limiting, perhaps through the binding of a *trans*-acting protein in the attenuator region. This regulatory mechanism is slightly different from those described for other *B. subtilis* operons (73), in which the default state is readthrough.

Additional evidence for the involvement of the A:B hairpin loop comes from studies of *azlA* mutants; as previously mentioned, the *ilv* and *leu* genes are constitutively expressed in these mutants (64). The nucleotide sequence of the DNA upstream of the *ilvB* gene

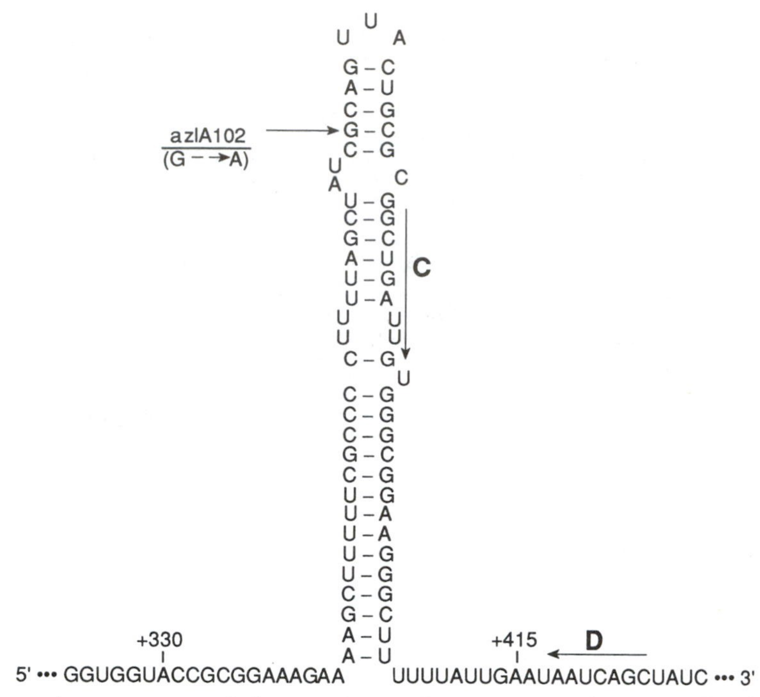

Figure 5. Predicted secondary structure of the terminator that occurs in the untranslated leader transcript upstream of the *B. subtilis ilv-leu* gene cluster. Structure shown is the hairpin A:B, the terminator, which was identified in Fig. 4. C and D represent inverted repeats (arrows) that are capable of base pairing and might interfere with the terminator structure. The base change that occurs in segment A of one *azlA* mutant, *azlA102*, is indicated.

from one such *azlA* mutant, *azlA102*, has been determined. In the mutant, there is a single base change, a G-to-A transition, in segment A of the hairpin loop shown in Fig. 5. This single base change apparently destabilizes the putative terminator structure and allows an excessive level of transcription to occur, i.e., about 10-fold higher than in a wild-type strain in the presence of leucine. Interestingly, if leucine is removed from the cells, the level of gene expression increases another fourfold, indicating that leucine still has some effect in regulating expression of the *ilv-leu* gene cluster in an *azlA* mutant (62).

Another gene that may play a role in regulating *ilv-leu* expression is the *ilvX* locus, which does not map near the *ilv-leu* cluster. In the presence of leucine, the *ilv* and *leu* genes are expressed at a level five times higher in a strain that carries a mutation called *ilvX1* than in wild-type cells; however, transcription of the *ilv-leu* cluster in this mutant is still regulated by leucine (62). The exact function of the *ilvX* product in regulating expression of the *ilv-leu* genes is not yet clear; perhaps it helps stabilize the terminator or destabilize antitermination mechanisms in the presence of leucine.

The only other *ilv* gene that has been examined at the DNA level in *B. subtilis* is *ilvA* (2, 3). The nucleotide sequence of the DNA upstream of the *ilvA* coding region and the sequence of the first part of the protein are presented in Fig. 6. It is apparent that there are no regions of potential dyad symmetry in the sequence that lies upstream of the coding region for threonine deaminase. It thus appears unlikely that an attenua-

```
5'...TTTACCGAGA AAGGATGTAT TCAATGAGAT TAATTAAATT AGAGCAGCCT
     AATGCAATCC AGTTAAAATG GTGTCCAATT ACTTAGAACA AGTAAATATT
     CAATTTGAGA CTGTTGACGT TACACAGGAC CAGAAGTAGC AGCAAGATTT
     GGTGGACATA GTCTGTATTT TTTTAATTAT TAAATAAAAT AATTGTCTTA
     TTTTTTGAAT ATTCATGTTA TAATGGCATA TTAAATAGGG ATTTAAAACA
     AGAAAGGAAT CTGTAC ATG AAA CCG TTG CTT AAA GAA AAC TCT
     .........    ..      Met Lys Pro Leu Leu Lys Glu Asn Ser

     CTC ATC CAA GTG AAA CAC ATT TTG AAG GCT CAC CAA AAC GTA
     Leu Ile Gln Val Lys His Ile Leu Lys Ala His Gln Asn Val

         >>> >>> >>> > >  >>      <<   <  << <<  <<< < <
     AAG GAC GTT GTC ATT CAC ACA CCT TTG CAG AGG ATT GAC AGA
     Lys Asp Val Val Ile His Thr Pro Leu Gln Arg Asn Asp Arg

     <
     CTT TCT GAG AGA TAT...3'
     Leu Ser Glu Arg Tyr...
```

Figure 6. Nucleotide sequence of the DNA upstream of the *B. subtilis ilvA* gene. Dots indicate the Shine-Dalgarno site preceding the ATG start codon. Arrowheads indicate a region of DNA that can potentially form a stem-and-loop structure. Codons in the amino-terminal end of the coding region that are specific for leucine are underlined.

tion mechanism similar to the one used to control *ilv-leu* gene expression is involved in controlling *ilvA* expression.

Does this *ilvA* gene have to be regulated? Lamb and Bott (33) reported that an excess of threonine or α-ketobutyrate inhibits growth and sporulation in *B. subtilis* because the cells starve for valine. Addition of isoleucine, which inhibits the activity of threonine deaminase (25), partially alleviates this threonine sensitivity. It appears, then, that an overexpressed *ilvA* gene may cause a disruption of the metabolic balance in the cells; hence, the gene should be under some type of regulatory control. Within the first 120 bp of the *ilvA* coding region, there is an area of potential dyad symmetry followed by a series of T residues (indicated by arrowheads in Fig. 6). In addition, a number of codons specific for leucine or valine lie in this part of the protein. The hairpin loop that this sequence would generate somewhat resembles a transcription terminator, but the calculated free energy of the structure is only −9.5 kcal/mol, which means that the structure is much less stable than other terminators (3). Cox et al. (11) reported that a stem-and-loop structure also appears in the 5′ end of the coding region in the *E. coli ilvA* gene, but they could not assign any functional significance to this structure. Whether the hairpin loop plays a role in regulating expression of the *B. subtilis ilvA* gene remains to be seen.

Since the genes involved in branched-chain amino acid biosynthesis are not all part of one operon in *B. subtilis*, a mechanism to coordinate expression of all of the genes must exist. Perhaps there is some *trans*-acting regulatory protein in *B. subtilis*, like the *ileR* or *leuJ* gene products of *E. coli*, that controls the activity of the *ilv* genes.

It is not clear what happens to *ilv* and *leu* gene expression during sporulation. A low level of expression of the genes may be required to supply the branched-chain intermediates needed for membrane biosynthesis. Lamb and Bott found that sporulation is inhibited when *B. subtilis* cells are starving for valine (33). Interestingly, it appears that the level of charging of two different valyl-tRNAs species changes during the early stages of sporulation (16) because of alterations in the valyl-tRNA synthetases (46). The precise role that valine plays during sporulation has not been thoroughly investigated, but it is clear that a certain level of valine is required for sporulation to proceed.

CONCLUSIONS

The biosynthesis of branched-chain amino acids has been well characterized for only a few gram-positive bacteria. In most cases, the pathways involved resemble those seen in gram-negative enteric bacteria. Some novel ways to synthesize isoleucine that do not involve the deamination of threonine have been described for several *Clostridium* species. This is probably a reflection of the fact that these anaerobic cells cannot carry out the oxidative-type reactions that other bacteria do.

The genetic organization and regulation of the *ilv* and *leu* genes are significantly different in gram-positive and gram-negative bacteria. As in *E. coli*, *B. subtilis* cells increase expression of the *ilv* and *leu* genes when there is a need for the branched-chain amino acids. In addition, there are at least two other levels of control over *ilv* and *leu* gene expression in *Bacillus* species: the cells need a supply of branched-chain keto acids, which are intermediates in the amino acid biosynthetic pathways, for membrane biosynthesis; and the cells must also determine when to commit to sporulation rather than activating biosynthetic genes. Several loci that have a regulatory role in expression of the *ilv* and *leu* genes have been identified in *B. subtilis*, but there are probably other such genes that have not yet been identified. Much more investigation is needed before the complete story of branched-chain amino acid biosynthesis in the gram-positive bacteria is known.

Acknowledgments. Studies of the *B. subtilis ilvA* gene were supported by an Academic Challenge Grant from the state of Ohio. I gratefully acknowledge the sharing of unpublished data by S. A. Zahler.

NOTE ADDED IN PROOF

Recently, Godon et al. (19a) characterized the branched-chain amino acid biosynthesis genes in *Lactococcus lactis* subsp. *lactis*. Nine structural genes are clustered in the order *leuABCD ilvDBNCA*. The authors suggest that regulation of the operon occurs by transcriptional attenuation, which is dependent on the translation of a leucine-rich leader peptide.

REFERENCES

1. **Alexander-Caudle, C., L. M. Latinwo, and J. H. Jackson.** 1990. Acetohydroxy acid synthase activity from a mutation at *ilvF* in *Escherichia coli* K-12. *J. Bacteriol.* **172:** 3060–3065.
2. **Armpriester, J. M., and P. S. Fink.** 1990. DNA sequence analysis of the *ilvA* gene from *Bacillus subtilis* 168, abstr. H-208, p. 189. Abstr. 90th Gen. Meet. Am. Soc. Microbiol. 1990.
3. **Armpriester, J. M., and P. S. Fink.** Submitted for publication.
4. **Bachmann, B. J.** 1990. Linkage map of *Escherichia coli* K-12, edition 8. *Microbiol. Rev.* **54:**130–197.
5. **Barat, M., C. Anagnostopoulos, and A.-M. Schneider.** 1965. Linkage relationships of genes controlling isoleucine, valine and leucine biosynthesis in *Bacillus subtilis. J. Bacteriol.* **90:**357–369.
6. **Berg, C. M., L. Liu, N. B. Vartak, W. A. Whalen, and B. Wang.** 1990. The branched chain amino acid transaminase genes and their products in *Escherichia coli*, p. 131–162. *In* Z. Barak, D. M. Chipman, and J. V. Schloss (ed.), *Biosynthesis of Branched Chain Amino Acids*. VCH Publishers, New York.
7. **Berg, C. M., W. A. Whalen, and L. B. Archambault.** 1983. Role of alanine-valine transaminase in *Salmonella typhimurium* and analysis of an *avtA*::Tn*5* mutant. *J. Bacteriol.* **155:**1009–1014.
8. **Buchanan, B. B.** 1969. Role of ferredoxin in the synthesis of α-ketobutyrate from propionyl coenzyme A and carbon dioxide by enzymes from photosynthetic and nonphotosynthetic bacteria. *J. Biol. Chem.* **244:**4218–4223.
9. **Calvo, J. M.** 1983. Leucine biosynthesis in prokaryotes, p. 267–284. *In* K. M. Herrmann and R. L. Somerville (ed.), *Amino Acid Biosynthesis and Genetic Regulation*. Addison-Wesley, London.
10. **Canard, B., and S. T. Cole.** 1989. Genome organization of the anaerobic pathogen *Clostridium perfringens*. *Proc. Natl. Acad. Sci. USA* **86:**6676–6680.

11. **Cox, J. L., B. J. Cox, V. Fidanza, and D. H. Calhoun.** 1987. The complete nucleotide sequence of the *ilvGMEDA* cluster of *Escherichia coli* K-12. *Gene* **56:**185–198.
12. **Daniel, J., E. Joseph, and A. Danchin.** 1984. Role of 2-ketobutyrate as an alarmone in *E. coli* K-12: inhibition of adenylate cyclase activity mediated by the phosphoenolpyruvate:glucose phosphotransferase transport system. *Mol. Gen. Genet.* **193:**467–472.
13. **Datta, P., T. Goss, J. Omnaas, and R. Patil.** 1987. Covalent structure of biodegradative threonine dehydratase of *Escherichia coli*: homology with other dehydratases. *Proc. Natl. Acad. Sci. USA* **84:**393–397.
14. **Dean, D. H., and D. R. Zeigler.** 1989. *Bacillus Genetic Stock Center Strains and Data*, 4th ed. *Bacillus* Genetic Stock Center, The Ohio State University.
15. **DeFelice, M., C. T. Lago, C. H. Squires, and J. M. Calvo.** 1982. Acetohydroxy acid synthase isozymes of *Escherichia coli* and *Salmonella typhimurium*. *Ann. Microbiol.* (Paris) **133A:**251–256.
16. **Doi, R. H., I. Kaneko, and R. T. Igarashi.** 1968. Pattern of valine transfer ribonucleic acid of *Bacillus subtilis* under different growth conditions. *J. Biol. Chem.* **243:**945–951.
17. **Efstathiou, I., and N. Traiffant.** 1986. Cloning of *Clostridium acetobutylicum* genes and their expression in *Escherichia coli* and *Bacillus subtilis*. *Mol. Gen. Genet.* **204:**317–321.
18. **Garbe, J. C., G. F. Hess, M. A. Franzen, and P. S. Vary.** 1984. Genetics of leucine biosynthesis in *Bacillus megaterium* QM B1551. *J. Bacteriol.* **157:**454–459.
19. **Gelfand, D. H., and R. A. Steinberg.** 1977. *Escherichia coli* mutants deficient in the aspartate and aromatic amino acid aminotransferases. *J. Bacteriol.* **130:**441–444.
19a. **Godon, J.-J., M.-C. Chopin, and S. D. Ehrlich.** 1992. Branched-chain amino acid biosynthesis genes in *Lactococcus lactis* subsp. *lactis*. *J. Bacteriol.* **174:**6580–6589.
20. **Goldstein, B. J., and S. A. Zahler.** 1976. Uptake of branched-chain α-keto acids in *Bacillus subtilis*. *J. Bacteriol.* **127:**667–670.
21. **Gottschalk, G.** 1979. *Bacterial Metabolism*, p. 138–139. Springer-Verlag, New York.
22. **Grandoni, J. A., J. M. Calvo, and S. A. Zahler.** 1991. Transcriptional regulation of the *ilv-leu* operon of *Bacillus subtilis*, abstr. H-70, p. 166. Abstr. 91st Gen. Meet. Am. Soc. Microbiol. 1991.
23. **Harms, E., E. Higgens, J. W. Chen, and H. E. Umbarger.** 1988. Translational coupling between the *ilvD* and *ilvA* genes of *Escherichia coli*. *J. Bacteriol.* **170:**4798–4807.
24. **Hatfield, G. W., and H. E. Umbarger.** 1970. Threonine deaminase from *Bacillus subtilis*. I. Purification of the enzyme. *J. Biol. Chem.* **245:**1736–1741.
25. **Hatfield, G. W., and H. E. Umbarger.** 1970. Threonine deaminase from *Bacillus subtilis*. II. The steady state kinetic properties. *J. Biol. Chem.* **245:**1742–1747.
25a. **Henkin, T. M., F. J. Grundy, W. L. Nicholson, and G. H. Chambliss.** 1991. *Mol. Microbiol.* **5:**575–584.
26. **Hercomb, J., G. Thierbach, S. Baumberg, and J. H. Parish.** 1987. Cloning, characterization and expression in *Escherichia coli* of a leucine biosynthetic gene from *Streptomyces rochei*. *J. Gen. Microbiol.* **133:**317–322.
27. **Holtzclaw, W. D., and L. F. Chapman.** 1975. Degradative acetolactate synthase of *Bacillus subtilis*: purification and properties. *J. Bacteriol.* **121:**917–922.
28. **Holtzclaw, W. D., and L. F. Chapman.** 1977. A new assay for transaminase C. *Anal. Biochem.* **83:**162–167.
29. **Ishii, K., T. Kudo, H. Honda, and L. Horikoshi.** 1983. Molecular cloning of β-isopropylmalate dehydrogenase from *Clostridium butyricum* M588. *Agric. Biol. Chem.* **43:**2313–2317.
30. **Johnson, D. I., and R. L. Somerville.** 1983. Evidence that repression mechanisms can exert control over the *thr*, *leu*, and *ilv* operons of *Escherichia coli* K-12. *J. Bacteriol.* **155:**49–55.
31. **Johnson, D. I., and R. L. Somerville.** 1984. New regulatory genes involved in the control of transcription initiation at the *thr* and *ilv* promoters in *Escherichia coli* K-12. *Mol. Gen. Genet.* **195:**70–76.
32. **Kaneda, T.** 1973. Biosynthesis of branched long-chain fatty acids from the related short-chain α-keto acid substrates by a cell-free system of *Bacillus subtilis*. *Can. J. Microbiol.* **19:**87–96.
33. **Lamb, D. H., and K. F. Bott.** 1979. Inhibition of *Bacillus subtilis* growth and sporulation by threonine. *J. Bacteriol.* **137:**213–220.
34. **Landick, R., and C. Yanofsky.** 1987. Transcription attenuation, p. 1276–1301. *In* F. C. Neidhardt, J. L. Ingraham, K. B. Low, B. Magasanik, M. Schaechter, and H. E. Umbarger (ed.), *Escherichia coli and Salmonella typhimurium: Cellular and Molecular Biology*, vol. 2. American Society for Microbiology, Washington, D.C.
35. **Lawther, R. P., and G. W. Hatfield.** 1980. Multivalent translational control of transcription termination at attenuator of *ilvGEDA* operon of *Escherichia coli* K-12. *Proc. Natl. Acad. Sci. USA* **77:**1862–1866.
36. **Lawther, R. P., R. C. Wek, J. M. Lopes, R. Periera, B. E. Taillon, and G. W. Hatfield.** 1987. The complete nucleotide sequence of the *ilvGMEDA* operon of *Escherichia coli* K-12. *Nucleic Acids Res.* **15:**2137–2155.
37. **Lessie, T. G., and C. B. Thorne.** 1976. Unusual mutations affecting branched-chain amino acid biosynthesis in *Bacillus licheniformis*, p. 91–100. *In* D. Schlessinger (ed.), *Microbiology—1976*. American Society for Microbiology, Washington, D.C.
38. **Lopes, J. M., and R. P. Lawther.** 1986. Analysis and comparison of the internal promoter, pE, of the *ilvGMEDA* operons of *Escherichia coli* K-12 and *Salmonella typhimurium*. *Nucleic Acids Res.* **14:**2779–2796.
39. **Lopes, J. M., and R. P. Lawther.** 1989. Physical identification of an internal promoter, *ilvAp*, in the distal portion of the *ilvGMEDA* operon. *Gene* **76:**255–269.
40. **Lowe, P. N., J. A. Hodgson, and R. N. Perham.** 1983. Dual role of a single multienzyme complex in the oxidative decarboxylation of pyruvate and branched-chain 2-oxo acids in *Bacillus subtilis*. *Biochem. J.* **215:**133–140.
41. **Luchansky, J. B., and P. A. Pattee.** 1984. Isolation of transposon Tn*551* insertions near chromosomal markers of interest in *Staphylococcus aureus*. *J. Bacteriol.* **159:** 894–899.
42. **Mackey, C. J., and S. A. Zahler.** 1982. Insertion of bacteriophage SPβ into the *citF* gene of *Bacillus subtilis* and specialized transduction of the *ilvBC-leu* genes. *J. Bacteriol.* **151:**1222–1229.
43. **Monticello, D. J., and R. N. Costilow.** 1982. Interconversion of valine and leucine by *Clostridium sporogenes*. *J. Bacteriol.* **152:**946–949.
44. **Monticello, D. J., R. S. Hadioetomo, and R. N. Costilow.** 1984. Isoleucine synthesis by *Clostridium sporogenes* from propionate or α-methylbutyrate. *J. Gen. Microbiol.* **130:**309–318.
45. **Nargang, F. E., C. S. Subrahmanyam, and H. E. Umbarger.** 1980. Nucleotide sequence of *ilvGEDA* operon attenuator region of *Escherichia coli*. *Proc. Natl. Acad. Sci. USA* **77:**1823–1827.
46. **Ohyama, K., I. Kaneko, T. Yamakawa, and T. Watanabe.** 1977. Alteration in two enzymatically active forms of valyl-tRNA synthetase during the sporulation of *Bacillus subtilis*. *J. Biochem.* **81:**1571–1574.
47. **Oultram, J. D., H. Peck, J. K. Brehm, D. E. Thompson, T. J. Swinfield, and N. P. Minton.** 1988. Introduction of genes from leucine biosynthesis from *Clostridium pasteurianum* into *C. acetobutylicum* by cointegrate conjugal transfer. *Mol. Gen. Genet.* **214:**177–179.
48. **Parsot, C.** 1986. Evolution of biosynthetic pathways: a common ancestor for threonine synthetase, threonine dehydratase and D-serine dehydratase. *EMBO J.* **5:**3013–3019.

49. **Pattee, P. A., T. Schutzbank, H. D. Kay, and M. H. Laughlin.** 1974. Genetic analysis of the leucine biosynthetic genes and their relationship to the *ilv* gene cluster. *Ann. N.Y. Acad. Sci.* **236:** 175–186.
50. **Perham, R. N., and P. N. Lowe.** 1988. Isolation and properties of the branched-chain 2-keto acid and pyruvate dehydrogenase multienzyme complex from *Bacillus subtilis*. *Methods Enzymol.* **166:**330–342.
51. **Piggot, P.** 1989. Revised genetic map of *Bacillus subtilis* 168, p. 1–41. *In* I. Smith, R. Slepecky, and P. Setlow (ed.), *Regulation of Prokaryotic Development*. American Society for Microbiology, Washington, D.C.
52. **Rudman, D., and A. Meister.** 1953. Transamination in *Escherichia coli*. *J. Biol. Chem.* **200:**591–604.
53. **Schildkraut, I., and S. Greer.** 1973. Threonine synthetase-catalyzed conversion of phosphohomoserine to α-ketobutyrate in *Bacillus subtilis*. *J. Bacteriol.* **115:**777–785.
54. **Shameshima, J. H., R. C. Wek, and G. W. Hatfield.** 1989. Overlapping transcription and termination of the convergent *ilvA* and *ilvY* genes of *Escherichia coli*. *J. Biol. Chem.* **264:**1224–1231.
55. **Skarstedt, M. T., and S. B. Greer.** 1973. Threonine synthetase of *Bacillus subtilis*: the nature of an associated dehydratase activity. *J. Biol. Chem.* **248:**1032–1044.
56. **Squires, C. H., M. DeFelice, J. Devereux, and J. M. Calvo.** 1983. Molecular structure of *ilvIH* and its evolutionary relationship to *ilvG* in *Escherichia coli* K-12. *Nucleic Acids Res.* **11:**5299–5313.
57. **Steiglitz, B. I., and J. M. Calvo.** 1974. Distribution of the isopropylmalate pathway to leucine among diverse bacteria. *J. Bacteriol.* **118:**935–941.
58. **Taillon, B. E., R. Little, and R. P. Lawther.** 1988. Analysis of the functional domains of biosynthetic threonine deaminase by comparison of the amino acid sequences of three wild-type alleles to the amino acid sequence of biodegradative threonine deaminase. *Gene* **63:**245–252.
59. **Temeyer, K. B., and L. F. Chapman.** 1983. Suppression of an *ilvA* mutation in *Bacillus subtilis*. *Mol. Gen. Genet.* **192:**198–203.
60. **Umbarger, H. E.** 1983. The biosynthesis of isoleucine and valine and its regulation, p. 245–266. *In* K. M. Herrmann and R. L. Sommerville (ed.), *Amino Acid Biosynthesis and Genetic Regulation*. Addison-Wesley, London.
61. **Umbarger, H. E., and B. Brown.** 1957. Threonine deamination in *Escherichia coli*. II. Evidence for two L-threonine deaminases. *J. Bacteriol.* **73:**105–112.
62. **Vandeyar, M. A.** 1987. Ph.D. thesis. Cornell University, Ithaca, N.Y.
63. **Ward, J. B., Jr., and S. A. Zahler.** 1973. Genetic studies of leucine biosynthesis in *Bacillus subtilis*. *J. Bacteriol.* **116:**719–726.
64. **Ward, J. B., Jr., and S. A. Zahler.** 1973. Regulation of leucine biosynthesis in *Bacillus subtilis*. *J. Bacteriol.* **116:**727–735.
65. **Wek, R. C., and G. W. Hatfield.** 1986. Nucleotide sequence and *in vivo* expression of the *ilvY* and *ilvC* genes in *Escherichia coli* K-12. Transcription from divergent overlapping promoters. *J. Biol. Chem.* **261:**2441–2450.
66. **Wek, R. C., and G. W. Hatfield.** 1986. Examination of the promoter P_e in the *ilvGMEDA* operon of *E. coli* K-12. *Nucleic Acids Res.* **14:**2763–2777.
67. **Wek, R. C., C. A. Hauser, and G. W. Hatfield.** 1985. The nucleotide sequence of the *ilvBN* operon of *Escherichia coli*: sequence homologies of the acetohydroxy acid synthase isozymes. *Nucleic Acids Res.* **13:**3995–4010.
68. **Wessler, S. R., and J. M. Calvo.** 1981. Control of *leu* operon expression in *Escherichia coli* by a transcription attenuation mechanism. *J. Mol. Biol.* **149:**579–597.
69. **Willecke, K., and A. B. Pardee.** 1971. Fatty acid requiring mutant of *Bacillus subtilis* defective in branched chain ketoacid dehydrogenase. *J. Biol. Chem.* **246:**5264–5272.
70. **Yanofsky, C.** 1988. Transcription attenuation. *J. Biol. Chem.* **263:**609–612.
71. **Zahler, S. A. (Cornell University).** 1982. Personal communication.
72. **Zahler, S. A., L. G. Benjamin, B. S. Glatz, P. F. Winter, and B. J. Goldstein.** 1976. Genetic mapping of *alsA*, *alsR*, *thyA*, *kauA*, and *citD* markers in *Bacillus subtilis*, p. 35–43. *In* D. Schlessinger (ed.), *Microbiology—1976*. American Society for Microbiology, Washington, D.C.
73. **Zalkin, H., and D. J. Ebbole.** 1988. Organization and regulation of genes encoding biosynthetic enzymes in *Bacillus subtilis*. *J. Biol. Chem.* **263:**1595–1598.

23. Biosynthesis of Riboflavin, Biotin, Folic Acid, and Cobalamin

JOHN B. PERKINS and JANICE G. PERO

This chapter reviews current knowledge of the biosynthetic pathways for four vitamins (riboflavin, biotin, folic acid, and cobalamin) in *Bacillus* spp. Biosynthesis of these vitamins has been studied in *Bacillus subtilis*, *Bacillus sphaericus*, and *Bacillus megaterium*. However, with the notable exception of riboflavin, information on the genes and enzymes involved in vitamin biosynthesis in these *Bacillus* species is much more limited than for the enteric bacteria. Whenever possible, the molecular biology of these pathways, including gene organization and regulation, will be emphasized. Only a brief summary of biosynthetic reactions and enzymes is presented, because in most cases, the enzymology of these pathways is similar to that found and extensively reviewed for *Escherichia coli* (3, 30, 45, 69).

RIBOFLAVIN

Riboflavin, also called vitamin B_2, is unique among the vitamins because more is known about the riboflavin (*rib*) biosynthetic genes in *B. subtilis* than is known for any other bacterium. The active forms of riboflavin in the cell are flavin mononucleotide and flavin adenine dinucleotide, which function as prosthetic groups of oxidation-reduction enzymes. The conversion of riboflavin into these coenzymes is controlled by genes encoding riboflavin kinase and flavin mononucleotide adenylyltransferase and will not be considered further. Here, we discuss the organization and regulation of the riboflavin biosynthetic genes and present results from our laboratory on how this information can be used to enhance vitamin production by *B. subtilis*.

Riboflavin Biosynthetic Enzymes and Intermediates in the Pathway

In both bacteria and fungi, riboflavin is synthesized in five enzymatic reactions beginning with the purine precursor GTP. As shown in Fig. 1, the steps in riboflavin biosynthesis are as follows. The imidazole ring of GTP is opened by GTP cyclohydrolase II to yield the pyrimidine intermediate 2,5-diamino-6-(ribosylamino)-4(3H)-pyrimidinone 5′-phosphate. This intermediate is converted to 5-amino-6-(ribosylamino)-2,4(1H,3H)-pyrimidinedione in sequential reactions by *rib*-specific deaminase and reductase enzymes. Condensation of this pyrimidinedione with a four-carbon moiety derived from the pentose pathway yields 6,7-dimethyl-8-ribityllumazine, which is then converted to riboflavin. These last two steps are mediated by the β and α subunits of riboflavin synthase, respectively. The only apparent difference between the prokaryotic and eukaryotic pathways is that the order of the second and third steps is reversed (for reviews, see references 3 and 30).

Table 1 lists the properties of the *B. subtilis* riboflavin biosynthetic enzymes that are currently known. These enzymes have characteristics similar to those described for the corresponding enzymes in other organisms, including *E. coli* and yeast cells. Although these developments have been extensively discussed in several recent reviews (3, 6, 30), descriptions of several key enzymes in the pathway deserve to be repeated here.

The first step in riboflavin synthesis is controlled by GTP cyclohydrolase II, which catalyzes the synthesis of 2,5-diamino-6-(ribosylamino)-4(3H)-pyrimidinone 5′-phosphate from GTP. This enzyme has been isolated from *E. coli* (31, 51) and is structurally unrelated to GTP cyclohydrolase I from the folic acid biosynthetic pathway (see Folic Acid below). GTP cyclohydrolase II is encoded by the *ribA* gene of *B. subtilis* and has a predicted molecular weight of 43,800. Recently, Bacher and his coworkers (117a) have sequenced the *E. coli* gene that encodes GTP cyclohydrolase II and found that the enteric form of this enzyme is substantially smaller (23,000 Da) than its *B. subtilis* counterpart. Comparison of the two predicted protein sequences revealed substantial identity (55%) of the entire *E. coli* enzyme to the COOH-terminal end of the *B. subtilis* protein. Interestingly, *ribA* in *B. subtilis* appears to encode a second enzymatic function for riboflavin synthesis. As discussed below, Bacher and his coworkers (118) have recently shown that the 5′ region of *ribA* encodes for 3,4-dihydroxy-2-butanone 4-phosphate synthase, which catalyzes the formation of the four-carbon unit from ribulose 5-phosphate.

Neither deaminase nor reductase enzymes have been isolated from *B. subtilis*. However, the *Bacillus* enzymes that control the last two steps of riboflavin synthesis have been extensively characterized because of their unusual catalytic and structural properties. In *B. subtilis*, two forms of riboflavin synthase can be isolated (5, 8, 9, 11): a "light" enzyme (70 kDa) that represents 80% of total enzyme activity and a "heavy" enzyme (1 MDa) that constitutes the remaining 20% (4). The light enzyme represents the "true" riboflavin synthase activity, which is composed of three identical α subunits (23,500 Da each) that catalyze the conversion of 6,7-dimethyl-8-ribityllumazine to ribo-

John B. Perkins and Janice G. Pero • OmniGene, Inc., 763D Concord Avenue, P.O. Box 9002, Cambridge, Massachusetts 02139-9002.

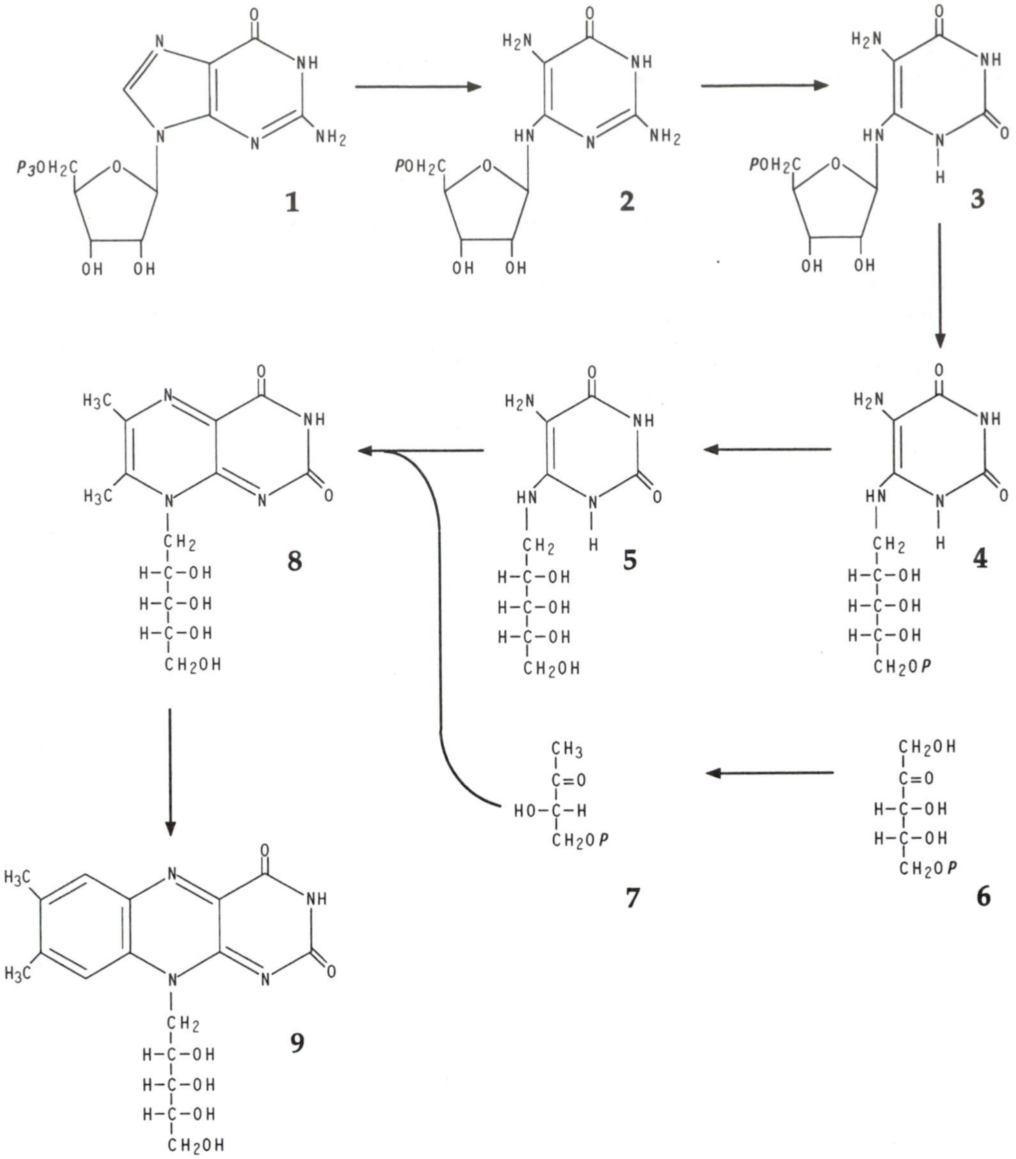

Figure 1. Bacterial riboflavin biosynthetic pathway. The corresponding intermediates shown are those produced by *E. coli* and *B. subtilis*. Structure 1, GTP; structure 2, 2,5-diamino-6-(ribosylamino)-4(3H)-pyrimidinone 5′-phosphate; structure 3, 5-amino-6-(ribosylamino)-2,4(1H,3H)-pyrimidinedione 5′-phosphate; structure 4, 5-amino-6-(ribitylamino)-2,4(1H,3H)-pyrimidinedione 5′-phosphate; structure 5, 5-amino-6-(ribitylamino)-2,4(1H,3H)-pyrimidinedione; structure 6, ribulose 5′-phosphate; structure 7, 3,4-dihydroxy-2-butanone 4-phosphate; structure 8, 6,7-dimethyl-8-ribityllumazine; structure 9, riboflavin. The enzymes that catalyze these reactions in *B. subtilis* are listed in Table 1. Structures are adapted from Bacher (3).

flavin by an unusual dismutation mechanism (4, 55, 102). The heavy enzyme consists of a complex of β and α subunits described below; the β subunit (16,200 Da), which is referred to a lumazine synthase, condenses 5-amino-6-(ribitylamino)-2,4(1H,3H)-pyrimidinedione with a four-carbon moiety, 3,4-dihydroxy-2-butanone 4-phosphate, to form this lumazine intermediate (Fig. 1; 102).

The complete amino acid sequence of the α subunit has been determined (121), and analysis suggests that there are two structurally similar domains per subunit. Because earlier in vitro ligand-binding studies also demonstrated two substrate-binding sites per subunit, it is likely that each domain provides one of these sites. Interestingly, the *ribB* gene that encodes this enzyme specifies a protein of 215 amino acids (96, 110), which is 13 amino acids longer than that determined from protein sequencing. Since the COOH-terminal sequence of the protein shows minor microheterogeneity, this discrepancy appears to be caused by nonspecific proteolytic attack on this region rather than by specific C-terminal processing (121).

Heavy riboflavin synthase (also referred to as lumazine synthase-riboflavin synthase complex) is a bifunctional enzyme complex with an unusual stoichiometry of 3 α subunits (riboflavin synthase) and 60 β

Table 1. Enzymes, genes, and regulatory elements of riboflavin synthesis in *B. subtilis*

Element (map position)	Gene[a]	Enzyme or function	Mol wt of: Gene product (no. of amino acids)	Mol wt of: Purified enzyme (no. of amino acids/ structure)	*E. coli* gene equivalent (map position)
Biosynthetic operon *ribOGBAHT* (209°)	*ribO*	Regulatory site			
	ribG	Deaminase	39,700 (361)	—[b]	*ribG* (9.5 min)[c]
	ribB	Riboflavin synthase (α subunit)	23,600 (215)	22,000[d]/23,500[e] (202/trimer)	*ribC* (40 min)
	ribA	GTP cyclohydrolase II and 3,4-dihydroxy-2-butanone 4-phosphate synthase	43,800 (398)		*ribA* (28 min) *ribB* (66 min)
	ribH	Lumazine synthase (β subunit)	16,900 (154)	16,200[d]/14,700[e] (154/icosohedron)	*ribH* (9.5 min)[c]
	ribT	Reductase	13,600 (124)		
Regulatory locus (145°)	*ribC*	Regulatory protein			

[a] Gene designations are based on Mironov et al. (96).
[b] —Enzyme may be a dimer, because the molecular weight of deaminase isolated from *E. coli* is 80,000 (32).
[c] Map position based on homology to the nucleotide sequence upstream from *E. coli nusB* (111, 129a).
[d] Calculated from the amino acid sequence (85, 121).
[e] Determined by sodium dodecyl sulfate-polyacrylamide electrophoresis (4).

subunits (lumazine synthase). In the absence of substrate, this enzyme complex is stable in only a narrow range of pH and dissociates into trimers of α subunits and aggregates of β subunits at pH levels above neutrality (4, 11). On the basis of analog binding studies and in vitro reconstruction experiments with purified subunits (10, 102), it was concluded that the β subunits are sufficient to catalyze formation of the lumazine intermediate depicted in Fig. 1. The complete amino acid sequence of this enzyme has also been determined (85) and reveals a molecular weight of 16,200, which is in good agreement with that predicted from nucleotide sequence data (*ribH*; 16,900 Da).

The structure of heavy riboflavin synthase shows several unusual features. Electron micrographic analysis of the purified enzyme revealed spherical particles with a diameter of approximately 15 nm (4, 8, 11). Fine-structure analysis of its quaternary form by X-ray diffraction of purified crystals and other analytical techniques revealed that the β subunits are arranged in a capsid structure with an icosohedral 532 symmetry (8, 78), a design that is intriguingly similar to that of capsid heads of bacterial viruses. More interestingly, the interior of this capsid contains a central pocket within which the α-trimer complex is sequestered (8, 11). This trimer is very similar to the subunits present in the light enzyme and appears to be enzymatically active.

The origin of the four-carbon unit, 3,4-dihydroxy-2-butanone 4-phosphate, which is used to generate the lumazine intermediate, has been the subject of lengthy investigations (3). In yeast cells (*Candida guilliermondii*), this compound is derived from ribulose 5-phosphate, a pentose pathway intermediate (12, 131), by an unusual rearrangement of the skeletal carbon atoms (7, 50, 81, 82, 104) catalyzed by 3,4-dihydroxy-2-butanone 4-phosphate synthase. Although the formation of this four-carbon unit has not been directly demonstrated in bacteria, ^{13}C-labeling studies with *B. subtilis* strongly implicate its formation from the pentose pool by a similar intramolecular reaction (79). With the recent discovery that the 3,4-dihydroxy-2-butanone 4-phosphate synthase coding region is within the *B. subtilis ribA* gene (see below), a better understanding of the biosynthesis of this compound is now possible.

In addition to these enzymes, it is also necessary to postulate an activity that dephosphorylates the pyrimidinedione intermediate prior to condensation with the four-carbon unit (Fig. 1), because this phosphorylated compound is not a substrate for lumazine synthase (58, 102). It remains unclear, however, whether this reaction is catalyzed by a *rib*-specific enzyme or a general phosphatase (3). Finally, it is also important to note that additional steps and intermediates in the *Bacillus* pathway have been postulated, in part on the basis of the accumulation of fluorescing compounds isolated from blocked mutants (25, 26, 99, 112). However, the role, if any, of these compounds in riboflavin synthesis has not been well documented, and some of the compounds may not be intermediates in the biosynthetic pathway.

Genetics of the Riboflavin Pathway

Descriptions of *B. subtilis rib* genes, their encoded enzymes, and other relevant parameters are given in Table 1. Stepanov, Bresler, and their colleagues showed that these genes mapped to two unlinked chromosomal loci (18, 22, 23, 52). All known biosynthetic genes are found at the single *rib* locus (described below) located adjacent to the *lys* marker at map position 209° (36, 60). A genetic map of the known biosynthetic mutations has existed for several years, although the postulated order of the genes has undergone several changes. The second locus, *ribC*, is thought to encode a riboflavin-activated repressor protein that negatively regulates expression of the biosynthetic genes (19, 20, 25, 34). This genetic arrangement is in stark contrast to that in *E. coli* and other gram-negative bacteria, in which the riboflavin biosynthetic genes are scattered around the chromo-

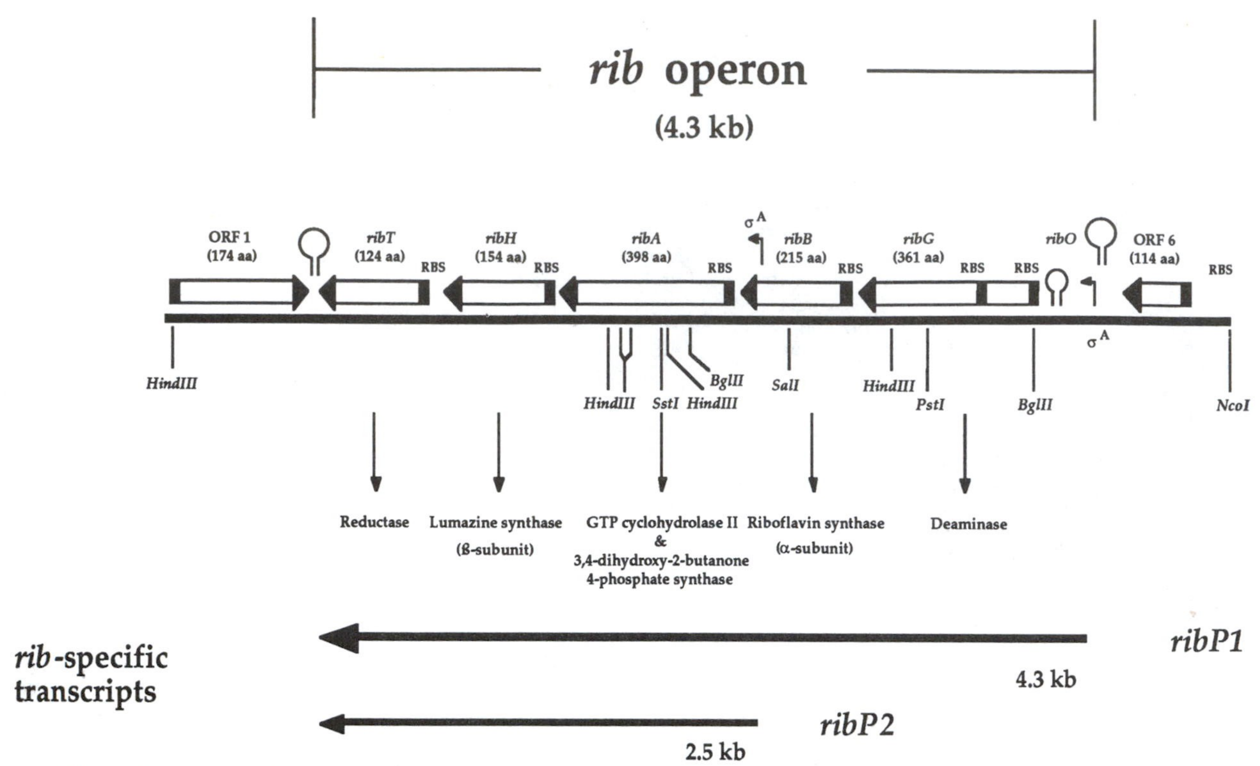

Figure 2. Structure of the riboflavin operon determined from the nucleotide sequence. Locations of the structural genes (*ribG, ribB, ribA, ribH,* and *ribT*), σ^A promoter regions (*rib*P_1 and *rib*P_2), and *ribO* regulatory site are shown in the upper diagram. Assignment of the *rib* genes to the indicated biosynthetic enzymes is described in the text. Tentatively identified genes ORF1 and ORF6 appear not to be involved in riboflavin synthesis. The bottom diagram indicates the *rib*-specific polycistronic RNA transcripts detected by Northern hybridization. Symbols: ▮, *Bacillus* ribosome-binding sites (RBS); ↰, start sites of transcription for σ^A-recognized promoters *rib*P_1 and *rib*P_2; ⚲, putative rho-independent transcription termination sites (the ⚲ located within *ribO* is postulated to be part of a transcriptional termination-antitermination regulatory mechanism). Not all restriction enzyme sites are shown. aa, amino acids.

some at four or five loci and no regulatory locus has been identified (13, 14, 123, 130).

A number of analog resistance mutations have been shown to cause deregulated riboflavin synthesis in *B. subtilis*. Selection for mutants resistant to the riboflavin analog roseoflavin (77, 92, 128) yields either *trans*-acting mutations within the *ribC* locus or *cis*-acting lesions in *ribO*, a putative binding site for *ribC* repressor located within the *rib* operon (19, 21, 23, 25). The *ribC* locus was originally located between the genes conferring resistance to streptomycin (*str*) and acriflavin (*acf*) at position 220° (34). However, its location has been recently remapped to position 145° on the other side of the chromosome (76). In addition to *ribO*, which is located at one end of the *rib* genetic map, a second operator site, *ribO*$_e$, has been postulated to control expression of the early biosynthetic genes, in particular the gene for GTP cyclohydrolase II (24, 112). Mutants with mutations in the putative *ribO*$_e$ region, however, have not been reported.

Finally, in addition to regulating riboflavin production, the *ribC* gene product has been proposed to control other flavinogenic pathways involved in the synthesis of flavin mononucleotide and flavin adenine dinucleotide (26, 27, 35). It is hypothesized that the *ribC* repressor can interact not only with riboflavin (and several of its biosynthetic intermediates) but also with flavin mononucleotide and flavin adenine dinucleotide.

Structure and Regulation of the *rib* Operon

Two laboratories have independently reported the nucleotide sequence of the riboflavin biosynthetic operon and have undertaken studies to elucidate how gene expression is regulated (95, 96, 110). These investigations have led to similar discoveries of coding regions, promoter and regulatory sites, and sizes and locations of transcription units, which are illustrated in Fig. 2. In general, riboflavin synthesis appears to be controlled by a single operon containing five biosynthetic genes that is flanked at either end by putative rho-independent transcriptional termination signals. Transcription of the structural genes is controlled by two σ^A-recognized promoters, *rib*P_1, which generates the primary transcript encompassing all five genes, and *rib*P_2, which generates a less-abundant transcript of the last three genes of the operon. Both transcripts appear to terminate at or near a strong rho-independent transcription termination sequence at the end of the operon. The *ribC* gene product and riboflavin appear to modulate the level of transcription from both promoters by a transcription termination-antitermination mechanism. These structural and regulatory features of the *rib* operon are described in detail below.

rib structural genes

Originally, the *rib* biosynthetic genes were cloned as an *Eco*RI fragment 10.5 kb long (117). Later physical mapping studies using *cat* insertional mutagenesis (110) or marker rescue transformation of *rib* mutants with subcloned fragments (37, 99) indicated that the genes were restricted to a region less than 5 kb long. The DNA sequence of this region revealed seven major open reading frames (ORFs), each with *B. subtilis*-specific ribosome-binding sites. Judging from the locations of transcriptional promoters and termination sites, five of these ORFs appear to be arranged in a single operon. These five ORFs were subsequently assigned to specific riboflavin biosynthetic enzymes by using various criteria, including amino acid sequence similarity to known *rib* biosynthetic enzymes or other enzymes with similar functions, alignment of the sequence to the published genetic map, and coupled in vitro transcription-translation assays to confirm the sizes of predicted gene products. For example, the lumazine synthase (β-subunit) and riboflavin synthase (α-subunit) genes were identified on the basis of almost perfect identity of their encoded products with published amino acid sequences (85, 121), with only a few minor differences noted (110). Similarly, the protein product of the first structural gene in the operon was tentatively identified as a *rib*-specific deaminase on the basis of significant similarity to the *E. coli* T_2 deoxycytidylate deaminase protein (39% identity in an 88-amino-acid overlap; 87, 110). Interestingly, no similarity to the *B. subtilis* chromosomal *cdd* gene product, cytidine/2′-deoxycytidine deaminase, was detected (126). The GTP cyclohydrolase II and reductase genes were originally identified by aligning the ORFs with the physical map of *rib* mutations determined by Morozov et al. (99). In addition, the putative proteins encoded by all the genes, except lumazine synthase, were identified by using coupled in vitro transcription-translation reactions with the cloned DNA as template (110). Finally, the gene products of ORFs 1 and 6, which lie outside the operon, could not be identified by using GenBank similarity searches. However, insertional mutagenesis studies have shown that expression of either gene is not necessary for riboflavin synthesis (111).

Several features of this operon also deserve special comment, because this information may be particularly useful in future studies of riboflavin biosynthesis. Unlike the *trp*, *pyr*, and *pur* operons of *B. subtilis* (41, 59, 116), the translational signals of the *rib* structural genes do not overlap. This fact suggests that translational coupling does not play an important role in the expression of this operon (134). However, other unique gene structures may play a role in controlling synthesis of the biosynthetic enzymes. As mentioned previously, Bacher and his coworkers (73, 117a) have determined that the *ribA* gene encodes both GTP cyclohydrolase II and the enzyme responsible for synthesis of the four-carbon unit. This conclusion is based on significant similarity of an N-terminal amino acid sequence of 3,4-dihydroxy-2-butanone 4-phosphate synthase isolated from *E. coli* to the 5′ end of *ribA* and similarity of the deduced protein sequence of *E. coli* GTP cyclohydrolase II to the 3′ end. A possible explanation of why the coding regions of these enzymes are condensed into a single gene is that synthesis of both enzymes must be coordinated to ensure efficient production of riboflavin. It is not known, however, whether both enzyme activities are contained on a single large polypeptide or whether it is necessary to process this polypeptide into two individual enzymes. In addition, there is some evidence that the deaminase coding region may contain a second, in-frame overlapping gene. Chen et al. (33) previously showed that the α and β subunits of *B. subtilis* aspartokinase II are translated from two different sets of in-frame ribosome-binding sites and Met start signals. Inspection of the deaminase gene reveals a similar structural motif, with the second set of translational signals located 114 amino acids from the beginning of the gene. The significance, if any, of this second initiation site has not yet been established.

rib promoters

Although examination of the nucleotide sequence reveals several sequences that potentially could be recognized by the various forms of *B. subtilis* RNA polymerase, the evidence supports the view that transcription of the *rib* genes is primarily controlled from two promoters recognized by the primary (σ^A-containing) form of the holoenzyme. These are *rib*P_1 (TTGC GT-N_{17}-TATAAT), which is located upstream from the first structural gene (deaminase) of the operon, and *rib*P_2 (TTGAAC-N_{17}-TACTAT), which is an internal promoter located upstream from the third structural gene (*ribA*). Evidence also indicates that transcription from both promoters is controlled by the *ribC* repressor and the availability of riboflavin.

Primer extension (110) and S1 mapping experiments (95) showed the start site of *rib*P_1 transcription as an A residue located 294 bp upstream from the first start codon. S1 mapping has also been used to show that this transcript terminates within the T-rich region following the stem-loop structure ($\Delta G = -16.5$ kcal/mol [1 cal = 4.184 J]) located at the end of the operon (95). Northern (RNA) blots have also detected a *rib*-specific polycistronic RNA message of the expected size (4.3 kb) that originates from this promoter (95, 110). Transcription from this promoter appears to be regulated by the *ribC* repressor and the availability of riboflavin, because the amount of transcript is appreciably reduced when cells are grown in the presence of exogenous riboflavin (95) but increases in cells harboring a roseoflavin resistance mutation (RoF^r) within the *ribC* gene (111). The effects of these factors on *rib*P_1 activity have been quantified by using a *lacZ* transcriptional fusion to *rib*P_1, which was integrated in single copy in the chromosome (110). In wild-type (*rib*$^+$) strains, β-galactosidase activity was increased 2-3-fold in the absence of riboflavin, whereas in RoF^r strains, 15-fold-higher constitutive synthesis was observed. Compared with the activities of other regulated *B. subtilis* biosynthetic promoters (e.g., *trp*, *pur*, *glt*, and *pyr*), *rib*P_1 activity appears to be weak even under deregulated conditions. Similar qualitative results were obtained with transcriptional fusions to a *cat* reporter gene (38, 100).

Although the *rib*P_2 promoter also appears to be regulated, the exact role of this promoter in riboflavin synthesis has not been resolved. This promoter activity was first confirmed by showing that restriction fragments containing *rib*P_2 could activate a promoterless *tet* gene in *E. coli* (38, 100). In addition, disruption of the *rib* operon by integration of *E. coli* plasmids with subcloned *rib*P_2-containing fragments had no effect on riboflavin prototrophy (111), as expected of an internal operon promoter. Single-copy *rib*P_2-*lacZ* transcriptional fusions using a 700-bp *Sal*I-*Bgl*II region showed weak but measurable β-galactosidase activity in wild-type (*rib*$^+$) cells that was approximately 50-fold less than that found with the *rib*P_1-*lacZ* fusion. Although it was not possible to determine whether riboflavin repressed activity, β-galactosidase activity could be increased approximately six- to eightfold when the *rib*P_2-*lacZ* fusion was introduced into cells containing a RoF^r mutation (111). Finally, Northern blots have detected a polycistronic RNA transcript of the appropriate size (2.5 to 2.8 kb) (111). Paradoxically, this transcript appears to be more abundant than what one would predict from a weak promoter. Although this discrepancy may simply relate to the stability of the mRNA, further investigation of this promoter may provide more insight into its role in riboflavin synthesis.

Finally, a third promoter, *rib*P_3, has been postulated by several investigators (38, 100) as controlling expression of *ribH* encoding lumazine synthase. No familiar promoter sequences were detected at the beginning of *ribH*; however a σ^A-like promoter sequence (TTGAAT-N_{18}-TAAAAA) was located within the 110-bp intercistronic gap between *ribH* and *ribT*. Although several transcripts close to the expected size can be detected by Northern blots (95, 111), it remains to be seen whether this promoter can be confirmed by other means and has any significant role in synthesis of the *rib*-specific reductase enzyme.

Model for regulation of the *rib* operon

In *B. subtilis*, a recurring pattern of gene organization and regulation for biosynthetic pathways has been observed by several investigators. The nucleotide sequences of the biosynthetic operon for tryptophan (59), the de novo purine nucleotide pathway (41–43), and the *sacPA* operon for sucrose utilization (40) each contain clustered genes transcribed as a polycistronic message and regulated at least in part by a transcription termination-antitermination mechanism involving an activated repressor protein that in some instances can be encoded by a gene unlinked to the biosynthetic operon (124, 134). It has been recently shown that these effector proteins are RNA-binding proteins that control termination by binding to specific sequences at or near a rho-independent terminator located before the first structural gene (40, 54). A fuller discussion of this regulatory mechanism is presented in chapters 11 and 20 of this book.

From the standpoint of common structural features, the *rib* biosynthetic operon and regulatory gene are strikingly similar to these examples and thus may be regulated by a similar mechanism. Examination of the nucleotide sequence immediately following the *rib*P_1 transcription initiation site reveals many of the features involved in this type of termination-antitermination regulation. These include an extended 5′ untranslated leader sequence and a strong putative rho-independent transcriptional terminator (ΔG = −26 kcal/mol) just upstream from the start of the first structural gene. In addition, sequencing of this region of the *rib* operon from *ribO* "operator" mutants revealed alterations of individual nucleotides within this leader region (97). Interestingly, however, these operator mutations were not located at or near the terminator but instead were localized to a 45-bp region approximately 40 bp downstream from the *rib*P_1 transcriptional start site. It will be interesting to learn the mechanism by which these mutations alter expression of the *rib* genes and if, in fact, regulation of the *rib* operon will be similar to that found for *trp* and other biosynthetic operons in *B. subtilis*.

In addition to the structures described above, Mironov et al. (97) and Perkins et al. (110) have noted the presence of a small (50-amino-acid) coding region within the 5′ leader that encompasses both the terminator stem-loop and the beginning of the *ribG* gene. Whether this element plays a role in regulation remains to be discovered.

Although regulation of the *rib*P_2 promoter also appears to involve the *ribC* repressor, the exact mechanism is unknown. Several nonoverlapping sequences with dyad symmetry can be detected around the promoter, but none of these structures resembles a rho-independent terminator. Therefore, a different *ribC*-dependent mechanism may regulate transcription from *rib*P_2. Since the *ribO*$_e$ operator site is proposed to control synthesis of GTP cyclohydrolase II and the other early enzymes (described above) and is physically dissociated from *ribO* (101), it will be interesting to see whether this site is part of the *rib*P_2 regulatory apparatus.

Commercial Production of Riboflavin

Stepanov et al. (129) demonstrated the feasibility of enhancing riboflavin production in *B. subtilis* by increasing the expression of the biosynthetic genes using recombinant DNA techniques. Riboflavin titers of 4.5 g/liter were achieved by ligating the *rib* operon to a low-copy-number *Bacillus* plasmid and then introducing this vector into a deregulated RoF^r *B. subtilis* strain containing additional purine analog resistance mutations. Resistance to purine analogs such as decoyinine (90, 91) and 8-azaguanine (75, 120) or the purine antagonist methionine sulfoxide (89) indirectly increases riboflavin levels by altering various stages of the de novo purine biosynthesis pathway to increase the level of the key riboflavin precursor GTP (47).

With the identification of the regulated *rib*P_1 and *rib*P_2 promoters within the *rib* operon, it has become possible to enhance further riboflavin production. For example, expression of the *rib* biosynthetic genes has been deregulated by replacing or bypassing the native promoters and regulatory sites with stronger, constitutive promoters derived from bacteriophage SPO1 of *B. subtilis* (110). In these constructions, the *rib*P_1

promoter and associated regulatory region were replaced with the SPO1-15 promoter (80), and a second copy of this promoter was inserted within the small intercistronic gap upstream from *ribA*, which encodes the first enzyme in the pathway, GTP cyclohydrolase II (the *rib*P_2 promoter cannot be removed because it is embedded within the *ribB* gene). When these engineered genes were then integrated and amplified in the chromosome by single reciprocal (Campbell) recombination at the *rib* locus, a dramatic increase in the level of riboflavin product was detected (110). In wild-type (*rib*$^+$) strains, these engineered *rib* genes caused the bacteria to overproduce riboflavin. However, a much stronger increase in riboflavin synthesis was observed when the same vector was integrated into RoF^r (deregulated) strains containing additional purine analog-resistant mutations, as described above. Fermentation of these deregulated strains containing the engineered *rib* operons was reported to yield 15 g of riboflavin per liter (110).

These experiments demonstrate the utility of *B. subtilis* for investigations of fundamental questions of gene regulation and for application of this knowledge in the development of new approaches for producing industrially important metabolites, including those that are currently produced by chemical synthesis.

BIOTIN

The molecular genetics and enzymology of the biosynthesis of biotin, also called vitamin B_8 or vitamin H, in bacteria have been studied extensively. In fact, more is known about the biotin (*bio*) genes and their regulation in *E. coli* than is known for any other vitamin (reviewed in references 39, 44, and 45). Most of our knowledge about biotin synthesis in *Bacillus* spp. comes from work on *B. sphaericus* (53, 63, 67).

Pathway for Biotin Biosynthesis

Early work in the 1960s by Ogata and his colleagues, who were studying biotin vitamers in *Bacillus* spp., was instrumental in establishing the enzymatic pathway for biotin biosynthesis, which appears to be similar in *E. coli* and *Bacillus* spp. Ogata's group (62) identified the biotin-vitamer produced from pimelic acid by a *Bacillus* sp. as 7-keto-8-aminopelargonic acid (7-KAP) and established that it was an intermediate in the pathway for dethiobiotin (DTB) synthesis. Rolfe and Eisenberg (119) identified similar compounds in *E. coli* and proposed the pathway for biotin synthesis depicted in Fig. 3. The steps in biosynthesis from pimelyl coenzyme A (pimelyl-CoA) and L-alanine to 7-KAP, 7,8-diaminopelargonic acid (DAPA), DTB, and finally biotin have been studied in considerable detail and are reviewed in reference 45.

All of the enzyme activities involved in the synthesis of DTB from pimelic acid, including pimelyl-CoA synthetase, 7-KAP synthetase, DAPA aminotransferase, and DTB synthetase, have been identified and characterized in *B. sphaericus* (63). The primary structure and properties of the latter three enzymes are similar to those described for the corresponding enzymes in *E. coli*. As in *E. coli*, 7-KAP synthetase requires pyridoxal 5′-phosphate as a cofactor in catalyzing the synthesis of 7-KAP from pimelyl-CoA and L-alanine (66, 68). 7-KAP synthetase is encoded by the *bioF* gene, which has been cloned and sequenced from *B. sphaericus* (53). This gene encodes a 42,000-Da protein that is homologous (65% similarity allowing conservative replacements) to the corresponding polypeptide from *E. coli*. The *B. sphaericus* enzyme has been purified to homogeneity from an *E. coli* overpro-

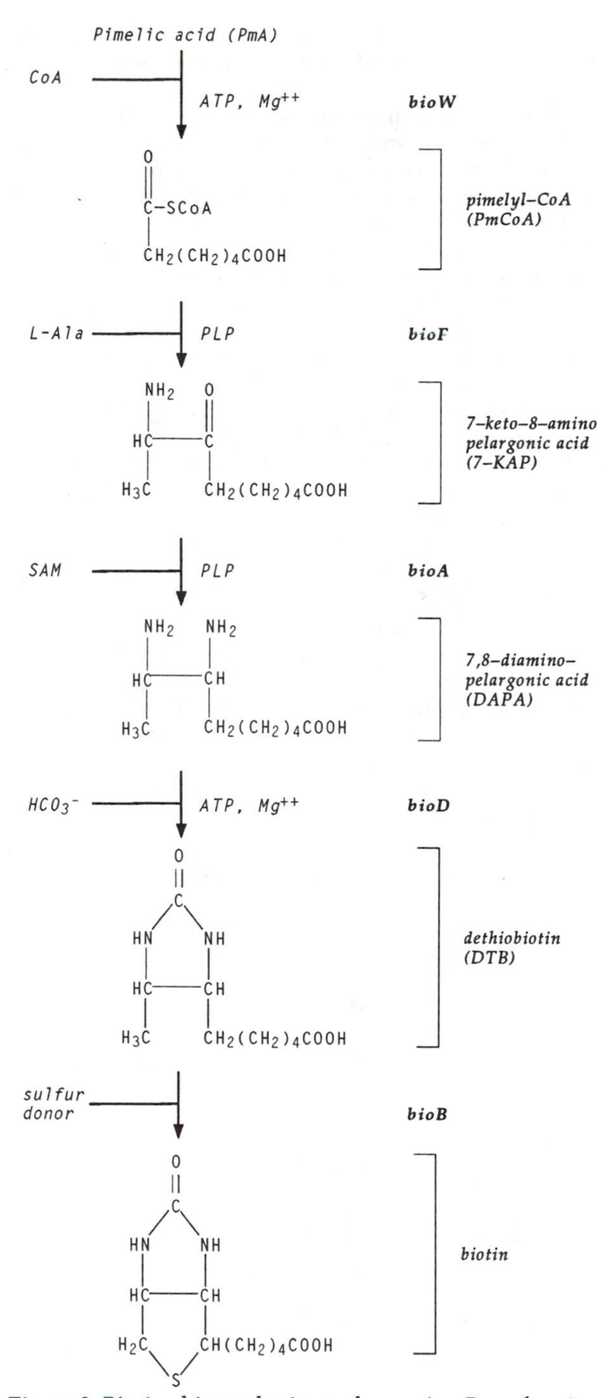

Figure 3. Biotin biosynthesis pathway in *B. sphaericus*. Adapted from Gloeckler et al. (53). SCoA, coenzyme A; Ala, alanine; PLP, pyridoxal 5′-phosphate; SAM, *S*-adenosylmethionine.

ducing strain, and the native enzyme has a molecular weight consistent with that of a monomeric structure (114).

DAPA aminotransferase is dependent on *S*-adenosyl-L-methionine and pyridoxal 5′-phosphate as well as 7-KAP for activity and has been shown to be the target of the biotin antagonist 5-(2-thienyl)-*n*-valeric acid (63). DAPA aminotransferase is the product of the *bioA* gene, which has also been cloned from *B. sphaericus* and shown to encode a polypeptide of 47,000 Da that shares 65% similarity with DAPA aminotransferase from *E. coli* (53).

DTB synthetase has been purified about 500-fold from *B. sphaericus* and, as with the *E. coli* enzyme, shown to be specific for the ureido ring synthesizing reaction and to require DAPA, HCO_3^- (CO_2), ATP, and Mg^{2+} for activity (63). DTB synthetase is the product of the *bioD* gene, which in *B. sphaericus* encodes a protein of 27,000 Da with 52% similarity to the equivalent *E. coli* enzyme (53).

As with *E. coli*, steps in the synthesis of biotin from DTB have not been resolved in *Bacillus* spp. (105). The enzyme catalyzing this reaction is called biotin synthetase, and its activity has been studied in intact cells (63); however, an in vitro enzymatic activity specific for biotin synthesis from DTB and the unknown sulfur donor has never been demonstrated. Biotin synthetase is encoded by the *bioB* gene, which specifies a protein of 37,000 Da in *B. sphaericus*, which again is similar to its *E. coli* counterpart (55%; 105). Also, biotin synthetase has been shown to be the target of actithiazic acid, a potent biotin antagonist (63).

As reviewed above, the enzymatic steps in the synthesis of biotin from pimelyl-CoA and alanine in *Bacillus* spp. appear to be similar to those described for *E. coli* (45). However, it is not clear that the same situation is true for the synthesis of pimelyl-CoA. Izumi et al. (65) first showed an activity in cell extracts of *B. sphaericus* and *B. megaterium* that could synthesize pimelyl-CoA from pimelic acid and CoA. In addition to pimelic acid and CoA, ATP and Mg^{2+} were required for this reaction. Later, this enzyme was purified 34-fold from *B. megaterium* and characterized as a new enzyme belonging to the family of acid-thiol ligases (EC 6.2.1) (64). Analysis of the cloned and sequenced *bio* genes from *B. sphaericus* led Gloeckler et al. (53) to suggest that pimelyl-CoA synthetase is encoded by *bioW*, a suggestion that was proven true by N-terminal sequencing of pimelyl-CoA synthetase purified to homogeneity from an *E. coli* strain engineered to overproduce the *bioW* gene product (115). Pimelyl-CoA synthetase is a dimer composed of two identical subunits of 28,000 Da each (115). These subunits have no obvious similarity to the products of the *bioC* (108) and *bioH* (107) genes, the two *E. coli* genes involved in steps preceding pimelyl-CoA synthesis. *bioX* is another gene of *B. sphaericus* thought to be involved in pimelate biosynthesis that again has no significant similarity to the *E. coli bioC* and *bioH* genes. It has a consensus phosphopantetheine attachment site sequence, suggesting that one function of the BioX protein requires the binding of an acyl group (53). These differences suggest that the products of the *bioX* and *bioW* genes of *B. sphaericus* probably do not

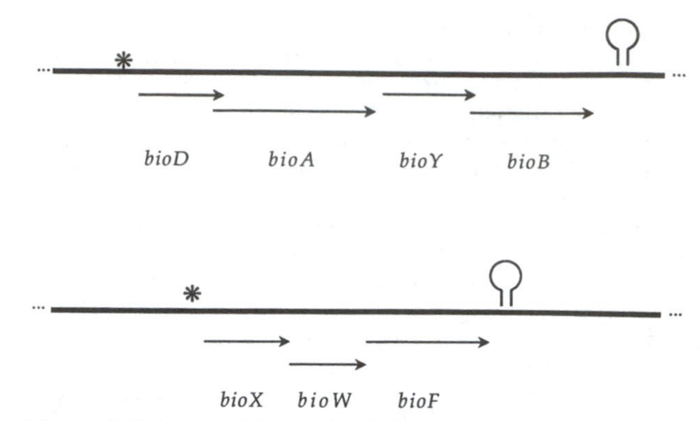

Figure 4. Schematic diagram of cloned and sequenced parts of the biotin operons of *B. sphaericus* (53). Putative transcription termination and operator sites are indicated by the symbols ⚲ and *, respectively (53, 127).

have the same functions in biotin synthesis as the products of *bioC* and *bioH* in *E. coli* (53).

Structure and Organization of Biotin Genes

In *E. coli*, the extensively studied biotin biosynthetic genes are located at three sites in the chromosome. Most of the *bio* genes are located in two divergently transcribed operons (reviewed in reference 44; DNA sequence is given in reference 108). The *bioA* gene is located in one operon, and the *bioBFCD* genes are located in the other operon. The *bioH* gene, involved in steps preceding pimelyl-CoA synthesis, is unlinked to the other *bio* genes (107, 119). Finally, the *birA* (or *bioR*) gene, whose product functions both as the enzyme (biotin holoenzyme synthetase) responsible for catalyzing the attachment of biotin to the lysine residue of target apoenzymes and as a repressor of the *bio* operon, is not linked to the other *bio* genes (reviewed in reference 39; nucleotide sequence is given in reference 61).

The organization of the *bio* genes in *B. sphaericus* is clearly different. By complementation of various *bio* mutants of *E. coli*, Gloeckler et al. (53) have isolated and characterized two DNA fragments from *B. sphaericus* that encode *bio* genes. One fragment contains an operon encoding the *bioD, bioA, bioY*, and *bioB* genes, and the other fragment contains an operon encoding the *bioX, bioW*, and *bioF* genes (Fig. 4). Unlike *bioF* in *E. coli*, *bioF* in *B. sphaericus* is in a different operon from the *bioB* and *bioD*, and the *bioA* gene is located in the same operon as *bioD* and *bioB* (Fig. 4). The *bioY, bioX*, and *bioW* genes of *B. sphaericus* have no known homologs in *E. coli*. *bioY* encodes a 21,000-Da protein of unknown function that may be anchored in the membrane (53). As discussed above, *bioW* encodes pimelyl-CoA synthetase (115), and *bioX* appears to be involved in steps leading to the synthesis of pimelyl-CoA, but its primary structure is completely unrelated to those of *E. coli bioC* and *bioH*.

The DNA fragments containing the *bio* operons from *B. sphaericus* do not overlap, and their locations on the *B. sphaericus* chromosome are not known (53). However, several *bio* mutants have been isolated and characterized in *B. subtilis* and shown to map to a single locus. Pai (109) showed that a collection of

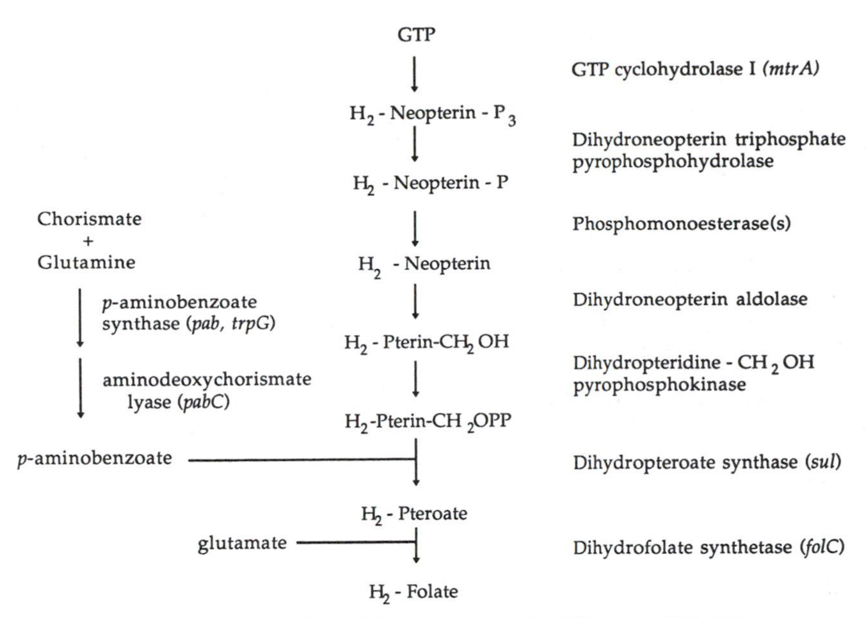

Figure 5. Folic acid biosynthesis pathway. Adapted from Brown and Williamson (30). Where known, the *B. subtilis* genes encoding the enzymes are indicated.

biotin auxotrophs fell into three groups based on nutritional requirements. One group, which grew on biotin only and excreted DTB, was clearly mutated in the *bioB* gene, whereas a second group, which was able to grow on biotin, DTB, or DAPA but not on 7-KAP and which excreted 7-KAP, was mutated in *bioA*. The third group appeared to be mutated in *bioF* (109), and more recent experiments have confirmed this result (53). The *bio* mutants of *B. subtilis* were mapped by transduction and linked to *aroG* in the order *bio–aroG–argA–leu-1*. All of the *bio* mutants mapped to the same location and were weakly linked (5%) to *aroG* by transformation (109; map position 268°).

Regulation of Biotin Biosynthesis

As in other bacteria, the synthesis of biotin in *Bacillus* spp. is tightly regulated. Biotin concentrations as low as 0.2 μg/ml are sufficient to repress completely the synthesis of biotin (67). Ogata and coworkers showed that this inhibition results from the repression of enzymes in the biotin biosynthetic pathway (64, 68). They found that biotin concentrations of 0.1 μg/ml prevented the synthesis of 7-KAP synthetase in *B. sphaericus* (68) and concentrations of 0.25 μg/ml completely repressed the synthesis of DTB synthetase in *B. megaterium* (64). Interestingly, they did not see any repression by biotin of pimelyl-CoA synthetase in *B. megaterium* even at concentrations of 1.0 μg/ml (64).

Unlike the situation in *E. coli*, for which extensive studies have shown that binding of an activated *birA* gene product to an operator located between the divergent *bioA* and *bioBFCD* operons regulates transcriptions (reviewed in reference 39), little is known about the mechanisms involved in biotin gene regulation in *Bacillus* spp. Experiments by Speck et al. (127) with *B. sphaericus* strains containing pUB110 plasmids with the regulatory region of the *bioDAYB* operon of *B. sphaericus* fused to a marker gene showed directly that biotin concentrations as low as 50 ng/ml prevent expression of the operon and that mutants in which this repression by biotin is overcome could be isolated. Two classes of mutants were isolated (127). One class represents chromosomal mutations that act in *trans* to allow constitutive expression of both *bio* operons. The locations of these mutations or other analog-resistant mutations (133) isolated for their abilities to produce increased amounts of biotin have not yet been reported. The second class of plasmid-borne *cis*-acting mutations all mapped to a 15-bp region common to the promoter region of both *bio* operons (127). Upstream of the first gene in both the *bioDAYB* and the *bioXWF* operons in *B. sphaericus* is an identical 15-bp sequence "located within sequences exhibiting a center of imperfect symmetry typical of control regions" (53). The isolation of four different constitutive mutations within this 15-bp sequence argues strongly that this 15-bp sequence is the operator site for biotin-mediated negative regulation (127). It will now be interesting to see whether the chromosomal mutations causing deregulation of the *bio* operons in *B. sphaericus* define a repressor gene similar to *birA* in *E. coli*.

FOLIC ACID

Enzymatic steps in the biosynthesis of folic acid in enteric bacteria have been extensively reviewed by Brown and Williamson (29, 30) and are outlined in Fig. 5. Much less work has been done on *Bacillus* species; however, the overall pathway is believed to be similar and will not be discussed further in this review. Rather, the focus will be on recently characterized *B. subtilis* genes known to be involved in folic acid biosynthesis.

H_2-folic acid is synthesized from the pteridine pre-

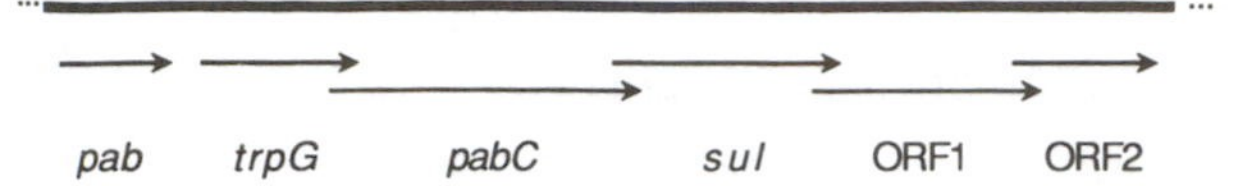

Figure 6. Schematic diagram of cloned and sequenced part of a folic acid operon of *B. subtilis* (125).

cursor 6-CH_2OH-7,8-H_2-pterin pyrophosphate (H_2-pterin-CH_2OPP), *p*-aminobenzoate, and glutamate (Fig. 5). Part of an apparent *B. subtilis* folic acid biosynthetic operon containing the three genes required for the synthesis of *p*-aminobenzoic acid and for the enzyme that catalyzes its condensation with H_2-pterin-CH_2OPP has been cloned and characterized by Slock et al. (125). In addition, a four-gene operon dedicated to folate biosynthesis has been cloned from *Streptococcus pneumoniae* and characterized (84).

Synthesis of *p*-Aminobenzoate

Three enzymes are involved in the synthesis of *p*-aminobenzoate in *B. subtilis*, and their genes (*pab*, *trpG*, and *pabC*) are located at adjacent sites on the chromosome (map position 10°; Fig. 6) (71, 125). The arrangement of these genes within a single operon in *B. subtilis* is much different than the situation in *E. coli*, in which the corresponding genes are located at three unlinked sites on the chromosome: *pabA* (74 min), *pabB* (40 min), and *pabC* (25 min) (56). The *pab* gene, which encodes a protein of 470 amino acids, shares a high degree of sequence similarity with the large subunit of *p*-aminobenzoate synthetase (PabB) from *E. coli* (125). The *trpG* gene, originally designated *trpX* and mapped between *pab* and *sul* (71), encodes a glutamine-binding protein of 194 residues with strong sequence similarity to the small subunit of *p*-aminobenzoate synthase (PabA) from *E. coli* and the *trpG* gene products from a number of bacteria (125). Unlike the situation in *E. coli*, *trpG* in *B. subtilis* and several other bacterial species encodes an amphibolic glutamine amindotransferase that functions as both a subunit of *p*-aminobenzoate synthase in the synthesis of *p*-aminobenzoate and a subunit of anthranilate synthase in the synthesis of anthranilate, an intermediate in tryptophan biosynthesis (71, 72; see chapter 19 in this book for a review of tryptophan biosynthesis and a discussion of *trpG* regulation). The *pabC* gene encodes a protein of 293 amino acids that is essential for *p*-aminobenzoate synthesis. This protein appears to correspond to aminodeoxychorismate lysase, an enzyme recently purified in *E. coli* (56). It is hypothesized that in *E. coli*, the two subunits of *p*-aminobenzoate synthase, PabA and PabB, bind chorismate and glutamine, respectively, to form 4-amino-4-deoxychorismate, which is then converted by aminodeoxychorismate lyase to *p*-aminobenzoate (56, 103). The strong sequence similarity among these enzymes indicates that a similar two-step process is likely to be involved in the synthesis of *p*-aminobenzoate in *B. subtilis*.

Synthesis of 7,8-Dihydropteroate

The enzyme dihydropteroate synthase catalyzes the formation of 7,8-dihydropteroate (H_2-pteroate) from *p*-aminobenzoate and H_2-pterin-CH_2OPP. Dihydropteroate synthase is the target of sulfonamides, which inhibit bacterial growth by competing with *p*-aminobenzoate as substrate for this enzyme, thereby blocking the synthesis of H_2-pteroate, the immediate precursor to folic acid. Bacterial sulfanilamide-resistant mutants (*sul*) generally carry a mutation in the dihydropteroate synthase gene. McDonald and Burke (93) originally cloned the *sul* gene of *B. subtilis* on a 4.9-kb DNA fragment, and it was from sequencing this fragment that Slock et al. (125) discovered that the *sul* gene was part of a folic acid biosynthetic operon containing the *p*-aminobenzoic acid biosynthetic genes located upstream from *sul* and a downstream gene (ORF1 or ORF2) that appears to encode H_2-pteridine-CH_2OH pyrophosphokinase (PPPK) (Fig. 6). The *sul* gene encodes a protein of 285 amino acids that has significant similarity to the dihydropteroate synthase of *S. pneumoniae* (125).

Lopez et al. (83, 84) have cloned and characterized a four-gene folic acid biosynthetic operon from *S. pneumoniae*. The first gene in this operon, *sulA*, encodes dihydropteroate synthase, and the last gene (*sulD*) encodes PPPK. The functions of the second and third genes, *sulB* and *sulC*, have not been determined, although *sulB* bears some similarity to *folC* of *E. coli* and may encode dihydrofolate synthase. Unlike the situation in *B. subtilis*, the genes involved in *p*-aminobenzoate synthesis in *S. pneumoniae* are not located in the same operon with other folic acid biosynthetic genes.

Other Genes Required for Folic Acid Biosynthesis

Of the four enzymes (Fig. 5) involved in synthesis of the precursor, H_2-pterin-CH_2OH, only the gene for GTP cyclohydrolase I, the enzyme for the first step in this pathway, has been reported in *B. subtilis*. Yanofsky and coworkers (2) recently reported that the previously unknown product of *mtrA* is GTP cyclohydrolase I. *mtrA* is part of a two-gene operon located at position 204° on the *B. subtilis* chromosome (54). The second gene in this operon, *mtrB*, encodes an 8,000-Da RNA-binding regulatory protein required for attenuation control of the *trp* operon (54). (The *mtrB* protein has also been proposed to regulate *trpG* expression at the level of translation [125]; see chapter 20 of this book.) GTP cyclohydrolase I purified from *E. coli* has a molecular mass of 210,000 and consists of eight identical polypeptide chains with a molecular mass of 24,873 each (73). The *mtrA* product of *B. subtilis* is 22,000 Da, a size consistent with the formation of a similar multimeric enzyme. *B. subtilis* mutants blocked in the other steps of pterine biosynthesis have not been reported.

In both *E. coli* (17) and *Corynebacterium* spp. (122), the protein that carries out the last step in folic acid biosynthesis, the conversion of H_2-pteroate and glutamate to H_2-folate (Fig. 5), has been shown to be a bifunctional enzyme catalyzing not only the reaction described above but also the synthesis of 10-formyl-H_4-pteroyldiglutamate (a folypolyglutamate). The gene for this enzyme (*folC*) has been cloned and sequenced in *E. coli* (16). The cloning and sequencing of a competence gene, *comC*, in *B. subtilis* has revealed an adjacent gene that appears by sequence similarity to be the *B. subtilis folC* gene (98). Resequencing of this region indicates that the *B. subtilis folC* gene encodes a protein similar in size to the *E. coli* folypolyglutamate synthetase-dihydrofolate synthetase with a high degree of sequence similarity over its entire length to the protein from *E. coli* (88). It appears from these results that *folC* (map position 258°) is unlinked to *mtrA* (map position 204°) or to other genes in the folic acid biosynthetic operon (map position 10°).

COBALAMIN (VITAMIN B_{12})

Vitamin B_{12}, a complex organic molecule, is the largest of all the vitamins. Most current knowledge of genetic pathways involved in vitamin B_{12} or cobalamin biosynthesis comes from studies of *Salmonella typhimurium* (reviewed in reference 69). Figure 7 outlines the steps in vitamin B_{12} biosynthesis as they are proposed to occur in most bacteria. One of the two enzymes in bacteria known to require vitamin B_{12} is ethanolamine ammonia-lyase, an adenosylcobalamin-dependent enzyme. *B. megaterium* contains this enzyme, and growth of *B. megaterium* on ethanolamine as the sole nitrogen source requires vitamin B_{12} (132). As a result, most of the work on vitamin B_{12} biosynthesis in *Bacillus* species has been carried out in *B. megaterium*. Recently, however, the genes involved in synthesis of uroporphyrinogen III (UroIII), a key precursor for cobalamin, have been cloned from *B. subtilis* (57, 113).

Genes Encoding the B_{12} Biosynthetic Enzymes

From analysis of blocked nutritional mutants of *S. typhimurium*, Jeter et al. (69, 70) divided the cobalamin pathway in bacteria into three major branches (*cobI*, *cobII*, and *cobIII*; Fig. 7). The *cobI* pathway defines the synthesis of cobinamide from dihydrosirohydrochlorin, which includes the corrinoid ring, cobalt, and the side chain (D-1-amino-2-propanol) to which 5,6-demethylbenzimidazole (DMB) becomes attached. Mutants belonging to this branch can synthesize B_{12} only if cobinamide is provided. The second branch, *cobII*, involves the synthesis of DMB; thus, vitamin production is restored only in the presence of DMB. *cobIII* encompasses the linking of the ribose moiety of nicotinic acid mononucleotide to DMB followed by the condensation of this intermediate to cobinamide to form B_{12}. The *cobIII* mutants cannot make cobalamin even in the presence of both DMB and cobinamide. Mapping studies indicate that in *S. typhimurium* each class of mutations defines a set of genes organized in a different operon, all of which are grouped together next to the *his* locus (70). The order of these operons is *cobI-cobIII-cobII*, and together they are estimated to contain 25 to 30 genes (*cobI*, 20 genes; *cobIII*, 5 genes; *cobII*, 3 genes). Expression of these genes is repressed by B_{12} and aerobic growth conditions and stimulated by cyclic AMP (48, 69). Recently, a new locus, *cobA*, that is unlinked to the other three operons has been identified. The *cobA* protein product is thought to play two key roles in *S. typhimurium* cobalamin biosynthesis: adenosylation of an early corrinoid intermediate in the *cobI* branch and adenosylation of exogenous corrinoids (e.g., B_{12}, cobinamide, or cobyric acid) (49). *cobA* appears to be functionally equivalent to the *E. coli btuR* gene that regulates synthesis of the outer membrane receptor of B_{12} coded for by the *btuB* locus (49, 86).

Although the *B. megaterium* pathway for cobalamin synthesis is likely to utilize enzymes similar to those described for *S. typhimurium*, less is known about the actual steps of B_{12} biosynthesis in *Bacillus* spp. Wolf and Brey (132) characterized a set of 34 *B. megaterium* mutants auxotrophic for B_{12} that could be divided into two major phenotypic groups: *cob* mutants are blocked in cobinamide synthesis after formation of dihydrosirohydrochlorin, and *cbl* mutants require B_{12} for growth and cannot be rescued by DMB. Thus, the *B. megaterium cob* and *cbl* genes appear to be equivalent to the genes within the *S. typhimurium cobI* and *cobIII* loci, respectively. The Cob mutants were further divided into six classes based on the accumulation of different cobalt-labeled corrinoid compounds. The Cbl mutants were similarly divided into three classes (132). In transduction mapping studies, Cob and Cbl mutants displayed strong linkage within each class, and as in *S. typhimurium*, the two loci were sufficiently close together on the chromosome to show weak linkage in two-factor crosses (132).

At least 11 of the estimated 20 to 30 genes of the *B. megaterium* B_{12} pathway have been cloned by complementation of various *cob* and *cbl* mutants using gene banks prepared with *B. subtilis* plasmid vectors (28). Since many of the recombinant plasmids were able to complement more than one mutation, it was possible to resolve these mutations into four complementation groups. For example, in one complementation group, at least six *cob* genes were detected on a 2.7-kb restriction fragment. Because many of these *cob*- and *cbl*-containing fragments did not overlap, it was not possible to determine whether these genes were organized into one or more operons. No further characterization of these genes has been reported. However, the *B. megaterium* gene that encodes the methyltransferase which catalyzes the formation of dihydrosirohydrochlorin from UroIII has been recently cloned and sequenced (118a).

Genes Encoding the Biosynthetic Enzymes for Key B_{12} Precursors

In addition to the similar primary biosynthetic steps, both *Salmonella* and *Bacillus* spp. appear to use similar pathways to synthesize key B_{12} precursors. The primary precursor of cobinamide is dihydrosirohydrochlorin (Fig. 7), which is directly formed from UroIII. UroIII represents a key metabolic branch point for the synthesis of other metal-containing tetrapyrrole compounds, such as heme and siroheme. In *B. subtilis*, *E. coli*, and *S. typhimurium*, UroIII is derived from 5-aminolevulinic acid, an intermediate that in turn is synthesized by the C_5 pathway. In this pathway, 5-aminolevulinic acid is generated from the carbon skelton of glutamate, involving at least three enzymatic reactions (1, 46, 106; see also chapter 14 of this volume). The *B. subtilis* genes responsible for UroIII synthesis are organized in a single operon

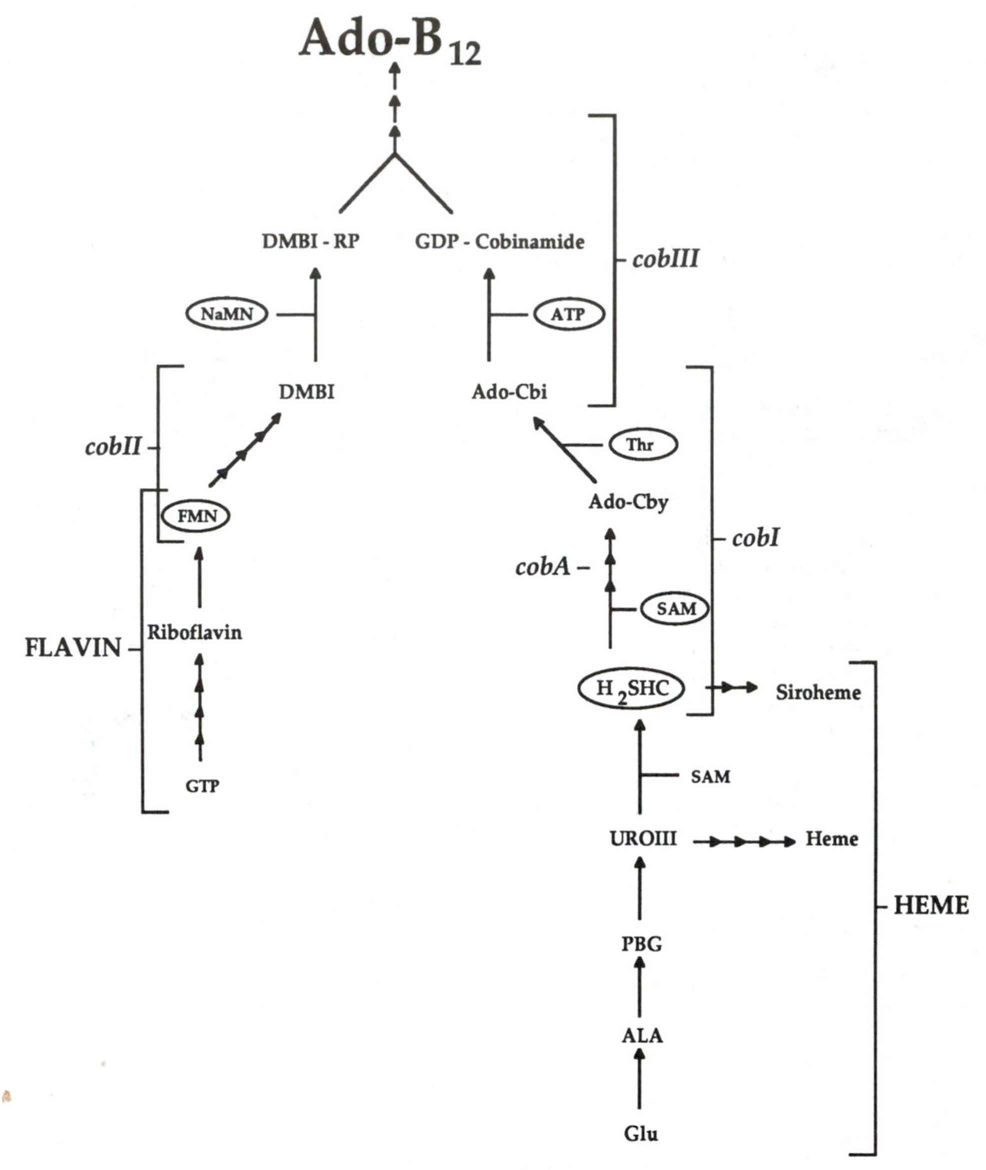

Figure 7. De novo synthesis of cobalamin. The corresponding intermediates shown are those produced by *S. typhimurium*, which are presumably the same as those produced by *B. subtilis*. Branch I of the pathway represents synthesis of the corrinoid ring, branch II represents synthesis of DMBI, and branch III represents assembly of the several parts to form the mature cobalamin molecule. *cobI*, *cobII*, and *cobIII* represents *S. typhimurium* mutations that block synthesis within each of these branches, respectively; *cobA* mutations block formation of adenosylated cobyric acid within the branch I pathway. Encircled compounds are those that make a direct contribution to the final structure. Abbreviations: ALA, 5-aminolevulinic acid; Ado-Cbi, adenosylated cobinamide; Ado-Cby, adenosylated cobyric acid; DMBI, 5,6-dimethylbenzimidazole; DMBI-RP, 1-α-D-ribofuranosido-DMBI; FMN, flavin mononucleotide; GDP-Cobinamide, guanosine diphosphocobinamide; Glu, glutamic acid; H_2SHC, dihydrosirohydrochlorin; NaMN, nicotinic acid mononucleotide; PBG, porphyrobilinogen; SAM; *S*-adenosylmethionine; Thr, L-threonine. Adapted from Jeter et al. (69) and Escalante-Semerena et al. (49).

located at 245° on the genetic map (15, 74, 94). Recently, this operon has been cloned and sequenced, revealing at least six genes (*hemAXCDBL*) (57, 113). Several of these genes encode enzymes with significant protein sequence similarity to their *E. coli* counterparts. It is not known how expression of the *hem* genes is regulated; however, since UroIII is an important metabolic branch point for tetrapyrrole synthesis, it would not be surprising if B_{12}, heme, and siroheme play influential roles.

Acknowledgments. We especially thank A. Bacher, Y. Lemoine, and C. Yanofsky for providing research results prior to publication; A. Sloma, A. Bacher, Y. Lemoine, D. Stahly, and J. Brenchley for their helpful comments on this review; and D. Potvin and A. Huot for preparation of the text and figures.

REFERENCES

1. **Avissar, Y. J., and S. I. Beale.** 1989. Identification of the enzymatic basis for δ-aminolevulinic acid auxotrophy

in a *hemA* mutant of *Escherichia coli*. *J. Bacteriol.* **171**:2919–2924.

2. **Babitzke, P., P. Gollnick, and C. Yanofsky.** 1992. The *mtrAB* operon of *Bacillus subtilis* encodes GTP cyclohydrolase I (MtrA), an enzyme involved in folic acid biosynthesis, and MtrB, a regulator of tryptophan biosynthesis. *J. Bacteriol.* **174**:2059–2064.
3. **Bacher, A.** 1991. Biosynthesis of flavins, p. 215–259. *In* F. Müller (ed.), *Chemistry and Biochemistry of Flavins*, vol. 1. Chemical Rubber Co., Boca Raton, Fla.
4. **Bacher, A., R. Baur, U. Eggers, H. Harders, M. K. Otto, and H. Schnepple.** 1980. Riboflavin synthases of *Bacillus subtilis*: purification and properties. *J. Biol. Chem.* **255**:632–637.
5. **Bacher, A., R. Baur, U. Eggers, H. Harders, and H. Schnepple.** 1976. Riboflavin synthases of *Bacillus subtilis*, p. 729–732. *In* T. P. Singer (ed.), *Flavins and Flavoproteins*. Biological and Medical Press, Amsterdam.
6. **Bacher, A., and R. Ladenstein.** 1991. The lumazine synthase/riboflavin synthase complex of *Bacillus subtilis*, p. 293–316. *In* F. Müller (ed.), *Chemistry and Biochemistry of Flavins*, vol. 1. Chemical Rubber Co., Boca Raton, Fla.
7. **Bacher, A., Q. LeVan, P. J. Keller, and H. G. Floss.** 1985. Biosynthesis of riboflavin. Incorporation of multiply ^{13}C-labeled precursors into the xylene ring. *J. Am. Chem. Soc.* **107**:6380–6385.
8. **Bacher, A., H. C. Ludwig, H. Schnepple, and Y. Ben-Shaul.** 1986. Heavy riboflavin synthase from *Bacillus subtilis*. Quaternary structure and reaggregation. *J. Mol. Biol.* **187**:75–86.
9. **Bacher, A., B. Mailänder, R. Baur, U. Eggers, H. Harders, and H. Schnepple.** 1976. Studies on the biosynthesis of riboflavin, p. 285–290. *In* W. Pfleiderer (ed.), *Chemistry and Biology of Pteridines*. Walter de Gruyter, Berlin.
10. **Bacher, A., G. Neuberger, and R. Volk.** 1986. Enzymatic synthesis of 6,7-dimethyl-8-ribityllumazine, p. 227–230. *In* B. A. Cooper and V. M. Whitehead (ed.), *Chemistry and Biology of Pteridines*. Walter de Gruyter, Berlin.
11. **Bacher, A., H. Schnepple, B. Mailänder, M. K. Otto, and Y. Ben-Shaul.** 1980. Structure and function of the riboflavin synthase complex of *Bacillus subtilis*, p. 579–586. *In* K. Yagi and T. Yamano (ed.), *Flavins and Flavoproteins*. Japan Scientific Societies Press, Tokyo.
12. **Bacher, A., R. Volk, P. J. Keller, H. G. Floss, Q. LeVan, W. Eisenreich, and B. Schwarzkopf.** 1988. Biosynthesis of flavins and deazaflavins, p. 431–440. *In* D. E. Edmondson and D. B. McCormick (ed.), *Flavins and Flavoproteins*. Walter de Gruyter, Berlin.
13. **Bachmann, B. J.** 1990. Linkage map of *Escherichia coli* K-12, edition 8. *Microbiol. Rev.* **54**:130–197.
14. **Bandrin, S. V., P. M. Rabinovich, and A. I. Stepanov.** 1983. Three linkage groups of genes involved in riboflavine biosynthesis in *Escherichia coli*. *Sov. Genet.* **19**: 1103–1109.
15. **Berek, I., A. Miczak, I. Kiss, G. Ivanovics, and I. Durkö.** 1975. Genetic and biochemical analysis of hemin dependent mutants of *Bacillus subtilis*. *Acta Microbiol. Acad. Sci. Hung.* **22**:157–167.
16. **Bognar, A. L., C. Osborne, and B. Shane.** 1987. Primary structure of the *Escherichia coli folC* gene and its folylpolyglutamate synthetase-dihydrofolate synthetase product and regulation of expression by an upstream gene. *J. Biol. Chem.* **262**:12337–12342.
17. **Bognar, A. L., C. Osborne, B. Shane, S. C. Singer, and R. Ferone.** 1985. Folylpoly-γ-glutamate synthetase-dihydrofolate synthetase. Cloning and high expression of the *Escherichia coli folC* gene and purification and properties of the gene product. *J. Biol. Chem.* **260**:5625–5630.
18. **Bresler, S. E., E. I. Cherepenko, T. P. Chernik, V. L. Kalinin, and D. A. Perumov.** 1970. Investigation of the operon of riboflavin synthesis in *Bacillus subtilis*. I. Genetic mapping of the linkage group. *Genetika* **6**:116–124.
19. **Bresler, S. E., E. I. Cherepenko, and D. A. Perumov.** 1970. Investigation of the operon of riboflavin synthesis in *Bacillus subtilis*. II. Biochemical study of regulator mutations. *Genetika* **6**:126–135.
20. **Bresler, S. E., I. E. Cherepenko, and D. A. Perumov.** 1971. Investigation of the operon of riboflavin biosynthesis in *Bacillus subtilis*. III. Production and properties of mutants with a complex regulatory genotype. *Genetika* **7**:117–123.
21. **Bresler, S. E., E. A. Glazunov, and D. A. Perumov.** 1972. Study of the operon of riboflavin biosynthesis in *Bacillus subtilis*. IV. Regulation of the synthesis of riboflavin synthetase. Investigation of riboflavin transport through the cell membrane. *Genetika* **8**:109–118.
22. **Bresler, S. E., E. A. Glazunov, D. A. Perumov, and T. P. Chernik.** 1977. Investigation of the operon of riboflavin biosynthesis in *Bacillus subtilis*. XIII. Genetic and biochemical investigation of mutants related to intermediate stages of biosynthesis. *Genetika* **13**:2007–2016.
23. **Bresler, S. E., G. F. Gorinchuk, T. P. Chernik, and D. A. Perumov.** 1978. Riboflavin operon in *Bacillus subtilis*. XV. Investigation of the mutants relating to initial steps of biosynthesis. The origin of ribityl side chain. *Genetika* **14**:2082–2090.
24. **Bresler, S. E., and D. A. Perumov.** 1979. Riboflavin operon in *Bacillus subtilis*. Regulation of GTP-cyclohydrolase synthesis in strains of different genotypes. *Genetika* **15**:967–971.
25. **Bresler, S. E., D. A. Perumov, T. P. Chernik, and E. A. Glazunov.** 1976. Investigation of the operon of riboflavin bisynthesis in *Bacillus subtilis*. X. Genetic and biochemical study of mutants that accumulate 6-methyl-7-(1′,2′-dihydroxyethyl)-8-ribityllumazine. *Genetika* **12**:83–91.
26. **Bresler, S. E., D. A. Perumov, E. A. Glazunov, T. N. Shevchenko, and T. P. Chernik.** 1977. Investigation of the operon of riboflavin biosynthesis in *Bacillus subtilis*. XII. Determination of the ATP: riboflavin-5′-phosphotransferase and riboflavin synthetase content in cells with different genotypes. *Genetika* **13**:880–887.
27. **Bresler, S. E., D. A. Perumov, A. P. Skvortsova, T. P. Chernik, and T. N. Shevchenko.** 1975. Study of the riboflavin operon of *Bacillus subtilis*. VIII. Genetic mapping of *ribC* markers in relation to the cluster of structural genes. *Genetika* **11**:95–100.
28. **Brey, R. N., C. D. B. Banner, and J. B. Wolf.** 1986. Cloning of multiple genes involved with cobalamin (vitamin B_{12}) biosynthesis in *Bacillus megaterium*. *J. Bacteriol.* **167**:623–630.
29. **Brown, G. M., and H. Williamson.** 1982. Biosynthesis of riboflavin, folic acid, thiamine, and pantothenic acid. *Adv. Enzymol. Relat. Areas Mol. Biol.* **53**:345–381.
30. **Brown, G. M., and J. M. Williamson.** 1987. Biosynthesis of folic acid, riboflavin, thiamine, and pantothenic acid, p. 521–538. *In* F. C. Neidhardt, J. L. Ingraham, K. B. Low, B. Magasanik, M. Schaechter, and H. E. Umbarger (ed.), *Escherichia coli and Salmonella typhimurium: Cellular and Molecular Biology*, vol. 1. American Society for Microbiology, Washington, D.C.
31. **Brown, G. M., J. Yim, Y. Suzuki, M. C. Heine, and F. Foor.** 1975. The enzymatic synthesis of pterins in *Escherichia coli*, p. 219–245. *In* W. Pfleiderer (ed.), *Chemistry and Biology of Pteridines*. Walter de Gruyter, Berlin.
32. **Burrows, R. B., and G. M. Brown.** 1978. Presence in *Escherichia coli* of a deaminase and a reductase involved in biosynthesis of riboflavin. *J. Bacteriol.* **136**: 657–667.
33. **Chen, N.-Y., F.-M. Hu, and H. Paulus.** 1987. Nucleotide

sequence of the overlapping genes for the subunits of *Bacillus subtilis* aspartokinase II and their control regions. *J. Biol. Chem.* **262:**8787–8798.

34. **Chernik, T. P., S. E. Bresler, V. V. Machkovsky, and D. A. Perumov.** 1979. Riboflavin-biosynthesis operon in *Bacillus subtilis*. XVI. Localization of group *ribC* markers on the chromosome. *Genetika* **15:**1569–1577.
35. **Chernik, T. P., A. P. Skvortsova, T. N. Shevchenko, D. A. Perumov, and S. E. Bresler.** 1974. Riboflavin operon in *Bacillus subtilis*. VI. Nature and properties of mutants constitutive in the presence of flavin mononucleotide. *Genetika* **10:**94–104.
36. **Chikindas, M. L., E. V. Luk'yanov, P. M. Rabinovich, and A. I. Stepanov.** 1986. Study of the 210° region of the *Bacillus subtilis* chromosome. *Mol. Genet. Mikrobiol. Virusol.* **2:**20–24.
37. **Chikindas, M. L., V. N. Mironov, E. V. Luk'yanov, Y. R. Boretskii, L. S. Artyunova, P. M. Rabinovich, and A. I. Stepanov.** 1987. Determination of the bounds of the *Bacillus subtilis* riboflavin operon. *Mol. Genet. Mikrobiol. Virusol.* **4:**22–26.
38. **Chikindas, M. L., G. I. Morozov, V. N. Mironov, E. V. Luk'yanov, V. V. Emel'yanov, and A. I. Stepanov.** 1988. Regulatory regions of the operon for biosynthesis of *Bacillus subtilis* riboflavin. *Dokl. Akad. Nauk SSSR* **298:** 997–1000.
39. **Cronan, J. E., Jr.** 1989. The *E. coli bio* operon: transcriptional repression by an essential protein modification enzyme. *Cell* **58:**427–429.
40. **Debarbouille, M., M. Arnaud, A. Fouet, A. Klier, and G. Rapoport.** 1990. The *sacT* gene regulating the *sacPA* operon in *Bacillus subtilis* shares strong homology with transcriptional antiterminators. *J. Bacteriol.* **172:**3966–3973.
41. **Ebbole, D. J., and H. Zalkin.** 1987. Cloning and characterization of a 12-gene cluster from *Bacillus subtilis* encoding nine enzymes for *de novo* purine nucleotide synthesis. *J. Biol. Chem.* **262:**8274–8287.
42. **Ebbole, D. J., and H. Zalkin.** 1988. Detection of *pur* operon-attenuated mRNA and accumulated degradation intermediates in *Bacillus subtilis*. *J. Biol. Chem.* **263:**10894–10902.
43. **Ebbole, D. J., and H. Zalkin.** 1989. Interaction of a putative repressor protein with an extended control region of the *Bacillus subtilis pur* operon. *J. Biol. Chem.* **264:**3553–3561.
44. **Eisenberg, M. A.** 1984. Regulation of the biotin operon. *Ann. N.Y. Acad. Sci.* **447:**335–349.
45. **Eisenberg, M. A.** 1987. Biosynthesis of biotin and lipoic acid, p. 544–550. *In* F. C. Neidhardt, J. L. Ingraham, K. B. Low, B. Magasanik, M. Schaechter, and H. E. Umbarger (ed.), *Escherichia coli and Salmonella typhimurium: Cellular and Molecular Biology*, vol 1. American Society for Microbiology, Washington, D.C.
46. **Elliott, T.** 1989. Cloning, genetic characterization, and nucleotide sequence of the *hemA-prfA* operon of *Salmonella typhimurium*. *J. Bacteriol.* **171:**3948–3960.
47. **Enei, H., K. Sato, Y. Anzai, and H. Okada.** August 1975. Fermentative production of riboflavine. U.S. patent 3,900,368.
48. **Escalante-Semerena, J. C., and J. R. Roth.** 1987. Regulation of cobalamin biosynthetic operons in *Salmonella typhimurium*. *J. Bacteriol.* **169:**2251–2258.
49. **Escalante-Semerena, J. C., S.-J. Suh, and J. R. Roth.** 1990. *cobA* function is required for both de novo cobalamin biosynthesis and assimilation of exogenous corrinoids in *Salmonella typhimurium*. *J. Bacteriol.* **172:**273–280.
50. **Floss, G. H., Q. LeVan, P. J. Keller, and A. Bacher.** 1983. Biosynthesis of riboflavin. An unusual rearrangement in the formation of 6,7-dimenthyl-8-ribityllumazine. *J. Am. Chem. Soc.* **105:**2493–2495.
51. **Foor, F., and G. M. Brown.** 1975. Purification and properties of guanosine triphosphate cyclohydrolase II from *Escherichia coli*. *J. Biol. Chem.* **250:**3545–3551.
52. **Glazunov, E. A., S. E. Bresler, and D. A. Perumov.** 1974. Investigation of the riboflavin operon of *Bacillus subtilis*. VII. Biochemical study of mutants relating to early stages of biosynthesis. *Genetika* **10:**83–92.
53. **Gloeckler, R., I. Ohsawa, D. Speck, C. Ledoux, S. Bernard, M. Zinsius, D. Villeval, T. Kisou, K. Kamogawa, and Y. Lemoine.** 1990. Cloning and characterization of the *Bacillus sphaericus* genes controlling the bioconversion of pimelate into dethiobiotin. *Gene* **87:** 63–70.
54. **Gollnick, P., S. Ishino, M. I. Kuroda, D. J. Henner, and C. Yanofsky.** 1990. The *mtr* locus is a two-gene operon required for transcription attenuation in the *trp* operon of *Bacillus subtilis*. *Genetics* **87:**8726–8730.
55. **Goodwin, T. W., and A. A. Horton.** 1961. Biosynthesis of riboflavin in cell-free systems. *Nature* (London) **191:** 772–774.
56. **Green, J. M., and B. P. Nichols.** 1991. *p*-Aminobenzoate biosynthesis in *Escherichia coli*. *J. Biol. Chem.* **266:** 12971–12975.
57. **Hanson, M., L. Rutberg, I. Schröder, and L. Hederstedt.** 1991. The *Bacillus subtilis hemAXCDBL* gene cluster, which encodes enzymes for the biosynthetic pathway from glutamate to uroporphyrinogen III. *J. Bacteriol.* **173:**2590–2599.
58. **Harzer, G., H. Rokos, M. K. Otto, A. Bacher, and S. Ghisla.** 1978. Biosynthesis of riboflavin. 6,7-Dimethyl-8-ribityllumazine 5′-phosphate is not a substrate for riboflavin synthase. *Biochim. Biophys. Acta* **540:**48–54.
59. **Henner, D. J., L. Band, and H. Shimotsu.** 1984. Nucleotide sequence of the *Bacillus subtilis* tryptophan operon. *Gene* **34:**169–177.
60. **Henner, D. J., and J. A. Hoch.** 1982. The genetic map of *Bacillus subtilis*, p. 1–33. *In* D. Dubnau (ed.), *The Molecular Biology of the Bacilli*. Academic Press, Inc., New York.
61. **Howard, P. K., J. Shaw, and A. J. Otsuka.** 1985. Nucleotide sequence of the *birA* gene encoding the biotin operon repressor and biotin holoenzyme synthetase functions of *Escherichia coli*. *Gene* **35:**321–331.
62. **Iwahara, S., M. Kikuchi, T. Tochikura, and K. Ogata.** 1966. Some properties of avidin-uncombinable unknown biotin-vitamer produced by *Bacillus* sp. and its role in biosynthesis of desthiobiotin. *Agric. Biol. Chem.* **30:**304–306.
63. **Izumi, Y., Y. Kano, K. Inagaski, N. Kawase, Y. Tani, and H. Yamada.** 1981. Characterization of biotin biosynthetic enzymes of *Bacillus sphaericus*: a desthiobiotin producing bacterium. *Agric. Biol. Chem.* **45:**1983–1989.
64. **Izumi, Y., H. Morita, Y. Tani, and K. Ogata.** 1974. The pimelyl-CoA synthetase responsible for the first step in biotin biosynthesis by microorganisms. *Agric. Biol. Chem.* **38:**2257–2262.
65. **Izumi, Y., H. Morita, K. Sato, Y. Tani, and K. Ogata.** 1972. Synthesis of biotin-vitamers from pimelic acid and coenzyme A by cell-free extracts of various bacteria. *Biochim. Biophys. Acta* **264:**210–213.
66. **Izumi, Y., H. Morita, Y. Tani, and K. Ogata.** 1973. Partial purification and some properties of 7-keto-8-aminopelargonic acid synthetase, an enzyme involved in biotin biosynthesis. *Agric. Biol. Chem.* **37:**1327–1333.
67. **Izumi, Y., and K. Ogata.** 1975. Some aspects of the microbial production of biotin. *Adv. Appl. Microbiol.* **22:**145–176.
68. **Izumi, Y., K. Sato, Y. Tani, and K. Ogata.** 1973. Distribution of 7-keto-8-aminopelargonic acid synthetase in bacteria and the control mechanism of the enzyme activity. *Agric. Biol. Chem.* **37:**1335–1340.
69. **Jeter, R., J. C. Escalante-Semerena, D. Roof, B. Olivera, and J. Roth.** 1987. Synthesis and use of vitamin B_{12}, p.

551–556. *In* F. C. Neidhardt, J. L. Ingraham, K. B. Low, B. Magasanik, M. Schaechter, and H. E. Umbarger (ed.), *Escherichia coli and Salmonella typhimurium: Cellular and Molecular Biology*, vol. 1. American Society for Microbiology, Washington, D.C.

70. **Jeter, R. M., B. M. Olivera, and J. R. Roth.** 1984. *Salmonella typhimurium* synthesizes cobalamin (vitamin B_{12}) de novo under anaerobic growth conditions. *J. Bacteriol.* **159:**206–213.
71. **Kane, J. F.** 1977. Regulation of a common amidotransferase subunit. *J. Bacteriol.* **132:**419–425.
72. **Kane, J. F., W. M. Holmes, and R. A. Jensen.** 1972. Metabolic interlock: the dual function of a folate pathway gene as an extraoperonic gene of tryptophan biosynthesis. *J. Biol. Chem.* **247:**1587–1596.
73. **Katzenmeier, G., C. Schmid, J. Kellermann, F. Lottspeich, and A. Bacher.** 1991. Sequence of GTP cyclohydrolase I from *Escherichia coli*. *Biol. Chem. Hoppe-Seyler* **372:**991–997.
74. **Kiss, I., I. Berek, and G. Ivanovics.** 1971. Mapping the δ-aminolevulinic acid synthetase locus in *Bacillus subtilis*. *J. Gen. Microbiol.* **66:**153–159.
75. **Konishi, S., and T. Shiro.** 1968. Fermentation production of guanosine by 8-azaguanine resistance of *Bacillus subtilis*. *Agric. Biol. Chem.* **32:**396–398.
76. **Kreneva, R. A., and D. A. Perumov.** 1990. Genetic mapping of regulatory mutations of *Bacillus subtilis* riboflavin operon. *Mol. Gen. Genet.* **222:**467–469.
77. **Kukanova, A. Y., V. G. Zhdanov, and A. I. Stepanov.** 1982. The roseoflavin-resistant mutants of *Bacillus subtilis*. *Genetika* **18:**319–321.
78. **Ladenstein, R., B. Meyer, R. Huber, H. Labischinski, K. Bartels, H. D. Bartunik, L. Bachmann, H. C. Ludwig, and A. Bacher.** 1986. Heavy riboflavin synthase from *Bacillus subtilis*. Particle dimensions, crystal packing and molecular symmetry. *J. Mol. Biol.* **187:**87–100.
79. **Le Van, Q., P. J. Keller, D. H. Brown, H. G. Floss, and A. Bacher.** 1985. Biosynthesis of riboflavin in *Bacillus subtilis*: origin of the four-carbon moiety. *J. Bacteriol.* **162:**1280–1284.
80. **Lee, G., and J. Pero.** 1981. Conserved nucleotide sequences in temporally-controlled phage promoters. *J. Mol. Biol.* **152:**247–265.
81. **Logvinenko, E. M., G. M. Shavlovsky, and N. Y. Tsarenko.** 1984. The role of iron in regulation of 6,7-dimethyl-8-ribityllumazine synthetase synthesis in flavinogenic yeasts. *Biokhimiya* **49:**45–50.
82. **Logvinenko, E. M., G. M. Shavlovsky, A. E. Zakal'sky, and I. V. Zakhodylo.** 1982. Biosynthesis of 6,7-dimethyl-8-ribityllumazine in yeast extracts of *Pichia guilliermondii*. *Biokhimiya* **47:**931–936.
83. **Lopez, P., M. Espinosa, B. Greenberg, and S. A. Lacks.** 1987. Sulfonamide resistance in *Streptococcus pneumoniae*: DNA sequence of the gene encoding dihydropteroate synthase and characterization of the enzyme. *J. Bacteriol.* **169:**4320–4326.
84. **Lopez, P., B. Greenberg, and S. A. Lacks.** 1990. DNA sequence of folate biosynthesis gene *sulD*, encoding hydroxymethyldihydropterin pyrophosphokinase in *Streptococcus pneumoniae*, and characterization of the enzyme. *J. Bacteriol.* **172:**4766–4774.
85. **Ludwig, H. C., F. Lottspeich, A. Henschen, R. Ladenstein, and A. Bacher.** 1987. Heavy riboflavin synthase of *Bacillus subtilis*: primary structure of the β subunit. *J. Biol. Chem.* **262:**1016–1021.
86. **Lundrigan, M. D., L. C. DeVeaux, B. J. Mann, and R. J. Kadner.** 1987. Separate regulatory systems for repression of *metE* and *btuB* by vitamin B_{12} in *Escherichia coli*. *Mol. Gen. Genet.* **206:**401–407.
87. **Maley, G. F., D. U. Guarino, and F. Maley.** 1983. Complete amino acid sequence of an allosteric enzyme, T2 bacteriophage deoxycytidylate deaminase. *J. Biol. Chem.* **258:**8290–8297.
88. **Margolis, P. S., A. Driks, and R. Losick.** 1993. Sporulation gene *spoIIB* from *Bacillus subtilis*. *J. Bacteriol.* **175:**528–540.
89. **Matsui, H., K. Sato, H. Enei, and Y. Hirose.** 1977. Mutation of an inosine-producing strain of *Bacillus subtilis* to DL-methionine sulfoxide resistance for guanosine production. *Appl. Environ. Microbiol.* **34:**337–341.
90. **Matsui, H., K. Sato, H. Enei, and Y. Hirose.** 1979. A guanosine-producing mutant of *Bacillus subtilis* with high productivity. *Agric. Biol. Chem.* **43:**393–394.
91. **Matsui, H., K. Sato, H. Enei, and Y. Hirose.** 1979. Production of guanosine by psicofuranine and decoyinine resistant mutants of *Bacillus subtilis*. *Agric. Biol. Chem.* **43:**1739–1744.
92. **Matsui, K., H.-C. Wang, T. Hirota, H. Matsukawa, S. Kasai, K. Shinagawa, and S. Otani.** 1982. Riboflavin production by roseoflavin-resistant strains of some bacteria. *Agric. Biol. Chem.* **46:**2003–2008.
93. **McDonald, K. O., and J. W. F. Burke.** 1982. Cloning of the *Bacillus subtilis* sulfanilamide resistance gene in *Bacillus subtilis*. *J. Bacteriol.* **149:**391–394.
94. **Miczak, A., B. Progai, and I. Berek.** 1979. Mapping the uroporphyrinogen III cosynthase locus in *Bacillus subtilis*. *Mol. Gen. Genet.* **174:**293–295.
95. **Mironov, V. N., M. L. Chikindas, A. S. Kraev, A. I. Stepanov, and K. G. Skryabin.** 1990. Operon organization of genes of riboflavin biosynthesis in *Bacillus subtilis*. *Dok. Akad. Nauk SSSR* **312:**237–240.
96. **Mironov, V. N., A. S. Kraev, B. K. Chernov, A. V. Ul'yanov, Y. B. Golva, G. E. Pozmogova, M. L. Simonova, V. K. Gordeev, A. I. Stepanov, and K. G. Skryabin.** 1989. Genes of riboflavin biosynthesis of *Bacillus subtilis*—complete primary structure and model of organization. *Dokl. Adad. Nauk SSSR* **305:**482–486.
97. **Mironov, V. N., D. A. Perumov, A. S. Kraev, A. I. Stepanov, and K. G. Skryabin.** 1990. Unusual structure in the regulation region of the *Bacillus subtilis* riboflavin biosynthesis operon. *Mol. Biol.* (Moscow) **24:**256–261.
98. **Mohan, S., J. Aghion, N. Guillen, and D. Dubnau.** 1989. Molecular cloning and characterization of *comC*, a late competence gene of *Bacillus subtilis*. *J. Bacteriol.* **171:**6043–6051.
99. **Morozov, G. I., P. M. Rabinovich, S. V. Bandrin, and A. I. Stepanov.** 1984. Organization of *Bacillus subtilis* riboflavin operon. *Mol. Genet. Mikrobiol. Virusol.* **7:**42–46.
100. **Morozov, G. I., P. M. Rabinovich, V. V. Emel'yanov, and A. I. Stepanov.** 1985. Operon for biosynthesis of *Bacillus subtilis* riboflavin contains additional promoters. *Mol. Genet. Mikrobiol. Virusol.* **12:**14–19.
101. **Morozov, G. I., P. M. Rabinovich, and A. I. Stepanov.** 1984. Operator position in the coupling group of structural genes of the *Bacillus subtilis* riboflavin operon. *Mol. Genet. Mikrobiol. Virusol.* **11:**11–16.
102. **Neuberger, G., and A. Bacher.** 1986. Biosynthesis of riboflavin. Enzymatic formation of 6,7-dimethyl-8-ribityllumazine by heavy riboflavin synthase from *Bacillus subtilis*. *Biochem. Biophys. Res. Commun.* **139:**1111–1116.
103. **Nichols, B. P., A. M. Seibold, and S. Z. Boktor.** 1989. *para*-Aminobenzoate synthesis from chorismate occurs in two steps. *J. Biol. Chem.* **264:**8587–8601.
104. **Nielsen, P., G. Neuberger, I. Fujii, D. H. Brown, P. J. Keller, H. G. Floss, and A. Bacher.** 1986. Biosynthesis of riboflavin. Enzymatic formation of 6,7-dimethyl-8-ribityllumazine from pentose phosphates. *J. Biol. Chem.* **261:**3661–3669.
105. **Ohsawa, I., D. Speck, T. Kisou, K. Hayakawa, M. Zinsius, R. Gloeckler, Y. Lemoine, and K. Kamogawa.** 1989. Cloning of the biotin synthetase gene from *Bacil-*

lus sphaericus and expression in *Escherichia coli* and *Bacilli*. *Gene* **80:**39–48.
106. **O'Neill, G. P., M.-W. Chen, and D. Söll.** 1989. δ-Aminolevulinic acid biosynthesis in *Escherichia coli* and *Bacillus subtilis* involves formation of glutamyl-tRNA. *FEMS Microbiol. Lett.* **60:**255–260.
107. **O'Regan, M., R. Gloeckler, S. Bernard, C. Ledoux, I. Ohsawa, and Y. Lemoine.** 1989. Nucleotide sequence of the *bioH* gene of *Escherichia coli*. *Nucleic Acids Res.* **17:**8004.
108. **Otsuka, A. J., M. R. Buoncristiani, P. K. Howard, J. Flamm, C. Johnson, R. Yamamoto, K. Uchida, C. Cook, J. Ruppert, and J. Matsuzaki.** 1988. The *Escherichia coli* biotin biosynthetic enzyme sequences predicted from the nucleotide sequence of the *bio* operon. *J. Biol. Chem.* **263:**19577–19585.
109. **Pai, C. H.** 1975. Genetics of biotin biosynthesis in *Bacillus subtilis*. *J. Bacteriol.* **121:**1–8.
110. **Perkins, J. B., J. G. Pero, and A. Sloma.** January 1991. Riboflavin overproducing strains of bacteria. European patent application 0405370.
111. **Perkins, J. B., A. Sloma, and J. Pero.** Unpublished data.
112. **Perumov, D. A., E. A. Glazunov, and G. F. Gorinchuk.** 1985. Riboflavin operon in *Bacillus subtilis*. XVII. Regulatory functions of intermediate products and their derivatives. *Genetika* **22:**748–754.
113. **Petricek, M., L. Rutberg, I. Schröder, and L. Hederstedt.** 1990. Cloning and characterization of the *hemA* region of the *Bacillus subtilis* chromosome. *J. Bacteriol.* **172:**2250–2258.
114. **Ploux, O., and A. Marquet.** 1992. The 8-amino-7-oxoate pelargon synthase from *Bacillus sphaericus*. Purification and preliminary characterization of the cloned enzyme overproduced in *Escherichia coli*. *Biochem. J.* **283:**327–331.
115. **Ploux, O., P. Soularue, A. Marquet, R. Gloeckler, and Y. Lemoine.** 1992. Investigation of the first step of biotin biosynthesis in *Bacillus sphaericus*. Purification and characterization of the pimeloyl-CoA synthase, and uptake of pimelate. *Biochem. J.* **287:**685–690.
116. **Quinn, C. L., B. T. Stephenson, and R. L. Switzer.** 1991. Functional organization and nucleotide sequence of the *Bacillus subtilis* pyrimidine biosynthetic operon. *J. Biol. Chem.* **266:**9113–9127.
117. **Rabinovich, P. M., M. Y. Beburov, Z. K. Linevich, and A. I. Stepanov.** 1978. Amplification of *Bacillus subtilis* riboflavin operon in *Escherichia coli*. *Genetika* **14:**1696–1705.
117a. **Richter, G., H. Ritz, G. Katzenmeier, R. Volk, A. Kohnle, F. Lottspeich, D. Allendorf, and A. Bacher.** Biosynthesis of riboflavin. Cloning, sequencing, mapping and expression of the gene coding for GTPcyclohydrolase II of *Escherichia coli*. Submitted for publication.
118. **Richter, G., R. Volk, C. Krieger, H.-W. Lahm, U. Röthlisberger, and A. Bacher.** 1992. Biosynthesis of riboflavin: cloning, sequencing, and expression of the gene coding 3,4-dihydroxy-2-butanone 4-phosphate synthase of *Escherichia coli*. *J. Bacteriol.* **174:**4050–4056.
118a. **Robin, C., F. Blance, L. Cauchois, B. Cameron, M. Couder, and J. Crouzet.** 1991. Primary structure, expression in *Escherichia coli*, and properties of *S*-adenosyl-L-methionine: uroporphyrinogen III methyltransferase from *Bacillus megaterium*. *J. Bacteriol.* **173:**4893–4896.
119. **Rolfe, B., and M. A. Eisenberg.** 1968. Genetic and biochemical analysis of the biotin loci of *Escherichia coli* K-12. *J. Bacteriol.* **96:**515–524.
120. **Saxild, H. H., and P. Nygaard.** 1987. Genetic and physiological characterization of *Bacillus subtilis* mutants resistant to purine analogs. *J. Bacteriol.* **169:**2977–2983.
121. **Schott, K., J. Kellermann, F. Lottspeich, and A. Bacher.** 1990. Riboflavin synthases of *Bacillus subtilis*: purification and amino acid sequence of the α subunit. *J. Biol. Chem.* **265:**4204–4209.
122. **Shane, B.** 1980. Pteroylpoly (γ-glutamate) synthesis by *Corynebacterium* species. Purification and properties of folylpoly(γ-glutamate) synthetase. *J. Biol. Chem.* **255:**5655–5662.
123. **Shavlovsky, G. M., G. E. Teslyar, and L. P. Strugovshchikova.** 1982. Regulation of flavinogenesis in riboflavin-dependent *Escherichia coli* mutants. *Mikrobiologiya* **51:**986–992.
124. **Shimotsu, H., M. Kuroda, C. Yanofsky, and D. J. Henner.** 1986. Novel form of transcription attenuation regulates expression of the *Bacillus subtilis* tryptophan operon. *J. Bacteriol.* **166:**461–471.
125. **Slock, J., D. P. Stahly, C.-Y. Han, E. W. Six, and I. P. Crawford.** 1990. An apparent *Bacillus subtilis* folic acid biosynthetic operon containing *pab*, an amphibolic *trpG* gene, a third gene required for synthesis of *para*-aminobenzoic acid, and the dihydropteroate synthase gene. *J. Bacteriol.* **172:**7211–7226.
126. **Song, B.-H., and J. Neuhard.** 1989. Chromosomal location, cloning and nucleotide sequence of the *Bacillus subtilis cdd* gene encoding cytidine/deoxycytidine deaminase. *Mol. Gen. Genet.* **216:**462–468.
127. **Speck, D., I. Ohsawa, R. Gloeckler, M. Zinsius, S. Bernard, C. Ledoux, T. Kisou, K. Kamogawa, and Y. Lemoine.** 1991. Isolation of *Bacillus sphaericus* mutants affected in their control of biotin synthesis: evidence for transcriptional regulation of the *bio* genes. *Gene* **108:**39–45.
128. **Stepanov, A. I., A. Y. Kukanova, E. A. Glazunov, and V. G. Zhdanov.** 1977. Analogs of riboflavin and lumiflavin and derivatives of alloxazine. II. Effect of roseoflavin on synthesis of 6,7-dimethyl-8-ribityllumazine and riboflavin synthetase and growth of *Bacillus subtilis*. *Genetika* **13:**490–495.
129. **Stepanov, A. I., V. G. Zhdanov, A. Y. Kukanova, M. Y. Khaikinson, P. M. Rabinovich, J. A. V. Iomantas, and Z. M. Galushkina.** December 1984. Method for preparing riboflavin. French patent application 3599355.
129a. **Taura, T., C. Ueguchi, K. Shiba, and K. Ito.** 1992. Insertional disruption of the *nusB* (*ssyB*) gene leads to cold-sensitive growth of *Escherichia coli* and suppression of the *secY24* mutation. *Mol. Gen. Genet.* **234:**429–432.
130. **Teslyar, G. E., and G. M. Shavlovsky.** 1983. Localization of the genes coding cyclohexrolase [cyclohydrolase] II and riboflavin synthase on the *Escherichia coli* K-12 chromosome. *Cytol. Genet.* **17:**57–59.
131. **Volk, R., and A. Bacher.** 1988. Biosynthesis of riboflavin. The structure of the four-carbon precursor. *J. Am. Chem. Soc.* **110:**3651–3653.
132. **Wolf, J. B., and R. N. Brey.** 1986. Isolation and genetic characterizations of *Bacillus megaterium* cobalamin biosynthesis-deficient mutants. *J. Bacteriol.* **166:**51–58.
133. **Yamada, H., M. Osakai, Y. Tani, and Y. Izumi.** 1983. Biotin overproduction by biotin analog-resistant mutants of *Bacillus sphaericus*. *Agric. Biol. Chem.* **47:**1011–1016.
134. **Zalkin, H., and D. J. Ebbole.** 1988. Organization and regulation of genes encoding biosynthetic enzymes in *Bacillus subtilis*. *J. Biol. Chem.* **26:**1595–1598.

24. De Novo Purine Nucleotide Synthesis

HOWARD ZALKIN

The de novo synthesis of purine nucleotides proceeds by a 14-step branched pathway via IMP (Fig. 1). Interrelationships with pathways for synthesis of thiamine and histidine are also shown in Fig. 1. Salvage reactions to reutilize purine bases provide alternative routes for synthesis of AMP and GMP. Purine nucleotide degradation and salvage reactions are covered in chapter 26. Although the de novo pathway is invariant in all organisms investigated, the genetic organization and regulation of expression differ between organisms. There are also significant structural differences in enzymes of the pathway in gram-positive and enteric bacteria as well as in higher eukaryotes. This chapter reviews our current understanding of de novo purine nucleotide synthesis in *Bacillus subtilis*, with emphasis on recent developments in understanding gene organization and regulation and on the distinctive structural features of two enzymes of the pathway.

GENE ORGANIZATION

Genes for de novo purine nucleotide synthesis are found in four positions on the *B. subtilis* chromosome (37). A 12-gene cluster, designated the *pur* operon, encoding all of the enzymes required for synthesis of IMP from 5-phosphoribosyl-1-pyrophosphate (PRPP) is located at 55°. Genes *guaA*, *guaB*, and *purA*, which are involved in the conversion of IMP to GMP and AMP, have been placed at 50°, 0°, and 355°, respectively. All of these genes have been cloned, and their nucleotide sequences have been determined.

The *pur* operon

The *B. subtilis pur* operon was cloned in a series of steps starting with *purF*. Gene *purF* was initially isolated by functional complementation in *Escherichia coli* (24). Approximately 14 kb of 5′ and 3′ DNA flanking *purF* was subsequently obtained by a series of plasmid rescue steps and was sequenced (8). Contiguous genes flanking *purF* were identified by complementation of *E. coli* mutants and sequence comparisons with homologous genes from *E. coli*, yeast cells, and *Drosophila melanogaster*. By this analysis, a 12-gene operon containing 13,080 nucleotides that encodes all of the enzymes required for de novo synthesis of IMP was identified. Several of the genes listed in Fig. 1 were cloned and assigned independently (42). The *pur* operon gene order is *purEKBC* (open reading frame [ORF]) *QLFMNHD*. Gene nomenclature follows that used for *E. coli* (3). To conform with *E. coli* nomenclature (2), the earlier designation *purH(J)* (8), for the gene encoding a bifunctional enzyme catalyzing steps 9 and 10, has been changed to *purH*. The relationship between genes and steps in the pathway is shown in Fig. 1. There is an unidentified ORF of 84 codons in the *pur* operon that encodes a 9- to 10-kDa protein when it is expressed from a multicopy plasmid in *E. coli*. Transcription of the *pur* operon is initiated from a σ^A-dependent promoter located 242 nucleotides upstream of *purE* and is terminated approximately 38 nucleotides downstream of *purD* (8).

The *pur* operon is organized into three sets of overlapping genes separated by intercistronic gaps: *purEKB*–73 bp–*purC*(ORF)*QLF*–101 bp–*purMNH*–15 bp–*purD*. The overlaps of the 3′ ends of one coding sequence with the 5′ end of the contiguous downstream sequence are as follows: *purE-purK*, 8 bp; *purK-purB*, 4 bp; *purC*-ORF, 8 bp; ORF-*purQ*, 4 bp; *purQ-purL*, 17 bp; *purL-purF*, 25 bp; *purM-purN*, 4 bp; *purN-purH*, 4 bp. Sites for translation initiation were inferred from the nucleotide sequence except for *purF*, for which NH_2-terminal amino acid sequence analysis of glutamine PRPP amidotransferase confirmed the translation initiation site (24, 51). However, all inferred sites for translation initiation are preceded by good ribosome-binding sequences having ΔGs of −12 to −19 kcal/mol (1 cal = 4.184 J) for complementary base pairing with the 3′ end of *B. subtilis* 16S rRNA. A possible role of gene overlaps is discussed below.

In contrast to the organization of the *B. subtilis pur* operon, genes required for synthesis of IMP are dispersed in seven operons on the *E. coli* chromosome (14). It appears that biosynthetic pathways in *Bacillus* species are coded for by clustered overlapping genes more often than is the case in enteric bacteria. Clustered overlapping genes are found for the *B. subtilis trp* (15), *pyr* (38), and *fol* (46) operons. Multiple, presumably contiguous genes for biosynthesis of arginine (32), cobalamin (5), branched-chain amino acids (22), and menaquinone (29) are also found in *Bacillus* species.

What is the likely role of the translational overlaps? Studies on overlapping *trpEG* genes in *E. coli* indicated that efficient translation of *trpG* was dependent on the translation of the distal segment of *trpE* (35). Oppenheim and Yanofsky introduced the term "translational coupling" to describe this phenomenon (35). Translational coupling refers to situations in which translation of a gene in a polycistronic operon is at least partially dependent on the translation of a contiguous upstream gene. However, the translation initiation region need not overlap with the upstream gene (36). Translational coupling provides one mechanism for coordinating the synthesis of functionally related proteins from a polycistronic mRNA (see dis-

Howard Zalkin • Department of Biochemistry, Purdue University, West Lafayette, Indiana 47907.

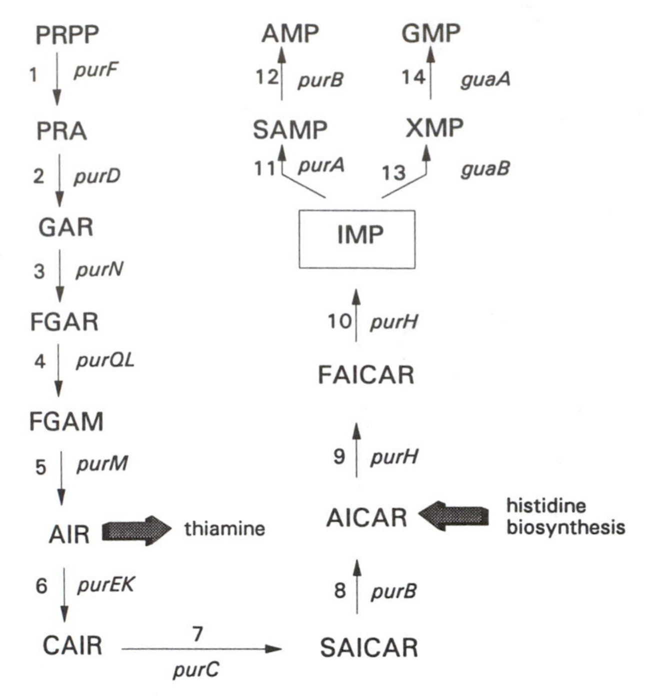

Figure 1. Pathway for de novo purine nucleotide synthesis. 5′-Phosphoribosyl-5-aminoimidazole (AIR) is a branch point for the synthesis of thiamine. Step 5 in the histidine biosynthetic pathway generates 5-phosphoribosyl-4-carboxamide-5-aminoimidazole (AICAR) as a by-product. Thick arrows designate multistep pathways. Abbreviations: GAR, 5′-phosphoribosyl-1-glycinamide; FGAR, 5′-phosphoribosylformylglycinamide; 5′-CAIR, 5′-phosphoribosyl-5-aminoimidazole-4-carboxylate; SAICAR, 5′-phosphoribosyl-4-(*N*-succinocarboxamide)-5-aminoimidazole; FAICAR, 5′-phosphoribosyl-4-carboxamide-5-formamidoimidazole; SAMP, adenylosuccinate.

cussion in reference 54). This coordination is often 1:1 in heterodimers. It has been proposed that enzyme synthesis from each cluster of overlapping genes in the *pur* operon, *purEKB*, *purC*(ORF)*QLF*, and *purMNH* is set to yield a 1:1 stoichiometry (54). Genes *purEK* and *purQL* are assumed to encode subunits of hetero-oligomeric enzymes.

The stoichiometry for expression of a gene from each of the three gene clusters was investigated by using *lacZ* fusions to *purE*, *purC*, and *purM* (11). β-Galactosidase activity from each *lacZ* fusion was corrected for small differences in specific activity in order to obtain true estimates of synthesis. Under conditions of derepression resulting from purine limitation, the average relative expression was 1.0:5.0:1.8 for *purE*:*purC*:*purM*. The factors responsible for these differences in expression remain to be determined.

Genes *guaB*, *guaA*, and *purA*

Gene *guaB* encoding IMP dehydrogenase was cloned by complementation of an *E. coli guaB* mutant with a library of *B. subtilis* DNA fragments constructed from a mutant strain selected for high guanosine production (31). A σ^A promoter sequence within 37 bp of the translational start was inferred (19); it requires confirmation by mapping of the mRNA 5′ end. Genes *guaA* and *purA* have recently been isolated and sequenced (26), and work is in progress to examine regulation.

REGULATION

In Vivo Experiments

Addition of purine compounds to the growth medium represses the levels of enzymes involved in de novo purine nucleotide synthesis. Adenosine or guanosine repress enzyme levels in the pathway to IMP in mutants unable to interconvert adenine and guanine nucleotides, whereas synthesis of *purA*-encoded adenylosuccinate synthetase and *guaB*-encoded IMP dehydrogenase is repressed specifically by adenosine and guanosine, respectively (33). Mutant strains in which levels of enzymes involved in the synthesis of IMP are insensitive to repression by adenosine but retain guanosine repression have been reported (18). The corresponding regulatory loci remain to be identified.

More recent work has confirmed the repression of enzyme synthesis by added purine compounds (43). Measurements of purine nucleotide pools provide insights into the purine regulatory molecules. ATP and guanine (or hypoxanthine) are implicated in the regulation of *pur* operon expression. The IMP dehydrogenase level is repressed by guanosine but not in the presence of adenine and is negatively correlated with the ratio of GTP/ATP pools. The adenylosuccinate synthetase level is repressed by adenine, increased by guanosine, and positively correlated with the ratio of GTP/ATP pools.

Transcriptional Regulation of the *pur* Operon

Nuclease S1 mapping experiments have provided evidence for transcription initiation from a σ^A-type promoter 242 bp upstream of *purE*, the first structural gene (8). Two species of RNA were detected: mRNA that extends into the coding region and a second mRNA that is transcribed from the same promoter but is prematurely terminated at approximately nucleotide 200 in the 5′ untranslated leader region. The truncated mRNA results from transcription termination at a secondary structure corresponding to a site for rho-independent termination. Transcription is subject to dual regulation by purine compounds, which may be bases, nucleosides, or nucleotides. For convenience, the regulatory molecules are arbitrarily referred to as nucleobases. Adenine represses transcription initiation and guanine regulates transcription attenuation in the leader mRNA.

On the basis of computer analysis of the 242-nucleotide leader mRNA, antiterminator ($\Delta G = -38$ kcal/mol) and terminator ($\Delta G = -32$ kcal/mol) secondary structures were proposed as regulating attenuation. This hypothesis is based on the attenuation model for regulation of the *B. subtilis trp* operon (44), in which a *trans*-acting regulatory protein, the *mtr* gene product, binds to untranslated leader mRNA and promotes formation of the secondary structure required for transcription termination (2a, 12, 35a). By analogy, a guanine-activated regulatory protein is visualized to bind to an mRNA site to inhibit formation of the

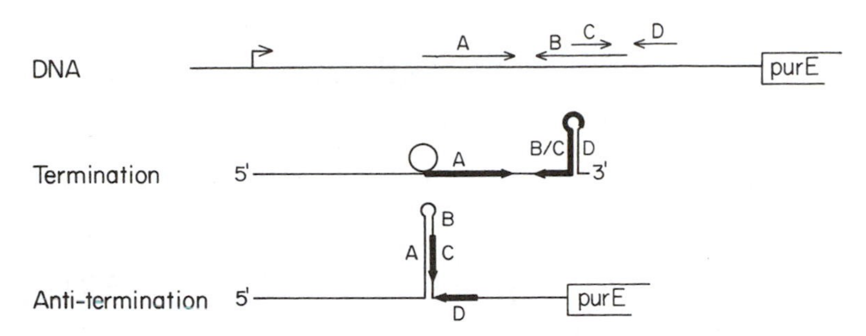

Figure 2. Schematic representation of a termination-antitermination model for regulation of the *B. subtilis pur* operon. The first line shows the region of DNA between the promoter (↱) and the first structural gene. Dyad symmetries are indicated by arrows and letters. The second line shows the prematurely terminated mRNA. Hairpin C:D followed by several uridylate residues is the terminator. A hypothetical guanine-activated regulatory protein, shown as a circle, is bound to segment A. The third line shows the antitermination secondary structure that leads to transcription of coding-length mRNA. Taken from reference 54 with permission.

antiterminator secondary structure, thus allowing the terminator stem-loop to form (Fig. 2). According to this model, when the pool of guanine is insufficient to activate the regulatory protein, binding to mRNA does not occur. In this case, the mutually exclusive antiterminator secondary structure forms and prevents formation of the transcription terminator. Evidence for a formally similar antitermination mechanism for the *E. coli bgl* operon (16) and for *B. subtilis sac* regulon genes (7) and the levanase operon (27) has been reported.

The 200-nucleotide attenuated *pur* operon mRNA accumulates in cells grown with excess guanosine (9). Addition of adenosine represses transcription initiation and prevents accumulation of the attenuated mRNA. Thus, adenine-mediated repression of transcription initiation and guanine-dependent attenuation are independent control mechanisms.

A *cis*-acting region extending from nucleotides −145 to −29 relative to the start of transcription is required for interaction of a protein and repression of transcription initiation (10). Deletions into the control site abolished repression by adenine but had little or no effect on expression or guanine-dependent attenuation. A protein that binds to the control region and is a candidate for a repressor was partially purified. Protein interaction was specific for the *pur* operon control site and exhibited a pattern of alternating protected and hypersensitive regions in a DNase I footprint. Protein-DNA binding was lost gradually as the control region was processively truncated from the 5′ end. Thus, specific subsites were not detected. This protein-DNA interaction is unlike that of the *E. coli pur* regulon repressor, with its 16-bp palindromic control site (40). Although the protein-DNA interaction was not dependent on an adenine compound, the binding conditions that were used are known to permit corepressor-independent binding of *E. coli* purine repressor to specific operator sites (39). The protein-DNA interaction has yet to be investigated under conditions (20) used to demonstrate corepressor dependence for *E. coli* purine repressor (39).

A number of experiments were conducted to quantitate the *pur* operon regulatory response. Regulation by adenine and guanine was determined from measurements of a *lacZ* reporter gene with *pur-lacZ* fusions integrated into the *pur* operon (11). Measurements of β-galactosidase from cells grown under conditions of purine limitation or excess indicated up to 400-fold regulation. The main conclusions that can be derived from these experiments are as follows. (i) No evidence for internal promoters was obtained. All *pur* operon transcription is derived from the σ^A-dependent promoter upstream of *purE*. (ii) In an experiment in which repression of *purM* by adenine plus guanine was 175-fold, regulation of *purE* and *purC* was comparable. Thus, the expectation of coordinate regulation of gene expression from the polycistronic mRNA was confirmed. However, the possibility for low-level transcriptional or translational modulation at an internal site remains open. (iii) Genes *purF* in the *purC QLF* cluster and *purM* in the *purMNH* cluster were closely coregulated over a 2- to 400-fold range. (iv) The individual contributions of adenine-dependent repression of transcription initiation and guanine-mediated attenuation are difficult to assess because of purine interconversions. However, the individual contributions of the two mechanisms can be estimated from analysis of a mutant in which the *cis* control site for repression was deleted (10). In the wild type, 72-fold regulation of *purC-lacZ* by adenine plus guanine was reduced to 12-fold when the control site for repression of transcription initiation was deleted. This suggests that 12-fold guanine-mediated attenuation and 6-fold adenine-dependent repression of transcription initiation account for the 72-fold overall regulation that was observed.

Accumulation and Degradation of Attenuated *pur* Operon mRNA

An apparently intact *pur* operon-attenuated mRNA of approximately 200 nucleotides was found to accumulate in cells grown with excess guanine (9). A combination of Northern (RNA) blot and RNA mapping experiments localized the 5′ end at nucleotides +1 to 4 and the 3′ end at nucleotides 200 to 203. This truncated mRNA thus likely results from transcription attenuation at the stretch of uracil residues at nucleotides 196 to 208, which is part of the rho-independent terminator in the *pur* operon leader. Accumulation of a well-defined truncated mRNA species permitted a detailed analysis of its degradation in

	-11 1 382 434
B. subtilis	MLAEIKGLNEECGV----PCFYGID---CGQCLACF---465
E. coli	MCGI----PNVYGID---QQFECSVF---503

Figure 3. Alignment of *B. subtilis* and *E. coli* glutamine PRPP amidotransferase segments. The alignment shows sequences at the NH_2 terminus and the four cysteinyl ligands to the 4Fe-4S cluster in the *B. subtilis* enzyme. Cys^1 is the NH_2-terminal residue in mature enzymes. Cysteinyl ligands are shaded in grey. Numbers for the COOH termini are at the right.

cells. Four major degradation intermediates accumulated and were characterized. A variety of secondary structures in the leader mRNA appears to stabilize the degradation intermediates and permit their accumulation. By using size and mapping data, it was possible to align the degradation intermediates with the nucleotide sequence, infer secondary structures, and deduce the order of degradation events. Accordingly, degradation of the ~200-nucleotide mRNA is initiated by an endonucleolytic cleavage following nucleotide 106 to yield left (L) and right (R) halves as follows.

$$\sim 200 \begin{array}{l} \nearrow [106L] \rightarrow 93L \rightarrow 88L \rightarrow \text{nucleotides} \\ \searrow 97R \rightarrow 58R \rightarrow \text{nucleotides} \end{array}$$

In this scheme, RNA species are designated by size. A 106L species does not accumulate but undergoes rapid 3′-5′ exonucleolytic cleavage in a short unpaired sequence to yield 93L. Endonucleolytic cleavage in a single-stranded looped-out region of 93L generates 88L. Further degradation intermediates of 88L were not detected. Intermediate 97R decays to 58R by an initial endonucleolytic cleavage followed by 3′-5′ exonuclease activity. Smaller intermediates were not detected for further decay of 58R. Half-lives for decay varied from 0.7 min for the ~200-nucleotide attenuated species to 3.1 and 3.6 min for 88L and 93L, respectively.

The main conclusions from this analysis of mRNA decay are as follows. (i) Each degradation step is initiated by an endonuclease cleavage followed by 3′-5′ exonuclease digestion. (ii) The 3′-5′ exonuclease digestion continues in non-base-paired regions and halts when secondary structure is encountered. (iii) Endonuclease cleavage sites are in short single-stranded loops that connect regions of secondary structure. (iv) The 12 nucleotides at the mRNA 5′ end, although apparently not involved in secondary structure, are retained in the 93L and 88L intermediates, supporting the conclusion that 5′-3′ exonuclease activity plays no role in mRNA decay in *B. subtilis*. (V) The ~200-nucleotide attenuated transcript and degradation intermediates are not generated from decay of full-length *pur* operon mRNA. Their formation is dependent on secondary structures that arise as a result of guanine-dependent attenuation.

RELATIONSHIP OF GUANINE NUCLEOTIDES TO SPORULATION

Limitation of the growth rate resulting from starvation for guanine nucleotides is known to initiate sporulation (21). Drugs that inhibit synthesis of purine nucleotides and decrease guanine nucleotide pools induce sporulation even in media with sufficient sources of carbon, nitrogen, and phosphate. Decoyinine, an inhibitor of GMP synthetase (Fig. 1, step 14); mycophenolic acid, an inhibitor of IMP dehydrogenase (step 13); and hadacidin, an inhibitor of adenylosuccinate synthetase (step 11) were all effective (30). With an expanded series of drugs, the GTP pool was reduced under all conditions that induced sporulation, whereas levels of all other nucleotides were noted to increase under some and decrease under other conditions (1). Formycin-A, which inhibited phosphorylation of GMP to GDP and GTP, caused a 20-fold increase in spore titer, indicating that GDP and/or GTP but not GMP are the signal molecules. Antibiotics inhibiting growth by direct inhibition of nucleic acid synthesis did not induce sporulation. The mechanism relating the GTP pool to sporulation remains to be determined.

ENZYMES OF PURINE NUCLEOTIDE SYNTHESIS

Information about the enzymes in the pathway for de novo purine nucleotide synthesis from *B. subtilis* is limited. Judging from deduced sequence similarities to *E. coli* genes, most enzymes in the pathway are likely similar to those from *E. coli*. There are, however, distinctive features for glutamine PRPP amidotransferase (Fig. 1, reaction 1) and 5′-phosphoribosyl-*N*-formylglycinamidine (FGAM) synthetase (reaction 4).

Glutamine PRPP Amidotransferase

Glutamine PRPP amidotransferase catalyzes the first committed step in de novo purine nucleotide synthesis:glutamine + PRPP → 5-phosphoribosyl-1-amine (PRA) + glutamate + PP_i. Similar to other glutamine amidotransferases, NH_3 can replace glutamine in vitro (28) and in vivo (25). Only the enzymes from *B. subtilis* (52) and *E. coli* (28) have been purified to homogeneity and characterized. Catalytic properties and feedback inhibition by purine nucleotides are similar for the enzymes from *B. subtilis* and *E. coli* (48), reflecting the 41% amino acid sequence identity (24). However, the amidotransferase from *B. subtilis* has two distinctive structural features not found in the *E. coli* enzyme. The *B. subtilis* enzyme is synthesized with an 11-amino-acid NH_2-terminal propeptide that must be cleaved to expose an NH_2-terminal active-site cysteine (Cys^1 in Fig. 3). To understand the role of Cys^1 in the mature enzyme, it is helpful to visualize the overall amidotransferase reaction in terms of two partial reactions (equations 1 and 2).

$$\text{glutamine} + H_2O \rightarrow \text{glutamate} + NH_3 \quad (1)$$

$$\text{PRPP} + NH_3 \rightarrow \text{PRA} + PP_i \quad (2)$$

$$\text{glutamine} + \text{PRPP} + H_2O \rightarrow \text{PRA} + \text{glutamate} + PP_i \quad (3)$$

Reaction 1 is a glutaminase activity, and reaction 2 is an NH_3-dependent synthesis of PRA. Reaction 3 is the sum of these reactions, i.e., glutamine-dependent synthesis of PRA. For the glutamine-dependent reaction (equation 3), it is not known whether the amide of glutamine is released (equation 1) and sequestered as enzyme-bound NH_3 prior to reaction with PRPP (equation 2) or is transferred directly to PRPP (equation 3). Regardless of the mechanism, Cys^1 in the mature enzyme forms a covalent glutaminyl intermediate that is required for the initial step of glutamine hydrolysis and/or amide transfer (25, 28, 51). Mutations that prevent propeptide processing abolish glutamine-dependent activity (47), leading to the conclusion that the NH_2-terminal position of the active-site cysteine residue is obligatory for glutamine utilization. In the *E. coli* enzyme, Met^{-1} is cleaved by methionine aminopeptidase to expose Cys^1.

Although the role of the 11-amino-acid propeptide is not understood, the peptide is not peculiar to the *B. subtilis* enzyme. Recent cloning and analyses of avian (55) and human (6) glutamine PRPP amidotransferase cDNAs indicate conservation of the 11-amino-acid propeptide. In addition, *E. coli* glutamate synthase, an Fe-S-containing glutamine amidotransferase, contains a comparable propeptide sequence (34, 49).

A 4Fe-4S cluster is the second distinctive structural feature in the *B. subtilis* amidotransferase (53). Cysteine residues 382, 434, 437, and 440 are ligands to the Fe-S cluster (Fig. 3) (23, 24). These cysteine residues are not conserved in the *E. coli* enzyme. Cys-382 is distal to a sequence motif at residues 344 to 354 that is conserved in all enzymes that utilize PRPP (17), suggesting proximity of the 4Fe-4S cluster to the catalytic site. However, the Fe-S cluster does not appear to have a direct role in catalysis (50). Present evidence supports a role for the 4Fe-4S cluster in a mechanism to shut down purine biosynthesis prior to sporulation (13). The 4Fe-4S cluster serves as a site for oxidative inactivation of the amidotransferase, the rate-limiting step in enzyme degradation in nutrient-starved cells. The *E. coli* amidotransferase does not contain a 4Fe-4S cluster and is not subject to oxidative inactivation.

FGAM Synthetase

FGAM synthetase is a glutamine amidotransferase that catalyzes step 4 in the de novo pathway. *E. coli purL* conforms to a fusion of genes *purL* and *purQ* from *B. subtilis* (8, 41). *B. subtilis* FGAM synthetase has not been studied. However, we infer that *B. subtilis* FGAM synthetase is a PurL PurQ heterodimer in which the 742-amino-acid PurL subunit has the capacity for NH_3-dependent synthesis of FGAM and the 227-amino-acid PurQ subunit confers the capacity to utilize the amide of glutamine in place of NH_3. These functional domains are fused in *E. coli purL*. This relationship of separated versus fused domains for a glutamine amidotransferase is also found in carbamoyl phosphate synthetase (45) and anthranilate synthase (4). Functional distinctions between multisubunit and fused-protein chains are not known.

PURINE NUCLEOTIDE SYNTHESIS IN OTHER GRAM-POSITIVE ORGANISMS

A computer search of nine biological science and biotechnology data bases with DIALOG did not identify references to genes or enzymes of purine nucleotide synthesis in other gram-positive bacteria for the period 1974 through 1992. A cluster of purine genes corresponding to a portion of the *B. subtilis pur* operon was recently cloned from *Lactobacillus casei* (13a).

NOTE ADDED IN PROOF

A recently completed X-ray structure of *B. subtilis* glutamine PRPP amidotransferase shows that Cys-236, Cys-382, Cys-437, and Cys-440 are ligands to the Fe-S cluster (46a).

Acknowledgments. I thank Per Nygaard for communicating information prior to publication. Research from my laboratory has been supported by Public Health Service grant GM24658 from the National Institutes of Health.

REFERENCES

1. **Abedin, Z., J. M. Lopez, and E. Freese.** 1983. Induction of bacterial differentiation by adenine- and adenosine-analogs and inhibitors of nucleic acid synthesis. *Nucleosides Nucleotides* **2:**257–274.
2. **Aiba, A., and K. Mizobuchi.** 1989. Nucleotide sequence analysis of genes *purH* and *purD* involved in the *de novo* purine nucleotide biosynthesis of *Escherichia coli*. *J. Biol. Chem.* **264:**21239–21246.

2a. **Babitzke, P., and C. Yanofsky.** 1992. Reconstitution of *Bacillus subtilis trp* attenuation *in vitro* with TRAP, the *trp* RNA-binding attenuation protein. *Proc. Natl. Acad. Sci. USA* **90:**133–137.

3. **Bachmann, B. J.** 1990. Linkage map of *Escherichia coli* K-12, edition 8. *Microbiol. Rev.* **54:**130–197.
4. **Bae, Y. M., E. Holmgren, and I. P. Crawford.** 1989. *Rhizobium meliloti* anthranilate synthase gene: cloning, sequence and expression in *Escherichia coli*. *J. Bacteriol.* **171:**3471–3478.
5. **Brey, R. N., C. D. B. Banner, and J. B. Wolf.** 1986. Cloning of multiple genes involved with cobalamin (vitamin B_{12}) biosynthesis in *Bacillus megaturium*. *J. Bacteriol.* **167:**623–630.
6. **Chen, Z., G. Zhou, A. Gavalas, J. E. Dixon, and H. Zalkin.** Unpublished data.
7. **Debarbouille, M., M. Arnaud, A. Fouet, A. Klier, and G. Rapoport.** 1990. The *sacT* gene regulating the *sacPA* operon in *Bacillus subtilis* shares strong homology with transcriptional antiterminators. *J. Bacteriol.* **172:**3966–3973.
8. **Ebbole, D. J., and H. Zalkin.** 1987. Cloning and characterization of a 12-gene cluster from *Bacillus subtilis* encoding nine enzymes for *de novo* purine nucleotide synthesis. *J. Biol. Chem.* **262:**8274–8287.
9. **Ebbole, D. J., and H. Zalkin.** 1988. Detection of *pur* operon-attenuated mRNA and accumulated degradation intermediates in *Bacillus subtilis*. *J. Biol. Chem.* **263:**10894–10902.
10. **Ebbole, D. J., and H. Zalkin.** 1989. Interaction of a putative repressor protein with an extended control region of the *Bacillus subtilis pur* operon. *J. Biol. Chem.* **264:**3553–3561.
11. **Ebbole, D. J., and H. Zalkin.** 1989. *Bacillus subtilis pur* operon expression and regulation. *J. Bacteriol.* **171:**2136–2141.
12. **Gollnick, P., S. Ishino, M. I. Kuroda, D. J. Henner, and C. Yanofsky.** 1990. The *mtr* locus is a two-gene operon required for transcription attenuation in the *trp* operon

of *Bacillus subtilis*. *Proc. Natl. Acad. Sci. USA* **87**:8726–8730.

13. **Grandoni, J. A., R. L. Switzer, C. A. Makaroff, and H. Zalkin.** 1989. Evidence that the iron-sulfur cluster of *Bacillus subtilis* glutamine phosphoribosylpyrophosphate amidotransferase determines stability of the enzyme to degradation *in vivo*. *J. Biol. Chem.* **264**:6058–6064.

13a. **Gu, Z.-M., D. W. Martindale, and B. H. Lee.** 1992. Isolation and complete sequence of the *purL* gene encoding FGAM synthase II in *Lactobacillus casei*. Gene **119**: 123–126.

14. **He, B., A. Shiau, K. Y. Choi, H. Zalkin, and J. M. Smith.** 1990. Genes of the *Escherichia coli pur* regulon are negatively controlled by a repressor-operator interaction. *J. Bacteriol.* **172**:4555–4562.

15. **Henner, D. J., L. Band, and H. Shimotsu.** 1984. Nucleotide sequence of the *Bacillus subtilis* tryptophan operon. *Gene* (Amsterdam) **34**:169–177.

16. **Houman, F., M. R. Diaz-Torres, and A. Wright.** 1990. Transcriptional antitermination in the *bgl* operon of *E. coli* is modulated by a specific RNA binding protein. *Cell* **62**:1143–1163.

17. **Hove-Jensen, B., K. W. Harlow, C. J. King, and R. L. Switzer.** 1986. Phosphoribosylpyrophosphate synthetase of *Escherichia coli*. Properties of the purified enzyme and primary structure of the *prs* gene. *J. Biol. Chem.* **261**: 6765–6771.

18. **Ishii, K., and I. Shiio.** 1972. Improved inosine production and derepression of purine nucleotide biosynthetic enzymes in 8-azaguanine resistant mutants of *Bacillus subtilis*. *Agric. Biol. Chem.* **38**:1511–1522.

19. **Kanzaki, N., and K. Miyagawa.** 1990. Nucleotide sequence of the *Bacillus subtilis* IMP dehydrogenase gene. *Nucleic Acids Res.* **18**:6710.

20. **Leirmo, S., C. Harrison, D. S. Cayley, R. R. Burgess, and M. T. Record, Jr.** 1987. Replacement of potassium chloride by potassium glutamate dramatically enhances protein-DNA interactions *in vitro*. *Biochemistry* **26**:2095–2101.

21. **Lopez, J. M., C. L. Marks, and E. Freeze.** 1979. The decrease of guanine nucleotides initiates sporulation of *Bacillus subtilis*. *Biochim. Biophys. Acta* **587**:238–252.

22. **Mackey, C. J., R. J. Warburg, H. O. Halvorson, and S. A. Zahler.** 1984. Genetic and physical analysis of the *ilvBC-leu* region in *Bacillus subtilis*. *Gene* (Amsterdam) **32**:49–56.

23. **Makaroff, C. A., J. L. Paluh, and H. Zalkin.** 1986. Mutagenesis of ligands to the [4Fe-4S] center of *Bacillus subtilis* glutamine phosphoribosylpyrophosphate amidotransferase. *J. Biol. Chem.* **261**:11416–11423.

24. **Makaroff, C. A., H. Zalkin, R. L. Switzer, and S. J. Vollmer.** 1983. Cloning of the *Bacillus subtilis* glutamine phosphoribosylpyrophosphate amidotransferase gene in *Escherichia coli*. Nucleotide sequence determination and properties of the plasmid-encoded enzyme. *J. Biol. Chem.* **258**:10586–10593.

25. **Mäntsälä, P., and H. Zalkin.** 1984. Glutamine amidotransferase function. Replacement of the active site cysteine in glutamine phosphoribosylpyrophosphate amidotransferase by site-directed mutagenesis. *J. Biol. Chem.* **259**:14230–14236.

26. **Mäntsälä, P., and H. Zalkin.** 1992. Cloning and sequence of *Bacillus subtilis purA* and *guaA* involved in the conversion of IMP to AMP and GMP. *J. Bacteriol.* **174**:1883–1890.

27. **Martin-Verstraete, I., M. Débarbouillé, A. Klier, and G. Rapoport.** 1990. Levanase operon of *Bacillus subtilis* includes a fructose-specific phosphotransferase system regulating the expression of the operon. *J. Mol. Biol.* **214**:657–671.

28. **Messenger, L. J., and H. Zalkin.** 1979. Glutamine phosphoribosylpyrophosphate amidotransferase from *Escherichia coli*. Purification and properties. *J. Biol. Chem.* **254**:3382–3392.

29. **Miller, P., A. Rabinowitz, and H. Taber.** 1988. Molecular cloning and preliminary genetic analysis of the *men* gene cluster of *Bacillus subtilis*. *J. Bacteriol.* **170**:2735–2741.

30. **Mitani, T., J. E. Heinze, and E. Freese.** 1977. Induction of sporulation in *Bacillus subtilis* by decoyinine or hadacidin. *Biochem. Biophys. Res. Commun.* **77**:1118–1125.

31. **Miyagawa, K., H. Kimura, K. Nakahama, M. Kikuchi, M. Doi, S. Akiyama, and Y. Nakao.** 1986. Cloning of the *Bacillus subtilis* IMP dehydrogenase gene and its application to increased production of guanosine. *Bio/Technology* **4**:225–228.

32. **Mountain, A., J. McChesney, M. C. M. Smith, and S. Baumberg.** 1986. Gene sequence encoding early enzymes of arginine synthesis within a cluster in *Bacillus subtilis*, as revealed by cloning in *Escherichia coli*. *J. Bacteriol.* **165**:1026–1028.

33. **Nishikawa, H., H. Momose, and I. Shiio.** 1967. Regulation of purine nucleotide synthesis in *Bacillus subtilis*. II. Specificity of purine derivatives for enzyme repression. *J. Biochem.* **62**:92–98.

34. **Oliver, G., G. Gosset, R. Sanchez-Pescador, E. Lozoya, C. M. Ku, N. Flores, B. Becerril, F. Valle, and F. Bolivar.** 1987. Determination of the nucleotide sequence for the glutamate synthase structural genes of *Escherichia coli* K-12. *Gene* **60**:1–11.

35. **Oppenheim, D., and C. Yanofsky.** 1980. Translational coupling during expression of the tryptophan operon of *Escherichia coli*. *Genetics* **95**:785–795.

35a. **Otridge, J., and P. Gollnick.** 1992. MtrB from *Bacillus subtilis* binds specifically to *trp* leader RNA in a tryptophan-dependent manner. *Proc. Natl. Acad. Sci. USA* **90**: 128–132.

36. **Petersen, C.** 1989. Long range translational coupling in the *rplJL-rpoBC* operon of *Escherichia coli*. *J. Mol. Biol.* **206**:323–332.

37. **Piggot, P. J., and J. Hoch.** 1985. Revised genetic linkage map of *Bacillus subtilis*. *Microbiol. Rev.* **49**:158–179.

38. **Quinn, C. L., B. T. Stephenson, and R. L. Switzer.** 1991. Functional organization and nucleotide sequence of the *Bacillus subtilis* pyrimidine biosynthetic operon. *J. Biol. Chem.* **266**:9113–9127.

39. **Rolfes, R. J., and H. Zalkin.** 1990. Purification of the *Escherichia coli* purine regulon repressor and identification of corepressors. *J. Bacteriol.* **172**:5637–5642.

40. **Rolfes, R. J., and H. Zalkin.** 1990. Autoregulation of *Escherichia coli purR* requires two control sites downstream of the promoter. *J. Bacteriol.* **172**:5758–5766.

41. **Sampei, G., and K. Mizobuchi.** 1989. The organization of the *purL* gene encoding 5′-phosphoribosylformylglycinamide amidotransferase of *Escherichia coli*. *J. Biol. Chem.* **264**:21230–21238.

42. **Saxild, H. H., and P. Nygaard.** 1988. Gene-enzyme relationships of the purine biosynthetic pathway in *Bacillus subtilis*. *Mol. Gen. Genet.* **211**:160–167.

43. **Saxild, H. H., and P. Nygaard.** 1991. Regulation of levels of purine biosynthetic enzymes in *Bacillus subtilis*: effects of changing purine nucleotide pools. *J. Gen. Microbiol.* **137**:2387–2394.

44. **Shimotsu, H., M. Kuroda, C. Yanofsky, and D. J. Henner.** 1986. Novel form of transcription attenuation regulates expression of the *Bacillus subtilis* tryptophan operon. *J. Bacteriol.* **166**:461–471.

45. **Simmer, J. P., R. E. Kelly, A. G. Rinker, Jr., J. L. Scully, and D. R. Evans.** 1990. Mammalian carbamyl phosphate synthetase (CPS). cDNA sequence and evolution of the CPS domain of the Syrian hamster multifunctional protein CAD. *J. Biol. Chem.* **265**:10395–10402.

46. **Slock, J., D. P. Stahly, C.-Y. Han, E. W. Six, and I. P. Crawford.** 1990. An apparent *Bacillus subtilis* folic acid biosynthetic operon containing *pab*, an amphibolic *trpG* gene, a third gene required for synthesis of *para*-amino-

benzoic acid, and the dihydropteroate synthase gene. *J. Bacteriol.* **172:**7211–7226.

46a. **Smith, J. L.** Personal communication.

47. **Souciet, J.-L., M. A. Hermodson, and H. Zalkin.** 1988. Mutational analysis of the glutamine phosphoribosylpyrophosphate amidotransferase pro-peptide. *J. Biol. Chem.* **263:**3323–3327.

48. **Switzer, R. L.** 1989. Regulation of bacterial glutamine phosphoribosylpyrophosphate amidotransferase, p. 129–151. *In* G. Hervé (ed.), *Allosteric Enzymes*. CRC Press, Boca Raton, Fla.

49. **Velázquez, L., L. Camarena, J. L. Reyes, and F. Bastarrachea.** 1991. Mutations affecting the Shine-Dalgarno sequences of the untranslated regions of the *Escherichia coli gltBDF* operon. *J. Bacteriol.* **173:**3261–3264.

50. **Vollmer, S. J., R. L. Switzer, and P. J. Debrunner.** 1983. Oxidation-reduction properties of the iron-sulfur cluster in *Bacillus subtilis* glutamine phosphoribosylpyrophosphate amidotransferase. *J. Biol. Chem.* **258:**14284–14293.

51. **Vollmer, S. J., R. L. Switzer, M. A. Hermodson, and H. Zalkin.** 1983. The glutamine-utilizing site of *Bacillus subtilis* glutamine phosphoribosylpyrophosphate amidotransferase. *J. Biol. Chem.* **258:**10582–10585.

52. **Wong, J. Y., D. A. Bernlohr, C. L. Turnbough, and R. L. Switzer.** 1981. Purification and properties of glutamine phosphoribosylpyrophosphate amidotransferase from *Bacillus subtilis*. *Biochemistry* **20:**5669–5674.

53. **Wong, J. Y., E. Meyer, and R. L. Switzer.** 1977. Glutamine phosphoribosylpyrophosphate amidotransferase from *Bacillus subtilis*. A novel iron-sulfur protein. *J. Biol. Chem.* **252:**7424–7426.

54. **Zalkin, H., and D. J. Ebbole.** 1988. Organization and regulation of genes encoding biosynthetic enzymes in *Bacillus subtilis*. *J. Biol. Chem.* **263:**1595–1598.

55. **Zhou, G., J. E. Dixon, and H. Zalkin.** 1990. Cloning and expression of avian glutamine phosphoribosylpyrophosphate amidotransferase. Conservation of a bacterial propeptide sequence supports a role for posttranslational processing. *J. Biol. Chem.* **265:**21152–21159.

25. De Novo Pyrimidine Nucleotide Synthesis

ROBERT L. SWITZER and CHERYL L. QUINN

PATHWAY AND ENZYMES OF PYRIMIDINE BIOSYNTHESIS

Overall Pathway

The pathway of biosynthesis of pyrimidine nucleotides de novo from central metabolic intermediates (Fig. 1) is the same in all prokaryotic and eukaryotic organisms in which it has been examined, but as will be seen, the functional organization of the enzymes of the pathway and the means adopted for regulation of pyrimidine biosynthesis vary greatly from species to species.

Enzymes of UMP Biosynthesis

The primary structures of all of the enzymes of de novo UMP biosynthesis from *Bacillus subtilis* are known because the genes encoding them have been sequenced (44, 85). General properties of each of the enzymes have also been determined in studies with crude extracts (72), but only aspartate transcarbamylase has been purified to homogeneity and characterized in detail (9, 10, 103). General properties of the enzymes and genes encoding them are summarized in Tables 1 and 2. Further detail, with emphasis on the properties that distinguish the *B. subtilis* enzymes from their homologs in other species, follows.

Carbamylphosphate synthetases (*pyrAA* and *pyrAB*)

B. subtilis resembles eukaryotic cells in possessing two carbamylphosphate synthetases that are specialized to serve either pyrimidine biosynthesis or arginine biosynthesis (73). The arginine-repressible carbamylphosphate synthetase is encoded by the *cpa* gene in the *B. subtilis* arginine biosynthetic operon, *argCAEBD-cpa-argF*, which maps at 102° on the chromosome (57). Synthesis of this isozyme is coregulated with that of the other arginine biosynthetic enzymes (66); no allosteric regulators of its activity have been identified (73). The pyrimidine-repressible carbamylphosphate synthetase is composed of two subunits, which are encoded by the *pyrAA* and *pyrAB* genes. These genes are the fourth and fifth cistrons of a nine-cistron pyrimidine biosynthetic cluster that maps at 139° on the chromosome (76, 77, 85). Carbamylphosphate pools produced by the two carbamylphosphate synthetases evidently commingle in vivo, because *cpa* mutants are not arginine auxotrophs, nor are *pyrA* mutants pyrimidine auxotrophs. Rather, these mutants have the phenotypes of pyrimidine-sensitive and arginine-sensitive growth, respectively (73, 80). These phenotypes are readily understood by assuming that both biosynthetic pathways can use carbamylphosphate provided by either enzyme, so that a mutant lacking one of the enzymes will be starved for arginine or pyrimidines when the remaining carbamylphosphate synthetase is repressed by the other end product. As expected, both *cpa* and *pyrA* mutants grow well when both arginine and a pyrimidine are added to the medium (73).

B. subtilis pyrimidine-repressible carbamylphosphate synthetase strongly resembles the corresponding enzyme from enteric bacteria. The glutaminase and catalytic subunit sequences are 46 and 48% identical to the *Escherichia coli carA*- and *carB*-encoded sequences, respectively (85). The sequence identity is especially pronounced in regions of the sequence that have been identified from studies of the *E. coli* enzyme's involvement in the catalytic mechanism. Specifically, a triad of cysteine, histidine, and glutamic acid residues identified by Amuro et al. (1) as involved in the glutaminase activity of the *trpG* class of amidotransferases is conserved in the *B. subtilis pyrAA* subunit (85). Likewise, glycine residues thought to form part of two distinct ATP-binding sites in the catalytic subunit (79) are conserved and lie in highly conserved sequences of the *B. subtilis pyrAB* subunit (85). The sequence similarities make it virtually certain that the catalytic mechanism of the *B. subtilis* carbamylphosphate synthetase is essentially the same as that of the much more extensively studied *E. coli* enzyme. A minimal molecular weight for the native enzyme is 158,000, but gel filtration analysis of the native enzyme in bacterial extracts provided an estimate of 90,000 to 100,000 (73). This may indicate that the subunits dissociate readily on gel filtration so that activity was observed only in the zone of overlap between the elution of the two subunits. Alternatively, the enzyme may have some affinity for the gel matrix used in the chromatography.

The resemblance between the enteric and *B. subtilis pyrA* carbamylphosphate synthetases extends to their kinetic properties. The *B. subtilis* enzyme absolutely requires high levels of K^+ ions (250 mM) for activity and stability; other monovalent cations do not substitute for K^+ and inhibit the K^+-supported reaction (73). Ammonium ions will substitute for glutamine, but much higher concentrations are required (Table 1). ATP saturation curves are sigmoid and are shifted dramatically by allosteric inhibitors and activators (73). The most important difference between the *E. coli* and *B. subtilis pyrA* carbamylphosphate synthetases is the complete insensitivity of the latter to

Robert L. Switzer • Department of Biochemistry, University of Illinois, Urbana, Illinois 61801. **Cheryl L. Quinn** • Molecular Immunology Group, Institute of Molecular Medicine, University of Oxford, John Radcliffe Hospital, Headington, Oxford OX3 9DU, England.

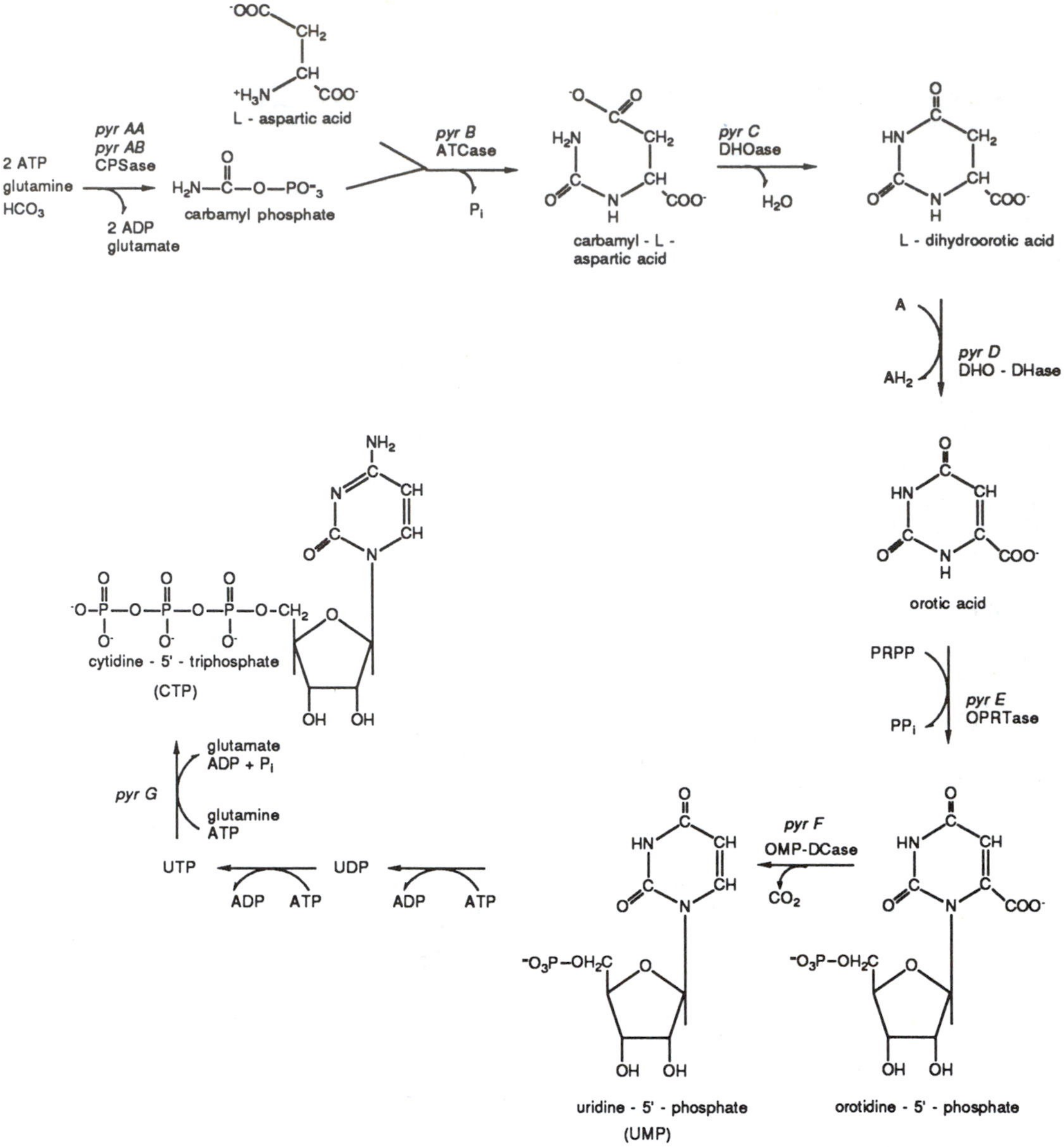

Figure 1. Pathway for de novo synthesis of pyrimidine nucleotides. Gene designations and commonly used abbreviations for the individual enzymes are shown. These are defined fully in the text and in Table 2. CPSase, carbamylphosphate synthetase; ATCase, aspartate transcarbamylase; DHOase, dihydroorotase; DHO-DHase, dihydroorotate dehydrogenase; A, electron acceptor; AH_2, reduced acceptor; OPRTase, orotate phosphoribosyltransferase; OMP-DCase, OMP decarboxylase.

allosteric effects of ornithine (or other intermediates of arginine biosynthesis) (73). Ornithine also has no effect on inhibition of the *B. subtilis* enzyme by UMP. This difference no doubt reflects the existence of the *cpa*-encoded carbamylphosphate synthetase in *B. subtilis*, which is repressed by arginine.

Aspartate transcarbamylase (*pyrB*)

The *B. subtilis* aspartate transcarbamylase has been extensively studied (9, 10, 44, 103). The native enzyme is a trimer of catalytic subunits and is not subject to allosteric regulation, as would be expected from its lack of regulatory subunits (10). Steady-state kinetic studies (10), sensitivity of the enzyme to the transition analog *N*-phosphonoacetyl-L-aspartate (11), and detailed comparison of the enzyme's amino acid sequence with that of the *E. coli* catalytic subunit (44) all indicate that the mechanism of catalysis and the overall polypeptide folding are the same for the two enzymes. Thus, the *B. subtilis* enzyme presents a useful object for study of aspartate transcarbamylase structure and catalysis separately from the complexities introduced by homotropic cooperativity of substrate binding and the effects of regulatory subunits. The *B. subtilis pyrB* gene has been introduced into an overproducing vector (42), the enzyme has been crystallized (9, 103), and X-ray diffraction studies of its structure have recently been published (103). The *B. subtilis* aspartate transcarbamylase is activated two- to fivefold by anions such as acetate, citrate, lactate,

Table 1. Properties of pyrimidine biosynthetic enzymes from *B. subtilis*

Enzyme	Mol wt		Substrate and K_m or $S_{0.5}$[a] (mM)	pH optimum	Activator(s)	Inhibitor(s)	Reference(s)
	Subunit	Native					
Carbamylphosphate synthetase		160,000	ATP, 10; glutamine, 0.36; HCO_3, 3.6; NH_4^+, 11	7.5	PRPP, GTP, GDP, GMP	UTP, UDP, UMP	73, 85
Glutaminase subunit	40,123						
Catalytic subunit	117,663			K^+			
Aspartate transcarbamylase	34,173	100,000	Aspartate, 7.0; carbamylphosphate, 0.11	8.5	Anions, such as acetate, citrate		9, 10, 44
Dihydroorotase	46,614	?	Dihydroorotate, 1.1	9.8	Zn^{2+}		72, 85, 95
Dihydroorotate dehydrogenase	33,094	?	Dihydroorotate, 0.2	7.8		Orotate	72, 85, 95
Orotate phosphoribosyltransferase	23,524	?	Orotate, ?; PRPP, ?	8–9	Mg^{2+}	Dihydroorotate	52, 72, 85
OMP decarboxylase	25,994	?	OMP, 0.0019	6.7–9.5	None identified	None identified	52, 72, 85
Pyrimidine ribonucleoside monophosphokinase	25,000	25,000	CMP, 0.04; UMP, 0.25; ATP, 0.04–0.4	6–9		Several ribonucleoside di- and triphosphates	115
Nucleoside diphosphokinase	?	100,000	UTP, 0.13; CTP, 0.28; ATP, 0.15	?	Mg^{2+}, Mn^{2+}		94
CTP synthetase	59,717	?	Glutamine, ?; UTP, ?; ATP, ?	?	?	?	3, 110

[a] $S_{0.5}$, concentration yielding half of the maximal velocity in cases where the saturation curve is not hyperbolic.

and phosphate (10). In other ways, the enzyme is remarkably similar to the isolated catalytic subunits of *E. coli* aspartate transcarbamylase in its kinetic properties.

Dihydroorotase (*pyrC*)

B. subtilis dihydroorotase has not been purified, but optimal conditions for its assay and general kinetic characteristics of the enzyme in crude extracts have been described elsewhere (72, 95) (Table 1). Like the dihydroorotase from *E. coli* (117), the *B. subtilis* enzyme is stimulated by Zn^{2+} ions. The amino acid sequences of dihydroorotase enzymes or domains from seven diverse species have been compared and found to show little sequence identity (85). Only 22 residues were conserved among all seven species compared. Five of the conserved residues are histidines; it is likely that two or three of these are involved in binding Zn^{2+} to the enzyme. Many other conserved residues are prolines, glycines, or hydrophobic amino acids, so they are probably important in maintaining the correct tertiary structure of the enzyme.

Dihydroorotate dehydrogenase (*pyrD*)

B. subtilis dihydroorotate dehydrogenase has been studied only in crude extracts (72, 95) and by sequencing of the *pyrD* gene (85). The enzyme is unstable in extracts but can be stabilized sufficiently for general characterization by inclusion of bovine serum albumin in the buffers (72) (Table 1). The enzyme can be assayed with ferricyanide as the electron acceptor (this assay has been validated radiochemically with [^{14}C]dihydroorotate), which suggests that, like the purified *E. coli* enzyme (41), *B. subtilis* dihydroorotate

Table 2. *B. subtilis* genes of pyrimidine biosynthesis

Gene	Map position	Enzyme encoded	Mutants known[a]	Gene cloned and sequenced	Reference(s)
pyrAA	139	Carbamylphosphate synthetase, glutaminase subunit	Yes[b]	Yes	43, 80, 81, 85
pyrAB	139	Carbamylphosphate synthetase, catalytic subunit	Yes[b]	Yes	43, 80, 81, 85
pyrB	139	Aspartate transcarbamylase	Yes	Yes	43, 44, 81, 85
pyrC	139	Dihydroorotase	Yes	Yes	43, 81, 85
pyrD	139	Dihydroorotate dehydrogenase	Yes	Yes	43, 81, 85
pyrE	139	Orotate phosphoribosyltransferase	No[c]	Yes	43, 81, 85
pyrF	139	OMP decarboxylase	Yes	Yes	43, 81, 85
pyrG (*ctrA*)	324	CTP synthetase	Yes	Yes	110
pmk[d]	?	Pyrimidine ribonucleoside monophosphokinase	No	No	
ndk	204	Nucleoside diphosphokinase	No	Yes	24a
prs	6–7	PRPP synthetase	No	Yes	64, 65

[a] Known mutants are available in the *Bacillus* Genetic Stock Center.
[b] *pyrA* mutants have been described, but it is not known whether they lie in *pyrAA* or *pyrAB*.
[c] Several mutants designated *pyrX* (81) are probably leaky *pyrE* mutants.
[d] *pmk* is suggested as a genetic designation for this gene.

dehydrogenase contains a flavin nucleotide. Comparison of the amino acid sequences of dihydroorotate dehydrogenases from three species showed only weak sequence identity but led to the identification of a probable flavin nucleotide binding sequence near the carboxyl terminus of the enzyme (85). The *B. subtilis* enzyme appears to be soluble rather than membrane bound (95). The enzyme is partially inhibited by orotate at millimolar concentrations, although such product inhibition is of doubtful physiological importance. Contrary to an early report (81), the enzyme is not inhibited by purine or pyrimidine nucleotides (72, 95).

Orotate phosphoribosyltransferase (*pyrE*)

A reliable radiochemical assay for *B. subtilis* orotate phosphoribosyltransferase in crude extracts has been developed. This assay is based on the use of [*carboxy*-^{14}C]orotate and coupling to excess orotidylate (OMP) decarboxylase (52, 72). The enzyme is quite unstable in extracts, but a few kinetic properties have been determined (Table 1). The enzyme requires Mg^{2+} ions and is inhibited by dihydroorotate. Sequencing of the *B. subtilis pyrE* gene and comparison of its deduced amino acid sequence with the sequences of corresponding enzymes and catalytic domains from other species revealed only weak sequence relatedness except in a region that has been identified as a probable 5-phosphoribosyl-1-pyrophosphate (PRPP)-binding sequence (85).

OMP decarboxylase (*pyrF*)

As for orotate phosphoribosyltransferase, *B. subtilis* OMP decarboxylase has been characterized only by radiochemical assays in crude extracts (52, 72). The enzyme displays a small K_m for OMP (1.9 μM; Table 1) and, when assayed with 0.5 mM OMP, is insensitive to a variety of purine and pyrimidine nucleotides or divalent cations (52, 72). A surprising result was complete inhibition by 1 mM phenylmethylsulfonyl fluoride. Genes encoding OMP decarboxylases from 20 diverse species have been sequenced and compared (36, 86). The amino acid sequences show little similarity except for 10 conserved residues that lie in four regions (36) and are probably critical for enzyme function.

Accessory Enzymes

Conversion of UMP to UTP

In *E. coli* and *Salmonella typhimurium*, UMP is known to be converted to UTP by two successive ATP-dependent phosphorylation steps. The first step is catalyzed by the UMP-specific UMP kinase encoded by the *pyrH* gene (27). Temperature-sensitive *pyrH* mutants have deficient UTP pools and are derepressed for enzymes of pyrimidine biosynthesis (27). The second step, conversion of UDP to UTP, is catalyzed in enteric bacteria by the nonspecific enzyme nucleoside diphosphokinase, which is encoded by the *ndk* gene (21, 22, 87). The corresponding pathway in *B. subtilis* has not been investigated by genetic approaches, but studies of purified enzyme activities in vitro strongly indicate that similar steps are involved. A pyrimidine ribonucleoside monophosphokinase from *B. subtilis* cells has been purified to homogeneity (115) (Table 1). The purified enzyme catalyzes the ATP-dependent phosphorylation of both UMP and CMP but has no activity with pyrimidine deoxyribonucleoside monophosphates or with purine ribonucleoside monophosphates. It has not been possible to separate activities that use UMP or CMP as specific substrates. These results indicate that in contrast to enteric bacteria, which produce specific UMP and CMP kinases, *B. subtilis* uses a single enzyme for conversion of both UMP and CMP to the corresponding diphosphates (61). A dCMP kinase that shows significant activity with CMP and UMP as substrates has also been purified from *B. subtilis* (58). Neuhard has suggested that this enzyme is the same as the CMP-UMP kinase previously purified and that it loses its dCMP kinase activity under some conditions (58).

A nucleoside diphosphokinase with broad substrate specificity for the reaction XTP + YDP = XDP + YTP has been purified to homogeneity from *B. subtilis* cells (94). The enzyme accepts both purine and pyrimidine nucleotides and both ribo- and deoxyribonucleotides as substrates. Kinetic studies indicate that the enzyme operates via a ping-pong mechanism, with formation of a phosphoenzyme intermediate. The intermediate has been demonstrated experimentally (94).

CTP synthetase

The CTP synthetase of *B. subtilis* has been identified only by means of molecular genetics. A mutant, called *ctrA1* and characterized by a requirement for cytidine for growth in the absence of ammonium ions, maps at 324° on the *B. subtilis* chromosome (76). The corresponding enzyme was identified by sequencing of chromosomal DNA adjacent to the *spo0F* locus, which lies very near *ctrA*, as CTP synthetase (110). *ctrA* is encoded by a 535-codon open reading frame (ORF) that encodes a sequence that is 53% identical to the amino acid sequence of *E. coli* CTP synthetase (118). The similarity increases to 73% if both identical residues and conservative substitutions are included. Functional residues involved in glutamine utilization by *E. coli* CTP synthetase are all conserved in the *B. subtilis* sequence and lie in very highly conserved regions. Disruption of the *B. subtilis ctrA* locus by introduction of a chloramphenicol resistance gene within the coding sequence led to the expected *ctrA* phenotype, except that the disrupted strain grew poorly without cytidine even in the presence of ammonium salts. This suggests that the original *ctrA1* allele was capable of CTP synthesis but defective in the utilization of glutamine as an amino group donor. Although the identity of the *ctrA*-encoded protein was not shown by direct enzymatic assay or by complementation of an *E. coli pyrG* mutant, there is little doubt that *ctrA* encodes *B. subtilis* CTP synthetase, and we propose that the gene be renamed *pyrG*. Even though the *B. subtilis* CTP synthetase has never been purified or characterized in vitro, its strong similarity in sequence to the well-studied *E. coli* CTP synthetase (37) suggests that the kinetic properties and allosteric regulation by GTP of the *B. subtilis* enzyme will resemble those of the *E. coli* CTP synthetase.

PRPP synthetase

PRPP is crucial for pyrimidine nucleotide biosynthesis both as a substrate for the orotate phosphoribosyltransferase step of the de novo pathway and as a substrate for the salvage pathways. *B. subtilis* PRPP synthetase is encoded by the *prs* gene, which maps at 6° to 7° on the chromosome (64). The *prs* gene has been cloned and sequenced (65), and the enzyme has been purified to near homogeneity (2). The enzyme subunit has a molecular weight of 34,696 (317 codons; the initiator methionine is removed from the NH_2 terminus [2]) and appears to be an octamer in the native state. *B. subtilis* PRPP synthetase resembles the enzymes from other species in requiring Mg^{2+} or Mn^{2+} for activity and P_i as an essential activator (2). The enzyme is subject to allosteric inhibition by ADP and GDP. In this respect, the *B. subtilis* enzyme resembles the mammalian PRPP synthetases, which are inhibited by both ADP and GDP or GMP; only ADP is an effective allosteric inhibitor of PRPP synthetase from enteric bacteria (2). Amino acid sequences of all known PRPP synthetases are quite strongly related (2). Expression of the *E. coli* and *S. typhimurium prs* genes is regulated to a modest extent (fivefold) by pyrimidines in the growth medium (26, 68, 119). Effects of pyrimidines on expression of *B. subtilis prs* have not been observed.

Formation of deoxythymidylate and dTTP

The pathway for formation of dTTP and dCTP from pyrimidine ribonucleotides in *B. subtilis* was reviewed by Møllgaard and Neuhard in 1983 (56), and little additional information has been published since then. Current knowledge of *B. subtilis* ribonucleotide reductase, which is thought to catalyze reduction of CTP and UTP to dCTP and dUTP, will be presented in chapter 26 of this volume. The precursor for thymidylate synthesis, dUMP, is formed in *B. subtilis* by two pathways, which appear from radiolabeling studies to contribute about equally (55). The first of these involves deamination of dCMP by dCMP deaminase, which has been purified and characterized (55). Interestingly, dCMP deaminase requires dCTP as a positive allosteric activator and is inhibited by dTTP. These properties appear to provide a mechanism for coordinating intracellular pools of dTTP and dCTP. The second pathway to dUMP probably involves hydrolysis of dUTP to dUMP plus PP_i, which is catalyzed by the enzyme deoxyuridine triphosphatase. This enzyme has been purified from *B. subtilis* and characterized (84); it is highly specific for dUTP, and no allosteric regulators have been identified. The properties of *B. subtilis dcd* mutants, which lack dCMP deaminase, indicate that either the deoxyuridine triphosphatase or the dCMP deaminase pathway can serve to provide precursors for dTTP synthesis but that dCMP deaminase is the only enzyme that does so from deoxycytidine nucleotides (55).

Reductive methylation of dUMP with 5,10-methylenetetrahydrofolate as the one-carbon donor is catalyzed in *B. subtilis*, as in other organisms, by thymidylate synthase. *B. subtilis* is unusual in that it produces two thymidylate synthases, A and B, encoded by the *thyA* and *thyB* genes, respectively (62). Thymidylate synthases A and B can be separated by ion-exchange chromatography, and the two enzymes have somewhat different kinetic properties. Thymidylate synthase A appears to play a dominant role in vivo, because *thyB* mutants are indistinguishable from wild-type cells except for their lack of thymidylate synthase B after chromatography to separate the isozymes (62). Thymidylate synthase B is temperature sensitive in vivo, and *thyA* mutants are temperature sensitive for growth. *thyA* mutants also have alterations in dUMP and dTTP pools that indicate decreased thymidylate synthase activity in vivo (62). Conversion of dTMP to dTTP involves sequential participation of two kinases: dTMP kinase, which has not been characterized genetically or enzymologically, and nucleoside diphosphokinase, which was described above.

ALLOSTERIC REGULATION OF PYRIMIDINE BIOSYNTHESIS

Carbamylphosphate Synthetase

The only clearly demonstrated site of allosteric regulation of de novo pyrimidine nucleotide biosynthesis in *B. subtilis* is the first step of the pathway, which is catalyzed by the pyrimidine-repressible carbamylphosphate synthetase. This enzyme has not been purified, but characterization of its properties in extracts clearly showed that it is a typical allosteric enzyme (73). ATP saturation curves are sigmoid and shifted toward hyperbolic form by the allosteric activators PRPP and GTP (Fig. 2). GMP and GDP are also activators, but GTP is more effective as a function of concentration. UTP, UDP, and UMP are all inhibitors; UMP is the most effective as a function of concentration. Cytidine nucleotides are not effective inhibitors at physiological concentration. The inhibiting nucleotides act by reducing both the apparent affinity of the enzyme for ATP and the maximal velocity (Fig. 2). As expected, activators and inhibitors effectively antagonize one another. These properties are consistent with the idea that pyrimidine biosynthesis is regulated by a combination of feedback inhibition by uridine nucleotides and activation by purine nucleotides. The latter may be seen as a means of providing balanced purine and pyrimidine nucleotide biosynthesis, since both pools are needed for nucleic acid synthesis. Both GTP, which is an allosteric activator, and the substrate ATP, which can overcome inhibition by uridine nucleotides at high concentrations (Fig. 2), participate in activation of carbamylphosphate synthetase.

Activation by PRPP can be seen as part of a global regulatory pattern. PRPP pools would not be expected to rise unless (i) metabolic energy, as measured by ATP/ADP, is abundant; (ii) carbon sources, as measured by ribose-5-phosphate, are abundant; and (iii) utilization of PRPP for nucleotide biosynthesis is sufficiently slow to allow PRPP accumulation (2). Thus, PRPP accumulation is an appropriate signal for activation of pyrimidine nucleotide biosynthesis. PRPP accumulation is also associated with purine starvation in *B. subtilis* (91), and PRPP activates purine nucleotide biosynthesis in *B. subtilis* through its effects on the first enzyme of that pathway (54). Failure of CTP or other cytidine nucleotides to regulate *B. subtilis* carbamylphosphate synthetase may indicate that the CTP pools are largely regulated by interactions with CTP synthetase, as is thought to be the case in enteric bacteria (37).

Other Steps

Kinetic studies with crude extracts have not identified any obvious allosteric regulatory interactions

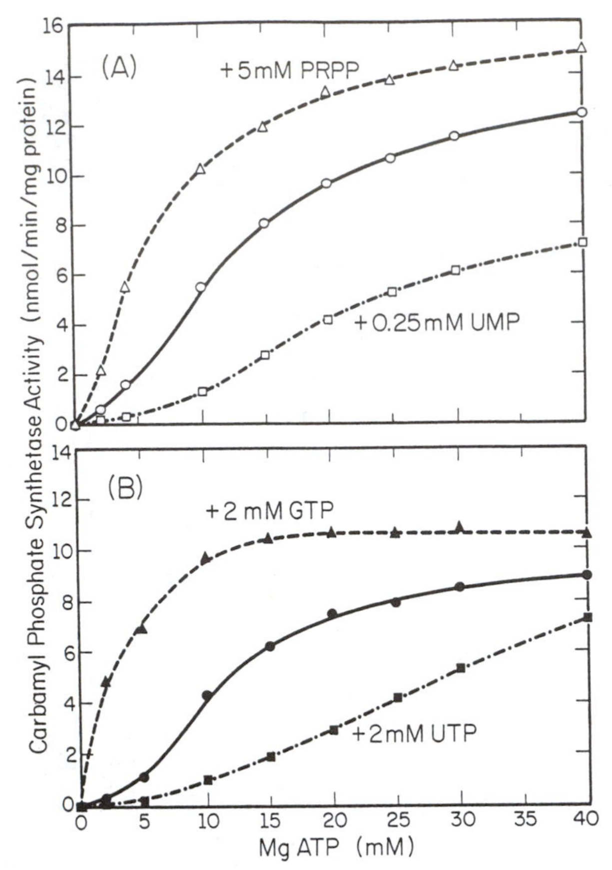

Figure 2. Allosteric regulation of the pyrimidine-repressible carbamylphosphate synthetase from *B. subtilis*. Effects of allosteric activators and inhibitors as a function of the concentration of the substrate Mg-ATP are shown. In all cases, $MgCl_2$ was maintained in 6 mM excess over the total ATP-plus-effector concentration. (A) ○, no effectors; △, plus 5 mM PRPP; □, plus 0.25 mM UMP. (B) ●, no effectors; ▲, plus 2 mM GTP; ■, plus 2 mM UTP. Reprinted from reference 73 with permission of copyright holder.

involving other *B. subtilis* enzymes of pyrimidine biosynthesis. Dihydroorotate dehydrogenase is inhibited by orotate (30% inhibition by 1 mM orotate at 5 mM dihydroorotate [72]), but this is probably simple product inhibition. Orotate phosphoribosyltransferase is inhibited by high concentrations of dihydroorotate (72), but this is probably due to competition for orotate at the active site. Similarly, inhibition of pyrimidine ribonucleotide monophosphokinase by ribonucleotide di- and triphosphates (115) is likely to result from competitive action at the substrate sites.

Properties of *B. subtilis* CTP synthetase, as determined from studies with crude extracts, have been briefly reported elsewhere (3). The enzyme is inhibited strongly by CTP and less effectively by CMP. Allosteric activation by GTP does not appear to have been tested, but we predict that it will be found because of the strong sequence similarity between *E. coli* and *B. subtilis* CTP synthetases. The report (3) of a small activation by GMP is also a hint that GTP may activate the *B. subtilis* enzyme. Activation of CTP synthetase by GTP and its substrate ATP is functionally analogous to the activation of carbamylphosphate synthetase by purine nucleotides. AMP was reported to inhibit *B. subtilis* CTP synthetase, but this may result from steric interactions at the active site.

GENETICS OF PYRIMIDINE BIOSYNTHESIS

The *pyr* Cluster

A collection of more than 50 Pyr^- mutant strains of *B. subtilis* was characterized by Potvin et al. (80, 81). The mutants were classified by assays for the activities of each of the six pyrimidine biosynthetic enzymes (Fig. 1). Mutants with lesions in five of the six genes were identified; the sixth gene, *pyrE*, was probably represented by a class of mutants called *pyrX* by Potvin et al. The genes, including *pyrX*, were shown by cotransformation analysis to map in a tight cluster on the *B. subtilis* chromosome. The gene order deduced was *pyrACBDFE*. This order was subsequently shown to be incorrect when the *pyr* genes were cloned and mapped on the basis of abilities of subclones to complement *E. coli pyr* mutants (43). The correct gene order is *pyrBCADFE*, a conclusion that has been confirmed by sequencing the entire *pyr* cluster (85). Representatives of the mutants characterized by Potvin et al. are available in the *Bacillus* Genetic Stock Center.

A substantial fraction of the mutants Potvin et al. characterized appear to be pleiotropic (81). It is not clear whether these are polar effects, as might be expected in a polycistronic operon, or whether the mutants actually carry multiple lesions that could result from the harsh mutagenesis used to generate most of them (81). In some cases, the enzymatic assays used to characterize the mutants may have been misleading, because it is now known that the assay conditions used were quite different from optimal conditions for the *B. subtilis* enzymes (72, 81). Many of the pleiotropic mutations were mapped in the *pyrD* locus (81). It is now known that *pyrD* is adjacent to an ORF encoding a protein of unknown function (85) (Fig. 3). It is possible that mutations in this unknown protein affect the expression of other *pyr* genes or the activities of enzymes they encode. Potvin et al. (81) suggested that the pyrimidine biosynthetic enzymes of *B. subtilis* are organized into a multienzyme complex in which dihydroorotate dehydrogenase plays a central role. Such a model could explain why so many of the pleiotropic mutants map in *pyrD*. The activities of carbamylphosphate synthetase, aspartate transcarbamylase, and dihydroorotase were shown to cosediment on sucrose density gradient centrifugation at a rate corresponding to a molecular weight of 130,000 (81). It is now clear, however, that these enzymes do not form a complex with one another and dihydroorotate dehydrogenase, because the minimal molecular weight for such a complex would be 340,000 (85).

Cloning, physical mapping, and determination of the nucleotide sequence of the *B. subtilis* chromosomal DNA encoding the *pyr* cluster have now provided a definitive picture of this system (85). As summarized in Fig. 3, the *B. subtilis pyr* cluster is organized in a single nine-cistron operon arranged in the following sequence: promoter-ORF1-*pyrB*-*pyrC*-*pyrAA*-*pyrAB*-ORF2-*pyrD*-*pyrF*-*pyrE*. The complete cluster is nearly 12 kbp long. ORF1 and ORF2 refer to ORFs comprising 673 and 256 codons, respectively, whose functions are unknown. With the exception of

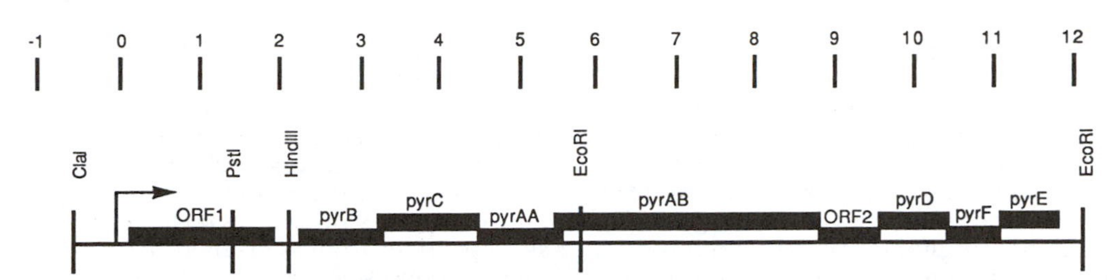

Figure 3. Organization of the *B. subtilis pyr* operon. Numbers represent the length of the DNA in kilobase pairs from the start of transcription, which is indicated by the arrow. Each thick dark line represents an ORF for each gene, with the overlap between lines depicting in exaggerated form the overlap between reading frames. Reprinted from reference 85 with permission of the copyright holder.

ORF1 and *pyrB*, which are separated by a 145-bp intercistronic region, the 3′ ends of all of the ORFs overlap the 5′ ends of the downstream cistrons by 1 to 32 bp. The overlapping of reading frames in the cistrons of biosynthetic operons is a common feature of gene organization in *B. subtilis* (126). The site of transcription initiation for the operon has been mapped to a G residue 151 bp upstream from the start codon of ORF1 (85). This site is preceded by a typical *B. subtilis* σ^A-dependent promoter sequence. Transcription from this promoter was shown to be strongly regulated by pyrimidines (85). A putative transcription termination sequence follows the 3′ end of the *pyrE* gene. Detailed comparisons of the deduced amino acid sequences of the *B. subtilis* pyrimidine biosynthetic enzyme with the sequences of the corresponding enzymes from numerous other species have been presented elsewhere (85). With the exception of the subunits of carbamylphosphate synthetase and the NH_2-terminal domain of aspartate transcarbamylase, the sequence similarity among these enzymes is very low. In general, the *B. subtilis* enzymes resemble the enzymes from gram-negative eubacteria no more closely than they resemble the enzymes from eukaryotes or even mammals. Although the *B. subtilis* pyrimidine biosynthetic enzymes do not strongly resemble their *E. coli* counterparts in their primary structures, it is noteworthy that all of the *B. subtilis* enzymes are able to complement the corresponding *E. coli pyr* mutants.

The available evidence indicates that the *B. subtilis pyr* genes are transcribed as a single polycistronic message. The gene cluster contains two rho-independent terminators, one located 87 to 119 bp downstream from the promoter and upstream of ORF1 and a second located in the ORF1-*pyrB* intercistronic region. There is evidence that transcriptional termination occurs at both terminators in vivo and that regulation of transcriptional termination versus readthrough is important in regulation of expression of the operon by pyrimidines. This will be discussed further in the next section.

pyrG (*ctrA*)

The identification of the *ctrA* locus as the structural gene for CTP synthetase (*pyrG*) and the molecular genetic characterization of this gene were described above.

prs

The *B. subtilis prs* gene, which encodes PRPP synthetase, was cloned by complementation of a corresponding *E. coli prs* mutant. Use of an integration vector containing both the *prs* gene and a chloramphenicol resistance gene allowed the chromosomal location of the *prs* gene to be mapped (64). A much more precise chromosomal location was obtained from sequencing the cloned *prs* gene and its flanking sequences (65). *prs* was shown to lie between the *tms* and *ctc* genes at 6° to 7° on the *B. subtilis* chromosome. The available evidence indicates that *prs* is the second cistron in a bicistronic transcription unit that also includes *tms*. The function of *tms* is unknown; *tms* mutants are temperature sensitive for growth. The promoter for *tms* and *prs* is a typical *B. subtilis* σ^A-dependent promoter and has been studied in some detail (34).

Other Genes

The *ndk* gene encoding nucleoside diphosphokinase has been located by sequencing of chromosomal DNA adjacent to the *mtr* and *trp* genes and was recognized by the strong similarity of its sequence to *ndk* sequences from other species (24a). The gene encoding pyrimidine ribonucleoside monophosphokinase, which we propose to name *pmk*, has not been located. Judging by their deduced amino acid sequences, it is unlikely that either ORF1 or ORF2 of the *pyr* cluster encodes this enzyme. In the case of ORF2, cells containing plasmids encoding and overexpressing ORF2 were assayed for nucleoside monophosphokinase; no evidence for elevated activity was found (102).

Five genetic loci (*furABCEF*) that render the cells resistant to 5-fluorouracil have been described in *B. subtilis*. All have been mapped, but none has been fully characterized biochemically. The *fur* mutants are briefly reviewed here because it is likely that they result from mutations in genes encoding enzymes of pyrimidine uptake, utilization, or biosynthesis or from mutations affecting regulation of these processes. Some speculations as to functions of specific *fur* genes are offered. One of the *fur* loci probably encodes uracil phosphoribosyltransferase (*upp*), since *upp* mutants have been isolated in *B. subtilis* by selection for fluorouracil resistance, although they have not been mapped (58). *furA* mutants confer resistance to 1 to 2 μg of fluorouracil per ml, map at 135° to 141°, and are linked by cotransformation to an unidentified *ura* (*pyr*) gene (25, 78). Recently, *furA* mutants were shown to be defective in uracil transport (67). *furC* and *furE* mutations are closely linked to one another and to *pyrG* (*ctrA*) (111); these mutations may be different alleles of the *pyrG* (CTP synthetase) gene that render the enzyme resistant to fluoro-UTP.

furB (125) and *furF* (16) mutants map at 41° to 49° and 154° to 170° on the *B. subtilis* chromosome, respectively. Both mutants are described as conferring resistance to 40 μg of fluorouracil per ml in the presence of 40 μg of uracil per ml.

B. subtilis cpa (*urs*) mutants may provide a valuable system for understanding repression of *pyr* genes by uracil. Because these mutants lack the arginine-repressible carbamylphosphate synthetase, repression of the *pyr*-encoded carbamylphosphate synthetase by uracil prevents *cpa* mutants from synthesizing arginine and growing. Spontaneous revertants of *cpa* mutants to uracil tolerance have been isolated (72). None of the mutants characterized had acquired arginine-repressible carbamylphosphate synthetase. Rather, all were insensitive to various degrees to repression of *pyr* gene expression by uracil. While none of the mutants have been mapped or characterized biochemically, some may have defects in *pyr* regulation. Some of the mutants are uracil tolerant but remain uridine sensitive, probably as the result of defects in the uptake of uracil or its conversion to UTP. Other revertants could very well be true *pyr* regulatory mutants and will be interesting to characterize.

REGULATION OF *PYR* GENE EXPRESSION

Identity of the Repressing Metabolite

In wild-type *B. subtilis*, the enzymes of pyrimidine biosynthesis are repressed equally well by growth in the presence of either uracil or cytidine. However, in a cytidine deaminase-deficient mutant strain, only uracil is capable of repressing *pyr* expression (3). In contrast, CTP synthetase expression is strongly repressed by cytidine in the cytidine deaminase mutant, and uracil does not repress *pyrG* expression (3). These results confirm that *pyrG* is regulated independently from the *pyr* operon, as might be expected from their different positions on the chromosome. They also indicate that the repressing metabolite for *pyrG* is a cytidine derivative, most likely CTP, whereas the *pyr* operon is specifically repressed by a uracil derivative, probably UTP.

Evidence for Regulation by Transcriptional Attenuation

Understanding of the regulation of the *B. subtilis pyr* operon at the molecular level is still very incomplete. Early studies (72) demonstrated that the six enzymes of UMP biosynthesis were coordinately regulated in *B. subtilis*, as expected if the *pyr* cluster constitutes an operon. More recently, analysis of the *pyr* cluster nucleotide sequence, mapping of *pyr* transcripts, and studies with promoter-indicator plasmids by Quinn et al. (85) have allowed a partial description to be obtained. Transcripts whose abundance is regulated by pyrimidines in vivo have been mapped by primer extension. The 5′ ends of these transcripts originate at a G residue located 151 nucleotide residues upstream from the initiator codon of ORF1 (Fig. 4). The abundance of these transcripts is reduced to below the detection limit by growth of *B. subtilis* cells in the presence of uracil. Spaced appropriately 5′ to the transcription start point is a promoter sequence, (−37)-TTGACA-N_{17}-AATAAT-(−7), which is an almost perfect consensus sequence for the σ^A-dependent form of *B. subtilis* RNA polymerase. Segments of *pyr* DNA that contain this region have been introduced into a promoter indicator plasmid, which expresses subtilisin (*aprE*) in a protease-deficient *B. subtilis* host when a promoter has been cloned into a multiple cloning site upstream from a promoterless *aprE* gene (116). These plasmids containing the putative *pyr* promoter expressed subtilisin in *B. subtilis*; expression was sharply reduced by growth of the cells with uracil. Strikingly, a plasmid in which the promoter region and residues 1 through 83 of the 5′ untranslated leader were inserted upstream of *aprE* still permitted the production of subtilisin, but expression was no longer reduced by uracil in the medium (85). This plasmid differed from the other plasmids tested in that it contained the *pyr* promoter but lacked a putative rho-independent terminator structure located at nucleotides 87 to 119 in the 5′ leader sequence (Fig. 4). Thus, it appears that this structure is crucial to pyrimidine regulation of *pyr* expression. Since this region contains no evident ORF until the start of ORF1 at residue 151, it does not seem likely that a coupled transcription-translation attenuation mechanism, such as has been well characterized for the *E. coli pyrBI* operon (40), is operative in the *B. subtilis pyr* operon. Rather, it is more likely that a factor-dependent attenuation mechanism, in which the ratio of transcriptional termination versus readthrough is regulated by a protein whose effects are modulated by intracellular UTP levels, will be found. Such a regulatory system would be analogous in some ways to the involvement of the *mtr* locus in regulation of the *B. subtilis trp* operon by transcription attenuation (23, 38, 96). However, we have not been able to identify an obvious antiterminator structure in the *pyr* leader sequence that would interfere with formation of the terminator, as is found in the *trp* operon (38, 96). There is also no direct evidence for a *pyr* regulatory protein at present. Clearly, many more experiments are required to test these suggestions. The possibility is not excluded that expression of the *pyr* operon is instead (or also) regulated by a repressor protein that binds in such a way as to block transcription initiation. However, the experiments described above appear to require that the operator site for such a repressor lie downstream from nucleotide 83 of the 5′ leader. No sequences containing the dyad symmetry typical of repressor-binding sites in other systems are evident in the regions flanking the *pyr* promoter.

A second region in the *pyr* operon sequence has also attracted attention as a possible site of regulation by transcriptional attenuation (44, 85). This is the intercistronic region between ORF1 and *pyrB* (Fig. 5). Here are found all of the elements required for a coupled transcription-translation attenuation mechanism as found in *E. coli pyrBI*: a putative rho-independent terminator sequence just upstream of the *pyrB* ORF, a ribosome-binding site preceding a short ORF that extends into the terminator sequence, and a T-rich region that could act as a transcription pause site. The structure of the region clearly would permit a mechanism of attenuation in which transcription would terminate after ORF1 when pyrimidine nucleotides are abundant and in which readthrough could occur

```
ACCATCCACGAACAGGAGAATATGTCGAATTTTGAAGCGTCGCGTTCCCGAGGATATGGCAGAATTAATCGAAAACCTCAGAAA
                       Spo0A box

      -35 .           .-10     +1
AAACGGTTGACAGAGGGTTTCTTTTCTGAAATAATAAACGAAGCTGAATAGATTCTTTAAAACAGTCCAGAGAGGCTGAGAAGG

                                                    rho-independent terminator?
ATAACGGATAGACGGGATGCGTGTATAGGCGCGCACCTTGTCCTAAAACCCCTCTATGCTCTGGCAGGAGGGGTTTTTTCTTCT

           RBS
ATATGAACTGTGAGGTGTCACACATTGAATCAAAAA...
                       MetAsnGlnLys...
                       ORF1------>
```

Figure 4. Sequence of the *B. subtilis pyr* promoter region. The sequence spans nucleotides −125 to +170 relative to the site of transcription initiation (+1). The −35 and −10 regions of the promoter, the initial sequence of ORF1, and its predicted ribosome-binding site (RBS) are shown. A putative rho-independent transcriptional terminator sequence, in which arrows denote sequences that could base pair to form a stable stem-loop structure in the *pyr* transcript, is shown. A consensus binding site for the Spo0A protein (104) 100 bp 5′ to the site of transcription initiation is indicated by double underlining. Reprinted with modifications from reference 85 with permission of the copyright holder.

to allow transcription of *pyrB* (and all of the downstream *pyr* genes) when pyrimidine nucleotides were scarce. However, no direct evidence that such a mechanism operates has been obtained as yet. S1 nuclease mapping of transcripts clearly established that both readthrough and terminated transcripts are found in cells grown on minimal medium without pyrimidines (85). Both kinds of transcripts were much less abundant in cells grown in the presence of uracil, but the relative amounts of terminated versus readthrough transcripts were not altered. Thus, there was no evidence that pyrimidines affected the frequency of transcription termination, as would be required by the coupled transcription-translation attenuation model. However, the structure of this intercistronic region is so strikingly similar to the 5′ region of *E. coli pyrBI* that it seems necessary to interpret these results with caution. It is possible that there remain conditions in which regulation of transcription by pyrimidines will be found. For example, the abundance of terminated and readthrough transcripts has not yet been determined in cells subjected to pyrimidine starvation.

The ORF1-*pyrB* intercistronic region also contains a sequence that could serve as a σ^A-dependent transcription promoter (Fig. 5). Experiments with the promoter-indicator plasmid system described above showed that transcription from this promoter could be observed in vivo, but the promoter was much weaker than the *pyr* promoter 5′ to ORF1, and no regulation of this promoter by pyrimidines was detectable (85). Furthermore, transcription from the intercistronic promoter could be detected only in a plasmid in which the downstream terminator sequence had been deleted. Attempts to locate transcripts originating from the intercistronic promoter by primer extension and S1 nuclease mapping were unsuccessful. We suggest that relatively little, if any, of *pyr* transcription in *B. subtilis* originates from this promoter in vivo.

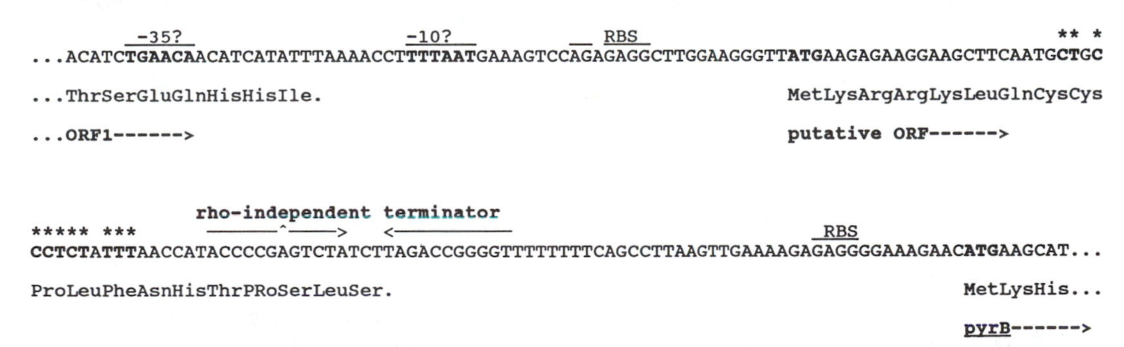

Figure 5. Features of the ORF1-*pyrB* intercistronic region of the *B. subtilis pyr* operon. The sequence is of the nontranscribed strand of *pyr* DNA from the 7 C-terminal codons of ORF1 through the 3 N-terminal codons of *pyrB*. A putative weak promoter sequence is designated by −35? and −10? Sequence elements that are consistent with a coupled transcription translation mechanism of attenuation control of expression of *pyrB* and downstream *pyr* genes are as follows: RBS, a possible ribosome-binding site and a short ORF for a putative leader peptide, whose coding sequence also includes a pyrimidine-rich transcription pause site (asterisks) and a potential rho-independent terminator (arrows indicate potential base-pairing sequences of the stem-loops). Reprinted from reference 85 with permission of the copyright holder.

Cessation of Transcription at the End of Exponential Growth

Pyrimidine biosynthesis in *B. subtilis* is subject to a level of regulation, which we term developmental regulation, that is not seen in many other species. When *B. subtilis* cells are starved for a carbon or nitrogen source, a condition that leads to sporulation, synthesis of enzymes of pyrimidine biosynthesis ceases, and enzyme activities disappear from the cells (50, 114). A similar result is obtained when an auxotrophic strain is starved for a required amino acid, a condition that generally does not permit efficient sporulation. These regulatory events have been studied most thoroughly with aspartate transcarbamylase, in which case it is clear that the loss of activity results from energy-dependent degradation of the enzyme (49, 114). Cessation of aspartate transcarbamylase synthesis in glucose-starved *B. subtilis* was first demonstrated by pulse-labeling with radioactive amino acids and immunoprecipitation (Fig. 6; 50). Synthesis of the enzyme drops to very low levels within 1 h after the end of exponential growth (Fig. 6). Measurements of nucleotide pools in cells under these conditions showed that pyrimidine nucleotide levels drop during this period, which indicated that the cessation of enzyme synthesis does not result simply from repression by end product nucleotides (50). More recently, we showed that *pyr* transcripts dropped to undetectable levels within 1 h after the end of exponential growth (85; unpublished data). Thus, some mechanism for shutting off *pyr* transcription in carbon-starved cells appears to exist. The nature of this mechanism is unknown at present. However, an intriguing possibility is raised by the observation that a consensus sequence for the binding of the Spo0A protein is found 100 nucleotides upstream from the *pyr* promoter (Fig. 4). The Spo0A protein has been shown to be involved in the regulation of several genes that regulate the transition of *B. subtilis* cells from exponential growth to the initiation of sporulation (104). The function of Spo0A as a gene regulator is in turn regulated by a phosphorelay signalling system that involves several *spo0* gene products (12). Indirect evidence for the involvement of the *spo0A* gene in the shutoff of aspartate transcarbamylase synthesis was obtained by Maurizi and Switzer (50), but direct measurements of the effects of *spo0* genes on *pyr* transcription are now possible.

The possibility that *pyr* transcription and possibly expression of other major biosynthetic pathways is shut off by the *spo0A* regulatory system as part of the multiple events that convert cellular metabolism from exponential growth to differentiation into endospores is an attractive one. Clearly, this is at present only a speculation and must be tested experimentally.

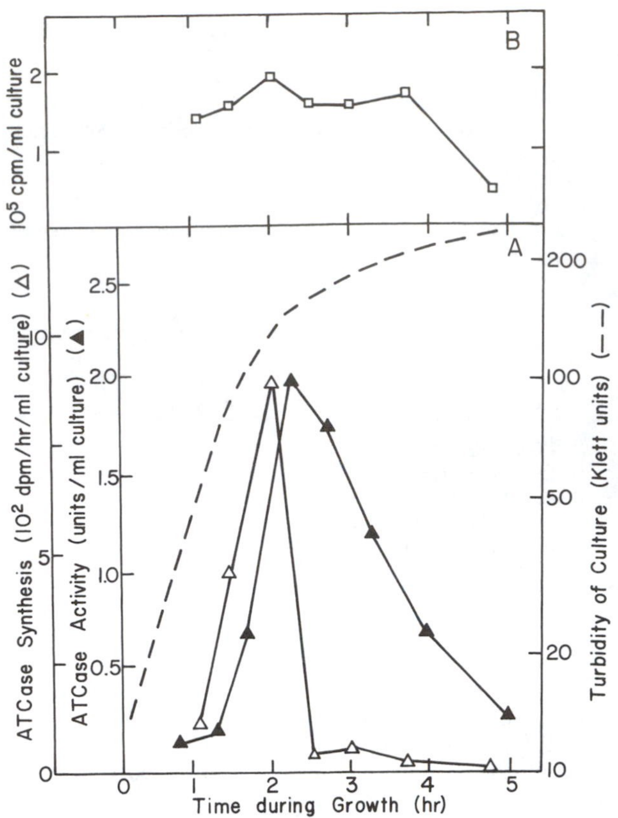

Figure 6. Cessation of aspartate transcarbamylase (ATCase) synthesis prior to its degradation during stationary phase in *B. subtilis* 168. The bacteria were grown on rich medium and pulse-labeled with [^{3}H]leucine for 30 min at the times shown. The amount of radioactivity in immunoprecipitated aspartate transcarbamylase (△) and bulk trichloroacetic acid-precipitable protein (□) from samples collected at various times during growth and stationary phase are shown. Total aspartate transcarbamylase activity per milliliter of culture was also determined (▲) along with the turbidity of the culture (– – –). Details are given in reference 50, from which this figure was reprinted with permission of the copyright holder.

INACTIVATION AND DEGRADATION OF PYRIMIDINE BIOSYNTHETIC ENZYMES IN STARVING CELLS

As described in the preceding section, enzymes of pyrimidine biosynthesis are inactivated in starving cells. Enzymes of purine and amino acid biosynthesis are also inactivated under the same conditions (24, 51, 63, 90, 106, 113). We believe that this is part of an adaptation that allows the cell to prevent wasteful synthesis of nucleotides and amino acids and assume a metabolism based on turnover of vegetative proteins and RNA for the synthesis of spore products. The degradation of aspartate transcarbamylase has been characterized in detail, but it has also been shown that the activities of *pyrA*-encoded carbamylphosphate synthetase (74) dihydroorotate dehydrogenase (95), orotate phosphoribosyltransferase (52), and OMP decarboxylase (52) are all lost from carbon-starved *B. subtilis* cells at rates similar to those at which aspartate transcarbamylase is lost. Dihydroorotase activity is lost much more slowly (95). Only in the case of aspartate transcarbamylase has it been shown directly by immunochemical means that inactivation is coincident with degradation of the enzyme protein (49).

The results of studies of the physiological regulation and mechanism of aspartate transcarbamylase degradation have been summarized elsewhere (107). Very briefly, activation and degradation require metabolic energy, as is common for intracellular proteolysis.

Degradation is initiated by starvation for glucose, a nitrogen source, or a required amino acid. Amino acids play an important role in regulation of the stability of aspartate transcarbamylase in vivo, because the enzyme is degraded even during exponential growth on glucose-containing medium if only ammonium ions are present as the nitrogen source (6). Amino acids in the medium cause the enzyme to be completely stable, however. The effects of amino acids may be mediated via the stringent response, because aspartate transcarbamylase degradation is sharply reduced in *relA* and *relC* mutants (7, 107). During amino acid starvation, there is a rough correlation between levels of pppGpp and ppGpp and the rate of aspartate transcarbamylase degradation, but under other conditions, such as glucose starvation of *relA* cells or after addition of the isoleucyl tRNA synthetase inhibitor pseudomonic acid, this correlation does not exist (7, 107). A biochemical model that describes the degradation of the enzyme and accounts for these observation has not yet been deduced.

COMPARISONS WITH OTHER SPECIES

Other Gram-Positive Bacteria

The genes and enzymes of pyrimidine biosynthesis have not been extensively investigated in other gram-positive species. Members of some genera, such as *Lactobacillus* and *Streptococcus*, do not produce carbamylphosphate synthetase; these organisms require arginine for growth and appear to obtain carbamylphosphate for pyrimidine biosynthesis from arginine via citrulline and a catabolic ornithine transcarbamylase (48). The aspartate transcarbamylase from *Streptococcus faecalis* has been purified to homogeneity and characterized (13). Its properties are generally similar to those of the *B. subtilis* aspartate transcarbamylase (10) except that its activation by anions is appreciably greater. On the basis of subunit and native molecular weight measurements, a tetrameric structure was proposed for the *S. faecalis* enzyme. In view of the conservation of trimeric subunit structures of aspartate transcarbamylases from other species and the fact that the active site is formed at subunit interfaces, however, we suggest that the native *S. faecalis* enzyme is also a trimer of catalytic subunits.

As of this writing, the cloning and sequencing of the *pyr* genes of the thermophile *Bacillus caldolyticus* was nearing completion in the laboratory of Jan Neuhard, University of Copenhagen (35). The *B. caldolyticus* genes were cloned in pUC vectors by complementation of *E. coli pyr* mutants. The genes are clustered on the chromosome in exactly the same order as in *B. subtilis*, i.e., ORF1-*pyrB*-*pyrC*-*pyrAA*-*pyrAB*-*pyrD*-ORF2-*pyrF*-*pyrE*. An intercistronic region lies between ORF1 and *pyrB* in *B. caldolyticus* as in *B. subtilis* but is 170 bp long instead of 145 bp. This region also carries the sequence of a putative rho-independent terminator. Unlike the case for *B. subtilis*, there is a second intercistronic region (about 300 bp) between *pyrF* and *pyrE*. Otherwise, the 5′ ends of the ORFs overlap the 3′ ends of the preceding genes as in *B. subtilis*. The deduced amino acid sequences of the *pyr*-encoded enzymes from *B. caldolyticus* show about 65% identity to their counterparts from *B. subtilis*. Studies of the regulation of the *B. caldolyticus pyr* cluster have yet to be conducted, but the sequence data obtained so far suggest that the two organisms use similar regulatory mechanisms.

Gram-Negative Bacteria and Archaebacteria

The molecular genetics and enzymology of pyrimidine nucleotide biosynthesis have been quite well studied in *E. coli* and *S. typhimurium* (61). In these species, the *pyr* genes are scattered around the chromosome and are not coordinately regulated. Gram-negative bacteria generally produce only a single carbamylphosphate synthetase whose synthesis and catalytic activity are regulated by both pyrimidines and metabolites of the arginine biosynthetic pathway, reflecting the function of the enzyme in providing precursors to both end products. Since aspartate transcarbamylase is the first enzyme of the pyrimidine biosynthetic branch derived from carbamylphosphate in gram-negative bacteria, this enzyme is generally subject to allosteric regulation by pyrimidine nucleotide inhibitors and purine nucleotide activators. Like gram-positive bacteria, gram-negative bacteria produce all pyrimidine biosynthetic enzymes as polypeptide products of separate cistrons. No polyfunctional proteins or multienzyme complexes have been found. Very little information about the genes and enzymes of pyrimidine biosynthesis in archaebacteria is yet available. A brief overview of pyrimidine biosynthesis in *E. coli* and *S. typhimurium*, with emphasis on points of difference with *B. subtilis*, follows.

The glutaminase and catalytic subunits of carbamylphosphate synthetase are encoded by the *carAB* operon, which is regulated by tandem arginine-repressible and pyrimidine-repressible promoters (75). The mechanism for pyrimidine-mediated repression has not been clarified, but a preliminary report (14) indicates that a DNA-binding factor, governed or encoded by the *carP* gene, binds to the pyrimidine-regulated promoter. Arginine repression is mediated through binding of the arginine repressor to a consensus sequence that overlaps the second promoter (75). *E. coli* carbamylphosphate synthetase is allosterically inhibited by the end products UMP and arginine and is activated by ornithine and IMP (37).

The *E. coli pyrBI* operon, which encodes the catalytic and regulatory subunits of aspartate transcarbamylase, has been extensively characterized (40). Of the approximately 300-fold derepression of *pyrBI* expression obtained in pyrimidine-starved cells, about 50-fold can be ascribed to regulation by a UTP-sensitive transcriptional attenuation mechanism, and the remaining 6-fold can be ascribed to mechanisms that affect transcriptional initiation but do not involve a specific repressor protein (45, 46). In a recent review, Landick and Turnbough (40) described the attenuation mechanism as follows. Low intracellular levels of UTP cause RNA polymerase to pause during transcription of uridine-rich segments of the leader transcript. This pausing permits ribosomes to bind to the transcript, to begin translation of a short leader polypeptide, and to follow the RNA polymerase so closely that when a transcriptional terminator sequence is transcribed, it is prevented from folding into an RNA stem-loop. Failure to form the stem-loop prevents transcriptional termination and permits readthrough into the ORFs of the *pyrBI* operon. High intracellular UTP levels do not cause transcriptional pausing, so that translation of the leader polypeptide does not prevent formation of the transcriptional terminator, and readthrough is greatly reduced. *E. coli* aspartate transcarbamylase has also received very extensive study as a prototypical allosteric enzyme (39, 61, 92). The enzyme displays cooperativity of aspartate binding and is allosterically inhibited by CTP and activated by ATP. The comparative biochemistry of aspartate transcarbamylases was recently summarized by Wild and Wales (120).

The genes encoding *E. coli* (122) and *S. typhimurium* (59) PyrC appear to be moncistronic and are regulated by both pyrimidines and purines. Control of *pyrC* does not appear to involve attenuation. Rather, the gene is regulated by two independent mechanisms. The quantitatively most important regulatory mechanism (10- to 15-fold) involves primarily cytidine (32, 93) and guanosine nucleotides (28, 30). On the basis of studies of *S. typhimurium pyrC* expression, a novel mechanism for the translational control of this gene has been proposed by Neuhard and his collaborators (33, 100). When CTP/GTP ratios are high (conditions that repress *pyrC*), *pyrC* transcripts begin at a C residue to generate a leader sequence that can base pair with a region that includes the *pyrC* ribosome-binding site, which sequesters this site and reduces *pyrC* translation. When CTP/GTP ratios are low, a different, slightly shorter transcript, which initiates at a downstream G residue, is formed. This transcript does not form a stem-loop with the ribosome-binding site, and *pyrC* translation is approximately 10-fold higher. Experimental support for a similar mechanism of regulation of *E. coli pyrC* has been presented (121), although the authors of the report concluded that it is the absolute concentrations of CTP and GTP rather than the ratio of CTP to GTP that regulate the selection of transcription start sites. A second mechanism by which purines regulate *pyrC* expression is quantitatively less important than the CTP/GTP-dependent mechanism just described. The effectors for the second mechanism are free hypoxanthine and guanine (53, 88), which act by preventing the binding of the *purR*-encoded purine repressor to a relatively weak *purR* binding site, which lies within the *pyrC* promoter. Repression of transcriptional initiation by this mechanism reduces *pyrC* expression by about twofold. This mechanism has been demonstrated in both *E. coli* (15, 123) and *S. typhimurium* (60).

Expression of *pyrD* in *E. coli* (41) and *S. typhimurum* (20) appears to be regulated by both pyrimidines and purines in the same two ways as *pyrC*. Although *pyrD* has not been studied as extensively as *pyrC*, both the sequence of the gene and the properties of regulatory mutants in the leader region suggest that a translational control mechanism like that proposed for *pyrC* regulates *pyrD* expression by pyrimidines and purines in *S. typhimurium* (20). A consensus PurR-binding site is found upstream of the *pyrD* promoter, and there is evidence that a normal *purR* gene is required to mediate a small (twofold) repression of *pyrD* by purines in *E. coli* (123).

Extensive study of the *E. coli pyrE* gene has shown that it is regulated by a UTP-modulated transcriptional attenuation mechanism similar to that described for the *pyrBI* operon (8, 29, 82, 83). *pyrE* is the second cistron of a bicistronic operon. Upstream of the *pyrE* ORF is a rho-independent terminator sequence, which is preceded by a pyrimidine-rich transcription pause site. Apparently, the upstream cistron (*orfE*), which was recently shown to encode RNase PH (71), plays the role of leader peptide in the *pyrBI* system, so that the extent of transcriptional readthrough into *pyrE* is regulated by UTP-dependent transcriptional attenuation at the *orfE-pyrE* junction.

The nucleotide sequences of *E. coli* (112) and *S. typhimurium* (108) *pyrF* genes reveal that in both cases, *pyrF* is the first gene of a bicistronic operon in which the second cistron (*orfF*) encodes an 11-kDa protein of unknown function. *pyrF* and *orfF* are coordinately regulated at the transcriptional level, and translation of the two genes is coupled (109). These observations might suggest a close linkage in function between the two genes, but OMP decarboxylase activity and regulation of *pyrF* expression by pyrimidines were normal in strains bearing a plasmid in which *orfF* was partially deleted compared with plasmids with intact *pyrF* and *orfF* genes (108, 109). Upstream sequence elements required for a transcriptional attenuation mechanism, such as those found for *pyrBI* or *pyrE*, are not evident in the *pyrF* sequence. This is surprising, because *pyrF* shares with *pyrBI* and *pyrE* the properties of specific repression by uridine nucleotides (93) and derepression by guanine limitation (28). The mechanisms governing *pyrF* expression have yet to be clarified.

Eukaryotes

The association of enzymatic activities of pyrimidine biosynthesis into polyfunctional proteins was first observed in eukaryotic organisms. In lower eukaryotes (yeasts, fungi), carbamylphosphate synthetase and aspartate transcarbamylase are encoded in a single bifunctional protein whose activity is allosterically regulated by UTP. The other enzymes of the pathway are encoded by unlinked genes and apparently do not associate to form complexes. A recent review of pyrimidine biosynthesis in *Saccharomyces cerevisiae*, which is the best studied and is typical of this group of eukaryotes, has been presented by Denis-Duphil (17). In higher eukaryotes (*Drosophila melanogaster*, mammals), the first three enzymes of the pathway are found in a single trifunctional protein, which in mammals is called the CAD protein (19). Dihydroorotate dehydrogenase is encoded by a separate gene and is located in the mitochondria. The last two enzymes of the pathway, orotate phosphoribosyltransferase and OMP decarboxylase, are again found in a single bifunctional protein, generally called UMP synthase (31, 105).

The complete structure of the hamster CAD protein is now known from cDNA sequencing (5, 97–99). The catalytic functions of this complex protein are arrayed in the following order from the NH_2 terminus to the COOH terminus: carbamylphosphate synthetase glutaminase domain, carbamylphosphate synthetase catalytic domain, dihydroorotase domain, and aspartate transcarbamylase domain, i.e., equivalent to the order *pyrAA-pyrAB-pyrC-pyrB*. The distinction between the trifunctional CAD protein of mammals and the bifunctional carbamylphosphate synthetase-aspartate transcarbamylase of lower eukaryotes may be more apparent than real, because the sequence of this bifunctional gene *ura2* from *Saccharomyces cerevisiae* (17, 101) shows the same arrangement of functional domains and considerable sequence identity to the CAD protein. Even the domain of *ura2* corresponding to the dihydroorotase region of the CAD protein shows sequence relatedness, even though it does not confer dihydroorotase activity. These relationships suggest that the trifunctional and bifunctional enzymes are derived from a common ancestor. On the other hand, the order of the corresponding genes in *B. subtilis*, *pyrB-pyrC-pyrAA-pyrAB*, is completely different from the order of the domains in the CAD protein. Thus, there is no reason to suppose that the multifunctional pyrimidine biosynthetic proteins arose from gene fusions within an ancestor that resembled *Bacillus* spp. in its genetic arrangement of *pyr* genes.

Our current understanding of pyrimidine biosynthesis in plants is less clear because the enzymes have not been purified to homogeneity, nor have the genes or cDNA encoding them been cloned and sequenced. Most reports indicate that plants possess a single glutamine-dependent carbamylphosphate synthetase, but the existence of a second carbamylphosphate synthetase cannot be excluded (69, 70, 89). Partial purification of the early enzymes of the pathway indicates that they are separate enzymes (89). Orotate phosphoribosyltransferase and OMP decarboxylase copurify through several steps (4), which suggests that these enzymes either reside in a tight complex or are part of a bifunctional UMP synthase protein. There is evidence that all of the enzymes of pyrimidine biosynthesis except dihydroorotate dehydrogenase are compartmentalized in the chloroplasts of plants (18). Regulation of plant carbamylphosphate synthetase in vitro (69, 70) resembles regulation of the *E. coli* enzyme in many ways. The plant enzyme is inhibited by uridine nucleotides and activated by the purine nucleotides IMP, ITP, GMP, and GTP. Ornithine antagonizes inhibition by uridine nucleotides. Plant aspartate transcarbamylase is reported to have a molecular weight of only 80 to 100 kDa, like the *B. subtilis* enzyme, but it is also subject to inhibition by uridine nucleotides (124). In fact, there is evidence from labeling studies in vivo that the primary site of end product inhibition of pyrimidine biosynthesis is at the aspartate transcarbamylase step rather than at the level of carbamylphosphate synthesis (47).

NOTE ADDED IN PROOF

Recent results with *B. subtilis* (113a) and *B. caldolyticus* (20a) *pyr* gene clusters have revealed that the region described in Fig. 3 as encoding ORF1 actually encodes two ORFs: ORF1A, which encodes a 21-kDa protein, and ORF2A, which encodes a 46-kDa protein. ORF1A is initiated at the codon previously described as the initiator codon for ORF1 (85). ORF1A and ORF1B are separated by a 170-nucleotide intercistronic region. The ability of plasmids encoding ORF1A to complement *E. coli upp* mutants suggests that this gene encodes uracil phosphoribosyltransferase. The sequence of ORF1B suggests that this gene encodes an integral membrane protein involved in uracil transport. More complete analysis of the sequence indicates that *pyr* transcripts could form stable, mutually exclusive antiterminator and terminator stem-loop structures at three sites: between the promoter and ORF1A (Fig. 4), between ORF1A and ORF1B, and between ORF1B and *pyrB* (Fig. 5). Thus, regulation of *pyr* gene expression in *Bacillus* may involve factor-dependent transcription attenuation at three separate sites. Analysis of *pyr* promoter-indicator plasmids and *pyr-lacZ* fusion integrants in *B. subtilis* supports this hypothesis. Remarkably, the results indicate that the protein encoded by ORF1A is required for mediation of transcription termination by exogenous pyrimidines. Thus, this protein may be bifunctional.

Acknowledgments. We thank C. J. Lovatt, J. Neuhard, P. Nygaard, and C. L. Turnbough, Jr., for helpful comments during preparation of this chapter. We also thank Marsha McCormack for keyboarding the manuscript. Research on this topic in our laboratory at the University of Illinois is supported by Public Health Service grant GM47112 (formerly AI11121).

REFERENCES

1. **Amuro, N., J. L. Paluh, and H. Zalkin.** 1985. Replacement by site-directed mutagenesis indicates a role for histidine 170 in the glutamine amide transfer function of anthranilate synthetase. *J. Biol. Chem.* **260:**14844–14849.
2. **Arnvig, K., B. Hove-Jensen, and R. L. Switzer.** 1990. Purification and properties of phosphoribosyl-diphosphate synthetase from *Bacillus subtilis*. *Eur. J. Biochem.* **192:**195–200.
3. **Asa, S., M. Doi, Y. Tsunemi, and S. Akiyama.** 1989. Regulation of pyrimidine nucleotide biosynthesis in cytidine deaminase-negative mutants of *Bacillus subtilis*. *Agric. Biol. Chem.* **53:**97–102.
4. **Ashihara, H.** 1978. Orotate phosphoribosyltransferase and orotidine-5′-monophosphate decarboxylase of black gram (*Phaseolus mungo*) seedlings. *Z. Pflanzenphysiol.* **87:**225–241.
5. **Bein, K., J. P. Simmer, and D. R. Evans.** 1991. Molecular cloning of a cDNA encoding the amino end of the mammalian multifunctional protein CAD and analysis of the 5′-flanking region of the CAD gene. *J. Biol. Chem.* **266:**3791–3799.
6. **Bond, R. W., A. S. Field, and R. L. Switzer.** 1983. Nutritional regulation of the degradation of aspartate transcarbamylase and of bulk protein in exponentially growing *Bacillus subtilis* cells. *J. Bacteriol.* **153:**253–258.
7. **Bond, R. W., and R. L. Switzer.** 1984. Degradation of aspartate transcarbamylase in *Bacillus subtilis* is deficient in *rel* mutants but is not mediated by guanosine polyphosphates. *J. Bacteriol.* **158:**746–748.
8. **Bonekamp, F., K. Clemmesen, O. Karlstrom, and K. F. Jensen.** 1984. Mechanism of UTP-modulated attenuation at the *pyrE* gene of *Escherichia coli*: an example of operon polarity control through the coupling of translation to transcription. *EMBO J.* **3:**2857–2861.
9. **Brabson, J. S., M. R. Maurizi, and R. L. Switzer.** 1985. Aspartate transcarbamylase from *Bacillus subtilis*. *Methods Enzymol.* **113:**627–635.
10. **Brabson, J. S., and R. L. Switzer.** 1975. Purification and properties of *Bacillus subtilis* aspartate transcarbamylase. *J. Biol. Chem.* **250:**8664–8669.
11. **Brofman, J. D.** 1979. The interaction of N-(phosphonoacetyl)-L-aspartate with *Bacillus subtilis* aspartate transcarbamylase. B.Sc. thesis. University of Illinois, Urbana.
12. **Burbulys, D., K. A. Trach, and J. A. Hoch.** 1991. The

initiation of sporulation in *Bacillus subtilis* is controlled by a multicomponent phosphorelay. *Cell* **64:**545–552.
13. **Chang, T.-Y., and M. E. Jones.** 1974. Aspartate transcarbamylase from *Streptococcus faecalis*. Purification, properties, and nature of an allosteric activator site. *Biochemistry* **13:**629–638.
14. **Charlier, D., M. Roovers, D. Gigot, A. Piérard, and N. Glansdorff.** 1990. Molecular interactions involved in the dual regulation of the *carAB* operon in *E. coli* K-12. *Abstr. Genes Enzymes Pyrimidine Biosynth.* American Society for Biochemistry and Molecular Biology.
15. **Choi, K. Y., and H. Zalkin.** 1990. Regulation of *Escherichia coli pyrC* by the purine regulon repressor protein. *J. Bacteriol.* **172:**3201–3207.
16. **Dean, D. R., J. A. Hoch, and A. I. Aronson.** 1977. Alteration of *Bacillus subtilis* glutamine synthetase results in overproduction of the enzyme. *J. Bacteriol.* **131:**981–987.
17. **Denis-Duphil, M.** 1989. Pyrimidine biosynthesis in *Saccharomyces cerevisiae*: the *ura2* cluster gene, its multifunctional product, and other structural or regulatory genes involved in *de novo* UMP synthesis. *Biochem. Cell. Biol.* **67:**612–631.
18. **Doremus, H. D., and A. T. Jagendorf.** 1985. Subcellular localization of the pathway of *de novo* pyrimidine nucleotide biosynthesis in pea leaves. *Plant Physiol.* **79:**856–861.
19. **Evans, D. R.** 1985. CAD, a chimeric protein that initiates *de novo* pyrimidine biosynthesis in eukaryotes, p. 283–331. *In* D. G. Hardie and J. R. Coggins (ed.), *Multidomain Proteins*. Elsevier Science Publishers, Amsterdam.
20. **Frick, M. M., J. Neuhard, and R. A. Kelln.** 1990. Cloning, nucleotide sequence and regulation of the *Salmonella typhimurium pyrD* gene encoding dihydroorotate dehydrogenase. *Eur. J. Biochem.* **194:**573–578.
20a. **Ghim, S. Y., P. Neilsen, and J. Neuhard.** Unpublished data.
21. **Ginther, C. L., and J. L. Ingraham.** 1974. Cold-sensitive mutant of *Salmonella typhimurium* defective in nucleoside diphosphokinase. *J. Bacteriol.* **118:**1020–1026.
22. **Ginther, C. L., and J. L. Ingraham.** 1974. Nucleoside diphosphokinase of *Salmonella typhimurium*. *J. Biol. Chem.* **249:**3406–3411.
23. **Gollnick, P., S. Ishino, M. I. Kuroda, D. J. Henner, and C. Yanofsky.** 1990. The *mtr* locus is a two-gene operon required for transcription attenuation in the *trp* operon of *Bacillus subtilis*. *Proc. Natl. Acad. Sci. USA* **87:**8726–8730.
24. **Graves, L. M., and R. L. Switzer.** 1990. Aspartokinase II from *Bacillus subtilis* is degraded in response to nutrient limitation. *J. Biol. Chem.* **265:**14947–14955.
24a. **Henner, D. J.** Unpublished data (GenBank accession number M80245).
25. **Hoch, J. A., and C. Anagnostopoulos.** 1970. Chromosomal location and properties of radiation sensitivity mutations in *Bacillus subtilis*. *J. Bacteriol.* **103:**295–301.
26. **Hove-Jensen, B., and P. Nygaard.** 1982. Phosphoribosylpyrophosphate synthetase of *Escherichia coli*. Identification of a mutant enzyme. *Eur. J. Biochem.* **126:** 327–332.
27. **Ingraham, J. L., and J. Neuhard.** 1972. Cold-sensitive mutants of *Salmonella typhimurium* defective in uridine monophosphate kinase (*pyrH*). *J. Biol. Chem.* **247:**6259–6265.
28. **Jensen, K. F.** 1979. Apparent involvement of purines in the control of expression of *Salmonella typhimurium pyr* genes: analysis of a leaky *guaB* mutant resistant to pyrimidine analogs. *J. Bacteriol.* **138:**731–738.
29. **Jensen, K. F.** 1988. Hyper-regulation of *pyr* gene expression in *Escherichia coli* cells with slow ribosomes. Evidence for RNA polymerase pausing *in vivo*? *Eur. J. Biochem.* **175:**587–593.
30. **Jensen, K. F.** 1989. Regulation of *Salmonella typhimurium pyr* gene expression: effect of changing both purine and pyrimidine nucleotide pools. *J. Gen. Microbiol.* **135:**805–815.
31. **Jones, M. E.** 1980. Pyrimidine nucleotide biosynthesis in animals: genes, enzymes, and regulation of UMP biosynthesis. *Annu. Rev. Biochem.* **49:**253–279.
32. **Kelln, R. A., J. J. Kinahan, K. F. Folderman, and G. A. O'Donovan.** 1975. Pyrimidine biosynthetic enzymes of *Salmonella typhimurium*, repressed specifically by growth in the presence of cytidine. *J. Bacteriol.* **124:** 764–774.
33. **Kelln, R. A., and J. Neuhard.** 1988. Regulation of *pyrC* expression in *Salmonella typhimurium*: identification of a regulatory region. *Mol. Gen. Genet.* **212:**287–294.
34. **Kenny, T. J., and C. P. Moran, Jr.** 1991. Genetic evidence for the interaction of σ^A with two promoters in *Bacillus subtilis*. *J. Bacteriol.* **173:**3282–3290.
35. **Kim, S. Y., P. Nielsen, and J. Neuhard.** Unpublished data.
36. **Kimsey, H. H., and D. Kaiser.** 1992. The orotidine 5′-monophosphate decarboxylase gene of *Myxococcus xanthus*: comparison to the OMP decarboxylase gene family. *J. Biol. Chem.* **267:**819–824.
37. **Koshland, D. E., Jr., and A. Levitzki.** 1974. CTP synthetase and related enzymes, p. 539–559. *In* P. D. Boyer (ed.), *The Enzymes*, vol. 9. Academic Press, Inc., New York.
38. **Kuroda, M. I., D. Henner, and C. Yanofsky.** 1988. *cis*-acting sites in the transcript of the *Bacillus subtilis trp* operon regulate expression of the operon. *J. Bacteriol.* **170:**3080–3088.
39. **Ladjimi, M. M., and E. R. Kantrowitz.** 1988. A possible model for the concerted allosteric transition in *Escherichia coli* aspartate transcarbamylase as deduced from site-directed mutagenesis studies. *Biochemistry* **27:**276–283.
40. **Landick, R., and C. L. Turnbough, Jr.** 1992. Transcriptional attenuation, p. 407–446. *In* S. L. McKnight and K. R. Yamamoto (ed.), *Transcriptional Regulation*. Cold Spring Harbor Laboratory, Cold Spring Harbor, N.Y.
41. **Larsen, J. N., and K. F. Jensen.** 1985. Nucleotide sequence of the *pyrD* gene of *Escherichia coli* and characterization of the flavoprotein dihyroorotate dehydrogenase. *Eur. J. Biochem.* **151:**59–65.
42. **Lerner, C. G., P. J. Hickman, E. E. Ferguson, and R. L. Switzer.** Unpublished data.
43. **Lerner, C. G., B. T. Stephenson, and R. L. Switzer.** 1987. Structure of the *Bacillus subtilis* pyrimidine biosynthetic (*pyr*) gene cluster. *J. Bacteriol.* **169:**2202–2206.
44. **Lerner, C. G., and R. L. Switzer.** 1986. Cloning and structure of the *Bacillus subtilis* aspartate transcarbamylase gene (*pyrB*). *J. Biol. Chem.* **261:**11156–11165.
45. **Liu, C., and C. L. Turnbough, Jr.** 1989. Multiple control mechanisms for pyrimidine-mediated regulation of *pyrBI* operon expression in *Escherichia coli* K-12. *J. Bacteriol.* **171:**3337–3342.
46. **Liu, C., and C. L. Turnbough, Jr.** Unpublished data.
47. **Lovatt, C. J., and A. H. Cheng.** 1984. Aspartate carbamyltransferase. Site of end-product inhibition of the orotate pathway in intact cells of *Cucurbita pepo*. *Plant Physiol.* **75:**511–515.
48. **Makoff, A. J., and A. Radford.** 1978. Genetics and biochemistry of carbamoylphosphate biosynthesis and its utilization in the pyrimidine biosynthetic pathway. *Microbiol. Rev.* **42:**307–328.
49. **Maurizi, M. R., J. S. Brabson, and R. L. Switzer.** 1978. Immunochemical studies of the inactivation of aspartate transcarbamylase by stationary phase *Bacillus subtilis* cells: evidence for selective, energy-dependent degradation. *J. Biol. Chem.* **253:**5585–5593.
50. **Maurizi, M. R., and R. L. Switzer.** 1978. Aspartate transcarbamylase synthesis ceases prior to activation

of the enzyme in *Bacillus subtilis*. *J. Bacteriol.* **135**:943–951.

51. **Maurizi, M. R., and R. L. Switzer.** 1980. Proteolysis in bacterial sporulation. *Curr. Top. Cell. Regul.* **16**:163–224.
52. **McGarry, T. J.** 1980. Coordinate expression and developmental fate of pyrimidine biosynthetic enzymes on *Bacillus subtilis*: orotidine 5′-monophosphate pyrophosphorylase and orotidine 5′-monophosphate decarboxylase. B.Sc. thesis. University of Illinois, Urbana.
53. **Meng, L. M., and P. Nygaard.** 1990. Identification of hypoxanthine and guanine as the corepressors for the purine regulon genes of *Escherichia coli*. *Mol. Microbiol.* **4**:2187–2192.
54. **Meyer, E., and R. L. Switzer.** 1979. Regulation of *Bacillus subtilis* glutamine phosphoribosylpyrophosphate amidotransferase activity by end products. *J. Biol. Chem.* **254**:5397–5402.
55. **Møllgaard, H., and J. Neuhard.** 1978. Deoxycytidine deaminase from *Bacillus subtilis*. Purification, characterization, and physiological function. *J. Biol. Chem.* **253**:3536–3542.
56. **Møllgaard, H., and J. Neuhard.** 1983. Biosynthesis of deoxythymidine triphosphate, p. 149–201. *In* A. Munch-Peterson (ed.), *Metabolism of Nucleotides, Nucleosides and Nucleobases in Microorganisms*. Academic Press, London.
57. **Mountain, A., J. McChesney, M. C. M. Smith, and S. Baumberg.** 1986. Gene sequence encoding early enzymes of arginine synthesis within a cluster in *Bacillus subtilis*, as revealed by cloning in *Escherichia coli*. *J. Bacteriol.* **165**:1026–1028.
58. **Neuhard, J.** 1983. Utilization of preformed pyrimidine bases and nucleosides, p. 95–148. *In* A. Munch-Petersen (ed.), *Metabolism of Nucleotides, Nucleosides and Nucleobases in Miroorganisms*. Acadmic Press, London.
59. **Neuhard, J., R. A. Kelln, and E. Stauning.** 1986. Cloning and structural characterization of the *Salmonella typhimurium pyrC* gene encoding dihydroorotase. *Eur. J. Biochem.* **157**:335–342.
60. **Neuhard, J., L. M. Meng, K. I. Sorensen, and R. A. Kelln.** 1990. Purine control of *pyrC* expression in *Salmonella typhimurium*. *Int. J. Purine Pyrimidine Res.* **1**:61–66.
61. **Neuhard, J., and P. Nygaard.** 1987. Purines and pyrimidines, p. 445–473. *In* F. C. Neidhardt, J. L. Ingraham, K. B. Low, B. Magasanik, M. Schaechter, and H. E. Umbarger (ed.), *Escherichia coli and Salmonella typhimurium: Cellular and Molecular Biology*, vol. 1. American Society for Microbiology, Washington, D.C.
62. **Neuhard, J., A. R. Price, L. Schack, and E. Thomassen.** 1978. Two thymidylate synthetases in *Bacillus subtilis*. *Proc. Natl. Acad. Sci. USA* **75**:1194–1198.
63. **Neway, J. O., and R. L. Switzer.** 1983. Degradation of ornithine transcarbamylase in sporulating *Bacillus subtilis* cells. *J. Bacteriol.* **155**:522–530.
64. **Nilsson, D., and B. Hove-Jensen.** 1987. Phosphoribosylpyrophosphate synthetase of *Bacillus subtilis*. Cloning, characterization and chromosomal mapping of the *prs* gene. *Gene* **53**:247–255.
65. **Nilsson, D., B. Hove-Jensen, and K. Arnvig.** 1989. Primary structure of the *tms* and *prs* genes of *Bacillus subtilis*. *Mol. Gen. Genet.* **218**:565–571.
66. **North, A. K., M. C. M. Smith, and S. Baumberg.** 1989. Nucleotide sequence of a *Bacillus subtilis* arginine regulatory gene and homology of its product to the *Escherichia coli* arginine repressor. *Gene* **80**:29–38.
67. **Nygaard, P.** Unpublished data.
68. **Olszowy, J., and R. L. Switzer.** 1972. Specific repression of phosphoribosylpyrophosphate synthetase by uridine compounds in *Salmonella typhimurium*. *J. Bacteriol.* **110**:450–451.
69. **O'Neal, T. D., and A. W. Naylor.** 1976. Some regulatory properties of pea leaf carbamoylphosphate synthetase. *Plant Physiol.* **57**:23–28.
70. **Ong, B. L., and J. F. Jackson.** 1972. Pyrimidine nucleotide biosynthesis in *Phaseolus aureus*. Enzymic aspects of the control of carbamoylphosphate synthesis and utilization. *Biochem. J.* **129**:583–593.
71. **Ost, K. A., and M. P. Deutscher.** 1991. *Escherichia coli orfE* (upstream of *pyrE*) encodes RNase PH. *J. Bacteriol.* **173**:5589–5591.
72. **Paulus, T. J., T. J. McGarry, P. G. Shekelle, S. Rosenzweig, and R. L. Switzer.** 1982. Coordinate synthesis of the enzymes of pyrimidine biosynthesis in *Bacillus subtilis*. *J. Bacteriol.* **149**:775–778.
73. **Paulus, T. J., and R. L. Switzer.** 1979. Characterization of pyrimidine-repressible and arginine-repressible carbamylphosphate synthetases from *Bacillus subtilis*. *J. Bacteriol.* **137**:82–91.
74. **Paulus, T. J., and R. L. Switzer.** 1979. Synthesis and inactivation of the carbamylphosphate synthetase isozymes of *Bacillus subtilis* during growth and sporulation. *J. Bacteriol.* **140**:769–773.
75. **Piette, J., H. Nyonoya, C. J. Lusty, R. Cunin, G. Weyens, M. Crabeel, D. Chalier, N. Glansdorff, and A. Pierard.** 1984. DNA sequence of the *carA* gene and the control region of *carAB*: tandem promoters, respectively controlled by arginine and the pyrimidines, regulate the synthesis of carbamoylphosphate synthetase in *Escherichia coli* K-12. *Proc. Natl. Acad. Sci. USA* **81**:4134–4138.
76. **Piggot, P. J., M. Amjad, J.-J. Wu, H. Sandoval, and J. Castro.** 1990. Genetic and physical maps of *Bacillus subtilis* 168, p. 493–540. *In* C. R. Harwood and S. M. Cutting (ed.), *Molecular Biology Methods for Bacillus*. John Wiley & Sons, Inc., Chichester.
77. **Piggot, P. J., and J. A. Hoch.** 1985. Revised genetic linkage map of *Bacillus subtilis*. *Microbiol. Rev.* **49**:158–179.
78. **Polsinelli, M.** 1965. Linkage between genes for amino acid or nitrogenous base biosynthesis and genes controlling resistance to structural correlated analogies. *Giorn. Microbiol.* **13**:99–110.
79. **Post, L. E., D. J. Post, and F. M. Raushel.** 1990. Dissection of the functional domains of *Escherichia coli* carbamylphosphate synthetase by site-directed mutagenesis. *J. Biol. Chem.* **265**:7742–7747.
80. **Potvin, B. W., and H. Gooder.** 1975. Carbamylphosphate synthesis in *Bacillus subtilis*. *Biochem. Genet.* **13**:125–143.
81. **Potvin, B. W., R. J. Kelleher, Jr., and H. Gooder.** 1975. Pyrimidine biosynthetic pathway of *Bacillus subtilis*. *J. Bacteriol.* **123**:605–615.
82. **Poulsen, P., F. Bonekamp, and K. F. Jensen.** 1984. Structure of the *Escherichia coli pyrE* operon and control of *pyrE* expression by a UTP modulated intercistronic attenuation. *EMBO J.* **3**:1783–1790.
83. **Poulsen, P., and K. F. Jensen.** 1987. Effect of UTP and GTP pools on attenuation at the *pyrE* gene of *Escherichia coli*. *Mol. Gen. Genet.* **208**:152–158.
84. **Price, A. R., and J. Frato.** 1975. *Bacillus subtilis* deoxyuridinetriphosphatase and its bacteriophage PBS2-induced inhibitor. *J. Biol. Chem.* **250**:8804–8811.
85. **Quinn, C. L., B. T. Stephenson, and R. L. Switzer.** 1991. Functional organization and nucleotide sequence of the *Bacillus subtilis* pyrimidine biosynthetic operon. *J. Biol. Chem.* **266**:9113–9127.
86. **Radford, A., and N. I. M. Dix.** 1988. Comparison of the orotidine 5′-monophosphate decarboxylase sequences of eight species. *Genome* **30**:501–505.
87. **Rodriguez, S. B., and J. L. Ingraham.** 1983. Location on the *Salmonella typhimurium* chromosome of the gene encoding nucleoside diphosphokinase (*ndk*). *J. Bacteriol.* **153**:1101–1103.
88. **Rolfes, R. J., and H. Zalkin.** 1990. Purification of the *Escherichia coli* purine regulon repressor and identification of corepressors. *J. Bacteriol.* **172**:5637–5642.
89. **Ross, C. W.** 1981. Biosynthesis of nucleotides, p. 169–

205. *In* A. Marcus (ed.), *The Biochemistry of Plants*, vol. 6. Academic Press, Inc., New York.
90. **Ruppen, M. E., and R. L. Switzer.** 1983. Degradation of *Bacillus subtilis* glutamine phosphoribosylpyrophosphate amidotransferase *in vivo*. *J. Biol. Chem.* **258:**2843–2851.
91. **Saxild, H. H., and P. Nygaard.** 1991. Regulation of levels of purine biosynthetic enzymes in *Bacillus subtilis*: effects of changing purine nucleotide pools. *J. Gen. Microbiol.* **137:**2387–2394.
92. **Schachman, H. K.** 1988. Can a simple model account for the allosteric transition of aspartate transcarbamoylase? *J. Biol. Chem.* **263:**18583–18586.
93. **Schwartz, M., and J. Neuhard.** 1975. Control of expression of *pyr* genes in *Salmonella typhimurium*: effects of variations in uridine and cytidine nucleotide pools. *J. Bacteriol.* **121:**814–822.
94. **Sedmak, J., and R. Ramaley.** 1971. Purification and properties of *Bacillus subtilis* nucleoside diphosphokinase. *J. Biol. Chem.* **246:**5365–5372.
95. **Shekelle, P. G.** 1978. The development fate of enzymes of pyrimidine biosynthesis during growth and sporulation of *Bacillus subtilis*: dihydroorotase and dihydroorotate dehydrogenase. B.Sc. thesis. University of Illinois, Urbana.
96. **Shimotsu, H., M. I. Kuroda, C. Yanofsky, and D. J. Henner.** 1986. Novel form of transcription attenuation regulates expression of the *Bacillus subtilis* tryptophan operon. *J. Bacteriol.* **166:**461–471.
97. **Simmer, J. P., R. E. Kelly, A. G. Rinker, Jr., J. L. Scully, and D. R. Evans.** 1990. Mammalian carbamylphosphate synthetase (CPS). cDNA sequence and evolution of the CPS domain of the Syrian hamster multifunctional protein CAD. *J. Biol. Chem.* **265:**10395–10402.
98. **Simmer, J. P., R. E. Kelly, A. G. Rinker, Jr., B. H. Zimmerman, J. L. Scully, H. Kim, and D. R. Evans.** 1990. Mammalian dihydroorotase: nucleotide sequence, peptide sequences, and evolution of the dihydroorotase domain of the multifunctional protein CAD. *Proc. Natl. Acad. Sci. USA* **87:**174–178.
99. **Simmer, J. P., R. E. Kelly, J. L. Scully, D. R. Grayson, A. G. Rinker, Jr., S. T. Bergh, and D. R. Evans.** 1989. Mammalian aspartate transcarbamylase (ATCase): sequence of the ATCase domain and interdomain linker in the CAD multifunctional polypeptide and properties of the isolated domain. *Proc. Natl. Acad. Sci. USA* **86:**4382–4386.
100. **Sørenson, K. I., and J. Neuhard.** 1991. Dual transcriptional initiation sites from the *pyrC* promoter control expression of the gene in *Salmonella typhimurium*. *Mol. Gen. Genet.* **225:**249–256.
101. **Souciet, J. L., M. Nagy, M. LeGouar, F. Lacroute, and S. Potier.** 1989. Organization of the yeast URA2 gene: identification of a defective dihydroorotase-like domain in the multifunctional carbamoylphophate synthetase-aspartate transcarbamoylase complex. *Gene* **79:**59–70.
102. **Stephenson, B. T.** Unpublished data.
103. **Stevens, R. C., K. M. Reinisch, and W. N. Lipscomb.** 1991. Molecular structure of *Bacillus subtilis* aspartate transcarbamoylase at 3.0Å resolution. *Proc. Natl. Acad. Sci. USA* **88:**6087–6091.
104. **Strauch, M., V. Webb, G. Spiegelman, and J. A. Hoch.** 1990. The *spo0A* protein of *Bacillus subtilis* is a repressor of the *abrB* gene. *Proc. Natl. Acad. Sci. USA* **87:**1801–1805.
105. **Suttle, D. P., B. Y. Bugg, J. K. Winkler, and J. J. Kanalas.** 1988. Molecular cloning and nucleotide sequence for the complete coding region of human UMP synthase. *Proc. Natl. Acad. Sci. USA* **85:**1754–1758.
106. **Switzer, R. L.** 1977. The inactivation of microbial enzymes *in vivo*. *Annu. Rev. Microbiol.* **31:**135–157.
107. **Switzer, R. L., R. W. Bond, M. E. Ruppen, and S. Rosenzweig.** 1985. Involvement of the stringent response in regulation of protein degradation in *Bacillus subtilis*. *Curr. Top. Cell. Regul.* **27:**373–385.
108. **Theisen, M., R. A. Kelln, and J. Neuhard.** 1987. Cloning and characterization of the *pyrF* operon of *Salmonella typhimurium*. *Eur. J. Biochem.* **164:**613–619.
109. **Theisen, M., and J. Neuhard.** 1990. Translational coupling in the *pyrF* operon. *Mol. Gen. Genet.* **222:**345–352.
110. **Trach, K., J. W. Chapman, P. Piggot, D. LeCoq, and J. A. Hoch.** 1988. Complete sequence and transcriptional analysis of the *spo0F* region of the *Bacillus subtilis* chromosome. *J. Bacteriol.* **170:**4194–4208.
111. **Trowsdale, J., S. M. H. Chen, and J. A. Hoch.** 1978. Genetic analysis of phenotypic revertants of *spo0A* mutants in *Bacillus subtilis*: a new cluster of ribosomal genes. *In* G. Chambliss, and J. C. Vary (ed.), *Spores VII*. American Society for Microbiology, Washington, D.C.
112. **Turnbough, C. L., Jr., K. H. Kerr, W. R. Funderberg, J. P. Donohue, and F. E. Powell.** 1987. Nucleotide sequence and characterization of the *pyrF* operon of *Escherichia coli* K-12. *J. Biol. Chem.* **262:**10239–10245.
113. **Turnbough, C. L., Jr., and R. L. Switzer.** 1975. Oxygen dependent inactivation of glutamine phosphoribosylpyrophosphate amidotransferase in stationary-phase cultures of *Bacillus subtilis*. *J. Bacteriol.* **121:**108–114.
113a. **Turner, R. J., Y. Lu, and R. L. Switzer.** Unpublished data.
114. **Waindle, L. M., and R. L. Switzer.** 1973. Inactivation of aspartic transcarbamylase in sporulating *Bacillus subtilis*: demonstration of a requirement for metabolic energy. *J. Bacteriol.* **144:**517–527.
115. **Waleh, N. S., and J. L. Ingraham.** 1976. Pyrimidine ribonucleoside monophosphokinase and the mode of RNA turnover in *Bacillus subtilis*. *Arch. Microbiol.* **110:**49–54.
116. **Wang, L.-F.** 1986. Gene organization and regulation of the *Bacillus subtilis* RNA polymerase major sigma operon. Ph.D. thesis. University of California, Davis.
117. **Washabaugh, M. W., and K. D. Collins.** 1984. Dihydroorotase from *Escherichia coli*. Purification and characterization. *J. Biol. Chem.* **259:**3293–3298.
118. **Weng, M., C. A. Makaroff, and H. Zalkin.** 1986. Nucleotide sequence of *Escherichia coli pyrG* encoding CTP synthetase. *J. Biol. Chem.* **261:**5568–5574.
119. **White, M. N., J. Olszowy, and R. L. Switzer.** 1971. Regulation and mechanism of phosphoribosylpyrophosphate synthetase: repression by end products. *J. Bacteriol.* **108:**122–131.
120. **Wild, J. R., and M. E. Wales.** 1990. Molecular evolution and genetic engineering of protein domains involving aspartate transcarbamoylase. *Annu. Rev. Microbiol.* **44:** 193–218.
121. **Wilson, H. R., C. D. Archer, J. Liu, and C. L. Turnbough, Jr.** 1992. Translational control of *pyrC* expression mediated by nucleotide-sensitive selective transcriptional initiation in *Escherichia coli*. *J. Bacteriol.* **174:**514–524.
122. **Wilson, H. R., P. T. Chan, and C. L. Turnbough, Jr.** 1987. Nucleotide sequence and expression of the *pyrC* gene of *Escherichia coli* K-12. *J. Bacteriol.* **169:**3051–3058.
123. **Wilson, H. R. and C. L. Turnbough, Jr.** 1990. Role of the purine repressor in the regulation of pyrimidine gene expression in *Escherichia coli* K-12. *J. Bacteriol.* **172:** 3208–3213.
124. **Yon, R. J.** 1972. Wheat germ aspartate transcarbamoylase. Kinetic behavior suggesting an allosteric mechanism of regulation. *Biochem. J.* **128:**311–320.
125. **Zahler, S. A.** 1978. An adenine-thiamin auxotrophic mutant of *Bacillus subtilis*. *J. Gen. Microbiol.* **107:**199–201.
126. **Zalkin, H., and D. J. Ebbole.** 1988. Organization and regulation of genes encoding biosynthetic enzymes in *Bacillus subtilis*. *J. Biol. Chem.* **263:**1595–1598.

26. Purine and Pyrimidine Salvage Pathways

PER NYGAARD

OVERVIEW

Many bacteria are able to produce their nucleotides both by synthesis de novo and from exogenously supplied bases and nucleosides, but other bacteria are incapable of synthesis de novo and require an exogenous supply of a purine compound, a pyrimidine compound, or both. Phosphoribosylpyrophosphate is an essential metabolite in the de novo synthesis of IMP and UMP and in the synthesis of nucleotides from bases.

The salvage pathways mediate a continuous influx into the nucleotide pools of nucleosides and bases arising from intracellular breakdown of nucleic acids and nucleotides, thereby preventing the loss of energy and valuable precursors. Because of active salvage, purine and pyrimidine bases and nucleosides are normally not found intracellularly. However, they may be formed in excess under some growth conditions and excreted into the culture medium. The natural habitat of bacteria contains highly variable levels of nucleic acid components that arise as degradation products from decaying cells or as excretion products. Most bacteria have evolved efficient transport mechanisms for the utilization of exogenously present nucleosides and bases, and uptake of these compounds is often coupled to the metabolic processes that convert them to nucleotides. Nucleotides in the growth medium cannot be taken up as such but must be dephosphorylated by nucleotidases located on the surfaces of the cells or excreted into the growth medium (20, 155). An important function of the ectonucleotidases is the assimilation of extracellular nucleotides.

The first complete purine nucleotide formed in the purine biosynthetic pathway is IMP, which is converted to AMP and GMP via two separate pathways (Fig. 1). These pathways are also important for the conversion of adenine to GMP and guanine to AMP and for the synthesis of AMP and GMP from xanthine and hypoxanthine. In the pyrimidine biosynthetic pathway, UMP is the precursor for all pyrimidine nucleotides, whether synthesized de novo or via the salvage pathways from pyrimidine bases or nucleosides (Fig. 2). Little is known about the enzymes that phosphorylate nucleoside monophosphates. An enzyme that specifically phosphorylates UMP and CMP has been purified from *Bacillus subtilis* (144). A single enzyme catalyzes the phosphorylation of both ribo- and deoxyribonucleoside diphosphates (184). Biosynthesis of deoxyribonucleotides is catalyzed by a single enzyme that converts all four ribonucleoside diphosphates (or triphosphates) to their corresponding 2′-deoxyribonucleotides (Fig. 3). The synthesis of dTTP proceeds via dUMP, which can be formed by alternative pathways that may differ among the various bacteria. Deoxyribonucleotides can also be synthesized from deoxyribonucleosides.

Bacteria exhibit major differences in their abilities to utilize and interconvert exogenously supplied purine and pyrimidine derivatives. Several species of lactic acid bacteria are known to require a purine and a pyrimidine base for growth (169), and some *Lactobacillus* species require a deoxyribonucleoside or thymine (8, 76, 153). *Lactococcus lactis*, on the other hand, is prototrophic for purines and pyrimidines (103). *Mycoplasma* and *Acholesplasma* species have a specific requirement for guanine, uracil, and thymine (81, 87). Within the genus *Mycobacterium*, *Mycobacterium leprae* is auxotrophic for purines, while other species are prototrophic (186).

The salvage pathways also make the pentose moieties of nucleosides and the free amino groups of adenine, guanine, and cytosine compounds available as sources of carbon, energy, and nitrogen, respectively. Under some conditions, the purine and pyrimidine rings can be degraded, and this degradation may be the major pathway by which some bacteria obtain carbon, energy, and nitrogen for growth (178). Consult reference 178 for a detailed description of the enzymic reactions involved in degradation of the purine and pyrimidine rings. The salvage pathways also define the sensitivity to purine and pyrimidine analogs that require conversion to nucleotides in order to be toxic. Inhibition of bacterial growth by naturally occurring purine bases and nucleosides has been observed in many species. The types of effects and apparent biochemical mechanisms associated with these effects are quite diverse. The extensive literature on this subject has been reviewed elsewhere (48, 49).

Among gram-positive bacteria, the salvage pathways have been studied in most detail in *B. subtilis* and *Bacillus cereus*. For previous reviews on purine and pyrimidine metabolism, see references 46, 59, 76, 93, 96, 99, and 107.

PURINE SALVAGE AND INTERCONVERSION

Utilization of Purine Bases and Nucleosides

The pathways by which purine bases and ribonucleosides are metabolized in *B. subtilis* are shown in Fig. 1. Different species of gram-positive bacteria show great variations in their abilities to metabolize bases and nucleosides. Our knowledge of purine salvage reactions and transport systems in *B. subtilis* is to a great extent based on studies of mutants defective

Per Nygaard • Institute of Biological Chemistry B, University of Copenhagen, DK-1307 Copenhagen K, Denmark.

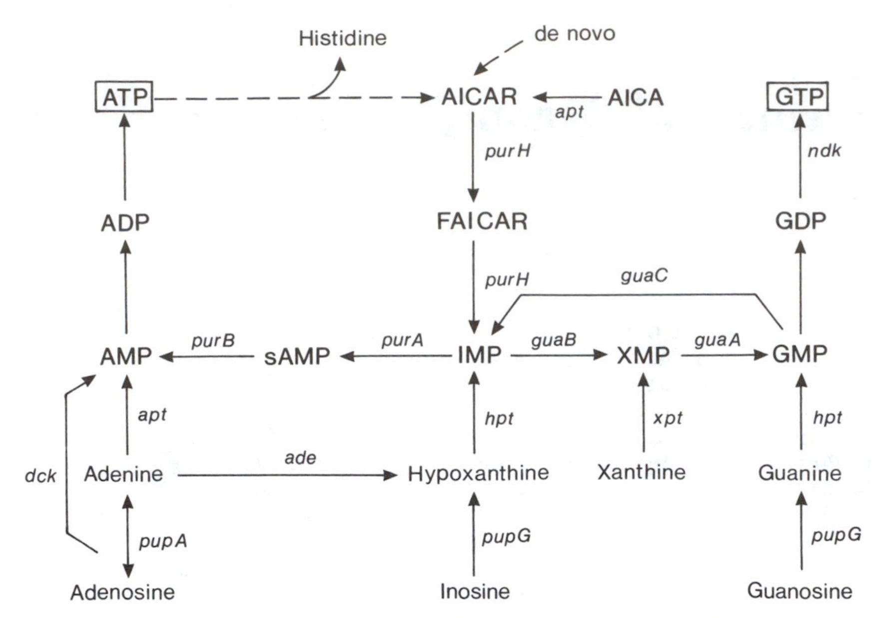

Figure 1. Pathways of ATP and GTP synthesis from purine bases and nucleosides in *B. subtilis*. Individual reactions are identified by gene symbols. AICA, aminoimidazolecarboxamide; AICAR, AICA ribonucleotide; FAICAR, formamidoimidazolecarboxamide ribonucleotide; sAMP, adenylosuccinate; *purH*, AICAR transformylase; *purA*, adenylosuccinate synthetase; *purB*, adenylosuccinate lyase; *guaB*, IMP dehydrogenase; *guaA*, GMP synthetase; *guaC*, GMP reductase; *apt*, adenine phosphoribosyltransferase; *xpt*, xanthine phosphoribosyltransferase; *hpt*, hypoxanthine (guanine) phosphoribosyltransferase; *dck*, deoxycytidine (adenosine) kinase; *pupA*, adenosine phosphorylase; *ade*, adenine deaminase; *pupG*, guanosine (inosine) phosphorylase.

in these reactions (30, 105, 137). Nucleotides present in the growth medium also serve as purine sources (Table 1). However, their utilization requires prior dephosphorylation to nucleosides by periplasmic or

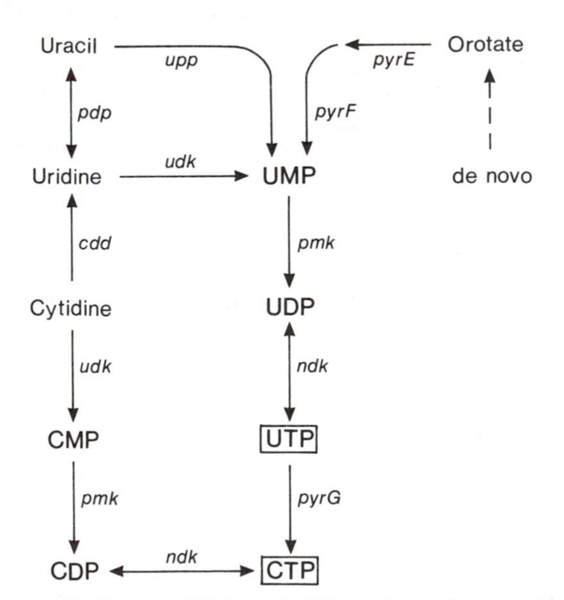

Figure 2. Pathways of UTP and CTP synthesis from pyrimidine bases and nucleosides in *B. subtilis*. Individual reactions are identified by the corresponding gene symbols, *pyrE*, orotate phosphoribosyltransferase; *pyrF*, orotidylate decarboxylase; *upp*, uracil phosphoribosyltransferase; *pdp*, pyrimidine nucleoside phosphorylase; *udk*, uridine (cytidine) kinase; *cdd*, cytidine deaminase; *pmk*, pyrimidine ribonucleoside monophosphate kinase; *pyrG*, CTP synthetase.

excreted nucleotidases (115, 155). The uptake and metabolism of purine bases and ribonucleosides are inhibited by the stringent response to amino acid deprivation (7). After removal of isoleucine from the culture medium of a *B. subtilis ile* mutant, the concentration of ppGpp increases dramatically and purine uptake is inhibited (Table 2). Salvaging of purine deoxyribonucleosides for the synthesis of deoxyribonucleotides will be dealt with in a separate section below.

Transport of bases and nucleosides

In *B. subtilis*, evidence for three purine transport systems has been obtained either by studying competition of uptake (7) or by selecting mutants defective in purine uptake (137) (Table 3). One transport system is specific for adenine. Another, encoded by the *pbuG* gene, is specific for hypoxanthine and guanine and ensures the utilization of hypoxanthine and guanine when they are present at low concentrations in the growth medium. Because xanthine uptake is not affected by the *pbuG* mutation and depends on xanthine phosphoribosyltransferase activity, a third transport system specific for xanthine must exist (137). Two nucleoside transport systems have been suggested, one specific for adenosine and another specific for guanosine and inosine (7). The transport systems require energy, as shown by the loss of transport capacity in the presence of energy poisons (36).

Metabolism of bases

Uptake of purines is strictly coupled to the intracellular formation of nucleotides and controlled by the nucleotide pools, probably by feedback mechanisms

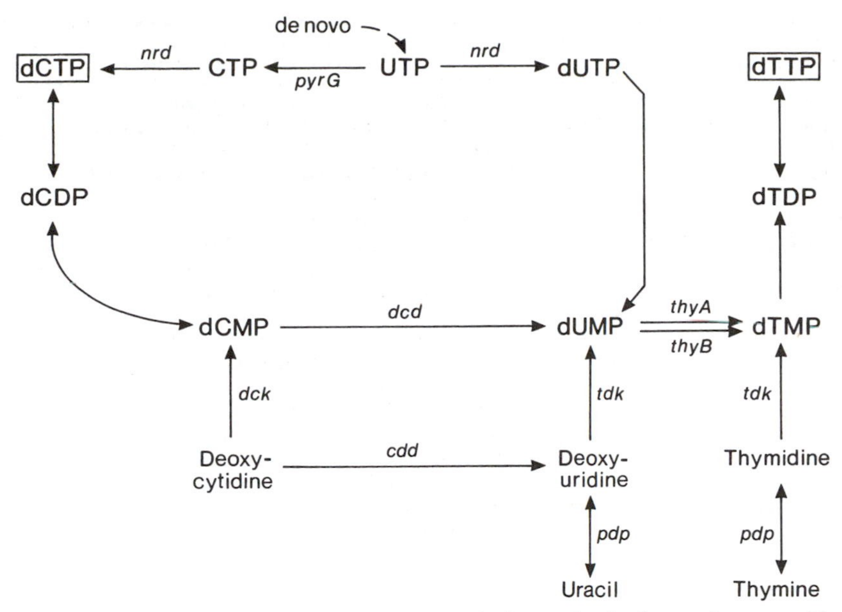

Figure 3. Pyrimidine deoxyribonucleotide synthesis in *B. subtilis*. Individual reactions are identified by gene symbols. *pyrG*, CTP synthetase; *nrd*, ribonucleotide reductase; *dcd*, dCMP deaminase; *thyA* and *thyB*, thymidylate synthase; *dck*, deoxycytidine kinase; *cdd*, cytidine (deoxycytidine) deaminase; *tdk*, thymidine kinase; *pdp*, pyrimidine nucleoside phosphorylase.

on the phosphoribosylation step (11). The four naturally occurring bases serve as sole purine source in mutants defective in IMP biosynthesis in *B. subtilis* (Table 1), but adenine and hypoxanthine are the best sources. Mutants defective in steps prior to aminoimidazolecarboxamide ribonucleotide (AICAR) (Fig. 1) can also utilize 4-amino-5-imidazolecarboxamide (AICA) in place of a purine base for growth (33, 137). AICA must be supplied in considerably higher concentrations than purine bases to support growth and is phosphoribosylated by adenine phosphoribosyltransferase to AICAR.

Adenine can be metabolized by two routes in *B. subtilis*, either directly to AMP, catalyzed by adenine phosphoribosyltransferase (the *apt* gene product), or via deamination to hypoxanthine, catalyzed by adenine deaminase (Fig. 1). Mutants defective in each of the two reactions have been isolated. The first reaction is essential in *purA* mutants, while adenine deamination is required when adenine serves as a general purine source. A *purB* mutant also requires adenine for growth, and both phosphoribosylation and deamination are essential in a *purB* mutant unless the requirement for guanine is met by including hypoxanthine or guanine compounds in the growth medium. During growth in adenine-supplemented media, 40% of the adenine taken up by wild-type cells is deaminated to hypoxanthine, which accumulates in the medium. Adenine phosphoribosyltransferase has been purified from *B. subtilis* (10).

Hypoxanthine and guanine are phosphoribosylated by the same enzyme. A mutant with a partial deficiency in hypoxanthine (guanine) phosphoribosyltransferase has been isolated (30). *B. subtilis* possesses a phosphoribosyltransferase specific for xanthine. A mutant defective in this enzyme can no longer metabolize xanthine. Mutants defective in IMP dehydrogenase (*guaB*) will grow on either xanthine or guanine,

Table 1. Growth of purine and pyrimidine auxotrophic strains of *B. subtilis* on different purine or pyrimidine compounds[a]

Strain genotype	Doubling time (min)	Purine or pyrimidine source
purE	42–47	Adenine, hypoxanthine, adenosine, inosine, deoxyadenosine, deoxyinosine, 2′-AMP (1 mM), 3′-AMP (1 mM)
	63–80	Guanine, guanosine, deoxyguanosine
	160–200	Xanthine, 5′-AMP (1 mM)
	240–250	AICA (2 mM), 2′-AMP 3′-AMP, 5′-AMP
	∞	Xanthosine, AICA riboside (1 mM)
pyrB	48–53	Uracil, uridine, cytidine, deoxyuridine, deoxycytidine, 2′-CMP, 3′-CMP
	58	Orotate (1 mM)
	>500	Orotate, 5′-CMP (1 mM)
	∞	Cytosine, dihydroorotate (1 mM)

[a] Determined in glucose minimal medium at 37°C. Purines and pyrimidines were added to a concentration of 0.1 mM if not otherwise stated.

Table 2. Effects of amino acid starvation on incorporation of bases and nucleosides in *B. subtilis* (*ile*)

Isoleucine present[a]	ppGpp concn[b]	Incorporation (nmol/min/mg [dry wt]) of[c]:				
		Adenine	Hypoxanthine	Guanosine	Uracil	Cytidine
+	0.01	1.47	1.43	0.90	0.74	0.75
–	0.50	0.03	0.01	0.04	0.09	0.09

[a] Cells were starved for isoleucine for 20 min.
[b] Nanomoles per milligram dry weight.
[c] ^{14}C-labeled compounds were added to cultures, and radioactivity accumulated within the cells was determined after 10 min.

Table 3. Selection of mutants resistant to purine and pyrimidine analogs in *B. subtilis*

Analog (concn)	Mutation selected	Reference(s)
2-Fluoroadenine (0.1 mM)	*apt* (adenine phosphoribosyltransferase)	137
2-Fluoroadenosine (2 mM) + hypoxanthine (0.5 mM)	*apt* (adenine phosphoribosyltransferase)	30
8-Azaguanine (0.5 mM)	*pbuG* (guanine [hypoxanthine] transport)	137
8-Azaxanthine (0.5 mM)	*xpt* (xanthine phosphoribosyltransferase)	137
	guaA (GMP synthetase)	137
8-Azaxanthine (2 mM)	*cpm* (pleiotropic mutation[a])	79
8-Azaguanine (0.15 mM) + adenine (0.6 mM)	*purR* (purine biosynthesis[b])	62
8-Thioguanosine (2 mM)	*hpt* (hypoxanthine [guanine] phosphoribosyltransferase)	30
5-Fluorouracil (10 μM)	*upp* (uracil phosphoribosyltransferase)	99
5-Fluorouracil (0.4 mM) + uracil (0.45 mM)	*furA* (uracil transport)	52, 106
5-Fluorouracil (0.4 mM) + uracil (0.45 mM)	*furB* (excretion of pyrimidine compounds)	106, 190
5-Fluorouracil (10 μM) + deoxyadenosine (0.1 mM)	*pupA* (adenosine phosphorylase[c])	
	pdp (pyrimidine nucleoside phosphorylase[c])	
	tdk (thymidine kinase[c])	99
5-Fluorouridine (50 μM)	*udk* (uridine [cytidine] kinase[c])	99
5-Fluorocytidine (10 μM)	*udk* (uridine [cytidine] kinase)	127
5-Azacytidine (0.5 mM)	*udk* (uridine [cytidine] kinase)	45
5-Fluorodeoxycytidine (10 μM)	*dck* (deoxycytidine [deoxyadenosine] kinase[d])	91, 127
	dcd (dCMP kinase[d])	92

[a] Required genetic background of *ade guaC*. See Fig. 1.
[b] Required genetic background of *ade purA*. See Fig. 1.
[c] Required genetic background of *upp*. See Fig. 2.
[d] Required genetic background of *cdd*. See Fig. 3.

while mutants defective in GMP synthetase (*guaA*) have a specific requirement for guanine. Both *guaA* and *guaB* mutants can use 2,6-diaminopurine as a guanine source (21). In a *guaA* mutant, competition between purine base and nucleoside utilization has been investigated. Hypoxanthine but not inosine inhibits growth on guanine. The inhibition of growth by hypoxanthine has been ascribed to a competition with guanine for the transport and phosphoribosylation steps. Adenine inhibits growth on guanine, and to do so, it must either be deaminated to hypoxanthine or phosphoribosylated to AMP. Phosphoribosylation of adenine causes a twofold reduction in the pool of 5-phosphoribosyl-1-pyrophosphate (PRPP) (Table 4) (139), and this might affect the phosphoribosylation of guanine.

The *apt* and *xpt* genes have been located on the chromosomal map of *B. subtilis* (118). Expression of the purine phosphoribosyltransferases in *B. subtilis* is not influenced by purines in the growth medium. Two purine phosphoribosyltransferases have been identified in *Lactobacillus casei*; one of them reacts with hypoxanthine and guanine, and the other reacts with xanthine and guanine (70).

Table 4. Intracellular amounts of nucleoside triphosphates and PRPP in *B. subtilis*[a]

Compound	Amt (nmol/mg [dry wt])	Compound	Amt (nmol/mg [dry wt])
ATP	4.8	CTP	1.0
GTP	1.4	UTP	1.6
dATP	0.24	dCTP	0.24
dGTP	0.15	dTTP	0.65
PRPP	1.5		

[a] Determined in cells growing exponentially in glucose minimal medium (139).

Metabolism of nucleosides

Both purine ribonucleosides and deoxyribonucleosides serve as purine sources in *B. subtilis* (Table 1). The nucleosides are converted to the corresponding ribonucleotide in two steps. The first is phosphorolytic cleavage to the free base and ribose 1-phosphate or deoxyribose 1-phosphate, and the second is phosphoribosylation of the base to the nucleotide. Two purine nucleoside phosphorylases have been characterized in *B. subtilis*, one specific for adenosine (deoxyadenosine) and another specific for guanosine, inosine, and their corresponding deoxyribonucleosides (64). A mutant with reduced guanosine (inosine) phosphorylase activity and another defective in adenosine phosphorylase have been isolated (30, 99). Guanosine (inosine) phosphorylase but not adenosine phosphorylase is induced in *B. subtilis* when cells are grown in the presence of inosine or adenosine (108). The utilization of guanosine in a *guaA* mutant is inhibited by inosine, which competes with guanosine for the transport system and for guanosine (inosine) phosphorylase (7). The uptake of nucleosides in *B. subtilis*, here exemplified by inosine (Fig. 4), parallels their incorporation into nucleotides and nucleic acids, because very little free base is excreted into the growth medium. This is in contrast to the situation in *Escherichia coli*, *Salmonella typhimurium* (107), and *B. cereus* (59, 172). In these organisms, nucleosides induce the synthesis of nucleoside-catabolizing enzymes, and as a consequence, nucleosides are rapidly taken up and metabolized. While the pentose moiety is being catabolized, excess purine base is excreted into the growth medium. In *B. cereus* cells grown in the presence of adenosine, inosine phosphorylase and adenosine deaminase are induced, while adenosine phosphorylase is constitutively synthesized. When induced cells are incubated in fresh medium containing adenosine, almost all adenosine is broken down to

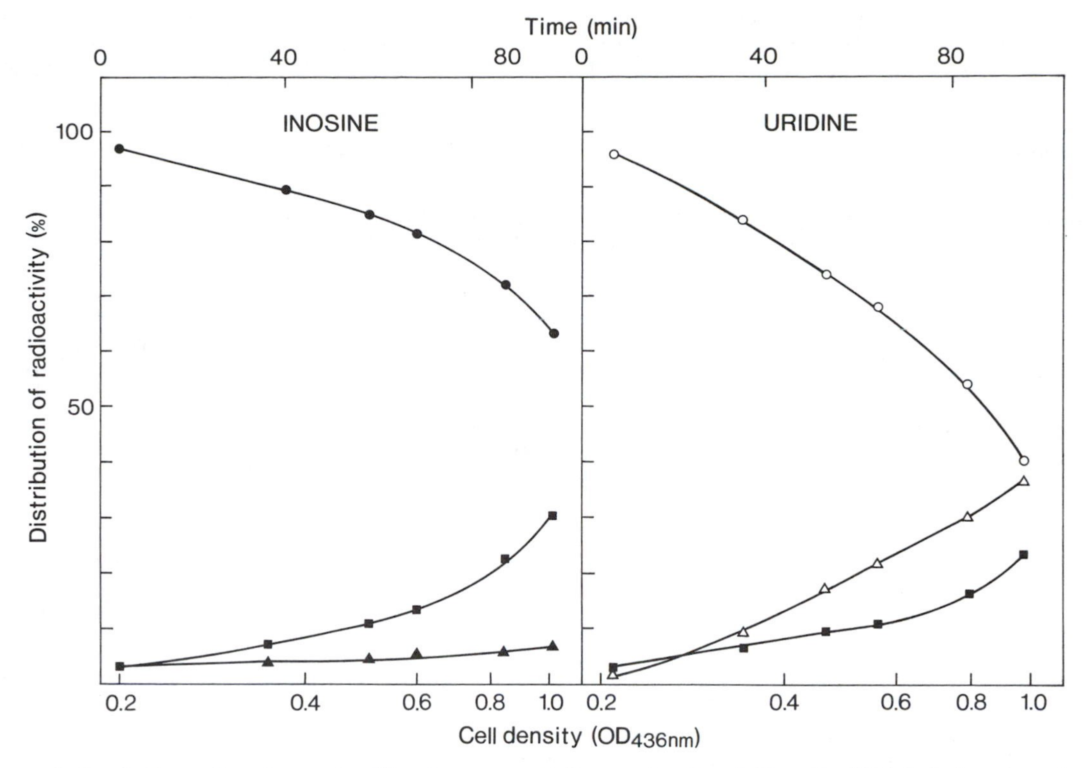

Figure 4. Metabolism of inosine and uridine in exponentially growing *B. subtilis* cells. ^{14}C-labeled nucleoside (100 nmol/ml) was added at time zero. At different times thereafter, samples of the culture were analyzed for the distribution of label in the culture. Symbols: ■, label incorporated into whole cells; ●, inosine; ▲, hypoxanthine; ○, uridine; △, uracil. OD_{436nm}, optical density at 436 nm.

hypoxanthine, indicating a preferential deamination of adenosine. In inosine-grown cells, inosine phosphorylase is induced, while adenosine deaminase is repressed; in a medium containing adenosine, equal amounts of adenine and hypoxanthine are formed, indicating that adenosine can be phosphorolytically cleaved when the level of adenosine deaminase is low. Two purine nucleoside phosphorylases have also been found in *Bacillus stearothermophilus*, one with specificity toward guanosine and inosine and another with specificity toward adenosine, guanosine, and inosine (53, 54).

B. subtilis does not contain specific ribonucleoside kinases (30, 137), but the *B. subtilis* deoxycytidine kinase possesses some adenosine kinase activity (91). This activity can be demonstrated in vivo in mutants defective in adenosine phosphorylase activity (137). Adenosine and inosine kinase activities have been determined in crude extracts of *B. cereus* (59). Nucleotidases may also catalyze the transfer of a phosphate group from a nucleoside monophosphate to a nucleoside (143). Since this reaction requires high concentrations of the nucleoside, it is probably not normally involved in nucleotide formation. Lactobacilli with a growth requirement for a deoxynucleoside do not contain purine nucleoside phosphorylase activity, but they all contain a ribonucleoside hydrolase that will hydrolyze adenosine, guanosine, and inosine (167). In these organisms, the free bases are phosphoribosylated by appropriate phosphoribosyltransferases or they are substrates for nucleoside deoxyribosyltransferases. Other lactic acid bacteria, which do not possess nucleoside deoxyribosyltransferase, however, have inosine phosphorylase activity (15). In coryneform bacteria, deoxyadenosine phosphorylase activity is present in *Micrococcus luteus* but not in *Brevibacterium ammoniagenes*, which cannot metabolize deoxyadenosine (4).

Conversion of IMP to ATP and GTP

The enzymic steps by which AMP and GMP are formed from IMP are shown in Fig. 1. The branching from IMP is a part of the de novo biosynthetic pathway but is also essential to the salvage pathways. When adenine compounds serve as purine sources, the branching from IMP to AMP becomes dispensable, and when guanine compounds serve as purine sources, the branching from IMP to GMP is no longer required. The details of the reactions that convert IMP to ATP and GTP and the regulation of enzyme synthesis and activity are found in chapter 24.

Interconversion of Adenine and Guanine Compounds

Adenine and guanine compounds are interconvertible through the common precursor IMP. This conversion uses special pathways, since the reactions leading from IMP to AMP and from IMP to GMP are virtually irreversible. The purine interconversion reactions occur under conditions of imbalance between the adenine and guanine nucleotide pools and are obligatory when adenine or guanine compounds serve as sole purine sources.

The histidine pathway

As a by-product of histidine biosynthesis de novo from ATP and PRPP, a molar equivalent amount of AICAR, and hence IMP, is formed. This pathway for guanine synthesis depends on the *purH*-encoded enzyme (Fig. 1) and is inhibited by histidine in the growth medium (105). The histidine pathway, however, does not alone account for the conversion of adenine to GMP.

The adenine deaminase pathway

In *B. subtilis*, adenine can be deaminated to hypoxanthine, and the hypoxanthine formed is phosphoribosylated to IMP, which in turn can be converted to GMP. This pathway is very efficient, and *B. subtilis* grows on adenine with a doubling time equivalent to that on hypoxanthine (Table 1). In mutants defective in adenine phosphoribosyltransferase and adenosine phosphorylase, it has been shown that the adenine deaminase-catalyzed reaction is an essential step in the conversion of adenine and adenosine to GMP (137). In agreement with this is the finding that a mutant defective in IMP biosynthesis and deficient for adenine deaminase fails to grow on adenine as sole purine (30, 106). The gene encoding adenine deaminase (*ade*) is located on the chromosomal map at 130°. The level of adenine deaminase is reduced twofold when both adenine and guanine are present in the growth medium (106). The reported adenosine and AMP deaminase activities in *B. subtilis* (30, 105) most likely can be ascribed to the combined action of 5′-nucleotidase, adenosine phosphorylase, and adenine deaminase. The existence and possible involvement of AMP deaminase in interconversion reactions in gram-positive bacteria remain to be established.

In *B. cereus*, *E. coli*, and *S. typhimurium* (57, 58, 100), the conversion of adenine proceeds via the intermediate formation of adenosine, inosine, and hypoxanthine catalyzed by adenosine phosphorylase, adenosine deaminase, and inosine phosphorylase. In this pathway, ribose 1-phosphate and P_i act catalytically, being continuously regenerated by inosine phosphorylase and adenosine phosphorylase in *B. cereus* and by purine nucleoside phosphorylase in *E. coli* and *S. typhimurium*. The latter enzyme uses both inosine and adenosine as substrates (66).

The *B. cereus* adenosine deaminase (37) is induced by addition of adenine or adenosine to the growth medium. The inosine formed from adenosine then causes the induction of inosine phosphorylase and has a repressing effect on adenosine deaminase. Adenosine phosphorylase is synthesized constitutively (172). Adenine and hypoxanthine will induce adenosine deaminase in *E. coli* but not in *S. typhimurium*, and inosine and adenosine will induce the synthesis of purine nucleoside phosphorylase in both organisms (107).

The intracellular formation of adenosine from AMP is less well characterized but must involve 5′-nucleotidase activity that is strictly controlled or that works only at high intracellular concentrations of AMP to avoid excess nucleotide degradation. In *E. coli*, AMP can be hydrolytically cleaved to adenine and ribose 5-phosphate by AMP nucleosidase, an enzyme that is activated by ATP and inhibited by P_i (72).

The GMP reductase pathway

The conversion of guanine compounds and xanthine to adenine nucleotides proceeds through IMP to AMP (Fig. 1). The only known enzyme catalyzing the conversion of guanine compounds to IMP is NADPH-dependent GMP reductase. *B. subtilis* mutants defective in GMP reductase (*guaC*) are unable to utilize guanine compounds as the sole purine source (30). The *guaC* gene has not been localized on the chromosomal map of *B. subtilis*. The growth of a *B. subtilis guaC* mutant is completely inhibited by guanine compounds but is restored by the addition of adenine compounds (94). Inhibition is best explained by feedback inhibition of PRPP amidotransferase; as a consequence of this inhibition, the cells are starved for adenine nucleotides. The level of GMP reductase is increased in *B. subtilis* grown in the presence of exogenous guanosine (106). Growth of a purine auxotroph (Table 1) on guanine and guanine nucleosides is slower than on hypoxanthine, indicating that the GMP reduction step is limiting. The failure of a purine-requiring mutant of *B. subtilis* to utilize 2,6-diaminopurine as sole purine source (21) most likely can be ascribed to the fact that the formation of guanine from 2,6-diaminopurine is too slow to allow accumulation of guanine nucleotides and hence stimulation of GMP reductase synthesis. GMP reductase from *E. coli* is inhibited by ATP and activated by GTP (77, 107), and the cellular level increases with increasing concentrations of guanine in the growth medium (102).

PYRIMIDINE SALVAGE: UTILIZATION OF PYRIMIDINE BASES AND NUCLEOSIDES

The pathways by which pyrimidine bases and nucleosides are converted to the nucleotide level vary among the different bacteria studied. A key enzyme is uracil phosphoribosyltransferase, which seems to be present in all gram-positive bacteria. The pyrimidine salvage pathways in *B. subtilis* are shown in Fig. 2. Each function has been established by studies of mutant strains with single or multiple blocks in the pathway (99, 127). Mutants defective in pyrimidine salvage enzymes and transport functions can be isolated by selecting for resistance to pyrimidine analogs (Table 3). Uptake of uracil, uridine, and cytidine in *B. subtilis* is reduced under conditions of stringent control (Table 2). Mutants defective in the pyrimidine biosynthetic pathway require an exogenous pyrimidine source for growth; this requirement can be met by several pyrimidine compounds (Table 1). Pyrimidine nucleotides are hydrolyzed to their nucleoside derivatives prior to uptake by ectonucleotidases. The enzymatic reactions involved in pyrimidine biosynthesis are described in chapter 25. The only intermediary compound of the UMP biosynthetic pathway that can be supplied from the outside is orotic acid. However, orotic acid will not support growth of mutants defective in orotate phosphoribosyltransferase or orotidylate decarboxylase activity. Because *B. subtilis* possesses two carbamylphosphate synthetases, a pyrimidine-repressible enzyme coded for by the *pyrA* gene and an arginine-repressible enzyme (117), a *pyrA* mutant has no pyrimidine requirement unless arginine is present in the growth medium and represses synthesis of the arginine-controlled enzyme. The pathways by which thymine and pyrimidine deoxyribonucleosides are metabolized (Fig. 3) will be dealt with in a special section below.

Metabolism and Transport of Bases

Uracil is either phosphoribosylated to UMP or converted to UMP through the concerted actions of uridine phosphorylase and uridine kinase. The uptake of uracil in *B. subtilis* depends on the intracellular pool size of GTP and is reduced when ppGpp accumulates (7). In *E. coli*, GTP activates uracil phosphoribosyltransferase activity, while ppGpp inhibits the enzyme (99). A similar regulatory mechanism may exist in *B. subtilis*. Pyrimidine starvation results in threefold-increased levels of uracil phosphoribosyltransferase activity, indicating that expression of the *upp* gene is controlled by a pyrimidine compound (106). Mutants defective in the *upp* gene still take up uracil and convert it to UMP via the intermediate formation of uridine catalyzed by pyrimidine nucleoside phosphorylase and uridine kinase. The mutant defective in the uracil transport function (Table 3) is not defective in the intracellular metabolism of uracil (106).

B. subtilis, several *Lactobacillus* species, *L. lactis*, and some coryneform bacteria lack the ability to deaminate cytosine (3, 99, 134). So far, cytosine deaminase activity has not been reported in gram-positive bacteria. No enzyme capable of converting cytosine directly to CMP has been described. Only bacteria like *E. coli* and *S. typhimurium* that possess cytosine deaminase activity can satisfy their pyrimidine requirements with cytosine. The same organisms are sensitive to 5-fluorocytosine.

Metabolism and Transport of Nucleosides

In *B. subtilis*, uridine and cytidine share a common transport system and pyrimidine nucleoside kinase (36, 68, 99). Uridine is converted to UMP by two routes, either directly by phosphorylation or via a phosphorolytic cleavage to uracil. When *B. subtilis* cells are grown in uridine-supplemented medium, significant amounts of uracil are excreted into the culture medium (Fig. 4), indicating a less strict coupling between uptake and salvage than is observed with inosine. Uridine will serve as the sole but poor carbon source in *B. subtilis*. Uridine is extensively cleaved to uracil in *Brevibacterium ammoniagenes* but will not serve as sole carbon source (3). However, in *E. coli* and *S. typhimurium*, the rapid cleavage of uridine makes the ribose moiety of uridine available as a good carbon source, while excess uracil appears in the growth medium (100).

A single enzyme catalyzes the phosphorolysis of pyrimidine ribonucleosides and deoxyribonucleosides in *B. subtilis*. Uridine, deoxyuridine, and thymidine are substrates, but cytidine and deoxycytidine are not (Fig. 2 and 3). Enzyme synthesis is induced by deoxyribonucleosides but not by ribonucleosides in the growth medium (99, 132). The *pdp* gene has been cloned (80). Pyrimidine nucleoside phosphorylase in *Bacillus thuringiensis* possesses activity toward both uridine and thymidine and is induced by deoxyribonucleosides (43, 44). Pyrimidine nucleoside phosphorylase has been purified from *B. stearothermophilus* (55, 136), and both uridine and thymidine phosphorylase activities reside in one protein; both uridine and thymidine act as inducers (136). *L. casei* contains two pyrimidine nucleoside phosphorylases, i.e., a uridine phosphorylase and a thymidine phosphorylase (6). Certain *Lactobacillus* species can hydrolytically degrade pyrimidine ribonucleosides but not pyrimidine deoxyribonucleosides. In these organisms, pyrimidine deoxyribonucleosides are substrates for nucleoside deoxyribosyltransferases (46, 99). In *E. coli* and *S. typhimurium*, synthesis of uridine phosphorylase and a number of other proteins involved in nucleoside transport and catabolism is negatively regulated by the CytR repressor and positively controlled by cyclic AMP and the cyclic-AMP receptor protein (46, 159). The effector is cytidine in *E. coli*, whereas both cytidine and uridine are effectors in *S. typhimurium* (100).

A single enzyme, coded for by the *udk* gene, catalyzes the phosphorylation of uridine and cytidine in *B. subtilis* and *E. coli*. Pyrimidine ribonucleoside kinase has been purified from *B. stearothermophilus*; both uridine and cytidine act as substrates, and the enzyme activity is inhibited by CTP (114). The expression of the *udk* gene in *E. coli* is derepressed during pyrimidine starvation (100).

Cytidine can be converted directly to CMP, but the major route involves deamination to uridine catalyzed by cytidine deaminase. Cytidine does not serve as sole pyrimidine source in mutants defective in cytidine deaminase (*cdd*) activity (127). Mutants defective in CTP synthetase have an absolute requirement for cytidine. The gene encoding cytidine (deoxycytidine) deaminase (*cdd*) in *B. subtilis* has been cloned, and the nucleotide sequence has been determined (162). The level of cytidine deaminase is not affected by pyrimidine compounds in the growth medium. A number of *Lactobacillus* species contain cytidine deaminase activity (134). The failure of pyrimidine-requiring strains of *Brevibacterium ammoniagenes* to grow on cytidine as pyrimidine source and the finding that wild-type strains are resistant to 5-fluorocytidine indicate that this organism is unable to metabolize cytidine (3). In *E. coli* and *S. typhimurium*, expression of the *cdd* gene is regulated by the CytR protein together with the *udp* gene as described above.

DEOXYRIBONUCLEOTIDE METABOLISM

De Novo Synthesis of Deoxyribonucleotides

Purine and pyrimidine 2′-deoxyribonucleoside phosphates may be synthesized via two different routes. The de novo pathway is initiated by reduction of the ribonucleotides to their corresponding deoxyribonucleotides. Thus, three of the DNA precursors, dATP, dGTP, and dCTP, are derived directly from their corresponding ribonucleotides. De novo synthesis of dTTP proceeds via the intermediate formation of dUMP, which may be derived from pathways starting with dUTP, dCTP, or both. The other routes by which deoxyribonucleotides can be synthesized involve salvaging of deoxyribonucleosides and thymine. Three deoxyribonucleoside kinases that can together phosphorylate deoxyadenosine, deoxyguanosine, deoxycytidine, thymidine, and deoxyuridine seem to be abundant in gram-positive bacteria. Nothing is known about enzymes that phosphorylate deoxyribonucleoside monophosphates to the diphosphate level. The last phosphorylation step is probably catalyzed by nucleoside diphosphokinase (Fig. 3). *Lactobacillus* species possess one or two nucleoside deoxyribosyltransferases that catalyze the synthesis of the five deoxyribonucleosides from a single deoxyribonucleoside and the appropriate bases. Only a few gram-positive organisms, such as certain *Lactobacillus* species, require thymine and a deoxyribonucleoside, while certain members of the genus *Mycoplasma* require thymine for growth (87, 99). A more extensive discussion of deoxyribonucleotide synthesis may be found in recent reviews (93, 126, 163).

Reduction of ribonucleotides

A single enzyme can catalyze the reduction of all four common ribonucleotides, but some organisms contain more than one enzyme with different cofactor requirements (28, 126, 163). Either ribonucleoside diphosphates or ribonucleoside triphosphates are substrates. Some evidence obtained with crude extracts indicates that ribonucleotide reduction occurs at the triphosphate level in *B. subtilis* and that the reduction is not dependent on coenzyme B_{12} (128). Ribonucleoside triphosphate reductase from *Lactobacillus leichmannii* is a single protein with an absolute requirement for 5′-deoxyadenosylcobalamin (coenzyme B_{12}). The ribonucleoside triphosphates are reduced to deoxyribonucleoside triphosphates concomitant with the oxidation of a thiol component (2, 41). The substrate specificity of the protein is governed by appropriate allosteric effectors (12, 99). Cells of *L. leichmannii* grown in the absence of B_{12} have a requirement for deoxyribonucleosides (40). The ribonucleotide reductase activity in *Bacillus megaterium* is stimulated by B_{12} (188), while ribonucleotide reductase activity in *Brevibacterium ammoniagenes* and *M. luteus* depends on manganese. During growth under conditions of manganese deficiency, DNA synthesis is specifically arrested and the cells have no detectable ribonucleotide reductase activity (142). The ribonucleotide reductase from *Brevibacterium ammoniagenes* represents a new type of ribonucleotide reductase composed of two subunits and having manganese as cofactor (187). The aerobic *E. coli* enzyme consists of two proteins and catalyzes reduction of the ribonucleoside diphosphate. Most of our information on the reduction of ribonucleotides in prokaryotes comes from studies of this enzyme (93, 126). Recently, an anaerobic ribonucleoside triphosphate reductase that requires *S*-adenosylmethionine as cofactor was identified in *E. coli* (28).

Biosynthesis of dTMP

B. subtilis possesses two thymidylate synthases, one encoded by the *thyA* gene and the other encoded by *thyB*. Only the *thyA*-encoded enzyme functions in cells growing at high temperatures (101). Both enzymes catalyze the synthesis of dTMP from dUMP (Fig. 3). A single and very specific enzyme, deoxyuridine triphosphatase, catalyzes the generation of dUMP and PP_i from dUTP in *B. subtilis* (124) and *E. coli* (157). The deamination of dCMP deaminase is not an essential reaction, because mutants defective in dCMP deaminase (*dcd*) grew normally but with twofold-increased dCTP pools and lowered dTTP pools (99). By using a *dcd* mutant, it was determined that 45% of the dUMP formed is derived from deamination of dCMP and that this reaction is the only one by which deoxycytidine phosphates can be deaminated in *B. subtilis* (92). The enzyme responsible for the synthesis of dCMP from dCTP is not known. Purified dCMP deaminase requires dCTP for activity and is inhibited by dTTP, suggesting that the level of dCTP is important for the activity in vivo also. *Lactobacillus acidophilus* contains dCMP deaminase with properties similar to those of the *B. subtilis* enzyme (149). *E. coli* and *S. typhimurium* do not contain dCMP deaminase activity. Instead, they possess activity by dCTP deaminase, an enzyme that shows cooperativity towards dCTP and is feedback inhibited by dTTP (99).

Utilization of Preformed DNA Precursors

Metabolism of thymine

Thymine enters *B. subtilis* cells by passive diffusion (129) and is metabolized through thymidine phosphorylase (Fig. 3). The reaction requires deoxyribose 1-phosphate, which normally is not present in wild-type cells. No phosphoribosyltransferase with specificity towards thymine has ever been found. When cells are supplied with a deoxyribosyl donor, such as deoxyadenosine, thymine can react with the deoxyribose 1-phosphate formed from the phosphorolysis of deoxyadenosine to form thymidine, which can then be converted to dTTP (Fig. 3). A thymine dependence phenotype can be observed only in strains with mutations in both the *thyA* and the *thyB* genes or in a *thyA* $thyB^+$ strain grown at high temperatures (101). In a double mutant, the pool of dUMP is increased dramatically. Thus, the increased level of dUMP may result in an increased availability of deoxyribose 1-phosphate and may thereby enhance the metabolism of thymine. Mutants with a low thymine requirement can be isolated; they are defective in the catabolism of deoxyribose phosphates (132), providing evidence for the role of deoxyribose 1-phosphate in thymine utilization. A *thyA* $thyB^+$ strain can utilize exogenous thymine because this mutant also contains increased dUMP pools (101). In *E. coli*, the thymine requirement is determined by the levels of thymidine phosphorylase and the deoxyribose 1-phosphate pool (65, 100). *L. leichmannii* and *L. casei* are unable to synthesize dTMP unless folate or thymine is added to the growth medium (5, 99).

Deoxyribonucleoside salvage routes

The five common deoxyribonucleosides are precursors of deoxyribonucleotides in *B. subtilis*. Conditional mutants defective in the reduction of ribonucleotides have been isolated in *B. subtilis*. At the nonpermissive temperature, these mutants have a minimum requirement for deoxyadenosine, deoxyguanosine, and deoxycytidine for growth (128). Mutants defective in cytidine (deoxycytidine) deaminase and dCMP deaminase have an additional requirement for thymidine or deoxyuridine (99). Three different deoxyribonucleoside kinases have been identified in *B. subtilis*: thymidine kinase with specificities toward thymidine and deoxyuridine, deoxyguanosine kinase, and deoxyadenosine (deoxycytidine) kinase (91). Mutants defective in thymidine kinase and deoxyadenosine (deoxycytidine) kinase have been isolated (Table 3). Synthesis of the deoxyribonucleoside kinases seems to be constitutive. Thymidine kinase is present almost universally; the only gram-positive organism in which this activity has been shown to be absent is *Brevibacterium ammoniagenes* (4). Three deoxyribonucleoside kinases have been purified from *L. acidophilus*, *L. leichmannii*, and *B. megaterium*, one with specificity towards deoxyadenosine and deoxycytidine, another with specificity towards deoxyguanosine and deoxyadenosine, and the third a thymidine (deoxyuridine) kinase (19, 56, 120, 181). The only deoxyribonucleoside kinase activity in *E. coli* and *S. typhimurium* is thymidine (deoxyuridine) kinase.

Nucleoside deoxyribosyltransferases are found in *Lactobacillus* and *Streptococcus* species and provide a route for the synthesis of any deoxyribonucleoside from a single deoxyribonucleoside. Interestingly, the same organisms are devoid of inosine phosphorylase activity (15). In the absence of acceptors, nucleoside deoxyribosyltransferases may catalyze the slow hydrolysis of deoxyribonucleosides (158). Nucleoside deoxyribosyltransferases, sometimes called *trans-N*-deoxyribosylases, catalyze two types of transfer of the deoxyribosyl moiety according to the nature of the donor and acceptor bases as follows.

$$\text{purine}_1 \text{ deoxyriboside} + \text{purine}_2 \rightarrow \text{purine}_2 \text{ deoxyriboside} + \text{purine}_1$$

$$\text{pyrimidine deoxyriboside} + \text{purine} \rightarrow \text{purine deoxyriboside} + \text{pyrimidine}$$

Lactobacillus helveticus contains an enzyme that catalyzes the purine-specific transfer and another that catalyzes the transfer of the deoxyribosyl moiety between purines and pyrimidines (13). It remains to be established whether there is a pyrimidine-specific enzyme. Nucleoside deoxyribosyltransferase from *L. leichmannii* has been purified, and the gene specifying the enzyme has been cloned. The enzyme has low specificity for both donor and acceptor substrates (16). Most *Lactobacillus* species have a growth requirement for a deoxyribonucleoside. The requirement for deoxyribonucleoside is absolute for some *Lactobacillus* species, but B_{12} can substitute for it in others (8, 40).

Catabolism of deoxyribonucleosides

The pathways by which deoxyribonucleosides in *B. subtilis* are catabolized involve the initial phosphorolytic cleavage to the free base and deoxyribose 1-phosphate. While a single enzyme catalyzes the cleavage of deoxyuridine and thymidine (Fig. 3), two enzymes are required for purine deoxyribonucleosides: (i) adenosine phosphorylase, with activity toward deoxyadenosine, and (ii) guanosine (inosine) phosphorylase, with activity towards deoxyguanosine and deoxyinosine (64). Deoxycytidine catabolism proceeds to deoxyuridine via deamination catalyzed by cytidine (deoxycytidine) deaminase. The deoxyribose 1-phosphate formed is converted to deoxyribose 5-phosphate, with phosphopentomutase (*drm*) as catalyst, and deoxyribose 5-phosphate is cleaved to acetaldehyde and glyceraldehyde 3-phosphate, with deoxyriboaldolase (*dra*) as catalyst. Through the reversal of these reactions, acetaldehyde can be converted to deoxyribose, but this "pathway" has never been shown to be an alternative route in deoxyribonucleotide synthesis. Unlike the situation in *E. coli*, in which the deoxyribose moiety is preferentially degraded even in glucose-rich media, deoxyribonucleosides are poor carbon sources in *B. subtilis*. The *pdp* and *dra* genes are closely linked on the *B. subtilis* chromosome, while the *drm* gene is located in another region (132, 166). The *pdp* and *dra* genes belong to the same regulon and are induced fivefold by addition of deoxyribonucleosides or acetaldehyde to the growth medium. The low-molecular-weight effector seems to be deoxyribose 5-phosphate (132). There is some evidence for a second pair of genes analogous to *dra* and *pdp* (80, 165). The two purine nucleoside phosphorylases, the phosphopentomutase and the cytidine (deoxycytidine) deaminase, are not induced by deoxyribonucleosides (46, 162). Mutants defective in the initial reactions of deoxyribonucleoside can be isolated by selecting for resistance to analogs (Table 3). Mutants defective in phosphopentomutase and deoxyriboaldolase can be isolated in thymine auxotrophs (requiring 20 μg of thymine per ml for growth) as clones possessing a low thymine requirement (1 μg/ml) (132). Such mutants are unable to catabolize the deoxyribose moiety of deoxyribonucleosides and therefore cannot grow on deoxyribonucleosides as carbon source. In contrast to wild-type cells, these cells contain deoxyribose-phosphates. Cells with mutations in the *dra* genes are extremely sensitive to deoxyribonucleosides because of accumulation of deoxyribose 5-phosphate, which is toxic to the cells (42, 43, 46).

Catabolism of deoxyribonucleosides in *B. cereus* and *B. thuringiensis* is induced by deoxyribonucleosides and occurs through the same reactions as in *B. subtilis* (58, 171). From thymine auxotrophs, *dra* and *drm* mutants have been isolated (42). Use of such mutants led to the conclusion that pyrimidine nucleoside phosphorylase is induced by deoxyribose 5-phosphate in the exponential growth phase and that deoxyribose 5-phosphate is involved in inactivation of the enzyme in the late growth phases (43, 44). Synthesis of thymidine phosphorylase is derepressed in thymine-starved cells of *L. casei* (6). *Lactobacillus* species with a growth requirement for a deoxyribonucleoside are unable to catabolize the deoxyribose moiety, and the deoxyribonucleosides are preferentially used for the synthesis of deoxyribonucleotides (46, 174), while it is likely that other *Lactobacillus* and *Lactococcus* species can catabolize deoxyribonucleosides (15, 46). Among coryneform bacteria, *Micrococcus luteus* can catabolize thymidine and deoxyadenosine, while these compounds are not metabolized at all by *Brevibacterium ammoniagenes* (4).

E. coli has evolved a highly regulated set of genes

governing proteins involved in the transport and catabolism of deoxyribonucleosides (46, 96, 100). The *deo* operon, which encodes four enzymes involved in catabolism of deoxyribonucleosides, is transcribed as a tetracistronic message from two promoters. The operon is induced by deoxyribose 5-phosphate, which binds to the DeoR protein, and by cytidine, which binds to the CytR protein. Regulation of gene expression occurs at the level of transcription initiation. The DeoR protein represses expression from both promoters, while the cyclic-AMP–cyclic-AMP receptor protein complex activates the second promoter and the CytR protein counteracts this activation (18, 159, 176).

Catabolism of deoxyribonucleotides

Deoxyribonucleotides are normally not degraded inside cells. However, certain growth conditions lead to a shortage of deoxyribosyl groups and induce degradation as described above. Uptake of foreign DNA, degradation of such DNA, and stress situations may lead to degradation of deoxyribonucleotides. Infection of *B. subtilis* with bacteriophages PBS1 and PBS2, which contain uracil instead of thymine in their DNA, results in the induction of phage-specific nucleotide-metabolizing enzymes such as dCTP deaminase, dTMP phosphohydrolase, dUMP kinase, and a protein that inhibits the cellular deoxyuridine triphosphatase (122–124, 170). These enzymes function to hydrolyze dTMP and thus prevent the accumulation of dTTP and allow the accumulation of dUTP (51). Infection with bacteriophages like SP8, which contain 5-hydroxymethyluracil in place of thymine in their DNA, results in the induction of phage-specific enzymes like dCMP deaminase (131), an enzyme that cleaves dTTP and dUTP to their respective monophosphates, and enzymes involved in the synthesis of 5-hydroxymethyl dUTP (78).

The unusual enzyme dGTP triphosphohydrolase, which cleaves dGTP to deoxyguanosine and inorganic tripolyphosphate, is found in members of the family *Enterobacteriaceae* but is not present in gram-positive bacteria (125).

MUTANTS DEFECTIVE IN PURINE AND PYRIMIDINE METABOLISM

Selection and Utilization of Mutants

The isolation and characterization of mutant strains defective in purine and pyrimidine metabolism have been powerful tools in the identification of the pathways of purine and pyrimidine metabolism and their regulation. Mutagenesis followed by penicillin counterselection has been used to isolate purine and pyrimidine auxotrophs and also to isolate salvage mutants (30, 71). Several analogs that can be taken up by gram-positive bacteria and that inhibit nucleotide metabolism exist. They might exert their toxic effect(s) either directly or after conversion to toxic compounds inside the cell. Mutants defective in purine and pyrimidine metabolism can be isolated by selecting for resistance to toxic analogs. Such selections are possible because most analogs are metabolized by the same enzymes that metabolize the natural compounds and are transported by the same transport system as their natural counterparts. Other types of mutants, namely, mutants with altered expression of purine or pyrimidine genes and mutants that synthesize enzymes with altered properties, have also been identified in this way. Mutants with some altered enzyme properties and defective for other enzymes are used in the production of purine compounds in the food industry. Nucleosides and nucleotides are also important materials in the synthesis of pharmaceuticals and compounds used in research. Interesting developments are the use of bacteria to synthesize nucleoside analogs (71, 113). The *L. helveticus* nucleoside deoxyribosyltransferases have been employed for synthesis of a variety of nucleoside analogs that have antileukemic and immunosuppressive activity (14). Many microorganisms produce purine- and pyrimidine-related antibiotics. For a comprehensive review, see reference 164.

Analogs

Purine analogs

B. subtilis mutants resistant to 2-fluoroadenine and 2-fluoroadenosine are defective in adenine phosphoribosyltransferase activity, indicating that phosphoribosylation of 2-fluoroadenine to 2-fluoro-AMP is required for toxicity. In agreement with this, mutants defective in adenosine phosphorylase activity are resistant to 2-fluoroadenosine (137). Different types of mutants resistant to 2-oxopurine analogs exist. Some are defective in salvaging or transport functions, while others have mutations that affect levels or activities of purine biosynthetic enzymes (Table 3). The *purR* mutation (identified by resistance to 8-azaguanine) eliminates control of the *pur* operon by guanine metabolites, while control by adenine remains unaffected (61, 62). The *cpm* mutation causes resistance to 8-azaxanthine, increased levels of the enzymes of the purine biosynthetic pathway, a reduced level of inosine (guanosine) phosphorylase, and increased levels of 5′-nucleotidases (79). The nature of the effect of this mutation on purine gene expression has not been investigated. The purine analog 6-mercaptopurine inhibits growth of *B. subtilis* and *E. coli* (34, 100). In *E. coli*, two types of resistance have been identified: a mutation in the *hpt* gene that prevents the formation of a toxic nucleotide analog and a mutation in the *purR* gene so that the mutant no longer produces the PurR protein, which negatively regulates expression of the *pur* genes (83, 130). 6-Mercaptopurine can bind to the PurR protein and thereby inhibit the expression of *pur* genes encoding the enzymes of the purine biosynthetic pathway (84). Two *guaA* mutants, one resistant to 8-azaxanthine (Table 3) and one resistant to decoyinine and psicofuranine, have been isolated (71). None of these mutants requires guanine for growth. These findings provide evidence for two inhibitory mechanisms by which either a nucleotide analog or the free base (nucleoside) might interfere with purine metabolism. A number of purine analogs and a few pyrimidine analogs induce sporulation in *B. subtilis* grown in the presence of excess glucose, ammonia, and phosphate (34, 86). The induction of sporulation can be prevented by adding the natural counterparts of the different analogs to the cultures. The interpretation of this observation led to the hypothesis that a partial inhibition of nucleotide synthesis induces sporulation (34). A result of this was the important finding that decoyinine, which specifically inhibits guanine nucleotide synthesis, causes efficient sporulation.

Pyrimidine analogs

The pyrimidine analog 5-fluorouracil seems to be toxic to most bacteria. To become toxic, it must be converted to 5-fluoro-UMP or 5-fluoro-dUMP. 5-Fluoro-dUMP is an exceedingly potent inhibitor of thymidylate synthetase (99). Mutants of *B. subtilis* resistant to 5-fluorouracil are defective in uracil phosphoribosyltransferase or uracil transport or they excrete pyrimidine compounds into the growth medium (Table 3). In *E. coli* and *S. typhimurium*, certain mutants resistant to 5-fluorouracil have reduced UMP kinase activities and exhibit derepressed levels of the pyrimidine biosynthetic enzymes, accumulate UMP, and excrete uracil (100). This type of mutation has not been found among the 5-fluorouracil-resistant mutants isolated in *B. subtilis* (106). In *upp* mutants, 5-fluorouracil can become toxic if the cells are grown in the presence of deoxyadenosine. Deoxyadenosine is phosphorolyzed to adenine and deoxyribose 1-phosphate by adenosine phosphorylase. 5-Fluorouracil can then react with deoxyribose 1-phosphate to form 5-fluorodeoxyuridine (a reaction catalyzed by pyrimidine nucleoside phosphorylase), and 5-fluorodeoxyuridine can be converted to 5-fluoro-dUMP by thymidine kinase. A mutation in one of the genes catalyzing one of the reactions described above will therefore lead to resistance (Table 3). 5-Fluorouridine and 5-fluorocytidine are toxic both to wild-type cells and in *upp* mutants. In *upp* mutants, they are converted to the nucleotide level by uridine (cytidine) kinase, and mutants defective in this step are resistant. 5-Fluorodeoxycytidine is not toxic in *cdd dcd* double mutants (cf. Fig. 3) (99). It appears that the deamination of 5-fluorodeoxycytidine to 5-fluorodeoxyuridine catalyzed by cytidine deaminase is required for toxicity, probably because 5-fluoro-dCMP is not toxic. In wild-type cells, 5-fluorodeoxycytidine is toxic because it can be converted to 5-fluoro-dUMP through reactions catalyzed by deoxycytidine kinase and dCMP deaminase; hence, mutations in genes encoding these enzymes lead to resistance.

Antifolates

Antifolate drugs such as trimethoprim, aminopterin, and methotrexate are toxic to many bacteria because they are potent inhibitors of dihydrofolate reductase, the enzyme that catalyzes the reduction of dihydrofolate to tetrahydrofolate. This reaction is the only way for dihydrofolate, formed in the synthesis of thymidylate, to be converted to tetrahydrofolate (183). A mutant enzyme that is not inhibited by trimethoprim has been found in *B. subtilis* (98). Mutants lacking thymidylate synthase activity can be isolated by a positive selection procedure on media containing thymine or thymidine and an antifolate drug. In *B. subtilis*, this procedure involves two events, because this organism possesses two thymidylate synthases. In this way, *thyA* and *thyB* mutants have been isolated from *B. subtilis* (101), and *thyA* mutants have been isolated from *B. thuringiensis* (42) and *B. megaterium* (180). A methotrexate-resistant *L. casei* mutant overproduces thymidylate synthase (17). It is interesting that the thymidylate synthase and dihydrofolate reductase genes are closely linked in *L. casei* (119) and that the *thyB*-encoded thymidylate synthase and *dfrA*-encoded dihydrofolate reductase form an operon in *B. subtilis* (63, 97). These findings suggest that the methotrexate resistance in *L. casei* originally ascribed to overproduction of thymidylate synthase is instead a result of high levels of dihydrofolate reductase.

Excretion of Purine and Pyrimidine Compounds

Wild-type cells of *B. subtilis* do not excrete purine and pyrimidine compounds during exponential growth. At a late growth stage and during sporulation, however, these compounds may occur in the growth medium. In certain mutants defective in reactions of the salvage pathways, some of the endogenously formed purine and pyrimidine bases or their nucleoside derivatives cannot be rescued and are then excreted. Adenine is excreted from a mutant defective in both adenine phosphoribosyltransferase (*apt*) and adenine deaminase (*ade*) but not in strains with single mutations in the *apt* or *ade* gene (106). Uracil is excreted in strains defective in uracil phosphoribosyltransferase (*upp*) (Table 3) and in mutants with derepressed synthesis of the pyrimidine biosynthetic enzymes. Such mutants may also excrete uridine (25). If a mutant of the latter type is defective, in addition, in pyrimidine nucleoside phosphorylase (*pdp*), only uridine is excreted (24). Strains auxotrophic for purines or pyrimidines accumulate and excrete intermediary compounds of the respective biosynthetic pathways when the purine or the pyrimidine source is exhausted from the growth medium. Under these conditions, the feedback control exerted by end products of the pathways is released and the repressing effects of the purine or pyrimidine compounds on the synthesis of the enzymes of the de novo pathway no longer operate (1, 35, 138).

Bacterial production of nucleotides and nucleosides is of industrial importance. IMP and GMP are used as flavor enhancers in food and are produced by bacteria that excrete these compounds or their nucleoside derivatives into the growth medium. Even though the cytoplasmic membrane is impermeable to exogenous nucleotides, it seems to allow these nucleotides to exit from cells when they tend to accumulate in the cytoplasm. The bacteria used for nucleoside and nucleotide production have multiple mutations that affect enzymes of the purine pathways. These strains have been only partially characterized; however, some important features of the mutants will be dealt with below. For previous reviews, readers are referred to references 71 and 112. *B. subtilis* and *Brevibacterium ammoniagenes* mutants that excrete IMP usually contain low levels of 5′-nucleotidase activity and often require adenine. Mutants that excrete inosine are defective in adenylosuccinate synthetase, have low levels of IMP dehydrogenase, and are defective in inosine phosphorylase activity. Guanosine-producing strains are characterized by low levels of adenylosuccinate synthetase, GMP reductase, and guanosine nucleoside phosphorylase, and adenosine-producing strains are defective in adenine deaminase and adenosine phosphorylase (47, 71, 89, 90). Recently, a pleiotropic mutation (*cpm*) affecting purine metabolism in *B. subtilis* was described (79). The mutant excretes inosine; has derepressed levels of purine biosynthetic enzymes, high levels of 5′-nucleotidases, and a strongly reduced level of inosine phosphorylase; and requires adenine and tyrosine for growth.

PHYSIOLOGICAL FUNCTION OF NUCLEOBASE AND NUCLEOSIDE METABOLISM

The enzymes catalyzing the intracellular metabolism of nucleobases and nucleosides serve two major functions: one is anabolic and ensures the reutilization of endogenously formed nucleosides and bases for nucleotide synthesis; the other is catabolic and makes nucleosides and bases available as carbon and nitrogen sources. When nucleosides are present at low concentrations in media rich in carbon and nitrogen sources, the nucleosides escape the catabolic enzymes and are preferentially phosphorylated directly to the corresponding nucleoside monophosphate. The activities of the anabolic enzymes are subjected to feedback control, and the synthesis of these enzymes is often derepressed when the cellular supply of nucleic acid precursors is reduced. The catabolic enzymes, on the other hand, are often inducible, and their activities are not controlled by feedback inhibitory mechanisms. In media deficient in carbon and/or nitrogen sources, degradation rather than salvaging of nucleosides and bases may become more important.

Purines and pyrimidines occur in large amounts in a number of ecosystems, and many gram-positive bacteria living in such habitats utilize these compounds as carbon, energy, nitrogen, and phosphate sources.

Salvaging of Nucleobases and Nucleosides

Purine and pyrimidine salvage in growing cells

Nucleosides and bases are continuously generated in cells from the degradation of nucleic acids, nucleotides, and other metabolites. For example, adenosine and adenine are formed from *S*-adenosylmethionine as by-products in methylation reactions and spermidine synthesis, respectively (107, 156). The rescuing of the intracellularly formed bases and nucleosides is usually very efficient. A *B. subtilis ade apt* mutant is unable to salvage endogenously formed adenine (Fig. 1) and therefore excretes adenine into the culture medium in amounts that correspond to about 10% of that synthesized de novo (106). The activities of various intracellular nucleotidases are under strict regulatory control to avoid useless degradation of nucleotides. One type of 5′-nucleotidase is inhibited by nucleoside triphosphates (32, 147); other types, because of their high K_ms, function only when nucleoside monophosphates accumulate (155). Nevertheless, some recycling of nucleotides from the degradation of mRNA and DNA gives rise to nucleosides, which are salvaged. A 3′-nucleoside monophosphate formed intracellularly can be reutilized only if it is dephosphorylated to the nucleoside level.

When purine and pyrimidine compounds are present in the growth medium, the contribution to nucleotide synthesis from the de novo pathways is greatly reduced. This suppression is accomplished by increased feedback inhibition of key enzymes of the de novo pathways and by repression of the synthesis of the de novo enzymes (27, 60, 95, 104, 116). Expression of the *pur* genes in *B. subtilis*, except for the *guaA* gene, is repressed when preformed purines are added to the growth medium (139). Even at low concentrations, exogenous purines are preferentially used and synthesis of the de novo enzymes is concomitantly reduced (Fig. 5). Adenine is efficiently used in the synthesis of both adenine and guanine nucleotides, while guanosine is mainly used for guanine nucleotide synthesis, indicating that the interconversion pathways operate with different efficiencies in wild-type cells (Fig. 5), as in purine auxotrophs (Table 1). A major effect of adenine addition is a twofold-smaller PRPP pool and larger adenine and guanine nucleotide pools, which together act through feedback inhibition of the first reaction of the biosynthetic pathway catalyzed by PRPP amidotransferase (85). Guanosine, on the other hand, enlarges the PRPP and guanine nucleotide pools, the PRPP amidotransferase activity is not feedback inhibited under these conditions. This explains why adenine nucleotides are synthesized de novo in guanosine-supplemented cultures. Regulatory mechanisms that couple nucleotide synthesize to the cell cycle may exist. The DnaA protein has been implicated in the regulation of *gua* gene expression through binding to DNA boxes in *E. coli* (82, 168). Because DNA boxes (189) are also present in the *guaB* gene of *B. subtilis* (67), this type of regulation might also occur in gram-positive bacteria.

Increasing concentrations of uracil during pyrimidine biosynthesis evoke a similar but more pronounced response to low concentrations of the preformed precursor (Fig. 6) with respect to both repression of enzyme synthesis and utilization of uracil for pyrimidine nucleotide synthesis. As a result of uracil addition, the UTP pool increases about 50%. However, no simple correlation between nucleotide pool size and *pyr* gene expression has ever been established (106). It is pertinent to recall here that all the genes encoding the enzymes of UMP biosynthesis are arranged in an operon and that none of the enzymes is required when uracil serves as pyrimidine source. The first enzyme of the UMP biosynthetic pathway, carbamylphosphate synthetase, is inhibited by uridine nucleotides and is activated by PRPP and guanosine nucleotides (117). The synthesis of CTP from UTP is inhibited by cytidine nucleotides. Addition of preformed pyrimidines has no effect on the level of CTP synthetase except in the case of cytidine deaminase-deficient mutants, in which cytidine represses the synthesis of CTP synthetase (1).

Purine and pyrimidine metabolism in starving cells

Bacteria growing in a rich medium will use the available metabolites, including amino acids, purines, and pyrimidines, rather than synthesize them de novo. To be able to adapt to a situation in which a

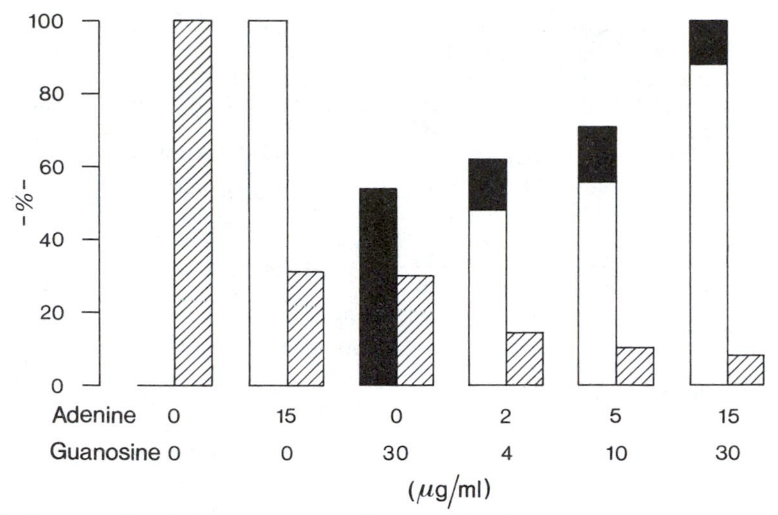

Figure 5. Effects of adenine and guanosine on purine salvage and levels of a purine biosynthetic enzyme in wild-type *B. subtilis*. Cells were grown with ^{14}C-labeled purines, and the contribution from the exogenous source to RNA nucleotides was calculated. Symbols: □, adenine; ■, guanosine; ▨, levels of glycinamide ribonucleotide synthetase. 100% ~ 24 nmol/min/mg of protein.

single compound becomes limiting, bacteria have evolved the stringent response, whereby synthesis of ppGpp and pppGpp decreases RNA synthesis (111). Starvation of an amino acid auxotrophic strain for the required amino acid induces the stringent response, reduces purine and pyrimidine salvage (Table 3), inhibits IMP dehydrogenase, and initiates sporulation (74, 111).

Partial purine or pyrimidine limitation decreases the intracellular level of GTP or UTP, respectively, resulting in decreased synthesis of tRNA and rRNA, while mRNA synthesis is less affected (177). Purine and pyrimidine limitations induce the synthesis of enzymes involved in the degradation of RNA, particularly 5′-nucleotidase and phosphodiesterase I, which leads to degradation of RNA to 5′-nucleoside monophosphates and nucleosides (23). This degradation may provide the basis for the resynthesis of nucleotides under starvation conditions.

Sporulation

In sporulating bacteria, net nucleic acid synthesis ceases at the end of the exponential growth phase, and PRPP amidotransferase and aspartate transcarbamylase are specifically inactivated through degradation (22, 173, 182). Synthesis of nucleotides during sporulation therefore depends on the turnover of preexisting nucleic acids and on the availability of purines and pyrimidines in the surroundings. When sporulation is initiated in *B. subtilis*, uptake of preformed nucleic acid precursors, in particular of uracil and guanine, decreases (7).

Several observations have indicated that purine

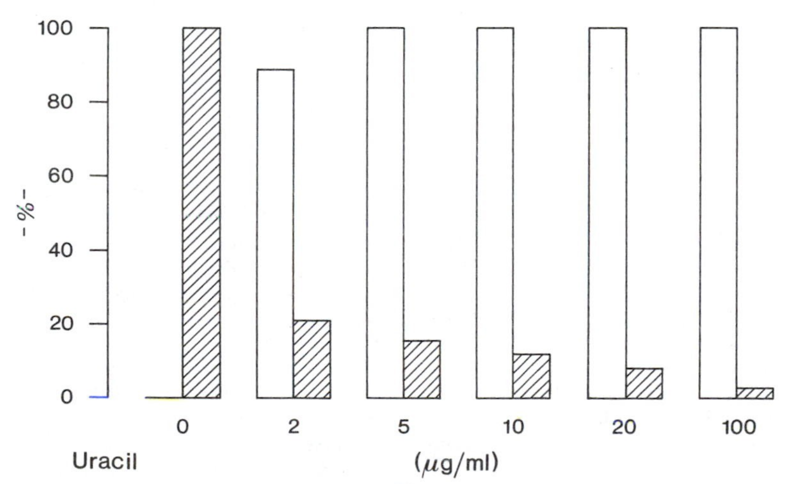

Figure 6. Effects of uracil on pyrimidine salvage and levels of a pyrimidine biosynthetic enzyme in wild-type *B. subtilis*. Cells were grown with ^{14}C-labeled uracil, and the contribution from the exogenous source to RNA nucleotides was calculated (□). ▨, levels of aspartate transcarbamylase. 100% ~ 77 nmol/min/mg of protein.

compounds are important for the initial processes of sporulation. Mutants of *B. cereus* (73), *B. megaterium* (29), and *B. subtilis* (35) defective in purine biosynthesis sporulate more frequently in excess glucose, ammonia, and amino acids than do purine prototrophs. It is somewhat surprising that a preformed purine precursor like inosine stimulates sporulation in *B. subtilis*, *B. cereus* (145), and *Clostridium perfringens* (133).

Sporulation of *B. subtilis* can be induced, even in rich media, by partial deprivation of the de novo synthesis of purines, specifically of guanine nucleotides (33–35, 74, 75, 111). The guanine nucleotide pools can be reduced by adding decoyinine to the growth medium, because decoyinine specifically inhibits GMP synthetase (35, 86). However, several other analogs that inhibit purine biosynthesis can also induce sporulation (35, 86, 191). Under most, if not all, sporulation conditions, there is a drop in the intracellular concentrations of GDP and GTP (74). The pool size of GTP seems to be a general signal for sporulation in other gram-positive bacteria such as *B. megaterium* (175), *Streptomyces griseus* (109), and *Clostridium acetobutylicum* (135). *B. subtilis* and *B. megaterium* contain several apparent GTP-binding proteins (88). However, where GTP exerts its effects is not known, and there may be several target sites (160, 161).

Accepting the concept that a decrease in the GTP pool is perceived by the cell as a signal to initiate sporulation leads to the question of which mechanisms reduce the GTP pool. Obviously, shortage of nitrogen, carbon, and phosphate will ultimately lead to smaller nucleotide pools and sporulation. Although ppGpp has been ruled out as a direct inducer of sporulation (111), it certainly has an inhibitory effect on GMP synthesis, whether de novo or via salvaging of preformed guanine compounds, by inhibiting IMP dehydrogenase (74, 111) and guanine and guanosine salvage (7) (Table 2). Inhibition of guanine salvage may be important for the regulation of GMP synthesis from guanine derived from degradation of RNA. Any 3′-GMP formed can be reutilized only after conversion to guanine, which must in turn be phosphoribosylated to GMP, a reaction inhibited by ppGpp. However, if 5′-GMP is a major degradation product, the control of GTP synthesis by ppGpp is circumvented, unless phosphorylation of GMP is inhibited by ppGpp.

A possible role for nucleosides in the induction of sporulation remains to be established. The level of inosine phosphorylase increases at the onset of sporulation in *B. cereus* (31, 38), and the level is reduced in a sporulation-deficient mutant of *B. subtilis* containing the pleiotropic mutation *cpm* (79). The equilibrium of the reaction catalyzed by nucleoside phosphorylase favors nucleoside synthesis and thus may provide a means by which a certain level of nucleoside can be maintained. Assuming that nucleosides are involved in the induction of sporulation, there are two target sites, GTP and *S*-adenosylmethionine synthesis (110, 179), that can be inhibited by adenosine and that when inhibited will induce sporulation. Like decoyinine, adenosine in combination with PP_i inhibits GMP synthetase (86, 192), and such inhibition may result in a smaller GTP pool. Adenosine is also a potent inhibitor of *S*-adenosylhomocysteine hydrolase (50, 69); inhibition of this enzyme causes *S*-adenosylhomocysteine accumulation, resulting in the inhibition of methylation reactions mediated by *S*-adenosylmethionine-utilizing methyl transferase.

Spore germination

Germination of spores of several species is stimulated by nucleosides (38, 39, 146). In *B. cereus* and *B. megaterium*, inosine enhances germination triggered by glucose or L-alanine (9, 121), while neither adenine nor hypoxanthine has any effect on spore germination (154). The stimulatory effect of inosine in *B. cereus* is not eliminated in a mutant strain containing only 3% of the wild-type level of inosine phosphorylase (38). In an attempt to identify the mechanism that underlies the stimulation, several nucleoside analogs with modifications in either the purine or the sugar moiety were tested on *B. cereus* spores (9, 154). The results of these studies indicate that germination-inducing activity is governed by the structural properties of the sugar rather than of the purine moiety. Only those analogs containing ribose or 2′-deoxyribose stimulate spore germination. This suggests that salvaging of the exogenous nucleoside is not required for stimulation of germination and that the nucleosides act, rather, as cofactors or allosteric effector molecules.

At the early germination stage, nucleotide biosynthesis is not possible in spores from *Bacillus* species because the necessary enzymes are absent. Most knowledge about nucleotide synthesis in germinating spores comes from studies with *B. megaterium*, in which nucleotide synthesis can be separated into two periods, i.e., an initial turnover stage lasting about 15 min and a subsequent biosynthetic stage. During the turnover stage, ribonucleoside triphosphates are formed from an endogenous pool of ribonucleoside 5′-monophosphates and from degradation of RNA, which in *B. subtilis* is degraded to ribonucleoside 5′-monophosphates that cannot be degraded further because of the absence of 5′-nucleotidases in spores (148). Nucleosides and bases formed endogenously or present in the germination medium may also provide a source (150–152). However, little is known about endogenous purine and pyrimidine salvage during germination, and no systematic studies have dealt with spore germination in mutants defective in salvage enzymes. At the biosynthetic stage, there is an increase in the ribonucleoside triphosphate pool due to nucleotide synthesis de novo and, as a result of this, increased RNA synthesis and decreased RNA breakdown (148, 151, 152).

The dormant spore contains all of the enzymes necessary for deoxyribonucleotide synthesis, but the deoxyribonucleotides themselves are absent and are first synthesized in the biosynthetic stage (150, 151) coincident with a 5- to 10-fold increase in the activity of ribonucleotide reductase.

Degradation of Nucleobases

The utilization of nucleobases as carbon or nitrogen sources appears to be of minor importance in *B. subtilis*, while other bacteria have evolved efficient and highly regulated enzyme systems for their degradation. However, the reactions involved in the interconversion of adenine and guanine compounds and deamination of cytosine compounds give rise to NH_3.

Purine degradation

Many bacteria degrade purine bases via xanthine to uric acid and allantoin. The aerobic degradation of allantoin gives rise to urea and glyoxylic acid. Bacteria belonging to several genera, including *Bacillus*, *Brevibacterium*, and *Mycobacterium*, can perform only some of these reactions and use purines as nitrogen sources. *Arthrobacter* species utilize purines as sole nitrogen and carbon sources. In the anaerobic pathways, xanthine is degraded through formyglycine to NH_3 and CO_2; other degradation products include formic acid, acetic acid, and glycine (178). These degradation pathways have been most extensively studied in bacteria belonging to the genus *Clostridium* (26), and some species are nutritionally restricted to the fermentation of purines and their degradation products (140, 141). The choice between purine salvage and degradation must be subjected to metabolic control. In a *Streptomyces* strain, purine degradation is promoted under conditions of nitrogen limitation, while purine salvage is inhibited (185).

Pyrimidine degradation

Degradation of uracil, orotic acid, and thymine occurs by two different pathways. One pathway involves the reductive degradation of uracil and thymine to β-alanine, CO_2, and NH_3 and is found in several genera, including *Mycobacterium*, *Corynebacterium*, and *Clostridium* (178). The oxidative pathway is found in the same genera. In this pathway, pyrimidines are degraded to barbituric acid, which is hydrolyzed to urea and malonic acid (178).

REFERENCES

1. **Asahi, S., M. Doi, Y. Tsunemi, and S. J. Akiyma.** 1989. Regulation of pyrimidine nucleotide biosynthesis in cytidine deaminase-negative mutants of *Bacillus subtilis*. *Agric. Biol. Chem.* **53:**97–102.
2. **Ashley, G. W., G. Harris, and J. A. Stubbe.** 1986. The mechanism of *Lactobacillus leichmannii* ribonucleotide reductase. Evidence for 3′ carbon-hydrogen bond cleavage and a unique role for coenzyme B_{12}. *J. Biol. Chem.* **261:**3958–3964.
3. **Auling, G., and B. Moss.** 1984. Metabolism of pyrimidine bases and nucleosides in the coryneform bacteria *Brevibacterium ammoniagenes*, and *Micrococcus luteus*. *J. Bacteriol.* **158:**733–736.
4. **Auling, G., H. Prelle, and H. Diekmann.** 1982. Incorporation of deoxyribonucleosides into DNA of coryneform bacteria and the relevance of deoxyribonucleoside kinases. *Eur. J. Biochem.* **121:**365–370.
5. **Avraham, Y., N. Grossowicz, and J. Yashphe.** 1991. Regulation of the synthesis and activity of thymidine phosphorylase in *Lactobacillus casei*. *FEMS Microbiol. Lett.* **82:**287–292.
6. **Avraham, Y., J. Yashphe, and N. Grossowicz.** 1988. Thymidine phosphorylase and uridine phosphorylase of *Lactobacillus casei*. *FEMS Microbiol. Lett.* **56:**29–34.
7. **Beaman, T. C., A. D. Hitchins, K. Ochi, N. Vasantha, T. Endo, and E. Freese.** 1983. Specificity and control of uptake of purines and other compounds in *Bacillus subtilis*. *J. Bacteriol.* **156:**1107–1117.
8. **Beck, W. S., and M. Levin.** 1963. Purification, kinetics, and repression control of bacterial trans-*N*-deoxyribosylase. *J. Biol. Chem.* **238:**702–709.
9. **Bédard, J., and G. M. Lefebvre.** 1989. L-Alanine and inosine enhancement of glucose triggering in *Bacillus megaterium* spores, p. 760–763. *In* H. O. Halvorson, R. Hanson, and L. L. Campbell (ed.), *Spores V*. American Society for Microbiology, Washington, D.C.
10. **Berlin, R. D.** 1969. Adenylate pyrophosphorylase: purification, reaction sequence, and inhibition by sodium ion. *Arch. Biochem. Biophys.* **134:**120–129.
11. **Berlin, R. D., and E. R. Stadtman.** 1966. A possible role of purine nucleotide pyrophosphorylases in the regulation of purine uptake by *Bacillus subtilis*. *J. Biol. Chem.* **241:**2679–2686.
12. **Blakley, R. L.** 1978. Ribonucleoside triphosphate reductase from *Lactobacillus leichmannii*. *Methods Enzymol.* **51:**246–259.
13. **Cardinaud, R.** 1978. Nucleoside deoxyribosyltransferase from *Lactobacillus helveticus*. *Methods Enzymol.* **51:**446–455.
14. **Carson, D. A., D. B. Wasson, and E. Beutler.** 1984. Antileukemic and immunosuppressive activity of 2-chloro-2′-deoxyadenosine. *Proc. Natl. Acad. Sci. USA* **81:**2232–2236.
15. **Chawdhri, R. F., D. W. Hutchinson, and A. O. Richards.** 1991. Nucleoside deoxyribosyltransferase and inosine phosphorylase activity in lactic acid bacteria. *Arch. Microbiol.* **155:**409–411.
16. **Cook, W. J., S. A. Short, and S. E. Ealick.** 1990. Crystallization and preliminary X-ray investigation of recombinant *Lactobacillus leichmannii* nucleoside deoxyribosyltransferase. *J. Biol. Chem.* **265:**2682–2683.
17. **Crusberg, T. C., R. Leary, and R. L. Kisliuk.** 1970. Properties of thymidylate synthetase from dichloromethotrexate-resistant *Lactobacillus casei*. *J. Biol. Chem.* **245:**5292–5296.
18. **Dandanell, G., and K. Hammer.** 1991. *deoP1* promoter and operator mutants in *Escherichia coli*: isolation and characterization. *Mol. Microbiol.* **5:**2371–2376.
19. **Deibel, M. R., Jr., and D. H. Ives.** 1978. Deoxynucleoside kinases from *Lactobacillus acidophilus* R-26. *Methods Enzymol.* **51:**346–354.
20. **Demain, A. L., and D. Hendlin.** 1967. Phosphohydrolases of a *Bacillus subtilis* mutant accumulating inosine and hypoxanthine. *J. Bacteriol.* **9:**66–74.
21. **Demain, A. L., and H. T. Shigeura.** 1968. Dependence of diaminopurine utilization on the mutational site of purine auxotrophy in *Bacillus subtilis*. *J. Bacteriol.* **95:** 555–571.
22. **Deutscher, M. P., and A. Kornberg.** 1968. Biochemical studies of bacterial sporulation and germination. VIII. Patterns of enzyme development during growth and sporulation of *Bacillus subtilis*. *J. Biol. Chem.* **243:**4653–4660.
23. **Dhariwal, K. R., N. Vasantha, and E. Freese.** 1982. Partial nucleotide limitation induces phosphodiesterase I and 5′-nucleotidase in *Bacillus subtilis*. *J. Bacteriol.* **149:**1146–1149.
24. **Doi, M., S. Asahi, Y. Tsunemi, and S. J. Akiyami.** 1989. Mechanism of uridine production by *Bacillus subtilis* mutants. *Appl. Microbiol. Technol.* **30:**234–238.
25. **Doi, M., Y. Tsunemi, S. Asahi, S. J. Akiyama, and Y. Nakao.** 1988. *Bacillus subtilis* mutants producing uridine in high yields. *Agric. Biol. Chem.* **52:**1479–1484.
26. **Dürre, P., and J. R. Andreesen.** 1983. Purine and glycine metabolism by purinolytic clostridia. *J. Bacteriol.* **154:** 192–199.
27. **Ebbole, D. J., and H. Zalkin.** 1987. Cloning and characterization of a 12-gene cluster from *Bacillus subtilis* encoding nine enzymes for *de novo* purine nucleotide synthesis. *J. Biol. Chem.* **262:**8274–8287.
28. **Eliasson, R., M. Fontecave, H. Jörnvall, M. Krook, E. Pontis, and P. Reichard.** 1990. The anaerobic ribonu-

cleoside triphosphate reductase from *Escherichia coli* requires *S*-adenosylmethionine as a cofactor. *Proc. Natl. Acad. Sci. USA* **87:**3314–3318.

29. **Elmerich, C., and J. P. Aubert.** 1975. Involvement of glutamine synthetase and the purine nucleotide pathway in repression of bacterial sporulation, p. 385–390. *In* P. Gerhardt, R. N. Costilow, and H. L. Sadoff (ed.), *Spores VI.* American Society for Microbiology, Washington, D.C.
30. **Endo, T., B. Uratani, and E. Freese.** 1983. Purine salvage pathways of *Bacillus subtilis* and effect of guanine on growth of GMP reductase mutants. *J. Bacteriol.* **155:**169–179.
31. **Engelbrecht, H. L.** 1972. Time course of purine nucleoside phosphorylase occurrence in sporulation of *Bacillus cereus. J. Bacteriol.* **111:**33–36.
32. **Felicioli, R. A., S. Senesi, F. Marmocchi, G. Falcone, and P. L. Ipata.** 1973. Nucleoside phosphomonoesterases during growth cycle of *Bacillus subtilis. Biochemistry* **12:**547–552.
33. **Freese, E.** 1989. Control of differentiation by GTP. *Nucleosides Nucleotides* **8:**975–978.
34. **Freese, E., J. Heinze, T. Mitani, and E. B. Freese.** 1978. Limitation of nucleotides induces sporulation, p. 277–285. *In* G. Chambliss and J. C. Vary (ed.), *Spores VII.* American Society for Microbiology, Washington, D.C.
35. **Freese, E., J. E. Heinze, and E. M. Galliers.** 1979. Partial purine deprivation causes sporulation of *Bacillus subtilis* in the presence of excess ammonia, glucose, and phosphate. *J. Gen. Microbiol.* **115:**193–205.
36. **Fucik, V., A. Kloudova, and A. Holý.** 1974. Transport of nucleosides in *Bacillus subtilis*: the effect of purine nucleosides on the cytidine-uptake. *Nucleic Acids Res.* **1:**639–644.
37. **Gabiellieri, E., S. Bernini, L. Piras, P. Cioni, E. Balestreri, G. Cerugnani, and R. Felicioli.** 1986. Purification, stability and kinetic properties of highly purified adenosine deaminase from *Bacillus cereus* NCIB 8122. *Biochim. Biophys. Acta* **884:**490–496.
38. **Gardner, R., and A. Kornberg.** 1967. Biochemical studies of bacterial sporulation and germination. V. Purine nucleoside phosphorylase of vegetative cells and spores of *Bacillus cereus. J. Biol. Chem.* **242:**2383–2388.
39. **Gould, G. W., and G. J. Dring.** 1972. Biochemical mechanisms of spore germination, p. 401–408. *In* H. O. Halvorson, R. Hanson, and L. L. Campbell (ed.), *Spores V.* American Society for Microbiology, Washington, D.C.
40. **Goulian, M., and W. S. Beck.** 1966. Variations of intracellular deoxyribosyl compounds in deficiencies of vitamin B_{12}, folic acid and thymine. *Biochim. Biophys. Acta* **129:**336–349.
41. **Goulian, M., and W. S. Beck.** 1966. Purification and properties of cobamide-dependent ribonucleotide reductase from *Lactobacillus leichmannii. J. Biol. Chem.* **242:**4233–4242.
42. **Grigorieva, T. M., Y. V. Smirnov, and V. V. Sukhodolets.** 1978. Thymine auxotrophs and secondary mutations decreasing the growth requirement for thymine in *Bacillus thuringiensis. Genetika* **14:**1310–1318.
43. **Grigorieva, T. M., and V. V. Sukhodolets.** 1979. Regulation of pyrimidine nucleoside phosphorylase activity in *Bacillus thuringiensis var. galleriae.* I. Change of the activity under the effect of exogenous nucleosides in thymine auxotrophs and mutants for *drm* and *dra* genes. *Genetika* **15:**1159–1168.
44. **Grigorieva, T. M., and V. V. Sukhodolets.** 1979. Regulation of pyrimidine nucleoside in *Bacillus thuringiensis var. galleriae.* II. The induction of the enzyme activity at different growth stages of bacterial cells. *Genetika* **15:**1169–1176.
45. **Guha, S.** 1984. Effects of 5-azacytidine on DNA methylation and on the enzymes of *de novo* pyrimidine biosynthesis in *Bacillus subtilis* Marburg strain. *Eur. J. Biochem.* **145:**99–106.
46. **Hammer-Jespersen, K.** 1983. Nucleoside catabolism, p. 203–258. *In* A. Munch-Petersen (ed.), *Metabolism of Nucleotides, Nucleosides and Nucleobases in Microorganisms.* Academic Press, London.
47. **Haneda, K., and R. Kodaira.** 1987. Mechanism of adenosine production by xanthine-requiring mutants derived from a *Bacillus* strain. *J. Ferment. Technol.* **65:**145–151.
48. **Henderson, J. F.** 1980. Inhibition of microbial growth by naturally-occurring purine bases and nucleosides. *Pharmacol. Ther.* **8:**605–627.
49. **Henderson, J. F., F. W. Scott, and J. K. Lowe.** 1980. Toxicity of naturally occurring purine deoxyribonucleosides. *Pharmacol. Ther.* **8:**573–604.
50. **Hersfield, M. S., and N. M. Kredich.** 1978. *S*-Adenosylhomocysteine hydrolase is an adenosine-binding protein: a target for adenosine toxicity. *Science* **202:**757–760.
51. **Hitzeman, R. A., A. R. Price, J. Neuhard, and H. Møllgaard.** 1978. Deoxyribonucleoside triphosphates and DNA polymerase in bacteriophage PBS1-infected *Bacillus subtilis*, p. 255–266. *In* J. Molineux and M. Kohiyama (ed.), *DNA Synthesis Present and Future.* Plenum Press, New York.
52. **Hoch, J. A., and C. Anagnostopoulos.** 1970. Chromosome location and properties of radiation sensitivity mutations in *Bacillus subtilis. J. Bacteriol.* **103:**295–301.
53. **Hori, N., M. Watanabe, Y. Yamazaki, and Y. Mikami.** 1989. Purification and characterization of thermostable purine nucleoside phosphorylase of *Bacillus stearothermophilus* JTS 859. *Agric. Biol. Chem.* **53:**2205–2210.
54. **Hori, N., M. Watanabe, Y. Yamazaki, and Y. Mikami.** 1989. Purification and characterization of second thermostable purine nucleoside phosphorylase of *Bacillus stearothermophilus* JTS 859. *Agric. Biol. Chem.* **53:**3219–3224.
55. **Hori, N., M. Watanabe, Y. Yamazaki, and Y. Mikami.** 1990. Purification and characterization of thermostable pyrimidine phosphorylase from *Bacillus stearothermophilus* JTS 859. *Agric. Biol. Chem.* **54:**763–768.
56. **Ikeda, S., and D. H. Ives.** 1985. Multisubstrate analogs for deoxynucleoside kinases. *J. Biol. Chem.* **260:**12659–12664.
57. **Ipata, P. L., S. Gini, and M. G. Tozzi.** 1985. In vitro 5-phosphoribosyl 1-pyrophosphate-independent salvage biosynthesis of ribo- and deoxyriboadenine nucleotides in *Bacillus cereus. Biochim. Biophys. Acta* **842:**84–89.
58. **Ipata, P. L., F. Sgarrella, R. Catalani, and M. G. Tozzi.** 1983. Induction of phosphoribomutase in *Bacillus cereus* growing on nucleosides. *Biochim. Biophys. Acta* **755:**253–256.
59. **Ipata, P. L., F. Sgarrella, and M. G. Tozzi.** 1985. Mechanisms of exogenous purine nucleotide utilization in *Bacillus cereus. Curr. Top. Cell. Regul.* **26:**419–432.
60. **Ishii, K., and I. Shiio.** 1970. Regulation of purine ribonucleotide synthesis by end product inhibition. III. Effect of purine nucleotides on succino-AMP synthetase of *Bacillus subtilis. J. Biochem.* **68:**171–176.
61. **Ishii, K., and I. Shiio.** 1972. Improved inosine production and derepression of purine nucleotide biosynthetic enzymes in 8-azaguanine resistant mutants of *Bacillus subtilis. Agric. Biol. Chem.* **36:**1511–1522.
62. **Ishii, K., and I. Shiio.** 1973. Regulation of purine nucleotide biosynthesis in *Bacillus subtilis. Agric. Biol. Chem.* **37:**287–300.
63. **Iwakura, M., M. Kawata, K. Tsuda, and T. Tanaka.** 1988. Nucleotide sequence of the thymidylate synthase B and dihydrofolate reductase genes contained in one *Bacillus subtilis* operon. *Gene* **64:**9–20.
64. **Jensen, K. F.** 1978. Two purine nucleoside phosphory-

lases in *Bacillus subtilis*. Purification and some properties of the adenosine-specific phosphorylase. *Biochim. Biophys. Acta* **525:**346–356.

65. **Jensen, K. F., J. C. Leer, and P. Nygaard.** 1973. Thymine utilization in *Escherichia coli* K12 on the role of deoxyribose 1-phosphate and thymidine phosphorylase. *Eur. J. Biochem.* **40:**345–354.
66. **Jensen, K. F., and P. Nygaard.** 1975. Purine nucleoside phosphorylase from *Escherichia coli* and *Salmonella typhimurium*. Purification and some properties. *Eur. J. Biochem.* **51:**253–265.
67. **Kanzaki, N., and K. Miyagawa.** 1990. Nucleotide sequence of the *Bacillus subtilis* IMP dehydrogenase gene. *Nucleic Acids Res.* **18:**6710.
68. **Kloudova, A., and V. Fucik.** 1974. Transport of nucleosides in *Bacillus subtilis*: characteristics of cytidine-uptake. *Nucleic Acids Res.* **1:**629–638.
69. **Kredich, N. M., and D. W. Martin, Jr.** 1977. Role of S-adenosylhomocysteine in adenosine-mediated toxicity in cultured mouse T lymphoma cells. *Cell* **12:**931–938.
70. **Krenitzky, T. A., S. M. Neil, and R. L. Miller.** 1970. Guanine and xanthine phosphoribosyltransfer activities of *Lactobacillus casei* and *Escherichia coli*. Their relationship to hypoxanthine and adenine phosphoribosyl transfer activities. *J. Biol. Chem.* **245:**2605–2611.
71. **Kuninaka, A.** 1986. Nucleic acids, nucleotides, and related compounds, p. 71–114. *In* H. J. Rehm and G. Reed (ed.), *Biotechnology*. VCH Verlagsgesellschaft, Weinheim, Germany.
72. **Leung, H. B., and V. L. Schramm.** 1980. Adenylate degradation in *Escherichia coli*. The role of AMP nucleosidase and properties of the purified enzyme. *J. Biol. Chem.* **255:**10867–10874.
73. **Levisohn, S., and A. I. Aronson.** 1967. Regulation of extracellular protease production in *Bacillus cereus*. *J. Bacteriol.* **93:**1023–1030.
74. **Lopez, J. M., A. Dromerick, and E. Freese.** 1981. Response of guanosine 5′-triphosphate concentration to nutritional changes and its significance for *Bacillus subtilis* sporulation. *J. Bacteriol.* **146:**1447–1449.
75. **Lopez, J. M., C. L. Marks, and E. Freese.** 1979. The decrease of guanine nucleotides initiates sporulation of *Bacillus subtilis*. *Biochim. Biophys. Acta* **587:**238–252.
76. **Magasanik, B.** 1962. Biosynthesis of purine and pyrimidine nucleotides, p. 295–334. *In* J. C. Gunsalus and R. Y. Stanier (ed.), *The Bacteria*, vol. 3. Academic Press, Inc., New York.
77. **Mager, J., and B. Magasanik.** 1960. Guanosine 5′-phosphate reductase and its role in the interconversion of purine nucleotides. *J. Biol. Chem.* **235:**1474–1478.
78. **Mathews, K. C. (ed.).** 1971. *Bacteriophage Biochemistry*. Van Nostrand Reinhold Co., New York.
79. **Maznitsa, I. I., A. A. Nudler, and G. I. Bourd.** 1990. A pleiotropic mutation affecting purine metabolism in *Bacillus subtilis*. *FEMS Microbiol. Lett.* **72:**173–176.
80. **Maznitsa, I. I., V. V. Sukhodolets, and L. S. Ukhabotina.** 1983. Cloning of *Bacillus subtilis* 168 genes compensating the defect of mutations for thymidine phosphorylase and uridine phosphorylase in *Escherichia coli* cells. *Genetika* **19:**881–887.
81. **McIvor, R. S., R. M. Wohlhueter, and P. G. W. Plagemann.** 1983. Uracil phosphoribosyltransferase from *Acholeplasma laidlawii*: partial purification and kinetic properties. *J. Bacteriol.* **156:**192–197.
82. **Mehra, R. K., and W. T. Drabble.** 1981. Dual control of the *gua* operon of *Escherichia coli* K12 by adenine and guanine nucleotides. *J. Gen. Microbiol.* **123:**27–37.
83. **Meng, L. M., M. Kilstrup, and P. Nygaard.** 1990. Autoregulation of PurR repressor synthesis and involvement of *purR* in the regulation of *purB*, *purC*, *purL*, *purMN* and *guaBA* expression in *Escherichia coli*. *Eur. J. Biochem.* **187:**373–379.
84. **Meng, L. M., and P. Nygaard.** 1990. Identification of hypoxanthine and guanine as the corepressors for the purine regulon genes of *Escherichia coli*. *Mol. Microbiol.* **4:**2187–2192.
85. **Meyer, E., and R. L. Switzer.** 1979. Regulation of *Bacillus subtilis* glutamine phosphoribosylpyrophosphate amidotransferase activity by end products. *J. Biol. Chem.* **254:**5397–5402.
86. **Mitani, T., J. E. Heinze, and E. Freese.** 1977. Induction of sporulation in *Bacillus subtilis* by deocyinine or hadacidin. *Biochem. Biophys. Res. Commun.* **77:**1118–1125.
87. **Mitchell, A., and L. R. Finch.** 1977. Pathways of nucleotide biosynthesis in *Mycoplasma mycoides* subsp. *mycoides*. *J. Bacteriol.* **130:**1047–1054.
88. **Mitchell, C., and J. C. Vary.** 1989. Proteins that interact with GTP during sporulation of *Bacillus subtilis*. *J. Bacteriol.* **171:**2915–2918.
89. **Miyagawa, K., N. Kanzaki, H. Kimura, Y. Sumino, S. Akiyama, and Y. Nakao.** 1989. Increased inosine production by a *Bacillus subtilis* xanthine-requiring mutant derived by insertional inactivation of the IMP dehydrogenase gene. *Biotechnology* **7:**821–824.
90. **Miyagawa, K., H. Kimura, K. Nakahama, M. Kikuchi, M. Doi, S. Akiyama, and Y. Nakao.** 1986. Cloning of the *Bacillus subtilis* IMP dehydrogenase gene and its application to increased production of guanosine. *Biotechnology* **4:**225–228.
91. **Møllgaard, H.** 1980. Deoxyadenosine/deoxycytidine kinase from *Bacillus subtilis*. *J. Biol. Chem.* **255:**8216–8220.
92. **Møllgaard, H., and J. Neuhard.** 1978. Deoxycytidylate deaminase from *Bacillus subtilis*. *J. Biol. Chem.* **253:** 3536–3542.
93. **Møllgaard, H., and J. Neuhard.** 1983. Biosynthesis of deoxythymidine triphosphate, p. 149–201. *In* A. Munch-Petersen (ed.), *Metabolism of Nucleotides, Nucleosides and Nucleobases in Microorganisms*. Academic Press, Inc., New York.
94. **Momose, H., H. Nishikawa, and N. Katsuya.** 1965. Genetic and biochemical studies on 5′-nucleotide fermentation. II. Repression of enzyme formation in purine nucleotide biosynthesis in *Bacillus subtilis* and derivation of derepressed mutants. *J. Gen. Appl. Microbiol.* **11:**211–220.
95. **Momose, H., H. Nishikawa, and I. Shiio.** 1966. Regulation of purine nucleotide synthesis in *Bacillus subtilis*. I. Enzyme repression by purine derivatives. *J. Biochem.* **59:**325–331.
96. **Munch-Petersen, A., and B. Mygind.** 1983. Transport of nucleic acid precursors, p. 259–305. *In* A. Munch-Petersen (ed.), *Metabolism of Nucleotides, Nucleosides and Nucleobases in Microorganisms*. Academic Press, Inc., New York.
97. **Myoda, T. T., and V. L. Funanage.** 1985. Coregulation of dihydrofolate reductase and thymidylate synthase B in *Bacillus subtilis*. *Biochim. Biophys. Acta* **824:**99–103.
98. **Myoda, T. T., S. V. Lowther, V. L. Funanage, and F. E. Young.** 1984. Cloning and mapping of the dihydrofolate reductase gene of *Bacillus subtilis*. *Gene* **29:**135–143.
99. **Neuhard, J.** 1983. Utilization of preformed pyrimidine bases and nucleosides, p. 95–148. *In* A. Munch-Petersen (ed.), *Metabolism of Nucleotides, Nucleosides and Nucleobases in Microorganisms*. Academic Press, Inc., New York.
100. **Neuhard, J., and P. Nygaard.** 1987. Purines and pyrimidines, p. 445–473. *In* F. C. Neidhardt, J. L. Ingraham, K. B. Low, B. Magasanik, M. Schaechter, and H. E. Umbarger (ed.), *Escherichia coli and Salmonella typhimurium: Cellular and Molecular Biology*, vol. 1. American Society for Microbiology, Washington, D.C.
101. **Neuhard, J., A. R. Price, L. Schack, and E. Thomassen.**

1978. Two thymidylate synthetases in *Bacillus subtilis*. *Proc. Natl. Acad. Sci. USA* **75**:1194–1198.

102. **Nijkamp, H. J. J., and P. G. DeHaan.** 1967. Genetic and biochemical studies of the guanosine 5′-monophosphate pathway in *Escherichia coli*. *Biochim. Biophys. Acta* **145**:31–40.
103. **Nilsson, D., and P. S. Andersen.** Unpublished observations.
104. **Nishikawa, H., H. Momose, and I. Shiio.** 1967. Regulation of purine nucleotide synthesis in *Bacillus subtilis*. II. Specificity of purine derivatives for enzyme repression. *J. Biochem.* **62**:92–98.
105. **Nishikawa, H., H. Momose, and I. Shiio.** 1968. Pathway of purine nucleotide synthesis in *Bacillus subtilis*. *J. Biochem.* **63**:149–155.
106. **Nygaard, P.** Unpublished observations.
107. **Nygaard, P.** 1983. Utilization of preformed purine bases and nucleosides, p. 27–93. *In* A. Munch-Petersen (ed.), *Metabolism of Nucleotides, Nucleosides and Nucleobases in Microorganisms*. Academic Press, Inc., New York.
108. **Nygaard, P., P. Duckert, and H. H. Saxild.** 1988. Purine gene organization and regulation in *Bacillus subtilis*, p. 57–61. *In* M. M. Zukowski, A. T. Ganesan, and J. A. Hoch (ed.), *Genetics and Biotechnology of Bacilli*, vol. 3. Academic Press, Inc., New York.
109. **Ochi, K.** 1987. Changes in nucleotide pools during sporulation of *Streptomyces griseus* in submerged culture. *J. Gen. Microbiol.* **133**:2787–2795.
110. **Ochi, K., and E. Freese.** 1982. A decrease in *S*-adenosylmethionine synthetase activity increases the probability of spontaneous sporulation. *J. Bacteriol.* **152**:400–410.
111. **Ochi, K., J. Kandala, and E. Freese.** 1982. Evidence that *Bacillus subtilis* sporulation induced by the stringent response is caused by a decrease in GTP and GDP. *J. Bacteriol.* **150**:704–711.
112. **Ogata, K.** 1975. The microbial production of nucleic acid-related compounds, p. 209–247. *In* D. Perlman (ed.), *Applied Microbiology*. Academic Press, Inc., New York.
113. **Ohno, M.** 1980. Nucleosides, p. 73–130. *In* H. Umezawa (ed.), *Anticancer Agents Based on Natural Product Models*. Academic Press, Inc., New York.
114. **Orengo, A., and S.-H. Kobayashi.** 1978. Uridine-cytidine kinase from Novikoff ascites rat tumor and *Bacillus stearothermophilus*. *Methods Enzymol.* **51**:299–305.
115. **Ozaki, H., and I. Shiio.** 1979. Two cytoplasmic 5′-nucleotidases of *Bacillus subtilis* K. *J. Biochem.* **85**:1083–1089.
116. **Paulus, T. J., T. J. McGarry, P. G. Shekelle, S. Rosenzweig, and R. L. Switzer.** 1982. Coordinate synthesis of the enzymes of pyrimidine biosynthesis in *Bacillus subtilis*. *J. Bacteriol.* **149**:775–778.
117. **Paulus, T. J., and R. L. Switzer.** 1979. Characterization of pyrimidine-repressible and arginine-repressible carbamylphosphate synthetase from *Bacillus subtilis*. *J. Bacteriol.* **137**:82–91.
118. **Piggot, P. J., M. Amjad, J.-J. Wu, H. Sandoval, and J. Castro.** 1990. Genetic and physical maps of *Bacillus subtilis* 168, p. 493–543. *In* C. R. Harwood and S. M. Cutting (ed.), *Molecular Biology Methods for Bacillus*. John Wiley & Sons, Ltd., Chichester, United Kingdom.
119. **Pinter, K., V. J. Davisson, and D. V. Santi.** 1988. Cloning, sequencing, and expression of the *Lactobacillus casei* thymidylate synthase gene. *DNA* **7**:235–241.
120. **Powell, J. W., and J. T. Wachsman.** 1973. Evidence for four deoxynucleoside kinase activities in extracts of *Lactobacillus leichmannii*. *Appl. Microbiol.* **25**:869–872.
121. **Preston, R. A., and H. A. Douthit.** 1988. Functional relationships between L- and D-alanine, inosine and NH_4Cl during germination of spores of *Bacillus cereus* T. *J. Gen. Microbiol.* **134**:3001–3010.
122. **Price, A. R.** 1976. Bacteriophage-induced inhibitor of a host enzyme, p. 290–294. *In* D. Schlessinger (ed.), *Microbiology—1976*. American Society for Microbiology, Washington, D.C.
123. **Price, A. R.** 1978. Deoxythymidylate phosphohydrolase from PBS2 phage-infected *Bacillus subtilis*. *Methods Enzymol.* **51**:185–290.
124. **Price, A. R., and J. Frato.** 1975. *Bacillus subtilis* deoxyuridinetriphosphatase and its bacteriophage PBS2-induced inhibitor. *J. Biol. Chem.* **250**:8804–8811.
125. **Quirk, S., and M. J. Bessman.** 1991. dGTP triphosphohydrolase, a unique enzyme confined to members of the family *Enterobacteriaceae*. *J. Bacteriol.* **173**:6665–6669.
126. **Reichard, P.** 1988. Interactions between deoxyribonucleotide and DNA synthesis. *Annu. Rev. Biochem.* **57**:349–374.
127. **Rima, B. K., and J. Takahashi.** 1977. Metabolism of pyrimidine bases and nucleosides in *Bacillus subtilis*. *J. Bacteriol.* **129**:574–579.
128. **Rima, B. K., and J. Takahashi.** 1978. Deoxyribonucleoside-requiring mutants of *Bacillus subtilis*. *J. Gen. Microbiol.* **107**:139–145.
129. **Rinehart, K. V., and J. C. Copeland.** 1973. Evidence that thymine is not a normal metabolite in wild-type *Bacillus subtilis*. *Biochim. Biophys. Acta* **294**:1–7.
130. **Rolfes, R. J., and H. Zalkin.** 1990. Purification of the *Escherichia coli* purine regulon repressor and identification of corepressors. *J. Bacteriol.* **172**:5637–5642.
131. **Roscoe, D. H., and R. G. Tucker.** 1966. The biosynthesis of 5-hydroxy-methyldeoxyuridylic acid in bacteriophage-infected *Bacillus subtilis*. *Virology* **29**:157–166.
132. **Rumyantseva, E. V., V. V. Sukhodolets, and Y. V. Smirnov.** 1979. Isolation and characterization of mutants for genes of nucleoside catabolism in *Bacillus subtilis*. *Genetika* **15**:594–604.
133. **Sacks, L. E., and P. A. Thompson.** 1975. Influence of methylxanthines on sporulation of *Clostridium perfringens* cells, p. 341–345. *In* P. Gerhardt, R. N. Costilow, and H. L. Sadoff (ed.), *Spores VI*. American Society for Microbiology, Washington, D.C.
134. **Sakai, T., T.-S. Yu, and S. Omata.** 1976. Distribution of enzymes related to cytidine degradation in bacteria. *Agric. Biol. Chem.* **40**:1893–1895.
135. **Santangelo, J. D., D. T. Jones, and D. R. Woods.** 1989. Comparison of nucleoside triphosphate levels in the wild-type strain with those in sporulation-deficient and solvent-deficient mutants of *Clostridium acetobutylicum* P262. *J. Gen. Microbiol.* **135**:711–719.
136. **Saunders, P. P., B. A. Wilson, and G. F. Saunders.** 1969. Purification and comparative properties of a pyrimidine nucleoside phosphorylase from *Bacillus stearothermophilus*. *J. Biol. Chem.* **244**:3691–3697.
137. **Saxild, H. H., and P. Nygaard.** 1987. Genetic and physiological characterization of *Bacillus subtilis* mutants resistant to purine analogs. *J. Bacteriol.* **169**:2977–2983.
138. **Saxild, H. H., and P. Nygaard.** 1988. Gene-enzyme relationships of the purine biosynthetic pathway in *Bacillus subtilis*. *Mol. Gen. Genet.* **211**:160–167.
139. **Saxild, H. H., and P. Nygaard.** 1991. Regulation of levels of purine biosynthetic enzymes in *Bacillus subtilis*: effects of changing purine nucleotide pools. *J. Gen. Microbiol.* **137**:2387–2394.
140. **Schiefer-Ullrich, H., and J. R. Andreesen.** 1985. *Peptostreptococcus barnesae* sp. nov., a gram-positive, anaerobic, obligately purine utilizing coccus from chicken feces. *Arch. Microbiol.* **143**:26–31.
141. **Schiefer-Ullrich, H., R. Wagner, P. Dürre, and J. R. Andreesen.** 1984. Comparative studies on physiology and taxonomy of obligately purinolytic clostridia. *Arch. Microbiol.* **138**:345–353.
142. **Schimpff-Weiland, G., H. Follmann, and G. Auling.** 1981. A new manganese-activated ribonucleotide re-

ductase found in gram-positive bacteria. *Biochem. Biophys. Res. Commun.* **102:**1276–1282.

143. **Schrader, W. P., and J. S. Anderson.** 1978. Membrane-bound nucleotidase of *Bacillus cereus. J. Bacteriol.* **133:** 576–583.
144. **Sedmak, J., and R. Ramaley.** 1971. Purification and properties of *Bacillus subtilis* nucleoside diphosphokinase. *J. Biol. Chem.* **246:**5365–5372.
145. **Sekar, V., S. P. Wilson, and J. H. Hageman.** 1981. Induction of *Bacillus subtilis* sporulation by nucleosides: inosine appears to be a sporogen. *J. Bacteriol.* **145:**489–493.
146. **Senesi, S., G. Cercignani, G. Freer, G. Batoni, S. Barnini, and F. Ota.** 1991. Structural and stereospecific requirements for the nucleoside-triggered germination of *Bacillus cereus* spores. *J. Gen. Microbiol.* **137:**399–404.
147. **Senesi, S., G. Falcone, P. L. Ipata, and R. A. Felicioli.** 1974. Inhibition of phosphodiesterases from *Bacillus subtilis* by nucleoside triphosphates. *Biochemistry* **13:** 5008–5011.
148. **Senesi, S., R. A. Felicioli, P. L. Ipata, and G. Falcone.** 1975. Regulation of polyribonucleotide turnover in vegative cells and spores of *Bacillus subtilis*, p. 265–270. *In* P. Gerhardt, R. N. Costilow, and H. L. Sadoff (ed.), *Spores VI*. American Society for Microbiology, Washington, D.C.
149. **Sergott, R. C., L. J. Debeer, and M. J. Bessman.** 1971. On the regulation of a bacterial deoxycytidylate deaminase. *J. Biol. Chem.* **246:**7755–7758.
150. **Setlow, P.** 1973. Deoxyribonucleic acid synthesis and deoxynucleotide metabolism during bacterial spore germination. *J. Bacteriol.* **114:**1099–1107.
151. **Setlow, P.** 1974. Energy and small-molecule metabolism during germination of *Bacillus* spores, p. 443–450. *In* P. Gerhardt, R. N. Costilow, and H. L. Sadoff (ed.), *Spores VI*. American Society for Microbiology, Washington, D.C.
152. **Setlow, P., and A. Kornberg.** 1970. Biochemical studies of bacterial sporulation and germination. XXIII. Nucleotide metabolism during spore germination. *J. Biol. Chem.* **245:**3645–3652.
153. **Sharpe, M. E.** 1981. The genus *Lactobacillus*, p. 1653–1679. *In* P. S. Mortimer, H. Stolp, H. G. Trüper, A. Balows, and H. G. Schlegel (ed.), *The prokaryotes*, vol. 2. Springer-Verlag, New York.
154. **Shibata, H., N. Ohnishi, K. Takeda, H. Fukunaga, K. Shimamura, E. Yasunobu, and I. Tani.** 1986. Germination of *Bacillus cereus* spores induced by purine ribosides and their analogs: effects of modification of base and sugar moieties of purine nucleosides on germination-inducing activity. *Can. J. Microbiol.* **32:**186–189.
155. **Shiio, I., and H. Ozaki.** 1978. Cellular distribution and some properties of 5′ nucleotidases in *Bacillus subtilis* K. *J. Biochem.* **83:**409–421.
156. **Shimizu, S., T. Abe, and H. Yamada.** 1988. Distribution of methylthioadenosine phosphorylase in eubacteria. *FEMS Microbiol. Lett.* **51:**177–180.
157. **Shlomai, J., and A. Kornberg.** 1978. Deoxyuridine triphosphatase of *Escherichia coli*. Purification, properties, and use as reagent to reduce uracil incorporation into DNA. *J. Biol. Chem.* **253:**3305–3312.
158. **Smar, M., S. A. Short, and R. Wolfenden.** 1991. Lyase activity of nucleoside 2-deoxyribosyltransferase: transient generation of ribal and its use in the synthesis of 2′-deoxynucleosides. *Biochemistry* **30:**7908–7912.
159. **Søgaard-Andersen, L., H. Pedersen, B. Holst, and P. Valentin-Hansen.** 1991. A novel function of the cAMP-CPR complex in *Escherichia coli*: cAMP-CRP functions as an adaptor for the CytR repressor in the *deo* operon. *Mol. Microbiol.* **5:**969–975.
160. **Sonenshein, A. L.** 1985. Recent progress in metabolic regulation of sporulation, p. 185–193. *In* J. A. Hoch and P. Setlow (ed.), *Molecular Biology of Microbial Differentiation*. American Society for Microbiology, Washington, D.C.
161. **Sonenshein, A. L.** 1989. Metabolic regulation of sporulation and other stationary-phase phenomena, p. 109–130. *In* I. Smith, R. A. Slepecky, and P. Setlow (ed.), *Regulation of Procaryotic Development*. American Society for Microbiology, Washington, D.C.
162. **Song, B. H., and J. Neuhard.** 1989. Chromosomal location, cloning and nucleotide sequence of the *Bacillus subtilis cdd* gene encoding cytidine/deoxycytidine deaminase. *Mol. Gen. Genet.* **216:**462–468.
163. **Stubbe, J. A.** 1990. Ribonucleotide reductases: amazing and confusing. *J. Biol. Chem.* **265:**5329–5332.
164. **Suhadolnik, R. J.** 1979. *Nucleosides as Biological Probes*. John Wiley & Sons, Inc., New York.
165. **Sukhodolets, V. V.** Personal communication.
166. **Sukhodolets, V. V., Y. Flyakh, and E. V. Rumyantseva.** 1983. Mapping of mutations in genes for nucleoside catabolism on the *Bacillus subtilis* chromosome. *Genetika* **19:**221–226.
167. **Takagi, Y., and B. L. Horecker.** 1957. Purification and properties of a bacterial riboside hydrolase. *J. Biol. Chem.* **225:**77–86.
168. **Tesfa-Selase, F., and W. T. Drabble.** 1990. Regulation of the *gua* operon of *Escherichia coli* by the DnaA protein. *Mol. Gen. Genet.* **231:**256–264.
169. **Teuber, M., and A. Geis.** 1981. The family *Streptococcaceae* (nonmedical aspects), p. 1614–1630. *In* P. S. Mortimer, H. Stolp, H. G. Trüper, A. Balows, and H. G. Schlegel (ed.), *The prokaryotes*, vol. 2. Springer-Verlag, New York.
170. **Tomati, F., and I. Takahashi.** 1976. Changes in enzyme activities in *Bacillus subtilis* infected with bacteriophage PBS1, p. 315–318. *In* D. Schlessinger (ed.), *Microbiology—1976*. American Society for Microbiology, Washington, D.C.
171. **Tozzi, M. G., F. Sgarrella, D. Barsacchi, and P. L. Ipata.** 1984. Induction of deoxyribose-5-phosphate aldolase of *Bacillus cereus* by deoxyribonucleosides. *Biochem. Int.* **9:**319–325.
172. **Tozzi, M. G., F. Sgarrella, and P. L. Ipata.** 1981. Induction and repression of enzymes involved exogenous purine compound utilization in *Bacillus cereus. Biochim. Biophys. Acta* **678:**460–466.
173. **Turnbough, C. L., Jr., and R. L. Switzer.** 1975. Oxygen-dependent inactivation of glutamine phosphoribosylpyrophosphate amidotransferase in stationary-phase cultures of *Bacillus subtilis. J. Bacteriol.* **121:**108–114.
174. **Uerkvitz, W.** 1974. Trans-*N*-deoxyribosylase from *Lactobacillus helveticus*. Crystallization and properties. *Eur. J. Biochem.* **23:**387–395.
175. **Váchová, L., M. Strnadová, H. Kucerova, and J. Chaloupka.** 1990. Effect of actinomycin D on viability, sporulation and nucleotide pool of *Bacillus megaterium*. Folia Microbiol. **35:**190–199.
176. **Valentin-Hansen, P., H. Aiba, and D. Schümperli.** 1982. The structure of tandem regulatory regions in the *deo* operon of *Escherichia coli* K12. *EMBO J.* **1:**317–322.
177. **Vasantha, N., E. M. Galliers, and J. N. Hansen.** 1984. Effect of purine and pyrimidine limitations on RNA synthesis in *Bacillus subtilis. J. Bacteriol.* **158:**884–889.
178. **Vogels, G. D., and C. van der Drift.** 1976. Degradation of purines and pyrimidines by microorganisms. *Bacteriol. Rev.* **40:**403–468.
179. **Wabiko, H., K. Ochi, D. M. Nguyen, E. R. Allen, and E. Freese.** 1988. Genetic mapping and physiological consequences of *metE* mutations of *Bacillus subtilis. J. Bacteriol.* **170:**2705–2710.
180. **Wachsman, J. T., S. Kemp, and L. Hogg.** 1964. Thymineless death in *Bacillus megaterium. J. Bacteriol.* **87:**1079–1086.
181. **Wachsman, J. T., and D. D. Morgan.** 1971. Deoxynucle-

oside kinases of *Bacillus megaterium. J. Bacteriol.* **105:** 787–792.

182. **Waindle, L. M., and R. L. Switzer.** 1973. Inactivation of aspartic transcarbamylase in sporulating *Bacillus subtilis*: demonstration of a requirement for metabolic energy. *J. Bacteriol.* **114:**517–527.
183. **Wainscott, V. J., and J. F. Kane.** 1976. Dihydrofolate reductase in *Bacillus subtilis*, p. 208–213. *In* D. Schlessinger (ed.), *Microbiology—1976*. American Society for Microbiology, Washington, D.C.
184. **Waleh, N. S., and J. L. Ingraham.** 1976. Pyrimidine ribonucleoside monophosphokinase and the mode of RNA turnover in *Bacillus subtilis*. *Arch. Microbiol.* **110:** 49–54.
185. **Watanabe, Y., O. Tatsuhiko, and Y. Tsujisaka.** 1976. Changes in the metabolic pathways of hypoxanthine in *Streptomyces*. *J. Gen. Appl. Microbiol.* **22:**13–23.
186. **Wheeler, P. R.** 1987. Biosynthesis and scavenging of purines by pathogenic mycobacteria including *Mycobacterium leprae*. *J. Gen. Microbiol.* **133:**2999–3011.
187. **Willing, A., H. Follmann, and G. Auling.** 1988. Ribonucleotide reductase of *Brevibacterium ammoniagenes* is a manganese enzyme. *Eur. J. Biochem.* **170:**603–611.
188. **Yau, S., and J. T. Wachsman.** 1973. The *Bacillus megaterium* ribonucleotide reductase: evidence for a B_{12} coenzyme requirement. *Mol. Cell. Biochem.* **1:**101–105.
189. **Yoshikawa, H., and N. Ogasawara.** 1991. Structure and function of DnaA and the DnaA-box in eubacteria: evolutionary relationships of bacterial replication origins. *Mol. Microbiol.* **5:**2589–2597.
190. **Zahler, S. A.** 1978. An adenine-thiamin auxotrophic mutant of *Bacillus subtilis*. *J. Gen. Microbiol.* **107:**199–201.
191. **Zain-ul-Abedin, J. M. Lopez, and E. Freese.** 1983. Induction of bacterial differentiation by adenine and adenosine analogs and inhibitors of nucleic acid synthesis. *Nucleoside Nucleotides* **2:**257–274.
192. **Zalkin, H., and C. D. Truit.** 1977. Characterization of the glutamine site of *Escherichia coli* guanosine 5′-monophosphate synthetase. *J. Biol. Chem.* **252:**5431–5436.

III. CELL ENVELOPE

27. Cell Wall Structure, Synthesis, and Turnover

A. R. ARCHIBALD, I. C. HANCOCK, and C. R. HARWOOD

Walls of gram-positive bacteria are dynamically variable and flexible structures that enclose and protect the underlying cytoplasmic membranes. They are intimately involved in cell growth and morphogenesis, cell division, genome segregation, interaction between the cell and its environment, and movement of materials into and out of the cell. During growth, the wall has to enlarge and change shape to accommodate the exponentially increasing volume of the cell and must divide to allow the formation of two daughter cells. The forces that drive wall growth and the ways in which growth is accomplished without compromising the wall's strength and ability to protect the protoplast are now understood in some detail (60, 157, 277). However, the mechanism of chromosome segregation and the way in which it is coordinated with wall assembly and cell division remain unclear (276).

The cell envelope of each member of the genus *Bacillus* consists of a cytoplasmic membrane, a wall, and in several strains, a proteinaceous surface layer (28, 267). Additional surface structures such as capsules, slimes, fimbriae, and flagella may also be present (25, 27). In electron micrographs of conventionally fixed preparations, the wall usually appears as a relatively amorphous structure between 20 and 50 nm thick that is tightly apposed to the underlying protoplast membrane (Fig. 1). Freeze-substituted walls show more structure and in a number of species reveal a fibrous matrix radiating outward 10 to 30 nm beyond the membrane (3, 59, 106, 241, 292). Walls of *Bacillus* species are composed of cross-linked peptidoglycans to which other, usually anionic, polymers are attached. Variable amounts of protein and lipoteichoic acid may also be present in the wall and at its surface. Unlike gram-negative bacteria, *Bacillus* species do not have outer membranes or membrane-enclosed periplasms (245).

The wall serves to protect the underlying protoplast, resist turgor, and maintain the shape of the cell. Physical studies with twisted chains (macrofibers) of *Bacillus subtilis* (209) indicate that hydrated walls could withstand turgor pressures of about 24 atm (ca. 2,431 kPa), which is of the same order as earlier estimates obtained indirectly (194). Walls of gram-positive bacteria can be up to 10 times thicker than the peptidoglycan layers in gram-negative bacteria. This extra thickness may be required to withstand the additional turgor in gram-positive bacteria (turgor in *Escherichia coli* has been estimated at 5 atm [ca. 500 kPa]). The stress in walls that is a result of turgor is important in determining cell shape (152, 277).

The importance of the influence of the wall in interactions between the cell and its environment is often underestimated. Because of its physical and chemical properties, in particular its limited porosity and high concentration of charged groups, the wall may block the unrestricted movement of materials between the cytoplasmic membrane and the environment. The wall is involved in cation binding and in maintaining an optimum ionic environment for membrane-bound enzyme systems. It is the site of immobilization of specific proteins (e.g., autolysins). Several phages adsorb to the wall as the first stage in their infective cycles, and the wall is the initial site of binding of DNA to competent cells during transformation.

Reviews of walls of gram-positive bacteria include those by Rogers et al. (245) and Shockman and Barrett (262). Walls of *Bacillus* spp. have been reviewed recently by Archibald (12).

WALL STRUCTURE AND ORGANIZATION

The wall in *B. subtilis*, like the walls in many other gram-positive bacteria, is composed mainly of peptidoglycan and one or more anionic polymers. These components are synthesized on identical anchor lipids (304), covalently attached to each other before or during insertion into the wall (197), processed through the wall, and finally released by turnover while still attached to each other. It is now clear that both types of polymer are essential for normal wall function and morphogenesis (199). While it remains convenient to consider the two components separately, as below, it is timely to discard older concepts that envisage the wall as a sacculus of peptidoglycan to which anionic polymers may be attached as components of only secondary significance.

In addition to containing peptidoglycan and anionic polymers, walls of gram-positive bacteria may contain substantial proportions of protein, held either covalently or noncovalently within the peptidoglycan–anionic-polymer complex, together with neutral polysaccharides, lipoteichoic acid, and the cations that form part of the polyelectrolyte gel structure (see below) of the wall complex (245, 262). Structured protein layers, polysaccharide capsules, and slimes coat the walls of many strains of gram-positive species (267).

Peptidoglycan

Peptidoglycans in gram-positive bacteria are built on the same basic structural plan as those in *E. coli*

A. R. Archibald, I. C. Hancock, and C. R. Harwood • Department of Microbiology, The Medical School, The University of Newcastle upon Tyne, Framlington Place, Newcastle upon Tyne NE2 4HH, United Kingdom.

Figure 1. Electron micrograph of a thin section of the cell wall of *B. subtilis* 168. A, outer leaf of the cytoplasmic membrane plus bound teichoic acid; B, less electron-opaque thin inner layer; C, electron-opaque heterogeneous outer layer (Jan A. Hobot).

and other gram-negative bacteria. Rigid glycan chains are cross-linked through flexible peptides to form a strong but elastic structure. Since each glycan chain carries several peptide substituents, it can become linked to more than one neighboring chain so that a "bag-shaped" macromolecule comprising all of the peptidoglycan and other covalently attached polymers in the wall is formed (308).

Glycan chains consist of alternating units of *N*-acetylglucosamine and *N*-acetylmuramic acid held together by $\beta 1 \rightarrow 4$ glycosidic linkages (Fig. 2). The average biosynthetic chain length of the glycan is about 20 disaccharide units in *Staphylococcus lactis* and *Staphylococcus aureus* (15, 284), 100 disaccharide units in *B. subtilis* 168, and 140 disaccharide units in an autolysin-deficient strain of *Bacillus licheniformis* (297). The reducing-terminal sugar of these glycan chains is *N*-acetylmuramic acid. However, in wild-type strains, the chains are subsequently hydrolyzed by an endo-β-*N*-acetylglucosaminidase (see below) to yield shorter chains with either *N*-acetylmuramic acid or *N*-acetylglucosamine at the reducing end. The terminal *N*-acetylglucosamine residues formed by this hydrolysis retain their free reducing groups and are not converted to the 1,6-anhydro derivatives characteristic of gram-negative bacteria.

The glycan chains are connected to short "stem" peptides through amide linkages between the carboxyl groups on muramyl residues and the terminal amino groups of the peptides. The stem peptides are linked through alternating L- and D-amino acid centers (Fig. 2 and 3a). In most strains of *Bacillus* (254), as in gram-negative bacteria, the stem peptide contains an *N*-terminal L-alanine linked through its carboxyl group to the amino group of D-glutamic acid. The γ (not the α) carboxyl group of this glutamic acid residue is linked to the amino group at the L center of *meso*-diaminopimelic acid, and the carboxyl group at this center is linked to D-alanyl-D-alanine.

This basic structural organization is well conserved, but there are substantial differences in detail among gram-positive bacteria. The compositions of the stem peptides, particularly the identities of the amino acids at position 3, vary in different organisms, though cysteine, methionine, arginine, histidine, proline, and aromatic amino acids are not found. Instead of direct cross-linkage, bridging peptides of various lengths may be present (Fig. 3b), and the location of the cross-linkage can vary. For example, peptidoglycans in *Bacillus pasteurii* and *Bacillus sphaericus* contain lysine (238) rather than diaminopimelic acid (even though the latter amino acid is present in the spore cortex peptidoglycan), and adjacent peptides are linked through bridge units of D-aspartic acid or L-alanyl-D-aspartic acid. Variations may also include the presence of more than one kind of cross-linkage. Thus, *B. megaterium* and *Bacillus cereus* contain small proportions of the DD isomer of diaminopimelic acid

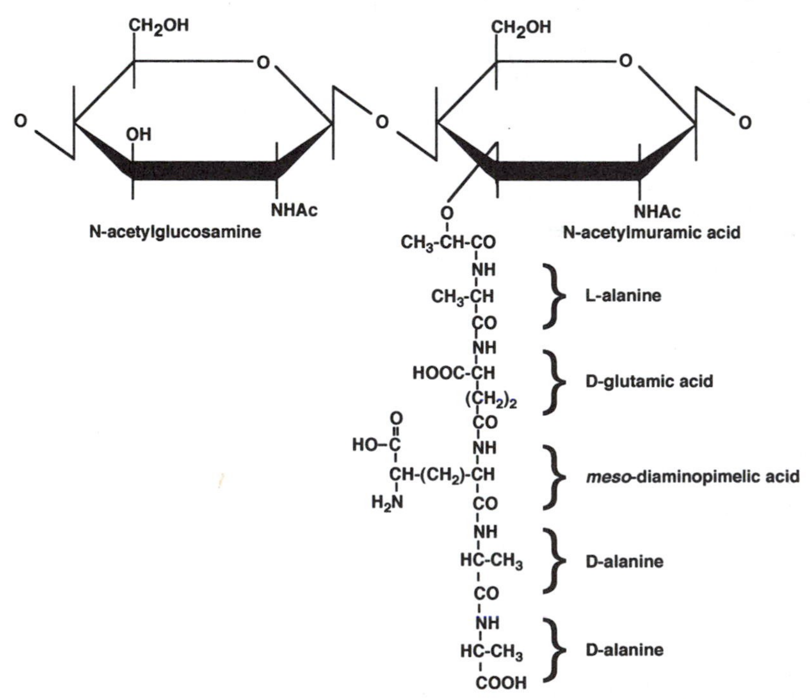

Figure 2. Disaccharide pentapeptide subunit of the peptidoglycan of *B. subtilis*.

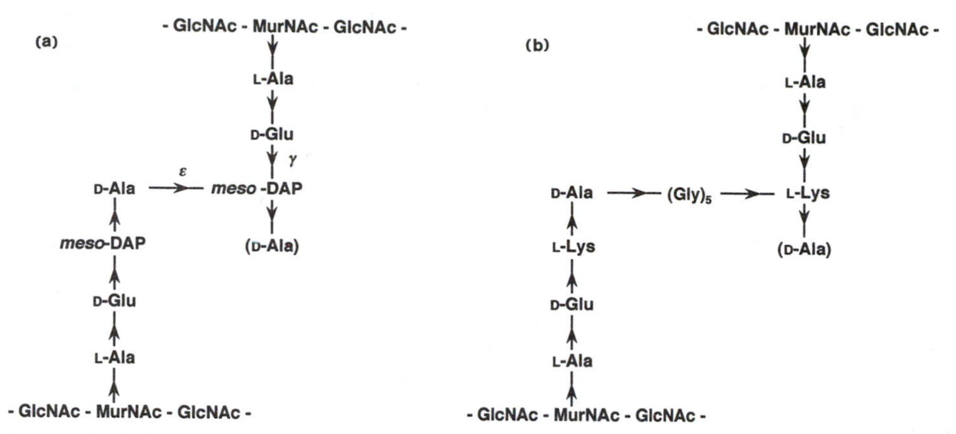

Figure 3. (a) Cross-linked peptidoglycan of the A1γ type found in *B. subtilis* and *E. coli*; (b) cross-linked peptidoglycan of the A3α type found in *S. aureus*. *meso*-DAP, *meso*-diaminopimelic acid.

that may act (314) as a bridge connecting stem peptides through their D-alanine residues at position 4.

The differing types of cross-linkage and the identity of the amino acid at position 3 have been used as chemotaxonomic markers, and various classification schemes have been described. The most widely used is the three-element scheme described by Schleifer and Kandler (254). An initial capital letter is used to designate the class of cross-linkage: A for cross-linkage between positions 3 and 4 of the peptide units or B for cross-linkage between positions 2 and 4. The arabic numeral following the capital letter designates the type of cross-linking, which can be direct (arabic numeral 1) or through a bridge peptide containing from one to several amino acids (arabic numerals 2 through 5). Finally, a Greek letter (α through γ) designates the amino acid at position 3 of the stem peptide. Thus, the peptidoglycan found in *E. coli* and *B. subtilis* (Fig. 3a) is classified as A1γ under this system, whereas *S. aureus* peptidoglycan (Fig. 3b) is described as A3α.

The detailed structure and composition of peptidoglycan in a given organism are more complex than implied by the description just given because of a number of variations and postsynthetic modifications that can alter both glycan and peptide components.

Variations in glycan structure include the presence of *N*-glycolyl substituents in *Nocardia* and *Mycobacterium* spp., the presence of 6-*O*-acetyl groups as a consequence of postsynthetic acetylation of a portion of the *N*-acetylmuramic acid units in *Streptococcus faecalis* (*Enterococcus hirae*) (2) and *S. aureus* (282), and the presence of unsubstituted amino groups of amino sugars in various *Bacillus* species as a consequence of the postsynthetic action of *N*-acetylase. In strains of *B. subtilis*, up to 35% of the glucosamine may be present in *N*-unsubstituted form (327). In *B. cereus*, up to 70% of the *N*-acetylglucosamine residues and a smaller proportion of the *N*-acetylmuramic acid residues may become de-*N*-acetylated (7, 310), rendering the peptidoglycan resistant to hen egg white lysozyme but more sensitive to its endogenous autolysins. In *Bacillus anthracis*, approximately 88% of the glucosamine residues and 34% of the muramic acid residues contain unsubstituted amino groups (327). The walls of these bacteria become sensitive to lysozyme following chemical N acetylation. A modification found in spore peptidoglycan is the occurrence of some of the muramyl residues in the form of internal lactams, which are formed by condensation between their own carboxyl and amino groups (303).

Variations in peptide structure include the amidation of the α carboxyl group of glutamic acid, as in *Bacillus licheniformis*, or of the second carboxyl group of *meso*-diaminopimelic acid, as in *B. subtilis* (281). Amidation neutralizes the acidic carboxyl groups and will reduce the charge density in the wall. Not all *Bacillus* strains have such amide groups, and their physiological significance is not clear. Stem peptides that have become incorporated into wall without having acted as donor in cross-linking reactions (see below) may retain their structures or may lose their terminal alanine residues (and their donor potential) through hydrolysis by D-alanine carboxypeptidase (see below) to give shortened tetrapeptide units. Tetra- and tripeptides may also be formed before incorporation into the wall. The distribution of tri-, tetra-, and pentapeptides differs from strain to strain, as does the extent of cross-linking.

Together with the presence of attachment sites for anionic polymers (183), the variations in peptidoglycan composition described above result in considerable structural complexity. Muramidase digestion of peptidoglycans from *B. subtilis* and *B. megaterium* gives a mixture of products at least as complex (95) as that for *E. coli*, which releases more than 80 different components (103).

Anionic Polymers

Anionic polymers constitute a substantial proportion of the weight of the walls of many gram-positive bacteria. They participate in the electrostatic interactions that determine the polyelectrolyte properties of the wall and are required for viability and normal morphogenesis (33, 91, 199). Anionic polymers fall into two classes: teichoic acids, in which a negative charge is provided by phosphodiester groups in the repeating units of the polymers, and teichuronic acids, in which carboxyl groups of uronic acid residues provide the negative charge. The types and composi-

Table 1. Teichoic acids in members of the genus *Bacillus*[a]

Species and strains	Repeating unit structure(s)	Reference
B. subtilis[b]		
NCTC 3610 (Marburg)	-1-*sn*-glycerol-3-phosphate-; partial 2-α-glucosyl[c]	75, 102
168	Also -(*N*-acetylgalactosamine, glucose-6-phosphate)-	29
W23 and S31	-1-D-ribitol-5-phosphate-[c], partial 4-β-glucosyl	20, 52
B. subtilis subsp. *niger* WM	-2-*sn*-glycerol-3-phosphate-; partial 1-β-glucosyl	68, 147
B. cereus AHU 1030[a]	As *B. subtilis* 168	311, 253
B. coagulans	-6-D-galactose-α1-2-*sn*-glycerol-3-phosphate-[c], with β-glucosyl substitution at C-1 of glycerol	146, 167
B. licheniformis	-1-*sn*-glycerol-3-phosphate-	
	-6-D-glucose-α1-1-*sn*-glycerol-3-phosphate-	
	-6-D-galactose-β1-1-*sn*-glycerol-3-phosphate-	42
	-2-D-galactose-α1-2-*sn*-glycerol-3-phosphate	147
B. pumilus AHU 1650	-6-D-*N*-acetylmannosamine-1-phosphate-	294
	-6-D-*N*-acetylglucosamine-1-phosphate-	166
B. stearothermophilus B65	-6-D-glucose-α1-1-*sn*-glycerol-3-phosphate-	68

[a] Teichoic acids are reported to be absent from all or most strains of *B. cereus* (311), *B. megaterium* (311), *B. polymyxa* (105), and *B. sphaericus* (132).
[b] For other *B. subtilis* strains, see reference 147.
[c] See text.

tions of the anionic polymers present in the wall may vary with growth conditions, but the overall provision of negative charge appears to be rather constant (177). Anionic polymers of both classes are covalently attached to peptidoglycan by a phosphodiester linkage to the hydroxyl group at C-6 of an *N*-acetylmuramyl residue in the glycan chain. In *B. licheniformis*, in which teichoic acid and teichuronic acid occur together, the two classes of anionic polymer are always linked to different glycan chains (300), suggesting an association of the synthetic complexes for peptidoglycan and anionic polymer (see below).

Endospores of *B. subtilis* are reported not to contain teichoic acid (52), but Nishikawa et al. (223) have detected a polymer of galactosamine phosphate in the spore coat of *B. megaterium*.

Anionic polymers in *Bacillus* spp.

Teichoic acids in walls of *B. subtilis* strains fall into two groups (Table 1). The Marburg strain (NCTC 3610) (102) and strain 168 (29) possess 1,3-linked glycerol teichoic acid (Fig. 4a) in which the hydroxyl group at C-2 of the glycerol residue may bear an α-glucosyl or a D-alanyl ester substituent. The extent of alanyl substitution at C-2 is affected by the extreme lability of the alanyl ester linkage, which has a very short half-life at pH 7 and above (256). Although not directly demonstrated in *B. subtilis*, in glycerol teichoic acids of other species in which C-2 is occupied by a glycosyl substituent, the sugar may carry an alanyl ester (312). In this position, the ester linkage is more stable. Strain 168 contains a second wall teichoic acid composed of *N*-acetylgalactosamine and glucose phosphate (75, 229). Depending on growth conditions, it may contribute between 10 and 30% of the total wall phosphate. Its synthesis is temperature sensitive, and the polymer is virtually absent from the walls of bacteria grown at 42°C (33).

In strains W23 and S31, the teichoic acids are ribitol phosphate polymers (Fig. 5) with D-alanyl ester substitutions at C-2 (Table 1). Some chains are fully β-glucosylated at C-4, while others are entirely

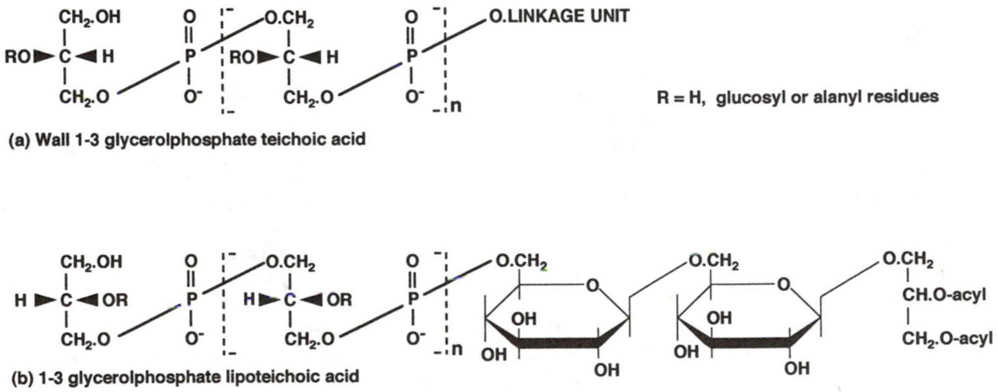

(a) Wall 1-3 glycerolphosphate teichoic acid

R = H, glucosyl or alanyl residues

(b) 1-3 glycerolphosphate lipoteichoic acid

Figure 4. Structures of glycerol teichoic acids from *B. subtilis*. (a) Wall teichoic acid; (b) lipoteichoic acid.

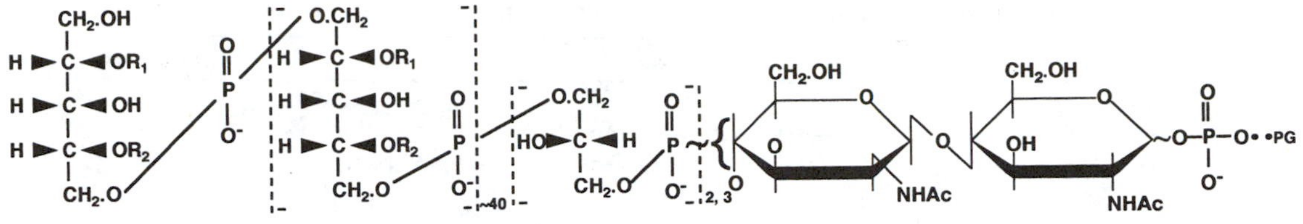

Figure 5. Mode of linkage of ribitol teichoic acid to peptidoglycan in *B. subtilis* W23. The polyribitolphosphate is attached to the linkage unit, which consists of glycerolphosphate and an *N*-acetylhexosamine-containing disaccharide 1-phosphate. PG, peptidoglycan; R_1, H or alanyl ester substituent; R_2, H or β glucosyl substituent.

nonglucosylated (52). The *N*-acetylgalactosamine–glucose phosphate polymer is absent in these strains. Teichoic acids found in *Bacillus* species other than *B. subtilis* are also listed in Table 1.

Early measurements underestimated the chain lengths of ribitol teichoic acids because the acid conditions used to extract the polymers from the wall caused cleavage of occasional phosphodiester linkages in the chains (14). More recent analyses of teichoic acids in intact walls of *B. subtilis* gave an average chain length of about 40 ribitol phosphate units. This value did not vary under different growth conditions and is similar to that reported for ribitol teichoic acid in other organisms (176). Biosynthetic evidence for ribitol teichoic acids indicates that chains are fairly homogeneous in length (83).

Teichuronic acids are found in significant amounts in phosphate-limited cells of most wild-type strains of *B. subtilis*, though a polymer containing both phosphate and mannosaminuronic acid occurs constitutively in strain AHU 1031 (324). The structure of the teichuronic acid from strain 168 has not been determined, but the structure for teichuronic acid in strain W23 (Fig. 6) is supported by biosynthetic data (110). Closely related polymers have been characterized in *B. licheniformis* NCTC 6346 and ATCC 9945 (180, 300), in which they are synthesized constitutively. Teichuronic acids (Table 2) have been reported to occur together with teichoic acids in *B. licheniformis* and as the sole anionic wall polymer of phosphate-limited cells of *B. subtilis* and *B. subtilis* subsp. *niger*. Teichuronic acids are the only anionic polymers in the walls of *B. megaterium* and several strains of *B. cereus* (311) even in conditions of phosphate excess. Yoneyama et al. (325) have demonstrated the presence of constitutive *N*-acetylmannosaminuronic acid-containing teichuronic acids in strains of *B. megaterium* and in *Bacillus polymyxa*.

Anionic polymers in other gram-positive bacteria

The range of teichoic acid types displayed by *Bacillus* spp. reflects the range found in gram-positive bacteria generally. Glycosylated glycerol teichoic acids have been found in coagulase-negative staphylococci (79, 85) and lactobacilli (259). Glycosylated ribitol teichoic acids are present in *Staphylococcus saprophyticus* and *S. aureus* (79), *Lactobacillus* spp. (258), and *Listeria* spp. (291). Alditol phosphate polymers in which a sugar forms an integral part of the main polymer chain, as in the teichoic acids of *B. licheniformis*, occur in *Lactobacillus* spp. (17) and in *Streptococcus pneumoniae* (239). Some strains of *Streptococcus sanguis* (202) and *Streptococcus pneumoniae* (293) contain anionic polymers consisting of oligosaccharide units linked by phosphodiester groups. Polymers in which only sugar phosphates occur, as in *Bacillus pumilus*, have been reported for some *Staphylococcus* species (19). Teichoic acids containing mannitol phosphate occur in *Brevibacterium* spp. (84).

A teichuronic acid of glucose and *N*-acetylmannosaminuronic acid has been described for *Micrococcus luteus*, with an additional *N*-acetylglucosamine-1-phosphate residue as linkage unit between the terminal sugar of the polymer and the peptidoglycan (121). Aminouronic acid-containing polymers also occur as capsules or microcapsules in *S. aureus* (145).

Evidence for the presence of polysaccharide as part of the wall of *B. anthracis* has been well documented (41, 136). The exact composition of the polysaccharide has been in dispute, although there is a general agreement that D-galactose and *N*-acetyl-D-glucosamine are constituents. A recent report (77) describes a new method for isolation of wall polysaccharides from *B. anthracis* and indicates that the polysaccharides contain galactose, *N*-acetylglucosamine, and *N*-acetylmannosamine in an approximate molar ratio of 3:2:1. The polysaccharide was not characterized further.

Attachment of anionic polymers to wall

Acid hydrolysis of walls yields muramic acid 6-phosphate (183), which originates at the site of attachment of anionic polymers to peptidoglycan (58). Chemical and biosynthetic studies (113, 164, 323) have shown that a "linkage unit" intervenes between the phosphate terminus of the main polyalditol phosphate chain of teichoic acid and an *N*-acetylmuramyl residue of the peptidoglycan. In *B. subtilis* W23, the linkage unit consists of two glycerolphosphate residues attached to *N*-acetylmannosaminyl-*N*-acetylglucosaminyl-phosphate (Fig. 5). The acid lability of the *N*-acetylglucosaminyl-phosphate accounts for the ease of extraction of wall teichoic acid with acid, while the phosphodiester linkage between glycerolphosphate and *N*-acetylmannosamine is unusually susceptible to alkaline hydrolysis (58).

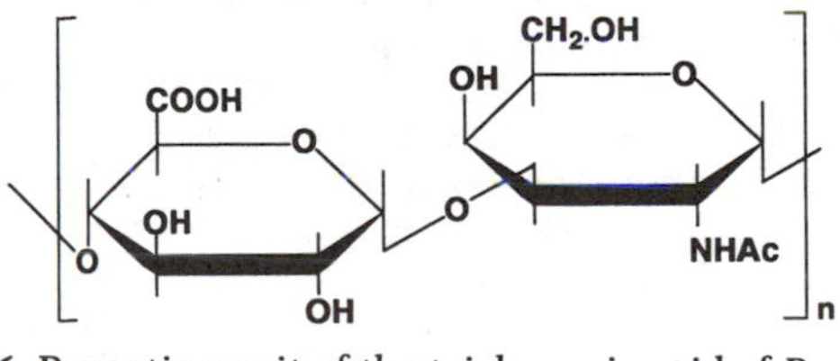

Figure 6. Repeating unit of the teichuronic acid of *B. subtilis* W23.

Table 2. Teichuronic acids in members of the genus *Bacillus*

Species and strain	Repeating unit structure[a]	Reference
B. subtilis		
W23	-4-GlcA-α1-3-GalNAc-α1-	316
AHU 1031	-6-Glc-α1-3/4-ManNAcA-β1-4-GlcNAc-β1- | Glc About 50% of Glc branches are esterified with glycerolphosphate.	324
B. licheniformis ATCC 9945	-4-GlcA-β1-4-GlcA-β1-3-GalNAc-β1-6-GalNAc-α1-	180
B. megaterium M46	-4-Glc-1-3-Rham-1-4-Rham-1- | GlcA	137
Alkalophilic *Bacillus* sp. strain C-125	Polymer containing polyglucuronic acid and poly-γ-L-glutamic acid	5

[a] Abbreviations: Glc, glucose; GlcA, glucuronic acid; GlcNAc, *N*-acetylglucosamine; GalNAc, *N*-acetylgalactosamine; ManNAcA, *N*-acetylmannosaminuronic acid; Rham, rhamnose.

Since in strains having glycerol teichoic acids the same precursor, CDP-glycerol, is required for synthesis of the main polymer chain and for glycerolphosphate residues in the linkage unit (see below), the two parts of the molecule will be indistinguishable chemically unless they differ in glycosylation. However, the same *N*-acetylhexosamine-containing disaccharide is found in strains W23 and 168 (164). The mode of attachment of the galactosamine-containing teichoic acid to the wall in strain 168 has not been characterized, but the susceptibility to acid of the attachment (75, 229) suggests that it, too, involves a sugar-1-phosphate linkage. The attachments of teichoic acids to walls in all other bacteria so far examined involve linkage units closely related in structure to that found in *B. subtilis* W23 (Fig. 5). This structure appears to be well conserved, despite the great diversity of teichoic acids. Species-specific variations of the teichoic acid linkage unit involve the number of glycerol phosphate residues (164, 322), replacement of *N*-acetylmannosamine by glucose (146), and replacement of *N*-acetylglucosamine by glucosamine (165).

There is no evidence for a distinct linkage unit between teichuronic acid and peptidoglycan as in the case of teichoic acids. In *B. licheniformis*, teichuronic acid is linked to the wall by a sugar-1-phosphate linkage between the terminal *N*-acetylgalactosaminyl residue of the teichuronic acid and the C-6 of a muramic acid residue in the peptidoglycan (300).

Roles of anionic polymers

There is now convincing evidence that the presence of an anionic polymer in the wall of *B. subtilis* is essential for normal cell division and hence for viability. Mutants constitutively deficient in all wall anionic polymers have not been isolated, and thermosensitive mutants of the *tag* locus, which encodes enzymes of the teichoic acid synthetic pathway, are severely disabled or not viable at the nonpermissive temperature (199). The cell's requirement appears to be for the presence of anionic groups in the wall rather than for teichoic acid specifically. Phosphate-limited bacteria in which teichuronic acid replaces teichoic acid appear to grow and divide normally, whereas under the same conditions, mutants blocked in teichuronic acid synthesis grow with abnormal morphology or die (78, 91). Growth of teichuronic acid-deficient mutants in near-limiting phosphate concentrations produces cells whose walls have diminished amounts of anionic polymer. As the total amount of anionic polymer decreases, the cells become shorter and fatter, suggesting that changes in these polymers differentially affect the incorporation of new material into cylindrical and septal walls (18b).

Physical studies have revealed the important contribution anionic polymers make to the polyelectrolyte gel structure of the wall (see below). They are linked to peptidoglycan throughout the thickness of the wall, where a large majority (>85% in *B. licheniformis*, for example) of peptidoglycan chains carry at least one covalently linked polyanion chain. The repulsion of like charges maintains the peptidoglycan network in an expanded state, and neutralization of these charges causes contraction of the network, with concomitant changes in wall porosity and density (193).

Glycerol teichoic acid isolated from walls of *B. subtilis* adopts a rigid, rod-type conformation in aqueous solutions of low ionic strength but changes to a random coil structure at high salt concentrations (74). Evidence for coiling of polyglycerolphosphate in lipoteichoic acid (see below) has also been obtained (173). Ribitol teichoic acids can also adopt ordered conformations in aqueous solution (227), but energy minimization studies (18a) show that these structures are dominated by charge interactions that are likely to be affected markedly by the ionic environment of the wall.

The presence of alanyl ester substituents on teichoic acids reduces the net negative charge in individual molecules and in the wall as a whole. The role of the alanyl substituents is unclear, but effects both at the level of gross wall properties and at the molecular level are likely. X-ray photoelectron spectroscopy (21) of lyophilized walls shows that the presence of alanyl ester substituents affects the interaction of bivalent cations with the phosphate groups of glycerol teichoic acid, confirming measurements of binding affinities for Mg^{2+} in wall suspensions (16, 175). ^{31}P nuclear magnetic resonance studies of solutions of purified lipoteichoic acid do not reveal any interaction between the alanyl amino groups and specific phosphate groups in the polymer chain (24), but such interactions might be affected by the steric effects of constraining the teichoic acid chains within the cross-linked matrix of the wall (124).

A variety of specific functions has been attributed to the wall anionic polymers. Their abilities to bind metal cations may protect the cell against toxic metals and could provide a buffer of essential bivalent cations such as Mg^{2+} (125, 131). There are strong, specific interactions between the major wall autolysin, *N*-acetylmuramoyl-L-alanine amidase (see below), and teichoic acid in the wall of *B. subtilis* (126, 127). Teichoic acid also acts as the binding site for a number of *B. subtilis* bacteriophages (9). A role in cell-to-cell adhesion has been demonstrated for the anionic wall polymers of oral streptococci, which interact with lectinlike surface proteins of *Actinomyces* and *Capnocytophaga* spp. (168).

Lipoteichoic Acid

Like most other gram-positive bacteria, *Bacillus* species contain substantial amounts (1 to 2% by weight) of lipoteichoic acid. It is mainly associated with the cytoplasmic membrane but may also be present, sometimes in a partially deacylated form, in the wall, at the surface, and excreted into the culture fluid. The chemistry and physiology of lipoteichoic acids have been thoroughly reviewed by Fischer (88).

Structure

In *B. subtilis*, lipoteichoic acid consists of a polyglycerolphosphate chain of between 24 and 33 phosphodiester-linked units attached through its terminal phosphate to 3-gentiobiosyl diglyceride (138). Maurer and Mattingly (200) have demonstrated considerable chain length heterogeneity in preparations of lipoteichoic acids from several strains of *Streptococcus*. However, in order to differentiate the different molecular species, it was necessary to deacylate the terminal glycolipids chemically. The alkaline conditions used to do this are likely to have caused hydrolysis of a portion of the phosphodiesters in the polyglycerolphosphate (122). Consequently, the range of chain lengths in the native polymer is still open to question.

The polyglycerolphosphate chain of lipoteichoic acid differs in an important respect from that of wall glycerol teichoic acids. Because of their biosynthetic origins, the glycerolphosphate groups are of opposite stereochemistry in the two polymer types (Fig. 4) (162). Though this will be most apparent at the glycerol-terminal end of the chain, which is *sn*-glycerol 1-phosphate in lipoteichoic acid and *sn*-glycerol 3-phosphate in wall teichoic acid (Fig. 4b), the whole polymer chain is asymmetric, and the two types of teichoic acid will be recognizably different over regions that include either end of the chain. The functional significance of this difference in stereochemistry is not yet clear.

In *B. subtilis* 168, between 40 and 60% of glycerolphosphates in the lipoteichoic acid chain are substituted with α-glucosyl or α-*N*-acetylglucosaminyl residues in about equal proportions. The remaining glycerolphosphates carry D-alanyl ester substituents (57, 87, 138), which display the same lability to alkaline conditions as those on wall teichoic acids. In *S. aureus* and *Lactobacillus fermentum*, substitution appears to be uniform along the chain (88). Lipoteichoic acids of other species of *Bacillus* differ in structure from that of *B. subtilis* (138). The lipoteichoic acids from *B. megaterium* and *Bacillus coagulans* appear to contain no alanine and are substituted to a small extent with α-galactosyl residues. Moreover, the terminal lipids are diglycerides rather than glycolipid. In one strain of *B. licheniformis*, alanine is present and the glycosyl substituent is α-*N*-acetylglucosamine (138).

Cellular location

Studies with *B. megaterium* (123) and *E. hirae* (269) showed that some lipoteichoic acid remains associated with cytoplasmic membrane during fractionation of disrupted cells and protoplasts, though the extent of this association depends on the concentration of bivalent cations, such as Mg^{2+}, in the surrounding fluid (131, 264). Immunoelectron microscopy shows lipoteichoic acid spread over the entire outer surfaces of protoplasts of *S. aureus* (1) and filling the space between the cytoplasmic membranes and the outer surfaces of the walls in thin sections of various gram-positive bacteria (88). Membrane-attached lipoteichoic acid may be partly responsible for the thin, electron-opaque layer seen underlying the wall in thin sections of acrylic resin-embedded *B. subtilis* stained with uranyl ions (Fig. 1) (129, 210).

It is likely that membrane-associated lipoteichoic acid is held in the membrane by intercalation of its acyl groups into the hydrophobic region of the membrane and is stabilized by bridging with divalent cations. Thus, deacylation, by an endogenous deacylating lipase, is associated with the release of lipoteichoic acid (149). Lipoteichoic acid is released by treatment with the metal-chelating agent EDTA (131) under conditions in which deacylation does not occur (191).

Lipoteichoic acid can be isolated from the walls and culture fluids of many bacteria, and it is probable that the wall material is simply a fraction of the exported polymer (88). There are, however, conflicting reports of the conditions under which release of lipoteichoic acid from the cell occurs, and no comparative studies have been carried out under controlled growth conditions. It was reported that no release of membrane-bound lipoteichoic acid occurred in growing *B. subtilis* (163), but turnover was detected in *Bacillus stearothermophilus* (45). Lipoteichoic acid-deacylating lipase activity has been demonstrated in several gram-positive bacteria, but the extent to which this enzyme regulates lipoteichoic acid export is not known (88).

In aqueous solution, lipoteichoic acid forms micelles, which in some cases are essential for its biological activity (62). Like other amphiphiles, lipoteichoic acid exhibits a well-defined critical micelle concentration below which it exists in solution as a monomer and above which it is almost entirely micellar (313). X-ray analysis has shown that the lipoteichoic acid of *S. aureus* forms micelles of fairly homogeneous size, each containing about 150 molecules (173). The hydrophilic exterior, consisting of the polyglycerolphosphate chains, appears to be only 0.34 nm thick, whereas a fully extended chain has a theoretical length of 0.8 nm. This length is unaffected by ionic strength or alanine substitution. These measurements indicate that the polyglycerolphosphate chain adopts a highly coiled conformation.

Other amphiphilic macromolecules in gram-positive bacteria

Not all gram-positive bacteria contain lipoteichoic acid. It is absent from alkalophilic *Bacillus* sp. strain A007 (163) and from several *Micrococcus* species. In the latter cases, it is replaced by an acidic lipomannan in which anionic groups are provided by ester-linked succinyl groups (181, 237). Analogous molecules appear to replace lipoteichoic acid in a variety of species, including *Streptococcus pneumoniae*, *Streptococcus sanguis*, *Bifidobacterium* spp., mycobacteria, and propionibacteria. These cases have been reviewed recently (274).

Proteins

Wall-bound proteins

Characteristically, the walls of gram-positive bacteria contain less protein than those of gram-negative bacteria, but substantial amounts of protein may nonetheless be present in some cases. Small amounts of protein or polypeptide (less than 1% of wall weight) are so firmly associated with the wall in *B. subtilis* (216) that they cannot be removed by treatment with detergent or phenol. Such treatment can, however, remove other wall-associated proteins. These proteins include enzymes involved in cross-linking and modifying peptidoglycan. Both autolytic enzymes and "modifier" proteins involved in the modulation of enzyme activities exhibit a functional association with wall anionic polymers (127, 240). In general, these proteins are present in small amounts, and >90% of the wall isolated even under fairly gentle conditions consists of peptidoglycan and anionic polymers.

Other gram-positive bacteria may contain substantial proportions of wall protein. For example, up to 16% of the weights of walls of streptococci (242) and up to 30% of the weights of walls of staphylococci (265) consist of covalently linked proteins. The M proteins in group A streptococci and protein A in *S. aureus* are important surface-associated virulence factors. Both of these proteins have distinctive C-terminal domains containing repeated proline-rich oligopeptide sequences that could span the wall (214).

Proteins exiting the cytoplasmic membrane in gram-positive bacteria may be affected by their interactions with the wall (63, 119, 120). Proteins such as autolysins and their regulator proteins, which interact strongly with the wall, are not generally found free in the culture supernatant. A second class of proteins is exported into the growth medium but delayed during passage through the wall, possibly because the wall charge or porosity limits their diffusion. A third class of proteins is translocated into the culture medium extremely rapidly, either because these proteins do not interact with wall components or because they have unique export channels. There is evidence that metal counterions, such as ferric and magnesium ions, may play a role in protein export, as may lipoteichoic acid (32, 45).

S-layer proteins

In a number of bacteria, substantial amounts of protein are found as envelope components in the form of a structured outer layer of wall. These surface layers (S layers) are composed of protein or glycoprotein subunits that are arranged in self-assembling hexagonal, square, or oblique arrays over the whole surface of the cell and are held together and to the underlying wall mainly by electrostatic and hydrophobic interactions (28, 267). Though their specific functions are not wholly clear, these paracrystalline arrays form an interface with the environment and so are responsible for the cell's surface properties. They presumably afford some protection to the underlying wall, possibly acting as a barrier to macromolecules. Pores formed between adjacent subunits in *B. stearothermophilus* strains have sharply defined exclusion cutoffs for molecules of molecular weights between 50,000 and 80,000 (251, 268). S layers may act like the outer membranes in gram-negative bacteria in trapping exoenzymes at a surface location where they can destroy potentially damaging or toxic components from the environment.

S layers have been found in several strains of *Bacillus* (105, 174, 221, 266). Among the best studied are the S layers of *B. sphaericus* NTCC 9602, which is composed of an acidic glycoprotein containing a single homogeneous peptide chain, and *Bacillus brevis* 47. *B. brevis* has a three-layered wall between 27 and 29 nm thick in which the thin innermost layer of peptidoglycan and teichoic acid is covered by a hexagonal array of subunits that have center-to-center spacings of 14.5 nm and are composed of 150-kDa proteins (289). On top of this is a second S layer composed of a serologically distinct 130-kDa protein. These proteins are shed during growth (225, 318) in amounts of up to 12 g of extracellular protein per liter of culture (215).

Sleytr and his colleagues have shown that the S layers of several thermophilic gram-positive bacteria consist of glycoproteins. The protein from the surface of a strain of *B. stearothermophilus* contains two glycan substituents, i.e., an asparagine-linked rhamnan and an oligosaccharide containing a diaminouronic acid (53, 212). The surface glycoproteins of several *Clostridium* species have also been characterized. That of *Clostridium symbiosum* HB 25 contains tetrasaccharides that are attached by phosphodiester linkages to the protein (211).

S-layer proteins in rod-shaped bacteria appear to cover the surface completely to give lattices that are well ordered and fairly uniform over the cylindrical surface but that have, at the poles and septation sites, numerous faults that arise because of their geometry. Analysis of the distribution of faults suggests that the constituent subunits are incorporated at specific sites and recrystallize during cell growth, maintaining a lowest-energy-state equilibrium (266). The S layer thus provides an ordered protein meshwork that requires a minimum of genetic information to ensure that it encloses the growing cell.

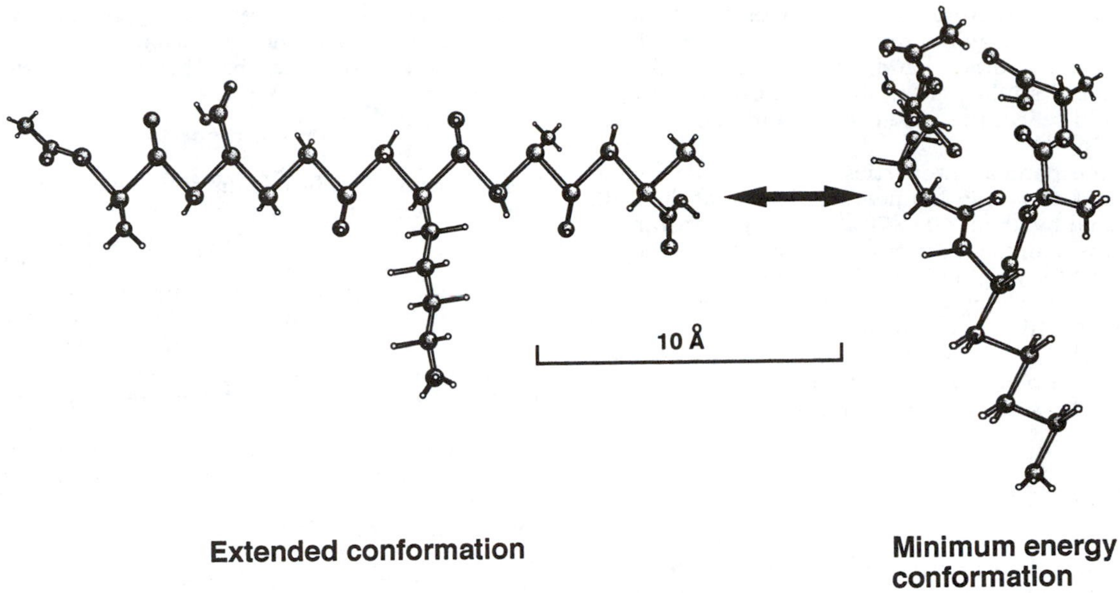

Figure 7. Diagrammatic representation of extended- and minimum-energy conformations of the stem peptide of *B. subtilis* peptidoglycan. 1 Å = 0.1 nm. (Redrawn from reference 171.)

Structural Organization of the Wall

Peptidoglycan

The strength of plant and fungal walls is due primarily to fibrils of cellulose or chitin that are formed by semicrystalline packing of large numbers of polysaccharide chains. Such fibrils are not present in bacterial walls, and the glycan chains in peptidoglycan, although chemically similar to chitin, have bulky peptide substituents that prevent close packing. Instead, the glycan chains in peptidoglycan are held together by cross-linkage through flexible peptide chains (Fig. 3). As a result, bacterial walls are elastic structures that can be stretched by turgor and by electrostatic repulsion between similarly charged groups. This elasticity has great significance for wall assembly and growth.

X-ray diffraction studies have confirmed the absence of semicrystalline structures in bacterial walls but indicate that the walls are not completely unordered (39, 40, 170, 219). The diffraction patterns are compatible with models (172) in which the wall contains several sheets of peptidoglycan in a layered structure. The energetically most favored conformations of glycan chains are those showing twists between adjacent sugars, giving extended chains from which peptide side chains emerge at different angles in an approximately helical arrangement (23, 172). Cross-linked peptides can extend between neighboring glycan chains in the same plane so that the glycans become interconnected through flexible peptides.

Infrared spectroscopy excludes a highly ordered structure for the peptides as well as for the glycan and indicates the absence of regular conformations such as alpha helices or beta pleated sheets (219). The peptides may adopt a minimum-energy ringlike conformation that could be unwound by mechanical forces to give more extended arrangements (Fig. 7) (22, 23). The glycan chains may thus form rigid elements within an elastic network of peptide.

Because of the twisted conformation of the glycan, peptides can extend outward above and below the plane of a layer so that cross-linkages between them can connect different layers within the wall. Labischinski et al. (172) calculate that a continuous cross-linked two-dimensional network is possible only in a structure in which murein strands are packed in parallel but staggered by half a helix pitch length. Modeling studies show that peptide dimers and possibly trimers but no higher oligomers may be formed within a single two-dimensional layer. Longer peptide oligomers can be formed within multilayered structures depending on the orientation of the layers with respect to each other and the lengths and flexibilities of the peptide side chains. Permitted arrangements (170) for the peptidoglycan in gram-positive cocci would be compatible with the isotropic shrinking and swelling behavior of staphylococci (171). A permitted arrangement for cylindrical wall, called topotype 4 by Labischinski et al. (172), is consistent with the low content of oligomers reported in *B. subtilis* (245). Topotype 4 allows a hooplike arrangement of glycan chains in a spiral or corkscrew winding at an angle to the short axis of the cell (295). This kind of arrangement would be consistent with the observation that swelling of *B. subtilis* occurs via variation in cell length without change in diameter (171). It is also compatible with an anisotropic disposition of peptidoglycan that may be deduced (208) from the helical morphology (206) shown by Lyt$^-$ mutants of *B. subtilis* (see below). A glycan chain comprising 100 disaccharide units would be approximately 0.1 μm in length, so about 16 such chains end to end would be needed to circumscribe the wall (297).

The physical existence of layers in cylindrical wall has not been demonstrated directly, and the concept may be an oversimplification. Thus, Koch has argued

(153) that the structure might more closely resemble a nonwoven fabric, with only partially oriented glycan fibrils. On the other hand, freeze-fracture electron microscopic studies (107) show a fibrous structure that could reflect the presence of bundles of oriented glycans.

The mechanical properties of bacterial threads made from *B. subtilis* filaments indicate that the walls can extend by about 60 to 70%. Although substantial, this is less than the maximum extension that would be calculated for ringlike cross-linked peptides oriented parallel to the long axis of the cell cylinder but is consistent with arrangements in which the peptides are oriented at an angle to the cell axis. Anionic polymers attached to the glycan chains may also influence the extensibility and organization of peptidoglycan (278).

Anionic polymers

The influence of anionic polymers on wall structure, assembly, and morphogenesis is not well understood, though it is clear that they make substantial contributions to the polyelectrolyte gel structure and mechanical properties of the wall (279). Stresses in wall produced by electrostatic repulsion between similarly charged groups may be of the same order of magnitude as those produced by turgor (278). Most of the charge in wall is due to the anionic polymers, and early support for the significance of this charge came from the finding that mutants deficient in anionic polymer were frequently irregular in shape (56). The importance of anionic polymers for normal growth (78, 91, 199) may reflect an essential contribution to morphogenesis by stresses produced in the wall through electrostatic repulsion between the charged groups of the polymers.

Anionic polymers are also responsible for specific surface properties, modulation of autolysin action (see above), cation binding, and surface charge. Estimation of the proportions of teichoic acid exposed at the outer surfaces of bacteria is difficult, since limits to the binding of adsorbed probes such as phages, antibody molecules, or lectins may be due to steric crowding of the adsorbed probes rather than to anionic-polymer density (8). It has been suggested that the outer wall region is a loosely packed, highly porous, and partly autolyzed gel that is freely permeable to macromolecular probes. The defined outer surface seen in stained thin section may represent collapsed or condensed material, and the "layer" of adsorbed concanavalin A observed by Birdsell et al. (29) may reflect the stabilization of a "glycocalyx" containing both teichoic acid and peptidoglycan. Underlying wall regions may be more tightly packed, less porous, and less accessible to macromolecular probes. The packing density at the pole may differ from that in the cylinder (271).

Individual glycan chains in *B. licheniformis* (300) are linked to only one type of anionic polymer, so teichoic acid and teichuronic acid could be present in separate domains in walls containing both polymers. So far, such segregation has been demonstrated only in bacteria that have undergone transitions affecting anionic-polymer synthesis (210), but it could also occur in bacteria growing under other conditions. It is possible also that the structurally distinct teichoic acids that occur together in various species may not be uniformly distributed in different wall regions.

Surface Properties

The bacterial surface is the region of physical interaction between the cell and its environment. It may encompass the wall and overlying components such as S layers and capsules. In general, the surface is electronegative at physiological pHs, and its charge density is higher than that of the surrounding environment. The charge density varies from organism to organism. For example, the polyglutamic acid capsule of *B. licheniformis* has a higher density of anionic groups than does the S layer at the surface of *B. stearothermophilus* (203, 250). Environmental pH and ionic composition affect not only the extent of ionization of surface components, and hence surface charge, but also the arrangement and disposition of molecules at the surface (141). Nevertheless, *B. subtilis* surface charge remains remarkably constant over the physiological pH range of 4 to 7, even in bacteria grown under conditions that lead to substantial differences in wall composition (11).

Binding of cations by surface components has long been observed to be an important property of a cell (13). The early suggestion (125) that both wall and membrane teichoic acids are involved in the binding of cations and in the provision of an optimum ionic environment for the activity of enzymes in the cytoplasmic membrane is supported by later work (131). Binding of cations also affects the charge and disposition of components in walls. The high and selective affinity of bacterial walls for metals is relevant both to natural mineralization of soluble environmental cations (26) and to the microbial recovery or removal of metals from industrial waste and the environment (111).

The presence of lipoteichoic acid at the surface of the wall (see above) contributes to both the anionic nature and the hydrophobicity of the wall. Ofek et al. suggested (224) that the association of M protein with virulence in group A streptococci may be due in part to an ionic interaction between the protein and lipoteichoic acid such that the lipid moiety of the teichoic acid is held at the surface so that it can facilitate bacterial attachment to host cell membrane.

BIOSYNTHESIS OF WALL POLYMERS

Peptidoglycan

Biosynthesis of peptidoglycan proceeds by polymerization of disaccharide-peptide subunits that are assembled on an undecaprenylphosphate anchor lipid to give a substituted glycan in which *N*-acetylmuramic acid is present at the reducing terminus of the chain, which is the site of attachment to the lipid. The pathway is outlined in Fig. 8. Mutations affecting synthesis of the pentapeptide can result in the formation of L forms in *Bacillus* spp. (298). Strains defective in diaminopimelic acid synthesis have also been identified among temperature-sensitive mutants of *B. subtilis* deficient in peptidoglycan synthesis (44), and

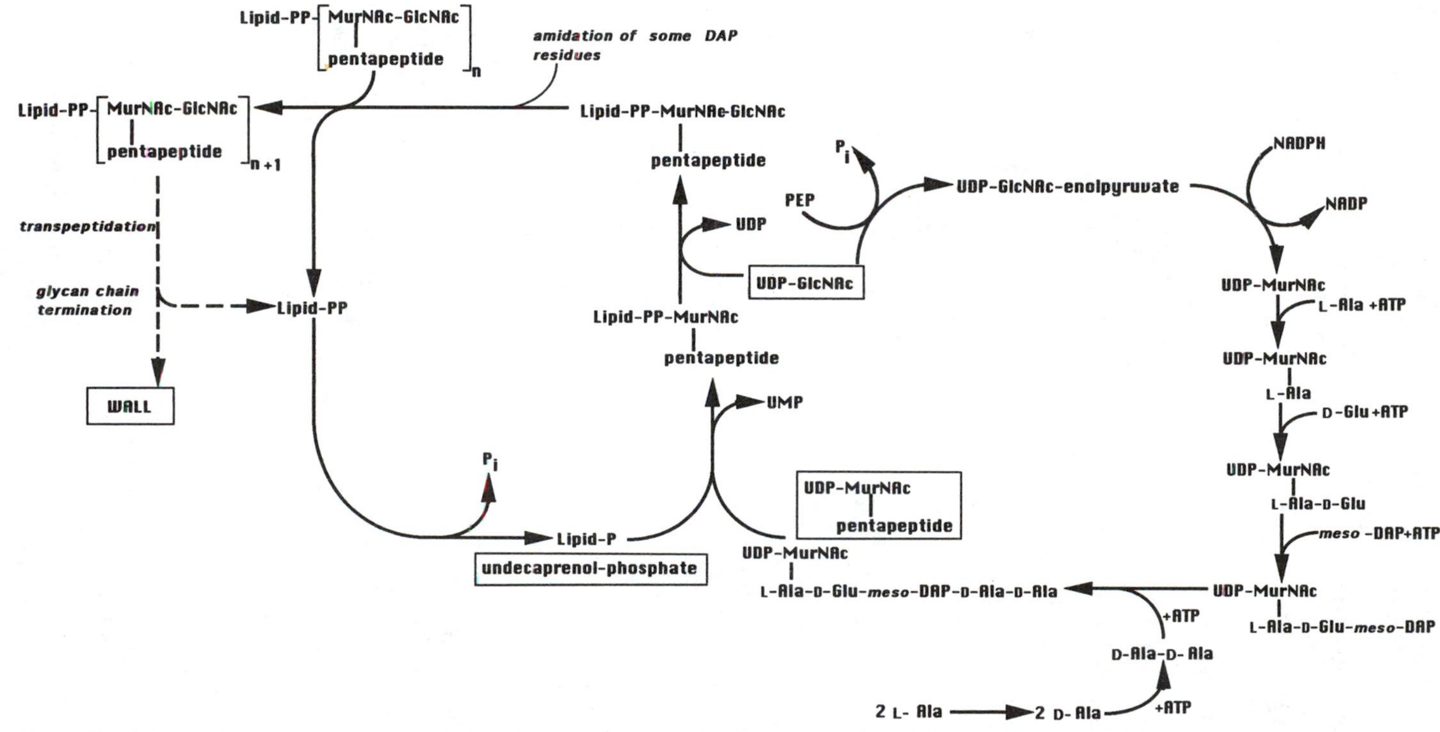

Figure 8. Biosynthesis of peptidoglycan in *B. subtilis*. Ala, alanine; DAP, diaminopimelic acid; Glu, glutamic acid; GlcNAc, *N*-acetylglucosamine; MurNAc, *N*-acetylmuramic acid; PP, pyrophosphate; PEP, phosphoenolpyruvate.

recent evidence indicates multiple isoenzymes in this pathway (248). The gene *murA*, which encodes the enzyme that catalyzes the first step in the synthesis of UDP-*N*-acetylmuramic acid from UDP-*N*-acetylglucosamine, has recently been identified (249). Mutants that are affected in this gene and that also have a temperature sensitivity phenotype are lytic at the nonpermissive temperature.

In the initial step of glycan synthesis, phospho-*N*-acetylmuramoylpentapeptide is transferred to undecaprenylphosphate by a membrane-bound "translocase" to give undecaprenylpyrophosphate-*N*-acetylmuramoylpentapeptide. Despite its name, there is no evidence that this enzyme is involved in a vectorial translocation across the membrane (see below). *N*-Acetylglucosamine is then transferred from UDP-*N*-acetylglucosamine to form the disaccharide-pentapeptide lipid (Fig. 8). This product is used directly for peptidoglycan synthesis in some bacteria, but in various species of *Bacillus*, it is first amidated by a membrane-bound enzyme at the free carboxyl group of either glutamic acid or diaminopimelic acid (182).

Polymerization of the disaccharide-pentapeptide repeating units is catalyzed by multiple membrane-bound transglycosylases. In a number of bacteria, some, but not all (228), of the transglycosylases are bifunctional enzymes that also have peptidoglycan transpeptidase activity and are therefore recognizable as penicillin-binding proteins (PBPs) (306). There has been no direct demonstration that this is the case in *Bacillus* species, but it would be expected by analogy with *E. coli* that the high-molecular-weight PBPs of *Bacillus* spp. are enzymes of this type. In *B. subtilis*, a mixture of PBP1, PBP2, and PBP4 catalyzed the transglycosylation reaction, though transpeptidase activity could not be detected (139). Pure PBP1 also showed some transglycosylase activity. A mixture of PBP1 through PBP4 of *B. megaterium* gave similar results, and a small amount of transpeptidase activity was also detected (139). Since in all these cases the transglycosylase activity is not penicillin sensitive, it is reasonable to assume that the penicillin-sensitive domain is a transpeptidase, as in *E. coli*. There is indirect evidence for this in PBP1 of *B. megaterium* (99) and PBP2 of *B. subtilis* (36). The multiple forms of transglycosylase-transpeptidase are to some extent interchangeable. A mutant of *B. megaterium* with only 10% of the normal level of PBP1 was still able to grow, albeit filamentously, apparently because of an increase in the amount of PBP3 (99). Similarly, PBP1 of *S. aureus* is not essential for growth, though PBP2 and PBP3 are, and mutants lacking PBP4 continue to grow with apparently normal morphology but with a low degree of peptidoglycan cross-linking (317).

During the transglycosylation reaction, both the growing chain and the incoming subunit are attached to anchor lipid molecules. Studies with membrane preparations from *B. licheniformis* (302) and *B. megaterium* (94) show that new units are incorporated at the lipid-bound end of the growing glycan chain by transfer of the nascent chain to the lipid-bound disaccharide peptide. The undecaprenylpyrophosphate so released must be converted to the monophosphate form by a pyrophosphatase (the reaction inhibited by bacitracin) so that it can rejoin the biosynthetic cycle. The chain termination mechanism is not known.

In wild-type strains, glycan chains are subsequently hydrolyzed by an endo-β-*N*-acetylglucosaminidase to yield shorter chains with *N*-acetylmuramic acid or *N*-acetylglucosamine at the reducing ends (see above). The peptide chains of the peptidoglycan are also susceptible to postsynthetic modification by carboxypeptidase activity (see below). The low-molecular-weight PBPs, such as PBP5, the predominant PBP in *B. subtilis* (30), are DD-carboxypeptidases that remove the terminal D-alanine from the pentapeptide. Unlike the transglycosylase-transpeptidases, these enzymes appear not to be essential for vegetative growth. Mutants of *B. subtilis* in which the gene encoding PBP5 (*dacA*) is inactivated grow without detectable alteration to morphology or peptidoglycan cross-linking (257). Thus, the precise role of these enzymes is unknown, although it is speculated that they are involved in the regulation of cross-linking. However, the *dacA* gene is strongly conserved across a range of *Bacillus* species (37) and appears to play an important role in cortex synthesis during sporulation (285).

In *Bacillus* spp., modification of nascent peptidoglycan structure also occurs as a result of removal of *N*-acetyl groups by an *N*-acetylase (see above).

Incorporation of nascent peptidoglycan into the wall is accomplished by transpeptidation reactions in which the terminal D-alanine is removed from the donor peptide while its penultimate D-alanine becomes linked to the free amino group of a bridge peptide or diamino acid in an adjacent acceptor peptide on another glycan chain (Fig. 3). There is evidence that this process of attachment to existing wall commences while the glycan chain is still in the process of elongating. Although inhibition of transpeptidation in vivo with a low concentration of penicillin leads to excretion of the nascent peptidoglycan in a linear, un-cross-linked form (287, 290, 307), in vitro systems capable of catalyzing the synthesis of peptidoglycan and its attachment to existing wall continue to synthesize some wall-linked peptidoglycan when transpeptidation is completely inhibited by β-lactams (301). Under these conditions, elongation of nascent, wall-linked peptidoglycan must be occurring by transglycosylation.

Anionic Polymers

Diverse mechanisms for anionic polymer synthesis in *Bacillus* spp. and other gram-positive bacteria have been described. They vary in the direction of polymer chain extension and the role of undecaprenylphosphate anchor lipid. In some cases, the polymer chain grows by the addition of units to the end attached to the membrane anchor, while in others, addition is to the end distant from the anchor. Undecaprenylphosphate lipid always serves to anchor the growing chain to the membrane but in some cases also acts as a carrier for the assembly of a polymer repeating unit before polymerization, as in the case of peptidoglycan synthesis. In a few cases, the lipid is also involved in glycosylation of the teichoic acid, while in other cases, glycosyl substituents are added directly from the

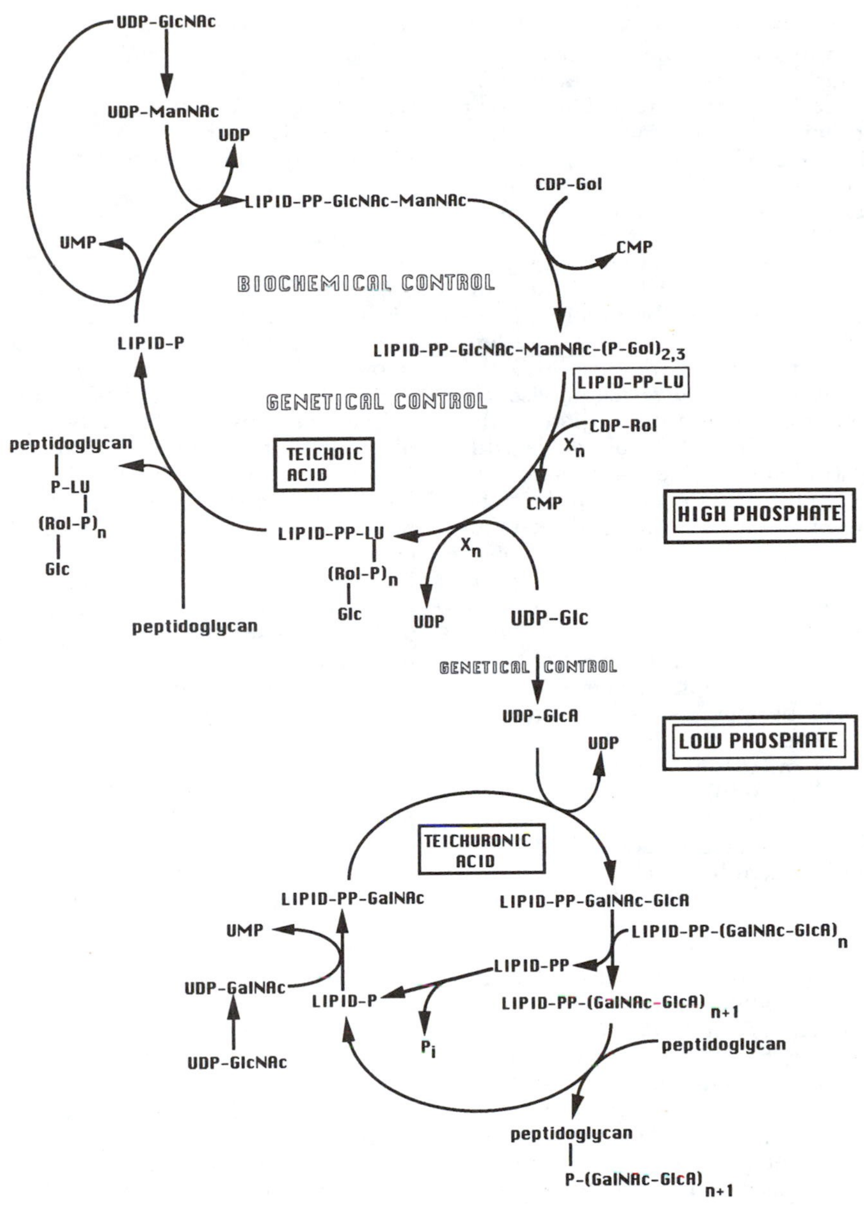

Figure 9. Biosynthesis of ribitol teichoic acid and teichuronic acid in *Bacillus* spp. GalNAc, *N*-acetylgalactosamine; Glc, glucose; GlcA, glucuronic acid; GlcNAc, *N*-acetylglucosamine; Gol, glycerol; Lipid-P, undecaprenylphosphate; LU, linkage unit; ManNAc, *N*-acetylmannosamine; PP, pyrophosphate; Rol, ribitol.

appropriate sugar nucleotide. At present, there is no unifying hypothesis to account for these variations.

Teichoic acids

The pathway for the biosynthesis of ribitol teichoic acid in *B. subtilis* W23 is as shown in Fig. 9 (109, 118, 147, 322). This strain is used to exemplify the specific role of CDP-glycerol in linkage unit synthesis. The same nucleotide is used as precursor for main-chain synthesis in strain 168.

The nucleotide precursors are synthesized in the cytoplasm (299). The activities of CTP-alditolphosphate cytidilyltransferases (CDP-alditol pyrophosphorylases) are regulated by the phosphate concentration of the growth medium (47, 48). CDP-glycerol pyrophosphorylase is also rapidly and irreversibly inactivated under conditions of phosphate limitation (110, 134).

Subsequent stages of polymer synthesis involve membrane-bound enzymes and an undecaprenylphosphate anchor lipid apparently identical with that involved in peptidoglycan synthesis (109, 322, 323). The polyalditolphosphate grows by consecutive additions of alditolphosphate residues from the nucleotide precursor to the end remote from the anchor lipid (133, 148), the reverse of the direction of peptidoglycan synthesis.

In strains such as 168, in which the main polymer chain is polyglycerolphosphate, the presence of distinct linkage unit glycerolphosphate residues would not be readily detectable. However, it seems likely that the initial glycerolphosphate addition, that to the lipid-bound disaccharide phosphate, is catalyzed by an enzyme that would be common to strains having glycerol and ribitol teichoic acids and distinct from that involved in chain elongation.

Glycosylation of the polyalditolphosphate chains of teichoic acids probably proceeds concomitantly with chain elongation (135). In *B. subtilis*, the enzymes phosphoglucomutase and UDP-glucose pyrophosphorylase are required for synthesis of UDP-glucose. Glucosyltransferase, a peripheral membrane protein encoded by the *gtaA* gene (35, 102), then transfers glucosyl residues directly from UDP-glucose to the polymer chain. However, in *B. coagulans*, polyisoprenylmonophosphate glucose is an intermediate in the glucosyl transfer process (261). Teichoic acids with sugars or sugar phosphates in the main chain are synthesized on lipid linkage unit anchors in the same way polyalditol phosphates are. A study of the synthesis of the teichoic acid of *B. licheniformis* ATCC 9945 indicated that the glucosylglycerolphosphate repeating unit of the polymer was assembled on a carrier lipid before being transferred to the growing polymer chain (112), but in *B. coagulans*, a teichoic acid consisting of a galactosylglycerolphosphate repeating unit was reported to be synthesized by alternate direct transfers of galactose and glycerolphosphate from nucleotide precursors to the growing chain (320).

Addition of D-alanyl ester substituents to wall teichoic acid has not been investigated in detail, but Koch et al. (161) have presented evidence that it can occur by transacylation from lipoteichoic acid.

Teichuronic acids

Biosynthesis of a teichuronic acid similar in structure to that found in the walls of phosphate-limited *B. subtilis* has been studied in *B. licheniformis* (300) (Fig. 9). The disaccharide repeating unit is assembled on an anchor lipid, probably undecaprenylphosphate, by sequential additions of *N*-acetylgalactosamine 1-phosphate from UDP-*N*-acetylgalactosamine and glucuronic acid from UDP-glucuronic acid. New disaccharide units are incorporated at the lipid-bound end of the growing polymer chain by transfer of the nascent chain to the lipid-bound disaccharide so that the direction of chain extension is the same as that of peptidoglycan. Each round of disaccharide transfer results in the formation of an undecaprenylpyrophosphate that must be converted to the monophosphate form before it can take part in another round of synthesis. Thus, teichuronic acid synthesis is also inhibited by bacitracin. The completed teichuronic acid chain is transferred to peptidoglycan together with its terminal 1-phosphate group, which forms the link to muramic acid in the peptidoglycan. Synthesis in *B. megaterium* of a teichuronic acid that has a trisaccharide repeating unit proceeds by a similar mechanism (6). These pathways resemble those for neutral wall polysaccharides in some *Bacillus* species (218) and for the O-antigen chains of lipopolysaccharide in gram-negative bacteria.

The synthesis of teichuronic acid of *M. luteus* is organized differently. Chains are initiated by transfer of *N*-acetylglucosamine 1-phosphate (247) from its uridine nucleotide to undecaprenylphosphate. Alternating additions of glucose and *N*-acetylmannosaminuronic acid from their uridine nucleotides are made by glycosyltransferases (128) to achieve chain extension. Thus, the direction of chain growth resembles that of teichoic acid rather than of *B. licheniformis* teichuronic acid.

Lipoteichoic acids

Studies with a variety of species, including *B. subtilis* (163), *B. stearothermophilus* (45), and *B. megaterium* (185, 186), have shown that biosynthesis of lipoteichoic acid proceeds by transfer of *sn*-glycerol 1-phosphate units from phosphatidylglycerol to a glycolipid and subsequently to the glycerol-terminal end of the growing chain (97, 275). The products of the reaction are the growing polymer and a 1,2-diglyceride that may then be phosphorylated to phosphatidic acid, a precursor of phosphatidylglycerol, thus recycling the diglyceride moiety of the phospholipid. Phosphatidylglycerol is present in *B. megaterium* in two distinct pools (185, 186), of which one is metabolically stable and the other undergoes rapid turnover, presumably into lipoteichoic acid.

Glycosylation of lipoteichoic acid appears to involve an undecaprenylphosphate lipid carrier. In several *Bacillus* strains, *N*-acetylglucosaminyl substituents are added from β-*N*-acetylglucosaminylmonophosphorylundecaprenol (260), while in *B. coagulans*, galactose is added from galactosylmonophosphorylprenol (321).

The mechanism of incorporation of alanine ester into lipoteichoic acid has not been studied in *Bacillus* spp., but in *Lactobacillus* spp. (50, 51) it proceeds by a two-step reaction in which D-alanine is first activated by reaction with ATP, forming D-alanyl-AMP, and is then transferred to a lipoteichoic acid acceptor in the membrane. This appears to take place during growth of the polymer chain, since lipid-soluble alanylated intermediates have been detected (50).

The D-alanyl esters can be transferred within lipoteichoic acid and from lipoteichoic acid to wall teichoic acid by inter- and intrachain transacylation reactions (161). Perturbation of the D-alanine ester content at one location in the envelope could be translated to another location by transacylation of the alanine. Neuhaus (222) has suggested that this could be a way of transmitting a signal from one location to another.

The amount of lipoteichoic acid in the cell may vary with growth rate (88). In *B. megaterium*, lipoteichoic acid synthesis ceases abruptly at the end of exponential growth (142), but there is a temporary resumption at about stage II of sporulation. A specific membrane pool of phosphatidylglycerol is the biosynthetic precursor of glycerolphosphate units in lipoteichoic acid (185, 186), and turnover of this lipid into both membrane-bound and extracellular lipoteichoic acids has been demonstrated in some species. While radioactively labeled glycerol from phosphatidylglycerol appeared in both lipid and lipoteichoic acid in the culture fluid in *B. stearothermophilus* (45), label travelled no further than membrane-bound lipoteichoic acid in *B. subtilis* W23 (163). Taron et al. (275) found that inorganic phosphate stimulates both the synthesis of phosphatidylglycerol and the rate of chain extension of lipoteichoic acid. This may partly explain the observation that phosphate limitation of growth reduces the lipoteichoic acid content of *B. licheniformis* (43) and of group B streptococci (220).

Regulation of anionic polymer synthesis

In strain *B. subtilis* W23, the only cellular role for CDP-glycerol is in synthesis of the linkage unit. Teichoic acid is assembled on a lipid-bound linkage unit, and CDP-glycerol pyrophosphorylase is a key regulatory enzyme. This enzyme appears to be regulated in the same way in strain 168 and in *B. licheniformis*, in which CDP-glycerol is also required for synthesis of the main polymer chain (134). During transition to phosphate limitation, synthesis of CDP-glycerol pyrophosphorylase is repressed and existing enzyme is inactivated. This may indicate that there are specific linkage unit glycerolphosphate groups in those strains as well as in strain W23, since synthesis of CDP-ribitol, the precursor of the main chain in that strain, is not regulated in the same way (47). Synthesis of CDP-ribitol pyrophosphorylase of strain W23 is repressed but the enzyme is not inactivated at low phosphate concentrations. Synthesis of the linkage unit disaccharide moiety on anchor lipid is rapidly inactivated under phosphate limitation (110) but recovers on addition of phosphate, even when protein synthesis is inhibited. Phosphate does not affect the activity of the enzyme in vitro (48), suggesting a role for a transmembrane signal transduction system in its regulation.

An understanding of the genetics of teichoic acid synthesis in *B. subtilis* 168 is beginning to emerge, mainly as a result of the work of Karamata and his coworkers. Since mutation of the glycerol teichoic acid synthetic genes is lethal (199), temperature-sensitive *tag* mutants, originally described by Boylan et al. (33), have had to be used. The genes concerned with synthesis of glucosyl polyglycerolphosphate appear to be organized in two adjacent, divergently transcribed operons at about 310° on the genetic map (198). The *tagABC* operon contains genes believed to be involved in polyglycerolphosphate synthesis and export (198) and includes the *tagB* (formerly *tag-1*) locus previously described by Boylan et al. (33). The deduced products of the *tagDEF* operon are very hydrophilic. *tagD* encodes CDP-glycerol pyrophosphorylase (235), while *tagF* (previously designated *rodC*) encodes a polyglycerolphosphate glycerolphosphotransferase (232). The latter enzyme, in its role as the main chain polymerase of teichoic acid synthesis, is membrane associated but does not contain motifs typical of membrane-bound enzymes (232).

The work of Young et al. (326) previously demonstrated the roles of three genes, which they designated *gtaABC*, in glucosylation of the glycerol teichoic acid of strain 168. *gtaA* (renamed *tagE* [198]) and *gtaB* map close to the *tag* locus, and their products have been identified as polyglycerolphosphate glucosyltransferase and UDP-glucose pyrophosphorylase (234). *gtaC* is located at 77° on the genetic map, distant from the *tag* locus, but is also essential for glucosylation since it encodes phosphoglucomutase, which is responsible for the synthesis of glucose 1-phosphate, the precursor of UDP-glucose (188). While the *gta* genes are not essential for growth, their mutation leads to synthesis of wall teichoic acid lacking glucosyl substituents. *gta* mutants are consequently more resistant to the action of the autolytic L-alanine amidase and exhibit various degrees of chaining due to their failure to separate after cell division (see below).

The genes encoding the membrane-bound regulatory enzymes of linkage unit synthesis have not yet been identified, and genetic analysis of the *tag* region has so far yielded little information about regulation at the genetic level. Wagner and Stewart (296) proposed that *tagE* and *tagF* possessed a vegetative (σ^A) promoter, but this could not be confirmed by other work (198). Mauel et al. mentioned the presence of putative Pho boxes in the *tag* locus, indicative of transcriptional regulation in response to deprivation of environmental P_i (198).

INCORPORATION OF ANIONIC POLYMERS AND PEPTIDOGLYCAN

Linkage of Anionic Polymers and Peptidoglycan

Teichoic acid is always linked to peptidoglycan synthesized at the same time (197). Though there has been no detailed study of the organization of wall synthetic enzymes at the molecular level, diverse investigations (4, 83, 178, 197, 304) lead to the conclusion that the cytoplasmic membrane contains ordered assemblies of the enzymes of anionic polymer and peptidoglycan synthesis together with shared anchor lipids. Attachment of anionic polymers to peptidoglycan almost certainly occurs at the outer surface of the membrane. Soluble, un-cross-linked peptidoglycan released from bacteria treated with penicillin does not have teichoic acid attached to it, although teichoic acid synthesis is not directly inhibited by the antibiotic (86, 305). The occurrence of teichoic acid and teichuronic acid on separate peptidoglycan chains in *B. licheniformis* (300) provides further evidence of a high level of organization within the biosynthetic complex.

Since the nucleotide precursors of the wall polymers are cytoplasmic components, wall synthesis must be a transmembrane process involving vectorial translocation with reactions at both faces of the membrane. Three potential mechanisms can be envisaged: (i) undecaprenylphosphate acts as a carrier for translocation of precursor subunits across the membrane, with polymerization at the outer surface of the membrane; (ii) polymerization occurs at the inner surface of the membrane, and the polymers are extruded through the membrane; and (iii) precursor subunits are translocated through the membrane via specific membrane permeases (gated pores), with polymerization at the outer surface of the membrane.

Physical studies of polyisoprenylphosphate lipids in lipid bilayer membranes have failed to detect any potential for flip-flop motion (115, 201, 309). The view that the undecaprenylphosphate acts as a translocating carrier is thus not supported by the available evidence. If polymerization occurred at the inner surface of the membrane, the hydrophilic polymer so formed would have to be extruded through the hydrophobic lipid barrier of the membrane. This would

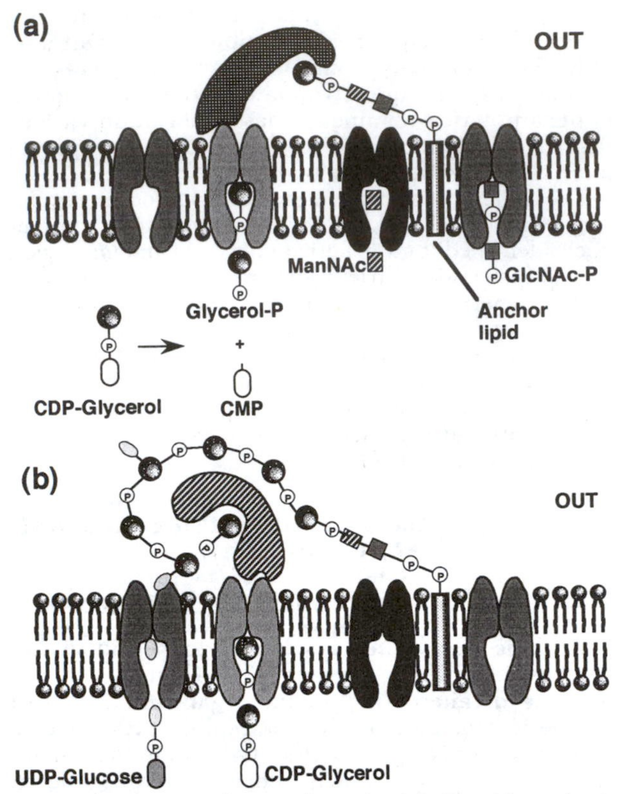

Figure 10. Transmembrane glycerol teichoic acid synthesis involving undecaprenylphosphate anchor lipid and the proposed gated-pore protein complexes. (a) Assembly of linkage unit; (b) glycerolphosphate chain extension and glycosylation.

presumably require an appropriate transmembrane protein complex. Un-cross-linked lipid-bound oligomers of peptidoglycan have been detected in *B. megaterium* and *E. coli* (94, 213). It has been proposed (262) that these are intermediates in the process of extrusion through the membrane, although they could equally well be oligomers being assembled at the outer surface of the membrane but not yet cross-linked into the wall.

Cross-linking can occur only at the outer surface of the membrane when new material is incorporated into the wall through the transpeptidase activity of high-molecular-weight PBPs. Since these are bifunctional enzymes, with both transglycosylase and transpeptidase activities, it seems likely that transglycosylation also occurs at the outer surface of the membrane. Direct support for this comes from the demonstration that enzymes involved in the polymerization of peptidoglycan and teichoic acid are present at the outer surfaces of the membranes in cells of *B. subtilis* W23 that have undergone partial wall autolysis (117).

We therefore think that wall polymer synthesis takes place at the outside surface of the membrane and that precursors must be transported to this location (Fig. 10). Since it is unlikely that the undecaprenylphosphate can transport precursors, a transmembrane protein complex is the more probable mediator of group translocation of the precursors of both anionic polymer and peptidoglycan. The undecaprenylphosphate acts as an acceptor of these units and anchors them and the growing chains to the enzyme complex in the membrane.

As the new material becomes linked to the wall, and so immobilized, the synthetic complex will be forced to move within the plane of the membrane so that the nascent chain can align with successive acceptor peptides at the inner surface of the wall (see below). For enzyme complexes underlying cylindrical wall regions, this movement must presumably follow a spiral path at an angle to the short axis of the cell (see above).

Incorporation of Peptidoglycan–Anionic-Polymer Complex into Wall

The final stage in the incorporation of peptidoglycan or peptidoglycan–anionic-polymer complex is accomplished by transpeptidation reactions (Fig. 11; see above). In *B. megaterium* (100) and *B. licheniformis* (301), incoming peptides on the nascent glycan-peptide act as donors and become linked to the ε-amino groups of diaminopimelic acid residues already in the wall. This same direction of cross-linking has been shown in *Salmonella typhimurium* (61) and *E. coli* (104), though the opposite direction has been reported in *Gaffkya homari*, in which the amino groups on the incoming new material act as acceptors in transpeptidation reactions involving pentapeptides present in older wall (108).

Attempts to gain an understanding of the kinetics of cross-link formation have been made by reference to the known kinetics of random and monomer addition polymerization reactions in solution. The size distribution of muropeptides isolated from muramidase-digested walls of *E. hirae* was closer to that predicted for a monomer addition polymerization mechanism (226), whereas similar analysis of *S. aureus* gave results considered to be in accord with an overall random polymerization process, possibly involving an initial monomer addition followed by a random secondary cross-linking process (270). An increase in cross-linking following incorporation has been reported for *S. aureus* (283), *B. megaterium* (90), *E. hirae* (71), and *E. coli* (70). Recent studies of cross-link formation in *B. megaterium* (96) have shown that the increase in cross-linking following incorporation could be explained on the basis that incorporation of new material must inevitably increase the number of cross-linkages associated with older material already in the wall. This last study also questioned the applicability of polymerization kinetics derived for homogeneous catalysis in solution, in which all oligomers present are equally accessible to monomer, to wall assembly processes, in which incoming monomers are able to link only to those peptides adjacent to the membrane. A model (Fig. 12) in which cross-links are formed only when new material is incorporated at the inner surface of the wall has been developed. The pattern of cross-link formation may reflect both enzymic and steric influences and could differ in different parts of the wall (i.e., in septal and cylindrical walls).

GROWTH AND CELL DIVISION

Growth of individual bacterial cells is approximately exponential (60, 195). The increase in cytoplas-

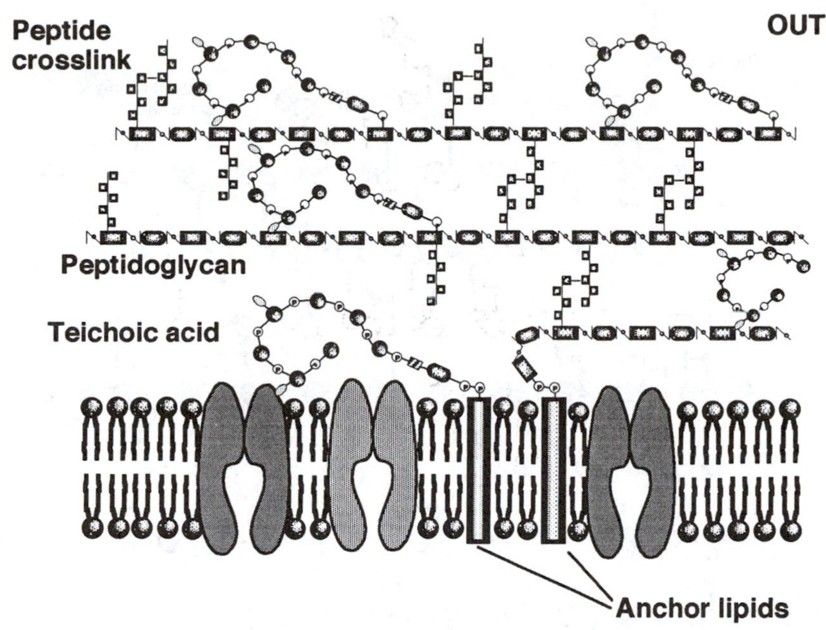

Figure 11. Formation of teichoic acid-peptidoglycan complex and its incorporation into the cell wall.

mic mass due to growth requires a corresponding increase in cell volume that has to be accommodated by an increase in wall. The surface-to-volume ratio changes during the cell cycle: as the cells get bigger, they change shape, and the rate of wall synthesis varies accordingly. Consequently, the ratio of wall synthesis to cytoplasm synthesis is not constant but can be understood in terms of the changing geometry of exponentially growing individual cells. It is now accepted that at least two processes, cylinder elongation and septation, are involved in wall assembly in rod-shaped bacteria. Cell shape reflects the balance between these processes (184).

The concept (152) that wall synthesis and growth are driven by stresses set up by turgor has greatly illuminated current understanding (116). Koch's hypothesis of surface stress (152) has been examined by Thwaites (277), who has carried out a stress analysis for the cylinder walls of rod-shaped cells and developed a model incorporating stresses due to turgor and electrostatic pressures that explains how wall thickness and diameter can remain constant during elongation. The recognition that new wall must be incorporated under stress-free conditions and the proposal (152) that old wall must be underlain or bridged by new material before it can be safely cut are consistent with experimental evidence and offer a unifying basis for the understanding of wall assembly for both gram-positive and gram-negative bacteria.

Wall Assembly in *Bacillus* spp.

The principal features of wall assembly in *Bacillus* spp. are as follows (Fig. 13).

(i) Cylindrical extension proceeds through the incorporation of new wall material at many sites (69) to form a layer covering the entire inner surface of the cylindrical wall (8). To allow an increase in cell volume during growth, this material is underlain by newer wall material, pushed outwards, and stretched, eventually reaching the surface, where it is removed by "turnover" (18, 80, 152, 231). Though experimental observations implicate autolysins in turnover, these enzymes might not be essential for elongation, since overstretched peptide cross-linkages in the outer wall layers could in principle be broken purely by physical forces. It is possible (157) that autolysins preferentially cut bonds that are already stretched, since this would reduce the activation energy of hydrolysis.

(ii) Septation proceeds from an annular zone to give a flat disk of wall material bisecting the cell. During cell separation, this disk is cleaved to form two new polar caps. The symmetrical cleavage of the septal disk implies a high degree of regulation of autolytic activity, though the mechanism governing this cleavage is not known. One suggestion is that proton gradients generated by the activated cell membranes (144) on either side of the septum (151) inhibit autolysins except in the region of lowest proton concentration equidistant from the two membranes. A supplementary form of control may involve "smart" autolysins (see below) able to cut stressed bonds that develop as poles form (159). The change in shape following cleavage is due to turgor, which stretches the wall material to form the new, nearly hemispherical polar caps. This change can take place even in the absence of further synthesis of wall material (154, 155).

(iii) New wall material is incorporated into polar caps by an inside-to-outside mechanism but more slowly than into the cylinder, so polar caps turn over less rapidly (18, 54, 217, 273).

Koch (152) proposed that the poles act as fixed supports to permit cylindrical elongation. However, poles are not completely conserved, and old wall is eventually removed after several generations (54). The forces driving longitudinal extension have been analyzed by Thwaites (277). Because the cylinder ends are

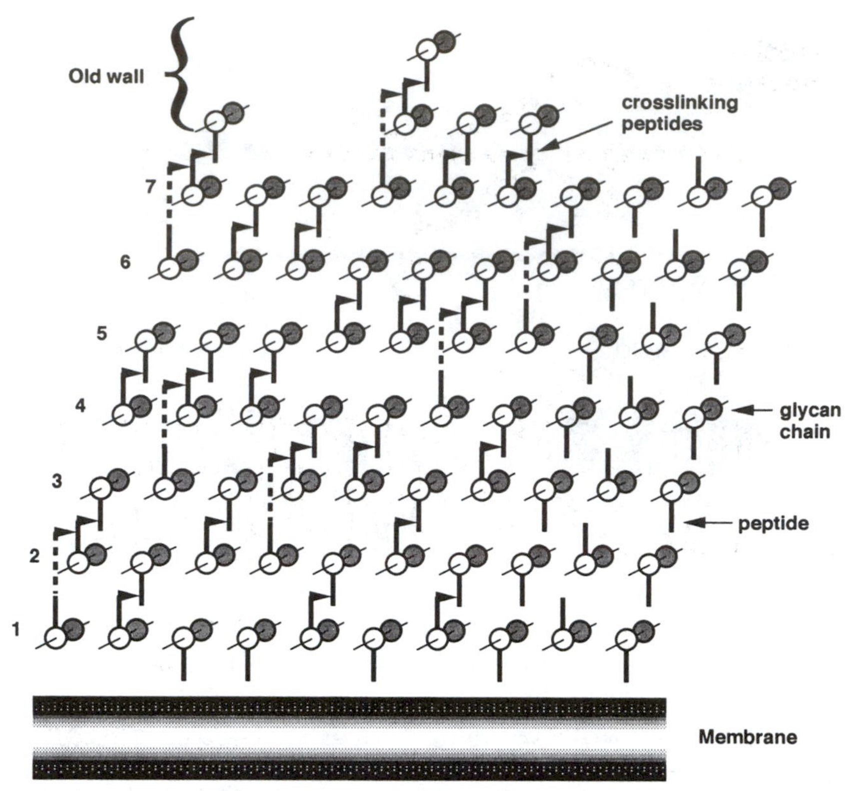

Figure 12. Pattern of cross-linking in a growing cell wall. Layer 1 shows a pattern of incorporation of new material in which, for every 10 peptide monomers incorporated, 6 remain as monomers, 3 cross-link to existing wall monomers to form dimers, and 1 cross-links to an existing wall dimer to form a trimer. When a further layer of material is incorporated by the same pattern of incorporation, layer 1 moves up to become the second layer. Three of the original six monomers are converted to dimers, and one of the original three dimers becomes a trimer. Consequently, the layer now contains three monomers and peptides forming part of five dimers and two trimers. When the next layer is incorporated, the layer moves up to become the third layer, and in the process, one of its dimers becomes a trimer, so that the now-mature layer contains three monomers and peptides forming part of four dimers and three trimers. Further incorporation of new material has no effect on the pattern of cross-linking involving this layer, and its pattern of cross-linking reflects that of the mature cell wall. The apparent increase in cross-linking is thus a direct consequence of the initial incorporation. To allow the process to be depicted clearly, cross-linkages are shown perpendicular to glycan sheets. In the wall, however, peptides radiate from the helically twisted glycan chain and so connect chains at various angles: these chains are unlikely to be arranged into the regular sheets shown, for simplicity, here.

closed by the poles, turgor produces both longitudinal and hoop (i.e., circumferential) stress components, while electrostatic repulsion produces longitudinal stress. Though the longitudinal and hoop stresses are of the same order, bacilli increase in length while their diameters remain constant. A likely explanation is that because of the orientations of the glycan chains (see above), the wall material is much stiffer in the hoop direction than longitudinally (277).

A consequence of these assembly processes is that all new wall material is incorporated into locations (the inner region of cylindrical wall, the septum, and the inner region of the pole) at which it does not experience stress until some time after it has been integrated into the cross-linked structure of the wall. This conclusion is supported by Mendelson's studies on mutants of *B. subtilis* that grow as twisted macrofibers because of their failure to separate following cell division (205, 206). Macrofibers treated with lysozyme undergo a sequence of relaxation motions, showing that the structures are under stress and that this stress is maintained by peptidoglycan. Twist reflects the anisotropy in the arrangement of peptidoglycan in the wall (see above) and can range from tight right-handed to tight left-handed according to growth conditions. Twist is also sensitive to agents that affect peptidoglycan synthesis (207, 208). Reversals of twist observed following changes in growth conditions (81) take place only after an interval that corresponds to the time taken for new material to move from the inner to the outer regions of the wall. Twist due to the anisotropic arrangement of peptidoglycan thus arises only when that peptidoglycan is in the older, i.e., stressed, region of the wall.

Wall Assembly in Gram-Positive Cocci

Wall growth in gram-positive cocci usually proceeds by septation, though an inside-to-outside replacement of old wall may take place in cocci that undergo wall turnover. Some cocci appear to retain a potential for lateral wall assembly that may be expressed when septum formation is inhibited (184).

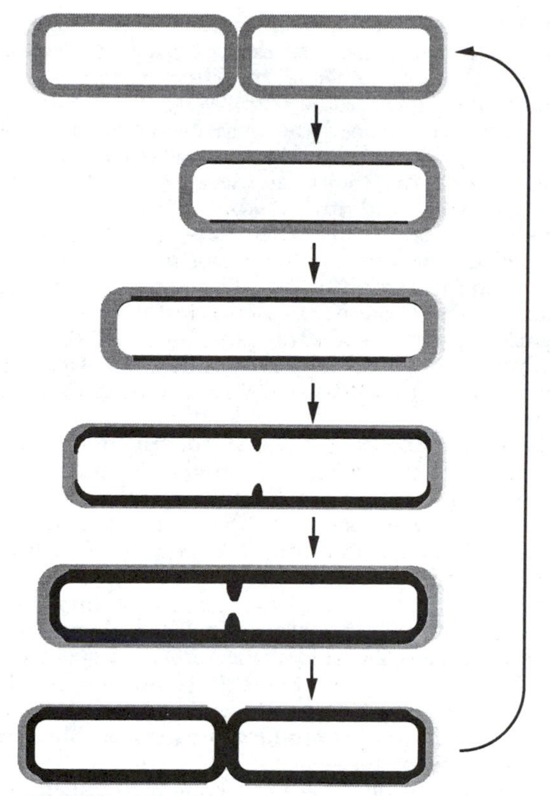

Figure 13. Cell wall growth in *B. subtilis*. Successive layers of cell wall are shown by progressively darker shading.

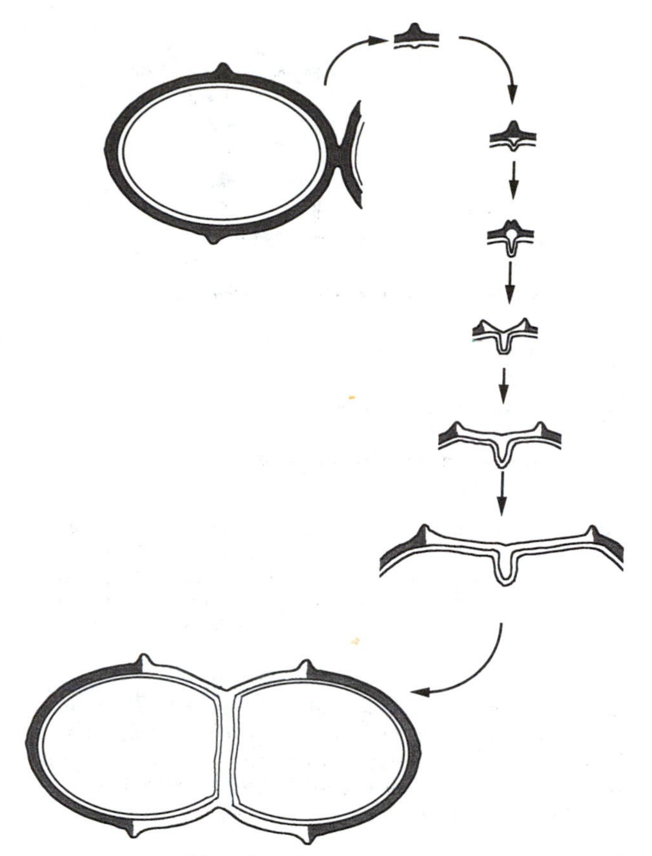

Figure 14. Assembly of cell wall in *E. hirae*. Older wall material is shaded. Modified from Harold (116).

The classic work by Cole and Hahn (55) showed that streptococci grow from an annular growth zone, so old and new areas of wall could be distinguished by using fluorescent antibody. Subsequent studies (187) have confirmed that gram-positive cocci grow by the zonal assembly of an internal septum that is ultimately split by controlled autolysis to give separate daughter cells. Cells of *E. hirae* are spheroid, with raised equatorial wall bands that mark the sites of the future septa. Shockman and his colleagues have shown that the wall bands split during growth (Fig. 14) and that a septum is formed between the daughter bands (65, 263). This septum is progressively split as the cell enlarges. New wall material is added mainly at the growing edge of the septal ring, but some new wall is also incorporated into the separating pole. The shapes of the resulting poles are consistent with Koch's theory of surface stress (158). Wall assembly thus takes place under stress-free conditions, and wall material has to resist turgor only after it is externalized by division.

Coordination of Cell Division and Chromosome Segregation

The mechanism of chromosome segregation in bacteria is not yet understood, and little information is available on the way in which it is coordinated with cell division. The replicon hypothesis (140) proposes that chromosome segregation is effected through the attachment of daughter chromosomes on either side of a membrane growth zone. Eberle and Lark (76) suggested that segregation of chromosomes in growing *Bacillus* spp. might involve their attachment to surface structures that are additional to the membrane and provide a degree of rigidity to facilitate segregation. Separation of chromosome attachment sites in the wall was a feature, and one of the principal attractions, of the early zonal-growth models proposed for these *Bacillus* spp. (252). The finding that only polar wall material is conserved (even partially) excluded these models, though it raised the possibility that the cell poles might be implicated in anchoring and segregating the chromosomes (160). Both origin and terminus regions of chromosomes are associated with wall in *B. subtilis* (73, 272), but segregation of such attachment sites would not necessarily serve to separate the chromosomes, and the forces responsible for this are not yet understood.

AUTOLYTIC ENZYMES AND WALL TURNOVER

The end of exponential growth in batch cultures of *B. subtilis* may be followed by marked lysis of the culture. This is due to the action of autolysins, enzymes that hydrolyze either the glycan or the peptide moieties of peptidoglycan. The widespread occurrence of such potentially lethal enzymes indicates a functional importance that is borne out by studies showing that cell separation is blocked in both rods and cocci defective in autolysin action so that mutants that lack or are resistant to autolysins (Lyt^{-}) grow as

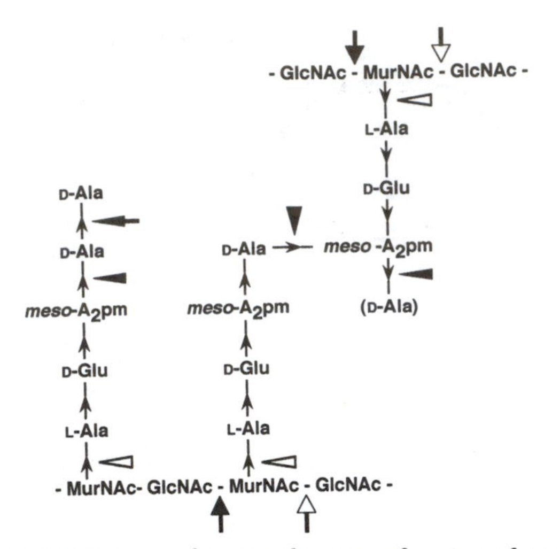

Figure 15. Diagram showing the sites of action of specific autolysins. Symbols: ◁, *N*-acetylmuranoyl-L-alanine amidase; ◀, carboxypeptidase; ↓, endo-β-*N*-acetylglucosaminidase; ⇩, endo-β-*N*-acetylmuramidase; ◀ and ▼, endopeptidase. *meso*-A_2pm, *meso*-diaminopimelic acid.

long multiseptate filaments (204, 280, 286) or as "packets" of coccal cells (319). Autolytic enzymes are involved in the development of competence, in sporulation and germination, and in the bactericidal effect of β-lactam antibiotics.

Autolytic Enzymes

A variety of different lysins have been found in bacteria (Fig. 15). The major autolysin in *B. subtilis* is the *N*-acetylmuramoyl-L-alanine amidase that splits the linkage between glycan and peptide. Amidase from *B. subtilis* 168 exists as a dimer with a 50-kDa subunit and a pH optimum of about 8.0. This enzyme occurs together with an 80-kDa modifier protein that interacts with the amidase and modulates its activity (126). Amidases from other strains differ in pH optima and molecular masses even though they hydrolyze similar linkages. In strain 168, the amidase occurs together with a 90-kDa monomeric glucosaminidase (126, 246).

Several classes of mutation produce phenotypically Lyt$^-$ cells in *B. subtilis*. Mutations causing a deficiency in autolytic enzyme production were originally described as *lyt-1* and *lyt-2* (82) but have now been shown to be alleles of *flaD* (233). These mutants grow as multiseptate filaments devoid of flagella. *flaD* has been cloned and sequenced by Sekiguchi et al. (255). It is identical to the *sin* gene (98), which in multiple copy suppresses sporulation and when mutated leads to hyperproduction of protease. *flaD* mutants are deficient in both amidase and *N*-acetylglucosaminidase. A homologous locus has been identified in *B. licheniformis* (255).

It is now clear that *flaD* (*sin*) does not in itself encode an autolytic enzyme (233). However, there have been three reports of the cloning and sequencing of structural genes for autolytic amidases. A mixed-oligonucleotide probe derived from the amino acid sequence of the major *B. subtilis* L-alanine amidase (169) has been used to identify a locus containing the *cwbA* and *cwlB* genes. The derived products resemble, respectively, the 80-kDa modifier protein and the major autolytic amidase (126). Margot and Karamata (189) have also cloned and sequenced a locus containing genes, which they designate *lytA*, *lytB*, and *lytC*, that encode the major amidase (*lytC*), its modifier protein (*lytB*), and an 11.2-kDa acidic protein of unknown function (*lytA*). They have also mapped and sequenced the gene for the major autolytic *N*-acetylglucosaminidase (*lytD*) (190). The deduced sequence of each of these proteins contains a putative N-terminal signal sequence, and each gene has a σ^D-dependent promoter. Inactivation of the amidase (*lytC*) gene in *B. subtilis* leads to a substantial reduction in total autolytic activity.

Sekiguchi and his colleagues identified a gene designated *cwlA* that, when expressed in *E. coli*, yielded a 30-kDa protein lytic for *B. subtilis* walls (169). *cwlA* shows no homology with the gene encoding muramoyl-L-alanine amidase in *S. pneumoniae*, but its derived C-terminal sequence is homologous with the N terminus of the D-carboxypeptidase of *Streptomyces albus* G. Considerable N-terminal homology with an autolysin from an unidentified *Bacillus* species was also shown (236). The 30-kDa protein has autolytic amidase activity (92), but insertional inactivation of the gene in *B. subtilis* produced no measurable change in overall autolytic activity (169).

Surprisingly, although insertional mutagenesis of the major amidase gene, *cwlB* (*lytC*), leads to loss of about 97% of the cell's autolytic amidase activity (189), it causes no detectable changes in growth or morphogenesis. Similar results have been reported for the 30-kDa amidase and the 90-kDa *N*-acetylglucosaminidase. Therefore, conclusions drawn about the roles of these enzymes in cell separation need to be treated with caution (93). It is possible that there are multiple wall-bound autolysins and that cell separation can be accomplished by any of a number of them. Recent work has demonstrated multiple autolysins in *B. subtilis* 168 (93), *B. subtilis* ATCC 6633, and *S. aureus* (179).

In *Bacillus* spp., some of these enzymes are likely to be developmentally controlled. Foster (93) has shown, for example, that a 30-kDa autolysin distinct from the *cwlA* gene product described above is particularly active toward spore cortex peptidoglycan and that its specific activity in the cell increases markedly during sporulation, just before mother cell lysis.

Autolysins generally show high affinity for homologous walls, and in several cases, the presence of a wall anionic polymer is required for hydrolysis of peptidoglycan. In *B. subtilis*, interaction of the major amidase with wall is affected by the modifier protein and the type of teichoic acid present (127). Autolysin in *B. licheniformis* binds only to walls that contain teichuronic acid. Mutants lacking this polymer (243) are resistant to autolysin and consequently grow in long, multiseptate filaments (244) (Fig. 16) similar to those found in *B. subtilis* (206, 280) (see above). A mutant of *M. luteus* deficient in wall teichuronic acid similarly grew in clusters of unseparated cells (319).

Perhaps the best studied and certainly one of the most striking examples of the interaction of autolysin and teichoic acid is that observed in *Streptococcus*

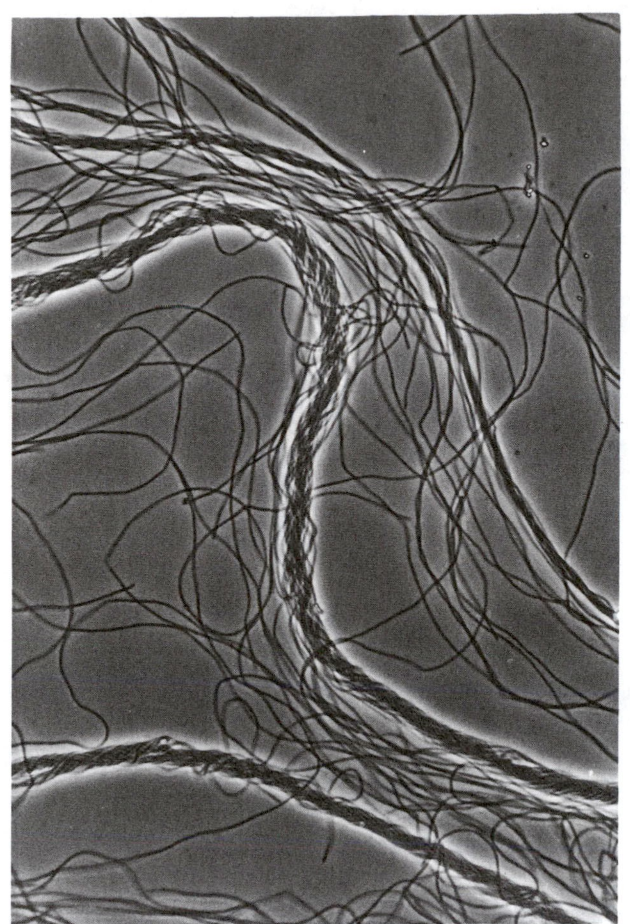

Figure 16. Micrograph of autolysin-deficient Nov-12 strain *B. licheniformis* organism, showing long multiseptate filaments.

pneumoniae (288), in which autolysis requires the presence of choline phosphate substituents in the wall teichoic acid. Bacteria grown in such a way that they incorporate ethanolamine rather than choline into their teichoic acid are resistant to autolysis, grow as long multiseptate chains, are not transformable, and do not lyse with penicillin or in the presence of deoxycholate. Interaction of autolysin with the choline phosphate substituents is required both to bind the autolysin to the wall (101, 192) and to activate it. Choline phosphate occupies an allosteric site at the C-terminal end of the autolysin and effects a conformational change that exposes the hydrolytic site (130). Activation of autolysin can also be effected by choline-containing lipoteichoic acid, and it is possible that this acid functions as the in vivo activator (34). When ethanolamine replaces choline in the lipoteichoic acid, inactive autolysin is retained in the cytoplasm, and it is possible (34) that export through the membrane and into the wall requires choline-lipoteichoic acid. Control of autolysin action might therefore consist of controlled export, activation, and localization, all of which are effected by choline attached to teichoic acid.

Charged amino groups in teichoic acids in other bacteria are provided by D-alanine rather than choline substituents, and it is possible that these also affect the interaction between teichoic acids and autolysin. Thus, in *S. aureus*, autolysin is inhibited by isolated lipoteichoic acid but only if the alanine ester content is low (89).

Role and Regulation of Autolytic Activity

Given their potential for destroying the cell, autolysins are likely to be tightly controlled in growing bacteria. Their activities can be affected by the ionic environment (49) and by secreted proteases (143) as well as by anionic polymers. In *B. subtilis*, an important overall regulatory control seems to be associated with the energized membrane (144). Lysis does not occur during growth, but conditions that do not permit energy generation (deprivation of oxygen or energy source, presence of metabolic poisons) result in rapid lysis of the cells.

The actions of autolysins are responsible both for the cleavage of the septum and for the excision of old material from the walls in growing bacteria. The mechanism and regulation of septal cleavage in *B. subtilis* have been explored by Koch and Burdett (154, 155; see above). The action of exogenous autolysins or lysozyme on filaments of autolysin-deficient mutants gives cells that have normally shaped poles (38, 155). Apparently, therefore, normal cleavage of the septum does not require production of autolysins at specific locations. It is possible that the septum contains structural discontinuities that mark the cleavage site, but no evidence for this is available, and cleavage sites may be determined in other ways. Koch has pointed out that protons extruded from the active membrane generate a pH gradient in the wall so that the midpoint of septal wall, equidistant from the underlying membranes of the two daughter cells, would be the region least affected by the inhibitory effect of low pH and thus the region most readily cleaved. Alternatively, septal cleavage may involve smart autolysins that cut stressed bonds that develop as the poles form (150).

Turnover

While cell separation is of sufficient importance on its own to have ensured the evolution of an appropriate mechanism, lysins may also function in wall growth or processing, particularly in rods, where the inside-to-outside mechanism requires stretching and eventual rupture of old peptidoglycan in cylindrical wall. Koch (159) has proposed that this stretching facilitates lysis, and he postulates that smart autolysins recognize (and hydrolyze) only stretched bonds. Whether smart or otherwise, lysins are involved in the shedding of wall material during growth. This process of turnover was first described in *B. megaterium* (46) and has been observed in growing cultures of various gram-positive bacteria and also in gram-negative bacteria, cyanobacteria, and yeasts (72). Turnover in *B. subtilis* is effected by *N*-acetylmuramoyl-L-alanine amidases (196, 197) and involves a more rapid release of material from cylindrical than polar regions (18, 217, 273), though poles do turn over (54, 150), as does wall material in *S. aureus* (31, 315). Turnover involves release of material from the outer region of cylindrical wall, i.e., that part of the wall in which material is stretched and is also distant from the active membrane. Factors similar to those responsible for controlling septal cleavage may thus also be invoked to explain how gram-positive rods maintain constant wall thickness. Turnover kinetics have been examined by several workers, and both mechanistic (67, 156, 230) and purely mathematical (10) models have been described.

In wild-type strains of *B. subtilis*, old wall material is shed at rates of about 50% per generation, and up to one-third of the total wall material present in growing cultures may be in the form of soluble turnover product. Release of such large proportions of wall in soluble form may have considerable significance for pathogens in relation both to the establishment of infection and to its consequences for the host (32). Wall turnover in encapsulated strains of *S. aureus* is accompanied by release of the capsular polysaccharide concomitantly with wall material (114).

Small amounts of cell-free autolysin can be detected in cultures of *B. subtilis*. The secreted autolysin is susceptible to inactivation by protease. Protease-deficient mutants demonstrate higher turnover rates than protease-sufficient strains (143) and have a tendency to lyse (64). In contrast, protease-hyperproducing strains show diminished turnover (143). Secreted lysins may give a competitive advantage to the producing organism. This was suggested in an early study by Davie and Brock (66), who showed that *Streptococcus zymogenes* secretes a lysin to which it is resistant, apparently because of the inhibitory effect of its alanylated teichoic acid. The corresponding teichoic acids from sensitive streptococci lacked alanyl ester residues.

REFERENCES

1. **Aasjord, P., and A. Grov.** 1980. Immunoperoxidase and electron microscopy studies of staphylococcal lipoteichoic acid. *Acta Pathol. Microbiol. Scand. Sect. B* **88:** 47–52.
2. **Abrams, A.** 1958. O-Acetyl groups in the cell wall of *Streptococcus faecalis*. *J. Biol. Chem.* **230:**949–959.
3. **Amako, K., and A. Takade.** 1985. The fine structure of *Bacillus subtilis* revealed by the rapid-freezing and substitution-fixation method. *J. Electron Microsc.* **34:** 13–17.
4. **Anderson, R. G., H. Hussey, and J. Baddiley.** 1972. The mechanism of wall synthesis in bacteria. *Biochem. J.* **127:**11–25.
5. **Aono, R.** 1989. Characterization of cell wall components of the alkalophilic *Bacillus* strain C-125: identification of a polymer composed of polyglutamate and polyglucuronate. *J. Gen. Microbiol.* **135:**265–271.
6. **Arakawa, H., and E. Ito.** 1986. Biosynthetic studies on N-acetylmannosaminuronic acid containing teichuronic acid in *Bacillus megaterium*. *Can. J. Microbiol.* **32:**822–825.
7. **Araki, Y., S. Oba, S. Araki, and E. Ito.** 1980. Enzymatic deacetylation of N-acetylglucosamine residues in cell wall peptidoglycan. *J. Biochem.* **88:**469–479.
8. **Archibald, A. R.** 1976. Cell wall assembly in *Bacillus subtilis*: development of bacteriophage-binding properties as a result of the pulsed incorporation of teichoic acid. *J. Bacteriol.* **127:**956–960.
9. **Archibald, A. R.** 1980. Phage receptors in gram positive bacteria, p. 5–26. *In* L. L. Randall and L. Philipson (ed.), *Receptors and Recognition*, series B, vol. 7. *Virus Receptors*. Chapman and Hall, London.
10. **Archibald, A. R.** 1987. Continuous culture, growth kinetics and wall turnover, p. 37–54. *In* I. C. Hancock and I. R. Poxton (ed.), *Bacterial Cell Surface Techniques*. John Wiley & Sons, London.
11. **Archibald, A. R.** 1988. Bacterial cell wall structure and the ionic environment, p. 159–173. *In* R. Whittenbury, G. W. Gould, J. G. Banks, and R. G. Board (ed.), *Homeostatic Mechanisms in Micro-Organisms*. Bath University Press, Bath, England.
12. **Archibald, A. R.** 1989. The *Bacillus* cell envelope, p. 217–254. *In* C. R. Harwood (ed.), *Bacillus*. Plenum Press, New York.
13. **Archibald, A. R., J. J. Armstrong, J. Baddiley, and J. B. Hay.** 1961. Teichoic acids and the structure of bacterial cell walls. *Nature* (London) **191:**570–572.
14. **Archibald, A. R., J. Baddiley, and N. L. Blumson.** 1968. The teichoic acids. *Adv. Enzymol.* **30:**223–245.
15. **Archibald, A. R., J. Baddiley, and J. E. Heckels.** 1973. Molecular arrangement of teichoic acid in the cell wall of *Staphylococcus lactis*. *Nature* (London) **241:**29–31.
16. **Archibald, A. R., J. Baddiley, and S. Heptinstall.** 1973. The alanine ester content and magnesium binding capacity of walls of *Staphylococcus aureus* H grown at different pH values. *Biochim. Biophys. Acta* **291:**629–634.
17. **Archibald, A. R., and H. E. Coapes.** 1971. The wall teichoic acids of *Lactobacillus plantarum* NIRD C106. Location of the phosphodiester groups and separation of the chains. *Biochem. J.* **124:**449–460.
18. **Archibald, A. R., and H. E. Coapes.** 1976. Bacteriophage SP50 as a marker for cell wall growth in *Bacillus subtilis*. *J. Bacteriol.* **125:**1195–1206.

18a. **Archibald, A. R., I. C. Hancock, C. R. Harwood, and P. Brown.** Unpublished data.

18b. **Archibald, A. R., I. C. Hancock, C. R. Harwood, and N. Prayitno.** Unpublished data.

19. **Archibald, A. R., and G. H. Stafford.** 1972. A polymer of N-acetylglucosamine 1-phosphate in the wall of *Staphylococcus lactis* 2102. *Biochem. J.* **130:**681–690.
20. **Armstrong, J. J., J. Baddiley, and J. G. Buchanan.** 1960. Structure of the ribitol teichoic acid from walls of *Bacillus subtilis*. *Biochem. J.* **76:**610–621.
21. **Baddiley, J., I. C. Hancock, and P. M. A. Sherwood.** 1973. X-ray photoelectron studies of magnesium ions bound to the cell walls of Gram-positive bacteria. *Nature* (London) **243:**43–45.
22. **Barnickel, G., H. Labischinski, H. Bradaczek, and P. Giesbrecht.** 1979. Conformational energy calculation of the peptide part of murein. *Eur. J. Biochem.* **95:**157–165.
23. **Barnickel, G., D. Naumann, H. Bradaczek, H. Labischinski, and P. Giesbrecht.** 1983. Computer-aided molecular modelling of the three-dimensional structure of bacterial peptidoglycan, p. 61–66. *In* R. Hakenbeck, J. V. Höltje, and H. Labischinski (ed.), *The Target of Penicillin*. Walter de Gruyter, Berlin.
24. **Batley, M., J. W. Redmond, and A. J. Wicken.** 1987. Nuclear magnetic resonance spectra of lipoteichoic acid. *Biochim. Biophys. Acta* **901:**127–133.
25. **Beveridge, T. J.** 1981. Ultrastructure, chemistry and function of the bacterial wall. *Int. Rev. Cytol.* **72:**229–317.
26. **Beveridge, T. J.** 1989. Role of cellular design in bacterial metal accumulation and mineralisation. *Annu. Rev. Microbiol.* **43:**147–171.
27. **Beveridge, T. J.** 1989. The structure of bacteria, p. 1–65. *In* J. S. Poindexter and E. R. Leadbetter (ed.), *Bacteria in Nature*. Plenum Publishing Corp., New York.
28. **Beveridge, T. J., and L. L. Graham.** 1991. Surface layers of bacteria. *Microbiol. Rev.* **55:**684–705.

29. **Birdsell, D. C., R. J. Doyle, and M. Morgenstern.** 1975. Organization of teichoic acid in the cell wall of *Bacillus subtilis*. *J. Bacteriol.* **121:**726–734.
30. **Blumberg, P. M., and J. L. Strominger.** 1974. Interaction of penicillin with the bacterial cell: penicillin-binding proteins and penicillin-sensitive enzymes. *Bacteriol. Rev.* **38:**291–335.
31. **Blumel, P., W. Uecker, and P. Giesbrecht.** 1979. Zero order kinetics of cell wall turnover in *Staphylococcus aureus*. *Arch. Microbiol.* **121:**103–110.
32. **Blumel, P., W. Uecker, and P. Giesbrecht.** 1981. *In vitro* studies on the possible role of cell wall turnover in *Staphylococcus aureus* during infection, p. 435–439. *In* J. Jeljasewicz (ed.), *Proceedings of the 4th International Conference of Staphylococci and Staphylococcal Infections*. Gustav Fischer Verlag, Stuttgart, Germany.
33. **Boylan, R. J., N. H. Mendelson, D. Brooks, and F. E. Young.** 1972. Regulation of the cell wall: analysis of a mutant of *Bacillus subtilis* defective in biosynthesis of teichoic acid. *J. Bacteriol.* **110:**281–290.
34. **Briese, T., and R. Hakenbeck.** 1985. Interaction of pneumococcal amidase with lipoteichoic acid and choline. *Eur. J. Biochem.* **146:**417–427.
35. **Brooks, D., L. L. Mays, Y. Hatefi, and F. E. Young.** 1971. Glucosylation of teichoic acid: solubilization and partial characterization of the uridine diphosphoglucose:polyglycerophosphate teichoic acid glucosyltransferase from membranes of *Bacillus subtilis*. *J. Bacteriol.* **107:**223–229.
36. **Buchanan, C. E.** 1987. Absence of penicillin-binding protein 4 from an apparently normal strain of *Bacillus subtilis*. *J. Bacteriol.* **269:**5301–5303.
37. **Buchanan, C. E., and M.-L. Ling.** 1992. Isolation and sequence analysis of *dacB*, which encodes a sporulation-specific penicillin-binding protein in *Bacillus subtilis*. *J. Bacteriol.* **174:**1717–1725.
38. **Burdett, I. D. J., and A. L. Koch.** 1984. Shape of nascent and completed poles of *Bacillus subtilis*. *J. Gen. Microbiol.* **130:**1711–1722.
39. **Burge, R. E., R. Adams, H. H. M. Balyuzi, and D. A. Reavely.** 1977. Structure of the peptidoglycan of bacterial cell walls. 1. *J. Mol. Biol.* **117:**955–974.
40. **Burge, R. E., A. G. Fowler, and D. A. Reavely.** 1977. Structure of the peptidoglycan of bacterial cell walls. 2. *J. Mol. Biol.* **117:**927–953.
41. **Burger, M.** 1950. *Bacterial Polysaccharides: Their Chemical and Immunological Aspects*, p. 121–134. Charles C Thomas, Publisher, Springfield, Ill.
42. **Burger, M. M., and L. Glaser.** 1966. The synthesis of teichoic acids. V. Poly(glucosyl glycerol phosphate) and poly(galactosylglycerol phosphate). *J. Biol. Chem.* **241:** 494–506.
43. **Button, D., M. K. Choudry, and N. C. Hemmings.** 1975. Lipoteichoic acid from *Bacillus licheniformis* and one of its mutants. *Proc. Soc. Gen. Microbiol.* **2:**45–46.
44. **Buxton, R. S., and J. B. Ward.** 1980. Heat sensitive lysis mutants of *Bacillus subtilis* 168 blocked at three different stages of peptidoglycan synthesis. *J. Gen. Microbiol.* **120:**283–293.
45. **Card, G. L., and D. J. Finn.** 1983. Products of phospholipid metabolism in *Bacillus stearothermophilus*. *J. Bacteriol.* **154:**294–303.
46. **Chaloupka, J., P. Kreckova, and L. Rihova.** 1962. The mucopeptide turnover in the cell walls of growing cultures of *Bacillus megaterium* KM. *Experientia* **18:** 362–364.
47. **Cheah, S. C., H. Hussey, and J. Baddiley.** 1981. Control of synthesis of wall teichoic acid in phosphate-starved cultures of *Bacillus subtilis* W23. *Eur. J. Biochem.* **118:** 497–500.
48. **Cheah, S. C., H. Hussey, I. C. Hancock, and J. Baddiley.** 1982. Control of synthesis of wall teichoic acid during balanced growth of *Bacillus subtilis* W23. *J. Gen. Microbiol.* **128:**593–599.
49. **Cheung, H.-Y., and E. Freese.** 1985. Monovalent cations enable cell wall turnover of the turnover-deficient *lyt-15* mutant of *Bacillus subtilis*. *J. Bacteriol.* **161:**1222–1225.
50. **Childs, W. C., and F. C. Neuhaus.** 1980. Biosynthesis of D-alanyl-lipoteichoic acid: characterization of ester linked alanine in the in vitro-synthesized product. *J. Bacteriol.* **143:**293–301.
51. **Childs, W. C., D. J. Taron, and F. C. Neuhaus.** 1986. Biosynthesis of D-alanyl lipoteichoic acid by *Lactobacillus casei*: interchain transacylation of D-alanyl ester residues. *J. Bacteriol.* **162:**1191–1195.
52. **Chin, T., M. M. Burger, and L. Glaser.** 1966. Synthesis of teichoic acids. VI. The formation of multiple wall polymers in *Bacillus subtilis* W23. *Arch. Biochem. Biophys.* **116:**358–367.
53. **Christian, R., G. Schultz, F. M. Unger, P. Messner, Z. Kupcu, and U. B. Sleytr.** 1986. Structure of a rhamnan from the surface layer glycoprotein of *Bacillus stearothermophilus* strain NRS 2004/3a. *Carbohydr. Res.* **150:** 265–272.
54. **Clarke-Sturman, A. J., A. R. Archibald, I. C. Hancock, C. R. Harwood, T. Merad, and J. A. Hobot.** 1989. Cell wall assembly in *Bacillus subtilis*: partial conservation of polar wall material and the effect of conditions on the pattern of incorporation of new material at the polar caps. *J. Gen. Microbiol.* **135:**657–665.
55. **Cole, R. M., and J. J. Hahn.** 1962. Cell wall replication in *Streptococcus pyogenes*: immunofluorescent methods applied during growth show that new wall is formed equatorially. *Science* **135:**722–724.
56. **Cole, R. M., T. S. Popkin, R. J. Boylan, and N. H. Mendelson.** 1970. Ultrastructure of a temperature-sensitive *rod* mutant of *Bacillus subtilis*. *J. Bacteriol.* **103:** 793–810.
57. **Coley, J., M. Duckworth, and J. Baddiley.** 1975. Extraction and purification of lipoteichoic acid from Gram positive bacteria. *Carbohydr. Res.* **40:**41–52.
58. **Coley, J., E. Tarelli, A. R. Archibald, and J. Baddiley.** 1978. The linkage between teichoic acid and peptidoglycan in bacterial cell walls. *FEBS Lett.* **88:**1–9.
59. **Cook, R. L., R. J. Harris, and G. Reid.** 1988. Effect of culture media and growth phase on the morphology of lactobacilli and on their ability to adhere to epithelial cells. *Curr. Microbiol.* **17:**159–166.
60. **Cooper, S.** 1991. *Bacterial Growth and Division*. Academic Press, Inc., New York.
61. **Cooper, S., M.-L. Hsieh, and B. Guenther.** 1988. Mode of peptidoglycan synthesis in *Salmonella typhimurium*: single-strand insertion. *J. Bacteriol.* **170:**3509–3512.
62. **Courtney, H. S., I. Ofek, W. A. Simpson, D. L. Hasty, and E. H. Beachey.** 1986. Binding of *Streptococcus pyogenes* to soluble and insoluble fibronectin. *Infect. Immun.* **53:**454–459.
63. **Coxon, R. D., A. R. Archibald, and C. R. Harwood.** 1989. Kinetics of protein export from *Bacillus subtilis*, p. 547–552. *In* L. O. Butler, C. R. Harwood, and B. E. B. Moseley (ed.), *Genetic Transformation and Expression*. Intercept Ltd., Andover, Hants, England.
64. **Coxon, R. D., C. R. Harwood, and A. R. Archibald.** 1991. Protein export during growth of *Bacillus subtilis*: the effect of extracellular protease deficiency. *Lett. Appl. Microbiol.* **12:**91–94.
65. **Daneo-Moore, L., and G. D. Shockman.** 1977. The bacterial cell surface in growth and division, p. 597–715. *In* G. Poste and G. L. Nicolson (ed.), *The Synthesis, Assembly and Turnover of Cell Surface Components*. Elsevier/North Holland Publishing Co., Amsterdam.
66. **Davie, J. M., and T. D. Brock.** 1966. Effect of teichoic acid on resistance to the membrane-lytic agent of *Streptococcus zymogenes*. *J. Bacteriol.* **92:**1623–1631.
67. **de Boer, W. R., F. J. Kruyssen, and J. T. M. Wouters.**

1981. Cell wall turnover in batch and chemostat cultures of *Bacillus subtilis*. *J. Bacteriol.* **145:**50–60.

68. **de Boer, W. R., J. T. M. Wouters, A. J. Anderson, and A. R. Archibald.** 1978. Further evidence for the structure of the teichoic acids from *Bacillus subtilis* var *niger*. *Eur. J. Biochem.* **85:**433–436.
69. **de Chastelier, C., R. Hellio, and A. Ryter.** 1975. Study of cell wall growth in *Bacillus megaterium* by high resolution autoradiography. *J. Bacteriol.* **123:**1184–1196.
70. **de Pedro, M. A., and U. Schwarz.** 1981. Heterogeneity of newly inserted and pre-existing murein in the sacculus of *Escherichia coli*. *Proc. Natl. Acad. Sci. USA* **78:**5856–5866.
71. **Dezelee, P., and G. D. Shockman.** 1975. Studies of the formation of peptide cross-links in the cell wall peptidoglycan of *Streptococcus faecalis*. *J. Biol. Chem.* **250:**6806–6816.
72. **Doyle, R. J., J. Chaloupka, and V. Vinter.** 1988. Turnover of cell walls in microorganisms. *Microbiol. Rev.* **52:**554–567.
73. **Doyle, R. J., A. L. Koch, and P. H. B. Carstens.** 1983. Cell wall-DNA association in *Bacillus subtilis*. *J. Bacteriol.* **153:**1521–1527.
74. **Doyle, R. J., M. L. McDannel, J. R. Helman, and U. N. Streips.** 1975. Distribution of teichoic acid in the cell wall of *Bacillus subtilis*. *J. Bacteriol.* **122:**152–158.
75. **Duckworth, M., A. R. Archibald, and J. Baddiley.** 1972. The location of N-acetylgalactosamine in the walls of *Bacillus subtilis* 168. *Biochem. J.* **130:**691–696.
76. **Eberle, H., and K. G. Lark.** 1966. Chromosome segregation in *Bacillus subtilis*. *J. Mol. Biol.* **22:**183–186.
77. **Ekwunife, F., J. Singh, K. Taylor, and R. J. Doyle.** 1991. Isolation and purification of cell wall polysaccharide of *Bacillus anthracis* (Δ Sterne). *FEMS Microbiol. Lett.* **82:**257–262.
78. **Ellwood, D. C., and D. W. Tempest.** 1969. Control of teichoic acid and teichuronic acid biosynthesis in chemostat cultures of *Bacillus subtilis* var *niger*. *Biochem. J.* **111:**1–5.
79. **Endl, J., H. P. Seidl, F. Fiedler, and K. H. Schleifer.** 1983. Chemical composition and structure of cell wall teichoic acids of staphylococci. *Arch. Microbiol.* **135:**215–223.
80. **Fan, D. P., B. E. Beckman, and H. L. Gardner-Eckstrom.** 1975. Mode of cell wall synthesis in gram-positive bacilli. *J. Bacteriol.* **123:**1157–1162.
81. **Favre, D., J. J. Thwaites, and N. H. Mendelson.** 1985. Kinetic studies of temperature-induced helix hand inversion in *Bacillus subtilis* macrofibers. *J. Bacteriol.* **164:**1136–1140.
82. **Fein, J. E., and H. J. Rogers.** 1976. Autolytic enzyme-deficient mutants of *Bacillus subtilis* 168. *J. Bacteriol.* **127:**1427–1442.
83. **Fiedler, F., and L. Glaser.** 1974. The synthesis of poly (ribitolphosphate). II. On the mechanism of poly(ribitolphosphate) polymerase. *J. Biol. Chem.* **249:**2690–2695.
84. **Fiedler, F., M. J. Schaffler, and E. Stackebrandt.** 1981. Biochemical and nucleic acid hybridisation studies on *Brevibacterium linens* and related strains. *Arch. Microbiol.* **129:**85–93.
85. **Fiedler, F., and E. Steber.** 1984. Structure and biosynthesis of teichoic acids in the cell walls of *Staphylococcus xylosus*. DSM 20266. *Arch. Microbiol.* **138:**321–328.
86. **Fischer, H., and A. Tomasz.** 1984. Production and release of peptidoglycan and wall teichoic acid polymers in pneumococci treated with beta-lactam antibiotics. *J. Bacteriol.* **157:**507–513.
87. **Fischer, W.** 1981. Glycerophosphoglycolipids: presumptive biosynthetic precursors of lipoteichoic acids, p. 209–228. *In* G. D. Shockman and A. J. Wicken (ed.), *Chemistry and Biological Activities of Bacterial Surface Amphiphiles*. Academic Press, Inc., New York.
88. **Fischer, W.** 1988. Physiology of lipoteichoic acids in bacteria. *Adv. Microb. Physiol.* **29:**233–303.
89. **Fischer, W., P. Rosel, and H. U. Koch.** 1981. Effect of alanine ester substitution and other structural features of lipoteichoic acids on their inhibitory activity against autolysins of *Staphylococcus aureus*. *J. Bacteriol.* **146:**467–475.
90. **Fordham, W. D., and C. Gilvarg.** 1974. Kinetics of crosslinking of peptidoglycan in *Bacillus megaterium*. *J. Biol. Chem.* **249:**2478–2482.
91. **Forsberg, C. W., P. B. Wyrick, J. B. Ward, and H. J. Rogers.** 1973. The effect of phosphate limitation on the morphology and wall composition of *Bacillus licheniformis* and its phosphoglucomutase-deficient mutants. *J. Bacteriol.* **113:**969–984.
92. **Foster, S. J.** 1991. Cloning, expression, sequence analysis and biochemical characterization of an autolytic amidase of *Bacillus subtilis* 168 *trpC2*. *J. Gen. Microbiol.* **137:**1987–1998.
93. **Foster, S. J.** 1992. Analysis of the autolysins of *Bacillus subtilis* 168 during vegetative growth and differentiation by using renaturing polyacrylamide gel electrophoresis. *J. Bacteriol.* **174:**464–470.
94. **Fuchs-Cleveland, E., and C. C. Gilvarg.** 1976. Oligomeric intermediates in peptidoglycan biosynthesis in *Bacillus megaterium*. *Proc. Natl. Acad. Sci. USA* **73:**4200–4204.
95. **Gally, D. L.** 1991. Cell wall assembly in Gram-positive bacteria. Ph.D. thesis, University of Newcastle upon Tyne, Newcastle upon Tyne, England.
96. **Gally, D. L., I. C. Hancock, C. R. Harwood, and A. R. Archibald.** 1991. Cell wall assembly in *Bacillus megaterium*: incorporation of new peptidoglycan by a monomer addition process. *J. Bacteriol.* **173:**2548–2589.
97. **Ganfield, M. C. W., and R. A. Peiringer.** 1980. The biosynthesis of nascent membrane lipoteichoic acid of *Streptococcus faecium* (*S. faecalis* ATCC 9790) from phosphatidyl-kojibiosyldiacylglycerol and phosphatidyl-glycerol. *J. Biol. Chem.* **255:**5164–5169.
98. **Gaur, N. K., K. Cabane, and I. Smith.** 1988. Structure and expression of the *Bacillus subtilis sin* operon. *J. Bacteriol.* **170:**1046–1053.
99. **Giles, A. F., and P. E. Reynolds.** 1979. *Bacillus megaterium* resistance to cloxacillin accompanied by a compensatory change in penicillin binding proteins. *Nature* (London) **280:**167–168.
100. **Giles, A. F., and P. E. Reynolds.** 1979. The direction of transpeptidation during cell wall peptidoglycan biosynthesis in *Bacillus megaterium*. *FEBS Lett.* **101:**244–248.
101. **Giudicelli, S., and A. Tomasz.** 1984. Attachment of pneumococcal autolysin to wall teichoic acids an essential step in enzymatic wall degradation. *J. Bacteriol.* **158:**1188–1190.
102. **Glaser, L., and M. M. Burger.** 1964. The synthesis of teichoic acids. III. Glycosylation of polyglycerolphosphate. *J. Biol. Chem.* **239:**3187–3191.
103. **Glauner, B., and U. Schwarz.** 1983. The analysis of murein composition with high pressure liquid chromatography, p. 29–34. *In* R. Hakenbeck, J. V. Höltje, and H. Labischinski (ed.), *The Target of Penicillin*. Walter de Gruyter, Berlin.
104. **Goodell, E. W., and U. Schwarz.** 1983. Cleavage and resynthesis of peptide cross-bridges in *Escherichia coli* murein. *J. Bacteriol.* **156:**136–140.
105. **Goundry, J., A. R. Archibald, J. Baddiley, and A. L. Davison.** 1967. The structure of the cell wall of *Bacillus polymyxa* NCIB 4747. *Biochem. J.* **104:**1–3.
106. **Graham, L. L., and T. J. Beveridge.** 1990. Effect of chemical fixatives on accurate preservation of *Escherichia coli* and *Bacillus subtilis* structure in cells prepared by freeze-substitution. *J. Bacteriol.* **172:**2150–2159.

107. **Graham, L. L., and T. J. Beveridge.** 1990. Evaluation of freeze-substitution and conventional embedding protocols for routine electron microscopic processing of eubacteria. *J. Bacteriol.* **172:**2141–2149.
108. **Hammes, W. P., and O. Kandler.** 1976. Biosynthesis of peptidoglycan in *Gaffkya homari*. The incorporation of peptidoglycan into the cell wall and its direction of transpeptidation. *Eur. J. Biochem.* **70:**97–106.
109. **Hancock, I. C.** 1981. The biosynthesis of ribitol teichoic acid by toluenised cells of *Bacillus subtilis* W23. *Eur. J. Biochem.* **119:**85–90.
110. **Hancock, I. C.** 1983. Activation and inactivation of secondary wall polymers in *Bacillus subtilis* W23. *Arch. Microbiol.* **134:**222–226.
111. **Hancock, I. C.** 1986. The use of gram positive bacteria for the removal of metals from aqueous solution, p. 25–43. *In* R. Thompson (ed.), *Trace Metal Removal from Aqueous Solution*. Royal Chemical Society, London.
112. **Hancock, I. C., and J. Baddiley.** 1972. Biosynthesis of the wall teichoic acid in *Bacillus licheniformis*. *Biochem. J.* **127:**27–37.
113. **Hancock, I. C., and J. Baddiley.** 1985. Biosynthesis of the bacterial envelope polymers teichoic acid and teichuronic acid, p. 279–307. *In* A. N. Martonosi (ed.), *The Enzymes of Biological Macromolecules*, vol. 2. Plenum Press, Inc., New York.
114. **Hancock, I. C., and C. M. Cox.** 1991. Turnover of cell surface-bound capsular polysaccharide in *Staphylococcus aureus*. *FEMS Microbiol. Lett.* **77:**25–30.
115. **Hanover, J. A., and W. J. Lennartz.** 1979. The topological orientation of N,N′-diacetylchitobiosyl pyrophosphoryl dolichol in artificial and natural membranes. *J. Biol. Chem.* **254:**9237–9246.
116. **Harold, F. M.** 1990. To shape a cell: an inquiry into the causes of morphogenesis of microorganisms. *Microbiol. Rev.* **54:**381–431.
117. **Harrington, C. R., and J. Baddiley.** 1983. Peptidoglycan synthesis by partly autolyzed cells of *Bacillus subtilis* W23. *J. Bacteriol.* **155:**776–792.
118. **Harrington, C. R., and J. Baddiley.** 1985. Biosynthesis of wall teichoic acids in *Staphylococcus aureus* H, *Micrococcus varians* and *Bacillus subtilis* W23: involvement of lipid. *J. Biochem.* **153:**539–545.
119. **Harwood, C. R.** 1992. *Bacillus subtilis* and its relatives: molecular biological and industrial workhorses. *Trends Biotechnol.* **10:**247–256.
120. **Harwood, C. R., R. D. Coxon, and I. C. Hancock.** 1990. The *Bacillus* cell envelope and secretion, p. 327–369. *In* C. R. Harwood and S. M. Cutting (ed.), *Molecular Biological Methods for Bacillus*. John Wiley & Sons, Chichester, England.
121. **Hase, S., and Y. Matsushima.** 1979. The structure of the branching point between acidic polysaccharide and peptidoglycan in *Micrococcus lysodeikticus* cell wall. *J. Biochem.* **81:**1181–1186.
122. **Hay, J. B.** 1970. Studies on the structure of bacterial cell walls and underlying regions. Ph.D. thesis, University of Newcastle upon Tyne, Newcastle upon Tyne, England.
123. **Hay, J. B., A. J. Wicken, and J. Baddiley.** 1963. The location of intracellular teichoic acids. *Biochim. Biophys. Acta* **71:**188–190.
124. **Heckels, J. E., P. A. Lambert, and J. Baddiley.** 1977. Binding of magnesium ions to cell walls of *Bacillus subtilis* W23 containing teichoic acid or teichuronic acid. *Biochem. J.* **162:**359–365.
125. **Heptinstall, S., A. R. Archibald, and J. Baddiley.** 1970. Teichoic acids and membrane function in bacteria. *Nature* (London) **225:**519–521.
126. **Herbold, D. R., and L. Glaser.** 1975. *Bacillus subtilis* N-acetylmuramic acid L-alanine amidase. *J. Biol. Chem.* **250:**1676–1680.
127. **Herbold, D. R., and L. Glaser.** 1975. Interaction of N-acetylmuramic acid L-alanine amidase with cell wall polymers. *J. Biol. Chem.* **250:**7231–7238.
128. **Hildebrandt, K. M., and J. S. Anderson.** 1990. Biosynthetic elongation of isolated teichuronic acid polymers via glucosyl- and *N*-acetylmannosaminuronosyltransferases from solubilized cytoplasmic membrane fragments of *Micrococcus luteus*. *J. Bacteriol.* **172:**5160–5164.
129. **Hobot, J. A.** 1990. New aspects of bacterial ultrastructure as revealed by modern acrylics for electron microscopy. *J. Struct. Biol.* **104:**169–177.
130. **Höltje, J. V., and A. Tomasz.** 1975. Specific recognition of choline residues in the cell wall teichoic acid by the N-acetylmuramyl-L-alanine amidase of pneumococcus. *J. Biol. Chem.* **250:**6072–6076.
131. **Hughes, A. H., I. C. Hancock, and J. Baddiley.** 1973. The function of teichoic acids in cation control in bacterial membranes. *Biochem. J.* **132:**83–93.
132. **Hungerer, K. D., and D. J. Tipper.** 1969. Cell wall polymers of *Bacillus sphaericus* 9602. *Biochemistry* **8:**3577–3587.
133. **Hussey, H., D. Brooks, and J. Baddiley.** 1969. Direction of chain extension during the synthesis of teichoic acid in bacterial cell walls. *Nature* (London) **221:**665–666.
134. **Hussey, H., S. Sueda, S. C. Cheah, and J. Baddiley.** 1978. Control of teichoic acid synthesis in *Bacillus licheniformis* ATCC 9945. *Eur. J. Biochem.* **82:**169–174.
135. **Ishimoto, N., and J. L. Strominger.** 1966. Polyribitolphosphate synthetase of *Staphylococcus aureus* strain Copenhagen. *J. Biol. Chem.* **241:**639–650.
136. **Ivanovics, G.** 1940. Untersuchungen uber das Polysaccharid der Milzbrandbazillen. *Z. Immunitaetsforsch. Exp. Ther.* **97:**402–423.
137. **Ivatt, R. J., and C. Gilvarg.** 1979. The primary structure of the teichuronic acid of *Bacillus megaterium*. *J. Biol. Chem.* **254:**2759–2765.
138. **Iwasaki, H., A. Shimada, and E. Ito.** 1986. Comparative studies of lipoteichoic acids from several *Bacillus* strains. *J. Bacteriol.* **167:**508–516.
139. **Jackson, G. E. D., and J. L. Strominger.** 1984. Synthesis of peptidoglycan by high molecular weight penicillin binding proteins of *Bacillus subtilis* and *Bacillus stearothermophilus*. *J. Biol. Chem.* **259:**1483–1490.
140. **Jacob, F., S. Brenner, and F. Cuzin.** 1963. On the regulation of DNA replication in bacteria. *Cold Spring Harbor Symp. Quant. Biol.* **28:**329–347.
141. **James, A. M., and J. E. Brewer.** 1968. Non-protein components of the cell surface of *Staphylococcus aureus*. *Biochem. J.* **107:**817–821.
142. **Johnstone, K., F. A. Simion, and D. J. Ellar.** 1982. Teichoic acid and lipid metabolism during sporulation of *Bacillus megaterium* KM. *Biochem. J.* **202:**459–467.
143. **Joliffe, L. K., R. J. Doyle, and U. N. Streips.** 1980. Extracellular proteases modify cell wall turnover in *Bacillus subtilis*. *J. Bacteriol.* **141:**1191–1208.
144. **Joliffe, L. K., R. J. Doyle, and U. N. Streips.** 1981. The energised membrane and cellular autolysis in *Bacillus subtilis*. *Cell* **25:**753–763.
145. **Karakawa, W. W., and W. F. Vann.** 1982. Capsular polysaccharides of *Staphylococcus aureus*, p. 285–293. *In* J. B. Robbins, J. C. Hill, and J. C. Sadoff (ed.), *Seminars in Infectious Diseases: Bacterial Vaccines*, vol. 4. Thieme Stratton Inc., New York.
146. **Kaya, S., Y. Araki, and E. Ito.** 1985. Structural studies on the linkage unit between polygalactosylglycerol phosphate and peptidoglycan in walls of *Bacillus coagulans*. *Eur. J. Biochem.* **147:**41–46.
147. **Kaya, S., K. Yokoyama, Y. Araki, and E. Ito.** 1984. *N*-Acetylmannosaminyl(1-4)*N*-acetylglucosamine, a linkage unit between glycerol teichoic acid and peptidoglycan in cell walls. *J. Bacteriol.* **158:**990–996.
148. **Kennedy, L. D., and D. R. D. Shaw.** 1968. Direction of

polyglycerolphosphate chain growth in *Bacillus subtilis*. *Biochem. Biophys. Res. Commun.* **32:**861–865.
149. **Kessler, R. E., I. Van de Rijn, and M. McCarty.** 1979. Characterisation and localisation of the enzymatic deacylation of lipoteichoic acid in group A streptococci. *J. Exp. Med.* **150:**1498–1509.
150. **Kirchner, G., M. A. Kemper, A. L. Koch, and R. J. Doyle.** 1988. Zonal turnover of cell poles of *Bacillus subtilis*. *Ann. Inst. Pasteur Microbiol.* **139:**645–654.
151. **Kirchner, G., A. L. Koch, and R. J. Doyle.** 1984. Energised membrane regulates pole formation in *Bacillus subtilis*. *FEMS Microbiol. Lett.* **24:**438–441.
152. **Koch, A. L.** 1983. The surface stress theory of microbial morphogenesis. *Adv. Microb. Physiol.* **24:**301–366.
153. **Koch, A. L.** 1988. Biophysics of bacterial walls viewed as stress-bearing fabric. *Microbiol. Rev.* **52:**337–353.
154. **Koch, A. L., and I. J. D. Burdett.** 1986. Biophysics of pole formation of Gram-positive rods. *J. Gen. Microbiol.* **132:**3451–3457.
155. **Koch, A. L., and I. J. D. Burdett.** 1986. Normal pole formation during total inhibition of wall synthesis of *Bacillus subtilis*. *J. Gen. Microbiol.* **132:**3441–3449.
156. **Koch, A. L., and A. J. Doyle.** 1985. Inside-to-outside growth and turnover of the wall of gram positive rods. *J. Theor. Biol.* **117:**137–157.
157. **Koch, A. L., and R. J. Doyle.** 1986. Growth strategy for the Gram positive rod. *FEMS Microbiol. Rev.* **32:**247–254.
158. **Koch, A. L., M. L. Higgins, and R. J. Doyle.** 1981. Surface tension-like forces determine bacterial shapes. *J. Gen. Microbiol.* **123:**151–161.
159. **Koch, A. L., G. Kirchner, R. J. Doyle, and I. D. J. Burdett.** 1985. How does a *Bacillus* split its septum right down the middle? *Ann. Inst. Pasteur Microbiol.* **136:**91–98.
160. **Koch, A. L., H. L. T. Mobley, R. J. Doyle, and U. N. Streips.** 1981. The coupling of wall growth and chromosome replication in Gram positive rods. *FEMS Microbiol. Lett.* **12:**201–208.
161. **Koch, H. U., R. Doker, and W. Fischer.** 1985. Maintenance of D-alanine ester substitution of lipoteichoic acid by reesterification in *Staphylococcus aureus*. *J. Bacteriol.* **164:**1211–1217.
162. **Koch, H. U., and W. Fischer.** 1978. Acyldiglucosyldiacylglycerol-containing lipoteichoic acid with a poly(3-O-galabiosyl 2-O galactosyl-*sn*-glycero-1-phosphate) chain from *Streptococcus lactis* Kiel 42172. *Biochemistry* **17:**5275–5281.
163. **Koga, Y., M. Nishihara, and H. Morii.** 1984. Products of phosphatidylglycerol turnover in two *Bacillus* strains with and without lipoteichoic acid in the cells. *Biochim. Biophys. Acta* **793:**86–94.
164. **Kojima, N., Y. Araki, and E. Ito.** 1985. Structure of the linkage units between ribitol teichoic acids and peptidoglycan. *J. Bacteriol.* **161:**299–306.
165. **Kojima, N., Y. Araki, and E. Ito.** 1985. Structural studies on the linkage unit of ribitol teichoic acid of *Lactobacillus plantarum*. *Eur. J. Biochem.* **148:**29–34.
166. **Kojima, N., J. Lida, Y. Araki, and E. Ito.** 1985. Structural studies on the linkage unit between poly(N-acetylglucosamine 1-phosphate) and peptidoglycan in cell walls of *Bacillus pumilis*. *Eur. J. Biochem.* **149:**331–336.
167. **Kojima, N., K. Uchikawa, Y. Araki, and E. Ito.** 1985. A common linkage saccharide unit between teichoic acids and peptidoglycan in cell walls of *Bacillus coagulans*. *J. Biochem.* **97:**1085–1092.
168. **Kolenbrander, P. E.** 1988. Intergeneric coaggregation among human oral bacteria and ecology of dental plaque. *Annu. Rev. Microbiol.* **42:**627–656.
169. **Kuroda, A., and J. Sekiguchi.** 1991. Molecular cloning and sequencing of a major *Bacillus subtilis* autolysin gene. *J. Bacteriol.* **173:**7304–7312.
170. **Labischinski, H., G. Barnickel, H. Bradaczek, and P. Giesbrecht.** 1979. On the secondary and tertiary structure of murein. *Eur. J. Biochem.* **95:**147–155.
171. **Labischinski, H., G. Barnickel, and D. Naumann.** 1983. The state of order of bacterial peptidoglycan, p. 49–54. *In* R. Hakenbeck, J. V. Holtje, and H. Labischinski (ed.), *The Target of Penicillin*. Walter de Gruyter, New York.
172. **Labischinski, H., G. Barnickel, D. Naumann, and P. Keller.** 1985. Conformational and topological aspects of the three-dimensional architecture of bacterial peptidoglycan. *Ann. Inst. Pasteur Microbiol.* **136A:**45–50.
173. **Labischinski, H., D. Naumann, and W. Fischer.** 1991. Small and medium-angle X-ray analysis of bacterial lipoteichoic acid phase structure. *Eur. J. Biochem.* **202:**1269–1274.
174. **Lablaw, L. W., and V. M. Mosley.** 1954. Periodic structure in the flagella and cell walls of a bacterium. *Biochim. Biophys. Acta* **15:**325–331.
175. **Lambert, P. A., I. C. Hancock, and J. Baddiley.** 1975. Influence of alanyl ester residues on the binding of magnesium ions to teichoic acids. *Biochem. J.* **151:**671–676.
176. **Lang, W. K., and A. R. Archibald.** 1982. Length of teichoic acid chains incorporated into walls of *Bacillus subtilis* grown under conditions of differing phosphate supply. *FEMS Microbiol. Lett.* **13:**93–97.
177. **Lang, W. K., K. Glassey, and A. R. Archibald.** 1982. Influence of phosphate supply on teichoic and teichuronic acid content of *Bacillus subtilis* walls. *J. Bacteriol.* **151:**367–375.
178. **Leaver, J., I. C. Hancock, and J. Baddiley.** 1981. Fractionation studies of the enzyme complex involved in teichoic acid synthesis. *J. Bacteriol.* **146:**847–852.
179. **Leclerc, D., and A. Asselin.** 1989. Detection of bacterial cell wall hydrolases after denaturing polyacrylamide gel electrophoresis. *Can. J. Microbiol.* **35:**749–753.
180. **Lifely, M. R., E. Tarelli, and J. Baddiley.** 1980. The teichuronic acid from walls of *Bacillus licheniformis* ATCC 9945. *Biochem. J.* **191:**305–318.
181. **Lim, S. H., and M. R. J. Salton.** 1985. Comparison of the chemical composition of lipomannan from *Micrococcus agilis* membranes with that of *Micrococcus luteus* strains. *FEMS Microbiol. Lett.* **27:**287–291.
182. **Linnett, P. E., and J. L. Strominger.** 1974. Amidation and cross-linking of the enzymatically synthesised peptidoglycan of *Bacillus stearothermophilus*. *J. Biol. Chem.* **249:**2489–2496.
183. **Liu, T. Y., and E. C. Gottschlick.** 1967. Muramic acid phosphate as a component of the mucopeptide of Gram-positive bacteria. *J. Biol. Chem.* **242:**471–476.
184. **Lleo, M. M., P. Canepari, and G. Satta.** 1990. Bacterial cell shape regulation: testing of additional predictions unique to the two-competing-sites model for peptidoglycan assembly and isolation of conditional rod-shaped mutants from some wild-type cocci. *J. Bacteriol.* **172:**3758–3771.
185. **Lombardi, S. J., S. L. Chen, and A. J. Fulco.** 1980. A rapidly metabolizing pool of phosphatidylglycerol as a precursor for phosphatidylethanolamine and diglyceride in *Bacillus megaterium*. *J. Bacteriol.* **141:**626–634.
186. **Lombardi, S. J., and A. J. Fulco.** 1980. Two distinct pools of phosphatidylglycerol in *Bacillus megaterium*. *J. Bacteriol.* **141:**618–625.
187. **Lopez, R., E. Garcia, P. Garcia, C. Ronda, and A. Tomasz.** 1982. Choline-containing bacteriophage receptors in *Streptococcus pneumoniae*. *J. Bacteriol.* **151:**1581–1590.
188. **Maino, V. C., and F. E. Young.** 1974. Regulation of glucosylation of teichoic acid. *J. Biol. Chem.* **249:**5169–5175.
189. **Margot, P., and D. Karamata.** 1992. Identification of the structural genes for N-acetylmuramoyl-L-alanine amidase and its modifier in *Bacillus subtilis* 168: inactivation of these genes by insertional mutagenesis has no

effect on growth or cell separation. *Mol. Gen. Genet.* **232:**359–366.

190. **Margot, P., C. Mauël, and D. Karamata.** 1991. The *Bacillus subtilis* N-acetylglucosaminidase is encoded by a monocistronic operon controlled by a σ^D dependent promoter, p. W-6. *Proceedings of 6th International Conference on Bacilli.* Stanford University Press, Stanford, Calif.
191. **Markham, J. L., K. W. Knox, A. J. Wicken, and M. J. Hewett.** 1975. Formation of extracellular lipoteichoic acid by oral streptococci and lactobacilli. *Infect. Immun.* **12:**378–385.
192. **Markiewicz, Z., and A. Tomasz.** 1990. Protein-bound choline is released from the pneumococcal autolytic enzyme during adsorption of the enzyme to cell wall particles. *J. Bacteriol.* **172:**2241–2244.
193. **Marquis, R. E.** 1973. Immersion refractometry of isolated bacterial cell walls. *J. Bacteriol.* **116:**1273–1279.
194. **Marquis, R. E., and E. L. Carstensen.** 1973. Electric conductivity and internal osmolality of intact bacterial cells. *J. Bacteriol.* **113:**1198–1206.
195. **Marr, A. G.** 1991. Growth rate of *Escherichia coli. Microbiol. Rev.* **55:**316–333.
196. **Mauck, J., L. Chin, and L. Glaser.** 1971. Turnover of cell wall of gram positive bacteria. *J. Biol. Chem.* **246:**1820–1827.
197. **Mauck, J., and L. Glaser.** 1972. On the mode of *in vivo* assembly of the cell wall of *Bacillus subtilis. J. Biol. Chem.* **247:**1180–1827.
198. **Mauël, C., M. Young, and D. Karamata.** 1991. Genes concerned with synthesis of poly(glycerol phosphate), the essential teichoic acid in *Bacillus subtilis* strain 168, are organized in two divergent transcription units. *J. Gen. Microbiol.* **137:**929–941.
199. **Mauël, C., M. Young, P. Margot, and D. Karamata.** 1989. The essential nature of teichoic acids in *Bacillus subtilis* as revealed by insertional mutagenesis. *Mol. Gen. Genet.* **215:**388–394.
200. **Maurer, J. J., and S. J. Mattingly.** 1991. Molecular analysis of lipoteichoic acid from *Streptococcus agalactiae. J. Bacteriol.* **173:**487–494.
201. **McCloskey, M. A., and F. A. Troy.** 1980. Paramagnetic isoprenoid lipid carrier. 1. Chemical synthesis and incorporation into model membrane. *Biochemistry* **19:**2056–2060.
202. **McIntire, F. C., C. A. Bush, S. S. Wu, S. C. Li, Y. T. Li, M. McNeil, S. S. Tjoa, and P. V. Fennesey.** 1987. Structure of a new hexasaccharide from the coaggregation polysaccharide of *Streptococcus sanguis. Carbohydr. Res.* **166:**133–143.
203. **McLean, R. J. C., D. Beauchemin, L. Chapman, and D. C. Beveridge.** 1990. Metal binding character of the γ-glutamylpolymer of *Bacillus licheniformis* ATCC 9945. *Appl. Environ. Microbiol.* **56:**3671–3677.
204. **Mendelson, N. H.** 1976. Helical growth of *Bacillus subtilis*: a new model of cell growth. *Proc. Natl. Acad. Sci. USA* **73:**1740–1744.
205. **Mendelson, N. H.** 1978. Helical *Bacillus* macrofibres; morphogenesis of a bacterial multicellular microorganism. *Proc. Natl. Acad. Sci. USA* **75:**2478–2482.
206. **Mendelson, N. H.** 1982. Bacterial growth and division: genes, structures, forces and clocks. *Microbiol. Rev.* **46:**341–375.
207. **Mendelson, N. H.** 1982. Dynamics of *Bacillus subtilis* helical macrofiber morphogenesis: writhing, folding, close packing and contraction. *J. Bacteriol.* **151:**438–449.
208. **Mendelson, N. H., D. Favre, and J. J. Thwaites.** 1984. Twisted states of *Bacillus subtilis* macrofibres reflect structural states of the cell wall. *Proc. Natl. Acad. Sci. USA* **81:**3562–3566.
209. **Mendelson, N. H., and J. J. Thwaites.** 1989. Cell wall mechanical properties as measured with bacterial thread made from *Bacillus subtilis. J. Bacteriol.* **171:**1055–1062.
210. **Merad, T., A. R. Archibald, I. C. Hancock, C. R. Harwood, and J. A. Hobot.** 1989. Cell wall assembly in *Bacillus subtilis*: visualisation of old and new wall material by electron microscopic examination of samples stained selectively for teichoic acid and teichuronic acid. *J. Gen. Microbiol.* **135:**645–655.
211. **Messner, P., K. Bock, R. Christian, G. Schultz, and U. B. Sleytr.** 1990. Characterization of the surface layer glycoprotein of *Clostridium symbiosum* HB25. *J. Bacteriol.* **172:**2576–2583.
212. **Messner, P., U. B. Sleytr, R. Christian, G. Schultz, and F. M. Unger.** 1987. Isolation and structural determination of a diacetamidodideoxyuronic acid-containing glycan chain from the S-layer glycoprotein of *Bacillus stearothermophilus* NRS 2004/3a. *Carbohydr. Res.* **168:**211–218.
213. **Mett, H., R. Bracha, and D. Mirelman.** 1980. Soluble nascent peptidoglycan in growing *Escherichia coli* cells. *J. Biol. Chem.* **255:**9584–9590.
214. **Miller, L., L. Gray, E. Beachey, and M. Kehoe.** 1988. Antigenic variation among group A streptococcal M proteins. *J. Biol. Chem.* **263:**5668–5673.
215. **Miyashiro, M. S., H. Enei, Y. Hirose, and S. Udaka.** 1980. Extracellular production of proteins. 5. Stimulating effect of inhibitors of cell wall synthesis on protein production by *Bacillus brevis. Agric. Biol. Chem.* **44:**2297–2303.
216. **Mobley, H. L. T., R. J. Doyle, and L. K. Joliffe.** 1983. Cell wall polypeptide complexes in *Bacillus subtilis. Carbohydr. Res.* **116:**113–125.
217. **Mobley, H. L. T., A. L. Koch, R. J. Doyle, and U. N. Streips.** 1984. Insertion and fate of the cell wall in *Bacillus subtilis. J. Bacteriol.* **158:**169–179.
218. **Murazumi, N., Y. Araki, and E. Ito.** 1986. Biosynthesis of the wall neutral polysaccharide in *Bacillus cereus. Eur. J. Biochem.* **161:**51–59.
219. **Nauman, D., G. Barnickel, H. Bradaczek, H. Labischinski, and P. Giesbrecht.** 1982. Infrared spectroscopy, a tool for probing bacterial peptidoglycan. *Eur. J. Biochem.* **125:**505–515.
220. **Nealon, T. J., and S. J. Mattingly.** 1984. Role of cellular lipoteichoic acids in mediating adherence of serotype III strains of group B streptococci to human embryonic, fetal and adult epithelial cells. *Infect. Immun.* **43:**523–530.
221. **Nermut, M. V., and R. G. E. Murray.** 1967. Ultrastructure of the cell wall of *Bacillus polymyxa. J. Bacteriol.* **93:**1949–1965.
222. **Neuhaus, F. C.** 1985. Interchain transacylation of D-alanine ester residues of lipoteichoic acid. A unique mechanism of membrane communication. *Biochem. Soc. Trans.* **13:**987–990.
223. **Nishikawa, J., M. Tada, Y. Takubo, T. Nishihara, and M. Kondo.** 1987. Occurrence in *Bacillus megaterium* of uridine 5′-diphospho-*N*-acetylgalactosamine and uridine 5′-diphosphogalactosamine, intermediates in the biosynthesis of galactosamine-6-phosphate polymer. *J. Bacteriol.* **169:**1338–1340.
224. **Ofek, I., W. A. Simpson, and E. H. Beachey.** 1982. Formation of molecular complexes between a structurally defined M protein and acylated or deacylated lipoteichoic acid of *Streptococcus. J. Bacteriol.* **149:**426–433.
225. **Ohmizu, H., N. Tsukagos, S. Udata, N. Kaneda, and K. Yagi.** 1983. Major proteins released by a protein producing bacterium, *Bacillus brevis* 47, are derived from the cell wall protein. *J. Bacteriol.* **94:**1077–1084.
226. **Oldmixon, E. H., P. Dezelee, M. C. Ziskin, and G. D. Shockman.** 1976. Monomer addition as a mechanism of forming peptide cross-links in the cell wall peptidogly-

can of *Streptococcus faecalis* ATCC 9790. *Eur. J. Biochem.* **68:**271–280.
227. **Pal, M. K., T. C. Ghosh, and J. K. Ghosh.** 1990. Studies on the conformation of and metal ion binding by teichoic acid of *Staphylococcus aureus*. *Biopolymers* **30:**273–277.
228. **Park, W., H. Seto, R. Hackenbeck, and M. Matsuhashi.** 1985. Major peptidoglycan transglycosylase activity in *Streptococcus pneumoniae* that is not a penicillin binding protein. *FEMS Microbiol. Lett.* **27:**45–48.
229. **Pavlik, H. G., and H. J. Rogers.** 1973. Selective extraction of polymers from cell walls of Gram-positive bacteria. *Biochem. J.* **131:**619–621.
230. **Pooley, H. M.** 1976. Layered distribution according to age within the cell wall of *Bacillus subtilis*. *J. Bacteriol.* **125:**1139–1147.
231. **Pooley, H. M.** 1976. Turnover and spreading of old wall during surface growth of *Bacillus subtilis*. *J. Bacteriol.* **125:**1127–1138.
232. **Pooley, H. M., F.-X. Abellan, and D. Karamata.** 1992. CDP-glycerol: poly(glycerophosphate) glycerophosphotransferase, which is involved in the synthesis of the major wall teichoic acid in *Bacillus subtilis* 168, is encoded by tagF (*rodC*). *J. Bacteriol.* **174:**646–649.
233. **Pooley, H. M., and H. Karamata.** 1984. Genetic analysis of autolysin-deficient and flagellaless mutants of *Bacillus subtilis*. *J. Bacteriol.* **160:**1123–1129.
234. **Pooley, H. M., D. Paschoud, and D. Karamata.** 1987. The *gtaB* marker in *Bacillus subtilis* 168 is associated with a deficiency in UDP-glucose pyrophosphorylase. *J. Gen. Microbiol.* **133:**3481–3493.
235. **Pooley, P. R., R.-X. Abellan, and D. Karamata.** 1991. A conditional-lethal mutant of *Bacillus subtilis* 168 with a thermosensitive glycerol-3-phosphate cytidylyltransferase, an enzyme specific for the synthesis of the major wall teichoic acid. *J. Gen. Microbiol.* **137:**921–928.
236. **Potvin, C., D. Leclerc, G. Tremblay, A. Asselin, and G. Bellemare.** 1988. Cloning, sequencing and expression of a *Bacillus* bacteriolytic enzyme in *Escherichia coli*. *Mol. Gen. Genet.* **214:**241–248.
237. **Powell, D. A., M. Duckworth, and J. Baddiley.** 1975. A membrane-associated lipomannan from *Micrococcus lysodeikticus*. *Biochem. J.* **151:**387–397.
238. **Powell, J. F., and R. E. Strange.** 1957. Diaminopimelic acid metabolism and sporulation in *Bacillus sphaericus*. *Biochem. J.* **65:**700–708.
239. **Poxton, I. R., E. Tarelli, and J. Baddiley.** 1978. The structure of C-polysaccharide from the walls of *Streptococcus pneumoniae*. *Biochem. J.* **175:**1033–1042.
240. **Priest, F. G.** 1977. Extracellular enzyme synthesis in the genus *Bacillus*. *Bacteriol. Rev.* **41:**711–753.
241. **Reid, G., R. L. Cook, R. J. Harris, J. D. Rousseau, and H. Lawford.** 1988. Development of a freeze substitution technique to examine the structure of *Lactobacillus casei* GR-1 grown in agar and under batch and chemostat culture conditions. *Curr. Microbiol.* **17:**151–158.
242. **Reis, K. J., E. M. Ayoub, and M. D. P. Boyle.** 1984. Streptococcal Fc receptors. I. Isolation and partial characterisation of the receptor from a group C *Streptococcus*. *J. Immunol.* **132:**3091–3097.
243. **Robson, R. L., and J. Baddiley.** 1977. Morphological changes associated with novobiocin resistance in *Bacillus licheniformis*. *J. Bacteriol.* **129:**1045–1050.
244. **Robson, R. L., and J. Baddiley.** 1977. Role of teichuronic acid in *Bacillus licheniformis*: defective autolysis due to deficiency of teichuronic acid in a novobiocin-resistant mutant. *J. Bacteriol.* **129:**1051–1058.
245. **Rogers, H. J., H. R. Perkins, and J. B. Ward.** 1980. *Microbial Cell Walls and Membranes*. Chapman & Hall, Ltd., London.
246. **Rogers, H. J., C. Taylor, S. Rayter, and J. B. Ward.** 1984. Purification and properties of autolytic endo-β-N-acetyl glucosaminidase and the N-acetyl muramyl-L-alanine amidase. *J. Gen. Microbiol.* **130:**2395–2402.
247. **Rohr, T. E., G. N. Levy, N. J. Stark, and J. S. Anderson.** 1977. Initial reactions in biosynthesis of teichuronic acid of *Micrococcus lysodeikticus* cell walls. *J. Biol. Chem.* **252:**3460–3465.
248. **Roten, C.-A. H., C. Brandt, and D. Karamata.** 1991. Genes involved in meso-diaminopimelate synthesis in *Bacillus subtilis*: identification of the gene encoding aspartokinase I. *J. Gen. Microbiol.* **137:**951–962.
249. **Roten, C.-A. H., C. Brandt, and D. Karamata.** 1991. Identification of *murA*, the structural gene of phosphoenolpyruvate:uridine-N-acetyl-glucosamine enolpyruvoyltransferase in *Bacillus subtilis*, p. W-8. *Proceedings of the 6th International Conference on Bacilli*. Stanford University, Stanford, Calif.
250. **Sara, M., and U. B. Sleytr.** 1987. Charge distribution on the S-layer of *Bacillus stearothermophilus* NRS 1536/3C and importance of charged groups for morphogenesis and function. *J. Bacteriol.* **169:**2804–2809.
251. **Sara, M., and U. B. Sleytr.** 1987. Molecular sieving through S-layers of *Bacillus stearothermophilus* strains. *J. Bacteriol.* **169:**4092–4098.
252. **Sargent, M. G.** 1978. Surface extension and the cell cycle in procaryotes. *Adv. Microb. Physiol.* **18:**106–176.
253. **Sasaki, Y., Y. Araki, and E. Ito.** 1980. Structure of the linkage region between glycerol teichoic acid and peptidoglycan in *Bacillus cereus* AHU cell walls. *Biochem. Biophys. Res. Commun.* **96:**529–534.
254. **Schleifer, K. H., and O. Kandler.** 1972. Peptidoglycan types of bacterial cell walls and their taxonomic implications. *Bacteriol. Rev.* **36:**407–477.
255. **Sekiguchi, J., H. Ohsu, A. Kuroda, H. Moriyama, and T. Akamatsu.** 1990. Nucleotide sequences of the *Bacillus subtilis flaD* locus and a *Bacillus licheniformis* homologue affecting autolysin level and flagellation. *J. Gen. Microbiol.* **136:**1223–1230.
256. **Shabarova, Z. A., N. A. Hughes, and J. Baddiley.** 1962. The influence of adjacent phosphate and hydroxyl groups on amino acid esters. *Biochem. J.* **83:**216–219.
257. **Sharpe, A., P. M. Blumberg, and J. L. Strominger.** 1974. D-Alanine carboxypeptidase and cell wall cross-linking in *Bacillus subtilis*. *J. Bacteriol.* **117:**926–927.
258. **Sharpe, M. E., A. L. Davison, and J. Baddiley.** 1964. Teichoic acids and group antigens in lactobacilli. *J. Gen. Microbiol.* **34:**333–340.
259. **Shaw, N., and J. Baddiley.** 1964. The teichoic acid from the walls of *Lactobacillus buchneri* NCIB8007. *Biochem. J.* **93:**317–321.
260. **Shimada, A., M. Ohta, H. Iwasaki, and E. Ito.** 1988. The function of β-N-acetylglucosaminyl monophosphoryl-undecaprenol in biosynthesis of lipoteichoic acids in a group of *Bacillus* strains. *Eur. J. Biochem.* **176:**559–565.
261. **Shimada, A., J. Tamatukuri, and E. Ito.** 1989. Function of α-D-glucosyl monophosphorylpolyprenol in biosynthesis of cell wall teichoic acid in *Bacillus coagulans*. *J. Bacteriol.* **171:**2835–2841.
262. **Shockman, G. D., and J. F. Barrett.** 1983. Structure, function and assembly of cell walls of Gram positive bacteria. *Annu. Rev. Microbiol.* **37:**501–527.
263. **Shockman, G. D., L. Daneo-Moore, and M. L. Higgins.** 1974. Problems of cell wall and membrane growth, enlargement and division. *Ann. N.Y. Acad. Sci.* **235:**161–196.
264. **Shockman, G. D., and H. D. Slade.** 1964. The cellular location of streptococcal group D antigens. *J. Gen. Microbiol.* **37:**297–305.
265. **Sjöquist, J., J. Moritz, I. B. Johansson, and H. Hjelm.** 1972. Localisation of protein A in the bacteria. *Eur. J. Biochem.* **30:**190–194.
266. **Sleytr, U. B.** 1983. Crystalline surface layers on bacteria. *Arch. Microbiol.* **37:**311–339.

267. **Sleytr, U. B., and P. Messner.** 1988. Crystalline surface layers in procaryotes. *J. Bacteriol.* **170:**2891–2897.
268. **Sleytr, U. B., and M. Sara.** 1986. Ultrafiltration membranes with uniform pores from crystalline bacterial cell wall layers. *Appl. Microbiol. Biotechnol.* **25:**83–90.
269. **Smith, D. G., and P. M. F. Shattock.** 1964. The cellular location of antigens in streptococci of groups D, N and Q. *J. Gen. Microbiol.* **34:**165–175.
270. **Snowden, M. A., H. R. Perkins, A. W. Wyke, M. W. Hayes, and J. B. Ward.** 1989. Cross-linking O-acetylation of newly synthesized peptidoglycan in *Staphylococcus aureus* H. *J. Gen. Microbiol.* **135:**3015–3022.
271. **Sonnenfeld, E. M., T. J. Beveridge, and R. J. Doyle.** 1985. Discontinuity of charge on cell wall poles of *Bacillus subtilis*. *Can. J. Microbiol.* **131:**875–877.
272. **Sonnenfeld, E. M., A. L. Koch, and R. J. Doyle.** 1985. Cellular location of origin and terminus in *Bacillus subtilis*. *J. Bacteriol.* **163:**895–899.
273. **Sturman, A. J., and A. R. Archibald.** 1978. Conservation of phage receptor material at the polar caps of *Bacillus subtilis* W23. *FEMS Microbiol. Lett.* **4:**255–259.
274. **Sutcliffe, I. C., and N. Shaw.** 1991. Atypical lipoteichoic acids of gram-positive bacteria. *J. Bacteriol.* **173:**7065–7069.
275. **Taron, D. J., W. C. Childs, and F. C. Neuhaus.** 1983. Biosynthesis of D-alanyl lipoteichoic acid: role of diglyceride kinase in the synthesis of phosphatidylglycerol for chain elongation. *J. Bacteriol.* **154:**1110–1116.
276. **Thomas, C. M., and G. Jagura-Burbzy.** 1992. Replication and segregation: the replicon hypothesis revisited, p. 45–80. *In* C. D. S. Hohan and J. A. Cole (ed.), *Prokaryotic Structure and Function: a New Perspective*. Cambridge University Press, Cambridge.
277. **Thwaites, J. J.** 1992. Growth and control of the cell wall: a mechanical model for *Bacillus subtilis*, p. 1–9. *In* M. A. De Pedro, J. V. Holtje, and W. Loffelhardt (ed.), *Bacterial Growth and Lysis: Metabolism and Structure of the Bacterial Sacculus*. Plenum Press, New York.
278. **Thwaites, J. J., and N. H. Mendelson.** 1991. Mechanical behaviour of bacterial cell walls. *Adv. Microb. Physiol.* **32:**173–222.
279. **Thwaites, J. J., U. C. Surana, and A. M. Jones.** 1991. Mechanical properties of *Bacillus subtilis* cell walls: effects of ions and lysozyme. *J. Bacteriol.* **173:**204–210.
280. **Tilby, M. J.** 1977. Helical shape and wall synthesis in bacteria. *Nature* (London) **266:**450–452.
281. **Tipper, D. J., W. Katz, J. L. Strominger, and J.-M. Ghuysen.** 1967. Substituents on the carboxyl group of D-glutamic acid in the peptidoglycan of several bacterial cell walls. *Biochem. J.* **6:**921–929.
282. **Tipper, D. J., and J. L. Strominger.** 1966. Isolation of 4-O-β-N-acetylmuramyl-N-acetylglucosamine and 4-O-β-N,6-O-diacetylmuramyl-N-acetylglucosamine and the structure of the cell wall polysaccharide of *Staphylococcus aureus*. *Biochem. Biophys. Res. Commun.* **22:**48–56.
283. **Tipper, D. J., and J. L. Strominger.** 1968. Biosynthesis of the peptidoglycan of bacterial cell walls. XII. Inhibition of crosslinking by penicillins and cephalosporins: studies in *Staphylococcus aureus*. *J. Biol. Chem.* **243:**3169–3175.
284. **Tipper, D. J., J. L. Strominger, and J. C. Ensign.** 1967. Structure of the cell wall of *Staphylococcus aureus* strain Copenhagen. VI. Mode of action of the bacteriolytic peptidase from *Myxobacter* and the isolation of intact cell wall polysaccharides. *Biochemistry* **6:**906–920.
285. **Todd, J. A., A. N. Roberts, K. Johnstone, P. J. Piggot, G. Winter, and D. J. Ellar.** 1986. Reduced heat resistance of mutant spores after cloning and mutagenesis of the *Bacillus subtilis* gene encoding penicillin-binding protein 5. *J. Bacteriol.* **167:**257–264.
286. **Tomasz, A.** 1968. Biological consequences of the replacement of choline by ethanolamine in the cell wall of pneumococcus: chain formation, loss of transformability, and loss of autolysis. *Proc. Natl. Acad. Sci. USA* **59:**86–93.
287. **Tomasz, A., M. McDonnell, H. Westphal, and E. Zanati.** 1975. Coordinated incorporation of nascent peptidoglycan and teichoic acid into pneumococcal cell walls during growth. *J. Biol. Chem.* **250:**337–341.
288. **Tomasz, A., and M. Westphal.** 1971. Abnormal autolytic enzyme in a pneumococcus with altered teichoic acid composition. *Proc. Natl. Acad. Sci. USA* **68:**2627–2630.
289. **Tsuboi, A., T. Norihiro, and S. Udaka.** 1982. Reassembly in vitro of hexagonal surface arrays in a protein-producing bacterium *Bacillus brevis*. *J. Bacteriol.* **151:**1485–1497.
290. **Tynecka, Z., and J. B. Ward.** 1975. Peptidoglycan synthesis in *Bacillus licheniformis*. The inhibition of cross-linking by benzylpenicillin and cephaloridine *in vitro*. *Biochem. J.* **146:**253–267.
291. **Uchikawa, K., I. Sekikawa, and I. Azuma.** 1986. Structural studies on teichoic acids in cell walls of several serotypes of *Listeria monocytogenes*. *J. Biochem.* **99:**315–329.
292. **Umeda, A., Y. Ueki, and K. Amako.** 1987. Structure of the *Staphylococcus aureus* cell wall determined by the freeze-substitution method. *J. Bacteriol.* **169:**2482–2487.
293. **Van Dam, J. E. G., J. Breg, R. Komen, J. P. Kamerling, and J. F. G. Vliegenthart.** 1989. Isolation and structural studies of phosphate-containing oligosaccharides from alkaline and acid hydrolysates of *Streptococcus pneumoniae* type 6B capsular polysaccharide. *Carbohydr. Res.* **187:**267–286.
294. **Vann, W. F., T.-Y. Liu, and J. B. Robbins.** 1976. *Bacillus pumilis* polysaccharide cross-reactive with meningococcal group A polysaccharide. *Infect. Immun.* **13:**1654–1662.
295. **Verwer, R. W. H., N. Nanninga, W. Keck, and U. Schwarz.** 1978. Arrangement of the glycan chains in the sacculus of *Escherichia coli*. *J. Bacteriol.* **136:**723–729.
296. **Wagner, P. M., and G. C. Stewart.** 1991. Role and expression of the *Bacillus subtilis rodC* operon. *J. Bacteriol.* **173:**4341–4346.
297. **Ward, J. B.** 1973. The chain length of glycans in bacterial cell walls. *Biochem. J.* **133:**395–398.
298. **Ward, J. B.** 1975. Peptidoglycan synthesis in L-phase variants of *Bacillus licheniformis* and *Bacillus subtilis*. *J. Bacteriol.* **124:**668–678.
299. **Ward, J. B.** 1981. Teichoic and teichuronic acids: biosynthesis, assembly and location. *Microbiol. Rev.* **45:**211–243.
300. **Ward, J. B., and C. A. M. Curtis.** 1982. The biosynthesis and linkage of teichuronic acid to peptidoglycan in *Bacillus licheniformis*. *Eur. J. Biochem.* **122:**125–132.
301. **Ward, J. B., and H. Perkins.** 1974. Peptidoglycan biosynthesis by preparations from *Bacillus licheniformis*: cross-linking of newly synthesised chains to preformed cell wall. *Biochem. J.* **139:**781–784.
302. **Ward, J. B., and H. R. Perkins.** 1973. The direction of glycan synthesis in a bacterial peptidoglycan. *Biochem. J.* **135:**721–728.
303. **Warth, A. D., and J. L. Strominger.** 1971. Structure of the peptidoglycan from vegetative cell walls of *Bacillus subtilis*. *Biochemistry* **10:**4349–4358.
304. **Watkinson, B. J., H. Hussey, and J. Baddiley.** 1971. Shared lipid phosphate carrier in the biosynthesis of teichoic acid and peptidoglycan. *Nature New Biol.* **229:**57–59.
305. **Waxman, D. J., D. M. Lindgren, and J. L. Strominger.** 1981. High molecular weight penicillin binding proteins from membranes of bacilli. *J. Bacteriol.* **148:**950–955.
306. **Waxman, D. J., and J. L. Strominger.** 1980. Penicillin-

binding proteins and the mechanism of action of beta-lactam antibiotics. *Annu. Rev. Biochem.* **52:**825–869.

307. **Waxman, D. J., W. Yu, and J. L. Strominger.** 1980. Linear uncrosslinked peptidoglycan secreted by penicillin treated *Bacillus subtilis. J. Biol. Chem.* **255:**11577–11587.
308. **Weidel, W., and H. Pelzer.** 1964. Bag shaped macromolecules. *Adv. Enzymol.* **16:**193–232.
309. **Weppner, L., and F. C. Neuhaus.** 1977. A fluorescent substrate for peptidoglycan synthesis. *J. Biol. Chem.* **252:**2296–2303.
310. **Westmacott, D., and H. R. Perkins.** 1979. Effects of lysozyme on *Bacillus cereus*: rupture of chains of bacteria and enhancement of sensitivity to autolysins. *J. Gen. Microbiol.* **115:**1–11.
311. **White, P. J.** 1977. A survey for the presence of teichuronic acid in walls of *Bacillus megaterium* and *Bacillus cereus. J. Gen. Microbiol.* **102:**435–439.
312. **Wicken, A. J., and J. Baddiley.** 1963. Structure of the intracellular teichoic acids from group D streptococci. *Biochem. J.* **87:**54–59.
313. **Wicken, A. J., J. D. Evans, and K. W. Knox.** 1986. Critical micelle concentrations of lipoteichoic acids. *J. Bacteriol.* **166:**72–77.
314. **Wickus, G. G., and J. L. Strominger.** 1972. Penicillin-sensitive transpeptidation during peptidoglycan biosynthesis in cell free preparations from *Bacillus megaterium. J. Biol. Chem.* **247:**5297–5306.
315. **Wong, W., F. E. Young, and A. N. Chatterjee.** 1974. Regulation of bacterial cell walls: turnover of cell wall in *Staphylococcus aureus. J. Bacteriol.* **120:**837–843.
316. **Wright, J., and J. E. Heckels.** 1975. The teichuronic acid of *Bacillus subtilis* W23 grown in a chemostat under phosphate limitation. *Biochem. J.* **147:**186–189.
317. **Wyke, A. W., J. B. Ward, and M. V. Hayes.** 1982. Synthesis of peptidoglycan *in vivo* in methicillin-resistant *Staphylococcus aureus. Eur. J. Biochem.* **127:**553–558.
318. **Yamada, H., N. Tsukagoshi, and S. Udaka.** 1981. Morphological alterations of cell wall concomitant with protein release in a protein producing bacterium, *Bacillus brevis* 47. *J. Bacteriol.* **148:**322–332.
319. **Yamada, M., A. Hirose, and M. Matsuhashi.** 1975. Association of lack of cell wall teichuronic acid with formation of cell packets of *Micrococcus lysodeikticus* (*luteus*) mutants. *J. Bacteriol.* **123:**678–686.
320. **Yokoyama, K., Y. Araki, and E. Ito.** 1987. Biosynthesis of poly(galactosylglycerolphosphate) in *Bacillus coagulans. Eur. J. Biochem.* **165:**47–53.
321. **Yokoyama, K., Y. Araki, and E. Ito.** 1988. The function of galactosyl phosphorylprenol in biosynthesis of lipoteichoic acid in *Bacillus coagulans. Eur. J. Biochem.* **173:**453–458.
322. **Yokoyama, K., G. N. La Mar, Y. Araki, and E. Ito.** 1986. Structure and functions of linkage unit intermediates in biosynthesis of ribitol teichoic acids in *Staphylococcus aureus* H and *Bacillus subtilis* W23. *Eur. J. Biochem.* **161:**479–489.
323. **Yokoyama, K., H. Mizuguchi, Y. Araki, S. Kaya, and E. Ito.** 1989. Biosynthesis of linkage units for teichoic acids in gram-positive bacteria: distribution of related enzymes and their specificities for UDP-sugars and lipid-linked intermediates. *J. Bacteriol.* **171:**940–946.
324. **Yoneyama, T., Y. Araki, and E. Ito.** 1984. The primary structure of teichuronic acid in *Bacillus subtilis* AHU 1031. *Eur. J. Biochem.* **141:**83–89.
325. **Yoneyama, T., Y. Koike, Y. Araki, H. Arakawa, K. Yokohama, Y. Sasaki, T. Kawamura, E. Ito, and S. Takao.** 1982. Distribution of mannosamine and mannosaminuronic acid among cell walls of *Bacillus* species. *J. Bacteriol.* **149:**15–21.
326. **Young, F. E., C. Smith, and B. E. Reilly.** 1969. Chromosomal location of genes regulating resistance to bacteriophage in *Bacillus subtilis. J. Bacteriol.* **98:**1087–1097.
327. **Zipperle, G. F., J. W. Ezzell, and R. J. Doyle.** 1984. Glucosamine substitution and muramidase susceptibility in *Bacillus anthracis. Can. J. Microbiol.* **30:**553–559.

28. Biosynthesis and Function of Membrane Lipids

DIEGO DE MENDOZA, ROBERTO GRAU, and JOHN E. CRONAN, JR.

Bacillus spp., especially *Bacillus subtilis*, have major advantages for the study of the function of membrane lipids in bacterial membrane physiology. Some properties such as genetic analysis, growth under chemically defined conditions, and ability to produce large numbers of cells for biochemical studies are shared with other bacteria (e.g., *Escherichia coli* and its relatives). However, a major advantage over gram-negative bacteria is the presence in *Bacillus* spp. of only a single membrane rather than the dual membranes (cytoplasmic and outer) found in gram-negative species. The interpretation of studies of lipid function in *E. coli* and other gram-negative bacteria is complicated by the presence of these two membranes. Although most such studies have been interpreted in terms of altered properties of only the inner membrane lipids, such interpretation is generally based only on assumptions and preconceptions concerning the physiological roles of the two membranes.

Despite the advantage of a single membrane, *Bacillus* spp. have been underutilized in membrane lipid research. One reason for this may be that the compositions of the membrane lipids differs markedly from those found in mammalian tissues, which lack both the branched-chain fatty acids and the major amounts of the glycolipid class found in *Bacillus* spp. (although such glycolipids are major components of plant lipids). A second reason is that genetic dissection of membrane lipid synthesis has been little exploited in *Bacillus* spp., very few mutants have been isolated, and no detailed genetic and molecular analyses of any mutant are available. Indeed, to date, no *Bacillus* lipid metabolic gene has been cloned.

Bacillus spp. offer two additional biological phenomena to membrane lipid research. The first is sporulation, during which membrane lipid phenomena accompanying differentiation could be studied. The second is temperature-induced control of the composition of the membrane lipids, in which detectable amounts of unsaturated fatty acids (UFAs) are synthesized by *Bacillus* spp. only at low growth temperatures. In contrast to the situation in other organisms, the *Bacillus* regulatory system seems to involve an "on or off" transcriptional regulatory system rather than only thermal modulation of the activity of a lipid biosynthetic enzyme. We believe major efforts should be directed at understanding these phenomena.

In this review, we focus on the aspects of lipid metabolism that differ from those of the better studied gram-negative bacteria. The primary focus will be *B. subtilis*, because of its advantages in genetic analysis, but data from other *Bacillus* species will be used when appropriate.

COMPLEX LIPID COMPOSITION

The phospholipids of *B. subtilis* consist largely of the common bacterial lipids, phosphatidylethanolamine and phosphatidylglycerol, which together constitute about 80% of the phospholipid fraction, with phosphatidylglycerol being about 75% of this total (4, 26). The remaining phospholipids consist mainly of the lysine ester of phosphatidylglycerol (in which the carboxyl group of lysine is esterified to the terminal hydroxyl group of the nonesterified glycerol moiety) and a small amount of cardiolipin (diphosphatidylglycerol). Major lipid components of *B. subtilis* include the glycolipids diglucosyl-diglyceride and monoglucosyl diglyceride and the neutral lipid 1,2-diglyceride. By analogy with model lipid membranes and other biological membranes (41), the lipids of *B. subtilis* are expected to be structured in a standard lipid bilayer (a lamellar phase).

COMPLEX LIPID BIOSYNTHESIS

The synthetic routes of phospholipid and glycolipid synthesis are assumed to follow the standard pathways documented in other organisms (Fig. 1 and 2). The few cases for which direct data on *Bacillus* spp. are available strengthen this assumption (38). A novel synthetic pathway is that of the synthesis of the lysine ester of phosphatidylglycerol (sometimes called lysylphosphatidylglycerol), in which the donor of the lysyl group is lysyl-tRNA rather than the free amino acid (38).

In bacilli, the synthesis of phospholipids seems to involve a membrane-bound pathway (38), whereas fatty acid synthesis is catalyzed by soluble enzymes in the cytosol (9). Phospholipid synthesis is thought to occur on the inner face of the cytoplasmic membrane, since the enzymes require access to cytosolic precursors (e.g., serine, ATP). Experiments with *Bacillus megaterium* (58, 59) indicate that newly synthesized phosphatidylethanolamine molecules first reside on the cytoplasmic face of the membrane lipid bilayer and are later somehow rotated ("flip-flopped") through the bilayer to become part of the lipid leaflet of the external face of the membrane. Phospholipid synthesis does not appear to be localized to distinct parts of the cell (e.g., poles or septum). In *B. subtilis*, Mindich (47) examined cells pulse-labeled with glyc-

Diego de Mendoza and Roberto Grau • Departamento de Microbiologia, Facultad de Ciencias Bioquimicas y Farmaceuticas, Universidad Nacional de Rosario, Suipacha 531, 2000 Rosario, Argentina. **John E. Cronan, Jr.** • Department of Microbiology and Department of Biochemistry, University of Illinois, Urbana, Illinois 61801.

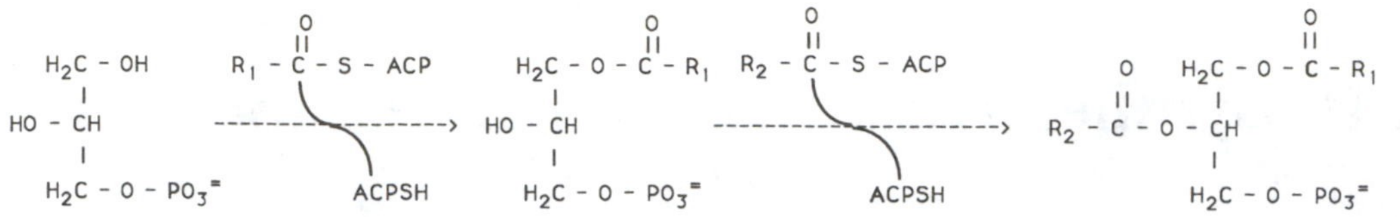

Figure 1. Acylation of *sn*-G3P by two successive transfers of acyl groups from acyl-ACP. This pathway has been demonstrated in *E. coli* and clostridia. *sn*-G3P is probably synthesized by reduction of a glycolytic intermediate such as dihydroxyacetone phosphate or glyceraldehyde 3-phosphate.

erol and found that incorporation of the label into lipid was a random unlocalized process.

FATTY ACID COMPOSITION

Iso and anteiso branched fatty acids are the dominant acyl components of the lipids of *Bacillus* species and are also formed as major constituents in several other species of gram-positive bacteria (34). The major components of *B. subtilis* are anteiso-$C_{15:0}$ and iso-$C_{17:0}$ (10, 31; see Table 1). Myristic and palmitic acids, the most common fatty acids in the majority of organisms, are generally minor constituents in the genus *Bacillus* (Table 1). The fatty acid compositions of *Bacillus* cultures depend on growth temperature. At 37°C, the phospholipids of a typical *B. subtilis* strain contain only saturated fatty acids (31, 33). However, cultures shifted from 37°C to 20°C synthesize a C_{16} monounsaturated fatty acid (24, 25). The synthesis of this UFA is a cold-inducible process and takes place by aerobic desaturation of saturated fatty acids (21, 25). Fatty acid composition can also be altered by supplementation of the growth medium with compounds that become primers for fatty acid synthesis (33, 34).

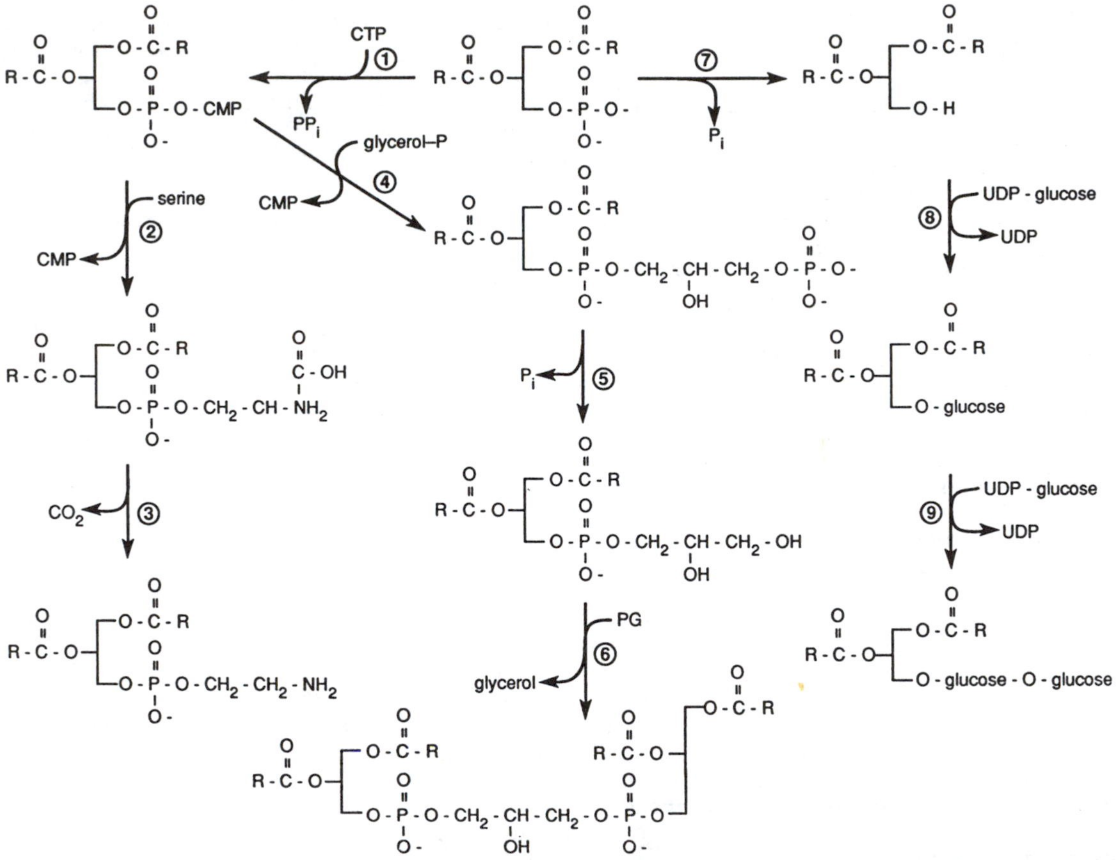

Figure 2. Synthesis of complex lipids. The three phospholipid species found in *E. coli* are synthesized from phosphatidic acid (top center structure) by a series of reactions catalyzed by six enzymes: 1, CDP-diglyceride synthase (phosphatidate cytidyltransferase); 2, phosphatidylserine synthase; 3, phosphatidylserine decarboxylase; 4, phosphatidylglycerol-phosphate synthase; 5, phosphatidylglycerol-phosphate phosphatase; 6, cardiolipin synthase. The same reactions are believed to occur in bacilli. In addition, diglyceride is synthesized by dephosphorylation of phosphatidic acid by phosphatidic acid phosphatase (reaction 7). A portion of the diglyceride is then glucosylated by one or two transfers of glucose from UDP-glucose (reaction 8 and 9). Reactions 7 through 9 have been detected in other organisms but not bacilli. Not shown is the amino acylation of phosphatidylglycerol (the product of reaction 5) by lysyl-tRNA to form lysylphosphatidylglycerol.

Table 1. Fatty acid composition of total membrane lipid extracts from *B. subtilis*[a]

Fatty acid	% of total fatty acids
Iso-C_{12}:	1
Iso-$C_{14:0}$	1
n-$C_{14:0}$	1
Iso-$C_{15:0}$	15
Anteiso-$C_{15:0}$	40
Iso-$C_{16:0}$	5
n-$C_{16:0}$	5
n-$C_{16:1}$	7
Iso-$C_{17:0}$	20
Anteiso-$C_{17:0}$	1
Iso-$C_{17:1}$	1
Anteiso-$C_{17:1}$	1
n-$C_{18:1}$	0.5

[a] Data were taken from reference 10 and are given as percentages by mass. *B. subtilis* BD99 was used for determination of the fatty acid composition. The strain was grown at 30°C in Spizizen salts medium supplemented with yeast extract, the required amino acids, and DL-malate as a carbon source (10).

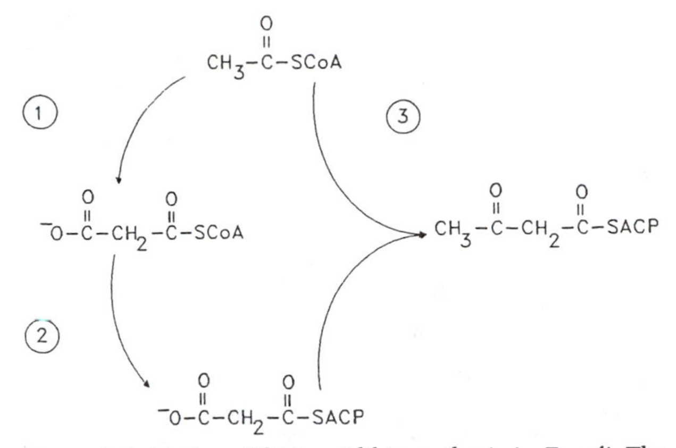

Figure 3. Initiation of fatty acid biosynthesis in *E. coli*. The initiation of new acyl chains is accomplished by the action of three enzymes: 1, acetyl-CoA carboxylase; 2, malonyl-CoA: ACP transacylase; 3, 3-ketoacyl-ACP synthase III.

Thus, *B. subtilis* can synthesize ω-alicyclic fatty acids if it is provided with the appropriate precursors (18, 34).

FATTY ACID BIOSYNTHESIS

Fatty acid synthesis (see Fig. 3 through 6) is the best-studied aspect of *Bacillus* lipid metabolism. We will first focus on the aspects of the pathway utilized in other bacteria and then discuss the product diversification reactions that result in the distinctive fatty acid compositions of bacilli.

OVERALL PATHWAY: ACP

The fatty acid systems of a variety of bacteria depend for their activity on acyl carrier proteins (ACPs) (for a recent review, see reference 64). ACPs are small, very acidic proteins that have a phosphopantetheine prosthetic group covalently linked to a serine residue via a phosphodiester bond (13, 64). Butterworth and Bloch (9) reported that the fatty acid synthetase activity of a 100,000 × *g* supernatant fraction from *B. subtilis* required a heat-stable protein that could be replaced by *E. coli* ACP. The ACP from *B. subtilis* was recently purified to homogeneity and found to be small (molecular weight, 9,000), thermostable, and very acidic (isoelectric point, 3.8) (48). The amino-terminal region (24 amino acids) of this protein has been determined and shows 50% homology with the corresponding sequence of *E. coli* ACP (48) (as does the overall amino acid composition). Moreover, the *B. subtilis* ACP was immunoprecipitated with antibodies against *E. coli* ACP (48) and functioned as an effective acceptor (at about half the activity of similar concentrations of *E. coli* ACP) for labeled palmitate in the reaction catalyzed by the ATP-requiring acylase (acyl-ACP synthetase) of *E. coli* (48). Approximately the same palmitate-accepting activity observed for the *B. subtilis* ACP has been shown for the ACPs of *Rhodobacter sphaeroides*, *Rhizobium meliloti*, and spinach (54).

The cloning and characterization of a gene encoding an ACP potentially involved in fatty acid biosynthesis in the gram-positive filamentous bacterium *Saccharopolyspora erythraea* was recently reported (55). The nucleotide sequence analysis of this gene provided evidence that the ACP gene is clustered with at least one other gene that would be required for fatty acid biosynthesis, i.e., that for a 3-ketoacyl-ACP synthase (55). It will be of considerable interest to clone the ACP gene of *B. subtilis* and determine whether it is located in a cluster containing other genes encoding components of the fatty acid synthase. The *E. coli* ACP gene is located in a gene cluster of six fatty acid synthetic proteins (64). It was recently reported that the *outG* locus from *B. subtilis*, originally identified by differential hybridization to RNA extracted from germinating spores, shows similarity to genes for fatty acid and polyketide antibiotic biosynthesis (1). The identity of this locus is particularly high around the cysteine residue of the active site of the 3-ketoacyl ACP synthase. However, the introduction of deletions into the two open reading frames of this locus did not show any lethal phenotype (1). Thus, it seems that *outG* may be involved in the synthesis of some nonessential secondary metabolite (1).

Synthesis of Acyl Chains

De novo synthesis of saturated fatty acids in bacteria is carried out by two different types of ACP-requiring fatty acid synthetases (FAS): straight chain and branched chain. Most of our knowledge of the biochemical and molecular aspects of the straight-chain FAS from bacteria is based on studies of *E. coli* (for reviews, see references 11, 13, and 64). The major saturated fatty acid of *E. coli*, palmitic acid, is formed by using acetyl coenzyme A (acetyl-CoA) as a primer and malonyl-ACP as the chain extender (11, 13, 60). The initial condensation of malonyl-ACP and acetyl-CoA (Fig. 3) is catalyzed by a recently described acetoacetyl-ACP synthase (28). The resulting 3-ketoester is reduced by an NADPH-dependent 3-ketoacyl-ACP reductase, and a water molecule is then removed by the 3-hydroxyacyl-ACP-dehydrase (Fig. 4). The last step is catalyzed by enoyl ACP reductase to form a saturated acyl-ACP that in turn can serve as the

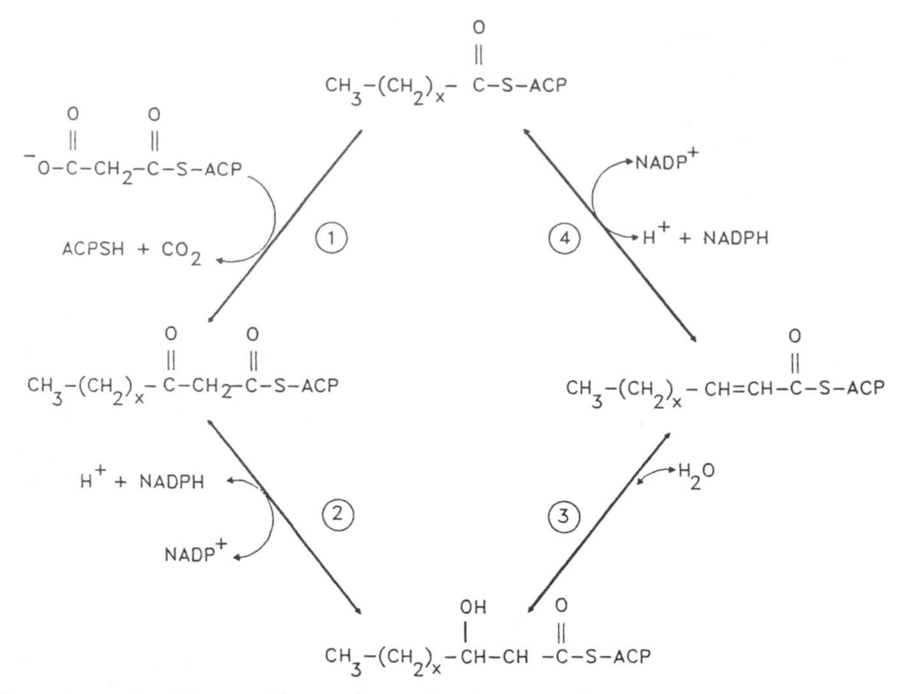

Figure 4. Elongation cycle of fatty acid biosynthesis. The elongation of a growing acyl chains is accomplished by the action of four enzymes: 1, 3-ketoacyl-ACP synthase; 2, 3-ketoacyl-ACP reductase; 3, 3-hydroxyacyl-ACP dehydrase; 4, enoyl reductase (*trans*-2-acyl-ACP reductase).

substrate for another condensation reaction (Fig. 4). Although branched-chain fatty acids are generally much more abundant than straight-chain acids (Table 1), straight-chain acids are abundant in some bacilli (particularly some thermophiles) (34).

The UFAs of bacilli are formed by desaturation of a finished acyl chain having either a straight or a branched chain (24). Desaturation can thus be considered a postsynthetic modification of the acyl chain rather than an aspect of acyl chain synthesis per se.

DIVERSIFICATION OF THE PATHWAY

The small portions of straight-chain even-numbered fatty acids present in *B. subtilis* are assumed to be synthesized from an acetyl-CoA primer for which the fatty acid synthase of *B. subtilis* shows a small but significant activity (9). Since it is known that acetyl-ACP but not acetyl-CoA serves as excellent primer for FAS from *B. subtilis* (9), Kaneda (34) proposed that acetyl-ACP could be formed by decarboxylation of malonyl-ACP. Therefore, malonyl-ACP may function both as a source of primer and as the chain extender. Kaneda also proposed that decarboxylation of malonyl-ACP and elongation of the primer could be catalyzed by a 3-ketoacyl-ACP-synthetase (34). In support of the proposed pathway, ACP-dependent decarboxylation and the ACP-dependent formation of a condensation product from malonyl-CoA has been detected in partially purified preparations of *B. subtilis* (34). Obviously, more experiments are required to support the proposed pathway of straight-chain fatty acid initiation synthesis in *B. subtilis*, especially in view of the report that *E. coli* contains a third condensing enzyme able to catalyze only limited elongation of acyl chains (29). This enzyme has been named acetoacetyl-ACP synthetase because the major product is a four-carbon acid (29). The acetyl donor is acetyl-CoA rather than acetyl-ACP (29). Thus, the substrates in the acetoacetyl-CoA synthetase reaction are acetyl-CoA and malonyl-ACP. The 3-ketoacyl-ACP synthetase and acetyl-CoA–to–ACP transacylase activities reside in the same protein molecule, and thus, fatty acid synthesis may not require the synthesis of acetyl-ACP (29). These findings suggest that *B. subtilis* may possess an enzyme that catalyzes the condensation of both branched short-chain CoAs and acetyl-CoA esters with malonyl-ACP. If this is the case, the ratio between branched- and straight-chain fatty acids produced could be due to the relative affinities of the condensing enzymes for acetyl-CoA or the branched short-chain acyl-CoA esters, respectively (Fig. 1).

Studies on the enzymes of branched-chain fatty acid biosynthesis in *B. subtilis* were initiated at about the same time as those involving palmitic acid synthesis, but work on branched-chain fatty acids has progressed very slowly (33–35). This situation seems to be due mainly to the low activity of the enzymes involved (33–35). It has been reported that the overall rate of fatty acid synthesis in *B. subtilis* extracts is only 2% that seen in comparable extracts of *E. coli* (35). This low activity of branched-chain FAS in cell-free systems has severely impeded study of the individual enzymes. However, the extensive studies performed by Kaneda and coworkers strongly support the idea that branched-chain fatty acids are synthesized by a mechanism very similar to that of straight-chain fatty acid synthesis in *E. coli*, including involvement of ACP in the synthetic reactions (for a recent review, see reference 34).

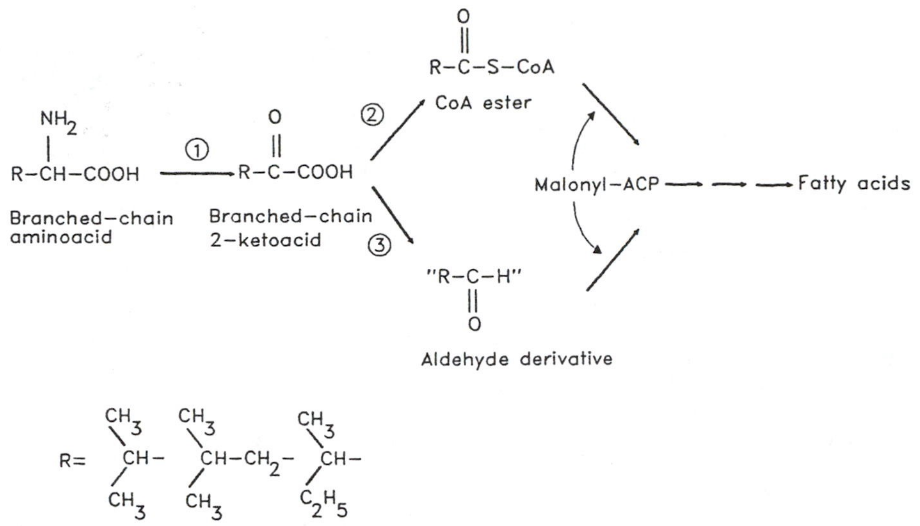

NH2 | R–CH–COOH ① → R–C(=O)–COOH ② → R–C(=O)–S–CoA (CoA ester) → Malonyl–ACP → → → Fatty acids

Branched–chain aminoacid; Branched–chain 2–ketoacid ③ → "R–C(=O)–H" Aldehyde derivative → Malonyl–ACP

R= $(CH_3)_2CH$–; $(CH_3)_2CH$–CH_2–; $CH_3(C_2H_5)CH$–

Figure 5. Proposed pathway for incorporation of branched-chain 2-keto acids into fatty acids in *B. subtilis*. Branched-chain amino acids are converted to branched-chain 2-keto acids by a branched-chain amino acid transaminase (reaction 1). The branched-chain 2-keto acid can then be converted into a CoA ester by a branched-chain 2-keto acid dehydrogenase (reaction 2) or an aldehyde derivative by a branched-chain 2-keto acid decarboxylase (reaction 3). It is assumed that the primers produced by reactions 2 and 3 could then be condensed with malonyl-ACP (see text and Fig. 6).

Since the branched-chain FAS and the straight-chain FAS use different primers, it appears that the two enzyme systems differ mainly in the specificities of their acyl-CoA–ACP transacylases (or 3-ketoacyl-ACP synthases; see above). *B. subtilis* FAS cannot efficiently convert acetyl-CoA to acetyl-ACP (9). If, however, acetyl-CoA is replaced by acetyl-ACP, the synthesis of straight-chain fatty acids from [^{14}C]malonyl-CoA is effectively primed (9). The *B. subtilis* FAS also synthesizes branched-chain fatty acids from branched short-chain acyl-CoA esters such as isobutyryl-CoA as a primer (9). However, if isobutyryl-ACP is provided to the *E. coli* synthase, it produces the appropriate branched-chain fatty acids but only at low rates (9). It is clear, however, that all branched intermediates are handled much less effectively by the *E. coli* enzymes than are those having straight chains (9).

In *B. subtilis*, branched-chain fatty acids are synthesized by using 2-keto acids derived from valine, leucine, and isoleucine as primer sources (32, 49). The major products, in this case, are iso and anteiso fatty acids with 14 to 17 carbons (32, 49). Extensive studies, mostly with *B. subtilis*, have been done to elucidate the pathway of branched-chain keto acid incorporation into fatty acids. Acyl-CoA esters with three to five carbons have been found to be good primers for the fatty acid synthase of *B. subtilis* (32, 33, 35). An NAD-CoA-dependent dehydrogenase that catalyzes the formation of acyl-CoA esters from the ketoacids of valine, leucine, and isoleucine has been detected in cell extracts of *B. subtilis* (50). These observations appear to indicate that CoA esters of isobutyrate, isovalerate, and 2-methylbutyrate should be the obligate primers for fatty acid synthesis from branched-chain 2-keto acid substrates (Fig. 5). However, it was reported that CoA and NAD, which are required for 2-keto acid dehydrogenase activity, are not required for the synthesis of fatty acids from 2-keto acid substrates; rather, they inhibit synthesis (32). Since decarboxylation of a 2-keto acid substrate should be essential for fatty acid synthesis, it was postulated that the primer for fatty acid synthesis could be produced by decarboxylation of 2-keto acids rather than by oxidative decarboxylation (Fig. 5), and the isolation of two keto acid decarboxylases from cell extracts of *B. subtilis* was reported (52). Immunoprecipitation experiments revealed that the branched-chain 2-keto acid decarboxylase but not the other decarboxylase (pyruvate decarboxylase) is essential to the incorporation of branched-chain 2-keto acids substrates into fatty acids (52). The two keto acid decarboxylases have similar ranges of substrate specificity, but branched-chain 2-keto acid decarboxylase has a much higher affinity for branched-chain substrates than does pyruvate decarboxylase (52). These experiments suggest that *B. subtilis* may possess two pathways for the supply of chain initiators of branched fatty acids: (i) from branched short-chain carboxylic acid CoA esters formed by the 2-keto acid dehydrogenase and (ii) from an aldehyde derivative (rather than acyl-CoA) generated by the branched-chain keto-acid decarboxylase (Fig. 5). However, both the exact chemical nature of the aldehyde derivative and the identity of the enzyme catalyzing the condensation of the aldehyde with malonyl ACP remain to be determined (Fig. 5).

GENETIC APPROACHES TO LIPID FUNCTION AND METABOLIC REGULATION

Two efforts that used *E. coli* as a model organism have contributed substantially to our understanding of membrane lipids in bacteria. The first was the isolation of a growing collection of well-characterized mutants affecting lipid metabolism (11, 13, 64). The second is the continuing progress in molecular cloning of the lipid metabolism genes of this organism (64). The availability of the cloned structural genes offers numerous advantages, including (i) structural analysis of the genes and their regulatory elements, (ii) determination of the deduced amino acid sequences of the gene products, (iii) rapid physical mapping of the genes on the chromosome, (iv) examination of the in vivo effects of enzyme overproduction, (v) use of overproducing strains for purification of large amounts of enzyme for structural and enzymological studies, (vi) localized mutagenesis of the gene to obtain additional mutants, and (vii) structural manipulation of chromosomal genes for physiological studies. The impact of the molecular approach on the understanding of lipid metabolism in *E. coli* has been recently reviewed (64). Unfortunately, parallel studies with bacilli are at an early stage.

Mutants Defective in Fatty Acid Synthesis

Although *B. subtilis* is genetically the best-characterized gram-positive organism, it has not been fully exploited for lipid metabolism and membrane research, and only two classes of mutants defective in branched fatty acids have been reported. One mutant (67) lacks the ability to synthesize branched-chain fatty acids because of a deficiency of the branched-chain 2-keto acid dehydrogenase (*bfmA*). The *bfmA* mutation could not be resolved from the *aceA* locus (which encodes the E1α subunit of the pyruvate dehydrogenase complex) by using transduction or transformation procedures. The inability to resolve the *bfmA* and *aceA* mutations was explained by the presence of pyruvate dehydrogenase and branched-chain 2-keto acid dehydrogenase activities in a single complex (40). It was suggested that the mutants *bfmA*, *aceB*, and *aceA* may be synonymous. The *bfmA* mutant can grow on short branched-chain fatty acids that are biosynthetically related to branched amino acids. Furthermore, it was reported that long branched-chain fatty acids and cyclic fatty acids can also serve as a growth supplement (67). The *bfmB* mutation of *B. subtilis* results in an acyl-CoA:ACP transacylase with low affinity for branched acyl-CoA precursor (but not for straight-chain acyl-CoA precursors) (6). The consequence of this mutation is a growth requirement for a branched-chain fatty acid precursor. In our hands, *bfmA* and *bfmB* strains displayed a high frequency of reversion when grown in minimal or rich media, even in the presence of branched-chain fatty acid precursors. The high reversion frequency precluded any further genetic manipulation.

A temperature-sensitive mutant of *B. subtilis* can grow at high temperatures (45°C) if supplemented with 14-methylhexadecanoic acid (a slight growth simulation with palmitic acid was also reported [27]). The mutation remains uncharacterized and to the best of our knowledge has not been investigated further.

Glycerol Auxotrophs

Glycerol auxotrophs are particularly useful in the study of membrane lipid synthesis for two reasons: (i) glycerol is a component of all the phospholipid species of the cell (Fig. 1 and 2), and (ii) the intracellular pools of glycerol or *sn*-glycerol 3-phosphate (G3P) are very small, thereby allowing phospholipid synthesis to be arrested when glycerol is removed from the medium. Glycerol auxotrophy has been shown to occur in two kind of *E. coli* mutants: *gpsA* mutants that lack the biosynthetic G3P dehydrogenase responsible for the synthesis of G3P from dihydroxyacetone phosphate (12) and *plsBX* mutants that owe their glycerol or G3P auxotrophy to a defect in G3P acyltransferase that results in an elevated K_m for G3P (3, 64).

A glycerol auxotroph of *B. subtilis* has been isolated and used in several investigations. The enzymatic defect in this mutant (called *glyc*) has not yet been determined, but it was demonstrated that lipid synthesis is dependent on externally supplied glycerol or G3P (43, 44, 46). Early researchers thought that some of the effects of glycerol deprivation observed with this mutant might result from blockage of the synthesis of the membrane teichoic acids. However, subsequent studies with glycerol auxotrophs of several different microorganisms showed that the results obtained were generally similar in all cases (14, 42), indicating that the effects observed are primarily those pertaining to lipid synthesis rather than to teichoic acid synthesis (43, 47). Moreover, in *B. subtilis*, no differences in response between strains having glycerol teichoic acids or ribitol teichoic acids as major cell wall constituents were observed (43).

Coupling of Fatty Acid and Phospholipid Syntheses

When *B. subtilis* glycerol auxotrophs are deprived of glycerol, net phospholipid synthesis stops immediately (43, 44, 46). Incorporation of radioactively labeled acetate or glucose into phospholipids stops abruptly after glycerol deprivation (46). However, a considerable incorporation of label (about 20 to 50% compared with supplemented cultures) into free fatty acids is seen (46). It has been shown that the free fatty acids are turning over rapidly and that the rate of fatty acid synthesis, as calculated from the observed rate of accumulation of free fatty acids and the known rate of turnover, is normal (46). The free fatty acids that accumulate have different chain length patterns than normal fatty acids (46). In *B. subtilis*, the normal branched-chain fatty acids are primarily C_{15} and C_{17}, but the accumulated free fatty acids present in glycerol-deprived cells primarily have 17 and 19 carbon atoms (46). These experiments argue that in *B. subtilis*, the rate of fatty acid synthesis either is not coupled to the rate of phospholipid synthesis or is controlled by a feedback regulation mechanism dependent on the synthesis of complex lipids. A different conclusion concerning the regulation of fatty acid synthesis by phospholipid synthesis in *E. coli* was first reported by Mindich (46), who found that fatty acid synthesis was completely inhibited when phospholipid synthesis was blocked in a *gpsA* strain. However, these experiments failed to account for the influence of fatty acid degradation. Subsequently, Cronan et al. (14) blocked fatty acid degradation and found that, similar to the situation in *B. subtilis*, free fatty acids with abnormally long chains accumulated in the absence of phospholipid synthesis. The conclusion that fatty acid and phospholipid syntheses are not tightly coupled was challenged by Nunn et al. (51), who concluded that fatty acid and phospholipid syntheses are coordinately inhibited during glycerol starvation. More recently, an estimation (57) of the acyl-ACP and acyl-CoA pools of *plsB* (or *gpsA*) pantothenate auxotrophs deprived of glycerol indicated that fatty acid synthesis proceeds in the absence of phospholipid synthesis at 20 to 30% of the normal rate. Thus, it seems that in both gram-positive and gram-negative bacteria, fatty acid and phospholipid synthescs are not tightly coupled.

It should be noted that although the results obtained with *B. subtilis glyc* mutants have been interpreted in terms of manipulation of phospholipid synthesis, the neutral lipids also contain glycerol (Fig. 2). Since these lipids (the glycolipids and diglyceride) are major components of the lipid fraction, inhibition of the synthesis of these compounds will also be blocked by glycerol deprivation. However, since phosphatidic acid, a key phospholipid precursor, is also the neutral lipid precursor (Fig. 2), the kinetics of inhibition probably follow those of phospholipid synthesis.

Phospholipid Synthesis and Insertion of Membrane Proteins

Most cells maintain a constant ratio of proteins (both whole cell and membrane) to phospholipids: neither membrane phospholipids nor proteins are overproduced. The mechanism responsible for this homeostasis is unknown. The simplest mechanism would be a strict coupling of the synthesis of membrane proteins and the synthesis of membrane phospholipids. It has been reported that concomitant phospholipid synthesis was required for assembly of the lactose transport system of the *E. coli* membranes (for reviews, see references 11 and 13). However, subsequent studies found the lactose transporter to be inserted normally in the absence of phospholipid synthesis (45, 66) and demonstrated that the previous results were due to invalid transport assays and extensive nonspecific damage to the cells (11, 13). In agreement with the *E. coli* results, it was demonstrated that in gram-positive bacteria, transport proteins were integrated into the membranes in the absence of phospholipid synthesis. Using *glyc* mutants of *B. subtilis* and *Staphylococcus aureus*, it was found that in the absence of lipid synthesis, the citrate transporter (66) and the lactose transporter (45) were normally integrated into membranes. Membranes isolated from *glyc* mutants of *B. subtilis* deprived of glycerol show an increase in protein content but no increase in lipid content, indicating that membrane proteins are made and integrated into the membrane in the absence of lipid synthesis (44). In *E. coli* mutants blocked in the first step of phospholipid synthesis, both the inner and the outer membranes became enriched with protein in the absence of phospholipid synthesis (42). These results demonstrate that the membranes of both gram-positive and gram-negative bacteria are not normally saturated with proteins and that the synthesis of phospholipids is not required for the synthesis and insertion of bulk membrane protein. The amount of lipid synthesized by *E. coli* can be increased twofold by overproduction of inner membrane proteins (13), indicating that the enzymes of lipid synthesis are all synthesized in functional excess. A similar study has not yet been reported for gram-positive bacteria.

UFA Synthesis: Regulation by Growth Temperature

It is well documented that bacteria adjust the fatty acid compositions of their phospholipids in response to growth temperature (13, 15, 16, 60). As the growth temperature decreases, the incorporation of proportionally more low-melting-point fatty acids into membrane lipids occurs. This adaptative response functions to lower the temperature of the order-disorder lipid-phase transition and thus optimizes membrane function at lower growth temperatures (13, 15, 16, 60).

The method used to alter the unsaturation of membrane lipids in response to growth temperature depends on the mechanism for synthesizing UFAs. In bacteria, either anaerobic or aerobic mechanisms are responsible for the synthesis of UFAs (reviewed in reference 16). The anaerobic pathway, elucidated in detail with *E. coli*, produces *cis*-UFA by the intervention of a specific 2,3-dehydrase acting at the C_{10} level (reviewed in reference 5). In certain bacteria (usually gram positive) and in all eukaryotic life-forms, double bonds are introduced into fatty acids by a common general mechanism in which the reaction is catalyzed by an oxygen-dependent desaturation system requiring the participation of a specific electron transport chain (16, 21).

The molecular mechanism of temperature regulation has been extensively studied in *E. coli* (reviewed in references 13 and 15). In this organism, as the temperature of growth decreases, the proportion of *cis*-vaccenate increases greatly. This regulatory response does not require induction of a protein (23), and the mechanism functions by thermal modulation of the soluble enzyme 3-keto acyl-acyl carrier protein synthase II (17).

A different mechanism of thermal regulation of UFA synthesis has been reported in a large number of *Bacillus* species that synthesize UFAs by the oxygen-dependent desaturation mechanism (21, 22). In these bacteria, there is an induction of a desaturating system when the cultures are grown at a low temperature (reviewed in reference 22). This adaptative response has been extensively characterized in *Bacillus megaterium* (20, 22) and more recently investigated in *B. subtilis* (24). *B. subtilis* growing at 37°C synthesizes only saturated fatty acids; however, when a culture growing at 37°C is transferred to 20°C, synthesis of a C_{16} monounsaturated fatty acid is induced (24, 25). Similar to *B. megaterium*, the proportion of UFAs initially formed in cultures of *B. subtilis* transferred from 37 to 20°C far exceed the levels of synthesis found in cultures grown for several generations at 20°C (22, 24, 25). Other bacilli synthesize diunsaturated acids in addition to monounsaturates in a temperature-dependent manner (22).

To explain the dramatic change in lipid composition of *Bacillus* spp. shifted from 37 to 20°C, it was proposed that transcription of the fatty acid desaturase gene(s) can occur only at low temperatures (20). To account for the unexpectedly large initial degree of unsaturation seen immediately after a downward temperature shift, the existence of a modulator protein whose synthesis also proceeds at lower temperatures but only following a brief delay was postulated (20, 22). Thus, the rapid desaturation taking place in freshly down-shifted cells would soon be moderated to a rate yielding the steady-state level of fatty acid unsaturation characteristic for that temperature (20, 22).

Recent results suggest that temperature-induced desaturation is more complex than the simple model given above. Mutants of *B. subtilis* selected for resistance to the protonophore carbonyl cyanide *m*-chlorophenylhydrazone (CCCP) (26) have decreased contents of monounsaturated straight-chain C_{16} fatty acids (ca. 50% of the normal levels). An analysis of these mutants found that they lack an NADH-dependent palmitoyl-CoA desaturase activity (19). Revertants of these strains regained palmitoyl-CoA desaturase activity and normal UFA contents, whereas temperature-sensitive revertants lacked desaturase activity only at high temperatures. These data argue strongly that fatty acid composition plays a causal role in protonophore resistance. However, the in vivo data seem only partially explained. First, although the CCCP-resistant mutants completely lack palmitoyl-CoA desaturase activity, the UFA content was decreased only by half (19). Second, cells grown at high temperatures contain appreciable desaturase activity, although no UFA were synthesized at this temperature. The thermolability of the enzyme in vitro also seemed to differ depending on the growth temperature of the cells examined. Palmitoyl-CoA desaturase activity required both membrane and cytosolic components and thus seems complex in both enzymatic and regulatory functions. These data indicate that *B. subtilis* contains an alternate UFA synthetic pathway accounting for the remaining UFA content of the CCCP-resistant mutants. It should be noted that *B. megaterium* seems to lack this alternate pathway. A CCCP-resistant mutant of this organism contains no UFA or detectable desaturase (19).

The alternate pathway of *B. subtilis* seems to be desaturation of fatty acyl groups of intact phospholipids. Strong evidence for this pathway has been obtained by examining the UFAs that accumulate in cultures of *B. subtilis* in which acyl transfer is blocked by starvation of glycerol auxotrophs (25). Starvation of these auxotrophs results in the accumulation of free fatty acids due to inhibition of acyl transfer caused by the lack of acyl acceptor. These free fatty acids are the product of de novo synthesis (46), and if fatty acyl-thioesters are used as desaturation substrates, synthesis of UFAs would be expected in these cells following temperature shift. However, no free UFAs were formed after glycerol deprivation (25). The inability of glycerol-starved cells to desaturate those fatty acyl moieties not present in phospholipids was not due to reduced induction of desaturation (i.e., resulting from the absence of phospholipid biosynthesis), since lipids labeled before glycerol starvation were readily desaturated (25). When the CoA pool was depleted by pantothenate starvation of pantothenate auxotrophs (2), prelabeled lipids of starved cells were desaturated

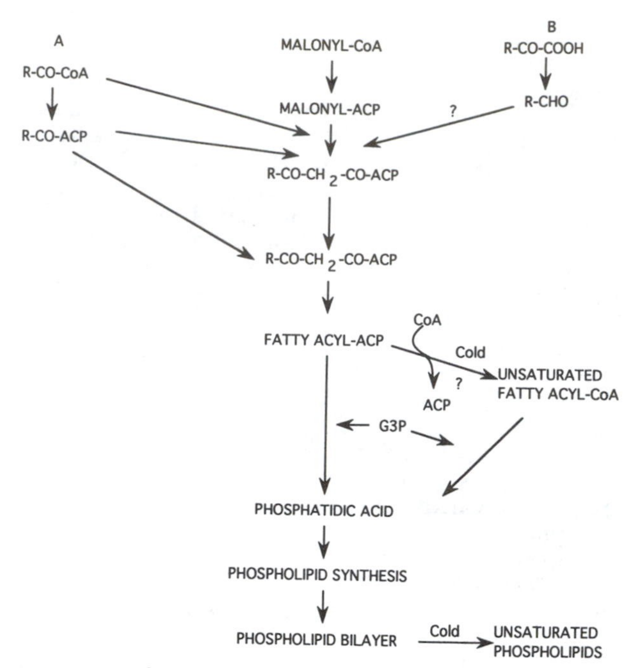

Figure 6. Pathways of branched-chain fatty acid and UFA syntheses in *B. subtilis*. (A) Pathway of synthesis from branched-chain acyl-CoA esters as primers. (B) The other proposed pathway (52) of synthesis from branched-chain 2-keto acids as primer sources. UFAs are formed by cold-induced desaturation of phospholipids (25) or acyl-CoAs (19).

upon transfer from 37 to 20°C (25). These results argue strongly that in vivo, the fatty acid moieties of membrane phospholipids (rather than fatty acid thioesters) are substrates of a cold-induced desaturation system of *B. subtilis* (Fig. 6). Clearly, a molecular understanding of the relationship between the two desaturation systems and the effects observed in vivo will require much further effort.

Function of Branched-Chain Fatty Acids

An important lesson learned from studies of the various mutants of *E. coli* blocked in fatty acid synthesis is that the organism tolerates a wide variation in the fatty acid composition of the membrane phospholipids. *E. coli* UFA auxotrophs (reviewed in reference 13) can grow if the medium is supplemented with any of a large number of fatty acids. Although saturated fatty acids will not support growth, a wide variety of *cis*-UFA (mono-, di-, or triunsaturated) support growth of UFA auxotrophs (13). Indeed, even UFAs with centrally located *trans* double bonds (a type of fatty acid not found in *E. coli* and very rarely found in nature) will suffice. It is clear that the double bond per se plays no chemical role in metabolism. The role of the double bond is only to decrease the temperature of the phase transition of the phospholipid in which it resides, since a number of fatty acids lacking double bonds also support growth. The acids that support growth (*cis*- or *trans*-cyclopropane, branched, centrally brominated) do, however, share with fatty acid double bonds the abilities to disrupt the close packing of phospholipid acyl chains and lower the temperature of the phase transition. Again, this property is purely physical. The presence of a substituent or a double bond in the middle of the hydrocarbon chain sterically disrupts strong hydrophobic interactions with other acyl chains (41). The finding that *E. coli fabA* and *fabB* mutants require an unsaturated (or equivalent) fatty acid for growth indicates that a membrane composed of phospholipids containing only saturated fatty acids is not functional (reviewed in references 11 and 13). Indeed, these mutants undergo cell lysis when deprived of the UFA supplement.

It seems that the physical states of the membrane lipids have a similar importance in bacilli. *B. subtilis* has membrane lipids that contain mainly iso-branched and anteiso-methyl-branched saturated acyl chains (Table 1). The physicochemical effect of a methyl branch in a long chain is similar to that of a *cis* double bond. Both alterations of the hydrocarbon chain cause a decrease in the melting point and an increase of the surface area in monolayer films formed from the corresponding free fatty acids or phosphoglycerides (41, 63). Iso-branched acyl phosphatidylcholines show gel-to-liquid crystalline phase transition temperatures about 20°C below those of the corresponding straight-chain phosphatidylcholines (63).

Studies with mutants defective in fatty acid synthesis support these model studies. A *bfmA* mutant of *B. subtilis* lacking 2-keto acid dehydrogenase was dependent on an exogenous supply of branched-chain fatty acids, whereas straight-chain fatty acids failed to support growth (67). These results were further confirmed by the use of the antibiotic cerulenin to block de novo fatty acid synthesis in *B. subtilis* (65). Upon cerulenin treatment, the bacteria ceased growth but remained completely viable. At 37°C, growth could be restored by supplementation with a mixture of 12-methyltetradecanoic acid and palmitic acid but not with a mixture of long straight-chain fatty acids (65). Moreover, *bfmB* mutants of *B. subtilis* grown in a medium supplemented with the required branched amino acids were very sensitive to inhibition by low concentrations of the straight-chain precursors butyrate, propionate, and pentanoate. The inhibition, which was accompanied by a large increase in the fraction of straight-chain fatty acids in membrane lipids, should result in an increased temperature transition. Both inhibition of growth and increased straight-chain fatty acid content were prevented by the addition of a branched fatty acid to the medium (6). Thus, straight-chain and branched-chain fatty acids are antagonists (as expected from the behavior of phospholipid model membranes).

Since spore formation in bacilli requires major changes in membrane structure, it seemed plausible that the fatty acid composition required for differentiation might differ from that required for growth. Differences between the fatty acid compositions of vegetative cells and spores of *Bacillus thuringiensis* (8), *B. megaterium* (62), and *B. subtilis* have been reported (56). However, it is not known whether these changes reflect the change in medium composition during the postexponential phase leading to sporulation or during sporulation per se. To address the question of whether certain branched-chain fatty acids are required for sporulation, the growth and sporulation of a *bfmB* mutant of *B. subtilis* growing in medium containing one of a wide range of fatty acid precursors were examined (7). The *bfmB* mutant could grow and sporulate in synthetic medium containing any branched or cyclic fatty acid primer (7). The fatty acid profiles of the lipids extracted from spores were similar to those seen in vegetative cells grown under the same conditions, i.e., the anteiso-fatty acids predominated in spores produced in the presence of isovalerate, and the iso-even fatty acids predominated in spores produced in the presence of isobutyrate (7). However, isolated spores contained only 10% or less of straight-chain fatty acids, whereas vegetative cells contained a much higher fraction of straight-chain fatty acids (7). An extreme result was found in cultures grown in 3-methylcrotonate: vegetative cells had 73% straight-chain fatty acids, whereas spores had only 9% (7). The highest straight-chain fatty acid content of spores was 28% for cultures grew in cyclobutanecarboxylate (7). These results demonstrate that sporulation in *B. subtilis* is independent of fatty acid composition as long as the fraction of bulkier fatty acids is sufficiently high (at least 70%) so that the phospholipid acyl chains are loosely packed and flexible.

Function of UFAs

The finding that an in vivo substrate for desaturation seems to be membrane lipids (25, 39) provides the cell with a rapid mechanism for decreasing the fluidity of preexisting membranes upon temperature decrease. The old lipids can be retailored to the new temperature. However, the membranes of *B. subtilis* (growing at 37°C) contain more than 90% methyl branched residues (33). Since the physicochemical effect of a methyl branch in a long chain is similar to that of a *cis* double bond (39), it seems possible that desaturation of lipids has a function(s) other than decreasing the temperature of transition of *B. subtilis* membrane lipids. Protonophore-resistant mutant strains of *B. subtilis* have increased saturated fatty acid/UFA ratios in the membrane phospholipids (26). The mutants exhibited elevated levels of membrane ATPase activity and synthesized ATP more effectively than the wild-type parent under conditions when the electrochemical proton gradient (μH^+) had been reduced by protonophores (26). Growth of mutants in the presence of palmitoleic acid restored the levels of UFA and resulted in a pronounced diminution in the protonophore resistance of growth and ATP synthesis (36). These findings support a hypothesis that specific changes in membrane lipid composition underlie the bioenergetic changes associated with protonophore resistance. However, more experiments are necessary to understand the mechanism by which a decrease in UFA leads to altered energy-coupling properties and specifically to the greater efficacy of ATP synthesis at low μH^+. Psychrophilic *Bacillus* species are known to be unusually rich in *cis*-5-monounsaturated acids (34), and it will be interesting to see whether these organisms are unusually sensitive to protonophores.

OUTLOOK

Study of the synthesis and functions of *B. subtilis* lipids is at an early stage. Despite the scanty and spotty nature of current knowledge, an understanding of the role of lipids in membrane function is beginning to emerge. A focused attack using the new armament of *B. subtilis* genetic tools should yield much more information.

Acknowledgments. This work was partially supported by grants from Fundacion Antorchas (Argentina) and Consejo Nacional de Investigaciones Científicas y Técnicas (CONICET). R.G. is a fellow from CONICET, and D.d.M. is a Career Investigator from the same institution.

REFERENCES

1. **Albertini, A. M., C. Scotti, and A. Galizzi.** 1991. The *Bacillus subtilis out*G locus shows similarity to genes for fatty acids and polyketide biosynthesis. Abstr. 6th Int. Conf. Bacilli 1991.
2. **Baigori, M., R. Grau, H. R. Morbidoni, and D. de Mendoza.** 1991. Isolation and characterization of *Bacillus subtilis* mutants blocked in the synthesis of pantothenic acid. *J. Bacteriol.* **173:**4240–4242.
3. **Bell, R. M.** 1975. Mutants of *Escherichia coli* defective in membrane phospholipid synthesis. Properties of wild type and Km defective *sn*-glycerol-3-phosphate acyltransferase activities. *J. Biol. Chem.* **250:**7147–7152.
4. **Bishop, D. G., L. Rutberg, and B. Samuelson.** 1967. The chemical composition of the cytoplasmic membrane of *Bacillus subtilis*. *Eur. J. Biochem.* **2:**448–453.
5. **Bloch, K.** 1970. β-Hydroxydecanoyl thioester dehydrase. *Enzymes* **5:**441–464.
6. **Boudreaux, D. P., E. Eisenstat, T. Ijima, and E. Freese.** 1981. Biochemical and genetic characterization of an auxotroph of *Bacillus subtilis* altered in the acyl CoA: acyl-carrier-protein-transacylase. *Eur. J. Biochem.* **115:** 175–181.
7. **Boudreaux, D. P., and E. Freese.** 1981. Sporulation in *Bacillus subtilis* is independent of membrane fatty acid composition. *J. Bacteriol.* **148:**480–486.
8. **Bulla, L. A., K. W. Nickerson, T. L. Mount, and J. J. Iandolo.** 1975. Biosynthesis of fatty acids during germination and outgrowth of *Bacillus thuringiensis* spores, p. 520–525. *In* P. Gerhardt, R. N. Costilow, and H. L. Sadof (ed.), *Spores VI*. American Society for Microbiology, Washington, D.C.
9. **Butterworth, P. H. W., and K. Bloch.** 1970. Comparative aspects of fatty acid synthesis in *Bacillus subtilis* and *Escherichia coli*. *Eur. J. Biochem.* **12:**496–501.
10. **Clejan, S., T. A. Krulwich, K. R. Mondrus, and D. Seto-Young.** 1986. Membrane lipid composition of obligately and facultatively alkalophilic strains of *Bacillus* spp. *J. Bacteriol.* **168:**334–340.
11. **Cronan, J. E., Jr.** 1978. Molecular biology of bacterial membrane lipids. *Annu. Rev. Biochem.* **47:**163–189.
12. **Cronan, J. E., Jr., and R. M. Bell.** 1974. Mutants of *Escherichia coli* defective in membrane phospholipid synthesis: mapping of the structural gene for L-glycerol-3-phosphate dehydrogenase. *J. Bacteriol.* **118:**598–605.
13. **Cronan, J. E., Jr., and C. O. Rock.** 1987. Biosynthesis of membrane lipids, p. 474–497. *In* F. C. Neidhart, J. L. Ingrahan, K. B. Low, B. Magasanik, M. Schaechter, and H. E. Umbarger (ed.), *Escherichia coli and Salmonella typhimurium: Cellular and Molecular Biology*, vol. 1. American Society for Microbiology, Washington, D.C.
14. **Cronan, J. E., Jr., L. J. Weisberg, and R. G. Allen.** 1975. Regulation of membrane lipid synthesis in *Escherichia coli*: accumulation of free fatty acids of abnormal length during inhibition of phospholipid synthesis. *J. Biol. Chem.* **250:**5835–5840.
15. **de Mendoza, D., and J. E. Cronan, Jr.** 1983. Thermal regulation of membrane lipid fluidity in bacteria. *Trends Biochem. Sci.* **8:**49–52.
16. **de Mendoza, D., and R. N. Farias.** 1988. Effect of fatty acid supplementation on membrane fluidity in microorganisms, p. 119–149. *In* R. C. Aloia, C. C. Curtain, and L. M. Gordon (ed.), *Physiological Regulation of Membrane Fluidity*, vol. 3. Alan R. Liss, Inc., New York.
17. **de Mendoza, D., A. Klages-Ulrich, and J. E. Cronan, Jr.** 1983. Thermal regulation of membrane fluidity in *Escherichia coli*: effects of overproduction of β-keto-acyl-acyl-carrier protein synthase I. *J. Biol. Chem.* **258:**2098–2101.
18. **Dreher, R., K. Poralla, and W. A. Koning.** 1976. Synthesis of ω-alicyclic fatty acids from cyclic precursors in *Bacillus subtilis*. *J. Bacteriol.* **127:**1136–1140.
19. **Dunkley, E. A., Jr., S. Clejan, and T. A. Krulwich.** 1991. Mutants of *Bacillus* species isolated on the basis of protonophore resistance are deficient in fatty acid desaturase activity. *J. Bacteriol.* **173:**7750–7755.
20. **Fujii, D. K., and A. J. Fulco.** 1977. Biosynthesis of unsaturated fatty acids by *Bacilli*: hyperinduction and modulation of desaturase synthesis. *J. Biol. Chem.* **252:** 3660–3670.
21. **Fulco, A. J.** 1969. The biosynthesis of unsaturated fatty acids by *Bacilli*. I. Temperature induction of the desaturation reaction. *J. Biol. Chem.* **244:**889–895.
22. **Fulco, A. J.** 1983. Fatty acid metabolism in bacteria. *Prog. Lipid Res.* **22:**133–160.
23. **Garwin, J. L., and J. E. Cronan, Jr.** 1980. Thermal modulation of fatty acid synthesis in *Escherichia coli* does not involve de novo synthesis. *J. Bacteriol.* **141:** 1457–1459.
24. **Grau, R., and D. de Mendoza.** 1993. Ph.D. thesis. University of Rosario, Rosario, Argentina.
25. **Grau, R., and D. de Mendoza.** Regulation of the synthesis of unsaturated fatty acids by growth temperature in *Bacillus subtilis*: evidence that phospholipids are substrates of a cold-inducible fatty acid desaturation pathway. *Mol. Microbiol.*, in press.
26. **Guffanti, A. A., S. Clejan, L. H. Falk, D. B. Hicks, and T. A. Krulwich.** 1987. Isolation and characterization of uncoupler resistant mutants of *Bacillus subtilis*. *J. Bacteriol.* **169:**4479–4485.
27. **Holgrem, E.** 1978. A mutant of *Bacillus subtilis* with temperature sensitive synthesis of fatty acids. *FEMS Microbiol. Lett.* **3:**327–329.
28. **Jackowski, S., C. M. Murphy, J. E. Cronan, Jr., and C. O. Rock.** 1989. Acetoacetyl-acyl carrier protein synthase: a target for the antibiotic thiolactomycin. *J. Biol. Chem.* **264:**7624–7629.
29. **Jackowski, S., and C. O. Rock.** 1987. Acetoacetyl-acyl carrier protein synthase as potential regulator of fatty acid biosynthesis in bacteria. *J. Biol. Chem.* **262:**7927–7931.
30. **Kaneda, T.** 1963. Biosynthesis of branched-chain fatty acids. I. Isolation and identification of fatty acids from *Bacillus subtilis* (ATCC 7059). *J. Biol. Chem.* **238:**1222–1228.
31. **Kaneda, T.** 1967. Fatty acids in the genus *Bacillus*. I. Iso- and anteiso-fatty acids as characteristic constituents of lipids in 10 species. *J. Bacteriol.* **93:**894–903.
32. **Kaneda, T.** 1973. Biosynthesis of branched long-chain fatty acids from the related short-chain α-keto acid substrates by a cell free system of *Bacillus subtilis*. *Can. J. Microbiol.* **19:**87–96.
33. **Kaneda, T.** 1977. Fatty acids of the genus *Bacillus*: an example of branched-chain preference. *Bacteriol. Rev.* **41:**391–418.
34. **Kaneda, T.** 1991. Iso- and anteiso-fatty acids in bacteria: biosynthesis, function, and taxonomic significance. *Microbiol. Rev.* **55:**288–302.
35. **Kaneda, T., and E. J. Smith.** 1980. Relationship of primer specificity of fatty acid *de novo* synthetase to fatty acid composition in ten species of bacteria and yeast. *Can. J. Microbiol.* **26:**893–898.
36. **Krulwich, T. A., S. Clejan, L. Falk, and A. A. Guffanti.**

1987. Incorporation of specific exogenous fatty acids into membrane lipids modulates protonophore resistance in *Bacillus subtilis*. *J. Bacteriol.* **169:**4479–4485.

37. **Krulwich, T. A., P. G. Quish, and A. A. Guffanti.** 1990. Uncoupler-resistant mutants of bacteria. *Microbiol. Rev.* **54:**52–65.
38. **Lennarz, W. J.** 1970. Bacterial lipids, p. 210–270. *In* J. J. Wakil (ed.), *Lipid Metabolism*. Academic Press, Inc., New York.
39. **Lombardi, F. J., and A. J. Fulco.** 1980. Temperature-mediated hyperinduction of fatty acid desaturation in pre-existing and newly formed fatty acid synthesized endogenously in *Bacillus megaterium*. *Biochim. Biophys. Acta* **618:**359–363.
40. **Lowe, P. N., J. A. Hodgson, and R. N. Perham.** 1983. Dual role of a single multienzyme complex in the oxidative decarboxylation of pyruvate and branched-chain 2-oxoacids in *Bacillus subtilis*. *Biochem. J.* **215:**133–140.
41. **McElhaney, R. N.** 1982. Effects of membrane lipids on transport and enzymic activities. *Curr. Top. Membr. Transp.* **17:**317–380.
42. **McIntyre, T. M., B. K. Chamberlain, R. E. Webster, and R. M. Bell.** 1977. Mutants of *Escherichia coli* defective in membrane phospholipid synthesis. Effects of cessation and reinitiation of phospholipid synthesis on macromolecular synthesis and phospholipid turnover. *J. Biol. Chem.* **252:**4487–4493.
43. **Mindich, L.** 1970. Membrane synthesis in *Bacillus subtilis*. I. Isolation and properties of strains bearing mutations in glycerol metabolism. *J. Mol. Biol.* **49:**415–432.
44. **Mindich, L.** 1970. Membrane synthesis in *Bacillus subtilis*. II. Integration of membrane proteins in the absence of lipid synthesis. *J. Mol. Biol.* **49:**433–439.
45. **Mindich, L.** 1971. Induction of *Staphylococcus aureus* lactose permease in the absence of glycerolipid synthesis. *Proc. Natl. Acad. Sci. USA* **68:**420–424.
46. **Mindich, L.** 1972. Control of fatty acid synthesis in bacteria. *J. Bacteriol.* **110:**96–102.
47. **Mindich, L.** 1975. Studies on bacterial membrane biogenesis using glycerol auxotrophs, p. 429–455. *In* A. Tzagolof (ed.), *Membrane Biogenesis: Mitochondria, Chloroplasts and Bacteria*. Plenum Press, New York.
48. **Morbidoni, R., M. Baigori, D. de Mendoza, and J. E. Cronan, Jr.** Unpublished results.
49. **Naik, D. N., and T. Kaneda.** 1974. Biosynthesis of branched long-chain fatty acids by species of *Bacillus*: relative activities of three α-keto acids substrates and factors affecting chain length. *Can. J. Microbiol.* **20:**1701–1708.
50. **Namba, Y., K. Yoshizawa, A. Ejima, T. Hayashi, and T. Kaneda.** 1969. Coenzyme A and nicotinamide adenine dinucleotide-dependent branched chain α-keto acid dehydrogenase. I. Purification and properties of the enzyme from *Bacillus subtilis*. *J. Biol. Chem.* **244:**4437–4447.
51. **Nunn, W. D., D. L. Kelley, and N. Y. Stumfall.** 1977. Regulation of fatty acid synthesis during glycerol starvation of glycerol auxotrophs of *Escherichia coli*. *J. Bacteriol.* **132:**526–531.
52. **Oku, H., and T. Kaneda.** 1988. Biosynthesis of branched-chain fatty acids in *Bacillus subtilis*. A decarboxylase is essential for branched-chain fatty acid synthetase. *J. Biol. Chem.* **263:**18386–18396.
53. **Op den Kamp, J. A. F., I. Redai, and L. L. M. van Deenen.** 1969. Phospholipid composition of *Bacillus subtilis*. *J. Bacteriol.* **99:**298–303.
54. **Platt, M. W., K. J. Miller, W. S. Lane, and E. P. Kennedy.** 1990. Isolation and characterization of the constitutive acyl carrier protein from *Rhizobium meliloti*. *J. Bacteriol.* **172:**5440–5444.
55. **Revill, W. P., and P. F. Leadlay.** 1991. Cloning, characterization, and high-level expression in *Escherichia coli* of the *Saccharopolyspora erythraea* gene encoding an acyl carrier protein potentially involved in fatty acid biosynthesis. *J. Bacteriol.* **173:**4379–4385.
56. **Rigomier, D., and B. Lubochinsky.** 1974. Metabolisme des phospholipides chez des mutants asporogenes de *Bacillus subtilis* au cours de la croissance exponentialle. *Ann. Microbiol.* **125B:**295–303.
57. **Rock, C. O., and S. Jackowski.** 1982. Regulation of phospholipid synthesis in *Escherichia coli*. Composition of the acyl-acyl carrier protein pool *in vivo*. *J. Biol. Chem.* **257:**10759–10765.
58. **Rothman, J. E., and E. P. Kennedy.** 1977. Asymmetrical distribution of phospholipids in the membrane of *Bacillus megaterium*. *J. Mol. Biol.* **110:**603–618.
59. **Rothman, J. E., and E. P. Kennedy.** 1977. Rapid transmembrane movement of newly synthesized phospholipids during membrane assembly. *Proc. Natl. Acad. Sci. USA* **74:**1821–1825.
60. **Russell, N. J.** 1984. Mechanisms of thermal adaptation in bacteria: blueprints for survival. *Trends Biochem. Sci.* **9:**108–112.
61. **Saito, K.** 1960. Chromatographic studies on bacterial fatty acids. *J. Biochem.* **47:**699–719.
62. **Scandella, C. J., and A. Kornberg.** 1969. Biochemical studies of bacterial sporulation and germination. XV. Fatty acids in growth and sporulation of *Bacillus megaterium*. *J. Bacteriol.* **98:**82–86.
63. **Silvius, J. R., and R. N. McElhaney.** 1979. Effects of phospholipid acyl chain structure on physical properties. I. Isobranched phosphatidylcholines. *Chem. Phys. Lipids* **24:**287–296.
64. **Vanden Boom, T., and J. E. Cronan, Jr.** 1989. Genetics and regulation of bacterial lipid metabolism. *Annu. Rev. Microbiol.* **43:**317–343.
65. **Wille, W., E. Eisenstand, and K. Willecke.** 1975. Inhibition of the *de novo* fatty acid biosynthesis by the antibiotic cerulenin in *Bacillus subtilis*: effect of citrate-Mg^{2+} transport and synthesis of macromolecules. *Antimicrob. Agents Chemother.* **8:**231–237.
66. **Willecke, K., and L. Mindich.** 1971. Induction of citrate transport in *Bacillus subtilis* during the absence of phospholipid synthesis. *J. Bacteriol.* **106:**514–518.
67. **Willecke, K., and A. B. Pardee.** 1971. Fatty acid requiring mutants of *Bacillus subtilis* defective in branched chain keto acid dehydrogenase. *J. Biol. Chem.* **246:**5264–5272.

IV. CHROMOSOME STRUCTURE

29. The Genetic Map of *Bacillus subtilis*

C. ANAGNOSTOPOULOS, PATRICK J. PIGGOT, and JAMES A. HOCH

This is the sixth edition in a succession of revised versions of the *Bacillus subtilis* 168 genetic map, in which at least one of the present authors was either the sole author or a coauthor (216, 217, 426, 427, 430). For the first time in this series, the map is drawn in linear form. The number of genes located on the map has increased in the last 2 years to the extent that a circular drawing would have been unaesthetic, hard to construct, and even more difficult to consult. The map (Fig. 1) is divided in degrees of a circle, starting from the origin of replication. Two physical maps of the *B. subtilis* chromosome were published last year (14, 246). They were constructed from data on the restriction of chromosomal DNA by the endonucleases *Sfi*I and *Not*I. The size of the *B. subtilis* genome estimated from these maps is 4,175 kb. One degree of the genetic map should therefore correspond to about 11.6 kb. The number of genetic loci positioned on the physical maps is at present too small to allow a correlation of significant value with the genetic map. We therefore did not attempt to align the two kinds of maps in this edition. Nucleotide sequence data of large areas of the chromosome are rapidly accumulating. When the location of these areas on the physical map is well established, a certain distortion of the genetic map may be revealed.

The readers and users of this review have to be reminded that the information reported is a compilation of the literature. They should consult original papers to decide whether the data are of the desired precision. This is particularly important for distances between genes. Most of the mapping in *B. subtilis* has been done by PBS1-mediated transduction. This method is sufficiently reliable for gene ordering in three-point crosses. However, several authors deduce gene order and distance from recombination values in two-point crosses. Interpretation of such results requires caution since the recombination frequencies measured are known to vary from one PBS1 lysate to another and also with the age of the lysates.

The genetic nomenclature in this map (Table 1), as in previous editions, follows the general rules adopted by the *Journal of Bacteriology*. Several changes of gene symbols have been made since the last edition. These were decided by the authors of the original papers themselves and sometimes by a consensus of all those working on the subject. When more precise information about the function controlled, and/or the protein encoded, by a gene is obtained, most authors have the tendency to propose a new designation. Examples: *kinA* (= protein kinase) replaced *spoIIF* and *spoIIJ*; *gga* replaced *pha-3*. In some cases the change involved whole clusters of genes. Example: the genes for flagellar proteins, symbols *flg*, *flh*, and *fli*, replace the previous *che* and *fla* symbols for the corresponding loci. In other cases the change does not concern the three-letter symbol but only the capital letter specifying the locus. For example, see the *pur*, *dna*, and *rec* genes. Very often the change was motivated by the desire to conform to the more precise nomenclature of *Escherichia coli* genetics, especially when determination of the nucleotide sequence of the gene and/or the amino acid sequences of the product revealed homology with the *E. coli* counterpart. We took particular care to dispel confusion due to these changes; although only the latest adopted symbols appear on the map, both old and new designations are listed in the Table, and cross-references clarify the point.

A special mention must be made in regard to the restriction and modification (R/M) system. The symbol *hsd* (= host-specific determinant, as in *E. coli*) replaces the previous *hsr* and *hsm*. This is proposed by Trautner and Noyer-Weidner in chapter 38 of this volume. However, we did not follow all the recommendations of these authors, which involve more complicated designations for individual genes and markers, including subscripts and superscripts, at variance with the generally adopted genetic nomenclature for bacteria. The complications arise from the fact that only one R/M system figures on the *E. coli* K-12 linkage map, indigenous to this strain, while in *B. subtilis*, besides the indigenous system of strain 168 (*hsdM*, for Marburg), five other systems (B, C, E, F, and R) have been introduced from other strains of the species. The more simplified nomenclature we adopted in this edition of the map is the following: the *hsd* symbol is followed by an italicized capital specifying the R/M system (B, C, E, F, M, or R) and then by the capital letter R or M, specifying the individual locus (R for restriction endonuclease and M for modification methyltransferase). Example: *hsdFM* is the gene for the methyltransferase of the F system of restriction and modification.

We believe we have included in this map all genes identified on the *B. subtilis* chromosome as of September 1, 1992. Table 1 gives information and references about the function controlled by each gene, the phenotype acquired by mutation, the product when known, and whether the gene has been cloned and sequenced. Open reading frames that have been discovered during the determination of the nucleotide sequence of a chromosome segment but are still of

C. Anagnostopoulos • Laboratoire de Génétique Microbienne, Institut National de la Recherche Agronomique, 78352 Jouy en Josas Cedex, France. **Patrick J. Piggot** • Department of Microbiology and Immunology, Temple University School of Medicine, Philadelphia, Pennsylvania 19140-5196. **James A. Hoch** • Division of Cellular Biology, Department of Molecular and Experimental Medicine, The Scripps Research Institute, La Jolla, California 92037-1093.

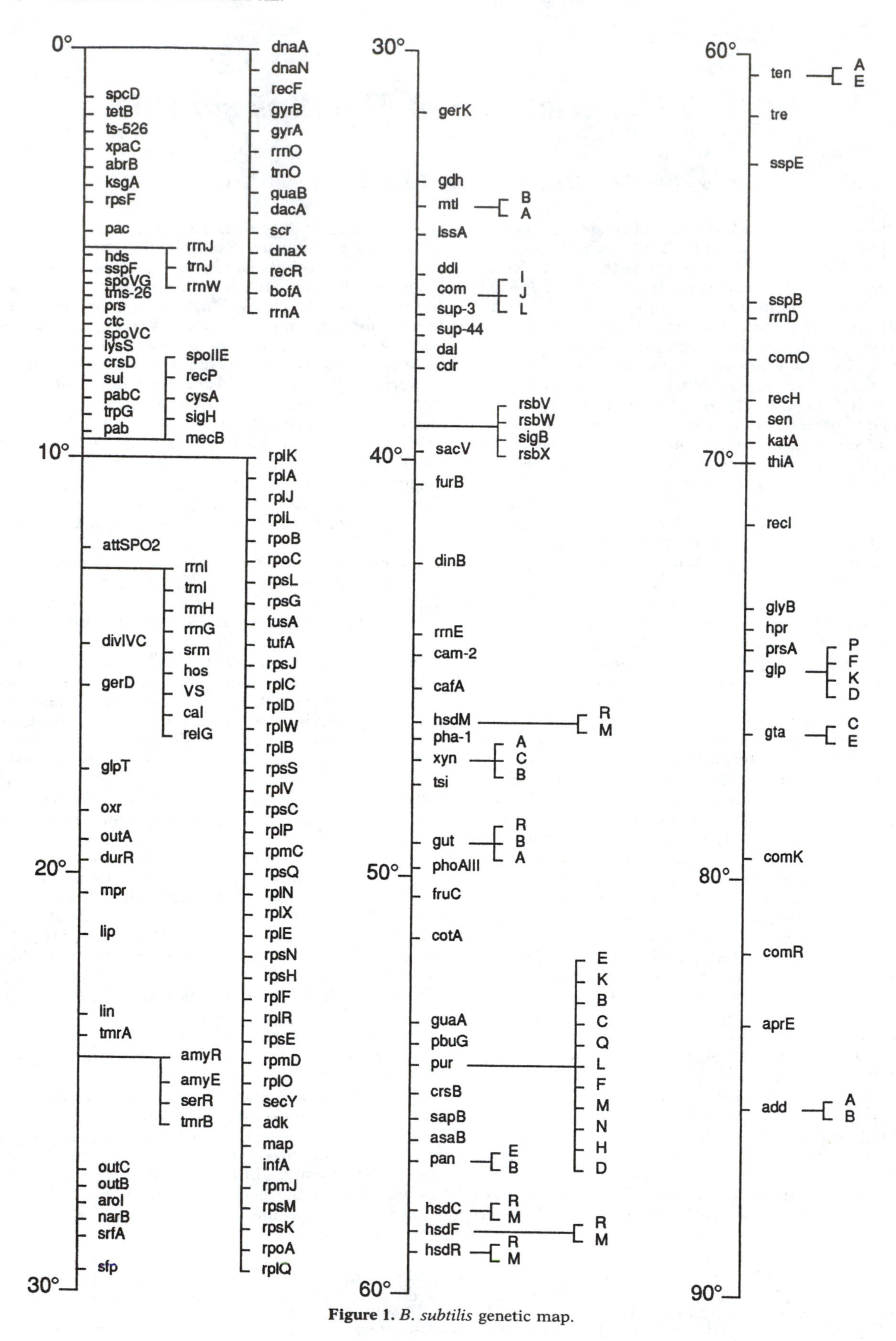

Figure 1. *B. subtilis* genetic map.

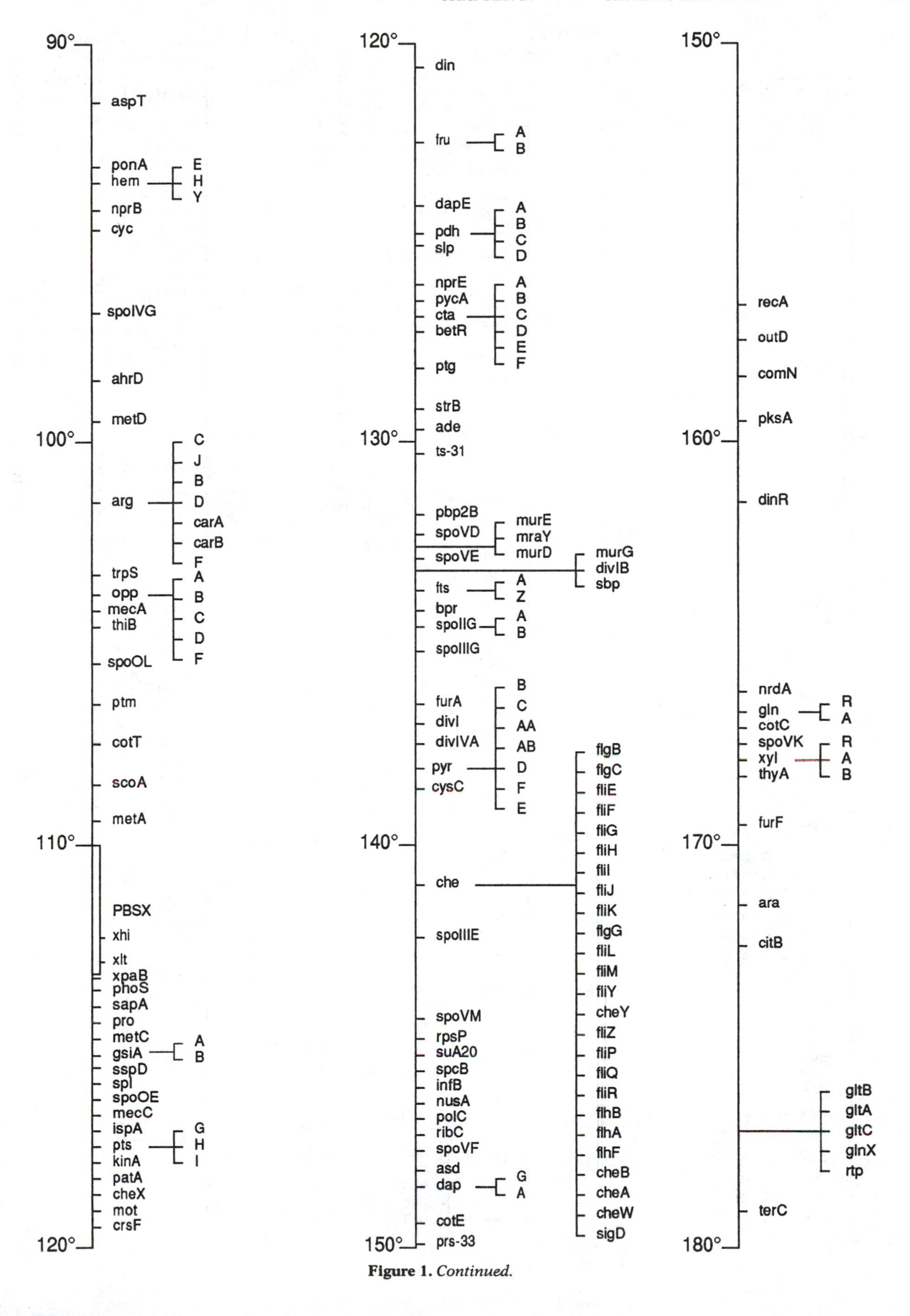

Figure 1. *Continued.*

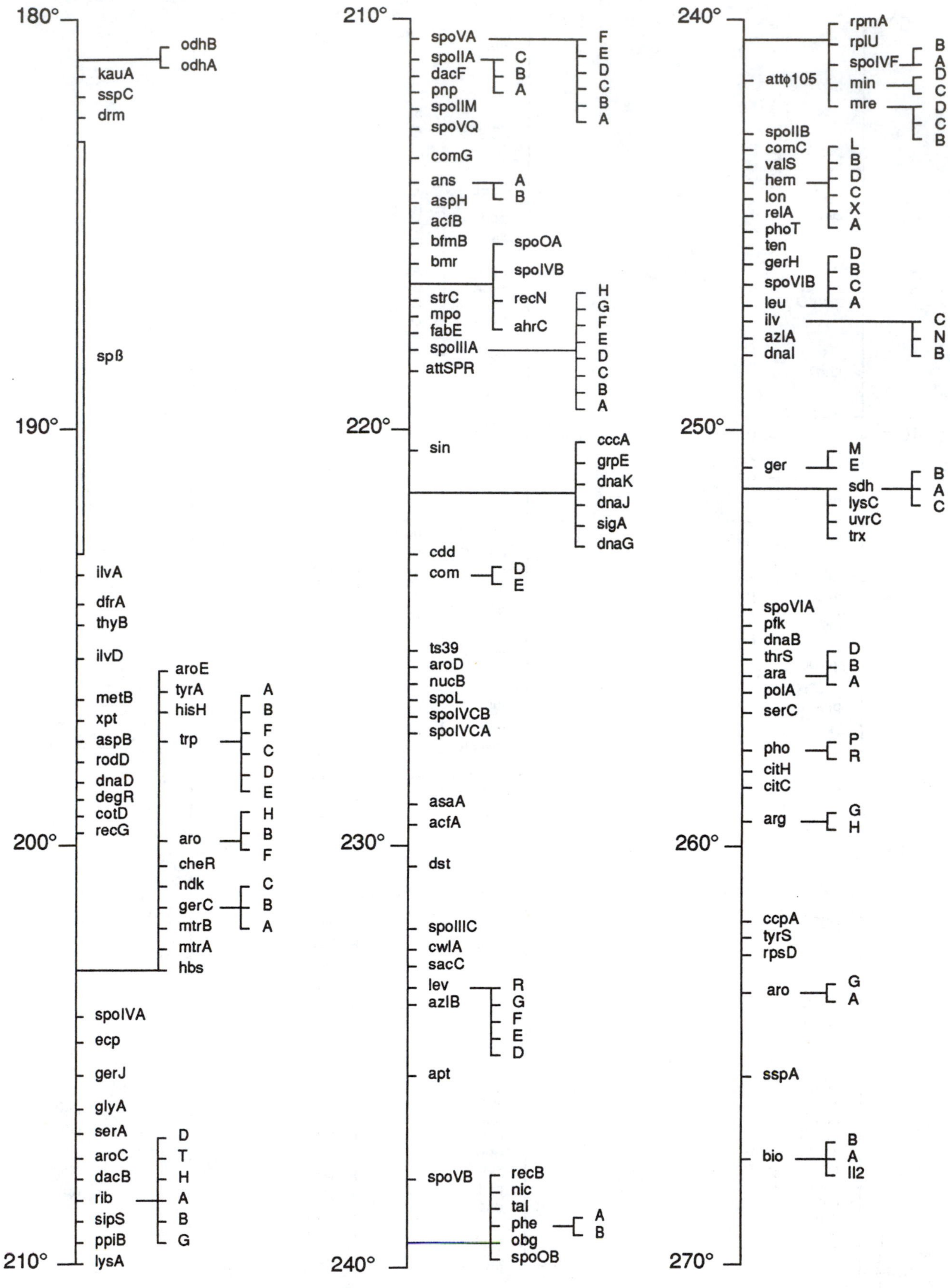

Figure 1. *Continued.*

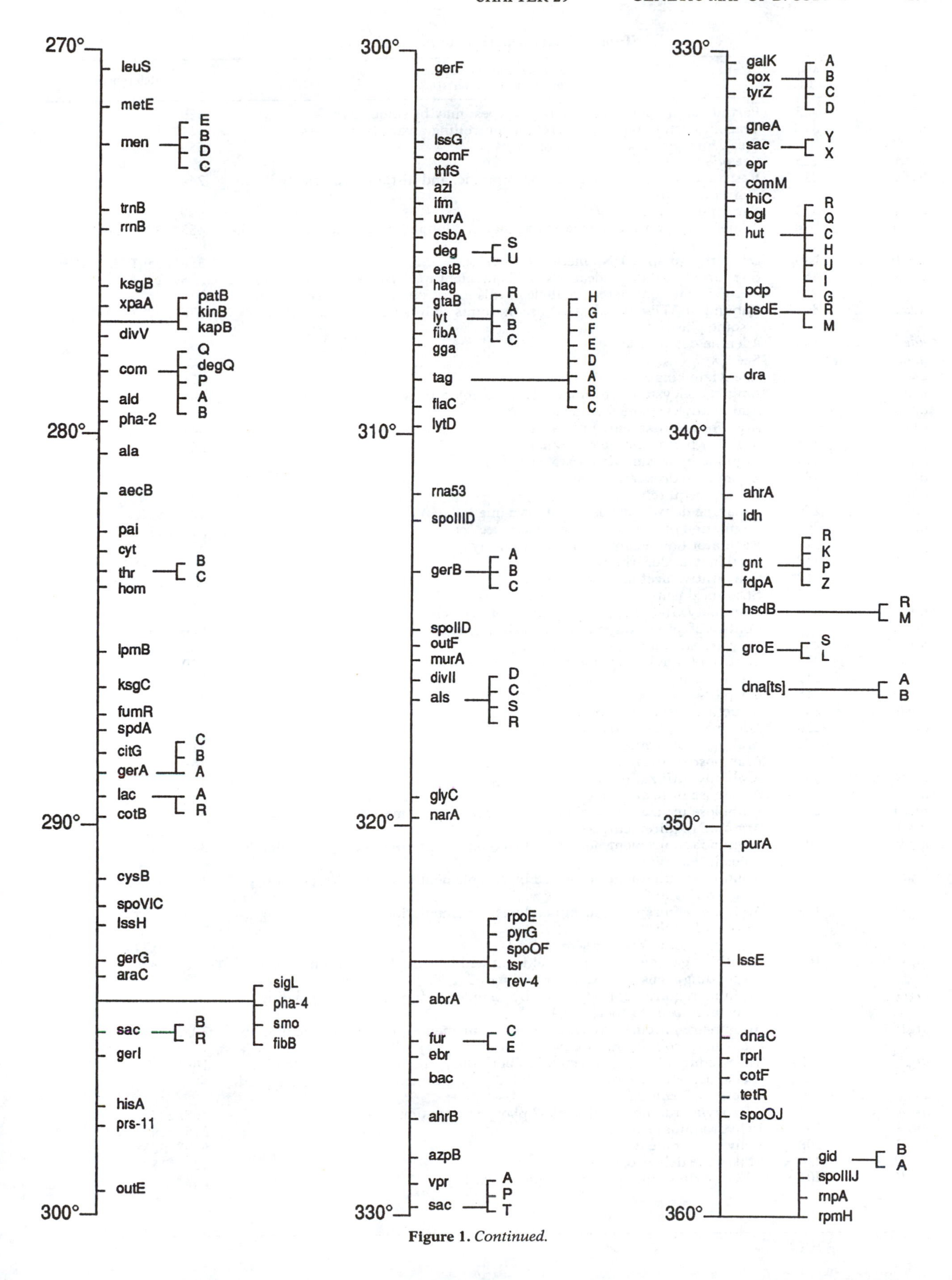

Figure 1. *Continued.*

Table 1. Genetic markers of *B. subtilis* 168

Gene symbol	Map position[a]	Phenotype, enzyme deficiency, or other characteristic[b]	References[c]
abrA	325	Partial suppressor of stage 0 phenotypes; may be same as *rev-4*	559
abrB	3	Transition state regulatory protein controlling a variety of genes	417*
abrC		Weak intragenic suppressors of *spo0A*	559
aceA	125	See *pdhA*	49
acfA	230	Resistance to acriflavin, ethidium bromide, and distamycin; sensitivity to streptomycin	244
acfB	215	Same phenotype as *acfA*	499
adaA	NM	Adaptive response to alkylating agents; methylphosphotriester-DNA methyltransferase	358*, 359*, 360
adaB	NM	O^6-methylguanine-DNA methyltransferase	358*, 359*, 360
addA	86	Subunit of ATP-dependent DNase; equivalent to *E. coli* RecBCD enzyme; *recE5* now considered an allele of this locus; possible helicase	276*, 277†
addB	86	Subunit of ATP-dependent DNase; mutants show increased stability for some plasmids	276*, 277†
ade	130	Adenine deaminase	391
adeF		See *apt*	
adk	11	Adenylate kinase	375*
aecA	252	Aminoethylcysteine resistance; aspartokinase II; see *lysC*	80*, 81, 82, 310, 420
aecB	282	Aminoethylcysteine resistance	333
ahrA	342	Arginine hydroxamate resistance	361
ahrB	328	Arginine hydroxamate resistance	361
ahrC	219	Arginine hydroxamate resistance	513†
ahrD	99	Arginine hydroxamate resistance	32
ala	281	Alanine requirement	333
ald	279	L-Alanine dehydrogenase; another name for *spoVN*	184*, 561
alsA	263	Regulation of acetolactate synthase; see *ccpA*	211, 623, 625
alsC	317	Regulator homologous to the *lysR* family	622
alsD	317	Acetolactate decarboxylase	622
alsR	317	Constitutive acetolactate synthase	623, 625
alsS	317	Structural gene for acetolactate synthase	622*
amt		3-Aminotyrosine resistance; part of, or very close to, *tyrA*	433
amyB		Control of amylase synthesis; identical to *degQ*	525, 605
amyE	25	Amylase structural gene; also called *amyA*	606*
amyR	25	Control of amylase synthesis; also called *amyH*	620
ansA	215	L-Asparaginase	541*
ansB	215	L-Aspartase	541*
aprE	84	Structural gene for subtilisin	521*
apt	236	Adenine phosphoribosyl transferase	482
ara	172	Arabinose utilization	411
araA	256	Arabinose utilization; L-arabinose isomerase	468†
araB	256	Arabinose utilization; L-ribulokinase	468†
araC	294	Arabinose utilization regulation	469
araD	256	Arabinose utilization; L-ribulose 5-phosphate 4-epimerase	468†
argA		Arginine requirement; see *argJ*	106*, 362†
argB	102	Arginine requirement; identified by complementation of corresponding locus in *E. coli*	362†
argC	102	Arginine requirement; identified by complementation of corresponding locus in *E. coli*	106*, 512*, 514*
argD	102	Arginine requirement; identified by complementation of corresponding locus in *E. coli*	106*, 362†
argE		Arginine requirement; see *argJ*	106*, 362†
argF	102	Ornithine carbamoyltransferase; identified by complementation of corresponding locus in *E. coli*; formerly *argC*	106*, 362, 363†
argG	260	Arginine requirement; identified by complementation of corresponding locus in *E. coli*; formerly *argA*	30, 362
argH	260	Arginine requirement; identified by complementation of corresponding locus in *E. coli*; formerly *argA*	30, 362
argJ	102	The product of this gene combines both the activities of the *argA* (amino acid acetyltransferase) and *argE* (acetylornithine deacetylase) gene products of *E. coli*	106*
aroA	264	3-Deoxy-D-arabinoheptalosonic-7-phosphate synthase	46*, 227
aroB	203	Dehydroquinate synthase	215, 314*
aroC	208	Dehydroquinate dehydratase	519*, 585†
aroD	226	Shikimate dehydrogenase	227, 515†
aroE	202	3-Enolpyruvylshikimate-5-phosphate synthase	213*

(Continued)

Table 1—*Continued*

Gene symbol	Map position[a]	Phenotype, enzyme deficiency, or other characteristic[b]	References[c]
aroF	203	Chorismate synthase	215*, 227
aroG	264	Chorismate mutase; isozyme 3	46*, 380
aroH	203	Chorismate mutase; isozymes 1 and 2	180*, 215
aroI	28	Shikimate kinase	404†
aroJ		Tyrosine and phenylalanine; see *hisH*	382
asaA	227	Arsenate resistance	2
asaB	57	Arsenate resistance (derived from *B. subtilis* W23)	1
asd	149	Aspartate semialdehyde dehydrogenase	410
ask		See *lysC*	
aspA		See *pycA*	
aspB	198	Aspartate aminotransferase	226
aspH	215	Aspartase, constitutive	237
aspT	92	Aspartate transport	593
ath	55	See *purM*	
attSPβ	190	Integration site for phage SPβ and related phages IG1, IG3, and IG4	100, 624
attSPO2	13	Integration site for phage SPO2	510
attSPR	218	Integration site for phage SPR	100
attϕ3T		Integration site for phage ϕ3T; probably the same as *attSPβ*	598, 622
attϕ105	242	Integration site for phage ϕ105	419
azc	25	Resistance to azetidine-2-carboxylic acid; *azc* 90% cotransformed with *aroI*	159
azi	304	Resistance to sodium azide	223
azlA	247	4-Azaleucine resistance; derepressed leucine biosynthetic enzymes	312†
azlB	234	4-Azaleucine resistance	587
azpA		Resistance to azopyrimidines; alteration of DNA polymerase III, see *polC*	587
azpB	329	Resistance to azopyrimidines	72
bac	327	Bacilysin biosynthesis	222
betR	128	Resistant to betacin produced by SPβ lysogens	207
bfmA	126	Lacks branched-chain-keto acid dehydrogenase; isolated in *aceA* strain, and not separated genetically from *aceA*, probably *pdhA*	49
bfmB	216	Requires branched-chain fatty acid, valine, or isoleucine; maps between *strC* and *lys*	49
bgl	335	Endo β1,3-1,4-glucanase (lichenase)	372*
bglC	NM	Endo β1,4-glucanase (CM cellulase)	311*
bioA	268	Biotin synthesis, 8-diaminopelargonate aminotransferase	405
bioB	268	Biotin synthetase	405
bio-112	268	Early defect in biotin synthesis	405
bmr	216	Bacterial multidrug resistance; homologous to *tet* gene of pBR322	385*
bofA	0	Activation of σ^K during sporulation	455*
bofB		See *spoIVF*	
bpf		Bacillopeptidase F; see *bpr*	
bpr	135	Bacillopeptidase F	123*, 505*, 601*
bry		Bryamycin (thiostrepton) resistance; maps in ribosomal protein cluster	510
bsr		Restriction, modification by *B. subtilis* R; see *hsdR*	557
but	NM	5-Bromouracil tolerance	41
cafA	46	Caffeine resistance	71
cafB	NM	Caffeine and nalidixic acid sensitivity; allele of *gyrA*	71
cal	12	Chalcomycin resistant	495
cam-2	45	Resistant to chloramphenicol	18
car	102	Carbamoylphosphate transferase-arginine; two cistrons, A and B; previously *cpa*	106*, 362†
catA		Hyperproduction of extracellular protease; can sporulate in presence of glucose; allele of *hpr*	
cccA	222	Gene for cytochrome c_{550}	576*
ccpA	263	Catabolite control protein involved in glucose regulation of several genes; *alsA* is allelic to *ccpA*	211*, 625
cdd	223	Deoxycytidine-cytidine deaminase	518*
cdr	38	Cadmium resistance	626
cel	NM	Cellulase	374*
che	142	Chemotaxis; includes 20 complementation groups, nomenclature modified; see *flg*, *flr* and *fih*; *che* designation kept for *A*, *B*, *D*, *W*, and *Y*	399, 401†, 627*, 628†
cheF	142	Chemotaxis; same as *fliJ*	611*, 612
cheR	204	Chemotactic methyltransferase	215*, 400
cheX	118	Chemotaxis	628
citA	NM	Citrate synthase I	515*
citB	173	Aconitase	108*
citC	259	Isocitrate dehydrogenase	163, 422†

(Continued)

Table 1—*Continued*

Gene symbol	Map position[a]	Phenotype, enzyme deficiency, or other characteristic[b]	References[c]
citF		See *sdh*	
citG	289	Fumarate hydratase	344*
citH	259	Malate dehydrogenase; same as *mdh*	163
citK		See *odhA*	
citL		See *pdhD*	
citM		See *odhB*	
citR	NM	Regulatory protein of the Lys family located upstream of *citA*	515
citZ	260	Citrate synthase II	515*
cml		Chloramphenicol resistance, caused by mutations in one of at least five 50S ribosomal proteins	402
comA	279	Competence; early block; *com-9* may map in *comA* or *comB*	112*
comB	279	Competence; early block	112*
comC	258	Competence	112*
comD	224	Competence	112*
comE	224	Competence	112*
comF	304	Competence	112
comG	213	Competence; may include *com-71*; DNA sequence contains seven open reading frames	112*
comI	37	Competence; codes for 17-kDa nuclease	577*
comJ	37	Competence; codes for 18-kDa protein that is isolated with the ComI protein	577*
comK	80	Competence; includes *com-30*	112, 566*, 567
comL	37	Competence	567†
comM	334	Competence; previously *com-104*	112, 133
comN	159	Competence	326
comO	68	Competence; previously *com-18*	326
comP	279	Competence: homologous to sensor class of signal transduction proteins	590*
comQ	279	Competence	364*
comR	82	Competence; previously *com-44*	31†, 326
com-9		See *comA*	133
com-30		See *comK*	133
com-71		See *comG*	133
com-104		See *comM*	133
cotA	52	Spore coat protein	110†
cotB	290	Spore coat protein	110†
cotC	168	Spore coat protein	110*
cotD	200	Spore coat protein	110*
cotE	150	Spore coat protein; peroxidase	105, 307*
cotF	356	Spore coat protein	93*, 393*
cotT	108	Spore coat protein	21*
cpa		See *car*	362†
cpsX		See *abrB*	
crk	NM	Cytidine kinase	456
crr	118	Phosphocarrier protein for glucose in the phosphotransferase system; now considered part of *ptsG*	177†, 523*
crsA	222	Carbon source-resistant sporulation; resistant to novobiocin and acridine orange during sporulation; mutation in *sigA*; has similar phenotype to *rvt* mutations	543
crsB	56	Requires high glucose for sporulation	543
crsC	217	Carbon source-resistant sporulation; maps close to or in *spo0A*	543
crsD	8	Carbon source-resistant sporulation	543
crsE	12	Carbon source-resistant sporulation; maps in *rpoBC* operon; *rfm-11* suppresses *crsE*	544
crsF	119	Carbon source-resistant sporulation	543
csaA	NM	Involved in protein secretion	370*
csbA	305	Gene controlled by sigma factor B	52*
css		Cysteine sensitivity; see *cysA*	258
ctaA	127	Required for expression of cytochrome aa_3	368*
ctaB	127	Cytochrome *a* oxidase assembly factor	478*
ctaC	127	Cytochrome aa_3 oxidase subunit II	478*
ctaD	127	Cytochrome aa_3 oxidase subunit I	478*
ctaE	127	Cytochrome aa_3 oxidase subunit III	478*
ctaF	127	Cytochrome aa_3 oxidase subunit IVB	478*
ctc	7	Expressed at the end of exponential growth under conditions in which the enzymes of the tricarboxylic acid cycle are repressed	236*, 388*
ctrA	324	CTP synthetase; requirement for cytidine in the absence of ammonium ion; see *pyrG*	553*

(Continued)

Table 1—*Continued*

Gene symbol	Map position[a]	Phenotype, enzyme deficiency, or other characteristic[b]	References[c]
cwbA		Modifier of *cwlB*; most probably identical to *lytB*	284
cwlA	233	Autolytic amidase; minor autolysin	145*, 285*, 286
cwlB		*N*-Acetyl muramoyl-L-alanine amidase; major autolysin, most probably identical to *lytC*	284*, 286*, 294
cyc	95	D-Cycloserine resistance	160
cym		Requirement for cysteine or methionine; see *cysA*	425
cysA	10	Cysteine requirement, serine *trans*-acetylase; original isolate contained four linked mutations, *cys* (cysteine sensitivity), *cym* (cysteine or methionine), *hts* (hydrogen sulfide excretion), and *cysA*	258
cysB	292	Cysteine requirement	114
cysC	139	Cysteine, methionine, sulfite, or sulfide requirement	616
cyt	284	Requires cytidine	229
dacA	0	D-Alanine carboxypeptidase; vegetative penicillin-binding protein 5	62, 549*
dacB	208	DD-carboxypeptidase penicillin-binding protein 5*	63*, 519*
dacF	211	Putative DD-carboxypeptidase; penicillin-binding protein	599*, 600*
dal	38	D-Alanine requirement; alanine racemase	136*
dapA	149	Dihydrodipicolinate synthase	410*
dapE	124	*N*-Acetyl-LL-diaminopimelic acid deacylase; previously *lssB*	68, 463
dapG	149	Aspartokinase I; previously *lssD*	410*, 463
dat	NM	Constitutive O^6-methylguanine-DNA alkyltransferase	358*
dcd	NM	dCMP deaminase	356
dciA		See *dpp*	
dck	NM	Deoxycytidine/deoxyadenosine kinase	355
ddd	NM	Deoxycytidine kinase	456
ddl	36	D-Alanyl-D-alanine ligase	68
dds		Deficient in division and sporulation; see *divIB*	36*, 123*
degQ	279	Degradation enzyme regulation; previously called *sacQ*	605*
degR	199	Degradation enzyme regulation; previously called *prtR*	607*
degS	306	Degradation enzyme regulation; part of *sacU*	219*, 283*
degU	306	Degradation enzyme regulation; part of *sacU*; also called *iep*	548*
deoA	NM	Thymidine phosphorylase	336
dfrA	195	Dihydrofolate reductase; coordinately regulated with *thyB*	249*
din	121	DNase inhibitor	440
dinA		DNA damage inducible; same as *uvrA*; homologous to *uvrB* of *E. coli*	82*, 169
dinB	43	DNA damage inducible	82*, 169
dinC		DNA damage inducible; see *tagC*	82*, 169, 335, 476
dinR	162	Regulation of DNA damage-inducible genes; product homologous to *lexA* of *E. coli*	451*
divI	138	Temperature-sensitive cell division; formerly *divD*	69, 562
divIB	135	Deficient in division and sporulation; also called *dds*	36*, 123*, 199*
divII	317	Temperature-sensitive cell division; formerly *divC*	562
divIVA	138	Minicell production	69, 452
divIVB	242	Minicell production; see *min* and *mre*	300*, 452
divIVC	15	Minicell production; formerly *divA*	562
divV	278	Temperature-sensitive cell division; formerly *divB*	562
dnaA	0	Same as *dnaA* of *E. coli*; formerly *dnaH* and *dnaK*; previous *dnaA* locus is now *nrd*	9, 394, 396*, 413†, 515†
dnaB	256	DNA synthesis; initiation of chromosome replication; probably more than one gene	395*
dnaC	355	DNA synthesis; homologous to *dnaB* of *E. coli*	17, 393*
dnaD	199	DNA synthesis, initiation of chromosome replication	262
dnaF		DNA synthesis; DNA polymerase III; see *polC*	
dnaG	224	DNA synthesis; primase; formerly *dnaE*	11, 584*
dnaI	248	DNA synthesis	262
dnaJ	223	Homologous to *dnaJ* of *E. coli*	592*
dnaK	223	Homologous to *dnaK* of *E. coli* and *Bacillus megaterium*	202*, 591*, 592*
dnaN	0	Previously *dnaG*	394*
dna(Ts)*A*	347	DNA synthesis	196
dna(Ts)*B*	347	DNA synthesis	196
dnaX	0	Homologous to *dnaZX* of *E. coli*; subunit of holoenzyme DNA polymerase III; formerly *dnaH* and *dna-8132*	10*, 534*, 537*
dpa	148	Requires dipicolinic acid for heat-resistant spores; same as *spoVF*	26
dpp	NM	Dipeptide permease; DNA sequence has five adjacent open reading frames; also called *dciA*	329*, 330*
dra	339	Deoxyriboaldolase	540
drm	182	Phosphodeoxyribomutase	540

(Continued)

Table 1—*Continued*

Gene symbol	Map position[a]	Phenotype, enzyme deficiency, or other characteristic[b]	References[c]
dst	231	Resistant to distamycin and acriflavin	499
durR	20	Duramycin resistance	381
D-*tyr*		Resistance to D-tyrosine; maps within the *tyrA* locus; see *tyrA*	78
ebr	326	Ethidium bromide resistance	41
ecp	204	Resistance to 2-amino-5-ethoxycarbonyl pyrimidine-4(3H)-one	450
efg		See *fusA*; elongation factor G	
eno	NM	Enolase	107*
epr	333	Minor extracellular protease	503*
ery		Erythromycin resistance; ribosomal protein L22; see *rplV*	547
estB	306	Esterase B defect	221
ethA	NM	Ethionine resistance	578
fabE	218	Fatty acid biosynthesis; homologous to *fabE* of *E. coli*	529*
fdpA	344	Fructose-bisphosphatase	152
ffs		Homologous to *ffs* of *E. coli*, but different size; see *scr*	536*
fibA	307	Macrofiber formation; weak linkage to *hisA*	481
fibB	295	Macrofiber formation	481
flaA	142	Defect in flagellar synthesis; autolysin deficient; part of the *che* operon; nomenclature changed, see *flg*, *fli*, *flh*, and *che* genes	6*, 201*, 437
flaB	142	Defect in flagellar synthesis; see *sigD*	321, 437
flaC	310	Defect in flagellar synthesis	437
flaD	221	Defect in flagellar synthesis; autolysin deficient; identical to *sin*	437, 490*
flg	142	Part of *che-fla* operon; three loci, *B*, *C*, and *G*; flagellar genes; homologous to the corresponding genes of *E. coli* and *S. typhimurium*	40*, 627*
flh	142	Part of the *che-fla* operon; three loci, *A*, *B*, and *F*; homologous to the corresponding genes of *E. coli* and *S. typhimurium*; flagella genes	40*, 627
fli	142	Part of *che-fla* operon; flagellar genes; 13 loci, E to K, P to R, and Z; homologous to the corresponding genes of *E. coli* and *S. typhimurium*	40*, 612, 627*
fruA	123	Fructose transport	166
fruB	123	Fructose-1-phosphate kinase	166
fruC	51	Fructokinase	165
ftr	UC	Fluorotryptophan resistance; maps near *aroC*	33
ftsA	134	Homologous to *E. coli* cell division gene *ftsA*; location of *spoIIN279*	35*, 123*, 265
ftsZ	134	Homologous to *E. coli* cell division gene *ftsZ*	35*, 123*
fumR	288	Regulation of fumarate hydratase	223
fun	12	Streptomycin-resistant oligosporogenous mutant; allele of *rpsL*	210
furA	137	Fluorouracil resistance	69, 114
furB	41	Fluorouracil resistance	621
furC	326	Resistance to 5-fluorouracil in the presence of uracil	622
furE	326	Resistance to 5-fluorouracil in the presence of uracil	622
furF	170	Resistance to 5-fluorouracil in the presence of uracil	101
fus		Fusidic acid resistance; elongation factor G; same as *efg*	272
galK	331	Galactokinase	170*
gap	NM	Glyceraldehyde-3-phosphate dehydrogenase	574*
gca	NM	L-Glutamine-D-fructose-6-phosphate aminotransferase	151
gdh	34	Structural gene for glucose dehydrogenase	291*
gerA	289	Germination defective; defective in germination response to alanine and related amino acids; consists of three genes	134*
gerB	314	Germination defective; defective in germination response to the combinations of glucose, fructose, asparagine, and KCl; consists of three genes	89, 353*, 354, 610†
gerC	204	Germination defective; temperature-sensitive germination in alanine; consists of three genes; *gerCB* product containing SAM binding site; *gerCC* is homologous to *cotE*; *gerC58* mutation is in *gerCC*	353a*, 354, 610†
gerD	16	Germination defective; defective germination in a range of germinants	354, 613*
gerE	251	Germination defective; defective germination in a range of germinants; transcriptional regulator of spore coat genes	91*
gerF	301	Germination defective; defective germination in a range of germinants	354
gerG	294	Germination defective; a mutant lacks phosphoglycerate kinase (*pgk*) activity; germinates poorly in alanine; sporulates poorly	354
gerH	246	Defective germination in a range of germinants	431
gerI	296	Defective germination in a range of germinants	431
gerJ	206	Defective germination in a range of germinants; allele *gerJ51* (also called *tzm*) is present in many laboratory strains	314†
gerK	32	Defective germination response to glucose	245
gerM	251	Defective germination and sporulation	473†
gga	308	= *pha-3*; membrane-bound enzymes involved in the biosynthesis of galactosamine-containing teichoic acids	130, 261*

(Continued)

Table 1—*Continued*

Gene symbol	Map position[a]	Phenotype, enzyme deficiency, or other characteristic[b]	References[c]
gid	360	Two genes homologous to *gidA* and *-B* of *E. coli*	398*
glnA	167	Glutamine synthetase structural gene	533*
glnR	167	Negative regulator of glutamine synthetase	142†, 515*
glnX	178	Glutamine biosynthesis	3*
glpD	75	Glycerol-3-phosphate dehydrogenase	230*
glpF	75	Shows homology to *E. coli* gene *glpF*, which is a glycerol facilitator	37*
glpK	75	Glycerol kinase	37*, 230*
glpP	75	Pleiotropic glycerol mutant; positively acting regulator protein	37*, 303
glpT	18	Fosfomycin-resistant, glycerol phosphate transport defect	267
gltA	177	Glutamate or aspartate requirement, glutamate synthase	42†, 515*
gltB	177	Glutamate synthase	42†, 515*
gltC	177	Positive regulator of *gltAB*	43*
gltX	NM	Glutamyl-tRNA synthetase	58*
glyA	207	Glycine requirement	266
glyB	74	Glycine requirement	197
glyC	320	Glycine requirement	72
gneA	332	UDP-*N*-acetylglucosamine 4-epimerase	130
gntK	344	Gluconate utilization; gluconate kinase	154*
gntP	344	Gluconate utilization; gluconate permease	154*
gntR	344	Gluconate utilization; repressor	153*
gntZ	344	Gluconate utilization; probably 6-phosphogluconate dehydrogenase	154*, 454
gpr	NM	Spore protein; germination protease for SASPs	545*
gra	25	Glucose-resistant amylase production; allele of *amyR*	386*
groEL	344	Phage growth; heat shock protein chaperonin	76, 301*, 484*
groES	344	Phage growth; heat shock protein chaperonin	76, 301*, 484*
grpE	222	Homologous to *grpE* of *E. coli*	591*
gsiA	115	Glucose starvation-inducible genes; two cistrons	365*, 366*
gsiB	NM	Glucose starvation-inducible genes	365*
gsiC		Same as *kinA*	366*
gsp10	UC	Spore outgrowth; maps near the *che* cluster	157*
gtaA		See *tagE*	
gtaB	308	Glucosylation of teichoic acid; UDP-glucose pyrophosphorylase	59†, 261*, 438
gtaC	77	Glucosylation of teichoic acid; lacks phosphoglucomutase	438
gtaD		See *tagE*	
gtaE	77	Glucosylation of teichoic acid	438
guaA	54	GMP synthetase; previously *guaB*	317*, 483
guaB	0	IMP dehydrogenase; previously *guaA*	259*, 483
guaC	NM	GMP reductase	121
guaF	NM	Hypoxanthine-guanine phosphoribosyltransferase	121
guaP	NM	Inosine-guanosine phosphorylase	121
gutA	50	D-Glucitol permease	165†
gutB	50	D-Glucitol hydrogenase	165†
gutR	50	Constitutive synthesis of D-glucitol permease and D-glucitol dehydrogenase; regulatory gene for glucitol catabolism	165†
gyrA	0	DNA gyrase (*nalA*)	357*
gyrB	0	DNA gyrase (*novA*)	357*
hag	307	Flagellin	292*, 347
hbs	204	DNA-binding protein HBsu; immediately upstream of *mtrA*, also called *dbpA*	215*, 341*
hds	6	Pleiotropic extragenic suppressors of *dna* mutations	500
hemA	244	glutamyl-tRNA reductase of the C-5 pathway for Δ-aminolevulinate synthetase	269, 421*, 486
hemB	244	Porphobilinogen synthase	38, 195*
hemC	244	Porphobilinogen deaminase	38, 421*
hemD	244	Uroporphyrinogen III cosynthase	195*, 343
hemE	94	Uroporphyrinogen decarboxylase	194*, 342
hemF	94	See *hemH*	194*, 342
hemG	94	See *hemY*	194*
hemH	94	Ferrochelatase	194*, 342
hemL	244	Glutamate-1-semialdehyde-2,1-aminotransferase	195*
hemX	244	Membrane protein of unknown function required for *hemL* synthesis	195*, 485
hemY	94	Coproporphyrinogen III oxidase and protoporphyrinogen IX oxidase	194
hisA	298	Histidine requirement; probable location of genes for all histidine enzymes except *hisH*	583†
hisH	202	Histidinol-phosphate aminotransferase, tyrosine and phenylalanine aminotransferase	213*

(Continued)

Table 1—*Continued*

Gene symbol	Map position[a]	Phenotype, enzyme deficiency, or other characteristic[b]	References[c]
hom	284	Threonine and methionine requirement; homoserine dehydrogenase; a deletion in this area covers this gene, the gene for homoserine dehydratase, and *thrB*; also see *tdm*	15, 409*
hos	12	Suppresses the temperature-sensitive phenotype of elongation factor G mutants	332
hpr	75	Overproduction of proteases; shown to be the same as *catA* and *scoC*	416*
hsdBM	345	DNA-methyltransferase *Bsu*B	239, 602; Trautner and Noyer-Weidner, this volume
hsdBR	345	Host-specificity determinant B of *B. subtilis* IAM 1247; restriction endonuclease *Bsu*B; previously *hsrB*	239, 602; Trautner and Noyer-Weidner, this volume
hsdCM	59	DNA-methyltransferase *Bsu*C	239; Trautner and Noyer-Weidner, this volume
hsdCR	59	Host specificity determinant C of *B. subtilis* IAM 1247; restriction endonuclease *Bsu*C; previously *hsrC*	239; Trautner and Noyer-Weidner, this volume
hsdEM	337	DNA methyltransferase *Bsu*E	156, 239; Trautner and Noyer-Weidner, this volume
hsdER	337	Host specificity determinant E of *B. subtilis* IAM 1231; restriction endonuclease *Bsu*E; previously *hsdE*	239; Trautner and Noyer-Weidner, this volume
hsdFM	59	DNA methyltransferase *Bsu*F	239, 580; Trautner and Noyer-Weidner, this volume
hsdFR	59	Host specificity determinant F of *B. subtilis* IAM 1231; restriction endonuclease *Bsu*F; isoschizomer of *Hpa*II; previously *hsrF*	239, 260*; Trautner and Noyer-Weidner, this volume
hsdMM	47	DNA methytransferase *Bsu*M	239, 471; Trautner and Noyer-Weidner, this volume
hsdMR	47	Host specificity determinant M of *B. subtilis* Marburg; identical to *nonB*; restriction endonuclease *Bsu*M; isosochizomer of *Xho*I; previously *hsrM*	239, 471; Trautner and Noyer-Weidner, this volume
hsdRM	59	DNA methyltransferase *Bsu*R	239, 268*, 557; Trautner and Noyer-Weidner, this volume
hsdRR	59	Host specificity determinant R of *B. subtilis*; restriction endonuclease *Bsu*R; isoschizomer of *Hae*III; previously *hsrR*	239, 268*, 557; Trautner and Noyer-Weidner, this volume
hts		Excretion of hydrogen sulfide; see *cysA*	258
hutC	335	Histidine utilization; control locus	256
hutG	335	Histidine utilization; formimino glutamate hydrolase	256
hutH	335	Histidine utilization; histidase	392*
hutI	335	Histidine utilization; imidazolone propionate hydrolase	256
hutP	335	Histidine utilization; positive control element	392*
hutR	335	Histidine utilization; control locus	256
hutU	335	Histidine utilization; urocanase	392*
idh	343	Inositol dehydrogenase	155*
iep		See *degU*	
ifm	304	Hypermotility; suppresses *flaA* and *flaD* mutations	437
igf	343	Deletion covering *iol*, *gnt*, *fdpA*, and *hsrB* (formerly *fdpA1*)	152
ilvA	194	Threonine dehydratase	16, 20*, 30, 141
ilvB	247	Condensing enzyme	312†
ilvC	247	α-Hydroxy-β-keto acid reductoisomerase	312†
ilvD	196	Dihydroxyacid dehydratase	16, 30
ilvN	246	Subunit of the *ahaS* enzyme (*ilvB*)	312†
infA	11	Initiation factor I	51*
infB	147	Initiation factor II	497*
inh		Inhibition by histidine; probably within *tyrA* locus; see *tyrA*	380
iol	343	Inability to grow on myoinositol; myoinositol dehydrogenase gene; see *idh*	152, 155*
ispA	117	Intracellular serine protease	273*

(Continued)

Table 1—*Continued*

Gene symbol	Map position[a]	Phenotype, enzyme deficiency, or other characteristic[b]	References[c]
kan	10	Kanamycin resistance; maps in the ribosomal protein cluster	174
kapB	277	Kinase B-associated protein	555*
katA	70	Catalase	305
kat-19	NM	Catalase; distinct from *katA*	45*
kauA	182	Branched-chain α-keto transport	172
kinA	118	Sporulation-specific protein kinase	19*, 412*
kinB	277	Sporulation-specific protein kinase	555*
kir		Probably mutation in the structural gene for elongation factor Tu; see *tuf*	508
ksgA	4	High-level kasugamycin resistance	551
ksgB	277	Low-level kasugamycin resistance	551
ksgC	287	Fumarate hydratase defective; kasugamycin resistance	223
lacA	290	Structural gene for beta-galactosidase	128
lacR	290	Regulatory gene for beta-galactosidase	128
leuA	247	α-Isopropylmalate synthase	373†
leuB	247	Isopropylmalate isomerase	373†
leuC	247	β-Isopropylmalate dehydrogenase	243*
leuD	247	Possibly a subunit of isopropylmalate isomerase	312†
leuS	271	Leucyl tRNA synthetase	568
levD	233	Enzyme III for fructose (PTS) transport system	325*
levE	233	Enzyme III for fructose (PTS) transport system	325*
levF	233	Enzyme II for fructose (PTS) transport system	325*
levG	233	Enzyme II for fructose (PTS) transport system	325*
levR	233	Regulator of levanase	103*
lin	24	Lincomycin resistance	173
lip	22	Lipase	99
lon	245	Homologous to *lon* of *E. coli*	307*
lpm		Lipiarmycin resistance; RNA polymerase; see *rpoC*	517
lpmB	286	Lipiarmycin resistance	516
lppX		Gene upstream of *cwbA*; lipoprotein of unknown function	284*
lssA	35	Thermosensitive lysis; linked to *ddl* by transformation	57
lssB	124	Thermosensitive lysis; see *dapE*	57, 463
lssC		Thermosensitive lysis; peptidoglycan precursor synthesis; same as *lssA*	57, 261
lssD		See *dapG*	
lssE	354	Thermosensitive lysis	57
lssF	316	Thermosensitive lysis; same as *murA*	57, 464
lssG	303	Thermosensitive lysis	57, 261†
lssH	293	Thermosensitive lysis	57
lysA	210	Lysine requirement, diaminopimelate decarboxylase	30, 253†, 519*, 606*
lysC	252	Aspartokinase II (formerly *ask* and *aecA*)	80*, 81, 310
lysS	8	Lysyl-tRNA synthetase	449
lyt		Autolytic enzymes; see *flaD*	135
lytA	307	Involved in the secretion of LytC (amidase)	294*
lytB	307	Modifier of LytC (amidase); see also *cwbA*	294*
lytC	307	*N*-Acetylmuramoyl-L-alanine amidase in vegetative growth; major autolysin; see also *cwlB*	294*, 319
lytD	310	*N*-Acetylglucosaminidase	320*
lytR	307	Regulator of LytC (amidase)	294*
map	11	Methionine aminopeptidase	375*
mdh	259	Malate dehydrogenase; same as *citH*	163
mecA	97	Medium-independent expression of competence	115
mecB	10	Medium-independent expression of competence	115
mecC	117	Medium-independent expression of competence	112
menB	273	Menaquinone deficient; multiple aminoglycoside resistance; dihydroxynaphthoate synthase	111*, 346†
menC,D	273	Menaquinone deficient; multiple aminoglycoside resistance; blocked in formation of *o*-succinylbenzoic acid from chorismic acid	346†
menE	273	Menaquinone deficient; multiple aminoglycoside resistance; *o*-succinylbenzoyl coenzyme A synthetase	111*, 346†
metA	110	Responds to methionine, cystathionine, or homocysteine	522
metB	197	Responds to methionine or homocysteine	16, 30
metC	115	Responds to methionine	114
metD	100	Responds to methionine	615
metE	272	SAM synthetase	578†
mic		Resistance to micrococcin; see *rplC*	511
min	242	Minicell production; two cistrons, *C* and *D*	300*
motA,B	118	Paralyzed flagella; an operon of two motility genes	348, 628

(Continued)

Table 1—*Continued*

Gene symbol	Map position[a]	Phenotype, enzyme deficiency, or other characteristic[b]	References[c]
mpo	218	Membrane protein overproduction, temperature-sensitive sporulation	331
mpr	21	Metalloprotease, extracellular	504*
mraY	133	Cell wall synthesis; homologous to *mraY* of *E. coli*	123*
mre	242	Three genes, *B*, *C*, and *D*; homologous to the *E. coli* shape determining genes *mre*; *divIVB* and *rodB* mutations map inside the *mre min* locus	300*
mtlA	34	Lacks mannitol transport; maps near *mtlB*	271
mtlB	34	Mannitol-1-phosphate dehydrogenase	79†
mtrA	204	Resistance to 5-methyl-tryptophan; GTP cyclohydrolase I, involved in folic acid biosynthesis	24, 175*
mtrB	204	Resistance to 5-methyl-tryptophan; regulator of tryptophan biosynthesis	24, 175*
murA	316	Thermosensitive lysis; phosphoenolpyruvate: uridine-*N*-acetylglucosamine enolpyruvoyl transferase; formerly *lssF*	57, 464
murD	134	Cell wall synthesis; homologous to *murD* of *E. coli*	123*, 220*
murE	133	Cell wall synthesis; homologous to *murE* of *E. coli*	123*, 220*
murG	139	Cell wall synthesis; homologous to *murG* of *E. coli*	123*, 220*, 241*
mutI		See *polC*	
nalA		Resistance to nalidixic acid; see *gyrA*	197
narA	320	Inability to use nitrate as a nitrogen source	298
narB	28	Inability to use nitrate as a nitrogen source	623
ndk	204	Nucleotide diphosphate kinase	215*
nea	UC	Neamine resistance; see ribosomal protein cluster	173
neo	UC	Neomycin resistance; see ribosomal protein cluster	198
nic	240	Nicotinic acid requirement	138†
nonA	UC	Permissive for bacteriophage SP10 and ϕNR2; closely linked to *rfm*	471
nonB		Permissive for bacteriophage SP10 and ϕNR2; see *hsdM*	471
novA	0	Resistance to novobiocin; see *gyrB*	198
nprB	95	Neutral protease	556*
nprE	127	Neutral metalloprotease	608*
nprR	127	Regulatory gene for neutral protease	550*
nrdA	167	Ribonucleoside diphosphate reductase; formerly *dnaA*	9, 395*
nucB	227	Extracellular DNase produced during sporulation	5, 566
nusA	147	Transcription termination	497*
obg	240	Essential GTP-binding protein of unknown function	554*
odhA	181	2-Oxoglutarate dehydrogenase, E1 subenzyme of the 2-oxoglutarate dehydrogenase complex; formerly *citK*	74†, 75*, 203
odhB	181	Dihydrolipoamide transsuccinylase, E2 subenzyme of the 2-oxoglutarate dehydrogenase complex; originally designated *citM*	74†, 75*
ole	UC	Oleandomycin resistance; see ribosomal protein cluster	173
opp	104	Oligopeptide permease; consists of five genes, *A* through *E*; also called *spo0K*	414*, 465*
ordA	NM	Ornithine-2-oxoacid aminotransferase	593
outA	20	Blocked in outgrowth after RNA, protein, and DNA syntheses have started; previous designation, *gspIV*	8
outB	28	Blocked in outgrowth before most macromolecular syntheses have started; previous designations, *gsp-81* and *tscBGH*	7*, 73
outC	28	Blocked in outgrowth after RNA and protein syntheses have started, but before DNA synthesis; previoius designation, *gsp-25*	8
outD	158	Blocked in outgrowth; protein and DNA syntheses reduced; previous designation, *gsp-1*	158
outE	300	Blocked in outgrowth; RNA synthesis normal; protein synthesis reduced; DNA synthesis prevented; previous designation, *gsp-42*	8
outF	316	Blocked in outgrowth; RNA and protein syntheses reduced; DNA synthesis prevented; previous designation, *gsp-4*	8
outG	160	Identified by outgrowth-specific transcript; see *pksA*	168†
outH	160	Identified by outgrowth-specific transcript; see *pksA*	168†
outI		Identified by outgrowth-specific transcript; part of chemotaxis operon; = *fla-che*	168†, 201*
oxr	19	Oxolinic acid resistance	573
pab	10	*p*-Aminobenzoic acid requirement; subunit A of *p*-aminobenzoate synthase	257, 502*
pabC	10	Enzyme required for *p*-aminobenzoate synthesis	502
pac	5	Resistance to pactamycin	560
pai	283	Regulation of sporulation and degradative enzyme production	233*
pal	126	Peptidoglycan-associated lipoprotein; see *slp*	205*
panB	57	Ketopantoate hydroxymethyltransferase	25
panE	58	Ketopantoic acid reductase	25
pap		Hyperproduction of proteases and amylase; see *degO*	525

(Continued)

Table 1—*Continued*

Gene symbol	Map position[a]	Phenotype, enzyme deficiency, or other characteristic[b]	References[c]
patA	118	Putative aminotransferase; other name, *uat*	19*, 412*, 555
patB	277	Putative aminotransferase	555*
pbp		Penicillin-binding protein; mutations affecting PBPs 2a, 2b, 3, and 4; probably four loci; see also *dac*	60, 498
pbp2B	133	Penicillin-binding protein; closely related to *pbpB* of *E. coli*	61, 123*, 498
pbuG	55	Purine base uptake	482
pcp	NM	Pyrrolidone carboxylpeptidase	22*
pdhA	125	Pyruvate dehydrogenase; E-1α (pyruvate decarboxylase) component; formerly *aceA*	49, 206*
pdhB	125	Pyruvate dehydrogenase component E-1β	206*
pdhC	125	Pyruvate dehydrogenase component E2	206*
pdhD	125	Pyruvate dehydrogenase; lipoamide dehydrogenase E-3 component of both pyruvate dehydrogenase and alpha-ketoglutarate dehydrogenase complexes, formerly *citL*	206*, 225
pdp	337	Pyrimidine nucleoside phosphorylase	467
pfk	255	Phosphofructokinase	165
pgk	NM	3′-Phosphoglycerolkinase	150
pha-1	47	Resistance to phage SPO1	298
pha-2	280	Resistance to phage SPP1	477
pha-3	308	Resistance to phage of group III; same as *gga*	129
pha-4	295	Resistance to phage	100
pheA	240	Phenylalanine requirement; prephenate dehydratase	554*
pheB	240	Chorismate mutase	554*
pheS	NM	Phe-tRNA synthetase subunit	53†, 54*
pheT	NM	Phe-tRNA synthetase subunit	53†, 54*
phl	12	Phleomycin resistance; mutator marker; probably allele of *rspL*	238
phoAIII	50	Alkaline phosphatase III	47*
phoP	258	Regulation of alkaline phosphatase and phosphodiesterase; homologous to *phoB* of *E. coli*	488*
phoR	258	Regulation of alkaline phosphatase and phosphodiesterase; homologous to *phoR* or *E. coli*	489*
phoS	114	Constitutive alkaline phosphatase	432
phoT	245	Constitutive alkaline phosphatase	424
pig		Sporulation-associated pigment; allele of *cotA*	459
pksA	160	Polyketide synthase; previously *outG*	157*, 168†
pnp	211	Putative purine nucleoside phosphorylase	600*
polA	257	DNA polymerase A	413†
polC	147	DNA polymerase III; azopyrimidine resistance; *mut-1*, *azh-12*, and *dnaF133* are alleles	193*, 475
ponA	93	Penicillin-binding protein; homologous to 1A of *E. coli* and *S. pneumoniae*	194*
ppiB	210	Putative protein-prolyl isomerase; 45% homologous to *ppiB* of *E. coli*	519*
pro	115	Proline requirement	67
prs	7	Phosphoribosylpyrophosphate synthetase	387†, 388*
prsA	75	Lipoprotein involved in protein secretion	274*, 275
prs-11	298	Protein secretion	275
prs-33	150	Protein secretion	275
prtR		See *degR*	
ptg	129	Peptidoglycan biosynthesis	68
ptm	107	Pyrithymine resistance	157
ptsG	118	An open reading frame with homology to the *E. coli ptsG* locus (glucose-specific enzyme II of the PTS) has been identified adjacent to *ptsH*	523*, 546*
ptsH	118	Phosphoenolpyruvate phosphotransferase, *Hpr*	178*
ptsI	118	Phosphoenolpyruvate phosphotransferase, enzyme I	178*
ptsX	118	Now considered part of *ptsG*	
pupA	NM	Adenosine phosphorylase	192
pupG	NM	Inosine-guanosine phosphorylase	391†
purA	350	Adenylsuccinate synthetase	317*, 434†, 483†
purB	55	Purine requirement; adenylosuccinate lyase; previously designated *purE*	119*
purC	55	Purine requirement; phosphoribosyl-aminoimidazole-succinocarboxamide synthetase	119*
purD	55	Purine requirement; GAR synthetase	119*
purE	55	Purine requirement; AIR carboxylase I	119*
purF	55	Purine requirement; amido-phosphoribosyltransferase; previously designated *purB*	315*

(Continued)

Table 1—*Continued*

Gene symbol	Map position[a]	Phenotype, enzyme deficiency, or other characteristic[b]	References[c]
purH	55	Purine requirement; phosphoribosylaminoimidazole-carboxamide formyltransferase/IMP cyclohydrolase	119*
purK	55	Purine requirement; AIR carboxylase II; previously designated *purD*	119*
purL	55	Purine requirement; FGAM synthetase II	119*
purM	55	AIR synthetase	119*
purN	55	Purine requirement; GAR formyltransferase	119*
purQ	55	Purine requirement; FGAM synthetase II	119*
pycA	127	Pyruvate carboxylase	367†
pyrA	139	Carbamyl phosphate synthetase	299†, 446*
pyrB	139	Aspartate carbamoyltransferase	299†, 446*
pyrC	139	Dihydroorotase	299†, 446*
pyrD	139	Dihydroorotate dehydrogenase	299†, 446*
pyrE	139	Orotate phosphoribosyltransferase; also called *pyrX*	299†, 446*
pyrF	139	Orotidine-5′-phosphate decarboxylase	299†, 446*
pyrG	323	Cytidine-5′-triphosphate synthase; same as *ctrA*	456, 553*
qox	331	Quinol oxidase $aa_{3\text{-}600}$; operon of four genes, *A*, *B*, *C*, and *D*	476*
recA	157	Genetic recombination and radiation resistance; homologous to *recA* of *E. coli*; the former *recE* mutations correspond to alleles of *recA*	113, 224, 322†, 532*, 609
recB	240	Genetic recombination and radiation resistance	224
recC		Genetic recombination; indirect effect of bacteriophage SPO2 lysogeny; see *attSPO2*	162
recD	0	Probably identical to *recR*	534
recE		Renamed *recA*	609
recF	0	Genetic recombination and radiation resistance	357*
recG	200	Genetic recombination and radiation resistance	337, 413†
recH	69	Genetic recombination and radiation resistance	338
recI	72	Genetic recombination and radiation resistance	338
recL	UC	Genetic recombination and radiation resistance; linked to *cysA*	113
recM		See *recR*	
recN	217	Genetic recombination, homologous to *E. coli recN*	565*
recP	11	Reduced recombination and high sensitivity to mitomycin C; prevents φ105 restriction in BsuR⁺ hosts; formerly *rec149*	12
recR	0	Genetic recombination and radiation resistance; homologous to *recR* of *E. coli*; formerly *recM*	9, 10*, 337
relA	245	ATP:GTP 3′-phosphotransferase; resistance to aminotriazole	466, 509
relC		See *tsp*, *rplK*	509
relG	13	Defect in glucose uptake	444
rev-4	324	Suppressor of some pleiotropic effects (but not asporogeny) of *spo0* mutations; suppresses effect on sporulation of various drug-resistant mutations	553*
rfm		Rifampin resistance; RNA polymerase; see *rpoB*	190
rgn	UC	Improved protoplast regeneration; maps near *cysA*	4
ribA	209	Riboflavin requirement	349*, 350*, 352, 447†
ribB	209	Riboflavin requirement	350*, 352, 447†
ribC	147	Riboflavin requirement	278
ribD	209	Riboflavin requirement	349*, 350*, 352, 447†
ribG	209	Riboflavin requirement	349*, 350*, 352, 448†
ribH	209	Riboflavin requirement	349*, 350*, 352, 448†
ribT	209	Riboflavin requirement	349*, 350*, 352, 448†
rna-53	312	Temperature-sensitive RNA synthesis	457
rnpA	360	Homologous to *rnpA* of *E. coli*	396*
rnpB	NM	RNase P, RNA component	336*, 453*
rodB	242	Cell wall defective; see *mre*	263, 300*
rodC	309	Cell wall defective; shown to be the same as *tagF*	231, 232*, 435
rodD	198	Cell wall defective	424
rplA	12	Ribosomal protein BL1, chloramphenicol resistance II	98
rplB	12	Ribosomal protein L2 (BL2)	507
rplC	12	Ribosomal protein L3 (BL3); probably micrococcin resistance	94
rplD	12	Ribosomal protein L4	441*
rplE	12	Ribosomal protein L5 (BL6)	212*
rplF	12	Ribosomal protein L6 (BL8)	94
rplJ	12	Ribosomal protein L10 (BL5)	98
rplK	12	Ribosomal protein L11 (BL11); thiostrepton resistance; (*relC*)	98
rplL	12	Ribosomal protein L12 (BL9); chloramphenicol resistance VI	98
rplN	12	Ribosomal protein L14	212*
rplO	11	Ribosomal protein L15; chloramphenicol resistance III	375*, 403, 441*

(Continued)

Table 1—*Continued*

Gene symbol	Map position[a]	Phenotype, enzyme deficiency, or other characteristic[b]	References[c]
rplP	12	Ribosomal protein L16	212*
rplQ	11	Ribosomal protein L17 (BL15)	51*
rplR	12	Ribosomal protein L18	539*
rplU	242	Ribosomal protein L21 (BL20)	95
rplV	12	Ribosomal protein L22 (BL17); erythromycin resistance	507
rplW	12	Ribosomal protein L23	441*
rplX	12	Ribosomal protein L24 (BL23)	212*, 494*
rpmA	241	Ribosomal protein L27 (BL24)	139*
rpmC	12	Ribosomal protein L29	212*
rpmD	11	Ribosomal protein L30 (BL27)	375*
rpmH	0	Ribosomal protein L34	357*
rpmJ	11	Ribosomal protein B	441*
rpoA	11	RNA polymerase α subunit	538*
rpoB	12	β subunit of RNA polymerase; rifampin resistance	442†
rpoC	12	β' subunit of RNA polymerase; streptolydigin resistance	442†
rpoD	224	RNA polymerase σ^{43} subunit; see *sigA*	443*
rpoE	325	δ subunit	290*
rprI	355	Putative protease	393*
rpsC	12	Ribosomal protein S3 (BS3)	507
rpsD	263	Ribosomal protein S4 (BS4)	185*
rpsE	11	Ribosomal protein S5; spectinomycin resistance	96, 441*, 614
rpsF	4	Ribosomal protein S6 (BS9)	97
rpsG	12	Ribosomal protein S7 (BS7)	98
rpsH	12	Ribosomal protein S8 (BS8)	212*
rpsI	12	Ribosomal protein S9 (BS10)	96
rpsJ	12	Ribosomal protein S10 (BS13) (*tetA*)	507
rpsK	12	Ribosomal protein S11 (BS11)	538*
rpsL	12	Ribosomal protein S12 (BS12) (*strA*); streptomycin resistance; also see *fun*	94
rpsM	11	Ribosomal protein S13	538*
rpsN	12	Ribosomal protein S14	212*
rpsP	145	Ribosomal protein S16 (BS17)	97
rpsQ	12	Ribosomal protein S17 (BS16)	212*
rpsS	12	Ribosomal protein S19 (BS19)	403
rpsT	12	Ribosomal protein S20 (BS20)	403
rrnA	0	rRNA operon	397*
rrnB	275	rRNA operon	181*
rrnD	67	rRNA operon	287†
rrnE	45	rRNA operon	287†
rrnF	NM	rRNA operon known to exist but not cloned or mapped	48
rrnG	15	rRNA operon	48†
rrnH	15	rRNA operon	48†
rrnI	15	rRNA operon	48†
rrnJ	6	rRNA operon	594†
rrnO	0	rRNA operon	397*
rrnW	6	rRNA operon	594†
rsbV	39	Positive regulator of σ^B-dependent gene expression	50*, 52
rsbW	39	Negative regulator of σ^B-dependent gene expression	50*, 52
rsbX	39	Negative regulator of σ^B-dependent gene expression	50, 52
rtp	178	Replication terminator protein	3*
rvt	NM	Mutations causing same phenotype as *rvtA* mutations but not mapping in the *rvtA* region	496
rvtA	217	Suppressor of sporulation defect in *spo0B*, *spo0E*, *spo0F*, and *spoIIA* mutants; allele of *spo0A*	496
sacA	330	β-Fructofuranosidase (sucrase)	146*
sacB	296	Levansucrase	524*
sacC	233	Levanase	323*
sacL		See *levD*, *levE*	
sacP	330	Sucrose transport; enzyme II of sucrose PTS	147*
sacQ	279	Hyperproduction of levansucrase and other extracellular enzymes; see *degQ*	297
sacR	296	Constitutive for levansucrase	526*
sacS	333	Regulation of β-fructofuranosidase; consists of two genes designated *sacX* and *sacY*	23†
sacT	330	Constitutive α-fructofuranosidase production	102*, 297
sacU		Regulatory gene for levansucrase and other extracellular enzymes; see *degS,U*	525

(Continued)

Table 1—*Continued*

Gene symbol	Map position[a]	Phenotype, enzyme deficiency, or other characteristic[b]	References[c]
sacV	40	Regulatory gene for levansucrase	324*
sacX	333	Sucrase regulation; see *sacS*	524*
sacY	333	Sucrase regulation; antiterminator; see *sacS*	629*
sapA	114	Mutations overcome the sporulation-phosphatase-negative phenotype of early blocked *spo* mutants	428
sapB	56	Mutations overcome sporulation-phosphatase-negative phenotype of early blocked *spo* mutations	432
sas	211	Weak intragenic suppressor mutations of *spoIIA*	619
sbp	134	Small basic protein	123*
scoA	109	Sporulation control; protease and phosphatase overproduction; delayed spore formation	345
scoB		Sporulation control; protease and phosphatase overproduction; delayed spore formation; allele of *kinA*	109
scoC		Sporulation control; see *hpr*	345
scoD		Sporulation control; the same as *kinA*	345
scr	0	Small cytoplasmic RNA; homologous to ffs of *E. coli*	10*, 534, 535*
sdhA	252	Flavoprotein subunit of succinate dehydrogenase; note that the *sdh* loci have been reorganized; originally designated *citF*	423*
sdhB	252	Iron protein subunit of succinate dehydrogenase	423*
sdhC	252	Cytochrome b_{558}; subunit of succinate dehydrogenase complex	313*
secA	303	Secretion of proteins; homologous to *secA* of *E. coli*; originally called *div-341, ts-341*; see *lssG*	470*
secY	11	Secretion of proteins; homologous to *secY* of *E. coli*	375*, 539*
sen	69	Enhances expression of genes for several extracellular proteins	582*
serA	207	Requirement for serine; probably glycerophosphate dehydrogenase	314, 519*
serC	257	Requirement for serine; defined by Tn*917* insertion	569
serR	25	Serine resistance	622
sfp	29	Surfactin production	179*, 376*
sigA	222	Major RNA polymerase sigma factor, σ^A or σ^{43}; locus also called *rpoD*	443, 583*
sigB	39	RNA polymerase sigma factor, σ^B or σ^{37}	39*, 50, 52, 118*
sigC	NM	RNA polymerase sigma factor, σ^C or σ^{32}	254
sigD	142	RNA polymerase sigma factor, σ^D or σ^{28}; autolytic enzymes, allelic to *flaB*	204*, 321
sigE	135	RNA polymerase sigma factor, σ^E or σ^{29}; locus the same as *spoIIGB*; see *spoIIG*	558*
sigF	211	RNA polymerase sigma factor, σ^F; locus the same as *spoIIAC*; see *spoIIA*	125*, 542*
sigG	135	RNA polymerase sigma factor, σ^G; locus the same as *spoIIIG*	328*, 542*
sigH	10	RNA polymerase sigma factor, σ^H or σ^{30}; locus the same as *spo0H*; see *spo0H*	117*
sigK	227	See *spoIVC* and *spoIIIC*; RNA polymerase sigma factor, σ^K; the gene is interrupted by an insertion excised in the mother cell during stage IV of sporulation; *spoIVCB* and *spoIIIC* code for, respectively, the N- and C-terminal halves of SigK	281, 531*
sigL	295	RNA polymerase sigma factor, σ^L	104*
sin	221	Sporulation inhibition at high copy; see *flaD*	164*
sipS	209	Signal peptidase I	519, 563*, 564
slp	125	Small lipoprotein	205*
smo	295	Smooth/rough colony morphology	167†
sof-1	217	Suppressor of sporulation defects in *spo0B*, *spo0E*, and *spo0F* mutants; mutation is an alteration in codon *12* of the *spo0A* gene	228*
spcA		Spectinomycin resistance; see *rpsE*	248
spcB	146	Spectinomycin resistance	308
spcD	2	Spectinomycin dependence; maps between *cysA* and *purA*	209
spdA	288	Sporulation derepressed; low pyruvate carboxylase activity	120
spe	NM	Endonuclease excising spore photoproducts (formerly *ssp-1*)	371
spg		Sporangiomycin resistance; 50S ribosomal alteration	34
spl	118	Spore photoproduct lyase	131*
spoCM	359	Stage 0 sporulation; allele of *spo0J*	55, 56
spo0A	217	Sporulation; mutants blocked at stage 0; transcription regulator; part of phosphorelay	140*
spo0B	240	Sporulation; mutants blocked at stage 0; part of phosphorelay	139*
spo0C	217	Stage 0 sporulation; mutations with less pleiotropic phenotypes known to be missense alterations in the *spo0A* gene product	228*
spo0D		Stage 0 sporulation; single allele resulting in stage 0 block of sporulation; shown to be an allele of *spo0B*	223
spo0E	115	Stage 0 sporulation; oligosporogenous mutations giving stage 0 block; less pleiotropic than *spo0A*, *spo0B*, or *spo0F* mutations	415*

(Continued)

Table 1—*Continued*

Gene symbol	Map position[a]	Phenotype, enzyme deficiency, or other characteristic[b]	References[c]
spo0F	324	Stage 0 sporulation; part of phosphorelay	552*
spo0G		Stage 0 sporulation; shown to be an allele of *spo0A*	223
spo0H	10	Stage 0 sporulation; locus codes for RNA polymerase sigma factor, H; also called *sigH*	117*
spo0J	359	Stage 0 sporulation; *spo-87* originally placed in this locus was found to be in a distinct one; named *spoIIIJ*	55*, 124*, 126†, 398*, 462*
spo0K	104	Stage 0 sporulation; see *opp* operon	429
spo0L	106	Stage 0 sporulation, uncharacterized allele giving *spo0* phenotype, maps near *spo0K* but genetically distinct	223
spoIIA	211	Stage II sporulation; DNA sequence has three adjacent open reading frames, *spoIIAA*, *spoIIAB*, and *spoIIAC*; *spoIIAC* also called *sigF*	125*, 144*, 600*
spoIIB	244	Stage II sporulation	429, 515*
spoIIC		Stage II sporulation; shown to be an allele of *spoIID*	429
spoIID	316	Stage II sporulation	306*
spoIIE	10	Stage II sporulation; *spoIIH* is an allele of *spoIIE*	187*, 618*
spoIIF		See *kinA*	429, 515*
spoIIG	135	Stage II sporulation; DNA sequence has two adjacent open reading frames, *spoIIGA* and *spoIIGB*; *spoIIGB* (also called *sigE*) codes for an RNA polymerase sigma factor, σ^E	123*, 327*, 462, 530*
spoIIH		See *spoIIE*	
spoIIJ		See *kinA*	
spoIIL		Defined by Tn*917* insertion; *spo0A* allele	176, 474
spoIIM	212	Defined by Tn*917* insertion; distinct from *spoIIA*	474, 618*
spoIIN		Defined by mutation *spo-279*; originally placed in *spoIIG*; maps near to, but separate from, *spoIIG*; *spoIIN279* is an allele of *ftsA*	265*, 296, 617
spoIIIA	218	Stage III sporulation; an eight-gene operon, *A* through *H*	126†, 186†, 242*
spoIIIB		Stage III sporulation; may be the same as *spoIIIE*	429
spoIIIC	232	Initially designated as stage III sporulation, but very similar phenotype to *spoIVC*; 3′ end of *sigK*	127*, 531*
spoIIID	312	Stage III sporulation	280*, 429
spoIIIE	142	Stage III sporulation	65*, 148*
spoIIIF		See *spoVB*	289, 439
spoIIIG	135	Stage III sporulation; codes for RNA polymerase sigma factor G; see also *sigG*	123*, 264*, 328*
spoIIIJ	360	Stage III sporulation; vegetatively expressed gene essential for σ^G activity at stage III; previously *spo0J87*	124*, 398*
spoIVA	204	Stage IV sporulation	314†, 458*, 528*
spoIVB	217	Stage IV sporulation; maps immediately upstream of *spo0A*	429, 565*
spoIVC	227	Stage IV sporulation; contains two cistrons; *spoIVCA* codes for a recombinase; *spoIVCB* codes for the N-terminal half of SigK	282*, 480*, 531*
spoIVD		Stage IV sporulation; allelic to *spoIIIC*	281, 429
spoIVE		Stage IV sporulation; allelic to *spoIIIC*	281, 429
spoIVF	241	Stage IV sporulation; other name, *bofB*	92*, 126†
spoIVG	97	Stage IV sporulation	429
spoVA	211	Stage V sporulation; contains six open reading frames, *spoVAA* through *spoVAF*	143*, 519*, 604*
spoVB	239	Stage V sporulation; the previous *spoIIIF* is part of *spoVB*; membrane protein involved in cortex biosynthesis	126†, 438†, 439*
spoVC	7	Stage V sporulation; the two mutations placed in this locus may be in separate loci	90, 236*
spoVD	133	Stage V sporulation; closely related to *pbpB* of *E. coli*	123*, 429
spoVE	134	Stage V sporulation; homologous to *ftsW* of *E. coli*	64*, 123*, 220, 240*, 255*
spoVF	148	Stage V sporulation; probably the same as *dpa*; mutants form octanol and chloroform-resistant, heat-sensitive spores; form heat-resistant spores in presence of dipicolinic acid	410*, 431
spoVG	6	Stage V sporulation; previously called 0.4-kb gene	234*, 487†
spoVH		Found to be an allele of *spoVA*	122
spoVJ		Found to be an allele of *spoVK*	126†, 132†, 149
spoVK	168	Defined by Tn*917* insertion	132, 474
spoVL		Defined by Tn*917* insertion; allele of *spoIVF*	90*, 474
spoVM	145	Defined by Tn*917* insertion	474
spoVN	279	Defined by Tn*917* insertion; same gene as *ald*	184*, 474
spoVP		Defined by Tn*917* insertion; same as *spoIVA*	314†, 458*, 474, 528*
spoVQ	213	Defined by Tn*917* insertion; less closely linked to *lys* than *spoVA*	474
spoVIA	255	Stage VI of sporulation	251

(Continued)

Table 1—*Continued*

Gene symbol	Map position[a]	Phenotype, enzyme deficiency, or other characteristic[b]	References[c]
spoVIB	247	Stage VI of sporulation	252
spoVIC	294	Stage VI of sporulation	250
spoL	227	"Decadent" sporulation	27
sprA	NM	Derepression of homoserine kinase, homoserine dehydrogenase, and the weak threonine dehydratase activity (*tdm*); see *hom*	570
sprB		Partial suppression of isoleucine requirement allows threonine dehydratase *sprA* mutants to grow in minimal medium; maps near the *tdm* locus; see *tdm* and *hom*	571
sprE		See *aprE*	
srfA	29	Surfactin production	377*, 378†
srfB		Surfactin production; identical to *comA*	379*
srm	12	Modifies resistance of *spcA* strains	70
ssa	217	Alcohol-resistant sporulation; maps close to, or in, *spo0A*; *rvt* mutations have same phenotype	44
sspA	266	One member of multigene family coding for small acid-soluble spore proteins	86*
sspB	66	Similar product to *sspA*	14, 85, 86*
sspC	182	Similar product to *sspA*	85, 87*
sspD	121	Similar product to *sspA*	85, 86*
sspE	62	Similar product to *sspA*	14, 188*
sspF	6	Similar product to *sspA*; same as "0.3-kb gene"	493, 527*
ssp-1		See *spe*	
std		Streptolydigin resistance; RNA polymerase; see *rpoC*	191
strA		Streptomycin resistance; see *rpsL*	173
strB	130	Streptomycin resistance	520
strC	219	Streptomycin uptake deficient, possible cytochrome oxidase regulator	340
sub	NM	Structural gene for peptide antibiotic subtilin; also called *spaS*	29*, 270*
sul	9	Sulfonilamide resistance; probably dihydropteroate synthase	339†, 502*
suA20	145	Suppressor of *recH* mutations with increased ATP-dependent DNase	288
sup-3	37	Suppressor, lysyl tRNA	218, 304, 369
sup-44	37	Suppressor, leucyl tRNA	304, 309
tagA	309	Polyglycerol phosphate assembly and export	335*
tagB	309	Polyglycerol phosphate assembly and export	335*
tagC	309	Polyglycerol phosphate assembly and export; probably identical to *dinC*	335*
tagD	309	Glycerol-3-phosphate cytidylyltransferase	335*, 436
tagE	309	= *gtaA* + *gtaD*	334
tagF	309	CDP-glycerol:poly(glycerolphosphate)glycerophosphotransferase; = *rodC*	435*
tagG	309	Teichoic acid biosynthesis; essential gene	293
tagH	309	Teichoic acid biosynthesis; essential gene	293
tal	240	Resistant to β-thienylalanine	84
tdm	284	Weak threonine dehydratase activity due to homoserine dehydratase; see *hom*	15, 570, 571
ten	245	Constitutive transfection enhancement of SP82 DNA; transformation defective	183
tenA	61	Transcription enhancement of *aprE*	407*
tenE	61	Reduction of *aprE* transcription	407*
terC	179	Terminus of replication	77*
tetB	3	Resistance to tetracycline	13*, 472*, 597
tetR	360	Tetracycline resistance	393
thfS	304	Tetrahydrofolate synthetase	296†
thiA	70	Thiamine requirement	266
thiB	105	Thiamine requirement	226
thiC	332	Thiamine requirement	622
thrB	284	Threonine requirement; homoserine kinase; previously designated *thrA*	408*
thrC	284	Threonine requirement; threonine synthase; previously called *thrB*	408*
thrS	256	Major threonyl-tRNA synthetase	445*
thrZ	UC	Minor threonyl-tRNA synthetase; maps near *sacA*	445*, 491
thyA	168	Thymidylate synthetase A	384
thyB	195	Thymidylate synthetase B	16, 249*
tkt	NM	Transketolase	479
tmp		Trimethoprim resistance; most probably the same as *dfrA*	249*, 579
tmrA	24	Tunicamycin resistance; hyperproductivity of extracellular α-amylase	390
tmrB	25	Tunicamycin resistance; ATP-binding membrane protein	389, 404†
tms-12		Temperature-sensitive cell division; allele of *divIB*	88, 199*
tms-26	7	Temperature-sensitive cell division	189†, 388*
tolB	NM	Tolerance to bacteriophage	247

(Continued)

Table 1—*Continued*

Gene symbol	Map position[a]	Phenotype, enzyme deficiency, or other characteristic[b]	References[c]
tre	62	Trehalose	298
trnA	0	Genes for Ile and Ala tRNAs located between 16S and 23S RNAs in the *rrnA* operon	397*
trnB	275	Linked set of transfer RNA genes distal to *rrnB* that contains transfer RNAs for Val, Thr, Lys, Leu, Gly, Leu, Arg, Pro, Ala, Met, Ile, Ser, fMet, Asp, Phe, His, Gly, Ile/Met, Asn, Ser, and Glu; formerly *trnE*	181*
trnI	14	Linked set of six tRNA genes including tRNAs for Ala, Pro, Arg, Gly, Thr, and Asn located between *rrnH* and *rrnI*; formerly *trnH* and *trnB*	588*
trnJ	6	Linked set of nine tRNA genes located between *rrnJ* and *rrnW*	182*
trnO	0	Same tRNA genes as *trnA* but located in *rrnO*	575
trnR	UC	Linked set of tRNA genes thought to be distal to *rrnR* that contains transfer RNAs for Asn, Ser, Glu, Val, Met, Asp, Phe, Thr, Tyr, Trp, His, Gln, Gly, Cys, Leu, and Leu; formerly called *trnD*	589*
trnY	NM	Linked set of tRNA genes for Lys, Glu, Asp, and Phe	603*
trpA	203	Tryptophan synthase α	214*
trpB	203	Tryptophan synthase β	214*
trpC	203	Indol-3-glycerolphosphate synthase	214*
trpD	203	Anthranilate phosphoribosyltransferase	28*
trpE	203	Anthranilate synthase	28*
trpF	203	Phosphoribosylanthranilate isomerase	214*
trpG	9	Glutamine aminotransferase	502*
trpS	104	Tryptophanyl-transfer RNA synthase	83*, 522
trpX	9	See *trpG*	257
trx	252	Thioredoxin	81*
ts-1	145	Temperature-sensitive division; allele of *ftsZ*	199*
ts-31	131	Temperature-sensitive division	351
ts-39	226	Temperature-sensitive synthesis of phosphatidylethanolamine	302
ts-341	312	Temperature-sensitive division; see *secA*; see *lssG*	351
ts-526	3	Temperature-sensitive division	351
tscB, G, H		Temperature-sensitive vegetative growth; alleles of *outB*	73, 137†
tsi	48	Temperature-sensitive induction of all known SOS functions	501
tsp		Thiostrepton resistance; 50S subunit; maps in ribosomal protein cluster	418
tsr	324	Temperature-sensitive RNA synthesis; fructose-1,6-biphosphate aldolase	553*, 572
tufA	10	Elongation factor Tu	116
tyrA	202	Tyrosine requirement; prephenate dehydrogenase	213*
tyrS	263	Tyrosyl-tRNA synthetase	185, 208*
tyrZ	332	Tyrosyl-tRNA synthetase, minor, expresses when *tyrS* is inactivated; formerly *tyrT*	171*
tzm		Tetrazolium reaction; see *gerJ*	586
uat		Unidentified aminotransferase; see *patA*	19*
udk	NM	Uridine kinase; also lacks cytidine kinase; mutant resistant to fluorouridine	383
udp	NM	Uridine phosphorylase	336†
unc	NM	Uncoupler resistant	279
upp	NM	Uracil phosphoribosyltransferase	383
urg	NM	*N*-Glucosidase	316
urs		Uracil sensitivity; arginine-specific carbamoylphosphate synthase; see *car*	
uvrA	305	Excision of UV light-induced pyrimidine dimers in DNA	82*, 224
uvrC	252	Excision of UV light-induced pyrimidine dimers in DNA; homologous to *uvrC* of *E. coli*; formerly *uvrB*	81*, 371
valS	242	Valyl-t-RNA synthetase	318*
vas		Valine sensitivity; maps within threonine dehydratase locus; see *ilvA*	295
vpr	330	Minor extracellular serine protease	506*
VS	12	Virginiamycin (VS component) resistance	460
xhi	112	Heat-inducible PBSX	66
xlt	112	Induced PBSX lacks tails	161
xpaA	277	Hydrolysis of 5-bromo 4-chloroindolyl phosphate (X-phos)	235†
xpaB	114	Hydrolysis of 5-bromo 4-chloroindolyl phosphate (X-phos)	235†
xpaC	2	Hydrolysis of 5-bromo 4-chloroindolyl phosphate (X-phos)	235†
xpt	198	Xanthine phosphoribosyltransferase	482
xre	NM	PBSX repressors	106
xre-1	NM	PBSX repressors	106
xylA	168	Xylose isomerase	492, 596*, 622
xylB	168	Xylose kinase	492, 595†, 622
xylR	168	Regulation of xylose regulon	200†, 622
xynA	48	Extracellular β-xylanase	406*, 461
xynB	48	Cell-associated β-xyloxidase	200†, 461, 622
xynC	48	Probably xyloside permease	200†, 622

[a] NM, not mapped; UC, map position not fully defined.
[b] Abbreviations: SASP, small acid-soluble spore protein; PTS, phosphotransferase system; SAM, *S*-adenosylmethionine; PRPP, 5-phosphoribosyl 1-pyrophosphate; GAR, phosphoribosylglycinamide; AIR, phosphoribosylaminoimidazole; IMP, inosinic acid; FGAM, 5′-phosphoribosyl-*N*-formylglycinamidine.
[c] *, locus sequenced; †, locus cloned.

unknown function were not listed in Table 1 or indicated on the map. Although more references from the early papers were included in this edition than in the two previous ones, the list does not cover all the genetic literature of *B. subtilis*. The reader is advised to consult the earliest editions and original papers for details on pioneering work.

REFERENCES

1. **Adams, A.** 1973. Transposition of the arsenate resistance locus of *Bacillus subtilis*. *Genetics* **74:**197–213.
2. **Adams, A., and M. Oishi.** 1972. Genetic properties of arsenate sensitive mutants of *Bacillus subtilis* 168. *Mol. Gen. Genet.* **118:**295–310.
3. **Ahn, K. S., and R. G. Wake.** 1991. Variations and coding features of the sequence spanning the replication terminus of *Bacillus subtilis* 168 and W23 chromosomes. *Gene* **98:**107–112.
4. **Akamatsu, T., and J. Sekiguichi.** 1983. Properties of regeneration mutants of *Bacillus subtilis*. *FEMS Microbiol. Lett.* **20:**425–428.
5. **Akrigg, A., and J. Mandelstam.** 1978. Extracellular manganese stimulated deoxyribonuclease as a marker event in sporulation of *Bacillus subtilis*. *Biochem. J.* **172:**63–67.
6. **Albertini, A. M., T. Caramori, W. D. Crabb, F. Scoffone, and A. Galizzi.** 1991. The *flaA* locus of *Bacillus subtilis* is part of a large operon coding for flagellar structures, motility functions and an ATPase-like polypeptide. *J. Bacteriol.* **173:**3573–3579.
7. **Albertini, A. M., T. Caramori, D. Henner, E. Ferrari, and A. Galizzi.** 1987. Nucleotide sequence of the *outB* locus of *Bacillus subtilis* and regulation of its expression. *J. Bacteriol.* **169:**1480–1484.
8. **Albertini, A. M., and A. Galizzi.** 1975. Mutant of *Bacillus subtilis* with a temperature-sensitive lesion in ribonucleic acid synthesis during germination. *J. Bacteriol.* **124:**14–25.
9. **Alonso, J. C.** Unpublished data.
10. **Alonso, J. C., K. Shirahige, and N. Ogasawara.** 1990. Molecular cloning, genetic characterization and DNA sequence analysis of the *recM* region of *Bacillus subtilis*. *Nucleic Acids Res.* **18:**6771–6777.
11. **Alonso, J. C., C. A. Stiege, R. H. Tailor, and J.-F. Viret.** 1988. Functional analysis of the *dna*(Ts) mutants of *Bacillus subtilis*: plasmid pUB110 replication as a model system. *Mol. Gen. Genet.* **214:**482–489.
12. **Alonso, J. C., R. H. Tailor, and G. Luder.** 1988. Characterization of recombination-deficient mutants of *Bacillus subtilis*. *J. Bacteriol.* **170:**3001–3007.
13. **Amano, H., C. L. Ives, K. F. Bott, and K. Shishido.** 1991. A limited number of *Bacillus subtilis* strains carry a tetracycline-resistance determinant at a site close to the origin of replication. *Biochim. Biophys. Acta* **1088:**251–258.
14. **Amjad, M., J. M. Castro, H. Sandoval, J.-J. Wu, M. Yang, D. J. Henner, and P. J. Piggot.** 1990. An SfiI restriction map of the *Bacillus subtilis* 168 genome. *Gene* **101:**15–21.
15. **Anagnostopoulos, C.** Unpublished data.
16. **Anagnostopoulos, C., and A. M. Schneider-Champagne.** 1966. Déterminisme génétique de l'éxigence en thymine chez certains mutants de *Bacillus subtilis*. *C. R. Acad. Sci. Ser. D* **262:**1311–1314.
17. **Anderson, J. J., and A. T. Ganesan.** 1975. Temperature-sensitive deoxyribonucleic acid replication in a *dnaC* mutant of *Bacillus subtilis*. *J. Bacteriol.* **121:**173–183.
18. **Anderson, L. M., T. M. Henkin, G. H. Chambliss, and K. F. Bott.** 1984. New chloramphenicol resistance locus in *Bacillus subtilis*. *J. Bacteriol.* **158:**386–388.
19. **Antoniewski, C., B. Savelli, and P. Stragier.** 1990. The *spoIIJ* gene, which regulates early developmental steps in *Bacillus subtilis*, belongs to a class of environmentally responsive genes. *J. Bacteriol.* **172:**86–93.
20. **Armpriester, J. M., Jr., and P. S. Fink.** Unpublished data.
21. **Aronson, A. I., H.-Y. Song, and N. Bourne.** 1989. Gene structure and precursor processing of a novel *Bacillus subtilis* spore coat protein. *Mol. Microbiol.* **3:**437–444.
22. **Awadé, A., P. Cleuziat, T. Gonzalès, and J. Robert-Baudouy.** 1992. Characterization of the *pcp* gene encoding the pyrrolidone carboxyl peptidase of *Bacillus subtilis*. *FEBS Lett.* **305:**67–73.
23. **Aymerich, S., and M. Steinmetz.** 1987. Cloning and preliminary characterization of the *sacS* locus from *Bacillus subtilis* which controls the regulation of the exoenzyme levansucrase. *Mol. Gen. Genet.* **208:**114–120.
24. **Babitzke, P., P. Gollnick, and C. Yanofsky.** 1992. The *mtrAB* operon of *Bacillus subtilis* encodes GTP cyclohydrolase I (*mtrA*), an enzyme involved in folic acid biosynthesis, and *mtrB*, a regulator of tryptophan biosynthesis. *J. Bacteriol.* **174:**2059–2064.
25. **Baigori, M., R. Grau, H. R. Morbidoni, and D. de Mendoza.** 1991. Isolation and characterization of *Bacillus subtilis* mutants blocked in the synthesis of pantothenic acid. *J. Bacteriol.* **173:**4240–4242.
26. **Balassa, G., P. Milhaud, E. Raulet, M. T. Silva, and J. C. F. Sousa.** 1979. A *Bacillus subtilis* mutant requiring dipicolinic acid for the development of heat-resistant spores. *J. Gen. Microbiol.* **110:**365–379.
27. **Balassa, G., P. Milhaud, J. C. F. Sousa, and M. T. Silva.** 1979. Decadent sporulation mutants of *Bacillus subtilis*. *J. Gen. Microbiol.* **110:**381–392.
28. **Band, L., H. Shimotsu, and D. J. Henner.** 1984. Nucleotide sequence of the *Bacillus subtilis trpE* and *trpD* genes. *Gene* **27:**55–65.
29. **Banerjee, S., and J. N. Hansen.** 1988. Structure and expression of a gene encoding the precursor of subtilin, a small protein antibiotic. *J. Biol. Chem.* **263:**9508–9514.
30. **Barat, M., C. Anagnostopoulos, and A.-M. Schneider.** 1965. Linkage relationships of genes controlling isoleucine, valine, and leucine biosynthesis in *Bacillus subtilis*. *J. Bacteriol.* **90:**357–369.
31. **Barberio, C.** Unpublished data.
32. **Baumberg, S., and A. Mountain.** 1984. *Bacillus subtilis* 168 mutants resistant to arginine hydroxamate in the presence of ornithine or citrulline. *J. Gen. Microbiol.* **130:**1247–1252.
33. **Bazzicalupo, M., E. Gallori, and M. Polsinelli.** 1980. Characterization of 5-fluoroindole and 5-fluorotryptophan resistant mutants in *Bacillus subtilis*. *Microbiologica* **3:**15–23.
34. **Bazzicalupo, M., B. Parisi, G. Pirali, M. Polsinelli, and F. Sala.** 1975. Genetic and biochemical characterization of a ribosomal mutant of *Bacillus subtilis* resistant to sporangiomycin. *Antimicrob. Agents Chemother.* **8:**651–656.
35. **Beall, B., M. Lowe, and J. Lutkenhaus.** 1988. Cloning and characterization of *Bacillus subtilis* homologs of *Escherichia coli* cell division genes *ftsZ* and *ftsA*. *J. Bacteriol.* **170:**4855–4864.
36. **Beall, B., and J. Lutkenhaus.** 1989. Nucleotide sequence and insertional inactivation of a *Bacillus subtilis* gene that affects cell division, sporulation, and temperature sensitivity. *J. Bacteriol.* **171:**6821–6834.
37. **Beijer, L., R.-P. Nilsson, C. Hulmberg, and L. Rutberg.** Nucleotide sequence of the *Bacillus subtilis glpP* gene encoding a regulatory protein and the *glpF* gene encoding the glycerol uptake facilitator and evidence for a *glpFK* operon. *J. Gen. Microbiol.*, in press.
38. **Berek, I., A. Miczak, and G. Ivanovics.** 1974. Mapping of the delta-aminolaevulinic acid dehydrase and por-

phobilinogen deaminase loci in *Bacillus subtilis*. *Mol. Gen. Genet.* **132:**233–239.

39. **Binnie, C., M. Lampe, and R. Losick.** 1986. Gene encoding the σ^{37} species of RNA polymerase σ factor from *Bacillus subtilis*. *Proc. Natl. Acad. Sci. USA* **83:**5943–5947.
40. **Bischoff, D., M. D. Weinreich, and G. W. Ordal.** 1992. Nucleotide sequence of *Bacillus subtilis* flagellar biosynthetic genes *fliP* and *fliQ* and identification of a novel flagellar gene, *fliZ*. *J. Bacteriol.* **174:**4017–4025.
41. **Bishop, P. E., and L. R. Brown.** 1973. Ethidium bromide-resistant mutant of *Bacillus subtilis*. *J. Bacteriol.* **115:**1077–1083.
42. **Bohannon, D. E., M. S. Rosenkrantz, and A. L. Sonenshein.** 1985. Regulation of *Bacillus subtilis* glutamate synthase genes by the nitrogen source. *J. Bacteriol.* **163:**957–964.
43. **Bohannon, D. E., and A. L. Sonenshein.** 1989. Positive regulation of glutamate biosynthesis in *Bacillus subtilis*. *J. Bacteriol.* **171:**4718–4727.
44. **Bohin, J.-P., and B. Lubochinsky.** 1982. Alcohol-resistant sporulation mutants of *Bacillus subtilis*. *J. Bacteriol.* **150:**944–955.
45. **Bol, D. K., and R. E. Yasbin.** 1991. The isolation, cloning and identification of a vegetative catalase gene from *Bacillus subtilis*. *Gene* **109:**31–37.
46. **Bolotin, A.** Unpublished data.
47. **Bookstein, C., C. W. Edwards, N. V. Kapp, and F. M. Hulett.** 1990. The *Bacillus subtilis* 168 alkaline phosphatase III gene: impact of a *phoAIII* mutation on total alkaline phosphatase synthesis. *J. Bacteriol.* **172:**3730–3737.
48. **Bott, K. F., G. C. Stewart, and A. G. Anderson.** 1984. Genetic mapping of cloned ribosomal RNA genes, p. 19–34. *In* A. T. Ganesan and J. A. Hoch (ed.), *Genetics and Biotechnology of Bacilli*. Academic Press, Inc., Orlando, Fla.
49. **Boudreaux, D. P., E. Eisenstadt, T. Iijima, and E. Freese.** 1981. Biochemical and genetic characterization of an auxotroph of *Bacillus subtilis* altered in the acyl-CoA: acyl-carrier protein transacylase. *Eur. J. Biochem.* **115:**175–181.
50. **Boylan, S. A., A. Rutherford, S. M. Thomas, and C. W. Price.** 1992. Activation of *Bacillus subtilis* transcription factor σ^B by a regulatory pathway responsive to stationary phase signals. *J. Bacteriol.* **174:**3695–3706.
51. **Boylan, S. A., J. W. Suh, S. M. Thomas, and C. W. Price.** 1989. Gene encoding the alpha core subunit of *Bacillus subtilis* RNA polymerase is cotranscribed with the genes for initiation factor 1 and ribosomal proteins B, S13, S11, and L17. *J. Bacteriol.* **171:**2553–2562.
52. **Boylan, S. A., M. D. Thomas, and C. W. Price.** 1991. Genetic method to identify regulons controlled by nonessential elements: isolation of a gene dependent on alternate transcription factor sigma^B of *Bacillus subtilis*. *J. Bacteriol.* **173:**7856–7866.
53. **Brakhage, A. A., H. Putzer, K. Shazand, R. J. Roschenthaler, and M. Grunberg-Manago.** 1989. *Bacillus subtilis* phenylalanyl-tRNA synthetase genes: cloning and expression in *Escherichia coli* and *B. subtilis*. *J. Bacteriol.* **171:**1228–1232.
54. **Brakhage, A. A., M. Wozny, and H. Putzer.** 1990. Structure and nucleotide sequence of the *Bacillus subtilis* phenylalanyl-tRNA synthetase genes. *Biochimie* **72:**725–734.
55. **Bramucci, M.** Unpublished data.
56. **Bramucci, M. G., K. M. Keggins, and P. S. Lovett.** 1977. Bacteriophage PMB12 conversion of the sporulation defect in RNA polymerase mutants of *Bacillus subtilis*. *J. Virol.* **24:**194–200.
57. **Brandt, C., and D. Karamata.** 1987. Thermosensitive *Bacillus subtilis* mutants which lyse at the non-permissive temperature. *J. Gen. Microbiol.* **133:**1159–1170.
58. **Breton, R., D. Watson, M. Yaguchi, and J. Lapointe.** 1990. Glutamyl-tRNA synthetases of *Bacillus subtilis* 168T and of *Bacillus stearothermophilus*. *J. Biol. Chem.* **265:**18248–18255.
59. **Briehl, M., H. M. Pooley, and D. Karamata.** 1989. Mutants of *Bacillus subtilis* 168 thermosensitive for growth and wall teichoic acid synthesis. *J. Gen. Microbiol.* **135:**1325–1334.
60. **Buchanan, C. E.** 1987. Absence of penicillin-binding protein 4 from an apparently normal strain of *Bacillus subtilis*. *J. Bacteriol.* **169:**5301–5303.
61. **Buchanan, C. E.** Unpublished data.
62. **Buchanan, C. E., and A. Gustafson.** 1991. Mapping of the gene for a major penicillin-binding protein to a genetically conserved region of the *Bacillus subtilis* chromosome and conservation of the protein among related species of *Bacillus*. *J. Bacteriol.* **173:**1807–1809.
63. **Buchanan, C. E., and M.-L. Ling.** 1992. Isolation and sequence analysis of *dacB* which encodes a sporulation-specific penicillin binding protein in *Bacillus subtilis*. *J. Bacteriol.* **174:**1717–1725.
64. **Bugaichuk, U. D., and P. J. Piggot.** 1986. Nucleotide sequence of the *Bacillus subtilis* developmental gene *spoVE*. *J. Gen. Microbiol.* **132:**1883–1890.
65. **Butler, P. D., and J. Mandelstam.** 1987. Nucleotide sequence of the sporulation operon, *spoIIIE*, of *Bacillus subtilis*. *J. Gen. Microbiol.* **133:**2359–2370.
66. **Buxton, R. S.** 1976. Prophage mutation causing heat inducibility of defective *Bacillus subtilis* bacteriophage PBSX. *J. Virol.* **20:**22–28.
67. **Buxton, R. S.** 1980. Selection of *Bacillus subtilis* 168 mutants with deletions of PBSX prophage. *J. Gen. Virol.* **46:**427–437.
68. **Buxton, R. S., and J. B. Ward.** 1980. Heat-sensitive lysis mutants of *Bacillus subtilis* 168 blocked at three different stages of peptidoglycan synthesis. *J. Gen. Microbiol.* **120:**283–293.
69. **Callister, H., and R. G. Wake.** 1981. Characterization and mapping of temperature-sensitive division initiation mutations of *Bacillus subtilis*. *J. Bacteriol.* **145:** 1042–1051.
70. **Cannon, J. G., and K. F. Bott.** 1980. Mutation affecting expression of spectinomycin resistance in *Bacillus subtilis*. *J. Bacteriol.* **141:**409–412.
71. **Canosi, U., M. Nolli, E. Ferrari, R. Marinone, and G. Mazza.** 1979. Genetic mapping of caffeine resistant and sensitive mutants of *B. subtilis*. *Microbiologica* **2:**167–172.
72. **Canosi, U., A. G. Siccardi, A. Falaschi, and G. Mazza.** 1976. Effect of deoxyribonucleic acid replication inhibitors on bacterial recombination. *J. Bacteriol.* **126:**108–121.
73. **Caramori, T., S. Calogero, A. M. Albertini, and A. Galizzi.** 1993. Functional analysis of the *outB* gene of *Bacillus subtilis*. *J. Gen. Microbiol.* **139:**31–37.
74. **Carlsson, P., and L. Hederstedt.** 1987. *Bacillus subtilis citM*, the structural gene for dihydrolipoamide transsuccinylase: cloning and expression in *Escherichia coli*. *Gene* **61:**217–224.
75. **Carlsson, P., and L. Hederstedt.** 1989. Genetic characterization of *Bacillus subtilis odhA* and *odhB*, encoding 2-oxoglutarate dehydrogenase and dihydrolipoamide transsuccinylase, respectively. *J. Bacteriol.* **171:**3667–3672.
76. **Carrascosa, J. L., J. A. Garcia, and M. Salas.** 1982. A protein similar to *Escherichia coli groEL* is present in *Bacillus subtilis*. *J. Mol. Biol.* **148:**731–737.
77. **Carrigan, C. M., J. A. Haarsma, M. T. Smith, and R. G. Wake.** 1987. Sequence features of the replication terminus of the *Bacillus subtilis* chromosome. *Nucleic Acids Res.* **15:**8501–8509.
78. **Champney, W. S., and R. A. Jensen.** 1969. D-Tyrosine as

a metabolic inhibitor of *Bacillus subtilis*. *J. Bacteriol.* **98:**205–214.

79. **Chaudhry, G. R., Y. S. Halpern, C. Saunders, N. Vasantha, B. J. Schmidt, and E. Freese.** 1984. Mapping of the glucose dehydrogenase gene in *Bacillus subtilis*. *J. Bacteriol.* **160:**607–611.
80. **Chen, N. Y., F. M. Hu, and H. Paulus.** 1987. Nucleotide sequence of the overlapping genes for the subunits of *Bacillus subtilis* aspartokinase II and their control regions. *J. Biol. Chem.* **262:**8787–8798.
81. **Chen, N. Y., J. J. Zhang, and H. Paulus.** 1989. Chromosomal location of the *Bacillus subtilis* aspartokinase II gene and nucleotide sequence of the adjacent genes homologous to *uvrC* and *trx* of *Escherichia coli*. *J. Gen. Microbiol.* **135:**2931–2940.
82. **Cheo, D. L., K. W. Bayles, and R. E. Yasbin.** 1991. Cloning and characterization of DNA damage-inducible promoter regions from *Bacillus subtilis*. *J. Bacteriol.* **173:**1696–1703.
83. **Chow, K. C., and J. T. Wong.** 1988. Cloning and nucleotide sequence of the structural gene coding for *Bacillus subtilis* tryptophanyl-tRNA synthetase. *Gene* **73:**537–543.
84. **Coats, J. H., and E. W. Nester.** 1967. Regulation reversal mutation: characterization of end product-activated mutants of *Bacillus subtilis*. *J. Biol. Chem.* **242:**4948–4955.
85. **Connors, M. J., B. Howard, J. Hoch, and P. Setlow.** 1986. Determination of the chromosomal location of four *Bacillus subtilis* genes which code for a family of small acid-soluble spore proteins. *J. Bacteriol.* **166:**412–416.
86. **Connors, M. J., J. M. Mason, and P. Setlow.** 1986. Cloning and nucleotide sequencing of genes for three small, acid-soluble proteins from *Bacillus subtilis* spores. *J. Bacteriol.* **166:**417–425.
87. **Connors, M. J., and P. Setlow.** 1985. Cloning of a small, acid-soluble spore protein gene from *Bacillus subtilis* and determination of its complete nucleotide sequence. *J. Bacteriol.* **161:**333–339.
88. **Copeland, J. C., and J. Marmur.** 1968. Identification of conserved genetic functions in *Bacillus* by use of temperature-sensitive mutants. *Bacteriol. Rev.* **32:**302–312.
89. **Corfe, B. M.** Unpublished data.
90. **Cutting, S.** Unpublished data.
91. **Cutting, S., and J. Mandelstam.** 1986. The nucleotide sequence and the transcription during sporulation of the *gerE* gene of *Bacillus subtilis*. *J. Gen. Microbiol.* **132:**3012–3024.
92. **Cutting, S., S. Roels, and R. Losick.** 1991. Sporulation operon *spoIVF* and the characterization of mutations that uncouple mother-cell from forespore gene expression in *Bacillus subtilis*. *J. Mol. Biol.* **221:**1237–1256.
93. **Cutting, S., L. Zheng, and R. Losick.** 1991. Gene encoding two alkali-soluble components of the spore coat from *Bacillus subtilis*. *J. Bacteriol.* **173:**2915–2919.
94. **Dabbs, E. R.** 1983. Arrangement of loci within the principal cluster of ribosomal protein genes of *Bacillus subtilis*. *Mol. Gen. Genet.* **192:**124–130.
95. **Dabbs, E. R.** 1983. A pair of *Bacillus subtilis* ribosomal protein genes mapping outside the principal ribosomal protein cluster. *J. Bacteriol.* **156:**966–969.
96. **Dabbs, E. R.** 1983. Mapping of the genes for *Bacillus subtilis* ribosomal proteins S9, S11 and BL27 by means of antibiotic resistant mutants. *Mol. Gen. Genet.* **191:**295–300.
97. **Dabbs, E. R.** 1983. Mapping of the genes for *Bacillus subtilis* ribosomal proteins S6 and S16: comparison of the chromosomal distribution of ribosomal protein genes in this bacterium with the distribution in *Escherichia coli*. *Mol. Gen. Genet.* **192:**386–390.
98. **Dabbs, E. R.** 1984. Order of ribosomal protein genes in the *rif* cluster of *Bacillus subtilis* is identical to that of *Escherichia coli*. *J. Bacteriol.* **159:**770–772.
99. **Dartois, V., A. Baulard, K. Schanck, and C. Colson.** 1992. Cloning, nucleotide sequence and expression in *Escherichia coli* of a lipase gene from *Bacillus subtilis* 168. *Biochim. Biophys. Acta* **1131:**253–260.
100. **de Lencastre, H.** Unpublished data.
101. **Dean, D. R., J. A. Hoch, and A. I. Aronson.** 1977. Alteration of the *Bacillus subtilis* glutamine synthetase results in overproduction of the enzyme. *J. Bacteriol.* **131:**981–987.
102. **Débarbouillé, M., M. Arnaud, A. Fouet, A. Klier, and G. Rapoport.** 1990. The *sacT* gene regulating the *sacPA* operon in *Bacillus subtilis* shares strong homology with transcriptional antiterminators. *J. Bacteriol.* **172:**3966–3973.
103. **Débarbouillé, M., I. Martin-Verstraete, A. Klier, and G. Rapoport.** 1991. The transcriptional regulator LevR of *Bacillus subtilis* has domains homologous to both σ^{54}- and phosphotransferase system-dependent regulators. *Proc. Natl. Acad. Sci. USA* **88:**2212–2216.
104. **Débarbouillé, M., I. Martin-Verstraete, F. Kunst, and G. Rapoport.** 1991. The *Bacillus subtilis sigL* gene encodes an equivalent of σ^{54} from gram-negative bacteria. *Proc. Natl. Acad. Sci. USA* **88:**9092–9096.
105. **Deits, T. A.** Unpublished data.
106. **Devine, K.** Unpublished data.
107. **Dhaese, P.** Unpublished data.
108. **Dingman, D. W., and A. L. Sonenshein.** 1987. Purification of aconitase from *Bacillus subtilis* and correlation of its N-terminal amino acid sequence with the sequence of the *citB* gene. *J. Bacteriol.* **169:**3062–3067.
109. **Dod, B., G. Balassa, E. Raulet, and V. Jeannoda.** 1978. Spore control (Sco) mutations in *Bacillus subtilis*. II. Sporulation and the production of extracellular proteases and alpha-amylases by Sco mutants. *Mol. Gen. Genet.* **163:**45–56.
110. **Donovan, W., L. Zheng, K. Sandman, and R. Losick.** 1987. Genes encoding spore coat polypeptides from *Bacillus subtilis*. *J. Mol. Biol.* **196:**1–10.
111. **Driscoll, J. R., and H. W. Taber.** 1992. Sequence analysis and regulation of the *Bacillus subtilis menBE* operon. *J. Bacteriol.* **174:**5063–5074.
112. **Dubnau, D.** 1991. Genetic competence in *Bacillus subtilis*. *Microbiol. Rev.* **55:**395–424.
113. **Dubnau, D., and C. Cirigliano.** 1974. Genetic characterization of recombination-deficient mutants of *Bacillus subtilis*. *J. Bacteriol.* **117:**488–493.
114. **Dubnau, D., C. Goldthwaite, I. Smith, and J. Marmur.** 1967. Genetic mapping in *Bacillus subtilis*. *J. Mol. Biol.* **27:**163–185.
115. **Dubnau, D., and M. Roggiani.** 1990. Growth medium-independent genetic competence mutants of *Bacillus subtilis*. *J. Bacteriol.* **172:**4048–4055.
116. **Dubnau, E., S. Pifko, A. Sloma, K. Cabane, and I. Smith.** 1976. Conditional mutations in the translational apparatus of *Bacillus subtilis*. *Mol. Gen. Genet.* **147:**1–12.
117. **Dubnau, E., J. Weir, G. Nair, L. Carter, III, C. Moran, Jr., and I. Smith.** 1988. *Bacillus* sporulation gene *spo0H* codes for sigma-30 (sigma-H). *J. Bacteriol.* **170:**1054–1062.
118. **Duncan, M. L., S. S. Kalman, S. M. Thomas, and C. W. Price.** 1987. Gene encoding the 37,000-dalton minor sigma factor of *Bacillus subtilis* RNA polymerase: isolation, nucleotide sequence, chromosomal locus, and cryptic function. *J. Bacteriol.* **169:**771–778.
119. **Ebbole, D. J., and H. Zalkin.** 1987. Cloning and characterization of a 12-gene cluster from *Bacillus subtilis* encoding nine enzymes for *de novo* purine nucleotide synthesis. *J. Biol. Chem.* **262:**8274–8287.
120. **Endo, T., H. Ishikawa, and E. Freese.** 1983. Properties of a *Bacillus subtilis* mutant able to sporulate continu-

ally during growth in synthetic medium. *J. Gen. Microbiol.* **129:**17–30.

121. **Endo, T., B. Uratani, and E. Freese.** 1983. Purine salvage pathways of *Bacillus subtilis* and effect of guanine on growth of GMP reductase mutants. *J. Bacteriol.* **155:**169–179.
122. **Errington, J.** Unpublished data.
123. **Errington, J.** Sporulation in *Bacillus subtilis*: regulation of gene expression and control of morphogenesis. Submitted for publication.
124. **Errington, J., L. Appleby, R. Daniel, H. Goodfellow, S. R. Partridge, and M. D. Yudkin.** 1992. Structure and function of the *spoIIIJ* gene of *Bacillus subtilis*: a vegetatively expressed gene that is essential for σ^G activity at an intermediate stage of sporulation. *J. Gen. Microbiol.* **138:**2609–2618.
125. **Errington, J., P. Fort, and J. Mandelstam.** 1985. Duplicated sporulation genes in bacteria: implications for simple developmental systems. *FEBS Lett.* **188:**184–188.
126. **Errington, J., and D. Jones.** 1987. Cloning in *Bacillus subtilis* by transfection with bacteriophage vector ϕ105J27: isolation and preliminary characterization of transducing phages for 23 sporulation loci. *J. Gen. Microbiol.* **133:**493–502.
127. **Errington, J., S. Rong, M. S. Rosenkrantz, and A. L. Sonenshein.** 1988. Transcriptional regulation and structure of the *Bacillus subtilis* sporulation locus *spoIIIC*. *J. Bacteriol.* **170:**1162–1167.
128. **Errington, J., and C. H. Vogt.** 1990. Isolation and characterization of mutations in the gene encoding an endogenous *Bacillus subtilis* β-galactosidase and its regulator. *J. Bacteriol.* **172:**488–490.
129. **Estrela, A. I., H. de Lencastre, and L. J. Archer.** 1986. Resistance of a *Bacillus subtilis* mutant to a group of temperate bacteriophages. *J. Gen. Microbiol.* **132:**411–415.
130. **Estrela, A. I., H. M. Pooley, H. de Lencastre, and D. Karamata.** 1991. Genetic and biochemical characterization in *Bacillus subtilis* 168 mutants specifically blocked in the synthesis of the teichoic acid poly(3-0-β-D-glucopyranosyl-*N*-acetylgalactosamine 1-phosphate): *gneA*, a new locus, is associated with UDP-*N*-acetylglucosamine 4-epimerase activity. *J. Gen. Microbiol.* **137:**943–950.
131. **Fajardo-Cavazos, P., C. Salazar, and W. L. Nicholson.** 1992. Molecular cloning and characterization of *spl*, the gene encoding spore photoproduct lyase, which is involved in repair of ultraviolet radiation-induced DNA damage in *Bacillus subtilis* spores. XI International Spores Conference (Spores XI), abstr. no. 133.
132. **Fan, N., S. Cutting, and R. Losick.** 1992. Characterization of the *Bacillus subtilis* sporulation gene *spoVK*. *J. Bacteriol.* **174:**1053–1056.
133. **Fani, R., G. Mastromei, and M. Polsinelli.** 1984. Isolation and characterization of *Bacillus subtilis* mutants altered in competence. *J. Bacteriol.* **157:**152–157.
134. **Feavers, I. M., J. S. Miles, and A. Moir.** 1985. The nucleotide sequence of a spore germination gene (*gerA*) of *Bacillus subtilis* 168. *Gene* **38:**95–102.
135. **Fein, J. E., and H. J. Rogers.** 1976. Autolytic enzyme-deficient mutants of *Bacillus subtilis* 168. *J. Bacteriol.* **127:**1427–1442.
136. **Ferrari, E., D. J. Henner, and M. Y. Yang.** 1985. Isolation of an alanine racemase gene from *Bacillus subtilis* and its use for plasmid maintenance in *B. subtilis*. *Bio/Technology* **3:**1003–1007.
137. **Ferrari, E., F. Scoffone, G. Ciarrocchi, and A. Galizzi.** 1985. Molecular cloning of a *Bacillus subtilis* gene involved in spore outgrowth. *J. Gen. Microbiol.* **131:**2831–2838.
138. **Ferrari, F. A., D. Lang, E. Ferrari, and J. A. Hoch.** 1982. Molecular cloning of the *spo0B* sporulation locus in bacteriophage lambda. *J. Bacteriol.* **152:**809–814.
139. **Ferrari, F. A., K. Trach, and J. A. Hoch.** 1985. Sequence analysis of the *spo0B* locus reveals a polycistronic transcription unit. *J. Bacteriol.* **161:**556–562.
140. **Ferrari, F. A., K. Trach, D. LeCoq, J. Spence, E. Ferrari, and J. A. Hoch.** 1985. Characterization of the *spo0A* locus and its deduced product. *Proc. Natl. Acad. Sci. USA* **82:**2647–2651.
141. **Fink, P. S.** Unpublished data.
142. **Fisher, S. H., M. S. Rosenkrantz, and A. L. Sonenshein.** 1984. Glutamine synthetase gene of *Bacillus subtilis*. *Gene* **32:**427–438.
143. **Fort, P., and J. Errington.** 1985. Nucleotide sequence and complementation analysis of a polycistronic sporulation operon, *spoVA* in *Bacillus subtilis*. *J. Gen. Microbiol.* **131:**1091–1105.
144. **Fort, P., and P. J. Piggot.** 1984. Nucleotide sequence of sporulation locus *spoIIA* in *Bacillus subtilis*. *J. Gen. Microbiol.* **130:**2147–2153.
145. **Foster, S. J.** 1991. Cloning, expression, sequence analysis and biochemical characterization of an autolytic amidase of *Bacillus subtilis* 168 *trpC2*. *J. Gen. Microbiol.* **137:**1987–1998.
146. **Fouet, A., M. Arnaud, A. Klier, and G. Rapoport.** 1987. *Bacillus subtilis* sucrose-specific enzyme II of the phosphotransferase system: expression in *Escherichia coli* and homology to enzymes II from enteric bacteria. *Proc. Natl. Acad. Sci. USA* **84:**8773–8777.
147. **Fouet, A., A. Klier, and G. Rapoport.** 1986. Nucleotide sequence of the sucrase gene of *Bacillus subtilis*. *Gene* **45:**221–225.
148. **Foulger, D., and J. Errington.** 1989. The role of the sporulation gene *spoIIIE* in the regulation of prespore-specific gene expression in *Bacillus subtilis*. *Mol. Microbiol.* **3:**1247–1255.
149. **Foulger, D., and J. Errington.** 1991. Sequential activation of dual promoters by different sigma factors maintains *spoVJ* expression during successive developmental stages of *Bacillus subtilis*. *Mol. Microbiol.* **5:**1363–1373.
150. **Freese, E.** Unpublished data.
151. **Freese, E. B., R. M. Cole, W. Klofat, and E. Freese.** 1970. Growth, sporulation, and enzyme defects of glucosamine mutants of *Bacillus subtilis*. *J. Bacteriol.* **101:**1046–1062.
152. **Fujita, Y., and T. Fujita.** 1983. Genetic analysis of a pleiotropic deletion mutation (delta-igf) in *Bacillus subtilis*. *J. Bacteriol.* **154:**864–869.
153. **Fujita, Y., and T. Fujita.** 1987. The gluconate operon *gnt* of *Bacillus subtilis* encodes its own transcriptional negative regulator. *Proc. Natl. Acad. Sci. USA* **84:**4524–4528.
154. **Fujita, Y., T. Fujita, Y. Miwa, J. Nihashi, and Y. Aratani.** 1986. Organization and transcription of the gluconate operon, *gnt*, of *Bacillus subtilis*. *J. Biol. Chem.* **261:**13744–13753.
155. **Fujita, Y., K. Shindo, Y. Miwa, and K.-I. Yoshida.** 1991. *Bacillus subtilis* inositol dehydrogenase-encoding gene (*idh*): sequence and expression in *Escherichia coli*. *Gene* **108:**121–125.
156. **Gaido, M. L., C. R. Prostko, and J. S. Strobl.** 1988. Isolation and characterization of BsuE methyltransferase, a CGCG specific DNA methyltransferase from *Bacillus subtilis*. *J. Biol. Chem.* **263:**4832–4836.
157. **Galizzi, A.** Unpublished data.
158. **Galizzi, A., F. Gorrini, A. Rollier, and M. Polsinelli.** 1973. Mutants of *Bacillus subtilis* temperature sensitive in the outgrowth phase of spore germination. *J. Bacteriol.* **113:**1482–1490.
159. **Gallori, E., M. Bazzicalupo, B. Parisi, G. Pedaggi, and M. Polsinelli.** 1978. Resistance to (L)-azetidine-2-car-

boxylic acid in *Bacillus subtilis*. *Biochem. Biophys. Res. Commun.* **85:**1518–1525.

160. **Gallori, E., and R. Fani.** 1983. Characterization of D-cycloserine resistant mutants in *Bacillus subtilis*. *Microbiologica* **6:**19–26.
161. **Garro, A. J., H. Leffert, and J. Marmur.** 1970. Genetic mapping of a defective bacteriophage on the chromosome of *Bacillus subtilis* 168. *J. Virol.* **6:**340–343.
162. **Garro, A. J., C. Sprouse, and J. G. Wetmur.** 1976. Association of the recombination-deficient phenotype of *Bacillus subtilis recC* strains with the presence of an SPO2 prophage. *J. Bacteriol.* **126:**556–558.
163. **Gass, K. B., and N. R. Cozzarelli.** 1973. Further genetic and enzymological characterization of the three *Bacillus subtilis* deoxyribonucleic acid polymerases. *J. Biol. Chem.* **248:**7688–7700.
164. **Gaur, N. K., K. Cabane, and I. Smith.** 1988. Structure and expression of the *Bacillus subtilis sin* operon. *J. Bacteriol.* **170:**1046–1053.
165. **Gay, P., H. Chalumeau, and M. Steinmetz.** 1983. Chromosomal localization of *gut*, *fruC*, and *pfk* mutations affecting genes involved in *Bacillus subtilis* D-glucitol catabolism. *J. Bacteriol.* **153:**1133–1137.
166. **Gay, P., and A. Delobbe.** 1977. Fructose transport in *Bacillus subtilis*. *Eur. J. Biochem.* **79:**363–373.
167. **Gay, P., D. LeCoq, M. Steinmetz, E. Ferrari, and J. A. Hoch.** 1983. Cloning structural gene *sacB*, which codes for exoenzyme levansucrase of *Bacillus subtilis*: expression of the gene in *Escherichia coli*. *J. Bacteriol.* **153:** 1424–1431.
168. **Gianni, M., and A. Galizzi.** 1986. Isolation of genes preferentially expressed during *Bacillus subtilis* spore outgrowth. *J. Bacteriol.* **165:**123–132.
169. **Gillespie, K., and R. E. Yasbin.** 1987. Chromosomal locations of three *Bacillus subtilis din* genes. *J. Bacteriol.* **169:**3372–3374.
170. **Glaser, P.** Unpublished data.
171. **Glaser, P., A. Danchin, F. Kunst, M. Débarbouillé, A. Vertes, and R. Dedonder.** 1991. A gene encoding a tyrosine tRNA synthetase is located near *sacS* in *Bacillus subtilis*. *DNA Seq.* **1:**251–261.
172. **Goldstein, B. J., and S. A. Zahler.** 1976. Uptake of branched-chain alpha-keto acids in *Bacillus subtilis*. *J. Bacteriol.* **127:**667–670.
173. **Goldthwaite, C., D. Dubnau, and I. Smith.** 1970. Genetic mapping of antibiotic resistance markers in *Bacillus subtilis*. *Proc. Natl. Acad. Sci. USA* **65:**96–103.
174. **Goldthwaite, C., and I. Smith.** 1972. Genetic mapping of amino-glycoside and fusidic acid resistant mutations in *Bacillus subtilis*. *Mol. Gen. Genet.* **114:**181–189.
175. **Gollnick, P., S. Ishino, M. I. Kuroda, D. J. Henner, and C. Yanofsky.** 1990. The *mtr* locus is a two-gene operon required for transcription attenuation in the *trp* operon of *Bacillus subtilis*. *Proc. Natl. Acad. Sci. USA* **87:**8726–8730.
176. **Gonzy-Treboul, G.** Unpublished data.
177. **Gonzy-Treboul, G., and M. Steinmetz.** 1987. Phosphoenolpyryvate:sugar phosphotransferase system of *Bacillus subtilis*: cloning of the region containing the *ptsH* and *ptsI* genes and evidence for the *crr*-like gene. *J. Bacteriol.* **169:**2287–2290.
178. **Gonzy-Treboul, G., M. Zagorec, M.-C. Rain-Guion, and M. Steinmetz.** 1989. Phosphoenolpyruvate:sugar phosphotransferase system of *Bacillus subtilis*: nucleotide sequence of *ptsX*, *ptsH* and the 5′-end of *ptsI* and evidence for a *ptsHI* operon. *Mol. Microbiol.* **3:**103–112.
179. **Grandi, G.** Unpublished data.
180. **Gray, J. V., G. Golinelli-Pimpaneau, and J. R. Knowles.** 1990. Monofunctional chorismate mutase from *Bacillus subtilis*: purification of the protein, molecular cloning of the gene, and overexpression of the gene product in *Escherichia coli*. *Biochemistry* **29:**376–383.
181. **Green, C. J., G. C. Stewart, M. A. Hollis, B. S. Vold, and K. F. Bott.** 1985. Nucleotide sequence of the *Bacillus subtilis* ribosomal RNA operon, *rrnB*. *Gene* **37:**261–266.
182. **Green, C. J., and B. S. Vold.** 1992. A cluster of nine tRNA genes between ribosomal gene operons in *Bacillus subtilis*. *J. Bacteriol.* **174:**3147–3151.
183. **Green, D. M.** Unpublished data.
184. **Grossman, A.** Unpublished data.
185. **Grundy, F. J., and T. M. Henkin.** 1990. Cloning and analysis of the *Bacillus subtilis rpsD* gene, encoding ribosomal protein S4. *J. Bacteriol.* **172:**6372–6379.
186. **Guerout-Fleury, A.-M., and P. Stragier.** 1992. Unexpected complexity of the *spoIIIA* locus. XI International Spores Conference (Spores XI), abstr. no. 151.
187. **Guzman, P., J. Westpheling, and P. Youngman.** 1988. Characterization of the promoter region of the *Bacillus subtilis spoIIE* operon. *J. Bacteriol.* **170:**1598–1609.
188. **Hackett, R. H., and P. Setlow.** 1987. Cloning, nucleotide sequencing, and genetic mapping of the gene for small, acid-soluble spore protein gamma of *Bacillus subtilis*. *J. Bacteriol.* **169:**1985–1992.
189. **Haldenwang, W. G., C. D. B. Banner, J. F. Ollington, R. Losick, J. A. Hoch, M. B. O'Connor, and A. L. Sonenshein.** 1980. Mapping a cloned gene under sporulation control by insertion of a drug resistance marker into the *Bacillus subtilis* chromosome. *J. Bacteriol.* **142:**90–98.
190. **Halling, S. M., and K. C. Burtis.** 1977. Reconstitution studies show that rifampicin resistance is determined by the largest polypeptide of *Bacillus subtilis* RNA polymerase. *J. Biol. Chem.* **252:**9024–9031.
191. **Halling, S. M., K. C. Burtis, and R. H. Doi.** 1978. Beta subunit of bacterial RNA polymerase is responsible for streptolydigin resistance in *Bacillus subtilis*. *Nature* (London) **272:**837–839.
192. **Hammer-Jespersen, K.** 1983. Nucleoside catabolism, p. 203–258. *In* A. Munch-Petersen (ed.), *Metabolism of Nucleotides, Nucleosides and Nucleobases in Microorganisms*. Academic Press, Inc., New York.
193. **Hammond, R. A., M. H. Barnes, S. L. Mack, J. A. Mitchener, and N. C. Brown.** 1991. *Bacillus subtilis* DNA polymerase III: complete sequence, overexpression, and characterization of the *polC* gene. *Gene* **98:**29–36.
194. **Hansson, M., and L. Hederstedt.** 1992. Cloning and characterization of the *Bacillus subtilis hemEHY* gene cluster, which encodes protoheme IX biosynthetic enzymes. *J. Bacteriol.* **174:**8081–8093.
195. **Hansson, M., L. Rutberg, I. Schroder, and L. Hederstedt.** 1991. The *Bacillus subtilis hemAXCDBL* gene cluster, which encodes enzymes of the biosynthetic pathway from glutamate to uroporphyrinogen III. *J. Bacteriol.* **173:**2590–2599.
196. **Hara, H., and H. Yoshikawa.** 1973. Asymmetric bidirectional replication of *Bacillus subtilis* chromosome. *Nature* (London) *New Biol.* **244:**200–203.
197. **Harford, N., J. Lepesant-Kejzlarova, J.-A. Lepesant, R. Hamers, and R. Dedonder.** 1976. Genetic circularity and mapping of the replication origin region of the *Bacillus subtilis* chromosome, p. 28–34. *In* D. Schlessinger (ed.), *Microbiology—1976*. American Society for Microbiology, Washington, D.C.
198. **Harford, N., and N. Sueoka.** 1970. Chromosomal location of antibiotic resistance markers in *Bacillus subtilis*. *J. Mol. Biol.* **51:**267–286.
199. **Harry, E. J., and R. G. Wake.** 1989. Cloning and expression of a *Bacillus subtilis* division initiation gene for which a homolog has not been identified in another organism. *J. Bacteriol.* **171:**6835–6839.
200. **Hastrup, S.** 1988. Analysis of the *Bacillus subtilis* xylose regulon, p. 79–83. *In* A. T. Ganesan and J. A. Hoch (ed.), *Genetics and Biotechnology of Bacilli*, vol. 2. Academic Press, Inc., San Diego, Calif.
201. **Hauser, P. M., W. D. Crabb, M. G. Fioria, F. Scoffone,**

and A. Galizzi. 1991. A genetic analysis of the *flaA* locus of *Bacillus subtilis*. *J. Bacteriol.* **173:**3580–3583.
202. **Hearne, C. M., and D. J. Ellar.** 1989. Nucleotide sequence of a *Bacillus subtilis* gene homologous to the *dnaK* gene of *Escherichia coli*. *Nucleic Acids Res.* **17:** 8373.
203. **Hederstedt, L.** Unpublished data.
204. **Helmann, J. D., L. M. Marquez, and M. J. Chamberlin.** 1988. Cloning, sequencing, and disruption of the *Bacillus subtilis* sigma-28 gene. *J. Bacteriol.* **170:**1568–1574.
205. **Hemila, H.** 1991. Sequence of a PAL-related lipoprotein from *Bacillus subtilis*. *FEMS Microbiol. Lett.* **82:**37–42.
206. **Hemila, H., A. Palva, L. Paulin, S. Arvidson, and I. Palva.** 1990. Secretory S complex of *Bacillus subtilis*: sequence analysis and identity to pyruvate dehydrogenase. *J. Bacteriol.* **172:**5052–5063.
207. **Hemphill, E. H., I. Gage, S. A. Zahler, and R. Z. Korman.** 1980. Prophage mediated production of bacteriocin-like substance by SP-beta lysogens of *Bacillus subtilis*. *Can. J. Microbiol.* **26:**1328–1333.
208. **Henkin, T.** Unpublished data.
209. **Henkin, T. M., K. M. Campbell, and G. H. Chambliss.** 1979. Spectinomycin dependence in *Bacillus subtilis*. *J. Bacteriol.* **137:**1452–1455.
210. **Henkin, T. M., and G. H. Chambliss.** 1984. Genetic analysis of a streptomycin-resistant oligosporogenous *Bacillus subtilis*. *J. Bacteriol.* **157:**202–210.
211. **Henkin, T. M., F. J. Grundy, W. L. Nicholson, and G. H. Chambliss.** 1991. Catabolite repression of α-amylase gene expression in *Bacillus subtilis* involves a trans-acting gene product homologous to the *Escherichia coli lacI* and *galR* repressors. *Mol. Microbiol.* **5:**575–584.
212. **Henkin, T. M., S. H. Moon, L. C. Mattheakis, and M. Nomura.** 1989. Cloning and analysis of the *spc* ribosomal protein operon of *Bacillus subtilis*: comparison with the *spc* operon of *Escherichia coli*. *Nucleic Acids Res.* **17:**7469–7486.
213. **Henner, D. J., L. Band, G. Flaggs, and E. Chen.** 1986. The organization and nucleotide sequence of the *Bacillus subtilis hisH*, *tyrA*, and *aroE* genes. *Gene* **49:**147–152.
214. **Henner, D. J., L. Band, and H. Shimotsu.** 1984. Nucleotide sequence of the *Bacillus subtilis* tryptophan operon. *Gene* **34:**169–177.
215. **Henner, D. J., P. Gollnick, and A. Moir.** 1990. Analysis of an 18 kilobase pair region of the *Bacillus subtilis* chromosome containing the *mtr* and *gerC* operons and the *aro-trp-aro* supraoperon, p. 657–665. *In* H. Heslot, J. Davies, J. Florent, L. Bohichou, G. Durand, and L. Penaasse (ed.), *6th International Symposium on Genetics of Industrial Microorganisms, Proceedings*, vol. II. Société Française de Microbiologie, Strasbourg.
216. **Henner, D. J., and J. A. Hoch.** 1980. The *Bacillus subtilis* chromosome. *Microbiol. Rev.* **44:**57–82.
217. **Henner, D. J., and J. A. Hoch.** 1982. The genetic map of *Bacillus subtilis*, p. 1–33. *In* D. Dubnau (ed.), *The Molecular Biology of the Bacilli*. Academic Press, Inc., New York.
218. **Henner, D. J., and W. Steinberg.** 1979. Genetic location of the *Bacillus subtilis sup-3* suppressor mutation. *J. Bacteriol.* **139:**668–670.
219. **Henner, D. J., M. Yang, and E. Ferrari.** 1988. Localization of *Bacillus subtilis sacU*(Hy) mutations to two linked genes with similarities to the conserved procaryotic family of two-component signalling systems. *J. Bacteriol.* **170:**5102–5109.
220. **Henriques, A. O., H. de Lencastre, and P. J. Piggot.** 1992. A *Bacillus subtilis* morphogene cluster that includes *spoVE* is homologous to the *mra* region of *Escherichia coli*. *Biochimie* **74:**735–748.
221. **Higerd, T. B.** 1977. Isolation of acetyl esterase mutants of *Bacillus subtilis* 168. *J. Bacteriol.* **129:**973–977.
222. **Hilton, M. D., N. G. Alaeddinoglu, and A. L. Demain.** 1988. *Bacillus subtilis* mutant deficient in the ability to produce the dipeptide antibiotic bacilysin: isolation and mapping of the mutation. *J. Bacteriol.* **170:**1018–1020.
223. **Hoch, J. A.** Unpublished data.
224. **Hoch, J. A., and C. Anagnostopoulos.** 1970. Chromosomal location and properties of radiation sensitivity mutations in *Bacillus subtilis*. *J. Bacteriol.* **103:**295–301.
225. **Hoch, J. A., and H. J. Coukoulis.** 1978. Genetics of the alpha-ketoglutarate dehydrogenase complex of *Bacillus subtilis*. *J. Bacteriol.* **133:**265–269.
226. **Hoch, J. A., and J. Mathews.** 1972. Genetic studies in *Bacillus subtilis*, p. 113–116. *In* H. O. Halvorson, R. Hanson, and L. L. Campbell (ed.), *Spores V*. American Society for Microbiology, Washington, D.C.
227. **Hoch, J. A., and E. W. Nester.** 1973. Gene-enzyme relationships of aromatic acid biosynthesis in *Bacillus subtilis*. *J. Bacteriol.* **116:**59–66.
228. **Hoch, J. A., K. Trach, F. Kawamura, and H. Saito.** 1985. Identification of the transcriptional suppressor *sof-1* as an alteration in the *spo0A* protein. *J. Bacteriol.* **161:**552–555.
229. **Hofemeister, J., M. Israeli-Reches, and D. Dubnau.** 1983. Integration of plasmid pE194 at multiple sites on the *Bacillus subtilis* chromosome. *Mol. Gen. Genet.* **189:** 58–68.
230. **Holmberg, C., L. Beijer, B. Rutberg, and L. Rutberg.** 1990. Glycerol catabolism in *Bacillus subtilis*: nucleotide sequence of the genes encoding glycerol kinase (*glpK*) and glycerol-3-phosphate dehydrogenase (*glpD*). *J. Gen. Microbiol.* **136:**2367–2375.
231. **Honeyman, A. L., and G. C. Stewart.** 1989. Identification of the protein encoded by *rodC*, a cell division gene from *Bacillus subtilis*. *Mol. Microbiol.* **2:**735–741.
232. **Honeyman, A. L., and G. C. Stewart.** 1989. The nucleotide sequence of the *rodC* operon of *Bacillus subtilis*. *Mol. Microbiol.* **3:**1257–1268.
233. **Honjo, M., A. Nakayama, K. Fukazawa, K. Kawamura, K. Ando, M. Hori, and Y. Furutani.** 1990. A novel *Bacillus subtilis* gene involved in negative control of sporulation and degradative-enzyme production. *J. Bacteriol.* **172:**1783–1790.
234. **Hudspeth, D. S. S., and P. S. Vary.** 1992. *spoVG* sequence of *Bacillus megaterium* and *Bacillus subtilis*. *Biochim. Biophys. Acta* **1130:**229–231.
235. **Hulett, M.** Unpublished data.
236. **Igo, M., M. Lampe, and R. Losick.** 1988. Structure and regulation of a *Bacillus subtilis* gene that is transcribed by the EσB form of RNA polymerase holoenzyme, p. 151–156. *In* A. T. Ganesan and J. A. Hoch (ed.), *Genetics and Biotechnology of Bacilli*, vol. 2. Academic Press, Inc., San Diego.
237. **Iijima, T., M. D. Diesterhaft, and E. Freese.** 1977. Sodium effect of growth on aspartate and genetic analysis of a *Bacillus subtilis* mutant with high aspartase activity. *J. Bacteriol.* **129:**1440–1447.
238. **Iijima, T., and Y. Ikeda.** 1970. Mutability of the phleomycin-resistant mutants of *Bacillus subtilis*. I. Isolation of genetically unstable mutants. *J. Gen. Appl. Microbiol.* **16:**419–427.
239. **Ikawa, S., T. Shibata, K. Matsumoto, T. Iijima, H. Saito, and T. Ando.** 1981. Chromosomal loci of genes controlling site-specific restriction endonucleases of *Bacillus subtilis*. *Mol. Gen. Genet.* **183:**1–6.
240. **Ikeda, M., T. Sato, M. Wachi, H. K. Jung, F. Ishino, Y. Kobayashi, and M. Matsuhashi.** 1989. Structural similarity among *Escherichia coli* FtsW and RodA proteins and *Bacillus subtilis* SpoVE protein, which function in cell division, cell elongation, and spore formation, respectively. *J. Bacteriol.* **171:**6375–6378.
241. **Ikeda, M., M. Wachi, H. K. Jung, F. Ishino, and M. Matsuhashi.** 1990. Homology among MurC, MurD, MurE and MurF proteins in *Escherichia coli* and that

between *E. coli* MurG and a possible MurG protein in *Bacillus subtilis*. *J. Gen. Appl. Microbiol.* **36:**179–187.

242. **Illing, N., and J. Errington.** 1991. The *spoIIIA* operon of *Bacillus subtilis* defines a new temporal class of mother-cell-specific sporulation genes under the control of the σ^E form of RNA polymerase. *Mol. Microbiol.* **5:**1927–1940.

243. **Imai, R., T. Sekiguchi, Y. Nosoh, and K. Tsuda.** 1987. The nucleotide sequence of 3-isopropylmalate dehydrogenase gene from *Bacillus subtilis*. *Nucleic Acids Res.* **15:**4988.

244. **Ionesco, H., J. Michel, B. Cami, and P. Schaeffer.** 1970. Genetics of sporulation in *Bacillus subtilis* Marburg. *J. Appl. Bacteriol.* **33:**13–24.

245. **Irie, R., T. Okamoto, and Y. Fujita.** 1982. A germination mutant of *Bacillus subtilis* deficient in response to glucose. *J. Gen. Appl. Microbiol.* **28:**345–354.

246. **Itaya, M., and T. Tanaka.** 1991. Complete physical map of the *Bacillus subtilis* 168 chromosome constructed by a gene-directed mutagenesis method. *J. Mol. Biol.* **220:**631–648.

247. **Ito, J.** 1973. Pleiotropic nature of bacteriophage tolerant mutants obtained in early-blocked asporogenous mutants of *Bacillus subtilis* 168. *Mol. Gen. Genet.* **124:**97–106.

248. **Itoh, T.** 1976. Amino acid replacement in the protein S5 from a spectinomycin resistant mutant of *Bacillus subtilis*. *Mol. Gen. Genet.* **144:**39–42.

249. **Iwakura, M., M. Kawata, K. Tsuda, and T. Tanaka.** 1988. Nucleotide sequences of the thymidylate synthase B and dihydrofolate reductase genes contained in one *Bacillus subtilis* operon. *Gene* **64:**9–20.

250. **James, W., and J. Mandelstam.** 1985. *spoVIC*, a new sporulation locus in *Bacillus subtilis* affecting spore coats, germination and the rate of sporulation. *J. Gen. Microbiol.* **131:**2409–2419.

251. **Jenkinson, H. F.** 1981. Germination and resistance defects in spores of *Bacillus subtilis* mutant lacking a coat polypeptide. *J. Gen. Microbiol.* **127:**81–91.

252. **Jenkinson, H. F.** 1983. Altered arrangement of proteins in the spore coat of a germination mutant of *Bacillus subtilis*. *J. Gen. Microbiol.* **129:**1945–1958.

253. **Jenkinson, H. F., and J. Mandelstam.** 1983. Cloning of the *Bacillus subtilis lys* and *spoIIIB* genes in phage ϕ105. *J. Gen. Microbiol.* **129:**2229–2240.

254. **Johnson, W. C., C. P. Moran, Jr., and R. Losick.** 1984. Two RNA polymerase sigma factors from *Bacillus subtilis* discriminate between overlapping promoters for a developmentally regulated gene. *Nature* (London) **302:**800–804.

255. **Joris, B., G. Dive, A. Henriques, P. J. Piggot, and J. M. Ghuysen.** 1990. The life-cycle proteins RodA of *Escherichia coli* and SpoVE of *Bacillus subtilis* have very similar primary structures. *Mol. Microbiol.* **4:**513–517.

256. **Kaminskas, E., Y. Kimhi, and B. Magasanik.** 1970. Urocanase and *N*-formimino-L-glutamate formiminohydrolase of *Bacillus subtilis*, two enzymes of the histidine degradation pathway. *J. Biol. Chem.* **245:**3536–3544.

257. **Kane, J. F.** 1977. Regulation of a common aminotransferase subunit. *J. Bacteriol.* **132:**419–425.

258. **Kane, J. F., R. L. Goode, and J. Wainscott.** 1975. Multiple mutations in *cysA14* mutants of *Bacillus subtilis*. *J. Bacteriol.* **121:**204–211.

259. **Kanzaki, N., and K. Miyagawa.** 1990. Nucleotide sequence of the *Bacillus subtilis* IMP dehydrogenase gene. *Nucleic Acids Res.* **18:**6710.

260. **Kapfer, W., J. Walter, and T. A. Trautner.** 1991. Cloning, characterization and evolution of the *BsuFI* restriction endonuclease gene of *Bacillus subtilis* and purification of the enzyme. *Nucleic Acids Res.* **19:**6457–6463.

261. **Karamata, D.** Unpublished data.

262. **Karamata, D., and J. D. Gross.** 1970. Isolation and genetic analysis of temperature-sensitive mutants of *Bacillus subtilis* defective in DNA synthesis. *Mol. Gen. Genet.* **108:**277–287.

263. **Karamata, D., M. McConnell, and H. J. Rogers.** 1972. Mapping of *rod* mutants of *Bacillus subtilis*. *J. Bacteriol.* **111:**73–79.

264. **Karmazyn-Campelli, C., C. Bonamy, B. Savelli, and P. Stragier.** 1989. Tandem genes encoding sigma-factors for consecutive steps of development in *Bacillus subtilis*. *Genes Dev.* **3:**150–157.

265. **Karmazyn-Campelli, C., L. Fluss, T. Leighton, and P. Stragier.** 1992. The spoIIN279 (ts) mutation affects the *ftsA* protein of *Bacillus subtilis*. *Biochimie* **74:**689–694.

266. **Kelly, M. S.** 1967. Physical and mapping properties of distant linkages between genetic markers in transformation of *Bacillus subtilis*. *Mol. Gen. Genet.* **99:**333–349.

267. **Kelly, M. S., and R. H. Pritchard.** 1963. Selection for linked loci in *Bacillus subtilis* by means of transformation. *Heredity* **17:**598–603.

268. **Kiss, A., G. Posfai, C. C. Keller, P. Venetianer, and R. J. Roberts.** 1985. Nucleotide sequence of the BsuRI restriction-modification system. *Nucleic Acids Res.* **13:**6403–6421.

269. **Kiss, I., I. Berek, and G. Ivanovics.** 1971. Mapping the delta-aminolaevulinic acid synthetase locus in *Bacillus subtilis*. *J. Gen. Microbiol.* **66:**153–159.

270. **Klein, C., C. Kaletta, N. Schnell, and K. D. Endian.** 1992. Analysis of genes involved in biosynthesis of the antibiotic subtilin. *Appl. Environ. Microbiol.* **58:**132–142.

271. **Klier, A. F., and G. Rapoport.** 1988. Genetics and regulation of carbohydrate catabolism in *Bacillus*. *Annu. Rev. Microbiol.* **42:**65–95.

272. **Kobayashi, H., K. Kobayashi, and Y. Kobayashi.** 1977. Isolation and characterization of fusidic acid-resistant, sporulation-defective mutants of *Bacillus subtilis*. *J. Bacteriol.* **132:**262–269.

273. **Koide, Y., A. Nakamura, T. Uozumi, and T. Beppu.** 1986. Cloning and sequencing of the major intracellular serine protease gene of *Bacillus subtilis*. *J. Bacteriol.* **167:**110–116.

274. **Kontinen, V. P., P. Saris, and M. Sarvas.** 1991. A gene (*prsA*) of *Bacillus subtilis* involved in a novel, late stage of protein export. *Mol. Microbiol.* **5:**1273–1283.

275. **Kontinen, V. P., and M. Sarvas.** 1988. Mutants of *Bacillus subtilis* defective in protein export. *J. Gen. Microbiol.* **134:**2333–2344.

276. **Kooistra, J., and G. Venema.** 1991. Cloning, sequencing, and expression of *Bacillus subtilis* genes involved in ATP-dependent nuclease synthesis. *J. Bacteriol.* **173:**3644–3655.

277. **Kooistra, J., B. Vosman, and G. Venema.** 1988. Cloning and characterization of a *Bacillus subtilis* transcription unit involved in ATP-dependent DNase synthesis. *J. Bacteriol.* **170:**4791–4797.

278. **Kreneva, R. A., and D. A. Perumov.** 1990. Genetic mapping of regulatory mutations of *Bacillus subtilis* riboflavin operon. *Mol. Gen. Genet.* **222:**467–469.

279. **Krulwich, T. A., S. Clejan, L. H. Falk, and A. A. Guffanti.** 1987. Incorporation of specific exogenous fatty acids into membrane lipids modulates protonophore resistance in *Bacillus subtilis*. *J. Bacteriol.* **169:**4479–4485.

280. **Kunkel, B., L. Kroos, H. Poth, P. Youngman, and R. Losick.** 1989. Temporal and spatial control of the mother-cell regulatory gene *spoIIID* of *Bacillus subtilis*. *Genes. Dev.* **3:**1735–1744.

281. **Kunkel, B., R. Losick, and P. Stragier.** 1990. The *Bacillus subtilis* gene for the developmental transcription factor sigma K is generated by excision of a dispensable DNA element containing a sporulation recombinase gene. *Genes Dev.* **4:**525–535.

282. **Kunkel, B., K. Sandman, S. Panzer, P. Youngman, and**

R. Losick. 1988. The promoter for a sporulation gene in the *spoIVC* locus of *Bacillus subtilis* and its use in studies of temporal and spatial control of gene expression. *J. Bacteriol.* **170:**3513–3522.

283. **Kunst, F., M. Débarbouillé, T. Msadek, M. Young, C. Mauel, D. Karamata, A. Klier, G. Rapoport, and R. Dedonder.** 1988. Deduced polypeptides encoded by the *Bacillus subtilis sacU* locus share homology with two-component sensor-regulator systems. *J. Bacteriol.* **170:** 5093–5101.
284. **Kuroda, A., M. H. Rashid, and J. Sekiguchi.** 1992. Molecular cloning and sequencing of the upstream region of the major *Bacillus subtilis* autolysin gene: a modifier protein exhibiting sequence homology to the major autolysing and the *spoIID* product. *J. Gen. Microbiol.* **138:**1067–1076.
285. **Kuroda, A., and J. Sekiguchi.** 1990. Cloning, sequencing and genetic mapping of a *Bacillus subtilis* cell wall hydrolase gene. *J. Gen. Microbiol.* **136:**2209–2216.
286. **Kuroda, A., and J. Sekiguchi.** 1991. Molecular cloning and sequencing of a major *Bacillus subtilis* autolysin gene. *J. Bacteriol.* **173:**7304–7312.
287. **LaFauci, G., R. L. Widom, R. L. Eisner, E. D. Jarvis, and R. Rudner.** 1986. Mapping of rRNA genes with integrable plasmids in *Bacillus subtilis. J. Bacteriol.* **165:**204–214.
288. **Lakomova, N. M., T. S. Tsurikova, and A. A. Prozorov.** 1980. Possible participation of RNA polymerase III in suppression of *recH* mutation of *Bacillus subtilis. Genetica* (USSR) **16:**583–587.
289. **Lamont, I. L., and J. Mandelstam.** 1984. Identification of a new sporulation locus, *spoIIIF*, in *Bacillus subtilis. J. Gen. Microbiol.* **130:**1253–1261.
290. **Lampe, M., C. Binnie, R. Schmidt, and R. Losick.** 1988. Cloned gene encoding the delta subunit of *Bacillus subtilis* RNA polymerase. *Gene* **67:**13–19.
291. **Lampel, K. A., B. Uratani, G. R. Chaudhry, R. F. Ramaley, and S. Rudikoff.** 1986. Characterization of the developmentally regulated *Bacillus subtilis* glucose dehydrogenase gene. *J. Bacteriol.* **166:**238–243.
292. **LaVallie, E. R., and M. L. Stahl.** 1989. Cloning of the flagellin gene from *Bacillus subtilis* and complementation studies of an in vitro-derived deletion mutation. *J. Bacteriol.* **171:**3085–3094.
293. **Lazarevic, V., and D. Karamata.** Unpublished data.
294. **Lazarevic, V., P. Margot, B. Suldo, and D. Karamata.** 1992. Sequencing and analysis of the *Bacillus subtilis lytRABC* divergon: a regulatory unit encompassing the structural genes of the *N*-acetylmuramoyl-L-alanine amidase and its modifier. *J. Gen. Microbiol.* **138:**1949–1961.
295. **Leibovici, J., and C. Anagnostopoulos.** 1969. Propriétés de la thréonine désaminase de la souche sauvage et d'un mutant sensible à la valine de *Bacillus subtilis. Bull. Soc. Chim. Biol.* **51:**691–707.
296. **Leighton, T.** Unpublished data.
297. **Lepesant, J.-A., F. Kunst, J. Lepesant-Kejzlarova, and R. Dedonder.** 1972. Chromosomal location of mutations affecting sucrose metabolism in *Bacillus subtilis* Marburg. *Mol. Gen. Genet.* **118:**135–160.
298. **Lepesant-Kejzlarova, J., J.-A. Lepesant, J. Walle, A. Billault, and R. Dedonder.** 1975. Revision of the linkage map of *Bacillus subtilis* 168: indications for circularity of the chromosome. *J. Bacteriol.* **121:**823–834.
299. **Lerner, C. G., B. T. Stephenson, and R. L. Switzer.** 1987. Structure of the *Bacillus subtilis* pyrimidine biosynthetic (*pyr*) gene cluster. *J. Bacteriol.* **169:**2202–2206.
300. **Levin, P. A., P. S. Margolis, and D. Sun.** 1992. Cloning and characterization of the *B. subtilis* homologs of the *Escherichia coli* cell division genes *minC* and *minD*. XI International Spores Conference (Spores XI), abstr. no. 54.
301. **Li, M., and S. L. Wong.** 1992. Cloning and characterization of the *groESL* operon from *Bacillus subtilis. J. Bacteriol.* **174:**3981–3992.
302. **Lindgren, V., E. Holmgren, and L. Rutberg.** 1977. *Bacillus subtilis* mutant with temperature-sensitive net synthesis of phosphatidylethanolamine. *J. Bacteriol.* **132:**473–484.
303. **Lindgren, V., and L. Rutberg.** 1974. Glycerol metabolism in *Bacillus subtilis*: gene-enzyme relationships. *J. Bacteriol.* **119:**431–442.
304. **Lipsky, R. H., R. Rosenthal, and S. A. Zahler.** 1981. Defective specialized SP-beta transducing bacteriophage of *Bacillus subtilis* that carry the *sup-3* or *sup-44* gene. *J. Bacteriol.* **148:**1012–1015.
305. **Loewen, P. C., and J. Switala.** 1987. Genetic mapping of *katA*, a locus that affects catalase 1 levels in *Bacillus subtilis. J. Bacteriol.* **169:**5848–5851.
306. **Lopez-Dias, I., S. Clarke, and J. Mandelstam.** 1986. *spoIID* operon of *Bacillus subtilis*: cloning and sequence. *J. Gen. Microbiol.* **132:**341–354.
307. **Losick, R.** Unpublished data.
308. **Love, E., J. D'Ambrosio, N. C. Brown, and D. Dubnau.** 1976. Mapping of the gene specifying DNA polymerase III of *Bacillus subtilis. Mol. Gen. Genet.* **144:**313–321.
309. **Lovett, P. S., N. P. Ambulos, Jr., W. Mulbry, N. Noguchi, and E. J. Rogers.** 1991. UGA can be decoded as tryptophan at low efficiency in *Bacillus subtilis. J. Bacteriol.* **173:**1810–1812.
310. **Lu, Y., N.-Y. Chen, and H. Paulus.** 1991. Identification of *aecA* mutations in *Bacillus subtilis* as nucleotide substitutions in the untranslated leader region of the aspartokinase II operon. *J. Gen. Microbiol.* **137:**1135–1143.
311. **MacKay, R. M., A. Lo, G. Willick, M. Zuker, S. Baird, M. Dove, F. Moranelli, and V. Seligy.** 1986. Structure of a *Bacillus subtilis* endo-β-1,4-glucanase gene. *Nucleic Acids Res.* **14:**9159–9170.
312. **Mackey, C. J., R. J. Warburg, H. O. Halvorson, and S. A. Zahler.** 1984. Genetic and physical analysis of the *ilvBC-leu* region in *Bacillus subtilis. Gene* **32:**49–56.
313. **Magnusson, K., M. K. Philips, J. R. Guest, and L. Rutberg.** 1986. Nucleotide sequence of the gene for cytochrome b_{558} of the *Bacillus subtilis* succinate dehydrogenase complex. *J. Bacteriol.* **166:**1067–1071.
314. **Mahler, I., R. Warburg, D. J. Tipper, and H. O. Halvorson.** 1984. Cloning of an unstable *spoIIA-tyrA* fragment from *Bacillus subtilis. J. Gen. Microbiol.* **130:**411–421.
315. **Makaroff, C. A., H. Zalkin, R. L. Switzer, and S. J. Vollmer.** 1983. Cloning of the *Bacillus subtilis* glutamine phosphoribosylpyrophosphate amidotransferase gene in *Escherichia coli. J. Biol. Chem.* **258:** 10586–10593.
316. **Makino, F., and N. Munakata.** 1977. Isolation and characterization of a *Bacillus subtilis* mutant with a defective *N*-glycosidase activity for uracil-containing deoxyribonucleic acid. *J. Bacteriol.* **131:**438–445.
317. **Mäntsälä, P., and H. Zalkin.** 1992. Cloning and sequence of *Bacillus subtilis purA* and *guaA*, involved in the conversion of IMP to AMP and GMP. *J. Bacteriol.* **174:**1883–1890.
318. **Margolis, P., A. Driks, and R. Losick.** 1993. Sporulation gene *spoIIB* from *Bacillus subtilis. J. Bacteriol.* **175:**528–546.
319. **Margot, P., and D. Karamata.** 1992. Identification of the structural gene for *N*-acetylmuramoyl-L-alanine amidase and its modifier in *Bacillus subtilis* 168: inactivation of these genes by insertional mutagenesis has no effect on growth or cell separation. *Mol. Gen. Genet.* **232:**359–366.
320. **Margot, P., C. Mauël, and D. Karamata.** 1991. The *Bacillus subtilis N*-acetylglucosaminidase is encoded by a monocistronic operon controlled by a sigma^{D} dependent promoter, abstr. W-6. 6th International Conference on Bacilli.
321. **Marquez, L. M., J. D. Helmann, E. Ferrari, H. M.**

Parker, G. W. Ordal, and M. J. Chamberlin. 1990. Studies of sigma D-dependent functions in *Bacillus subtilis*. *J. Bacteriol.* **172:**3435–3443.

322. **Marrero, R., and R. E. Yasbin.** 1988. Cloning of the *Bacillus subtilis recE*$^+$ gene and functional expression of *recE*$^+$ in *B. subtilis*. *J. Bacteriol.* **170:**335–344.

323. **Martin, I., M. Débarbouillé, E. Ferrari, A. Klier, and G. Rapoport.** 1987. Characterization of the levanase gene of *Bacillus subtilis* which shows homology to yeast invertase. *Mol. Gen. Genet.* **208:**177–184.

324. **Martin, I., M. Débarbouillé, A. Klier, and G. Rapoport.** 1987. Identification of a new locus, *sacV*, involved in the regulation of levansucrase synthesis in *Bacillus subtilis*. *FEMS Microbiol. Lett.* **44:**39–43.

325. **Martin-Verstraete, I., M. Débarbouillé, A. Klier, and G. Rapoport.** 1990. Levanase operon of *Bacillus subtilis* includes a fructose-specific phosphotransferase system regulating the expression of the operon. *J. Mol. Biol.* **214:**657–671.

326. **Mastromei, G., C. Barberio, S. Pistolesi, and M. Polsinelli.** 1989. Isolation of *Bacillus subtilis* transformation-deficient mutants and mapping of competence genes. *Genet. Res.* **54:**1–5.

327. **Masuda, E. S., H. Anaguchi, T. Sato, M. Takeuchi, and Y. Kobayashi.** 1990. Nucleotide sequence of the sporulation gene *spoIIGA* from *Bacillus subtilis*. *Nucleic Acids Res.* **18:**657.

328. **Masuda, E. S., H. Anaguchi, K. Yamada, and Y. Kobayashi.** 1988. Two developmental genes encoding sigma factor homologs are arranged in tandem in *Bacillus subtilis*. *Proc. Natl. Acad. Sci. USA* **85:**7637–7641.

329. **Mathiopoulos, C., J. P. Mueller, F. J. Slack, C. G. Murphy, S. Patankar, G. Bukusoglu, and A. L. Sonenshein.** 1991. A *Bacillus subtilis* dipeptide transport system expressed early during sporulation. *Mol. Microbiol.* **5:**1903–1913.

330. **Mathiopoulos, C., and A. L. Sonenshein.** 1989. Identification of *Bacillus subtilis* genes expressed early during sporulation. *Mol. Microbiol.* **3:**1071–1081.

331. **Matsuzaki, S., and Y. Kobayashi.** 1984. New mutation affecting the synthesis of some membrane proteins and sporulation in *Bacillus subtilis*. *J. Bacteriol.* **159:**228–232.

332. **Matsuzaki, S., and Y. Kobayashi.** 1985. Genetic heterogeneity in the *cysA-fus* region of the *Bacillus subtilis* chromosome: identification of the *hos* gene. *J. Bacteriol.* **163:**1336–1338.

333. **Mattioli, R., M. Bazzicalupo, G. Federici, E. Gallori, and M. Polsinelli.** 1979. Characterization of mutants of *Bacillus subtilis* resistant to *S*-(2-aminoethyl) cysteine. *J. Gen. Microbiol.* **114:**223–225.

334. **Mauël, C., and D. Karamata.** Unpublished data.

335. **Mauël, C., M. Young, and D. Karamata.** 1991. Genes concerned with synthesis of poly-(glycerol phosphate), the essential teichoic acid in *Bacillus subtilis* strain 168, are organized on two divergent transcription units. *J. Gen. Microbiol.* **137:**929–941.

336. **Maznitsa, I. I., V. V. Sukhodolets, and L. S. Ukhabotina.** 1983. Cloning of *Bacillus subtilis* 168 genes compensating the defect of mutations for thymidine phosphorylase and uridine phosphorylase in *Escherichia coli* cells. *Genetica* (USSR) **19:**881–887.

337. **Mazza, G., A. Forunato, E. Ferrari, U. Canosi, A. Falaschi, and M. Polsinelli.** 1975. Genetic and enzymic studies on the recombination process in *Bacillus subtilis*. *Mol. Gen. Genet.* **136:**9–30.

338. **Mazza, G., and A. Galizzi.** 1978. The genetics of DNA replication, repair and recombination in *Bacillus subtilis*. *Microbiologica* **1:**111–135.

339. **McDonald, K. O., and W. F. Burke, Jr.** 1982. Cloning of the *Bacillus subtilis* sulfanilamide resistance gene in *Bacillus subtilis*. *J. Bacteriol.* **149:**391–394.

340. **McEnroe, A. S., and H. Taber.** 1984. Correlation between cytrochrome aa_3 concentration and streptomycin accumulation in *Bacillus subtilis*. *Antimicrob. Agents Chemother.* **26:**507–512.

341. **Micka, B., N. Groch, U. Heinemann, and M. Marahiel.** 1991. Molecular cloning, nucleotide sequence, and characterization of the *Bacillus subtilis* gene encoding the DNA-binding protein HBsu. *J. Bacteriol.* **173:**3191–3198.

342. **Miczak, A., I. Berek, and G. Ivanovics.** 1976. Mapping the uroporphyrinogen decarboxylase, coproporphyinogen oxidase and ferrochelatase loci in *Bacillus subtilis*. *Mol. Gen. Genet.* **146:**85–87.

343. **Miczak, A., B. Pragai, and I. Berek.** 1979. Mapping the uroporphyrinogen III cosynthase locus in *Bacillus subtilis*. *Mol. Gen. Genet.* **174:**293–295.

344. **Miles, J. S., and J. R. Guest.** 1985. Complete nucleotide sequence of the fumarase gene *citG* of *Bacillus subtilis* 168. *Nucleic Acids Res.* **13:**131–140.

345. **Milhaud, P., G. Balassa, and J. Zucca.** 1978. Spore control (Sco) mutations in *Bacillus subtilis*. 1. Selection and genetic mapping of Sco mutations. *Mol. Gen. Genet.* **163:**35–44.

346. **Miller, P., A. Rabinowitz, and H. Taber.** 1988. Molecular cloning and preliminary genetic analysis of the *men* gene cluster of *Bacillus subtilis*. *J. Bacteriol.* **170:**2735–2741.

347. **Mirel, D. B., and M. J. Chamberlin.** 1989. The *Bacillus subtilis* flagellin gene (*hag*) is transcribed by the sigma-28 form of RNA polymerase. *J. Bacteriol.* **171:**3095–3101.

348. **Mirel, D. B., V. M. Lustre, and M. J. Chamberlin.** 1992. An operon of *Bacillus subtilis* motility genes transcribed by the σ^D form of RNA polymerase. *J. Bacteriol.* **174:**4197–4204.

349. **Mironov, V. N., M. L. Chikindas, A. S. Kraev, A. I. Stepanov, and K. G. Skryabin.** 1990. Operon organization of the riboflavin biosynthesis genes of *Bacillus subtilis*. *Dokl. Akad. Nauk. SSSR* **312:**237–240.

350. **Mironov, V. N., A. S. Kraev, B. K. Chernov, A. B. Vlyanov, Y. B. Golova, G. E. Pozmogova, M. L. Simonova, and K. G. Skryabin.** 1989. Genes of riboflavin biosynthesis of *Bacillus subtilis*: complete primary structure and model of organization. *Dokl. Akad. Nauk. SSSR* **305:**482–486.

351. **Miyakawa, Y., and T. Komano.** 1981. Study on the cell cycle of *Bacillus subtilis* using temperature-sensitive mutants. I. Isolation and genetic analysis of the mutants defective in septum formation. *Mol. Gen. Genet.* **181:**207–214.

352. **Mizonov, V. N.** Unpublished data.

353. **Moir, A.** Unpublished data.

353a. **Moir, A.** Personal communication.

354. **Moir, A., E. Lafferty, and D. A. Smith.** 1979. Genetic analysis of spore germination mutants of *Bacillus subtilis* 168: the correlation of phenotype with map location. *J. Gen. Microbiol.* **111:**165–180.

355. **Mollgaard, H.** 1980. Deoxyadenosine/deoxycytidine kinase from *Bacillus subtilis*. Purification, characterization, and physiological function. *J. Biol. Chem.* **255:**8216–8220.

356. **Mollgaard, H., and J. Neuhard.** 1978. Deoxycytidylate deaminase from *Bacillus subtilis*. Purification, characterization and physiological function. *J. Biol. Chem.* **253:**3536–3542.

357. **Moriya, S., N. Ogasawara, and H. Yoshikawa.** 1985. Structure and function of the region of the replication origin of the *Bacillus subtilis* chromosome. III. Nucleotide sequence of some 10,000 base pairs in the origin region. *Nucleic Acids Res.* **13:**2251–2265.

358. **Morohoshi, F., K. Hayashi, and N. Munakata.** 1989. *Bacillus subtilis* gene coding for constitutive *O*-6 methylguanine-DNA alkyltransferase. *Nucleic Acids Res.* **17:**6531–6543.

359. **Morohoshi, F., K. Hayashi, and N. Munakata.** 1990. *Bacillus subtilis ada* operon encodes two DNA alkyltransferases. *Nucleic Acids Res.* **18:**5473–5480.
360. **Morohoshi, F., K. Hayashi, and N. Munakata.** 1991. Molecular analysis of *Bacillus subtilis ada* mutants deficient in the adaptive response to simple alkylating agents. *J. Bacteriol.* **173:**7834–7840.
361. **Mountain, A., and S. Baumberg.** 1980. Map locations of some mutations conferring resistance to arginine hydroxamate in *Bacillus subtilis* 168. *Mol. Gen. Genet.* **178:**691–701.
362. **Mountain, A., J. McChesney, M. C. M. Smith, and S. Baumberg.** 1986. Gene sequence encoding early enzymes of arginine synthesis within a cluster in *Bacillus subtilis*, as revealed by cloning in *Escherichia coli*. *J. Bacteriol.* **165:**1026–1028.
363. **Mountain, A., M. C. M. Smith, and S. Baumberg.** 1990. Nucleotide sequence of the *Bacillus subtilis argF* gene encoding ornithine carbamoyltransferase. *Nucleic Acids Res.* **18:**4594.
364. **Msadek, T., F. Kunst, A. Klier, and G. Rapoport.** 1991. DegS-DegU and ComP-ComA modulator-effector pairs control expression of the *Bacillus subtilis* pleiotropic regulatory gene *degQ*. *J. Bacteriol.* **173:**2366–2377.
365. **Mueller, J. P., G. Bukusoglu, and A. L. Sonenshein.** 1992. Transcriptional regulation of *Bacillus subtilis* glucose starvation-inducible genes: control of *gsiA* by the ComP-ComA signal transduction system. *J. Bacteriol.* **174:**4361–4373.
366. **Mueller, J. P., and A. L. Sonenshein.** 1992. Role of the *Bacillus subtilis gsiA* gene in regulation of early sporulation gene expression. *J. Bacteriol.* **174:**4374–4383.
367. **Mueller, J. P., and H. W. Taber.** 1988. Genetic regulation of cytochrome aa_3 in *Bacillus subtilis*, p. 91–95. *In* A. T. Ganesan and J. A. Hoch (ed.), *Genetics and Biotechnology of Bacilli*, vol. II. Academic Press, Inc., San Diego.
368. **Mueller, J. P., and H. W. Taber.** 1989. Isolation and sequence of *ctaA*, a gene required for cytochrome aa_3 biosynthesis and sporulation in *Bacillus subtilis*. *J. Bacteriol.* **171:**4967–4978.
369. **Mulbry, W. W., N. P. Ambulos, Jr., and P. S. Lovett.** 1989. *Bacillus subtilis* mutant allele *sup-3* causes lysine insertion at ochre codons: use of *sup-3* in studies of translational attenuation. *J. Bacteriol.* **171:**5322–5324.
370. **Muller, J.** Unpublished data.
371. **Munakata, H., and Y. Ikeda.** 1968. Mutant of *Bacillus subtilis* producing ultraviolet-sensitive spores. *Biochem. Biophys. Res. Commun.* **33:**469–475.
372. **Murphy, N., D. J. McConnell, and B. A. Cantwell.** 1984. The DNA sequence of the gene and genetic control sites for the excreted *B. subtilis* enzyme β-glucanase. *Nucleic Acids Res.* **12:**5355–5367.
373. **Nagahari, K., and K. Sakaguchi.** 1978. Cloning of *Bacillus subtilis* leucine A, B and C genes with *Escherichia coli* plasmids and expression of the *leuC* gene in *E. coli*. *Mol. Gen. Genet.* **158:**263–270.
374. **Nakamura, A., T. Uozumi, and T. Beppu.** 1987. Nucleotide sequence of a cellulase gene of *Bacillus subtilis*. *Eur. J. Biochem.* **164:**317–320.
375. **Nakamura, K., A. Nakamura, H. Takamatsu, H. Yoshikawa, and K. Yamane.** 1990. Cloning and characterization of a *Bacillus subtilis* gene homologous to *E. coli secY*. *J. Biochem.* (Tokyo) **107:**603–607.
376. **Nakano, M. M., N. Corbell, J. Besson, and P. Zuber.** 1992. Isolation and characterization of *sfp*: a gene that functions in the production of the lipopeptide biasurfactant, surfactin, in *Bacillus subtilis*. *Mol. Gen. Genet.* **232:**313–321.
377. **Nakano, M. M., R. Magnuson, A. Myers, J. Curry, A. D. Grossman, and P. Zuber.** 1991. *srfA* is an operon required for surfactin production, competence development, and efficient sporulation in *Bacillus subtilis*. *J. Bacteriol.* **173:**1770–1778.
378. **Nakano, M. M., M. A. Marahiel, and P. Zuber.** 1988. Identification of a genetic locus required for biosynthesis of the lipopeptide antibiotic surfactin in *Bacillus subtilis*. *J. Bacteriol.* **170:**5662–5668.
379. **Nakano, M. M., and P. Zuber.** 1989. Cloning and characterization of *srfB*, a regulatory gene involved in surfactin production and competence in *Bacillus subtilis*. *J. Bacteriol.* **171:**5347–5353.
380. **Nasser, D., and E. W. Nester.** 1967. Aromatic amino acid biosynthesis: gene-enzyme relationships in *Bacillus subtilis*. *J. Bacteriol.* **94:**1706–1714.
381. **Navarro, J., J. Chabot, K. Sherrill, R. Aneja, S. A. Zahler, and E. Racker.** 1985. Interaction of duramycin with artificial and natural membranes. *Biochemistry* **24:**4645–4650.
382. **Nester, E. W., and A. L. Montoya.** 1976. An enzyme common to histidine and aromatic amino acid biosynthesis in *Bacillus subtilis*. *J. Bacteriol.* **126:**699–705.
383. **Neuhard, J.** 1983. Utilization of preformed pyrimidine bases and nucleosides, p. 95–148. *In* A. Munch-Petersen (ed.), *Metabolism of Nucleotides, Nucleosides, and Nucleobases in Microorganisms*. Academic Press, Inc., New York.
384. **Neuhard, J., A. R. Price, L. Schack, and E. Thomassen.** 1978. Two thymidylate synthetases in *Bacillus subtilis*. *Proc. Natl. Acad. Sci. USA* **75:**1194–1198.
385. **Neyfakh, A. A., V. E. Bidnenko, and L. B. Chen.** 1991. Efflux-mediated multidrug resistance in *Bacillus subtilis*: similarities and dissimilarities with the mammalian system. *Proc. Natl. Acad. Sci. USA* **88:**4781–4785.
386. **Nicholson, W. L., and G. H. Chambliss.** 1986. Molecular cloning of *cis*-acting regulatory alleles of the *Bacillus subtilis amyR* region by using gene conversion transformation. *J. Bacteriol.* **165:**663–670.
387. **Nilsson, D., and B. Hove-Jensen.** 1987. Phosphoribosylpyrophosphate synthetase of *Bacillus subtilis*: cloning, characterization and chromosomal mapping of the *prs* gene. *Gene* **53:**247–255.
388. **Nilsson, D., B. Hove-Jensen, and K. Arnvig.** 1989. Primary structure of the *tms* and *prs* genes of *Bacillus subtilis*. *Mol. Gen. Genet.* **218:**565–571.
389. **Noda, Y., K. Yoda, A. Takatsuki, and M. Yamasaki.** 1992. TmrB protein, responsible for tunicamycin resistance of *Bacillus subtilis*, is a novel ATP-binding membrane protein. *J. Bacteriol.* **174:**4302–4307.
390. **Nomura, S., K. Yamane, T. Sasaki, M. Yamasaki, G. Tamura, and B. Maruo.** 1978. Tunicamycin-resistant mutants and chromosomal locations of mutational sites in *Bacillus subtilis*. *J. Bacteriol.* **136:**818–821.
391. **Nygaard, P., P. Duckert, and H. H. Saxild.** 1988. Purine gene organization and regulation in *Bacillus subtilis*, p. 57–61. *In* A. T. Ganesan and J. A. Hoch (ed.), *Genetics and Biotechnology of Bacilli*, vol. II. Academic Press, Inc., San Diego.
392. **Oda, M., A. Sugishita, and K. Furukawa.** 1988. Cloning and nucleotide sequences of histidase and regulatory genes in the *Bacillus subtilis hut* operon and positive regulation of the operon. *J. Bacteriol.* **170:**3199–3205.
393. **Ogasawara, N.** Unpublished data.
394. **Ogasawara, N., S. Moriya, P. Mazza, and H. Yoshikawa.** 1986. A *Bacillus subtilis dnaG* mutant harbours a mutation in a gene homologous to the *dnaN* gene of *Escherichia coli*. *Gene* **45:**227–231.
395. **Ogasawara, N., S. Moriya, P. G. Mazza, and H. Yoshikawa.** 1986. Nucleotide sequence and organization of *dnaB* gene and neighboring genes on the *Bacillus subtilis* chromosome. *Nucleic Acids Res.* **14:**9989–9999.
396. **Ogasawara, N., S. Moriya, K. von Meyenburg, F. G. Hansen, and H. Yoshikawa.** 1985. Conservation of genes and their organization in the chromosomal rep-

lication origin region of *Bacillus subtilis* and *Escherichia coli*. *EMBO J.* **4**:3345–3350.
397. **Ogasawara, N., S. Moriya, and H. Yoshikawa.** 1983. Structure and organization of rRNA operons in the region of the replication origin of the *Bacillus subtilis* chromosome. *Nucleic Acids Res.* **11**:6301–6318.
398. **Ogasawara, N., and H. Yoshikawa.** 1992. Genes and their organization in the replication origin region of the bacterial chromosome. *Mol. Microbiol.* **6**:629–634.
399. **Ordal, G. W.** Unpublished data.
400. **Ordal, G. W., D. O. Nettleton, and J. A. Hoch.** 1983. Genetics of *Bacillus subtilis* chemotaxis: isolation and mapping of mutations and cloning of chemotaxis genes. *J. Bacteriol.* **154**:1088–1097.
401. **Ordal, G. W., H. M. Parker, and J. R. Kirby.** 1985. Complementation and characterization of chemotaxis mutants of *Bacillus subtilis*. *J. Bacteriol.* **164**:802–810.
402. **Osawa, S., R. Takata, K. Tanaka, and M. Tamaki.** 1973. Chloramphenicol resistant mutants of *Bacillus subtilis*. *Mol. Gen. Genet.* **127**:163–173.
403. **Osawa, S., and A. Tuki.** 1978. Mapping by interspecies transformation experiments of several ribosomal protein genes near the replication origin of *Bacillus subtilis* chromosome. *Mol. Gen. Genet.* **164**:113–129.
404. **Otozai, K., Y. Takeichi, A. Nakayama, K. Yamane, T. Tanimoto, M. Yamasaki, G. Tamura, S. Nomura, F. Kawamura, and H. Saito.** 1984. Cloning of the *AROI*$^{\pm}$ gene regions of *Bacillus subtilis* chromosomal DNAs by *B. subtilis* temperate phage *rho11* and *Escherichia coli* vector systems, and a comparison of physical maps of the gene regions. *J. Gen. Appl. Microbiol.* **30**:15–25.
405. **Pai, C. H.** 1975. Genetics of biotin biosynthesis in *Bacillus subtilis*. *J. Bacteriol.* **121**:1–8.
406. **Paice, M. G., R. Bourbonnais, M. Desrochers, L. Jurasek, and M. Yaguchi.** 1986. A xylanase gene from *Bacillus subtilis*: nucleotide sequence and comparison with *B. pumilus* gene. *Arch. Microbiol.* **144**:201–206.
407. **Pang, A. S.-H., S. Nathoo, and S.-L. Wong.** 1991. Cloning and characterization of a pair of novel genes that regulate production of extracellular enzymes in *Bacillus subtilis*. *J. Bacteriol.* **173**:46–54.
408. **Parsot, C.** 1986. Evolution of biosynthetic pathways: a common ancestor for threonine synthase, threonine dehydratase and D-serine dehydratase. *EMBO J.* **5**: 3013–3019.
409. **Parsot, C., and G. N. Cohen.** 1988. Cloning and nucleotide sequence of the *Bacillus subtilis hom* gene coding for homoserine dehydrogenase. *J. Biol. Chem.* **263**: 14654–14660.
410. **Paulus, H.** Unpublished data.
411. **Paveia, M. H., and L. J. Archer.** 1980. Location of genes for arabinose utilization in the *Bacillus subtilis* chromosome. *Broteria Genetica* (Lisbon) **1**:169–176.
412. **Perego, M., S. P. Cole, D. Burbulys, K. Trach, and J. A. Hoch.** 1989. Characterization of the gene for a protein kinase which phosphorylates the sporulation-regulatory proteins Spo0A and Spo0F of *Bacillus subtilis*. *J. Bacteriol.* **171**:6187–6196.
413. **Perego, M., E. Ferrari, M. T. Bassi, A. Galizzi, and P. Mazza.** 1987. Molecular cloning of *Bacillus subtilis* genes involved in DNA metabolism. *Mol. Gen. Genet.* **209**:8–14.
414. **Perego, M., C. F. Higgins, S. R. Pearce, M. P. Gallagher, and J. A. Hoch.** 1991. The oligopeptide transport system of *Bacillus subtilis* plays a role in the initiation of sporulation. *Mol. Microbiol.* **5**:173–185.
415. **Perego, M., and J. A. Hoch.** 1987. Isolation and sequence of the *spo0E* gene: its role in initiation of sporulation in *Bacillus subtilis*. *Mol. Microbiol.* **1**:125–132.
416. **Perego, M., and J. A. Hoch.** 1988. Sequence analysis and regulation of the *hpr* locus, a regulatory gene for protease production and sporulation in *Bacillus subtilis*. *J. Bacteriol.* **170**:2560–2567.
417. **Perego, M., G. B. Spiegelman, and J. A. Hoch.** 1988. Structure of the gene for the transition state regulator, *abrB*: regulator synthesis is controlled by the *spo0A* sporulation gene in *Bacillus subtilis*. *Mol. Microbiol.* **2**:689–699.
418. **Pestka, S., D. Weiss, R. Vince, B. Wienen, G. Stoffler, and I. Smith.** 1976. Thiostrepton-resistant mutants of *Bacillus subtilis*: localization of resistance to 50S subunit. *Mol. Gen. Genet.* **144**:235–241.
419. **Peterson, A. M., and L. Rutberg.** 1969. Linked transformation of bacterial and prophage markers in *Bacillus subtilis* 168 lysogenic for bacteriophage phi-105. *J. Bacteriol.* **98**:874–877.
420. **Petricek, M., L. Rutberg, and L. Hederstedt.** 1989. The structural gene for aspartokinase II in *Bacillus subtilis* is closely linked to the *sdh* operon. *FEMS Microbiol. Lett.* **61**:85–88.
421. **Petricek, M., L. Rutberg, I. Schroder, and L. Hederstedt.** 1990. Cloning and characterization of the *hemA* region of the *Bacillus subtilis* chromosome. *J. Bacteriol.* **172**:2250–2258.
422. **Phang, C. H., and K. Jeyaseelan.** 1988. Isolation and characterization of *citC* gene of *Bacillus subtilis*, p. 97–100. *In* A. T. Ganesan and J. A. Hoch (ed.), *Genetics and Biotechnology of Bacilli*, vol. 2. Academic Press, Inc., San Diego.
423. **Phillips, M. K., L. Hederstedt, S. Hasnain, L. Rutberg, and J. R. Guest.** 1987. Nucleotide sequence encoding the flavoprotein and iron-sulfur protein subunits of the *Bacillus subtilis* PY79 succinate dehydrogenase complex. *J. Bacteriol.* **169**:864–873.
424. **Piggot, P. J.** Unpublished data.
425. **Piggot, P. J.** 1975. Characterization of a *cym* mutant of *Bacillus subtilis*. *J. Gen. Microbiol.* **89**:371–374.
426. **Piggot, P. J.** 1989. Revised genetic map of *Bacillus subtilis* 168, p. 1–41. *In* I. Smith, R. A. Slepecky, and P. Setlow (ed.), *Regulation of Prokaryotic Development*. American Society for Microbiology, Washington, D.C.
427. **Piggot, P. J., M. Amjad, J.-J. Wu, H. Sandoval, and J. Castro.** 1990. Genetic and physical maps of *Bacillus subtilis* 168, p. 494–532. *In* C. R. Harwood and S. M. Cutting (ed.), *Molecular Biology Methods for Bacillus*. John Wiley and Sons, Ltd., London.
428. **Piggot, P. J., and R. S. Buxton.** 1982. Bacteriophage PBSX-induced deletion mutants of *Bacillus subtilis* 168 constitutive for alkaline phosphatase. *J. Gen. Microbiol.* **128**:663–669.
429. **Piggot, P. J., and J. G. Coote.** 1976. Genetic aspects of bacterial endospore formation. *Bacteriol. Rev.* **40**:908–962.
430. **Piggot, P. J., and J. A. Hoch.** 1985. Revised genetic linkage map of *Bacillus subtilis*. *Microbiol. Rev.* **49**:158–179.
431. **Piggot, P. J., A. Moir, and D. A. Smith.** 1981. Advances in the genetics of *Bacillus subtilis* differentiation, p. 29–39. *In* H. S. Levinson, A. L. Sonenshein, and D. J. Tipper (ed.), *Sporulation and Germination*. American Society for Microbiology, Washington, D.C.
432. **Piggot, P. J., and S. Y. Taylor.** 1977. New types of mutation affecting formation of alkaline phosphatase by *Bacillus subtilis*. *J. Gen. Microbiol.* **102**:69–80.
433. **Polsinelli, M.** 1965. Linkage relationship between genes for amino acid or nitrogenous base biosynthesis and genes controlling resistance to structurally correlated analogues. *J. Gen. Microbiol.* **13**:99–110.
434. **Poluektova, E. U., N. M. Lakomova, T. S. Belova, and A. A. Prozorov.** 1984. Cloning of *purA16* locus in Rec$^+$ cells of *Bacillus subtilis*. *Genetika* **20**:943–948.
435. **Pooley, H. M., F.-X. Abellan, and D. Karamata.** 1992. CDP-glycerol:poly(glycerophosphate) glycerophosphotransferase, which is involved in the synthesis of the

major wall teichoic acid in *Bacillus subtilis* 168, is encoded by *tagF* (*rodC*). *J. Bacteriol.* **174:**646–649.

436. **Pooley, H. M., F. X. Abellan, and D. Karamata.** 1991. A conditional lethal mutant of *Bacillus subtilis* 168 with a thermosensitive glycerol-3-phosphate cytidylyltransferase, an enzyme specific for the synthesis of the major wall teichoic acid. *J. Gen. Microbiol.* **137:**921–928.
437. **Pooley, H. M., and D. Karamata.** 1984. Genetic analysis of autolysin-deficient and flagellaless mutants of *Bacillus subtilis. J. Bacteriol.* **160:**1123–1129.
438. **Pooley, H. M., D. Paschoud, and D. Karamata.** 1987. The *gtaB* marker in *Bacillus subtilis* 168 is associated with a deficiency in UDPglucose pyrophosphorylase. *J. Gen. Microbiol.* **133:**3481–3493.
439. **Popham, D. L., and P. Stragier.** 1991. Cloning, characterization, and expression of the *spoVB* gene of *Bacillus subtilis. J. Bacteriol.* **173:**7942–7949.
440. **Porter, A. C. G., and J. Mandelstam.** 1982. A mutant of *Bacillus subtilis* secreting a DNase inhibitor during sporulation. *J. Gen. Microbiol.* **128:**1903–1914.
441. **Price, C.** Unpublished data.
442. **Price, C., S. Boylan, M. Duncan, S. Kalman, J. W. Suh, S. Thomas, and B. Van Hoy.** 1988. Use of lambda-gt11 and antibody probes to isolate genes encoding RNA polymerase subunits from *Bacillus subtilis*, p. 183–188. *In* A. T. Ganesan and J. A. Hoch (ed.), *Genetics and Biotechnology of Bacilli*, vol. 2. Academic Press, Inc., San Diego.
443. **Price, C. W., M. A. Gitt, and R. H. Doi.** 1983. Isolation and physical mapping of the gene encoding the major sigma-factor of *Bacillus subtilis* RNA polymerase. *Proc. Natl. Acad. Sci. USA* **80:**4074–4078.
444. **Price, V. L., and J. A. Gallant.** 1983. *Bacillus subtilis relG* mutant: defect in glucose uptake. *J. Bacteriol.* **153:**270–273.
445. **Putzer, H., A. A. Brakhage, and M. Grunberg-Manago.** 1990. Independent genes for two threonyl-tRNA synthetases in *Bacillus subtilis. J. Bacteriol.* **172:**4593–4602.
446. **Quinn, C. L., B. T. Stephenson, and R. L. Switzer.** 1991. Functional organization and nucleotide sequence of the *Bacillus subtilis* pyrimidine biosynthetic operon. *J. Biol. Chem.* **266:**9113–9127.
447. **Rabinovich, P. M., M. Yu. Beburov, Z. K. Linevich, and A. I. Stepanov.** 1978. Amplification of *Bacillus subtilis* riboflavin operon genes in *Escherichia coli* cells. *Genetica* (USSR) **14:**1696–1705.
448. **Rabinovich, P. M., Yu. V. Yomantas, M. Ya. Haykinson, and A. I. Stepanov.** 1984. Cloning of genetic material in bacilli, p. 297–308. *In* A. T. Ganesan and J. A. Hoch (ed.), *Genetics and Biotechnology of Bacilli*. Academic Press, Inc., Orlando, Fla.
449. **Racine, F. M., and W. Steinberg.** 1974. Genetic location of two mutations affecting the lysyltransfer ribonucleic acid synthetase of *Bacillus subtilis. J. Bacteriol.* **120:**384–389.
450. **Raugei, G., M. Bazzicalupo, G. Federici, and E. Gallori.** 1981. Effect of a new pyrimidine analog on *Bacillus subtilis* growth. *J. Bacteriol.* **145:**1079–1081.
451. **Raymond-Denise, A., and N. Guillen.** 1991. Identification of *dinR*, a DNA damage-inducible regulator gene of *Bacillus subtilis. J. Bacteriol.* **173:**7084–7091.
452. **Reeve, J. N., N. H. Mendelson, S. I. Coyne, L. L. Hallock, and R. M. Cole.** 1973. Minicells of *Bacillus subtilis. J. Bacteriol.* **114:**860–873.
453. **Reich, C., K. J. Gardiner, G. J. Olsen, B. Pace, T. L. Marsh, and N. R. Pace.** 1986. The RNA component of the *Bacillus subtilis* RNase P. Sequence, activity, and partial secondary structure. *J. Biol. Chem.* **261:**7888–7893.
454. **Reizer, A., J. Deutscher, M. H. Saier, Jr., and J. Reizer.** 1991. Analysis of the gluconate (*gnt*) operon of *Bacillus subtilis. Mol. Microbiol.* **5:**1081–1089.
455. **Ricca, E., S. Cutting, and R. Losick.** 1992. Characterization of *bofA*, a gene involved in intercompartmental regulation of pro-σ^K processing during sporulation in *Bacillus subtilis. J. Bacteriol.* **174:**3177–3184.
456. **Rima, B. K., and I. Takahashi.** 1978. Synthesis of thymidine nucleotides in *Bacillus subtilis. Can. J. Biochem.* **56:**158–160.
457. **Riva, S., G. Villani, G. Mastromei, and G. Mazza.** 1976. *Bacillus subtilis* mutant temperature sensitive in the synthesis of ribonucleic acid. *J. Bacteriol.* **127:**679–690.
458. **Roels, S., A. Driks, and R. Losick.** 1992. Characterization of *spoIVA*, a sporulation gene involved in coat morphogenesis in *Bacillus subtilis. J. Bacteriol.* **174:**575–585.
459. **Rogolsky, M.** 1968. Genetic mapping of a locus which regulates the production of pigment associated with spores of *Bacillus subtilis. J. Bacteriol.* **95:**2426–2427.
460. **Ron, E. Z., M.-P. de Bethune, and C. G. Cocito.** 1980. Mapping of virginiamycin S resistance in *Bacillus subtilis. Mol. Gen. Genet.* **180:**639–640.
461. **Roncero, M. I. G.** 1983. Genes controlling xylan utilization by *Bacillus subtilis. J. Bacteriol.* **156:**257–263.
462. **Rong, S., and A. L. Sonenshein.** 1992. Mutations in the precursor region of a *Bacillus subtilis* sporulation sigma factor. *J. Bacteriol.* **174:**3812–3817.
463. **Roten, C.-A. H., C. Brandt, and D. Karamata.** 1991. Genes involved in *meso*-diaminopimelate synthesis in *Bacillus subtilis*: identification of the gene encoding aspartokinase I. *J. Gen. Microbiol.* **137:**951–962.
464. **Roten, C.-A. H., C. Brandt, and D. Karamata.** 1991. Identification of *murA*, the structural gene of phosphoenolpyruvate:uridine-*N*-acetyl-glucosamine enolpyruvoyl transferase in *Bacillus subtilis*, abstr. W-8. 6th International Conference on Bacilli.
465. **Rudner, D. Z., J. R. Ladeaux, K. Breton, and A. D. Grossman.** 1991. The *spo0K* locus of *Bacillus subtilis* is homologous to the oligopeptide permease locus and is required for sporulation and competence. *J. Bacteriol.* **173:**1388–1398.
466. **Rudner, R.** Unpublished data.
467. **Rumyantseva, E. V., V. V. Sukhodolets, and Yu. V. Smirnov.** 1979. Isolation and characterization of mutants for genes of nucleoside catabolism in *Bacillus subtilis. Genetica* (USSR) **15:**594–604.
468. **Sa-Nogueira, I., and H. de Lencastre.** 1989. Cloning and characterization of *araA*, *araB*, and *araD*, the structural genes for L-arabinose utilization in *Bacillus subtilis. J. Bacteriol.* **171:**4088–4091.
469. **Sa-Nogueira, I., H. Paveia, and H. de Lencastre.** 1988. Isolation of constitutive mutants for L-arabinose utilization in *Bacillus subtilis. J. Bacteriol.* **170:**2855–2857.
470. **Sadaie, Y., H. Takamatsu, K. Nakamura, and K. Yamane.** 1991. Sequencing reveals similarity of the wild-type div^+ gene of *Bacillus subtilis* to the *Escherichia coli secA* gene. *Gene* **98:**101–105.
471. **Saito, H., T. Shibata, and T. Ando.** 1979. Mapping of genes determining nonpermissiveness and host-specific restriction to bacteriophages in *Bacillus subtilis* Marburg. *Mol. Gen. Genet.* **170:**117–122.
472. **Sakaguchi, R., H. Amano, and K. Shishido.** 1988. Nucleotide sequence homology of the tetracycline-resistance determinant naturally maintained in *Bacillus subtilis* Marburg 168 chromosome and the tetracycline-resistance gene of *B. subtilis* plasmid pNS1981. *Biochim. Biophys. Acta* **950:**441–444.
473. **Sammons, R. L., G. M. Slynn, and D. A. Smith.** 1987. Genetical and molecular studies on *gerM*, a new developmental locus of *Bacillus subtilis. J. Gen. Microbiol.* **133:**3299–3312.
474. **Sandman, K., R. Losick, and P. Youngman.** 1987. Genetic analysis of *Bacillus subtilis spo* mutations generated by Tn917-mediated insertional mutagenesis. *Genetics* **117:**603–617.

475. **Sanjanwala, B., and A. T. Ganesan.** 1991. Genetic structure and domains of DNA polymerase III of *Bacillus subtilis. Mol. Gen. Genet.* **226:**467–472.
476. **Santana, M., F. Kunst, M. F. Hullo, G. Rapoport, A. Danchin, and P. Glaser.** 1992. Molecular cloning sequencing and physiological characterization of the *qox* operon from *Bacillus subtilis* encoding the aa3-600 quinol oxidase. *J. Biol. Chem.* **267:**10225–10231.
477. **Santos, M. A., H. de Lencastre, and L. J. Archer.** 1983. *Bacillus subtilis* mutation blocking irreversible binding of bacteriophage SSP1. *J. Gen. Microbiol.* **129:**3499–3504.
478. **Saraste, M., T. Metso, T. Nakari, T. Jalli, M. Lauraeus, and J. van der Oost.** 1991. The *Bacillus subtilis* cytochrome-*c* oxidase: variations on a conserved protein theme. *Eur. J. Biochem.* **195:**517–525.
479. **Sasajima, K., and T. Kumada.** 1983. Deficiency of flagellation in *Bacillus subtilis* pleiotropic mutant lacking transketolase. *Agric. Biol. Chem.* **47:**1375–1376.
480. **Sato, T., Y. Samori, and Y. Kobayashi.** 1990. The *cisA* cistron of *Bacillus subtilis* sporulation gene *spoIVC* encodes a protein homologous to a site-specific recombinase. *J. Bacteriol.* **172:**1092–1098.
481. **Saxe, C. L., and N. H. Mendelson.** 1984. Identification of mutations associated with macrofiber formation in *Bacillus subtilis. Genetics* **107:**551–561.
482. **Saxild, H. H., and P. Nygaard.** 1987. Genetic and physiological characterization of *Bacillus subtilis* mutants resistant to purine analogs. *J. Bacteriol.* **169:**2977–2983.
483. **Saxild, H. H., and P. Nygaard.** 1988. Gene-enzyme relationships of the purine biosynthetic pathway in *Bacillus subtilis. Mol. Gen. Genet.* **211:**160–167.
484. **Schmidt, A., M. Schiesswohl, U. Völker, M. Hecker, and W. Schumann.** 1992. Cloning, sequencing, mapping and transcriptional analysis of the *groESL* operon from *Bacillus subtilis. J. Bacteriol.* **174:**3993–3999.
485. **Schroeder, I., and L. Hederstedt.** Unpublished data.
486. **Schroeder, I., L. Hederstedt, and S. Gough.** Unpublished data.
487. **Segall, J., and R. Losick.** 1977. Cloned *Bacillus subtilis* DNA containing a gene that is activated early during sporulation. *Cell* **11:**751–761.
488. **Seki, T., H. Yoshikawa, H. Takahashi, and H. Saito.** 1987. Cloning and nucleotide sequence of *phoP*, the regulatory gene for alkaline phosphatase and phosphodiesterase in *Bacillus subtilis. J. Bacteriol.* **169:**2913–2916.
489. **Seki, T., H. Yoshikawa, H. Takahashi, and H. Saito.** 1988. Nucleotide sequence of the *Bacillus subtilis phoR* gene. *J. Bacteriol.* **170:**5935–5938.
490. **Sekiguchi, J., H. Ohsu, A. Kuroda, H. Moriyama, and T. Akamatsu.** 1990. Nucleotide sequences of the *Bacillus subtilis flaD* locus and a *B. licheniformis* homologue affecting the autolysin level and flagellation. *J. Gen. Microbiol.* **136:**1223–1230.
491. **Serror, P.** Unpublished data.
492. **Serror, P., and S. D. Ehrlich.** Unpublished data.
493. **Setlow, P.** Unpublished data.
494. **Sharp, P. M., N. C. Nolan, N. N. Cholmain, and K. M. Devine.** 1992. DNA sequence variability at the *rplX* locus of *Bacillus subtilis. J. Gen. Microbiol.* **138:**39–45.
495. **Sharrock, R. A., and T. Leighton.** 1981. Intergenic suppressors of temperature-sensitive sporulation in *Bacillus subtilis. Mol. Gen. Genet.* **183:**532–537.
496. **Sharrock, R. A., S. Rubenstein, M. Chan, and T. Leighton.** 1984. Intergenic suppression of *spo0* phenotype by the *Bacillus subtilis* mutation *rvtA. Mol. Gen. Genet.* **194:**260–264.
497. **Shazand, K., J. Tucker, R. Chiang, K. Stansmore, H. U. Sperling-Petersen, M. Grunberg-Manago, J. C. Rabinowitz, and T. Leighton.** 1990. Isolation and molecular genetic characterization of the *Bacillus subtilis* gene (*infB*) encoding protein synthesis initiation factor 2. *J. Bacteriol.* **172:**2675–2687.
498. **Shohayeb, M., and I. Chopra.** 1987. Mutations affecting penicillin-binding proteins 2a, 2b and 3 in *Bacillus subtilis* alter cell shape and peptidoglycan metabolism. *J. Gen. Microbiol.* **133:**1733–1742.
499. **Siccardi, A. G., E. Lanza, E. Nielsen, A. Galizzi, and G. Mazza.** 1975. Genetic and physiological studies on the site of action of distamycin A. *Antimicrob. Agents Chemother.* **8:**370–376.
500. **Siccardi, A. G., S. Ottolenghi, A. Fortunato, and G. Mazza.** 1976. Pleiotropic, extragenic suppression of *dna* mutations in *Bacillus subtilis. J. Bacteriol.* **128:**174–181.
501. **Siegel, E. C., and J. Marmur.** 1969. Temperature-sensitive induction of bacteriophage in *Bacillus subtilis* 168. *J. Virol.* **4:**610–618.
502. **Slock, J., D. P. Stahly, C.-Y. Han, E. W. Six, and I. P. Crawford.** 1990. An apparent *Bacillus subtilis* folic acid biosynthetic operon containing *pab*, an amphibolic *trpG* gene, a third gene required for synthesis of *para*-aminobenzoic acid, and the dihydropteroate synthase gene. *J. Bacteriol.* **172:**7211–7226.
503. **Sloma, A., A. Ally, D. Ally, and J. Pero.** 1988. Gene encoding a minor extracellular protease in *Bacillus subtilis. J. Bacteriol.* **170:**5557–5563.
504. **Sloma, A., C. F. Rudolph, G. A. Rufo, Jr., B. J. Sullivan, K. A. Theriault, D. Ally, and J. Pero.** 1990. Gene encoding a novel extracellular metalloprotease in *Bacillus subtilis. J. Bacteriol.* **172:**1024–1029.
505. **Sloma, A., G. A. Rufo, Jr., C. F. Rudolph, B. J. Sullivan, K. A. Theriault, and J. Pero.** 1990. Bacillopeptidase F of *Bacillus subtilis*: purification of the protein and cloning of the gene. *J. Bacteriol.* **172:**1470–1477. (Erratum, **172:**5520–5521.)
506. **Sloma, A., G. A. Rufo, Jr., K. A. Theriault, M. Dwyer, S. W. Wilson, and J. Pero.** 1991. Cloning and characterization of the gene for an additional extracellular serine protease of *Bacillus subtilis. J. Bacteriol.* **173:**6889–6895.
507. **Smith, I.** 1982. The translational apparatus of *Bacillus subtilis*, p. 111–145. *In* D. A. Dubnau (ed.), *Molecular Biology of the Bacilli*. Academic Press, Inc., New York.
508. **Smith, I., and P. Paress.** 1978. Genetic and biochemical characterization of kirromycin resistance mutations in *Bacillus subtilis. J. Bacteriol.* **135:**1101–1117.
509. **Smith, I., P. Paress, K. Cabane, and E. Dubnau.** 1980. Genetics and physiology of the *rel* system of *Bacillus subtilis. Mol. Gen. Genet.* **178:**271–279.
510. **Smith, I., and H. Smith.** 1973. Location of the SPO2 attachment site and the bryamycin resistance marker on the *Bacillus subtilis* chromosome. *J. Bacteriol.* **114:**1138–1142.
511. **Smith, I., D. Weiss, and S. Pestka.** 1976. A micrococcin-resistant mutant of *Bacillus subtilis*: localization of resistance to the 50S subunit. *Mol. Gen. Genet.* **144:**231–233.
512. **Smith, M. C. M., L. Czaplewsi, A. K. North, S. Baumberg, and P. G. Stockley.** 1989. Sequences required for regulation of arginine biosynthesis promoters are conserved between *Bacillus subtilis* and *Escherichia coli. Mol. Microbiol.* **3:**23–28.
513. **Smith, M. C. M., A. Mountain, and S. Baumberg.** 1986. Cloning in *Escherichia coli* of a *Bacillus subtilis* arginine repressor gene through its ability to confer structural stability on a fragment carrying genes of arginine biosynthesis. *Mol. Gen. Genet.* **205:**176–182.
514. **Smith, M. C. M., A. Mountain, and S. Baumberg.** 1990. Nucleotide sequence of the *Bacillus subtilis argC* gene encoding *N*-acetylglutamate-gamma-semialdehyde dehydrogenase. *Nucleic Acids Res.* **18:**4595.
515. **Sonenshein, A. L.** Unpublished data.
516. **Sonenshein, A. L., and H. B. Alexander.** 1979. Initiation of transcription *in vitro* is inhibited by lipiarmycin. *J. Mol. Biol.* **127:**55–72.

517. **Sonenshein, A. L., H. B. Alexander, D. M. Rothstein, and S. H. Fisher.** 1977. Lipiarmycin-resistant ribonucleic acid polymerase mutants of *Bacillus subtilis*. *J. Bacteriol.* **132:**73–79.
518. **Song, B. H., and J. Neuhard.** 1989. Chromosomal location, cloning and nucleotide sequence of the *Bacillus subtilis cdd* gene encoding cytidine/deoxycytidine deaminase. *Mol. Gen. Genet.* **216:**462–468.
519. **Sorokin, A., and S. D. Ehrlich.** Unpublished data.
520. **Staal, S. P., and J. A. Hoch.** 1972. Conditional dihydrostreptomycin resistance in *Bacillus subtilis*. *J. Bacteriol.* **110:**202–207.
521. **Stahl, M. L., and E. Ferrari.** 1984. Replacement of the *Bacillus subtilis* subtilisin structural gene with an in-vitro-derived deletion mutation. *J. Bacteriol.* **158:**411–418.
522. **Steinberg, W., and C. Anagnostopoulos.** 1971. Biochemical and genetic characterization of a temperature-sensitive tryptophanyl-transfer ribonucleic acid synthetase mutant of *Bacillus subtilis*. *J. Bacteriol.* **105:**6–19.
523. **Steinmetz, M.** Unpublished data.
524. **Steinmetz, M., S. Aymerich, G. Gonzy-Treboul, and D. LeCoq.** 1988. Levansucrase induction by sucrose in *Bacillus subtilis* involves an antiterminator. Homology with the *Escherichia coli bgl* operon, p. 11–15. *In* A. T. Ganesan and J. A. Hoch (ed.), *Genetics and Biotechnology of Bacilli*, vol. 2. Academic Press, Inc., San Diego.
525. **Steinmetz, M., F. Kunst, and R. Dedonder.** 1976. Mapping of mutations affecting synthesis of exocellular enzymes in *Bacillus*. Identity of *sacU*, *amyB* and *pap* mutations. *Mol. Gen. Genet.* **148:**281–285.
526. **Steinmetz, M., D. LeCoq, S. Aymerich, G. Gonzy-Treboul, and P. Gay.** 1985. The DNA sequence of the gene for the secreted *Bacillus subtilis* enzyme levansucrase and its genetic control sites. *Mol. Gen. Genet.* **200:**220–228.
527. **Stephens, M. A., N. Lang, K. Sandman, and R. Losick.** 1984. A promoter whose utilization is temporally regulated during sporulation in *Bacillus subtilis*. *J. Mol. Biol.* **176:**333–348.
528. **Stevens, C. M., R. Daniel, N. Illing, and J. Errington.** 1992. Characterization of a sporulation gene, *spoIV A*, involved in spore coat morphogenesis in *Bacillus subtilis*. *J. Bacteriol.* **174:**586–594.
529. **Stragier, P.** Unpublished data.
530. **Stragier, P., J. Bouvier, C. Bonamy, and J. Szulmajster.** 1984. A developmental gene product of *Bacillus subtilis* homologous to the sigma factor of *Escherichia coli*. *Nature* (London) **312:**376–378.
531. **Stragier, P., B. Kunkel, L. Kroos, and R. Losick.** 1989. Chromosomal rearrangement generating a composite gene for a developmental transcription factor. *Science* **243:**507–512.
532. **Stranathan, M. C., K. W. Bayles, and R. E. Yasbin.** 1990. The nucleotide sequence of the *recE*$^+$ gene of *Bacillus subtilis*. *Nucleic Acids Res.* **18:**4249.
533. **Strauch, M. A., A. I. Aronson, S. W. Brown, H. J. Schreier, and A. L. Sonenshein.** 1988. Sequence of the *Bacillus subtilis* glutamine synthetase gene region. *Gene* **71:**257–265.
534. **Struck, J. C. R., J. C. Alonso, H. Y. Toschka, and V. A. Erdmann.** 1990. The *Bacillus subtilis* small cytoplasmic RNA gene and '*dnaX*' map near the chromosomal replication origin. *Mol. Gen. Genet.* **222:**470–472.
535. **Struck, J. C. R., R. K. Hartmann, H. Y. Toschka, and V. A. Erdmann.** 1989. Transcription and processing of *Bacillus subtilis* small cytoplasmic RNA. *Mol. Gen. Genet.* **215:**478–482.
536. **Struck, J. C. R., R. A. Lempicki, H. Y. Toschka, V. A. Erdmann, and M. J. Fournier.** 1990. *Escherichia coli* 4.5S RNA gene function can be complemented by heterologous bacterial RNA genes. *J. Bacteriol.* **172:**1284–1288.
537. **Struck, J. C. R., D. W. Vogel, N. Ulbrich, and V. A. Erdmann.** 1988. A *dnaZX*-like open reading frame downstream from the *Bacillus subtilis scRNA* gene. *Nucleic Acids Res.* **16:**2720.
538. **Suh, J. W., S. A. Boylan, and C. W. Price.** 1986. Gene for the alpha subunit of *Bacillus subtilis* RNA polymerase maps in the ribosomal protein gene cluster. *J. Bacteriol.* **168:**65–71.
539. **Suh, J. W., S. A. Boylan, S. M. Thomas, K. M. Dolan, D. B. Oliver, and C. W. Price.** 1990. Isolation of a *secY* homologue from *Bacillus subtilis*: evidence for a common protein export pathway in eubacteria. *Mol. Microbiol.* **4:**305–314.
540. **Sukhodolets, V. V., Ya V. Flyakh, and E. V. Rumyantseva.** 1983. Mapping of mutations in genes for nucleoside catabolism on the *Bacillus subtilis* chromosome. *Genetica* (USSR) **19:**221–226.
541. **Sun, D., and P. Setlow.** 1991. Cloning, nucleotide sequence, and expression of the *Bacillus subtilis ans* operon, which codes for L-asparaginase and L-aspartase. *J. Bacteriol.* **173:**3831–3845.
542. **Sun, D., P. Stragier, and P. Setlow.** 1989. Identification of a new sigma-factor involved in compartmentalized gene expression during sporulation of *Bacillus subtilis*. *Genes Dev.* **3:**141–149.
543. **Sun, D., and I. Takahashi.** 1982. Genetic mapping of catabolite-resistant mutants of *Bacillus subtilis*. *Can J. Microbiol.* **28:**1242–1251.
544. **Sun, D., and I. Takahashi.** 1984. A catabolite-resistance mutation is localized in the *rpo* operon of *Bacillus subtilis*. *Can. J. Microbiol.* **30:**423–429.
545. **Sussman, M. D., and P. Setlow.** 1991. Cloning, nucleotide sequence, and regulation of the *Bacillus subtilis gpr* gene, which codes for the protease that initiates degradation of small, acid-soluble proteins during spore germination. *J. Bacteriol.* **173:**291–300.
546. **Sutrina, S. L., P. Reddy, M. H. Saier, Jr., and J. Reizer.** 1990. The glucose permease of *Bacillus subtilis* is a single polypeptide chain that functions to energize the sucrose permease. *J. Biol. Chem.* **265:**18581–18589.
547. **Tanaka, K., M. Tamaki, S. Osawa, A. Kimura, and R. Takata.** 1973. Erythromycin resistant mutants of *Bacillus subtilis*. *Mol. Gen. Genet.* **127:**157–161.
548. **Tanaka, T., and M. Kawata.** 1988. Cloning and characterization of *Bacillus subtilis iep*, which has positive and negative effects on production of extracellular proteases. *J. Bacteriol.* **170:**3593–3600.
549. **Todd, J. A., A. N. Roberts, K. Johnstone, P. J. Piggot, G. Winter, and D. J. Ellar.** 1986. Reduced heat resistance of mutant spores after cloning and mutagenesis of the *Bacillus subtilis* gene encoding penicillin-binding protein 5. *J. Bacteriol.* **167:**257–262.
550. **Toma, S., M. Del Bue, A. Pirola, and G. Grandi.** 1986. *nprRI* and *nprR2* regulatory regions for neutral protease expression in *Bacillus subtilis*. *J. Bacteriol.* **167:**740–743.
551. **Tominaga, A., and Y. Kobayashi.** 1978. Kasugamycin-resistant mutations of *Bacillus subtilis*. *J. Bacteriol.* **135:**1149–1150.
552. **Trach, K., J. W. Chapman, P. J. Piggot, and J. A. Hoch.** 1985. Deduced product of the stage 0 sporulation gene spo0F shares homology with the Spo0A, OmpR and SfrA proteins. *Proc. Natl. Acad. Sci. USA* **82:**7260–7264.
553. **Trach, K., J. W. Chapman, P. Piggot, D. LeCoq, and J. A. Hoch.** 1988. Complete sequence and transcriptional analysis of the *spo0F* region of the *Bacillus subtilis* chromosome. *J. Bacteriol.* **170:**4194–4208.
554. **Trach, K., and J. A. Hoch.** 1989. The *Bacillus subtilis spo0B* stage 0 sporulation operon encodes an essential GTP-binding protein. *J. Bacteriol.* **171:**1362–1371.
555. **Trach, K. A., and J. A. Hoch.** Multisensory activation of

the phosphorelay initiating sporulation in *Bacillus subtilis*: identification and sequence of the protein kinase of the alternate pathway. *Mol. Microbiol.*, in press.
556. **Tran, L., X.-C. Wu, and S.-L. Wong.** 1991. Cloning and expression of a novel protease gene encoding an extracellular neutral protease from *Bacillus subtilis. J. Bacteriol.* **173:**6364–6372.
557. **Trautner, T. A., B. Pawlek, S. Bron, and C. Anagnostopoulos.** 1974. Restriction and modification in *Bacillus subtilis*: biological aspects. *Mol. Gen. Genet.* **131:**181–191.
558. **Trempy, J. E., C. Bonamy, J. Szulmajster, and W. G. Haldenwang.** 1985. *Bacillus subtilis* sigma factor sigma 29 is the product of the sporulation-essential gene *spoIIG. Proc. Natl. Acad. Sci. USA* **82:**4189–4192.
559. **Trowsdale, J., S. M. H. Chen, and J. A. Hoch.** 1978. Genetic analysis of phenotype revertants of *spo0A* mutants in *Bacillus subtilis*: a new cluster of ribosomal genes, p. 131–135. *In* G. Chambliss and J. C. Vary (ed.), *Spores VII*. American Society for Microbiology, Washington, D.C.
560. **Trowsdale, J., S. M. H. Chen, and J. A. Hoch.** 1979. Genetic analysis of a class of polymyxin resistant partial revertants of stage 0 sporulation mutants of *Bacillus subtilis*: map of the chromosome region near the origin of replication. *Mol. Gen. Genet.* **173:**61–70.
561. **Trowsdale, J., and D. A. Smith.** 1975. Isolation, characterization, and mapping of *Bacillus subtilis* 168 germination mutants. *J. Bacteriol.* **123:**85–95.
562. **Van Alstyne, D., and M. I. Simon.** 1971. Division mutants of *Bacillus subtilis*: isolation and PBS1 transduction of division-specific markers. *J. Bacteriol.* **108:**1366–1379.
563. **Van der Oost, C., C. Von Wachenfeld, L. Hederstedt, and M. Saraste.** 1991. *Bacillus subtilis* cytochrome oxidase mutants: biochemical analysis and genetic evidence for two aa_3-type oxidases. *Mol. Microbiol.* **5:**2063–2072.
564. **Van Dijl, J. M., A. de Jong, J. Vehmaanperà, G. Venema, and S. Bron.** 1992. Signal peptidase I of *Bacillus subtilis*: patterns of amino acids on prokaryotic and eukaryotic type I signal peptidases. *EMBO J.* **11:**2819–2828.
565. **Van Hoy, B. E., and J. A. Hoch.** 1990. Characterization of the *spoIVB* and *recN* loci of *Bacillus subtilis. J. Bacteriol.* **172:**1306–1311.
566. **van Sinderen, D.** Unpublished data.
567. **van Sinderen, D., S. Withoff, H. Boels, and G. Venema.** 1990. Isolation and characterization of *comL*, a transcription unit involved in competence development of *Bacillus subtilis. Mol. Gen. Genet.* **224:**396–404.
568. **Vander Horn, P. B., and S. A. Zahler.** 1992. Cloning and nucleotide sequence of the leucyl-tRNA synthetase gene of *Bacillus subtilis. J. Bacteriol.* **174:**3928–3935.
569. **Vandeyar, M. A., and S. A. Zahler.** 1986. Chromosomal insertions of Tn*917* in *Bacillus subtilis. J. Bacteriol.* **167:**530–534.
570. **Vapnek, D., and S. Greer.** 1971. Suppression by derepression in threonine dehydratase-deficient mutants of *Bacillus subtilis. J. Bacteriol.* **106:**615–625.
571. **Vapnek, D., and S. Greer.** 1971. Minor threonine dehydratase encoded within the threonine synthetic region of *Bacillus subtilis. J. Bacteriol.* **106:**983–993.
572. **Vary, J.** Unpublished data.
573. **Vazquez-Ramos, J. M., and J. Mandelstam.** 1981. Oxolinic acid-resistant mutants of *Bacillus subtilis. J. Gen. Microbiol.* **127:**1–9.
574. **Viaene, A., and P. Dhaese.** 1989. Sequence of the glyceraldehyde-3-phosphate dehydrogenase gene from *Bacillus subtilis. Nucleic Acids Res.* **17:**1251.
575. **Vold, B. S.** 1985. Structure and organization of genes for transfer RNA in *Bacillus subtilis. Microbiol. Rev.* **49:**71–80.
576. **von Wachenfeldt, C., and L. Hederstedt.** 1990. *Bacillus subtilis* 13-kilodalton cytochrome c-550 encoded by *cccA* consists of a membrane-anchor and a heme domain. *J. Biol. Chem.* **265:**13939–13948.
577. **Vosman, B., G. Kuiken, J. Kooistra, and G. Venema.** 1988. Transformation in *Bacillus subtilis*: involvement of the 17-kilodalton DNA-entry nuclease and the competence-specific 18-kilodalton protein. *J. Bacteriol.* **170:** 3703–3710.
578. **Wabiko, H., K. Ochi, D. M. Nguyen, E. R. Allen, and E. Freese.** 1988. Genetic mapping and physiological consequences of *metE* mutations of *Bacillus subtilis. J. Bacteriol.* **170:**2705–2710.
579. **Wainscott, V. J., and J. F. Kane.** 1976. Dihydrofolate reductase in *Bacillus subtilis*, p. 208–213. *In* D. Schlessinger (ed.), *Microbiology—1976*. American Society for Microbiology, Washington, D.C.
580. **Walter, J., M. Noyer-Weidner, and T. A. Trautner.** 1990. The amino acid sequence of the CCGG recognizing DNA methyltransferase M.*Bsu*F1: implications for the analysis of sequence recognition by cytosine DNA methyltransferases. *EMBO J.* **9:**1007–1013.
581. **Walton, D. A., A. Moir, R. Morse, I. Roberts, and D. A. Smith.** 1984. The isolation of a lambda phage carrying DNA from the histidine and isoleucine-valine regions of the *Bacillus subtilis* chromosome. *J. Gen. Microbiol.* **130:**1577–1586.
582. **Wang, L.-F., and R. H. Doi.** 1990. Complex character of *senS*, a novel gene regulating expression of extracellular-protein genes of *Bacillus subtilis. J. Bacteriol.* **172:** 1939–1947.
583. **Wang, L. F., and R. H. Doi.** 1986. Nucleotide sequence and organization of *Bacillus subtilis* RNA polymerase major sigma (σ^{43}). *Nucleic Acids Res.* **14:**4293–4307.
584. **Wang, L. F., C. W. Price, and R. H. Doi.** 1985. *Bacillus subtilis dnaE* encodes a protein homologous to DNA primase of *Escherichia coli. J. Biol. Chem.* **260:**3368–3372.
585. **Warburg, R. J., I. Mahler, D. J. Tipper, and H. O. Halvorson.** 1984. Cloning the *Bacillus subtilis* 168 *aroC* gene encoding dehydroquinase. *Gene* **32:**57–66.
586. **Warburg, R. J., and A. Moir.** 1981. Properties of a mutant of *Bacillus subtilis* 168 in which spore germination is blocked at a late stage. *J. Gen. Microbiol.* **124:** 243–253.
587. **Ward, J. B., Jr., and S. A. Zahler.** 1973. Genetic studies of leucine biosynthesis in *Bacillus subtilis. J. Bacteriol.* **116:**719–726.
588. **Wawrousek, E. F., and J. N. Hansen.** 1983. Structure and organization of a cluster of six tRNA genes in the space between tandem ribosomal RNA gene sets in *Bacillus subtilis. J. Biol. Chem.* **258:**291–298.
589. **Wawrousek, E. F., N. Narasimhan, and J. N. Hansen.** 1984. Two large clusters with thirty-seven transfer RNA genes adjacent to ribosomal RNA gene sets in *Bacillus subtilis*: sequence and organization of *trrnD* and *trrnE* gene clusters. *J. Biol. Chem.* **259:**3694–3702.
590. **Weinrauch, Y., R. Penchev, E. Dubnau, I. Smith, and D. Dubnau.** 1990. A *Bacillus subtilis* regulatory gene product for genetic competence and sporulation resembles sensor protein members of the bacterial two-component signal-transduction systems. *Genes Dev.* **4:**860–872.
591. **Wetzstein, M., and W. Schumann.** 1990. Nucleotide sequence of a *Bacillus subtilis* gene homologous to the *grpE* gene of *E. coli* located immediately upstream of the *dnaK* gene. *Nucleic Acids Res.* **18:**1289.
592. **Wetzstein, M., V. Völker, J. Dedio, S. Löbau, P. Zuber, M. Schisesswohl, C. Herget, M. Hecker, and W. Schumann.** 1992. Cloning, sequencing, and molecular analysis of the *dnaK* locus from *Bacillus subtilis. J. Bacteriol.* **174:**3300–3310.
593. **Whiteman, P., C. Marks, and E. Freese.** 1980. The

sodium effect of *Bacillus subtilis* growth on aspartate. *J. Gen. Microbiol.* **119:**493–504.

594. **Widom, R. L., E. D. Jarvis, G. LaFauci, and R. Rudner.** 1988. Instability of rRNA operons in *Bacillus subtilis. J. Bacteriol.* **170:**605–610.
595. **Wilhelm, M.** 1984. The cloning of *Bacillus subtilis* xylose isomerase xylulokinase in *Escherichia coli* genes by IS5-mediated expression. *EMBO J.* **3:**2555–2560.
596. **Wilhelm, M., and C. P. Hollenberg.** 1985. Nucleotide sequence of the *Bacillus subtilis* xylose isomerase gene: extensive homology between the *Bacillus* and *Escherichia coli* enzyme. *Nucleic Acids Res.* **13:**5717–5722.
597. **Williams, G., and I. Smith.** 1979. Chromosomal mutants causing resistance to tetracycline in *Bacillus subtilis. Mol. Gen. Genet.* **177:**23–29.
598. **Williams, M. T., and F. E. Young.** 1977. Temperate *Bacillus subtilis* bacteriophage phi-3T: chromosomal attachment site and comparison with temperate bacteriophage phi-105 and SPO2. *J. Virol.* **21:**522–529.
599. **Wu, J.-J., and P. J. Piggot.** 1990. Regulation of late expression of the *Bacillus subtilis spoIIA* locus: evidence that it is cotranscribed with the gene for a putative penicillin-binding protein, p. 321–327. *In* A. T. Ganesan and J. A. Hoch (ed.), *Genetics and Biotechnology of Bacilli*, vol. 3. Academic Press, Inc., New York.
600. **Wu., J.-J., R. Schuch, and P. J. Piggot.** 1992. Characterization of a *Bacillus subtilis* sporulation operon that includes genes for a RNA polymerase σ-factor and for a putative DD-carboxypeptidase. *J. Bacteriol.* **174:**4885–4892.
601. **Wu, X.-C., S. Nathoo, A. S.-H. Pang, T. Carne, and S.-L. Wong.** 1990. Cloning, genetic organization, and characterization of a structural gene encoding bacillopeptidase F from *Bacillus subtilis. J. Biol. Chem.* **265:**6845–6850.
602. **Xu, G.** Unpublished data.
603. **Yamada, Y., M. Ohki, and H. Ishikura.** 1983. The nucleotide sequence of *Bacillus subtilis* tRNA genes. *Nucleic Acids Res.* **11:**3037–3045.
604. **Yamamoto, J., M. Shimizu, and K. Yamane.** 1991. Molecular cloning and analysis of nucleotide sequence of the *Bacillus subtilis lysA* gene region using *B. subtilis* phage vectors and a multi-copy plasmid, pUB110. *Agric. Biol. Chem.* **55:**1615–1626.
605. **Yang, M., E. Ferrari, E. Chen, and D. J. Henner.** 1986. Identification of the pleiotropic *sacQ* gene of *Bacillus subtilis. J. Bacteriol.* **166:**113–119.
606. **Yang, M., A. Galizzi, and D. Henner.** 1983. Nucleotide sequence of the amylase gene from *Bacillus subtilis. Nucleic Acids Res.* **11:**237–249.
607. **Yang, M., H. Shimotsu, E. Ferrari, and D. J. Henner.** 1987. Characterization and mapping of the *Bacillus subtilis prtR* gene. *J. Bacteriol.* **169:**434–437.
608. **Yang, M. Y., E. Ferrari, and D. J. Henner.** 1984. Cloning of the neutral protease gene of *Bacillus subtilis* and the use of the cloned gene to create an in vitro-derived deletion mutation. *J. Bacteriol.* **160:**16–21.
609. **Yasbin, R. E., D. Cheo, and K. W. Bayles.** 1991. The SOB system of *Bacillus subtilis*: a global regulon involved in DNA repair and differentiation. *Res. Microbiol.* **142:**885–892.
610. **Yazdi, M. A., and A. Moir.** 1990. Characterization and cloning of the *gerC* locus of *Bacillus subtilis* 168. *J. Gen. Microbiol.* **136:**1335–1342.
611. **Ying, C., and G. W. Ordal.** 1989. Nucleotide sequence and expression of *cheF*, an essential gene for chemotaxis in *Bacillus subtilis. J. Bacteriol.* **171:**1631–1637.
612. **Ying, C., F. Scoffone, A. M. Albertini, A. Galizzi, and G. W. Ordal.** 1991. Properties of the *Bacillus subtilis* chemotaxis protein CheF, a homolog of the *Salmonella typhimurium* flagellar protein FliJ. *J. Bacteriol.* **173:**3584–3586.
613. **Yon, J. R., R. L. Sammons, and D. A. Smith.** 1989. Cloning and sequencing of the *gerD* gene of *Bacillus subtilis. J. Gen. Microbiol.* **135:**3431–3445.
614. **Yoshikawa, H., and R. H. Doi.** 1990. Sequence of the *Bacillus subtilis* spectinomycin resistance gene region. *Nucleic Acids Res.* **18:**1647.
615. **Young, F. E., D. Smith, and B. E. Reilly.** 1969. Chromosomal location of genes regulating resistance to bacteriophage in *Bacillus subtilis. J. Bacteriol.* **98:**1087–1097.
616. **Young, M.** 1975. Genetic mapping of sporulation operons in *Bacillus subtilis* using a thermosensitive sporulation mutant. *J. Bacteriol.* **122:**1109–1116.
617. **Young, M.** 1976. Use of temperature-sensitive mutants to study gene expression during sporulation in *Bacillus subtilis. J. Bacteriol.* **126:**928–936.
618. **Youngman, P.** Unpublished data.
619. **Yudkin, M. D., and L. Turley.** 1980. Suppression of asporogeny in *Bacillus subtilis*: allele-specific suppression of a mutation in the *spoIIA* locus. *J. Gen. Microbiol.* **121:**69–78.
620. **Yuki, S.** 1975. The chromosomal location of the structure gene for amylase in *Bacillus subtilis. Jpn. J. Genet.* **50:**155–157.
621. **Zahler, S. A.** 1978. An adenine-thiamin auxotrophic mutant of *Bacillus subtilis. J. Gen. Microbiol.* **107:**199–201.
622. **Zahler, S. A.** Unpublished data.
623. **Zahler, S. A., L. G. Benjamin, B. S. Glatz, P. F. Winter, and B. J. Goldstein.** 1976. Genetic mapping of *alsA*, *alsR*, *thyA*, *kauA*, and *citD* markers in *Bacillus subtilis*, p. 35–43. *In* D. Schlessinger (ed.), *Microbiology—1976*. American Society for Microbiology, Washington, D.C.
624. **Zahler, S. A., R. Z. Korman, R. Rosenthal, and H. E. Hemphill.** 1977. *Bacillus subtilis* bacteriophage SPβ: localization of the prophage attachment site, and specialized transduction. *J. Bacteriol.* **129:**556–558.
625. **Zahler, S. A., N. Najimudin, D. S. Kessler, and M. A. Vandeyar.** 1990. Alpha-acetolactate synthesis by *Bacillus subtilis*, p. 25–42. *In* Z. Barak, D. M. Chipman, and J. V. Schloss (ed.), *Biosynthesis of Branched Chain Amino Acids*. Verlags-gesellschaft, Weinheim, Germany.
626. **Zeigler, D. R., B. E. Burke, R. M. Pfister, and D. H. Dean.** 1987. Genetic mapping of cadmium resistance mutations in *Bacillus subtilis. Curr. Microbiol.* **16:**163–165.
627. **Zuberi, A. R., C. Ying, D. S. Bischoff, and G. W. Ordal.** 1991. Gene-protein relationships in the flagellar hook-basal body complex of *Bacillus subtilis*: sequences of the *flgB*, *flgC*, *flgG*, *fliE*, and *fliF* genes. *Gene* **101:**23–31.
628. **Zuberi, A. R., C. Ying, H. M. Parker, and G. W. Ordal.** 1990. Transposon Tn*917lacZ* mutagenesis of *Bacillus subtilis*: identification of two new loci required for motility and chemotaxis. *J. Bacteriol.* **172:**6841–6848.
629. **Zukowski, M., L. Miller, P. Cogswell, and K. Chen.** 1988. Inducible expression system based on sucrose metabolism genes of *Bacillus subtilis*, p. 17–22. *In* A. T. Ganesan and J. A. Hoch (ed.), *Genetics and Biotechnology of Bacilli*, vol. 2. Academic Press, Inc., San Diego.

30. Physical Map of the *Bacillus subtilis* 168 Chromosome

MITSUHIRO ITAYA

Since the construction of a physical map of the chromosome of *Escherichia coli* K-12 in the pioneering work of Kohara et al. (17) and Smith et al. (29), physical maps of the chromosomes of a variety of bacterial species have been determined (19). A standard protocol for chromosome mapping is the use of pulsed-field gel electrophoresis (PFGE) (7, 26) to separate very large segments of DNA generated by restriction enzymes that cleave the bacterial chromosome infrequently (28). Recently, PFGE in combination with a new chromosome mapping strategy based on the genetic manipulation of restriction sites in the chromosome has been used to generate a detailed physical map of the chromosome of *Bacillus subtilis* (15). The merging of this physical map with the extensive genetic map of the *B. subtilis* chromosome (see chapter 29) is expected to provide important insights into the organization and rearrangement of genes involved in the processes of secretion, competence-mediated genetic recombination, and sporulation and in other distinctive aspects of the biology of this gram-positive bacterium. Here, I summarize information on the physical and genetic maps of the chromosome, the integration of the physical and genetic maps, and the unique features of the *B. subtilis* genome.

CONSTRUCTION OF THE *B. SUBTILIS* PHYSICAL MAP

Basic Strategy for *B. subtilis* Genome Map Construction

Two basic physical maps of bacterial genomes have been developed: (i) a long-range restriction map generated by sequence-specific restriction endonucleases that cleave the bacterial chromosome infrequently and (ii) a detailed restriction map derived from the assembly of overlapping cloned DNA segments (contiguous DNA libraries, or "contig library"). These methods have been briefly summarized elsewhere (15, 19, 28). The physical map of the *B. subtilis* 168 chromosome has been constructed by the first method (15). Because the strategy and ideas for its construction are unique and rather specific to *B. subtilis* 168, they will be described briefly.

B. subtilis 168 possesses the remarkable characteristic of being competent for DNA uptake followed by homologous recombination between the invading DNA and homologous DNA residing on the chromosome (see chapter 39). Because of the extremely high efficiency and fidelity of recombination, the probability of cotransformation of two unlinked genetic markers is remarkably high. This phenomenon, known as "congression," has been widely used to transfer genetic markers between *B. subtilis* strains (see chapter 39). An extremely powerful and useful variation of classical congression occurs when one or a few specific chromosomal segments are used rather than the entire chromosomal DNA. A novel protocol based on this idea has been developed; this protocol can be used to change any particular cloned *B. subtilis* sequence (14). As outlined in Fig. 1, three transformation steps are required to accomplish this process. In the example shown, two hypothetical DNA segments are used. The first is the hypothetical target gene *pay*. A selectable drug resistance marker (e.g., the chloramphenicol acetyltransferase [*cat*] gene) is cloned into the *pay* gene on a plasmid vector. The first transformation (step I) is simply to obtain a strain whose *pay* gene is marked with the *cat* gene. A second hypothetical gene, *tax*, has been cloned and modified to carry the neomycin resistance gene, *neo*, to yield *tax*::*neo*. Transformation number two (step II) is performed by using both *tax*::*neo* DNA and *pay* mutant DNA (*pay* mutant DNA contains the desired mutational change, e.g., a new restriction enzyme site, in the *pay* gene). Transformants are selected for resistance to neomycin and screened for chloramphenicol sensitivity. Chloramphenicol-sensitive neomycin-resistant cells have the genotype *pay tax*::*neo*. Step III is performed to change the genotype to *pay tax*$^+$, leaving the single change at the *pay* locus as the only difference between the new strain and the original parental strain. If the hypothetical *tax* gene were in actuality a gene such as *leuB*, the *tax*::*neo* strains would require leucine for growth. A positive selection could be used in step III, in which *leuB*$^+$ (*tax*$^+$) DNA would be used for transformation of the *pay tax*::*neo* strain and transformants could be selected as leucine prototrophs.

This procedure has been termed "gene-directed mutagenesis" (14). Other than the fact that only a single mutation distinguishes the final strain from the parental strain, a second important feature of the procedure is that the process can go through many cycles to continually modify a particular strain. It is always possible to use the same drug resistance markers, since they are removed during steps II and III. Changes in the migration pattern of DNA fragments in gels after removal of one, two, or three restriction enzyme sites are shown in Fig. 2.

As depicted in Fig. 3, specific nucleotide sequences for *Not*I and *Sfi*I restriction enzymes have been elim-

Mitsuhiro Itaya • Mitsubishi Kasei Institute of Life Sciences, 11, Minamiooya, Machida-shi, Tokyo 194, Japan.

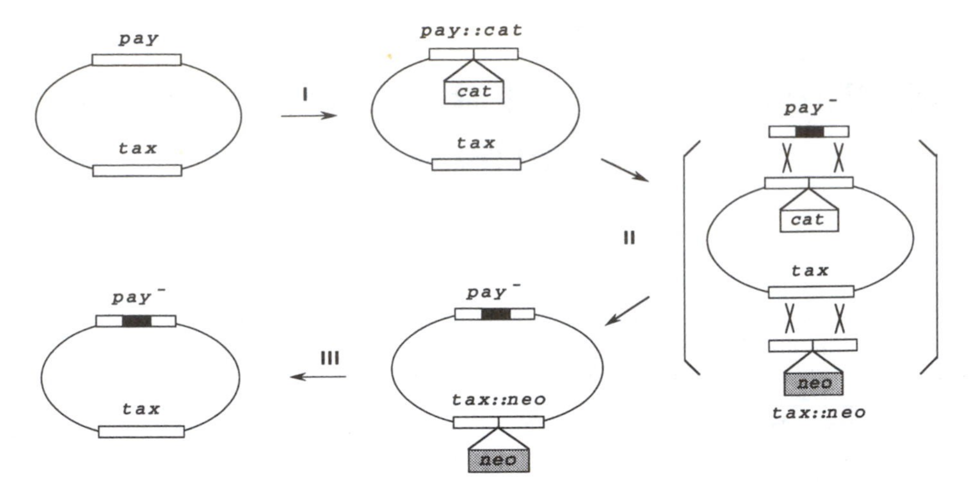

Figure 1. Conversion of nucleotide sequences on the *B. subtilis* chromosome. Circles represent the *B. subtilis* chromosome. Two hypothetical genes, *pay* and *tax*, are in boxed regions, and *cat* and *neo* denote the drug resistance gene cassettes. The DNA segments used in transformation steps I through III are prepared in *E. coli* and linearized. × × indicates a double crossover followed by homologous recombination. A net change that the last strain acquires is only the *pay* mutations. Details are given in the text.

inated in the chromosome of *B. subtilis* 168 to aid in establishing a detailed restriction map for the entire genome. This type of linkage of certain fragments already provides some order to the map. To construct many of these *Not*I or *Sfi*I changes, it was necessary to isolate so-called *Not*I- or *Sfi*I-linking clones (cloned DNA fragments carrying *Not*I or *Sfi*I sites). A total of 13 *Sfi*I-linking clones and 45 *Not*I-linking clones were obtained. The sum of the chromosomal segments in these linking clones was about 100 kb of the chromosome of *B. subtilis* 168. These clones were used as either target (*pay*-like) DNA or catalyst (*tax*-like) DNA in gene-directed mutagenesis.

I will not repeat the details used for constructing the physical map of *B. subtilis* 168, but the map we generated differs from other current maps (as summarized in reference 19). Two other groups have reported partial *Not*I (33) or *Sfi*I (2) maps. Kobayashi and co-workers (16) are working to generate a physical map for sites of restriction enzymes that employ six-base recognition sequences. Also, Ehrlich and co-workers are constructing a physical map by utilizing yeast artificial chromosome vector (YAC vector; 6, 27; chapter 31). In this chapter, I will use the map constructed by Itaya and Tanaka (15) because it is com-

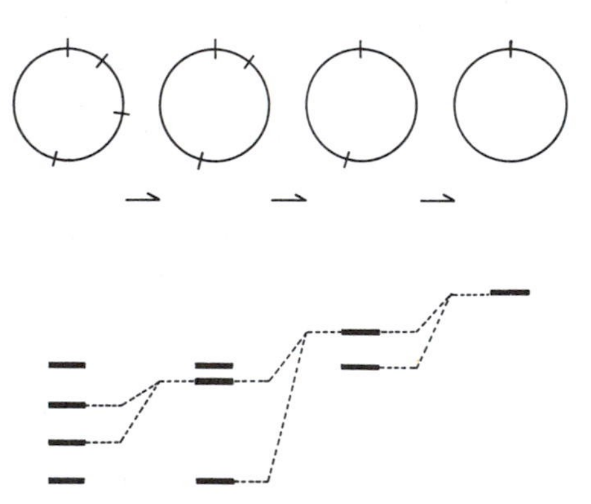

Figure 2. Strategy for determination of restriction enzyme sites on the chromosome. Circles represent the *B. subtilis* chromosome, and short bars around the circles represent recognition sites of restriction enzymes. The sites can be eradicated, i.e., converted to other sequences, from the chromosome by using the gene-directed mutagenesis method described in the legend to Fig. 1. The method can be sequentially applied, and the number of recognition sites can be decreased progressively from left to right. Patterns of fragments after enzyme digestion and gel electrophoresis are shown at the bottom. Dotted lines indicate changes of fragments resulting from the loss of the recognition site. Details are described in reference 15.

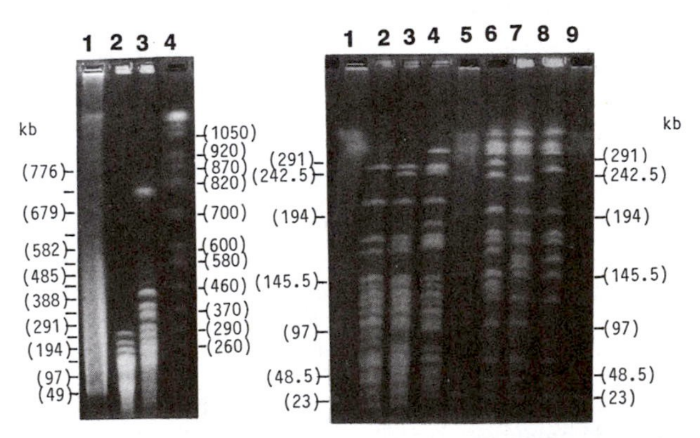

Figure 3. Separation of *B. subtilis* chromosome DNA by PFGE after *Sfi*I or *Not*I digestion. (Left) Lane 1, concatemeric lambda DNA size markers. Chromosomal DNA from CU741, a wild-type *B. subtilis*, was digested with *Not*I (lane 2), *Sfi*I (lane 3), or DNA from *Saccharomyces cerevisiae* (lane 4). (Right) Lanes 1, 5, and 9 contain concatemeric lambda DNA plus lambda DNA digested with *Hin*dIII as size standards. *Not*I-digested chromosomal DNA was from CU741 (lane 2), BEST4041 (lane 3), or BEST4101 (lane 4). *Sfi*I-digested chromosomal DNA was from OA101 (lanc 6), BEST3015 (lane 7), or BEST3019 (lane 8). Sizes (in kilobase pairs) are in parentheses. *Sfi*I or *Not*I generates 26 or more than 80 fragments. Designations for each fragment are given in Table 1. Derivatives constructed by progressive removal of *Not*I (BEST4041 and BEST4101) or *Sfi*I (BEST3015 and BEST3019) are described in referencc 15.

plete and sufficiently accurate for the discussions which follow.

Choice of Restriction Enzymes

The primary requirement for choosing a particular enzyme is the frequency of appearance of its recognition site in the genome, which is reflected in the number of fragments observed in PFGE. Most six-base cutters generate too many fragments to be resolved on agarose gels (21). When used in conjunction with cloned DNA pieces, six-base cutters can be useful (e.g., Kohara-type mapping [17]). Recently, an 18-base sequence recognition endonuclease, omeganuclease I-*Sce*I(TAGGGATAACAGGGTAAT) (referred to here as I-*Sce*I) became commercially available (32). A site for this nuclease is absent in the *B. subtilis* 168 genome (13). As shown in Fig. 3, digestion with *Sfi*I(GGCCNN NNNGGCC) produces 26 fragment, whereas digestion with *Not*I(GCGGCCGC) yields more than 80 DNA fragments. Other rare cutters [*Pac*I(TTAATTAA), *Avr*II (CCTAGG), and *Sse*8387I(CCTGCAGG)] generate about 60, 60, and 40 DNA fragments, respectively, from the *B. subtilis* chromosome (11). One unusual observation is illustrated in Fig. 4: *Sfi*I cleaves two sites weakly, probably because of variable nucleotide sequences within and/or near the *Sfi*I recognition site (. . .NNGGCCNNNNNGGCCNN. . .; 22). We have observed this phenomenon repeatedly, and it has been reported by others (2).

The absence of a site for I-*Sce*I (18-base recognition site) turns out to be quite useful. Integration of a site for I-*Sce*I into the *B. subtilis* 168 chromosome permits the accurate measurement of distances between known genes or sequences. Two examples of this (11, 13) have been reported. In one case, an I-*Sce*I site was inserted at the *recF* locus, near the *oriC* region, and a second site was inserted into the region for type II DNA membrane binding (13). The 49-kb fragment produced after digestion of DNA from this doubly modified strain unequivocally gives the physical distance between these two loci. Similarly, the disparity in lengths of the two regions between *oriC* and *terC* observed with the *Sfi*I-*Not*I-generated physical map (15) has been confirmed by insertion of I-*Sce*I sites at appropriate positions on the chromosome (11; see below for a detailed description of the relative locations of *oriC* and *terC*). In the future, it should be possible to analyze *B. subtilis* chromosomes whose structures are dramatically altered either by unclear events (e.g., the *trpE26* mutant change described by Anagnostopoulos [3]) or by some of the manipulations to be described below.

Genome Size

The total genome of the *B. subtilis* 168 chromosome is estimated to be 4,188 kb, which is slightly larger than the original estimate of 4,165 kb (see reference 15 and Table 1). The difference between these two values is due to a 23-kb deletion found in the strain originally used for construction of the map. Amjad et al. also reported a value of 4,188 kb for a *B. subtilis* strain (2). The precision with which we know the size of the chromosome is quite good. Size resolution of each fragment separated by PFGE is about 1% (11). Also, the sums of the sizes of the *Not*I fragments (4, 123 plus 23 kb) and the *Sfi*I fragments (4,165 plus 23 kb) agree well (11, 15; Table 1). Of course, the exact size of the chromosome awaits determination of the complete nucleotide sequence. I use the originally determined size (4,165 kb) as the genome size throughout this article to avoid confusion.

A size of 4.165 Mbp places the *B. subtilis* chromosome in the middle of the range of 1 to 9 Mbp reported for sizes of chromosomes from other bacteria (19). It is somewhat smaller than the *Escherichia coli* chromosome (4.7 Mbp [17, 29]) and also smaller than that of *Bacillus cereus* (5.2 Mbp [18]). A number of strains of *B. subtilis* have been shown to contain rearrangements or deletions (11). To date, however, no variation of more than 4% of the standard genome has been observed except under specific growth conditions (see below).

Correlation between the Physical and Genetic Maps

The loci of 11 known genes (*abrB*, *amyR*, *cotA*, *aprE*, *spoIIG*, *terC*, *cotD*, *leuB*, *sacQ*, *degU*, and *sacP*) have been precisely placed on the physical map (15). These sites were located on the physical map by creation of additional *Sfi*I and *Not*I sites by inserting a neomycin resistance gene cassette carrying both restriction sites (12, 15). This initial result showed a good correlation between the physical and genetic maps. A relationship of 11.57 kb/degree (4,165 kb/360°) was derived. In addition to the 11 genes described above, a number of others have been localized to particular *Sfi*I and/or *Not*I DNA fragments (2, 11, 15, 33). These results are summarized in Table 2. Recent data on the locations of some genes are included in Fig. 4. All of the recent data show the same overall linearity between the physical and genetic maps except in at least three areas (Fig. 5). Part of the mismatch has also been suggested by Amjad et al. (2). The significance of these differences may become clearer as the number of mapped genes increases.

It is expected that the availability of a physical map for *B. subtilis* will facilitate mapping of any gene by simple Southern hybridization. Unpublished results in Table 2 obtained in my laboratory suggest that it is rather easy to determine to which *Not*I and/or *Sfi*I fragment any DNA hybridizes. Table 2 presents a format for describing the location of DNA mapped in this manner.

FEATURES OF THE *B. SUBTILIS* CHROMOSOME

Origins of Initiation and Termination of DNA Replication Are Not Separated by 180°

As with other bacteria, unique origins of initiation (*oriC*) and termination (*terC*) of DNA replication have been identified (see chapter 36). It has generally been assumed that following initiation of replication at *oriC*, the replication of the chromosome would proceed at an equal rate in both directions and the site for termination would be 180°C from the point of initiation (4; see chapter 29). Traditionally, then, these sites

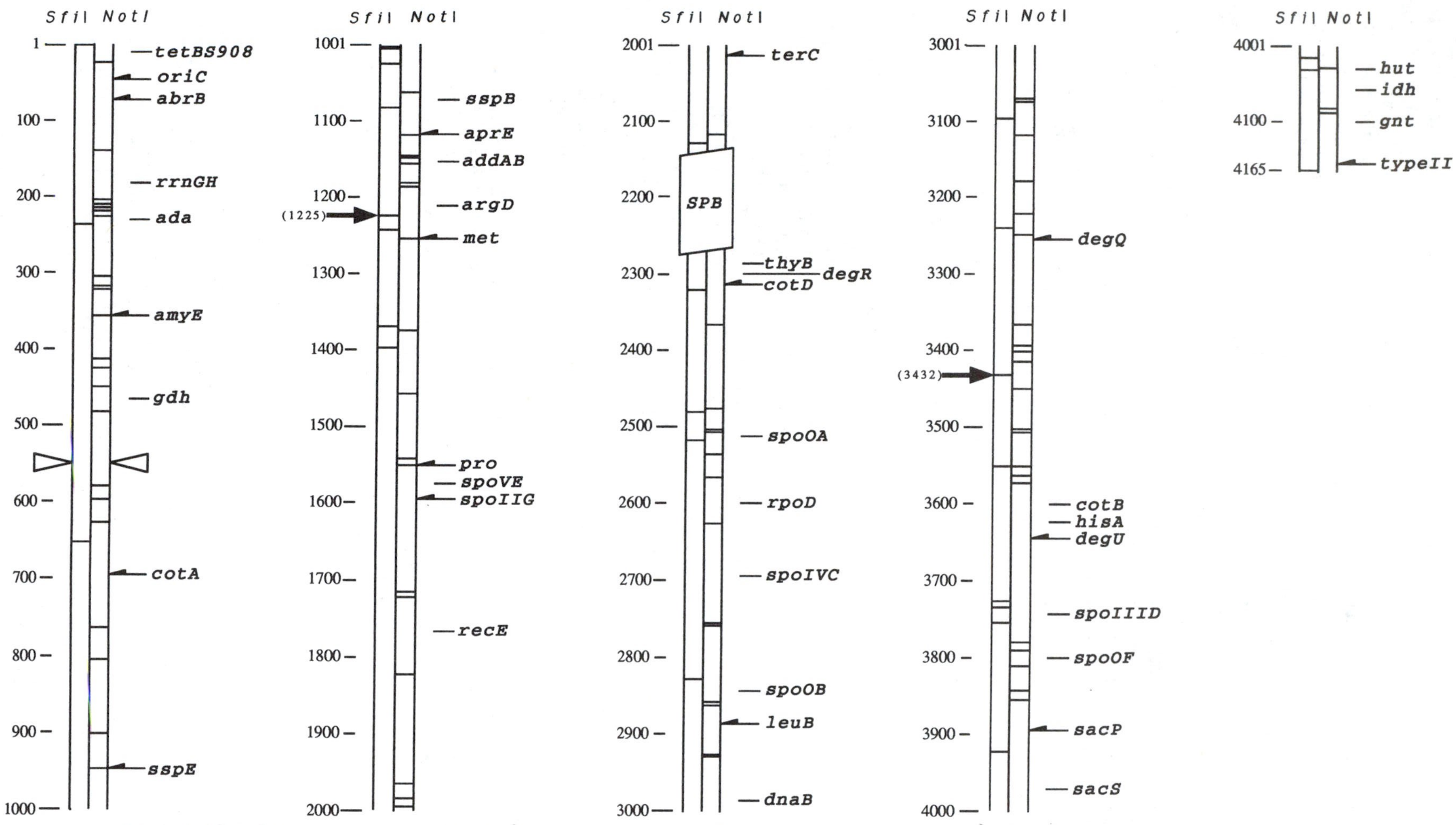

Figure 4. *Sfi*I and *Not*I physical map of the *B. subtilis* 168 chromosome. The map was redrawn on the basis of data in reference 15. Total genome size is 4,165 kb, which is divided into five linearized parts (1,000 kb each). Short horizontal bars across the vertical lines mark *Sfi*I or *Not*I sites. Physical distances (in kilobases) are indicated on the left. Locations of genes at arrowheads had been precisely determined and were used as landmarks on the physical map (Table 2 and text). Genes whose locations on both *Sfi*I and *Not*I fragments have been determined are included beside the *Not*I segments (Table 2). Open triangles indicate 23-kb deletions found in the standard strain, CU741, used for map construction (11; see text). Two weakly cleavable *Sfi*I sites are indicated by boldface arrows. The region of the SPβ prophage is boxed.

Table 1. Assignment of *Not*I and *Sfi*I fragments of *B. subtilis* CU741 DNA to the physical map[a]

*Sfi*I fragment	Size (kb)	Location (kb)	*Not*I fragment	Size (kb)	Location (kb)	*Not*I fragment	Size (kb)	Location (kb)
AS	730	1399–2129	1N	265	2118–2368		35	322–357
BS	414	238–652	2N	207	3573–3780		35	3415–3450
CS	350	652–1002	3N	174	3855–4029		33	450–483
DS	310	2519–2829	4N	165	1551–1716		32	3811–3843
ES	268	2829–3097	5N	143	1823–1966		30	598–628
FS	238	0–238	6N	140	2930–3070		30	2507–2537
GS	192	2129–2321	7N	135	628–763		30	2537–2567
HS	192	3240–3432	8N	129	2627–2756		27	3367–3394
IS	175	3551–3726	9N	127	13–140		26	2477–2503
JS	168	3754–3922	10N	122	1996–2118		25	1119–1144
KS	160	2321–2481	11N	120	1255–1375		25	1156–1181
LS	143	1082–1225	12N	118	3249–3367		23	427–450
MS	143	3097–3240	13N	114	947–1061		20	1966–1986
NS	133	4032–4165	14N	110	4089–13		20	3791–3811
OS	126	1244–1370	15N	109	2368–2477		18	580–598
PS	119	3432–3551	16N	100	1723–1823		14	3401–3415
QS	95	3922–4017	17N	100	2759–2859		13	414–427
RS	57	1025–1082	18N	97	483–580		12	306–318
SS	38	2481–2519	19N	96.5	804–901		12	3551–3563
TS	29	1370–1399	20N	85	1459–1544		12	3843–3855
US	20	1005–1025	21N	84	1375–1459		11	3780–3791
VS	20	3734–3754	22N	78	228–306		10	1986–1996
WS	19	1225–1244	23N	67	1188–1255		8	220–228
XS	15	4017–4032	24N	65	140–205		8	1148–1156
YS	8	3726–3734	25N	63	2863–2926		7	1181–1188
ZS	3	1002–1005	26N	60	2567–2627		7	1544–1551
			27N	60	3119–3179		7	1716–1723
			28N	58	1061–1119		7	3394–3401
			29N	57	357–414		6.6	4082–4089
			30N	53	4029–4082		6	205–211
			31N	53	3450–3503		5	3503–3508
			32N	46	901–947		4	216–220
			33N	45	3074–3119		3.7	2859–2863
				43	3179–3222		3.7	2926–2930
				43	3508–3551		3.5	3070–3074
				42	763–805		2	1146–1148
Total	4,165					Total	4,138	

[a] All *Sfi*I fragments and 33 *Not*I fragments were given designations. Physical distances used for locating each segment are from the 0-kb site (the *Sfi*I site between fragments FS and NS; Fig. 4). Details are in reference 15.

have been placed at 0 and 180° on the genetic map. However, the locations of *oriC* at 39 kb (13) and *terC* at 2,012 kb on the physical map (15) indicate that the clockwise and counterclockwise segments differ in length by about 170 kb, or nearly 4% of the total length of the chromosome (Fig. 6). These locations have been confirmed by inserting I-*Sce*I sequence recognition sites at *oriC* and *terC* and showing that I-*Sce*I cleaves the chromosome into two fragments that differ in size by about 170 kb (11). The disparity in size of the two segments is unrelated to the variations in certain regions of the chromosome discussed in the next section.

To account for the differences in lengths of the two segments, it has been suggested that replication proceeds at the same rate in each direction. When the clockwise replication fork reaches the *terC* region, synthesis is arrested, and then the counterclockwise replication process has additional time to reach the *terC* region. A relevant biological role for the disparity in length of the two segments is unclear and may be subtle, particularly since the *terC* region can be displaced or deleted without altering the viability of the cells (see chapter 36). Because the *B. subtilis* genome, unlike that of *E. coli* (10), has a single *terC* region, it might be a simpler system for the study of regulation of termination of replication (see chapter 36).

DNA Rearrangements on the *B. subtilis* 168 Chromosome

Diversity of the *B. subtilis* genome based on the physical map can be used to evaluate the frequency of DNA rearrangements (e.g., inversions, deletions, and insertions). Macro-restriction fragment length polymorphisms can result from DNA rearrangements, acquisition of new *Not*I or *Sfi*I sites, or eradication of preexisting *Not*I or *Sfi*I sites. Among the altered *Not*I and *Sfi*I fragments, three regions were frequently found to have deletions (11; regions, I, II, and III in Fig. 6). A 5-kb deletion was observed for region I. This region corresponds to the DNA from *rrnH* to *rrnG* and results from a spontaneous homologous recombination event between these two rRNA cistrons, as has been reported by Widom et al. (34). A second deletion region (region II) has been discovered in laboratory stocks representing several strains. The minimum

Table 2. Designation of genes on physical map

Genetic map location[a]	Physical map data[b]				Reference(s)
	Gene, plasmid, or phage	Location (kb)[b]	Fragment designation		
			*Sfi*I	*Not*I (kb)[c]	
0	*oriC*	39	FS	9N (127)	13
0	*recF*	41	FS	9N (127)	2, 13
0	*gyrB*	43	FS	9N (127)	2, 13
0	*gyrA*	45	FS	9N (127)	2, 13
3	*abrB*	73	FS	9N (127)	2, 15
11	*spo0H*	?	FS	?	2
15	*rrnGH*	140–205	FS	24N (65)	11
ND	*ada*	228–238	FS	22N (78)	11
25	*amyE*	358	BS	29N (57)	2
25	*amyR*	359	BS	29N (57)	15
34	*gdh*	?	BS	N (40)	2, 33
40	*sigB*	?	BS	18N (97)	2
52	*cotA*	695	CS	7N (135)	2, 15, 33
65	*sspE*	947	CS	32N (46)*13N (114)	2, 15
65	*sspB*	1061–1082	RS	28N (58)	2
74	*glyB*	?	RS	?	2
84	*aprE*	1119	LS	28N (58)*N (27)	2, 15
86	*addAB*	1146–1156	LS	N (16)	11
102	*argCAEBD*	1188–1255	LS	23N (67)	2, 33
102	*cpaF*	?	LS	?	2
104	*trpS*	?	LS	?	2
100	*met*	1255	OS	23N (67)*11N (120)	15
?	pbsx	?	OS	?	2
115	*pro*	1551	AS	N(7)*4N (165)	15
120	*spo0E*	?	AS	?	2
121	*sspD*	?	AS	?	2
133	*spoVE*	1551–1595	AS	4N (165)	2, 11
135	*spoIIG*	1595	AS	4N (165)	15
157	*recE*	1723–1823	AS	16N (100)	15
173	*citB*	?	AS	?	2
180	*gltA*	1996–2118	AS	10N (122)	2, 33
180	*terC*	2012	AS	10N (122)	11,15
182	*sspC*	?	GS	1N (265)	2
190	*SPβ*	2129–2314	GS	1N (265)	2, 11, 33
198	*asp*	?	GS	1N (265)	2
200	*thyB*	2129–2314	GS	1N (265)	15
200	*cotD*	2314	GS	1N (265)	2, 15
200	*degR*	2129–2314	GS	1N (265)	11
203	*trpDEC*	?	KS	?	2
204	*spoVA*	?	KS	?	2
211	*spoIIA*	?	KS	?	2
217	*spo0A*	2507–2519	SS	N (30)	2, 15
219	*ahrC*	?	SS	?	2
224	*rpoD*	2567–2627	DS	26N (60)	2, 15
226	*aroD*	?	DS	?	2
227	*spoIIIC*	?	DS	?	2
227	*spoIVC*	2627–2756	DS	8N (129)	11, 33
240	*phe*	?	ES	?	2
241	*spo0B*	2829–2859	ES	17N (100)	2, 11, 33
243	*attO105*	?	ES	?	2
247	*leuB*	2886	ES	25N (63)	15, 33
256	*dnaB*	2930–3070	ES	6N (140)	2, 15
266	*sspA*	?	ES	?	2
273	*menCDBE*	?	MS	?	2
275	*rrnB*	?	MS	?	2
279	*degQ*	3256	HS	12N (118)	2, 15
284	*hom*	?	HS	?	2
289	*gerA*	?	HS	?	2
290	*cotB*	3573–3726	IS	2N (207)	2, 15
298	*hisA*	?	IS	2N (207)	2
306	*degU*	3643	IS	2N (207)	15
302	*spoIIID*	3734–3754	VS	2N (207)	2
316	*spoIID*	?	JS	?	2
323	*spo0F*	3791–3811	JS	N (20)	2, 15, 33
324	*ctrA*	?	JS	?	2

(Continued)

Table 2—*Continued*

Genetic map location[a]	Physical map data[b]				Reference(s)
	Gene, plasmid, or phage	Location (kb)[b]	Fragment designation		
			*Sfi*I	*Not*I (kb)[c]	
324	*tsr*	?	JS	?	2
324	*rev*	?	JS	?	2
325	*rpoE*	?	JS	?	2
330	*sacP*	3895	JS	3N (174)	15
330	*sacS*	3922–4017	QS	3N (174)	2
335	*hut*	4029–4032	XS	30N (53)	11
343	*idh*	4032–4082	NS	30N (53)	11
344	*gnt*	4089–4165	NS	14N (110)	2, 15
352	*spo0J*	?	FS	14N (110)	2
ND	*Type II*	4155	FS	14N (110)	13
0	*tetBS908*	?	FS	14N (110)	1

[a] Data are from Anagnostopoulos et al. (chapter 29). ND, not determined.
[b] Location on the physical map or minimum region determined by both *Sfi*I and *Not*I segments. Physical distance data used are the same as those in reference 15.
[c] * indicates linking clones. Genes with both *Not*I and *Sfi*I physical location data are indicated in Fig. 4.

deletion we have observed in region II is 23 kb (11; Fig. 4 and 6). No good explanation for deletions in region II has yet been made. A third deletion at region III was the largest observed (130 kb; Fig. 4 and 6). It is known that an SPβ prophage resides in this region, and the presence or absence of this prophage is responsible for the difference in sizes observed for region III in various strains. Somewhat unexpectedly, this prophage can excise spontaneously during normal growth or can be found in strains that have undergone transformation. Not all prophage excise under the same conditions. PBSX (24) seems to be stably maintained under conditions in which SPβ is excised. If the number of *B. subtilis* strains examined by the current methods increases, or if more new and sensitive methods are employed, other differences of DNA structure might be observed. The overall maintenance of genome structure has been expected for *B. subtilis* and *E. coli* (8, 23), in contrast to the case for *Streptomyces* spp. (5).

Several important DNA rearrangements have been observed under circumstances very different from normal propagation of cells. Although it is important to know of these changes, they need to be thought of in a slightly different manner from the changes observed during normal laboratory growth and strain manipulations. Examples of special DNA rearrangements are (i) deletion of a 44-kb segment that occurs during sporulation (20, 25, 30), (ii) changes resulting from diploid formation following fusion of two cells (31), (iii) amplification of a particular segment of the chromosome in response to selection conditions (e.g., the *tmr* [9] or *tet* [1] locus), or (iv) formation or maintenance of merodiploids (3).

Origins and Diversity of Strains Derived from *B. subtilis* 168

The most widely studied strain of *B. subtilis* is *B. subtilis* 168. A *B. subtilis* Marburg stock was mutated, and certain auxotrophic markers were selected to produce the original *B. subtilis* 168 strain. Because of its ability to be readily transformed, many investigators have used this strain and introduced other markers, leading to distinctly different strains. As described in the previous section, relatively large segments of the genome are deleted in some *B. subtilis* 168 derivatives. If DNA from a strain quite unrelated to *B. subtilis* 168 is used for transformation, then deletions, insertions, and other DNA rearrangements might also be introduced either at the locus being intentionally altered or at other loci as a result of congression. In fact, some unpublished data from my laboratory confirm that when derivatives of *B. subtilis* 168 are transformed with DNA derived from *B. subtilis* W23 or other *B. subtilis* species, they acquire new DNA arrangements distinct from those described above. I believe that many strains have been selected after transformation of *B. subtilis* 168-derived strains by DNA from other strains than *B. subtilis trpC2* (BGSC stock 1A1). If it is important for interpretation that certain experiments be based on isogenic strains, it may not be possible to use two separate laboratory strains that are *B. subtilis* 168-derived strains.

FUTURE PERSPECTIVES

How Large or Small Can the *B. subtilis* Genome Be?

Extreme sizes of the *B. subtilis* genome have yet to be determined. Anagnostopoulos has constructed a strain in which several hundred kilobases of DNA are duplicated (3). However, this increased genome size may not have reached the limit of the additional amount of DNA that *B. subtilis* can maintain. In turn, one can ask, how small can the genome be? A complete answer to this question may come only after we understand the various networks involved in cell growth as well as the number of genes required for growth. The fraction of genes required for growth in the laboratory under defined conditions has not yet been determined, but an estimate can be made from the work I have reported on changes in *Sfi*I and *Not*I sequences. Changes in 3 of 57 *Not*I and *Sfi*I sites resulted in a lethal phenotype. If we assume that this frequency is typical throughout the entire genome, only about 5% of the DNA is essential (3 of 57 sites).

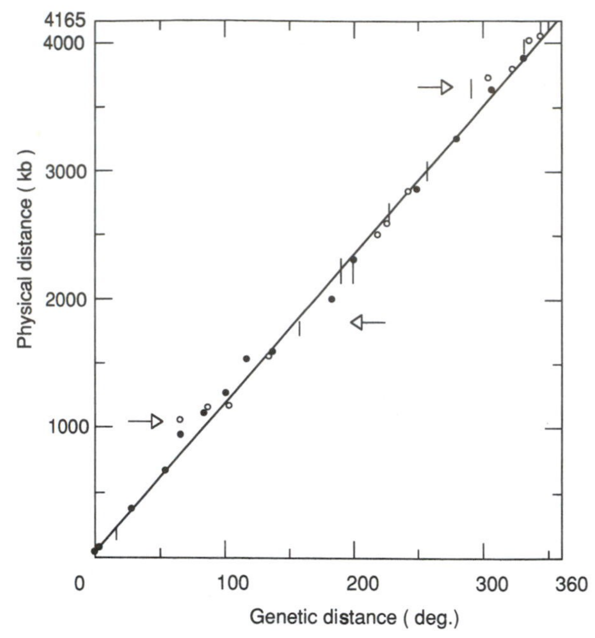

Figure 5. Correlation between the genetic and physical maps. Genetic map information given in Table 2 is included. Closed circles indicate gene loci whose precise physical locations have been determined. Vertical bars or open circles represent DNA regions to which cloned genes have been assigned by the results of Southern hybridization. Open arrows indicate regions where the physical location is slightly displaced from that on the genetic map.

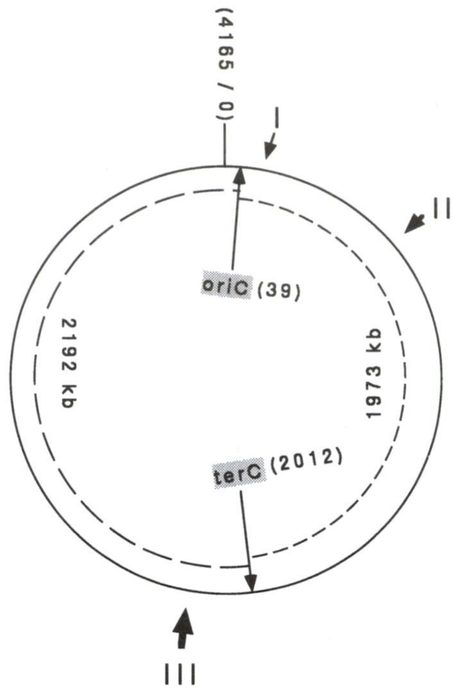

Figure 6. Stability and asymmetry of the *B. subtilis* 168 chromosome. Locations of the *oriC* site (at 39 kb) and the *terC* site (at kb 2012) and regions for naturally occurring DNA deletions on the *B. subtilis* 168 chromosome (regions I, II, and III) are indicated by arrows. The genome size for *B. subtilis* 168 (strain CU741) was estimated to be 4,165 kb (see details in the text and Fig. 4). The broken circle indicates the lengths for the clockwise segment (1,973 kb) and the counterclockwise segment (2,192 kb).

Alteration in seven other *Not*I or *Sfi*I sites resulted in obvious phenotypes in the strain, including those concerning nutritional requirements. About 10 to 15% (7 of 57 sites) of the DNA would fall into this classification. It may be possible to develop a strain of *B. subtilis* in which all but the essential genes are deleted.

Use of *B. subtilis* Genome as a Cloning Vector and a Repository for Cloned DNA

One rather remarkable feature of the transformation process of *B. subtilis* 168 is the ability to incorporate DNA from bacteria that are somewhat distantly related to *B. subtilis* 168 (e.g., *B. subtilis* W23 and *Bacillus natto*). This ability may be due in part to the fact that *B. subtilis* 168 has only a weak restriction-modification system (see chapter 38). The relative stability of the *B. subtilis* genome probably reflects the fact that uptake of DNA does not necessarily lead to integration of the DNA without nucleotide sequence homology. I have recently been able to integrate the complete phage lambda (48.5 kb) into the *B. subtilis* chromosome. Having foreign DNA in the *B. subtilis* genome should permit introduction of other DNA via homologous recombination. Since it has already been shown (3) that *B. subtilis* can accommodate large amounts of additional DNA, *B. subtilis* might be an excellent vector for many types of recombinant DNA studies. *B. subtilis* can form spores that are refractory to many types of damaging agents. This property could be used as a means of long-term storage of specific types of DNA at room temperature or for transportation of DNAs over great distances such as would be encountered in space travel.

REFERENCES

1. **Amano, H., L. I. Catherine, K. F. Bott, and K. Shishido.** 1991. A limited number of *Bacillus subtilis* strains carry a tetracycline-resistance determinant at a site close to the origin of replication. *Biochim. Biophys. Acta* **1088:** 251–258.
2. **Amjad, M., J. M. Castro, H. Sandoval, J.-J. Wu, M. Yang, D. J. Henner, and P. J. Piggot.** 1991. An *Sfi*I restriction map of the *Bacillus subtilis* 168 genome. *Gene* **101:**15–21.
3. **Anagnostopoulos, C.** 1990. Genetic rearrangements in *Bacillus subtilis*, p. 361–371. *In* K. Drlica and M. Riley (ed.), *The Bacterial Chromosome*. American Society for Microbiology, Washington, D.C.
4. **Bachmann, B. J.** 1972. Pedigrees of some mutant strains of *Escherichia coli* K-12. *Bacteriol. Rev.* **36:**525–557.
5. **Birch, A., A. Hausler, and R. Hutter.** 1990. Genome rearrangement and genetic instability in *Streptomyces* spp. *J. Bacteriol.* **172:**4138–4142.
6. **Burke, D. T., G. F. Carle, and M. V. Olson.** 1987. Cloning of large segments of exogenous DNA into yeast by means of artificial chromosome vector. *Science* **236:**806–812.
7. **Chu, G., D. Vollrath, and R. W. Davis.** 1986. Separation of large DNA molecules by contour-clamped homogeneous electric fields. *Science* **234:**1582–1585.
8. **Daniels, D. L.** 1990. The complete Avr II restriction map of the *Escherichia coli* genome and comparisons of several laboratory strains. *Nucleic Acids Res.* **18:**2649–2651.

9. **Furusato, T., J.-I. Takano, K. Yamane, K.-I. Hashiguchi, A. Tanimoto, M. Mori, K. Yoda, M. Yamasaki, and G. Tamura.** 1986. Amplification and deletion of the *amyE-tmrB* gene region in a *Bacillus subtilis* recombination-phage genome by the *tmrA7* mutation. *J. Bacteriol.* **165:** 549–556.
10. **Hidaka, M., T. Kobayashi, and T. Horiuchi.** 1991. A newly identified DNA replication terminus site, *terE*, on the *Escherichia coli* chromosome. *J. Bacteriol.* **173:**391–393.
11. **Itaya, M.** 1993. Stability and asymmetric replication of the *Bacillus subtilis* 168 chromosome structure. *J. Bacteriol.* **175:**741–749.
12. **Itaya, M., K. Kondo, and T. Tanaka.** 1989. A neomycin resistance gene cassette selectable in a single copy state in the *Bacillus subtilis* chromosome. *Nucleic Acids Res.* **17:**4410.
13. **Itaya, M., J. Laffan, and N. Sueoka.** 1992. Physical distance between the site of type II binding to the membrane and *oriC* on the *Bacillus subtilis* 168 chromosome. *J. Bacteriol.* **174:**5466–5470.
14. **Itaya, M., and T. Tanaka.** 1990. Gene-directed mutagenesis on the chromosome of *Bacillus subtilis* 168. *Mol. Gen. Genet.* **223:**268–272.
15. **Itaya, M., and T. Tanaka.** 1991. Complete physical map of the *Bacillus subtilis* 168 chromosome constructed by a gene-directed mutagenesis method. *J. Mol. Biol.* **220:** 631–648.
16. **Kobayashi, Y., et al.** Personal communication.
17. **Kohara, Y., K. Akiyama, and K. Isono.** 1987. The physical map of the whole *E. coli* chromosome: application of a new strategy for rapid analysis and sorting of a lambda genomic library. *Cell* **50:**495–508.
18. **Kolsto, A.-B., A. Gronstad, and H. Oppegaard.** 1990. Physical map of the *Bacillus cereus* chromosome. *J. Bacteriol.* **172:**3821–3825.
19. **Krawiec, S., and M. Riley.** 1990. Organization of the bacterial chromosome. *Microbiol. Rev.* **54:**502–539.
20. **Kunkel, B., R. Losick, and P. Stragier.** 1990. The *Bacillus subtilis* gene for the developmental transcription recombinase gene. *Genes Dev.* **4:**525–535.
21. **McClell, M., R. Jones, Y. Patel, and M. Nelson.** 1987. Restriction endonucleases for pulsed field mapping of bacterial genomes. *Nucleic Acids Res.* **15:**5985–6005.
22. **Qiang, B.-Q., and I. Schildkraut.** 1984. A type II restriction endonuclease with an eight nucleotide specificity from *Streptomyces fimbriatus*. *Nucleic Acids Res.* **12:** 4507–4516.
23. **Riley, M., and S. Krawiec.** 1987. Genome organization, p. 967–981. *In* F. C. Neidhardt, J. L. Ingraham, K. B. Low, M. Magasanik, M. Schaechter, and H. E. Umbarger (ed.), *Escherichia coli* and *Salmonella typhimurium: Cellular and Molecular Biology*, vol. 2. American Society for Microbiology, Washington, D.C.
24. **Rutberg, L.** 1982. Temperate bacteriophages of *Bacillus subtilis*, p. 247–268. *In* D. A. Dubnau (ed.), *The Molecular Biology of the Bacilli*. Academic Press, Inc., New York.
25. **Sato, T., Y. Samori, and Y. Kobayashi.** 1990. The *cisA* cistron of *Bacillus subtilis* sporulation gene *spoIVC* encodes a protein homologous to a site-specific recombinase. *J. Bacteriol.* **172:**1092–1098.
26. **Schwartz, D., J. Saffran, R. Welsh, M. Haas, M. Goldberg, and C. Cantor.** 1983. New techniques for purifying large DNAs and studying their properties and packaging. *Cold Spring Harbor Symp. Quant. Biol.* **47:**189–195.
27. **Serror, P.** Personal communication.
28. **Smith, C. L., and G. Condemine.** 1990. New approaches for physical mapping of small genomes. *J. Bacteriol.* **172:**1167–1172.
29. **Smith, C. L., J. G. Econome, A. Scutt, S. Klco, and C. R. Cantor.** 1987. A physical map of the *Escherichia coli* K-12 genome. *Science* **236:**1448–1453.
30. **Stragier, P., B. Kunkel, L. Kroos, and R. Losick.** 1989. Chromosome rearrangement generating a composite gene for a developmental transcription factor. *Science* **243:**507–512.
31. **Thaler, D. S., J. R. Roth, and L. Hirshbein.** 1990. Imprinting as a mechanism for the control of gene expression, p. 445–456. *In* K. Drlica and M. Riley (ed.), *The Bacterial Chromosome*. American Society for Microbiology, Washington, D.C.
32. **Thierry, A., A. Perrin, J. Boyer, C. Fairhead, B. Dujon, B. Frey, and G. Schmitz.** 1991. Cleavage of yeast and bacteriophage T7 genomes at a single site using the rare cutter endonuclease I-*Sce* I. *Nucleic Acids Res.* **19:**189–190.
33. **Ventra, L., and A. S. Weiss.** 1989. Transposon-mediated restriction mapping of the *Bacillus subtilis* chromosome. *Gene* **78:**29–36.
34. **Widom, R. L., E. D. Jarris, G. LaFauci, and R. Rudner.** 1988. Instability of rRNA operons in *Bacillus subtilis*. *J. Bacteriol.* **170:**605–610. (Erratum, **170:**2003.)

31. An Ordered Collection of *Bacillus subtilis* DNA Segments in Yeast Artificial Chromosomes

PASCALE SERROR, VASCO AZEVEDO, and S. DUSKO EHRLICH

A collection of *Bacillus subtilis* chromosomal DNA fragments inserted in yeast artificial chromosomes (YACs) is represented in Fig. 1. It is composed of 59 YACs which carry segments of *B. subtilis* DNA with an average size of 115 kb. It was constructed as follows.

YACs were obtained by random cloning of *B. subtilis* DNA, partially digested with *Eco*RI, in the pYAC4 vector (2). A total of 677 YACs carried inserts of >40 kb, as deduced by pulsed-field gel electrophoresis. The total length of cloned DNA was about 40 Mb or about 12 *B. subtilis* genome equivalents. The structure of inserts from 13 YACs, carrying a total of about 1.2 Mb of *B. subtilis* DNA, was compared with the structure of the corresponding chromosomal region by Southern blotting. Twelve inserts were colinear with the chromosome and one was not, which indicates that >90% of the cloned *B. subtilis* DNA did not rearrange in yeast.

To order the YACs, we first used 66 previously cloned *B. subtilis* genes as hybridization probes. Next, the ends of the inserts carrying these genes were isolated. Cloning in *Escherichia coli* and inverse polymerase chain reaction (3) were used for the left and the right end, respectively (orientation corresponds to vector arms, as defined in reference 2). The ends were used as probes to identify clones with overlapping inserts. Iteration of the procedure resulted in the ordered collection. The identity of several segments was ascertained by transformation of *B. subtilis* mutants.

Despite the relatively high redundancy of the random collection, four gaps remain in the ordered collection. Maximum gap size was estimated by hybridizing *B. subtilis* DNA cleaved with several restriction enzymes with the ends of segments bordering the gaps. The total gap size is <82 kb, indicating that 98% of the *B. subtilis* chromosome is represented in the ordered collection.

Ordering of the collection led us to suggest different map positions than those previously reported for certain genes (5): (i) *sspD* and *pts* should be inverted; (ii) *xyl* maps near *thyA*, as also found by Zahler (8); (iii) *thrZ* (6) maps between *spoIID* and *sacA*; and (iv) *sspB* is close to 90°, as already suggested by Amjad et al. (1).

The ordered collection should be useful for genetic studies in *B. subtilis*. It has already been used for mapping of the β-galactosidase gene (7) and is being used for mapping of the repeated stable RNA genes (4). It was also used for identification of λ clones derived from a particular chromosomal region and should be useful for clone ordering. It will be used as a source of DNA for cloning and sequencing large regions of the *B. subtilis* genome.

Acknowledgments. E. Alvarez, G. Damiani, and E. Zumstein participated in the creation of the collection. We thank V. Sgaramella for initial encouragement and guidance, C. Anagnostopoulos for continuing interest and many useful discussions, numerous members of the *B. subtilis* community for providing hybridization probes, and F. Haimet for the artwork. V.A. was a recipient of a Brazilian government fellowship. This work was supported, in part, by EEC SCIENCE grant SC1-0211.

REFERENCES

1. **Amjad, M., J. M. Castro, H. Sandoval, J. J. Wu, M. Yang, D. J. Henner, and P. J. Piggot.** 1990. An *Sfi*I restriction map of the *B. subtilis* 168 genome. *Gene* **101:**15–21.
2. **Burke, D. T., G. F. Carle, and M. V. Olson.** 1887. Cloning of large segments of exogenous DNA into yeast by means of artificial chromosome vectors. *Science* **236:**806–812.
3. **Ochman, H., A. S. Gerber, and D. L. Hartl.** 1988. Genetic applications of an inverse polymerase chain reaction. *Genetics* **120:**621–623.
4. **Okamoto, K., P. Serror, V. A. Azevedo, and B. Vold.** 1992. The use of yeast artificial chromosomes and PCR to map ribosomal and tRNA gene regions of *B. subtilis*. The Eleventh International Spores Conference, Woods Hole, Mass., 9–13 May 1992.
5. **Piggot, P. J.** 1989. Revised genetic map of *Bacillus subtilis* 168, p. 1–41. *In* I. Smith, R. A. Slepecky, and P. Setlow (ed.), *Regulation of Prokaryotic Development*. American Society for Microbiology, Washington, D.C.
6. **Putzer, H., A. A. Brakhage, and M. Grunberg-Manago.** 1990. Independent genes for two threonyl-tRNA synthetases in *Bacillus subtilis*. *J. Bacteriol.* **172:**4593–4602.
7. **Zagorec, M., and M. Steinmetz.** 1991. Construction of a derivative of *Tn917* containing an outward-directed promoter and its use in *B. subtilis*. *J. Gen. Microbiol.* **137:**107–112.
8. **Zahler, S. A.** Personal communication.

(Figure 1 on following page)

Pascale Serror, Vasco Azevedo, and S. Dusko Ehrlich • Laboratoire de Génétique Microbienne, Institut National de la Recherche Agronomique, Domaine de Vilvert, 78352 Jouy en Josas Cedex, France.

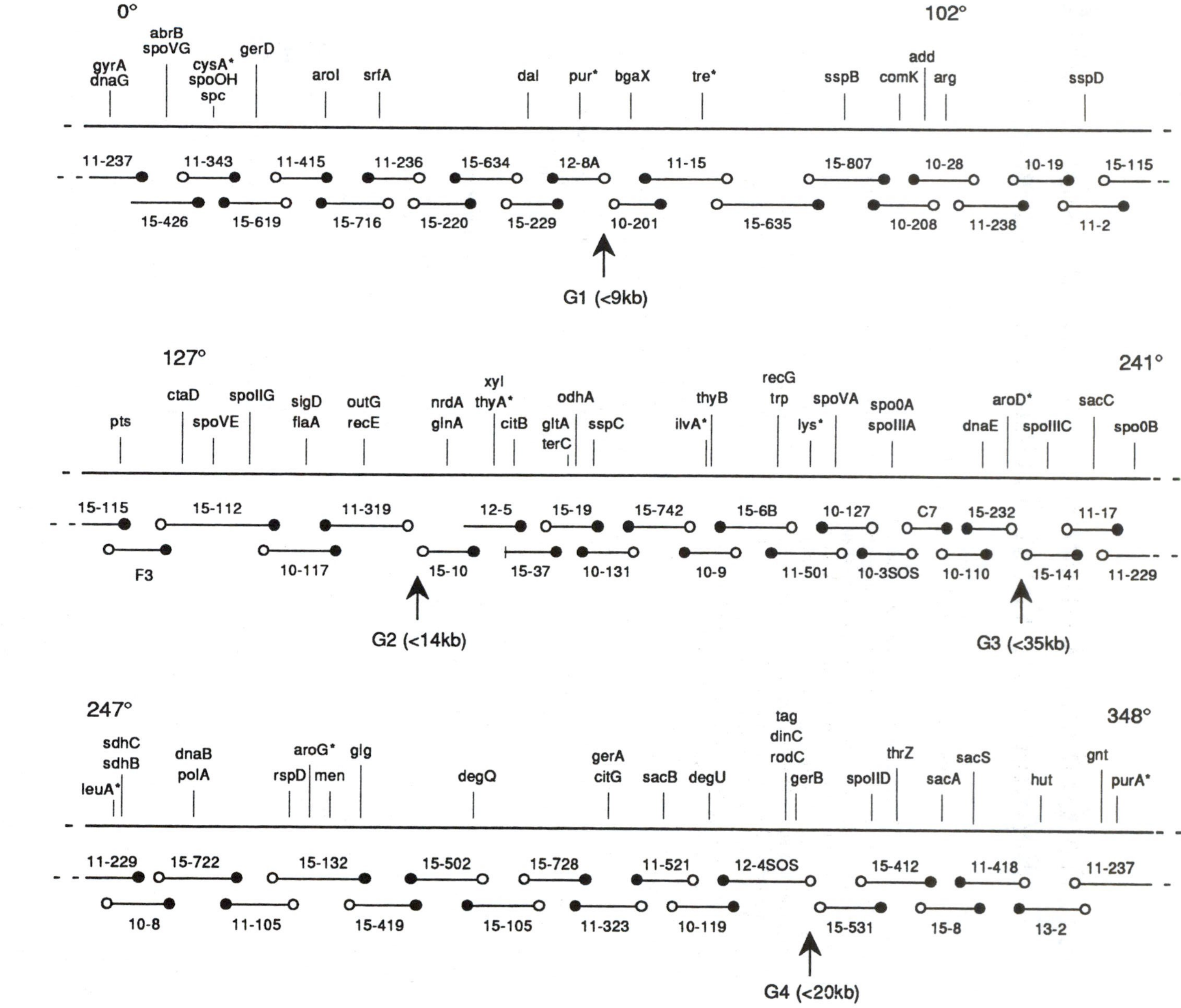

Figure 1. The ordered YAC clones containing *B. subtilis* DNA. The segments carried in individual YAC clones are represented by horizontal lines. They are placed according to their relationships with the neighboring clones and the *B. subtilis* genes they contain, but the extent of overlap between the segments has not been determined. YAC accession numbers are also mentioned. *B. subtilis* genes which hybridized with the corresponding YACs are indicated. Genetic position (in degrees) is given for *dnaG*, *add*, *ctaD*, *spo0B*, *sdh*, and *gnt* loci. Stars indicate markers which were identified in YAC clones by transformation of *B. subtilis* mutants. Filled and open circles represent left and right ends of inserts, respectively, which were (i) successfully isolated and (ii) shown to hybridize with the overlapping insert (except for inserts bordering the gaps). We failed to isolate one end of the insert carried on YAC 12-5 (carrying the *thyA* gene), possibly because of their toxicity for *E. coli* and aberrant amplification in vitro. One end of the insert from YAC 15-37 (close to *terC*), signalled by a vertical line, hybridized with the insert carried by YAC 15-634 (close to *dal*), which shows that it is composed of nonadjacent *B. subtilis* DNA segments. However, the neighboring clone, YAC 15-10, is likely to contain the segment adjacent to *thyA*. The positions of four gaps present in the collection are indicated by vertical arrows and letters G1 to G4. The numbers in parentheses refer to gap size.

32. The Genetic Map of *Bacillus megaterium*

PATRICIA S. VARY

Bacillus megaterium is a species of *Bacillus* that has been studied for many years because of its interesting physiology and its ability to sporulate at great efficiency. It was also the organism in which lysogeny was discovered by Lwoff (31). The species produces a variety of unusual compounds, many of which are useful for industry, and it is a proficient industrial cloning host for foreign proteins. Most strains harbor several plasmids. Strain QM B1551 contains at least seven plasmids ranging from 5.4 to 166 kb in size (25). Over the last few years, my laboratory has developed the genetics of this organism using strain QM B1551. Many multiply marked strains are now available (over 600 mutants from this laboratory alone), including a mapping kit for further genetic mapping (63). We reported the first transduction in *B. megaterium* by phage MP13 in 1979 (58). At least 14 other related phages with different host ranges on *B. megaterium* strains also transduce (59) and are available for mapping in strains that MP13 does not transduce (61, 62). Over 50 loci have been mapped, spanning approximately 75% of the chromosome (60). Moreover, protoplast transformation of plasmids has also been developed in the laboratories of Alfoldi and Carlton (64, 66), and most plasmids used in *Bacillus*, *Streptococcus*, and *Staphylococcus* function similarly in *B. megaterium* (25, 63). In addition, we have introduced transposon Tn*917* and derivatives carrying *lacZ* and *cat* genes (1, 53) and have exploited the transposon technology developed by Youngman et al. (70) to isolate and characterize several sporulation mutants and to clone sporulation genes of the *spoIIA* and *spoVA* operons (53, 54). It is also fitting at this time to compile the mapping data available for *B. megaterium* since its role in industrial production is expanding. Plasmidless derivatives are presently being used as hosts for foreign protein production by industries in the United States and Europe (16, 35, 44). The major impetus for their use is the now well-documented (44, 47) absence of the alkaline proteases that have hampered protein production by *Bacillus subtilis* and thus production of quality protein. In addition to its protein stability, *B. megaterium* also has the advantage of higher plasmid stability (25, 35). Its genetics and industrial uses have recently been reviewed (60). In the last 5 years, over 46 patents have been issued using this species, at least 40 genes have been cloned, and 34 genes have been sequenced. We have recently reported a codon usage analysis based on the sequences of 29 of these genes (60).

This report presents the mapping data for *B. megaterium* and compares these data with genetic mapping data compiled for *B. amyloliquefaciens*, *B. licheniformis*, *B. stearothermophilus*, *B. subtilis*, and *B. thuringiensis*.

DETAILED MAP

Figure 1 presents the detailed linkage map for *B. megaterium* constructed from our laboratory data. The map is not as yet circular. The linkage groups have been assigned letters A through F, which are then used in Table 1 to designate map position and have been placed in positions corresponding to the position of similar genes in *B. subtilis*. No attempt has been made here to assign a more specific map unit since the final placement of linkage groups has not been made. The regions near *leu-ilv* and *trp-his* have been mapped in detail through three-factor crosses (4, 15). The enzyme deficiencies in the *leu* genes as well as in some of the *trp* genes have also been determined (4, 15). Two-factor crosses that are reasonably unambiguous have generated the data in groups B, E, and F. Linkage group A has been very difficult to map. Several of the resistance markers are only in an approximate position and cannot be ordered definitively. However, the *cysB* and *cysC* genes appear to be on that fragment, unlike *B. subtilis* (see below). A partial physical map has been produced (37) but is not complete. *Not*I cuts the *B. megaterium* genome into at least 24 fragments. *Sfi*I cuts it into at least seven, two of which are greater than 1,100 kb. Loci *hisD* and *nic* have been mapped only on *Not*I restriction fragments that also contain other genes of known position, so these have been included in the figure. Mapping was done by probing transposition mutants with a Tn*917* probe. When the sizes of the *Not*I fragments of over 16 mutants and the wild type are totaled, a minimum size for the *B. megaterium* genome of 4.67 Mb is found. This is probably an underestimate, since some fragments observed on gels probably are doublets as found by Kølsto et al. (27), who have reported a genome size for *B. cereus* of 5.7 Mb. In accordance with the *B. subtilis* map, the sporulation genes have been placed inside the circle.

Two other groups of loci have been shown to be linked, but their position relative to other genes is not known. The *cob* and *cbl* genes, important in cobalamin synthesis, have been mapped relative to each other using phage MP13 in strain ATCC 10778 by Wolf and Brey (69). They also cloned the genes and established four physical linkage groups and 12 complementation groups (2). Several other cloned fragments have contained interesting linked genes. For example, all seven genes coding for the proteins found in both

Patricia S. Vary • Department of Biological Sciences, Northern Illinois University, DeKalb, Illinois 60115.

Table 1. Genetic markers of *B. megaterium*

Gene	Map position[a]	Phenotype or genotype	Reference(s)[b]
amyA	NM	Amylase, 29-amino acid signal sequence, not *amyE***	36
amyB	NM	Amylase, not homologous to *amyE*(Bs)*	26
argA	F, N184	Requires arginine	51
argC	NM	Requires arginine or citrulline	59
argO	B	Requires arginine, citrulline, or ornithine	51
atpA	NM	Alpha subunit of F_1 ATPase**	3
atpB	NM	A subunit of F_0 ATPase**	3, 18
atpC	NM	Epsilon subunit of F_1 ATPase**	3, 18, 19
atpD	NM	Beta subunit**	3, 19
atpE	NM	c subunit of F_0 ATPase**	3
atpF	NM	b subunit of F_0 ATPase**	3
atpG	NM	Gamma subunit of F_1 ATPase**	3
atpH	NM	Delta subunit of F_1 ATPase**	3
atpI	NM	Gene with *atp* operon, unknown function	3
azi	F	Resistant to 2 mM sodium azide	4
bio-4	NM	Requires biotin	59
cbl	NM	X–XII complementation groups, cobalamine synthesis**	2, 69
cml	A, N413	Resistant to chloramphenicol	39
cob	NM	I–VI complementation groups, cobinamide synthesis**	2, 69
cysB	A, S584	Requires cysteine	39
cysC	A, S584	Requires cysteine	39
cyt	NM	Cytochrome monooxygenase**	43, 68
divIVA	C	Filamentous, asporogenous, overproduces PHB[c]	28
dnaK	NM	Heat shock protein**	50
gdhA	NM	Glucose dehydrogenase**, probably two genes	21
gdhB	NM	Glucose dehydrogenase*	21, 32, 35
ger-1	NM	Germination on glucose, proline, and leucine*	52
ger-2	NM	Proline germination receptor	42
ger-28	NM	Germination deficient on glucose	29
ger-37	NM	Germination deficient on KBr	29
ger-44	NM	Germination deficient on proline	29
ger-45	NM	Germination deficient on leucine	29
ger(Ts)	NM	Temperature sensitive for germination	57
gerP	Plasmid pVY109	Germination deficient on leucine, proline, glucose, and KBr; on 166-kb plasmid	49
glt	NM	Requires glutamate	29
gly	NM	Requires glycine	29
gpr	E	*ssp* germination protease**	14, 51
guaB	S486	Requires guanine	37
gyrA	A, S584	Resistant to nalidixic acid	39
hem-1	F	Requires hemin, is not *hemA* or *hemB*, produces porphobilinogens	15
hemA	B	Requires hemin, grows on δ-aminolevulinic acid	15, 51
hisA	F	Requires histidine	4
hisD	N184	Requires histidine, cannot grow on histidinol	4
hisH	E, N339	Requires histidine, inhibited by *p*-fluorophenylalanine	4
ilvA	D	Requires isoleucine and valine	39, 59
ilvB/C	F	Requires isoleucine and valine	15
kan	A, S584	Resistant to kanamycin	39
lac	NM	Cannot utilize lactose	53
ldh	NM	Lactic dehydrogenase**	67, 71
leuA	F, N184	α-Isopropylmalate synthase	15
leuB	F, N184	Isopropylmalate isomerase	15
leuC	F, N184	β-Isopropylmalate dehydrogenase	15
megA	Plasmid pBM309	Megacin A, of strain QM B1551 and ATCC 19213 and immunity genes, all plasmid borne*	64, 65
metA	NM	Requires methionine, homocystine, or cystine	39
metC/D	B, N37	Requires methionine	51
mitC	NM	Sensitive to mitomycin C, not *uvr* or *rec*	8
neo	A	Resistant to neomycin	39
neo-6	NM	Resistant to neomycin, ATPase reduced	39
nic	N184	Requires nicotinic acid	37
nov-1	A	Resistant to novobiocin	39
npr	NM	Neutral protease negative	63
pam	NM	Penicillin amidase*	33, 34, 48
pbp	D	Homologous to *pbp*(Bs)** partial clone	54
pheA	E	Requires phenylalanine	15

(Continued)

Table 1—*Continued*

Gene	Map position[a]	Phenotype or genotype	Reference(s)[b]
pigA	NM	White on minimal medium	59
pigB	NM	Brown-pigmented colonies	29
pigY	NM	Bright yellow-pigmented colonies	59
purA	A, N413	Requires adenine	39
purB	A, S75	Requires adenine, guanine, or hypoxanthine	39
purC/D	A, S75	Requires adenine or hypoxanthine	39
pyrB	C	Requires uracil, or any intermediate except carbamyl phosphate	28
pyrC	C	Requires uracil, will not grow on carbamyl aspartate or orotidine monophosphate	28
pyrD	C	Requires uracil or orotic acid	28
pyrF	C	Requires uracil	28
rec-1	NM	Recombination deficient	8
rec-3	NM	Recombination deficient, rough morphology	8
rib-20	E	Requires riboflavin	53
rplV	A	Resistant to erythromycin	39
rpoB	A, N413	Resistant to rifampin	39
rpsL	A, S584	Resistant to streptomycin	39, 56
ser	NM	Requires serine	59
spo-57	E, N339	Tn*917-lacZ cat* insertion, linked to *spoI* expressed at t_0	53
spo0H	NM	Sigma H homolog	38
spoI54	E, N339	*spoI* Tn*917-lacZ cat* fusion, slight asymmetric septation, expressed at t_0	53
spoIIAA	E, N339	Homologous to *spoIIAA*(Bs), t_0**	54
spoIIAB	E, N339	Homologous to *spoIIAB*(Bs), t_0**	54
spoIIAC	E, N339	Homologous to *spoIIAC*(Bs), t_0, probably sigma F**	53, 54
spoVAA	E, N339	Homologous to *spoVAA*(Bs)**	54
spoVAB	E, N339	Homologous to *spoVAB*(Bs)**, partial clone	54
spoVG	NM	Homologous to *spoVG*(Bs)**	23
sspA	F	Major small acid-soluble protein A**	10, 11, 51
sspB	B	Major small acid-soluble protein C**	6, 10, 46, 51
sspC1	NM	Small acid-soluble protein C-1, homology to *sspB***	13
sspC2	NM	Small acid-soluble protein C-2**, homology to *sspB*	13
sspC5	NM	Small acid-soluble protein C-5**	11
sspD	B	Small acid-soluble protein C-3**	7, 12, 51
sspE	NM	Small acid-soluble protein B	17
sspF	B	Small acid-soluble protein C-4**	11, 51
thr	NM	Requires threonine	59
thyA	D	Requires thymine, trimethoprin resistant	59
thyB	NM	Requires thymine, trimethoprin resistant, with *thyA mtn*	59
trpA	E, N339	Complements *trpA* gene of *B. subtilis**	33
trpB	E, N339	Tryptophan synthase*	4, 33
trpC	E, N339	Indoleglycerol phosphate synthase**	4, 22, 33
trpD	E, N339	Anthranylphosphoribosyltransferase	4, 33
trpE	E, N339	Anthranilate synthase	4
trpF	E, N339	Complements *trpF* of *B. subtilis*	33
tyr-21	N200	Requires tyrosine	37
ura	E, N339	Requires uracil	37
uvr	NM	UV sensitive, recombination proficient	8
val	NM	Requires valine	59
xylA	NM	Xylose isomerase**	44, 45
xylB	NM	Xylulokinase**	44, 45
xylR	NM	*xyl* repressor**	44, 45

[a] NM, not mapped.
[b] One asterisk (*) means the gene has been cloned; two asterisks (**), it has been sequenced.
[c] PHB, polyhydroxy butyrate.

the F_1 and F_0 subunits of *B. megaterium* QM B1551 ATPase are linked as in *Escherichia coli* and have been sequenced and ordered as *atpI-atpB-atpE-atpF-atpH-atpA-atpG-atpD-atpC* (3, 19). The genes show great homology to the *E. coli unc* genes, but have been named *atp* in accordance with most other systems. The linked genes *pbp-spoIIA-spoVA* have been cloned and sequenced on one 3.34-kb fragment (54). Three other early *spo* mutations map in the same region, including an interesting *spoI* mutation, but have not been unambiguously ordered (53). The *spoVG* gene has been cloned, but not as yet mapped (23). A clone carrying the *trp* genes was isolated by McCullough (33) and was found to complement all the *B. subtilis trp* genes except *trpE*. The *trpC* gene has recently been cloned and sequenced (22). Genes have been named to be consistent with *B. subtilis* and *E. coli* when applicable.

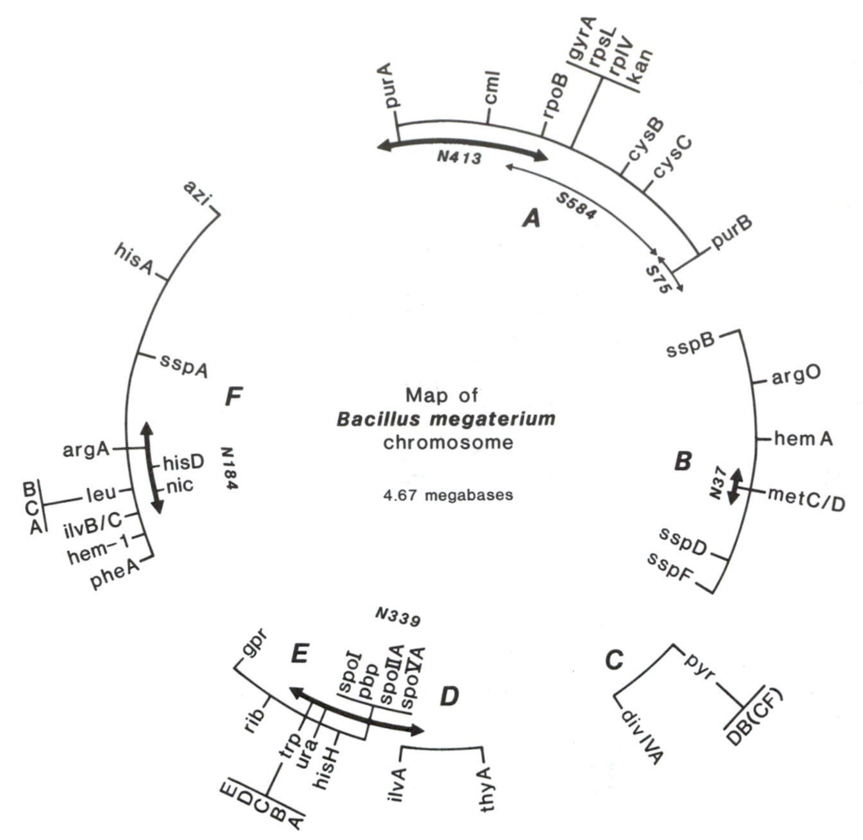

Figure 1. Genetic map of *B. megaterium*. Shown are the six linkage groups A through F. The map is depicted as circular by analogy with *B. subtilis*, *B. stearothermophilus*, and *B. cereus*. Restriction fragments from *Not*I (N) or *Sfi*I (S) digestion, with their corresponding size in kilobases, are also shown (37). Figure is from Vary (60).

This is the first compilation of genes for *B. megaterium*. I have attempted to include as many genes as possible, including several unmapped but cloned genes so as to present a complete current picture of the genetics of the organism.

COMPARISON WITH OTHER BACILLI

Considerable genetic mapping data have now been reported for several other bacilli. These data are presented, some in abbreviated form, in Fig. 2. Those with incomplete maps have been ordered in relation to *B. subtilis*. In an impressive set of experiments, Welker and coworkers have recently reported the results of mapping *B. stearothermophilus* both by transduction and protoplast fusion (5, 55), a technique that has not proven useful for *B. megaterium* or *B. subtilis* (9). They have been able to demonstrate a circular genetic chromosome, only the second *Bacillus* species to have this demonstrated. However, Kølsto et al. (27) have recently assembled a physical circular map for *B. cereus* and have reported considerable variation in restriction maps within the species. Luckily, *B. cereus* yielded only 11 *Not*I fragments, in contrast to the 24 for *B. megaterium* and over 40 for *B. subtilis*.

As can be seen in Fig. 2, many of the genes are in the same regions in all the species. There are, however, exceptions. In *B. megaterium*, the *leu* and *ilv* genes are inverted in relation to *phe*, the *hemA* gene (mutants can grow on δ-aminolevulinic acid) is near *metC* and not near *leu*, *argO* and *hem* are inverted, and *cysB* and *cysC* are in the ribosomal protein region. Although none of the *cys* mutants produces H_2S, characteristic of *cysA* mutants, and all five mutants mapped have been found to map only in this region (59), at this time it cannot be ruled out that the mutations may all be in a complex *cysA* locus as in *B. subtilis*. Comparison shows that the *ssp* gene positions in both organisms are strikingly similar even though the surrounding genes have undergone more rearrangement (51). Some negative results suggest other rearrangements. For example, *ser* cannot be cotransduced with *rib* or *trp* (59). In *B. stearothermophilus*, there is a *serB* locus near *pyr* that possibly could be a locus not mapped in *B. subtilis*, and the *leu* and *ilv* genes are in the same order as in *B. megaterium*. *B. licheniformis* appears to have the same order for *leu* and *ilv* as *B. subtilis*. In *B. thuringiensis*, *lys* is between *his* and *cys*, rather than near *trp*. This seems to also be true in *B. amyloliquefaciens*.

From the small amount of data presented, one cannot make too many conclusions. However, the data suggest that the maps of these six bacilli are roughly similar, with species more distantly related to *B. subtilis* exhibiting more variation in genetic loci. One therefore cannot assume that the maps are identical when projecting the site of a specific gene, but there should be a high probability that it will be in the same region.

Bacillus megaterium

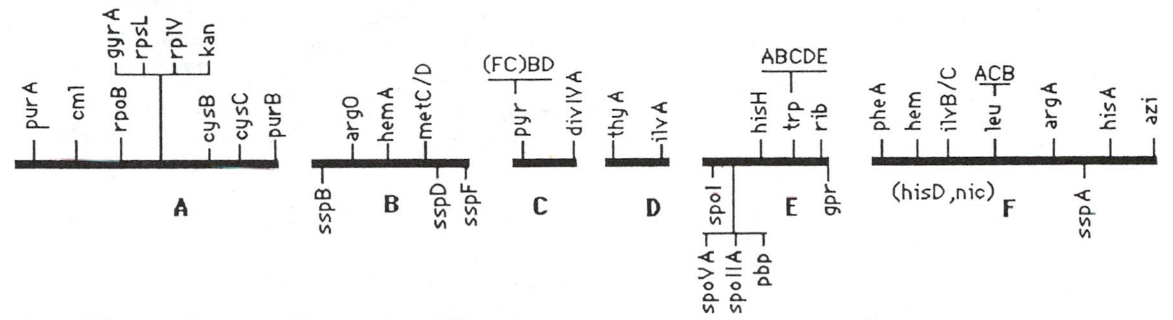

Bacillus stearothermophilus

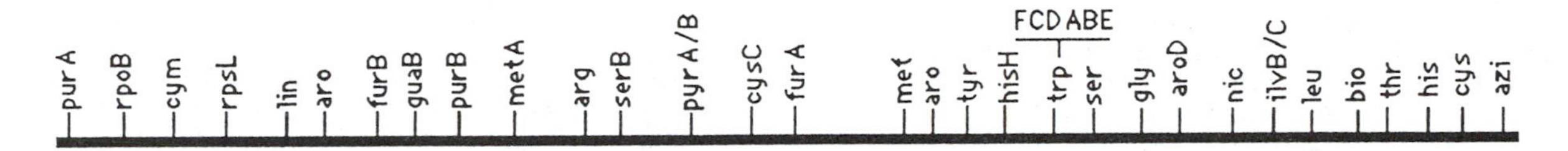

Bacillus thuringiensis

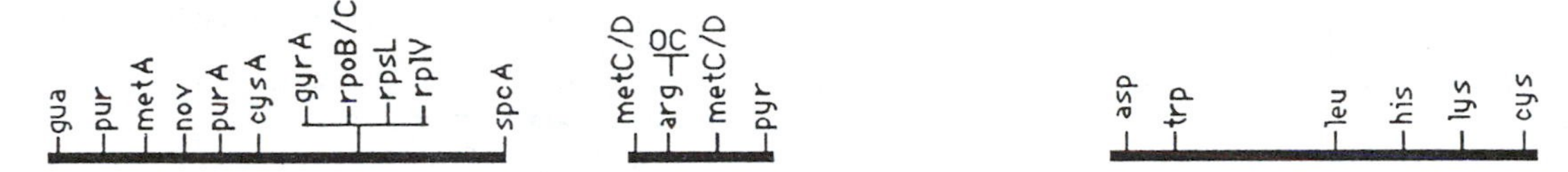

Bacillus amyloliquefaciens

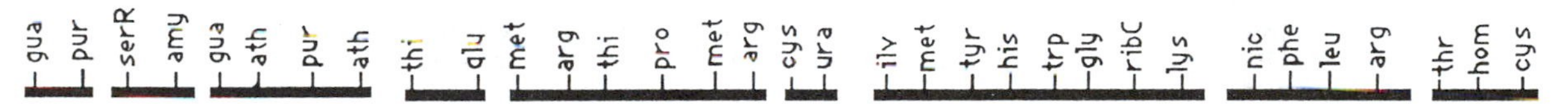

Bacillus licheniformis

Bacillus subtilis

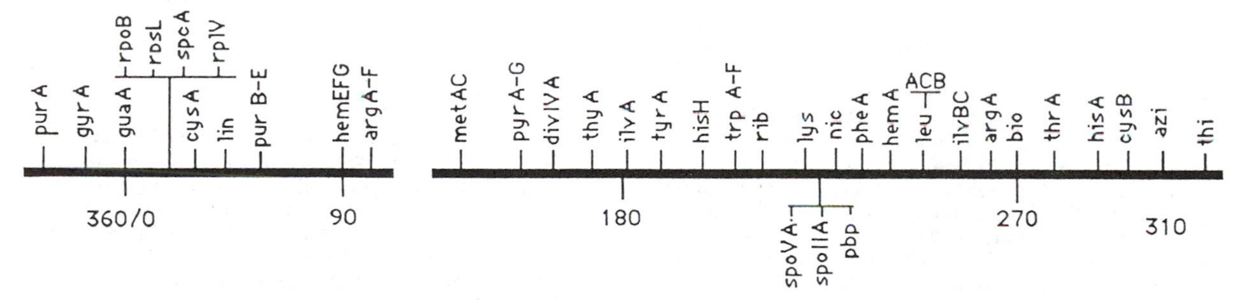

Figure 2. Comparison of the genetic maps of six bacilli. The maps have been aligned when possible, but are not to scale. References: *B. megaterium* (see Table 1), *B. amyloliquefaciens* (24), *B. stearothermophilus* (5, 55), *B. thuringiensis* (20, 30), *B. licheniformis* (40), and *B. subtilis* (41).

Acknowledgments. This review was supported in part by a grant from Abbott Laboratories. I thank M. M. Lecadet for unpublished data for *B. thuringiensis*.

REFERENCES

1. **Bohall, N. A., Jr., and P. S. Vary.** 1986. Transposition of Tn*917* in *Bacillus megaterium*. *J. Bacteriol.* **167:**716–718.
2. **Brey, R. N., C. D. B. Bonner, and F. B. Wolf.** 1986. Cloning of multiple genes involved with cobalamin (vitamin B_{12}) biosynthesis in *Bacillus megaterium*. *J. Bacteriol.* **167:**623–630.
3. **Brusilow, W. S. A., M. A. Scarpetta, C. A. Hawthorne, and W. P. Clark.** 1989. Organization and sequence of the gene coding for the proton-translocating ATPase of *Bacillus megaterium*. *J. Biol. Chem.* **264:**1528–1533.
4. **Callahan, J. P., I. P. Crawford, G. F. Hess, and P. S. Vary.** 1983. Cotransductional mapping of the *trp-his* region of *Bacillus megaterium*. *J. Bacteriol.* **154:**1455–1458.
5. **Chen, Z., S. F. Wojcik, and N. E. Welker.** 1986. Genetic analysis of *Bacillus stearothermophilus* by protoplast fusion. *J. Bacteriol.* **165:**994–1001.
6. **Curiel-Quesada, E., B. Setlow, and P. Setlow.** 1983. Cloning of the gene for C protein, a low molecular weight spore-specific protein from *Bacillus megaterium*. *Proc. Natl. Acad. Sci. USA* **80:**3250–3254.
7. **Curiel-Quesada, E., and P. Setlow.** 1984. Cloning of a new low-molecular-weight spore-specific protein gene from *Bacillus megaterium*. *J. Bacteriol.* **157:**751–757.
8. **English, J. D., and P. S. Vary.** 1986. Isolation of recombination defective and UV sensitive mutants of *Bacillus megaterium*. *J. Bacteriol.* **163:**155–160.
9. **Fleischer, E. R., and P. S. Vary.** 1985. Genetic analysis of fusion recombinants and presence of noncomplementary diploids in *Bacillus megaterium*. *J. Gen. Microbiol.* **131:** 919–926.
10. **Fliss, E. R., M. J. Conners, C. A. Loshon, E. Curiel-Quesada, B. Setlow, and P. Setlow.** 1985. Small, acid-soluble spore proteins of *Bacillus*: products of a sporulation-specific multigene family, p. 60–66. *In* J. A. Hoch and P. Setlow (ed.), *Molecular Biology of Microbial Differentiation*. American Society for Microbiology, Washington, D.C.
11. **Fliss, E. R., C. A. Loshon, and P. Setlow.** 1986. Genes for *Bacillus megaterium* small, acid-soluble spore proteins, cloning and nucleotide sequence of three additional genes from this multigene family. *J. Bacteriol.* **165:**467–473.
12. **Fliss, E. R., and P. Setlow.** 1984. *Bacillus megaterium* spore protein C-3: nucleotide sequence of its gene. *Gene* **30:**167–172.
13. **Fliss, E. R., and P. Setlow.** 1984. Complete nucleotide sequence and start sites for transcription and translation of the *Bacillus megaterium* protein C gene. *J. Bacteriol.* **158:**809–813.
14. **Fliss, E. R., and P. Setlow.** 1985. Genes for *Bacillus megaterium* small, acid soluble spore proteins: nucleotide sequence of two genes and their expression during sporulation. *Gene* **35:**151–157.
15. **Garbe, J. C., M. A. Franzen, G. F. Hess, and P. S. Vary.** 1984. Genetics of leucine biosynthesis in *Bacillus megaterium* QM B1551. *J. Bacteriol.* **157:**454–459.
16. **Ginsburgh, C., D. Spaulding, G. Robey, A. Shivakumar, O. McCall, R. Vanags, and L. Katz.** 1989. Sporulation promoter *spoVG* controlled expression of pp42 gene of HIV-1 in *Bacillus megaterium*. Abstr. V International Conference on AIDS, Montreal, abstr. no. 1262.
17. **Hackett, R. H., B. Setlow, and P. Setlow.** 1986. Cloning and nucleotide sequence of the *Bacillus megaterium* gene coding for the small, acid-soluble spore protein B. *J. Bacteriol.* **168:**1023–1025.
18. **Hawthorne, C. A., and W. S. A. Brusilow.** 1986. Complementation of mutants in the *Escherichia coli* proton-translocating ATPase by cloned DNA from *Bacillus megaterium*. *J. Biol. Chem.* **261:**45–48.
19. **Hawthorne, C. A., and W. S. A. Brusilow.** 1988. Sequence of the genes for the beta and epsilon subunits of the ATP synthase of *Bacillus megaterium* QM B1551. *Biochem. Biophys. Res. Commun.* **151:**926–931.
20. **Heierson, A., R. Landen, and H. G. Boman.** 1983. Transductional mapping of nine linked chromosomal genes in *Bacillus thuringiensis*. *Mol. Gen. Genet.* **192:**118–123.
21. **Heilmann, H. J., H. J. Margert, and H. G. Gassen.** 1988. Identification and isolation of glucose dehydrogenase genes of *Bacillus megaterium* M1286 and their expression in *E. coli*. *J. Biochem.* **174:**485–490.
22. **Hudspeth, D. S., P. Pun, and P. S. Vary.** Cloning and sequencing of the *trpC* gene of *Bacillus megaterium*. In preparation.
23. **Hudspeth, D. S., and P. S. Vary.** 1992. Sequencing of the *spoVG* genes from *B. megaterium* and *B. subtilis*. *Biochim. Biophys. Acta.* **1130:**229–231.
24. **Jomantas, J. A. V., J. A. Fiodorova, E. G. Abalakina, and Y. I. Kozlov.** 1991. Genetics of *Bacillus amyloliquefaciens*. 6th International Conference on Bacilli, Stanford, Calif., abstr. no. T7.
25. **Kieselburg, M. K., M. Weickert, and P. S. Vary.** 1984. Analysis of resident and transformant plasmids in *Bacillus megaterium*. *Bio/Technology* **2:**254–259.
26. **Klug, C., P. Pun, and P. S. Vary.** 1988. Cloning of a starch degrading gene of *B. megaterium*, abstr. H-134, p. 167. *Abstr. 89th Annu. Meet. Am. Soc. Microbiol. 1988*. American Society for Microbiology, Washington, D.C.
27. **Kølsto, A. B., A. Gronstad, and H. Oppegaard.** 1990. Physical map of the *Bacillus cereus* chromosome. *J. Bacteriol.* **172:**3821–3825.
28. **Lach, D. A., V. K. Sharma, and P. S. Vary.** 1990. Isolation and characterization of a unique division mutant of *Bacillus megaterium*. *J. Gen. Microbiol.* **136:**545–553.
29. **Lach, D. A., and P. S. Vary.** Unpublished data.
30. **Lecadet, M. M.** Personal communication.
31. **Lwoff, A., and A. Gutmann.** 1950. Recherches sur *Bacillus megatherium* lysogene. *Ann. Inst. Pasteur* **78:**711–713.
32. **Makino, Y., S. Negoro, I. Urabe, and H. Okada.** 1989. Stability-increasing mutants of glucose dehydrogenases from *Bacillus megaterium* IWG3. *J. Biol. Chem.* **264:**6381–6385.
33. **McCullough, J. E.** 1983. Gene cloning in bacilli related to enhanced penicillin acylase production. *Bio/Technology* **1:**879–882.
34. **Meevootisom, V., and J. R. Saunders.** 1986. Cloning and expression of penicillin acylase genes from overproducing strains of *Escherichia coli* and *Bacillus megaterium*. *Appl. Microbiol. Biotechnol.* **25:**372–378.
35. **Meinhardt, F., U. Stahl, and W. Ebeling.** 1989. Highly efficient expression of homologous and heterologous genes in *Bacillus megaterium*. *Appl. Microbiol. Biotechnol.* **30:**343–350.
36. **Metz, R. J., L. N. Allen, T. M. Cao, and N. W. Zeman.** 1988. Nucleotide sequence of an amylase gene from *Bacillus megaterium*. *Nucleic Acids Res.* **16:**5203.
37. **Muse, W. B.** 1990. The development of a partial physical map of *Bacillus megaterium*. M.S. thesis. Northern Illinois University, DeKalb.
38. **Naumann, G., K.-D. Neitzke, J. Ribbe, W. Honerlage, and P. Fortnagel.** 1990. The *Bacillus megaterium spo0H* gene. Abstr. Gene Manipulations in Bacilli, abstr. no. 53. Second Workshop on Gene Manipulations in Bacilli. Academy of Sciences of the GDR, Holzhau/Erzgebirge, German Democratic Republic, April, 1990.
39. **Palm, S., and P. S. Vary.** Unpublished data.
40. **Perlak, F. J., and C. B. Thorne.** 1981. Genetic map of *Bacillus licheniformis*, p. 79–82. *In* H. S. Levinson, A. L. Sonenshein, and D. J. Tipper (ed.), *Sporulation and Germination*. American Society for Microbiology, Washington, D.C.
41. **Piggot, P. J.** 1989. Revised genetic map of *Bacillus*

subtilis 168, p. 1–41. *In* I. Smith, R. A. Slepecky, and P. Setlow (ed.), *Regulation of Prokaryotic Development: Structural and Functional Analysis of Bacterial Sporulation and Germination*. American Society for Microbiology, Washington, D.C.

42. **Rossignol, D. P., and J. C. Vary.** 1979. L-proline site for triggering *Bacillus megaterium* spore germination. *Biochem. Biophys. Res. Commun.* **89:**547–551.
43. **Ruettinger, R. T., L. P. Wen, and A. J. Fulco.** 1989. Coding nucleotide, 5′ regulatory, and deduced amino acid sequences of P-450BM-3, a single peptide cytochrome P-450. *J. Biol. Chem.* **264:**10987–10995.
44. **Rygus, T., and W. Hillen.** 1991. Inducible high-level expression of heterologous genes in *Bacillus megaterium* using the regulatory elements of the xylose-utilization operon. *Appl. Microbiol. Biotechnol.* **35:**594–599.
45. **Rygus, T., A. Scheler, R. Allmansberger, and W. Hillen.** 1991. Molecular cloning, structure, promoters and regulatory elements for transcription of the *Bacillus megaterium* encoded regulon for xylose utilization. *Arch. Microbiol.* **155:**535–542.
46. **Setlow, P., and J. Ozols.** 1980. Covalent structure of protein C: a second major low molecular weight protein degraded during germination of *Bacillus megaterium* spores. *J. Biol. Chem.* **255:**8413–8416.
47. **Shivakumar, A. G., R. I. Vanags, D. R. Willcox, L. Katz, P. S. Vary, and J. L. Fox.** 1989. Gene dosage effect on the expression of the delta-endotoxin genes of *Bacillus thuringiensis* subsp. kurstaki in *Bacillus subtilis* and *Bacillus megaterium*. *Gene* **79:**21–31.
48. **Son, H.-J., T.-I. Mheen, B.-L. Seong, and M. H. Han.** 1982. Studies on microbial penicillin amidase (IV). The production of penicillin amidase from a partially constitutive mutant of *Bacillus megaterium*. *J. Gen. Appl. Microbiol.* **28:**281–291.
49. **Stevenson, D. M., D. Lach, and P. S. Vary.** A gene required for germination in *Bacillus megaterium* is plasmid-borne. *In* E. Balla and G. Berencie (ed.), *Proceedings of the 11th European Meeting on Genetic Transformation*. INTERCEPT, in press.
50. **Sussman, M. D., and P. Setlow.** 1987. Nucleotide sequence of a *Bacillus megaterium*, gene homologous to the *dnaK* gene of *Escherichia coli*. *Nucleic Acids Res.* **15:**3923.
51. **Sussman, M. D., P. S. Vary, C. Hartman, and P. Setlow.** 1988. Integration and mapping of *Bacillus megaterium* genes which code for small, acid-soluble spore proteins and their protease. *J. Bacteriol.* **170:**4942–4945.
52. **Tani, K., M. Kawanishi, J. Nishikawa, M. Sasaki, Y. Takubo, T. Nishihara, and M. Kondo.** 1990. Identification of a germination gene of *Bacillus megaterium*. *Biochem. Biophys. Res. Commun.* **167:**402–406.
53. **Tao, Y.-P., and P. S. Vary.** 1991. Isolation and characterization of sporulation *lacZ* fusion mutants of *Bacillus megaterium*. *J. Gen. Microbiol.* **137:**797–806.
54. **Tao, Y.-P., and P. S. Vary.** 1992. Cloning and sequencing of the *spoIIA* operon of *Bacillus megaterium*. *Biochimie* **74:**695–704.
55. **Vallier, H., and N. E. Welker.** 1990. Genetic map of the *Bacillus stearothermophilus* NUB36 chromosome. *J. Bacteriol.* **172:**793–801.
56. **Vary, J. C.** 1973. Germination of *Bacillus megaterium* spores after various extraction procedures. *J. Bacteriol.* **116:**797–802.
57. **Vary, J. C., and A. Kornberg.** 1970. Biochemical studies of bacterial sporulation and germination. *J. Bacteriol.* **101:**327–329.
58. **Vary, P. S.** 1979. Transduction in *Bacillus megaterium*. *Biochem. Biophys. Res. Commun.* **88:**1119–1124.
59. **Vary, P. S.** Unpublished data.
60. **Vary, P. S.** 1992. Development of genetic engineering in *Bacillus megaterium*: an example of the versatility and potential of industrially important bacilli, p. 253–311. *In* R. Doi (ed.), *Biology of the Bacilli, Applications in Industry*. Butterworth Heinemann, San Diego.
61. **Vary, P. S., J. C. Garbe, M. Franzen, and E. W. Frampton.** 1982. MP13, a generalized transducing bacteriophage for *Bacillus megaterium*. *J. Bacteriol.* **149:**1112–1119.
62. **Vary, P. S., and W. F. Halsey.** 1980. Isolation and partial characterization of several bacteriophages for *Bacillus megaterium* QM B1551. *J. Gen. Virol.* **51:**137–146.
63. **Vary, P. S., and Y.-P. Tao.** 1988. Development of genetic methods in *Bacillus megaterium*, p. 403–407. *In* A. T. Ganesan and J. A. Hoch (ed.), *Genetics and Biotechnology of Bacilli*, vol. 2. Academic Press, Inc., New York.
64. **Von Tersch, M. A., and B. C. Carlton.** 1983. Megacinogenic plasmids of *Bacillus megaterium*. *J. Bacteriol.* **155:**872–877.
65. **Von Tersch, M. A., and B. C. Carlton.** 1984. Molecular cloning of structural and immunity genes for megacins A-216 and A-19213 in *Bacillus megaterium*. *J. Bacteriol.* **160:**854–859.
66. **Vorobjeva, I. P., I. A. Khmel, and L. Alfoldi.** 1980. Transformation of *Bacillus megaterium* protoplasts by plasmid DNA. *FEMS Microbiol. Lett.* **7:**261–263.
67. **Waldvogel, S., H. Weber, and H. Zuber.** 1987. Structure and function of L-lactate dehydrogenase from thermophilic and mesophilic bacteria. VII. Nucleotide sequence of the lactate dehydrogenase gene from the mesophilic bacterium *Bacillus megaterium*. *Hoppe-Seyler's Z. Physiol. Chem.* **368:**1391–1399.
68. **Wen, L. P., and A. J. Fulco.** 1987. Cloning of the gene encoding a catalytically self-sufficient cytochrome P-450 fatty acid monooxygenase induced by barbiturates in *Bacillus megaterium* and its functional expression and regulation in heterologous (*Escherichia coli*) and homologous (*Bacillus megaterium*) hosts. *J. Biol. Chem.* **262:**6676–6682.
69. **Wolf, J. B., and R. N. Brey.** 1986. Isolation and genetic characterizations of *Bacillus megaterium* cobalamin biosynthesis-deficient mutants. *J. Bacteriol.* **166:**51–58.
70. **Youngman, P., H. Poth, B. Green, K. York, G. Olmedo, and K. Smith.** 1989. Methods for genetic manipulation, cloning, and functional analysis of sporulation genes in *Bacillus subtilis*, p. 65–88. *In* I. Smith, R. A. Slepecky, and P. Setlow (ed.), *Regulation of Prokaryotic Development: Structural and Functional Analysis of Bacterial Sporulation and Germination*. American Society for Microbiology, Washington, D.C.
71. **Zuelli, F., S. Waldvogel, R. W. Schneiter, and H. Zuber.** 1988. Structure and function of L-lactate-dehydrogenases (LDH) from thermophilic and mesophilic bacteria. *Experientia* **44:**A37.

33. The Genetic Map of *Bacillus stearothermophilus* NUB36

NEIL E. WELKER

Organisms adapted to high temperature have evolved unique solutions to the biochemical problems imposed by this environment. Considerable information has been obtained about the ecology, physiology, metabolic capabilities, and biochemical and physical properties of cellular components of thermophiles. It is clear that the ability to grow at high temperature cannot be explained by a single mechanism but, rather, that thermophiles appear to use a variety of mechanisms. Insights into the molecular mechanisms of thermophily have been hampered because little is known about the genetic and molecular basis of thermophily. This gap in our knowledge is directly related to a lack of genetic exchange systems for the manipulation and genetic analysis of a thermophilic species. Although some progress has been made in the development of genetic technology for *Bacillus stearothermophilus* (2, 3), strains of the genus *Thermus* (4), and *Methanobacterium thermoautotrophicum* (5, 11), the lack of efficient and reliable genetic exchange systems, a repertoire of mutants, or plasmids that express useful genetic markers has hampered the exploitation of these organisms for basic and applied research. Recent progress in the development of genetic techniques for *B. stearothermophilus* NUB36 (1, 9, 12) makes it possible to elucidate the molecular and genetic mechanisms of thermophily in this organism. The genetic characterization of the *B. stearothermophilus* NUB36 genome is the first step in attaining this goal.

GENETIC MAP

Figure 1 presents the first linkage map of the *B. stearothermophilus* NUB36 chromosome. The map was constructed using the linkages reported by Vallier and Welker (8). The genetic markers listed in Table 1 contain a number of new markers that were not included in the previous investigation (8) and two loci (*glnQ* and *glnH*) that have been sequenced (13).

Preliminary studies (1, 9) indicated that the markers on the *B. stearothermophilus* chromosome would exhibit the same phenotype and would be located in the same relative position as the analogous markers on the *Bacillus subtilis* chromosome. This facilitated the selection of markers used in linkage analysis experiments. Nine linkage groups were established by using two-factor crosses with transducing phage TP-42C (9). The order of each linkage group was established by protoplast fusion (1). The auxotrophic fusion parents were either streptomycin resistant (Str^r) or chloramphenicol resistant (Cml^r), and Str^r Cml^r recombinants were selected from the fusion products. Each pair of nonselected markers was assigned a coinheritance frequency. The coinheritance frequency for a specific pair of markers is defined as the percentage of Str^r Cml^r recombinants that have either parental phenotype for that specific pair of markers. Coinheritance frequency values between 60% and 90% were used to establish linkage between two markers. The genetic linkage map presented in Fig. 1 contains 71 loci. By using characteristics such as growth response to intermediates or end products of metabolism, cross-feeding, and the relative order within a linkage group or between two linkage groups, each of 46 loci was tentatively assigned a cognate *B. subtilis* gene. The map location, in degrees, of the analogous *B. subtilis* loci was used to position these landmark loci on the *B. stearothermophilus* chromosome. Thus, the 360° scale brings the *B. stearothermophilus* NUB36 map into conformity with the *B. subtilis* map. The selection of a cognate *B. subtilis* gene for nine other loci was based on similarities of the growth response to nutrients and cross-feeding analyses or resistance phenotype. Four of the loci exhibited linkages with adjacent loci that are different from their linkages on the *B. subtilis* map, and the relative positions of five loci on the *B. stearothermophilus* map could not be determined by transduction or protoplast fusion. The tentative location of the other loci was determined by their linkage to adjacent landmark loci.

The position of most loci was determined by two-factor crosses (transduction or protoplast fusion or both). However, the most reliable mapping data come from three-factor crosses. The order of the loci in the *aro-trp* region was verified by three-factor crosses with transducing phage TP-42C (10). Thus, the order of the genes in this region is likely to be that presented in Fig. 1. The tentative order of the genes in the *B. stearothermophilus aro* region is *aspB-aroBAFEC-tyrA-hisH-(trp)*, whereas it is *aspB-aroE-tyrA-hisH-(trp)-aroHBF* in *B. subtilis*. The tentative order of genes in the *trp* operon of *B. stearothermophilus* is *trpFCDABE*, whereas it is *trpABFCDE* in *B. subtilis*. The genetic map of *B. stearothermophilus* NUB36 may be similar to the *B. subtilis* 168 map; however, the identity of the genes in *B. stearothermophilus* with the putative analogous genes of *B. subtilis* must be rigorously established by biochemical and molecular biological techniques. The relative order of *pur-6*, *gua-11*, *met-1*, and

Neil E. Welker • Department of Biochemistry, Molecular Biology, and Cell Biology, Northwestern University, Evanston, Illinois 60208.

Table 1. Genetic markers of *B. stearothermophilus* NUB36

Gene symbol[a]	Map position (degrees)[b]	Phenotype, enzyme deficiency, or other characteristic	Reference[c]
aec-1	287‡	Resistance to *S*-(2-aminoethyl)-L-cysteine	8
aec-2	284† (*aecB*, 282)	Resistance to *S*-(2-aminoethyl)-L-cysteine	8
aec-3	252* (*aecA*, 252)	Resistance to *S*-(2-aminoethyl)-L-cysteine	8
agr-1	NM	Aspartate, asparagine, glutamate, or glutamine requirement; defined by Tn*917* insertion	10
ahr-1	207† (*ahrC*, 219)	Resistance to arginine hydroxamate	8
ahr-2	99§ (*ahrD*, 99)	Resistance to arginine hydroxamate	8
ala-1	281* (*ala*, 281)	Alanine requirement	8
arg-9	117* (*arg-342*, 117)	Arginine, ornithine, or citrulline requirement	8
arg-13	260§ (*argG* or *argH*, 260)	Arginine requirement	8
arg-14	102§ (*arg*, *argA*, *argB*, *argC*, *argD*, or *argE*, 102)	Arginine, ornithine, or citrulline requirement	8
aro-1	203† (*aroA*, 264)	Shikimate or chorismate or phenylalanine, tyrosine, and tryptophan requirement	8
aro-2	203* (*aroB*, 203)	Shikimate or chorismate or phenylalanine, tyrosine, and tryptophan requirement	8
aro-4	206* (*aroC*, 206)	Shikimate or chorismate or phenylalanine, tyrosine, and tryptophan requirement	8
aro-5	226* (*aroD*, 226)	Shikimate or chorismate or phenylalanine, tyrosine, and tryptophan requirement	8
aro-7	26* (*aroI*, 26)	Chorismate or phenylalanine, tyrosine, and tryptophan requirement	8
aro-8	202* (*aroE*, 202)	Chorismate or phenylalanine, tyrosine, and tryptophan requirement	8
aro-10	203* (*aroF*, 203)	Chorismate or phenylalanine, tyrosine, and tryptophan requirement	8
aro-12	262§ (*aroG*, 262)	Phenylalanine or tyrosine requirement	8
asp-1	198* (*aspB*, 198)	Aspartate requirement	8
asp-3	209‡	Aspartate or fumarate requirement	8, 10
azi-2	304* (*azi*, 304)	Resistance to sodium azide	8
bio-2	268* (*bioA*, *bioB*, or *bio-112*, 268)	Biotin requirement	8
caf-1	46* (*cafA*, 46)	Resistance to caffeine	8
cml-1	11* (*cml*, 11)	Resistance to chlormaphenicol	8
cys-1	140* (*cysC*, 140)	Cysteine, methionine, cystathionine, sulfide, or sulfite requirement	8
cys-2	11* (*cysA*, 11)	Cysteine, methionine, cystathionine, or sulfide requirement	8
cys-3	292* (*cysB*, 292)	Cysteine requirement	8
egr-1	NM	Enhanced growth at 70°C	10
fur-1	41* (*furB*, 41)	Resistance to 5-fluorouracil	8
fur-2	141* (*furA*, 141)	Resistance to 5-fluorouracil	8
fut-1	NM	Fumarate utilization	10
glh-1	NM	Resistance to γ-glutamylhydrazide	10
gln-1	NM‖	Glutamine or glutamate requirement	10
glnH	NM	L-Glutamine transport; 26-kDa cytoplasmic membrane protein	13*
glnQ	NM	L-Glutamine transport; 27.4-kDa cytoplasmic membrane protein	13*
glu-1	180§ (*gltA* or *gltB*, 180)	Glutamate or aspartate requirement	8
gly-1	207‡	Glycine or serine requirement	8
gly-4	207* (*glyA*, 207)	Glycine requirement	8
gua-4	0/360* (*guaB*, 0/360)	Guanine requirement	8
gua-6	54* (*guaA*, 54)	Guanine requirement	8
gua-11	4‡	Guanine requirement	8
his-1	202* (*hisH*, 202)	Histidine requirement	8
his-3	287† (*hisA*, 298)	Histidine requirement	8
hom-1	284* (*hom*, 284)	Homoserine or threonine and methionine requirement	8
hsmI	NM	Modification methylase of the *Bsm*I system	1
hsrI	NM	Restriction endonuclease of the *Bsm*I system	1
ilv-2	247* (*ilvB* or *ilvC*, 247)	Isoleucine and valine requirement	8
ksg-1	277* (*ksgB*, 277)	Resistance to kasugamycin	8
ksg-2	287* (*ksgC*, 287)	Resistance to kasugamycin	8
ksg-3	4* (*ksgA*, 4)	Resistance to kasugamycin	8
leu-2	247* (*leuA*, *leuB*, or *leuC*, 247)	Leucine requirement	8
lin-1	24* (*lin*, 24)	Resistance to lincomycin	8
met-1	11‡	Methionine, homocysteine, cystathionine, sulfide, or sulfite requirement	8

(Continued)

Table 1—*Continued*

Gene symbol[a]	Map position (degrees)[b]	Phenotype, enzyme deficiency, or other characteristic	Reference[c]
met-2	11‡	Methionine, homocysteine, cystathionine, sulfide, or sulfite requirement	8
met-3	141–198¶	Methionine, homocysteine, or cystathionine requirement	8
met-4	141–198¶	Methionine requirement	8
met-5	110* (*metA*, 110)	Methionine, homocysteine, or cystathionine requirement	8
mut-1	NM	Mannitol utilization	10
nic-2	240* (*nic*, 240)	Nicotinate requirement	8
pig-1	NM	Pigment produced by cells growing on minimal glucose agar plates	1
pro-1	NM	Proline requirement	10
pur-5	55‡	Adenine or guanine requirement	8
pur-6	348* (*purA*, 348)	Adenine requirement	8
pur-21	55‡	Guanine or hypoxanthine requirement	8
pur-22	55* (*purB*, 55)	Adenine, guanine, or hypoxanthine requirement	8
pyr-2	139* (*pyrA*, 139)	Carbamyl phosphate, *N*-carbamyl-DL-aspartate, orotate, thymine, uracil, or cytosine requirement	8
pyr-3	139* (*pyrB*, 139)	Carbamyl phosphate, orotate, thymine, uracil, or cytosine requirement	8
pyr-4	NM	Orotate, uracil, cytosine, or thymine requirement	10
pyr-5	NM	Uracil, cytosine, or thymine requirement	10
rec-1	NM	Genetic recombination and radiation resistance	10
rfm-1	12* (*rfm*, 12)	Resistance to rifampin	8
rib-2	209‡	Riboflavin requirement	8
rib-3	209* (*ribO*, 209)	Riboflavin requirement	8
ser-1	207* (*serA*, 207)	Serine requirement	8
sin-1	NM	Sugar inhibition; sensitivity to glucose and decreased utilization of sucrose	10
sin-2	NM	Sugar inhibition; sensitivity to glucose	10
sin-3	NM	Sugar inhibition; sensitivity to sucrose	10
sin-5	NM	Sugar inhibition; sensitivity to glucose and sucrose	10
str-1	12* (*strA*, 12)	Resistance to streptomycin	8
str-3	130* (*strB*, 130)	Resistance to streptomycin	8
str-4	219* (*strC*, 219)	Resistance to streptomycin	8
sut-1	NM	Sugar utilization, sucrose	10
sut-2	NM	Sugar utilization, galactose and glucose; decreased maltose utilization. Partial resistance to 2-deoxyglucose and 2-deoxygalactose	10†
sut-3	NM	Sugar utilization, glucose; decreased maltose utilization. Resistance to 2-deoxyglucose and 2-deoxygalactose	10†
sut-4	NM	Sugar utilization; decreased maltose and glucose utilization. Resistance to 2-deoxyglucose and 2-deoxygalactose	10†
sut-5	NM	Sugar utilization, glycerol. Resistance to 2-deoxyglucose and 2-deoxygalactose	10†
sut-6	NM	Sugar utilization, galactose and sucrose; decreased glucose utilization. Partial resistance to 2-deoxyglucose and 2-deoxygalactose	10†
sut-7	NM	Sugar utilization, galactose, glucose, sucrose, and trehalose. Resistance to 2-deoxyglucose and 2-deoxygalactose and partial resistance to 5-thioglucose	10†
sut-8	NM	Sugar utilization, glucose; decreased maltose utilization and increased galactose utilization. Partial resistance to 2-deoxyglucose	10†
sut-9	NM	Sugar utilization, glucose; decreased maltose utilization and increased galactose utilization. Resistance to 2-deoxyglucose	10†
sut-10	NM	Sugar utilization, glucose; increased galactose utilization. Resistance to 2-deoxyglucose, 2-deoxygalactose, and 5-thioglucose	10†
thi-1	NM	Thiamine requirement	8
thr-1	284* (*thrB*, 284)	Threonine requirement	8
thy-1	NM	Thymine requirement	10
tmr-1	24–25* (*tmrA* or *tmrB*, 24–25)	Resistance to tunicamycin	8
trp-1	203* (*trpA*, 203)	Tryptophan or indole requirement	8
trp-2	203* (*trpB*, 203)	Tryptophan requirement	8
trp-4	203* (*trpE*, 203)	Tryptophan, indole, or anthranilate requirement	8
trp-6	203* (*trpD*, 203)	Tryptophan or indole requirement	8
trp-7	203* (*trpF*, 203)	Tryptophan or indole requirement	8
trp-13	203* (*trpC*, 203)	Tryptophan or indole requirement	8

(Continued)

Table 1—*Continued*

Gene symbol[a]	Map position (degrees)[b]	Phenotype, enzyme deficiency, or other characteristic	Reference[c]
tsg-3	NM	Temperature-sensitive growth at 70°C	10
tsg-6	NM	Temperature-sensitive growth at 60°C and at 70°C	8
tyr-1	202* (*tyrA*, 202)	Tyrosine requirement	

[a] Genotype designation of loci conforms to the *B. subtilis* system (7). The nomenclature of *glnQ* and *glnH* conforms to the *Escherichia coli* system (6). Other genotype designations are taken from the phenotype characteristics in column 3.

[b] NM, not mapped by transduction or by protoplast fusion. *, loci that were assigned a cognate *B. subtilis* gene by using phenotypic characteristics, such as growth response to intermediates or end products of metabolism, cross-feeding, accumulation of intermediates, or sensitivity to antibiotics and other compounds, and their position relative to other loci, as established by transduction or protoplast fusion or both. The map position, in degrees, of the analogous *B. subtilis* loci was used to position these landmark loci on the *B. stearothermophilus* chromosome. †, loci that were assigned a cognate *B. subtilis* gene by using phenotypic characteristics and their location on the chromosome as established by transduction or protoplast fusion or both. The relative position of these loci to adjacent loci, however, was different from the relative positions of the cognate loci on the *B. subtilis* chromosome. The map position of these loci was estimated from transduction and protoplast fusion analyses. ‡, loci that could not be assigned a cognate *B. subtilis* gene. The map position of these loci was estimated from transduction and protoplast fusion analyses. §, loci that were assigned a cognate *B. subtilis* gene by using phenotypic characteristics but not mapped by transduction or protoplast fusion. The map position of the analogous *B. subtilis* loci was used to position these loci on the *B. stearothermophilus* chromosome. ||, *gln-1* is linked to *glnQH* by transduction. ¶, the relative position of *met-4* and *met-3* to *asp-1*, *aro-2*, and *his-1* was determined by two-factor transduction crosses. The position of the two loci relative to *fur-2* was not determined. The analogous *B. subtilis* genes and their map location on the *B. subtilis* chromosome are in parentheses.

[c] *, locus sequenced; †, mutants were kindly provided by T. Ferenci. Numbers refer to references.

tmr-1; *gua-6*, *pur-5*, and *pur-21*; *asp-1* and *his-1*; *gly-4* and *asp-3*; and *hom-1* and *his-4*, established by two-factor transduction crosses, was verified by protoplast fusion.

The availability of phages that transduce chromosomal DNA fragments that cover a range of sizes (9) will make it possible to identify linkage groups and to determine the order of closely linked markers. Protoplast fusion can be used to order transduction linkage groups, verify the order of genes in a linkage group, and locate individual genes on the genetic map. Two methods that have been used to identify mutations in *B. subtilis* have been developed for use in *B. stearothermophilus* NUB36. The first is an integrational vector

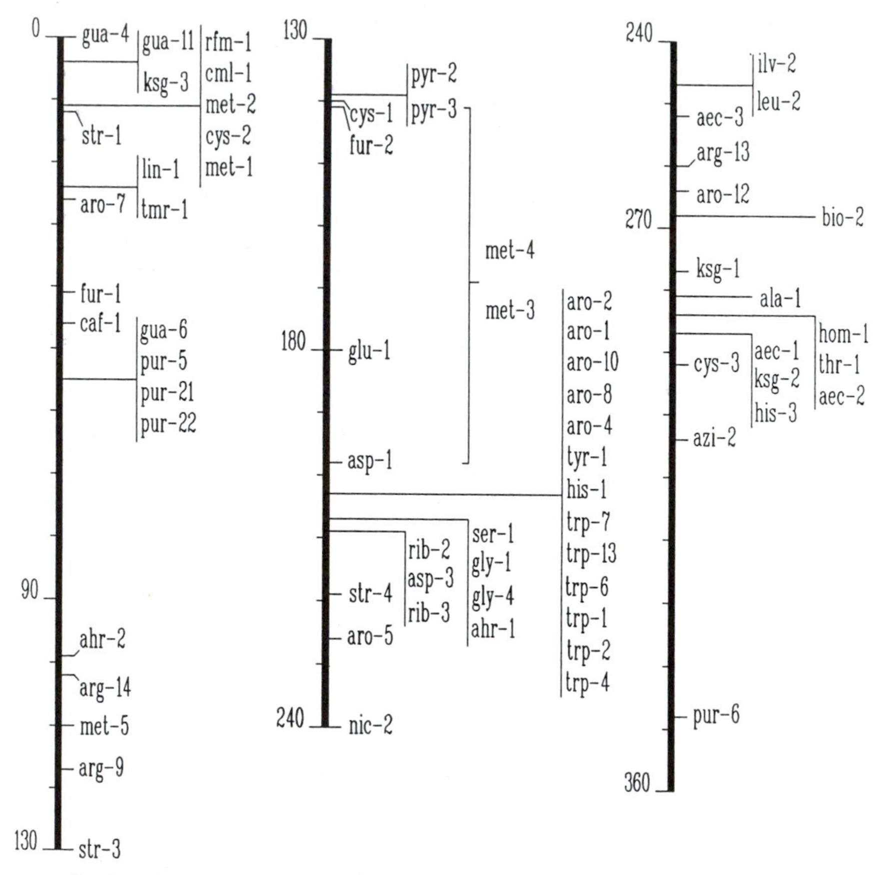

Figure 1. Linear scale drawings representing the circular linkage map of the *B. stearothermophilus* NUB36 chromosome. The scale of 360°, beginning with zero at the *gua-4* locus, is based on the *B. subtilis* map (7). The genetic symbols used in this figure are defined in Table 1. The map position of the loci was determined as described in Table 1, footnote *b*.

carrying cloned *B. stearothermophilus* DNA (13). The second is the use of transposon Tn*917* (10). In addition, we have constructed six isogenic reference strains that carry markers that span the chromosome. Since loci that are separated by 20 to 60° on the *B. stearothermophilus* genetic map can be linked by protoplast fusion, this will be the main method of mapping new markers.

Acknowledgments. This work was supported in part by grant DE-FG02-84ER13204 from the U.S. Department of Energy. I thank M. Heaton for preparing the linkage map, G. Munley for the isolation of the *gln-1* mutant, M. Hinkle for the isolation of the *tsg* mutants, and M. Hinkle, M. Kraujalis, and A. Carr, Jr., for their help in developing the Tn*917* system.

REFERENCES

1. **Chen, Z.-F., S. F. Wojcik, and N. E. Welker.** 1986. Genetic analysis of *Bacillus stearothermophilus* by protoplast fusion. *J. Bacteriol.* **165:**994–1001.
2. **Hoshino, T., T. Ikeda, H. Narushima, and N. Tomizuka.** 1985. Isolation and characterization of antibiotic-resistance plasmids in thermophilic bacilli. *Can. J. Microbiol.* **31:**339–345.
3. **Imanaka, T., M. Fujii, E. Aramori, and S. Aiba.** 1982. Transformation of *Bacillus stearothermophilus* with plasmid DNA and characterization of shuttle plasmids between *Bacillus stearothermophilus* and *Bacillus subtilis*. *J. Bacteriol.* **149:**824–830.
4. **Koyama, Y., T. Hoshino, N. Tomizuka, and K. Furukawa.** 1986. Genetic transformation of the extreme thermophile *Thermus thermophilus* and of other *Thermus* spp. *J. Bacteriol.* **166:**338–340.
5. **Meile, L., P. Abendshein, and T. Leisinger.** 1990. Transduction in the archaebacterium *Methanobacterium thermoautotrophicum* Marburg. *J. Bacteriol.* **172:**3507–3508.
6. **Nohno, T., T. Saito, and J.-S. Hong.** 1986. Cloning and complete nucleotide sequence of the *Escherichia coli* glutamine permease operon (*gln*HPQ). *Mol. Gen. Genet.* **205:**260–269.
7. **Piggot, P. J.** 1989. Revised genetic map of *Bacillus subtilis* 168, p. 1–41. *In* I. Smith, R. A. Slepecky, and P. Setlow (ed.), *Regulation of Procaryotic Development*. American Society for Microbiology, Washington, D.C.
8. **Vallier, H., and N. E. Welker.** 1990. Genetic map of the *Bacillus stearothermophilus* NUB36 chromosome. *J. Bacteriol.* **172:**793–801.
9. **Welker, N. E.** 1988. Transduction in *Bacillus stearothermophilus*. *J. Bacteriol.* **170:**3761–3764.
10. **Welker, N. E.** Unpublished data.
11. **Worrell, V. E., D. P. Nagle, Jr., D. McCarthy, and A. Eisenbraun.** 1988. Genetic transformation system in the archaebacterium *Methanobacterium thermoautotrophicum* Marburg. *J. Bacteriol.* **170:**653–656.
12. **Wu, L., and N. E. Welker.** 1989. Protoplast transformation of *Bacillus stearothermophilus* NUB36 by plasmid DNA. *J. Gen. Microbiol.* **135:**1315–1324.
13. **Wu, L., and N. E. Welker.** 1991. Cloning and characterization of a glutamine transport operon of *Bacillus stearothermophilus* NUB36: effect of temperature on regulation of transcription. *J. Bacteriol.* **173:**4877–4888.

34. The Genetic Map of *Staphylococcus aureus*

PETER A. PATTEE

The first report of genetic linkage involving the chromosome of *Staphylococcus aureus* came from a transduction analysis of L-tryptophan-dependent mutants performed by Harry L. Ritz, working in the laboratory of Jack N. Baldwin, in 1962 (84). This report was followed by similar studies that showed linkage among mutations affecting individual biosynthetic activities (3, 4, 39, 40, 74, 81, 96). Because genetic analyses were done by generalized transduction with phages with headful capacities of about 40 kbp (105), the linkage relationships among these individual regions remained undefined. Only when Lindberg and coworkers (47) reported transformation in *S. aureus* NCTC 8325 (which is Ps47, the propagating strain for phage 47 of the International Typing Series), and when some functionally unrelated genes were shown to be cotransformable (72), was it feasible to attempt to develop a genetic map of the *S. aureus* chromosome. The first maps consisted of individual linkage groups, defined by transformation, that consisted primarily of genes involved in biosynthetic and resistance phenotypes (66, 70, 72). Invaluable to these early studies were diverse silent chromosomal insertions of Tn*551* isolated several years earlier by Richard Novick using UV-irradiated transducing phage carrying the pI258 plasmid, on which Tn*551* was originally observed (60, 75). By use of a thermosensitive mutant of pI258 developed in the Novick laboratory, auxotrophic mutants resulting from insertional inactivation of various biosynthetic genes (66) and Tn*551* insertions adjacent to markers of interest (49) were isolated and their chromosomal loci were identified. The ability to isolate transposons inserted near target markers has been valuable in strain constructions, in the development of readily portable mutations, and for the extension of the known linkage groups. As increasing numbers of mutations were mapped, it became evident that the three linkage groups defined by transformation represented a major portion of what was assumed to be a single circular chromosome. This view was confirmed by Mark Stahl, who used protoplast fusion between multiply marked parental strains to determine the most probable relationship among the known linkage groups on a circular chromosome map (101). Transformation with high-molecular-weight DNA confirmed the orientation and linkage of two of the major linkage groups previously characterized by transformation (102). However, continuity of these linkage groups on a single circular genetic map, inferred by protoplast fusion, remained unsubstantiated. Protoplast fusion analysis also proved to be an efficient means of estimating the approximate location of previously unmapped mutations (20, 67, 102). However, the chromosome map based on these genetic methods of analysis was deficient in that neither the size of the chromosome, nor the relationship between genetic and physical distances, was known. Furthermore, the genetic map was based almost solely on NCTC 8325, a phage group III strain of *S. aureus*.

THE MAP

The availability of methods for the construction of physical maps of prokaryotic chromosomes based on pulsed-field agarose gel electrophoresis (PFGE) and DNA hybridization (97, 98), combined with existing libraries of genetically mapped insertions of transposons Tn*551*, Tn*917*, Tn*4001*, and Tn*916*, allowed the construction of a physical map of NCTC 8325 that was correlated with the genetic map (Fig. 1; 68). These studies show that the chromosome of *S. aureus* NCTC 8325 is about 2,800 kb and show physical continuity of the circular map. The distance between *pan* and Ω1252[Tn*4001*] (at 4 o'clock; Fig. 1) is presumably beyond that detectable by transformation, although the *Sgr*A1-A PFGE fragment (whose ends are defined by *coa* and Ω1043[Tn*551*]) clearly defines linkage across this region (68a).

With the exception of the *Sau*3AI endonuclease and methylase genes (89), the genetic nomenclature used with respect to the chromosomal markers of *S. aureus* is that of Demerec et al. (24), modified (19) with reference to transposons and similar determinants (Table 1). The silent insertions of transposition elements are retained in the current map because they serve as useful chromosomal markers representing regions otherwise deficient in more conventional markers. They also have been located on the genetic map and therefore provide a means of rapid identification (by DNA hybridization) of individual restriction fragments used to construct the physical map. As shown in Table 1, the Ω number of a silent chromosomal insertion is indicative of the identity of that transposon.

The restriction fragments shown in Fig. 1 have been drawn approximately to scale as determined by PFGE analysis relative to bacteriophage lambda DNA multimers and *Saccharomyces cerevisiae* chromosomes used as molecular weight standards. The locations of individual markers relative to one another on any one fragment are estimates based on cotransformation frequencies; although the order of markers shown is believed to be correct, the distances separating individual markers are estimates only, based on cotrans-

Peter A. Pattee • Department of Microbiology, Immunology and Preventive Medicine, Iowa State University, Ames, Iowa 50011.

Table 1. Genetic markers of the chromosome of *S. aureus*

Gene symbol[a]	Phenotype	Map location[b]	Reference(s)
aacA	Aminoglycoside acetyltransferase–aminoglycoside phosphotransferase bifunctional enzyme (AAC6′-APH2″)	Unknown	26
agr[c] (*hla*)	Accessory gene regulator of several exoproteins and toxins	F	16, 51, 54, 55, 76, 83
ala[c]	L-Alanine requirement	A	52, 66, 70
aphA	Aminoglycoside phosphotransferase (APH3′-III)	A	26
atr	Aminotriazole resistance	Unknown	18
attϕ11	Prophage ϕ11 integration site	F	75
attϕ12	Prophage ϕ12 integration site	A	46
attϕ13	Prophage ϕ13 integration site (in *hlb*)	F	21, 22
attϕL54a	Prophage ϕL54a integration site (in *geh*)	E	42, 43, 68a
att554	Primary integration site for Tn*554*	A	61, 77
bla (*pen*)	β-Lactamase production	F	75, 79
cfxB	Fluoroquinolone resistance	A	110
coa	Coagulase	E	37, 68a, 78
cyt	Cytosine requirement	Unknown	85
dna	Temperature-sensitive DNA replication	G	68a, 109
ermB	Impaired erythromycin resistance by Tn*551*	In Tn*551*	73
ery	Erythromycin resistance	Unknown	52
eta	Exfoliative toxin serotype A	Unknown	44, 63
etb	Exfoliative toxin serotype B	Unknown	32, 44
femA[c] (Ω2003)	Factor essential for expression of methicillin resistance; see Ω2004	A	6–9
fnb	Fibronectin binding protein	Unknown	28
fus	Fusidic acid resistance	D	101, 102
geh[d]	Glycerol ester hydrolase (lipase)	E	42, 43, 68a
gly	Glycine requirement	Unknown	52
gyrA	DNA gyrase; ciprofloxacin resistance	G	100, 110
his	L-Histidine requirement (*hisEABCDG*)	G or M	39, 40, 52, 72
hla (*hly*)	α-Toxin structural gene	B	27, 38, 62, 67
hlb[d]	β-Hemolysin	F	21, 22, 68a
hld	Delta hemolysin	F	33
hlg	Gamma hemolysin	C	23, 68a
ilv-leu[c]	L-Isoleucine, L-valine, L-leucine requirement (*ilvABCD-leuABCD*)	F	52, 72, 74, 96
ilvR	Resistance to D-leucine	F	65, 74
lac[c]	Phospho-β-galactosidase	H	15, 68a
lin	Lincomycin resistance	Unknown	52
lip[c]	Lipoic acid requirement	A	68a
lys[c]	L-Lysine requirement (*lysOABFG*)	A	3, 4, 72
lytA	*N*-Acetylmuramyl-L-alanine amidase (peptidoglycan hydrolase)	F	34, 68a, 111
mec	Methicillin resistance	G	25, 41, 103, 104
metA	L-Methionine requirement	K	50
metC[c]	L-Methionine requirement; β-cystathionase	A	52, 88
mit	Enhanced sensitivity to mitomycin C, nitrosoguanidine, and UV light	G	68a, 107, 112
mtl	Mannitol catabolism	Unknown	57, 58
ngr	Apurinic endonuclease deficiency	A	108
norA	Hydrophilic quinolone resistance	D	110, 113
nov	Novobiocin resistance	G	52, 69, 72
nuc	Staphylococcal thermostable nuclease deficiency	Unknown	92, 94
ofxC	Fluorquinolone resistance	A	110
ole	Oleandomycin resistance	Unknown	52
pan[c]	Pantothenate requirement	C	68a
pdx	Pyridoxal requirement	Unknown	66
phe	L-Phenylalanine requirement	Unknown	52
pig	Golden-yellow pigment deficiency	I	65
purA	Adenine requirement	G	72
purB[c]	Adenine + guanine requirement	F	72
purC[c]	Purine requirement	B	102
purD[c]	Guanine requirement	B	68a
pur-140	Purine requirement	B	49
recA	Homologous recombination deficiency	Unknown	112
rib[c]	Riboflavin requirement	J	66, 70
rif	Rifampin resistance	D	52, 56
rpoB	RNA polymerase β-subunit	Unknown	56

(Continued)

Table 1—*Continued*

Gene symbol[a]	Phenotype	Map location[b]	Reference(s)
rsv	Rifamycin SV resistance	Unknown	56
sak[e]	Staphylokinase production	Unknown	5, 22, 86
sau3AIR	*Sau*3AI restriction endonuclease	Unknown	89
sau3AIM	*Sau*3AI methyltransferase	Unknown	89
sea[e] (*entA*)	Enterotoxin A production	F	10–12, 22, 51, 70, 91
seb (*entB*)	Enterotoxin B production	Unknown	12, 35, 82, 90
sec (*entC*)	Enterotoxin C production	Unknown	12–14, 31
sed (*entD*)	Enterotoxin D production	Unknown	12, 17
see (*entE*)	Enterotoxin E production	Unknown	12
ser	L-Serine requirement	Unknown	52
sezA	Silent enterotoxin structural gene	Unknown	99
spa	Staphylococcal protein A	G	48, 64, 93
stl	Streptolydigin resistance	Unknown	56
str	Streptomycin resistance	Unknown	48
stv	Streptomaricin resistance	Unknown	56
tet	Tetracycline resistance	E	1, 65, 101, 102
thrA[c]	L-Lysine + L-methionine + L-threonine requirement; failure to convert aspartate to aspartic β-semialdehyde	A	66, 70, 88
thrB[c]	L-Threonine requirement; homoserine kinase deficiency	A	52, 66, 72, 88
thrC[c]	L-Threonine requirement; threonine synthetase deficiency	A	88
thy	Thymine requirement	A	47, 52, 72
tmn	Tetracycline and minocycline resistance	K	1, 65, 87
tofA	Temperature-sensitive osmotically remedial cell wall synthesis; D-glutamate addition defect	B	30, 59
trp[c]	L-Tryptophan requirement (*trpABFCDE*)	A	52, 72, 81, 84, 88
tsr	Temperature-sensitive growth	Unknown	106, 109
tst	Toxic shock syndrome toxin 1 structural gene (see Ω401 and Ω402)	A (2 loci)	20
tyrA[c]	L-Tyrosine requirement	A	66, 88
tyrB[c]	L-Tyrosine requirement	A	66, 88
uraA	Uracil requirement	C	52, 72
uraB[c]	Uracil requirement	B	66
uvr (*uvs*)	Enhanced sensitivity to UV light	Near or in N[f]	29, 68a, 112
Ω1–Ω100; Ω1000–Ω1099	Silent insertions of Tn*551*	Various	49, 60, 66
Ω401 and Ω402	Insertion sites of Hi*555*[g], the heterologous insertion carrying *tst*; Ω401 maps immediately adjacent to *trp* and causes a Trp⁻ phenotype; Ω402 maps near *tyrB*	A	20
Ω420 (Ω42)	Insertion site of pI258	F	60, 75
Ω500 (Ω50)	Insertion site of pI258	F	60, 75
Ω1100–Ω1199	Silent insertions of Tn*916*	Various	36, 114
Ω101–Ω120; Ω1200–Ω1299	Silent insertions of Tn*4001*	Various	50
Ω2004	Insertion site of Tn*551* that impairs Mec; may affect penicillin-binding proteins; see *femA*	A	6–9

[a] Gene designations in parentheses are old designations.
[b] Refers to specific *Sma*I restriction fragments in Fig. 1. Genes identified on a specific fragment but not shown in Fig. 1 have only been mapped by PFGE and DNA hybridization. Other genes (see Fig. 1 legend) have only been mapped by multifactorial transformation analyses.
[c] Identifies a phenotype for which insertional inactivation with Tn*551*, Tn*917*, and/or Tn*4001* is known.
[d] Controlled by negative phage conversion.
[e] Controlled by positive phage conversion.
[f] Chromosomal location of *uvr* is based on physical and genetic map data for Ω1073 and Ω1074, which exhibit greater than 50% cotransduction with *uvr*.
[g] Provisional designation for an element that exhibits some of the characteristics of a transposable element, but for which definitive evidence is lacking.

formation frequencies. Experience has shown that map units based on transformation are only approximations. While they may be relatively reproducible for pairs of point mutations, they often lack reciprocal agreement (particularly when working with insertion mutations), and considerable caution must be exercised in their interpretation. Some linkage data have been obtained by transduction, and an estimate of the relationship between transduction and transformation linkage values has been made (88).

PERSPECTIVES FOR THE FUTURE

At present, the most efficient way to identify the chromosomal locus occupied by a new marker of

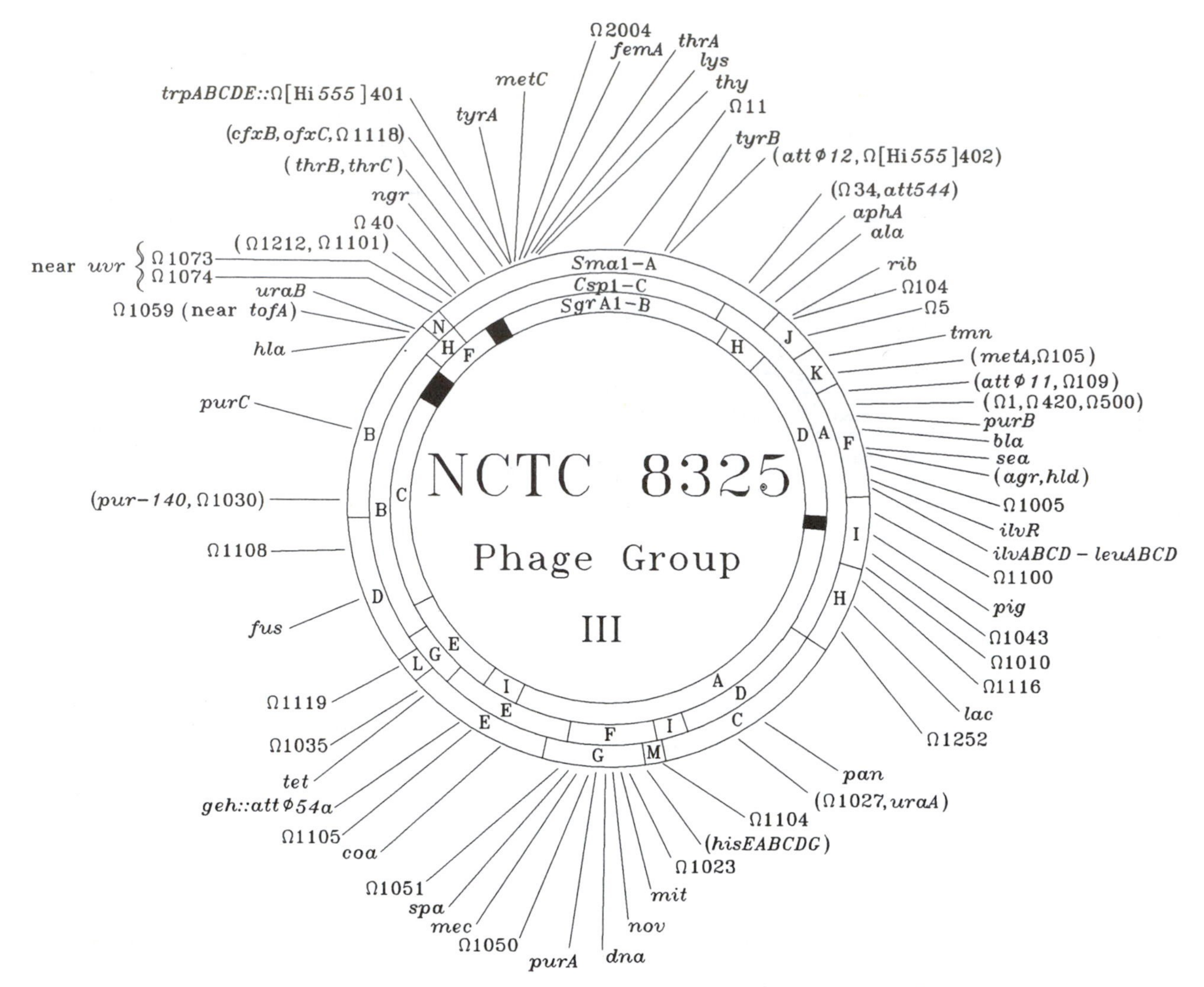

Figure 1. Chromosome map of *S. aureus*. The identity of individual markers is in Table 1. The order of the majority of the markers shown is based on genetic analysis and DNA hybridization analysis of *Sma*I, *Csp*I, and *Sgr*AI restriction fragments resolved by PFGE. Among these markers, *lys, thy,* Ω11, *att*ϕ*12,* Ω402, *aphA, att*ϕ*11,* Ω420, Ω500, *purB, bla, sea, hld,* Ω1005, *ilvR, leu, pig, uraA, his, mit, nov, dna, purA, fus, pur-140, tofA, uvr, ngr, thrC, cfxB, ofxC,* Ω401, *tyrA, metC,* Ω2004, and *femA,* have not been physically mapped, and their map locations are based only on multifactorial transformation analyses. The silent insertions of Tn*551* shown to be near *tofA* and near *uvr* (on the *Sgr*AI F fragment) are readily cotransducible with *tofA* and *uvr*, respectively. The order of markers shown within parentheses is not known. The entire chromosome is about 2,800 kb, and the individual fragments are drawn approximately to scale. The gaps in the *Sgr*AI map reflect regions that may be occupied by smaller fragments not located on the *Sgr*AI physical map (*Sgr*AI-G, -J, and -K, representing a total of about 300 kb). There are at least two *Csp*I fragments representing about 100 kb, and three *Sma*I fragments representing about 50 kb, that also have not been located on the map.

interest is by a combination of physical mapping by PFGE with *Sma*I-, *Csp*I-, and *Sgr*AI-digested chromosomal DNA and DNA hybridization with an appropriate probe. As additional details are added to the physical map, greater precision of these data will be possible. Because there is considerable heterogeneity in the distribution of restriction endonuclease recognition sites among different strains, which is most pronounced between strains in different phage groups (71), the marker of interest should be moved into a strain where this information is known (i.e., at this writing, strains 112 [45], Ps55 [2], and NCTC 8325, representing phage groups I, II, and III, respectively). Markers whose phenotypes permit direct selection by transduction present little problem in this regard. However, markers encoding exoprotein synthesis, for example, cannot be selected for during transfer. In these instances, isolation of a transposon inserted adjacent to the marker of interest provides the means of moving that marker by cotransfer (49, 59). Because their patterns of insertion differ, and because delivery plasmids for them are available, transposons Tn*551* and Tn*917* are the most useful for this purpose. Both encode macrolide resistance, however, which constrains their utility in some instances. Tn*4001* (encoding gentamicin resistance) and Tn*916* (encoding tetracycline resistance) exhibit diversity of chromosomal insertions in *S. aureus*; however, populations of chromosomal insertions of Tn*4001* that are relatively free of the delivery plasmid integrated into the chromosome can only be obtained in a Rec⁻ background (50), and an efficient means of generating large populations of independent chromosomal insertions of Tn*916* in *S. aureus* is not available (36, 114). For chromosomal markers for which a probe may not be available, the isolation of a transposon inserted very near that marker allows DNA hybridization analysis of the PFGE fragments with a transposon-specific probe to facilitate mapping; because close proximity of transposon and the target marker is important, transduction is preferable to transformation for isolating the insert near the marker of interest. Plasmid pTV20, which contains Tn*917* sequences (115), can be used to probe for either Tn*551* or Tn*917*. Tn*917 lac*, which contains an internal *Sma*I recognition site, is useful for physical mapping and can be delivered using the thermosensitive plasmid pTV32Ts (115); this provides the means of more exactly determining the location of a transposon within a *Sma*I restriction fragment based on the sizes of the two newly created *Sma*I fragments.

While it is apparent that there is reasonable conservation of the gene order among the very limited numbers of strains examined in phage groups I, II, and III (71), it is clear that some diversity exists. The best documented example of variability of the *S. aureus* chromosome occurs in methicillin-resistant clinical isolates. In these strains, considerable instability exists in the region of the chromosome surrounding the *mec* determinant (53, 80, 95). There is no reason to believe that other similarly variable regions of the chromosome of *S. aureus* will not come to light as additional isolates are examined in this context.

Dedication. This chapter is dedicated to the memory of Jack N. Baldwin.

Acknowledgments. Sincere thanks are extended to the many investigators who have generously supplied probes, mutants, and other materials without which the continued development of this map would not be possible. The expert technical assistance of Robyn L. Hottman, Lillian McLeod, and Judson Guericke is gratefully acknowledged. The unpublished data from this laboratory were obtained with support from the National Science Foundation.

REFERENCES

1. **Asheshov, E. H.** 1975. The genetics of tetracycline resistance in *Staphylococcus aureus*. *J. Gen. Microbiol.* **88:**132–140.
2. **Bannantine, J. P.** 1991. M.S. thesis. Iowa State University, Ames.
3. **Barnes, I. J., A. Bondi, and K. S. Fuscaldo.** 1971. Genetic analysis of lysine auxotrophs of *Staphylococcus aureus*. *J. Bacteriol.* **105:**553–555.
4. **Barnes, I. J., A. Bondi, and A. G. Moat.** 1969. Biochemical characterization of lysine auxotrophs of *Staphylococcus aureus*. *J. Bacteriol.* **99:**169–174.
5. **Behnke, D., and D. Gerlach.** 1987. Cloning and expression in *Escherichia coli*, *Bacillus subtilis*, and *Streptococcus sanguis* of a gene for staphylokinase—a bacterial plasminogen activator. *Mol. Gen. Genet.* **210:**528–534.
6. **Berger-Bachi, B.** 1983. Insertional inactivation of staphylococcal methicillin resistance by Tn*551*. *J. Bacteriol.* **154:**479–487.
7. **Berger-Bachi, B.** 1983. Increase in transduction efficiency of Tn*551* mediated by the methicillin resistance marker. *J. Bacteriol.* **154:**533–535.
8. **Berger-Bachi, B., L. Barberis-Maino, A. Strassle, and F. H. Kayser.** 1989. *femA*, a host-mediated factor essential for methicillin resistance in *Staphylococcus aureus*: molecular cloning and characterization. *Mol. Gen. Genet.* **219:**263–269.
9. **Berger-Bachi, B., and M. L. Kohler.** 1983. A novel site on the chromosome of *Staphylococcus aureus* influencing the level of methicillin resistance: genetic mapping. *FEMS Microbiol. Lett.* **20:**305–309.
10. **Betley, M. J., and J. J. Mekalanos.** 1985. Staphylococcal enterotoxin A is encoded by phage. *Science* **229:**185–187.
11. **Betley, M. J., and J. J. Mekalanos.** 1988. Nucleotide sequence of the type A staphylococcal enterotoxin gene. *J. Bacteriol.* **170:**34–41.
12. **Betley, M. J., P. M. Schlievert, M. S. Bergdoll, G. A. Bohach, J. J. Iandolo, S. A. Khan, P. A. Pattee, and R. R. Reiser.** 1990. Staphylococcal gene nomenclature. *ASM News* **56:**182. (Letter.)
13. **Bohach, G. A., and P. M. Schlievert.** 1987. Expression of staphylococcal enterotoxin C_1 in *Escherichia coli*. *Infect. Immun.* **55:**428–432.
14. **Bohach, G. A., and P. M. Schlievert.** 1987. Nucleotide sequence of the staphylococcal enterotoxin C1 gene and relatedness to other pyrogenic toxins. *Mol. Gen. Genet.* **209:**15–20.
15. **Breidt, F., Jr., W. Hengstenberg, U. Finkeldei, and G. C. Stewart.** 1987. Identification of the genes for the lactose-specific components of the phosphotransferase system in the *lac* operon of *Staphylococcus aureus*. *J. Biol. Chem.* **262:**16444–16449.
16. **Brown, D. R., and P. A. Pattee.** 1980. Identification of a chromosomal determinant of alpha-toxin production in *Staphylococcus aureus*. *Infect. Immun.* **30:**36–42.
17. **Brown, R. C., and A. H. A. Bingham.** 1991. Expression of *Staphylococcus aureus* enterotoxin type D in *Escherichia coli* X1776. *FEMS Microbiol. Lett.* **80:**299–304.
18. **Burke, M. E., and P. A. Pattee.** 1972. Histidine biosynthetic pathway in *Staphylococcus aureus*. *Can. J. Microbiol.* **18:**569–576.
19. **Campbell, A., P. Starlinger, D. E. Berg, D. Botstein, E. M. Lederberg, R. P. Novick, and W. Szybalski.** 1979. Nomenclature of transposable elements in prokaryotes. *Plasmid* **2:**466–473.
20. **Chu, M. C., B. N. Kreiswirth, P. A. Pattee, R. P. Novick, M. E. Melish, and J. F. James.** 1988. Association of toxic shock toxin-1 determinant with a heterologous insertion at multiple loci in the *Staphylococcus aureus* chromosome. *Infect. Immun.* **56:**2702–2708.
21. **Coleman, D. C., J. P. Arbuthnott, H. M. Pomeroy, and T. H. Birkbeck.** 1986. Cloning and expression in *Escherichia coli* and *Staphylococcus aureus* of the beta-lysin determinant from *Staphylococcus aureus*: evidence that bacteriophage conversion of beta-lysin activity is caused by insertional inactivation of the beta-lysin determinant. *Microb. Pathogen.* **1:**549–564.
22. **Coleman, D. C., D. S. Sullivan, R. J. Russell, J. P. Arbuthnott, B. F. Carey, and H. M. Pomeroy.** 1989. *Staphylococcus aureus* bacteriophages mediating the simultaneous lysogenic conversion of β-lysin, staphylokinase and enterotoxin A: molecular mechanism of triple conversion. *J. Gen. Microbiol.* **135:**1679–1697.
23. **Cooney, J., M. Mulvey, J. P. Arbuthnott, and T. J. Foster.** 1988. Molecular cloning and genetic analysis of the determinant for gamma-lysin, a two-component toxin of *Staphylococcus aureus*. *J. Gen. Microbiol.* **134:**2179–2188.
24. **Demerec, M., E. A. Adelberg, A. J. Clark, and P. E. Hartman.** 1966. A proposal for a uniform nomenclature in bacterial genetics. *Genetics* **54:**61–76.

25. **Dornbusch, K., H. O. Hallander, and F. Lofquist.** 1969. Extrachromosomal control of methicillin resistance and toxin production in *Staphylococcus aureus*. *J. Bacteriol.* **98:**351–358.
26. **El Solh, N., N. Moreau, and S. D. Ehrlich.** 1986. Molecular cloning and analysis of *Staphylococcus aureus* chromosomal aminoglycoside resistance genes. *Plasmid* **15:**104–118.
27. **Fairweather, N., S. Kennedy, T. J. Foster, M. Kehoe, and G. Dougan.** 1983. Expression of a cloned *Staphylococcus aureus* α-hemolysin determinant in *Bacillus subtilis* and *Staphylococcus aureus*. *Infect. Immun.* **41:** 1112–1117.
28. **Flock, J.-I., G. Froman, K. Jonsson, B. Guss, C. Signas, B. Nilsson, G. Raucci, M. Hook, T. Wadstrom, and M. Lindberg.** 1987. Cloning and expression of the gene for a fibronectin-binding protein from *Staphylococcus aureus*. *EMBO J.* **6:**2351–2357.
29. **Goering, R. V., and P. A. Pattee.** 1971. Mutants of *Staphylococcus aureus* with increased sensitivity to ultraviolet radiation. *J. Bacteriol.* **106:**157–161.
30. **Good, G. M., and P. A. Pattee.** 1970. Temperature-sensitive osmotically fragile mutants of *Staphylococcus aureus*. *J. Bacteriol.* **104:**1401–1403.
31. **Hovde, C. J., S. P. Hackett, and G. A. Bohack.** 1990. Nucleotide sequence of the staphylococcal enterotoxin C3 gene: sequence comparison of all three type C staphylococcal enterotoxins. *Mol. Gen. Genet.* **220:**329–333.
32. **Jackson, M. P., and J. J. Iandolo.** 1986. Cloning and expression of the exfoliative toxin B gene from *Staphylococcus aureus*. *J. Bacteriol.* **166:**574–580.
33. **Janzon, L., S. Lofdahl, and S. Arvidson.** 1989. Identification and nucleotide sequence of the delta-lysin gene, *hld*, adjacent to the accessory gene regulator (*agr*) of *Staphylococcus aureus*. *Mol. Gen. Genet.* **219:**480–485.
34. **Jayaswal, R. K., Y.-I. Lee, and B. J. Wilkinson.** 1990. Cloning and expression of a *Staphylococcus aureus* gene encoding a peptidoglycan hydrolase activity. *J. Bacteriol.* **172:**5783–5788.
35. **Johns, M. B., Jr., and S. A. Khan.** 1988. Staphylococcal enterotoxin B gene is associated with a discrete genetic element. *J. Bacteriol.* **170:**4033–4039.
36. **Jones, J. M., S. C. Yost, and P. A. Pattee.** 1987. Transfer of the conjugal tetracycline resistance transposon Tn*916* from *Streptococcus faecalis* to *Staphylococcus aureus* and identification of some insertion sites in the staphylococcal chromosome. *J. Bacteriol.* **169:**2121–2131.
37. **Kaida, S., T. Miyata, Y. Yoshizawa, H. Igarashi, and S. Iwanaga.** 1989. Nucleotide and deduced amino acid sequence of staphylocoagulase gene from *Staphylococcus aureus* strain 213. *Nucleic Acids Res.* **17:**8871.
38. **Kehoe, M., J. Duncan, T. Foster, N. Fairweather, and G. Dougan.** 1983. Cloning, expression, and mapping of the *Staphylococcus aureus* α-hemolysin determinant in *Escherichia coli* K-12. *Infect. Immun.* **41:**1105–1111.
39. **Kloos, W. E., and P. A. Pattee.** 1965. A biochemical characterization of histidine-dependent mutants of *Staphylococcus aureus*. *J. Gen. Microbiol.* **39:**185–194.
40. **Kloos, W. E., and P. A. Pattee.** 1965. Transduction analysis of the histidine region in *Staphylococcus aureus*. *J. Gen. Microbiol.* **39:**195–207.
41. **Kuhl, S. A., P. A. Pattee, and J. N. Baldwin.** 1978. Chromosome map location of the methicillin resistance determinant in *Staphylococcus aureus*. *J. Bacteriol.* **135:** 460–465.
42. **Lee, C. Y., and J. J. Iandolo.** 1986. Lysogenic conversion of staphylococcal lipase is caused by insertion of the bacteriophage L54a genome into the lipase structural gene. *J. Bacteriol.* **166:**385–391.
43. **Lee, C. Y., and J. J. Iandolo.** 1988. Structural analysis of staphylococcal bacteriophage ϕ11 attachment sites. *J. Bacteriol.* **170:**2409–2411.
44. **Lee, C. Y., J. J. Schmidt, A. D. Johnson-Winegar, L. Spero, and J. J. Iandolo.** 1987. Sequence determination and comparison of the exfoliative toxin A and toxin B genes from *Staphylococcus aureus*. *J. Bacteriol.* **169:** 3904–3909.
45. **Lee, H.-C.** 1991. M.S. thesis. Iowa State University, Ames.
46. **Limpa-Amara, Y.** 1978. M.S. thesis. Iowa State University, Ames.
47. **Lindberg, M., J.-E. Sjostrom, and T. Johansson.** 1972. Transformation of chromosomal and plasmid characters in *Staphylococcus aureus*. *J. Bacteriol.* **109:**844–847.
48. **Lofdahl, S., B. Guss, M. Uhlen, L. Philipson, and M. Lindberg.** 1983. Gene for staphylococcal protein A. *Proc. Natl. Acad. Sci. USA* **80:**697–701.
49. **Luchansky, J. B., and P. A. Pattee.** 1984. Isolation of transposon Tn*551* insertions near chromosomal markers of interest in *Staphylococcus aureus*. *J. Bacteriol.* **159:**894–899.
50. **Mahairas, G. G., B. R. Lyon, R. A. Skurray, and P. A. Pattee.** 1989. Genetic analysis of *Staphylococcus aureus* with Tn*4001*. *J. Bacteriol.* **171:**3968–3972.
51. **Mallonee, D. H., B. A. Glatz, and P. A. Pattee.** 1982. Chromosomal mapping of a gene affecting enterotoxin A production in *Staphylococcus aureus*. *Appl. Environ. Microbiol.* **43:**397–402.
52. **Martin, S. M., S. C. Shoham, M. Alsup, and M. Rogolsky.** 1980. Genetic mapping in phage group 2 *Staphylococcus aureus*. *Infect. Immun.* **27:**532–541.
53. **Matthews, P. R., and P. R. Stewart.** 1988. Amplification of a section of chromosomal DNA in methicillin-resistant *Staphylococcus aureus* following growth in high concentrations of methicillin. *J. Gen. Microbiol.* **134:** 1455–1464.
54. **McClatchy, J. K., and E. D. Rosenblum.** 1966. Biological properties of α-toxin mutants of *Staphylococcus aureus*. *J. Bacteriol.* **92:**575–579.
55. **McClatchy, J. K., and E. D. Rosenblum.** 1966. Genetic recombination between α-toxin mutants of *Staphylococcus aureus*. *J. Bacteriol.* **92:**580–583.
56. **Morrow, T. O., and S. A. Harmon.** 1979. Genetic analysis of *Staphylococcus aureus* RNA polymerase mutants. *J. Bacteriol.* **137:**374–383.
57. **Murphey, W. H., and E. D. Rosenblum.** 1964. Mannitol catabolism by *Staphylococcus aureus*. *Arch. Biochem. Biophys.* **107:**292–297.
58. **Murphey, W. H., and E. D. Rosenblum.** 1964. Genetic recombination by transduction between mannitol-negative mutants of *Staphylococcus aureus*. *Proc. Soc. Exp. Biol. Med.* **116:**544–548.
59. **Nieuwlandt, D. T., and P. A. Pattee.** 1989. Transformation of a conditionally peptidoglycan-deficient mutant of *Staphylococcus aureus* with plasmid DNA. *J. Bacteriol.* **171:**4906–4913.
60. **Novick, R. P., E. Edelman, M. D. Schwesinger, A. D. Gruss, E. C. Swanson, and P. A. Pattee.** 1979. Genetic translocation in *Staphylococcus aureus*. *Proc. Natl. Acad. Sci. USA* **76:**400–404.
61. **Novick, R. P., S. A. Khan, E. Murphy, S. Iordanescu, I. Edelman, J. Krolewski, and M. Rush.** 1981. Hitchhiking transposons and other mobile genetic elements and site-specific recombination systems in *Staphylococcus aureus*. *Cold Spring Harbor Symp. Quant. Biol.* **45:**67–76.
62. **O'Reilly, M., J. C. S. de Azavedo, S. Kennedy, and T. J. Foster.** 1986. Inactivation of the alpha-haemolysin gene of *Staphylococcus aureus* 8325-4 by site-directed mutagenesis and studies on the expression of its haemolysins. *Microb. Pathogen.* **1:**125–138.
63. **O'Toole, P. W., and T. J. Foster.** 1986. Molecular cloning and expression of the epidermolytic toxin A gene of *Staphylococcus aureus*. *Microb. Pathogen.* **1:**583–594.

64. **Patel, A. H., T. J. Foster, and P. A. Pattee.** 1989. Physical and genetic mapping of the protein A gene in the chromosome of *Staphylococcus aureus* 8325-4. *J. Gen. Microbiol.* **135:**1799–1807.
65. **Pattee, P. A.** 1976. Genetic linkage of chromosomal tetracycline resistance and pigmentation to a purine auxotrophic marker and the isoleucine-valine-leucine structural genes in *Staphylococcus aureus*. *J. Bacteriol.* **127:**1167–1172.
66. **Pattee, P. A.** 1981. Distribution of Tn*551* insertion sites responsible for auxotrophy on the *Staphylococcus aureus* chromosome. *J. Bacteriol.* **145:**479–488.
67. **Pattee, P. A.** 1986. Chromosomal map location of the alpha-hemolysin structural gene in *Staphylococcus aureus* NCTC 8325. *Infect. Immun.* **54:**593–596.
68. **Pattee, P. A.** 1990. Genetic and physical mapping of the chromosome of *Staphylococcus aureus* NCTC 8325, p. 163–169. *In* K. Drlica and M. Riley (ed.), *The Bacterial Chromosome*. American Society for Microbiology, Washington, D.C.
68a. **Pattee, P. A.** Unpublished data.
69. **Pattee, P. A., and J. N. Baldwin.** 1961. Transduction of resistance to chlortetracycline and novobiocin in *Staphylococcus aureus*. *J. Bacteriol.* **82:**875–881.
70. **Pattee, P. A., and B. A. Glatz.** 1980. Identification of a chromosomal determinant of enterotoxin A production in *Staphylococcus aureus*. *Appl. Environ. Microbiol.* **39:** 186–193.
71. **Pattee, P. A., H.-C. Lee, and J. P. Bannantine.** 1990. Genetic and physical mapping of the chromosome of *Staphylococcus aureus*, p. 41–58. *In* R. P. Novick and R. A. Skurray (ed.), *Molecular Biology of the Staphylococci*. VCH Publishers, Inc., New York.
72. **Pattee, P. A., and D. S. Neveln.** 1975. Transformation analysis of three linkage groups in *Staphylococcus aureus*. *J. Bacteriol.* **124:**201–211.
73. **Pattee, P. A., C. J. Schroeder, and M. L. Stahl.** 1983. Erythromycin-sensitive mutations of transposon Tn*551* in *Staphylococcus aureus*. *Iowa State J. Res.* **58:**175–180.
74. **Pattee, P. A., T. Schutzbank, H. D. Kay, and M. H. Laughlin.** 1974. Genetic analysis of the leucine biosynthetic genes and their relationship to the *ilv* gene cluster. *Ann. N.Y. Acad. Sci.* **236:**175–186.
75. **Pattee, P. A., N. E. Thompson, D. Haubrich, and R. P. Novick.** 1977. Chromosomal map locations of integrated plasmids and related elements in *Staphylococcus aureus*. *Plasmid* **1:**38–51.
76. **Peng, H.-L., R. P. Novick, B. Kreiswirth, J. Kornblum, and P. Schlievert.** 1988. Cloning, characterization, and sequencing of an accessory gene regulator (*agr*) in *Staphylococcus aureus*. *J. Bacteriol.* **170:**4365–4372.
77. **Phillips, S., and R. P. Novick.** 1979. Tn*554*: a site-specific repressor-controlled transposon in *Staphylococcus aureus*. *Nature* (London) **278:**476–478.
78. **Phonimdaeng, P., M. O'Reilly, P. W. O'Toole, and T. J. Foster.** 1988. Molecular cloning and expression of the coagulase gene of *Staphylococcus aureus* 8325-4. *J. Gen. Microbiol.* **134:**75–83.
79. **Poston, S. M.** 1966. Cellular location of the genes controlling penicillinase production and resistance to streptomycin and tetracycline in a strain of *Staphylococcus aureus*. *Nature* (London) **210:**802–804.
80. **Poston, S. M., and F.-L. L. S. Hee.** 1991. Genetic characterisation of resistance to metal ions in methicillin-resistant *Staphylococcus aureus*: elimination of resistance to cadmium, mercury and tetracycline with loss of methicillin resistance. *J. Med. Microbiol.* **34:**193–201.
81. **Proctor, A. R., and W. E. Kloos.** 1970. The tryptophan gene cluster of *Staphylococcus aureus*. *J. Gen. Microbiol.* **64:**319–327.
82. **Ranelli, D. M., C. R. Jones, M. B. Johns, G. J. Mussey, and S. A. Khan.** 1985. Molecular cloning of staphylococcal enterotoxin B gene in *Escherichia coli* and *Staphylococcus aureus*. *Proc. Natl. Acad. Sci. USA* **82:**5850–5854.
83. **Recsei, P. A., B. Kreiswirth, M. O'Reilly, S. Schlievert, A. Gruss, and R. P. Novick.** 1986. Regulation of exoprotein gene expression in *Staphylococcus aureus* by *agr*. *Mol. Gen. Genet.* **202:**58–61.
84. **Ritz, H. L., and J. N. Baldwin.** 1962. A transduction analysis of complex loci governing the synthesis of tryptophan by *Staphylococcus aureus*. *Proc. Soc. Exp. Biol. Med.* **110:**667–671.
85. **Rudin, L., J.-E. Sjostrom, M. Lindberg, and L. Philipson.** 1974. Factors affecting competence for transformation in *Staphylococcus aureus*. *J. Bacteriol.* **118:**155–164.
86. **Sako, T., S. Sawaki, T. Sakurai, S. Ito, Y. Yoshizawa, and I. Kondo.** 1983. Cloning and expression of the staphylokinase gene of *Staphylococcus aureus* in *Escherichia coli*. *Mol. Gen. Genet.* **190:**271–277.
87. **Schaefler, S., W. Francois, and C. L. Ruby.** 1976. Minocycline resistance in *Staphylococcus aureus*: effect on phage susceptibility. *Antimicrob. Agents Chemother.* **9:**600–613.
88. **Schroeder, C. J., and P. A. Pattee.** 1984. Transduction analysis of transposon Tn*551* insertions in the *trp-thy* region of the *Staphylococcus aureus* chromosome. *J. Bacteriol.* **157:**533–537.
89. **Seeber, S., C. Kessler, and F. Gotz.** 1990. Cloning, expression and characterization of the *Sau*3AI restriction and modification genes in *Staphylococcus carnosus* TM300. *Gene* **94:**37–43.
90. **Shafer, W. M., and J. J. Iandolo.** 1978. Chromosomal locus for staphylococcal enterotoxin B. *Infect. Immun.* **20:**273–278.
91. **Shafer, W. M., and J. J. Iandolo.** 1978. Staphylococcal enterotoxin A: a chromosomal gene product. *Appl. Environ. Microbiol.* **36:**389–391.
92. **Shortle, D.** 1983. A genetic system for analysis of staphylococcal nuclease. *Gene* **22:**181–189.
93. **Shuttleworth, H. L., C. J. Duggleby, S. A. Jones, T. Atkinson, and N. P. Minton.** 1987. Nucleotide sequence analysis of the gene for protein A from *Staphylococcus aureus* Cowan 1 (NCTC8530) and its enhanced expression in *Escherichia coli*. *Gene* **58:**283–295.
94. **Sjostrom, J.-E., M. Lindberg, and L. Philipson.** 1973. Competence for transfection in *Staphylococcus aureus*. *J. Bacteriol.* **113:**576–585.
95. **Skinner, S., B. Inglis, P. R. Matthews, and P. R. Stewart.** 1988. Mercury and tetracycline resistance genes and flanking repeats associated with methicillin resistance on the chromosome of *Staphylococcus aureus*. *Mol. Microbiol.* **2:**289–292.
96. **Smith, C. D., and P. A. Pattee.** 1967. Biochemical and genetic analysis of isoleucine and valine biosynthesis in *Staphylococcus aureus*. *J. Bacteriol.* **93:**1832–1838.
97. **Smith, C. L., J. G. Econome, A. Schutt, S. Klco, and C. R. Cantor.** 1987. A physical map of the *Escherichia coli* K-12 genome. *Science* **236:**1448–1453.
98. **Smith, C. L., P. E. Warburton, A. Gaal, and C. R. Cantor.** 1986. Analysis of genome organization and rearrangements by pulsed field gradient gel electrophoresis, p. 45–70. *In* J. K. Setlow and A. Hollaender (ed.), *Genetic Engineering. Principles and Methods*, vol. 8. Plenum Publishing Corp., New York.
99. **Soltis, M. T., J. J. Mekalanos, and M. J. Betley.** 1990. Identification of a bacteriophage containing a silent staphylococcal variant enterotoxin gene (*sezA*$^+$). *Infect. Immun.* **58:**1614–1619.
100. **Sreedharan, S., M. Oram, B. Jensen, L. R. Peterson, and L. M. Fisher.** 1990. DNA gyrase *gyrA* mutations in ciprofloxacin-resistant strains of *Staphylococcus aureus*: close similarity with quinolone resistance mutations in *Escherichia coli*. *J. Bacteriol.* **172:**7260–7262.
101. **Stahl, M. L., and P. A. Pattee.** 1983. Computer-assisted

chromosome mapping by protoplast fusion in *Staphylococcus aureus. J. Bacteriol.* **154:**395–405.

102. **Stahl, M. L., and P. A. Pattee.** 1983. Confirmation of protoplast fusion-derived linkages in *Staphylococcus aureus* by transformation with protoplast DNA. *J. Bacteriol.* **154:**406–412.
103. **Stewart, G. C., and E. D. Rosenblum.** 1980. Genetic behavior of the methicillin resistance determinant in *Staphylococcus aureus. J. Bacteriol.* **144:**1200–1202.
104. **Stewart, G. C., and E. D. Rosenblum.** 1980. Transduction of methicillin resistance in *Staphylococcus aureus*: recipient effectiveness and beta-lactamase production. *Antimicrob. Agents Chemother.* **18:**424–432.
105. **Stewart, P. R., H. G. Waldron, J. S. Lee, and P. R. Matthews.** 1985. Molecular relationships among serogroup B bacteriophages of *Staphylococcus aureus. J. Virol.* **55:**111–116.
106. **Summers, D. K., and K. G. H. Dyke.** 1982. Characterization of a temperature-sensitive DNA replication mutant of *Staphylococcus aureus. J. Gen. Microbiol.* **128:** 1735–1741.
107. **Tam, J. E.** 1985. Ph.D. thesis. Iowa State University, Ames.
108. **Tam, J. E., and P. A. Pattee.** 1986. Characterization and genetic mapping of a mutation affecting apurinic endonuclease activity in *Staphylococcus aureus. J. Bacteriol.* **168:**708–714.
109. **Thomas, C. M., and K. G. H. Dyke.** 1978. Isolation and transduction analysis of temperature-sensitive mutants of *Staphylococcus aureus* defective in DNA replication. *J. Gen. Microbiol.* **106:**41–47.
110. **Trucksis, M., J. S. Wolfson, and D. C. Hooper.** 1991. A novel locus conferring fluoroquinolone resistance in *Staphylococcus aureus. J. Bacteriol.* **173:**5854–5860.
111. **Wang, X., B. J. Wilkinson, and R. K. Jayaswal.** 1991. Sequence analysis of a *Staphylococcus aureus* gene encoding a peptidoglycan hydrolase activity. *Gene* **102:** 105–109.
112. **Wyman, L., R. V. Goering, and R. P. Novick.** 1974. Genetic control of chromosomal and plasmid recombination in *Staphylococcus aureus. Genetics* **76:**681–702.
113. **Yoshida, H., M. Bogaki, S. Nakamura, K. Ubukata, and M. Konno.** 1990. Nucleotide sequence and characterization of the *Staphylococcus aureus norA* gene, which confers resistance to quinolones. *J. Bacteriol.* **172:**6942–6949.
114. **Yost, S. C., J. M. Jones, and P. A. Pattee.** 1988. Sequential transposition of Tn*916* among *Staphylococcus aureus* protoplasts. *Plasmid* **19:**13–20.
115. **Youngman, P.** 1987. Plasmid vectors for recovering and exploiting Tn917 transpositions in Bacillus and other Gram-positive bacteria, p. 79–103. *In* K. G. Hardy (ed.), *Plasmids: A Practical Approach.* IRL Press, Oxford.

35. The Chromosome Map of *Streptomyces coelicolor* A3(2)

DAVID A. HOPWOOD, HELEN M. KIESER, and TOBIAS KIESER

Genetic mapping of the chromosome of *Streptomyces coelicolor* A3(2) began with the analysis of haploid recombinants selected from matings (44). Two separate linkage groups were first recognized, but later they were joined into a single circular map (46). This conclusion was particularly significant because it was the first demonstration of map circularity in a gram-positive bacterium (and only *Escherichia coli* K-12 and the related *Salmonella typhimurium* LT2 provided precedents among gram-negative organisms). Useful information on the linkage map also came from the analysis of heteroclones. This was the name given to heterozygous colonies that grew when selection was made for progeny of a cross inheriting two closely linked markers, one from each parent (85). Heteroclones were deduced to arise from partially diploid plating units (51). Further rounds of recombination took place as the colonies developed so that the spores on individual heteroclones represented a population of haploid recombinant progeny from which linkage relationships could be deduced by nonselective analysis. These studies led to more accurate mapping within the two well-marked quadrants of the map that corresponded to the two original linkage groups (45).

Later, a large population of heteroclones from a particular seven-factor cross were analyzed simply for the presence or absence of each parental allele. It was deduced that each heteroclone contained a complete chromosome from one parent and a genomic fragment representing an incomplete chromosome from the other, and that the latter could represent any arc of the circular linkage map (48). This result established that the chromosome itself could not be a linear structure with constant ends (a situation that could have been compatible with a circular linkage map if even-numbered, and not odd-numbered, crossovers yielded viable recombinants; 88), but a circularly permuted genome, rather than a truly circular one, remained a possibility.

The data from this experiment were also used to calculate the relative lengths of the intervals between the seven loci by a maximum likelihood analysis (24, 48), on the assumption that the incomplete genomic segment in each heteroclone arose from a complete parental chromosome by random breakage. By this approach it was possible to estimate the lengths not only of the intervals in the two well-marked segments of the map, but also those of the two long opposite "silent regions" of the linkage map that were almost devoid of marker genes. The relative positions of the seven markers studied in this experiment were kept as constant reference points for all subsequent editions of the linkage map (50) until it was possible to construct a physical map of the genome.

The finding of good agreement between interval lengths derived from the heteroclone experiment and those previously deduced by conventional analysis of recombination in the well-marked quadrants of the chromosome did not mean that both accurately reflected physical distances, because heteroclone formation might require events that are also involved in recombination (48). Therefore the long silent regions of the linkage map might not have corresponded to long stretches of DNA, but could have contained recombinational "hot spots" that would have inflated their apparent length. In the absence of a means—such as interrupted mating or transduction of overlapping genomic segments—to relate genetic to physical map lengths in *S. coelicolor* (or any other actinomycete), there seemed no possibility of resolving the problem of the silent regions. The advent of physical genome mapping of *E. coli* by pulsed-field gel electrophoresis of large restriction fragments (86) or the construction of ordered clone libraries (61) has changed all that.

A physical map of the *S. coelicolor* chromosome has recently been constructed, using pulsed-field gel electrophoresis of *Ase*I and *Dra*I fragments (60). These two enzymes recognize sites (ATTAAT and TTTAAA, respectively) that, because of the unusually high average G+C content of *Streptomyces* DNA (~74%) (93), are quite rare: there are 17 and 8 sites, respectively, on the *S. coelicolor* chromosome. The physical and genetic maps were correlated by assigning mapped and cloned genes and the integrated genetic elements SLP1 and IS*117* to restriction fragments using Southern hybridization and by deducing the sites of integration of prophages VP5 and ϕC31, insertion element IS*110*, and integrating plasmid SCP1 from the increases in length of specific restriction fragments that occurred on integration of the elements.

The total size of the *S. coelicolor* A3(2) chromosome turned out to be about 8 Mb, some 75% larger than that of *E. coli* K-12 (61, 86). (The genome size of strains of *Streptomyces ambofaciens* has also been estimated to be ~8 Mb [64].)

The genetic map was found to be reasonably congruent with the physical map over the well-marked quadrants. It was also apparent that the silent regions are indeed long stretches of DNA. In fact, we argued (60) that the "3 o'clock" segment—which is consider-

David A. Hopwood, Helen M. Kieser, and Tobias Kieser • John Innes Institute, John Innes Centre, Norwich NR4 7UH, United Kingdom.

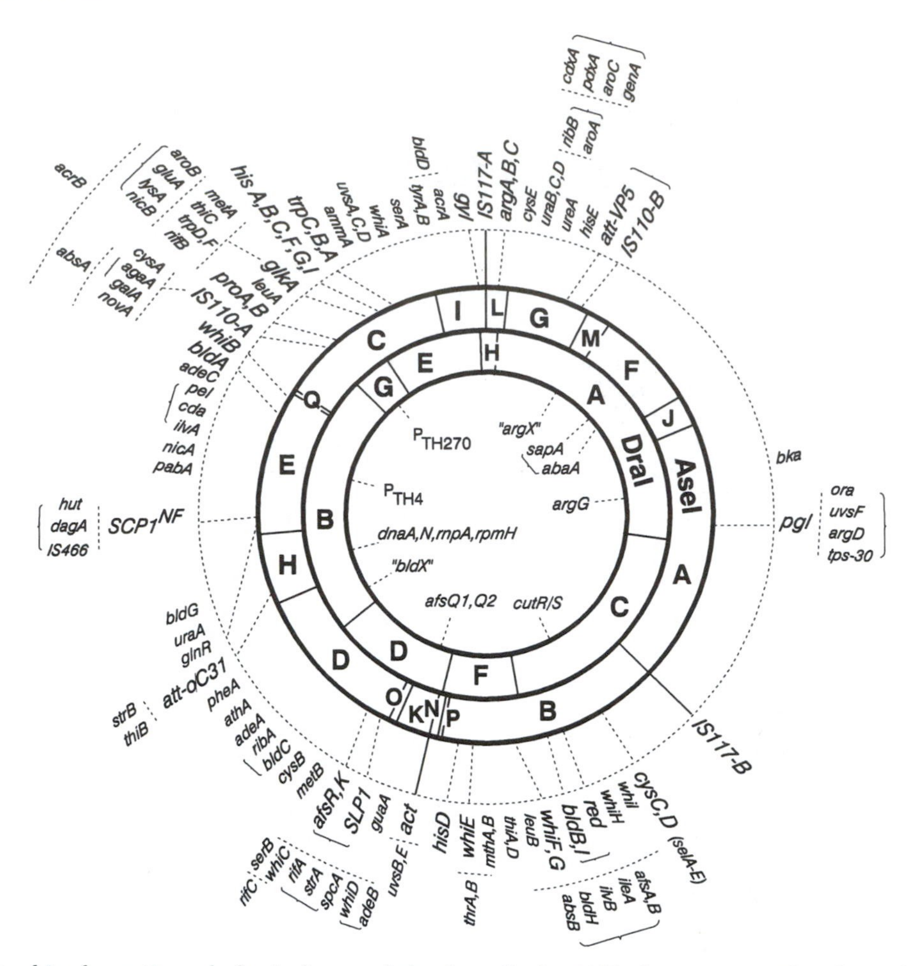

Figure 1. Combined genetic and physical map of the *S. coelicolor* A3(2) chromosome. The physical map for the restriction enzymes *Ase*I and *Dra*I (60) is shown as outer and inner bold circles, respectively. Restriction fragment sizes are given in Table 2 (which also includes the sizes of the unmapped *Ssp*I fragments). Evidence on which the order of fragments is based (including linking clones) is in Kieser et al. (60). On the inside are markers located physically but not genetically; their positions are indicated at the midpoint of the restriction fragment to which they have been assigned (sometimes a notional *Ase*I-*Dra*I fragment). The outside circle shows the genetic map, corrected by expansion of the 3 o'clock silent region (between the loci of *argA,B,C* and *cysC,D*) by one-third (see text). The spacing of loci on the genetic map is approximate only. Dotted lines connect physically mapped loci (shown in large type) to their approximate positions on the physical map; the three continuous lines indicate the precise positions of IS*117*-A, IS*117*-B, and the *act* cluster (which contain *Ase*I sites). Braces indicate groups of loci whose order has not been determined. Loci outside of short dotted lines have not been ordered with respect to loci inside such lines.

ably longer than the "9 o'clock" segment on both physical and genetic maps—was actually underestimated by the heteroclone experiment. By increasing the length of the 3 o'clock segment on the genetic map (between the closest well-mapped loci, *argA* and *cysC/D*) by about one-third, the two maps were made essentially superimposable; this correction has been applied in the map in Fig. 1. Table 1 describes all of the loci or gene clusters mapped genetically and/or physically.

Having established that the silent regions of the chromosome consist of long stretches of DNA, it is highly significant that they are almost devoid of typical classes of genetic markers found elsewhere on the map. These include the 60 or so genes involved in the biosynthesis of amino acids, purines, pyrimidines, or vitamins (except for *argG*; see below); 10 genes that mutated to give resistance to antibiotics and other inhibitors (*acr, gen, nov, rif, str*); the *act, cda,* and *red* gene clusters for antibiotic biosynthetic pathways and the 9 *aba, abs, afs,* and *ora* loci involved in the regulation of antibiotic biosynthesis; the 16 *bld* and *whi* loci for morphological differentiation; the 6 *uvs* genes for radiation repair; and, even more telling, the set of temperature-sensitive lethal mutations that identified 37 indispensable genes, none of which mapped to the silent regions (47). (Of these, only *tps-30*, which was one of the seven key markers in the heteroclone experiment discussed above, is shown in Fig. 1.) On the other hand, there is a suggestion that genes for the utilization of various carbon sources might be arranged more randomly on the map, since *dagA* maps at 9 o'clock (42) and the *bka* and *hut* loci were located near 3 o'clock and 9 o'clock, respectively, on the genetic map (58), whereas *agaA, galA,* and *gyl* map in the well-marked regions (Fig. 1).

It is a feature of streptomycetes that long segments of genomic DNA can be deleted from many laboratory

Table 1. Mapped genes of *S. coelicolor* A3(2)

Locus	Role	Reference(s)
abaA	Locus of at least two ORFs[a] which causes overproduction of actinorhodin when cloned in *S. lividans*; disruption of at least one ORF in *S. coelicolor* abolishes or reduces production of three antibiotics.	29
absA,B	Required for production of all four *S. coelicolor* antibiotics	1
acrA,B	Resistance to acriflavine	49
act	Cluster of at least 20 genes (ORFs) for actinorhodin biosynthesis and export, covering ~22 kb. Five ORFs (*act*I-ORF1, -2, and -3, *act*III, and *act*VII) encode components of the type II polyketide synthase; 11 ORFs (*act*IV, *act*VA-ORF2, -3, -4, -5, and -6, *act*VB, and *act*VI-ORF1, -2, -3, and -4) encode later biosynthetic steps; *act*II-ORF4 is a transcriptional activator; *act*II-ORF1, -2, and -3 and probably *act*VA-ORF1 are concerned with antibiotic export. The *act* cluster contains two *Ase*I sites.	9, 10, 27, 28, 35, 70, 71, 73, 77, 78, 81
adeA,C	Biosynthesis of purines	49
adeB	Biosynthesis of adenine	49
afsA	Production of A factor	37
afsB	Production of A factor and antibiotics	37
afsK	Encodes a protein kinase that phosphorylates the product of *afsR*	53
afsQ1,Q2	An operon of two genes whose products resemble bacterial two-component sensor-regulator systems. Causes overproduction of actinorhodin and undecylprodigiosin when cloned in *S. lividans*	56
afsR	Gene that causes overproduction of actinorhodin and undecylprodigiosin when cloned in *S. lividans*; disruption of *afsR* in *S. coelicolor* diminishes their production.	54
agaA	Utilization of agar as carbon source	49
ammA	Utilization of nitrate (mutants grow on ammonium)	49
argA,C,D	Biosynthesis of arginine (mutants can use ornithine)	49
argB	Biosynthesis of arginine	49
argG	Biosynthesis of arginine (mutants can use ornithine) (argininosuccinase)	5, 60
"argX"	Biosynthesis of arginine (mutants can use ornithine or proline) (ornithine acetyltransferase) (provisional designation)	18, 60
aroA,B,C	Biosynthesis of aromatic amino acids	49
athA	Biosynthesis of purines and thiamine	49
attVP5	Attachment site for phage VP5	22
attC31	Attachment site for phage ϕC31	67
bka	Utilization of *p*-hydroxybenzoate	58
bldA	tRNA gene for the rare leucine codon UUA. Mutations prevent the normal pattern of antibiotic synthesis and aerial mycelium formation.	62, 65, 75, 79
bldB	Encodes a small protein; mutants delayed or deficient in aerial mycelium formation and antibiotic production	38, 75
bldC	Mutant lacks aerial mycelium and produces little actinorhodin or red pigment on most media.	75
bldD,G	Mutants phenotypically resemble *bldA* mutants.	12, 75
bldH	Mutants lack aerial mycelium and antibiotic production on most media.	12
bldI	Mutants resemble *bldB* mutants.	38
"bldX"	*S. coelicolor* homolog of an *S. griseus* gene needed for aerial mycelium formation (provisional designation)	4, 60
cda	Production of calcium-dependent antibiotic	52
cdxA	Mutants require carbon dioxide.	91
cutR/S	Pair of genes whose products resemble bacterial two-component sensor-regulator systems; involved in copper metabolism	60, 90
cysA	Biosynthesis of cysteine	49
cysB	Biosynthesis of cysteine (mutants can use S_2O_3)	49
cysC	Biosynthesis of cysteine (mutants can use S_2O_3, or S_2O_4)	49
cysD (*selC*)	Biosynthesis of cysteine (mutants can use S_2O_3, S_2O_4, or S_2O_5); resistance to selenate and chromate	49, 69
cysE	Biosynthesis of cysteine (mutants can use S_2O_3)	49
dagA	Encodes diffusible agarase	42, 57
dnaA	Encodes DNA replication initiator protein	11, 32
dnaN	Encodes β-subunit of DNA polymerase III	11
galA	Utilization of galactose	83
genA	Resistance to gentamicin	49
glkA	Encodes glucose kinase; transcribed from its own promoter and also cotranscribed with an upstream ORF of unknown function. *glkA* mutants were isolated by resistance to 2-deoxyglucose and most cannot use glucose and are defective in glucose repression.	3, 30, 55
glnR	Regulatory gene for glutamine biosynthesis	92
gluA	Biosynthesis of glutamic acid	49
guaA	Biosynthesis of guanine	49

(Continued)

Table 1—*Continued*

Locus	Role	Reference(s)
gyl	*gylA,B,X* constitute an operon of three genes for glycerol utilization: *gylA*, glycerol kinase; *gylB*, glycerol-3-phosphate dehydrogenase; *gylX*, function not known. *gylR* is a regulator gene for *gylA,B,X* on a separate upstream transcript	84, 87
hisA	Biosynthesis of histidine (histidinol dehydrogenase) (equivalent to *hisD* of *E. coli*)	49, 66
hisB	Biosynthesis of histidine (imidazoleglycerol-phosphate dehydrase) (equivalent to *hisBd* of *E. coli*)	49, 66
hisC,F,I	Biosynthesis of histidine (isomerase, amidotransferase, or cyclase) (equivalent to *hisH* and/or *hisA* of *E. coli*)	49, 66
hisD	Biosynthesis of histidine (histidinol-phosphate phosphatase) (equivalent to *hisBpx* of *E. coli*)	49
hisE	Biosynthesis of histidine (phosphoribosyl-AMP 1,6-cyclohydrolase)	49
hisG	Biosynthesis of histidine (histidinol phosphate aminotransferase) (equivalent to *hisC* of *E. coli*)	49, 66
hut	Utilization of histidine (?formylglutamate iminohydrolase)	58
ileA	Biosynthesis of isoleucine	49
ilvA,B	Biosynthesis of isoleucine and valine	49
IS*110*-*A,B*	Inserted copies of mobile element IS*110*	15
IS*117*-*A,B*	Inserted copies of mobile element IS*117* (the minicircle). Each copy contains an *Ase*I site.	68
IS*466*	Two closely linked inserted copies of mobile element IS*466*	19, 36, 57
leuA,B	Biosynthesis of leucine (mutants can use α-ketoisocaproic acid)	49
lysA	Biosynthesis of lysine	49
metA	Biosynthesis of methionine	49
metB	Biosynthesis of methionine (mutants can use homocysteine)	49
mthA	Biosynthesis of methionine and threonine	49
mthB	Biosynthesis of methionine and threonine (mutants can use homoserine)	49
nicA,B	Biosynthesis of nicotinamide	49
novA	Resistance to novobiocin	59
ora	(Over)production of an orange pigment	80
P_{TH4},P_{TH270}	Two promoters for unknown genes, transcriptionally dependent on *whiG*	89
pabA	Biosynthesis of *p*-aminobenzoic acid	49
pdxA	Biosynthesis of pyridoxine	49
pel	Resistance to phage VP11	49
pgl	Sensitivity to plaque formation by φC31	60, 76
pheA	Biosynthesis of phenylalanine	49
proA,B	Two closely linked but separately transcribed genes for glutamylphosphate dehydrogenase and glutamyl kinase, respectively (Note: *proA* in the *Streptomyces* literature is equivalent to *E. coli proB*.)	41
red	Cluster of genes for biosynthesis of the red-pigmented antibiotic complex (at least four prodiginines, the major compound being undecylprodigiosin) covering not more than 34 kb of DNA and including *redD*, for a transcriptional activator	25, 26, 72, 82
ribA,B	Biosynthesis of riboflavin	49
rifA	Resistance to high levels of rifampin; subunit of core RNA polymerase	14
rifB,C	Resistance to low levels of rifampin	14
rnpA	Encodes RNase P	11
rpmH	Encodes ribosomal protein L34	11
sapA	Encodes one of the spore-associated proteins (13 kDa)	34
$SCP1^{NF}$	Integrated copy of giant linear plasmid SCP1. Integration was associated with (or followed by) deletion of ~40 kb of the chromosome, one chromosomal copy of IS*466*, and ~80 kb of SCP1.	36, 60
selA	Resistance to selenate; biosynthesis of cysteine (ATP sulfurylase)	69
selB	Resistance to selenate; biosynthesis of cysteine	69
selC (*cysD*)	Resistance to selenate and chromate; biosynthesis of cysteine (sulfate permease)	69
selD	Resistance to selenate and chromate; biosynthesis of cysteine	69
selE	Resistance to selenate and chromate	69
serA,B	Biosynthesis of serine/glycine	49
SLP1	Inserted copy of the SLP1 element	6
spcA	Resistance to spectinomycin	49
strA	Resistance to streptomycin (high level)	49
strB	Resistance to streptomycin (low level)	49
thiA	Biosynthesis of thiamine (mutants do not cross-feed *thiB* mutants)	49
thiB	Biosynthesis of thiamine (mutants can use thiazole)	49
thiC,D	Biosynthesis of thiamine (mutants cross-feed *thiB* mutants)	49
thrA	Biosynthesis of threonine	49

(Continued)

Table 1—*Continued*

Locus	Role	Reference(s)
thrB	Biosynthesis of threonine (mutants also grow on serine)	49
tps-30	Mutants fail to grow at 37°C (temperature sensitive)	47
trpA,B	Encode the two subunits of tryptophan synthetase	41
trpC	Encodes indoleglycerolphosphate synthetase	41
trpD	Encodes anthranilate phosphoribosyltransferase	41
trpF	Encodes phosphoribosyl anthranilate isomerase	41
tyrA	Biosynthesis of tyrosine and phenylalanine	49
tyrB	Biosynthesis of tyrosine	49
uraA,B,C	Biosynthesis of uracil	49
uraD	Biosynthesis of uracil and arginine	49
ureA	Encodes urease	49
uvsA,C	UV repair (mutants are Hcr^- for phage VP11)	39, 40
uvsB,D	UV repair (mutants are Hcr^+ for phage VP11)	39, 40
uvsE	UV repair	39, 40
uvsF	Mutant enhances UV sensitivity of *uvsA*, *uvsC*, and *uvsD* mutants.	39, 40
whiA,H,I	Mutants have white colonies because of early block in sporulation.	13
whiB	As *whiA*; encodes a small, highly charged, cysteine-rich protein	13, 21
whiC	Mutant has white colonies because few spores are formed.	13
whiD	Mutant has white colonies with thin-walled spores.	13
whiE	Cluster of at least seven genes for polyketide spore pigment; ORFIII, -IV, -V, and -VI are homologous to *act*I-ORF1, -2, and -3 and *act*VII, respectively	13, 20
whiF	Mutant has white colonies with rod-shaped (rather than ellipsoidal) spores. Mutation is very closely linked to *whiG* and may possibly be a special *whiG* allele.	13
whiG	Encodes homolog of sigma factors involved in motility in other bacteria. Required for sporulation but not for growth. Mutants therefore have white aerial mycelium devoid of spores.	13, 16, 74

[a] ORF, open reading frame.

stocks, and other segments can be amplified many times, with only limited phenotypic effects (reviewed in reference 7). The deletable segments exceed 800 kb in *Streptomyces glaucescens*, where their loss leads to streptomycin sensitivity, lack of production of dihydrostreptomycin, and/or lack of melanin production (8); in *S. ambofaciens*, deletable DNA covers 2 Mb (63). The deletions and amplifications have not been related to a detailed genetic map in these two species. In *S. coelicolor* A3(2) (and its close relative *Streptomyces*

Table 2. Restriction fragments of the chromosome of *S. coelicolor* M145

*Ase*I (ATTAAT)		*Dra*I (TTTAAA)		*Ssp*I (AATATT)	
Fragment	Size (kb)	Fragment	Size (kb)	Fragment	Size (kb)
A	1,500	A	2,100	A	960
B	1,300	B	1,770	B	605
C	960	C	1,650	C	605
D	900	D	820	D	590
E	820	E	690	E	500
F	610	F	530	F	450
G	480	G	280	G	425
H	380	H	115	H	420
I	290			I	400
J	200			J	370
K	190			K	370
L	140			L	350
M	110			M	310
N	45			N	300
O	16			O	232
P	11			P	232
Q	10			Q	200
				R	160
				S	150
				T	120
				U	80
				V	60
				W	50
				X	30
				Y	15
17	7,962	8	7,955	25	7,984

lividans 66) deletions up to at least 40 kb have been detected, and these appear to be located in the silent regions (5, 17, 31, 60). Some of the deletions are associated with chloramphenicol sensitivity or arginine auxotrophy (*argG* mutations) or both, phenotypes that have been recognized to be highly unstable in extensive studies (2, 5, 23), but otherwise the deletions are largely silent phenotypically.

Much more needs to be done to determine the extent and map locations of the deletable and amplifiable segments of *S. coelicolor* (and *S. lividans*) DNA, for which the genomic map and the availability of materials stemming from its construction, such as isolated genomic fragments and, eventually, ordered clone libraries, will be invaluable. It certainly appears at present that wild-type streptomycetes carry considerable stretches of DNA (which is not merely highly repetitive DNA as judged by the renaturation kinetics of extracted chromosomal DNA samples, as discussed in reference 60) that presumably has selective value in natural habitats—otherwise it would not have persisted—but has only subtle effects on phenotype when deleted in laboratory variants. It will be interesting to find out the role(s) of this DNA and to determine whether its sequence has diverged between species at the same or at a different rate compared with the DNA in the well-marked segments of the map. The arrangement of genes in the well-marked segments seems to be extensively conserved even between streptomycetes that are not very closely related (33), as in the well-known case of *E. coli* K-12 and *S. typhimurium* LT2, or in pseudomonads, which show extensive map similarities for groups of genes of primary metabolism but not for segments determining specialized functions peculiar to individual species (43). Are the silent regions of streptomycete genomes "genetic fossils," or are they rapidly evolving segments of DNA ready to take on new functions?

REFERENCES

1. **Adamidis, T., P. Riggle, and W. Champness.** 1990. Mutations in a new *Streptomyces coelicolor* locus which globally block antibiotic biosynthesis but not sporulation. *J. Bacteriol.* **172:**2962–2969.
2. **Altenbuchner, J., and J. Cullum.** 1984. DNA amplification and an unstable arginine gene in *Streptomyces lividans* 66. *Mol. Gen. Genet.* **195:**134–138.
3. **Angell, S. M., E. Schwarz, and M. J. Bibb.** 1992. The glucose kinase gene of *Streptomyces coelicolor* A3(2): its nucleotide sequence, transcriptional analysis and role in glucose repression. *Mol. Microbiol.* **6:**2833–2844.
4. **Babcock, M. J., and K. E. Kendrick.** 1988. Cloning of DNA involved in sporulation of *Streptomyces griseus*. *J. Bacteriol.* **170:**2802–2808.
5. **Betzler, M., P. Dyson, and H. Schrempf.** 1987. Relationship of an unstable *argG* gene to a 5.7-kilobase amplifiable DNA sequence in *Streptomyces lividans* 66. *J. Bacteriol.* **169:**4804–4810.
6. **Bibb, M. J., J. M. Ward, T. Kieser, S. N. Cohen, and D. A. Hopwood.** 1981. Excision of chromosomal DNA sequences from *Streptomyces coelicolor* forms a novel family of plasmids detectable in *Streptomyces lividans*. *Mol. Gen. Genet.* **184:**230–240.
7. **Birch, A., A. Häusler, and R. Hütter.** 1990. Genome rearrangement and genetic instability in *Streptomyces* spp. *J. Bacteriol.* **172:**4138–4142.
8. **Birch, A., A. Häusler, M. Vögtli, W. Krek, and R. Hütter.** 1989. Extremely large chromosomal deletions are intimately involved in genetic instability and genomic rearrangements in *Streptomyces glaucescens*. *Mol. Gen. Genet.* **217:**447–458.
9. **Caballero, J. L., F. Malpartida, and D. A. Hopwood.** 1991. Transcriptional organization and regulation of an antibiotic export complex in the producing *Streptomyces* culture. *Mol. Gen. Genet.* **228:**372–380.
10. **Caballero, J. L., E. Martinez, F. Malpartida, and D. A. Hopwood.** 1991. Organisation and functions of the *act*VA region of the actinorhodin biosynthetic gene cluster of *Streptomyces coelicolor*. *Mol. Gen. Genet.* **230:**401–412.
11. **Calcutt, M. J., and F. J. Schmidt.** 1992. Conserved gene arrangement in the origin region of the *Streptomyces coelicolor* chromosome. *J. Bacteriol.* **174:**3220–3226.
12. **Champness, W. C.** 1988. New loci required for *Streptomyces coelicolor* morphological and physiological differentiation. *J. Bacteriol.* **170:**1168–1174.
13. **Chater, K. F.** 1972. A morphological and genetic mapping study of white colony mutants of *Streptomyces coelicolor*. *J. Gen. Microbiol.* **72:**9–28.
14. **Chater, K. F.** 1974. Rifampicin-resistant mutants of *Streptomyces coelicolor* A3(2). *J. Gen. Microbiol.* **80:**277–290.
15. **Chater, K. F., C. J. Bruton, S. G. Foster, and I. Tobek.** 1985. Physical and genetic analysis of IS*110*, a transposable element of *Streptomyces coelicolor* A3(2). *Mol. Gen. Genet.* **200:**235–239.
16. **Chater, K. F., C. J. Bruton, K. A. Plaskitt, M. J. Buttner, C. Mendez, and J. D. Helmann.** 1989. The developmental fate of *S. coelicolor* hyphae depends upon a gene product homologous with the motility σ factor of *B. subtilis*. *Cell* **59:**133–143.
17. **Chater, K. F., D. J. Henderson, M. J. Bibb, and D. A. Hopwood.** 1988. Genome flux in *Streptomyces coelicolor* and other streptomycetes and its possible relevance to the evolution of mobile antibiotic resistance determinants, p. 7–42. *In* A. J. Kingsman, K. F. Chater, and S. M. Kingsman (ed.), *Transposition*. University Press, Cambridge.
18. **Chou, C.-F., and C. W. Chen.** Personal communication.
19. **Cullum, J.** Personal communication.
20. **Davis, N. K., and K. F. Chater.** 1990. Spore colour in *Streptomyces coelicolor* A3(2) involves the developmentally regulated synthesis of a compound biosynthetically related to polyketide antibiotics. *Mol. Microbiol.* **4:**1679–1691.
21. **Davis, N. K., and K. F. Chater.** 1992. The *Streptomyces coelicolor whiB* gene encodes a small transcription factor-like protein dispensable for growth but essential for sporulation. *Mol. Gen. Genet.* **232:**351–358.
22. **Dowding, J. E., and D. A. Hopwood.** 1973. Temperate bacteriophages for *Streptomyces coelicolor* A3(2) isolated from soil. *J. Gen. Microbiol.* **78:**349–359.
23. **Dyson, P., and H. Schrempf.** 1987. Genetic instability and DNA amplification in *Streptomyces lividans* 66. *J. Bacteriol.* **169:**4796–4803.
24. **Edwards, A. W. F.** 1966. Linkage on a circular map. *Genetics* **54:**1185–1187.
25. **Feitelson, J. S., and D. A. Hopwood.** 1983. Cloning a *Streptomyces* gene for an O-methyltransferase involved in antibiotic biosynthesis. *Mol. Gen. Genet.* **190:**394–398.
26. **Feitelson, J. S., F. Malpartida, and D. A. Hopwood.** 1985. Genetic and biochemical characterization of the *red* gene cluster of *Streptomyces coelicolor* A3(2). *J. Gen. Microbiol.* **131:**2431–2441.
27. **Fernández-Moreno, M. A., J. L. Caballero, D. A. Hopwood, and F. Malpartida.** 1991. The *act* cluster contains regulatory and antibiotic export genes, direct targets for translational control by the *bldA* tRNA gene of *Streptomyces*. *Cell* **66:**769–780.
28. **Fernández-Moreno, M. A., E. Martínez, L. Boto, D. A. Hopwood, and F. Malpartida.** 1992. Nucleotide sequence

and deduced functions of a set of co-transcribed genes of *Streptomyces coelicolor* A3(2) including the polyketide synthase for the antibiotic actinorhodin. *J. Biol.* **267:** 19278–19290.

29. **Fernández-Moreno, M. A., E. Martínez, J. Niemi, H. M. Kieser, D. A. Hopwood, and F. Malpartida.** 1992. A new locus in *Streptomyces coelicolor* which controls the biosynthesis of antibiotics, but not sporulation. *J. Bacteriol.* **174:**2958–2967.
30. **Fisher, S. H., C. J. Bruton, and K. F. Chater.** 1987. The glucose kinase gene of *Streptomyces coelicolor* and its use in selecting spontaneous deletions for desired regions of the genome. *Mol. Gen. Genet.* **206:**35–44.
31. **Flett, F., and J. Cullum.** 1987. DNA deletions in spontaneous chloramphenicol-sensitive mutants of *Streptomyces coelicolor* A3(2) and *Streptomyces lividans* 66. *Mol. Gen. Genet.* **207:**499–502.
32. **Flett, F., and C. P. Smith.** Personal communication.
33. **Friend, E. J., and D. A. Hopwood.** 1971. The linkage map of *Streptomyces rimosus. J. Gen. Microbiol.* **68:**187–197.
34. **Guijarro, J., R. Santamaria, A. Schauer, and R. Losick.** 1988. Promoter determining the timing and spatial localization of transcription of a cloned *Streptomyces coelicolor* gene encoding a spore-associated polypeptide. *J. Bacteriol.* **170:**1895–1901.
35. **Hallam, S. E., F. Malpartida, and D. A. Hopwood.** 1988. DNA sequence, transcription and deduced function of a gene involved in polyketide antibiotic synthesis in *Streptomyces coelicolor. Gene* **74:**305–320.
36. **Hanafusa, T., and H. Kinashi.** 1992. The structure of an integrated copy of the giant linear plasmid SCP1 in the chromosome of *Streptomyces coelicolor* 2612. *Mol. Gen. Genet.* **231:**363–368.
37. **Hara, O., S. Horinouchi, T. Uozumi, and T. Beppu.** 1983. Genetic analysis of A-factor synthesis in *Streptomyces coelicolor* A3(2) and *Streptomyces griseus. J. Gen. Microbiol.* **129:**2939–2944.
38. **Harasym, M., L.-H. Zhang, K. Chater, and J. Piret.** 1990. The *Streptomyces coelicolor* A3(2) *bldB* region contains at least two genes involved in morphological development. *J. Gen. Microbiol.* **136:**1543–1550.
39. **Harold, R. J., and D. A. Hopwood.** 1970. Ultraviolet-sensitive mutants of *Streptomyces coelicolor.* II. Genetics. *Mut. Res.* **10:**439–448.
40. **Harold, R. J., and D. A. Hopwood.** 1972. A rapid method for complementation testing of ultraviolet-sensitive (*uvs*) mutants of *Streptomyces coelicolor. Mut. Res.* **16:**27–34.
41. **Hodgson, D. A.** Personal communication.
42. **Hodgson, D. A., and K. F. Chater.** 1991. A chromosomal locus controlling extracellular agarase production by *Streptomyces coelicolor* A3(2), and its inactivation by chromosomal integration of plasmid SCP1. *J. Gen. Microbiol.* **124:**339–348.
43. **Holloway, B. W.** Personal communication.
44. **Hopwood, D. A.** 1959. Linkage and the mechanism of recombination in *Streptomyces coelicolor. Ann. N.Y. Acad. Sci.* **81:**887–898.
45. **Hopwood, D. A.** 1965. New data on the linkage map of *Streptomyces coelicolor. Genet. Res.* **6:**248–262.
46. **Hopwood, D. A.** 1965. A circular linkage map in the actinomycete *Streptomyces coelicolor. J. Mol. Biol.* **12:** 514–516.
47. **Hopwood, D. A.** 1966. Non-random location of temperature-sensitive mutants on the linkage map of *Streptomyces coelicolor. Genetics* **54:**1169–1176.
48. **Hopwood, D. A.** 1966. Lack of constant genome ends in *Streptomyces coelicolor. Genetics* **54:**1177–1184.
49. **Hopwood, D. A., K. F. Chater, J. E. Dowding, and A. Vivian.** 1973. Advances in *Streptomyces coelicolor* genetics. *Bacteriol. Rev.* **37:**371–405.
50. **Hopwood, D. A., and T. Kieser.** 1990. The *Streptomyces* genome, p. 147–162. *In* K. Drlica and M. Riley (ed.), *The Bacterial Chromosome.* American Society for Microbiology, Washington, D.C.
51. **Hopwood, D. A., and G. Sermonti.** 1962. The genetics of *Streptomyces coelicolor. Adv. Genet.* **11:**273–342.
52. **Hopwood, D. A., and H. M. Wright.** 1983. CDA is a new chromosomally-determined antibiotic from *Streptomyces coelicolor* A3(2). *J. Gen. Microbiol.* **129:**3575–3579.
53. **Horinouchi, S., and T. Beppu.** 1992. Regulation of secondary metabolism and cell differentiation in *Streptomyces*: A-factor as a microbial hormone and the AfsR protein as a component of a two-component regulatory system. *Gene* **115:**167–172.
54. **Horinouchi, S., M. Kito, M. Nishiyama, K. Furuya, S.-K. Hong, K. Miyake, and T. Beppu.** 1990. Primary structure of AfsR, a global regulatory protein for secondary metabolite formation in *Streptomyces coelicolor* A3(2). *Gene* **95:**49–56.
55. **Ikeda, H., E. T. Seno, C. J. Bruton, and K. F. Chater.** 1984. Genetic mapping, cloning and physiological aspects of the glucose kinase gene of *Streptomyces coelicolor. Mol. Gen. Genet.* **196:**501–507.
56. **Ishizuka, H., S. Horinouchi, H. M. Kieser, D. A. Hopwood, and T. Beppu.** 1992. A putative two-component regulatory system involved in secondary metabolism in *Streptomyces. J. Bacteriol.* **174:**7585–7594.
57. **Kendall, K., and J. Cullum.** 1986. Identification of a DNA sequence associated with plasmid integration in *Streptomyces coelicolor* A3(2). *Mol. Gen. Genet.* **202:**240–245.
58. **Kendrick, K. E., and M. L. Wheelis.** 1982. Histidine dissimilation in *Streptomyces coelicolor. J. Gen. Microbiol.* **128:**2029–2040.
59. **Kieser, H. M., and D. A. Hopwood.** Unpublished data.
60. **Kieser, H. M., T. Kieser, and D. A. Hopwood.** 1992. A combined genetic and physical map of the *Streptomyces coelicolor* A3(2) chromosome. *J. Bacteriol.* **174:**5496–5507.
61. **Kohara, Y., K. Akiyama, and K. Isono.** 1987. The physical map of the whole *E. coli* chromosome: application of a new strategy for rapid analysis and sorting of a large genomic library. *Cell* **50:**495–508.
62. **Lawlor, E. J., H. A. Bayliss, and K. F. Chater.** 1987. Pleiotropic morphological and antibiotic deficiencies result from mutations in a gene encoding a tRNA-like product in *Streptomyces coelicolor* A3(2). *Genes Dev.* **1:**1305–1310.
63. **Leblond, P., P. Demuyter, J.-M. Simonet, and B. Decaris.** 1991. Genetic instability and associated genome plasticity in *Streptomyces ambofaciens*: pulsed-field gel electrophoresis evidence for large DNA alterations in a limited genomic region. *J. Bacteriol.* **173:**4229–4233.
64. **Leblond, P., F. X. Francou, J.-M. Simonet, and B. Decaris.** 1990. Pulsed-field gel electrophoresis analysis of the genome of *Streptomyces ambofaciens* strains. *FEMS Microbiol. Lett.* **72:**79–88.
65. **Leskiw, B. K., E. J. Lawlor, J. M. Fernandez-Abalos, and K. F. Chater.** 1991. TTA codons in some genes prevent their expression in a class of developmental, antibiotic-negative, *Streptomyces* mutants. *Proc. Natl. Acad. Sci. USA* **88:**2461–2465.
66. **Limauro, D., A. Avitabile, C. Cappellano, A. M. Puglia, and C. B. Bruni.** 1990. Cloning and characterization of the histidine biosynthetic gene cluster of *Streptomyces coelicolor* A3(2). *Gene* **90:**31–41.
67. **Lomovskaya, N. D., L. K. Emeljanova, and S. I. Alikhanian.** 1971. The genetic location of prophage on the chromosome of *Streptomyces coelicolor. Genetics* **68:** 341–347.
68. **Lydiate, D. J., A. M. Ashby, D. J. Henderson, H. M. Kieser, and D. A. Hopwood.** 1989. Physical and genetic characterization of chromosomal copies of the *Streptomyces coelicolor* mini-circle. *J. Gen. Microbiol.* **135:**941–955.
69. **Lydiate, D. J., C. Mendez, H. M. Kieser, and D. A.**

Hopwood. 1988. Mutation and cloning of clustered *Streptomyces* genes essential for sulphate metabolism. *Mol. Gen. Genet.* **211:**415–423.
70. **Malpartida, F., and D. A. Hopwood.** 1984. Molecular cloning of the whole biosynthetic pathway of a *Streptomyces* antibiotic and its expression in a heterologous host. *Nature* (London) **309:**462–464.
71. **Malpartida, F., and D. A. Hopwood.** 1986. Physical and genetic characterisation of the gene cluster for the antibiotic actinorhodin in *Streptomyces coelicolor* A3(2). *Mol. Gen. Genet.* **205:**66–73.
72. **Malpartida, F., J. Niemi, R. Navarrete, and D. A. Hopwood.** 1990. Cloning and expression in a heterologous host of the complete set of genes for biosynthesis of the *Streptomyces coelicolor* antibiotic undecylprodigiosin. *Gene* **93:**91–99.
73. **Martinez, E., M. A. Fernández-Moreno, J. L. Caballero, D. A. Hopwood, and F. Malpartida.** Unpublished data.
74. **Mendez, C., and K. F. Chater.** 1987. Cloning of *whiG*, a gene critical for sporulation of *Streptomyces coelicolor* A3(2). *J. Bacteriol.* **169:**5715–5720.
75. **Merrick, M. J.** 1976. A morphological and genetic mapping study of bald colony mutants of *Streptomyces coelicolor. J. Gen. Microbiol.* **96:**299–315.
76. **Mkrtumian, N. M., and N. D. Lomovskaya.** 1972. A mutation affecting the ability of temperate actinophage ϕC31 of *Streptomyces coelicolor* to lyse and lysogenize. *Genetika* **8:**135–141.
77. **Parro, V., D. A. Hopwood, F. Malpartida, and R. P. Mellado.** 1991. Transcription of genes involved in the earliest steps of actinorhodin biosynthesis in *Streptomyces coelicolor. Nucleic Acids Res.* **19:**2623–2627.
78. **Passantino, R., A.-M. Puglia, and K. F. Chater.** 1991. Additional copies of the *act*II regulatory gene induce actinorhodin production in pleiotropic *bld* mutants of *Streptomyces coelicolor* A3(2). *J. Gen. Microbiol.* **137:** 2059–2064.
79. **Piret, J. M., and K. F. Chater.** 1985. Phage-mediated cloning of *bldA*, a region involved in *Streptomyces coelicolor* morphological development, and its analysis by genetic complementation. *J. Bacteriol.* **163:**965–972.
80. **Rudd, B. A. M.** 1987. Genetics of pigmented secondary metabolites in *Streptomyces coelicolor*. Ph.D. thesis. University of East Anglia, Norwich, United Kingdom.
81. **Rudd, B. A. M., and D. A. Hopwood.** 1979. Genetics of actinorhodin biosynthesis by *Streptomyces coelicolor* A3(2). *J. Gen. Microbiol.* **114:**35–43.
82. **Rudd, B. A. M., and D. A. Hopwood.** 1980. A pigmented mycelial antibiotic in *Streptomyces coelicolor*: control by a chromosomal gene cluster. *J. Gen. Microbiol.* **119:**333–340.
83. **Seno, E. T.** 1982. Biochemical and genetic studies of the catabolism of glycerol and other carbon sources in *Streptomyces coelicolor* A3(2). Ph.D. thesis, University of East Anglia, Norwich, United Kingdom.
84. **Seno, E. T., C. J. Bruton, and K. F. Chater.** 1984. The glycerol utilization operon of *Streptomyces coelicolor*: genetic mapping of *gyl* mutations and the analysis of cloned *gyl* DNA. *Mol. Gen. Genet.* **193:**119–128.
85. **Sermonti, G., A. Mancinelli, and I. Spada-Sermonti.** 1960. Heterogeneous clones ("heteroclones") in *Streptomyces coelicolor* A3(2). *Genetics* **45:**669–672.
86. **Smith, C. L., J. G. Econome, A. Schutt, S. Klco, and C. R. Cantor.** 1987. A physical map of the *Escherichia coli* K12 genome. *Science* **236:**1448–1453.
87. **Smith, C. P., and K. F. Chater.** 1988. Structure and regulation of controlling sequences for the *Streptomyces coelicolor* glycerol operon. *J. Mol. Biol.* **204:**569–580.
88. **Stahl, F. W.** 1967. Circular genetic maps. *J. Cell. Physiol.* **70:**1–12.
89. **Tan, H.** 1991. Molecular genetics of developmentally regulated promoters in *Streptomyces coelicolor* A3(2). Ph.D. thesis. University of East Anglia, Norwich, United Kingdom.
90. **Tseng, H.-C., and C. W. Chen.** 1991. A cloned *ompR*-like gene of *Streptomyces lividans* 66 suppresses defective *melC1*, a putative copper-transfer gene. *Mol. Microbiol.* **5:**1187–1196.
91. **Vivian, A., and H. P. Charles.** 1970. The occurrence and genetics of some CO_2 mutants in *Streptomyces coelicolor. J. Gen. Microbiol.* **61:**263–271.
92. **Wray, L. V., M. R. Atkinson, and S. H. Fisher.** 1991. Identification and cloning of the *glnR* locus, which is required for transcription of the *glnA* gene in *Streptomyces coelicolor* A3(2). *J. Bacteriol.* **173:**7351–7360.
93. **Wright, F., and M. J. Bibb.** 1992. Codon usage in the G+C-rich *Streptomyces* genome. *Gene* **113:**55–65.

V. CHROMOSOME REPLICATION, MODIFICATION, AND REPAIR

36. Initiation and Termination of Chromosome Replication

H. YOSHIKAWA and R. G. WAKE

INTRODUCTION

Our current understanding of the overall cyclic process of bacterial chromosome replication, encompassing the phases of initiation, elongation, and termination, has relied heavily on studies using both *Bacillus subtilis* and *Escherichia coli*. In several instances, crucial aspects have been elucidated primarily and most convincingly with the former. Thus, in the 1960s, Cairns (21) showed that the circular *E. coli* chromosome replicated as a θ structure, but it was the classic work of Yoshikawa and Sueoka with *B. subtilis* (139) that first demonstrated that replication commenced from a unique origin, *oriC*, on the chromosome. The features that made *B. subtilis* particularly suitable for studies on chromosome replication were its efficient transformation system and the ease with which a synchronous cycle of replication could be achieved by the use of germinating spores. The initial success in defining *oriC* in *B. subtilis* has been followed by very substantial contributions to our understanding of the structure and evolution of the origin region and the mechanism and control of initiation of replication. In more recent years, the process of sporulation in *B. subtilis* has been exploited to expose the termination process to detailed investigation at the molecular level (122). In this chapter, the emphasis is on the processes of initiation and termination of replication of the *B. subtilis* chromosome. As an introduction to the discussion of these topics, a brief account is given of the work that has afforded a general picture of the topology of the *B. subtilis* chromosome through the three phases of the replication cycle. This account will be followed by a description of currently identified DNA replication genes in *B. subtilis* and a summary of what is known about the enzymology of the elongation phase of DNA replication in this organism.

Historical Overview

Marker frequency analysis and sequential replication from a unique chromosomal origin

Cairns' autoradiographic studies in 1963 (21) showed that the *E. coli* chromosome consisted of a single piece of DNA with no free end. In other words, it was circular. During replication, this molecule splits into two arms over part of its length. Cairns interpreted the split region, or "eye," to represent the replicated portion of the chromosome. From a consideration of the lengths of DNA replicated upon pulse-labeling with [^{3}H]thymidine, the length of the whole chromosome, and the generation time of the bacterial culture, Cairns concluded that "one or at the most two regions of the chromosome are being duplicated at any moment." For reasons that are not entirely clear, he preferred the apparently simpler situation of only one replication region (fork) per chromosome.

At the same time that Cairns was doing his illuminating studies on *E. coli*, Yoshikawa and Sueoka were using a different approach to investigate the process of bacterial chromosome replication in *B. subtilis* (139). They wanted to know whether there was a fixed order of replication of genetic markers during the replication cycle. From a comparison of the relative frequencies of various markers in DNA obtained from exponentially growing and stationary cultures of strain W23 and assayable by transformation of strain 168, they concluded that there was a fixed order of replication. Of the markers analyzed at that time, it was considered that *purB* was replicated very early in the cycle and *metB* was replicated very late. Other markers were replicated at intermediate stages. The simplest model to explain such an ordered replication, and the one proposed by Yoshikawa and Sueoka, was that replication always started at the end (the origin) of the chromosome that was close to *purB* and proceeded in a continuous fashion towards the other end (the terminus), which was close to *metB* (at that time, circularity of the *B. subtilis* chromosome had not been established, and a linear structure was the simplest to assume). However, they made it quite clear that their data did not rule out a model in which the whole chromosome was replicated as a series of discrete segments (or subunits). Bidirectional replication of a circular chromosome (to be established later) in which DNA synthesis occurred simultaneously in two separate segments would fit into the "subunit model." But because of Cairns's preference for only one replication fork in the replicating *E. coli* chromosome, the subunit model was considered unlikely.

The replication order map constructed from measurements of the relative frequencies of various markers was confirmed by direct determination of the time of replication of markers (transfer from unreplicated to replicated DNA in a density shift approach) during the synchronous germination of spores (140). (It had already been established that spores contained completed chromosomes.) In effect, Yoshikawa and Sueoka established what it was not possible to establish by autoradiography: that replication of the bac-

H. Yoshikawa • Department of Genetics, Osaka University Medical School, Osaka 565, Japan. **R. G. Wake** • Department of Biochemistry, University of Sydney, Sydney, New South Wales 2006, Australia.

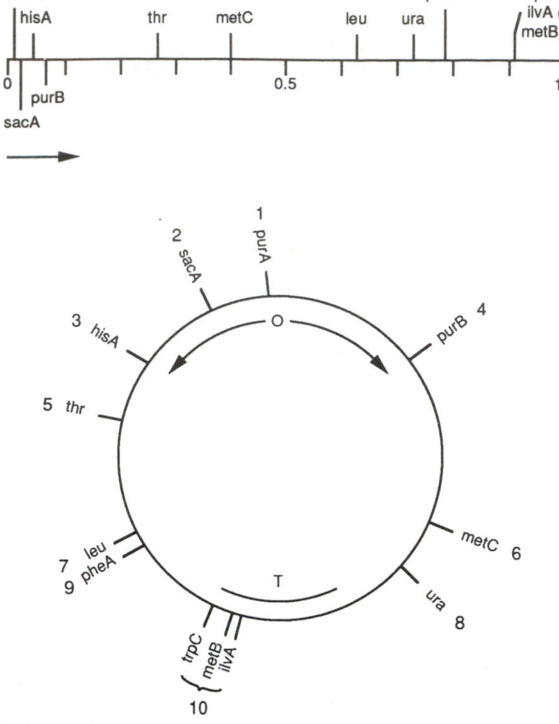

Figure 1. Gene replication order map of *B. subtilis* as the basis for the circular genetic linkage map. (A) Chromosome depicted as a linear structure with replication starting at position 0 (origin) and proceeding sequentially through the markers (genes) indicated towards 1.0 (terminus) (94). (B) The same genes, with their order of replication (genes 1 through 10) on the Harford-Dedonder circular linkage map (54).

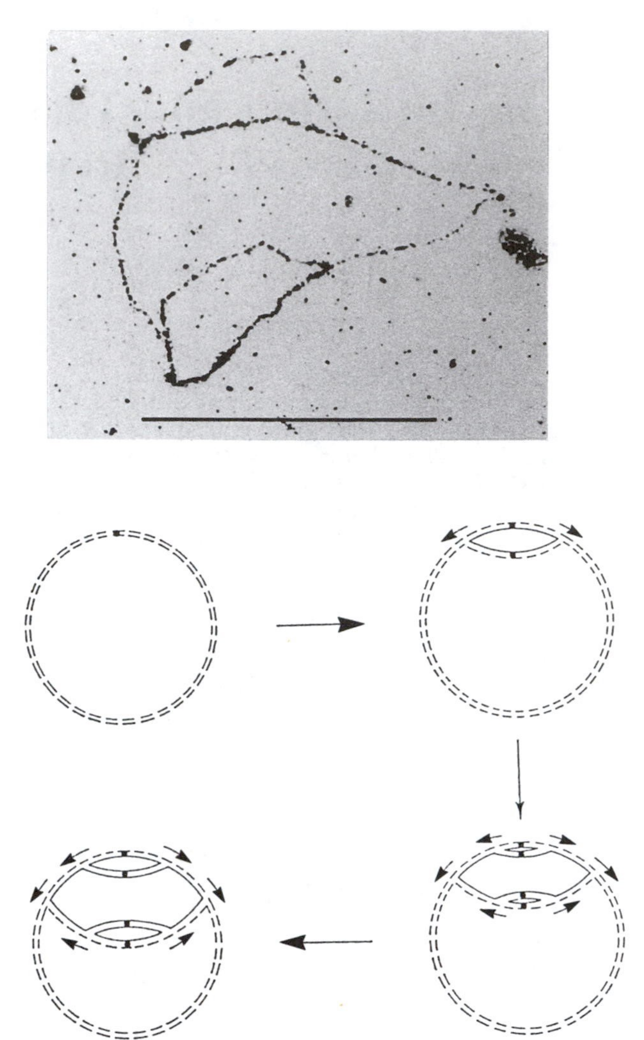

Figure 2. Autoradiographic visualization of the replicated portion of a reinitiated chromosome and suggestion of bidirectional replication to explain the result obtained. (A) The autoradiograph shows two smaller (reinitiated) loops contained within a larger loop. (B) Scheme for the formation of the structure seen by autoradiography from a circular chromosome containing a single *oriC* site (filled circle). Arrows indicate directions of movement of replication forks. The broken line represents the parental strand, and the solid line represents the newly synthesized one (taken from reference 120).

terial chromosome started at a unique site, now called *oriC*. Also, the replication order map that was constructed and subsequently extended established the framework for the current map of the circular *B. subtilis* chromosome (Fig. 1).

Reinitiated chromosomes, physical circularity, and bidirectional replication

In another major contribution to our knowledge of the process of chromosome replication in bacteria, Sueoka and his colleagues in 1964 showed by marker frequency analysis that in a rich medium, reinitiation of chromosome replication at *oriC* occurred before the round in progress was complete (138). This was one of the first suggestions that chromosome replication was controlled at the level of initiation. While circularity of the *B. subtilis* chromosome was not definitively established until 1973 (121), many researchers before this time considered that circularity was likely. The demonstration of reinitiation of chromosome replication and the generally held view of unidirectional replication posed problems with respect to the possible structure of a reinitiated circular chromosome. The θ structure allowed two origins to be connected to a single terminus. How could four origins be accommodated? The problem was resolved in 1972 by again taking advantage of chromosome replication following spore germination. When spores were germinated, chromosomes were induced to undergo reinitiation (dichotomous replication) in the presence of [^{3}H] thymine, and replicated portions of such chromosomes were examined by autoradiography, an unexpected result was obtained (120). Many of the structures were displayed as fully closed multiforked forms, actually large loops split into two arms at two positions (Fig. 2A). The interpretation put forward to account for the general forms of these structures was that replication involved the initial movement of replication forks in both directions away from *oriC* (Fig. 2B). The second round of replication starting soon afterwards (i.e., reinitiation) in the same bidirectional manner in the replicated segments would generate the types of reinitiated structures visualized by autoradiography. This was the first substantial evidence to suggest that replication of the *B. subtilis* chromosome was a bidirectional process. Also, it made it more likely that the chromosome was a circular structure, as had been shown for the *E. coli* chromosome 10 years earlier. Subsequent autoradiographic studies using spores germinated and growing out in an unenriched medium that did not induce reinitiation of replication established that the chromosome was a physically circular structure and that it replicated in a completely bidirectional manner (36).

The realization that replication was bidirectional necessitated a reinvestigation of the genetic map of *B. subtilis* 168, which was based on the assumption of unidirectional replication. This was achieved by Kejzlarova-Lepesant et al. after a detailed study of the order of replication of many genetic markers and the accumulation of new data to connect certain sections of the genetic linkage map (54). The Harford-Dedonder map is the basis for the current map of the circular *B. subtilis* 168 chromosome (Fig. 1; 98).

A unique terminus of replication

After it had been established that the *B. subtilis* chromosome was a circular structure and that replication from *oriC* was a bidirectional process, 8 years elapsed before the question of the nature of the termination site or region was addressed. In the intervening period, studies with other bidirectionally replicating circular replicons such as phage lambda (117) and simian virus 40 (58) chromosomes showed that the two forks did not meet at a defined location. It was reasonable to assume that the same situation existed in the bacterial chromosome. Furthermore, there was no obvious way of labeling the bacterial chromosome in its terminus region, which was known to be located approximately opposite *oriC* on the genetic map. In 1978, Dunn et al. (26) investigated the relationship between DNA replication and the induction of sporulation in a starvation medium. Their data suggested that the DNA replication inhibitor 6-(*para*-hydroxy-phenylazo)uracil had no effect on the level of sporulation achieved if it was added after all cycles of replication proceeding in the starvation medium had terminated. Their conclusion was in accord with the fact that spores contain only completed (i.e., nonreplicating) chromosomes. Using this information, two independent groups devised an approach involving radioactive labeling of DNA after suspension of cells in a starvation (sporulation) medium and addition of 6-(*para*-hydroxy-phenylazo)uracil shortly afterwards to incorporate radioactivity into only the terminus region of the spore chromosome (1, 102). Subsequent restriction fragment analysis of the region labeled for progressively shorter periods prior to termination showed convincingly that the fusion of approaching replication forks occurred at a very precise site, called *terC* (125).

Thus, the picture that emerges for replication of the circular *B. subtilis* chromosome (approximately 4,000 kb long; see chapter 30) can be summarized as follows. (i) Initiation at the unique *oriC* site yields two replication forks. (ii) The replication forks move away from each other during the elongation phase (DNA chain growth) and eventually approach each other toward the end of the cycle in the terminus region. (iii) The approaching forks finally meet and fuse at the defined site, *terC*.

Genes for DNA Replication and Their Products

Temperature-sensitive mutations that affect largely DNA replication in *B. subtilis* were first identified around the late 1960s by Gross and his coworkers (53, 77). They were distributed among nine linkage groups (A through I). There was no evidence for more than one gene in each linkage group, and the genes have since been called *dnaA*, etc. They were classified as affecting the initiation and/or elongation phase of the replication cycle. Since that time, other workers have described additional temperature-sensitive mutations. In many cases, the original designations of genes in which these mutations occur have been changed to conform with those used for equivalent genes in *E. coli* and for which more extensive information on the role of the protein products in replication is available (4), but it should be remembered that in some cases, the same designation persists for quite different genes in the two organisms. Thus, *dnaB* of *B. subtilis* codes for a protein involved in initiation of replication (132), while *dnaB* of *E. coli* codes for the helicase operating to unwind DNA at the chromosomal replication fork (76). Not all *dna* genes in *B. subtilis* were originally identified through temperature-sensitive mutations. For example, what is now called *dnaA* in *B. subtilis* was established by sequence homology with *dnaA* of *E. coli* (82). Table 1 presents a summary of the currently known genes whose products have direct roles in DNA replication in *B. subtilis*. For the sake of clarity, many of the original designations do not appear in the first column of Table 1. The mutations first described by Karamata and Gross (53) as falling into group A are now known to be in a gene encoding ribonucleotide reductase, and the *dnaA* designation has subsequently been assigned to one that codes for a protein that is involved in initiation of replication and is homologous to the DnaA protein of *E. coli* (81). No attempt has been made here to list or compare the numerous mutations that have been described for individual genes. These have been covered well in the review by Winston and Sueoka (132). Also, it is clear that there are many *dna* genes yet to be identified.

Enzymology of DNA Replication

Far more is known about the molecular aspects of DNA replication in *E. coli* than in *B. subtilis*. This reflects the relatively early success by Kornberg and coworkers in developing a soluble in vitro system for DNA replication in *E. coli* based on viral single-strand DNA templates and the subsequent use of *oriC* plasmids as the basis of an in vitro system for initiation of replication (76). By comparison, the less well defined systems available for *B. subtilis*, such as toluene-treated cells (74), have not been very suitable for establishing the roles of various proteins in the replication process. Recently, a more promising in vitro system comprising a membrane-associated DNA com-

Table 1. DNA replication genes of *B. subtilis* 168

Gene	Map position (°)[a]	Protein and function	Previous designation	Reference(s)
dnaA	360	DnaA; initiation of replication cycle, homology to *E. coli* DnaA protein	*dnaJ*	81, 82, 143
dnaB[b]	255	DnaB; initiation of replication cycle		47, 53, 88
dnaC[c]	357	Unknown; DNA elongation		5, 53
dnaD	201	Unknown; initiation of replication cycle		53
dnaG	233	Primase; initiation of Okazaki fragments, homology to primase of *E. coli*	*dnaE*	53, 123
dnaH	4	Unknown; probably involved in DNA elongation, assumed to encode subunit of DNA polymerase III holoenzyme		4, 53
dnaI	251	Unknown; probably involved in DNA elongation		4, 53
dnaN	0	β subunit of DNA polymerase III holoenzyme; DNA elongation, homology to *E. coli* β subunit	*dnaG, dnaK*	4, 53, 82, 143
dnaX	143	τ and γ subunits of DNA polymerase III holoenzyme; DNA elongation, homology to *E. coli* τ and γ subunits	*dnaZX*	63, 114
gyrA	2	DNA gyrase α subunit; initiation of replication cycle and DNA elongation	*nalA*	59, 82, 116
gyrB	2	DNA gyrase β subunit; initiation of replication cycle and DNA elongation	*novA*	59, 82, 116
polA	255	DNA polymerase I; possibly needed for DNA elongation[d]		34
polC	143	DNA polymerase III α subunit; initiation of replication cycle and DNA elongation, contains 3' → 5' exonuclease (proofreading) activity	*dnaF, dnaP*	8, 10, 13, 37, 96, 101
rtp	168	Replication terminator protein (RTP); clockwise-fork arrest at *terC*		67

[a] In most cases, these are given as positions listed in reference 143.
[b] May be part of a four-gene operon.
[c] Two other groups of temperature-sensitive DNA replication mutations (*tsA* and *tsB*) map very close to *dnaC*, but the relationship between *tsA*, *tsB*, and *dnaC* is not clear (39, 132).
[d] There is doubt as to whether DNA polymerase I is a replication enzyme in *B. subtilis* (34).

plex has been developed (57). It has yielded information on the likely role of a membrane-associated protein in the regulation of replication. The lack of a soluble in vitro system for *B. subtilis* for identifying replication proteins has led to alternative approaches such as analysis of the effect of temperature shifts on the replication of the pUB110 plasmid present in various *dna*(Ts) mutants (4). Through such an approach and in conjunction with information on homology relationships of gene products with those of *E. coli*, it has been possible to define the likely roles of several proteins in the reactions at the replication fork in *B. subtilis*. In this regard, there is direct evidence that the process of DNA chain growth in *B. subtilis*, like that in *E. coli*, is semidiscontinuous, reflecting leading- and lagging-strand synthesis, but some data suggest that chain growth is discontinuous in both strands (132). Presumably, there are in *B. subtilis* counterparts of many if not all of the established *E. coli* proteins.

Table 2 lists the known essential proteins required for DNA chain growth accompanying replication fork movement in *E. coli* (the elongation phase) and the equivalent proteins in *B. subtilis*. (A more detailed consideration of the proteins involved in the initiation and termination phases is given below.) In the case of *E. coli*, the DnaB helicase, which migrates along the 5' → 3' strand complementary to the lagging strand, is considered to be largely responsible for the unwinding of DNA at the fork junction (76). There have been no reports of studies on helicases in *B. subtilis*, but such helicases must exist. There is some evidence to suggest that the DnaC protein of *E. coli* is involved in both the initiation and elongation phases in this organism (119), but recent data point to a role largely, if not exclusively, in initiation (3a). In most models, DnaC has a role in initiation and then dissociates from its complex with DnaB. As yet, there is no known equivalent in *B. subtilis* of this protein. So far, a chromosomally encoded single-stranded binding protein for *B. subtilis* needs to be identified, but a phage-encoded single-stranded binding protein has recently been described (35). *dnaG*, encoding primase, has been identified in *B. subtilis* through protein homology with *E. coli* (123). The DNA polymerase III holoenzyme of *E. coli* contains many subunits, for which all genes have been identified. The *polC* gene and its encoded α subunit (DNA polymerase) of the *B. subtilis* holoenzyme have been characterized in some detail (10, 37, 96, 101). Unlike the analogous protein in *E. coli*, the α subunit contains 3' → 5' exonuclease (proofreading) activity within a domain homologous to that of the ε subunit of the *E. coli* holoenzyme (13, 71, 101). The equivalent product of *dnaN* (β subunit) in *B. subtilis* was identified through protein homology (82), and the two subunits (τ and γ) encoded by a single DNA sequence (*dnaX*) were identified in both organisms (114). The genes for the δ, δ', χ, ψ, and θ subunits of the *E. coli* DNA polymerase III holoenzyme (72) have only recently been identified (Table 2), and as yet, there are no reported homologs in *B. subtilis*. It has been suggested that the product of *dnaH* is a subunit of the *B. subtilis* holoenzyme (4), so it could be equivalent to one of these *E. coli* subunits.

Table 2. Essential proteins (and genes) for DNA chain growth and replication fork movement (elongation phase) in *E. coli* and equivalent proteins and genes in *B. subtilis*

E. coli[a]		*B. subtilis*[b]	
Protein and subunit	Gene	Protein and subunit	Gene
DnaB helicase	*dnaB*	Unknown	Unknown
DnaC protein	*dnaC*	Unknown	Unknown
Single-stranded binding protein	*ssb*	Unknown	Unknown
Primase	*dnaG*	Primase (assumed)[c]	*dnaG*
DNA polymerase III holoenzyme		DNA polymerase III holoenzyme	
α (DNA polymerase)	*polC* (*dnaE*)	α (DNA polymerase + 3′ → 5′ exonuclease)	*polC*
β	*dnaN*	β (assumed)[c]	*dnaN*
ε (3′ → 5′ exonuclease)	*dnaQ* (*mutD*)	ε[d]	Unknown
τ	*dnaX*	τ (assumed)[c]	*dnaX*
γ	*dnaX*	γ (assumed)[c]	*dnaX*
δ	*holA*[e]	Unknown	Unknown
δ′	*holB*[e]	Unknown	Unknown
χ	*holC*[e]	Unknown	Unknown
ψ	*holD*[e]	Unknown	Unknown
θ	*holE*[e]	Unknown	Unknown
Gyrase		Gyrase	
α	*gyrA*	α	*gyrA*
β	*gyrB*	β	*gyrB*
DNA polymerase I	*polA*	DNA polymerase I[f]	*polA*
DNA ligase	*lig*	Unknown	Unknown

[a] Most of the data for *E. coli* have been taken from the review by McMacken et al. (76). The data for the δ, δ′, χ, ψ, and θ subunits of DNA polymerase III holoenzyme have not yet been published (see footnote *e* below).
[b] See Table 1 for references.
[c] These proteins have not been identified by either isolation or assay. They are assumed to exist because of homology with *E. coli* proteins.
[d] This might not exist as a separate subunit because of the presence of a domain within the α subunit with homology to the ε subunit of *E. coli* (13, 101).
[e] Reports on the identification of these genes were made recently by separate groups led by C. McHenry and M. O'Donnell, who have agreed on this terminology (86a).
[f] There is doubt was to whether DNA polymerase I is an essential replication enzyme in *B. subtilis* (132).

DNA gyrase from *B. subtilis* and the genes (*gyrA*, *gyrB*) defining its subunits have been investigated (59, 82, 116). DNA polymerase I, encoded by *polA*, was the first DNA polymerase to be purified from *B. subtilis*, and it was reported to contain no significant exonuclease activity, unlike DNA polymerase I of *E. coli* (92). At least part of the *polA* gene has been cloned, but there is no sequence information available for comparisons of homology (97). Thus, there must remain some doubt about its possible role in excision of RNA primers in the lagging strand of replication forks in *B. subtilis* (4, 132). *B. subtilis* DNA polymerase III has been reported to contain a low level of RNase activity (71), and this activity may be involved in removal of RNA primers (132). Neither DNA ligase of *B. subtilis* nor its gene has been identified. It is possible that the protein products of the *B. subtilis* genes *dnaC* and *dnaI* encode one or more of the "unknown" *B. subtilis* proteins listed in Table 1.

INITIATION OF CHROMOSOME REPLICATION

Initiation of bacterial chromosome replication was first defined as the step that requires synthesis of new proteins, while elongation proceeds to completion without concomitant synthesis of proteins. Genetic (139) and autoradiographic (21) studies had shown that bacterial chromosomes had a fixed origin of replication. On the basis of these observations, Jacob et al. proposed that initiation of replication was controlled by a replicon-specific protein factor (initiator) acting on a specific site (replicator) in the replicon (52). The replicon hypothesis provided a guide for studying the molecular mechanism of initiation of replication of bacterial chromosomes. Fifteen years after the proposal of the hypothesis, the DnaA protein and its binding sequence, the DnaA box, were shown to be the two elements essential for initiation of replication of the *E. coli* chromosome (32, 118). Later, studies on chromosomes of *B. subtilis* and other bacteria revealed that these two elements were commonly conserved in most eubacteria and played central roles in initiation of replication (137). In this review, the emphasis is on specific features of initiation of chromosome replication as observed in *B. subtilis* and based on the universal mechanism commonly found in other bacterial species, particularly *E. coli*.

Structure of the Origin Region

Identification of the origin region

The replication origin region of the *B. subtilis* chromosome was first identified by marker frequency analysis as described above (139). Obviously, it was not an accurate measurement of the origin, as the method was limited by the availability of genetic markers defined by mutant cells. *purA* was demonstrated to be the first gene replicated; it is now known to be located about 40 kb away from the origin (unpublished data). The second approach was to isolate the origin region of the chromosome by taking advantage of the fact that *purA* was enriched in the membrane fraction (134). Purification of a specific origin fragment from such an origin-membrane fraction was unsuccessful, although a clear, unique pattern of restriction fragments was reproducibly observed (135, 141). After cloning of DNA became possible, the isolation of autonomously replicating fragments from the chromosome was attempted in a number of laboratories without success. As will be described below, these initial attempts failed because of the strong inhibition against cell growth exerted by sequences in the origin region of the chromosome (80). Eventually, isolation of the active origin became possible only after the structure and function of the origin region were more thoroughly understood (79).

A more precise method for identifying the origin of replication was to characterize the first replicating fragments during synchronous initiation of replication. This method was first used successfully to define the *E. coli oriC* by using mutant cells temperature-sensitive for initiation for synchronization (73). Seiki et al. have applied this method to *B. subtilis* by using spore germination to synchronize the initiation of the first replication cycle (105). The very first replicating fragments were labeled by [^{3}H]thymine or [^{3}H]bromodeoxyuridine after germination of a thymine-requiring mutant in the absence of thymine. Analysis of the labeled DNA by fluorography after digestion with appropriate restriction enzymes and electrophoresis allowed Seiki et al. to construct a restriction map of some 50 kb in the origin region (105). Later, these fragments were cloned to confirm the physical map (82). Furthermore, Ogasawara et al. could identify the origin to within 1 kb by hybridization of the various cloned fragments with the first Okazaki fragments synthesized immediately after the synchronous initiation of chromosomal replication during germination of thymine-requiring spores (87). The region coincided with the *dnaA* and DnaA box region, recently established as the *oriC* of the *B. subtilis* chromosome (79).

Using a similar approach, Levine et al. observed a second origin from which initiation occurred preferentially during synchronous initiation in *dnaB37* cells following prior completion at the nonpermissive temperature and reinitiation at the permissive temperature in the presence of novobiocin (64). The possibility of the second origin in the *gyrA* rRNA gene (*rrnO*) region remains unproved until established by isolation as an active origin. The original observation should also be reexamined with a more direct and more sensitive method such as the detection of eye forms with two-dimensional gel electrophoresis (19a).

Gene organization of the origin region

Over the past several years, Ogasawara and Yoshikawa have cloned approximately 50 kb of DNA from the replication origin region and sequenced approximately 20 kb, of which 16 kb constitutes a continuous stretch (Fig. 3; 90). Altogether, 15 open reading frames (ORFs) were identified either by mutation or by structural homology with known genes in *E. coli*. It was very significant to find a gene highly homologous with *E. coli dnaA*. Furthermore, the *dnaA* gene was flanked by noncoding regions containing multiple repeats of the DnaA-binding sequence (DnaA box: TTATCCACA). In addition, there was a third noncoding region containing DnaA boxes 1.5 kb left of *dnaA*. These three regions were named DnaA box regions and included DnaA box regions L (left), C (center), and R (right) (Fig. 3; 137). Characteristic of the gene arrangement in this region was the symmetric orientation of most genes around DnaA box region C. Thus, all genes to the right of the DnaA box region were transcribed from left to right, while genes to the left of the DnaA box region were transcribed in the opposite direction. As a consequence of this organization, the direction of replication would be the same as the direction of transcription if replication started from DnaA box region C and proceeded bidirectionally. Genes in the origin region whose functions are either known by mutation or deduced from their homologous genes in *E. coli* are important for activities included in the cell's basic metabolism, such as DNA or RNA metabolism (*dnaA*, *dnaN*, *recF*, *gyrAB*, and *rnpA*), protein synthesis (*rmpH*), sporulation (*orf-282* and *orf-261*), and possibly cell division (*gidAB*). The product of gene *50K* belongs to a GTPase family (14), and its *E. coli* counterpart is known to be essential for cell growth (38a).

Conservation of genes and their organization in the origin region of eubacteria

There is a remarkable conservation of genes and their arrangements between the origin region of the *B. subtilis* chromosome and the *dnaA* region of the *E. coli* chromosome (Fig. 3; 89, 137). Of 11 ORFs from *50K* to *gyrB*, 8 are similar in structure (similarity becomes significantly higher if amino acid sequences are compared) and identical in relative locations to those in the *E. coli* chromosome. Thus, the genes are, from left to right, *50K*, *orf-261* (partly similar to *60K* of *E. coli*), *rpmH*, *rnpA*, *dnaA*, noncoding region (DnaA box C in *B. subtilis*), *dnaN*, *recF*, and *gyrB* in both organisms. Three additional genes defined by ORFs (*orf-208*, *orf-71*, and *orf-52*) and two DnaA box regions (L and R) are found in *B. subtilis*. The symmetrical arrangement of gene orientation relative to DnaA box region C is also conserved in *E. coli*. However, there is no DnaA box region in *E. coli* (only one DnaA box is found in the corresponding noncoding region); hence, this region does not serve as an origin of replication. Instead, the corresponding DnaA box region is located about 40 kb away from the *dnaA* region and acts as the *oriC* of the *E. coli* chromosome. To determine which of the two, *B. subtilis* type or *E. coli* type, is the ancestral type of the origin region, the gene organization of the *dnaA* region of another gram-negative bacterium, *Pseudomonas putida*, was investigated (Fig. 3; 28, 90). Surprisingly, the genes and their arrangement in the *Pseudomonas dnaA* region are more like those of the distantly related *B. subtilis* than those of the closely related *E. coli*. At least 12 genes and two DnaA box regions (L and C) are conserved between *B. subtilis* and *P. putida*. Comparative analysis suggests strongly that an approximately 50-kb segment of DNA containing a DnaA box region corresponding to the DnaA box L regions of *B. subtilis* and *P. putida* has been inverted during evolution of the *E. coli* chromosome. A few additional genes in the origin region are conserved between *B. subtilis* and *P. putida*. Particularly inter-

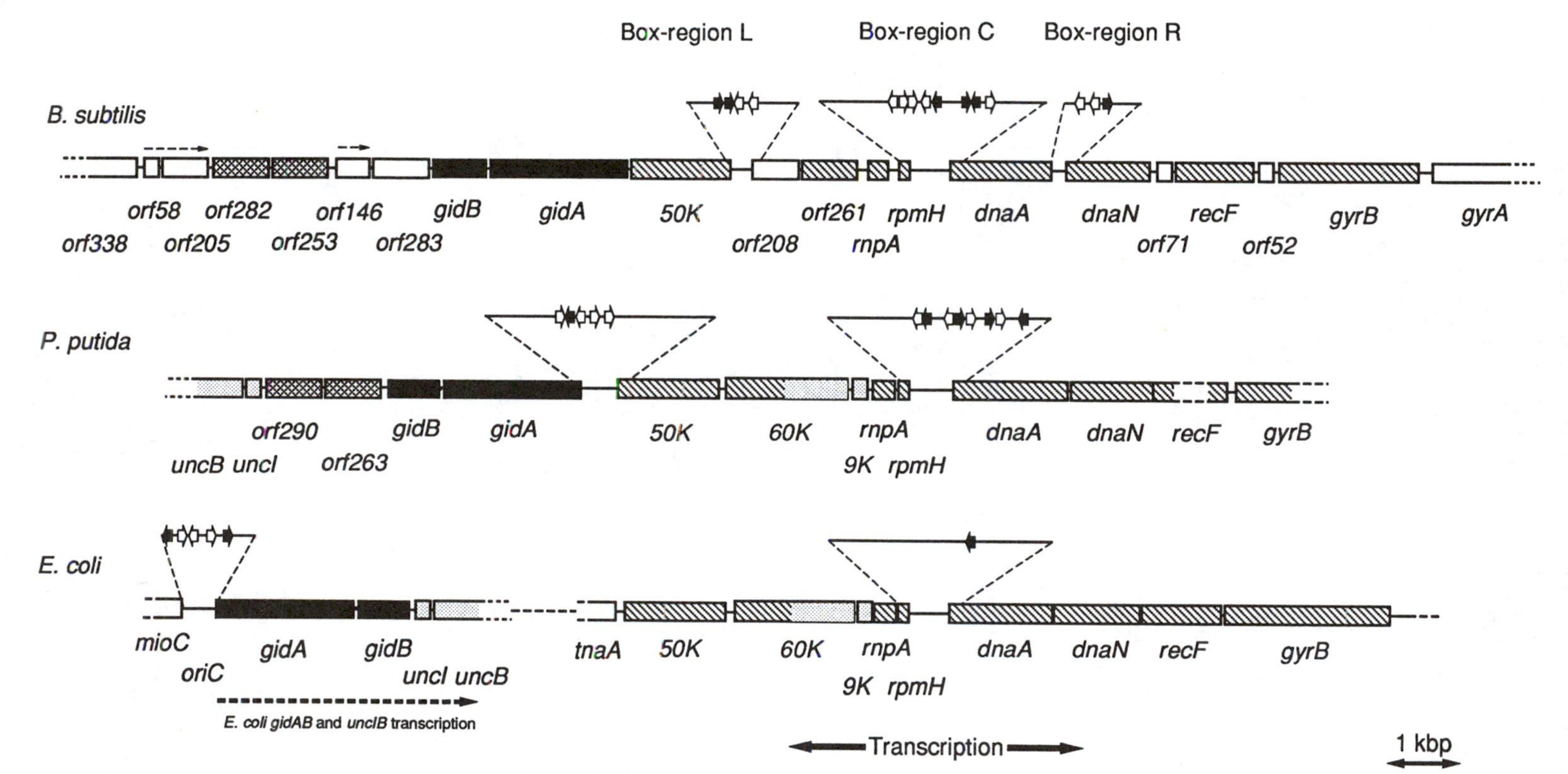

Figure 3. Conservation and variations of genes and their organization in the replication origin regions of bacterial chromosomes. Locations of ORFs (deduced from nucleotide sequences) are shown for three species. ORFs conserved between two or more species are indicated by various shadings. Open boxes are unique for one species. Sizes of ORFs are roughly to scale. Dotted lines in ORF boxes show the regions not yet sequenced. ORFs whose functions have not been identified by mutations are named either with ORF followed by the number of amino acids or with the molecular weight of the gene product. DnaA box regions are shown on a magnified scale with filled arrows for consensus boxes and open arrows for boxes different from the consensus by one base. The *E. coli oriC* region contains a region (dotted) of approximately 45 kb. Directions of transcription are indicated by solid arrows. Exceptions from the general rule are shown by dotted arrows.

esting was the conservation of two genes located at the left of the conserved gene *rnpA*. They have been identified as two *spo0J* genes in *B. subtilis* (85, 90). It would be interesting to examine the function of the homologous genes in *P. putida*, which apparently has no sporulation ability. These results clearly show that the gene arrangement common to *B. subtilis* and *P. putida* is an ancestral type and represents the replication origin region of ancestral bacteria.

DnaA and the DnaA Box in Initiation

Conservation and evolutionary relationships in eubacteria

The *dnaA* gene was first identified as a gene essential for initiation of chromosomal replication in *E. coli* (118). Several temperature-sensitive replication genes of *B. subtilis* have also been isolated from mutants temperature sensitive for growth (53, 132). Among these mutants, the *dnaB* mutant showed phenotypic properties very similar to those of *E. coli dnaA* mutants (60, 83). However, when the gene was cloned and sequenced, there was no similarity in predicted amino acid sequence between the two gene products (47, 88). The presence of the *dnaA* gene in *B. subtilis* was demonstrated through determination of the nucleotide sequence of the origin region of the chromosome (82). Discovery of a number of 9-mer sequences identical to the *E. coli* DnaA box in the two noncoding regions flanking the *B. subtilis dnaA* gene strongly suggested that *dnaA* is also essential for initiation of chromosomal replication in *B. subtilis*. This was proved by the isolation of a *dnaA*(Ts) mutant (81) and autonomously replicating sequences (*ars*) consisting of DnaA box regions as essential elements for *ars* activity (79). Before we consider the functional aspects of *dnaA* and DnaA boxes in *B. subtilis*, the conservation of *dnaA* and DnaA box regions among eubacteria and possible evolutionary relationships will be discussed.

As mentioned above, *P. putida* contains the *dnaA* gene and two DnaA box regions in the replication origin region. Recently, these two DnaA box regions in two pseudomonads, *P. putida* and *Pseudomonas aeruginosa*, were found to be active as *ars* elements (5a, 136). The comparative analysis has been extended to two other gram-positive bacteria with extreme chromosomal base compositions, *Micrococcus luteus* (75% G+C) (29) and *Mycoplasma capricolum* (25% G+C) (30). The *dnaA* gene and DnaA box regions are well conserved in these bacteria, although their sequence differences are greater, reflecting the strong mutation pressure under which the genomes of these bacteria have evolved. Comparison of the DnaA amino acid sequences from five eubacteria (three gram-positive bacteria and two gram-negative bacteria) clearly revealed four domains differing in levels of similarity (30, 137). The most conservative domains, domains III and IV, contain ATP-binding sites (107) and DNA-binding sites (established by partial proteolytic digestion of the protein; unpublished observation), respectively, that are known to be essential for the function of DnaA in the activation of *E. coli oriC* (16). Diversity of the DnaA protein is obvious in the least conserved domain, II, in which even the number of amino acids is widely variable. It is interesting to note that species with a high genomic G+C content contain a larger number of the amino acids Ala and Pro coded for by codons rich in G and C (30).

Conservation of the DnaA box sequence is very obvious except in species of extreme genomic G+C content. Thus, in *M. luteus*, the consensus sequence was deduced as TTGTCCACA (the third A is replaced by G) (29), and in *Mycoplasma capricolum*, one of each of the three Cs in the consensus sequence was replaced by either A or T (30). On the other hand, no similarity in the number of DnaA boxes and their relative locations within the DnaA box regions was observed in gram-positive and gram-negative bacteria. However, the structures of the DnaA box regions in relatively closely related species, *E. coli* and *P. putida*, are similar and can be superimposed to align most of the DnaA boxes within the DnaA box regions (90). These results are consistent with the observation that not only are sequence-specific elements like the DnaA boxes essential for *oriC* function in *E. coli* (91), but also the relative distance between neighboring DnaA boxes is important, probably for cooperative binding of DnaA protein to the DnaA box region.

Comparison of the structures of the origin regions, of the *dnaA* genes and DnaA box regions in particular, in various bacterial chromosomes allowed deduction of the possible evolutionary relationships of bacterial replication origins as shown in Fig. 4 (137). The replication origin region of the prototype ancestral bacterium consists of a *dnaA* gene and a DnaA box region surrounded by at least 12 genes that are conserved between *B. subtilis* and *P. putida*. Duplication of the DnaA box region may be the first evolutionary step (as in pseudomonads) and may be followed by translocation of the duplicated DnaA box region in *E. coli* and triplication of the DnaA box region in *B. subtilis* and other gram-positive organisms. In addition, deletion of DnaA boxes from their original locations occurred in *E. coli*, so that the second and translocated DnaA box now constitutes the replication origin of the chromosome. In contrast, DnaA boxes were added in *B. subtilis*, resulting in tighter regulation of both initiation from this region and transcription of the *dnaA* gene. Mutations found in *Mycoplasma* and *Micrococcus* spp. seem to be more recent events. Functional aspects accompanying multiplication of DnaA box regions as it occurs in *B. subtilis* will be discussed below.

Function of DnaA in initiation

The DnaA protein has been identified as a protein essential for the initiation of chromosomal replication in *E. coli* both in vivo and in vitro (118). Activation of *oriC* by interaction of DnaA with DnaA boxes and subsequent formation of the primosome have been established categorically in *E. coli* (16). Studies of DnaA in *B. subtilis* have not been as extensive as studies of *E. coli* DnaA because the *B. subtilis dnaA* gene was discovered only very recently.

Mutations in the *dnaA* gene of *B. subtilis* were not isolated until the cloned gene was mutagenized in vitro and introduced into the cell for selection of a temperature-sensitive mutant (81). The mutant obtained carried a single amino acid substitution in

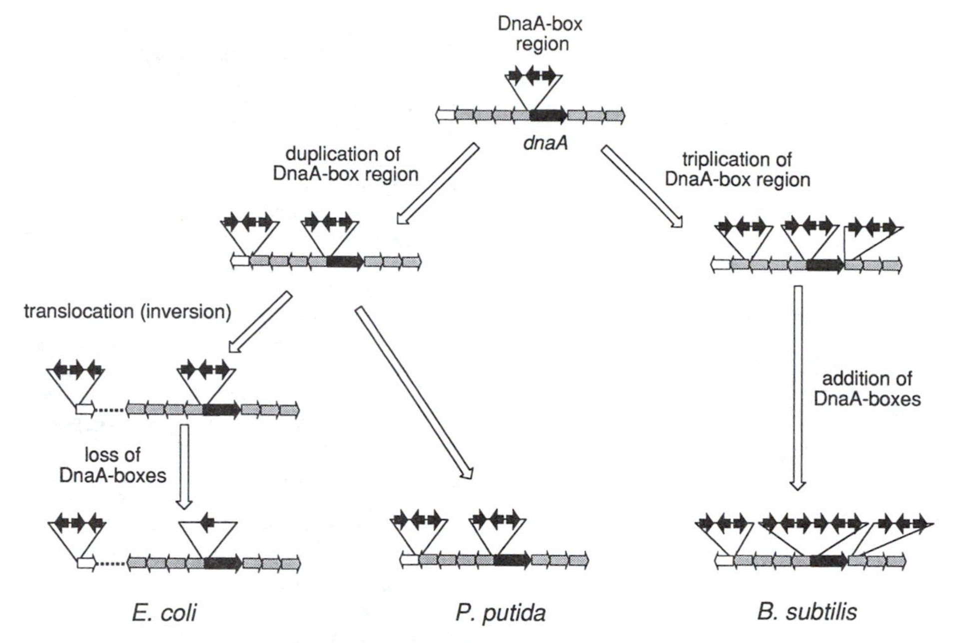

Figure 4. Evolutionary relationships of replication origins. Hypothetical evolutionary relationships between three bacterial origins are shown schematically. The main filled arrow (*dnaA*), shaded arrows (other conserved genes flanking *dnaA*), and open arrow (*gidA*) show the organization and orientations of genes in the replication origin region. The numbers and orientations of DnaA boxes (small filled arrows) are arbitrary, showing that there are multiple repeats of DnaA boxes. The *E. coli* DnaA box region linked to *dnaA* is exceptional because there is only one DnaA box. Reprinted from Yoshikawa and Ogasawara (137) with permission.

domain IV of DnaA that caused a defect specifically in initiation of chromosomal replication at the nonpermissive temperature. The mutant protein is inactive and turns over (observed in the presence of rifampin) rapidly, with a half-life of 7 min at 49°C. However, the cell overproduced the inactive protein and accumulated it during incubation at the nonpermissive temperature in spite of the constant rapid turnover. Upon a temperature shift-down to the permissive temperature, the cell reinitiated chromosomal replication without synthesis of new proteins, indicating that inactivation of the DnaA protein was reversible and could be utilized upon renaturation at the permissive temperature. Moriya et al. (81) measured the amounts of DnaA protein accumulated at 49°C for various periods as well as the cell's potential for initiating rounds of replication without new protein synthesis after the temperature shift-down and found a linear relationship between the amount of DnaA protein and the cell's initiation potential (Fig. 5; 81).

B. subtilis DnaA protein has been produced on a large scale in mutant *E. coli* cells in which the endogenous *dnaA* gene was destroyed (31). The purified enzyme binds preferentially to DNA fragments containing multiple repeats of DnaA boxes (31). The binding may be cooperative, because not only the numbers but also the relative locations of DnaA boxes are important in binding. By using various DnaA box-containing fragments derived from the two DnaA box regions flanking *dnaA*, it was demonstrated that the consensus sequence TTATCCACA was strongest in binding followed by those sequences differing from the consensus by one base. Sequences differing from the consensus by two bases bound the protein only when those sequences were located adjacent to a consensus sequence. The protein bound both ATP and ADP with high affinity (31). The ATP form was as stable as the ADP form but was changed rapidly to the ADP form by the addition of DNA. No sequence specificity was observed with respect to the requirement for DNA in the DNA-dependent ATPase activity of the DnaA protein. The anti-DnaA antibody has allowed estimation of the amount of DnaA protein in the whole-cell lysate of exponentially growing cells to be 200 to 300 molecules for each chromosomal origin (unpublished data).

DnaA box regions act as autonomously replicating sequences (*ars*)

Moriya et al. have recently succeeded in isolating fragments from the origin region of the chromosome containing *ars* activity (79). The two DnaA box regions flanking the *dnaA* gene are required simultaneously in *cis* configuration in order to exert *ars* activity (Fig. 6). Thus, the minimum chromosomal segment essential for *ars* spans some 2.3 kb including the *dnaA* gene. When inserted into a plasmid vector and introduced into *B. subtilis* as *oriC* plasmids, the coding sequence of *dnaA* could be deleted without affecting the *ars* activity. The shortest distance between the two DnaA box regions so far obtained has been 274 bases. Deletion analysis from the ends of the two DnaA box regions revealed that the AT-rich 16-mer repeats in the 5′-terminal portion of DnaA box region C upstream from *dnaA* and an extremely AT-rich stretch (60 bp that were 86% A+T) in the 3′-terminal portion of the downstream DnaA box region R were both

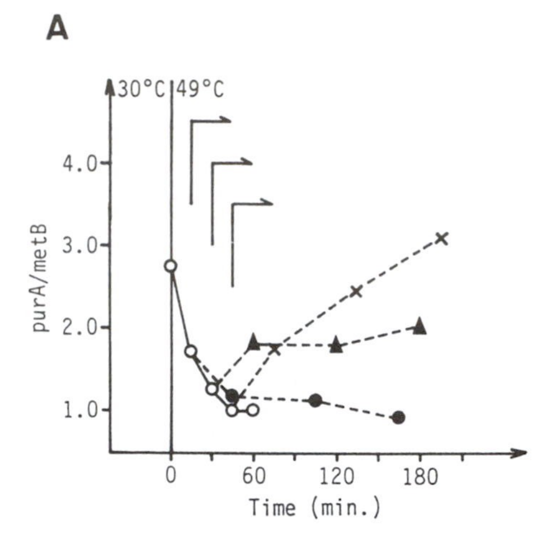

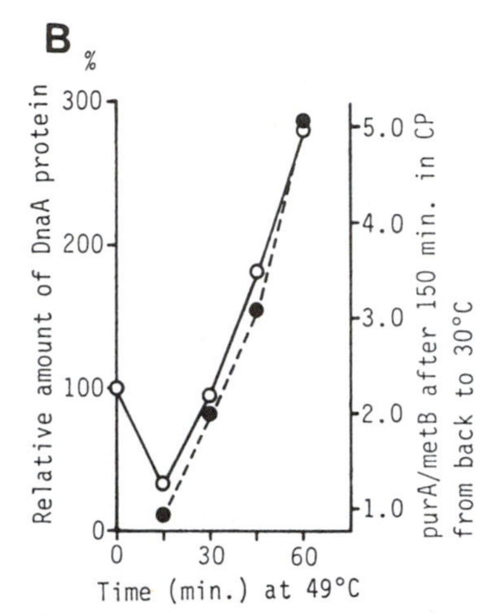

Figure 5. Correlation between the cell's capacity for initiation and the amount of DnaA protein analyzed in a *dnaA* mutant cell. Cells of a *dnaA1* mutant of *B. subtilis* were grown in brain heart infusion medium for about six generations at 30°C, and then the culture was shifted to 49°C. At various times during incubation at 49°C, aliquots were shifted to 30°C in the presence of chloramphenicol (CP). At times indicated at both 49 and 30°C, *purA/metB* ratios of the chromosomal DNAs were determined. In parallel, the relative amounts of DnaA protein per unit volume of the culture incubated for indicated times at 49°C were determined by Western blotting (immunoblotting) using an anti-DnaA antibody. (A) Normalized *purA/metB* ratios are plotted against the time of incubation at 49°C (○) and at 30°C following incubation at 49°C for 15 min (●), 30 min (▲), and 45 min (×). Bent arrows indicate times of shift-down to 30°C. (B) Data in panel A were replotted together with data for DnaA content of the cell. The *purA/metB* ratios at 150 min after temperature shift-down to 30°C are plotted against the incubation period at 49°C before the shift-down (●). The value for 60 min was taken from the data in a separate experiment. Datum points for DnaA content are averages of two experiments (○). Reprinted from Moriya et al. (81) with permission.

essential for *ars* activity (79). It is known that in *E. coli oriC*, the AT-rich region adjacent to the repeat of DnaA boxes is the area first unwound as a consequence of the binding of the DnaA protein to the DnaA boxes (17). It seems that both DnaA box regions are activated simultaneously in the case of *B. subtilis oriC*. Interaction between the two DnaA box regions through interaction with DnaA protein bound to each region may be involved in such a simultaneous activation. The function of DnaA box region L is not known at the moment. Like the other two DnaA box regions, it did not show *ars* activity by itself. Using *E. coli* as a host cell, cloning of fragments containing DnaA box region L in combination with either one or both of the other two DnaA box regions has not been successful.

ars fragments of various constructs are all, regardless of size, unstable as *oriC* plasmids and tend to be lost or integrated into the chromosome even in *recE* mutant cells (79). The copy number of a relatively stable plasmid was estimated as one per chromosomal replication origin. In retrospect, the instability and the low copy numbers of *oriC* plasmids were probably the major cause of earlier failures to isolate *ars* fragments. In addition, the *dnaA* gene segment intervening between the two DnaA box regions contains many sites cleaved by restriction enzymes conventionally used in random cloning of *ars* fragments.

Incompatibility caused by interaction of DnaA with DnaA box regions

The *B. subtilis oriC* plasmids are unstable and of low copy number because they are barely compatible with *oriC* on the chromosome. Moriya et al. had shown previously that DNA fragments from the replication origin inhibited cell growth when they were introduced in multiple copies carried on plasmids (80). DnaA box regions turned out to be responsible for the inhibition. AT-rich stretches within DnaA box regions C and R, which are essential for *ars* activity, did not show the inhibition. The three DnaA box regions can be divided into four segments containing four or five DnaA boxes exhibiting different levels of inhibition (Fig. 6). The *incB* region immediately upstream of *dnaA* is the strongest and could not transform cells even when carried on a low-copy-number plasmid (four or five copies per cell). *incC* and *incD* are intermediate and could transform cells when carried on a low-copy-number plasmid but not on a high-copy-number plasmid (30 to 40 copies per cell). The *incA* region, which contains five DnaA boxes (one consensus and four sequences differing by 1 base from the consensus), did not show inhibition even when carried on a high-copy-number plasmid. When part of the *incC* region was introduced on the high-copy-number plasmid, transformants grew with a doubling time twice that of the parental cells. The amount of DNA per cell and the *purA/metB* ratios of the transformed cells indicated that the increase in doubling time resulted from a decreased frequency of initiation of chromosomal replication. Spontaneous revertants in the growth rate contained the same plasmid with a mutation in DnaA box sequences (80). These results suggest strongly that the growth inhibition was caused by incompatibility between extra copies of the DnaA boxes and chromosomal origins, the phenomenon being similar to incompatibility between two plasmids sharing the same replication origin and initiator protein. The incompatibility may be caused by competition between the two plasmid origins against the common initiator protein or, alterna-

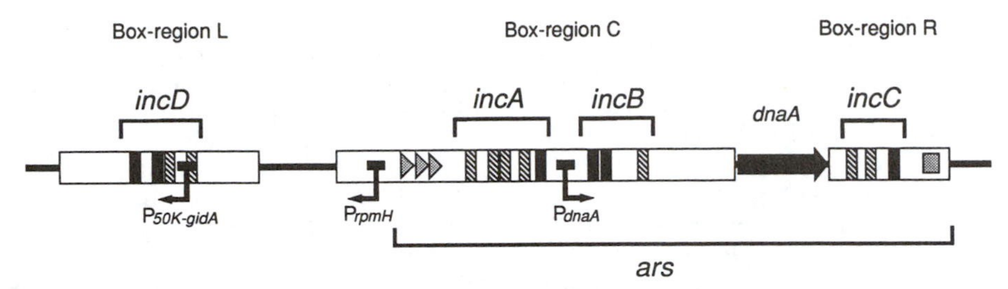

Figure 6. Functional elements in DnaA box regions of the *B. subtilis* chromosome. Structural elements that show various functions in initiation of chromosomal replication and its regulation are shown schematically. DnaA box regions are to scale, while the *dnaA* gene is reduced. Within the DnaA box regions, DnaA boxes are shown as filled vertical strips for consensus boxes and shaded vertical strips for boxes different from the consensus by one base. AT-rich 16-mers are indicated by triangles, and an AT-rich stretch is indicated by a shaded square. Promoters and directions of transcription are indicated by bent arrows. Regions responsible for *inc* activity and required for *ars* activity are indicated.

tively, could result from interaction between the origin and *inc* regions through interaction of initiator proteins bound to each sequence (86). Extra DnaA boxes may inhibit chromosomal replication by titration of the limited amount of DnaA protein in the cell. However, >10-fold overproduction of the DnaA protein did not diminish inhibition by the *inc* fragments (unpublished data), and the copy number of *oriC* was not affected by the presence of the intact *dnaA* gene in the *oriC* plasmids (79). These results suggest strongly that interaction of DnaA box regions through DnaA protein bound to DnaA box regions occurs in *B. subtilis* and plays a role in the control of initiation of replication.

Other Factors in Initiation

dnaB and other initiation genes

The *dnaB* gene was one of the initiation genes most extensively studied before the development of DNA cloning (132). Many mutations between *polA* and *citF* were isolated and mapped (50), and all of them except one (*dnaB19*) were very closely linked. Furthermore, *dnaB19* affected only the initiation of chromosomal replication, while other mutations affected plasmid replication as well (132). Therefore, it was once proposed that *dnaB* covered two cistrons. However, it was shown later by cloning and sequencing that the *dnaB* mutations were located in two separate portions of a single ORF (47, 88). Some of the *dnaB* mutants were reversible and could reinitiate chromosomal replication when cells treated for a certain time at the nonpermissive temperature were shifted down to the permissive temperature (61, 83). The DnaB protein seems to be involved in the early stage of initiation, because RNA synthesis was required for reinitiation of the new round of replication dependent on the DnaB protein accumulated during the incubation of the mutant cell at the nonpermissive temperature (61, 83). However, the function of the *dnaB* gene does not seem to be replicon specific, because the replication of pUB110 is also affected by *dnaB* mutations (132). Mutations affecting initiation of chromosomal replication and induction of prophage SPO2 were isolated by first screening for mutants that did not induce the phage by hydroxyphenylazouracil at high temperature and then selecting for temperature-sensitive growth (84). All mutants mapped in *dnaB*, indicating that the DnaB protein is also involved in the early phase of prophage induction. The DnaB protein may be involved in binding of DNA to the cell membrane at or near the replication origin regions of both chromosome and plasmid DNAs (131). The binding of DNA to the membrane may be a common step in the first stage in replication of chromosome, plasmid, and bacteriophage. The effect of a *dnaB* mutation on the two types of membrane-DNA complexes will be described below. The sequence data show that *dnaB* lies within an operon that also contains three downstream genes defined by ORF311, ORF213, and ORF281. The amino acid sequence suggests that ORF311 is a nucleotide-binding protein and ORF213 is a hydrophobic-residue-rich membrane protein (88). The function of these genes should be examined in relation to the function of DnaB both in initiation and membrane binding of the chromosome. The DnaB protein has been partially purified, but no function such as binding to DNA or membrane has been demonstrated, probably because of DnaB's strong tendency to self-aggregate (unpublished data). Neither the sequences of action of the DnaA and DnaB proteins nor the possible interaction between the two proteins has been studied. It should be noted that the *B. subtilis dnaB* gene has no relationship with *E. coli dnaB* (see Introduction). In fact, no homolog of *B. subtilis dnaB* has been identified in *E. coli*.

In addition to the two major classes of initiation genes (*dnaA* and *dnaB*), *dnaD* (53) and *dna-199*(Ts) (*tsB*; 39) have been reported to map at other locations on the chromosome. Although temperature-sensitive mutants of these genes show phenotypes very similar to that of *dnaA* or *dnaB*, no follow-up studies have been performed to further characterize the function of the protein products of these genes.

Involvement of the membrane in initiation

Since the proposal of the replicon hypothesis by Jacob et al. (52), involvement of the cell membrane has been a major issue in relation to replication and partitioning of the bacterial chromosome. Ryter et al. demonstrated clearly by electron microscopy that the *B. subtilis* chromosome is attached to the cell membrane at a limited number of sites (100). Isolation of membrane fractions by various means (sucrose gradient [133], sucrose-CsCl gradient [115], Renografin [46], and Mg^{2+}-Sarkosyl complex [40]) has also shown that the whole chromosome can be isolated in association with the membrane. Surprisingly, such association was resistant to high salt, like 6M CsCl. Breakage of the chromosome by hydrodynamic shear (134) or restriction enzyme digestion (103) released most of the chromosomal DNA from the membrane fraction, leaving 5 to 15% of the DNA behind, depending on the degree of breakage (134). Pulse-labeled replicating DNA was preferentially concentrated in the latter fraction, suggesting that newly replicated DNA at or near the replication machinery is associated with the cell membrane (33). By taking advantage of the transformation assay of genetic markers remaining in the membrane fractions, two regions on the chromosome, one near the replication origin and the other near the terminus, were shown to be associated with the membrane (95, 134). The association with the membrane is resistant to high salt, EDTA, and nonionic detergent but sensitive to a strong ionic detergent such as Sarkosyl (134). In addition to a stable membrane complex (M complex), the markers near the replication origin form a salt-sensitive complex (S complex) containing DNA, RNA, and proteins (134). A definite restriction fragment pattern was reproducibly obtained from the S complex (135, 141). Later, Sargent and Bennett prepared membrane-bound DNA following treatment with a restriction enzyme and subsequently cloned a specific DNA sequence that mapped close to the *purA* marker (103). Formation of the S complex paralleled formation of the cell's capacity to initiate chromosomal replication during spore germination (141). Markers near the replication terminus did not form the S complex. These results suggest that both the origin and the terminus regions bind to membrane but in two different forms, one (origin region) being changeable from one state (M complex) to the other (S complex) during one cell cycle, and the other (terminus region) being stable throughout the cell cycle. Origin-membrane association may be involved in initiation, and terminus-membrane association may be involved in partitioning of the chromosome. It is interesting to note that the origin region but not the terminus was attached to the membrane in a stable L-form cell in which segregation of daughter chromosomes was irregular (46). That attachment of origin to the membrane has a role in initiation is supported by the recent report that the association of the *purA* marker in both the M complex and the S complex was affected by mutation in the initiation protein, DnaB (131). These observations, however, should be reevaluated in the light of recent knowledge of the structure and function of the replication origin of the chromosome. *purA*, which had been used exclusively as an origin marker, is now known to be located some 40 kb away from *oriC*.

A functional role for the cell membrane in initiation of chromosomal replication was suggested by the treatment of cells with toluene (74). The permeabilized cells thus prepared lost the ability to initiate replication of DNA permanently without any significant effect on elongation of replication (130). A more direct demonstration of the role of the membrane was achieved by in vitro replication of chromosomal DNA in a membrane preparation (12). Sensitivity to drugs, dependency on the *dnaB* gene, and identification of the labeled origin fragments by hybridization suggested that initiation of a round of chromosomal replication occurred in the membrane preparation in vitro (11). Furthermore, Laffan and Firshein have identified a 64-kDa protein that binds preferentially to origin DNA (57). Antibody against this protein stimulated in vitro DNA synthesis by the membrane preparation, suggesting a role for the 64-kDa protein in negative regulation of initiation in vivo. However, more detailed analyses are required to establish that the DNA synthesis observed in these experiments reflects origin-dependent replication of the chromosome.

Roles of RNA and RNA synthesis in initiation

The involvement of RNA or RNA synthesis directly in initiation of chromosome replication in *B. subtilis* was demonstrated by the use of *dnaB* mutant cells (61, 83). After treatment at the nonpermissive temperature for a certain period, such mutant cells reinitiated replication upon the shift to the permissive temperature in the absence of protein synthesis but still required RNA synthesis. Extensive studies on the kinetics of the RNA synthesis required for the reinitiation revealed that the RNA species were short lived (half-life of 2 min) at the nonpermissive temperature when functional DnaB was not available. They were more stable (half-life of 12 min) at the permissive temperature (83). RNA specific for initiation may be involved in forming an initiation complex, or alternatively, transcription per se in the region of the origin may be involved in creating a conformation of DNA appropriate for the initiation proteins to act on. In *E. coli*, genetic suppression studies suggest an interaction between DnaA protein and RNA polymerase (6, 99). More stable RNA, covalently linked to DNA, has been shown to be synthesized at a specific time during an early phase of the initiation of replication of the *B. subtilis* chromosome (42). Neither the structure of the RNA-linked DNA nor its location on the chromosome has been described.

Mechanism and Control of Initiation of Replication in *B. subtilis*

The molecular mechanism of initiation of replication based on interaction of the DnaA protein with the DnaA box region (*oriC*) of the *E. coli* chromosome has been well documented (16). It is assumed that the molecular mechanism of the initiation stage that activates *oriC* is essentially the same in *B. subtilis*, since both DnaA and the DnaA box region are conserved in structure and function. However, the formation of a primosome complex as a whole may be entirely different in *B. subtilis* and *E. coli*. At least two DnaA box regions 1.5 kb apart are required simultaneously for initiation. The third DnaA box region, located symmetrically to the left, may also participate in initiation. All of these elements may constitute a large, complex superstructure (initiosome) together with DnaA, DnaB, and other yet unknown accessory proteins. Binding of chromosomal DNA to the membrane at the specific sequence about 40 kb away from the *oriC* region may form an anchor to assemble such an initiosome in close association with the cell membrane.

Both the frequency and the timing of initiation of chromosome replication are strictly controlled within the cell division cycle in bacteria. Initiation cell mass is the key factor in the regulation of both frequency and timing of initiation (25). In *E. coli*, the amount of DnaA protein is thought to determine the initiation cell mass (69). In addition, hemimethylation of the *oriC* region is responsible for preventing premature initiation of the newly replicated *oriC*, thus resulting in more precise timing of initiation (9, 15, 78). The DnaA protein also determines the frequency of initiation of chromosomal replication in *B. subtilis* when other components of the initiation-elongation machinery are available in excess (81). The linear relationship between the amount of DnaA and the cell's capacity to initiate rounds of replication revealed that the amount of DnaA in exponentially growing cells corresponded to the amount of DnaA just sufficient to initiate one round of replication (81). However, overproduction of the DnaA protein, which caused extra initiations in *E. coli* (7), did not induce extra initiations in *B. subtilis* (unpublished data), suggesting that other factors are also limiting for initiation in exponentially growing *B. subtilis* cells. Expression of the *dnaA* gene is autoregulated both in *E. coli* (18, 124) and in *B. subtilis* (unpublished data). In *B. subtilis*, the autogenous suppression is much stronger than in *E. coli*, and the expression is coupled with initiation of replication, i.e., replication of the *dnaA* region itself (89). These observations suggest that DnaA is synthesized only at the early phase of the cell cycle when the newly replicated promoter region of the *dnaA* gene is not completely bound by the DnaA protein. In other words, DnaA protein synthesis is completed before the cell acquires the proper initiation cell mass (Fig. 7). It is therefore necessary to look for other factors that can measure initiation cell mass and control the timing of initiation in *B. subtilis*. The DnaB protein and assembly of the initiosome through DnaB in association with the cell membrane may contribute to the controlling agency. When the initiosome structure is completed, the *oriC* region may take a specific conformation by using a large number of DnaA molecules and multiple DnaA box regions to activate *oriC* (Fig. 7). After initiation takes place, the initiosome decomposes, releasing the DnaA protein from the *oriC* region to inactivate *oriC*. Premature initiation of the newly replicated *oriC* is prevented by titration of free DnaA protein by many newly replicated DnaA boxes in the *oriC* region. Inhibition of premature initiation by hemimethylation of the *oriC* region is a recently developed device found only in *E. coli* and other enteric bacteria. Clusters of GATC sites for *dam* methylation in *E. coli* are not found in other bacteria, including pseudomonads, which are evolutionarily fairly closely related to *E. coli*. The evolution of the methylation device seems to be related to the loss of DnaA box regions from the *oriC* region in *E. coli* (137). Extra DnaA box regions introduced into *B. subtilis* cells inhibit initiation of chromosomal replication because they interfere with formation of a specific active conformation of the *oriC* region by interacting with DnaA already bound to DnaA box regions. Identification of the molecular entities in a putative initiosome

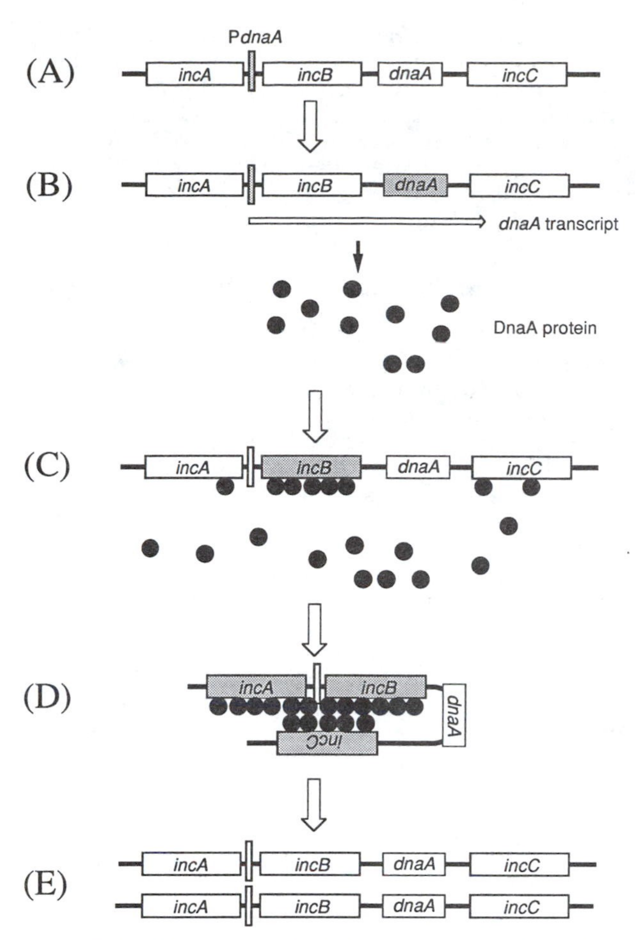

Figure 7. Model for control of the initiation cycle and for expression and function of DnaA protein. (A) DnaA box regions release DnaA protein temporarily after they are replicated. (B) *dnaA* gene expression is derepressed and actively transcribed to produce DnaA protein (●). (C) DnaA protein preferentially binds to the *incB* region of the DnaA box region to repress *dnaA* gene expression. (D) DnaA box regions form the initiation complex (initiosome); DnaB protein and binding to the cell membrane may be involved in this stage. Preformed DnaA protein now binds to all *inc* regions within at least two DnaA box regions to form the active *oriC* conformation. (E) Initiation and consequent replication of *oriC* take place. The relative sizes of the *inc* regions and the *dnaA* gene are as shown in Fig. 3 and 6.

is essential for understanding the molecular mechanism and control of initiation of chromosomal replication.

TERMINATION OF CHROMOSOME REPLICATION

The cycle of chromosome replication commences with initiation. Toward the end of the second (elongation) phase, the approaching forks meet and fuse. Termination is defined here as the meeting and fusion of the forks to yield two separate and continuous double-stranded segments of DNA spanning the site of fusion. In the overall division cycle, this fusion is followed (or accompanied) by segregation of the two chromosomes. In the case of *E. coli* at least, it is possible that termination leaves the two daughter chromosomes twined around one another, presumably in their terminus regions (113). Decatenation or resolution of such a structure into two separate chromosomes can be achieved by the action of DNA gyrase (113). It appears that another enzyme, topiosomerase IV, could also be involved in decatenation to allow segregation of daughter chromosomes (53a).

A Specific Terminus and the Manner of Replication Fork Approach

Reference has already been made to the early difficulty in identifying the terminus region of the bacterial chromosome and the use of the sporulation process in *B. subtilis* to overcome this problem (1, 102). What this approach provided was a means of incorporating radioactivity specifically into that portion of the chromosome replicated toward the end of the cycle and just before termination. Analysis of spore DNA labeled in this way identified a unique collection of restriction fragments whose number decreased in a defined order with progressively shorter times of labeling (125). Upon construction of a restriction map of ~250 kb of the labeled terminus region, it became clear from the order of fragment replication that over the last few (<5) minutes of the cycle of replication, only the anticlockwise fork was moving (48, 126). To account for this, it was suggested that the other (clockwise) fork had already been arrested at a site, called *terC*, toward which the anticlockwise fork continued to progress during the final few minutes. These studies showed convincingly that termination of replication in *B. subtilis* occurred at a unique site on the chromosome, a situation quite different from that established for the case of the simian virus 40 (58) and lambda (117) chromosomes.

Clockwise-Fork Arrest at *terC* as First Stage in Termination

Could the proposed arrested clockwise fork at *terC* be detected? It would be expected to exist transiently around the time of termination and was predicted to occur within a specific *Bam*HI fragment. Germinated and outgrowing spores were used to provide synchrony in a cycle of replication, and the arrested fork was identified at the expected time by hybridization to a specific probe as a slowly migrating species on agarose gel electrophoresis (127). Its forked structure and predictable dimensions (arm and stem regions) were confirmed by electron microscopy and analysis of the single strands (128).

Could the same forked molecule be observed during exponential growth? The merodiploid strain GSY1127 described by Anagnostopoulos and coworkers (104) permitted such observation. Due to a nontandem duplication of part of the anticlockwise segment of the chromosome in this strain, *terC* would be expected to be offset (relative to *oriC*) to a grossly asymmetric location. This would cause the clockwise fork to remain arrested at *terC* for a much longer portion of the replication cycle in this strain than in the wild type. The predicted enhanced level of the arrested fork was readily observed (127). Thus, it was established that clockwise-fork arrest at *terC*, the first stage in the termination process, is common to the situations of sporulation, outgrowth following spore germination, and exponential growth. O'Sullivan and Anagnostopoulos (93) independently used the GSY1127 merodiploid strain to show that there is a preferred chromosomal site for termination of replication. In subsequent work, use of the merodiploid strain has provided a convenient assay for the functioning of *terC*, at least as a replication fork arrest site (109, 110). It remains to be established definitively whether or not there is a complete rather than a partial block to clockwise fork movement at *terC*. Marker frequency analysis experiments have given results at least consistent with a complete block (38). However, the recently achieved relocation of the *terC* region to asymmetric sites on the wild-type chromosome (23; see below) will permit a more rigorous examination of this aspect of fork movement associated with termination.

terC Region Sequence and Requirements for Fork Arrest

rtp gene and the IRR

The DNA sequence spanning *terC* established the existence of features that have subsequently been shown to be the major and probably sole determinants of the termination-associated clockwise-fork arrest (22, 110). These are summarized in Fig. 8A. The relevant features are a gene (*rtp*) for a small basic protein (replication terminator protein; 122 amino acids, 14.5 kDa) and, just upstream of this, an inverted repeat (IR) region (IRR) made up of two imperfect IRs (IRI and IRII; 47 and 48 bp, respectively) separated by 59 bp. The clockwise fork enters this region of the chromosome from the right and is blocked in its movement at approximately the position labeled *terC*. Deletion experiments confirmed the importance of these features (110). When up to 80 bp to the left of IRI was deleted, there was no effect on fork arrest. When a further 130 bp including IRI was deleted, fork arrest was no longer observed. Removal of the bulk of the ORF defining *rtp* or disruption of this ORF also abolished fork arrest. These results suggested a possible role in fork arrest for both the IRR and the product of the *rtp* gene. They also established that the normal termination mechanism involving fork arrest at *terC* is not essential for cell viability. This is discussed further below.

The sequence features of the *terC* region just discussed relate to strain 168 of *B. subtilis*. To examine the generality of these features, the situation in the related *B. subtilis* strain, W23, has also been examined (68). The extent of overall DNA sequence homology between the W23 and 168 strains has been reported to be 67 to 89% (70, 106). Strain 168 *terC* region sequence features are largely conserved in strain W23. Significantly, the amino acid sequence of the *rtp* gene product (RTP) is 100% conserved in spite of 22 base differences at the DNA level. IRI and IRII are largely conserved. Clockwise-fork arrest has been established to occur in W23 in the same manner as in 168. Presumably, the sequence features common to both 168 and W23 are operative in effecting fork arrest.

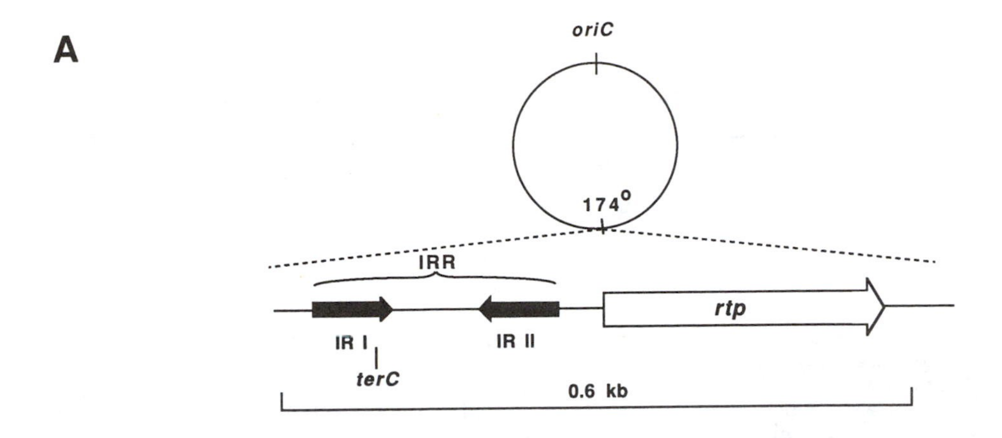

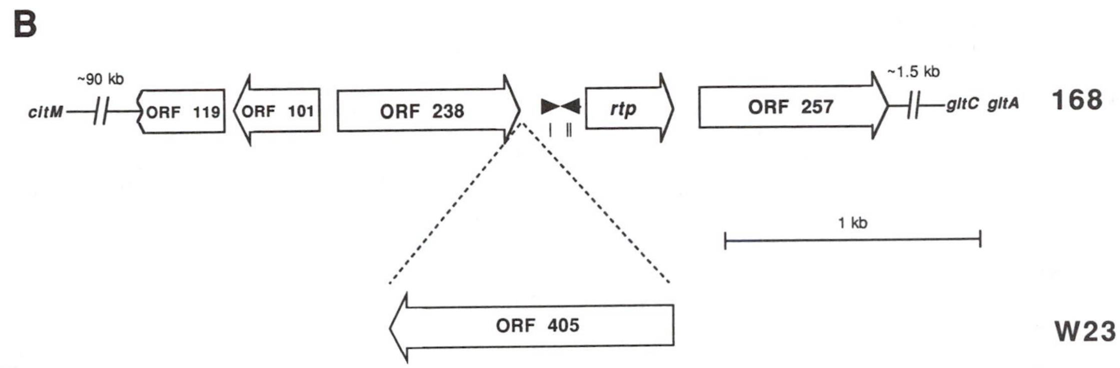

Figure 8. Sequence features of the *terC* region of the *B. subtilis* chromosome. (A) The region located at 174° on the physical map (51) and slightly offset in relation to the origin (*oriC*) comprises the gene for *rtp* and the upstream IRR, which is made up of IRI plus IRII. The clockwise-fork arrest site is shown as *terC* (see also Fig. 9). (B) Arrangement of sequences (ORFs) over a more extended chromosomal segment (~3 kb) spanning the *terC* region (IRR + *rtp*) (3). Strain W23 contains an additional ORF (ORF405) that strain 168 does not contain, which is inserted just left of the IRR. The *gltC* and *gltA* genes lie ~1.5 kb to the right of the chromosomal segment sequenced, and the *citM* gene lies ~90 kb to the left of it.

Recently, a comparison over a more extensive region (~3 kb) spanning *terC* in the two strains has been made (3). The findings are summarized in Fig. 8B, which shows the organization of sequences (ORFs) in the terminus region of strain 168. A significant difference in W23 results from the insertion of an ORF comprising 405 amino acids just to the left of IRI. ORF405 is a member of the cytochrome P-450 family. It is not known whether any of the ORFs identified in Fig. 8B (other than *rtp*) encode functional proteins.

Fork arrest requires *rtp* expression and involves RTP-IRR interactions

That clockwise-fork arrest requires the *rtp* gene product was established by showing that arrest in vivo is dependent on *rtp* expression. A strain in which *rtp* expression was placed under the control of the isopropyl-β-D-thiogalactopyranoside (IPTG)-inducible *spac-1* promoter was constructed (111). Replication fork arrest, monitored by the appearance of a new forked DNA molecule of predicted dimensions, was shown to be dependent on IPTG-induced expression of *rtp* in this strain. Also, the very low levels of IPTG required to induce fork arrest suggested that relatively little RTP was needed.

It had been suggested that RTP binds directly to the IRR to effect arrest (110). That such binding does occur was established by in vitro studies using purified RTP. Band retardation experiments showed RTP to be a DNA-binding protein with specific affinity for sequences within the IRR (67). When assays were performed with the whole IRR over a range of RTP concentrations, a series of four retarded species could be identified. This suggested that there were four RTP-binding sites within the IRR. Further experiments showed that RTP bound to two sites in IRI and two sites in IRII (66). DNase I footprinting studies using the IRR and a segment of IRI containing a single RTP-binding site confirmed that RTP bound specifically to sequences within IRI and IRII and identified the regions involved (66). Furthermore, it was clear that RTP had a greater affinity for the IRI within the IRR than it did for IRII. The overall affinity of RTP for the IRR (K_d = 1.2 × 10^{-11} M in 50% glycerol) was at least as high as that of the *E. coli trp* repressor for its operator.

RTP exists as a tight dimer of 29 kDa in solution at neutral pH, and direct measurement of the stoichiometry of the RTP-IRR interaction established that a dimer of RTP binds to each of four sites (two each in

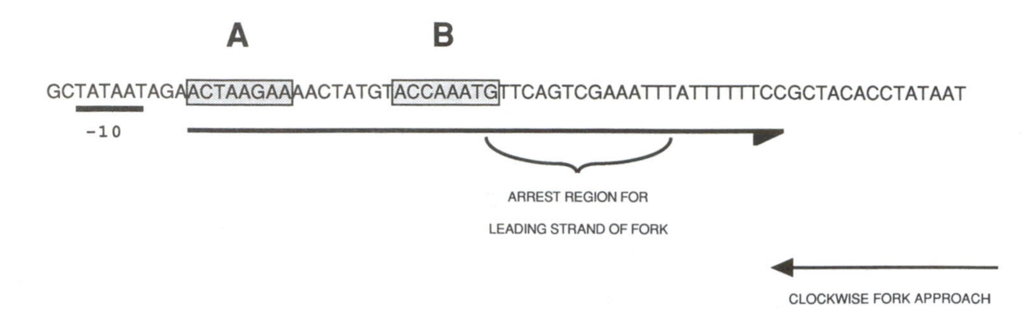

Figure 9. Putative RTP recognition sequences within the functional IRI of the *B. subtilis* chromosome. The two 8-bp segments (A and B, shaded) proposed as recognition sequences have been identified from the relative positionings of DNase I-protected regions and comparison of sequences within a total of 8 IRs. The heavy arrow defines IRI as originally described (22). The −10 region of the *rtp* promoter is located just to the left of A. The region of arrest of the leading strand of the clockwise replication fork is based on the work of Jannière and coworkers (20).

IRI and IRII) within the IRR (66). It is likely that the binding of RTP to adjacent sites within each IR is cooperative.

DNA recognition sequences for RTP and precise location of *terC*

While it is known that IRI and IRII each contain two binding sites for RTP, the exact nature of the recognition sequences has not yet been established. However, examination of the sequences that are protected from DNase I cleavage by binding of RTP to a single and adjacent sites within an IRR raise the possibility that a largely conserved 8-bp sequence (consensus ACTAAATA) is involved in defining the recognition sequence of each site for the RTP dimer (66). Figure 9 shows the location of these 8-bp sequences, labeled A and B, within IRI of the 168 strain. From in vivo studies, it has been shown that the leading strand of the clockwise fork of the chromosome approaching from the right in Fig. 9 can pass through IRII (not shown) but not through IRI (shown) (129). The arrest site has now been mapped more precisely by Jannière and coworkers (20) by the use of a *terC* region-containing derivative of the *Streptococcus faecalis* plasmid pAMβ1 replicating in *B. subtilis*. The majority of the arrested leading strands were distributed within the region indicated by the bracket in Fig. 9. Clearly, this finding is consistent with RTP binding to the 8-bp segments effecting arrest of the leading strand of the clockwise fork.

Model for Fork Arrest and Fusion at *terC*

Termination of chromosome replication in *E. coli* also involves arrest of replication forks, which is effected by protein-DNA interactions analogous to those observed in *B. subtilis* (56). The terminator protein in this case has been named Tus (terminus utilization substance), and the DNA sequences to which it binds have been referred to as terminators. Five terminators have been found in *E. coli*; they are distributed over a terminus region of ~600 kb and are oriented such that two are opposed to the other three (43, 56). There appears to be significant sequence homology between the *E. coli* terminators and segments of the inverted repeats (IRI and IRII) of the *B. subtilis* IRR (44), although no similarity between Tus and RTP has been identified. The main differences between *B. subtilis* and *E. coli* therefore appear to be in the number, arrangement, and spacing of terminators. All of the data accumulated so far for *B. subtilis* point to the existence of a single termination region centered around the close-together and opposed IRI and IRII. This is similar to the situation in *E. coli* plasmid R6K, in which there are two opposed terminators, highly homologous in sequence to those in the *E. coli* chromosome itself and separated by only 73 bp (45). It is known that an individual *E. coli* or R6K terminator with its bound Tus protein can arrest a fork (44). But, significantly, the effect is polar in the sense that only one of the two possible orientations of the DNA terminator relative to the direction of approach of a replication fork is functional in causing arrest. It is also known that this polar effect is brought about by a polar inhibition of the DnaB helicase that unwinds the two strands at the apex of the fork (55, 62).

It is now known that IRI and IRII in *B. subtilis* can also function as polar terminators, at least in plasmids (112). The fact that the clockwise replication fork traverses IRII but not IRI is consistent with such a situation. IRI would thus be the functional terminator in the chromosome, and its orientation relative to the approaching clockwise fork is consistent with the known active orientation of the homologous *E. coli* terminator sequence (44).

A model describing clockwise-fork arrest in the *terC* region of the *B. subtilis* chromosome is shown in Fig. 10. It incorporates recent information on the autoregulation of *rtp* (2), which was proposed as a possibility some time ago (110). The clockwise fork enters the region (IRR + *rtp*) from the right. IRI and IRII each contain two binding sites for RTP. The promoter for *rtp* lies just to the left of IRI. The −10 region of this promoter (Fig. 9) is protected from DNase I cleavage when RTP binds to IRI (66). Following expression of *rtp* (Fig. 10A), the RTP protein binds to IRI and IRII (panel B). It is not known whether sufficient RTP is produced in vivo to saturate both IRs as shown, but it is clear that RTP binds preferentially to IRI (66) and that this binding shuts down expression of *rtp*. Around the time of termination, the approaching clockwise fork passes through IRII, presumably releasing any bound RTP. RTP does not bind single-stranded terminators (unpublished observations). The clockwise fork is then arrested by the appropriately oriented IRI-RTP

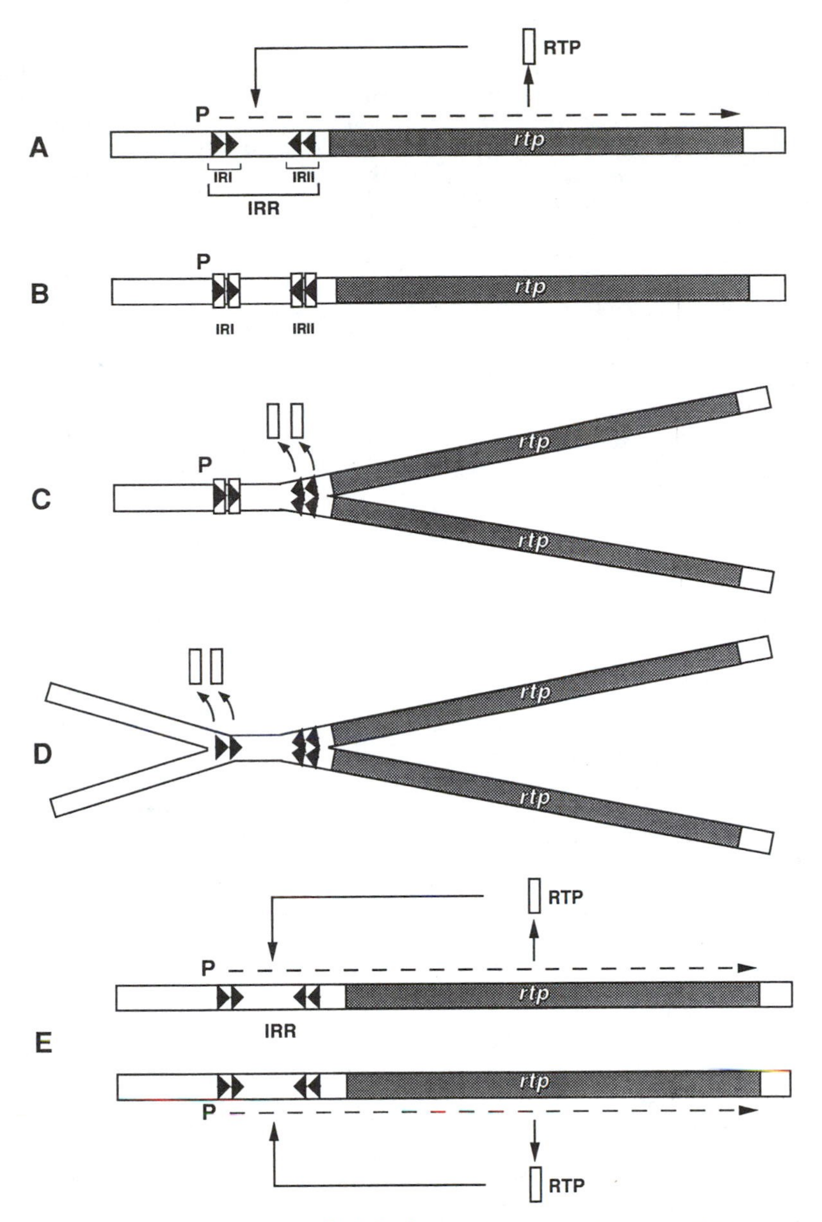

Figure 10. Model for clockwise-fork arrest and fork fusion within the *terC* region (IRR + *rtp*) of the *B. subtilis* chromosome. See text for explanation of model. ▶, ◀, binding sites for RTP; □, RTP; P, promoter.

complex (panel C). It is likely that IRII at this stage cannot bind RTP because of the presence of a stalled replisome complex in its vicinity, its partial single-stranded nature, and/or a limited amount of RTP. The orientation of IRI is such that upon arrival of the anticlockwise fork a few minutes later, the fork passes through IRI, releasing RTP (panel D). Upon fusion with the arrested fork (panel E) and regeneration of two new double-stranded IRRs, *rtp* would be expressed again. (This last aspect of the overall model has not yet been tested.) As yet, there is no definitive information on how RTP bound to its terminator blocks fork movement, but it is highly likely that, as in the case of *E. coli*, it specifically inhibits unwinding at the fork by the major replicative helicase (not yet identified; Table 2).

The essence of the model just described is that the two replication forks meet and fuse within a very restricted region of the chromosome. In the terminology of Kuempel et al. (56), the opposed IRI and IRII terminators provide a replication fork trap. In this context, if the anticlockwise fork arrived in the *terC* region ahead of the clockwise fork, its movement would be expected to be arrested by IRII. However, it appears that in vivo, IRII as part of the IRR in the chromosome is not an effective terminator, possibly because of a limiting amount of RTP (see below).

Dispensability of the *terC* Region and Relocation to New Chromosomal Sites

Strains with deleted *terC* regions are viable and can sporulate

The *terC* region is located on the chromosome between *citM* (originally referred to as *citK*) and *gltA* (48;

Fig. 8B). In 1982, Zahler (142) reported the isolation of three strains of *B. subtilis* (CU1693, CU1694, and CU1695) whose auxotrophic requirements were consistent with a deletion spanning at least the *citM-gltA* segment. Hybridization experiments using SPβc2 DNA and cloned DNA from the terminus region provided more detailed information on the extent of the deletions and confirmed directly that the *terC* region (*rtp* + IRR) was missing from each of these strains (49). Thus, the normal termination mechanism involving clockwise-fork arrest at *terC* is not essential for viability. The Zahler strains do not sporulate, but this is not because they lack the normal termination mechanism. It presumably reflects the deletion of the *cit* genes coding for the citric acid cycle enzyme α-ketoglutarate dehydrogenase, which is needed for sporulation (27). A smaller *terC* region deletion than those of the Zahler strains had no marked effect on the level of sporulation (49).

It is likely that termination via the normal mechanism involving the *terC* region confers a significant advantage on the cell. This is suggested by the 100% conservation of the amino acid sequence of RTP in the 168 and W23 strains of *B. subtilis* in spite of 22 base differences at the DNA sequence level. What could the advantage be? The highly preferred coorientation of replication fork movement and direction of transcription through the clockwise and anticlockwise replicating segments of the chromosome of *B. subtilis* (144) (and also of *E. coli* [19]) raises the possibility that the maintenance of this more efficient situation, made possible by the existence of a replication fork trap at *terC*, is achieved via the termination mechanism that has been uncovered. Another possibility is that termination involving fork fusion within a specific domain of the chromosome allows more efficient segregation, perhaps requiring prior decatenation, of sister chromosomes. Clearly, the precise advantage given to a *B. subtilis* cell by the presence of a specific terminus and termination mechanism has yet to be uncovered.

That one can delete the *terC* region with no grossly adverse effect on the symmetric division associated with vegetative growth or the asymmetric division associated with sporulation (49) is consistent with the earlier conclusion that in *B. subtilis* at least, termination of replication is not the trigger for cell division (75). However, this topic is still a point of contention (24) and perhaps should be reexamined by a much more direct approach now possible because of knowledge of exactly where termination occurs in the *B. subtilis* chromosome and the ease of establishing whether or not termination has occurred by using DNA hybridization probes.

Polarity of action of the *terC* region at new chromosomal sites

In the model presented in Fig. 10, the *terC* region is shown to contain two terminators, IRI and IRII. IRI is depicted as the functional terminator, and this reflects the clockwise fork reaching *terC* ahead of the anticlockwise fork. The earlier arrival of the clockwise fork is probably the direct result of *terC* being slightly offset on the physical map of the chromosome, making the clockwise-replicating segment the shorter of the two by ~140 kb (51). (It is significant that for a chromosome of 4,165 kb [51], a replication cycle time of ~50 min at 37°C [24] would result in an approximately 3-min difference in arrival of the two forks at *terC*, assuming that both travel at the same rate. This is consistent with what has been observed experimentally [126]). Until recently, there was no direct evidence that IRII could function as a terminator, although it was known to bind RTP in a manner similar to IRI. To examine this aspect of terminator function within the chromosome, alternate orientations of the *terC* region (IRR + *rtp*) were relocated to grossly asymmetric sites, *pyr* at 139° and *metD* at 100°, making the clockwise fork arrive at *terC* much earlier than normal in the replication cycle (23). To achieve this, Tn*917* was inserted at the relocation sites. Surprisingly, clockwise-fork arrest was effected only when IRI was operative; IRII was essentially ineffective. It has now been established that both IRI and IRII can function separately as terminators in a plasmid system, although IRII is less effective than IRI when they are present together within the IRR (112). A likely explanation for the nonfunctioning of IRII in the chromosome is a limiting amount of RTP and its preferential binding to IRI (66). Apparently, because in the wild-type 168 strain the clockwise fork always arrives first at *terC* (IRR), only IRI is utilized, and IRII has degenerated to become largely ineffective.

The ability to relocate the *terC* region to potentially any site on the chromosome will allow the exploration of several interesting aspects and effects of the termination process. For example, is the block to clockwise-fork movement at new asymmetric sites complete, and what might be the consequence of the anticlockwise fork proceeding through a chromosomal segment in a direction opposite to that in which fork movement normally proceeds? What are the consequences of relocating the *terC* region such that the anticlockwise fork reaches it first, and how close to *oriC* can the *terC* region be placed? How do *terC* region relocations affect coordination between the replication and division cycles, and what are the consequences with respect to the asymmetric division associated with sporulation? *terC* region relocations could give unique insights into the relationship between chromosome replication and other processes that are closely related to or dependent on it.

Arrest of Replication Forks, Postinitiation Control of Replication, and the Stringent Response

Several years ago, Séror and her colleagues (108) showed that the stringent response involves an effect on DNA replication at or near the initiation stage. The effect was observable through the use of a temperature-sensitive DNA initiation mutant of the *dnaB* class. In an extension of this work, it was established that when the temperature-sensitive *dnaB37* mutant was returned to the permissive temperature following a period of nonpermissive conditions, replication was restricted in the immediate term to a region of 120 to 175 kb on either side of *oriC* (41). Replication forks were presumably stalled for a significant time on either side of the origin at the limits of the overreplicated segment. This was referred to as postinitiation control, a level of control in addition to that at *oriC* exerted in order to limit replication under conditions of unbalanced growth. It has now been shown that stalling (or arrest) of forks at what appear to be the same sites (subsequently more precisely estimated as ~200 kb from *oriC* on each side) can be brought on by the stringent response (65). In contrast, the stringent response in *E. coli* blocks initiation of DNA replication at *oriC* itself. The nature of these special sites on either side of *oriC* in *B. subtilis* at which replication forks can be arrested has not yet been explored. It will be interesting to know whether they are related in any way to the DNA terminators (IRI and IRII) identified in the *terC* region. Probably of more significance is the precise role of this secondary control of DNA replication in *B. subtilis*. The possibility that postinitiation control is an important aspect of growth and sporulation control has been raised (65).

Acknowledgments. We gratefully acknowledge the assistance of Peter J. Lewis in preparation of the tables.

REFERENCES

1. **Adams, R. T., and R. G. Wake.** 1980. Highly specific labeling of the *Bacillus subtilis* chromosome terminus. *J. Bacteriol.* **143:**1036–1038.
2. **Ahn, K. S., M. S. Malo, M. T. Smith, and R. G. Wake.** Submitted for publication.
3. **Ahn, K. S., and R. G. Wake.** 1991. Variations and coding features of the sequence spanning the replication terminus of *Bacillus subtilis* 168 and W23 chromosomes. *Gene* **98:**107–112.

3a. **Allen, G. C., and A. Kornberg.** 1991. Fine balance in the regulation of DnaB helicase by DnaC protein in replication in *Escherichia coli. J. Biol. Chem.* **266:**22096–22101.

4. **Alonso, J. C., C. A. Stiege, R. H. Tailor, and J.-F. Viret.** 1988. Functional analysis of the *dna*(Ts) mutants of *Bacillus subtilis*. Plasmid pUB110 replication as a model system. *Mol. Gen. Genet.* **214:**482–489.
5. **Andersen, J., and A. T. Ganesan.** 1975. Temperature-sensitive deoxyribonucleic acid replication in a *dnaC* mutant of *Bacillus subtilis. J. Bacteriol.* **121:**173–183.

5a. **Atlung, T.** Personal communication.

6. **Atlung, T.** 1984. Allele-specific suppression of *dnaA*(ts) mutations in *Escherichia coli. Mol. Gen. Genet.* **197:**125–128.
7. **Atlung, T., A. Lobner-Olesen, and F. G. Hansen.** 1987. Overproduction of DnaA protein stimulates initiation of chromosome and minichromosome replication in *Escherichia coli. Mol. Gen. Genet.* **206:**51–59.
8. **Attolini, C., G. Mazza, A. Fortunato, G. Ciarrocchi, G. Mastromei, S. Riva, and A. Falaschi.** 1976. On the identity of *dnaP* and *dnaF* genes of *Bacillus subtilis. Mol. Gen. Genet.* **148:**9–17.
9. **Bakker, A., and D. W. Smith.** 1989. Methylation of GATC sites is required for precise timing between rounds of DNA replication in *Escherichia coli. J. Bacteriol.* **171:**5738–5742.
10. **Barnes, M. H., R. A. Hammond, K. A. Foster, J. A. Michener, and N. C. Brown.** 1989. The cloned *polC* gene of *Bacillus subtilis*: characterization of the azp12 mutation and controlled in vitro synthesis of active DNA polymerase III. *Gene* **85:**177–186.
11. **Benjamin, P., and W. Firshein.** 1983. Initiation of DNA replication in vitro by a DNA-membrane complex extracted from *Bacillus subtilis. Proc. Natl. Acad. Sci. USA* **80:**6214–6218.
12. **Benjamin, P., P. Strumph, M. Kenny, and W. Firshein.** 1982. DNA synthesis in purified DNA-membrane complexes extracted from a *Bacillus subtilis polA* mutant. *Nature* (London) **298:**769–771.
13. **Bernad, A., L. Blanco, J. M. Lazaro, G. Martin, and M. Salas.** 1989. A conserved 3′ → 5′ exonuclease active site in prokaryotic and eukaryotic DNA polymerases. *Cell* **59:**219–228.
14. **Bourne, H. R., D. A. Sanders, and F. McCormick.** 1991. The GTPase superfamily: conserved structure and molecular mechanisms. *Nature* (London) **349:**117–127.
15. **Boye, E., and A. Løbner-Olesen.** 1990. The role of *dam* methyltransferase in the control of DNA replication in *E. coli. Cell* **62:**981–989.
16. **Bramhill, D., and A. Kornberg.** 1988. A model for initiation at origins of DNA replication. *Cell* **54:**915–918.
17. **Bramhill, D., and A. Kornberg.** 1988. Duplex opening by dnaA protein at novel sequences in initiation of replication at the origin of the *E. coli* chromosome. *Cell* **52:**743–755.
18. **Braun, R. E., K. O'Day, and A. Wright.** 1985. Autoregulation of the DNA replication gene *dnaA* in *E. coli* K-12. *Cell* **40:**159–169.
19. **Brewer, B. J.** 1988. When polymerases collide: replication and the transcriptional organization of the *Escherichia coli* chromosome. *Cell* **53:**679–686.

19a. **Brewer, B. J., and W. L. Fangman.** 1987. The localization of replication origins on ARS plasmids in *S. cerevisiae. Cell* **51:**463–471.

20. **Bruand, C., S. D. Ehrlich, and L. Jannière.** 1991. Unidirectional theta replication of the structurally stable *Enterococcus faecalis* plasmid pAMβ1. *EMBO J.* **10:**2171–2178.
21. **Cairns, J.** 1963. The bacterial chromosome and its manner of replication as seen by autoradiography. *J. Mol. Biol.* **6:**208–213.
22. **Carrigan, C. M., J. A. Haarsma, M. T. Smith, and R. G. Wake.** 1987. Sequence features of the replication terminus of the *Bacillus subtilis* chromosome. *Nucleic Acids Res.* **15:**8501–8509.
23. **Carrigan, C. M., R. A. Pack, M. T. Smith, and R. G. Wake.** 1991. The normal *terC*-region of the *Bacillus subtilis* chromosome acts in a polar manner to arrest the clockwise replication fork. *J. Mol. Biol.* **22:**197–207.
24. **Cooper, S.** 1991. *Bacterial Growth and Division*. Academic Press, Inc., San Diego, Calif.
25. **Donachie, W. D.** 1968. Relationship between cell size and time of initiation of DNA replication. *Nature* (London) **219:**1077–1079.
26. **Dunn, G., P. Jeffs, N. H. Mann, D. M. Torgerson, and M. Young.** 1978. The relationship between DNA replication and the induction of sporulation in *Bacillus subtilis. J. Gen. Microbiol.* **108:**189–195.
27. **Freese, E., P. Fortnagel, R. Schmitt, W. Klofat, E. Chappelle, and G. Picciolo.** 1969. Biochemical genetics of initial sporulation stages, p. 82–101. *In* L. L. Campbell (ed.), *Spores IV*. American Society for Microbiology, Bethesda, Md.
28. **Fujita, M. Q., H. Yoshikawa, and N. Ogasawara.** 1989. Structure of the *dnaA* region of *Pseudomonas putida*. Conservation among three bacteria, *Bacillus subtilis, Escherichia coli* and *Pseudomonas putida. Mol. Gen. Genet.* **215:**381–387.
29. **Fujita, M. Q., H. Yoshikawa, and N. Ogasawara.** 1990. Structure of the *dnaA* region of *Micrococcus luteus*. Conservation and variation among eubacteria. *Gene* **93:**73–78.
30. **Fujita, M. Q., H. Yoshikawa, and N. Ogasawara.** 1992. Structure of the *dnaA* and DnaA-box region in the *Mycoplasma capricolum* chromosome: conservation and variation in the course of evolution. *Gene* **110:**17–23.
31. **Fukuoka, T., S. Moriya, H. Yoshikawa, and N. Ogasawara.** 1990. Purification and characterization of an

initiation protein for chromosomal replication, DnaA, in *Bacillus subtilis*. *J. Biochem.* **107:**732–739.
32. **Fuller, R., B. E. Funnell, and A. Kornberg.** 1984. The dnaA protein complex with the *E. coli* chromosomal origin (*oriC*) and other DNA sites. *Cell* **38:**889–900.
33. **Ganesan, A. T., and J. Lederberg.** 1965. A cell-membrane bound fraction of bacterial DNA. Biochem. Biophys. Res. Commun. **18:**824–835.
34. **Gass, K. B., and N. R. Cozzarelli.** 1973. Further genetic and enzymological characterization of the three *Bacillus subtilis* deoxyribonucleic acid polymerases. *J. Biol. Chem.* **248:**7688–7700.
35. **Gutierrez, C., G. Martin, J. M. Sogo, and M. Salas.** 1991. Mechanism of stimulation of DNA replication by bacteriophage ϕ29 single-strand DNA-binding protein p5. *J. Biol. Chem.* **266:**2104–2111.
36. **Gyurasits, E. B., and R. G. Wake.** 1973. Bidirectional chromosome replication in *Bacillus subtilis*. *J. Mol. Biol.* **73:**55–63.
37. **Hammond, R. A., M. H. Barnes, S. L. Mack, J. A. Mitchener, and N. C. Brown.** 1991. *Bacillus subtilis* DNA polymerase III: complete sequence, overexpression and characterization of the *polC* gene. *Gene* **98:**29–36.
38. **Hanley, P. J. B., C. M. Carrigan, D. B. Rowe, and R. G. Wake.** 1987. Breakdown and quantitation of the forked termination of replication intermediate of *Bacillus subtilis*. *J. Mol. Biol.* **196:**721–727.
38a. **Hansen, F.** Personal communication.
39. **Hara, H., and H. Yoshikawa.** 1973. Asymmetric bidirectional replication of *Bacillus subtilis* chromosome. *Nature New Biol.* **244:**200–203.
40. **Harmon, J. M., and H. W. Taber.** 1977. Some properties of a membrane-deoxyribonucleic acid complex isolated from *Bacillus subtilis*. *J. Bacteriol.* **129:**789–795.
41. **Henckes, G., F. Harper, A. Levine, F. Vannier, and S. J. Séror.** 1989. Overreplication of the origin region in the *dnaB37* mutant of *Bacillus subtilis*: postinitiation control of chromosomal replication. *Proc. Natl. Acad. Sci. USA* **86:**8660–8664.
42. **Henckes, G., F. Vannier, A. Buu, and L. S. Seror.** 1982. Possible involvement of DNA-linked RNA in the initiation of *Bacillus subtilis* chromosome replication. *J. Bacteriol.* **149:**79–91.
43. **Hidaka, M., T. Kobayashi, and T. Horiuchi.** 1991. A newly identified DNA replication terminus site. *TerE*, on the *Escherichia coli* chromosome. *J. Bacteriol.* **173:**391–393.
44. **Hill, T. M., A. J. Pelletier, M. L. Tecklenburg, and P. J. Kuempel.** 1988. Identification of the DNA sequence from the *E. coli* terminus region that halts replication forks. *Cell* **55:**459–466.
45. **Horiuchi, T., and M. Hidaka.** 1988. Core sequence of two separable terminus sites of the R6K plasmid that exhibit polar inhibition of replication is a 20bp inverted repeat. *Cell* **54:**515–523.
46. **Horowitz, S., R. J. Doyle, F. E. Young, and U. N. Streips.** 1979. Selective association of the chromosome with membrane in a stable L-form of *Bacillus subtilis*. *J. Bacteriol.* **138:**915–922.
47. **Hoshino, T., T. McKenzie, S. Schmidt, T. Tanaka, and N. Sueoka.** 1987. Nucleotide sequence of *Bacillus subtilis dnaB*. A gene essential for DNA replication initiation and membrane attachment. *Proc. Natl. Acad. Sci. USA* **84:**653–657.
48. **Iismaa, T. P., M. T. Smith, and R. G. Wake.** 1984. Physical map of the *Bacillus subtilis* replication terminus region: its confirmation, extension and genetic orientation. *Gene* **32:**171–180.
49. **Iismaa, T. P., and R. G. Wake.** 1987. The normal replication terminus of the *Bacillus subtilis* chromosome, *terC*, is dispensable for vegetative growth and sporulation. *J. Mol. Biol.* **195:**299–310.
50. **Imada, S., L. E. Carroll, and N. Sueoka.** 1980. Genetic mapping of a group of temperature-sensitive *dna* initiation mutants in *Bacillus subtilis*. *Genetics* **94:**809–823.
51. **Itaya, M., and T. Tanaka.** 1991. Complete physical map of the *Bacillus subtilis* 168 chromosome constructed by a gene-directed mutagenesis method. *J. Mol. Biol.* **220:**631–648.
52. **Jacob, F., S. Brenner, and F. Cuzin.** 1963. On the regulation of DNA replication in bacteria. *Cold Spring Harbor Symp. Quant. Biol.* **28:**329–348.
53. **Karamata, D., and J. D. Gross.** 1970. Isolation and genetic analysis of temperature-sensitive mutants of *B. subtilis* defective in DNA synthesis. *Mol. Gen. Genet.* **108:**277–287.
53a. **Kato, J., Y. Nishimura, R. Imamura, H. Niki, S. Hiraga, and H. Suzuki.** 1990. New topoisomerase essential for chromosome segregation in *E. coli*. *Cell* **63:**393–404.
54. **Kejzlarova-Lepesant, J., N. Harford, J.-A. Lepesant, and R. Dedonder.** 1975. Revised genetic map for *Bacillus subtilis* 168, p. 592–595. *In* P. Gerhardt, R. N. Costilow, and H. L. Sadoff (ed.), *Spores VI*. American Society for Microbiology, Washington, D.C.
55. **Khatri, G. S., T. MacAllister, P. R. Sista, and D. Bastia.** 1989. The replication terminator protein of *E. coli* is a DNA sequence-specific contra-helicase. *Cell* **59:**667–674.
56. **Kuempel, P. L., A. J. Pelletier, and T. M. Hill.** 1989. Tus and the terminators: the arrest of replication in prokaryotes. *Cell* **59:**581–583.
57. **Laffan, J., and W. Firshein.** 1988. Origin specific DNA binding membrane-associated protein may be involved in repression of initiation in *Bacillus subtilis*. *Proc. Natl. Acad. Sci. USA* **85:**7452–7456.
58. **Lai, C.-J., and D. Nathans.** 1975. Non-specific termination of simian virus 40 DNA replication. *J. Mol. Biol.* **97:**113–118.
59. **Lampe, M. F. and K. F. Bott.** 1985. Genetic and physical organization of the cloned *gyrA* and *gyrB* genes of *Bacillus subtilis*. *J. Bacteriol.* **162:**78–84.
60. **Laurent, S. J.** 1973. Initiation of deoxyribonucleic acid replication in a temperature-sensitive mutant of *B. subtilis*. Evidence for a transcriptional step. *J. Bacteriol.* **116:**141–145.
61. **Laurent, S. J., and F. S. Vannier.** 1973. Temperature-sensitive initiation of chromosome replication in a mutant of *Bacillus subtilis*. *J. Bacteriol.* **114:**474–484.
62. **Lee, E. H., A. Kornberg, M. Hidaka, T. Kobayashi, and T. Horiuchi.** 1989. The *E. coli* replication termination protein impedes the action of helicases. *Proc. Natl. Acad. Sci. USA* **86:**9104–9108.
63. **Lee, S., P. Kanda, R. C. Kennedy, and J. R. Walker.** 1987. Relation of the *Escherichia coli dnaX* gene to its two products—the τ and δ subunits of DNA polymerase III holoenzyme. *Nucleic Acids Res.* **15:**7663–7675.
64. **Levine, A., G. Henckes, F. Vannier, and S. J. Seror.** 1987. Chromosomal initiation in *Bacillus subtilis* may involve two closely linked origins. *Mol. Gen. Genet.* **208:**37–44.
65. **Levine, A., F. Vannier, M. Dehbi, G. Henckes, and S. J. Séror.** 1991. The stringent response blocks DNA replication outside the *ori* region in *Bacillus subtilis* and at the origin in *Escherichia coli*. *J. Mol. Biol.* **219:**605–613.
66. **Lewis, P. J., G. B. Ralston, R. I. Christopherson, and R. G. Wake.** 1990. Identification of the replication terminator protein binding sites in the terminus region of the *Bacillus subtilis* chromosome and stoichiometry of the binding. *J. Mol. Biol.* **214:**73–84.
67. **Lewis, P. J., M. T. Smith, and R. G. Wake.** 1989. A protein involved in termination of chromosome replication in *Bacillus subtilis* binds specifically to the *terC* site. *J. Bacteriol.* **171:**3564–3567.
68. **Lewis, P. J., and R. G. Wake.** 1989. DNA and protein sequence conservation at the replication terminus in

Bacillus subtilis 168 and W23. *J. Bacteriol.* **171:**1402–1408.

69. **Løbner-Olesen, A., K. Skarstad, F. G. Hansen, K. von Meyenburg, and E. Boye.** 1989. The DnaA protein determines the initiation mass of *Escherichia coli* K-12. *Cell* **57:**881–889.
70. **Lovett, P. S., and F. E. Young.** 1969. Identification of *Bacillus subtilis* NRRL B-3275 as a strain of *Bacillus pumilis. J. Bacteriol.* **100:**658–661.
71. **Low, R. L., S. A. Rashbaum, and N. R. Cozzarelli.** 1976. Purification and characterization of DNA polymerase III from *Bacillus subtilis. J. Biol. Chem.* **251:**1311–1325.
72. **Maki, H., S. Maki, and A. Kornberg.** 1988. DNA polymerase III holoenzyme of *Escherichia coli.* IV. The holoenzyme is an asymmetric dimer with twin active sites. *J. Biol. Chem.* **263:**6570–6578.
73. **Marsh, R. C., and A. Worcel.** 1977. A DNA fragment containing the origin of replication of the *Escherichia coli* chromosome. *Proc. Natl. Acad. Sci. USA* **74:**2720–2724.
74. **Matsushita, T., K. White, and N. Sueoka.** 1971. Chromosome replication in toluenized *Bacillus subtilis* cells. *Nature New Biol.* **232:**111–114.
75. **McGinness, T., and R. G. Wake.** 1979. Division septation in the absence of chromosome termination in *Bacillus subtilis. J. Mol. Biol.* **134:**251–264.
76. **McMacken, R., L. Silver, and C. Georgopoulos.** 1987. DNA replication, p. 564–612. *In* F. C. Neidhardt, J. L. Ingraham, K. B. Low, B. Magasanik, M. Schaechter, and H. E. Umbarger (ed.), *Escherichia coli and Salmonella typhimurium*: Cellular and Molecular Biology, vol. 1. American Society for Microbiology, Washington, D.C.
77. **Mendelson, N. H., and J. D. Gross.** 1967. Characterization of a temperature-sensitive mutant of *Bacillus subtilis* defective in deoxyribonucleic acid replication. *J. Bacteriol.* **94:**1603–1608.
78. **Messer, W., and M. Noyer-Weidner.** 1988. Timing and targeting: the biological functions of Dam methylation in *E. coli. Cell* **54:**735–737.
79. **Moriya, S., T. Atlung, F. G. Hansen, H. Yoshikawa, and N. Ogasawara.** 1992. Cloning of autonomously replicating sequence (ars) from the *Bacillus subtilis* chromosome. *Mol. Microbiol.* **6:**309–315.
80. **Moriya, S., T. Fukuoka, N. Ogasawara, and H. Yoshikawa.** 1988. Regulation of initiation of the chromosomal replication by DnaA-boxes in the origin region of the *Bacillus subtilis* chromosome. *EMBO J.* **7:**2911–2917.
81. **Moriya, S., K. Kato, H. Yoshikawa, and N. Ogasawara.** 1990. Isolation of a *dnaA* mutant of *Bacillus subtilis* defective in initiation of replication: amount of DnaA protein determines cells' initiation potential. *EMBO J.* **9:**2905–2910.
82. **Moriya, S., N. Ogasawara, and H. Yoshikawa.** 1985. Structure and function of the region of the replication origin of the *Bacillus subtilis* chromosome. III. Nucleotide sequence of some 10,000 base pairs in the origin region. *Nucleic Acids Res.* **13:**2251–2265.
83. **Murakami, S., N. Inuzuka, M. Yamaguchi, K. Yamaguchi, and H. Yoshikawa.** 1976. Initiation of DNA replication in *Bacillus subtilis.* III. Analysis of molecular events involved in the initiation using a temperature-sensitive *dna* mutant. *J. Mol. Biol.* **108:**683–704.
84. **Murakami, S., and H. Yoshikawa.** 1976. Gene that controls initiation of chromosome replication and prophage induction in *Bacillus subtilis. Nature* (London) **259:**215–218.
85. **Mysliwiec, T. H., J. Errington, A. B. Vaidya, and M. G. Bramucci.** 1991. The *Bacillus subtilis spo0J* gene: evidence for involvement in catabolite repression of sporulation. *J. Bacteriol.* **173:**1911–1919.
86. **Nordström, K.** 1990. Control of plasmid replication—how do DNA iterons set the replication frequency? *Cell* **63:**1121–1124.

86a. **O'Donnell, M.** Personal communication.

87. **Ogasawara, N., S. Mizumoto, and H. Yoshikawa.** 1984. Replication origin of the *Bacillus subtilis* chromosome determined by hybridization of the first-replicating DNA with cloned fragments from the replication origin region of the chromosome. *Gene* **30:**173–182.
88. **Ogasawara, N., S. Moriya, P. G. Mazza, and H. Yoshikawa.** 1986. Nucleotide sequence and organization of *dnaB* gene and neighbouring genes on the *Bacillus subtilis* chromosome. *Nucleic Acids Res.* **14:**9989–9999.
89. **Ogasawara, N., S. Moriya, K. von Meyenburg, F. G. Hansen, and H. Yoshikawa.** 1985. Conservation of genes and their organization in the chromosomal replication origin region of *Bacillus subtilis* and *Escherichia coli. EMBO J.* **4:**3345–3350.
90. **Ogasawara, N., and H. Yoshikawa.** Genes and their organization in replication origin region of bacterial chromosome. *Mol. Microbiol.*, in press.
91. **Oka, A., K. Sugimoto, M. Takanami, and Y. Hirota.** 1980. Replication origin of the *Escherichia coli* K-12 chromosome. *J. Mol. Biol.* **176:**443–458.
92. **Okazaki, R., and A. Kornberg.** 1964. Enzymatic synthesis of deoxyribonucleic acid. XV. Purification and properties of a polymerase from *Bacillus subtilis. J. Biol. Chem.* **239:**259–268.
93. **O'Sullivan, A., and C. Anagnostopoulos.** 1982. Replication terminus of the *Bacillus subtilis* chromosome. *J. Bacteriol.* **151:**135–143.
94. **O'Sullivan, A., and N. Sueoka.** 1967. Sequential replication of the *Bacillus subtilis* chromosome. IV. Genetic mapping by density transfer experiment. *J. Mol. Biol.* **27:**349–368.
95. **O'Sullivan, M. A., and N. Sueoka.** 1972. Membrane attachment of the replication origins of a multifork (dichotomous) chromosome in *Bacillus subtilis. J. Mol. Biol.* **69:**237–248.
96. **Ott, R. W., M. H. Barnes, N. C. Brown, and A. T. Ganesan.** 1986. Cloning and characterization of the *polC* region of *Bacillus subtilis. J. Bacteriol.* **165:**951–957.
97. **Perego, M., E. Ferrari, M. T. Bassi, A. Galizzi, and P. Mazza.** 1987. Molecular cloning of *Bacillus subtilis* genes involved in DNA metabolism. *Mol. Gen. Genet.* **209:**8–14.
98. **Piggot, P. J.** 1990. Genetic map of *Bacillus subtilis* 168, p. 107–146. *In* K. Drlica and M. Riley (ed.), *The Bacterial Chromosome.* American Society for Microbiology, Washington, D.C.
99. **Rasmussen, K. V., T. Atlung, G. Kerszman, G. E. Hansen, and F. G. Hansen.** 1983. Conditional change of DNA replication control in an RNA polymerase mutant of *Escherichia coli. J. Bacteriol.* **154:**443–451.
100. **Ryter, A., Y. Hirota, and F. Jacob.** 1968. DNA-membrane complex and nuclear segregation in bacteria. *Cold Spring Harbor Symp. Quant. Biol.* **33:**669–676.
101. **Sanjanwala, B., and A. T. Ganesan.** 1989. DNA polymerase III gene of *Bacillus subtilis. Proc. Natl. Acad. Sci. USA* **86:**4421–4424.
102. **Sargent, M. G.** 1980. Specific labeling of the *Bacillus subtilis* chromosome terminus. *J. Bacteriol.* **143:**1033–1035.
103. **Sargent, M. G., and M. F. Bennett.** 1985. Amplification of a major membrane-bound DNA sequence of *Bacillus subtilis. J. Bacteriol.* **161:**589–595.
104. **Schneider, A.-M., M. Gaisne, and C. Anagnostopoulos.** 1982. Genetic structure and internal rearrangements of stable merodiploids from *Bacillus subtilis* strains carrying the *trpE26* mutation. *Genetics* **101:**189–210.
105. **Seiki, M., N. Ogasawara, and H. Yoshikawa.** 1979. Structure of the region of the replication origin of the

Bacillus subtilis chromosome. *Nature* (London) **281:** 699–701.

106. **Seki, T., T. Oshima, and Y. Oshima.** 1975. Taxonomic study of *Bacillus* by deoxyribonucleic acid-deoxyribonucleic acid hybridization and interspecific transformation. *Int. J. Syst. Bacteriol.* **25:**258–270.
107. **Sekimizu, K., D. Bramhill, and A. Kornberg.** 1987. ATP activates dnaA protein in initiating replication of plasmids bearing the origin of the *E. coli* chromosome. *Cell* **50:**259–265.
108. **Séror, S. J., F. Vannier, A. Levine, and G. Henckes.** 1986. Stringent control of chromosomal replication in *Bacillus subtilis. Nature* (London) **321:**709–710.
109. **Smith, M. T., C. Aynsley, and R. G. Wake.** 1985. Cloning and localization of the *Bacillus subtilis* chromosome replication terminus, *terC. Gene* **38:**9–17.
110. **Smith, M. T., and R. G. Wake.** 1988. DNA sequence requirements for replication fork arrest at *terC* in *Bacillus subtilis. J. Bacteriol.* **170:**4083–4090.
111. **Smith, M. T., and R. G. Wake.** 1989. Expression of the *rtp* gene of *Bacillus subtilis* is required for replication fork arrest at the chromosome terminus. *Gene* **85:**187–192.
112. **Smith, M. T., and R. G. Wake.** 1992. Definition and polarity of action of DNA replication terminators in *Bacillus subtilis. J. Mol. Biol.* **227:**648–657.
113. **Steck, T. R., and K. Drlica.** 1984. Bacterial chromosome segregation: evidence for DNA gyrase involvement in decatenation. *Cell* **36:**1081–1088.
114. **Struck, J. C. R., D. W. Vogel, N. Ulbrich, and V. A. Erdmann.** 1986. A *dnaZX*-like open reading frame downstream from the *Bacillus subtilis* scRNA gene. *Nucleic Acids Res.* **16:**2720.
115. **Sueoka, N., and W. G. Quinn.** 1968. Membrane attachment of the chromosome replication origin in *Bacillus subtilis. Cold Spring Harbor Symp. Quant. Biol.* **33:**695–705.
116. **Sugino, A., and K. F. Bott.** 1980. *Bacillus subtilis* deoxyribonucleic acid gyrase. *J. Bacteriol.* **141:**1331–1339.
117. **Valenzuela, M. S., D. Freifelder, and R. B. Inman.** 1976. Lack of a unique termination site for the first round of bacteriophage lambda DNA replication. *J. Mol. Biol.* **102:**569–589.
118. **von Meyenburg, K., and F. G. Hansen.** 1987. Regulation of chromosome replication, p. 1555–1577. *In* F. C. Neidhardt, J. L. Ingraham, K. B. Low, B. Magasanik, M. Schaechter, and H. E. Umbarger (ed.), *Escherichia coli and Salmonella typhimurium: Cellular and Molecular Biology*, vol. 2. American Society for Microbiology, Washington D.C.
119. **Wahle, E., R. S. Lasken, and A. Kornberg.** 1989. The dnaB-dnaC replication protein complex of *Escherichia coli*. I. Formation and properties. *J. Biol. Chem.* **264:** 2463–2468.
120. **Wake, R. G.** 1972. Visualization of reinitiated chromosomes in *Bacillus subtilis. J. Mol. Biol.* **68:**501–509.
121. **Wake, R. G.** 1973. Circularity of the *Bacillus subtilis* chromosome and further studies on its bidirectional replication. *J. Mol. Biol.* **77:**569–575.
122. **Wake, R. G., P. J. Lewis, and M. T. Smith.** 1990. The *rtp* gene and termination of chromosome replication in *Bacillus subtilis*, p. 99–108. *In* M. Zukowski, A. T. Ganesan, and J. A. Hoch (ed.), *Genetics and Biotechnology of Bacilli*, vol. 3. Academic Press, Inc., San Diego, Calif.
123. **Wang, L.-F., C. W. Price, and R. H. Doi.** 1985. *Bacillus subtilis dnaE* encodes a protein homologous to DNA primase of *Escherichia coli. J. Biol. Chem.* **260:**3368–3372.
124. **Wang, Q., and M. Kaguni.** 1987. Transcriptional repression of the dnaA gene of *Escherichia coli* by dnaA protein. *Mol. Gen. Genet.* **209:**518–525.
125. **Weiss, A. S., I. K. Hariharan, and R. G. Wake.** 1981. Analysis of the terminus region of the *Bacillus subtilis* chromosome. *Nature* (London) **293:**673–675.
126. **Weiss, A. S., and R. G. Wake.** 1983. Restriction map of DNA spanning the replication terminus of the *Bacillus subtilis* chromosome. *J. Mol. Biol.* **171:**119–137.
127. **Weiss, A. S., and R. G. Wake.** 1984. A unique DNA intermediate associated with termination of chromosome replication in *Bacillus subtilis. Cell* **39:**683–689.
128. **Weiss, A. S., R. G. Wake, and R. B. Inman.** 1986. An immobilized fork as a termination of replication intermediate in *Bacillus subtilis. J. Mol. Biol.* **188:**199–205.
129. **Williams, N. K., and R. G. Wake.** 1989. Sequence limits of DNA strands in the arrested replication fork at the *Bacillus subtilis* chromosome terminus. *Nucleic Acids Res.* **17:**9947–9956.
130. **Winston, S., and T. Matsushita.** 1975. Permanent loss of chromosome initiation in toluene-treated *Bacillus subtilis* cells. *J. Bacteriol.* **123:**921–927.
131. **Winston, S., and N. Sueoka.** 1980. DNA-membrane association is necessary for initiation of chromosomal and plasmid replication in *Bacillus subtilis. Proc. Natl. Acad. Sci. USA* **77:**2834–2838.
132. **Winston, S., and N. Sueoka.** 1982. DNA replication in *Bacillus subtilis*, p. 35–69. *In* D. A. Dubnau (ed.), *The Molecular Biology of the Bacilli*, vol. 1. *Bacillus subtilis*. Academic Press, Inc., New York.
133. **Yamaguchi, K., S. Murakami, and H. Yoshikawa.** 1971. Chromosome-membrane association in *Bacillus subtilis*. I. DNA release from membrane fraction. *Biochem. Biophys. Res. Commun.* **44:**1559–1565.
134. **Yamaguchi, K., and H. Yoshikawa.** 1973. Topography of chromosome membrane junction in *Bacillus subtilis. Nature New Biol.* **244:**204–206.
135. **Yamaguchi, K., and H. Yoshikawa.** 1977. Chromosome-membrane association in *Bacillus subtilis*. III. Isolation and characterization of a DNA-protein complex carrying replication origin markers. *J. Mol. Biol.* **110:**219–253.
136. **Yee, Y. W., and D. W. Smith.** 1990. Pseudomonas chromosomal replication origins: a bacterial class distinct from *Escherichia coli* type origins. *Proc. Natl. Acad. Sci. USA* **87:**1278–1282.
137. **Yoshikawa, H., and N. Ogasawara.** 1991. Structure and function of DnaA and the DnaA-box in eubacteria: evolutionary relationships of bacterial replication origin. *Mol. Microbiol.* **5:**2589–2597.
138. **Yoshikawa, H., A. O'Sullivan, and N. Sueoka.** 1964. Sequential replication of the *Bacillus subtilis* chromosome. III. Regulation of initiation. *Proc. Natl. Acad. Sci. USA* **52:**973–980.
139. **Yoshikawa, H., and N. Sueoka.** 1963. Sequential replication of *Bacillus subtilis* chromosome. I. Comparison of marker frequencies in exponential and stationary growth phases. *Proc. Natl. Acad. Sci. USA* **49:**555–566.
140. **Yoshikawa, H., and Sueoka.** 1963. Sequential replication of the *Bacillus subtilis* chromosome. II. Isotopic transfer experiments. *Proc. Natl. Acad. Sci. USA* **49:**806–813.
141. **Yoshikawa, H., K. Yamaguchi, M. Seiki, N. Ogasawara, and H. Toyoda.** 1979. Organization of the replication-origin region of the *Bacillus subtilis* chromosome. *Cold Spring Harbor Symp. Quant. Biol.* **43:**569–576.
142. **Zahler, S. A.** 1982. Specialized transduction in *Bacillus subtilis*, p. 269–305. *In* D. A. Dubnau (ed.), *Molecular Biology of the Bacilli*. Academic Press, Inc., New York.
143. **Zeigler, D. R.** 1989. Genetic map of *Bacillus subtilis* 168, p. 4.1–4.29. *In* D. R. Zeigler and D. H. Dean (ed.), *Bacillus Genetic Stock Center Strains and Data*, 4th ed. Ohio State University, Columbus.
144. **Zeigler, D. R., and D. H. Dean.** 1990. Orientation of genes in the *Bacillus subtilis* chromosome. *Genetics* **125:**703–708.

37. DNA Repair Systems

RONALD E. YASBIN, DAVID CHEO, and DAVID BOL

All living cells are constantly exposed to chemical and physical agents that have the ability to alter the primary structure of DNA. Such alterations, if not corrected, could result in mutations. The generation of such mutations is in itself not necessarily deleterious to the organism. In fact, the accumulation of mutations increases the genetic variability of a species as long as the number of mutations does not reach a critical level (genetic load) at which point the species could no longer exist. Thus, it would seem obvious that living systems must maintain mechanisms for repairing DNA damage. It would also seem obvious that these same systems must balance the removal of DNA damage with the accumulation of a finite number of mutations.

Escherichia coli has served as the principal model for investigations into DNA repair mechanisms. The DNA repair systems identified in this paradigm have also been discovered in most other organisms studied. While differences do exist, it is clear that DNA repair systems have been conserved in both prokaryotic and eukaryotic systems. For instance, lesions can be removed from damaged DNA through the action of such systems as photoreactivation (41, 79), nucleotide excision repair (NER; 24, 42, 80), base excision repair (BER; 23, 24), the adaptive response (42, 85), and the oxidative stress response (11).

On the other hand, bulky noncoding lesions in the DNA that are not removed before being encountered by the replication machinery constitute a block to further DNA synthesis (84). This blockage in DNA replication can be circumvented if the DNA polymerase dissociates from the DNA upon encountering a noncoding lesion and then reinitiates replication on the other side of the lesion (24, 78). The resulting gaps in the newly synthesized daughter strand can then be "repaired" by a process(es) that has been variously called postreplication repair, daughter strand gap repair, and recombination repair (25, 79, 95). Regardless of its name, this type of mechanism exemplifies tolerance of DNA damage rather than a true repair process since the actual damage is never physically removed but is diluted out by subsequent DNA replication.

While postreplication repair illustrates DNA damage tolerance via a discontinuous mode of DNA synthesis, DNA damage tolerance could also occur via a continuation of DNA synthesis opposite a noncoding lesion without any gap formation. This is termed translesion DNA synthesis (79). Translesion DNA synthesis is one of a myriad of coordinately induced cellular responses observed in *E. coli* that collectively compose the SOS regulon (46, 74, 95, 99).

All of the systems and regulons that have been mentioned so far have been reviewed recently, at least with respect to how they function in *E. coli*. In contrast, this review attempts to look at these repair systems with respect to differentiation processes and developmental biology as studied in the gram-positive bacterium *Bacillus subtilis*. In fact, one of the most fascinating and significant aspects of studying the biological sciences has been the obscuring of the boundaries that have defined established disciplines. Originally, DNA repair systems were considered integral parts of an organism's ability to survive the effects of environmental insults and metabolic processes. However, as the molecular characterization of these repair systems proceeded it became obvious that in addition to determining mutation frequency and cell survival, DNA repair systems also play important roles in viral activation, DNA replication, genetic recombination, metabolism, and cancer (34). In order to investigate systematically the interrelationship(s) between DNA repair systems and these other phenomena, an appropriate model system must be identified. In this case, *E. coli* is not the appropriate organism since it lacks readily identifiable developmental cycles. However, *B. subtilis* has emerged as the paradigm for the study of developmental processes.

DNA REPAIR SYSTEMS

Photoreactivation

The first DNA repair mechanism discovered in *E. coli* was photoreactivation (41, 79). Photoreactivation reduces the deleterious effects of UV irradiation (200 to 300 nm) by means of a light-dependent process in which the *cis-syn* cyclobutyl pyrimidine dimers are enzymatically monomerized by a photolyase (24). Interestingly, *B. subtilis* and all other naturally competent eubacteria lack this type of DNA repair system (102).

Nucleotide Excision Repair

Bulky, noncoding lesions that produce a block to DNA replication can be removed from damaged DNA through the action of an NER system (24, 42, 80). A noncoding lesion constitutes some alteration of a nucleotide(s) contained within the DNA such that the replication machinery of the cell can no longer proceed with accuracy or efficiency (24). The UVR system of *E. coli* (42, 92) is the best characterized example of nucleotide excision repair. In this NER system, a

Ronald E. Yasbin, David Cheo, and David Bol • Department of Biological Sciences and The Program in Molecular and Cell Biology, University of Maryland Baltimore County, Baltimore, Maryland 21045.

variety of bulky lesions are recognized and endonucleolytic cleavage of the DNA is initiated, both 3′ and 5′ to the lesion, by the activities of the UvrA, UvrB, and UvrC proteins (29, 42, 94). The UvrD helicase and DNA polymerase I then play critical, but not essential, roles in the further processing of lesions and the resynthesis of the DNA (42). The distinction between critical and essential roles for all of these gene products has been based upon the fact that strains carrying mutations in either the *uvrD* or the *polA* gene are less sensitive to certain DNA-damaging agents than are strains carrying mutations in *uvrA*, *uvrB*, or *uvrC* (24, 42).

In *B. subtilis* an NER system phenotypically similar to the one in *E. coli* has been identified (31, 86, 102). Essentially, following UV irradiation or following treatment with other DNA-damaging agents, lesions are removed from the DNA of repair-proficient strains of this bacterium (30, 64). Two genes, *uvrA* and *uvrB*, have been shown to be absolutely required for excision repair (64, 102). Interestingly, DNA sequence comparisons have shown that the UvrA and UvrB proteins of *B. subtilis* are homologous to the UVrB and UvrC proteins of *E. coli*, respectively (7, 8). Although a third Uvr gene has not been identified in *B. subtilis*, analysis of *uvrA* mutations has resulted in the suggestion that this locus might actually represent an operon consisting of the *uvrA* and the *uvrC* genes (102). Similar analysis has demonstrated that the *uvrA* gene and possibly the *uvrB* gene are components of the SOS regulon of *B. subtilis* (11; see below). At the present time, the enzymology of the *B. subtilis* NER system has not been elucidated.

Base Excision Repair

BER is a second method by which bulky, noncoding lesions as well as inappropriate bases can be removed from DNA (23, 24). BER differs from NER in that damaged or incorrect bases are excised as free bases rather than nucleotides or oligonucleotides. Total removal of the DNA lesion requires a two-step process. First, there is hydrolysis of the *N*-glycosylic bond that links the inappropriate base or the lesion to the deoxyribose-phosphate backbone of DNA. This is performed by the action of a class of DNA repair enzymes called DNA glycosylases. Once a damaged base or incorrect base is recognized by a specific glycosylase, the *N*-glycosylic bond is cut, leaving an apurinic or apyrimidinic (AP) site in the DNA (23, 24, 43, 44). The second step in BER is removal of this AP site via the action of one or more nucleases. Abasic sites in DNA are specifically recognized by enzymes known as AP endonucleases. Repair synthesis and ligation in BER proceed as in NER.

BER mechanisms have been identified in *B. subtilis*. Specifically, glycosylase and AP endonuclease activities have been isolated from this organism (23, 37) and other gram-positive bacteria (3, 72). Evidence has also been presented indicating that the W-reactivation phenomenon (21) and possibly the transitory germinative excision repair function (98) of *B. subtilis* are specific for pyrimidine dimers (see below). This type of repair specificity might indicate the involvement of a glycosylase. In addition, the enzyme(s) responsible for the phenomenon of transfection enhancement (19, 56, 73) may be an additional example of a base-specific excision repair protein.

Spore-Specific DNA Repair

The spores of *B. subtilis* and other *Bacillus* species are as much as 10- or 20-fold more resistant to the cytotoxic effects of UV radiation than are their growing vegetative cells. Upon spore germination, UV resistance increases to even higher levels before declining to the level of resistance of vegetative cells (90). The nature of these dramatic differences in UV resistance between vegetative cells and spores of *Bacillus* species has been the subject of much investigation and debate.

Early studies of the UV photochemistry of DNA in spores of *Bacillus* species led to the discovery that the *cis-syn* cyclobutane-type pyrimidine dimers, of which the thymine-thymine (TT) dimer represents the primary photoproduct found in the DNA of UV-irradiated vegetative cells, was not contained in the DNA of UV-irradiated spores (15). In contrast to vegetative cells, a different series of UV photoproducts was identified within the DNA of spores. The primary photoproduct of UV-irradiated spores is a 5-thyminyl-5,6-dihydrothymine adduct (SP; 93).

Lesions within the DNA of *Bacillus* spores are efficiently repaired by at least two independent mechanisms (65, 66). These repair systems function only upon germination and not within the dormant spore. As mentioned previously, the NER system is a general repair system that will recognize, excise, and repair bulky lesions in DNA including both SP and TT. In addition there is a repair system that acts specifically to monomerize SP to thymine residues (65). The mechanism of action of this SP-specific repair system is not known. However, it is dependent upon the endonuclease that is encoded by the *spe* (*spl*) gene. Strains that are deficient in the products of either the *uvrA* or *spe* genes alone produce spores that are no more sensitive to UV irradiation than are the spores produced by repair-proficient strains. However, *uvrA spe* double mutants produce spores that are much more sensitive to UV irradiation than are the spores isolated from the repair-proficient strains (63, 67). These results suggest that excision repair and SP-specific repair are independent mechanisms that are both involved in the ability of spores to remove UV-induced DNA damage. It also appears that either one of the systems must be functional in order for spores to remain resistant to the damaging effects of UV light. Recently, Fajardo-Cavazos et al. have cloned a DNA segment that complements the defect in UV resistance that exists in *spl* (*spe*) mutants (69).

The small acid-soluble spore proteins (SASPs) of the α and β types are synthesized in *B. subtilis* specifically during stage III of sporulation (see chapter 55, this volume). These proteins accumulate in spores to approximately 5 to 12% of total protein and are rapidly degraded during spore germination (83). The degradation of SASPs provides the germinating spore with amino acids for protein synthesis and metabolism. Expression of SASP genes occurs just before the developing spore becomes resistant to UV irradiation and correlates with the change in UV photochemistry of its DNA from the vegetative cell type to spore type: the A conformation (83). The SASP genes (*ssp*) of several *Bacillus* species have been cloned and their protein products have been purified (83). The α/β type SASP genes make up a multigene family of at least seven members that are highly conserved among *Bacillus* species (83). Deletion of α/β type SASP genes results in a loss of UV resistance and the generation of TT within spore DNA rather than SP (82). The SASPs have thus been shown to play an important role in the resistance of spores to UV irradiation (54). In vitro and in vivo data have demonstrated that the α/β type SASPs play a major role in determining the structure of DNA in the spore. As mentioned previously, the α/β type SASPs bind DNA and change the conformation of DNA from the B to A conformation (58, 70). UV irradiation of DNA complexed with α/β type SASPs in vitro generates no TT but does generate SP. Thus the binding of spore DNA by α/β type SASPs is primarily responsible for the novel UV photochemistry of their DNA.

Although the types of photoproducts formed from α/β-type SASP/DNA complexes in vitro and in vivo are the same, the yield of SP generated in vitro is more than 10-fold lower than the yield of SP generated in vivo (71). Thus there are most likely other factors that influence DNA structure and photochemistry in vivo. These other factors may involve, for example, the state of hydration, ionic conditions, and pH within the spore, as well as other proteins and molecules that interact with DNA within the spore. For instance, *Bacillus* spores contain high levels (up to approximately 10% of the dry weight of spores) of dipicolinic acid. Dipicolinic acid is found in the central core of the spore, where it may associate with spore DNA. Spores with decreased levels of dipicolinic acid demonstrate higher levels of UV resistance, thus implicating dipicolinic acid in the sensitization of DNA to UV (25). Upon germination, the uptake of water and loss of dipicolinic acid are complete prior to the degradation of the α/β-type SASPs.

Postreplication Repair

C. T. Hadden and colleagues (14) established that postreplication repair occurs in *B. subtilis*. Mutations that eliminated this type of repair included alleles of the *recA*, *recF*, and *recM* genes (1, 14). Dodson and Hadden also suggested that the molecular mechanisms accomplishing postreplication repair play important roles in other types of recombination, especially transductional recombination. A more thorough review of the *B. subtilis* recombination mechanisms can be found in chapter 39 (this volume). On the basis of the results of Dodson and Hadden, it was suspected that competent *B. subtilis* cells would be more resistant to DNA-damaging agents than the noncompetent cells within the same culture. This suggestion was made because of the increased recombination frequency of competent cells and because of the increased amount of RecA protein (formerly called RecE protein) found in competent cells (50, 105). However, initial results indicated that competent cells were more sensitive to DNA-damaging agents than were the noncompetent bacteria (101). Later, it was discovered that this increased sensitivity to DNA-damaging agents among competent cells was the result of the enhanced induction of endogenous prophage(s) in these cells (106). Accordingly, "prophage-cured" competent cells of *B. subtilis* were found to be more resistant to DNA-damaging agents than were the noncompetent bacteria within the same culture (104, 106). Furthermore, the demonstration that prophage were selectively induced from competent cells led to the hypothesis that an SOS system existed in *B. subtilis* and that this regulon was spontaneously induced during the development of the competent state (48, 101).

SOS Regulon

To determine whether or not an SOS system existed in *B. subtilis*, genetic studies were initiated using the various mutations that had been shown to be involved in DNA repair, recombination, or both (49). During these investigations it became apparent that there existed a great deal of confusion concerning the names and map locations of several "*rec*" genes. A principal point of confusion concerned the gene that codes for the major recombination protein. Lovett and Roberts (51) had purified the major recombination protein of *B. subtilis* by its activity and by the fact that it cross-reacted with antibodies raised against the RecA protein of *E. coli*. Because of the extreme sensitivity to DNA-damaging agents and the lack of recombination in strains carrying the *recE4* mutation, it had been suggested that the *recE* gene encoded the major recombination protein (the RecA analog) of *B. subtilis*. This suggestion was later substantiated following the cloning and characterization of the *recE* gene by Marrero and Yasbin (53).

Previously, two genes that play roles in DNA repair and recombination (*recE* and *recA*) were mapped close to each other on the *B. subtilis* chromosome. After the wild-type *recE* allele was sequenced, as well as a few mutant *recE* and *recA* alleles, it was demonstrated that the *recE* and *recA* mutations actually represent different alleles of the same gene (107). Accordingly, this gene has been designated *recA*.

Using a set of isogenic strains that differed only in the presence of specific mutations in genes that were thought to be involved in DNA recombination and repair, it was demonstrated that an SOS-like system did exist in *B. subtilis* (49). Phenomena shown to be associated with the SOS-like system of *B. subtilis* are: W-reactivation, error-prone repair, prophage induction, induction of a DNA methylase activity, induction of DNA damage-inducible (*din*) genes, induction of the *recA* gene, and filamentous growth or filamentation (49, 103). In actuality, the SOS system of *B. subtilis* appears to represent more than one global regulon, since four types of phenomena can be distinguished (see review in reference 103).

Type 1 phenomena

Type 1 SOS phenomena are those RecA-dependent events that are induced following DNA damage as well as during the development of the competent state. The phenomena of the type 1 category include the induction of certain prophage (phage such as ϕ105 and SPO2), W-reactivation, error-prone repair, DNA methylase activity (in strains carrying phage from the SPβ family of viruses), and DNA damage inducibility or *din* gene induction. These phenomena do not occur in the absence of a functional RecA protein and are spontaneously induced during the development of the competent state (103).

Type 2 phenomenon

The SOS type 2 phenomenon is a RecA-independent event that is induced following DNA damage but is not induced during the development of the competent state. Following treatment of *B. subtilis* with certain DNA-damaging agents, the bacteria begin to form long filaments. This type of filamentous growth is also part of the SOS response in *E. coli*. While in the *E. coli* model this filamentation is dependent upon the presence of a functional RecA protein, filamentation of *B. subtilis* does not require a functional RecA gene product.

Type 3 phenomenon

The SOS type 3 phenomenon is represented by the regulation of the *recA* gene. The *recA* gene is under more than one type of regulation. Following UV irradiation (or treatment with other DNA-damaging agents), *B. subtilis* does not induce the transcription of the *recA* gene unless there is a functional RecA protein present (50, 105). On the other hand, during the development of the competent state, a functional RecA protein is not required for induction of the *recA* gene. However, functional RecA product is required for the competence state induction of transcription of *dinA*, *dinB*, and *dinC* as well as for the induction of prophage, W-reactivation, and error-prone repair (103). Further analysis has revealed that the competence-specific induction of the *recA* gene is dependent upon functional *spo0A*, *spo0H*, and *degU* gene products (10) while DNA damage induction of the *recA* gene is not dependent upon any of these gene products.

Type 4 phenomenon

Finally, the SOS type 4 phenomenon is a RecA-dependent event that is induced following DNA damage but is not induced during the development of the competent state. At the present time this category includes only induction of the SPβ prophage family. It has been hypothesized that there are two types of temperate *Bacillus* bacteriophage: "smart" and "naive" (103). The smart phages (the SPβ family) distinguish between activation of the SOS system following DNA damage and activation of this system when cells differentiate into their competent state. On the other hand, the naive prophages are induced whenever the SOS system is activated (see SOS type 1 phenomena).

Regulation of type I phenomena

The nucleotide sequences of the *din* promoter regions were determined and the consensus sequence GAAC-N4-GTTC, common to all *din* promoter regions, has been identified (8, 10). In addition to the presence of this putative operator for the SOS genes (termed a "Cheo" box) in the suspected promoter regions of all of the *din* genes including *recA*, additional copies of this consensus sequence are found further upstream in the *dinB*, *dinC*, and *recA* loci as well as within the *recA* coding sequence (8). Copies of the Cheo box were also found in the promoter region of the *uvrB* gene of *B. subtilis* (8), as well as in the promoter of the *dinR* gene (demonstrated to be an SOS-inducible gene; 75). Removal of the Cheo box from the *dinA* promoter region resulted in constitutive transcription of this gene (9, 10). Similar deletion studies on the operator/promoter region of the *recA* gene (9, 10) again strongly indicate that Cheo boxes are involved in *din* gene regulation. Thus, it is envisioned that in the absence of DNA damage a SOS cellular repressor(s) binds to the Cheo box(es) in the operator/promoter region of a *din* gene and the transcription of this DNA damage-inducible gene is repressed. It is also postulated that this repressor is inactivated, following the introduction of DNA damage and/or the abnormal cessation of DNA replication, by a mechanism dependent upon a functional RecA protein. It is important to note that at present there is no conclusive evidence demonstrating that the *B. subtilis* RecA protein is involved directly with the putative SOS cellular repressor. However, due to the phenotypic similarities between the SOS systems of *B. subtilis* and *E. coli* it is assumed that following the introduction of DNA damage into the cell the *B. subtilis* RecA protein is allosterically altered so that it now has apoprotease activity. This activated form of the RecA protein is then believed to be involved in the autoproteolytic cleavage of the cellular SOS repressor. At the present time the best candidate for this cellular repressor is the DinR protein recently identified by Raymond-Denise and Guillen (75).

The SOS System and the Development of the Competent State

Although type 1 SOS phenomena are all induced during the development of the competent state, the interesting relationship that exists between the SOS regulon and differentiation in *B. subtilis* is best examined by studying the phenomena that represent SOS types 2, 3, and 4. Hints about the regulation of these phenomena come primarily from studies on the RecA-independent induction of transcription of the *recA* gene itself during the development of the competent state.

A deletion of 20 bp (10) centered around position −150 with respect to the putative transcription start site of the *recA* gene has been shown to eliminate competence-specific induction but not DNA damage-dependent induction of this promoter. Within this region is the base sequence TTTAGAGCAA, which is followed by the similar sequence TTTTCAGCAC (10). This region upstream of the *recA* gene (which can be classified as a late competence gene using the guidelines of Dubnau [chapter 39; 18]) shares significant similarity with regions upstream of the late competence genes *comC* and *comG*. Furthermore, these regions appear to be likely binding sites for the putative competence activation transcription factor (chapter 39; 18). Based on the results obtained from the deletion studies, it has been proposed that the competence activation transcription factor binds to the operator of *recA* during the development of the competent state to induce transcription of this gene. The presence of this transcription activator bound to the appropriate operator region of the *recA* gene causes the SOS repressor protein to be displaced. Once the competence-induced transcription of the *recA* gene begins, the induction of the type 1 SOS phenomena follows when the RecA protein is activated by the presence of single-strand regions in chromosomal DNA (35). An accumulation of such single-stranded regions of the DNA has been demonstrated to occur as the cells become competent (35). Through this type of a molecular mechanism (or some related mechanism), *B. subtilis* has ensured itself that an increased amount of RecA protein will be present in competent cells. This protein is then available for its essential role in recombination and/or repair.

Adaptive response

An inducible repair system that protects against the lethal and mutagenic effects of alkylation damage has been detected in a variety of different organisms. This repair system has been termed the adaptive response, due to its particular mode of functioning (for reviews see references 42 and 85). Specifically, cultures exposed to low levels of an alkylating agent such as methyl methane sulfonate, and then subsequently challenged by a larger dose of an alkylating agent, are able to withstand both the cytotoxic and mutagenic effects of such an exposure. Adaptation has been shown to involve the enhanced production of proteins that are capable of accepting methyl groups from the DNA or are capable of removing damaged bases from DNA (45). In *E. coli* the *ada* gene possesses two distinct DNA methyltransferase activities and it is also the transcriptional activator of the adaptive response (42, 85).

In *B. subtilis*, six different mutant strains have been identified that are deficient in this response (32, 59–61). Analysis delineated two distinct groups within these mutant strains (60). The first group was totally deficient in the adaptive response, while the second group was deficient only in the induction of O^6-methylguanine-DNA methyltransferase synthesis. Three species of methyl-accepting proteins were detected by Morohoshi and Munakata (61). A 20-kDa protein was constitutively expressed in wild-type cells and in all of the strains deficient in the adaptive response. This protein has O^6-methylguanine-DNA methyltransferase activity. A 22-kDa inducible protein was found only in wild-type cells. This protein also had O^6-methylguanine-DNA methyltransferase activity. In addition, a 27-kDa protein was identified in wild-type cells and in those mutant strains that were deficient only in the induction of the synthesis of an O^6-methylguanine-DNA methyltransferase activity. This protein was designated as the methylphosphotriester-DNA methyltransferase. The 22-kDa and 27-kDa proteins are encoded by the genes *adaA* and *adaB*, respectively, which lie in an operon and overlap by 11 bp (59). Interestingly, the separate functions of the AdaA and AdaB proteins of *B. subtilis* are handled by a single Ada protein of *E. coli* (85). Another difference between the adaptive responses of these two organisms is that approximately 10 times as much O^6-methylguanine-DNA methyltransferase activity is expressed constitutively in *B. subtilis* as compared to *E. coli* (42, 85). As a result of this difference *B. subtilis* is much more resistant to certain alkylating agents than is *E. coli*. Furthermore, *polA* mutants of *B. subtilis* are able to adapt to both the mutagenic and lethal effects of alkylating agents, while in *E. coli* such mutants are adapted for mutagenicity but not for lethality (42, 85). Finally, the spectrum of alkylating agents that can be used to induce the adaptive response, as well as those agents for which the cells can be adapted, differs between the two systems (85).

Oxidative Stress Response

The enzymes necessary to allow cells to tolerate oxidative stress can be classified into two functional categories. First, harmful agents can be removed from the environment before they disrupt essential cell functions. Examples of this type of protection are enzymes such as superoxide dismutase and catalase, which prevent damage caused by O_2* and H_2O_2, respectively. Second, mechanisms exist that actually remove deleterious lesions in cellular components (22, 33). Alkylhydroperoxidase, glutathione reductase, and DNA repair enzymes are examples of this type of repair action. Accordingly, genes that are responsible for repairing the DNA damage resulting from H_2O_2 exposure, such as *recA*, *polA*, and *xth* have been identified in *E. coli* (12, 13, 97). In addition, exposure of aerobically grown *E. coli* to H_2O_2 results in the coordinate induction of nine genes that compose the *oxyR* global regulon (11, 96). Transcription of these H_2O_2-inducible genes requires the positive action of the *oxyR* gene product.

Two catalase enzymes are known to be produced in *E. coli* (47). The *katG* gene encodes a H_2O_2-inducible catalase, HPI, that is part of the *oxyR* regulon (91). On the other hand, HPII catalase, the product of the *katE* gene, is expressed as cultures of *E. coli* enter stationary phase, or when a carbon source other than glucose is utilized for growth (76). The expression of the *katE* gene is proposed to be regulated by an alternative sigma factor, the product of the *katF* gene (39, 57, 81).

Based on observations made with the *E. coli* paradigm, a model for the mechanisms of resistance to oxidative stress has been proposed. This model is composed of cellular components required to remove H_2O_2 from the environment and to repair the lesions resulting from exposure to H_2O_2. These functions appear to be conserved throughout the diverse species of organisms.

The presence of H_2O_2 can induce two independent systems that aid *B. subtilis* in its ability to handle oxidative stress (5, 16, 17). Specifically, both the SOS regulon and an oxidative stress-specific system are induced following exposure of this bacterium to hydrogen peroxide (5). Of specific importance to the oxidative stress system is the inducible gene *kat-19*, which encodes a vegetative catalase of 65,830 kDa (6). This catalase is expressed during vegetative growth of *B. subtilis* and is H_2O_2 inducible as well as growth-stage regulated (5, 6). Interestingly, comparison of the amino acid sequence of the *B. subtilis* catalase enzyme encoded by the *kat-19* gene with the amino acid sequence of catalases from other organisms revealed greater similarity to enzymes isolated from eukaryotic organisms than to either of the two catalases of *E. coli* (6).

As it does for the SOS response, *B. subtilis* regulates components of the oxidative stress system in coordination with developmental and differentiation cycles. This type of regulation has also been observed for the temporally regulated *E. coli* catalase gene *katE*. However, the mechanism by which *katE* is regulated is significantly different from the developmental coordination demonstrated by the *kat-19* gene. Transcription of the *kat-19* gene during the transition from exponential growth to stationary phase occurs from a σ^A promoter (4). Thus, the vegetative "housekeeping" machinery is utilized, and other regulatory components that are active at this stage of growth function to control *kat-19* gene expression. Some candidates for the temporal regulation of the *kat-19* gene can be deduced from the dependence on *spo0A* for the temporal expression of *kat-19* and from the abolition of this dependence in the absence of a functional *abrB* gene product. Since the *abrB* gene product is a positive regulator of the gene that encodes the repressor Hpr, and since the *spo0A* gene product represses the *abrB* gene, it has been postulated that Hpr may repress the expression of the *kat-19* gene (4, 16). Furthermore, the protein encoded by the *hpr* gene has been shown to bind specific DNA sequences in the operator/promoter regions of some temporally regulated genes in *B. subtilis* (40). Interestingly, three copies of the A-T-rich Hpr "consensus" recognition sequence [(A/G)ATAntat(t/c)] are found near the promoter for the *kat-19* gene (4). Thus, in the *spo0A* mutant strain, reduced expression of *kat-19* could be due to overexpression of *hpr*. Further support for the involvement of *hpr* in the regulation of the oxidative stress system has been demonstrated by the hyperresistance to H_2O_2 of *B. subtilis* strains carrying the *hpr* mutation (16). Presumably, the *hpr* mutation alone may result in constant derepression of the oxidative stress system.

The complexity evident in the control of expression of the *kat-19* gene emphasizes the difference between the oxidative stress system of *B. subtilis* and that of the gram-negative organisms. Most obvious is the difference in the regulation of catalase enzyme production. While *E. coli* expresses catalase enzymes in response to H_2O_2 and at the transition phase of the growth cycle, it does so by expressing two different genes. In *B. subtilis* the *kat-19* gene is the only catalase gene expressed during vegetative growth and during the transition to stationary phase.

DNA REPAIR SYSTEMS OF OTHER GRAM-POSITIVE BACTERIA

Before the elegant enzymatic characterization of the Uvr system of *E. coli*, the best characterized DNA repair system (with respect to enzyme activity) was the pyrimidine dimer specific BER isolated from *Micrococcus luteus* (24). Subsequently, from this organism AP-endonucleases, endonucleases with activity against photoalkylated DNA, 3-methyladenine glycosylase, uracil glycosylase, and enzymes involved in an NER system have been isolated and characterized (for review see reference 24; 52, 68). Recently, a significant degree of gene homology has been identified between the UvrB protein of *E. coli* and proteins with similar activities that have been isolated from *M. luteus* and *Streptococcus pneumoniae* (87, 88). Mutants of *Enterococcus faecalis* and *Staphylococcus aureus* that are sensitive to DNA-damaging agents have been isolated and *recA* genes have been cloned out of *Streptococcus mutans* and *S. aureus* (2, 27, 28, 36, 77, 100). Collectively these results indicate that DNA repair systems in general, and certain specific proteins associated with the repair of DNA (i.e., UvrB and RecA), have been conserved during the evolutionary process.

An extreme example of the ability of bacteria to survive DNA-damaging agents can be found among the genus *Deinococcus*. Few organisms approach the ability of this genus and the related genus *Deinobacter* to be resistant to far-UV and ionizing radiation (for a recent review, see reference 89). These two genera are in the family *Deinobacteriaceae* and are able to repair large amounts of radiation-induced DNA damage by using nonmutagenic repair systems. Within the *Deinobacteriaceae* are gram-positive cocci and gram-negative bacilli. Two independent excision repair systems have been identified in *Deinococcus radiodurans* for the repair of UV damage, and one of these systems is also required to repair the damage produced by the bifunctional alkylating agent mitomycin C (20, 62). In addition to these excision repair systems, enzymes associated with BER and with oxidative stress systems have also been identified in these extremely radiation-resistant organisms (38, 55, 62). Although these systems have been identified, a complete characterization has not yet occurred. Such a characterization is essential for determining whether these radiation-resistant bacteria obtain their resistance through the previously identified systems described above.

SUMMARY AND CONCLUSIONS

All of the major DNA repair systems and stress tolerance mechanisms that have been identified and characterized for *E. coli* have also been observed among the gram-positive bacteria. While the phenotypic similarities exist, essential differences have been observed in the regulation as well as in the activities of these repair systems among these organisms. This is especially true for the DNA repair systems of *B. subtilis*. These systems are generally linked in their expression and activity with one or more of the developmental states that have been identified for this bacterium. It is in the elucidation of this linkage that *B. subtilis* becomes a critical model for the understanding of how organisms respond at the molecular level to stressful situations. Similar characterizations of these repair systems in other gram-positive bacteria, especially among the extremely resistant bacteria, will determine whether or not *B. subtilis* represents a paradigm for gram-positive bacteria or for bacteria that have distinct developmental cycles.

REFERENCES

1. **Alonso, J. E., G. Luder, and R. H. Tailor.** 1991. Characterization of *Bacillus subtilis* recombinational pathways. *J. Bacteriol.* **173:**3977–3980.
2. **Bayles, K. W.** 1991. Cloning of the *Staphylococcus aureus recA* gene and generation and characterization of a *recA* mutant. *Plasmid* **25:**154–161.
3. **Bibor, V., and W. G. Verly.** 1978. Purification and properties of the endonuclease specific for apurinic sites of *Bacillus stearothermophilus*. *J. Biol. Chem.* **253:** 850–855.
4. **Bol, D. K.** 1991. *Bacillus subtilis* responses and mechanisms of resistance to hydrogen peroxide. Ph.D. thesis. University of Maryland Baltimore County, Baltimore.
5. **Bol, D. K., and R. E. Yasbin.** 1990. Characterization of an inducible oxidative stress system in *Bacillus subtilis*. *J. Bacteriol.* **172:**3503–3506.
6. **Bol, D. K., and R. E. Yasbin.** 1991. The isolation and identification of a vegetative catalase gene from *Bacillus subtilis*. *Gene* **109:**31–37.
7. **Chen, N.-Y., J.-J. Zhang, and H. Paulus.** 1989. Chromosomal location of the *Bacillus subtilis* aspartokinase II gene and nucleotide sequence of the adjacent genes homologous to *uvrC* and *trx* of *Escherichia coli*. *J. Gen. Microbiol.* **135:**2931–2940.
8. **Cheo, D. L., K. W. Bayles, and R. E. Yasbin.** 1991. Cloning and characterization of DNA damage-inducible promoter regions from *Bacillus subtilis*. *J. Bacteriol.* **173:**1696–1703.
9. **Cheo, D. L., K. W. Bayles, and R. E. Yasbin.** 1992. Molecular characterization of regulatory elements controlling expression of the *Bacillus subtilis recA*$^+$ gene. *Biochimie* **74:**755–762.
10. **Cheo, D. L., K. W. Bayles, and R. E. Yasbin.** Elucidation of regulatory elements that control damage induction and competence-induction of the *Bacillus subtilis* SOS system. Submitted for publication.
11. **Christman, M. F., R. W. Morgan, F. S. Jacobson, and B. N. Ames.** 1985. Positive control of a regulon for defense against oxidative stress and some heat shock proteins in *Salmonella typhimurium*. *Cell* **41:**753–762.
12. **Demple, B., J. Halbrook, and S. Linn.** 1983. *Escherichia coli xth* mutants are hypersensitive to hydrogen peroxide. *J. Bacteriol.* **153:**1079–1082.
13. **Demple, B., A. Johnson, and D. Fung.** 1986. Exonuclease III and endonuclease IV remove 3′ blocks from DNA synthesis primers in H_2O_2-damaged *E. coli*. *Proc. Natl. Acad. Sci. USA* **83:**7731–7735.
14. **Dodson, L. A., and C. T. Hadden.** 1980. Capacity for postreplication repair correlated with transducibility in Rec$^-$ mutants of *Bacillus subtilis*. *J. Bacteriol.* **144:** 608–615.
15. **Donnellan, J. E., Jr., and R. B. Setlow.** 1965. Thymine photoproducts but not thymine dimers are found in ultraviolet irradiated bacterial spores. *Science* **149:** 308–310.
16. **Dowds, B. C. A., and J. A. Hoch.** 1991. Regulation of the oxidative stress response by the *hpr* gene in *Bacillus subtilis*. *J. Gen. Microbiol.* **137:**1121–1125.
17. **Dowds, B. C. A., P. Murphy, D. J. McConnell, and K. M. Devine.** 1987. Relationship among oxidative stress, growth cycle, and sporulation in *Bacillus subtilis*. *J. Bacteriol.* **169:**5771–5775.
18. **Dubnau, D.** 1991. Genetic competence in *Bacillus subtilis*. *Microbiol. Rev.* **55:**395–424.
19. **Epstein, H. T., and I. Mahler.** 1968. Mechanism of enhancement of SP82 transfection. *J. Virol.* **2:**710–715.
20. **Evans, D. M., and B. E. B. Moseley.** 1983. Roles of *uvsC*, *uvsD*, *uvsE*, and *mtcA* genes in the two pyrimidine dimer excision repair pathways of *Deinococcus radiodurans*. *J. Bacteriol.* **156:**576–583.
21. **Fields, P. I., and R. E. Yasbin.** 1980. Involvement of deoxyribonucleic acid polymerase III in W-reactivation in *Bacillus subtilis*. *J. Bacteriol.* **144:**473–475.
22. **Fridovich, I.** 1978. The biology of oxygen radicals. *Science* **201:**875–900.
23. **Friedberg, E. C.** 1981. Base excision repair of DNA, p. 77–83. *In* E. Seeberg and K. Kleppi (ed.), *Chromosome Damage and Repair.* Plenum Press, New York.
24. **Friedberg, E. C.** 1985. *DNA Repair*, p. 141–211. W. H. Freeman and Co., New York.
25. **Ganesan, A. K.** 1974. Persistence of pyrimidine dimers during post-replication repair in ultraviolet light-irradiated *Escherichia coli* K-12. *J. Mol. Biol.* **87:**102–119.
26. **Germaine, G. R., and W. G. Murrel.** 1973. Effect of dipicolinic acid on the ultraviolet radiation resistance of *Bacillus cereus* spores. *Photochem. Photobiol.* **17:**145–153.
27. **Goering, R. V.** 1979. Mutants of *Staphylococcus aureus* deficient in recombinational repair. Improved isolation by selecting for mutants exhibiting concurrent sensitivity to ultraviolet radiation and N-methyl-N′-nitro-N-nitrosoguanidine. *Mutat. Res.* **60:**279–289.
28. **Goering, R. V., and P. A. Pattee.** 1971. Mutants of *Staphylococcus aureus* with increased sensitivity to ultraviolet radiation. *J. Bacteriol.* **106:**157–161.
29. **Grossman, L., and A. T. Yeung.** 1990. The UvrABC endonuclease system of *Escherichia coli*: a view from Baltimore. *Mutat. Res.* **236:**213–221.
30. **Hadden, C. T.** 1979. Gap-filling repair synthesis induced by ultraviolet light in a *Bacillus subtilis* Uvr$^-$ mutant. *J. Bacteriol.* **139:**239–246.
31. **Hadden, C. T.** 1979. Pyrimidine dimer excision in a *Bacillus subtilis* Uvr$^-$ mutant. *J. Bacteriol.* **139:**247–255.
32. **Hadden, C. T., R. S. Foote, and S. Mitra.** 1983. Adaptive response of *Bacillus subtilis* to *N*-methyl-*N*′-nitro-*N*-nitrosoguanidine. *J. Bacteriol.* **153:**756–762.
33. **Hagensee, M., and R. E. Moses.** 1989. Multiple pathways for repair of H_2O_2-induced damage in *Escherichia coli*. *J. Bacteriol.* **171:**991–995.
34. **Hanawalt, P. C.** 1989. Concepts and models for DNA repair: from *E. coli* to mammalian cells. *Environ. Mol. Mutagen.* **14**(Suppl. 16):90–98.
35. **Harris, W. J., and G. C. Barr.** 1971. Mechanism of transformation in *Bacillus subtilis*. *Mol. Gen. Genet.* **113:**316–330.
36. **Inoue, M., H. Oshima, T. Okubo, and S. Mitsuhashi.** 1972. Isolation of the *rec* mutants in *Staphylococcus aureus*. *J. Bacteriol.* **112:**1169–1176.
37. **Inoue, T., and T. Kada.** 1978. Purification and properties of a *Bacillus subtilis* endonuclease specific for apurinic sites in DNA. *J. Biol. Chem.* **253:**8559–8563.
38. **Juan, J.-Y., S. N. Keeney, and E. M. Gregory.** 1991. Reconstitution of the *Deinococcus radiodurans* aposuperoxide dismutase. *Arch. Biochem. Biophys.* **286:**257–263.
39. **Kaasen, I., P. Falkenberg, O. B. Styrvold, and A. R. Strøm.** 1992. Molecular cloning and physical mapping of the *otsBA* genes, which encode the osmoregulatory trehalose pathway of *Escherichia coli*: evidence that transcription is activated by KatF (AppR). *J. Bacteriol.* **174:**889–898.
40. **Kallio, P. T., J. E. Fagelson, J. A. Hoch, and M. A. Strauch.** 1991. The transition state regulator Hpr of *Bacillus subtilis* is a DNA-binding protein. *J. Biol. Chem.* **266:**13411–13417.
41. **Kelner, A.** 1949. Effect of visible light on the recovery of *Streptomyces griseus* conidia from ultraviolet irradiation injury. *Proc. Natl. Acad. Sci. USA* **35:**73–79.
42. **Kushner, S. R.** 1987. DNA repair, p. 1044–1053. *In* F. C. Neidhardt, J. L. Ingraham, K. B. Low, B. Magasanik, M. Schaechter, and H. E. Umbarger (ed.), *Escherichia coli and Salmonella typhimurium: Cellular and Molecular Biology*. American Society for Microbiology, Washington, D.C.

43. **Lindahl, T.** 1979. DNA glycosylases, endonucleases, endonucleases for apurinic/apyrimidinic sites, and base excision repair. *Prog. Nucleic Acids Res. Mol. Biol.* **22:** 135–192.
44. **Lindahl, T.** 1982. DNA repair enzymes. *Annu. Rev. Biochem.* **51:**61–87.
45. **Lindahl, T., B. Sedgwick, M. Sekiguchi, and Y. Nakabeppu.** 1988. Regulation and expression of the adaptive response to alkylating agents. *Annu. Rev. Biochem.* **57:**133–157.
46. **Little, J. W., and D. W. Mount.** 1982. The SOS regulatory system of *Escherichia coli. Cell* **29:**11–22.
47. **Loewen, P. C., J. Switala, and B. L. Triggs-Riane.** 1985. Catalases HPI and HPII in *E. coli* are induced independently. *Arch. Biochem. Biophys.* **243:**144–149.
48. **Love, P. E., M. J. Lyle, and R. E. Yasbin.** 1985. DNA damage inducible (DIN) loci are transcriptionally activated in competent *Bacillus subtilis. Proc. Natl. Acad. Sci. USA* **82:**6201–6205.
49. **Love, P. E., and R. E. Yasbin.** 1984. Genetic characterization of the inducible "SOS-like" system of *Bacillus subtilis. J. Bacteriol.* **160:**910–920.
50. **Lovett, C. M., Jr., P. E. Love, and R. E. Yasbin.** 1989. Competence-specific induction of the *Bacillus subtilis* RecA protein analog: evidence for dual regulation of a recombination protein. *J. Bacteriol.* **171:**2318–2322.
51. **Lovett, C. M., Jr., and J. W. Roberts.** 1985. Purification of a RecA protein analogue from *Bacillus subtilis. J. Biol. Chem.* **260:**3305–3313.
52. **Malvy, C., J. Pierre, M. Lefrancois, and J. Markovits.** 1990. Low concentrations of acridine dimers inhibit *Micrococcus* AP endonuclease through interaction with apurinic sites in DNA. *Chem-Biol. Interact.* **73:**249–260.
53. **Marrero, R., and R. E. Yasbin.** 1988. Cloning of the *Bacillus subtilis recE*$^+$ gene and functional expression of *recE*$^+$ in *B. subtilis. J. Bacteriol.* **170:**335–344.
54. **Mason, J. M., and P. Setlow.** 1986. Essential role of small acid-soluble spore proteins in resistance of *Bacillus subtilis* spores to UV light. *J. Bacteriol.* **167:**174–178.
55. **Masters, C. I., B. E. B. Moseley, and K. W. Minton.** 1991. AP endonuclease and uracil DNA glycosylase activities in *Deinococcus radiodurans. Mutat. Res.* **254:** 263–272.
56. **McAllister, W. T., and D. M. Green.** 1972. Bacteriophage SP82G inhibition of intracellular deoxyribonucleic acid inactivation process of *Bacillus subtilis. J. Virol.* **10:**51–59.
57. **McCann, M. P., J. P. Kidwell, and A. Matin.** 1991. The putative sigma factor KatF has a central role in development of starvation-mediated general resistance in *Escherichia coli. J. Bacteriol.* **173:**4188–4194.
58. **Mohr, S. C., N. V. H. A. Sokolov, C. He, and P. Setlow.** 1991. Binding of small acid-soluble spore proteins from *Bacillus subtilis* changes the conformation of DNA from B to A. *Proc. Natl. Acad. Sci. USA* **88:**77–81.
59. **Morohoshi, F., K. Hayashi, and N. Munakata.** 1991. Molecular analysis of *Bacillus subtilis ada* mutants deficient in the adaptive response to simple alkylating agents. *J. Bacteriol.* **173:**7834–7840.
60. **Morohoshi, F., and N. Munakata.** 1986. Two classes of *Bacillus subtilis* mutants deficient in the adaptive response to simple alkylating agents. *Mol. Gen. Genet.* **202:**200–206.
61. **Morohoshi, F., and N. Munakata.** 1987. Multiple species of *Bacillus subtilis* DNA alkyltransferase involved in the adaptive response to simple alkylating agents. *J. Bacteriol.* **169:**587–592.
62. **Moseley, B. E. B.** 1983. Photobiology and radiobiology of *Micrococcus* (*Deinococcus*) *radiodurans. Photochem. Photobiol. Rev.* **7:**223–274.
63. **Munakata, N.** 1974. Ultraviolet sensitivity of *Bacillus subtilis* spores upon germination and outgrowth. *J. Bacteriol.* **120:**59–65.
64. **Munakata, N.** 1977. Mapping of the genes controlling excision repair of pyrimidine photoproducts in *Bacillus subtilis. Mol. Gen. Genet.* **156:**49–54.
65. **Munakata, N., and C. S. Rupert.** 1972. Genetically controlled removal of "spore photoproduct" from deoxyribonucleic acid of ultraviolet-irradiated *Bacillus subtilis* spores. *J. Bacteriol.* **111:**192–198.
66. **Munakata, N., and C. S. Rupert.** 1974. Dark repair of DNA containing "spore photoproduct" in *Bacillus subtilis. Mol. Gen. Genet.* **130:**239–250.
67. **Munakata, N., and C. S. Rupert.** 1975. Effects of DNA-polymerase-defective and recombination-deficient mutations on the ultraviolet sensitivity of *Bacillus subtilis* spores. *Mutat. Res.* **27:**157–169.
68. **Nakayama, H., S. Shiota, and K. Umezu.** 1992. UV endonuclease-mediated enhancement of UV survival in *Micrococcus luteus*: evidence revealed by deficiency in the Uvr homolog. *Mutat. Res.* **273:**43–48.
69. **Nicholson, W.** Personal communication.
70. **Nicholson, W. L., B. Setlow, and P. Setlow.** 1990. Binding of DNA in vitro by a small, acid-soluble spore protein and its effect on DNA topology. *J. Bacteriol.* **172:**6900–6906.
71. **Nicholson, W. L., B. Setlow, and P. Setlow.** 1991. Ultraviolet irradiation of DNA complexed with alpha/beta-type small, acid-soluble proteins from spores of *Bacillus* or *Clostridium* species makes spore photoproduct but not thymine dimers. *Proc. Natl. Acad. Sci. USA* **88:**8288–8292.
72. **Pierre, J., and J. Laval.** 1980. *Micrococcus luteus* endonucleases for apurinic/apyrimidinic sites in deoxyribonucleic acid. I. Purification and general properties. *Biochemistry* **19:**5018–5024.
73. **Radany, E. H., G. Malanoski, N. Ambulos, E. C. Friedberg, and R. E. Yasbin.** 1988. Transfection enhancement of bacteriophage DNA may reflect a novel repair pathway for UV-irradiated DNA in *Bacillus subtilis. J. Cell. Biochem.* (Suppl.) **12A:**323.
74. **Radman, M.** 1974. Phenomenology of an inducible mutagenic DNA repair pathway in *Escherichia coli*: SOS repair hypothesis, p. 128–142. *In* S. Prakash, F. Sherman, M. Miller, C. Lawrence, and H. W. Tabor (ed.), *Molecular and Environmental Aspects of Mutagenesis.* Charles C. Thomas, Publisher, Springfield, Ill.
75. **Raymond-Denise, A., and N. Guillen.** 1991. Identification of *dinR*, a DNA damage-inducible regulator gene of *Bacillus subtilis. J. Bacteriol.* **173:**7084–7091.
76. **Richter, H. E., and P. C. Loewen.** 1981. Induction of catalase in *E. coli* by ascorbic acid involves hydrogen peroxide. *Biochem. Biophys. Res. Commun.* **100:**1039–1046.
77. **Roca, A. I., and M. M. Cox.** 1990. The RecA protein: structure and function. *Crit. Rev. Biochem.* **25:**415–456.
78. **Rupp, W. D., and P. Howard-Flanders.** 1968. Discontinuities in the DNA synthesized in an excision-defective strain of *Escherichia coli* following ultraviolet irradiation. *J. Mol. Biol.* **31:**291–304.
79. **Sancar, A., K. A. Franklin, and G. B. Sancar.** 1984. *Escherichia coli* photolyase stimulates UvrABC excision nuclease *in vitro. Proc. Natl. Acad. Sci. USA* **81:**7397–7401.
80. **Sancar, A., and W. D. Rupp.** 1983. A novel repair enzyme: UVRABC excision nuclease of *Escherichia coli* cuts a DNA strand on both sides of the damaged region. *Cell* **33:**249–260.
81. **Schellhorn, H. E., and V. L. Stones.** 1992. Regulation of *katF* and *katE* in *Escherichia coli* K-12 by weak acids. *J. Bacteriol.* **174:**4769–4776.
82. **Setlow, B., and P. Setlow.** 1987. Thymine-containing dimers as well as spore photoproducts are found in ultraviolet-irradiated *Bacillus subtilis* spores that lack small, acid-soluble proteins. *Proc. Natl. Acad. Sci. USA* **84:**421–423.

83. **Setlow, P.** 1988. Small acid-soluble, spore proteins of *Bacillus* species: structure, synthesis, genetics, function and degradation. *Annu. Rev. Microbiol.* **42:**319–338.
84. **Setlow, R. B., P. A. Swenson, and W. L. Carrier.** 1963. Thymine dimers and inhibition of DNA synthesis by ultraviolet irradiation of cells. *Science* **142:**1464–1466.
85. **Shevell, D. E., B. M. Friedman, and G. C. Walker.** 1990. Resistance to alkylation damage in *Escherichia coli*: role of the Ada protein in induction of the adaptive response. *Mutat. Res.* **223:**53–72.
86. **Shuster, R. C.** 1967. Fate of thymine-containing dimers in the deoxyribonucleic acid of ultraviolet-irradiated *Bacillus subtilis*. *J. Bacteriol.* **93:**811–815.
87. **Sicard, N., and A. M. Estevenon.** 1990. Excision-repair capacity in *Streptococcus pneumoniae*: cloning and expression of a uvr-like gene. *Mutat. Res.* **235:**195–201.
88. **Sicard, N., J. Oreglia, and A.-M. Estevenon.** 1992. Structure of the gene complementing *uvr-402* in *Streptococcus pneumoniae*: homology with *Escherichia coli uvrB* and the homologous gene in *Micrococcus luteus*. *J. Bacteriol.* **174:**2412–2415.
89. **Smith, M. D., C. I. Masters, and B. E. B. Moseley.** 1992. Molecular biology of radiation-resistant bacteria, p. 258–280. *In* R. A. Herbert and R. J. Sharp (ed.), *Molecular Biology and Biotechnology of Extremophiles*. Blackie & Son Limited, Glasgow.
90. **Stafford, R. S., and J. E. Donnellan, Jr.** 1968. Photochemical evidence for conformation changes in DNA during germination of bacterial spores. *Proc. Natl. Acad. Sci. USA* **59:**822–828.
91. **Tartaglia, L. A., G. Storz, and B. N. Ames.** 1989. Identification and molecular analysis of *oxyR*-regulated promoters important for the bacterial adaptation to oxidative stress. *J. Mol. Biol.* **209:**709–719.
92. **van Houten, B.** 1990. Nucleotide excision repair in *Escherichia coli*. *Microbiol. Rev.* **54:**18–51.
93. **Varghese, A. J.** 1970. 5-Thyminyl-5,6-dihydrothymine from DNA irradiated with ultraviolet light. *Biochem. Biophys. Res. Commun.* **38:**484–490.
94. **Vissc, R., M. de Ruijter, and J. Brouwer.** 1991. Uvr excision repair protein complex of *Escherichia coli* binds to the convex side of a cisplatin-induced kink in the DNA. *J. Biol. Chem.* **266:**7609–7617.
95. **Walker, G. C.** 1984. Mutagenic and inducible responses to deoxyribonucleic acid damage in *Escherichia coli*. *Microbiol. Rev.* **48:**60–93.
96. **Walkup, L. K. B., and T. Kogoma.** 1989. *Escherichia coli* proteins inducible by oxidative stress mediated by the superoxide radical. *J. Bacteriol.* **171:**1476–1484.
97. **Wallace, S. S.** 1988. AP endonucleases and DNA glycosylases that recognize oxidative DNA damage. *Environ. Mol. Mutagen.* **12:**431–477.
98. **Wang, T.-C. V., and C. S. Rupert.** 1977. Transitory germinative excision repair in *Bacillus subtilis*. *J. Bacteriol.* **129:**1313–1319.
99. **Witkin, E. M.** 1976. Ultraviolet mutagenesis and inducible DNA repair in *Escherichia coli*. *Bacteriol. Rev.* **40:**869–907.
100. **Yagi, Y., and D. B. Clewell.** 1980. Recombination-deficient mutant of *Streptococcus faecalis*. *J. Bacteriol.* **143:**966–970.
101. **Yasbin, R. E.** 1977. DNA repair in *Bacillus subtilis*. II. Activation of the inducible system in competent bacteria. *Mol. Gen. Genet.* **153:**219–225.
102. **Yasbin, R. E.** 1985. DNA repair in *Bacillus subtilis*, p. 33–52. *In* D. Dubnau (ed.), *Molecular Biology of the Bacilli*, vol. II. Academic Press, Inc., New York.
103. **Yasbin, R. E., D. L. Cheo, and K. W. Bayles.** 1992. Inducible DNA repair and differentiation in *Bacillus subtilis*: interactions between global regulons. *Mol. Microbiol.* **6:**1263–1270.
104. **Yasbin, R. E., P. I. Fields, and B. J. Andersen.** 1980. Properties of *Bacillus subtilis* 168 derivatives freed of their natural prophages. *Gene* **12:**155–159.
105. **Yasbin, R. E., P. Love, J. Jackson, and C. M. Lovett, Jr.** 1988. Evolutionary divergence of the SOS-like (SOB) system of *Bacillus subtilis*, p. 485–490. *In* E. C. Friedberg and P. Hanawalt (ed.), *Mechanisms and Consequences of DNA Damage Processing*. Alan R. Liss, Inc., New York.
106. **Yasbin, R. E., and R. Miehl.** 1980. Deoxyribonucleic acid repair in *Bacillus subtilis*: development of competent cells into a tester for carcinogens. *Appl. Environ. Microbiol.* **39:**854–858.
107. **Yasbin, R. E., M. Stranathan, and K. W. Bayles.** 1991. The *recE(A)*$^+$ gene of *B. subtilis* and its gene product: further characterization of this universal protein. *Biochimie* **73:**245–250.

38. Restriction/Modification and Methylation Systems in *Bacillus subtilis*, Related Species, and Their Phages

THOMAS A. TRAUTNER and MARIO NOYER-WEIDNER

Several restriction/modification (R/M) systems have been identified in *Bacillus subtilis* and related bacteria and will be described here. Accepting the view that R/M systems have evolved to defend bacteria effectively against attack by bacterial viruses, we shall discuss the question of what mechanisms bacteriophages have developed to overcome barriers provided by host R/M systems. To what extent do R/M systems affect other inter- and intraspecific transport of DNA? In this connection, we shall discuss the usefulness of restriction systems, with their high substrate specificities for double-stranded DNA, in understanding the processing of free or packaged DNA during uptake into *B. subtilis* cells or subcellular structures. Recent progress in the characterization of genes encoding restriction endonucleases (ENases) and modification methyltransferases (MTases) from a wide range of organisms has made such systems per se interesting paradigms for the study of the evolution of highly specific DNA-binding proteins. In particular, our interest is focused here on the phylogenetic relationship among the various ENases and MTases of *Bacillus* and other bacterial species. Furthermore, the requirement for coexistence of an ENase with an MTase represents an interesting case of obligatory coevolution of two genes. The study of multispecific MTases, discovered so far in only some temperate *B. subtilis* and *Bacillus amyloliquefaciens* phages, has made significant contributions to our present understanding of the nature, function, and evolution of R/M systems and particularly of their MTases. The interesting biochemical aspects of the action of ENases and MTases will not be covered here. Such discussions are included in a recent review (72).

R/M SYSTEMS

Nomenclature

In this review, we shall follow the convention of Smith and Nathans (90) in describing R/M systems. In addition, we propose that for *B. subtilis*, the gene descriptions as used for *Escherichia coli* (3) be adopted. The present nomenclature, as described, e.g., in Piggot et al. (74), does not take into account recent refinements in the genetic analysis of R/M systems. Here we propose that the locus encoding a R/M system be termed *hsd* (for host specificity determinant). The specificity of the system is then indicated by an index (e.g., hsd_{RI} for genes carrying wild-type alleles of the *Bsu*RI R/M system). Mutant alleles of the contributing genes would be indicated by R^- or M^-. Thus, $hsd_{RI}R^-$, $hsd_{RI}M^-$, and $hsd_{RI}R^-M^-$ represent the possible mutant forms of the hsd_{RI} locus. This nomenclature will be used throughout this review.

Description of R/M Systems

In *B. subtilis*, the phenomenon of restriction and modification was first observed by Trautner et al. (104) with phage SPP1 and a newly isolated strain, *B. subtilis* R, and by Shibata and Ando (85) with phage ϕ105c and bacterial strains *B. subtilis* 168 and *B. amyloliquefaciens* N and H. Later, Uozumi et al. (105) screened their laboratory collection of some 80 different *B. subtilis*-related bacteria and described a further 13 strains that showed restriction and modification of phage ϕ105c. Altogether, this analysis has led to the identification of six different R/M systems in *B. subtilis*. These and well-characterized R/M systems from the related species *B. amyloliquefaciens* (15, 16, 68), *Bacillus sphaericus* (76, 96), and *Bacillus aneurinolyticus* (62) are described in Table 1. A list of all *Bacillus* R/M systems identified so far has been presented by Wilson (113). According to the classification used by Bickle (9), all systems characterized are of type II and are encoded chromosomally. Neither type I nor type III R/M systems have been identified in *B. subtilis* or other gram-positive bacteria.

The commonly used strain *B. subtilis* 168 has only one inherent R/M system, *Bsu*MI (85), whose genes have been mapped at coordinate 47 (Fig. 1). Phage ϕ105c, grown on *B. amyloliquefaciens* N, has an efficiency of plating of 10^{-3} to 10^{-2} on the *Bsu*MI-proficient strain 168 (85). Plaques grown on this strain plate on 168 strains with an efficiency of plating of 1. This capacity is again lost when such phages are passaged on nonmodifying *B. amyloliquefaciens* strain N. A mutant strain of 168, RM125, which has lost the hsd_{MI} genes, has been identified by Uozumi et al. (105).

In addition to performing like a classical R/M system, the *Bsu*MI system of strain 168 is involved in intricate controls of permissiveness of the strain for the growth of other phages. Phages SP10 and ϕNR2 do not grow on the wild-type 168 strain of *B. subtilis* (80, 100a). However, a double mutation in *B. subtilis* 168, *nonA nonB*, eliminates this nonpermissiveness. The *nonB* mutation thus defined is localized within the hsd_{MI} locus, also generating a *Bsu*MI restriction-deficient phenotype. No phenotype could be associated with the *nonA* gene, which has been localized close to

Thomas A. Trautner and Mario Noyer-Weidner • Max-Planck-Institut für Molekulare Genetik, Ihnestrasse 73, DW-1000 Berlin 33, Germany.

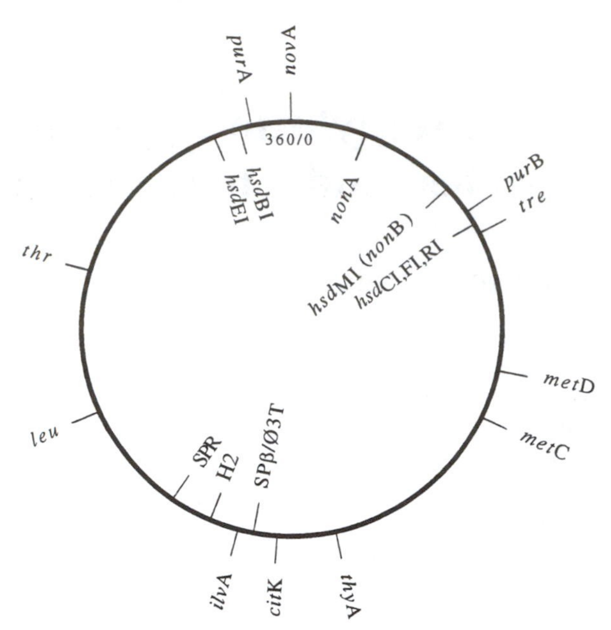

Figure 1. Genetic map of *B. subtilis* 168 (adapted from Piggot et al. [74]). Outside circle shows several physiological markers as reference. Inner circle shows positions of *hsd* loci and attachment sites of MTase-encoding phages (the attachment site of phage H2, originally isolated from *B. amyloliquefaciens*, has been determined in *B. subtilis* lysogens of this phage [121]; the attachment sites of phages $\rho 11_B$ and $\rho 11_s$ have not yet been determined). The inner circle also shows the position of the *nonA* locus, which, in interplay with the hsd_{MI} locus, determines nonpermissiveness of *B. subtilis* 168 for some phages.

the *rfm* gene of *B. subtilis* (80). SP10 could, however, grow in $nonB^+$ cells when these were preinfected with phage SP18. SP18 infection was found to antagonize the degradation of SP10 DNA, which is the cause of restriction of SP10 in a $nonB^+$ background (114). Nonpermissiveness of strain 168 to SP10 infection could also be eliminated by exposing the cells to heat shock prior to infection (35). Nonpermissiveness of growth of the related phages ϕ15 and PZA is also determined by the *Bsu*MI system (28). Fucik et al. (28) further reported that introducing an early sporulation defect (*spo0A*) into *B. subtilis* 168 abolishes the *Bsu*MI restriction capacity (but not the modification potential) and at the same time renders the cells permissive for growth of ϕ15 and PZA. This permissiveness in the *spo0A* background can be eliminated by making the cells streptomycin resistant. These interactions demonstrate the close interrelationship between the activity of the *Bsu*MI R/M system and other functions encoded by the genomes of 168-type strains, although molecular interpretations of these phenomena have not yet been provided.

All other R/M systems studied have been introduced into *B. subtilis* 168 from various *B. subtilis* strains by transformation and were mapped by PBS1 transduction (42, 43, 104), defining four locations of *hsd* loci in the *B. subtilis* chromosome (Table 1, Fig. 1). No pair of the *Bsu*CI, *Bsu*RI, and *Bsu*FI systems has been found to coexist within one cell. Apparently, the corresponding genes occupy one and the same location within the chromosome, a fact that is consistent with mapping experiments (42). The location of both the hsd_{RI} and hsd_{FI} genes within an identical nucleotide sequence environment was shown in later experiments (108). In line with the mapping data, any one R/M system encoded at this locus can coexist with all other *hsd* genes. In fact, a strain with a fourfold R/M capacity comprising the *Bsu*RI system and three other R/M systems encoded at the other loci has been constructed (42).

In all transfer experiments, transformation of an *hsd* system with its R/M genes had the characteristics of single-gene transformation, pointing to the close linkage of the two genes. The specific locations of the various *hsd* genes introduced in the strain 168 chromosome were reproduced in independently isolated transformants (43). Those authors attributed the specificity of insertion as a consequence of recombination between DNA flanking the *hsd* genes of the donor and homologous DNA of the recipient chromosome rather than to recombination with "inert" *hsd* genes in strain 168. In support of this notion are unpublished experiments from our laboratory, which showed an absence of DNA-DNA homology when the strain 168 chromosome was probed with DNA of identified *hsd* systems other than *Bsu*MI. No systematic search regarding the interesting question of whether a given *hsd* system could be integrated at any region of the chromosome that would accommodate foreign DNA has been reported. It would also be desirable to determine experimentally whether cells with an R/M system in which the cognate ENase and MTase genes with appropriate promoters occupy unlinked positions within the chromosome could be constructed and maintained. As derived from cloning experiments in *E. coli* and *B. amyloliquefaciens*, linkage of the two genes constituting an R/M system is not obligatory for their functioning, since cells with a proficient R/M system have been constructed by placing the R/M genes on two different compatible plasmids (15, 48, 51). With the DNA sequence of many *B. subtilis hsd* systems established and the possibility of integrating foreign DNA at defined locations within the bacterial chromosome, such investigations have become feasible. Results of such experiments would contribute to the understanding of the selective forces that have universally brought ENase and MTase genes of the same R/M system into the closest possible proximity irrespective of their relative transcriptional orientations.

B. subtilis and *B. amyloliquefaciens* temperate phages SPR, ϕ3T, SPβ, $\rho 11_B$, $\rho 11_s$ and H2 contain genes that encode monospecific and/or multispecific MTases in the absence of cognate ENases. The established chromosomal locations in lysogenic cells of these phages are given in Fig. 1.

The target sites recognized by the *B. subtilis* R/M systems and of phage-encoded MTases are listed in Tables 1 and 2. Following partial or complete purification of a new ENase, targets were frequently first determined indirectly by comparing restriction patterns generated by this new enzyme with that obtained with ENases of known specificities. Cleavage sites were then often verified by biochemical methods. By analogy, methylation sites within the targets could be predicted by differential treatment of DNA modified by the MTases investigated with restriction ENases whose sensitivity to methylation had been established. Biochemical analyses were frequently applied to confirm such indirect predictions. Of particular use in the identification of locations of 5mC residues were Maxam-Gilbert-type sequence analyses with relevant fragments of 5mC containing DNA (99, 108). The hydrazine-mediated cleavage, which is specific for C (or N4mC), is strongly reduced in sequences containing 5mC (73), causing a gap in the sequence ladder.

Methylation of C to produce 5mC is the most prominent modification type in *B. subtilis* (Table 1). A–N-6 methylation caused by M.*Bsu*BI (Table 1; 116) represents the only other type of methylation described for *B. subtilis* or in *B. subtilis* phages. C–N-4 methylation has not been reported in *B. subtilis*. It is, however, the modification produced by M.*Bam*HI (16) and possibly also by M.H2II (16, 23).

MTases encoded by various temperate bacteriophages of *B. subtilis* and *B. amyloliquefaciens* are expressed only following induction of the prophage. They represent "early" functions. Expression of these genes leads to self-methylation of the induced phage DNA and of other DNAs residing in or invading the induced cells (1, 117, 118).

No site-specific ENases or MTases functionally unrelated to R/M systems, e.g., the Mrr, McrA, and McrBC restriction activities (37, 69, 77), or M.*Dam* or M.*Dcm* (65, 67) modification activities of *E. coli* have been identified in *B. subtilis* so far. The presence of a C-5 MTase associated with the state of competence has been reported by Ganesan (30). This activity was also observed in SPβ-cured *B. subtilis* cells and is therefore not encoded by this temperate phage (29a).

PHAGE-R/M SYSTEM INTERACTION

The interactions of bacteriophages and their hosts, analyzed according to the categories of bacterial R/M systems and the mechanisms that bacteriophages have evolved to overcome restriction, represent an interesting case of coevolution of two competing organisms. A review by Krüger and Bickle (54) describes this subject.

B. subtilis phages, like those of *E. coli* and of *Salmonella* and *Shigella* spp., have developed several mechanisms that alone or in combination permit them to overcome barriers set by R/M systems. A mechanism observed in *B. subtilis* and *B. amyloliquefaciens* phages is self-methylation by phage-encoded MTases. To be efficient, self-methylation requires that the methylation specificity encoded by the phage modify the DNA target recognized by the resident ENase. Such a direct relationship obtains in the case of *B. amyloliquefaciens* phage H2, which encodes an MTase, M.H2II, that methylates the target sequence GGATCC of the host's (*Bam*HI) R/M system (23) (Table 1). The methylation potential of the ϕ3T- and $\rho 11_s$-encoded monospecific MTases affecting TCGA (99) could also overcome the restriction potential of the *Bsu*MI system (PyTCGAPu) (10, 47). Similarly, the methylation specificities of many temperate *B. subtilis* phages directed against the sequence GGCC relates to a matching R/M modification system (*Bsu*RI). With the exception of the target of the *Bsu*FI system (CCGG), which is also recognized by M.SPR, no other methylation specificity identified in the phage MTases corresponds to known host R/M specificities. It might very well be that these methylation potentials are directed against phage-sensitive bacteria encoding unidentified R/M systems with matching specificities. This would in turn imply that the genes or domains encoding such MTases represent dispensable genetic information under laboratory conditions. The *B. subtilis* phage Z (38) demonstrates the dispensability of such genetic information, as it carries a natural deletion eliminating an MTase gene (70, 101). Other phages in which mutations destroying methylation activity have been introduced are also not affected in their growth on various *B. subtilis* strains in comparison with methylation-proficient phages (110). Apparently, there is no selective pressure on the phage genome to dispose of dispensable DNA. This conclusion is further supported by the recent discovery of inert target-recognizing "pseudodomains" that differ from active domains by only a few amino acid exchanges (56, 57). Self-methylation has also been observed in phages of other species, e.g., *E. coli* phage T1, which encodes an MTase modifying the M.*Dam* target site GATC (82). Here, the function of methylation is unknown. It is, however, not used to provide protection from restriction (82, 84). A similar situation holds for phages T2 and T4, which encode GATC-specific MTases (36, 83).

The presence of unusual bases in phage DNA has also been shown to provide resistance to restriction. Examples of such *B. subtilis* phages are SPO1, SP82G, H1, SP8, and ϕe, whose DNAs contain hydroxymethyl uracil (HMU) instead of thymine (7, 39), and PBS2, whose thymine is replaced by uracil (8). In vitro, such DNAs are generally cleaved less rapidly than the corresponding thymine-containing DNAs by restriction ENases whose target sequences have A/T pairs (8, 25), whereas cleavage by restriction enzymes recognizing only G/C-containing targets is not affected. These rules alone, however, do not necessarily explain in vivo phenomena. Phage SPO1, e.g., whose DNA contains five GGCC cleavage sites (61), is neither restricted nor modified following infection of hsd_{RI} R^+M^+ cells (104), although such DNA can be both restricted and methylated in vitro (78). It has not been clarified whether this insensitivity in vivo can be attributed to a phage-encoded activity, which would actively interfere with the functioning of the *Bsu*RI R/M system, or to a more general property of HMU-containing DNA, which would make it less accessible to R/M enzymes.

The best-studied cases of *B. subtilis* phage-encoded inhibitors of restriction are described by Makino et al. (63, 64) for phages ϕNR2 and ϕ1. In the case of ϕ1, they discovered a restriction-resistant phage mutant, ϕ1rH, that, after infection, produces an inhibitor protein of 20 kDa that specifically and reversibly binds to R.*Bam*N but neither to the isoschizomeric ENase R.*Ava*II nor to any other type II restriction enzyme. Makino et al. (64) found that phage ϕ1rH DNA recovered from *Bam*Nx-proficient cells was as sensitive to R.*Bam*Nx cleavage in vitro as the ϕ1 wild-type DNA, showing that the inhibitor of restriction also interfered with the MTase. From an evolutionary point of view, the observations of Makino et al. (63, 64) indicate that wild-type phages ϕ1 and ϕNR2 maintain a latent potential to overcome *Bam*Nx restriction that becomes manifest through selection of mutants by

Table 1. R/M systems of *B. subtilis*, *B. amyloliquefaciens*, and some other *Bacillus* strains

Organism and system (reference)	Genomic location (reference)[a]	DNA target[b] (reference[s])	Prototype R/M system	Phages (reference[s])	
				Restricted	Nonrestricted
B. subtilis					
*Bsu*MI hsd_{MI} = *nonB* (80)	47 (80)	Py ↓ TĊGAPu (10, 31, 47)	*Xho*I-*Asu*II	φ105c (85)	φ29, SPP1 (85), φ2 (46) SP18 (114), PZE (28)
				SP10, φNR2 (80)	φ15, PZA plate only on Spo⁻ cells (28); SP10, φNR2 plate on *nonA nonB* (hsd_{MI} R^- M^-) 168 strains (80)
*Bsu*RI hsd_{RI}	59 (43)	GG ↓ ĊC (13, 33, 34, 50)	*Hae*III	SPP1, φ105c, SP02 (104); SP8 (33); Z (71)	φ29, SP82, SP50, H1 (12); PBS1, SP01 (104); SPR, φ3T, $\rho11_B$, $\rho11_s$, SPβ, H2 (71, 109)
*Bsu*FI hsd_{FI}	59 (43)	Ċ ↓ CGG (48, 108)	*Msp*I	φ105c (85), SPP1 (48)	
*Bsu*CI hsd_{CI}	59 (43)	ND	ND	φ105c (85)	
*Bsu*EI hsd_{EI}	337 (43)	ĊGCG	*Bst*UI	φ105c (85), SPP1 (47)	
*Bsu*BI hsd_{BI}	345 (43)	CTGCȦ ↓ G (88, 116)	*Pst*I	φ105c (85)	SPP1 (116)
B. amyloliquefaciens					
*Bam*HI	ND	G ↓ GATĊC (16, 79); modified base is N4mC (16)	*Bam*HI	φ105c (85)	φ1 (63)
*Bam*NI	ND	G ↓ GATCC (87)	*Bam*HI	φ105c (85, 86), φ1 (63), φNR2 (63)	φ29, SPP1 (85)
*Bam*Nx	ND	G ↓ GACC (41)	*Ava*II	φ105c (85, 86), φ1 (63), φNR2 (63)	φ1rH (64) φNR2rH (63)
B. aneurinolyticus					
*Ban*I	ND	G ↓ GPyPuCC (95)	*Ban*I		
B. sphaericus					
*Bsp*RI	ND	GG ↓ ĊC (53)	*Hae*III		

[a] Coordinates are taken from reference 74. ND, not done.
[b] Including cleavage point (↓) and methylated base (˙).
[c] This description follows the convention of Wilson (113). Arrows give directions of transcription. Filled, open, and hatched symbols represent ENase and MTase genes and regulatory sequences. Numbers of amino acids of the restriction and modification gene products are indicated above the genes. The modification type generated by the MTases is indicated below the gene.

exposure to selective pressure following infection of R.*Bam*Nx-proficient cells.

Finally, sequence biases in phage DNA against targets recognized by restriction enzymes represent a widespread way of overcoming host restriction. Thus, phages φ29 and SPP1 do not contain target sites of the *Bsu*MI system in their DNA and therefore plate with equal efficiency on *B. subtilis* 168 strains containing or lacking the *Bsu*MI system. Additionally, the absence of the sequence GGCC in φ29 DNA makes the phage insensitive to *Bsu*RI restriction (45). It is of interest that the *Bsu*RI system is the one against which most antirestriction activities of phages are directed. When this activity is generated by an absence or paucity of GGCC sites in the phage DNAs, it represents a remarkable sequence bias. For example, phage SPO1 DNA has only 5 GGCC sites (and only 28 CCGG sites, the potential targets of the *Bsu*FI system) (61), with about 300 sites of each anticipated on the basis of a random base distribution. Similarly, 23 GGCC sites were anticipated in φ29 DNA (45). An example of the effects of a combination of sequence bias and the presence of unusual bases is provided by phage H1, which is not restricted by *Bsu*RI (12). The DNA of this HMU-containing phage is not degraded by the *Bsu*RI ENase in vitro under standard conditions of *Bsu*RI restriction. Only after the use of nonstandard conditions could 4 of an anticipated 250 sites be detected. Replacing HMU by T in a DNA fragment containing one of these four sites rendered the fragment sensitive to restriction under all conditions. HMU-T substitutions in 11% of the H2 genome occurring pari passu with shotgun cloning did not reveal additional *Bsu*RI sites.

Table 1—*Continued*

M_r (kDa) of enzymes derived from sequence of (reference):		Gene arrangement[c]	Comments
Enase	MTase		
ND	ND	ND	*nonA* is located at coordinate 12 (80).
66.314 (51)	49.642 (51)	576 436 5mC	hsd_{RI}, hsd_{FI}, and hsd_{CI} are allelic (42, 108). *Bsu*RI ENase is the only restriction ENase active as a monomer.
45.546 (48)	46.918 (108)	395 409 5mC	
ND	ND	ND	
ND	42.000 (29)	ND	M_r of the MTase was determined biochemically (gel filtration).
36.174 (116)	57.179 (116)	501 316 N6mA	M.*Bsu*BI is the only MTase in *B. subtilis* that has been shown to be an A–N-6 MTase. Modification and restriction genes overlap by 17 bp.
24.570 (16)	49.547 (16.106)	213 423 N4mC	ENase and Mtase genes are separated by open reading frame encoding 102-amino-acid protein that is involved in regulation of expression of these genes (16, 68).
ND	ND	ND	Two activities described were observed within one strain, *B. amyloliquefaciens* N, which was later termed *B. subtilis* N (86, 87). Phage mutants of ϕ1 and ϕNR2 overcome *Bam*Nx restriction (63, 64).
ND	ND	ND	
39.841 (62)	42.637 (62)	354 428 5mC	
ND	48.264 (76)	424 5mC	

EFFECTS OF R/M SYSTEMS ON DNA PROCESSING FOLLOWING UPTAKE

The most obvious effect of R/M systems is the restriction of phage growth and the transient acquisition of a "modification" by phages surviving such an infection. The latter extends the host range of phages to include the original R/M cells. Biochemically, these biological phenomena could be attributed to the concerted action of site-specific restriction ENases and modification MTases on DNA. A relevant biological question, particularly with an organism like *B. subtilis*, with its many genetic exchange processes involving cells or subcellular structures in various phases, is how these processes are affected by R/M systems. On the other hand, the stringent requirement of the enzymes of R/M systems for double-stranded DNA has permitted conclusions to be drawn about the mechanism of processing restriction-sensitive DNA when DNA transport phenomena in R/M-proficient and -deficient strains were compared. Altogether, such studies have pointed to the prominent role played in the processing of "outside" DNA by sequence homology between donor and recipient DNAs (11). In Table 3, we summarize the effects of R/M on various DNA transport processes and provide information on the substrate nature of incoming DNA, the requirement for *rec* proficiency of the recipient cell, and the number of DNA molecules required for the biological process under study.

In R/M-proficient cells, infection by bacteriophage, transduction of chromosomal or plasmid DNA, and transformations of protoplasts derived from such cells are sensitive to restriction provided no antirestriction mechanisms are operative. In these cases, DNA is taken up as a double-stranded molecule, which is a substrate for ENases. The efficiency of restriction depends on the number of restriction sites present in the DNA taken up. One molecule of DNA is sufficient to give the biological effect.

Transformation of competent cells with homologous chromosomal DNA is insensitive to R/M (104).

Table 2. DNA MTases of *B. subtilis* and *B. amyloliquefaciens* phages

Organism and MTase	Genomic location of prophage (reference)[a]	DNA target(s) (reference[s])[b]	Prototype modification	M_r (kDa) of MTase derived from sequence (reference)	Gene arrangement[c]	Comments
B. subtilis						
M.SPR	210/226 (81)	GGĊC (17) ĊCGG (17, 108) CĊ(A/T)GG (32)	*Hae*III *Msp*I *Eco*RII	49.844 (17)	439 → 5mC	M.SPR has strong immunological cross-reaction with all other multispecific MTases but not with monospecific MTases (102).
M.ϕ3TI M.ρ11$_B$I M.SPβI	190 (111) ND 190 (120)	GGĊC (71) GĊNGC (71)	*Hae*III *Fnu*4HI	50.507 (101)		Nucleotide sequences of genes encoding these MTases are identical (71, 101).
M.ϕ3TII	190 (111)	TĊGA (99)		36.402 (99)	326 → 5mC; 443 → 5mC	Structural gene of M.ϕ3TII is located 84 bp upstream of M.ϕ3TI gene. M.ϕ3TII sequence has significant similarity to that of M.*Taq*I, which is an A–N-6 MTase.
M.ρ11$_s$I	ND	GGĊC (5) G C GAGCAC (5) T T	*Hae*III *Bsp* 1286I	57.166 (5)	326 → 5mC; 503 → 5mC	M.ρ11$_s$I carries two pseudodomains that can be activated by mutagenesis to give the enzyme capacities to methylate *Eco*RII and/or *Fnu*4HI targets (56, 57).
M.ρ11$_s$II	ND	TĊGA (99)		36.402 (99)		M.ρ11$_s$II is identical to M.ϕ3TII.
B. amyloliquefaciens						
M.H2I	ND[d]	GGĊC (57, 121) GĊNGC (57, 121) G C GAGCAC (57, 121) T T	*Hae*III *Fnu*4HI *Bsp* 1286I	57.166 (57)	503 → 5mC	M.H2I carries a pseudodomain that can be activated by mutagenesis to give the enzyme capacity to methylate *Eco*RII target (56, 57).
M.H2II	ND[d]	GGATCC (23)	*Bam*HI	30.979 (23)	265 → N4mC	Unlike genes of M.ϕ3TI and M.ϕ3TII, genes of M.H2I and M.H2II are not adjacent to each other.

[a] ND, not done.
[b] Dot indicates methylated base.
[c] Directions of transcription and numbers of amino acids are indicated.
[d] When transferred to *B. subtilis*, phage H2 becomes integrated near coordinate 199 (121).

Table 3. Effect of restriction and modification on processing of invading DNA in *B. subtilis*

Process(es)	Mode of processing[a]	Response to restriction[b]	No. of molecules needed	Dependence on RecE[c]	Comments
Phage infection, transduction, protoplast transformation	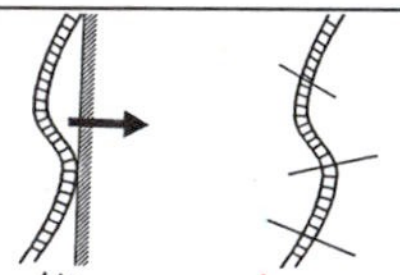	S	1	I	
Transformation with homologous DNA	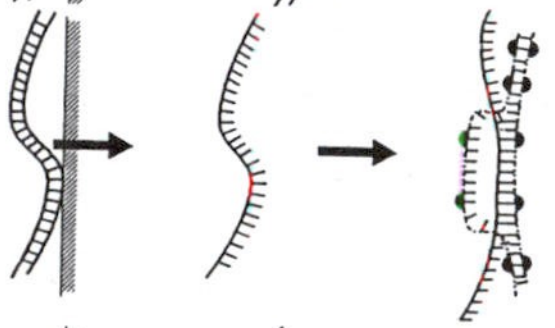	R	1	D	
Transfection	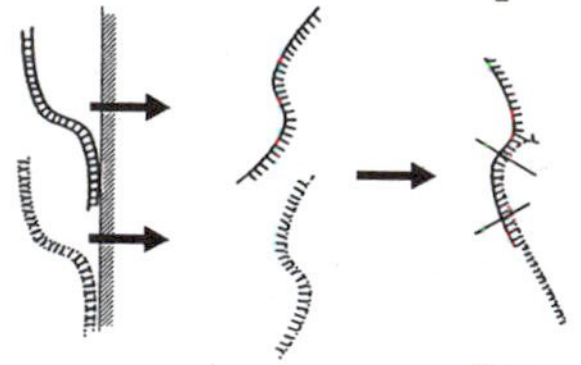	S	≥2	I	
Transfection under marker rescue conditions	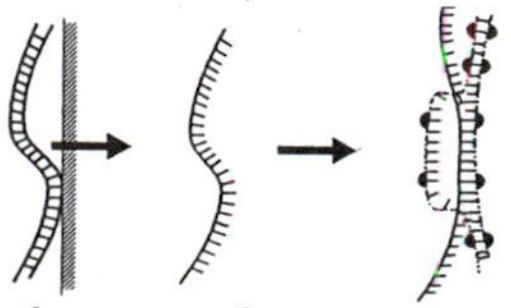	R	1	?	Two components contribute to insensitivity of this process to restriction: (i) small target size of DNA required for marker rescue transfection; (ii) synapsis between resident (modified) and transfected DNA.
Plasmid transformation No homology of plasmid DNA to resident DNA	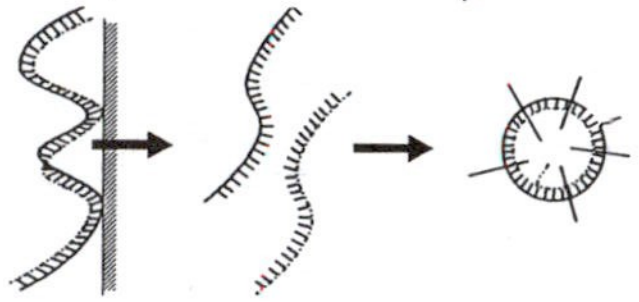	S	>2	I	In *B. subtilis*, requirement for presence of more than one genome in transformation is provided by uptake of different components of concatemeric DNA complex.
Homology to resident plasmid	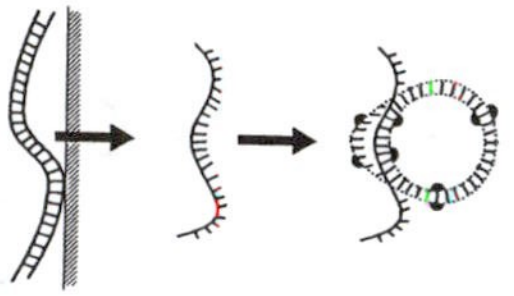	R	1	?	Arguments for insensitivity of this process to restriction are analogous to those described for transfection under marker rescue conditions
Monomeric plasmid transformation (partial homology to chromosome)	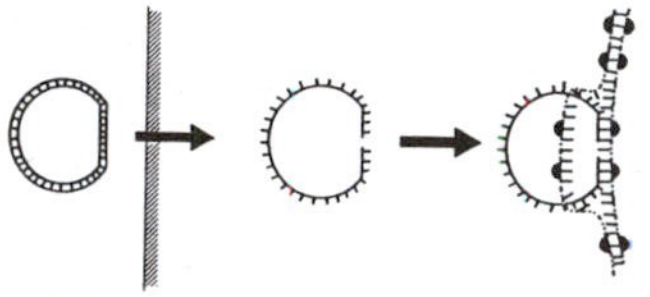	R/S (see comments)	1	D	Only synaptic structure of donor/resident DNA is shown. Restriction is observed only when restriction sites are present in nonsynapsed plasmid portion. Under Rec^- conditions also, restriction sites within plasmid part, which is homologous to chromosome, contribute to sensitivity to restriction.

[a] Half-circles represent methyl groups; lines across DNA strands represent targets for ENase action.
[b] S, sensitive; R, resistant.
[c] I, independent; D, dependent.

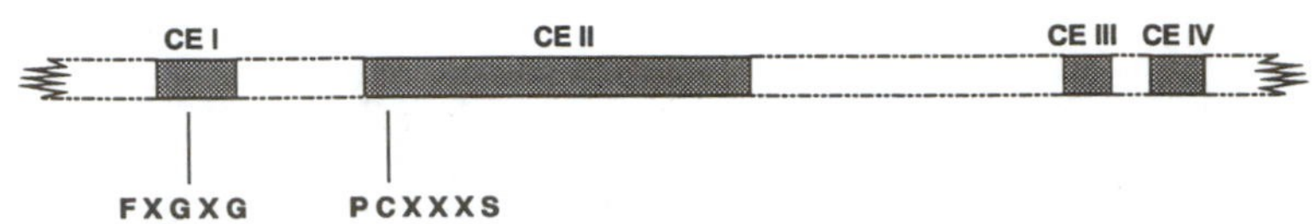

Figure 2. Building plan of C-5 MTases. The primary sequence of a typical C-5 MTase is shown as a double line. Four conserved elements (CEs I to IV) identifiable in all C-5 MTases (60) are indicated as shaded boxes. CEs I and II contain the amino acid motifs F-X-G-X-G and P-C-X-X-X-S, thought to be involved in adenosylmethionine binding and methyl group transfer. The CEs are flanked and spaced by regions variable in size (indicated by tattered ends and dotted lines, respectively) and amino acid composition. TRDs have been allocated to the variable region separating CEs II and III.

Irrespective of the R/M proficiency of recipient cells, transformation is *rec* dependent. The DNA is converted during uptake to a single-stranded form (2, 26), which is a poor substrate for ENases. The donor DNA synapses with homologous, modified recipient DNA to form a segment of hemimethylated DNA within the chromosome, which would be both resistant to restriction and rapidly converted to fully methylated DNA by the preexisting modification MTase (11). For single marker transformation, the relationship between the number of transformants obtained and the concentration of transforming DNA is linear, indicating that one DNA molecule is sufficient to produce a transformant.

Transfection of competent R/M cells is sensitive to restriction (104). As in transformation, the transfecting DNA is also converted during uptake to a single-stranded form, which would reduce the sensitivity of the DNA to ENases. However, such DNA must obligatorily hybridize to a complementary phage DNA strand provided by a second, independent uptake event of phage DNA. The newly formed double-stranded DNA, composed of complementary nonmodified DNA, is a substrate for restriction. Transfection with nonmodified DNA, performed under conditions of marker rescue, is analogous to transformation by homologous chromosomal DNA and hence is insensitive to restriction. The frequency of transfection under these conditions depends linearly on the concentration of DNA (102). Similarly, transfection with nonmodified phage DNA of lysogenic cells having the same DNA as a prophage is insensitive to restriction (11).

Plasmid tranformation in *B. subtilis* requires multimeric DNA molecules (20). Again, complementary single DNA strands, which in this case derive from separated uptake events to include different portions of the multimer, have to synapse to form a double-stranded replicon (27). Hence, plasmid transformation is sensitive to restriction (10, 19). Analogous to the marker rescue situation described for transfection, plasmid transformation of R/M-proficient cells carrying appropriately marked identical plasmids is not sensitive to restriction (24).

Canosi et al. (18) have reported that inserting fragments of chromosomal *B. subtilis* DNA into plasmid DNA also facilitated the transformation of monomeric plasmid DNA. The strong dependence of this process on recombination proficiency of the competent cells suggested that in this situation, monomeric single-stranded molecules linearized within the region of homology would synapse with their chromosomal inserts at the homologous chromosomal locations. Subsequent DNA synthesis would then convert the single-stranded synapsed plasmid molecules to double-stranded replicons (6, 40). The extent of restriction in such a constellation is determined only by the number of restriction sites present in the plasmid moiety, which cannot synapse with the bacterial chromosome (18). These are the sites that DNA synthesis converts to a double-stranded nonmodified DNA. Plasmid restriction sites within the synapsed region are in a hemimethylated configuration that, as in transformation, is not accessible to restriction. If synapsis is prevented by the use of a *rec*-deficient recipient cell that is proficient in R/M, transformation with such plasmids is drastically reduced, since all restriction sites, including those that would be protected by synapsis formation in the *rec*$^+$ situation, could then be recognized.

SEQUENCE COMPARISONS: IMPLICATIONS FOR ENZYME FUNCTION AND EVOLUTION

In eight *Bacillus* R/M systems (*Bam*HI, *Ban*I, *Bcn*I, *Bgl*I, *Bsp* 6I, *Bsu*BI, *Bsu*FI, and *Bsu*RI), the sequences of both the restriction and the modification genes have been established. Since the results of sequence analyses of the *Bcn*I, *Bgl*I, and *Bsp* 6I R/M genes, summarized by Wilson (113), have not yet been published in detail, these systems will not be treated here (Table 1). For the *Bsp*RI R/M system, only the modification gene has been sequenced. In addition, the sequences of six mono- or multispecific MTases encoded by a group of related temperate *B. subtilis* and *B. amyloliquefaciens* phages have been reported. The modification genes of these phages are solitary; i.e., they are not accompanied by restriction genes (Table 2).

Intramolecular Comparisons

In the few cases analyzed (58), short repetitive sequence motifs have been recognized within the primary sequences of MTases encoded by *B. subtilis* and its phages. The presence and spacing of these motifs support the idea that gene duplication (several rounds) has occurred in the evolution of MTases (58). Similarly, single or multiple sequence duplications have also occurred in the evolution of at least some *B. subtilis* ENases, including R.*Bsu*FI and R.*Bsu*RI (14, 48, 51, 58).

Intermolecular Comparisons

Comparison of ENases and MTases of individual R/M systems

In general, ENases and MTases of the same type II R/M system, including those of *Bacillus* spp., do not show significant similarities in their primary sequences

(16, 22, 48, 51, 62, 116). This suggests that ENases and MTases, although recognizing the same DNA sequences with high selectivity, have evolved independently and use different strategies to interact with identical targets. This is probably a reflection of the different reaction types catalyzed by these enzymes.

Analysis of C-5 MTases

As described above (see also Tables 1 and 2), all kinds of type II MTases have been identified in *Bacillus* spp. Since C-5 MTases and C–N-4/A–N-6 MTases show considerable differences with respect to biochemistry and evolution, we will discuss them separately.

C-5 MTases, the only class of MTases also widespread among eukaryotes, have very similar primary sequences (Fig. 2). In all C-5 MTases, including those encoded by *Bacillus* species and their phages, at least 4 conserved elements (CEs) can be identified (60) (the description of 6 [55, 91] or 10 [75] CEs mainly results from further subdivision of these 4 elements). The four CEs whose sequential arrangements are identical in all C-5 MTases are flanked and interrupted by short segments that are variable in length and amino acid composition (60, 75). Some of these variable elements may show similarities in pairwise comparisons, indicating a relatedness of the corresponding C-5 MTases beyond that generally observed among C-5 MTases.

The preservation of CEs in all C-5 MTases suggests that CEs are responsible for functional or structural properties that are common to all C-5 MTases. Two such functions, binding of *S*-adenosylmethionine and covalent binding of MTases to the C-6 of the cytosine to be methylated, are apparently mediated by distinct amino acid motifs. These are F-X-G-X-G for *S*-adenosylmethionine binding and P-C-X-X-X-S for cytosine–C-6 binding (44, 52, 60, 75, 115). These motifs are part of CEs I and II, respectively. No distinct functions have yet been attributed to CEs III and IV. Uniformly, the largest and most heterogeneous variable region among MTases with different DNA target specificities is that separating CEs II and III (60, 75). This region has been proposed to carry the enzymes' target-recognizing domains (TRDs), which are responsible for the enzymes' characteristic capacities to interact with defined DNA sequences (60, 75). Experimental proof for this assumption was provided in the case of the multispecific C-5 MTases encoded by various temperate *B. subtilis* phages (4, 102, 103, 110). From a comparison of several TRDs identified in these MTases, a consensus sequence characterizing TRDs could be derived. Similarity to this consensus sequence was also found in the variable region separating CEs II and III of monospecific MTases (60), supporting the concept that this region is generally responsible for DNA target recognition by C-5 MTases. Experimental evidence for this assumption has recently been provided for the monospecific MTases M.*Hpa*II and M.*Hha*I (51a).

The multispecificity of the phage-encoded C-5 MTases is due to the presence of several TRDs within this region (110). These TRDs represent contiguous segments consisting of about 40 amino acids each. They are arranged consecutively, do not overlap, and are not separated by one or more "linker" amino acids (102). TRDs can be rearranged within and exchanged among multispecific phage C-5 MTases without functional loss (107). They represent functionally independent "modules" whose activities do not depend on a particular sequence context. Some multispecific phage MTases (M.$\rho11_s$I and M.H2I) carry elements that resemble active TRDs of other MTases but do not contribute to the enzymes' methylation potentials (56, 57). These inert TRDs can be functionally activated by directed mutagenesis (56). The maintenance of inactive TRDs in the absence of positive selective pressure may have played a role in development of the multispecificity of phage enzymes.

Comparing the putative TRD regions of the monospecific enzyme M.*Bsu*FI with those of M.*Msp*I and M.*Hpa*II, which all recognize CCGG sites (21, 108), showed that the three MTases share a highly conserved enzyme core. However, extensive similarity of the TRDs is found only between M.*Bsu*FI and M.*Msp*I (75% amino acid identity), which both modify the outer cytosine residue of the CCGG target. The TRD region of M.*Hpa*II, which methylates the inner cytosine of the CCGG sequence, is substantially different from those present in M.*Bsu*FI and M.*Msp*I. This suggests that TRDs serve not only in recognizing a target but also in directing methylation to a defined cytosine residue within the target when the target contains several cytosines.

Comparative analyses have also revealed interesting aspects concerning the phylogeny of C-5 MTases encoded by *Bacillus* species and their phages.

(i) There is no general correlation between the degree of sequence similarity of C-5 MTases and the phylogenetic distance between the organisms encoding them. In some cases, both parameters correlate, e.g., in the case of the isomethylomeric enzymes M.*Bsu*RI and M.*Bsp*RI (about 84% amino acid identity) (51, 60) encoded by the related species *B. subtilis* and *B. sphaericus* or in the case of the multispecific C-5 MTases (up to 96% amino acid identity) (57) encoded by a related group of temperate *B. subtilis* and *B. amyloliquefaciens* phages. M.*Bsu*FI, on the other hand, has more similarity to C-5 MTases encoded by gram-negative bacteria (M.*Msp*I, M.*Hpa*II, M.*Eco*RII, and M.*Hha*I) than to the MTase of the allelic *Bsu*RI R/M system or to the multispecific MTases of *B. subtilis* phages (108). The extensive similarity of the core sequence of M.*Bsu*FI to those of M.*Msp*I and M.*Hpa*II (56 and 45% amino acid and 62 and 51% nucleotide identities, respectively) (108) suggests that the M.*Bsu*FI, M.*Msp*I, and M.*Hpa*II genes derive from a common ancestor that might have been distributed by horizontal gene transfer among different bacterial species.

(ii) The recently identified monospecific C-5 MTase M.ϕ3T II, which recognizes the target TCGA, not only exhibits motifs characteristic for C-5 MTases but also shows significant similarities to the A–N-6 MTase M.*Taq*I (99). M.*Taq*I, encoded by the gram-negative species *Thermus aquaticus*, also recognizes TCGA targets but modifies adenine residues of this sequence. The similarities of M.ϕ3TII and M.*Taq*I suggest that C-5 MTases and A–N-6 MTases, although in general quite different (see below), have a common evolutionary origin (58).

(iii) There is no particularly pronounced sequence similarity between multispecific C-5 MTases encoded by

Bacillus phages and monospecific C-5 MTases encoded by their hosts. Thus, M.SPR is more similar to an enzyme from a gram-negative organism (M.*Eco*RII; about 48% amino acid homology) than to an isomethylomeric enzyme of its gram-positive host (M.*Bsu*RI; about 37% amino acid homology) (60). This finding suggests that the phage-borne genes encoding multispecific C-5 MTases are not directly derived from host genes encoding C-5 MTases, in contrast to the tight phage-host interrelationship demonstrated by the close relatedness of the thymidylate-synthetase genes carried by some of the phages discussed here and the *B. subtilis thyA* gene (93, 94).

(iv) The highly related multispecific MTases encoded by *Bacillus* phages can be subdivided into three groups. Group I consists of the apparently identical MTases M.ϕ3TI, M.SPβ, and M.$\rho 11_B$ (70, 101). This group is distinguished from the other multispecific MTases by a 33-amino-acid "insert" within the conserved N terminus whose sequence resembles a protein zinc finger (101). Group II consists of M.$\rho 11_s$I and M.H2I, whose core and TRD sequences are nearly identical (96% overall amino acid identity) (57). The only major functional difference between these two enzymes, the capacity (M.H2I) or inability (M.$\rho 11_s$I) to methylate GCNGC sites, could be attributed to a single amino acid exchange (56) defining a pseudodomain in M.$\rho 11_s$I. Group III is represented by M.SPR, the only multispecific MTase with a CCGG-specific TRD. Although the three groups show differences in the degree of similarity of their conserved sequences (unpublished results), deriving a phylogenetic tree would be highly speculative in view of the extensive ability of the phages encoding these enzymes to recombine with each other (71, 100, 119).

Analysis of A–N-6 and C–N-4 MTases

In general, A–N-6 and C–N-4 MTases show much higher sequence diversity among themselves than do C-5 MTases. In overall comparisons (22, 52, 58, 59, 89), only two motifs were revealed to be common to all A–N-6 MTases. One resembles the F-X-G-X-G motif also found in C-5 MTases and other *S*-adenosylmethionine-binding proteins (44). The second motif, (ND)-P-P-(YF), may be involved in the transfer of methyl groups to adenine residues of the target sequence (22, 52). Primary sequences of C–N-4 MTases show similarities with A–N-6 MTases (16, 52, 98). These include the F-X-G-X-G motif and a sequence similar to the (ND)-P-P-(YF) motif of A–N-6 MTases that is presumably responsible for methyl group transfer (98). No homologies with C-5 MTases could be detected.

The only A–N-6 MTase so far identified in *B. subtilis* is M.*Bsu*BI. M.*Bsu*BI is an isomethylomer of M.*Pst*I that modifies adenine residues of CTGCAG target sites (116). The sequences of the two MTases (M.*Bsu*BI, 501 amino acids; M.*Pst*I, 507 amino acids) can be aligned throughout their entire lengths, showing about 40% amino acid identity (116). The similarity of the two enzymes, which is also reflected at the level of the corresponding genes (58% nucleotide identity), suggests a common ancestor for the M.*Bsu*BI and M.*Pst*I genes that might have been horizontally transmitted between gram-positive and gram-negative species.

The *B. amyloliquefaciens*-encoded enzyme M.*Bam*HI, which recognizes the target GGATCC, is a C–N-4 MTase (16). Comparison of its sequence revealed extensive similarities prominent in five regions, with two A–N-6 MTases (M.*Hin*fI and M.*Dpn*A) and two C–N-4 MTases (M.*Cfr*9I and M.*Pvu*II) (16). This similarity also transgresses the phylogenetic separation into gram-positive and gram-negative bacteria. Conservation of the same amino acid sequences, although not as extensive as observed with the bacterial MTases, were also detected when M.*Bam*HI was compared with M.H2II, an MTase encoded by *B. amyloliquefaciens* phage H2 (16). Because of the presence in its sequence of the S-P-P-Y motif, which has been identified in several enzymes of this class (52), this enzyme is also presumed to be a C–N-4 MTase (16, 23). Furthermore, M.H2II methylation protects phage H2 against *Bam*HI restriction (109), which would not be mediated by methylation of the adenine residue of the *Bam*HI target site GGATCC (66). Consideration of the overall similarities of the enzymes discussed here suggests that, as observed in with phage- and host-encoded *B. subtilis* C-5 MTases (see above), the C–N-4 MTases encoded by *B. amyloliquefaciens* phage and host genes are not directly derived from each other.

Comparison of ENases

The primary sequences of ENases, including those of *B. subtilis* and *B. amyloliquefaciens*, are in general much more diverse than those of MTases (22, 58). They do not share any generally conserved sequence motif. This applies to both hetero- and isoschizomeric enzymes. However, in comparisons of a few pairs of isoschizomers, extensive similarities have been observed (48, 92, 116). Interestingly, the members of such pairs of enzymes may derive from gram-positive (*B. subtilis*) and gram-negative (*Providencia stuartii* or *Moraxella* sp.) bacteria. Thus, the sequence of R.*Bsu*BI, consisting of 316 amino acids, can be aligned over its entire length with that of R.*Pst*I (326 amino acids), revealing some 42% overall amino acid identity (116). R.*Msp*I (262 amino acids) can be aligned over its entire length with the COOH part of R.*Bsu*FI (395 amino acids) (48). This alignment reveals 45% overall similarity. Intramolecular comparisons of R.*Bsu*FI suggested that its larger size derived from duplication of a limited gene segment (48). Since the MTases of the pairs of the *Bsu*BI-*Pst*I and *Bsu*FI-*Msp*I R/M systems are also highly related, the complete R/M systems apparently have a direct evolutionary connection. This situation is different from that of the isospecific R/M pair *Eco*RI-*Rsr*I, where similarity is observed only between the ENases (49, 92).

COMPARISON OF GENE ARRANGEMENTS

Arrangement of Bacterial Restriction and Modification Genes

Genes encoding MTases and ENases of one and the same type II R/M system are uniformly located close to each other (112, 113). Beyond this generality, however, no further conservation in the arrangement of restriction and modification genes can be observed. Most frequently, genes are transcribed codirectionally. In other cases, restriction and modification genes show opposite orientations that lead to either divergent or convergent transcription (113).

The closely linked restriction and modification genes of *B. subtilis* and *B. amyloliquefaciens* type II R/M systems also show this diversity of arrangement (Table 1). Within systems transcribed codirectionally, the 3′ end of the *Bsu*BI modification gene shows a 17-bp overlap with the 5′ end of the restriction gene (116; Table 1). The high frequency of overlaps observed with codirectionally transcribed type II restriction and modification genes (113) indicates that such overlaps are of functional importance, e.g., in regulating the coordinated expression of these genes. On the other extreme is the *Bsu*RI system, in which the restriction and modification genes are separated by a spacer of exceptional length (780 bp) (51). These genes are transcribed by their own promoters (51).

The great similarity of primary sequences of the *Msp*I-*Bsu*FI or *Pst*I-*Bsu*BI R/M systems suggested that their gene pairs derived from common ancestors (see above). However, during the establishment of these R/M systems in the gram-positive or gram-negative background, their composite genes came into different transcriptional orientations. Thus the *Msp*I R/M genes are transcribed divergently, while the *Bsu*FI R/M genes are oriented convergently (48). The *Pst*I R/M genes are transcribed divergently, while the *Bsu*BI R/M genes are transcribed codirectionally, with the modification gene overlapping the restriction gene (116). The different arrangements of the *Bsu*FI-*Msp*I and *Bsu*BI-*Pst*I R/M genes reflect either independent integration of their composite genes into the different host chromosomes or rearrangements after transmission of the complete set of ancestral restriction and modification genes. Irrespective of their orientations, the genes have been maintained in close proximity. Such close linkage is apparently of biological significance, perhaps in the transfer of type II R/M systems to other cells or in the regulation of the activity of these systems.

At this time, little is known about the regulation of type II R/M systems. Recently, however, a related family of genes whose products are apparently involved in the regulation of type II R/M systems with either divergent or convergent transcription has been described. These genes, termed "C" for "controller," have been found in association with the *Bam*HI system (16, 68) (Table 1) and several systems (*Pvu*II, *Eco*RV, *Eco* 72I, and *Sma*I) encoded by gram-negative species (97, 113). The C genes are uniformly codirectional with the restriction genes which they precede and in most cases overlap (97). The product of the C gene connected with the *Bam*HI R/M system apparently regulates expression of the M.*Bam*HI gene (68). In contrast, in the *Pvu*II system, the C-gene product is required for expression of restricting activity without affecting the expression of the methylation activity (97). While the biological significance of this regulatory potential is still unclear (in the *Pvu*II system it may delay [deleterious] expression of the restriction gene compared with the modification gene after transfer of the *Pvu*II system into new cells [97]), it is most likely exerted at the level of transcription. This possibility is suggested by the observation that the C proteins carry a helix-turn-helix motif related to that present in the λ *c*I repressor and several other transcriptional regulators (16, 97).

Arrangement of Modification Genes in Phage Genomes

The *B. subtilis* and *B. amyloliquefaciens* phages ϕ3T, $\rho 11_s$, and H2 code for multispecific (M.ϕ3TI, M.$\rho 11_s$I, and M.H2I) and monospecific (M.ϕ3TII, M.$\rho 11_s$II, and M.H2II) MTases (Table 2). The arrangement of the genes encoding mono- and multispecific MTases is different in the ϕ3T-$\rho 11_s$ and H2 genomes. The genes encoding M.ϕ3TII and M.$\rho 11_s$II directly precede the 5′ flank of the M.ϕ3TI and M.$\rho 11_s$I genes (99). As they are transcribed codirectionally with the M.ϕ3TI and M.$\rho 11_s$I genes, the two MTase genes might actually form an operon (99). In contrast, the M.H2II gene is remote from the M.H2I gene (57). Although the M.$\rho 11_s$ and M.H2I genes are highly similar (97.8% nucleotide identity) (57), the sequence environments in which they are found in their respective genomes are quite different. Therefore, these genes must have been established in their contexts by a mechanism different from recombination of homologous regions bracketing the MTase genes.

Acknowledgments. We thank J. Walter and H. Zabin and other members of our group for helpful comments and assistance during the preparation of this manuscript. We acknowledge thoughtful criticism of the manuscript by S. Zahler, C. Anagnostopoulos, A. Kiss, and an anonymous referee from Cold Spring Harbor, N.Y. Part of the work reported from our laboratory was supported by Deutsche Forschungsgemeinschaft (SFB 344) and the European Community [Contract SCI-CT90-0472 (TSTS)].

REFERENCES

1. **Arwert, F., and L. Rutberg.** 1974. Restriction and modification in *Bacillus subtilis*. Induction of a modifying activity in *Bacillus subtilis* 168. *Mol. Gen. Genet.* **133:** 175–177.
2. **Arwert, F., and G. Venema.** 1973. Transformation in *Bacillus subtilis*. Fate of newly introduced transforming DNA. *Mol. Gen. Genet.* **123:**185–198.
3. **Bachmann, B. J.** 1987. Linkage map of *Escherichia coli* K-12, p. 807–876. *In* F. C. Neidhardt, L. Ingraham, K. B. Low, B. Magasanik, M. Schaechter, and H. E. Umbarger (ed.), *Escherichia coli and Salmonella typhimurium: Cellular and Molecular Biology*, vol. 2. American Society for Microbiology, Washington, D.C.
4. **Balganesh, T. S., L. Reiners, R. Lauster, M. Noyer-Weidner, K. Wilke, and T. A. Trautner.** 1987. Construction and use of chimeric SPR/ϕ3T DNA methyltransferases in the definition of sequence recognizing enzyme regions. *EMBO J.* **6:**3543–3549.
5. **Behrens, B., M. Noyer-Weidner, B. Pawlek, R. Lauster, T. S. Balganesh, and T. A. Trautner.** 1987. Organization of multispecific DNA methyltransferases encoded by temperate *Bacillus subtilis* phages. *EMBO J.* **6:**1137–1142.
6. **Bensi, G., A. Iglesias, U. Canosi, and T. A. Trautner.** 1981. Plasmid transformation in *Bacillus subtilis*. The significance of partial homology between plasmid and recipient cell DNAs. *Mol. Gen. Genet.* **184:**400–404.
7. **Berkner, K. L., and W. R. Folk.** 1977. *Eco*RI cleavage and methylation of DNAs containing modified pyrimidines in the recognition sequence. *J. Biol. Chem.* **252:** 3185–3193.
8. **Berkner, K. L., and W. R. Folk.** 1979. The effects of substituted pyrimidines in DNAs on cleavage by sequence-specific endonucleases. *J. Biol. Chem.* **254:**2551–2560.

9. **Bickle, T. A.** 1987. DNA restriction and modification systems, p. 692–696. *In* F. C. Neidhardt, L. Ingraham, K. B. Low, B. Magasanik, M. Schaechter, and H. E. Umbarger (ed.), *Escherichia coli and Salmonella typhimurium: Cellular and Molecular Biology*, vol. 1. American Society for Microbiology, Washington, D.C.
10. **Bron, S., L. Jannière, and S. D. Ehrlich.** 1988. Restriction and modification in *Bacillus subtilis* Marburg 168: target sites and effects on plasmid transformation. *Mol. Gen. Genet.* **211:**186–189.
11. **Bron, S., E. Luxen, and T. A. Trautner.** 1980. Restriction and modification in *B. subtilis*. The role of homology between donor and recipient DNA in transformation and transfection. *Mol. Gen. Genet.* **179:**111–117.
12. **Bron, S., E. Luxen, and G. Venema.** 1983. Resistance of bacteriophage H1 to restriction and modification by *Bacillus subtilis* R. *J. Virol.* **46:**703–708.
13. **Bron, S., and K. Murray.** 1975. Restriction and modification in *B. subtilis*. Nucleotide sequence recognized by restriction endonuclease R.*Bsu*R from strain R. *Mol. Gen. Genet.* **143:**25–33.
14. **Bron, S., K. Murray, and T. A. Trautner.** 1975. Restriction and modification in *B. subtilis*. Purification and general properties of a restriction endonuclease from strain R. *Mol. Gen. Genet.* **143:**13–23.
15. **Brooks, J. E., J. S. Benner, D. F. Heiter, K. S. Silber, L. A. Sznyter, T. Jager-Quinton, L. S. Moran, B. E. Slatko, G. G. Wilson, and D. O. Nwankwo.** 1989. Cloning the *Bam*HI restriction modification system. *Nucleic Acids Res.* **17:**979–997.
16. **Brooks, J. E., P. D. Nathan, D. Landry, L. A. Sznyter, P. Waite-Rees, C. L. Ives, L. S. Moran, B. E. Slatko, and J. S. Benner.** 1991. Characterization of the cloned *Bam*HI restriction modification system: its nucleotide sequence, properties of the methylase, and expression in heterologous hosts. *Nucleic Acids Res.* **19:**841–850.
17. **Buhk, H.-J., B. Behrens, R. Tailor, K. Wilke, J. J. Prada, U. Günthert, M. Noyer-Weidner, S. Jentsch, and T. A. Trautner.** 1984. Restriction and modification in *Bacillus subtilis*: nucleotide sequence, functional organization and product of the DNA methyltransferase gene of bacteriophage SPR. *Gene* **29:**51–61.
18. **Canosi, U., A. Iglesias, and T. A. Trautner.** 1981. Plasmid transformation in *Bacillus subtilis*: effects of insertion of *Bacillus subtilis* DNA into plasmid pC194. *Mol. Gen. Genet.* **181:**434–440.
19. **Canosi, U., G. Lüder, T. A. Trautner, and S. Bron.** 1981. Restriction and modification in *B. subtilis*: effects on plasmid transformation, p. 179–187. *In* M. Polsinelli and G. Mazza (ed.), *Proceedings of the 5th European Meeting on Bacterial Transformation and Transfection*. Cotswold Press, Oxford.
20. **Canosi, U., G. Morelli, and T. A. Trautner.** 1978. The relationship between molecular structure and transformation efficiency of some *S. aureus* plasmids isolated from *B. subtilis*. *Mol. Gen. Genet.* **166:**259–267.
21. **Card, C. O., G. G. Wilson, K. Weule, J. Hasapes, A. Kiss, and R. J. Roberts.** 1990. Cloning and characterization of the *Hpa*II methylase gene. *Nucleic Acids Res.* **18:**1377–1383.
22. **Chandrasegaran, S., and H. O. Smith.** 1987. Amino acid sequence homologies among twenty-five restriction endonucleases and methylases, p. 149–156. *In* R. H. Sarma and M. H. Sarma (ed.), *Structure and Expression*. Adenine Press, Guilderland, N.Y.
23. **Connaughton, J. F., W. D. Kaloss, P. G. Vanek, G. A. Nardone, and J. G. Chirikjian.** 1990. The complete sequence of the *Bacillus amyloliquefaciens* proviral H2, *Bam*HI methylase gene. *Nucleic Acids Res.* **18:**4002.
24. **Contente, S., and D. Dubnau.** 1979. Marker rescue transformation by linear plasmid DNA in *Bacillus subtilis*. *Plasmid* **2:**555–571.
25. **Cregg, J. M., and C. R. Steward.** 1978. *Eco*RI cleavage of DNA from *Bacillus subtilis* phage SPO1. *Virology* **85:**601–605.
26. **Davidoff-Abelson, R., and D. Dubnau.** 1973. Conditions affecting the isolation from transformed cells of *B. subtilis* of high-molecular-weight single-stranded DNA of donor origin. *J. Bacteriol.* **116:**146–153.
27. **deVos, W., G. Venema, U. Canosi, and T. A. Trautner.** 1981. Plasmid transformation in *Bacillus subtilis*: fate of plasmid DNA. *Mol. Gen. Genet.* **181:**424–433.
28. **Fucik, V., H. Grünnerová, and S. Zadrazil.** 1982. Restriction and modification in *Bacillus subtilis* 168. Regulation of *hsr* (*nonB*) expression in *spoA* mutants and effects on permissiveness of ϕ15 and ϕ105 phages. *Mol. Gen. Genet.* **186:**118–121.
29. **Gaido, M. L., C. R. Prostko, and J. S. Strobl.** 1988. Isolation and characterization of *Bsu*E methyltransferase, a CGCG specific DNA methyltransferase from *Bacillus subtilis*. *J. Biol. Chem.* **263:**4832–4836.
29a. **Ganesan, A. T.** Personal communication.
30. **Ganesan, A. T.** 1979. Genetic recombination during transformation in *Bacillus subtilis*: appearance of a deoxyribonucleic acid methylase. *J. Bacteriol.* **139:**270–279.
31. **Guha, S.** 1985. Determination of DNA sequences containing methylcytosine in *Bacillus subtilis* Marburg. *J. Bacteriol.* **163:**573–579.
32. **Günthert, U., and L. Reiners.** 1987. *Bacillus subtilis* phage SPR codes for a DNA methyltransferase with triple sequence specificity. *Nucleic Acids Res.* **15:**3689–3702.
33. **Günthert, U., K. Storm, and R. Bald.** 1978. Restriction and modification in *Bacillus subtilis*. Localization of the methylated nucleotide in the *Bsu*RI recognition sequence. *Eur. J. Biochem.* **90:**581–583.
34. **Günthert, U., J. Stutz, and G. Klotz.** 1975. Restriction and modification in *B. subtilis*. *Mol. Gen. Genet.* **142:**185–191.
35. **Gwinn, D. D., and W. D. Lawton.** 1968. Alteration of host specificity in *Bacillus subtilis*. *Bacteriol. Rev.* **32:**297–301.
36. **Hattman, S.** 1978. Sequence specificity of the wild-type (*dam*$^+$) and mutant (*dam*^{h}) forms of bacteriophage T2 DNA adenine methylase. *J. Mol. Biol.* **119:**361–376.
37. **Heitman, J., and P. Model.** 1987. Site-specific methylases induce the SOS DNA repair response in *Escherichia coli*. *J. Bacteriol.* **169:**3243–3250.
38. **Hemphill, H. E., I. Gage, S. A. Zahler, and R. Z. Korman.** 1980. Prophage-mediated production of a bacteriocinlike substance by SPβ lysogens of *Bacillus subtilis*. *Can. J. Microbiol.* **26:**1328–1333.
39. **Hemphill, H. E., and H. R. Whiteley.** 1975. Bacteriophages of *Bacillus subtilis*. *Bacteriol. Rev.* **39:**257–315.
40. **Iglesias, A., G. Bensi, U. Canosi, and T. A. Trautner.** 1981. Plasmid transformation in *Bacillus subtilis*: alterations introduced into the recipient-homologous DNA of hybrid plasmids can be corrected in transformation. *Mol. Gen. Genet.* **184:**405–409.
41. **Ikawa, S., T. Shibata, and T. Ando.** 1979. Recognition sequence of endonuclease R.*Bam*Nx from *Bacillus amyloliquefaciens* N. *Agric. Biol. Chem.* **43:**873–875.
42. **Ikawa, S., T. Shibata, T. Ando, and H. Saito.** 1980. Genetic studies on site-specific endodeoxyribonucleases in *B. subtilis*: multiple modification and restriction systems in transformants of *Bacillus subtilis* 168. *Mol. Gen. Genet.* **177:**359–368.
43. **Ikawa, S., T. Shibata, K. Matsumoto, T. Iijima, H. Saito, and T. Ando.** 1981. Chromosomal loci of genes controlling site-specific restriction endonucleases of *Bacillus subtilis*. *Mol. Gen. Genet.* **183:**1–6.
44. **Ingrosso, D., A. V. Fowler, J. Bleibaum, and S. Clarke.** 1989. Sequence of the D-aspartyl/L-isoaspartyl protein methyltransferase from human erythrocytes. *J. Biol. Chem.* **264:**20131–20139.

45. **Ito, J., and R. J. Roberts.** 1979. Unusual base sequence arrangements in phage ϕ29 DNA. *Gene* **5:**1–7.
46. **Ito, J., and J. Spizizen.** 1971. Abortive infection of sporulating *Bacillus subtilis* 168 by ϕ2 bacteriophage. *J. Virol.* **7:**515–523.
47. **Jentsch, S.** 1983. Restriction and modification in *Bacillus subtilis*: sequence specificities of restriction/modification systems *Bsu*M, *Bsu*E, and *Bsu*F. *J. Bacteriol.* **156:**800–808.
48. **Kapfer, W., J. Walter, and T. A. Trautner.** 1991. Cloning, characterization and evolution of the *Bsu*FI restriction endonuclease gene of *Bacillus subtilis* and purification of the enzyme. *Nucleic Acids Res.* **19:**6457–6463.
49. **Kaszubska, W., C. Aiken, C. D. O'Connor, and R. I. Gumport.** 1989. Purification, cloning and sequence analysis of *Rsr*I DNA methyltransferase: lack of homology between two enzymes, *Rsr*I and *Eco*RI, that methylate the same nucleotide in identical recognition sequences. *Nucleic Acids Res.* **17:**10403–10425.
50. **Kiss, A., and F. Baldauf.** 1983. Molecular cloning and expression in *Escherichia coli* of two modification methylase genes of *Bacillus subtilis*. *Gene* **21:**111–119.
51. **Kiss, A., G. Posfai, C. C. Keller, P. Venetianer, and R. J. Roberts.** 1985. Nucleotide sequence of the *Bsu*RI restriction-modification system. *Nucleic Acids Res.* **13:** 6403–6421.
51a. **Klimasauskas, S., J. L. Nelson, and R. J. Roberts.** 1991. The sequence specificity domain of cytosine-C5 methylases. *Nucleic Acids Res.* **22:**6183–6190.
52. **Klimasauskas, S., A. Timinskas, S. Menkevicius, D. Butkienè, V. Butkus, and A. Janulaitis.** 1989. Sequence motifs characteristic of DNA[cytosine-N4]methyltransferases: similarity to adenine and cytosine-C5 DNA-methylases. *Nucleic Acids Res.* **17:**9823–9832.
53. **Koncz, C., A. Kiss, and P. Venetianer.** 1978. Biochemical characterization of the restriction-modification system of *Bacillus sphaericus*. *Eur. J. Biochem.* **89:**523–529.
54. **Krüger, D. H., and T. A. Bickle.** 1983. Bacteriophage survival: multiple mechanisms for avoiding the deoxyribonucleic acid restriction systems of their hosts. *Microbiol. Rev.* **47:**345–360.
55. **Kupper, D., J.-G. Zhou, P. Venetianer, and A. Kiss.** 1989. Cloning and structure of the *Bep*I modification methylase. *Nucleic Acids Res.* **17:**1077–1088.
56. **Lange, C., A. Jugel, J. Walter, M. Noyer-Weidner, and T. A. Trautner.** 1991. 'Pseudo' domains in phage-encoded DNA methyltransferases. *Nature* (London) **352:** 645–648.
57. **Lange, C., M. Noyer-Weidner, T. A. Trautner, M. Weiner, and S. A. Zahler.** 1991. M.H2I, a multispecific 5C-DNA methyltransferase encoded by *Bacillus amyloliquefaciens* phage H2. *Gene* **100:**213–218.
58. **Lauster, R.** 1989. Evolution of type II DNA methyltransferases. A gene duplication model. *J. Mol. Biol.* **206:**313–321.
59. **Lauster, R., A. Kriebardis, and W. Guschlbauer.** 1987. The GATATC-modification enzyme *Eco*RV is closely related to the GATC-recognizing methyltransferases *Dpn*II and *dam* from *E. coli* and phage T_4. *FEBS Lett.* **220:**167–176.
60. **Lauster, R., T. A. Trautner, and M. Noyer-Weidner.** 1989. Cytosine-specific type II DNA methyltransferases. A conserved enzyme core with variable target-recognizing domains. *J. Mol. Biol.* **206:**305–312.
61. **Lawrie, J. M., J. S. Downard, and H. R. Whiteley.** 1978. *Bacillus subtilis* bacteriophages SP82, SP01 and ϕe: a comparison of DNAs and peptides synthesized during infection. *J. Virol.* **27:**725–737.
62. **Maekawa, Y., H. Yasukawa, and B. Kawakami.** 1990. Cloning and nucleotide sequences of the *Ban*I restriction-modification genes in *Bacillus aneurinolyticus*. *J. Biochem.* **107:**645–649.
63. **Makino, O., J. Kawamura, H. Saito, and Y. Ikada.** 1979. Inactivation of restriction endonuclease *Bam*Nx after infection with phage ϕNR2. *Nature* (London) **277:**64–66.
64. **Makino, O., H. Saito, and T. Ando.** 1980. *Bacillus subtilis*-phage ϕ1 overcomes host-controlled restriction by producing *Bam*Nx inhibitor protein. *Mol. Gen. Genet.* **179:**463–468.
65. **Marinus, M. G.** 1984. Methylation of prokaryotic DNA, p. 81–109. *In* A. Razin, H. Cedar, and A. D. Riggs (ed.), *DNA Methylation. Biochemistry and Biological Significance*. Springer-Verlag, New York.
66. **McClelland, M., and M. Nelson.** 1988. The effect of site specific DNA modification methyltransferases—a review. *Gene* (Amsterdam) **74:**291–304.
67. **Messer, W., and M. Noyer-Weidner.** 1988. Timing and targeting: the biological functions of Dam methylation in *E. coli*. *Cell* **54:**735–737.
68. **Nathan, P. D., and J. E. Brooks.** 1988. Characterization of clones of the *Bam*HI methyltransferase gene. *Gene* **74:**35–36.
69. **Noyer-Weidner, M., R. Diaz, and L. Reiners.** 1986. Cytosine-specific DNA modification interferes with plasmid establishment in *Escherichia coli* K12: involvement of *rgl*B. *Mol. Gen. Genet.* **205:**469–475.
70. **Noyer-Weidner, M., S. Jentsch, J. Kupsch, M. Bergbauer, and T. A. Trautner.** 1985. DNA methyltransferase genes of *Bacillus subtilis* phages: structural relatedness and gene expression. *Gene* **35:**143–150.
71. **Noyer-Weidner, M., S. Jentsch, B. Pawlek, U. Günthert, and T. A. Trautner.** 1983. Restriction and modification in *Bacillus subtilis*: DNA methylation potential of the related bacteriophages Z, SPR, SPβ, ϕ3T and q11. *J. Virol.* **46:**446–453.
72. **Noyer-Weidner, M., and T. A. Trautner.** 1993. DNA methylation in prokaryotes, p. 39–108. *In* J. D. Jost and H. P. Saluz (ed.), *DNA Methylation: Molecular Biology and Biological Significance*. Birkhäuser Verlag, Basel.
73. **Ohmori, H., J. Tomizawa, and G. Maxam.** 1978. Detection of 5-methylcytosine in DNA sequences *Nucleic Acids Res.* **5:**1479–1485.
74. **Piggot, P. J., M. Amjad, J.-J. Wu, H. Sandoval, and J. Castro.** 1990. Genetic and physical maps of *Bacillus subtilis* 168, p. 493–543. *In* C. R. Harwood and S. M. Cutting (ed.), *Molecular Biological Methods for Bacillus*. John Wiley & Sons, Chichester, United Kingdom.
75. **Pósfai, J., A. S. Bhagwat, G. Pósfai, and R. J. Roberts.** 1989. Predictive motifs derived from cytosine methyltransferases. *Nucleic Acids Res.* **17:**2421–2435.
76. **Pósfai, G., A. Kiss, S. Erdei, J. Pósfai, and P. Venetianer.** 1983. Structure of the *Bacillus sphaericus* R modification methylase gene. *J. Mol. Biol.* **170:**597–610.
77. **Raleigh, E. A., and G. Wilson.** 1986. *Escherichia coli* K-12 restricts DNA containing 5-methylcytosine. *Proc. Natl. Acad. Sci. USA* **83:**9070–9074.
78. **Reeve, J. N., E. Amann, R. Tailor, U. Günthert, K. Scholz, and T. A. Trautner.** 1980. Unusual behaviour of SP01 DNA with respect to restriction and modification enzymes recognizing the sequence 5'-G-G-C-C. *Mol. Gen. Genet.* **178:**229–331.
79. **Roberts, R. J., G. A. Wilson, and F. E. Young.** 1977. Recognition sequence of specific endonuclease *Bam*HI from *Bacillus amyloliquefaciens* H. *Nature* (London) **265:**82–84.
80. **Saito, H., T. Shibata, and T. Ando.** 1979. Mapping of genes determining nonpermissiveness and host-specific restriction to bacteriophages in *Bacillus subtilis* Marburg. *Mol. Gen. Genet.* **170:**117–122.
81. **Santos, I., and H. deLencastre.** Unpublished results.
82. **Scherzer, E., B. Auer, and M. Schweiger.** 1987. Identification, purification, and characterization of *Escherichia coli* virus T1 DNA methyltransferase. *J. Biol. Chem.* **262:**15225–15231.
83. **Schlagman, S., and S. Hattman.** 1983. Molecular clon-

ing of a functional *dam*+ gene coding for phage T4 DNA adenine methylase. *Gene* **22:**139–156.
84. **Schneider-Scherzer, E., B. Auer, E. J. deGroot, and M. Schweiger.** 1990. Primary structure of a DNA (N^6-adenine)-methyltransferase from *Escherichia coli* virus T1. *J. Biol. Chem.* **265:**6086–6091.
85. **Shibata, T., and T. Ando.** 1974. Host controlled modification and restriction in *Bacillus subtilis*. *Mol. Gen. Genet.* **131:**275–280.
86. **Shibata, T., and T. Ando.** 1975. *In vitro* modification and restriction of phage φ105C DNA with *Bacillus subtilis* N cell-free extract. *Mol. Gen. Genet.* **138:**269–279.
87. **Shibata, T., and T. Ando.** 1976. The restriction endonucleases in *Bacillus amyloliquefaciens* N strain. Substrate specificities. *Biochim. Biophys. Acta* **442:**184–196.
88. **Shibata, T., S. Ikawa, Y. Komatsu, T. Ando, and H. Saito.** 1979. Introduction of host-controlled modification and restriction systems of *Bacillus subtilis* IAM1247 into *Bacillus subtilis* 168. *J. Bacteriol.* **139:**308–310.
89. **Smith, H. O., T. M. Annau, and S. Chandrasegaran.** 1990. Finding sequence motifs in groups of functionally related proteins. *Proc. Natl. Acad. Sci. USA* **87:**826–830.
90. **Smith, H. O., and D. Nathans.** 1973. A suggested nomenclature for bacterial host modification and restriction systems and their enzymes. *J. Mol. Biol.* **81:**419–423.
91. **Som, S., A. S. Bhagwat, and S. Friedman.** 1987. Nucleotide sequence and expression of the gene encoding the *Eco*RII modification enzyme. *Nucleic Acids Res.* **15:**313–323.
92. **Stephenson, F. H., B. T. Ballard, H. W. Boyer, J. M. Rosenberg, and P. J. Greene.** 1989. Comparison of the nucleotide and amino acid sequences of the *Rsr*I and *Eco*RI restriction endonucleases. *Gene* **85:**1–13.
93. **Stroynowski, I. T.** 1981. Distribution of bacteriophage φ3T homologous deoxyribonucleic acid sequences in *Bacillus subtilis* 168, related bacteriophages, and other *Bacillus* species. *J. Bacteriol.* **148:**91–100.
94. **Stroynowski, I. T.** 1981. Integration of bacteriophage φ3T-coded thymidylate synthetase gene into the *Bacillus subtilis* chromosome. *J. Bacteriol.* **148:**101–108.
95. **Sugisaki, H., Y. Maekawa, S. Kanazawa, and M. Takanami.** 1982. New restriction endonucleases from *Acetobacter aceti* and *Bacillus aneurinolyticus*. *Nucleic Acids Res.* **10:**5747–5752.
96. **Szomolanyi, E., A. Kiss, and P. Venetianer.** 1980. Cloning the modification methylase gene of *Bacillus sphaericus* R in *Escherichia coli*. *Gene* **10:**219–225.
97. **Tao, T., J. C. Bourne, and R. M. Blumenthal.** 1991. A family of regulatory genes associated with type II restriction-modification systems. *J. Bacteriol.* **173:**1367–1375.
98. **Tao, T., J. Walter, K. J. Brennan, M. M. Cotterman, and R. M. Blumenthal.** 1989. Sequence, internal homology and high-level expression of the gene for a DNA-(cytosine *N4*)-methyltransferase, M.*Pvu*II. *Nucleic Acids Res.* **17:**4161–4175.
99. **Terschüren, P. A.** Unpublished results.
100. **Terschüren, P.-A., M. Noyer-Weidner, and T. A. Trautner.** 1987. Recombinant derivatives of *Bacillus subtilis* phage Z containing the DNA methyltransferase genes of related methylation-proficient phages. *J. Gen. Microbiol.* **133:**945–952.
100a. **Thorne, C. B.** 1962. Transduction in *Bacillus subtilis*. *J. Bacteriol.* **83:**106–111.
101. **Tran-Betcke, A., B. Behrens, M. Noyer-Weidner, and T. A. Trautner.** 1986. DNA methyltransferase genes of *Bacillus subtilis* phages: comparison of their nucleotide sequences. *Gene* **42:**89–96.
102. **Trautner, T. A.** Unpublished results.
103. **Trautner, T. A., T. S. Balganesh, and B. Pawlek.** 1988. Chimeric multispecific DNA methyltransferases with novel combination of target recognition. *Nucleic Acids Res.* **16:**6649–6657.
104. **Trautner, T. A., B. Pawlek, S. Bron, and C. Anagnostopoulos.** 1974. Restriction and modification in *B. subtilis*. Biological aspects. *Mol. Gen. Genet.* **131:**181–191.
105. **Uozumi, T., T. Hoshino, K. Miwa, S. Horinouchi, T. Beppu, and K. Arima.** 1977. Restriction and modification in *Bacillus* species. Genetic transformation of bacteria with DNA from different species, part I. *Mol. Gen. Genet.* **152:**65–69.
106. **Vanek, P. G., J. F. Connaughton, W. D. Kaloss, and J. G. Chirikjian.** 1991. The complete sequence of the *Bacillus amyloliquefaciens* strain H, cellular *Bam*HI methylase gene. *Nucleic Acids Res.* **18:**6145.
107. **Walter, J., T. A. Trautner, and M. Noyer-Weidner.** 1992. High plasticity of multispecific DNA methyltransferases in the region carrying DNA target recognizing enzyme modules. *EMBO J.* **11:**4445–4450.
108. **Walter, J., M. Noyer-Weidner, and T. A. Trautner.** 1990. The amino acid sequence of the CCGG recognizing DNA methyltransferase M.*Bsu*FI: implications for the analysis of sequence recognition by cytosine DNA methyltransferases. *EMBO J.* **9:**1007–1013.
109. **Weiner, M. P.** 1986. Characterization of bacteriophage H2. Ph.D. thesis. Cornell University, Ithaca, N.Y.
110. **Wilke, K., E. Rauhut, M. Noyer-Weidner, R. Lauster, B. Pawlek, B. Behrens, and T. A. Trautner.** 1988. Sequential order of target-recognizing domains in multispecific DNA-methyltransferases. *EMBO J.* **7:**2601–2609.
111. **Williams, M. T., and F. E. Young.** 1977. Temperate *Bacillus subtilis* bacteriophage φ3T: chromosomal attachment site and comparison with temperate bacteriophages φ105 and SPO2. *J. Virol.* **21:**522–529.
112. **Wilson, G. G.** 1988. Type II restriction-modification systems. *Trends Genet.* **4:**314–318.
113. **Wilson, G. G.** 1991. Organization of restriction-modification systems. *Nucleic Acids Res.* **19:**2539–2566.
114. **Witmer, H., and M. Franks.** 1981. Restriction and modification of bacteriophage SP10 DNA by *Bacillus subtilis* Marburg 168: stabilization of SP10 DNA in restricting hosts preinfected with a heterologous phage SP18. *J. Virol.* **37:**148–155.
115. **Wu, J. C., and D. V. Santi.** 1987. Kinetic and catalytic mechanism of *Hha*I methyltransferase. *J. Biol. Chem.* **262:**4776–4786.
116. **Xu, G.-L., W. Kapfer, J. Walter, and T. A. Trautner.** 1992. *Bsu*BI—an isospecific restriction and modification system of *Pst*I: characterization of the *Bsu*BI genes and enzymes. *Nucleic Acids Res.* **20:**6517–6523.
117. **Yasbin, R.** 1977. DNA repair in *Bacillus subtilis*. I. The presence of an inducible system. *Mol. Gen. Genet.* **153:**211–218.
118. **Yasbin, R.** 1977. DNA repair in *Bacillus subtilis*. II. Activation of the inducible system in competent cells. *Mol. Gen. Genet.* **153:**219–225.
119. **Zahler, S. A.** 1982. Specialized transduction in *Bacillus subtilis*, p. 269–360. *In* D. Dubnau (ed.), *The Molecular Biology of the Bacilli*, vol. I. *Bacillus subtilis*. Academic Press, Inc., New York.
120. **Zahler, S. A., R. Z. Korman, R. Rosenthal, and H. E. Hemphill.** 1977. *Bacillus subtilis* bacteriophage SPβ: localization of the prophage attachment site, and specialized transduction. *J. Bacteriol.* **129:**556–558.
121. **Zahler, S. A., R. Z. Korman, C. Thomas, P. S. Fink, M. P. Weiner, and J. M. Odebralski.** 1987. H2, a temperate bacteriophage isolated from *Bacillus amyloliquefaciens* strain H. *J. Gen. Microbiol.* **133:**2937–2944.

VI. GENETIC EXCHANGE AND GENETIC ENGINEERING

39. Genetic Exchange and Homologous Recombination

DAVID DUBNAU

After *Escherichia coli*, *Bacillus subtilis* is genetically the best-characterized bacterium. Homologous recombination has been an essential tool in the genetic analysis of *B. subtilis*, and in turn, the availability of powerful genetic tools has facilitated investigation of the exchange of DNA-encoded information in this organism. This review will briefly summarize the present state of knowledge concerning several modes of homologous genetic exchange and will place greater emphasis on natural competence, about which much is known, and one section will discuss our understanding of the genetics and biochemistry of homologous recombination per se. This review will not touch on site-specific recombination, such as the integration of temperate phages, transposition, and related phenomena. For reasons of space and because it has been recently and ably reviewed for gram-positive organisms (122), little attention will be paid to the mechanics of homologous recombination except for the case of transformation by replacement.

MODES OF GENETIC EXCHANGE

Several modes of homologous genetic exchange in *B. subtilis* have been described. This section will provide a brief overview of the essential properties of these systems.

Transformation

Transformation may be defined as a process in which exogenous DNA is introduced into the bacterial cell so as to permanently alter its heredity. Transformation in *B. subtilis* may be accomplished in two ways. Transformation via natural competence is the principal focus of this review and will be dealt with in detail below. Artificial transformation is carried out either by electroporation, in which a transient high-voltage electric field serves to permeabilize the cell to DNA in solution, or by another procedure in which protoplasts are transformed with the aid of polyethylene glycol.

Electroporation of plasmid DNA into *B. subtilis* has been reported to occur with a frequency of 10^4 transformants per μg of DNA under optimized conditions (29). Electroporation has been useful for introducing plasmids into otherwise untransformable bacilli (see, for instance, references 19, 142, and 242) and might also be useful for the transformation of competence-deficient mutants of *B. subtilis*.

Protoplast transformation of *B. subtilis* was first reported by Chang and Cohen (34). The frequency of this transformation event was quite high (up to more than 10^7 transformants per μg of DNA). Unlike natural transformation with plasmid DNA (see below), linear monomeric molecules could be used in this case. In both the protoplast and the electroporation systems, no successful transformation for chromosomal markers has been reported. Transformation appears to be limited to DNA molecules that are intact replicons, such as plasmids and phage genomes. The reason for this limitation is not clear. It is interesting, however, that the artificial (Ca^{2+} and heat shock-induced) transformation of *E. coli* also proceeds inefficiently for chromosomal markers unless *recBC* mutants are used as recipients (42, 169), presumably to avoid exonucleolytic damage to incoming strands. No comparable experiments appear to have been carried out in *B. subtilis*.

Transduction

Several transducing phages for *B. subtilis* have been described. The most widely used to date are PBS1 (227) and the closely related AR9 (20). These phages present a major advantage for genetic manipulation, since they are large and a single transducing particle encompasses roughly 250 kbp, or about 7% of the host genome (105). This facilitates rapid genetic mapping to the *B. subtilis* chromosome (47, 236). A sometimes troublesome disadvantage of these transducing vectors is that they attach to the bacterial cell by binding to flagella. This requires that the donor and recipient strains be flagellated, and in practice, it is advisable to test the strains for motility before a transductional cross is attempted.

Little is known about the recombination mechanisms involved in *B. subtilis* transduction. Both types of transduction utilize *recA*-dependent pathways, although PBS1 transduction differs from transformation in its stricter dependence on base sequence homology and its lower dependence on the *recA1* allele (67). Probably as a consequence of the largeness of the PBS1-transducing fragment, the distance over which linkage can be measured with PBS1 transduction is greater and the frequency of recombination between relatively close markers is lower than with transformation. However, when pains are taken to isolate very-high-molecular-weight transforming DNA, linkage patterns similar to those observed with transduction are obtained (114). Transformation is generally resistant to restriction, whereas transduction is not (74, 234). This has been reasonably interpreted as indicating that whereas transformation proceeds via

David Dubnau • Department of Microbiology, Public Health Research Institute, 455 First Avenue, New York, New York 10016.

the uptake of single-stranded DNA (see below), transduction involves the exchange of duplex DNA (see chapter 38).

Some work has been devoted to characterizing transducing particles and their mechanism of formation. In the case of PBS1, plaque-forming and transducing phage particles have differing buoyant densities, and it appears that they contain only phage and bacterial DNAs, respectively (263). Nothing is known about the mechanism of packaging used by this phage. However, plasmid transduction by the virulent phage SPP1 has been better characterized. The DNA within the transducing particles consists exclusively of plasmid DNA in the form of concatemers possessing the molecular weight of phage DNA (31, 48). It has been reported that infection by SPP1 causes the formation of plasmid concatemers by two mechanisms (25). In the first (homology dependent), phage-plasmid interaction results in the formation of a chimeric molecule, which then replicates to form a concatemer. In the second mechanism (homology independent), a phage-encoded product triggers synthesis of multigenome-length plasmid molecules. Phage genome-length concatemers are then packaged in SPP1 capsids. These investigations have yielded several interesting and potentially useful results. When any fragment of SPP1 DNA is inserted into a plasmid of interest, the plasmid transduction frequency increases 100- to 1,000-fold (48). Also, insertion of an SPP1 *pac* site into the chromosome led to highly efficient packaging of nearby chromosomal DNA when the strain was infected by SPP1. These procedures could likely be exploited to yield high-efficiency specialized transduction systems for genes and plasmids of interest.

Conjugation

Several transposons and plasmids that appear to mediate conjugative transfer either intraspecifically or interspecifically among gram-positive bacteria and even between gram-negative and gram-positive bacteria have been described (21, 37, 92, 115). By conjugative transfer is meant an exchange of genetic information that is not inhibited by DNase and appears to require cell-cell contact. Transfer involving *B. subtilis* occurs at a relatively low frequency (10^{-4} to 10^{-5}) and takes place on a solid interface, often when the donor and recipient cells impinged on each other on a filter or are spread on an agar surface. The conjugative plasmid or transposon itself is transferred, and in addition, several small plasmids can be mobilized for transfer. This low-frequency transfer appears to be distinct from the more efficient conjugative transfer that takes place between strains of *Enterococcus faecalis* (37).

Particularly interesting is a study by Torres et al. (232), who demonstrated that the *E. faecalis* tetracycline resistance transposon Tn*925* not only will mediate its own transfer in *B. subtilis* but also will mobilize the transfer of chromosomal genes. Using multiply marked strains in which only one strain carried a copy of Tn*925* inserted in the chromosome, the genetic compositions of exconjugants produced during filter matings were examined. Extensive recombination between the parental chromosomes occurred at a fairly low frequency. A rough estimate suggested about one recombination event per several hundred base pairs. The data were not compatible with the unidirectional transfer of markers from a fixed origin, as in the case of conjugation in *E. coli*. Instead, it suggested extensive interaction between the parental chromosomes, perhaps over their entire lengths. Torres et al. (232) proposed the interesting notion that the zygotes formed as a result of cell-cell interaction in this system resemble the diploid products of protoplast fusion events (see below). It is also noteworthy that some of the recombinants scored in these crosses were tetracycline sensitive and probably had not inherited the Tn*925* element. In fact, the distribution of recombinant classes appears to be consistent with the idea that Tn*925* segregates in these crosses like an ordinary chromosomal marker. If these inferences are correct, it would be wrong to think in terms of recipient and donor cells. Instead, we should regard this type of cross as occurring between parents that make equal contributions to the generation of haploid progeny.

These various conjugative systems have several potential uses. They can be used to mediate transposon transfers between strains that cannot ordinarily exchange genetic information either because they are mutationally altered (e.g., competence-deficient or nonmotile strains) or because they are members of different species. Such transfer is useful, because elements like Tn*916* and Tn*925* can be used as mutagens following introduction into a strain that is otherwise refractory to transposon mutagenesis and because they can be engineered to carry genes of interest. Second, these elements can be used to introduce plasmids into otherwise nontransformable or nontransducible strains or species. And finally, the observations of Torres et al. (232) suggest that Tn*925*, at least, will prove useful for the transfer of chromosomal markers into such strains.

Protoplast Fusion

The fusion of *B. subtilis* protoplasts to yield transiently diploid cells provides another mode of genetic exchange (195), which is reviewed in reference 109. Protoplasts of suitably marked parental strains are prepared by lysozyme treatment in hypertonic medium, mixed, and treated with polyethylene glycol to induce fusion. The fused cells are then allowed to regenerate cell walls, and the genetic outcome of the fusion events can then be monitored on appropriate selective media. The frequency of successful fusion, as measured by complementation assays, usually amounts to a few percent of the parental population. Genetic tests indicate that the parental chromosomes make equal contributions to the fusants (124). In addition, complementation experiments with lysogenic and nonlysogenic parents suggest that cytoplasmic mixing accompanies fusion (195).

Two expected outcomes have been reported following the fusion event: it is possible to recover haploid parental or recombinant clones. One of the most extensive studies of fusion-induced recombination is that of Ftouhi and Guillen (80). Multiply marked strains were fused, and the recombinant haploid progeny were scored. Most of the colonies contained only a single class of recombinants, although some contained reciprocal recombinant types. Recombination was detected in all 10 chromosomal intervals tested. These intervals covered the entire chromosome and ranged in size from 1 to 1,265 kb. Interestingly, the regions that contained the chromosomal replication origin exhibited an enhanced frequency of recombination. The recombination observed in these experiments was generally dependent on the product of *recA*, the *B. subtilis* homolog of the *E. coli recA* gene (see below). However, recombination in the *ori*-containing regions was largely *recA* independent.

Other unanticipated outcomes of fusion crosses have also been described: complementing (125) and noncomplementing (108) diploids. These consist of clones that appear to propagate relatively stably. The former type exhibits a prototrophic phenotype when parental auxotrophs are used, and the latter exhibits one of the parental phenotypes. Both of these diploid classes infrequently segregate parental clones. The noncomplementing diploid clones are considered to consist of stable diploids in which one parental chromosome is completely silent. Recently, the existence of stable diploids, both complementing and noncomplementing, has been questioned (101). Reconstruction experiments, measurements of the DNA to dry weight ratios in presumed diploids, and self-fusion and hybridization experiments have led to the conclusion that the presumed complementing and noncomplementing diploids are in reality haploid and that their phenotypic behavior can be explained by the presence of mixed cultures and cross feeding.

NATURAL GENETIC COMPETENCE

Unlike the various forms of "artificial" competence discussed above, natural competence (referred to below simply as competence) is a physiologically and genetically determined property of a particular strain. In *B. subtilis*, it results from the growth of a bacterial culture under defined growth conditions. In several organisms, including *B. subtilis*, competence has provided a principal means of genetic exchange, which has been used for genetic analysis. As such, an improved understanding of the regulation and mechanism of competence promises to enhance the usefulness and facilitate the interpretation of transformational crosses. Of more intrinsic interest are two additional aspects.

In *B. subtilis*, competence is usually expressed postexponentially. It constitutes an experimentally amenable global regulatory system and is one of several postexponential systems in *B. subtilis* (including competence, sporulation, motility, degradative enzyme expression, and antibiotic production) that are currently of considerable interest. Competent cells are able to efficiently bind, process, and internalize exogenous high-molecular-weight DNA. This poses interesting problems concerning macromolecular transport. Earlier work focused on the alterations in DNA molecular weight and strandedness that accompany the transformation process. Recently, the characterization of gene products that are involved in DNA uptake has begun, allowing investigators to pose new questions concerning this process. In this review, emphasis is on the newer work that concerns the regulation of competence and the nature of transformation-specific gene products.

Competence is widespread in both gram-positive and gram-negative bacteria. The best-studied systems are those of *Streptococcus pneumoniae*, *Streptococcus sanguis*, *Haemophilus influenzae*, *Neisseria gonorrhoeae*, and *B. subtilis*. Competence and genetic transformation have been the subjects of several recent reviews, to which the interested reader is referred (55–58, 68, 84, 113, 204, 216).

Properties of Competent Cultures

Competent cultures of *B. subtilis* are heterogeneous, a fact first suggested by the analysis of transformation for unlinked pairs of markers (165). The conclusion from this indirect analysis was then substantiated by the resolution of competent cultures on isopycnic gradients of Renografin into a majority noncompetent population of high buoyant density and a minority subpopulation of more buoyant competent cells (95, 100). An independent study revealed that competent cells could also be resolved on the basis of sedimentation velocity in sucrose gradients (202).

Further study revealed that competent cells were in an altered metabolic state (52, 146, 165). They were found to be relatively dormant with respect to most forms of macromolecular synthesis and were reported to have completed a round of chromosomal replication with the result that their chromosomes were "lined up" in a terminated configuration. It has also been suggested that competent cells may exhibit a block in deoxyribonucleotide synthesis (134). Relatively little has been reported concerning the ultrastructure of competent cells of *B. subtilis*, although Vermuelen and Venema have suggested that they contain a higher density of mesosome structures than do noncompetent cells (243). Although for the most part these early studies have not been confirmed and extended, they serve to underscore the metabolic and morphological uniqueness of the competent state.

THE TRANSFORMATION PATHWAY

Much of the earlier work characterizing the transformation pathway traced the fate of transforming DNA during binding, processing, and uptake by the competent cell. The result of these studies was a description of the molecular weight and strandedness of transforming DNA at various stages of the transformation process but relatively little understanding of the gene products involved in transformation or of the processing of the recipient chromosome. A detailed description of the experiments that have led to this descriptive pathway is beyond the scope of this review, and instead, a relatively brief summary of the results of these studies will be presented. First, however, it may be useful to describe the various types of transformation that have been studied.

Transformation Classified by Type of Donor DNA

The transformation process may be subdivided according to the nature of the donor molecule. For instance, the donor may consist of phage DNA (transfection), plasmid DNA, or bacterial chromosomal DNA. Although the outcome of the transformation process will differ in each case, the uptake of DNA probably proceeds by a common pathway. This underscores an important feature of the *B. subtilis* competence system, namely, that the binding and uptake of DNA are not base sequence specific. In transfection, a successful event results in a burst of phage particles, usually detected as an infectious center on a lawn of indicator cells. In plasmid transformation, a clone in which the donor plasmid has been established as an autonomously replicating element results. In chromosomal transformation, the donor DNA is rescued by recombination with a homologous resident DNA segment, and the transformation event is usually detected by permanent genetic alteration of the recipient.

A form of plasmid transformation in which donor DNA can recombine with a homologous plasmid replicon has also been described (41, 149). This form of transformation has been used to increase the efficiency of molecular cloning experiments in *B. subtilis* (91) and has served as a model for chromosomal transformation (41, 254, 255).

Not only does the fate of the DNA differ in these various types of transformation experiments, but so does the efficiency of the various events. For instance, nearly one transformant results from each molecule of chromosomal DNA taken up (23, 60, 209). However, in the cases of transfection and plasmid transformation, several hundred molecules are taken up for each successful transformation event, measured as a colony or a plaque on a petri dish. The major consequence of this difference lies in the inapplicability of experiments that determine the biochemical fates of bulk plasmid or bacteriophage DNA in order to draw inferences concerning the pathways of these types of transformation, since only a very small proportion of the DNA taken up participates in a transfection or plasmid transformation event. In spite of this difficulty, models to account for the various forms of transformation have been advanced, but they will not be discussed further here (64, 122, 244).

Stages during Transformation

Several stages have been discerned during the processing of transforming DNA: binding, fragmentation, uptake, and in the case of chromosomal transformation, integration and resolution.

Binding

Binding is usually measured by the wash-resistant association of radioactive DNA with competent cells, although a filter-binding assay may also be used (53). Physiologically noncompetent cells exhibit little or no DNA binding. The attachment of DNA to the competent cell proceeds with no discernible lag and is saturable (54, 60, 127). Two studies have concluded that about 50 binding sites are present on the surface of the average competent cell (60, 201). The mass of DNA bound at saturation is proportional to the molecular weights of the donor DNA molecules (54). This suggests that each binding site is occupied by a fixed number of DNA molecules (presumably one) rather than by a fixed mass of DNA. Immediately after binding, the DNA is sensitive to hydrodynamic shear and to exogenous DNase (60, 127). It is therefore extended into the aqueous phase and must be attached to the cell surface at relatively few points per molecule. This inference has been confirmed by electron microscopy (55, 176). All or nearly all of the bound DNA constitutes a precursor form for the generation of chromosomal transformants, since the addition of heterologous DNA to prevent further binding by competition permits the nearly quantitative conversion of previously bound radioactive DNA either to the integrated form or to acid-soluble material (46, 60). The binding of DNA is not accompanied by immediate double-strand cleavage, since intact donor molecules can be recovered from the cell surface by treatment with phenol and detergent (54). It is possible that binding does occur concomitantly with nicking; this question has not been addressed in *B. subtilis*. The fact that bound DNA can be removed by phenol-detergent treatment indicates that its attachment is noncovalent. The precise locations of the binding sites on the cell surface are not known; they may be localized on the cell wall, on an exposed portion of the membrane, or on a specialized structure that traverses both. Although a careful quantitative study has not been reported, there appears to be little if any base sequence specificity in binding. All DNA samples tested appear to be bound (and taken up) with approximately equal frequency. This is evident when DNA from *B. subtilis*, *E. coli*, phage T7, or a variety of other phages and plasmids is used (60, 72, 77, 87, 208). On the other hand, it has been reported that glucosylated DNA, double-stranded RNA, and various synthetic polymers bind poorly if at all (39, 208).

Fragmentation

Bound DNA rapidly suffers double-strand cleavage (15, 54, 61). The cleaved molecules are relatively resistant to removal by hydrodynamic shear, possibly because they are not as extended into the liquid phase as the intact molecules initially bound. However, they are still completely susceptible to exogenous DNase and are thus exposed on the cell surface (60, 61).

The pattern of cleavage has been studied by using radiolabeled phage T7 DNA as well as chromosomal DNA of known average molecular size (54). This was possible because bound DNA before and after fragmentation could be quantitatively recovered in soluble form by treatment of the DNA-cell complexes with Sarkosyl in the presence of phenol. The T7 molecules each underwent an average of one double-strand cleavage at an essentially random position. This resulted in a halving of the initial number average molecular mass to about 12.5×10^6 Da. The only exception to the random location of the cleavage was an apparent bias against cleavage near the ends of the molecule. When chromosomal DNA with a starting weight average molecular weight of 125×10^6 was used, a broad distribution of recovered-fragment molecular weights was observed, with some starting molecules suffering more than one double-strand break in a time-dependent manner. An accumulation of recovered material was observed at the low end of the distribution, with the same weight average molecular weight as the distribution of cleaved T7 DNA fragments, together with polydispersed material of intermediate size. These observations can be at least partially rationalized by assuming that cleavage occurs at fixed points on the cell surface. These points may most simply be assumed to correspond to binding sites. A characteristic distribution of cleavage sizes would then be observed, depending on the stiffness of the DNA, the average distance between the binding-cleavage sites on the cell surface, and other factors such as the preferred range of angles at which a DNA molecule emerges from the binding-cleavage site. The average size of the fragmented molecules held on the competent cell surface may therefore be determined in part by the distribution of binding-cleavage sites on the cell surface and the probability that, given the stiffness of DNA, the end-to-end distance of any segment will coincide with the distance between two sites at least some of the time. As noted above, the observed number average lower-molecular-mass limit of DNA recovered from the cell surface was 12.5×10^6 Da, which corresponds to a contour length of 6.4 μm. When DNA is free in solution, this length corresponds to an average end-to-end distance of about 0.74 μm (32). The range of molecular weights recovered from bound T7 DNA was 5×10^6 to 21×10^6 Da, which corresponds to end-to-end distances of 0.47 to 0.96 μm. Although these values appear to be reasonable and of the same order of magnitude as the dimensions of a bacillus, it is not possible with our current understanding to precisely predict the observed distribution of sizes. The geometric ideas proposed to account for the fragmentation of donor DNA therefore remain plausible but unproven.

Uptake

As mentioned above, donor material is attached to the cell surface in a form that is completely accessible to exogenous DNase. Beginning at 1 to 2 min following the addition of DNA to the competent cells, transformation becomes resistant to incubation with DNase (127). The lag preceding this resistance increases with decreasing temperature. The precise interpretation of DNase resistance is uncertain. It may represent transport across the cell membrane or penetration into a compartment in which the DNA is protected by the cell wall from access to the added nuclease. It was observed that transformation becomes cyanide resistant with a distinct time course following the acquisition of DNase resistance (223). This was interpreted as suggesting that cyanide resistance reflects transport while DNase resistance measures some prior step. However, steps following transport, such as integration, might easily be cyanide sensitive. In fact, cyanide sensitivity appears at 6 to 8 min, which is when integration occurs. It seems prudent to use a relatively neutral term such as "uptake" to describe the step resulting in DNase resistance and to defer a more precise interpretation.

Strauss (221, 222) has used the kinetics of DNase resistance to study the uptake of single and linked pairs of transforming markers. He has shown clearly that pairs of markers are taken up with a longer lag time than are single markers and that the length of this lag time is proportional to the genetic distance between the markers. This was interpreted as indicating that uptake is linear. Since several single markers were taken up with identical kinetics, it appears that the competent cell does not distinguish the "right" and "left" ends of transforming molecules, a conclusion that is in accord with the observation that there is no apparent base sequence specificity in the binding and uptake steps. The lag observed in the uptake of linked marker pairs was used to infer that about 55 bp are taken up per s at 28°C. Since the chromosomal-DNA region used for these studies has now been completely sequenced (104), it is possible to arrive at a more reliable estimate. The extreme markers studied (*aroE1* and *mtr*) are separated by about 15 kb. Since the time required for entry of this marker pair is 1.4 min, DNA enters at a rate of about 179 bp/s, at least for this segment.

Donor DNA is recoverable from transformed cells as single-stranded fragments (174). The weight average molecular weight of this material is 3×10^6 to 5×10^6 Da (46). Kinetic studies have shown that these single strands, which are resistant to the action of exogenous DNase, are derived from surface-localized double-stranded fragments and are precursors of integrated donor DNA (46). The first appearance of single-stranded material coincides with the development of DNase-resistant transformation. In addition, the appearance of single-stranded DNA is coincident in time with the release into the cell medium of acid-soluble products derived from donor DNA (46, 61, 112). The acid-soluble and single-stranded products each make up about half of the material initially bound to the cell surface, suggesting that one strand may be completely degraded. When the transforming DNA was labeled with [^{3}H]thymidine, this acid-soluble material was recovered as 5′-TMP, thymidine, and thymine (61). It is tempting to conclude that these three events (the appearance of acid-soluble material, single-stranded DNA, and DNase-resistant transformants) all reflect a single concerted process. Similar reasoning (based on data obtained with *S. pneumoniae*) has led Rosenthal and Lacks to propose a model in which a nuclease asymmetrically located in association with a membrane pore for DNA transport degrades one strand (185). This would result in the release of acid-soluble material into the medium and could drive concomitant transport of the complementary strand. This model will be discussed further below.

Recently, Vagner et al. (235) have investigated the chemical polarity of uptake. A variety of DNA molecules with radioactive labels located on one strand, either centrally or near the 5′ or 3′ terminus, were constructed. After 15 min at 32°C, the uptake of radioactivity by the competent cells was measured and found to be quite similar for the various donor DNA samples. This similarity was interpreted as indicating either that single strands are taken up with no preferential polarity or that initial uptake is of double-stranded DNA, with subsequent intracellular degradation of one strand. Although neither of these possibilities is ruled out by other data, the former appears to be unlikely, since a single strand entering with 5′-to-3′ or 3′-to-5′ polarities would present very different aspects to the uptake machinery, requiring two uptake systems. The possibility that uptake of an entire strand took place in these experiments was also considered. Such uptake would obscure the underlying polarity of uptake. This possibility was rejected, basically because *B. subtilis* cells introduce double-strand breaks into DNA from which uptake is thought to be initiated, and this was assumed to prevent the uptake of an entire strand. However, two factors require a reconsideration of these interesting experiments. First, the donor molecules used were derived from M13 phage and had molecular weights of about 4.5×10^6. The smallest molecules used in studies of double-strand cleavage on the competent cell surface had molecular weights of about 2×10^7 (54, 61). It is uncertain whether molecules as small as those used in the study of Vagner et al. undergo efficient cleavage. Second, the samples were taken for determination of uptake after 15 min. By this time, uptake has usually been completed, and even if cleavage did occur, it is possible that an entire strand was taken up in segments, one from a preexisting molecular terminus and the other from the point of cleavage. It would be better to carry out such studies as a function of time. Finally, it should be noted that in *S. pneumoniae*, which exhibits a transformation pathway similar to that in *B. subtilis*, a 3′-to-5′ polarity of uptake has been indicated (147).

Energetics of uptake

The transport of DNA into the competent cell is extremely efficient. About half of the mass of donor DNA initially bound to the cell surface is found as DNase-resistant single strands, with the remainder accounted for as acid-soluble material. Obviously, a DNA-specific transport mechanism must exist in competent cells. As mentioned above, Rosenthal and Lacks (185) have suggested that transport occurs through a membrane channel that is associated with an asymmetrically located nuclease. Asymmetry of the nuclease active site ensures that only one strand is degraded during transport, and it is implicit in the model that the driving force for transport is provided by the action of the nuclease. In support of this model are two observations. In *S. pneumoniae* as in *B. subtilis*, uptake of single-stranded DNA occurs after the binding of double-stranded DNA. Second, Rosenthal and Lacks (185) have identified a membrane-localized nuclease required for transformation and DNA entry in *S. pneumoniae*. This is an attractively simple model that is consistent with the data so far presented for the *B. subtilis* system.

Grinius (88) has used uncoupling agents in a variety of transformation experiments and has concluded that DNA uptake in *B. subtilis* requires both the pH and the electrical components of the proton motive force. Poisoning with arsenate led Grinius to suggest that uptake is not driven directly by ATP. He proposed instead that the transforming DNA binds to proteins on the cell surface, that this complex acquires a net positive charge by binding protons, and that the entire complex then electrophoreses through a water-filled membrane channel. This electrogenic mechanism explains the need for both components of the proton motive force. Van Nieuwenhoven et al. (237) have criticized aspects of the Grinius model, suggesting that the ΔpH alone provides the driving force. Theirs is an electroneutral proton symport mechanism, and its continued operation would require pumps to prevent intracellular accumulation of protons. Neither model would explicitly explain the requirement for a membrane nuclease, and neither relates the transport process to the appearance of single strands. Both of these studies measured ΔpH and $\Delta\psi$, and Grinius's study also measured the ATP pool in bulk competent cells. Since the competent cell fraction is physiologically distinct and makes up about 10% of the culture, it would be important to repeat such measurements in the competent cell subfraction. A further highly speculative model that is consistent with all of the existing data has been proposed (56). This proposal was suggested by the anion-exchange model of sugar phosphate transport (8, 9, 138). Perhaps duplex DNA enters a water-filled membrane channel, where it encounters a nuclease, with the result that one residue is cleaved and expelled, possibly by reorientation of a membrane exchange center. If the ΔpH is appropriate, the incoming base pair, consisting of two molecules of a monovalent nucleotide, may release two protons into the cytoplasm. If a single divalent nucleotide is expelled, electroneutrality would be maintained but the system would be continually coupled to proton circulation. Clearly, more data are needed to distinguish between these and additional models for the energetics of DNA transport.

Integration and resolution

As stated above, single-stranded donor molecules of 3×10^6 to 5×10^6 Da are recoverable from transformed cells. These are direct precursors of integrated material (donor-recipient complex); about 72% of the single-stranded DNA mass eventually is integrated (46).

The product of recombination is a heteroduplex consisting of paired donor and recipient strands (22, 59, 65, 224). The average size of the integrated donor segment is 2.8×10^6 (weight average), or ca. 8.5 kb, in one study that used physical measurements (59) and 2.9×10^6 to 3.9×10^6 in another study that used electron microscopy (79). On the basis of what was known about transformation in *B. subtilis*, it was predicted (60) that the integrated segments would be clustered. This follows from the fact that a donor DNA molecule is fragmented on the cell surface and that uptake and integration of the resulting fragments proceed efficiently. This should result in a number of clustered but independent integration events from each molecule initially bound. This prediction was confirmed by both studies.

Early during the integration process, a noncovalent association of donor and recipient DNA from which the donor moiety could be liberated at pH 12 was detected (14, 63). The freed donor single-stranded moiety had a weight average molecular weight of 2.5 $\times 10^6$ to 3×10^6, in good agreement with the measured sizes of both the precursor single strands and the final integrated moiety. Arwert and Venema (14) have suggested that nicks rather than gaps separate the donor and recipient moieties of the joint molecules, since DNA ligase alone increases the transforming activity of the joint molecule preparation as much as does ligase plus DNA polymerase. The structure of the joint molecules is not known; they may consist of a branched triple-stranded molecule, a completely base-paired molecule, or some other form. Also, little is known about the mechanics of recombination in this system, although it is tempting to think in terms of a strand assimilation mechanism, since RecA, the *B. subtilis* analog of the *E. coli* RecA protein, is required for integration (66).

Several studies have concluded that DNA replication is not required for the integration step (62, 126, 255). However, it is possible that some limited synthesis is required for repair at the ends of the donor segment prior to ligation of the segment to the adjacent recipient strands. This question has been addressed by using a model system in which transforming DNA was used in recombination with a resident plasmid molecule (255). This study permitted the approximation that no more than about 300 bases were synthesized for each integration event, and this number may be a considerable overestimate. Thus, the extent of new synthesis appears to be quite limited.

Kinetic analysis of the dependence of transformation on length of the donor-homologous segment has been carried out in two systems: ordinary chromosomal transformation (153) and the plasmid recombination system referred to above (41, 254). These studies resulted in strikingly similar conclusions for the two systems. First, transformation behaves as though a portion of the donor segment is degraded or otherwise made unavailable for integration. This so-called excluded length is equivalent to about 400 to 500 bases per donor segment and appears to represent an absolute molecular weight limit below which no replacement integration can occur. A similar value for the excluded length was previously obtained with *S. pneumoniae* (93). A second conclusion from the kinetic analysis concerns the recombination of switch frequency during transformation. It may be estimated that during transformation of *B. subtilis*, about 1×10^{-4} to 3×10^{-4} genetic exchanges per base occur (41, 254). Since the average size of an integrated segment is about 9 kb, one to three exchanges per integrated segment are predicted. Since replacement recombination of a continuous DNA segment requires exactly two exchanges, the estimate of exchange frequency appears to be reasonable. The rather low transformational exchange frequency is consistent with the genetic evidence; multiple exchanges are rare in *B. subtilis* (45, 255).

The final heteroduplex molecule formed by the integration step can be resolved by replication or by mismatch repair to form a homoduplex. Mismatch repair is known to occur following transformation of *B. subtilis*, and the probability of repair depends on the nature of the mismatch (28, 213). A precise description of this dependence and information about the probability of cocorrection of linked markers as a function of distance have not been reported for *B. subtilis*.

GENETICS OF HOMOLOGOUS RECOMBINATION

Several loci that appear to play roles in homologous recombination have been identified in *B. subtilis*. As in *E. coli*, these loci are also involved in the repair of DNA damage. In the section that follows, the aspects most directly pertinent to recombination will be briefly reviewed. A more comprehensive summary of these genes and of the *B. subtilis* SOS system in general can be found in chapter 37 (this volume). The nomenclature of the *B. subtilis rec* genes has been rationalized, and with the exceptions noted below, this review utilizes the now-standard nomenclature of Mazza and Galizzi (144).

recA Gene

As in *E. coli*, the *recA* gene in *B. subtilis* plays a central role. This *B. subtilis* locus was previously known as *recE* but has been renamed in conformance with *E. coli* nomenclature. Certain recombination mutations previously referred to as *recA* (e.g., *recA1*) have been shown to be alleles of the newly named locus, and their designations have therefore been retained (266). Other alleles of *recA*, notably *recA4* and *recA6*, an insertional knockout of the cloned gene, exhibit complete deficiencies in homologous recombination (33, 66). Thus, it appears that *recA* is absolutely required for this process. The *B. subtilis recA* gene has been sequenced (219) and exhibits 60% amino acid sequence similarity to its *E. coli* counterpart. Furthermore, the *E. coli* gene can complement the *recA4* allele for recombinational activity (but not for induction of lysogenic prophage) (50, 133), and the two gene products are immunologically cross-reactive (136, 141).

In view of the predicted amino acid similarity between the *E. coli* and *B. subtilis* proteins, it is virtually certain that the latter is directly involved in recombination as well as in induction of the SOS system. This conclusion is consistent with the older observation that during transformation, a *recA4* mutant can bind, process, and internalize high-molecular-weight DNA but fails to exhibit association of the donor single strands with recipient DNA (66). Little can be said concerning the nature of the involvement of RecA in *B. subtilis* recombination except to suppose that it resembles the well-characterized strand assimilatory activities of the *E. coli* protein (179).

Other Recombination Loci

Several additional *rec* loci in *B. subtilis* have been identified (145). In most cases, the gene products have not been extensively characterized, and little is known concerning their roles in recombination. In several cases, the gene products appear to be related in amino acid sequence to *rec* genes of *E. coli*, permitting potentially important inferences to be drawn and suggesting that the DNA repair and genetic recombination mechanisms in these two organisms have some basic similarities. These similar gene products include the products of *recF* (168) and *recR* (6), which correspond to their similarly named *E. coli* counterparts. (The *B. subtilis recR* gene was previously named *recM*.) The *B. subtilis recL* gene appears to be the counterpart of *E. coli recO* (4). Also, the *B. subtilis addAB* genes are apparently the equivalents of *E. coli recBCD* (118). This equivalence is based largely on the similar in vitro activities exhibited by the corresponding enzymes, including ATP-dependent double-stranded DNA exonuclease activity, DNA-dependent ATPase activity (51), and ATP-dependent helicase activity (199). In addition, the predicted gene product of the *addA* gene shows limited sequence similarity to the *E. coli* RecB protein. An interesting difference in the organization of the enzymes from the two organisms is the apparent absence of a RecD protein in *B. subtilis*. Although in *E. coli* the inactivation of *recD* by mutation results in the loss of exonuclease activity, the *B. subtilis addA* and *addB* genes appear to be sufficient for the expression of this activity (118).

Kooistra et al. report that mutations in either *addA* or *addB* cause severe deficiencies in transformability by chromosomal DNA (118), while others report only minor effects (5). The latter finding is in accordance with conclusions concerning both gram-negative and gram-positive systems that the ATP-dependent nucleases are not required for transformation (246, 262). It is conceivable that some of the mutations used in these studies affected the equivalent of the *E. coli recD* gene, which is required for DNA repair but not for recombination, if indeed such an equivalent exists in the gram-positive systems (see above). It is also possible that the discrepancy among the *B. subtilis* reports lies in differences in the genetic backgrounds used. For instance, the studies of Kooistra et al. (119) were carried out with strains that had been repeatedly mutagenized with nitrosoguanidine and were not cured of the defective bacteriophage that inhabits wild-type strains of *B. subtilis*. In any event, this is an important point that deserves to be resolved. For instance, if a major role for the products of *recBCD* in *E. coli* is to generate a single-stranded DNA substrate for recombination (229), it would be reasonable that the *B. subtilis* equivalent is not needed for transformation, since single strands are generated by the uptake mechanism.

Unlike the case of *recA*, which is absolutely required for homologous recombination, and aside from the question of *addA* and *addB* dealt with above, mutations in the other *rec* genes have only moderate effects on recombination frequencies (5, 7, 66). Alonso et al. (5) have shown that certain combinations of *rec* mutations depress recombination more severely than do either of the single mutations. These data permitted Alonso et al. to classify the genes into three epistatic groups. Group α contains *recB*, *recD*, *recF*, *recG*, *recL*, and *recR*. Group β consists of *addA* and *addB*. Finally, group γ contains *recH* and *recP*. When mutations within a group were combined, little additional deficiency in recombination was noted compared with that of either of the single mutants. It appears that *B. subtilis*, like *E. coli* (137), possesses several alternative recombinational pathways and that *recA* is required for all of them. It would certainly be premature to speculate further on the natures of these pathways in *B. subtilis* in the absence of further information concerning the biochemical activities of the various gene products and some understanding of the preferred substrates for each of the putative pathways.

LATE COMPETENCE GENE PRODUCTS

Types of Gene Products

Several approaches have identified gene products required for transformation. These products can be classified into two groups. One type of product plays a regulatory role and is needed for appropriate expression of other competence-specific proteins. These regulatory proteins have also been referred to as "early competence products." Since some of them are not expressed constitutively but only in certain media and at certain growth stages, the term "regulatory product" is preferable. Other proteins either are required directly for the binding and processing of DNA or are needed for proper assembly of the competence machinery. These have been called "late competence proteins." Since it is not yet possible to distinguish those that are directly involved in transformation and those that play a morphogenetic role in the assembly of the competence apparatus, we will retain the term "late proteins." The regulatory and late genes have usually been distinguished by using genetic tests for the dependencies of expression of one gene on another, most often with fusions of target gene promoters to the *E. coli lacZ* marker. All of the known late-gene products have been identified by characterization of competence-deficient mutations followed by cloning and sequencing of the cognate gene. In a number of cases, the proteins encoded by these genes have been studied directly.

Late Competence Gene Products

Competence nuclease

Two potential late competence gene products were studied biochemically and genetically by Smith et al. (205–207). These include (i) a 75-kDa membrane-localized protein complex that contains a 17-kDa nuclease and (ii) an 18-kDa protein that seems to limit the activity of the nuclease. The genes encoding these two proteins (*comI* and *comJ*, respectively) have been cloned and sequenced, and a null mutation at least in *comI* was reported to be partially competence deficient (245). However, recent evidence has called this conclusion into question (240); a null mutant of the nuclease gene exhibits normal or near normal transformability. Therefore, no *B. subtilis* analog of the *Streptococcus* entry nuclease (185) has been clearly identified to date. However, it has been pointed out by van Sinderen and Venema (240) that in view of the report by Vagner et al. that uptake may occur with either polarity (235), an alternative entry nuclease may exist in *B. subtilis*. If this is true, inactivation of only one of these nucleases would have a minor effect on transformability.

The *div-341* (*secA*) locus

An interesting mutation (*div-341*) that exhibits a pleiotropic phenotype has been described (190, 192). This mutation, which maps near the *degS-degU* genes (Fig. 1), results in filamentous growth above 45°C. At an intermediate temperature (37°C), competence and sporulation were strongly inhibited, and exoenzyme production was also limited. At a permissive temperature (30°C), competence, sporulation, and exoenzyme production were normal. These pleiotropic effects seem to suggest classification of the *div-341* locus as regulatory. However, cloning (188) and sequencing (193) of the *div-341* gene have revealed the presence of an open reading frame (ORF) with striking similarity

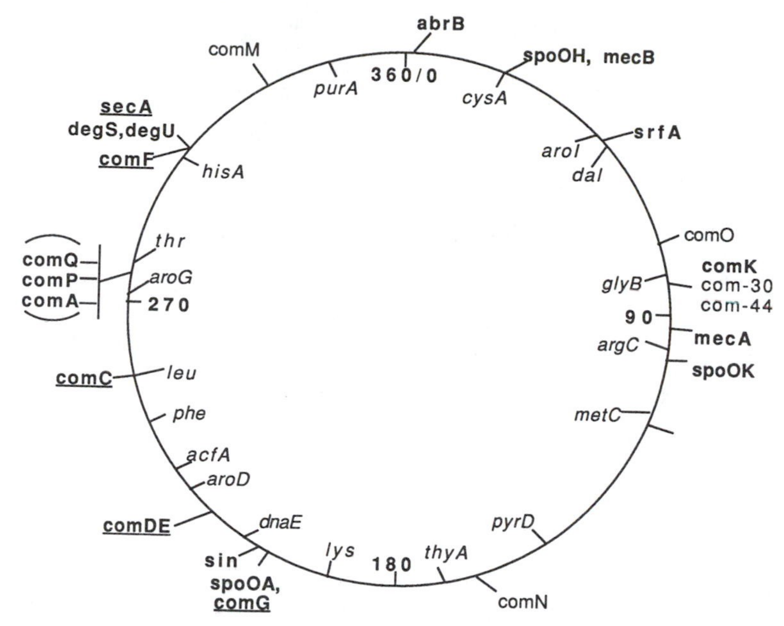

Figure 1. Genetic map of *B. subtilis* competence genes. Known competence loci are indicated on the outside of the circle. Italicized markers on the inside are included for reference. Boldface type without underlining indicates regulatory loci. Boldface underlined type indicates late competence loci. Normal type indicates loci that may include either regulatory or late competence genes. The order of the *comQ-comP-comA* cluster relative to flanking markers is not known.

on both the DNA and amino acid sequence levels to that of the *E. coli secA* gene. The *div-341* mutation results from a single amino acid change in this ORF (189). This finding provides an obvious explanation for the exoenzyme deficiency of the *div-342* mutants, since *secA* lies on the main secretion pathway in *E. coli* and presumably in *B. subtilis* as well. The deficiencies in sporulation and competence may result from the failure to secrete products essential for these systems. A further experiment lends support to this notion (189). The mutant strain was grown to late exponential phase at 30°C and then transferred to a second medium for 90 min to allow the development of competence. Competence was then measured by a third incubation with transforming DNA for 30 min. Transformability was high when competence was allowed to develop at the permissive temperature, whether incubation with transforming DNA was at the permissive (30°C) or the nonpermissive (37°C) temperature. When the second incubation occurred at the nonpermissive temperature, however, almost no transformants were obtained. This experiment is consistent with the idea that the *div-341* (*secA*) gene plays a role during the development of competence rather than during the binding and uptake of DNA. However, these data may be interpreted in several ways. SecA may play either a regulatory or a morphogenetic role during the development of competence. It is also possible that SecA plays a direct role during transformation but that the *div-341* mutation prevents the synthesis or proper folding rather than the activity of SecA. This would restrict the window during which incubation at the nonpermissive temperature can inhibit transformation to the period of formation of the competence machinery. It would be useful to determine the dependencies of late-gene expression on SecA and the effect of *div-341* on the assembly of known late competence proteins on the surfaces of competent cells. It is also possible that the effect of the *secA*(Ts) mutation on competence is indirect. For instance, the intracellular accumulation of precursor proteins or the effect of the SecA lesion on the regulation of other Sec proteins (196) may interfere with the development of competence.

Genetic Identification of Additional Late Competence Genes

Several competence loci have been identified following the isolation of *com* mutants. In some cases, it is not possible to firmly classify these mutants as having alterations in regulatory or late genes. For instance, Fani et al. (73) used nitrosoguanidine to isolate four *com* mutants. These four mutants were reduced in transformation by three to five orders of magnitude but were not reduced in transduction or protoplast transformation. Three of the mutations cannot yet be classified as defining regulatory or late genes. Judging from its map position (Fig. 1), one of these mutations (*comM104*) defines a new gene. The fourth, *com-30*, caused deficient uptake, but DNA binding was at nearly wild-type levels. Therefore, the latter probably defines a late competence gene. Its reported map position is near that of *comK*, a regulatory gene (241). An additional mutation, *com-44*, also maps near *comK* but is not an allele of the latter (18).

Mastromei et al. (143) used transposon Tn*917* to identify at least four distinct *com* genes. Of these, two can be said to be distinct from other previously identified loci. One of these two (*comO18*) was strongly blocked in DNA binding and was depressed in transformation about 10,000-fold. The other (*comN114*) was depressed about 200-fold and took up about two-thirds as much DNA as did the wild type. The approximate map locations of *comO* and *comN* are shown in Fig. 1. These are candidates for late competence genes, but their roles in competence cannot be specified with certainty.

The powerful modifications to Tn*917* now available (267–269) have permitted the isolation of a number of *com* mutations as gene fusions to a promoterless *lacZ* element and the ready cloning of the *com* loci so identified (3, 96). The availability of *lacZ* fusions has permitted examination of the expression patterns of these new *com* determinants and classification of them as regulatory or late genes. Several of the late *com* loci have been characterized by sequencing and transcriptional mapping, and the latter aspect will now be described.

Transcriptional Organization and Phenotypic Properties of Late Competence Loci

Identification of late *com* genes for further study

By the following criteria, *comG*, *comC*, and *comDE* have been identified as late *com* transcription units. First, they are expressed only at the time of transition to the postexponential state and in competence medium. Second (with one exception to be discussed below), they are not required for the resolution of populations in Renografin gradients into competent and noncompetent subpopulations. Third, expression of these genes is restricted to the competent Renografin-resolved subpopulation. Fourth, expression of these genes is dependent on several other determinants identified as regulatory genes but not on one another. Finally, the late *com* mutants appeared otherwise normal in growth, colony morphology, and other late growth properties, such as sporulation. This is in contrast to certain of the regulatory mutants, which exhibit pleiotropic phenotypes. All but the third of these criteria are known to apply as well to the *comF* locus (131) (previously referred to as *com-524* [96]). Expression of *comF* in Renografin-separated cells has not yet been tested. The transcriptional organization of each of these loci and the phenotypes conferred by their cognate mutations will now be described.

comG. Several closely linked insertions of the transposon Tn*917-lacZ* that resulted in a complete loss of transformability were isolated; competence was reduced by at least 6 orders of magnitude (1–3, 96). The mutant *com* strains grown through the usual competence regimen were unable to bind transforming DNA. Transcription of the mutant locus occurred normally as judged by β-galactosidase production from the *lacZ* fusion construct and by S1 nuclease and primer extension measurement of mRNA synthesis. The region of DNA in which the transposons had inserted was cloned, sequenced, transcriptionally mapped, and shown to consist of an operon (*comG*) with seven ORFs (1, 2). The end of ORF3 and the beginning of ORF4, the end of ORF4 and the beginning of ORF5, and the end of ORF5 and the beginning of ORF6 overlap to various extents. This configuration suggests that these four ORFs may be translationally coupled, as if their products are synthesized in fixed proportions. The map location of *comG* is shown in Fig. 1.

Although evidence for at least one and possibly two minor downstream promoters was obtained, these were not able to support the expression of more than very limited transformability, and it is clear that a single major promoter drives the great bulk of *comG* expression. Upstream from the transcriptional start site, identified by S1 nuclease mapping and primer extension analysis, was a sequence that resembles a vegetative *B. subtilis* promoter. Downstream from the final ORF was a dyad symmetry element that resembles a rho factor-independent terminator.

Although Tn*917* insertions in five of the seven *comG* ORFs were identified, it is possible that not all of the ORFs are required for transformation, since Tn*917* insertion would be expected to be polar on downstream genes. However, at least ORFs 1, 3, and 7 are essential competence genes. Inactivation of ORF1 results in a competence-deficient strain that fails to resolve into two cell populations on gradients of Renografin (1). Inactivation of ORF2 does not prevent resolution in Renografin. Complementation in *trans* of a strain with a transposon insertion in ORF1 by a plasmid carrying only *comG* ORF1 restores Renografin separation but not competence. Thus, it appears that ORF1 is essential. ORF7 is also needed for transformability, since it is the most distal gene of the *comG* operon and an ORF7 insertion is competence deficient. Finally, an in-frame deletion of ORF3 has been isolated and shown to completely eliminate transformability (27). This deletant can be complemented by ORF3 in *trans*.

comC. Other insertions of Tn*917-lacZ* defined the *comC* locus, which consists of a single ORF (3, 96, 151, 152). *comC* mutants are completely deficient in competence and cannot bind transforming DNA. The location of *comC* is shown in Fig. 1. The transcriptional start site was identified by primer extension and is located downstream from a sequence that resembles a vegetative promoter. *comC*, like *comG*, is followed by an apparent transcriptional terminator element.

comDE. Four Tn*917-lacZ* insertions were used to identify a locus that we provisionally call *comDE* (3, 96). Mutations in this locus result in a complete loss of competence. Their map positions are shown in Fig. 1. The sequence of the *comDE* region has recently been completed (99); it contains four ORFs, three of which are read in one direction, almost certainly as an operon. The Tn*917* insertion in the third (most distal) ORF results in the complete loss of competence and in a mutant strain that binds DNA but cannot take it up. Insertion in the most promoter-proximal ORF (ORF1) also results in the complete loss of competence but abolishes the ability to bind DNA. This loss is presumably not due to a polar effect of the Tn*917* insertion on ORF3 because of the distinct phenotype conferred by ORF3 inactivation. Since Tn*917* is polar on downstream genes, however, it is not certain whether ORF1 or ORF2 or both are essential competence genes. The promoter of this operon has been localized by low-resolution S1 mapping to a region about 1 kb upstream from the start of ORF1 translation. Contained within this leader sequence is a large ORF that would be read in a direction opposite to that of ORFs 1 through 3. It is not known what role, if any, is played by this reverse ORF or even whether it is expressed. Interestingly, whereas transcription in the forward direction increases when a culture reaches stationary phase in competence medium, transcription in the reverse direction decreases at this time (97).

comF. A single Tn*917-lacZ* insertion defines the *comF* locus (96). This consists of three ORFs (131). Two overlapping reading frames are located downstream from *comF* ORF1, and all three are driven by one major promoter. The transposon insertion is located in ORF1 and lowers competence by a factor of 10^4. An in-frame deletion in ORF1 lowers competence by 10^3, and antibiotic resistance cassettes inserted in the downstream ORFs lower competence by a factor of about 10.

Some Properties of Late *com* Proteins

The amino acid sequences of the seven *comG* proteins inferred from the DNA sequence demonstrated that all but ComG ORF1 possess at least one highly hydrophobic, probably transmembrane segment (1). The ORF1 sequence exhibits one hydrophobic region with a marginal probability of being classified as a transmembrane segment. The ORF2 protein (323 residues) has three predicted transmembrane segments. The remaining ORFs (3 through 7) are each predicted to contain a single N-terminally located transmembrane segment and consist of 98, 115, 165, and 124 amino acid residues, respectively.

ComG ORF1 is predicted to be a protein of 356 residues. Located within ORF1 is a sequence that resembles a nucleotide-binding site, similar to those found in other proteins such as AMP kinase, bovine ATPase, and the RecA protein (247). Antiserum against a hydrophilic synthetic peptide derived from the ComG ORF1 sequence was prepared and used to study the subcellular location of this protein (26). The ComG ORF1 signal, detected on Western blots (immunoblots), was localized to the membrane fraction of competent cells. It could be solubilized from the membrane by extraction with 0.1 M NaOH (187) and was thus considered a peripheral membrane protein. Probing of membrane and protoplast preparations with protease suggested that the ORF1 protein was located on the cytoplasmic face of the membrane.

Antipeptide antiserum against ComG ORF3 was also prepared (27) and used to demonstrate that in competent cells, the protein is in part an intrinsic membrane protein because of its insolubility in 0.1 M NaOH. All of the signal detected on Western blots could be eliminated by prior treatment of protoplasts or even of whole competent cells with protease. These observations, together with the single predicted transmembrane domain located near the N terminus of the ORF3 protein, suggested that about half of the protein in competent cells was situated with the N terminus in the cytoplasm and the C-terminal moiety exposed at the exterior of the membrane. This was also true when ORF3 was expressed under control of the regulatable *Pspac* promoter in noncompetent cells. However, an intriguing difference between the localizations of ORF3 in competent and noncompetent cells was noted. In the latter case, none of the signal was removed from the membrane by 0.1 M NaOH, whereas in competent-cell membranes, about half of the signal was solubilized. Since all of the signal could be eliminated by protease treatment of protoplasts, it was concluded that in the competent cell, a portion of the ComG ORF3 protein was on the exterior of the membrane and the remainder was intrinsically associated with the membrane. This points to a specific organization of ORF3 protein in the surface layers of competent cells. No biochemical evidence concerning localization or properties of the remaining ComG proteins (ORFs 2, 4, 5, 6, and 7) is yet available, although as noted above, these proteins are probably intrinsic membrane components.

ComC is predicted to be a polytopic membrane protein of 248 residues containing four or five transmembrane segments (151). The ORF3 protein of the *comDE* operon is also likely to be a polytopic membrane protein (99). This gene product, 776 amino acid residues long, is highly hydrophobic and contains about eight predicted transmembrane segments. The ComDE ORF1 and ORF2 proteins are hydrophilic except that ORF1 has a single predicted N-terminal transmembrane region.

The *comF* ORF1 protein is predicted to be highly hydrophilic and, like *comG* ORF1, to contain an ATP-binding site (131).

The Transformation Machine

The presence of at least some late competence products in association with the membrane and differences in the dispositions of ComG ORF3 in noncompetent and competent cells support the existence of a surface-localized multicomponent transformation machine. It is likely that such a machine includes a DNA-binding component, a channel through the peptidoglycan layer, and an aqueous pore that extends through the cell membrane. Since the transport process is apparently coupled to an energy source (see above), the machine probably includes proteins that are required for this coupling. Finally, a nuclease capable of processively degrading one strand of donor DNA may also be present in the machine. The existence of such a multicomponent transformation machine, in which the individual components are in intimate contact, has not been demonstrated. It is, however, an attractive hypothesis that has heuristic value, since it is testable biochemically as well as genetically.

Similarities of Late *com* Genes to Other Proteins

Identification of protein similarities

We have noted interesting similarities of several *com* gene products to other proteins in the data base (57, 131). Some of the similarities discussed below have been discovered independently by Whitchurch et al. (261).

Certain *com* proteins exhibit similarity to proteins of the pullulanase secretion system of *Klebsiella pneumoniae* (reviewed in reference 177). This gram-negative organism secretes the enzyme pullulanase across the inner and outer membranes and into the extracellular space. In order to secrete pullulanase, the product of *pulA*, at least 12 gene products are specifically

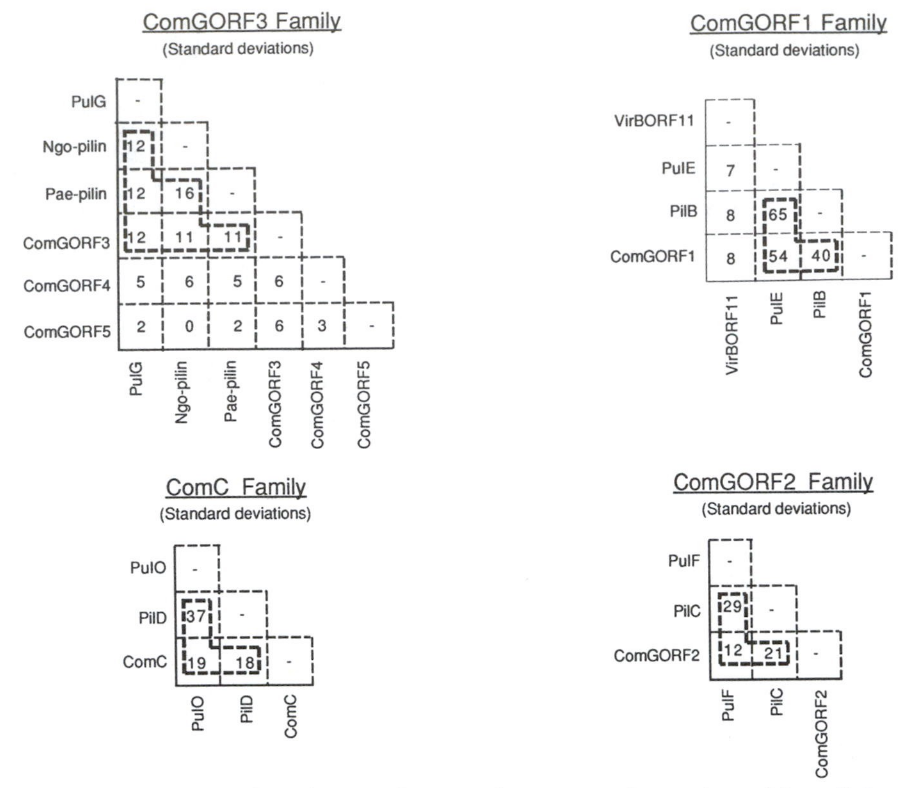

Figure 2. Numerical representation of similarities of *com* products to protein members of the pullulanase secretion, pilin assembly and synthesis, and *A. tumefaciens* Ti plasmid systems. The numbers are *Z* values obtained by using the RDF program (129) with 100 randomizations of the Com amino acid sequences. *Z* values in excess of 10 standard deviations are regarded as highly significant, and values of 5 to 10 standard deviations are regarded as possibly significant.

required. Of these, PulE, PulF, PulG, and PulO are similar to ComG ORF1, ComG ORF2, ComG ORF3, and ComC, respectively (Fig. 2). In addition, similarity of these Com proteins to several gene products required for the processing of the *Pseudomonas* strain PAK pilin and the assembly of this protein into a pilus structure (167) as well as to the primary sequence of the pilin protein itself (1) is evident. Specifically, PilB, PilC, and PilD are similar to ComG ORF1, ComG ORF2, and ComC, while ComG ORF3 resembles the pilin protein (Fig. 2). In Fig. 2, the *Z* values for these and other similarities are shown in tabular form. These values indicate that resemblances of the *com* products to those of the *pul* and *pil* systems are quite significant. In addition to participating in the assembly of pili, PilD is also required for the export of several extracellular enzymes in *Pseudomonas aeruginosa* PAK, and a nearly identical protein (XcpA) is similarly required in *P. aeruginosa* PAO (17, 226). Other proteins required for extracellular secretion in *Pseudomonas* spp. (XcpY and XcpZ) exhibit similarity to PulL and PulM (76). Secretion systems like those encoded by the *xcp* and *pul* genes may be widespread in gram-negative organisms. For instance, the *out* system of the plant pathogen *Erwinia chrysanthemi* is required for the export of various cell wall degradative enzymes, and several *out* and *pul* products are quite similar (102). Finally, a protein required in *P. aeruginosa* for twitching motility (PilT) is similar to PilB, PulE, and ComG ORF1 (261).

In addition to these similarities, which will be discussed further below, we have noted a weaker resemblance between ComG ORF1 and the VirB ORF11 protein of *Agrobacterium tumefaciens* (1, 200, 249, 250) (Fig. 2). The *virB* operon is thought to play a role in the transport of T-DNA from the bacterium to the plant cell. The purified VirB ORF11 protein is localized on the cytoplasmic face of the cell membrane (as is ComG ORF1), and it possesses both ATPase and autophosphorylating activities (38). The similarities among ComG ORF1, VirB ORF11, PilB, PilT, and PulE are most marked in the region corresponding to the nucleotide-binding site proposed above.

The relationships among ComG ORFs 3, 4, and 5; PulG; and the *Pseudomonas*-type pilin molecules are restricted to the N-terminal portions of these molecules and correspond approximately to a portion of pilin that is conserved among the so-called *N*-methylphenylalanine pilins. The latter proteins, found in *Neisseria*, *Moraxella*, and *Bacteroides* spp. in addition to *Pseudomonas* spp., are major subunits of pili, which are filamentous extracellular appendages. The pilins are processed by the removal of a few amino acid residues from their N termini followed by N methylation of the terminal Phe residue. Nonpiliated variants of *Neisseria* and *Moraxella* spp. have been reported to be competence deficient (24, 197, 212), as are strains deficient in ComG ORF3 (27). It is remarkable also that ComG ORFs 4, 5, 6, and 7 each exhibit some

```
orf3comG:1     M N E K G F T L V E M L I V L F I I G I L L L I T I P N V - T K H N Q T I Q K K
orf4comG:24    N E E K G F T L L E S L L V L S L A S I L L V A V F T T L P P A Y D N T A V R Q
orf5comG:3     R E N K G F S T I E T M S A L S L W L F V L L T V V P L W - D K L M A D E K M A
Ngopil:3       T L Q K G F T L I E L M I V I A I V G I L A A V A L P A Y - Q D Y T A R A Q V S
Paepil:2       K A Q K G F T L I E L M I V V A I I G I L A A I A I P Q Y - Q N Y V A R S E G A
pulG:2         Q R Q R G F T L L E I M V V I V I L G V L A S I V V P N L - M G N K E K A D R Q

orf3comG:40    G C E G L Q N M V K A Q M T A F E L D H E G Q T P S L A D L Q S E G Y V K K D A
orf4comG:64    A A S Q L K N D I M L T Q Q T A I S R Q Q R T K I L F H K K E Y Q L V I G D T V
orf5comG:42    E S R E I G Y Q M M N E S I S K Y V M S G E G A A S K T I T K N N H I Y A M K W
Ngopil:42      E A I L L A E G Q K S A V T E Y Y L N H G K W P E N N T S A G V A S P P S D I K
Paepil:41      S A L A S V N P L K T T V E E S L S R G W S V K S G T G T E D A T K K E V P L G
pulG:41        K V V S D L V A L E G A L D M Y K L D N S R Y P T T E Q G L Q A L V S A P S A E

orf3comG:80    V C P N G K R I I I T G G E V K V E H * * *
orf4comG:104   I E R P Y A T G L S I E L L T L K E H * * *
orf5comG:82    E E E G E Y Q N V C I K A A A Y K E K S F C L S I L Q T E W L H A S * * *
Ngopil:82      G K Y V K E V E V K N G V V T A T M L S S G V N N E I K G K K L S L W A R R E N
Paepil:81      V A A D A N K L G T I A L K P D P A D G T A D I T L T F T M G G A G P K N N K G
pulG:81        P H A R N Y P E G G Y I R R L P Q D P W G S D Y Q L L S P G Q H G Q V D I F S L

Ngopil:122     G S V K W F C G Q P V T R T D D D T V A D A K D G K E I D T K H L P S T C R D K
Paepil:121     K I I T L T R T A A D G L W K C T S D Q D E Q F I P K G C S R * * *
pulG:121       G P D G V P E S N D D I G N W T I G K K * * *

Ngopil:162     A S D A K * * *
```

Figure 3. Similarities of ComG ORF3, ORF4, and ORF5 (1) to protein members of the pullulanase secretion system (181) and to type IV pilins from *N. gonorrhoeae* (148) and *P. aeruginosa* (194). Amino acid identities are indicated by shading whenever at least three amino acids in the same position are identical except in the last few lines, in which two identical amino acids are indicated. The residue number of the first amino acid on each line is indicated. Type IV pilins are processed by cleavage between the conserved Gly (residues 6 and 7) and Phe (residues 7 and 8).

resemblance to ComG ORF3. All are small proteins (98 to 165 residues), all have hydrophobic N-terminal segments and very similar hydropathy profiles (1), and there appears to be some conservation of amino acid sequence within the N-terminal segments of ORFs 3, 4, and 5. Finally, it is worth noting that in addition to the resemblance of PulG to pilin and ComG ORF3 already noted, the N-terminal region of the PulG protein also resembles those of PulI and PulJ (261). Thus, both the Pul and the Com systems seem to encode several small proteins belonging to the pilin-like family.

Strom and Lory (225) examined structural requirements for the processing of pilin derived from *P. aeruginosa* PAK-NP and discovered that at least the Gly residue immediately preceding the conserved Phe (Fig. 3) is required for cleavage. These and other residues surrounding the pilin cleavage sites are conserved in *comG* ORFs 3, 4, and 5. Possibly, the pilin-like Com proteins are also processed (although no direct evidence for this exists), and the sequence similarity may reflect requirements for the processing site. The possibility that ComG ORFs 3, 4, and 5 are processed is supported by the observation that the ComC protein closely resembles PilD (see Fig. 5 and 6), and the latter is specifically required for the processing of *Pseudomonas* pilin (167). Most convincingly, Pugsley (178) has found that when *comC* is expressed in *E. coli*, it can mediate the processing of pilin from *N. gonorrhoeae*. In addition to providing cleavage sites, the conserved sequences may reflect requirements for assembly of the pilinlike proteins into a higher-order structure.

Finally, *comF* ORF1 closely resembles a large class of ATP-dependent helicases found in both prokaryotic and eukaryotic systems (131). These include PriA, a 3′→ 5′ helicase and DNA translocase that is a component of the *E. coli* primosome and is required for its assembly (211).

What do the protein similarities mean?

The striking amino acid sequence similarities noted above establish the relationships between the competence system of *B. subtilis* and the pilus assembly and protein export systems of *Pseudomonas*, *Xanthomonas*, *Erwinia*, and *Klebsiella* spp. Two types of roles for the proteins involved in these relationships may be considered. First, the *pul*, *pil*, *out*, and *xcp* products may be directly required for protein secretion in the gram-negative systems, while their homologs in the *B. subtilis com* system may be required for the uptake of DNA. This hypothesis therefore proposes that many of these late competence proteins are directly involved in the transport of DNA. Alternatively, certain of these *com* products may fulfill morphogenetic roles required for the transport or assembly of components of the transformation machine without direct involvement in DNA processing and transport. The similarities of *com* products to those of the *pil*, *pul*, *xcp*, and *out* systems are certainly suggestive of such a morphogenetic role and strengthen the second hypothesis. The last four systems are ultimately required for the export of specific extracellular proteins through the inner and outer membranes of *Klebsiella*, *Xanthomonas*, *Erwinia*, and *Pseudomonas* spp. These export systems also require *sec* products, presumably for inner membrane transport. The latter requirement may be true of the *B. subtilis com* system as well in view of the probable role of *secA* in competence. However, *B. subtilis* does not possess an outer membrane. What, then, would be the role of morphogenetic Com proteins that are similar to those of the gram-negative systems? It seems likely that certain of the *pul* products are needed to assemble a cell surface apparatus that exports pullulanase. The Com proteins may function analogously, assembling components of the DNA transport machine.

At this time, it is not possible to rigorously distinguish between a direct or a morphogenetic role for any given *com* protein. In those assembly systems for which a final structure has been defined and for which components have been described, it is possible to classify a gene product as fulfilling a morphogenetic role if it is required for assembly but is not a component of the structure. Such a classification appears to be possible in the case of the *Pseudomonas* pili; PilB, PilC, and PilD are in this sense morphogenetic proteins, while the product of *pilA* is not. It is also the case with certain bacteriophages, in which morphogenetic proteins that are not themselves part of the completed virion have been identified. However, in the *com* system, although it is likely that a multiprotein surface-localized machine is involved in the binding, processing, and transport of DNA, this complex has not been demonstrated, and its composition is certainly not understood. These ambiguities can be illustrated by the case of ComG ORF1, a probable ATP-binding protein. It is possible that this protein serves to couple DNA uptake to ATP hydrolysis. Alternatively, it may be required for the transport of another protein component of the competence apparatus, either as an energy coupler or as molecular chaperone. For instance, PapJ is a likely nucleotide-binding protein that appears to play a role as such a chaperone in the assembly of the P pili of *E. coli* (230). Although we cannot rigorously classify any of the late competence products either as morphogenetic or as directly involved in the uptake of DNA, we can present some reasonable inferences.

The product of *comE* ORF3 appears to be required specifically for DNA uptake as opposed to binding (see above) and may be directly involved in this process. The ComG ORF3 product, which resembles the structural protein pilin, may play a direct role as part of the competence apparatus. This is consistent with the evidence reviewed above that suggests a specific structural organization of ComG ORF3 in competent cells (27). It is possible that the products of *comG* ORFs 4, 5, 6, and 7, which, as noted above, resemble ComG ORF3 in some respects, also play direct roles as part of the competence apparatus. The related N-terminal hydrophobic sequences of ComG ORFs 3, 4, and 5 resemble the conserved N-terminal sequences of the *N*-methylphenylalanine pilins, which have been implicated in the protein-protein interactions involved in the assembly of pili (251). We therefore propose that the five small ComG proteins assemble to form part of a cell surface-associated structure for the binding and uptake of transforming DNA (1, 27). Consistent with this idea is the overlap of the *comG* ORF3, -4, -5, and -6 reading frames noted above, which suggests that these products may be translationally coupled to ensure their synthesis in appropriate relative amounts. It is noteworthy that the published model for the structure of *Pseudomonas* pili proposes a helical array of pilin subunits with fivefold symmetry and a central pore of 1.2 nm (78). A tubular structure may contribute to the formation of a channel for passage of transforming DNA through the cell wall and membrane.

The *comF* ORF1 product may function as a helicase-DNA translocation protein during transformation. Interestingly, recent evidence suggests that this protein is required for DNA uptake and not for binding (131).

On the other hand, the ComG ORF1, ComG ORF2, and ComC proteins, which are similar to PilB, PilC, and PilD, may very well be morphogenetic, since the last three products are required for the assembly of pili but not for the synthesis of pilin or for its transport into the inner or outer membranes of *Pseudomonas* spp. (167). PilD is needed for the proteolytic processing of *Pseudomonas* pilin (167). It will be interesting to determine whether ComG ORFs 3, 4, and 5 are processed by a ComC-dependent mechanism. Another candidate for a role in assembly rather than for direct involvement in the DNA uptake process is the *div-341* (*secA*) locus described above (188, 190, 192, 193). One interpretation of the requirement for this *secA* homolog is that the export of one or more late competence proteins requires the operation of the major (*sec*) secretion pathway in addition to more specialized gene products.

Much of the preceding discussion is speculative. Clearly, experiments designed to describe the postulated competence machinery and enumerate its protein components are needed.

REGULATION OF COMPETENCE

Three Modes of Competence Regulation in *B. subtilis*

Competence in *B. subtilis* is subject to three types of control. In the usual glucose–minimal-salts-based competence medium supplemented with amino acids, competence develops postexponentially. The time of transition from exponential growth is defined as t_0. Competence is maximal about 2 h later, at t_2 (3). This period reflects the first mode of competence regulation, i.e., growth stage dependence. Transformability develops to extremely low levels in complex media. When glucose is replaced by glycerol in the usual competence medium or when the glucose medium is supplemented with glutamine, the expression of competence is also low (3). These observations indicate a second aspect of control, i.e., dependence on medium. Finally, as noted above, only about 10% of the cells in the culture achieve competence, and these cells are metabolically and morphologically distinct (52, 95, 100, 146, 165, 202). Thus, an event that results in the development of at least two cell types, with expression of competence and late competence genes (3) in one of them, occurs.

The *B. subtilis* SOS System

E. coli responds to the stress of DNA damage by inducing the expression of a series of genes, including several devoted to repair of this damage (248). The signal for induction is thought to involve single-stranded DNA, which interacts with the RecA protein (in *E. coli*) and results in cleavage of the LexA repressor and consequent derepression of the SOS regulon. A similar pathway appears to exist in *B. subtilis* (see chapter 37). In addition to this mode of regulation, however, the *B. subtilis* SOS system is at least partially derepressed as cells reach competence (132, 265), as can be shown by using transcriptional fusions of *lacZ* to a set of genes that can be induced by DNA damage (*din* genes). These genes are also induced in the competent state and specifically in the competent cell subpopulation. Competence-specific induction of the *din* genes, like damage induction, requires a functional RecA product. Also, as part of the competence response, the *recA* gene itself is strongly derepressed (135). However, the competence induction pathway for *recA* expression, unlike the DNA damage-inducible pathway, does not depend on the presence of a functional RecA gene product. Also unlike the DNA damage-induced pathway, competence-dependent induction of the SOS-like response depends on the *degU*, *spo0A*, and *spo0H* gene products (265, 266), as does competence itself (see below). It thus appears that induction of the SOS response by DNA damage and induction during the development of competence differ and that the latter regulatory mode overlaps regulation of the competence regulon. It would be of great interest to determine whether this form of SOS-like induction depends on all the competence regulatory genes described below. The implications of competence-linked induction of DNA repair genes and a possible mechanism for competence-linked induction of the SOS system will be further discussed later in this review.

Regulatory Genes

Most investigations of the control mechanisms involved in competence have been concerned with identifying and characterizing regulatory genes. More than a dozen such genetic elements have been identified, and their relationships have been explored. As a result, a good deal is known about the regulatory pathways involved, although it is clear that much more remains to be understood. What follows is a catalog of these loci and a brief description of the phenotypes associated with each locus. This initial description will make little attempt to explore the mechanisms of action of these genes or how they interact to form a regulatory network; those tasks will be attempted later in this review. The locations of the regulatory loci on the *B. subtilis* genetic map are shown in Fig. 1.

Mutations of each of the regulatory loci have been tested for their effects on expression of other regulatory genes and on the late competence genes. In most cases, this testing was done by using *lacZ* transcriptional fusions to the target genes; in some cases, testing was by direct measurement of steady-state mRNA concentration. These experiments have generally revealed that the mutations have profound effects on transcription of the late genes but no detectable effects on one another. Exceptions to this generalization will be noted below.

Competence Regulatory Loci

spo0H

Null mutations in *spo0H* are asporogenic and are about 20-fold depressed in transformation efficiency (3, 191). Mutants exhibit decreased expression of all of the known late *com* genes, as is shown by using transcriptional fusions to *lacZ* (3). The *spo0H* locus encodes σ^H, an alternative σ factor that is required for reading essential sporulation genes (70). The *com* phenotype of *spo0H* null mutants implies that at least one essential *com* gene is transcribed by $E\sigma^H$; the identity of the $E\sigma^H$ target(s) is unknown.

abrB

Null mutants in *abrB* are depressed about 20- to 50-fold in competence and fail to express late competence genes at the wild-type level (70). *abrB* encodes a so-called transition state regulator that prevents the inappropriate expression of various genes that are normally expressed postexponentially and acts as a DNA-binding protein (172, 220). The positive effect of *abrB* on the expression of *com* genes may be due to a positive effect of AbrB on expression of at least one essential gene in the competence pathway or to repression of a negatively acting competence control element. The identity of this target gene is not known.

spo0A

spo0A null mutants are depressed 500- to 1,000-fold in competence, are severely depressed in the expression of late competence genes (3, 191, 214), and are completely deficient in sporulation (107). Null mutants in this locus exhibit a highly pleiotropic phenotype, reflecting the central role played by the Spo0A protein in a variety of forms of postexponential expression. One of the functions of Spo0A is to down-regulate the expression of *abrB* on the level of transcription; *spo0A* mutants overexpress AbrB (172). A *spo0A abrB* double null mutant exhibits the same level of competence as does a null mutant in *abrB* alone (3). It thus appears that the sole role for Spo0A in the competence system is to prevent the overexpression of AbrB, which must be capable of playing a negative as well as a positive role in the development of competence. A mutation of *spo0A* that abolishes the ability of Spo0A to act as a transcriptional activator but does not interfere with its function as a negative regulator of *abrB* has recently been described (173). This mutation abolishes sporulation. However, it exhibits normal regulation of competence (182). This further supports the conclusion that Spo0A functions in competence only to prevent the accumulation of excessive levels of AbrB. Further support will be presented below in the section on *mec* mutants.

The amino acid sequence of the N-terminal domain of Spo0A is similar to that of a class of proteins known as response regulators (75, 217). In most cases, these are transcriptionally active molecules that can be phosphorylated by cognate proteins collectively referred to as histidine kinases. Phosphorylation of the conserved N-terminal domain of a response regulator is thought to cause a conformational change that alters the activity of the C-terminal DNA-binding domain. Phosphorylation of Spo0A increases its binding affinity for a sequence upstream from the *abrB* promoter about 20-fold (233), presumably facilitating its action as a repressor. A protein with homology to the histidine kinases, which can phosphorylate Spo0A (via a phosphorylation cascade), is the SpoIIJ (KinA) protein (12, 30, 170).

This and other evidence clearly indicate that phosphorylation of Spo0A is a critical event in the initiation of sporulation. Is phosphorylation a signaling event for competence? Null mutations of *spoIIJ* have no obvious effect on competence (253, 259). It has recently been shown that SpoIIJ acts by phosphorylating the Spo0F protein and that the latter transfers the phosphate to Spo0B, which in turn phosphorylates Spo0A (30). Null mutations in the *spo0F* or *spo0B* genes, like those in *spoIIJ*, have no significant effects on competence (3). It is tempting therefore to conclude that signaling for the development of competence does not proceed via the phosphorylation of Spo0A. Since, as noted above, AbrB plays both positive and negative roles in competence, perhaps the affinity of the unphosphorylated Spo0A protein for the upstream *abrB* site is sufficient to prevent excessive expression of *abrB* but not so great as to drive expression below the required level. However, it is likely that other histidine kinases can phosphorylate Spo0A by what is often referred to as cross-talk. Perhaps one or more of these kinases are required to activate Spo0A for purposes of competence. It is also possible that the increase in *spo0A* expression that occurs at t_0 is involved in signaling the onset of competence. However, this increase is glucose repressed (264) and may not occur in competence medium. Further light on the role of Spo0A can be shed by the study of *spo0A* mutations. Green and Youngman (86) have altered the conserved aspartate residue of Spo0A that is thought to be the site of phosphorylation. This change completely abolishes sporulation but results in an intermediate level of competence about 10- to 100-fold higher than that of a *spo0A* null mutant. This result implies at least that phosphorylation of Spo0A is not absolutely required for the development of competence. The partial competence deficiency may be due to a potentiating effect of phosphorylation or to interference of the mutation with some aspect of Spo0A function other than phosphorylation. Green and Youngman have combined this mutation with a second alteration in Spo0A: a deletion in the protein that appears to render sporulation independent of Spo0A phosphorylation. This combination of mutations restores a nearly wild-type level of competence. This finding is consistent with the first notion stated above, i.e., that phosphorylation of Spo0A by a histidine kinase other than SpoIIJ plays a potentiating role in competence. Perhaps nonphosphorylated Spo0A can partially control *abrB* expression but not sufficiently to permit full expression of competence. Phosphorylation of Spo0A (by a SpoIIJ-independent pathway) would result in a more optimal concentration of AbrB and hence in a higher expression of competence.

spo0K

The only known mutation in this locus, until recently, was *spo0K141* (175). This mutation causes an oligosporogenic phenotype and decreases competence several hundredfold (191). It is required for the expression of the late competence genes (183). Recently, two laboratories have determined the complete sequence of *spo0K* and shown it to consist of five ORFs, all of which are highly similar to the similarly organized oligopeptide permease (*opp*) operons of *E. coli* and *Salmonella typhimurium* (171, 186). This permease is related to several of the bacterial periplasmic permease systems (10). The promoter-proximal ORF encodes a peptide-binding protein that is periplasmically located in gram-negative bacteria and is probably covalently anchored to the exterior face of the gram-positive cell membrane by fatty acid acylation (171). The second and third ORFs encode integral membrane proteins, and the fourth and fifth encode hydrophilic proteins that possess ATPase activity and probably directly couple ATP hydrolysis to transport (10). At least the fifth ORF of *spo0K* (ORFE) is required for competence (186), but its disruption has only a minimal effect on sporulation and on transport of at least one peptide (171, 186).

sin

Null mutations in the *sin* locus depress competence several hundredfold (82) and lower the expression of late competence genes (94). *sin* was originally identified as an ORF that inhibits sporulation when overexpressed (82). It is known to encode a DNA-binding protein that exerts negative effects on the expression of several genes involved in sporulation and also on the expression of various degradative enzymes that normally exhibit increased postexponential expression in *B. subtilis* (83, 139). It is not known whether Sin acts positively or negatively during the development of competence; like the AbrB protein, it may activate expression of an essential competence gene or it may repress expression of a repressor. *sin* is apparently transcribed by an $E\sigma^A$ promoter and is expressed throughout growth at an approximately constant rate (81).

degU

Null mutations in *degU* are severely deficient in transformation (228) and prevent the expression of the known late competence genes (131, 183). *degU* was originally identified because it is needed for the expression of certain degradative enzymes, mostly extracellular, that are normally synthesized at an enhanced rate postexponentially (16, 106, 120, 121). Upstream from *degU* is the *degS* gene, which is also required for degradative enzyme synthesis but not for competence (106, 120, 155, 156, 183). The DegS and DegU products are histidine kinase and response regulator proteins, as suggested originally by their amino acid sequences and confirmed by biochemical tests on the purified proteins (44, 106, 120, 158).

In addition to null mutations in these genes, point mutations that result in hyperproduction of the degradative enzymes have been described and characterized on the sequence level (155, 215). These so-called *degS*(Hy) and *degU*(Hy) mutations profoundly interfere with development of competence (155, 215) and with expression of the known late competence genes (183). They have usually been interpreted as alterations in the proteins that result in enhanced activation (presumably via phosphoryl-group transfer) of DegU by DegS. Inactivation of *degS* restores competence to a *degU*(Hy) mutant (155). These observations have led to the suggestion that the "unactivated" (presumably unphosphorylated) form of DegU is specifically required for competence, whereas the activated form is needed for the expression of degradative enzymes. Strong support for this hypothesis was derived from the *degU146* mutation, which converts the

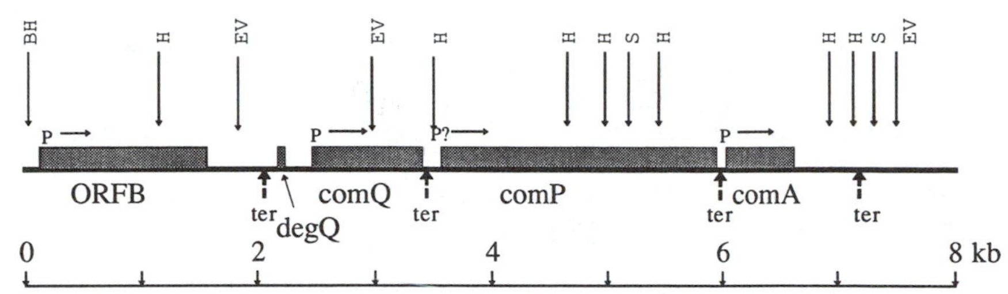

Figure 4. Genetic and physical maps of the *comA-comP* region of the *B. subtilis* chromosome. The ORFB determinant was previously called *comB* (94) but is now known not to be a competence gene (57). The DNA sequence of the entire 8-kb region has been determined. References are given in the text. BH, *Bam*HI; H, *Hin*dIII; EV, *Eco*RV; S, *Sst*I. The positions of promoters (P) and terminators (ter) are given. The *degQ* determinant encodes a 46-amino-acid peptide and is not drawn to scale.

conserved aspartate residue that is the likely target for phosphorylation to an asparagine residue (155). In addition, preliminary experiments using crude extracts suggest that the *degU146* allele encodes a protein that is no longer phosphorylated by the DegS protein kinase (44). This mutation decreases degradative enzyme production but was reported to have little if any effect on competence. However, we have recently noted that in a different genetic background from the one used by Dahl et al. (44), the *degU146* mutation severely depresses competence (97). Thus, the role of DegU phosphorylation may be complex.

comA and *comP*

Null mutations in *comA* or *comP* reduce competence about 400-fold and interfere with late *com* gene expression (94, 162, 256, 259). *comA* was independently discovered as a locus needed for expression of the peptide antibiotic surfactin and was named *srfB* (162). *comA* and *comP* are adjacent genes (Fig. 4). Judged by their predicted amino acid sequences, they appear to encode response regulator and histidine kinase proteins, respectively. More recently, ComA has been purified from an overproducing *E. coli* clone and shown to accept phosphate from the *E. coli* NRII (NtrB) protein, a histidine kinase, in vitro (231). In addition to the resemblance of the ComA N-terminal domain to other response regulators, the C-terminal domain of ComA exhibits similarity to the C-terminal sequences of several known DNA-binding proteins and to DegU (Fig. 5). These similarities, plus the effects of *comA* and *degU* mutations on expression of a number of genes, implies that the two response regulators function as transcription factors. The predicted

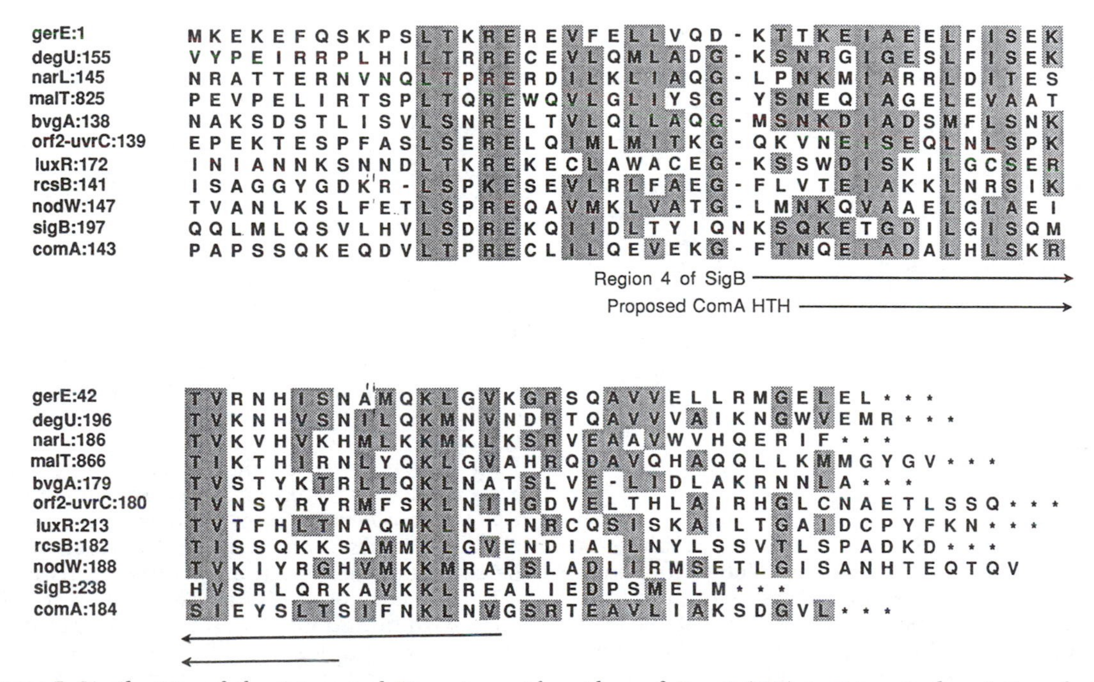

Figure 5. Similarities of the C-terminal 71 amino acid residues of ComA (256) to C-terminal moieties of other suspected and known transcription factors. Amino acid similarities are indicated by shading when six or more of them occur at the same position. For this purpose, the amino acids are grouped as follows: M, L, I, and V; H, K, and R; D and E; Q and N; A, G, S, and T; F, W, and Y; P; C. Locations of the region 4 helix-turn-helix (103) of SigB (71), based on comparison of σ factors, and of the proposed helix-turn-helix of ComA are indicated. Additional protein sequences and their sources are GerE (43), DegU (106, 120, 228), NarL (166), MalT (40), BvgA (13), ORF2-UvrC (198), LuxR (49), RcsB (218), and NodW (85).

ComP amino acid sequence exhibits eight regions of pronounced hydrophobicity near the N terminus, and a specific model has been proposed for the organization of ComP as a polytopic membrane protein (259). Thus, although the signal to which ComP responds is not known (see below), this signal is presumably extracellular or mediated via other membrane components.

lacZ fusions to *comA* have shown that the expression of this gene increases about twofold during growth, reaching a maximum after t_0 (259). High-resolution primer extension has revealed a start site upstream from *comA*, near a probable Eσ^A promoter. However, disruption of *comP* prevents the twofold increase in *comA* expression, suggesting that this increase is due to readthrough from a promoter upstream of *comP*. The predicted readthrough transcript was detected by S1 nuclease analysis. The significance of this is not known, nor is the location of the promoter responsible for transcription of *comP*.

comA and *comP* are also required for starvation-induced expression of certain other loci, such as *degQ* (156) and *gsiA* (157), neither of which is required for competence.

comQ

Upstream from *comP* (Fig. 4) is an ORF, identified by sequencing, that encodes a predicted protein of 328 amino acids (257). Disruption of this ORF by insertion of antibiotic resistance cassettes resulted in competence deficiency to the same residual level as that seen in *comA* or *comP* null mutants and also eliminated expression of the known late competence genes. These disruptions were not polar on *comA* fusions to *lacZ*, and they could be fully complemented for competence by a single copy of the ORF integrated at an ectopic chromosomal site. This ORF has been named *comQ* and appears to be a new competence locus.

High-resolution primer extension analysis identified a transcriptional start site for *comQ* appropriately positioned downstream from a sequence suggestive of a vegetative Eσ^A promoter (257). Downstream from *comQ* and between it and *comP* is a palindrome that resembles a rho factor-independent transcriptional terminator, and S1 nuclease mapping confirmed the existence of a termination site near this element. As judged by primer extension analysis and by study of a translational fusion of *comQ* to *lacZ*, the expression of this gene in competence medium was relatively high 2 h before t_0 (t_{-2}) and declined to an unmeasurable level after about t_{-1}.

srfA

A Tn*917-lacZ* insertion (*csh-293*) that resulted in both competence deficiency (about 400-fold) and a subtle oligosporogenic phenotype was identified (110). This mutation also depressed expression of the late competence genes (183). A locus that was needed for competence was identified independently and called *comL* (241). Finally, in a search for genes required for postexponential expression of the extracellular peptide antibiotic surfactin, a locus known as *srfA* was characterized (160). It has been shown that *csh-293*, *comL*, and *srfA* mutations identify the same locus, and we propose here to retain the name *srfA* (159). *srfA* consists of a large operon (greater than 20 kb), of which only the promoter-proximal portion (about 15 kb) appears to be required for the development of competence (159, 240, 241). This portion of the operon encodes two ORFs with predicted amino acid sequences similar to those of tyrocidine synthetases I and II (140, 150, 159, 163, 239, 241, 252). These enzymes are required for the nonribosomal assembly of peptide antibiotics via the stepwise amino acyl activation and polymerization of amino acids (130). It has been pointed out that the involvement of this operon is particularly interesting in view of the possible involvement of a soluble peptide factor in signaling for competence development (159).

srfA is expressed at a low level throughout growth, but its expression increases sharply just before t_0 (110, 160). This kinetic pattern suggests that *srfA* may play an intermediate role in the competence signaling pathway, a prediction that is supported by further evidence. The increase in *srfA* expression noted in competence medium at t_0, is dependent on expression of *comQ* (258), *comP* (98), *comA* (98, 162, 241), and *spo0K* (89, 98, 161) and is partially dependent on *spo0A* and *spo0H* (98, 110, 160). *srfA* expression is not measurably dependent on *abrB*, *sin*, and *degU*. This suggests a hierarchical arrangement of gene products in the competence signaling pathway, a point that will be further discussed below. However, one may ask at this point whether the increase in *srfA* expression noted at t_0 is important for competence. It is possible that it is not and that the dependence of the increased *srfA* expression on several competence regulatory genes has nothing to do with the dependence of competence on these genes. Nakano et al. have constructed a strain in which the expression of *srfA* has been placed under control of the regulatable *Pspac* promoter (161). In this strain, competence develops only in the presence of an inducer of *Pspac* such as isopropylthiogalactose (IPTG). The level of competence achieved under these conditions is lower than in the wild-type strain, presumably because the *Pspac* promoter is weaker than the normal *srfA* promoter (163). It is interesting that this level of competence is expressed throughout growth when the strain is grown continuously in the presence of IPTG (98, 164). This observation suggests that the growth stage-specific signal may normally act prior to the increased expression of *srfA* at t_0. When the *srfA-Pspac* construct is combined with several of the null mutations in other competence regulatory genes, the level of transformability reached is extremely low in the absence of IPTG, as it is in the original *Pspac-srfA* strain. In the cases of the *Pspac-srfA* strains carrying *comP*, *comA*, *comQ*, or *spo0K* mutations, growth in the presence of inducer restores competence to the level achieved in the construct carrying the *srfA-Pspac* construct alone in the presence of IPTG (98, 164). When *srfA* is combined with mutations in *degU* or *abrB* mutations, the addition of IPTG does not elevate the level of competence (98). In other words, mutations in *comP*, *comA*, *comQ*, and *spo0K* seem to be bypassed for their effects on competence by expression of *srfA* under *Pspac* control. These observations suggest that some of the regulatory gene products (*comA*, *comP*, *comQ*, and *spo0KORFE*) are needed only to turn on *srfA* expression in response to environmental signals, whereas others (*degU*, *abrB*, and *sin*) are not and act either later than or in parallel to *srfA* for competence signaling.

comK

comK was identified by van Sinderen et al. (241). It has been shown to increase its expression at t_0 and to be required for the expression of the late competence genes (241). In addition, *comK* expression is dependent on that of *comA*, *srfA* (*comL*), *abrB*, and *sin* (97, 238, 241). Van Sinderen and Venema (240) also found that although it was impossible to achieve the expression of ComK on a high-copy-number plasmid, moderate overexpression appeared to result in the elevated expression of competence during exponential growth. This suggests that the ComK product may act positively in turning on competence at t_0. The *comK* locus has been cloned and sequenced (240), and its predicted product bears no striking relationship to any known protein.

mec Loci

As noted above, the development of competence and the expression of late competence genes is poor in complex media and maximal in glucose–minimal-salts-based media. To further our understanding of the regulatory loci involved in the response to nutritional environment, mutants that were capable of expressing late competence genes (and competence itself) in complex media were isolated following ethyl methanesulfonate mutagenesis (69). The mutations responsible for this phenotype were called *mec* (medium-independent expression of competence). They were mapped to at least two loci, *mecA* and *mecB* (Fig. 1) (69). Although mutations in *mecA* and *mecB* permitted the development of competence and the expression of late competence genes in complex media, they exhibited enhanced expression at t_0, as for the mec^+ strain. Thus, it might appear, as we originally inferred, that growth stage-related regulation is independent of the *mec* mutations. However, the *mecA* and *mecB* mutants did express the late competence genes prior to t_0 at a rate greater than that of the wild type, so that interpretation of this point is somewhat ambiguous (69).

mecA is closely linked by transduction to the *spo0K* locus (183). Recently, I have observed that it is linked as well to *comK* and is located between these two loci (182). *mecA* has been cloned, and the *mecA42* allele has been shown to be recessive to the wild type (117). *mecA*, therefore, probably encodes a negative regulator of competence. *mecB* is closely linked by transformation to rifampin, streptolydigin, and lipiarmycin resistance alleles (69) that were previously shown to be in or near the *rpoB* gene (128, 210). The latter encodes the β subunit of RNA polymerase. The close transformational linkage observed is consistent with the hypothesis that the two *mecB* mutations studied are alleles of *rpoB*. Other than these comments, we can say little about the nature of the *mecA* and *mecB* loci and their respective mutations.

The *mecA42*, *mecB31*, and *mecB23* mutations bypass all of the competence regulatory genes described above except for *spo0A*, which is partially bypassed (116, 183, 184), and *comK* (116, 238). For example, the *mec* mutations permit the development of competence and the expression of late competence genes even in genetic backgrounds carrying inactivating mutations in the *comA*, *comP*, *degU*, *abrB*, *spo0K*, *srfA*, and *sin* genes. Taken at face value, the suppression of all of these regulatory mutations implies that the *mec* products act later than the products of the other regulatory genes during competence development.

Partial suppression of the *spo0A* mutant would imply that the negative effect of AbrB (which, as described above, is overproduced in the *spo0A* background) must be exerted at multiple points both before and after the point of action of *mec*. Some recent experiments are consistent with this interpretation of the relationship of *mecA* to *spo0A* and *abrB* (182). A *spo0A abrB* double null mutant was prepared in a *mecA42* background that also carried a *lacZ* fusion to *comG*. In this strain, essentially normal levels of *comG*-driven β-galactosidase were observed. This is as predicted, since the *abrB* mutation was able to suppress the effect of the *spo0A* mutation, and *mecA42* bypassed the effect of the *abrB* null mutation.

The ability of the *mec* mutations to bypass the early gene requirements suggests that these loci are all required for a signaling process that converges at some later point. Since the *mec* mutations allow expression of competence and late competence genes in complex media, in the absence of glucose, and in the presence of a high glutamine concentration (69) (all treatments that ordinarily repress competence), it appears that the signals involved are at least in part nutritional.

A further inference is that the early regulatory gene products, several of which are known or thought to be regulators of transcription (ComA, DegU, Sin, AbrB, Spo0A, and Spo0H), probably do not act directly on the late-gene upstream sequences to modulate their expression. This follows from our interpretation of the *mec* suppression data, since the Mec products appear to act later than the other regulatory products.

Finally, it appears that the role of the *mec* genes is not restricted to the competence system. Mueller and Sonenshein (157) have identified two genes (*gsiA* and *gsiC*) that are induced to express themselves at elevated rates under conditions of glucose starvation. *gsiA* is not required for competence but is dependent for its expression on *comA* and *comP*. It is not dependent on *degU*, *degS*, *sin*, *srfA*, or *spo0K*, at least in complex medium. However, the *comA*-*comP* dependence of *gsiA* expression is bypassed by mutations in *mecA*, *mecB*, or *gsiC*. In addition to presenting additional complexities, these observations suggest a wider role for the *mec*, *comA*, and *comP* genes and imply that the set of *mec*-bypassable genes constitutes a pathway with more than one branch. Additional support for this notion derives from unpublished results of Msadek and Kunst (154), who have observed that the dependence of expression of levansucrase on *degU* is bypassed by mutations in *mecA* and *mecB*. This implies that DegU may not interact directly with targets near the degradative enzyme promoters and again suggests that the *mec*-bypassable pathway contains several branches.

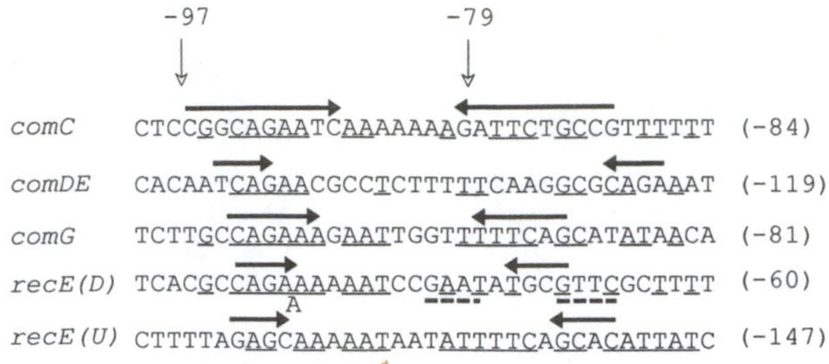

Figure 6. Comparison of proposed CTF-binding sites upstream from *comC*, *comDE*, *comG*, and *recA*. Identities in at least three of the four sequences are indicated by underlining. Arrows indicate partial dyad symmetries. A deletion extending from the left to position −97 has no effect on expression of *comC*, whereas a deletion extending to −79 reduces *comC* expression (152). Positions of the centers of proposed dyad symmetries relative to transcriptional start sites (+1) are indicated to the right. Dotted lines indicate a *din* consensus site upstream from the *recA* promoter (35).

What Turns On Late-Gene Expression?

This question of what turns on late-gene expression is raised acutely by the inference that early regulatory products do not act directly on late competence gene promoters. A clue to the answer was provided by observations that DNA fragments derived from the promoter regions of the *comG* and *comC* transcription units were capable of repressing expression of the known late competence genes when many copies of these fragments were present (1, 152). This observation suggested that a limiting transcription factor was being titrated by the promoter fragments and that the action of this factor was general, since the *comC* fragment repressed *comG* expression and vice versa.

The upstream sequences required for expression of *comC* were explored by the analysis of deletions (152). It was found that sequences located between −97 and −79 were required for this expression. Centered at −84 was a sequence with partial dyad symmetry. Upstream from the *comG* and *comDE* promoters, similar sequences containing weak dyads with centers of symmetry at −81 and −119, respectively, were noted (Fig. 6). When various *comC* fragments were placed in multicopy in a wild-type *B. subtilis* strain, the intact fragment depressed competence as expected, as did the fragment deleted to −97 from upstream. However, the −79 deletion that removed the sequence found to be essential for *comC* expression did not inhibit transformability. It thus appears that these sequences define a binding site for a factor required for the expression of at least some of the late competence genes. My coworkers and I have named the factor defined by these in vivo experiments competence transcription factor (CTF).

Gel shift experiments have served to identify an activity that has many of the properties expected of CTF (152). This activity binds specifically to a fragment carrying the sequences defined by the in vivo experiments described above as responsible for CTF binding. The activity binds to the fragment that is deleted up to −97 but not to the fragment deleted to −79 (116). The binding activity is detectable only in extracts of the wild-type strain prepared after t_0 and from competence medium (152). It is detectable in *mecA42* extracts prepared from complex medium as well. No expression of the binding activity is detectable in null mutants for *comA*, *degU*, *spoOA*, *srfA*, *spoOH*, *comK*, or *spoOK* (116, 152). The *comA* dependence of the gel shift activity was bypassed by the *mecA42* mutation. This in vitro activity therefore appears to be under competence control and is likely, although not proven, to represent the CTF activity as defined by in vivo experiments.

CTF may be required for expression of the late competence genes that have been tested, i.e., *comG*, *comC*, and *comDE*. This is inferred from the in vivo titration experiments (152) and is consistent with the presence of similar sequences upstream of these genes (Fig. 6). The simplest interpretation is that CTF is not only necessary but also sufficient for the expression of competence. If this hypothesis is correct, then the entire regulatory cascade may be needed only to express or activate CTF. This possibility will be testable when *ctf* is cloned and characterized.

At this time, it is appropriate to return to the question of competence-linked induction of the SOS regulon. Cheo et al. (35) have sequenced the promoter regions of several DNA damage-inducible genes from *B. subtilis* and have identified sequences necessary for damage-inducible regulation of the *din* genes. Within these essential regions were conserved sequences that were interpreted as representing recognition elements for the binding of a presumed SOS-specific regulatory protein, possibly analogous to the LexA product of *E. coli*. Several of the *din* gene regulatory regions contained more than one of these putative recognition elements. Upstream from one of the three copies of this element associated with the *recA* gene, we have identified a sequence that resembles the putative CTF-binding element (Fig. 6). The SOS recognition sequence is centered at position −50, and the CTF recognition element is centered at about −60. Yasbin and coworkers (36) have identified an additional sequence upstream from *recA* that also resembles a CTF-binding site (Fig. 6). This suggests a model for dual regulation of the *recA* determinant in which binding of a LexA-like protein represses expression, and this repression can be relieved in two ways: by DNA damage and RecA-dependent cleavage of LexA and by binding of CTF, possibly resulting in displacement of LexA. The promoter regions of the *din* genes other than *recA* do not seem to possess CTF-binding sequences. This is consistent with the finding of Love et al. (132) that whereas the induction of *recA* in the competent state does not require a functional RecA product, the competence-specific induction of the *din* genes does require this gene product. Presumably, the *din* gene induction is a secondary consequence of *recA* induction and/or activation.

The Regulatory Cascade

The information outlined above suggests a pathway of information flow involved in signaling the onset of competence. Various forms of this pathway have been suggested before, and they differ slightly from one another (57, 58, 68, 183, 241). Hopefully, these differences reflect the evolution of our understanding. They also serve to emphasize the provisional and incom-

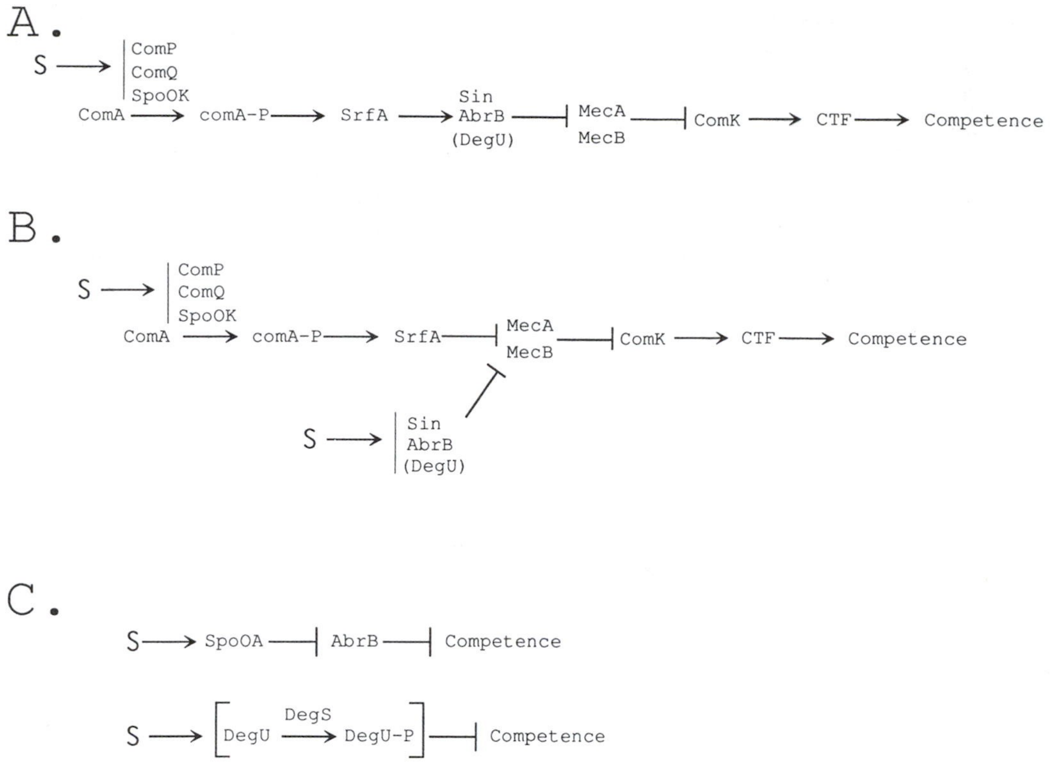

A.

S ⟶ ComP / ComQ / SpoOK

ComA ⟶ comA-P ⟶ SrfA ⟶ Sin / AbrB / (DegU) ⊣ MecA / MecB ⊣ ComK ⟶ CTF ⟶ Competence

B.

S ⟶ ComP / ComQ / SpoOK

ComA ⟶ comA-P ⟶ SrfA ⊣ MecA / MecB ⊣ ComK ⟶ CTF ⟶ Competence

S ⟶ Sin / AbrB / (DegU)

C.

S ⟶ SpoOA ⊣ AbrB ⊣ Competence

S ⟶ [DegU —DegS⟶ DegU-P] ⊣ Competence

Figure 7. Regulation of competence: schemes of signal transduction and information flow. Each boldface S indicates a likely point at which signals might be received by the regulatory apparatus. Arrows and lines terminated by perpendiculars indicate positive and negative regulation, respectively. DegU is in parentheses because its location in the scheme is not certain. (A and B) Alternative arrangements that are consistent with the available data as outlined in the text. The schemes differ in one respect: whether the *srfA* signal is transduced through *sin* and/or *abrB* (A) or whether an independent signal is relayed through some combination of the products of these genes that then acts directly or indirectly on *mecA* in combination with the *srfA* signal (B). (C) Summary of two possible pathways of negative control. Spo0A is required to downregulate the expression of *abrB*, which can act negatively as well as positively on competence (see panels A and B). Since activation of DegU may act negatively on competence, DegS may be considered a negative regulator. The latter mechanism is speculative, since the exact roles of phosphorylated and unphosphorylated DegU in competence are unclear (see text).

plete nature of much of our knowledge and the correspondingly provisional nature of our schematic attempts to integrate this information.

The various genetic dependencies suggest a hierarchical ordering of the known regulatory gene products (Fig. 7). The dependence of *srfA* expression on several gene products but not on others suggests either that some of the regulatory proteins act later than *srfA* (Fig. 7A) or that they lie on a separate branch of the pathway (Fig. 7B). The dependence of *comK* expression on *srfA* and the dependence of late-gene expression on *comK* suggest that the *comK* product acts later than *srfA* and before the turning on of late genes (241). The failure of *mecA42* to bypass *comK* (116, 238) suggests the order for these elements shown in Fig. 7. The positions of *spo0H*, *degU*, *sin*, and *abrB* in Fig. 7A are based on unpublished data establishing the dependence of *comK* expression on these genes (97, 240). Also shown in Fig. 7C are two potential negative circuit elements in the competence scheme. As noted above, DegS is not required for the development of competence. However, since the *degS*(Hy) and *degU*(Hy) mutations confer competence deficiency, it may be that under certain conditions, DegS transfers phosphate to DegU, thus acting as a negative regulator. Also, as noted above, an excess of AbrB acts negatively on competence, explaining the requirement for the Spo0A protein. It is possible that Spo0A acts at t_0 to downregulate *abrB* expression and that AbrB is a normal negative regulator of competence.

The position of the *spo0H* product in these schemes is still somewhat ambiguous. Although the *srfA* mutation *csh-293* was identified on the basis of its *spo0H* dependency (110), we have noted only a two- to three-fold effect of *spo0H* deficiency on *srfA* expression in competence medium (98). The bypass of *spo0H* by *mecA42* suggests that σ^H is required prior to *mecA*. Similarly ambiguous are the points of negative action of AbrB. As noted above, overproduction of this product seems to be the sole reason for the competence deficiency of *spo0A* mutants. This deficiency is partially bypassed by *mec* mutants, implying multiple sites of this negative AbrB activity both before and after the point of *mec* action. As noted above, a double *abrB spo0A* mutant is completely bypassed by *mecA42*, since in this strain, the AbrB protein is no longer available to act negatively and is no longer required to act positively because of the *mecA* mutation.

Competence Signaling: How Does the Cascade Function?

The formal schemes presented in Fig. 7 tell us little about mechanism. What can be inferred concerning the modes of action of the various regulatory elements? Four known loci are required to turn on *srfA* expression: *comP*, *comA*, *comQ*, and *spo0K*. The observation reported above, i.e., that *srfA* under *Pspac* control bypasses the requirements for *comP*, *comA*, *comQ*, and *spo0K*, suggests that these genes are not required for any purpose other than the activation of *srfA* expression.

As noted above, ComA and ComP are response regulator and histidine kinase members, respectively, of the two-component signal transduction systems. It is likely that ComP serves to activate ComA, both because of the genetic proximity of these two genes and because overexpression of ComA bypasses the competence deficiency of a *comP* mutant (259). This has been observed in the case of other cognate two-component regulators (203, 260). Presumably, this activation involves phosphate transfer. Little is known of the role of ComQ. However, the extents of competence deficiency in null mutants of *comP*, *comA*, and *comQ* as well as of mutants with multiple mutations in these genes are identical, suggesting that the corresponding gene products may act in concert. Also, overexpression of ComA bypasses a *comQ* knockout mutant (253), and *comQ* is located immediately upstream from *comP* on the genetic map (Fig. 4). Perhaps ComQ functions as an auxiliary protein facilitating or regulating the transfer of phosphate to ComA or the stability of the phosphorylated protein.

The role of *spo0K* is particularly intriguing. Rudner et al. (186) have shown that at least *spo0K* ORFE is required for competence. We have found that the requirement for this ORF can be partially bypassed by overexpression of ComP or ComQ (253). This appears to be the case for overexpression of ComA as well, although the *spo0KORFE* strain carrying *comA* in multicopy grew poorly and was difficult to work with. As noted above, ComP was predicted to be a membrane protein, possessing eight transmembrane segments in its N-terminal domain (259). The *spo0K* proteins, which closely resemble the gram-negative oligopeptide permease proteins, are almost certainly associated with the membrane. Perhaps the *spo0K* products and ComP "talk" to one another (see reference 186 and the discussion below), transducing a signal via ComA and possibly ComQ.

It has been suggested (186) that the Spo0K-dependent signal transduction system may be sensitive to the concentration of peptide pheromones, such as the soluble sporulation factor of *B. subtilis* (90) or the postulated soluble competence factor (111). It has also been proposed that the *spo0K* system may respond to cell wall-derived peptides, as the homologous *opp* operon is thought to do in *Salmonella* spp. (171). Of course, these two hypotheses are not mutually exclusive, since cell wall-derived peptides may diffuse into the extracellular space and act as soluble factors.

As a result of signaling events involving *comQ*, *comA*, *comP*, and *spo0K*, the expression of *srfA* is activated. This operon contains genes required for the stepwise assembly of peptides (159, 241). It is tempting to propose, as has been done elsewhere (159, 241), that one of the intermediates in surfactin synthesis or a derivative of such an intermediate serves as an extracellular competence factor for *B. subtilis*. Although this is indeed a plausible notion, it is well to recall that the pathway responsible for nonribosomal peptide synthesis involves the formation of aminoacyl adenylates and their coupling to phosphopantetheine "arms" on a multicomponent synthetic complex (130). There are several opportunities for interaction of these intermediates with other pathways, and perhaps these interactions provide an intracellular signal for competence development. If *srfA* is needed for the synthesis of a peptide competence factor, then it might seem less likely that *spo0K* is a sensor for such a peptide, since it acts prior to *srfA*. On the other hand, we can envisage an autocatalytic amplification system in which a low concentration of the factor is sensed by the *spo0K-comP-comQ-comA* system, resulting in activation of *srfA* and increased production of the factor. If this type of "autocrine" signaling cycle operates, then the observation (97) that expression of *srfA* under *Pspac* control bypasses the competence requirement for *spo0K* implies that a sensor for the postulated SrfA peptide that does not involve *spo0K* but acts later than *srfA* must exist.

The thoughts and observations presented in the preceding paragraphs suggest the following working hypothesis. A low-level synthesis of the postulated competence factor may be constitutively produced, since *srfA* is expressed at a low but measurable level throughout growth (97, 110, 160). This implies that as growth proceeds, the extracellular concentration of this factor would increase to a critical level and interact with *spo0K* products, which in turn would transmit a signal to the *comPAQ* proteins, thus activating ComA by phosphorylation and initiating an increase in *srfA* transcription. This increase in transcription would result in an increase in synthesis of the factor, which in turn would trigger further *srfA* expression by the same mechanism. An autocatalytic "switch" responding to cell density would thus be operative. In order for the postulated extracellular factor to interact with the cell and potentiate the development of competence, a receptor-signal transduction system must also operate downstream of *srfA*. The ability of *srfA* expression under *Pspac* control to bypass a *spo0K* disruption suggests that the Spo0K proteins are not needed for this downstream step.

These postulated roles for *spo0K* in competence-specific signal transduction and the proposed interactions of *spo0K* with the ComA-ComP two-component regulators are strikingly reminiscent of the workings of the *E. coli pst* operon during phosphate regulation (180). *pst* and *spo0K* encode related ATP-driven, membrane-associated transport systems. Both operons encode substrate-specific binding proteins, two hydrophobic membrane-associated components, and a less hydrophobic protein that contains an ATP-binding consensus sequence. Both are required for control of global regulons. If the hypothesis outlined above is correct, regulation occurs in both cases by interaction with two-component regulators (*comP-comA* and *phoR-phoB*). Both operons include a fifth ORF (*phoU* and *spo0KORFE*) not found in various otherwise similar ATP-driven transport systems, and in both cases, this ORF appears to be required for interaction with the two-component regulatory pathways but not for transport of P_i or oligopeptides (171). Thus, the two global regulation systems may well be similarly organized and may have mechanistic features in common. For instance, the *pho* system does not appear to be regulated by a sensing of the intracellular level of P_i or the flux of P_i across the membrane; instead, the Pst proteins appear to function as part of a transmembrane signaling device (180). Possibly, the products of *spo0K* act similarly downstream of *srfA* without actually internalizing the postulated competence factor.

Clearly, the development of competence responds to signals that are at least in part nutritional. A typical medium used for the development of competence consists of salts, glucose, amino acids, and yeast extract (11). The yeast extract is dispensable, serving only to increase the growth rate and yield of competent cells. Replacement of glucose by glycerol in this medium has little or no effect on growth but decreases the yield of transformants and the expression of late competence genes about 10-fold (3). In this medium, as stated above, competence develops postexponentially, although an initial minor wave of competence is almost always detected at about t_{-3} to t_{-2} (3). When amino acids are omitted (except for those required for growth), competence develops to the usual level but is expressed throughout growth, with a further increase at t_0 (182). Addition of amino acids in the form of casein hydrolysate to either 0.02 or 2% final concentration have the same effect: the expression of competence is repressed during exponential growth but is induced beginning at t_0. This result suggests that repression by amino acids is overridden at the time of transition to stationary phase by some specific postexponential mechanism other than exhaustion of amino acids. The regulatory genes required for exponential expression of competence in the absence of amino acids as opposed to the postexponential form of expression are not known.

Some inferences concerning other forms of nutritional signaling may be drawn. The expression of *srfA* is delayed about 1 h when glucose is replaced by glycerol (182). This suggests that at least some of the competence glucose signaling is exercised prior to the onset of *srfA* expression. In a *comP* null mutant bypassed for competence by overexpression of ComA (259), competence and the expression of late competence genes are no longer as dependent on the presence of glucose as in the wild type, nor are they inhibited by the addition of glutamine. This suggests that the component of glucose and glutamine signaling that occurs prior to *srfA* expression requires ComP. Msadek et al. (156) have observed that *degQ* expression is dependent on *comA* and *comP* (and also on *comQ*). *degQ* is present in the *comA* gene cluster (Fig. 4) but is not required for the development of competence. In fact, the overproduction of DegQ has a slight negative effect on competence (156). The induction of *degQ* expression can be triggered by amino acid starvation with a *comA*- and *comP*-dependent mechanism. Perhaps one of the signals detected by *comP* and *comA* is the availability of amino acids, although the relationship of observations in my laboratory on the development of competence in the presence and absence of amino acid supplements to observations by Msadek et al. is not yet clear. These fragmentary observations indicate that the nutritional signals detected by the ComP mechanism are likely to be complex but may involve, at least in part, amino acid starvation-dependent and glucose-dependent signals. A major unanswered question involves the relationship of nutritional signaling to the sensing of a competence factor. Perhaps the proposed interaction of *comP* and *spo0K* permits integration of these two types of signals.

Another nutritional signal is received following *srfA* expression and prior to the point at which the *mec* genes function, since the sensitivity of competence development to complex media is bypassed by *mecA* and *mecB* mutants (69) but not when *srfA* is driven by the *Pspac* promoter (98, 164). Perhaps the *abrB-sin-degU* genes are involved in receiving this signal (Fig. 7).

It is difficult to even posit a plausible hypothesis concerning the signals and mechanisms responsible for cell type-specific regulation of competence without resorting to fantasy. Perhaps the heterogeneity of competent cultures is determined by titration of a limiting substance (for instance, a competence pheromone) or by the proportion of cells in a critical window of the cell cycle when a specific signal is received. It is also conceivable that a cell division event occurs with the daughter cells enjoying alternative fates (as in sporulation). The noncompetent daughter might then undergo about three divisions to yield the observed ratio of competent to noncompetent cells. One interesting observation is that ComG ORF1 is required for separation in Renografin (1). It is not known whether this ORF is also needed for heterogeneity in gene expression of the competent culture or whether it is required only for the density difference, as appears plausible.

POSTEXPONENTIAL EXPRESSION: THE SIGNAL TRANSDUCTION NETWORK

When *B. subtilis* reaches the transition from exponential to stationary phase, several forms of expression may be activated. In addition to competence and sporulation, the cells may release antibiotics, increase the synthesis of a variety of degradative enzymes, and increase in motility. Information as to the relationship among these various forms of expression is sketchy. Clearly, sporulation is an ultimate response to stress, since it seems to preclude other responses. Sporulation is inhibited by high concentrations of nutrients, particularly glucose. Competence, on the other hand, is stimulated by the presence of glucose. In fact, to a certain extent, these seem to be alternative responses when viewed genetically as well as nutritionally. For instance, Sin is required for competence, but overexpression of this protein inhibits sporulation (82). Conversely, ComA is required for competence, but overexpression inhibits sporulation (256). *degU*(Hy) and *degS*(Hy) mutations enable sporulation to occur in the presence of glucose but prevent the development of competence (155, 183, 215). It is possible that the signaling requirements for competence and sporulation are arranged so that these two responses occur sequentially; perhaps the exhaustion of nutrients that favor the former leads to the initiation of sporulation. Certainly, competent cells can go on to sporulate, presumably without an intervening period of cell division (123).

These genetic relationships between sporulation and competence serve to introduce an interesting generalization. Few if any of the early regulatory genes discussed in this review are specific for competence. In fact, most of the early regulatory genes that act on a given form of postexponential expression seem to affect others as well. This is true of *spoOK*, *comA*, *comP*, *spo0A*, *spo0H*, *abrB*, *sin*, *degU*, *degS*, *srfA*, *mecA*, and *mecB* at least. In the cases of *spo0H*, *spo0A*, and *spo0K*, this pleiotropy is obvious. As noted above, overexpression of Sin results in asporogeny (82). A loss of *sin* function results in competence deficiency, nonmotility, and overexpression of certain degradative enzymes. *degS*(Hy) and *degU*(Hy) mutants, in addition to the phenotypes described above, are nonmotile and, of course, overproduce degradative enzymes. ComP deficiency also affects sporulation (259). *mecB* mutants are oligosporogenic, and *mecA* mutations may exhibit an altered response of sporulation to the presence of glucose (69). Also, as noted above, *mec* mutations affect regulation of the *gsiA* (157) and degradative enzyme (154) systems. Thus, for each of these loci, either overexpression or particular alleles affect more than one form of expression. This has led to the notion of a "signal transduction network" that regulates the expression of a postexponential super regulon (58). This network is presumably capable of gathering information from various signal molecules and integrating this information to ensure that the appropriate response occurs at the proper time. Particular combinations of signals may therefore lead to competence, competence plus degradative enzyme synthesis, or sporulation, etc.

Acknowledgments. I am deeply appreciative of the many investigators who have communicated unpublished results and contributed to this review with many stimulating discussions: F. Breidt, Y. Chung, M. K. Dahl, J. Dubnau, N. K. Gaur, G. Grandi, B. Green, A. Grossman, J. Hahn, D. Henner, M. Hobbs, G. Inamine, K. Jaacks-Siranosian, L. Kong, V. Krishnapillai, F. Kunst, T. Leighton, S. P. Livingston, A. Londoño, R. Magnuson, J. S. Mattick, I. Mandic-Mulec, T. Msadek, J. Mueller, M. Nakano, A. Ninfa, A. Pugsley, G. Rapoport, M. Roggiani, Y. Sadaie, D. van Sinderen, I. Smith, L. Sonenshein, G. Venema, Y. Weinrauch, C. B. Whitchurch, P. Youngman, and P. Zuber. Work in this laboratory was supported by NIH grants AI10311, GM37137, and GM43756.

REFERENCES

1. **Albano, M., R. Breitling, and D. Dubnau.** 1989. Nucleotide sequence and genetic organization of the *Bacillus subtilis comG* operon. *J. Bacteriol.* **171:**5386–5404.
2. **Albano, M., and D. Dubnau.** 1989. Cloning and characterization of a cluster of linked *Bacillus subtilis* late competence mutants. *J. Bacteriol.* **171:**5376–5385.
3. **Albano, M., J. Hahn, and D. Dubnau.** 1987. Expression of competence genes in *Bacillus subtilis*. *J. Bacteriol.* **169:**3110–3117.
4. **Alonso, J. C. (Max Planck Institut für Molekulare Genetik, Berlin, Germany).** 1991. Personal communication.
5. **Alonso, J. C., G. Lüder, and R. H. Taylor.** 1991. Characterization of *Bacillus subtilis* recombinational pathways. *J. Bacteriol.* **173:**3977–3980.
6. **Alonso, J. C., K. Shirahige, and N. Ogasawara.** 1990. Molecular cloning, genetic characterization and NA sequence analysis of the *recM* region of *Bacillus subtilis*. *Nucleic Acids Res.* **18:**6771–6777.
7. **Alonso, J. C., R. H. Tailor, and G. Lüder.** 1988. Characterization of recombination-deficient mutants of *Bacillus subtilis*. *J. Bacteriol.* **170:**3001–3007.
8. **Ambudkar, S. V., T. J. Larson, and P. C. Maloney.** 1986. Reconstitution of sugar phosphate transport systems of *Escherichia coli*. *J. Biol. Chem.* **261:**9083–9086.
9. **Ambudkar, S. V., L. A. Sonna, and P. C. Maloney.** 1986. Variable stoichiometry of phosphate-linked anion exchange in *Streptococcus lactis*: implications for the mechanism of sugar phosphate transport by bacteria. *Proc. Natl. Acad. Sci. USA* **83:**280–284.
10. **Ames, G. F.-L.** 1990. Energy coupling in bacterial periplasmic permeases. *J. Bacteriol.* **172:**4133–4137.
11. **Anagnostopoulos, C., and J. Spizizen.** 1961. Requirements for transformation in *Bacillus subtilis*. *J. Bacteriol.* **81:**741–746.
12. **Antoniewski, C., B. Savelli, and P. Stragier.** 1990. The *spoIIJ* gene, which regulates early developmental steps in *Bacillus subtilis*, belongs to a class of environmentally responsive genes. *J. Bacteriol.* **172:**86–93.
13. **Aricó, B., J. F. Miller, C. Roy, S. Stibitz, D. Monack, S. Falkow, R. Gross, and R. Rappuoli.** 1989. Sequences required for expression of *Bordetella pertussis* virulence factors share homology with prokaryotic signal transduction proteins. *Proc. Natl. Acad. Sci. USA* **86:**6671–6675.
14. **Arwert, F., and G. Venema.** 1973. Evidence for a noncovalently bonded intermediate in recombining during transformation in *Bacillus subtilis*, p. 203–214. *In* L. J. Archer (ed.), *Bacterial Transformation*. Academic Press, Inc., New York.
15. **Arwert, F., and G. Venema.** 1973. Transformation in *Bacillus subtilis*. Fate of newly introduced transforming DNA. *Mol. Gen. Genet.* **123:**185–198.
16. **Ayusawa, D., Y. Yoneda, K. Yamane, and B. Maruo.** 1975. Pleiotropic phenomena in autolytic enzyme(s) content, flagellation, and simultaneous hyperproduction of extracellular α-amylase and protease in a *Bacillus subtilis* mutant. *J. Bacteriol.* **124:**459–469.
17. **Bally, M., G. Ball, A. Baudere, and A. Lazdunski.** 1991. Protein secretion in *Pseudomonas aeruginosa*: the *xcpA* gene encodes an integral inner membrane protein homologous to *Klebsiella pneumoniae* secretion function protein PulO. *J. Bacteriol.* **173:**479–486.
18. **Barberio, C. (Università degli studi di Firenze, Florence, Italy).** 1991. Personal communication.
19. **Belliveau, B. H., and J. T. Trevors.** 1989. Transformation of *Bacillus cereus* vegetative cells by electroporation. *Appl. Environ. Microbiol.* **55:**1649–1652.
20. **Belyaeva, N. N., and R. R. Azizbekyan.** 1968. Fine structure of new *Bacillus subtilis* phage AR9 with complex morphology. *Virology* **34:**176–179.
21. **Bertram, J., M. Stratz, and P. Durre.** 1991. Natural transfer of conjugative plasmid Tn*916* between gram-positive and gram-negative bacteria. *J. Bacteriol.* **173:**443–448.
22. **Bodmer, W., and A. T. Ganesan.** 1964. Biochemical and genetic studies of integration and recombination in Bacillus subtilis transformation. *Genetics* **50:**717–738.
23. **Bodmer, W. F.** 1966. Integration of deoxyribonuclease-treated DNA in *Bacillus subtilis* transformation. *J. Gen. Physiol.* **49:**233–258.
24. **Bovre, K., and L. O. Froholm.** 1972. Competence in

genetic transformation related to colony type and fimbriation in three species of *Moraxella*. *Acta Pathol. Microbiol. Scand.* **80:**649–659.
25. **Bravo, A., and J. C. Alonso.** 1990. The generation of concatemeric plasmid DNA in *Bacillus subtilis* as a consequence of bacteriophage SPP1 infection. *Nucleic Acids Res.* **18:**4651–4657.
26. **Breidt, F., M. Roggiani, and D. Dubnau.** Unpublished data.
27. **Breitling, R., and D. Dubnau.** 1990. A pilin-like membrane protein is essential for DNA binding by competent *Bacillus subtilis*. *J. Bacteriol.* **172:**1499–1508.
28. **Bresler, S. E., R. A. Kreneva, and V. V. Kushev.** 1968. Correction of molecular heterozygotes in the course of transformation. *Mol. Gen. Genet.* **102:**257–268.
29. **Brigidi, P., E. D. Rossi, M. L. Bertarini, G. Riccardi, and D. Matteuzzi.** 1990. Genetic transformation of intact cells of *Bacillus subtilis* by electroporation. *FEMS Microbiol. Lett.* **67:**135–138.
30. **Burbulys, D., K. A. Trach, and J. A. Hoch.** 1991. Initiation of sporulation in *B. subtilis* is controlled by a multicomponent phosphorelay. *Cell* **64:**545–552.
31. **Canosi, U., G. Luder, and T. A. Trautner.** 1978. SPP1-mediated plasmid transformation. *J. Virol.* **44:**431–436.
32. **Cantor, C. R., and P. R. Schimmel.** 1980. *Biophysical Chemistry*, Part III. *The Behavior of Biological Macromolecules*. W. H. Freeman & Co., San Francisco.
33. **Ceglowski, P., G. Lüder, and J. C. Alonso.** 1990. Genetic analysis of *recE* activities in *Bacillus subtilis*. *Mol. Gen. Genet.* **222:**441–445.
34. **Chang, S., and S. N. Cohen.** 1979. High frequency transformation of *Bacillus subtilis* protoplasts by plasmid DNA. *Mol. Gen. Genet.* **168:**111–115.
35. **Cheo, D. L., K. W. Bayles, and R. E. Yasbin.** 1991. Cloning and characterization of DNA damage-inducible promoter regions from *Bacillus subtilis*. *J. Bacteriol.* **173:**1696–1703.
36. **Cheo, D. L., K. W. Bayles, and R. E. Yasbin.** 1992. Molecular characterization of regulatory elements controlling expression of the *Bacillus subtilis* *recA*$^+$ gene. *Biochimie* **74:**755–762.
37. **Christie, P. J., R. Z. Korman, S. A. Zahler, J. C. Adsit, and G. M. Dunny.** 1987. Two conjugation systems associated with *Streptococcus faecalis* plasmid pCF10: identification of a conjugative transposon that transfers between *S. faecalis* and *Bacillus subtilis*. *J. Bacteriol.* **169:**2529–2536.
38. **Christie, P. J., J. J. E. Ward, M. P. Gordon, and E. W. Nester.** 1989. A gene required for transfer of T-DNA to plants encodes an ATPase with autophosphorylating activity. *Proc. Natl. Acad. Sci. USA* **86:**9677–9681.
39. **Ciferri, O., S. Barlati, and J. Lederberg.** 1970. Uptake of synthetic polynucleotides by competent cells of *Bacillus subtilis*. *J. Bacteriol.* **104:**684–688.
40. **Cole, S. T., and O. Raibaud.** 1986. The nucleotide sequence of the *malT* gene encoding the positive regulator of the *Escherichia coli* maltose regulon. *Gene* **42:**201–208.
41. **Contente, S., and D. Dubnau.** 1979. Marker rescue transformation by linear plasmid DNA in *Bacillus subtilis*. *Plasmid* **2:**555–571.
42. **Cosloy, S., and M. Oishi.** 1973. The nature of the transformation process in *Escherichia coli* K12. *Mol. Gen. Genet.* **124:**1–10.
43. **Cutting, S., and J. Mandelstam.** 1986. The nucleotide sequence and the transcription during sporulation of the *gerE* gene of *Bacillus subtilis*. *J. Gen. Microbiol.* **132:**3013–3024.
44. **Dahl, M. K., T. Msadek, F. Kunst, and G. Rapoport.** 1991. Mutational analysis of the *Bacillus subtilis* DegU regulator and its phosphorylation by the DegS protein kinase. *J. Bacteriol.* **173:**2539–2547.
45. **Darlington, A. J., and W. F. Bodmer.** 1968. Events occurring at the site of integration of a DNA molecule in *Bacillus subtilis* transformation. *Genetics* **60:**681–684.
46. **Davidoff-Abelson, R., and D. Dubnau.** 1973. Kinetic analysis of the products of donor deoxyribonucleate in transformed cells of *Bacillus subtilis*. *J. Bacteriol.* **116:**154–162.
47. **Dedonder, R. A., J. Lepesant, J. Lepesant-Kejzlarova, A. Billault, M. Steinmetz, and F. Kunst.** 1977. Construction of a kit of reference strains for rapid genetic mapping in *Bacillus subtilis* P 168. *Appl. Environ. Microbiol.* **33:**989–993.
48. **Deichelbohrer, I., J. C. Alonso, G. Luder, and T. A. Trautner.** 1985. Plasmid transduction by *Bacillus subtilis* bacteriophage SPP1: effects of DNA homology between plasmid and bacteriophage. *J. Bacteriol.* **162:**1238–1243.
49. **Devine, J. H., G. S. Shadel, and T. O. Baldwin.** 1989. Identification of the operator of the *lux* regulon from the *Vibrio fischeri* strain ATCC7744. *Proc. Natl. Acad. Sci. USA* **86:**5688–5692.
50. **deVos, W. M., S. C. deVries, and G. Venema.** 1983. Cloning and expression of the *Escherichia coli recA* gene in *Bacillus subtilis*. *Gene* **25:**301–308.
51. **Doly, J., and C. Anagnostopoulos.** 1976. Isolation, subunit structure and properties of the ATP-dependent deoxyribonuclease of *Bacillus subtilis*. *Eur. J. Biochem.* **71:**309–316.
52. **Dooley, D. C., C. T. Hadden, and E. W. Nester.** 1971. Macromolecular synthesis in *Bacillus subtilis* during development of the competent state. *J. Bacteriol.* **108:**668–679.
53. **Dubnau, D.** Unpublished data.
54. **Dubnau, D.** 1976. Genetic transformation of *Bacillus subtilis*: a review with emphasis on the recombination mechanism, p. 14–27. *In* D. Schlessinger (ed.), *Microbiology—1976*. American Society for Microbiology, Washington, D.C.
55. **Dubnau, D.** 1982. Genetic transformation in *Bacillus subtilis*, p. 148–178. *In* D. Dubnau (ed.), *The Molecular Biology of the Bacilli*, vol. I. *Bacillus subtilis*. Academic Press, Inc., New York.
56. **Dubnau, D.** 1989. The competence regulon of *Bacillus subtilis*, p. 147–166. *In* I. Smith, R. A. Slepecky, and P. Setlow (ed.), *Regulation of Procaryotic Development*. American Society for Microbiology, Washington, D.C.
57. **Dubnau, D.** 1991. Genetic competence in *Bacillus subtilis*. *Microbiol. Rev.* **55:**395–424.
58. **Dubnau, D.** 1991. The regulation of genetic competence in *Bacillus subtilis*. *Mol. Microbiol.* **5:**11–18.
59. **Dubnau, D., and C. Cirigliano.** 1972. Fate of transforming deoxyribonucleic acid after uptake by competent *Bacillus subtilis*: size and distribution of the integrated donor sequences. *J. Bacteriol.* **111:**488–494.
60. **Dubnau, D., and C. Cirigliano.** 1972. Fate of transforming DNA following uptake by competent *Bacillus subtilis*. IV. The endwise attachment and uptake of transforming DNA. *J. Mol. Biol.* **64:**31–46.
61. **Dubnau, D., and C. Cirigliano.** 1972. Fate of transforming DNA following uptake by competent *Bacillus subtilis*. III. Formation and properties of products isolated from transformed cells which are derived entirely from donor DNA. *J. Mol. Biol.* **64:**9–29.
62. **Dubnau, D., and C. Cirigliano.** 1973. Fate of transforming deoxyribonucleic acid after uptake by competent *Bacillus subtilis*: nonrequirement of deoxyribonucleic acid replication for uptake and integration of transforming deoxyribonucleic acid. *J. Bacteriol.* **113:**1512–1514.
63. **Dubnau, D., and C. Cirigliano.** 1973. Fate of transforming DNA following uptake by competent *Bacillus subtilis*. VI. Non-covalent association of donor and recipient DNA. *Mol. Gen. Genet.* **120:**101–106.
64. **Dubnau, D., S. Contente, and T. J. Gryczan.** 1980. On

the use of plasmids for the study of genetic recombination in *Bacillus subtilis*, p. 365–386. *In* S. Zadrazil and J. Sponar (ed.), *DNA-Recombination Interactions and Repair*. Pergamon Press, Oxford.
65. **Dubnau, D., and R. Davidoff-Abelson.** 1971. Fate of transforming DNA following uptake by competent *Bacillus subtilis*. I. Formation and properties of the donor-recipient complex. *J. Mol. Biol.* **56:**209–221.
66. **Dubnau, D., R. Davidoff-Abelson, B. Scher, and C. Cirigliano.** 1973. Fate of transforming deoxyribonucleic acid after uptake by competent *Bacillus subtilis*: phenotypic characterization of radiation-sensitive recombination-deficient mutants. *J. Bacteriol.* **114:**273–286.
67. **Dubnau, D., R. Davidoff-Abelson, and I. Smith.** 1969. Transformation and transduction in *Bacillus subtilis*: evidence for separate modes of recombinant formation. *J. Mol. Biol.* **45:**155–179.
68. **Dubnau, D., J. Hahn, L. Kong, M. Roggiani, and Y. Weinrauch.** 1991. Genetic competence as a post-exponential global response. *Semin. Dev. Biol.* **2:**3–12.
69. **Dubnau, D., M. Roggiani, and J. Hahn.** 1990. Growth medium-independent genetic competence mutants of *Bacillus subtilis*. *J. Bacteriol.* **172:**4048–4055.
70. **Dubnau, E., J. Weir, G. Nair, L. Carter III, C. Moran, Jr., and I. Smith.** 1988. *Bacillus* sporulation gene *spoOH* codes for σ^{30} (σ^{H}). *J. Bacteriol.* **170:**1054–1062.
71. **Duncan, M. L., S. S. Kalman, S. M. Thomas, and C. W. Price.** 1987. Gene encoding the 37,000-dalton minor sigma factor of *Bacillus subtilis* RNA polymerase: isolation, nucleotide sequence, chromosomal locus, and cryptic function. *J. Bacteriol.* **169:**771–778.
72. **Ehrlich, S. D.** 1977. Replication and expression of plasmids from *Staphylococcus aureus* in *Bacillus subtilis*. *Proc. Natl. Acad. Sci. USA* **74:**1680–1682.
73. **Fani, R., G. Mastromei, M. Polsinelli, and G. Venema.** 1984. Isolation and characterization of *Bacillus subtilis* mutants altered in competence. *J. Bacteriol.* **157:**153–157.
74. **Ferrari, E., U. Canosi, A. Gallizi, and G. Mazza.** 1978. Studies on transduction processes by SPP1 phage. *J. Gen. Virol.* **41:**563–572.
75. **Ferrari, F. A., K. Trach, D. LeCoq, J. Spence, E. Ferrari, and J. A. Hoch.** 1985. Characterization of the *spoOA* locus and its deduced product. *Proc. Natl. Acad. Sci. USA* **82:**2647–2651.
76. **Filloux, A., M. Bally, G. Ball, M. Akrim, J. Tommassen, and A. Lazdunski.** 1990. Protein secretion in gram-negative bacteria: transport across the outer membrane involves common mechanisms in different bacteria. *EMBO J.* **9:**4323–4329.
77. **Foldes, J., and T. A. Trautner.** 1964. Infectious DNA from a newly isolated *B. subtilis* phage. Z. Vererbungs. **95:**57–65.
78. **Folkhard, W., D. A. Marvin, T. H. Watts, and W. Paranchych.** 1981. Structure of polar pili from *Pseudomonas aeruginosa* strains K and O. *J. Mol. Biol.* **149:**79–93.
79. **Fornilli, S. L., and M. S. Fox.** 1977. Electron microscope visualization of the products of *Bacillus subtilis* transformation. *J. Mol. Biol.* **113:**181–191.
80. **Ftouhi, N., and N. Guillen.** 1990. Genetic analysis of fusion recombinants in *Bacillus subtilis*: function of the *recE* gene. *Genetics* **126:**487–496.
81. **Gaur, N. K., K. Cabane, and I. Smith.** 1988. Structure and expression of the *Bacillus subtilis sin* operon. *J. Bacteriol.* **170:**1046–1053.
82. **Gaur, N. K., E. Dubnau, and I. Smith.** 1986. Characterization of a cloned *Bacillus subtilis* gene which inhibits sporulation in multiple copies. *J. Bacteriol.* **168:**860–869.
83. **Gaur, N. K., J. Oppenheim, and I. Smith.** 1991. The *Bacillus subtilis sin* gene, a regulator of alternate developmental processes, codes for a DNA-binding protein. *J. Bacteriol.* **173:**678–686.
84. **Goodgal, S. H.** 1982. DNA uptake in *Haemophilus* transformation. *Annu. Rev. Genet.* **16:**169–192.
85. **Gottfert, M., P. Grob, and H. Hennecke.** 1990. Proposed regulatory pathway encoded by *nodV* and *nodW* genes, determinants of host specificity in *Bradyrhizobium japonicum*. *Proc. Natl. Acad. Sci. USA* **87:**2680–2684.
86. **Green, B., and P. Youngman (University of Georgia, Athens).** 1991. Personal communication.
87. **Green, D. M.** 1964. Infectivity of DNA isolated from *Bacillus subtilis* bacteriophage, SP82. *J. Mol. Biol.* **10:** 438–451.
88. **Grinius, L.** 1982. Energetics of gene transfer into bacteria. Sov. *Sci. Rev. Sec. D* **3:**115–165.
89. **Grossman, A. D., K. Ireton, E. F. Hoff, J. R. LeDeaux, D. Z. Rudner, R. Magnuson, and K. A. Hicks.** 1991. Signal transduction and the initiation of sporulation in *Bacillus subtilis*. *Semin. Dev. Biol.* **2:**31–36.
90. **Grossman, A. D., and R. Losick.** 1988. Extracellular control of spore formation in *Bacillus subtilis*. *Proc. Natl. Acad. Sci. USA* **85:**4369–4373.
91. **Gryczan, T., S. Contente, and D. Dubnau.** 1980. Molecular cloning of heterologous chromosomal DNA by recombination between a plasmid vector and a homologous resident plasmid in *Bacillus subtilis*. *Mol. Gen. Genet.* **177:**459–467.
92. **Guffanti, A. A., P. G. Quirk, and T. A. Krulwich.** 1991. Transfer of Tn*925* and plasmids between *Bacillus subtilis* and alkaliphilic *Bacillus firmus* OF4 during Tn*925*-mediated conjugation. *J. Bacteriol.* **173:**1686–1689.
93. **Guild, W. R., A. Cato, and S. Lacks.** 1968. Transformation and DNA size: two controlling parameters and the efficiency of the single strand intermediate. *Cold Spring Harbor Symp. Quant. Biol.* **33:**643–645.
94. **Guillen, N., Y. Weinrauch, and D. Dubnau.** 1989. Cloning and characterization of the regulatory *Bacillus subtilis* competence genes, *comA* and *comB*. *J. Bacteriol.* **171:**5354–5361.
95. **Hadden, C., and E. W. Nester.** 1968. Purification of competent cells in the *Bacillus subtilis* transformation system. *J. Bacteriol.* **95:**876–885.
96. **Hahn, J., M. Albano, and D. Dubnau.** 1987. Isolation and characterization of competence mutants in *Bacillus subtilis*. *J. Bacteriol.* **169:**3104–3109.
97. **Hahn, J., and D. Dubnau.** Unpublished data.
98. **Hahn, J., and D. Dubnau.** 1991. Growth stage signal transduction and the requirements for srfA induction in the development of competence. *J. Bacteriol.* **173:**7275–7282.
99. **Hahn, J., Y. Kozlov, and D. Dubnau.** Unpublished data.
100. **Haseltine-Cahn, F., and M. S. Fox.** 1968. Fractionation of transformable bacteria from competent cultures of *Bacillus subtilis* on Renografin gradients. *J. Bacteriol.* **95:**867–875.
101. **Hauser, P., and D. Karamata (Institut de Genetique et de Biologie Microbiennes, Lausanne, Switzerland).** 1991. Personal communication.
102. **He, S. Y., M. Lindberg, A. K. Chatterjee, and A. Collmer.** 1991. Cloned *Erwinia chrysanthemi out* genes enable *Escherichia coli* to selectively secrete a diverse family of heterologous proteins to its milieu. *Proc. Natl. Acad. Sci. USA* **88:**1079–1083.
103. **Helmann, J. D., and M. J. Chamberlin.** 1988. Structure and function of bacterial sigma factors. *Annu. Rev. Biochem.* **57:**839–872.
104. **Henner, D. (Genentech Corp., San Francisco, Calif.).** 1991. Personal communication.
105. **Henner, D. J., and J. A. Hoch.** 1980. The *Bacillus subtilis* chromosome. *Microbiol. Rev.* **44:**57–82.
106. **Henner, D. J., M. Yang, and E. Ferrari.** 1988. Localization of *Bacillus subtilis sacU*(Hy) mutations to two linked genes with similarities to the conserved procar-

yotic family of two-component signalling systems. *J. Bacteriol.* **170:**5102–5109.

107. **Hoch, J. A., and J. Spizizen.** 1969. Genetic control of some early events in sporulation of *Bacillus subtilis* 168, p. 112–120. *In* L. L. Campbell (ed.), *Spores IV*. American Society for Microbiology, Washington, D.C.
108. **Hotchkiss, R. D., and M. Gabor.** 1980. Biparental products of bacterial protoplast fusion showing unequal parental chromosome expression. *Proc. Natl. Acad. Sci. USA* **77:**3553–3557.
109. **Hotchkiss, R. D., and M. H. Gabor.** 1985. Protoplast fusion in *Bacillus* and its consequences, p. 109–149. *In* D. Dubnau (ed.), *The Molecular Biology of the Bacilli*. Academic Press, Inc., New York.
110. **Jaacks, K. J., J. Healy, R. Losick, and A. D. Grossman.** 1989. Identification and characterization of genes controlled by the sporulation-regulatory gene *spoOH* in *Bacillus subtilis*. *J. Bacteriol.* **171:**4121–4129.
111. **Joenje, H., M. Gruber, and G. Venema.** 1972. Stimulation of the development of competence by culture fluids in *Bacillus subtilis* transformation. *Biochim. Biophys. Acta* **262:**189–199.
112. **Joenje, H., and G. Venema.** 1975. Different nuclease activities in competent and non-competent cells of *Bacillus subtilis*. *J. Bacteriol.* **122:**25–33.
113. **Kahn, M. E., and H. O. Smith.** 1984. Transformation in Haemophilus: a problem in membrane biology. *J. Membr. Biol.* **81:**89–103.
114. **Kelly, M. S., and R. H. Pritchard.** 1965. Unstable linkage between genetic markers in transformation. *J. Bacteriol.* **89:**1314–1321.
115. **Koehler, T. M., and C. B. Thorne.** 1987. *Bacillus subtilis* (*natto*) plasmid pLS20 mediates interspecies plasmid transfer. *J. Bacteriol.* **169:**5271–5278.
116. **Kong, L., and D. Dubnau.** Unpublished data.
117. **Kong, L., K. Jaacks-Siranosian, A. G. Grossman, and D. Dubnau.** Unpublished data.
118. **Kooistra, J., and G. Venema.** 1991. Cloning, sequencing, and expression of *Bacillus subtilis* genes involved in ATP-dependent nuclease synthesis. *J. Bacteriol.* **173:** 3644–3655.
119. **Kooistra, J., B. Vosman, and G. Venema.** 1988. Cloning and characterization of a *Bacillus subtilis* transcription unit involved in ATP-dependent DNase synthesis. *J. Bacteriol.* **170:**4791–4797.
120. **Kunst, F., M. Debarbouille, T. Msadek, M. Young, C. Mauel, D. Karamata, A. Klier, G. Rapoport, and R. Dedonder.** 1988. Deduced polypeptides encoded by the *Bacillus subtilis sacU* locus share homology with two-component sensor-regulator systems. *J. Bacteriol.* **170:** 5093–5101.
121. **Kunst, F., M. Pascal, J. Lepesant-Kejzlarova, J.-A. Lepesant, A. Billault, and R. Dedonder.** 1974. Pleiotropic mutations affecting sporulation conditions and the synthesis of extracellular enzymes in *Bacillus subtilis* 168. *Biochimie* **56:**1481–1489.
122. **Lacks, S. A.** 1988. Mechanisms of genetic recombination in gram-positive bacteria, p. 43–86. *In* R. Kucherlapti and G. R. Smith (ed.), *Genetic Recombination*. American Society for Microbiology, Washington, D.C.
123. **Lencastre, H. D., and P. J. Piggot.** 1979. Identification of different sites of expression for *spo* loci by transformation of *Bacillus subtilis*. *J. Gen. Microbiol.* **114:**377–389.
124. **Lévi, C., C. Sanchez-Rivas, and P. Schaeffer.** 1977. Further genetic studies on the fusion of bacterial protoplasts. *FEMS Microbiol. Lett.* **2:**323–326.
125. **Lévi-Meyrueis, C., C. Sanchez-Rivas, and P. Schaeffer.** 1980. Formation de bactéries diploïdes stables par fusion de protoplastes de *Bacillus subtilis* et effet de mutations *rec*$^-$ sur les produits de fusion formes. *C.R. Acad. Sci. Ser. D* **291:**67–70.
126. **Levin, B. C., and O. E. Landman.** 1973. DNA synthesis inhibition by 6-(p-hydroxyphenylazo)-uracil in relation to uptake and integration of transforming DNA in *Bacillus subtilis*, p. 217–240. *In* L. J. Archer (ed.), *Bacterial Transformation*. Academic Press, Inc., New York.
127. **Levine, J. S., and N. Strauss.** 1965. Lag period characterizing the entry of transforming deoxyribonucleic acid into *Bacillus subtilis*. *J. Bacteriol.* **89:**281–287.
128. **Linn, T., R. Losick, and A. L. Sonenshein.** 1975. Rifampin resistance mutation of *Bacillus subtilis* altering the electrophoretic mobility of the beta subunit of ribonucleic acid polymerase. *J. Bacteriol.* **122:**1387–1390.
129. **Lipman, D. J., and W. R. Pearson.** 1985. Rapid and sensitive protein similarity searches. *Science* **227:**1435–1441.
130. **Lipmann, F.** 1980. Bacterial production of antibiotic polypeptides by thiol-linked synthesis on protein templates. *Adv. Microb. Physiol.* **21:**227–266.
131. **Londoño, A., and D. Dubnau.** Unpublished data.
132. **Love, P. E., M. J. Lyle, and R. E. Yasbin.** 1985. DNA-damage-inducible (*din*) loci are transcriptionally activated in competent *Bacillus subtilis*. *Proc. Natl. Acad. Sci. USA* **82:**6201–6205.
133. **Love, P. E., and R. E. Yasbin.** 1986. Induction of the *Bacillus subtilis* SOS-like response by *Escherichia coli* RecA protein. *Proc. Natl. Acad. Sci. USA* **83:**5204–5208.
134. **Loveday, K. S.** 1978. DNA synthesis in competent *Bacillus subtilis* cells. *J. Bacteriol.* **135:**1158–1161.
135. **Lovett, C. M., Jr., P. E. Love, and R. E. Yasbin.** 1989. Competence-specific induction of the *Bacillus subtilis* RecA protein analog: evidence for dual regulation of a recombination protein. *J. Bacteriol.* **171:**2318–2322.
136. **Lovett, C. M., Jr., and J. W. Roberts.** 1985. Purification of a RecA protein analogue from *Bacillus subtilis*. *J. Biol. Chem.* **260:**3305–3313.
137. **Mahajan, S. K.** 1988. Pathways of homologous recombination in *Escherichia coli*, p. 87–140. *In* R. Kucherlapati and G. R. Smith (ed.), *Genetic Recombination*. American Society for Microbiology, Washington, D.C.
138. **Maloney, P. C.** 1987. Coupling to an energized membrane: role of ion-motive gradients in the transduction of metabolic energy, p. 222–243. *In* F. C. Neidhardt, J. L. Ingraham, K. B. Low, B. Magasanik, M. Schaechter, and H. E. Umbarger (ed.), *Escherichia coli and Salmonella typhimurium: Cellular and Molecular Biology*, vol. 1. American Society for Microbiology, Washington, D.C.
139. **Mandic-Mulec, I., N. K. Gaur, and I. Smith (Public Health Research Institute, New York, N.Y.).** 1991. Personal communication.
140. **Marahiel, M. A., M. Krause, and H. Sharpeid.** 1985. Cloning of the tyrocidine synthetase I gene from *Bacillus brevis* and its expression in *Escherichia coli*. *Mol. Gen. Genet.* **201:**231–236.
141. **Marrero, R., and R. Yasbin.** 1988. Cloning of the *Bacillus subtilis recE*$^+$ gene and functional expression of *recE*$^+$ in *B. subtilis*. *J. Bacteriol.* **170:**335–344.
142. **Masson, L., G. Prefontaine, and R. Brousseau.** 1989. Transformation of *Bacillus thuringiensis* vegetative cells by electroporation. *FEMS Microbiol. Lett.* **51:**273–277.
143. **Mastromei, G., C. Barberio, S. Pistolesi, and M. Poisinelli.** 1989. Isolation of transformation-deficient mutants and mapping of competence genes. *Genet. Res.* **54:**1–5.
144. **Mazza, G., and A. Galizzi.** 1978. The genetics of DNA replication, repair and recombination in *Bacillus subtilis*. *Microbiologica* **1:**111–135.
145. **Mazza, G., and A. Galizzi.** 1989. Revised genetics of DNA metabolism in *Bacillus subtilis*. *Microbiologica* **12:**157–179.
146. **McCarthy, C., and E. W. Nester.** 1967. Macromolecular

synthesis in newly transformed cells of *Bacillus subtilis*. *J. Bacteriol.* **94:**131–140.

147. **Mejean, V., and J. Claverys.** 1988. Polarity of DNA entry in transformation of *Streptococcus pneumoniae*. *Mol. Gen. Genet.* **213:**444–448.
148. **Meyer, T. F., E. Billyard, R. Haas, S. Storzbach, and M. So.** 1984. Pilus genes of *Neisseria gonorrhoeae*: chromosomal organization and DNA sequence. *Proc. Natl. Acad. Sci. USA* **81:**6110–6114.
149. **Michel, B., B. Niaudet, and S. D. Ehrlich.** 1983. Intermolecular recombination during transformation of *Bacillus subtilis* competent cells by monomeric and dimeric plasmids. *Plasmid* **10:**1–10.
150. **Mittenhuber, G., R. Weckerman, and M. A. Marahiel.** 1989. Gene cluster containing the genes for tyrocidine synthetases 1 and 2 from *Bacillus brevis*: evidence for an operon. *J. Bacteriol.* **171:**4881–4887.
151. **Mohan, S., J. Aghion, N. Guillen, and D. Dubnau.** 1989. Molecular cloning and characterization of *comC*, a late competence gene of *Bacillus subtilis*. *J. Bacteriol.* **171:** 6043–6051.
152. **Mohan, S., and D. Dubnau.** 1990. Transcriptional regulation of *comC*: evidence for a competence-specific factor in *Bacillus subtilis*. *J. Bacteriol.* **172:**4064–4071.
153. **Morrison, D. A., and W. R. Guild.** 1972. Activity of deoxyribonucleic acid fragments of defined size in *Bacillus subtilis* transformation. *J. Bacteriol.* **112:**220–223.
154. **Msadek, T., and F. Kunst (Institut Pasteur, Paris, France).** 1991. Personal communication.
155. **Msadek, T., F. Kunst, D. Henner, A. Klier, G. Rapoport, and R. Dedonder.** 1990. Signal transduction pathway controlling the synthesis of a class of degradative enzymes in *Bacillus subtilis*: expression of the regulatory genes and analysis of mutations in *degS* and *degU*. *J. Bacteriol.* **172:**824–834.
156. **Msadek, T., F. Kunst, A. Klier, and G. Rapoport.** 1991. DegS-DegU and ComP-ComA modulator-effector pairs control expression of the *Bacillus subtilis* pleiotropic regulatory gene *degQ*. *J. Bacteriol.* **173:**2366–2377.
157. **Mueller, J. P., and A. L. Sonenshein (Tufts University, Boston, Mass.).** 1991. Personal communication.
158. **Mukai, K., M. Kawata, and T. Tanaka.** 1990. Isolation and phosphorylation of the *Bacillus subtilis degS* and *degU* gene products. *J. Biol. Chem.* **265:**20000–20006.
159. **Nakano, M. M., R. Magnusson, A. Myers, J. Curry, A. D. Grossman, and P. Zuber.** 1991. *srfA* is an operon required for surfactin production, competence development, and efficient sporulation in *Bacillus subtilis*. *J. Bacteriol.* **173:**1770–1778.
160. **Nakano, M. M., M. A. Marahiel, and P. Zuber.** 1988. Identification of a genetic locus required for the biosynthesis of the lipopeptide antibiotic surfactin in *Bacillus subtilis*. *J. Bacteriol.* **170:**5662–5668.
161. **Nakano, M. M., L. Xia, and P. Zuber.** 1991. Transcription initiation region of the *srfA* operon which is controlled by the *comP-comA* signal transduction system in *Bacillus subtilis*. *J. Bacteriol.* **173:**5487–5493.
162. **Nakano, M. M., and P. Zuber.** 1989. Cloning and characterization of *srfB*, a regulatory gene involved in surfactin production and competence in *Bacillus subtilis*. *J. Bacteriol.* **171:**5347–5353.
163. **Nakano, M. M., and P. Zuber (Louisiana University, Shreveport).** 1991. Personal communication.
164. **Nakano, M. N., and P. Zuber.** 1991. The primary role of ComA in establishment of the competent state in *Bacillus subtilis* is to activate the expression of *srfA*. *J. Bacteriol.* **173:**7269–7274.
165. **Nester, E. W., and B. A. D. Stocker.** 1963. Biosynthetic latency in early stages of deoxyribonucleic acid transformation in *Bacillus subtilis*. *J. Bacteriol.* **86:**785–796.
166. **Nohno, T., S. Noji, S. Taniguchi, and T. Saito.** 1989. The *narX* and *narL* genes encoding the nitrate-sensing regulators of *Escherichia coli* are homologous to a family of prokaryotic two-component regulatory genes. *Nucleic Acids Res.* **17:**2947–2957.
167. **Nunn, D., S. Bergman, and S. Lory.** 1990. Products of three accessory genes, *pilB*, *pilC*, and *pilD*. *J. Bacteriol.* **172:**2911–2919.
168. **Ogasawara, N., S. Moriya, K. von Meyenburg, F. G. Hansen, and H. Yoshikawa.** 1985. Conservation of genes and their organization in the chromosomal replication region of *Bacillus subtilis* and *Escherichia coli*. *EMBO J.* **4:**3345–3350.
169. **Oishi, M., and S. Cosloy.** 1972. The genetic and biochemical basis of transformability of *Escherichia coli* K12. *Biochem. Biophys. Res. Commun.* **49:**1568–1572.
170. **Perego, M., S. P. Cole, D. Burbulys, K. Trach, and J. A. Hoch.** 1989. Characterization of the gene for a protein kinase which phosphorylates the sporulation-regulatory proteins SpoOA and SpoOF of *Bacillus subtilis*. *J. Bacteriol.* **171:**6187–6196.
171. **Perego, M., C. F. Higgins, S. R. Pearce, M. P. Gallagher, and J. A. Hoch.** 1991. The oligopeptide transport system of *Bacillus subtilis* plays a role in the initiation of sporulation. *Mol. Microbiol.* **5:**173–185.
172. **Perego, M., G. B. Spiegelman, and J. A. Hoch.** 1988. Structure of the gene for the transition state regulator *abrB*: regulator synthesis is controlled by the *spoOA* sporulation gene in *Bacillus subtilis*. *Mol. Microbiol.* **2:**689–699.
173. **Perego, M., J.-J. Wong, G. B. Spiegelman, and J. A. Hoch.** 1991. Mutational dissociation of the positive and negative regulatory properties of the SpoOA sporulation transcription factor of *Bacillus subtilis*. *Gene* **100:** 207–212.
174. **Piechowska, M., and M. S. Fox.** 1971. Fate of transforming deoxyribonucleate in *Bacillus subtilis*. *J. Bacteriol.* **108:**680–689.
175. **Piggot, P., and J. G. Coote.** 1976. Genetic aspects of bacterial endospore formation. *Bacteriol. Rev.* **40:**908–962.
176. **Poindexter, J., and D. Dubnau.** Unpublished data.
177. **Pugsley, A., C. d'Enfert, I. Reyss, and M. G. Kornacker.** 1990. Genetics of extracellular protein secretion by gram-negative bacteria. *Annu. Rev. Genet.* **24:**67–90.
178. **Pugsley, T. (Institut Pasteur, Paris, France).** 1991. Personal communication.
179. **Radding, C. M.** 1988. Homologous pairing and strand exchange promoted by *Escherichia coli* RecA protein, p. 193–230. *In* R. Kucherlapti and G. R. Smith (ed.), *Genetic Recombination*. American Society for Microbiology, Washington, D.C.
180. **Rao, N. N., and A. Torriani.** 1990. Molecular aspects of phosphate transport in *Escherichia coli*. *Mol. Microbiol.* **4:**1083–1090.
181. **Reyss, I., and A. P. Pugsley.** 1990. Five additional genes in the *pulC-O* operon of the gram-negative bacterium *Klebsiella oxytoca* UNF5023 which are required for pullulanase secretion. *Mol. Gen. Genet.* **222:**176–184.
182. **Roggiani, M., and D. Dubnau.** Unpublished data.
183. **Roggiani, M., J. Hahn, and D. Dubnau.** 1990. Suppression of early competence mutations in *Bacillus subtilis* by *mec* mutations. *J. Bacteriol.* **172:**4056–4063.
184. **Roggiani, M., L. Kong, and D. Dubnau.** Unpublished data.
185. **Rosenthal, A. L., and S. D. Lacks.** 1980. Complex structure of the membrane nuclease of *Streptococcus pneumoniae* revealed by two dimensional electrophoresis. *J. Mol. Biol.* **141:**133–146.
186. **Rudner, D. Z., J. R. LeDeaux, K. Ireton, and A. D. Grossman.** 1991. The *spoOK* locus of *Bacillus subtilis* is homologous to the oligopeptide permease locus and is required for sporulation and competence. *J. Bacteriol.* **173:**1388–1398.
187. **Russel, M., and P. Model.** 1982. Filamentous phage

pre-coat is an integral membrane protein: analysis by a new method of membrane preparation. *Cell* **28:**177–184.
188. **Sadaie, Y.** 1989. Molecular cloning of a *Bacillus subtilis* gene involved in cell division, sporulation, and exoenzyme secretion. *Jpn. J. Genet.* **64:**111–119.
189. **Sadaie, Y. (National Institute of Genetics, Mishima, Japan).** 1991. Personal communication.
190. **Sadaie, Y., and T. Kada.** 1983. Effect of septum-initiation mutations on sporulation and competent cell formation in *Bacillus subtilis. Mol. Gen. Genet.* **190:**176–178.
191. **Sadaie, Y., and T. Kada.** 1983. Formation of competent *Bacillus subtilis* cells. *J. Bacteriol.* **153:**813–821.
192. **Sadaie, Y., and T. Kada.** 1985. *Bacillus subtilis* gene involved in cell division, sporulation, and exoenzyme secretion. *J. Bacteriol.* **163:**648–653.
193. **Sadaie, Y., H. Takamatsu, K. Nakamura, and K. Yamane.** 1991. Sequencing reveals similarity of the wild type *div*$^+$ gene of *Bacillus subtilis* to the *Escherichia coli secA* gene. *Gene* **98:**101–105.
194. **Sastry, P. A., B. L. Pasloske, W. Paranchych, J. R. Pearlstone, and L. B. Smillie.** 1985. Comparative studies of the amino acid and nucleotide sequences of pilin derived from *Pseudomonas aeruginosa* PAK and PAO. *J. Bacteriol.* **164:**571–577.
195. **Schaeffer, P., B. Cami, and R. D. Hotchkiss.** 1976. Fusion of bacterial protoplasts. *Proc. Natl. Acad. Sci. USA* **77:**2151–2155.
196. **Schatz, P., and J. Beckwith.** 1990. Genetic analysis of protein export in *Escherichia coli. Annu. Rev. Genet.* **24:**215–248.
197. **Seifert, H. S., R. S. Ajioka, C. Marchal, P. F. Sparling, and M. So.** 1988. DNA transformation leads to pilin antigenic variation in *Neisseria gonorrhoeae. Nature* (London) **336:**392–395.
198. **Sharma, S., T. F. Stark, W. G. Beattie, and R. E. Moses.** 1986. Multiple control elements for the *uvrC* gene unit of *Escherichia coli. Nucleic Acids Res.* **14:**2301–2318.
199. **Shemyakin, F. M., A. A. Grepachevsky, and A. V. Chestukhin.** 1979. Properties of *Bacillus subtilis* ATP-dependent deoxyribonuclease. *Eur. J. Biochem.* **98:**417–423.
200. **Shirasu, K., P. Morel, and C. I. Kado.** 1990. Characterization of the *virB* operon of an *Agrobacterium tumefaciens* Ti plasmid: nucleotide sequence and protein analysis. *Mol. Microbiol.* **4:**1153–1163.
201. **Singh, R. N.** 1972. Number of deoxyribonucleic acid uptake sites in competent cells of *Bacillus subtilis. J. Bacteriol.* **110:**266–272.
202. **Singh, R. N., and M. P. Pitale.** 1967. Enrichment of *Bacillus subtilis* transformants by zonal centrifugation. *Nature* (London) **213:**1262–1263.
203. **Slauch, J. M., S. Garrett, D. E. Jackson, and T. J. Silhavy.** 1988. *envZ* functions through *ompR* to control porin gene expression in *Escherichia coli* K-12. *J. Bacteriol.* **170:**439–441.
204. **Smith, H., D. B. Danner, and R. A. Deich.** 1981. Genetic transformation. *Annu. Rev. Biochem.* **50:**41–68.
205. **Smith, H., K. Wiersma, S. Bron, and G. Venema.** 1983. Transformation in *Bacillus subtilis*: purification and partial characterization of a membrane-bound DNA-binding protein. *J. Bacteriol.* **156:**101–108.
206. **Smith, H., K. Wiersma, S. Bron, and G. Venema.** 1984. Transformation in *Bacillus subtilis*: a 75,000-dalton protein complex is involved in binding and entry of donor DNA. *J. Bacteriol.* **157:**733–738.
207. **Smith, H., K. Wiersma, G. Venema, and S. Bron.** 1985. Transformation in *Bacillus subtilis*: further characterization of a 75,000-dalton protein complex involved in binding and entry of donor DNA. *J. Bacteriol.* **164:**201–206.
208. **Soltyk, A., D. Shugar, and M. Piechowska.** 1975. Heterologous deoxyribonucleic acid uptake and complexing with cellular constituents in competent *Bacillus subtilis. J. Bacteriol.* **124:**1429–1438.
209. **Somma, S., and M. Poisinelli.** 1970. Quantitative autoradiographic study of competence and deoxyribonucleic acid incorporation in *Bacillus subtilis. J. Bacteriol.* **101:**851–855.
210. **Sonenshein, A. L., H. B. Alexander, D. M. Rothstein, and S. H. Fisher.** 1977. Lipiarmycin-resistant ribonucleic acid polymerase mutants of *Bacillus subtilis. J. Bacteriol.* **132:**73–79.
211. **Soo-Lee, M., and K. J. Marians.** 1990. Differential ATP requirements distinguish the DNA translocation and DNA unwinding activities of the *Escherichia coli* PRI A protein. *J. Biol. Chem.* **265:**17078–17083.
212. **Sparling, P. F.** 1966. Genetic transformation of *Neisseria gonorrhoeae* to streptomycin resistance. *J. Bacteriol.* **92:**1364–1371.
213. **Spatz, C. H., and T. A. Trautner.** 1970. One way to do experiments on gene conversion? Transfection with heteroduplex SPP1 DNA. *Mol. Gen. Genet.* **109:**84–106.
214. **Spizizen, J.** 1965. Analysis of asporogenic mutants in *Bacillus subtilis* by genetic transformation, p. 125–137. *In* L. L. Campbell and H. O. Halvorsen (ed.), *Spores III*. American Society for Microbiology, Washington, D.C.
215. **Steinmetz, M., F. Kunst, and R. Dedonder.** 1976. Mapping of mutations affecting synthesis of exocellular enzymes in *Bacillus subtilis. Mol. Gen. Genet.* **148:**281–285.
216. **Stewart, G. J., and C. A. Carlson.** 1986. The biology of natural transformation. *Annu. Rev. Microbiol.* **40:**211–235.
217. **Stock, J. B., A. J. Ninfa, and A. M. Stock.** 1989. Protein phosphorylation and the regulation of adaptive responses in bacteria. *Microbiol. Rev.* **53:**450–490.
218. **Stout, V., and S. Gottesman.** 1990. RcsB and RscC: a two-component regulator of capsule synthesis in *Escherichia coli. J. Bacteriol.* **172:**659–669.
219. **Stranathan, M. C., K. W. Bayles, and R. E. Yasbin.** 1990. The nucleotide sequence of the *recE*$^+$ gene of *Bacillus subtilis. Nucleic Acids Res.* **18:**4249.
220. **Strauch, M. A., G. B. Spiegelman, M. Perego, W. C. Johnson, D. Burbulys, and J. A. Hoch.** 1989. The transition state transcription regulator *abrB* of *Bacillus subtilis* is a DNA binding protein. *EMBO J.* **8:**1615–1621.
221. **Strauss, N.** 1965. Configuration of transforming deoxyribonucleic acid during entry into *Bacillus subtilis. J. Bacteriol.* **89:**288–293.
222. **Strauss, N.** 1966. Further evidence concerning the configuration of transforming deoxyribonucleic acid during entry into *Bacillus subtilis. J. Bacteriol.* **91:**702–708.
223. **Strauss, N.** 1970. Early energy-dependent step in the entry of transforming deoxyribonucleic acid. *J. Bacteriol.* **101:**35–37.
224. **Strauss, N.** 1970. Transformation of *Bacillus subtilis* using hybrid DNA molecules constructed by annealing resolved complementary strands. *Genetics* **66:**583–593.
225. **Strom, M. S., and S. Lory.** 1991. Amino acid substitutions in pilin of *Pseudomonas aeruginosa*. Effect on leader peptide cleavage, amino terminal methylation, and pilus assembly. *J. Biol. Chem.* **266:**1656–1664.
226. **Strom, M. S., D. Nunn, and S. Lory.** 1991. Multiple roles of the pilus biogenesis protein PilD: involvement of PilD in excretion of enzymes from *Pseudomonas aeruginosa. J. Bacteriol.* **173:**1175–1180.
227. **Takahashi, I.** 1963. Transducing phages for *Bacillus subtilis. J. Gen. Microbiol.* **31:**211–217.
228. **Tanaka, T., and M. Kawata.** 1988. Cloning and characterization of *Bacillus subtilis iep*, which has positive and negative effects on production of extracellular proteases. *J. Bacteriol.* **170:**3593–3600.
229. **Taylor, A. F.** 1988. RecBCD enzyme of *Escherichia coli*, p. 231–263. *In* R. Kucherlapati and G. R. Smith (ed.), *Genetic Recombination*. American Society for Microbiology, Washington, D.C.
230. **Tennent, J. M., F. Lindberg, and S. Normark.** 1990. Integrity of *Escherichia coli* P pili during biogenesis: properties and role of PapJ. *Mol. Microbiol.* **4:**747–758.

231. **Tieng, E., M. Roggiani, and D. Dubnau.** Unpublished data.
232. **Torres, O. R., R. Z. Korman, S. A. Zahler, and G. M. Dunny.** 1991. The conjugative transposon Tn*925*: enhancement of conjugative transfer by tetracycline in *Enterococcus faecalis* and mobilization of chromosomal genes in *Bacillus subtilis* and *E. faecalis*. *Mol. Gen. Genet.* **225:**395–400.
233. **Trach, K., D. Burbulys, J. Wu, R. Jonas, J. Day, P. Kallio, M. Strauch, M. Perego, G. Spiegelman, C. Fogher, and J. Hoch.** 1990. Control of the initiation of sporulation, p. 13. Abstr. Int. Conf. *Bacillus subtilis* Genome, Paris.
234. **Trautner, T. A., B. Pawlek, S. Bron, and C. Anagnostopoulos.** 1974. Restriction and modification in *B. subtilis*. Biological aspects. *Mol. Gen. Genet.* **131:**181–191.
235. **Vagner, V., J.-P. Claverys, S. D. Ehrlich, and V. Mejean.** 1990. Direction of DNA entry in competent cells of *Bacillus subtilis*. *Mol. Microbiol.* **4:**1785–1788.
236. **Vandeyar, M. A., and S. A. Zahler.** 1986. Chromosomal insertions of Tn*917* in *Bacillus subtilis*. *J. Bacteriol.* **167:**530–534.
237. **van Nieuwenhoven, M. H., K. J. Hellingwerf, G. Venema, and W. N. Konings.** 1982. Role of proton motive force in genetic transformation of *Bacillus subtilis*. *J. Bacteriol.* **151:**771–776.
238. **van Sinderen, D. (University of Groningen, Groningen, The Netherlands).** 1991. Personal communication.
239. **van Sinderen, D., and G. Grandi (University of Groningen, Groningen, The Netherlands).** 1991. Personal communication.
240. **van Sinderen, D., and G. Venema (University of Groningen, Groningen, The Netherlands).** 1991. Personal communication.
241. **van Sinderen, D., S. Withoff, H. Boels, and G. Venema.** 1990. Isolation and characterization of *comL*, a transcription unit involved in competence development in *Bacillus subtilis*. *Mol. Gen. Genet.* **224:**396–404.
242. **Vehmaanpera, J.** 1989. Transformation of *Bacillus amyloliquefaciens* by electroporation. *FEMS Microbiol. Lett.* **52:**165–169.
243. **Vermuelen, C. A., and G. Venema.** 1974. Electron microscope and autoradiographic study of ultrastructural aspects of competence and deoxyribonucleic acid absorption in *Bacillus subtilis*: ultrastructure of competent and noncompetent cells and cellular changes during development of competence. *J. Bacteriol.* **118:**334–341.
244. **Vos, W. D., G. Venema, U. Canosi, and T. A. Trautner.** 1981. Plasmid transformation in *Bacillus subtilis*: fate of plasmid DNA. *Mol. Gen. Genet.* **181:**424–433.
245. **Vosman, B., G. Kuiken, and G. Venema.** 1988. Transformation in *Bacillus subtilis*: involvement of the 17-kilodalton DNA-entry nuclease and the competence-specific 18-kilodalton protein. *J. Bacteriol.* **170:**3703–3710.
246. **Vovis, G.** 1973. Adenosine triphosphate-dependent deoxyribonuclease from *Diplococcus pneumoniae*: fate of transforming deoxyribonucleic acid. *J. Bacteriol.* **113:** 718–723.
247. **Walker, G. C.** 1985. Inducible DNA repair systems. *Annu. Rev. Biochem.* **54:**425–457.
248. **Walker, G. C.** 1987. The SOS response of *Escherichia coli*, p. 1346–1357. *In* F. C. Neidhardt, J. L. Ingraham, K. B. Low, B. Magasanik, M. Schaechter, and H. E. Umbarger (ed.), *Escherichia coli and Salmonella typhimurium: Cellular and Molecular Biology*, vol. 2. American Society for Microbiology, Washington, D.C.
249. **Ward, J. E., D. E. Akiyoshi, D. Regier, A. Datta, M. P. Gordon, and E. W. Nester.** 1988. Characterization of the *virB* operon from an *Agrobacterium tumefaciens* Ti plasmid. *J. Biol. Chem.* **263:**5804–5814.
250. **Ward, J. E., D. E. Akiyoshi, D. Regier, A. Datta, M. P. Gordon, and E. W. Nester.** 1990. Characterization of the *virB* operon from an *Agrobacterium tumefaciens* Ti plasmid (a correction). *J. Biol. Chem.* **265:**4768.
251. **Watts, T. H., C. M. Kay, and W. Paranchych.** 1983. Spectral properties of three quaternary arrangements of *Pseudomonas* pilin. *Biochemistry* **22:**3640–3646.
252. **Weckerman, R., R. Furbass, and M. A. Marahiel.** 1988. Complete nucleotide sequence of the *tycA* gene coding the tyrocidine synthetase I from *Bacillus brevis*. *Nucleic Acids Res.* **16:**11841.
253. **Weinrauch, Y., and D. Dubnau.** Unpublished data.
254. **Weinrauch, Y., and D. Dubnau.** 1983. Plasmid marker rescue transformation in *Bacillus subtilis*. *J. Bacteriol.* **154:**1077–1087.
255. **Weinrauch, Y., and D. Dubnau.** 1987. Plasmid marker rescue transformation proceeds by breakage-reunion in *Bacillus subtilis*. *J. Bacteriol.* **169:**1205–1211.
256. **Weinrauch, Y., N. Guillen, and D. Dubnau.** 1989. Sequence and transcription mapping of *Bacillus subtilis* competence genes *comB* and *comA*, one of which is related to a family of bacterial regulatory determinants. *J. Bacteriol.* **171:**5362–5375.
257. **Weinrauch, Y., T. Msadek, F. Kunst, and D. Dubnau.** Unpublished data.
258. **Weinrauch, Y., T. Msadek, F. Kunst, and D. Dubnau.** 1991. Sequence and properties of *comQ*, a new competence regulatory gene of *Bacillus subtilis*. *J. Bacteriol.* **173:**5685–5693.
259. **Weinrauch, Y., R. Penchev, E. Dubnau, I. Smith, and D. Dubnau.** 1990. A *Bacillus subtilis* regulatory gene product for genetic competence and sporulation resembles sensor protein members of the bacterial two-component signal-transduction systems. *Genes Dev.* **4:**860–872.
260. **Weston, L. A., and R. J. Kadner.** 1987. Identification of UhP polypeptides and evidence for their role in exogenous induction of the sugar phosphate transport system of *Escherichia coli* K-12. *J. Bacteriol.* **169:**3546–3555.
261. **Whitchurch, C. B., M. Hobbs, S. P. Livingston, V. Krishnapillai, and J. S. Mattick.** 1990. Characterization of a *Pseudomonas aeruginosa* twitching motility gene and evidence for a specialised protein export system widespread in eubacteria. *Gene* **101:**33–34.
262. **Wilcox, K. W., and H. O. Smith.** 1975. Isolation and characterization of mutants of *Haemophilus influenzae* deficient in an adenosine 5′-triphosphate-dependent deoxyribonuclease activity. *J. Bacteriol.* **122:**443–453.
263. **Yamagishi, H., and I. Takahashi.** 1968. Transducing particles of PBS 1. *Virology* **36:**639–645.
264. **Yamashita, S., F. Kawamura, H. Yoshikawa, H. Takahashi, Y. Kobayashi, and H. Saito.** 1989. Dissection of the expression signals of the *spoOA* gene of *Bacillus subtilis*: glucose represses sporulation-specific expression. *J. Gen. Microbiol.* **135:**1335–1345.
265. **Yasbin, R., J. Jackson, P. Love, and R. Marrero.** 1988. Dual regulation of the *recE* gene, p. 109–114. *In* A. T. Ganesan and J. A. Hoch (ed.), *Genetics and Biotechnology of Bacilli*, vol. 2. Academic Press, Inc., San Diego, Calif.
266. **Yasbin, R. E., M. Stranathan, and K. W. Bayles (University of Maryland Baltimore County, Baltimore).** 1991. Personal communication.
267. **Youngman, P., J. B. Perkins, and R. Losick.** 1984. A novel method for the rapid cloning in *Escherichia coli* of *Bacillus subtilis* chromosomal DNA adjacent to Tn917 insertions. *Mol. Gen. Genet.* **195:**424–433.
268. **Youngman, P., J. B. Perkins, and K. Sandman.** 1985. Use of Tn*917*-mediated transcriptional gene fusions to *lacZ* and *cat-86* for the identification and study of *spo* genes in *Bacillus subtilis*, p. 47–54. *In* J. A. Hoch and P. Setlow (ed.), *Molecular Biology of Microbial Differentiation*. American Society for Microbiology, Washington, D.C.
269. **Youngman, P., P. Zuber, J. B. Perkins, K. Sandman, M. Igo, and R. Losick.** 1985. New ways to study developmental genes in spore-forming bacteria. *Science* **228:** 285–291.

40. Transposons and Their Applications

PHILIP YOUNGMAN

Although transposons indigenous to *Bacillus subtilis* have yet to be discovered, it has proved possible to introduce into this species insertion elements from other gram-positive bacteria and to exploit these elements for a wide range of genetic manipulations (88). By far the most extensively used transposon in *B. subtilis* is the *Enterococcus faecalis* element Tn*917* (78), which will be the main subject of this chapter. Numerous derivatives of Tn*917* have been developed to facilitate the cloning and functional analysis of regulated genes identified by insertional mutations, and delivery vectors for obtaining Tn*917* insertions function effectively in several other gram-positive species (87). Also described in this chapter is a new family of insertion elements based on derivatives of the *Escherichia coli* transposon Tn*10* that were specifically engineered for use in *B. subtilis* (65). The performance characteristics of these Tn*10* derivatives have not yet been extensively evaluated, and specialized versions designed to facilitate cloning and the recovery of insertion-mediated gene fusions have not yet been developed, but available information suggests that Tn*10* might ultimately prove to be superior to Tn*917* as a general-purpose insertional mutagen in *B. subtilis*. Although vectors developed for the delivery of Tn*917* and Tn*10* in *B. subtilis* function well in some other gram-positive bacteria, they apparently cannot be used in several genera of interest and importance to both basic researchers and industrial scientists, including the genera *Streptomyces* and *Clostridium*. For *Streptomyces* spp., this lack has been overcome by the development of a completely independent set of vectors and transposons derived entirely from insertion elements, plasmids, and phages indigenous to *Streptomyces* species. The insertional mutagenesis systems now available for *Streptomyces* spp. will be discussed here briefly, primarily as a model for the establishment of such systems in gram-positive species in which barriers to the use of nonindigenous elements exist.

INSERTIONAL MUTAGENESIS IN *B. SUBTILIS* WITH Tn*917*

Tn*917* was first identified as a 5.3-kb translocatable segment of *E. faecalis* plasmid pAD2 that encoded inducible erythromycin resistance (Em^r) (78). It is now recognized as a member of the Tn*3* family of transposable elements, very similar in sequence and genetic organization to Tn*551* of *Staphylococcus aureus* (3, 63, 71). Like other members of the Tn*3* family, Tn*917* generates a 5-bp duplication of target sequences upon insertion (63) and exhibits a very low frequency of precise excision (88). Although direct mapping of transcripts has not been carried out, the three major open reading frames of Tn*917* are apparently transcribed in the same direction (Fig. 1). When viewed such that transcription would proceed from left to right (Fig. 1), the leftmost gene is *erm*, the erythromycin resistance determinant. The two other major open reading frames, *tnpR* and *tnpA*, exhibit substantial homology to their Tn*3* counterparts and thus are presumed to encode, respectively, the resolvase and transposase functions of Tn*917*. It was noted by Tomich and colleagues (78) that both expression of *erm* and levels of Tn*917* transposition are significantly higher in the presence of subselective levels of erythromycin. Conceivably, this reflects induced transcriptional readthrough of *tnpA* and *tnpR* from the *erm* promoter. Of particular importance for understanding the origins and properties of most of the Tn*917* derivatives discussed below is the fact that the interval between *erm* and the nearest terminal inverted repeat (Fig. 1) consists of nonessential DNA that may be modified without interfering with transposition (89).

The most effective method currently available for delivering Tn*917* to the *B. subtilis* chromosome involves the use of temperature-sensitive plasmids derived from the *S. aureus* replicon pE194 (37). Older vectors, such as pTV1 (88), were constructed by using the wild-type replication origin of pE194, which fails to function at temperatures above 45°C. Newer versions of these vectors incorporate a mutation that significantly enhances the temperature-sensitive replication defect of pE194 (81). These vectors were generally given the suffix Ts (e.g., pTV1Ts) to distinguish them from structurally similar constructions that do not incorporate the temperature sensitivity mutation (86). Using the Ts vectors, chromosomal insertions are readily obtained by 1:100 dilution of mid-log-phase cultures into fresh broth prewarmed to 40°C and then overnight aeration at that temperature (86, 87). A similar protocol is effective in at least some other gram-positive species in which pE194-derived vectors can replicate (7, 9, 17).

Although Tn*917* insertions are not perfectly random in their distribution over the *B. subtilis* chromosome (88), it is apparent that insertions may be obtained in any chromosomal region and perhaps within the coding sequence of any gene. More than 100 chromosomal insertions have been characterized thus far; most of these are displayed on the map in Fig. 2 and listed in Table 1. The following is a brief summary of the kinds of insertions included.

Phenotypically Cryptic Mutations

Vandeyar and Zahler (80) characterized a collection of 20 phenotypically cryptic chromosomal Tn*917* in-

Philip Youngman • Department of Genetics, University of Georgia, Athens, Georgia 30602.

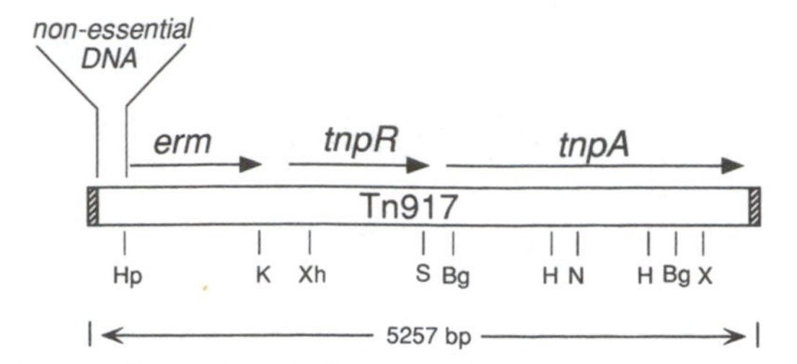

Figure 1. Physical and functional map of Tn*917*. Arrows indicate locations and orientations of transposon-associated genes: *erm*, gene encoding inducible erythromycin resistance; *tnpA* and *tnpR*, genes homologous to Tn*3* genes with the same designations (32). Hatched boxes at the ends represent 38-bp inverted repeats. The region labeled nonessential DNA consists of a 176-bp interval of DNA between the *erm* gene and the nearest inverted repeat that can be deleted or altered without interfering with transposition (89). The nucleotide sequence of this 176-bp interval is given in reference 63. Hp, *Hpa*I restriction sites; N, *Nco*I restriction sites. Other restriction sites are identified in footnote *a* to Table 2. The complete nucleotide sequence of Tn*917* is given in reference 71, but it contains several errors, some of which (in the *tnpA* region) were corrected in a subsequent publication (3).

sertions that they found scattered over most chromosomal regions. Insertions in regions spanning 11° to 54° and 140° to 177° (Fig. 2) were relatively less abundant, however, and as in previous studies (88), the immediate vicinity of *gltA* (165° to 170°) appeared to be a "hot spot" where insertions were disproportionately frequent. Although most of the cryptic insertions in that study were obtained without bias for position, it was also demonstrated that PBS1-mediated linkage to a selectable marker (*purA*) could be used to target cryptic insertions to a specified chromosomal location. Zahler and colleagues subsequently characterized 11 more cryptic insertions (94), which are also shown in Fig. 2 and listed in Table 1. Sandman et al. (69) demonstrated that for each of 12 randomly chosen selectable markers, it was possible to recover a Tn*917* insertion linked by cotransformation, which stands as perhaps the most persuasive indication that insertions can occur in any chromosomal interval. Twelve phenotypically cryptic insertions from that study are also shown in Fig. 2 and listed in Table 1. Additional well-characterized cryptic insertions (e.g., near *gltA* [64] and *terC* [34]) were omitted because of space limitations.

Auxotrophic Mutations

Again, the largest collection of Tn*917*-mediated insertional auxotrophs comes from the study of Vandeyar and Zahler (80) and from subsequent unpublished work by Zahler and colleagues (94). A total of 29 insertions from that collection are shown in Fig. 2 and listed in Table 1. One additional insertion (*trpE*::Tn*917*lacΩHU114) derives from the work of Perkins and Youngman (64). Analysis of these insertional mutations reinforced several general observations concerning the behavior of Tn*917* as a chromosomal mutagen: insertions are relatively random, precise excision of insertions is extremely rare, and deletions accompany the insertion process about 10% of the time (69, 80).

Sporulation-Related Mutations

Tn*917* has been exploited extensively to facilitate the characterization of sporulation genes. In the study by Sandman et al. (69), a library of chromosomal insertions was screened to recover 20 distinct insertional *spo* mutations distributed over 18 distinct genetic loci. Several of these mutations have been instrumental in the cloning and characterization of the mutated loci (e.g., *spoIIE*::Tn*917*ΩHU181 [27] and *spoIVC*::Tn*917*ΩHU215 [43]). Moreover, at least five of the insertional mutations from this collection (*spoIIJ*::Tn*917*ΩHU19, *spoVM*::Tn*917*ΩHU324, *spoIIM*::Tn*917*ΩHU287, *spoVQ*::Tn*917*ΩHU195, and *spoVN*::Tn*297*ΩHU297) are apparently different from previously identified *spo* loci, and these likewise have proved valuable for cloning the mutated genes (e.g., *spoIIJ*::Tn*917*ΩHU19 [4, 61] and *spoIIM*::Tn*917*ΩHU287 [72]). Other sporulation-related Tn*917* mutations exploited for molecular analysis of the disrupted locus include *abrB*::Tn*917* (62, 66), *spoOK418*::Tn*917-lac* (67), *gerM*::Tn*917* (68), and *gerD*::Tn*917* (85).

Identification of *din* Loci

Love and colleagues (47) were successful in identifying several DNA damage-inducible loci by screening a large library of Tn*917-lac* insertions for insertion-mediated fusions that exhibited increased expression when exposed to mitomycin C, a study modeled after the earlier work of Kenyon and Walker (39). After further genetic analysis, these insertions were found to define at least three distinct loci, represented in Fig. 2 and Table 1 by *dinA76*::Tn*917lac*, *dinB7*::Tn*917lac*, and *dinC1*::Tn*917lac* (10, 25). The library of Tn*917lac* chromosomal insertions generated for this work has been made available by R. Yasbin and colleagues to several other investigators and has proved to be of great value in the recovery of other types of insertions, as noted below.

Insertional *com* Mutations

Transformation competence in *B. subtilis* has been of long-standing interest both as a mechanism of genetic exchange and as a model system for understanding developmentally regulated gene activation programs (22). Recent progress in identifying genes whose products participate in the competence process has been dramatic, and this progress has been made possible in part by the recovery of transposon-mediated insertional mutations that reduce or abolish transformation (29, 51). These mutations were found to cluster in several chromosomal regions and apparently define at least nine distinct genetic loci, represented in Fig. 2 and listed in Table 1 as *comA124*::Tn*917-lac* (26), *comC530*::Tn*917-lac* (54), *comD413*::Tn*917-lac* (26), *comE518*::Tn*917-lac* (26), *comF524*::Tn*917-lac* (45), *comG12*::Tn*917-lac* (1), *comK44*::Tn*917* (51), *comM67*::Tn*917* (51), *comN114*::Tn*917* (51), and *comO18*::Tn*917* (51). Because many of these mutations were produced with Tn*917-lac*, useful information about the expression patterns and regulation of the disrupted genes was obtained even before they were cloned (2). Techniques developed to facili-

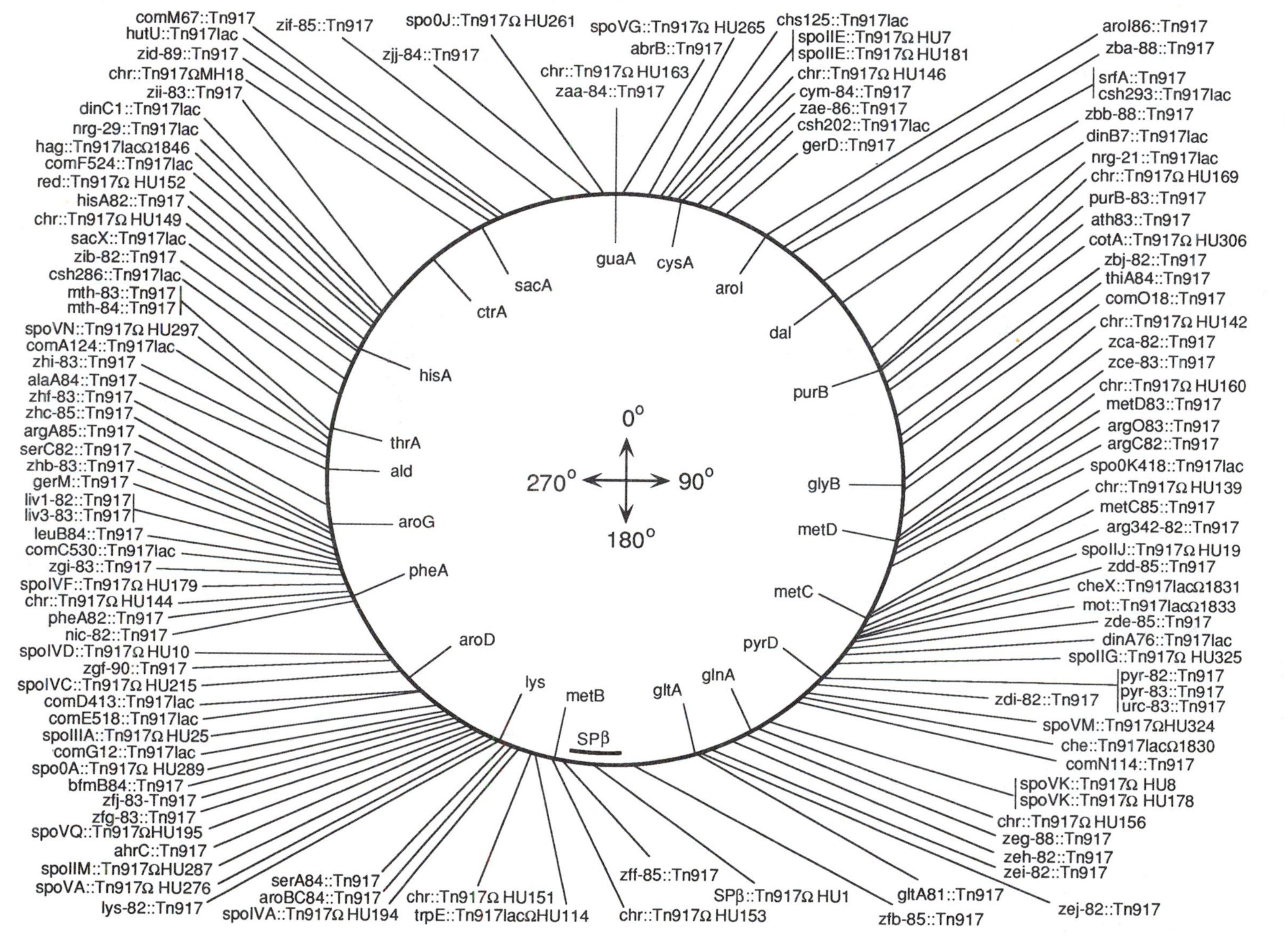

Figure 2. Genetic map of the *B. subtilis* chromosome showing locations of characterized Tn*917* insertions relative to several commonly used chromosomal markers. Most of these insertions were not positioned by three-factor crosses, so except for insertional mutations that correspond to carefully mapped loci, locations are approximate and the relative positions of some closely spaced insertions are arbitrary. Additional information concerning specific insertions is given in Table 1.

Table 1. Catalog of Tn*917* insertions

Insertion	Approx map position (°)[a]	Phenotype and comments[b]	Reference(s)
chr::Tn*917*ΩHU163	0	Cryptic; 41% cotransformation with *guaA*	69
zaa-84::Tn*917*	0	Cryptic; 95% cotransduction with *guaA*	80
abrB::Tn*917*	6	Partial suppressor of *spoOA* mutations	66
spoVG::Tn*917*ΩHU265	7	Sporulation block at stage V	69
csh-125::Tn*917-lac*	8	Expression controlled by σ^H	38
spoIIE::Tn*917*ΩHU7	9	Sporulation block at stage II	27
spoIIE::Tn*917*ΩHU181	9	Sporulation block at stage II	27
chr::Tn*917*ΩHU146	11	Cryptic; 86% cotransformation with *cysA*	69
cym-84::Tn*917*	11	Requires cysteine or methionine	80
zae-86::Tn*917*	18	Cryptic (linkage relationships unpublished)	94
csh-202::Tn*917-lac*	19	Expression controlled by σ^H	38
gerD::Tn*917*	20	Defective germination	85
aroI86::Tn*917*	26	Requires phenylalanine, tyrosine, and tryptophan	94
zba-88::Tn*917*	33	Cryptic (linkage relationships unpublished)	94
srfA::Tn*917*	35	Defect in surfactin biosynthesis, competence; same locus identified by *csh-293*::Tn*917-lac*	38, 57
zbb-88::Tn*917*	44	Cryptic (linkage relationships unpublished)	94
dinB7::Tn*917-lac*	48	Expression induced by DNA damage-competence	25
nrg-21::Tn*917-lac*	55	Expression induced by nitrogen limitation	5
chr::Tn*917*ΩHU169	62	Cryptic; 23% cotransformed with *purB33*	69
purB83::Tn*917*	65	Requires purines	80
ath-83::Tn*917*	66	Requires adenine and thiamine	80
cotA::Tn*917*ΩHU306	67	Defect in spore coat, sporulation pigmentation	21
zbj-82::Tn*917*	75	Cryptic; 38% cotransformed with *tre*	80
thiA84::Tn*917*	84	Requires thiamine	80
comO18::Tn*917*	88	Defect in transformation competence	51
chr::Tn*917*ΩHU142	93	Cryptic; 84% cotransformed with *glyB*	69
zca-82::Tn*917*	94	Cryptic; 62% cotransformed with *glyB*	80
comK44::Tn*917*	96	Defect in transformation competence	51
zce-83::Tn*917*	98	Cryptic; 73% cotransduction with *metD*	80
chr::Tn*917*ΩHU160	100	Cryptic; 81% cotransformed with *metD*	69
metD83::Tn*917*	101	Requires methionine	80
argO83::Tn*917*	102	Requires arginine	80
argC82::Tn*917*	103	Requires arginine	80
spoOK418::Tn*917-lac*	107	Sporulation block at stage 0; competence defect	67
chr::Tn*917*ΩHU139	116	Cryptic; 38% cotransformed with *metC*	69
metC85::Tn*917*	118	Requires methionine	80
arg342-82::Tn*917*	119	Requires arginine	80
spoIIJ::Tn*917*ΩHU19	121	Sporulation block at stage 0; same locus as kinA; encodes sporulation-related protein kinase	61
zdd-85::Tn*917*	125	Cryptic; 62% cotransduced with *metC*	80
cheX::Tn*917-lac*Ω1831	127	Chemotaxis defect	95
mot::Tn*917-lac*Ω1833	128	Motility defect	95
zde-85::Tn*917*	129	Cryptic; 36% cotransduced with *metC*	80
dinA76::Tn*917-lac*	130	Expression induced by DNA damage, competence	25
spoIIG::Tn*917*ΩHU325	132	Sporulation block at stage II	69
pyr-82::Tn*917*	134	Requires pyrimidines	80
pyr-83::Tn*917*	134	Requires pyrimidines	80
urc-83::Tn*917*	134	Requires uracil and either cysteine or methionine	80
zdi-82::Tn*917*	140	Cryptic; 97% cotransduced with *cysC*	80
spoVM::Tn*917*ΩHU324	141	Sporulation block at stage V	69
che::Tn*917-lac*Ω1830	143	One of several insertions in major *che* locus	95
spoVK::Tn*917*ΩHU8	152	Sporulation block at stage V; same locus as spoVJ	69
spoVK::Tn*917*ΩHU178	152	Sporulation block at stage V	69
chr::Tn*917*ΩHU156	155	Cryptic; 19% cotransformed with *glnA*	69
zeg-88::Tn*917*	160	Cryptic (linkage relationships unpublished)	94
zeh-82::Tn*917*	162	Cryptic (linkage relationships unpublished)	94
zei-82::Tn*917*	165	Cryptic; 40% cotransformed with *gltB*	80
zej-82::Tn*917*	167	Cryptic; 98% cotransduced with *gltB*	80
gltA81::Tn*917*	168	Requires glutamate	80
zfb::Tn*917*	175	Cryptic; 80% cotransduced with *citK*	80
SPβ::Tn*917*ΩHU1	185	Cryptic; prophage nondefective	88
zff-85::Tn*917*	194	Cryptic; 95% cotransduced with *ilvA*	80
chr::Tn*917*ΩHU153	195	Cryptic; 77% cotransformed with *metB*	69
trpE::Tn*917-lac*ΩHU114	200	Requires tryptophan	64
chr::Tn*917*ΩHU151	202	Cryptic; 36% cotransformed with *trpC*	69
spoIVA::Tn*917*ΩHU194	203	Sporulation block at stage IV; formerly spoVP	69
aroBC84::Tn*917*	204	Requires shikimic acid	80

(Continued)

Table 1—*Continued*

Insertion	Approx map position (°)[a]	Phenotype and comments[b]	Reference(s)
serA84::Tn*917*	205	Requires serine	80
lys-82::Tn*917*	206	Requires lysine	80
spoVA::Tn*917*ΩHU276	206	Sporulation block at stage V	69
spoIIM::Tn*917*ΩHU287	207	Sporulation block at stage II	69
ahrC::Tn*917*	208	Arginine hydroxamate resistance	58
spoVQ::Tn*917*ΩHU195	209	Sporulation block at stage V	69
zfg-83::Tn*917*	210	Cryptic; 75% cotransduced with *lys*	80
zfj-83::Tn*917*	211	Cryptic (linkage relationships unpublished)	94
bfmB84::Tn*917*	212	Requires branched-chain fatty acids	94
spoOA::Tn*917*ΩHU289	213	Sporulation block at stage 0; formerly *spoIIL*	69
comG12::Tn*917-lac*	215	Defect in transformation competence	1
spoIIIA::Tn*917*ΩHU25	217	Sporulation block at stage III; formerly *spoIIIB*	69
comE518::Tn*917-lac*	220	Defect in transformation competence	29
comD413::Tn*917-lac*	222	Defect in transformation competence	29
spoIVC::Tn*917*ΩHU215	235	Sporulation block at stage IV	69
zgf-90::Tn*917*	236	Cryptic (linkage relationships unpublished)	94
spoIVD::Tn*917*ΩHU10	237	Sporulation block at stage IV	69
nic-82::Tn*917*	246	Requires nicotinic acid	80
pheA82::Tn*917*	247	Requires phenylalanine	80
chr::Tn*917*ΩHU144	247	Cryptic; 99% cotransformation with *pheA*	69
spoIVF::Tn*917*ΩHU179	247	Sporulation block at stage IV; formerly *spoVL*	69
zgi-83::Tn*917*	250	Cryptic; 84% cotransduced with *pheA*	80
comC530::Tn*917*	251	Defect in transformation competence	54
leuB84::Tn*917*	251	Requires leucine	80
liv1-82::Tn*917*	251	Requires leucine, isoleucine, and valine	80
liv3-83::Tn*917*	251	Requires leucine, isoleucine, and valine	80
gerM::Tn*917*	252	Defective in spore germination	68
zhb-83::Tn*917*	257	Cryptic; 61% cotransduced with *argA*	80
serC82::Tn*917*	257	Requires serine	80
argA85::Tn*917*	259	Requires arginine	80
zhc-85::Tn*917*	260	Cryptic; 93% cotransformed with *argA*	80
zhf-83::Tn*917*	271	Cryptic (linkage relationships unpublished)	94
alaA84::Tn*917*	275	Requires alanine	80
zhi-83::Tn*917*	277	Cryptic: 85% cotransduced with *ald*	80
comA124::Tn*917-lac*	278	Defect in transformation competence	26
spoVN::Tn*917*ΩHU297	279	Sporulation block at stage V	69
mth-83::Tn*917*	280	Requires threonine and methionine	80
mth-84::Tn*917*	280	Requires threonine and methionine	80
csh-286::Tn*917*	285	Expression controlled by σ^H	38
zib-82::Tn*917*	287	Cryptic; 83% cotransduced with *cysB*	80
sacX::Tn*917-lac*	295	Regulation of levansucrase expression	46
chr::Tn*917*ΩHU149	298	Cryptic; 37% cotransformed with *hisA*	69
hisA82::Tn*917*	299	Requires histidine	80
red::Tn*917*ΩHU152	302	Overproduction of pulcheriminic acid	88
comF524::Tn*917-lac*	303	Defect in transformation competence	29, 45
hag::Tn*917*Ω1846	304	Defect in motility	95
nrg-29::Tn*917-lac*	306	Expression induced by nitrogen limitation	5
dinC1::Tn*917-lac*	309	Expression induced by DNA damage	47
zii-83::Tn*917*	317	Cryptic; 58% cotransduced with *ctrA*	80
chr::Tn*917*ΩMH18	327	Cryptic; closely linked to *bac-1*	33
zid-89::Tn*917*	337	Cryptic (linkage relationships unpublished)	94
hutU::Tn*917-lac*	339	Defect in histidine utilization	6
comM67::Tn*917*	343	Defect in transformation competence	51
zif-85::Tn*917*	345	Cryptic; 64% cotransduced with *purA*	80
zjj-84::Tn*917*	359	Cryptic (linkage relationships unpublished)	94
spoOJ::Tn*917*ΩHU261	359	Sporulation block at stage 0	69

[a] Positions were fixed with reference to map coordinates established by Dean and Ziegler (20).
[b] Cotransduction refers to PBS1-mediated generalized transduction.

tate the cloning of genes identified by Tn*917* insertions (90) have helped accelerate the further physical and functional characterization of these *com* loci, and in several cases, tentative functional or regulatory roles can be assigned to certain Com gene products (22). The highly informative collection of insertional *com* mutations characterized by D. Dubnau and colleagues was derived from the Yasbin library of Tn*917-lac* insertions.

Chemotaxis-Related Tn*917* Insertions

Zuberi and colleagues (95) have used insertional mutagenesis with Tn*917-lac* to generate a collection of *mot* and *che* mutations. Four of these were located in the previously identified major *che* operon of *B. subtilis* (96) and are represented in Fig. 2 and Table 1 by *che*::Tn*917-lac*Ω1830. Several other insertions cluster near the *pstI* locus and apparently define at least two novel *che-mot* transcription units (e.g., *cheX*::Tn*917-lac*Ω1831 and *mot*::Tn*917-lac*Ω1833). An additional insertion disrupts the *hag* locus (*hag*::Tn*917-lac*Ω 1846). The fact that many of these insertions also produced *lacZ* fusions made it possible to infer which *che* and *mot* loci were transcribed by (or are under the regulation of) σ^D-associated RNA polymerase.

Genes Controlled by σ^H

Jaacks et al. (38) employed a novel approach to identifying 18 *B. subtilis* genes whose expression is controlled by σ^H, a sigma factor required for sporulation initiation. A library of Tn*917-lac* insertions was introduced by transformation into a strain containing the *sigH* coding sequence under control of the isopropyl-β-D-thiogalactopyranoside (IPTG)-inducible P_{spac} promoter, and transformants were screened for increased expression in the presence of IPTG. Four of the so-called *csh* insertions characterized in this study are shown in Fig. 2 and listed in Table 1. Again, the source of the Tn*917-lac* insertions used in this work was the Yasbin insertion library referred to above.

Nitrogen Regulation

Tn*917-lac* insertions have also been exploited for the investigation of nitrogen regulation in *B. subtilis*. In one study, an insertional mutation in the *hut* operon was used to analyze transcriptional regulation of this operon (6). In another study (5), a library of Tn*917-lac* insertions was screened for enhanced expression under conditions of nitrogen limitation. This screen yielded two insertions (*nrg-21*::Tn*917-lac* and *nrg-29*::Tn*917-lac*) that showed dramatic induction in nitrogen-poor media (Fig. 2, Table 1).

DELIVERY VECTORS FOR OBTAINING CHROMOSOMAL Tn*917* INSERTIONS

As explained above, delivery systems for generating chromosomal Tn*917* insertions involve the use of temperature-sensitive plasmid replicons derived from pE194 or pE194Ts (81). Other recent reviews provide detailed information concerning the origins, structures, and properties of these vectors (87, 91). Table 2 schematically displays the important features of the most widely used Tn*917* derivatives. The recently described derivative Tn*917PF1* of Zagorec and Steinmetz (93) was not included in previous reviews. This novel construction contains a strong, outward-facing promoter associated with the *aphA* gene of Tn*1545* (79) inserted near the *erm*-proximal end of Tn*917*. When Tn*917PF1* is inserted in the appropriate orientation upstream from a cryptic chromosomal gene, this transposon derivative is capable of activating expression of the gene. Tn*917PF1* is thus conceptually similar to Tn*5-tac-1* of Chow and Berg (11), which was designed for generating conditional insertional mutations in *E. coli*. Zagorec and Steinmetz (93) demonstrated the utility of Tn*917PF1* by using it to isolate insertions that activated expression of a cryptic *B. subtilis* gene encoding a β-galactosidase activity.

Some of the newer transposon derivatives listed in Table 2 supersede older vectors still in wide use. For example, the Tn*917* derivative in pLTV1 (9) generates *lacZ* transcriptional fusions in the same way as the derivatives in pTV32 and pTV51 but also contains ColE1-derived replication functions (Table 2). Thus, pLTV1 would be superior to pTV32 or pTV51 for many applications. In addition, as an inadvertant consequence of the way it was constructed, the transposon derivative in pLTV1 has a significantly elevated transposition frequency in *B. subtilis*. Thus, pLTV1 would be much better for generating libraries of Tn*917* insertional mutations even when the feature of creating *lacZ* fusions is irrelevant (87).

SELECTION FOR Tn*917* INSERTIONS INTO *B. SUBTILIS* PLASMIDS

For Tn*917*-mediated mutagenesis of plasmid DNA, Kopec et al. (40a) have devised an effective method that is based on the observation that phage SPO2 can very efficiently transduce any plasmid containing an SPO2 *cos* site. Thus, when the transduction donor contains both a chromosomal copy of Tn*917* and a *cos*-bearing plasmid, a selection for SPO2-mediated transfer of transposon-associated antibiotic resistance to a new strain often recovers mutants with transpositions into plasmid sequences. When this kind of experiment was carried out by inducing lysis of an SPO2 lysogen containing a *cos*-bearing derivative of pC194, transductants containing plasmids with Tn*917* insertions were recovered at a frequency of better than 10^{-8}/PFU, and insertions were distributed with a high degree of randomness within cloned sequences carried by the pC194 derivative (50). Although this method has not been widely exploited, it appears to be a useful strategy for functional analysis of cloned genes in *B. subtilis*.

USE OF Tn*917* IN OTHER GRAM-POSITIVE BACTERIA

The observation by Kuramitsu and Casadaban (44) that Tn*917* derivatives can actually function in *E. coli* makes it apparent that this transposon has a very broad potential host range. This wide host range has been most readily exploited in other gram-positive species in which the standard delivery vectors developed for *B. subtilis* can replicate. For example, using pTV1, pTV32Ts, and pLTV1 in *Bacillus megaterium*, Vary and colleagues have obtained several insertions that block sporulation or cause auxotrophic phenotypes (7, 76). Hartley and Paddon have likewise generated several different kinds of insertional mutations by using pTV1 in *Bacillus amyloliquefaciens* (31), where they find indications that Tn*917* insertions are perhaps even more randomly distributed than in *B. subtilis*. In the extreme alkalophiles, which present a combination of highly interesting biology and nonexistent genetics, Krulwich and colleagues have demonstrated the likelihood that Tn*917* will be a very useful tool (42). For example, in preliminary studies with the

Table 2. Tn*917* Derivatives

Vector	Schematic structure of transposon derivative[a]	Properties or applications	Reference
pTV1Ts	erm tnpR tnpA	Insertional mutagenesis	88
pTV8	erm tnpR tnpA; Sm B Sm	Insertional mutagenesis; contains cloning site in nonessential DNA	89
pTV20	erm tnpR cat bla {tnpA}; K Xh Sp E ColE1	Transposition functions inactive; used for recovery of cloned sequences adjacent to sites of existing insertions	90
pTV21	erm tnpR bla cat {tnpA}; ColE1 E Sp H Bg X	Transposition functions inactive; used for recovery of cloned sequences adjacent to sites of existing insertions	90
pTV21Δ2	bla cat {tnpA}; ColE1 E Sp S H Bg X	Transposition functions inactive; used for recovery of cloned sequences adjacent to sites of existing insertions	90
pTV24	cat erm tnpR tnpA; Sm B S	Insertional mutagenesis; a selection for Cm^r may be used to recover transpositions	86
pTV32Ts	(lacZ) erm tnpR tnpA; Sm B	Insertional mutagenesis; creates transcriptional *lacZ* fusions	64
pTV51Ts	(lacZ) erm tnpR tnpA; B Sm	Insertional mutagenesis; creates transcriptional *lacZ* fusions; cloning site allows further modification without disrupting fusions	86
pTV52	(cat-86) erm tnpR tnpA; Sm B S P	Insertional mutagenesis; creates transcriptional *cat* fusions	86
pTV53	(lacZ) (cat-86) erm tnpR tnpA; B P	Insertional mutagenesis; creates simultaneous *lac-cat* transcriptional fusions	92
pTV55	(lacZ) cat {tnpR} {tnpA}; B Sm B	Transposition functions inactivated; used to generate *lacZ* fusions by recombination at the sites of existing insertions	87
pLTV1Ts	(lacZ) cat bla erm tnpR tnpA; M13 ColE1 E Ss K Sm B X Sp Xh H	Insertional mutagenesis; contains ColE1-derived replication functions to facilitate cloning DNA adjacent to sites of insertions; exhibits enhanced frequency of transposition	9
pLTV3	(lacZ) cat neo ble erm tnpR tnpA; M13 ColE1 Sm X Ss K B Sp Xh H	Similar to pLTV1 but contains *neo* and *ble* genes in place of *bla*; for insertional mutagenesis in pathogenic bacteria	9
pTnPF1	aph' erm tnpR tnpA	Contains outward-directed promoter for activation of cryptic genes	93

[a] Open blocks, sequences derived from Tn*917*; thin lines, foreign DNA inserted into Tn*917*; hatched region of TnPF1, the *aph'* insert; arrows, location and orientation of transposon-associated genes; { }, partially deleted or inactivated genes; ( ), promoterless genes; *erm*, erythromycin gene of Tn*917*; *tnpA* and *tnpR*, genes of Tn*917* homologous to genes with same designation in Tn*3* (32); *cat*, a gene of gram-positive origin encoding chloramphenicol resistance; *bla*, a gene of gram-negative origin encoding ampicillin resistance; *lacZ*, a promoterless copy of the *E. coli* β-galactosidase gene; *cat-86*, a promoterless copy of a *B. pumilus* chloramphenicol acetyltransferase gene (55); *neo* and *ble*, the neomycin resistance and bleomycin resistance determinants of Tn*5* (52); *aph'*, a kanamycin resistance gene from Tn*1545* minus its transcription terminator (79); -⊖- *ColE1*, a plasmid ColE1-derived replication origin; -⊖- *M13*, a phage M13-derived origin of replication; B, *Bam*HI; Sm, *Sma*I; S, *Sal*I; P, *Pst*I; E, *Eco*RI; Sp, *Sph*I; K, *Kpn*I; Xh, *Xho*I; Bg, *Bgl*II; X, *Xba*I.

alkalophilic species *Bacillus firmus*, it was possible to select for Tn*917*-generated mutants that can no longer tolerate extreme alkaline conditions (42). Good prospects for the use of Tn*917* for insertional mutagenesis in *Lactobacillus plantarum* (17) and *Listeria monocytogenes* (18) also exist.

Several laboratories have exploited Tn*917* for physical and functional analyses of conjugative plasmids in streptococci, enterococci, and staphylococci. These analyses were particularly successful with the large, functionally complex plasmids of *E. faecalis*, for which intensive insertional mutagenesis was used to identify regions specifying plasmid transfer, pheromone response, congugative transposition, drug resistance, and hemolysin-bacteriocin activity. In the work of Dunny and colleagues, more than 50 Tn*917* insertions allowed the construction of a fine-structure physical map of plasmid pCF-10, which was shown to be a composite replicon containing two distinct functional sets of conjugation genes, one of which is associated with a Tn*916*-like element (12–14). Clewell and colleagues carried out a similar analysis of the pheromone-responsive plasmid pAD1, which included the use of Tn*917-lac* insertions to construct a map of conjugation-associated transcription units; subsequent work has exploited insertional mutations to clone and further characterize the disrupted genes (23, 35, 83, 84). Krah and Macrina used similar strategies to identify and study transfer functions in the broad-host-range streptococcal plasmid pIP501 (41), as did Thomas and Archer with the staphylococcal conjugative plasmid pGO1 (77).

One problem encountered in the adaptation of Tn*917*-related methodology to other species concerns the cloning of chromosomal sequences adjacent to the sites of insertions. The methodology developed for *B. subtilis* took advantage of the natural transformation competence of that species. Techniques were devised for recombinational integration of ColE1-derived replication functions into existing insertional mutations; sequences adjacent to insertions could then easily be rescued into *E. coli* (90). To permit the rescue of adjacent sequences in species in which recombinational integration of cloning vectors into existing insertions might be difficult or impossible, Tn*917* derivatives that already contained ColE1-derived replicons and antibiotic resistance markers selectable in *E. coli* yet still capable of transposition were developed (9). Two such derivatives (in pLTV1Ts and pLTV3Ts) are shown in Table 2. Both of these derivatives also generate transcriptional *lacZ* fusions, and one of them (in pLTV3Ts) contains selectable markers other than *bla* for use in pathogenic bacteria for which β-lactam antibiotics are important therapeutic agents. Portnoy and colleagues have made extensive use of pLTV3Ts for insertional mutagenesis in *L. monocytogenes* (9, 75).

USE OF OTHER TRANSPOSONS IN *B. SUBTILIS*

Although two other transposons of gram-positive origin, Tn*916* (24) and Tn*1545* (19), have been introduced into *B. subtilis* and shown to transpose, only Tn*917* has seen extensive use as an insertional mutagen. A new alternative to Tn*917* was recently made available, however, with the construction of a mini-Tn*10* derivative that was specifically designed for use in *B. subtilis* (65). This derivative consists essentially of a segment of DNA containing the *cat* gene of pC194 (37) flanked by 307-bp fragments derived from IS*10* that include complete inside ends of the insertion sequence. To generate transpositions, this mini-Tn*10* construction is carried on a plasmid derived from pE194Ts that also contains a copy of the Tn*10* transposase gene engineered for efficient expression in *B. subtilis* (e.g., pHV1249 in Fig. 3). The *cat* gene of pC194 is selectable in single copy in *B. subtilis* (30), and the pE194Ts-derived replication functions are identical to those incorporated into vectors such as pTV1Ts. Thus, transpositions can be obtained with these mini-Tn*10* derivatives in much the same way as with the Tn*917* derivatives listed in Table 2. In addition, the mini-Tn*10* derivatives offer two potential advantages over Tn*917*. First, in a direct comparison of levels of transposition achieved with pTV1Ts and pHV1249, the

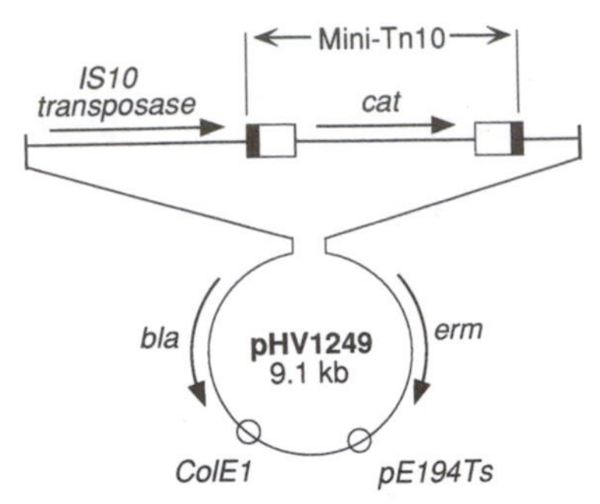

Figure 3. Delivery vector designed for selection of mini-Tn*10* derivatives in *B. subtilis* (65). Arrows with solid arrowheads represent locations and orientations of relevant genes: *cat*, chloramphenicol resistance gene of gram-positive origin and selectable in single copy in *B. subtilis*; IS*10* transposase, copy of the IS*10* transposase gene provided with a ribosome-binding site appropriate for efficient translation in *B. subtilis*; *bla*, β-lactamase gene conferring ampicillin resistance in *E. coli*; *erm*, ribosome methyltransferase gene conferring resistance to erythromycin in *B. subtilis*. The segment marked Mini-Tn10 consists of the *cat* gene flanked by 307-bp inverted repeats that include the inside ends of IS*10* (40). ColE1, replication origin derived from pBR322 (8); pE194Ts, replication functions derived from the temperature-sensitive gram-positive replicon pE194ts (81). Most of the transposition-related elements of pHV1249 were constructed from pNK1250 (82).

mini-Tn*10* derivative was reported to transpose at a significantly higher frequency (65). Second, and perhaps of even more importance, there were indications that the mini-Tn*10* derivatives might be more random with respect to target site.

USE OF OTHER TRANSPOSONS IN OTHER GRAM-POSITIVE BACTERIA

Although transposons indigenous to other gram-positive bacteria, such as Tn*551* (59, 60) and Tn*4001* (48, 49) of *S. aureus*, have been utilized effectively for insertional mutagenesis, only in the case of *Streptomyces* insertion elements have efforts been made to alter natural transposons to produce derivatives more useful for genetic analysis. Tn*4556* is a 6.8-kb transposon related to Tn*3* and discovered in an isolate of *Streptomyces fradiae* (15). It contains no known selectable marker but was modified by Chung (16) to carry a gene encoding resistance to viomycin (Table 3), which is selectable in most *Streptomyces* species. A barrier to the widespread use of Tn*4556* for insertional mutagenesis has been the absence until recently of an effective delivery system that allowed selection for recovery of chromosomal insertions. Although the transposon was available on a temperature-sensitive replicon, curing of this replicon at the nonpermissive temperature was not reliable in most species. This problem has apparently been overcome by Schauer and colleagues (70), who have described a method for using conjugation as a basis for counterselecting against plasmid-borne copies of the transposon. This approach was used to recover several mutants of *Streptomyces coelicolor* altered in morphological development.

Perhaps the most effective system for transposon-

Table 3. *Streptomyces* transposons

Transposon (vector)	Schematic structure of transposon derivative[a]	Properties or applications	Reference
Tn*4563* (pUC1172)	*tnpA* *tnpR* *Vm*^{r}	Insertional mutagenesis; derived from Tn*4556*; affords selection for viomycin resistance to recover transpositions	70
Tn*5096* (pCZA159) (pRHB126)	*OrfA* *Am*^{r} *OrfB*; Sp B E	Insertional mutagenesis; derived from IS*493*; affords selection for apramycin resistance to recover transpositions	73
Tn*5098* (pCZA172)	*OrfA* *tylJ* *tylF* *Am*^{r} *OrfB*; Sp B; E	Derived from IS*493*; carries two genes involved in tylosin biosynthesis	73
Tn*5099* (pXH106)	**xylE** *Hm*^{r} *OrfA* *OrfB*; Su D; D E X Hp A B N Nt Ss As Sp Sy	Insertional mutagenesis; derived from IS*493*; confers hygromycin resistance; creates transcriptional *xylE* fusions	28

[a] Open blocks, sequences derived from original insertion element; thin lines, foreign DNA inserted into element; arrows, location and orientation of transposon-associated genes; *tnpA* and *tnpR*, genes of Tn*4556* origin homologous to genes with same designation in Tn3 (32); *Vm*^{r}, a viomycin resistance gene; *OrfA* and *OrfB*, open reading frames of IS*493* presumably encoding transposition-related functions (74); *tylJ* and *tylF*, genes involved in tylosin biosynthesis; *Am*^{r}, a gene encoding apramycin resistance; ***xylE***, a promoterless copy of a catecholdioxygenase gene from *P. putida*; *Hm*^{r}, a gene encoding hygromycin resistance; stem-loop symbol, a synthetic linker containing the *korB* transcription terminator; Su, *Stu*I restriction sites; D, *Dra*I restriction sites; A, *Avr*II restriction sites; Nt, *Not*I restriction sites; As, *Ase*I restriction sites; Sy, *Sty*I restriction sites; other restriction sites identified as in the legends or footnotes to other tables and figures.

mediated insertional mutagenesis in *Streptomyces* spp. is based on the use of elements derived from IS*493*, a 1.6-kb insertion sequence identified in an isolate of *Streptomyces lividans* (74). It is interesting to note that IS*493* was actually discovered by using a simple genetic screen devised specifically to detect the presence of indigenous elements that transposed at a high enough frequency to compete with other kinds of spontaneous mutations that can inactivate cloned genes. IS*493* contains two open reading frames (encoding transposition-related functions?) separated by a unique *Sty*I site. Material inserted into this *Sty*I site has no effect on ability of the element to transpose, which has made it easy to construct several useful derivatives (Table 3), including one that generates transcriptional fusions to a promoterless copy of the *Pseudomonas putida xylE* gene (28, 73). The *xylE* gene product, a catechol dioxygenase, converts colorless catechol to an intensely yellow oxidation product and has thus been widely exploited as a reporter gene (97). It represents perhaps the best reporter gene system available in *Streptomyces* spp., in which the more commonly used alternatives, such as *lacZ*, are relatively ineffective (36). Two different methods for the selection of chromosomal insertions of IS493 derivatives exist. One approach (73) makes use of a low-copy-number, temperature-sensitive vector related to the SCP2* replicon (56). The transposon derivatives carried on these vectors are introduced by transformation and propagated in liquid culture at the permissive temperature (29 to 30°C). The resulting mycelium suspension is then homogenized and plated to produce approximately 20 to 100 colonies per standard petri plate, initially also at the permissive temperature. After 2 days of growth, the plates are shifted to the nonpermissive temperature (39°C) and then subjected to about 10 additional days of incubation under continued selection for the resistance marker carried by the transposon. After the additional incubation period, products of transposition are apparent as rapidly growing sectors emanating from the primary colonies. Another approach (53) employs temperature-sensitive transposon-carrying vectors that also contain cloned segments of the broad-host-range phage FP43 DNA that include its *pac* site. The presence in these vectors of the FP43 *pac* site enables the phage to transduce them with high efficiency into a wide array of *Streptomyces* species, and they can then be selected at the nonpermissive temperature for chromosomal transpositions.

Acknowledgments. I gratefully acknowledge S. A. Zahler, D. Dubnau, A. Grossman, R. Losick, P. Vary, T. Krulwich, D. Clewell, M. Young, G. Ordal, M. Zagorec, R. G. Wake, and I. Smith for personal communications and for help with assembling the information required for this chapter. Work from my laboratory was supported by NIH grant GM35495.

REFERENCES

1. **Albano, M., R. Breitling, and D. A. Dubnau.** 1989. Nucleotide sequence and genetic organization of the *Bacillus subtilis comG* operon. *J. Bacteriol.* **171:**5386–5404.
2. **Albano, M., J. Hahn, and D. Dubnau.** 1987. Expression of competence genes in *Bacillus subtilis*. *J. Bacteriol.* **169:** 3110–3117.
3. **An, F. Y., and D. B. Clewell.** 1991. Tn*917* transposase sequence correction reveals a single open reading frame corresponding to the *tnpA* determinant of Tn3-family elements. *Plasmid* **25:**121–124.
4. **Antoniewski, C., B. Savelli, and P. Stragier.** 1990. The *spoIIJ* gene, which regulates early developmental steps in *Bacillus subtilis*, belongs to a class of environmentally responsive genes. *J. Bacteriol.* **172:**86–93.
5. **Atkinson, M. R., and S. H. Fisher.** 1991. Identification of genes and gene products whose expression is activated during nitrogen-limited growth in *Bacillus subtilis*. *J. Bacteriol.* **173:**23–27.
6. **Atkinson, M. R., L. J. Wray, and S. H. Fisher.** 1990. Regulation of histidine and proline degradation enzymes by amino acid availability in *Bacillus subtilis*. *J. Bacteriol.* **172:**4758–4765.

7. **Bohall, N. J., and P. S. Vary.** 1986. Transposition of Tn*917* in *Bacillus megaterium*. *J. Bacteriol.* **167:**716–718.
8. **Bolivar, F., R. L. Rodriguez, P. J. Greene, M. C. Betlach, H. L. Heyneker, and H. W. Boyer.** 1977. Construction and characterization of new cloning vehicles. II. A multipurpose cloning system. *Gene* **2:**95–113.
9. **Camilli, A., A. Portnoy, and P. Youngman.** 1990. Insertional mutagenesis of *Listeria monocytogenes* with a novel Tn*917* derivative that allows direct cloning of DNA flanking transposon insertions. *J. Bacteriol.* **172:**3738–3744.
10. **Cheo, D. L., K. W. Bayles, and R. E. Yasbin.** 1991. Cloning and characterization of DNA damage-inducible promoter regions from *Bacillus subtilis*. *J. Bacteriol.* **173:**1696–1703.
11. **Chow, W. Y., and D. E. Berg.** 1988. Tn*5tac1*, a derivative of Tn*5* that generates conditional mutations. *Proc. Natl. Acad. Sci. USA* **85:**6468–6472.
12. **Christie, P. J., and G. M. Dunny.** 1986. Identification of regions of the *Streptococcus faecalis* plasmid pCF-10 that encode antibiotic resistance and pheromone response functions. *Plasmid* **15:**230–241.
13. **Christie, P. J., S. M. Kao, J. C. Adsit, and G. M. Dunny.** 1988. Cloning and expression of genes encoding pheromone-inducible antigens of *Enterococcus* (*Streptococcus*) *faecalis*. *J. Bacteriol.* **170:**5161–5168.
14. **Christie, P. J., R. Z. Korman, S. A. Zahler, J. C. Adsit, and G. M. Dunny.** 1987. Two conjugation systems associated with *Streptococcus faecalis* plasmid pCF10: identification of a conjugative transposon that transfers between *S. faecalis* and *Bacillus subtilis*. *J. Bacteriol.* **169:** 2529–2536.
15. **Chung, S.-T.** 1987. Tn*4556*, a 6.8-kilobase-pair transposable element of *Streptomyces fradiae*. *J. Bacteriol.* **169:** 4436–4441.
16. **Chung, S.-T.** 1988. Transposition of Tn*4556* in *Streptomyces*. *Dev. Ind. Microbiol.* **29:**81–88.
17. **Cosby, W. M., L. T. Axelsson, and W. J. Dobrogosz.** 1989. Tn*917* transposition in *Lactobacillus plantarum* using the highly temperature-sensitive plasmid pTV1Ts as a vector. *Plasmid* **22:**236–243.
18. **Cossart, P., M. F. Vicente, J. Mengaud, F. Baquero, D. J. Perez, and P. Berche.** 1989. Listeriolysin O is essential for virulence of *Listeria monocytogenes*: direct evidence obtained by gene complementation. *Infect. Immun.* **57:** 3629–3636.
19. **Courvalin, P., and C. Carlier.** 1987. Tn*1545*: a conjugative shuttle transposon. *Mol. Gen. Genet.* **206:**259–264.
20. **Dean, D. H., and D. R. Ziegler.** 1989. *Bacillus Genetic Stock Center Strains and Data*, 4th ed. Ohio State University, Columbus.
21. **Donovan, W., L. Zheng, K. Sandman, and R. Losick.** 1987. Genes encoding spore coat polypeptides from *Bacillus subtilis*. *J. Mol. Biol.* **196:**1–10.
22. **Dubnau, D.** 1991. Genetic competence in *Bacillus subtilis*. *Microbiol. Rev.* **55:**395–424.
23. **Ehrenfeld, E. E., and D. B. Clewell.** 1987. Transfer functions of the *Streptococcus faecalis* plasmid pAD1: organization of plasmid DNA encoding response to sex pheromone. *J. Bacteriol.* **169:**3473–3481.
24. **Franke, A. E., and D. B. Clewell.** 1981. Evidence for a chromosome-borne resistance transposon (Tn*916*) in *Streptococcus faecalis* that is capable of conjugal transfer in the absence of a conjugative plasmid. *J. Bacteriol.* **145:**494–502.
25. **Gillespie, K., and R. E. Yasbin.** 1987. Chromosomal locations of three *Bacillus subtilis din* genes. *J. Bacteriol.* **169:**3372–3374.
26. **Guillen, N., Y. Weinrauch, and D. A. Dubnau.** 1989. Cloning and characterization of the regulatory *Bacillus subtilis* competence genes *comA* and *comB*. *J. Bacteriol.* **171:**5354–5361.
27. **Guzmán, P., J. Westpheling, and P. Youngman.** 1988. Characterization of the promoter region of the *Bacillus subtilis spoIIE* operon. *J. Bacteriol.* **170:**1598–1609.
28. **Hahn, D. R., P. J. Solenberg, and R. H. Baltz.** 1991. Tn*5099*, a *xylE* promoter probe transposon for *Streptomyces* spp. *J. Bacteriol.* **173:**5573–5577.
29. **Hahn, J., M. Albano, and D. Dubnau.** 1987. Isolation and characterization of Tn*917lac*-generated competence mutants of *Bacillus subtilis*. *J. Bacteriol.* **169:**3104–3109.
30. **Haldenwang, W. G., C. D. B. Banner, J. F. Ollington, R. Losick, J. A. Hoch, M. B. O'Connor, and A. L. Sonenshein.** 1980. Mapping a cloned gene under sporulation control by insertion of a drug resistance marker into the *Bacillus subtilis* chromosome. *J. Bacteriol.* **142:**90–98.
31. **Hartley, R. W., and C. J. Paddon.** 1986. Use of plasmid pTV1 in transposon mutagenesis and gene cloning in *Bacillus amyloliquefaciens*. *Plasmid* **16:**45–51.
32. **Heffron, F., B. J. McCarthy, H. Ohtsubo, and I. Ohtsubo.** 1979. DNA sequence analysis of the transposon Tn*3*: three genes and other sites involved in transposition. *Cell* **18:**1153–1163.
33. **Hilton, M. D., N. G. Alaeddinoglu, and A. L. Demain.** 1988. *Bacillus subtilis* mutant deficient in the ability to produce the dipeptide antibiotic bacilysin: isolation and mapping of the mutation. *J. Bacteriol.* **170:**1018–1020.
34. **Iismaa, T. P., M. T. Smith, and R. G. Wake.** 1984. Physical map of the *Bacillus subtilis* replication terminus region: its confirmation, extension and genetic orientation. *Gene* **32:**171–180.
35. **Ike, Y., D. B. Clewell, R. A. Segarra, and M. S. Gilmore.** 1990. Genetic analysis of the pAD1 hemolysin/bacteriocin determinant in *Enterococcus faecalis*: Tn*917* insertional mutagenesis and cloning. *J. Bacteriol.* **172:**155–163.
36. **Ingram, C., M. Brawner, P. Youngman, and J. Westpheling.** 1989. *xylE* functions as an efficient reporter gene in *Streptomyces* spp.: use for the study of *galP1*, a catabolite-controlled promoter. *J. Bacteriol.* **171:**6617–6624.
37. **Iordanescu, S.** 1976. Three distinct plasmids originating in the same *Staphylococcus aureus* strain. *Arch. Roum. Pathol. Exp. Microbiol.* **35:**111–118.
38. **Jaacks, K. J., J. Healy, R. Losick, and A. D. Grossman.** 1989. Identification and characterization of genes controlled by the sporulation-regulatory gene *spoOH* in *Bacillus subtilis*. *J. Bacteriol.* **171:**4121–4129.
39. **Kenyon, C. J., and G. C. Walker.** 1980. DNA-damaging agents stimulate gene expression at specific loci in *Escherichia coli*. *Proc. Natl. Acad. Sci. USA* **77:**2819–2823.
40. **Kleckner, N.** 1989. Transposon Tn*10*, p. 227–268. *In* D. M. Berg and M. M. Howe (ed.), *Mobile DNA*. American Society for Microbiology, Washington, D.C.

40a. **Kopec, L. K., R. E. Yasbin, and R. Marrero.** 1985. Bacteriophage SPO2-mediated plasmid transduction in transpositional mutagenesis within the genus *Bacillus*. *J. Bacteriol.* **164:**1283–1287.

41. **Krah, E., and F. L. Macrina.** 1989. Genetic analysis of the conjugal transfer determinants encoded by the streptococcal broad-host-range plasmid pIP501. *J. Bacteriol.* **171:**6005–6012.
42. **Krulwich, T. A.** Personal communication.
43. **Kunkel, B., K. Sandman, S. Panzer, P. Youngman, and R. Losick.** 1988. The promoter for a sporulation gene in the *spoIVC* locus of *Bacillus subtilis* and its use in studies of temporal and spatial control of gene expression. *J. Bacteriol.* **170:**3513–3522.
44. **Kuramitsu, H. K., and M. J. Casadaban.** 1986. Transposition of the gram-positive transposon Tn*917* in *Escherichia coli*. *J. Bacteriol.* **167:**711–712.
45. **Landono, A., and D. Dubnau.** Personal communication.
46. **Le Coq, D., S. Aymerich, and M. Steinmetz.** 1991. Dual effect of a Tn*917* insertion into the *Bacillus subtilis sacX* gene. *J. Gen. Microbiol.* **137:**101–106.
47. **Love, P. E., M. J. Lyle, and R. E. Yasbin.** 1985. DNA-damage-inducible (*din*) loci are transcriptionally acti-

vated in competent *Bacillus subtilis*. *Proc. Natl. Acad. Sci. USA* **82:**6201–6205.

48. **Lyon, B. R., J. W. May, and R. A. Skurray.** 1984. Tn*4001*: a gentamycin and kanamycin resistance transposon in *Staphylococcus aureus*. *Mol. Gen. Genet.* **193:**554–556.
49. **Mahairas, G. G., B. R. Lyon, R. A. Skurray, and P. A. Pattee.** 1989. Genetic analysis of *Staphylococcus aureus* with Tn*4001*. *J. Bacteriol.* **171:**3968–3972.
50. **Marrero, R., and R. E. Yasbin.** 1988. Cloning of the *Bacillus subtilis recE*$^+$ gene and functional expression of *recE*$^+$ in *B. subtilis*. *J. Bacteriol.* **170:**335–344.
51. **Mastromei, G., C. Barberio, S. Pistolesi, and M. Polsinelli.** 1989. Isolation of *Bacillus subtilis* transformation-deficient mutants and mapping of competence genes. *Genet. Res.* **54:**1–5.
52. **Mazodier, P., P. Cossart, E. Giraud, and F. Gassner.** 1985. Completion of the nucleotide sequence of the central region of Tn*5* reveals the presence of three resistance genes. *Nucleic Acids Res.* **13:**195–205.
53. **McHenney, M. A., and R. H. Baltz.** 1991. Transposition of Tn*5096* from a temperature-sensitive transducible plasmid in *Streptomyces* spp. *J. Bacteriol.* **173:**5578–5581.
54. **Mohan, S., J. Aghion, N. Guillen, and D. Dubnau.** 1989. Molecular cloning and characterization of *comC*, a late competence gene of *Bacillus subtilis*. *J. Bacteriol.* **171:** 6043–6051.
55. **Mongkolsuk, S., Y.-W. Chiang, R. B. Reynolds, and P. S. Lovett.** 1983. Restriction fragments that exert promoter activity during postexponential growth of *Bacillus subtilis*. *J. Bacteriol.* **155:**1399–1406.
56. **Muth, G., B. Nussbaumer, W. Wolleben, and A. Pühler.** 1989. A vector system with temperature-sensitive replication for gene disruption and mutational cloning in streptomycetes. *Mol. Gen. Genet.* **219:**341–348.
57. **Nakano, M. M., R. Magnuson, A. Myers, J. Curry, A. D. Grossman, and P. Zuber.** 1991. *srfA* is an operon required for surfactin production, competence development, and efficient sporulation in *Bacillus subtilis*. *J. Bacteriol.* **173:** 1770–1778.
58. **North, A. K., M. C. Smith, and S. Baumberg.** 1989. Nucleotide sequence of a *Bacillus subtilis* arginine regulatory gene and homology of its product to the *Escherichia coli* arginine repressor. *Gene* **80:**29–38.
59. **Novick, R. P., I. Edelman, M. D. Schwesinger, A. D. Gruss, E. C. Swanson, and P. A. Pattee.** 1979. Genetic translocation in *Staphylococcus aureus*. *Proc. Natl. Acad. Sci. USA* **76:**400–404.
60. **Pattee, P. A.** 1981. Distribution of Tn*551* insertion sites responsible for auxotrophy on the *Staphylococcus aureus* chromosome. *J. Bacteriol.* **145:**479–488.
61. **Perego, M., S. P. Cole, D. Burbulys, K. Trach, and J. A. Hoch.** 1989. Characterization of the gene for a protein kinase which phosphorylates the sporulation-regulatory proteins SpoOA and SpoOF of *Bacillus subtilis*. *J. Bacteriol.* **171:**6187–6196.
62. **Perego, M., G. B. Spiegelman, and J. A. Hoch.** 1988. Structure of the gene for the transition state regulator *abrB*: regulator synthesis is controlled by the *spoOA* sporulation gene in *Bacillus subtilis*. *Mol. Microbiol.* **2:**689–699.
63. **Perkins, J. B.,20and P. J. Youngman.** 1984. A physical and functional analysis of Tn*917*, a *Streptococcus* transposon in the Tn*3* family that functions in *Bacillus*. *Plasmid* **12:**119–138.
64. **Perkins, J. B., and P. J. Youngman.** 1986. Construction and properties of Tn*917*-lac, a transposon derivative that mediates transcriptional gene fusions in *Bacillus subtilis*. *Proc. Natl. Acad. Sci. USA* **83:**140–144.
65. **Petit, M.-A., C. Bruand, L. Jannière, and S. D. Ehrlich.** 1990. Tn*10*-derived transposons active in *Bacillus subtilis*. *J. Bacteriol.* **172:**6736–6740.
66. **Robertson, J. B., M. Gocht, M. A. Marahiel, and P. Zuber.** 1989. AbrB, a regulator of gene expression in *Bacillus*, interacts with the transcription initiation regions of a sporulation gene and an antibiotic biosynthesis gene. *Proc. Natl. Acad. Sci. USA* **86:**8457–8461.
67. **Rudner, D. Z., J. R. LeDeaux, K. Ireton, and A. D. Grossman.** 1991. The *spoOK* locus of *Bacillus subtilis* is homologous to the oligopeptide permease locus and is required for sporulation and competence. *J. Bacteriol.* **173:**1388–1398.
68. **Sammons, R. L., G. M. Slynn, and D. A. Smith.** 1987. Genetical and molecular studies on *gerM*, a new developmental locus of *Bacillus subtilis*. *J. Gen. Microbiol.* **133:**3299–3312.
69. **Sandman, K., R. Losick, and P. Youngman.** 1987. Genetic analysis of *Bacillus subtilis spo* mutations generated by Tn*917*-mediated insertional mutagenesis. *Genetics* **117:**603–617.
70. **Schauer, A. T., A. D. Nelson, and J. B. Daniel.** 1991. Tn*4563* transposition in *Streptomyces coelicolor* and its application to isolation of new morphological mutants. *J. Bacteriol.* **173:**5060–5067.
71. **Shaw, J. H., and D. B. Clewell.** 1985. Complete nucleotide sequence of macrolide-lincosamide-streptogramin B-resistance transposon Tn*917* in *Streptococcus faecalis*. *J. Bacteriol.* **164:**782–796.
72. **Smith, K., and P. Youngman.** Unpublished data.
73. **Solenberg, P. J., and R. H. Baltz.** 1991. Transposition of Tn*5096* and other IS*493* derivatives in *Streptomyces griseofuscus*. *J. Bacteriol.* **173:**1096–1104.
74. **Solenberg, P. J., and S. G. Burgett.** 1989. Method for selection of transposable DNA and characterization of a new insertion sequence, IS*493*, from *Streptomyces lividans*. *J. Bacteriol.* **171:**4807–4813.
75. **Sun, A. N., A. Camilli, and D. A. Portnoy.** 1990. Isolation of *Listeria monocytogenes* small-plaque mutants defective for intracellular growth and cell-to-cell spread. *Infect. Immun.* **58:**3770–3778.
76. **Tao, Y.-P., and P. S. Vary.** 1991. Isolation and characterization of sporulation *lacZ* fusion mutants of *Bacillus megaterium*. *J. Gen. Microbiol.* **137:**797–806.
77. **Thomas, W. J., and G. L. Archer.** 1989. Identification and cloning of the conjugative transfer region of *Staphylococcus aureus* plasmid pGO1. *J. Bacteriol.* **171:**684–691.
78. **Tomich, P. K., F. Y. An, and D. B. Clewell.** 1980. Properties of erythromycin-inducible transposon Tn*917* in *Streptococcus faecalis*. *J. Bacteriol.* **141:**1366–1374.
79. **Trieu-Cuot, P., A. Klier, and P. Courvalin.** 1985. DNA sequences specifying the transcription of the streptococcal kanamycin resistance gene in *E. coli* and *B. subtilis*. *Mol. Gen. Genet.* **198:**348–352.
80. **Vandeyar, M. A., and S. A. Zahler.** 1986. Chromosomal insertions of Tn*917* in *Bacillus subtilis*. *J. Bacteriol.* **167:** 530–534.
81. **Villafane, R., D. H. Bechhofer, C. S. Narayanan, and D. Dubnau.** 1987. Replication control genes of plasmid pE194. *J. Bacteriol.* **169:**4822–4829.
82. **Way, J. C., M. A. Davis, D. Morisato, D. E. Roberts, and N. Kleckner.** 1984. New Tn*10* derivatives for transposon mutagenesis and for construction of *lacZ* operon fusions by transposition. *Gene* **32:**369–379.
83. **Weaver, K. E., and D. B. Clewell.** 1988. Regulation of the pAD1 sex pheromone response in *Enterococcus faecalis*: construction and characterization of *lacZ* transcriptional fusions in a key control region of the plasmid. *J. Bacteriol.* **170:**4343–4352.
84. **Weaver, K. E., and D. B. Clewell.** 1990. Regulation of the pAD1 sex pheromone response in *Enterococcus faecalis*: effects of host strain and *traA*, *traB*, and C region mutants on expression of an E region pheromone-inducible *lacZ* fusion. *J. Bacteriol.* **172:**2633–2641.
85. **Yon, J. R., R. L. Sammons, and D. A. Smith.** 1989. Cloning and sequencing of the *gerD* gene of *Bacillus subtilis*. *J. Gen. Microbiol.* **135:**3431–3445.
86. **Youngman, P.** 1987. Plasmid vectors for recovering and

exploiting Tn*917* transpositions in *Bacillus* and other gram-positive bacteria, p. 79–104. *In* K. Hardy (ed.), *Plasmids: a Practical Approach*. IRL Press, Oxford.

87. **Youngman, P.** 1990. Use of transposons and integrational vectors for mutagenesis and construction of gene fusions in *Bacillus* species, p. 221–266. *In* S. M. C. C. R. Harwood (ed.), *Molecular Biological Methods for Bacillus*. John Wiley & Sons Ltd., Chichester, United Kingdom.
88. **Youngman, P. J., J. B. Perkins, and R. Losick.** 1983. Genetic transposition and insertional mutagenesis in *Bacillus subtilis* with *Streptococcus faecalis* transposon Tn*917*. *Proc. Natl. Acad. Sci. USA* **80:**2305–2309.
89. **Youngman, P., J. B. Perkins, and R. Losick.** 1984. Construction of a cloning site near one end of Tn*917* into which foreign DNA may be inserted without affecting transposition in *Bacillus subtilis* or expression of the transposon-borne *erm* gene. *Plasmid* **12:**1–9.
90. **Youngman, P., J. B. Perkins, and R. Losick.** 1984. A novel method for the rapid cloning in *Escherichia coli* of *Bacillus subtilis* chromosomal DNA adjacent to Tn*917* insertions. *Mol. Gen. Genet.* **195:**424–433.
91. **Youngman, P., H. Poth, B. D. Green, K. York, G. Olmedo, and K. Smith.** 1989. Methods for genetic manipulation, cloning, and functional analysis of sporulation genes in *Bacillus subtilis*, p. 65–88. *In* I. Smith, R. A. Slepecky, and P. Setlow (ed.), *Regulation of Procaryotic Development*. American Society for Microbiology, Washington, D.C.
92. **Youngman, P., P. Zuber, J. B. Perkins, K. Sandman, M. Igo, and R. Losick.** 1985. New ways to study developmental genes in bacteria. *Science* **228:**285–291.
93. **Zagorec, M., and M. Steinmetz.** 1991. Construction of a derivative of Tn*917* containing an outward-directed promoter and its use in *Bacillus subtilis*. *J. Gen. Microbiol.* **137:**107–112.
94. **Zahler, S. A.** Personal communication.
95. **Zuberi, A. R., C. Ying, H. M. Parker, and G. W. Ordal.** 1990. Transposon Tn*917lacZ* mutagenesis of *Bacillus subtilis*: identification of two new loci required for motility and chemotaxis. *J. Bacteriol.* **172:**6841–6848.
96. **Zuberi, A. R., C. Ying, M. R. Weinreich, and G. W. Ordal.** 1990. Transcriptional organization of a cloned chemotaxis locus of *Bacillus subtilis*. *J. Bacteriol.* **172:**1870–1876.
97. **Zukowski, M. M., D. F. Gaffney, D. Speck, M. Kauffmann, A. Findeli, A. Wisecup, and J. P. Lecocq.** 1983. Chromogenic identification of genetic regulatory signals in *Bacillus subtilis* based on expression of a cloned pseudomonas gene. *Proc. Natl. Acad. Sci. USA* **80:**1101–1105.

41. Conjugative Transposons

JUNE R. SCOTT

DEFINITION AND IMPORTANCE

Among medically important gram-positive bacteria, antibiotic resistance is often determined by genes present in a class of transposable elements that Don Clewell named conjugative transposons (29). The name derives from the observation that these elements confer on their gram-positive hosts the ability to act as conjugative donors. During this mating event, the transposon is transferred to a new location in the genome of the recipient strain. As will be explained below, the basic mechanism of transposition of these elements is fundamentally different from that of the better-studied transposons commonly found in *Escherichia coli*. Recent reviews of conjugative transposons include those by Clewell et al. (16, 20, 21) and Scott (56, 56a). Conjugative elements in *Bacteroides* spp. have also been described (32, 51, 55, 64), but these will not be included in this review because although little is known about them, they seem to be significantly different.

Conjugative transposons were first isolated from streptococcal strains associated with human disease. The best studied include Tn*916*, which was found in a strain of *Enterococcus faecalis* (25); Tn*3701*, which came from *Streptococcus pyogenes* (group A streptococcus; 43); and Tn*1545* (10, 14, 22) and the element originally called omega *cat-tet* (59, 60, 68) and now renamed Tn*5253* (2), which were both isolated from *Streptococcus pneumoniae* strains. Conjugative transposons have been found in many natural bacterial isolates (listed in reference 18). All conjugative transposons isolated to date carry *tetM* (8) or a closely related gene for tetracycline resistance (61) that appears to be expressed in all gram-positive and gram-negative bacteria. Unlike other *tet* markers, *tetM* causes resistance to minocycline, a semisynthetic tetracycline derivative (4), as well as to tetracycline.

The smallest conjugative transposon described is Tn*916*, which is about 16.4 kb. This is a "simple" transposon in the sense that it has been observed to transpose only as a unit. Complex conjugative transposons, which are much larger (they may be in excess of 60 kb), may be composed of several transposable units. The best studied of these are Tn*3701* (reviewed by Le Bouguenec et al. [43] and Horaud et al. [33]) and Tn*5253* (reviewed by Vijayakumar et al. [67]).

Complex conjugative transposons with homology to Tn*3701* are widespread in nature, as is suggested by examination of a collection of clinical streptococcal isolates (44). In addition to tetracycline resistance, the larger simple and complex elements often encode resistance to antibiotics like chloramphenicol, kanamycin, and the MLS group, which includes the macrolide erythromycin as well as lincosamide and streptogramin.

Disease-associated gram-positive bacterial strains from which conjugative transposons have been isolated (especially the streptococci) frequently contain no detectable plasmids. Ninety percent of the group A, B, C, D, F, and G streptococcal isolates and the oral streptococci examined by Horodniceanu et al. (34, 35) and Buu-Hoi and Horaud (9) fall into this category. Therefore, it appears likely that conjugative transposons are a major source of spread of antibiotic resistance, especially among the clinically important gram-positive bacteria. These transposons represent an important medical problem, and a molecular understanding of them is highly desirable.

Within Tn*3701* is a region homologous to Tn*916* that has an erythromycin resistance marker inserted near its right end. This segment is called Tn*3703*, since in crosses with an *E. faecalis* recipient, it is sometimes found inserted unlinked to the rest of the element. Tn*3703* has been seen to transpose from a plasmid to the chromosome but not to undergo independent conjugative transposition (33, 43). The lack of autonomous transposition may result from interruption of the sequence of a gene required for conjugation (see below) by the *erm* marker. The *erm* marker and adjacent DNA to its right are often lost by spontaneous deletion (43), suggesting that *erm* may be on a separate element (nonconjugative) that inserted later in the evolution of the complex transposon.

The composite element Tn*5253*, on the other hand, appears to be composed of two complete, independent, simple transposable elements (2). It contains a Tn*916*-like internal *tet*-containing element called Tn*5251* that excises from the larger element spontaneously and undergoes conjugative transposition on its own. When Tn*5251* excises from Tn*5253*, the remaining element, called Tn*5252*, is also able to conjugatively transpose. Some complex elements may have arisen by transposition of one simple element into another. The presence of DNA homology between Tn*3701* and Tn*5252* (outside of the Tn*916*-like element) suggests that these two transposons may be closely related (44). Comparison of the restriction maps of these elements implies that these transposons differ in the orientation of the Tn*916*-like element within them. This suggests that the smaller element was inserted into each of the larger elements separately and that the larger element existed first. In this light, it should be remembered that the small element Tn*1545* (which carries Em Km Tc) originally cotrans-

June R. Scott • Department of Microbiology and Immunology, Emory University Health Sciences Center, Atlanta, Georgia 30322.

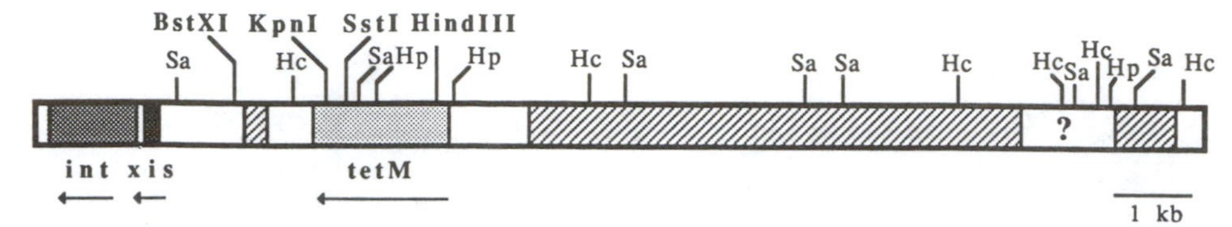

Figure 1. Map of Tn*916*. Locations of the genes *tetM*, *int*, and *xis* are shown, and their transcription directions are indicated by arrows beneath the map. There are 173 bases between the stop codon of *int* and the end of the transposon. The region required for conjugative transfer (58) is shaded. The question mark represents a region whose function is unknown because no mutants with insertions into it were isolated. Restriction enzyme sites are indicated above the line: S, *Sau*3A; Hc, *Hin*cII; Hp, *Hpa*II. There are no recognition sites for *Bam*HI, *Bgl*II, *Cla*I, *Eco*RI, *Hae*III, *Nco*I, *Pst*I, *Sal*I, *Sph*I, *Xba*I, or *Xho*I (30) or for *Ava*I or *Pvu*II (6a).

ferred with a Cm marker and may have been part of a larger composite conjugative transposon (10, 44).

As a consequence of their mechanism of transposition, the better-studied nonconjugative transposons all cause duplications of the target DNA into which they insert. The discovery that conjugative transposons do not cause such duplications indicated that their mechanism of transposition was completely different from that of the other types of transposable elements. The idea originally suggested by Gawron-Burke and Clewell (29), that conjugative transposons, like lambda prophage, excise to form a circular molecule that then inserts into its target, has turned out to be correct (see below).

The smaller conjugative transposons all seem to be very closely related (Fig. 1 shows a map of Tn*916*). The sequences of the ends (about 200 bp) of Tn*1545* and Tn*916* are almost identical (11, 19; GenBank accession number M20864). (By convention, however, the end designated "left" for Tn*916* is called the "right" end of Tn*1545*. In this review, I use the Tn*916* designations.) Furthermore, the two open reading frames of Tn*1545* that encode products involved in transposition (52; EMBL accession number X61025) are almost completely conserved (one amino acid difference in their predicted proteins) in Tn*916* (see below). The only transposon ends known to differ from those of Tn*916* (20; GenBank accession number M37184) are those of the element reported in *Clostridium difficile* (46) and the outside ends of the large composite transposon Tn*5253* (67). Since the large elements have not yet been studied in detail, it is possible that they and the *C. difficile* elements transpose by a mechanism different from that used by Tn*916*. All of the small elements, however, are likely to transpose by the same molecular mechanism.

BROAD HOST RANGE

The host range of conjugative transposons is extraordinarily extensive, and in all the hosts containing these elements, resistance to tetracycline is expressed. As mentioned above, these elements appear to play a significant role in the spread of antibiotic resistance, especially among gram-positive bacteria. Transposition of conjugative transposons has been observed in both gram-positive and gram-negative bacteria.

The mating stimulated by the presence of conjugative transposons appears to be extremely promiscuous, since at least among gram-positive bacteria, the two parents need not be of the same species or even the same genus. Recently, conjugative transposition from gram-positive to gram-negative bacteria was described (6). An *E. faecalis* donor was reported to transfer Tn*916* to *Alcaligenes eutrophus*, *Citrobacter freundii*, and *E. coli*. To distinguish tetracycline-resistant transconjugants from parents, counterselection with penicillin was used. The transposon was present on a plasmid in the donor, and in the transconjugants, the transposon appeared to have formed a cointegrate between the chromosome of the recipient and the entire Tn*916*-containing plasmid, since the sizes of the transposon-host junction fragments were the same in the donor and the transconjugants. This type of cointegrate formation has not been described in other work, and the results would have been more convincing if controls to distinguish transconjugants from penicillin-resistant donors (which arise from *E. faecalis* at the concentration of penicillin used to counterselect in these experiments) had been presented.

When the gram-negative phototrophic recipients (*Rhodobacter* and *Rhodospirillum* spp.) were crossed with the same *E. faecalis* strain carrying Tn*916* on a plasmid, no transconjugants were detected. However, in these crosses, in addition to the use of a different recipient, a different counterselection was used (selective malate medium instead of penicillin). Since no negative result was reported in transfer from *E. faecalis* to any gram-negative organism when penicillin counterselection was used, it remains possible that the apparent transfer of Tn*916* from *E. faecalis* to gram-negative bacteria is an artifact. Further work is needed to clarify the results of these crosses. If the strains resistant to penicillin and tetracycline are true transconjugants, the results suggest that some type of conjugal transfer occurs in the absence of transposon excision, and it remains to be determined whether this transfer is dependent on the presence of the transposon.

The same authors reported transfer of a conjugative transposon from *E. coli* to the gram-positive recipients *Bacillus subtilis*, *Clostridium acetobutylicum*, *E. faecalis*, and *Streptococcus lactis* subsp. *diacetylactis* (6). Because the Tn*916* transferred from *E. coli* appears to have integrated at different sites in the transconjugants, the transfer from gram-negative to gram-positive strains is more convincing. However, other investigators have been unable to obtain such transfer, and this finding awaits further confirmation.

It appears that the host range of conjugative transposons is not affected by DNA restriction. Guild et al. (31) showed that in *S. pneumoniae*, although transfer

Table 1. Frequency of Tc^r transconjugants when *B. subtilis* CKS102 (57) is used as donor in plate matings

Recipient strain	Marker selected	Reference	Frequency[a]	No. repeats of expt
B. subtilis BD1590	*cat*	1	$1.0 \times 10^{-6} \pm 8 \times 10^{-7}$	4
E. faecalis JH2-2	*fus rif*	38	6×10^{-6}	1
E. faecalis BM4110	*str*	22	$3 \times 10^{-7} \pm 2 \times 10^{-7}$	2
L. lactis subsp. *lactis* MG1363	*str*	28	$2.0 \times 10^{-4} \pm 5 \times 10^{-5}$	4
S. pneumoniae R800-str41	*str-41*	15	$1.0 \times 10^{-5} \pm 5 \times 10^{-6}$	3
S. pyogenes JRS75	*aphA3 str*	49	7×10^{-7}	1

[a] Number of transconjugants divided by number of parents. The parent used for this calculation is the one present in a limiting amount in each mating and is the donor for all but the mating with *S. pneumoniae*. Data are from reference 7a.

by mating of the conjugative plasmid pIP501 was restricted by a *Dpn*II$^+$ recipient compared with a *Dpn*II$^-$ strain, conjugative transposition of the complex element Tn*5253* from the same donor to the *Dpn*II$^+$ and *Dpn*$^-$ recipient strains occurred at the same frequency. In vitro, the transposon was sensitive to *Dpn*II, so this experiment indicates that this transposable element is not subject to restriction during conjugation.

Although Guild's study was done with a large complex transposon whose ends may differ from those of the better-studied small elements like Tn*916*, I suspect that the 16-kb element Tn*916* is also not subject to restriction after conjugative transposition. For this element, the frequency of formation of transconjugants is not related to the genetic distance between the two parents. From the same *B. subtilis* donor carrying a single Tn*916* in its chromosome, transconjugants are obtained at frequencies greater than or equal to that for an isogenic *B. subtilis* recipient. This was true with recipient strains of *E. faecalis*, *S. pyogenes*, *S. pneumoniae*, or *Lactococcus lactis* (7a; Table 1). This is surprising, because DNA from organisms as evolutionarily distant as these would be expected to be restricted on entry into the foreign host.

Lack of restriction of the element transferred in conjugative transposition would occur if the transposon DNA were transferred in mating as a single strand, as in F-plasmid-promoted matings in *E. coli* (see below). Alternatively, the transposon may avoid restriction either by assuming a protective conformation or by encoding a general system for protection of its DNA against restriction. Systems for defense against restriction have been found in two phages that have broad host ranges: Mu, which encodes the *mom* system (39), and P1, which has the *dar* system (36). In the former, acetamide modification of adenosines occurs so that restriction enzymes whose recognition sites include A cannot act. The *dar* system of P1 involves two phage-encoded proteins that are closely associated and injected with DNA packaged into a phage particle. In a *cis*-specific manner, this protein protects the DNA with which it is associated from a particular group of restriction enzymes (36). Although an understanding of the mechanism involved must await further experimentation, the apparent abilities of conjugative transposons to avoid restriction are an obvious advantage for broadening their host ranges.

MECHANISM OF TRANSPOSITION

Excision Forms Covalently Closed Circular Transposon Intermediate

Gawron-Burke and Clewell (29) originally suggested that transposition of Tn*916* proceeds through formation of an excised intermediate that they thought might be circular. This idea was based largely on their finding that in overnight filter matings between strains of *E. faecalis*, a conjugative transposon present on a conjugative plasmid in the donor was often found unlinked to the plasmid in the transconjugants. Furthermore, in *E. coli*, excision of Tn*916* is very frequent and usually leads to loss of the element (17, 30). In an overnight culture of an *E. coli* strain containing pAM120, a multicopy plasmid with a cloned region from the genome of an *E. faecalis* strain that contains Tn*916*, about 90% of the cells are tetracycline sensitive (13a). These cells have lost the conjugative transposon. Very rarely, conjugative transposons are found to have transposed from the plasmid to the *E. coli* chromosome (20, 23, 30).

We identified the predicted circular transposon form in *E. coli* (57). Tn*916* was used as an insertional mutagen in *S. pyogenes*, and the region of streptococcal chromosomal DNA containing the transposon was then cloned on a cosmid in *E. coli*. Analysis of agarose gels of this cosmid led to the identification of a ca. 16-kb covalently closed circular molecule that hybridized with Tn*916* but not with the vector. This molecule was digested with several restriction enzymes known to cut the transposon and was not sensitive to others that do not cut this element.

To demonstrate that the covalently closed circular transposon could act as an intermediate in conjugative transposition, it was purified from an agarose gel and used to transform *B. subtilis* protoplasts (57). The frequency of transformation with the free circular form was about 150-fold greater than that of the Tn*916*-containing cosmid from which it was derived. Furthermore, the products of the insertion of the circular intermediate into the *B. subtilis* chromosome had all of the expected characteristics: the transposon inserted into different chromosomal sites as determined by DNA hybridization analysis (Southern blot), and its presence caused the *B. subtilis* strain to become a conjugative donor of the Tn*916* element. There is no evidence to suggest that the circular transposon can replicate.

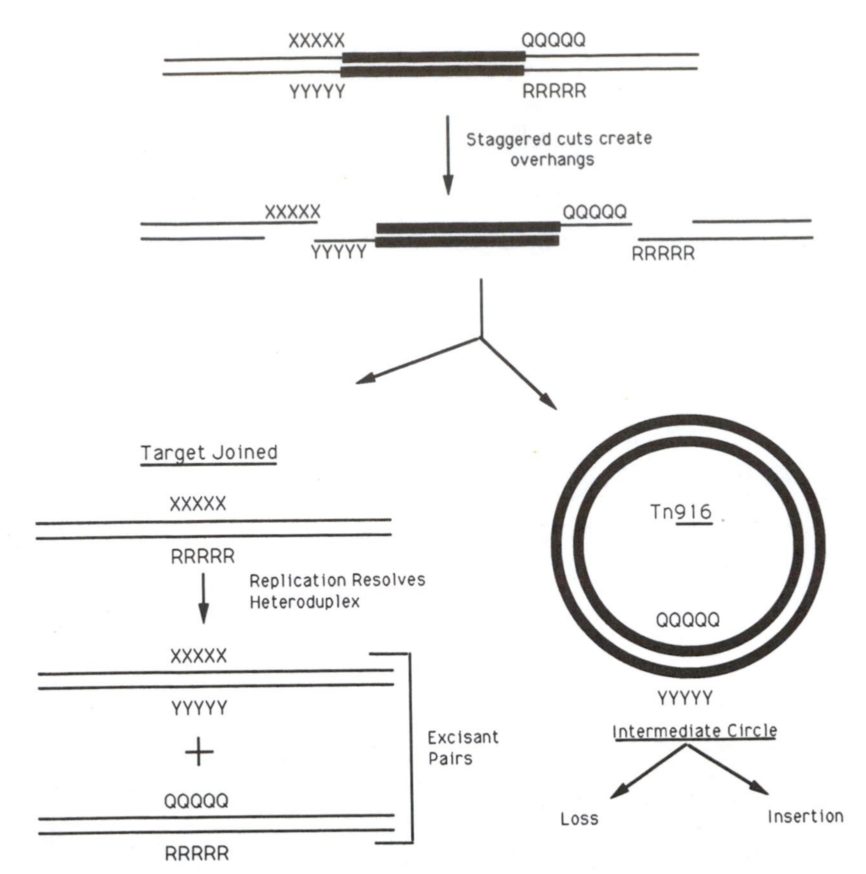

Figure 2. Model for excision of Tn*916*. Thick lines represent Tn*916*, and thin lines represent DNA adjacent to the transposon. Coupling sequences are indicated by the hypothetical nucleotide pairs X-Y and Q-R. A staggered cleavage of the phosphodiester backbone on the 3′ side of the coupling sequence on both strands (first line) generates molecules with 3′ single-stranded ends (second line). The target and transposon sequences are joined by their 3′ single-stranded regions to generate an excisant molecule and the transposon circle. Because there is no apparent requirement for homology between the two coupling sequences, both the excisant and the transposon circle contain heteroduplexes consisting of the base pairs originally present in the coupling sequences. Semiconservative replication resolves the heteroduplex in the excisant and generates a pair of molecules (excisant pairs), of which one has the left coupling sequence at the site of excision and the other has the right coupling sequence. (Reproduced from reference 12 with permission.)

Coupling Sequences and Mismatches at the Joint of the Excised Element

Analysis of the sequence of a target from which the transposon has excised (the excisant) revealed the presence of unexpected bases (11, 19, 52). These bases were not present in the target prior to insertion of the transposon, and their generation remained a mystery for several years. However, when it was recognized that two different excisant forms can result from each inserted molecule and when their sequences were compared, it appeared that these bases must have been inserted with the transposon (12). We called these 5 bases "coupling sequences," because they are not permanently attached either to the transposon or to the target. A model for Tn*916* excision consistent with the data is illustrated in Fig. 2.

According to this model, staggered nicks occur on each DNA strand during excision. One nick is adjacent to the transposon end, and the other is displaced into the target DNA by the 5 bases, which correspond to the coupling sequences. (In Fig. 2, the nick is arbitrarily shown 3′ to the coupling sequences.) If the excised element were to circularize directly, these 5-base coupling sequences would have to pair and the junction they form would be ligated closed. However, because the coupling sequences need not be related and are not usually homologous, they produce a heteroduplex at the closure joint of the circular transposon. Both by indirect analysis of the cloned joint region of these molecules (progeny of the chimeric plasmid containing the cloned joint were sequenced) and by direct testing of the circular transposons for resistance to a restriction enzyme that should recognize a correctly paired joint sequence, such covalently closed circular excised transposons were shown to often have mismatches at the joint. When purified, these circular molecules served as intermediates in transposition (see above; 57).

It seems unusual that the mismatches at the joint are not repaired in the excised transposon in *E. coli*. We have suggested that they may be protected from repair enzymes by the presence of a protein or proteins (12). Mismatched bases might stimulate the recombination with target DNA required for transpo-

```
                              TARGET

                ......CCTCTTTTAAAAT*AAAAAGAG.......
                ......GGAGAAAATTTTA*TTTTTCTC.......

                          CLASS I EXCISION

pJRS1006:

                           L              R
                      --------Tn916---------
 ......CCTCTTTT AAAAT AAACAAAGT:::CTATTTTT AACTA AAAAAGAG...
 ......GGAGAAAA TTTTA TTTGTTTCT:::GATAAAAA TTGAT TTTTTCTC...

Excisants:                          Closed transposon circle joint:

                                              R           L
                                         -----------   --------
   ..CCTCTTTT AAAAT AAAAAGAG.. +      ::CTATTTTT AACTA AAACAAAGT::
   ..GGAGAAAA TTGAT TTTTTCTC..        ::GATAAAAA TTTTA TTTGTTTCT::

  After replication:

1. ..CCTCTTTT AAAAT AAAAAGAG..
   ..GGAGAAAA TTTTA TTTTTCTC..

   And:

2. ..CCTCTTTT AACTA AAAAAGAG..
   ..GGAGAAAA TTGAT TTTTTCTC..
                          CLASS II EXCISION

pJRS1006:
                            L             R
                       --------Tn916------- -
  .......CCTCTTTT AAAAT AAACAAAGT:::CTATTTTT TAACT AAAAAAGAG...
  .......GGAGAAAA TTTTA TTTGTTTCA:::GATAAAAA ATTGA TTTTTTCTC...

Excisants:                          Closed transposon circle joint:

                                              R            L
                                         ----------    ---------
   ..CCTCTTTT AAAAT AAAAAAGAG.. +     ::CTATTTTT TAACT AAACAAAGT::
   ..GGAGAAAA ATTGA TTTTTTCTC..       ::GATAAAAA TTTTA TTTGTTTCA::

 After replication:

1. ..CCTCTTTT AAAAT AAAAAAGAG..
   ..GGAGAAAA TTTTA TTTTTTCTC..

   And:

2. ..CCTCTTTT TAACT AAAAAAGAG..
   ..GGAGAAAA ATTGA TTTTTTCTC..
```

Figure 3. Excision products from pJRS1006. The transposon sequence is delineated by dotted lines above the bases; coupling regions, which are in boldface type, are isolated by spaces except in the target. * (in the target), site of insertion of Tn*916*; :, bases internal to the transposon; ·, bases internal to target DNA. Underlined bases in class II excisants are derived from the coupling sequence that was brought in with the transposon. L and R, left and right ends of Tn*916*, respectively. Sequences of the transposon circle joints, target, and class I excisants after replication are from Caparon and Scott (12). Sequences of the other excisants are predicted.

sition, but it is not known whether a heteroduplex joint region is required for transposition of Tn*916*.

Of the nine excisants we analyzed from a single insertion (12), seven behave as described above (class I; Fig. 3). The two exceptions (class II) behave as if the enzyme responsible for creating the staggered nick made an error in recognition of the last T in the string of six Ts at the right end of Tn*916* and instead nicked after the fifth T. In these cases, 6 instead of 5 bases are left behind as the coupling sequence in the host (excisant) chromosome. The transposon circle then has only five Ts at its right end. All three laboratories working on the sequences at the ends of conjugative transposons have reported that there are sometimes five and sometimes six Ts at the right end of Tn*916* (11, 12, 19).

In summary, during excision, three different excision products have been found (Fig. 3; 12). First, for about half the class I excision events analyzed in this laboratory, the original target sequence is restored in the excisant. Second, for the rest of the class I events, there is a substitution of 5 coupling-sequence bases that were brought in with the transposon for 5 bases that were originally present in the target prior to insertion of the element. This substitution may produce a detectable mutation or may look like "precise excision" at the phenotypic level because the change affects a maximum of 5 bases, which may or may not affect the amino acid sequence, and if the protein sequence is changed, this change may or may not alter the activity of the gene product. Third, the excision may be of the class II type, in which the excisant has

6 bases instead of 5. One of these bases (the A underlined in Fig. 3) came from the 3′ end of the coupling sequence on the right of the transposon, and two different excisants may result from this class II event. This, of course, alters the reading frame if the transposon had inserted into a gene. Therefore, when a conjugative transposon is used as a mutagen, it is not safe to assume that the excisant produced after cloning into *E. coli* will have the wild-type sequence in the region that served as the transposon target. The frequency at which each class of excision event occurs is probably determined by the specific coupling sequence involved and/or by the DNA surrounding this sequence.

The question of the polarity of the nicking event (shown 3′ to the coupling sequences in the figures) has not been directly addressed. When the purified circular Tn*916* intermediate was introduced into *B. subtilis*, one coupling sequence that had been located 3′ of the transposon in the parental insertion was 3′ of the transposon in one transformant but was shifted to 5′ of the transposon in another. This might suggest a surprising lack of polarity of the nicking enzyme(s). However, because the transposon DNA was deproteinated prior to artificial introduction into a recipient by transformation, repair of mismatches that are normally protected by a protein might be responsible for this result. A direct test of the polarity of the coupling sequences with relation to the transposon is needed to answer the question of polarity of the nicking enzyme.

The generation of both excision products (excisant and transposon circle) need not occur in a single event. Anecdotal evidence (56a) suggests that recovery of these products might be exclusive, since it appears that when the circular transposon is plentiful, the excisant is scarce and vice versa. We envision the possibility that in one excision reaction, only one of the two nicked molecules is ligated. The form chosen for closure may depend on the specific bases present in the coupling sequences or on other factors.

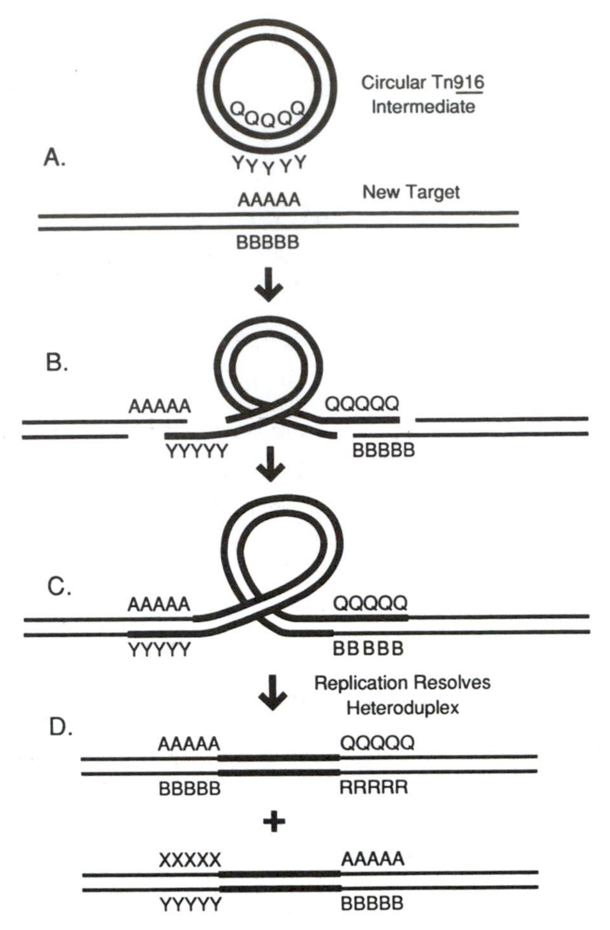

Figure 4. Model for insertion of Tn*916*. Symbols are the same as in Fig. 2. (A and B) Complementary bases of the target. Staggered cleavage at the 3′ ends of the new target and the heteroduplex-containing joint of the transposon circle (A) generates molecules with 3′ single-stranded ends (B). The transposon circle and target are covalently joined by their single-stranded ends to create a new insertion with a heteroduplex at each end (C). Replication resolves the heteroduplexes and generates a pair of molecules in which one member is flanked by the target sequence at one end and a coupling sequence from the previous insertion at its other end. The second member of the pair would be flanked by the target sequence on the side opposite its location in the first member and would have at its other end the second of the two coupling sequences of the first insertion (D). The two forms would segregate at the first replication event. (Reproduced from reference 12 with permission.)

Insertion of the Transposon into Its Target

Figure 4 diagrams the current understanding of how Tn*916* inserts into its target by a reversal of the process used for excision (12). Staggered nicks (shown 3′ to the coupling sequences in Fig. 4) are made in the circular transposon molecule that serves as an intermediate, and the nonhomologous coupling sequences are then simply ligated to a target that has been nicked in a similar manner to give staggered ends with a 5-base overhang. However, because the target is a replicon, the mismatches formed by this insertion process are resolved by replication. Two different types of daughter molecules with the transposon flanked by different sequences are then expected. Each will have only one of the two coupling sequences present in the circular intermediate form, and this sequence will be on a different end of each daughter. Analysis of the transformants obtained from introduction of the purified circular transposon into *B. subtilis* protoplasts showed that this was the case. Thus, the coupling sequences have been traced directly from the first transposon target through the circular intermediate to the second transposon target, validating the model as diagrammed in Fig. 2 and 4.

Ends of the Transposon

Because of the coupling sequences and the imprecision of cutting that results in a variable number of Ts at the right end of the transposon, the ends of the element have to be defined arbitrarily. For the purposes of explaining excision and insertion by the simplest possible model, I defined ends that include six Ts at the right. The transposon has an inverted repeat (IR) element in which 20 of 26 nucleotides are

Table 2. Insertion targets and transposon circle joint[a]

Right		Left	Location of Tn	Reference
GCGGATAACTAGATTTTTATGCTATTTTT	NNNNN	AAACAAAGTATAAATTTCTAATTATCTTTT	Transposon circle	12
			No selection or screen	
CAAATATTGAATGTAAAAATGCGCTTTTTT	CTAGG	AAAAAAAAGCGCATTTTTAACCTGCATTATA	CKS101[b]	13a
ATGGCAATCTTT	AGTTA	GACACAAAGTATATTTG	CKS103	13a
TCTCTTAACCAGGTCAATAGAATTATTTTT	GATAT	AAAAATAGAATACTATGCTCACAAAACCCG	CKS107	13a
ACGAAAGAGATACTGAAGAAATCGGTTTTT	CTTTT	AAAAAGAGCTTAGGACAGAATTTCTTAATT	CKS109.3	13a
			Selected for phenotype	
ATAAAAAAAAATATAAATTAAGACCTCTTT	AAAAT′	AAAAAGAGCATAAATGGCATAAAAAATGGGT	pJRS1006	12
ATGATAATATTATT	AAACT	AAAGAAATATCTTTT	pAM160	19
AGAGAGCTATTTTT	AATAA	AAAAACCAATACTGT	pAM120	26
ATAAGTTAATTCTTTT	TTTGG	AAAAACTAGT	*L. monocytogenes* internalin ustream	26

[a] All sequences are from Tn*916* except for *L. monocytogenes* internalin upstream sequence, which is from Tn*1545* and for which the sequence of the target with the transposon is not available. The sequences from Courvalin's group working on Tn*1545* excision may not fit this model, but the data in the literature (11) have been corrected (52) and are not complete enough to be fit to a model different from theirs. They have defined the ends of the transposon differently and believe there are 6 or 7 bases in the coupling sequence.
[b] Obtained by transformation of *B. subtilis* with purified circular Tn*916*.

complementary (19). The left IR ends at the left terminus of Tn*916*, and the right IR is at the right terminus if this transposon end has only five Ts. The sixth T, when present, is beyond the last base in the IR. The importance of this motif or of several direct repeats that occur within 200 bases of the ends of Tn*916* (19) has not been experimentally determined. There appears also to be a potential promoter reading out from the right end of Tn*916* (19). A partial restriction map of Tn*916* showing some of its features is presented in Fig. 1.

Target

Although insertion of Tn*916* and its relatives does not appear to be site specific in most organisms (7), it is not random, either (see below). No studies have focused on the DNA sequences of sites that serve as targets for conjugative transposons. However, sequences of some sites into which these elements have inserted are available (Table 2), mostly from studies in which conjugative transposons were used as mutagens. In these cases, it should be remembered that these are not random isolates of inserted elements but, rather, isolates that have been selected on the basis of phenotype. Other target sequences listed in Table 2 are from transformation of *B. subtilis* protoplasts with purified circular transposons (57) and therefore represent unselected sites.

The genomes of the organisms in which insertion sites of conjugative transposons have been sequenced are highly A+T rich. It appears that next to the end of the transposon that has a string of Ts (the right end of Tn*916*), the target is usually A rich, and next to the left end of the transposon, which is A rich, the target is usually T rich. It is not clear whether there are enough unselected data at this time to draw a general conclusion about this or other suggested target consensus sequences (53).

Transposon Genes

Sequence analysis of the end of Tn*1545* revealed two open reading frames (Fig. 1; 53). The larger, originally called ORF2, has been renamed *int*, and the smaller is called *xis*. The deduced protein sequence for Int suggests that it belongs to the family of proteins that includes lambda integrase. Using a minitransposon constructed in vitro by joining fragments of the ends of Tn*1545*, including these open reading frames, and inserting a drug resistance determinant between them, Poyart-Salmeron et al. (53) demonstrated that the *int* gene is required for excision of mini-Tn*1545* in *E. coli*. Furthermore, they showed that excision occurred when a plasmid containing this *int* gene cloned under the *lac* promoter was present in *trans*. By linking an *E. coli* replicon to a fragment of *E. faecalis* DNA containing Tn*1545*, Poyart-Salmeron et al. (53) showed that Int is the only transposon function needed for integration in *E. coli* when it is provided in *trans* from the *lac* promoter on a high-copy-number plasmid.

To show that Int is required for conjugative transposition, the *int* gene was inactivated by insertion of an Erm marker (63). This Tn*916 int-1* mutant did not transfer from *B. subtilis* to *E. faecalis*. However, the mutation was complemented by a wild-type *int* gene cloned in a plasmid (together with the 3′ end of *xis*) under the Spac-1 promoter (70) if the plasmid was present in both the donor and recipient. In this laboratory, we have recently found (7a) that if *int* is supplied in adequate amounts, it need be provided only in the donor in this mating (see below for discussion of conjugational transfer). Thus, the transposon integrase seems to be necessary for conjugational transposition, and if it is provided in adequate amounts (see below), it is the only transposon function required for the excision step (which occurs in the donor in conjugation [see below]).

The smaller open reading frame was not required for excision of the minitransposon but seemed to enhance the frequency of this reaction, at least when it was expressed from the *lac* promoter in *E. coli* (53). Its role in integration has not been examined. This open reading frame was originally called ORF1 and is now designated *xis* on the basis of the weak homology of its deduced amino acid sequence with the N-terminal region of the Xis protein of P22 (6 of 16 amino acids identical and an additional 5 conservative substitutions) and the site-specific invertase Pin from *E. coli* (16 of 57 identical residues and an additional 7 conservative substitutions).

At present, there is no information about regulation or expression of *int* and *xis*, and no promoters for either of these genes have been demonstrated. Although both genes are read in the same direction on the basis of sequence, a potential promoter has been noted between them (20).

Equilibrium between Insertion and Excision

Insertion and excision of conjugative transposons are the results of a reversible reaction. The direction of this equilibrium seems to differ for gram-negative and gram-positive bacteria.

In gram-positive bacteria, the reaction appears to be skewed toward insertion. In *B. subtilis* CKS102, it has not proved possible to detect the excised circular transposon by polymerase chain reaction with primers that read across the joint (6a). This failure suggests that any excised forms are rapidly inserted and therefore are not detectable by the usual assays.

In *E. coli*, on the other hand, excision seems to be more frequent than insertion. Enough of the circular excised transposon was produced from a cosmid-cloned fragment from *S. pyogenes* containing Tn*916* to be detected on an agarose gel (57). Furthermore, in *E. coli*, excision is often followed by loss of the element instead of by insertion (30). In an overnight *E. coli* culture grown without antibiotic selection, 50 to 90% of the cells lost the transposon.

Gawron-Burke and Clewell (30) suggested that this discrepancy in the direction of the insertion-excision equilibrium between gram-positive organisms and *E. coli* might be caused by an imbalance in the controls for transposition in *E. coli*. In support of this, Poyart-Salmeron et al. (53) reported that in *E. coli*, the stability of Tn*1545* depends on the copy number of the vector on which it resides. Attempts to clone mini-Tn*1545* on a high-copy-number pUC-derived vector led to recovery of the excisant only, indicating that excision was so frequent that it precluded isolation of the inserted form of the element. However, the element was successfully cloned on the lower-copy-number vector pSC101 (53). This suggests that the amount of transposon-encoded proteins required for transposition or the ratio of these proteins to those encoded by the host may play a major role in determining the direction of the excision-insertion equilibrium. The nature of the coupling sequences may play a role as well.

There is also evidence that the amount of integrase plays a significant role in determining the excision rate in gram-positive organisms. Complementation of Tn*916 int-1* for transfer from the chromosome of *B. subtilis* to *E. faecalis* was not detected if the integrase was supplied only in the donor from plasmid pAT145 (a pAM beta-1 derivative) when the *int* gene had been cloned under the pSpac promoter (63). (In this construction, a fragment of the Xis protein that may negatively complement the wild-type form may also be produced.) However, if the int^{+} gene is on a transfer-deficient Tn*916* mutant (Tra^{-} mutant) in the *B. subtilis* chromosome, transfer of the erythromycin-resistant Tn*916 int-1* is as efficient as that from the control carrying only the tetracycline-resistant wild-type Tn*916* (7a). This suggests that the amount of integrase is important in determining the rate of excision of conjugative transposons.

These indications that the relative amounts of the proteins involved in transposition of these elements determine the direction of the excision-insertion equilibrium make it difficult to interpret results of complementation experiments in which products are supplied from foreign promoters in unknown quantity.

A second factor that is likely to influence the outcome of this reversible reaction is the nature of the coupling sequence and adjacent sequences in the case being studied. It is likely that the nicking and closing enzyme(s) has substrate preferences that influence the frequency of each reaction. Indirect evidence to support this idea comes from the wide variation in frequency of isolation of transconjugants from different conjugative-transposon-carrying donors of the same strain. Even when there is a single copy of the element in the donor, the frequency of conjugative transposition may differ by 10^{4}-fold in matings with the same recipient (7). Since the only difference in these crosses is the location of the transposon in the donor chromosome, this location must have a major effect on the frequency of conjugative transposition. Thus, the coupling sequences seem likely to play a major role in determining the excision frequency, and it is possible that DNA adjacent to the coupling sequences is important as well.

Host Function(s) Needed

L. lactis is unable to act as a donor for conjugative transposition, although when Tn*916* is transferred from *B. subtilis* or *E. faecalis* to *L. lactis*, transposition occurs. An *L. lactis* strain carrying a conjugative transposon cannot serve as a transposition donor either to another *L. lactis* strain or to a strain of *B. subtilis*, *E. faecalis*, or *S. pyogenes*. Thus, excision of Tn*916* does not seem to occur in *L. lactis* (7). This indicates that a host function is required for excision and that this function is absent from *L. lactis*. The function may alter the expression of transposon genes that are essential for the excision process, or it may act directly in the excision event. Other transposons and prophages require IHF (integration host factor) for transposition or excision, and there is a putative IHF-binding site near the left end of Tn*916* (7). It is possible, therefore, that an IHF-like factor is needed for excision and that *L. lactis* has no such factor. However, it is equally likely that the function missing from *L. lactis* affects expression of the transposon integrase. To test this, we are attempting in my laboratory to complement Tn*916* for excision in *L. lactis* with an integrase gene provided in *trans*.

Comparison with Phage Lambda

In its general outline, transposition of conjugative transposons appears similar to excision and insertion by lambda prophage in a host lacking the specific chromosomal attachment site. In both systems, a circular intermediate is formed in a reversible process (54, 57). However, the overlap region of lambda formed by joining the single-strand overhangs on the left and right of the prophage does not include mismatches. In fact, if mismatches are experimentally introduced into this region, further processing and resolution are very inefficient and if they occur at all are often incorrect (3, 42). For Tn*916*, on the other hand, mismatches of 2 or 3 bases in the 5-base coupling sequence are very common and may even be required for insertion into a new target.

It was originally suggested that transfer of conjugative transposons leads to a "burst" of transposition by a zygotic-induction-like mechanism (29). For example, when Tn*916* is on a conjugative plasmid in the donor, it is often found in the chromosome in the transconjugants (see above). However, because a transposon present in the recipient does not reduce the frequency of conjugative transposition, no lambdalike repressor is present (20, 50). The alternative that conjugation stimulates excision is discussed below.

The conjugative transposon integrase was so named because it has a region with similarity to the active site of the lambdoid integrase proteins. Little is known about the mode of action of the transposon protein, so the similarity to lambda is still limited to the level of amino acid sequence. It seems likely, however, that there is a functional similarity as well.

Excision of lambda requires a second phage-encoded protein called *xis*. There is a transposon protein called *xis* that has very weak homology with the P22 Xis protein (52). The transposon protein stimulates transposon excision in certain experiments in *E. coli* but is not required for the excision event (52).

Lambda also requires the host protein IHF to facilitate folding and optimal binding of the phage enzymes needed in the reaction (42). There is a sequence near the left end of Tn*916* that has 8 of the 9 conserved bases for an IHF-binding site and also has the conserved spacing (7), but a possible role for IHF in conjugative transposition has not yet been investigated.

To understand the extent of the analogy between insertion and excision of conjugative transposons and lambdoid prophages, further careful quantitation and biochemical work on the former are needed.

CONJUGATION

Functions Required for Conjugation

The best-understood conjugation system is that promoted by the F plasmid in *E. coli* (reviewed by Willetts and Wilkins [69] and Ippen-Ihler and Minkley [37]). The F plasmid contains genes required for synthesis of pili, which project from the surface of the *E. coli* cell and mediate cell-cell contact of mating pairs. Formation of these F^+/F^- mating pairs triggers a reaction that leads to unwinding of the double-stranded circular F plasmid after it has been nicked. This initiates DNA replication at *oriT*, the transfer origin on F. The nicked strand of F is transferred to the recipient, beginning at its 5′ end at *oriT*. With the strand of DNA, the primase enzyme is transferred to the recipient, and this enzyme is responsible for synthesis of the complementary F-plasmid DNA strand in the transconjugant cell. Many plasmid-encoded functions, including those for synthesis of pili and those for processing of the DNA, are required for these events. Although the mating process may be quite different, it appears that Tn*916* has enough DNA to encode many functions for conjugation and transfer.

An analysis of the functions of Tn*916* required for conjugative transfer has been started by isolating mutants deficient in this process (58). Mutants of Tn*916* into which Tn*5* had inserted were obtained in *E. coli* and scored for ability to transfer between strains of *E. faecalis* in filter matings and for ability to transform *E. faecalis* protoplasts. From these experiments, it appears that most of the transposon DNA to the "right" of the *tetM* gene (as Tn*916* is usually drawn) is needed for successful conjugation (Fig. 1).

Some of these transfer-deficient mutants (Tra^-) were complemented by the conjugative plasmid pAD1 for transfer to the *E. faecalis* recipient. This phenotype was called Tn^+ by Senghas et al. (58) because it was believed to represent transfer of the Tra^- transposon to pAD1. However, in these cases, over 90% of the transconjugants had the Tn*916* element on the chromosome of the recipient unlinked to pAD1. Some transconjugants contained the transposon in the plasmid, and these may be interpreted as evidence for insertion of the Tn*916* element into the plasmid prior to conjugative transfer, as proposed by Senghas et al. (58). Alternatively, the plasmid may promote conjugative transfer of Tn*916* in *trans* (see below), and the transposon may occasionally insert into the plasmid, presumably after the mating event. No data on whether the transconjugants with the element on the plasmid have one or more additional copies in the chromosome were presented. Further analysis of these conjugation-deficient Tn*5* insertion mutants should lead to significant advances in our understanding of conjugation mediated by these transposable elements.

Is the Transposon Transferred as a Single or a Double Strand?

The method by which DNA is transferred during conjugation between gram-positive organisms is not yet understood. Furthermore, it has been suggested, at least for the pneumococcus, that the transfer mechanism used by conjugative plasmids may differ from that of conjugative transposons, since the former are subject to DNA restriction in crosses in which the latter are not (31; see above). The interpretation of the differential restriction is that both plasmid and transposon transfer a single DNA strand but that in the recipient cell, the complementary strand of the plasmid is quickly synthesized. The double-stranded form is then sensitive to restriction endonuclease digestion. By analogy with conjugative plasmids transferred among gram-negative bacteria, it is assumed that the primase that primes synthesis of the complementary DNA strand is transferred with the strand of plasmid DNA. In this model, the transposon is presumably maintained longer as a single strand in the transconjugant than is the plasmid because the transposon is not a replicon and so does not have a primase associated with it. However, if conjugative transposons have a special restriction protection system (see above), this argument becomes moot. Therefore, conclusions about the number of strands of the transposon that are transferred await more direct experimentation.

Nature of a Mating Pair

One method by which mating between strains of *B. subtilis* and possibly between strains of *E. faecalis* appears to proceed is a mechanism that resembles cell fusion and results in formation of a fairly complete zygote (65). This was demonstrated by scoring unlinked markers in crosses between a *B. subtilis* donor carrying the small conjugative transposon Tn*925* and a closely related multiply marked *B. subtilis* recipient strain that had a *comG* mutation (which prevents transformation). Transconjugants selected for one marker were found to inherit linked markers from the donor, even when these markers were very distant from the site of the transposon and the transposon was not present in the transconjugants. These observations strongly suggest that the diploidy of the zygote is extensive. However, because the frequency of formation of transconjugants selected in this experiment has not been compared with the frequency of formation of transconjugants selected for a marker on the transposon, the cell fusion type of mating may be rare among transposon-promoted conjugation events. No evidence was presented to indicate whether transposition accompanies this type of mating.

If a true zygote is the first product of conjugative transposition, markers unlinked to the conjugative transposon should be cotransferred. Experiments to test this have scored cotransfer of nonconjugative plasmids, and the evidence from these experiments is mixed. In crosses between *B. subtilis*::Tn*916* containing either pC194, pUB110, or pE194 and *B. thuringiensis*, selection for transconjugants containing pC194 or pUB110 was successful, but pE194 was not found in the transconjugants (47). It is difficult to interpret these results, but it should be noted that a different selective marker was used in each case. It therefore remains possible that mobilization of pE194 can be detected if a selective marker other than erythromycin is used. It is also possible that if pE194 is present at a low copy number and is geographically constrained in the cell, it is not coinherited with the transposon in descendants of the zygote even if it is present in the zygote itself.

More recently, Clewell et al. (20) showed that in *E. faecalis* matings, when the Tn*916* element is located on a nonconjugative plasmid near an erythromycin resistance marker, the transposon does not mobilize the plasmid marker and transfer of the plasmid is not detectable. It therefore appears that complete zygotes are not formed in all cases of Tn*916*-mediated conjugation.

When a *B. subtilis* donor with a Tn*916 int-1* mutation (which contains an *erm* determinant inserted into the essential transposon gene *int*; 63) was crossed with an *E. faecalis* recipient, transconjugants were obtained only when the wild-type transposon gene was provided from a plasmid present both in the donor and in the recipient strain. This would not be expected if complete cell fusion occurred. However, when my colleagues and I recently repeated these crosses but provided the essential transposon function from another transposon in the donor instead of from a foreign promoter on a multicopy plasmid, complementation was obtained when the wild-type gene was present only in the donor (7a). These results are therefore consistent with formation of a complete diploid intermediate during conjugative transposition.

In summary, at this time, the nature of the mating pair and of the DNA transferred during conjugative transposition remains unclear.

EXCISION AND CONJUGATION

Conjugation May Trigger Excision

Excision and loss of conjugative transposons, which are very frequent in *E. coli*, are rarely seen in gram-positive organisms. It was originally suggested by Clewell and Gawron-Burke (21) that excision of these elements is rate limiting in the process of conjugative transposition in gram-positive bacteria. This implies that when the element excises, it is usually transferred in a mating event. However, data collected more recently make it seem to me more likely that excision in gram-positive bacteria is itself stimulated by formation of appropriate cell-cell contacts during mating. In this view, mating-pair formation is rate limiting and triggers excision of these conjugative transposons. It is possible that in *E. coli*, excision of conjugative transposons is more frequent because stimulation by effective mating-pair formation is not required. Perhaps, instead, the excision process is always induced in *E. coli*.

This hypothesis predicts the existence of a class of transfer-deficient transposon mutants that is not complemented by a conjugative plasmid (Tra$^-$ Tn$^-$ of Senghas et al. [58]) and does not excise in gram-positive bacteria. Such a mutant should excise in *E. coli*, so the circular form of the transposon would be present in a DNA preparation made from a strain of *E. coli*. Therefore, the test used by Senghas et al. (58) for transformation of *E. faecalis* protoplasts that uses DNA from *E. coli* should give a positive result. The predicted mutant would thus have a Tra$^-$ Tn$^-$ PT$^+$ phenotype in the scheme of Senghas et al. (58). Such a class of mutants was identified and has not yet been tested for excision in gram-positive organisms.

As mentioned above, the presence of a conjugative transposon in the recipient does not inhibit the entry of a second element by conjugative transposition, indicating that there is no repressor to prevent this (50). Therefore, since there is nothing inhibiting formation of mating pairs among different bacteria within one strain, transposition within a strain should also occur by a conjugative mechanism. Because transposition is a low-frequency event (10^{-4} to 10^{-9}, with the frequency probably depending on the coupling sequences adjacent to the element; see above), transposition between two locations on the chromosome of one strain is difficult to observe. However, we have recently isolated an insertion of Tn*916* into a gene encoding chloramphenicol resistance (7c). Excision of the transposon generates chloramphenicol-resistant survivors, so it is now possible to score excision frequency accurately. Work in progress should allow estimation of the frequencies of excision, transposition, and loss of the transposon in this strain.

The following sections present evidence supporting the hypothesis that effective mating-pair formation triggers excision of conjugative transposons in gram-positive bacteria.

Excision Occurs in the Donor

It appears that conjugative transposons are excised in the donor before they are transferred to the recipient cell. When Tn*916* is located on a nonconjugative plasmid, the plasmid markers are not found in the transconjugants selected for a transposon marker, and no transconjugants are found when the plasmid marker is selected (20, 24; see above). Thus, the conjugative transposon does not mobilize the plasmid DNA in which it resided in the donor, and excision of the transposon occurs prior to transfer.

It has also been observed that when a conjugative transposon is present in a conjugative plasmid in the donor, transfer of the transposon is independent of that of the plasmid. When a plasmid marker was selected, 43 to 90% of the transconjugants received the transposon marker, and of these, many no longer had the transposon on the plasmid (20, 24, 66). This would be predicted if the plasmid-stimulated mating event induces transposon excision in the donor prior to transfer. The same result is also obtained when a Tn*916* element (with a different marker) is present in the recipient as well. On the other hand, when the same plasmid containing Tn*916* is introduced into the same recipient by protoplast transformation instead of mating, linkage between the transposon and the plasmid is maintained as expected (20). The latter two experiments rule out zygotic induction as the explanation of genetic dissociation between the transposon and plasmid in the transconjugants. A simple interpretation is that mating-pair formation stimulates excision of Tn*916* in the donor and the transposon is then transferred independently of the plasmid to the recipient. Introduction of the DNA by nonconjugative methods would not lead to excision.

Additional evidence that excision of conjugative transposons occurs in the donor in a mating event comes from our work with *L. lactis*. This organism seems to be defective in excision of conjugative transposons (see above). Although it cannot act as a donor in conjugative transposition, *L. lactis* can act as a recipient. This implies that the transposon excises in the donor prior to transfer (7).

In addition, my colleagues and I (7a) have recently found that in a mating between *B. subtilis*::Tn*916–int-1* and *E. faecalis*, the *int* mutant can be complemented by an appropriate source of the Int protein (which is required for transposon excision; see above) in the donor alone (see above). It is not necessary to provide the protein in the recipient. This suggests either that no integrase is needed in the gram-positive recipient and therefore that excision does not take place there or that the required integrase is transferred together with the transposon during mating. When the integrase was present only in the recipient, *E. faecalis* transconjugants were not found (7a). This implies that excision takes place prior to transfer of the conjugative transposon.

Transposon Transfer Is Stimulated by Conjugative Plasmid in Same Donor Cell

The presence of a conjugative plasmid in the donor has been reported to stimulate the frequency of transfer of an unlinked chromosomally located conjugative transposon in *E. faecalis* by several orders of magnitude (25). It was suggested that this stimulation results from cointegrate formation. The hypothesis was that the transposon first inserts into the plasmid and that the cointegrate is then transferred to the recipient strain (21). However, cointegrates are rarely observed in the transconjugants, so if they occur, they must quickly resolve in the recipient. Furthermore, for the large element Tn*5253*, Guild et al. showed that when the donor *Pneumococcus* strain with a transposon in the chromosome carries the conjugative plasmid pIP501, inheritance of the plasmid is restricted by a *Dpn*II$^+$ recipient, while that of the transposon is not (31). At least for large composite transposons, this strongly suggests that the plasmid and transposon are not transferred as one molecule. By analogy with lambda, the term "zygotic induction" was used to describe segregation of the transposon and the conjugative plasmid in the recipient. However, since conjugative transposition of Tn*916* into a recipient that contains a Tn*916* element occurs with the same frequency as into one that does not, nothing analogous to lambda prophage repression appears to occur (24, 50).

An alternative interpretation for stimulation of the frequency of conjugative transposition by a coresident conjugative plasmid is that the plasmid in the donor increases the frequency with which mating pairs are formed and that formation of the pairs either allows or stimulates conjugative transfer of the transposon from the chromosome of the donor to the chromosome of the recipient. If this is correct, it implies that mating-pair formation stimulates excision and that excision is rate limiting for conjugative transposition. It will be interesting to see the results of an experimental test of this issue.

Transfer of Unlinked Conjugative Transposons Frequently Occurs in One Mating Event

Very recently, Clewell's group (20, 24) reported that when there are two differently marked copies of Tn*916* in the donor chromosome, transconjugants selected for having acquired one of these elements often contain the unlinked transposon too. This result again suggests that the conjugation event stimulates excision of the second transposon and that since a mating pair already exists, the unselected element is often found in the transconjugants as well.

USE AS MUTAGENS

Introduction

Insertional mutagens are valuable tools in all organisms, and those that can be used in gram-positive bacteria are limited. Tn*917* has been exploited effectively, largely because of the useful derivatives constructed by Youngman and his colleagues (71 and chapter 40, this volume). However, in many hosts, including *B. subtilis*, Tn*917* displays strong site specificity. To facilitate selection of mutations at other sites, Tn*916* and its relatives can be employed. The conjugative elements are also valuable for mutagenesis of hosts that do not undergo transformation and for which no electroporation system has been developed (13). Thus, Tn*916* has been a favorite mutagen for the streptococci (13) and has been valuable also for *Listeria monocytogenes* mutagenesis (27, 40, 45). It is also being used in *C. acetobutylicum* (5) and the pathogenic neisseriae (40, 41, 48).

Target Specificity

To be an ideal mutagen, a transposon must show little or no target specificity. At first, conjugative transposons were thought to be ideal in this way. However, more recent work with *B. subtilis* and *L. lactis* indicates that these transposons do show target preferences. Among 18 independent *B. subtilis* transformants into which Tn*916* was introduced from pAM120 grown in *E. coli*, we found by Southern blot analysis that insertion had occurred at only 10 different sites (or regions). The transposon appeared to have inserted into five sites in several different strains (6c). Thus, transposition from this donor is not random in *B. subtilis*. Further, in an effort to isolate an auxotrophic mutant, this laboratory screened >6,000 transformants and transconjugants of *B. subtilis* into which Tn*916* had inserted and never found a single such mutant! Since the frequency of such mutations should be about 1 in 100, these results again indicate that transposition by this element is not random. Finally, for conjugation between *B. subtilis* CKS102 (which has one copy of Tn*916*) and *L. lactis*, 24 transconjugants were analyzed by Southern blots. Among these transconjugants, at least two types of insertions (and possibly as many as six types) were isolated more than once (7c). In summary, although insertion of Tn*916* is not random, there are many sites into which it inserts.

Choice of Donor

The target range of a conjugative transposon may be determined at least in part by the coupling sequences that surround the element in the donor DNA. If this is true, then the randomness of mutagenesis by Tn*916* would be increased by introducing the transposon from different locations; i.e., instead of using pAM120 as the only source of the transposon, other sources should be employed as well. Introduction of the transposon from the chromosome of other organisms by mating is a quick and simple method of mutagenesis as long as counterselection to kill the donor strain is effective.

One additional cautionary note is provided by the case of *L. lactis* (7). This organism cannot act as a conjugative transposon donor because excision does not occur in it. Some strains of *L. lactis* carry fertility factors and are able to transfer the transposon like any chromosomal marker (6b). In crosses between *L. lactis* strains when one strain happens to carry this factor, the transposon will be found in transconjugants in the absence of transposition, so no mutagenesis will occur. However, *L. lactis* may be mutagenized with conjugative transposons as long as they are introduced by conjugation from a strain of a different species. There may be other species of bacteria defective in excision of conjugative transposons that will also not serve as donors of these elements.

Availability of Mini-Tn*1545*

To study a gene in a gram-positive organism following transposon mutagenesis, it is often desirable to clone the DNA including the transposon into *E. coli*. With the large conjugative transposons (which are 16.4 kb or larger), this can be done by using a cosmid vector. However, to make cloning and manipulation more convenient, Poyart-Salmeron et al. (52) have developed a small version of Tn*1545* (3.4 kb). The element, called Tn*1545*delta3, consists of the kanamycin resistance determinant of Tn*1545* (AphA3) cloned between the left and right ends of the transposon. Both the *int* and the *xis* genes are present in this construction. This minitransposon is capable of insertion and excision and can be used for mutagenesis, as shown by the work of Nassif et al. with the pathogenic neisseriae (48). It is available on a pSC101-derived vector.

Occurrence of Multiple Transposon Insertions

When conjugative transposons are introduced, especially in mating events, it is common for the transconjugants to contain more than one copy of the transposon. Many explanations of this are possible. There may be more than one transposon in the donor, and in the long matings performed (usually overnight), multiple rounds of conjugation may occur. To avoid working with mutants bearing multiple insertions, it is very important to move the mutation to an unmutagenized strain. Depending on the organism being studied, transduction, conjugation, or electroporation with the mutated DNA followed by selection for tetracycline-resistant recipients should allow the isolation of single-insertion mutations (13).

Outward-Reading Promoter from Tn*916*

It cannot be assumed that insertion of Tn*916* will inactivate downstream genes in an operon. In fact, there is a potential promoter near the right end of the element that will transcribe into adjacent target DNA. This was first observed when Tn*916* inserted into the hemolysin gene of a plasmid in *E. faecalis* (19). In addition to the nonhemolytic mutants, there was a large class of insertions that produced a larger zone of hemolysis and were called hyperhemolytic. Further analysis showed that in these insertions, the transposon was always at the same site and in the orientation that would permit transcription originating within the transposon to continue into the hemolysin gene.

The asymmetrically located *Hin*dIII site of Tn*916*, which is within the *tetM* gene, provides a convenient way of determining the orientation of the element in

Southern blots. The outward-reading promoter will be in a Tn*916* fragment that is larger than 11 kb (the right end of Tn*916*; Fig. 1).

Excisants Are Not Always Wild Type

It was originally reported that in *E. coli*, excision of a conjugative transposon restores the wild type, which was expected to make the conjugative transposon very convenient as a mutagen (30). To obtain the wild-type form of the mutant allele, it was believed to be necessary only to clone the mutant in *E. coli* and allow excision to occur by removing the selection pressure. However, on closer examination of sequences of excisants, it became clear that this would not always work (19). Often, the wild-type phenotype is restored but the coupling sequence that entered with the transposon has replaced that which was present in the unmutagenized parental DNA (12; see above and Fig. 3). It is also possible to have a class II excision event in which the reading frame is changed because the excisant contains 6 bases instead of the 5 originally present (12). Use of the term "precise excision" for conjugative transposons is confusing, because although excision occurs at a precise location in class I events, i.e., at one end of each coupling sequence, it does not always restore the wild-type sequence. It is important, therefore, not to assume that the excisant generated in *E. coli* has the wild-type sequence. Instead, the cloned mutant can be used to identify the wild type by hybridization to a gene library made from the parental strain (13).

In addition, conjugative transposons have been observed to generate deletions that may extend into host DNA in some situations (50, 62). There is not enough information about these deletions to be able to guess how they are formed, but their existence provides a note of caution for analysis of conjugative transposon mutations.

USE AS SHUTTLE BETWEEN *E. COLI* AND GRAM-POSITIVE BACTERIA

Introduction of Cloned DNA from *E. coli*

In some gram-positive bacteria, there is no convenient shuttle system for introduction of DNA that has been cloned or altered in *E. coli*. In other cases in which high-copy-number shuttle plasmids are available, it is often important to maintain the gene of interest at the same copy number as the chromosome. In both situations, conjugative transposons are attractive shuttle vector systems.

The cloned DNA to be studied can be inserted into Tn*916* at the *Bst*XI site without altering the function of this conjugative transposon (49). The chimeric transposon is then introduced into the chromosome of a gram-positive organism by the easiest means that will not lead to rearrangement of large incoming DNA molecules. Electroporation is often the method of choice, although in the case of *B. subtilis*, transformation of protoplasts is effective. The transposon will then insert into the host chromosome, taking the cloned gene with it.

If it is difficult to introduce DNA directly into the host to be studied, an intermediate host can be used (49). For this purpose, *B. subtilis* has been effective, but any gram-positive organism in which excision of conjugative transposons occurs can be used. The intermediate host is transformed with the DNA of an *E. coli* plasmid containing Tn*916* with the gene of interest cloned into it. The transposon with the cloned DNA will be located in the chromosome of the transformant. This strain will then act as a conjugative donor to any other gram-positive strain, so the transposon containing the cloned gene can be transferred to the desired host strain by mating on a solid surface. The presence of a single copy of the vector transposon in the chromosome can be confirmed by Southern blot analysis. In my laboratory, we have employed this approach successfully in our work on the M protein of *S. pyogenes*, with *B. subtilis* as the intermediate host (49).

Allele Replacement

Allele replacement can be used to replace a chromosomal allele with one engineered in vitro and cloned into *E. coli* (49). In this procedure, either the allele to be replaced or the incoming allele must be readily scorable. If the cloned DNA has no selectable phenotype, an antibiotic resistance marker can be cloned downstream of the allele used for replacement so that direct selection is possible (51a).

If the recipient chromosome has a region of homology with the DNA cloned into the conjugative transposon, the transposon will insert into this DNA by homologous recombination at some low frequency. In *S. pyogenes*, in over 95% of the transconjugants selected for the presence of Tn*916* (tetracycline resistance), the element had inserted into the chromosome by transposition (49). However, in the other 5%, the chimeric transposon had recombined into the homologous chromosomal region. (It is now possible to use an *int* mutant transposon as a vector to eliminate the class of transconjugants in which the transposon inserts by transposition.) In the transconjugants that arose by homologous recombination, the transposon was no longer present and the resident allele had been replaced by the new one (Fig. 5). Two homologous recombination events must occur to obtain this product. These events may occur at the same time, replacing the resident allele with a linear fragment bracketed by regions homologous to the chromosome, or sequentially. In the latter case, there should be an unstable intermediate with the entire transposon inserted at a region of homology.

In my laboratory, we used this strategy successfully to replace the *emm6* allele encoding type 6 M protein in group A streptococcus strain D471 with a kanamycin resistance gene (49; Fig. 5). In this case, recombinants containing the entering allele could be selected with kanamycin. Recombinants resulting from integration of the whole transposon were tetracycline resistant, and those that had resulted from a double crossover in which the kanamycin resistance gene replaced the resident *emm6* allele were tetracycline sensitive (Fig. 5). Thus, even though the desired replacement recombinants were a minority class, the

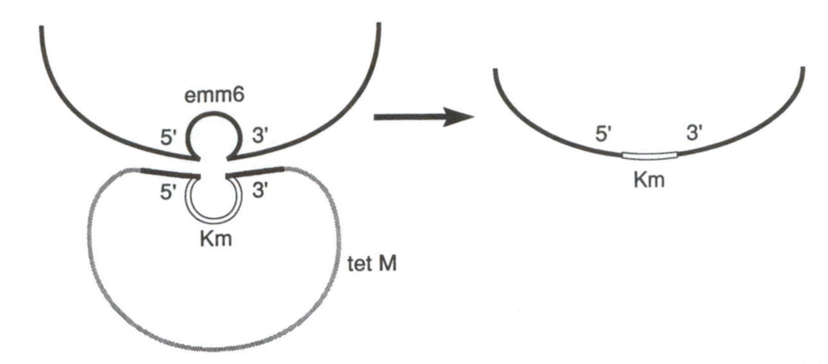

Figure 5. Insertion of shuttle transposon. Allelic replacement of *emm6* by *aphA3* (kanamycin resistance gene [Km]). The dark line represents chromosomal DNA of *emm6*$^+$ *S. pyogenes* JRS4. The gray circle represents the covalently closed circular form of Tn*916*, which is introduced into JRS4 by mating with a *B. subtilis* strain and contains the sandwich cloned at the *Bst*XI site. The sandwich contains the *aphA3* gene (represented by a white box) cloned between the region 5′ of *emm6* and the region 3′ of *emm6*. The homologous regions 5′ and 3′ of *emm6* on the transposon and chromosome of the recipient *S. pyogenes* strain pair with each other, and recombination leads to substitution of the kanamycin resistance gene for *emm6* and loss of the rest of the transposon, including the *tetM* gene.

presence of a selectable marker made it possible to isolate them.

If no selectable marker is available, one can be placed downstream of the desired allele in the chimeric transposon chimera. For example, an altered form of the *emm* gene can be cloned between the region 5′ of *emm* that is homologous to the chromosome and the kanamycin resistance marker. This chimeric transposon can then be transformed into *B. subtilis*, and from there, it can be mated into *S. pyogenes*. The desired transconjugants will be kanamycin resistant and tetracycline sensitive as a result of replacement of the resident *emm6* allele with the DNA cloned between the 5′ and 3′ regions of chromosomal homology.

Complementation with a Single Copy of a Cloned Gene

The presence of a resident copy of Tn*916* in the recipient does not decrease the frequency of introduction of a second copy by conjugation (50). Thus, Tn*916* can be used as a low-copy-number shuttle to study complementation either of an allele present on a transposon previously introduced into the host or of a chromosomal allele (13). The gene of interest is cloned into Tn*916* (at, for example, the *Bst*XI site), and the chimeric transposon is introduced into the strain to be studied. In the large majority of the transconjugants or transformants that receive the transposon, it inserts by transposition instead of recombination into the resident homologous allele.

It is necessary to perform Southern blot analysis to be sure that there is only one copy of the new Tn*916* element in the chromosome. If there are more copies, the element can be moved by transformation or transduction to a new host (13).

Vectors for Insertion of Cloned DNA into Tn*916* on *B. subtilis* Chromosome

There are several technical difficulties in using Tn*916* as a shuttle. First, the large size of this element makes in vitro cloning manipulations difficult. Second, these manipulations are further complicated by the necessity of using *Bst*XI, an enzyme that is optimally active at high temperature and that generates ambiguous ends. Third, in *E. coli*, Tn*916* is excised and lost at high frequency, so analysis of the potential clone constructed is difficult. Fourth, to introduce the chimeric Tn*916* into *B. subtilis*, the tedious and fastidious protoplast transformation system must be used. Fifth, the site of insertion of the transposon into the chromosome cannot be predicted. For efficient conjugative transfer of the element to the recipient strain of choice, several *B. subtilis* transformants must be screened for donor potential. In addition, complementation analysis might be influenced by the location of the Tn*916*, because it might inactivate an operon into which it inserts or activate an operon to its right because of the strong promoter located at this end of the element (19; see above).

To avoid these problems, Caparon has constructed the pVIT (vector for insertion into Tn*916*) plasmids (11a, 13). These pUC-based vectors contain a fragment of Tn*916* that includes the *Bst*XI site and contains over 1 kb of Tn*916* DNA on each side of the site (Fig. 1). In these vectors (Fig. 6, pVIT130), the *Bst*XI site has been converted to a *Bam*HI site that is unique in the plasmid (Fig. 6). The gene to be cloned (*aphA3* in pVIT121 in Fig. 6) is inserted into this *Bam*HI site along with a marker selectable in *B. subtilis*. The chimeric plasmid is linearized with a restriction enzyme that recognizes a site located exclusively in the pUC portion of the plasmid. The linear DNA is then introduced by transformation of competent cells into a strain of *B. subtilis* containing a single copy of Tn*916* at a known location (Fig. 6, bottom). If further conjugative transfer of the element is desired, the recipient strain should have already been characterized as a high-frequency conjugative transposon donor, and for complementation studies, the recipient strain should have been tested to determine that the resident Tn*916* is at a site that does not affect the phenotype to be investigated.

Transformants in which the inserted DNA replaces

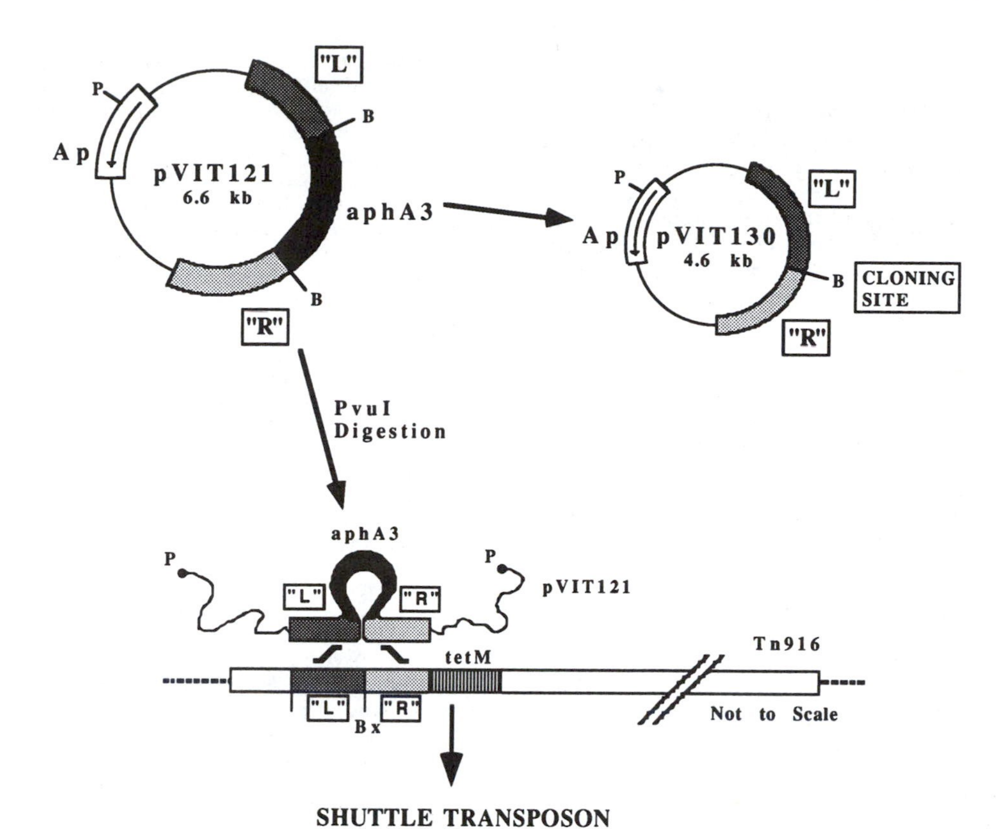

Figure 6. Vectors for construction of chimeric shuttle transposons. The plasmid pVIT121 is a derivative of a pUC plasmid that replicates only in *E. coli*. It contains a DNA fragment that was originally located immediately to the left (Fig. 1) of the *Bst*XI site of Tn*916* (labeled "L" in this figure) and a fragment originally located immediately to the right ("R") of this site. The *aphA3* gene (for kanamycin resistance) is inserted between the L and R fragments. Digestion of pVIT121 with *Bam*HI followed by ligation removed *aphA3* and generated pVIT130, which contains a *Bam*HI site that can be used for insertion of the DNA fragments to be studied. The pVIT vectors can be used to generate a shuttle transposon following the introduction of linearized DNA into a gram-positive host that contains a copy of Tn*916* in its chromosome (pVIT121 linearized by digestion with *Pvu*I is shown at the bottom of the figure). Homologous recombination between the L and R segments of both the pVIT vector and the transposon resident in the chromosome result in integration of the cloned DNA into Tn*916*. The shuttle transposon can then be introduced into the desired host by conjugation. Restriction enzyme sites: B, *Bam*HI; Bx, *Bst*XI; P, *Pvu*I. Ap is the vector *bla* gene encoding ampicillin resistance.

the *Bst*XI site region of the resident transposon by homologous recombination are selected by using the marker cloned into the pVIT vector along with the gene of interest (Fig. 6, bottom; 11a, 13). Crossover events within both homology regions of the linearized Tn*916* chimeric DNA and the resident Tn*916* element in the chromosome generate the recombinant shuttle transposon containing the gene to be studied. This transposon can then be transferred by conjugation to the gram-positive host of choice.

CONCLUSION

Conjugative transposons represent an important class of movable drug resistance elements. Their importance results both from their involvement in the spread of antibiotic resistance among human pathogens and their potential use as molecular biological tools.

Conjugative transposons represent a new insertional mutagen with specificities different from those of other types of transposon, and they should be of especially great importance in gram-positive organisms, for which the variety of insertional mutagens is limited. Although these elements apparently do not insert totally randomly, they do insert at a large number of different sites. Because they cause gram-positive hosts to be conjugational donors and because they show no apparent sensitivity to restriction barriers, their transfer from one gram-positive host to another is technically simple.

For gram-positive organisms in which no electroporation system is yet available, a conjugative transposon may possibly be used as a shuttle vector for analysis of DNA cloned in *E. coli*. The DNA of interest can be cloned into a nonessential site in the transposon, and the chimeric element can then be transformed into a *B. subtilis* strain as an intermediate host. This strain can then act as a conjugational donor to the gram-positive host of choice.

Conjugative transposons are also important as models for basic studies of the mechanisms of genetic recombination. It seems possible at this time that the enzyme(s) involved in excision and insertion of conjugative transposons may have some important fundamental differences from excision enzymes previously studied. These possible differences include the ability to efficiently process coupling sequences containing mismatches in the DNA. Further molecular analysis of the mechanism of transposition of these elements and of the mechanisms of action of what may be a new class of recombination enzyme should therefore lead to a better understanding of the basic steps of recombination. If the rate of development of this relatively new field continues, we will have a much better understanding of conjugative transposons in the near future.

Acknowledgments. I am happy to acknowledge many enjoyable discussions about conjugative transposons with F. Bringel and M. Caparon. I am grateful to M. Caparon for Fig. 1 and 6 and to F. Bringel, M. Caparon, G. Churchward, and C. Moran for critical reading of the manuscript. The cited unpublished work in my laboratory was supported by grant MV-500 from the American Cancer Society.

REFERENCES

1. **Albano, M., and D. A. Dubnau.** 1989. Cloning and characterization of a cluster of Linked *Bacillus subtilis* late competence mutations. *J. Bacteriol.* **171:**5376–5385.
2. **Ayoubi, P., A. O. Kilic, and M. N. Vijayakumar.** 1991. Tn*5253*, the pneumococcal omega (*cat tet*) BM6001 element, is a composite structure of two conjugative transposons, Tn*5251* and Tn*5252*. *J. Bacteriol.* **173:**1617–1622.
3. **Bauer, C. E., J. F. Gardner, R. I. Gumport, and R. A. Weisberg.** 1989. The effect of attachment site mutations on strand exchange in bacteriophage lambda site-specific recombination. *Genetics* **122:**727–735.
4. **Bentorche, F., G. De Cespedes, and T. Horaud.** 1991. Tetracycline resistance heterogeneity in *Enterococcus faecium*. *Antimicrob. Agents Chemother.* **35:**808–812.
5. **Bertram, J., A. Kuhn, and P. Durre.** 1990. Tn*916*-induced mutants of *Clostridium acetobutylicum* defective in regulation of solvent formation. *Arch. Microbiol.* **153:**373–377.
6. **Bertram, J., M. Stratz, and P. Durre.** 1991. Natural transfer of conjugative transposon Tn*916* between gram-positive and gram-negative bacteria. *J. Bacteriol.* **173:** 443–448.

6a. **Bringel, F.** Unpublished data.

6b. **Bringel, F., G. L. Van Alstine, M. G. Caparon, and J. R. Scott.** Unpublished data.

7. **Bringel, F., G. L. Van Alstine, and J. R. Scott.** 1991. A host factor absent from *Lactococcus lactis* subspecies *lactis* MG1363 is required for conjugative transposition. *Mol. Microbiol.* **5:**2983–2993.

7a. **Bringel, F., G. L. Van Alstine, and J. R. Scott.** 1992. Conjugative transposition of Tn*916*: the transposon *int* gene is required only in the donor. *J. Bacteriol.* **174:**4036–4041.

7b. **Bringel, F., G. L. Van Alstine, and J. R. Scott.** 1992. Transfer of Tn*916* between *Lactococcus lactis* subsp. *lactis* strains is nontranspositional: evidence for a chromosomal fertility function in strain MG1363. *J. Bacteriol.* **174:**5840–5847.

7c. **Bringel, F., G. L. Van Alstine, and J. R. Scott.** Unpublished data.

8. **Burdett, V., J. Inamine, and S. A. Rajagopalan.** 1982. Heterogeneity of tetracycline resistance determinants in *Streptococcus*. *J. Bacteriol.* **149:**995–1004.
9. **Buu-Hoi, A., and T. Horaud.** 1985. Genetic basis of antibiotic resistance in group A, C, and G streptococci, p. 231–232. *In* Y. Kimura, S. Kotami, and Y. Shiokawa (ed.), *Recent Advances in Streptococci and Streptococcal Diseases*. Reedbooks, Bracknell, England.
10. **Buu-Hoi, A., and T. Horodniceanu.** 1980. Conjugative transfer of multiple antibiotic resistance markers of *Streptococcus pneumoniae*. *J. Bacteriol.* **143:**313–320.
11. **Caillaud, F., and P. Courvalin.** 1987. Nucleotide sequence of the ends of the conjugative shuttle transposon Tn*1545*. *Mol. Gen. Genet.* **209:**110–115.

11a. **Caparon, M. G.** Unpublished data.

12. **Caparon, M. G., and J. R. Scott.** 1989. Excision and insertion of the conjugative transposon Tn*916* involves a novel recombination mechanism. *Cell* **59:**1027–1034.
13. **Caparon, M. G., and J. R. Scott.** 1991. Genetic manipulation of pathogenic streptococci. *Methods Enzymol.* **204:** 556–586.

13a. **Caparon, M. G., and J. R. Scott.** Unpublished data.

14. **Carlier, C., and P. Courvalin.** 1982. Resistance of streptococci to aminoglycoside-aminocyclitol antibiotics, p. 162–166. *In* D. Schlessinger (ed.), *Microbiology—1982*. American Society for Microbiology, Washington, D.C.
15. **Claverys, J. P., H. Prats, H. Vasseghi, and M. Gherardi.** 1984. Identification of *Streptococcus pneumoniae* mismatch repair genes by an additive transformation approach. *Mol. Gen. Genet.* **196:**91–96.
16. **Clewell, D. B.** 1990. Movable genetic elements and antibiotic resistance in enterococci. *Eur. J. Clin. Microbiol. Infect. Dis.* **9:**90–102.
17. **Clewell, D. B., G. F. Fitzgerald, L. Dempsey, L. E. Pearce, F. Y. An, B. A. White, Y. Yagi, C. Gawron-Burke.** 1985. Streptococcal conjugation: plasmids, sex pheromones, and conjugative transposons, p. 194–203. *In* S. E. Mergenhagen and B. Rosan (ed.), *Molecular Basis of Oral Microbial Adhesion*. American Society for Microbiology, Washington, D.C.
18. **Clewell, D. B., and S. E. Flannagan.** The conjugative transposons of gram positive bacteria. *In* D. B. Clewell (ed.), *Bacterial Conjugation*, in press. Plenum Press, New York.
19. **Clewell, D. B., S. E. Flannagan, Y. Ike, J. M. Jones, and C. Gawron-Burke.** 1988. Sequence analysis of termini of conjugative transposon Tn*916*. *J. Bacteriol.* **170:**3046–3052.
20. **Clewell, D. B., S. E. Flannagan, L. A. Zitzow, Y. A. Su, P. He, E. Senghas, and K. E. Weaver.** 1991. Properties of conjugative transposon Tn*916*, p. 39–44. *In* G. M. Dunny, P. P. Cleary, and L. L. McKay (ed.), *Genetics and Molecular Biology of Streptococci, Lactococci, and Enterococci*. American Society for Microbiology, Washington, D.C.
21. **Clewell, D. B., and C. Gawron-Burke.** 1986. Conjugative transposons and the dissemination of antibiotic resistance in streptococci. *Annu. Rev. Microbiol.* **40:**635–659.
22. **Courvalin, P., and C. Carlier.** 1986. Transposable multiple antibiotic resistance in *Streptococcus pneumoniae*. *Mol. Gen. Genet.* **205:**291–297.
23. **Courvalin, P., and C. Carlier.** 1987. Tn*1545*: a conjugative shuttle transposon. *Mol. Gen. Genet.* **206:**259–264.
24. **Flannagan, S. E., and D. B. Clewell.** 1991. Conjugative transfer of Tn*916* in *Enterococcus faecalis*: *trans*-activation of homologous transposons. *J. Bacteriol.* **173:**7136–7141.
25. **Franke, A. E., and D. B. Clewell.** 1981. Evidence for a chromosome-borne resistance transposon (Tn*916*) in *Streptococcus faecalis* that is capable of conjugal transfer in the absence of a conjugative plasmid. *J. Bacteriol.* **145:**494–502.
26. **Gaillard, J.-L., P. Berche, C. Frehel, E. Gouin, and P. Cossart.** 1991. Entry of *L. monocytogenes* into cells is mediated by internalin, a repeat protein reminiscent of surface antigens from gram-positive cocci. *Cell* **65:**1127–1141.
27. **Gaillard, J. L., P. Berche, and P. Sansonetti.** 1986. Transposon mutagenesis as a tool to study the role of hemolysin in the virulence of *Listeria monocytogenes*. *Infect. Immun.* **52:**50–55.
28. **Gasson, M.** 1983. Plasmid complements of *Streptococcus*

lactis NCDO 712 and other lactic streptococci after protoplast-induced curing. *J. Bacteriol.* **154:**1–9.

29. **Gawron-Burke, C., and D. B. Clewell.** 1982. A transposon in *Streptococcus faecalis* with fertility properties. *Nature* (London) **300:**281–284.
30. **Gawron-Burke, C., and D. B. Clewell.** 1984. Regeneration of insertionally inactivated streptococcal DNA fragments after excision of transposon Tn*916* in *Escherichia coli*: strategy for targeting and cloning genes from gram-positive bacteria. *J. Bacteriol.* **159:**214–221.
31. **Guild, W. R., M. D. Smith, and N. B. Shoemaker.** 1982. Conjugative transfer of chromosomal R determinants in *Streptococcus pneumoniae*, p. 88–92. *In* D. Schlessinger (ed.), *Microbiology—1982*. American Society for Microbiology, Washington, D.C.
32. **Halula, M., and F. L. Macrina.** 1990. Tn*5030*: a conjugative transposon conferring clindamycin resistance in Bacteriodes species. *Rev. Infect. Dis.* **12:**S235–S242.
33. **Horaud, T., G. De Cespedes, D. Clermont, F. David, and F. Delbos.** 1991. Variability of chromosomal genetic elements in streptococci, p. 16–20. *In* G. M. Dunny, P. P. Cleary, and L. L. McKay (ed.), *Genetics and Molecular Biology of Streptococci, Lactococci, and Enterococci*. American Society for Microbiology, Washington, D.C.
34. **Horodniceanu, T., L. Bougueleret, and G. Bieth.** 1981. Conjugative transfer of multiple-antibiotic resistance markers in beta-hemolytic group A, B, F, and G streptococci in the absence of extrachromosomal deoxyribonucleic acid. *Plasmid* **5:**127–187.
35. **Horodniceanu, T., C. Le Bouguenec, A. Buu-Hoi, and G. Bieth.** 1982. Conjugative transfer of antibiotic resistance in beta-hemolytic streptococci in the presence and absence of plasmid DNA, p. 105–108. *In* D. Schlessinger (ed.), *Microbiology—1982*. American Society for Microbiology, Washington, D.C.
36. **Iida, S., M. B. Streiff, T. A. Bickle, and W. Arber.** 1987. Two DNA antirestriction systems of bacteriophage P1 *dar*A, and *dar*B: characterization of *dar*A$^-$ phages. *Virology* **157:**156–166.
37. **Ippen-Ihler, K. A., and E. G. Minkley, Jr.** 1986. The conjugation system of F, the fertility factor of *Escherichia coli*. *Annu. Rev. Genet.* **20:**593–624.
38. **Jacob, A. E., and S. J. Hobbs.** 1974. Conjugal transfer of plasmid-borne multiple antibiotic resistance in *Streptococcus faecalis* var. *zymogenes*. *J. Bacteriol.* **117:**360–372.
39. **Kahmann, R., and S. Hattman.** 1987. Regulation and expression of the *mom* gene, p. 93–109. *In* N. Symonds, A. Toussaint, P. van de Putte, and M. M. Howe (ed.), *Phage Mu*. Cold Spring Harbor Laboratory, Cold Spring Harbor, N.Y.
40. **Kathariou, S., P. Metz, H. Hof, and W. Gobel.** 1987. Tn*916*-induced mutations in the hemolysin determinant affecting virulence of *Listeria monocytogenes*. *J. Bacteriol.* **169:**1291–1297.
41. **Kathariou, S., D. S. Stephens, P. Spellman, and S. A. Morse.** 1990. Transposition of Tn*916* to different sites in the chromosome of *Neisseria meningitidis*: a genetic tool for meningococcal mutagenesis. *Mol. Microbiol.* **4:**729–735.
42. **Landy, A.** 1989. Dynamic, structural and regulatory aspect of lambda site-specific recombination. *Annu. Rev. Biochem.* **58:**913–949.
43. **Le Bouguenec, C., G. De Cespedes, and T. Horaud.** 1988. Molecular analysis of a composite chromosomal conjugative element (Tn*3701*) of *Streptococcus pyogenes*. *J. Bacteriol.* **170:**3930–3936.
44. **Le Bouguenec, C., G. De Cespedes, and T. Horaud.** 1990. Presence of chromosomal elements resembling the composite structure Tn*3701* in streptococci. *J. Bacteriol.* **172:**727–734.
45. **Mengaud, J., J. Chenevert, C. Geoffroy, J. L. Gaillard, and P. Cossart.** 1987. Identification of the structural gene encoding the SH-activated hemolysin of *Listeria monocytogenes*: listeriolysin O is homologous to streptolysin O and pneumolysin. *Infect. Immun.* **55:**3225–3227.
46. **Mullany, P., M. Wilks, and S. Tabaqchali.** 1991. Transfer of Tn*916* and Tn*916*deltaE into *Clostridium difficile*: demonstration of a hot-spot for these elements in the *C. difficile* genome. *FEMS Microbiol. Lett.* **79:**191–194.
47. **Naglich, J. G., and R. E. Andrews.** 1988. Tn*916*-dependent conjugal transfer of PC194 and PUB110 from *Bacillus subtilis* into *Bacillus thuringiensis* subsp. *israelensis*. *Plasmid* **20:**113–126.
48. **Nassif, X., D. Puaoi, and M. So.** 1991. Transposition of Tn*1545*-Δ3 in the pathogenic neisseriae: a genetic tool for mutagenesis. *J. Bacteriol.* **173:**2147–2154.
49. **Norgren, M., M. G. Caparon, and J. R. Scott.** 1989. A method for allelic replacement that uses the conjugative transposon Tn*916*: deletion of the *emm*6.1 allele in *Streptococcus pyogenes* JRS4. *Infect. Immun.* **57:**3846–3850.
50. **Norgren, M. G., and J. R. Scott.** 1991. Presence of the conjugative transposon Tn*916* in the recipient strain does not impede transfer of a second copy of the element. *J. Bacteriol.* **173:**319–324.
51. **Odelson, D. A., J. L. Rasmussen, C. J. Smith, and F. L. Macrina.** 1987. Extrachromosomal systems and gene transmission in anaerobic bacteria. *Plasmid* **17:**87–109.
51a. **Perez-Casal, J., M. G. Caparon, and J. R. Scott.** 1992. Introduction of the *emm*6 gene into an *emm*-deleted strain of *Streptococcus pyogenes* restores its ability to resist phagocytosis. *Res. Microbiol.* **143:**549–558.
52. **Poyart-Salmeron, C., P. Trieu-Cuot, C. Carlier, and P. Courvalin.** 1989. Molecular characterization of two proteins involved in the excision of the conjugative transposon Tn*1545*: homologies with other site-specific recombinases. *EMBO J.* **8:**2425–2433.
53. **Poyart-Salmeron, C., P. Trieu-Cuot, C. Carlier, and P. Courvalin.** 1990. The integration-excision system of the conjugative transposon Tn*1545* is structurally and functionally related to those of lambdoid phages. *Mol. Microbiol.* **4:**1513–1521.
54. **Rothman, J. L.** 1965. Transduction studies on the relation between prophage and host chromosome. *J. Mol. Biol.* **12:**892–912.
55. **Salyers, A. A., N. B. Shoemaker, and E. P. Guthrie.** 1987. Recent advances in *Bacteroides* genetics. *Crit. Rev. Microbiol.* **14:**49–71.
56. **Scott, J. R.** 1991. Mechanism of transposition of conjugative transposons, p. 28–33. *In* G. M. Dunny, P. P. Cleary, and L. L. McKay (ed.), *Genetics and Molecular Biology of Streptococci, Lactococci, and Enterococci*. American Society for Microbiology, Washington, D.C.
56a. **Scott, J. R.** 1992. Sex and the single circle: conjugative transposition. *J. Bacteriol.* **174:**6005–6010.
56b. **Scott, J. R., M. G. Caparon, and M. Norgren.** Unpublished data.
57. **Scott, J. R., P. A. Kirchman, and M. G. Caparon.** 1988. An intermediate in transposition of the conjugative transposon Tn*916*. *Proc. Natl. Acad. Sci. USA* **85:**4809–4813.
58. **Senghas, E., J. M. Jones, M. Yamamoto, C. Gawron-Burke, and D. B. Clewell.** 1988. Genetic organization of the bacterial conjugative transposon Tn*916*. *J. Bacteriol.* **170:**245–249.
59. **Shoemaker, N. B., M. D. Smith, and W. R. Guild.** 1979. Organization and transfer of heterologous chloramphenicol and tetracycline resistance genes in pneumococcus. *J. Bacteriol.* **139:**432–441.
60. **Shoemaker, N. B., M. D. Smith, and W. R. Guild.** 1980. DNase-resistant transfer of chromosomal *cat* and *tet* insertions by filter mating in pneumococcus. *Plasmid* **3:**80–87.
61. **Smith, M. D., S. Hazum, and W. R. Guild.** 1981. Homology among *tet* determinants in conjugative elements of streptococci. *J. Bacteriol.* **148:**232–240.
62. **Stephens, D. S., J. S. Swartley, S. Kathariou, and S. A.**

Morse. 1991. Insertion of Tn*916* in *Neisseria meningitidis* resulting in loss of group B capsular polysaccharide. *Infect. Immun.* **59:**4097–4102.

63. **Storrs, M. J., C. Poyart-Salmeron, P. Trieu-Cuot, and P. Courvalin.** 1991. Conjugative transposition of Tn*916* requires the excisive and integrative activities of the transposon-encoded integrase. *J. Bacteriol.* **173:**4347–4352.
64. **Tally, F. P., M. J. Shimell, G. R. Carson, and M. H. Malamy.** 1981. Chromosomal and plasmid-mediated transfer of clindamycin resistance in *Bacteroides fragilis*, p. 51. *In* S. B. Levy and R. C. Clowes (ed.), *Molecular Biology, Pathogenicity and Ecology of Bacterial Plasmids*. Plenum Press, New York.
65. **Torres, O. R., R. Z. Korman, S. A. Zahler, and G. M. Dunny.** 1991. The conjugative transposon Tn*925*: enhancement of conjugal transfer by tetracycline in *Enterococcus faecalis* and mobilization of chromosomal genes in *Bacillus subtilis* and *E. faecalis*. *Mol. Gen. Genet.* **225:**395–400.
66. **Trieu-Cuot, P., C. Poyart-Salmeron, C. Carlier, and P. Courvalin.** 1991. Molecular dissection of the transposition mechanism of conjugative transposons from gram-positive cocci, p. 21–27. *In* G. M. Dunny, P. P. Cleary, and L. L. McKay (ed.), *Genetics and Molecular Biology of Streptococci, Lactococci, and Enterococci*. American Society for Microbiology, Washington, D.C.
67. **Vijayakumar, M. N., P. Ayoubi, and A. O. Kilic.** 1991. Organization of Tn*5253*, the pneumococcal Ω (*cat tet*) BM6001 element, p. 49–53. *In* G. M. Dunny, P. P. Cleary, and L. L. McKay (ed.), *Genetics and Molecular Biology of Streptococci, Lactococci, and Enterococci*. American Society for Microbiology, Washington, D.C.
68. **Vijayakumar, M. N., S. D. Priebe, G. Pozzi, J. M. Hageman, and W. R. Guild.** 1986. Cloning and physical characterization of chromosomal conjugative elements in streptococci. *J. Bacteriol.* **166:**972–977.
69. **Willetts, N., and B. Wilkins.** 1984. Processing of plasmid DNA during bacterial conjugation. *Microbiol. Rev.* **48:** 24–41.
70. **Yansura, D. C., and D. J. Henner.** 1984. Use of the *Escherichia coli lac* repressor and operator to control gene expression in *Bacillus subtilis*. *Proc. Natl. Acad. Sci. USA* **81:**439–443.
71. **Youngman, P., J. B. Perkins, and K. Sandman.** 1985. Use of Tn*917*-mediated transcriptional gene fusions to *lacZ* and *cat-86* for the identification and study of *spo* genes in *Bacillus subtilis*, p. 47–54. *In* J. A. Hoch and P. Setlow (ed.), *Molecular Biology of Microbial Differentiation* 1985. American Society for Microbiology, Washington, D.C.

42. Integrational Vectors for Genetic Manipulation in *Bacillus subtilis*

MARTA PEREGO

INTEGRATIONAL VECTORS

The development of integrational vectors has provided a versatile system for the advance of molecular genetic and mutagenic studies in *Bacillus subtilis*. Over the past decade, the identification of genes of interest and the subsequent study of their structures and regulation have been greatly facilitated by the application of techniques for using integrational vectors. This chapter will describe these techniques and the strategies used and vectors devised to investigate the mechanisms of gene function and regulation in *B. subtilis*.

First Generation

One distinctive biological property of *B. subtilis* competent cells is the highly recombinogenic nature of DNA uptake. Once introduced into the cell by transformation, DNA fragments carrying regions of the *B. subtilis* chromosome can recombine with the host chromosome; the site of recombination is determined by the region of homology shared by the fragment and the *B. subtilis* chromosome. In fact, it was shown that heterologous DNA flanked on both sides by sequences homologous with the *B. subtilis* chromosome inserts efficiently into the *B. subtilis* chromosome (14). Heterologous DNA carried on plasmids was also shown to integrate into the *B. subtilis* chromosome when linked to sequences homologous to the chromosome (6). In one instance, a construct like this was used as a "mapping vehicle" for cloned *B. subtilis* DNA (13). This recombinant plasmid was unable to replicate in *B. subtilis* but was able to integrate into the chromosome by recombination at the site of homology with the cloned DNA inserted into the plasmid vector. Furthermore, such a construct carried a chloramphenicol resistance determinant that provided the selective marker in transformation and/or transduction experiments.

On the basis of these findings, some integrational vectors carrying additional features were constructed. The first generally useful integrational vectors described were plasmids pHV32 (32) and pJH101 (9) (Fig. 1). They are both pBR322 derivatives that carry the ColE1 replication functions, the β-lactamase gene, and the tetracycline resistance gene (2). Both these antibiotic resistances are expressed only in *Escherichia coli*. In addition, pHV32 and pJH101 each carry a *cat* gene of gram-positive origin obtained from *Staphylococcus aureus* plasmid pC194 (17) that confers chloramphenicol resistance on both *E. coli* and *B. subtilis*. The presence in these plasmids of unique restriction sites allows the easy cloning of DNA fragments, and the detection of insertion can be monitored by disruption of the ampicillin or tetracycline resistance markers.

pHV32 and pJH101 do not contain origins of replication for *B. subtilis* and cannot replicate in this host. For this reason, they cannot transform *B. subtilis* competent cells to chloramphenicol resistance. However, if any DNA fragment providing a region of homology with the *B. subtilis* chromosome is inserted into these plasmids, they acquire the ability to transform *B. subtilis* competent cells to chloramphenicol resistance at high frequency. This cloned fragment provides the basis for the recombinational integration of the entire plasmid into the chromosome. The integration occurs via a Campbell-type recombination event that results in a duplication of the cloned region at either end of the vector (3) (Fig. 2). Reversal of the integration event is a less efficient process that occurs at a frequency of 10^{-4} to 10^{-5} per bacterial generation (46, 49) and leads to precise excision of the plasmid and restoration of gene structure on the chromosome.

Improved Vectors (Second Generation)

Several modifications have been introduced into pJH101 in order to improve vector performance. For example, plasmid pJAB1 (37) is a chimera of the positive cloning vectors pTR262 (38) and pJH101. Cloning in its unique *Bcl*I or *Hin*dIII sites inactivates the lambda repressor, thus allowing expression of tetracycline resistance and positive selection of recombinant clones. Additional drug resistance genes that can be selected for in single copy can be useful for genetic manipulation in *B. subtilis*. Thus, introduction of the *erm* gene for macrolide-lincosamide-streptogramin B resistance in pJH101 was described by Kenney and Moran (24).

In more recent years, a new series of integrational vectors has been developed by several laboratories. Added improvements include such features as polylinkers, detection of inserts by disruption of the *lac* alpha complementation, origins of replication for coliphage M13 or f1 to produce single-stranded DNA, and T7 or SP6 promoter sites for generating in vitro transcripts through cloned inserts.

Marta Perego • Istituto di Tecnica Farmaceutica, Università degli Studi di Parma, Via M. D'Azeglio 85, 43100 Parma, Italy, and Dipartimento di Genetica e Microbiologia, Università degli Studi di Pavia, Via Abbiategrasso 207, 27100 Pavia, Italy.

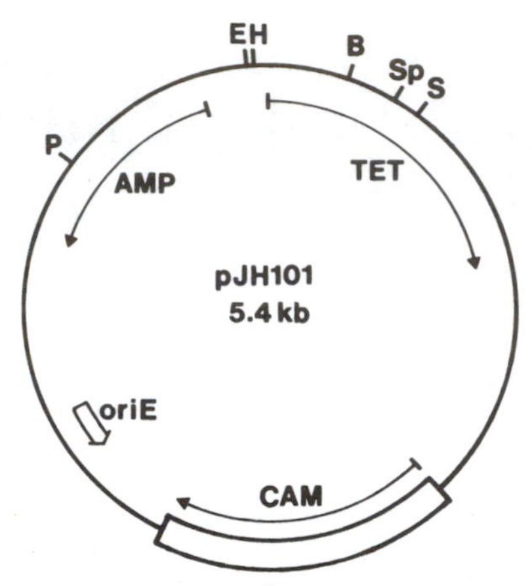

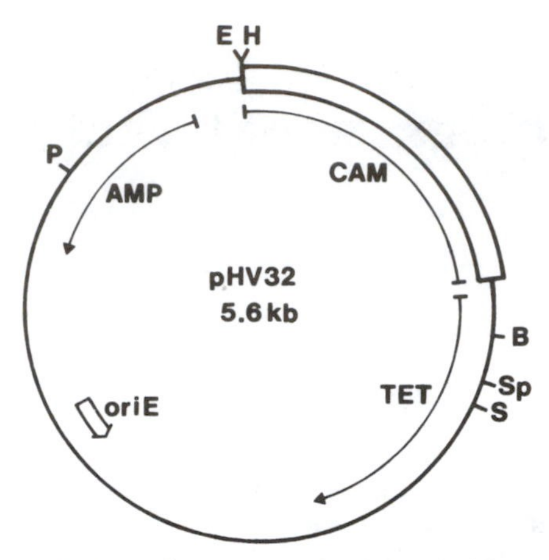

Figure 1. First generation of integrative vectors for *B. subtilis*. pJH101 and pHV32 are both derived from pBR322. The fragment carrying the *cat* gene is from pC194 (17). Restriction sites: B, *Bam*HI; E, *Eco*RI; H, *Hin*dIII; P, *Pst*I; S, *Sal*I; Sp, *Sph*I. *oriE* is the ColE1 origin of replication for *E. coli*. AMP, ampicillin; TET, tetracycline; CAM, chloramphenicol.

Extending the tradition started with pJH101, two plasmids, pJM102 and pJM103 (34a), were constructed in Hoch's laboratory (Fig. 3). They are derivatives of pUC18 and pUC19, respectively, and carry the *cat* gene from pC194. Insertion of fragments in pJM102 and pJM103 can be detected by inactivation of the Lac phenotype in *E. coli*. The presence of the pUC18 or pUC19 multiple cloning site (MCS) facilitates cloning into the vector and excision of the vector and adjacent sequences from the chromosome (see below). Two more plasmids, pJM112 and pJM113, were constructed by using a different antibiotic resistance gene (34a) (Fig. 3). They are, respectively, pUC18 and pUC19 derivatives carrying the kanamycin resistance gene originally found in the *Streptococcus faecalis* plasmid pJH1 (45). pJM112 and pJM113 confer resistance to 20 μg of kanamycin per ml in *E. coli* and 2 μg/ml in *B. subtilis*. Similarly, plasmid pSGMU2 (11) (Fig. 3) is based on the *E. coli* plasmid pUC13 (28), into which the original chloramphenicol resistance marker from pC194 has been inserted in the *Nar*I site.

Plasmid pBG6 (50) (Fig. 4) carries an additional feature: it contains the M13 origin of replication that allows production of single-stranded DNA virions and so facilitates additional manipulations such as generation of unidirectional deletions or oligonucleotide-mediated mutagenesis and characterization by sequencing with the M13 universal primer. The same advantages were offered by another kind of integrational vector completely based on the single-stranded filamentous coliphage M13. Vector M13mp19cat (12) is an M13mp19 phage vector carrying the *cat* gene originally from pC194 inserted in the *Ava*II site. Because the *Ava*II site in M13mp19 is in nonessential DNA, the construct is fully viable in *E. coli*, it produces blue plaques in the presence of 5-bromo-4-chloro-3-indolyl-β-D-galactopyranoside (X-Gal), and the *cat* gene does not impair phage growth or stability of inserts cloned into the phage. In order to transform *B. subtilis* competent cells, the double-stranded RFI form of this vector is prepared from *E. coli*-infected cells and used as conventional plasmid-derived vectors.

The ability to generate in vitro transcripts through cloned inserts has been introduced in the pGEM-3Zf(+)cat-1 vector (50). This vector is a derivative of the Promega pGEM-3Zf(+) phagemid that contains the PT7 promoter for the *E. coli* T7 phage and the PSP6 promoter for *Salmonella typhimurium* phage SP6 flanking the pUC19 polylinker. This introduction makes possible the synthesis of RNA corresponding to either the coding or the noncoding strand of cloned DNA from a single plasmid construct. pGEM-3Zf(+)cat-1 also contains the f1 single-stranded phage origin of replication, which allows the induction of single-stranded DNA when bacterial cells containing the recombinant plasmid are infected with an appropriate helper phage (5). The remainder of the plasmid is pUC19.

A *B. subtilis* integrational vector that can be maintained at low or high copy number in *E. coli* has been described by Hasnain and Thomas (15). pCT571 confers Km^r, Tc^r, and Cm^r in *E. coli* and Cm^r in *B. subtilis*. It has seven unique restriction sites that allow the screening of recombinant clones by insertional inactivation of the Km^r or Tc^r markers. It contains the

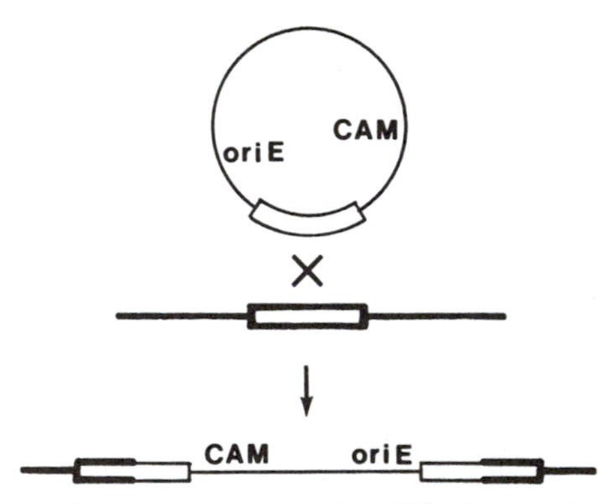

Figure 2. Schematic representation of the integrative recombination event by the Campbell-type mechanism, which results in duplication of the region cloned into the plasmid. Bold lines represent the chromosome, and thin lines represent the plasmid; the open box represents the region of homology between plasmid and chromosome. CAM, chloramphenicol.

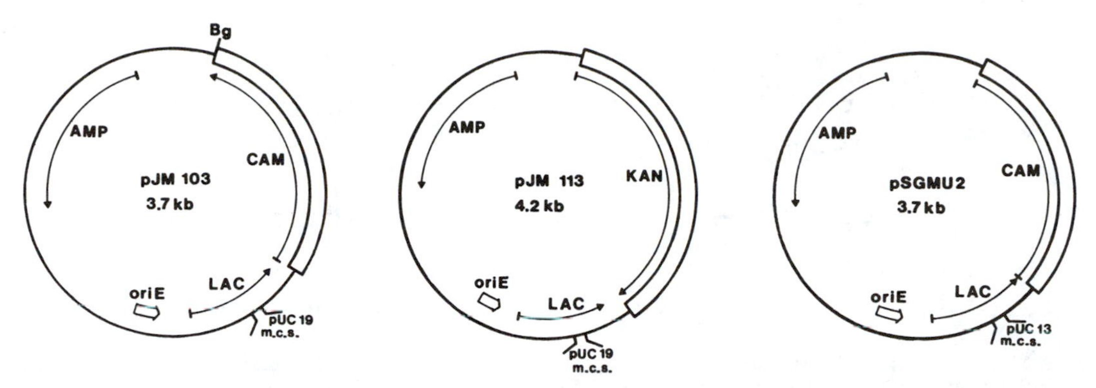

Figure 3. Second generation of integrative vectors. pJM103 and pJM113 (34a) are derivatives of pUC19 (47), while pSGMU2 (11) derives from pUC13 (28). The fragment carrying the *cat* gene is from pC194, and the one carrying the kanamycin resistance gene is from pJH1 (45). Plasmids pJM102 and pJM112, described in the text, differ from pJM103 and pJM113, respectively, in that they are derivatives of pUC18. The pUC18-19 MCS contains the following restriction sites: *Eco*RI, *Sst*I, *Kpn*I, *Sma*I, *Xma*I, *Bam*HI, *Xba*I, *Sal*I, *Acc*I, *Hinc*II, *Pst*I, *Sph*I, and *Hin*dIII. The pUC13 MCS, compared with the pUC18-19 MCS, is missing the *Kpn*I and *Sph*I sites. Bg, *Bgl*II. AMP, ampicillin; CAM, chloramphenicol; KAN, kanamycin.

origin of replication of pSC101 (42) and replicates normally at six to eight copies per chromosome in *E. coli*. It also contains the oriVRK2 origin of replication (43), which when supplied with the product of the *trfA* gene of RK2 in *trans*, allows pCT571 to replicate at 35 to 40 copies per chromosome. This plasmid is particularly useful for the construction of *B. subtilis* genomic libraries and the cloning of genes whose products at high copy numbers are deleterious to *E. coli*.

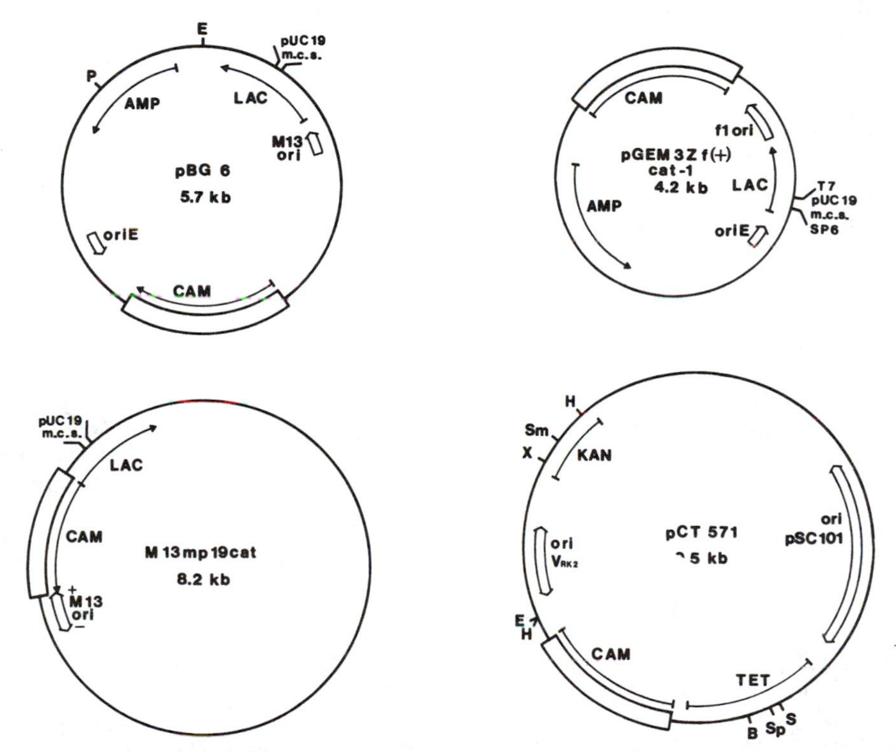

Figure 4. Integrative vectors carrying additional features. pBG6 (50) derives from pJH101, in which the *Cla*I-*Bal*I region has been replaced by the *Cla*I-*Bal*I fragment from M13mp19 that contains the M13 origin of replication, the *lacZ* gene, and the MCS. M13mp19cat contains the *cat* gene from pC194 in the *Ava*II site of M13mp19. pGEM-3Zf(+)cat-1 (50) was obtained from pGEM-3Zf(+) (Promega) by inserting a 1-kb fragment containing the *cat* gene originally associated with pC194 into the unique *Nde*I site filled in with Klenow polymerase; f1 ori is the f1 phage origin for single-stranded replication; PT7 is the promoter for *E. coli* phage T7, while PSP6 is the promoter for *S. typhimurium* phage SP6. pCT571 (15) is a low- to high-copy-number integrative vector: the low-copy-number state is controlled by the pSC101 replicon, while oriVRK2, when supplied with the *trfA* gene in *trans*, allows the plasmid to replicate at an elevated copy number. For the pUC19 MCS, see Fig. 3. X, *Xho*I; Sm, *Sma*I. All other restriction sites are as in Fig. 1. AMP, ampicillin; CAM, chloramphenicol; KAN, kanamycin; TET, tetracycline.

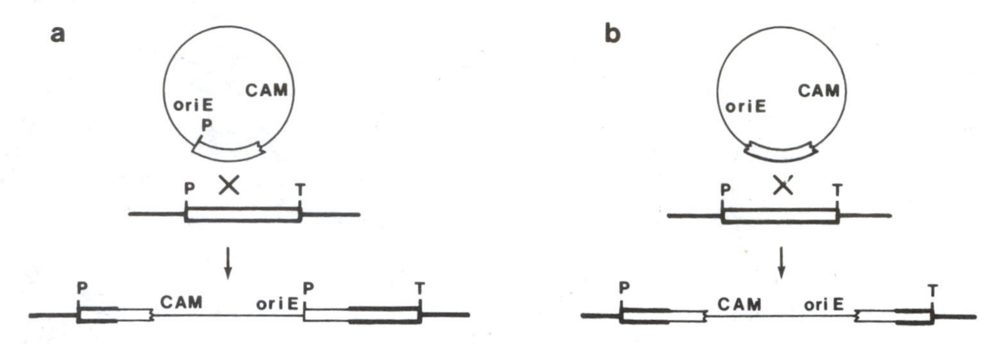

Figure 5. Schematic representation of the use of integrative vectors in gene function analysis. The boundaries of the transcriptional unit are indicated by P for promoter and T for terminus. (a) When the cloned region contains one end of the transcriptional unit, integration gives rise to a heterologous but functional unit. (b) When the cloned region is internal to the transcriptional unit, integration results in gene disruption. See also the legend to Fig. 2. CAM, chloramphenicol.

Recently, more drug resistance markers that can be selected for in single copy in *B. subtilis* have been described. One is the gene from the *S. aureus* plasmid pUB110 that confers resistance to antibiotics such as bleomycin (Bm) or phleomycin (Pm) in both *E. coli* and *B. subtilis* (39). A plasmid carrying the bleomycin resistance gene in the pUC18 polylinker pUC18-ble-1 has been also described by Youngman et al. (50). A second antibiotic gene derives from the pUB110 neomycin resistance marker whose promoter has been modified so that Nm^r transformants can be isolated after integration of plasmid pBEST501 or pBEST502 in a single copy into the *B. subtilis* chromosome (18). The last single-copy resistance gene described for *B. subtilis* is the blasticidin S gene (21). These bring to six the number of drug resistance genes available for genetic manipulation of the *B. subtilis* chromosome by means of integrational vectors.

Applications in Gene Mapping and Functional Analysis

Integrational vectors have proved useful for a wide range of applications in *B. subtilis* molecular genetic studies. The first vectors, pJH101 and pHV32, were described as valuable tools for mapping studies. The chromosomal locations of cloned DNA fragments with no inherent phenotypes can be analyzed with relative ease. In fact, since the plasmid contains an expressible gene for antibiotic resistance, once the plasmid becomes integrated in the region of homology with the *B. subtilis* chromosome, it can be located in genetic crosses by the resistance phenotype (8, 13).

As previously mentioned, integration of the recombinant plasmid into the chromosome by the single-crossover Campbell-type mechanism gives rise to a nontandem duplication of the cloned region. When the cloned fragment contains a functional transcription unit of the gene to be analyzed, then complementation and dominance analysis can be undertaken, with the advantage over the use of replicative vectors that it is possible to analyze the effect of only one extra copy of the gene instead of an excess of one of the two alleles (7).

The discovery that in *B. subtilis*, as in several prokaryotes, the duplicated region flanking the Campbell-type integrated plasmid represents an amplifiable structure (1, 23) has led to the development of procedures for evaluating the effects of copy number on gene function and interactions (4). Starting from the nontandem duplication obtained by single-crossover integration, amplification is easily achieved in a $RecE^+$ background by growing the bacteria in increasing concentrations of the antibiotic to which the plasmid confers resistance. Amplification occurs only for the sequence present on the transforming plasmid, which means that a gene dosage effect can be expected only if a functional transcriptional unit is present on the cloned fragment. When the integrational plasmid carries only a portion of a transcriptional unit, then integration necessarily interrupts that unit. If the vector contains either boundary of the transcriptional unit, then after integration, a complete new transcriptional unit consisting of cloned DNA recombined with chromosomal DNA is created (Fig. 5a). On the other hand, integration of vectors containing fragments internal to the transcriptional unit results in disruption of the unit and loss of function, possibly producing a mutant phenotype. This phenomenon is called integrational mutagenesis (Fig. 5b). Therefore, this integration mechanism allows definition of the functional boundaries of a cloned gene or operon as first described by Piggot et al. (37).

Integrative gene disruption is a powerful tool for the functional analysis of cloned genes. The method described above, which uses an integrational vector that includes sequences internal to the gene, may sometimes have the disadvantage of being unstable unless antibiotic selection is maintained in order to avoid reversion by Campbell excision. However, permanent chromosomal deletions can be constructed by following the strategy described by Stahl and Ferrari (41). This involves, first, in vitro construction of a gene deletion in an integrational vector. The deletion-containing plasmid is then inserted into the chromosome by homologous recombination with selection for antibiotic resistance. Transformants are analyzed in the absence of selective pressure for the generation of antibiotic-sensitive segregants resulting from a Campbell excision, which may leave behind a permanent deletion of the gene.

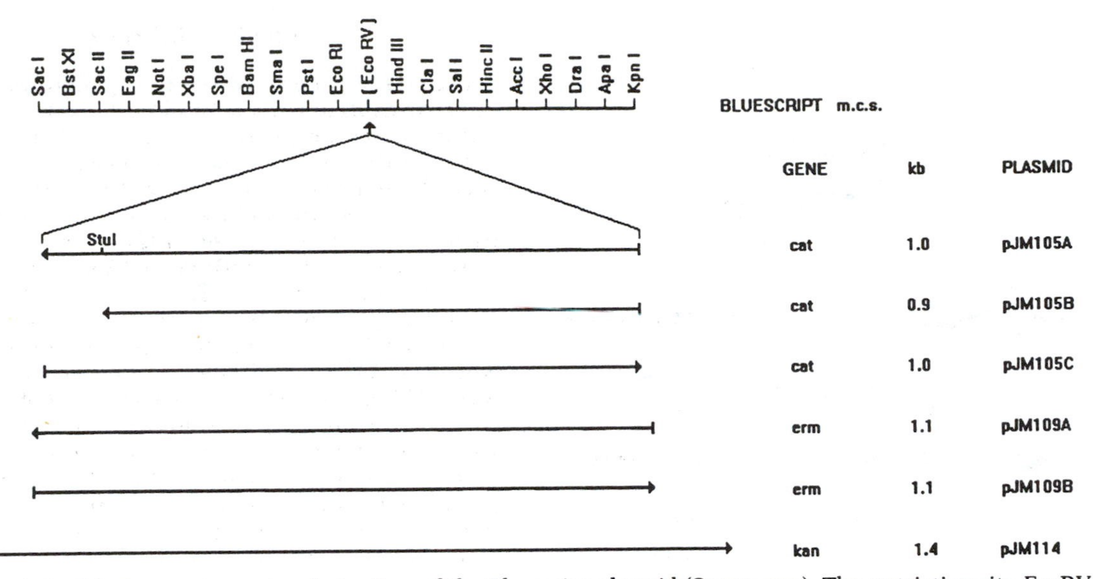

Figure 6. Antibiotic cassette vector derivatives of the Bluescript plasmid (Stratagene). The restriction site *Eco*RV used to construct the cassettes is pointed out by the arrow and is in parentheses to indicate its loss, which was due to the cloning. Directions of transcription of the antibiotic genes are indicated by arrows. The *cat* gene in pJM105A and pJM105C was a *Sau*3A-*Hpa*II fragment from pC194 blunted with Klenow polymerase. The truncated *cat* gene in pJM105B was obtained as *Hin*dIII-*Stu*I fragment from pJM105A recloned in a *Hin*dIII-*Eco*RV-cut Bluescript plasmid. The macrolide-lincosamide-streptogramin B resistance gene *ermG* was a *Hae*III-*Bst*NI Klenow-blunted fragment from pBD370 (30). The kanamycin resistance gene was recovered from plasmid pJH1 (45) as a *Cla*I fragment whose ends were filled in with Klenow.

ANTIBIOTIC RESISTANCE CASSETTE VECTORS FOR GENE INACTIVATION

Alternative ways to mutagenize genes in *B. subtilis* are offered by variants of the typical integrational vectors, also called "antibiotic resistance cassette vectors." Cloning of an antibiotic resistance gene in a *B. subtilis* chromosomal fragment allows stable integration of the construct in the recipient chromosome of competent cells via a double-crossover event. Stable insertion or deletion-insertion mutations can be created by using cassettes from plasmids such as pJM105A or pJM105C, which carry the *cat* gene from pC194 in the *Eco*RV site internal to the multiple cloning site of the Bluescript plasmid (Stratagene) (35). A similar construct is present in pJM105B, although the *cat* fragment is missing 115 bp at the 3′ end that code for the last six carboxy-terminal amino acids of the *cat* protein and the *cat* transcriptional terminator. This truncated *cat* gene is still able to confer Cm^r on *B. subtilis* and *E. coli* and can be placed ahead of essential genes in operons because transcription of the essential downstream genes is not terminated by the *cat* gene terminator (44) (Fig. 6). Antibiotic resistance cassette vectors pJM109A and pJM109B were constructed by following the same strategy with the *Bacillus sphaericus ermG* resistance gene from pBD370 (30) or with the kanamycin resistance gene, pJM114, from *Streptococcus faecalis* (45) (Fig. 6).

These plasmids tremendously increase the possibilities for gene manipulation. Recently, antibiotic cassette vectors have been used to construct a complete physical map of the *B. subtilis* chromosome by the method called "gene-directed mutagenesis" (19, 20).

CONSTRUCTION OF GENOMIC LIBRARIES

Several strategies have been successfully used to prepare gene banks of the *B. subtilis* genome in *E. coli* plasmid by using fragments cloned in integrational vectors. Integrational libraries present several advantages over other libraries (for example, Tn*917* or lambda) in the screening procedures. In fact, four different strategies based on the recombinational property of integrative plasmids can be applied in the screening procedures, depending on the genetic and molecular situations.

First, by using libraries made of plasmid pools, it is possible to clone directly any selectable locus by transformation of the *B. subtilis* mutant strain. Double selection for the wild-type phenotype and antibiotic resistance should yield transformants with the integrated plasmid. In this case, transformation requires the presence on the plasmid of either an entire wild-type copy of the gene or a fragment overlapping the site of mutation on the chromosome and one end of the transcription unit.

In a second approach, a selectable marker can also be transformed directly to the wild-type phenotype with a pool of plasmids obtaining transformants via marker replacement recombination due to a double-crossover event. A third screening procedure can be applied to clone genes whose inactivation gives rise to a mutant phenotype. This depends on the insertional mutagenesis property of integrational vectors. After transformation of a wild-type *B. subtilis* strain with a library of integrational plasmids, antibiotic-resistant transformants must be analyzed for the mutant phenotype.

In all these screening procedures, the recombinant

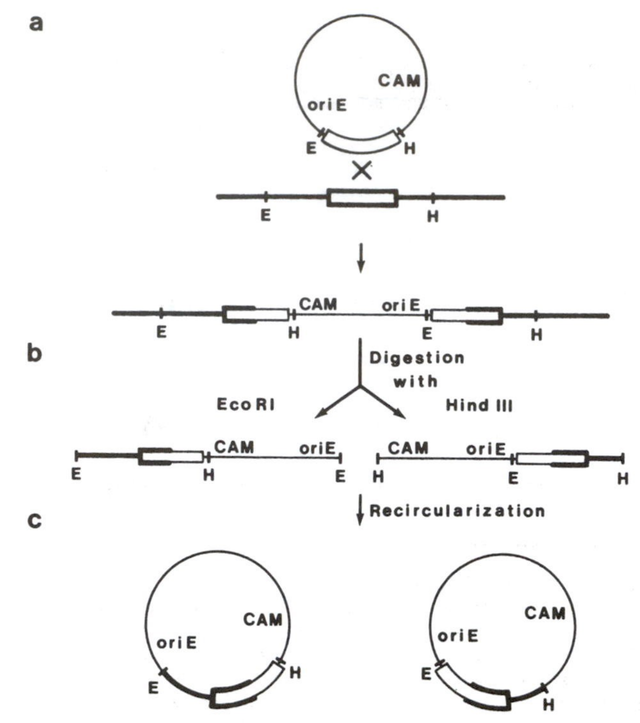

Figure 7. Plasmid rescue after chromosomal integration and cloning of adjacent sequences. (a) The *Eco*RI and *Hin*dIII sites on the plasmid belong to the vector MCS. (b) After plasmid integration, chromosomal DNA is extracted and then digested with a restriction enzyme that is unique in the plasmid in order to create fragments carrying the entire vector and adjacent sequences. (c) Digested chromosomal DNA is used at 10 μg/ml in a ligation mixture and then transformed into *E. coli* competent cells in order to recover the recircularized plasmid. See also the legend to Fig. 2. CAM, chloramphenicol; H, *Hin*dIII; E, *Eco*RI.

plasmid carrying the fragment of interest can be easily isolated from the library pools (34). With the first and third procedures, one of the advantages of having a plasmid integrated into the chromosome is the fact that the plasmid carrying the gene of interest can be rescued together with adjacent chromosomal sequences and recloned in *E. coli* in a straightforward manner. Chromosomal DNA from the *B. subtilis* strain carrying the integrated plasmid is digested with appropriate restriction enzymes, ligated in order to recircularize the plasmid, and then used to transform *E. coli* cells (Fig. 7).

A fourth screening procedure involves the preparation of a library of chromosomal insertions and depends on the great efficiency of transformation-mediated genetic exchange, which promotes incorporation of linked mutations in *B. subtilis*. After transformation of *B. subtilis* with plasmids from an integrational library, antibiotic-resistant transformants are pooled, and the chromosomal DNA is extracted and used to transform the appropriate *B. subtilis* mutant. This strategy is particularly useful in obtaining an integrated plasmid closely linked to a gene of interest for which a direct selection cannot be applied or that cannot be insertionally inactivated. A plasmid located very close to the gene can be cloned by rescuing chromosomal sequences adjacent to the insertion with the "walking" technique.

Cloning Efficiency

The choice of enzyme to use for generating chromosomal DNA fragments to be cloned in an integrational library may depend on the projected use of the library. If the intention is to generate a collection of gene disruptions, then a restriction enzyme with a common recognition sequence in *B. subtilis* should be used. For example, a *Taq*I library and a *Pst*I library were constructed in plasmid pJM102, and their efficiencies for insertional mutagenesis were tested (34a). For the *Taq*I library, which contained fragments ranging in size from 0.5 to 2 kb, the yield of auxotrophs was 5% and the yield of sporulation-deficient mutants was 25% among the Cm^r transformants obtained from a *B. subtilis* wild-type strain. When the *Pst*I library, which contained fragments bigger than 2 kb, was used, the yield of auxotrophs was 3% and the yield of sporulation mutants was 10%. The frequency of insertional mutagenesis when this technique is used depends on the average size of the cloned fragments: the smaller the average fragment, the higher the efficiency of the library in insertional mutagenesis. With small fragments, the efficiency of recombination is somewhat reduced (but not by more than a factor of 10 to 50) until the insert is less than 200 bp in size.

The frequencies of specific genetic markers were checked in a library constructed in pJH101 with *Sau*3A fragments ranging from 1 to 10 kb in size. In a study aimed at identifying genes involved in DNA metabolism, five genes were isolated by marker rescue from 17 mutants tested (34). A *Sau*3A library constructed in pCT571 with *B. subtilis* chromosomal fragments ranging in size from 4 to 20 kb was shown to contain 14 different nutritional markers in the 17 auxotrophs tested (15).

The major disadvantage of *B. subtilis* libraries constructed in integrational vectors is the difficulty often encountered in cloning sequences that are toxic to *E. coli* when present at high copy numbers. However, this problem can be circumvented either by cloning in the *E. coli pcnB* mutant, in which the copy number of ColE1 derivatives is reduced by a factor of 10 to 15 (27), or by using the low-copy-number plasmid pCT571 previously described.

TRANSCRIPTION OR GENE FUSION CONSTRUCTIONS

The need to analyze the expression of regulated genes in *B. subtilis* has motivated the development of several approaches for constructing fusions to specific cloned genes or generating libraries of fusions at random chromosomal locations. The advantage of using integrational vectors is that obtaining the fusions does not necessarily result in gene disruption, and thus the pattern of regulation of essential genes can be monitored. Although integrational vectors using a promoterless copy of the *cat* gene or the *amyE* gene as indicators have been described (29, 33), the *E. coli lacZ* gene is the indicator of choice for studies in *B. subtilis* because of the ease in assaying the β-galactosidase activity of its gene product. Typically, *lacZ* integrational fusion vectors are derivatives of ColE1 plasmids carrying antibiotic resistance markers selectable in *B. subtilis* and promoterless copies of the

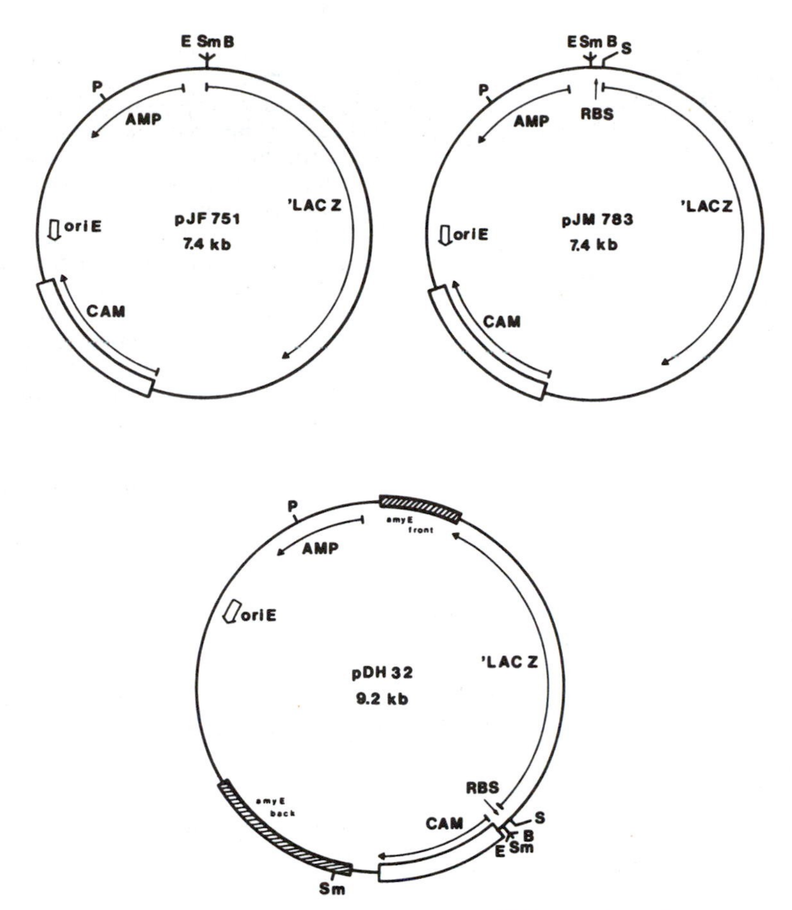

Figure 8. Integrational vectors for *lacZ* fusion constructions. pJF751 and pJM783 promote *lacZ* fusions for, respectively, translational and transcriptional analyses after integration into the *B. subtilis* chromosome by the Campbell-type mechanism. pDH32, a derivative of pBGtrp (40), integrates via a double-crossover event in the *amyE* region and promotes *lacZ* transcriptional fusions. The *lacZ* gene in each of these vectors is a promoterless truncated version of the *E. coli lacZ* gene. The *spoVG* ribosome-binding sites (RBS) in pJM783 and pDH32 come from pTV32 (36) and are preceded by one translational stop codon in each of the three possible frames and by a small MCS. Restriction sites: E, *Eco*RI; Sm, *Sma*I; B, *Bam*HI; S, *Sal*I; P, *Pst*I. AMP, ampicillin; CAM, chloramphenicol.

lacZ gene preceded by multiple cloning sites. One such plasmid, pJF751, was described by Ferrari et al. (10). pJF751 mediates the generation of translational fusions to β-galactosidase and is a derivative of pJH101, in which the tetracycline gene has been replaced by the promoterless *lac* gene preceded by an *Eco*RI-, *Sma*I-, and *Bam*HI-containing MCS (Fig. 8). An integrative vector, pJM783 (34a) (Fig. 8), which was designed to generate transcriptional fusions, was derived from pJF751 by introducing the ribosome-binding site from the *spoVG* gene of *B. subtilis*, which provides a translational initiation signal for *lacZ* preceded by three stop codons in the three possible frames (36). A similar construction is present in plasmid pVG34 described by Youngman et al. (50), while plasmid pKSV3 (50) differs by the presence of the bleomycin resistance gene derived from pUB110.

The study of the transcriptional potential of a specific DNA fragment is best accomplished if the DNA fragment is placed in front of *lacZ* in a transcriptionally neutral region of the chromosome. Plasmid pDH32 (15a) (Fig. 8) is one good example of a vector for this purpose. Transcriptional fusions to *lacZ* are placed in the chromosome in the amylase gene, opposite to its direction of transcription, by a double-crossover event. This avoids the problem of amplification observed for vectors that integrate by a single-crossover event. A similar vector, pGV34 (50), offers the advantage of ease of transfer from strain to strain, since it resides in the SPβ prophage. Such plasmids are carried by the phage after induction (7). There are several advantages offered by this approach. First, the sequence at the normal chromosomal location of the gene used to create the fusion remains intact. Second, comparative studies with the fusions in the wild-type position allow genetic analysis for the identification, for example, of upstream regulatory sequences or autogenous control of gene expression. Third, such constructs are highly stable in the chromosome, since excision cannot occur, and easily transferred from strain to strain by transformation or transduction in order to analyze the behavior of the fusions in different backgrounds.

A plasmid, pIS112 (26), in which *lacZ* translational fusions can be made in *E. coli* and analyzed in *B. subtilis* in three contexts (multicopy during propagation of the plasmid, single copy integrated into the chromosome at the wild-type locus, and single copy

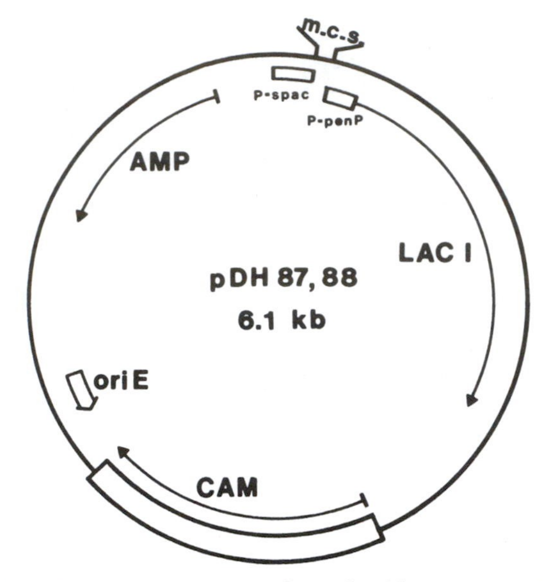

Figure 9. Integrative vectors for inducible systems. pDH87 and pDH88 derive from pSI1 (48). They have origins of replication and ampicillin genes from pBR322, while the *cat* gene is from pC194. The multiple cloning site in pDH87 contains the following restriction sites: *Hin*dIII, *Xba*I, *Sal*I, *Pst*I, and *Sph*I. The MCS of pDH88 is made of the following sites: *Hin*dIII, *Sma*I, *Xba*I, *Hpa*I, *Bgl*II, *Cla*I, and *Sph*I. *Pst*I in pDH87 and *Hpa*I in pDH88 are not unique (16).

integrated in the chromosome at a heterologous locus) has been described.

INDUCIBLE SYSTEMS

Manipulation of gene expression represents an important feature in genetic studies on gene regulation in every organism. In *B. subtilis*, the homologous recombination of integrative plasmids offers a simple way to put any gene under the control of a regulated promoter integrated into the chromosome. The first inducible gene expression system in *B. subtilis* in the replicative vector pSI1, was described by Yansura and Henner (48). This system, obtained by a modification of an *E. coli* inducible promoter system, contains the *lacI* gene coding for the *lac* repressor with modified transcriptional and translational control signals in order to ensure constitutive expression in *B. subtilis*. The expression promoter, called Pspac-1, consists of a hybrid construct containing the RNA polymerase recognition sequence from the *B. subtilis* phage SPO1 and the *lac* operator. Two integrative vectors, pDH87 and pDH88 (16), were derived from pSI1 (Fig. 9). In these two plasmids, the presence of the *cat* gene from pC194 allows selection in *B. subtilis*, while the replication origin and the ampicillin resistance gene from pBR322 allow genetic manipulation in *E. coli*. Placing a promoterless gene downstream from Pspac-1 results in induction of gene transcription in the presence of the inducer isopropyl-β-D-thiogalactopyranoside (IPTG).

Recently, these plasmids or similar constructs derived from them, like pAG58 (22), were used to investigate the regulatory mechanisms controlling sporulation gene expression. By placing a truncated copy of the *spo0A* gene (4a) or the *spo0H* gene (22) behind the spac promoter, the investigators were able to identify genes whose expression is regulated by the *trans*-acting regulatory factor Spo0A or by the sigma factor σ^H. Other applications for these expression vectors or similar versions were described in studies aimed at investigating the genetic or physiological role of *B. subtilis* gene products such as the vegetative σ^A factor (25) or the competence factor SrfA (31). Two of the major limitations of the spac system are leakiness in the control of repression, which results in a low level of expression of the cloned gene even in the absence of inducer, and reversibility of the integration process, which can lead to loss of the integrated plasmid unless selective pressure is maintained.

CONCLUSIONS

Improvement of genetic manipulation in *B. subtilis* has been greatly facilitated by integrational vectors: they are important factors contributing to the innovative interest in studying gene regulation and developmental processes in this organism. Furthermore, they contributed to the development of biological properties that are unique to *B. subtilis*, thus eliminating the cultural and methodological gap between *E. coli* and *B. subtilis*.

Acknowledgments. I express my gratitude to J. A. Hoch for revising the manuscript and for his help and patience, to A. Galizzi and A. M. Albertini for their support in the preparation of this contribution, and to F. Scoffone for art work. The work conducted in J. A. Hoch's laboratory for the construction of the pJM plasmids described in this chapter was supported by Public Health Service grant GM-19416.

REFERENCES

1. **Albertini, A. M., and A. Galizzi.** 1985. Amplification of a chromosomal region in *Bacillus subtilis*. *J. Bacteriol.* **162:**1203–1211.
2. **Bolivar, F., R. L. Rodriguez, P. J. Greene, M. C. Betlach, H. L. Heyneker, H. W. Boyer, J. H. Crosa, and S. Falkow.** 1977. Construction and characterization of new cloning vehicles. II. A multipurpose cloning system. *Gene* **2:**95–113.
3. **Campbell, A.** 1962. Episomes. *Adv. Genet.* **11:**101–145.
4. **Chapman, J. W., and P. J. Piggot.** 1987. Analysis of the inhibition of sporulation of *Bacillus subtilis* caused by increasing the number of copies of the *spo0F* gene. *J. Gen. Microbiol.* **133:**2079–2088.

4a. **Devine, K.** Personal communication.

5. **Dotto, G. P., K. Horiuchi, and N. D. Zinder.** 1984. The functional origin of bacteriophage f1 DNA replication: its signals and domains. *J. Mol. Biol.* **172:**507–521.
6. **Duncan, C. H., G. A. Wilson, and F. E. Young.** 1978. Mechanism of integrating foreign DNA during transformation of *Bacillus subtilis*. *Proc. Natl. Acad. Sci. USA* **75:**3664–3668.
7. **Ferrari, E., and J. A. Hoch.** 1983. A single copy, transducible system for complementation and dominance analysis in *Bacillus subtilis*. *Mol. Gen. Genet.* **189:**321–325.
8. **Ferrari, F. A., and J. A. Hoch.** 1982. Chromosomal location of *Bacillus subtilis* DNA fragment uniquely transcribed by sigma-28 containing DNA polymerase. *J. Bacteriol.* **152:**780–785.
9. **Ferrari, F. A., A. Nguyen, D. Lang, and J. A. Hoch.** 1983.

Construction and properties of an integrable plasmid for *Bacillus subtilis*. *J. Bacteriol.* **154:**1513–1515.
10. **Ferrari, F. A., K. Trach, and J. A. Hoch.** 1985. Sequence analysis of the *spo0B* locus reveals a polycistronic transcription unit. *J. Bacteriol.* **161:**556–562.
11. **Fort, P., and J. Errington.** 1985. Nucleotide sequence and complementation analysis of a polycistronic sporulation operon, *spoVA* in *Bacillus subtilis*. *J. Gen. Microbiol.* **131:**1091–1105.
12. **Guzman, P., J. Westpheling, and P. Youngman.** 1988. Characterization of the promoter region of the *Bacillus subtilis spoIIE* operon. *J. Bacteriol.* **170:**1598–1609.
13. **Haldenwang, W. G., C. D. B. Banner, J. F. Ollington, R. Losick, J. A. Hoch, M. B. O'Connor, and A. L. Sonenshein.** 1980. Mapping a cloned gene under sporulation control by insertion of a drug resistance marker into the *Bacillus subtilis* chromosome. *J. Bacteriol.* **142:**90–98.
14. **Harris-Warrick, R. M., and J. Lederberg.** 1978. Interspecies transformation in *Bacillus subtilis*: mechanism of heterologous intergenote transformation. *J. Bacteriol.* **133:**1246–1253.
15. **Hasnain, S., and C. M. Thomas.** 1986. Construction of a novel gene bank of *Bacillus subtilis* using a low copy number vector in *Escherichia coli*. *J. Gen. Microbiol.* **132:**1863–1874.
15a. **Henner, D.** Unpublished data.
16. **Henner, D. J.** 1990. Inducible expression of regulatory genes in *Bacillus subtilis*. *Methods Enzymol.* **185:**223–228.
17. **Horinouchi, S., and B. Weisblum.** 1982. Nucleotide sequence and functional map of pC194, a plasmid that specifies inducible chloramphenicol resistance. *J. Bacteriol.* **150:**815–825.
18. **Itaya, M., K. Kondo, and T. Tanaka.** 1989. A neomycin resistance gene cassette selectable in a single copy state in the *Bacillus subtilis* chromosome. *Nucleic Acids Res.* **17:**4410.
19. **Itaya, M., and T. Tanaka.** 1990. Gene-directed mutagenesis on the chromosome of *Bacillus subtilis* 168. *Mol. Gen. Genet.* **223:**268–272.
20. **Itaya, M., and T. Tanaka.** 1991. Complete physical map of the *Bacillus subtilis* 168 chromosome constructed by a gene-directed mutagenesis method. *J. Mol. Biol.* **220:**631–648.
21. **Itaya, M., I. Yamaguchi, K. Kobayashi, T. Endo, and T. Tanaka.** 1990. The blasticidin S resistance gene (*bsr*) selectable in a single copy state in the *Bacillus subtilis* chromosome. *J. Biochem.* (Tokyo) **107:**799–801.
22. **Jaacks, K. J., J. Healey, R. Losick, and A. D. Grossman.** 1989. Identification and characterization of genes controlled by the sporulation-regulatory gene *spo0H* in *Bacillus subtilis*. *J. Bacteriol.* **171:**4121–4129.
23. **Janniere, L., B. Niaudet, E. Pierre, and S. D. Ehrlich.** 1985. Stable gene amplification in the chromosome of *Bacillus subtilis*. *Gene* **40:**47–55.
24. **Kenney, T. J., and C. P. Moran, Jr.** 1987. Organization and regulation of an operon that encodes a sporulation-essential sigma factor in *Bacillus subtilis*. *J. Bacteriol.* **169:**3329–3339.
25. **Kenney, T. J., K. York, P. Youngman, and C. P. Moran, Jr.** 1989. Genetic evidence that RNA polymerase associated with sigma A factor uses a sporulation-specific promoter in *Bacillus subtilis*. *Proc. Natl. Acad. Sci. USA* **86:**9109–9113.
26. **Lewandoski, M., and I. Smith.** 1988. Use of a versatile *lacZ* vector to analyze the upstream region of the *Bacillus subtilis spo0F* gene. *Plasmid* **20:**148–154.
27. **Lopilato, J., S. Bortner, and J. Beckwith.** 1986. Mutations in a new chromosomal gene of *Escherichia coli* K-12, pcnB, reduce plasmid copy number of pBR322 and its derivatives. *Mol. Gen. Genet.* **205:**285–290.
28. **Messing, J.** 1983. New M13 vectors for cloning. *Methods Enzymol.* **101:**20–78.
29. **Mongkolsuk, S., Y. W. Chiang, R. B. Reynolds, and P. S. Lovett.** 1983. Restriction fragments that exert promoter activity during postexponential growth in *Bacillus subtilis*. *J. Bacteriol.* **155:**1399–1406.
30. **Monod, M., S. Mohan, and D. Dubnau.** 1987. Cloning and analysis of *ermG*, a new macrolide-lincosamide-streptogramin B resistance element from *Bacillus sphaericus*. *J. Bacteriol.* **169:**340–350.
31. **Nakano, M. M., and P. Zuber.** 1991. The primary role of ComA in establishment of the competent state in *Bacillus subtilis* is to activate expression of *srfA*. *J. Bacteriol.* **173:**7269–7274.
32. **Niaudet, B., A. Goze, and S. D. Ehrlich.** 1982. Insertional mutagenesis in *Bacillus subtilis*: mechanism and use in gene cloning. *Gene* **19:**277–284.
33. **O'Kane, C., M. A. Stephens, and D. McConnell.** 1986. Integrable alpha-amylase plasmid for generating random transcriptional fusions in *Bacillus subtilis*. *J. Bacteriol.* **168:**973–981.
34. **Perego, M., E. Ferrari, M. T. Bassi, A. Galizzi, and P. Mazza.** 1987. Molecular cloning of *Bacillus subtilis* genes involved in DNA metabolism. *Mol. Gen. Genet.* **209:**8–14.
34a. **Perego, M., and J. Hoch.** Unpublished data.
35. **Perego, M., G. B. Spiegelman, and J. A. Hoch.** 1988. Structure of the gene for the transition state regulator, *abrB*: regulator synthesis is controlled by the *spo0A* sporulation gene in *Bacillus subtilis*. *Mol. Microbiol.* **2:**689–699.
36. **Perkins, J. B., and P. J. Youngman.** 1986. Construction and properties of Tn917-lac, a transposon derivative that mediates transcriptional gene fusions in *Bacillus subtilis*. *Proc. Natl. Acad. Sci. USA* **83:**140–144.
37. **Piggot, P. J., C. A. M. Curtis, and H. DeLencastre.** 1984. Use of integrational plasmid vectors to demonstrate the polycistronic nature of a transcriptional unit (*spoIIA*) required for sporulation of *Bacillus subtilis*. *J. Gen. Microbiol.* **130:**2123–2136.
38. **Roberts, T. M., S. L. Swanberg, A. Poteete, G. Riedel, and K. Backman.** 1980. A plasmid cloning vehicle allowing a positive selection for inserted fragments. *Gene* **12:**123–127.
39. **Semon, D., N. R. Movva, T. F. Smith, M. E. Alama, and J. Davies.** 1987. Plasmid determined bleomycin resistance in *Staphylococcus aureus*. *Plasmid* **17:**46–53.
40. **Shimotsu, H., and D. J. Henner.** 1986. Construction of a single-copy integration vector and its use in analysis of regulation of the *trp* operon of *Bacillus subtilis*. *Gene* **43:**85–94.
41. **Stahl, M. L., and E. Ferrari.** 1984. Replacement of the *Bacillus subtilis* subtilisin structural gene with an in vitro-derived deletion mutation. *J. Bacteriol.* **158:**411–418.
42. **Stoker, N. G., N. F. Fairweather, and B. G. Spratt.** 1982. Versatile low copy-number plasmid vectors for cloning in *Escherichia coli*. *Gene* **18:**335–341.
43. **Thomas, C. M., D. M. Stalker, and D. R. Helinski.** 1981. Replication and incompatibility properties of segments of the origin of replication of the broad host range plasmid RK2. *Mol. Gen. Genet.* **181:**1–7.
44. **Trach, K., and J. A. Hoch.** 1989. The *Bacillus subtilis spo0B* stage 0 sporulation operon encodes an essential GTP-binding protein. *J. Bacteriol.* **171:**1362–1371.
45. **Trieu-Cuot, P., and P. Courvalin.** 1983. Nucleotide sequence of the Streptococcus faecalis plasmid gene encoding the 3'5'-aminoglycoside phosphotransferase type III. *Gene* **23:**331–341.
46. **Vagner, V., and S. D. Ehrlich.** 1988. Efficiency of homologous DNA recombination varies along the *Bacillus subtilis* chromosome. *J. Bacteriol.* **170:**3978–3982.
47. **Yanisch-Perron, C., J. Vieira, and J. Messing.** 1985. Improved M13 phage cloning vectors and host strains: nucleotide sequences of the m13mp18 and pUC10 vectors. *Gene* **33:**103–119.
48. **Yansura, D. G., and D. J. Henner.** 1984. Use of the

Escherichia coli lac repressor and operator to control gene expression in *Bacillus subtilis*. *Proc. Natl. Acad. Sci. USA* **81:**439–443.

49. **Young, M., and S. D. Ehrlich.** 1989. Stability of reiterated sequences in the *Bacillus subtilis* chromosome. *J. Bacteriol.* **171:**2653–2656.

50. **Youngman, P., H. Poth, B. Green, K. York, G. Olmedo, and K. Smith.** 1989. Methods for genetic manipulation, cloning, and functional analysis of sporulation genes in *Bacillus subtilis*, p. 65–87. *In* I. Smith, R. Slepecky, and P. Setlow (ed.), *Regulation of Procaryotic Development*. American Society for Microbiology, Washington, D.C.

43. Plasmids

LAURENT JANNIÈRE, ALEXANDRA GRUSS, and S. DUSKO EHRLICH

Laboratory strains of *Bacillus subtilis*, which are derived from strain 168, are devoid of endogenous plasmids. Only a few plasmids have been isolated from other, mostly industrial *B. subtilis* strains. However, many plasmids from other *Bacillus* spp. and from other genera of gram-positive bacteria, such as staphylococci and streptococci, have been introduced, studied, and used in *B. subtilis* 168. Studies conducted in *B. subtilis* as well as in natural plasmid hosts and in *Escherichia coli* are presented in this chapter, as they have all contributed to our understanding of plasmids from gram-positive bacteria.

An essential feature of any plasmid is its capacity to replicate. Two types of replication have been observed for plasmids from gram-positive bacteria. One is called rolling-circle replication (RCR), and the other is called theta replication. The first is used by small plasmids, and the second is used by large plasmids: the average size of some 40 RCR plasmids is only 4 kb (Table 1), while the average size of 24 theta replicating plasmids is 31 kb (see Tables 2 and 3). This correlation is not fortuitous, as is discussed more fully below. It is noticeable that in most cases, both types of replication require only two plasmid-encoded elements, an origin and a cognate protein, which often function in a remarkably broad range of hosts, extending sometimes even across the Gram stain barrier.

RCR plasmids have been studied much more extensively than theta replicating plasmids and have been reviewed recently in depth (78, 143, 144). We focus, therefore, on newly found aspects of the replication mechanism, copy control, and interaction with host functions that help explain the life cycles of these plasmids. In contrast, theta replicating plasmids have not been previously reviewed in detail (see, however, references 27 and 60) and are discussed more extensively here.

ROLLING-CIRCLE PLASMIDS

It is remarkable that many plasmids isolated from gram-positive bacteria use RCR (Fig. 1) (176, 177). RCR is initiated by a site-specific nick, and the replication intermediate forms a σ-shaped molecule. Since replication products include free single-stranded DNA (ssDNA) circles, these plasmids are referred to as ssDNA plasmids or RCR plasmids. The vast number of plasmids using RCR in gram-positive bacteria (Table 1) contrasts sharply with the exclusively theta-type replicons found in *E. coli*, which are initiated by the unwinding of two strands in the origin region and a priming reaction on one or both strands (see below). Only two RCR plasmids have been isolated from gram-negative bacteria (Table 1) (101, 193). However, several of the RCR plasmids isolated from gram-positive bacteria are replicative in *E. coli*.

Although no plasmids of *E. coli* are known to replicate by RCR, numerous bacteriophages use a similar replication mechanism. The replication mechanism of phage ϕX174 most closely resembles that of the ssDNA plasmids (75). One might think that the so-called ssDNA plasmids are in fact phage. This is not likely, however, since they remain intracellular, encode no encapsidation functions, and perhaps most importantly, regulate their copy numbers. The last feature would not be necessary if the object of the replicon was to escape from the cell.

Why RCR rather than θ replication? As described below, RCR provides greater chance of error, which may increase diversity and accelerate dissemination. The close similarities between some 40 sequenced ssDNA plasmids found in 20 microorganisms attest to their relatedness and spread (78).

Plasmid Families

Four classes of ssDNA plasmids have been distinguished according to distinct plus-origin functions (78) and are represented by pE194, pC194, pT181, and pSN2 (Table 1) (78). A fifth class has recently been reported (120a). There is now much information on the replication functions of the first three classes. Comparisons of plasmid overall structure indicate that the plasmids are assembled from cassettes, since the same characteristics are linked with various plasmid types. Three commonly found cassette units are the antibiotic resistance gene, the plus origin and Rep, and the minus origin. In addition, cassettes that specify functions involved in plasmid mobilization have been identified (156, 158). Exchange of cassettes could arise by cointegrate formation between two RCR plasmids at short repeat DNA sequences, with resolution at other short repeats (78).

Plasmid Replication

Monomeric products

The RCR plasmid encodes its own replication protein (Rep). As part of its nicking and closing activities (53, 103), Rep creates the site-specific nick that initiates synthesis from the plus origin (Fig. 1). The Rep protein is (where examined) covalently linked to the 5′ end created by the nick. The strand is displaced with the help of a host-encoded helicase (95a, 188), while a new strand is synthesized by extension from the 3′ OH

Laurent Jannière, Alexandra Gruss, and S. Dusko Ehrlich • Laboratoire de Génétique Microbienne, Institut National de la Recherche Agronomique, 78352 Jouy en Josas Cedex, France.

Table 1. Rolling-circle plasmids and their hosts of origin[a]

Replicon	Size (kb)	Original host	Reference or review
pT181	4.4	*S. aureus*	143
pC221	4.6	*S. aureus*	143
pC223	4.6	*S. aureus*	143
pS194	4.4	*S. aureus*	143
pUB112	4.1	*S. aureus*	143
pCW7	4.2	*S. aureus*	143
pHD2	2.1	*Bacillus thuringiensis*	129a
pC194	2.9	*S. aureus*	78
pUB110	4.5	*S. aureus*	78
pOX6	3.2	*S. aureus*	143
pLS11	8.4	*S. aureus*	78
pTA1060	8.4	*B. subtilis*	78
pBAA1	6.8	*B. subtilis*	78
pBS2	2.3	*B. subtilis*	49
pUH1	5.7	*B. subtilis*	80
pFTB14	8.2	*Bacillus amyloliquefaciens*	136
pBC16	4.6	*Bacillus cereus*	78
pBC1	1.6	*Bacillus coagulans*	54
pCB101	6.0	*Clostridium butyricum*	78
pLP1	2.1	*Lactobacillus plantarum*	21
pIJ101[b]	8.8	*Streptomyces lividans*	78
pC30il	2.1	*L. plantarum*	167
pTD1	2.6	*Treponema denticola*	129
pKYM[c]	2.1	*Shigella sonnei*	192
ϕX174[c,d]	5.3	*E. coli*	78
pLAB1000	3.3	*Lactobacillus hilgardii*	100
pWGB32	2.4	*S. aureus*	74
pVA380-1	4.2	*Streptococcus ferus*	113a
pRF1	4.2	*Plectonema*	151a
pE194	3.7	*S. aureus*	78
pMV158 (pLS1)	5.5	*Streptococcus agalactiae*	78
pWV01	3.3	*L. lactis* subsp. *cremoris*	116
pSH71	2.1	*L. lactis*	78
pFX2	2.5	*L. lactis*	195
pLB4	3.5	*L. plantarum*	8
pA1	2.8	*L. plantarum*	187a
pADB201	1.7	*Mycoplasma mycoides*	78
pKMK1	1.9	*M. mycoides*	100a
pHPK255[c]	1.5	*Helicobacter pylori*	101
pSN2	1.3	*S. aureus*	78
pE12	2.2	*S. aureus*	143
pE5	2.1	*S. aureus*	78
pT48	2.1	*S. aureus*	143
pTCS1	1.3	*S. aureus*	143
pNE131	2.1	*Staphylococcus epidermidis*	78
pIM13	2.1	*B. subtilis*	78
pTKX14	7.5	*B. thuringiensis*	120a

[a] Plasmids are grouped according to homologies in plus-origin replication functions.
[b] pIJ101 generates ssDNA. It is tentatively placed with the other plasmids in this group on the basis of amino acid homology of its Rep protein in the region thought to correspond to the enzymatic active site.
[c] Isolated from gram-negative bacteria.
[d] ssDNA phage of *E. coli*.

(also formed from the nick). This replication intermediate has the form of a sigma with a short (less-than-monomer-length) tail (44a, 79a). Rep creates a second nick at the origin to terminate a round of replication and religates the displaced ssDNA monomeric circular plasmid. The signals recognized by Rep for initiation and termination are overlapping but not identical (76). The predominant products of RCR are monomeric plasmids, but dimers and oligomers are present at low levels, presumably because of inefficient nicking at termination. Indeed, certain plasmids are naturally present as oligomers (157), and a point mutation mapping in Rep can shift the plasmid product from monomeric to oligomeric form, probably by affecting the capacity of the protein to terminate (13a).

As described above, an RCR plasmid-encoded protein (Rep) initiates replication by a nick at the plus origin. The nature of the interaction was suspected because of similarities between plasmid pC194 and bacteriophage ϕX174 (75). An amino acid stretch similar to the active site of the ϕX174 Rep protein was identified in the pC194 Rep protein sequence, providing a clue that the proteins have analogous functions. This is partly true; as in ϕX174, DNA nicking involves a tyrosine present on the Reps of both pC194 (75) and pT181 (178). However, unlike ϕX174, the Rep of pC194 cannot reinitiate after one round of replication (76); thus, each time Rep binds to the origin, it can effect one single round of replication, whereas once bound, ϕX174 Rep effects multiple rounds. This property of plasmid replication is vital to the maintenance of a copy control system.

For the pT181 plasmid family, origin recognition and nicking functions are located at distinct domains of a protein that binds as a dimer (188). Binding of Rep to the origin is facilitated by a supercoiled substrate (178). In the presence of Rep, origin sequences show enhanced sensitivity to ssDNA nucleases and other agents that cleave cruciforms (seen for pT181; 139).

The ssDNA generated during plus-strand replication (Fig. 1) serves as the template for minus-strand replication; although it is clear that plus- and minus-strand syntheses are discontinuous, it is not known whether initiation occurs during plus-strand synthesis or after the ssDNA circle is released (79). Efficient initiation requires a specific minus replication origin. Three kinds of minus origins have been identified on the ssDNA plasmids (19, 55, 79). All of them rely on RNA polymerase to prime DNA synthesis on the ssDNA template. Its action was demonstrated in vivo by the addition of rifampin to cultures; this addition induced the accumulation of ssDNA (19). This has also been observed in vitro when the pT181 replicon was used (15). Since ssDNA circular monomers do accumulate in the cell, the conversion of ssDNA to double-stranded DNA (dsDNA) is probably not completely efficient. ssDNA circles are particularly abundant if a functional minus origin is not present in the plasmid, which is often the case in a heterologous host. In *B. subtilis*, RNA polymerase-dependent nonspecific initiation (19) is efficient enough to allow maintenance of plasmid copy number, whereas in *Staphylococcus aureus*, it is not, and the copy number is substantially reduced (79).

HMW: the other RCR replication product

The replication cycle represented in Fig. 1 predicts monomeric products if the components function without any hitches. This is not always the case. The Rep protein sometimes initiates but does not terminate

(Fig. 1, right) (46). Without termination, a σ molecule with a long tail is one of the replication products. The ssDNA tail is converted to dsDNA, and the linear end is susceptible to degradation by exonuclease; this replication product would thus be eliminated. However, if a major host exonuclease (like RecBCD in *E. coli*; 175) is mutated (47, 187) or if the plasmid carries a DNA segment containing an exonuclease-blocking sequence (45–47, 77), a high-molecular-weight linear multimeric ssDNA plasmid product (called HMW) does appear. The tail of the nonterminated σ contains tens and perhaps even hundreds of tandem plasmid copies (44a).

Foreign inserts induce HMW formation

The foreign DNA segments that block the degradation of σ forms and thus result in accumulation of HMW vary according to the bacterial host. In *B. subtilis*, commonly cloned sequences such as pBR322 provoke HMW formation, thus increasing total plasmid copy number by as much as 10-fold (45, 77). This might affect recombination and explain, in part, high recombination frequencies observed in experiments involving the RCR plasmids as vectors (99). Theta replicons with or without foreign inserts were reported to generate little or no HMW (45, 46, 77). However, recent evidence shows that certain derivatives in pAMβ1 do produce some HMW, albeit quantitatively less than RCR plasmids (see below).

CHI sequence on RCR plasmids is responsible for HMW

The foreign DNA sequences on RCR plasmids cause HMW formation in *E. coli* because they contain CHI (5′GCTGGTGG3′) in an appropriate orientation. An *E. coli* strain in which RecBCD is mutated only for CHI recognition (1) does not accumulate HMW, regardless of whether the insert is CHI or foreign DNA (47). Presumably, in other hosts, CHI-like sequences also underlie HMW production. An insert inducing HMW in one host may not do so in another, indicating that the CHI sequence recognized by the RecBCD of each host is not the same (47). The correspondence between CHI and HMW formation may be an important tool for identification of the CHI sequence in other organisms; using RCR-induced HMW formation as the criterion, one can predict which DNA segment contains a CHI site and in which orientation the site is active in a given organism.

CHI is known to stimulate recombination in *E. coli* (171). As mentioned above, recombination frequencies from ssDNA plasmids have been reported as being 1,000 times greater per chromosome equivalent than those of theta replicons (98, 99). Since most of these experiments were performed with plasmids that produce HMW, the elevated recombination frequency could be due in part to the presence of CHI.

Copy Number Control

Two ssDNA plasmids, the pT181 type (143) and pLS1 (pE194 type) (51), have been analyzed extensively for their copy number control systems. As for all known *E. coli* plasmids, control is exerted at the level of Rep protein expression and/or origin accessibility, which together regulate frequency of initiation of plasmid DNA synthesis. In the case of pT181, control is mediated by countertranscripts and origin configuration. When the copy number of the plasmid is high, countertranscripts block the transcription of further Rep message by causing premature termination of the mRNA (145). Plasmid pLS1 encodes a countertranscript and a repressor protein (51). The protein is proposed to activate or repress Rep transcription by bending DNA around the promoter controlling both Rep and repressor protein synthesis (51, 126, 151).

pT181 may have a second means of controlling replication, i.e., by structural changes in the origin that affect its recognition. In vitro relaxed plasmid DNA is less readily recognized than supercoiled DNA (178). Thus, plasmid configuration may be an important factor in the frequency of initiation of plasmid replication (139). The *cmp* locus, identified on plasmid pT181, maps outside the minimal replicon yet seems to affect origin configuration (69); a *cmp* deletion or mutation results in slower origin recognition so that *cmp* plasmids cannot compete with the wild-type plasmid in the same strain. A *cmp*-like locus also exists on pUB110 (13a, 154a).

The repressor system should keep plasmid copy number within a certain range by limiting Rep synthesis. However, the plasmid has two means of escaping this system of control. One is "overshoot," and the other is HMW production. Overshoot might be related to the likely absence of a partitioning system from RCR plasmids; if segregation gives rise to a cell with a low plasmid copy number, an excess of Rep is synthesized and the copy number increases above the average (87), as shown in a model system in which a thermosensitive (Ts) plasmid was used to reduce copy number. This phenomenon, shown in *S. aureus*, escapes the usual Rep limitation and copy control.

As mentioned above, plasmid copy number is regulated by the frequency with which plasmid synthesis is initiated. This means that a single initiation can give rise to many plasmid copies in concatemer form. Copy number regulation may thus be circumvented by multimerization. Not surprisingly, plasmids producing HMW are unstable, possibly because of their toxicity and outgrowth of plasmid-free cells.

Segregational Stability

RCR plasmids have no known functions that ensure equipartitioning into daughter cells. Their stability may be due to elevated copy numbers and random partitioning; furthermore, overshoot replication in daughter cells with few plasmid copies could correct partition errors. The generally small size of RCR plasmids would preclude the existence of plasmid-encoded proteins required for stable maintenance; if partition functions exist, they are more likely to correspond to noncoding sequences. The absence of a functional minus replication origin is known to affect plasmid stability (28a) without substantially reducing copy number in *B. subtilis* (79) by a mechanism that is not understood. Insertion of foreign DNA segments was also reported to affect segregational stability (28), probably by inducing HMW synthesis. In some cases, a reduced monomer plasmid copy number was ob-

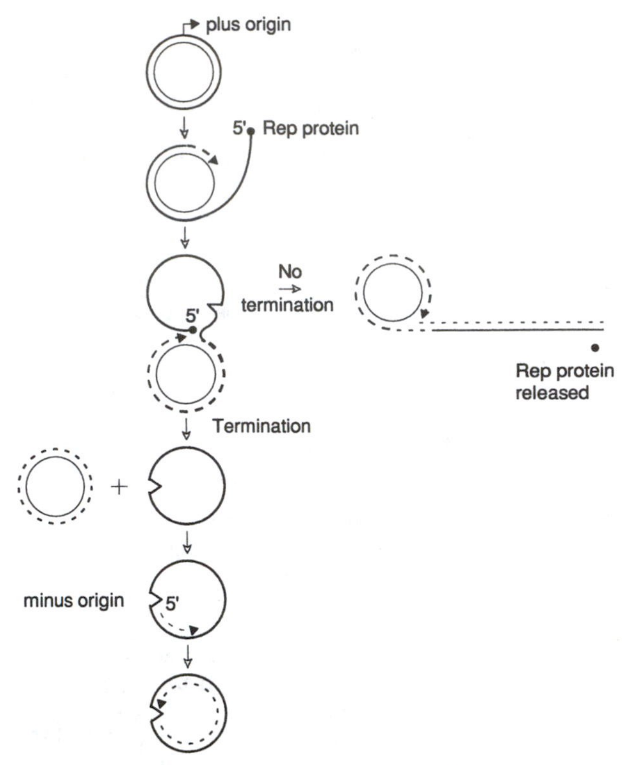

Figure 1. Unidirectional replication of rolling-circle replicons. Heavy lines, plus strands; light lines, minus strands; continuous lines, parental DNA; discontinuous lines, newly synthesized DNA. "Plus origin" represents the terminus–plus-origin overlapping region, and the bent arrow indicates direction of replication. <, secondary structure at the minus origin. Steps are as follows. (i) Initiation. Nick by Rep at the plus origin. Copy number control is exerted at this step. (ii) Displacement of the plus strand and elongation from the 3' OH nick to synthesize a new plus strand. (iii) Termination of plus-strand synthesis. Rep renicks at the plus origin to release one complete double-stranded plasmid that is then free to reenter the replication pool. Or, (iii) if termination does not occur because of, for example, plus-strand breakage or Rep protein release, high-molecular-weight linear concatemers (HMW) will form. (iv) Religation of the displaced plus strand to form a free ssDNA replication intermediate. (v) Conversion synthesis of ssDNA to dsDNA. Initiation at the minus origin (utilizing host factors) is followed by elongation to synthesize a new minus strand. (vi) Termination of conversion synthesis, releasing a second double-stranded molecule for reentry into the replication pool.

served (79), and the growth rate of the host cell was decreased (unpublished observations). Both events are expected to facilitate the emergence of a plasmid-free bacterial population.

Structural Stability

RCR plasmids are notoriously unstable (59, 61), probably because their replication generates intermediates that are susceptible to rearrangements. There are two reasons for this instability. First, the ssDNA nick used for initiation can be recombinogenic (6, 76, 132). Second, since plus- and minus-strand syntheses are discontinuous, free ssDNA is generated, and this ssDNA appears to elevate intraplasmid recombination frequencies (150).

Nick errors

A sequence resembling the plus origin may be present in an insert and may be recognized for initiation or termination. Such events would give rise to plasmid deletions (78) and would provide a consistent explanation for the endpoints found in deletions induced by pC194 replication (6, 132). In addition, there is a certain amount of cross recognition by Rep within a plasmid type (75); the pC194 Rep protein can recognize the pUB110 origin to terminate replication (75, 76). The two origins have in common 18 bp that appear to be sufficient for termination (76). This degenerate feature of Rep recognition has been exploited to map the nick sites of pUB110 (19) and pE194 (170). Errors of this type are a possible source of plasmid DNA rearrangements. In addition, nick sites can be deletion hot spots in which the second endpoint is a replication pause site with no homology with the origin (14).

ssDNA and stimulation of recombination

RCR plasmids show a high level of recombination between long homologous sequences (138) or short direct repeats (99). The latter are attributed in part to the accumulation of ssDNA (i.e., absence of a plasmid minus origin). Furthermore, the presence of an active replicon integrated in the chromosome stimulates replication forks in the neighboring region. When the replicon is an RCR plasmid, a nick and unidirectional replication may be initiated. In *B. subtilis*, if repeated segments are in the vicinity of the RCR replicon, significant stimulation of both deletion (140) and amplification (154) of the DNA between the repeats is observed (20 to 160 times). The amplification is seen by using an RCR but not a theta integrated replicon (133a). Recombination is stimulated whether repeats are to one side or flanking the replicon, although the stimulation is greatest when the repeats are downstream from the origin in the direction of replication. This property has been successfully used to conditionally amplify the cellulase gene inserted in the *B. subtilis* chromosome, yielding increased expression (153).

Although the effects of a CHI sequence in the repeats have not been explored, it is notable that the repeat sequences used in these experiments do contain a CHI-equivalent site active in *B. subtilis* (47). CHI sequences can stimulate recombination if there is a free dsDNA end (175), which might result from initiation of RCR by conversion of the displaced strand to a double-stranded form.

Cloning Vectors and Transposon Delivery Systems

The most popular vectors in *B. subtilis* employ replicons originally isolated from *S. aureus* (e.g., pC194, pUB110, and pT181). The plasmids isolated directly from *B. subtilis* were found to have close relatives in other species. However, relatively little systematic engineering has been done to make good cloning RCR vectors, possibly because of segregational (28) and structural (98) instabilities observed in

Table 2. pAMβ1 plasmid family[a]

Plasmid[b]	Size (kb)	Host range (no. of genera[c])	Original host	Reference
pAMβ1	26.5	11	*E. faecalis*	40
pIP501	30	7	*S. agalactiae*	92
pAC1	25.5	2	*Streptococcus pyogenes*	37
pRI405	25.5	2	*E. faecalis*	185
pIP612	34	ND	*S. agalactiae*	62
pIP613	26	3	*E. faecalis*	42
pIP646	27	3	*Streptococcus equissimilis*	22
pIP920	30	1	*E. faecium*	22
pMV103	26.5	2	*S. agalactiae*	85
pMV141	25	2	*S. agalactiae*	85
pSF9400	ND	ND	*E. faecalis*	65
p43	60	1	*B. thuringiensis*	8
pSM19035	27	2	*S. pyogenes*	122a
pERL1	26	2	*S. pyogenes*	124
pSM10419	22.5	1	*S. pyogenes*	125

[a] Most of the data are from reference 90. For pAMβ1 and pIP501, see also references 17a, 22, 86, 112, 149, 159, 165, 174, 182, and 189. ND, not determined.
[b] Plasmids are grouped according to absence (pAMβ1) or presence (pSM19035) of long inverted repeats. Classification of pSF9400 and p43 is tentative, since the absence of inverted repeats has not been ascertained. All plasmids except pSF9400, p43, and pSM10419 were shown to be conjugative.
[c] Number of genera into which a plasmid has been introduced.

plasmids carrying foreign inserts. The induction of HMW upon cloning of foreign segments (77) is now understood to be due to the presence of CHI sites on the foreign segments (45, 46). Although the exact sequence of CHI recognized in *B. subtilis* is not known, it is clear that the sequence is present in pBR322 sequences, which are included in most *E. coli-B. subtilis* shuttle vectors, as well as in many of the cloned DNAs. This sequence probably affects the stabilities of vectors and clones. Nevertheless, some RCR derivatives are exceptionally useful and do not have theta replicating equivalents.

Numerous Ts RCR plasmids have been isolated (95). pE194 is the most widely used, in part for its convenience of selection (erythromycin resistance) (186). It is Ts above 39°C, which is convenient for *B. subtilis*. A Ts variant of another plasmid of the pE194 replicon family, pWV01, has recently been isolated (121). It is Ts above 35°C, has multiple cloning sites, and may be useful in experiments in which the temperature range is limited. Both Ts plasmids are successfully used for transposon insertion into the chromosome: Tn*917* has been widely developed for use in *B. subtilis* and other gram-positive hosts (195), with the Ts pE194 used as vector. Tn*10* derivatives have been adapted for use in *B. subtilis* with both the Ts pE194 vector (152) and Ts pWV01 (120b, 121). The formation of HMW by these constructs seems to be a handy way of increasing transposition frequency by increasing plasmid copy number; 100-fold-higher frequencies were observed when pBR322 sequences were present on the delivery vector (33, 152).

THETA REPLICATING PLASMIDS

RCR plasmids are small and are unable to stably maintain large DNA inserts, which suggests that RCR might be too error prone to be used by large plasmids (60). Not surprisingly, studies of one family of large plasmids, represented by pAMβ1, have revealed a different mode of replication, named theta. It is likely that large plasmids that belong to other families use a similar replication mode.

Plasmid Families

Restriction and hybridization data indicate that numerous large plasmids (22 to 34 kb) isolated from streptococci and enterococci are closely related to pAMβ1 (Table 2; see reference 90 for a review). The following observations suggest that these plasmids carry highly conserved replication regions. First, restriction analysis revealed that the replication regions of plasmids pAMβ1, pIP501, pSM19035, and pSM10419, which have been located in 2- to 3-kb segments by deletions generated in vivo and in vitro, have similar restriction maps (10, 12, 113, 123). Second, hybridization experiments showed that plasmids pIP501, pSM10419, pERL1, and pSF9400 carry sequences highly homologous to the replication region of pSM19035 (71). Third, sequence determination demonstrated that the replication regions of pAMβ1, pIP501, and pSM19035 are >80% homologous and carry large open reading frames (ORFs) which encode a highly (>98% identity) conserved Rep protein (24, 25, 169, 173) (Fig. 2). The functional organization of these regions is also conserved, since chimeric replicons pAMβ1-pIP501 and pIP501-pSM19035 are viable (24).

Sequences flanking the replication regions are also highly conserved (Fig. 2). At the 5′ ends, homology was detected by restriction and (partial) sequence analysis (10, 12, 23a, 24, 69, 113, 123, 173). Deletion analysis revealed that this region negatively controls plasmid copy number (10, 12, 98, 160, 166, 173). At the 3′ end, a highly homologous 1-kb sequence involved in resolution of multimers was found in pAMβ1 and pIP501 (98, 159a, 172). Such a function is probably

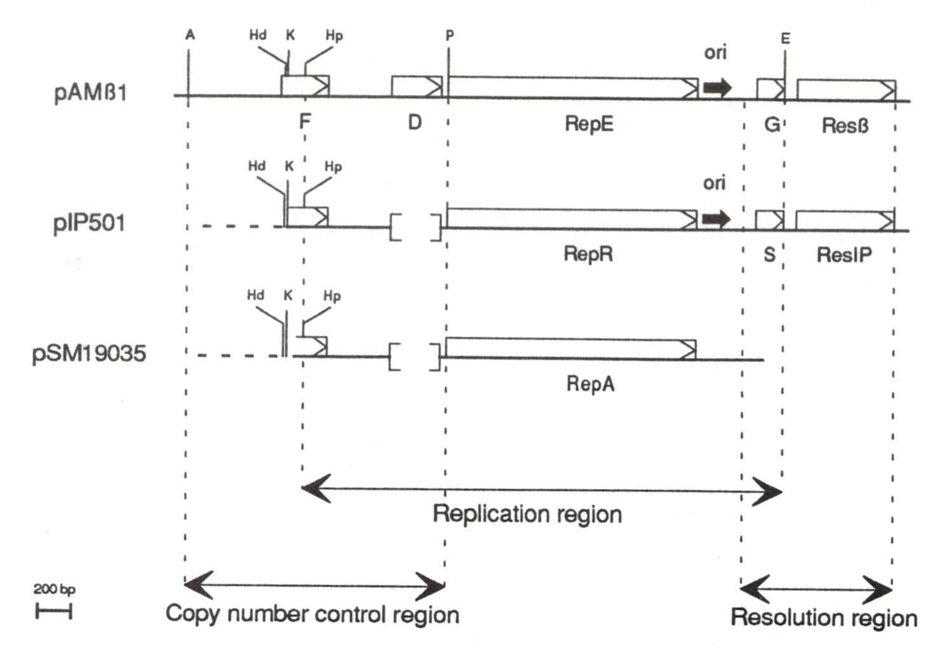

Figure 2. Regions required for copy control, replication, and multimer resolution in three plasmids from the pAMβ1 family. Continuous line, sequenced regions; broken line, unsequenced regions; boxes, ORFs; arrows within the boxes, direction of transcription. The region lacking in pIP501 and pSM19035 is shown in brackets. Replication origin and direction of replication are indicated by heavy arrows. Several restriction sites that suggest sequence homology outside of the common sequenced regions are represented. Rep and Res stand for the replication protein and resolvase, respectively. A, *Acc*I; Hd, *Hin*dIII; K, *Kpn*I; Hp, *Hpa*I; P, *Pvu*I; E, *Eco*RI.

also present downstream of the pSM19035 replication region (168a; see below).

The pAMβ1 plasmid family can be divided into two classes, one of plasmids that do not contain long inverted repeats, represented by pAMβ1, and one of plasmids that do contain such repeats, represented by pSM19035 (Table 2; see reference 90 for a review). The repeats represent 40, 60, and 80% of the genome in pERL1, pSM10419, and pSM19035, respectively (11, 20, 73). Interestingly, they include the plasmid replication region (12, 73), which directs unidirectional theta replication (see below). Initiation at both origins within the same molecule should lead to fork collision before the entire plasmid is replicated. To prevent such events, it is possible that only one of the replication regions is functional or that initiation at one origin precludes initiation at the other.

Plasmids of the pAMβ1 family have a large host range and are autotransferable (Table 2; see reference 90 for a review). This could explain their dissemination among streptococci and enterococci. They might also be carried by other bacterial genera, as is suggested by the observation that the large (60-kb) *Bacillus thuringiensis* plasmid p43 encodes a replication protein 21 to 25% homologous with the Rep proteins of pAMβ1, pIP501, and pSM19035 (8, 9).

Plasmids listed in Table 3 do not belong to the pAMβ1 family. Replication regions of all of them except pI524 were cloned and sequenced (8, 9, 26a, 66, 67, 82, 83, 92, 93, 94, 97a, 117, 118). Their sizes vary between 1 and 2.3 kb, and all but pTH1030 contain one long ORF, which probably corresponds to a *rep* gene, since its alteration abolishes plasmid transformation ability. The encoded proteins vary between 312 and 406 amino acids. Rep proteins of pTB19 (RepA) and p60 are 31.5% identical, and those of pCI305, pSK11L, pSL2, and pWV02 are 52 to 76% identical, which indicates the existence of two families of theta replicating plasmids other than pAMβ1. The pCI305 family includes a number of other plasmids from lactococci (26b, 83). Interestingly, the replication region of pSK11L and pTB19 (RepA) present 51% of homology at the DNA level (91), which indicates that the two plasmid families might be related.

Several of the plasmids listed in Table 3 are close to 10 kb or smaller, which indicates that the θ replication mode is not associated only with large plasmids.

Table 3. θ replicating plasmids which do not belong to the pAMβ1 family[a]

Plasmid	Size (kb)	Original host	Reference
pTB19	26.5	*Bacillus stearothermophilus*	93
p60	95	*B. thuringiensis*	9
pCI305	8.7	*L. lactis* subsp. *cremoris*	82
pSK11L	47.3	*L. lactis* subsp. *cremoris*	64
pWV02	3.8	*L. lactis* subsp. *cremoris*	26a
pSL2	7.8	*L. lactis* subsp. *lactis*	97a
p44	69	*B. thuringiensis*	9
pHT1030	15	*B. thuringiensis*	117
pIP404	10.2	*Clostridium perfringens*	26
pI524	31.8	*S. aureus*	142

[a] Plasmids are grouped according to homologies in the replication functions.

In contrast, RCR has never been found in large plasmids. The host range of θ-replicating plasmids varies; it appears to be narrow for pCI305 and broad for pIP404 (66, 82).

Replication of pAMβ1 Family

Most studies of θ replicons of gram-positive bacteria are based on plasmids of the pAMβ1 family and are therefore emphasized here. Replication of plasmids that belong to other families is treated separately.

Mode of replication

Several observations suggest that the pAMβ1 family uses a unidirectional theta mode of replication. First, pAMβ1 derivatives do not generate ssDNA during replication (98). Second, pAMβ1 and pIP501 replication intermediates are bubble shaped, with one fork at a constant position (the origin) and the other at a variable position, as is shown by two-dimensional gel electrophoretic analysis (31, 114, 115). Third, insertion of the *B. subtilis* terminus (a signal that causes arrest of the chromosomal replication at *terC*; 34, 35, 119) in a pAMβ1 derivative leads to accumulation of bubble-shaped replication intermediates having one fork at the pAMβ1 origin and the other at *terC* (30, 31). These characteristics are not observed in RCR plasmids (30, 79a).

Two-dimensional gel analysis revealed that the pAMβ1 and pIP501 origins lie close to the 3′ end of the gene encoding the Rep protein and that the replication fork progresses in the direction of transcription of this gene (Fig. 2; 31, 115). A biochemical analysis of the intermediates accumulated upon insertion of the replication terminus in pAMβ1 derivatives confirmed these conclusions. It also showed that leading-strand synthesis starts 27 nucleotides downstream from the stop codon of the replication protein (at position 4631 of pAMβ1 [coordinates according to reference 173]) and that lagging-strand synthesis stops 13 to 15 nucleotides further (between positions 4644 and 4646), which results in a short single-stranded region (31). These conclusions probably extend to pIP501 and pSM19035 (23c, 24, 173). Interestingly, the structures frequently observed in origins of large *E. coli* plasmids and bacterial chromosomes (short direct repeats, *dnaA* boxes, and AT-rich regions; 23, 104, 147) were not found in the vicinity of the initiation site of DNA synthesis. The only structures detected were a pair of inverted and a pair of direct repeats, which are, however, not essential for pAMβ1 replication (see below). These observations suggest that pAMβ1 replication might initiate in a unique way.

Primary replication functions

Two elements were shown to be necessary and sufficient for replication of pAMβ1 or pIP501 in *B. subtilis* and are therefore considered primary plasmid replication functions. The first is the protein encoded by ORFE or ORFR (RepE or RepR, respectively). It operates as a positive, rate-limiting replication effector since (i) frameshift mutations within these ORFs abolish the plasmid's ability to transform *B. subtilis* (24, 114, 173), (ii) RepR activates the cognate origin in *trans* (23c), (iii) overproduction of RepE measured in vitro correlates with an increase of pAMβ1 copy number measured in vivo (173), and (iv) the activity of a foreign promoter used to control the synthesis of RepE or RepR determines plasmid copy number (23b, 114, 115).

The second element required for replication is the replication origin, which is located at the 3′ end of the replication region as shown by biochemical analysis of pAMβ1 and pIP501 replication intermediates. In pIP501 this element was mapped to a 72-bp segment (between positions 2108 and 2180, coordinates according to reference 24), which contains the putative initiation site of leading-strand synthesis (position 2138 [23c]). In pAMβ1 the right boundary of the origin was mapped to a 13-bp stretch (coordinates 4632 to 4649 [114]) which begins 1 bp downstream from the initiation site. Thus, in both plasmids the origins appear to be carried by the same sequence. However, deletion of the right branch of the inverted repeats which flank the pAMβ1 initiation site does not affect plasmid replication (114), while removal of either branch of the corresponding pIP501 repeats abolishes replication (23c). This discrepancy is not understood but might be due to the differences in experimental procedures used to test origin activity in the two cases rather than to a real difference of the initiation of DNA replication in pAMβ1 and pIP501.

The two primary replication functions confer a broad host range upon pAMβ1 and pIP501 (10, 48, 114, 133, 166, 182). However, the host ranges of pAMβ1 derivatives in which the natural promoters are replaced by foreign promoters vary (114). This suggests that the pAMβ1 host range depends on the promiscuity of the signals that ensure RepE synthesis. A low requirement for plasmid-encoded elements in plasmid replication is also observed for the promiscuous rolling-circle ssDNA plasmids from gram-positive bacteria and theta replicating IncP plasmids from gram-negative bacteria (see above; 179).

Secondary replication functions

Two functions that affect but are not strictly required for replication of pAMβ1 were detected downstream from the plasmid replication origin. The first is named *ssiA*, for single-strand initiation, and the second is named anti-HMW, since it prevents formation of high-molecular-weight multimers.

ssiA. *ssiA* is localized ≈100 bp downstream from the initiation site of leading-strand synthesis (between coordinates 4712 and 4856) within a 145-bp segment that is relatively GC rich (50% rather than 32% for the replication region) and carries inverted repeats (Fig. 3; 29a). Its role is presumably to facilitate lagging-strand synthesis by stimulating formation of a primosome *priA*-dependent complex, as suggested by the following evidence. First, insertion of *ssiA* into an RCR plasmid lacking an active minus origin strongly reduces the amount of ssDNA accumulating in *B. subtilis*. This reduction is orientation specific. The polarity shows that *ssiA* is carried on the lagging-strand template of pAMβ1. Second, *ssiA* activity is resistant to rifampin, which indicates that the activity is not mediated by RNA polymerase. This shows that *ssiA* differs from the three types of RNA polymerase-dependent single-stranded initiation sites in RCR plasmids described so far (minus origins; 15, 19, 27).

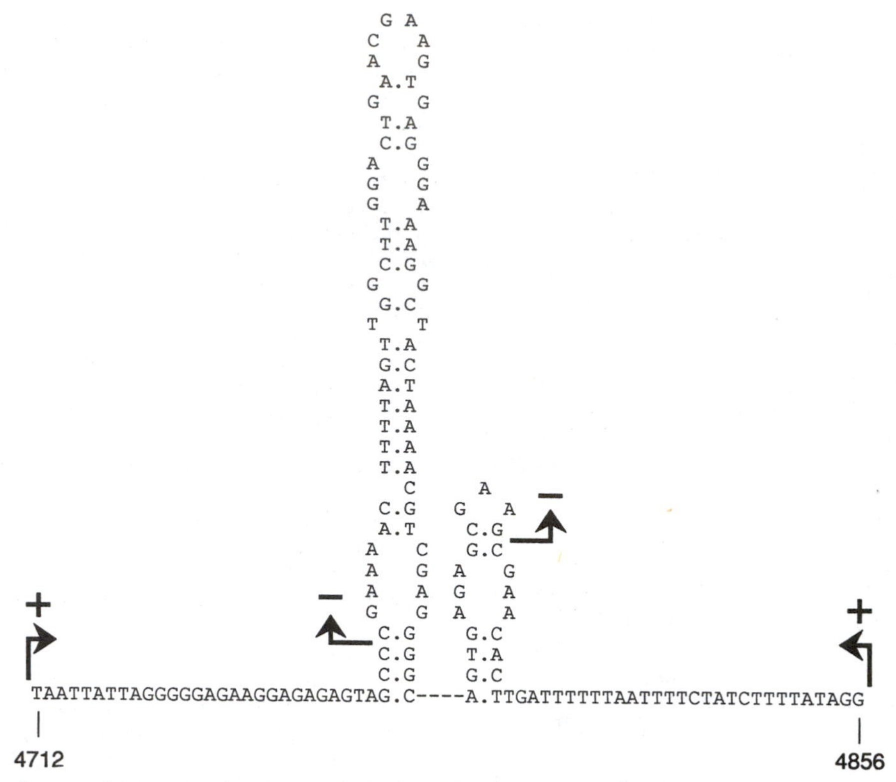

Figure 3. Single-strand initiation (*ssiA*) site of pAMβ1. The sequence of the active strand is shown. Endpoints of deletions generated by exonuclease III are indicated by arrows; + and − refer to deletions that preserve or abolish *ssiA* activity.

Thus, *ssiA* represents a new signal for DNA replication in gram-positive bacteria. Third, *ssiA* carries inverted repeats, like *priA*-dependent primosome assembly sites of *E. coli* (*pas*), and has some homology (in the region 4812-4823; Fig. 3) with *pas* of φX174, R100 (*ssiB*), and F (*ssiC*) (128, 140a), which suggests that *ssiA* operates as *pas* does. Sequence comparisons suggest that pIP501 and pSM19035 also carry *ssiA* downstream of their replication origins on their lagging-strand templates.

Since deletion of *ssiA* does not abolish replication of pAMβ1 in *B. subtilis*, alternative modes of initiation of lagging-strand synthesis must exist. Synthesis could be initiated by a nonspecific, RNA polymerase-dependent mechanism previously observed in this host with RCR plasmids (19) or by a specific DnaA-dependent mechanism similar to that described for *E. coli* (127, 163). Interestingly, several *dnaA* boxes are present in the available sequence of pAMβ1 (134; unpublished data).

Anti-HMW. The other secondary replication function, anti-HMW, was detected a few nucleotides downstream from *ssiA* within a 250-bp region that contains a short ORF, ORFG, preceded by a ribosome-binding site and two putative promoters (unpublished data). Deletion of this region in high-copy-number plasmids results in accumulation of high-molecular-weight plasmid multimers (HMW). This accumulation is not observed in *B. subtilis* hosts affected in homologous recombination (*recE* mutants). Unlike the RCR plasmids, no foreign DNA insert is required to induce HMW formation in plasmids derived from pAMβ1 missing the anti-HMW. The multimers are composed of head-to-tail monomers and appear to be linear. Up to 25% (by mass) of plasmid can be present as HMW (HMW of RCR plasmids accounts for up to 90% of plasmid mass). Deletion of the region in low-copy-number plasmids renders them segregationally instable (173). The instability might be due to HMW formation, as has also been suggested for RCR plasmids (see above). The anti-HMW activity appears to be independent from the position and orientation of the 250-bp region, and it is not known whether it is mediated by the ORFG product. Interestingly, in some *Lactococcus lactis* strains, the large (46-kb) plasmid pRT2030 exists exclusively in the form of HMW (88). So far, this property has not been associated with any plasmid or host trait.

HMW formation is likely to require unending RCR and concomitant protection of the extremity of the nascent σ-shaped molecule from exonuclease degradation, as was discussed more fully above. An anti-HMW function could interfere with the establishment of the unending replication or facilitate degradation of the σ-shaped molecules. Such a function has been observed in plasmid pTX143 (120a).

Interestingly, Rep proteins of pAMβ1, pIP501, pSM19305, and two other theta replicating plasmids (pCI305 and R1) contain motifs resembling the active sites of Rep proteins of RCR plasmids (83, 169, 178a) and might therefore be endowed with a nicking and closing activity. We speculate that this activity could initiate unscheduled RCR and therefore stimulate HMW formation, which could be limited in turn by a plasmid-encoded anti-HMW function.

Initiation of replication

Replication origins of bacterial chromosomes and of many theta replicating plasmids from gram-negative bacteria contain particular structures, such as *dnaA* boxes and direct AT-rich repeats (23; see reference 104 for a review). These structures play a central role in initiation of replication, since binding of DnaA protein to its boxes in conjunction with the Rep protein in plasmid systems melts the repeats, which allows loading of a DNA helicase and primase and subsequent synthesis of primers for DNA synthesis (see reference 104 for a review). Such structures are not present in the vicinity of the leading-strand initiation site of the pAMβ1 plasmid family, which indicates that replication of these plasmids is initiated in a different way (23c, 31). An alternative initiation mechanism is that of the ColE1 plasmid family, in which the primer for leading-strand synthesis is generated by RNase H cleavage of a transcript at the origin. The primer is then extended by DNA polymerase I, generating a D-loop structure of about 400 bp. This activates a *pas* sequence that is carried 180 nucleotides downstream from the origin on the lagging-strand template. Upon activation, *pas* directs primosome assembly, which allows initiation of lagging-strand synthesis and probably also replacement of DNA polymerase I by DNA polymerase III (see references 109 and 196 for reviews). The following observations indicate that plasmids of the pAMβ1 family use an initiation mechanism related to that of ColE1-type replicons (29a, 114). First, pAMβ1, pIP501, and pSM19035 do not transform *polA* mutants of *B. subtilis*. Second, DNA polymerase I initiates pAMβ1 leading-strand synthesis. Third, pAMβ1, pIP501, and pSM19035 carry a *pas*-like sequence on the lagging-strand template, ~100 bp downstream of the initiation site of the leading-strand synthesis.

However, several features differentiate replication machineries of plasmids from the ColE1 and pAMβ1 families. First, ColE1 does not encode a protein required for its replication, whereas pAMβ1 does. Second, there is no homology between replication regions of ColE1- and pAMβ1-type plasmids. Third, replication of ColE1 is not promiscuous, and that of pAMβ1 is (see references 90 and 109 for reviews). An elegant way to reconcile these differences would be to attribute the RNase H activity, which is host encoded in the case of ColE1 plasmids, to Rep proteins of the pAMβ1 family. Rep would thus be able to cleave its own mRNA at the origin, which is situated very close to the *rep* stop codon, and therefore generate a primer for leading-strand synthesis. Plasmid promiscuity would depend only on the ability of the Rep promoter to function in numerous hosts, as is suggested by promoter replacement experiments in pAMβ1 (114).

Replication of Other Plasmids

Several plasmids that do not belong to the pAMβ1 family (Table 3) seem to use theta replication, as suggested by the following evidence. First, small plasmids derived from p60, p44, pCI305, pIP404, and pHT1030 or from the RepA replication region of pTB19 (this plasmid contains, in addition to the RepA determinant, two inactive RCR replicons; 135, 148, 184) do not accumulate ssDNA during replication (9, 66, 83, 98, 118). Second, bubble-shaped pWV02 replication intermediates were detected by two-dimensional gel electrophoresis and electron microscopy (26a). Third, pI524 replicates bidirectionally, as indicated by pulse-labeling (141a).

All the sequenced replication regions except that of pHT1030 carry repeats that either are AT rich or are localized close to AT-rich sequences and are found within (in pTB19 RepA and p60), upstream from (in pCI305, pKSL11, pSL2, pWV02, and p44), or downstream from (pIP404) the *rep* gene (9, 26a, 66, 83, 91, 94). A DnaA box is present upstream from the repeats in pTB19 RepA (94). These structures resemble replication origins of numerous plasmids from gram-negative bacteria (see reference 104 for a review) and might function as such, since they appear to confer autonomous replication on plasmids derived from pTB19 RepA and pCI305 in the presence of the *trans*-encoded cognate Rep protein (82, 83, 94). Presumably, the mechanism of replication of these plasmids resembles that of their counterparts from gram-negative bacteria. As in the pAMβ1 family, only two plasmid-encoded elements (the Rep protein and the replication origin) are required for replication.

The replication region of pHT1030 resembles that of ColE1 in several aspects (118). First, it does not carry an ORF. Second, its activity depends on a promoter situated at one extremity of the replication region and oriented toward the other extremity. Third, it contains two long imperfect inverted repeats, one of which carries a short inverted repeat. However, protein neosynthesis, which is not required for ColE1 replication (56, 97), seems to be required for replication of pTH1030, since this plasmid is not amplified in *B. subtilis* upon chloramphenicol addition (118).

Copy Control

Replication proteins of pAMβ1 and pIP501 are rate-limiting positive effectors, which implies that their activities must be strictly controlled to keep plasmid copy number within convenient limits. This control could be exercised at the level of protein synthesis or protein stability. Evidence for the first level of control is described below. There is presently no direct evidence for the second level.

One system that regulates Rep synthesis appears to operate by a countertranscript-driven transcriptional attenuation mechanism similar to that described for the RCR plasmid pT181 (145), as deduced from analyses of the region localized 5′ to pAMβ1 ORFE and pIP501 ORFR (23a, 25a, 113b). The key elements of the putative system of pAMβ1 (Fig. 4A) are two promoters, P_{DE} and P_{CT}, which face each other, and two pairs of partially homologous inverted repeats, of which one pair (repeats I and II) is between the two promoters and the other pair (repeats III and IV) is behind P_{CT} and is followed by a series of thymines (Fig. 4B). P_{DE} drives the synthesis of the mRNA of ORFD and ORFE, and P_{CT} ensures synthesis of countertranscript

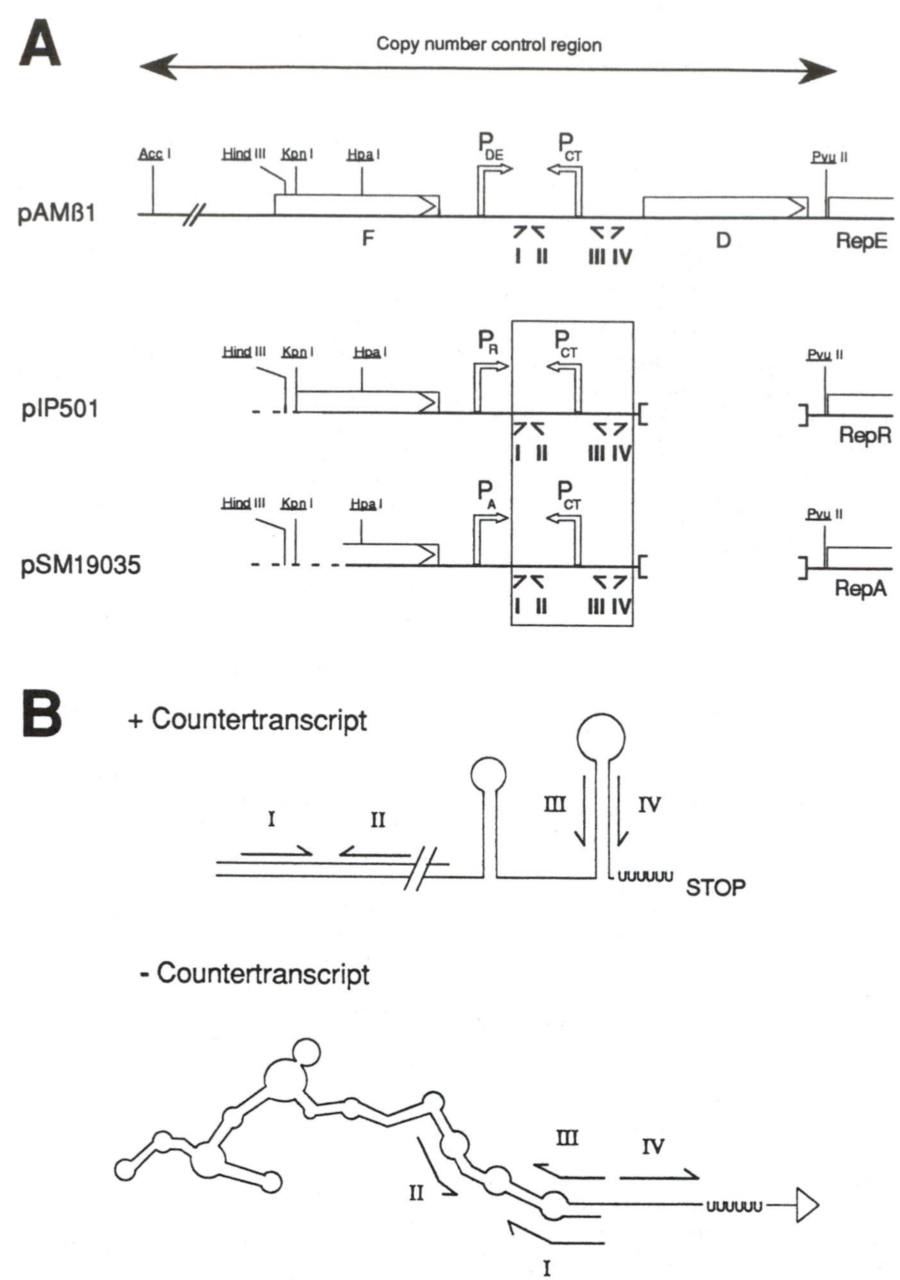

Figure 4. Copy control regions of three plasmids of the pAMβ1 family and putative control mechanisms. (A) Critical promoters are represented by bent arrows, which indicate directions of transcription; inverted repeats are labeled I to IV. The region diverging to almost 50% between pAMβ1 and two other plasmids is boxed. Other symbols are explained in the legend to Fig. 2. (B) Hypothetical secondary structures generated upon interaction of mRNA and countertranscript (top) and folding of the 5′ end of the mRNA (bottom).

RNA. The system might operate as follows. When the nascent mRNA pairs with the countertranscript, repeats III and IV can anneal and form a Rho-independent transcription terminator, which impedes further mRNA synthesis (Fig. 4B). When this pairing does not occur, the 5′ end of the nascent mRNA can form a complex structure involving interaction of repeats I and III, which prevents formation of the terminator (Fig. 4B) and allows synthesis of full-length mRNA. Sequence analysis suggests that plasmid pSM19035 carries the same key elements (Fig. 4A), which argues for a similar control of expression of its Rep protein (114).

A possible difference between regulation of pT181 and pAMβ1 copy number is that the synthesis of the countertranscript is constitutive in the former plasmid, whereas it might be regulated in the latter (23a, 113b). Preliminary observations indicate that the product of ORFF, which is known to negatively regulate plasmid copy number (173), stimulates synthesis of the countertranscript, while that of ORFD might repress it. A similar type of control, involving the countertranscript and the ORFF equivalent, was revealed in pIP501.

Another system of Rep regulation, observed in pIP501 and pAMβ1, possibly operates by interfering with transcription from the *rep* promoter (23a, 26a). This interference involves a region containing direct and inverted repeats which partially overlap the promoter. Mutation or deletion within the repeats can cause a dramatic increase in the plasmid copy number (up to 50-fold; 23a). How the two control systems coordinately regulate copy number of pAMβ1 family plasmids is not known.

Yet another type of copy number control might also be present in pAMβ1. It involves binding of the resolvase (RES) of pAMβ1 (Resβ) to the *res* site was observed to result in a sharp decrease of pAMβ1 copy number. Two models could be proposed to explain this phenomenon. First, the complex Resβ-*res*, which maps 200 bp downstream from the pAMβ1 initiation site of DNA synthesis, might inactivate the origin by modifying its topology. Second, the complex Resβ-*res* might arrest DNA synthesis, which is mediated by DNA polymerase I in this region of the pAMβ1 genome (see above).

Elements regulating replication were shown to exist in plasmids from other families, such as pIP404 and pTH1030, by isolation of high-copy-number mutants (2, 66). These mutants were studied only in pIP404 and found to be localized upstream of the *rep* gene in a region encoding a protein of 198 amino acids (66). A deletion of the 3′ end of the coding region increased plasmid copy number five times, which suggests that the protein, named Cop, is a negative copy number regulator. However, the hydrophobic nature of the protein and its probable membrane localization argue against such a role (67). A countertranscript homologous to the 3′ region of the Rep mRNA and localized closely to the putative origin was also detected (66). Its role was not investigated, but it may regulate replication by affecting either synthesis of the Rep protein or activity of the origin.

Experimental evidence for regulation of replication of other plasmids is not available, but sequence analysis reveals suggestive features. For instance, the putative promoters of the *rep* genes of several plasmids (pTB19 RepA, p44, pCI305, pSL2, and pSKL11) are localized in the vicinities of inverted repeats, which suggests that plasmid copy number might be regulated by controlling Rep synthesis (9, 83, 91, 94). This control might involve the Rep protein in pCI305, pSK11L, and pTB19 (RepA), since the inverted repeats upstream of the *rep* gene are homologous with the direct repeats present in the presumptive plasmid origin. On the other hand, a region of pIP404 thought to correspond to the plasmid origin contains a large number of two types of direct repeats that have the structure X_6Y_{16} and might regulate replication by sequestering the Rep protein or inactivating the origin by steric hindrance (141). Finally, it is likely that copy number control of pTB19 RepA, pCI305, and pHT1030 does not involve a repressor-like protein, since the regions that allow autonomous replication and confer a copy number similar to that of the parental plasmid do not contain ORFs other than *rep* (83, 94, 118).

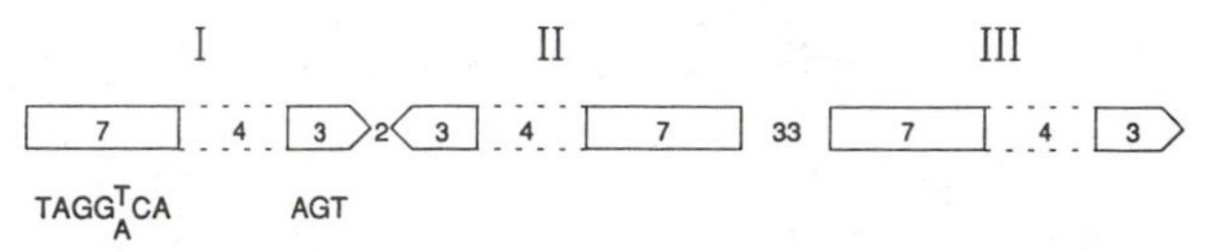

Figure 5. Structure of the resolution sites of pAMβ1, pIP501, and pSM19035. The three imperfect inverted repeats (I through III) are represented by arrows, matching regions are indicated by a continuous line, and nonhomologous regions are indicated by a broken line. The numbers refer to lengths (in base pairs) of different regions and distances between the repeats. The sequence shown beneath the box is present in the nine repeats from the three plasmids.

Segregational Stability

Only a few functions that improve segregational stability of theta replicating plasmids from gram-positive bacteria are known. Nevertheless, such functions are probably frequent, since (i) the stability of theta replicating plasmids is often affected by deletions (63, 66, 82, 91, 98, 113, 118, 160, 166, 173), (ii) a putative resolution function has been detected by sequence analysis of pIP404 (68), and (iii) several regions involved in plasmid stabilization have been found in pAMβ1, pSK11L, and pHT1030 (91, 118, 172). In addition, a *par*-like function was reported for a 15-kb plasmid from *Corynebacterium glutamicum* (110). However, there is no indication besides size that the plasmid uses the theta mode of replication. Elements improving the stability of pAMβ1, pIP404, and pHT1030, which have been characterized to some extent, are discussed in this section.

pAMβ1 family

A function that increases the segregational stability of high-copy-number derivatives of pAMβ1 has been found within a 1-kb region located downstream from the replication origin (positions ≈4800 to 5800; 172; unpublished results). This function has a broad host range, since it is active in *B. subtilis* and *Clostridium acetobutylicum* (172). It operates by maintaining high-copy-number derivatives in a predominantly monomeric state and has thus been termed resolution function (172).

The resolution function is composed of two elements, a *trans*-active protein named Resβ, encoded by ORFH, and a *cis*-active site named *res*β, localized within a ≈100-bp segment that maps upstream from ORFH, between *ssi* and ORFG (coordinates 4841 to 4951; Fig. 5; unpublished results). This segment contains three imperfect repeats forming 1.5 inverted repeats (Fig. 5) and is active independently of its position and orientation within plasmids derived from pAMβ1.

Sequence comparisons indicated that the N-terminal domain of Resβ is highly homologous to the corresponding region of site-specific recombinases of the invertase (INV)-RES class (172; for a review, see reference 164). Surprisingly, a more detailed analysis detected a weak but significant homology between the C-terminal domains of Resβ and site-specific recombinases of the integrase (INT; 4) class (Fig. 6). Specifically, the amino acids involved in transitory covalent binding of INV-RES or INT recombinases to DNA (Ser or Tyr, respectively; see references 44, 81, and 164 for

```
ResD         NH2 -- 200 H t f R H S y A m h M l y a G I - p L k v L Q s L M G H k - s I s s T - e v Y - t k  29 -- COOH
Cre          NH2 -- 287 H s a R v g a A r d M a r a G V - s I p e I m q a g G w t n - V n I v - m n Y - i r  17 -- COOH
XerC         NH2 -- 238 H k L R H S f A T h M l e s s - g d L r g V Q e L L G H a n - L s t T - q i Y - t H  21 -- COOH
TnpITn4430   NH2 -- 232 H q L R H f f c T n a i e k G - f s I h e V a n q a G H s n - I h t T - l l Y - t -  14 -- COOH
FLP          NH2 -- 303 H i g R H l m t S f L s m k G L t e L t n V v g n w s d k r a s a V a r t t Y - t H  78 -- COOH
Rci          NH2 -- 233 H d L R H e a i S r f f e l G s l n V m e I a a I s G H r - s M n M l - k r Y - t H 111 -- COOH
FimB         NH2 -- 139 H m L R H S c g f a L a n m G I - d t r l I Q d y L G H r n - I r h T - v w Y - t -  23 -- COOH
FimE         NH2 -- 134 H m L R H a c g y e L a e r G - a d t r l I Q d y L G H r n - I r h T - v r Y - t -  26 -- COOH
Intλ         NH2 -- 306 H e L R s l s A - r L y e k q I - s d k f a Q h L L G H k - s d t M a - s q Y r - -  13 -- COOH
IntP2        NH2 -- 267 H a L R H S f A T h f m i n G - g s I i t L Q r I L G H t - r I e q T - m v Y - a H  31 -- COOH
Int186       NH2 -- 275 H v L R H T f A S h f m m n G - g n I l v L Q r V L G H t d - I k M T - m r Y - a H  22 -- COOH
IntP22       NH2 -- 312 H d L R H T w A S w L v q a G V - p I s v L Q e M g G w e - s I e M v - r r Y - a H  36 -- COOH
IntΦ80       NH2 -- 350 H d M R r T i A T n L s e l G - c p p h v I e k L L G H q m - V g V m - a h Y n l H  27 -- COOH
IntP4        NH2 -- 346 H g f R t m a r g a L g e s G L w s d d a I e r q L s H s e r n n V - r a a Y - i H  53 -- COOH
IntpSAM2     NH2 -- 325 r e L R H S f v S l L s d r G V - p L e e I s r L V G H s g t a v t e - e v Y r - k  23 -- COOH
IntpSE211    NH2 -- 377 H d a R H T a A T v L l v l G V - p d r v V m e L M G w s - s V t M - k q r Y - m H  21 -- COOH
TnpITn21     NH2 -- 275 H t L R H S f A T a L l r s G - y d I r t V Q d L L G H s d - V s t T - m i Y - t H  23 -- COOH
ORF2Tn1545   NH2 -- 341 H s L R H T f c T n y a n a G M - n p k a L Q y I M G H a n - I a M T - l n Y y a H  24 -- COOH
TnpATn554    NH2 -- 300 H m L R H T h A T q L i r e G - w d V a f V Q k r L G H a - h V q t T l n t Y - v H  21 -- COOH
TnpBTn554    NH2 -- 463 H a f R H T v g T r M i n n G M - p q h i V Q k f L G H e s p - e M T - s r Y - a H 128 -- COOH

CONSENSUS               H - L R H 1 - A 1 - L - - - G 2 - - 2 - - 2 Q - 2 L G H - - - 2 - 2 T - - - Y - - H

Resβ         NH2 -- 150 n d p R l q h A f d L f l n G L - s d k e V e e q t G - i n r - - r T f r r Y r s r  16 -- COOH
ResIP        NH2 -- 150 n d p R l k h A f d L f l n G L - s d k e V e e q t G - i n r - - r T f r r Y r a r  16 -- COOH
```

Figure 6. Comparison of conserved C-terminal domains of INT recombinases and Resβ and ResIP proteins. The amino acid sequences are from previously published data (4, 18, 29, 41, 102, 107, 111, 122, 130, 137, 155, 159a, 172). Alignments are based on those reported previously (4, 18, 29, 41, 57, 107, 122, 130, 137, 155). Positions within the INT recombinases, where at least 11 residues are related, are shown by uppercase letters (related amino acids are considered to be ST, ILMV, DE, QN, FY, and C; 50). Positions at which at least 11 residues are identical are indicated by boldface uppercase letters. Resβ and ResIP amino acids matching the consensus are shown. The arrow indicates the tyrosine likely to be transiently linked to DNA during recombination is close to the C-terminal part of the conserved domain.

reviews) are conserved in Resβ (172; Fig. 6) suggesting that both domains of Resβ might be functional. This is the first case of homology with the two classes of site-specific recombinases, which suggests that Resβ might be an ancestral enzyme from which the two classes have subsequently evolved or might result from a recent fusion of recombinases belonging to the two classes. In spite of this dual relationship, Resβ might be functionally related to the INT recombinases, as is suggested by the following observations. First, Resβ seems to have a low substrate topology requirement in vivo, as do INT but not INV-RES recombinases (see references 44 and 162 for reviews). This conclusion is based on the observation that Resβ is active on supercoiled and relaxed molecules and catalyzes deletions, inversions, and possibly genome fusions (82a, 172; unpublished results). Second, *res*β resembles recombination sites of INT recombinases such as Cre, FLP, and Int (*lox*, FRT, and *attB*, respectively), since it contains 1.5 inverted repeats, while the typical sites of the INV-RES class of recombinases contain either 2.5 to 3 inverted repeats (RES subclass) or 1 inverted repeat and an enhancer (INV subclass; see references 44, 72, and 164 for reviews).

The appearance of plasmids of the pAMβ1 family that carry inverted repeats could be due to the low substrate topology requirement of Resβ. Intermolecular recombination between *res* sites in opposite orientations would generate a head-to-head dimer molecule, which could evolve by deletion-insertion rearrangements into plasmids of the pSM19305 type.

Sequence comparisons indicated that pIP501 carries a resolution system similar to that of pAMβ1 (159a). The two systems cross-react, which shows that they are functionally related (159a). It is likely that pSM19035 (and possibly all plasmids of the pAMβ1 family) carries a similar system, since a putative *res* site was found 3′ of the *rep* gene of this plasmid by sequence analysis and a stabilization function was mapped downstream from this gene by genetic analysis (168a, 169). Comparison of the *res* sites of the three plasmids revealed two highly conserved sequences within the imperfect repeats (Fig. 5). One, TAGG(A/T)CA, might be the Res-binding site; the other, GTA, is located close to the center of the inverted repeat, where strand exchange occurs in other systems (see reference 44 for a review). The importance of the center of the inverted repeat was indicated by the observation that alteration of the center abolishes resolution activity.

In addition to the resolution activity, Resβ when overproduced has a strong ($\geq$10-fold) copy number reduction (CNR) activity on pAMβ1 and pIP501 derivatives carrying *res* sites (unpublished results). However, resolution and CNR activities have different requirements, since a point mutation at the top of the palindrome abolishes resolution but does not affect CNR, while displacing the *res*β 3.5 kb downstream from the origin does not affect resolution but abolishes CNR activity. These observations suggest a possible mechanism for CNR. Binding of Resβ to *res*, independent of the sequence at the top of the palindrome, could interfere with replication from a nearby but not a distant origin either by affecting initiation of replication by modifying the topology at the origin or by impeding replication fork movement solely in the early replication intermediates. Saturation of *res* sites at high Resβ concentration could then result in a decreased copy number. Interestingly, a twofold decrease of copy number is observed upon insertion of the *res*β gene into a high-copy-number pAMβ1 derivative, which suggests that the CNR system might be part of the regular plasmid copy number control (98).

It is interesting that in the pAMβ1 family, replication and resolution functions are closely linked and even intermingled, which suggests that they evolved in concert to form a cassette ensuring efficient plasmid maintenance. Association of replication and INT-dependent site-specific recombination functions has been observed in other genomes as well. For instance, in *E. coli* plasmid F and *Saccharomyces cerevisiae* plasmid 2μm, the recombination systems ResD-*rfsF* and FLP-FRT, respectively, map close to the replication origins (4, 43, 111, 146, 168). In the *E. coli* chromosome, the *dif* site recognized by XerC maps close to *terC*, which is a major pause site of the DNA replication initiated at *oriC* (16, 17, 36, 41, 52, 108). Possibly, this association is not fortuitous and points to involvement of the site-specific recombination INT-dependent systems in initiation and/or termination of DNA replication of the theta type.

Other plasmids

A RES belonging to the INV-RES family and relatively closely related to the RES of transposon Tn*917* was revealed by sequence analysis of pIP404 (68). According to transcription analysis, its expression is directed by a weak promoter situated 140 bp from the start codon. The promoter is flanked by four partially homologous inverted repeats, which might form the resolution site. It was proposed that this region maintains plasmid population in the monomeric state and thus increases pIP404 segregational stability.

A RES-independent stability system active in *B. subtilis* and *B. thuringiensis* has been found in pHT1030 (118). It is the first system that confers stability not only during vegetative cell growth but also during sporulation and germination. It is localized within a 1-kb region neighboring the replication origin and is composed of two elements, a 15-kDa polypeptide named SpbA and an upstream sequence containing dyad symmetry. The polypeptide has no homology with any of the proteins contained in the data banks. The system is not active in *trans*, and its mode of action is not known. However, the polypeptide might have a lethal effect on the cells, which suggests that the system resembles *ccd* or Kil-Kor of F or R1, respectively (70, 116a).

Structural Stability

Studies with plasmids derived from pAMβ1 and pTB19 RepA determinant and carried out in *B. subtilis* indicate that the theta mode of replication confers much higher stability than the RCR mode. First, recombination between short (9- to 27-bp) and long (several-kilobase) direct repeats is much less frequent (up to 1,000-fold) in theta than in RCR genomes (27a, 99, 138). Second, a theta replicating plasmid inserted into the *B. subtilis* chromosome stimulated 100 times less recombination between long direct repeats placed in their vicinity than did an RCR plasmid (133a, 140, 194). Third, much larger segments can generally be cloned in theta replicating than in RCR plasmids; in a shotgun experiment, the average insert sizes were about 10 and 1 kb, with ranges of 3 to 17 and 0.1 to 1.7 kb in the theta- and RCR-derived vectors, respectively (98). These observations have stimulated development of a number of theta replicating cloning vectors (2, 8, 82, 98, 133, 149, 161, 166, 169). Their successful use in organisms other than *B. subtilis*, such as *B. thuringiensis*, *L. lactis*, *S. aureus*, and *C. acetobutylicum*, indicates that high structural stability might be a general property of theta replicating plasmids from gram-positive bacteria.

Structural instability is a poorly understood phenomenon, in spite of its importance for evolution, medicine, and biotechnology, and the reasons for the difference between theta and RCR plasmids are not entirely clear. However, several features that might lead to high structural instability, such as (i) uncoupling of leading- and lagging-strand syntheses, which generates single-stranded molecules active in recombination (61, 61a); (ii) nicking and closing activities of the Rep proteins, which might be deletogenic (131, 132); and (iii) the propensity to generate HMW upon insertion of foreign DNA (45, 77), which might select for plasmids having lost the insert, characterize RCR replication and are absent from theta replication.

Conjugational Transfer

Large plasmids (>25 kb) from gram-positive bacteria are frequently autotransferable. In general, the transfer is efficient on solid media but not in liquid. However, in response to a pheromone synthesized by the potential recipient bacteria, intraspecific transfer of certain *Enterococcus faecalis* plasmids becomes very efficient in liquid because of induction of genes that encode a substance conferring aggregation properties on donor bacteria. This phenomenon, recently reviewed elsewhere (38, 39), is not discussed here.

Transfer on solid media allows DNA exchange between very divergent bacteria, since it can occur interspecifically and even heterogramically (181, 182). It seems to be involved in the horizontal dissemination of adaptive genes, such as antibiotic resistance and substrate utilization, and is widely used in the laboratory to introduce foreign genetic information into otherwise nontransformable organisms (84, 183). In spite of the importance of such transfer, its mechanism, especially in gram-positive bacteria, is poorly understood (see references 96, 190, and 191 for a review of transfer of plasmids originating from gram-negative bacteria).

Generally, the determinants for conjugal transfer of plasmids isolated from gram-positive bacteria are specified by large (15- to 20-kb) regions, as has been shown for pAMβ1, pIP501, pAD1, pGO1, pSK41, pUW3626, and pCRG1690 (5, 32, 58, 105, 113, 180) (however, theta replication has not been demonstrated for all these plasmids). This indicates involvement of numerous genes, as previously observed for conjugative plasmids in gram-negative bacteria. It is possible that plasmid-borne transfer functions are simpler in streptomycetes, since the functions are carried on a much smaller region in SCP2 (120). However, additional functions could be encoded on the host chromosome (89).

Three regions were shown to be involved in conjugal transfer of pIP501. Two of these, named A and B, which measure 7.5 and 8 kb, respectively, are sufficient for efficient plasmid transfer between strains of *E. faecalis*, while the third region (4.5 kb) is necessary for transfer to *Streptococcus sanguis* (105, 106). Five genes were detected in region A, but only three, named *cnjA*, *cnjB*, and *cnjC*, which are consecutive and are transcribed in the same direction, appear to be involved in conjugation. The functions of the other two genes, which are also consecutive and are transcribed in the opposite direction, are not known (105). Transfer genes of pGO1 are also organized in several transcription units (3).

Nonmobilizable plasmids containing a specific segment of pIP501 are transmitted at high frequencies among *L. lactis* subsp. *lactis* strains if transfer functions are provided in *trans* by a pIP501 derivative (111a). The DNA segment containing the mobilization (*mob*) function is characterized by three palindromes and a 202-amino-acid ORF of unknown function. The smallest segment conferring high mobilization frequency was localized to a 1-kb segment extending between pIP501 coordinates 3.6 and 4.6. This segment carries only one of the three palindromes, named *palI*, with a predicted ΔG of −27 kcal/mol and no ORF. A second sequence nearly identical to *palI* is present on pIP501, upstream of the plasmid copy control region. Further homologies with *palI* are found on pAMβ1 (13). The *mob* region maps outside of the previously defined transfer region and is inactive in *recA L. lactis* subsp. *lactis* strains, which suggests that it might act as a hot spot for cointegrate formation by homologous recombination.

Acknowledgments. We thank V. Bidnenko, A. Blanchard, S. Bron, C. Bruand, P. Dabert, D. Halpern, P. Langella, E. Le Chatelier, E. Maguin, F. Morel, M. F. Noirot-Gros, R. P. Novick, J. Polak, C. Pujol, A. V. Sorokin, and C. M. Thomas, who communicated unpublished results to us, and F. Haimet, who did the artwork. Work from our laboratory was supported in part by grants from the Commission of the European Community (reference BIOT-CT91-0268) and from the Fondation pour la Recherche Médicale.

REFERENCES

1. **Amundsen, S. K., A. M. Neiman, S. M. Thibodeaux, and G. R. Smith.** 1990. Genetic dissection of the biochemical activities of RecBCD enzyme. *Genetics* **126:**25–40.
2. **Arantes, O., and D. Lereclus.** 1991. Construction of cloning vectors for *Bacillus thuringiensis*. *Gene* **108:**115–119.
3. **Archer, G. L., and W. D. Thomas, Jr.** 1990. Conjugative transfer of antimicrobial resistance genes between staphylococci, p. 115–122. *In* R. Novick (ed.), *Molecular Biology of the Staphylococci*. VCH Press, New York.
4. **Argos, P., A. Landy, K. Abremski, J. B. Egan, E. Haggard-Ljungquist, R. H. Hoess, M. L. Kahn, B. Kalionis, S. V. L. Narayana, L. S. Pierson III, N. Sternberg, and J. M. Leong.** 1986. The integrase family of site-specific recombinases: regional similarities and global diversity. *EMBO J.* **5:**433–440.
5. **Asch, D. K., R. V. Goering, and E. A. Ruff.** 1984. Isolation and preliminary characterization of a plasmid

mutant derepressed for conjugal transfer in *Staphylococcus aureus*. *Plasmid* **12**:197–202.

6. **Ballester, S., P. Lopez, M. Espinosa, J. C. Alonso, and S. A. Lacks.** 1989. Plasmid structural instability associated with pC194 replication functions. *J. Bacteriol.* **171**: 2271–2277.
7. **Bates, E. E., and H. J. Gilbert.** 1989. Characterization of a cryptic plasmid from *Lactobacillus plantarum*. *Gene* **85**:253–258.
8. **Baum, J. A., D. M. Coyle, M. P. Gilbert, C. S. Jany, and C. Gawron-Burke.** 1990. Novel cloning vectors for *Bacillus thuringiensis*. *Appl. Environ. Microbiol.* **56**:3420–3428.
9. **Baum, J. A., and M. P. Gilbert.** 1991. Characterization and comparative sequence analysis of replication origins from three large *Bacillus thuringiensis* plasmids. *J. Bacteriol.* **173**:5280–5289.
10. **Behnke, D., and M. S. Gilmore.** 1981. Location of antibiotic resistance determinants, copy control, and replication functions on the double-selective streptococcal cloning vector pGB301. *Mol. Gen. Genet.* **184**: 115–120.
11. **Behnke, D., V. I. Golubkov, H. Malke, A. S. Boitsov, and A. A. Totolian.** 1979. Restriction endonuclease analysis of group a streptococcal plasmids determining resistance to macrolides, lincosamides and streptogramin-B antibiotics. *FEMS Microbiol. Lett.* **6**:5–9.
12. **Behnke, D., and S. Klaus.** 1983. Double or triple sets of replication functions as inverted and direct repeats on *in vitro* reconstructed streptococcal MLS resistance plasmids. *Z. Allg. Mikrobiol.* **23**:539–547.
13. **Bidnenko, V.** Personal communication.

13a. **Bidnenko, V. E., A. Gruss, and S. D. Ehrlich.** Mutation in the plasmid pUB110 Rep protein affects its ability to terminate rolling circle replication. Submitted for publication.

14. **Bièrne, H., S. D. Ehrlich, and B. Michel.** 1991. The replication termination signal *terB* of the *Escherichia coli* chromosome is a deletion hot spot. *EMBO J.* **10**: 2699–2705.
15. **Birch, P., and S. A. Khan.** 1992. Replication of single-stranded plasmid pT181 DNA *in vitro*. *Proc. Natl. Acad. Sci. USA* **89**:290–294.
16. **Black, D. S., A. J. Kelly, M. J. Mardis, and H. S. Moyed.** 1991. Structure and organization of *hip*, an operon that affects lethality due to inhibition of peptidoglycan or DNA synthesis. *J. Bacteriol.* **173**:5732–5739.
17. **Blakely, G., S. Colloms, G. May, M. Burke, and D. Sherratt.** 1991. *Escherichia coli* XerC recombinase is required for chromosomal segregation at cell division. *New Biol.* **3**:789–798.

17a. **Blanchard, A.** Personal communication.

18. **Boccard, F., T. Smokvina, J.-L. Pernodet, A. Friedmann, and M. Guérineau.** 1989. The integrated conjugative plasmid pSAM2 of *Streptomyces ambofaciens* is related to temperate bacteriophages. *EMBO J.* **8**:973–980.
19. **Boe, L., M.-F. Gros, H. te Riele, S. Ehrlich, and A. Gruss.** 1989. Replication origins of single-stranded DNA plasmid pUB110. *J. Bacteriol.* **171**:3366–3372.
20. **Boistov, A. S., V. I. Golubkov, I. M. Iontova, E. N. Zaitsev, H. Malke, and A. A. Totolian.** 1979. Inverted repeats on plasmids determining resistance to MLS antibiotics in group A streptococci. *FEMS Microbiol. Lett.* **6**:11–14.
21. **Bouia, A., F. Bringel, L. Frey, B. Kammerer, A. Belarbi, A. Guyonvarch, and J.-C. Hubert.** 1989. Structural organization of pLP1, a cryptic plasmid from *Lactobacillus plantarum* CCM 1904. *Plasmid* **22**:185–192.
22. **Bourgueleret, L., G. Bieth, and T. Horodniceanu.** 1981. Conjugative R plasmids in group C and G streptococci. *J. Bacteriol.* **145**:1102–1105.
23. **Bramhill, D., and A. Kornberg.** 1988. A model for initiation at origins of DNA replication. *Cell* **54**:915–918.

23a. **Brantl, S., and D. Behnke.** 1991. Copy number control of the streptococcal plasmid pIP501 occurs at three levels. *Nucleic Acids Res.* **20**:395–400.

23b. **Brantl, S., and D. Behnke.** 1992. The amount of RepR protein determines the copy number of plasmid pIP501 in *Bacillus subtilis*. *J. Bacteriol.* **174**:5475–5478.

23c. **Brantl, S., and D. Behnke.** 1992. Characterization of the minimal origin required for replication of the streptococcal plasmid pIP501 in *Bacillus subtilis*. *Mol. Microbiol.* **6**:3501–3510.

24. **Brantl, S., D. Behnke, and J. C. Alonso.** 1990. Molecular analysis of the replication region of the conjugative *Streptococcus agalactiae* plasmid pIP501 in *Bacillus subtilis*. Comparison with plasmids pAMβ1 and pSM19035. *Nucleic Acids Res.* **18**:4783–4789.
25. **Brantl, S., A. Nowak, D. Behnke, and J. C. Alonso.** 1989. Revision of the nucleotide sequence of the *Streptococcus pyogenes* plasmid pSM19035 *repS* gene. *Nucleic Acids Res.* **17**:10110.

25a. **Brantl, S., B. Nuez, and D. Behnke.** 1992. *In vitro* and *in vivo* analysis of transcription within the replication region of plasmid pIP501. *Mol. Gen. Genet.* **234**:105–112.

26. **Brefort, G., M. Magot, H. Ionesco, and M. Sebald.** 1977. Characterization and transferability of *Clostridium perfringens* plasmids. *Plasmid* **1**:52–66.

26a. **Bron, S.** Personal communication.

26b. **Bron, S.** Unpublished data.

27. **Bron, S.** 1990. Plasmids, p. 75–138. *In* C. R. Harwood, and S. M. Cutting (ed.), *Molecular Biology Methods for Bacillus*, vol. 3. John Wiley & Sons, Ltd., London.

27a. **Bron, S., S. Holsappel, G. Venema, and B. Peeters.** 1991. Plasmid deletion formation between short, direct repeats in *Bacillus subtilis* is stimulated by single-stranded rolling-circle replication intermediates. *Mol. Gen Genet.* **226**:88–96.

28. **Bron, S., and E. Luxen.** 1985. Segregational instability of pUB110-derived recombinant plasmids in *Bacillus subtilis*. *Plasmid* **14**:235–244.

28a. **Bron, S., E. Luxen, and P. Swart.** 1988. Instability of recombinant pUB110 plasmids in *Bacillus subtilis*: plasmid-encoded stability function and effects of DNA inserts. *Plasmid* **19**:231–241.

29. **Brown, D. P., K. B. Idler, and L. Katz.** 1990. Characterization of the genetic elements required for site-specific integration of plasmid pSE211 in *Saccharopolyspora erythraea*. *J. Bacteriol.* **172**:1877–1888.

29a. **Bruand, C.** Personal communication.

30. **Bruand, C., S. D. Ehrlich, and L. Jannière.** 1990. A method for detecting unidirectional theta replication in *Bacillus subtilis* plasmids, p. 123–129. *In* M. M. Zukowski, A. T. Ganesan, and J. A. Hoch (ed.), *Genetics and biotechnology of Bacilli*, vol. 3. Academic Press, Inc., San Diego, Calif.
31. **Bruand, C., S. D. Ehrlich, and L. Jannière.** 1991. Unidirectional theta replication of the structurally stable *Enterococcus faecalis* plasmid pAMβ1. *EMBO J.* **10**: 2171–2177.
32. **Byrne, M. E., M. T. Gillepsie, and R. A. Skurray.** 1990. Molecular analysis of a gentamicin resistance transposonlike element on plasmids isolated from North American *Staphylococcus aureus* strains. *Antimicrob. Agents Chemother.* **34**:2106–2113.
33. **Camilli, A., D. A. Portnoy, and P. Youngman.** 1990. Insertional mutagenesis of *Listeria monocytogenes* with a novel Tn*917* derivative that allows direct cloning of DNA flanking transposon insertions. *J. Bacteriol.* **172**: 3738–3744.
34. **Carrigan, C. M., J. A. Haarsma, M. T. Smith, and R. G. Wake.** 1987. Sequence features of the replication termi-

nus of the *Bacillus subtilis* chromosome. *Nucleic Acids Res.* **15:**8501–8509.
35. **Carrigan, C. M., R. A. Pack, M. T. Smith, and R. G. Wake.** 1991. Normal *terC*-region of the *Bacillus subtilis* chromosome acts in a polar manner to arrest the clockwise replication fork. *J. Mol. Biol.* **222:**197–207.
36. **Clerget, M.** 1991. Site-specific recombination promoted by short DNA segment of plasmid R1 and by homologous segment in the terminus region of the *Escherichia coli* chromosome. *New Biol.* **3:**780–788.
37. **Clewell, D. B., and A. E. Franke.** 1974. Characterization of a plasmid determining resistance to erythromycin, lincomycin, and vernamycin B_α in a strain of *Streptococcus pyogenes*. *Antimicrob. Agents Chemother.* **5:**534–537.
38. **Clewell, D. B., L. T. Pontius, K. E. Weaver, F. Y. An, Y. Ike, A. Suzuki, and J. Nakayama.** 1990. *Enterococcus faecalis* hemolysin/bacteriocin plasmid pAD1: regulation of the pheromone response, p. 3–8. *In* G. M. Dunny, P. P. Cleary, and L. L. McKay (ed.), *Genetics and Molecular Biology of Streptococci, Lactococci and Enterococci*. American Society for Microbiology, Washington, D.C.
39. **Clewell, D. B., and K. E. Weaver.** 1989. Sex pheromones and plasmid transfer in *Enterococcus faecalis*. *Plasmid* **21:**175–184.
40. **Clewell, D. B., Y. Yagi, G. M. Dunny, and S. K. Schultz.** 1974. Characterization of three plasmid deoxyribonucleic acid molecules in a strain of *Streptococcus faecalis*: identification of a plasmid determining erythromycin resistance. *J. Bacteriol.* **117:**283–289.
41. **Colloms, S. D., P. Sykora, G. Szatmari, and D. J. Sherratt.** 1990. Recombination at ColE1 requires the *Escherichia coli xerC* gene product, a member of the lambda integrase family of site-specific recombinases. *J. Bacteriol.* **172:**6973–6980.
42. **Courvalin, P. M., C. Carlier, and Y. A. Chabbert.** 1972. Plasmid-linked tetracycline and erythromycin resistance in group D *"Streptococcus."* *Ann. Inst. Pasteur* (Paris) **123:**755–759.
43. **Cox, M. M.** 1989. DNA inversion in the 2μm plasmid of *Saccharomyces cerevisiae*, p. 661–670. *In* D. E. Berg and M. M. Howe (ed.), *Mobile DNA*. American Society for Microbiology, Washington, D.C.
44. **Craig, N. L.** 1988. The mechanism of conservative site-specific recombination. *Annu. Rev. Genet.* **22:**77–105.
44a. **Dabert, P.** Personal communication.
45. **Dabert, P., S. D. Ehrlich, and A. Gruss.** 1992. High-molecular-weight linear multimer formation by single-stranded DNA plasmids in *Escherichia coli*. *J. Bacteriol.* **174:**173–178.
46. **Dabert, P., S. D. Ehrlich, and A. Gruss.** 1992. Chi sequence protects against RecBCD degradation of DNA *in vivo*. *Proc. Natl. Acad. Sci. USA* **89:**12073–12077.
47. **Dabert, P., S. D. Ehrlich, and A. Gruss.** Effects of chi on ssDNA plasmids and high-molecular-weight linear plasmid multimer formation. Submitted for publication.
48. **Dao, M. L., and J. J. Ferretti.** 1985. *Streptococcus-Escherichia coli* shuttle vector pSA3 and its use in the cloning of streptococcal genes. *Appl. Environ. Microbiol.* **49:**115–119.
49. **Darabi, A., R. Forough, G. Bhardwaj, M. Watabe, G. Goodarazi, S. C. Gross, and K. Watabe.** 1989. Identification and nucleotide sequence of the minimal replicon of the low-copy-number plasmid pBS2. *Plasmid* **22:**281–286.
50. **Dayhoff, M. O., R. M. Schwartz, and B. C. Orcutt.** 1978. A model of evolutionary change in proteins, p. 345. *In* M. O. Dayhoff (ed.), *Atlas of Protein Sequence and Structure*, vol. 5, suppl. 3. National Biomedical Research Foundation, Washington, D.C.
51. **del Solar, G., and M. Espinosa.** 1992. The copy number of plasmid pLS1 is regulated by two *trans*-acting plasmid products: the antisense RNAII and the repressor protein, RepA. *Mol. Microbiol.* **6:**83–94.
52. **de Massy, B., S. Béjar, J. Louarn, and J.-M. Louarn.** 1987. Inhibition of replication forks exiting the terminus region of the *Escherichia coli* chromosome occurs at loci separated by 5 min. *Proc. Natl. Acad. Sci. USA* **84:**1759–1763.
53. **Dempsey, L. A., P. Birch, and S. A. Khan.** 1992. Uncoupling of the DNA topoisomerase and replication activities of an initiator protein. *Proc. Natl. Acad. Sci. USA* **89:**3083–3087.
54. **de Rossi, E., A. Milano, P. Brigidi, F. Brini, and G. Riccardi.** 1992. Structural organization of pBC1, a cryptic plasmid from *Bacillus coagulans*. *J. Bacteriol.* **174:**638–642.
55. **Devine, K., S. Hogan, D. Higgins, and D. McConnell.** 1989. Replication and segregational stability of the *Bacillus* plasmid pBAA1. *J. Bacteriol.* **171:**1166–1172.
56. **Donoghue, D. J., and P. A. Sharp.** 1978. Replication of colicin E1 plasmid DNA in vivo requires no plasmid-encoded proteins. *J. Bacteriol.* **133:**1287–1294.
57. **Dorman, C. J., and C. F. Higgins.** 1987. Fimbrial phase variation in *Escherichia coli*: dependence on integration host factor and homologies with other site-specific recombinases. *J. Bacteriol.* **169:**3840–3843.
58. **Ehrenfeld, E. E., and D. B. Clewell.** 1987. Transfer functions of the *Streptococcus faecalis* plasmid pAD1: organization of plasmid DNA encoding response to sex pheromone. *J. Bacteriol.* **169:**3473–3481.
59. **Ehrlich, S. D.** 1989. Illegitimate recombination in bacteria, p. 799–832. *In* D. E. Berg and M. M. Howe (ed.), *Mobile DNA*. American Society for Microbiology, Washington, D.C.
60. **Ehrlich, S. D., C. Bruand, S. Sozhamannan, P. Dabert, M.-F. Gros, L. Jannière, and A. Gruss.** 1991. Plasmid replication and structural stability in *Bacillus subtilis*. *Res. Microbiol.* **142:**869–873.
61. **Ehrlich, S. D., P. Noirot, M. A. Petit, L. Jannière, B. Michel, and H. te Riele.** 1986. Structural instability of *Bacillus subtilis* plasmids, p. 71–83. *In* J. K. Setlow and A. Hollaender (ed.), *Genetic Engineering*, vol. 8. Plenum Press, New York.
61a. **Ehrlich, S. D., H. te Riele, M. A. Petit, L. Jannière, P. Noirot, and B. Michel.** 1986. DNA recombination in plasmids and the chromosome of *Bacillus subtilis*, p. 27–34. *In* A. T. Ganesan and J. Hoch (ed.), *Bacillus Molecular Genetics and Biotechnology Applications*. Academic Press, Inc., Orlando, Fla.
62. **El-Solh, N., D. H. Bouanchaud, T. Horodniceanu, A. Roussel, and Y. A. Chabbert.** 1978. Molecular studies and possible relatedness between R plasmids from groups B and D streptococci. *Antimicrob. Agents Chemother.* **14:**19–23.
63. **Evans, R. P., Jr., and F. L. Macrina.** 1983. Streptococcal R plasmid pIP501: endonuclease site map, resistance determinant location, and construction of novel derivatives. *J. Bacteriol.* **154:**1347–1355.
64. **Feirtag, J. M., J. P. Petzel, E. Pasalodos, K. A. Baldwin, and L. L. McKay.** 1991. Thermosensitive plasmid replication, temperature-sensitive host growth, and chromosomal plasmid integration conferred by *Lactococcus lactis* subsp. *cremoris* lactose plasmids in *Lactococcus lactis* subsp. *lactis*. *Appl. Environ. Microbiol.* **57:**539–548.
65. **Forbes, B. A.** 1979. Ph.D. thesis. University of Oklahoma Health Sciences Center, Oklahoma City.
66. **Garnier, T., and S. T. Cole.** 1988. Identification and molecular genetic analysis of replication functions of the bacteriocinogenic plasmid pIP404 from *Clostridium perfringens*. *Plasmid* **19:**151–160.
67. **Garnier, T., and S. T. Cole.** 1988. Complete nucleotide

sequence and genomic organization of the bacteriocinogenic plasmid, pIP404, from *Clostridium perfringens*. *Plasmid* **19:**134–150.

68. **Garnier, T., W. Saurin, and S. T. Cole.** 1987. Molecular characterization of the resolvase gene, *res*, carried by a multicopy plasmid from *Clostridium perfringens*: common evolutionary origin for prokaryotic site-specific recombinases. *Mol. Microbiol.* **1:**371–376.
69. **Gennaro, M. L.** 1990. DNA replication and its regulation in the pT181 plasmid family, p. 183–195. *In* R. P. Novick (ed.), *Molecular Biology of the Staphylococci*. VCH Press, New York.
70. **Gerdes, K., L. K. Poulsen, T. Thisted, A. K. Nielsen, J. Martinussen, and P. H. Andreasen.** 1990. The *hok* killer gene family in gram-negative bacteria. *New Biol.* **2:**946–956.
71. **Gilmore, M. S., D. Behnke, and J. J. Ferretti.** 1982. Evolutionary relatedness of MLS resistance and replication function sequences on streptococcal antibiotic resistance plasmids, p. 174–176. *In* D. Schlessinger (ed.), *Microbiology—1982*. American Society for Microbiology, Washington, D.C.
72. **Glasgow, A. C., K. T. Hughes, and M. I. Simon.** 1989. Bacterial DNA inversion systems, p. 637–659. *In* D. E. Berg and M. M. Howe (ed.), *Mobile DNA*. American Society for Microbiology, Washington, D.C.
73. **Golubkov, V. I., W. Reichardt, A. S. Boistov, I. M. Iontova, H. Malke, and A. A. Totolian.** 1982. Sequence relationships between plasmids associated with conventional MLS resistance and zonal lincomycin resistance in *Streptococcus pyogenes*. *Mol. Gen. Genet.* **187:**310–315.
74. **Grinius, L., G. Dreguniene, E. B. Goldberg, C.-H. Liao, and S. J. Projan.** 1992. A staphylococcal multidrug resistance gene product is a member of a new protein family. *Plasmid* **27:**119–129.
75. **Gros, M. F., H. te Riele, and S. D. Ehrlich.** 1987. Rolling circle replication of the single-stranded plasmid pC194. *EMBO J.* **6:**3863–3869.
76. **Gros, M.-F., H. te Riele, and S. D. Ehrlich.** 1989. Replication origin of a single-stranded DNA plasmid pC194. *EMBO J.* **8:**2711–2716.
77. **Gruss, A., and S. D. Ehrlich.** 1988. Insertion of foreign DNA into plasmids from gram-positive bacteria induces formation of high-molecular-weight plasmid multimers. *J. Bacteriol.* **170:**1183–1190.
78. **Gruss, A., and S. D. Ehrlich.** 1989. The family of highly interrelated single-stranded deoxyribonucleic acid plasmids. *Microbiol. Rev.* **53:**231–241.
79. **Gruss, A., H. Ross, and R. Novick.** 1987. Functional analysis of a palindromic sequence required for normal replication of several staphylococcal plasmids. *Proc. Natl. Acad. Sci. USA* **84:**2165–2169.
79a. **Halpern, D.** Personal communication.
80. **Hara, T., S. Nagatomo, S. Ogata, and S. Ueda.** 1991. Molecular structure of the replication origin of a *Bacillus subtilis* (*natto*) plasmid, pUH1. *Appl. Environ. Microbiol.* **57:**1838–1841.
81. **Hatfull, G. F., and N. D. F. Grindley.** 1988. Resolvases and DNA-invertases: a family of enzymes active in site-specific recombination, p. 357–396. *In* R. Kucherlapati and G. R. Smith (ed.), *Genetic Recombination*. American Society for Microbiology, Washington, D.C.
82. **Hayes, F., C. Daly, and G. F. Fitzgerald.** 1990. Identification of the minimal replicon of *Lactococcus lactis* subsp. *lactis* UC317 plasmid pCI305. *Appl. Environ. Microbiol.* **56:**202–209.
82a. **Hayes, F., C. Daly, and G. F. Fitzgerald.** 1990. High-frequency, site-specific recombination between lactococcal and pAMβ1 plasmid DNAs. *J. Bacteriol.* **172:**3485–3489.
83. **Hayes, F., P. Vos, G. F. Fitzgerald, W. M. de Vos, and C. Daly.** 1991. Molecular organization of the minimal replicon of novel, narrow-host-range, lactococcal plasmid pCI305. *Plasmid* **25:**16–26.
84. **Heinemann, J. A.** 1991. Genetics of gene transfer between species. *Trends Genet.* **7:**181–185.
85. **Hershfield, V.** 1979. Plasmids mediating multiple drug resistance in group B streptococcus: transferability and molecular properties. *Plasmid* **2:**137–149.
86. **Hespell, R. B., and T. R. Whitehead.** 1991. Conjugal transfer of Tn*916*, Tn*916*ΔE, and pAMβ1 from *Enterococcus faecalis* to *Butyrivibrio fibrisolvens* strains. *Appl. Environ. Microbiol.* **57:**2703–2709.
87. **Highlander, S., and R. P. Novick.** 1987. Plasmid repopulation kinetics in *Staphylococcus aureus*. *Plasmid* **17:**210–221.
88. **Hill, C., L. A. Miller, and T. R. Klaenhammer.** 1991. The bacteriophage resistance plasmid pTR2030 forms high-molecular-weight multimers in lactococci. *Plasmid* **25:**105–112.
89. **Hopwood, D. A., D. J. Lydiate, F. Malpartida, and H. M. Wright.** 1984. Conjugative sex plasmids of *Streptomyces*, p. 615–634. *In* D. R. Helinski, S. N. Cohen, D. B. Clewell, D. A. Jackson, and A. Hollaender (ed.), *Plasmids in Bacteria*. Plenum Press, New York.
90. **Horaud, T., C. Le Bouguenec, and K. Pepper.** 1985. Molecular genetics of resistance to macrolides, lincosamides and streptogramin B (MLS) in streptococci. *J. Antimicrob. Chemother.* **16:**111–135.
91. **Horng, J. S., K. M. Polzin, and L. L. McKay.** 1991. Replication and temperature-sensitive maintenance functions of lactose plasmid pSK11L from *Lactococcus lactis* subsp. *cremoris*. *J. Bacteriol.* **173:**7573–7581.
92. **Horodniceanu, T., D. H. Bouanchaud, G. Bieth, and Y. A. Chabbert.** 1976. R plasmids in *Streptococcus agalactiae* (group B). *Antimicrob. Agents Chemother.* **10:**795–801.
93. **Imanaka, T., M. Fujii, and S. Aiba.** 1981. Isolation and characterization of antibiotic resistance plasmids from thermophilic bacilli and construction of deletion plasmids. *J. Bacteriol.* **146:**1091–1097.
94. **Imanaka, T., H. Ishikawa, and S. Aiba.** 1986. Complete nucleotide sequence of the low copy number plasmid pRAT11 and replication control by the RepA protein in *Bacillus subtilis*. *Mol. Gen. Genet.* **205:**90–96.
95. **Iordanescu, S.** 1976. Temperature-sensitive mutant of a tetracycline resistant staphylococcal plasmid. *Arch. Roum. Pathol. Exp. Microbiol.* **35:**257–264.
95a. **Iordanescu, S., and R. Basheer.** 1991. The *Staphylococcus aureus* mutation pcrA3 leads to the accumulation of pT181 replication initiation complexes. *J. Mol. Biol.* **221:**1183–1189.
96. **Ippen-Ihler, K. A., and E. G. Minkley, Jr.** 1986. The conjugation system of F, the fertility factor of *Escherichia coli*. *Annu. Rev. Genet.* **20:**593–624.
97. **Itoh, T., and J. Tomizawa.** 1978. Initiation of replication of plasmid ColE1 DNA by RNA polymerase, ribonuclease H, and DNA polymerase I. Cold Spring Harbor Symp. Quant. Biol. **43:**409–417.
97a. **Jahns, A., A. Schäfer, A. Geis, and M. Teuber.** 1991. Identification, cloning and sequencing of the replication region of *Lactococcus lactis* ssp. *lactis* biovar. diacetylactis Bu2 citrate plasmid pSL2. *FEMS Microbiol. Lett.* **80:**253–258.
98. **Jannière, L., C. Bruand, and S. D. Ehrlich.** 1990. Structurally stable *Bacillus subtilis* cloning vectors. *Gene* **87:**53–61.
99. **Jannière, L., and S. D. Ehrlich.** 1987. Recombination between short repeat sequences is more frequent in plasmids than in the chromosome of *Bacillus subtilis*. *Mol. Gen. Genet.* **210:**116–121.
100. **Josson, K., P. Soetaert, F. Michiels, H. Joos, and J. Mahillon.** 1990. *Lactobacillus hilgardii* plasmid pLAB1000 consists of two functional cassettes com-

monly found in other gram-positive organisms. *J. Bacteriol.* **172:**3089–3099.

100a. **King, K. W., and K. Dybvig.** 1992. Nucleotide sequence of *Mycoplasma mycoides* subspecies *mycoides* plasmid pKMK1. *Plasmid* **28:**86–91.

101. **Kleanthous, H., C. L. Clayton, and S. Tabaqchali.** 1991. Characterization of a plasmid from *Helicobacter pylori* encoding a replication protein common to plasmids in gram-positive bacteria. *Mol. Microbiol.* **5:**2377–2389.

102. **Klemm, P.** 1986. Two regulatory *fim* genes, *fimB* and *fimE*, control the phase variation of type 1 fimbriae in *Escherichia coli. EMBO J.* **5:**1389–1393.

103. **Koepsel, R., R. Murray, W. Rosenblum, and S. Khan.** 1985. The replication initiator protein of plasmid pT181 has sequence-specific endonuclease and topoisomerase-like activities. *Proc. Natl. Acad. Sci. USA* **82:**6845–6849.

104. **Kornberg, A., and T. A. Baker.** 1991. *DNA Replication*, 2nd ed. W. H. Freeman & Co., New York.

105. **Krah, E. R., III, and F. L. Macrina.** 1989. Genetic analysis of the conjugal transfer determinants encoded by the streptococcal broad-host-range plasmid pIP501. *J. Bacteriol.* **171:**6005–6012.

106. **Krah, E. R., III, and F. L. Macrina.** 1991. Identification of a region that influences host range of the streptococcal conjugative plasmid pIP501. *Plasmid* **25:**64–69.

107. **Kubo, A., A. Kusukawa, and T. Komano.** 1988. Nucleotide sequence of the *rci* gene encoding shufflon-specific DNA recombinase in the IncI1 plasmid R64: homology to the site-specific recombinases of integrase family. *Mol. Gen. Genet.* **213:**30–35.

108. **Kuempel, P. L., J. M. Henson, L. Dircks, M. Tecklenburg, and D. F. Lim.** 1991. *dif*, a *recA*-independent recombination site in the terminus region of the chromosome of *Escherichia coli. New Biol.* **3:**799–811.

109. **Kües, U., and U. Stahl.** 1989. Replication of plasmids in gram-negative bacteria. *Microbiol. Rev.* **53:**491–516.

110. **Kurusu, Y., M. Inui, K. Kohama, M. Kobayashi, M. Terasawa, and H. Yukawa.** 1991. Identification of plasmid partition function in coryneform bacteria. *Appl. Environ. Microbiol.* **57:**759–764.

111. **Lane, D., R. de Feyter, M. Kennedy, S.-H. Phua, and D. Semon.** 1986. D protein on miniF plasmid acts as a repressor of transcription and as a site-specific resolvase. *Nucleic Acids Res.* **14:**9713–9728.

111a. **Langella, P.** Personal communication.

112. **Langella, P., and A. Chopin.** 1989. Effect of restriction-modification systems on transfer of foreign DNA into *Lactococcus lactis* subsp. *lactis. FEMS Microbiol. Lett.* **59:**301–306.

113. **Leblanc, D. J., and L. N. Lee.** 1984. Physical and genetic analyses of streptococcal plasmid pAMβ1 and cloning of its replication region. *J. Bacteriol.* **157:**445–453.

113a. **LeBlanc, D. J., L. N. Lee, and A. Abu-Al-Jaibat.** 1992. Molecular, genetic, and functional analysis of the basic replicon of pVA380-1, a plasmid of oral streptococcal origin. *Plasmid* **28:**130–145.

113b. **Le Chatelier, E.** Personal communication.

114. **Le Chatelier, E., C. Bruand, S. D. Ehrlich, and L. Jannière.** A novel class of theta-replicating plasmids: the pAMβ1 family from gram-positive bacteria. Submitted for publication.

115. **Le Chatelier, E., S. D. Ehrlich, and L. Jannière.** 1993. Biochemical and genetic analysis of the unidirectional theta replication of the *S. agalactiae* plasmid pIP501. *Plasmid* **29:**50–56.

116. **Leenhouts, K., J. B. Tolner, S. Bron, J. Kok, G. Venema, and J. F. M. L. Seegers.** 1991. Nucleotide sequence and characterization of the broad-host-range lactococcal plasmid pWV01. *Plasmid* **26:**55–66.

116a. **Lereclus, D.** Personal communication.

117. **Lereclus, D., S. Guo, V. Sanchis, and M.-M. Lecadet.** 1988. Characterization of two *Bacillus thuringiensis* plasmids whose replication is thermosensitive in *B. subtilis. FEMS Microbiol. Lett.* **49:**417–422.

118. **Lereclus, D., and O. Orantes.** 1992. *spbA* locus ensures the segregational stability of pHT1030, a novel type of gram-positive replicon. *Mol. Microbiol.* **6:**35–46.

119. **Lewis, P. J., G. B. Ralston, R. I. Christopherson, and R. G. Wake.** 1990. Identification of the replication terminator protein binding sites in the terminus region of the *Bacillus subtilis* chromosome and stoichiometry of the binding. *J. Mol. Biol.* **214:**73–84.

120. **Lydiate, D. J., F. Malpartida, and D. A. Hopwood.** 1985. The *Streptomyces* plasmid SCP2: its functional analysis and development into useful cloning vectors. *Gene* **35:**223–235.

120a. **Madsen, S. M., L. Andrup, and L. Boe.** Fine mapping and DNA sequence of replication functions of *Bacillus thuringiensis* plasmid pTX14-3. *Plasmid*, in press.

120b. **Maguin, E.** Personal communication.

121. **Maguin, E., P. Duwat, T. Hege, D. Ehrlich, and A. Gruss.** 1992. New thermosensitive plasmid for gram-positive bacteria. *J. Bacteriol.* **174:**5633–5638.

122. **Mahillon, J., and D. Lereclus.** 1988. Structural and functional analysis of Tn*4430*: identification of an integrase-like protein involved in the co-integrate-resolution process. *EMBO J.* **7:**1515–1526.

122a. **Malke, H.** 1974. Genetics of resistance to macrolide antibiotics and lincomycin in natural isolates of *Streptococcus pyogenes. Mol. Gen. Genet.* **135:**349–367.

123. **Malke, H., and S. E. Holm.** 1982. Streptococcal DNA cloning vehicles derived from a plasmid associated with zonal lincomycin resistance, p. 233–235. *In* S. E. Holm and P. Christensen (ed.), *Basic Concepts of Streptococci and Streptococcal Diseases*. Reedbooks, Chertsey, England.

124. **Malke, H., H. E. Jacob, and K. Störl.** 1976. Characterization of the antibiotic resistance plasmid ERL1 from *Streptococcus pyogenes. Mol. Gen. Genet.* **144:**333–338.

125. **Malke, H., W. Reichardt, M. Hartmann, and F. Walter.** 1981. Genetic study of plasmid-associated zonal resistance to lincomycin in *Streptococcus pyogenes. Antimicrob. Agents Chemother.* **19:**91–100.

126. **Martin, J. P., G. H. del Solar, R. Lurz, A. G. de la Campa, B. Dobrinski, and M. Espinosa.** 1989. Induced bending of plasmid pLS1 DNA by the plasmid-encoded protein RepA. *J. Biol. Chem.* **264:**21334–21339.

127. **Masai, H., N. Nomura, and K.-I. Arai.** 1990. The ABC-primosome: a novel priming system employing dnaA, dnaB, dnaC and primase on a hairpin containing a dnaA box sequence. *J. Biol. Chem.* **265:**15134–15144.

128. **Masai, H., N. Nomura, Y. Kubota, and K.-I. Arai.** 1990. Roles of ϕX174 type primosome- and G4 type primase-dependent primings in initiation of lagging and leading strand syntheses of DNA replication. *J. Biol. Chem.* **265:**15124–15133.

129. **McDougall, J., D. Margarita, and I. Saint Girons.** 1992. Homology of a plasmid from the spirochete *Treponema denticola* with the single-stranded DNA plasmid. *J. Bacteriol.* **174:**2724–2728.

129a. **McDowell, D. G., and N. H. Mann.** 1991. Characterization and sequence analysis of a small plasmid from *Bacillus thuringiensis* var. *kurstaki* strain HD1-DIPEL. *Plasmid* **25:**113–120.

130. **Mercier, J., J. Lachapelle, F. Couture, M. Lafond, G. Vézina, M. Boissinot, and R. C. Levesque.** 1990. Structural and functional characterization of *tnpI*, a recombinase locus in Tn*21* and related β-lactamase transposons. *J. Bacteriol.* **172:**3745–3757.

131. **Michel, B., and S. D. Ehrlich.** 1986. Illegitimate recombination at the replication origin of the bacteriophage M13. *Proc. Natl. Acad. Sci. USA* **83:**3386–3390.

132. **Michel, B., and S. D. Ehrlich.** 1986. Illegitimate recombination occurs between the replication origin of plas-

mid pC194 and a progressing replication fork. *EMBO J.* **5:**3691–3696.

133. **Minton, N. P., T.-J. Swinfield, J. K. Brehm, S. M. Whelan, and J. D. Oultram.** 1991. Vectors for use in *Clostridium acetobutylicum*, p. 120–140. *In* M. Sebald (ed.), *Genetics and Molecular Biology of Anaerobic Bacteria*. Springer Verlag, New York.

133a. **Morel, F.** Personal communication.

134. **Moriya, S., T. Fukuoka, N. Ogasawara, and H. Yoshikawa.** 1988. Regulation of initiation by DnaA-boxes in the origin region of the *Bacillus subtilis* chromosome. *EMBO J.* **7:**2911–2917.

135. **Muller, R. E., T. Ano, T. Imanaka, and S. Aiba.** 1986. Complete nucleotide sequences of *Bacillus* plasmids pUB110dB, pRBH1 and its copy mutants. *Mol. Gen. Genet.* **202:**169–171.

136. **Murai, M., H. Miyashita, H. Araki, T. Seki, and Y. Oshima.** 1987. Molecular structure of the replication origin of a *Bacillus liquefaciens* plasmid pFTB14. *Mol. Gen. Genet.* **210:**92–100.

137. **Murphy, E.** 1990. Properties of the site-specific transposable element Tn*554*, p. 123–135. *In* R. P. Novick (ed.), *Molecular Biology of the Staphylococci*. VCH Press, New York.

138. **Niaudet, B., L. Jannière, and S. D. Ehrlich.** 1984. Recombination between repeated DNA sequences occurs more often in plasmids than in the chromosome of *Bacillus subtilis. Mol. Gen. Genet.* **197:**46–54.

139. **Noirot, P., and R. P. Novick.** 1990. Initiation of rolling-circle replication in pT181 plasmid: initiator protein enhances cruciform extrusion at the origin. *Proc. Natl. Acad. Sci. USA* **87:**8560–8564.

140. **Noirot, P., M.-A. Petit, and S. D. Ehrlich.** 1987. Plasmid replication stimulates DNA recombination in *Bacillus subtilis. J. Mol. Biol.* **196:**39–48.

140a. **Nomura, N., H. Masai, M. Inuzuka, C. Miyazaki, E. Ohtsubo, T. Itoh, S. Sasamoto, M. Matsui, R. Ishizaki, and K.-I. Arai.** 1991. Identification of eleven single-strand initiation sequences (*ssi*) for priming of DNA replication in the F, R6K, R100 and ColE2 plasmids. *Gene* **108:**15–22.

141. **Nordström, K.** 1990. Control of plasmid replication—how do DNA iterons set the replication frequency? *Cell* **63:**1121–1124.

141a. **Novick, R. P.** Personal communication.

142. **Novick, R. P.** 1963. Analysis by transduction of mutations affecting penicillinase formation in *Staphylococcus aureus. J. Gen. Microbiol.* **33:**121–136.

143. **Novick, R. P.** 1989. Staphylococcal plasmids and their replication. *Annu. Rev. Microbiol.* **43:**537–565.

144. **Novick, R. P.** 1991. Genetic systems in staphylococci. *Methods Enzymol.* **204:**587–636.

145. **Novick, R. P., S. Iordanescu, S. J. Projan, J. Kornblum, and I. Edelman.** 1989. pT181 plasmid replication is regulated by a countertranscript-driven transcriptional attenuator. *Cell* **59:**395–404.

146. **O'Connor, M. B., J. J. Kilbane, and M. H. Malamy.** 1986. Site-specific and illegitimate recombination in the *oriV1* region of the F factor: DNA sequences involved in recombination and resolution. *J. Mol. Biol.* **189:**85–102.

147. **Ogasawara, N., M. Q. Fujita, S. Moriya, T. Fukuoka, M. Hirano, and H. Yoshikawa.** 1990. Comparative anatomy of *oriC* of eubacteria, p. 287–295. *In* K. Drlica and M. Riley (ed.), *The Bacterial Chromosome*. American Society for Microbiology, Washington, D.C.

148. **Oskam, L., D. J. Hillenga, G. Venema, and S. Bron.** 1991. The large *Bacillus* plasmid pTB19 contains two integrated rolling-circle plasmids carrying mobilization functions. *Plasmid* **26:**30–39.

149. **Oultram, J. D., and M. Young.** 1985. Conjugal transfer of plasmid pAMβ1 from *Streptococcus lactis* and *Bacillus subtilis* to *Clostridium acetobutylicum. FEMS Microbiol. Lett.* **27:**129–134.

150. **Peeters, B., J. de Voer, S. Bron, and G. Venema.** 1988. Structural plasmid instability in *Bacillus subtilis*: effect of direct and inverted repeats. *Mol. Gen. Genet.* **212:** 450–458.

151. **Pérez-Martin, J., and M. Espinosa.** 1991. The RepA repressor can act as a transcriptional activator by inducing DNA bends. *EMBO J.* **10:**1375–1382.

151a. **Perkins, D. R., and S. R. Barnum.** 1992. DNA sequence and analysis of a cryptic 4.2 kb plasmid from the filamentous cyanobacterium, *Plectonema* sp. strain pCC6402. *Plasmid* **28:**170–176.

152. **Petit, M.-A., C. Bruand, L. Jannière, and S. D. Ehrlich.** 1990. Tn*10*-derived transposons active in *Bacillus subtilis. J. Bacteriol.* **172:**6736–6740.

153. **Petit, M.-A., G. Joliff, J. M. Mesas, A. Klier, G. Rapoport, and S. D. Ehrlich.** 1990. Hypersecretion of a cellulase from *Clostridium thermocellum* in *Bacillus subtilis* by induction of chromosomal DNA amplification. *Bio/Technology* **8:**559–563.

154. **Petit, M.-A., J. M. Mesas, P. Noirot, F. Morel, and S. D. Ehrlich.** 1992. Induction of DNA amplification in the *Bacillus subtilis* chromosome. *EMBO J.* **11:**1317–1326.

154a. **Polak, J.** Personal communication.

155. **Poyart-Salmeron, C., P. Trieu-Cuot, C. Carlier, and P. Courvalin.** 1989. Molecular characterization of two proteins involved in the excision of the conjugative transposon Tn*1545*: homologies with other site-specific recombinases. *EMBO J.* **8:**2425–2433.

156. **Projan, S., S. Moghazeh, and R. Novick.** 1988. Nucleotide sequence of pS194, a streptomycin-resistance plasmid from *Staphylococcus aureus. Nucleic Acids Res.* **16:**2179–2187.

157. **Projan, S., M. Monod, C. Narayanan, and D. Dubnau.** 1987. Replication properties of pIM13, a naturally occurring plasmid found in *Bacillus subtilis*, and of its close relative pE5, a plasmid native to *Staphylococcus aureus. J. Bacteriol.* **169:**5131–5139.

158. **Projan, S., and R. Novick.** 1988. Comparative analysis of five related staphylococcal plasmids. *Plasmid* **19:** 203–221.

159. **Pucci, M. J., M. E. Monteschio, and C. L. Kemker.** 1988. Intergeneric and intrageneric conjugal transfer of plasmid-encoded antibiotic resistance determinants in *Leuconostoc* spp. *Appl. Environ. Microbiol.* **54:**281–287.

159a. **Pujol, C., S. D. Ehrlich, and L. Jannière.** The promiscuous plasmids pIP501 and pAMβ1 from gram-positive bacteria encode cross-reacting resolution functions. Submitted for publication.

160. **Rabinovich, P. M., M. Y. Haykinson, L. S. Arutyunova, Y. V. Yomantas, and A. I. Stepanov.** 1985. The structure and source of plasmid DNA determine the cloning properties of vectors for *Bacillus subtilis*, p. 635–652. *In* D. R. Helinski, S. N. Cohen, D. B. Clewell, D. A. Jackson, and A. Hollaender (ed.), *Plasmids in Bacteria*. Plenum Press, New York.

161. **Rood, J. I., and S. T. Cole.** 1991. Molecular genetics and pathogenesis of *Clostridium perfringens. Microbiol. Rev.* **55:**621–648.

162. **Sadowski, P.** 1986. Site-specific recombinases: changing partners and doing the twist. *J. Bacteriol.* **165:**341–347.

163. **Seufert, W., and W. Messer.** 1987. DnaA protein binding to the plasmid origin region can substitute for primosome assembly during replication of pBR322 in vitro. *Cell* **48:**73–78.

164. **Sherratt, D.** 1989. Tn*3*-related transposable elements: site-specific recombination and transposition, p. 163–184. *In* D. E. Berg and M. M. Howe (ed.), *Mobile DNA*. American Society for Microbiology, Washington, D.C.

165. **Shrago, A. W., and W. J. Dobrogosz.** 1988. Conjugal transfer of group B streptococcal plasmids and como-

bilization of *Escherichia coli-Streptococcus* shuttle plasmids to *Lactobacillus plantarum. Appl. Environ. Microbiol.* **54:**824–826.
166. **Simon, D., and A. Chopin.** 1988. Construction of a vector plasmid family and its use for molecular cloning in *Streptococcus lactis. Biochimie* **70:**559–566.
167. **Skaugen, M.** 1989. The complete nucleotide sequence of a small cryptic plasmid from *Lactobacillus plantarum. Plasmid* **22:**175–179.
168. **Smith, H. O., A. M. Annau, and S. Chandrasegaran.** 1990. Finding sequence motifs in groups of functionally related proteins. *Proc. Natl. Acad. Sci. USA* **87:**826–830.
168a. **Sorokin, A. V.** Personal communication.
169. **Sorokin, A. V., and V. E. Khazak.** 1987. Structure of pSM19035 replication region and MLS-resistance gene, p. 269–281. *In* L. O. Butler, C. Hartwood, and B. E. B. Moseley (ed.), *Genetic Transformation and Expression.* Intercept Ltd., Andover, United Kingdom.
170. **Sozhamannan, S., P. Dabert, V. Moretto, S. D. Ehrlich, and A. Gruss.** 1990. Plus-origin mapping of single-stranded DNA plasmid pE194 and nick site homologies with other plasmids. *J. Bacteriol.* **172:**4543–4548.
171. **Stahl, F. W., J. M. Crasemann, and M. M. Stahl.** 1975. Rec-mediated recombinational hot spot activity in bacteriophage lambda. *J. Mol. Biol.* **94:**203–212.
172. **Swinfield, T. J., L. Jannière, S. D. Ehrlich, and N. P. Minton.** 1991. Characterization of a region of the *Enterococcus faecalis* plasmid pAMβ1 which enhances the segregational stability of pAMβ1-derived cloning vectors in *Bacillus subtilis. Plasmid* **26:**209–221.
173. **Swinfield, T. J., J. D. Oultram, D. E. Thompson, J. K. Brehm, and N. P. Minton.** 1990. Physical characterization of the replication region of the *Streptococcus faecalis* plasmid pAMβ1. *Gene* **87:**79–90.
174. **Tannock, G. W.** 1987. Conjugal transfer of plasmid pAMβ1 in *Lactobacillus reuteri* and between lactobacilli and *Enterococcus faecalis. Appl. Environ. Microbiol.* **53:**2693–2695.
175. **Taylor, A. F.** 1988. RecBCD enzyme in *Escherichia coli*, p. 231–263. *In* R. Kucherlapati and G. R. Smith (ed.), *Genetic Recombination.* American Society for Microbiology, Washington, D.C.
176. **te Riele, H., B. Michel, and S. Ehrlich.** 1986. Are single-stranded circles intermediates in plasmid DNA replication? *EMBO J.* **5:**631–637.
177. **te Riele, H., B. Michel, and S. Ehrlich.** 1986. Single-stranded plasmid DNA in *Bacillus subtilis* and *Staphylococcus aureus. Proc. Natl. Acad. Sci. USA* **83:**2541–2545.
178. **Thomas, C. D., D. Balson, and W. V. Shaw.** 1990. *In vitro* studies of the initiation of staphylococcal replication. *J. Biol. Chem.* **265:**5519–5530.
178a. **Thomas, C. M.** Personal communication.
179. **Thomas, C. M., and C. A. Smith.** 1987. Incompatibility group P plasmids: genetics, evolution, and use in genetic manipulation. *Annu. Rev. Microbiol.* **41:**77–101.
180. **Thomas, W. D., Jr., and G. L. Archer.** 1989. Identification and cloning of the conjugative transfer region of *Staphylococcus aureus* plasmid pGO1. *J. Bacteriol.* **171:** 684–691.
181. **Trieu-Cuot, P., C. Carlier, and P. Courvalin.** 1988. Conjugative plasmid transfer from *Enterococcus faecalis* to *Escherichia coli. J. Bacteriol.* **170:**4388–4391.
182. **Trieu-Cuot, P., C. Carlier, P. Martin, and P. Courvalin.** 1987. Plasmid transfer by conjugation from *Escherichia coli* to gram-positive bacteria. *FEMS Microbiol. Lett.* **48:**289–294.
183. **Trieu-Cuot, P., C. Carlier, C. Poyart-Salmeron, and P. Courvalin.** 1991. Shuttle vectors containing a multiple cloning site and a *lacZα* gene for conjugal transfer of DNA from *Escherichia coli* to gram-positive bacteria. *Gene* **102:**99–104.
184. **van der Lelie, D., S. Bron, G. Venema, and L. Oskam.** 1989. Similarity of minus origins of replication and flanking open reading frames of plasmids pUB110, pTB913 and pMV158. *Nucleic Acids Res.* **17:**7283–7294.
185. **van Embden, J. D. A., N. Soedirman, and H. W. B. Engel.** 1978. Transferable drug resistance to group A and group B streptococci. *Lancet* **i:**655–656.
186. **Villafane, R., D. Bechhofer, C. Narayanan, and D. Dubnau.** 1987. Replication control of genes of plasmid pE194. *J. Bacteriol.* **169:**4822–4829.
187. **Viret, J.-F., and J. C. Alonso.** 1987. Generation of linear multigenome-length plasmid molecules in *Bacillus subtilis. Nucleic Acids Res.* **15:**6349–6367.
187a. **Vujcic, M., and L. Topisirovic.** 1993. Molecular analysis of the rolling-circle replicating plasmid pA1 of *Lactobacillus plantarum* A112. *Appl. Environ. Microbiol.* **59:**274–280.
188. **Wang, P.-Z., S. Projan, V. Henriquez, and R. P. Novick.** 1992. Specificity of origin recognition by replication initiator protein in plasmids of the pT181 family is determined by a six amino acid residue element. *J. Mol. Biol.* **223:**145–158.
189. **West, C. A., and P. J. Warner.** 1985. Plasmid profiles and transfer of plasmid-encoded antibiotic resistance in *Lactobacillus plantarum. Appl. Environ. Microbiol.* **50:**1319–1321.
190. **Willetts, N., and R. Skurray.** 1987. Structure and function of the F factor and mechanism of conjugation, p. 1110–1133. *In* F. C. Neidhardt, J. L. Ingraham, K. B. Low, B. Magasanik, M. Schaechter, and H. E. Umbarger (ed.), *Escherichia coli and Salmonella typhimurium: Cellular and Molecular Biology.* American Society for Microbiology, Washington, D.C.
191. **Willetts, N., and B. Wilkins.** 1984. Processing of plasmid DNA during bacterial conjugation. *Microbiol. Rev.* **48:**24–41.
192. **Xu, F., L. E. Pierce, and P.-L. Yu.** 1989. Genetic analysis of a lactococcal plasmid replicon. *Mol. Gen. Genet.* **227:**33–39.
193. **Yakusawa, H., T. Hase, A. Sakai, and Y. Masamune.** 1991. Rolling-circle replication of the plasmid pKYM isolated from a gram-negative bacterium. *Proc. Natl. Acad. Sci. USA* **88:**10282–10286.
194. **Young, M., and S. D. Ehrlich.** 1989. Stability of reiterated sequences in the *Bacillus subtilis* chromosome. *J. Bacteriol.* **171:**2653–2656.
195. **Youngman, P.** 1987. Plasmid vectors for recovering and exploiting Tn917 transpositions in *Bacillus* and other gram-positive bacteria, p. 79–103. *In* K. Hardy (ed.), *Plasmids, a Practical Approach.* IRL Press, Oxford.
196. **Zavitz, K. H., and K. J. Marians.** 1991. Dissecting the functional role of PriA protein-catalysed primosome assembly in *Escherichia coli* DNA replication. *Mol. Microbiol.* **5:**2869–2873.

44. Temperate Phage Vectors

J. ERRINGTON

Temperate phages have been widely used as tools for genetic manipulation and analysis of *Bacillus subtilis*. Until recently, the vast majority of successful cloning experiments were based on the use of phage vectors, particularly ϕ105 but also, more recently, SPβ. Infectivity and the formation of stable lysogens with the prophage integrated in single copy also provided a convenient means of carrying out various kinds of genetic experiments such as complementation analysis and measurement of the expression of gene fusions in different genetic backgrounds. It has recently become possible to use bacteriophage vectors for the overexpression of genes encoding heterologous or genetically manipulated proteins. Although the problem of plasmid instability now seems to have been overcome (mainly by avoiding those plasmids that replicate by rolling-circle replication; reviewed in reference 21), there are still potential advantages, particularly the single copy of the prophage and its relative stability in the absence of selective pressure, in phage cloning methods.

Probably the most important application for phage vectors in *B. subtilis* has been in gene cloning. Cloning directly in *B. subtilis* is advantageous in two types of situation: first, where the gene of interest can be selected directly by complementation of a specific mutation; and second, where there is a convenient screening method based on a phenotypic property of the desired recombinant. To a certain extent, phage cloning systems have been superseded by the application of powerful "chromosome walking" techniques based on the excision of integrated plasmids (reviewed by Youngman [47]), but these techniques are convenient only when the gene of interest is known to lie close to a previously cloned gene or a transposon insertion. They may also be impeded by intervening regions of DNA that cause instability of the recombinant plasmid in *Escherichia coli*.

Two temperate phages, ϕ105 and SPβ, have proved to be particularly useful for cloning and gene manipulation in *B. subtilis*. The biology of these phages is considered in detail in chapter 57, but a brief recapitulation may be useful. ϕ105 is somewhat similar to coliphage lambda in terms of its size (39.2 kb) (1, 3, 30), morphology (2), and genome organization (chapter 57). After infection, the DNA circularizes via single-stranded cohesive termini (8). During lysogeny of a susceptible host, the circularized phage integrates at a unique chromosomal attachment site near *pheA* (39) via a phage attachment site not far from the midpoint of the vegetative phage DNA (9, 31). Thus, as in phage lambda, the prophage map is a circularly permuted version of the vegetative map. Again, as in lambda, a region of the phage DNA to one side of the attachment site, i.e., the side opposite that of the immunity region, is dispensable (18). Most of the useful vectors based on ϕ105 are derived from a deletion mutant, ϕ105DI:1t, which has a 4-kbp deletion in this region (18). Other deletions are known to abolish phage tail synthesis or assembly (22, 27) or to result in the prevention of host cell lysis (17). Mutations conferring temperature inducibility (*cts*) first defined the immunity region of the phage (39). Mutations preventing RecE-dependent induction (*ind*; 16, 20) probably lie in the same gene, which has been cloned and extensively characterized (see chapter 57).

Phage SPβ is much larger (120 kb; 41), and less is known about its general organization. However, there are again temperature-inducible mutants (38) and deletions defining a large (27-kb) nonessential region of DNA (41). Thus, although its larger size makes SPβ DNA more difficult than ϕ105 to characterize directly, SPβ is likely to have a much larger capacity for heterologous DNA. Most of the genetic strains of *B. subtilis* are SPβ lysogens, as this prophage was present in the original Marburg strain. However, strains carrying the temperature-inducible mutation *c2* can be readily cured by growth at high temperature (38).

CLONING VECTORS

Cloning in ϕ105 by Direct Transfection

There are two very different approaches to cloning in temperate phage vectors in *B. subtilis*. The first approach, direct transfection (10), is similar to the familiar shotgun cloning methods used in *E. coli*. The vector ϕ105J106 (Fig. 1) has a cloning capacity of 5 to 6 kbp and unique cloning sites for *Bam*HI, *Sal*I, and *Xba*I (6). The *Sal*I site is particularly useful for cloning by the partial-fill-in method of Zabarovsky and Allikmets (48). This method minimizes the formation of recircularized vector molecules, and the insertion of multiple fragments of target DNA is prevented. The target DNA is prepared by partial digestion with *Mbo*I, partial end fill, and gradient fractionation to the desired size range (for detailed protocols for this and other cloning methods, see reference 15). Recombinant molecules are amplified by transfection of protoplasts followed by plaquing in a lawn of sensitive bacteria. The library is generated by harvesting the progeny phage as a pool and screened by infecting a suitable recipient strain with a sample of the lysate and then plating the strain on a suitable selective or indicator medium. A small proportion of the infected cells will form lysogens. Cloning the gene of interest depends on being able to select or detect a recombinant carrying the desired gene.

J. Errington • Sir William Dunn School of Pathology, South Parks Road, Oxford OX1 3RE, United Kingdom.

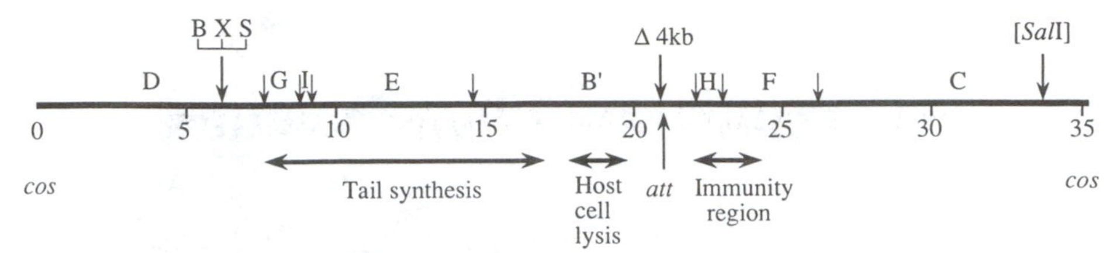

Figure 1. Schematic drawing of the structural and functional organization of phage cloning vector ϕ105J106. The vegetative phage genome is represented by the central bar, with distances (in kilobase pairs) from its left end noted below the bar. Immediately above the bar are the positions of *Eco*RI restriction sites. The fragments generated are labeled with standard one-letter designations (see reference 15 for a detailed restriction map). The modifications of wild-type ϕ105 incorporated into vector ϕ105J106 are shown at the top. The polylinker inserted into a nonessential site in the phage contains unique restriction sites for *Bam*HI (B), *Xba*I (X), and *Sal*I (S), the last having been made unique by removal of a *Sal*I site to the right of the genome (denoted [*Sal*I]). A deletion removing 4 kbp of nonessential phage DNA (Δ 4kb) allows for the insertion of exogenous DNA fragments without exceeding the packaging limits of the phage. Approximate positions of regions of the phage genome devoted to particular functions are indicated below the bar. *att* denotes the site at which the phage DNA integrates into the host chromosome by site-specific recombination (9, 31). The vegetative genome has single-stranded complementary termini (*cos*) that allow circularization of the genome prior to integration into the host chromosome (8).

This approach has been very successful, particularly for cloning sporulation genes. At least 23 distinct *spo* loci have been isolated (6, 10, 16a). In addition, this system has been used to clone several genes involved in spore germination (25, 46) and the biosynthesis of small molecules (10, 13) and to clone the *recE* gene (32).

A disadvantage of the transfection system is that the cloning capacity of vectors such as ϕ105J106 is relatively small (5 to 6 kb; 17, 46). In principle, the capacity could be increased by deleting the phage immunity region, as in many of the lambda vectors, resulting in an additional capacity of at least 2.5 kb (18). However, the advantage of being able to screen the recombinants by complementation in stable lysogens would be lost. ϕ105 derivatives with much larger deletions have been isolated (13, 26), but they are invariably defective, being capable of infection only in the presence of wild-type helper phage (22, 27). Unfortunately, more extensively deleted defective vectors cannot be used for cloning by direct transfection, because this procedure requires that the recombinant phages be amplifiable by plaque formation. However, it is possible to use a rather different delivery system called "prophage transformation" (29) in conjunction with severely deleted prophages.

Prophage Transformation

Prophage transformation is an ingenious cloning system that is unique to *B. subtilis* (29) and takes advantage of the efficient transformation of this organism with linear DNA fragments. The principles underlying this method are illustrated in Fig. 2. Restriction fragments of the DNA to be cloned are ligated at a relatively high DNA concentration in an approximately 1-to-1 molar ratio with fragments of phage vector DNA. The resultant linear concatemeric DNA is introduced into a host strain carrying a prophage. Homologous recombination can result in the insertion of passenger DNA into the prophage genome. Although early experiments based on either ρ11 or ϕ105 met with some success, the method suffered from several disadvantages (reviewed in detail previously [14]), but these have mainly been overcome now. The use of *ind cts* double mutants (16, 20) overcomes the otherwise poor transformation efficiency of strains lysogenic for ϕ105 (44, 45). The use of a plasmid containing defined regions of homology with the prophage greatly improves the efficiency of generating recombinants. Finally, the use of a selectable marker placed between one of the fragments of phage DNA and the target DNA allows for the generation of genomic libraries containing random insertions of target DNA. The original protocol was based on selection for recombinants with a particular insertion at the initial transformation step. It was thus suitable only for the cloning of genes for which a direct selection was available, and the procedure had to be repeated for each gene to be cloned.

The last two improvements have also been incorporated into a prophage transformation system based on SPβ (37). This system has the additional advantage that it incorporates devices that facilitate the subcloning of inserts into plasmids in either *B. subtilis* or *E. coli*. Again, detailed protocols for using the prophage transformation systems are described in reference 15.

EXPRESSION VECTORS

Vectors for Detecting Products of Cloned Genes

In general, vectors in which the products of a cloned gene are expressed are useful for two purposes. First, they can be used to identify the product of the cloned gene. DNA sequencing studies often highlight open reading frames that probably encode protein products. Confirmation that such products are actually made in vivo is not always straightforward. Plasmid-encoded products may be visualized by using "minicell" methods in *B. subtilis* (34). However, an alternative method, and one that is particularly convenient for use in conjunction with the ϕ105 cloning system described above, is the use of phage infection of maxicells. We fortuitously discovered that the cloning site used in most of the successful ϕ105 transfection vectors actually lies within a phage transcription unit (6). Moreover, for reasons that are still not clear, it seems that most, if not all, of the insertions that result in viable recombinant phages have the cloned DNA

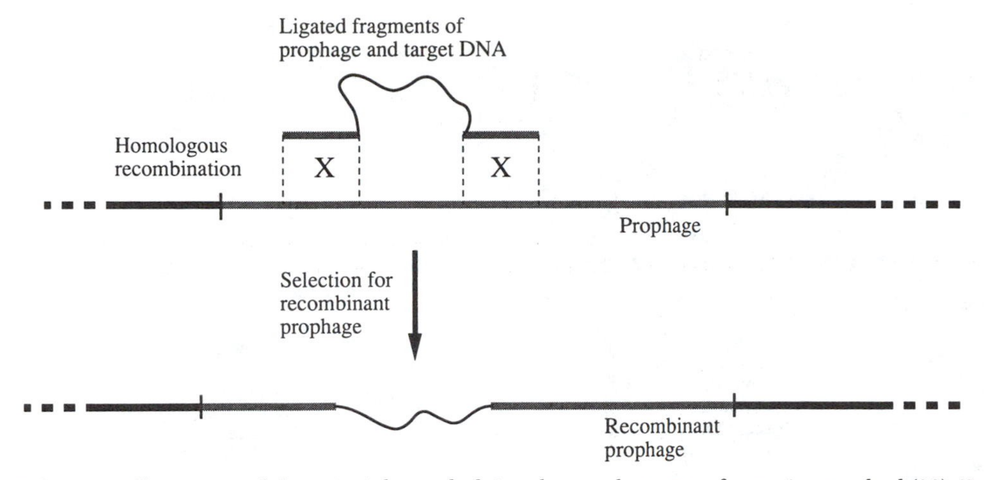

Figure 2. Schematic illustration of the principles underlying the prophage transformation method (29). Fragments of phage DNA and the target DNA to be cloned are ligated at a high DNA concentration and an approximately 1:1 molar ratio to form concatemeric molecules. Some of the molecules will have the structure shown at the top. Recombination between the fragments of phage DNA and their homologous sequences in the host prophage leads to insertion of target DNA in place of a segment of prophage DNA. Recombinants with the appropriate genotype can either be selected directly, as in this instance, or be generated at random if a selectable marker is placed at the inner margin of one of the fragments of phage DNA. In the original procedure, phage DNA fragments were derived by digestion of total phage DNA, sometimes generating many fragments, only a proportion of which would generate viable recombinant prophages. It is now more usual to provide specific fragments of cloned phage DNA flanking dispensable regions of the prophage genome.

oriented such that the cloned genes are transcribed from the phage promoter (7). Thus, the maxicell method for visualizing phage-encoded proteins can be used to visualize the additional proteins encoded by recombinant phages (6). Presumably, some essential phage-encoded proteins lie downstream from the phage cloning site, and their expression is disrupted by insertions in the "incorrect" orientation but not in the "correct" orientation. The recipient cells are UV irradiated to destroy the host DNA, so that after a period of incubation, host protein synthesis is virtually abolished. However, the cells retain the capacity to synthesize phage proteins after infection and injection of phage DNA. By adding labeled amino acids, these newly synthesized proteins can be specifically labeled and visualized by autoradiography. Recombinant phages produce additional proteins corresponding to the products of the cloned genes.

Vectors for Protein Production

The second application for phage expression vectors lies in the overexpression of cloned genes for purposes of protein purification. Although, again, a number of plasmid expression vectors are available, the prophage-based vectors have the advantages of increased stability because of their chromosomal locations and convenient regulation provided by the phage immunity system. Moreover, phage induction, which can be controlled by temperature shift in appropriate mutants, results not only in expression from strong phage promoters but also in an increased copy number through phage DNA replication.

Expression vectors have been developed from two phages, ϕ105 and a defective phage, PBSX, which is, like SPβ, present in most of the genetic strains of *B. subtilis* (40). In both cases, the development of an expression vector required the incorporation of several modifications to the prophage: (i) a mutation rendering the prophage temperature inducible, (ii) the identification of a suitable cloning site within a strongly induced transcription unit, and (iii) (optional) a mutation preventing lysis of the host cell. Clearly, for a secreted gene product, it is desirable that the host cell does not lyse so that the product can be purified from the cell supernatant. Figures 3 and 4 illustrate two of the vectors that have been developed. One of the systems has been derived by manipulation of the PBSX prophage of *B. subtilis* 168 (33). Campbell-type integration introduces a plasmid carrying the gene to be expressed into the PBSX prophage and downstream from a strong prophage promoter (36). The prophage carries a mutation, *xhi-1479* (4), that allows thermoinduction of prophage development, and the plasmid insertion blocks lysis of the host cell because the genes needed lie downstream from the same promoter (43). The construction is relatively stable and is maintained in single copy during the growth phase. Transcription and DNA amplification are activated by thermoinduction.

Similar systems have been developed from phage ϕ105. One vector utilizing the phage promoter that directs transcription through the ϕ105J106 cloning site, and mutations preventing host cell lysis and allowing thermoinduction, gave a relatively modest yield of a secreted heterologous protein (about 20 μg/ml [20a]). Vectors giving substantially greater yields have been developed more recently, mainly by the identification and use of a stronger ϕ105 promoter. ϕ105MU209 was isolated as a colony exhibiting extremely strong expression of β-galactosidase among transformants generated by random insertion of a *lacZ* reporter gene into a temperature-inducible ϕ105 prophage (11). The structure of the insertion in

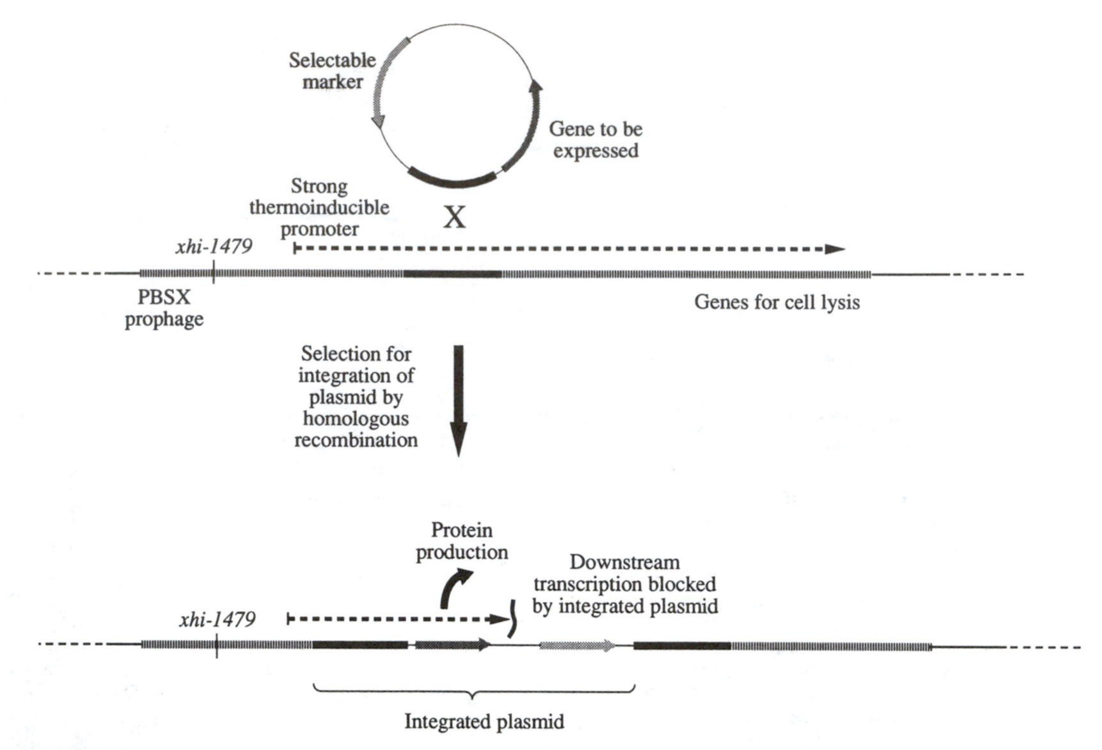

Figure 3. Schematic illustration of an expression system based on the defective *B. subtilis* prophage PBSX. The circular plasmid at the top of the diagram can be grown and manipulated in *E. coli* but cannot replicate autonomously in *B. subtilis*. Transformants of *B. subtilis* selected on the basis of the plasmid-borne resistance marker arise by Campbell-type integration of the plasmid into the PBSX prophage via its region of homology (indicated by the black box). The prophage carries the *xhi-1479* mutation, which allows thermoinduction of the strong phage promoter. After plasmid integration, this promoter drives expression of the cloned gene. In addition, a transcription terminator in the plasmid blocks transcription of downstream phage genes, including genes required for host cell lysis.

ϕ105MU209 is complex and has not yet been fully resolved. However, it is clear that the insertion has not only conferred strong inducible expression of *lacZ* following temperature induction but has also somehow blocked host cell lysis. Despite the fact that the structure of the insertion is not fully known, it has been possible to use homologous recombination to replace part of the coding region of *lacZ* with heterologous genes, which are then overexpressed following thermoinduction. This system has been used to produce 6,000 to 9,000 Miller units of β-galactosidase (approximately 5 to 10% of total cellular protein; 7) and wild-type and mutant versions of the secreted *Bacillus cereus* β-lactamase I, with yields of up to 400 μg/ml (42). Again, it seems that the constructions are relatively stable in the absence of selection (before induction), and it seems likely that phage DNA replication and the resultant increase in copy number contribute to the high level of expression.

Clearly, these are first-generation vectors, and there is considerable potential for further development leading to improved yields of protein.

OTHER APPLICATIONS

Phage vectors have also been exploited as general tools for the analysis of gene structure and regulation in *B. subtilis* in much the same way that phage lambda has been used in *E. coli*. There are two main types of application. In the first, complementation analysis, the specialized transducing phage provides a convenient means of generating partially diploid cells (e.g., see references 19 and 23). In the second, "portable" gene fusions, prophages have often been manipulated to carry reporter gene constructs. A transducing lysate provides a convenient means of introducing a gene fusion into strains with different genetic backgrounds. This has been particularly useful in deducing the pattern of interactions between the sporulation genes (see references 5 and 24 for recent examples).

In both cases, the phage derivatives needed can be generated by direct transfection (e.g., see reference 35) or by homologous recombination between a wild-type copy of the gene on the prophage and its allele on the chromosome. The latter can be achieved either by the homologous recombination that occurs between incoming phage DNA and host chromosome during or after infection (23) or by transformation of a strain carrying a prophage with DNA carrying the mutant allele or inserted reporter gene (12, 19, 49). As in the case of cloning by prophage transformation, the efficiency of such manipulations involving ϕ105 prophages is greatly facilitated by incorporation of the *ind* and *cts* mutations.

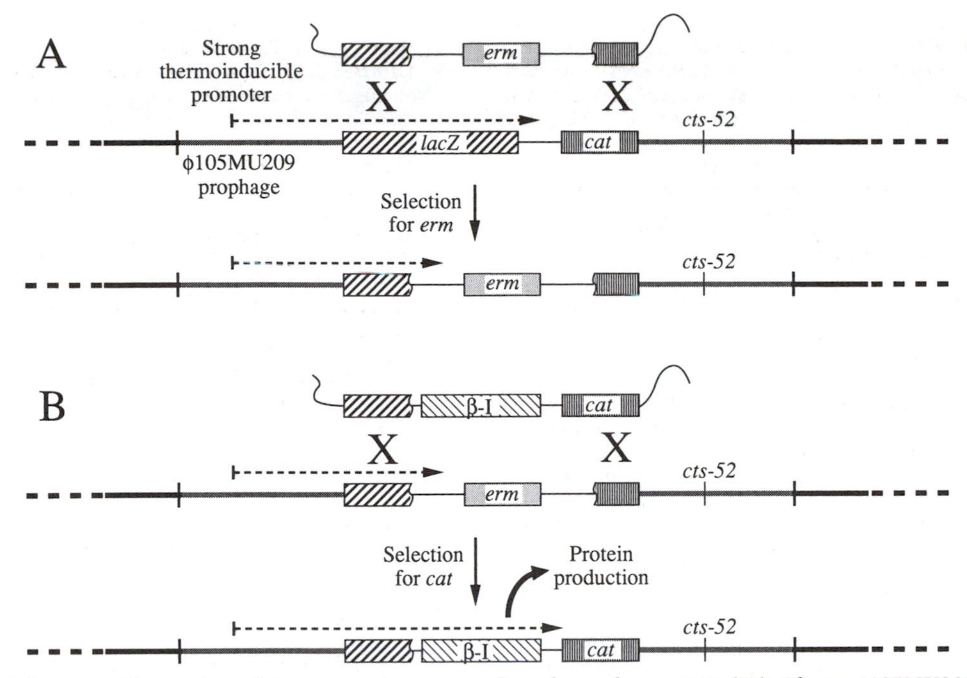

Figure 4. Schematic illustration of an expression system based on phage ϕ105 (42). Phage ϕ105MU209 has an insertion consisting of a reporter gene *lacZ* followed by a selectable chloramphenicol resistance gene. The structure of the insertion is complex and not yet fully resolved. However, *lacZ* lies downstream from a strong phage promoter that can be induced about 100-fold by thermoinduction of the prophage, which carries the *cts-52* mutation. The insertion also blocks the host cell lysis that normally follows phage induction. To express other proteins in this system, the *lacZ* gene and part of the chloramphenicol resistance gene are first replaced with an erythromycin resistance gene (A). The latter is then replaced by the gene to be expressed (in this case, β-lactamase I [β-I] from *B. cereus*) by selection for the full-length chloramphenicol resistance gene (B).

CONCLUSIONS

Bacteriophages have provided a range of useful tools for gene cloning and genetic analysis in *B. subtilis*, and they will probably continue to do so for the foreseeable future. However, there are several areas in which improvements in the efficiency or facility of such systems would be desirable. It is also clear on reading the review of temperature phages elsewhere in this volume (chapter 57) that there are a number of interesting phages in *B. subtilis* but that progress in understanding their biology lags far behind our understanding of the equivalent phages of *E. coli*. Some of the motivation for the progress that has been made has stemmed from the practical need to clone, express, and manipulate genes in *B. subtilis*. However, without detailed knowledge of the underlying biology, progress beyond basic vector systems is slow and unpredictable.

Acknowledgments. Work in this laboratory was supported by the Science and Engineering Research Council and the EC SCIENCE program.

REFERENCES

1. **Anaguchi, H., S. Fukui, and Y. Kobayashi.** 1984. Revised restriction map of *Bacillus subtilis* bacteriophage ϕ105 DNA. *J. Bacteriol.* **159:**1080–1082.
2. **Birdsell, D. C., G. M. Hathaway, and L. Rutberg.** 1969. Characterization of temperate *Bacillus* bacteriophage ϕ105. *J. Virol.* **4:**264–270.
3. **Bugaichuk, U. D., M. Deadman, J. Errington, and D. Savva.** 1984. Restriction enzyme analysis of *Bacillus subtilis* bacteriophage ϕ105 DNA. *J. Gen. Microbiol.* **130:** 2165–2167.
4. **Buxton, R. S.** 1976. Prophage mutation causing heat inducibility of defective *Bacillus subtilis* bacteriophage PBSX. *J. Virol.* **20:**22–28.
5. **Cutting, S., S. Panzer, and R. Losick.** 1989. Regulatory studies on the promoter for a gene governing synthesis and assembly of the spore coat in *Bacillus subtilis*. *J. Mol. Biol.* **207:**393–404.
6. **East, A. K., and J. Errington.** 1989. A new bacteriophage vector for cloning in *Bacillus subtilis* and the use of ϕ105 for protein synthesis in maxicells. *Gene* **81:**35–43.
7. **East, A. K., and J. Errington.** Unpublished data.
8. **Ellis, D. M., and D. H. Dean.** 1985. Nucleotide sequence of the cohesive single-stranded ends of *Bacillus subtilis* temperate bacteriophage ϕ105. *J. Virol.* **55:**513–515.
9. **Ellis, D. M., and D. H. Dean.** 1986. Location of the *Bacillus subtilis* temperature bacteriophage ϕ105 *attP* attachment site. *J. Virol.* **5:**223–224.
10. **Errington, J.** 1984. Efficient *Bacillus subtilis* cloning system using bacteriophage vector ϕ105J9. *J. Gen. Microbiol.* **130:**2615–2628.
11. **Errington, J.** 1986. Gene cloning and expression vectors based on *Bacillus subtilis* bacteriophage ϕ105, p. 217–227. *In* A. T. Ganesan and J. A. Hoch (ed.), *Bacillus Molecular Genetics and Biotechnology Applications*. Academic Press, Inc., Orlando, Fla.
12. **Errington, J.** 1986. A general method for fusion of the *Escherichia coli lacZ* gene to chromosomal genes in *Bacillus subtilis*. *J. Gen. Microbiol.* **132:**2953–2966.

13. **Errington, J.** 1987. New vectors for gene cloning in *Bacillus subtilis*, p. 71–80. *In* M. Alacevic, D. Hranueli, and Z. Toman (ed.), *Proceedings of the Fifth International Symposium on the Genetics of Industrial Microorganisms, 1986.* Ognjen Prica Printing Works, Karlovac, Yugoslavia.
14. **Errington, J.** 1988. Generalized cloning vectors for *Bacillus subtilis*, p. 345–362. *In* R. L. Rodriguez and D. T. Denhardt (ed.), *Vectors: a Survey of Molecular Cloning Vectors and Their Uses.* Butterworths, Boston.
15. **Errington, J.** 1990. Gene cloning techniques, p. 175–220. *In* C. R. Harwood and S. M. Cutting (ed.), *Molecular Biological Methods for Bacillus*. John Wiley & Sons, Ltd., Chichester, United Kingdom.
16. **Errington, J.** 1991. Unpublished data.

16a. **Errington, J., and D. Jones.** 1987. Cloning in *Bacillus subtilis* by transfection with bacteriophage ϕ105: isolation and characterization of transducing phages for 23 sporulation loci. *J. Gen. Microbiol.* **133:**493–502.

17. **Errington, J., and N. Pughe.** 1987. Upper limit for DNA packaging by *Bacillus subtilis* bacteriophage ϕ105: isolation of phage deletion mutants by induction of oversize prophages. *Mol. Gen. Genet.* **210:**347–351.
18. **Flock, J.-I.** 1977. Deletion mutants of temperate *Bacillus subtilis* bacteriophage ϕ105. *Mol. Gen. Genet.* **155:**241–247.
19. **Fort, P., and J. Errington.** 1985. Nucleotide sequence and complementation analysis of a polycistronic sporulation operon, *spoVA*, in *Bacillus subtilis*. *J. Gen. Microbiol.* **131:**1091–1105.
20. **Garro, A. J., and M.-F. Law.** 1974. Relationship between lysogeny, spontaneous induction, and transformation efficiencies in *Bacillus subtilis*. *J. Bacteriol.* **120:**1256–1259.

20a. **Gibson, R. M., and J. Errington.** 1992. A novel *Bacillus subtilis* cloning vector based on bacteriophage ϕ105. *Gene* **121:**137–142.

21. **Gruss, A., and S. D. Ehrlich.** 1989. The family of highly interrelated single-stranded deoxyribonucleic acid plasmids. *Microbiol. Rev.* **53:**231–241.
22. **Iiima, T., F. Kawamura, H. Saito, and Y. Ikeda.** 1980. A specialized transducing phage constructed from *Bacillus subtilis* phage ϕ105. *Gene* **9:**115–126.
23. **Ikeuchi, T., J. Kudoh, and K. Kurahashi.** 1985. Genetic analysis of *spo0A* and *spo0C* mutants of *Bacillus subtilis* with a ϕ105 prophage merodiploid system. *J. Bacteriol.* **163:**411–416.
24. **Illing, N., and J. Errington.** 1991. The *spoIIIA* operon of *Bacillus subtilis* defines a new temporal class of mother cell-specific sporulation genes under the control of the σ^E form of RNA polymerase. *Mol. Microbiol.* **5:**1927–1940.
25. **James, W., and J. Mandelstam.** 1985. Protease production during sporulation of germination mutants of *Bacillus subtilis* and the cloning of a functional *gerE* gene. *J. Gen. Microbiol.* **131:**2421–2430.
26. **Jenkinson, H. F., and M. Deadman.** 1984. Construction and characterization of recombinant phage ϕ105d(Cm^r met) for cloning in *Bacillus subtilis*. *J. Gen. Microbiol.* **130:**2155–2164.
27. **Jenkinson, H. F., and J. Mandelstam.** 1983. Cloning of the *Bacillus subtilis lys* and *spoIIIB* genes in phage ϕ105. *J. Gen. Microbiol.* **129:**2229–2240.
28. **Jones, D., and J. Errington.** 1987. Construction of improved bacteriophage ϕ105 vectors for cloning by transfection in *Bacillus subtilis*. *J. Gen. Microbiol.* **133:**483–492.
29. **Kawamura, F., H. Saito, and Y. Ikeda.** 1979. A method for construction of specialized transducing phage ρ11 of *Bacillus subtilis*. *Gene* **5:**87–91.
30. **Lampel, J. S., D. M. Ellis, and D. H. Dean.** 1984. Reorienting and expanding the physical map of temperate *Bacillus subtilis* bacteriophage ϕ105. *J. Bacteriol.* **160:** 1178–1180.
31. **Marrero, R., and R. E. Yasbin.** 1986. Evidence for circular permutation of the prophage genome of *Bacillus subtilis* bacteriophage ϕ105. *J. Virol.* **57:**1145–1148.
32. **Marrero, R., and R. E. Yasbin.** 1988. Cloning of the *Bacillus subtilis* recE$^+$ gene and functional expression of recE$^+$ in *B. subtilis*. *J. Bacteriol.* **170:**335–344.
33. **McConnell, D. J.** Personal communication.
34. **Moran, C. P., Jr.** 1990. Measuring gene expression in Bacillus, p. 267–293. *In* C. R. Harwood and S. M. Cutting (ed.), *Molecular Biological Methods for Bacillus*. John Wiley & Sons, Ltd., Chichester, United Kingdom.
35. **Mueller, J. P., and H. W. Taber.** 1989. Isolation and sequence of *ctaA*, a gene required for cytochrome aa_3 biosynthesis and sporulation in *Bacillus subtilis*. *J. Bacteriol.* **171:**4967–4978.
36. **O'Kane, C., M. A. Stephens, and D. McConnell.** 1986. Integrable α-amylase plasmid for generating random transcriptional fusions in *Bacillus subtilis*. *J. Bacteriol.* **168:**973–981.
37. **Poth, H., and P. Youngman.** 1988. A new cloning system for *Bacillus subtilis* comprising elements of phage, plasmid and transposon vectors. *Gene* **73:**215–226.
38. **Rosenthal, R., P. A. Toye, R. Z. Korman, and S. A. Zahler.** 1979. The prophage of SPβ*c2dcitK*$_1$, a defective specialized transducing phage of *Bacillus subtilis*. *Genetics* **92:**721–739.
39. **Rutberg, L.** 1969. Mapping of a temperate bacteriophage active on *Bacillus subtilis*. *J. Virol.* **3:**38–44.
40. **Seaman, E., E. Tarmy, and J. Marmur.** 1964. Inducible phages of *Bacillus subtilis*. *Biochemistry* **3:**607–613.
41. **Spancake, G. A., and H. E. Hemphill.** 1985. Deletion mutants of *Bacillus subtilis* bacteriophage SPβ. *J. Virol.* **55:**39–44.
42. **Thornewell, S. J., A. K. East, and J. Errington.** Manuscript in preparation.
43. **Wood, H. E., M. T. Dawson, K. M. Devine, and D. J. McConnell.** 1990. Characterization of PBSX, a defective prophage of *Bacillus subtilis*. *J. Bacteriol.* **172:**2667–2674.
44. **Yasbin, R. E., G. A. Wilson, and F. E. Young.** 1973. Transformation and transfection in lysogenic strains of *Bacillus subtilis* 168. *J. Bacteriol.* **113:**540–548.
45. **Yasbin, R. E., G. A. Wilson, and F. E. Young.** 1975. Transformation and transfection in lysogenic strains of *Bacillus subtilis*: evidence for selective induction of prophage in competent cells. *J. Bacteriol.* **121:**296–304.
46. **Yasdi, M. A., and A. Moir.** 1990. Characterization and cloning of the *gerC* locus of *Bacillus subtilis* 168. *J. Gen. Microbiol.* **136:**1335–1342.
47. **Youngman, P.** 1990. Use of transposons and integrational vectors for mutagenesis and construction of gene fusions in Bacillus species, p. 221–266. *In* C. R. Harwood and S. M. Cutting (ed.), *Molecular Biological Methods for Bacillus*. John Wiley & Sons, Ltd., Chichester, United Kingdom.
48. **Zabarovsky, E. R., and R. L. Allikmets.** 1986. An improved technique for the efficient construction of gene libraries by partial filling-in of cohesive ends. *Gene* **42:**119–123.
49. **Zuber, P., and R. Losick.** 1983. Use of a *lacZ* fusion to study the role of the *spo0* genes of *Bacillus subtilis* in developmental regulation. *Cell* **35:**275–283.

VII. TRANSCRIPTION AND TRANSLATION MACHINERY

45. RNA Polymerase and Transcription Factors

CHARLES P. MORAN, JR.

Gene expression in bacteria is regulated primarily at the level of transcription. Transcription and its regulation have been studied extensively in several gram-negative bacteria, especially *Escherichia coli*, but mostly neglected in gram-positive bacteria. The major exception is *Bacillus subtilis*, which because of its amenability to genetic analysis has served as a paradigm for understanding transcription in gram-positive bacteria. The enzymes that catalyze transcription in *B. subtilis* are structurally and functionally similar to those in *E. coli* and other bacteria. Furthermore, the factors that regulate gene transcription in these bacteria are similar. For example, several families of structurally related transcription factors have been identified in gram-negative bacteria. Two major families of sigma factors bind to RNA polymerase to form holoenzymes that bind and utilize specific promoters. A large family of proteins homologous to NtrC of enteric bacteria are phosphorylated and bind to specific DNA sequences to activate specific promoters. Other types of activators as well as repressors of promoter activity have been described in gram-negative bacteria. *B. subtilis* and other gram-positive bacteria contain members of most of the major families of positive and negative regulators of transcription that have been found in *E. coli*. However, the specific physiological problems that these regulators are employed to solve are sometimes different than in *E. coli*.

The mechanisms of transcriptional regulation for many genes in gram-positive bacteria are described throughout this volume in the context of the specific physiological processes discussed in each chapter. Therefore, this review attempts to provide an overview of the roles of the major families of proteins required for transcription and its regulation in *B. subtilis*. Emphasis is placed on comparing the roles of these factors in *B. subtilis* and gram-negative bacteria.

COMPOSITION OF RNA POLYMERASE

RNA polymerase in *B. subtilis*, like that in *E. coli*, is composed of several protein subunits. The core enzyme is composed of α, β, and β' in a stoichiometry of 2:1:1, respectively (reviewed in reference 30). The core polymerase is capable of transcribing DNA into RNA, but it initiates transcription inefficiently at random sites. Like its counterpart in *E. coli*, association of a sigma subunit with the core polymerase forms the holoenzyme, which is able to bind and utilize specific promoters (reviewed in reference 91).

The core subunits of *B. subtilis* RNA polymerase are structurally and functionally similar to those in *E. coli*. The structural gene for the α subunit, *rpoA*, has been cloned and sequenced (11). The sequence of *rpoA* indicates that it encodes a 314-amino-acid peptide that is highly homologous to the α subunit from *E. coli*. The amino acid sequences of the two peptides contain 46% identical residues. The nucleotide sequences of the structural genes for the β and β' subunits have not been determined. However, the proteins are probably very similar to those of *E. coli*, since the amino acid sequences of the large subunits of RNA polymerase are highly conserved in *E. coli* and eukaryotic organisms (3, 8). Furthermore, heterologous but functional holoenzymes have been reconstituted in vitro with sigma factors from *B. subtilis* and core polymerase from *E. coli* as well as a sigma factor from *E. coli* and core polymerase from *B. subtilis* (119). The core RNA polymerases in other gram-positive bacteria appear to be composed of similar-sized subunits (Table 1).

Two copies of an additional subunit, ω, are often isolated with RNA polymerase core from *E. coli*. This factor is not essential for reconstituted activity in vitro, and no phenotype is associated with disruption of its structural gene (45). Therefore, the role of this factor in *E. coli* is unknown. A similar-sized protein is associated with RNA polymerase from *B. subtilis* (cited in reference 135), but its role is also unknown.

Another protein, δ, has been found associated with RNA polymerase from *B. subtilis*. In vitro, δ binds to the core enzyme and a variety of holoenzymic forms and inhibits transcription of nonspecific templates (135). Disruption of the structural gene for δ was not associated with a phenotype; the role of this subunit therefore remains unknown (83).

The *nusA* product from *E. coli* associates with RNA polymerase after initiation of transcription and remains associated with the elongation complex (reviewed in reference 39). A *nusA* homolog has not been identified in gram-positive bacteria.

SIGMA FACTORS

Role in Promoter Utilization

The sigma subunit of RNA polymerase determines the specificity of promoter utilization. There are at least 10 different sigma factors in *B. subtilis*, each of which directs RNA polymerase to a different set of promoters (Table 2). Two regions in most sigma factors in bacteria (excluding σ^{54} [109] and its homologs) probably determine the specificity of promoter utilization by making sequence-specific contacts at two regions of promoters located approximately 10 and 35 bp upstream from the start point of transcription, i.e.,

Charles P. Moran, Jr. • Department of Microbiology and Immunology, Emory University School of Medicine, Atlanta, Georgia 30322.

Table 1. Subunits of core RNA polymerase in gram-positive bacteria

Organism	Subunit	Apparent mass	Reference
Bacillus subtilis	β	140,000	29
	β'	130,000	29
	α	34,815[a]	11
Streptomyces coelicolor	β	150,000	14
	β'	140,000	14
	α	47,000	14
Clostridium perfringens	β	145,000	43
	β'	135,000	43
	α	44,000	43
Clostridium acetobutylicum	A	128,000	101
	B	117,000	101
	D	42,000	101
Lactobacillus curvatus	β	151,000	122
	β'	145,000	122
	α	42,500	122

[a] Estimated from the sequence of the structural gene. All other masses are estimated from mobility during electrophoresis.

in the −10 and −35 regions, respectively (84; reviewed recently in references 54 and 91). This model is based primarily on the results of genetic suppression experiments with four different sigma factors (σ^{70} from *E. coli* [42, 120, 137] and σ^{A} [73, 74], σ^{H} [26, 146], and σ^{E} [133] from *B. subtilis*). In these experiments, single amino acid substitutions in the sigma factors were found to suppress the effect of specific base pair substitutions in their cognate promoters.

The primary sigma factor in growing *B. subtilis*, σ^{A}, is homologous to σ^{70} in *E. coli* (48). Utilization of promoters by RNA polymerase containing σ^{A} appears to be governed primarily by the same sequences at the −35 and −10 regions that signal recognition of promoters by σ^{70} in *E. coli* (5′-TTGACA-3′ and 5′-TATA AT-3′, respectively) (Table 2) (57, 73, 92). Suppression studies with σ^{70} identified two amino acids that probably interact with the −10 region of promoters; the glutamine at position 437 of σ^{70} interacts with the first base pair of the −10 region hexamer (TATAAT) (137), and the threonine at position 440 may interact

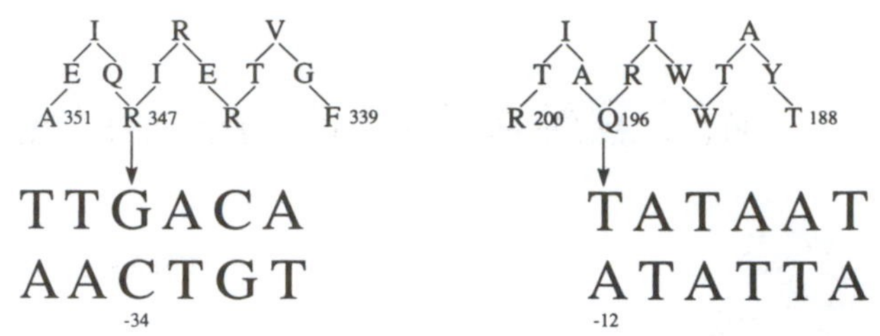

Figure 1. Model of interactions of σ^{A} with a consensus promoter. Nucleotide sequences are the canonical hexameric sequences found at the −10 and −35 regions of consensus σ^{A}-dependent promoters. Above the promoter sequences are the amino acid sequences (one-letter code) in σ^{A} that may interact with the −35 and −10 regions of σ^{A} promoters. Used with permission from the publisher (73).

with the same base pair (120). A second region of σ^{70} interacts with the −35 region of promoters. Suppression experiments demonstrated genetic interactions between the arginine at position 588 of σ^{70} and the third base pair of the −35 region hexamer sequence (42), while the arginine at position 584 interacts with the fifth base pair of the −35 region hexamer (120). The amino acid sequence of σ^{A} is highly conserved compared with that of σ^{70}, although σ^{A} is missing a region containing 245 amino acids found near the N terminus of σ^{70} (48). The sequences of amino acids in σ^{A} that interact with the −10 and −35 regions of promoters appears to be identical to those of σ^{70}. Suppression studies with four promoters showed that the glutamine at position 196 of σ^{A}, which is homologous to the glutamine at 437 of σ^{70}, interacts with the first base pair of the −10 region hexamer (73, 74, 142), and the arginine at position 347 of σ^{A}, which is homologous to the arginine at position 588 of σ^{70}, interacts with the third base pair of the −35 region hexamer in two promoters in *B. subtilis* (73) (Fig. 1).

The allele specificity of genetic interactions between the promoters and sigma factors suggests that these amino acid substitutions define positions of the sigma factors that directly contact base pairs of the promoter. Two types of allele-specific interactions have

Table 2. Sigma factors in *B. subtilis*

Sigma factor	Primary function	Cognate promoter consensus sequence[a]		Reference(s)
		−35	−10	
σ^{A}	Housekeeping	TTGACA	TATAAT	92
σ^{B}	*ctc* transcription[b]	AGGTTT	GGGTAT	12, 134
σ^{C}	Unknown[c]	AAATC	TANTGNTTNTA	66
σ^{D}	Flagellar synthesis	CTAAA	CCGATAT	47
σ^{E}	Stage II sporulation	KMATATT	CATACA-T	107
σ^{F}	Stage II sporulation	Similar to that of σ^{G}		129
σ^{G}	Forespore specific	YGHATR	CAHWHTAH	95
σ^{H}	Stage 0 sporulation	AGGA--T	GAATT	132
σ^{K}	Mother cell specific	GKMACA	CATANNNT	38
σ^{L}	Levanase operon	CTGGA[d]	TTGCA[e]	28

[a] Nontranscribed strands are shown 5′ to 3′, left to right. H = A, C, or T; R = A or G; W = A or T; Y = C or T; K = G or T; M = C or A.
[b] σ^{B} directs transcription of several promoters (7, 12, 61, 104); however, the role of σ^{B} in the life cycle of *B. subtilis* remains unknown (9, 34).
[c] The structural gene for σ^{C} is unknown (66).
[d] −24 region.
[e] −12 region.

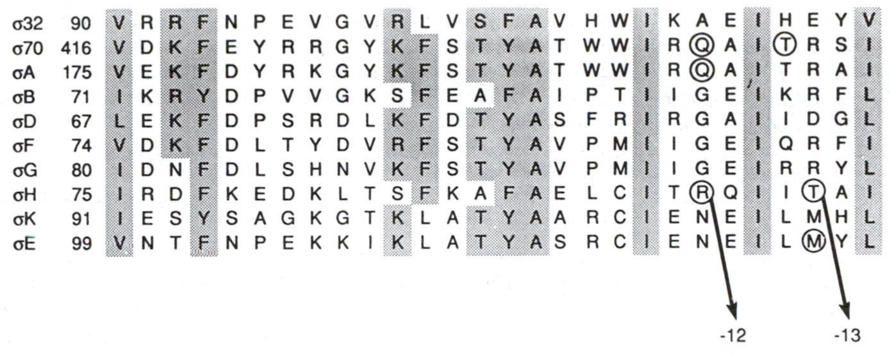

Figure 2. The −10 recognition regions of sigma factors. Amino acid sequences are those of the regions in sigma factors that may interact with the −10 regions of promoters. Sigma factors are from *B. subtilis* except for two from *E. coli* (σ^{70} and σ^{32}). Circles indicate positions at which amino acid substitutions specifically suppress the effects of mutations in the −10 regions of promoters, σ^{70} (120, 137), σ^{A} (73, 74), σ^{H} (26, 146), and σ^{E} (133). Conserved regions are shaded. Used with permission of the publisher (133).

been found: base pair specificity and position specificity. In the case of σ^{H}, a secondary sigma factor in *B. subtilis*, substitution of isoleucine at position 100 of σ^{H} suppressed the effect of G · C to A · T substitution at position −13 in the *spoVG* promoter (146). This amino acid substitution did not suppress the effects of substitutions at other positions of the *spoVG* promoter (position specificity) or of other base pair substitutions at position −13 (base pair specificity). The model that holds that this specificity is due to a direct interaction between the side chain to the amino acid at position 100 of σ^{H} and the base pair at position −13 of the *spoVG* promoter is further strengthened by the effect of a loss-of-contact substitution at position 100 of σ^{H}. Substitution of alanine at position 100 of σ^{H}, which essentially removes the amino acid side chain normally present at this position, resulted in a sigma factor that was unable to discriminate between *spoVG* promoter derivatives that differed only by single base pair substitutions at position −13 (26).

The amino acid sequences in the regions of σ^{70}, σ^{A}, and σ^{H} that interact with the −10 and −35 regions of their cognate promoters are partially conserved among most sigma factors (reviewed in reference 54) (Fig. 2 and 3). Most sigma factors may use similar structural motifs to interact with the −10 and −35 regions of their cognate promoters. This model is supported by the observations of the effects of amino acid substitutions in σ^{E} and σ^{F}, secondary sigma factors that are produced during sporulation in *B. subtilis*. Recently, substitution of alanine for the methionine at position 124 of σ^{E}, which lies within the region of σ^{E} that is similar to the −10 region recognition domains of the other sigmas (Fig. 2), was shown to suppress the effect of a base pair substitution at position −13 of the *spoIIID* promoter (133). Since this suppression was position specific, it is likely that the methionine at position 124 of σ^{E} interacts with the −10 region of promoters.

The sequences of amino acids in the regions of σ^{70} and σ^{A} that are implicated by suppression experiments to interact with the −35 region of promoters are similar to the sequences found in a region in most sigma factors in *B. subtilis* (Fig. 3). These regions in the other sigma factors probably interact with the −35 regions of their cognate promoters. Although this model has not been tested directly, recent evidence indicates that this conserved region in other sigma factors plays a role in the specificity of promoter utilization. An amino acid substitution in this region of σ^{F}, i.e., substitution of alanine for valine at position 233, was shown to change the specificity of promoter utilization so that σ^{F} could direct transcription from a promoter that is normally used by σ^{G} RNA polymerase (113). Moreover, substitution of methionine at this position in σ^{F} changed the specificity such that the altered σ^{F} directed transcription from a promoter that is used normally by σ^{B} RNA polymerase (87).

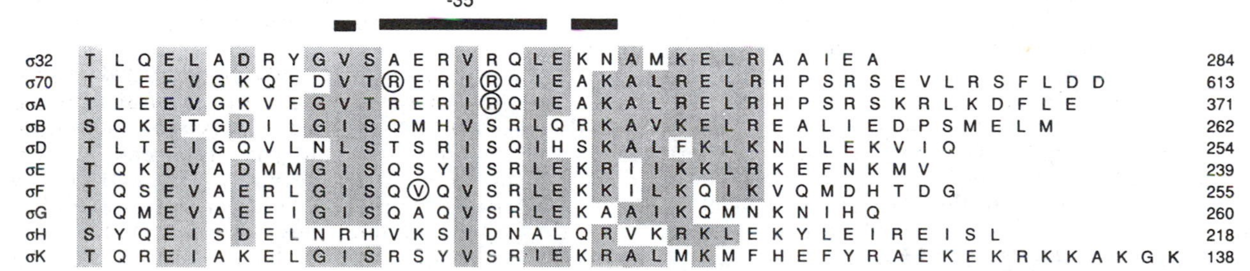

Figure 3. The −35 recognition regions of sigma factors. Amino acid sequences are those of the regions in sigma factors that may interact with the −35 regions of promoters. Sigma factors are from *B. subtilis* except for two from *E. coli* (σ^{70} and σ^{32}). Circles indicate positions at which amino acid substitutions specifically suppress the effects of mutations in the −35 regions of promoters, σ^{70} (42, 120), and σ^{A} (73) or in σ^{F} to change the specificity to σ^{G}- and σ^{B}-like (87, 113). Sequences are aligned and conserved regions are shaded according to Stragier et al. (125).

Evidently, the specificity of promoter utilization is directed by two regions of most sigma factors that make sequence-specific contacts at two regions of the promoters. The sigma subunit of RNA polymerase is required for specific binding of the enzyme to promoters. However, promoter utilization involves several distinct steps after the initial binding of RNA polymerase to the promoter. These steps include unwinding of the DNA, incorporation of the initial ribonucleotides into the transcript, and finally, release of the sigma factor so that the polymerase can elongate the transcript (reviewed in reference 88). It is not known whether sigma factors participate directly in the steps of promoter utilization that follow the initial binding. However, the recognition of limited amino acid sequence similarities in sigma factors and single-stranded nucleic acid-binding proteins has led to the suggestion that sigma factors may be directly involved in unwinding the DNA at the promoter (54). Recently, an amino acid substitution in the -10 recognition domain of σ^E, i.e., substitution of arginine for the cysteine at position 117, produced a sigma factor that associated with RNA polymerase and trapped the polymerase in a stable complex with promoter DNA in which the polymerase was unable to initiate transcription (68). This cysteine at position 117 of σ^E is located only 7 amino acid residues from the methionine at position 124, which was implicated in the suppression experiments described above to interact with the -10 region of promoters (Fig. 2). Therefore, the cysteine at position 117 probably is located near the -10 region of the promoter in the polymerase-promoter complex. The simplest interpretation of the finding that substitution of an arginine for the cysteine at position 124 allows promoter binding by polymerase but prevents transcription initiation is that this substitution inactivates a function of the sigma factor that is required during a step of promoter utilization that follows the initial binding of polymerase to the promoter. Other, more complex explanations are possible; therefore, additional work is needed to determine whether sigma factors participate directly in multiple steps of promoter utilization.

In addition to the insights into promoter utilization that have been provided by analyses of sigma structure, these analyses have also provided important tools for the study of the role of specific sigma factors and other key regulatory factors in the life cycle of *B. subtilis*. For example, suppression experiments such as those in which single amino acid substitutions in σ^A, σ^H, and σ^E have been found to specifically suppress base pair substitutions in promoters provide compelling evidence that transcription from specific promoters is directed by specific sigma factors. During sporulation of *B. subtilis*, several sigma factors are produced, and these factors play a central role in controlling gene expression. The role of sigma factors during sporulation will be discussed in more detail below; however, many of the conclusions depend on the knowledge that sigma factors direct transcription of specific genes. The suppression experiments have been particularly helpful in determining the roles of some sigma factors. For example, allele-specific suppression of promoter mutations by a single amino acid substitution in σ^A provided strong evidence that transcription of two key operons, *spoIIG* and *spoIIE*, is directed by σ^A after the onset of sporulation (74, 143). Other examples of the use of altered specificity mutations in sigma factors to determine the role of these factors during sporulation will be described below.

Use in Control of Gene Expression

Sigma factors play a variety of roles in controlling gene expression. At one extreme, the production of a secondary sigma factor can be primarily responsible for the activation of specific promoters. Alternatively, a secondary sigma factor can be necessary for transcription from a specific promoter; however, the activity of the promoter may be regulated primarily by an ancillary factor such as a repressor or an activator. The production of sigma factors that activate transcription from specific promoters plays a major role in controlling gene expression during endospore formation in *B. subtilis*. Several aspects of sporulation are summarized in other chapters in this volume. However, the sporulation process is reviewed briefly here to illustrate the role of sigma factors in this process.

In response to nutrient depletion, *B. subtilis* undergoes a morphologically complex differentiation that culminates in production of endospores (Fig. 4). This differentiation requires the expression of a large number of genes (probably about 100) (reviewed in reference 121). Transcription of many of these genes is activated after the onset of sporulation, and regulation of this transcription is governed principally by the production of four sigma factors during sporulation. The activities of these sigma factors appear sequentially during sporulation in the order σ^F, σ^E, σ^G, and σ^K, and each sigma activates transcription of a different set of genes (reviewed in references 85, 91, and 126). One role for these sigma factors is control of the temporal pattern of transcription during sporulation. The transcription of each set of genes that is directed by each sigma factor can occur only during that developmental period in which the sigma factor is present. However, as will be described below, the activity of each sigma factor is regulated, and transcription from some promoters is subject to regulation by promoter-specific DNA-binding proteins. Therefore, the sequential production of sigma factors ensures that specific sets of genes are transcribed during specific developmental periods. However, within a specific sigma factor-dependent set, the temporal patterns of expression of specific genes are differentially regulated by promoter-specific regulators.

Sigma factors control not only the temporal pattern of gene expression during sporulation but also the spatial distribution of gene expression. Endospore formation involves two cell types (reviewed in references 85 and 126). At an early stage, the cell divides asymmetrically to produce daughter cells with different developmental fates (Fig. 4). The smaller cell, known as the forespore, is destined to become the spore. The larger cell, or mother cell, engulfs the forespore during an intermediate stage of sporulation. The mother cell becomes a terminally differentiated cell that produces specialized products for the developing endospore and ultimately lyses to release the

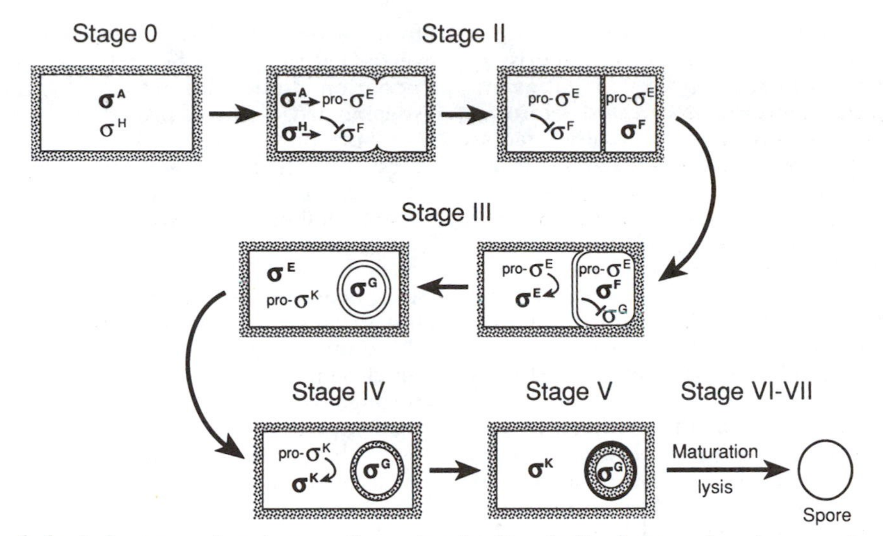

Figure 4. Morphological stages of endospore formation in *B. subtilis*. Stages of endospore development are described in the text. The vegetative cell begins endospore development during stage 0. Completion of the asymmetric cell division marks stage II. Engulfment of the forespore protoplast is completed in stage III. The spore coat begins to accumulate in stage IV. Subsequent maturation of the spore during the late stages culminates in mother cell lysis and release of the mature spore. Accumulation and activities of the RNA polymerase sigma factors during the endospore development are also indicated. σ^A and σ^H are active during the earliest stage of sporulation. The concentration of σ^H increases fourfold at the onset of sporulation (active σ factors are indicated with boldface type). σ^F is shown in both the mother cell and the forespore; however, it is active predominantly in the forespore. The symbol —⊣ indicates that σ^F is held inactive until after septation and that σ^G is held inactive until after engulfment of the forespore. σ^G is produced predominantly in the forespore, and σ^K is produced predominantly in the mother cell. This figure is based on results described in the text and a similar figure in reference 85.

mature endospore. Although each cell, mother cell and forespore, contains a copy of the genome, different sets of genes are transcribed in the two cell types. This cell type-specific transcription results, at least in part, from the production of cell type-specific sigma factors. σ^G is produced predominantly in the forespore and directs the transcription of forespore-specific genes, such as the *ssp* family (130). σ^K is produced exclusively in the mother cell and directs the transcription of mother cell-specific genes, such as those that produce the proteins that coat the endospore (125). The establishment of cell type-specific production of sigma factors, σ^G, and σ^K, appears to result from differential activities of the sigma factors σ^F and σ^E in the forespore and mother cell, respectively.

σ^F is produced before the asymmetric cell division (46); however, it remains inactive until after septation (87). Electron microscopy of immunogold-labeled cells demonstrated that transcription from a σ^F-dependent promoter is restricted to the forespore compartment (87). Since a mutant derivative of σ^F that utilized the *ctc* promoter, which is not normally active during sporulation, resulted in the forespore-specific transcription of *ctc*, it is likely that σ^F is active exclusively in the forespore (87). The protein that may be responsible for repressing σ^F activity in the predivisional cell and in the mother cell after septation is encoded by *spoIIAB* (87, 113), which is in the same operon as the structural gene for σ^F, *spoIIAC*. It is not known how the antagonism of σ^F activity by SpoIIAB is relieved in the forespore compartment, but this relief appears to require both the products of *spoIIAA* and the *spoIIE* operon (87, 113). In one plausible but completely hypothetical model, the differential accumulation of an ion caused by the greater surface-to-volume ratio of the smaller forespore compartment could signal the inactivation of SpoIIAB by SpoIIAA. The mechanism by which SpoIIAB antagonizes σ^F activity also is not known, but several results suggest that SpoIIAB may act directly on σ^F or RNA polymerase containing σ^F rather than acting as a DNA-binding protein that represses transcription from σ^F-dependent promoters. For example, a mutant derivative of σ^F with an altered specificity is able to use the *ctc* promoter, which is not normally active during sporulation. The utilization of the *ctc* promoter by the altered σ^F is repressed by SpoIIAB. It is unlikely that SpoIIAB binds to the *ctc* promoter; therefore, it is likely that SpoIIAB controls σ^F activity directly. Furthermore, SpoIIAB is homologous to at least one other protein in *B. subtilis*. This protein, the product of *orfW*, is encoded in the *sigB* operon adjacent to the structural gene for σ^B (69). The results of genetic analyses of *orfW* function are consistent with the model that the *orfW* product antagonizes σ^B function (7, 102). Therefore, the products of *spoIIAB* and *orfW* may be sigma factor antagonists. This type of negative regulator of transcription has not yet been described in gram-negative bacteria.

σ^F directs transcription of the structural gene for σ^G, *spoIIIG* (113). Since σ^F activity is restricted to the forespore compartment, σ^G is produced predominantly in the forespore. The forespore-specific production of σ^G is amplified to some degree, because σ^G also directs transcription from the *spoIIIG* promoter (20, 70, 103). The production of σ^G in the forespore results in the transcription of forespore-specific genes; however, several observations remain unexplained. For example, not all σ^G-dependent promoters are activated simultaneously; i.e., there are apparently at least two temporal subclasses of σ^G-dependent genes. For example, *sspE* transcription is activated before *0.3* gene transcription (99). Presumably, another factor controls the temporal expression of these genes. It has also been noted that σ^G is not active until after engulfment of the forespore (reviewed in reference 85). It is not known what factor delays σ^G activation or how activation of σ^G is coupled to engulfment of the forespore. SpoIIAB can antagonize σ^G activity (74a, 103), but if it is SpoIIAB that holds σ^G inactive until after engulfment, the repression of σ^F and σ^G activities by SpoIIAB must be mediated differently, since σ^F is activated earlier than σ^G.

σ^K is produced exclusively in the mother cell and directs the expression of mother cell-specific genes such as the *cot* genes, which encode the proteins that coat the endospore (76). σ^K controls the compartment-specific expression of these genes and is required for their temporal regulation. One of the σ^K-dependent genes, *gerE*, encodes a small sequence-specific DNA-binding protein that is required for activation of a subset of σ^K-dependent promoters (24, 145). Therefore, there are at least two temporal classes of σ^K-dependent promoters, a *gerE*-independent class and a *gerE*-dependent class.

σ^K is produced exclusively in the mother cell primarily because transcription of *sigK*, the structural gene for σ^K, is restricted to the mother cell (reviewed in reference 85). σ^K directs transcription from the *sigK* promoter; therefore, after the accumulation of active σ^K in the mother cell is established, its synthesis is amplified in an autocatalytic loop similar to that of the σ^G-directed production of σ^G in the forespore. Establishment of *sigK* transcription in the mother cell depends on two factors, the product of *spoIIID* and σ^E. The product of *spoIIID* is a sequence-specific DNA-binding protein that stimulates transcription from several promoters, including σ^K- and σ^E-directed transcription from the *sigK* promoter (76, 79). *spoIIID* transcription is restricted to the mother cell, which accounts in part for the restriction to the mother cell, which accounts in part for the restriction of *sigK* transcription to the mother cell (77). *spoIIID* transcription is directed by σ^E (77, 123, 133).

The activity of σ^E may also be restricted to the mother cell (31, 62). This model is supported by the observation that in electron micrographs of immunogold-labeled cells containing *lacZ* fused to a σ^E-dependent promoter, *spoIID*, the β-galactosidase accumulated predominantly in the mother cell (31). As noted above, σ^E also directs the expression of *spoIIID*, which is transcribed exclusively in the mother cell. It is not known whether the activity of all σ^E-dependent promoters is restricted to the mother cell. (A change-of-specificity mutant of σ^E has not been used in experiments analogous to those described above for σ^F to study the compartment-specific activity of σ^E.) However, since the activity of several promoters used by σ^E is restricted to the mother cell, it is likely that σ^E is active predominantly in the mother cell (62). It is not known how σ^E activity is restricted to the mother cell, since the structural gene for σ^E, *sigE*, is transcribed before septation. σ^E is activated by proteolytic removal of several N-terminal amino acids from a pro-form of σ^E (82); therefore, this processing could be restricted to the mother cell (31). However, cell fractionation studies did not demonstrate compartment-specific processing of pro-σ^E (18).

σ^K production is controlled at two additional levels. First, the structural gene for *sigK* is produced by a deletion of about 40 kb that joins the N-terminal coding half of the gene to the C-terminal coding half of this gene (125). This deletion is mediated by a specific recombinase and the *spoIIID* product (78, 110, 125). This deletion occurs exclusively in the mother cell, but the sporulation proficiency of a mutant in which the intervening DNA has been deleted to produce a functional *sigK* gene demonstrated that the DNA rearrangement is not an essential component for establishing cell type-specific expression of σ^K (23). Moreover, the intervening sequence is not present in the structural gene for the *sigK* homolog in *Bacillus thuringiensis* (1).

The activation of σ^K requires proteolytic removal of 20 N-terminal amino acids from a precursor or pro-σ^K. A mutant in which the coding sequences for the proregion of pro-σ^K had been deleted produced σ^K but formed spores that germinated inefficiently (23). Analysis of this mutant showed that σ^K-directed transcription of the *cotA* gene occurred prematurely. Since almost morphologically normal spores were formed by this mutant, the compartment-specific expression of σ^K probably was not affected by deletion of the prosequence; rather, the processing of pro-σ^K seems to be required to delay activation of σ^K-dependent gene transcription in the mother cell (23). Evidently, premature activation of σ^K results in premature deposition of the coat proteins, which produces germination-defective spores. The primary role of the pro-σ^K processing step may be to provide a mechanism by which to couple σ^K-activated gene expression to attainment of a specific stage of forespore development; i.e., a morphological landmark or checkpoint signals the processing step (23).

Although the forespore- and mother cell-specific programs of gene expression have been discussed as two parallel pathways, expression of the genes in these two compartments is highly coordinated. The proteolytic processing of pro-σ^K depends on forespore-specific gene expression (23). Mutations in the structural gene for the forespore-specific sigma factor σ^G prevent transcription of σ^K-dependent genes in the mother cell. This dependency of σ^K function on σ^G can be bypassed by the deletion mutation that removes the proregion from pro-σ^K (23). Therefore, forespore-specific events are required for processing of σ^K in the mother cell.

Some of the gene products required for the coupling of σ^K activation to forespore-specific gene expression have been identified. *spoIVF* probably encodes the specific protease that removes the proregion from pro-σ^K or a regulator of its activity (23). *spoIVB* is expressed in the forespore and is required to signal the *spoIVF*-dependent processing of pro-σ^K (22).

Mother cell-specific gene expression also effects forespore gene expression. As noted earlier, the activation of σ^G in the forespore is delayed by some unknown mechanism until the forespore is engulfed. Engulfment requires σ^E, which probably is active predominantly in the mother cell. This so-called "crisscross" regulation of cell type-specific gene expression (85), in which forespore-specific gene expression affects mother cell-specific gene expression and mother cell gene expression affects forespore gene expression, occurs even at the early stages of development. σ^E is processed from a precursor protein, pro-σ^E. The processing of pro-σ^E requires σ^F activity (67, 82), which is restricted to the forespore (87). It is not known how σ^F affects pro-σ^E processing, but presumably, σ^F directs the transcription of a gene whose product signals the processing of σ^E. The crisscross regulation probably maintains the coordination of development in the two cell types, mother cell and forespore (85).

Two other sigma factors, σ^H and σ^A, function during sporulation. σ^H is necessary for the initiation of sporulation (33). Inactivation of the structural gene for σ^H, *spo0H*, blocks the initiation of sporulation but has no apparent effect on growth. σ^H directs transcription from several promoters that are activated within minutes after the onset of sporulation. These include the promoter for the *spoIIA* operon (140), which encodes σ^F. However, σ^H is present in growing cells before initiation of sporulation and activation of the *spoIIA* promoter (53, 132). The σ^H present in growing cells evidently is active, since it directs transcription from the major promoter for *citG*, the structural gene for the tricarboxylic acid cycle enzyme fumarase (132). The concentration of σ^H increases about fourfold at the onset of sporulation (53). It is not known what role, if any, the increase in σ^H concentration plays in the activation of σ^H-dependent promoters. The activities of some σ^H-dependent promoters are regulated by DNA-binding proteins. Transcription from the σ^H-dependent promoter *spoVG* is repressed by *abrB* (106, 147), whereas transcription from *spoIIA* probably is activated by Spo0A (136). These activators and repressors will be described below.

Many genes transcribed by σ^H-RNA polymerase do not encode products that function exclusively during endospore formation (e.g., *citG* [134], *rpoD* [19], and others [65]). Moreover, σ^H does not direct all of the transcription of sporulation-essential genes during the earliest stages of sporulation. The primary sigma factor in growing cells, σ^A, directs the transcription of at least two key operons, *spoIIE* and *spoIIG*, which includes *sigE*, the structural gene for σ^E, at the onset of sporulation. σ^A was shown to interact with these promoters in genetic suppression experiments in which the effects of single-base-pair substitutions in the *spoIIG* promoter (74) and the *spoIIE* promoter (142) were suppressed by a single-amino-acid substitution in σ^A. The *spoIIG* and *spoIIE* promoters are used by RNA polymerase containing the primary sigma factor in growing cells, yet these promoters are silent during growth and are activated in stationary phase after the onset of sporulation. The activities of these promoters, like that of the σ^H-dependent promoter *spoIIA*, are probably regulated by the sequence-specific DNA-binding protein Spo0A (111, 112, 136). The role of this transcriptional activator will be discussed below.

The sigma factors that act during sporulation illustrate two roles for sigma factors in the determination of promoter activity. In several cases, the sigma factors are essential for utilization of specific promoters, but ancillary factors regulate promoter activity. In other cases, the accumulation of sigma factor activates specific promoters. The same types of roles are expected of the additional sigma factors in *B. subtilis*, σ^B, σ^C, σ^D, and σ^L. However, there is less information about the regulation of specific promoters used by these factors.

σ^D and σ^L are described in more detail in other chapters in this volume but are described briefly here to illustrate the roles of sigma factors that have and have not been conserved between *Bacillus* spp. and the enteric bacteria during evolution. σ^D in *B. subtilis* is required for motility (55). It directs the transcription of *hag*, the structural gene for flagellin (89), and of several chemotaxis genes (148). *Salmonella* spp. (96) and *E. coli* (5) contain homologs of σ^D. The sequence of this sigma factor (σ^F), encoded by *fliA* in *Salmonella* spp., is 37% homologous with σ^D; 84 of the 239 amino acids of σ^F from *Salmonella* spp. are identical to those of σ^D from *B. subtilis* (96). Like its homolog in *B. subtilis*, σ^F directs the transcription of *Salmonella* genes required for motility (96).

σ^L in *B. subtilis* is most similar to σ^{54} from gram-negative bacteria (28). The σ^{54} homologs compose a distinct family of sigma factors not structurally related to the other sigma factors. The sequences of promoters that signal their recognition by RNA polymerase containing σ^{54} are located around positions -12 and -24 rather than in the -10 and -35 regions, which signal recognition of promoters by RNA polymerase containing the other types of sigma factors (reviewed in references 50 and 109). In gram-negative bacteria, activation of promoters used by RNA polymerase containing σ^{54} requires site-specific DNA-binding proteins (e.g., NtrC or NifA) that bind to sites far upstream from the promoters (reviewed in reference 50). Promoter utilization by RNA polymerase containing σ^L in *B. subtilis* probably follows similar rules, since the σ^L-dependent promoter for the levanase operon contains sequences at positions -12 and -24 that are similar to those conserved among σ^{54} promoters (28). Moreover, utilization of this promoter requires sequences located 130 bp upstream from the transcription start point and requires LevR, a protein that is homologous to NtrC and NifA (27). Many of the σ^{54}-dependent promoters in enteric bacteria direct transcription of genes required for nitrogen assimilation, e.g., *glnA*, which encodes glutamine synthetase (59, 60). Disruption of the σ^L structural gene in *B. subtilis* prevents growth on arginine, ornithine, valine, or isoleucine as nitrogen sources (28). However, the structural gene for glutamine synthetase is transcribed by RNA polymerase containing σ^A (114) (see chapter 20, this volume). σ^L directs transcription of the levanase operon in *B. subtilis* (28). In spite of the similarities between the σ^L and σ^{54} of gram-negative bacteria, σ^L has been recruited to direct transcription of a set of genes in *B. subtilis* that are different from those dependent on σ^{54} in enteric bacteria.

PHOSPHORYLATED ACTIVATORS OF PROMOTERS

Responses to many environmental signals by bacteria in several genera are mediated by two-component systems that are composed of a sensor protein and a

Table 3. Examples of transcription activators in gram-positive bacteria[a]

Organism(s)	Transcription factor	Family type	Activator
B. subtilis	Spo0A	Phosphorylated two-component response regulator	Phosphorelay (kinA, Spo0B, Spo0F)
	ComA		ComP
	DegU		DegS
	PhoP		PhoR
	LevR		Unknown
Streptococcus spp.	Mry		Unknown
B. subtilis	GltC	LysR family	Unknown

[a] See text for references.

response regulator protein (for reviews, see references 108 and 124). In several instances, the response regulator has been shown to be a sequence-specific DNA-binding protein that activates transcription from specific promoters. Furthermore, the response regulator is activated when it is phosphorylated by the sensor, which is an autophosphorylating histidine kinase. Other less well characterized members of these families probably function by a similar mechanism, because the structures of these factors are highly conserved. Several of these types of two-component regulatory systems from *B. subtilis* have been characterized to various degrees (Table 3).

The product of *spo0A*, Spo0A, is homologous to the response regulator class of signal transducers (36). Spo0A is essential for the initiation of sporulation, development of competence for DNA uptake, and other stationary-growth-phase-associated processes (see chapters 9, 39, 51, and 54, this volume). Several lines of evidence demonstrate the importance of phosphorylation of Spo0A. Spo0A can be phosphorylated in vitro by kinases isolated from *B. subtilis* (13) and by heterologous kinases from *E. coli*, CheA (97) and NRII (6). Mutations in *spo0A* (*coi*) that increase the rate of phosphorylation of Spo0A in vitro cause increased sporulation in cultures growing in the presence of excess glucose and GTP (97). Hoch and his colleagues (13) have presented evidence from in vitro studies that the product of *spoIIJ* (*kinA*), a member of the histidine kinase family, begins a cascade of phosphotransfers that culminates in the phosphorylation of Spo0A. In this model, the phosphate moiety is transferred from KinA to Spo0A through intermediary phosphotransferases Spo0F and Spo0B.

Spo0A is a sequence-specific DNA-binding protein. Binding of Spo0A to the *abrB* promoter represses its activity in vitro (127), and phosphorylation of Spo0A increases its affinity for the *abrB* promoter in vitro (136). Therefore, during the early stages of sporulation, phosphorylation probably causes Spo0A to bind to the *abrB* promoter and repress its activity.

Spo0A binding also probably directly activates several promoters. Within the first hour after the onset of sporulation, the *spoIIG* and *spoIIE* promoters are activated (52, 72). These promoters are used by RNA polymerase containing σ^A, and their activation is dependent on Spo0A (52, 71, 72).

Spo0A binds to two sites at the *spoIIG* promoter that are located approximately 87 and 37 bp upstream from the start point of transcription (111). Single-base-pair substitutions at these sites reduce *spoIIG* promoter activity in vivo, and several of these mutations have been shown to reduce binding to Spo0A to these sites in vitro (111, 112). Moreover, Spo0A stimulates transcription from the *spoIIG* promoter by σ^A-RNA polymerase in vitro (112). These results strongly support the model that Spo0A binding to the *spoIIG* promoter activates its activity.

Since *spoIIG* promoter activity in vivo is also dependent on those products that are believed to function in the cascade that leads to phosphorylation of Spo0A (i.e., SpoIIJ, Spo0B, and Spo0F), it is tempting to assume that phosphorylation of Spo0A increases the affinity of the protein for the *spoIIG* promoter and results in activation of the promoter. However, this model has not been tested, and these results could be explained by alternative models. For example, the dependency of *spoIIG* promoter activity on phosphorylation of Spo0A could reflect the fact that Spo0A stimulates its own synthesis, which is dependent on phosphorylation (141). Therefore, the nonphosphorylated form of Spo0A could act on the *spoIIG* promoter during the early stage of sporulation when the concentration of Spo0A increases. The *spoIIE* promoter is probably activated by a similar mechanism, since Spo0A binds specifically to four sites in the upstream region of *spoIIE* (142). Evidently, binding of Spo0A also activates the *spoIIA* promoter, which is used by σ^H-RNA polymerase (136).

ComA is another member of the response regulator class and, together with ComP, a histidine kinase, is essential for the development of competence for DNA uptake (32). Like Spo0A, ComA affects several stationary-phase-specific responses (e.g., expression of *degO* [see chapter 50, this volume], *gsiA* [chapter 9], and surfactin [chapter 61]). ComA has been phosphorylated in vitro by using NRII, a kinase from *E. coli*; therefore, it is likely that it is phosphorylated in vivo, probably by ComP (32). The amino acid sequence of the carboxy-terminal region of ComA is similar to that of several DNA-binding proteins; therefore, it is likely that phosphorylation of ComA and binding of ComA to its target sequences stimulate transcription of some competence-essential genes. However, the target of ComA has not been identified.

DegU is phosphorylated by DegS in vitro and probably in vivo (25, 93). Although originally defined as a gene product affecting expression of extracellular degradative enzymes, DegU is also required for the development of competence (reviewed in chapter 50). The similarity of its sequence to those of other DNA-binding proteins suggests that DegU binds DNA (58,

Table 4. Some examples of transcription repressors in *Bacillus* spp.[a]

Repressor	Target	Regulation of repressor action
GntR	Gluconate operon	Repressor inactivated by gluconate
PenI	*penP* (β-lactamase) promoter	Inactivation of PenI by penicillin-mediated membrane protein and cytoplasmic factor
GlnR	*glnRA* operon, which encodes glutamine synthetase	Corepressors glutamine, GlnA, and unknown factor
AbrB	Stationary-phase-induced genes: *tycA*, *spoVG spo0E*	Synthesis repressed by Spo0A during stationary phase

[a] See text for references.

80) but that its target sequences have not been defined. It has not been shown to directly stimulate transcription.

The sequence similarity of PhoP from *B. subtilis* to PhoB from *E. coli* and the requirement of PhoP for alkaline phosphatase expression in *B. subtilis* supports the model that PhoP functions as a PhoB homolog (115) (see chapter 17). However, there is no direct evidence that PhoP binds to DNA, can be phosphorylated, or can stimulate transcription.

LevR is required for expression of the levanase operon in *B. subtilis* (reviewed in chapter 50). A central 200-amino-acid domain of LevR is similar to domains in other response regulators (27). Expression from the *B. subtilis* levanase operon promoter in *E. coli* is stimulated by LevR and is dependent on σ^{54} (27). In *B. subtilis*, the LevR-dependent expression of the levanase promoter requires sequences located 130 bp upstream from the start point of transcription (27), although direct binding to this target by LevR has not been reported. Since σ^{L} from *B. subtilis*, a σ^{54} homolog, is also required for levanase promoter activity, it is likely that LevR stimulates RNA polymerase containing σ^{L} to use the levanase promoter by a mechanism similar to that of NifA and NtrC stimulation of σ^{54}-dependent promoters in gram-negative bacteria (28).

Since the phosphorylated activators of transcription are found in many gram-negative species as well as in *B. subtilis*, these factors will probably be found in all gram-positive bacteria. Recently, one such activator has been identified in *Streptococcus* spp. (17, 105).

OTHER ACTIVATORS OF TRANSCRIPTION INITIATION

Another large family of promoter activator proteins in bacteria is homologous to LysR (56). These proteins may bind to DNA via a helix-turn-helix motif (56). No common type of coactivator has been identified for these proteins; however, several of these proteins are involved in controlling amino acid biosynthesis. The LysR family originally included only proteins from gram-negative bacteria, but the sequence of GltC from *B. subtilis* revealed that this protein, too, is a member of the LysR family (10). GltC is an activator of *gltA* and *gltB* transcription. *gltA* and *gltB* encode the subunits of glutamate synthase (see chapter 20).

The cyclic-AMP (cAMP) receptor protein (CAP) (reviewed in reference 2) and fumarate nitrate regulator (FNR) (35) are structurally related promoter activators that recognize similar target sequences in enteric bacteria. However, these activators are used for very different physiological functions. In enteric bacteria, CAP binds to and activates promoters in response to elevated cAMP levels, thereby playing a central role in catabolite control (2). FNR activates transcription of genes required for anaerobic respiration (35). If the progenitor of FNR and CAP evolved before the split between gram-negative and gram-positive bacteria, then it would be surprising if the apparently versatile FNR-CAP structure had not been recruited for a role in gram-positive bacteria. However, similar proteins have not been described in gram-positive bacteria. If proteins that are similar to CAP are found in *B. subtilis*, they will play different physiological roles, since catabolite control in *B. subtilis* does not involve cAMP (116) (see chapter 15).

REPRESSORS OF TRANSCRIPTION INITIATION

A relatively large number of sequence-specific DNA-binding proteins that repress promoter activity in *B. subtilis* have been described (Table 4). Regulation of the *gnt* operon by its repressor appears to be among the simplest and best characterized mechanisms (40, 41, 90). The gluconate operon consists of four genes: *gntR*, which encodes the repressor; *gntK* and *gntP*, which encode gluconate kinase and permease, respectively; and *gntZ*, whose function is unknown (40). These products allow *B. subtilis* to grow on gluconate. The *gntK* product converts gluconate to gluconate 6-phosphate, which is metabolized through the pentose cycle. The operon is transcribed from a single promoter in the order *gntR*, *gntK*, *gntP*, *gntZ*. The inducer, gluconate or glucono-δ-lactone, inactivates the repressor. In the absence of inducers, the repressor binds to a single operator at the start point of transcription of the *gnt* promoter (41). This model is supported by genetic evidence that shows that constitutive *gnt* synthesis results from mutations in *gntR* and by in vitro experiments that show that homodimers of the repressor bind to the operator and repress transcription (90). Moreover, the inducers prevent repressor binding in vitro (90).

In the case of the *gnt* operon, it is clear which signal induces the operon (e.g., gluconate) and that the inducer acts directly by inactivation of the repressor GntR. In other cases, the regulatory pathway is not as clearly defined. For example, *penP* from *Bacillus licheniformis* encodes an inducible penicillinase (β-lactamase). *penP* promoter activity is regulated by a well-characterized repressor, PenI (139), and is inducible by penicillins. However, unlike the *gnt* operon, the inducer does not appear to act directly, since addition of inducer does not relieve repression of *penP* promoter activity by PenI in vitro. Mutations in an unlinked gene, *penR1*, cause an uninducible phenotype (75). Since the amino acid sequence of the *penR1* product is similar to penicillin-binding proteins from *E. coli*, it has been suggested that PenR1 binds the inducer, penicillin, and transmits a signal that inactivates the repressor PenI (75). Since another gene product, PenR2, is also required for induction, it has been suggested that PenR2 may transduce the signal from the transmembrane protein PenR1 to the repressor PenI (63, 75).

Operons that are repressible by environmental signals or metabolites are also controlled by repressors. Transcription of the structural gene for glutamine synthetase (*glnA*) in *B. subtilis* is repressed by excess preferred nitrogen sources (114). This repression is mediated in part by a repressor (GlnR), which is encoded by the promoter proximal gene in the *glnRA* operon (114) (see chapter 20). GlnR has been shown to bind to and repress transcription from the *glnRA* promoter in vitro (114). However, regulation of *glnRA* promoter activity appears to be somewhat more complicated than expected, since mutations in *glnA* also cause transcription of *glnRA* in the presence of excess glutamine (51). Evidently, GlnR is not able to sense and respond to nitrogen availability without the aid of glutamine synthetase. It is not known how glutamine synthetase affects changes in GlnR activity.

There are also repressors in *Bacillus* spp. that appear to respond to growth phase-specific signals. *abrB* encodes a DNA-binding protein that represses transcription from several promoters (100, 106); see chapter 52, this volume. DNase footprinting experiments have been used to show that the *abrB* product binds specifically to the promoters of *aprE* (128), the structural gene for the protease subtilisin (*tycA*), an antibiotic biosynthetic gene (106), and sporulation genes *spoVG* and *spo0E* (106, 128). The activity of each of these promoters is normally induced at the end of the exponential growth phase. Mutation of *abrB* causes constitutive expression from the *tycA* promoter (106). Evidently, AbrB represses transcription from each of these promoters during the growth phase. At the end of exponential growth phase, i.e., the onset of sporulation, *abrB* expression is repressed by Spo0A, resulting in a decrease of AbrB concentration and derepression of the stationary-phase-specific promoters (127). Although *abrB* is necessary for the activity of some promoters (e.g., *hpr*), it is not known whether AbrB binds and activates these promoters directly or whether it represses transcription of a gene whose product represses transcription from these *abrB*-dependent promoters.

Derepression of *abrB*-repressed genes after the end of exponential growth phase apparently results because AbrB synthesis is repressed by Spo0A (127). In this model, Spo0A (or its modifiers) and not AbrB senses the growth phase or metabolic state of the cell and regulates the synthesis of AbrB. In contrast, another repressor of stationary-growth-phase-specific promoters, Sin, evidently is inactivated at the end of the exponential growth phase (see chapter 54, this volume). Sin is a DNA-binding protein that has been shown to bind to an operator at the *aprE* promoter (44). Overproduction of Sin inhibits sporulation gene expression; therefore, Sin may also bind to and repress the activity of some promoters that are essential for sporulation (see chapter 54). Western blot (immunoblot) experiments show that Sin continues to accumulate after the onset of sporulation; therefore, it is probably inactivated to derepress transcription from the promoters that are negatively regulated by Sin (86).

There are probably multiple repressors involved in catabolite control of promoter activity in *Bacillus* spp. (see chapter 15). In enteric bacteria, carbon catabolite control is mediated in part by cAMP and its receptor protein, CAP, which binds to and activates specific promoters (reviewed in reference 2). Carbon catabolite control in *B. subtilis* is regulated by a different mechanism, since cAMP does not accumulate in *B. subtilis* (116). Moreover, genetic evidence such as the identification of operator sites at which mutations cause glucose-resistant promoter activity and evidence that proteins bind to these sites (e.g., *amyE* [94] and *citB* [37]) suggest that several repressors rather than a global activator control the activity of promoters in response to carbon availability. However, in no case has a catabolite-responsive repressor protein been characterized (see chapters 11 and 15).

ATTENUATION AND ANTITERMINATION

As in gram-negative bacteria, not all transcriptional regulation in *B. subtilis* occurs at the level of promoter activity. Studies of several systems have revealed regulation by attenuation or antitermination of transcription. These mechanisms are not identical in all cases to those that have been studied in gram-negative bacteria.

Transcription of the tryptophan biosynthetic operon on *B. subtilis* is regulated by attenuation in response to tryptophan (81, 117, 118). The first structural gene in the *B. subtilis trp* operon is preceded by a 204-base leader sequence. Unlike the regulation of the *trp* operons in gram-negative bacteria, transcription through the leader of the *B. subtilis trp* operon in response to tryptophan limitation does not involve translation of the leader (118). Since expression of the leader sequence from a multicopy plasmid causes constitutive expression of a chromosomal copy of the *trp* operon, it is thought that the leader region transcripts bind a factor that terminates transcription. Presumably, this factor requires tryptophan for its activity. This titratable factor may be encoded by the *mtr* locus, since mutations at this locus prevent attenuation of *trp* operon transcription (49). The *mtr* locus encodes two proteins, MtrA and MtrB (49). MtrB shows homology with the RNA-binding protein of *E. coli* phage T4, RegA (49). Therefore, MtrB may bind directly to the *trp* leader region transcript to prevent termination of transcription (49).

Antitermination of transcription also plays a role in regulation of inducible operons. Transcription of *sacB*, the structural gene for levansucrase, is regulated by an antiterminator (21). A palindromic sequence located between the promoter and the *sacB* coding sequence causes termination of transcription. Termination is prevented by the *sacY* product in response to the inducer sucrose. Mutations at two other genes cause constitutive transcription of *sacB*. These genes are *ptsI*, which encodes enzyme I of the phosphoenolpyruvate-dependent phosphotransferase system (PTS), and *sacX*, which appears to be a sucrose-specific permease of the PTS (21). These enzymes probably inactivate the *sacY* product in the absence of sucrose. Sucrose interferes with the inactivation of SacY, so in the presence of sucrose, SacY is available to prevent termination of the *sacB* transcript. SacX may inactivate SacY by phosphorylation, since the PTS enzymes are known to transfer phosphate groups. Moreover, SacY is homologous to an antiterminator from *E. coli*, BglG, which has been shown to be phosphorylated by the PTS component enzyme II^{Bgl} (4). Other antitermination mechanisms may be used to regulate other operons in *B. subtilis*, for example, the purine biosynthetic operon (144) (see chapter 24).

TRANSCRIPTION IN *B. SUBTILIS*: A PARADIGM FOR OTHER GRAM-POSITIVE BACTERIA?

Although the current picture of transcription machinery in other gram-positive organisms is fragmented, the available evidence supports the use of *B. subtilis* as a paradigm. Multiple RNA polymerase sigma factors have been identified and characterized in *Streptomyces* spp. (15, 16, 131, 138), and the gene for a sigma factor from *Staphylococcus aureus* has been sequenced and shown to be most homologous to σ^A from *B. subtilis* (64). A putative member of the two-component family of transcription activators has been identified in *Streptococcus* spp. (17, 105), and there is also at least one characterized transcriptional repressor from *Staphylococcus* spp. (98).

The lessons learned from the study of transcription in *B. subtilis* lead to two important expectations concerning transcriptional regulation in other gram-positive bacteria. On the one hand, members of most of the major families of transcription factors are likely to be found in all gram-positive bacteria. On the other hand, it is likely that some of these factors have been adapted to solve different physiological problems in each species. Therefore, the results of studies of *B. subtilis* provide a good starting point for the exploration of other gram-positive bacteria. However, the study of other gram-positive bacteria will undoubtedly reveal important new insights into the control of transcription in bacteria.

Acknowledgments. I thank E. M. Bryan, T. J. Kenney, J. R. Scott, A. L. Sonenshein, K. M. Tatti, and P. Kirchman for helpful suggestions on the manuscript. Work in my laboratory has been supported by Public Health Service grants AI20319 and GM39917 and by a Research Career Development Award.

REFERENCES

1. **Adams, L. F., K. L. Brown, and J. R. Whitely.** 1991. Molecular cloning and characterization of two genes encoding sigma factors that direct transcription from a *Bacillus thuringiensis* crystal protein gene promoter. *J. Bacteriol.* **173:**3846–3854.
2. **Adhya, S., and S. Garges.** 1990. Positive control. *J. Biol. Chem.* **265:**10797–10800.
3. **Allison, L. A., M. Moyle, M. Shales, and C. J. Ingles.** 1985. Extensive homology among the largest subunits of eukaryotic and prokaryotic RNA polymerases. *Cell* **42:**599–610.
4. **Amster-Choder, O., F. Houman, and A. Wright.** 1989. Protein phosphorylation regulates transcription of the β-glucoside utilization operon in E. coli. *Cell* **58:**847–855.
5. **Arnosti, D. N., and M. J. Chamberlin.** 1989. Secondary σ factor controls transcription of flagellar and chemotaxis genes in Escherichia coli. *Proc. Natl. Acad. Sci. USA* **86:**830–834.
6. **Baldus, J., and C. P. Moran, Jr.** 1992. Unpublished data.
7. **Benson, A. K., and W. G. Haldenwang.** 1992. Characterization of a regulatory network that controls σ^B expression in *Bacillus subtilis. J. Bacteriol.* **174:**749–757.
8. **Biggs, J., L. L. Searles, and A. L. Greenleaf.** 1985. Structure of the eukaryotic transcription apparatus: features of the gene for the largest subunit of Drosophila RNA polymerase II. *Cell* **42:**611–621.
9. **Binnie, C., M. Lampe, and R. Losick.** 1986. Gene encoding the σ37 species of RNA polymerase σ factor from *Bacillus subtilis. Proc. Natl. Acad. Sci. USA* **83:**5943–5947.
10. **Bohannon, D. E., and A. L. Sonenshein.** 1989. Positive regulation of glutamate biosynthesis in *Bacillus subtilis. J. Bacteriol.* **171:**4718–4727.
11. **Boylan, S. A., J.-W. Suh, S. M. Thomas, and C. W. Price.** 1989. Gene encoding the alpha core subunit of *Bacillus subtilis* RNA polymerase is cotranscribed with the genes for initiation factor 1 and ribosomal proteins B, S13, S11, and L17. *J. Bacteriol.* **171:**2553–2562.
12. **Boylan, S. A., M. D. Thomas, and C. W. Price.** 1991. Genetic method to identify regulons controlled by nonessential elements: isolation of a gene dependent on alternate transcription factor σB of *Bacillus subtilis. J. Bacteriol.* **173:**7856–7866.
13. **Burbulys, D., K. A. Trach, and J. A. Hoch.** 1991. Initiation of sporulation in *B. subtilis* is controlled by a multicomponent phosphorelay. *Cell* **64:**545–552.
14. **Buttner, M. J., and N. L. Brown.** 1985. RNA polymerase-DNA interactions in Streptomyces. *J. Mol. Biol.* **185:**177–188.
15. **Buttner, M. J., K. F. Chater, and M. J. Bibb.** 1990. Cloning, disruption, and transcriptional analysis of three RNA polymerase sigma factor genes of *Streptomyces coelicolor* A3(2). *J. Bacteriol.* **172:**3367–3378.
16. **Buttner, M. J., A. M. Smith, and M. J. Bibb.** 1988. At least three different RNA polymerase holoenzymes direct transcription of the agarose gene (*dag*A) of *Streptomyces coelicolor* A3(2). *Cell* **52:**599–607.
17. **Caparon, M. G., and J. R. Scott.** 1987. Identification of a gene that regulates expression of M protein, the major virulence determinant of group A streptococci. *Proc. Natl. Acad. Sci. USA* **84:**8677–8681.
18. **Carlson, H. C., and W. G. Haldenwang.** 1989. The σE subunit of *Bacillus subtilis* RNA polymerase is present in both forespore and mother cell compartments. *J. Bacteriol.* **171:**2216–2218.
19. **Carter, H. L., III, L. F. Wang, R. H. Doi, and C. P. Moran, Jr.** 1988. *rpoD* operon promoter used by σ^H-RNA polymerase in *Bacillus subtilis. J. Bacteriol.* **170:** 1617–1621.
20. **Coppolecchia, R., H. DeGrazia, and C. P. Moran, Jr.** 1991. Deletion of *spoIIAB* blocks endospore formation in *Bacillus subtilis* at an early stage. *J. Bacteriol.* **173:** 6678–6685.
21. **Crutz, A., M. Steinmetz, S. Aymerich, R. Richter, and D. LeCoq.** 1990. Induction of levansucrase in *Bacillus subtilis*: an antitermination mechanism negatively con-

trolled by the phosphotransferase system. *J. Bacteriol.* **172:**1043–1050.

22. **Cutting, S., A. Driks, R. Schmidt, B. Kunkel, and R. Losick.** 1991. Forespore-specific transcription of a gene in the signal transduction pathway that governs pro-σK processing in Bacillus subtilis. *Genes Dev.* **5:**456–466.
23. **Cutting, S., V. Oke, A. Driks, R. Losick, S. Lu, and L. Kroos.** 1990. A forespore checkpoint for mother cell gene expression during development in B. subtilis. *Cell* **62:**239–250.
24. **Cutting, S., S. Panzer, and R. Losick.** 1989. Regulatory studies on the promoter for a gene governing synthesis and assembly of the spore coat in *Bacillus subtilis. J. Mol. Biol.* **207:**393–404.
25. **Dahl, M. K., T. Msadek, F. Kunst, and G. Rapoport.** 1991. Mutational analysis of the *Bacillus subtilis* DegU regulator and its phosphorylation by the DegS protein kinase. *J. Bacteriol.* **173:**2539–2547.
26. **Daniels, D., P. Zuber, and R. Losick.** 1990. Two amino acids in an RNA polymerase σ factor involved in the recognition of adjacent base pairs in the -10 region of a cognate promoter. *Proc. Natl. Acad. Sci. USA* **87:**8075–8079.
27. **Debarbouille, M., I. Martin-Verstraete, A. Klier, and G. Rapoport.** 1991. The transcriptional regulator LevR of Bacillus subtilis has domains homologous to both σ54- and phosphotransferase system-dependent regulators. *Proc. Natl. Acad. Sci. USA* **88:**2212–2216.
28. **Debarbouille, M., I. Martin-Verstraete, F. Kunst, and G. Rapoport.** 1991. The Bacillus subtilis sigL gene encodes an equivalent of σ54 from gram-negative bacteria. *Proc. Natl. Acad. Sci. USA* **88:**9092–9096.
29. **Doi, R. H., T. Kudo, and C. Dickel.** 1981. RNA polymerase forms in vegetative and sporulating cells of *Bacillus subtilis*, p. 219–223. *In* H. S. Levinson, A. L. Sonenshein, and D. J. Tipper (ed.), *Sporulation and Germination.* American Society for Microbiology, Washington, D.C.
30. **Doi, R. H., and L.-F. Wang.** 1986. Multiple procaryotic ribonucleic polymerase sigma factors. *Microbiol. Rev.* **50:**227–243.
31. **Driks, A., and R. Losick.** 1991. Compartmentalized expression of a gene under the control of sporulation transcription factor σE in Bacillus subtilis. *Proc. Natl. Acad. Sci. USA* **88:**9934–9938.
32. **Dubnau, D.** 1989. The competence regulon of *Bacillus subtilis*, p. 147–166. *In* I. Smith, R. A. Slepecky, and P. Setlow (ed.), *Regulation of Procaryotic Development.* American Society for Microbiology, Washington, D.C.
33. **Dubnau, E., J. Weir, G. Nair, H. L. Carter III, C. P. Moran, Jr., and I. Smith.** 1988. *Bacillus* sporulation gene *spo0H* codes for σ^{30} (σ^{H}). *J. Bacteriol.* **170:**1054–1062.
34. **Duncan, M. L., S. S. Kalman, S. M. Thomas, and C. W. Price.** 1987. Gene encoding the 37,000-dalton minor sigma factor of *Bacillus subtilis* RNA polymerase: isolation, nucleotide sequence, chromosomal locus, and cryptic function. *J. Bacteriol.* **169:**771–778.
35. **Eiglmeier, K., N. Honore, S. Luchi, E. C. C. Lin, and S. T. Cole.** 1989. Molecular genetic analysis of FNR-dependent promoters. *Mol. Microbiol.* **3:**869–878.
36. **Ferrari, F. A., K. Trach, D. LeCoq, J. Spence, E. Ferrari, and J. A. Hoch.** 1985. Characterization of the *spo0A* Locus and its deduced product. *Proc. Natl. Acad. Sci. USA* **82:**2647–2651.
37. **Fouet, A., and A. L. Sonenshein.** 1990. A target for carbon source-dependent negative regulation of the *citB* promoter of *Bacillus subtilis. J. Bacteriol.* **172:**835–844.
38. **Foulger, D., and J. Errington.** 1991. Sequential activation of dual promoters by different sigma factors maintains spoVJ expression during successive developmental stages of Bacillus subtilis. *Mol. Microbiol.* **5:**1363–1373.
39. **Friedman, D. I., E. R. Olson, C. Georgopoulos, K. Tilly, I. Herskowitz, and F. Banuett.** 1984. Interactions of bacteriophage and host macromolecules in the growth of bacteriophage. *Microbiol. Rev.* **48:**299–325.
40. **Fujita, Y., and T. Fujita.** 1987. The gluconate operon *gnt* of *Bacillus subtilis* encodes its own transcriptional negative regulator. *Proc. Natl. Acad. Sci. USA* **84:**4524–4528.
41. **Fujita, Y., and Y. Miwa.** 1989. Identification of an operator sequence for the *Bacillus subtilis gnt* operon. *J. Biol. Chem.* **264:**4201–4206.
42. **Gardella, T., H. Moyle, and M. M. Susskind.** 1989. A mutant *E. coli* σ^{70} subunit of RNA polymerase with altered promoter specificity. *J. Mol. Biol.* **206:**579–590.
43. **Garnier, T., and S. T. Cole.** 1988. Studies of UV-inducible promoters from Clostridium perfringens in vivo and in vitro. *Mol. Microbiol.* **2:**607–614.
44. **Gaur, N. K., J. Oppenheim, and I. Smith.** 1991. The *Bacillus subtilis sin* gene, a regulator of alternate developmental processes, codes for a DNA-binding protein. *J. Bacteriol.* **173:**678–686.
45. **Gentry, D. R., and R. R. Burgess.** 1986. The cloning and sequence of the gene encoding the omega subunit of Escherichia coli RNA polymerase. *Gene* **48:**33–40.
46. **Gholamhoseinian, A., and P. J. Piggot.** 1989. Timing of *spoII* gene expression relative to septum formation during sporulation of *Bacillus subtilis. J. Bacteriol.* **171:** 5747–5749.
47. **Gilman, M. Z., J. L. Wings, and M. J. Chamberlin.** 1981. Nucleotide sequence of two *Bacillus subtilis* promoters used by *Bacillus subtilis* sigma-28 RNA polymerase. *Nucleic Acids Res.* **9:**5991–6000.
48. **Gitt, M. A., L. F. Wang, and R. H. Doi.** 1985. A strong sequence homology exists between RNA polymerase sigma factors of *Bacillus subtilis* and *Escherichia coli. J. Biol. Chem.* **260:**7178–7185.
49. **Gollnick, P., S. Ishino, M. I. Kuroda, D. J. Henner, and C. Yanofsky.** 1990. The mtr locus is a two-gene operon required for transcription attenuation in the trp operon of Bacillus subtilis. *Proc. Natl. Acad. Sci. USA* **87:**8726–8730.
50. **Gralla, J. D.** 1991. Transcriptional control—lessons from an *E. coli* promoter data base. *Cell* **66:**415–418.
51. **Gutowski, J. C., and H. J. Schreier.** 1992. Interaction of the *Bacillus subtilis glnRA* repressor with operator and promoter sequences in vivo. *J. Bacteriol.* **174:**671–681.
52. **Guzman, P., J. Westpheling, and P. Youngman.** 1988. Characterization of the promoter region of the *Bacillus subtilis spoIIE* operon. *J. Bacteriol.* **170:**1598–1609.
53. **Healy, J., J. Weir, I. Smith, and R. Losick.** 1991. Post-transcriptional control of a sporulation regulatory gene encoding transcription factor σH in *Bacillus subtilis. Mol. Microbiol.* **5:**477–487.
54. **Helman, J. D., and M. J. Chamberlin.** 1988. Structure and function of bacterial sigma factors. *Annu. Rev. Biochem.* **57:**839–879.
55. **Helman, J. D., L. M. Marquez, and J. J. Chamberlin.** 1988. Cloning, sequencing and disruption of the *Bacillus subtilis* σ28 gene. *J. Bacteriol.* **170:**1568–1574.
56. **Henikoff, S., G. W. Haughn, J. M. Calvo, and J. C. Wallace.** 1988. A large family of bacterial activator proteins. *Proc. Natl. Acad. Sci. USA* **85:**6602–6606.
57. **Henkin, T. M., and A. L. Sonenshein.** 1987. Mutations of the *Escherichia coli lacUV5* promoter resulting in increased expression in *Bacillus subtilis. Mol. Gen. Genet.* **209:**467–474.
58. **Henner, D. J., M. Yang, and E. Ferrari.** 1988. Localization of *Bacillus subtilis sacU*(Hy) mutations to two linked genes with similarities to the conserved family of two-component signalling systems. *J. Bacteriol.* **170:** 5102–5109.
59. **Hirschman, J., P. K. Wong, K. Sei, J. Keener, and S. Kustu.** 1985. Products of nitrogen regulatory genes *ntrA*

and *ntrC* of enteric bacteria activate *glnA* transcription *in vitro*: evidence that the *ntrA* product is a sigma factor. *Proc. Natl. Acad. Sci. USA* **82:**7525–7529.
60. **Hunt, T. P., and B. Magasanik.** 1985. Transcription of glnA by purified *Escherichia coli* components: core RNA polymerase and the products of *glnF, glnG* and *glnL*. *Proc. Natl. Acad. Sci. USA* **82:**8453–8457.
61. **Igo, M., M. Lampe, C. Ray, W. Schafer, C. P. Moran, Jr., and R. Losick.** 1987. Genetic studies of a secondary RNA polymerase sigma factor in *Bacillus subtilis. J. Bacteriol.* **169:**3464–3469.
62. **Illing, N., and J. Errington.** 1991. Genetic regulation of morphogenesis in *Bacillus subtilis*: roles of σE and σF in prespore engulfment. *J. Bacteriol.* **173:**3159–3169.
63. **Imanaka, T., T. Himeno, and S. Aiba.** 1987. Cloning and nucleotide sequence of the penicillinase antirepressor gene *penJ* of *Bacillus licheniformis. J. Bacteriol.* **169:**3867–3872.
64. **Iordanescu, S.** 1991. The Staphylococcus aureus chromosomal gene plaC, identified by mutations amplifying plasmid pT181, encodes a sigma factor. *Nucleic Acids Res.* **19:**4921–4924.
65. **Jaacks, K. J., J. Healy, R. Losick, and A. Grossman.** 1989. Identification and characterization of genes controlled by the sporulation-regulatory gene *spo0H* in *Bacillus subtilis. J. Bacteriol.* **171:**4121–4129.
66. **Johnson, W. C., C. P. Moran, Jr., and R. Losick.** 1983. Two RNA polymerase σ from *Bacillus subtilis* discriminate between overlapping promoters for a developmentally regulated gene. *Nature* (London) **302:**800–804.
67. **Jonas, R. M., and W. G. Haldenwang.** 1989. Influence of *spo* mutations of σE synthesis in *Bacillus subtilis. J. Bacteriol.* **171:**5226–5228.
68. **Jones, C. H., and C. P. Moran, Jr.** 1991. A mutant sigma factor blocks the transition between promoter binding and initiation of transcription. *Proc. Natl. Acad. Sci. USA* **89:**1958–1962.
69. **Kalman, S., M. L. Duncan, S. M. Thomas, and C. W. Price.** 1990. Similar organization of the *sigB* and *spoIIA* operons encoding alternate sigma factor of *Bacillus subtilis* RNA polymerase. *J. Bacteriol.* **172:**5575–5585.
70. **Karmazyn-Campelli, C., C. Bonamy, B. Savelli, and P. Stragier.** 1989. Tandem gene encoding σ factors for consecutive steps of development in *Bacillus subtilis. Genes Dev.* **3:**150–157.
71. **Kenney, T. J., P. A. Kirchman, and C. P. Moran, Jr.** 1988. Gene encoding σE is transcribed from a σA-like promoter in *Bacillus subtilis. J. Bacteriol.* **170:**3058–3064.
72. **Kenney, T. J., and C. P. Moran, Jr.** 1987. Organization and regulation of an operon that encodes a sporulation-essential sigma factor in *Bacillus subtilis. J. Bacteriol.* **169:**3329–3339.
73. **Kenney, T. J., and C. P. Moran, Jr.** 1991. Genetic evidence for interaction of σA with two promoters in *Bacillus subtilis. J. Bacteriol.* **173:**3282–3290.
74. **Kenney, T. J., K. York, P. Youngman, and C. P. Moran, Jr.** 1989. Genetic evidence that RNA polymerase associated with σA uses a sporulation-specific promoter in *Bacillus subtilis. Proc. Natl. Acad. Sci. USA* **86:**9109–9113.
74a. **Kirchman, P., and C. P. Moran.** Unpublished data.
75. **Kobayashi, T., Y. F. Zhu, N. J. Nicholls, and J. O. Lampen.** 1987. A second regulatory gene, blaR1, encoding a potential penicillin-binding protein for induction of β-lactamase in *Bacillus licheniformis. J. Bacteriol.* **169:**3873–3878.
76. **Kroos, L., B. Kunkel, and R. Losick.** 1989. Developmental regulatory protein from *Bacillus subtilis. Science* **243:**526–528.
77. **Kunkel, B., L. Kroos, H. Poth, P. Youngman, and R. Losick.** 1989. Temporal and spatial control of the mother-cell regulatory gene spoIIID of *Bacillus subtilis. Genes Dev.* **3:**1735–1744.
78. **Kunkel, B., R. Losick, and P. Stragier.** 1990. The *Bacillus subtilis* gene for the developmental transcription factor σK is generated by excision of a dispensable DNA element containing a sporulation recombinase gene. *Genes Dev.* **4:**525–535.
79. **Kunkel, B., K. Sandman, S. Panzer, P. Youngerman, and R. Losick.** 1988. Identification of the promoter for the *Bacillus subtilis* sporulation locus *spoVC* and its use in studies of temporal and spatial control of gene expression. *J. Bacteriol.* **170:**3515–3522.
80. **Kunst, F., M. Debarbouille, T. Msadek, M. Young, C. Mauel, D. Karamata, A. Klier, G. Rapoport, and R. Dedonder.** 1988. Deduced polypeptides encoded by the *Bacillus subtilis sacU* locus share homology with two-component sensor-regulator systems. *J. Bacteriol.* **170:** 5093–5101.
81. **Kuroda, M. I., D. Henner, and C. Yanofsky.** 1988. *cis*-acting sites in the transcript of the *Bacillus subtilis trp* operon regulate expression of the operon. *J. Bacteriol.* **170:**3038–3088.
82. **LaBell, T. J., J. E. Trempy, and W. G. Haldenwang.** 1987. Sporulation-specific σ factor σ29 of *Bacillus subtilis* is synthesized from a precursor protein, P31. *Proc. Natl. Acad. Sci. USA* **84:**1784–1788.
83. **Lampe, M., C. Binnie, R. Schmidt, and R. Losick.** 1988. Cloned gene encoding the delta subunit of Bacillus subtilis RNA polymerase. *Gene* **67:**13–19.
84. **Losick, R., and J. Pero.** 1981. Cascades of sigma factors. *Cell* **25:**582–584.
85. **Losick, R., and P. Stragier.** 1992. Crisscross regulation of cell-type-specific gene expression during development in Bacillus subtilis. *Nature* (London) **355:**601–604.
86. **Mandic-Mulec, I., N. Gaur, U. Bai, and I. Smith.** 1992. Sin, a stage-specific repressor of cellular differentiation. *J. Bacteriol.* **174:**3561–3569.
87. **Margolis, P., A. Driks, and R. Losick.** 1991. Establishment of cell type of compartmentalized activation of a transcription factor. *Science* **254:**562–565.
88. **McClure, W.** 1985. *Annu. Rev. Biochem.* **54:**171–204.
89. **Mirel, D. B., and M. J. Chamberlin.** 1989. The *Bacillus subtilis* flagellin gene (*hag*) is transcribed by the σ28 form of RNA polymerase. *J. Bacteriol.* **171:**3095–3101.
90. **Miwa, Y., and Y. Fujita.** 1988. Purification and characterization of a repressor for the *Bacillus subtilis gnt* operon. *J. Biol. Chem.* **263:**13252–13257.
91. **Moran, C. P., Jr.** 1989. Sigma factors and the regulation of transcription, p. 167–184. *In* I. Smith, R. Slepecky, and P. Setlow (ed.), *Regulation of Procaryotic Development.* American Society for Microbiology, Washington, D.C.
92. **Moran, C. P., Jr., N. Lang, S. F. J. LeGrice, G. Lee, M. Stephens, A. L. Sonnenshein, J. Pero, and R. Losick.** 1982. Nucleotide sequences that signal the initiation of transcription and translation in *Bacillus subtilis. Mol. Gen. Genet.* **186:**339–346.
93. **Mukai, K., M. Kawata, and T. Tanaka.** 1990. Isolation and phosphorylation of the Bacillus subtilis degS and degU gene products. *J. Biol. Chem.* **265:**20000–20006.
94. **Nicholson, W. L., Y. Park, T. M. Henkin, M. Won, M. J. Weikert, J. A. Gaskell, and G. H. Chambliss.** 1987. Catabolite repression-resistant mutations of the *Bacillus subtilis* alpha-amylase promoter affect transcription levels and are in an operator-like sequence. *J. Mol. Biol.* **198:**609–618.
95. **Nicholson, W. L., D. Sun, B. Setlow, and P. Setlow.** 1989. Promoter specificity of σG-containing RNA polymerase from sporulating cells of *Bacillus subtilis*: identification of a group of forespore-specific promoters. *J. Bacteriol.* **171:**2708–2718.
96. **Ohnishi, H., K. Kusukake, H. Suzuki, and T. Iino.** 1990. Gene fliA encodes an alternative sigma factor specific

for flagellar operons in Salmonella typhimurium. *Mol. Gen. Genet.* **221:**139–147.

97. **Olmedo, G., E. G. Ninfa, J. Stock, and P. Youngman.** 1990. Novel mutations that alter the regulation of sporulation in Bacillus subtilis. Evidence that phosphorylation of regulatory protein Spo0A controls the initiation of sporulation. *J. Mol. Biol.* **215:**359–372.
98. **Oskouian, B., and G. C. Stewart.** 1990. Repression and catabolite repression of the lactose operon of *Staphylococcus aureus. J. Bacteriol.* **172:**3804–3812.
99. **Panzer, S., R. Losick, D. Sun, and P. Setlow.** 1989. Evidence for an additional temporal class of gene expression in the forespore compartment of sporulating *Bacillus subtilis. J. Bacteriol.* **171:**561–564.
100. **Perego, M., G. B. Spiegelman, and J. A. Hoch.** 1988. Structure of the gene for the transition state regulator, *abrB*: regulator synthesis is controlled by the *spo0A* sporulation gene in *Bacillus subtilis. Mol. Microbiol.* **2:**689–699.
101. **Pich, A., and H. Bahl.** 1991. Purification and characterization of the DNA-dependent RNA polymerase from *Clostridium acetobutylicum. J. Bacteriol.* **173:**2120–2124.
102. **Price, C. W.** 1992. Personal communication.
103. **Rather, P. N., R. Coppolecchia, H. DeGrazia, and C. P. Moran, Jr.** 1990. Negative regulator of σG-controlled gene expression in stationary-phase *Bacillus subtilis. J. Bacteriol.* **172:**709–715.
104. **Ray, C., R. E. Hay, H. L. Carter III, and C. P. Moran, Jr.** 1985. Mutations that affect utilization of a promoter in stationary-phase *Bacillus subtilis. J. Bacteriol.* **163:**610–614.
105. **Robbins, J. C., J. G. Spanier, S. J. Jones, W. J. Simpson, and P. P. Cleary.** 1987. *Streptococcus pyrogenes* type 12 M protein gene regulation by upstream sequences. *J. Bacteriol.* **169:**5633–5640.
106. **Robertson, J. B., M. Gocht, M. A. Marahiel, and P. Zuber.** 1989. AbrB, a regulator of gene expression in *Bacillus*, interacts with the transcription initiation regions of a sporulation gene and an antibiotic biosynthesis gene. *Proc. Natl. Acad. Sci. USA* **86:**8457–8461.
107. **Roels, S., A. Driks, and R. Losick.** 1992. Characterization of *spoIVA*, a sporulation gene involved in coat morphogenesis in *Bacillus subtilis. J. Bacteriol.* **174:** 575–585.
108. **Ronson, C. W., B. T. Nixon, and F. M. Ausubel.** 1987. Conserved domains in bacterial regulatory proteins that respond to environmental stimuli. *Cell* **49:**579–581.
109. **Sasse-Dwight, S., and J. D. Gralla.** 1990. Role of eukaryotic-type functional domains found in the prokaryotic enhancer receptor factor σ54. *Cell* **62:**945–954.
110. **Sato, T., Y. Samori, and Y. Kobayashi.** 1990. The *cisA* cistron of *Bacillus subtilis* gene *spoIVC* encodes a protein homologous to a site-specific recombinase. *J. Bacteriol.* **172:**1092–1098.
111. **Satola, S., P. Kirchman, and C. P. Moran, Jr.** 1991. Spo0A binds to a promoter used by σA RNA polymerase during sporulation in *Bacillus subtilis. Proc. Natl. Acad. Sci. USA* **88:**4533–4537.
112. **Satola, S. W., J. M. Baldus, and C. P. Moran, Jr.** 1992. Binding of Spo0A stimulates *spoIIG* promoter activity in *Bacillus subtilis. J. Bacteriol.* **174:**1448–1453.
113. **Schmidt, R., P. Margolis, L. Duncan, R. Coppolecchia, C. P. Moran, Jr., and R. Losick.** 1990. Control of developmental transcription factor σF by sporulation regulatory proteins SpoIIAA and SpoIIAB in Bacillus subtilis. *Proc. Natl. Acad. Sci. USA* **87:**9221–9225.
114. **Schreier, H. J., S. W. Brown, K. D. Hirschi, J. F. Nomellini, and A. L. Sonenshein.** 1989. Regulation of Bacillus subtilis glutamine synthetase gene expression by the product of the glnR gene. *J. Mol. Biol.* **210:**51–63.
115. **Seki, T., H. Yoshikawa, H. Takahashi, and H. Saito.** 1987. Cloning and nucleotide sequence of *phoP*, the regulatory gene for alkaline phosphotase and phosphodiesterase in *Bacillus subtilis. J. Bacteriol.* **169:**2913–2916.
116. **Setlow, P.** 1973. Inability to detect cAMP in vegetative or sporulating cells or dormant spores of *Bacillus megaterium. Biochem. Biophys. Res. Commun.* **52:**365–372.
117. **Shimotsu, H., and D. J. Henner.** 1984. Characterization of the *Bacillus subtilis* tryptophan promoter region. *Proc. Natl. Acad. Sci. USA* **81:**6315–6319.
118. **Shimotsu, H., M. I. Kuroda, C. Yanofsky, and D. J. Henner.** 1986. Novel form of transcription attenuation regulates expression of the *Bacillus subtilis* tryptophan operon. *J. Bacteriol.* **166:**461–471.
119. **Shorenstein, R. G., and R. Losick.** 1973. Comparative size and properties of sigma subunits of ribonucleic acid polymerase from Bacillus subtilis and Escherichia coli. *J. Biol. Chem.* **248:**6170–6173.
120. **Siegle, D. A., J. C. Hu, W. A. Walter, and C. A. Gross.** 1989. Altered promoter recognition by mutant forms of the σ70 subunit of *Escherichia coli* RNA polymerase. *J. Mol. Biol.* **206:**591–603.
121. **Smith, I., R. Slepecky, and P. Setlow (ed.).** 1989. *Regulation of Procaryotic Development*. American Society for Microbiology, Washington, D.C.
122. **Stetter, K. O., and W. Zillig.** 1974. Transcription in Lactobacillaceae. *Eur. J. Biochem.* **48:**527–540.
123. **Stevens, C. M., and J. Errington.** 1990. Differential gene expression during sporulation in Bacillus subtilis: structure and regulation of the spoIIID gene. *Mol. Microbiol.* **4:**543–551.
124. **Stock, J. B., A. J. Ninfa, and A. M. Stock.** 1989. Protein phosphorylation and regulation of adaptive responses in bacteria. *Microbiol. Rev.* **53:**450–490.
125. **Stragier, P., B. Kunkel, L. Kroos, and R. Losick.** 1989. Chromosomal rearrangement generating a composite gene for a developmental transcription factor. *Science* **243:**507–512.
126. **Stragier, P., and R. Losick.** 1990. Cascades of sigma factors revisited. *Mol. Microbiol.* **4:**1801–1806.
127. **Strauch, M., V. Webb, G. Spiegelman, and J. A. Hoch.** 1990. The Spo0A protein of *Bacillus subtilis* is a repressor of the abrB gene. *Proc. Natl. Acad. Sci. USA* **87:**1801–1805.
128. **Strauch, M. A., G. B. Spiegelman, M. Perego, W. C. Johnson, D. Burbulys, and J. A. Hoch.** 1989. The transition state transcription regulator *abrB* of *Bacillus subtilis* is a DNA binding protein. *EMBO J.* **8:**1615–1621.
129. **Sun, D., P. Fajardo-Cavazos, D. Sussman, F. Tovar-Rojo, R. Cabera-Martinez, and P. Setlow.** 1991. Effect of chromosome location of *Bacillus subtilis* forespore genes on their *spo* gene dependence and transcription by EσF: identification of features of good EσF-dependent promoters. *J. Bacteriol.* **173:**7867–7874.
130. **Sun, D., P. Stragier, and P. Setlow.** 1989. Identification of a new σ-factor involved in compartmentalized gene expression in *Bacillus subtilis. Genes Dev.* **3:**141–149.
131. **Tanaka, K., T. Shiina, and H. Takahashi.** 1991. Nucleotide sequence of genes hrdA, hrdC, and hrdD from Streptomyces coelicolor A3(2) having similarity to rpoD genes. *Mol. Gen. Genet.* **229:**334–340.
132. **Tatti, K. M., H. L. Carter III, A. Moir, and C. P. Moran, Jr.** 1989. Sigma H-directed transcription of *citG* in *Bacillus subtilis. J. Bacteriol.* **171:**5928–5932.
133. **Tatti, K. M., C. H. Jones, and C. P. Moran, Jr.** 1991. Genetic evidence for interaction of σE with the *spoIIID* promoter in *Bacillus subtilis. J. Bacteriol.* **173:**7828–7833.
134. **Tatti, K. M., and C. P. Moran, Jr.** 1984. Promoter recognition by sigma-37 RNA polymerase from *Bacillus subtilis. J. Mol. Biol.* **175:**285–297.
135. **Tjian, R., R. Losick, J. Pero, and A. Hinnenbush.** 1977. Purification and comparative properties of the delta

and sigma subunits of RNA polymerase from Bacillus subtilis. *Eur. J. Biochem.* **74:**149–154.

136. **Trach, K., D. Burbulys, M. Strauch, J.-J. Wu, N. Dhillon, R. Jonas, C. Hanstein, P. Kallio, M. Perego, T. Bird, G. Spiegelman, C. Fogher, and J. A. Hoch.** 1991. Control of the initiation of sporulation in Bacillus subtilis by a phosphorelay. *Res. Microbiol.* 142:815–823.
137. **Waldburger, C., T. Gardella, R. Wong, and M. M. Susskind.** 1990. Changes in conserved region 2 of Escherichia coli σ70 affecting promoter recognition. *J. Mol. Biol.* **215:**267–276.
138. **Westpheling, J., M. Ranes, and R. Losick.** 1985. RNA polymerase heterogeneity in Streptomyces coelicolor. *Nature* (London) **313:**22–27.
139. **Wittman, V., and H. C. Wong.** 1988. Regulation of the penicillinase genes of *Bacillus licheniformis*: interaction of the *pen* repressor with its operators. *J. Bacteriol.* **170:**3206–3212.
140. **Wu, J.-J., P. J. Piggot, K. M. Tatti, and C. P. Moran, Jr.** 1991. Transcription of the *Bacillus subtilis spoIIA* locus. *Gene* **101:**113–116.
141. **Yamashita, H., H. Yoshikawa, F. Kawamura, H. Takahashi, T. Yamamoto, Y. Kobayashi, and H. Saito.** 1986. The effect of *spo0* mutations on the expression of *spo0A*- and *spo0F-lacZ* fusions. *Mol. Gen. Genet.* **205:**28–33.
142. **York, K., T. J. Kenney, S. Satola, C. P. Moran, Jr., H. Poth, and P. Youngman.** 1991. Spo0A controls the σ^A-dependent activation of *Bacillus subtilis* sporulation-specific transcription unit *spoIIE*. *J. Bacteriol.* **174:** 2648–2658.
143. **Yudkin, M. D.** 1987. Structure and function in a *Bacillus subtilis* sporulation-specific sigma factor: molecular nature of mutations in spoIIAC. *J. Gen. Microbiol.* **133:** 475–481.
144. **Zalkin, H,. and D. J. Ebbole.** 1988. Organization and regulation of genes encoding biosynthetic enzymes in Bacillus subtilis. *J. Biol. Chem.* **263:**1595–1598.
145. **Zheng, L., and R. Losick.** 1990. Cascade regulation of spore coat gene expression in *Bacillus subtilis*. *J. Mol. Biol.* **212:**645–660.
146. **Zuber, P., J. Healy, H. L. Carter III, S. Cutting, C. P. Moran, Jr., and R. Losick.** 1989. Mutation changing the specificity of an RNA polymerase sigma factor. *J. Mol. Biol.* **206:**605–614.
147. **Zuber, P., and R. Losick.** 1983. Use of a *lacZ* fusion to study developmental regulation by the *spo0* genes of *Bacillus subtilis*. *Cell* **35:**275–283.
148. **Zuberi, A. R., C. Ying, M. R. Weinreich, and G. Ordal.** 1990. Transcriptional organization of a cloned chemotaxis locus of *Bacillus subtilis*. *J. Bacteriol.* **172:**1870–1876.

46. Ribosomal Structure and Genetics

TINA M. HENKIN

The machinery of protein synthesis is of ancient origin, and conservation of ribosome structure and function is very high. Bacterial cells devote a large proportion of their resources to the synthesis and maintenance of the translational apparatus. It is therefore critical that the expression of translational genes be tightly controlled to maintain an appropriate balance. This is a difficult problem because of the complexity of this machinery, which encompasses rRNA and ribosomal proteins, translation factors, tRNAs, and aminoacyl-tRNA synthetases. Although a great deal has been learned in recent years about ribosome structure, function, and gene regulation in *Escherichia coli*, considerably less is known about other bacterial systems. The goal of this chapter is to summarize current knowledge about *Bacillus* ribosomes and translation factors; tRNA and tRNA synthetase genes are discussed in chapter 47. Much of the information concerning *Bacillus* ribosome structure and function has been obtained from work with *Bacillus stearothermophilus*, while most of the genetic analysis has been carried out with *Bacillus subtilis*. Available information about ribosome structure and genetics in other gram-positive systems will also be summarized.

RIBOSOME STRUCTURE AND ASSEMBLY

Bacillus ribosomes, like those of *E. coli*, are 70S in size and are composed of two subunits that are 50S and 30S. The larger subunit contains one molecule each of 23S rRNA, 5S rRNA, and approximately 30 proteins; L7 and L12 are differently modified versions of the same protein, and four copies are present. The smaller subunit contains one molecule each of 16S rRNA and approximately 20 proteins. The structure in general is very similar to that of the *E. coli* ribosome, as is the case for all eubacterial ribosomes characterized thus far.

Ribosome Morphology

Several experimental approaches have been used in the analysis of ribosome morphology. Transmission and scanning transmission electron microscopy (10) and three-dimensional electron microscopy (176) have been applied most extensively to *E. coli* ribosomes, while X-ray crystallography has been utilized more successfully in the analysis of *B. stearothermophilus* ribosomes because of the greater stability of ribosomal particles from thermophilic organisms (190). The resolution of the crystallography data is still fairly low, but the data have revealed features predicted from studies using other techniques as well as a clear definition of the intersubunit space and a tunnel through the 50S subunit, which is presumed to be the region through which the nascent polypeptide chain passes (189).

Immunoelectron microscopy has been used for both *E. coli* and *B. stearothermophilus* to localize proteins within the ribosomal subunits, demonstrating that protein S4 is located in the small lobe of the 30S subunit (162), L1 is located at the tip of a protuberance of the 50S subunit, L6 is at the base of the L7-L12 stalk, L23 is at the base of the subunit, L29 is close to L23, and L2, which is part of the peptidyl transferase center, is located on the interface side of the 50S subunit between the central and L1 protuberances (59, 60, 163). Lactoperoxidase-catalyzed iodination (113, 114) and limited proteolysis (97) have been used in both systems to identify proteins accessible on the ribosomal surface. The results from all of these studies indicate that the overall morphology of the ribosome and the arrangement of proteins within the ribosomal subunits are highly conserved, permitting incorporation of data obtained from analysis of either system into a single model.

rRNA Sequence Data

B. subtilis contains 10 rRNA operons (106, 116, 161). The sequences of the *rrnB* and *rrnO* operons have been determined (52, 127, 129); they show an organizational pattern identical to that of the *E. coli* rRNA operons, containing genes for 16S, 23S, and 5S rRNA in that order. Several *rrn* operons are associated with tRNA genes (see below). Conservation compared with the *E. coli* rRNA sequence is approximately 75%. Analysis of rRNA sequences has proved to be a useful tool in the definition of phylogenetic relationships. This has been exploited recently in analysis of the genus *Bacillus* (4, 146, 158) as the basis for its division into at least four groups: a *B. subtilis*-*B. stearothermophilus* group; a *Bacillus brevis*-*Bacillus laterosporus* group; a *Bacillus alvei* group, including *Bacillus macerans*, *Bacillus macquariensis*, and *Bacillus polymyxa*; and a more distant *Bacillus cycloheptanicus* branch, including *Bacillus acidocaldarius* and *Bacillus acidoterrestris* (43, 185). Additional members of the *B. subtilis* group recently characterized include *Bacillus cereus*, *Bacillus thuringiensis*, *Bacillus anthracis*, and *Bacillus mycoides*, all of which proved to be very closely interrelated (3). Although the diversity within a given group is as expected for members of a genus,

Tina M. Henkin • Department of Biochemistry and Molecular Biology, Albany Medical College, Albany, New York 12208.

conservation between members of two different groups is much lower. In fact, conservation between members of the *B. subtilis* group and members of other genera such as *Enterococcus* and *Lactobacillus* is higher than that between the *B. subtilis* group and the other "*Bacillus*" groups. For this reason, it has been suggested that the genus *Bacillus* is extremely divergent phylogenetically and should be subdivided (146). The accumulation of a larger collection of rRNA sequences will facilitate this subdivision; this work is in progress (4, 42).

rRNA sequences from other gram-positive organisms, including species of *Lactobacillus*, *Mycoplasma*, *Mycobacterium*, *Enterococcus*, *Streptomyces*, and *Clostridium*, have also recently been determined (58, 118). The available data have led to subdivision of the gram-positive group into two subgroups based on G+C content, with *Bacillus* species in the low-G+C group (186). This set of data is likely to grow rapidly in the near future and will be extremely valuable in improving our understanding of the evolutionary history of this diverse group.

Although little detailed information about rRNA processing is available, signals for processing and maturation of rRNA were found to be similar to those identified in *E. coli* and include a tripartite stem-loop structure with the 16S, 23S, and 5S rRNA sequences located within the loops (108, 127). Processing sites within the stem of this structure have been identified and are conserved in *rrn* operons of *Mycoplasma*, *Streptomyces*, and *Clostridium* spp. (47, 173). An RNase III-like enzyme that acts at these sites has been identified in *B. subtilis* (140). RNase P, which is responsible for cleavage at the 5′ end of tRNA precursors (45), probably also plays a role in rRNA processing in rRNA-tRNA transcriptional units. Maturation of 5S rRNA has been characterized in most detail in *B. subtilis*, in which it is mediated by RNase M5 (138), which acts on a complex of precursor 5S rRNA and ribosomal protein L18 (BL16) to release mature 5S rRNA. The requirement for the ribosomal protein can be fulfilled by incubation in the presence of dimethyl sulfoxide, indicating that the recognition determinants for the enzyme are entirely within the RNA substrate and that the role of the protein is likely to be in maintenance of the appropriate conformation. Processing of rRNA-tRNA precursors is discussed in more detail in chapter 47.

Ribosomal-Protein Sequence Data

Initial comparisons of ribosomal proteins from *E. coli* and *Bacillus* spp. relied on electrophoretic migration patterns and immunological analysis (38, 48, 77) as well as functional correspondence in heterologous ribosomal reconstitution assays (72, 124). Proteins were originally named on the basis of polyacrylamide gel electrophoresis patterns and have been renamed when the homology to *E. coli* proteins became apparent after sequence determination (134, 135). The correspondence to *E. coli* proteins of each of the *B. subtilis* (73) and *B. stearothermophilus* (188) 30S proteins has been determined on the basis of amino-terminal protein sequence analysis, and these proteins are named accordingly (S2, S3, etc.). The 50S proteins are less well characterized; 50S proteins for which correspondence to *E. coli* nomenclature is known are named according to that nomenclature (L1, L2, etc.), while proteins for which correspondence is unknown retain the BL nomenclature. This confusion will be remedied as the remaining sequence data become available. In this chapter, use of protein names without the B prefix indicates correspondence to the *E. coli* protein bearing the same name.

A large number of *B. subtilis* and *B. stearothermophilus* protein sequences have been determined either from direct protein sequencing or from gene cloning and DNA sequence determination. The proteins for which sequence data are now available are listed in Table 1. In all cases, correspondence between the *E. coli* and *Bacillus* proteins is unambiguous, although the level of sequence conservation varies from approximately 30 to 75%.

The most striking difference in the protein profiles of *Bacillus* and *E. coli* ribosomes is the absence in *Bacillus* spp. of a homolog for protein S1 (61, 76). This protein plays a major role in translation initiation in *E. coli*, and its absence in *Bacillus* spp. and certain other gram-positive bacteria has been suggested as a factor in the difference in translation start site selection in these organisms, which exhibit a more stringent Shine-Dalgarno sequence requirement (6, 110, 144). Members of the high-G+C gram-positive group, including *Streptomyces* and *Micrococcus* spp., apparently do contain an S1 homolog and exhibit low-stringency translation initiation (40).

The motivation for much of the protein sequence analysis in *B. stearothermophilus* comes from the greater ease with which ribosomal proteins from this thermophilic organism can be crystallized for detailed structural studies. A number of proteins, including S5, L6, L9, L30, and a complex of L7-L12-L10, have now been crystallized, and structural analysis is in progress (1, 102, 175, 181). Ribosomal proteins of the extreme thermophile *Bacillus caldolyticus* have also been characterized in the hope of more efficient crystallization properties (7); two-dimensional electrophoretic profiles of ribosomal proteins from this organism are very similar to those of proteins from *B. stearothermophilus*, although several proteins exhibit alterations in charge.

Characterization of ribosomal-protein profiles and protein sequences of other gram-positive organisms, including *Streptomyces* (125), *Micrococcus* (130, 131), and *Mycoplasma* (132) spp., has recently been initiated. These studies will increase our pool of knowledge of ribosomal-protein structural features and will permit broader conclusions about the significance of conserved elements.

Ribosome Assembly

In vitro ribosome assembly using purified components has been accomplished for both *Bacillus* and *E. coli* systems. Reconstitution of the 30S subunit was accomplished earlier for both systems (65, 124, 172), but 50S subunit reconstitution proved to be more difficult. It was first successful with ribosomes from *B. stearothermophilus* because of the greater heat stability of the components (22, 123) but was later carried

Table 1. *Bacillus* ribosomal-protein and translation factor sequences

Protein	Gene	Reference for sequence data		*B. subtilis* map location (°) (reference)[c]
		Protein[a,b]	Gene[b]	
S2	*rpsB*	2		
S3	*rpsC*		95 (BST)	12
S4	*rpsD*	2	55 (BSU), 57 (BST)	263 (69)
S5	*rpsE*	83	191 (BSU), 143 (BST)	12
S6	*rpsF*			4 (26)
S7	*rpsG*		84 (BST)	12
S8	*rpsH*	2	115 (BSU)	12
S9	*rpsI*	85		13 (27)
S10	*rpsJ*			12 (183)
S11	*rpsK*	90	11 (BSU)	12 (27)
S12	*rpsL*	88	84 (BST)	12 (49, 63)
S13	*rpsM*	12	165 (BSU)	12
S14	*rpsN*		70 (BSU)	12
S15	*rpsO*	2		
S16	*rpsP*	2		145 (26)
S17	*rpsQ*	71	70 (BSU)	12
S18	*rpsR*	109		
S19	*rpsS*	74	95 (BST)	12
S20	*rpsT*	2		13 (136)
S21	*rpsU*	71		
L1	*rplA*	89		11 (29)
L2	*rplB*	91	95 (BST)[d]	
L3	*rplC*			12 (25)
L4	*rplD*			
L5	*rplE*	82	70 (BSU)	12
L6	*rplF*	92	166 (BSU), 143 (BST)	12
L8	*rplH*			
L9	*rplI*	87	143, 175 (BST)	
L10	*rplJ*			11 (29)
L11	*rplK*			11 (29)
L12 (L7)	*rplL*	46 (BST), 79 (BSU)		11 (29)
L13	*rplM*			13[e] (137)
L14	*rplN*	89	70 (BSU)	12
L15	*rplO*	89	191 (BSU)	12
L16	*rplP*		70 (BSU)[d], 95 (BST)[d]	12
L17	*rplQ*	85	11 (BSU)	12
L18	*rplR*	82	166 (BSU), 143 (BST)	12
L19	*rplS*			
L20	*rplT*		141 (BST)	
L21	*rplU*		23 (BSU)	241 (28)
L22	*rplV*		95 (BST)	12 (171)
L23	*rplW*	89		
L24	*rplX*	89	70 (BSU)	12
L25	*rplY*			
L26	*rplZ*			
L27	*rpmA*	85	41 (BSU)	241 (28)
L28	*rpmB*	96	105 (BSU)	145
L29	*rpmC*	89	70 (BSU)	12
L30	*rpmD*	83	191 (BSU)	12 (27)
L31	*rpmE*			
L32	*rpmF*	168		
L33	*rpmG*	96		
L34	*rpmH*	96	117 (BSU)	0
L35	*rpmI*		141 (BST)	
L36	*rpmJ*	168	11 (BSU)	12
IF1	*infA*		11 (BSU)	12
IF2	*infB*		149 (BSU), 13 (BST)	145
IF3	*infC*	86	141 (BST)	
EF-G	*efg*		84 (BST)	12 (33)
EF-Tu	*tuf*			12 (33, 154)

[a] Sequences were determined for *B. stearothermophilus* protein unless otherwise noted.
[b] BSU, *B. subtilis*; BST, *B. stearothermophilus*.
[c] No reference is listed where map location is derived only from DNA sequence analysis.
[d] Partial sequence.
[e] Map location suggested only by interspecies transformation experiments.

out for *E. coli*, *B. subtilis*, and *Bacillus licheniformis* as well (37, 121). The overall assembly map is very similar in the *Bacillus* and *E. coli* systems. Proteins isolated from *B. stearothermophilus* could interact appropriately with *E. coli* 16S rRNA to assemble active 30S subunits (124), indicating that major features of the assembly process are conserved. Similarly, 30S proteins from *E. coli* could interact with *B. stearothermophilus* 16S rRNA (72). However, in studies of binding of *E. coli* S4 to 16S rRNA of various organisms, binding to rRNA from *B. subtilis* and *B. brevis* was less efficient than to that of *B. stearothermophilus* (170), indicating that there may be differences in rRNA and protein structure that affect ribosome assembly when heterologous components are used. Efficient reconstitution of active *B. stearothermophilus* 50S subunits is possible when heterologous 5S rRNA derived from *E. coli*, *Caulobacter* spp., or archaebacteria is used, while incorporation of 5S rRNA from eukaryotic sources such as yeast or rat is impaired; eukaryotic 5.8S rRNA exhibited very low incorporation (64). The heat resistance of ribosomes from thermophilic *B. stearothermophilus* was shown to be primarily dependent on the source of the ribosomal proteins in hybrid reconstitution experiments with components from *B. stearothermophilus* and the mesophilic *B. licheniformis* (37).

Heterologous reconstitution systems have also been useful in identification of functional homology of ribosomal proteins. Single-protein-omission studies were used to determine the abilities of *Bacillus* ribosomal proteins to replace proteins omitted from *E. coli* subunit reconstitution mixes (72). Ribosomes from thiostrepton-resistant mutants of *B. subtilis* and *Bacillus megaterium* lack protein L11; *E. coli* L11 was shown to replace the *Bacillus* protein, demonstrating functional homology (182). A similar approach took advantage of the existence of a number of *E. coli* mutants missing individual ribosomal proteins (30). The missing protein could be incorporated into mutant ribosomal particles, so that addition of ribosomal proteins isolated from *B. subtilis* to 50S subunits from an *E. coli* mutant lacking protein L27 permitted identification of the L27 homolog (27).

In Vitro Translation Systems

Systems for in vitro translation systems using *Bacillus* cell extracts have been established in several laboratories and include RNA-directed systems (19, 100) and DNA-directed coupled transcription-translation systems (101). A cell-free translation system for *Micrococcus luteus* was also recently described (40). The production of high levels of proteases by *Bacillus* cells, especially in stationary-phase cultures, requires the use of careful precautions against proteolytic degradation; this problem may be circumvented in the future by the use of mutant strains defective in protease production. These systems are useful both in the characterization of cloned genes and in the analysis of specificity of translation.

RIBOSOMAL GENETICS

A large number of ribosomal determinants have been localized on the *B. subtilis* chromosome; their arrangement is shown in Fig. 1. These determinants have been identified by a combination of classical and molecular genetic techniques, and the remaining genes are likely to be localized in the near future. For those ribosomal genes for which the chromosomal orientation has been determined, the majority have been shown to be oriented so that transcription proceeds in the same direction as DNA replication, presumably to avoid interference from the high transcriptional activity of these genes (192); this pattern is also found in *E. coli* (81).

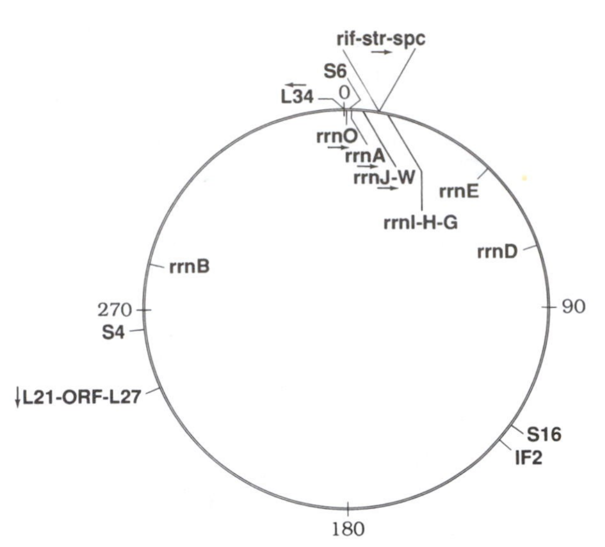

Figure 1. Location of ribosomal genes on the *B. subtilis* chromosome. Map coordinates (0 to 360°) are indicated on the outside of the circle. The origin of chromosomal replication is at 0°. rRNA operons (*rrn*) are shown on the inside of the circle. Ribosomal-protein and translation factor genes as well as other genes known to be located within ribosomal gene clusters are shown on the outside of the circle. Arrows indicate the direction of transcription, where known. The *rif-str-spc* cluster at 12° includes genes for the following: L11-L1-L10-L12-ORF-β-β'-S12-S7-EFG-EFTu-S10-(L3,L4, L23?)-L2-S19-L22-S3-L16-L29-S17-L14-L24-L5-S14-S8-L6-L18-S5-L30-L15-SecY-Adk-Map-IF1-L36-S13-S11-α-L17-ORF--S9-(L13?)--S20, where ORF is an open reading frame. Parentheses indicate uncertainty in gene localization. Additional genes are also likely to be present in this region. The gene for protein L28 is also located near 145°.

rRNA Genes

B. subtilis was found by hybridization analysis to contain 10 rRNA operons (106, 161). All 10 operons have now been cloned, and the *rrnB* and *rrnO* operons have been sequenced. Only two of the operons, *rrnA* and *rrnO*, both of which are located near the origin of chromosomal replication, contain tRNA genes ($tRNA^{Ile}$ and $tRNA^{Ala}$) between the 16S and 23S rRNA genes (106); in *E. coli*, all seven *rrn* operons contain intervening tRNA genes, three with $tRNA^{Ile}$ and $tRNA^{Ala}$ and four with $tRNA^{Glu}$ (81). Immediately downstream from the *rrnB* operon is a very large cluster of tRNA genes containing 21 genes (53, 179). A second large cluster of 16 tRNA genes is located downstream from another rRNA operon, probably *rrnW* (147, 184); this operon also contains a minor 5S rRNA gene (174). No such large clusters of tRNA genes are found in *E. coli*. There is a smaller six-tRNA-gene cluster between the *rrnI* and *rrnH* operons and a nine-gene cluster between *rrnJ* and *rrnW*, while $tRNA^{Met}$ and $tRNA^{Asp}$ genes are located at the 3′ ends of *rrnE* and *rrnD* (54, 178).

Initial studies of rRNA gene localization by heteroduplex and density transfer analyses indicated that the majority (80%) of rRNA genes are located near the origin of chromosomal replication, while the remaining genes were predicted to be located near the terminus (20, 21, 152). The chromosomal loci of all 10 rRNA operons were identified by integrational mapping (80, 98). As predicted from the earlier studies, the majority are located very close to the origin (*rrnO*, 1°; *rrnA*, 6°; *rrnJ-rrnW*, 10°; *rrnI-rrnH-rrnG*, 14°); the remainder are more distant (*rrnE*, 44°, *rrnD*, 70°; *rrnB*, 270°), although not near the terminus of replication. In *E. coli*, the rRNA operons are not grouped in tandem arrays, as is the case for *B. subtilis rrnI-rrnH-rrnG* and *rrnJ-rrnW*, and they exhibit less clustering near the origin of replication (36). In *B. subtilis*, it is possible to identify spontaneously arising isolates in which one rRNA operon has been deleted without a major effect on viability (50, 107). These deletion events appear to occur at high frequency, and it has been suggested that the arrangement of rRNA operons in tandem arrays increases instability. Deletion of one rRNA operon in *E. coli* by genetic manipulation also had no effect on viability (35).

Clostridium perfringens was also found to contain 10 rRNA operons, with similar clustering near the origin of chromosomal replication (47). *Streptomyces coelicolor* contains six rRNA operons (173), *Lactococcus* spp. contain six (9), and *Streptococcus pneumoniae* contains four (5).

Ribosomal-Protein Genes

Ribosomal-protein gene characterization was initiated by the isolation and mapping of mutations conferring resistance to antibiotics that were known to target the translational machinery. Some of these antibiotic resistance mutations were shown to be associated with alterations in specific ribosomal proteins. Mutations conferring resistance to streptomycin (*strA*, S12; 78), spectinomycin (*spcA*, S5; 49, 63), thiostrepton (*tsp*, L11; 182), erythromycin (*ery*, L22; 49, 171), chloramphenicol (*cam*, L1, L13, L17; 136), tetracycline (*tetA*, S10; 183), and kasugamycin (*ksg*, S9; 27) all proved to map in a single region near the origin of replication, at 12°. Other mutations mapping in this region with which no ribosomal-protein alteration was associated included resistance to micrococcin (*mic*, upstream of *strA*; 49), virginiamycin (between *efg* and *ery*; 145), and neomycin (*neo*; 49). Also within this cluster were genes conferring resistance to antibiotics known to affect RNA polymerase, such as rifampin (*rfm*, β; 104), streptolydigin (*std*, β'; 62), and lipiarmycin (*lpm*, β'; 157) as well as elongation factors such as fusidic acid and kirromycin, which target translation factors EF-G (*fus*; 93) and EF-Tu (*kir*; 153, 154), respectively; temperature-sensitive mutants with alterations in EF-G and EF-Tu were also isolated (33). The *strR* mutation, which in combination with the *rpsL2* (protein S12) mutation confers resistance to high levels of streptomycin, was found to map to the far end of the ribosomal gene cluster; this phenotype was associated with the 30S subunit in cell-free translation assays, but no ribosomal-protein alteration was detected (68). The order of antibiotic resistance markers in this region, starting from the outside marker *cysA*, is *cysA-rfm(rpoB)-std, lpm(rpoC)-fus(efg)-kir(tuf)-mic*(50S?) - *strA(rpsL)* - *tet(rpsH)* - *ery(rplV)* - *spcA(rpsE)* - *tsp(rplK)-ksg(rpsI)--strR*(30S?) (49, 63, 151). Additional antibiotic resistance mutations, including resistance to pactamycin (*pac*, 5°; 63), lincomycin (*lin*, 24°; 49, 63) and kasugamycin (*ksgA*, 4° [184] and *ksgB*, 277° [139]), mapped to other regions of the chromosome. A mutation resulting in dependence on spectinomycin for growth was located near 2°, but no ribosomal protein alteration was detected (66). A set of *B. stearothermophilus* antibiotic-resistant and -dependent mutants with alterations in proteins S4, S5, L4, BL14, and BL19 or missing L11 has recently been isolated (148).

A second approach to identification of ribosomal-protein genes in the major cluster involved the use of interspecies transformation experiments (137). DNAs isolated from related *Bacillus* species such as *B. licheniformis*, *Bacillus amyloliquefaciens*, and *Bacillus niger* were introduced into competent *B. subtilis* cells with selection for major cluster region antibiotic resistance markers. From these experiments, the genes encoding proteins S3, S5, S8, S17, S19, L2, L4, L5, L6, L13, L17, L22, L23, L24, and L29 were suggested to be located in the major ribosomal cluster, while the genes for proteins S4, S6, S16, S20, L19, L21, L27, and L32 were believed to be absent; the remaining proteins could not be assayed because their electrophoretic migration patterns were not sufficiently different from those of the equivalent *B. subtilis* proteins. Most of these assignments were confirmed by other studies, although the gene for protein S20 has been reported to be located at the far end of the cluster (133).

Another approach to this problem involved the selection for second-site mutations that suppressed the effects of other ribosomal alterations. Revertants of the spore-minus phenotype of a strain containing two ribosomal mutations, *rpsL2* and *strR*, which together conferred a streptomycin-resistant, oligosporogenous phenotype, were found to contain alterations in a variety of different ribosomal proteins, including S17 and S5, which were located in the major cluster, S4, which mapped to a new site at 263°, and BL17, which did not map to the major cluster and whose location is unknown (67).

A similar approach involved selection for antibiotic-resistant or -dependent strains of *B. subtilis* and screening for protein alterations (24). Mutants containing changes in the electrophoretic mobilities of a variety of ribosomal proteins were identified, resulting in localization of the genes for proteins S7, S8, S11, L1, L3, L5, L6, L10, L11, L12, L24, and L30 to the major cluster (12°), for S6 to 4°, for S16 to 135°, and for L21 and L27 to 241°.

Most recently, the technique of choice for characterization of ribosomal-protein genetics involves the cloning and DNA sequence analysis of ribosomal-protein genes. A number of genes including many in the major cluster at 12° and several at other positions, have now been cloned (Fig. 1 and Table 1). In all cases, the localization and arrangement of ribosomal protein genes determined by molecular techniques have proved to be consistent with the results of classical genetic analyses and interspecies transformation experiments.

A large proportion of the ribosomal-protein genes (as well as genes encoding RNA polymerase subunits and translation factors) in *E. coli* are organized into operons, and a number of operons are clustered together in the *rif* and *str-spc* clusters; these two clusters are at separate positions in the *E. coli* chromosome (81). In *B. subtilis*, the regions equivalent to the *rif* and *str-spc* clusters are adjacent to each other in the 12° region. The apparent order of genes in this cluster indicates that the homologs of the L11 and β operons are followed by the *str*, S10, *spc*, and α homologs. There is also evidence for the presence of the genes encoding proteins S9 and S20 at the far end of the cluster, and the gene for protein L13, which is adjacent to the gene for S9 in *E. coli*, is predicted from interspecies transformation experiments to be located in the major cluster. The S9-L13 pair is located downstream of the end of the *str-spc* cluster in *E. coli*, while the gene for S20 is not located near other ribosomal genes. Analysis of mutants with alterations in *rif* cluster ribosomal proteins indicated that these genes are organized as in *E. coli* (L11-L1-L10-L12-β; 29). Analysis of antibiotic-resistant mutants indicated that *rpoC* (β′) is downstream from *rpoB* (β) (62, 157). The DNA sequence of the *B. subtilis* β cluster has been partially determined (34); it includes an additional open reading frame predicted to encode a polypeptide 23 kDa in length between *rplL* (L7-L12) and *rpoB* (β). The beginning of the *str* operon equivalent of *B. stearothermophilus* has been characterized; it contains the genes for S12, S7, and EF-G (84). The *tuf* gene, which encodes EF-Tu, is downstream from *efg* (154). DNA sequencing is complete for the *B. subtilis* region extending from the last part of the S10 cluster through the *spc* and α clusters (11, 70, 115, 166, 191), while the adjoining region containing additional S10 cluster genes from *B. stearothermophilus* was characterized (L2-S19-L22-S3-L16; 95). The sequence of the beginning of the S10 region has not yet been determined. In *E. coli*, the gene order in this region is S10-L3-L4-L23-L2 (81). Genetic mapping of a tetracycline resistance mutation localized the S10 gene between *tuf* and *ery* (L17; 183). On the basis of mapping of a mutation causing altered L3 electrophoretic mobility, the gene for protein L3 is also predicted to be in this region (25), while the genes for proteins L4 and L23 were localized to the major cluster by interspecies transformation experiments (137). It therefore seems likely that the organization of the S10 cluster is conserved, although determination of the DNA sequence of the remaining portion of this region is necessary to confirm this possibility. The gene orders for the *spc* and α regions are generally consistent with those of the equivalent regions in *E. coli*, with these exceptions: the *adk* (adenylate kinase), *map* (methionine aminopeptidase), and *infA* (initiation factor 1 [IF-1]) genes are present in *B. subtilis* DNA in the region between the *secY* (secretion) and *rpmJ* (ribosomal protein L36) genes at the end of the *spc* cluster (11, 166); and the *rpsD* gene, which is the third gene of the *E. coli* α operon, is absent from the major cluster (165) and is located instead at 263° (55, 69).

Several ribosomal-protein genes have been found to map at sites other than the major cluster. These include the genes for proteins S4 (263°; 69), S6 (4°; 26), S16 (145°; 26), L21 and L27 (241°; 23, 41), L34 (0°; 128), and L28 (145°; 105). Except for the gene for protein S4, the genes equivalent to these are also absent from the major clusters in *E. coli*, while the S4 gene is located in the α operon in *E. coli*. The gene for protein S21, which is in the *rpoD* operon in *E. coli*, is not in this position in *B. subtilis* (177), and its chromosomal location is unknown. The L21 and L27 genes are in a cluster that includes a third open reading frame, whose function is unknown; these genes are between the *spoIVF* and *spo0B* genes, in the order *spoIV-FAB*-L21-ORF-L27-*spo0B* (23, 41). The gene for L28 is adjacent to *spoVM*, at 145° (105); the exact position of this locus relative to the genes for S16 and IF2, which are also in this region, is not known. The genes for L20 and L35 are clustered with the *infC* gene, encoding initiation factor IF-3 in *B. stearothermophilus* (141), as in *E. coli*, but the position of this cluster on the *B. subtilis* map has not been determined.

Preliminary analysis of ribosomal-gene organization in other gram-positive organisms has indicated that the patterns observed in *B. subtilis* are well conserved. In *M. luteus*, a member of the high-G+C group, the organization of the *str* cluster is identical to that found in *Bacillus* spp. and *E. coli* (S12-S7-EFG-EFTu; 131). This gene order is also conserved in cyanobacteria (15, 111) and is similar in the genome of cyanelles, which are primitive plastids in certain lower plants, except that the gene for EF-G is absent and the gene for S10 is located at the end of the *str* operon rather than at the beginning of the "S10" cluster (119). The *spc* gene cluster in *M. luteus* also exhibits a gene order similar to that found in *B. subtilis* and *E. coli* except that the genes for S14 and L34 are absent (130). In *Mycoplasma capricolum*, a low-G+C gram-positive bacterium evolutionarily distant from *Bacillus* spp., the S10-*spc* gene order is also similar to that found in *B. subtilis* and *E. coli* except that the genes for L30 and L34 are absent (132). In the cyanelle and higher-plant chloroplast genomes, a number of S10-*spc* genes are arranged as in *B. subtilis* and *E. coli*, although several genes are missing from this cluster (112). The *infA* gene, which encodes IF-1, is located at the end of the *spc* cluster in *B. subtilis*, *Lactococcus lactis*, and higher-plant chloroplasts but not in *E. coli*, and in both *B. subtilis* and chloroplasts, the *rpsD* gene, which encodes ribosomal protein S4, is absent from the α cluster (11, 94, 150). In contrast, in *Euglena gracilis* chloroplasts, the genes for S4 and S11 are in a dicistronic operon in reverse order from that found in *E. coli* (159). These data on ribosomal-gene organization, though preliminary, are consistent with other results suggesting an evolutionary linkage between gram-positive and photosynthetic bacteria (187) and indicate that numerous rearrangements in ribosomal-gene organization have occurred in the course of evolution.

Translation Factors

The gene for IF-1 (*infA*) was identified by DNA sequence analysis near the end of the *spc* gene cluster, at 12° (11), and is apparently cotranscribed with the *spc* and α cluster genes. This gene is not located near other translational genes in *E. coli* (81). The *infB* gene, which encodes IF-2 (α and β), has been isolated from both *B. subtilis* (149) and *B. stearothermophilus* (13) and was found to map at 145°. The genetic organization of the region surrounding *infB* has apparently been conserved and includes the homologs of *nusA*, the product of which is involved in transcription elongation, upstream of *infB*, and two downstream open reading frames (149); in *E. coli*, the *rpsO* gene, which encodes ribosomal S15, is also in this region, but complete DNA sequence information is not yet available for *Bacillus* spp. The *infC* gene, which encodes IF-3, has been isolated from *B. stearothermophilus* and was found to be located next to the genes for ribosomal proteins L35 and L20 (141), as is also the case in *E. coli*. The position of this cluster on the *B. subtilis* genome has not yet been determined. In *E. coli*, these genes are located immediately downstream from *thrS*, which encodes threonyl-tRNA synthetase; in *B. subtilis*, *thrS* is at 250° (142), but this gene has not yet been found in the region upstream from *infC* in the cloned *B. stearothermophilus* DNA (39). Genes encoding elongation factors EF-G and EF-Tu were localized to the major cluster of ribosomal-protein genes by characterization of temperature-sensitive and antibiotic-resistant mutants. Mutations in EF-G confer resistance to fusidic acid (93), while mutations in EF-Tu confer resistance to kirromycin (154). These genes are located downstream of *rpsL* (ribosomal protein S12) in the order *rpsL-efg-tuf*. This gene organization is also found in *E. coli*, in which *rpsL* is followed by the S7 gene to form the *str* operon (81). This region has been isolated from *B. stearothermophilus* and shows the expected *rpsL-rpsG-efg* order in the segment sequenced thus far (84).

Ribosomal Mutants and Sporulation

A number of mutants that have alterations in components of the translational apparatus and exhibit defects in sporulation have been identified (18). These include mutants with altered sensitivity to antibiotics that act on the ribosome and temperature-sensitive mutants that result in spore-minus or spore-conditional phenotypes. In several cases, the effect of these mutations on sporulation could be an indirect result of membrane alterations, since membrane-stabilizing agents such as glycerol or Tween 80 suppress the sporulation defect (180). The spore-minus phenotypes of a number of these mutants as well as mutants with alterations in RNA polymerase were also suppressed by mutations in the *rev*, *rpoD* (*sigA*), and *spo0A* genes (120). These results suggest the possibility of interactions of the transcription and translation machinery with components of the sensory systems responsible for triggering the initiation of sporulation. Further studies will be necessary to determine whether these interactions are part of the normal roles of these systems or whether they represent a separate regulatory phenomenon.

RIBOSOMAL-GENE REGULATION

Most of the information concerning regulation of ribosome biosynthesis comes from the analysis of *E. coli*. Analysis of this regulatory scheme in other bacterial systems may provide valuable information as to which aspects of the regulatory pathway are universally conserved (and therefore likely to be important) and which are more divergent. The extensive information already available about *B. subtilis* ribosomal-gene organization and the availability of powerful tools for genetic analysis make this the ideal system for such a study.

rRNA Gene Regulation

Data on rRNA gene regulation in *B. subtilis* are limited. *E. coli rrn* operons contain two tandem promoters; the upstream promoter (P_1) exhibits higher transcriptional activity and is subject to growth rate-dependent control, while the downstream promoter (P_2) is less active and does not respond to growth rate (51). *B. subtilis rrn* operons also contain tandem promoters (160) and exhibit growth rate-dependent regulation (99). At least some of the tRNA genes associated with *rrn* operons are cotranscribed with the rRNA genes (see chapter 47).

Examination of the expression of *B. subtilis rrnB-cat* fusions in *E. coli* indicated that in this case, the downstream P_2 promoter rather than P_1 was more active and was regulated in response to growth rate (32). These studies suggest that the signals important for growth rate response in *E. coli* are present in the *B. subtilis* DNA, but the mechanism for this response in *B. subtilis* has not been explored. Growth rate-dependent expression of *rrn-lacZ* fusions integrated into the *B. subtilis* genome has also been demonstrated (147). Differential expression of different rRNA operons was also observed in these studies, but the basis for the variation in expression is not known.

Synthesis of rRNA, measured by incorporation of radioactive uracil into ribosomal subunits and 16S and 23S rRNAs, was found to decrease early during sporulation, presumably because of the shift in transcriptional activity from vegetative to sporulation-associated genes (75). This effect was blocked in a rifampin-resistant, spore-minus mutant. Expression of *rrn-lacZ* fusions was also found to be greatly reduced by 1 h after the initiation of sporulation (147).

These initial experiments on rRNA gene expression in *B. subtilis* demonstrate that the tools for a detailed examination of the regulation of this system are now available. It is likely that much more information will be available in the near future.

Ribosomal-Protein Gene Regulation

Very little is known about the transcriptional organization and regulation of the ribosomal-protein genes in the major cluster. Preliminary studies have failed to locate a transcriptional initiation site for either the α or the *spc* gene clusters (115, 166); it appears that transcription of this region originates further upstream and that this set of genes is part of a very large transcriptional unit. In the *E. coli str-spc* cluster, there are promoter and terminator sites at the beginning and end of each of the operons within the cluster; however, significant readthrough transcription from P_{spc} into the α operon was detected (103).

In *E. coli*, the *spc*, *str*, and α operons are each regulated by binding of a ribosomal protein encoded within the operon to a target site on the mRNA, resulting in turnoff of operon expression by translational repression and mRNA degradation. For the *spc* operon, S8, the product of the fifth gene in the operon, is the regulatory protein; its target site is a region of the mRNA at the start of the L5 coding sequence that resembles the binding site for S8 on 16S rRNA (17). Inspection of the equivalent region of the *B. subtilis* sequence revealed no sequence similar to the *E. coli* regulatory target site, suggesting that the regulatory mechanism may differ (70). In the case of the *E. coli str* operon, the S7 protein acts as the translational regulator, and the target site has been proposed to be the mRNA region from the end of the S12 coding sequence through the start of the S7 coding sequence (31). This region exhibits structural similarity to the binding site for S7 on 16S rRNA. This structural element is conserved in the *str* operon of the cyanobacterium *Anacystis nidulans* (111), suggesting the possibility that the regulatory mechanism is also conserved, but is not present in the equivalent regions of *B. stearothermophilus* or *M. luteus*, in which the S12-S7 intergenic regions are much shorter (84, 131). Similarly, the *E. coli* α operon is regulated by binding of protein S4 to a target site at the beginning of the S13 coding sequence; in this case, the target site does not greatly resemble the S4 binding site on 16S rRNA at the primary sequence level and has been found to fold into a pseudoknot structure (169). No sequence resembling this target site is found in this region in *B. subtilis* (11). No other information about regulation of ribosomal-protein genes in the major cluster is currently available.

In *E. coli*, S4 is encoded in the α operon and acts as the regulator for α operon ribosomal-gene expression. The location of the *B. subtilis rpsD* gene, which encodes protein S4, at a position distant from the major cluster suggested that the regulatory pattern might differ from that of the *E. coli* system. Studies of *rpsD* regulation in *B. subtilis* indicate that this gene is autogenously regulated and that the target for regulation is the mRNA leader region, since mutations in this region result in derepression (56). The regulatory target site at the start of the *rpsD* gene is not related at the primary sequence level to the target site for S4 in the *E. coli* α operon mRNA; however, it has the potential to form a structure similar to the *E. coli* α operon pseudoknot. Phylogenetic and mutagenic analyses of the *rpsD* leader region failed to support a pseudoknot model and suggest that a simpler structure containing two stem-loops is more likely (57). Replacement of the *rpsD* promoter region with a derivative of the *E. coli lac* promoter had no effect on regulation, indicating that regulation occurs post-transcription initiation, but the molecular mechanism for repression has not yet been characterized.

Preliminary analysis of the *B. stearothermophilus infC* operon, which includes the genes for IF-3, L35, and L20, suggests that the three genes form a single transcriptional unit (39). IF-3 is produced at a lower level than are the ribosomal proteins. In *E. coli*, transcription of this region is complex, and the presence of a transcription initiation point for L35 and L20 within *infC* may be responsible in part for the observed differential expression, while translation autoregulation of *infC*, mediated by an unusual AUU translation initiation codon, is also involved (14). There is no evidence for differential transcription in *B. stearothermophilus*, and it appears likely that regulation occurs primarily at the translational level, since an AUU-to-AUG mutation in *B. stearothermophilus infC* results in a 30-fold increase in expression (39).

Stringent Response

In a number of bacterial systems, the synthesis of ribosomal components has been found to decrease in response to amino acid starvation. This effect, called the stringent response, is mediated by an increase in levels of the phosphorylated nucleotides ppGpp and pppGpp, which are synthesized by the product of the *relA* gene (stringent factor) in response to binding of uncharged tRNA to the ribosomal A site. This system has been examined in *B. subtilis* and was shown to be very similar to that of *E. coli* (44, 122). Relaxed mutants defective in *relA* have been identified, and the *relA* gene has been localized to 235° (155, 167). Thiostrepton-resistant mutants lacking ribosomal protein L11 also exhibit a relaxed phenotype; these mutations, designated *relC*, are in *rplK*, the structural gene for L11 (182). There is also evidence for a ppGpp degradation activity analogous to that of the *E. coli spoT* product (8). In *Streptomyces* spp., amino acid starvation was shown to result in (p)ppGpp synthesis and inhibition of rRNA transcription; this effect was blocked in *relC* mutants with alterations in ribosomal protein L11 (126, 164).

These results indicate that the stringent-response systems of *B. subtilis* and *Streptomyces* spp. are very similar to that of *E. coli*. However, no information concerning the mechanisms for response to (p)ppGpp levels is available. In *E. coli*, sequences in the promoter regions of genes sensitive to the stringent response have been identified, and it has been proposed that (p)ppGpp acts directly as an effector in modulating promoter recognition by RNA polymerase (16). Conservation of sequence elements in promoters of *B. subtilis* rRNA and tRNA operons has also been noted (127), but no mutational analyses have yet been employed to test the importance of these elements. Expression of a fusion of the *B. subtilis rrnO* promoter to *lacZ* was repressed by induction of the stringent response by the addition of serine hydroxamate in wild-type strains but not in a *relA*-minus mutant (147). This type of *rrn-lacZ* construct should permit identification of *cis*-acting sequence elements in the *rrnO* promoter region involved in this response.

FUTURE DIRECTIONS

A great deal of progress has been made in recent years in the analysis of the structure and function of the bacterial ribosome. Characterization of the translational apparatus of organisms other than *E. coli* will continue to play a major role in increasing the breadth of our understanding of this very complicated system by drawing our attention to those features that have been maintained through the course of evolution as well as those that vary in interesting ways.

The identification and characterization of the genetic determinants for ribosomal components in *Bacillus* species have progressed rapidly. A complete picture of ribosomal-gene organization should be available in the near future. Initial studies indicate remarkable conservation of ribosomal-gene organization in prokaryotes, and comparisons with archaebacteria and chloroplast gene organization in conjunction with rRNA sequence analysis may provide clues to the evolutionary history of these groups.

At this time, our only concept of the mechanisms for regulation of ribosome biosynthesis in prokaryotes is based on detailed studies of this system in *E. coli*. Studies of ribosomal-gene expression in *B. subtilis* are just beginning and so far reveal both similarities to and differences from expression in *E. coli*. Autogenous control has been demonstrated in one case, but it is not known whether this is a general feature of ribosomal-protein gene regulation in *B. subtilis*, as it is in *E. coli*. The mechanisms for coupling ribosomal-protein synthesis to rRNA levels have also not been determined in this system, nor do we understand the regulation of rRNA synthesis. The tools for genetic manipulation and gene expression studies are now in hand; the next few years should provide a burst of new information in this area.

Acknowledgments. I thank G. Fox, R. Losick, C. Price, and R. Rudner for generously providing data prior to publication and F. Grundy for critically reading the manuscript. The work in my laboratory is supported by NIH grant GM40650.

REFERENCES

1. **Appelt, K., S. W. White, and K. S. Wilson.** 1983. Proteins of the *Bacillus stearothermophilus* ribosome. Crystallization of proteins L30 and S5. *J. Biol. Chem.* **258:** 13328–13330.
2. **Arndt, E., T. Scholzen, W. Kroner, T. Hatakeyama, and M. Kimura.** 1991. Primary structures of ribosomal proteins from the archaebacterium *Halobacterium marismortui* and the eubacterium *Bacillus stearothermophilus*. *Biochimie* **73:**657–668.
3. **Ash, C., J. A. E. Farrow, M. Dorsch, E. Stackebrandt, and M. D. Collins.** 1991. Comparative analysis of *Bacillus anthracis*, *Bacillus cereus*, and related species on the basis of reverse transcriptase sequencing of 16S rRNA. *Int. J. Syst. Bacteriol.* **41:**343–346.
4. **Ash, C., J. A. E. Farrow, S. Wallbanks, and M. D. Collins.** 1991. Phylogenetic heterogeneity of the genus *Bacillus* revealed by comparative analysis of small-subunit-ribosomal RNA sequences. *Lett. Appl. Microbiol.* **13:**202–206.
5. **Bacot, C. M., and R. H. Reeves.** 1991. Novel tRNA gene organization in the 16S-23S intergenic spacer of *Streptococcus pneumoniae* rRNA gene cluster. *J. Bacteriol.* **173:**4234–4236.
6. **Band, L., and D. J. Henner.** 1984. *Bacillus subtilis* requires a "stringent" Shine-Dalgarno region for gene expression. *DNA* **3:**17–21.
7. **Beck, J. A., J. Dijk, and R. Reinhardt.** 1987. Ribosomal proteins and DNA-binding protein II from the extreme thermophile *Bacillus caldolyticus*. *Biol. Chem. Hoppe-Seyler* **368:**121–130.
8. **Belitsky, B. R., and R. S. Shakulov.** 1982. Functioning of *spoT* gene product in *Bacillus subtilis* cells. *FEBS Lett.* **138:**226–228.
9. **Beresford, T., and S. Condon.** 1991. Cloning and partial characterization of genes for ribosomal ribonucleic acid in *Lactococcus lactis* subsp. *lactis*. *FEMS Microbiol. Lett.* **78:**319–324.
10. **Boublik, M., V. Mandiyan, and S. Tumminia.** 1990. Potential for electron microscopic techniques for structural analysis of ribosomes, p. 114–122. *In* W. E. Hill, A. Dahlberg, R. A. Garrett, P. B. Moore, D. Schlessinger, and J. R. Warner (ed.), *The Ribosome: Structure, Function, and Evolution*. American Society for Microbiology, Washington, D.C.
11. **Boylan, S. A., J.-W. Suh, S. M. Thomas, and C. W. Price.** 1989. Gene encoding the alpha core subunit of *Bacillus subtilis* RNA polymerase is cotranscribed with the genes for initiation factor 1 and ribosomal proteins B, S13, S11, and L17. *J. Bacteriol.* **171:**2553–2562.
12. **Brockmoller, J., and R. M. Kamp.** 1988. Cross-linked amino acids in the protein pair S13-S19 and sequence analysis of protein S13 of *Bacillus stearothermophilus* ribosomes. *Biochemistry* **27:**3372–3381.
13. **Brombach, M., C. O. Gualerzi, Y. Nakamura, and C. L. Pon.** 1986. Molecular cloning and sequence of the *Bacillus stearothermophilus* translational initiation factor IF2 gene. *Mol. Gen. Genet.* **205:**97–102.
14. **Brombach, M., and C. L. Pon.** 1987. The unusual translation initiation codon AUU limits the expression of the *infC* (initiation factor IF3) gene of *Escherichia coli*. *Mol. Gen. Genet.* **208:**94–100.
15. **Buttarelli, F. R., R. A. Calogero, O. Tiboni, C. O. Gualerzi, and C. L. Pon.** 1989. Characterization of the *str* operon genes from *Spirulina pastensis* and their evolutionary relationship to those of other prokaryotes. *Mol. Gen. Genet.* **217:**97–104.
16. **Cashel, M., and K. E. Rudd.** 1987. The stringent response, p. 1410–1438. *In* F. C. Neidhardt, J. L. Ingraham, K. B. Low, B. Magasanik, M. Schaechter, and H. E. Umbarger (ed.), *Escherichia coli and Salmonella typhimurium: Cellular and Molecular Biology*. American Society for Microbiology, Washington, D.C.
17. **Cerretti, D. P., L. Mattheakis, K. R. Kearney, L. Vu, and M. Nomura.** 1988. Translational regulation of the *spc* operon in *Escherichia coli*. Identification and structural analysis of the target site for S8 repressor protein. *J. Mol. Biol.* **204:**309–325.
18. **Chambliss, G. H.** 1980. Ribosomes and sporulation in *Bacillus subtilis*, p. 781–794. *In* G. Chambliss, G. R. Craven, J. Davies, K. Davis, L. Kahan, and M. Nomura (ed.), *Ribosomes: Structure, Function and Genetics*. University Park Press, Baltimore.
19. **Chambliss, G. H., T. M. Henkin, and J. M. Leventhal.** 1983. Bacterial in vitro protein synthesizing systems. *Methods Enzymol.* **101:**598–605.
20. **Chilton, M. D., and B. J. McCarthy.** 1969. Genetic and base sequence homologies in bacilli. *Genetics* **62:**697–710.
21. **Chow, L. T., and N. Davidson.** 1973. Electron microscope mapping of the distribution of ribosomal genes of the *Bacillus subtilis* chromosome. *J. Mol. Biol.* **75:**265–279.
22. **Cohlberg, J. A., and M. Nomura.** 1976. Reconstitution of *Bacillus stearothermophilus* 50S ribosomal subunits from purified molecular components. *J. Biol. Chem.* **251:**209–221.
23. **Cutting, S., S. Roels, and R. Losick.** 1991. Sporulation operon *spoIVF* and the characterization of mutations that uncouple mother-cell from forespore gene expression in *Bacillus subtilis*. *J. Mol. Biol.* **221:**1237–1256.
24. **Dabbs, E. R.** 1982. Selection in *Bacillus subtilis* giving rise to strains with mutational alterations in a variety of ribosomal proteins. *Mol. Gen. Genet.* **187:**297–301.

25. **Dabbs, E. R.** 1983. Arrangement of loci within the principal cluster of ribosomal protein genes of *Bacillus subtilis*. *Mol. Gen. Genet.* **192:**124–130.
26. **Dabbs, E. R.** 1983. Mapping of the genes for *Bacillus subtilis* ribosomal proteins S6 and S16: comparison of the chromosomal distribution of ribosomal protein genes in this bacterium with the distribution in *Escherichia coli*. *Mol. Gen. Genet.* **192:**386–390.
27. **Dabbs, E. R.** 1983. Mapping of the genes for *Bacillus subtilis* ribosomal protein S9, protein S11 and protein BL27 by means of antibiotic resistant mutants. *Mol. Gen. Genet.* **191:**295–300.
28. **Dabbs, E. R.** 1983. A pair of *Bacillus subtilis* ribosomal protein genes mapping outside the principal ribosomal protein cluster. *J. Bacteriol.* **156:**966–969.
29. **Dabbs, E. R.** 1984. Order of ribosomal protein genes in the Rif cluster of *Bacillus subtilis* is identical to that of *Escherichia coli*. *J. Bacteriol.* **159:**770–772.
30. **Dabbs, E. R.** 1991. Mutants lacking individual ribosomal proteins as a tool to investigate ribosomal properties. *Biochimie* **73:**639–645.
31. **Dean, D., J. L. Yates, and M. Nomura.** 1981. Identification of ribosomal protein S7 as a repressor of translation within the *str* operon of *E. coli*. *Cell* **24:**413–419.
32. **Deneer, H. G., and G. B. Spiegelman.** 1987. *Bacillus subtilis* rRNA promoters are growth rate regulated in *Escherichia coli*. *J. Bacteriol.* **169:**995–1002.
33. **Dubnau, E., S. Pifko, A. Sloma, K. Cabane, and I. Smith.** 1976. Conditional mutations in the translational apparatus of *Bacillus subtilis*. *Mol. Gen. Genet.* **147:**1–12.
34. **Duncan, M., and C. W. Price.** Personal communication.
35. **Ellwood, M., and M. Nomura.** 1980. Deletion of a ribosomal nucleic acid operon in *Escherichia coli*. *J. Bacteriol.* **143:**1077–1080.
36. **Ellwood, M., and M. Nomura.** 1982. Chromosomal locations of the genes for rRNA in *Escherichia coli* K-12. *J. Bacteriol.* **149:**458–468.
37. **Fahnestock, S. R.** 1977. Reconstruction of active 50S ribosomal subunits from *Bacillus licheniformis* and *Bacillus subtilis*. *Arch. Biochem. Biophys.* **182:**497–505.
38. **Fahnestock, S. R., W. A. Strycharz, and D. M. Marquis.** 1981. Immunochemical evidence of homologies among 50S ribosomal proteins of *Bacillus stearothermophilus* and *Escherichia coli*. *J. Biol. Chem.* **256:**10111–10116.
39. **Falconi, M., M. Brombach, C. O. Gualerzi, and C. L. Pon.** 1991. In vivo transcriptional pattern of the *infC* operon of *Bacillus stearothermophilus*. *Mol. Gen. Genet.* **227:**60–64.
40. **Farwell, M. A., and J. C. Rabinowitz.** 1991. Protein synthesis in vitro by *Micrococcus luteus*. *J. Bacteriol.* **173:**3514–3522.
41. **Ferrari, F. A., K. Trach, and J. A. Hoch.** 1985. Sequence analysis of the *spo0B* locus reveals a polycistronic transcription unit. *J. Bacteriol.* **161:**556–562.
42. **Fox, G. E.** Personal communication.
43. **Fox, G. E., J. D. Wisotzkey, and P. Jurtshuk, Jr.** 1992. How close is close: 16S rRNA sequence identity may not be sufficient to guarantee species identity. *Int. J. Syst. Bacteriol.* **42:**166–170.
44. **Gallant, J., and G. Margason.** 1972. Amino acid control of messenger ribonucleic acid synthesis in *Bacillus subtilis*. *J. Biol. Chem.* **247:**2289–2294.
45. **Gardiner, K., and N. R. Pace.** 1980. RNase P from *Bacillus subtilis* has an RNA component. *J. Biol. Chem.* **255:**7507–7509.
46. **Garland, W. G., K. A. Louie, A. T. Matheson, and A. Liljas.** 1987. The complete amino acid sequence of the ribosomal "A" protein (L12) from *Bacillus stearothermophilus*. *FEBS Lett.* **220:**43–46.
47. **Garnier, T., B. Canard, and S. T. Cole.** 1991. Cloning, mapping, and molecular characterization of the rRNA operons of *Clostridium perfringens*. *J. Bacteriol.* **173:**5431–5438.
48. **Geisser, M., G. W. Tischendorf, and G. Stoffler.** 1973. Comparative immunological and electrophoretic studies on ribosomal proteins of Bacillaceae. *Mol. Gen. Genet.* **127:**129–145.
49. **Goldthwaite, C., D. Dubnau, and I. Smith.** 1970. Genetic mapping of antibiotic resistance markers in *Bacillus subtilis*. *Proc. Natl. Acad. Sci. USA* **65:**96–103.
50. **Gottlieb, P., G. LaFauci, and R. Rudner.** 1985. Alterations in the number of rRNA operons within the *Bacillus subtilis* genome. *Gene* **33:**259–268.
51. **Gourse, R. L., H. A. deBoer, and M. Nomura.** 1986. DNA determinants of rRNA synthesis in *E. coli*: growth rate dependent regulation, feedback inhibition, upstream activation, antitermination. *Cell* **14:**197–205.
52. **Green, C. J., G. C. Stewart, M. A. Hollis, B. S. Vold, and K. F. Bott.** 1985. Nucleotide sequence of the *Bacillus subtilis* ribosomal RNA operon, *rrnB*. *Gene* **37:**261–266.
53. **Green, C. J., and B. S. Vold.** 1983. Sequence analysis of a cluster of twenty-one tRNA genes in *Bacillus subtilis*. *Nucleic Acids Res.* **11:**5763–5774.
54. **Green, C. J., and B. S. Vold.** 1992. A cluster of nine tRNA genes between ribosomal gene operons in *Bacillus subtilis*. *J. Bacteriol.* **174:**3147–3151.
55. **Grundy, F. J., and T. M. Henkin.** 1990. Cloning and analysis of the *Bacillus subtilis rpsD* gene, encoding ribosomal protein S4. *J. Bacteriol.* **172:**6372–6379.
56. **Grundy, F. J., and T. M. Henkin.** 1991. The *rpsD* gene, encoding ribosomal protein S4, is autogenously regulated in *Bacillus subtilis*. *J. Bacteriol.* **173:**4595–4602.
57. **Grundy, F. J., and T. M. Henkin.** 1992. Characterization of the *Bacillus subtilis rpsD* regulatory target site. *J. Bacteriol.* **174:**6763–6770.
58. **Gutell, R. R., and G. E. Fox.** 1988. A compilation of large subunit RNA sequences presented in a structural format. *Nucleic Acids Res.* **16:**r175–r269.
59. **Hackl, W., and M. Stoffler-Meilicke.** 1988. Immunoelectron microscopic localisation of ribosomal proteins from *Bacillus stearothermophilus* that are homologous to *Escherichia coli* L1, L6, L23 and L29. *Eur. J. Biochem.* **174:**431–435.
60. **Hackl, W., M. Stoffler-Meilicke, and G. Stoffler.** 1988. Three-dimensional location of ribosomal protein BL2 from *Bacillus stearothermophilus*, a key component of the peptidyltransferase center. *FEBS Lett.* **233:**119–123.
61. **Hahn, V., and P. Stiegler.** 1986. An *Escherichia coli* S1-like ribosomal protein is immunologically conserved in Gram-negative bacteria, but not in Gram-positive bacteria. *FEMS Microbiol. Lett.* **36:**293–297.
62. **Halling, S. M., K. C. Burtis, and R. H. Doi.** 1978. β' subunit of bacterial RNA polymerase is responsible for streptolydigin resistance in *Bacillus subtilis*. *Nature* (London) **272:**837–839.
63. **Harford, N., and N. Sueoka.** 1970. Chromosomal location of antibiotic resistance markers in *Bacillus subtilis*. *J. Mol. Biol.* **51:**267–286.
64. **Hartmann, R. K., D. W. Vogel, R. T. Walker, and V. A. Erdmann.** 1988. In vitro incorporation of eubacterial, archaebacterial and eukaryotic 5S rRNAs into large ribosomal subunits of *Bacillus stearothermophilus*. *Nucleic Acids Res.* **16:**3511–3524.
65. **Held, W. A., S. Mizushima, and M. Nomura.** 1973. Reconstitution of *Escherichia coli* 30S ribosomal subunits from purified molecular components. *J. Biol. Chem.* **248:**5720–5730.
66. **Henkin, T. M., K. M. Campbell, and G. H. Chambliss.** 1979. Spectinomycin dependence in *Bacillus subtilis*. *J. Bacteriol.* **137:**1452–1455.
67. **Henkin, T. M., K. M. Campbell, and G. H. Chambliss.** 1982. Revertants of a streptomycin-resistant, oligosporogenous mutant of *Bacillus subtilis*. *Mol. Gen. Genet.* **186:**347–354.

68. **Henkin, T. M., and G. H. Chambliss.** 1984. Genetic analysis of a streptomycin-resistant, oligosporogenous mutant of *Bacillus subtilis*. *J. Bacteriol.* **157:**202–210.
69. **Henkin, T. M., and G. H. Chambliss.** 1984. Genetic mapping of a mutation causing an alteration in *Bacillus subtilis* ribosomal protein S4. *Mol. Gen. Genet.* **193:**364–369.
70. **Henkin, T. M., S. H. Moon, L. C. Mattheakis, and M. Nomura.** 1989. Cloning and analysis of the *spc* ribosomal protein operon of *Bacillus subtilis*: comparison with the *spc* operon of *Escherichia coli*. *Nucleic Acids Res.* **17:**7469–7486.
71. **Herfurth, E., H. Hirano, and B. Wittmann-Liebold.** 1991. The amino-acid sequences of the *Bacillus stearothermophilus* ribosomal proteins S17 and S21 and their comparison to homologous proteins of other ribosomes. *Biol. Chem. Hoppe-Seyler* **372:**955–961.
72. **Higo, K., W. Held, L. Kahan, and M. Nomura.** 1973. Functional correspondence between 30S ribosomal proteins of *Escherichia coli* and *Bacillus stearothermophilus*. *Proc. Natl. Acad. Sci. USA* **70:**944–948.
73. **Higo, K., E. Otaka, and S. Osawa.** 1982. Purification and characterization of 30S ribosomal proteins from *Bacillus subtilis*: correlation to *Escherichia coli* 30S proteins. *Mol. Gen. Genet.* **184:**239–244.
74. **Hirano, H., K. Eckart, M. Kimura, and B. Wittmann-Liebold.** 1987. Semi-preparative HPLC purification of ribosomal proteins from *Bacillus stearothermophilus* and sequence determination of the highly conserved protein S19. *Eur. J. Biochem.* **170:**149–157.
75. **Hussey, C., R. Losick, and A. L. Sonenshein.** 1971. Ribosomal RNA synthesis is turned off during sporulation of *Bacillus subtilis*. *J. Mol. Biol.* **57:**59–70.
76. **Isono, K., and S. Isono.** 1976. Lack of ribosomal protein S1 in *Bacillus stearothermophilus*. *Proc. Natl. Acad. Sci. USA* **73:**767–770.
77. **Isono, K., S. Isono, G. Stoffler, L. P. Visentin, M. Yaguchi, and A. T. Matheson.** 1973. Correlation between 30S ribosomal proteins of *Bacillus stearothermophilus* and *Escherichia coli*. *Mol. Gen. Genet.* **127:**191–195.
78. **Itoh, T., H. Kosugi, K. Higo, and S. Osawa.** 1975. Ribosomal proteins from streptomycin-resistant and dependent mutants, and revertants from streptomycin-dependence to independence in *Bacillus subtilis*. *Mol. Gen. Genet.* **139:**293–301.
79. **Itoh, T., and B. Wittmann-Liebold.** 1980. The primary structure of *Bacillus subtilis* acidic protein BL-9. *J. Biochem.* **87:**1185–1198.
80. **Jarvis, E. D., R. L. Widom, G. LaFauci, Y. Setoguchi, I. R. Richter, and R. Rudner.** 1988. Chromosomal organization of rRNA operons in *Bacillus subtilis*. *Genetics* **120:**625–635.
81. **Jinks-Robertson, S., and M. Nomura.** 1987. Ribosomes and tRNA, p. 1358–1385. *In* F. C. Neidhardt, J. L. Ingraham, K. B. Low, B. Magasanik, M. Schaechter, and H. E. Umbarger (ed.), *Escherichia coli and Salmonella typhimurium*: *Molecular and Cellular Biology*. American Society for Microbiology, Washington, D.C.
82. **Kimura, J., and M. Kimura.** 1987. The complete amino acid sequences of the 5S rRNA binding proteins L5 and L18 from the moderate thermophile *Bacillus stearothermophilus* ribosome. *FEBS Lett.* **210:**85–90.
83. **Kimura, M.** 1984. Proteins of the *Bacillus stearothermophilus* ribosome. The amino acid sequences of proteins S5 and L30. *J. Biol. Chem.* **259:**1051–1055.
84. **Kimura, M.** 1991. The nucleotide sequences of *Bacillus stearothermophilus* ribosomal protein S12 and S7 genes: comparison with the *str* operon of *Escherichia coli*. *Agric. Biol. Chem.* **55:**207–213.
85. **Kimura, M., and C. K. Chow.** 1984. The complete amino acid sequences of ribosomal proteins L17, L27, and S9 from *Bacillus stearothermophilus*. *Eur. J. Biochem.* **139:**225–234.
86. **Kimura, M., H. Ernst, and K. Appelt.** 1983. The primary structure of initiation factor IF3 from *Bacillus stearothermophilus*. *FEBS Lett.* **160:**78–81.
87. **Kimura, M., J. Kijk, and I. Heiland.** 1980. The primary structure of protein BL17 isolated from the large subunit of the *Bacillus stearothermophilus* ribosome. *FEBS Lett.* **121:**323–326.
88. **Kimura, M., and J. Kimura.** 1987. The complete amino acid sequence of ribosomal protein S12 from *Bacillus stearothermophilus*. *FEBS Lett.* **210:**91–96.
89. **Kimura, M., J. Kimura, and K. Ashman.** 1985. The complete primary structure of ribosomal proteins L1, L14, L15, L23, L24 and L29 from *Bacillus stearothermophilus*. *Eur. J. Biochem.* **150:**491–497.
90. **Kimura, M., J. Kimura, and T. Hatakeyama.** 1988. Amino acid sequences of ribosomal proteins S11 from *Bacillus stearothermophilus* and S19 from *Halobacterium marismortui*: comparison of the ribosomal protein S11 family. *FEBS Lett.* **240:**15–20.
91. **Kimura, M., J. Kimura, and K. Watanabe.** 1985. The primary structure of ribosomal protein L2 from *Bacillus stearothermophilus*. *Eur. J. Biochem.* **153:**289–297.
92. **Kimura, M., N. Rawlings, and K. Appelt.** 1981. The amino acid sequence of protein BL10 from the 50S subunit of the *Bacillus stearothermophilus* ribosome. *FEBS Lett.* **136:**58–64.
93. **Kobayashi, H., K. Kobayashi, and Y. Kobayashi.** 1977. Isolation and characterization of fusidic acid-resistant, sporulation-defective mutants of *Bacillus subtilis*. *J. Bacteriol.* **132:**262–269.
94. **Koivula, T., and H. Hemila.** 1991. Nucleotide sequence of a *Lactococcus lactis* gene cluster encoding adenylate kinase, initiation factor 1 and ribosomal proteins. *J. Gen. Microbiol.* **137:**2595–2600.
95. **Kromer, W., T. Hatakeyama, and M. Kimura.** 1990. Nucleotide sequence of *Bacillus stearothermophilus* ribosomal protein genes: part of the ribosomal S10 operon. *Biol. Chem. Hoppe-Seyler* **371:**631–636.
96. **Kruft, V., U. Kapp, and B. Wittmann-Liebold.** 1991. Characterization and primary structure of proteins L28, L33 and L34 from *Bacillus stearothermophilus* ribosomes. *Biochimie* **73:**855–860.
97. **Kruft, V., and B. Wittmann-Liebold.** 1991. Determination of peptide regions on the surface of the eubacterial and archaebacterial ribosome by limited proteolytic digestion. *Biochemistry* **30:**11781–11787.
98. **LaFauci, G., R. L. Widom, R. L. Eisner, E. D. Jarvis, and R. Rudner.** 1986. Mapping of rRNA genes with integrable plasmids in *Bacillus subtilis*. *J. Bacteriol.* **165:**204–214.
99. **Leduc, E., M. Hoekstra, and G. B. Spiegelman.** 1982. Relationship between synthesis of ribosomes and RNA polymerase in *Bacillus subtilis*. *Can. J. Microbiol.* **28:**1280–1288.
100. **Legault-Demare, L., and G. H. Chambliss.** 1974. Natural messenger ribonucleic acid-directed cell-free protein-synthesizing system of *Bacillus subtilis*. *J. Bacteriol.* **120:**1300–1307.
101. **Leventhal, J. M., and G. H. Chambliss.** 1979. DNA-directed cell-free protein-synthesizing system of *Bacillus subtilis*. *Biochim. Biophys. Acta* **564:**162–171.
102. **Liljas, A., and M. E. Newcomer.** 1981. Purification and crystallization of a protein complex from *Bacillus stearothermophilus* ribosomes. *J. Mol. Biol.* **153:**393–398.
103. **Lindahl, L., F. Sor, R. H. Archer, M. Nomura, and J. M. Zengel.** 1990. Transcriptional organization of the S10, *spc* and α operons of *Escherichia coli*. *Biochim. Biophys. Acta* **1050:**337–342.
104. **Linn, T., R. Losick, and A. L. Sonenshein.** 1975. Rifampin resistance mutation of *Bacillus subtilis* altering the electrophoretic mobility of the beta subunit of

ribonucleic acid polymerase. *J. Bacteriol.* **122:**1387–1390.

105. **Losick, R.** Personal communication.
106. **Loughney, K., E. Lund, and J. E. Dahlberg.** 1982. tRNA genes are found between the 16S and 23S rRNA genes in *Bacillus subtilis. Nucleic Acids Res.* **10:**1607–1624.
107. **Loughney, K., E. Lund, and J. E. Dahlberg.** 1983. Deletion of a rRNA gene set in *Bacillus subtilis. J. Bacteriol.* **154:**529–532.
108. **Loughney, K., E. Lund, and J. E. Dahlberg.** 1983. Ribosomal RNA precursors of *Bacillus subtilis. Nucleic Acids Res.* **11:**6709–6721.
109. **McDougall, J., T. Choli, V. Kruft, U. Kapp, and B. Wittman-Liebold.** 1989. The complete amino acid sequence of ribosomal protein S18 from the moderate thermophile *Bacillus stearothermophilus. FEBS Lett.* **245:**253–260.
110. **McLaughlin, J. R., C. L. Murray, and J. C. Rabinowitz.** 1981. Unique features in the ribosome binding site sequence of the Gram-positive *Staphylococcus aureus* β-lactamase gene. *J. Biol. Chem.* **256:**11283–11291.
111. **Meng, B. Y., K. Shinozaki, and M. Sugiura.** 1989. Genes for the ribosomal proteins S12 and S7 and elongation factors EF-G and EF-Tu of the cyanobacterium *Anacystis nidulans*: structural homology between 16S rRNA and S7 mRNA. *Mol. Gen. Genet.* **216:**25–30.
112. **Michalowski, C. B., B. Pfanzagl, W. Loffelhardt, and H. J. Bohnert.** 1990. The cyanelle S10-*spc* ribosomal protein gene operon from *Cyanophora paradoxa. Mol. Gen. Genet.* **224:**222–231.
113. **Michalski, C. J., and B. H. Sells.** 1975. Molecular morphology of ribosomes. Iodination of *Escherichia coli* ribosomes with solid-state lactoperoxidase. *Eur. J. Biochem.* **52:**385–389.
114. **Miller, H. M., S. M. Friedman, D. J. Litman, and C. R. Cantor.** 1976. Surface topography of the *Bacillus stearothermophilus* ribosome. *Mol. Gen. Genet.* **144:**273–280.
115. **Moon, S. H., and T. M. Henkin.** Unpublished observations.
116. **Moran, C. P., Jr., and K. F. Bott.** 1979. Organization of transfer and ribosomal ribonucleic acid genes in *Bacillus subtilis. J. Bacteriol.* **140:**742–744.
117. **Moriya, S., N. Ogasawara, and H. Yoshikawa.** 1985. Structure and function of the region of the replication origin of the *Bacillus subtilis* chromosome. III. Nucleotide sequence of some 10,000 base pairs in the origin region. *Nucleic Acids Res.* **13:**2251–2265.
118. **Neefs, J.-M., Y. Van de Peer, P. DeRijk, A. Goris, and R. DeWachter.** 1991. Compilation of small ribosomal subunit RNA sequences. *Nucleic Acids Res.* **19:**1987–2015.
119. **Neumann-Spallart, C., J. Jakowitsch, M. Kraus, M. Brandtner, H. J. Bohnert, and W. Loffehardt.** 1991. *rps10*, unreported for plastid DNAs, is located on the cyanelle genome of *Cyanophora paradoxa* and is cotranscribed with the *str* operon genes. *Curr. Genet.* **19:**313–315.
120. **Ng, C., C. Buchanan, A. Leung, C. Ginther, and T. Leighton.** 1991. Suppression of defective-sporulation phenotypes by mutations in transcription factor genes of *Bacillus subtilis. Biochimie* **73:**1163–1170.
121. **Nierhaus, K. H., and F. Dohme.** 1974. Total reconstruction of functionally active 50S ribosomal subunits from *Escherichia coli. Proc. Natl. Acad. Sci. USA* **71:**4713–4717.
122. **Nishino, T., J. Gallant, P. Shalit, L. Palmer, and T. Wehr.** 1979. Regulatory nucleotides involved in the *rel* function of *Bacillus subtilis. J. Bacteriol.* **140:**671–679.
123. **Nomura, M., and V. A. Erdmann.** 1970. Reconstruction of 50S subunits from dissociated molecular components. *Nature* (London) **228:**744–748.
124. **Nomura, M., P. Traub, and H. Bechmann.** 1968. Hybrid 30S ribosomal particles reconstituted from components of different bacterial origins. *Nature* (London) **219:**793–799.
125. **Ochi, K.** 1989. Heterogeneity of ribosomal proteins among *Streptomyces* species and its application to identification. *J. Gen. Microbiol.* **135:**2635–2642.
126. **Ochi, K.** 1990. *Streptomyces relC* mutants with an altered ribosomal protein ST-L11 and genetic analysis of a *Streptomyces griseus relC* mutant. *J. Bacteriol.* **172:**4008–4016.
127. **Ogasawara, N., S. Moriya, and H. Hoshikawa.** 1983. Structure and organization of rRNA operons in the region of the replication origin of the *Bacillus subtilis* chromosome. *Nucleic Acids Res.* **11:**6301–6318.
128. **Ogasawara, N., S. Moriya, K. von Meyenburg, F. G. Hansen, and H. Hoshikawa.** 1985. Conservation of genes and their organization in the chromosomal replication origin region of *Bacillus subtilis* and *Escherichia coli. EMBO J.* **4:**3345–3350.
129. **Ogasawara, N., M. Seiki, and H. Yoshikawa.** 1983. Replication origin region of *Bacillus subtilis* chromosome contains two rRNA operons. *J. Bacteriol.* **154:**50–57.
130. **Ohama, T., A. Muto, and S. Osawa.** 1989. Spectinomycin operon of *Micrococcus luteus*: evolutionary implications of organization and novel codon usage. *J. Mol. Evol.* **29:**381–395.
131. **Ohama, T., F. Yamao, A. Muto, and S. Osawa.** 1987. Organization and codon usage of the streptomycin operon in *Micrococcus luteus*, a bacterium with a high genomic G+C content. *J. Bacteriol.* **169:**4770–4777.
132. **Ohkubo, S., A. Muto, Y. Kawauchi, F. Yamao, and S. Osawa.** 1987. The ribosomal protein gene cluster of *Mycoplasma capricolum. Mol. Gen. Genet.* **210:**314–322.
133. **Osawa, S.** 1976. Gene locus of a 30S ribosomal protein S20 of *Bacillus subtilis. Mol. Gen. Genet.* **144:**49–51.
134. **Osawa, S.** 1982. Ribosomal genes in *Bacillus subtilis*: comparison with *Escherichia coli*, p. 19–21. *In* D. Schlessinger (ed.), Microbiology—1982. American Society for Microbiology, Washington, D.C.
135. **Osawa, S., and H. Hori.** 1980. Molecular evolution of ribosomal components, p. 333–355. *In* G. Chambliss, R. Craven, J. Davies, L. Kahan, and M. Nomura (ed.), *Ribosomes: Structure, Function and Genetics*. University Park Press, Baltimore.
136. **Osawa, S., R. Takata, K. Tanaka, and M. Tamaki.** 1973. Chloramphenicol resistant mutants of *Bacillus subtilis. Mol. Gen. Genet.* **127:**163–173.
137. **Osawa, S., A. Tokui, and H. Saito.** 1978. Mapping by interspecies transformation experiments of several ribosomal protein genes near the replication origin of *Bacillus subtilis* chromosome. *Mol. Gen. Genet.* **164:**113–129.
138. **Pace, N. R., and B. Pace.** 1990. Ribosomal RNA terminal maturase: ribonuclease M5 from *Bacillus subtilis. Methods Enzymol.* **181:**366–374.
139. **Pai, Y. L., and R. Dabbs.** 1981. Conditional lethal mutants of *Bacillus subtilis* dependent on kasugamycin for growth. *Mol. Gen. Genet.* **183:**478–483.
140. **Panganiban, A. T., and H. R. Whiteley.** 1983. Purification and properties of a new *Bacillus subtilis* RNA processing enzyme. Cleavage of phage SP82 mRNA and *Bacillus subtilis* precursor rRNA. *J. Biol. Chem.* **258:**12487–12493.
141. **Pon, C. L., M. Brombach, S. Thamm, and C. O. Gualerzi.** 1989. Cloning and characterization of a gene cluster from *Bacillus stearothermophilus* comprising *infC*, *rpmI* and *rplT*. *Mol. Gen. Genet.* **218:**355–357.
142. **Putzer, H., A. A. Brakhage, and M. Grunberg-Manago.** 1990. Independent genes for two threonyl-tRNA synthetases in *Bacillus subtilis. J. Bacteriol.* **172:**4593–4602.
143. **Ramakrishnan, V., and S. E. Gerchman.** 1991. Cloning, sequencing and overexpression of genes for ribosomal

proteins from *Bacillus stearothermophilus*. *J. Biol. Chem.* **266**:880–885.

144. **Roberts, M. W., and J. C. Rabinowitz.** 1989. The effect of *Escherichia coli* ribosomal protein S1 on the translational specificity of bacterial ribosomes. *J. Biol. Chem.* **264**:2228–2235.
145. **Ron, E. Z., M.-P. deBethune, and C. G. Cocito.** 1980. Mapping of virginiamycin S resistance in *Bacillus subtilis*. *Mol. Gen. Genet.* **180**:639–640.
146. **Rossler, D., W. Ludwig, K. H. Schleifer, C. Lin, T. J. McGill, J. D. Wisotzkey, P. Jurtskuk, Jr., and G. E. Fox.** 1991. Phylogenetic diversity in the genus *Bacillus* as seen by 16S rRNA sequencing studies. *Syst. Appl. Microbiol.* **14**:266–269.
147. **Rudner, R.** Personal communication.
148. **Schnier, J., H.-S. Gewitz, S.-E. Behrens, A. Lee, C. Ginther, and T. Leighton.** 1990. Isolation and characterization of *Bacillus stearothermophilus* 30S and 50S ribosomal protein mutations. *J. Bacteriol.* **172**:7306–7309.
149. **Shazand, K., J. Tucker, R. Chiang, K. Stansmore, H. V. Sperling-Peterson, M. Grunberg-Manago, J. C. Rabinowitz, and T. Leighton.** 1990. Isolation and molecular genetic characterization of the *Bacillus subtilis* gene (*infB*) encoding protein synthesis initiation factor 2. *J. Bacteriol.* **172**:2675–2687.
150. **Sijben-Muller, G., R. B. Hallick, J. Alt, P. Westloff, and R. G. Hermann.** 1986. Spinach plastid genes coding for initiation factor IF1, ribosomal protein S11 and RNA polymerase α subunit. *Nucleic Acids Res.* **14**:1029–1044.
151. **Smith, I.** 1982. The translational apparatus of *Bacillus subtilis*, p. 111–145. *In* D. Dubnau (ed.), *The Molecular Biology of the Bacilli*, vol. 1. Academic Press, Inc., New York.
152. **Smith, I., D. Dubnau, P. Morell, and J. Marmur.** 1968. Chromosomal location of DNA base sequences complementary to transfer RNA and to 5S, 16S and 23S ribosomal RNA in *Bacillus subtilis*. *J. Mol. Biol.* **33**:123–140.
153. **Smith, I., C. Goldthwaite, and D. Dubnau.** 1969. The genetics of ribosomes in *Bacillus subtilis*. *Cold Spring Harbor Symp. Quant. Biol.* **34**:85–89.
154. **Smith, I., and P. Paress.** 1978. Genetic and biochemical characterization of kirromycin resistance mutations in *Bacillus subtilis*. *J. Bacteriol.* **135**:1107–1117.
155. **Smith, I., P. Paress, K. Cabane, and E. Dubnau.** 1980. Genetics and physiology of the *rel* system of *Bacillus subtilis*. *Mol. Gen. Genet.* **178**:271–279.
156. **Smith, I., P. Paress, and S. Pestka.** 1978. Thiostrepton-resistant mutants exhibit relaxed synthesis of RNA. *Proc. Natl. Acad. Sci. USA* **75**:5993–5997.
157. **Sonenshein, A. L., H. B. Alexander, D. M. Rothstein, and S. H. Fisher.** 1977. Lipiarmycin-resistant ribonucleic acid polymerase mutants of *Bacillus subtilis*. *J. Bacteriol.* **132**:73–79.
158. **Stackebrandt, E., W. Ludwig, M. Weizenegger, S. Dorn, McGill, G. E. Fox, C. R. Woese, W. Schubert, and K. H. Schleifer.** 1987. Comparative 16S rRNA oligonucleotide analyses and murein types of round sporeforming bacilli and non-sporeforming relatives. *J. Gen. Microbiol.* **133**:2523–2529.
159. **Stevenson, J. K., R. G. Drager, D. W. Copertino, D. A. Christopher, K. P. Jenkins, G. Yepiz-Plascencia, and R. B. Hallick.** 1991. Intercistronic group III introns in polycistronic ribosomal protein operons of chloroplasts. *Mol. Gen. Genet.* **228**:183–192.
160. **Stewart, G. C., and K. F. Bott.** 1983. DNA sequence of the tandem ribosomal RNA promoter for *B. subtilis* operon *rrnB*. *Nucleic Acids Res.* **11**:6289–6300.
161. **Stewart, G. C., F. E. Wilson, and K. F. Bott.** 1982. Detailed physical mapping of the ribosomal RNA genes of *Bacillus subtilis*. *Gene* **19**:153–162.
162. **Stoffler-Meilicke, M., B. Epe, P. Woolley, M. Lotti, J. Littlechild, and G. Stoffler.** 1984. Location of protein S4 on the small ribosomal subunit of *E. coli* and *B. stearothermophilus* with protein-specific and hapten-specific antibodies. *Mol. Gen. Genet.* **197**:8–18.
163. **Stoffler-Meilicke, M., and G. Stoffler.** 1990. Topography of the ribosomal proteins from *Escherichia coli* within the intact subunits as determined by immunoelectron microscopy and protein-protein cross-linking, p. 123–133. *In* W. E. Hill, A. Dahlberg, R. A. Garrett, P. B. Moore, D. Schlessinger, and J. R. Warner (ed.), *The Ribosome: Structure, Function and Evolution*. American Society for Microbiology, Washington, D.C.
164. **Strauch, E., E. Takano, H. A. Baylis, and M. J. Bibb.** 1991. The stringent response in *Streptomyces coelicolor* A3(2). *Mol. Microbiol.* **5**:289–298.
165. **Suh, J.-W., S. A. Boylan, and C. W. Price.** 1986. Gene for the alpha subunit of *Bacillus subtilis* RNA polymerase maps in the ribosomal protein gene cluster. *J. Bacteriol.* **168**:65–71.
166. **Suh, J.-W., S. A. Boylan, S. M. Thomas, K. M. Dolan, D. B. Oliver, and C. W. Price.** 1990. Isolation of a *secY* homologue from *Bacillus subtilis*: evidence for a common protein export pathway in eubacteria. *Mol. Microbiol.* **4**:305–315.
167. **Swanton, M., and G. Edlin.** 1972. Isolation and characterization of an RNA-relaxed mutant of *B. subtilis*. *Biochem. Biophys. Res. Commun.* **46**:583–588.
168. **Tanaka, I., M. Kumura, J. Kimura, and J. Dijk.** 1984. The amino acid sequence of two small ribosomal proteins from *Bacillus stearothermophilus*. *FEBS Lett.* **166**:343–346.
169. **Tang, C. K., and D. E. Draper.** 1989. Unusual mRNA pseudoknot structure is recognized by a protein translational repressor. *Cell* **57**:531–536.
170. **Thurlow, D. L., and R. A. Zimmermann.** 1978. Conservation of ribosomal protein binding sites in prokaryotic 16S RNAs. *Proc. Natl. Acad. Sci. USA* **75**:2859–2863.
171. **Tipper, D. J., C. W. Johnson, C. L. Ginther, T. Leighton, and H. G. Wittmann.** 1977. Erythromycin resistant mutations in *Bacillus subtilis* cause temperature sensitive sporulation. *Mol. Gen. Genet.* **150**:147–159.
172. **Traub, P., and M. Nomura.** 1969. Structure and function of *Escherichia coli* ribosomes. VI. Mechanism of assembly of 30S ribosomes studied in vitro. *J. Mol. Biol.* **40**:391–413.
173. **van Wezel, G. P., E. Vijgenboom, and L. Bosch.** 1991. A comparative study of the ribosomal RNA operons of *Streptomyces coelicolor* A3(2) and sequence analysis of *rrnA*. *Nucleic Acids Res.* **19**:4399–4403.
174. **Vold, B. S., C. J. Green, N. Narasimhan, M. Strem, and J. N. Hansen.** 1988. Transcriptional analysis of *Bacillus subtilis* rRNA-tRNA operons. II. Unique properties of an operon containing a minor 5S rRNA gene. *J. Biol. Chem.* **263**:14485–14490.
175. **Vorgias, C. E., A. J. Kingswell, Z. Dauter, and K. S. Wilson.** 1991. Cloning, overexpression, purification and crystallisation of ribosomal protein L9 from *Bacillus stearothermophilus*. *FEBS Lett.* **286**:204–208.
176. **Wagenknecht, T., J. M. Carazo, M. Radermacher, and J. Frank.** 1989. Three-dimensional reconstruction of the ribosome from *Escherichia coli*. *Biophys. J.* **55**:465–477.
177. **Wang, L.-F., and R. H. Doi.** 1986. Nucleotide sequence and organization of *Bacillus subtilis* RNA polymerase major sigma (σ^{43}) operon. *Nucleic Acids Res.* **14**:4293–4307.
178. **Wawrousek, E. F., and J. N. Hansen.** 1983. Structure and organization of a cluster of six tRNA genes in the space between tandem ribosomal RNA gene sets in *Bacillus subtilis*. *J. Biol. Chem.* **258**:291–298.
179. **Wawrousek, E. F., N. Narasimhan, and J. N. Hansen.** 1984. Two large clusters with thirty-seven transfer RNA genes adjacent to ribosomal RNA gene sets in *Bacillus subtilis*. *J. Biol. Chem.* **259**:3694–3702.

180. **Wayne, R. R., and T. Leighton.** 1981. Physiological suppression of *Bacillus subtilis* conditional sporulation phenotypes—RNA polymerase and ribosomal mutations. *Mol. Gen. Genet.* **183:**550–552.
181. **White, S. W., K. Appelt, J. Dijk, and K. S. Wilson.** 1983. Proteins of the *Bacillus stearothermophilus* ribosome: a 5Å structural analysis of protein S5. *FEBS Lett.* **163:**73–75.
182. **Wienen, B., R. Ehrlich, M. Stoffler-Meilicke, G. Stoffler, I. Smith, D. Weiss, R. Vince, and S. Pestka.** 1979. Ribosomal protein alterations in thiostrepton- and micrococcin-resistant mutants of *Bacillus subtilis. J. Biol. Chem.* **254:**8031–8041.
183. **Williams, G., and I. Smith.** 1979. Chromosomal mutations causing resistance to tetracycline in *Bacillus subtilis. Mol. Gen. Genet.* **177:**23–29.
184. **Wilson, F. E., J. A. Hoch, and K. Bott.** 1981. Genetic mapping of a linked cluster of ribosomal ribonucleic acid genes in *Bacillus subtilis. J. Bacteriol.* **148:**624–628.
185. **Wisotzkey, J. D., P. Jurtshuk, Jr., G. E. Fox, G. Deinhard, and K. Poralla.** 1992. Comparative sequence analyses on the 16S rRNA (rDNA) of *Bacillus acidocaldarius, Bacillus acidoterrestris,* and *Bacillus cycloheptanicus* and proposal for creation of a new genus, *Alicyclobacillus* gen. nov. *Int. J. Syst. Bacteriol.* **42:**263–269.
186. **Woese, C. R.** 1987. Bacterial evolution. *Microbiol. Rev.* **51:**221–271.
187. **Woese, C. R., B. A. Debrunner-Vossbrinck, H. Oyaizu, E. Stackebrandt, and W. Ludwig.** 1984. Gram-positive bacteria: possible photosynthetic ancestry. *Science* **229:**761–765.
188. **Yaguchi, M., A. T. Matheson, and L. P. Visentin.** 1974. Procaryotic ribosomal proteins: N-terminal sequence homologies and structural correspondence of 30S ribosomal proteins from *Escherichia coli* and *Bacillus stearothermophilus. FEBS Lett.* **46:**296–300.
189. **Yonath, A., W. Bennett, S. Weinstein, and H. G. Wittmann.** 1990. Crystallography and image reconstructions of ribosomes, p. 134–147. *In* W. E. Hill, A. Dahlberg, R. A. Garrett, P. B. Moore, D. Schlessinger, and J. R. Warner (ed.), *The Ribosome: Structure, Function and Evolution.* American Society for Microbiology, Washington, D.C.
190. **Yonath, A., M. A. Saper, I. Makowski, J. Mussig, J. Piefte, H. D. Bartunik, K. S. Bartels, and H. G. Wittmann.** 1986. Characterization of single crystals of the large ribosomal particles from *B. stearothermophilus. J. Mol. Biol.* **187:**633–636.
191. **Yoshikawa, H., and R. H. Doi.** 1990. Sequence of the *Bacillus subtilis* spectinomycin resistance gene region. *Nucleic Acids Res.* **18:**1647.
192. **Zeigler, D. R., and D. H. Dean.** 1990. Orientation of genes in the *Bacillus subtilis* chromosome. *Genetics* **125:**703–708.

47. tRNA, tRNA Processing, and Aminoacyl-tRNA Synthetases

CHRISTOPHER J. GREEN and BARBARA S. VOLD

Most of the research on tRNA, tRNA processing, and tRNA synthetases in bacteria has been performed with *Escherichia coli*. Because of this, *E. coli* has often been considered the prototypical bacterium. However, even among the eubacteria, there is a large amount of diversity. Therefore, our understanding of bacterial metabolism and genomic organization should not be based solely on what is known from one gram-negative bacterium. In this chapter, we give an overview of tRNA, tRNA processing, and tRNA synthetases in *Bacillus subtilis*, emphasizing the areas in which there are distinct differences from *E. coli*. We also discuss the general characteristics of tRNA in the gram-positive bacteria. Despite the common belief that most bacteria are similar to *E. coli*, the tRNA gene organization and anticodon complement in *B. subtilis*, and probably in most gram-positive bacteria, are remarkably different from those in *E. coli*.

tRNA

Mature tRNAs

Sequences

tRNAs were among the first nucleic acids to be sequenced because of their small size and ease of purification. In recent years, the pace of sequencing of new mature tRNA species has declined considerably, and sequencing at the gene level is now more common. This has resulted in a lack of information on the modified nucleoside composition of individual tRNA species in *B. subtilis*.

The sequences of only 12 different mature *B. subtilis* tRNAs have been reported, i.e., those for the amino acids Ala, Arg, Gly, Leu, Lys, Met, fMet, Phe, Pro, Thr, Tyr, and Val, not including modification variants (130). Four tRNA sequences are known from *Bacillus stearothermophilus*, i.e., those for the amino acids Leu, Phe, Tyr, and Val (130). All 29 tRNAs with 28 different anticodons from *Mycoplasma capricolum* have been sequenced (3), making this bacterium's tRNAs the best characterized of the gram-positive group. The mycoplasmas (class *Mollicutes*) have no cell walls but are grouped with the other low-G+C gram-positive bacteria on the basis of rRNA sequence comparisons (151). The mycoplasma tRNAs have been of particular interest because of their unusually degenerate codon recognition capabilities compared with those of *E. coli*, which allows them to reduce their total number of tRNA species. In many cases in *Mycoplasma* spp., only one tRNA is needed to read all four codons in a family of codons that differ at their third positions (3, 118, 119).

Isoaccepting tRNA species

Chromatography on RPC-5 columns indicates that *B. subtilis* has 42 different isoaccepting tRNA species (139). RPC-5 chromatography is very sensitive to modification differences (28); therefore, the number of primary tRNA sequences is probably much smaller. Modification differences in a single *B. subtilis* tRNA gene transcript have been shown to account for multiple isoacceptors for Lys, Trp, and Tyr (140). Comparison of RPC-5 chromatographic profiles of *B. subtilis* and *E. coli* tRNAs suggests that the gram-positive bacterium has a smaller number of tRNA isoacceptors (70, 139). For example, although *E. coli* has four different species of $tRNA^{Arg}$ (70), *B. subtilis* has only two RPC-5-resolvable isoacceptors (139). This may indicate that the *B. subtilis* tRNA population is similar to that of *Mycoplasma* spp. in having relatively fewer tRNA species with more flexible codon-reading rules. The unconventional codon reading seen in *Mycoplasma* spp. may actually be typical of the gram-positive bacteria.

Is there a tRNA that can read the UGA "termination" codon?

There is evidence that *B. subtilis* has a tRNA that can read the UGA termination codon as a tryptophan, normally coded by UGG. No opal mutations (UGA termination codon) have ever been found in *B. subtilis*. A mutated chloramphenicol acetyltransferase gene (*cat-86*) with an internal UGA termination codon has been shown to confer resistance to chloramphenicol in a *B. subtilis* host. The gene product was produced at a relative efficiency of 6% compared with the wild-type product by the substitution of a tryptophan residue. Replacement of the UGA codon with a UAA termination codon blocked expression of the *cat* gene in *B. subtilis* (79). A single $tRNA^{Trp}$ gene has been found in a 16-tRNA gene cluster that has the expected CCA anticodon sequence (150). The sequence of the mature $tRNA^{Trp}$ has not been determined. RPC-5 chromatography of *B. subtilis* $tRNA^{Trp}$ shows that there is one isoacceptor in exponentially growing cells but that a second $tRNA^{Trp}$ is present in spores, probably because of modification differences (140). This putative UGA codon-reading capability might be due to unconventional translation by the product of the known $tRNA^{Trp}$ gene. It is also possible that *B. subtilis* con-

Christopher J. Green • SRI International, 333 Ravenswood Avenue, Menlo Park, California 94025. **Barbara S. Vold** • Syva Company, P.O. Box 10058, Palo Alto, California 94303.

tains a minor $tRNA^{Trp}$ with a UCA anticodon that has not yet been discovered that can read the UGA codon. It is interesting that *M. capricolum* contains just such a tRNA species (3, 92, 162). It has been proposed that the existence of this type of tRNA in *Mycoplasma* spp. is evidence for the relatively recent evolution of some new codons in these bacteria (93, 101). The mechanism of this apparent UGA codon translation in *B. subtilis* may have important implications for understanding how the genetic code can change during evolution.

Methionine initiator tRNA

Like *E. coli*, *B. subtilis* has an initiator $tRNA^{Met}$ (159) as well as an elongator $tRNA^{Met}$ (158). While most tRNAs that recognize codons beginning with an A have a modified A residue immediately 3′ to the anticodon (13), the initiator tRNAs have an unmodified A at that position. This lack of modification in the initiator may cause the codon translation rules to be relaxed, allowing for initiation of translation at UUG and GUG codons in a relatively small number of cases in *E. coli* (72). In *Bacillus* spp. and other gram-positive bacteria, initiation at UUG and GUG codons is more common, even in highly expressed proteins (1, 51, 86, 88, 109, 161). It is not yet clear whether this difference in initiation is due to differences in initiator $tRNA^{Met}$, ribosomes, or context differences in the mRNAs.

Modified nucleosides

The total array of modified bases in *B. subtilis* tRNA is generally similar to that of *E. coli*. Approximately three or four residues are present as modified bases in the average tRNA molecule. Although in general the types of modifications are the same, there are some significant differences. One unusual difference is the presence of 1-methyladenosine in *B. subtilis* and other gram-positive tRNAs (3, 137). This modified nucleotide is common in eukaryotes but not in *E. coli* tRNAs (131). Another significant difference is the relatively small amount of 4-thiouridine found in *B. subtilis* tRNA compared with that in *E. coli* (24). Almost all of the *E. coli* tRNAs have this modified base at position 8, but only 2 of the 12 known *B. subtilis* tRNA sequences have it (131). The overall level of thiolation for *B. subtilis* is between 0.36 and 0.44 mol of total thionucleoside per mol of tRNA, i.e., three to four times lower than that in *E. coli* tRNAs (144). Nevertheless, there are at least five types of thiolated nucleosides in *B. subtilis*, and approximately one-third exist as thiomethyl groups at position 2 of an adenosine or modified adenosine residue (144). The levels of one thiomethylated nucleoside, N^6-(Δ^2-isopentenyl)-2-methylthioadenosine (ms^2i^6A), has been reported to change during sporulation (68, 140). In fact, while ms^2i^6A constitutes about 90% of the total hydrophobic nucleosides at all growth stages in *E. coli*, *B. subtilis* tRNAs have N^6-(Δ^2-isopentenyl)adenosine (i^6A) as the predominant hydrophobic nucleoside in exponential growth and ms^2i^6A as the predominant nucleoside in stationary phase (140). The i^6A-ms^2i^6A modification occurs only in the anticodon loop. Nine of the 11 tRNA isoacceptors that showed chromatographic alterations during development translate codons that begin with A or U. The reason for this is not known, but for tyrosyl-tRNA (68) and lysyl-tRNA (127), the difference in modification in the anticodon loop results in a change in translational efficiency.

A very significant difference is in the biosynthetic pathways for methylation of 5-methyluridine (ribothymidine). Originally, there was a controversy about whether or not the tRNA of *B. subtilis* contained 5-methyluridine (5, 110). This was resolved by the observation that the methyl donor for 5-methyluridine in *B. subtilis* did not involve *S*-adenosylmethionine as it did in *E. coli* but a tetrahydrofolate derivative instead (6, 52). The tetrahydrofolate pathway is also used by *Streptococcus faecalis* (30); however, *B. stearothermophilus* (2) apparently uses the *S*-adenosylmethionine pathway, so this difference is not reflected across all the gram-positive lines.

tRNAs and Sporulation

Modification

Early experiments indicated that tRNAs and rRNAs from spores and vegetative cells were similar in total base composition, size, and conformation (31). Melting-temperature curves for tRNAs from logarithmically growing cells, young spores, and mature spores showed similarity hyperchromicity, and spore tRNAs could be aminoacylated (145). In vivo and in vitro methylation studies of exponentially growing cells, sporulation cells, and spores showed that tRNAs from exponentially growing cells are slightly undermethylated and that tRNAs from stationary cells or spores are more completely modified. In general, the types of methylated nucleosides are preserved with fidelity even in the dormant spore, although changes in specific isoacceptors were not examined in this study (126).

Are new tRNA species made in sporulating cells?

Chromatographic comparisons of vegetative cell and sporulating cell isoaccepting species by RPC-5 showed (i) no alterations in elution profile of the tRNA species for Phe, Ile, Met, fMet, Val, Ala, Asp, His, or Pro; (ii) changes in relative amounts of isoaccepting tRNA species for Tyr, Leu, Ser, Lys, Thr, Asp, Glu, Gly, and Arg; and (iii) the appearance of a new species for Trp-tRNA (140). The pairs of isoacceptors for Lys-, Tyr-, and Trp-tRNAs showing differences in relative levels during sporulation were studied for composition of modified nucleosides and two-dimensional electrophoretic fingerprint analysis of oligonucleotides. The results suggested that the differences among these pairs of isoaccepting species were in the degree of posttranscriptional modifications of the anticodon loop regions. The nucleosides involved were thiomethylations of i^6A in Tyr- and Trp-tRNAs and differences in an unknown nucleoside K, which occurred in a position analogous to *N*-[(9-β-D-ribofuranosyl)purin-6-yl]carbamoylthreonine (t^6A) (140). By antibody recognition and mass spectrometry, this nucleoside was subsequently identified as N^6-[(9-β-D-ribofuranosyl-2-methylthiopurine-6-yl)carbamoyl]threonine (ms^2t^6A) (144). Therefore, changes in modifications of modified nucleosides are important aspects of the sporulation phenomenon. All of the tRNA groups investigated expressed different temporal patterns of change in isoaccepting species beginning at t_0 and continuing through the sporulation process (139).

Another method of studying changes in tRNA species is that of Henner and Steinberg (55), who used two-dimensional polyacrylamide gel electrophoresis for the separation of in vivo-^{32}P-labeled RNA. They resolved *B. subtilis* tRNAs into 32 well-separated spots, with the relative abundances ranging from 0.9 to 17% of the total. All of the tRNA species resolved by this gel system were synthesized at every stage examined, including vegetative growth, sporulation, and outgrowth. No evidence of preferential transcription, degradation, or compartmentation of unique tRNA species during spore formation was found. They concluded that posttranscriptional modifications were being used to change the complement of tRNAs rather than the expression of new tRNA genes.

Loss of CCA end

The most dramatic difference between tRNAs from exponentially growing cells and spores is the relative loss of the 3′-terminal adenylate and cytidylate residues that has been observed in both *Bacillus megaterium* (124) and *B. subtilis* (136). In *B. subtilis*, approximately 2.3% of the tRNA in exponentially growing cells lacks the 3′-terminal adenosine, 3.0% lacks it at sporulation stages I and II, 10.6% lack it at stage III, and 19.3 to 23.9% lack it in spores. In *B. megaterium*, the loss occurred after the appearance of refractile spores (124).

Changes in aminoacyl-tRNA synthetases during sporulation

Aminoacyl-tRNA synthetase and methionyl-tRNA transformylase activity have been found at various growth stages and in spores (138). No major changes in synthetase activity or in methionyl-tRNA transformylase activity during the sporulation cycle was demonstrated. However, there was a burst of activity of aminoacyl-tRNA synthetases during exponential growth. Dormant spores appeared to contain the same kinds of aminoacyl-tRNA synthetase activities as vegetative cells. Thus, the capacities to aminoacylate tRNAs and formylate fMet-tRNA are present through sporulation in *B. subtilis*.

RNA turnover during sporulation

Numerous changes in RNA metabolism occur during sporulation in bacilli. Although a net accumulation of RNA ceases at the beginning of the sporulation cycle, turnover of rRNA, mRNA, and tRNA provides for continuing synthesis (8, 16, 134). Radioactive nucleosides are incorporated into tRNA and rRNA during the sporulation process. Both tRNA and rRNA in dormant spores contained radioactive label added late in sporulation (31). Perhaps because of differences in degradation rates, the tRNA/rRNA ratio during sporulation in *B. subtilis* changes from 0.20 in vegetative cells to 0.36 to 0.43 in dormant spores (31).

The RNA turnover is caused by an increase in RNase activity that begins during the first hour of sporulation. The rise in extracellular RNase activity is a biochemical marker for stage I of sporulation (81). This increase in RNase makes characterization of RNA from sporulation cells and spores difficult and creates havoc for those of us interested in that field.

rRNA transcription during sporulation

Since the majority of *B. subtilis* tRNA genes follow rRNA operons and appear to be part of their transcriptional units, transcription of rRNA genes is an integral part of tRNA gene transcription. Genes for stable RNA in eubacteria represent very actively transcribed genes, and their expression is governed by a variety of control mechanisms. In the early 1970s, a controversy arose about rRNA gene transcription during sporulation in *B. subtilis*. One view was that rRNA synthesis was turned off during sporulation (58), while another view was that rRNAs continued to be synthesized during sporulation, although at a somewhat lower rate than during vegetative growth (16). Pulse-chase labeling experiments have indicated that by the early stages of sporulation, relative amounts of rRNA synthesis are reduced severalfold (104, 134). It has been suggested that a modification of RNA polymerase may play a role in this regulation of rRNA expression during sporulation (104), and this is certainly the case for some other genes. However, we do not know the mechanism by which the rRNA transcription decreases, and we do not know whether tRNA transcription falls under the same control mechanism.

RNA synthesis during germination

Studies of RNA, protein, and DNA synthesis during synchronized spore germination have indicated that RNA synthesis precedes synthesis of protein and DNA. There is an emphasis on rRNA and tRNA synthesis at the earliest stages of germination, with an increasing amount of mRNA synthesis as germination continues (4, 82, 125). Thus, even though the spore contains tRNAs and ribosomes and the ^{32}P-labeling studies of Henner and Steinberg (55) suggested that the tRNAs made during spore outgrowth were the same as those synthesized at other stages of the life cycle, the first commitment upon germination is to produce tRNA and rRNA. Even though tRNA and ribosomes in spores may prove functional in laboratory experiments, dormant spores in the real world are meant to survive very long periods; thus, long-term damage is likely, and a renewal of the translational scaffolding may be necessary to ensure translational fidelity.

tRNA Gene Organization in *B. subtilis*

Comparison with *E. coli* tRNA genes

Transfer RNA gene organization in *Bacillus* spp. differs significantly from that in *E. coli* in that the majority of the gram-positive bacterium's tRNA genes are clustered (141). In *E. coli*, 78 tRNA genes have been sequenced and mapped. They are scattered around the chromosome at 40 different loci (70). *E. coli* tRNA genes have been found as separate transcriptional units and parts of ribosomal operons; they have also been found linked to mRNAs (37). As parts of rRNA operons, the most common site for tRNA genes in *E. coli* is in the spacer region between the 16S and 23S rRNA genes. Three of the seven *E. coli* rRNA operons have genes coding for $tRNA^{Ile}$ and $tRNA^{Ala}$ at this position. The other four rRNA operons have

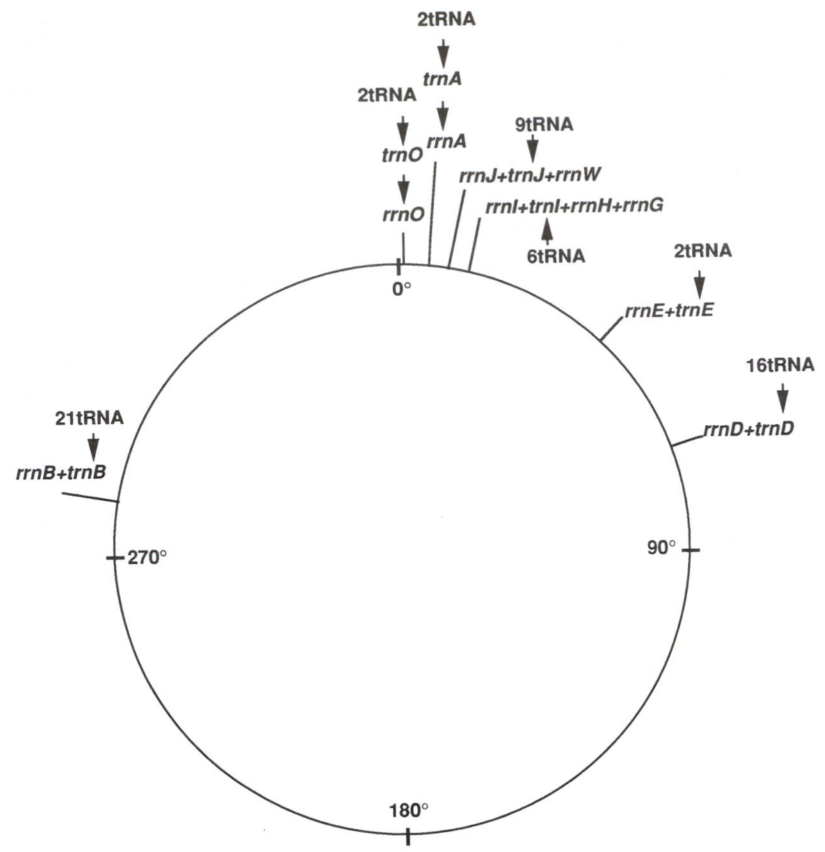

Figure 1. Known map positions of rRNA and tRNA operons on the *B. subtilis* chromosome divided into 360°. Numbers of genes in each tRNA gene cluster are indicated by arrows.

tRNA^{Glu} genes in the same location (91). Some rRNA operons have tRNAs distal to the 23S rRNA gene. One rRNA operon has two tRNA genes, for tRNA^{Asp} and tRNA^{Trp}, at the 3′ end (90). Two other rRNA operons each have a single tRNA gene at the same place (32, 61). The largest tRNA gene cluster in *E. coli* is an independent group of seven in a single transcriptional unit that includes three pairs of repeated genes (94). There are also two clusters, each with four genes, all coding for different tRNAs (57, 76, 89).

Most tRNA genes of *B. subtilis* are closely linked to rRNA operons

Since almost all of the *B. subtilis* tRNA genes are associated with rRNA operons, the organization of rRNA genes is important to the discussion of tRNA gene organization. *B. subtilis* has 10 rRNA operons. All have been mapped (Fig. 1) (64), and one has been completely sequenced (42). *E. coli* and *Salmonella typhimurium* both have only seven rRNA operons (65). The basic rRNA operon organization in *B. subtilis* is the same closely linked 16S-23S-5S pattern found in *E. coli*. Seven of the *B. subtilis* genes are within 1° to 14° on the genomic map (64). This concentration of rRNA operons near the replication origin of the *B. subtilis* chromosome is more extreme than that found in *E. coli*, in which five of the seven operons are in approximately 20% of the genome near the origin of replication (65).

Three *B. subtilis* rRNA operons (*rrnI*, *rrnH*, and *rrnG*) are closely linked and map to 14° on the chromosome. All are transcribed in the same direction (64, 153). The first two are separated by only 794 bp, while the last is only 186 bp from the two upstream rRNA operons (45). The closely linked ribosomal operons *rrnJ* and *rrnW* map to 10° (64, 153) and are only 1,040 bp apart.

Although all *E. coli* rRNA operons have tRNA genes in the 16S-23S spacer region, this is uncommon for *B. subtilis* rRNA operons. Pairs of tRNA genes, for tRNA^{Ile} and tRNA^{Ala}, are found in the 344-bp spacer region between the 16S and 23S rRNA genes in the *B. subtilis* rRNA *rrnO* and *rrnA* operons found near the origin of replication (78, 98), similar to the pattern often seen in *E. coli*. However, Southern hybridization studies indicate that these two are probably the only *B. subtilis* rRNA operons with such spacer tRNA genes. All of the other operons have 16S-23S spacers of approximately 164 bp with no tRNA genes (78).

Clustered tRNA genes in *B. subtilis*

It had long been suspected that the tRNA genes in *B. subtilis* are clustered. Hybridization experiments had shown that the majority of tRNA in *B. subtilis* was transcribed from two areas of the genome (99, 128). Early transcriptional mapping studies had shown that the tRNA genes of *B. subtilis* were clustered into large polycistronic groups (14).

A total of 62 *B. subtilis* tRNA gene sequences in eight clusters coding for 29 different tRNAs have now been reported (Fig. 2). These tRNA gene clusters are usually associated with rRNA genes and are usually named after the preceding rRNA operons (105). The majority of the tRNA genes are in two large groups, one containing 21 and the other containing 16 tRNA genes, immediately following the 3′ ends of ribosomal gene operons (43, 150). The tRNA genes of the largest group have intergenic spacers between 2 and 37 bases long. This tRNA gene cluster (*trnB*) is part of the *rrnB* operon that has been completely sequenced (42) and mapped to 280° on the *B. subtilis* chromosome (18). The 16-tRNA gene cluster has been tentatively mapped to the 3′ end of the *rrnD* operon at 70° (117) and therefore is designated *trnD*.

Another cluster of six tRNA genes (*trnI*) is found in the 894-bp space between the two rRNA operons *rrnI* and *rrnH* (18, 64, 149, 164). The 1040-bp section between *rrnJ* and *rrnW* has recently been found to contain nine tRNA genes (*trnJ*), while the 186-bp sequence between *rrnH* and *rrnG* has none (45). An independent cluster of four tRNA genes (*trnY*) that is unlinked to rRNA operons has been found but not yet mapped (160). In addition, the *rrnE* operon at map position 44° (74) has recently been found to have genes for $tRNA^{Met}$ and $tRNA^{Asp}$ at its 3′ end (117). Therefore, 9 of the 10 rRNA operons in *B. subtilis* have now been found to be immediately adjacent to tRNA genes.

The 62 *B. subtilis* tRNA genes that have been sequenced probably account for almost all such genes in this bacterium's genome. Southern analysis has shown that there are nine *Eco*RI restriction fragments of *B. subtilis* genomic DNA that hybridize to tRNA (149). With the sequencing of the *rrnJ-rrnW* spacer containing nine tRNA genes (45), the six most intensely hybridizing bands have been identified. There remain three other relatively lightly hybridizing *Eco*RI bands that have not yet been sequenced. It is likely that these contain a small number of as-yet-unidentified tRNA genes.

tRNAs and the anticodon complement of *B. subtilis*

An examination of the sequences of the known *B. subtilis* tRNA genes indicates that at least one acceptor for each amino acid has been found (Table 1). However, the known *B. subtilis* tRNA genes do not correspond to all of the *E. coli* anticodons. Determining whether certain tRNA gene products can read multiple codons based on unusual "wobbling" is difficult, since the information on the modified bases at the first anticodon position of *B. subtilis* tRNAs is far from complete. The most extreme type of flexible codon reading in bacteria has been found in *M. capricolum*, a cell wall-less bacterium related to the gram-positive branch, in which most of the four codon sets for a given amino acid can be read by a single tRNA (3, 119). Assuming that modified bases on many *Bacillus* tRNA species can allow for a similar promiscuous pairing at the third position of the codons, the only conventional codons that cannot be read by the gene products of the known tRNA genes are the AGA and AGG arginine codons. Although these arginine codons are used very rarely in *E. coli*, they are used frequently in *B. subtilis* (96). A $tRNA^{Arg}$ with a UCU anticodon capable of translating such codons has been found in *M. capricolum* (3). It is likely that a similar $tRNA^{Arg}$ will be found among the few remaining undiscovered *B. subtilis* tRNA genes.

Although the sequencing of *B. subtilis* tRNA genes is not quite complete, it is obvious that this bacterium's total anticodon repertoire is smaller than that found in *E. coli* (Table 1). It appears that the number of *B. subtilis* anticodons is similar to that for *M. capricolum*. The 78 tRNA genes of *E. coli* (70) and the 30 tRNA genes of *M. capricolum* (92) show that these bacteria have 41 and 28 anticodons, respectively. The 62 known tRNA genes of *B. subtilis* have 28 anticodons. Four anticodons found in *B. subtilis* are not found in *M. capricolum*, i.e., those for Gly (GCC), Leu (CAG), Ser (GGA), and Thr (GGU). Four anticodons found in *M. capricolum* have not yet been found in *B. subtilis*, i.e., those for Arg (UCU), Lys (CUU), Thr (AGU), and Trp (UCA). As discussed above, the missing $tRNA^{Arg}$ is likely to be found in the remaining unsequenced tRNA genes. The $tRNA^{Lys}$ with the unusual CUU anticodon in *M. capricolum* may not actually be used in translation because of some unusual features (3). The *Mycoplasma* $tRNA^{Trp}$ with the UCA anticodon is capable of translating the UGA termination codon. Such a tRNA may explain the ability of *B. subtilis* to translate such codons (79) as discussed above, but its presence in this bacterium has not yet confirmed.

The *M. capricolum* anticodon set is likely to be only slightly smaller than that of *B. subtilis*, perhaps because of the evident selective pressure toward a very high A/T content in *Mycoplasma* spp. that has allowed codons rich in G/C to be greatly reduced in number or even eliminated. The reduced complement of anticodons found in *M. capricolum* compared with *E. coli* may not be due to the small genome size but may instead be typical of gram-positive bacteria. Perhaps the larger anticodon complement of *E. coli* should be considered unusual by comparison with other known bacteria.

$tRNA^{Ile}$ gene with a CAU anticodon

In the 21-tRNA gene cluster, the methionine anticodon CAU is found in the 10th, 11th, and 13th genes in the cluster. The 10th gene codes for the $tRNA^{Met}$ known from tRNA sequencing to be the methionine elongator tRNA (159). The 13th tRNA gene in the cluster codes for the methionine initiator tRNA (160). The 11th gene also has a CAU anticodon but was classified as an isoleucine receptor (43) on the basis of the following. Only two methionine isoacceptors have been found by high-pressure liquid chromatography of *B. subtilis* tRNAs (160). The presence of an A-U base pair in the third position of the aminoacyl stem of this predicted tRNA rules out the tRNA being recognized by the methionyl-tRNA synthetase (123). Phage T4, *E. coli*, and spinach chloroplasts each have an unknown modified nucleoside in the third position of the anticodon of a similar $tRNA^{Ile}$ that may have originated from a C in the original transcript that allows it to recognize AUA codons (38, 67, 73). More recently, it has been shown that a homologous $tRNA^{Ile}$ gene with the same CAU anticodon in *M. capricolum* is posttranscriptionally modified to a lysidine, N^2-(5-amino-5-carboxypentyl)cytidine, residue at the first position so that it can recognize AUA codons (3, 92).

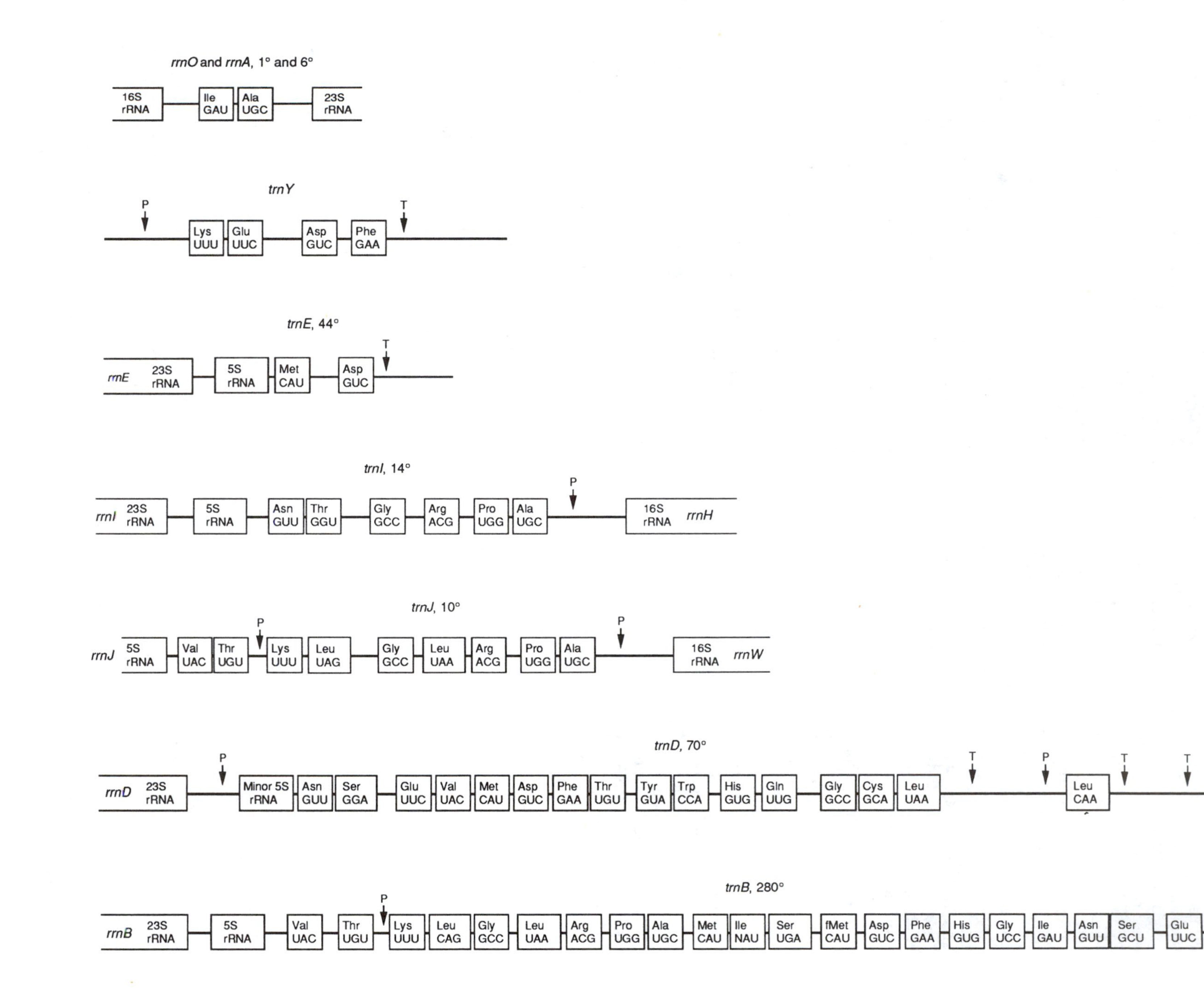
rrnO and rrnA, 1° and 6°
16S rRNA
Ile GAU
Ala UGC
23S rRNA
trnY
P
Lys UUU
Glu UUC
Asp GUC
Phe GAA
T
trnE, 44°
rrnE
23S rRNA
5S rRNA
Met CAU
Asp GUC
T
trnI, 14°
rrnI
23S rRNA
5S rRNA
Asn GUU
Thr GGU
Gly GCC
Arg ACG
Pro UGG
Ala UGC
P
16S rRNA
rrnH
trnJ, 10°
rrnJ
5S rRNA
Val UAC
Thr UGU
P
Lys UUU
Leu UAG
Gly GCC
Leu UAA
Arg ACG
Pro UGG
Ala UGC
P
16S rRNA
rrnW
trnD, 70°
rrnD
23S rRNA
P
Minor 5S rRNA
Asn GUU
Ser GGA
Glu UUC
Val UAC
Met CAU
Asp GUC
Phe GAA
Thr UGU
Tyr GUA
Trp CCA
His GUG
Gln UUG
Gly GCC
Cys GCA
Leu UAA
T
P
Leu CAA
T
T
trnB, 280°
rrnB
23S rRNA
5S rRNA
Val UAC
Thr UGU
P
Lys UUU
Leu CAG
Gly GCC
Leu UAA
Arg ACG
Pro UGG
Ala UGC
Met CAU
Ile NAU
Ser UGA
fMet CAU
Asp GUC
Phe GAA
His GUG
Gly UCC
Ile GAU
Asn GUU
Ser GCU
Glu UUC
TT

Histidine tRNA

In both eukaryotes and prokaryotes, the $tRNA^{His}$ has an extra base, usually a G residue, at the 5′ end that increases the length of the aminoacyl stem from the normal 7 to 8 bp (131). In eukaryotes, this extra base is added posttranscriptionally (27). However, in *E. coli*, the extra G residue is encoded in the gene (57). In both the 16- and the 21-tRNA gene clusters of *B. subtilis*, the $tRNA^{His}$ genes also encode the extra base (43, 150). In both *E. coli* (100) and *B. subtilis* (44), the extra G residue is left on by an unusual cleavage by RNase P, the enzyme that removes the 5′ leader sequence from tRNA precursors, leaving the 8-bp aminoacyl stem found only on this tRNA.

3′-terminal CCA

Most tRNA genes of eubacteria encode the 3′-terminal CCA sequence to which the aminoacyl residue is attached, whereas the tRNA genes found in eukaryotes, including mitochondria and chloroplasts, usually do not encode this feature (131). The CCA sequence can be added posttranscriptionally by a nucleotidyltransferase during the maturation process. All known *E. coli* tRNA genes encode the CCA sequence (131). However, in the 62 known *B. subtilis* tRNA genes, 14 do not encode the CCA end. The lack of the CCA sequence may make processing of the tRNA precursors by RNase P less efficient, since in *E. coli*, the processing of phage T4 tRNA precursors lacking CCA is inhibited (121). In vitro RNase P processing of synthetic *B. subtilis* $tRNA^{His}$ precursors with and without the CCA have shown that the latter substrate has a K_m at least 1 order of magnitude higher than that of the former (44).

tRNA Gene Promoters

Sequencing of many of the *B. subtilis* tRNA genes has revealed a number of possible promoters for σ^A-containing RNA polymerase. σ^A is the principal sigma species in *B. subtilis* and is the species most like *E. coli* σ^{70} in terms of promoter specificity (77). The cluster of four tRNAs in *trnY* (Fig. 2) has the typical sequences for a −35 and −10 promoter region that probably initiates transcription approximately 84 bases 5′ to the first tRNA. A putative terminator consisting of a base-paired stem-loop preceding a T-rich region is present immediately following the fourth tRNA gene. Therefore, this cluster is probably a separate transcriptional unit (160). The $tRNA^{Ile}$-$tRNA^{Ala}$ gene pair between the 16S and 23S genes of two rRNA operons has no obvious promoter or terminator sequences other than those belonging to the ribosomal operon. It is likely that these tRNAs are made as part of a large combined rRNA-tRNA precursor (78, 98).

The rRNA–21-tRNA gene cluster has what appears to be a σ^A promoter between the second and third tRNA genes. The −35 region for this promoter is inside the sequence of the second gene. The only large terminator sequence found in this cluster is after the last tRNA gene (43, 150). The transcription of the rRNA–16-tRNA gene cluster of *B. subtilis* also may be regulated by multiple promoters. The first terminatorlike hairpin structure in the rRNA-tRNA complex is found between the 15th and 16th tRNA genes. A second putative terminator is present at the end of the 16th tRNA gene. This last gene, for $tRNA^{Leu}$CAA, has a good consensus sequence for a σ^A promoter, presumably allowing independent expression. There is another putative promoter immediately 5′ to the 5S gene. This particular gene encodes the minor 5S rRNA (150) originally found from RNA sequencing (111). Southern analysis indicates that this is the only copy of the minor 5S gene in the *B. subtilis* genome (143).

The tRNA gene clusters in *trnI* and *trnJ* found between rRNA sets also have a complex organization. There is an obvious σ^A promoter between the last tRNA gene of each cluster and the second ribosomal cluster. There is also a possible promoter 5′ to the first tRNA gene of *trnI*, although it has a much poorer consensus promoter sequence. A possible attenuator sequence with a short hairpin followed by a T-rich region is found between the second and third tRNA genes of *trnI*. The entire dual rRNA-tRNA cluster may constitute a single operon that can be transcribed from the promoter of the first ribosomal gene set. The putative attenuator of *trnI* may reduce transcription of the last four tRNAs and the second ribosomal cluster (149). The *trnJ* 9-tRNA gene cluster (45) seems to be a combination of the *trnI* and *trnB* clusters. The gene order for the nine tRNAs in *trnJ* is the same as in the first 9 genes of the 21 genes of *trnB* except that the gene for $tRNA^{Leu}$CAG of the latter has been converted to a $tRNA^{Leu}$UAG by six base changes and one deletion in the variable loop. This tRNA gene cluster also has the same promoter as *trnB*; the promoter is located between the second and third tRNA genes. The sequence from the last three tRNA genes, including the intergenic spacers and the 3′-distal rRNA promoter of *trnJ*, is almost identical to that of *trnI*. This indicates a recent evolutionary origin, perhaps from a fusion of the 3′ half of a gene cluster similar to *trnI* with the 5′ half of a *trnB*-like operon.

Simple visual inspections of DNA sequences can easily miss promoters that do not conform to known consensus sequences. This is especially true in *B. subtilis*, which has many types of promoters. The cloning of the *B. subtilis trnB* 21-tRNA gene cluster into *E. coli* has demonstrated that the internal promoter between the second and third tRNA genes can direct transcription of the downstream genes. The *E. coli* host is also capable of correctly processing these *B. subtilis* tRNAs into functional form. S1 nuclease mapping with RNAs from *B. subtilis* and from an *E. coli* host bearing a plasmid with the cloned *B. subtilis* 21-tRNA cluster demonstrated that the putative promoter was active in both and initiated transcription at a position 14 or 15 nucleotides 5′ to the third tRNA. Readthrough from the ribosomal promoter also oc-

Figure 2. Gene orders of known *B. subtilis* tRNA gene clusters. Each tRNA gene is represented by a rectangle enclosing its cognate amino acid and anticodon. Map positions, where known, are listed above each operon. Putative promoters and terminators are labeled P and T, respectively.

Table 1. *E. coli* tRNA anticodons compared with *B. subtilis* and *M. capricolum* anticodons

E. coli tRNA	Anti-codon	Loci of corresponding *B. subtilis* tRNA operons[a]	Found in *M. capricolum*
Ala	GGC	—	No
	UGC	*rrnO, rrnA, trnB, trnI, trnJ*	Yes
Arg	ACG	*trnB, trnI, trnJ*	Yes
	CCG	—	No
	CCU	—	No
	UCU	—	Yes
Asn	GUU	*trnB, trnD, trnI*	Yes
Asp	GUC	*trnB, trnD, trnE, trnY*	Yes
Cys	GCA	*trnD*	Yes
Gln	CUG	—	No
	UUG	*trnD*	Yes
Glu	UUC	*trnB, trnD, trnY*	Yes
Gly	CCC	—	No
	GCC	*trnB, trnD, trnI, trnJ*	No
	UCC	*trnB*	Yes
His	GUG	*trnB, trnD*	Yes
Ile	GAU	*rrnO, rrnA, trnB*	Yes
	NAU	*trnB*	Yes
Leu	CAA	*trnD*	Yes
	CAG	*trnB*	No
	GAG	—	No
	UAA	*trnB, trnD, trnJ*	Yes
	UAG	*trnJ*	Yes
Lys	UUU	*trnB, trnJ, trnY*	Yes
Met	CAU	*trnB, trnD, trnE*	Yes
fMet	CAU	*trnB*	Yes
Phe	GAA	*trnB, trnD, trnY*	Yes
Pro	CGG	—	No
	GGG	—	No
	UGG	*trnB, trnI, trnJ*	Yes
Ser	CGA	—	No
	GCU	*trnB*	Yes
	GGA	*trnD*	No
	UGA	*trnB*	Yes
Thr	CGU	—	No
	GGU	*trnI*	No
	UGU	*trnB, trnD, trnJ*	Yes
Trp	CCA	*trnD*	Yes
Tyr	GUA	*trnD*	Yes
Val	GAC	—	No
	UAC	*trnB, trnD, trnJ*	Yes
Se-Cys	UCA	—	No

[a] —, codon not yet found.

curs in *B. subtilis*, indicating that the entire cluster can be transcribed as part of the ribosomal operon, with additional independent transcription of the last 19 tRNA genes (142, 146).

Transcriptional mapping of the rRNA–16-tRNA gene cluster from *B. subtilis* proved that both the minor 5S promoter and the promoter for the last tRNA gene in the cluster are active (143). The minor 5S gene and the downstream tRNA genes can be transcribed from the promoter immediately upstream of the 5S and by readthrough, presumably from the promoter at the beginning of the ribosomal operon. The minor 5S promoter was active during exponential growth and sporulation stages. The promoter for the last tRNA gene, $tRNA^{Leu}$CAA, also was active during both exponential growth and sporulation. A considerable amount of transcription from upstream also occurred, indicating that the putative terminator 5′ to this promoter was not completely effective. These data indicate that the entire rRNA-tRNA cluster can be transcribed as a single precursor, although processing probably begins before the transcript is finished. Additional transcription from the minor 5S promoter can independently increase the amount of the minor 5S and downstream tRNAs. The $tRNA^{Leu}$-CAA gene at the end of the cluster can be transcribed independently and as part of larger precursors. Therefore, this last tRNA gene can be transcribed from at least three different promoters (143).

Evolution of tRNA Gene Clusters

Bacterial tRNA gene sequences are highly conserved, even between gram-positive organisms and the purple bacterium-pseudomonad group, but the tRNA gene order may be useful for evolutionary comparisons of related bacterial species. Recent sequencing of tRNA gene clusters from mycoplasmas (class *Mollicutes*) has revealed some interesting homologies. These cell wall-less bacteria have extremely small genomes 600 to 1,200 kb in size and have been grouped with low-G+C gram-positive bacteria on the basis of their 16S and 5S rRNA sequences (115, 151). In *Spiroplasma melliferum* (116), *Mycoplasma mycoides* (118), and *M. capricolum* (92), tRNA gene clusters containing isoacceptors for Arg, Pro, Ala, Met, Ile, Ser, fMet, Asp, and Phe exactly duplicate the gene order found in a portion of the *B. subtilis* 21-tRNA gene cluster. These clusters are apparently not associated with rRNA operons, and the *Spiroplasma* cluster is preceded by a $tRNA^{Cys}$ gene not found in the *B. subtilis* 21-tRNA gene cluster (116). Virtually all of the tRNAs and tRNA genes have now been sequenced for *M. capricolum*, the first bacterium for which this has been accomplished (3, 92). This bacterium has only 30 genes coding for 29 different tRNA species with 28 different anticodons, largely because of very relaxed codon recognition requirements in *Mycoplasma* spp. similar to those found in mitochondria and chloroplasts (119). *M. capricolum* has several other small clusters of tRNA genes that appear to have preserved the gene orders seen in *B. subtilis* (92). The sequences of the tRNA genes themselves are approximately 85% identical to those of the corresponding *B. subtilis* genes, although there is no similarity in the spacer regions. All of the known *Mycoplasma* tRNA genes encode the 3′-terminal CCA sequence (3, 92, 116, 118).

This conservation of tRNA gene order indicates that the *Mycoplasma* class of bacteria may have at least partially preserved the tRNA gene clustering found in a common ancestor of many of the other gram-positive bacteria. It is less clear whether *Bacillus* spp. have preserved their extensive tRNA gene clustering from a distant ancestor or whether there has been selective pressure to further increase this type of clustering in the *Bacillus* line. In either case, it is evident that powerful evolutionary constraints maintain tRNA gene orders in the gram-positive bacteria.

Genomic mapping of *Clostridium perfringens* indicates that this anaerobic spore-forming gram-positive bacterium has nine ribosomal operons with seven grouped within one third of the genome. The tRNA genes of this bacterium are also closely linked to the ribosomal operons, perhaps indicating that it has tRNA gene clustering similar to that of *B. subtilis* (23).

Recent restriction enzyme mapping of the more distantly related high-G+C gram-positive *Mycobacteria* spp. of the actinomycete group has indicated that these bacteria also have clustered tRNA genes associated with rRNA genes (11). DNA sequencing of another actinomycete, *Micrococcus luteus*, has shown that many of its tRNA genes are on separate transcriptional units; however, one group of four genes in one operon has been found. Most of these tRNA genes do not encode the 3′-terminal CCA sequence (59, 60).

Not all of the gram-positive bacteria have similar tRNA gene orders. The four ribosomal gene operons of *Streptococcus pneumoniae* have a single $tRNA^{Ala}$ in the spacer region between the 16S and 23S genes compared with the $tRNA^{Ile}$-$tRNA^{Ala}$ genes known in two such operons on *B. subtilis* (7). A six-member tRNA gene cluster has been found at the 3′ end of an rRNA operon in *Lactobacillus bulgaricus* (also known as *Lactobacillus delbrueckii*) (106). This tRNA gene cluster has the isoacceptor gene order Asn, Pro, Gly, Arg, Val, Asp. None of the genes encodes the CCA 3′-terminal sequence. There is no obvious similarity with the tRNA gene orders of the known clusters in *B. subtilis*. A transcriptional unit for a single gene of $tRNA^{Ser}$CGA has also been found in *L. bulgaricus* (56). No such tRNA gene has yet been found in *B. subtilis*.

A very different type of tRNA gene clustering has been found in *Photobacterium phosphoreum*, a member of the *Vibrionaceae* family. This gram-negative enteric marine bacterium has a cluster consisting of six $tRNA^{Pro}$ and two $tRNA^{His}$ genes interspersed by $tRNA^{Pro}$ pseudogenes. A second, smaller cluster that was apparently derived from the larger cluster is also present (40).

The reason for the extensive tRNA gene clustering in *B. subtilis* is not known. Although the clustering may provide a convenient way of coordinating transcription of tRNA genes or of processing tRNA precursors, it is obvious that *E. coli* does not need the same type of organization. One possible role of the clusters may be in spore germination, when a rapid production of the protein synthesis machinery may be required. If this is the case, then an analysis of tRNA gene organizations from other more closely related gram-positive bacteria, both spore forming and nonspore forming, may explain the situation. Since at least part of the tRNA gene order of the 21-gene cluster of *B. subtilis* has been preserved in *Mycoplasma* spp. despite the large evolutionary distance between these organisms, it is likely that many other more closely related gram-positive bacteria have similar clusters. Tracking this tRNA gene cluster through bacterial phylogenies may provide information on its function.

tRNA PROCESSING

RNase P in Gram-Positive Bacteria

The enzymatic activity known as RNase P is responsible for the 5′-end maturation of tRNA precursors by endonucleolytic cleavage of the leader sequence, leaving a 3′-hydroxyl and a 5′-phosphate. This enzyme from both *E. coli* and *B. subtilis* is a complex of a protein and an RNA approximately 400 residues long (39, 69, 132). *B. subtilis* cells contain approximately 20 to 50 copies of this ribonuclear enzyme (113). The enzymes from both bacteria are inactivated by micrococcal nuclease, indicating that the RNA component was essential for the enzyme's activity (39, 132). At first, it was suggested that the RNA subunit of RNase P was involved only in recognizing the tRNA precursor (112). However, it was later found that the RNA component from both the *E. coli* and the *B. subtilis* RNase P enzymes (M1 RNA and P-RNA, respectively) was the actual catalyst for the reaction (48). This was the first demonstration of an RNA molecule acting as a catalyst in an in vivo reaction.

The gene coding for the RNase P protein subunit (*rnpA*) maps to the replication origin region in both *E. coli* and *B. subtilis*, an area with significant similarities in gene organization in the two bacteria (53, 97). Both genes code for highly basic 119-amino-acid residue peptides; however, they have only 24.6% sequence similarity (97). The *rnpB* gene of *B. subtilis* has not yet been mapped (105). The same gene from *E. coli* has been mapped to 68 min on the chromosome (71).

The RNase P protein components of both *E. coli* and *B. subtilis* enable the holoenzyme to cleave its substrates at low-ionic-strength conditions of 60 mM NH_4Cl and 10 mM $MgCl_2$ (48). The RNase P protein has no catalytic activity by itself. Without the protein, the *B. subtilis* P-RNA requires 800 mM NH_4Cl and 100 mM $MgCl_2$ for optimal activity against most substrates. Under these high-ionic-strength conditions, although the P-RNA alone has catalytic activity, the cleavage rate (k_{cat}) is approximately 5% that of the holoenzyme because of a slower dissociation of the cleaved substrate from the catalyst (114). It appears that the high-ionic-strength buffer serves to screen the anionic repulsion that would normally be expected to occur between negatively charged substrate and catalyst. However, this ionic screening by the buffer is nonspecific, resulting in the cleaved reaction product loitering longer in the binding site. The protein subunit of the holoenzyme may be able to provide specific screening of anionic repulsion in certain areas while leaving other areas exposed that allow for a more rapid ejection of the product after cleavage (102, 114). Another possible role of the RNase P protein may be to stabilize an active conformation of the catalytic RNA subunit. This is suggested by the fact that many mutations in M1 RNA that eliminate or reduce RNA-only catalysis can be compensated for by the presence of the RNase P protein (80).

The *B. subtilis* P-RNA is slightly larger, at 401 nucleotides, than the *E. coli* M1 RNA, which is 377 nucleotides. Sequence comparisons of the RNase P-RNA subunits from *E. coli* and *B. subtilis* revealed only 43% similarity (112, 113). This was not sufficient for reliable identification of conserved residues. In order to perform a phylogenetic comparison to predict secondary structures for the P-RNA, sequences from seven different bacterial species, including the gram-positive organisms *B. subtilis*, *B. stearothermophilus*, *B. megaterium*, *Bacillus brevis*, and members of the "purple bacteria group" to which *E. coli* belongs, were compared (63, 103). The RNase P-RNA sequences could all be folded up into a similar structure, with phylum-specific stems and loops inserted at various locations. To prove that this model was correct, a synthetic, simplified, composite P-RNA of 263 nucleotides that combined the conserved features found in the gram-positive bacteria (using *B. megaterium* sequences) and the *E. coli* RNA and deleted any other structures that seemed to be specific to either group was constructed. This catalytic RNA had a specificity identical to those of the native ribozymes but a K_m approximately 2 orders of magnitude higher and a requirement for a higher-ionic-strength buffer for maximal activity. The synthetic catalytic RNA's activity was not affected by the presence of the RNase P protein (147).

Secondary Structure of RNase P

The secondary structural model for RNase P-RNA has been further refined by comparison of 16 different eubacterial sequences (22, 50, 63, 103) and comparisons of the catalytic efficiencies of other synthetic composite RNAs (29). The latest model for the catalytic RNAs from a large number of bacteria indicates that the gram-positive catalytic RNAs are unusual in that they lack an important pseudoknot structure found in all other bacterial RNase P-RNAs (50). Even a recently sequenced archaebacterial RNase P-RNA may have this same feature seen in RNase Ps from eubacteria other than gram-positive organisms (95). Mutational analyses have shown that the lack of this pseudoknot was probably responsible for most of the increase in the K_m of the synthetic *B. megaterium-E. coli* RNase P-RNA hybrid molecule (29). Since the gram-positive P-RNAs have K_ms comparable to those of the other eubacterial catalytic RNAs, other parts of their structures may compensate. Because of other structural features present, it is possible that gram-positive-bacterium RNA in this region has a tertiary structure that is very similar to that of the pseudoknot in the catalytic RNAs from other bacteria (50).

Although in vitro mixing of the catalytic RNAs and RNase P proteins can help demonstrate that there is a basic conservation of RNase P function across a large evolutionary distance, in vivo experiments are required to discover more subtle differences that may be due to group-specific inserts in the basic secondary structure model. Waugh and Pace (148) have constructed a strain of *E. coli* that has an *rnpB* chromosomal deletion and also has an *rnpB* gene on a plasmid that is temperature sensitive for replication. The conditionally lethal strain could be complemented by plasmids containing *rnpB* genes from several other bacteria, including *Alcaligenes euthrophus*, *B. subtilis*, and *Chromatium vinosum*. However, the RNase P-RNA genes from *Agrobacterium tumefaciens*, *Pseudomonas fluorescens*, *B. brevis*, *B. megaterium*, and *B. stearothermophilus* could not complement the *rnpB* deletion. Apparently, in some of these combinations, the lack of phylum-specific stems and loops in the catalytic RNA was lethal, and in others, the lack was at least marginally functional. These results imply that phylum-specific RNase P-RNA stems and loops may not be absolutely required for at least some in vivo functioning in all cases. Perhaps differences in the catalytic RNA sequences in some cases are responsible only for relatively minor changes in tRNA precursor processing efficiency or for interaction with other enzymes.

RNase P Recognition Site

E. coli M1-RNA has been shown to cleave modified tRNA-like substrates that consist of only the amino acid acceptor stem and the T-stem and loop (85). Chemical modifications of precursor tRNA substrates block cleavage by the catalytic RNase P-RNAs from either *E. coli* or *B. subtilis* if the lesions are at positions on the amino acid acceptor stem and the T-stem and loop (135). This indicates that the catalytic RNA binds to the upper portion of the inverted L-shaped tRNA tertiary structure, the most highly conserved structure of tRNAs formed from the aminoacyl stem and the ribothymidine stem and loop. This explains how RNase P can recognize all of its tRNA precursor substrates and contrasts with the aminoacyl-tRNA synthetases that recognize many different nucleotide positions scattered throughout the tRNA (120).

RNase P Reaction Mechanism

Little is known about the exact cleavage mechanism of RNase P. However, it is clear that the mechanism is different from the self-splicing intron reaction in *Tetrahymena* rRNA or the self-cleavage seen in many plant virusoids. Although the self-splicing intron also generates 5'-phosphate and 3'-hydroxyl ends, it requires the 3'-hydroxyl of a guanosine cofactor (163). RNase P has no cofactor requirement, and the periodate oxidation of the 3' end of the RNase P RNA and its precursor tRNA substrates does not affect the reaction (83). The self-cleaving virusoid RNAs generate 5'-hydroxyls and 2',3'-phosphates (36). It has been proposed that the catalytic RNA of RNase P activates a water molecule by abstracting a proton with one of the nucleotide bases or one of the phosphodiester groups (102).

Why Is RNase P a Catalytic RNA?

The explanation of why RNase P uses a catalytic RNA can lead to several interesting speculations. One is that this enzyme is a remnant of an "RNA world" in early evolution during which many enzymatic functions were carried out by nucleic acids. In the bacterial RNase P molecules known, the RNA catalyst has never been replaced by a purely protein-based enzyme, perhaps because of significant evolutionary constraints. However, this explanation is unsatisfactory, since the RNase P-RNA sequences are so poorly conserved, especially when compared with those of rRNA, a molecule of undeniably ancient origin. Although RNase P has a requirement to recognize a wide range of similar but not identical substrates, this problem has been solved by many other protein-based enzymes. In general, it is obvious either that the RNase P must perform its catalysis more efficiently than any hypothetical protein-based nuclease or that this RNA-based enzyme has other functions besides the obvious role in performing a single type of cleavage. In vivo, enzymes do not exist in isolation. The processing of the tRNA precursors may involve a multienzyme complex. If RNase P is part of such a complex, perhaps it functions as a scaffold to which other protein-based enzymes are associated. In vivo interactions with other proteins, perhaps a complex of other tRNA precursor processing enzymes, may explain the presence of phylum-specific inserts in the otherwise conserved secondary structure of this catalytic RNA.

AMINOACYL-tRNA SYNTHETASES

Aminoacyl-tRNA synthetases catalyze the charging of amino acids to tRNA. This two-step reaction is performed by using ATP to create an aminoacyl adenylate-enzyme complex that is then used to attach the amino acid residue to the terminal 3′-adenosine on the specific tRNA (35). The synthetases naturally play an important role in protein metabolism and cell growth. There is a large amount of information on the regulation of expression and genetics of the aminoacyl-tRNA synthetases in prokaryotes. However, most of this is from *E. coli* (46, 120).

There are significant functional homologies between the aminoacyl-tRNA synthetases of *Bacillus* spp. and *E. coli*, as demonstrated by several examples of the cloning by complementation of *Bacillus* synthetase genes into *E. coli* hosts with temperature-sensitive mutations (9, 19, 20, 66, 88). It is also apparent that there is a high degree of structural similarity, from the amino acid sequence to the overall three-dimensional structure, between aminoacyl-tRNA synthetases for the same amino acid from all species (120). The four basic types of aminoacyl synthetase quaternary structures are a monomer, a homodimer, a homotetramer, and a tetramer consisting of two types of subunits (120). The subunit length for prokaryotic synthetases ranges from 303 to 939 amino acid residues. There are no significant differences in size between the *E. coli* and *B. subtilis* charging enzymes for the same amino acid (120). The most unusual difference between the aminoacyl synthetases of *B. subtilis* and *E. coli* is that the former lacks an enzyme capable of directly charging a tRNA with glutamine (154).

X-Ray Crystal Structure of Aminoacyl-tRNA Synthetases

Several aminoacyl-tRNA synthetases have been crystallized and analyzed by X-ray diffraction. This work is often difficult because of problems in obtaining good-quality crystals. One of the easiest synthetases to crystallize has been the tyrosyl-tRNA synthetase from *B. stearothermophilus*. Consequently, this is one of the synthetases whose higher-order structure is best understood. It is a dimer with two identical peptides of 419 residues, and its sequence is 56% similar to that of the corresponding enzyme from *E. coli* (157).

X-ray crystallographic data indicate that the *B. stearothermophilus* tyrosyl-tRNA synthetase has a three-dimensional structure that is very similar to that of the well-studied *E. coli* methionyl-tRNA synthetase. These two enzymes have a similarly placed "mononucleotide-binding fold," even though their amino acid sequences have only slight similarities in this region (12, 15). However, these similarities do not extend to all of the other aminoacyl-tRNA synthetases (157).

Aminoacyl-tRNA Synthetase Genes

Most of the aminoacyl-tRNA synthetase genes from *E. coli* have been cloned, sequenced, and mapped (46). Because aminoacyl-tRNA synthetases are so critical to the cell's metabolism, understanding their genetic regulation is especially important. Interestingly, these genes in *E. coli* appear to have several different mechanisms by which to regulate their expression. The *E. coli* alanyl-tRNA synthetase has a negative feedback regulation by the protein binding near the transcription initiation site of its gene (107). The phenylalanyl-tRNA synthetase from the same bacterium is regulated by an attenuation mechanism (34). The *E. coli* threonyl-tRNA synthetase can bind to its own mRNA to inhibit translation. The enzyme apparently recognizes a region 5′ to the coding sequence of the mRNA, which has a secondary structure that mimics the anticodon stem and loop of $tRNA^{Thr}$ (129).

Since this class of genes in *E. coli* is so rich in regulatory mechanisms, it is likely that much can be learned about analogous mechanisms in gram-positive bacteria by the study of aminoacyl-tRNA synthetase genes. Unfortunately, work in this area is just beginning, but we can expect that more data will be available soon.

The *B. subtilis* aminoacyl-tRNA synthetase genes for glutamate, threonine, tryptophan, and tyrosine have been cloned and sequenced (21, 26, 41, 47, 108). Both of the *B. subtilis* phenylalanyl-tRNA synthetase subunit genes have been cloned by complementation of a temperature-sensitive *E. coli* mutant (19). Five synthetase genes from *B. stearothermophilus* (those for glutamate, methionine, tryptophan, tyrosine, and valine) have been sequenced (10, 17, 21, 87, 157). In *Bacillus caldotenax*, the tyrosyl-tRNA synthetase gene has been cloned and sequenced (66). For the *Bacillus* aminoacyl-tRNA synthetase genes that have been sequenced, the predicted amino acid similarity with the corresponding *E. coli* peptides ranges from 32 to 56%.

Aminoacyl-tRNA synthetases have been divided into two classes on the basis of the presence of several different stretches of amino acid sequences (33). Class I synthetases include those for the amino acids arginine, glutamine, glutamate, isoleucine, leucine, methionine, tryptophan, tyrosine, and valine. This grouping is based on the presence of an at least partially conserved histidine-isoleucine-glycine-histidine (HIGH) sequence and a lysine-methionine-serine-lysine-serine (KMSKS) sequence. Class II includes the aminoacyl-tRNA synthetases for aspartate, asparagine, histidine, lysine, phenylanine, proline, serine, and threonine. Enzymes in this class lack the distinguishing sequences of class I and have several less conserved consensus stretches of amino acid sequences. The *Bacillus* aminoacyl-tRNA synthetases also appear to fit into these two classes, with the class I enzymes usually having some variation of the conserved HIGH and KMSKS sequences. Site-directed mutagenesis and crystallography of the *B. stearothermophilus* tyrosyl-tRNA synthetase (a class I enzyme) indicate that the first histidine residue in the HIGH sequence binds to the γ-phosphate of ATP during its hydrolysis. The second histidine residue in the HIGH sequence may bind to the ribose of the ATP (35). Mutagenizing this position to an asparagine in the *B. stearothermophilus* enzyme does not affect its kinetic parameters. The *B. caldotenax* tyrosyl-tRNA synthetase naturally has an arginine in the same position, indicating that either histidine or asparagine can function at this point (66). This amino acid residue may not function similarly in all class I enzymes, however, since in both the *E. coli* and *B. stearothermophilus* glutamyl-tRNA synthetases, the characteristic HIGH sequence is replaced with a histidine-isoleucine-glycine-glycine sequence. Work with the *E. coli* methionyl-tRNA synthetase has shown that the second lysine in the conserved KM-SKS sequence stabilizes the transition state during formation of the aminoacyl adenylate (87).

Threonyl-tRNA synthetase

B. subtilis has two distinct threonyl-tRNA synthetase genes, *thrS* and *thrZ*, that map at 250° and 344° on the chromosome. Only the first synthetase gene is normally expressed. These two genes were originally known as *thrSv* and *thrS2*, respectively (108). The second gene can be induced by blocking the expression of the first. Both *thrS* and *thrZ* mutant strains appear to grow and sporulate equally well. The two proteins have only 51.5% sequence similarity, just slightly better than the 42 to 47% similarity with the *E. coli* threonyl synthetase (108). This indicates that the divergence of these two forms of *B. subtilis* threonyl synthetases is quite ancient, but at present, the role of the *thrZ* gene product is not known.

A similar situation may exist with the *B. subtilis* tyrosyl-tRNA synthetase genes. There are two genes, *tyrS* and *tyrZ*, for different forms of this enzyme (41, 47). The *tyrS* gene maps to 263° on the *B. subtilis* chromosome (47). The *tyrZ* gene is not expressed in vegetative cells unless the *tyrS* gene is inactivated. Expression of tyrosyl-tRNA synthetase can be induced by starvation for either tyrosine or tryptophan, indicating that the aminoacyl-synthetase genes for these two amino acids may be coordinately regulated by a transcription antitermination mechanism (54).

Lack of Glutaminyl-tRNA Synthetase in Gram-Positive Bacteria

The most unusual aspect of the amino acid-tRNA synthetases of gram-positive bacteria compared with those of *E. coli* is the complete lack of an enzyme capable of charging a tRNA with glutamine (154). In *B. subtilis*, *B. megaterium*, and *Lactobacillus acidophilus*, $tRNA^{Gln}$ is first charged with glutamate, which is subsequently converted to glutamine by an amidotransferase while it is attached to the tRNA (155, 156). Although this $tRNA^{Gln}$-charging mechanism seems unusual when compared with what happens in *E. coli* or the cytoplasm of eukaryotes, it apparently also takes place in a wide range of different cells and organelles, such as the cyanobacterium genus *Synechocystis* (122), archaebacteria (49, 152), the chloroplasts of the green alga *Chlamydomonas reinhardtii* (25, 62), the chloroplasts of higher plants (122), and mammalian (122) and yeast (84) mitochondria. The *B. subtilis* amidotransferase that converts the glutamyl-$tRNA^{Gln}$ to glutaminyl-$tRNA^{Gln}$ requires ATP as a cofactor and favors free glutamine over asparagine as a nitrogen donor. This activity is distinct from that of the *B. subtilis* glutamine synthetase that converts free glutamate into glutamine (133). In *B. subtilis*, a single glutamyl-tRNA synthetase is responsible for aminoacylating both $tRNA^{Glu}$ and $tRNA^{Gln}$ with glutamate. The same *Bacillus* enzyme can also efficiently attach glutamate to *E. coli* $tRNA_1^{Gln}$ in vitro but not to *E. coli* $tRNA^{Glu}$ and $tRNA_2^{Gln}$ (75). Sequence analysis of the glutamyl-tRNA synthetase gene from both *B. subtilis* and *B. stearothermophilus* indicates that there is a region of significant similarity with the *E. coli* glutaminyl-tRNA synthetase in an area that is believed to bind to the anticodon region of their tRNA substrates. This may account for the ability of the *Bacillus* glutamyl-tRNA synthetases to recognize $tRNA^{Gln}$ (21).

NOTE ADDED IN PROOF

We have recently discovered a large rRNA-tRNA operon in *Staphylococcus aureus* that has 27 tRNA genes immediately 3′ to an rRNA operon (Green and Vold, unpublished results). Using PCR on a yeast artificial chromosome library of *B. subtilis* genomic DNA, the *trnY* and *rnpB* genes of *B. subtilis* have been located near the origin and at approximately 200°, respectively (personal communication, K. Okamoto and B. Vold).

Acknowledgments. This work was supported by National Science Foundation grant DMB-91-96116 and National Institutes of Health grant GM29231-10A1.

REFERENCES

1. **Adachi, T., H. Yamagata, N. Tsukagoshi, and S. Udaka.** 1990. Use of both initiation sites of the middle wall protein gene in *Bacillus brevis*. *J. Bacteriol.* **172:**511–513.
2. **Agris, P. F., H. Koh, and D. Söll.** 1973. The effect of growth temperature on the *in vivo* ribose methylation of *Bacillus stearothermophilus* transfer RNA. *Arch. Biochem. Biophys.* **154:**277–282.
3. **Andachi, Y., F. Yamao, A. Muto, and S. Osawa.** 1989. Codon recognition patterns as deduced from sequences of the complete set of transfer RNA species in *Mycoplasma capricolum*. *J. Mol. Biol.* **209:**37–54.
4. **Armstrong, R. L., and N. Sueoka.** 1968. Phase transitions in ribonucleic acid synthesis during germination of *Bacillus subtilis* spores. *Proc. Natl. Acad. Sci. USA* **59:**153–160.
5. **Arnold, H., and H. Kersten.** 1973. The occurrence of ribothymidine, 1-methyladenosine, methylated guanosines and the corresponding methyltransferases in *E. coli* and *Bacillus subtilis*. *FEBS Lett.* **36:**34–38.
6. **Arnold, H. J., W. Schmidt, and H. Kersten.** 1975. Occurrence and biosynthesis of ribothymidine in tRNA's of *B. subtilis*. *FEBS Lett.* **52:**62–65.
7. **Bacot, C. M., and R. H. Reeves.** 1991. Novel tRNA gene organization in the 16S-23S intergenic spacer of *Streptococcus pneumoniae* rRNA gene cluster. *J. Bacteriol.* **173:**4234–4236.
8. **Balassa, G.** 1966. Renouvellement des ARN et des proteines au cours de la sporulation de *Bacillus subtilis*. *Ann. Inst. Pasteur* **110:**316–346.
9. **Barker, D.** 1982. Cloning and amplified expression of the tyrosyl-tRNA synthetase genes of *Bacillus stearo-*

thermophilus and *Escherichia coli*. *Eur. J. Biochem.* **125:**357–360.

10. **Barstow, D. A., A. F. Sharman, T. Atkinson, and N. P. Minton.** 1986. Cloning and complete nucleotide sequence of the *Bacillus stearothermophilus* tryptophanyl tRNA synthetase gene. *Gene* **46:**37–45.
11. **Bhargava, S., A. K. Tyagi, and J. S. Tyagi.** 1990. tRNA genes in mycobacteria: organization and molecular cloning. *J. Bacteriol.* **172:**2930–2934.
12. **Bhat, T. N., D. M. Blow, and P. Brick.** 1982. Tyrosyl-tRNA synthetase forms a mononucleotide-binding fold. *J. Mol. Biol.* **158:**699–709.
13. **Björk, G. R., J. U. Ericson, C. E. D. Gustafsson, T. G. Hagervall, Y. H. Jönsson, and P. M. Wikström.** 1987. Transfer RNA Modification. *Annu. Rev. Biochem.* **56:** 263–287.
14. **Bleyman, M., M. Kondo, N. Hecht, and C. Woese.** 1969. Transcriptional mapping: functional organization of the ribosomal and transfer ribonucleic acid cistrons in the *Bacillus subtilis* genome. *J. Bacteriol.* **99:**535–543.
15. **Blow, D. M., T. N. Bhat, A. Metcalfe, J. L. Risler, S. Brunie, and C. Zelwer.** 1983. Structural homology in the amino-terminal domains of two aminoacyl-tRNA synthetases. *J. Mol. Biol.* **171:**571–576.
16. **Bonamy, C., L. Hirschbein, and J. Szulmajster.** 1973. Synthesis of ribosomal ribonucleic acid during sporulation of *Bacillus subtilis*. *J. Bacteriol.* **113:**1296–1306.
17. **Borgford, T. J., N. J. Brand, T. E. Gray, and A. R. Fersht.** 1987. The valyl-tRNA synthetase from *Bacillus stearothermophilus* has considerable sequence homology with the isoleucyl-tRNA synthetase from *Escherichia coli*. *Biochemistry* **26:**2480–2486.
18. **Bott, K. F., G. C. Stewart, and A. G. Anderson.** 1984. Genetic mapping of cloned ribosomal RNA genes, p. 19–34. *In* A. T. Ganesan and J. A. Hoch (ed.), *Genetics and Biotechnology of the Bacilli*. Academic Press, Inc., New York.
19. **Brakhage, A. A., H. Putzer, H. K. Shazand, R. J. Röschenthaler, and M. Grunberg-Manago.** 1989. *Bacillus subtilis* phenylalanyl-tRNA synthetase genes: cloning and expression in *Escherichia coli* and *B. subtilis*. *J. Bacteriol.* **171:**1228–1232.
20. **Brand, N. J., and A. R. Fersht.** 1986. Molecular cloning of the gene encoding the valyl-tRNA synthetase from *Bacillus stearothermophilus*. *Gene* **44:**139–142.
21. **Breton, R., D. Watson, M. Yaguchi, and J. Lapointe.** 1990. Glutamyl-synthetases of *Bacillus subtilis* 168T and of *Bacillus stearothermophilus*: cloning and sequencing of the *gltX* genes and comparison with other aminoacyl-tRNA synthetases. *J. Biol. Chem.* **265:**18248–18255.
22. **Brown, J. W., E. S. Haas, B. D. James, D. A. Hunt, and N. R. Pace.** 1991. Phylogenetic analysis and evolution of RNase P RNA in proteobacteria. *J. Bacteriol.* **173:**3855–3863.
23. **Canard, B., and S. T. Cole.** 1989. Genome organization of the anaerobic pathogen *Clostridium perfringens*. *Proc. Natl. Acad. Sci. USA* **86:**6676–6680.
24. **Cerutti, P., J. W. Holt, and N. Miller.** 1968. Detection and determination of 5,6-dihydrouridine and 4-thiouridine in transfer ribonucleic acid from different sources. *J. Mol. Biol.* **34:**505–518.
25. **Chen, M. W., D. Jahn, A. Schön, G. P. O'Neill, and D. Söll.** 1990. Purification and characterization of *Chlamydomonas reinhardtii* chloroplast glutamyl-tRNA synthetase, a naturally misacylating enzyme. *J. Biol. Chem.* **265:**4054–4057.
26. **Chow, K., and J. T. Wong.** 1988. Cloning and nucleotide sequence of the structural gene coding for *Bacillus subtilis* tryptophanyl-tRNA synthetase. *Gene* **73:**537–543.
27. **Cooley, L., B. Appel, and D. Söll.** 1982. Post-transcriptional nucleotide addition is responsible for the formation of the 5′ terminus of histidine. *Proc. Natl. Acad. Sci. USA* **79:**6475–6479.
28. **Cortese, R. R., R. Landsberg, R. A. Von der Haar, H. E. Umbarger, and B. N. Ames.** 1974. Pleiotropy of *his*T mutants blocked in pseudouridine synthesis in tRNA: leucine and isoleucine-valine operons. *Proc. Natl. Acad. Sci. USA* **71:**1857–1861.
29. **Darr, S. C., K. Zito, D. Smith, and N. R. Pace.** 1992. Contributions of phylogenetically variable structural elements to the function of the ribozyme ribonuclease P. *Biochemistry* **31:**328–333.
30. **Delk, A. S., and J. C. Rabinowitz.** 1975. Biosynthesis of ribosylthymine in the transfer RNA of *Streptococcus faecalis*: a folate-dependent methylation not involving S-adenosylmethionine. *Proc. Natl. Acad. Sci. USA* **72:** 528–530.
31. **Doi, R. H.** 1969. Changes in nucleic acids during sporulation, p. 125–166. *In* G. W. Gould and A. Hurst (ed.), *The Bacterial Spore*. Academic Press, London.
32. **Duester, G. L., and W. M. Holmes.** 1980. The distal end of the ribosomal RNA operon *rrn*D of *Escherichia coli* contains a tRNA^{Thr}1 gene, two 5S rRNA genes and a transcription terminator. *Nucleic Acids Res.* **8:**3793–3807.
33. **Eriani, G., M. Delarue, O. Poch, J. Gangloff, and D. Moras.** 1990. Partition of tRNA synthetases into two classes based on mutually exclusive sets of sequence motifs. *Nature* (London) **347:**203–206.
34. **Fayat, G., J. F. Mayaux, C. Sacerdot, M. Fromant, M. Springer, M. Grunberg-Manago, and S. Blanquet.** 1983. *Escherichia coli* phenylalanyl-tRNA synthetase operon region. Evidence for an attenuation mechanism. Identification of the gene for the ribosomal protein L20. *J. Mol. Biol.* **260:**10063–10068.
35. **Fersht, A. R.** 1987. Dissection of the structure and activity of the tyrosyl-tRNA synthetase by site-directed mutagenesis. *Biochemistry* **26:**8031–8037.
36. **Forster, A. C., and R. H. Symons.** 1987. Self-cleavage of plus and minus RNAs of a virusoid and a structural model for the active sites. *Cell* **49:**211–220.
37. **Fournier, M. J., and H. Ozeki.** 1985. Structure and organization of the tRNA genes of *Escherichia coli* K-12. *Microbiol. Rev.* **49:**379–397.
38. **Fukada, K., and J. Abelson.** 1980. DNA sequence of a T4 transfer RNA gene cluster. *J. Mol. Biol.* **139:**377–391.
39. **Gardiner, K., and N. R. Pace.** 1980. RNase P from *Bacillus subtilis* has an RNA component. *J. Biol. Chem.* **255:**7507–7509.
40. **Giroux, S., J. Beaudet, and R. Cedergren.** 1988. Highly repetitive tRNA^{Pro}-tRNA^{His} gene cluster from *Photobacterium phosphoreum*. *J. Bacteriol.* **170:**5601–5606.
41. **Glaser, P., A. Danchin, F. Kunst, M. DéBarboullé, A. Vertés, and R. Dedonder.** 1990. A gene encoding a tyrosine tRNA synthetase is located near *sac*S in *Bacillus subtilis*. *Sequence-J. DNA Mapping Sequencing* **1:**251–261.
42. **Green, C. J., G. C. Stewart, M. A. Hollis, B. S. Vold, and K. F. Bott.** 1985. Nucleotide sequence of the *Bacillus subtilis* ribosomal RNA operon, *rrn*B. *Gene* **37:**261–266.
43. **Green, C. J., and B. S. Vold.** 1983. Sequence analysis of a cluster of twenty-one tRNA genes in *Bacillus subtilis*. *Nucleic Acids Res.* **11:**5763–5774.
44. **Green, C. J., and B. S. Vold.** 1988. Structural requirements for the processing of synthetic tRNA^{His} precursors by the catalytic RNA component of RNase P. *J. Biol. Chem.* **263:**652–657.
45. **Green, C. J., and B. S. Vold.** 1992. A cluster of nine tRNA genes between ribosomal gene operons in *Bacillus subtilis*. *J. Bacteriol.* **174:**3147–3151.
46. **Grunberg-Manago, M.** 1987. Regulation of the expression of aminoacyl-tRNA synthetases and translation factors, p. 1386–1409. *In* F. C. Neidhardt, J. L. Ingraham, K. B. Low, B. Magasanik, M. Schaechter, and

H. E. Umbarger (ed.), *Escherichia coli and Salmonella typhimurium: Cellular and Molecular Biology*. American Society for Microbiology, Washington, D.C.

47. **Grundy, F. J., and T. M. Henkin.** 1987. Cloning and analysis of the *Bacillus subtilis rpsD* gene, encoding ribosomal protein S4. *J. Bacteriol.* **172:**6372–6379.
48. **Guerrier-Takada, C., K. Gardiner, T. Marsh, N. Pace, and S. Altman.** 1983. The RNA moiety of ribonuclease P is the catalytic subunit of the enzyme. *Cell* **35:**849–857.
49. **Gupta, R.** 1984. *Halobacterium volcanii* tRNAs: identification of 41 tRNAs covering all amino acids, and the sequences of 33 class I tRNAs. *J. Biol. Chem.* **259:**9461–9471.
50. **Haas, E. S., D. P. Morse, J. W. Brown, F. J. Schmidt, and N. R. Pace.** 1991. New long-range structure in ribonuclease P RNA. *Science* **254:**853–856.
51. **Hager, P. W., and J. C. Rabinowitz.** 1985. Translational specificity in *Bacillus subtilis*, p. 1–32. *In* D. A. Dubnau (ed.), *The Molecular Biology of the Bacilli*, vol. 2. Academic Press, Inc., New York.
52. **Hall, R. H.** 1971. *The Modified Nucleosides in Nucleic Acids*. Columbia University Press, New York.
53. **Hansen, F. G., E. B. Hansen, and T. Atlung.** 1985. Physical mapping and nucleotide sequence of the *rnpA* gene that encodes the protein component of ribonuclease P in *Escherichia coli*. *Gene* **38:**85–93.
54. **Henkin, T.** Personal communication.
55. **Henner, D. J., and W. Steinberg.** 1979. Transfer ribonucleic acid synthesis during sporulation and spore outgrowth in *Bacillus subtilis* studied by two-dimensional polyacrylamide gel electrophoresis. *J. Bacteriol.* **140:**555–566.
56. **Hottinger, H., T. Ohgi, M.-C. Zwahlen, S. Dhamija, and D. Söll.** 1987. Allele-specific complementation of an *Escherichia coli leuB* mutation by a *Lactobacillus bulgaricus* tRNA gene. *Gene* **60:**75–83.
57. **Hsu, L. M., H. J. Klee, J. Zagorski, and M. J. Fournier.** 1984. Structure of an *Escherichia coli* tRNA operon containing linked genes for arginine, histidine, leucine, and proline tRNAs. *J. Bacteriol.* **158:**934–942.
58. **Hussey, C., R. Losick, and A. L. Sonenshein.** 1971. Ribosomal RNA synthesis is turned off during sporulation of *Bacillus subtilis*. *J. Mol. Biol.* **57:**59–70.
59. **Ikeda, R., T. Ohama, A. Muto, and S. Osawa.** 1990. Nucleotide sequences of nine tRNA genes from *Micrococcus luteus*. *Nucleic Acids Res.* **18:**7154.
60. **Ikeda, R., T. Ohama, A. Muto, and S. Osawa.** 1990. Nucleotide sequences of two tRNA gene clusters from *Micrococcus luteus*. *Nucleic Acids Res.* **18:**7155.
61. **Ikemura, T., and M. Nomura.** 1977. Expression of spacer tRNA genes in ribosomal RNA transcription units by hybrid ColE1 plasmids in *E. coli*. *Cell* **11:**779–793.
62. **Jahn, D., Y.-C. Kim, Y. Ishino, M.-W. Chen, and D. Söll.** 1990. Purification and functional characterization of the Glu-tRNA^{Gln} amidotransferase from *Chlamydomonas reinhardtii*. *J. Biol. Chem.* **265:**8059–8064.
63. **James, B. D., G. J. Olsen, J. Liu, and N. R. Pace.** 1988. The secondary structure of a ribonuclease P RNA, the catalytic element of a ribonucleoprotein enzyme. *Cell* **52:**19–26.
64. **Jarvis, E. D., R. L. Widom, G. LaFauci, Y. Setoguchi, I. R. Richter, and R. Rudner.** 1988. Chromosomal organization of rRNA operons in *Bacillus subtilis*. *Genetics* **120:**625–635.
65. **Jinks-Robertson, S., and M. Nomura.** 1987. Ribosomes and tRNA, p. 1358–1385. *In* F. C. Neidhardt, J. L. Ingraham, K. B. Low, B. Magansanik, M. Schaechter, and H. E. Umbarger (ed.), *Escherichia coli and Salmonella typhimurium: Cellular and Molecular Biology*. American Society for Microbiology, Washington, D.C.
66. **Jones, M. D., D. M. Lowe, T. Borgford, and A. R. Fersht.** 1986. Natural variations of tyrosyl-tRNA synthetase and comparison with engineered mutants. *Biochemistry* **25:**1887–1891.
67. **Kashdan, M., and B. Dudock.** 1982. The gene for a spinach chloroplast isoleucine tRNA has a methionine anticodon. *J. Biol. Chem.* **257:**11191–11194.
68. **Keith, G., H. Rogg, G. Dirheimer, B. Menichi, and T. Heyman.** 1976. Post-transcriptional modification of tyrosine tRNA as a function of growth in *Bacillus subtilis*. *FEBS Lett.* **61:**120–122.
69. **Kole, R., and S. Altman.** 1979. Reconstitution of RNase P activity from inactive RNA and protein. *Proc. Natl. Acad. Sci. USA* **76:**3795–3799.
70. **Komine, Y., T. Adachi, H. Inokuchi, and H. Ozeki.** 1990. Genomic organization and physical mapping of the transfer RNA genes in *Escherichia coli* K12. *J. Mol. Biol.* **212:**579–598.
71. **Komine, Y., and H. Inokuchi.** 1991. Precise mapping of the *rnpB* gene encoding the RNA component of RNase P in *Escherichia coli* K-12. *J. Bacteriol.* **173:**1813–1816.
72. **Kozak, M.** 1983. Comparison of initiation of protein synthesis in protein synthesis in procaryotes, eucaryotes, and organelles. *Microbiol. Rev.* **47:**1–45.
73. **Kuchino, Y., S. Watanabe, F. Harada, and S. Nishimura.** 1980. Primary structure of AUA specific isoleucine tRNA from *Escherichia coli*. *Biochemistry* **19:**2085–2089.
74. **LaFauci, G., R. L. Widom, R. L. Eisner, E. D. Jarvis, and R. Rudner.** 1986. Mapping of rRNA genes with integrable plasmids in *Bacillus subtilis*. *J. Bacteriol.* **165:**204–214.
75. **Lapointe, J., L. Duplan, and M. Proulx.** 1986. A single glutamyl-tRNA synthetase aminoacylates tRNA^{Glu} and tRNA^{Gln} in *Bacillus subtilis* and efficiently misacylates *Escherichia coli* tRNA^{Gln}1 in vitro. *J. Bacteriol.* **165:**88–93.
76. **Lee, J. S., G. An, and J. D. Friesen.** 1981. Location of the *tufB* promoter of *E. coli*: cotranscription of *tufB* with four transfer RNA genes. *Cell* **25:**251–258.
77. **Losick, R., P. Youngman, and P. J. Piggot.** 1986. Genetics of endospore formation in *Bacillus subtilis*. *Annu. Rev. Genet.* **20:**625–669.
78. **Loughney, K., E. Lund, and J. E. Dahlberg.** 1982. tRNA genes are found between the 16S and 23S rRNA genes in *Bacillus subtilis*. *Nucleic Acids Res.* **10:**1607–1624.
79. **Lovett, P. S., N. P. Ambulos, Jr., W. Mulbry, N. Noguchi, and E. J. Rogers.** 1991. UGA can be decoded as tryptophan at low efficiency in *Bacillus subtilis*. *J. Bacteriol.* **173:**1810–1812.
80. **Lumelsky, N., and S. Altman.** 1988. Selection and characterization of randomly produced mutants in the gene coding for M1 RNA. *J. Mol. Biol.* **202:**443–454.
81. **Mandelstam, J.** 1969. Regulation of bacterial spore formation, p. 377–401. *In* P. Meadow and S. J. Pirt (ed.), *Microbial Growth*. Cambridge University Press, London.
82. **Margulies, L., Y. Setoguchi, and R. Rudner.** 1978. Asymmetric transcription during post-germinative development of *Bacillus subtilis* spores. I. Hybridization patterns. *Biochim. Biophys. Acta* **521:**708–718.
83. **Marsh, T. L., and N. R. Pace.** 1985. Ribonuclease P catalysis differs from ribosomal RNA self-splicing. *Science* **229:**79–81.
84. **Martin, N. C., M. Rabinowitz, and H. Fukuhara.** 1977. Yeast mitochondrial DNA specifies tRNA for 19 amino acids. Deletion mapping of the tRNA genes. *Biochemistry* **16:**4672–4677.
85. **McClain, W. H., C. Guerrier-Takada, and S. Altman.** 1987. Model substrates for an RNA enzyme. *Science* **238:**527–530.
86. **McLaughlin, J. R., C. L. Murray, and J. C. Rabinowitz.** 1981. Unique features in the ribosome binding sequence of the Gram-positive *Staphylococcus aureus* beta-lactamase gene. *J. Biol. Chem.* **256:**11283–11291.

87. **Mechulam, Y., E. Schmitt, M. Panvert, J. Schmitter, M. Lapadat-Tapolsky, T. Meinnel, P. Dessen, S. Blanquet, and G. Fayat.** 1991. Methionyl-tRNA synthetase from *Bacillus stearothermophilus*: structural and functional identities with the *Escherichia coli* enzyme. *Nucleic Acids Res.* **19:**3673–3681.
88. **Mézes, P. S. F., R. W. Blacher, and J. O. Lampen.** 1985. Processing of *Bacillus cereus* 569/H beta-lactamase I in *Escherichia coli* and *Bacillus subtilis*. *J. Biol. Chem.* **260:**1218–1223.
89. **Miyajima, A., M. Shibuya, Y. Kuchino, and Y. Kaziro.** 1981. Transcription of the *E. coli tuf*B gene: cotranscription with four tRNA genes and inhibition by guanosine-5′-diphosphate-3′-diphosphate. *Mol. Gen. Genet.* **189:**13–19.
90. **Morgan, E. A., T. Ikemura, L. Lindahl, A. M. Fallon, and M. Nomura.** 1978. Some rRNA operons in *E. coli* have tRNA genes at their distal ends. *Cell* **13:**335–344.
91. **Morgan, E. A., T. Ikemura, and M. Nomura.** 1977. Identification of spacer tRNA genes in individual ribosomal RNA transcription units of *Escherichia coli*. *Proc. Natl. Acad. Sci. USA* **74:**2710–2714.
92. **Muto, A., Y. Andachi, H. Yuzawa, F. Yamao, and S. Osawa.** 1990. The organization and evolution of transfer RNA genes in *Mycoplasma capricolum*. *Nucleic Acids Res.* **18:**5037–5043.
93. **Muto, A., T. Ohama, Y. Andachi, F. Yamao, R. Tanaka, and S. Osawa.** 1991. Evolution of codons and anticodons in eubacteria, p. 179–193. *In* M. Kimura and N. Takahata (ed.), *New Aspects of the Genetics of Molecular Evolution*. Japan Scientific Society Press, Tokyo.
94. **Nakajima, N., H. Ozeki, and Y. Shimura.** 1981. Organization and structure of an *E. coli* tRNA operon containing seven tRNA genes. *Cell* **23:**239–249.
95. **Nieuwlandt, D. T., E. S. Haas, and C. J. Daniels.** 1991. The RNA component of RNase P from the archaebacterium *Haloferax volcanii*. *J. Biol. Chem.* **266:**5689–5695.
96. **Ogasawara, N.** 1985. Markedly unbiased codon usage in *Bacillus subtilis*. *Gene* **40:**145–150.
97. **Ogasawara, N., S. Moriya, K. von Meyenburg, F. G. Hansen, and H. Yoshikawa.** 1985. Conservation of genes and their organization in the chromosomal replication origin region of *Bacillus subtilis* and *Escherichia coli*. *EMBO J.* **4:**3345–3350.
98. **Ogasawara, N., S. Moriya, and H. Yoshikawa.** 1983. The structure and organization of rRNA operons at the region of the replication origin of the *B. subtilis* chromosome. *Nucleic Acids Res.* **11:**6301–6318.
99. **Oishi, M., A. Oishi, and N. Sueoka.** 1966. Location of genetic loci of soluble RNA on *Bacillus subtilis* chromosome. *J. Mol. Biol.* **55:**1095–1103.
100. **Orellana, O., L. Cooley, and D. Söll.** 1986. The additional guanylate at the 5′-end of *Escherichia coli* $tRNA^{His}$ is the result of unusual processing by RNase P. *Mol. Cell. Biol.* **6:**525–529.
101. **Osawa, S., A. Muto, T. H. Jukes, and T. Ohama.** 1990. Evolutionary changes in the genetic code. *Proc. R. Soc. Lond. Sect. B* **241:**19–28.
102. **Pace, N. R., C. Reich, B. D. James, G. J. Olsen, B. Pace, and D. S. Waugh.** 1987. Structure and catalytic function in ribonuclease P. *Cold Spring Harbor Symp. Quant. Biol.* **52:**239–248.
103. **Pace, N. R., D. K. Smith, G. J. Olsen, and B. D. James.** 1989. Phylogenetic comparative analysis and the secondary structure of ribonuclease P RNA—a review. *Gene* **82:**65–75.
104. **Pero, J., J. Nelson, and R. Losick.** 1975. In vitro and in vivo transcription by vegetative and sporulating *Bacillus subtilis*, p. 202–212. *In* P. Gerhardt, R. N. Costilow, and H. L. Sadoff (ed.), *Spores VI*. American Society for Microbiology, Washington, D.C.
105. **Piggot, P. J., M. Amjad, J.-J. Wu, H. Sandoval, and J. Castro.** 1990. Genetic and physical maps of *Bacillus subtilis* 168, p. 493–534. *In* C. R. Harwood and S. M. Cutting (ed.), *Molecular Biology Methods for Bacillus*. John Wiley & Sons Ltd., London.
106. **Pittet, A. C., and H. Hottinger.** 1989. Sequence of a hexameric tRNA gene cluster associated with rRNA genes in *Lactobacillus bulgaricus*. *Nucleic Acids Res.* **17:**4873.
107. **Putney, S. D., and P. Schimmel.** 1981. An aminoacyl tRNA synthetase binds to a specific DNA sequence and regulates its gene transcription. *Nature* (London) **291:** 632–635.
108. **Putzer, H., A. A. Brakhage, and M. Grunberg-Manago.** 1990. Independent genes for two threonyl-tRNA synthetases in *Bacillus subtilis*. *J. Bacteriol.* **172:**4593–4602.
109. **Ramakrishna, N., E. Dubnau, and I. Smith.** 1984. The complete DNA sequence and regulatory regions of the *Bacillus licheniformis spo*OH gene. *Nucleic Acids Res.* **12:**1779–1790.
110. **Randerath, E., L.-L. S. Y. Chia, H. P. Morris, and K. Randerath.** 1974. Base analysis of RNA by ^3H postlabeling—a study of ribothymidine content and degree of base methylation of 4S RNA. *Biochim. Biophys. Acta* **366:**159–167.
111. **Raue, H. A., and R. J. Planta.** 1977. Heterogeneity of the genes coding for 5S RNA in three related strains of the genus *Bacillus*. *Mol. Gen. Genet.* **156:**185–193.
112. **Reed, R. E., M. F. Baer, C. Guerrier-Takada, H. Donis-Keller, and S. Altman.** 1982. Nucleotide sequence of the gene coding for the RNA subunit (M1 RNA) of ribonuclease P from *Escherichia coli*. *Cell* **30:**627–636.
113. **Reich, C., K. J. Gardiner, G. J. Olsen, B. Pace, T. L. Marsh, and N. R. Pace.** 1986. The RNA component of the *Bacillus subtilis* RNase P. *J. Biol. Chem.* **261:**7888–7893.
114. **Reich, C., G. J. Olsen, B. Pace, and N. R. Pace.** 1988. The role of the protein moiety of ribonuclease P, a ribonucleoprotein enzyme. *Science* **239:**178–181.
115. **Rogers, M. J., J. Simmons, R. T. Walker, W. G. Weisburg, C. R. Woese, R. S. Tanner, I. M. Robinson, D. A. Stahl, G. Olsen, R. H. Leach, and J. Maniloff.** 1985. Construction of the mycoplasma evolutionary tree from 5S rRNA sequence data. *Proc. Natl. Acad. Sci. USA* **82:**1160–1164.
116. **Rogers, M. J., A. A. Steinmetz, and R. T. Walker.** 1987. Organization and structure of tRNA genes in *Spiroplasma melliferum*. *Isr. J. Med. Sci.* **23:**357–360.
117. **Rudner, R.** Personal communication.
118. **Samuelsson, T., P. Elias, F. Lustig, and Y. S. Guindy.** 1985. Cloning and nucleotide sequence analysis of transfer RNA genes from *Mycoplasma mycoides*. *Biochem. J.* **232:**223–228.
119. **Samuelsson, T., Y. S. Guindy, F. Lustig, T. Borén, and U. Lagerkvist.** 1987. Apparent lack of discrimination in the reading of certain codons in *Mycoplasma mycoides*. *Proc. Natl. Acad. Sci. USA* **84:**3166–3170.
120. **Schimmel, P.** 1987. Aminoacyl tRNA synthetases: general scheme of structure-function relationships in the polypeptides and recognition of transfer RNAs. *Annu. Rev. Biochem.* **56:**125–158.
121. **Schmidt, F. J., and W. H. McClain.** 1978. Alternate orders of processing by RNase P occur *in vitro* but not *in vivo*. *J. Biol. Chem.* **253:**4730–4734.
122. **Schön, A., C. G. Kannangara, S. Gough, and D. Söll.** 1988. Protein biosynthesis in organelles requires misaminoacylation of tRNA. *Nature* (London) **331:**187–190.
123. **Schulman, L.** 1979. Chemical approaches to the study of protein-tRNA recognition, p. 311–324. *In* P. R. Schimmel, D. Söll, and J. N. Abelson (ed.), *Transfer RNA: Structure, Properties and Recognition*. Cold Spring Harbor Laboratory, Cold Spring Harbor, N.Y.
124. **Setlow, P., G. Primuw, and M. P. Deutscher.** 1974. Absence of 3′-terminal residues from transfer ribonucleic acid of dormant spores of *Bacillus megaterium*. *J. Bacteriol.* **117:**127–132.

125. **Setoguchi, Y., L. Margulies, and R. Rudner.** 1978. Asymmetric transcription during post-germinative development of *Bacillus subtilis* spores. II. Hybrid competition analyses. *Biochim. Biophys. Acta* **521:**719–725.
126. **Singhal, R. P., and B. Vold.** 1976. Changes in transfer ribonucleic acids of *Bacillus subtilis* during different growth phases. *Nucleic Acids Res.* **3:**1249–1261.
127. **Smith, D. W. E., A. L. McNamara, and B. S. Vold.** 1982. Lysine tRNAs from *Bacillus subtilis* 168: function of the isoacceptors in a rabbit reticulocyte cell-free protein-synthesizing system. *Nucleic Acids Res.* **10:**3117–3123.
128. **Smith, I., D. Dubnau, P. Morell, and J. Marmur.** 1968. Chromosome location of DNA base sequences complementary to transfer RNA and to 5S, 16S and 23S ribosomal RNA in *Bacillus subtilis. J. Mol. Biol.* **33:**123–140.
129. **Springer, M., M. Graffe, J. S. Butler, and M. Grunberg-Manago.** 1986. Genetic definition of the translational operator of the threonyl-tRNA synthetase gene in *E. coli. Proc. Natl. Acad. Sci. USA* **83:**4384–4388.
130. **Sprinzl, M., N. Dank, S. Nock, and A. Schön.** 1991. Compilation of tRNA sequences and sequences of tRNA genes. *Nucleic Acids Res.* **19**(Suppl.):2127–2171.
131. **Sprinzl, M., T. Hartman, J. Weber, J. Blank, and R. Zeidler.** 1989. Compilation of tRNA sequences and sequences of tRNA genes. *Nucleic Acids Res.* **17:**r1–r172.
132. **Stark, B. C., R. Kole, E. J. Bowman, and S. Altman.** 1978. Ribonuclease P: an enzyme with an essential RNA component. *Proc. Natl. Acad. Sci. USA* **75:**3717–3721.
133. **Strauch, M. A., H. Zalkin, and A. I. Aronson.** 1988. Characterization of the glutamyl-tRNA^{Gln}-to-glutaminyl-tRNA^{Gln} amidotransferase reaction of *Bacillus subtilis. J. Bacteriol.* **170:**916–920.
134. **Testa, D., and R. Rudner.** 1975. Synthesis of ribosomal RNA during sporulation in *Bacillus subtilis. Nature* (London) **254:**630–632.
135. **Thurlow, D. L., D. Shilowski, and T. L. Marsh.** 1991. Nucleotides in precursor tRNAs that are required intact for catalysis by RNase P RNAs. *Nucleic Acids Res.* **19:**885–891.
136. **Vold, B.** 1974. Degree of completion of 3′-terminus of transfer ribonucleic acids of *Bacillus subtilis* 168 at various developmental stages and asporogenous mutants. *J. Bacteriol.* **117:**1361–1362.
137. **Vold, B.** 1976. Modified nucleosides of *Bacillus subtilis* transfer ribonucleic acids. *J. Bacteriol.* **127:**258–267.
138. **Vold, B. S.** 1973. Variations in activity of aminoacyl-tRNA synthetases as a function of development in *Bacillus subtilis. Arch. Biochem. Biophys.* **154:**691–695.
139. **Vold, B. S.** 1973. Analysis of isoaccepting transfer ribonucleic acid species of *Bacillus subtilis*: chromatographic differences between transfer ribonucleic acids from spores and cells in exponential growth. *J. Bacteriol.* **113:**825–833.
140. **Vold, B. S.** 1978. Post-transcriptional modifications of the anticodon loop region: alterations in isoaccepting species of tRNA's during development in *Bacillus subtilis. J. Bacteriol.* **135:**124–132.
141. **Vold, B. S.** 1985. Structure and organization of genes for transfer ribonucleic acid in *Bacillus subtilis. Microbiol. Rev.* **49:**71–80.
142. **Vold, B. S., and C. J. Green.** 1986. Expression in *Escherichia coli* of *Bacillus subtilis* tRNA genes from a promoter within the tRNA gene region. *J. Bacteriol.* **166:**306–312.
143. **Vold, B. S., C. J. Green, N. Narasimhan, M. Strem, and J. N. Hansen.** 1988. Transcriptional analysis of *Bacillus subtilis* rRNA-tRNA operons. Unique properties of an operon containing a minor 5S rRNA gene. *J. Biol. Chem.* **263:**14485–14490.
144. **Vold, B. S., M. E. Longmire, and D. E. Keith, Jr.** 1981. Thiolation and 2-methylthio-modification of *Bacillus subtilis* transfer ribonucleic acids. *J. Bacteriol.* **148:**869–876.
145. **Vold, B. S., and S. Minatogawa.** 1972. Characterization of changes in transfer ribonucleic acids during sporulation in *Bacillus subtilis*, p. 254–263. *In* H. O. Halvorson, R. Hanson, and L. L. Campbell (ed.), *Spores V.* American Society for Microbiology, Washington, D.C.
146. **Vold, B. S., K. O. Okamoto, B. J. Murphy, and C. J. Green.** 1988. Transcriptional analysis of *Bacillus subtilis* rRNA-tRNA operons. The tRNA gene cluster of *rrnB* has an internal promoter. *J. Biol. Chem.* **263:**14480–14484.
147. **Waugh, D. S., C. J. Green, and N. R. Pace.** 1989. The design and catalytic properties of a simplified ribonuclease P RNA. *Science* **244:**1569–1571.
148. **Waugh, D. S., and N. R. Pace.** 1990. Complementation of an RNase P RNA (*rnpB*) gene deletion in *Escherichia coli* by homologous genes from distantly related eubacteria. *J. Bacteriol.* **172:**6316–6322.
149. **Wawrousek, E. F., and J. N. Hansen.** 1983. Structure and organization of a cluster of six tRNA genes in the space between tandem ribosomal RNA gene sets in *Bacillus subtilis. J. Biol. Chem.* **258:**291–299.
150. **Wawrousek, E. F., N. Narasimhan, and J. N. Hansen.** 1984. Two large clusters with thirty-seven transfer RNA genes adjacent to ribosomal RNA gene sets in *Bacillus subtilis*: sequence and organization of trrnD and trrnE clusters. *J. Biol. Chem.* **259:**3694–3702.
151. **Weisburg, W. G., J. G. Tully, D. L. Rose, J. P. Petzel, H. Oyaizu, D. Yang, L. Mandelco, J. Sechrest, T. G. Lawrence, J. van Etten, J. Maniloff, and C. R. Woese.** 1989. A phylogenetic analysis of the mycoplasmas: basis for their classification. *J. Bacteriol.* **171:**6455–6467.
152. **White, B. N., and S. T. Bayley.** 1972. Further codon assignments in an extremely halophilic bacterium using a cell-free protein-synthesizing system and a ribosomal binding assay. *Can. J. Biochem.* **50:**601–609.
153. **Widom, R. L., E. D. Jarvis, G. LaFauci, and R. Rudner.** 1988. Instability of rRNA operons in *Bacillus subtilis. J. Bacteriol.* **170:**605–610.
154. **Wilcox, M.** 1969. γ-Phosphoryl ester of Glu-tRNA^{Gln} as an intermediate in *Bacillus subtilis* glutaminyl-tRNA synthesis. *Cold Spring Harbor Symp. Quant. Biol.* **34:**521–528.
155. **Wilcox, M.** 1969. γ-Glutamyl phosphate attached to glutamine-specific tRNA. *Eur. J. Biochem.* **11:**405–412.
156. **Wilcox, M., and M. Nirenberg.** 1968. Transfer RNA as a cofactor coupling amino acid synthesis with that of protein. *Proc. Natl. Acad. Sci. USA* **61:**229–236.
157. **Winter, G., G. L. E. Koch, B. S. Hartley, and D. G. Barker.** 1983. The amino acid sequence of the tyrosyl-tRNA synthetase from *Bacillus stearothermophilus. Eur. J. Biochem.* **132:**383–387.
158. **Yamada, Y., and H. Ishikura.** 1980. Nucleotide sequence of non-initiator methionine tRNA from *Bacillus subtilis. Nucleic Acids Res.* **8:**4517–4520.
159. **Yamada, Y., Y. Kuchino, and H. Ishikura.** 1980. Nucleotide sequence of initiator tRNA from *Bacillus subtilis. J. Biochem.* **87:**1261–1269.
160. **Yamada, Y., M. Ohki, and H. Ishikura.** 1983. The nucleotide sequence of *Bacillus subtilis* tRNA genes. *Nucleic Acids Res.* **11:**3037–3045.
161. **Yamagata, H., T. Adachi, A. Tsuboi, M. Takao, T. Sasaki, N. Tsukagoshi, and S. Udaka.** 1987. Cloning and characterization of the 5′ region of the cell wall protein gene operon in *Bacillus brevis* 47. *J. Bacteriol.* **169:**1239–1245.
162. **Yamao, F., A. Muto, Y. Kawauchi, M. Iwami, S. Iwagami, Y. Azumi, and S. Osawa.** 1985. UGA is read as tryptophan in *Mycoplasma capricolum. Proc. Natl. Acad. Sci. USA* **82:**2306–2309.
163. **Zaug, A. J., P. J. Grabowski, and T. R. Cech.** 1983. Autocatalytic cyclization reaction of an excised intervening sequence RNA is a cleavage-ligation reaction. *Nature* (London) **301:**578–583.
164. **Zuber, P. A.** 1982. Ph.D. thesis. University of Virginia, Charlottesville.

48. Translation and Its Regulation

ROBERT LUIS VELLANOWETH

Much of our current knowledge concerning the process of translation in prokaryotes comes from extensive studies of one organism, *Escherichia coli*. These investigations have provided detailed insights into many aspects of protein synthesis, including the mechanism and kinetics of initiation; the cyclical reactions involving transfer RNAs, protein factors, GTP, and ribosomal proteins in elongation; and the fine structure and relationships of proteins and RNA in the ribosomal machinery itself (for recent reviews, see references 55, 68, and 101). Results obtained through the use of *E. coli* have provided the basis of the genetic code, the recognition of specific mRNA signals directing ribosomes to initiation points, and insight into subtle translational regulatory mechanisms. Although this wealth of knowledge has unraveled many of the mysteries concerning the transfer of the one-dimensional information inherent in DNA to three-dimensional active molecules, biochemical evidence suggests that some of the specific conclusions cannot be extrapolated to all other bacteria. For example, not all bacteria contain the same number of ribosomal proteins. One ribosomal protein (S1) missing in many bacteria is known to play a critical role in initiation in *E. coli*, suggesting that fundamental processes in translation may differ among the bacteria. Another indication of such differences is that mRNAs cannot be freely exchanged among all bacteria and still result in efficient recognition and translation. Lodish was the first to observe that ribosomes from *E. coli* and *Bacillus stearothermophilus* do not translate f2 coliphage RNA into three products equally well (81). Not only was overall incorporation by *B. stearothermophilus* only 5% that of *E. coli*, but the products were made in vastly different relative amounts by the two ribosomes. It was further demonstrated that the inability of *B. stearothermophilus* ribosomes to efficiently translate the f2 message was a function of the small ribosomal subunit; neither the source of crude initiation factors nor that of large ribosomal subunit had any effect on the observed translational selection (82).

Around the same time as these discoveries, J. C. Rabinowitz and his coinvestigators were interested in the biosynthesis of the small, clostridial iron-sulfur-containing protein ferredoxin. Specifically, they set out to ascertain the mechanism and pathway of incorporation of the iron-sulfur groups into the apoprotein during biosynthesis. A cell-free protein-synthesizing system that efficiently translated crude *Clostridium pasteurianum* mRNA and incorporated Phe in response to poly(U) but was unexpectedly inactive in response to f2 RNA was thus developed (58). A similar system derived from *E. coli*, on the other hand, was active on both *C. pasteurianum* mRNA and f2 RNA. This phenomenon of translational species specificity was expanded in a subsequent study to include several other organisms (128). Three gram-negative organisms (*E. coli*, *Pseudomonas fluorescens*, and *Azotobacter vinelandii*) were found to be nonspecific with respect to message selection, while five gram-positive organisms (*Bacillus subtilis*, *Clostridium acidiurici*, *Clostridium tetanomorphum*, *Streptococcus faecalis*, and *Peptococcus aerogenes*) were selective in that only gram-positive mRNAs were translated. The general conclusion from these and other studies has been that ribosomes from gram-positive organisms can efficiently translate only mRNAs from gram-positive organisms and their phages, whereas ribosomes from gram-negative organisms can efficiently translate both gram-negative and gram-positive mRNAs (for a review, see reference 49).

This division of organisms with respect to their translational characteristics on the basis of their reaction to the Gram stain was valid for the limited number of organisms studied but was certainly an oversimplification. At the time of the gram-negative–gram-positive hypothesis of translational specificity, all of the gram-negative organisms tested fell into a group termed the "purple bacteria," and the gram-positive organisms formed a subset of the clostridial branch of bacteria (30). However, at least two members of the high-G+C-content actinomycete branch of gram-positive bacteria have been shown to efficiently translate *E. coli* mRNA in vivo and in vitro (12, 29, 120, 140) and thus can be considered translationally nonspecific gram-positive organisms. The evolutionary relationships among these groups of bacteria and the translational characteristics of those that have been studied are shown in Fig. 1. Thus, while the actinomycetes are gram positive, their relatively nonspecific translational characteristics, including the presence on the 30S subunit of a counterpart to ribosomal protein S1, are more closely related to those of the gram-negative purple bacteria. Only those species making up the clostridial branch are selective in the translation of mRNAs and, perhaps important for translational specificity (see below), lack a functional counterpart to ribosomal protein S1.

The focus of this chapter will be a review of the current status of translation in *B. subtilis* since publication of a paper on translational specificity by Hager and Rabinowitz (49). I will describe the important features of a *B. subtilis* ribosome binding site (RBS) as distinct from an *E. coli* RBS and suggest a subtly different mechanism for initiation, taking into ac-

Robert Luis Vellanoweth • Department of Cellular and Structural Biology, University of Texas Health Science Center, 7703 Floyd Curl Drive, San Antonio, Texas 78284-7762.

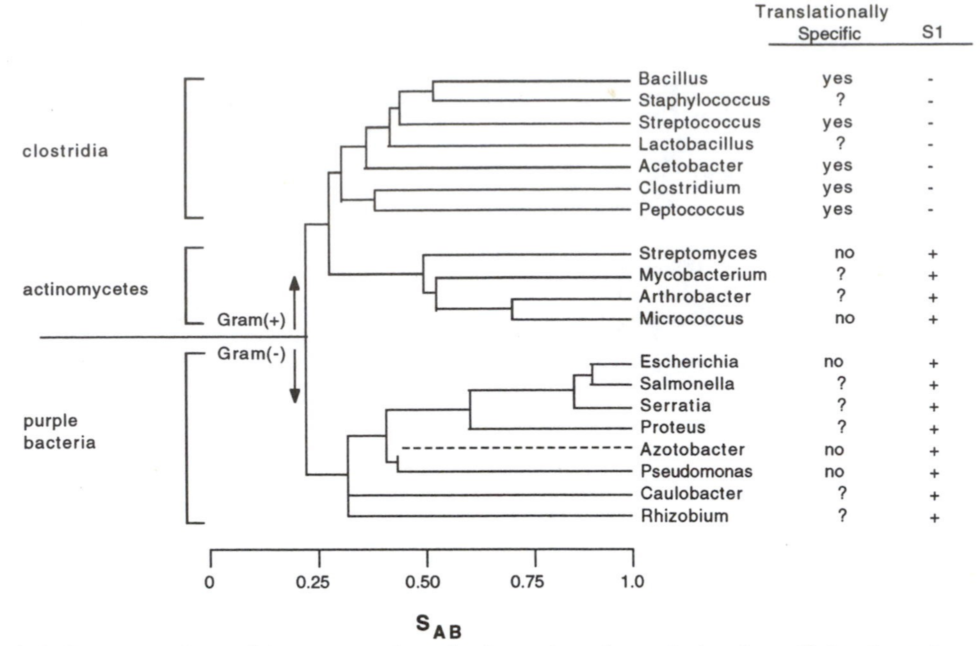

Figure 1. Phylogenetic relationships among selected eubacteria and translational specificity. Groupings are based on binary association coefficients (S_{AB}s) (30) of 16S rRNA sequences.

count the known requirements for efficient initiation. From these features, potential sites for translational regulation governing the overall yield of protein product will be proposed, and a discussion of known translational regulatory mechanisms in *B. subtilis* will follow. Finally, I will compare and contrast the translational characteristics of other gram-positive organisms, including those from the high-G+C-content actinomycete group.

TRANSLATION INITIATION IN *B. SUBTILIS*

Given the considerable conservation of rRNA structures, ribosomal proteins, and initiation factors between *E. coli* and *B. subtilis* (with the notable exception of ribosomal protein S1), the basic steps in the translation initiation pathway in *E. coli* are likely to be replicated in *B. subtilis*. Initiation, as proposed by Gualerzi et al. (45) on the basis of kinetic studies (17, 44–47), involves the association of an mRNA containing a Shine-Dalgarno (SD) sequence (125) and initiation codon, the 30S ribosomal subunit plus initiation factors and GTP, and the fMet-tRNA$_f$ into a preinitiation ternary complex. A first-order rearrangement, thought to position the initiator tRNA properly with respect to the initiation codon, is followed by binding of the 50S subunit and formation of the 70S initiation complex. Gold and Stormo (38) described features of an mRNA RBS that could influence the rates of the individual reactions leading up to the ternary complex and, by consequence, overall translational yield. For example, an SD sequence sequestered by secondary structure would be expected to slow the association of the mRNA and the 30S subunit, while a strong, unstructured SD sequence would increase the rate of association. The rate of dissociation of the 30S-mRNA complex would also depend on the strength of the SD sequence. The spacing between the SD sequence and the initiation codon as well as whether the AUG was bound up in a secondary structure would influence the rate of first-order rearrangement into the initiation complex. The rate of formation of the 70S initiation complex may be related to the nature of the second codon or even sequences further downstream from the RBS. Subtle differences between the *E. coli* and *B. subtilis* ribosome in the effects these features have on the rates of individual reactions of the initiation pathway are probably responsible for the observed translational specificity.

SD Sequence Interaction

The free 3′ end of 16S rRNA plays a critical role in the initiation of protein synthesis by serving as the site for Watson-Crick base pairing with the complementary SD sequence upstream of the initiation codon of eubacterial mRNAs. Considerable evidence supports the role of this interaction in directing *E. coli* ribosomes to appropriate translation start points and has been reviewed several times (36, 131). The primary difference between *B. subtilis* and gram-negative organisms centers on the nature of this interaction. McLaughlin et al. (87) first pointed out that gram-positive mRNAs contain SD sequences exhibiting more-extensive complementarity with the 16S rRNA and proposed that this was a requirement responsible for the observed inability of the *B. subtilis* ribosome to translate gram-negative messages. When put forth, the hypothesis was based on only a limited number of sequenced gram-positive genes but has been continuously supported by further sequencing of *B. subtilis* and other "clostridial" genes. From an analysis of these gene sequences, it was found that the average free energy of binding of a *B. subtilis* SD sequence to the 16S rRNA is −16.7 kcal/mol (1 cal = 4.184 J), while that of an average *E. coli* SD sequence is −11.7 kcal/mol (49). More significantly, while *E. coli* SD sequences showed a distribution from −2 to −22 kcal/mol, no *B. subtilis* SD sequence had a free energy of formation (ΔG_f) less stable than −12 kcal/mol.

Numerous examples in the literature demonstrate the importance of a strong SD for efficient translation in *B. subtilis*. For example, Band and Henner (6) constructed a series of plasmids with the leukocyte interferon gene containing SD sequences ranging from −9.4 to −29.4 kcal/mol. In *B. subtilis*, the yield of interferon product correlated directly with the strength of the SD sequence interaction. In *E. coli*, on the other hand, no direct correlation was observed, indicating the greater importance of the SD sequence in *B. subtilis*. Rabinowitz and I recently found (142) that in a collection of *lacZ* clones whose RBSs were generated randomly (34a), only those containing a sequence resembling a polypurine-rich SD sequence gave any detectable β-galactosidase activity in *B. subtilis*, while all gave at least some activity in *E. coli*. Also, the amount of product formed in *B. subtilis* correlated with the strength of the SD sequence interaction, while such was not the case in *E. coli*. In the same study, when clones containing SD sequences of AAGGA or AAGGAGG in identical contexts were used, an increase in ΔG_f from −11.8 to −19.8 kcal/mol resulted in a 9-fold increase in β-galactosidase activity in *E. coli*, but in *B. subtilis*, the difference was 114-fold. Analysis of mutations affecting cytochrome b_{558} synthesis revealed that a G-to-A transition in the SD sequence, which reduced the stability from −14.4 to −7.2 kcal/mol, resulted in a 10-fold decrease in yield in *B. subtilis* (31). The same mutation had no effect in *E. coli*. Similarly, a mutation of the poor SD sequence of the *Streptomyces lavendulae* streptothricin acetyltransferase gene to the canonical AAGGAGG sequence increased the resistance of *B. subtilis* cells to streptothricin twofold, apparently reflecting increased translation of the streptothricin acetyltransferase message (61). In a study analyzing the mRNA sequence determinants for induction of the *B. pumilus cat-86* gene in *B. subtilis*, Harwood et al. (53) isolated a deletion mutant that removed half of the SD sequence, leaving a ΔG_f of −9.2 kcal/mol. This construct still gave rise to wild-type levels of chloramphenicol acetyltransferase (CAT) expression, and thus, those authors questioned whether a stringent SD sequence was really necessary for efficient synthesis in *B. subtilis*. However, as they pointed out (53), a new SD sequence with a ΔG_f of −16.2 kcal/mol and a spacing of 6 bases could be drawn slightly downstream. This spacing is well within the limits for expression in *B. subtilis* (see below) and probably explains the absence of an effect of this deletion. Similarly, Hung et al. (64) have suggested that an SD sequence with a ΔG_f of −9.4 kcal/mol functions in *B. subtilis*, but again, another SD sequence with a ΔG_f of −12 kcal/mol and a spacing of 10 can be proposed slightly upstream. Mountain (93) examined the SD regions of the genes for seven highly abundant intracellular proteins and observed that the average number of bases available for base pairing was 8.7, which is much greater than the average of 3 to 4 in *E. coli*.

Zaghloul and Doi analyzed the specific circumstances in which the *E. coli* Tn*9*-derived CAT gene (*cat*) could be expressed in *B. subtilis* (148). Under normal conditions, the *cat* message is not translated by *B. subtilis* ribosomes. However, several plasmid constructions that allow expression have been isolated. In one instance (148), the CAT protein was produced not from the native *E. coli* RBS but rather from a fusion of the *cat* coding sequence to a *B. subtilis* gene directed by its own strong SD sequence. Two specialized cases in which *B. subtilis* utilizes a poor *E. coli* SD sequence have been reported. One appears to result from reinitiation of a ribosome terminating from a *B. subtilis* coding sequence onto the adjacent *E. coli* SD sequence (149). A second, more interesting case involves the duplication of the *E. coli* RBS that resulted in significant CAT production in *B. subtilis* without an alteration of the SD sequence (80). Those authors proposed that the increased expression was due to an mRNA stem-loop structure that placed the SD and the initiation codon in the loop. In fact, deletion of the upstream SD sequence did not affect translation, but removal of sequences involved in formation of the stem reduced expression nearly 130-fold. Although the maximal levels of CAT expression in *B. subtilis* were 14% of that found in *E. coli*, these results suggest that the requirement for a strong SD sequence in *B. subtilis* may not be absolute but rather may depend on the nature of the total RBS. Whether such structures exist in natural *B. subtilis* messages remains to be seen.

Initiation Codon

Another major difference between *B. subtilis* and the gram-negative organisms lies in the choice of initiation codon. While 91% of the sequenced *E. coli* genes start with AUG (36), nearly 30% of *B. subtilis* and other clostridial branch genes start with UUG or GUG (49). Recently, CUG has also been shown to function as a start codon in *B. subtilis* (5). Mutations of an AUG initiation codon to GUG or UUG have generally resulted in decreased expression in *B. subtilis*, as has been found in *E. coli* (110), with the possible exception of the β subunit of aspartokinase II. Although the results were not quantitated, it appears that mutating the initiation codon from AUG to GUG or even AUC had little effect on the translational yield of aspartokinase in *B. subtilis* (18). Relative translational yield from the three initiation codons may differ between the two organisms. Ambulos et al. (5) found the order AUG > UUG > GUG in *B. subtilis* which contrasts with the relative order AUG > GUG > UUG in *E. coli* (110), a result recently confirmed by us in vivo (142). These orders of preference correlate with the relative abundance of genes containing the different initiation codons in both organisms (36, 49), but the significance of their differential use is unclear.

In *E. coli*, it is apparent that the use of a non-AUG initiation codon serves to limit expression of a gene product at the translational level (110). Since a much higher percentage of *B. subtilis* genes contain non-AUG codons, it is unlikely that the synthesis of those gene products is also tightly regulated by the same mechanism. We have analyzed this problem both in vitro (29a) and in vivo (142) by comparing relative translational efficiency with each of the three initiation codons as a function of SD-sequence strength. In both studies, large differences in product formation were obtained in the presence of a weak SD sequence. However, as the strength of the SD sequence was increased, differences between the three initiation codons moderated considerably in both *B. subtilis* and *E. coli*. Thus, since *B. subtilis* genes contain strong SD sequences, changes in the initiation codon can be tolerated without significantly affecting translational yield, again implying the larger relative importance of a strong SD sequence in translation initiation in *B. subtilis*.

At least one gene, *infC*, which codes for initiation factor 3 (IF-3) and contains an unusual AUU initiation codon, has been reported in *B. subtilis* (106). In *E. coli*, the counterpart IF-3 gene also contains AUU as the start codon. Given the role of IF-3 in the initiation of protein synthesis, Gold et al. (39) proposed an unusual scheme for IF-3-independent translation of the IF-3 message, including base pairing of regions of the IF-3 RBS to complementary sequences of the 16S rRNA outside of the anti-SD-sequence region. The mechanism was proposed to explain the use of the AUU start codon and the observed autoregulation of IF-3 synthesis. Since the *B. subtilis* IF-3 message also contains the AUU initiator, it seemed possible that a mechanism similar to that proposed for *E. coli* was operative. However, a search for sequences in the *B. subtilis* 16S rRNA (42) that would base pair in a manner analogous to that proposed (39) was unsuccessful, suggesting either that the proposal of Gold et al. is questionable or that the mechanism in *B. subtilis* for IF-3-independent translation of the *infC* message is different.

Spacing between the SD Sequence and the Initiation Codon

Spacing between the SD sequence and the initiation codon would not be expected to affect binding of the message to the 30S subunit but would affect the rate of the first-order rearrangement to an initiation complex with the start codon base paired to the anticodon of the tRNA. Thus, longer spacings would require some looping out of intervening sequences, while shorter spacings might even involve melting of the SD–16S-rRNA duplex. Gold has described a message with a spacing of 2 bases (counting from the most 3′ base that can base pair to the 16S rRNA, to the first position of the initiation codon) that is cold sensitive for translation in *E. coli*, suggesting just such a scenario (37). Spacing effects have been tested experimentally in *E. coli* (34, 136, 147), but an interpretation of the results in many cases is complicated by differences in sequence context and potential secondary structures. Hartz et al. (52) constructed a series of plasmids containing the SD sequence at various positions from the initiation codon separated by all As to minimize secondary-structure effects. They found that in *E. coli*, spacings in the range of 6 to 10 nucleotides had a minor effect on translational yield. Using the same constructs, Rabinowitz and I found similar results for *B. subtilis* (142). The optimum spacing in both organisms was around 7 to 9 bases, which is consistent with the average spacing found in sequenced genes (49, 93, 130). With a poor SD sequence (AAGGA), however, *B. subtilis* exhibited an optimum of 9 to 12 bases. Furthermore, regardless of the strength of the SD sequence, *B. subtilis* tended to discriminate more severely against messages with spacings shorter than optimal than did *E. coli*. This suggests, perhaps, that the P-site-to-CCUCC distance on the *B. subtilis* ribosome is longer than that in *E. coli*. Although the sequences are different, the number of bases between the CCUCC sequence on the 16S rRNA and the first stem in the secondary structure is the same in the two organisms (42). Thus, it is more likely that the discrimination simply reflects the requirement for a stronger interaction of the ribosome with the SD sequence.

Other Nucleotides in the RBS

Statistical analyses of *E. coli* genes have revealed that the entire 35-base region bound by an initiating ribosome contains a nonrandom distribution of nucleotides (117, 118, 131). This distribution has been used to distinguish true initiation sites from pseudosites containing an AUG codon downstream of an SD-like sequence (131). The presence of preferred nucleotides at certain positions of the RBS suggests that additional contacts exist between the ribosome and the message besides those involving the SD sequence and initiation codon, but the nature of these contacts is unknown. It is possible that these additional determinants of an RBS differ in *B. subtilis* from those in *E. coli*, since the preferred nucleotides are not the same in the two organisms (49). In support of this view, Hager and Rabinowitz showed that a strong SD sequence on an *E. coli* mRNA, such as those of the T7 coliphage, was not sufficient for significant translation by the *B. subtilis* ribosome (48).

An alternative view (22) concerning the existence of base preferences in natural RBSs is that these preferences reflect the necessity for an unstructured binding site for the 30S subunit. Several mRNAs are known to be highly structured except in the region of the RBS (20, 94, 97). Ganoza et al. (33) carried out a statistical analysis of the possible secondary structures in 123 RBSs and compared the results with structures surrounding noninitiator AUGs. The noninitiation sites were found to form stable secondary structures much more frequently than the true initiation sites. The differences, then, in the preferred RBS nucleotides outside the SD sequence and initiation codon between *B. subtilis* and *E. coli* may simply reflect an alternate evolutionary pathway to decrease secondary structure around RBSs rather than different specific ribosome-mRNA contacts.

Secondary-Structure Effects

The most complete analysis to date of the influence of mRNA secondary structure on translation initiation in prokaryotes was reported by de Smit and van Duin (22, 23). By systematically altering the stability of the MS2 coat gene hairpin, they found a linear relationship between hairpin stability and coat protein yield in *E. coli*. Each 2-kcal/mol decrease in the stability of the hairpin resulted in a roughly 10-fold increase in expression. Numerous other reports demonstrate the effect of secondary structure in the RBS region on translation in *E. coli* (51, 71, 88, 126, 129, 135). In most cases, secondary structure decreases the level of translation. In *B. subtilis*, expression of the *cat* and *ermC* gene products provides well-known examples of control by secondary structure (see below). A G-to-T mutation in the SD region of the *repU* gene of plasmid pUB110 reduces copy number and increases segregational stability (78). While not altering the strength of the SD sequence, the mutation did increase the stability of a hairpin involving the SD sequence, implying a role for secondary structure in decreased translation of the *repU* message. In the Rabinowitz laboratory, we have recently investigated the effect of three base changes in the spacer region of a synthetic RBS that introduced a stem-loop downstream of the SD sequence and included part of the initiation codon (142). The changes caused a decrease in translational yield of 98% in *B. subtilis* but only a 2-fold decrease in yield in *E. coli*. It thus appears that *B. subtilis* ribosomes are less able to tolerate secondary structure in the RBS than are *E. coli* ribosomes. This difference may reflect the absence in *B. subtilis* of an S1-like protein, known to contain a helix-destabilizing activity (7, 73, 134, 137, 138, 141).

Implications of the Absence of S1

B. subtilis and all of the clostridial gram-positive organisms that have been surveyed lack a counterpart to the largest *E. coli* ribosomal protein, S1 (56, 57, 67, 95, 111, 119). All of the gram-negative organisms surveyed as well as the gram-positive organisms that make up the actinomycete branch contain S1-like proteins as part of their ribosomal-protein makeups (111). The absence of protein S1 correlates with translational specificity, while organisms containing S1 are translationally nonspecific (Fig. 1). Higo et al. (57) first suggested that the absence of S1 may account for species specificity in translation. Roberts and Rabinowitz tested this idea by preparing S1 from *E. coli* ribosomes and determining the effect of its addition to *B. subtilis* ribosomes in in vitro translation reactions (109, 111). They found that while S1 would bind to *B. subtilis* 30S subunits and stimulate their translation of poly(U), the addition did not affect the translation of natural *E. coli* messages by *B. subtilis* ribosomes. More interestingly, *E. coli* ribosomes depleted of S1 preferentially translated *B. subtilis* mRNAs over homologous messages. A more thorough analysis of the translational characteristics of S1-depleted *E. coli* ribosomes revealed that they are dependent on a strong SD sequence, much as *B. subtilis* ribosomes are (29a). Although S1 addition did not overcome the translational specificity of *B. subtilis* ribosomes, the results suggest a different mechanism for initiation than that exhibited by *E. coli*. S1 is thought to bind RNA nonspecifically and bring it into the decoding site of the 30S subunit, where proper positioning of the SD sequence and initiation codon signals can take place (22, 132, 133). This additional step in binding mRNA may somewhat alleviate the necessity of a strong SD sequence interaction for binding. In the absence of S1, the *B. subtilis* 30S subunit and the S1-depleted *E. coli* 30S subunit probably rely exclusively on a strong interaction between the SD sequence and the 16S rRNA for efficient binding of mRNA.

Definition of an Optimal *B. subtilis* RBS

Of utmost importance for high-level initiation of protein synthesis in *B. subtilis* is the presence of a strong SD sequence. The most highly expressed gene products contain 7 to 10 complementary bases (93), but at a minimum, 6 bases, including the canonical GGAGG sequence, should be present. Seven- to 9-base spacings between the SD sequence and the initiation codon appear to be about optimal; fewer than 6 nucleotides severely diminishes translational yield, while spacings of 12 bases decrease yields by about 40%. AUG appears to be the codon of choice; GUG or UUG reduces translation, but the magnitude of the reduction depends on the strength of the SD sequence. The entire RBS region must be free of secondary structure for efficient recognition by the *B. subtilis* ribosome, although it appears in one instance that enhanced translation from a poor SD sequence can be attained by placing the SD sequence and the initiation codon in the loop of a stem. The spacer region should be free of Cs and Gs, since As and Us (but especially As) are more conducive to high-level expression. Preferred nucleotides at specific positions include Us at −20, −19, −6, +5, and +16; As at −18 and from +6 to +15; and Cs at +12 and +23 (49). In *E. coli*, nucleotides upstream of the defined RBS appear important for maximal translation. While a few examples of *E. coli* messages starting with an AUG or with very little upstream sequence exist (105, 107), most mRNAs require a leader extending further upstream than that bound by the ribosome. *B. subtilis* most likely also requires additional sequences upstream of the tailing edge of a bound ribosome for efficient expression; a transcript beginning at the SD sequence is not translated in *B. subtilis* (108).

TRANSLATIONAL CONTROL MECHANISMS

Potential Means of Translational Control

Given the known mRNA sequence requirements for efficient initiation of translation, slight sequence differences can easily result in a vast range of protein yields. With equivalent mRNA levels, increasing or decreasing the strength of the SD sequence can lead to differences in yield of at least 2 and probably 3 orders of magnitude. Thus, the expression of proteins that may be deleterious to the *B. subtilis* cell at high levels can be controlled simply by decreasing the strength of the SD interaction. The range of SD strengths in *B. subtilis* genes is −12 to −22 kcal/mol, and these differences could account for much of the variation of individual protein levels found in the cell, ignoring, for the moment, differential promoter strengths. Short spacings may also serve to limit expression at the translational level, but natural *Bacillus* genes with spacings of less than 7 bases are uncommon.

The use of a non-AUG initiation codon in the adenylate cyclase gene of *E. coli* serves to limit the amount of expression of the enzyme, which at high levels is toxic to the cell (110). Non-AUG codons are much more prevalent in *B. subtilis* than in *E. coli* (30 versus 9%), so it is possible that expression of that 30% of *B. subtilis* genes is also restricted at the translational level. However, the influence of a non-AUG initiator depends on the strength of the SD sequence interaction (29a, 142). Thus, messages with strong SD sequences, as found in natural *B. subtilis* genes, are less affected by the presence of a non-AUG start codon, especially GUG, than are messages with weak SD sequences, as are normally found in *E. coli*. While the use of UUG or GUG still diminishes translational yield in *B. subtilis*, such use is probably not as important as a means of controlling gene expression as it is in *E. coli*.

In my studies, secondary structures involving the RBS have the most drastic effect on translation in *B. subtilis*. The *B. subtilis* ribosome appears to be unable to initiate on structured messages to any significant extent and initiates much less than an *E. coli* ribosome in the same situation. This inability may be related to the absence of an S1-like protein in the initiation process. Even secondary structures in the open reading frame of a *B. subtilis* message appear to be inhibitory to translation (74), a phenomenon not observed in *E. coli* (20, 94, 97). As in *E. coli* (35, 63), inhibition of translation by secondary structure in *B. subtilis* may be related more to the exact nature of the structure than to simply the presence of a duplex in the RBS (80). Thus, the formation of alternate secondary structures can serve as an absolute means of controlling expression at the level of translation, as has been found with antibiotic-inducible genes (see below).

In a number of organisms, a statistical examination of coding sequences shows that synonymous codon usage is not random (65, 96). The degree of bias in codon selection correlates with the level of expression of a given gene product in *E. coli* (40, 121), *Saccharomyces cerevisiae* (11, 122), and *Neurospora crassa* (70). Highly expressed genes in these species contain codons that correspond to the most abundant of the isoaccepting tRNA species, while genes expressed at low levels tend to contain an abundance of rare codons. Thus, the use of codon bias can serve as a means of translationally controlling the level of a gene product. The first examination of codon usage in *B. subtilis* genes suggested that relatively little bias exists in this organism (102). In a more extensive analysis, Shields and Sharp (123) determined the synonymous codon usage of a set of 56 *B. subtilis* genes. They found by correspondence analysis that highly expressed genes, such as the ribosomal-protein genes and the small, acid-soluble spore protein (SASP) genes, showed a relatively high degree of bias, while poorly expressed genes exhibited a usage pattern least like those of the former. In the high-bias genes, certain codons were avoided, suggesting that their presence might limit expression, although cases of high expression levels being achieved even with nonoptimal codons are known (123). It is important to note that the preferred codons in *B. subtilis* are generally different from those in *E. coli* or *S. cerevisiae*, and this fact may be more relevant to the heterologous expression of foreign genes than to translational control in *B. subtilis* per se.

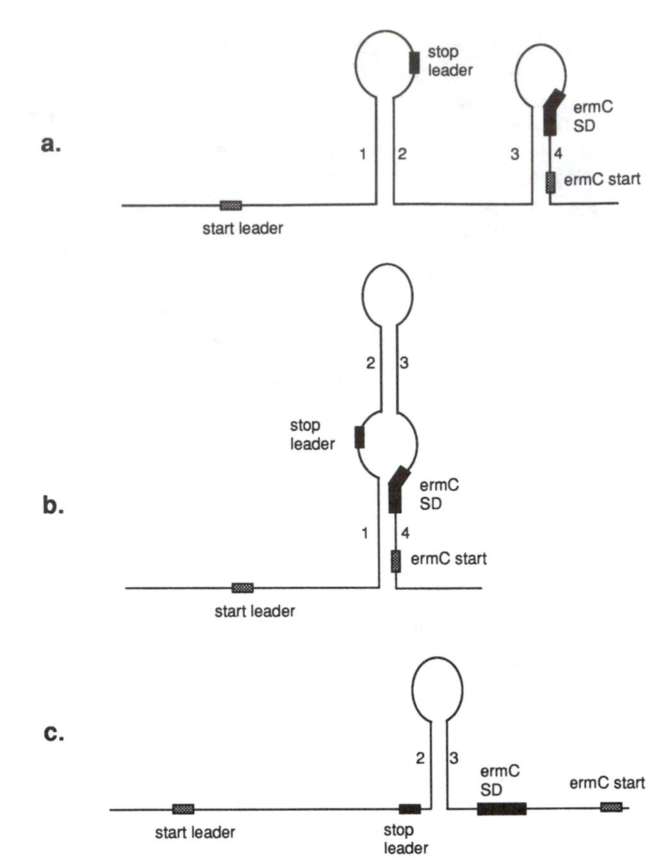

Figure 2. Schematics of alternative secondary structures involved in translational attenuation of the *ermC* gene. See text for discussion.

Translational Attenuation of *erm* Genes

Resistance to the macrolide-lincosamide-streptogramin B group of antibiotics involves dimethylation of a specific A residue on the 23S rRNA by an rRNA methyltransferase. Methylated ribosomes exhibit a markedly reduced affinity for the macrolide-lincosamide-streptogramin B antibiotics. The *ermC* gene, which codes for a 29,000-Da methylase that confers resistance to erythromycin, was first discovered on the *Staphylococcus aureus* plasmid pE194 and later introduced into *B. subtilis* by transformation. The gene is constitutively transcribed but leads to very little synthesis of the methylase unless the cells are exposed to erythromycin. At subinhibitory concentrations of the antibiotic, cells carrying the *ermC* gene acquire a resistance to inhibitory levels of erythromycin within 1 h (143, 144) without an increase in *ermC* transcription. When the sequence of the *ermC* gene was reported simultaneously by two groups (43, 62), a model was put forth to explain the posttranscriptional induction of methylase synthesis in the presence of erythromycin. In Fig. 2 are schematic representations of the 5′ region of the *ermC* gene. The methylase gene is preceded by an open reading frame coding for a translatable leader peptide of 19 amino acids. Within this region are three alternative secondary structures. Two of these structures are inhibitory for methylase synthesis because of the inaccessibility of the methylase SD sequence and the initiation codon (Fig. 2a and b), and one structure is active for methylase synthesis (Fig. 2c). It was proposed (43, 62) that translation of the leader in the presence of erythromycin would cause a stall of the ribosome, resulting in the destabilization of stem 1-4 or 3-4 and freeing the methylase RBS for high-level initiation.

Numerous experiments have demonstrated the roles of both stalling and secondary structure in *ermC* translational attenuation (reviewed by Bechhofer [8] and de Smit and van Duin [22]). Translation of the leader is necessary for induction, since both a nonsense codon at position 2 (100) and oligonucleotides complementary to the leader RBS (99) inhibited methylase synthesis. An extensive analysis of amino acid substitutions in the leader peptide (86) showed that the nature of residues 5 through 9 and not that of the nucleotide sequence is important for induction. It was concluded that an erythromycin-bound ribosome stalls with position 8 in the P site and position 9 in the A site because of the interaction of erythromycin with residues 5 through 9 of the leader peptide. In vitro nuclease protection analysis (99) suggested that the stall site occurred at positions 12 through 15, although de Smit and van Duin (22) interpreted the data to suggest that the stall occurs around position 10. Given the size of a ribosome, a stall at this position would only melt the first 5 bp of stem 1-4 (or 1-2), but the destabilization may be enough to cause isomerization to 2-3 pairing and enable a 30S subunit to initiate at the methylase RBS (86). By probing the structure of the leader region with enzymes and chemical reagents, two groups confirmed the secondary-structure predictions of the induction model (85, 98), although the two alternative inhibitory structures (Fig. 2a and b) could not be distinguished.

Methylase synthesis is further controlled by a methylation-mediated feedback loop and autoregulation. In the first case, the induction of *ermC* requires the presence of sensitive ribosomes for erythromycin-induced stalling and resistant ribosomes for *ermC* translation (100). As the degree of ribosome methylation increases following induction, the fraction of sensitive ribosomes decreases and induction ceases. Initial *ermC* translation is provided for by a low basal level of methylase production (50). Methylase synthesis is also autoregulated through an interaction of the methylase protein with its own mRNA. Although the *ermC* mRNA was initially proposed to resemble the 23S rRNA methylation site (21), further analysis suggested that the overall structure of the 5′ region is important for recognition by the methylase (13). One final means of posttranscriptional control involves the stabilization of *ermC* mRNA and the homologous, inducible *ermA* transcript as a result of stalling (9, 115). Stabilization appears to result from protection of the transcript from a novel 5′-to-3′ exonuclease or a processive endonuclease (10, 116).

Control of *cat* Gene Expression

Chloramphenicol resistance in bacteria involves acetylation of the drug by the enzyme CAT. In gram-negative organisms, *cat* genes are nearly always expressed constitutively, but in the gram-positive species *Bacillus* and *Staphylococcus*, expression is inducible by chloramphenicol and other antibiotics. The *cat-86* gene of *Bacillus pumilus*, since transferred into a promoter-probe plasmid in *B. subtilis* (146), is inducible by chloramphenicol independently of the promoter used (91, 92) and is observed in the presence of rifampin (27), implying a posttranscriptional mode of induction. The sequences of most of the inducible *cat* genes reveal the presence of a short upstream open reading frame overlapping a hairpin structure containing the *cat* SD sequence (28). Further analysis has demonstrated a translational attenuation mode of induction similar to that described for the *ermC* gene (reviewed by Lovett [83]) and including stabilization of the *cat* mRNA (25).

Ambulos et al. (4) showed by deletion analysis that although two leader sequences are found upstream of *cat-86*, only the most 5′ proximal is necessary for chloramphenicol induction. This leader sequence codes for a 9-amino-acid peptide that terminates just before the loop of the hairpin. Destruction of the hairpin by deletion (4, 15, 53, 77) resulted in constitutive expression of the *cat* gene, demonstrating a role for secondary structure in the inhibition of *cat* translation. Translation of the leader is required for induction, since replacements of leader codon positions 2 through 5 with nonsense codons or starvation for the amino acids encoded by codons 2 through 5 resulted in a loss of inducibility (2, 14, 24). Both ochre replacements and amino acid starvation experiments implicate codon 6 in the A site as the site at which the ribosome stalls (2, 24, 26). Thus, codons 2 through 5 specify a stall sequence (113) that interacts with a chloramphenicol-bound ribosome to induce a translational pause at a site that disrupts the inhibitory duplex and allows entry of initiating ribosomes into the *cat* RBS. By analogy with *ermC* induction, the peptide itself may cause the ribosome to stall, but specific nucleotides in this region may also interact directly with the ribosome to facilitate stalling (83, 113, 114).

Each stall sequence of six inducible *cat* genes contains a 12-base region, termed the *crb* box (112), complementary to sequences between positions 1294 and 1305 of the 16S rRNA. These *crb* boxes are 50 to 83% homologous to the rRNA sequence, but the homology increases to 75 to 88% if a more limited 8-base region is considered (83). Rogers et al. (112) tested the role of this sequence element in the stalling of ribosomes in growth rate experiments. They hypothesized that in the presence of chloramphenicol, high-level expression of the *crb* box from a multicopy plasmid would remove enough of the ribosome pool through stalling to retard the growth rates of the cells. Growth rate reductions dependent on the presence of the *crb* box were indeed observed. Changes in the sequence of the *crb* box that decreased its complementarity to the 16S rRNA but retained the amino acid sequence restored the normal growth rate. The minimum stability of the *crb* box–16S-rRNA interaction necessary for growth rate reduction was −11.8 kcal/mol (wild-type stability of *cat-86* is −14.2 kcal/mol), although it is unclear in this model why some chloramphenicol-inducible *cat* genes have stabilities much lower than this minimum (112).

Control in the *trp* Operon

The control of *trp* operon expression in *B. subtilis* provides another example of translational control at the level of mRNA secondary structure. *trp* operon expression is controlled primarily by transcriptional attenuation involving alternate secondary structures that produce either a transcriptional terminator or an antiterminator. In contrast to *trp* attenuation in *E. coli* (76), where the translation rate of a leader determines the formation of alternate structures, in *B. subtilis*, a *trans*-acting regulatory factor, a product of the *mtr* gene (59), binds specifically to a region of the leader mRNA in the presence of tryptophan to induce termination of transcription (124). However, the fact that attenuation by *trp* resulted in 1,000-fold regulation, when only about 100-fold regulation was evident in other examples of transcriptional attenuation, suggested to Kuroda et al. (75) that an additional means of control was operative. They found by comparing transcriptional and translational fusions of the *trpE* sequence to *lacZ* that *mtr* control also involved an inhibition of translation of the escaped *trpE* messages. Binding of the *mtr* protein to its mRNA binding site in the leader was proposed to enable the formation of a secondary structure sequestering the *trpE* SD sequence, thus exerting an additional 14-fold decrease in *trpE* expression.

Coupling Phenomena

Translational coupling, or the reinitiation of a terminating ribosome on an adjacent RBS, is a mechanism observed in several *E. coli* operons that controls the synthesis of individual members of the operons (37). Most examples of translational coupling in *B. subtilis* are artificial *B. subtilis-E. coli* gene constructs on plasmids. The general phenomenon observed is that a poor *E. coli* SD sequence can be recognized by the *B. subtilis* ribosome when the sequence is placed near a termination codon of an efficiently translated *B. subtilis* gene, while a disruption of translation through the upstream coding sequence causes a loss of expression of the downstream gene (32, 104, 149). Although each of these constructs is unnatural, such a mechanism may occur in at least one operon in *Bacillus licheniformis*. Resistance to β-lactams in this organism is inducible by subinhibitory concentrations of the antibiotic (19). Control of penicillinase expression appears to involve both a repressor and an antirepressor, which modulate the activity of the penicillinase gene promoter. The sequence of the antirepressor gene (*penJ* or *blaR1*) was published simultaneously by two groups (66, 72). The sequence reveals that the antirepressor gene follows the repressor gene (*penI*) in a manner very similar to that of the coupled genes alluded to above. Furthermore, the SD sequence for *penJ* (*blaR1*) is rather poor and may be sequestered in a hairpin structure (72). Thus, it is possible that translation of the antirepressor requires translation through *penI* for the ribosome to disrupt the secondary structure and reinitiate at the adjacent, poor antirepressor SD sequence.

TRANSLATION IN OTHER GRAM-POSITIVE ORGANISMS

All of the clostridial gram-positive organisms whose translational characteristics have been analyzed display essentially the same features as *B. subtilis* (111, 128). None contain an S1-like ribosomal protein, and thus, they initiate translation with an emphasis on the SD–16S-rRNA interaction. Even the high-A+T *Clostridium* spp. contain very strong SD sequences in their genes (41, 145). Gene regulation at the translational level probably takes advantage of the same features found in *B. subtilis*, especially in light of the fact that the *cat-86* and *ermC* genes originally came from *B. pumilus* and *S. aureus*, respectively.

Organisms of the actinomycete group have a number of features that distinguish them from other gram-positive organisms. Their genomes generally have very high G+C contents (66 to 73%), which is immediately reflected in the codon usage of the known genes (69) and is apparently responsible for the much more frequent use of GUG as an initiation codon in *Micrococcus luteus* and *Streptomyces* spp. than in *B. subtilis*, even in highly expressed genes (60, 103). The average SD sequence binding strength of the known *M. luteus* genes is −14.5 kcal/mol (29), which is intermediate between those of *E. coli* (−11.7) and *B. subtilis* (−16.7) (49). The SD sequence binding strengths of the three genes constituting the *Streptomyces lividans gal* operon range from −9 to −13.5 kcal/mol (1). The aminoglycoside phosphotransferase gene of *Streptomyces fradiae* contains no apparent SD sequence in the upstream region (139), while the streptothricin acetyltransferase mRNA from *S. lavendulae* begins right at the AUG initiation codon (61). The entire RBS regions of the known actinomycete genes contain a preponderance of Cs and Gs, especially at −5 and −6 (60), again probably reflecting the high G+C content but very different from the selected exclusion of those bases from *B. subtilis* and *E. coli* mRNAs. In that preferred nucleotides in the RBS region of *B. subtilis* and *E. coli* genes reflect either additional ribosome-message contacts or a need to diminish secondary structures, such contacts or RBS structural features in the actinomycetes may be fundamentally different from the *B. subtilis* paradigm. Like the gram-negative organisms but unlike the clostridial gram-positive organisms, all the actinomycetes contain an S1-like ribosomal protein (Fig. 1). Thus, these organisms are apparently capable of initiating translation in a manner less dependent on the strength of the SD sequence interaction, as observed in *E. coli*. It is interesting, however, that the *Streptomyces aureofaciens* S1 protein (SS1) is only partially homologous to *E. coli* S1 either structurally or functionally (90).

It was noted early on that antibiotic resistance genes from *E. coli* confer resistance on *S. lividans* (120). Bibb and Cohen found that the *E. coli cat* gene was expressed in *S. lividans*, and they correctly suggested that the barrier to heterologous expression was not simply a gram-positive–gram-negative distinction, as initially proposed by McLaughlin et al. (87), but rather was species specific (12). In vitro experiments have confirmed that *S. lividans* ribosomes can efficiently translate *E. coli* mRNAs (140). Similarly, *M. luteus* ribosomes are nonspecific in their selection of mRNAs, translating messages from *M. luteus*, *B. subtilis*, and *E. coli* (29). Thus, ribosomes of the actinomycete species are different from those of the clostridial gram-positive organisms in having an S1-like protein and being able to efficiently translate gram-negative mRNAs.

The use of an in vitro translation system has allowed investigations into the nature of resistance to translational inhibitors in *Streptomyces* spp. In one study, resistance to the aminoglycosides kanamycin, apramycin, and sisomicin was dependent on methylation of the 16S rRNA (127). More recently, a fractionated, coupled transcription-translation system was used to investigate the nature of resistance to panctamycin, a potent inhibitor of translation in both prokaryotes and eukaryotes, in panctamycin-producing *Streptomyces* spp. (16). Through mixing experiments with ribosomal subunits from *S. lividans* and *Streptomyces panctum*, the 30S subunit was found to be responsible for the resistance. An interesting finding in this survey was that some antibiotic-producing species apparently do not constitutively produce a resistance determinant that can be isolated with the ribosome. No attempts to induce the resistance phenotype were made, but it will be interesting to see what the mode of induction is, whether transcriptional or, as with the *cat* and *erm* genes, translational.

Recently, Leskiw et al. proposed (79) that the very rare TTA leucine codon in *Streptomyces* genes might play a role in developmentally regulated gene expression during differentiation. It had been known that mutations in the structural gene for $tRNA_{TTA}$, while having no effect on vegetative growth, prevent the emergence of aerial hyphae and the production of antibiotics. In 100 *Streptomyces* genes analyzed, the distribution of TTA codons was confined chiefly to genes required for aerial mycelium formation or antibiotic production (79). Thus, these species may use a rare codon to specifically control the expression of genes at different developmental stages through a translational mechanism.

CONCLUSION

While transcriptional control through alternative sigma factors and variable promoter strengths plays a major role in gene regulation, the examples discussed above highlight the importance of translational control as a means of setting the final yield of a protein product in *B. subtilis*. SD sequence strengths, the initiation codon, spacing differences, and the nature of nucleotide distribution within the RBS all determine the efficiency with which a RBS is recognized by the *B. subtilis* ribosome. Given the extremely inhibitory influence of mRNA secondary structure within the RBS, it is apparent why these structures play such an important role in translational regulation in the cases described.

Although this discussion has focused on control at the level of initiation of translation, other means determine the translatability of a message in *B. subtilis*. Variation in mRNA half-life is an important aspect in control of the expression of SASP and *sdh* genes (84, 89). The production of an antisense RNA transcript controls the synthesis of RepH and, in turn, the replication of pC194 (3). Simple promoter switching in the *sigA* operon controls the synthesis of DNA primase by producing a transcript no longer functional for translation of the primase mRNA (108). The induction of σ^H upon sporulation appears not to occur at the level of transcription but rather through a stimulation of translation of the σ^H mRNA by some unknown mechanism (54). Continued analysis of these and other mechanisms of translational control will lead to a more complete understanding of how *B. subtilis* fully controls the synthesis of its genes.

Acknowledgments. I thank Thomas W. Okita of the IBC at Washington State University, in whose laboratory this chapter was written; the McKnight Foundation for postdoctoral fellowship support; and Jesse C. Rabinowitz for comments.

REFERENCES

1. **Adams, C. W., J. A. Fornwald, F. J. Schmidt, M. Rosenberg, and M. E. Brawner.** 1988. Gene organization and structure of the *Streptomyces lividans gal* operon. *J. Bacteriol.* **170:**203–212.
2. **Alexieva, Z., E. J. Duvall, N. P. Ambulos, Jr., U. J. Kim, and P. S. Lovett.** 1988. Chloramphenicol induction of *cat-86* requires ribosome stalling at a specific site in the leader. *Proc. Natl. Acad. Sci. USA* **85:**3057–3061.
3. **Alonso, J. C., and R. H. Tailor.** 1989. Initiation of plasmid pC194 replication and its control in *Bacillus subtilis*. *Mol. Gen. Genet.* **210:**476–484.
4. **Ambulos, N. P., Jr., E. J. Duvall, and P. S. Lovett.** 1986. Analysis of the regulatory sequences needed for induction of the chloramphenicol acetyltransferase gene *cat-86* by chloramphenicol and amecetin. *J. Bacteriol.* **167:**842–849.
5. **Ambulos, N. P., Jr., T. Smith, W. Mulbry, and P. S. Lovett.** 1990. CUG as a mutant start codon for *cat-86* and *xylE* in *Bacillus subtilis*. *Gene* **94:**125–128.
6. **Band, L., and D. J. Henner.** 1984. *Bacillus subtilis* requires a "stringent" Shine-Dalgarno region for gene expression. *DNA* **3:**17–21.
7. **Bear, D. G., R. Ng, D. van Deever, N. P. Johnson, G. Thomas, T. Schleich, and H. F. Noller.** 1976. Alteration of polynucleotide secondary structure by ribosomal protein S1. *Proc. Natl. Acad. Sci. USA* **73:**1824–1828.
8. **Bechhofer, D. H.** 1990. Triple post-transcriptional control. *Mol. Microbiol.* **4:**1419–1423.
9. **Bechhofer, D. H., and D. Dubnau.** 1987. Induced mRNA stability in *Bacillus subtilis*. *Proc. Natl. Acad. Sci. USA* **84:**498–502.
10. **Bechhofer, D. H., and K. Zen.** 1989. Mechanism of erythromycin-induced *ermC* mRNA stability in *Bacillus subtilis*. *J. Bacteriol.* **171:**5803–5811.
11. **Bennetzen, J. L., and B. D. Hall.** 1982. Codon selection in yeast. *J. Biol. Chem.* **257:**3026–3031.
12. **Bibb, M. J., and S. N. Cohen.** 1982. Gene expression in *Streptomyces*: construction and application of promoter-probe plasmid vectors in *Streptomyces lividans*. *Mol. Gen. Genet.* **187:**265–277.
13. **Breidt, F., and D. Dubnau.** 1990. Identification of *cis*-acting sequences required for translational autoregulation of the *ermC* methylase. *J. Bacteriol.* **172:**3661–3668.
14. **Bruckner, R., T. Dick, H. Matzura, and E. Zyprian.** 1988. Regulation of inducible *Staphylococcus aureus* CAT gene by translational attenuation, p. 263–266. *In* A. T. Ganesan and J. A. Hoch (ed.), *Genetics and Bio-*

technology of Bacilli. Academic Press, Inc., San Diego, Calif.

15. **Bruckner, R., and H. Matzura.** 1985. Regulation of the inducible chloramphenicol acetyltransferase gene of the *Staphylococcus aureus* plasmid pUB110. *EMBO J.* **4:**2295–2300.
16. **Calcutt, M. J., and E. Cundliffe.** 1989. Use of a fractionated, coupled transcription-translation system in the study of ribosomal resistance mechanisms in antibiotic-producing *Streptomyces*. *J. Gen. Microbiol.* **135:**1071–1081.
17. **Canonaco, M. A., R. A. Calogero, and C. O. Gualerzi.** 1986. Mechanism of translational initiation in procaryotes. Evidence for a direct effect of IF2 on the activity of the 30S ribosomal subunit. *FEBS Lett.* **207:**198–204.
18. **Chen, N.-Y., and H. Paulus.** 1988. Mechanism of expression of the overlapping genes of *Bacillus subtilis* aspartokinase II. *J. Biol. Chem.* **263:**9526–9532.
19. **Collins, J. R.** 1979. The *Bacillus licheniformis* β-lactamase system, p. 351–368. *In* M. T. Hamilton-Miller and J. T. Smith (ed.), *Beta-Lactamases*. Academic Press Ltd., London.
20. **Cone, K. C., and D. A. Steege.** 1985. Messenger RNA conformation and ribosome selection of translational reinitiation sites in the lac repressor mRNA. *J. Mol. Biol.* **186:**725–732.
21. **Denoya, C. D., D. H. Bechhofer, and D. Dubnau.** 1986. Translational autoregulation of *ermC* 23S rRNA methyltransferase expression in *Bacillus subtilis*. *J. Bacteriol.* **168:**1133–1141.
22. **de Smit, M. H., and J. van Duin.** 1990. Control of procaryotic translational initiation by mRNA secondary structure. *Prog. Nucleic Acid Res. Mol. Biol.* **38:**1–35.
23. **de Smit, M. H., and J. van Duin.** 1990. Secondary structure of the ribosome binding site determines translational efficiency. *Proc. Natl. Acad. Sci. USA* **87:**7668–7672.
24. **Dick, T., and H. Matzura.** 1988. Positioning ribosomes on leader mRNA for translational activation of the message of an inducible *Staphylococcus aureus cat* gene. *Mol. Gen. Genet.* **214:**108–111.
25. **Dreher, J., and H. Matzura.** 1991. Chloramphenicol-induced stabilization of *cat* messenger RNA in *Bacillus subtilis*. *Mol. Microbiol.* **5:**3025–3034.
26. **Duvall, E. J., N. P. Ambulos, Jr., and P. S. Lovett.** 1987. Drug-free induction of a chloramphenicol acetyltransferase gene in *Bacillus subtilis* by stalling ribosomes in a regulatory leader. *J. Bacteriol.* **169:**4235–4241.
27. **Duvall, E. J., and P. S. Lovett.** 1986. Chloramphenicol induces translation of the mRNA for a chloramphenicol-resistance gene in *Bacillus subtilis*. *Proc. Natl. Acad. Sci. USA* **83:**3939–3943.
28. **Duvall, E. J., D. M. Williams, P. S. Lovett, C. Rudolph, N. Vasantha, and M. Guyer.** 1983. Chloramphenicol-inducible gene expression in *Bacillus subtilis*. *Gene* **24:**170–177.
29. **Farwell, M. A., and J. C. Rabinowitz.** 1991. Protein synthesis in vitro by *Micrococcus luteus*. *J. Bacteriol.* **173:**3514–3522.

29a. **Farwell, M. A., M. W. Roberts, and J. C. Rabinowitz.** 1992. The effect of ribosomal protein S1 from *Escherichia coli* and *Micrococcus luteus* on protein synthesis in vitro by *E. coli* and *Bacillus subtilis*. *Mol. Microbiol.* **6:**3375–3383.

30. **Fox, G. E., E. Stackenbrandt, R. B. Hespell, J. Gibson, J. Maniloff, T. A. Dyer, R. S. Wolfe, W. E. Balch, R. S. Tanner, L. J. Magrum, L. B. Zablen, R. Blakemore, R. Gupta, L. Bonen, B. J. Lewis, D. A. Stahl, K. R. Luehrsen, K. N. Chen, and C. R. Woese.** 1980. The phylogeny of procaryotes. *Science* **209:**457–463.
31. **Friden, H., L. Rutberg, K. Magnusson, and L. Hederstedt.** 1987. Genetic and biochemical characterization of *Bacillus subtilis* mutants defective in expression and function of cytochrome b-558. *Eur. J. Biochem.* **168:**695–701.
32. **Fujiwara, S., N. Tsubokura, Y. Kurusu, K. Minami, and Y. Kobayashi.** 1990. Heat-inducible translational coupling in *Bacillus subtilis*. *Nucleic Acids Res.* **18:**739–744.
33. **Ganoza, M. C., E. C. Kofold, P. Marliere, and B. G. Louis.** 1987. Potential secondary structure at translation-initiation sites. *Nucleic Acids Res.* **15:**345–360.
34. **Gheysen, D., D. Iserentant, C. Derom, and W. Fiers.** 1982. Systematic alteration of the nucleotide sequence preceding the translation initiation codon and the effects on bacterial expression of the cloned SV40 small-t antigen gene. *Gene* **17:**55–63.

34a. **Gold, L.** Unpublished data.

35. **Gold, L.** 1988. Posttranscriptional regulatory mechanisms in *Escherichia coli*. *Annu. Rev. Biochem.* **57:**199–233.
36. **Gold, L., D. Pribnow, T. Schneider, S. Shinedling, B. S. Singer, and G. Stormo.** 1981. Translational initiation in procaryotes. *Annu. Rev. Microbiol.* **35:**365–403.
37. **Gold, L., and G. Stormo.** 1987. Translational initiation, p. 1302–1307. *In* F. C. Neidhardt, J. L. Ingraham, K. B. Low, B. Magasanik, M. Schaechter, and H. E. Umbarger (ed.), *Escherichia coli and Salmonella typhimurium: Cellular and Molecular Biology*, vol. 2. American Society for Microbiology, Washington, D.C.
38. **Gold, L., and G. D. Stormo.** 1990. High-level translation initiation. *Methods Enzymol.* **185:**89–93.
39. **Gold, L., G. Stormo, and R. Saunders.** 1984. *Escherichia coli* translational initiation factor IF3: a unique case of translational regulation. *Proc. Natl. Acad. Sci. USA* **81:**7061–7065.
40. **Gouy, M., and C. Gautier.** 1982. Codon usage in bacteria: correlation with gene expressivity. *Nucleic Acids Res.* **10:**7055–7074.
41. **Graves, M. C., G. T. Mullenbach, and J. C. Rabinowitz.** 1985. Cloning and nucleotide sequence determination of the *Clostridium pasteurianum* ferredoxin gene. *Proc. Natl. Acad. Sci. USA* **82:**1653–1657.
42. **Green, C. J., G. C. Stewart, M. A. Hollis, B. S. Vold, and K. F. Bott.** 1985. Nucleotide sequence of the *Bacillus subtilis* ribosomal RNA operon, *rrnB*. *Gene* **37:**261–266.
43. **Gryczan, T. J., G. Grandi, J. Hahn, R. Grandi, and D. Dubnau.** 1980. Conformational alteration of mRNA structure and the posttranscriptional regulation of erythromycin-induced drug resistance. *Nucleic Acids Res.* **8:**6081–6097.
44. **Gualerzi, C., G. Risuleo, and C. L. Pon.** 1977. Initial rate kinetic analysis of the mechanism of initiation complex formation and the role of initiation factor IF3. *Biochemistry* **16:**1684–1689.
45. **Gualerzi, C. O., R. A. Calogero, M. A. Canonaco, M. Brombach, and C. L. Pon.** 1987. Selection of mRNA by ribosomes during procaryotic translation initiation, p. 317–330. *In* M. F. Tuite, M. Picard, and M. Bolotin-Fukuhara (ed.), *Genetics of Translation: New Approaches*. Springer-Verlag, Heidelberg, Germany.
46. **Gualerzi, C. O., and C. L. Pon.** 1981. Protein biosynthesis in procaryotic cells: mechanism of 30S initiation complex formation in *Escherichia coli*, p. 805–826. *In* M. Balaban (ed.), *Structural Aspects of Recognition and Assembly in Biological Macromolecules*. Int. Sci. Serv., Rehovot, Israel.
47. **Gualerzi, C. O., C. L. Pon, R. T. Pawlik, M. A. Canonaco, M. Paci, and W. Wintermeyer.** 1986. Role of the initiation factors in *Escherichia coli* translational initiation, p. 621–641. *In* B. Hardesty and G. Kramer (ed.), *Structure, Function and Genetics of Ribosomes*. Springer-Verlag, New York.
48. **Hager, P. W., and J. C. Rabinowitz.** 1985. Inefficient translation of T7 late mRNA by *Bacillus subtilis* ribosomes. *J. Biol. Chem.* **260:**15163–15167.

49. **Hager, P. W., and J. C. Rabinowitz.** 1985. Translational specificity in *Bacillus subtilis*, p. 1–29. *In* D. Dubnau (ed.), *The Molecular Biology of the Bacilli*. Academic Press, Inc., New York.
50. **Hahn, J., G. Grandi, T. J. Gryczan, and D. Dubnau.** 1982. Translational attenuation of *ermC*: a deletion analysis. *Mol. Gen. Genet.* **186:**204–216.
51. **Hall, M. N., J. Gabay, M. Debarbouille, and M. Schwartz.** 1982. A role for mRNA secondary structure in the control of translation initiation. *Nature* (London) **295:**616–618.
52. **Hartz, D., D. S. McPheeters, and L. Gold.** 1991. Influence of mRNA determinants on translation initiation in *Escherichia coli*. *J. Mol. Biol.* **218:**83–97.
53. **Harwood, C. R., D. E. Bell, and A. R. Winston.** 1987. The effects of deletions in the leader sequence of *cat-86*, a chloramphenicol-resistance gene isolated from *Bacillus pumilus*. *Gene* **54:**267–273.
54. **Healy, J., J. Weir, I. Smith, and R. Losick.** 1991. Post-transcriptional control of a sporulation regulatory gene encoding transcription factor σ^H in *Bacillus subtilis*. *Mol. Microbiol.* **5:**477–487.
55. **Hershey, J. W. B.** 1987. Protein synthesis, p. 613–647. *In* F. C. Neidhardt, J. L. Ingraham, K. B. Low, B. Magasanik, M. Schaechter, and H. E. Umbarger (ed.), *Escherichia coli and Salmonella typhimurium: Cellular and Molecular Biology*, vol. 1. American Society for Microbiology, Washington, D.C.
56. **Higo, K.** 1973. Functional correspondence between 30S ribosomal proteins of *Escherichia coli* and *Bacillus stearothermophilus*. *Proc. Natl. Acad. Sci. USA* **70:**944–948.
57. **Higo, K., E. Otaka, and S. Osawa.** 1982. Purification and characterization of 30S ribosomal proteins from *Bacillus subtilis*: correlation to *Escherichia coli* 30S proteins. *Mol. Gen. Genet.* **185:**239–244.
58. **Himes, R. H., M. R. Stallcup, and J. C. Rabinowitz.** 1972. Translation of synthetic and endogenous messenger ribonucleic acid in vitro by ribosomes and polyribosomes from *Clostridium pasteurianum*. *J. Bacteriol.* **112:**1057–1069.
59. **Hoch, S. O., C. W. Roth, I. P. Crawford, and E. W. Nester.** 1971. Control of tryptophan biosynthesis by the methyltryptophan resistance gene in *Bacillus subtilis*. *J. Bacteriol.* **105:**38–45.
60. **Hopwood, D. A., M. J. Bibb, K. F. Chater, G. R. Janssen, F. Malpartida, and C. P. Smith.** 1986. Regulation of gene expression in antibiotic-producing Streptomyces, p. 251–276. *In* I. R. Booth and C. F. Higgins (ed.), *Regulation of Gene Expression. 25 Years On*. Cambridge University Press, Cambridge.
61. **Horinouchi, S., K. Furuya, M. Nishiyama, H. Suzuki, and T. Beppu.** 1987. Nucleotide sequence of the streptothricin acetyltransferase gene from *Streptomyces lavendulae* and its expression in heterologous hosts. *J. Bacteriol.* **169:**1929–1937.
62. **Horinouchi, S., and B. Weisblum.** 1980. Posttranscriptional modification of mRNA conformation: mechanism that regulates erythromycin-induced resistance. *Proc. Natl. Acad. Sci. USA* **77:**7079–7083.
63. **Huang, W. M., S.-Z. Ao, S. Casjens, R. Orlandi, R. Zeikus, R. Weiss, D. Winge, and M. Fang.** 1988. A persistent untranslated sequence within bacteriophage T4 DNA topoisomerase gene 60. *Science* **239:**1005–1012.
64. **Hung, A., J. Thillet, and R. Pictet.** 1989. In vivo selected promoter and ribosome binding site up-mutations: demonstration that the *Escherichia coli bla* promoter and Shine-Dalgarno region with low complementarity to the 16 S ribosomal RNA function in *Bacillus subtilis*. *Mol. Gen. Genet.* **219:**129–136.
65. **Ikemura, T.** 1985. Codon usage and tRNA content in unicellular and multicellular organisms. *Mol. Biol. Evol.* **2:**13–34.
66. **Imanaka, T., T. Himeno, and S. Aiba.** 1987. Cloning and nucleotide sequence of the penicillinase antirepressor gene *penJ* of *Bacillus licheniformis*. *J. Bacteriol.* **169:**3867–3872.
67. **Isono, K., and S. Isono.** 1976. Lack of ribosomal protein S1 in *Bacillus stearothermophilus*. *Proc. Natl. Acad. Sci. USA* **73:**767–770.
68. **Jinks-Robertson, S., and M. Nomura.** 1987. Ribosomes and tRNA, p. 1358–1385. *In* F. C. Neidhardt, J. L. Ingraham, K. B. Low, B. Magasanik, M. Schaechter, and H. E. Umbarger (ed.), *Escherichia coli and Salmonella typhimurium: Cellular and Molecular Biology*, vol. 2, American Society for Microbiology, Washington, D.C.
69. **Kano, A., Y. Andachi, T. Ohama, and S. Osawa.** 1991. Novel anticodon composition of transfer RNAs in *Micrococcus luteus*, a bacterium with a high genomic G+C content. Correlation with codon usage. *J. Mol. Biol.* **221:**387–401.
70. **Kinnaird, J. H., and P. A. Burns.** 1991. An apparent rare-codon effect on the rate of translation of a *Neurospora* gene. *J. Mol. Biol.* **221:**733–736.
71. **Knight, J. A., L. W. Hardy, D. Rennel, D. Herrick, and A. R. Poteete.** 1987. Mutations in an upstream regulatory sequence that increase expression of the bacteriophage T4 lysozyme gene. *J. Bacteriol.* **169:**4630–4636.
72. **Kobayashi, T., Y. F. Zhu, N. J. Nicholls, and J. O. Lampen.** 1987. A second regulatory gene, *blaR1*, encoding a potential penicillin-binding protein required for induction of β-lactamase in *Bacillus licheniformis*. *J. Bacteriol.* **169:**3873–3878.
73. **Kolb, A., J. M. Hermosa, J. O. Thomas, and W. Szer.** 1977. Nucleic acid helix unwinding properties of ribosomal protein S1 and the role of S1 in mRNA binding to ribosomes. *Proc. Natl. Acad. Sci. USA* **74:**2379–2383.
74. **Kubo, M., and T. Imanaka.** 1989. mRNA secondary structure in an open reading frame reduces translation efficiency in *Bacillus subtilis*. *J. Bacteriol.* **171:**4080–4082.
75. **Kuroda, M. I., D. Henner, and C. Yanofsky.** 1988. *cis*-acting sequences in the transcript of the *Bacillus subtilis trp* operon regulate the expression of the operon. *J. Bacteriol.* **170:**3080–3088.
76. **Landick, R., and C. Yanofsky.** 1987. Transcription attenuation, p. 1276–1301. *In* F. C. Neidhardt, J. L. Ingraham, K. B. Low, B. Magasanik, M. Schaechter, and H. E. Umbarger (ed.), *Escherichia coli and Salmonella typhimurium: Cellular and Molecular Biology*, vol. 2. American Society for Microbiology, Washington, D.C.
77. **Laredo, J., V. Wolff, and P. S. Lovett.** 1988. Chloramphenicol acetyltransferase specified by *cat-86*: gene and protein relationships. *Gene* **73:**209–214.
78. **Leonhardt, H.** 1990. Identification of a low copy number mutation within the pUB110 replicon and its effect on plasmid stability in *Bacillus subtilis*. *Gene* **94:**121–124.
79. **Leskiw, B. K., M. J. Bibb, and K. F. Chater.** 1991. The use of a rare codon specifically during development? *Mol. Microbiol.* **5:**2861–2867.
80. **Lin, C.-K., D. S. Goldfarb, R. H. Doi, and R. L. Rodriguez.** 1985. Mutations that affect the translation efficiency of Tn9-derived *cat* gene in *Bacillus subtilis*. *Proc. Natl. Acad. Sci. USA* **82:**173–177.
81. **Lodish, H. F.** 1969. Species specificity of polypeptide chain initiation. *Nature* (London) **224:**867–870.
82. **Lodish, H. F.** 1970. Specificity in bacterial protein synthesis: role of initiation factors and ribosomal subunits. *Nature* (London) **226:**705–707.
83. **Lovett, P. S.** 1990. Translational attenuation as the regulator of inducible *cat* genes. *J. Bacteriol.* **172:**1–6.
84. **Mason, J. M., P. Fajardo-Cavazos, and P. Setlow.** 1988. Levels of mRNAs which code for small, acid-soluble

spore proteins and their *LacZ* gene fusions in sporulating cells of *Bacillus subtilis*. *Nucleic Acids Res.* **16:**6567–6583.

85. **Mayford, M., and B. Weisblum.** 1989. Conformational alterations in the *ermC* transcript in vivo during induction. *EMBO J.* **8:**4307–4314.
86. **Mayford, M., and B. Weisblum.** 1989. *ermC* leader peptide. Amino acid sequence critical for induction by translational attenuation. *J. Mol. Biol.* **206:**69–79.
87. **McLaughlin, J. R., C. L. Murray, and J. C. Rabinowitz.** 1981. Unique features of the ribosomes binding site sequence of the Gram-positive *Staphylococcus aureus* β-lactamase gene. *J. Biol. Chem.* **256:**11283–11291.
88. **McPheeters, D. S., A. Christensen, E. T. Young, G. Stormo, and L. Gold.** 1986. Translational regulation of expression of the bacteriophage T4 lysozyme gene. *Nucleic Acids Res.* **14:**5813–5826.
89. **Melin, L., L. Rutberg, and A. von Gabain.** 1989. Transcriptional and posttranscriptional control of the *Bacillus subtilis* succinate dehydrogenase operon. *J. Bacteriol.* **171:**2110–2115.
90. **Mikulik, K., J. Smardova, A. Jiranova, and P. Branny.** 1986. Molecular and functional properties of protein SS1 from small ribosomal subunits of *Streptomyces aureofaciens*. *Eur. J. Biochem.* **155:**557–563.
91. **Mongkolsuk, S., N. P. Ambulos, Jr., and P. S. Lovett.** 1984. Chloramphenicol-inducible gene expression in *Bacillus subtilis* is independent of the chloramphenicol acetyltransferase structural gene and its promoter. *J. Bacteriol.* **160:**1–8.
92. **Mongkolsuk, S., Y.-W. Chiang, R. B. Reynolds, and P. S. Lovett.** 1983. Restriction fragments that exert promoter activity during postexponential growth of *Bacillus subtilis*. *J. Bacteriol.* **155:**1399–1406.
93. **Mountain, A.** 1989. Gene expression systems for *Bacillus subtilis*, p. 73–114. *In* C. R. Harwood (ed.), *Bacillus*. Plenum Press, New York.
94. **Movva, N. R., K. Nakamura, and M. Inouye.** 1980. Gene structure of the *ompA* protein, a major surface protein of *Escherichia coli* required for cell-cell interaction. *J. Mol. Biol.* **143:**317–328.
95. **Muralikrishna, P., and T. Suryanarayana.** 1985. Comparison of ribosomes from Gram-positive and Gram-negative bacteria with respect to the presence of protein S1. *Biochem. Int.* **11:**691–699.
96. **Murayama, T., T. Gojobori, S. Aota, and T. Ikemura.** 1986. Codon usage tabulated from the GenBank genetic sequence data. *Nucleic Acids Res.* **14:**r151–r197.
97. **Nakamura, K., R. M. Pirtle, I. L. Pirtle, K. Takeish, and M. Inouye.** 1980. Messenger ribonucleic acid of the lipoprotein of the *Escherichia coli* outer membrane. II. The complete nucleotide sequence. *J. Biol. Chem.* **255:** 210–216.
98. **Narayanan, C. S., and D. Dubnau.** 1985. Evidence for the translational attenuation model: ribosome-binding studies and structural analysis with an in vitro run off transcript of *ermC*. *Nucleic Acids Res.* **13:**7307–7326.
99. **Narayanan, C. S., and D. Dubnau.** 1987. Demonstration of erythromicin-dependent stalling of ribosome on the *ermC* leader transcript. *J. Biol. Chem.* **262:**1766–1771.
100. **Narayanan, C. S., and D. Dubnau.** 1987. An in vitro study of the translational attenuation model of *ermC* regulation. *J. Biol. Chem.* **262:**1756–1765.
101. **Noller, H. F., and M. Nomura.** 1987. Ribosomes, p. 104–125. *In* F. C. Neidhardt, J. L. Ingraham, K. B. Low, B. Magasanik, M. Schaechter, and H. E. Umbarger (ed.), *Escherichia coli and Salmonella typhimurium: Cellular and Molecular Biology*, vol. 1. American Society for Microbiology, Washington, D.C.
102. **Ogasawara, N.** 1985. Markedly unbiased codon usage in *Bacillus subtilis*. *Gene* **40:**145–150.
103. **Ohama, T., A. Muto, and S. Osawa.** 1989. Spectinomycin operon of *Micrococcus luteus*: evolutionary implications of organization and novel codon usage. *J. Mol. Evol.* **29:**381–395.
104. **Peijnenburg, A. A. C. M., G. Venema, and S. Bron.** 1990. Translational coupling in a *penP-lacZ* gene fusion in *Bacillus subtilis* and *Escherichia coli*: use of AUA as a restart codon. *Mol. Gen. Genet.* **221:**267–272.
105. **Pirotta, V.** 1979. Operators and promoters in the O_R region of phage 434. *Nucleic Acids Res.* **6:**1495–1508.
106. **Pon, C. L., M. Brombach, S. Thamm, and C. O. Gualerzi.** 1989. Cloning and characterization of a gene cluster from *Bacillus stearothermophilus* comprising *infC*, *rpmI* and *rplT*. *Mol. Gen. Genet.* **218:**355–357.
107. **Ptashne, M., K. Backman, M. Z. Humayun, A. Jeffrey, R. Mauer, B. Meyer, and R. T. Sauer.** 1976. Autoregulation and function of a repressor in bacteriophage lambda. *Science* **194:**156–161.
108. **Qi, F.-X., and R. H. Doi.** 1990. Localization of a second SigH promoter in the *Bacillus subtilis sigA* operon and regulation of *dnaE* expression by the promoter. *J. Bacteriol.* **172:**5631–5636.
109. **Rabinowitz, J. C., and M. Roberts.** 1986. Translational barriers limiting expression of *E. coli* genes in Bacillus and other Gram-positive organisms, p. 297–312. *In* S. B. Levy and R. P. Novick (ed.), *Antibiotic Resistance Genes: Ecology, Transfer, and Expression. Banbury Report 24.* Cold Spring Harbor Laboratory, Cold Spring Harbor, N.Y.
110. **Reddy, P., A. Peterkofsky, and K. McKenney.** 1985. Translational efficiency of the *Escherichia coli* adenylate cyclase gene: mutating the UUG initiation codon to GUG or AUG results in increased gene expression. *Proc. Natl. Acad. Sci. USA* **82:**5656–5660.
111. **Roberts, M. W., and J. C. Rabinowitz.** 1989. The effect of *Escherichia coli* ribosomal protein S1 on the translational specificity of bacterial ribosomes. *J. Biol. Chem.* **264:**2228–2235.
112. **Rogers, E. J., N. P. Ambulos, Jr., and P. S. Lovett.** 1990. Complementarity of *Bacillus subtilis* 16S rRNA with sites of antibiotic-dependent ribosome stalling in *cat* and *erm* leaders. *J. Bacteriol.* **172:**6282–6290.
113. **Rogers, E. J., U. J. Kim, N. P. Ambulos, Jr., and P. S. Lovett.** 1990. Four codons in the *cat-86* leader define a chloramphenicol-sensitive ribosome stall sequence. *J. Bacteriol.* **172:**110–115.
114. **Rogers, E. J., and P. S. Lovett.** 1990. Erythromycin induces expression of the chloramphenicol acetyltransferase gene *cat-86*. *J. Bacteriol.* **172:**4694–4695.
115. **Sandler, P., and B. Weisblum.** 1988. Erythromycin-induced stabilization of *ermA* messenger RNA in *Staphylococcus aureus* and *Bacillus subtilis*. *J. Mol. Biol.* **203:**905–915.
116. **Sandler, P., and B. Weisblum.** 1989. Erythromycin-induced ribosome stall in the *ermA* leader: a barricade to the 5′-to-3′ nucleotide cleavage of the *ermA* transcript. *J. Bacteriol.* **171:**6680–6688.
117. **Scherer, G. F. E., M. D. Walkinshaw, S. Arnott, and D. J. Morre.** 1980. The ribosome binding sites recognized by *E. coli* ribosomes have regions with signal character in both the leader and protein coding segments. *Nucleic Acids Res.* **8:**3895–3907.
118. **Schneider, T. D., G. D. Stormo, L. Gold, and A. Ehrenfeucht.** 1986. Information content of binding sites on nucleotide sequences. *J. Mol. Biol.* **188:**415–431.
119. **Schnier, J., and G. Faist.** 1985. Comparative studies on the structural gene for the ribosomal protein S1 in ten bacterial species. *Mol. Gen. Genet.* **200:**476–481.
120. **Schottel, J. L., M. J. Bibb, and S. N. Cohen.** 1981. Cloning and expression in *Streptomyces lividans* of antibiotic resistance genes derived from *Escherichia coli*. *J. Bacteriol.* **146:**360–368.
121. **Sharp, P. M., and W.-H. Li.** 1986. Codon usage in regulatory genes in *Escherichia coli* does not reflect

selection for 'rare' codons. *Nucleic Acids Res.* **14:**7737–7749.

122. **Sharp, P. M., T. M. F. Touhy, and K. R. Mosurski.** 1986. Codon usage in yeast: cluster analysis clearly differentiates highly and lowly expressed genes. *Nucleic Acids Res.* **14:**5125–5143.
123. **Shields, D. C., and P. M. Sharp.** 1987. Synonymous codon usage in *Bacillus subtilis* reflects both translational selection and mutational biases. *Nucleic Acids Res.* **15:**8023–8040.
124. **Shimotsu, H., M. I. Kuroda, C. Yanofsky, and D. J. Henner.** 1986. Novel form of transcription attenuation regulates expression of the *Bacillus subtilis* tryptophan operon. *J. Bacteriol.* **166:**461–471.
125. **Shine, J., and L. Dalgarno.** 1974. The 3′-terminal sequence of *Escherichia coli* 16S ribosomal RNA: complementarity to nonsense triplets and ribosome binding sites. *Proc. Natl. Acad. Sci. USA* **71:**1342–1346.
126. **Simons, R. W., and N. Kleckner.** 1983. Translational control of IS10 transposition. *Cell* **34:**683–691.
127. **Skeggs, P. A., J. Thompson, and E. Cundliffe.** 1985. Methylation of 16S ribosomal RNA and resistance to aminoglycoside antibiotics in clones of *Streptomyces lividans* carrying DNA from *Streptomyces tenjimariensis*. *Mol. Gen. Genet.* **200:**415–421.
128. **Stallcup, M. R., W. J. Sharrock, and J. C. Rabinowitz.** 1976. Specificity of bacterial ribosomes and messenger ribonucleic acids in protein synthesis reaction *in vitro*. *J. Biol. Chem.* **251:**2499–2510.
129. **Steitz, J. A.** 1969. Polypeptide chain initiation: nucleotide sequence of the three ribosome binding sites of R17 phage RNA. *Nature* (London) **224:**957–964.
130. **Stormo, G. D.** 1986. Translation initiation, p. 195–224. *In* W. Reznikoff and L. Gold (ed.), *Maximizing Gene Expression*. Butterworths, Mass.
131. **Stormo, G. D., T. D. Schneider, and L. M. Gold.** 1982. Characterization of translational initiation sites in *E. coli*. *Nucleic Acids Res.* **10:**2971–2996.
132. **Subramanian, A. R.** 1983. Structure and functions of ribosomal protein S1. *Prog. Nucleic Acids Res. Mol. Biol.* **28:**101–142.
133. **Subramanian, A. R.** 1984. Structure and functions of the largest *Escherichia coli* ribosomal protein. *Trends Biochem. Sci.* **9:**491–494.
134. **Szer, W., J. M. Hermosa, and M. Boublik.** 1976. Destabilization of the secondary structure of RNA by ribosomal protein S1 of *Escherichia coli*. *Biochem. Biophys. Res. Commun.* **70:**957–964.
135. **Tessier, L.-H., P. Sondermeyer, T. Faure, D. Dreyer, A. Benavente, D. Villeral, M. Courtney, and J.-P. Lecocq.** 1984. The influence of mRNA primary and secondary structure on human IFN-γ gene expression in *Escherichia coli*. *Nucleic Acids Res.* **12:**7663–7675.
136. **Thomas, D. Y., G. Dubuc, and S. Narang.** 1982. *Escherichia coli* plasmid vectors containing synthetic translational initiation sequences and ribosome binding sites fused with the *lacZ* gene. *Gene* **19:**211–219.
137. **Thomas, J. O., A. Kalb, and W. Szer.** 1978. Structure of single-stranded nucleic acids in the presence of ribosomal protein S1. *J. Mol. Biol.* **123:**163–176.
138. **Thomas, J. O., and W. Szer.** 1982. RNA helix destabilizing proteins. *Prog. Nucleic Acid Res. Mol. Biol.* **27:** 157–187.
139. **Thompson, C. J., and G. S. Gray.** 1983. Nucleotide sequence of a streptomycete aminoglycoside phosphotransferase gene and its relationship to phosphotransferases encoded by resistance plasmids. *Proc. Natl. Acad. Sci. USA* **80:**5190–5194.
140. **Thompson, J., S. Rae, and E. Cundliffe.** 1984. Coupled transcription-translation in extracts of *Streptomyces lividans*. *Mol. Gen. Genet.* **195:**39–43.
141. **van Dieijen, G., P. H. van Knippenberg, and J. van Duin.** 1976. The role of ribosomal protein S1 in the recognition of native phage RNA. *Eur. J. Biochem.* **64:**511–518.
142. **Vellanoweth, R. L., and J. C. Rabinowitz.** 1992. The influence of ribosome binding site elements on translational efficiency in *Bacillus subtilis* and *Escherichia coli in vivo*. *Mol. Microbiol.* **6:**1105–1114.
143. **Weaver, J. R., and P. A. Pattee.** 1964. Inducible resistance to erythromycin in *Staphylococcus aureus*. *J. Bacteriol.* **88:**574–580.
144. **Weisblum, B., C. Siddhikol, C.-J. Lai, and V. Demohn.** 1971. Erythromycin-inducible resistance in *Staphylococcus aureus*: requirements for induction. *J. Bacteriol.* **106:**835–847.
145. **Whitehead, T. R., and J. C. Rabinowitz.** 1988. Nucleotide sequence of the *Clostridium acidiurici* (*"Clostridium acidi-urici"*) gene for 10-formyltetrahydrofolate synthetase shows extensive amino acid homology with the trifunctional enzyme C_1-tetrahydrofolate synthetase from *Saccharomyces cerevisiae*. *J. Bacteriol.* **170:**3255–3261.
146. **Williams, D. M., E. J. Duvall, and P. S. Lovett.** 1981. Cloning restriction fragments that promote expression of a gene in *Bacillus subtilis*. *J. Bacteriol.* **146:**1162–1165.
147. **Wood, C. R., M. A. Boss, T. P. Patel, and J. S. Emtage.** 1984. The influence of messenger-RNA secondary structure on expression of an immunoglobulin heavy-chain in *Escherichia coli*. *Nucleic Acids Res.* **12:**3937–3950.
148. **Zaghloul, T. I., and R. H. Doi.** 1987. In vitro expression of a Tn*9*-derived chloramphenicol acetyltransferase gene fusion by using a *Bacillus subtilis* system. *J. Bacteriol.* **169:**1212–1216.
149. **Zaghloul, T. I., F. Kawamura, and R. H. Doi.** 1985. Translational coupling in *Bacillus subtilis* of a heterologous *Bacillus subtilis-Escherichia coli* gene fusion. *J. Bacteriol.* **164:**550–555.

49. Protein Secretion

VASANTHA NAGARAJAN

The basic mechanism by which proteins are transported across membranes appears to be universal, with important features conserved between bacteria and eukaryotes. Among the eubacteria, the mechanism of protein secretion has been extensively studied in *Escherichia coli* by using a variety of genetic and biochemical techniques (17, 102, 137). Some of the characteristics of protein secretion in *E. coli* and *Bacillus subtilis* are similar. However, certain variations exist, probably because of differences in the cell envelope. *B. subtilis* is amenable to a variety of biochemical and genetic manipulations and represents an alternative eubacterium in which to study protein transport.

Because they can secrete certain proteins in large quantities into the growth medium, *Bacillus* species are used for the industrial production of enzymes such as amylases and proteases. Most of the interest in protein secretion studies in *B. subtilis* stems from the possibility of using this bacterium for the production of commercially useful proteins. However, the protein transport mechanism in *B. subtilis* has not been extensively studied. Rather, the exoenzymes produced by *B. subtilis* have been used traditionally as biochemical markers for sporulation.

Protein secretion across the *B. subtilis* cell envelope is a complex process. The following steps are envisioned on the basis of current knowledge of protein export in bacteria (102, 137). These steps include insertion of the precursor protein into the membrane and translocation of the protein across the membrane. Signal peptide cleavage can occur prior to or after completion of the translocation process (co- or post-translational). Translocation is followed by folding of the mature protein, release from the membrane, and passage through the cell wall, resulting in secretion of the protein into the growth medium. Thus, protein transport involves an interaction between the exported protein and the cellular secretion factors.

The first part of this chapter discusses some of the structural features of the *B. subtilis* exoproteins and their role in protein secretion. This discussion is followed by an outline of the approaches taken to define the components of *B. subtilis* secretion machinery and of our current knowledge of protein secretion in *B. subtilis*. In this discussion, the term "translocation" refers to the transfer of protein across the membrane. The terms "protein secretion" and "export" include translocation, signal peptide processing, release, and passage of mature protein across the cell wall.

SIGNAL PEPTIDES

Signal peptides, which are usually present at the N termini of the precursors of exoproteins, have a critical cellular function and are necessary for the export of proteins across biological membranes (7, 21). Signal peptides have similar structural features but do not share extensive primary amino acid sequence homology. Features in the N-terminal region, the middle hydrophobic core region, and the C-terminal region are conserved in signal peptides derived from a wide variety of organisms (21).

The genes for several of the exoproteins from *Bacillus* species have been sequenced, and the deduced signal peptide sequences are shown in Table 1. The average length of a *B. subtilis* signal peptide is 29 amino acids, and some of the salient structural features are outlined here.

N-Terminal Region

The number of positively charged residues in the N-terminal region varies from 2 to 7. The average net of 3.1 positively charged residues is slightly higher than that observed for the majority of *E. coli* signal peptides. The positively charged N terminus may interact with the negatively charged phospholipid layer of the cell membrane (40). Thus, the higher number of positively charged residues might be due to the nature of the phospholipid in *B. subtilis* membrane.

Hydrophobic Core Region

The length of the hydrophobic core was calculated by counting residues between the N-terminal region and the end of the core. The core termination was defined by residues such as glycine or proline that were predicted to favor a β turn. The average length of the hydrophobic cores of known *Bacillus* signal peptides is 17 residues. The hydrophobic core region of the signal peptide is thought to span the cell membrane and facilitate the initiation of translocation. The molar distribution of the various residues in the hydrophobic core is as follows: L, 24%; S, 13%; F, 10%; I, 10%; A, 10%; V, 10%; T, 10%; M, 4%; G, 4%; and C, W, Y, P, Q, and N combined, 5%. Non-*Bacillus* signal peptides contain 37% L, 15% A, 10% F, 10% V, 5% T, 3% S, 7% I, 3% M, 2% G, and 8% C, W, Y, P, Q, and N combined (85). The significance of a higher frequency of serine, threonine, and glycine and a lower frequency of leucine in *B. subtilis* signal peptides compared with non-*Bacillus* signal peptides is not clear.

Vasantha Nagarajan • Central Research and Development Division, E. I. du Pont de Nemours & Company, Wilmington, Delaware 19880-0228.

Table 1. Signal peptide sequences[a]

Organism and protein	Sequence	Reference
Bacillus subtilis		
APR	MRSKKLWISLLFALTLIFTMAFSNMSAQA A	114
AMY	MFAKRFKTSLLPLFAGFLLLFYLVLAGPAAASA S	75
BPR	MRKKTKNRLISSVLSTVVISSLFPGAAGA S	109
CEL	MKRSISIFITCLLITVLTMGGLQASPASAGTKTPA A	66
EPR	MKNMSCKLVVSVTLFFSFLTIGPLAHA Q	107
GLU	MKRSISIFITCLLITLLTMGGMIASPASA A	51
GLU (β)	MPYLKRVLLLLVTGLFMSLFAVTATASA K	61
GLU (1-4)	MMRRRKRSDMKRSIISIFITCLLIAVLTMGGLLPSPASA A	95
LVS	MNIKKFAKQATVLTFTTALLAGGATQAFA K	115
LVN	MKKRLIQVMIMFTLLLTMAFSADA A	53
MPR	MKLVPRFRKQWFAYLTVLCLALAAAVSFGVPAKA A	107
NPR	MGLGKKLSVRVAASFMSLSISLPGVQA A	139
NPRB	MRNLTKTSLLLAGLCTAAQMVFVTHASA E	125
PhoAIII	MKKFPKKLLPIAVLSSIAFSSLASGSVPEASA Q	32
PBP5*	MRIFKKAVFVIMISLIATVNVNTAHA A	7a
PBP (DacF)	MKRLLSTLLIGIMLLTFAPSAAFA K	138
VPR	MKKGIIRFLLVSFVLFFALSTGITGVQA A	109
XYN	MFKFKKNFLVGLSAALMSISLFSATASA A	80
Bacillus stearothermophilus		
AMY	MLTFHRIIRKGWMFLLAFLLTALLFCPTGQPAKA A	65
NPRT	MNKRAMLGAIGLAFGLLAAPIGASA K	119
NPRM	MKRKMKMKLRSFGVAAGLAAQVFLPYNRLAST D	45
NPR	MNKRAMLGAIGLAFGLMAWPFGASA K	129
NPRS	MKRKMKMKLVRFGLAAGVAAQVFFLPYNALAST E	74
Bacillus licheniformis		
AMY	MKQQKRLYARLLTLLFALIFLLPHSAAAA A	116
APR	MMRKKSFWLGMLTAFMLVFTMAFSDSASA A	39
BLA	MKNKRMLKIGICVGILGLSITSLEAFTGESLQVEAKEKTGQV K	57
Bacillus amyloliquefaciens		
APR	MRGKKVWISLLFALALIFTMAFGSTSSAQA A	132
AMY	MIQKRKRTVSIFRLVLMCTLLFVSLPITKTSA V	122
GLU	MKRVLLILVTGLFMSLCGITSSVSA Q	29
BAR	MKKRLSWISVCLLVLVSAAGMLFSTA A	79
NPR	MGLGKKLSSAVAASFMSLTISLPGVQA A	104
NPR	MGLGKKLSVAVAASFMSLTISLPGVQA A	132
LVS	MNIKKIVKQATVLFTTALLAGGATQAFA K	123
Bacillus polymyxa		
AMY	MTLYRSLWKKGCMLLLSLVLSLTAFIGSPSNTASA A	128
NPR	MKKVWFSLLGGAMLLGSVASGASA E A	121
Bacillus pumilus		
XYN	MNLRKLRLLFVMCIGLTLILTAVPAHA R	19
Bacillus cereus		
PEN	MKNKRMLKIGICVGILGLSITSLEAFTGESLNVEAKEKTGQV K	57
PEN (5b)	MIVLKNKKMLKIGMCVGILGLSITSLVTFTGGALQVEAKEKTGQV K	46
Bacillus brevis		
MWP	MKKVVNSVLASALALTVAPMAFA A	126
OWP	MNKKVVLSVLSTTLVASVAASAFA A	127
Staphylococcus aureus		
SPA	MKKKNIYSIRKLGVGIASVTLGTLLISGGVTPAAN A	50
SAK	MLKRSLLFLTVLLLLFSFSSITNEVSA S	2

[a] Signal peptide sequences were deduced from DNA sequences. Amino acid sequences of random signal peptides and staphylococcal nuclease have not been included (44, 111).

C-Terminal Region

A majority of *Bacillus* signal peptides contain a glycine or a proline residue predicted to favor a β turn at the end of the hydrophobic core at positions −5 to −7 from the C terminus. (+1 refers to the N-terminal amino acid of the mature protein.) The signal peptide cleavage site has not been determined precisely for a majority of *Bacillus* exoproteins by N-terminal amino acid sequencing. Rather, it is based on similarity to a variety of prokaryotic and eukaryotic signal peptides (133, 134). Thus, alanine is the predominant residue (>90%) at the −1 and −3 positions of the *Bacillus* signal peptides. As shown in Table 1, a wide variety of residues are observed at the +1 position; however, alanine at the +1 position is deduced from the DNA sequences in 50% of the *B. subtilis* exoproteins.

Lipoproteins

The hydrophobic cores of the lipoprotein signal peptides are shorter than those of the exoproteins (Table 2). The signal peptide processing site has leucine and cysteine residue at the −3 and +1 posi-

Table 2. Lipoprotein signal sequences[a]

Organism and protein	Sequence	Reference
B. subtilis		
Slp	MRYRAVFPMLIIVFALSG CT	28
PrsA	MKKIAIAAAITATSILALSA CS	42
B. licheniformis		
PenP 749C	MKLWFSTLKLKKAAAVLLFSCVALAG CA	70
B. cereus		
PenP	MFVLNKFFTNSHYKKIVPVVLLSCATLIG CS	34
S. aureus		
PenP	MKKLIFLIVIALVLSA CN	56

[a] Amino acid sequences of lipoprotein signal peptides that function in *B. subtilis* have been compiled from DNA sequences. Reference 27 contains an extensive compilation of bacterial-lipoprotein signal peptide sequences.

tions, respectively, which is similar to other bacterial lipoproteins (27).

A majority of the signal peptides that function in *B. subtilis* have the structural features mentioned above, but certain exceptions are known. These include the staphylococcal nuclease signal peptide and some of the random *Bacillus* signal peptides isolated by Smith et al. (44, 111).

BACILLUS SIGNAL SEQUENCE MUTANTS

Signal sequence mutants in *E. coli* not only have helped elucidate the structure-function relationship of signal peptides but also have been used as genetic tools for identifying secretion genes (16, 20, 102). A majority of the mutations in *E. coli* signal sequences were isolated by genetic selection. The advent of site-directed mutagenesis technology has resulted in sets of designed *E. coli* signal sequence mutations. A series of signal sequence mutations in *Bacillus amyloliquefaciens* levansucrase (LVS), *B. amyloliquefaciens* alkaline protease, and *B. subtilis* α-amylase have been constructed by design (see below). This was necessitated in part by a lack of genetic methods for selecting for spontaneous signal sequence mutations in *B. subtilis*.

LVS Signal Sequence Mutants

Signal sequence mutants spanning the entire signal peptide were constructed (6). The LVS signal peptide contains three lysine residues (K4, K5, and K8) at the N terminus. Mutant I4I5K8, which contained a single lysine residue (+1), was transported at a slightly lower rate than the wild type. Mutant D4I5K8, with a net charge of 0 but with two charged amino acids (aspartic acid and lysine), had wild-type kinetics. However, mutant I4I5N8, also with a net charge of 0 (lacking any charged residues), was completely blocked in secretion. Thus, the secretion defect does not seem to be due to the net charge but rather to a specific requirement for the charged residues at the N terminus.

Deletion of five or more residues in the hydrophobic core of the LVS signal peptide resulted in a secretion defect. Introduction of glutamic acid in the middle of the hydrophobic core completely blocked secretion. The only leaky signal sequence mutant contained a deletion of three amino acids [Δ(14–16)] in the middle of the hydrophobic core. The kinetics of transport suggested that the translocation was slower, occurring exclusively posttranslationally, and that only 35% of the precursor was processed (5). However, Δ(14–16) precursor was completely processed in *E. coli* at a slower rate than the wild type (4).

Analysis of mutants altered at the LVS signal peptide processing site showed that only small amino acids were tolerated at the −1 and −3 positions and that a wide variety of amino acids were tolerated at the +1 position. In addition, the −1 position was more sensitive to substitution than the −3 position. The half-life of the defective LVS precursor as determined by radiolabeling was dependent on the nature of the signal sequence defect. Signal peptide processing site mutant precursors were stable, while precursors with defective cores were completely degraded. If it is assumed that the processing site mutations allow translocation of the precursor across the membrane but that core mutations prevent insertion into the membrane, then these results are consistent with the hypothesis that the LVS precursor is unable to fold into a stable conformation intracellularly.

α-Amylase Signal Sequence Mutants

Signal sequence mutations were constructed in the *B. subtilis* α-amylase signal peptide to determine the effect of negatively charged residues in the C terminus (67). Glutamic acid but not lysine at the −6 position blocked secretion in both *E. coli* and *B. subtilis*. The effect on secretion of increasing the length of the signal peptide from 33 to 36 residues was also studied. Introduction of two or three leucine residues at the end of the hydrophobic core resulted in a partial secretion defect, and this defect was more severe in *E. coli* than in *B. subtilis*. The behavior of α-amylase signal sequence mutants was studied by measuring either extracellular amylase or β-lactamase activity in *B. subtilis*. Thus, it is not clear whether these mutants are blocked in translocation across the membrane or in the signal peptide processing.

Alkaline Protease Signal Sequence Mutants

A set of charged substitutions in the hydrophobic core of the *B. amyloliquefaciens* alkaline protease signal peptide has been constructed (83). Analysis of these mutants revealed that the phenotype of any given mutation was dependent on both the position and the species of substituted amino acid. For example, the effect of lysine substituted at position 15 or 16 of the signal peptide was more severe than that of lysine at position 14 or 17. In addition, four different charged residue substitutions at position 17 had different effects. Arginine at position 17 completely blocked secretion, while aspartic acid or glutamic acid had little effect. These results are difficult to rationalize with the current knowledge of signal peptides. However, a similar position effect has been observed in *E. coli* (113).

Wheat α-Amylase Signal Sequence Mutants

The targeting signal for transport of proteins across the endoplasmic reticulum in eukaryotes is similar to bacterial signal peptides. The signal sequence mutants described below were constructed to improve the secretion efficiency of the wheat α-amylase signal sequence in *B. subtilis* (94). The wheat α-amylase signal peptide can transport LVS and *E. coli* alkaline phosphatase from *B. subtilis*, but the kinetics of transport were slower than those observed with *Bacillus* signal peptides. The amylase signal peptide is 24 amino acids long with a 10-amino-acid hydrophobic core; it is shorter than a typical *Bacillus* signal peptide. An attempt to improve the efficiency of wheat α-amylase signal peptide was made by increasing the length of the hydrophobic core. Introduction of two (S and T) or four (S, T, L, and L) amino acids in the core slightly increased the rate of transport. However, insertion of five (S, T, L, L, and I) or six (S, T, L, L, I, and L) residues resulted in a drastic decrease in secretion as judged by both the activity and the rate of signal peptide removal. The LVS precursors containing these mutant signal peptides are stable, which suggests that the defect may be in the signal peptide processing. These results suggest that there is an optimal length at which the hydrophobic core is functional.

Biochemical analyses of various signal sequence mutants have shown that a functional signal peptide in *B. subtilis* requires the presence of charged residues at the N terminus and that a hydrophobic core is necessary for efficient translocation (5, 6). The ability to form a β turn in the beginning of the C region is not necessary for secretion, as demonstrated in the case of LVS mutants (6). An increase in the hydrophobic core length probably does not prevent translocation but instead prevents signal peptide processing.

PRO PEPTIDE

The DNA sequences of several *Bacillus* exoproteins have revealed the presence of an open reading frame (ORF) coding for a "pro" peptide between the signal and the mature coding sequence (114, 132, 139). The pro peptide sequence may be unique to gram-positive exoproteins and has not been reported in periplasmic or outer membrane proteins of *E. coli*. The length of this pro peptide varies from 8 to 206 residues. Pro peptides contain a high percentage of charged residues. Several roles can be envisioned for the pro peptide: (i) promoting efficient insertion of the precursor into the membrane, (ii) serving as a temporary anchor to the membrane after signal peptide cleavage, (iii) interacting with cations in the cell wall, (iv) increasing the solubility of the mature protein, (v) participating in folding of the mature protein, and (vi) protecting the N terminus of mature protein from proteolysis in the growth medium.

Pro peptides of α-amylase and barnase are 8 and 13 residues long, respectively, and are not needed for secretion or activity (24, 78, 100, 120). However, the pro peptide seems to increase the stability of α-amylase in growth medium (100). Pro peptides are essential for in vivo production of active subtilisin and neutral protease (37, 118). Deletion of part or all of the pro peptide resulted in loss of activity of subtilisin E (*B. subtilis* alkaline protease) from *E. coli*, even though subtilisin was secreted (37). The deletion of pro peptide (77 amino acids) from subtilisin BPN′ (*B. amyloliquefaciens* alkaline protease) or replacement with the neutral protease pro peptide (196 residues) prevented the production of active subtilisin from *B. subtilis* (61a). The role of pro peptide in the production of active neutral protease (NprT) has been studied (118). The NprT pro peptide is 200 amino acids long. Deletions of 3 (His-91 to Val-93) and 33 (Asp-77 to Met-109) residues in the pro peptide resulted in 26 and 85% decreases in extracellular protease activity, respectively. Insertion of two amino acids (Gly-88 and His-89) resulted in a total loss of activity. Thus, it appears that the function of the pro peptide may depend on its sequence. However, the sequences of protease pro peptides are more divergent than their cognate mature sequences (39).

The role of pro peptide in the folding of subtilisin has been studied in detail (36, 37). The pro peptide of subtilisin E was able to catalyze in vitro refolding of mature subtilisin E, subtilisin BPN′, and Carlsberg in *trans* (141). Prosubtilisin produced intracellularly in *E. coli* as inclusion bodies could be converted into active subtilisin, and this renaturation process was not prevented by phenylmethylsulfonyl fluoride, which suggests that it is an intramolecular autocatalytic process (36). A similar autocatalytic process has been postulated to be the major method of conversion of preprosubtilisin to subtilisin BPN in vivo in *B. subtilis* (90).

Experiments to determine the behavior of pro peptide when it is fused to heterologous proteins have been performed. When the pro peptide of subtilisin BPN′ is fused to heterologous proteins, it can be secreted into the growth medium, demonstrating that the pro peptide can cross the *B. subtilis* cell wall (131). A membrane-bound intermediate with the pro peptide of subtilisin E fused to a heterologous protein has been reported. This suggests that the pro peptide might serve as a temporary anchor to the membrane (15).

The pro peptides that were originally observed for the major extracellular proteases from *B. amyloliquefaciens* and *B. subtilis* seem to be common among proteases from other bacteria. It is not clear whether proteases have evolved with a folding pathway that is different from those of other exoproteins. The pro peptide may help maintain the precursor protein in vivo in a translocation-competent state. The kinetics of signal peptide removal in the presence and absence of the pro peptide in vivo in *B. subtilis* have not been determined.

ROLE OF THE MATURE PROTEIN SEQUENCE IN SECRETION

Experiments in several laboratories have shown that *Bacillus* signal peptides are interchangeable in *B. subtilis* and that several heterologous proteins can be secreted, albeit with various efficiencies, when they are fused to these signals (99). Thus, signal peptides are sufficient for targeting heterologous proteins in the secretory pathway, but the differences in efficiency suggest that the mature sequence plays a specific role.

The kinetics of signal peptide removal of 15 different precursors were compared in *B. subtilis* by pulse-chase analysis (64). The hybrid precursors consisted of four different mature sequences (*E. coli* alkaline phosphatase, staphylococcal protein A, *B. amyloliquefaciens* barnase, and LVS) fused to several *B. amyloliquefaciens* signal peptides (alkaline protease, neutral protease, barnase, and LVS). The rate of signal peptide removal of any given precursor was dependent on both the signal peptide and the mature sequence and was not determined by either one alone (11a). This observation underlines an important role played by the mature sequence in protein secretion.

Studies have shown that the phenotype of any given signal sequence mutation depends on the mature sequence. The effect of most point mutations in the alkaline protease signal peptide was dependent on whether the mature sequence was alkaline protease or *E. coli* alkaline phosphatase (83). The effect of extending the length of *B. subtilis* α-amylase signal peptide was dependent on whether the mature sequence was β-lactamase or α-amylase (67, 75).

The N-terminal amino acids of the mature protein may be especially important in determining the efficiency of signal peptide processing in *B. subtilis*. The effects of various substitutions at the N-terminal second (+2) and third (+3) positions of mature human parathyroid hormone have been studied (101). The signal peptide processing was more efficient when residues that favored a β turn were located at the +2 and +3 positions of the mature protein.

A deletion in the *Bacillus licheniformis* subtilisin mature sequence resulted in the accumulation of precursor protein in the membrane fraction (103). Experimental results have not determined whether the signal peptide has been removed from the precursor, and hence, the secretion defect could be at the level of translocation or of release from the membrane. However, a role for the mature sequence in protein export was inferred from these studies.

The signal peptide has been shown to modulate folding of the mature protein in *E. coli* (48, 81). Randall and Hardy demonstrated that rapid folding of a precursor protein into a stable conformation results in an export-incompetent precursor (93). Thus, the effect of mature sequence on secretion observed in *B. subtilis* might also be due to differences in the rate of folding. An equally plausible explanation is the influence of the mature protein on signal recognition and insertion into the membrane. In addition, one can also envision a role played by the mature sequence in release from the membrane after the signal peptide has been removed. *B. subtilis* LVS mutants with alterations in residue 366 are not released rapidly from the membrane, even though the signal peptide removal is not affected (3).

VECTORS FOR STUDYING PROTEIN TRANSPORT IN *B. SUBTILIS*

Vectors for isolating random sequences that can function as signal peptides in *B. subtilis* have been reported (110). Smith et al. constructed vectors containing *E. coli* β-lactamase and *B. licheniformis* α-amylase that lacked a signal peptide (110). The β-lactamase vectors were able to confer ampicillin resistance on *E. coli* when a functional signal peptide was fused to β-lactamase. A set of clones containing putative signal peptide coding regions from *B. subtilis* was isolated (111). These hybrid protein fusions were used to demonstrate differences in the signal peptide cleavage site between *E. coli* and *B. subtilis*. One of the hybrid fusions was translocated in both *E. coli* and *B. subtilis*, but the signal peptide was processed only in *B. subtilis*. This difference in phenotype between *E. coli* and *B. subtilis* was useful in cloning the *B. subtilis* leader peptidase gene, as discussed below in the section on *sipS*.

E. coli alkaline phosphatase has been useful for determining the membrane topologies of several integral membrane proteins, because the phosphatase activity correlates with translocation of the protein (52, 59). The alkaline phosphatase activity correlates with export in *B. subtilis* also (84). This vector can be used to determine secretion-incompatible sequences in *B. subtilis*.

A set of modular vectors based on LVS, *E. coli* alkaline phosphatase, and staphylococcal protein A has been constructed for protein secretion studies in *B. subtilis* (62). These vectors contain compatible restriction sites and can be used to isolate novel signal sequences. One of the limitations of studying protein secretion in *B. subtilis* has been the lack of multiple reporter proteins for studying interference caused by a protein secretion defect. For example, overproduction of a defective precursor protein might affect a certain class of proteins and not others. Thus, the availability of multiple reporter proteins will aid in characterization of secretion mutants.

SECRETION GENES

Several of the *B. subtilis* mutations that had a pleiotropic effect on the accumulation of exoenzymes have turned out to be transcriptional regulators (see chapter 50). Thus, novel approaches are needed for isolating *B. subtilis* secretion mutants. Characterization of secretion mutations in *E. coli* has identified the following genes as essential for protein transport: *secA* (*prlD*), *secD*, *secE* (*prlG*), *secF*, *secY* (*prlA*), *lep*, and *lspA* (102, 137). An obvious method of isolating the *B. subtilis sec* homologs is based on the use of *E. coli sec* genes as probes for hybridization, and this approach has been used with limited success. The *B. subtilis* genes for *secA*, *secY*, and *sipS* have been isolated, and their properties are described below.

secA

A *B. subtilis* temperature-sensitive septum initiation mutant designated *div-341* and having the following properties was isolated (96). Sporulation efficiency, competence development, and α-amylase production were reduced at 37°C, but vegetative growth was unaffected. Protease production was abolished even at the permissive temperature. The pleiotropic nature of the defect led Sadaie and Kada (96) to conclude that *div-341* was blocked in stage 0 of sporulation. Sadaie et al. isolated the div^+ gene on a phage vector, and the DNA sequence showed homology to *E. coli secA* (97). The div^+ gene codes for an ORF encoding 841 amino acids (97). This ORF shares 50% identity with the *E. coli* SecA protein, and a homology of 71.4% was observed at the N terminus between residues 34 and 335. The majority of the temperature-sensitive mutations in *E. coli secA* are in the N terminus, which contains the ATP-binding domain (55). In the *secA* operon in *E. coli*, the *secA* gene is preceded by "gene *X*" (76). The ORF preceding div^+ does not share any homology with gene *X*, and it appears that *B. subtilis secA* may be the first gene of the operon. Thus, there may be differences between the organizations of the *secA* operons of *E. coli* and *B. subtilis*.

Overhoff et al. have attempted independently to isolate the *B. subtilis secA* gene on the basis of homology to *E. coli secA* (77). This approach has resulted in cloning a fragment coding for 364 N-terminal residues of *B. subtilis* SecA. This fragment was able to complement the growth defect of an *E. coli secA*(Ts) mutant at 42°C and partially suppress the transport defect.

Driessen and Freudl's laboratories have recently characterized the *B. subtilis* SecA protein (41a, 129a). The *B. subtilis* SecA protein is functional in *E. coli* in an in vitro translocation system. *B. subtilis* SecA has a lower affinity for *E. coli* membrane than does *E. coli* SecA. Site-directed mutagenesis has identified lysine 106 as part of the catalytic ATP-binding site (41a). It has been proposed that the GKT motif in the N-terminal domain of SecA is part of the catalytic ATP-binding site (129a). The SecA protein is an ATPase, and it couples the hydrolysis of ATP to the releases of bound precursor proteins to allow proton motive force-driven translocation across the cytoplasmic membrane.

secY

The *E. coli secY* (*prlA*) gene is located in the ribosomal-protein *spc* operon (38). The *B. subtilis secY* gene was identified as part of the *B. subtilis spc* operon (117). *secY* gene organization in the *spc* operon shares certain similarities and differences in *E. coli* and *B. subtilis*. Whereas the genes that are 5′ to *secY* are conserved, the 3′ genes are not. The *B. subtilis secY* codes for an ORF consisting of 431 residues that shares 41% identity with the *E. coli secY*. This level of identity has been observed between *E. coli* and *B. subtilis* in the components of transcriptional and translational apparatuses. *E. coli* SecY, an integral membrane protein, has 10 putative membrane-spanning segments (38). On the basis of the sequence of *B. subtilis secY* ORF, Su et al. (117) predict that SecY protein also has 10 membrane-spanning segments. The degree of similarity with *E. coli* SecY is greater in sequences containing the membrane-spanning and cytoplasmic domains and is limited in periplasmic-domain sequences. This is not surprising, because the milieu of the *E. coli* periplasm is likely to be quite different from that of *B. subtilis*. Nakamura et al. have also independently cloned the *B. subtilis secY* gene by using *E. coli secY* as a hybridization probe (68).

The expression of *B. subtilis secY* was unable to suppress a temperature-sensitive growth defect of an *E. coli secY24* mutant, but the protein transport defect was suppressed (69, 117). However, *B. subtilis* SecY also interfered with *E. coli* protein export and caused a slight increase in the level of SecA in *E. coli* (117) (*E. coli secA* is autoregulated, and secretion defects result in enhanced expression of *secA*).

sipS

The gene for the *B. subtilis* signal peptidase was recently isolated by van Dijl et al. by taking advantage of the difference in phenotype of a hybrid β-lactamase precursor that was cleaved in *B. subtilis* but not in *E. coli* (130). Shotgun cloning of *B. subtilis* DNA in *E. coli* resulted in identification of clones that were able to process the defective precursor and export β-lactamase. The DNA sequence has revealed a 21-kDa protein with homology to both the *E. coli* and the *Salmonella typhimurium* signal peptidases I. It also shares homology to the mitochondrial inner membrane protease I of *Saccharomyces cerevisiae*. The overproduction of *B. subtilis sipS* does not complement an *E. coli lep*(Ts) mutant for growth. Overproduction of *sipS* gene product in *B. subtilis* seems to increase the rate of signal peptide processing of at least one precursor protein, which suggests that the *sipS* gene product may be rate limiting for protein export. The *sipS* gene is not essential for cell viability, which suggests the possibility of a second signal peptidase in *B. subtilis* (130).

prsA

Kontinen and Sarvas have isolated a set of *B. subtilis* mutants (*prs*) based on the inability to transport *B. amyloliquefaciens* α-amylase (43). The various *prs* mutations mapped in four distinct loci in the *B. subtilis* chromosome. The *prs* mutants had a pleiotropic effect on different exoproteins, and the level of defectiveness was dependent on the mutation. Some of the *prs* mutants affected the secretion of both the lipoproteins and the exoproteins. One of the genes (*prsA*) that complements *prs-3* and *prs-29* mutations has been cloned (42). The *prs-3* and *prs-29* mutants have wild-type penicillinase activity and are normal for competence development and sporulation. However, they have a severe defect in the transport of α-amylase and a partial defect in the transport of protease. The *prsA* ORF codes for a protein with an N-terminal lipoprotein signal peptide followed by a 271-amino-acid mature protein sequence. The amino acid sequence does not contain any putative membrane-spanning domain and shares 30% identity with PrtM protein from *Lactococcus lactis*. PrtM, a lipoprotein, is involved in the maturation of an extracellular protease, PrtP. Kontinen et al. (42) speculate that PrsA may be needed to fold and release α-amylase from *B. subtilis* and that it thus might be an extracellular "molecular chaperone."

GENETIC METHODS OF ISOLATING *B. SUBTILIS* SECRETION MUTANTS

The progress made in understanding the mechanism of protein secretion in *E. coli* is due to the availability of secretion mutants (102, 105, 137). Isolation of *B. subtilis* secretion mutants would facilitate understanding the mechanism of protein transport, but efforts in this area have been rather limited. Two genetic approaches were used to identify secretion genes in *E. coli* (*prl* and *sec*). As discussed below, these two complementary approaches have identified mutations within the same structural genes, underlining the importance of these genes in protein export.

Suppressor of Signal Sequence Mutants (*prl*)

One of the powerful methods of isolating secretion mutants in *E. coli* was originally developed by Silhavy and coworkers and was based on the isolation of extragenic suppressors of signal sequence mutations (16). This method relies on isolating a dominant secretion mutant such that the altered secretion machinery can interact with a defective precursor and restore its translocation. For example, an *E. coli* strain containing a defective signal peptide for LamB cannot transport maltodextrins and has a Dex^- phenotype. Analysis of Dex^+ pseudorevertants led to the identification of some of the *prl* class of secretion genes such as *prlA* (*secY*), *prlD* (*secA*), and *prlG* (*secE*) (16, 102).

A similar approach to isolating *B. subtilis* secretion mutants is based on the secretion of LVS. A Δ*sacA* and Δ*sacB B. subtilis* strain has a Suc^- phenotype. This phenotype can be converted to Suc^+ when mature LVS fused to a functional signal sequence is introduced into this strain. However, when the LVS precursor contains a defective signal peptide, it is rapidly degraded and LVS secretion is completely blocked, resulting in a Suc^- phenotype (63). A series of Suc^+ clones have been isolated from one such strain, and the mutants are being characterized. An extensive set of LVS signal sequence mutants is available, and suppressors directed towards specific regions in the signal peptide can thus be isolated by using the above method.

There are advantages in using several defective precursor proteins to isolate suppressor mutants. Once a collection of these mutants is available, they can be classified on the basis of allele specificity. A majority of *Bacillus* exoenzymes can be detected by powerful screens, and identification of a single clone producing these proteins in a bacterial lawn is feasible. Thus, one can isolate suppressors of signal sequence mutants by using any of the exoenzymes for which there are existing signal sequence mutants.

Internalization of β-Galactosidase (*sec*)

Beckwith and coworkers developed a selection method for isolating *E. coli sec* mutants that is complementary to the method described above. This selection is based on a phenotype that depends on internalization of a precursor protein due to secretion defects. *E. coli* strains containing β-galactosidase fused to a functional signal peptide are Lac^- because of the targeting of β-galactosidase to the membrane (102, 105). In addition, overproduction of some of the fusion proteins interfered with the secretion of periplasmic and outer membrane proteins, resulting in a lethal phenotype (102). Thus, Lac^+ mutants were isolated, and extensive characterization of these mutants has identified the *E. coli secA*, *secB*, *secD*, and *secF* genes.

A similar approach to target β-galactosidase to the *B. subtilis* membrane has been attempted. Hastrup and Jacobs constructed an *apr-lacZ* gene fusion in which the promoter of *B. licheniformis apr* was replaced with a xylose-inducible promoter (25). A *B. subtilis* strain containing a *xyn-apr-lacZ* gene fusion on a multicopy plasmid has a Lac^- phenotype in the absence of xylose. In the presence of 0.2% xylose, filamentous growth and a weak Lac^+ phenotype were observed. However, at 2% xylose, growth was inhibited. Mutants that did not show any growth inhibition at a high level of xylose (2%) were isolated (4). Characterization of these mutants has shown that a majority were plasmid borne and were unrelated to secretion.

Two *sacB-lacZ* translational fusions consisting of the N-terminal 191 and 216 residues of LVS fused to β-galactosidase were integrated into the *B. subtilis* chromosome and characterized by Zagorec and Steinmetz (140). A secretion defect and growth inhibition were observed only when the host strain carried a *sacU*(Hy) mutation that resulted in enhanced expression of *sacB-lacZ*. The fusion containing 216 residues of LVS was more lethal than the fusion containing 191 residues. Since the lethal effect was not observed in the wild-type strain but only in the *sacU*(Hy) strain, this selection method has a limitation, and it is conceivable that most of the mutants that suppress the growth defect contain down mutations in the *sacU* locus or in sucrose regulation.

BIOCHEMISTRY OF PROTEIN SECRETION

A membrane fraction enriched with ribosomes has been observed in *B. subtilis* and *B. amyloliquefaciens*. It was postulated that these polysomes are the sites of synthesis of exported proteins (30, 31, 54, 112, 124). Membrane-bound lipases, amylase, protease, and LVS have been observed in *B. subtilis*, and these are postulated as intermediates of the exoenzymes (41, 86, 124). Immunoelectron microscopy of *B. subtilis* cells secreting human interferon or staphylokinase showed a clustering of cells and an enrichment of gold particles on the cell envelope, suggesting that preferential regions for secretion sites may exist (135).

Muren and Randall (60) demonstrated that the dissipation of electrochemical membrane potential resulted in the accumulation of α-amylase precursor in *B. amyloliquefaciens*. When the disruption of membrane potential was accomplished without alteration in the intracellular ATP pool, it still resulted in the blockage of secretion, suggesting that the inhibition was due to the requirement for proton motive force. It has been feasible to slow the rate of transport and/or accumulate precursors in *B. subtilis* in the presence of energy uncouplers such as carbonyl cyanide *m*-chlorophenylhydrazone (CCCP) or phenylethyl alcohol (6, 64, 101).

Biochemical studies on the secretion of amylase and proteases by *B. amyloliquefaciens* protoplasts suggest that the polypeptide chains are extruded through the membrane in an unfolded state (98). The presence of a high concentration of Mg^+, Ca^+, or spermidine in the protoplast medium was required for the production of active protease but not of amylase. The active enzymes produced by protoplasts were resistant to trypsin. However, addition of trypsin at the time of protoplast formation resulted in the degradation of amylase and protease. Thus, the proteins released from the protoplast were in a different conformation (trypsin sensitive) than the tertiary structure of amylase or protease (98).

The role of metals in the folding and release of LVS from *B. subtilis* has been extensively characterized (10, 11, 86, 88). The half-time for conversion of precursor LVS to mature LVS is 4 to 5 s. However, the disappearance of the membrane-bound species in the growth medium occurs in 35 s. The elution profile of membrane-bound mature LVS on a hydroxyapatite column was different from that of exolevansucrase. The activity of the membrane-bound form could be recovered only in the presence of iron. An elegant set of in vitro folding studies of LVS have shown that refolding is promoted by the addition of Ca^{2+} or Fe^{3+} (11). A set of LVS variants that are altered in Gly-366 have helped to understand the correlation between the roles of iron in protein folding and secretion (3). LVS variants containing Asp-366 or Val-366 were not released into the growth medium efficiently, even though the rate of the signal peptide removal was similar to that of wild type. In vitro refolding studies showed that Fe^{3+} was unable to promote the refolding of Asp-366 or Val-366 variants. However, Fe^{3+} was able to promote the refolding of another variant, Ser-366, which was efficiently secreted into the growth medium. Thus, Ca^{2+} or Fe^{3+} may be involved in vivo in the folding of LVS. This observation suggests that secreted proteins must assume a stable tertiary structure before they can be released from the membrane. The requirement of cobalt for the secretion of *B. licheniformis* alkaline phosphatase has been reported (33). The teichoic acids in *B. subtilis* cell wall may concentrate metals and thus indirectly assist in the folding and release of the mature protein (14).

Role of Membrane Lipids

Lipids are thought to play a role in protein secretion, but the precise mechanism is poorly understood (40). Some recent in vitro experiments concerning protein translocation in *E. coli* suggest an important role for acidic lipids in the binding of SecA to the membrane (137). Membrane fluidity, which is regulated in *B. subtilis* by altering the level of unbranched fatty acids, may play a major role in protein transport (see chapter 28).

The role played by lipids in protein transport is further substantiated by the inhibition of exoenzyme secretion by a variety of membrane-modifying agents in *Bacillus* species (18). Inhibition of the secretion of LVS and α-amylase has been observed in the presence of cerelunin (9, 82, 87). It is proposed that cerelunin and other amphiphilic compounds alter the membrane fluidity and thus inhibit secretion.

Role of Cell Wall

Gram-positive bacteria have complex cell wall structures (discussed in chapter 27). Upon being released from the membrane, exoproteins have to cross the cell wall and may remain associated with the cell wall before being released into the growth medium. A washed cell suspension of *B. amyloliquefaciens* released α-amylase and proteases for 15 min (22). This release was energy independent but temperature dependent and occurred in the absence of protein synthesis.

The cell wall is not a rigid structure, and its architecture can vary depending on the physiological status of the cell. Teichoic acid, a major component of the *B. subtilis* cell wall, has a rigid structure in distilled water and assumes a random coil structure in a polyelectrolyte solution (14). Interaction of metals (Mn, Ni, Ca, Na, and Li) with the *B. subtilis* cell wall has been studied. The rigidity of the cell wall may be altered by growth of *B. subtilis* in 0.2 M NaCl, a condition that does not alter the growth rate. Cheung and Freese showed that 0.2 M NaCl was necessary to observe cell wall turnover of a *lyt-15* (turnover-deficient) mutant (12). This requirement for 0.2 M NaCl for turnover was due to the low amounts of teichoic acid in this strain. Some of the cell wall-bound enzymes may be released into the growth medium by alterations in the cation concentration.

The recent observation of the role of metals in the folding of LVS and a possible role for them in the production of proteases suggests that cell wall may have an active role in the release of proteins from membranes after translocation and that this role may be the provision of metals (88, 98).

PHYSIOLOGY OF PROTEIN SECRETION

The morphological and biochemical changes that accompany sporulation of *B. subtilis* offer interesting challenges for protein transport. Upon forespore engulfment, the two membranes are in apposition, with the outer surfaces of both the inner and outer forespore membranes facing each other. The space between the two is topologically equivalent to the extracellular space and is filled with cortex. Thus, enzymes responsible for cortex synthesis presumably will have to be targeted. Proteins made by the mother cell can be targeted to either the mother cell membrane or the outer forespore membrane. The importance of secretion in sporulation is reflected in the fact that the only mutation in a secretion gene (*div-341* [*secA*]) in *B. subtilis* was isolated as a temperature-sensitive sporulation initiation mutant (96).

One of the major sporulation-specific penicillin-binding proteins (PBP5*) appears to be localized in the outer forespore membrane, where it might participate in cortex synthesis (8). DNA sequence analysis of the cloned PBP5* has revealed the presence of an N-terminal signal peptide, and PBP5* appears to be expressed in the mother cell (7a).

LIPOPROTEINS

A class of proteins in *Bacillus* species that is similar to the lipoproteins of *E. coli* is modified by lipids upon translocation across the membrane (27). The mechanism of the lipid modification is conserved between *E. coli* and *B. subtilis*. The N-terminal cysteine residue is covalently modified with diacyl glycerol through a thioether linkage and contains an amide-linked fatty acid. The signal peptide cleavage of lipoproteins is catalyzed by signal peptidase II, which is distinguished by its sensitivity to globomycin (27). Among the gram-positive lipoproteins, penicillinases have been extensively studied by Lampen and coworkers (46, 71–73). The membrane-bound lipid-modified penicillinase is cleaved nonspecifically by proteases, resulting in the release of exopenicillinase to the growth medium in stationary phase. A consensus L x x C sequence seems to be observed at cleavage sites in lipoprotein signal peptides. Deletion of these residues in the *B. licheniformis* penicillinase results in the secretion of penicillinase, suggesting that the covalently modified intermediate is not essential for secretion or activity of the penicillinase (58). Gram-positive lipoproteins are postulated to be analogous to the periplasmic proteins of *E. coli* (73).

The behavior of the *E. coli* lipoprotein (Lpp) signal peptide in *B. subtilis* has been studied (26). *B. subtilis* cells processed the Lpp precursor, and the lipid modification of the *E. coli* lipoprotein was observed. An additional precursor that contained a 45-amino-acid N-terminal peptide fused to the Lpp signal peptide was also characterized. *B. subtilis* cells were able to transport and covalently modify this hybrid lipoprotein, demonstrating that the signal peptide need not always be at the N terminus and that it can tolerate at least 45 additional N-terminal amino acids.

ALTERNATIVE SECRETORY PATHWAY

Gram-negative bacteria can accomplish extracellular secretion of proteins into the growth medium by using an alternative secretory pathway (91). Some of the recent studies on the export of subtilin from *B. subtilis* suggest the existence of an alternative secretory pathway (13). Subtilin, a peptide antibiotic, contains several unusual amino acids. The gene for subtilin has been cloned from *B. subtilis* ATCC 6633, and DNA sequence analysis reveals that the gene is synthesized with an additional 24-amino-acid N-terminal extension (1, 23). This extension, which may function as a signal peptide, is hydrophilic and does not resemble other *Bacillus* signal peptides. A *B. subtilis* 168 strain has been converted to subtilin production by competent-cell transformation with the chromosomal DNA from *B. subtilis* ATCC 6633 (49). DNA sequence analysis of the subtilin operon has identified an ORF, SpaB, that would encode a protein that shares homology to *E. coli* HlyB protein (13). The HlyB protein participates in the transport of hemolysin through an alternative export pathway in *E. coli*. Further characterization of the subtilin operon may help elucidate the mechanism of the alternative protein export pathway in *B. subtilis*.

ROLE OF CHAPERONES IN PROTEIN TRANSPORT

Some of the recent studies on protein transport in both prokaryotes and eukaryotes suggest that the conformational state of the precursor protein plays a major role in the rate of protein transport (137). Proteins competent for export are thought to exist in a structure that is different from their fully folded tertiary structures, and a variety of cellular factors seem to bind to the precursor proteins to maintain these proteins in an export-competent conformation (137). These cellular factors include *E. coli* SecB and the heat shock proteins GroEL, GroES, and DnaK (89, 137). The biochemical data strongly suggest that homologs for chaperones will be present in *B. subtilis*. The genes for *B. subtilis* homologs of *groEL*, *groES*, and *dnaK* have been cloned and sequenced (47, 102a, 136). Thus, the role of heat shock proteins in protein secretion in *B. subtilis* can be studied.

COMPARISON WITH *E. COLI*

Current knowledge of protein secretion in *E. coli* is the result of a decade of genetic and biochemical studies. Thus, *E. coli* serves as a paradigm for studies on protein secretion in bacteria. Protein transport across biological membranes requires an interplay between two major components, the exported protein and the cellular factors, for optimal efficiency. The role played by the exported protein is likely to be similar in both *E. coli* and *B. subtilis*. However, the physiology of the organism might dictate certain constraints on the host factors, and thus, differences are anticipated.

Bacillus signal peptides are longer than *E. coli* signal peptides. *Bacillus* signal peptides are functional in *E. coli*, whereas *E. coli* signal peptides have not been extensively tested in *B. subtilis*. Signal peptide processing sites for several of the *Bacillus* signal peptides are different in *E. coli* (111). In addition, *E. coli* appears to require a shorter hydrophobic core than *B. subtilis* (67, 75). To be functional, *Bacillus* signal peptides require the presence of positively charged residues at the N terminus (6). *E. coli* signal sequence mutants lacking any charged residues at the N terminus are still functional (35, 92).

A majority of *Bacillus* exoproteins contain a pro peptide in addition to the signal peptide. A pro peptide has not been reported for *E. coli* periplasmic and outer membrane proteins. The presence of the pro peptide in *Bacillus* exoproteins may be related to the differences in cell envelope between *E. coli* and *B. subtilis*. The outer membrane of *E. coli* is a barrier to the periplasmic proteins, and thus, a high concentration of periplasmic proteins is available to the producing organism. In contrast, *B. subtilis* does not contain an outer membrane, and the secreted proteins are released into the growth medium and may not be available to the producing organism. The pro peptide might serve as a temporary anchor to the membrane or cell wall, resulting in a higher concentration of the exoprotein.

At least some of the cellular factors involved in protein secretion seem to be conserved between *E. coli* and *B. subtilis*. The DNA sequences of *B. subtilis secA*, *secY*, and *sipS* (*lep*) are similar to those of their homologs in *E. coli*. The SecA protein in *E. coli* plays a key role in protein export and has to interact with the precursor protein, cytoplasmic membrane, and various Sec proteins such as SecB, SecY, and SecE (137). The ability of the *B. subtilis* SecA protein to function in an *E. coli* in vitro system further underscores the conservation of the protein export pathway in gram-positive and gram-negative bacteria (41a, 129a). The location of the *secY* gene in the *spc* operon is conserved between *E. coli* and *B. subtilis*. The organization of the *secA* operon in *B. subtilis* appears to be different from that of *E. coli*.

PERSPECTIVES

Most of the exciting genetic and biochemical studies in *B. subtilis* have revolved around sporulation. The ability of *Bacillus* species to secrete proteins has been known for a long time, and these proteins were useful markers for sporulation. However, the biochemical tools necessary for studying protein secretion have not been available.

In the last few years, a set of genetic and biochemical tools for studying protein transport in *B. subtilis* has been developed. A set of signal sequence mutants has been constructed and characterized. Three of the secretion genes, namely, *secA* (*div*), *secY*, and *sipS* (*lep*), have been identified and sequenced. Thus, the study of protein secretion in *B. subtilis* enters an exciting phase in which earlier progress in the study of *E. coli* serves as a guide but in which surprises are bound to be encountered.

Acknowledgments. I thank Stephen Fahnestock for stimulating discussions and critical reading of the manuscript. I also thank Don Oliver, Tom Silhavy, and Ethel Jackson for critical comments. I thank C. Buchanan, A. Driessen, N. Hansen, and J. van Dijl and S. Wong for sending preprints.

REFERENCES

1. **Banerjee, S., and J. N. Hansen.** 1988. Structure and expression of a gene encoding the precursor of subtilin, a small protein antibiotic. *J. Biol. Chem.* **263:**9508–9514.
2. **Behnke, D., and D. Gerlach.** 1987. Cloning and expression in *Escherichia coli*, *Bacillus subtilis*, and *Streptococcus sanguis* of a gene for staphylokinase—a bacterial plasminogen activator. *Mol. Gen. Genet.* **210:**528–534.
3. **Benyahia, F., R. Chambert, and M. F. Petit-Glatron.** 1988. Levansucrase of *Bacillus subtilis*: effects on the secretion process of single amino acid substitutions in the mature part of the protein. *J. Gen. Microbiol.* **134:** 3259–3268.
4. **Borchert, T. V.** 1991. A genetic approach in the study of protein secretion in *B. subtilis*. Ph.D. thesis. The Technical University of Denmark, Copenhagen.
5. **Borchert, T. V., and V. Nagarajan.** 1990. Structure-function studies on the *Bacillus amyloliquefaciens* levansucrase signal peptide, p. 171–177. *In* M. Zukowski, A. T. Ganesan, and J. Hoch (ed.), *Genetics and Biotechnology of Bacilli*, Academic Press, Inc., New York.
6. **Borchert, T. V., and V. Nagarajan.** 1991. Effect of signal sequence alteration on export of levansucrase in *Bacillus subtilis*. *J. Bacteriol.* **173:**276–282.
7. **Briggs, M. S., and L. M. Gierasch.** 1986. Molecular mechanisms of protein secretion: the role of the signal sequence. *Adv. Prot. Chem.* **38:**110–180.

7a. **Buchanan, C. E., and M. Ling.** 1992. Isolation and sequence analysis of *dacB*, which encodes a sporulation-specific penicillin-binding protein in *Bacillus subtilis*. *J. Bacteriol.* **174:**1717–1725.

8. **Buchanan, C. E., and S. L. Neyman.** 1986. Correlation of penicillin-binding protein composition with different functions of two membranes in *Bacillus subtilis* forespores. *J. Bacteriol.* **165:**498–503.
9. **Caulfield, M. P., R. C. W. Berkeley, E. A. Pepper, and J. Melling.** 1979. Export of extracellular levansucrase by *Bacillus subtilis*: inhibition by cerulenin and quinacrine. *J. Bacteriol.* **138:**345–351.
10. **Chambert, R., F. Benyahia, and M. F. Petit-Glatron.** 1990. Secretion of *Bacillus subtilis* levansucrase Fe(III) could act as a cofactor in efficient coupling of folding and translocation processes. *Biochem. J.* **265:**375–382.
11. **Chambert, R., and M. F. Petit-Glatron.** 1990. Reversible thermal unfolding of *Bacillus subtilis* levansucrase is modulated by Fe^{3+} and Ca^{2+}. *FEBS Lett.* **275:**61–64.

11a. **Chen, M., and V. Nagarajan.** The role of signal peptide and mature protein in RNase (Barnase) export from *Bacillus subtilis*. *Mol. Gen. Genet.*, in press.

12. **Cheung, H. Y., and E. Freese.** 1985. Monovalent cations enable cell wall turnover of the turnover-deficient *lyt-15* mutant of *Bacillus subtilis*. *J. Bacteriol.* **161:**1222–1225.
13. **Chung, Y. J., M. T. Steen, and J. N. Hansen.** The subtilin gene of *Bacillus subtilis* ATCC 6633 is encoded in an operon that contains a homolog of the hemolysin B transport mechanism. Submitted for publication.
14. **Doyle, R.** 1989. How cell walls of gram-positive bacteria interact with metal ions, p. 275–293. *In* T. J. Beveridge and R. J. Doyle (ed.), *Metal Ions in the Bacteria*. John Wiley & Sons, Inc., New York.
15. **Egnell, P., and J.-I. Flock.** 1991. The subtilisin Carlsberg pro-region is a membrane anchorage for two fusion proteins produced in *Bacillus subtilis*. *Gene* **97:** 49–54.
16. **Emr, S. D., S. Hanley-Way, and T. J. Silhavy.** 1981. Suppressor mutations that restore export of a protein with a defective signal sequence. *Cell* **23:**79–88.
17. **Fandl, J., and P. C. Tai.** 1990. Biochemical characterization of genetically defined translocation components. *J. Bioenerg. Biomembr.* **22:**369–388.
18. **Fishman, Y., S. Rottem, and N. Citri.** 1980. Preferential suppression of normal exoenzyme formation by membrane-modifying agents. *J. Bacteriol.* **141:**1435–1438.
19. **Fukusaki, E., W. Panbangred, A. Shinmyo, and H. Okada.** 1984. The complete nucleotide sequence of the xylanase gene (*xynA*) of *Bacillus pumilus*. *FEBS Lett.* **171:**197–201.
20. **Gennity, J., J. Goldstein, and M. Inouye.** 1990. Signal peptide mutants of *Escherichia coli*. *J. Bioenerg. Biomembr.* **22:**233–270.
21. **Gierasch, L. M.** 1989. Signal sequences. *Biochemistry* **28:**923–930.
22. **Gould, A. R., B. K. May, and W. H. Elliott.** 1975. Release of extracellular enzymes from *Bacillus amyloliquefaciens*. *J. Bacteriol.* **122:**34–40.
23. **Hansen, J. N.** 1992. The molecular biology of nisin and its structural analogs. *In* D. Hoover and L. Steenson

(ed.), *Bacteriocins of Lactic Acid Bacteria*, in press. Academic Press, Inc., New York.

24. **Hartley, R. W.** 1988. Barnase and barstar. Expression of its cloned inhibitor permits expression of a cloned ribonuclease. *J. Mol. Biol.* **202:**913–915.
25. **Hastrup, S., and M. F. Jacobs.** 1990. Lethal phenotype conferred by xylose-induced overproduction of an Apr-lacZ fusion protein, p. 33–41. *In* M. Zukowski, A. T. Ganesan, and J. Hoch (ed.), *Genetics and Biotechnology of Bacilli*, vol. 3. Academic Press, Inc., New York.
26. **Hayashi, S., S. Y. Chang, S. Chang, C. Z. Giam, and H. C. Wu.** 1985. Modification and processing of internalized signal sequences of prolipoprotein in *Escherichia coli* and in *Bacillus subtilis*. *J. Biol. Chem.* **260:** 5753–5759.
27. **Hayashi, S., and H. C. Wu.** 1990. Lipoproteins in bacteria. *J. Bioenerg. Biomembr.* **22:**451–472.
28. **Hemila, H.** 1991. Sequence of a PAL related lipoprotein from *Bacillus subtilis*. *FEMS Microbiol. Lett.* **82:**37–42.
29. **Hofemeister, J., A. Kurtz, R. Borriss, and J. Knowles.** 1986. The β-glucanase gene from *Bacillus amyloliquefaciens* shows extensive homology with that of *Bacillus subtilis*. *Gene* **49:**177–187.
30. **Horiuchi, S., D. Marty-Mazars, P. C. Tai, and B. D. Davis.** 1983. Localization and quantitation of proteins characteristic of the complexed membrane of *Bacillus subtilis*. *J. Bacteriol.* **154:**1215–1221.
31. **Horiuchi, S., P. C. Tai, and B. D. Davis.** 1983. A 64-kilodalton membrane protein of *Bacillus subtilis* covered by secreting ribosomes. *Proc. Natl. Acad. Sci. USA* **80:**3287–3291.
32. **Hulett, F. M., E. E. Kim, C. Bookstein, N. V. Kapp, C. W. Edwards, and H. W. Wyckoff.** 1991. *Bacillus subtilis* alkaline phosphatases III and IV. *J. Biol. Chem.* **172:**1077–1084.
33. **Hulett, F. M., K. Stuckmann, D. Spencer, and T. Sanopoulou.** 1986. Purification and characterization of the secreted alkaline phosphatase of *Bacillus licheniformis* MC14: identification of a possible precursor. *J. Gen. Microbiol.* **132:**2387–2395.
34. **Hussain, M., F. I. J. Pastor, and J. O. Lampen.** 1987. Cloning and sequencing of the *blaZ* gene encoding β-lactamase III. A lipoprotein of *Bacillus cereus* 569/H. *J. Bacteriol.* **169:**579–586.
35. **Iino, T., M. Takahashi, and T. Sako.** 1987. Role of amino-terminal positive charge on signal peptide in staphylokinase export across the cytoplasmic membrane of *Escherichia coli*. *J. Biol. Chem.* **262:**7412–7417.
36. **Ikemura, H., and M. Inouye.** 1988. *In vitro* processing of pro-subtilisin produced in *Escherichia coli*. *J. Biol. Chem.* **263:**12959–12963.
37. **Ikemura, H., H. Takagis, and M. Inouye.** 1987. Requirement of pro-sequence for the production of active subtilisin E in *Escherichia coli*. *J. Biol. Chem.* **262:**7859–7864.
38. **Ito, K.** 1990. Structure, function and biogenesis of SecY, an integral membrane protein involved in protein export. *J. Bioenerg. Biomembr.* **22:**353–368.
39. **Jacobs, M., M. Eliasson, M. Uhlen, and J.-I. Flock.** 1985. Cloning, sequencing and expression of subtilisin Carlsberg from *Bacillus licheniformis*. *Nucleic Acids Res.* **13:**8913–8926.
40. **Jones, J. D., C. J. Macknight, and L. Gierasch.** 1990. Biophysical studies of signal peptides: implications for signal sequence functions and involvement of lipid in protein export. *J. Bioenerg. Biomembr.* **22:**213–232.
41. **Kennedy, M. B., and W. J. Lennarz.** 1979. Characterization of the extracellular lipase of *Bacillus subtilis* and its relationship to a membrane-bound lipase found in a mutant strain. *J. Biol. Chem.* **254:**1080–1089.

41a. **Klose, M., K. Schimz, J. van der Wolk, A. Driessen, and R. Freudl.** Lysine[106] of the putative catalytic ATP-binding site of the *B. subtilis* SecA protein is required for functional complementation of *Escherichia coli* secA mutants *in vivo*. Submitted for publication.

42. **Kontinen, V., P. Saris, and M. Sarvas.** 1991. A gene (*prsA*) of *Bacillus subtilis* involved in a novel, late stage of protein export. *Mol. Microbiol.* **5:**1273–1283.
43. **Kontinen, V., and M. Sarvas.** 1988. Mutants of *B. subtilis* defective in protein export. *J. Gen. Microbiol.* **134:**2333–2344.
44. **Kovacevic, S., L. E. Veal, H. M. Hsiung, and J. R. Miller.** 1985. Secretion of staphylococcal nuclease by *Bacillus subtilis*. *J. Bacteriol.* **162:**521–528.
45. **Kubo, M., and T. Imanaka.** 1988. Cloning and nucleotide sequence of the highly thermostable neutral protease gene from *Bacillus stearothermophilus*. *J. Gen. Microbiol.* **134:**1883–1892.
46. **Lampen, J. O., W. Wang, P. S. F. Mezes, and Y. Yang.** 1984. β-Lactamases of bacilli: nature and processing, p. 129–140. *In* A. T. Ganesan and J. Hoch (ed.), *Genetics and Biotechnology of Bacilli*. Academic Press, Inc., New York.
47. **Li, M., and S. L. Wong.** 1992. Cloning and characterization of *groESL* operon from *Bacillus subtilis*. *J. Bacteriol.* **174:**3981–3992.
48. **Liu, G., T. B. Topping, and L. L. Randall.** 1989. Physiological role during export for the retardation of folding by the leader peptide of maltose-binding protein. *Proc. Natl. Acad. Sci. USA* **86:**9213–9217.
49. **Liu, W., and J. N. Hansen.** 1991. Conversion of *Bacillus subtilis* 168 to a subtilin producer by competence transformation. *J. Bacteriol.* **173:**7387–7390.
50. **Lofdahl, S., B. Guss, M. Uhlen, L. Philpson, and M. Lindenberg.** 1983. Gene for staphylococcal protein A. *Proc. Natl. Acad. Sci. USA* **80:**697–701.
51. **MacKay, R. M., A. Lo, G. Willick, M. Zuker, S. Baird, M. Dove, F. Moranelli, and Verner Seligy.** 1986. Structure of a *Bacillus subtilis* endo-β-1,4-glucanase gene. *Nucleic Acids Res.* **14:**9159–9170.
52. **Manoil, C., and J. Beckwith.** 1985. TnphoA: a transposon probe for protein export signals. *Proc. Natl. Acad. Sci. USA* **82:**8129–8133.
53. **Martin, I., M. Debarbouille, E. Ferrari, A. Klier, and G. Rapoport.** 1987. Characterization of the levanase gene of *Bacillus subtilis* which shows homology to yeast invertase. *Mol. Gen. Genet.* **208:**177–184.
54. **Marty-Mazars, D., S. Horiuchi, P. C. Tai, and B. D. Davis.** 1983. Proteins of ribosome-bearing and free-membrane domains in *Bacillus subtilis*. *J. Bacteriol.* **154:**1381–1388.
55. **Matsuyama, S., E. Kimura, and S. Mizushima.** 1990. Complementation of two overlapping fragments of SecA, a protein translocation ATPase of *Escherichia coli*, allow ATP binding to its amino-terminal region. *J. Biol. Chem.* **265:**8760–8765.
56. **McLaughlin, J. R., C. L. Murray, and J. C. Rabinowitz.** 1981. Unique features in the ribosome binding site sequence of the gram-positive *Staphylococcus aureus* β-lactamase gene. *J. Biol. Chem.* **256:**11283–11291.
57. **Mezes, P. S. F., R. H. Blacher, and J. O. Lampen.** 1985. Processing of *Bacillus cereus* 569/H β-lactamase I in *Escherichia coli* and *Bacillus subtilis*. *J. Biol. Chem.* **260:**1218–1223.
58. **Mezes, P. S. F., W. Wang, E. C. H. Yeh, and J. O. Lampen.** 1983. Construction of penP-1, *Bacillus licheniformis* 749/C β-lactamase lacking site for lipoprotein modification. *J. Biol. Chem.* **258:**11211–11218.
59. **Michealis, S., H. Inouye, D. Oliver, and J. Beckwith.** 1983. Mutations that alter the signal sequence of alkaline phosphatase in *Escherichia coli*. *J. Bacteriol.* **154:** 366–374.
60. **Muren, E. M., and L. Randall.** 1985. Export of α-amylase by *Bacillus amyloliquefaciens* requires proton motive force. *J. Bacteriol.* **164:**712–716.
61. **Murphy, N., D. J. McConnell, and B. A. Cantwell.** 1984.

The DNA sequence of the gene and genetic control sites for the excreted *B. subtilis* enzyme b-glucanase. *Nucleic Acids Res.* **12:**5355–5367.

61a. **Nagarajan, V.** Unpublished data.

62. **Nagarajan, V., H. Albertson, M. Chen, and J. Ribbe.** 1992. Modular vectors for protein expression in *Bacillus subtilis*. *Gene* **114:**121–126.

63. **Nagarajan, V., and T. V. Borchert.** 1991. Levansucrase—a tool to study protein secretion in *Bacillus subtilis*. *Res. Microbiol.* **142:**787–792.

64. **Nagarajan, V., and M. Chen.** Unpublished data.

65. **Nakajima, R., T. Imanaka, and S. Aiba.** 1985. Nucleotide sequence of the *Bacillus stearothermophilus* α-amylase Gene. *J. Bacteriol.* **163:**401–406.

66. **Nakamura, A., T. Uozumi, and T. Beppu.** 1987. Nucleotide sequence of a cellulase gene of *Bacillus subtilis*. *Eur. J. Biochem.* **164:**317–320.

67. **Nakamura, K., Y. Fujita, Y. Itoh, and K. Yamane.** 1989. Modification of length, hydrophobic properties and electric charge of *Bacillus subtilis* α amylase signal peptide and their effects on the production of secretory proteins in *B. subtilis* and *Escherichia coli* cells. *Mol. Gen. Genet.* **216:**1–9.

68. **Nakamura, K., A. Nakamura, H. Takamatsu, H. Yoshikawa, and K. Yamane.** 1990. Cloning and characterization of a *Bacillus subtilis* gene homologous to *E. coli secY*. *J. Biochem.* **107:**603–607.

69. **Nakamura, K., H. Takamatsu, Y. Akiyama, K. Ito, and K. Yamane.** 1990. Complementation of the protein transport defect of an *Escherichia coli secY* mutant (*secY24*) by *Bacillus subtilis secY* homologue. *FEBS Lett.* **273:**75–78.

70. **Neugebauer, K., R. Sprengel, and H. Schaller.** 1981. Penicillinase from *Bacillus licheniformis*: nucleotide sequence of the gene and implications for the biosynthesis of a secretory protein in a Gram-positive bacterium. *Nucleic Acids Res.* **9:**2577–2588.

71. **Nielsen, J. B. K., M. P. Caufield, and J. O. Lampen.** 1981. Lipoprotein nature of *Bacillus licheniformis* membrane penicillinase. *Proc. Natl. Acad. Sci. USA* **78:**3511–3515.

72. **Nielsen, J. B. K., and J. O. Lampen.** 1982. Membrane-bound penicillinase in gram-positive bacteria. *J. Biol. Chem.* **257:**4490–4495.

73. **Nielsen, J. B. K., and J. O. Lampen.** 1982. Glyceride-cysteine lipoproteins and secretion by gram-positive bacteria. *J. Bacteriol.* **152:**315–322.

74. **Nishiya, Y., and T. Imanaka.** 1990. Cloning and nucleotide sequences of the *Bacillus stearothermophilus* neutral protease gene and its transcriptional activator gene. *J. Bacteriol.* **172:**4861–4869.

75. **Ohmura, K., K. Nakamura, H. Yamazaki, T. Shiroza, K. Yamane, Y. Jigami, H. Tanaka, K. Yoda, M. Yamasaki, and G. Tamura.** 1984. Length and structural effect of signal peptides derived from *Bacillus subtilis* a-amylase on secretion of *Escherichia coli* β-lactamase in *B. subtilis* cells. *Nucleic Acids Res.* **12:**5307–5319.

76. **Oliver, D. B., R. J. Cabelli, and G. P. Jarosik.** 1990. SecA protein: autoregulated initiator of secretory precursor protein translocation across the *E. coli* plasma membrane. *J. Bioenerg. Biomembr.* **22:**311–336.

77. **Overhoff, B., M. Klein, M. Spies, and R. Freudl.** 1991. Identification of a gene fragment which codes for the 364 amino terminal amino acid residues of a SecA homologue from *Bacillus subtilis*: further evidence for the conservation of the protein export apparatus in gram-positive and gram-negative bacteria. *Mol. Gen. Genet.* **228:**417–423.

78. **Paddon, C. J., and R. W. Hartley.** 1987. Expression of *Bacillus amyloliquefaciens* extracellular ribonuclease (barnase) in *Escherichia coli* following an inactivating mutation. *Gene* **53:**11–19.

79. **Paddon, C. J., N. Vasantha, and R. W. Hartley.** 1989. Translation and processing of *Bacillus amyloliquefaciens* extracellular RNase. *J. Bacteriol.* **171:**1185–1187.

80. **Paice, M. G., R. Bourbonnais, M. Deschrochers, L. Jurasek, and M. Yaguchi.** 1986. A xylanase gene from *Bacillus subtilis*: nucleotide sequence and comparison with *B. pumilus* gene. *Arch. Microbiol.* **144:**201–206.

81. **Park, S., G. Liu, T. B. Topping, W. H. Cover, and L. Randall.** 1988. Modulation of folding pathways of exported proteins by leader sequences. *Science* **239:**1033–1035.

82. **Paton, J. C., B. K. May, and W. H. Elliot.** 1980. Cerulenin inhibits production of extracellular proteins but not membrane proteins in *Bacillus amyloliquefaciens*. *J. Gen. Microbiol.* **118:**179–187.

83. **Payne, M. S., A. Griveson, and E. N. Jackson.** Unpublished data.

84. **Payne, M. S., and E. N. Jackson.** 1991. The use of alkaline phosphatase fusions to study protein secretion in *Bacillus subtilis*. *J. Bacteriol.* **173:**2278–2282.

85. **Perlman, D., and H. O. Halvorson.** 1983. A putative signal peptidase recognition site and sequence in eukaryotic and prokaryotic signal peptides. *J. Mol. Biol.* **167:**391–409.

86. **Petit-Glatron, M. F., F. Benyahia, and R. Chambert.** 1987. Secretion of *Bacillus subtilis* levansucrase: a possible two-step mechanism. *Eur. J. Biochem.* **163:**379–387.

87. **Petit-Glatron, M. F., and R. Chambert.** 1981. Levansucrase of *Bacillus subtilis*. Conclusive evidence that its production and export are unrelated to fatty-acid synthesis but modulated by membrane-modifying agents. *Eur. J. Biochem.* **119:**603–611.

88. **Petit-Glatron, M. F., I. Monteil, F. Benyahia, and R. Chambert.** 1990. *Bacillus subtilis* levansucrase: amino acid substitutions at one site affect secretion efficiency and refolding kinetics mediated by metals. *Mol. Microbiol.* **4:**2063–2070.

89. **Philips, G. J., and T. J. Silhavy.** 1990. Heat-shock proteins dnaK and groEL facilitate export of LacZ hybrid proteins in *E. coli*. *Nature* (London) **344:**882–884.

90. **Powers, S. D., R. M. Adams, and J. A. Wells.** 1986. Secretion and autoproteolytic maturation of subtilisin. *Proc. Natl. Acad. Sci. USA* **83:**3096–3100.

91. **Pugsley, A. P., C. d'Enfert, I. Reyss, and M. G. Kornacker.** 1990. Genetics of extracellular protein secretion by gram-negative bacteria. *Annu. Rev. Genet.* **24:** 67–90.

92. **Puziss, J. W., J. D. Fikes, and P. J. Bassford, Jr.** 1989. Analysis of mutational alterations in the hydrophilic segment of the maltose-binding protein signal peptide. *J. Bacteriol.* **171:**2303–2311.

93. **Randall, L., and S. J. Hardy.** 1986. Correlation of competence for export with lack of tertiary structure of the mature species: a study *in vivo* of maltose-binding protein in *E. coli*. *Cell* **46:**921–928.

94. **Ribbe, J., and V. Nagarajan.** 1992. Characterization of secretion efficiency of a plant signal peptide in *Bacillus subtilis*. *Mol. Gen. Genet.* **235:**333–339.

95. **Robson, L. M., and G. H. Chambliss.** 1987. Endo-β-1,4-glucanase gene of *Bacillus subtilis* DLG. *J. Bacteriol.* **169:**2017–2025.

96. **Sadaie, Y., and T. Kada.** 1985. *Bacillus subtilis* gene involved in cell division, sporulation, and exoenzyme secretion. *J. Bacteriol.* **163:**648–653.

97. **Sadaie, Y., H. Takamatsu, K. Nakamura, and K. Yamane.** 1991. Sequencing reveals similarity of the wild-type div^{+} gene of *Bacillus subtilis* to the *Escherichia coli secA* gene. *Gene* **98:**101–105.

98. **Sanders, R. L., and B. K. May.** 1975. Evidence for extrusion of unfolded extracellular enzyme polypeptide

chains through membranes of *Bacillus amyloliquefaciens*. *J. Bacteriol.* **123:**806–814.

99. **Sarvas, M.** 1988. Protein secretion in bacilli. *Curr. Top. Microbiol. Immunol.* **125:**103–126.

100. **Sasamoto, H., K. Nakazawa, K. Tsutsumi, K. Takase, and K. Yamane.** 1989. Signal peptide of *Bacillus subtilis* α-amylase. *J. Biochem.* **106:**376–382.

101. **Saunders, C. W., J. A. Pedroni, and P. Monahan.** 1991. Secretion from B. subtilis of a 34 amino acid residue fragment of human parathyroid hormone: effect of sequences adjacent to the signal sequence cleavage. *Gene* **102:**277–282.

102. **Schatz, P. J., and J. Beckwith.** 1990. Genetic analysis of protein export in *Escherichia coli*. *Annu. Rev. Genet.* **24:**215–249.

102a. **Schmidt, A., M. Schiesswohl, U. Volker, M. Hecker, and W. Schuman.** 1992. Cloning, sequencing, mapping and transcriptional analysis of the groESL operon from *Bacillus subtilis*. *J. Bacteriol.* **174:**3993–3999.

103. **Schulein, R., J. Kreft, S. Gonski, and W. Goebel.** 1991. Preprosubtilisin Carlsberg processing and secretion is blocked after deletion of amino acids 97–101 in the mature part of the enzyme. *Mol. Gen. Genet.* **227:**137–143.

104. **Shimada, H., M. Honjo, I. Mita, A. Nakayama, A. Akaoka, K. Manabe, and Y. Furutani.** 1985. The nucleotide sequence and some properties of the neutral protease gene of *Bacillus amyloliquefaciens*. *J. Bacteriol.* **2:**75–85.

105. **Silhavy, T., and J. R. Beckwith.** 1985. Use of *lac* fusion for study of biological problems. *Microbiol. Rev.* **49:** 398–418.

106. **Sloma, A., A. Ally, D. Ally, and J. Pero.** 1988. Gene encoding a minor extracellular protease in *Bacillus subtilis*. *J. Bacteriol.* **170:**5557–5563.

107. **Sloma, A., C. F. Rudolph, G. A. Rufo, Jr., B. J. Sullivan, K. A. Theriault, D. Ally, and J. Pero.** 1990. Gene encoding a novel extracellular metalloprotease in *Bacillus subtilis*. *J. Bacteriol.* **172:**1024–1029.

108. **Sloma, A., G. A. Rufo, Jr., C. F. Rudolph, B. J. Sullivan, K. A. Theriault, D. Ally, and J. Pero.** 1990. Bacillopeptidase F of *Bacillus subtilis*: purification of the protein and cloning of the gene. *J. Bacteriol.* **172:**1470–1477.

109. **Sloma, A., G. A. Rufo, Jr., K. A. Theriault, M. Dwyer, S. W. Wilson, and J. Pero.** 1991. Cloning and characterization of the gene for an additional extracellular serine protease of *Bacillus subtilis*. *J. Bacteriol.* **173:**6889–6895.

110. **Smith, H., S. Bron, J. Vanee, and G. Venema.** 1987. Construction and use of signal sequence selection vectors in *Escherichia coli* and *Bacillus subtilis*. *J. Bacteriol.* **169:**3321–3328.

111. **Smith, H., A. D. Jong, S. Bron, and G. Venema.** 1988. Characterization of signal-sequence-coding regions selected from the *Bacillus subtilis* chromosome. *Gene* **70:**351–361.

112. **Smith, W. P., P.-C. Tai, and B. D. Davis.** 1978. Interaction of secreted nascent chains with surrounding membrane in *Bacillus subtilis*. *Proc. Natl. Acad. Sci. USA* **75:**5922–5925.

113. **Stader, J., S. A. Benson, and T. J. Silhavy.** 1986. Kinetic analysis of lamB mutants suggests that signal sequence plays multiple roles in protein export. *J. Biol. Chem.* **262:**15075–15080.

114. **Stahl, M. L., and E. Ferrari.** 1984. Replacement of the *Bacillus subtilis* subtilisin structural gene with an in vitro-derived deletion mutation. *J. Bacteriol.* **158:**411–418.

115. **Steinmetz, M., D. L. Coq., S. Aymerich, G. Gonzy-Treboul, and P. Gay.** 1985. The DNA sequence of the gene for the secreted *Bacillus subtilis* enzyme levansucrase and its genetic control sites. *Mol. Gen. Genet.* **200:**220–228.

116. **Stephens, M. A., S. A. Ortlepp, J. F. Ollington, and D. McConnell.** 1984. Nucleotide sequence of the 5′ region of the *Bacillus licheniformis* α-amylase gene: comparison with the *B. amyloliquefaciens* gene. *J. Bacteriol.* **158:**369–372.

117. **Su, J.-W., A. Boylan, S. M. Thomas, K. M. Dolan, D. B. Oliver, and C. W. Price.** 1990. Isolation of secY homologue from *Bacillus subtilis*: evidence for a common protein export pathway in eubacteria. *Mol. Microbiol.* **4:**305–314.

118. **Takagi, M., and T. Imanaka.** 1989. Role of the pre-pro-region of neutral protease in secretion in *Bacillus subtilis*. *J. Ferment. Bioeng.* **67:**71–76.

119. **Takagi, M., T. Imanaka, and S. Aiba.** 1985. Nucleotide sequence and promoter region for the neutral protease gene from *Bacillus stearothermophilus*. *J. Bacteriol.* **163:** 824–831.

120. **Takase, K., H. Mizuno, and K. Yamane.** 1988. NH_2-terminal processing of *Bacillus subtilis* α-amylase. *J. Biol. Chem.* **263:**11548–11553.

121. **Takekawa, S., N. Uozumi, N. Tsukagoshi, and S. Udaka.** 1991. Protease involved in generation of β and α amylases from large amylase precursor in *Bacillus polymyxa*. *J. Bacteriol.* **173:**6820–6825.

122. **Takkinen, K., R. F. Pettersson, N. Kalkkinen, I. Palva, H. Soderlund, and L. Kaariainen.** 1983. Amino acid sequence of α-amylase from *Bacillus amyloliquefaciens* deduced from the nucleotide sequence of the cloned gene. *J. Biol. Chem.* **298:**1007–1013.

123. **Tang, L., R. Lenstra, T. V. Borchert, and V. Nagarajan.** 1990. Isolation and characterization of levansucrase encoding gene from *Bacillus amyloliquefaciens*. *Gene* **96:**89–93.

124. **Thirunavukkarasu, M., and F. G. Priest.** 1983. Synthesis of α-amylase and α-glucosidase by membrane bound ribosomes from *Bacillus licheniformis*. *Biochem. Biophys. Res. Commun.* **114:**677–683.

125. **Tran, L., X. Wu, and S. Wong.** 1991. Cloning and expression of a novel protease gene encoding an extracellular neutral protease from *Bacillus subtilis*. *J. Bacteriol.* **173:**6364–6372.

126. **Tsuboi, A., R. Uchihi, T. Adachi, T. Sasaki, S. Hayakawa, H. Yamagata, N. Tsukagoshi, and S. Udaka.** 1988. Characterization of the genes for the hexagonally arranged surface layer proteins in protein-producing *Bacillus brevis* 47: complete nucleotide sequence of the middle wall protein gene. *J. Bacteriol.* **170:**935–945.

127. **Tsukagoshi, N., R. Tabata, T. Takemura, H. Yamagata, and S. Udaka.** 1984. Molecular cloning of a major cell wall protein gene from protein-producing *Bacillus brevis* 47 and its expression in *Escherichia coli* and *Bacillus subtilis*. *J. Bacteriol.* **158:**1054–1060.

128. **Uozumi, N., K. Sakurai, T. Sasaki, S. Takekawa, H. Yamagata, N. Tsukagoshi, and S. Udaka.** 1989. A single gene directs synthesis of a precursor protein with β- and α-amylase activities in *Bacillus polymyxa*. *J. Bacteriol.* **171:**375–382.

129. **Van Den Burg, B., H. G. Enequist, M. E. Van Der Haar, V. G. H. Eijsink, B. K. Stulp, and G. Venema.** 1991. A highly thermostable neutral protease from *Bacillus caldolyticus*: cloning and expression of the gene in *Bacillus subtilis* and characterization of the gene product. *J. Bacteriol.* **173:**4107–4115.

129a. **van der Wolk, J., M. Klose, E. Breukink, R. A. Demel, B. Kruijff, R. Freudl, and A. J. M. Driessen.** Characterization of a *Bacillus subtilis secA* mutant protein deficient in translocation ATPase and release from the membrane. Submitted for publication.

130. **van Dijl, J. M., A. de Jong, G. Venema, and S. Bron.** 1992. Signal peptidase I of *Bacillus subtilis*: patterns of conserved amino acids in prokaryotic and eukaryotic type I signal peptidases. *EMBO J.* **11:**2819–2828.

131. **Vasantha, N., and L. D. Thompson.** 1986. Fusion of pro

region of subtilisin to staphylococcal protein A and its secretion by *Bacillus subtilis. Gene* **49:**23–28.

132. **Vasantha, N., L. D. Thompson, C. Rhodes, C. Banner, J. Nagle, and D. Filpula.** 1984. Genes for alkaline protease and neutral protease from *Bacillus amyloliquefaciens* contain a large open reading frame between the regions coding for signal sequences and mature protein. *J. Bacteriol.* **159:**811–819.
133. **von Heijne, G.** 1983. Patterns of amino acids near signal sequence cleavage sites. *Eur. J. Biochem.* **133:**17–21.
134. **von Heijne, G.** 1986. A new method for predicting signal sequence cleavage sites. *Nucleic Acids Res.* **14:** 4683–4690.
135. **Wagner, B., M. Wagner, L. Wollweber, and D. Behnke.** 1989. Immunoelectron microscopy of *Bacillus subtilis* cell secreting human interferon a1 or staphylokinase. *FEMS Microbiol. Lett.* **65:**327–332.
136. **Wetzstein, M., J. Dedio, and W. Schumann.** 1990. Complete nucleotide sequence of the *Bacillus subtilis dnaK* gene. *Nucleic Acids Res.* **18:**2172.
137. **Wickner, W., A. J. M. Driessen, and F. Hartl.** 1991. The enzymology of protein translocation across the *Escherichia coli* plasma membrane. *Annu. Rev. Biochem.* **60:**101–124.
138. **Wu, J. J., and P. J. Piggot.** 1989. Regulation of late expression of the *Bacillus subtilis spoIIA* locus: evidence that it is cotranscribed with the gene for a putative penicillin binding protein, p. 321–328. *In* M. Zukowski, A. T. Ganesan, and J. Hoch (ed.), *Genetics and Biotechnology of Bacilli*. Academic Press, Inc., New York.
139. **Yang, M. Y., E. Ferrari, and D. J. Henner.** 1984. Cloning of the neutral protease gene of *Bacillus subtilis* and use of the cloned gene to create an in vitro-derived deletion mutation. *J. Bacteriol.* **160:**15–21.
140. **Zagorec, M., and M. Steinmetz.** 1990. Expression of levansucrase-β-galactosidase hybrids inhibits secretion and is lethal in *Bacillus subtilis. J. Gen. Microbiol.* **136:**1137–1143.
141. **Zhu, X., Y. Ohta, F. Jordan, and M. Inouye.** 1989. Pro-sequence of subtilisin can guide the refolding of denatured subtilisin in an intramolecular process. *Nature* (London) **339:**483–484.

VIII. POSTEXPONENTIAL-PHASE PHENOMENA

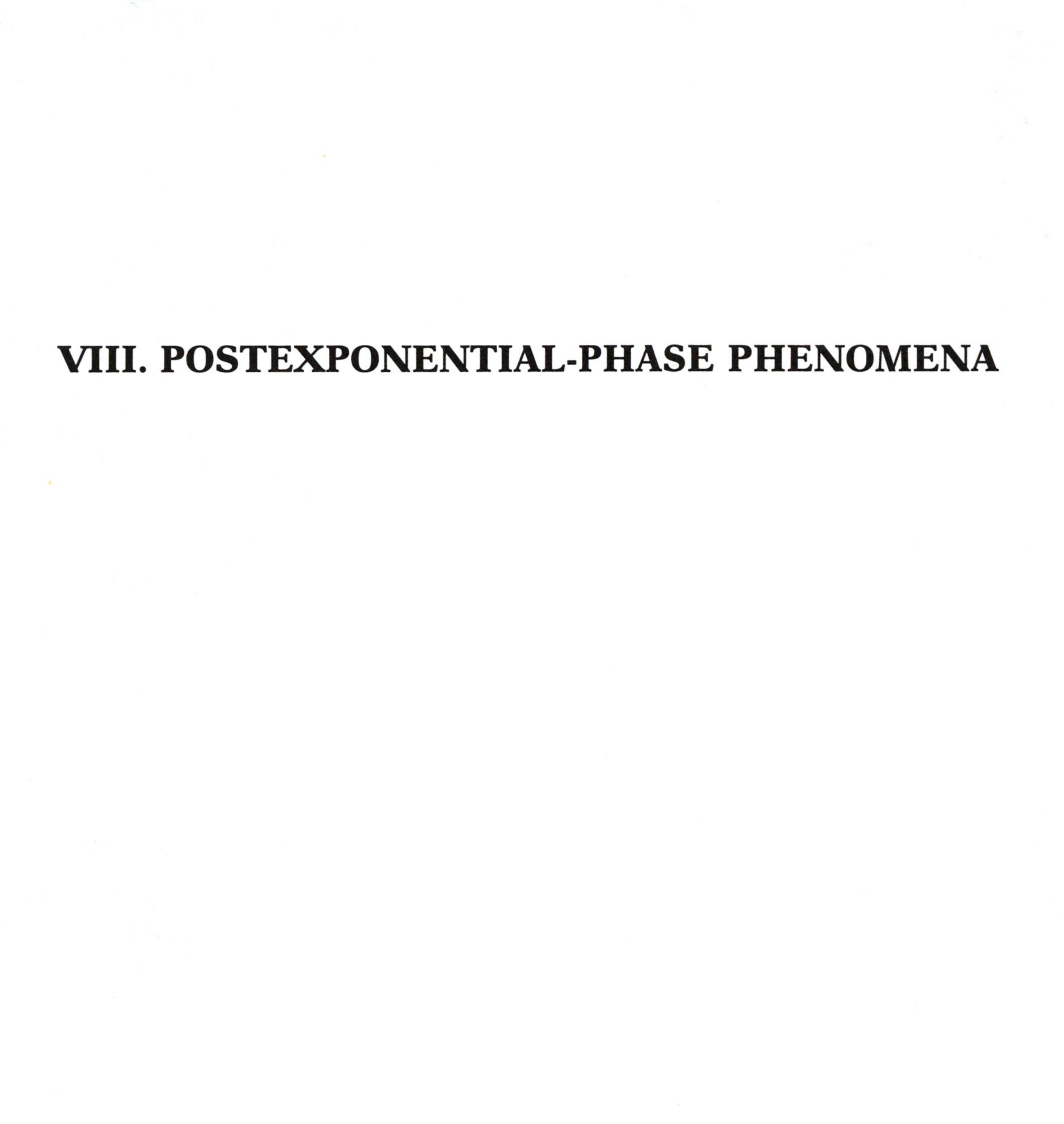

50. Two-Component Regulatory Systems

TAREK MSADEK, FRANK KUNST, and GEORGES RAPOPORT

INTRODUCTION

Two-Component System: a Paradigm for Regulation of Genetic Expression

Requirements for bacterial survival include both the continuous monitoring of external conditions and the ability to respond rapidly to extracellular and cytoplasmic signals reflecting changes in the environment. This dual process, termed signal transduction, generally involves modifications of gene expression, giving bacteria the flexibility to adapt to a wide variety of stimuli. During the past few years, a rapidly growing body of genetic and biochemical evidence has introduced a novel paradigm for the regulation of genetic expression: bacterial signal transduction through so-called two-component systems (71, 117, 126). Indeed, a large number of bacterial responses involve the interaction of two regulatory proteins, one of which controls the activity of the other. These paired regulators make up two families of homologous proteins and have been the subject of several reviews (3, 18, 19, 47, 71, 95, 126, 146, 147).

Typically, the first component, a histidine protein kinase, receives an extracellular signal directly or indirectly, possibly through its amino-terminal domain, which often includes transmembrane sequences. This signal is then transduced via the carboxy-terminal domain of the kinase to the second component, the response regulator, which is generally a transcriptional activator. The common mechanism underlying this signal transduction process involves a now-classic phosphotransfer reaction between the two proteins. The protein kinase is first autophosphorylated at a conserved histidine residue in an ATP-dependent reaction. In a second step, the phosphoryl group is transferred to an aspartate residue in the amino-terminal domain of the response regulator. Several members of the histidine protein kinase family also act as phosphatases, catalyzing dephosphorylation of the associated response regulator. Regulation may therefore take place by modulating either the kinase activity or the phosphatase activity of the protein. However, for reasons of simplicity, members of this family will be designated only as "kinases" within this review.

Originally characterized for nitrogen regulation (NtrB-NtrC) (86, 87), chemotaxis (CheA-CheY-CheB) (18, 19, 119, 145–147), osmoregulation (EnvZ-OmpR) (64, 100), and regulation of phosphate uptake (PhoR-PhoB) (88–90) in *Escherichia coli* and *Salmonella typhimurium*, the family of two-component systems has rapidly expanded, and new examples continue to emerge. More than 30 of these regulatory pairs have been identified in both gram-negative and gram-positive bacteria; they control functions ranging from virulence gene expression in *Agrobacterium tumefaciens* (VirA-VirG) (80, 93, 166), *S. typhimurium* (PhoQ-PhoP) (96), and *Bordetella pertussis* (BvgS-BvgA) (9, 130) to complex developmental pathways such as sporulation (SpoIIJ-Spo0F-Spo0A) (8, 40, 121, 156, 173) and competence for transformation by exogenous DNA (ComP-ComA and DegS-DegU) in *Bacillus subtilis* (57, 76, 150, 163, 165).

Table 1 lists the known and potential two-component regulatory pairs identified in gram-positive bacteria up to March 1992. These pairs are the focus of this chapter (see below). For specific details regarding two-component systems in gram-negative bacteria, the reader is referred to several excellent reviews (3, 18, 19, 47, 95, 146, 147).

Domain Structure and Phosphorylation of Two-Component Regulators

The term "two component" is somewhat misleading, since several of these systems involve additional proteins acting as the true environmental sensor (Che and Ntr systems in *E. coli*) (87, 145) and one system forms a multicomponent phosphorelay (sporulation in *B. subtilis*; see chapter 51, this volume). However, this terminology does underline the central importance of the two sets of proteins involved in these systems, i.e., protein kinases and response regulators.

Protein kinases

The protein kinase family is characterized by a highly conserved carboxy-terminal domain approximately 250 amino acids long and containing at least four invariant amino acid residues (47, 71, 146). Figure 1 shows an alignment of the conserved regions within the carboxy-terminal domains of protein kinases from gram-positive bacteria. A conserved histidine among the first residues of this region has been identified as the site of autophosphorylation of the EnvZ (His-243), NtrB (His-139), and VirA (His-474) protein kinases (42, 66, 114). In the case of the *E. coli* CheA protein, however, the phosphorylated residue is His-48, which is located within the nonconserved amino-terminal domain, and the site of phosphorylation is the N-3 position of the imidazole ring (58, 145, 146). Proteolytic cleavage of phosphorylated CheA releases an 18-kDa amino-terminal fragment that remains fully capable of transferring its phosphate group to CheY or CheB, the cognate response regula-

Tarek Msadek, Frank Kunst, and Georges Rapoport • Unité de Biochimie Microbienne, Centre National de la Recherche Scientifique, URA 1300, Institut Pasteur, 25 rue du Docteur Roux, 75724 Paris Cedex 15, France.

Table 1. Two-component regulatory pairs of gram-positive bacteria

Organism	Histidine protein kinase	Response regulator	Adaptive response	Reference(s)
B. subtilis	CheA	CheY and CheB	Chemotaxis	16, 44
B. subtilis	ComP	ComA	Competence	163, 165
B. subtilis	DegS	DegU	Degradative enzyme production, competence	57, 76, 150
B. subtilis	PhoR	PhoP	Phosphate assimilation	133, 134
B. subtilis	SpoIIJ (KinA) and KinB	Spo0F and Spo0A	Sporulation	8, 23, 40, 121, 155, 156, 173
B. subtilis	Urf-1	Urf-2	?	102
B. subtilis	OrfX-18	OrfX-17	?	140
B. megaterium	?	Orf-1	?	99
S. aureus	AgrB	AgrA	Extracellular enzyme production	73, 120
S. lividans	CutS	CutR	Melanin production, copper metabolism	157
E. faecium	VanS	VanR	Synthesis of peptidoglycan precursors	10

tors (58). However, this fragment can no longer be autophosphorylated, which suggests that the conserved carboxy-terminal region may define a catalytic domain involved in ATP binding and in transfer of the γ-phosphoryl group from ATP to a histidine side chain rather than in phosphotransfer from the kinase to the response regulator.

Phosphoramidates such as the phosphohistidine group of members of the protein kinase family exhibit a characteristic instability in acid as opposed to alkali. Although no nucleotide-binding site has been identified within the conserved amino-terminal domain of histidine protein kinases, one may speculate that the glycine-rich sequences within this domain (regions III and IV, Fig. 1) play the role of the Rossmann ATP-binding motif of eukaryotic protein kinases (GXGXXG) (50, 69, 127, 153). In addition to their kinase activities, histidine protein kinases such as EnvZ and NtrB also act as phosphoprotein phosphatases, accelerating the dephosphorylation of their cognate response regulators, OmpR or NtrC, in an ATP-dependent fashion (1, 2, 63, 68, 115, 154). The combined kinase and phosphatase activities of the histidine protein kinases thus act to control the amount of the phosphorylated form of the response regulator present in the cell in response to environmental signals.

The protein kinases can be divided into two groups: membrane-associated proteins such as EnvZ or PhoR of *E. coli* (27, 90, 101) and cytoplasmic proteins such as NtrB of *E. coli* and *Klebsiella pneumoniae* (22, 36, 85, 98) and CheA of *E. coli* and *S. typhimurium* (72, 109, 143). The amino-terminal transmembrane and periplasmic domains of the EnvZ and VirA protein kinases are presumably directly involved in sensing environmental signals (41, 92, 154). Cytoplasmic protein kinases are thought to respond to intracellular signals. Most of the protein kinases identified in gram-positive bacteria have membrane-spanning segments in their amino-terminal domains, suggesting that they are membrane-associated proteins: ComP, PhoR, KinB, and OrfX18 of *B. subtilis* (23, 134, 140, 155, 165); VanS of *Enterococcus faecium* (10); AgrB of *Staphylococcus aureus* (73); and CutS of *Streptomyces lividans* (157). The *B. subtilis* DegS, SpoIIJ (KinA), and CheA kinases appear to be cytoplasmic proteins (8, 44, 57, 76, 121).

Two of the histidine protein kinases from gram-positive bacteria, SpoIIJ (KinA) and DegS of *B. subtilis*, have been purified and characterized. Both proteins are labeled with ^{32}P when incubated in the presence of [γ-^{32}P]ATP (23, 30, 108, 121) (see chapter 51 and below).

Response regulators

Proteins in the response regulator family share a highly conserved amino-terminal domain that is approximately 120 amino acids long (3, 47, 71, 146). Figure 2 shows an alignment of this conserved region within the response regulators of gram-positive bacteria. This domain contains at least three invariant amino acid residues that correspond to residues Asp-11, Asp-56, and Lys-106 in the *B. subtilis* DegU amino acid sequence (57, 76, 146, 150). In addition, most of the response regulators have potential helix-turn-helix DNA-binding motifs in their carboxy-terminal domains.

The *E. coli* and *S. typhimurium* CheY proteins consist entirely of conserved amino-terminal response regulator domains, and their three-dimensional structures have been determined (144, 161). CheY is a single-domain protein with a doubly wound five-stranded parallel β sheet surrounded by five α helices (144, 161). Within this structure, aspartate residues Asp-12, Asp-13, and Asp-57 are clustered to form an acidic pocket, which constitutes the phosphorylation site of the protein (144). One of these conserved aspartate residues is the phosphoacceptor site for several response regulators: Asp-57 for CheY of *E. coli* and *S. typhimurium* (20, 129), Asp-54 for NtrC of *S. typhimurium* (cited in reference 18), and Asp-52 for VirG of *A. tumefaciens* (65). Divalent metal ions are essential for the transfer of phosphate from CheA to CheY, and the acidic pocket formed by residues Asp-12, Asp-13, and Asp-57 in CheY acts as a magnesium ion-binding site (83). Within the active site of the CheY molecule, the Lys-109 residue also plays a functionally important role (81, 161). The ε-amino group of the Lys-109 side chain is bound to one of the oxygens of the β-carbox-

Figure 1. Alignment of conserved regions within the carboxy-terminal domains of histidine protein kinases from gram-positive bacteria. Multiple protein sequence alignments were carried out by using the CLUSTAL V program (59, 60). Protein sequences are from the references in Table 1. Homologous residues are indicated by black boxes, stars indicate invariant residues, and numbers indicate positions in the amino acid sequences of the respective proteins. Accepted conservative substitutions are as follows: I, L, V, and M; K and R; S and T; D and E; F and Y; N and Q; G and A. *Bs*, *B. subtilis*; *Sa*, *S. aureus*; *Ef*, *E. faecium*.

ylate side chain of Asp-57 in CheY (161). In its unphosphorylated conformation, the *E. coli* CheY protein is thought to be inactive. Phosphorylation of Asp-57 would cause a local adjustment of the Asp-57 carboxyl side chain and a repositioning of the Lys-109 side chain (161). Phosphorylation-induced rearrangements within the active site are thought to give rise to larger-scale conformational changes within the CheY protein (81, 145, 161). Phosphorylated CheY would thus have an active conformation, allowing interaction with the flagellar switch proteins.

Acyl phosphate bonds such as those of phosphorylated response regulators tend to be more stable in acid than in alkali. Although the phosphorylated histidine protein kinases are relatively stable, the phosphorylated response regulators vary widely in stability, with half-lives ranging from 10 s for CheY (168) to approximately 1 h for OmpR or VirG (63, 65). These differences reflect the presence or absence of intrinsic autophosphatase activities, which, together with phosphatase activities of associated protein kinases or of additional proteins such as CheZ (Che system of *E. coli*), regulate the time scale response of the signal transduction system. It has been suggested that the role of phosphorylation in activation of response regulators is indirect, in that it generates or stabilizes a conformational change in the protein (18). Indeed, a modified CheY protein in which the Asp-13 residue is replaced by lysine is not phosphorylated by CheA in vitro, yet its in vivo phenotype is that of wild-type phosphorylated CheY, suggesting that it is locked in an active conformation (18, 20).

The phosphotransfer reaction in bacterial signal transduction regulatory pathways between protein phosphoramidates and protein acyl phosphates appears to be catalyzed by the response regulator alone. Indeed, by using phosphoramidate as a phosphodonor, both the CheY and CheB response regulators have been shown to be phosphorylated in vitro in the absence of CheA or any other auxiliary protein (82).

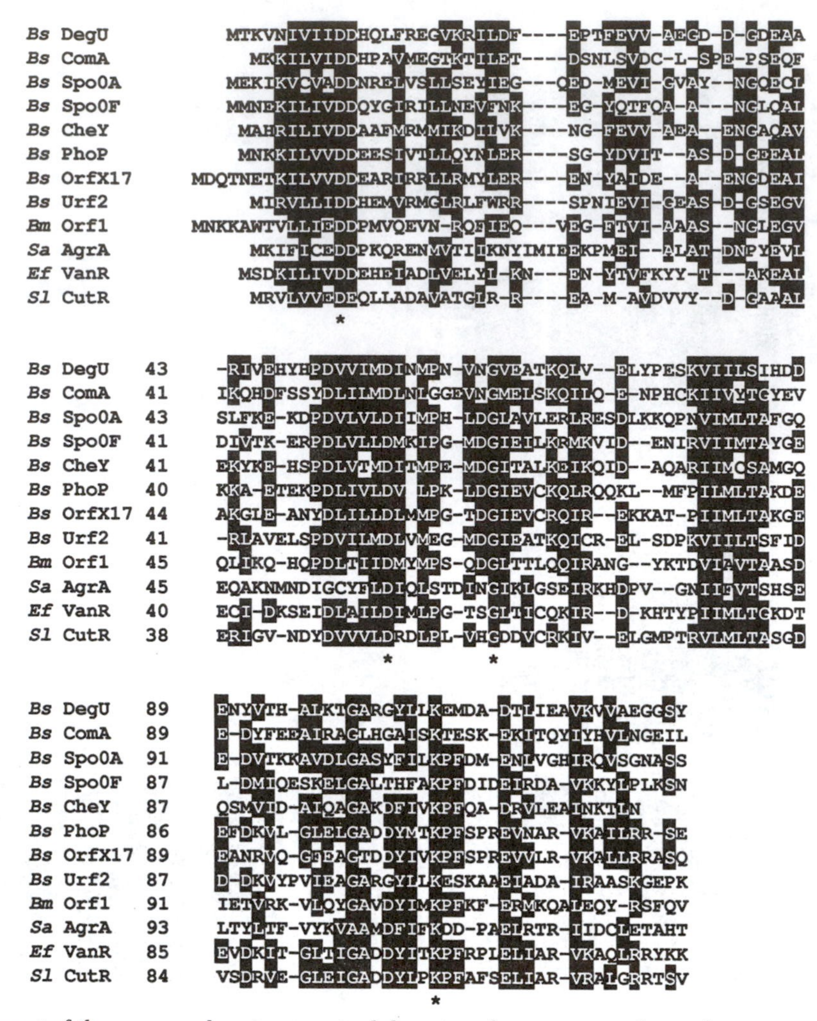

Figure 2. Alignment of the conserved amino-terminal domains of response regulators from gram-positive bacteria. References for protein sequences are listed in Table 1. Numbers correspond to positions in the respective amino acid sequences. Comparisons were carried out as indicated in the legend to Fig. 1. *Bs*, *B. subtilis*; *Bm*, *B. megaterium*; *Sa*, *S. aureus*; *Ef*, *E. faecium*; *Sl*, *S. lividans*.

CheY can also use intermediary metabolites such as acetyl phosphate or carbamoyl phosphate as phosphodonors (82). It was therefore suggested that low-molecular-weight phosphodonors such as acetyl phosphate may also be involved in regulating signal transduction pathways in vivo (82).

Amino acid sequence similarities between the carboxy-terminal domains of the response regulators allow the distinction of several subfamilies. In *E. coli*, five such groups have been defined: (i) proteins lacking a carboxy-terminal domain, such as CheY; (ii) the OmpR subfamily; (iii) the UhpA subfamily; (iv) response regulators of the NtrC subfamily that share both a central domain involved in ATP binding and interaction with σ^{54}-RNA polymerase holoenzyme and a carboxy-terminal DNA-binding domain; and (v) proteins such as CheB, whose carboxy-terminal domains are not significantly similar to those of any of the other response regulators (146).

In gram-positive bacteria, the *B. subtilis* CheY (16) and Spo0F (156, 173) proteins consist entirely of amino-terminal response regulator domains. This is consistent with their role in protein-protein interaction rather than direct regulation of genetic expression. Indeed, phosphorylated CheY is thought to interact directly with the FliM, FliN, and FliG flagellar motor components in *E. coli* (146), and Spo0F is involved in a multicomponent phosphorelay, interacting with both the SpoIIJ (KinA) protein kinase and the Spo0B phosphotransferase (23, 155) (see chapter 51 and below).

A response regulator with a carboxy-terminal domain homologous to that of NtrC has not yet been identified in gram-positive bacteria. However, both σ^{54} and an NifA-type regulator with the conserved central domain homologous to that of NtrC (LevR) have recently been identified in *B. subtilis* (32, 33).

The amino acid sequences of the *B. subtilis* Spo0A (40) and *S. aureus* AgrA (73, 120) proteins show no similarities in their carboxy-terminal domains to each other or to any other response regulators.

The remaining known response regulators of gram-positive bacteria can be divided into two groups on the basis of carboxy-terminal domain sequence simi-

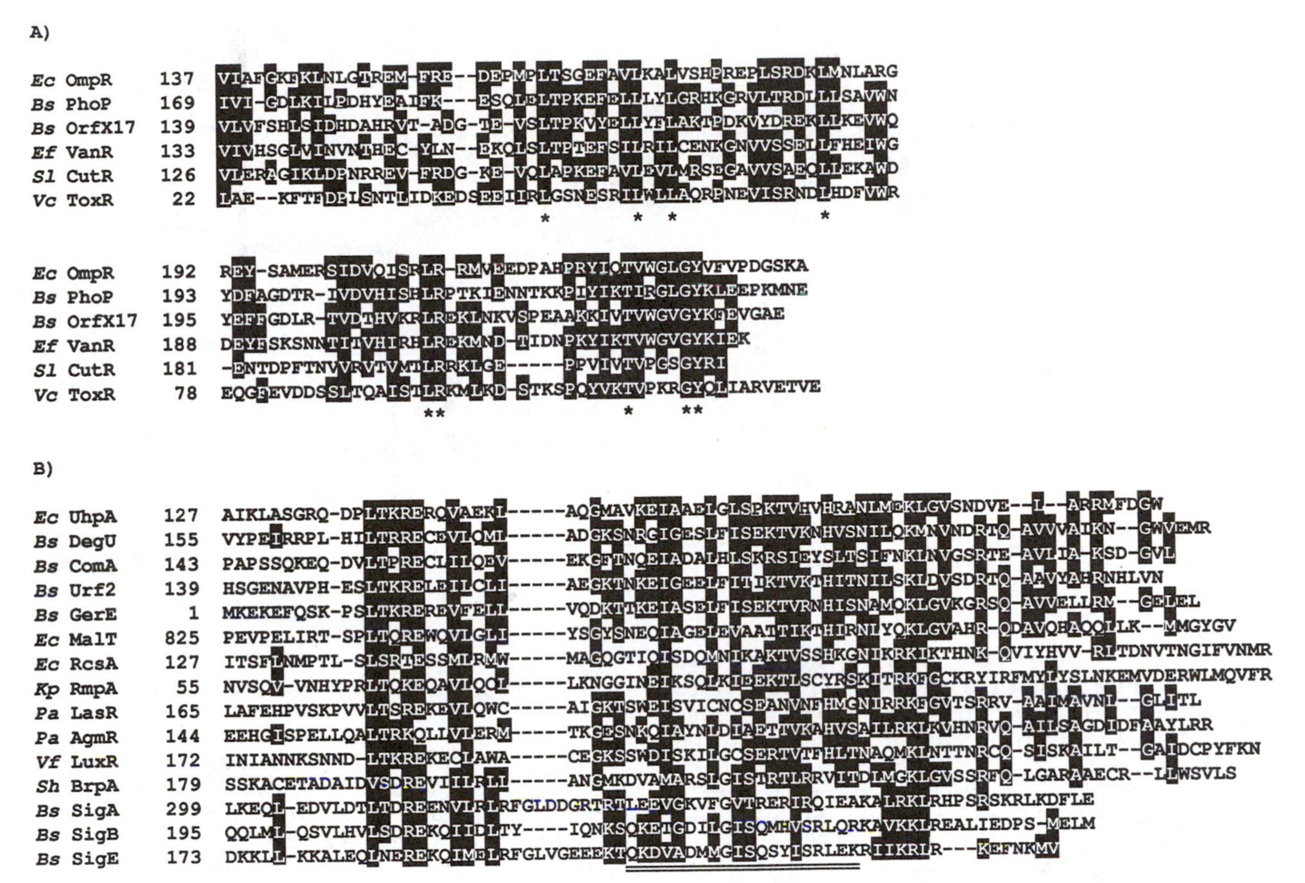

Figure 3. Conserved carboxy-terminal domains of response regulators from gram-positive bacteria. Numbers correspond to positions in the respective amino acid sequences. References for the response regulators from gram-positive bacteria are listed in Table 1. Comparisons were carried out as indicated in the legend to Fig. 1. (A) Alignment of the OmpR-PhoP subfamily with OmpR of *E. coli* (27, 167) and ToxR of *V. cholerae* (97). *Ec*, *E. coli*; *Bs*, *B. subtilis*; *Ef*, *E. faecium*; *Sl*, *S. lividans*; *Vc*, *V. cholerae*. (B) Alignment of the UhpA-DegU subfamily with the regulatory proteins RcsA and MalT of *E. coli* (25, 43, 148); RmpA of *K. pneumoniae* (*Kp*) (113, 160); LasR and AgmR of *P. aeruginosa* (*Pa*) (45, 132); LuxR of *V. fischeri* (*Vf*) (35, 39); BrpA of *S. hygroscopicus* (*Sh*) (123); and GerE (28) and sigma factors σ^A (46), σ^B (15), and σ^E (149) of *B. subtilis*. A potential DNA-binding helix-turn-helix motif within this region is underlined.

larities, as shown in Fig. 3. The *B. subtilis* PhoP (133) and OrfX17 (140), *E. faecium* VanR (10), and *S. lividans* CutR (157) proteins belong to the OmpR subfamily (Fig. 3A), whereas the *B. subtilis* DegU (57, 76, 150), ComA (163), and Urf-2 (102) response regulators are part of the UhpA subfamily (43, 146) (Fig. 3B).

Interestingly, both of these conserved carboxy-terminal domains are also found in proteins that are not members of the response regulator family, since they lack the conserved amino-terminal domain. The OmpR subfamily carboxy-terminal region is also found in the *Vibrio cholerae* ToxR protein, which controls cholera toxin gene expression (97, 146) (Fig. 3A). The domain shared by members of the UhpA subfamily contains a potential helix-turn-helix DNA-binding motif (43) and is present in a wide spectrum of bacterial regulatory proteins that are not members of the response regulator family (53, 57, 67, 76, 104, 148) (Fig. 3B). Included among these proteins are MalT of *E. coli* (25, 124); GerE of *B. subtilis* (28, 61); RcsA of *K. pneumoniae*, *E. coli*, *Erwinia amylovora*, and *Erwinia stewartii* (4, 13, 26, 122, 148); RmpA of *K. pneumoniae* (113, 160); LasR of *Pseudomonas aeruginosa* (45); LuxR of *Vibrio fischeri* (35, 39); BrpA of *Streptomyces hygroscopicus* (123); and several bacterial σ factors (67) (Fig. 3B).

The *B. subtilis* GerE protein binds to the promoter regions of two spore coat protein genes, *cotB* and *cotC*, and stimulates transcription by σ^K-RNA polymerase holoenzyme (74). In this respect, the GerE protein is particularly interesting, since it consists entirely of the conserved carboxy-terminal domain of the UhpA response regulator subfamily, suggesting an independent function for this region (28, 47, 61, 67). This has been convincingly demonstrated for one member of the UhpA subfamily, the *Rhizobium meliloti* FixJ response regulator. A truncated FixJ protein, consisting only of the conserved carboxy-terminal domain, can activate transcription from the *nifA* promoter independently of the FixL histidine protein kinase (67). Independent activity of the conserved carboxy-terminal domain was also demonstrated for the *V. fischeri* LuxR protein, which is not a member of the response regulator family (24).

By analogy with the *E. coli* and *S. typhimurium* CheB response regulators, in which conserved amino-

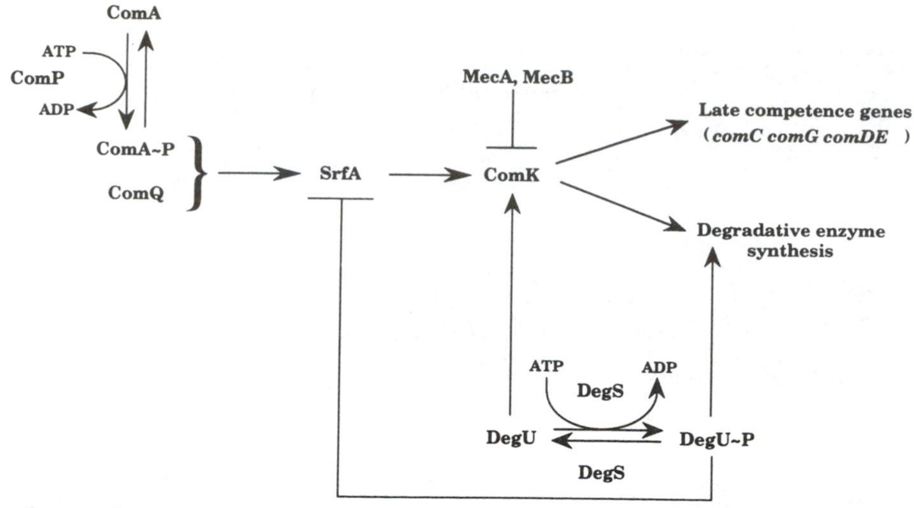

Figure 4. Regulation of competence gene expression in *B. subtilis*. Arrows indicate positive regulation, and perpendicular bars indicate negative regulation. The ComP-ComA and DegS-DegU two-component systems form two parallel pathways controlling competence gene expression (37).

terminal regions inhibit the methylesterase activities of the carboxy-terminal domains (84, 136, 142), it was proposed that response regulator activity may be modulated through an intramolecular negative interaction between the two halves of the protein (67). Phosphorylation by the cognate protein kinase would relieve this inhibition, presumably by inducing a transient conformational change. This type of modular structure, in which each domain is endowed with an independent function, suggests that response regulators such as DegU or ComA of *B. subtilis* may be viewed as composite proteins, with an amino-terminal Spo0F-type module controlling the activity of a carboxy-terminal GerE-type module that is probably involved in DNA binding and regulation of transcription.

TWO-COMPONENT REGULATORY SYSTEMS OF *B. SUBTILIS*

Soil bacteria such as *B. subtilis* are subject to drastic variations in environmental conditions such as temperature, humidity, and nutrient source availability. Faced with a depletion of essential nutrients at the onset of stationary phase, *B. subtilis* can adopt several responses, including induction of chemotaxis and motility, synthesis of macromolecule-degrading enzymes, competence for genetic transformation, antibiotic production, and finally, sporulation (138, 139). Each of these postexponential-phase responses is controlled by at least one two-component regulatory system, the components of which interact to form a signal transduction network. To date, eight potential histidine kinase proteins (CheA, ComP, DegS, PhoR, SpoIIJ [KinA], KinB, Urf-1, and OrfX-18) that are thought to interact (directly or indirectly) with nine cognate response regulators (CheY, CheB, ComA, DegU, PhoP, Spo0F, Spo0A, Urf-2, and OrfX-17) have been identified in *B. subtilis*.

Regulation of Genetic Competence

Competence in *B. subtilis* is a natural physiological state allowing the uptake of exogenous DNA molecules. The products of approximately 24 loci have been identified as playing a part in competence development (37) (see chapter 39, this volume). These loci can be divided into two classes, one of which encodes regulatory proteins that are required for the expression of genes in the second class. Examples of this second class of genes, referred to as late competence genes, since they are expressed only during the post-exponential growth phase, include *comC*, *comG*, and *comDE* (37). Specific regulatory proteins controlling the expression of these genes include the products of the *comP*, *comA*, *comQ*, *srfA* (*comL*), and *comK* genes (37). Other proteins controlling the expression of late competence genes include DegS, DegU, Spo0A, Sin, and AbrB (37). The regulatory pathway controlling competence gene expression is shown in Fig. 4. The *comP* and *comA* genes encode histidine kinase and response regulator proteins, respectively (Fig. 1, 2, and 3B) (163, 165). ComA (214 amino acids) has been shown to be phosphorylated in vitro by using the heterologous *E. coli* NtrB histidine protein kinase (37). ComQ is the product of a gene lying just upstream from the *comP* and *comA* genes (105, 164). In response to an as-yet-unidentified signal, ComA would be phosphorylated by ComP, a membrane-associated protein kinase (749 amino acids), and act in conjunction with ComQ to allow expression of *srfA*, an operon required for biosynthesis of the lipopeptide antibiotic surfactin (Fig. 4). This appears to be the major role of the ComP, ComA, and ComQ proteins in the development of competence (48, 112).

Two mutations, designated *mecA* and *mecB*, that allow a complete bypass of null mutations in *comP*, *comA*, *comQ*, and *srfA* for late competence gene expression have been identified (37, 38, 125) (see chapter 39).

Table 2. Mutations in *degS* and *degU* genes and associated phenotypes

Mutation	Associated modification	Phenotype		References
		In vivo	In vitro	
degS100(Hy)	DegS V236M	Degradative-enzyme hyperproduction, competence deficiency	Reduced DegU phosphatase activity	57, 151
degS200(Hy)	DegS G218E	Degradative-enzyme hyperproduction, competence deficiency	Reduced DegU phosphatase activity	31, 57, 151
degS42	DegS E300K	Deficiency in degradative-enzyme production	Loss of autophosphorylation activity	104, 151
degS220	DegS A193V	Deficiency in degradative-enzyme production	Deficiency in phosphotransfer to DegU	31, 104
degU24(Hy)	DegU T98I	Degradative-enzyme hyperproduction, competence deficiency	Increased DegU phosphorylation	30, 104
degU31(Hy)	DegU V131L	Degradative-enzyme hyperproduction, competence deficiency	Increased DegU phosphorylation	30, 104
degU32(Hy)	DegU H12L	Degradative-enzyme hyperproduction, competence deficiency	Decreased DegU dephosphorylation	31, 57
degU146	DegU D56N	Deficiency in degradative-enzyme production	Deficiency in DegU phosphorylation	30, 31, 104

Regulation of Degradative Enzyme Synthesis and Genetic Competence

Production in *B. subtilis* of a class of degradative enzymes, including an intracellular protease and several secreted enzymes (levansucrase, alkaline and metalloproteases, α-amylase, and β-glucanase[s]), is controlled at the transcriptional level by the products of the *degS* and *degU* genes (11, 54, 128, 135) and by at least two accessory regulatory genes, *degQ* and *degR*, which encode small polypeptides of 46 and 60 amino acids, respectively (7, 110, 152, 171, 172). Although both DegS and DegU are required for degradative enzyme production (57, 76, 150), the products of *degQ* and *degR* appear to be dispensable (70, 171, 172) (see chapter 54).

The *degS* and *degU* genes form an operon encoding a two-component system (57, 76, 104, 150). Indeed, similarities were found between DegS (385 amino acids) and the histidine protein kinase family and between DegU (229 amino acids) and the response regulator proteins, suggesting that DegS may modify DegU through phosphorylation (Fig. 1 and 2). Just upstream from the conserved carboxy-terminal histidine protein kinase domain in DegS lies a region sharing strong similarities with the catalytic domains of eukaryotic protein kinases (50, 104, 153). The DegS protein appears to be a cytoplasmic protein, since it does not contain any significant hydrophobic domains. However, the nature of the signal detected directly or indirectly by DegS is not known. Carboxy-terminal amino acid sequence similarities place DegU within the UhpA response regulator subfamily (Fig. 3B).

Two classes of mutations that have been identified in both the *degS* and the *degU* genes lead either to deficiency of degradative-enzyme production or to a pleiotropic Hy phenotype, which includes hyperproduction of degradative enzymes, ability to sporulate in the presence of glucose, decreased genetic competence, and loss of flagella (12, 57, 78, 104). Several of these mutations have been characterized at the molecular level, and the associated amino acid modifications in DegS or DegU are shown in Table 2. Missense mutations in *degS* and *degU* that abolish degradative-enzyme production do not affect the competence pathway, while those that lead to hyperproduction of degradative enzymes result in a lowered transformation frequency. Apparently, transformation frequency is high when degradative-enzyme production is low and vice versa (104).

Interpretation of these phenotypes led to the hypothesis that the DegU response regulator could act as a molecular switch allowing either degradative-enzyme synthesis or expression of genetic competence. Since degradative-enzyme production requires both the DegS protein kinase and the DegU response regulator, we postulated that the phosphorylated form of DegU may be necessary for this process (57, 76, 104, 150). Deletion of the *degS-degU* operon or disruption of the *degU* gene leads to a deficiency in both degradative-enzyme production and genetic competence (104, 150). However, a deletion of the *degS* gene alone, which also leads to a deficiency in degradative-enzyme synthesis, has no significant effect on the development of genetic competence (104, 105).

These results strongly suggest that the unphosphorylated form of DegU is necessary for competence gene expression. Additional support for this hypothesis came from the characterization of a mutation leading to a D-to-N change at position 56 in the DegU protein (DegU D56N) (Table 2). Originally identified as an intragenic suppressor of the pleiotropic Hy phenotype of a DegU H12L mutant, this mutation restores competence and leads to a deficiency in degradative-enzyme production in the double mutant DegU H12L D56N, presumably by inactivating the phosphorylation site of DegU through replacement of the conserved aspartate residue at position 56 by an

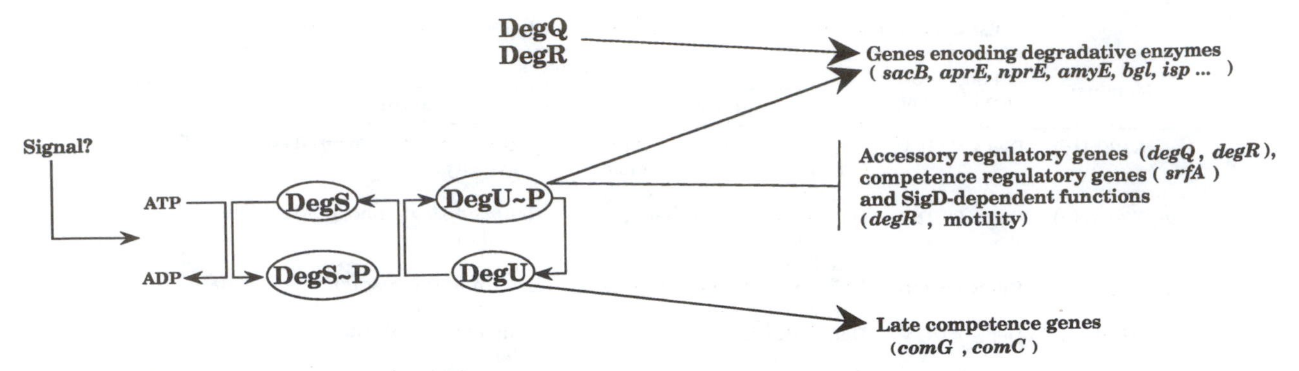

Figure 5. Pleiotropic regulation by the DegS-DegU signal transduction pathway controlling degradative enzyme synthesis and competence gene expression in *B. subtilis*. Arrows indicate positive regulation, and perpendicular bars indicate negative regulation. Regulation by DegS-DegU, DegQ, and DegR may be indirect.

asparagine (104). Substitutions leading to deficient degradative-enzyme synthesis, such as DegS E300K or DegU D56N, are thought to lead to accumulation of the unphosphorylated form of DegU (104). It is interesting that both forms of DegU seem to be required for two distinct functions and apparently act as positive regulators. Roggiani et al. have shown that the presence of a *degS*(Hy) or *degU*(Hy) mutation or a disruption of the *degU* gene results in decreased expression of late competence genes such as *comG*, whereas a disruption of the *degS* gene has little or no effect on competence gene expression (125). The effect of Hy mutations on competence appears to be twofold. On one hand, these mutations favor accumulation of the phosphorylated form of DegU, leading to a dearth of the unphosphorylated form of the response regulator, which is required for the expression of late competence genes. On the other hand, the phosphorylated form of DegU appears to act as a repressor of essential components of the competence development pathway encoded by the *srfA* operon. Indeed, although DegU is not required for *srfA* expression, this expression is markedly reduced in a DegU H12L (Hy) strain (37, 48, 158).

DegU appears to be the first response regulator described as having two active conformations: a phosphorylated form that is necessary for degradative enzyme production and a nonphosphorylated form that is required for the expression of genetic competence (Fig. 5). The in vivo equilibrium between the two forms of DegU is presumably regulated by DegS in response to an as-yet-unidentified environmental signal. The *degS*(Hy) and *degU*(Hy) mutations favor accumulation of the phosphorylated form of DegU by increasing the phosphorylation rate of the response regulator or by enhancing the stability of the phosphorylated protein by decreasing its dephosphorylation rate. Mutations in *degS* or *degU* that lead to a deficiency of degradative enzyme synthesis promote accumulation of the unphosphorylated form of DegU (Fig. 5). Biochemical evidence has provided strong support for this model. Both the DegS and DegU proteins have been purified. DegS has been shown to be autophosphorylated when incubated in the presence of [γ-^{32}P]ATP and to transfer its phosphate group to the DegU response regulator (30, 108). The pH stabilities of the phosphorylated proteins are consistent with the existence of a histidinyl-phosphate group for DegS and an aspartyl-phosphate group for DegU (108). Sequence similarities to other two-component systems suggest the conserved His-189 residue of the DegS protein kinase and Asp-56 residue of the DegU response regulator as likely candidates for the respective phosphorylation sites of the two proteins (104, 108).

In vitro phosphorylation experiments using modified DegS and DegU proteins have provided further evidence for this model. Proteins were purified from strains carrying *degS* or *degU* mutations that lead to a deficiency in degradative-enzyme synthesis. The DegS E300K protein is no longer autophosphorylated, whereas the DegS A193V protein is still autophosphorylated but has a strongly diminished ability to phosphorylate DegU (31, 151). In both cases, the in vivo effect is accumulation of the unphosphorylated form of DegU, which leads to a deficiency in degradative-enzyme synthesis. The reduced DegU phosphorylation activity of the DegS A193V protein could explain why the corresponding mutation was originally isolated as an extragenic suppressor of the Hy phenotype of the DegU H12L (Hy) mutant (78, 104). Indeed, we showed that, in contrast to wild-type DegS, the DegS A193V protein is unable to phosphorylate the DegU H12L response regulator (29). The DegU D56N response regulator, which is modified within the putative phosphorylation site, was also purified, and we showed that it could no longer be phosphorylated by DegS, suggesting that the Asp-56 residue is indeed the phosphorylation site of the DegU protein (31). We also showed that a *comG'-'lacZ* fusion was expressed at the same level in strains carrying wild-type *degS* and *degU* genes, a deletion of the *degS* gene, or the DegU D56N modification, but that this expression was abolished in strains carrying the DegU H12L (Hy) modification or a deletion of both the *degS* and *degU* genes (29). These results confirm the essential role in late competence gene expression of the unphosphorylated form of the DegU response regulator.

The effects of *degS*(Hy) and *degU*(Hy) mutations were also examined by using in vitro phosphorylation experiments. T. Tanaka and coworkers, who used modified DegS V236M and DegS G218E proteins, showed that these Hy mutations led to a decreased rate of dephosphorylation and thus to an increased stability of DegU~P (151). Using an in vitro phosphorylation assay, which consists of incubating purified DegS and radioactive ATP with crude extracts containing either wild-type or mutant DegU(Hy) proteins, we were able to show that a stronger phosphorylation signal was obtained with the DegU T98I and DegU V131L proteins than with the wild-type DegU protein. These modified proteins may therefore present higher rates of phosphorylation than the wild-type protein does. We also purified the modified DegU H12L (Hy) response regulator and showed that this protein is phosphorylated by DegS and has a significantly lower rate of dephosphorylation than the wild-type DegU protein, with a half-life of 120 min rather than 18 min. This strongly increased stability is consistent with the Hy phenotype of the DegU H12L mutant (31).

Finally, by comparing the stability of the purified phosphorylated DegU response regulator in the presence and absence of DegS, we demonstrated that DegS acts as a DegU phosphatase (31), as previously suggested by T. Tanaka and coworkers (151). We also showed that both the modified DegS G218E (Hy) and DegS A193V proteins have lost this DegU phosphatase activity (29).

In addition to the DegS-DegU two-component system, several accessory regulatory genes such as *degQ* and *degR* are also involved in controlling degradative-enzyme production (75). The *degQ36* mutation, a single base change in the promoter at position −10 that leads to overexpression of the *degQ* gene, gave a Hy phenotype similar to that of the *degS*(Hy) and *degU*(Hy) mutations (7, 78, 79, 171).

A number of reports have shown that Hy mutations in the *degU* or *degQ* genes lead to an increase in the level of *sacB* and *aprE* mRNAs (11, 54, 135). Thus, in every case examined, these pleiotropic mutations appear to increase the rate of transcription initiation of their target genes. The sites at which these pleiotropic mutations stimulate the transcription of the *sacB* and *aprE* promoters were localized by deletion analysis. The stimulation effects of the *degU32*(Hy) (Table 2) and *degQ36*(Hy) mutations require regions located between positions −164 and −141 upstream from the *aprE* transcription start site (54–56). Although comparison of the corresponding regions of the *sacB* and *aprE* promoters showed that 27 of 48 bases are identical in these upstream promoter regions, no obvious consensus sequence could be established (54, 55). In addition, no biochemical evidence for specific binding of DegU~P to these target sites yet exists. Some as-yet-unidentified intermediates may be involved in the regulation of degradative-enzyme synthesis. Since the same region was involved in stimulation by either the *degU32*(Hy) mutation or the *degQ36*(Hy) mutation for two of the target genes, *sacB*, and *aprE* (54–56), it remained to be determined whether the products of the two regulatory genes act independently or upon each other. Expression of the *degS* and *degU* genes was not affected by deletion of the *degQ* gene (104, 106). Strains carrying nonfunctional *degS* or *degU* genes are deficient for degradative-enzyme synthesis. The presence of the *degQ* gene on a multicopy plasmid or of the *degQ36* mutation in such strains does not lead to a hyperproduction phenotype (7, 105). It seems, therefore, that DegS and DegU are required for DegQ to act upon the target genes.

The expression of *degQ* is subject to growth phase regulation (171). This expression increased under conditions of carbon, nitrogen, or phosphate source limitation. This expression was reduced under all the conditions tested in a strain with the *degS* and *degU* genes deleted or carrying the DegU H12L (Hy) substitution (104, 105). Expression of *degQ* was shown to be controlled by both the DegS-DegU and ComP-ComA two-component systems. Separate regulatory targets for DegS-DegU and ComP-ComA were localized by deletion analysis upstream from the *degQ* gene at positions −393 to −186 and positions −78 to −40, respectively (105). A comparison of the sequence of the *degQ* ComP-ComA target (positions −78 to −40) with the upstream region of two other genes controlled by ComP-ComA, *srfA* and *gsiA*, a glucose starvation-inducible gene, allowed the identification of a 16-bp region of imperfect dyad symmetry (TTGCGGNNTC CCGCAN) that is thought to be the ComA target site (107, 111) (see chapter 39).

Both the ComP-ComA and DegS-DegU two-component systems control the expression of late competence genes. They seem to act through two different branches in the competence regulatory pathway. Indeed, the ComP-ComA branch involves at least one intermediate regulatory locus, *srfA*, while the DegS-DegU regulatory pair is not required for the expression of *srfA* (37). However, these two branches partially overlap. The DegS-DegU and ComP-ComA pathways may have several intermediate regulatory genes in common, such as the *comK* gene, whose expression requires both ComA and DegU (49, 158, 159) (Fig. 4). In addition, the *mec* mutations allow a bypass of both ComP-ComA and DegS-DegU for competence gene expression. We have recently shown that the *mec* mutations also bypass the requirement of DegS-DegU for degradative-enzyme synthesis (77; cited in 37). This result suggests that *mec* gene products play a central role in this regulatory network, affecting both competence gene expression and degradative-enzyme synthesis (Fig. 4).

Regulation of Sporulation Initiation

Sporulation in *B. subtilis* is a paradigm for prokaryotic developmental regulation, and well over 100 genes affecting this process have been identified. Sporulation and the development of competence represent two alternative developmental pathways available to *B. subtilis* at the onset of stationary phase. The initiation of sporulation occurs at the time of transition from exponential to stationary phase, when cell growth becomes limited by the depletion of essential nutrients, and involves several members of the histidine protein kinase-response regulator family, which form a multicomponent phosphorelay (see chapter 51).

At least five proteins, including the products of the *spoIIJ* (*kinA*), *kinB*, *spo0F*, *spo0B*, and *spo0A* genes, are involved in this multicomponent regulatory system (23, 155). The *spoIIJ* gene (also called *kinA*) encodes a 606-amino-acid cytoplasmic histidine protein kinase (8, 121) (Fig. 1). KinB is a membrane-bound histidine protein kinase (23, 155) (see chapter 51). The Spo0F protein (123 amino acids), is similar to the CheY regulator (see below and Fig. 2) in that it consists only of the conserved N-terminal response regulator domain (156, 173). Spo0A (267 amino acids) is a response regulator whose carboxy-terminal domain has no similarities to those of the other response regulators (40). The product of the *spo0B* gene (192 amino acids) has been shown to act as a phosphotransferase (21, 23, 40).

The sporulation phosphorelay is one of the best studied examples of histidine protein kinase-response regulator systems. The in vitro studies with purified proteins that provided the first biochemical evidence for such a phosphorelay can be summarized as follows. The initial event in the cascade is the phosphorylation of Spo0F by the SpoIIJ kinase. The key enzyme in the phosphorelay is Spo0B, a phosphotransferase that catalyzes the concurrent dephosphorylation of Spo0F and the phosphorylation of Spo0A (23). This is the only known example of such a phosphotransfer between two response regulators. The end product of the phosphorelay is the phosphorylated form of Spo0A, whose accumulation is essential for triggering subsequent steps in the sporulation process. Genetic evidence suggests that more than one histidine protein kinase may be involved in the phosphorylation of Spo0F, and one likely candidate is KinB (23, 141, 155) (see chapter 51).

Regulation of Phosphate Assimilation

B. subtilis produces phosphate-repressible alkaline phosphatase and phosphodiesterase during the vegetative growth phase. The synthesis of these enzymes is controlled by *phoS*, *phoT*, and two closely linked genes, *phoP* and *phoR* (94, 170). The *phoP* and *phoR* genes were cloned and sequenced (133, 134). Both gene products belong to the family of two-component systems. PhoR of *B. subtilis* (579 amino acids), presumably a membrane-associated protein, is thought to be the sensor histidine kinase for the PhoP response regulator (241 amino acids) (Fig. 1, 2, and 3A). Although there are no biochemical data on phosphorylation of the PhoR and PhoP proteins, the phosphate-regulated gene systems nevertheless appear to be organized in a similar fashion in gram-positive and gram-negative bacteria. It was recently demonstrated that alkaline phosphatases in *B. subtilis* are encoded by a multigene family and synthesized both in phosphate-starved vegetative cells and in sporulating cells (62). The PhoP and PhoR regulators have been shown to be required for vegetative alkaline phosphatase production during phosphate starvation but not for *phoB* transcription during sporulation (23a) (see chapter 17).

Regulation of Chemotaxis

Although the central regulators of chemotaxis in *B. subtilis* are similar to those in *E. coli*, the overall mechanism of chemotaxis differs in the two organisms (see chapter 53). The genes encoding CheA and CheY of *B. subtilis* (formerly designated CheN and CheB, respectively) have recently been cloned and sequenced (16, 44). Sequence similarities to other regulatory proteins suggest that CheA (671 amino acids) and CheY (120 amino acids) of *B. subtilis* form a histidine kinase-response regulator pair (Fig. 1 and 2). No biochemical evidence yet exists for phosphotransfer between the two proteins, but the strong homology with the corresponding *E. coli* proteins suggests that such transfer is more than likely. However, the chemotactic mechanism in *B. subtilis* is not the same as that in *E. coli*. Indeed, null *cheA* or *cheY* mutants of *B. subtilis* tumble, while those of *E. coli* exhibit smooth swimming (16, 17, 44). It has been suggested that phosphorylated CheY in *B. subtilis* may interact with the flagellar switch to cause smooth swimming rather than tumbling as in *E. coli* (see chapter 53).

Potential Two-Component Systems of Unknown Function in *B. subtilis*

Amino acid sequence similarities suggest the existence of two potential histidine kinase-response regulator pairs of unknown function in *B. subtilis* (Fig. 1, 2, and 3B).

The first of these regulatory pairs was found by E. H. Kemp and A. Moir (102). The DNA sequence of the region downstream from the *B. subtilis gerA* operon revealed the presence of two open reading frames transcribed in the opposite direction to *gerA* and apparently organized as an operon. These two open reading frames, designated Urf-1 and Urf-2, could encode a putative histidine kinase and response regulator, respectively (Fig. 1 and 2). The sequence of Urf-1 is incomplete, so that whether the deduced protein is membrane associated or not remains to be determined. Urf-2 presumably encodes a 211-amino-acid protein that is a member of the UhpA-DegU subfamily within the response regulators (Fig. 3B). Disruption of the corresponding genes had no effect on sporulation or germination in *B. subtilis* (102).

The second of these putative two-component systems was found by A. Sorokin and S. D. Ehrlich within the framework of the European *B. subtilis* genome-sequencing project (140). The sequence of a 22-kb DNA fragment extending from the *lys* gene and including the riboflavin biosynthesis operon was determined. Two open reading frames, designated ORFX-17 and ORFX-18, were located downstream from the *rib* operon. Amino acid similarities suggest that ORFX-18 could encode a putative membrane-associated protein kinase (589 amino acids) that shows strong homology with the *B. subtilis* PhoR protein (Fig. 1). ORFX-17 apparently encodes a response regulator (241 amino acids) that belongs to the OmpR-PhoP subfamily (Fig. 2 and 3A). Amino acid sequence similarities with the PhoR-PhoP regulatory pair and the presence of putative "Pho-boxes" upstream from the ORFX-17–ORFX-18 operon suggest that these proteins may be involved in regulation by phosphate sources in *B. subtilis* (140). Preliminary evidence suggests that mutations in these two genes reduce the cell's ability to produce alkaline phosphatase during phosphate limitation and to form resistant spores (23b).

SIGNAL TRANSDUCTION NETWORK IN *B. SUBTILIS*

Given the strong sequence similarities between two-component systems, it is not surprising that a certain amount of cross talk between heterologous histidine protein kinase-response regulator pairs has been shown to exist both in vivo and in vitro.

In vivo cross talk has been shown to occur between the *E. coli* Che and Ntr systems (116); the *E. coli* PhoR-PhoB and PhoM-PhoM ORF2 systems (162, 169); and the *E. coli* EnvZ-OmpR, *B. pertussis* Bvg, and *P. aeruginosa* Alg systems (14, 34, 47). Under in vitro conditions, it seems likely that any histidine kinase may interact with any response regulator to some degree. Transfer of phosphate has been shown to occur in vitro between *E. coli* heterologous regulatory pairs, from the PhoM kinase to the heterologous response regulator PhoB, from CheA to NtrC or OmpR, from EnvZ to NtrC, and from NtrB to CheY (6, 63, 116). In vitro phosphotransfer has also been demonstrated by using *E. coli* histidine protein kinases and *B. subtilis* response regulators (CheA-Spo0A, NtrB-ComA) (37, 118).

In *B. subtilis*, the different two-component systems controlling postexponential-phase responses appear to interconnect to a large extent to form a sensory transduction network. This network appears to be involved in a hierarchy of environmental signal responses, involving a choice between competence gene expression, degradative-enzyme production, and finally, sporulation.

The ComP-ComA and DegS-DegU histidine protein kinase-response regulator pairs are both required for late competence gene expression. These two systems have several target genes in common, including the *srfA* and *degQ* genes, and involve common intermediates such as the *comK* gene. Both ComP and DegS have been proposed to also affect sporulation to a minor extent, especially in the absence of the SpoIIJ (KinA) kinase, possibly through cross talk with the Spo0A phosphorelay (137, 165). Signals involved in the initiation of sporulation include the depletion of carbon, nitrogen, or phosphate sources (131, 139). The existence of multiple kinases that could channel information to the Spo0A phosphorelay may thus allow both interaction with the Com and Deg systems and specific responses to different sets of environmental conditions that trigger sporulation. In addition, as suggested for the *E. coli* Che system (82), intermediary metabolites such as acetyl phosphate or carbamoyl phosphate may also be involved in regulating inputs to the Com-Deg-Spo signal transduction network, providing further links between these systems and the nutritional state of the cell. In strains carrying *degS*(Hy) or *degU*(Hy) mutations, which lead to accumulation of the phosphorylated DegU response regulator, sporulation is no longer repressed by glucose (57, 78, 104). These mutations also affect several σ^D-dependent functions in *B. subtilis*, such as flagellar synthesis and expression of the *degR* gene, which encodes a 60-amino-acid accessory regulatory peptide (51, 52, 91, 110, 152, 172). Indeed, although deletion of the *degS* and *degU* genes has no effect on these functions, *degR* expression is abolished in a strain carrying a DegU H12L (Hy) substitution, and the cells are devoid of flagella and grow as long filaments (56, 78, 103).

Both the DegS-DegU two-component system and the Spo0A phosphorelay control the expression of the *aprE* and *nprE* genes encoding the major *B. subtilis* proteases (see chapters 51 and 52). The Spo0A and AbrB regulatory proteins are also required for the expression of late competence genes (37). The ComP-ComA regulatory pair also controls degradative-enzyme production through *degQ* gene expression and has been shown to be required for expression of the *degQ36*(Hy) phenotype (105).

At least three regulatory systems involving histidine kinase-response regulator proteins appear to form a signal transduction network in *B. subtilis*. At the onset of stationary phase, when growth is limited by the depletion of nitrogen sources such as amino acids but glucose is still present as a carbon source, the ComP-ComA system would allow competence gene expression, and the DegS-DegU regulatory pair could be involved in the choice between producing degradative enzymes or expressing competence genes. Finally, when glucose is no longer present and no alternative nutrient sources are accessible as substrates for degradative enzymes, the Spo0A phosphorelay would trigger sporulation initiation.

TWO-COMPONENT SYSTEMS OF OTHER GRAM-POSITIVE BACTERIA

On the basis of sequence similarities, several histidine kinase-response regulator pairs have been proposed to exist in other gram-positive bacteria. They control a wide variety of functions ranging from synthesis of exoproteins and cell wall precursors to copper metabolism.

Regulation of Exoprotein Synthesis in *S. aureus*

Mutations in the *agr* (accessory gene regulator) locus of *S. aureus* prevent postexponential-growth-phase synthesis of several exoproteins, including serine protease, nuclease, α-hemolysin, β-hemolysin, enterotoxin, and toxic shock syndrome toxin-1 (73, 120) (see chapter 2). The DNA sequence of the *agr* locus revealed six open reading frames, two of which apparently encode a two-component system (73). AgrB (423 amino acids) is a putative membrane-associated histidine protein kinase thought to interact with AgrA (241 amino acids), the corresponding response regulator (Fig. 1 and 2) (73, 120). Assuming that the amino acid sequence similarities to other known two-component systems reflect the existence of a phosphotransfer between AgrB and AgrA, AgrB is presumed to be autophosphorylated at the end of the exponential growth phase and would then transfer its phosphate group to AgrA. Phosphorylated AgrA would act as a transcriptional activator, leading to the accumulation of a regulatory RNA molecule, RNAIII, which controls the expression of genes encoding exoproteins (73). The association of the RNAIII regulatory molecule with a two-component system is a unique feature of the *S. aureus agr* regulon (see chapter 2).

Regulation of Cell Wall Precursor Synthesis Conferring Vancomycin Resistance on *E. faecium*

Inducible resistance to vancomycin in *E. faecium* involves plasmid-mediated synthesis of depsipeptide-containing peptidoglycan precursors that bind the antibiotic with reduced affinity. Plasmid pIP816 of *E. faecium* carries a five-gene cluster (*vanR*, *vanS*, *vanH*, *vanA*, and *vanX*) (10). The last three genes encode enzymes involved in synthesis of the depsipeptide precursors. The first two genes, *vanR* and *vanS*, encode a putative two-component system that controls the expression of *vanH*, *vanA*, and *vanX* at the transcriptional level in response to the presence of vancomycin in the external medium (10). VanS (384 amino acids) is a potential membrane-associated histidine protein kinase that shares strong amino acid sequence similarities with EnvZ and PhoR of *E. coli* (Fig. 1). VanR (231 amino acids) shares the characteristic response regulator amino-terminal domain and the OmpR subfamily carboxy-terminal region (Fig. 2 and 3A) (10). Sequence similarities strongly suggest that a phosphotransfer mechanism is involved in regulation by VanS-VanR.

Regulation of Copper Metabolism and Melanin Pigmentation in *S. lividans*

Melanin formation in *Streptomyces lividans* is catalyzed by an extracellular tyrosinase, encoded by the *melC2* gene, whose activity requires cupric ions. The product of *melC1*, which is located upstream from *melC2*, is thought to be a copper transfer protein that is essential for tyrosinase activity. A DNA fragment that allows the phenotypic suppression of *melC1* mutations in *S. lividans* was isolated (157). DNA sequence analysis revealed the presence of two open reading frames, designated *cutR* and *cutS*, whose predicted products seem to form a two-component system. The sequence of the gene encoding CutS is incomplete, but the truncated open reading frame could encode a putative membrane-associated histidine protein kinase with strong similarities to the *E. coli* EnvZ protein. CutR (217 amino acids) strongly resembles the OmpR subfamily of response regulators (Fig. 2 and 3A) and is thought to be a transcription regulator acting with CutS in response to changes in osmolarity or cupric ion concentration to regulate biosynthesis of the membrane proteins of a copper transport or efflux system (157). The presence of the *cutR* gene on a multicopy plasmid causes deregulated uptake or efflux of cupric ions, resulting in an aberrantly high intracellular copper concentration, which in turn complements the *melC1* defect (157).

Putative Response Regulator of Unknown Function in *Bacillus megaterium*

A search for sequence similarities between the *B. subtilis* DegU response regulator and proteins in the NBRF/PIR data base was carried out by using the BLAST alignment program (5). This allowed us to suggest that a previously identified open reading frame of *B. megaterium* could encode a response regulator. This open reading frame, *orf1*, is located upstream from the *gdh1* glucose dehydrogenase isozyme gene of *B. megaterium* IAM1030 (99). *orf1* encodes a putative 228-amino-acid protein of unknown function that has strong similarities to members of the response regulator family (Fig. 2).

SUMMARY

A large number of proteins belonging to the histidine protein kinase-response regulator family have been characterized in gram-positive bacteria. Those of *B. subtilis* are the genetically best studied among these, and extensive biochemical data are available for the systems controlling sporulation initiation and degradative-enzyme production. Regulation by the *B. subtilis* two-component systems involves several original features that have not been described for those of gram-negative bacteria.

Among these features is the possibility that the unphosphorylated form of the response regulator is active and controls specific functions. Biochemical and genetic evidence has shown that the DegU response regulator acts as a molecular switch controlling two alternative functions: the phosphorylated form of the protein is essential for degradative-enzyme production, while the unphosphorylated form allows expression of late competence genes (31, 104).

Another unique feature of this type of regulation in *B. subtilis* is the association of several histidine protein kinase and response regulator proteins in a multicomponent phosphorelay. Thus, several kinases are thought to phosphorylate the Spo0F protein, which would act as a funnel for different environmental signals. Spo0B is a unique protein that acts as a phosphotransferase, catalyzing phosphotransfer between two response regulators, Spo0F and Spo0A (see chapter 51).

Finally, it appears that a considerable amount of overlap exists between the different systems controlling postexponential-phase responses in *B. subtilis*. Thus, DegS-DegU, ComP-ComA, and Spo0A are all involved in controlling competence gene expression, and the Spo0A phosphorelay and DegS-DegU both control protease gene expression. Under certain conditions, the ComP and DegS kinases and the DegU response regulator can also affect the sporulation process to a minor extent, presumably through cross talk. The ComP-ComA and DegS-DegU systems appear to be closely related, as they share several common targets, including *degQ*, *srfA*, and late competence genes.

Acknowledgments. We thank M. Arthur, R. S. Chesnut, M. K. Dahl, D. Dubnau, S. D. Ehrlich, J. Hahn, J. A. Hoch, F. M. Hulett, E. H. Kemp, I. Mandic-Mulec, A. Moir, J. P. Mueller, M. Nakano, R. Novick, G. Ordal, H. Saito, A. L. Sonenshein, A. Sorokin, I. Smith, F. Vandenesch, D. van Sinderen, Y. Weinrauch, H. Yoshikawa, and P. Zuber for helpful discussion and for providing us with unpublished information. We are grateful to Christine Dugast for expert secretarial assistance and for help with the illustrations. The work from our laboratory reported in this review was supported by funds from Institut Pasteur, Centre National de la Recherche Scientifique, Université Paris 7, and Fondation pour la Recherche Médicale.

REFERENCES

1. **Aiba, H., T. Mizuno, and S. Mizushima.** 1989. Transfer of phosphoryl group between two regulatory proteins involved in osmoregulatory expression of the *ompF* and *ompC* genes in *Escherichia coli*. *J. Biol. Chem.* **264:**8563–8567.

2. **Aiba, H., F. Nakasai, S. Mizushima, and T. Mizuno.** 1989. Evidence for the physiological importance of the phosphotransfer between the two regulatory components, EnvZ and OmpR, in osmoregulation in *Escherichia coli. J. Biol. Chem.* **264:**14090–14094.
3. **Albright, L. M., E. Huala, and F. M. Ausubel.** 1989. Prokaryotic signal transduction mediated by sensor and regulator pairs. *Annu. Rev. Genet.* **23:**311–336.
4. **Allen, P., C. A. Hart, and J. R. Saunders.** 1987. Isolation from *Klebsiella* and characterization of two *rcs* genes that activate colanic acid capsular biosynthesis in *Escherichia coli. J. Gen. Microbiol.* **133:**331–340.
5. **Altschul, S. F., W. Gish, W. Miller, E. W. Myers, and D. J. Lipman.** 1990. Basic local alignment search tool. *J. Mol. Biol.* **215:**403–410.
6. **Amemura, M., K. Makino, H. Shinagawa, and A. Nakata.** 1990. Cross talk to the phosphate regulon of *Escherichia coli* by PhoM protein: PhoM is a histidine protein kinase and catalyzes phosphorylation of PhoB and PhoM-open reading frame 2. *J. Bacteriol.* **172:**6300–6307.
7. **Amory, A., F. Kunst, E. Aubert, A. Klier, and G. Rapoport.** 1987. Characterization of the *sacQ* genes from *Bacillus licheniformis* and *Bacillus subtilis. J. Bacteriol.* **169:**324–333.
8. **Antoniewski, C., B. Savelli, and P. Stragier.** 1990. The *spoIIJ* gene, which regulates early development steps in *Bacillus subtilis*, belongs to a class of environmentally responsive genes. *J. Bacteriol.* **172:**86–93.
9. **Aricó, B., J. F. Miller, C. Roy, S. Stibitz, D. Monack, S. Falkow, R. Gross, and R. Rappuoli.** 1989. Sequences required for expression of *Bordetella pertussis* virulence factors share homology with prokaryotic signal transduction proteins. *Proc. Natl. Acad. Sci. USA* **86:**6671–6675.
10. **Arthur, M., C. Molinas, and P. Courvalin.** 1992. The VanS-VanR two-component regulatory system controls synthesis of depsipeptide peptidoglycan precursors in *Enterococcus faecium* strain BM4147. *J. Bacteriol.* **173:** 2582–2591.
11. **Aymerich, S., G. Gonzy-Tréboul, and M. Steinmetz.** 1986. 5′-noncoding region *sacR* is the target of all identified regulation affecting the levansucrase gene in *Bacillus subtilis. J. Bacteriol.* **166:**993–998.
12. **Ayusawa, D., Y. Yoneda, K. Yamane, and B. Maruo.** 1975. Pleiotropic phenomena in autolytic enzyme(s) content, flagellation, and simultaneous hyperproduction of extracellular α-amylase and protease in a *Bacillus subtilis* mutant. *J. Bacteriol.* **124:**459–469.
13. **Bernhard, F., K. Poetter, K. Geider, and D. L. Coplin.** 1990. The *rcsA* gene from *Erwinia amylovora*: identification, nucleotide sequence, and regulation of exopolysaccharide biosynthesis. *Mol. Plant Microbe Interact.* **3:**429–437.
14. **Berry, A., J. D. DeVault, and A. M. Chakrabarty.** 1989. High osmolarity is a signal for enhanced *algD* transcription in mucoid and nonmucoid *Pseudomonas aeruginosa* strains. *J. Bacteriol.* **171:**2312–2317.
15. **Binnie, C., M. Lampe, and R. Losick.** 1986. Gene encoding the σ^{37} species of RNA polymerase σ factor from *Bacillus subtilis. Proc. Natl. Acad. Sci. USA* **83:**5943–5947.
16. **Bischoff, D. S., and G. W. Ordal.** 1991. Sequence and characterization of *Bacillus subtilis* CheB, a homolog of *Escherichia coli* CheY, and its role in a different mechanism of chemotaxis. *J. Biol. Chem.* **266:**12301–12305.
17. **Bischoff, D. S., and G. W. Ordal.** 1992. *Bacillus subtilis* chemotaxis: a deviation from the *Escherichia coli* paradigm. *Mol. Microbiol.* **6:**23–28.
18. **Bourret, R. B., K. A. Borkovich, and M. I. Simon.** 1991. Signal transduction pathways involving protein phosphorylation in prokaryotes. *Annu. Rev. Biochem.* **60:** 401–441.
19. **Bourret, R. B., J. F. Hess, K. A. Borkovich, A. A. Pakula, and M. I. Simon.** 1989. Protein phosphorylation in chemotaxis and two-component regulatory systems of bacteria. *J. Biol. Chem.* **264:**7085–7088.
20. **Bourret, R. B., J. F. Hess, and M. I. Simon.** 1990. Conserved aspartate residues and phosphorylation in signal transduction by the chemotaxis protein CheY. *Proc. Natl. Acad. Sci. USA* **87:**41–45.
21. **Bouvier, J., P. Stragier, C. Bonamy, and J. Szulmajster.** 1984. Nucleotide sequence of the *spo0B* gene of *Bacillus subtilis* and regulation of its expression. *Proc. Natl. Acad. Sci. USA* **81:**7012–7016.
22. **Buikema, W. J., W. W. Szeto, P. V. Lemley, W. H. Orme-Johnson, and F. M. Ausubel.** 1985. Nitrogen fixation specific regulatory genes of *Klebsiella pneumoniae* and *Rhizobium meliloti* share homology with the general nitrogen regulatory gene *ntrC* of *K. pneumoniae. Nucleic Acids Res.* **13:**4539–4555.
23. **Burbulys, D., K. A. Trach, and J. A. Hoch.** 1991. Initiation of sporulation in *B. subtilis* is controlled by a multicomponent phosphorelay. *Cell* **64:**545–552.

23a. **Chesnut, R. S., C. Bookstein, and F. M. Hulett.** 1991. Separate promoters direct expression of *phoAIII*, a member of the *Bacillus subtilis* alkaline phosphatase multigene family, during phosphate starvation and sporulation. *Mol. Microbiol.* **5:**2181–2190.

23b. **Chesnut, R. S., and F. M. Hulett.** 1993. Personal communication.

24. **Choi, S. H., and E. P. Greenberg.** 1991. The C-terminal region of the *Vibrio fischeri* LuxR protein contains an inducer-independent *lux* gene activating domain. *Proc. Natl. Acad. Sci. USA* **88:**11115–11119.
25. **Cole, S. T., and O. Raibaud.** 1986. The nucleotide sequence of the *malT* gene encoding the positive regulator of the *Escherichia coli* maltose regulon. *Gene* **42:**201–208.
26. **Coleman, M., R. Pearce, E. Hitchin, F. Busfield, J. W. Mansfield, and I. S. Roberts.** 1990. Molecular cloning, expression and nucleotide sequence of the *rcsA* gene of *Erwinia amylovora*, encoding a positive regulator of capsule expression: evidence for a family of related capsule activator proteins. *J. Gen. Microbiol.* **136:**1799–1806.
27. **Comeau, D. E., K. Ikenaka, K. Tsung, and M. Inouye.** 1985. Primary characterization of the protein products of the *Escherichia coli ompB* locus: structure and regulation of synthesis of the OmpR and EnvZ proteins. *J. Bacteriol.* **164:**578–584.
28. **Cutting, S., and J. Mandelstam.** 1986. The nucleotide sequence and the transcription during sporulation of the *gerE* gene of *Bacillus subtilis. J. Gen. Microbiol.* **132:**3013–3024.
29. **Dahl, M. K., T. Msadek, and F. Kunst.** 1992. Unpublished results.
30. **Dahl, M. K., T. Msadek, F. Kunst, and G. Rapoport.** 1991. Mutational analysis of the *Bacillus subtilis* DegU regulator and its phosphorylation by the DegS protein kinase. *J. Bacteriol.* **173:**2539–2547.
31. **Dahl, M. K., T. Msadek, F. Kunst, and G. Rapoport.** 1992. The phosphorylation state of the DegU response regulator acts as a molecular switch allowing either degradative enzyme synthesis or expression of genetic competence in *Bacillus subtilis. J. Biol. Chem.* **267:** 14509–14514.
32. **Débarbouillé, M., I. Martin-Verstraete, A. Klier, and G. Rapoport.** 1991. The levanase regulator LevR of *Bacillus subtilis* has domains homologous to both σ^{54}- and PTS-dependent regulators. *Proc. Natl. Acad. Sci. USA* **88:**2212–2216.
33. **Débarbouillé, M., I. Martin-Verstraete, F. Kunst, and G. Rapoport.** 1991. The *Bacillus subtilis sigL* gene encodes an equivalent of σ^{54} from Gram-negative bacteria. *Proc. Natl. Acad. Sci. USA* **88:**9092–9096.

34. **DeVault, J. D., A. Berry, T. K. Misra, A. Darzins, and A. M. Chakrabarty.** 1989. Environmental sensory signals and microbial pathogenesis: *Pseudomonas aeruginosa* infection in cystic fibrosis. *Bio/Technology* **7:**352–357.
35. **Devine, J. H., C. Countryman, and T. O. Baldwin.** 1988. Nucleotide sequence of the *luxR* and *luxI* genes and structure of the primary regulatory region of the *lux* regulon of *Vibrio fischeri* ATCC 7744. *Biochemistry* **27:**837–842.
36. **Drummond, M., P. Whitty, and J. Wootton.** 1986. Sequence and domain relationships of *ntrC* and *nifA* from *Klebsiella pneumoniae*: homologies to other regulatory proteins. *EMBO J.* **5:**441–447.
37. **Dubnau, D.** 1991. Genetic competence in *Bacillus subtilis*. *Microbiol. Rev.* **55:**395–424.
38. **Dubnau, D., and M. Roggiani.** 1990. Growth medium-independent genetic competence mutants of *Bacillus subtilis*. *J. Bacteriol.* **172:**4048–4055.
39. **Engebrecht, J., and M. Silverman.** 1987. Nucleotide sequence of the regulatory locus controlling expression of bacterial genes for bioluminescence. *Nucleic Acids Res.* **15:**10455–10467.
40. **Ferrari, F. A., K. Trach, D. LeCoq, J. Spence, E. Ferrari, and J. A. Hoch.** 1985. Characterization of the *spo0A* locus and its deduced product. *Proc. Natl. Acad. Sci. USA* **82:**2647–2651.
41. **Forst, S., D. Comeau, S. Norioka, and M. Inouye.** 1987. Localization and membrane topology of EnvZ, a protein involved in osmoregulation of OmpF and OmpC in *Escherichia coli*. *J. Biol. Chem.* **262:**16433–16438.
42. **Forst, S., J. Delgado, and M. Inouye.** 1989. Phosphorylation of OmpR by the osmosensor EnvZ modulates expression of the *ompF* and *ompC* genes in *Escherichia coli*. *Proc. Natl. Acad. Sci. USA* **86:**6052–6056.
43. **Friedrich, M. J., and R. J. Kadner.** 1987. Nucleotide sequence of the *uhp* region of *Escherichia coli*. *J. Bacteriol.* **169:**3556–3563.
44. **Fuhrer, D. K., and G. W. Ordal.** 1991. *Bacillus subtilis* CheN, a homolog of CheA, the central regulator of chemotaxis in *Escherichia coli*. *J. Bacteriol.* **173:**7443–7448.
45. **Gambello, M. J., and B. H. Iglewski.** 1991. Cloning and characterization of the *Pseudomonas aeruginosa lasR* gene, a transcriptional activator of elastase expression. *J. Bacteriol.* **173:**3000–3009.
46. **Gitt, M. A., L.-F. Wang, and R. H. Doi.** 1985. A strong sequence homology exists between the major RNA polymerase σ factors of *Bacillus subtilis* and *Escherichia coli*. *J. Biol. Chem.* **260:**7178–7185.
47. **Gross, R., B. Aricó, and R. Rappuoli.** 1989. Families of bacterial signal-transducing proteins. *Mol. Microbiol.* **3:**1661–1667.
48. **Hahn, J., and D. Dubnau.** 1991. Growth stage signal transduction and the requirements for *srfA* induction in development of competence. *J. Bacteriol.* **173:**7275–7282.
49. **Hahn, J., and D. Dubnau.** 1992. Personal communication.
50. **Hanks, S. K., A. M. Quinn, and T. Hunter.** 1988. The protein kinase family: conserved features and deduced phylogeny of the catalytic domains. *Science* **241:**42–52.
51. **Helmann, J. D.** 1991. Alternative sigma factors and the regulation of flagellar gene expression. *Mol. Microbiol.* **5:**2875–2882.
52. **Helmann, J. D., L. M. Márquez, V. L. Singer, and M. J. Chamberlin.** 1988. Cloning and characterization of the *Bacillus subtilis* sigma-28 gene, p. 189–193. *In* A. T. Ganesan, and J. A. Hoch (ed.), *Genetics and Biotechnology of Bacilli*, vol. 2. Academic Press, Inc., San Diego.
53. **Henikoff, S., J. C. Wallace, and J. P. Brown.** 1990. Finding protein similarities with nucleotide sequence databases. *Methods Enzymol.* **183:**111–132.
54. **Henner, D. J., E. Ferrari, M. Perego, and J. A. Hoch.** 1988. Location of the targets of the *hpr-97*, *sacU32*(Hy), and *sacQ36*(Hy) mutations in upstream regions of the subtilisin promoter. *J. Bacteriol.* **170:**296–300.
55. **Henner, D. J., E. Ferrari, M. Perego, and J. A. Hoch.** 1988. Upstream activating sequences in *Bacillus subtilis*, p. 3–9. *In* A. T. Ganesan and J. A. Hoch (ed.), *Genetics and Biotechnology of Bacilli*, vol. 2. Academic Press, Inc., San Diego.
56. **Henner, D. J., M. Yang, L. Band, H. Shimotsu, M. Ruppen, and E. Ferrari.** 1987. Genes of *Bacillus subtilis* that regulate the expression of degradative enzymes, p. 81–90. *In* M. Alacevic, D. Hranueli, and Z. Toman (ed.), *Genetics of Industrial Microorganisms. Proceedings of the Fifth International Symposium on the Genetics of Industrial Microorganisms.* Pliva, Zagreb, Yugoslavia.
57. **Henner, D. J., M. Yang, and E. Ferrari.** 1988. Localization of *Bacillus subtilis sacU*(Hy) mutations to two linked genes with similarities to the conserved procaryotic family of two-component signaling systems. *J. Bacteriol.* **170:**5102–5109.
58. **Hess, J. F., R. B. Bourret, and M. I. Simon.** 1988. Histidine phosphorylation and phosphoryl group transfer in bacterial chemotaxis. *Nature* (London) **336:**139–143.
59. **Higgins, D. G., and P. M. Sharp.** 1988. CLUSTAL: a package for performing multiple sequence alignment on a microcomputer. *Gene* **73:**237–244.
60. **Higgins, D. G., and P. M. Sharp.** 1989. Fast and sensitive multiple sequence alignments on a microcomputer. *Comput. Appl. Biosci.* **5:**151–153.
61. **Holland, S. K., S. Cutting, and J. Mandelstam.** 1987. The possible DNA-binding nature of the regulatory proteins, encoded by *spoIID* and *gerE*, involved in the sporulation of *Bacillus subtilis*. *J. Gen. Microbiol.* **133:**2381–2391.
62. **Hulett, F. M., E. E. Kim, C. Bookstein, N. V. Kapp, C. W. Edwards, and H. W. Wyckoff.** 1991. *Bacillus subtilis* alkaline phosphatases III and IV. Cloning, sequencing and comparisons of deduced amino acid sequence with *Escherichia coli* alkaline phosphatase three-dimensional structure. *J. Biol. Chem.* **266:**1077–1084.
63. **Igo, M. M., A. J. Ninfa, J. B. Stock, and T. J. Silhavy.** 1989. Phosphorylation and dephosphorylation of a bacterial transcriptional activator by a transmembrane receptor. *Genes Dev.* **3:**1725–1734.
64. **Igo, M. M., J. M. Slauch, and T. J. Silhavy.** 1990. Signal transduction in bacteria:kinases that control gene expression. *New Biol.* **2:**5–9.
65. **Jin, S., R. K. Prusti, T. Roitsch, R. G. Ankenbauer, and E. W. Nester.** 1990. Phosphorylation of the VirG protein of *Agrobacterium tumefaciens* by the autophosphorylated VirA protein: essential role in biological activity of VirG. *J. Bacteriol.* **172:**4945–4950.
66. **Jin, S., T. Roitsch, R. G. Ankenbauer, M. P. Gordon, and E. W. Nester.** 1990. The VirA protein of *Agrobacterium tumefaciens* is autophosphorylated and is essential for *vir* gene regulation. *J. Bacteriol.* **172:**525–530.
67. **Kahn, D., and G. Ditta.** 1991. Modular structure of FixJ: homology of the transcriptional activator domain with the −35 binding domain of sigma factors. *Mol. Microbiol.* **5:**987–997.
68. **Keener, J., and S. Kustu.** 1988. Protein kinase and phosphoprotein phosphatase activities of nitrogen regulatory proteins NTRB and NTRC of enteric bacteria: roles of the conserved amino-terminal domain of NTRC. *Proc. Natl. Acad. Sci. USA* **85:**4976–4980.
69. **Kemp, B. E., and R. B. Pearson.** 1990. Protein kinase recognition sequence motifs. *Trends Biochem. Sci.* **15:**342–346.
70. **Klier, A., T. Msadek, and G. Rapoport.** 1992. Positive

regulation in the Gram-positive bacterium: *Bacillus subtilis*. *Annu. Rev. Microbiol.* **46:**429–459.

71. **Kofoid, E. C., and J. S. Parkinson.** 1988. Transmitter and receiver modules in bacterial signaling proteins. *Proc. Natl. Acad. Sci. USA* **85:**4981–4985.
72. **Kofoid, E. C., and J. S. Parkinson.** 1991. Tandem translation starts in the *cheA* locus of *Escherichia coli*. *J. Bacteriol.* **173:**2116–2119.
73. **Kornblum, J., B. N. Kreiswirth, S. J. Projan, H. Ross, and R. P. Novick.** 1990. Agr: a polycistronic locus regulating exoprotein synthesis in *Staphylococcus aureus*, p. 373–402. *In* R. P. Novick (ed.), *Molecular Biology of the Staphylococci*. VCH, New York.
74. **Kroos, L.** 1991. Gene regulation in the mother-cell compartment of sporulating *Bacillus subtilis*. *Semin. Dev. Biol.* **2:**63–71.
75. **Kunst, F., A. Amory, M. Débarbouillé, I. Martin, A. Klier, and G. Rapoport.** 1988. Polypeptides activating the synthesis of secreted enzymes, p. 27–31. *In* A. T. Ganesan and J. A. Hoch (ed.), *Genetics and Biotechnology of Bacilli*, vol. 2. Academic Press Inc., San Diego.
76. **Kunst, F., M. Débarbouillé, T. Msadek, M. Young, C. Mauël, D. Karamata, A. Klier, G. Rapoport, and R. Dedonder.** 1988. Deduced polypeptides encoded by the *Bacillus subtilis sacU* locus share homology with two-component sensor-regulator systems. *J. Bacteriol.* **170:**5093–5101.
77. **Kunst, F., and T. Msadek.** 1991. Unpublished results.
78. **Kunst, F., M. Pascal, J. Lepesant-Kejzlarová, J.-A. Lepesant, A. Billault, and R. Dedonder.** 1974. Pleiotropic mutations affecting sporulation conditions and the synthesis of extracellular enzymes in *Bacillus subtilis* 168. *Biochimie* **56:**1481–1489.
79. **Lepesant, J.-A., F. Kunst, J. Lepesant-Kejzlarová, and R. Dedonder.** 1972. Chromosomal location of mutations affecting sucrose metabolism in *Bacillus subtilis* Marburg. *Mol. Gen. Genet.* **118:**135–160.
80. **Leroux, B., M. F. Yanofsky, S. C. Winans, J. E. Ward, S. F. Ziegler, and E. W. Nester.** 1987. Characterization of the *virA* locus of *Agrobacterium tumefaciens*: a transcriptional regulator and host range determinant. *EMBO J.* **6:**849–856.
81. **Lukat, G. S., B. H. Lee, J. M. Mottonen, A. M. Stock, and J. B. Stock.** 1991. Roles of the highly conserved aspartate and lysine residues in the response regulator of bacterial chemotaxis. *J. Biol. Chem.* **266:**8348–8354.
82. **Lukat, G. S., W. R. McCleary, A. M. Stock, and J. B. Stock.** 1992. Phosphorylation of bacterial response regulator proteins by low molecular weight phospho-donors. *Proc. Natl. Acad. Sci. USA* **89:**718–722.
83. **Lukat, G. S., A. M. Stock, and J. B. Stock.** 1990. Divalent metal ion binding to the CheY protein and its significance to phosphotransfer in bacterial chemotaxis. *Biochemistry* **29:**5436–5442.
84. **Lupas, A., and J. Stock.** 1989. Phosphorylation of an N-terminal regulatory domain activates the CheB methylesterase in bacterial chemotaxis. *J. Biol. Chem.* **264:**17337–17342.
85. **MacFarlane, S. A., and M. Merrick.** 1985. The nucleotide sequence of the nitrogen regulation gene *ntrB* and the *glnA-ntrBC* intergenic region of *Klebsiella pneumoniae*. *Nucleic Acids Res.* **13:**7591–7606.
86. **Magasanik, B.** 1982. Genetic control of nitrogen assimilation in bacteria. *Annu. Rev. Genet.* **16:**135–168.
87. **Magasanik, B.** 1988. Reversible phosphorylation of an enhancer binding protein regulates the transcription of bacterial nitrogen utilization genes. *Trends Biochem. Sci.* **13:**475–479.
88. **Makino, K., H. Shinagawa, M. Amemura, T. Kawamoto, M. Yamada, and A. Nakata.** 1989. Signal transduction in the phosphate regulon of *Escherichia coli* involves phosphotransfer between PhoR and PhoB proteins. *J. Mol. Biol.* **210:**551–559.
89. **Makino, K., H. Shinagawa, M. Amemura, and A. Nakata.** 1986. Nucleotide sequence of the *phoB* gene, the positive regulatory gene for the phosphate regulon of *Escherichia coli* K-12. *J. Mol. Biol.* **190:**37–44.
90. **Makino, K., H. Shinagawa, M. Amemura, and A. Nakata.** 1986. Nucleotide sequence of the *phoR* gene, a regulatory gene for the phosphate regulon of *Escherichia coli*. *J. Mol. Biol.* **192:**549–556.
91. **Márquez, L. M., J. D. Helmann, E. Ferrari, H. M. Parker, G. W. Ordal, and M. J. Chamberlin.** 1990. Studies of σ^D-dependent functions in *Bacillus subtilis*. *J. Bacteriol.* **172:**3435–3443.
92. **Melchers, L. S., T. J. G. Regensburg-Tuïnk, R. B. Bourret, N. J. A. Sedee, R. A. Schilperoort, and P. J. J. Hooykaas.** 1989. Membrane topology and functional analysis of the sensory protein VirA of *Agrobacterium tumefaciens*. *EMBO J.* **8:**1919–1925.
93. **Melchers, L. S., D. V. Thompson, K. B. Idler, R. A. Schilperoort, and P. J. J. Hooykaas.** 1986. Nucleotide sequence of the virulence gene *virG* of the *Agrobacterium tumefaciens* octopine Ti plasmid: significant homology between *virG* and the regulatory genes *ompR*, *phoB* and *dye* of *E. coli*. *Nucleic Acids Res.* **14:**9933–9942.
94. **Miki, T., Z. Minami, and Y. Ikeda.** 1965. The genetics of alkaline phosphatase formation in *Bacillus subtilis*. *Genetics* **52:**1093–1100.
95. **Miller, J. F., J. J. Mekalanos, and S. Falkow.** 1989. Coordinate regulation and sensory transduction in the control of bacterial virulence. *Science* **243:**916–922.
96. **Miller, S. I., A. M. Kukral, and J. J. Mekalanos.** 1989. A two-component regulatory system (*phoP phoQ*) controls *Salmonella typhimurium* virulence. *Proc. Natl. Acad. Sci. USA* **86:**5054–5058.
97. **Miller, V. L., R. K. Taylor, and J. J. Mekalanos.** 1987. Cholera toxin transcriptional activator ToxR is a transmembrane DNA binding protein. *Cell* **48:**271–279.
98. **Miranda-Ríos, J., R. Sánchez-Pescador, M. Urdea, and A. A. Covarrubias.** 1987. The complete nucleotide sequence of the *glnALG* operon of *Escherichia coli* K12. *Nucleic Acids Res.* **15:**2757–2770.
99. **Mitamura, T., R. V. Ebora, T. Nakai, Y. Makino, S. Negoro, I. Urabe, and H. Okada.** 1990. Structure of isozyme genes of glucose dehydrogenase from *Bacillus megaterium* IAM1030. *J. Ferment. Bioeng.* **70:**363–369.
100. **Mizuno, T., and S. Mizushima.** 1990. Signal transduction and gene regulation through the phosphorylation of two regulatory components: the molecular basis for the osmotic regulation of the porin genes. *Mol. Microbiol.* **4:**1077–1082.
101. **Mizuno, T., E. T. Wurtzel, and M. Inouye.** 1982. Osmoregulation of gene expression. II. DNA sequence of the *envZ* gene of the *ompB* operon of *Escherichia coli* and characterization of its gene products. *J. Biol. Chem.* **257:**13692–13698.
102. **Moir, A., and E. H. Kemp.** 1991. Personal communication.
103. **Msadek, T., and F. Kunst.** 1991. Unpublished results.
104. **Msadek, T., F. Kunst, D. Henner, A. Klier, G. Rapoport, and R. Dedonder.** 1990. Signal transduction pathway controlling synthesis of a class of degradative enzymes in *Bacillus subtilis*: expression of the regulatory genes and analysis of mutations in *degS* and *degU*. *J. Bacteriol.* **172:**824–834.
105. **Msadek, T., F. Kunst, A. Klier, and G. Rapoport.** 1991. DegS-DegU and ComP-ComA modulator-effector pairs control expression of the *Bacillus subtilis* pleiotropic regulatory gene *degQ*. *J. Bacteriol.* **173:**2366–2377.
106. **Msadek, T., F. Kunst, A. Klier, G. Rapoport, and R. Dedonder.** 1990. The Deg signal transduction pathway: mutations and regulation of expression of *degS*, *degU* and *degQ*, p. 245–255. *In* M. M. Zukowski, A. T. Gane-

san, and J. A. Hoch (ed.), *Genetics and Biotechnology of Bacilli*, vol. 3. Academic Press, Inc., San Diego.

107. **Mueller, J. P., G. Bukusoglu, and A. L. Sonenshein.** 1992. Transcriptional regulation of *Bacillus subtilis* glucose starvation inducible genes: control of *gsiA* by the ComP-ComA signal transduction system. *J. Bacteriol.* **174:**4361–4373.
108. **Mukai, K., M. Kawata, and T. Tanaka.** 1990. Isolation and phosphorylation of the *Bacillus subtilis degS* and *degU* gene products. *J. Biol. Chem.* **265:**20000–20006.
109. **Mutoh, N., and M. I. Simon.** 1986. Nucleotide sequence corresponding to five chemotaxis genes in *Escherichia coli. J. Bacteriol.* **165:**161–166.
110. **Nagami, Y., and T. Tanaka.** 1986. Molecular cloning and nucleotide sequence of a DNA fragment from *Bacillus natto* that enhances production of extracellular proteases and levansucrase in *Bacillus subtilis. J. Bacteriol.* **166:**20–28.
111. **Nakano, M. M., L. A. Xia, and P. Zuber.** 1991. Transcription initiation region of the *srfA* operon, which is controlled by the *comP-comA* signal transduction system in *Bacillus subtilis. J. Bacteriol.* **173:**5487–5493.
112. **Nakano, M. M., and P. Zuber.** 1991. The primary role of ComA in establishment of the competent state in *Bacillus subtilis* is to activate expression of *srfA. J. Bacteriol.* **173:**7269–7274.
113. **Nassif, X., N. Honoré, T. Vasselon, S. T. Cole, and P. J. Sansonetti.** 1989. Positive control of colanic acid synthesis in *Escherichia coli* by *rmpA* and *rmpB*, two virulence-plasmid genes of *Klebsiella pneumoniae. Mol. Microbiol.* **3:**1349–1359.
114. **Ninfa, A. J., and R. L. Bennett.** 1991. Identification of the site of autophosphorylation of the bacterial protein kinase/phosphatase NRII. *J. Biol. Chem.* **266:**6888–6893.
115. **Ninfa, A. J., and B. Magasanik.** 1986. Covalent modification of the *glnG* product, NR_I, by the *glnL* product, NR_{II}, regulates the transcription of the *glnALG* operon in *Escherichia coli. Proc. Natl. Acad. Sci. USA* **83:**5909–5913.
116. **Ninfa, A. J., E. G. Ninfa, A. N. Lupas, A. Stock, B. Magasanik, and J. Stock.** 1988. Crosstalk between bacterial chemotaxis signal transduction proteins and regulators of transcription of the Ntr regulon: evidence that nitrogen assimilation and chemotaxis are controlled by a common phosphotransfer mechanism. *Proc. Natl. Acad. Sci. USA* **85:**5492–5496.
117. **Nixon, B. T., C. W. Ronson, and F. M. Ausubel.** 1986. Two-component regulatory systems responsive to environmental stimuli share strongly conserved domains with the nitrogen assimilation regulatory genes *ntrB* and *ntrC. Proc. Natl. Acad. Sci. USA* **83:**7850–7854.
118. **Olmedo, G., E. Gottlin Ninfa, J. Stock, and P. Youngman.** 1990. Novel mutations that alter the regulation of sporulation in *Bacillus subtilis*: evidence that phosphorylation of regulatory protein Spo0A controls the initiation of sporulation. *J. Mol. Biol.* **215:**359–372.
119. **Parkinson, J. S.** 1988. Protein phosphorylation in bacterial chemotaxis. Cell **53:**1–2.
120. **Peng, H.-L., R. P. Novick, B. Kreiswirth, J. Kornblum, and P. Schlievert.** 1988. Cloning, characterization, and sequencing of an accessory gene regulator (*agr*) in *Staphylococcus aureus. J. Bacteriol.* **170:**4365–4372.
121. **Perego, M., S. P. Cole, D. Burbulys, K. Trach, and J. A. Hoch.** 1989. Characterization of the gene for a protein kinase which phosphorylates the sporulation-regulatory proteins Spo0A and Spo0F of *Bacillus subtilis. J. Bacteriol.* **171:**6187–6196.
122. **Poetter, K., and D. L. Coplin.** 1991. Structural and functional analysis of the *rcsA* gene from *Erwinia stewartii. Mol. Gen. Genet.* **229:**155–160.
123. **Raibaud, A., M. Zalacain, T. G. Holt, R. Tizard, and C. J. Thompson.** 1991. Nucleotide sequence analysis reveals linked *N*-acetyl hydrolase, thioesterase, transport, and regulatory genes encoded by the bialaphos biosynthetic gene cluster of *Streptomyces hygroscopicus. J. Bacteriol.* **173:**4454–4463.
124. **Richet, E., and O. Raibaud.** 1989. MalT, the regulatory protein of the *Escherichia coli* maltose system, is an ATP-dependent transcriptional activator. *EMBO J.* **8:**981–987.
125. **Roggiani, M., J. Hahn, and D. Dubnau.** 1990. Suppression of early competence mutations in *Bacillus subtilis* by *mec* mutations. *J. Bacteriol.* **172:**4056–4063.
126. **Ronson, C. W., B. T. Nixon, and F. M. Ausubel.** 1987. Conserved domains in bacterial regulatory proteins that respond to environmental stimuli. *Cell* **49:**579–581.
127. **Rossmann, M. G., D. Moras, and K. W. Olsen.** 1974. Chemical and biological evolution of a nucleotide-binding protein. *Nature* (London) **250:**194–199.
128. **Ruppen, M. E., G. L. Van Alstine, and L. Band.** 1988. Control of intracellular serine protease expression in *Bacillus subtilis. J. Bacteriol.* **170:**136–140.
129. **Sanders, D. A., B. L. Gillece-Castro, A. M. Stock, A. L. Burlingame, and D. E. Koshland, Jr.** 1989. Identification of the site of phosphorylation of the chemotaxis response regulator protein, CheY. *J. Biol. Chem.* **264:** 21770–21778.
130. **Scarlato, V., A. Prugnola, B. Aricó, and R. Rappuoli.** 1990. Positive transcriptional feedback at the *bvg* locus controls expression of virulence factors in *Bordetella pertussis. Proc. Natl. Acad. Sci. USA* **87:**6753–6757.
131. **Schaeffer, P., J. Millet, and J.-P. Aubert.** 1965. Catabolic repression of bacterial sporulation. *Proc. Natl. Acad. Sci. USA* **54:**704–711.
132. **Schweizer, H. P.** 1991. The *agmR* gene, an environmentally responsive gene, complements defective *glpR*, which encodes the putative activator for glycerol metabolism in *Pseudomonas aeruginosa. J. Bacteriol.* **173:** 6798–6806.
133. **Seki, T., H. Yoshikawa, H. Takahashi, and H. Saito.** 1987. Cloning and nucleotide sequence of *phoP*, the regulatory gene for alkaline phosphatase and phosphodiesterase in *Bacillus subtilis. J. Bacteriol.* **169:**2913–2916.
134. **Seki, T., H. Yoshikawa, H. Takahashi, and H. Saito.** 1988. Nucleotide sequence of the *Bacillus subtilis phoR* gene. *J. Bacteriol.* **170:**5935–5938.
135. **Shimotsu, H., and D. J. Henner.** 1986. Modulation of *Bacillus subtilis* levansucrase gene expression by sucrose and regulation of the steady-state mRNA level by *sacU* and *sacQ* genes. *J. Bacteriol.* **168:**380–388.
136. **Simms, S. A., M. G. Keane, and J. Stock.** 1985. Multiple forms of the CheB methylesterase in bacterial chemosensing. *J. Biol. Chem.* **260:**10161–10168.
137. **Smith, I., E. Dubnau, M. Predich, and R. Rudner.** 1992. Early *spo* gene expression in *Bacillus subtilis*: the role of interrelated signal transduction systems. *Biochimie* **74:** 669–678.
138. **Smith, I., I. Mandic-Mulec, and N. Gaur.** 1991. The role of negative control in sporulation. *Res. Microbiol.* **142:** 831–839.
139. **Sonenshein, A. L.** 1989. Metabolic regulation of sporulation and other stationary-phase phenomena, p. 109–130. *In* I. Smith, R. A. Slepecky, and P. Setlow (ed.), *Regulation of Procaryotic Development: Structural and Functional Analysis of Bacterial Sporulation and Germination.* American Society for Microbiology, Washington, D.C.
140. **Sorokin, A., and S. D. Ehrlich.** 1991. Personal communication.
141. **Spiegelman, G., B. Van Hoy, M. Perego, J. Day, K. Trach, and J. A. Hoch.** 1990. Structural alterations in the *Bacillus subtilis* Spo0A regulatory protein which suppress mutations at several *spo0* loci. *J. Bacteriol.* **172:**5011–5019.
142. **Stewart, R. C., and F. W. Dahlquist.** 1988. N-terminal

half of CheB is involved in methylesterase response to negative chemotactic stimuli in *Escherichia coli*. *J. Bacteriol.* **170:**5728–5738.

143. **Stock, A., T. Chen, D. Welsh, and J. Stock.** 1988. CheA protein, a central regulator of bacterial chemotaxis, belongs to a family of proteins that control gene expression in response to changing environmental conditions. *Proc. Natl. Acad. Sci. USA* **85:**1403–1407.
144. **Stock, A. M., J. M. Mottonen, J. B. Stock, and C. E. Schutt.** 1989. Three-dimensional structure of CheY, the response regulator of bacterial chemotaxis. *Nature* (London) **337:**745–749.
145. **Stock, J. B., G. S. Lukat, and A. M. Stock.** 1991. Bacterial chemotaxis and the molecular logic of intracellular signal transduction networks. *Annu. Rev. Biophys. Biophys. Chem.* **20:**109–136.
146. **Stock, J. B., A. J. Ninfa, and A. M. Stock.** 1989. Protein phosphorylation and regulation of adaptive responses in bacteria. *Microbiol. Rev.* **53:**450–490.
147. **Stock, J. B., A. M. Stock, and J. M. Mottonen.** 1990. Signal transduction in bacteria. *Nature* (London) **344:** 395–400.
148. **Stout, V., A. Torres-Cabassa, M. R. Maurizi, D. Gutnick, and S. Gottesman.** 1991. RcsA, an unstable positive regulator of capsular polysaccharide synthesis. *J. Bacteriol.* **173:**1738–1747.
149. **Stragier, P., J. Bouvier, C. Bonamy, and J. Szulmajster.** 1984. A developmental gene product of *Bacillus subtilis* homologous to the sigma factor of *Escherichia coli*. *Nature* (London) **312:**376–378.
150. **Tanaka, T., and M. Kawata.** 1988. Cloning and characterization of *Bacillus subtilis iep*, which has positive and negative effects on production of extracellular proteases. *J. Bacteriol.* **170:**3593–3600.
151. **Tanaka, T., M. Kawata, and K. Mukai.** 1991. Altered phosphorylation of *Bacillus subtilis* DegU caused by single amino acid changes in DegS. *J. Bacteriol.* **173:** 5507–5515.
152. **Tanaka, T., M. Kawata, Y. Nagami, and H. Uchiyama.** 1987. *prtR* enhances the mRNA level of the *Bacillus subtilis* extracellular proteases. *J. Bacteriol.* **169:**3044–3050.
153. **Taylor, S. S.** 1989. cAMP-dependent protein kinase. *J. Biol. Chem.* **264:**8443–8446.
154. **Tokishita, S., A. Kojima, H. Aiba, and T. Mizuno.** 1991. Transmembrane signal transduction and osmoregulation in *Escherichia coli*—functional importance of the periplasmic domain of the membrane-located protein kinase, EnvZ. *J. Biol. Chem.* **266:**6780–6785.
155. **Trach, K., D. Burbulys, G. Spiegelman, M. Perego, B. Van Hoy, M. Strauch, J. Day, and J. A. Hoch.** 1990. Phosphorylation of the Spo0A protein: a cumulative environsensory activation mechanism, p. 357–365. *In* M. M. Zukowski, A. T. Ganesan, and J. A. Hoch (ed.), *Genetics and Biotechnology of Bacilli*, vol. 3. Academic Press, Inc., San Diego.
156. **Trach, K. A., J. W. Chapman, P. J. Piggot, and J. A. Hoch.** 1985. Deduced product of the stage 0 sporulation gene *spo0F* shares homology with Spo0A, OmpR, and SfrA proteins. *Proc. Natl. Acad. Sci. USA* **82:**7260–7264.
157. **Tseng, H.-C., and C. W. Chen.** 1991. A cloned *ompR*-like gene of *Streptomyces lividans* 66 suppresses defective *melC1*, a putative copper-transfer gene. *Mol. Microbiol.* **5:**1187–1196.
158. **van Sinderen, D.** 1991. Personal communication.
159. **van Sinderen, D., S. Withoff, H. Boels, and G. Venema.** 1990. Isolation and characterization of *comL*, a transcription unit involved in competence development of *Bacillus subtilis*. *Mol. Gen. Genet.* **224:**396–404.
160. **Vasselon, T., P. J. Sansonetti, and X. Nassif.** 1991. Nucleotide sequence of *rmpB*, a *Klebsiella pneumoniae* gene that positively controls colanic biosynthesis in *Escherichia coli*. *Res. Microbiol.* **142:**47–54.
161. **Volz, K., and P. Matsumura.** 1991. Crystal structure of *Escherichia coli* CheY refined at 1.7-Å resolution. *J. Biol. Chem.* **266:**15511–15519.
162. **Wanner, B. L., and P. Latterell.** 1980. Mutants affected in alkaline phosphatase expression: evidence for multiple positive regulators of the phosphate regulon in *Escherichia coli*. *Genetics* **96:**353–366.
163. **Weinrauch, Y., N. Guillen, and D. A. Dubnau.** 1989. Sequence and transcription mapping of *Bacillus subtilis* competence genes *comB* and *comA*, one of which is related to a family of bacterial regulatory determinants. *J. Bacteriol.* **171:**5362–5375.
164. **Weinrauch, Y., T. Msadek, F. Kunst, and D. Dubnau.** 1991. Sequence and properties of *comQ*, a new competence regulatory gene of *Bacillus subtilis*. *J. Bacteriol.* **173:**5685–5693.
165. **Weinrauch, Y., R. Penchev, E. Dubnau, I. Smith, and D. Dubnau.** 1990. A *Bacillus subtilis* regulatory gene product for genetic competence and sporulation resembles sensor protein members of the bacterial two-component signal-transduction systems. *Genes Dev.* **4:**860–872.
166. **Winans, S. C., P. R. Ebert, S. E. Stachel, M. P. Gordon, and E. W. Nester.** 1986. A gene essential for *Agrobacterium* virulence is homologous to a family of positive regulatory loci. *Proc. Natl. Acad. Sci. USA* **83:**8278–8282.
167. **Wurtzel, E. T., M. Y. Chou, and M. Inouye.** 1982. Osmoregulation of gene expression. I. DNA sequence of the *ompR* gene of the *ompB* operon of *Escherichia coli* and characterization of its gene product. *J. Biol. Chem.* **257:**13685–13691.
168. **Wylie, D., A. Stock, C.-Y. Wong, and J. Stock.** 1988. Sensory transduction in bacterial chemotaxis involves phosphotransfer between Che proteins. *Biochem. Biophys. Res. Commun.* **151:**891–896.
169. **Yamada, M., K. Makino, M. Amemura, H. Shinegawa, and A. Nakata.** 1989. Regulation of the phosphate regulon of *Escherichia coli*: analysis of mutant *phoB* and *phoR* genes causing different phenotypes. *J. Bacteriol.* **171:**5601–5606.
170. **Yamane, K., and B. Maruo.** 1987. Alkaline phosphatase possessing alkaline phosphodiesterase activity and other phosphodiesterases in *Bacillus subtilis*. *J. Bacteriol.* **134:**108–114.
171. **Yang, M., E. Ferrari, E. Chen, and D. J. Henner.** 1986. Identification of the pleiotropic *sacQ* gene of *Bacillus subtilis*. *J. Bacteriol.* **166:**113–119.
172. **Yang, M., H. Shimotsu, E. Ferrari, and D. J. Henner.** 1987. Characterization and mapping of the *Bacillus subtilis prtR* gene. *J. Bacteriol.* **169:**434–437.
173. **Yoshikawa, H., J. Kazami, S. Yamashita, T. Chibazakura, H. Sone, F. Kawamura, M. Oda, M. Isaka, Y. Kobayashi, and H. Saito.** 1986. Revised assignment for the *Bacillus subtilis spo0F* gene and its homology with *spo0A* and with two *Escherichia coli* genes. *Nucleic Acids Res.* **14:**1063–1072.

51. *spo0* Genes, the Phosphorelay, and the Initiation of Sporulation

JAMES A. HOCH

A large fraction of the gram-positive bacilli have adopted sporulation as a means of survival when environmental conditions are less than optimal for growth. The initiation of sporulation begins at the end of exponential growth, occurs coincidentally with the onset of stationary phase in cultures grown in the laboratory, and is under genetic control. Sporulation involves an energy-intensive pathway and a commitment to the production of a complex morphological structure, and therefore is not a process that the cell undertakes frivolously. The cell responds to signals which it continually monitors in the environment, and when these environmental signals reach a point that indicates it is time to shut down division and begin sporulation, the initiation process begins. Thus, sporulation is the antithesis of division, and some of the same signals may be used for both, but in an opposite manner. The metabolic and environmental information that is monitored by the cell for these processes remains a mystery. Substantial progress has been made, however, in understanding the mechanisms by which this information is recognized and the signal transduction system that transmits variations in environmental and metabolic signals to the transcription machinery.

The signal transduction system, the phosphorelay (4), for the initiation of sporulation in *Bacillus subtilis* revolves around the transcription factor Spo0A. The role of the phosphorelay is to activate the Spo0A protein by means of phosphorylation to allow it to initiate or repress transcription at the promoters that it controls. Recognition of the mechanism of the phosphorelay evolved from studies of sporulation mutants blocked at the very early stages of sporulation, stage 0. Through mapping studies, *spo0* mutations were placed in several loci, and through cloning and sequencing efforts the composition of each of these loci has been determined. We shall explore in this chapter the genetic and chemical structure of the *spo0* loci and their products, along with other components of the signal transduction system, and try to provide an historic and mechanistic perspective of the phosphorelay and the function of the Spo0A transcription factor.

GENES FOR COMPONENTS OF THE PHOSPHORELAY

The genes for the component enzymes of the phosphorelay are scattered on the chromosome and, for the most part, individually regulated. The genes, their promoters and terminators, and the locations of relevant *cis* regulating sequences are shown in Fig. 1.

spo0A Locus

The *spo0A* locus consists of a single gene preceded by two promoters (11, 20). A σ^A promoter functions as a low-level promoter to produce a maintenance level of the Spo0A protein during exponential growth (50). Downstream from the σ^A promoter is a σ^H promoter, which is required for induction of the protein after the end of exponential growth and during the initiation process (11, 50). The gene is followed by a large hairpin loop that probably acts as a terminator. The *spo0A* locus is located at 216° on the genetic map.

The protein encoded by the *spo0A* locus is a 29,691-Da transcription factor. The amino-terminal portion of the Spo0A protein (up to approximately amino acid residue 120) is homologous to response regulators of procaryotic two-component regulatory systems (35), and it is this portion of the molecule that is phosphorylated to activate its transcription functions (4). The carboxy half of the protein, from approximately residue 120 until the end of the protein at residue 267, is unique among known carboxy-terminal portions of response regulator proteins. This portion of the molecule may contain the DNA-binding specificity of the protein and, presumably, that portion of the molecule that interacts with the transcription machinery (30).

spo0B Locus

The *spo0B* locus consists of a single transcription unit with several open reading frames (3). The first gene encodes a protein of 22,542 Da in which all the *spo0B* alleles reside (10); this gene is followed immediately downstream by the *obg* gene, coding for a protein of 47,668 Da (44). It is clear from gene inactivation studies that the *spo0B* and *obg* genes are in the same transcript since attempts to inactivate *spo0B* with a terminator containing a *cat* cassette have been unsuccessful, whereas the same cassette lacking a terminator can be used to inactivate *spo0B* (44). Presumably the terminator prevents transcription into the *obg* gene. In addition, it is likely that the *spo0B* transcription unit continues through the *pheB* and *pheA* genes downstream of *obg*. Transcription of this locus seems to occur from a single promoter (3, 10).

James A. Hoch • Division of Cellular Biology, Department of Molecular and Experimental Medicine, The Scripps Research Institute, 10666 North Torrey Pines Road, La Jolla, California 92037-1093.

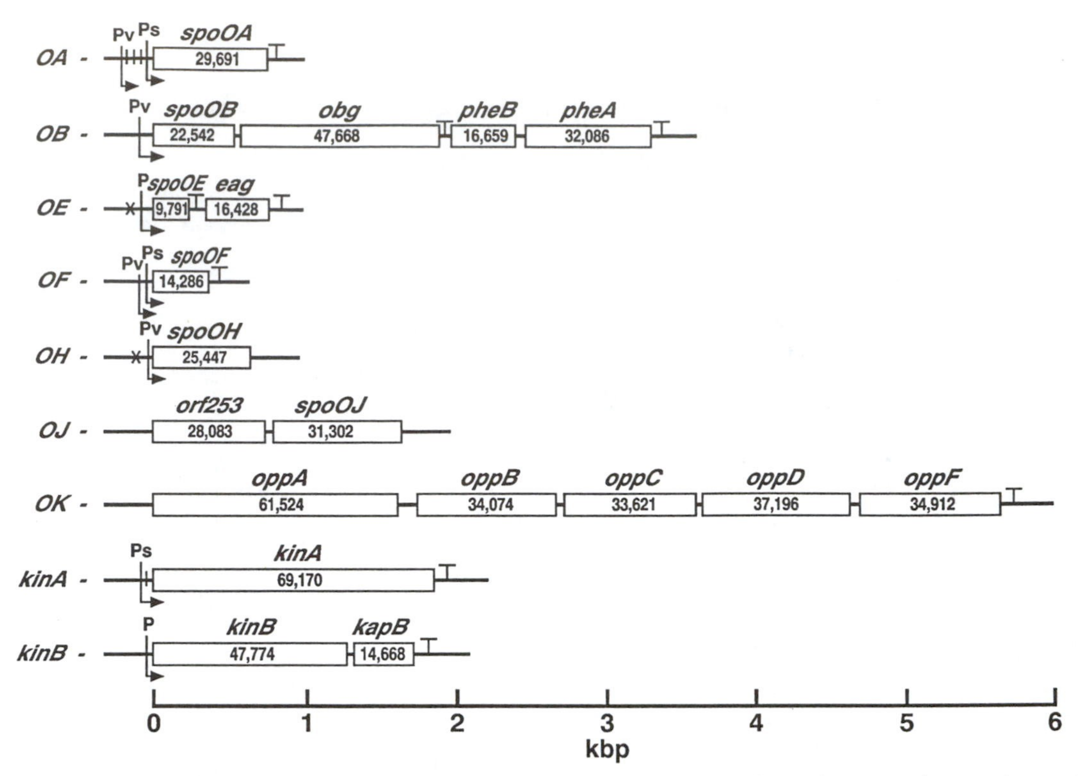

Figure 1. Structure of the *spo0* and *kin* genes. The complete transcription unit for each gene is shown. The numbers in each box are the molecular weights of the gene products. P or Pv indicates the start site of a vegetative promoter, whereas Ps indicates the location of a sporulation (σ^{H}) dependent promoter. Vertical lines in promoters indicate sites of 0A boxes, and an X indicates the approximate location of AbrB binding sites. Terminators are shown by a T.

The uncertainty in the length of the transcription unit is due to the fact that the *obg* gene is followed by what could be a transcription terminator loop, but the 5′ end of the transcript for the *pheB* and *pheA* genes that follow this gene was found in the middle of this loop. It is not certain whether all *pheB* and *pheA* transcription initiates from the *spo0B* promoter (in which case the loop would be a processing site) or whether there is another promoter for *pheB-pheA* partially contained in the *obg* gene itself.

The product of the *spo0B* gene is the phosphoprotein phosphotransferase of the phosphorelay (4). The enzymatic function of the Spo0B protein is to transfer phosphate from Spo0F-P to Spo0A, producing Spo0A-P. This is a unique feature of the phosphorelay not found in other two-component regulatory systems, and Spo0B is a unique type of enzyme, having never been described before. Its kinetic and physical parameters have not yet been examined in great detail.

The *obg* gene encodes an essential protein with GTP-binding capacity. This has been shown in vitro and confirms interpretations from the primary structure of the protein, which shows a clear GTP-binding motif (44). Further studies with purified protein have shown that Obg is a GTPase and has the ability to be phosphorylated by [γ-^{32}P]GTP (45). These activities are now being studied in detail but are reminiscent of the properties of G-proteins involved in signal transduction in higher organisms. The actual role of the Obg protein in the phosphorelay signal transduction system is completely unknown, and it is possible that the *spo0B* and *obg* genes are simply coincidentally transcribed in the same operon and have no interactive role.

The *spo0B* locus is located at 240° on the genetic map.

spo0E Locus

The *spo0E* locus codes for a protein of 9,791 Da (27). Transcription of this gene occurs from a σ^{A} promoter, and two start sites have been observed differing by only 3 bp. The *spo0E* gene is followed by a gene for a potential small membrane protein, *eag*, and these genes are separated by what appears to be a transcription terminator. Mapping the 5′ end of the *eag* transcript, however, showed that this transcript initiates in a hairpin loop between the two genes. It is unknown whether, like the *pheA* question in the *spo0B* operon, *eag* is transcribed from its own promoter, or whether the hairpin loop serves as a processing site for a longer transcript. *eag* transcription is low-level constitutive and does not correspond to that seen for the *spo0E* gene, suggesting that its control is different from that for *spo0E*. In addition, inactivation of *eag* appears to have no effect on either growth or sporulation of *B. subtilis*.

Spo0^{-} lesions in the *spo0E* gene were found to be

nonsense mutations concentrated in the carboxy-terminal one-third of the gene. However, recently it has been found that deletions of the *spo0E* gene are Spo^+ (29). Thus, the $Spo0^-$ phenotype of nonsense mutations in this gene is probably related to the production of a carboxy-truncated *spo0E* gene product.

The exact function of the *spo0E* gene in the phosphorelay is not certain. This protein has been exceedingly difficult to produce in expression vectors, and the product of the gene has not been isolated and tested for its biochemical properties. Studies with the deletion mutant, however, suggest that the *spo0E* gene product plays a negative role in the phosphorelay, where it may serve as a means to control the flow of phosphate through this signal transduction system (29).

The *spo0E* locus is located at 116° on the genetic map.

spo0F Locus

The *spo0F* locus appears to consist of a single gene (42, 43) preceded by a weak σ^A promoter and a strong σ^H promoter (21). The σ^A transcript presumably is used to maintain a low level of *spo0F* during exponential growth, whereas the σ^H promoter appears to be used to induce the synthesis of this enzyme once exponential growth ceases (31, 51).

The product of the *spo0F* gene is a 124-amino acid, 14,286-Da protein with strong similarity to the response regulators in two-component regulatory systems (42). This protein differs from most other response regulators in that it lacks an additional carboxy-terminal domain. In this regard, it is similar to the CheY protein of coliforms (35), although it serves a different role than CheY. The function of *spo0F* is to accept phosphate from the activating kinases for the phosphorelay (see below) and serve as a substrate for the *spo0B* phosphoprotein phosphotransferase. Although it shares primary sequence homology with the CheY response regulator protein, solution nuclear magnetic resonance studies (5) have shown that the arrangements of the helices are somewhat different from that of the CheY structure determined by X-ray crystallography (34). The *spo0F* locus is located at 323° on the genetic map.

spo0H Locus

The *spo0H* gene codes for a protein of 27,447 Da that functions as a sigma factor, σ^H (8). Transcription of this gene appears to occur exclusively from a σ^A promoter and no other genes are in an operon structure with it.

σ^H is the sigma factor used for high-level transcription of the *spo0A*, *spo0F*, and *kinA* genes, among others, and serves an important role in the transcription of genes for components of the phosphorelay during the initial stages of sporulation (31). This gene product and its specificity for various promoters are described in another chapter of this book.

The *spo0H* (*sigH*) locus is located at 9° on the chromosome.

spo0J Locus

The *spo0J* locus is dicistronic. The *orf-282* gene, defined by the *spo0J93* allele, codes for a protein of 282 amino acids (23, 24) and is paired with another gene, *orf-253*. This pair shares homology with the *sopB-sopA*, *parB-parA*, and *korB-incC* pairs in *Escherichia coli* (24). These latter gene pairs are suggested to function in chromosome partition or anchoring. The transcription and regulation of the *spo0J* genes have not been studied in great detail. Until recently it was thought that the *spo0J* locus was defined by two alleles, *spo0J93* and *spo0J87*. However, it has now been shown that *spo0J87* is actually a stage III locus that is in a different transcription unit from the *spo0J93* allele (23). The products of the *spo0J* locus have not been isolated, and their function is unknown.

spo0K Locus

The *spo0K* locus is an operon of genes coding for the oligopeptide transport system (26, 32). The operon consists of five genes, *oppA*, *oppB*, *oppC*, *oppD*, and *oppF*. This complex of genes makes up a membrane transport system with specificity for the uptake of peptides up to 5 amino acids long. The *oppA* gene product is located on the outer surface of the membrane and serves as a specificity determinant for transport of peptides. OppB and OppC are integral membrane proteins presumably making up the core of the transport channel, and OppD and OppF are both ATP-binding proteins that serve to energize the transport of peptides across the membrane. Mutations in the first four genes appear to completely abolish peptide transport through this system, whereas mutations in the *oppF* locus do not affect the transport of peptides (such as the antibiotic bialaphos). Similarly, *oppF* mutants are not sporulation defective but are defective in competence (32). The OppF protein must have some role in peptide transport, but not necessarily in the uptake of all peptides recognized by this system.

The function of the *opp* system in transport has been well documented, although its role in sporulation is uncertain. Some investigators believe that the transport of cell wall peptides across the membrane is an important initiating signal for sporulation (26), whereas others have postulated that internalization of sporulation-inducing peptides requires the *opp* system (14, 32). These hypotheses are not mutually exclusive. The competence-deficient phenotype of *oppF* mutants is probably due to an inability to transport the surfactin molecule which is intimately involved in the initiation of the physiological events that lead to competence (15).

The *opp* operon is located at 104° on the genetic map.

kinA Locus

The *kinA* locus contains a single gene which codes for a protein of 69,170 Da showing strong homology to the transmitter kinases of two-component regulatory systems (1, 25). The *kinA* gene is transcribed from a σ^H promoter (31). The KinA protein is capable of auto-

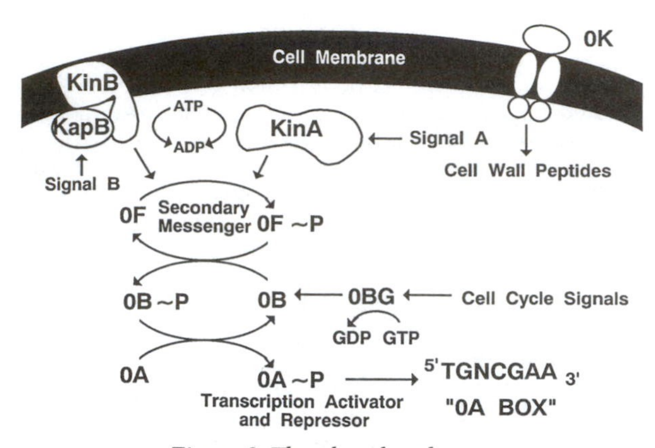

Figure 2. The phosphorelay.

phosphorylation in the presence of ATP and mediates the transfer of phosphate to both the Spo0A and Spo0F response regulator proteins (25). In vivo, however, its activity is probably confined to the Spo0F protein. Former names for this locus include *spoIIJ* and *spoIIF*. The *kinA* locus is located at 118° on the genetic map.

kinB Locus

The *kinB* locus consists of a transcription unit containing two open reading frames (46). Transcription of this locus appears to occur from a σ^A promoter and is terminated by a typical terminator. The first gene, *kinB*, codes for a 47,774-Da protein, showing strong similarity to the transmitter class of histidine protein kinases. The second gene in the operon codes for a protein, KapB, of 14,668 Da with no homology to any known protein, and especially no homology to the response regulator family of proteins. The function of this second protein is unknown, but it is postulated to be either required for the activity of KinB or required for the expression of this operon. The *kinB* locus maps at 278°.

THE PHOSPHORELAY SIGNAL TRANSDUCTION SYSTEM

The genes described above code for the components of the phosphorelay signal transduction system (Fig. 2). The role of the phosphorelay is to produce Spo0A-P, the activated form of the transcription factor responsible for transcription of genes involved in the initial stages of sporulation. Phosphorylation of Spo0A is accomplished through the phosphorelay initiated by the activity of two kinases, KinA and KinB. Both of these kinases may be involved in interpreting environmental signals and transducing this information into the formation of an autophosphorylated kinase protein. These kinases then transfer phosphate to the Spo0F protein, generating Spo0F-P. Spo0F serves as a secondary messenger in this system, being a substrate for KinA and KinB, and perhaps other kinases, and also serving as a substrate for the Spo0B protein. The enzymatic activity of the Spo0B enzyme is unique to the phosphorelay, and its function is to transfer phosphate from Spo0F-P to Spo0A, producing Spo0A-P. Thus, Spo0B is a phosphoprotein phosphotransferase acting in the role of a kinase to phosphorylate Spo0A. This series of reactions is somewhat more complex than a simple two-component regulatory system, perhaps reflecting the seriousness of the decision to sporulate, and its complexity may provide multiple levels of control on the production of Spo0A-P at the level both of transcription and of phosphate flow in the pathway.

REGULATION OF INFORMATION FLOW IN THE PHOSPHORELAY

The multiple points of control of the flow of phosphate through the phosphorelay begin with the kinases, KinA and KinB. Very little is known about the environmental or metabolic effectors that serve to activate either of these kinases. KinA is a cytoplasmic enzyme, with no obvious membrane-spanning regions in its primary structure, and is produced in *E. coli* expression vectors as a soluble enzyme. After a long series of experiments attempting to find either activators or inhibitors of this enzyme, it was discovered that *cis*-unsaturated fatty acids were inhibitory to the autophosphorylation activity of KinA (38). The most potent inhibitors found had at least one unsaturated double bond in the *cis* configuration and a chain length of 16 to 20 carbon atoms. *trans*-Isomers were not inhibitory. Furthermore, fatty acids, when released from the phospholipids isolated from *B. subtilis*, were also inhibitory, although the particular fatty acid causing inhibition was not characterized. It is possible that the concentration of certain specific unsaturated fatty acids may act as a signal linking the initiation of sporulation to the status of membrane synthesis and septation, or to some other membrane-associated activity. There is no evidence to prove this contention or to show that such fatty acids inhibit this enzyme in vivo. Thus, at present we are in the same position as those studying eukaryotic protein kinase C inhibition by fatty acids (9), in that the observation is solid but the physiological significance is unclear. No activators for KinA have been found.

When it was discovered that KinA mutants still sporulated although their sporulation was delayed, it was clear that there must be at least one other kinase able to activate the phosphorelay (25). Through experiments designed to identify additional histidine protein kinases, a gene for a second kinase, KinB, was uncovered which, when inactivated in a KinA background, dropped the sporulation frequency to practically zero (46). The product of the KinB gene is clearly a histidine protein kinase with very high similarity to KinA in its carboxy-terminal region. The amino-terminal portions of these two enzymes, however, are completely different, and the KinB primary structure shows six membrane-spanning regions, suggesting that the enzyme is an integral membrane protein. The *kinB* gene is associated in an operon with a gene for a small protein, KapB, of unknown function. Inactivation of *kapB* gene in a *kinA* background gives the same phenotype as inactivation of *kinB* in this background, suggesting that KapB either serves to control the expression of the *kinB* gene or is required for the function of KinB, perhaps acting as an effector binding domain (46). As with KinA, no effectors for KinB have been found and the mechanism of its activation is completely unknown.

One caveat with regard to the interpretation of the role of KinB is that it is not absolutely proven that Spo0F is the substrate for its phosphotransferase activity. Since KinB is tightly integrated into the membrane and recalcitrant to solubilization, at least so far, its enzymatic activity cannot be assayed in vitro. From genetic studies it is certain that KinB activation of the Spo0A protein requires the presence of both Spo0F and Spo0B proteins. Thus, KinB does not act directly on Spo0A but rather must have Spo0F in the chain to cause phosphate flow through the phosphorelay.

The substrate for KinA is clearly Spo0F. This kinase and Spo0F display interesting and complex kinetics; the kinase activity is activated by low levels of Spo0F and inhibited by high levels (13). Since the enzyme autophosphorylates poorly in the absence of Spo0F, it is possible that the availability of Spo0F, and perhaps the absence of free *cis*-unsaturated fatty acids, regulates its activity. Thus the possibility exists that KinA might not be subject to activation by an effector ligand acting as a signal from the environment, but rather KinA might be functionally inactive in autophosphorylation in the absence of Spo0F. This would place responsibility for the initiation of phosphate flow in the phosphorelay on transcriptional control of the *spo0F* gene.

KinA has a labile phosphatase activity, and upon purification many preparations lose this activity. It is unknown what activates the phosphatase activity of KinA on Spo0F-P, but control of this activity is another possible point of control of phosphate transfer through the phosphorelay. The phosphorylated form of Spo0F has the essential role of serving as a substrate for the Spo0B protein. Genetic studies suggest that it has no other critical role. For example, *spo0F* deletion mutants can be completely suppressed to sporulation proficiency by mutations in Spo0A that bypass the phosphorelay (17, 19). The available evidence suggests that Spo0F simply acts as a secondary messenger accumulating phosphate from these two kinases. Spo0F is subject to cross-talk at a low level by other kinases, however. In a *kinA-kinB* mutant where sporulation is completely eliminated, the regulation of the *abrB* gene shows normal kinetics. Since this regulation of *abrB* is probably the most sensitive indicator of low levels of Spo0A-P in the cell, the result suggests that Spo0F or Spo0A is subject to phosphorylation by kinases other than KinA and KinB. In a *spo0F* or *spo0B* mutant, *abrB* transcription is not regulated, indicating that in these mutants no Spo0A-P can be produced and therefore the cross-talk into this system must occur through the Spo0F response regulator.

No controls have been demonstrated on the Spo0B phosphoprotein phosphotransferase activity. Spo0B appears to be a simple enzyme transferring phosphate from Spo0F-P to Spo0A. However, its gene exists in a transcription unit with the *obg* gene, which codes for a protein clearly related to G-proteins. Obg has been shown to bind GTP, and further studies of purified Obg protein have shown that it is a GTPase and is phosphorylated by GTP but not ATP (45). Since Obg is essential for growth of the cell, it has not been possible to use mutant analysis to determine whether it has any role in the sporulation process. The association of Spo0B and Obg is strictly genetic and may be circumstantial, but it seems possible that Obg serves to regulate in some manner the activity of Spo0B or vice versa.

The role of the *spo0E* gene has been an enigma for many years. Cloning of the *spo0E* gene revealed that it coded for a protein of 9,791 Da. The mutant alleles of this gene which were sequenced were shown to be translation terminator mutations in the carboxyl third of the gene. Naturally, the assumption was made from these data that the *spo0E* gene coded for a protein whose function was required for sporulation (27). However, when a *spo0E* deletion mutation was constructed and inserted into the chromosome, there was no observable sporulation defect in such a strain (29). This surprising result suggested that the sporulation defect observed in the *spo0E* mutants was due to the carboxy-truncated Spo0E protein produced. Close observation of colonies of strains containing the *spo0E* deletion revealed that such strains segregated strains with spontaneous secondary mutations. The strains bearing these secondary mutations had the phenotype of *spo0* mutants, which implied that deletion of the *spo0E* gene relieved a negative regulatory function with the effect of activation of the phosphorelay at inappropriate times. The physiological consequences of activating the phosphorelay are detrimental to growth; therefore, secondary mutations in phosphorelay components occur which prevent these effects. Thus, Spo0E must have some negative regulatory role, but it is not clear at what point in the phosphorelay Spo0E acts. This conclusion is consistent with the observation that multiple copies of the *spo0E* gene, presumably leading to overproduction of its gene product, result in a sporulation-defective phenotype (29). Determination of the site of action of Spo0E has been hampered by inability to produce this protein in any quantity in either *E. coli* or *B. subtilis* vectors. We speculate that Spo0E is a phosphatase for the Spo0F-P molecule and that the effect of carboxy truncation is to render the phosphatase constitutively active. This interpretation labors under the premise that the carboxy terminus has some role in the regulation of phosphatase activity, perhaps in binding yet another effector molecule. This, of course, is pure speculation.

We probably have only touched the surface of the possible points of flow regulation in the phosphorelay. Still unknown are the effectors that activate either KinA or KinB, and so it is not clear what environmental signals, if any, activate the phosphorelay. Oligopeptide transport mutants, *opp*, are sporulation defective, suggesting that some peptide transported into the cell from outside has a role in either control of phosphate flow or transcription. Perhaps there are peptides, or molecules with some peptide structure that can be recognized by the Opp system, that are involved in both cell-cell communication and sporulation as has been suggested (14, 26, 32). How Spo0E activity is controlled remains a mystery, and perhaps other controls exist on Spo0B or even directly on Spo0A-P. Environmental signal input may also regulate the activity of transition state regulators, such as

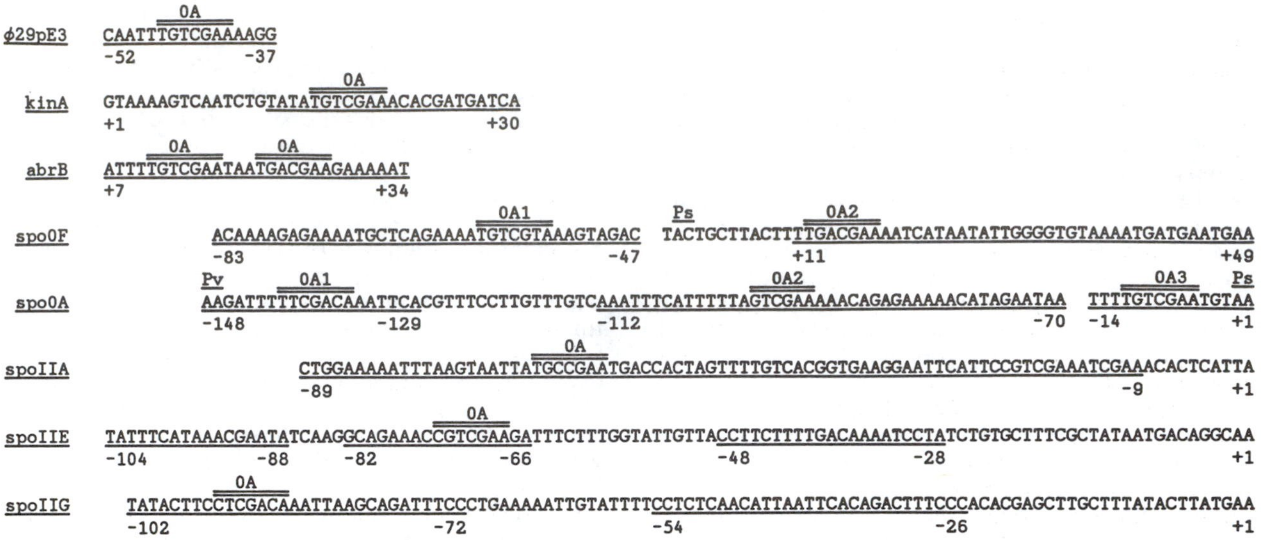

```
φ29pE3   CAATTTGTCGAAAAGG
         -52 ... -37
kinA     GTAAAAGTCAATCTGTATATGTCGAAACACGATGATCA
         +1 ... +30
abrB     ATTTTGTCGAATAATGACGAAGAAAAAT
         +7 ... +34
spo0F    ACAAAAGAGAAAATGCTCAGAAAATGTCGTAAAGTAGAC   TACTGCTTACTTTTGACGAAAATCATAATATTGGGGTGTAAAATGATGAATGAA
         -83 ... -47                               +11 ... +49
spo0A    AAGATTTTTCGACAAATTCACGTTTCCTTGTTTGTCAAATTTCATTTTTAGTCGAAAAACAGAGAAAAACATAGAATAA   TTTTGTCGAATGTAA
         -148 ... -129 ... -112 ... -70                                                  -14 ... +1
spoIIA   CTGGAAAAATTTAAGTAATTATGCCGAATGACCACTAGTTTTGTCACGGTGAAGGAATTCATTCCGTCGAAATCGAAACACTCATTA
         -89 ... -9 ... +1
spoIIE   TATTTCATAAACGAATATCAAGGCAGAAACCGTCGAAGATTTCTTTGGTATTGTTACCTTCTTTTGACAAAATCCTATCTGTGCTTTCGCTATAATGACAGGCAA
         -104 ... -88 ... -82 ... -66 ... -48 ... -28 ... +1
spoIIG   TATACTTCCTCGACAAATTAAGCAGATTTCCCTGAAAAATTGTATTTTCCTCTCAACATTAATTCACAGACTTTCCCACACGAGCTTGCTTTATACTTATGAA
         -102 ... -72 ... -54 ... -26 ... +1
```

Labels above sequences: φ29pE3: 0A; kinA: 0A; abrB: 0A, 0A; spo0F: 0A1, Ps, 0A2; spo0A: Pv, 0A1, 0A2, 0A3, Ps; spoIIA: 0A; spoIIE: 0A; spoIIG: 0A.

Figure 3. Spo0A binding sites in promoters. The underlined regions are the extent of Spo0A protection from DNase I treatment in footprint experiments. Double-overlined regions designate 0A boxes that are equal or close to the consensus site, TGNCGAA. Ps and Pv are sporulation (σ^H) dependent and vegetative (σ^A) dependent promoters, respectively.

Hpr, AbrB, and Sin, that have transcriptional effects on the phosphorelay. What is certain is that regulation of this pathway is very complex and very finely tuned to a variety of possible effectors derived from the environment and the metabolism of the cell.

AUTOREGULATION AND POSITIVE FEEDBACK LOOPS IN TRANSCRIPTIONAL CONTROL OF THE PHOSPHORELAY

Transcriptional control of the genes for components of the phosphorelay plays an important role in the control of phosphate flow through this system. The product of this signal transduction pathway can be thought of as Spo0A~P, which serves as a major transcription controller for the genes encoding the components of the pathway in a variation of autoregulation. Under conditions of nutrient excess where the cells are growing exponentially and there is no reason to initiate sporulation, the level of Spo0A~P in the cell is thought to be exceedingly low. This results in high-level expression of Spo0A~P repressed genes including the transition state regulator gene, *abrB*, which acts as a negative regulator of many genes, but in this context its most important effects are on the *spo0E* gene (29) and on the *spo0H* gene coding for σ^H (49). Thus, production of σ^H, which is necessary, but not sufficient for, the transcription of a variety of genes, including the *spo0A*, *spo0F*, and *kinA* genes (31), is directly controlled by the *abrB* concentration in the cell, which, in turn, reflects the Spo0A~P concentration. Maintenance levels of *spo0A* and *spo0F* transcription are provided by σ^A promoters insensitive to the effects of Spo0A~P.

As nutritional conditions in the culture deteriorate and environmental alterations dictate the end of exponential growth, low levels of Spo0A~P are produced from the maintenance level of phosphorelay components present. The promoter most sensitive to this low level of Spo0A~P is that of the *abrB* gene. Transcription of *abrB* is repressed by Spo0A~P (36), causing the AbrB concentration to fall and releasing repression of many genes including the *spo0E* and *spo0H* genes. This mechanism provides the sigma factor required for high-level transcription of *spo0F*, *spo0A*, and *kinA* during the transition state (31).

Transcription of the *spo0A* gene appears to occur by the following sequence. During exponential growth *spo0A* transcription occurs from the σ^A, or P_V, promoter. Accumulation of low levels of Spo0A~P inhibit transcription from this promoter, presumably due to Spo0A~P binding at the *spo0A* box that is located just downstream of the start site of this promoter (0A1; Fig. 3). This is a direct form of autoregulation where Spo0A~P controls the synthesis of Spo0A. High-level synthesis of Spo0A seems to be dependent on the P_S, σ^H-dependent promoter located downstream of P_V (6). This promoter cannot function during vegetative growth, presumably because it is catabolite repressed (6, 48, 50) and σ^H levels are low, although several other repressive mechanisms may also play a role here. As these vegetative repressive mechanisms are lifted from the P_S promoter, transcription from P_V is repressed by the accumulating Spo0A~P. At the same time, activation of transcription occurs from P_S. This is believed to be due to binding of Spo0A~P to one or the other, or both (0A1 and 0A2; Fig. 3), of the *spo0A* boxes upstream of the P_S promoter (39). In addition, there exists an 0A box covering the P_S promoter itself (0A3; Fig. 3) from approximately −10 to −30. We believe that this 0A box serves as a second repression site where Spo0A~P exerts a repressive effect when its concentration reaches a sufficiently high level (39). Thus, the *spo0A* promoters appear to be subject to autoregulatory controls superimposed on outside controls by catabolite repression and probably other vegetative regulators.

The cell can't produce Spo0A~P without the help of

Spo0F, and Spo0A~P controls *spo0F* transcription. In the *spo0F* promoter, transcription from the P_S (σ^H) promoter requires an upstream *spo0A* binding site (0A1; Fig. 3) (40). Spo0A~P is a positive regulator of this gene, and this activation appears to be modulated by an 0A-binding site downstream of the start site of transcription (0A2; Fig. 3). Thus, Spo0A~P regulates its own synthesis through a positive feedback loop by producing more of the components required for its own synthesis (40).

The genes for the other components of the phosphorelay are less dramatic in their regulation. The *spo0B* operon is derepressed in *spo0A* or *spo0B* mutants, although no detailed studies of its regulation have been undertaken (10). The *kinA* gene is induced at the end of growth (1, 31), and the presence of an 0A box downstream of the start site of transcription where Spo0A binds in vitro (47) suggests that Spo0A~P may modulate this induction (Fig. 3). Nothing is known of the regulation of *kinB* other than that it is transcribed from a σ^A-type promoter.

Although the role of Spo0A~P is emphasized here, one should not lose sight of an abundance of other possible controls on transcription. Catabolite repression is very important in the regulation of the *spo0A* gene, and this may occur by regulatory proteins binding at the P_S promoter (6, 48, 50). Transition state regulators Hpr and Sin are known to be inhibitors of sporulation when overexpressed, and both of these proteins bind specifically to promoters (12, 18, 28). It would not be surprising to find these proteins involved in the regulation of phosphorelay component genes.

DNA-BINDING SPECIFICITY OF Spo0A~P

Spo0A binds to the promoters it regulates by recognizing a 7-bp "0A box," TGNCGAA (where N can be any base, but usually is A or T [36]). The 0A box exists in all promoters known to be controlled by Spo0A~P and within the areas that are footprinted by this protein in vitro (Fig. 3). Although both the phosphorylated and the nonphosphorylated forms of Spo0A bind to promoters containing the 0A box, in the case of the *abrB* gene it has been shown that the phosphorylated form has an apparent 20- to 50-fold higher affinity of binding than was observed for the nonphosphorylated form (41). The AbrB promoter is very sensitive to extremely low levels of Spo0A~P concentration within the cell. It is possible that the tandem repeat of 0A boxes separated by one helical turn of the DNA duplex that is unique to this promoter may sequester more than one Spo0A~P molecule, and perhaps protein-protein interactions help to stabilize the complex of Spo0A~P and DNA. This argument could explain the sensitivity of the *abrB* promoter to Spo0A~P without invoking contextual arguments or the existence of an additional regulator. However, there is no experimental evidence that Spo0A~P forms a stable dimer in solution in the presence or absence of DNA.

Several mutants in the three aspartic residues known to make up the aspartic pocket where the Spo0A protein is phosphorylated have been made and tested for their ability to be phosphorylated in vitro. Mutations D10N and/or D56N completely abolish the capacity of the protein to be phosphorylated directly by KinA and by Spo0B in the phosphorelay reaction (4). Purified preparations of such mutant proteins bind and footprint as well as unphosphorylated Spo0A to the *abrB* promoter (39). Both of these mutants have been shown to be ineffective as repressors of the *abrB* gene in vivo, suggesting strongly that even though Spo0A itself binds to the promoters and to 0A boxes in vitro, it is only the phosphorylated form that is the effective species in vivo (39). There may be substantial context differences between the types of promoters in which Spo0A represses transcription and those in which it activates (Fig. 3). Binding of Spo0A to the AbrB promoter is clean, and small regions are footprinted in vitro (36). The same is true for the ϕ29pE3 promoter, where Spo0A footprints on an exceedingly tight region of the promoter (37). On the other hand, promoters that are activated by the presence of Spo0A~P show larger, more diffuse regions of binding. Three examples are the SpoIIA (41), SpoIIE (52), and SpoIIG (33) promoters. In all of these promoters there are very large regions of the promoter that are protected from DNase I in in vitro footprinting studies. In the *spoIIE* and *spoIIG* promoters, Spo0A binding to the −35 region may facilitate interaction of σ^A containing RNA polymerase with the promoters (33, 52). Footprint analyses with the combination of Spo0A~P, RNA polymerase, and *spoIIG* promoter are consistent with this conclusion, since a stable footprint is obtained with the ternary complex of DNA-Spo0A~P-RNA polymerase, whereas only weak and variable footprints are found with DNA and RNA polymerase (2). Furthermore, the large differences in apparent affinity between phosphorylated and nonphosphorylated Spo0A are not observed with the *spoIIG* promoter, consistent with a requirement for the ternary complex to obtain maximal binding (2). It has been shown that mutations in the carboxy-terminal end of the Spo0A protein, which appear not to affect its ability to repress *abrB*, nevertheless do prevent activation of the *spoIIA* promoter, suggesting that the carboxy-terminal portion of the Spo0A protein is involved in interaction with the transcription complex (30).

PERSPECTIVES ON SPORULATION INITIATION

Growth and sporulation are simply two means by which *Bacillus* and other sporeformers maintain their competitive edge and ensure survival under all environmental conditions. In ordinary laboratory media, we observe both growth and sporulation and are able to quantify both processes. We name the period in between the "transition state" and often consider it only a gear-shifting period before sporulation. In fact, the transition state in nature plays a much more important role and should be viewed as being as important as sporulation in conferring a competitive advantage to the organism. It is during the transition state that the organism excretes copious quantities of proteases, amylases, and other enzymes to degrade complex substrates in order to continue growth as long as possible, albeit at a slower rate. At the same time, the production of antibiotics and other bioactive secondary metabolites by *B. subtilis* ensures that other bacteria or fungi are prevented from invading its ecological niche. It's likely that *B. subtilis* and its relatives in their natural environment spend more time in the transition state than in exponential growth.

It should be clear that many regulatory pathways play some role, important or tangential, in the cellular conversion from exponential growth to transition state and in control of the initiation of sporulation. We are capable of observing what pathways are important only in the laboratory under artificial conditions. Thus, the *kinA* pathway to phosphorylate Spo0F appears to predominate in the laboratory, but could be insignificant under some natural conditions. Mutations are double-edged swords. The *spo0* mutants have allowed us to define the signal transduction pathway to transcription activation for both the transition state and the onset of sporulation. On the other hand, is the phenotype of a *spo0A* mutant representative of any growth phase of the organism in nature? Are conditions ever so favorable that the cellular level of Spo0A~P approaches zero? The answer may be yes, but probably only in early exponential growth at low cell densities in rich media in the laboratory. Several other regulators such as AbrB, Hpr, Sin, Sen, Pai, and others may have important roles and are described in other chapters of this book. Overproduction of the Spo0E, Hpr, or Sin proteins prevents sporulation. Are these regulators ever this active in nature? Maybe, but we don't know for sure.

It seems most reasonable to view sporulation as a failed transition state. The activation of genes (*spoIIA* and *spoIIE*) for sigma factors required in the morphological processes unique to sporulation (7, 22) is an indication that the cell has sensed a hostile environment and has given up on growth. The role of transition state regulators such as AbrB, Hpr, Sin, and probably others, is to prevent this from happening as long as the compounds that regulate them are still in sufficient quantity to signal growth rather than sporulation. Again, one or another of these regulators may be important only under certain limited environmental states. Thus, control of the flow of phosphate through the phosphorelay to Spo0A may be exceedingly complex, utilizing all the aforementioned regulators, or the redundancy of regulators may simply reflect a strategic position to take advantage of all possible means to promote growth and prevent sporulation.

Finally, the mechanism by which the cell coordinates chromosome replication and segregation with septation and sporulation remains a mystery. These physical and morphological constraints to the transcription of sporulation genes must be interpreted in some manner, possibly by control of the phosphorelay and, ultimately, the cellular level of Spo0A~P. One possible cell cycle regulator of this type might be the essential G-protein Obg (4) or others as yet undiscovered.

The discovery that the *spo0J* locus codes for two proteins with homology to proteins implicated in control of chromosome segregation may be providing the first insight into this problem (23, 24). The sequence of the origin region of the chromosome (24) has revealed 0A boxes in the origin and in the promoter for the *dnaA* gene (16). It seems possible that one role of increased Spo0A~-P concentration during the initial stages of sporulation might be to prevent reinitiation of chromosome replication, which could be accomplished by preventing *dnaA* transcription and/or binding at the origin to inhibit the reinitiation of DNA replication. On the other hand, maybe it is chromosome segregation rather than initiation that is affected by Spo0A~P. It is not difficult to imagine Spo0A~P serving as a cosegregation factor directing chromosome segregation to the forespore, either by facilitating interaction with sporulation-specific partition functions or by preventing interaction with similar vegetative functions.

Acknowledgments. This manuscript was supported in part by grant GM-19416 from the National Institute of General Medical Sciences. This is paper 7599-MEM from the Department of Molecular and Experimental Medicine, The Scripps Research Institute.

REFERENCES

1. **Antoniewski, C., B. Savelli, and P. Stragier.** 1990. The *spoIIJ* gene, which regulates early developmental steps in *Bacillus subtilis*, belongs to a class of environmentally responsive genes. *J. Bacteriol.* **172:**86–93.
2. **Bird, T., and G. Spiegelman.** Unpublished data.
3. **Bouvier, J., P. Stragier, C. Bonamy, and J. Szulmajster.** 1984. Nucleotide sequence of the *spo0B* gene of *Bacillus subtilis* and regulation of its expression. *Proc. Natl. Acad. Sci. USA* **81:**7012–7016.
4. **Burbulys, D., K. A. Trach, and J. A. Hoch.** 1991. The initiation of sporulation in *Bacillus subtilis* is controlled by a multicomponent phosphorelay. *Cell* **64:**545–552.
5. **Cavanagh, J., N. Skelton, T. Tucker, J. A. Hoch, and J. M. Whiteley.** Unpublished data.
6. **Chibazakura, T., F. Kawamura, and H. Takahashi.** 1991. Differential regulation of *spo0A* transcription in *Bacillus subtilis*: glucose represses promoter switching at the initiation of sporulation. *J. Bacteriol.* **173:**2625–2632.
7. **Driks, A., and R. Losick.** 1991. Compartmentalized expression of a gene under the control of sporulation transcription factor sigma^{E} in *Bacillus subtilis*. *Proc. Natl. Acad. Sci. USA* **88:**9934–9938.
8. **Dubnau, E., J. Weir, G. Nair, L. Carter III, C. Moran, Jr., and I. Smith.** 1988. *Bacillus* sporulation gene *spo0H* codes for sigma-30 (sigma-H). *J. Bacteriol.* **170:**1054–1062.
9. **El Touny, S., W. Kahn, and Y. Hannun.** 1990. Regulation of platelet protein kinase C by oleic acid. *J. Biol. Chem.* **265:**16437–16443.
10. **Ferrari, F. A., K. Trach, and J. A. Hoch.** 1985. Sequence analysis of the *spo0B* locus reveals a polycistronic transcription unit. *J. Bacteriol.* **161:**556–562.
11. **Ferrari, F. A., K. Trach, D. LeCoq, J. Spence, E. Ferrari, and J. A. Hoch.** 1985. Characterization of the *spo0A* locus and its deduced product. *Proc. Natl. Acad. Sci. USA* **82:**2647–2651.
12. **Gaur, N. K., J. Oppenheim, and I. Smith.** 1991. The *Bacillus subtilis sin* gene, a regulator of alternate developmental processes, codes for a DNA-binding protein. *J. Bacteriol.* **173:**678–686.
13. **Grimshaw, C. E., C. G. Hanstein, M. A. Strauch, D. Burbulys, J. A. Hoch, and J. M. Whiteley.** Unpublished data.
14. **Grossman, A. D., and R. Losick.** 1988. Extracellular control of spore formation in *Bacillus subtilis*. *Proc. Natl. Acad. Sci. USA* **85:**4369–4373.
15. **Hahn, J., and D. Dubnau.** 1991. Growth stage signal

transduction and the requirements for *srfA* induction in development of competence. *J. Bacteriol.* **173:**7275–7282.
16. **Hoch, J. A.** Unpublished data.
17. **Hoch, J. A., K. Trach, F. Kawamura, and H. Saito.** 1985. Identification of the transcriptional suppressor *sof-1* as an alteration in the *spo0A* protein. *J. Bacteriol.* **161:**552–555.
18. **Kallio, P. T., J. E. Fagelson, J. A. Hoch, and M. A. Strauch.** 1991. The transition state regulator Hpr of *Bacillus subtilis* is a DNA-binding protein. *J. Biol. Chem.* **266:**13411–13417.
19. **Kawamura, F., and H. Saito.** 1983. Isolation and mapping of a new suppressor mutation of an early sporulation gene *spo0F* mutation in *Bacillus subtilis. Mol. Gen. Genet.* **192:**330–334.
20. **Kudoh, J., T. Ikeuchi, and K. Kurahashi.** 1985. Nucleotide sequences of the sporulation gene *spo0A* and its mutant genes of *Bacillus subtilis. Proc. Natl. Acad. Sci. USA* **82:**2665–2668.
21. **Lewandoski, M., E. Dubnau, and I. Smith.** 1986. Transcriptional regulation of the *spo0F* gene of *Bacillus subtilis. J. Bacteriol.* **168:**870–877.
22. **Losick, R., and P. Stragier.** 1992. Crisscross regulation of cell-type-specific gene expressing during development in *B. subtilis. Nature* (London) **355:**601–604.
23. **Mysliwiec, T. H., J. Errington, A. B. Vaidya, and M. G. Bramucci.** 1991. The *Bacillus subtilis spo0J* gene: evidence for involvement in catabolite repression of sporulation. *J. Bacteriol.* **173:**1911–1919.
24. **Ogasawara, N., and H. Yoshikawa.** 1992. Genes and their organization in the replication origin region of the bacterial chromosome. *Mol. Microbiol.* **6:**629–634.
25. **Perego, M., S. P. Cole, D. Burbulys, K. Trach, and J. A. Hoch.** 1989. Characterization of the gene for a protein kinase which phosphorylates the sporulation-regulatory proteins Spo0A and Spo0F of *Bacillus subtilis. J. Bacteriol.* **171:**6187–6196.
26. **Perego, M., C. F. Higgins, S. R. Pearce, M. P. Gallagher, and J. A. Hoch.** 1991. The oligopeptide transport system of *Bacillus subtilis* plays a role in the initiation of sporulation. *Mol. Microbiol.* **5:**173–185.
27. **Perego, M., and J. A. Hoch.** 1987. Isolation and sequence of the *spo0E* gene: its role in initiation of sporulation in *Bacillus subtilis. Mol. Microbiol.* **1:**125–132.
28. **Perego, M., and J. A. Hoch.** 1988. Sequence analysis and regulation of the *hpr* locus, a regulatory gene for protease production and sporulation in *Bacillus subtilis. J. Bacteriol.* **170:**2560–2567.
29. **Perego, M., and J. A. Hoch.** 1991. Negative regulation of *Bacillus subtilis* sporulation by the *spo0E* gene product. *J. Bacteriol.* **173:**2514–2520.
30. **Perego, M., J.-J. Wu, G. B. Spiegelman, and J. A. Hoch.** 1991. Mutational dissociation of the positive and negative regulatory properties of the Spo0A sporulation transcription of *Bacillus subtilis. Gene* **100:**207–212.
31. **Predich, M., G. Nair, and I. Smith.** 1992. *Bacillus subtilis* early sporulation genes *kinA*, *spo0F*, and *spo0A* are transcribed by the RNA polymerase containing σ^H. *J. Bacteriol.* **174:**2771–2778.
32. **Rudner, D. Z., J. R. Ladeaux, K. Breton, and A. D. Grossman.** 1991. The *spo0K* locus of *Bacillus subtilis* is homologous to the oligopeptide permease locus and is required for sporulation and competence. *J. Bacteriol.* **173:**1388–1398.
33. **Satola, S. W., J. M. Baldus, and C. P. Moran, Jr.** 1992. Binding of Spo0A stimulates *spoIIG* promoter activity in *Bacillus subtilis. J. Bacteriol.* **174:**1448–1453.
34. **Stock, A. M., J. M. Mottonen, J. B. Stock, and C. E. Schutt.** 1989. Three-dimensional structure of CheY, the response regulator of bacterial chemotaxis. *Nature* (London) **337:**745–749.
35. **Stock, J. B., A. J. Ninfa, and A. M. Stock.** 1989. Protein phosphorylation and regulation of adaptive response in bacteria. *Microbiol. Rev.* **53:**450–490.
36. **Strauch, M., V. Webb, G. Spiegelman, and J. A. Hoch.** 1990. The Spo0A protein of *Bacillus subtilis* is a repressor of the *abrB* gene. *Proc. Natl. Acad. Sci. USA* **87:**1801–1805.
37. **Strauch, M. A.** Unpublished data.
38. **Strauch, M. A., D. de Mendoza, and J. A. Hoch.** 1992. *cis*-Unsaturated fatty acids specifically inhibit a signal-transducing protein kinase required for initiation of sporulation in *Bacillus subtilis. Mol. Microbiol.* **6:**2909–2917.
39. **Strauch, M. A., K. Trach, J. Day, and J. A. Hoch.** 1992. Spo0A activates and represses its own synthesis by binding at its dual promoters. *Biochimie* **74:**619–626.
40. **Strauch, M. A., J.-J. Wu, R. H. Jonas, and J. A. Hoch.** A positive feedback loop controls transcription of the *spo0F* gene, a component of the sporulation phosphorelay in *Bacillus subtilis. Mol. Microbiol.*, in press.
41. **Trach, K., D. Burbulys, M. Strauch, J.-J. Wu, N. Dhillon, R. Jonas, C. Hanstein, P. Kallio, M. Perego, T. Bird, G. Spiegelman, C. Fogher, and J. A. Hoch.** 1991. Control of the initiation of sporulation in *Bacillus subtilis* by a phosphorelay. *Res. Microbiol.* **142:**815–823.
42. **Trach, K., J. W. Chapman, P. J. Piggot, and J. A. Hoch.** 1985. Deduced product of the stage 0 sporulation gene *spo0F* shares homology with the Spo0A, OmpR and SfrA proteins. *Proc. Natl. Acad. Sci. USA* **82:**7260–7264.
43. **Trach, K., J. W. Chapman, P. Piggot, D. LeCoq, and J. A. Hoch.** 1988. Complete sequence and transcriptional analysis of the *spo0F* region of the *Bacillus subtilis* chromosome. *J. Bacteriol.* **170:**4194–4208.
44. **Trach, K., and J. A. Hoch.** 1989. The *Bacillus subtilis spo0B* stage 0 sporulation operon encodes an essential GTP-binding protein. *J. Bacteriol.* **171:**1362–1371.
45. **Trach, K. A., C. Folger, and J. A. Hoch.** Unpublished data.
46. **Trach, K. A., and J. A. Hoch.** Multisensory activation of the phosphorelay initiating sporulation in *Bacillus subtilis*: identification and sequence of the protein kinase of the alternate pathway. *Mol. Microbiol.*, in press.
47. **Trach, K. A., M. Strauch, and J. A. Hoch.** Unpublished data.
48. **Weickert, M. J., and G. H. Chambliss.** 1990. Site-directed mutagenesis of a catabolite repression operator sequence in *Bacillus subtilis. Proc. Natl. Acad. Sci. USA* **87:**6238–6242.
49. **Weir, J., M. Predich, E. Dubnau, G. Nair, and I. Smith.** 1991. Regulation of *spo0H*, a gene coding for the *Bacillus subtilis* σ^H factor. *J. Bacteriol.* **173:**521–529.
50. **Yamashita, S., F. Kawamura, H. Yoshikawa, H. Takahashi, Y. Kobayashi, and H. Saito.** 1989. Dissection of the expression signals of the *spo0A* gene of *Bacillus subtilis*: glucose represses sporulation-specific expression. *J. Gen. Microbiol.* **135:**1335–1345.
51. **Yamashita, S., H. Yoshikawa, F. Kawamura, H. Takahashi, T. Yamamoto, Y. Kobayashi, and H. Saito.** 1986. The effect of *spo0* mutations on the expression of *spo0A*- and *spo0F-lacZ* fusions. *Mol. Gen. Genet.* **205:**28–33.
52. **York, K., T. J. Kenney, S. Satola, C. P. Moran, Jr., H. Poth, and P. Youngman.** 1992. Spo0A controls the σ^A-dependent activation of *Bacillus subtilis* sporulation-specific transcription unit *spoIIE. J. Bacteriol.* **174:**2648–2658.

52. AbrB, a Transition State Regulator

MARK A. STRAUCH

The end of exponential growth in a bacterial culture is a time when many genes whose expression was repressed by conditions conducive to growth become induced. Several of these genes code for pathways of carbon and nitrogen utilization that will dominate the energy metabolism of the cell during a protracted period in stationary phase. The transition state between growth and stationary phases may be viewed as a crossroads where the cell is still expressing some growth-related functions but has also begun to express new gene products necessary for its survival in an increasingly hostile environment.

In *Bacillus* spp., some of the functions normally expressed only during the transition state include development of competence for the uptake of DNA molecules, production of antibiotics, synthesis of flagella, and synthesis of a wide variety of extracellular enzymes. While most of these transition state functions are deficient in mutants blocked in the earliest stage of sporulation by *spo0* mutants, many other regulatory genes (reviewed in reference 38 and chapter 54) have pleiotropic effects on these functions and so have been termed "transition state regulators." None of the known mutations in these regulators lead to a significant s/porulation defect, but the regulators are involved in coordinating and transmitting signals received by the sporulation-sensing network (i.e., *spo0* genes) to bring about regulation of the transition state functions. One of these regulators is the product of the *abrB* gene, which not only controls transition state genes that are unnecessary for the sporulation response but also regulates essential sporulation genes. In this manner, the *abrB* gene functions to coordinate the sporulation pathway with other transition state processes.

ISOLATION AND CHARACTERIZATION OF *abrB* MUTANTS

Initial efforts to dissect the pleiotropic nature of *spo0* mutations were focused on the isolation of mutants in which any or all of the phenotypes were reverted. By selection for resistance to polymyxin or to the antibiotic produced by wild-type cells (traits lost by *spo0* mutants), suppressors that, while not restoring the ability to sporulate, did cause reversion of several of the other phenotypes examined (such as transformability, protease production, regulation of nitrate reductase, and antibiotic production) were isolated (15). The vast majority of these suppressors were found to map at a single locus, termed *cpsX* (14), that was distinct from any known *spo0* gene. Similar suppressors obtained by selection for resistance to antibiotic or to infection by bacteriophages ϕ2 and ϕ15 were also obtained and were called *abs*, *tol*, and *abr* (18, 19, 45). Genetic analysis of *abr* mutations revealed that they mapped at four distinct loci, with the majority being located near the origin of chromosome replication at the locus termed *abrB* (46). In the same study, the *cpsX*, *abs*, and *tol* mutations were also found to map at the *abrB* locus. These findings suggest that the product of the *abrB* locus is a major determinant of the pleiotropic effects of *spo0* mutations. When this locus was cloned and sequenced (see below), it was found to consist of a single polypeptide-encoding gene, and the nature of the *abrB* mutations indicated that they resulted in loss of function of the protein.

None of the known *abrB* mutations had a detectable phenotype in spo^+ cells. Rather, their effects were seen only in the presence of *spo0* mutations. While it had been established that *abrB* mutations suppressed many *spo0*-associated phenotypes, nothing was known concerning the mechanisms by which they did so. Studies of isolated ribosomes from *spo0* versus *spo0 abrB* mutants indicated that *abrB* led to alterations or deficiencies in one or more ribosomal proteins (32, 47). This suggested that *spo0* mutants might not be able to translate certain mRNAs but could regain the ability to do so because the *abrB* mutations altered the ribosomes. However attractive this hypothesis was, studies examining the expression of the *spoVG* and *tycA* genes (22, 51) proved that AbrB acted primarily at the transcriptional level. In both cases, *abrB* mutations were able to restore the block in transcription caused by *spo0* mutations, and it was postulated that the role of the *spo0* gene products was to somehow inactivate the activity or synthesis of the AbrB protein. (What effect, if any, the *abrB*-induced alteration of ribosome structure has on the cell remains a mystery.) To elucidate the molecular mechanisms involved in AbrB-mediated regulation of transcription and the role of *spo0* gene products in its inactivation, the transcription of other *abrB* targets was examined, the *abrB* gene was cloned, and the gene's protein product was purified.

THE *abrB* LOCUS

Nucleotide sequence analysis revealed that the *abrB* locus consists of a single gene coding for a protein of 96 amino acids with a calculated molecular weight of 10,773 (29). The first three residues of the predicted protein are Met-Phe-Met, but another study (40) of the AbrB protein produced from an expression vector in *Escherichia coli* revealed that the amino terminus of the purified protein begins at the second methionine

Mark A. Strauch • Division of Cellular Biology, Department of Molecular and Experimental Medicine, The Scripps Research Institute, 10666 North Torrey Pines Road, La Jolla, California 92037.

residue. While this seems to indicate that translation initiates at the second Met, the possibility remains that the protein begins at the first Met and is post-translationally processed by the *E. coli* host. It is not known at which Met codon the protein initiates when it is translated in *Bacillus subtilis*. The deduced amino acid sequence of AbrB is not significantly similar to those of any other known proteins.

The *abrB* gene is transcribed by two promoters whose start sites are separated by 14 bp (29). The significance of dual promoters is unclear, but as will be discussed below, the two promoters are regulated differentially. Transformation analysis using defined fragments of the *abrB* region was used to separate *abrB* mutations into two classes (29). Approximately half of those tested were found to be located upstream of the coding sequence, with the other half either within the coding region or downstream of it. Four of the former class were analyzed in detail and found to contain mutations in the region of the dual promoters. One such promoter region mutation (*abrB15*) was shown to eliminate transcription from the downstream promoter but not from the upstream one. This suggested that the *abrB* phenotype could arise from lowering the level of *abrB* expression. To date, two mutations of the second class have been sequenced. One (*abrB4*) changes the Cys residue at position 54 to a Tyr; the other, *abrB703*, (50, 51) is a frameshift mutation produced by a single-base deletion that fuses the first 57 codons of *abrB* to as many as 53 downstream codons before the message is terminated. Thus, the AbrB phenotype is the result either of a lowering of the expression of the wild-type protein or of the production of an altered, presumably nonfunctional protein.

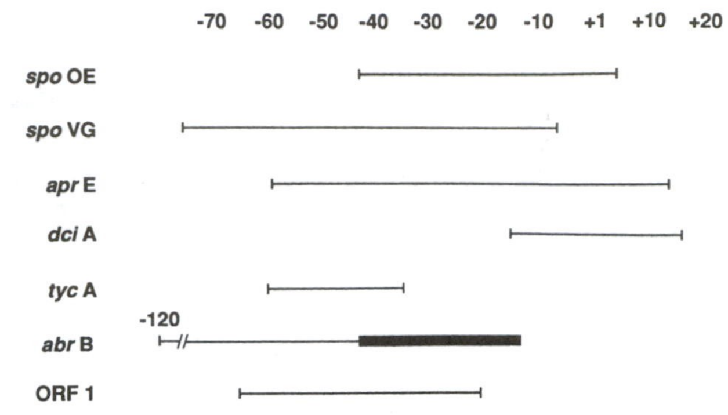

Figure 1. AbrB binding regions on seven negatively controlled genes. The extents and locations of AbrB-afforded protection from DNase I cleavage are shown relative to the start points of transcription (+1). In the cases of *abrB* and ORF1, two promoters transcribe the gene. Binding on *abrB* is shown relative to the P2 promoter; the P1 start site would be at −14 relative to P2. Binding on ORF1 is shown relative to the P2 promoter; the P1 start site in this case would be at −71. There are two transcriptional start sites located 3 bp apart for the *spo0E* gene, but it is not known whether these have common −10 and −35 elements (28). The heavy line from −14 to −43 in the *abrB*-protected region indicates a higher-affinity binding site (40). An additional binding region on the *tycA* gene from +169 to +231 is required for AbrB binding to and repression of the promoter-located site (10).

DNA-BINDING PROPERTIES OF AbrB PROTEIN

Purified AbrB is a hexamer of identical subunits (39, 40) that specifically binds the promoter regions of genes it is known to control (30, 40). AbrB contains a region of slight similarity to the helix-turn-helix motif found in the binding regions of several regulatory proteins (5). However, introduction of mutations expected to disrupt the structure of this putative helix-turn-helix motif resulted in mutant proteins whose abilities to bind a target DNA fragment in vitro were not altered (10). Additionally, AbrB does not contain any other known DNA-binding motif. Nevertheless, AbrB binding to DNA is specific and highly cooperative (9, 40). This cooperativity no doubt reflects the need for the cell to keep the AbrB concentration at a low but effective regulatory level (see below). In addition, AbrB appears to bind in discrete steps, since for any given subsaturating concentration of protein, only a single species of bound DNA is observed in gel retardation experiments. Each step may represent a partially occupied binding region (although this has not been proven) and seems to suggest that binding is occurring progressively and cooperatively from an initial nucleation site. However, DNase I protection (footprinting) experiments do not reveal a progressive increase in the length of the region protected with increasing AbrB concentration. Rather, the binding is all or none in appearance over a relatively large region (30 to 120 nucleotides, depending on the promoter; Fig. 1).

A possible explanation for the apparent discrepancy between the results of gel retardation and DNase footprinting experiments is that AbrB initially binds with similar affinity to several independent sites but that subsequent binding of additional AbrB molecules is cooperative. Each step in the mobility shift experiments might represent a given number of AbrB molecules bound per DNA, although the locations would vary from DNA molecule to DNA molecule. To see complete protection of a given region of DNA, the majority of molecules in the sample population must have the protein bound at that site. If not all the DNA molecules have the protein bound at the same site, then only a partial protection at any given site will be seen when the population as a whole is examined (as it is in footprinting procedures). The partial protection would extend over the same length as the fully protected region if the region were composed of multiple identical sites. This is basically what has been observed for AbrB binding, although the transitions from unprotected to partially protected to fully protected are quite abrupt (possibly due to cooperativity of binding) and not easily distinguishable in footprinting experiments. However, the critical feature of this hypothesis, the existence of multiple, identical binding sites in the footprint regions, has not yet been proven.

The binding of AbrB to the promoter regions of the *aprE*, *spo0E*, *abrB*, *dciA*, *spoVG*, *tyrA*, and *sin* operons has been demonstrated by DNase I footprinting (9, 33, 37, 40). As mentioned above, relatively large regions were protected. In each case, these regions include one or more crucial promoter elements (−35, −10, and +1). However, the locations of the upstream or downstream boundaries of the binding regions vary from promoter to promoter (Fig. 1). With the exception of the *tycA* promoter (9), only a single protected region was observed, no matter how large. The *tycA* promoter (from *Bacillus brevis*) has two adjacent protected sites well downstream of the transcription start site (+169 to +199 and +207 to +231) in addition to a protected site in the promoter region from −35 to −60. This last site appeared to have less affinity for AbrB binding in vitro and apparently depended on the integrity of the +169 to +231 binding region. In fact, deletion of the downstream sites abolished AbrB repression of *tycA* transcription in vivo (9). It is not known whether binding to the downstream sites alone is sufficient to cause repression. It is possible that binding at these sites serves to facilitate binding at the promoter site (i.e., by loop formation) and that the AbrB bound at the site causes repression. Nevertheless, in each of the cases examined, the location of a binding site within the promoter region itself implies that AbrB interferes with the interaction of RNA polymerase with these promoters. In fact, studies with the *abrB* (36, 37) and *dciA* (35) promoters have shown that AbrB can repress transcription in vitro.

An examination of the protected regions does not reveal any obvious candidate for a sequence motif that is recognized by AbrB. Each of the regions is relatively AT rich, and in one study that used a hydroxyl-radical footprinting technique on the AbrB-*spoVG* and -*tycA* interactions (9), it was found that in some areas known to be protected against DNase I attack (because of AbrB binding), short regions of AT-rich sequences regularly spaced appeared to be preferentially protected. However, in other areas of the DNase I footprint regions, candidate AT motifs were not protected from hydroxyl-radical cleavage, even though they appeared to be spaced properly in relation to the motifs that were protected.

Perhaps AbrB binds to a specific DNA structure and this structure is assumed by a variety of base sequences (40). A similar hypothesis has been proposed to explain the properties of the type II DNA-binding protein TF1 of bacteriophage SPO1 (12). However, the similarity between AbrB and TF1 seems to end at this point, since they share no amino acid similarity and AbrB does not exhibit the nonspecific binding properties that TF1 does (13). Because it binds with high specificity, AbrB must recognize a relatively unusual structure that is not found at random throughout the genome and particularly not in other promoter regions. It is not known what structure AbrB recognizes, but runs of A and T that are characteristic of AbrB-binding sites can produce a structure distinct from the usual B-DNA and can cause DNA bending (21, 25, 27). In fact, at least three of the promoters to which AbrB binds, *spoVG*, *tycA* (9), and *aprE* (7, 37), are known to be in (or near) intrinsically bent DNA regions (as judged by in vitro methods).

Examination of regions of the *aprE* promoter in which AbrB protects guanines from methylation (40) revealed high conformity to the 8-bp sequence TGNU WNNA (in which U is a purine, W is A or T, and N is any base). Sequences around the nonprotected guanines showed only 16% having seven or eight matches to this consensus sequence. Only 24% of the sequences around guanines in promoter regions to which AbrB does not bind matched in seven or eight positions. Statistical analysis indicated that the similarity around the protected guanines was not due to chance. Thus, the TGNUWNNA sequence could contribute to the formation of the structure recognized by AbrB. Interestingly, a point mutation in an AT-rich sequence upstream of the −35 region of the *spoVG* promoter overcomes AbrB repression in vivo (51) and shows impaired affinity for AbrB binding in vitro (30). The wild-type sequence of the transcribed strand is TGAAAAAA; the mutant sequence is TAAAAAAA. Note that the former can be aligned perfectly with the consensus sequence around protected guanines found in the *aprE* promoter (TGNUWNNA), while the latter is missing the guanine, which may be a critical structural element and a point of close AbrB contact in the *spoVG* promoter as well.

Methylation protection and hydroxyl-radical footprinting studies revealed an intriguing aspect of how AbrB binds to its target sequences. All of the AbrB-protected guanines occur on one face of the DNA helix, a finding that implies that binding involves stacking of the protein along one side of the helix and that recognition determinants are repeated with a 10-bp periodicity (40). In the hydroxyl-radial studies, the preferentially protected AT-rich sequences were also spaced about one helical turn apart, a fact pointing to the same conclusion (9). Binding of AbrB to one face of the helix may facilitate the regulation of certain promoters (such as *aprE*) that are controlled by multiple regulatory proteins (see below).

The AbrB protein evidently must form a hexamer in order to bind DNA. This conclusion is based on studies with a purified mutant AbrB protein (AbrB4) in which the unique cysteine is replaced by a tyrosine. Because the mutant protein forms a hexamer in solution (39), oligomerization does not depend on the formation of disulfide bonds. The AbrB4 protein binds poorly to DNA, and in subunit-mixing experiments, the presence of only one or two mutant subunits abolishes the ability of the heterogeneous protein to bind DNA effectively (39). Thus, a single amino acid change in only one or two subunits must radically alter the overall three-dimensional shape of the hexamer or the interaction between subunits that is required for DNA binding. Because no region of the AbrB monomer possesses significant similarity to known DNA-binding motifs (6, 10, 37, 42), the DNA-binding site of AbrB could conceivably be formed by two or more of the subunits.

AbrB REGULATION OF GENE EXPRESSION

AbrB participates in the regulation of a wide variety of processes that are associated with the end of the exponential phase of cell growth (Table 1). Although AbrB appears to play three different roles (repressor, "preventer," and activator), the overall purpose of AbrB-mediated regulation is to prevent the expression of postexponential-phase genes at inappropriate times, such as during vegetative growth on good nutrient sources. Once the cells enter the transition state, AbrB's down regulation is lifted, leading to the expression of a battery of genes concerned with the cell's survival and developmental options.

AbrB is a classical repressor of certain genes in the

Table 1. Transition state events controlled by AbrB

Process	Target gene(s)	Function of target	Mode of AbrB control[a]	Reference
Sporulation	*spo0E*	Negative regulator	R	28, 40
	spo0H	RNA polymerase sigma factor	R	49
	spoVG	?	P	51
Nutrient utilization	*aprE*	Alkaline protease (subtilisin)	P	40, 48
	nprE	Neutral protease	U	20, 48
	dciA	Dipeptide transport	P	33
	hut	Histidase	A	Fisher[b]
	?	Inducibility of nitrate reductase	U	15
	?	"Esterase(s)"	U	15
Competence	?	Regulator(s)?	U	Dubnau[b]
Motility	?	Flagellin production (?)	U	26
Antibiotic production	?	*B. subtilis* antibiotic	U	15, 45
	tycA	Tyrocidine synthetase I (*B. brevis*)	R	22
Oxidative stress response	*katA*	Catalase	U	3
Phage tolerance, antibiotic resistance	?	Cell membrane, envelope components?	U	18, 19, 45
Transition state regulation	*abrB*		*	40
	hpr	Regulator of proteases (and others)	A	20, 40
Unknown	ORF1$_{(Sin)}$[c]	Regulation of Sin?	P	34

[a] R, repressor; P, preventer; U, unclear, indirect, or multifaceted; A, activator; *, autoregulation.
[b] Fisher, chapter 16, this volume; Dubnau, chapter 39, this volume.
[c] First open reading frame of the *sin* operon.

sense that their expression is rendered constitutive (during vegetative growth) in *abrB* mutants. The expression of the tyrocidine synthetase I (*tycA*) gene of *B. brevis* is a striking example of this type of control (22). Two essential sporulation genes (*spo0E* and *spo0H*) are also subject to this type of regulation (28, 40, 49). Because *spo0H* codes for an RNA polymerase sigma factor (σ^H) required for the initial stage of sporulation (see chapter 45), the release of AbrB repression during the transition state makes intuitive sense for this gene. Not so obvious is the cell's rationale for having AbrB as the sole regulator (at least under the cultural conditions examined) of *spo0E*, whose product is a negative regulator of sporulation (28). Transcription of the *spo0E* gene can initiate from either of two start sites located 3 bp apart. Transcripts from the downstream start site (P1) are present at low constitutive levels under all conditions examined (28). Transcripts from the upstream site (P2) are subject to repression by AbrB and become induced during the transition state or are produced at elevated, constitutive levels during vegetative growth in an *abrB* mutant background. The location of AbrB binding to *spo0E* (Fig. 1), as determined by footprinting, should include the −10 and −35 elements for both of these start sites, so it is puzzling that transcription from only one is subject to AbrB repression. Several possibilities exist and have been pointed out previously (28). Since AbrB binding might involve stacking along one face of the helix, then while the recognition determinants for P2 are masked, the P1 determinants may lie enough on the opposite helix face that a transcriptional complex is able to recognize them. Since it is not known what forms of RNA polymerase holoenzymes are responsible for each transcript, other possibilities include AbrB being an effective binding competitor (or blockage) for one form but not the other or involvement of AbrB in the repression of some component of the form that initiates at P2. Nevertheless, the question remains: why should the cell want to lift AbrB repression of a negative regulator of sporulation just when the cell is most likely to decide to sporulate? A definitive answer awaits elucidation of the molecular mechanism of *spo0E* action, but it is possible that at this point, the cell must place the final decision-making process entirely under the control of genes (*spo0* genes) that are more directly responsive to the exact metabolic and environmental conditions encountered (see chapter 51).

AbrB also appears to be an activator of some genes. The product of the *hpr* locus is another transition state regulator that has been shown to be a DNA-binding protein (20). Results (17, 29) of examinations of the expression of *hpr* that used a *hpr-lacZ* fusion in various genetic backgrounds led to the conclusion that AbrB activates transcription of the *hpr* gene. However, attempts (37, 40) to demonstrate AbrB binding to the *hpr* promoter via footprinting procedures have been unsuccessful, although AbrB appears to bind, according to gel retardation assays. The meaning of this apparent incongruity is unknown, but perhaps the binding of AbrB at promoters it activates is intrinsically different from its binding at promoters it negatively regulates (and the former is somehow recalcitrant to analysis by footprinting methods). In addition to *hpr*, AbrB may activate some component or regulator of the competence pathway (1) (see chapter 39) and the enzymes involved in histidine utilization (see chapter 16).

A third role that AbrB plays is that of what can be thought of as a preventer. This role can be illustrated by the example of subtilisin (*aprE*) expression. *spo0* mutants are deficient in *aprE* expression during the transition state, but *abrB* mutations repair this defect. However, the *abrB* mutations do not render subtilisin production constitutive during exponential growth: other temporal controls are still intact. The repressive effect of AbrB is seen only in those *spo0* mutations that result in elevation of AbrB levels (see below) during the transition state. For genes subject to this type of control (Table 1), AbrB's role is to assist in preventing inappropriate expression during vegetative growth, and there must be other preventers acting on them (the appearance of activators during the transition state may also play a role). Other transition state regulators (see chapter 54) such as Hpr and Sin have been shown to serve in preventing expression of some of these genes. In fact, AbrB, Hpr, and Sin are all known to bind to the *aprE* promoter (11, 20). Interestingly, one of the Hpr-binding sites on *aprE* is contained within the AbrB binding region, and another partially overlaps that region (20). It has not yet been determined whether AbrB precludes Hpr binding or vice versa (37). Since AbrB appears to bind along one face of the helix, the Hpr determinants may lie along the opposite face, and binding of these two regulators could occur simultaneously. If so, perhaps the AbrB and Hpr proteins also interact physically in some manner to achieve more than just an additive effect. This scenario is even more intriguing when it is remembered that at least one other preventer (Sin) binds to *aprE*. A physical interaction of bound AbrB with other transition state regulators may be a hallmark of the regulatory mechanism at "prevented" genes.

Where the exact target genes of AbrB action are not known or have yet to be examined in vitro, it is uncertain as to which of these roles AbrB plays. For complex, multistep processes such as the development of competence (see chapter 39), AbrB may play a multifaceted role, i.e., activating some steps while repressing (preventing) others. It is clear that AbrB regulation observed in vivo need not be due to direct action of AbrB binding at the promoter. The production of neutral protease (product of the *nprE* gene) is under AbrB control (48), but AbrB does not bind to the *nprE* promoter (37). Hpr protein, however, does bind to *nprE* (20). Thus, it appears that AbrB's effect on neutral protease is indirect, i.e., through its regulation of Hpr. AbrB may also participate in the regulation of the Sin protein. The *sin* gene is preceded by an open reading frame, *orf-1*, in a two-gene operon (see chapter 54). It is believed that the ORF1 protein may play an important role in regulating expression or activity of the Sin protein (34). Footprinting studies have revealed that Hpr (20), AbrB (37), and Spo0A (37) bind to the *orf-1* promoter at distinct locations. This evidence suggests an even broader role for AbrB in transition state gene expression through its regulation of Sin functioning.

How might AbrB binding to DNA affect RNA polymerase's ability to transcribe? The simplest explanation in the case of negative control is that the bound AbrB sterically prevents the polymerase from interacting with its determinants on the DNA. The location of AbrB binding (Fig. 1) implies that this may be the case. But what of AbrB-mediated activation? Perhaps Abr binding to these promoters enhances a certain step such as formation of the closed or open complex. This could entail an AbrB-polymerase contact functionally analogous to the NtrC-polymerase contact that enhances formation of an open complex at the *glnA* operon of enteric bacteria (43). As in that case, AbrB might bind at some distance from the promoter and form a contact with bound but inactive polymerase. Unfortunately, we do not yet have evidence as to where AbrB binds to genes it activates, but the possibility exists that the location of AbrB binding in relation to the promoter dictates whether AbrB activates or represses.

Another explanation is that AbrB binding alters or stabilizes a DNA conformation necessary for proper polymerase recognition. AbrB binding at negatively controlled promoters might result in the formation (stabilization?) of the region into a structure that cannot be properly recognized by the transcriptional machinery. AbrB binding at positively controlled promoters would then be postulated to result in a structure recognizable by polymerase: in effect, AbrB would "present" these promoters to the polymerase. In this model, it is not the AbrB protein per se that affects the RNA polymerase interaction but rather the effect AbrB binding has on the localized structure of the promoter. Such situations are not without precedent: Hu (the major chromosomal-associated protein of *E. coli*) affects the binding abilities of regulatory proteins through its effect on DNA structure (8). Considering that AbrB binding appears to occur by stacking on one face of the helix, it is possible that RNA polymerase could "read" the other side of the helix and, in the case of positive control, displace bound AbrB if the promoter were presented in the proper conformation.

The effect of localized structure on a promoter's efficiency, particularly the effects of activating sequences that are curved or that become bent due to binding of activator proteins, is becoming well documented. Of interest to a consideration of AbrB's action are results regarding the activating nature of the AT-rich sequence upstream of the −35 region of the *spoVG* gene (4) and the evidence (23, 24) that curvature upstream of −35 can have dramatic effects on promoter utilization in *B. subtilis*. When the AT-rich *spoVG* sequence, known to be curved or bent (2), is placed upstream of the *E. coli gal* gene in place of the normal −35 region and the cyclic AMP receptor protein-binding site, it effectively substitutes for the activating effect of cyclic AMP receptor protein-cyclic AMP in vivo. The integrity of the *spoVG* AT-rich sequence is necessary for AbrB control in vivo (51) and AbrB binding in vitro (30). These results imply that the curvature of this region is necessary for promoter utilization and that AbrB binding acts to change the curvature so that the region cannot function as an activating sequence. In essence, AbrB's negative control of *spoVG* could act via masking of a DNA structure necessary for promoter usage, and its positive effect at other promoters might be through inducing the formation of a DNA conformation that activates transcription.

REGULATION OF *abrB* EXPRESSION

A complete understanding of how AbrB regulates transition state gene expression requires understanding how AbrB itself is regulated. The *abrB* gene is transcribed by two promoters located 14 bp apart, but it is not known which form(s) of RNA polymerase recognizes each promoter. mRNA quantization and primer extension analysis have shown that while both promoters are subject to autoregulation, the downstream one (P2) is also regulated by *spo0A* (29). Purified AbrB binds to a region encompassing critical elements of both promoters. Additional in vitro studies (41) revealed that Spo0A protein bound downstream of the start sites of the promoters and was independent of AbrB binding. During vegetative growth, *abrB* is controlled primarily by autoregulation, but at the onset of the transition state, Spo0A represses transcription regardless of AbrB levels (39). In vitro studies have confirmed that both AbrB (36) and Spo0A (41) can prevent RNA polymerase from transcribing the *abrB* gene.

This knowledge of the transcription control of *abrB* allows us to tie many findings together. The cooperativity of AbrB binding means that AbrB-mediated control can be modulated by slight increases or decreases in the intracellular AbrB concentration. Negative autoregulation during vegetative growth ties *abrB* gene expression to this intracellular concentration and serves to maintain an effective AbrB level that nonetheless will be rapidly sensitive to any sudden decreases. During the transition state, AbrB-mediated control needs to be abolished, and this is accomplished by having the Spo0A protein rapidly repress *abrB* transcription without regard to AbrB levels. The AbrB concentration drops below its threshold regulatory levels, resulting in the derepression or deactivation of the genes it controls.

What causes the Spo0A protein to commence repression of *abrB* during the transition state? Spo0A lies at the end of a phosphotransfer cascade that is initiated by metabolic signals indicating nutrient deprivation and the imminent cessation of vegetative growth (see chapter 51). The phosphorylated Spo0A protein has at least 20-fold-higher affinity for binding to the *abrB* promoter than does the unphosphorylated form (44). (Unphosphorylated Spo0A may also participate in *abrB* regulation during vegetative growth, as evidenced by a slight [two- to threefold] increase in *abrB* transcription during vegetative growth in *spo0A* mutants [39].) Presumably, the increased binding affinity of phosphorylated Spo0A is directly correlated with increased repression of *abrB* during the transition state. The binding activity of AbrB may also be regulated, but there is no evidence that a covalent modification of AbrB occurs, and despite an intensive search, no small effector molecule or ion has been found (37). At present, the only known mechanism for abolishing AbrB control is Spo0A-phosphate repression of *abrB* transcription. However, given the complexity of the circuits AbrB controls, it would not be surprising to discover regulation at the level of AbrB-binding activity as well.

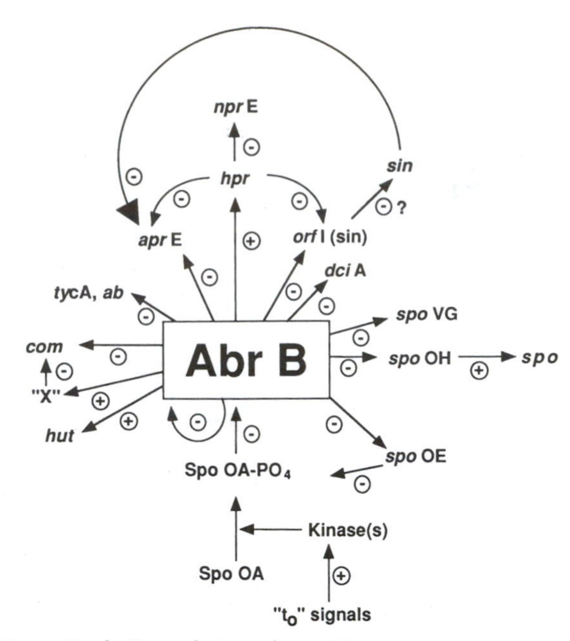

Figure 2. AbrB regulation of transition state gene expression. +, positive regulation; −, negative regulation. The conversion of the Spo0A protein to a form that is active in repressing *abrB* transcription occurs at the end of exponential growth and involves phosphorylation. t_o, time zero.

SUMMARY

The current model of the role AbrB plays in the regulation of transition state events is depicted in Fig. 2. During vegetative growth, AbrB prevents inappropriate functions from being expressed by (i) direct repression, (ii) action in concert with other regulators (preventers), or (iii) activation of other transition state repressors-preventers (such as Hpr). AbrB might also play a role in activating the expression of certain genes during vegetative growth under suboptimal conditions. (An example of this type of action is the apparent positive regulation by AbrB of the expression of histidase when the cells are growing on a poor carbon source such as arabinose [see chapter 16].) When signals that indicate the approaching end of vegetative growth are received, the Spo0A protein becomes phosphorylated and represses *abrB* transcription. The intracellular AbrB level (and possibly AbrB activity) drops, and the transition state functions under AbrB control are expressed because of direct derepression, "deprevention," or deactivation of other negative regulators.

Transition state regulators such as AbrB are likely to exist in other microbial cells that undergo a differentiation process. AbrB analogs may also play a role in any bacterial transition from active growth to stationary phase. Perhaps AbrB-like proteins also exist in eukaryotic cells (such as the pluripotential stem cells of the hemopoietic system) that are poised to respond in various developmental ways depending on the stimulus. It is interesting that although they share no significant amino acid homology, AbrB and eukaryotic homeotic selector (homeodomain) proteins have functional similarities (16, 31). Like AbrB, these are transcription regulators, some capable of both positive and negative effects. They are DNA-binding proteins with somewhat unclear binding specificities. Some regulate other homeotic selector genes (just as AbrB regulates other transition state regulators), and some are subject to autoregulation. The activation of their specific targets represents a developmental transition. Perhaps AbrB is a prokaryotic prototype of these eukaryotic regulators, each having evolved convergently.

Much remains to be learned regarding AbrB's role and functioning. For example, what DNA structure does AbrB recognize? How does it interact with RNA polymerase or other bound proteins? Is its binding activity subject to control by other gene products or small molecules? Whatever the answers to these questions are, it is clear that AbrB is a key regulator of the transition state. Although under laboratory conditions the transition state is usually quite transient, in nature, a *Bacillus* cell may be in this stage of growth for long periods because of the limited availability of nutrients in the soil. In any case, it is reasonable to assume that the cell has evolved to optimize all possible growth states it encounters. AbrB's role in coordinating the expression of certain sporulation genes with the expression of protective functions and those involved in searching for and utilizing alternative nutrients appears to be a critical feature of evolution's answer to the decision (to sporulate or not) *Bacillus* cells must make when they enter the transition state.

REFERENCES

1. **Albano, M. J. Hahn, and D. Dubnau.** 1987. Expression of competence genes in *Bacillus subtilis. J. Bacteriol.* **169:** 3110–3117.
2. **Banner, C. D. B., C. P. Moran, Jr., and R. Losick.** 1983. Deletion analysis of a complex promoter for a developmentally regulated gene from *Bacillus subtilis. J. Mol. Biol.* **168:**351–365.
3. **Bol, D., and R. Yasbin.** Personal communication.
4. **Bracco, L., D. Kottarz, A. Kolb, S. Diekmann, and H. Buc.** 1989. Synthetic curved DNA sequences can act as transcriptional activators in *Escherichia coli. EMBO J.* **8:**4289–4296.
5. **Brennan, R. G., and B. W. Matthews.** 1989. The helix-turn-helix DNA binding motif. *J. Biol. Chem.* **264:**1903–1906.
6. **Churchill, M. E. A., and A. A. Travers.** 1991. Protein motifs that recognize structural features of DNA. *Trends Biochem. Sci.* **16:**92–97.
7. **Ferrari, E.** Personal communication.
8. **Flashner, Y., and J. D. Gralla.** 1988. DNA dynamic flexibility and protein recognition: differential stimulation by bacterial histone-like protein HU. *Cell* **54:**713–721.
9. **Furbass, R., M. Gocht, P. Zuber, and M. A. Marahiel.** 1991. Interaction of AbrB, a transcriptional regulator from *Bacillus subtilis*, with the promoter of the transition state-activated genes *tycA* and *spoVG. Mol. Gen. Genet.* **225:**347–354.
10. **Furbass, R., and M. A. Marahiel.** 1991. Mutant analysis of interaction of the *Bacillus subtilis* transcription regulator AbrB with the antibiotic biosynthesis gene *tycA. FEBS Lett.* **287:**153–156.
11. **Gaur, N. K., J. Oppenheim, and I. Smith.** 1991. The *Bacillus subtilis sin* gene, a regulator of alternate development processes, codes for a DNA-binding protein. *J. Bacteriol.* **173:**678–686.
12. **Greene, J. R., S. M. Brennan, D. J. Andrew, C. C. Thompson, S. H. Richards, R. L. Heinrickson, and E. P. Geiduschek.** 1984. Sequence of the bacteriophage SPO1 gene coding for transcription factor 1, a viral homologue of the bacterial type II DNA-binding proteins. *Proc. Natl. Acad. Sci. USA* **81:**7031–7035.
13. **Greene, J. R., L. M. Morrissey, L. M. Foster, and E. P. Geiduschek.** 1986. DNA binding by the bacteriophage SPO1-encoded type II DNA-binding protein, transcription factor 1. *J. Biol. Chem.* **261:**12820–12827.
14. **Guespin-Michel, J. F.** 1971. Phenotypic reversion in some early blocked sporulation mutants of *Bacillus subtilis*. Genetic studies of polymyxin-resistant partial revertants. *Mol. Gen. Genet.* **112:**243–254.
15. **Guespin-Michel, J. F.** 1971. Phenotypic reversion in some early blocked sporulation mutants of *Bacillus subtilis*: isolation and phenotype identification of partial revertants. *J. Bacteriol.* **109:**241–247.
16. **Hayashi, S., and M. P. Scott.** 1990. What determines the specificity of action of Drosophila homeodomain proteins? *Cell* **63:**883–894.
17. **Hoch, J. A.** Personal communication.
18. **Ito, J.** 1973. Pleiotropic nature of bacteriophage tolerant mutants obtained in early-blocked asporogenous mutants of *Bacillus subtilis* 168. *Mol. Gen. Genet.* **124:**97–106.
19. **Ito, J., G. Mildner, and J. Spizizen.** 1971. Early blocked asporogenous mutants of *Bacillus subtilis* 168. I. Isolation and characterization of mutants resistant to antibiotic(s) produced by sporulating *Bacillus subtilis* 168. *Mol. Gen. Genet.* **112:**104–109.
20. **Kallio, P. T., J. E. Fagelson, J. A. Hoch, and M. A. Strauch.** 1991. The transition state regulator Hpr of *Bacillus subtilis* is a DNA-binding protein. *J. Biol. Chem.* **266:**13411–13417.
21. **Koo, H.-S., H.-M. Wu, and D. M. Crothers.** 1986. DNA bending at adenine-thymine tracts. *Nature* (London) **320:** 501–506.
22. **Marahiel, M. A., P. Zuber, G. Czekay, and R. Losick.** 1987. Identification of the promoter for a peptide antibiotic biosynthesis gene from *Bacillus brevis* and its regulation in *Bacillus subtilis. J. Bacteriol.* **169:**2215–2222.
23. **McAllister, C. F., and E. C. Achberger.** 1988. Effect of polyadenine-containing curved DNA on promoter utilization in *Bacillus subtilis. J. Biol. Chem.* **263:**11743–11749.
24. **McAllister, C. F., and E. C. Achberger.** 1989. Rotational orientation of upstream curved DNA affects promoter function in *Bacillus subtilis. J. Biol. Chem.* **264:**10451–10456.
25. **Milton, D. L., M. L. Casper, N. M. Wills, and R. F. Gesteland.** 1990. Guanine tracts enhance sequence directed DNA bends. *Nucleic Acids Res.* **18:**817–820.
26. **Mirel, D., and M. Chamberlain.** Personal communication.
27. **Nelson, H. C. M., J. T. Finch, B. F. Luisi, and A. Klug.** 1987. The structure of an oligo(dA)-oligo(dT) tract and its biological implications. *Nature* (London) **330:**221–226.
28. **Perego, M., and J. A. Hoch.** 1991. Negative regulation of *Bacillus subtilis* sporulation by the *spo0E* gene product. *J. Bacteriol.* **173:**2514–2520.
29. **Perego, M., G. B. Spiegelman, and J. A. Hoch.** 1988. Structure of the gene for the transition state regulator, *abr*B: regulator synthesis is controlled by the *spo*0A sporulation gene in *Bacillus subtilis. Mol. Microbiol.* **2:**689–699.
30. **Robertson, J. B., M. Gocht, M. A. Marahiel, and P. Zuber.** 1989. AbrB, a regulator of gene expression in *Bacillus*, interacts with the transcription initiation regions of a sporulation gene and an antibiotic biosynthesis gene. *Proc. Natl. Acad. Sci. USA* **86:**8457–8461.
31. **Scott, M. P., J. W. Tamkun, and G. W. Hartzell III.** 1989. The structure and function of the homeodomain. *Biochim. Biophys. Acta* **989:**25–48.
32. **Shiflett, M. A., and J. A. Hoch.** 1978. Alterations of ribosomal proteins causing changes in the phenotype of *spo0A* mutants of *Bacillus subtilis*, p. 136–138. *In* G. Chambliss and J. C. Vary (ed.), *Spores VII*. American Society for Microbiology, Washington, D.C.
33. **Slack, F. J., J. P. Mueller, M. A. Strauch, C. Mathiopoulos, and A. L. Sonenshein.** 1991. Transcriptional regulation of a *Bacillus subtilis* dipeptide transport operon. *Mol. Microbiol.* **5:**1915–1925.
34. **Smith, I.** Personal communication.
35. **Sonenshein, A. L.** Personal communication.

36. **Spiegelmann, G. B.** Personal communication.
37. **Strauch, M. A.** Unpublished data.
38. **Strauch, M. A., and J. A. Hoch.** 1991. Control of post-exponential gene expression by transition state regulators, p. 105–121. *In* R. H. Doi (ed.), *Biology of Bacilli—Applications to Industry*. Butterworths, Stoneham, Mass.
39. **Strauch, M. A., M. Perego, D. Burbulys, and J. A. Hoch.** 1989. The transition state transcription regulator AbrB of *Bacillus subtilis* is autoregulated during vegetative growth. *Mol. Microbiol.* **3:**1203–1210.
40. **Strauch, M. A., G. B. Spiegelman, M. Perego, W. C. Johnson, D. Burbulys, and J. A. Hoch.** 1989. The transition state transcription regulator *abr*B of *Bacillus subtilis* is a DNA binding protein. *EMBO J.* **8:**1615–1621.
41. **Strauch, M. A., V. Webb, G. Spiegelman, and J. A. Hoch.** 1990. The Spo0A protein of *Bacillus subtilis* is a repressor of the *abr*B gene. *Proc. Natl. Acad. Sci. USA* **87:**1801–1805.
42. **Struhl, K.** 1989. Helix-turn-helix, zinc-finger, and leucine-zipper motifs for eucaryotic transcriptional regulatory proteins. *Trends Biochem. Sci.* **14:**137–140.
43. **Su, W., S. Porter, S. Kustu, and H. Echols.** 1990. DNA-looping and enhancer activity: association between DNA-bound NtrC activator and RNA polymerase at the bacterial *gln*A promoter. *Proc. Natl. Acad. Sci. USA* **87:** 5504–5508.
44. **Trach, K., D. Burbulys, M. Strauch, J.-J. Wu, R. Jonas, N. Dhillon, C. Hanstein, P. Kallio, M. Perego, T. Bird, G. Spiegelman, C. Fogher, and J. A. Hoch.** 1991. Control of the initiation of sporulation in *Bacillus subtilis* by a phosphorelay. *Res. Microbiol.* **142:**815–823.
45. **Trowsdale, J., S. M. H. Chen, and J. A. Hoch.** 1978. Genetic analysis of phenotype revertants of *spo0A* mutants in *Bacillus subtilis*: a new cluster of ribosomal genes, p. 131–135. *In* G. Chambliss and J. C. Vary (ed.), *Spores VII*. American Society for Microbiology, Washington, D.C.
46. **Trowsdale, J., S. M. H. Chen, and J. A. Hoch.** 1979. Genetic analysis of a class of polymyxin-resistant partial revertants of stage 0 sporulation mutants of *Bacillus subtilis*; map of the chromosome region near the origin of replication. *Mol. Gen. Genet.* **173:**61–70.
47. **Trowsdale, J., M. Shiflett, and J. A. Hoch.** 1978. New cluster of ribosomal genes in *Bacillus subtilis* with regulatory role in sporulation. *Nature* (London) **272:**179–180.
48. **Valle, F., and E. Ferrari.** 1989. Subtilisin: a redundantly temporally regulated gene?, p. 131–146. *In* I. Smith, R. A. Slepecky, and P. Setlow (ed.), *Regulation of Procaryotic Development*. American Society for Microbiology, Washington, D.C.
49. **Weir, J., M. Predich, E. Dubnau, G. Nair, and I. Smith.** 1991. Regulation of *spo0H*, a gene coding for the *Bacillus subtilis* σ^H factor. *J. Bacteriol.* **173:**521–529.
50. **Zuber, P.** Personal communication.
51. **Zuber, P., and R. Losick.** 1987. Role of AbrB in Spo0A- and Spo0B-dependent utilization of a sporulation promoter in *Bacillus subtilis*. *J. Bacteriol.* **169:**2222–2230.

53. Motility and Chemotaxis

GEORGE W. ORDAL, LETICIA MÀRQUEZ-MAGAÑA, and MICHAEL J. CHAMBERLIN

GENES AFFECTING FLAGELLAR SYNTHESIS, MOTILITY, AND CHEMOTAXIS

Overview

The proteins that make up the bacterial sensory system for motility and chemotaxis represent a highly evolved and coordinated functional pathway. The motor that drives the bacterium is a complex, multicomponent organelle, the flagellum. It is not surprising that as information has become available about the expression of the genes that specify these proteins, they, too, have been found to be closely coordinated and regulated by a wide variety of environmental stimuli. In fact, synthesis of all of the proteins in this pathway represents a substantial burden on the bacterial cell in terms of energy expenditure, a burden justified only under certain conditions in which cell survival may be tied to a functional sensory system.

We will refer to the structural and regulatory genes that control this pathway as flagellar-chemotaxis-motility genes (*fla-che-mot* genes) and to the reactions and structures that they determine as flagellar-chemotaxis-motility functions. In this review, we consider first the genetics and mapping of *fla-che-mot* genes, then the mechanisms of motility and chemotaxis, and finally, what is known about regulation of gene expression in the *fla-che-mot* system.

Complexities of Genetic Studies of *fla-che-mot* Functions

The study of mutants defective in flagellar synthesis, motility, and chemotaxis has provided important insights in understanding these processes in *Bacillus subtilis* (32, 46, 56, 89, 110, 117, 150–153). In addition, such mutations have led to isolation, mapping, and structural analysis of many structural and regulatory genes that make up the *fla-che-mot* system. A compendium of genes implicated in this sensory pathway or in the regulation of genes in that pathway is given in Table 1.

Because of the complexity of the sensory pathway itself and the many levels at which *fla-che-mot* gene expression is regulated, it is often difficult to interpret the behavior of a particular mutation in molecular terms without a great deal of study. Thus, the isolation of a given mutation by criterion screening for one of these processes has often led to the identification of a genetic locus more directly involved in one of the other processes or in the identification of a locus involved not only in all three processes but also in additional regulatory systems. For example, several lesions originally identified as causing defects in chemotaxis have now been shown to encode components of the flagellar structure (5, 147, 150, 151), while several mutations originally identified as affecting flagellum synthesis have been shown to encode transcriptional regulators that affect not only expression of motility functions but also expression of autolytic enzymes and/or development of competence for transformation (36, 57, 73, 89, 119). Similarly, mutants defective in flagellum synthesis have often been found to have decreased levels of autolytic enzyme expression (31, 32, 89); hence, the isolation of mutants defective in autolytic activity has also led to the identification of loci required for flagellar synthesis (117).

Nomenclature

In discussing the isolation and properties of mutants that affect flagellar structure, chemotaxis, and motility, it is important to have a clear understanding of the nomenclature that has been used to designate different kinds of mutations. Important classes of mutations include the following.

Nonflagellate or partially flagellate (Fla$^-$ or *fla*) mutants

The Fla$^-$ phenotype is defined as any substantial diminution of the flagellar apparatus (81), and the term *fla* is used as a collective symbol for flagellar genes encoding structural proteins that make up the flagellar apparatus (60). In the enteric bacteria, specific genes whose gene products have been shown to be required for flagellar assembly fall into four regions within the bacterial chromosome and have therefore been designated *flg*, *flh*, *fli*, or *flj* genes and given an alphabetical addition according to their genomic order (60). For example, the first gene in the *flg* region is *flgA*, and the second gene is *flgB*. Although the same genomic order is not retained exactly in *B. subtilis* (see below), the same designation has been given to a *B. subtilis* gene if the predicted protein product is significantly similar in sequence to a flagellar gene product in the enteric bacteria (see chapter 29). Unlike the other *fla* genes, the gene for the flagellin protein is *hag* in *B. subtilis*, and mutations within this gene are designated *hag* alleles (the designation *hag* is derived from H antigen, originally defined in *Salmonella typhimurium*). Mutations in regulatory genes that give rise to a Fla$^-$ phenotype are

George W. Ordal • Department of Biochemistry, College of Medicine, University of Illinois, Urbana, Illinois 61801.
Leticia Màrquez-Magaña and Michael J. Chamberlin • Division of Biochemistry and Molecular Biology, University of California, Berkeley, California 94720.

Table 1. *B. subtilis* genes and gene products involved in motility and chemotaxis or its regulation

Gene[a]	Other allele	Region	Enteric homolog	% Identity[b]	Function of gene product[c]	Reference(s)
motA	*mot*	120	*motA*	27	(Motor rotation)	96, 152
motB	*mot*	120	*motB*	27	(Required for motor rotation)	96, 152
cheX		120			Required for chemotaxis to sugars	152
flgB		140	*flgB*	25	(Rod protein)	151
flgC		140	*flgC*	37	(Rod protein)	151
fliE		140	*fliE*	30	(Basal-body protein)	84a, 151
fliF		140	*fliF*	23	(Basal-body M-ring protein)	151
fliG	*flaA15*	140	*fliG*	36	(Flagellar switch protein)	5, 51, 151
fliH		140	*fliH*	23	(Flagellum-specific export?)	5
fliI		140	*fliI*	49	(Flagellum-specific export?)	5
fliJ	*cheF*	140	*fliJ*	21	(Required for formation of basal body)	5, 147
orf-6		140			(Required for flagellar formation)	5
fliK		140	*fliK*	22	(Hook length control)	5
orf-8					Unknown	5, 51
flgG		140	*flgG*	39	(Rod protein)	5, 51
fliL		140	*fliL*	21	(Required for flagellar formation)	5, 51, 103a
fliM		140	*fliM*	29	(Flagellar switch)	150
fliY	*cheD*	140	(*fliN*)		Flagellar switch	150
cheY	*cheB*	140	*cheY*	36	(Modulation of flagellar switch bias)	13
fliZ	*orf-219*	140			Required for flagellar formation	103a
fliP		140	*fliP*	48	Required for flagellar formation	14
fliQ		140	*fliQ*	49	Required for flagellar formation	14
fliR		140	*fliR*	27	Required for flagellar formation	84a, 103a
flhB		140	*flhB*	36	Required for flagellar formation	84a, 103a
flhA		140	*flhA*	45	Required for flagellar formation	24b, 84a
flhF		140	*flhF*		Unknown	24a, 103a
orf-298		140			Unknown	103a
cheB	*cheL*	140	*cheB*	39	Receptor-demethylating enzyme	103a
cheA	*cheN*	140	*cheA*	34	(Phosphorylation of CheY and CheB)	33
cheW		140	*cheW*	27	(Modulation of CheA activity)	48, 49
orfA		140			Required for receptor demethylation	88, 103a
orfB		140			Required for receptor demethylation	88, 103a
sigD	*flaB2*	140	*fliA*	37	Alternate sigma factor	25, 56
cheR		200	*cheR*	30	Receptor-methylating enzyme	57
orf-1		215			Can affect motility and autolysin expression	34
sin	*flaD2*	215			Can affect motility and autolysin expression	34
orf-3		215			Unknown	34
lytA		310			Unknown	77
lytB		310			Modifier of amidase activity	77
lytC		310			Structural gene for amidase	77
degS	*sacU*(Hy)	310			Regulatory sensor (kinase)	57, 73
degU	*sacU*(Hy)	310			Regulatory transducer	57, 73
flgM		310		27	(Repressor of σ^F function)	95a
flgK		310		32	(Hook-associated protein)	95a
hag		310	*fliC*	40	Flagellin protein	76, 95

[a] Suggested designation for the gene. Many of these are based on amino acid sequence homologies between *B. subtilis* and enteric bacterial gene products (see text).
[b] Percent amino acid sequence identity between *B. subtilis* and enteric gene products. NK, not known.
[c] Functions known or suggested for *B. subtilis* gene products. Where the *B. subtilis* function is not yet established, functions are given in parentheses.

named for their gene products (i.e., the gene for σ^D is *sigD*, and the gene for the Sin protein is *sin*).

Nonmotile or paralyzed (Mot⁻ or *mot*) mutants

In Mot⁻ mutants, the flagellar apparatus appears to be morphologically intact, but the flagellar filament fails to rotate. The *mot* designation has been used for two genes in *B. subtilis* whose gene products are homologous to the gene products of the *motA* and *motB* genes of the enteric bacteria (96); these enteric genes encode the structural proteins that make up the flagellar motors in these organisms (28, 126). Disruption of *B. subtilis motA* or *motB* gives a Mot⁻ phenotype (96).

Chemotaxis-defective (Che⁻ or *che*) mutants

Mutants defective in chemotaxis have functional flagella but fail to give a chemotactic response. The *B. subtilis* chemotactic response is described in detail in the first half of this chapter. In *B. subtilis*, several loci originally identified as *che* genes (110, 111) have subsequently been shown to be *fla* genes (5, 147, 150, 151). Despite the effect of these genes on chemotaxis, we suggest that these loci be referred to as *fla* genes and that the *che* alleles within them be designated *fla*(Che) mutations in accordance with the usage of Macnab (82).

Increased frequency of motility (Ifm⁻ or *ifm*) mutants

Mutants bearing *ifm* alleles have been operationally defined as strains exhibiting a zone of motility on semisolid plates with a diameter approaching four times that produced by the motile parent strain (117). Additionally, *ifm* mutations suppress specific lesions, giving rise to a motile phenotype, whereas the original lesion resulted in a nonmotility phenotype (46, 117). *ifm* strains show considerably increased autolysin levels (117).

Autolysis-deficient (Lyt⁻ or *lyt*) mutants

Lyt⁻ mutants grow as long filaments, since they are unable to produce the autolytic enzymes responsible for hydrolysis of the cell wall after cell division. Mutants defective in the expression of autolysins often lack the flagellar apparatus and have therefore been designated *fla* or regulatory protein mutations and not *lyt* mutations. The *lyt* designation is reserved for structural genes encoding the autolytic activity itself and its modifiers. Several Lyt⁻ mutants have been isolated, and the lesions within them were designated *fil* alleles (3). We suggest that where possible, these alleles be renamed *fla* or *lyt* alleles in keeping with the definitions described above.

Early Mutant Isolation and Studies

The isolation of mutants defective in the synthesis of flagella has been very useful in our understanding of the gene products required for not only flagellar synthesis but also for motility, chemotaxis, and autolytic enzyme expression. In 1969, a screen for temperature-sensitive mutations that resulted in the lack of assembled flagellin protein allowed for the identification of several *flaTS* loci (46). These mutations were not within the *hag* structural gene but mapped near it, close to the *hisA* locus (46). Additionally, the first *ifm* mutations were identified and found to map to this region as well, suggesting that many of the genes required for flagellar functions are clustered in this region of the chromosome (46). While the mapping of *hag*, *ifm*, and one *flaTS* (*flaTS-51*, renamed *flaC51* in reference 117) mutation was found to be correct, subsequent work has determined that the mapping of two other lesions (*flaTS4* and *flaTS2*, renamed *flaA4* and *flaB2*, respectively, in reference 117) was incorrect and that their respective genes are found on the opposite side of the bacterial chromosome between *pyrD* and *thyA* (117). The functions of the genes bearing the *flaA4* and *flaB2* mutations will be discussed in the next section. Additionally, it was determined that the *flaA4* and *flaB2* mutations were not temperature sensitive but that the original isolates also contained an unlinked *ifm*-like suppressor mutation that led to the temperature sensitivity phenotype (117).

Several mutations have been isolated on the basis of their abilities to reduce the level of autolytic enzyme expression (32). These strains were obtained by minimal mutagenesis and found to contain a single mutation that gave rise to a Lyt⁻ Fla⁻ phenotype (32). The determination that the pleiotropic phenotype was due to a single mutation is important, since several researchers have proposed that a single mutation gives rise to the hyperproduction of α-amylase and protease and to a decrease in autolytic enzyme expression, a lack of flagella, and in one case, the development of competence (7, 148). These latter studies involved extensive mutagenesis, however, and it is unclear what the number and true nature of the mutation(s) in the hyperproducing strains are. Nonetheless, linkage of the Fla⁻ and Lyt⁻ phenotypes has been observed by many researchers (31, 32, 56, 89, 117, 152). In fact, it has been suggested that a decrease in autolytic enzyme expression results in lack of flagellar assembly because an autolytic activity is required for the localized peptidoglycan hydrolysis to allow for the insertion of the basal-body structures of the flagellar apparatus into or through the gram-positive cell wall (31).

A paper by Pooley and Karamata in 1984 (117) was a pivotal work. In it, the authors not only corrected the mapping data of the earlier study but also demonstrated that several *lyt* mutations were identical in nature to the *flaA4* and *flaB2* mutations and were therefore appropriately renamed *fla*. In particular, the *lyt-1*, *lyt-2* (31, 32), and *lyt-15* alleles are identical in nature to the *flaA4* and *flaB2* mutations. Since *lyt-1* and *lyt-2* map with each other but separately from the previously identified *fla* mutations, they were renamed *flaD1* and *flaD2*; the *lyt-15* lesion, however, maps near *flaA4* and was therefore renamed *flaA15* (117). In this way, four *fla* loci were mapped to three regions of the chromosome. The *flaA* and *flaB* loci are located between *pyrD* and *thyA* on the bacterial chromosome (117), *flaC* is located near *hag* and *hisA* (46, 117), and *flaD* maps between *aroD* and *lys* (117). Additionally, the *ifm* mutation maps between *hisA* and *hag* and is able to suppress the *flaA4* and *flaD1* mutations (117); testing of suppression of the remaining *fla* alleles was not done. The *flaA*, *flaB*, *flaC*, and *flaD* loci are very important in that they define gene products and regions of the bacterial chromosome required for flagellar function (see next section).

Recent Studies of Mutants

Most of the work done recently on motility functions within *B. subtilis* has focused on understanding the process of chemotaxis (see above). Toward this end, a screen for chemotaxis mutants was carried out and a collection of chemotaxis mutants was obtained (110). A mutation in the gene specifying the chemotactic methyltransferase (*cheR*) maps very close to *aroF*, while the remainder of the mutations map between *pyrD* and *thyA* on the bacterial chromosome (110). Two lambda clones, together containing >23 kb of *B. subtilis* genomic DNA, were found to contain overlapping inserts that complement most of the mutations that map between *pyrD* and *thyA* (110). Studies done on this collection of mutants identified at least 10 complementation groups within this region of DNA (111); these groups were further localized to specific restriction fragments of the lambda inserts by recombination experiments (153). The genes encoded by this region of DNA have recently been sequenced, and many of the predicted protein products have been shown to be homologous to structural proteins that form part of the flagellar apparatus in the enteric bacteria (5, 147, 150, 151). The original *che* mutations

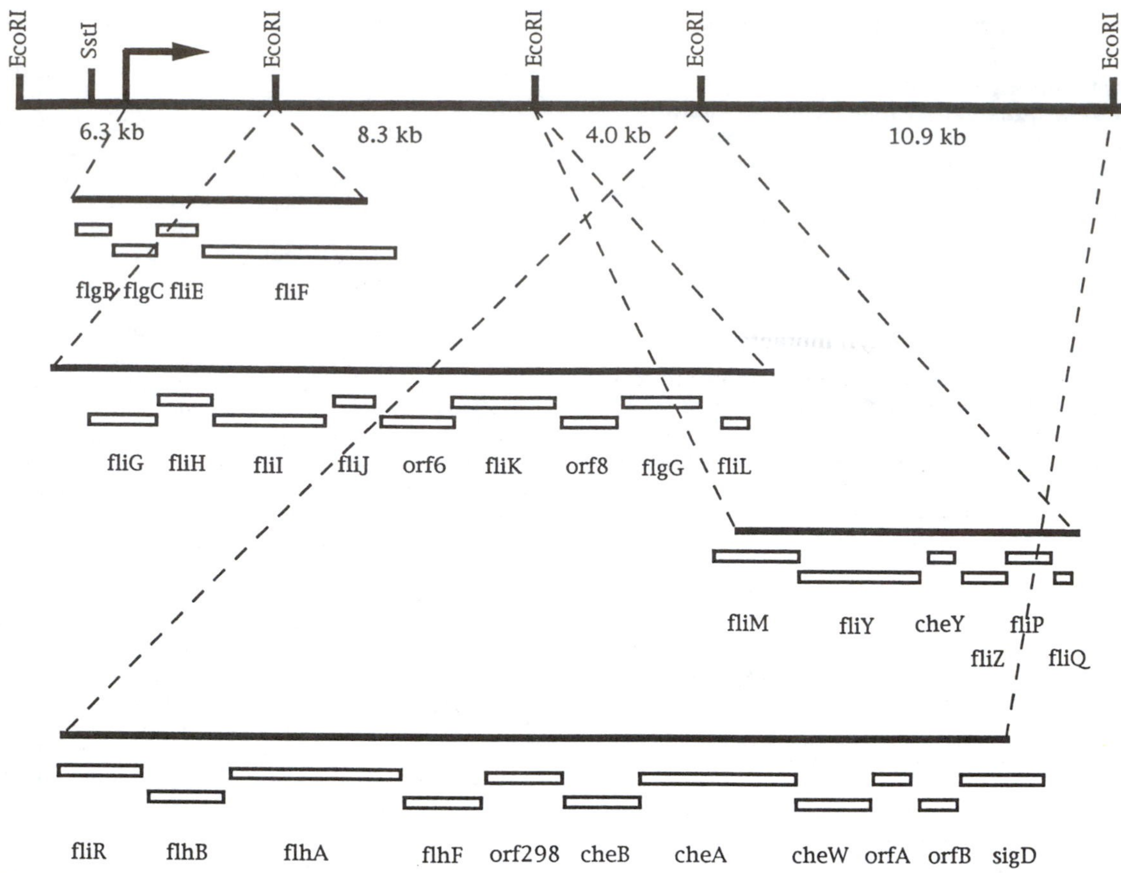

Figure 1. Map of *fla-che* region 140, originally designated the *che* operon. The heavy dark line at the top represents the restriction map of this region, showing the five *Eco*RI sites that span the region. The major promoter is shown as an arrow. Dashed lines lead to magnified maps of each region that show the individual open reading frames with their designations (Table 1). Data are from references 5 and 151. Adapted from reference 13.

will therefore be renamed in agreement with their enteric homologs (Fig. 1 and Table 1).

Interestingly, the *flaA4* and *flaA15* mutations have been mapped within this region of DNA by marker rescue (51). The *flaA15* (originally identified as *lyt-15*) lesion maps within *fliG*, the gene for a switch protein, while the *flaA4* mutation maps within a restriction fragment containing three genes, including the gene that encodes a rod protein (51). Although the original screen for *che* mutations actually led to identification of several *fla* genes, it also led to the identification of five genes encoding proteins homologous to the central regulators of chemotaxis in the enteric bacteria, i.e., CheY, CheR, CheA, CheB, and CheW (12, 33, 49, 58, 67a; Table 1). In fact, this large cluster of *fla-che* genes (region 140) probably contains a large fraction of all the genes required for flagellar assembly and chemotaxis in *B. subtilis* (Fig. 1).

Genes and Genomic Organization

Many of the genes originally identified by mutant studies as affecting flagellar, chemotaxis, and motility functions have now been cloned and sequenced. Along with mapping analyses, these studies have given us a great deal of information on the genomic and genetic organization of the *fla*, *mot*, and *che* loci (Fig. 1 and 2) as well as on the predicted functions of their gene products (Table 1). These genes map in five different regions of the chromosome and are clustered in two regions approximately 180° apart (Fig. 2). Genetic studies have shown that within these regions, several of the genes located physically adjacent to one an-

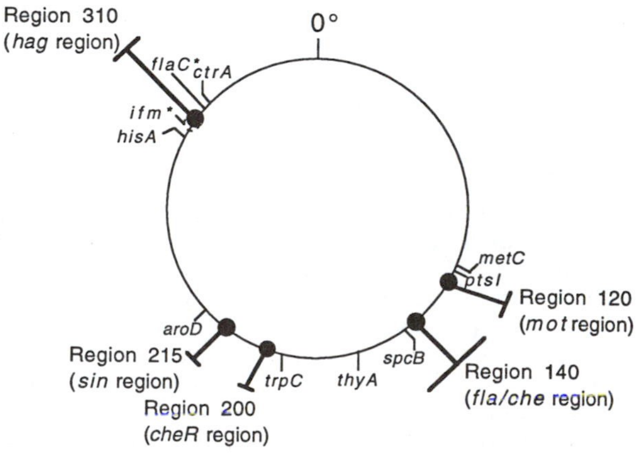

Figure 2. Genetic map of *B. subtilis* showing locations of genes required for flagellar assembly, motility, and chemotaxis.

other form operons. In one case, the operon appears to be larger than 25 kbp (88, 153; Fig. 1, region 140). In this section, we will describe the genes identified to date that are implicated as structural or regulatory elements in the signaling pathways of *B. subtilis*.

Region 120 (the *mot* region)

The *mot* region encodes two genes, *motA* and *motB*, whose predicted protein products are homologous to the structural proteins required for flagellar rotation in the enteric bacteria (96, 152). Disruption of either the *motA* or the *motB* gene results in a Mot$^-$ phenotype (96), supporting the notion that these are functionally homologous to their enteric counterparts. The *motAB* operon is transcribed by the σ^D form of RNA polymerase. An additional complementation group, implicated in chemotaxis to sugars, has been identified near the *motAB* operon and is named *cheX* (152). The *cheX* locus is thought to be transcribed from a promoter recognized by the major form of RNA polymerase (152).

Region 140 (the *fla-che* region)

The *fla-che* genes in region 140 appear to be expressed as a single, very large operon of over 25 kb (88, 153) transcribed clockwise (CW) with respect to the bacterial chromosome map. Many of the genes encoded in this region are homologous to enteric genes required for flagellar assembly and chemotaxis. The genes throughout this region generally encode homologs to the structural proteins that make up the basal-body structure (5, 147, 150, 151). Within these genes, however, are several genes specifically required for chemotaxis: *cheY*$_B$, *cheB*$_B$ (103a), *cheA*$_B$, *cheW*$_B$, *orfA*, and *orfB* (12, 33, 49, 88). Near the end of the region is the structural gene for σ^D (88). Analysis of the partial sequence obtained downstream from *sigD* demonstrates the existence of the 5′ end of another open reading frame, indicating that this cluster of genes extends further than the *sigD* gene (56). In fact, genes downstream from *sigD* have been shown to be required for taxis towards amino acids but not sugars (153).

Interestingly, the gene organization within *fla-che* region 140 is similar in some respects to the organization found within the enteric bacteria. In the latter case, the *flgBCDEFGHIJ* genes are cotranscribed as an operon, as are *fliFGHIJK* and *fliLMNOPQR* (84). In *B. subtilis*, the *flgB* gene is upstream from the *flgC* gene and is promoter proximal; these genes are immediately followed by the *fliF*, *fliG*, *fliH*, *fliI*, and *fliJ* genes. These, in turn, are followed by the *fliK*, *fliL*, *fliM*, *fliP*, *fliQ*, and *fliR* genes, although these last genes are interspersed with other coding sequences (Fig. 1).

Region 200 (the *cheR* region)

The *cheR* gene appears to be monocistronic and to be transcribed from P_A promoter element. *cheR* and *cheX* are the only known *che* genes not part of *fla-che* region 140 (58, 110).

Region 215 (the *sin* region)

The *sin* region contains the *sin* gene as well as two flanking open reading frames. However, only the open reading frame upstream from *sin* appears to be part of the same operon, since transcription terminators have been identified immediately downstream from *sin*, and an RNA species originating upstream of *orf-1* extends through *sin* and ends at the termination sequences (34). *orf-1* and *sin* are discoordinately expressed from three promoters (34). Expression from the two promoters upstream of *orf-1* increases during logarithmic growth and is maximal during early stationary phase, whereas expression from the promoter upstream of *sin* is constitutively expressed throughout growth (34). The two promoters upstream of *orf-1* are *spo0A* and *spo0H* dependent and are thought to be transcribed by the σ^E and σ^H forms of RNA polymerase; the promoter upstream from *sin* is independent of *spo0* regulation and is thought to be transcribed by the major holoenzyme (34).

Region 310 (the *hag* region)

The genetic organization of the *hag* region has recently been determined by consolidating the physical maps obtained by several groups involved in cloning genes found within this area of the chromosome (57, 73, 95, 149). All of the genes identified to date in this region are transcribed counterclockwise (CCW) with respect to the bacterial chromosome. Furthest upstream (closest to the chromosomal origin) are the *lyt* genes encoding the structural proteins for the amidase autolytic enzyme and its modifier protein (77, 87). These appear to be transcribed in part by the σ^D holoenzyme (77) from a promoter previously designated P_{D-7} (40, 122). They must also be transcribed from another promoter, since a strain lacking the σ^D factor continues to express 30% of the wild-type levels of this activity (89).

The *degSU* operon is approximately 15 kb downstream from the *lyt* operon. It is not known what genes are encoded in the intervening sequence between the two operons. The *degSU* operon may be transcribed by the major σ^A holoenzyme, since a strong consensus characteristic of σ^A promoter is located upstream from *degS* (57, 73).

Approximately 4 kb downstream from *degU* lies one end of a cluster of σ^D-dependent promoters. The first promoter, P_{D-1}, is followed by 4 to 5 kb of DNA that contains an operon having at least four open reading frames including some similar in amino acid sequence to the enteric FlgM (*orfB*) and FlgK (*orfD*) proteins (Table 1; 94). The next gene in this region is the well-characterized flagellin gene *hag*. The *hag* gene is monocistronic and is transcribed by σ^D RNA polymerase from the P_{D-6} promoter (95). Immediately downstream from the transcription terminator for *hag* is another promoter, P_{D-8}, that is followed by at least one open reading frame that has not yet been fully sequenced (95).

The *flaC* and *ifm* mutations originally isolated by Grant and Simon (46) map to the *hag* region as well. Although the *flaC* mutation is thought to map further upstream of the *lyt* operon (86a), the *ifm* allele maps between *hag* and *hisA*.

MECHANISMS OF MOTILITY AND CHEMOTAXIS

Overview

Most bacteria have efficient sensory systems coupled with motility mechanisms that permit taxis toward or away from different stimuli. In chemotaxis, binding of an attractant to a cell receptor produces a signal favoring movement toward increasing concentrations of that substance, while repellents have an opposite effect. *B. subtilis* has long been known to show motility and chemotaxis, and these processes provide important mechanisms by which this organism is able to grow in environments in which nutrients are scarce. However, it is only in the past decade that detailed studies of the genetics and molecular mechanisms controlling motility and chemotaxis have been undertaken.

The paradigm for bacterial motility and chemotaxis has been established through extensive studies of the enteric bacteria *Escherichia coli* and *S. typhimurium* (82). For these cells, motility is based on the presence of flagella, which are extracellular organelles that consist of a helical filament joined to the bacterial cell surface through a complex structure called a hook and basal body (HBB) (81, 83, 84). These flagella can be rotated by interaction with a molecular motor either in a CCW or a CW direction. Rotation CCW leads the bacterium to swim smoothly, while CW rotation leads to tumbling and a random change in the orientation of the bacterial cell (75). In a normal, homogeneous medium, periods of smooth swimming are interrupted by tumbling, permitting random exploration of all regions of the medium (10).

Chemotaxis in enteric bacteria is mediated through sets of cellular receptors that bind specific attractants (82, 86). Such binding initiates a signal that is transduced through a series of protein intermediates by specific methylation-demethylation and phosphorylation-dephosphorylation. The targets of this pathway, the motor proteins, cause rotation of the flagellum CCW or CW, depending on the nature of the signal. Increasing concentrations of attractants lead to CCW rotation and permit the bacterium to swim smoothly. Repellents, or decreasing concentrations of attractants, lead to CW rotation to bring about tumbling so that the bacteria can begin to move in the desired direction (92, 142, 143).

The *B. subtilis* motility-chemotaxis system appears to function overall in a manner reminiscent of the enteric paradigm. Fundamentally, both systems utilize a phosphorylation cascade for excitation and removal (or addition) of methyl groups from methyl-esterified receptors for adaptation. There are significant degrees of amino acid sequence homology between many *B. subtilis* proteins implicated in motility-chemotaxis and enteric proteins that make up the flagellar structures and signal pathways. This homology has led those in the field to surmise that at least some of these proteins play homologous functional roles in the two sensory systems, although this has been shown directly for only a few such proteins. Finally, the genes for these motility-chemotaxis proteins make up a coordinately controlled regulon in enteric bacteria, and there are some similarities in the regulatory mechanisms employed in *B. subtilis*. Hence, it is often useful to use the enteric paradigm for motility-chemotaxis in thinking about the parallel process in *B. subtilis*. However, it is already clear that there are important differences in the two systems, and the number of such differences is likely to increase as the detailed mechanisms of *B. subtilis* motility and chemotaxis become better understood.

The two foremost differences are the inversion of the excitatory reaction and a more complex adaptation mechanism in *B. subtilis*. In *E. coli*, repellent activates an autophosphorylating kinase (CheA), which subsequently phosphorylates CheY. CheY-P binds to a switch to cause tumbling (CW rotation of the flagella). In *B. subtilis*, attractant is believed to activate CheA to make CheY-P to cause smooth swimming (CCW rotation of the flagella). In *E. coli*, adaptation (return to prestimulus behavior) to repellent is caused by demethylation of the receptors with production of methanol. In *B. subtilis*, adaptation to attractant is caused in part by transfer of methyl groups from the receptors to another protein. The demethylation of this protein seems to occur at the flagellum, and methanol is produced.

Attractants, Repellents, and Their Receptors

All 20 common amino acids and many sugars are attractants for *B. subtilis* (107, 115). In general, attractants have unique binding sites for use in chemotaxis, suggesting that there are specific binding sites for these compounds on the membrane receptors or specific binding proteins that interact with receptors. In at least one case, the mechanism of sensing attractant concentration is separate from that for transport of attractant across the membrane. In this review, we use the term "receptors" to refer to the homologs of the methyl-esterified integral membrane proteins that are the direct receptors for amino acids in *E. coli*. When the receptors do not directly bind an amino acid, we use the term "binding protein" to refer to the protein that does directly bind the amino acid.

Many compounds serve as repellents for *B. subtilis*, and these repellents also seem to have unique binding sites, suggesting that there may be membrane receptors for them as well.

Amino acids as attractants

Amino acid attractants can be identified in a microscope assay by their abilities to cause smooth swimming in the suspensions of bacteria they are added to (Table 2). In a more sensitive assay, attractant is placed in a small capillary tube whose distal end is sealed and whose proximal end sits in a bacterial suspension. Attractant diffuses out of the capillary into the suspension so that a gradient is formed, and bacteria travel up the gradient and enter the capillary. After some time, the contents of the capillary are analyzed, usually by plating and counting colonies. The capillary assay allows accurate measurement of threshold values, i.e., minimum concentrations of attractant for which taxis is detected. Experiments with *E. coli* have shown that under certain conditions accumulation is proportional to the net change in fraction of receptor bound and thus can be used as a

Table 2. Capillary and microscope assays of thresholds for amino acid taxis by *B. subtilis*[a]

Amino acid	Threshold (M)[b]	
	Microscope assay	Capillary assay
Alanine	1.0×10^{-7}	3.0×10^{-9}
Proline	3.2×10^{-7}	7.0×10^{-9}
Cysteine	5.6×10^{-8}	1.0×10^{-8}
Threonine	1.8×10^{-6}	1.6×10^{-8}
Asparagine	1.8×10^{-6}	8.0×10^{-8}
Methionine	3.2×10^{-6}	9.0×10^{-8}
Valine	1.8×10^{-6}	1.3×10^{-7}
Leucine	5.6×10^{-6}	1.9×10^{-7}
Serine	3.2×10^{-6}	6.0×10^{-7}
Glycine	1.8×10^{-5}	6.0×10^{-7}
Isoleucine	5.6×10^{-6}	6.0×10^{-7}
Glutamine	5.6×10^{-6}	1.4×10^{-6}
Tryptophan	2.8×10^{-4}	2.0×10^{-6}
Arginine	5.6×10^{-5}	3.0×10^{-6}
Lysine	1.0×10^{-4}	3.0×10^{-6}
Phenylalanine	3.2×10^{-5}	4.0×10^{-6}
Tyrosine	1.0×10^{-4}	1.6×10^{-5}
Aspartate	1.0×10^{-4}	3.0×10^{-5}
Histidine	5.6×10^{-4}	3.0×10^{-5}
Glutamate	3.2×10^{-4}	1.0×10^{-4}

[a] Data are taken from Ordal and Gibson (107).
[b] Threshold values are minimum molar concentrations that give an effect different from buffer. For the microscope assay, values are minimum concentrations that cause smooth swimming in a suspension of bacteria. For the capillary assay, values are minimum concentrations that bring about accumulation of bacteria in a capillary tube.

Table 3. Apparent K_ds for amino acids as attractants of *B. subtilis*[a]

Amino acid	Apparent K_d (M)		
	Addition	Removal	Modified capillary assay
Alanine	1.9×10^{-7}	3.6×10^{-7}	3.2×10^{-7}
Arginine	8.8×10^{-5}	8.1×10^{-5}	1.0×10^{-4}
Asparagine	9.3×10^{-5}	7.3×10^{-5}	5.6×10^{-5}
Aspartate	2.9×10^{-2}	4.4×10^{-2}	1.0×10^{-2}
Cysteine	2.2×10^{-5}	1.7×10^{-5}	1.0×10^{-5}
Glutamate	6.1×10^{-2}	4.5×10^{-2}	3.3×10^{-2}
Glutamine	8.4×10^{-5}	7.6×10^{-5}	3.2×10^{-2}
Glycine	2.4×10^{-4}	1.8×10^{-4}	3.2×10^{-4}
Histidine	5.1×10^{-3}	5.4×10^{-3}	3.2×10^{-3}
Isoleucine	1.4×10^{-4}	2.7×10^{-4}	1.0×10^{-4}
Leucine	3.0×10^{-5}	1.3×10^{-5}	3.2×10^{-5}
Lysine			1.0×10^{-3}
Methionine	5.7×10^{-5}	8.7×10^{-5}	1.0×10^{-4}
Phenylalanine			3.2×10^{-4}
Proline	1.4×10^{-6}	1.6×10^{-6}	1.0×10^{-6}
Serine	6.5×10^{-5}	5.1×10^{-5}	3.2×10^{-5}
Threonine			5.6×10^{-6}
Tryptophan			3.2×10^{-3}
Tyrosine			3.2×10^{-4}
Valine	6.7×10^{-5}	3.6×10^{-5}	3.2×10^{-5}

[a] Data are from references 42 and 113. ND, not determined.

means of "measuring" the apparent K_d (20, 93). This is also assumed to be true for *B. subtilis* (113).

In a second kind of assay for measuring apparent K_ds, cells are sheared and then fixed to a coverslip by using antiflagellum antibody. In the shearing, some cells lose all of their flagella but retain a fragment of one of them, and this fragment can be used to tether its cell to the coverslip with antibody. Since the flagella rotate, the effect of tethering is to make the cell body rotate, and the behavior of rotating cells can be quantitated. The flagella rotate CCW for smooth swimming and CW for tumbling. Addition of attractant causes CCW rotation of the flagella, and removal causes CW rotation.

Experiments with enteric bacteria have shown that the period of smooth swimming caused by adding attractant is proportional to receptor occupancy (11, 125). Similar experiments were performed on *B. subtilis*, and the receptor K_ds inferred from these experiments agree fairly well with those calculated from the capillary assays (Table 3; 42).

It should be noted that these experiments reveal only the highest-affinity receptors. Many of the amino acids have multiple receptors that function over quite different concentration ranges. For instance, in *B. subtilis*, threonine appears to have two receptors, whose K_ds are approximately 10 μM and 7 mM (113).

Uniqueness of receptors

To determine whether the putative receptors were distinct, as are, for example, the serine and aspartate receptors of *E. coli*, competition experiments were undertaken. If two amino acids share the same binding site, they should reciprocally interfere with ("jam") each other's chemotaxis. Furthermore, other amino acids might jam both amino acids' chemotaxes or neither.

The outcome of these experiments suggested that all of the receptors—or all of the binding sites, to be more precise—were distinct. There were many instances of lack of jamming, 51 instances of nonreciprocal jamming, and 35 instances of reciprocal jamming. However, in the last case, there were always other amino acids that jammed one taxis but not the other. Interestingly, two amino acids, glutamate and aspartate, jammed taxes to all other amino acids but were themselves jammed by no other amino acids except each other (113). These amino acids also have the greatest effect on distribution of methyl groups on the methylated chemotaxis proteins (MCPs) (see below; 44).

We may speculate about the origin of the jamming that was observed. In *E. coli* and *S. typhimurium*, galactose is sensed by the galactose-binding protein (52), which then interacts with Trg protein to bring about smooth swimming (50, 53). Ribose is sensed by the ribose-binding protein, and the complex interacts with Trg. Ribose has no affinity for the galactose-binding protein, nor has galactose any affinity for the ribose-binding protein. Nevertheless, when the ribose-binding protein is induced, ribose can interfere with galactose taxis. This competition is presumed to be due to interactions at Trg (130). In *B. subtilis*, the MCPs are not specific for different amino acid attractants (see below; 136), and in any case, too many separate sites of interaction with the different amino acids must exist, judging from the foregoing experiments (113). Thus, it is tempting to speculate that individual amino acids interact with different binding proteins and the complex interacts with MCPs. The

Table 4. K_is for proline analogs as inhibitors of proline chemotaxis and transport[a]

Compound	K_i (M)	
	Chemotaxis	Transport
Azetidine-2-COOH	2.3×10^{-6}	2.3×10^{-5}
2,4-Dinitrophenylproline	$\sim 1.0 \times 10^{-6}$	1.1×10^{-2}
Glutamylproline	$> 1.0 \times 10^{-2}$	3.1×10^{-3}
4-Hydroxyproline	2.2×10^{-4}	2.3×10^{-4}
Pipecolinate	3.2×10^{-2}	1.05×10^{-2}
Proline	1.0×10^{-6}	2.3×10^{-6}
Proline amide	2.8×10^{-3}	1.7×10^{-4}
Prolylglutamate	3.2×10^{-3}	3.1×10^{-3}
Prolylhydroxyproline	3.2×10^{-4}	5.9×10^{-3}
Prolylproline	5.8×10^{-4}	2.6×10^{-4}

[a] Data are from reference 114.

ability of one amino acid to jam taxis toward another would then reflect the complex of one binding protein with its associated amino acid blocking access of the other binding protein with its associated amino acid to MCPs. However, there is no direct evidence for these amino acid-binding proteins.

Relation of chemotaxis and transport

Judging from experiments with proline taxis and transport, it appears that chemotaxis and transport are independent in *B. subtilis*. Various analogs of proline were tested as inhibitors of proline chemotaxis and inhibitors of proline transport (Table 4). Two of the analogs inhibited proline transport much more effectively than proline chemotaxis, and two inhibited it much less effectively. Furthermore, mutants that showed considerable shifts in K_m for proline transport but no change in proline chemotaxis were identified. Therefore, unlike galactose transport and chemotaxis in *E. coli*, proline transport and chemotaxis in *B. subtilis* are independent (114). Subsequently, by use of direct binding assays, aspartate chemotaxis and transport in *S. typhimurium* and serine chemotaxis and transport in *E. coli* were found to be independent (26, 54).

Sugars as attractants

Sugar taxis in *B. subtilis* is considerably weaker than amino acids taxis (Table 5). As a result, it was not possible to conduct experiments with sugar attractants in *B. subtilis* that were directly parallel to those conducted with amino acids. Competition experiments revealed that most of the sugar receptors are distinct, but there is a mechanism by which taxis to one sugar is substantially inhibited in the presence of another sugar, even when the sugars do not share a common receptor (115). Many instances of reciprocal jamming, nonreciprocal jamming, and lack of jamming were found. Usually, but not always, other sugars that jammed one member of the mutual jamming pair but not the other were identified. Some sugars that were good jammers were also good attractants. This group included 2-deoxyglucose, gentiobiose, gluconate, glucose, maltose, melibiose, α-methylglucoside, β-methylglucoside, sorbose, and xylose. However, gluconate and melibiose were poor attractants but good jammers. Finally, other sugars, includ-

Table 5. Capillary assays of sugar taxis by *B. subtilis*[a]

Attractant	Concn (M)[b]	
	Peak	Threshold
D-Glucosamine	10^{-6}	10^{-8}
D-Cellobiose	10^{-5}	10^{-6}–10^{-7}
D-Mannitol	3.2×10^{-5}	10^{-8}–10^{-9}
N-Acetyl-D-glucosamine	3.2×10^{-5}	10^{-7}–10^{-8}
D-Fructose	3.2×10^{-5}	10^{-7}–10^{-8}
D-Mannose	10^{-4}	10^{-7}–10^{-8}
α-Methyl-D-glucoside	10^{-4}	10^{-7}–10^{-8}
Sucrose	10^{-4}	10^{-7}–10^{-8}
Trehalose	10^{-4}	10^{-7}
β-Methyl-D-glucoside	10^{-4}	10^{-6}
D-Glucose	3.2×10^{-4}	10^{-7}
D-Sorbitol	10^{-3}	10^{-7}
Raffinose	10^{-3}	10^{-5}–10^{-6}
Maltose	3.2×10^{-3}	10^{-6}–10^{-7}
N-Acetyl-β-D-mannosamine	3.2×10^{-3}	10^{-5}–10^{-6}
Gentiobiose	10^{-2}	10^{-6}
α-Methyl-D-mannoside	10^{-2}	10^{-4}
α-Glycerophosphate	10^{-2}	10^{-3}
L-Sorbose	3.2×10^{-2}	10^{-5}
2-Deoxy-D-glucose	3.2×10^{-2}	10^{-4}–10^{-5}
D-Gluconate (10 mM)	10^{-1}	10^{-4}
DL-Glyceraldehyde[c]	10^{-1}	10^{-3}–10^{-4}
D-Xylose	10^{-1}	10^{-3}
Melibiose	10^{-1}	10^{-2}–10^{-3}

[a] Data are from reference 115.
[b] Threshold values are minimum molar concentrations that give an effect different from buffer in a capillary assay. Peak concentrations are molar concentrations of attractant for which maximal accumulation of cells occurs in a capillary assay.
[c] Absent from growth medium due to toxicity.

ing fructose, mannitol, mannose, α-methylmannoside, *N*-acetylglucosamine, *N*-acetylmannosamine, sorbitol, sucrose, and trehalose, were good attractants but poor jammers.

Repellents and possible receptors

A diverse catalog of substances are repellents of *B. subtilis* (108; see Table 6). Many of them are readily classified on the basis of their use in other contexts.

There are several indications that receptors for many repellents may exist. These indications come from two lines of experimentation. First, a repellent desensitized cells to further additions of itself but not to other chemicals, a result consistent with binding at different sites. However, there are exceptions to this rule (Table 7). In contrast, structurally similar repellents usually mutually desensitize; for instance, tetracaine, procaine, and lidocaine mutually interfere, suggesting a common receptor (104).

Second, when three uncouplers of oxidative phosphorylation (FCCP, TCSA, and PCP) were tested as inhibitors of glycine or proline transport, they showed uncompetitive inhibition in two instances, competitive inhibition in two, and noncompetitive inhibition in two (Table 8). These results have been interpreted to mean that (i) the transport substrate has to bind to make a site available for the inhibitor (uncompetitive); (ii) if one binds, the other cannot (competitive); or (iii) inhibition always occurs if the inhibitor binds (noncompetitive). It should be noted that in six instances where inhibition occurred, there were individ-

Table 6. Repellents of *B. subtilis*[a]

Compound[b]	Threshold concn	Reference
Uncouplers of oxidative phosphorylation		
FCCP	10 nM	109
CCCP	0.18 μM	109
CCP	1.8 μM	109
PCP	0.1 μM	109
3,4,5-Trichlorophenol	32 μM	112
3,5-Dichlorophenol	18 μM	112
2,4,6-Trichlorophenol	32 μM	112
3,3′,4′,5-Tetrachlorosalicylanalide	18 μM	109
2,6-Dibromophenol	6.3 μM	109
2,4-Dinitrophenol	36 μM	109
Inhibitors of electron transport		
Sodium azide	10 mM	109
2-Heptyl-4-hydroxyquinoline-*N*-oxide	4 μM	109
Sodium cyanide	56 μM	109
Amytal	0.5 mM	109
Local anesthetics		
Tetracaine	9.4 μM	109
Lidocaine	1.0 mM	109
Procaine	0.58 μM	109
Drugs		
Chlorpromazine	2.8 μM	104
Promazine	4 μM	99
Phenobarbital	40 μM	99
Diallylbarbital	5 μM	99
Hexobarbital	0.4 μM	99
Pentabarbital	40 μM	99
Primidone	5 μM	99
Glutethimide	9 μM	99
Diazepam (Valium)	7 μM	99
Flurazepam (dalmane)	3 μM	99
Librium	3 μM	99
Quinacrine	5.6 μM	109

[a] Threshold concentrations are minimal molar concentrations that cause tumbling of cells in a microscope assay. Butyrate, indole, and *p*-hydroxybenzoate are also repellents, but the threshold concentrations have not been determined (136).
[b] FCCP, trifluoromethoxycarbonylcyanidephenylhydrazone; CCCP, *m*-chlorocarbonylcyanidephenylhydrazone; CCP, carbonylcyanidephenylhydrazone; PCP, pentachlorophenol.

ual differences. This result implies that inhibition of action of membrane proteins might be a rather widespread phenomenon, and each receptor protein might have various sites on or within it at which different inhibitors can bind (21). It is not unreasonable to imagine that these repellent substances might also bind to proteins, like MCPs, that affect chemotactic signal transduction.

These results were further extended by analyzing the effects of PCP and analogs both as inhibitors of amino acid transport and as repellents. Less highly substituted chlorophenols, which are weaker acids than PCP, are also weaker inhibitors, although the mode of inhibition (noncompetitive for proline and uncompetitive for glycine) is the same. Pentachlorothiophenol, which is insoluble, was also a potent inhibitor (Table 9). These results are consistent with the inhibiting species being an anion binding to the transport proteins within the membrane (101). Their actions as repellents reflect the same order of potency (Table 9) (112). This supports the idea that these substances might be acting as repellents by binding as charged forms to some membrane-spanning receptor protein. The MCPs of *E. coli* appear to be the loci of repellent actions in that organism (29), and further work might reveal the same to be true in *B. subtilis*.

MOLECULAR MECHANISMS OF SIGNALING

Homologies of Structure and Function between *B. subtilis* and Enteric Bacterial Signaling Proteins

Our most detailed information about the mechanisms involved in sensory transduction comes from studies of enteric bacteria, in which the majority of the genes involved in structural and regulatory roles are probably known and have been sequenced (81, 82, 84). In addition, many of the proteins in the sensory pathway have been studied biochemically, and their major functions have been identified. Fewer genes involved in signaling have been identified and sequenced in *B. subtilis*, and the exact biochemical role of these genes is defined in only a few cases. However, a large number of *B. subtilis* genes implicated in signaling show significant protein sequence homology to enteric signaling proteins (usually from 25 to 40% amino acid sequence identity; Table 1). This suggests that these genes have evolved from common prokaryotic ancestors and have experienced a comparable evolutionary drift.

In many instances, the *B. subtilis* homologs of enteric flagellar, motility, and chemotaxis genes are

Table 7. Effect of preincubation with one repellent on postadaptive thresholds of others in *B. subtilis*[a]

Preincubation reagent[b]	Relative change in threshold for newly added reagent							Concn of preincubation reagent (M)
	NaCN	FCCP	PCP	TCSA	TPB	Chlorpromazine	Tetracaine	
NaCN	1.8	1	1	1	1	1	1	5.6×10^{-4}
FCCP	0.56	*18*	1.8	1	1.8	1	1.8	1.8×10^{-7}
PCP	1.8	1	*56*	1	1	1.8	*3.2*	3.2×10^{-6}
TCSA	0.56	1.8	1	*0.32*	1	1.8	*3.2*	3.2×10^{-8}
TPB	*3.2*	1.8	1.8	0.56	*18*	*0.32*	1	5.6×10^{-4}
Chlorpromazine	1.8	1	0.56	0.56	1	*10*	1.8	9.0×10^{-6}
Tetracaine	0.56	*5.6*	1	1.8	1	1.8	*32*	2.9×10^{-4}

[a] Data are from reference 104. True postadaptive threshold values, i.e., the molar concentrations needed to cause tumbling behavior in the microscope assay, can be calculated by multiplying the relative changes in threshold by the unperturbed threshold values for each repellent (given in Table 6). Numbers in italics indicate shifts of postadaptive threshold by 3.2-fold or more.
[b] TCSA, 3,3′,4′,5-tetrachlorosalicylanalide; TPB, tetraphenylboron. For other compounds, see Table 6, footnote *b*.

Table 8. Summary of effect of uncouplers on amino acid transport in *B. subtilis*[a]

Inhibitor[b]	Proline			Glycine		
	Nature	K (μM)	Time of complete inhibition	Nature	K (μM)	Time of complete inhibition
FCCP	Uncompetitive	0.25	Immediate	Competitive	0.21	~2 min
TCSA	Noncompetitive	0.020	Immediate	Competitive	0.037	~2 min
PCP	Noncompetitive	3.3	Immediate	Uncompetitive	1.4	Immediate

[a] Data are from reference 21.
[b] For abbreviations, see Table 6, footnote *b*, and Table 7, footnote *b*.

known to be involved in signaling. However, homologous functions for homologous proteins have been shown biochemically only for the flagellin genes (*hag-fliC*), the secondary sigma factor genes (*sigD-fliA*), *fliY-fliN*, *cheR*, and *cheB* (see below). It is important to stress that we cannot always assume that conservation of protein sequences between two genes signifies conservation of protein function. There is every likelihood that proteins playing a particular role in motility and chemotaxis in *E. coli* have evolved into somewhat different roles in *B. subtilis*.

Even so, many of the *B. subtilis* genes implicated in signaling have been given designations that correspond to those of their enteric homologs, as suggested by Albertini et al. (5; Table 1). Where it is appropriate in discussing the functions of these proteins, we will also use a subscript E or B to indicate the origin of the protein, e.g., $CheA_E$ for the enteric CheA protein, and $CheA_B$ for its *B. subtilis* counterpart.

The Enteric Paradigm

In enteric bacteria, amino acid attractants bind at receptors that also function as MCPs. These MCPs span the membrane twice. The chemoeffector binding region is located between the membrane-spanning regions in the N-terminal part of the protein; it lies in the periplasmic space. The much larger C-terminal part of the molecule, where sensory transduction takes place, lies in the cytoplasm (7, 97) and contains four to six glutamates that may be methyl-esterified (63–66; 133, 134). Here the interactions with other chemotaxis proteins, including the autophosphorylating kinase, $CheA_E$, presumably occur.

Of the four MCPs, one, Tar, binds aspartate and related amino acids (26) and another, Tsr, binds serine and related amino acids (54). These MCPs also mediate chemotaxis to certain repellents (123). A methyltransferase, $CheR_E$, uses *S*-adenosylmethionine (AdoMet) to methylate the MCPs (124), and a methylesterase, $CheB_E$, hydrolyses these methyl groups (128).

Table 9. Values of chlorophenols as inhibitors of proline and glycine transport in *B. subtilis*[a]

Inhibitor	pK_a	pK_i	
		Proline	Glycine
Pentachlorophenol	4.7	5.64	6.16
3,4,5-Trichlorophenol	6.2	5.51	5.94
3,5-Dichlorophenol	7.1	4.41	4.52
2,4,6-Trichlorophenol	7.5	3.28	3.74
2,4-Dichlorophenol	8.2	3.49	3.40

[a] Data are from reference 101.

Besides these proteins, four others—$CheA_E$, $CheY_E$, $CheW_E$, and $CheZ_E$—play important roles in mediating sensory transduction through protein phosphorylation (16, 17). $CheA_E$ is an autophosphorylating kinase that phosphorylates $CheY_E$ and $CheB_E$ (59). The phosphorylated form of $CheB_E$ ($CheB_E$-P) is much more active than $CheB_E$ (80, 127). $CheZ_E$ dephosphorylates $CheY_E$ (59). $CheW_E$ allows modulation of $CheA_E$ activity to reflect binding of attractant or repellent at the MCPs (16, 79). $CheY_E$-P binds at the switch at the base of each flagellum to cause clockwise rotation (18, 61, 85). The switch consists of FliM, FliG, and FliN (67).

Addition of attractant causes a reduced rate of $CheA_E$ kinase activity and hence a deficiency of $CheY_E$-P and $CheB_E$-P (16). Due to lack of $CheY_E$-P, the bacteria swim smoothly. Due to lack of $CheB_E$-P, increased methylation of the MCP occurs, especially of the MCP binding the attractant, and the rate of demethylation is low (65, 66, 129, 140). In the course of time, this increased methylation brings $CheA_E$ activity back to normal and adaptation occurs (45). Conversely, addition of repellent would be expected to enhance $CheA_E$ activity and hence levels of $CheY_E$-P and $CheB_E$-P. The $CheY_E$-P causes tumbling but the $CheB_E$-P demethylates the MCPs so that $CheA_E$ activity soon returns to normal. $CheZ_E$ constantly dephosphorylates $CheY_E$-P; thus the levels of $CheY_E$-P would reflect "current" $CheA_E$ activity, rather than remaining high.

Signal Transduction in *B. subtilis*

Except for CheZ, homologs of the chemotaxis proteins that mediate sensory transduction in enteric bacteria are all present in *B. subtilis*. They include $CheR_B$, $CheB_B$, $CheA_B$, $CheY_B$, and $CheW_B$. A plausible pathway for the functioning of this signal pathway is diagrammed in Fig. 3. In this formulation, attractant binds to MCPs, probably indirectly via specific binding proteins, to activate $CheA_B$ to autophosphorylate. $CheA_B$-P phosphorylates $CheY_B$ and $CheB_B$. $CheY_B$-P binds to the switch to cause smooth swimming. The switch consists of FliM, FliG, and FliY. FliY is homologous to FliN but is 28 kDa larger. $CheB_B$-P brings about adaptation through methyl transfer to an unidentified "regulator." In *B. subtilis*, it appears that CheY-P causes smooth swimming, whereas in *E. coli*, it causes tumbling. The methylated regulator may

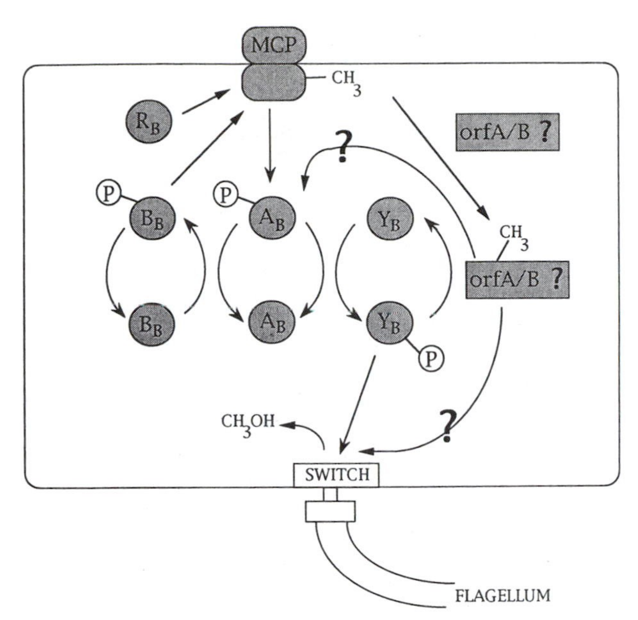

Figure 3. Plausible model for the signal pathway for chemotaxis in *B. subtilis*. Adapted from reference 13.

play the role of CheZ in *E. coli*, that is, catalyze dephosphorylation of $CheY_B$-P so that its levels do not stay high longer than desirable, and/or it may interact with the MCPs and $CheA_B$ to help restore $CheA_B$ activity to its unstimulated level. The methylated regulator may be demethylated at FliY. In the remainder of this section, some of the unique properties of *B. subtilis* chemotaxis that give rise to this model are delineated.

Turnover of Methyl Groups on MCPs

In contrast to what is found in *E. coli*, in *B. subtilis*, little change in total methylation occurs following the addition of attractant. Instead, there is a period of enhanced turnover of methyl groups on the MCPs that diminishes with time. Both amino acid and sugar attractants cause this effect. All MCPs are affected regardless of the attractant (135, 136, 138). The attractants aspartate, glutamate, and, to a lesser extent, histidine cause a substantial change in the distribution of methyl groups on the MCPs. They cause increased methylation of MCP H1, as evidenced by methylation of bands in the H1 region on a sodium dodecyl sulfate (SDS)-polyacrylamide gel (44). There is also increased methylation of MCP H2, apparently by methyl transfer from H1 (9). Overall, however, there is little net change in the number of methyl groups on the MCPs (136, 138). The other amino acids cause much more minor changes in MCP methylation, and the sugars cause none (44, 135, 136).

The effects of repellent on methylation and demethylation of the MCPs are complex. At low concentrations, repellent appears to cause decreased turnover of methyl groups on the MCPs. Both indole and butyrate have been analyzed. However, in the case of butyrate at high concentrations, turnover starts to increase as a function of concentration, as though an attractant were being added simultaneously. In a capillary assay, on the other hand, butyrate acts as a repellent throughout the whole range of concentrations (48). More work is needed to understand this puzzle.

Methyl Transfer beyond the MCPs

In *E. coli*, methyl groups on the MCPs are directly hydrolyzed to produce methanol. In *B. subtilis*, they are apparently transferred to another protein(s), possibly to form a methylated regulator (137). The evidence for this transfer can be summarized as follows. First, if attractant is added to cells previously labeled with [*methyl*-^{3}H]methionine, there is an immediate (within 5 s) delabeling of the MCPs. However, methanol appears in the medium much more slowly during the adaptation period. Thus, there is a kinetic discrepancy between the loss of radioactive methyl groups from the MCPs and the appearance of radioactive methanol (138).

Second, if the attractant is removed with excess nonradioactive methionine still present, the MCPs become relabeled. Thus, the radioactive methyl groups must have remained fixed to some receptor in the system and not have been significantly released as methanol as the result of the original addition of attractant (137).

Third, cells may be radiolabeled and then transferred to a nitrocellulose (Millipore) membrane in a flow cell. If buffer containing nonradioactive methionine with and then without attractant is alternately pumped past the cells, radioactive methanol will be released both when the attractant is added and when it is removed. The major release of radioactive methanol occurs on the third or fourth cycle, not the first. By contrast, in a similar experiment using the repellent leucine for *E. coli*, the major release occurs on the first cycle of addition, as expected if the MCPs are the direct source of methanol (137).

Methanol Formation

Although there is no direct release of methanol from the MCPs in *B. subtilis*, methanol is ultimately released from the protein(s) to which it is transferred. The addition or removal of high concentrations of attractant or repellent causes methanol to form (136). Interestingly, the addition of small amounts of attractant gives little or no methanol, and the addition of small amounts of repellent, which inhibits methyl transfer, also fails to release methanol. Thus, there may exist a process that allows adaptation to small stimuli but does not involve methanol formation.

Involvement of the Motor in Sensory Transduction

As discussed above, many chemotaxis mutants were isolated, divided into complementation groups, and subsequently mapped to region 140 on the *B. subtilis* chromosome. These mutants were motile but showed defects in chemotaxis both on swarm plates and in capillary assays. Hence, these complementation groups were originally designated *cheA-cheO* (111, 153). Subsequent sequencing has shown that the majority of these genes are homologous to enteric genes that specify flagellar structural or assembly proteins, and these open reading frames have been renamed in keeping with this homology (5, 150, 151; Fig. 1 and Table 1).

These results are most easily interpreted if it is assumed that the morphology and functioning of the flagellar motor can also play a role in signal transduction and methanol release (147). That is, these "*che*" mutants synthesize intact but defective flagellar structures that are blocked in signal transduction. Thus, for example, *cheE*, *cheG*, *cheJ*, and *cheK* mutants mostly tumble, and *cheI* mutants swim smoothly (110, 111). Furthermore, *cheE*, *cheF*, *cheG*, and *cheH* mutants show abnormal release of methanol upon the four stimuli of addition and removal of attractant and repellent (103a, 146). It is possible that this methanol release occurs at FliY, which is homologous to FliN except for having an additional 28 kDa of polypeptide. In *E. coli*, the methanol is released directly from the MCPs, which are located over the entire surfaces of the cells, not just near the motors (30). Thus, in this fundamental way, the two organisms appear to be different in the functioning of their chemotactic pathways.

Role of Calcium Ion

Over the years, there have been tantalizing suggestions that Ca^{2+} plays a role in chemotaxis in *B. subtilis*. In 1976 and 1977 Ordal found that in the presence of A23187, a Ca^{2+} ionophore, and 0.1 μM Ca^{2+}, the bacteria tumble perpetually and do not swim smoothly even when 10 mM alanine, an attractant, is added (105, 106). On this basis, he concluded that Ca^{2+} regulates chemotactic behavior in bacteria (106), but that conclusion is almost certainly incorrect. However, the possibility that Ca^{2+} plays some role was revived by the experiments of Matsushita et al. (91), who showed that Ca^{2+} channel blockers inhibit chemotaxis but not growth and motility in *B. subtilis*. This idea received further impetus from the purification of a 38-kDa Ca^{2+}-binding protein from *B. subtilis* that binds Ca^{2+} even in the presence of excess Mg^{2+} (141). This protein appears to be missing from a mutant lacking most flagellar, motility, and chemotactic genes (141a). To define the role of Ca^{2+}, it will be helpful to clone and inactivate the gene for this protein and characterize the corresponding mutant.

Comparison with *Halobacterium halobium*

As mentioned above, all stimuli (both addition and removal of attractant and repellent) can cause methanol production (137). In *E. coli*, only negative stimuli, such as addition of repellent, cause methanol production; positive stimuli inhibit it (137, 139, 140). Interestingly, all stimuli of *H. halobium*, one of the archebacteria, also cause methanol formation, as in *B. subtilis* (4). As mentioned in the next section, the MCPs of *B. subtilis* are larger than those of *E. coli* and more in line with those of *H. halobium*. It is tempting to speculate that the underlying mechanisms of chemotaxis in *H. halobium* and *B. subtilis* might both involve a methyl transfer process that might reflect properties of the ancestral bacterium from which the bacterial and archebacterial superkingdoms descend.

FUNCTIONAL CHARACTERISTICS OF SPECIFIC CHEMOTAXIS PROTEINS

MCPs of *B. subtilis*

As in *E. coli*, the MCPs of *B. subtilis* are integral membrane proteins that are methyl esterified on glutamate side chains as the result of methyl transfer from *S*-adenosylmethionine (2, 68, 145). This reaction is catalyzed by the chemotactic methyltransferase CheR. In *B. subtilis*, methylation of MCPs has been observed both in vivo and in vitro (144).

Some of the MCPs have been partially purified from *B. subtilis*. The purification was carried out on MCPs from a mutant lacking $CheR_B$ so that the MCPs were unmethylated and could be more easily obtained in homogeneous form. Methylation changes the charges on the MCPs and causes increased mobility on SDS-polyacrylamide gels. There are two groups of MCPs. The major MCPs in terms of susceptibility to be methylated have been designated H1 (97 kDa), H2 (86 kDa), and H3 (77 kDa) (49a). The *E. coli* MCPs are 58 to 60 kDa (15, 19, 72, 118), and *B. subtilis* has a couple of poorly methylated MCPs of about this size also (49a).

$CheR_B$ Methyltransferase

The $CheR_B$ methyltransferase has been purified to homogeneity. It transfers methyl groups from *S*-adenosylmethionine to glutamate residues on the MCPs. It is activated by divalent cations and has a K_m for *S*-adenosylmethionine of 5 μM. It is inhibited by the product of the reaction with *S*-adenosyl homocysteine; the K_i is 0.2 μM (24). It also methylates MCPs from *E. coli*, and reciprocally, $CheR_E$ methylates MCPs from *B. subtilis* (23), providing direct evidence for homologous functions.

Studies of a null mutant, made with an integration plasmid, have revealed very interesting behavior. First, the mutant is somewhat tumbly, unlike the *E. coli* counterpart, which swims smoothly. Second, when the cells are tethered by a single flagellum and exposed to attractant, the cells excite but adapt only partially (67a).

A plasmid carrying $CheR_B$ complements a $cheR_E$ mutant on a swarm plate. The methylation of *E. coli* MCPs carried out in this complemented strain is not completely normal, but the (subtle) differences have not yet been elucidated (67a). Comparison of gel filtration and SDS-polyacrylamide gel experiments with $CheR_B$ show that CheB acts as a 30-kDa monomer (24). The gene $cheR_B$ has recently been sequenced (58, 103a). It encodes a protein that is 31% identical to $CheR_E$ in amino acid sequence.

$CheB_B$ Methylesterase

The $CheB_B$ methylesterase has been purified to homogeneity. $CheB_B$ was assayed by measuring transfer of radioactive label from MCPs previously radiolabeled in vitro by using $CheR_B$ and *S*-adenosyl-[*methyl*-^3H]methionine. The apparent K_m for methylated MCPs was about 10 nM. Additional unmethylated MCPs do not inhibit this reaction (41). Other experiments indicate that $CheB_B$ has virtually no affinity for membranes containing unmethylated MCPs (99). The enzyme requires glycerol and Mg^{2+} ion for activity in vitro. The rate of demethylation was enhanced on addition of attractants. $CheB_B$ also demethylated *E. coli* MCPs, but this demethylation was inhibited by addition of attractants, and similarly, in vivo, attractants inhibit demethylation of MCPs. Finally, 1 mM Mg^{2+} ion was needed to demethylate the *B. subtilis* MCPs but not the *E. coli* MCPs (41, 43, 100).

Gel filtration of a cytoplasmic extract revealed activities at apparent molecular masses of 41 and 66 kDa. This result suggests that $CheB_B$ binds to a protein of 25 kDa. The identity of this protein is not yet known (99).

Studies of a null mutant of *cheB*$_B$ have been very revealing. A null mutant is very slightly biased toward smooth swimming compared with wild type. It excites normally and only partially adapts. Therefore, $CheB_B$ is not required for excitation but is required for complete adaptation. The null mutant shows only 10% normal methylation of MCPs in vivo. This degree of methylation reflects the natural rate of spontaneous hydrolysis of methyl groups from the MCPs (27a). Thus, $CheB_B$ is required for methyl transfer away from the MCPs. These experiments are the origin of the hypothesis that the purpose of methyl transfer, which is enhanced by addition of attractants, is adaptation. The protein to which methyl groups are transferred has not been identified.

We hypothesize that in vivo, $CheB_E$ carries out methanol release but $CheB_B$ does not. This hypothesis is based on a considerable body of evidence that suggests that methyl groups on the MCPs are transferred to an unidentified intermediate in *B. subtilis* but not in *E. coli*. However, $CheB_E$ and $CheB_B$ can both cause methanol formation in vitro from both *E. coli* and *B. subtilis* MCPs. Furthermore, complementation experiments indicate that expression of $CheB_B$ in *E. coli* complements a *cheB*$_E$ mutant, providing direct evidence for homologous functions of the two proteins. Additional work is needed to clarify these contradictory findings.

$CheA_B$ Autophosphorylating Kinase

According to an analysis of the *cheA*$_B$ gene sequence, $CheA_B$ is 34% identical to $CheA_E$ (Table 1), and $CheA_B$ appears to be a member of a family of autophosphorylating kinases. A number of invariant sequences in family members are also conserved in $CheA_B$, and the phosphorylated residue of $CheA_E$ (His-48) is conserved in $CheA_B$ as His-46. A nonpolar null mutant, made by inserting a *cat* gene into *cheA*$_B$, did not respond to addition of attractant or repellent and did not show any stimulation of methanol formation (33).

The phenotype of the $CheA_B^-$ null mutant is tumbly. This is the opposite of the phenotype of $CheA_E^-$ strains, which are smooth swimming, probably because of the deficiency of $CheY_E$-P. However, the $CheA_B^-$ mutants of *B. subtilis* are not just "opposite" those of *E. coli*. During the course of the original mutant isolation and characterization, six mutants (in *cheN*$_B$, later renamed *cheA*$_B$) were identified. Four were tumbly, one was random, and one swam smoothly. These differences notwithstanding, it seems very likely, on the basis of homology with $CheA_E$, that $CheA_B$ initiates the biochemical events that mediate chemotactic sensory transduction (33, 110, 111).

$CheY_B$ Regulator

cheY$_B$ (formerly named *cheB*) is located beside the putative "switch" genes *fliM*$_B$ and *fliY* (see below) and a considerable distance from the remaining chemotaxis genes (Fig. 1; 12). With only one exception, *cheY*$_E$ mutants of *E. coli* are smooth swimming. In that exceptional mutant, Asp-13 is changed to Lys-13 and the mutant becomes a tumbler, as though $CheY_E$ were "permanently phosphorylated" (18). In *B. subtilis*, all of the original five *cheY*$_B$ mutants are smooth swimming (13, 110, 111). However, null mutants produced by nonpolar insertion or deletion were tumbly (12). Recently, a mutant in which Asp-10 (equivalent to $CheY_B$ Asp-13) was changed to Lys-10 was found to be tumbly also; that is, unlike in *E. coli*, in which this mutation produced the opposite phenotype from the null mutant, it produced in *B. subtilis* the same phenotype as the null mutant (13). The fact that the *cheY*$_B$ and *cheA*$_B$ null mutants are both tumbly whereas the *E. coli* null mutants are both smooth swimming suggests that $CheY_B$-P in *B. subtilis* causes smooth swimming, whereas $CheY_E$-P in *E. coli* causes tumbling.

Work has been done on the smooth-swimming mutant bearing the *cheY2* allele. It was possible to partially complement this mutation on a swarm plate by introducing the wild-type gene on a plasmid. Furthermore, the bacteria were still almost exclusively smooth swimming and did not tumble when the repellent acetate was added. Thus, the mutant was found to be partially dominant. One interpretation of these results is that $CheY_B$ binds more tightly to the switch than does $CheY_E$, and it is much easier to get mutants in which $CheY_B$ binds strongly in the conformation that promotes smooth swimming. Presumably, this is the $CheY_B$-P conformation.

$CheW_B$

The *cheW*$_B$ gene was also identified during sequencing of the *che* operon. It encodes a protein that is 27% identical to its *E. coli* counterpart (49). The gene was inactivated by a nonpolar *cat* insertion. Interestingly, the mutant was as sensitive as wild type to attractant-induced increased turnover of methyl groups on the MCPs and repellent-induced decreased turnover. Therefore, unlike the *E. coli* homolog, $CheW_B$ is not needed to couple events at the receptors to subsequent biochemical events (48).

Other Potential Chemotaxis Proteins

Two other proteins encoded by genes in the "chemotaxis region" of the major *che-fla* operon may play roles in chemotaxis (Fig. 1 and Table 1). The respective genes, *orfB* and *orfA*, are immediately upstream from *sigD*. Inactivation of these genes makes the MCPs considerably less susceptible to methyl group turnover in vivo (117a), as was found for inactivation of *cheB*$_B$. These results are consistent with the defect in all these strains being in removal of methyl groups from the MCPs. Thus, during an in vivo methylation experiment, the turnover, which is needed to cycle radioactive methyl groups into the MCPs, is reduced. The *orfA* and *orfB* null mutants tumble incessantly and do not respond to addition of attractants (67a). Neither OrfA nor OrfB is homologous to any known protein, although OrfA has a near-consensus ATP-binding site. It is tempting to speculate that OrfA, OrfB, and CheB$_B$ interact and cooperate to carry out the methyl transfer process.

Mutants with mutations in *orf-298* are also being sought. The corresponding protein is 24% identical to the iron protein of nitrogenase, which is responsible for transferring electrons from ferredoxin or flavodoxin to the FeMo protein that reduces N_2 to NH_3 (22). Orf-298 also has a consensus ATP-binding site.

Switch Proteins

As noted above, the switch that controls the direction of flagellar rotation in *E. coli* consists of FliG, FliM, and FliN (67). FliN (14.4 kDa) is a short protein, about half the size of FliG (36.8 kDa) and FliM (37.8 kDa). *B. subtilis* has FliG and FliM and, instead of FliN, FliY, a much larger protein (42 kDa). Expression of FliY on a plasmid can complement *S. typhimurium* mutants in *fliN* (but not in *fliM*) for motility but not chemotaxis. The complemented bacteria swim smoothly. The N terminus of FliY shows 33% identity with the first 122 amino acids of FliM$_E$, whereas the C terminus of FliY has 52% identity with the last 30 amino acids of FliN$_E$. The middle 60% of FliY is not significantly homologous to any known protein (12, 13, 111). It is tempting to speculate that the increased turnover of methyl groups on the MCPs leads to methylation of a regulator, which binds to the nonhomologous part of FliY. The purpose of this methylation may be to facilitate dephosphorylation of CheY$_B$-P, built up following addition of attractant so that it does not persist longer than is desirable. *B. subtilis* apparently lacks CheZ, which fulfills the function of dephosphorylation of CheY$_E$-P in *E. coli* and may interact with the switch (116). Since release of methanol occurs at the motor (see above), it is not unlikely that it occurs at FliY, the unique part of the *B. subtilis* switch. Since FliN and FliY are related, we imagine that during evolution, FliY was truncated to make FliN at the same time that the methylation system was simplified.

REGULATION OF EXPRESSION OF *fla-che-mot* GENES

Expression of the genes of the sensory pathway is regulated in both *B. subtilis* and enteric bacteria by the composition of the growth medium (1, 71, 96, 121), by the stage of cell growth (70, 95, 102), and by the expression of other *fla-che-mot* genes needed for a functional sensory system (74, 150; for a review, see reference 83). The molecular mechanisms that coordinate and regulate *fla-che-mot* gene expression in these diverse pathways are still poorly understood even in the best-characterized systems and will probably prove to be quite complex.

The most clearly defined regulatory system for *fla-che-mot* genes is that of enteric bacteria (*E. coli* and *S. typhimurium*), in which a majority of the genes in the sensory pathway appear to have been identified. The genes form a regulatory hierarchy of at least three classes, designated I, II, and III. Genes in class I, *flhC* and *flhD*, are required for expression of all other *fla-che-mot* genes (69, 74). The expression of these "master regulatory genes" is regulated by growth medium, particularly by the presence of cyclic AMP (71, 121), and also by expression of "stress regulons" that repress *flhC-flhD* expression and consequently suppress *fla-che-mot* gene expression during many situations involving environmental stress (120).

Genes in classes II and III absolutely depend on expression of class I genes and, in addition, form a complex regulatory hierarchy of their own (74). One of the class II genes, designated *fliA* (103), encodes a transcriptional regulatory factor originally called σ^F (6) that programs the bacterial RNA polymerase to transcribe class III genes, which possess a conserved promoter sequence recognized by this factor (8, 55). The σ^F factor appears to be a member of a large class of related σ factors that are present in many different bacterial species, for which the general designation σ^{28} family has been suggested (96). These σ factors are implicated in flagellar gene expression in a wide variety of bacteria, and there is a homologous factor, σ^D, with a nearly identical promoter specificity and function, in *B. subtilis* (6, 39, 40).

However, expression of the *fliA*-σ^F regulatory factor does not lead to an immediate explanation for some important regulatory features of the *fla-che-mot* hierarchy. In particular, failure to express any of the class II genes leads to repression of class III gene expression (74). It seems that the entire flagellar HBB assembly must be completed to ensure normal synthesis of class III genes, including the flagellin and hook-associated protein genes, which are added on after completion of HBB assembly (62, 132).

Recent studies by Gillen and Hughes (37) have led to the identification of a negative regulator of class III gene expression, designated FlgM, that may be involved in the coupling of this expression to flagellar assembly. The *flgM* gene itself is transcribed by the *fliA* (σ^F) form of RNA polymerase.

Regulation of *fla-che-mot* gene expression in *B. subtilis* shows evidence of many similar kinds of control. However, elucidation of specific regulatory mechanisms has been handicapped by the fact that identification and mapping of specific *fla-che-mot* genes and studies of their regulation have lagged behind studies of their enteric counterparts. There are gene families in *B. subtilis* that roughly correspond in structure to the enteric class II and class III genes, and there are even some similarities in operon structure. A large cluster of genes homologous to enteric flagellar assembly and structural genes for the HHB (class II genes) is found in region 140 of the chromosome (see below;

Fig. 1 and Table 1), and these genes are apparently coordinately expressed in a very large operon.

Expression of a group of genes similar to enteric class III, which includes *hag* and the *motAB* operon (95, 96), is controlled by the σ^D factor, which is required for recognition of their specific promoters. The σ^D factor is homologous in structure and function to the enteric FliA protein, as confirmed by *sigD* complementation of *fliA* mutations in *E. coli* (25). In addition, a *B. subtilis* gene encoding a homolog to the enteric FlgM regulator has recently been found to lie in an operon controlled by the σ^D factor, mapping in region 310 near *hag* (94; Table 1). This operon is transcribed from the P_{D-1} promoter and contains at least one other flagellar gene.

There is currently no information as to whether *B. subtilis* has regulatory genes that correspond to those in the enteric class I master regulatory operon *flhD-flhC*. Flagellar synthesis in *B. subtilis* is not sensitive to glucose repression (94). There is evidence that a number of different kinds of mutations, including those that affect autolysin synthesis (*lyt*) and protease secretion (*deg*), can block flagellar synthesis, but it is not known at what level these effects are mediated.

Two well-known global regulators in *B. subtilis* are products of the *abrB* and *spo0A* genes, which act to control expression of vegetative and early postexponential genes (see chapters 51 and 52). Expression of at least two *B. subtilis* σ^D-dependent promoters has been reported to be dependent on *spo0A* expression (38). In addition, the rate of synthesis of *hag* RNA is increased significantly in *abrB spo0A* double mutants, suggesting that *abrB* may negatively regulate *hag* transcription (94a). However, recent studies (88, 90) show high levels of flagellin protein expression in *spo0A* mutants, and informal discussion with many workers suggests that most *spo0A* mutants are motile. Hence, the question of *spo0A* regulation of this system remains unresolved.

In addition to the structural similarities between class II and III enteric genes and the *B. subtilis fla-che-mot* equivalents, several kinds of regulatory phenomena that affect the expression of these genes appear quite similar in the two systems. These include (i) regulation of *B. subtilis* "class III" genes such as *hag* and *mot* by σ^D (95, 96), (ii) growth state regulation of the level of flagellin expression (94a, 95), and (iii) the control of *B. subtilis* class III gene expression by expression of flagellar structural genes that are homologs of enteric class II genes (88, 152). The first of these has already been considered.

Growth stage regulation of the level of flagellin expression is seen as an exponential increase in the amount of flagellin mRNA per cell as cells in rich media go from early to late exponential growth (94, 95). After exponential growth halts, flagellin mRNA synthesis reaches a peak and then decays rapidly in the next 2 h. This same pattern of expression is seen for all σ^D-dependent transcription units that have been studied (38, 122). A similar kind of increase in flagellin levels is seen for enteric bacteria (1, 70, 131). The mechanism for this kind of regulation is not known, but in *B. subtilis*, the increase in flagellin expression is paralleled by a corresponding increase in σ^D protein levels (88).

Expression of flagellin from the *B. subtilis hag* gene is shut off almost completely by mutations or insertions in the *fla-che* operon genes that specify HBB functions (88, 152). This effect is not due to a polar effect on σ^D expression, since σ^D levels remain high enough to account for normal levels of flagellin expression when compared with other constructions. Hence, it appears that there is a blockage of σ^D-σ^F function brought about by deficiencies in HBB expression or assembly that is similar to the blockage of FliA (σ^F) function in enteric bacteria brought about by HBB deficiency. It is tempting to think that the regulatory mechanisms that are involved in these last two phenomena are also homologous between *B. subtilis* and enterics, but these mechanisms are not yet defined in either system.

Genetic studies on the biological role of σ^D (56, 89), a sigma factor expressed in vegetative cells of *B. subtilis* (38), have led to additional insights into *fla-che-mot* gene expression. Disruption of the *sigD* gene, the structural gene for the alternate sigma factor, results in a lack of flagellin expression and filamentous growth by the null mutant bearing the disruption, whereas reduced levels of the σ factor give rise to defects in motility and/or chemotaxis (56, 89). In fact, the *hag*, *motA*, and *motB* genes of *B. subtilis* have been shown to be dependent on the σ^D factor for transcription (95, 96, 152). These results demonstrate that σ^D plays a significant role in the control of gene expression for flagellar assembly, motility, and autolytic functions. The *sigD* gene is allelic to *flaB2* (89), which explains the original phenotype observed in the mutant bearing the *flaB2* mutation (46, 117).

Two other regulatory loci, *sin* and *degS-degU*, at which mutant alleles can affect motility functions, autolytic enzyme expression, competence for transformation, and extracellular enzyme expression have been identified (35, 47, 73, 119). One of these gene products, the Sin protein, also inhibits sporulation when overexpressed in the cell (35). Sin is a DNA-binding protein that binds specifically to the upstream region of *aprE*, the structural gene for alkaline protease, and thereby represses its expression (36). Sin has also been shown to bind the *spoIIA* promoter and to repress its expression as well as the expression of several other stage II sporulation genes, including the structural gene for σ^E (36). Not only does Sin function to repress gene expression, but it is also required for the positive activation of several genes and physiological functions. It has been shown to be required for the expression of late competency (*com*) genes (47), for motility (35), and for autolytic enzyme synthesis (119). In fact, the *flaD2* (originally identified as *lyt-2*) lesion has been localized within the *sin* gene (119). However, *sin* is not essential for flagellin synthesis in *B. subtilis*, since disruptions of this gene show normal synthesis of *hag* mRNA (90).

Genes of the *degS-degU* locus can also affect motility functions. The wild-type *degS* and *degU* gene products appear to be required for competence as well as for control of extracellular enzyme production. Mutations designated *degS*(Hy) and *degU*(Hy) [originally *sacU*(Hy)] are missense mutations within the *degS* and *degU* genes (57) that result in a strain that lacks flagella, is poorly transformable, overexpresses extracellular enzymes, and sporulates in the presence of normally inhibitory amounts of glucose (78). However, disruptions of *degS* or *degU* do not affect motility; hence, these genes are not directly required for expression of motility functions (98).

The predicted protein products of the *degS* and *degU* genes are similar to the two-component signaling systems found in many prokaryotes (57, 73). The *degS* gene product is similar to the sensor class of proteins, which monitor the external environment, and is a protein kinase. The *degU* gene product is similar to the transducer-regulator class of proteins and is phosphorylated by DegS (27). It is possible that the lack of flagella in the *degS*(Hy) and *degU*(Hy) mutants results from cross talk between these mutant regulators and some components that directly control expression of motility functions.

REFERENCES

1. **Adler, J., and Templeton.** 1967. The effect of environmental conditions on the motility of *E. coli. J. Gen. Microbiol.* **46:**175–184.
2. **Ahlgren, J. A., and G. W. Ordal.** 1983. Methyl esterification of glutamic acid residues of methyl-accepting chemotaxis proteins in *Bacillus subtilis. Biochem. J.* **213:**759–763.
3. **Akamatsu, T., and J. Sekiguchi.** 1987. Genetic mapping and properties of filamentous mutations in *Bacillus subtilis. Agric. Biol. Chem.* **51:**2901–2909.
4. **Alam, M., M. Lebert, D. Oesterhelt, and G. L. Hazelbauer.** 1989. Methyl-accepting chemotaxis proteins in Halobacterium halobium. *EMBO J.* **8:**631–639.
5. **Albertini, A. M., T. Caramori, W. D. Crabb, F. Scoffone, and A. Galizzi.** 1991. The *flaA* locus of *Bacillus subtilis* is part of a large operon coding for flagellar structures, motility functions, and an ATPase-like polypeptide. *J. Bacteriol.* **173:**3573–3579.
6. **Arnosti, D. N., and M. J. Chamberlin.** 1989. A secondary sigma factor controls transcription of flagellar and chemotaxis genes in *E. coli. Proc. Natl. Acad. Sci. USA* **86:**830–834.
7. **Ayusawa, D., Y. Yoneda, K. Yamane, and B. Maruo.** 1975. Pleitropic phenomenon in autolytic enzyme content, flagellation, and simultaneous hyperproduction of extracellular α-amylase and protease in a *Bacillus subtilis* mutant. *J. Bacteriol.* **124:**459–469.
8. **Bartlett, D., B. Frantz, and P. Matsumura.** 1988. Flagellar transcriptional activators FlbB and FlaI: gene sequences and 5′ consensus sequences of operons under FlbB and FlaI control. *J. Bacteriol.* **170:**1575–1581.
9. **Bedale, W. A., D. O. Nettleton, C. S. Sopata, M. S. Thoelke, and G. W. Ordal.** 1988. Evidence for methyl-group transfer between the methyl-accepting chemotaxis proteins in *Bacillus subtilis. J. Bacteriol.* **170:**223–227.
10. **Berg, H. C., and D. A. Brown.** 1972. Chemotaxis in *Escherichia coli* analysed by three-dimensional tracking. *Nature* (London) **239:**500–504.
11. **Berg, H. C., and P. M. Tedesco.** 1975. Transient response to chemotactic stimuli in *Escherichia coli. Proc. Natl. Acad. Sci. USA* **72:**3235–3239.
12. **Bischoff, D. S., and G. W. Ordal.** 1991. Sequence and characterization of *B. subtilis* CheB, a homolog of *E. coli* CheY, and its role in a different mechanism of chemotaxis. *J. Biol. Chem.* **266:**12301–12305.
13. **Bischoff, D. S., and G. W. Ordal.** 1992. *Bacillus subtilis* chemotaxis: a deviation from the *Escherichia coli* paradigm. *Mol. Microbiol.* **6:**23–28.

13a. **Bischoff, D. S., and G. W. Ordal.** 1992. Identification and characterization of FliY, a novel component of the *Bacillus subtilis* flagellar switch complex. *Mol. Microbiol.* **6:**2715–2723.

14. **Bischoff, D. S., M. R. Weinreich, and G. W. Ordal.** 1992. Nucleotide sequences of *Bacillus subtilis* flagellar biosynthetic genes *fliP* and *fliQ*, and identification of a novel flagellar gene, *fliZ. J. Bacteriol.* **174:**4017–4025.
15. **Bollinger, J., C. Park, S. Harayama, and G. L. Hazelbauer.** 1984. Structure of the Trg protein: homologies with and differences from other sensory transducers of *Escherichia coli. Proc. Natl. Acad. Sci. USA* **81:**3287–3291.
16. **Borkovich, K. A., and M. I. Simon.** 1990. The dynamics of protein phosphorylation in bacterial chemotaxis. *Cell* **63:**1339–1348.
17. **Bourret, R. B., K. A. Borkovich, and M. I. Simon.** 1991. Signal transduction pathways involving protein phosphorylation in prokaryotes. *Annu. Rev. Biochem.* **60:** 401–444.
18. **Bourret, R. B., J. F. Hess, and M. I. Simon.** 1990. Conserved aspartate residues and phosphorylation in signal transduction by the chemotaxis protein CheY. *Proc. Natl. Acad. Sci. USA* **87:**41–45.
19. **Boyd, A., K. Kendall, and M. I. Simon.** 1983. Structure of the serine chemoreceptor in *Escherichia coli. Nature* (London) **301:**623–626.
20. **Brown, D. A., and H. C. Berg.** 1974. Temporal stimulation of chemotaxis in *Escherichia coli. Proc. Natl. Acad. Sci. USA* **71:**1388–1392.
21. **Brummet, T. B., and G. W. Ordal.** 1977. Inhibition of amino acid transport in *Bacillus subtilis* by uncouplers of oxidative phosphorylation. *Arch. Biochem. Biophys.* **178:**368–372.
22. **Burgess, B. K.** 1990. The iron-molybdenum cofactor of nitrogenase. *Chem. Rev.* **90:**1377–1406.
23. **Burgess-Cassler, A., and G. W. Ordal.** 1982. Functional homology of *Bacillus subtilis* methyltransferase II and *Escherichia coli* cheR protein. *J. Biol. Chem.* **257:**12835–12838.
24. **Burgess-Cassler, A., A. H. J. Ullah, and G. W. Ordal.** 1982. Purification and characterization of *Bacillus subtilis* methyl-accepting chemotaxis protein methyltransferase II. *J. Biol. Chem.* **257:**8412–8417.

24a. **Carpenter, P. B., D. W. Hanlon, and G. W. Ordal.** 1992. *flhF*, a *Bacillus subtilis* flagellar gene that encodes a putative GTP-binding protein. *Mol. Microbiol.* **6:**2705–2713.

24b. **Carpenter, P. B., and G. W. Ordal.** *Bacillus subtilis* FlhA: a flagellar protein related to a new family of signal-transducing receptors. *Mol. Microbiol.*, in press.

25. **Chen, Y. F., and J. D. Helmann.** 1992. Restoration of motility to an *Escherichia coli fliA* flagellar mutant by a *Bacillus subtilis* sigma factor. *Proc. Natl. Acad. Sci. USA* **89:**5123–5127.
26. **Clarke, S., and D. E. Koshland, Jr.** 1979. Membrane receptors for aspartate and serine in bacterial chemotaxis. *J. Biol. Chem.* **254:**9695–9702.
27. **Dahl, M. K., T. Msadek, F. Kunst, and G. Rapoport.** 1991. Mutational analysis of the *Bacillus subtilis* DegU regulator and its phosphorylation by the DegS protein kinase. *J. Bacteriol.* **173:**2539–2547.

27a. **Dahlquist, F.** Personal communication.

28. **Dean, G. E., R. M. Macnab, J. Stader, P. Matsumura, and C. Burks.** 1984. Gene sequence and predicted amino acid sequence of the *motA* protein, a membrane-associated protein required for flagellar rotation in *Escherichia coli. J. Bacteriol.* **159:**991–999.
29. **Eisenbach, M., C. Constantinous, H. Aloni, and M. Shinitsky.** 1990. Repellents for *Escherichia coli* operate neither by changing membrane fluidity nor by being sensed by periplasmic receptors during chemotaxis. *J. Bacteriol.* **172:**5218–5224.
30. **Engstrom, P., and G. L. Hazelbauer.** 1982. Methyl-accepting chemotaxis proteins are distributed in the membrane independently from basal ends of bacterial flagella. *Biochim. Biophys. Acta* **686:**19–26.

31. **Fein, J. E.** 1979. Possible involvement of bacterial autolytic enzymes in flagellar morphogenesis. *J. Bacteriol.* **137:**933–946.
32. **Fein, J. E., and H. J. Rogers.** 1976. Autolytic enzyme-deficient mutants of *Bacillus subtilis* 168. *J. Bacteriol.* **127:**1427–1442.
33. **Fuhrer, D. K., and G. W. Ordal.** 1991. *Bacillus subtilis* CheN, a homolog of CheA, the central regulator of chemotaxis in *Escherichia coli*. *J. Bacteriol.* **173:**7443–7448.
34. **Gaur, N. K., K. Cabane, and I. Smith.** 1988. Structure and expression of the *Bacillus subtilis sin* operon. *J. Bacteriol.* **170:**1046–1053.
35. **Gaur, N. K., E. Dubnau, and I. Smith.** 1986. Characterization of a cloned *Bacillus subtilis* gene that inhibits sporulation in multiple copies. *J. Bacteriol.* **168:**860–869.
36. **Gaur, N. K., J. Oppenheim, and I. Smith.** 1991. The *Bacillus subtilis sin* gene, a regulator of alternate developmental processes, codes for a DNA-binding protein. *J. Bacteriol.* **173:**678–686.
37. **Gillen, K. L., and K. T. Hughes.** 1991. Molecular characterization of *flgM*, a gene encoding a negative regulator of flagellin synthesis in *Salmonella typhimurium*. *J. Bacteriol.* **173:**6453–6459.
38. **Gilman, M. Z., and M. J. Chamberlin.** 1983. Developmental and genetic regulation of *Bacillus subtilis* genes transcribed by sigma-28 containing RNA polymerase. *Cell* **35:**285–293.
39. **Gilman, M. Z., J. S. Glenn, V. L. Singer, and M. J. Chamberlin.** 1984. Isolation of sigma-28 specific promoters from *Bacillus subtilis* DNA. *Gene* **32:**11–20.
40. **Gilman, M. Z., J. L. Wiggs, and M. J. Chamberlin.** 1981. Nucleotide sequences of two *Bacillus subtilis* promoters used by *B. subtilis* sigma-28 RNA polymerase. *Nucleic Acids Res.* **9:**5991–6000.
41. **Goldman, D. J., D. O. Nettleton, and G. W. Ordal.** 1984. Purification and characterization of chemotactic methylesterase from *Bacillus subtilis*. *Biochemistry* **23:**675–680.
42. **Goldman, D. J., and G. W. Ordal.** 1981. Sensory adaptation and deadaptation in *Bacillus subtilis*. *J. Bacteriol.* **147:**267–270.
43. **Goldman, D. J., and G. W. Ordal.** 1984. *In vitro* methylation and demethylation of methyl-accepting chemotaxis proteins in *Bacillus subtilis*. *Biochemistry* **23:** 2600–2606.
44. **Goldman, D. J., S. W. Worobec, R. B. Siegel, R. V. Hecker, and G. W. Ordal.** 1982. Chemotaxis in *Bacillus subtilis*: effects of attractants on the level of MCP methylation and the role of demethylation in the adaptation process. *Biochemistry* **21:**915–920.
45. **Goy, M. F., M. S. Springer, and J. Adler.** 1977. Sensory transduction in *Escherichia coli*: role of a protein methylation reaction in sensory adaptation. *Proc. Natl. Acad. Sci. USA* **74:**4964–4968.
46. **Grant, G. F., and M. I. Simon.** 1969. Synthesis of bacterial flagella. II. PBS1 Transduction of flagella-specific markers in *Bacillus subtilis*. *J. Bacteriol.* **99:**116–124.
47. **Guillen, N., Y. Weinrauch, and D. A. Dubnau.** 1989. Cloning and characterization of the regulatory *Bacillus subtilis* competence genes *comA* and *comB*. *J. Bacteriol.* **171:**5354–5361.
48. **Hanlon, D. W., P. B. Carpenter, and G. W. Ordal.** 1992. Influence of attractants and repellents on methyl group turnover on methyl-accepting chemotaxis proteins of *Bacillus subtilis* and the role of CheW. *J. Bacteriol.* **174:**4218–4222.
49. **Hanlon, D. W., L. M. Marquez-Magaña, P. B. Carpenter, M. J. Chamberlin, and G. W. Ordal.** 1992. Sequence and characterization of *Bacillus subtilis* CheW. *J. Biol. Chem.* **267:**12055–12060.

49a. **Hanlon, D. W., and G. Ordal.** Unpublished data.
50. **Harayama, S., P. Engstrom, H. Wolf-Watz, T. Iino, and G. L. Hazelbauer.** 1982. Cloning of *trg*, a gene for a sensory transducer in *Escherichia coli*. *J. Bacteriol.* **152:**372–383.
51. **Hauser, P. M., W. D. Crabb, M. G. Fiora, F. Scoffone, and A. Galizzi.** 1991. Genetic analysis of the *flaA* locus of *Bacillus subtilis*. *J. Bacteriol.* **173:**3580–3583.
52. **Hazelbauer, G. L., and J. Adler.** 1971. Role of the galactose binding protein in chemotaxis of *Escherichia coli* toward galactose. *Nature* (London) *New Biol.* **230:** 101–104.
53. **Hazelbauer, G. L., and S. Harayama.** 1979. Mutants in transmission of chemotactic signals from two independent receptors of *E. coli*. *Cell* **16:**617–625.
54. **Hedblom, M. L., and J. Adler.** 1980. Genetic and biochemical properties of *Escherichia coli* mutants with defects in serine chemotaxis. *J. Bacteriol.* **144:**1048–1060.
55. **Helmann, J., and M. Chamberlin.** 1987. DNA sequence analysis suggests that expression of flagellar and chemotaxis genes in *Escherichia coli* and *Salmonella typhimurium* is controlled by an alternative σ factor. *Proc. Natl. Acad. Sci. USA* **84:**6422–6424.
56. **Helmann, J., L. Márquez, and M. J. Chamberlin.** 1988. Cloning, sequencing, and disruption of the *Bacillus subtilis* σ^{28} gene. *J. Bacteriol.* **170:**1568–1574.
57. **Henner, D., M. Yang, and E. Ferrari.** 1988. Localization of *Bacillus subtilis sacU*(Hy) mutations to two linked genes with similarities to the conserved procaryotic family of two-component signalling systems. *J. Bacteriol.* **170:**5102–5109.
58. **Henner, D. J., P. Gollnick, and A. Moir.** 1990. An analysis of an 18 kilobase pair region of *Bacillus subtilis* chromosome containing *Mtr* and *GerC* operons and the *Aro-Trp-Aro* supraoperon, p. 657–666. *In* H. Heslot (ed.), *Proceedings of the 6th International Symposium on Genetics of Industrial Microorganisms*. Société Française de Microbiologie, Strasbourg, France.
59. **Hess, J. F., K. Oosawa, N. Kaplan, and M. I. Simon.** 1988. Phosphorylation of three proteins in the signaling pathway of bacterial chemotaxis. *Cell* **53:**79–87.
60. **Iino, T., Y. Komeda, K. Kutsukake, R. Macnab, P. Matsumura, J. S. Parkinson, M. I. Simon, and S. Yamaguchi.** 1988. New unified nomenclature for the flagellar genes of *Escherichia coli* and *Salmonella typhimurium*. *Microbiol. Rev.* **52:**533–535.
61. **Ishihara, A., J. E. Segall, S. M. Block, and H. C. Berg.** 1983. Coordination of flagella on filamentous cells of *Escherichia coli*. *J. Bacteriol.* **155:**228–237.
62. **Jones, C. L., and R. M. Macnab.** 1990. Flagellar assembly in *Salmonella typhimurium*: analysis with temperature-sensitive mutants. *J. Bacteriol.* **172:**1327–1339.
63. **Kehry, M. R., M. W. Bond, M. W. Hunkpiller, and F. W. Dahlquist.** 1983. Enzymatic deamidation of methyl-accepting chemotaxis proteins in *Escherichia coli* catalyzed by the cheB gene product. *Proc. Natl. Acad. Sci. USA* **80:**3599–3603.
64. **Kehry, M. R., and F. W. Dahlquist.** 1982. The methyl-accepting chemotaxis proteins of *Escherichia coli*. Identification of the multiple methylation sites on methyl-accepting chemotaxis protein I. *J. Biol. Chem.* **257:** 10378–10386.
65. **Kehry, M. R., T. G. Doak, and F. W. Dahlquist.** 1984. Stimulus-induced changes in methylesterase activity during chemotaxis in *Escherichia coli*. *J. Biol. Chem.* **259:**11828–11835.
66. **Kehry, M. R., T. G. Doak, and F. W. Dahlquist.** 1985. Sensory adaptation in bacterial chemotaxis: regulation of demethylation. *J. Bacteriol.* **163:**983–990.
67. **Kihara, M., M. Homma, K. Kutsukake, and R. M. Macnab.** 1989. Flagellar switch of *Salmonella typhimu-*

rium: gene sequences and deduced protein sequences. *J. Bacteriol.* **171:**3247–3257.

67a. **Kirsch, M., and G. W. Ordal.** Unpublished data.

68. **Kleene, S. J., M. L. Toews, and J. Adler.** 1977. Isolation of glutamic acid methyl ester from an *Escherichia coli* membrane protein involved in chemotaxis. *J. Biol. Chem.* **252:**3214–3218.

69. **Komeda, Y.** 1986. Transcriptional control of flagellar genes in *Escherichia coli* K-12. *J. Bacteriol.* **168:**1315–1318.

70. **Komeda, Y., and T. Iino.** 1979. Regulation of expression of flagellin (*hag*) in *E. coli* K12: analysis of *hag-lacZ* fusions. *J. Bacteriol.* **168:**1315–1318.

71. **Komeda, Y., H. Suzuki, J. Ishidsu, and T. Iino.** 1975. The role of cAMP in flagellation of *S. typhimurium*. *Mol. Gen. Genet.* **142:**289–298.

72. **Krikos, A., N. Mutoh, A. Boyd, and M. I. Simon.** 1983. Sensory transducers of *E. coli* are composed of discrete structural and functional domains. *Cell* **33:**615–622.

73. **Kunst, F., M. Debarbouille, T. Msadik, M. Young, C. Mauël, D. Karamata, A. Klier, G. Rapoport, and R. Dedonder.** 1988. Deduced polypeptides encoded by the *Bacillus subtilis sacU* locus share homology with two-component sensor-regulator systems. *J. Bacteriol.* **170:**5093–5101.

74. **Kutsukake, K., Y. Ohya, and T. Iino.** 1990. Transcriptional analysis of the flagellar regulon of *Salmonella typhimurium*. *J. Bacteriol.* **172:**741–747.

75. **Larsen, S. H., R. W. Reader, E. N. Kort, W.-W. Tso, and J. Adler.** 1974. Change in direction of flagellar rotation is the basis of the chemotactic response in *Escherichia coli*. *Nature* (London) **249:**74–77.

76. **LaVallie, E. R., and M. L. Stahl.** 1989. Cloning of the flagellin gene from *Bacillus subtilis* and complementation studies of an in vitro-derived deletion mutation. *J. Bacteriol.* **171:**3085–3094.

77. **Lazarevic, V., P. Margot, and D. Karamata.** 1991. *Bacillus subtilis* N-acetylmuramoyl-L-alanine amidase and its modifier are encoded by a three ORF operon called *lytABC*, abstr. W4. Sixth Int. Conf. Bacilli, July 28–31, Stanford University.

78. **Lepesant, J. A. F. Kunst, J. Lepesant-Kejzalarova, and R. Dedonder.** 1972. Chromosomal location of mutations affecting sucrose metabolism in *Bacillus subtilis* Marburg. *Mol. Gen. Genet.* **188:**135–160.

79. **Liu, J. D., and J. S. Parkinson.** 1989. Role of CheW protein in coupling membrane receptors to the intracellular signalling system of bacterial chemotaxis. *Proc. Natl. Acad. Sci. USA* **86:**8703–8707.

80. **Lupas, A., and J. Stock.** 1989. Phosphorylation of an N-terminal regulatory domain activates the CheB methylesterase in bacterial chemotaxis. *J. Biol. Chem.* **264:**17337–17342.

81. **Macnab, R. M.** 1987. Flagella, p. 70–83. *In* F. C. Neidhardt, J. L. Ingraham, K. B. Low, B. Magasanik, M. Schaechter, and H. E. Umbarger (ed.), *Escherichia coli and Salmonella typhimurium: Cellular and Molecular Biology*, vol. 1. American Society for Microbiology, Washington, D.C.

82. **Macnab, R. M.** 1987. Motility and chemotaxis, p. 732–759. *In* F. C. Neidhardt, J. L. Ingraham, K. B. Low, B. Magasanik, M. Schaechter, and H. E. Umbarger (ed.), *Escherichia coli and Salmonella typhimurium: Cellular and Molecular Biology*, vol. 1. American Society for Microbiology, Washington, D.C.

83. **Macnab, R. M.** 1990. Genetics, structure, and assembly of the bacterial flagellum, p. 77–106. *In* J. F. Armitage and J. M. Lackie (ed.), *Biology of the Chemotactic Response*. Cambridge University Press, Cambridge.

84. **Macnab, R. M.** 1992. Genetics and biogenesis of bacterial flagella. *Annu. Rev. Genet.* **26:**131–158.

84a. **Macnab, R. M.** Personal communication.

85. **Macnab, R. M., and D. P. Han.** 1983. Asynchronous switching of flagellar motors on a single bacterial cell. *Cell* **32:**109–117.

86. **Macnab, R. M., and D. E. Koshland, Jr.** 1972. The gradient-sensing mechanism in bacterial chemotaxis. *Proc. Natl. Acad. Sci. USA* **69:**2509–2512.

86a. **Margot, P., and D. Karamata.** Personal communication.

87. **Margot, P., C. Mauël, and D. Karamata.** 1991. The *Bacillus subtilis* N-acetylglucosaminidase is encoded by a monocistronic operon controlled by a σ^D dependent promoter, abstr. W6. Sixth Int. Conf. Bacilli, July 28–31, Stanford University.

88. **Màrquez, L.** 1991. Function and regulation of expression of the *Bacillus subtilis* sigma-D factor. Ph.D. thesis. University of California, Berkeley.

89. **Màrquez, L. M., J. D. Helmann, E. F. Ferrari, H. M. Parker, G. W. Ordal, and M. J. Chamberlin.** 1990. Studies of σ^D-dependent functions in *Bacillus subtilis*. *J. Bacteriol.* **172:**3435–3443.

90. **Màrquez-Magaña, L., D. B. Mirel, and M. J. Chamberlin.** Role of *spo0*, *abrB*, and *sin* gene products in regulation of σ^D expression and activity. Submitted for publication.

91. **Matsushita, T., H. Hirata, and I. Kusaka.** 1988. Calcium channel blockers inhibit bacterial chemotaxis. *FEBS Lett.* **236:**437–440.

92. **Mesibov, R., and J. Adler.** 1972. Chemotaxis toward amino acids in *Escherichia coli*. *J. Bacteriol.* **112:**315–326.

93. **Mesibov, R., G. W. Ordal, and J. Adler.** 1973. The range of attractant concentrations for bacterial chemotaxis and the threshold and size of response over this range. Weber law and related phenomena. *J. Gen. Physiol.* **62:**203–223.

94. **Mirel, D. B.** 1992. The *Bacillus subtilis* sigma-D regulon. Ph.D. thesis. University of California, Berkeley.

94a. **Mirel, D. B.** Unpublished data.

95. **Mirel, D. B., and M. J. Chamberlin.** 1989. The *Bacillus subtilis* flagellin gene (*hag*) is transcribed by the σ^{28} form of RNA polymerase. *J. Bacteriol.* **171:**3095–3101.

96. **Mirel, D. B., V. M. Lustre, and M. J. Chamberlin.** 1992. An operon of *Bacillus subtilis* motility genes transcribed by the σ^D form of RNA polymerase. *J. Bacteriol.* **174:**4197–4204.

97. **Mowbray, S. L., D. L. Foster, and D. E. Koshland, Jr.** 1985. Proteolytic fragments identified with domains of the aspartate chemoreceptor. *J. Biol. Chem.* **260:**11711–11718.

98. **Msadek, T., F. Kunst, D. Henner, A. Klier, G. Rapoport, and R. Dedonder.** 1990. Signal transduction pathway controlling synthesis of a class of degradative enzymes in *Bacillus subtilis*: expression of the regulatory genes and analysis of mutations in *degS* and *degU*. *J. Bacteriol.* **172:**824–834.

99. **Nettleton, D. O.** 1986. Chemotactic methylation in *Bacillus subtilis*. Ph.D. thesis. University of Illinois, Urbana.

100. **Nettleton, D. O., and G. W. Ordal.** 1989. Functional homology of chemotactic methylesterases from *Bacillus subtilis* and *Escherichia coli*. *J. Bacteriol.* **171:**120–123.

101. **Nicholas, R. A., and G. W. Ordal.** 1978. Inhibition of bacterial transport by uncouplers of oxidative phosphorylation: effect of pentachlorophenol and analogs in *Bacillus subtilis*. *Biochem. J.* **176:**639–647.

102. **Nishihara, T., and E. Freese.** 1975. Motility of *Bacillus subtilis* during growth and sporulation. *J. Bacteriol.* **123:**366–371.

103. **Ohnishi, K., K. Kutsukake, H. Suzuki, and T. Iino.** 1990. Gene *fliA* encodes an alternative sigma factor specific for flagellar operons in *Salmonella typhimurium*. *Mol. Gen. Genet.* **221:**139–147.

103a. **Ordal, G. W.** Unpublished data.

104. **Ordal, G. W.** 1976. Recognition sites for repellents of *Bacillus subtilis. J. Bacteriol.* **126:**72–79.
105. **Ordal, G. W.** 1976. Control of tumbling in bacterial chemotaxis by divalent cation. *J. Bacteriol.* **126:**706–711.
106. **Ordal, G. W.** 1977. Calcium ion regulates chemotactic behavior in bacteria. *Nature* (London) **270:**66–67.
107. **Ordal, G. W., and K. J. Gibson.** 1977. Chemotaxis toward amino acids by *Bacillus subtilis. J. Bacteriol.* **129:**151–155.
108. **Ordal, G. W., and D. J. Goldman.** 1975. Chemotaxis away from uncouplers of oxidative phosphorylation in *Bacillus subtilis. Science* **189:**802–805.
109. **Ordal, G. W., and D. J. Goldman.** 1976. Chemotactic repellents of *Bacillus subtilis. J. Mol. Biol.* **100:**103–108.
110. **Ordal, G. W., D. O. Nettleton, and J. A. Hoch.** 1983. Genetics of *Bacillus subtilis* chemotaxis: isolation and mapping of mutations and cloning of chemotaxis genes. *J. Bacteriol.* **154:**1088–1097.
111. **Ordal, G. W., H. M. Parker, and J. R. Kirby.** 1985. Complementation and characterization of chemotaxis mutants of *Bacillus subtilis. J. Bacteriol.* **164:**802–810.
112. **Ordal, G. W., and D. P. Villani.** 1980. Action of uncouplers of oxidative phosphorylation as repellents of *Bacillus subtilis. J. Gen. Microbiol.* **115:**471–478.
113. **Ordal, G. W., D. P. Villani, and K. J. Gibson.** 1977. Amino acid chemoreceptors of *Bacillus subtilis. J. Bacteriol.* **129:**156–165.
114. **Ordal, G. W., D. P. Villani, R. A. Nicholas, and F. G. Hamel.** 1978. Independence of proline chemotaxis and transport in *Bacillus subtilis. J. Biol. Chem.* **253:**4916–4919.
115. **Ordal, G. W., D. P. Villani, and M. S. Rosendahl.** 1979. Chemotaxis toward sugars by *Bacillus subtilis. J. Gen. Microbiol.* **118:**471–478.
116. **Parkinson, J. S., and S. R. Parker.** 1979. Interaction of the *cheC* and *cheZ* gene products is required for chemotactic behavior in *Escherichia coli. Proc. Natl. Acad. Sci. USA* **76:**2390–2394.
117. **Pooley, H. M., and D. Karamata.** 1984. Genetic analysis of autolysin-deficient and flagellaless mutants of *Bacillus subtilis. J. Bacteriol.* **160:**1123–1129.
117a. **Rosario, M., D. Bochar, and G. Ordal.** Unpublished data.
118. **Russo, A. F., and D. E. Koshland, Jr.** 1983. Separation of signal transduction and adaptation functions of the aspartate receptor in bacterial sensing. *Science* **220:**1016–1020.
119. **Sekiguchi, J., H. Ohsu, A. Kuroda, H. Moriyama, and T. Akamatsu.** 1990. Nucleotide sequences of the *Bacillus subtilis flaD* locus and a *B. licheniformis* homologue affecting the autolysis level and flagellation. *J. Gen. Microbiol.* **136:**1223–1230.
120. **Shi, W., Y. Zhou, J. Wild, J. Adler, and C. A. Gross.** 1992. DnaK, DnaJ, and GrpE are required for flagellum synthesis in *Escherichia coli. J. Bacteriol.* **174:**6256–6263.
121. **Silverman, M., and M. Simon.** 1974. Characterization of *Escherichia coli* flagellar mutants that are insensitive to catabolite repression. *J. Bacteriol.* **120:**1196–1203.
122. **Singer, V.** 1987. Characterization of promoters and genes controlled by *Bacillus subtilis* sigma-28 RNA polymerase. Ph.D. thesis. University of California, Berkeley.
123. **Springer, M. S., M. F. Goy, and J. Adler.** 1977. Sensory transduction in *Escherichia coli*: two complementary pathways of information processing that involve methylated proteins. *Proc. Natl. Acad. Sci. USA* **74:**3312–3316.
124. **Springer, W. R., and D. E. Koshland, Jr.** 1977. Identification of a protein methyltransferase as the cheR gene product in the bacterial sensing system. *Proc. Natl. Acad. Sci. USA* **74:**533–537.
125. **Spudich, J. A., and D. E. Koshland, Jr.** 1975. Quantitation of the sensory response in bacterial chemotaxis. *Proc. Natl. Acad. Sci. USA* **72:**710–713.
126. **Stader, J., P. Matsumura, D. Vacante, G. E. Dean, and R. M. Macnab.** 1986. Nucleotide sequence of the *Escherichia coli motB* gene and site-limited incorporation of its product into the cytoplasmic membrane. *J. Bacteriol.* **166:**244–252.
127. **Stewart, R. C., A. Roth, and F. W. Dahlquist.** 1990. Mutations that affect control of the methylesterase activity of CheB, a component of the chemotaxis adaptation system in *Escherichia coli. J. Bacteriol.* **172:**3388–3399.
128. **Stock, J. B., and D. E. Koshland, Jr.** 1978. A protein methylesterase involved in bacterial sensing. *Proc. Natl. Acad. Sci. USA* **75:**3659–3663.
129. **Stock, J. B., and D. E. Koshland, Jr.** 1981. Changing reactivity of receptor carboxyl groups during bacterial sensing. *J. Biol. Chem.* **256:**10826–10833.
130. **Strange, P. G., and D. E. Koshland, Jr.** 1976. Receptor interactions in a signalling system: competition between ribose receptor and galactose receptor in the chemotaxis response. *Proc. Natl. Acad. Sci. USA* **73:**762–766.
131. **Suzuki, T., and T. Iino.** 1973. In vitro synthesis of phase-specific flagellin of *Salmonella. J. Mol. Biol.* **81:**57–70.
132. **Suzuki, T., and Y. Komeda.** 1981. Incomplete flagellar structure in *Escherichia coli* mutants. *J. Bacteriol.* **145:**1036–1041.
133. **Terwilliger, T. C., E. Bogonez, E. A. Wang, and D. E. Koshland, Jr.** 1983. Sites of methyl esterification in the aspartate receptor involved in bacterial chemotaxis. *J. Biol. Chem.* **258:**9608–9611.
134. **Terwilliger, T. C., and D. E. Koshland, Jr.** 1984. Sites of methyl esterification and deamination on the aspartate receptor involved in chemotaxis. *J. Biol. Chem.* **259:**7719–7725.
135. **Thoelke, M. S., J. M. Casper, and G. W. Ordal.** 1990. Methyl transfer in chemotaxis toward sugars in *Bacillus subtilis. J. Bacteriol.* **172:**1148–1150.
136. **Thoelke, M. S., J. M. Casper, and G. W. Ordal.** 1990. Methyl group turnover on methyl-accepting chemotaxis proteins during chemotaxis by *Bacillus subtilis. J. Biol. Chem.* **265:**1928–1932.
137. **Thoelke, M. S., J. R. Kirby, and G. W. Ordal.** 1989. Novel methyl transfer during chemotaxis in *Bacillus subtilis. Biochemistry* **27:**8453–8457.
138. **Thoelke, M. S., H. M. Parker, E. A. Ordal, and G. W. Ordal.** 1988. Rapid attractant-induced changes in methylation of methyl-accepting chemotaxis proteins in *Bacillus subtilis. Biochemistry* **27:**8453–8457.
139. **Toews, M. L., and J. Adler.** 1979. Methanol formation in vivo from methylated chemotaxis proteins in *Escherichia coli. J. Biol. Chem.* **254:**1761–1764.
140. **Toews, M. L., M. F. Goy, M. S. Springer, and J. Adler.** 1979. Attractants and repellents control demethylation of methylated chemotaxis proteins in *Escherichia coli. Proc. Natl. Acad. Sci. USA* **76:**5544–5548.
141. **Tozzi, M. G., U. D'Arcangelo, A. Del Corso, and G. W. Ordal.** 1991. Identification and purification of a calcium-binding protein from *Bacillus subtilis. Biochim. Biophys. Acta* **1080:**160–164.
141a. **Tozzi, M. G., and G. W. Ordal.** Unpublished data.
142. **Tsang, N., R. M. Macnab, and D. E. Koshland, Jr.** 1973. Common mechanism for repellents and attractants in bacterial chemotaxis. *Science* **181:**60–63.
143. **Tso, W.-W., and J. Adler.** 1974. Negative chemotaxis in *Escherichia coli. J. Bacteriol.* **118:**560–576.
144. **Ullah, A. H. J., and G. W. Ordal.** 1981. In vivo and in vitro chemotactic methylation in *Bacillus subtilis. J. Bacteriol.* **145:**958–965.
145. **Van der Werf, P., and D. E. Koshland, Jr.** 1977. Identi-

fication of α-glutamyl methyl ester in bacterial membrane protein involved in chemotaxis. *J. Biol. Chem.* **252:**2793–2795.

146. **Ying, C., and G. W. Ordal.** 1988. Cloning and expression of a chemotaxis gene in *Bacillus subtilis,* p. 75–78. *In* A. T. Ganesan and J. A. Hoch (ed.), *Genetics and Biotechnology of Bacilli,* vol. 2. Academic Press, Inc., New York.
147. **Ying, C., F. Scoffone, A. M. Albertini, A. Galizzi, and G. W. Ordal.** 1991. Properties of the *Bacillus subtilis* chemotaxis protein CheF, a homolog of the *Salmonella typhimurium* flagellar protein FliJ. *J. Bacteriol.* **173:** 3584–3586.
148. **Yoneda, Y., and B. Maruo.** 1975. Mutation of *Bacillus subtilis* causing hyperproduction of α-amylase and protease, and its synergistic effect. *J. Bacteriol.* **124:**48–54.
149. **Young, M., C. Mauël, P. Margot, and D. Karamata.** 1989. Pseudo-allelic relationship between non-homologous genes concerned with biosynthesis of polyglycerol phosphate and polyribitol phosphate teichoic acids in *Bacillus subtilis* strains 168 and W23. *Mol. Microbiol.* **3:**1805–1812.
150. **Zuberi, A. R., D. S. Bischoff, and G. W. Ordal.** 1991. Nucleotide sequence and characterization of a *Bacillus subtilis* gene encoding a flagellar switch protein. *J. Bacteriol.* **173:**710–719.
151. **Zuberi, A. R., C. Ying, D. S. Bischoff, and G. W. Ordal.** 1991. Gene-protein relationships in the flagellar hook-basal body complex of *Bacillus subtilis*: sequences of the *flgB, flgC, flgG, fliE,* and *fliF* genes. *Gene* **101:**23–31.
152. **Zuberi, A. R., C. Ying, H. M. Parker, and G. W. Ordal.** 1990. Transposon Tn*917lacZ* mutagenesis of *Bacillus subtilis*: identification of two new loci required for motility and chemotaxis. *J. Bacteriol.* **172:**6841–6848.
153. **Zuberi, A. R., C. Ying, M. R. Weinreich, and G. W. Ordal.** 1990. Transcription organization of a cloned chemotaxis locus of *Bacillus subtilis. J. Bacteriol.* **172:** 1870–1876.

54. Regulatory Proteins That Control Late-Growth Development

ISSAR SMITH

Upon encountering nutrient deprivation, *Bacillus* and other soil organisms (e.g., those in the genus *Streptomyces*) initiate a series of responses that allow survival in the hostile environment. Among these responses are synthesis and secretion of degradative enzymes, production of antibiotics, development of motility and competence (in *Bacillus subtilis*), and finally, appearance of spores. These dormant, resistant life forms will germinate and start a new round of vegetative growth when exposed to adequate food supplies.

These temporally regulated processes show overlapping modes of control, but they can also be differentiated on the basis of genetic requirements and response to alternative starvation signals. For example, even though the development of competence and sporulation is temporally regulated and the two processes have certain gene products in common, the former process requires glucose, whereas the latter is repressed by this carbon source. One of the most actively studied areas in *Bacillus* physiology and molecular biology is the decision-making process whereby the cell faced with nutrient stress chooses one of the alternative late-growth pathways.

Several proteins that affect these pathways have been described. The purpose of this review will be to discuss the specific functions of these proteins, the regulation of their synthesis, and their roles in the circuitry controlling late-growth development in *B. subtilis* and closely related bacteria, where these processes are understood. This review will be restricted to proteins that have pleiotropic effects, i.e., those that can regulate more than one late-growth-regulated process. For this reason, there will be no coverage of regulators that are dedicated to one process or that affect phenomena that are not temporally controlled, e.g., SacV, LevR, SacX, and SacY, etc. (8, 41, 72).

In recent years, several genes that affect late-growth development have been cloned and characterized, and it has been demonstrated, usually by creating mutations in the chromosomal gene or by disrupting the gene on a multicopy plasmid, that the protein products of these genes are the functional agents for their effects. Among these proteins are AbrB, ComA, DegQ, DegR, DegU, DegT, Hpr, Pai (ORF1 and ORF2), SinR (and SinI), Sen, Spo0A, TenA, and TenI. The best-characterized proteins in terms of their functions and control of their synthesis are AbrB, ComA, DegU, Spo0A, Hpr, and SinR. Because the first four proteins are discussed elsewhere in this volume (see chapters 39, 50, 51, and 52), they will be mentioned only briefly in this review. The other regulatory factors are less characterized, since in most cases, the protein has not been purified and/or inactivation of the genetic determinant has no apparent phenotypic effect. Therefore, their mechanisms of action and physiological roles have not been determined. However, the control of expression of some of their genetic determinants, e.g., *degQ*, provides valuable insight into the networks regulating late growth development. For this reason and for the sake of completeness, salient information pertaining to the less well described proteins will be presented. The review will be organized in the following manner. It will begin with those proteins with no known physiological role in wild-type cells; i.e., when their structural genes are disrupted, there is no apparent phenotype. These proteins are usually activators of the processes they modulate. Next will be considered proteins that have physiological roles inferred from the phenotypes resulting from disruption of their genetic determinants; the major focus will be on SinR (formerly known as Sin), a protein with known functions and an exceedingly complex mode of regulation at both the transcriptional and the posttranslational levels. Interestingly, this latter group, which also includes the response regulators (RR) ComA, DegU, and Spo0A, consists of proteins that can be repressors or activators of many late-growth functions. Properties of these proteins and their regulation are summarized in Table 1. Finally, there will be a discussion of the networks controlling the transition from vegetative growth to stationary phase and how the proteins described in the review play roles in this developmental process.

Sen

A DNA fragment from *Bacillus natto* cloned on a multicopy plasmid in *B. subtilis* caused a two- to threefold elevation of production of alkaline and neutral proteases, α-amylase, and alkaline phosphatase (89). The gene contained on this fragment, *senN*, was used to clone its homolog in *B. subtilis*, *senS* (85). SenS and SenN, the proteins encoded by the two genes, are essentially identical. The former contains 65 amino acids, and the latter contains 60 residues. The proteins are highly charged (44% are charged, 29% are basic, and 18% are lysines), and no significant similarity to any other proteins was observed. Mapping of *senS* (97% linked to *thiA* at 70°) on the *B. subtilis* chromosome indicated that this gene was unrelated to previously described regulatory loci.

The mechanism of Sen action is not yet known,

Issar Smith • Department of Microbiology, The Public Health Research Institute, New York, New York 10016.

Table 1. Properties of proteins affecting late-growth development and regulation of their genes

Regulatory protein (size in amino acids)	Gene map position (°)[a]	Regulation of gene[b]	Function	Reference(s)
AbrB (96)	3	Expression constitutive and repressed by Spo0A and AbrB	DNA-binding protein represses *spo0H*, *spoVG*, *spo0E*, *tycA*, *aprE*, *nprE*, and *mot* regulons; essential for competence but inhibits when overproduced; activates *comK* and *hpr*	10, 12, 16, 56, 73–75, 82, 83, 95
TenA (236)	62	Not known; possibly cotranscribed with *tenI*	Activates *aprE*, *nprE*, *sacB*, and *phoA* when overproduced; requires DegS-DegU for effect; mechanism of action not known; null mutations have no phenotype	53
TenI (205)	62	See *tenA*	Inhibits activation effect of TenA	53
SenS (65)	70	Constitutively expressed	Activates *aprE*, **nprE**, and *amyE* when overproduced; mechanism of action not known; null mutations have no phenotype	85, 86, 89
HpR (119)	75	Expression growth regulated; positively regulated by AbrB	DNA-binding protein represses *nprE*, *aprE*, and *sin*; unknown role in glucose repression of sporulation and oxidative stress response	5, 25, 29, 57
DegR (60)	200	Expression constitutive; *degSU*(Hy) mutations lower expression; read in vitro by E-σ^D	Activates *aprE*, *nprE*, and *sacB* when overproduced; effect requires DegS-DegU; mechanism of action unknown; null mutations have no phenotype	24, 45, 78, 80, 91
Spo0A (267)	217	Dual promoters; P1 read by E-σ^A and transcribed in vegetative growth; P2 read by E-σ^H, induced at t_0, and repressed by glucose	DNA-binding protein of RR class; phosphorylated by phosphorelay and then represses *abrB* and activates *spoIIA*, *spoIIE*, and *spoIIG*; mutations in ORF allow glucose-insensitive sporulation and also bypass mutations in early *spo* genes	6, 26, 52, 58, 59, 62, 68, 73, 82, 92
SinR (111)	221	Second gene of two-gene *sin* operon; two major promoters, both read by E-σ^A; P1 is induced at t_0, repressed by HPr, Sin, and glucose, and requires *spo0A* and *spo0H*; P3 is read constitutively	DNA-binding protein represses *aprE*, *nprE*, *spoIIA*, *spoIIE*, *spoIIG*, and *sin* P1; essential for motility, competence, and autolysin production; activates *hag* and *comK*; not known whether positive effects of *comK* are direct	1, 11, 17–19, 39, 64, 65, 71
SinI (57)	221	First gene of *sin* operon; transcribed by *sin* P1	Antagonizes SinR function at level of protein	17, 18
DegQ (46)	279	Gene expression is growth regulated; repressed by glucose and amino acids; requires DegS-DegU, ComP-ComA, and comQ	Activates *aprE*, *nprE*, and *amyE* when overproduced; competence and motility are inhibited; activation needs DegS-DegU; mechanism of action unknown; null mutations have no phenotype	2, 25, 32, 33, 36, 45–47, 66, 69, 90
ComA (214)	280	Is in two-gene operon with upstream *comP*; has secondary promoter; expression constitutive	RR essential for late *com* genes and for *srfA*, *degQ*, *srfA*, and *gsiA* expression; not known whether effects are direct; requires ComP, an HK; is believed to be activated by phosphorylation, which then enables it to act as DNA-binding protein; mechanism of action not demonstrated yet; represses sporulation if overproduced	11, 47, 48, 50, 87
pai ORF1 (172)	284	First gene of two-gene *pai* operon; regulation unknown	Inhibits septation, sporulation, *aprE*, *nprE*, *sacB*, and *phoA* when overproduced; effect (except for septation) requires ORF2; disruption of chromosomal ORF1 gives glucose- and glutamine-insensitive sporulation; mechanism of action not known; sequence similar to those of two transacetylases (RimI and STAT[c])	27, 28, 93
pai ORF2 (207)	284	Second gene of *pai* operon; may have secondary promoter	Required for all *pai* overproduction effects save filamentous growth; may be essential for growth	27

(Continued)

Table 1—*Continued*

Regulatory protein (size in amino acids)	Gene map position (°)[a]	Regulation of gene[b]	Function	Reference(s)
DegU (229)	306	Expression constitutive; cotranscribed with *degS*; has secondary promoter	RR of degradative-enzyme synthesis; requires phosphorylation by DegS for activity; essential for *sacB* and *degQ* expression; unphosphorylated DegU essential for competence; this effect at level of *comK*; *degSU*(Hy) mutations activate *aprE*, **nprE**, and *amyE*, inhibit flagellar formation and competence, and cause glucose-insensitive sporulation; believed to be DNA-binding protein but direct targets not found; *mec* mutations bypass *sacB* dependence on DegS-DegU	7, 22, 25, 33, 45, 46, 78, 79
DegT (*B. stearothermophilus*) (372)		Regulation unknown	Activates aprE and sacB when overproduced; this also causes glucose-resistant sporulation and lowers competence, autolysin production, and flagellar formation; very high sequence similarity to several *Streptomyces* antibiotic biosynthetic proteins; mechanism of action unknown	76, 77

[a] Map position refers to the location of the gene on the *B. subtilis* chromosome.
[b] t_0, time zero (beginning of stationary phase).
[c] STAT, streptothricin acetyltransferase.

because the protein has not been purified and mutagenesis of its open reading frame (ORF) has not been attempted. In vitro, the *sen* genes, encode proteins of the expected molecular weights. In vivo, reporter gene assays for *aprE* expression indicate that Sen increases transcription of this target gene, a result that is consistent with the hypothesis that it is a DNA-binding regulatory protein, but there is no evidence that Sen acts directly at the *aprE* promoter (85). Temporal regulation of *aprE* is unchanged when Sen is overproduced.

Null mutations of *senS* had no affect on extracellular enzyme synthesis and did not cause any other discernible phenotype. Thus, the physiological role of Sen (when *sen* is present in single copy) is not yet known.

senS, as determined with gene reporter assays, is constitutively expressed during growth and early sporulation. The presence of an apparent transcription terminator between the promoter and the ribosome-binding site suggests that attenuation-antitermination may play a regulatory role in *sen* expression (86).

DegR

degR (formerly known as *prtR*) was originally cloned from *B. natto* (49) and then from *B. subtilis*, in which it was mapped near *metB*, which is at 200° on the *B. subtilis* chromosome (91). *degR* codes for a 60-amino-acid protein, DegR, that enhances the production of alkaline and neutral proteases and levansucrase when overproduced. The effect on protease levels is 40- to 400-fold, while the increase in levansucrase activity is 2 orders of magnitude, and *sacB-lacZ* reporter gene assays showed a 10- to 15-fold stimulation of *sacB* expression when *degR* was present in multiple copies (91). Similar quantitative results were obtained with *aprE* RNA levels (80). In the latter studies, *aprE* mRNA stability was not affected by DegR, indicating that the enhancement effect most likely occurs at the level of transcription initiation. In all cases, the temporal regulation of target genes was unaffected by elevated levels of DegR. *degR* deletion analyses, the creation of functional translational fusions between the *degR* ORF and reporter genes, and the actual purification of DegR (80, 91) indicate that the protein is made in vivo and is responsible for the phenotype when DegR is overproduced. Examination of the *degR* ORF shows no sequence similarity to that of other late-growth regulatory proteins or other proteins in various data banks. The fact that it is strongly hydrophilic, containing approximately 40% charged amino acids, suggests that it could be a DNA-binding protein; however, in vitro assays have not been performed with the purified protein, and the ORF has not been subjected to mutational studies. It has been reported that the DegR enhancement effect requires the DegS-DegU system (78). The physiological role of DegR is unknown, since null mutations of *degR* have no apparent phenotype, as is the case with *senS*.

Reporter gene studies indicate that *degR* expression, similar to that of *senS*, is constitutive. A sequence very similar to promoters recognized by RNA polymerase containing σ^D (20) was observed upstream of the transcriptional start site of this gene, and in vitro transcription studies showed that *degR* was a good template for E-σ^D (5, 24). It is not known whether this polymerase transcribes *degR* in vivo, since the epistatic effects of null mutations of *sigD* on *degR* expression have not been studied. A possible relationship between *degR* expression and other late-growth phenomena comes from the observation that the *degU32*(Hy) mutation eliminates *degR* expression (45).

The pleiotropic *degU*(Hy) allele also results in an absence of flagella (33). Perhaps E-σ^D, which transcribes the *hag* gene, the genetic determinant for flagellin (44), does transcribe *degR*, and *degU*(Hy) may regulate some step upstream from *hag* and *degR* transcription.

DegT

degT, a gene from *Bacillus stearothermophilus*, enhances production of alkaline protease and levansucrase when it is cloned on a multicopy plasmid in *B. subtilis* (77). On the other hand, overproduction of *degT* causes decreased autolysin activity, lowered competence, loss of flagella, filamentous growth, abnormal cell growth, and glucose-resistant sporulation. These phenotypes are similar to those observed in the *degSU*(Hy) and *sin* loss-of-function mutations discussed below.

The *degT* ORF is predicted to encode a protein containing 372 amino acids. Sequence analysis of the ORF shows a very hydrophobic N-terminal region I (residues 51 to 160) with several possible membrane-spanning domains. Region II (amino acids 161 to 372) is very hydrophilic and contains a possible DNA-binding domain with consensus amino acids for a helix-turn-helix at positions 160 to 179. The protein has not yet been purified, but deletion analysis of the gene on a multicopy plasmid indicates that the ORF causes the pleiotropic phenotypes. DegT shows very high similarity to several *Streptomyces* proteins that are involved in antibiotic synthesis (76). Among these proteins are DnrJ, which regulates daunorubicin synthesis in *Streptomyces peucetius*, and EryC1, a regulatory protein and/or enzyme in the erythromycin biosynthetic pathway of *Streptomyces erythraeus* (new name, *Saccharopolyspora erythraea*). The similarity between DegT and DnrJ is extremely high, with 39% identity in a 358-amino-acid overlap, and an optimized FASTA (55) score of 692 (69a). It has been postulated that DnrJ may be a protein kinase and part of a kinase-RR pair along with DnrI, another regulatory protein of daunorubicin synthesis (76). Interestingly, DnrJ and DegT show similarity at both the N-terminal hydrophobic and the postulated central DNA-binding domains. Specific hypotheses concerning the function of DegT must await biochemical information on the protein and the physiological results of inactivating the *degT* gene, especially its *B. subtilis* homolog.

TenA AND TenI

Shotgun cloning of *B. subtilis* chromosomal fragments, a technique used to clone many of the genes for the regulatory proteins described in this review, was employed for the isolation of the *tenA-tenI* operon, since multiple copies of this operon cause enhanced production of alkaline protease (53). DNA sequence analysis of the 2.8-kb DNA fragment showed the existence of two complete, partially overlapping ORFs, named *tenA* and *tenI*, and translational *lacZ* fusions with the *ten* operon indicate that both ORFs are translated in vivo. *tenA*, the left-most ORF, has coding capacity for a protein of 236 amino acids and is responsible for the enhanced production of extracellular enzymes. In a strain containing a multicopy plasmid with both *tenA* and *tenI*, the levels of secreted alkaline protease, neutral protease, and levansucrase showed increases of approximately 10-fold over those in strains with a plasmid vector alone. When a plasmid containing only *tenA* was used for these studies or if *tenI* was disrupted by a frameshift lesion in a plasmid containing both *tenA* and *tenI*, an additional enhancement of fivefold for alkaline protease and threefold for levansucrase activity was observed. This suggests that the 205-amino-acid protein encoded by *tenI* could be an inhibitor of *tenA* function. TenA does not stimulate production of α-amylase, xylanase, or cellulase and thus has a pattern of stimulation similar to that of DegR. Reporter gene assays indicate that TenA exerts its affects at the transcriptional level but does not change temporal regulation of its target genes. TenA may be a transcriptional regulator-activator. However, no similarity between TenA or TenI and any of the other late-growth regulatory proteins of *Bacillus* spp. or any other protein sequences in data banks has been observed.

Null mutations in both *tenA* and *tenI* have no effect on growth in rich or synthetic media or on the production of extracellular enzymes. A delay in sporulation was observed with both types of lesions, but after 20 h, wild-type levels of spores were observed. Mapping studies have shown that *ten* is closely linked to *tre*, which is at 62°.

Little is known about the regulation of the *ten* operon except that as is the case with DegR action, the DegS-DegU system is required for the stimulatory affects of TenA. In addition, the removal of a stem-loop structure between the transcriptional start site of the *ten* operon and the *tenA* Shine-Dalgarno sequence results in a twofold increase in *ten* promoter activity. This suggests an antitermination-attenuation mechanism for *ten* expression such as has been demonstrated for *sacB* (32, 66) and *trp* (67) and postulated for *senS* (85).

DegQ

The *degQ* gene (originally called *sacQ*) was first defined by the *sacQ36* mutation, which phenotypically resembles those of *sacU*(Hy) type. These two types of mutations were originally characterized by overproduction of levansucrase, but they mapped at separate genetic loci: *degQ* at 279° and *sacU* (renamed *degS degU*) at 306° (36). The Hy alleles of both *degU* and *degQ* cause elevated production of alkaline protease, neutral protease, xylanase, β-glucanase, and α-amylase (2, 33). In addition, both types of Hy alleles cause loss of motility and lowered competence (33, 45, 78). The cloning of *degQ* from *Bacillus amyloliquefaciens* then allowed the isolation of the *B. subtilis* counterpart by homology (90). A homologous gene was also isolated from *Bacillus licheniformis* (2). The derived amino acid sequences of the three *degQ* ORFs show strong similarity.

The presence of a multicopy plasmid containing the *B. licheniformis degQ* gene causes a 70-fold elevation of alkaline protease levels and a 50-fold enhancement of levansucrase activity. The *B. subtilis degQ* gene codes for a polypeptide of 46 amino acids, of which 40% are charged, and *degQ* is translated in vivo (90). Many data indicate that the DegQ protein causes the Hy phenotype: (i) sequencing of the *degQ36* mutation shows that a C-to-T transition at the −10 position of *degQ* (90) makes this promoter one of the strongest in *B. subtilis* and results in much higher expression of its gene product (47, 90); (ii) placing the *degQ* ORF behind an inducible promoter, i.e., under LacI control, results in the DegQ(Hy) phenotype when the inducer isopropyl-β-D-thiogalactopyranoside (IPTG) is added (2); and (iii) a fortuitous cloning accident that modified the C-terminal end of the *B. amyloliquefaciens* DegQ (the first 36 amino acids of the original protein were fused to a heterologous 28-amino-acid peptide) resulted in a protein, DegQ*, with much higher enhancement activity for exoenzyme production than the 46-amino-acid wild-type protein (69). In this last case, the presence of DegQ* on a multicopy plasmid raised the levels of exoprotease 50- to 100-fold over those found in wild-type strains.

Strains overproducing DegQ show increased transcription of *aprE* (66), and deletion mapping of *aprE* upstream regulatory sequences indicates that the site of DegQ action, whether direct or indirect, is in the −141 to −164 region of the promoter (25). Similar analyses of *sacB* indicate that the site of DegQ action is just upstream of the promoter (32), and examination of DNA sequences in the regions of both promoters that are essential for DegQ stimulation show a 30-nucleotide identity among 45 residues (25). The DegQ overexpression phenotype requires a functional DegS-DegU system, and deletion mapping experiments of the target genes *aprE* and *sacB* indicate that there is overlap in the sites of DegQ and DegS-DegU activation (25, 32). As is the case with most of the accessory proteins, overexpression of DegQ does not alter the temporal regulation of its target genes.

The foregoing data suggest that DegQ may be a transcription factor. However, the protein has not been purified, and mutational analyses of its ORF and putative binding sites have not been performed. The absence of such information and the fact that inactivation of chromosomal *degQ* has no apparent phenotype (90) make it difficult to discuss the role of DegQ in the regulation of its target genes. This is also true for the regulatory proteins discussed above. Unlike the genes discussed above, however, much is known about the regulation of *degQ* expression. *degQ* is temporally (90), nutritionally, and epistatically (46, 47) controlled. *degQ* expression, as measured by *lacZ* reporter assays, increases at the end of vegetative growth, is repressed by glucose in minimal salts media but stimulated by decoyinine under these conditions, and is also stimulated by growth in poor carbon sources, by amino acid deprivation, and by phosphate starvation. Interestingly, these are all conditions that cause the initiation of sporulation. *degQ* expression also requires a functional DegS-DegU system, which form a histidine kinase (HK)-RR pair (see below). Its expression also requires the presence of ComP and ComA, the competence HK-RR pair, and the newly described competence gene, *comQ*, which maps between *comPA* and *degQ* (47). Deletion analyses of *degQ* promoter sequences showed that the region from −393 to −186 was the site of DegU action and that ComA activation required positions −78 to −40. It was further shown that amino acid regulation required ComA and that phosphate starvation, catabolite repression, and decoyinine stimulation involved sequences downstream from position −78 and were independent of both two-component systems (HK-RR pairs).

DNA sequences upstream of the transcriptional start site of the *B. subtilis degQ* gene indicate a potential E-σ^A promoter (90), and alignment of the *B. licheniformis* and the *B. amyloliquefaciens degQ* upstream regions with the corresponding sequence from *B. subtilis* demonstrates conservation of the −35 and −10 hexamer sequences, with the major exception of the C at position −10 in the *B. subtilis* promoter (2). Interestingly, the *degQ36* mutation replaces the C with a T that is normally found at this position in *degQ* genes of the two other bacilli, and this fact could partially explain the lower stimulation of target genes by multiple copies of the wild-type *B. subtilis* gene compared with that of its homologs (2). In vitro transcription studies of *degQ* to confirm the nature of the RNA polymerase involved have not been performed. Inactivation of the *spo0H* gene, which codes for σ^H (12), causes a fivefold decrease in *degQ* expression (45), but it is not known whether the effect is direct. *aprE*, which is transcribed by E-σ^A (54), also requires *spo0H* for its maximum expression (13), and a *spo0H* requirement is also found for in vivo expression of the E-σ^A-transcribed *spoIIG* (31) and *spoIIE* (17, 21). Unlike these other developmentally regulated genes, which require *spo0A* (13, 62, 92), *degQ* expression is unaffected by the inactivation of this gene (45). The intriguing network of *degQ* regulation is more fully discussed in other articles in this volume (see chapters 39 and 50).

Pai

With the discussion of Pai, we enter the realm of demonstrated physiological significance of the protein under normal physiological conditions, which for the purposes of this review is defined by an altered phenotype when the gene in question is disrupted. The *pai* operon of *B. subtilis* was cloned on a multicopy plasmid by virtue of the phenotype of repression of extracellular protease production (27). In addition, lower levels of secreted alkaline protease, neutral protease, levansucrase, α-amylase, and alkaline phosphatase and greatly reduced sporulation (a reduction of 5 orders of magnitude) were observed with multicopy *pai*. Filamentous growth was also found under these conditions. Levels of *nprE* RNA were lower in strains with multicopy *pai*, suggesting an effect at the transcriptional level. Sequencing of the 2.5-kb DNA fragment revealed two ORFs, the left-most (ORF1) coding for a 21-kDa protein and the right-most (ORF2) coding for a 24-kDa protein, and there was some indication from the DNA sequence that the two genes may be in one transcription unit. Disruption of either of the ORFs eliminates the multicopy phenotype with regard to protease levels and sporulation frequency, whereas ORF2 is not required for the filamentous phenotype. Overexpression of the *pai* operon in *Escherichia coli* and partial purification of the gene products has shown that the two ORFs can be translated in vivo, at least in *E. coli*.

Chromosomal mapping of *pai* by using antibiotic cassettes and integrative vectors has placed the operon close to *thrB*, which is at 284°. Disruption of ORF1 in the chromosome results in sporulation and extracellular protease production that is completely resistant to normally repressing levels of glucose (2%). This level of glucose in wild-type strains lowers levels of total protease by 100-fold and of sporulation by approximately 6 log orders. No enhancement of protease production is observed in disrupted-ORF1 strains when glucose is absent. ORF2 disruptions in the chromosome were unobtainable, suggesting that this gene may be essential for growth. It may have a secondary promoter and could thus be transcribed even if ORF1 was disrupted.

Analysis of the two ORFs shows that both code for highly charged proteins that contain approximately 40% charged amino acids. Examination of various data bases shows no relationship between ORF2 and any other protein. However, a similar comparison with ORF1 indicated a reasonable similarity with two proteins (69a). ORF1 shows 30% identity in a 75-amino-acid overlap with the 161-codon *rimI* ORF of *E. coli*, giving a FASTA score of 146. *rimI* codes for an enzyme that catalyzes N-terminal acetylation of ribosomal protein S18 (93). A similar relationship was observed between ORF1 and *sta*, which encodes the 184-amino-acid streptothricin acetyltransferase of *Streptomyces lavendulae* (28). There was 30% identity in an 84-amino-acid overlap, giving an optimized FASTA score of 126. Interestingly, the identities are all localized in the C-terminal regions of the three proteins. Exactly how the ORF1 and ORF2 proteins are involved in the glucose repression of several late-growth processes must await their purification and further characterization. However, an intriguing observation concerning ORF1 function with regard to sporulation and exoprotease production has been made. As discussed above, cells with disruptions of *pai* ORF1 show essentially wild-type levels of sporulation and protease production in the presence of normally inhibitory levels of glucose. The addition of glutamine as well as glucose represses these phenomena in wild-type strains to a much greater extent than glucose alone. The expression of early sporulation genes is similarly effected (11a). However, disrupted-ORF1 strains are equally resistant to the combined effects of glucose and glutamine for the above-mentioned parameters (54a). This is not always the case with mutations causing glucose-insensitive sporulation (see below).

Hpr

The *hpr* locus at 75°, mapping between *glyB* and *glpDKP*, was originally defined by mutations that caused overproduction (16- to 37-fold) of alkaline and neutral proteases. Other genetic lesions, *catA* and *scoC*, which were initially isolated on the basis of their glucose-insensitive sporulation and subsequently were shown to cause elevated exoprotease production, also were mapped to this area of the *B. subtilis* chromosome. The cloning and characterization of the wild-type *hpr* gene and the subsequent DNA sequencing of several *hpr*, *catA*, and *scoC* mutations have now shown that all reside in the same gene and that all cause a loss of Hpr function (57). This result suggested that Hpr is a repressor of protease production and sporulation. Further evidence for this idea came from the gain-of-function phenotype when the *hpr* gene was present in multiple copies. Sporulation was reduced 3 to 4 log orders, and exoprotease activity was severely inhibited as judged by halo size on casein-containing solid media (57).

The *hpr* ORF codes for a protein with 119 amino acids and a calculated molecular mass of 23,718 kDa. Mutational analysis indicates that the protein and its absence are responsible for the gain-of-function and loss-of-function phenotypes, respectively. No significant similarity between the *hpr* ORF and any protein sequence in data banks has been observed. The protein has now been purified (29) and shown to bind to the upstream regulatory regions of *nprE* and *aprE*. Since previous studies had shown that *hpr* loss-of-function mutations caused an increase in levels of *aprE* mRNA (25), binding studies indicate that Hpr is a direct repressor of its target genes. Footprinting experiments further showed that there were four Hpr binding sites in the *aprE* upstream region and two in the *nprE* regulatory sequences. The Dnase I-protected regions were 20 to 30 bp long and were found at positions −324 to −295, −292 to −267, −74 to −59, and −35 to −14 on the former promoter and at positions −107 to −82 and −26 to −4 on the latter. No cooperativity was observed in the binding of Hpr to these sites. Deletions in the *aprE* promoter region that removed sequences upstream of −200 eliminated the *hpr* (loss-of-function) stimulation effect on *aprE* expression (25), indicating the physiological significance of the two Hpr upstream binding sites. Similarly, a deletion from −156 to −90 in the *nprE* promoter caused an increase in the levels of *nprE* RNA (81), suggesting that Hpr binding to this region is important for in vivo regulation of the gene. As is frequently the case with other modulators of late-growth gene expression, the absence of Hpr does not change the temporal expression of genes it represses, indicating that other factors play a role in the timing of target gene expression.

Hpr has been reported to be involved in the regulation of other late-growth functions, i.e., alkaline phosphatase production, motility, and the control of sporulation (57), but is not known whether the effect is direct. The effect on sporulation, as discussed above, is manifested by the inability of glucose to inhibit this process in strains with *hpr* loss-of-function mutations. Interestingly, the presence of glutamine with glucose still inhibits sporulation in these mutant strains (54a), unlike the case with strains containing null mutations of the *pai* ORF1 (see above). In addition to its late-growth role, Hpr is involved in at least one exponential-growth-phase function. *spo0A* mutant strains are resistant to normally toxic levels of hydrogen peroxide. Mutations in *abrB* or *hpr* eliminate this elevated resistance to oxidative stress (9). The end result of the *spo0A* mutation would be expected to be an overproduction of Hpr via the derepression of *abrB*, which occurs when its repressor, Spo0A, is not present (see below).

Clearly, Hpr is an important regulator of late-growth function. Of interest, then, are relationships between this protein and other factors functioning during late growth. Examination of the upstream regions of the *sin* operon, which codes for another late-growth regulatory protein, SinR (discussed below), showed the presence of two regions with a 4-base consensus, (G/A)ATA, that is found in all six of the Hpr-binding sites in *aprE* and *nprE*. It was then shown that Hpr binds to the *sin* promoter at these regions (positions −86 to −57 and −10 to +14) (29). This interaction is presumably of physiological importance, because strains with *hpr* loss-of-function mutations show approximately fivefold-elevated levels of *sin* operon expression (39a). SinR also binds to the *aprE* promoter, inhibiting expression of this gene (19). However, there is no cooperativity between the binding of SinR and Hpr on the *aprE* gene, even though their binding sites are adjacent to each other (19, 29).

The regulation of *hpr* expression also reflects the interdependence of the pleiotropic regulatory proteins. The late-growth regulator AbrB (discussed below) is a positive activator of the *hpr* gene and binds to the promoter region of *hpr* (75).

AbrB

The DNA-binding protein AbrB is an extremely important factor in the transition between vegetative growth and stationary phase (75). Since AbrB is described at length in this volume (see chapter 52), this section will briefly summarize the interactions of this protein with its target genes. This is especially relevant, because AbrB affects many processes that are regulated by the other late-growth regulatory proteins.

AbrB represses *aprE* (which codes for the major secreted alkaline protease) (83), *spoVG* (94), *spo0E* (56), and *spo0H* (88, 95), the last three genes controlling various aspects of early sporulation. *tycA* (an antibiotic synthesis gene [60]) and *abrB* itself (74) are also repressed by AbrB. It is important to stress that AbrB repression was demonstrated in vivo only by removing Spo0A, a repressor of *abrB*. The effect of overexpression of AbrB on its target genes in a SpoA$^+$ strain has not been demonstrated. However, the 66,000-molecular-weight AbrB protein, which is composed of six identical monomers of 11,000 molecular weight each, has been shown to bind to *tycA*, *spoVG* (16), *aprE*, *spo0E*, and *abrB* (75). Recent experiments have shown that AbrB binds to the *spo0H* promoter (89a). There seems to be no clear consensus binding sequence, but runs of As and Ts may play a role in the protein DNA interaction. Some of the genes repressed by AbrB that are temporally regulated show constitutive expression when the *abrB* gene is inactivated. This suggests that these genes, i.e., *tycA*, *spo0H*, and *spo0E*, are solely regulated by AbrB.

In addition to its repressor functions, AbrB is a positive regulator of *hpr*, presumably by binding to the promoter region (75). AbrB is also required for full expression of competence genes (11), and epistatic experiments have shown that AbrB acts to positively regulate expression of *comK*, an early competence regulatory gene, but it is not known whether the effect is direct (22).

The regulation of *abrB* is complex, as is true for many pleiotropic genes that affect late-growth processes. In addition to AbrB, Spo0A, the master regulator for most of these temporally regulated phenomena (see below and chapter 51), represses expression of *abrB* by directly binding to the promoter and blocking transcription (73).

ComA

ComA, DegU, and Spo0A are RRs for competence, extracellular enzyme synthesis, and sporulation regulons (see chapters 39, 50, and 51). It could be argued that discussing the three RRs in a chapter on accessory proteins involved in late-growth development is unnecessary, since the former are described more completely in the chapters just mentioned. However, the RRs show pleiotropic effects, they are involved in the control of many late-growth phenomena in addition to those for which they were first described, and they interact with many of the other regulatory proteins and their genes that have been described in this chapter.

ComA is an essential regulatory protein for the temporally and nutritionally controlled competence regulon (see chapter 39) (11). This protein is believed to be activated by phosphorylation at a conserved aspartate residue found in all RRs. ComP, an HK whose gene is in an operon with *comA*, is the leading candidate for the HK that phosphorylates ComA, as mutations in *comP* that are nonpolar on *comA* also lower competence and expression of late competence genes (11). ComA and its activator, ComP, are also necessary for expression of *degQ* (47) and the *srfA* operon (51). This latter locus is important for sporulation and competence as well as for biosynthesis of the antibiotic surfactin (50). The ComA-ComP system is also required for the expression of *gsiA*, a gene induced by glucose starvation (48).

Elevated levels of ComA inhibit sporulation, and inactivation of *comP* lowers sporulation frequency in the absence of the major sporulation HK, KinA-SpoIIJ (87). The block in sporulation caused by overexpression of ComA occurs at an early stage, as measured by the inhibition of early *spo* gene expression. The presence of ComP at equally high levels reverses the ComA inhibition (70). This suggests that unphosphorylated ComA may repress early steps in sporulation, but the physiological significance of the effect is not known.

While ComA has a C-terminal domain resembling certain DNA-binding proteins (11) and is believed to be a transcriptional activator, as are most of the RRs, it has not been shown to act directly at the gene level.

DegU

The RR protein DegU is a major factor in the control of extracellular enzyme synthesis, as the previous parts of this chapter amply demonstrate. It and its cognate HK, DegS, are essential for the expression of *sacB*, which is the structural gene for the secreted levansucrase, and *degQ*, which also regulates extracellular enzyme synthesis (46). While *degSU* loss-of-function mutations have negligible effects on the expression of *amyE*, *nprE*, and *aprE*, which code for α-amylase, neutral protease, and alkaline protease, respectively, the *degSU*(Hy) mutations dramatically increase expression of these genes. As discussed above in the section on DegQ, there are DNA sequences in the upstream regions of *sacB*, *aprE*, and *degQ* that are required for DegS-DegU action. However, DegU has not been shown to directly interact with its target genes.

degS loss-of-function mutations abolish the positive effect of *degU* in degradative-enzyme synthesis, and biochemical analysis of the nature of the *degSU*(Hy) mutations indicate that they generally result in more phosphorylation of DegU by DegS or cause more stable phosphorylation at the level of DegU (see chapter 50) (7, 79). These data indicate that DegU-phosphate is the form of the protein that activates the target genes discussed above.

DegU is also essential for competence, and recent data suggest that this protein is required to activate expression of *comK* (22). Unlike the situation in which DegU-phosphate stimulates the expression of extracellular protein genes, much evidence points to the need for unphosphorylated DegU as an activator of competence. Nonpolar *degS* loss-of-function mutations have no effect on competence, and *degSU*(Hy) mutations drastically lower its frequency (see chapters 39 and 50). Perhaps the strongest evidence for the need for unphosphorylated DegU in the competence pathway comes from analysis of the *degU146* mutation. The essential aspartic residue at position 56 was changed to an asparagine. The mutated DegU, which is unable to accept a phosphate residue, cannot function as an activator for degradative-enzyme synthesis, but it is still able to function, even though at a level much lower than that of wild type, in the competence pathway (see chapter 50). The close relationship between the competence regulon and the DegS-DegU pathway is also illustrated by the phenotype of *mec* mutations. These genetic lesions allow induction of competence in rich medium and bypass mutations in early competence genes (61). The *mec* mutations also suppress the negative effects of *degSU* mutations on *sacB* expression (45).

The *degSU*(Hy) mutations have other pleiotropic effects, e.g., lack of flagella and glucose-insensitive sporulation, a phenotype that is similar to that found when levels of certain other regulatory proteins are altered, e.g., when high levels of DegT occur (see above).

Spo0A

Spo0A is a key regulator for most late-growth functions (see chapter 51). It functions in a dual manner, acting both as a positive and a negative control element. It represses *abrB* when phosphorylated, thus allowing early events in the initiation of sporulation, i.e., derepression of *spo0H*, *spo0E*, and *spoVG*, and Spo0A controls the temporally regulated *aprE* and *tycA* in the same indirect manner (see chapter 51). Spo0A is also required for the development of competence through its control of the levels of AbrB. AbrB is essential for competence, but it can inhibit this process if overexpressed, which presumably occurs when Spo0A is absent (see chapter 39). Since competence, unlike the processes discussed above, is unaffected by mutations in *kinA-spoIIJ*, *spo0F*, or *spo0B* (see chapter 39), it is not clear whether Spo0A must be phosphorylated for its role in competence.

Spo0A also functions as a positive regulator of sporulation, binding to the upstream regulatory regions of three stage II genes, *spoIIA*, *spoIIE*, and *spoIIG* (62, 82, 92). Interestingly, Spo0A interacts with two different RNA polymerases, E-σ^A and E-σ^H, in its action as a positive regulator of transcription. Presumably, Spo0A must be phosphorylated by the SpoIIJ-Spo0F-Spo0B phosphorelay for its positive role in transcription (82). Mutational analysis has shown that alterations of the C-terminal region of Spo0A eliminate the positive activation of the protein without affecting its ability to down regulate *abrB* (58).

Genetic data also highlight the key role of Spo0A in the initiation of sporulation. *spo0A* is absolutely required for this process, since *spo0A* loss-of-function mutations, unlike the case with all other early sporulation genes (save *spo0H*, which codes for the earliest acting sporulation-specific σ factor), cannot be suppressed. In addition, various mutations in the *spo0A* ORF allow sporulation to proceed either in the absence of SpoIIJ, Spo0B, or SpoIIF or in the presence of normally inhibitory levels of glucose (26, 52, 68).

As would be expected from its role as a master late-growth regulator, *spo0A* itself is regulated in a complex manner. It is transcribed from two promoters, an upstream E-σ^A-like P1 and a downstream E-σ^H-like P2 (6). These RNA polymerase assignments have been confirmed by in vitro transcription studies (59). Transcription at these two promoters is differentially regulated temporally and nutritionally. *spo0A* P1 is transcribed during vegetative growth but is not functional as cells enter stationary phase. Conversely, *spo0A* P2 is poorly transcribed during vegetative growth and shows elevated levels of transcription initiation as cells stop growing (6, 59). Glucose represses transcription from P2 and prevents the down regulation of P1 usually observed at the end of exponential growth.

SinR AND SinI

Several aspects of late-growth accessory proteins and their regulation have been discussed in the preceding sections. Among these are pleiotropy of function, interaction between regulatory proteins, dual (positive and negative) regulatory roles, complex regulation of genetic determinants, and in some cases, the presence of a second cotranscribed ORF (the *pai* and *ten* operons). Mechanisms and/or functions for most of these phenomena are unknown. Studies of the *B. subtilis* SinR protein (formerly known as Sin) and its regulation, on the other hand, have provided much information on the themes discussed above, and this section will focus on the role of SinR in late-growth development.

Negative SinR Functions

A chromosomal fragment containing the *sinR* gene was originally cloned by means of a gain-of-function phenotype of sporulation inhibition (up to 3 log orders) when it was present on a multicopy plasmid (18). Alkaline and neutral exoprotease production was also inhibited under these conditions, so that less than 10% residual protease activity was observed. DNA sequencing and deletion analyses of the cloned DNA segment indicated that the *sinR* gene, the second of two cotranscribed ORFs and also containing a secondary promoter, was responsible for the Sin^+ (sporulation inhibition) phenotype. The 111-amino-acid SinR protein, the product of the *sinR* gene, was the active element in the Sin^+ phenotype, as demonstrated by the following: (i) point mutations in the *sinR* ORF destroy its ability to inhibit sporulation and exoprotease production (18); (ii) the *sinR* ORF is translated in vivo, and sporulation and protease inhibition are correlated with higher levels of SinR (17, 19); (iii) strains carrying a chromosomal copy of *sinR* under LacI-regulated pSPAC control demonstrate a Sin^+ phenotype in the presence of IPTG (59a); and (iv) SinR binds in vitro to certain target genes whose in vivo expression is inhibited by high levels of SinR (see below).

Sequence analysis of the *sinR* ORF indicated that it encoded a DNA-binding protein, a conclusion supported by the tight binding to and subsequent purification of the protein on heparin-Sepharose columns. Since high levels of SinR inhibit the expression of *aprE* and the deletion of sequences 200 bp upstream from the transcriptional start site of this gene prevents the inhibition phenotype, it was postulated that SinR exerts its repressive effects by directly binding to these sequences. Gel retardation and footprinting experiments have confirmed this hypothesis (19, 29). SinR specifically binds to the region from −219 to −268 in *aprE*, which is adjacent to one of the Hpr binding sites at −267 to −292.

Recent experiments have shown that SinR represses sporulation in a manner analogous to its effect on *aprE*. The expression of three stage II genes, *spoIIA*, *spoIIE*, and *spoIIG*, but not of earlier acting *spo* genes is almost completely inhibited when SinR is overproduced and SinR protein binds to the promoter regions of these target genes (39). Gel retardation studies have localized the apparent Sin binding to the region from −75 to −30 of the *spoIIA* promoter (39a). However, reproducible DNase footprinting of SinR to the SpoIIA promoter region has not been obtained. SinR also autogenously regulates its own synthesis, and exonuclease III footprinting experiments have shown that SinR binds to its gene in the region +1 to +43, which would presumably block transcription from *sin* P1 and P2 (19a, 71).

Inactivation of the chromosomal *sinR* gene provides further evidence for the role of SinR as a repressor of sporulation and certain other late-growth processes. The stage II operons that are inhibited by excess SinR are expressed at a threefold-higher level than in wild-type cells (*sinR* in single copy) when SinR is absent. Expression of these genes is also detectable up to 90 min earlier when *sinR* is inactivated. Significantly, *sinR* loss-of-function mutations suppress the sporulation-negative phenotype caused by null mutations of *spoIIJ* and missense mutations in *ftsA* (35, 39). The *sinR* operon and *aprE* are also severalfold overexpressed when *sinR* is inactivated (18, 19a, 39). Elimination of the SinR binding site in the region from −219 to −268 of *aprE* also raises the level of *aprE* expression in a strain that has wild-type levels of SinR and prevents the inhibition of *aprE* when SinR is overproduced (19).

Expression of *amyE* (18) and *sacB* (45) is also raised threefold when SinR is absent, but it is not known whether these genes are direct SinR targets. There is no change in the patterns of expression of SinR target genes when SinR is absent, outside of the earlier induction of stage II genes. Thus, other regulatory elements must play a role in their temporal regulation. This will be discussed in the concluding section of this review.

Positive SinR Functions

Disruption of the chromosomal *sinR* gene has many other ramifications besides the derepression of its negative target genes. In a *sinR* null background, cells become filamentous (but septate), nonmotile, autolysin negative, and competence deficient (3 log orders less than wild-type strains) and their colonial morphologies change to an extremely rough appearance. These pleiotropic effects that characterize the SinR loss-of-function phenotype are similar to those described for mutations in *flaD*, which maps in the same region of the chromosome at 221° (1), and it has recently been shown that *flaD* and *sinR* are identical (64, 65).

Thus, SinR plays a positive role in competence, motility, and autolysin production, and in the case of the first two phenomena, it is known to work at the level of gene expression. *comK*, a regulatory gene for competence, requires SinR for its expression, while *srfA*, which is directly epistatic on *comK*, is unaffected by null mutations in *sinR* (see chapter 39). Lower levels of flagellin are observed in cells lacking SinR, and the expression of the *hag* gene, which codes for flagellin, is severely inhibited (59a). However, SinR does not bind in vitro to *comK* or *hag* promoters. Thus, it does not seem that the SinR interaction with these genes is direct. It is also not known whether the SinR requirement for autolysin production is at the level of gene expression, and preliminary experiments (19b) have shown that SinR does not bind the promoter of *cwlB*, which codes for the major autolysin *N*-acetylmuramoyl-L-alanine amidase (34).

Mechanism of SinR Action

Comparison of known SinR binding sites gives a core consensus binding sequence of GNCNCGAAat ACA (71), but this is based on relatively few genes, and a definitive binding sequence awaits the isolation of more SinR operator sites. The nature of the SinR-DNA operator site interaction is unknown and is currently being studied by means of a mutational analysis of SinR, but amino acid similarities suggest that its amino-terminal domain is important for the binding of SinR to its operator sites. The *B. subtilis* bacteriophage ϕ105 repressor and SinR share extensive similarity in their N-terminal domains (22 of 42 amino acids are identical, and an additional 9 are conserved), especially in their helix-turn-helix DNA-binding sequences (71). In addition, the consensus sequence for ϕ105R operator sites is very similar to the putative SinR binding sequence, containing all of the conserved bases listed above. However, overexpression of SinR has no affect on the ϕ105 lytic response in vivo, and high levels of ϕ105 repressor do not mimic the Sin$^+$ phenotype (69a). This is not too surprising, since position 1 in the DNA recognition helix, a key determinant of DNA-binding specificity, differs in the two proteins. The ϕ105 repressor contains arginine at position 1 (84), and lysine is the corresponding amino acid in SinR. The significance of the strong similarity between the two proteins is not known, outside of a possible common evolutionary history.

In addition to the nature of the interaction between SinR and its DNA target sequences, there are other unanswered questions concerning the mechanisms of repression and activation of genes in the Sin regulon. The best-described example of SinR-DNA interaction is in the regulation of the *aprE* gene (19). However, it is not clear why the binding of SinR to a region almost 200 bp upstream from the −35 region should affect transcription, especially when these DNA sequences are not necessary for in vivo expression of *aprE*. There may be a very weak binding site for SinR close to the +1 position, and it is tempting to think that SinR causes looping between the −200 and +1 regions, which could block RNA polymerase and/or positive regulator access to the *aprE* promoter. However, attempts to footprint this putative binding site have been unsuccessful, and its physiological significance remains uncertain. SinR is a tetramer in solution, like LacI and GalR, which do cause looping of target DNA sequences (14, 38). However, there is even less evidence for looping in any other SinR-DNA interaction.

Less is known about the mechanism of SinR repression of stage II genes, but it does bind to an upstream region of *spo0IIA* (−75 to −30) (39, 39a), which also contains a binding site for Spo0A, the positive activator of *spoIIA*, *spoIIE*, and *spoIIG* (62, 82, 92). This suggests a possible competition between SinR and Spo0A at the level of DNA binding. However, this possibility remains untested, as it has not been possible to get reproducible SinR footprints on the SpoIIA promoter.

Regulation of SinR

SinR and its genetic determinant are regulated at several different levels, i.e., transcriptionally, translationally, and posttranslationally. The two-gene *sinR* operon has three promoters (P1, P2, and P3), and steady-state levels of RNA initiated at these sites show different patterns relative to each other during growth and sporulation (17). P1 RNA initiates 120 nucleotides from the translational start site of the upstream ORF1 (now known as *sinI*; see below), P2 RNA starts 50 nucleotides from this point, and P3 transcripts initiate 15 nucleotides upstream from the first codon of the downstream *sinR* gene. The transcriptional start of P3 RNA overlaps its ribosome-binding site, and the initiating codon of the *sin* protein is UUG. These two features of P3 RNA could explain its apparently low level of translation (see below). P1 and P3 show close resemblance to E-σ^A consensus promoters and are transcribed in vitro by this RNA polymerase species (49a). P3 RNA levels are high throughout growth and through the first 2 h of sporulation, while P1 RNA levels are low during growth and increase at the beginning of stationary phase, but at all time points, P3 RNA is 10 times more abundant than P1-initiated transcripts. P2 resembles E-σ^E promoters, which is consistent with the observed appearance of the P2 transcript 2 h after commencement of the stationary phase. In vitro transcription studies have not been performed with this promoter, and it is possible that the putative P2 transcripts, which are of low abundance relative to those initiated at P1, actually represent processed molecules that are initiated at P1. Most if not all of the *sin* operon transcripts end downstream from the last codon of *sinR* at a ρ-independent terminator that functions well both in vivo and in vitro (49a).

Detailed mutational analyses have not been performed on any *sin* operon promoters in the chromosome, but circumstantial evidence indicates that P1 and P2 are important for *sin* regulation. As discussed above, Hpr and SinR repress *sin* operon expression in vivo (19a, 39a). Since both of these proteins bind to the *sin* P1 and P2 promoter region in vitro, these promoters most likely play an important role in the initiation of *sin* operon transcription. The role of P3 is more problematic. Deletion experiments starting from the 5′ end of the *sin* operon carried on multicopy plasmids showed that P3 was sufficient for sporulation inhibition, and insertions of antibiotic resistance genes into the chromosomal *sinI* allow production of the downstream *sin* gene. Since the antibiotic cassette disruptions separate P1 and P2 from the *sin* gene but do not affect P3, the inference is that chromosomal P3 transcripts are translated.

Additional evidence suggests that differential regulation occurs in the *sin* operon. Translational *lacZ* fusions have been made with the upstream *sinI*, which is regulated by P1 and P2, and also with the downstream *sinR* gene, putatively controlled by P1, P2, and P3 (17). Studies with these fusions integrated into the chromosome have shown that both genes are translated at similar low levels during vegetative growth, but at the end of this stage, *sinI* expression shows a 5- to 10-fold increase, while *sinR* expression increases only very slightly. In addition, the rise in *sinI* expression is prevented by mutations in the *spo0A* and *spo0H* genes and by the presence of high levels of glucose. Thus, *sinI* seems to be under sporulation control. *sinR* expression is unaffected by these conditions.

Levels of SinR, as measured by Western immunoanalyses, do not decrease when the SinR stage II target genes show increased transcription (39). Thus, there was no apparent explanation for the down regulation of SinR function. Recent experiments have provided an answer to some of these unexplained phenomena. The facts that *sinI* expression increases at the end of exponential growth and that SinR levels do not decrease when the repressive effects of SinR on stage II genes are being overcome suggested that the protein encoded by *sinI* might be countering the action of SinR at a posttranslational level. Several lines of evidence indicate that this is indeed the case (3a). Greater sporulation inhibition was observed when plasmids contained only the *sinR* gene rather than the entire *sin* operon. Following this early observation, it was found that overproduction of SinI, whether this was achieved by a multicopy gene effect or by placing a chromosomal copy of the *sinI* gene under LacI-pSPAC control, gave a phenotype identical to that observed when the *sin* gene was inactivated. Among these characteristics are filamentous growth, rough colonial morphology, loss of competence, and overexpression of the SinR target stage II sporulation genes. The raising SinI levels had no effect on cellular SinR concentration, indicating that this protein does not lower SinR expression or cause its disappearance.

These results suggested that SinI was antagonizing SinR activity. To provide more evidence for this idea, strains lacking *sinI* but expressing SinR were constructed. A nonpolar in-frame deletion that destroyed the multicopy *sinI* phenotype was made in *sinI*. When this deletion was integrated into the chromosome by gene conversion, the cell behaved as though SinR was overexpressed; i.e., cells appeared smooth and sporulated at approximately 1 to 10% of wild-type levels. This phenotype was also observed when *sinI* was disrupted by insertion of an antibiotic resistance cassette (see above). In addition, the expression of *aprE* and stage II genes was inhibited. However, SinR levels remained unchanged. This pseudo "Sin$^+$" phenotype was eliminated by reintegrating a wild-type *sinI* gene at the region of homology or by providing SinI in *trans*.

Thus, the 57-amino-acid SinI protein controls the activity of Sin at a posttranslational level. Protein data base analyses reveal that of all proteins screened, SinI shows greatest similarity to SinR. In a 39-amino-acid overlap consisting of the central part of SinI and the C-terminal region of SinR, there is 28% identity and 52% conservation of amino acids. While the interaction between the two proteins could be some type of SinI-mediated covalent modification of Sin, my current hypothesis is that a protein-protein interaction initiated by SinI sequesters SinR in the form of nonfunctional heteromultimers and prevents it from binding to its target genes. The formation of these heteromultimers with SinI could be at the SinR middle and C-terminal regions, which are believed to be part of the multimerization domain of SinR. This would be similar to the NF$\kappa\beta$-I$\kappa\beta$ interaction in mammalian cells (3). Analogous types of protein-protein associations have been proposed for the regulation of σ^B and σ^F functions in *B. subtilis* (4, 30, 63). It is also possible that SinI competes with SinR in binding to DNA target sequences or interacts with Sin bound to DNA in a manner analogous to the "squelching" phenomenon observed with GAL4 and GAL80 in yeast cells (37). However, as described above, essentially all of the phenomena affected by alteration of Sin levels also respond in a reciprocal manner to the changing of SinI concentrations. It seems more economical to propose one bimolecular interaction between two proteins instead of several different trimolecular (SinI-SinR-DNA target) associations. In addition, SinI does not behave like a DNA-binding protein during its purification. The actual mechanism of SinI down regulation of SinR function is now being studied in vitro. Recent experiments have indicated that addition of purified SinI to SinR prevents the latter protein from binding to the *aprE* gene. Adding SinI to SinR after DNA binding is initiated has no effect on the SinR-*aprE* interaction (3a). This suggests that SinI interacts with SinR before the latter binds to its target. In addition, immunoprecipitation studies have shown that SinR and SinI can form stable complexes in vitro (3a).

ROLE OF PLEIOTROPIC REGULATORY PROTEINS IN LATE GROWTH DEVELOPMENT: CONCLUSIONS AND PERSPECTIVES

Why do *B. subtilis* and related bacteria go to the trouble of making several proteins with redundant, overlapping, and in some cases antagonistic physiological functions that affect late-growth-associated processes? A free-living *B. subtilis* cell often faces nutrient deprivation, and upon approaching the stationary phase, it has the option of becoming motile, secreting extracellular enzymes and antibiotics, becoming competent, or ultimately sporulating. Depending on the environmental signal (type of nutrient stress, etc.), one or more of these alternative processes can be stimulated or inhibited. Sensory-effector mechanisms are needed to integrate the environmental cues and channel the cell into the appropriate response. The end results of the late-growth pathways frequently are complex structures, e.g., flagellae, the DNA uptake machinery, the spore septum, the cortex and coat, etc., whose synthesis and assembly must be closely coordinated, both in a temporal and a spatial context. Since the synthesis of elaborate subcellular structures not required for vegetative growth is costly in terms of energy expended, negative control mechanisms are also required. This is especially true for sporulation, since a cell's premature entry into the dormant state would be disadvantageous, in a population sense, since its nonsporulating peers and competing species would be able to continue growth or resume growth more rapidly. Indirect evidence for this idea comes from chemostat (steady-state) growth studies, in which sporulation is not necessary for survival. Wild-type cultures gave rise to *spo* mutants, usually *spo0A*, which overgrew Spo$^+$ cells in less-than-optimal growth media (42, 43).

Thus, the proteins described in this review may be required to act as fail-safe regulators that can prevent commitment to unnecessary differentiation, to channel cells into alternative developmental fates, and possibly, to synchronize complex regulatory and structural pathways. Some of the proteins, e.g., AbrB, SinR, the RRs ComA, DegU, and Spo0A, etc., can act in more than one of these ways. It is not surprising that the control of expression of these multifaceted regulators is also exceedingly complex and that there are interdependent relationships controlling their synthesis.

Space limitations and a lack of knowledge do not allow a complete discussion of how each protein described in this review fits into the networks controlling late-growth development. By concentrating on a few of these proteins and the processes they control, however, one can attempt to form a coherent picture of how the proteins function in regard to the roles listed above, and perhaps this simplified model can

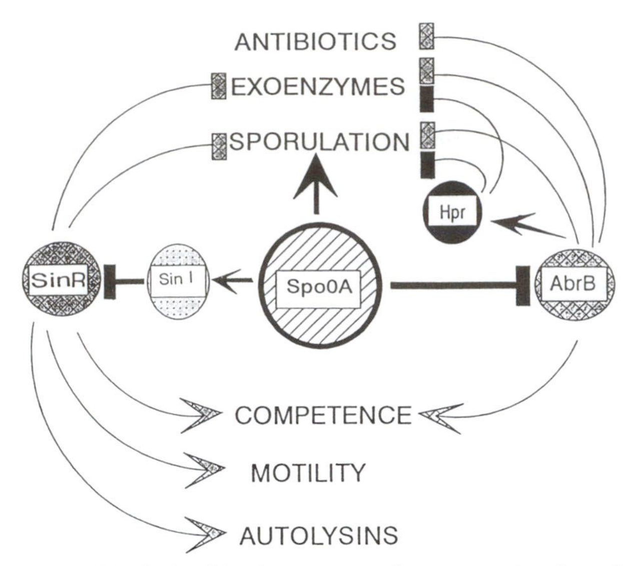

Figure 1. Relationships between regulatory proteins that affect late-growth processes, showing interactions between Spo0A, AbrB, Hpr, Sin, and SinI and the stationary-phase-activated processes they control. References and experimental details for this figure are in the text. For clarity, several elements of late-growth control, e.g., activation of Spo0A by the phosphorelay system and down regulation of *sin* by Hpr, etc., are not shown. The specific steps in the processes that are affected by the regulatory proteins are also not shown; they are described in the text. Positive interactions are indicated by an arrow, and negative ones are indicated by a bar. The figure does not differentiate between transcriptional and translational interactions.

provide a framework for understanding the actions of the other proteins. Figure 1 summarizes some of these relationships.

The regulation of *aprE* provides the clearest example of redundant control in late-growth development. Hpr, SinR, and AbrB are direct repressors of *aprE* expression. It is clear why inactivation of AbrB, Hpr, or SinR does not change temporal regulation of this gene, since there is a multiple fail-safe system for the repression of *aprE*. This redundant control system could explain why gain of function mutations in positive activators like DegU and overproduction of DegR, DegQ, DegT, TenA, and Sen do not alter the timing of *aprE* transcription, even though they result in severalfold stimulation of its expression. It is not clear why *aprE* is regulated in such an intricate fashion, since it is not essential for other late-growth processes. Of course, the regulation of production of subtilisin and other exoenzymes may be much more important for the survival of *B. subtilis* in the wild than in the laboratory. In any case, it would be important to try to alter the temporal control of *aprE* expression by sequentially inactivating all of the genes coding for its repressors.

It is of interest to compare the regulation of *spo0H* with that of *aprE*. The former gene, which codes for σ^H, the earliest acting sporulation-specific σ factor, is controlled negatively at the level of transcription by AbrB. Phosphorylated Spo0A, produced early in the response to nutrient stress, represses *abrB* (82). Unlike the situation with *aprE* expression, however, *abrB* loss-of-function mutations allow constitutive expression of *spo0H*. Thus, there is no backup system to regulate *spo0H* transcription. However, cellular concentrations of σ^H are controlled at the posttranscriptional level, because the protein becomes more stable when sporulation is initiated (23). The uncharacterized mechanism of protein stabilization thus acts as a second fail-safe system to control activation of the E-σ^H regulon, which is necessary for early stages of sporulation. The end results of higher levels of E-σ^H activity are higher levels of Spo0A-phosphate and, indirectly, more σ^H via the continued down regulation of *abrB* (59). These events are required for the transcription of three stage II genes, *spoIIA*, *spoIIE*, and *spoIIG* (62, 82, 92).

In addition to repressing *aprE*, SinR directly prevents the expression of these three stage II genes, and as previously discussed, there may be a competition between SinR and Spo0A for specific DNA-binding sites (39, 71). Thus, SinR acts as a stage-specific repressor, exerting its negative function after the action of AbrB has occurred. The existence of two points at which sporulation can be repressed could be another example of a fail-safe or redundant control, as is the case with *aprE*. It may also reflect different levels of commitment in the sense that reversal of sporulation by the addition of fresh growth medium is possible after the initiation process is begun. However, cells are irreversibly committed to sporulate once stage II, the formation of the spore septum, is completed (15). Nutrient stress may initiate the early Spo0A cascade, which is marked by the down regulation of *abrB*, but the existence of another checkpoint at stage II could allow the cell one last opportunity to decide on the appropriateness of sporulation. A third possibility is that SinR also acts to coordinate expression of its stage II target genes so that correct levels of their gene products (σ^E and its processing machinery and σ^F and its activation-inhibition loop) would be available for the next sporulation stage. A further refinement of these ideas concerns Spo0A and its interactions with AbrB and SinR. Spo0A represses *abrB* and indirectly controls SinR function through its control of *sinI* expression. As discussed above, *sinI* codes for SinI, a protein that inhibits the negative and positive functions of SinR at the posttranslational level. The fact that Spo0A controls the functional levels of both AbrB and SinR could reflect another way the environment, through its control of Spo0A synthesis and activation, influences late-growth events, even after the development process has been initiated. Spo0A control would also provide a way to coordinate the transition from stage 0 to stage II, since AbrB and SinR could also be viewed as modulators and not only repressors of sporulation.

At some point after the expression of *srfA* and before the expression of *comK*, both AbrB and SinR are required for competence (see chapter 39), but it is not known whether the dependencies are direct and whether the same step is affected by these regulators. These proteins can be viewed as developmental switches, since they can affect the cell's choice of developmental pathways. The role of AbrB is unusual in that AbrB, unlike SinR, can also repress competence when it is overproduced in *spo0A* mutants. This could represent a role of AbrB in the timing or other aspect of early competence control that would depend on the environment through the latter's effect on Spo0A synthesis and activation. In fact, the only role for Spo0A in competence (KinA-SpoIIJ, Spo0F, and Spo0B are not required) is to control AbrB levels (10). Presumably, there are sufficient levels of Spo0A for this task, because *spo0A* P1, the E-σ^A-dependent, glucose-insensitive promoter, can function in the high-glucose-containing media required for competence (6).

As discussed in the previous paragraph, Spo0A is also involved in the down regulation of Sin through the SinI protein. Since *sinI* expression is inhibited by glucose and is consequently poorly expressed in competence media (18a), the negative effect of Spo0A via ORF1 expression on SinR function is minimized in the competence pathway.

Thus, AbrB and SinR function in similar though not identical ways to repress genes in two late-growth pathways (extracellular protease production and sporulation) and to positively regulate a third (competence). Spo0A works to counter or modulate the activity of SinR and AbrB in these regulons, and this work provides another level of control at which the environment, through its affects on Spo0A, can determine pathways of development.

However, SinR and AbrB do not always work on the same target genes. The E-σ^D-transcribed gene *hag* (44) requires SinR in vivo for maximum transcription, even though the activation is not direct (59a). Other genes affecting motility and autolysin production and requiring SinR (1) are also believed to be in the E-σ^D regulon (40). However, there is no evidence that AbrB affects genes in the E-σ^D-dependent pathway, and it has recently been shown that mutations in *spo0A* and/or *abrB* have no effect on *hag* expression (59a).

This last aspect of regulation of late-growth pathways provides an informative example of multiple control. SinR and AbrB function together in many ways but probably not in the regulation of the E-σ^D regulon. This is reminiscent of the control of gene expression in eukaryotes, in which the combinatorial interaction of repressors and activators permits a tremendous flexibility in the choice of target genes and the quantitative control of their transcription. This combinatorial type of gene regulation could allow nutrient-starved *B. subtilis* to choose from a large number of developmental pathways. We hope to understand more about these decision mechanisms as we learn more about the functions of all the regulatory proteins affecting late-growth development in *B. subtilis*.

Acknowledgments. I especially thank David Dubnau, Frank Kunst, and Tarek Msadek for invaluable discussions, suggestions, and unpublished information. I thank many colleagues, including Michael Chamberlin, Roy Doi, Mark Strauch, and Charles Hutchinson, for information and preprints. Members of the Dubnau laboratory, especially Jeanette Hahn, Manuela Roggiani, and Yvette Weinrauch, were most helpful in explaining the intricacies of the regulation of competence. The work from my laboratory as well as many of the concepts discussed in this review come from my gifted and creative present and former laboratory associates Uma Bai, Jeanie Dubnau, Nand Gaur, Ines Mandic-Mulec, Gopal Nair, Seung Hwan Park, Mima Predich, and Matko Zelic. Annabel Howard performed invaluable secretarial assistance. Ines Mandic-Mulec kindly drew Fig. 1. Work from my laboratory that is discussed here was supported by Public Health Service grants GM-19693 and GM-32651 from the National Institutes of Health. Computer analyses were performed on a VAX 11/750 purchased with funds from NSF grant PCM-8313516 awarded to the Public Health Research Institute.

REFERENCES

1. **Akamatsu, T., and J. Sekiguchi.** 1987. Genetic mapping and properties of filamentous mutations in *Bacillus subtilis*. *Agric. Biol. Chem.* **51:**2901–2909.
2. **Amory, A. F., E. Kunst, A. Aubert, and G. Rapoport.** 1987. Characterization of the *sacQ* genes from *Bacillus licheniformis* and *Bacillus subtilis*. *J. Bacteriol.* **169:**324–333.
3. **Baeuerle, P. A., and D. Baltimore.** 1988. I-$\kappa\beta$: a specific inhibitor of the NF-$\kappa\beta$ transcriptional factor. *Science* **242:**540–546.

3a. **Bai, U., I. Mandic-Mulec, and I. Smith.** 1993. SinI modulates the activity of SinR, a developmental switch protein of *Bacillus subtilis*, by protein-protein interaction. *Genes Dev.* **7:**139–148.

4. **Benson, A. K., and W. G. Haldenwang.** 1992. Characterization of a regulatory network that controls σ^B expression in *Bacillus subtilis*. *J. Bacteriol.* **174:**749–757.
5. **Chamberlin, M.** Personal communication.
6. **Chibazakura, T., F. Kawamura, and H. Takahashi.** 1991. Differential regulation of *spo0A* transcription in *Bacillus subtilis*: glucose represses promoter switching at the initiation of sporulation. *J. Bacteriol.* **173:**2625–2632.
7. **Dahl, M. K., T. Msadek, F. Kunst, and G. Rapoport.** 1991. Mutational analysis of the *Bacillus subtilis* DegU regulator and its phosphorylation by the DegS protein kinase. *J. Bacteriol.* **173:**2539–2547.
8. **Debarbouille, M., I. Martin-Verstraete, A. Klier, and G. Rapoport.** 1991. The transcriptional regulator LevR of *Bacillus subtilis* has domains homologous to both σ^{54}- and phosphotransferase system-dependent regulators. *Proc. Natl. Acad. Sci. USA* **88:**2212–2216.
9. **Dowds, B. C. A., and J. A. Hoch.** 1991. Regulation of the oxidative stress response by the *hpr* gene in *Bacillus subtilis*. *J. Gen. Microbiol.* **137:**1121–1125.
10. **Dubnau, D.** Personal communication.
11. **Dubnau, D., J. Hahn, L. Kong, M. Roggiani, and Y. Weinrauch.** 1991. Genetic competence as a post-exponential global response. *Semin. Dev. Biol.* **2:**3–11.

11a. **Dubnau, E., S. H. Park, and I. Smith.** Unpublished data.

12. **Dubnau, E., J. Weir, G. Nair, L. Carter III, C. Moran, Jr., and I. Smith.** 1988. *Bacillus* sporulation gene *spo0H* codes for σ^{30} (σ^H). *J. Bacteriol.* **170:**1054–1062.
13. **Ferrari, E., D. J. Henner, M. Perego, and J. A. Hoch.** 1988. Transcription of *Bacillus subtilis* subtilisin and expression of subtilisin in sporulation mutants. *J. Bacteriol.* **170:**289–295.
14. **Flashner, Y., and J. D. Gralla.** 1988. Dual mechanism of repression at a distance in the *lac* operon. *Proc. Natl. Acad. Sci. USA* **85:**8968–8972.
15. **Freese, E., W. Klofat, and E. Galliers.** 1970. Commitment to sporulation and induction of glucose-phosphoenolpyruvate-transferase. *Biochim. Biophys. Acta* **222:**265–289.
16. **Furbass, R., M. Gocht, P. Zuber, and M. A. Marahiel.** 1991. Interaction of AbrB, a transcription regulator of *Bacillus subtilis*, with the promoters of the transition state activated genes *tycA* and *spoVG*. *Mol. Gen. Genet.* **225:**347–354.
17. **Gaur, N. K., K. Cabane, and I. Smith.** 1988. Structure and expression of the *Bacillus subtilis sin* operon. *J. Bacteriol.* **170:**1046–1053.
18. **Gaur, N. K., E. Dubnau, and I. Smith.** 1986. Character-

ization of a cloned *Bacillus subtilis* gene that inhibits sporulation in multiple copies. *J. Bacteriol.* **168:**860–869.
18a. **Gaur, N. K., I. Mandic-Mulec, and I. Smith.** Unpublished data.
19. **Gaur, N. K., J. Oppenheim, and I. Smith.** 1991. The *Bacillus subtilis sin* gene, a regulator of alternate developmental processes, codes for a DNA-binding protein. *J. Bacteriol.* **173:**678–686.
19a. **Gaur, N. K., and I. Smith.** Unpublished data.
19b. **Gaur, N. K., I. Smith, and D. Karamata.** Unpublished data.
20. **Gilman, M., and M. Chamberlin.** 1983. Developmental and genetic regulation of *Bacillus subtilis* genes transcribed by σ^{28}-RNA polymerase. *Cell* **35:**285–293.
21. **Guzman, P., J. Westpheling, and P. Youngman.** 1988. Characterization of the promoter region of the *Bacillus subtilis spoIIE* operon. *J. Bacteriol.* **170:**1598–1609.
22. **Hahn, J., and D. Dubnau.** Personal communication.
23. **Healy, J., J. Weir, I. Smith, and R. Losick.** 1991. Post-transcriptional control of a sporulation regulatory gene encoding transcription factor σ^{H} in *Bacillus subtilis*. *Mol. Microbiol.* **5:**477–487.
24. **Helmann, J. D., L. M. Marquez, and M. J. Chamberlin.** 1988. Cloning, sequencing, and disruption of the *Bacillus subtilis* σ^{28} gene. *J. Bacteriol.* **170:**1568–1574.
25. **Henner, D. J., E. Ferrari, M. Perego, and J. A. Hoch.** 1988. Location of the targets of the *hpr-97*, *sacU32*(Hy), and *sacQ36*(Hy) mutations in upstream regions of the subtilisin promoter. *J. Bacteriol.* **170:**296–300.
26. **Hoch, J. A., K. Trach, F. Kawamura, and H. Saito.** 1985. Identification of the transcriptional suppressor *sof-1* as an alteration in the *spo0A* protein. *J. Bacteriol.* **161:**552–555.
27. **Honjo, M., A. Nakayama, K. Fukazawa, K. Kawamura, K. Ando, M. Hori, and Y. Furutani.** 1990. A novel *Bacillus subtilis* gene involved in negative control of sporulation and degradative-enzyme production. *J. Bacteriol.* **172:**1783–1790.
28. **Horinouchi, S., K. Furuya, M. Nishiyama, H. Suzuki, and T. Beppu.** 1987. Nucleotide sequence of the streptothricin acetyltransferase gene from *Streptomyces lavendulae* and its expression in heterologous hosts. *J. Bacteriol.* **169:**1929–1937.
29. **Kallio, P. T., J. E. Fagelson, J. A. Hoch, and M. A. Strauch.** 1991. The transition state regulator Hpr of *Bacillus subtilis* is a DNA-binding protein. *J. Biol. Chem.* **266:**13411–13417.
30. **Kalman, S., M. L. Duncan, S. M. Thomas, and C. W. Price.** 1990. Similar organization of the *sigB* and *spoIIA* operons encoding alternate sigma factors of *Bacillus subtilis* RNA polymerase. *J. Bacteriol.* **172:**5575–5585.
31. **Kenney, T. J., K. York, P. Youngman, and C. P. Moran, Jr.** 1989. Genetic evidence that RNA polymerase associated with σ^{A} factor uses a sporulation-specific promoter in *Bacillus subtilis*. *Proc. Natl. Acad. Sci. USA* **86:**9109–9113.
32. **Klier, A., A. Fouet, M. Debarbouille, F. Kunst, and G. Rapoport.** 1987. Distinct control sites located upstream from the levansucrase gene of *Bacillus subtilis*. *Mol. Microbiol.* **1:**233–241.
33. **Kunst, F., M. Pascal, J. Lepesant-Kejzlarova, J. Lepesant, A. Billault, and R. Dedonder.** 1974. Pleiotropic mutations affecting sporulation conditions and the synthesis of extracellular enzymes in *Bacillus subtilis* 168. *Biochimie* **56:**1481–1489.
34. **Kuroda, A., and J. Sekiguchi.** 1991. Molecular cloning and sequencing of a major *Bacillus subtilis* autolysin gene. *J. Bacteriol.* **173:**7304–7312.
35. **Leighton, T.** Personal communication.
36. **Lepesant, J. A., F. Kunst, J. Lepesant Kejzlarova, and R. Dedonder.** 1972. Chromosomal location of mutations affecting sucrose metabolism in *Bacillus subtilis* Marburg. *Mol. Gen. Genet.* **118:**135–160.
37. **Lue, N., D. Chasman, A. Buchman, and R. Kornberg.** 1987. Interaction of *GAL4* and *GAL80* gene regulatory proteins in vitro. *Mol. Cell. Biol.* **7:**3446–3451.
38. **Mandal, N., W. Su, R. Haber, S. Adhya, and H. Echols.** 1990. DNA looping in cellular repression of transcription of the galactose operon. *Genes Dev.* **4:**410–418.
39. **Mandic-Mulec, I., N. Gaur, U. Bai, and I. Smith.** 1992. Sin, a stage-specific repressor of cellular differentiation. *J. Bacteriol.* **174:**3561–3569.
39a. **Mandic-Mulec, I., and I. Smith.** Unpublished data.
40. **Marquez, L. M., J. D. Helmann, E. Ferrari, H. M. Parker, G. W. Ordal, and M. J. Chamberlin.** 1990. Studies of σ^{D}-dependent functions in *Bacillus subtilis*. *J. Bacteriol.* **172:**3435–3443.
41. **Martin, I., M. Debarbouille, A. Klier, and G. Rapoport.** 1987. Identification of a new locus, *sacV*, involved in the regulation of levansucrase synthesis in *Bacillus subtilis*. *FEMS Microbiol. Lett.* **44:**39–43.
42. **Michel, J. F., B. Cami, and P. Schaeffer.** 1968. Selection de mutants de *Bacillus subtilis* bloques au debut de la sporulation. II. Selection par adaptation a une nouvelle source de carbone et par viellissement de cultures sporulees. *Ann. Inst. Pasteur* (Paris) **114:**21–27.
43. **Michel, J. F., B. Cami, and P. Schaeffer.** 1968. Selection de mutants de *Bacillus subtilis* bloques au debut de la sporulation. I. Mutants asporogenes pleiotropes selectionnees par croissance en milieu au nitrate. *Ann. Inst. Pasteur* (Paris) **114:**11–20.
44. **Mirel, D. B., and M. J. Chamberlin.** 1989. The *Bacillus subtilis* flagellin gene (*hag*) is transcribed by the σ^{28} form of RNA polymerase. *J. Bacteriol.* **171:**3095–3101.
45. **Msadek, T.** Personal communication.
46. **Msadek, T., F. Kunst, D. Henner, A. Klier, G. Rapoport, and R. Dedonder.** 1990. Signal transduction pathway controlling synthesis of a class of degradative enzymes in *Bacillus subtilis*: expression of the regulatory genes and analysis of mutations in *degS* and *degU*. *J. Bacteriol.* **172:**824–834.
47. **Msadek, T., F. Kunst, A. Klier, and G. Rapoport.** 1991. DegS-DegU and ComP-ComA modulator-effector pairs control expression of the *Bacillus subtilis* pleiotropic regulatory gene *degQ*. *J. Bacteriol.* **173:**2366–2377.
48. **Mueller, J. L., and A. L. Sonenshein.** Personal communication.
49. **Nagami, Y., and T. Tanaka.** 1986. Molecular cloning and nucleotide sequence of a DNA fragment from *Bacillus natto* that enhances production of extracellular proteases and levansucrase in *Bacillus subtilis*. *J. Bacteriol.* **166:**20–28.
49a. **Nair, G., and I. Smith.** Unpublished data.
50. **Nakano, M. M., R. Magnuson, A. Myers, J. Curry, A. D. Grossman, and P. Zuber.** 1991. *srfA* is an operon required for surfactin production, competence development, and efficient sporulation in *Bacillus subtilis*. *J. Bacteriol.* **173:**1770–1778.
51. **Nakano, M. M., L. Xia, and P. Zuber.** 1991. Transcription initiation region of the *srfA* operon, which is controlled by the *comP-comA* signal transduction system in *Bacillus subtilis*. *J. Bacteriol.* **173:**5487–5493.
52. **Olmedo, G., E. G. Ninfa, J. Stock, and P. Youngman.** 1990. Novel mutations that alter the regulation of sporulation in *Bacillus subtilis*. Evidence that phosphorylation of regulatory protein Spo0A controls the initiation of sporulation. *J. Mol. Biol.* **215:**359–372.
53. **Pang, A. S., S. Nathoo, and S. Wong.** 1991. Cloning and characterization of a pair of novel genes that regulate production of extracellular enzymes in *Bacillus subtilis*. *J. Bacteriol.* **173:**46–54.
54. **Park, S., S. Wong, L. Wang, and R. H. Doi.** 1989. *Bacillus subtilis* subtilisin gene (*aprE*) is expressed from a σ^{A} (σ^{43}) promoter in vitro and in vivo. *J. Bacteriol.* **171:**2657–2665.
54a. **Park, S. H., and I. Smith.** Unpublished data.

55. **Pearson, W. R., and D. J. Lipman.** 1988. Improved tools for biological sequence comparison. *Proc. Natl. Acad. Sci. USA* **85:**2444–2448.
56. **Perego, M., and J. A. Hoch.** 1987. Isolation and sequence of the *spo0E* gene: its role in initiation of sporulation in *Bacillus subtilis. Mol. Microbiol.* **1:**125–132.
57. **Perego, M., and J. A. Hoch.** 1988. Sequence analysis and regulation of the *hpr* locus, a regulatory gene for protease production and sporulation in *Bacillus subtilis. J. Bacteriol.* **170:**2560–2567.
58. **Perego, M., J. Wong, G. B. Spiegelman, and J. A. Hoch.** 1991. Mutational dissociation of the positive and negative regulatory properties of the Spo0A sporulation transcription factor of *Bacillus subtilis. Gene* **100:**207–212.
59. **Predich, M., G. Nair, and I. Smith.** 1992. *Bacillus subtilis* early sporulation genes *kinA*, *spo0F*, and *spo0A* are transcribed by the RNA polymerase containing σ^H. *J. Bacteriol.* **174:**2771–2778.

59a. **Predich, M., and I. Smith.** Unpublished data.

60. **Robertson, J. B., M. Gocht, M. A. Marahiel, and P. Zuber.** 1989. AbrB, a regulator of gene expression in *Bacillus*, interacts with the transcription initiation regions of a sporulation gene and an antibiotic biosynthesis gene. *Proc. Natl. Acad. Sci. USA* **86:**8457–8461.
61. **Roggiani, M., J. Hahn, and D. Dubnau.** 1990. Suppression of early competence mutations in *Bacillus subtilis* by *mec* mutations. *J. Bacteriol.* **172:**4056–4063.
62. **Satola, S., P. A. Kirchman, and C. P. Moran, Jr.** 1991. Spo0A binds to a promoter used by σ^A RNA polymerase during sporulation in *Bacillus subtilis. Proc. Natl. Acad. Sci. USA* **88:**4533–4537.
63. **Schmidt, R., P. Margolis, L. Duncan, R. Coppolecchia, C. P. Moran, Jr., and R. Losick.** 1990. Control of developmental transcription factor σ^F by sporulation regulatory proteins *Spo*IIAA and *Spo*IIAB in *Bacillus subtilis. Proc. Natl. Acad. Sci. USA* **87:**9221–9225.
64. **Sekiguchi, J., B. Ezaki, K. Kodama, and T. Akamatsu.** 1988. Molecular cloning of a gene affecting the autolysin level and flagellation in *Bacillus subtilis. J. Gen. Microbiol.* **134:**1611–1621.
65. **Sekiguchi, J., H. Ohsu, A. Kuroda, H. Moriyama, and T. Akamatsu.** 1990. Nucleotide sequences of the *Bacillus subtilis flaD* locus and a *B. licheniformis* homologue affecting the autolysin level and flagellation. *J. Gen. Microbiol.* **136:**1223–1230.
66. **Shimotsu, H., and D. J. Henner.** 1986. Modulation of *Bacillus subtilis* levansucrase gene expression by sucrose and regulation of the steady-state mRNA level by *sacU* and *sacQ* genes. *J. Bacteriol.* **168:**380–388.
67. **Shimotsu, H., M. I. Kuroda, C. Yanofsky, and D. J. Henner.** 1986. Novel form of transcription attenuation regulates expression of the *Bacillus subtilis* tryptophan operon. *J. Bacteriol.* **166:**461–471.
68. **Shoji, K., S. Hiratsuka, F. Kawamura, and Y. Kobayashi.** 1988. New suppressor mutation *sur0B* of *spo0B* and *spo0F* mutations in *Bacillus subtilis. J. Gen. Microbiol.* **134:**3249–3257.
69. **Sloma, A., D. Pawlyk, and J. Pero.** 1988. Development of an expression and secretion system in *Bacillus subtilis* utilizing *sacQ*, p. 23–26. *In* A. T. Ganesan and J. A. Hoch (ed.), *Genetics and Biotechnology of Bacilli*, vol. 2. Academic Press, Inc., San Diego, Calif.

69a. **Smith, I.** Unpublished data.

70. **Smith, I., E. Dubnau, M. Predich, U. Bai, and R. Rudner.** 1992. Early *spo* gene expression in *Bacillus subtilis*: the role of interrelated signal transduction systems. *Biochimie* **74:**669–678.
71. **Smith, I., I. Mandic-Mulec, and N. Gaur.** 1991. The role of negative control in sporulation. *Res. Microbiol.* **142:** 831–839.
72. **Steinmetz, M., and S. Aymerich.** 1990. The *Bacillus subtilis sac-deg* constellation: how and why?, p. 303–311. *In* M. M. Zukowski, A. T. Ganesan, and J. A. Hoch (ed.), *Genetics and Biotechnology of Bacilli*, vol. 3. Academic Press, Inc., San Diego, Calif.
73. **Strauch, M., V. Webb, G. Spiegelman, and J. A. Hoch.** 1990. The Spo0A protein of *Bacillus subtilis* is a repressor of the *abrB* gene. *Proc. Natl. Acad. Sci. USA* **87:**1801–1805.
74. **Strauch, M. A., M. Perego, D. Burbulys, and J. A. Hoch.** 1989. The transition state transcription regulator AbrB of *Bacillus subtilis* is autoregulated during vegetative growth. *Mol. Microbiol.* **3:**1203–1209.
75. **Strauch, M. A., G. B. Spiegelman, M. Perego, W. C. Johnson, D. Burbulys, and J. A. Hoch.** 1989. The transition state transcription regulator *abrB* of *Bacillus subtilis* is a DNA binding protein. *EMBO J.* **8:**1615–1621.
76. **Stutzman-Engwall, K. J., S. L. Otten, and C. R. Hutchinson.** 1992. Regulation of secondary metabolism in *Streptomyces* spp. and overproduction of daunorubicin in *Streptomyces peucetius. J. Bacteriol.* **174:**144–154.
77. **Takagi, M., H. Takada, and T. Imanaka.** 1990. Nucleotide sequence and cloning in *Bacillus subtilis* of the *Bacillus stearothermophilus* pleiotropic regulatory gene *degT. J. Bacteriol.* **172:**411–418.
78. **Tanaka, T., and M. Kawata.** 1988. Cloning and characterization of *Bacillus subtilis iep*, which has positive and negative effects on production of extracellular proteases. *J. Bacteriol.* **170:**3593–3600.
79. **Tanaka, T., M. Kawata, and K. Mukai.** 1991. Altered phosphorylation of *Bacillus subtilis* DegU caused by single amino acid changes in DegS. *J. Bacteriol.* **173:** 5507–5515.
80. **Tanaka, T., M. Kawata, M. Saitoh, and Y. Nagami.** 1988. Enhancement of mRNA level by *prtR*, p. 33–37. *In* A. T. Ganesan and J. A. Hoch (ed.), *Genetics and Biotechnology of Bacilli*, vol. 2. Academic Press, Inc., San Diego, Calif.
81. **Toma, S., M. D. Bue, A. Pirola, and G. Grandi.** 1986. *nprR1* and *nprR2* regulatory regions for neutral protease expression in *Bacillus subtilis. J. Bacteriol.* **167:**740–743.
82. **Trach, K., D. Burbulys, M. Strauch, J. Wu, N. Dhillon, R. Jonas, C. Hanstein, P. Kallio, M. Perego, T. Bird, G. Spiegelman, C. Fogher, and J. A. Hoch.** 1991. Control of the initiation of sporulation in *Bacillus subtilis* by a phosphorelay. *Res. Microbiol.* **142:**815–823.
83. **Valle, F., and E. Ferrari.** 1989. Subtilisin: a redundantly temporally regulated gene?, p. 131–146. *In* I. Smith, R. A. Slepecky, and P. Setlow (ed.), *Regulation of Procaryotic Development*. American Society for Microbiology, Washington, D.C.
84. **Van Kaer, L., Y. Gansemans, M. Van Montagu, and P. Dhaese.** 1988. Interaction of the *Bacillus subtilis* phage ϕ105 repressor with operator DNA: a genetic analysis. *EMBO J.* **7:**859–866.
85. **Wang, L., and R. H. Doi.** 1990. Complex character of *senS*, a novel gene regulating expression of extracellular-protein genes of *Bacillus subtilis. J. Bacteriol.* **172:**1939–1947.
86. **Wang, L., and R. H. Doi.** 1990. *senS*, a novel regulatory gene with complex structure and partial homology to sigma factors of *Bacillus subtilis*, p. 385–391. *In* M. M. Zukowski, A. T. Ganesan, and J. A. Hoch (ed.), *Genetics and Biotechnology of Bacilli*, vol. 3. Academic Press, Inc., San Diego, Calif.
87. **Weinrauch, Y., R. Penchev, E. Dubnau, I. Smith, and D. Dubnau.** 1990. A *Bacillus subtilis* regulatory gene product for genetic competence and sporulation resembles sensor protein members of the bacterial two-component signal-transduction systems. *Genes Dev.* **4:**860–872.
88. **Weir, J., M. Predich, E. Dubnau, G. Nair, and I. Smith.** 1991. Regulation of *spo0H*, a gene coding for the *Bacillus subtilis* σ^H factor. *J. Bacteriol.* **173:**521–529.
89. **Wong, S. L., L. F. Wang, and R. H. Doi.** 1988. Cloning and nucleotide sequence of senN, a novel *Bacillus natto* (*B. subtilis*) gene that regulates expression of extracellular protein genes. *J. Gen. Microbiol.* **134:**3264–3276.

89a. **Worman, D., M. Marahiel, M. Predich, and I. Smith.** Unpublished data.

90. **Yang, M., E. Ferrari, E. Chen, and D. J. Henner.** 1986. Identification of the pleiotropic *sacQ* gene of *Bacillus subtilis*. *J. Bacteriol.* **166:**113–119.

91. **Yang, M., H. Shimotsu, E. Ferrari, and D. J. Henner.** 1987. Characterization and mapping of the *Bacillus subtilis prtR* gene. *J. Bacteriol.* **169:**434–437.

92. **York, K., T. J. Kenney, S. Satola, C. P. Moran, Jr., H. Poth, and P. Youngman.** 1992. Spo0A controls the σ^A-dependent activation of *Bacillus subtilis* sporulation-specific transcription unit *spoIIE*. *J. Bacteriol.* **174:**2648–2658.

93. **Yoshikawa, A., S. Isono, A. Sheback, and K. Isono.** 1987. Cloning and nucleotide sequencing of the genes *rimI* and *rimJ* which encode enzymes acetylating ribosomal proteins S18 and S5 of *Escherichia coli* K12. *Mol. Gen. Genet.* **209:**481–488.

94. **Zuber, P., and R. Losick.** 1987. Role of *abrB* in *spo0A*- and *spo0B*-dependent utilization of a sporulation promoter in *Bacillus subtilis*. *J. Bacteriol.* **169:**2223–2230.

95. **Zuber, P., M. Marahiel, and J. Robertson.** 1988. Influence of *abrB* on the transcription of the sporulation-associated genes *spoVG* and *spo0H* in *Bacillus subtilis*, p. 123–127. *In* A. T. Ganesan and J. A. Hoch (ed.), *Genetics and Biotechnology of Bacilli*, vol. 2. Academic Press, Inc., New York.

55. Spore Structural Proteins

PETER SETLOW

Dormant spores of *Bacillus subtilis* have a number of properties quite different from those of growing cells. Included among these differences are increased resistance of spores to chemicals, heat, mechanical disruption, UV irradiation, and enzymes such as lysozyme (21). The morphological appearance of dormant spores is also very different from that of growing cells (56); included among the morphological differences are the presence of at least one structure in spores (the spore coat) that has no counterpart in growing cells and a greatly expanded peptidoglycan layer (the spore cortex) in spores. In addition, the structure of the spore's nucleoid appears to be different from that of the growing cell, judging from the nucleoid's appearance in the electron and light microscopes and from photochemistry of the DNA (3, 44, 48).

It seems highly likely that the spore's unique structural features play key roles in its physiological properties. Indeed, it has been suggested (i) that the spore coat has an important role in spore resistance to enzymes, chemicals, and mechanical disruption (1); (ii) that the spore cortex is essential in bringing about and maintaining the spore core dehydration needed to provide heat resistance (and possibly metabolic dormancy) (18, 21); and (iii) that the novel spore nucleoid structure is important in the regulation of forespore gene expression as well as in the resistance of spores to UV light (44, 48). While the full details of spore cortex structure, including its biosynthesis and cross-linking, are not yet available, much more is known about spore coat and nucleoid structures. Not surprisingly, these two structures are formed by a group of proteins unique to the spore stage of the *B. subtilis* life cycle. Each structure contains a large amount of protein, with spore coats comprising ~50% and spore nucleoid proteins ≥5% of total *B. subtilis* spore protein (1, 19, 26, 27, 39, 49); spores of other *Bacillus* species have comparable amounts of protein in these two structures (1, 49). Thus, both spore coat and spore nucleoid components are major spore structural proteins; both types of protein consist of multiple species coded for by monocistronic genes scattered around the *B. subtilis* chromosome. In addition, both spore coat and nucleoid proteins appear unique, since analogous proteins have not been found in other organisms except in spores from related species. It is hoped that understanding the structure and function of their individual protein components will give insight into spore coat and nucleoid structures. This in turn may give insight into how these structures determine spore properties.

SPORE COAT PROTEINS

Studies of spore coat structure by electron microscopy have shown that the spore coat lies immediately under the exosporium, which is quite small in *B. subtilis* (1, 56, 58). The spore coat has two distinct layers, a thick, more electron-dense outer coat and a less electron-dense inner coat that appears to be composed of a number of layers (Fig. 1) (1). The inner coat is underlain by (i) a somewhat granular zone whose origin (inner coat, outer forespore membrane, or cortex) is not clear and then (ii) the spore cortex (Fig. 1). The majority of the *B. subtilis* spore coat appears to be composed of protein (~50% of total spore protein), although small amounts of lipid should be present in spore coat preparations, since the outer forespore membrane is often extracted with spore coat proteins. The proteins in spore coats are quite insoluble, requiring harsh conditions (0.1 M NaOH for 15 min at 4°C or 1% sodium dodecyl sulfate [SDS] plus 50 mM dithiothreitol [DTT] for 30 min at 65°C) for their extraction (1, 14, 19, 26). Spores generally remain viable after these treatments. The NaOH treatment appears less harsh than SDS and DTT extraction, as the latter procedure solubilizes significantly more protein than the former. The coat proteins extracted by either procedure precipitate when brought to neutral pH in the absence of solubilizing agents (SDS, urea, etc.). This insolubility has rendered purification of individual coat protein species difficult other than by SDS-polyacrylamide gel electrophoresis. Even a combination of the harsh extraction conditions noted above fails to extract ~30% of total spore coat protein. The reason for the insolubility of this remaining protein is unclear. While it has been reported that the insoluble coat protein may be cross-linked via *o,o*-dityrosine residues (39), this observation could not be confirmed by another group (19). SDS-polyacrylamide gel electrophoresis of the proteins in extracts of *B. subtilis* spore coats has shown the presence of 10 to 20 distinct proteins, although some of them are only minor components (14, 26) (Table 1). These proteins range from 5 to 65 kDa in size (Table 1). Analysis of spore coat proteins in other *Bacillus* species has generally revealed fewer proteins, most of which have low molecular weights (<20 kDa) (1). The difficulties in purification of individual coat protein species and concerns about proteolysis during extraction initially made relationships between the many different coat proteins unclear. Thus, a number of the proteins of ~12 kDa were initially suggested to be derived from much larger precursors (40 to 60 kDa)

Peter Setlow • Department of Biochemistry, University of Connecticut Health Center, Farmington, Connecticut 06030-3305.

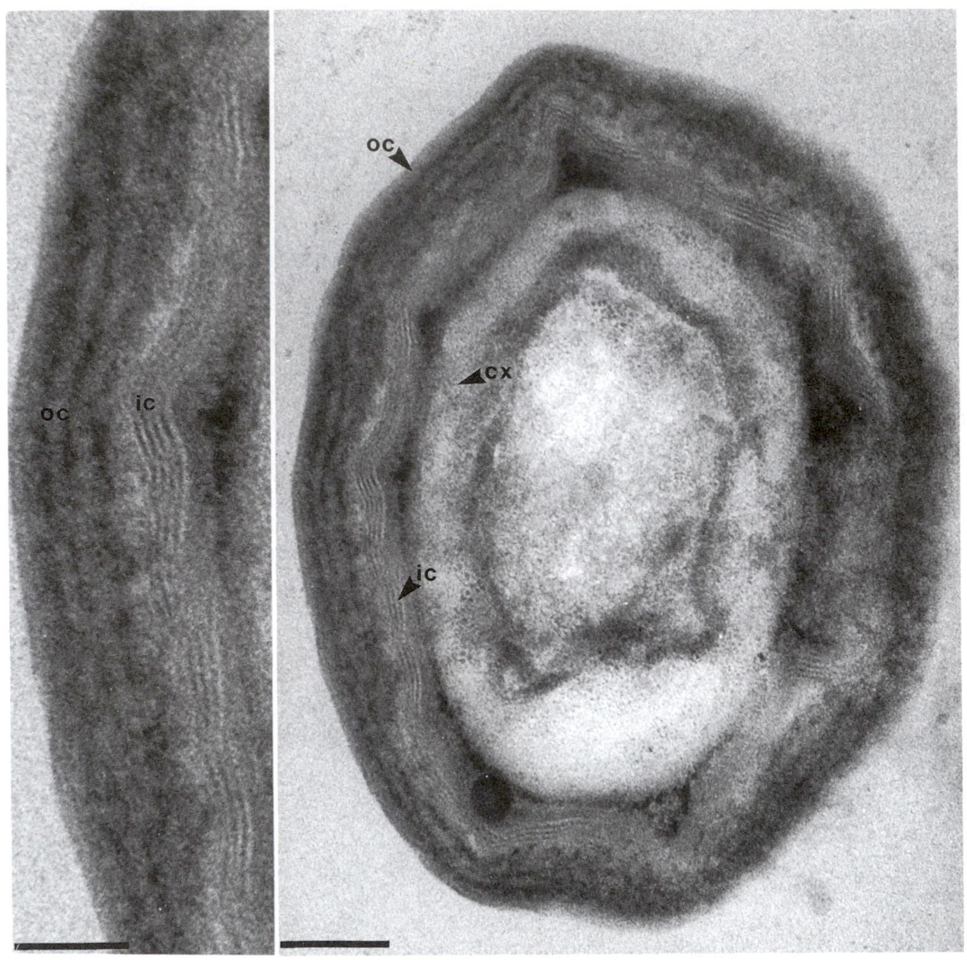

Figure 1. Electron micrograph of a dormant spore of *B. subtilis* PY79. Sporulation was caused by nutrient exhaustion, and spores were taken at t_{20} for fixation. Methods for fixation and staining were as described in reference 9. The outer spore coat (oc), inner spore coat (ic), and spore cortex (cx) layers are noted. Bar for entire spore, 100 nm; bar for enlargement of the coat region, 50 nm.

(1, 20). However, more recent work analyzing purified coat proteins and the relevant coat protein genes (termed *cot* genes) has shown that most coat proteins are unique species and are not derived from much larger precursors.

Spore coat proteins are synthesized only in the mother cell compartment beginning at h 3 to 4 of sporulation (t_{3-4}) and in a defined temporal order (12, 26, 42, 58, 59). This has been best established by studies of the expression of *cot* genes. These studies have identified three groups of *cot* genes that show different patterns of expression during sporulation (12, 29, 58, 59). The only member of the first group, whose expression begins at about t_{3-4}, is the *cotE* gene, which is transcribed by RNA polymerase containing σ^E (E-σ^E). Transcription of *cotE* takes place from two different promoters: initially from the P1 promoter, which is recognized by E-σ^E alone, and subsequently from the P2 promoter, whose recognition by E-σ^E requires the SpoIIID protein for activation of its expression (59). Transcription of *cotE* is followed by expression of *cot* genes transcribed by E-σ^K alone (*cotA, cotD, and cotF*) and finally by expression of those transcribed by E-σ^K in conjunction with the GerE gene product (*cotB* and *cotC*). The enzyme transcribing the *cotT* gene has not been definitively established, but it may be E-σ^K alone (2, 4, 59).

While the order and regulation of expression of a number of *cot* genes have been well established, the mechanism(s) whereby the individual gene products assemble into the mature spore coat is not well understood. However, assembly of the outer spore coat requires the product of *cotE* (CotE), and spores lacking the outer coat lack CotA, CotB, CotC, and the 34- and 38-kDa proteins (although at least some of the last five proteins are synthesized in *cotE* mutants) (58). Since the outer-coat-deficient spores retain CotD and CotT, these proteins may be in the inner spore coat (4, 12, 58), and spores of a *cotT* null mutant appear to be deficient in inner coat layers (4). While CotE is found in mature spores and appears to be a morphogenic protein required for complete spore coat assembly, its precise function is not clear. Although only low to moderate levels of CotE are found in spore coat extracts, it is possible that CotE levels in the spore coat are actually high but that this protein is extracted inefficiently. Alternatively, CotE levels in the spore coat may indeed be only low to moderate, and

Table 1. Properties of spore coat proteins[a]

Protein	Mol wt (kDa)[b]	Coat location[c]	Coding gene	Chromosomal location (°)	Remarks
CotA	65	Outer	*cotA*	52	Minor protein
CotB	59	Outer	*cotB*	290	Minor protein
38 kDa	38[d]	Outer	?	?	Major protein
34 kDa	34[e]	Outer	? (could be *cotB*)	?	Major protein; could be derived from CotB
CotE	24	?	*cotE*[f]	150	Probably minor protein; morphogenic protein for outer coat assembly; may be tyrosine peroxidase
18 kDa	18	?	?	?	Minor protein; similar but not identical to CotC
CotT	12.5	Inner	*cotT*[f]	108	Major protein; 14-kDa precursor processed to 12.5-kDa form
CotC	12	Outer	*cotC*[f]	168	Major protein
CotD	11	Inner	*cotD*[f]	200	Major protein
CotF	8 and 5	?	*cotF*[f]	349	19-kDa precursor processed to 5- and 8-kDa forms

[a] Data are from references 2, 4, 12, 13, 14, and 58.
[b] Values determined on SDS-polyacrylamide gels. For some of the proteins, their small sizes and/or unusual amino acid compositions may result in molecular weights on SDS-polyacrylamide gels different from weights predicted from gene sequences.
[c] Inferred from effect of *cotE* mutation on deposition of individual proteins on spores (58).
[d] Probably the 36-kDa protein referred to in reference 26.
[e] Probably the 33-kDa protein referred to in reference 26.
[f] Coding sequence determined.

this suggests that CotE might act catalytically. Indeed, a limited similarity between the CotE amino acid sequence and those of heme-containing peroxidases has recently been noted, and preliminary work suggests that the CotE protein has heme-dependent peroxidase activity (13). Since *o,o*-dityrosine can be formed by peroxidases and has been reported in spore coats as noted above, formation of this type of protein-protein cross-link may be the role of CotE in spore coat morphogenesis. As noted below, a number of spore coat proteins are extremely rich in tyrosine.

A second gene required for proper spore coat assembly is *spoIVA*, which encodes a 492-residue protein (41, 52). *spoIVA* mutants appear to express *cot* genes normally and to assemble a spore coat. However, the coat is not deposited on the forespore but misassembles as swirls within the mother cell (41). This observation suggests that individual spore coat proteins have at least some ability to self-assemble into spore coat-like structures. Mutations in at least two other loci (*spoVK* and *spoVM*) also result in a phenotype similar to that of *spoIVA* mutations (9). The product of the *spoIVA* gene is not a known spore coat protein, and its precise function is unknown. It is possible that SpoIVA is not involved in spore assembly directly but rather plays some role in spore cortex formation and that proper assembly of the spore coat on the forespore depends on some feature of the spore cortex. However, at least one mutant (*spoVE*) that is defective in spore cortex formation shows proper spore coat assembly (1, 53). Unfortunately, the precise relationship between spore cortex and spore coat formation is unclear.

With the exception of the *cotE* null mutant, whose spores lack outer spore coat proteins, spores of other *cot* gene mutants have a normal complement of other spore coat proteins. Consequently, the roles of most individual proteins in assembly of the spore coat are unclear. Similarly, there is generally no significant phenotypic effect upon loss of individual coat proteins from spores, again making it difficult to assign specific functions to individual species. The major exception to this is a *cotE* mutant whose spores lack the outer coat, as noted above. These *cotE* mutant spores are refractile and heat and chemical resistant but lysozyme sensitive. Thus, the outer coat (or at least one of its components) may play an important role in preventing access of lysozyme to the peptidoglycan of the spore cortex. Indeed, *spoVIA* mutant spores that apparently lack only a single outer coat protein (probably the 38-kDa species) are lysozyme sensitive (23). However, the nature of the gene in which the *spoVIA* mutation lies is not clear (i.e., does it code for a spore coat protein?). Spores of *cotE* and *cotT* null mutants also exhibit slight defects in spore germination. However, these are relatively small effects, and their meaning is not clear. Spores of *gerE* null mutants also lack a number of spore coat proteins, since GerE is a positive effector for transcription of some *cot* genes by E-σ^K, as noted above (2, 4, 11, 16, 24, 25, 59). Thus *gerE* mutant spores lack CotB, CotC, and mature CotT, although large amounts of CotT precursor are present (2, 11). The inner spore coat is also reported to be defective in *gerE* mutants. These mutant spores are lysozyme sensitive and show a severe germination defect (5, 33). There is also increased synthesis of a number of other coat proteins (i.e., CotA and CotT) in *gerE* mutants, since GerE may be a negative effector of transcription of some other *cot* genes (4, 11, 58). Two other loci in which mutations alter spore coat formation are *spoVIB* and *spoVIC* (22, 24). Spores of a *spoVIB* mutant have a normal complement of coat proteins, but the coat is misarranged, as at least one coat protein normally exposed on the spore surface is buried. The *spoVIB* mutant spores germinate more slowly than wild-type spores. Spores of *spoVIC* mutants have lowered levels of some coat proteins, and the coat layers appear disorganized upon examination in the electron microscope. The *spoVIC* mutant spores also germinate more slowly than wild-type spores. To date, the identity and function of the *spoVIB* and *spoVIC* gene products have not been determined.

Much of what we know about individual coat proteins has come from the cloning and analysis of *cot* genes. This work has led to a number of general conclusions. (i) *cot* genes are scattered around the *B. subtilis* genetic map and are monocistronic. (ii) Most coat proteins are products of unique genes. However, at least one and possibly two coat protein pairs exhibit a high degree of sequence identity, suggesting that there may be multiple related genes coding for a few coat protein species. (iii) Most smaller coat proteins (12 to 24 kDa) are not derived by proteolysis from large precursors. However, some of the smallest ones (5 to 8 kDa) are generated by proteolysis of 12- to 19-kDa species. (iv) Comparison of deduced amino acid sequences of spore coat proteins with sequences in available data bases has yet to reveal similar proteins in other organisms (except presumably in spores of other *Bacillus* or *Clostridium* species). Properties of individual coat proteins are given below.

CotA

The 65-kDa CotA protein is a minor spore coat component (presumably in the outer coat), most of which can be extracted from intact spores by alkali alone (14). The *cotA* gene is identical to a gene called *pig* that is responsible for the brown pigmentation of *B. subtilis* spores. However, the identity of this pigment and how CotA is responsible for the pigmentation are not clear. The protein loses its N-terminal methionine after synthesis; possible C-terminal processing has not been studied, and the determination of the *cotA* coding sequence is not yet complete. Spore coats of *cotA* null mutants lack only the 65-kDa protein and show no obvious phenotype.

CotB

The 59-kDa CotB protein is also a minor spore coat component (presumably from the outer coat) that can be completely extracted with alkali alone (14). CotB loses only the N-terminal methionine at its amino terminus, but possible C-terminal processing has not been studied. Again, the determination of the *cotB* coding sequence is not yet complete. Spore coats of null mutants in *cotB* lack only the 59-kDa protein and have no obvious phenotype.

CotC

The 12-kDa CotC protein is a major species extracted from the spore coat (presumably the outer coat) with alkali alone. Partial CotC sequence determination and the sequence of the *cotC* gene suggest that CotC undergoes no posttranslational processing and has an extremely unusual amino acid composition (14). Among its 66 residues, this protein has 20 tyrosines, 19 lysines, and 12 aspartates; the sequence includes five tyrosyl-tyrosine sequences and 9 lysyl-lysine sequences. It appears that there is a second coat protein related to CotC, since an 18-kDa coat protein has an identical N-terminal sequence save for one conservative (tryptophan-for-valine) change. Since spores of a *cotC* null mutant lack only the 12-kDa protein, the 18-kDa protein is presumably coded for by a second gene. The spores of a *cotC* mutant have no obvious phenotype.

CotD

The 11-kDa CotD protein is a major spore coat component (presumably from the inner coat) extracted from spores primarily with SDS and DTT, although a small amount is removed with alkali alone (14). This 75-residue protein appears to undergo no posttranslational processing and contains 17 histidine and 5 cysteine residues, with two cysteinyl-cysteine sequences. Spore coats of *cotD* null mutants lack only CotD and exhibit no striking phenotype, although a slight germination defect has been reported (14). Overproduction of CotD also results in spores with a slight germination defect (5).

CotE

The 24-kDa CotE protein is a minor spore coat component extracted from spores in large part with dilute alkali alone. Its amino acid composition and sequence are not remarkable, and it appears to undergo no posttranslational processing other than loss of the N-terminal methionine (58). Spores of *cotE* null mutants lack the outer spore coat and proteins CotA, CotB, CotC, and CotE but not CotD or CotT (the status of the *cotF*-derived gene products has not yet been reported). However, much of the CotT in *cotE* mutant spores exists as its precursor form (4). The 18-, 34-, and 38-kDa proteins, whose genes have not been cloned, are also absent in *cotE* mutants (58). CotE is thought to be a morphogenic protein required for outer coat assembly, since at least the *cotA* and *cotC* genes are expressed at normal levels in a *cotE* mutant, but the CotA and CotC proteins are not deposited on the spore (13, 58). CotE mutant spores are lysozyme sensitive and germinate slower and less efficiently than wild-type spores. Recent work (13) suggests that CotE may be a heme-containing tyrosine peroxidase involved in generating *o,o*-dityrosine cross-links between spore coat proteins.

CotF

The product of the *cotF* gene is predicted to be a 19-kDa protein (12). However, this species has not been found in spore coats, since the *cotF* gene product is processed to two proteins of 5 and 8 kDa, both of which are major spore components readily extracted from spores with alkali alone (12). The N-terminal amino acid sequences of the 5- and 8-kDa species indicate that the 5-kDa protein begins at the fifth residue of the *cotF* coding sequence and the 8-kDa protein begins at residue 77. The 8-kDa protein is tyrosine rich, with ~10% tyrosine residues. For both species, the putative cleavage site is C terminal to an arginyl residue (arginyl-arginine for the 5-kDa protein and arginyl-alanine for the 8-kDa protein). However, it has not been established that these are sites of endoproteolytic cleavage, nor has processing at the C terminus of the two small proteins been ruled out. Spores of *cotF* mutants lack the 5- and 8-kDa species but no other coat proteins and have no obvious phenotype.

CotT

The 12.5-kDa CotT protein is a major spore coat component and requires SDS and DTT for its extraction (presumably from the inner spore coat) (2, 4, 5, 19). The *cotT* gene codes for a protein with 19 additional residues at the N terminus. Interestingly, the putative cleavage site for the CotT precursor is also C terminal to an arginine residue (arginyl-glutamine), as was found for the *cotF*-derived proteins. The CotT precursor is present in sporulating cells, and small amounts are present in spores. While the amino acid sequence in the region removed in generating mature CotT is not remarkable, the 63-residue mature protein contains 22 tyrosines, 19 prolines, and 11 glycines, with five Pro-$(Tyr)_{2-3}$-Pro sequences. Spores of *cotT* mutants lack the 12.5-kDa species, and the only obvious phenotype of these mutant spores is slightly slower germination on some germinants. The *cotT* mutant spores also appear to have less inner spore coat upon examination in the electron microscope. Interestingly, overproduction of CotT results in spores with a rather severe germination defect (5).

Other Spore Coat Proteins

Spore coats contain two major proteins of 34 and 38 kDa that are extracted only with SDS and DTT and appear to be in the outer coat (14, 26). The N-terminal sequence of the 38-kDa species indicates that it is different from all other coat proteins (14). However, its coding gene has not been cloned. The N-terminal sequence of the 34-kDa protein is identical to that of CotB and could be derived from the latter by proteolysis. While the 34-kDa protein is present in spores of a *cotB* insertion mutant, this insertion could have permitted the synthesis of a 34-kDa N-terminal fragment of CotB (14). As noted above, spore coats also contain an 18-kDa protein extracted with alkali alone that is very similar but not identical in primary sequence to CotC (14).

It has been reported that *B. subtilis* spore coats contain protein species with some similarities to keratins (28). However, this assignment was based only on X-ray diffraction analysis of whole isolated spore coats, and no keratinlike proteins have been identified in spores. It is, of course, possible that such keratinlike proteins are in the spore coat protein insoluble in SDS and DTT. Indeed, this latter fraction has a rather high (~4%) cysteine-cystine content (19).

Table 2. Properties of *B. subtilis* SASP[a]

SASP	Coding gene	Chromosomal position (°)	Protein length (no. of residues[b])	Remarks[c]
α/β Type				
α	*sspA*	266	68	Major SASP (2.5)
β	*sspB*	65	66	Major SASP (1.5)
SspC	*sspC*	182	71	Minor SASP (0.5)
SspD	*sspD*	121	63	Minor SASP (<0.3)
SspF	*sspF*[d]	3	60	Not yet detected, but very likely minor SASP[e]
γ	*sspE*	65	73	Major SASP (4)

[a] Data are from references 27, 34, 38, 49, 51, and 57.
[b] Without the N-terminal methionine.
[c] Values in parentheses are the approximate percentages of total spore protein for this SASP.
[d] Previously called *0.3 kb* (38).
[e] This protein has not been detected by polyacrylamide gel analysis of acid extracts of spores lacking the three major SASP (α, β, and γ) (45).

SPORE NUCLEOID PROTEINS

Analysis by fluorescence microscopy of sporulating cells of *B. subtilis* stained with the fluorescent DNA stain 4′,6-diamidino-2-phenylindole (DAPI) has shown that the forespore nucleoid becomes significantly condensed relative to the mother cell nucleoid at $\sim t_1$ of sporulation (44). The mechanism for this forespore nucleoid condensation is not clear, but it could involve one or more forespore nucleoid-specific proteins. It has been suggested that forespore nucleoid condensation could play a role in regulation of gene expression in the forespore (44). However, this has by no means been proven.

Approximately 2 to 3 h later in sporulation (t_{3-4}), the appearance of forespore DNA as seen in the electron microscope changes further, to a more fibrillar state (3). This morphological change is associated with a change in forespore DNA's UV photochemistry such that UV irradiation produces a thyminyl-thymine adduct (termed spore photoproduct [SP]) in forespore DNA rather than the cyclobutane-type thymine dimers (TT) that are the most abundant photoproduct formed in the DNA of growing cells or the mother cell (3, 48). This latter change in forespore nucleoid properties is associated with synthesis of α/β-type small, acid-soluble spore proteins (SASP) in the forespore (48, 49). These proteins are major forespore nucleoid components and are double-stranded-DNA-binding proteins (46).

α/β-Type SASP

Approximately 5% of the protein of dormant spores of *B. subtilis* is composed of a group of α/β-type SASP that are 63 to 71 residues in length (49) (Table 2). There are two major α/β-type SASP in *B. subtilis* spores, termed α and β. SASP-α and -β are almost identical in primary sequence, and at least two other minor α/β-type SASP have been identified (Fig. 2) (49, 57). Judging from results with other *Bacillus* species, there are at least seven α/β-type SASP in *B. subtilis* (49). Genes for four α/β-type SASP (*sspABCD*) have

sspA: MANNNSGNSNNLLVPGAAQAIDQMKLEIASEFGVNLGA--DTTSRANGSVGGEITKRLVSFAQQNMGGGQF

sspB: MANQNSSNDLLVPGAAQAIDQMKLEIASEFGVNLGA--DTTSRANGSVGGEITKRLVSFAQQQMGGRVQ

sspC: MAQQSRSRSNNNNDLLIPQAASAIEQMKLEIASEFGVNLGA--ETTSRANGSVGGEITKRLVRLAQQNMGGQFH

sspD: MASRNKLVVPGVEQALDQFKLEVASEFGVNLGS--DTVARANGSVGGEMTKRLVQQAQSQLNGTTK

sspF: MGRRRGVMSDEFKYELAKDLGFYDTVKNGGWGEIRARDAGNMVKRAIEIAEQQMAQNQNNR

Figure 2. Primary sequence of *B. subtilis* α/β-type SASP. Known amino acid sequences of α/β-type SASP from *B. subtilis* and the protein specified by the *sspF* gene (previously *0.3 kb*) are shown in one-letter code and are taken from references 49 and 51. The N-terminal methionine in the coding gene is included but is undoubtedly removed posttranslationally. The *sspA* gene codes for SASP-α, *sspB* codes for SASP-β, *sspC* codes for SspC, and *sspD* codes for SspD (49). For the proteins specified by the *sspABCD* genes, residues conserved in all known α/β-type SASP from the *Bacillus* line of sporeformers are in boldface type; residues that are similar in these SASP are underlined. A gap has been introduced in these sequences to maximize the alignment with SspF. The large downward-pointing arrow indicates the site of cleavage by the SASP-specific protease (49). The two upward-pointing arrowheads indicate residues whose alteration in SspC (Gly→Ala or Lys→Gln) reduces or abolishes the ability of this protein to bind DNA in vivo or in vitro (57). For the protein specified by the *sspF* gene, residues identical to those conserved in other α/β-type SASP are in boldface; residues similar to those in other α/β-type SASP are underlined.

been cloned and sequenced, allowing determination of the primary sequences of their proteins (49). Primary sequences have also been determined (either directly or indirectly) for 19 α/β-type SASP from other *Bacillus* and *Clostridium* species (6, 7, 29, 49). None of these proteins exhibits significant sequence similarity to any other protein in available data bases. However, the sequences of the four α/β-type SASP from *B. subtilis* are extremely similar, with 35 identical residues and 11 similar ones (Fig. 2). The primary sequences of α/β-type SASP have also been very highly conserved across species, as 27 residues are conserved exactly in 19 α/β-type SASP from various *Bacillus* species and 11 residues are similar (Fig. 2). The great majority of the residues conserved in α/β-type SASP of *Bacillus* species are also conserved in those of *Clostridium* species (6, 7). In general, the amino- and carboxy-terminal regions of these proteins exhibit the least sequence conservation. One region of highly conserved sequence surrounds the site recognized by the SASP-specific protease (large downward arrow in Fig. 2) responsible for initiating degradation of these proteins during spore germination. A second more carboxy-terminal region is even more highly conserved; alteration of residues in this latter region by site-directed mutagenesis reduces or abolishes α/β-type SASP binding to DNA (57; upward arrows in Fig. 2).

All *ssp* genes are monocistronic and are scattered around the chromosome (49). The *ssp* genes are expressed only in the forespore during sporulation; their expression is regulated at the transcriptional level, with transcription carried out by RNA polymerase containing the forespore-specific factor σ^G (E-σ^G) (37, 54). In contrast to some spore coat proteins, SASP appear to undergo no posttranslational processing other than removal of N-terminal methionine residues (49). α/β-Type SASP synthesis takes place at t_{3-4} of sporulation, in parallel with expression of other E-σ^G-dependent genes such as *gdh* and *spoVA*. Also in parallel with SASP synthesis are acquisition of spore resistance to UV light, a change in forespore DNA's UV photochemistry as noted above, and an increase in the average number of negative supertwists in forespore plasmid DNA (3, 36, 48, 49). The magnitude of these changes is greatly reduced during sporulation of strains lacking SASP-α and -β (36, 48). However, it appears likely that sufficient levels of any α/β-type SASP will bring about these changes in forespore DNA properties (48, 57). Studies using several different techniques have shown that the α/β-type SASP are associated with spore DNA in vivo (17, 48). When synthesized in *E. coli*, these proteins also associate specifically with the cell's nucleoid (43). As discussed below, α/β-type SASP are double-stranded-DNA-binding proteins.

During spore germination, α/β-type SASP are rapidly degraded to amino acids, which can support much of the protein synthesis during this period of development (49). This is of particular importance, since a number of amino acid biosynthetic enzymes are lost during spore formation and then resynthesized when the germinated spore returns to vegetative growth; only then does amino acid biosynthesis begin. Not surprisingly, spores lacking SASP-α and -β as well as a third major SASP (γ) (see below) return to vegetative growth poorly if at all in amino acid-free medium (49). Spores lacking only SASP-α and -β ($\alpha^-\beta^-$) or only SASP-γ (γ^-) return to vegetative growth in amino acid-free medium at rates intermediate between those of wild-type and $\alpha^-\beta^-\gamma^-$ spores. SASP degradation during spore germination is initiated by endoproteolytic cleavage by a spore-specific endoprotease that requires an extended region (at least five residues) of SASP sequence for recognition and cleavage (49). How this enzyme becomes active in the first minutes of spore germination is not yet clear.

As noted above, the DNA of $\alpha^-\beta^-$ spores exhibits a number of differences from the DNA of wild-type spores. Thus, $\alpha^-\beta^-$ spores are much more sensitive to UV irradiation than are wild-type spores (48). This difference in UV sensitivity is due to a difference in the DNA's UV photochemistry. While UV irradiation of wild-type spores labeled with [^{3}H]thymidine generates only SP and no thymine dimers, UV irradiation of $\alpha^-\beta^-$ spores generates significant numbers of thymine dimers and only half as much SP as in wild-type spores. Plasmids from $\alpha^-\beta^-$ spores also have 25 to 50% fewer negative supertwists than plasmids from wild-type spores (36). Both the DNA photochemistry and the plasmid supercoiling of $\alpha^-\beta^-$ spores are restored to wild-type values by the synthesis of sufficient levels of any one of several α/β-type SASP (36, 48, 57). However, mutant α/β-type SASP that no longer bind DNA, as judged by in vitro assays, are ineffective in this regard (57).

As noted above, α/β-type SASP are double-stranded-DNA-binding proteins but show no binding to single-stranded DNA or double- or single-stranded RNA (34, 46). DNAs from a variety of sources, including bacterial, animal, and plasmid, are bound. The only significant requirement for the DNA is that it must be capable of adopting an A-like conformation; consequently, poly(dA) · poly(dT) is not bound (46). Binding of α/β-type SASP to relaxed but covalently closed plasmid DNA results in the introduction of a large number of negative supertwists (~1/150 bp) (34). In addition, SASP-DNA binding changes DNA's UV photochemistry in vitro such that UV irradiation produces SP but neither cyclobutane-type pyrimidine dimers nor 6-4 bipyrimidine photoadducts (15, 35). α/β-Type SASP binding also provides almost complete resistance to specific and nonspecific DNase cleavage and to chemical cleavage of the DNA backbone and blocks in vitro transcription of most genes (46). However, guanine methylation is not blocked (46).

Studies of α/β-type SASP-DNA binding by physical techniques have indicated that both the DNA and the protein undergo a large conformational change upon their interaction (31, 32). The protein changes from a rather loose, unstructured form in the free protein to a much more rigid form with a considerable amount of α-helical character when it is bound to DNA. In concert with the change in the protein, the DNA adopts a conformation very much like that of A-DNA (32). Although detailed structural analysis of the α/β-type SASP-DNA complex is not complete, it has been noted that an A-like structure of the DNA in such a complex could explain the effects of α/β-type SASP binding on DNA topology (34, 36). In addition, it appears that the structure of DNA in the α/β-type SASP-DNA complex in vivo is similar to that in vitro, as the DNA in dormant spores is very likely in an A-like conformation (50). The existence of the chromosome in an A-like structure in vivo is a unique property of spores, undoubtedly related to their dormancy and hence the absence of transcription. α/β-Type SASP are unique among known proteins in being able to promote a B→A conformational change in DNA in aqueous solution.

Minor Spore SASP

In addition to the minor α/β-type SASP, spores contain a number of other minor SASP (~10 to 20% the level of SASP-α or -β) that are not α/β-type (47, 49). These minor SASP have been studied most in *Bacillus megaterium* spores, although analogous proteins appear in *B. subtilis* spores (27, 47, 57). At least one of these minor SASP (termed E in *B. megaterium*) is made at about the time of forespore nucleoid condensation. However, no causal relationship between these events has been established, nor is it known whether this SASP is a DNA-binding protein (47). A number of the other minor SASP are synthesized in parallel with α/β-type SASP, and at least two of these appear to be DNA-binding proteins (8, 47). However, the relevant genes have not yet been cloned, nor are primary sequence data available.

One good candidate for a gene encoding a minor SASP is the gene termed *0.3 kb*, which encodes a rather stable, small mRNA that could direct synthesis of a 60-residue protein (Table 2) (38, 51). Expression of the *0.3 kb* gene is restricted to the forespore, appears to be due to transcription by E-σ^G, and takes place ~1 h later than expression of the *sspABCD* genes (40, 51). While the *0.3 kb* gene's mRNA is very abundant, translation of this mRNA may be low, since little β-galactosidase is generated from a translational *0.3 kb-lacZ* fusion (40, 51). Recent analysis of the predicted amino acid sequence of the *0.3 kb* gene product indicates that this protein has significant similarity to α/β-type SASP, although this similarity is by no means as great as that between SASP-α, SASP-β, SspC, and SspD (Fig. 2) (30). However, the similarity to α/β-type SASP does suggest that the *0.3 kb* gene product is a SASP and therefore that the *0.3 kb* gene should be renamed *sspF*. As noted above, the *sspF* mRNA is translated quite poorly, and the SspF protein has not been identified in spores, despite efforts to do so (45). Similarly, the function of SspF in spores (if any) is not known.

SASP-γ

In addition to the major SASP-α and -β, *B. subtilis* spores contain a third major SASP, termed γ, whose level is higher than that of either SASP-α or -β (27, 49). However, this protein is not associated with the nucleoid (17). SASP-γ is coded for by the *sspE* gene, which is unique; i.e., there is only one γ-type SASP. The monocistronic *sspE* gene is also transcribed by E-σ^G in parallel with genes for α/β-type SASP. Like α/β-type SASP, SASP-γ is rapidly degraded to amino acids early in spore germination, with its degradation initiated by the same endoprotease that acts on SASP-α and -β. The primary sequence of the 73-residue SASP-γ is quite different from that of α/β-type SASP and has been much less well conserved in evolution than those of α/β-type SASP (49). Indeed, γ-type SASP appear to be absent from spores of *Clostridium* species (6). γ-Type SASP show no significant sequence similarity to any other protein in available data bases (49).

Null mutants in SASP-γ display no phenotype, except that they return to vegetative growth more slowly upon germination in an amino acid-free medium than do wild-type spores (49). This phenotype is even more extreme when the γ^- mutation is coupled with α^- and β^- mutations. Thus, the only known function of SASP-γ is to serve as a reservoir of amino acids for protein synthesis during spore germination and outgrowth.

CONCLUSIONS

Dormant spores of *B. subtilis* contain a group of proteins associated with at least several of the unique or novel structural features of the spore. The proteins associated with the spore coat and nucleoid are unique both to the spore stage of the *B. subtilis* life cycle and also to this group of organisms. A number of the spore coat proteins have highly unusual amino acid compositions that may be important in their specific functions in the complex spore coat. α/β-Type SASP, on the other hand, do not have unusual amino acid compositions but have properties in vitro that clearly indicate how these proteins function in determining forespore nucleoid structure and properties in vivo. Studies of the assembly, functions, and structures of these two unique groups of proteins present fascinating problems whose investigation at many different levels continues to provide surprises.

Acknowledgments. Research in my laboratory has been supported by Public Health Service grant GM19698 and grants from the Army Research Office. I am grateful to T. Deits, S. Cutting, R. Losick, and P. Margolis for helpful conversations and communication of results prior to publication and to A. Driks for the electron micrograph in Fig. 1.

REFERENCES

1. **Aronson, A. I., and P. C. Fitz-James.** 1976. Structure and morphogenesis of the bacterial spore coat. *Bacteriol. Rev.* **40:**360–402.
2. **Aronson, A. I., H.-Y. Song, and N. Bourne.** 1989. Gene structure and precursor processing of a novel *Bacillus subtilis* spore coat protein. *Mol. Microbiol.* **3:**437–444.
3. **Baille, E., G. R. Germaine, W. G. Murrell, and D. F. Ohye.** 1974. Photoreactivation, photoproduct formation, and deoxyribonucleic acid state in ultraviolet-irradiated sporulating cultures of *Bacillus cereus*. *J. Bacteriol.* **120:** 516–523.
4. **Bourne, N., P. C. FitzJames, and A. I. Aronson.** 1991. Structural and germination defects of *Bacillus subtilis* spores with altered contents of a spore coat protein. *J. Bacteriol.* **173:**6618–6625.
5. **Bourne, N., T.-S. Huang, and A. I. Aronson.** 1990. Properties of *Bacillus subtilis* spores with alterations in spore coat structure, p. 329–338. *In* M. M. Zukowski, A. T. Ganesan, and J. A. Hoch (ed.), *Genetics and Biotechnology of Bacilli*, vol. 3. Academic Press, Inc., New York.
6. **Cabrera-Martinez, R., J. M. Mason, B. Setlow, W. M. Waites, and P. Setlow.** 1989. Purification and amino acid sequence of two small, acid-soluble proteins from *Clostridium bifermentans* spores. *FEMS Microbiol. Lett.* **61:** 139–144.
7. **Cabrera-Martinez, R. M., and P. Setlow.** 1991. Cloning and nucleotide sequence of three genes coding for small, acid-soluble proteins of *Clostridium perfringens* spores. *FEMS Microbiol. Lett.* **77:**127–132.
8. **Carillo, Y., and P. Setlow.** Unpublished results.
9. **Cutting, S.** Personal communication.
10. **Cutting, S., A. Driks, R. Schmidt, B. Kunkel, and R. Losick.** 1991. Forespore-specific transcription of a gene in the signal transduction pathway that governs Pro-σ^k processing in *Bacillus subtilis*. *Genes Dev.* **5:**456–466.
11. **Cutting, S., S. Panzer, and R. Losick.** 1989. Regulatory studies on the promoter for a gene governing synthesis and assembly of the spore coat in *Bacillus subtilis*. *J. Mol. Biol.* **207:**393–404.
12. **Cutting, S., L. Zheng, and R. Losick.** 1991. Gene encoding two alkali-soluble components of the spore coat from *Bacillus subtilis*. *J. Bacteriol.* **173:**2915–2919.
13. **Deits, T.** Personal communication.
14. **Donovan, W., L. Zheng, K. Sandman, and R. Losick.** 1987. Genes encoding spore coat polypeptides from *Bacillus subtilis*. *J. Mol. Biol.* **196:**1–10.
15. **Fairhead, H., and P. Setlow.** 1992. Binding of DNA to α/β-type small, acid-soluble proteins from spores of *Bacillus* or *Clostridium* species prevents formation of cytosine dimers, cytosine-thymine dimers, and bipyrimidine photoadducts after UV irradiation. *J. Bacteriol.* **174:**2874–2880.
16. **Feng, P., and A. I. Aronson.** 1986. Characterization of a *Bacillus subtilis* germination mutant with pleiotropic alterations in spore coat structure. *Curr. Microbiol.* **13:** 221–226.
17. **Francesconi, S. C., T. J. MacAlister, B. Setlow, and P. Setlow.** 1988. Immunoelectron microscopic localization of small, acid-soluble spore proteins in sporulating cells of *Bacillus subtilis*. *J. Bacteriol.* **170:**5963–5967.
18. **Gerhardt, P., and R. E. Marquis.** 1989. Spore thermoresistance mechanisms, p. 43–63. *In* I. Smith, R. A. Slepecky, and P. Setlow (ed.), *Regulation of Prokaryotic Development*. American Society for Microbiology, Washington, D.C.
19. **Goldman, R. C., and D. J. Tipper.** 1978. *Bacillus subtilis* spore coats: complexity and purification of a unique polypeptide component. *J. Bacteriol.* **135:**1091–1106.
20. **Goldman, R. C., and D. J. Tipper.** 1981. Coat protein synthesis during sporulation of *Bacillus subtilis*: immunological detection of soluble precursors to the 12,200-dalton spore coat protein. *J. Bacteriol.* **147:**1040–1048.
21. **Gould, G. W.** 1983. Mechanisms of resistance and dormancy, p. 173–209. *In* A. Hurst and G. W. Gould (ed.), *The Bacterial Spore*, vol. 2. Academic Press, Inc., London.
22. **James, W., and J. Mandelstam.** 1985. *spoVIC*, a new sporulation locus in *Bacillus subtilis* affecting spore coats, germination and the rate of sporulation. *J. Gen. Microbiol.* **131:**2409–2419.
23. **Jenkinson, H. F.** 1981. Germination and resistance defects in spores of a *Bacillus subtilis* mutant lacking a coat polypeptide. *J. Gen. Microbiol.* **127:**81–91.
24. **Jenkinson, H. F.** 1983. Altered arrangement of proteins in the spore coat of a germination mutant of *Bacillus subtilis*. *J. Gen. Microbiol.* **129:**1945–1958.
25. **Jenkinson, H. F., and H. Lord.** 1983. Protease deficiency and its association with defects in spore coat structure, germination and resistance properties in a mutant of *Bacillus subtilis*. *J. Gen. Microbiol.* **129:**2727–2737.
26. **Jenkinson, H. F., W. D. Sawyer, and J. Mandelstam.** 1981. Synthesis and order of assembly of spore coat proteins in *Bacillus subtilis*. *J. Gen. Microbiol.* **123:**1–16.
27. **Johnson, W. C., and D. J. Tipper.** 1981. Acid-soluble spore proteins of *Bacillus subtilis*. *J. Bacteriol.* **146:**972–982.
28. **Kadota, H., K. Iijima, and A. Uchida.** 1965. The presence of keratin-like substance in spore coat of *Bacillus subtilis*. *Agric. Biol. Chem.* **29:**870–875.
29. **Magill, N. G., C. A. Loshon, and P. Setlow.** 1990. Small, acid-soluble, spore proteins and their genes from two species of *Sporosarcina*. *FEMS Microbiol. Lett.* **72:**293–298.
30. **Margolis, P.** Personal communication.
31. **Mohr, S., B. Setlow, and P. Setlow.** Unpublished results.
32. **Mohr, S. C., N. V. H. A. Sokolov, C. He, and P. Setlow.** 1991. Binding of small acid-soluble spore proteins from *Bacillus subtilis* changes the conformation of DNA from B to A. *Proc. Natl. Acad. Sci. USA* **88:**77–81.
33. **Moir, A.** 1981. Germination properties of a spore coat-defective mutant of *Bacillus subtilis*. *J. Bacteriol.* **146:** 1106–1116.
34. **Nicholson, W. L., B. Setlow, and P. Setlow.** 1990. Binding of DNA in vitro by a small, acid-soluble spore protein

and its effect on DNA topology. *J. Bacteriol.* **172:**6900–6906.

35. **Nicholson, W. L., B. Setlow, and P. Setlow.** 1991. Ultraviolet irradiation of DNA complexed with α/β-type small, acid-soluble proteins from spores of *Bacillus* or *Clostridium* species makes spore photoproduct but not thymine dimers. *Proc. Natl. Acad. Sci. USA* **88:**8288–8292.
36. **Nicholson, W. L., and P. Setlow.** 1990. Dramatic increase in the negative superhelicity of plasmid DNA in the forespore compartment of sporulating cells of *Bacillus subtilis. J. Bacteriol.* **172:**7–14.
37. **Nicholson, W. L., D. Sun, B. Setlow, and P. Setlow.** 1989. Promoter specificity of sigma-G-containing RNA polymerase from sporulating cells of *Bacillus subtilis*: identification of a group of forespore-specific promoters. *J. Bacteriol.* **171:**2708–2718.
38. **Ollington, J. F., and R. Losick.** 1981. A cloned gene that is turned on at an intermediate stage of spore formation in *Bacillus subtilis. J. Bacteriol.* **147:**443–451.
39. **Pandey, N. K., and A. I. Aronson.** 1979. Properties of the *Bacillus subtilis* spore coat. *J. Bacteriol.* **137:**1208–1218.
40. **Panzer, S., R. Losick, D. Sun, and P. Setlow.** 1989. Evidence for an additional temporal class of gene expression in the forespore compartment of sporulating *Bacillus subtilis. J. Bacteriol.* **171:**561–564.
41. **Roels, S., A. Driks, and R. Losick.** 1992. Characterization of *spoIVA*, a sporulation gene involved in coat morphogenesis in *Bacillus subtilis. J. Bacteriol.* **174:**575–585.
42. **Sandman, K., L. Kroos, S. Cutting, P. Youngman, and R. Losick.** 1988. Identification of the promoter for a spore coat protein gene in *Bacillus subtilis* and studies on the regulation of its induction at a late stage of sporulation. *J. Mol. Biol.* **200:**461–473.
43. **Setlow, B., A. R. Hand, and P. Setlow.** 1991. Synthesis of a *Bacillus subtilis* small, acid-soluble spore protein in *Escherichia coli* causes cell DNA to assume some characteristics of spore DNA. *J. Bacteriol.* **173:**1642–1653.
44. **Setlow, B., N. Magill, P. Febbroriello, L. Nakhimovsky, D. E. Koppel, and P. Setlow.** 1991. Condensation of the forespore nucleoid early in sporulation of *Bacillus* species. *J. Bacteriol.* **173:**6270–6278.
45. **Setlow, B., and P. Setlow.** Unpublished results.
46. **Setlow, B., D. Sun., and P. Setlow.** 1992. Interaction between DNA and α/β-type small, acid-soluble spore proteins: a new class of DNA-binding protein. *J. Bacteriol.* **174:**2312–2322.
47. **Setlow, P.** 1978. Purification and characterization of additional low-molecular-weight basic proteins degraded during germination of *Bacillus megaterium* spores. *J. Bacteriol.* **136:**331–340.
48. **Setlow, P.** 1988. Resistance of bacterial spores to ultraviolet light. *Comments Mol. Cell. Biophys.* **5:**253–264.
49. **Setlow, P.** 1988. Small acid-soluble, spore proteins of *Bacillus* species: structure, synthesis, genetics, function and degradation. *Annu. Rev. Microbiol.* **42:**319–338.
50. **Setlow, P.** 1992. DNA in dormant spores of *Bacillus* species is in an A-like conformation. *Mol. Microbiol.* **6:**563–567.
51. **Stephens, M. A., N. Lang, K. Sandman, and R. Losick.** 1984. A promoter whose utilization is temporally regulated during sporulation in *Bacillus subtilis. J. Mol. Biol.* **176:**333–348.
52. **Stevens, C. M., R. Daniel, N. Illing, and J. Errington.** 1992. Characterization of a sporulation gene, *spoIVA*, involved in spore coat morphogenesis in *Bacillus subtilis. J. Bacteriol.* **174:**586–594.
53. **Stragier, P.** 1989. Temporal and spatial control of gene expression during sporulation: from facts to speculation, p. 243–254. *In* I. Smith, R. A. Slepecky, and P. Setlow (ed.), *Regulation of Procaryotic Development*. American Society for Microbiology, Washington, D.C.
54. **Sun, D., P. Stragier, and P. Setlow.** 1989. Identification of a new σ-factor involved in compartmentalized gene expression during sporulation of *Bacillus subtilis. Genes Dev.* **3:**141–149.
55. **Sussman, M. D., and P. Setlow.** 1991. Cloning, nucleotide sequence, and regulation of the *Bacillus subtilis gpr* gene which codes for the protease that initiates degradation of small, acid-soluble, proteins during spore germination. *J. Bacteriol.* **173:**293–300.
56. **Tipper, D. J., and J. J. Gauthier.** 1972. Structure of the bacterial endospore, p. 3–12. *In* H. O. Halvorson, R. Hanson, and L. L. Campbell (ed.), *Spores V*. American Society for Microbiology, Washington, D.C.
57. **Tovar-Rojo, F., and P. Setlow.** 1991. Analysis of the effects of mutant small, acid-soluble spore proteins from *Bacillus subtilis* on DNA in vivo and in vitro. *J. Bacteriol.* **173:**4827–4835.
58. **Zheng, L., W. P. Donovan, P. C. Fitz-James, and R. Losick.** 1988. Gene encoding a morphogenic protein required in the assembly of the outer coat of the *Bacillus subtilis* endospore. *Genes Dev.* **2:**1047–1054.
59. **Zheng, L., and R. Losick.** 1990. Cascade regulation of spore coat gene expression in *Bacillus subtilis. J. Mol. Biol.* **212:**645–660.

IX. BACTERIOPHAGES

56. SPO1 and Related Bacteriophages

CHARLES R. STEWART

SPO1, SP82, ϕe, 2C, SP8, H1, and SP5C constitute a family of large, virulent bacteriophages of *Bacillus subtilis* whose distinguishing feature is the complete replacement of thymine by hydroxymethyluracil (hmUra) in their DNA. Although isolated independently from soil in various locations in Japan, Europe, and the United States, their striking similarities suggest descent from a common ancestor. To the extent that they have been observed, they are similar with respect to size and structure of virions and antigenic specificity; size and CsCl buoyant density of both native and denatured DNAs; nucleotide sequence, as determined by DNA-DNA hybridization and by genetic and restriction maps; species of RNA and polypeptides synthesized; and regulation of the sequence of transcription and translation. However, although clearly related, they are by no means identical and show various degrees of divergence from each other (3–5, 15, 26, 27, 52, 54, 63, 67, 69, 73, 75, 76, 84, 87, 88, 91, 93, 99, 110, 113, 116, 130, 141).

This chapter will focus on SPO1, about which the most is known, but will also discuss data for the other phages that can illuminate areas not studied with SPO1. Section II is a discussion of several areas in which the study of SPO1 has made unique contributions to our understanding of important biological principles. Section III surveys a variety of interesting phenomena that have been observed in SPO1 infection but whose mechanisms are not understood and that thus offer opportunities for future research. Section IV is a compilation of practical information that should be useful to those doing research on SPO1.

NOTEWORTHY ELEMENTS OF SPO1 BIOLOGY

The Sigma Cascade

SPO1 infection involves a complex sequence of gene actions. By using hybridization competition, Gage and Geiduschek (42) distinguished six classes of transcripts on the basis of their times of synthesis. These classes can subdivided into additional classes, as discussed below. However, all transcripts can be grouped into three broad categories, i.e., early, middle, and late, on the basis of the time at which their synthesis begins. The mechanism by which SPO1 regulates this sequential onset of early, middle, and late transcription may be regarded as the paradigm of the use of a cascade of sigma factors to regulate sequential gene action (85).

Pero et al. (103, 104) identified three different RNA polymerases (A, B, and C), which become active at early, middle, and late times, respectively, and are thus responsible for transcription of the three categories of genes. This relationship can be seen in hybridization of pulse-labeled transcripts to Southern blots of SPO1 *Eco*RI* digests (136). RNA labeled early (1 to 4 min), midway (7 to 12 min), or late (later than 18 min) during infection produces three substantially different patterns, and the same three patterns are produced by RNA synthesized by the three polymerases in vitro.

Each of the three RNA polymerases recognizes a different promoter sequence (19, 78, 79, 138). Promoters specific for each polymerase were found by assaying for binding of each polymerase to restriction fragments. They were located within specific fragments by electron microscopy of polymerase-DNA complexes, subdivision of specific fragments with other restriction enzymes, size analysis of in vitro transcripts, and S1 mapping. Sequences were determined for 4 early, 13 middle, and 5 late promoters (12, 18, 19, 55, 57, 78, 79, 101, 111, 122, 133, 137, 138). Each of the three types shows a different consensus sequence in the −35 and −10 regions, as indicated in Table 1. The four early promoters (including P_E4, P_E5, and P_E6 in Fig. 3) are identical in the regions shown. The 13 middle promoters (including $P_{MII}3$, -6, -7, -8, -9, -10, -11, and -12 in Fig. 4 and the four shown in Fig. 5) are nearly identical for the −10 region, but there is some variation in the −35 region, with $P_{MII}8$ diverging strikingly from the others. The late promoters (shown in Fig. 6) show some variation in the −10 region and more variation in the −35 region.

RNA polymerase A, which has σ^A, is the predominant host polymerase, and the consensus sequences for the early promoters are similar to those for host promoters recognized by σ^A (90). In RNA polymerase B, σ^A has been replaced by a sigma factor specified by SPO1 gene 28 (30, 31, 36, 103). RNA polymerase C has, instead, the products of SPO1 genes 33 and 34 (35, 139). These replacements are responsible for the changes in specificity of transcription, since gene 33 and 34 mutations prevent the transcription of late genes, while gene 28 mutations block both middle and late transcription (40, 136). The sequences of genes 28 and 34 show significant similarity to those specifying other sigma factors (16, 18, 60).

Thus, host RNA polymerase transcribes early genes, which include gene 28 (62, 110). Gene product 28 (gp28) substitutes for the host sigma factor, changing the specificity of the RNA polymerase so that it now transcribes the middle genes, which include genes 33 and 34 (37, 104). The products of these genes then modify the RNA polymerase so that it transcribes the late genes.

Charles R. Stewart • Department of Biochemistry and Cell Biology, Rice University, Houston, Texas 77251.

Table 1. Consensus sequences of SPO1 promoters

Type of promoter	Sequence[a]	
	−35 Region	−10 Region
Early	TTGACT	ATAAT
Middle	AGGAGA--A-T	TTT-TTT
Late	CGTTAGA	GATATT

[a] Hyphens indicate nucleotide positions for which there is no consensus. T represents hmUra. Sequences are from references 19 and 78.

TF1

TF1 is an SPO1-specified DNA-binding protein that is synthesized in very large quantities during SPO1 infection (65, 120). It binds preferentially to hmUra-containing DNA (66) and to specific sites within SPO1 DNA (57–59). The binding of TF1 to restriction fragments carrying one of these sites causes DNA bending (as shown by differential gel retardation with various positions of the binding site on the fragment [125]) and can facilitate negative supercoiling (as shown by an increase in negative supercoiling when such a fragment is self-ligated in the presence of TF1 [58]).

The TF1 gene has been cloned and sequenced (55), and the inferred amino acid sequence shows extensive similarity to those of other prokaryotic type II DNA-binding proteins, such as the HU and IHF proteins of *Escherichia coli*. The complete biological roles of these proteins remain to be established, but they seem to be involved in the wrapping of DNA into higher-order "chromatinlike" structures and in the facilitation of reactions that require DNA bending, such as site-specific recombination and DNA loop-facilitated repression (38, 45, 105).

TF1 is essential to successful SPO1 infection, since a temperature-sensitive mutation in the TF1 gene prevents the normal production of progeny (120). TF1 plays an essential role, either directly or indirectly, in several of the processes that occur during infection, most prominently in the shutoff of middle genes. This shutoff cannot be explained by the disappearance of RNA polymerase B, which remains active late in infection, as shown by the continued activity of several middle promoters (122). Apparently, the substitution by gp33 and gp34 occurs on only a fraction of the RNA polymerase molecules present. Other middle promoters assayed in the same way are substantially turned off, implying the existence of an independent shutoff mechanism. TF1 appears to play an essential role in this mechanism, since the *ts* mutation (and other mutations in the TF1 gene) prevents the shutoff of some of the middle genes (120, 121). A plausible guess is that bending caused by TF1 either interferes with normal transcription of these genes or facilitates the activity of a separate repressor, but those details remain to be established.

The TF1 mutant is also deficient in processing of the major head protein (120). Although little is known about SPO1 morphogenesis, this encourages the speculation that TF1 participates in head assembly, possibly by promoting appropriate folding of the DNA or by participating in the reaction that generates mature DNA molecules from the concatemeric replicating form. IHF plays a role in the comparable reaction for lambda (38).

Since the TF1 gene is located in a cluster of genes essential for SPO1 DNA replication and is expressed at the same time as those genes, one might speculate about its involvement in the formation or maintenance of DNA structures optimal for replication. When replication was analyzed by incorporation of radioactive precursors, the initial rate during infection by the TF1 mutant was indistinguishable from that of wild type, but total accumulation of labeled DNA by the mutant was somewhat less (120). Thus, TF1 is certainly not needed for polymerization, but a more subtle structural role remains possible.

The *ts* mutant is also deficient in expression of several late genes (120). It is conceivable that the activation of certain late genes is TF1's only direct role, since the other activities affected by the mutation either are or could be dependent on the activity of late gene products. However, analysis of the pattern of TF1 binding suggests that other roles are likely. Several of its specific binding sites have been analyzed by DNA footprinting, and they show no particular relationship to either middle or late promoters (57–59).

Several of the binding sites do overlap early promoters, and TF1 inhibits transcription from such promoters in vitro (145). However, in vivo, the shutoff of early transcription is accomplished without the activity of middle genes (40) and thus without TF1. Perhaps TF1 would shut off those genes if they were still on when TF1 became active.

Although the sequences of several of the preferred binding sites have common features, such as closely juxtaposed blocks of alternating purines and pyrimidines, they have not revealed any consensus sequence for TF1 binding (59). TF1 binds as a dimer to the cores of these binding sites. As the TF1 concentration increases, additional dimers bind to sites contiguous to the core (57, 124).

The Intron

SPO1 (and those of its relatives that have been tested) contains the only introns known in gram-positive bacteria, which are among the few known in prokaryotes of any kind (50, 51). The SPO1 intron is a self-splicing intron of 882 nucleotides whose sequence and structure are congruent with the conserved features of group I introns found in eukaryotic organelles and bacteriophage T4 (49, 50). Sequencing of the products of in vitro splicing, combined with sequencing of the region surrounding the DNA sites probed by such products, showed that the intron is located within SPO1 gene 31, which specifies the DNA polymerase essential for SPO1 replication. This intron includes an open reading frame whose inferred amino acid sequence shows no obvious similarity to similarly located open reading frames found in T4 or eukaryotes. Preliminary experiments have indicated that the open reading frame specifies a DNA-binding protein with a specificity for the surrounding exons.

The occurrence of very similar introns in both SPO1 and T4, which have no known overlaps in host range, along with the recent discovery of similar introns in cyanobacteria (147), argues that introns existed in ancient evolutionary history before the divergence of gram-positive from gram-negative bacteria, although the argument is not absolute, since the possibility of lateral transfer cannot be excluded (6). Assuming that the argument is correct and that prokaryotes have lost most of their introns over the course of evolution, the preservation of these few introns in T4 and SPO1 suggests a significant selective advantage. Since two of the T4 introns also interrupt genes involved in DNA replication, it is tempting to speculate that the regulation of splicing has found a useful role in the regulation of DNA synthesis, but there is no direct evidence in support of such a role (50).

Special Features Resulting from the Presence of hmUra

The hmUra phages specify an array of enzymes whose apparent purpose is the substitution of hydroxymethyl dUTP (HMdUTP) for dTTP in the DNA precursor pool of infected cells. These include dCMP deaminase (92, 117), dUMP hydroxymethylase (1, 117), HMdUMP kinase (68), dTTPase-UTPase (32, 107, 115), dTMPase (4), and an inhibitor of thymidylate synthetase (61). One might ask what benefit is provided by the substitution of hmUra for thymine in the DNA that justifies this investment of energy and materials to accomplish it.

It has been proposed that this substitution permits ready distinction between phage and host DNAs, permitting host macromolecular syntheses to be shut off without interfering with comparable phage processes. Such an explanation may be applicable to T4, which substitutes hydroxymethyl cytosine for cytosine by processes quite analogous to those used by SPO1 (71) and which specifies enzymes that completely degrade the cytosine-containing *E. coli* DNA while having no effect on hydroxymethyl-cytosine-containing DNA (129). However, any such distinction made by SPO1 must be considerably more subtle, since host DNA is not degraded during SPO1 infection (148). Host DNA synthesis and gene action are shut off by SPO1 infection, but there is no evidence that it is the presence of thymine that identifies the host DNA as the target for such shutoffs. Indeed, some (thymine-containing) host genes, including those specifying rRNA and ribosomal protein, are not shut off (42, 126, 143). Thus, while it is possible that the presence of thymine is necessary to identify a shutoff target, it is certainly not sufficient.

The presence of hmUra does cause at least two biologically significant differences in the behavior of DNA. First, in vitro transcription of middle promoters by RNA polymerase B proceeds much more efficiently from natural SPO1 DNA fragments (containing hmUra) than from cloned versions of the same fragments (containing thymine) (14, 77, 112). However, since there is no reason to expect middle promoters to be present in host DNA in significant numbers, this may be, at best, a redundant control. Early promoters are transcribed equally well from both types of DNA (77).

The other known difference is that TF1 binds preferentially to hmUra-containing DNA (66). Thus, the hmUra permits TF1 to distinguish between host and phage DNAs. In view of the large quantity of TF1 made, its extensive binding to the SPO1 genome, the probability of its pleiotropic effects, and the heterogeneity in nucleotide sequence of TF1-binding sites, this may well be a necessary distinction and thus a sufficient justification for the presence of hmUra.

It is not even established that the presence of hmUra is necessary for successful infection. ϕe mutants deficient in dTTPase-dUTPase produce normal burst sizes, with thymine replacing up to 20% of the hmUra in the progeny DNA (33, 86). Thus, the total replacement of thymine by hmUra is not necessary for normal function. SPO1 mutants deficient in dUMP hydroxymethylase or HMdUMP kinase are completely deficient for DNA replication (48, 68, 92), which might suggest an absolute requirement for hmUra. However, these mutants have been tested only under conditions in which HMdUTP is the only triphosphate available with that hydrogen-bonding specificity (i.e., they have not been tested in the absence of the dTTPase), so it remains unproven that the presence of hmUra is essential.

Resolution of Concatemers

Because all known DNA polymerases require primers and polymerize only in the 5′- to -3′ direction, an additional mechanism must be available for synthesis of the 5′ ends of linear DNA molecules. SPO1 provides a test for one such mechanism, which Watson proposed (142) to account for the formation of concatemers by phages with linear, terminally redundant genomes. According to this hypothesis, concatemers would be formed by overlap of the complementary protruding 3′ ends left by each round of replication. The concatemers would be resolved by staggered cleavage of the two strands, leaving protruding 5′ ends that could be filled in by normal polymerization.

SPO1 is the type of phage to which Watson's hypothesis is applicable. It has a linear genome with terminal redundancy (22), it forms concatemers up to 20 genomes long as estimated from sedimentation-velocity experiments (83), and restriction digests of concatemers reveal a single copy of the terminal redundancy at each junction between unit genomes (22).

Watson's hypothesis predicts that when the first dimer has formed, the entire genome will have replicated except for the terminal redundancy involved in the junction. Thus, genetic markers in the rest of the genome will be completely replicated, whereas half of those in the terminal redundancy will remain unreplicated. This was precisely the result obtained for SPO1. When replication was arrested at the end of the first round by temperature shift with temperature-sensitive replication initiation mutants, and when the extents of replication of different genetic markers were assayed by density shift, markers in the terminal redundancy had been replicated to only half the extent of those in the rest of the genome (22, 46). The SPO1 terminal redundancy is much longer than necessary for this purpose, and its length makes the process more complicated. However, alternative explanations of this result seem even more improbable.

The temperature-sensitive initiation mutations used for that experiment are in SPO1 gene 32. Apparently, the gene 32 product, in addition to its role in initiation, is needed for the resolution of concatemers. There must be a way of resolving concatemers without initiating another round of replication, since a gene 21 mutant also prevents initiation of the second round but leaves all markers equally replicated (46). Resolution of concatemers apparently requires a late function, since concatemers accumulate during *sus-34* infection, but does not require packaging, since it takes place in a mutant that fails to make heads (82).

PROMISING AREAS FOR FUTURE RESEARCH

The SPO1 life cycle is considerably more complex than is apparent merely from consideration of the processes described above. Most notably, the regulation of the sequence of gene action involves many additional events. In this section, I will discuss a variety of potentially interesting phenomena that have been described before but whose mechanisms have not been elucidated and that thus offer potential for future research.

Additional Mechanisms for Regulation of Phage Gene Action

Section II described the three general categories of early, middle, and late genes, which are distinguished by the time at which their transcription begins and by the sigma factors responsible for recognizing their promoters. Each of these categories may be divided into subcategories of genes that are turned on or off at different times and/or under different controls, each implying the existence of regulatory mechanisms beyond the well-established sigma cascade.

Shutoff of early transcription

sus-28 infection permits shutoff of early transcription (40, 100), showing that that shutoff is accomplished by a mechanism independent of the substitution of gp28 for σ^A.

At least two mechanisms for shutoff of early genes

The early genes have been divided into two classes, *e* and *em*, that are turned off at the end of early and middle times, respectively (42). The middle-time transcription of some of the *em* genes is from their early promoters, since it occurs even in *sus-28* infection (100). Thus, there must be two separate mechanisms for shutting off different early promoters at the two different times.

Heterogeneity in time of onset of early transcription

Transcription of different early sequences begins at different times after infection, with some transcripts not appearing until middle times, even in *sus-28* infection (11, 12). In at least some cases, the difference is caused by a difference in the time at which transcription was initiated, as shown by the time at which rifampin resistance was achieved (100), and the difference cannot be explained by differences in the strength of promoters as measured in vitro (111). Thus, there must be additional regulatory factors that cause the onset of certain early transcripts to be delayed in comparison with others.

Mechanisms for the shutoff of middle genes

I have discussed above the evidence that TF1 plays an essential role in the mechanism responsible for shutoff of certain *m* genes. There must be at least one additional factor involved in middle gene shutoff, since TF1 is specified by a middle gene, but the shutoff of many middle transcripts requires gp34, the late-time-specific sigma factor (40, 136). In addition, at least one middle gene is shut off independently of TF1 (120), suggesting the existence of a second mechanism.

DNA replication and late transcription

Several replication-negative SPO1 mutants permitted synthesis of many late-time-specific transcripts, as measured by hybridization to Southern blots (56, 136), and of most late-time-specific proteins, as measured by polyacrylamide gel electrophoresis (PAGE) (62). Thus, DNA replication is not generally required for late gene activity. However, hybridization competition (43) showed that one subset of late transcripts (among those designated class 1 [42]), was dramatically underrepresented in infection by a replication-negative mutant. Although the effect of preventing replication by other means was not tested, leaving open the possibility that this mutation affects transcription independently of its effect on replication, this suggested that replication may be needed for normal expression of that subset in a way in which it is not needed for expression of the other late genes.

Distinction between the roles of genes 33 and 34

Mutations in genes 33 and 34, which specify the two polypeptides responsible for the promoter recognition specificity of RNA polymerase C, do not always have identical effects. Some middle transcripts are turned off in *sus-33* infection but not in *sus-34* infection (56, 136). In minicells, some late proteins are not made, and the synthesis of those that are takes place in *sus-33* but not in *sus-34* infection (110). Thus, there appear to be regulatory events that can be mediated by gp34 in the absence of gp33.

The following is the simplest, although far from the only, interpretation of these data. gp34 binds to RNA polymerase independently of gp33 and causes certain promoters to be recognized. These promoters are distinguished from other late promoters in that the others require both gp33 and gp34 and cannot be expressed in minicells. Among the gene products produced from these promoters is one whose activity is to repress certain middle genes.

As noted above, gene 34 but not gene 33 shows similarity to other sigma factors. gp34 may function as a normal sigma, with gp33 an accessory needed for some but not all promoters. There is a 4-bp overlap between the terminus of gene 33 and the origin of gene 34, suggesting a translational coupling that may ensure the presence of equimolar quantities of the two proteins (16).

Gene 27 product

A mutant defective in gene 27 shows a deficiency in late gene action (56, 62). The pattern of late transcription, as shown by hybridization to Southern blots, is significantly different from that shown by *sus-33* or *sus-34*, suggesting that the role of gp27 is different from those of gp33 and gp34.

The gene 27 mutant is also deficient in DNA replication (92, 131), and neither deficiency appears to be an indirect result of the other. Thus, gp27 may play a direct role in both replication and late transcription, although it remains possible that either deficiency is the result of a polar effect on a downstream gene (the mutation is a nonsense mutation near the beginning of gene 27 [17]).

We do not know the mechanism by which gp27 affects either replication or transcription. The gene has been sequenced (17), showing the product to be a basic protein with a high density of proline residues but not revealing its function. If gp27 does affect both processes directly, it would be similar to T4 gp45 (146). gp45 is an integral part of the T4 replication complex (128) and also interacts directly with RNA polymerase (109), but no comparable role for gp27 has been shown. The nucleotide sequences of the two genes show no extensive homologies.

Translational control mechanisms

Posttranscriptional processing of early RNAs occurs for both SPO1 and SP82. Early RNAs produced during infection are substantially smaller than those produced by in vitro transcription of phage DNA, and the in vitro transcripts can be converted into RNAs of the same size as several of the in vivo RNAs by an enzyme similar to *E. coli* RNase III that was purified from uninfected *B. subtilis* (29, 94, 95). By S1 mapping, the RNAs formed by in vitro processing were shown to be the same as those formed in vivo (95). Three of the cleavage sites have similar nucleotide sequences, each capable of forming a stem and a loop, and the cleavage site in each is at the 5′ side of an adenylate residue within the loop (95).

Thus, there is a processing activity that is potentially capable of regulating translation. However, no evidence that it does so has been obtained. In vitro cleavage of one of the SP82 transcripts into three subfragments causes no detectable changes in the proteins translated from that RNA in vitro (94).

There is, however, evidence suggesting that the translation of several of the RNA molecules formed by that processing requires the activity of a protein synthesized early during infection. In minicells, *sus-28* infection permits synthesis of several proteins that are not synthesized if chloramphenicol is present during the first 30 min of infection (and is then replaced by rifampin to permit protein synthesis but prevent further RNA synthesis). Several of these proteins are synthesized if RNA made in the presence of chloramphenicol is used to program in vitro translation. Thus, certain RNA molecules that are present are not translated (110). The simplest explanation is translational repression overcome by a process requiring protein synthesis.

Another mechanism available for translational control is suggested by the fact that during late times, one of the RNA classes (m_1l) decreases in abundance, although its relative rate of synthesis remains constant, suggesting a differential rate of degradation (43).

Shutoff of Host Functions

Host DNA, RNA, and protein syntheses are shut off during SPO1 infection. Since SPO1 macromolecular syntheses proceed, these shutoffs must be specific to the host systems and not merely the result of general disruption of the infected cells. Very little is known about the mechanisms of host shutoff in any viral system, so elucidation of these mechanisms should be particularly useful.

Host DNA synthesis

The high CsCl buoyant density of SPO1 DNA, which results from the presence of hmUra, makes it possible readily to differentiate phage from host DNA synthesis in infected cells and to show that host DNA synthesis is shut off virtually completely by 5 to 7 min after infection (126, 144). This shutoff takes place in *sus-28* infection (133), so it does not require any of the known modifications of RNA polymerase. However, it remains completely sensitive to inhibition by rifamycin until 4 min after infection (43), suggesting that at least one of the necessary transcriptions is delayed beyond the normal time for early transcription.

Strong arguments have been provided against several of the obvious hypotheses as to the mechanism by which this shutoff occurs. There is no detectable degradation of the host genome (148). The TTPase induced during infection might inhibit replication by limiting substrate. However, the shutoff of host replication is maintained in extracts of infected cells and is not relieved by excess TTP (148). A φe mutant deficient in TTPase was able to shut off host replication (114). Also in φe, the possibility that shutoff is caused by the phage-induced inhibitor of thymidylate synthetase was eliminated by showing that shutoff still occurs in a strain in which the requirement for thymidylate synthetase is circumvented (114).

Host RNA and protein synthesis

The synthesis of most host proteins is reduced by SPO1 infection to about 10% of its normal level (126). A substantial decrease in the intensity of pulse-labeled host protein bands on PAGE may be seen by 1 to 4 min after infection (62, 80, 110, 126), and most bands have reached their minima by 6 to 8 min. Host RNA synthesis also decreases during infection (42), and the rate of decrease in protein synthesis after infection is comparable to that accomplished by treatment of uninfected cells with actinomycin (126). Thus, the shutoff of protein synthesis may be entirely explained by the complete shutoff of most mRNA synthesis within a few minutes after infection, although the possibility that there is also a direct effect at the translational level has not been ruled out. Presumably, these shutoffs are caused by the products of an early gene or genes, since substantial effects occur before middle genes become active.

The shutoff mechanism discriminates between different host transcripts, permitting continued synthesis of rRNA and of the mRNA for ribosomal proteins (42, 126). However, even rRNA synthesis is not totally unaffected. The rate of rRNA synthesis decreases late in infection (42), and the normal increase in its rate of synthesis in response to a shift to enriched media does not occur during SPO1 infection (143).

None of the SPO1 genes responsible for shutoff of host functions has been identified, although a φe mutant deficient in shutoff of host replication has been isolated (86). Curran and Stewart (24) identified several SPO1 restriction fragments that may carry active host shutoff genes. These fragments are refractory to cloning, as would be expected of fragments carrying host shutoff genes, and alternative reasons for unclonability, such as the disruption of vector function by overactive promoters, were eliminated. Several of these fragments are located in the terminal redundancy, which, as discussed below, includes at least 11 early genes with unknown functions.

DNA Replication

Regulation of Replication

Infecting DNA is modified soon after infection so that if extracted, it is inactive in transfection (82). This modification is correlated in time with the introduction of a small number of single-strand breaks. Recovery from this modification takes place at about the time that replication begins, and there is no evidence as to which, if either, of these events causes the other. However, initiation of the replication of an infecting genome is hastened if the culture has been preinfected with another SPO1 genome (20), so it is not the time of residence of the infecting DNA within the infected cell that determines the time at which replication begins.

Incorporation of labeled precursors into SPO1 DNA begins about 10 min after infection (144). The capacity to replicate becomes insensitive to rifamycin at about 8 min (43), which is consistent with the time of synthesis of those DNA synthesis enzymes that have been tested, including dCMP deaminase (144), dUMP hydroxymethylase (1), DNA polymerase (148), gp30, and gp32 (62, 110). Thus, the time of initiation may be determined simply by the time at which a sufficient concentration of the replication machinery has been synthesized.

The origins and directions of replication were determined by infecting cells with light phage in heavy medium and following the shift in density of genetic markers in different portions of the genome. For the first round of replication, SPO1 DNA replicates from at least two origins, using at least three growing points. Most of the genome is replicated right to left from an origin near gene 32 to a terminus between genes 2 and 10. Replication of the region to the left of gene 3 requires at least one other origin. The region to the right of gene 32 may either use a third origin or be replicated rightward from the origin near gene 32 (24, 47).

In a mixed infection of SPO1 and SP82, SP82 predominates in the burst, and the site on the genome that determines this difference is close to the right-hand replication origin (134). Nothing is known about the mechanism of this predominance, but it is conceivable that such predominance reflects a negative regulation of replication. The precise positions of the replication origins remain to be established.

Roles of gene products in replication

At least 10 SPO1 gene products (genes 21a, 21b, 22, 23, 27, 28, 29, 30, 31, and 32) are essential for replication, as shown by the replication-negative phenotypes of mutants affected in those genes. Genes 23 and 29 specify HMdUMP kinase and dUMP hydroxymethylase (68, 92, 133), enzymes needed for nucleotide synthesis. Genes 32 and 21a and/or 21b are required specifically for initiation of replication at each origin, as shown by the fact that temperature-sensitive mutations in these genes permit completion of an ongoing round of replication after shift to the restrictive temperature but do not permit initiation of a new round (46). Nothing is known of the mechanisms by which initiation is accomplished. Genes 22, 30, and 31 are directly involved in polymerization, judging from the fact that replication is stopped abruptly by shifting temperature-sensitive mutants affected in those genes to the restrictive temperature. gp29 may also be directly involved, since one of the gene 29 *ts* mutations also causes abrupt termination (48). Others do not, as is expected for genes involved in precursor synthesis, suggesting for this gene product a multiple role possibly similar to that suggested for the analogous T4 enzyme (13, 140). Gene 31 specifies the DNA polymerase (28, 50). The nucleotide sequence of gene 30 shows similarities to those of lambda gene P and of T4 gene 46 (123), but nothing further is known of its activity. Nothing is known of the activity of gp22 or of the structure or function of a replisome.

The reason for the need for a phage-specific DNA polymerase has not been established, and evidence regarding the role of host DNA polymerases is ambiguous. Host mutants temperature sensitive for DNA polymerase III, the polymerase essential for *B. subtilis* replication, are also temperature sensitive for replication of φe or SPO1 DNA (74). This observation is confusing, since 6-(*p*-hydroxyphenylazo)-uracil (HPUra) inhibits DNA polymerase III but does not inhibit replication of the hmUra phages. At present, the resolution of this apparent conflict requires introduction of an ad hoc hypothesis, such as that polymerase III may have two activities, both of which are inactivated in the *ts* mutant and only one of which is inhibited by HPUra. A host mutation affecting DNA polymerase I had no effect on SPO1 infection (148).

DNA gyrase appears to play an essential role, since gyrase inhibitors nalidixic acid (41) and novobiocin (118) both inhibit SPO1 replication. Host mutants resistant to the two antibiotics make SPO1 infection resistant to novobiocin but not nalidixic acid, while an SPO1 mutation provides resistance to nalidixic acid (2). The host gyrase includes the products of two different genes, *gyrA* and *gyrB*, which specify resistance to nalidixic acid and novobiocin, respectively (106, 135), so these data suggest that the SPO1 genome specifies a product that modifies or substitutes for the bacterial *gyrA* product but that SPO1 uses the host *gyrB* product. To my knowledge, none of the mapped SPO1 mutants has been tested for gyrase activity, nor has the nalidixic acid-resistant mutation been mapped.

Inhibition of SPO1 replication by nalidixic acid requires higher concentrations than does inhibition of host replication, offering a means of inhibiting host replication while permitting SPO1 replication to proceed. The same facility is offered by HPUra (10, 74, 86).

Six SPO1 genes (genes 2, 3, 11, 12, 13, and 14) have

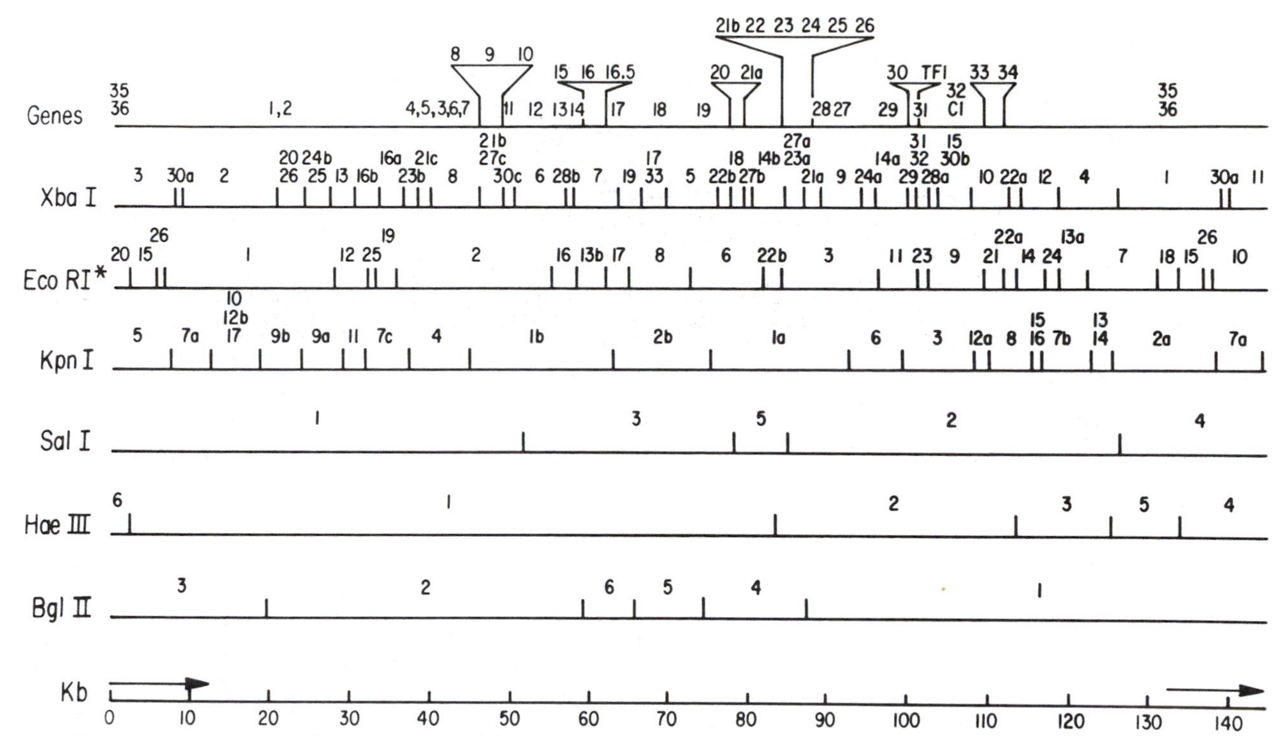

Figure 1. Restriction and genetic maps of the SPO1 genome (taken from references 24, 92, and 102). Arrows indicate positions of the terminal redundancy. The *Eco*RI* site between *Eco*RI*-3 and *Eco*RI*-22b is not there in some strains of SPO1 (e.g., see reference 102). Where the order of two or more adjacent genes or fragments with respect to each other has not been determined, their numbers are placed one above the other. The positions of genes 6, 7, 10, 15, 22, 23, 24, 25, and 26 have been determined only by genetic mapping; those of all other genes have been determined from the positions of the restriction fragments on which they are located (24). In most cases, the latter positions are the same as those determined by genetic mapping, but when those determinations differ, the positions given are those determined from the restriction map. For several pairs of genes (genes 1 and 2, 8 and 9, and 20 and 21a), their positions with respect to the rest of the genome were determined from the restriction map, but their positions with respect to each other were determined only from the genetic map. Mutation C1 was mapped adjacent to gene 32 but was not shown to be in a separate gene (92). Eleven additional genes have been located to the right of genes 35 and 36 in the terminal redundancy by in vitro expression from specific restriction fragments, although none of them has been defined by mutation (100). Their positions are shown in an expanded map of the terminal redundancy in Fig. 3. All locations are determined with respect to the *Eco*RI restriction map (102), using fragment sizes from reference 24. This map was published previously (132) and has been reproduced with permission from Plenum Press.

conditional lethal mutations that completely prevent phage multiplication but cause only partial deficiency in DNA replication (92). The nature and extent of the replication deficiency differ between mutants affected in the same gene, so the significance of these deficiencies is uncertain, and nothing further is known about the function of any of these genes.

Genomic organization of replication genes

Two clusters of replication genes, genes 21 to 23 and 29 to 32, behave differently in a variety of seemingly unrelated ways. (i) Cold-sensitive mutations tend to occur in genes 21 to 23, while heat-sensitive mutations occur in genes 29 to 32 (48). (ii) Of 34 genes tested, 7 could not be cloned on any available restriction fragments. Six of these are in a cluster that includes genes 21 to 23 (24). (iii) Proteins specified by genes in the second cluster can readily be detected on polyacrylamide gels; those specified by genes in the first cluster cannot (110). (iv) Mutations in the two clusters have been reported to have different effects on late transcription (108). (v) In sequential complementation experiments, mutants affected in the second cluster could be rescued more effectively by infection with a second mutant than could mutants affected in the first cluster (133).

There are a few exceptions to these general distinctions, and not all of the genes were tested in each case. However, clearly, general differences between these two clusters exist and would be interesting if they could be understood. The two clusters are separated by several genes not necessary for replication and by genes 27 and 28, whose behavior does not place them in either group.

Structure and Morphogenesis

An SPO1 particle consists of an icosahedral head; a contractile tail consisting of sheath, tube, and base plate; and a neck that attaches the head to the tail. The head diameter is about 87 nm, and the dimensions of sheath and base plate are about 19 by 140 nm and 25 by 60 nm, respectively. When contraction of the tail is induced in vitro, the configuration of the

Table 2. Summary of genes of SPO1

Gene[a]	Phenotype of mutant (reference)[b]	Activity of gene product (reference)	Reference for: Sequence determined	Reference for: Gene product purified or identified by PAGE	Restriction fragment(s) on which gene is located (reference)[c]
Mapped					
1	D$^+$ (92), deficient virion assembly (39)			110	*Kpn*I-9b (24, 25)
2	D$^+$ or DD (92); deficient virion assembly[d]			110	*Kpn*I-9b (24, 25)
4	D$^+$ or DA (92); deficient head formation[d]				*Kpn*I-4, *Xba*I-23b (24, 25)
5	D$^+$ (92); polyheads formed; deficient processing of gp6 (97)				*Kpn*I-4, *Xba*I-21c (24, 25)
3	D-int or DA (92); deficient virion assembly[d]				*Kpn*I-4, *Xba*I-8 (24, 25)
6	D$^+$ (92); deficient head formation (39, 81)	Part of head structure (127)		127	*Eco*RI-2 (21)
7	D$^+$ (92); deficient tail formation (39)				*Eco*RI-2 (21)
8	D$^+$ (92); deficient tail formation (98)				*Xba*I-21b (24, 25)
9	D$^+$ (92); deficient tail formation (39)				*Xba*I-21b (24, 25)
10	D$^+$ (92); deficient tail formation[d]				*Eco*RI-2 (21)
11	D$^+$ or D-int (92); deficient tail formation (39)				*Xba*I-30c, *Eco*RI-2Δ (24, 25)
12	D$^+$ or D-int (92); deficient tail formation (39)				*Eco*RI 2Δ (24, 25)
13	D$^+$ or D-int (92); deficient tail formation[d]				*Eco*RI 16 (24, 25)
14	D$^+$ or D-int (92); deficient tail formation[d]				*Eco*RI-16, *Xba*I-7 (24, 25)
15	D$^+$ (92); deficient tail formation (39)				*Xba*I-7 (24, 25)
16	D$^+$ (92); deficient tail formation (39)				*Xba*I-7, *Eco*RI-13b, *Bgl*II-6 (24, 25)
16.5	D$^+$ (131)				*Bgl*II-6, *Eco*RI-13b, *Xba*I (24, 25)
17	D$^+$ (92); deficient tail formation (39)				*Eco*RI-17, *Bgl*II-6, *Xba*I-19 (24, 25)
18	D$^+$ (92); deficient tail formation[d]				*Xba*I-8 (24, 25)
19	D$^+$ (92); deficient tail formation[d]; deficient lytic enzyme				*Xba*I-6 (24, 25)
20	D$^+$ (92); deficient tail formation (39)				*Xba*I-18 (24, 25)
21a and 21b	DO (92); deficient initiation of replication (46)[e]	Necessary for initiation of replication but not for elongation (46)[e]			*Xba*I-18 (gene 21a only) (24, 25)
22	DO (92)	Necessary for elongation stage of replication (48)			*Eco*RI-3 (21)
23	DO (92)	HMdUMP kinase[h]			*Eco*RI-3 (21)
24	D$^+$ (92)				*Eco*RI-3 (21)
25	D$^+$ (92)				*Eco*RI-3 (21)
26	D$^+$ (92)				*Eco*RI-3 (21)
28	DO (92); deficient transcription of middle and late genes (40)	σ factor that confers middle gene specificity on host RNA polymerase (30, 36, 136)	18	12, 36, 110	*Xba*I-21a, *Xba*I-9 (18, 24, 25)
27	DO (92, 131); deficient transcription of late genes (56)		17	12, 17	*Xba*I-9, *Kpn*I-1a (17, 24, 25)
29	DO; HMase$^-$ (92)	dUMP hydroxymethylase (92); necessary for elongation stage of replication (48)		1, 72	*Eco*RI-11 (21)
30	DO[f] (92)	Necessary for elongation stage of replication (48)	123	110	*Xba*I-14a, *Xba*I-29, *Eco*RI-11, *Kpn*I-6, *Kpn*I 3 (24, 25, 121)

TF1	D+; deficient middle gene shutoff and head morphogenesis (120)	Binds specifically to hmUra-containing DNA; causes DNA bending (66, 125)	55	65	*Xba*I-29, *Xba*I-31, *Eco*RI-11, *Kpn*I 3 (55)
31	DO (92)	DNA polymerase (28, 50)	50	62, 110	*Eco*RI-11, *Eco*RI-23, *Eco*RI-9, *Xba*I-31, *Xba*I-32, *Xba*I-28a, *Xba*I-15, *Kpn*I-3 (24, 25, 50)
32	DO (92); deficient initiation of replication (46)	Necessary for initiation of replication but not elongation (46)		110	*Xba*I-15 (24, 25)
C1	Clear plaques with distinct edges (92)				
33	D+ (92); deficient transcription of late genes (40)	Confers late gene specificity on RNA polymerase (35, 40)	16	35	*Eco*RI-21 (24, 25)
34	D+ (92); deficient transcription of late genes (40)	Confers late gene specificity on RNA polymerase (35, 40)	16	35	*Eco*RI-21 (24, 25)
35	D+ (92); deficient virion assembly[d]				*Eco*RI-18 (24, 25)
36	D+				*Eco*RI-18 (24, 25)
e9	—[g]			100	
e6	—[g]			100	
e18	—[g]			100	
e4	—[g]			100	
e15	—[g]			100	
e3	—[g]			100	
e16	—[g]			100	
e20	—[g]			100	
e7	—[g]			100	
e21	—[g]			100	
e12	—[g]			100	
Unmapped mutations					
bh	Plaques with black halo (91)				
hd	Forms plaques on host mutant on which wild type does not form plaques (91)				
Nal^r	Resistant to nalidixic acid (2)				
Unknown[i]					

[a] Genes are listed in map order, which differs from numerical order in two places.
[b] Abbreviations for replication phenotypes: D+, replication positive; DD, replication delayed; DA, replication arrested prematurely; D-int, partial but not normal replication; DO, replication negative. HMase, hydroxymethylase.
[c] Fragments shown carry at least part of the gene in question, most often shown by the presence of marker rescue activity with respect to mutations in the gene. Fragments underlined carry entire genes, judging by the fact that they are able to complement mutants affected in the gene, they have marker rescue activity for genes on both sides of the gene in question on the map, or their nucleotide sequence is entirely contained within the sequence of that fragment.
[d] These data have never been published, and the original data are no longer accessible (33a). The data were used by Geiduschek and Ito in writing an earlier review (44), and this information is taken from that review.
[e] Mutants N6 and F2 are affected in two different cistrons, 21a and 21b, since the two mutations are located on different restriction fragments at least 2.8 kb apart and since the gene 21a product is expressed from its restriction fragment (24, 25). However, the two mutants frequently do not complement each other or the mutant cs 21-1, which was used to identify the replication initiation phenotype. Because of the inconsistency of complementation patterns, it is not clear which of these cistrons is responsible for the replication initiation phenotype. Each has a DO phenotype.
[f] Three different mutations in gene 30 showed weak complementation of each other, and the gene was initially subdivided into 3 subcistrons (92). However, sequencing has now shown that all three mutations are in one gene (123).
[g] No mutation isolated. Gene was identified physically (100).
[h] Gene 23 mutants fail to complement SP82 mutant H20, which is deficient in HMdUMP kinase (68, 133).
[i] Activities of four genes are known, but the SPO1 genes have not been identified. Products are inhibitor of thymidylate synthetase (61), dTTase-UTPase (32, 33, 106, 113), dCMP deaminase (92, 116), and deoxythymidylate-5′-nucleotidase (4).

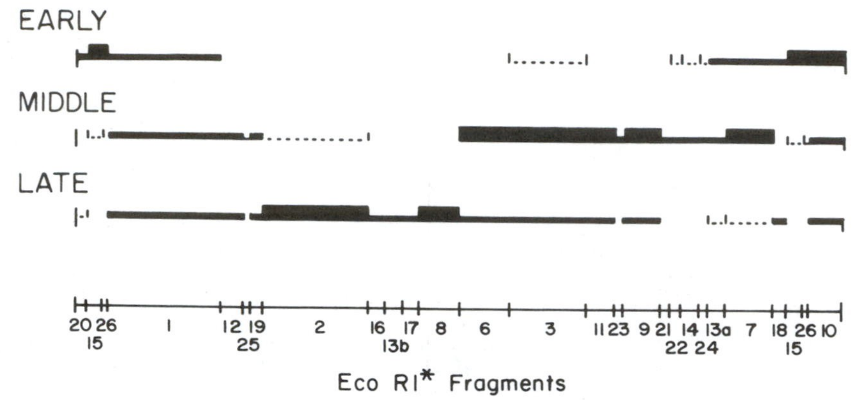

Figure 2. General locations of early, middle, and late sequences on the SPO1 genome. Thick, thin, and dotted lines represent relative amounts of early, middle, and late labeled RNA that hybridize to each of the indicated *Eco*RI* restriction fragments. *Eco*RI* fragments 13a and 13b and fragments 14 and 15 were not resolved on these Southern blots, so the hybridized RNAs for these bands were arbitrarily assigned to one or both members of each pair of DNAs on the basis of the amount of RNA hybridizing to restriction fragments neighboring each member of the pair. Data are from references 102 and 136, and the figure has been reproduced with permission from the American Society for Microbiology.

base plate is altered, and the sheath contracts to about 27 by 63 nm, exposing the tube (9 by 142 nm) (91, 96, 98, 141). Fifty-three polypeptides were identified by PAGE of the subassemblies of the bacteriophage particle (97). Twenty-eight of these are in the base plate, 6 are in the sheath and tube, 3 are in the neck, and 16 are in the head. The head contains a double-stranded DNA molecule of 140 to 145 kbp, including a 12.4-kbp terminal redundancy (21, 24, 27, 75, 98, 102, 141).

A number of mutations affect head or tail structure (Table 2), but there has been little integration of the genetic and physical analyses. Only one gene product has been identified as a structural protein (gp6 is part of the head [127]), and only one step in morphogenesis has been analyzed (the activities of gp5 and TF1 are necessary for proteolytic processing to generate the mature form of the major head protein [97, 120]). Thus, virtually the entire morphogenetic process of this complex structure remains to be worked out.

TOOLS FOR FUTURE RESEARCH

Maps

SPO1 genome

Figures 1 and 2 show restriction, genetic, and transcription maps of the whole SPO1 genome, while Fig. 3 through 6 show more detailed maps of the most intensely studied regions. Thirty-nine genes have been identified and mapped by means of conditional lethal mutations (24, 92). Another 11 have been identified and located by in vitro transcription and translation of specific restriction fragments from the terminal redundancy (100). Genes with related functions tend to be clustered, with genes 35 and 1 to 3 affecting virion assembly, genes 4 to 6 affecting head formation, genes 7 to 20 affecting tail formation, genes 21 to 23 and 27 to 32 affecting DNA synthesis, and genes 33 and 34 affecting regulation of gene action. Table 2 summarizes the information available about each of the known genes.

If the well-studied regions are representative, genes are packed tightly together in the SPO1 genome (Fig. 3 and 4). The average intergene distance, at the six places at which the intergene sequences are known, is 69 bp. In two cases, there is actually a short overlap between adjacent genes (16, 123), and three of the other four intergenes are occupied by promoters. The molecular weights of 53 structural proteins and of 11 proteins specified by in vitro expression of restriction fragments have been estimated by gel electrophoresis (97,100), permitting two very different estimates of the average size of an SPO1 gene (1.15 and 0.63 kb, respectively). Each estimate is undoubtedly biased against very small genes, whose products may not have been detected on the gels. However, the average of those figures (0.89 kb) may be used as a very crude estimate of the average size of all the other genes in the SPO1 genome. Assuming an average intergene distance of 69 bp, a total of 140 genes in the SPO1 genome is predicted.

About 60 SPO1-specific proteins have been identified by PAGE of extracts of infected cells. These are divided about equally among early, middle, and late gene products, on the basis of their dependence on gp28, gp33, and gp34 (62, 110). This is, of course, an underestimate of the number of active genes, since it includes only proteins that are clearly visible and resolvable on one-dimensional gels. For instance, the late proteins probably include at least all 53 of the proteins found in the mature virion (97). If the early and middle proteins were similarly underrepresented on the gels of infected-culture extracts, the total number of active genes would be more than 140.

Transcription mapping

The approximate sites at which early, middle, and late transcription takes place have been located on the restriction map by hybridizing RNA, pulse-labeled at various times, to Southern blots of *Eco*RI* digests of SPO1 DNA (102, 136). These data are summarized in Fig. 2. There is a tendency for the different categories of transcripts to be clustered, with the major early cluster in the terminal redundancy, the major middle

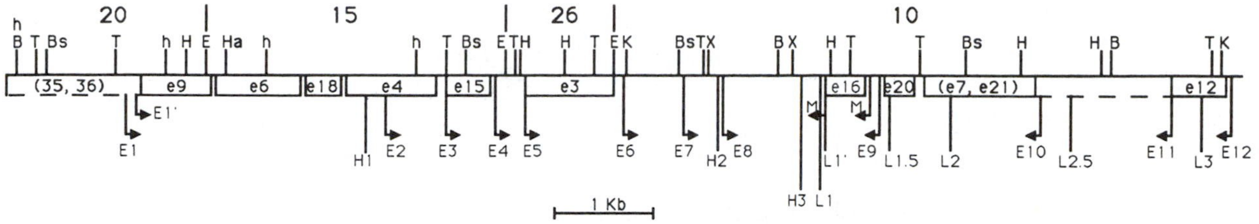

Figure 3. Map of the SPO1 terminal redundancy. Data for the restriction map are from references 8, 24, 100, 102, and 111 and from E. P. Geiduschek (43a). Small adjustments have been made to reconcile differences in positions reported by different groups. Restriction sites: B, *Bst*NI; Bs, *Bst*EII; E, *Eco*RI; H, *Hae*II; h, *Hpa*II; Ha, *Hae*III; K, *Kpn*I; T, *Taq*I; X, *Xba*I. The numbers of the *Eco*RI fragments are indicated above the line. Positions of early promoters (8, 111) are indicated by arrows labeled E1 through E12, corresponding to $P_{E}1$ through $P_{E}12$. Positions of two possible middle promoters are indicated by arrows labeled M. The leftmost of these is a middle promoter sequence (9) not known to be functional, and the rightmost is inferred from Perkus and Shub's (100) conclusion that gene e16 is expressed from a middle promoter. Its position is shown just to the right of that of e16, but it could also be farther to the right. (Some evidence suggests the presence of more than one middle promoter in this region [43a].) The direction of transcription driven by each promoter is indicated by an arrow. The positions of transcription terminators (8, 9) are indicated by straight lines descending from the map. Those designated H are heavy-strand terminators that terminate rightward transcription; the light-strand terminators, designated L, terminate leftward transcription. The efficient terminators in the central termination region are emphasized with longer lines. Transcription from each rightward promoter is terminated at the efficient terminator H3, and that from each leftward promoter is terminated at L1, with intermediate terminators functioning less efficiently or under particular conditions. The gene positions are from references 24 and 100. Placements are based on the best arguments available, but in some cases, those arguments are not conclusive, so placements should be considered approximations. Where there is no basis for determining the order of two genes, they are placed in parentheses. The length of the box for each gene represents the size of the gene calculated from the molecular weight of the gene product as estimated by gel electrophoresis (except that e4 has been shortened by about 18% to permit consistency with the restriction enzyme inactivation data without invoking overlapping genes). In the e9 to e15 and e16 to e20 regions, the indicated genes occupy nearly all the space available, assuming no overlapping. There is an indication of another open reading frame just downstream of $P_{E}4$ (133), so the space between e15 and e3 is probably occupied as well. In the e7-e21-e12 region, the dashed box represents space unaccounted for by known genes. Any of those three genes could be anyplace within that dashed region instead of at their indicated positions, with the only constraints being that e12 is to the right of e7 and e21 and that e21 cannot overlap a *Hae*II or *Bst*EII site. Markers in genes 35 and 36 are located in *Eco*RI* fragment 18 (24). That in gene 35 is in the terminal redundancy (22), and that in gene 36 behaves as if it is, too (46), so they are probably both in *Eco*RI*-20, which is the portion of fragment 18 that is in the redundant region. It is unlikely that either of these genes is e9, since gene 36 apparently is expressed from *Eco*RI*-18 (25), whereas e9 is not expressed from *Eco*RI*-20 (100), and since e9 is an early gene, whereas gene 35 is required for viral assembly (39) and is therefore probably a late gene. Therefore, genes 35 and 36 are placed to the left of e9. There is no basis for estimating the size of genes 35 and 36, so the dashed box merely indicates the region within which they are found.

cluster in the right half of the unique region, and the major late cluster in the left half of the unique region. The clustering is not absolute: smaller quantities of each category are scattered about the genome. These results from analyses of in vivo transcription are generally consistent with those from in vitro transcription of, or RNA polymerase binding to, separated restriction fragments (8, 12, 111, 137).

In several regions of the genome, individual transcription units have been mapped precisely. Most of this mapping has employed in vitro assays, although whenever they were available, in vivo assays have confirmed in vitro results (e.g., see references 9 and 111). These transcription units will be discussed individually below, in sections that analyze the most intensively studied regions of the SPO1 genome.

Terminal redundancy

All major early promoters are located in the terminal redundancy, as determined either by binding RNA polymerase to specific restriction fragments (137) or by the formation of ternary complexes of restriction fragments, RNA polymerase, and nascent RNA (111). Several weaker early promoters were detected outside the terminal redundancy by R-loop mapping and in vitro transcription (12, 111) and will be discussed below. No early promoters were detected in the left half of the unique region of the genome.

By electron microscopy of complexes of RNA polymerase and restriction fragments, 12 early promoters were mapped to specific sites within the terminal redundancy (111), and a 13th was detected by in vitro transcription (8). Transcription from the nine leftmost of these promoters proceeds rightward, and that from the four rightmost proceeds leftward. All transcripts terminate in a central termination region, which prevents transcription in one direction from overlapping that in the other. In addition, before reaching the central terminator, several of the transcripts reach other terminators, at which they may be terminated under certain conditions or with lesser efficiency (7–9). Figure 3 shows the positions of these promoters and terminators on the restriction map of the terminal redundancy. The genes specifying 11 of the early proteins have been mapped within the terminal redundancy by in vitro transcription and translation of restriction fragments (100), and the positions of these and the two previously known genes are also shown in Fig. 3. The transcripts necessary for synthesis of most of these proteins can be provided by one or more of the

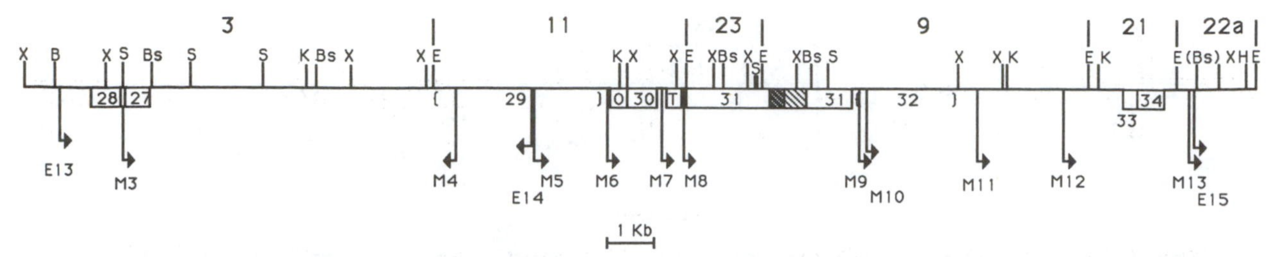

Figure 4. Map of gene 28 to 34 region. All of the known genes in this region are indicated by numbers immediately below the line (except that the TF1 gene is indicated by the letter T, and the unidentified open reading frame next to gene 30 is indicated by the letter O). Those genes enclosed in boxes have been completely sequenced (16–18, 50, 55, 121a, 123). Genes 29 and 32 have not been located precisely. Brackets enclose regions within which at least part of those genes must be located according to marker rescue activity associated with specific restriction fragments (21, 24). The shaded region between the two halves of gene 31 represents the intron. The more lightly shaded portion is the open reading frame within the intron. Restriction sites are indicated by letters above the line: B, *Bgl*II; Bs, *Bst*EII; E, *Eco*RI; H, *Hae*III; K, *Kpn*I; S, *Sca*I; X, *Xba*I. The numbers of the *Eco*RI fragments are given above the lines. The locations of restriction sites within the sequenced regions have been determined precisely from the sequence. Others have been determined by restriction mapping and are not as precise (12, 16, 24, 102, 133). The *Sca*I map is incomplete, as most of the unsequenced region has not been tested for *Sca*I sites. The *Sca*I site indicated within *Eco*RI fragment 23 is actually two sites 78 bp apart. Parentheses around one Bs indicate that the location of that *Bst*EII site is not precise, so its relationship to M13 and E15 is not known. The locations of all of the known promoters in the region are shown below the line, with directions of transcription indicated by arrows. M3 through M13 are middle promoters $P_{MII}3$ through $P_{MII}13$, and E13 through E15 are early promoters P_E13 through P_E15. Their locations were determined by Chelm et al. (12), with additional information about certain promoters coming from sequence information and from Scarlato et al. (122). All locations were determined with reference to the *Eco*RI* restriction map (102), using fragment sizes from reference 24.

13 promoters shown. However, the expression of e6, e18, and e4 from *Eco*RI* fragment 15 cannot be accounted for by known promoters, suggesting either that anomalous expression is occurring in the in vitro transcription-translation system or that there is at least one additional promoter not identified by the other techniques.

Many of these early promoters are among the strongest promoters known (111). It has been suggested (44) that the purpose of this concentration of strong promoters is to permit the SPO1 early genes to compete effectively with host genes for RNA polymerase. We do not know the roles of any of these early gene products, but they may include activities to shut off host functions, as discussed above.

Evidence for the presence of two middle promoters within the terminal redundancy has also been presented. The gene specifying protein e16 is expressed in vitro by RNA polymerase A or B, and its expression in vivo is enhanced by gp28 (100). The promoter from which this gp28-dependent transcription takes place has not been identified by other means, although a tentative location is indicated in Fig. 3. A middle promoter sequence has been found in the major termination region (9), but no gene product has been observed, and its position is not appropriate for e16.

Several of the transcription terminators have been sequenced (9). They show sequences similar to those of *E. coli* terminators, with a region of dyad symmetry followed by a stretch of hmUra residues. There are only slight differences between the sequences of the efficient terminators and those of the two less efficient ones that have been sequenced. One of the latter has a greater-than-usual distance between the two regions mentioned above. The other has a C within its stretch of hmUras. The inefficiency of the latter and of two others that have not been sequenced can be overcome by decreasing the ribonucleoside triphosphate concentration in the transcription mixture (7).

Gene 28 to 34 region

The gene 28 to 34 region is the most thoroughly studied region of the SPO1 genome, including all of the nine genes that have been completely sequenced. Among these are the five regulatory genes discussed

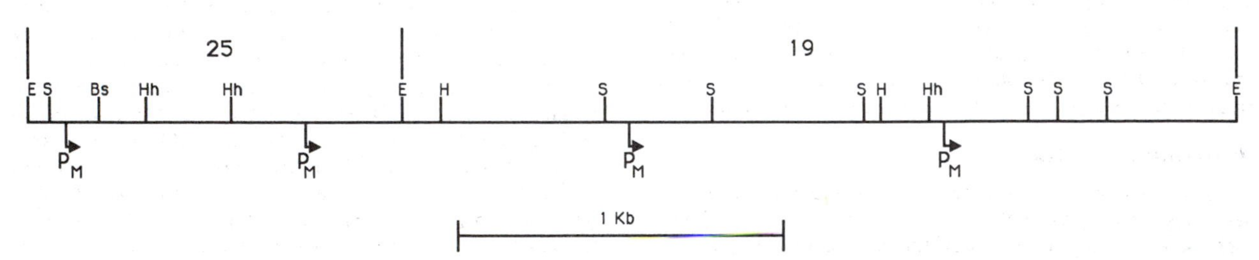

Figure 5. Map of a region of middle transcription. All information was taken from Lee and Pero (78). Restriction sites: Bs, *Bst*EII; E, *Eco*RI; H, *Hpa*II; Hh, *Hha*I; S, *Sau*3A. The numbers of the *Eco*RI fragments are given above the line. The four middle promoters are designated P_M. Arrows indicate directions of transcription driven by each promoter. No genes on these fragments have been identified.

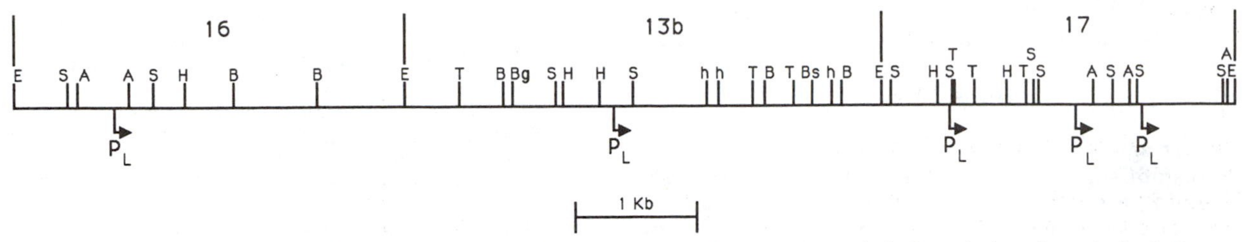

Figure 6. Map of a region of late transcription. All data are taken from Costanzo (19). Restriction sites: A, *Ava*I; B, *Bst*NI; Bs, *Bst*EII; Bg, *Bgl*II; E, *Eco*RI; H, *Hpa*II; h, *Hin*dIII; S, *Sau*3A; T, *Taq*I. The numbers of the *Eco*RI fragments are given above the line. Five late promoters are indicated by P_L, with arrows showing directions of transcription. Genes known from marker rescue experiments to be (at least partially) on these fragments include genes 13 and 14 on *Eco*RI fragment 16, genes 16 and 16.5 on fragment 13b, and gene 17 on fragment 17 (24). Since their locations are not known more precisely, the genes are not shown.

above (genes 27, 28, 33, 34, and TF1), two genes (genes 30 and 31) that are specifically required for SPO1 DNA replication, the open reading frame within the intron of gene 31, and another open reading frame whose function remains to be determined. Figure 4 shows the locations of these genes on the restriction map of this region as well as the 3 early and 11 middle promoters found there. (Figure 2 shows that this region is most active during the middle period. The three early promoters are all quite weak in comparison with those in the terminal redundancy.) The specific roles of several of the promoters can be inferred from their locations immediately before specific genes. Each of the early transcripts includes a region that would also be expected to be transcribed from a middle promoter, which could account for some of the genes that are transcribed at both early and middle times. For instance, gene 27 is downstream from both P_E13 and $P_{MII}3$, and there is no transcription terminator to stop either transcription from reaching gene 27 (12, 18). Hybridization competition shows that RNA transcribed in vitro from another early promoter, P_E15, is effectively competed by either early or middle in vivo RNAs (12).

Fragment 25 to 19 region

The fragment 25 to 19 region is also a region of middle gene activity. Figure 5 shows the locations of four middle promoters on its restriction map (78). None of the known genes is located in this region.

Fragment 16 to 17 region

The fragment 16 to 17 region is a region of late gene activity. Figure 6 shows the locations of five late promoters on its restriction map (19). There are also five known genes in the region, but there is no one-to-one correspondence of promoters with known genes, since at least four of the genes are located upstream of at least three of the promoters.

Cloning

Most of the SPO1 genome is available as fragments cloned in plasmids that replicate in *B. subtilis* (24). Although hmUra is inhibitory to certain restriction enzymes (64), most enzymes work well in its presence, as does DNA ligase. The thymine-containing cloned fragments are biologically functional and can both complement and recombine with superinfecting SPO1 mutants (23, 25), although the activity of at least some cloned middle genes is probably substantially lower than that of the same genes when present in the SPO1 genome (77, 119). Several SPO1 fragments are not clonable in either *B. subtilis* or *E. coli*, apparently because of the expression of deleterious genes on those fragments, since the active promoters on those fragments clone readily when the rest of the fragment has been either mutated or deleted (24, 133).

SPO1 can also serve as a cloning vehicle. Sayre and Geiduschek (119) showed that at least 5.6 kb of exogenous DNA could be inserted into the SPO1 genome by in vivo Campbell-mode integration of a plasmid carrying only 654 bp of homology to the SPO1 genome.

Recombination and Mutagenesis

Recombination occurs at high frequency during SPO1 infection. For instance, mixed infection with mutants F14 and F4, whose mutations are 359 bp apart in genes 33 and 34 (16), produced recombinants at a frequency of 5×10^{-3} (92). The efficiency is high enough that genetic information from small DNA fragments can be introduced into the SPO1 genome by in vivo recombination. Cloned fragments as small as 233 bp (123), with one end as little as 12 bp (123) or 16 bp (16) from the relevant marker, are effective in marker rescue experiments, thus permitting the use of such fragments for fine structure genetic mapping.

More importantly, the high efficiency of recombination permits mutations constructed in vitro to be introduced readily into the SPO1 genome. For instance, Sayre and Geiduschek (119) introduced site-specific mutations into the TF1 gene, cloned the mutant genes on a 654-bp fragment in *B. subtilis*, and permitted the cloned fragment to recombine with superinfecting SPO1. Mutants were identified among the progeny by plaque-filter hybridization at a frequency of 5×10^{-4}.

Recombination is essential for transfection by the DNA of hmUra phages. As shown primarily for SP82, transfecting DNA is susceptible to a host nuclease that causes sufficient damage that recombination between several genomes is necessary to form a single intact genome (53). Infection results in inhibition of this nuclease, which thus causes no problem for normal infection (70, 89). The effect of the nuclease on transfection can be diminished by the presence of UV-damaged DNA either exogenously added or generated by irradiation of recipient cells (34).

SUMMARY

SPO1 and its relatives have had important roles in the establishment of several biological principles, including the role of sigma cascades in gene regulation, the occurrence of introns in gram-positive microorganisms, and the functions of type II DNA-binding proteins. They also offer a fertile field for further investigation, with a variety of interesting phenomena observed but not elucidated.

Acknowledgments. I am grateful to E. P. Geiduschek, P. Hoet, V. Scarlato, and D. Shub for providing information prior to publication and to E. P. Geiduschek for a critical reading of the manuscript. Unpublished work from this laboratory discussed herein was supported by research grant DMB-8801703 from the National Science Foundation. Certain segments of this chapter have been adapted, with the permission of Plenum Press, from a previous article (132).

REFERENCES

1. **Alegria, A. H., F. M. Kahan, and J. Marmur.** 1968. A new assay for phage hydroxymethylases and its use in *Bacillus subtilis* transfection. *Biochemistry* **7:**3179–3186.
2. **Alonso, J. C., A. N. Sarachu, and O. Grau.** 1981. DNA gyrase inhibitors block development of *Bacillus subtilis* bacteriophage SPO1. *J. Virol.* **39:**855–860.
3. **Aposhian, H. V.** 1965. A dTMPase found after infection of *Bacillus subtilis* with phage SP5C. *Biochem. Biophys. Res. Commun.* **18:**230–235.
4. **Aposhian, H. V., and G. Y. Tremblay.** 1966. Deoxythymidylate 5′-nucleotidase. Purification and properties of an enzyme found after infection of *B. subtilis* with phage SP5C*. *J. Biol. Chem.* **241:**5095–5101.
5. **Arwert, F., and G. Venema.** 1974. Transfection of *Bacillus subtilis* with bacteriophage H1 DNA: fate of transfecting DNA and transfection enhancement in *B. subtilis* uvr$^+$ and uvr$^-$ strains. *Mol. Gen. Genet.* **128:**55–72.
6. **Belfort, M.** 1991. Self-splicing introns in prokaryotes: migrant fossils? *Cell* **64:**9–11.
7. **Brennan, S. M.** 1984. Ribonucleoside triphosphate concentration-dependent termination of bacteriophage SPO1 transcription *in vitro* by *B. subtilis* RNA polymerase. *Virology* **135:**555–560.
8. **Brennan, S. M., B. K. Chelm, J. M. Romeo, and E. P. Geiduschek.** 1981. A transcriptional map of the bacteriophage SPO1 genome. II. The major early transcription units. *Virology* **111:**604–628.
9. **Brennan, S. M., and E. P. Geiduschek.** 1983. Regions specifying transcriptional termination and pausing in the bacteriophage SPO1 terminal repeat. *Nucleic Acids Res.* **11:**4157–4175.
10. **Brown, N. C.** 1970. 6-(*p*-Hydroxyphenylazo)-uracil: a selective inhibitor of host DNA replication in phage-infected *Bacillus subtilis*. *Proc. Natl. Acad. Sci. USA* **67:**1454–1461.
11. **Chelm, B. K., J. J. Duffy, and E. P. Geiduschek.** 1982. Interaction of *Bacillus subtilis* RNA polymerase core with two specificity-determining subunits. *J. Biol. Chem.* **257:**6501–6508.
12. **Chelm, B. K., J. M. Romeo, S. M. Brennan, and E. P. Geiduschek.** 1981. A transcriptional map of the bacteriophage SPO1 genome. III. A region of early and middle promoters (the gene 28 region). *Virology* **112:** 572–588.
13. **Chiu, C.-S., P. K. Tomich, and G. R. Greenberg.** 1976. Simultaneous initiation of synthesis of bacteriophage T4 DNA and of deoxyribonucleotides. *Proc. Natl. Acad. Sci. USA* **73:**757–761.
14. **Choy, H. A., J. M. Romeo, and E. P. Geiduschek.** 1986. Activity of a phage-modified RNA polymerase at hybrid promoters: effects of substituting thymine for hydroxymethyluracil in a phage SPO1 middle promoter. *J. Mol. Biol.* **191:**59–73.
15. **Coene, M., P. Hoet, and C. Cocito.** 1983. Physical map of phage 2C DNA: evidence for the existence of large redundant ends. *Eur. J. Biochem.* **132:**69–75.
16. **Costanzo, M., L. Brzustowicz, N. Hannett, and J. Pero.** 1984. Bacteriophage SPO1 genes 33 and 34. Location and primary structure of genes encoding regulatory subunits of *Bacillus subtilis* RNA polymerase. *J. Mol. Biol.* **180:**533–547.
17. **Costanzo, M., N. Hannett, L. Brzustowicz, and J. Pero.** 1983. Bacteriophage SPO1 gene 27: location and nucleotide sequence. *J. Virol.* **48:**555–560.
18. **Costanzo, M., and J. Pero.** 1983. Structure of a *Bacillus subtilis* bacteriophage SPO1 gene encoding a RNA polymerase σ factor. *Proc. Natl. Acad. Sci. USA* **80:**1236–1240.
19. **Costanzo, M. C.** 1983. Bacteriophage SPO1 regulatory genes. Ph.D. thesis. Harvard University, Cambridge, Mass.
20. **Cregg, J. M., and C. R. Stewart.** 1977. Timing of initiation of DNA replication in SPO1 infection of *Bacillus subtilis*. *Virology* **80:**289–296.
21. **Cregg, J. M., and C. R. Stewart.** 1978. EcoRI cleavage of DNA from *Bacillus subtilis* phage SPO1. *Virology* **85:** 601–605.
22. **Cregg, J. M., and C. R. Stewart.** 1978. Terminal redundancy of "high frequency of recombination" markers of *Bacillus subtilis* phage SPO1. *Virology* **86:**530–541.
23. **Curran, J. F., and C. R. Stewart.** 1982. Recombination and expression of a cloned fragment of the DNA of *Bacillus subtilis* bacteriophage SPO1. *Virology* **120:**307–317.
24. **Curran, J. F., and C. R. Stewart.** 1985. Cloning and mapping of the SPO1 genome. *Virology* **142:**78–97.
25. **Curran, J. F., and C. R. Stewart.** 1985. Transcription of *Bacillus subtilis* plasmid pBD64 and expression of bacteriophage SPO1 genes cloned therein. *Virology* **142:**98–111.
26. **Davison, P. F.** 1963. The structure of bacteriophage SP8. *Virology* **21:**146–151.
27. **Davison, P. F., D. Freifelder, and B. W. Holloway.** 1964. Interruptions in the polynucleotide strands in bacteriophage DNA. *J. Mol. Biol.* **8:**1–10.
28. **DeAntoni, G. L., N. E. Besso, G. E. Zanassi, A. N. Sarachu, and O. Grau.** 1985. Bacteriophage SPO1 DNA polymerase and the activity of viral gene 31. *Virology* **143:**16–22.
29. **Downard, J. S., and H. R. Whiteley.** 1981. Early RNAs in SP82- and SPO1-infected *Bacillus subtilis* may be processed. *J. Virol.* **37:**1075–1078.
30. **Duffy, J. J., and E. P. Geiduschek.** 1977. Purification of a positive regulatory subunit from phage SPO1-modified RNA polymerase. *Nature* (London) **270:**28–32.
31. **Duffy, J. J., R. L. Petrusek, and E. P. Geiduschek.** 1975. Conversion of *Bacillus subtilis* RNA polymerase activity *in vitro* by a protein induced by phage SPO1. *Proc. Natl. Acad. Sci. USA* **72:**2366–2370.
32. **Dunham, L. T., and A. R. Price.** 1974. Deoxythymidine triphosphate-deoxyuridine triphosphate nucleotidohydrolase induced by *Bacillus subtilis* bacteriophage ϕe. *Biochemistry* **13:**2667–2672.
33. **Dunham, L. T., and A. R. Price.** 1974. Mutants of *Bacillus subtilis* bacteriophage ϕe defective in dTTP-dUTP nucleotidohydrolase. *J. Virol.* **14:**709–712.

33a. **Eiserling, F.** Personal communication.

34. **Epstein, H. T., and I. Mahler.** 1968. Mechanisms of enhancement of SP82 transfection. *J. Virol.* **2:**710–715.
35. **Fox, T. D.** 1976. Identification of phage SPO1 proteins coded by regulatory genes 33 and 34. *Nature* (London) **262:**748–753.
36. **Fox, T. D., R. Losick, and J. Pero.** 1976. Regulatory gene 28 of bacteriophage SPO1 codes for a phage-induced subunit of RNA polymerase. *J. Mol. Biol.* **101:**427–433.
37. **Fox, T. D., and J. Pero.** 1974. New phage SPO1-induced polypeptides associated with *Bacillus subtilis* RNA polymerase. *Proc. Natl. Acad. Sci. USA* **71:**2761–2765.
38. **Friedman, D. I.** 1988. Integration host factor: a protein for all reasons. *Cell* **55:**545–554.
39. **Fujita, D. J.** 1971. Studies on conditional lethal mutants of bacteriophage SPO1. Ph.D. thesis. University of Chicago, Chicago, Ill.
40. **Fujita, D. J., B. M. Ohlsson-Wilhelm, and E. P. Geiduschek.** 1971. Transcription during bacteriophage SPO1 development: mutations affecting the program of viral transcription. *J. Mol. Biol.* **57:**301–317.
41. **Gage, L. P., and D. J. Fujita.** 1969. Effect of nalidixic acid on deoxyribonucleic acid synthesis in bacteriophage SPO1-infected *Bacillus subtilis. J. Bacteriol.* **98:** 96–103.
42. **Gage, L. P., and E. P. Geiduschek.** 1971. RNA synthesis during bacteriophage SPO1 development: six classes of SPO1 RNA. *J. Mol. Biol.* **57:**279–300.
43. **Gage, L. P., and E. P. Geiduschek.** 1971. RNA synthesis during bacteriophage SPO1 development. II. Some modulations and prerequisites of the transcription program. *Virology* **44:**200–210.
43a. **Geiduschek, E. P.** Personal communication.
44. **Geiduschek, E. P., and J. Ito.** 1982. Regulatory mechanisms in the development of lytic bacteriophages in *Bacillus subtilis*, p. 203–245. *In* D. A. Dubnau (ed.), *The Molecular Biology of the Bacilli*, vol. 1. Academic Press, Inc., New York.
45. **Geiduschek, E. P., G. J. Schneider, and M. H. Sayre.** 1990. TF1, a bacteriophage-specific DNA-binding and DNA-bending protein. *J. Struct. Biol.* **104:**84–90.
46. **Glassberg, J., M. Franck, and C. R. Stewart.** 1977. Initiation and termination mutants of *Bacillus subtilis* bacteriophage SPO1. *J. Virol.* **21:**147–152.
47. **Glassberg, J., M. Franck, and C. R. Stewart.** 1977. Multiple origins of replication for *Bacillus subtilis* phage SPO1. *Virology* **78:**433–441.
48. **Glassberg, J., R. A. Slomiany, and C. R. Stewart.** 1977. Selective screening procedure for the isolation of heat- and cold-sensitive, DNA replication-deficient mutants of bacteriophage SPO1 and preliminary characterization of the mutants isolated. *J. Virol.* **21:**54–60.
49. **Goodrich, H. A., J. M. Gott, M.-Q. Xu, V. Scarlato, and D. A. Shub.** 1989. A group I intron in *Bacillus subtilis* bacteriophage SPO1, p. 59–66. *In* T. R. Cech (ed.), *Molecular Biology of RNA*. Alan R. Liss, Inc., New York.
50. **Goodrich-Blair, H., V. Scarlato, J. M. Gott, M.-Q. Xu, and D. A. Shub.** 1990. A self-splicing group I intron in the DNA polymerase gene of *Bacillus subtilis* bacteriophage SPO1. *Cell* **63:**417–424.
51. **Goodrich-Blair, H., and D. A. Shub.** 1991. Introns in HMU bacteriophage: intron structure and open reading frame function. *J. Cell Biochem.* **15**(Suppl. D):66.
52. **Green, D. M.** 1964. Infectivity of DNA isolated from *Bacillus subtilis* bacteriophage SP82. *J. Mol. Biol.* **10:** 438–451.
53. **Green, D. M.** 1966. Intracellular inactivation of infective SP82 bacteriophage DNA. *J. Mol. Biol.* **22:**1–13.
54. **Green, D. M., and D. Laman.** 1972. Organization of gene function in *Bacillus subtilis* bacteriophage SP82G. *J. Virol.* **9:**1033–1046.
55. **Greene, J. R., S. M. Brennan, D. J. Andrew, C. C. Thompson, S. H. Richards, R. L. Heinrikson, and E. P. Geiduschek.** 1984. Sequence of bacteriophage SPO1 gene coding for transcription factor 1, a viral homologue of the bacterial type II DNA-binding proteins. *Proc. Natl. Acad. Sci. USA* **81:**7031–7035.
56. **Greene, J. R., B. K. Chelm, and E. P. Geiduschek.** 1982. SPO1 gene 27 is required for viral late transcription. *J. Virol.* **41:**715–720.
57. **Greene, J. R., and E. P. Geiduschek.** 1985. Site-specific DNA binding by the bacteriophage SPO1-encoded type II DNA binding protein. *EMBO J.* **4:**1345–1349.
58. **Greene, J. R., and E. P. Geiduschek.** 1985. Interaction of a virus-coded type II DNA binding protein, p. 255–269. *In* R. Calendar (ed.), *Sequence Specificity in Transcription and Translation*. Alan R. Liss, Inc., New York.
59. **Greene, J. R., L. M. Morrissey, L. M. Foster, and E. P. Geiduschek.** 1986. DNA binding by the bacteriophage SPO1-encoded type II DNA-binding protein, TF1: formation of nested complexes at a selective binding site. *J. Biol. Chem.* **261:**12820–12827.
60. **Gribskov, M., and R. R. Burgess.** 1986. Sigma proteins from *E. coli, B. subtilis*, phage SPO1 and T4 are homologous proteins. *Nucleic Acids Res.* **14:**6745–6763.
61. **Haslam, E. A., D. H. Roscoe, and R. G. Tucker.** 1967. Inhibition of thymidylate synthetase in bacteriophage-infected *Bacillus subtilis. Biochim. Biophys. Acta* **134:** 312–326.
62. **Heintz, N., and D. A. Shub.** 1982. Transcriptional regulation of bacteriophage SPO1 protein synthesis in vivo and in vitro. *J. Virol.* **42:**951–962.
63. **Hoet, P., M. Coene, and C. Cocito.** 1983. Comparison of the physical maps and redundant ends of the chromosomes of phages 2C, SPO1, SP82 and ϕe. *Eur. J. Biochem.* **132:**63–67.
64. **Ito, J., F. Kawamura, and J. J. Duffy.** 1975. Susceptibility of non-thymine-containing DNA to bacterial restriction nucleases. *FEBS Lett.* **55:**278–281.
65. **Johnson, G. G., and E. P. Geiduschek.** 1972. Purification of the bacteriophage SPO1 transcription factor 1. *J. Biol. Chem.* **247:**3571–3578.
66. **Johnson, G. G., and E. P. Geiduschek.** 1977. Specificity of the weak binding between the phage SPO1 transcription-inhibitory protein, TF1, and SPO1 DNA. *Biochemistry* **16:**1473–1485.
67. **Kahan, E.** 1966. A genetic study of temperature-sensitive mutants of the *subtilis* phage SP82. *Virology* **30:** 650–660.
68. **Kahan, E.** 1971. Early and late gene function in bacteriophage SP82. *Virology* **46:**634–637.
69. **Kallen, R. G., M. Simon, and J. Marmur.** 1962. The occurrence of a new pyrimidine base replacing thymine in a bacteriophage DNA: 5-hydroxymethyl uracil. *J. Mol. Biol.* **5:**248–250.
70. **King, J. J., and D. M. Green.** 1977. Inhibition of nuclease activity in *Bacillus subtilis* following infection with bacteriophage SP82G. *Biochem. Biophys. Res. Commun.* **74:**492–498.
71. **Kornberg, A., and T. A. Baker.** 1992. *DNA Replication*, 2nd ed. W. H. Freeman & Co., San Francisco.
72. **Kunitani, M. G., and D. V. Santi.** 1980. On the mechanism of 2′-deoxyuridylate hydroxymethylase. *Biochemistry* **19:**1271–1275.
73. **Lannoy, N. N., P. P. Hoet, and C. G. Cocito.** 1985. Cloning of DNA segments of phage 2C, which allows autonomous plasmid replication in *B. subtilis. Eur. J. Biochem.* **152:**137–142.
74. **Lavi, U., A. Nattenberg, A. Ronen, and M. Marcus.** 1974. *Bacillus subtilis* DNA polymerase III is required for the replication of virulent bacteriophage ϕe. *J. Virol.* **14:**1337–1342.
75. **Lawrie, J. M., J. S. Downard, and H. R. Whiteley.** 1978. *Bacillus subtilis* bacteriophages SP82, SPO1 and ϕe: a comparison of DNAs and of peptides synthesized during infection. *J. Virol.* **27:**725–737.

76. **Lawrie, J. M., and H. R. Whiteley.** 1977. A physical map of bacteriophage SP82 DNA. *Gene* **2:**233–250.
77. **Lee, G., N. M. Hannett, A. Korman, and J. Pero.** 1980. Transcription of cloned DNA from *Bacillus subtilis* phage SPO1. Requirement for hydroxymethyluracil-containing DNA by phage-modified RNA polymerase. *J. Mol. Biol.* **139:**407–422.
78. **Lee, G., and J. Pero.** 1981. Conserved nucleotide sequences in temporally controlled bacteriophage promoters. *J. Mol. Biol.* **152:**247–265.
79. **Lee, G., C. Talkington, and J. Pero.** 1980. Nucleotide sequence of a promoter recognized by *Bacillus subtilis* RNA polymerase. *Mol. Gen. Genet.* **180:**57–65.
80. **Levinthal, C., J. Hosoda, and D. Shub.** 1967. The control of protein synthesis after phage infection, p. 71–87. *In* J. S. Colter and W. Paranchych (ed.), *The Molecular Biology of Viruses*. Academic Press, Inc., New York.
81. **Levner, M. H.** 1972. Replication of viral DNA in SPO1-infected *Bacillus subtilis*. II. DNA maturation during abortive infection. *Virology* **48:**417–429.
82. **Levner, M. H.** 1972. Eclipse of viral DNA infectivity in SPO1-infected *Bacillus subtilis*. *Virology* **50:**267–272.
83. **Levner, M. H., and N. R. Cozzarelli.** 1972. Replication of viral DNA in SPO1-infected *Bacillus subtilis*. I. Replicative intermediates. *Virology* **48:**402–416.
84. **Liljemark, W. F., and D. L. Anderson.** 1970. Structure of *Bacillus subtilis* bacteriophage ϕ25 and ϕ25 deoxyribonucleic acid. *J. Virol.* **6:**107–113.
85. **Losick, R., and J. Pero.** 1981. Cascades of sigma factors. *Cell* **25:**582–584.
86. **Marcus, M., and M. C. Newlon.** 1971. Control of DNA synthesis in *Bacillus subtilis* by ϕe. *Virology* **44:**83–93.
87. **Marmur, J., and C. M. Greenspan.** 1963. Transcription *in vivo* of DNA from bacteriophage SP8. *Science* **142:** 387–389.
88. **Marmur, J., C. M. Greenspan, E. Palecek, F. M. Kahan, J. Levine, and M. Mandel.** 1963. Specificity of the complementary RNA formed by *Bacillus subtilis* infected with bacteriophage SP8. *Cold Spring Harbor Symp. Quant. Biol.* **28:**191–199.
89. **McAllister, W. T., and D. M. Green.** 1972. Bacteriophage SP82G inhibition of intracellular deoxyribonucleic acid inactivation process in *Bacillus subtilis*. *J. Virol.* **10:**51–59.
90. **Moran, C. P.** 1989. Sigma factors and the regulation of transcription, p. 167–184. *In* I. Smith, R. A. Slepecky, and P. Setlow (ed.) *Regulation of Procaryotic Development*. American Society for Microbiology, Washington, D.C.
91. **Okubo, S., B. Strauss, and M. Stodolsky.** 1964. The possible role of recombination in the infection of competent *Bacillus subtilis* by bacteriophage deoxyribonucleic acid. *Virology* **24:**552–562.
92. **Okubo, S., T. Yanagida, D. J. Fujita, and B. M. Ohlsson-Wilhelm.** 1972. The genetics of bacteriophage SPO1. *Biken J.* **15:**81–97.
93. **Panganiban, A. T., and H. R. Whiteley.** 1981. Analysis of bacteriophage SP82 major "early" in vitro transcripts. *J. Virol.* **37:**372–382.
94. **Panganiban, A. T., and H. R. Whiteley.** 1983. Purification and properties of a new *Bacillus subtilis* RNA processing enzyme. *J. Biol. Chem.* **258:**12487–12493.
95. **Panganiban, A. T., and H. R. Whiteley.** 1983. *Bacillus subtilis* RNAase III cleavage sites in phage SP82 early mRNA. *Cell* **33:**907–913.
96. **Parker, M. L., and F. A. Eiserling.** 1983. Bacteriophage SPO1 structure and morphogenesis. I. Tail structure and length regulation. *J. Virol.* **46:**239–249.
97. **Parker, M. L., and F. A. Eiserling.** 1983. Bacteriophage SPO1 structure and morphogenesis. III. SPO1 proteins and synthesis. *J. Virol.* **46:**260–269.
98. **Parker, M. L., E. J. Ralston, and F. A. Eiserling.** 1983. Bacteriophage SPO1 structure and morphogenesis. II. Head structure and DNA size. *J. Virol.* **46:**250–259.
99. **Pene, J., and J. Marmur.** 1964. Infectious DNA from a virulent bacteriophage active on transformable *Bacillus subtilis*. *Fed. Proc.* **23:**318.
100. **Perkus, M. E., and D. A. Shub.** 1985. Mapping the genes in the terminal redundancy of bacteriophage SPO1 with restriction endonucleases. *J. Virol.* **56:**40–48.
101. **Pero, J.** 1983. A procaryotic model for the developmental control of gene expression, p. 227–233. *In* S. Subtelny and F. C. Kafatos (ed.), *Gene Structure and Regulation in Development*. Alan R. Liss, Inc., New York.
102. **Pero, J., N. M. Hannett, and C. Talkington.** 1979. Restriction cleavage map of SPO1 DNA: general location of early, middle, and late genes. *J. Virol.* **31:**156–171.
103. **Pero, J., J. Nelson, and T. D. Fox.** 1975. Highly asymmetric transcription by RNA polymerase containing phage SPO1-induced polypeptides and a new host protein. *Proc. Natl. Acad. Sci. USA* **72:**1589–1593.
104. **Pero, J., R. Tjian, J. Nelson, and R. Losick.** 1975. *In vitro* transcription of a late class of phage SPO1 genes. *Nature* (London) **257:**248–251.
105. **Pettijohn, D. E.** 1988. Histone-like proteins and bacterial chromosome structure. *J. Biol. Chem.* **263:**12793–12796.
106. **Piggot, P. J., and J. A. Hoch.** 1985. Revised genetic linkage map of *Bacillus subtilis*. *Microbiol. Rev.* **49:**158–179.
107. **Price, A. R., L. F. Dunham, and R. L. Walker.** 1972. Thymidine triphosphate nucleotidohydrolase and deoxyuridylate hydroxymethylase induced by mutants of *Bacillus subtilis* bacteriophage SP82G. *J. Virol.* **10:**1240–1241.
108. **Rabussay, D., and E. P. Geiduschek.** 1977. Regulation of gene action in the development of lytic bacteriophages, p. 1–196. *In* H. Fraenkel-Conrat and R. R. Wagner (ed.), *Comprehensive Virology* 8. Plenum Press, New York.
109. **Ratner, D.** 1974. The interaction of bacterial and phage proteins with immobilized *Escherichia coli* RNA polymerase. *J. Mol. Biol.* **88:**373.
110. **Reeve, J. N., G. Mertens, and E. Amann.** 1978. Early development of bacteriophages SPO1 and SP82G in minicells of *Bacillus subtilis*. *J. Mol. Biol.* **120:**183–207.
111. **Romeo, J. M., S. M. Brennan, B. K. Chelm, and E. P. Geiduschek.** 1981. A transcriptional map of the bacteriophage SPO1 genome. I. The major early promoters. *Virology* **111:**588–603.
112. **Romeo, J. H., J. R. Greene, S. H. Richards, and E. P. Geiduschek.** 1986. The phage SPO1-specific RNA polymerase, E. gp28, recognizes its cognate promoters in thymine-containing DNA. *Virology* **153:**46–52.
113. **Romig, W. R., and A. M. Brodetsky.** 1961. Isolation and preliminary characterization of bacteriophages for *Bacillus subtilis*. *J. Bacteriol.* **82:**135–141.
114. **Roscoe, D. H.** 1969. Synthesis of DNA in phage-infected *B. subtilis*. *Virology* **38:**527–537.
115. **Roscoe, D. H.** 1969. Thymidine triphosphate nucleotidohydrolase: a phage-induced enzyme in *Bacillus subtilis*. *Virology* **38:**520–526.
116. **Roscoe, D. H., and R. G. Tucker.** 1964. The biosynthesis of a pyrimidine replacing thymine in bacteriophage DNA. *Biochem. Biophys. Res. Commun.* **16:**106–110.
117. **Roscoe, D. H., and R. G. Tucker.** 1966. The biosynthesis of 5-hydroxymethyl-deoxyuridylic acid in bacteriophage-infected *Bacillus subtilis*. *Virology* **29:**157–166.
118. **Sarachu, A. N., J. C. Alonso, and O. Grau.** 1980. Novobiocin blocks the shutoff of SPO1 early transcription. *Virology* **105:**13–18.
119. **Sayre, M. H., and E. P. Geiduschek.** 1988. TF1, the bacteriophage SPO1-encoded type II DNA-binding pro-

tein, is essential for viral multiplication. *J. Virol.* **62:** 3455–3462.

120. **Sayre, M. H., and E. P. Geiduschek.** 1990. Construction and properties of a temperature-sensitive mutation in the gene for the bacteriophage SPO1 DNA-binding protein TF1. *J. Bacteriol.* **172:**4672–4681.
121. **Sayre, M. H., and E. P. Geiduschek.** 1990. Effects of mutations at amino acid 61 in the arm of TF1 on its DNA-binding properties. *J. Mol. Biol.* **216:**819–833.

121a. **Scarlato, V.** Personal communication.

122. **Scarlato, V., J. R. Greene, and E. P. Geiduschek.** 1991. Bacteriophage SPO1 middle transcripts. *Virology* **180:** 716–728.
123. **Scarlato, V., and M. H. Sayre.** 1992. The structure of the bacteriophage SPO1 gene 30. *Gene* **114:**115–119.
124. **Schneider, G. J., and E. P. Geiduschek.** 1990. Stoichiometry of DNA binding by the bacteriophage SPO1-encoded type II DNA-binding protein TF1. *J. Biol. Chem.* **265:**10198–10200.
125. **Schneider, G. J., M. H. Sayre, and E. P. Geiduschek.** 1991. The DNA-bending properties of TF1. *J. Mol. Biol.* **221:**777–794.
126. **Shub, D. A.** 1966. Functional stability of messenger RNA during bacteriophage development. Ph.D. thesis. Massachusetts Institute of Technology, Cambridge.
127. **Shub, D. A.** 1975. Nature of the suppressor of *Bacillus subtilis* HA101B. *J. Bacteriol.* **122:**788–790.
128. **Sinha, N. K., C. F. Morris, and B. M. Alberts.** 1980. Efficient *in vitro* replication of double-stranded DNA templates by purified T4 bacteriophage replication system. *J. Biol. Chem.* **255:**4290–4303.
129. **Snustad, D. P., L. Snyder, and E. Kutter.** 1983. Effects on host genome structure and expression, p. 40–55. *In* C. K. Mathews, E. M. Kutter, G. Mosig, and P. B. Berget (ed.), *Bacteriophage T4*. American Society for Microbiology, Washington, D.C.
130. **Spiegelman, G. B., and H. R. Whiteley.** 1974. *In vivo* and *in vitro* transcription by ribonucleic acid polymerase from SP82-infected *Bacillus subtilis*. *J. Biol. Chem.* **249:**1483–1489.
131. **Stewart, C. R.** 1984. Dissection of HA20, a double mutant of bacteriophage SPO1. *J. Virol.* **49:**300–301.
132. **Stewart, C. R.** 1988. Bacteriophage SPO1, p. 477–515. *In* R. Calendar (ed.), *The Bacteriophages*, vol. I. Plenum Press, New York.
133. **Stewart, C. R.** Unpublished data.
134. **Stewart, C. R., and M. Franck.** 1981. Predominance of bacteriophage SP82 over bacteriophage SPO1 in mixed infections of *Bacillus subtilis*. *J. Virol.* **38:**1081–1083.
135. **Sugino, A., and K. F. Bott.** 1980. *Bacillus subtilis* deoxyribonucleic acid gyrase. *J. Bacteriol.* **141:**1331–1339.
136. **Talkington, C., and J. Pero.** 1977. Restriction fragment analysis of temporal program of bacteriophage SPO1 transcription and its control by phage-modified RNA polymerases. *Virology* **83:**365–379.
137. **Talkington, C., and J. Pero.** 1978. Promoter recognition by phage SPO1-modified RNA polymerase. *Proc. Natl. Acad. Sci. USA* **75:**1185–1189.
138. **Talkington, C., and J. Pero.** 1979. Distinctive nucleotide sequences of promoters recognized by RNA polymerase containing a phage-coded "σ-like" protein. *Proc. Natl. Acad. Sci. USA* **76:**5465–5469.
139. **Tjian, R., and J. Pero.** 1976. Bacteriophage SPO1 regulatory proteins directing late gene transcription *in vitro*. *Nature* (London) **262:**753–757.
140. **Tomich, P. K., C.-S. Chiu, M. G. Wovcha, and G. R. Greenberg.** 1974. Evidence for a complex regulating the *in vivo* activation of early enzymes induced by bacteriophage T4. *J. Biol. Chem.* **249:**7613–7622.
141. **Truffaut, N., B. Revet, and M. O. Soulie.** 1970. Etude comparative des DNA de phages 2C, SP8*, SP82, φe, SPO1 et SP50. *J. Biochem.* **15:**391–400.
142. **Watson, J. D.** 1972. Origin of concatemeric T7 DNA. *Nature New Biol.* **239:**197–201.
143. **Webb, V. B., and G. B. Spiegelman.** 1984. Ribosomal RNA synthesis in uninfected and SPO1am 34 infected *B. subtilis*. *Mol. Gen. Genet.* **194:**98–104.
144. **Wilson, D. L., and L. P. Gage.** 1971. Certain aspects of SPO1 development. *J. Mol. Biol.* **57:**297–300.
145. **Wilson, D. L., and E. P. Geiduschek.** 1969. A template-selective inhibitor of *in vitro* transcription. *Proc. Natl. Acad. Sci. USA* **62:**514–520.
146. **Wu, R., and E. P. Geiduschek.** 1975. The role of replication proteins in the regulation of bacteriophage T4 transcription. *J. Mol. Biol.* **96:**513–538.
147. **Xu, M.-Q., S. D. Kathe, H. Goodrich-Blair, S. A. Nierzwicki-Bauer, and D. A. Shub.** 1990. Bacterial origin of a chloroplast intron: conserved self-splicing group I introns in cyanobacteria. *Science* **250:**1566–1570.
148. **Yehle, C. O., and A. T. Ganesan.** 1972. Deoxyribonucleic acid synthesis in bacteriophage SPO1-infected *Bacillus subtilis*. I. Bacteriophage deoxyribonucleic acid synthesis and fate of host deoxyribonucleic acid in normal and polymerase-deficient strains. *J. Virol.* **9:**263–272.

57. Temperate Bacteriophages

STANLEY A. ZAHLER

Reviews of *Bacillus subtilis* phages (59, 91, 97, 100, 112, 148, 149) and in particular of temperate *B. subtilis* phages (97, 148, 149) are available and useful. Use them to learn about the historical background that is mostly ignored in this chapter. Chapter 44 in this book describes the uses of various *B. subtilis* temperate phages for gene cloning.

B. subtilis is susceptible to a variety of bacteriophages, both virulent (lytic) and temperate (capable of forming lysogens). On the whole, the phages seem to be less exotic than those that infect *Escherichia coli*. Several *B. subtilis* virulent phages contain unusual bases in their DNAs (for reviews, see references 59 and 112), but there are no reports of RNA phages, single-stranded DNA phages, or transposonlike phages in *B. subtilis*. All of the temperate phages reported for *B. subtilis* contain linear double-stranded DNA, and all of the ones that have been studied sufficiently insert that DNA into the chromosome of the host at a particular favored site, the prophage attachment site (*att*). No case of a plasmidlike prophage genome (as occurs in the *E. coli* phage P1) is known in *B. subtilis*. However, the physical state of the prophage of phage SP16 has not been studied.

The adsorption of phages to *B. subtilis* often occurs at the glucosylated teichoic acids that make up much of the cell surface. Reference 38a is a useful introduction to the literature on that subject.

The known temperate phages of *B. subtilis* have been classified into four groups, I to IV, originally on the basis of serology, immunity, host range, and adsorption site (25, 135). More recently (149), group III has been expanded to include new phages related to other members of the group by DNA homology and physical structure, despite differences in the other characteristics. Table 1 outlines the physical and chemical characteristics of a typical member of each of the four groups.

Many, perhaps all, strains of *B. subtilis* and related *Bacillus* spp. when isolated from nature carry a defective bacteriophage genome. The one carried by *B. subtilis* 168 is called PBSX; it is a typical representative of group V temperate phages. When induced to lyse by exposure to DNA-damaging treatments, these bacilli release large numbers of phage particles (PBSX, in the case of strain 168) that contain in their heads DNA from the bacterial chromosome rather than phage DNA. The phage particles (or their tails alone) can adsorb to other strains of *B. subtilis* or related species of *Bacillus*, which they kill without injecting their DNA; the virions or their tails alone act as bacteriocins. Physical characteristics of group V phages are also given in Table 1.

B. subtilis 168 and most of its derivatives are lysogenic for a group III phage, SPβ, as well as for the defective group V phage PBSX. Yasbin et al. (140) have constructed useful bacterial strains that have lost the SPβ prophage and are unable to express PBSX.

B. subtilis strains lysogenic for ϕ105 (group I) or SPO2 (group II) are poorly transformable (86, 142–145), apparently because the regimens usually used to bring about competence cause induction of the prophages and lysis of the cells. *B. subtilis* lysogens for ϕ105 or SPO2 that carry the *recA4* mutation (formerly *recE4*) are poorly inducible by DNA-damaging treatments (for references, see reference 74).

GROUP I PHAGES: ϕ105, ρ6, ρ10, AND ρ14

The group I phages typified by ϕ105 are small phages with genomes consisting of linear double-stranded DNA of 38.5 to 40.1 kb, which is about 80% as much DNA as coliphage λ (25, 94). They are all homoimmune (i.e., lysogeny for one of them prevents lysis by any other member of the group) and serologically related. Bacterial mutants resistant to one member of the group are resistant to all members, suggesting that they all adsorb to a common adsorption site on the bacterial surface. DNA homology between pairs of members of the group is 80 to 90%. For ϕ105, the icosahedral heads are 50 to 52 nm in diameter, and the flexible noncontractile tails are 177 to 220 nm long. The burst size is 100 to 200, and the latent period at 37°C is 40 min (8). DNA content is about 39.2 kb (37).

Phage ϕ105 DNA has constant (not permuted) complementary 3′ single-stranded cohesive ends ($cos_{\phi105}$) 7 bases long (36). Their sequences are as follows:

```
5′-GCGCTCC-3′
3′-CGCGAGG-5′
```

This sequence has no obvious similarity to the sequence of the 12-base 5′ single-stranded complementary cohesive ends of coliphage λ, although both are rich in G-C pairs. The linear ϕ105 DNA becomes a circle after being injected into sensitive bacteria and recombines into the chromosomal attachment site at a phage site located about 64% of the genome's length from the left end of the standard ϕ105 genetic map.

The life cycle of phage ϕ105 is probably quite similar to that of coliphage λ. Complementation tests have identified 11 cistrons in which temperature-sensitive or suppressible (nonsense) mutations were isolated (3, 95). Deletion mutations are also known;

Stanley A. Zahler • Section of Genetics and Development, Division of Biological Sciences, Cornell University, Ithaca, New York 14853-2703.

Table 1. Characteristics of *B. subtilis* phages[a]

Group	Typical phage	Size			G+C content (mol%)	Prophage chromosome site (°)
		Head (nm)	Tail (nm)	DNA (kb)		
I	ϕ105	52 × 52 (8)	10 × 220 (8)	40 (16)	43.5 (8)	244 (82, 89)
II	SPO2	50 × 50 (10)	10 × 180 (10)	40 (16)	43 (91)	15 (91, 105)
III	SPβ	72 × 82 (129)	12 × 358 (129)	126 (129)	31 (129)	188 (87, 151)
IV	SP16	61 × 61 (77)	12 × 192 (77)	60 (85)	37.8 (77)	?
V	PBSX	45 × 45 (111)	20 × 200 (111)	13 (1)	43 (1)	112 (1, 87)

[a] Numbers in parentheses are references.

they occur in phage regions that are not essential for the lytic cycle. Some deletion mutations prevent lysogenization by the phage. Genetic maps of the phage and prophage (including some restriction endonuclease targets) can be found in references 13, 37, and 149.

The functions of most of the essential genes of ϕ105 are unknown. Induction of a temperature-sensitive prophage mutant (*c*ts23, probably mutated in the repressor gene) results in the initiation of several rounds of replication of prophage DNA before excision of the prophage from the chromosome occurs. This replication does not stop at the prophage ends; the nearby chromosomal genes *leu* and *phe* are replicated as well, and they remain linked genetically to prophage markers (4, 96). A similar phenomenon occurs with coliphage λ (60). During phage replication in induced lysogens, concatenated phage DNA is found in the cells. There is no evidence that ϕ105 encodes an RNA polymerase or its own DNA polymerase (79, 84).

Phage ϕ105 Repressor and Control Region

Mutations of ϕ105 that prevent lysogenization or the maintenance of lysogeny can be isolated. They lie in at least three different complementation groups (22, 23) and were named *cI*, *cII*, and *cIII* in conscious reflection of the coliphage λ immunity system (for a review, see reference 138). The *cI* gene is also called $c_{\phi105}$ and encodes the ϕ105 repressor.

Two laboratories cloned the $c_{\phi105}$ gene almost simultaneously (23, 30) by using the ability of the multicopy gene to protect *B. subtilis* from lysis by ϕ105 or its clear-plaque mutants. (A second DNA fragment of the phage that also protected bacteria from infection was cloned [28]. The nature of the fragment's activity has not been clarified.) The repressor gene encodes a protein of 144 amino acids that binds to DNA sequences, i.e., to the ϕ105 operators, to prevent transcription of most of the ϕ105 genome (23, 29, 30, 83, 125–127).

Two phage promoters, both located very close to the $c_{\phi105}$ gene, are subject to control by the repressor. They are transcribed divergently. The leftward promoter, P_M, is used to transcribe two genes: the repressor gene $c_{\phi105}$ and a second, unidentified open reading frame. The repressor protein stimulates transcription from the P_M promoter. The rightward promoter, P_R, is believed to transcribe the late genes of ϕ105. The repressor protein inhibits transcription from the P_R promoter. This is strongly reminiscent of the P_M-P_R operator region of coliphage λ (for reviews, see references 55 and 88).

The region contains six binding sites (operators) for repressor protein (127). Figure 1 shows their arrangement. Three of them (O_R1, O_R2, and O_R3) are 14-mers with the following sequence:

```
5'-GACGGAAATACAAG-3'
3'-CTGCCTTTATGTTC-5'
```

Two others, O_R4 and O_R5, differ from these sequences in two nucleotide pairs, and the sixth, O_R6, differs from the canonical sequence in five nucleotide pairs. O_R4 and O_R5 bind repressor much less strongly than do O_R1, O_R2, and O_R3. Surprisingly, O_R6 binds repressor almost as well as does the canonical sequence.

Five of the operators lie between the divergent transcription start sites for P_M and P_R, a distance of about 176 nucleotide pairs (Fig. 1). From left to right, their order is O_R4, O_R1, O_R2, O_R5, O_R6. There is no overlap between binding sites. The other operator, O_R3, lies several hundred nucleotides to the right, within the first gene (ORF3) transcribed by the P_R promoter. All of the operators except O_R5 face from left to right. O_R4 overlaps the −35 sequence of P_M. O_R6 overlaps the −10 sequence of P_R and in fact protects DNA (from DNase I digestion, in footprinting experiments) that includes part of the −35 region of P_R as well.

A synthetic fragment of DNA with the canonical operator sequence works very well as an operator or as a transcription stop in the presence of repressor (126). Mutations in the operator or in the repressor sequence alter binding in ways that agree with evidence that the repressor interacts with the canonical operator sequence with specific amino acid-nucleotide interactions (31, 125, 126). Surprisingly, the operator sequence has no obvious twofold rotational symmetry (126). Nevertheless, the repressor protein is probably present in cells as a tetramer (127).

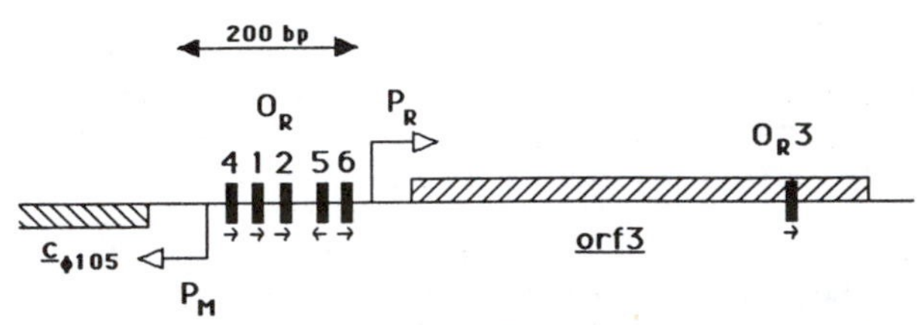

Figure 1. ϕ105 operator region. The six known sites to which the ϕ105 repressor binds are indicated by dark bars. The P_M and P_R promoters, the gene that encodes the repressor protein ($c_{\phi105}$), and the first gene of the late operon (*orf-3*) are shown. The direction of each operator sequence is indicated by short arrows beneath the bars. See reference 127 for further details.

Specialized Transduction by Phage φ105

The φ105 prophage lies between the *B. subtilis leu* and *phe* genes; the best estimate (127) places it at about 243°, perhaps between the *rodB* and *hemD* genes. Dean and his coworkers (24) showed that the linkage between *leuB* and *pheA* as measured by phage PBS1 transduction was decreased from 53% in φ105 nonlysogens to 20% in lysogens.

Mitomycin C-induced lysates of φ105 lysogens contain phage particles that can transduce auxotrophic bacteria to $IlvBC^+$, $LeuBC^+$, Nic^+, or $PheA^+$ but cannot transduce markers further away (103). These specialized transductions are much more efficient if the recipient auxotroph is lysogenic for φ105, at least in part because nonlysogens are often lysed by the free phage present in the lysates. There is about one transductant per 10^6 plaque-forming particles for *leuC*, the closest marker to $att_{\phi 105}$ that was tested.

High-frequency transducing (HFT) lysates could not be prepared, as can be done with coliphage λ (and with some of the group III *B. subtilis* phages). Perhaps the most likely explanation is that even *leuC* is too far from $att_{\phi 105}$ to permit the inclusion of $cos_{\phi 105}$ and *leuC* in a single phage particle, a necessity for the production of HFT lysates. A reexamination of the situation looking for the transduction of markers closer to $att_{\phi 105}$ might permit the isolation of HFT-producing strains for *hem*, for example.

GROUP II PHAGE: SPO2

SPO2 has a linear genome about 40 kb long with constant (not permuted) ends. It inserts into the *B. subtilis* chromosome in a permuted manner at an attachment site near the ribosomal-protein gene cluster (61, 105). The best estimate of the exact site is at 13° on the chromosome, perhaps between the *rps* genes and *relG* (87). A linear genetic map with 17 cistrons is available (146); a circular map with the SPO2 *cos* site indicated is also available (147). A map correlating the *sus* mutants of SPO2 with a phage restriction map has been published (49). Deletion mutants that complement all of the *sus* mutations can be isolated (50); they lie in a 3.5-kb region of the phage.

An unusual incompatibility phenomenon involving prophage SPO2 and the *Staphylococcus aureus* plasmid pC194, which is often used in *B. subtilis*, has been described (72, 73). If an SPO2 lysogen is transformed with pC194 and selection for the chloramphenicol resistance marker of the plasmid is maintained, the SPO2 prophage is lost from the culture; eventually, only nonlysogens remain. Explanations offered for this phenomenon are not very satisfactory.

The results of a study (46) of UV light inactivation of SPO2 and of its DNA (and comparison with two other phages) suggest that SPO2 may encode a protein(s) that carries out DNA repair reactions.

Relationship to Phage φ105

Physically, SPO2 and φ105 are quite similar (Table 1). There is some homology between their DNAs (9, 10, 16, 17) and some serological relationship as well (25). SPO2's adsorption site on the *B. subtilis* surface is different from that of φ105 (25) but may be identical to that of the group III phages φ3T and ρ11. SPO2 and φ105 are heteroimmune (10, 25).

The DNA homology between the two phages is restricted to a small region making up about 14% of each genome (5.7 kb) and lying in the central region of each phage genome (16). Homology in that region is only partial. Complementation of φ105 *sus* mutations by SPO2 was confined to one of the 11 cistrons tested (98), and that one was complemented inefficiently.

SPO2 DNA Polymerase

Rutberg and his coworkers (98) showed that phage φ105 uses host DNA polymerase III for the synthesis of DNA during the phage lytic cycle but that phage SPO2 does not. Rather, SPO2 encodes its own DNA polymerase, which has been cloned (99) and sequenced (89). These studies were facilitated by the sensitivity of *B. subtilis* DNA polymerase III to the inhibitor 6-(*p*-hydroxyphenylazo)uracil and the resistance of the phage enzyme to the inhibitor.

The amino acid sequence of the SPO2 DNA polymerase has some similarity to those of the carboxyl ends of *E. coli* DNA polymerase I and coliphage T7 DNA polymerase in the region that is believed to encode the DNA-binding domain (27, 63) and to various other motifs in the Klenow fragment of *E. coli* DNA polymerase I and other polymerases (eukaryotic DNA polymerases α and β, RNA-dependent RNA polymerases of viruses, reverse transcriptases, and DNA-dependent RNA polymerases of coliphage T3 and mitochondria [27]). It is likely that some of the motifs detected are nucleotide-binding sites. None of these observations seems to make it possible to determine evolutionary relatedness among the enzymes.

GROUP III PHAGES: SPβ, φ3T, ρ11, SPR, Z, IG1, IG3, IG4, AND H2

The large temperate phages of group III have similar appearances. They have icosahedral heads and long, flexible tails with complex but undefined structures at the tail termini. They contain single linear DNA molecules 110 to 129 kb long. The phage particles probably contain seven major proteins (106). Restriction endonuclease maps are available for φ3T (20, 106), ρ11 (64, 106), SPβ (42, 43 [with correction; see reference 149], 76, 106, 107), and H2 (67, 152). The relationships among group III phages have been examined by tests of DNA homologies (106, 131). The amount of DNA in these phages is about 2.5 times more than that in coliphage λ. A comparable increase in complexity is expected.

It has recently been shown (41a) that the IG1 prophage can be induced efficiently by 6-(*p*-hydroxyphenylazo)uracil, which is a known inhibitor of *B. subtilis* DNA polymerase III. This implies that IG1 and probably the other group III phages encode their own DNA polymerases that are resistant to the inhibitor.

It is probable that the chromosome of *B. subtilis* 168 contains relics of group III prophages, just as the chromosome of *E. coli* K-12 contains relics of lambdoid phages. (i) DNA cloned from near the terminus of

Table 2. DNA methyltransferases of group III phages[a]

Phage	Gene name	Endonuclease protected against	Target sequences	Reference(s)
SPR	*psmSPR*	*Hae*III	GGCC	122
		*Hpa*II	CCGG	
		*Eco*RII	CC(A/T)GG	
SPβ	*psmSPβ*	*Hae*III	GGCC	120
		*Fnu*4HI	GCNGC	
ϕ3T	*psmϕ3T*	*Hae*III	GGCC	120
		*Fnu*4HI	GCNGC	
$\rho 11_B$	*psm$\rho 11_B$*	*Hae*III	GGCC	67
		*Fnu*4HI	GCNGC	
$\rho 11_S$	*psm$\rho 11_S$*	*Hae*III	GGCC	67
		Bsp 1286	G(A/G/T)GC(T/C/A)C	
H2	*psm*$H2_I$	*Hae*III	GGCC	67
		*Fnu*4HI	GCNGC	
		Bsp 1286	G(A/G/T)GC(T/C/A)C	
H2	*bam*M_2	*Bam*HI	GGATCC	18, 152

[a] Any nucleotide within parentheses makes a suitable target for the restriction enzyme. Cytidylates known to be methylated by the phage enzyme are underlined. The methylated cytidylate of the *Bsp* 1286 target sequence is not known.

replication of *B. subtilis* has strong homology with the DNA of phage SPβ (93). There is also strong homology between the phage ϕ3T chromosome and *B. subtilis* chromosomal DNA (114). (ii) Spontaneous deletions ("*citD*" deletions) start within the SPβ prophage and extend to close to the replication terminus, between *odh* (=*citK*) and *gltAB* (the latter lies at the terminus) (148). The deletions may represent recombinations between prophage DNA and the DNA of the homologous region described by Rowe et al. (93). Other spontaneous deletions extend from the prophage past *gltAB* (93, 132, 148) and may represent recombinations between the prophage of SPβ and another region of homology that lies between *gltAB* and *citB*. (iii) Insertions of an integration-deficient mutant (*int-5*) of SPβ may frequently occur in regions of homology on the chromosome. Such insertions occur near *gltAB*, between *thyA* and *glnA*, near the *dal* region of the chromosome, and in the *sdh* gene cluster (47, 71, 148).

Sargent et al. (101) showed that protoplasts of SPβ lysogens degrade their own DNA during osmotic stress at least 10 times as rapidly as do nonlysogens. They suggested that a phage-encoded protein, possibly induced by protoplasting conditions, was responsible.

Isolation

The first isolation of a member of group III was of phage ϕ3T (124). It was isolated from soil and formed plaques on strain NCTC3610 of *B. subtilis*, the presumed parent from which the tryptophan auxotroph strain 168 was derived. Tucker (124) showed that ϕ3T converted Thy^- *B. subtilis* strains to Thy^+ and caused the production of thymidylate synthetase in the cells. All new isolates of group III phages are able to convert Thy^- hosts to Thy^+. (The only known group III phage without this ability is SPβ, the endogenous prophage of strain 168 and, presumably, of strain NCTC3610.) Phage ρ11 was also isolated on a derivative of strain 168 and is similar to ϕ3T (26). Two different group III phages, $\rho 11_S$ and $\rho 11_B$, are labeled "ρ11" in various laboratories (7); no systematic study of how they differ has been made, but they encode different DNA methyltransferases and can be differentiated by that characteristic (Table 2; see below).

Warner et al. (129) first showed that *B. subtilis* 168 is lysogenic for phage SPβ, a member of group III. This demonstration was made possible by the discovery of a *B. subtilis* strain, CU1050 (= su^{+3} = BGSC1A459), that had spontaneously lost the SPβ prophage and could therefore be used as a plating host for SPβ. Strain CU1050 also carries a mutation, *pla-1* (152), that permits phage SPβ to form plaques efficiently on it; bacteria that carry *pla*$^+$ give only 1% as many plaques.

Other members of group III were isolated on CU1050 or its derivatives that lack the SPβ prophage. These include SPR (123), Z (58), IG1, IG3, IG4 (39, 40), and H2 (152, 153). Phage SPR is found in the prophage state in *B. subtilis* R. Phage SPR was originally misidentified as SPβ (123), and some early papers confused the two. Phage H2 is found in the prophage state in *Bacillus amyloliquefaciens* H. Other temperate phages of *B. amyloliquefaciens* may include members of group III (152, 153).

Phage Immunity and Chromosome Attachment Sites

Phages SPβ, Z, SPR, IG1, IG3, IG4, and H2 form an immunity group: strains lysogenic for one cannot be lysed by another. Phages ϕ3T and ρ11 can make plaques on SPβ lysogens and form a second immunity group. However, the relationships are not reciprocal: SPβ cannot form plaques on ϕ3T or ρ11 lysogens (25, 40).

The chromosomal attachment sites of SPβ (151), ϕ3T (80, 134), IG1, IG3, and IG4 (41) lie between the *ilvA* and *kauA* genes on the *B. subtilis* chromosome map and about 132 kb from the chromosome's replication terminus (93). The relationship among the attachment sites is not clear; they may be identical. It is difficult to cite a precise position for *attSPβ*, since the chromosome map (87) contains prophage SPβ; the prophage extends approximately from 183° to 192°. The attachment site for the H2 prophage on the *B. subtilis* chromosome is a few kilobases farther from the chromosome terminus, between the *metB* and *tyrA* genes (152). Nothing is known of its attachment site in its normal host, *B. amyloliquefaciens*.

Host Ranges and Serology

The host ranges of most group III phages are restricted to *B. subtilis* strains, although none can infect strain W23. φ3T and SPβ can form plaques on some strains of *Bacillus globigii* (25). IG4 can infect a strain of *Bacillus pumilis* (40).

Youngman et al. (150) and Weiner and Zahler (131) isolated plaque-forming group III phages carrying transposon Tn*917*. The antibiotic-resistant markers of the transposon (erythromycin-lincomycin or, in some derivatives, chloramphenicol) allowed host range tests of the phages without demanding that phage replication and plaque formation take place. Phage H2 was able to infect and convert to antibiotic resistance the seven strains of *B. amyloliquefaciens* that were tested; neither SPβ nor SPR could do so. All three phages could infect the strain of *B. pumilis* that was tested. *B. subtilis* strains already carrying group III prophages were converted to antibiotic resistance with high efficiency, even if they were Rec$^-$.

One can isolate mutants of nonlysogenic derivatives of strain 168 that are resistant to clear-plaque mutants of group III phages. Estrela et al. (38) showed that a mutant isolated as resistant to φ3T*c* became resistant simultaneously to ρ11, Z, IG1, IG3, and IG4 but not to SPR. The bacterial mutation responsible for the resistance, *pha-3*, was mapped to a position close to the *gta* genes (141), in which mutations give rise to resistance to several phages because of alterations in the cells' glycosolated teichoic acids. Mutations in *pha*, *gtaB*, *gtaC*, and *gneA* (38a) caused resistance to φ3T. Weiner (130) showed that SPβ::Tn*917* could still infect the *pha-3* mutant and make it antibiotic resistant, although at lowered efficiency.

Antiserum against SPβ inactivates SPβ, φ3T, ρ11, IG1, IG3, and IG4 but not SPR (40) or H2 (131). Antiserum against SPR inactivates only SPR (79). Antiserum against H2 inactivates H2 but not SPβ or SPR (130).

Betacin Production

Two of the group III phages, SPβ and Z, cause bacteria in which they are present as prophages to excrete a bacteriocinlike protein (betacin) that kills nearby nonlysogenic bacteria on agar plates (58). A mutant of SPβ lacking this ability, SPβ*bet-1*, has been isolated. Lysogens of SPβ and of Z are not killed by betacin; this implies that there exists on the prophage a tolerance gene ("*tol*") as well. Both *bet* and *tol* are expressed in the prophage state.

Thymidylate Synthetase

Each of the group III phages except SPβ carries a gene for thymidylate synthetase, which is expressed by the prophage and converts thymine auxotrophs of *B. subtilis* to thymine prototrophy. Genes for thymidylate synthetase have been cloned from φ3T (gene *thyP3* [32–34, 48, 65, 108, 114, 115]) and from ρ11 (gene *thyP11* [48]). The *thyP3* gene has been sequenced (65). The cloned genes carry their own promoters and are expressed in *B. subtilis* and *E. coli*. They make Thy$^-$ auxotrophs of either species prototrophic. The *thyP3* gene is strongly homologous to the *B. subtilis thyA* gene, and the flanking DNAs of phage and bacterium are also homologous. (Unlike all other known organisms, *B. subtilis* has two thymidylate synthetases encoded by two unlinked genes, *thyA* and *thyB* [78]. A φ3T lysogen, then, has three *thy* genes, and a φ3T-Z double lysogen has four.)

The homology between *thyP3* and *thyA* permits the transformation of *thyA* mutants of *B. subtilis* by DNA from *thyP3*. If the thymine auxotroph is lysogenic for SPβ, two kinds of transformants are observed (115). If the *thyP3* gene is contained on a small DNA fragment, the bacterial *thyA* gene is usually replaced by *thyP3*. If the *thyP3* gene lies on a larger fragment of DNA, the *thyP3* gene is more likely to replace the region of the SPβ prophage that is homologous to φ3T regions that surround the *thyP3* gene. This results in the production of an SPβ prophage that carries the *thyP3* gene. Spancake et al. (108) have studied the *thyP3*-carrying SPβ phages (which they call SPβT) released by such transformants. The *thyP3* gene lies at the right end of the SPβ prophage and is approximately central in the DNAs of SPβT phage particles. The SPβT particles carry extensive DNA deletions.

Mutations of Group III Phages

Temperature-sensitive mutants of phage SPβ have been isolated by H. E. Hemphill and in my laboratory (149). None has been characterized physiologically. One cannot isolate nonsense mutations of SPβ by using the usual plating bacterium, CU1050, because that strain carries an ochre nonsense suppressor mutation, *sup-3*. We have isolated a derivative of CU1050, strain CU3069, that has lost the suppressor mutation and several other markers carried by CU1050 (153). It still carries the *pla-1* mutation that permits high-efficiency plaque formation. Using this strain as host, we have succeeded in isolating SPβ mutants with *sus* (suppressible) mutations in essential genes (unpublished results).

Certain mutants, φ1*m*, of the virulent phage φ1 are unable to grow in lysogens of SPβ, although they grow well in nonlysogens (90). Lysogens of an SPβ mutant called SPβ*mpi* no longer prevent productive infection by φ1*m*. This system is formally analogous to the relationship between the *rII* genes of phage T4 and the *rex* gene of coliphage λ (for a review, see reference 19).

Clear-plaque mutants of group III phages have been observed often. SPβ*c1*, a clear-plaque mutant of SPβ, is unable to lysogenize *B. subtilis* (129), as is SPβ*c10* (76).

Rosenthal et al. (92) described an SPβ mutant that seems to have a temperature-sensitive repressor. Lysogens carrying SPβ*c2* can be induced to lyse by brief treatment with elevated temperature. The c^+ and *c2* alleles can complement the *c1* clear-plaque mutation, leading to the conclusion that all three are alleles. Bacteria lysogenic for SPβ*c2* lose the prophage when incubated at 50°C. (Phage SPβ cannot replicate above about 45°C.)

Several laboratories have tried to clone the *c* (repressor) gene of phage SPβ without success. One attempt (76) resulted in the cloning and sequencing of another SPβ gene, called *d*, that protects nonlysogens from infection with SPβ when the gene is present on a multicopy plasmid. When the cloned *d* gene was inactivated in vitro and returned to the SPβ phage genome, a clear-plaque phenotype resulted; SPβ*d2* phage cannot lysogenize *B. subtilis*. The relationship between the *d* gene and the *c* gene is not at all clear.

Zahler (148) described an SPβ mutation, *int-5*, that greatly decreases the ability of SPβ to lysogenize *B. subtilis* at the normal bacterial attachment site (*attSPβ*) between *ilvA* and *kauA*. SPβ*c1* can complement the *int-5* mutation, permitting the insertion of SPβ*c2int-5* into *attSPβ*. Such a lysogen cannot be induced efficiently by mitomycin C or heat; the *int-5* mutation prevents proper excision of the prophage as well as insertion. Bacteria lysogenic for SPβ*c2int-5* are killed by attempts to grow them at 50°C. Many of the rare survivors carry large deletions that include much of the SPβ prophage and bacterial DNA extending through the *kauA* and *citK* genes (148) and occasionally through the *gltAB* genes as well (93). Apparently, the SPβ prophage has a gene ("*kil*") that is lethal to the host if it is expressed, as is true of coliphage λ (51).

Many deletion mutants of group III have been isolated. Deletion mutants of ρ11*c3*, a clear-plaque mutant of ρ11, that have lost as much as 9 kb of DNA have been described (64). Long deletions of SPβ have also been described (42, 107). Most of the mutants described in reference 107 fell into one of three classes that had lost 11.8, 14.0, or 14.2 kb of DNA. These deletions fall within a 27-kb contiguous region that lies near the center of the SPβ genome. There are no essential genes in the region, but the region is transcribed both in the prophage state and during phage replication. The phage attachment site (*attP*) at which the circular vegetative phage DNA opens during integration as prophage lies within this region and is deleted in some of the mutants.

DNA Methyltransferases

Several of the group III phages (but not phage Z, IG1, IG3, or IG4) encode DNA methyltransferases that methylate the position 5 carbons of cytidylic acids in DNA. The enzymes are expressed only during vegetative growth of the phages. The particular cytidylic acids affected differ from phage to phage and lie within the sequences corresponding to one or another restriction endonuclease target (Table 2). Early publications that gave evidence suggesting such enzyme activities include references 5, 11, 12, 20, and 53. Most of the restriction endonucleases mentioned in Table 2 have isoschizomers made by some strains of *Bacillus*. Presumably, the methylations protect phage DNAs from bacterial endonucleases. The phages themselves are not known to encode restriction endonucleases.

The first clear demonstrations that group III phages have DNA methyltransferase activity (21, 123) showed that phage φ3T or SPR DNA contains *Hae*III targets that are protected from the endonuclease activity of *Hae*III or its *Bacillus* isoschizomer, *Bsu*RI. The protection was shown to be due to the methylation of *Hae*III targets during phage replication.

Trautner and his colleagues have carried out an extensive series of studies on the DNA methylation systems of various members of group III phages (for a review of the early literature, see reference 149; 6, 7, 52, 54, 66–69, 120, 122, 128, 133). Several of the phages encode a DNA methyltransferase that recognizes two or three different DNA target sequences and methylates a cytidylic residue in each target (Table 2). These multispecific enzymes are intrinsically interesting. They have molecular weights of about 50,000 and extensive sequence homologies to each other. (They also have homology to some monospecific bacterial methyltransferases, parts of restriction-modification systems from various bacteria, and a mammalian DNA methyltransferase [68, 128].)

The general structure of the sequences of the multispecific enzymes is as follows. Toward the amino end of the polypeptide are at least two highly conserved motifs. Toward the carboxyl end are two other conserved motifs. It is likely that one motif is involved in binding the enzyme to DNA nonspecifically; one is involved in binding to *S*-adenosylmethionine, which is the methyl donor in all of the reactions; and one (or more) is involved in carrying out transmethylation. In the central portion of each enzyme's amino acid sequence is a variable region of a few hundred amino acids. Within the variable regions of the multispecific enzymes are target-specific domains (TSDs, or modules) that encode recognition sites for particular DNA targets. These target-recognizing domains, each about 50 amino acids long, determine precisely which DNA sequences the enzyme can methylate. Within each domain are some conserved amino acids that probably help define the general structure of the target-recognizing region and other variable amino acids that probably determine the specificity of the domain (66–69).

Mutations in the genes encoding multispecific phage methylases sometimes inactivate only one of the modules and hence prevent recognition of one DNA target site (52, 120, 133). Some of the enzymes have inactive TSDs ("pseudo" domains), sequences that differ by one or a few amino acids from active TSDs of other phage enzymes (66). Site-directed mutagenesis of the DNA that encodes an inactive TSD can add specificity to the phage enzyme (66). The TSDs can be moved about by genetic engineering (6, 121). The *pms* genes that encode them can be moved by transfection and recombination into the genome of phage Z, which has no methyltransferase gene of its own (116).

Sequence comparisons of multispecific DNA methyltransferases are being studied to shed light on the evolution of these enzymes (7, 52, 66–68, 120, 122). As is common in such studies, the speculations are fascinating, but convincing evidence is in short supply.

Two of the group III phages encode not only a multispecific DNA methyltransferase but also a second, monospecific DNA methyltransferase (67, 130, 131). Phage H2 encodes a monospecific DNA methyltransferase not closely linked physically to *pmsH2*$_I$ that recognizes the *Bam*HI DNA sequence and methylates one of its Cs (Table 2; 18, 67, 130). Phage $\rho 11_S$ encodes a monospecific DNA methyltransferase that recognizes a sequence that has not been completely identified (67).

Specialized Transduction

The group III phages, like coliphage λ, can carry out specialized transduction (for reviews, see references 148 and 149). The improper excision of prophage DNA results in the inclusion on defective phage genomes of bacterial genes that are linked to the bacterial *att* site. The packaged defective phage carry the defective genomes to new strains of bacteria, transducing them if there are selectable markers on the phage. Specialized transduction of genes linked to the normal *att* site has been demonstrated for SPβ (44, 92, 151), ϕ3T (80), H2 (152), and IG1, IG3, and IG4 (41). Only one nondefective specialized transducing phage (SPβ carrying the *ilvA* gene) has been reported (45).

If the prophage lies at an unusual site on the bacterial chromosome rather than at the normal attachment site, nearby bacterial genes can again be carried on defective prophages. Specialized transduction with SPβ or its derivatives has been demonstrated for bacterial genes in the *dal–sup-3–sup-44–ddl* chromosome region (70), *glnA* (47), and the *sdh-ilvBC-leu* region (71).

The availability of an SPβ derivative carrying the *c2* mutation (heat-sensitive repressor), the *int-5* mutation (inability to lysogenize at the normal *attSP*β site), the *del-2* deletion (a 10-kb deletion of nonessential DNA), and Tn*917cat* (a chloramphenicol-resistant derivative of transposon Tn*917*) has made it possible to select for insertion of the phage into Tn*917* transposons located anywhere on the chromosomes of SPβ nonlysogens of *B. subtilis*. Specialized transduction of bacterial genes close to the Tn*917* insert can sometimes be found after heat-induced lysis of the bacteria (unpublished results).

In vitro methods for constructing specialized transducing phages by using phage ρ11 were pioneered by H. Saito and his colleagues (64) and are described elsewhere in this volume (see chapter 44).

GROUP IV PHAGE: SP16

No publications concerning phage SP16 have appeared since 1986 (for a review, see reference 149). It is the only temperate phage known to infect *B. subtilis* W23, on which it was first isolated (77, 117). It also infects strain 168 and some strains of *B. amyloliquefaciens* and *Bacillus licheniformis*. It is unique in its immunity, antigenic properties, and host range (25, 85, 117).

Parker and Dean (85) have presented evidence that SP16 particles contain a DNA molecule of about 52.8 kb (measured by adding the sizes of restriction fragments) or 60.0 kb (measured by electron microscopy as the length of spread molecules). Their evidence also suggests that SP16 DNA has a terminal redundancy of 10 or 15% and that the phage DNA is circularly permuted; that is, the ends of the molecules are not all identical. This is similar to the DNA molecules of coliphage P1 (for a review, see reference 139), which may imply that SP16, like P1, also replicates as a plasmid. Phage SP16 does not carry out generalized transduction.

If *B. subtilis* lysogens for SP16 are used as plating bacteria for the group II phage SPO2, the efficiency of plating is reduced by a factor of 10^{-4} (117). This phenomenon is not due to a restriction-modification system carried by the SP16 prophage or to phage immunity. The rare SPO2 plaques found on 168(SP16) lysogens are due to mutants of SPO2; the mutants are not altered in their abilities to adsorb to host cells.

GROUP V DEFECTIVE PHAGES: PBSX, PBSZ, ETC.

When *B. subtilis* 168 cells are exposed to treatments (such as UV light or mitomycin C) that induce the SOS (=SOB) response, the cells lyse after incubation of 1 h and release particles of a defective bacteriophage called PBSX (102), SPα (35), μ (62), ϕ3610 (113), or PBSH (56, 57). Several hundred particles are released per cell; they can be detected by electron microscopy. The particles adsorb to and kill certain strains of related bacilli, in particular *B. subtilis* W23, thus acting like a bacteriocin. The phage particles are complex; at least 26 different proteins are present (75).

PBSX is defective in several interesting ways. (i) In the electron microscope, the particles look like phage particles with small heads and large, complex tails, but the heads do not contain phage DNA. Instead, each particle contains a randomly selected 13-kb fragment of bacterial chromosome (1, 81, 82, 104). The structural proteins of the phage particles (75) and the nonstructural proteins synthesized following induction of PBSX (118) require at least 54 kb of coding DNA. (ii) The DNA contained in the PBSX heads is not injected into bacteria to which the particles adsorb (81). Therefore, PBSX cannot transduce markers to new strains of bacteria. Furthermore, it does not adsorb to *B. subtilis* 168. (iii) Under some conditions, e.g., low Mg^{2+} concentrations (81, 111), the contractile tail sheaths contract in a manner quite different from that of other phages like coliphage T4: the distal end of the sheath stays attached to the tail core, while the head-proximal portion detaches from the core, increases in diameter, and decreases in length (109, 111, 113).

B. subtilis W23, on the other hand, produces a slightly different defective phage called PBSZ that pari passu adsorbs to and kills *B. subtilis* 168. Similar defective phages have been described in at least five other species of *Bacillus* (111). At least five different morphological types can be distinguished by the lengths of their tails, although their heads are all small, too small to contain the amount of DNA needed to encode the phage head and tail proteins. The phage heads are all about 45 nm in diameter; tails are 20 nm wide and range between 185 and 265 nm long for different phages. The tail of PBSX is 200 nm long; that of PBSZ is 255 nm long.

PBSX Prophage

The prophage of PBSX lies on the *B. subtilis* chromosome between the *metA* and *phoS* markers. Mutations that affect the expression of PBSX are located in this region: *xin* prevents the production of PBSX (119), *xhi* is heat inducible for PBSX (14), *xtl* prevents assembly of the tail (119), *xhd* results in the production of defective heads (119), and *xki* causes the production of PBSX particles that look normal but cannot kill strain W23 (14, 110). The last particles lack a particular protein that may be responsible for the killing effect. Finally, Buxton has isolated deletions that enter the PBSX prophage (15) and may remove it entirely.

A fragment of DNA from the PBSX region of the chromosome is capable of serving as a replication origin in *B. subtilis* when cloned onto plasmids that cannot otherwise replicate in *B. subtilis* (2). Presumably, the cloned DNA includes the replication origin for the historically original, nondefective phage precursor of PBSX. The cloned DNA has homology with chromosomal DNA from *B. subtilis* W23, presumably with part of the genome of its group V defective prophage, PBSZ.

Wood et al. (136) cloned and sequenced DNA that complements the temperature-sensitive *xhi* mutation. It encodes a repressor for PBSX. They renamed the gene *xre* (PBSX repressor). The repressor's deduced protein sequence shows some homology with the *c2* repressor of *Salmonella* phage P22, the coliphage P1 repressor, the $c_{\phi105}$ repressor, and other helix-turn-helix DNA-binding proteins.

More than 33 kb of PBSX DNA fragments have been cloned (137). The cloned DNA includes 21 kb of a single operon, apparently the "late operon" of the phage, transcribed in the direction *metA* → *phoS*. The genes for a number of proteins produced by PBSX have been located on the cloned DNA.

The widespread distribution of group V defective prophages in strains of *Bacillus* has led to speculation that the prophages may carry out some essential function in the lives of the bacteria (137). It will be useful to learn whether the Buxton deletions (14, 15) or others that can be isolated by his methods have left behind fragments of PBSX DNA that are required for bacterial survival.

Acknowledgments. I thank H. E. Hemphill for reading the manuscript and making useful suggestions. This work was supported by grants from the National Institutes of Health.

REFERENCES

1. **Anderson, L. M., and K. F. Bott.** 1985. DNA packaging by the *Bacillus subtilis* defective bacteriophage PBSX. *J. Virol.* **54:**773–780.
2. **Anderson, L. M., H. E. Ruley, and K. F. Bott.** 1982. Isolation of an autonomously replicating DNA fragment from the region of defective bacteriophage PBSX of *Bacillus subtilis*. *J. Bacteriol.* **150:**1280–1286.
3. **Armentrout, R. W., and L. Rutberg.** 1970. Mapping of prophage and mature deoxyribonucleic acid from temperate *Bacillus* bacteriophage φ105 by marker rescue. *J. Virol.* **6:**760–767.
4. **Armentrout, R. W., and L. Rutberg.** 1971. Heat induction of φ105 in *Bacillus subtilis*: replication of the bacterial and bacteriophage genomes. *J. Virol.* **8:**455–468.
5. **Arwert, F., and L. Rutberg.** 1974. Restriction and modification in *Bacillus subtilis*. Induction of a modifying activity in *Bacillus subtilis* 168. *Mol. Gen. Genet.* **133:**175–177.
6. **Balganesh, T. S., L. Reiners, R. Lauster, M. Noyer-Weidner, K. Wilke, and T. A. Trautner.** 1987. Construction and use of chimeric SPR/φ3T DNA methyltransferases in the definition of sequence recognizing enzyme regions. *EMBO J.* **6:**3543–3549.
7. **Behrens, B., M. Noyer-Weidner, B. Pawlek, R. Lauster, T. S. Balganesh, and T. A. Trautner.** 1987. Organization of multispecific DNA methyltransferases encoded by temperate *Bacillus subtilis* phages. *EMBO J.* **6:**1137–1142.
8. **Birdsell, D. C., G. M. Hathaway, and L. Rutberg.** 1969. Characterization of temperate *Bacillus* bacteriophage φ105. *J. Virol.* **4:**264–270.
9. **Boice, L., F. A. Eiserling, and W. R. Romig.** 1969. Structure of *Bacillus subtilis* phage SPO2 and its DNA: similarity of *Bacillus subtilis* phages SPO2, φ105 and SPP1. *Biochem. Biophys. Res. Commun.* **34:**398–403.
10. **Boice, L. B.** 1969. Evidence that *Bacillus subtilis* bacteriophage SPO2 is temperate and heteroimmune to bacteriophage φ105. *J. Virol.* **4:**47–49.
11. **Bron, S., and K. Murray.** 1975. Restriction and modification in *B. subtilis*. Nucleotide sequence recognized by restriction endonuclease R.*Bsu*R from strain R. *Mol. Gen. Genet.* **143:**25–33.
12. **Bron, S., K. Murray, and T. A. Trautner.** 1975. Restriction and modification in *B. subtilis*. Purification and general properties of a restriction endonuclease from strain R. *Mol. Gen. Genet.* **143:**13–23.
13. **Bugaichuk, U. D., M. Deadman, J. Errington, and D. Savva.** 1984. Restriction enzyme analysis of *Bacillus subtilis* bacteriophage φ105 DNA. *J. Gen. Microbiol.* **130:**2165–2167.
14. **Buxton, R. S.** 1976. Prophage mutation causing heat inducibility of defective *Bacillus subtilis* bacteriophage PBSX. *J. Virol.* **20:**22–28.
15. **Buxton, R. S.** 1980. Selection of *Bacillus subtilis* 168 mutants with deletions of the PBSX prophage. *J. Gen. Virol.* **46:**427–437.
16. **Chow, L. T., L. Boice, and N. Davidson.** 1972. Map of the partial sequence homology between DNA molecules of *Bacillus subtilis* bacteriophages SPO2 and φ105. *J. Mol. Biol.* **68:**391–400.
17. **Chow, L. T., and N. Davidson.** 1973. Electron microscope study of the structures of the *Bacillus subtilis* prophages SPO2 and φ105. *J. Mol. Biol.* **75:**257–264.
18. **Connaughton, J. F., W. D. Kaloss, P. G. Vanek, G. A. Nardone, and J. G. Chirikjian.** 1990. The complete sequence of the *Bacillus amyloliquefaciens* proviral H2, *Bam*HI methylase gene. *Nucleic Acids Res.* **18:**4002.
19. **Court, D., and A. B. Oppenheim.** 1983. Phage lambda's accessory genes, p. 251–277. *In* R. W. Hendrix, J. W. Roberts, F. W. Stahl, and R. A. Weisberg (ed.), *Lambda II*. Cold Spring Harbor Laboratory, Cold Spring Harbor, N.Y.
20. **Cregg, J. M., and J. Ito.** 1979. A physical map of the genome of temperate phage φ3T. *Gene* **6:**199–219.
21. **Cregg, J. M., A. H. Nguyen, and J. Ito.** 1980. DNA modification induced during infection of *Bacillus subtilis* by phage φ3T. *Gene* **12:**17–24.
22. **Cully, D. F., and A. J. Garro.** 1980. Expression of superinfection immunity to bacteriophage φ105 by *Bacillus subtilis* cells carrying a plasmid chimera of pUB110 and *Eco*RI fragment F of φ105 DNA. *J. Virol.* **34:**789–791.
23. **Cully, D. F., and A. J. Garro.** 1985. Nucleotide sequence of the immunity region of *Bacillus subtilis* bacteriophage φ105: identification of the repressor gene and its mRNA and protein products. *Gene* **38:**153–164.
24. **Dean, D. H., M. Arnaud, and H. O. Halvorson.** 1976. Genetic evidence that *Bacillus* bacteriophage φ105 integrates by insertion. *J. Virol.* **20:**339–341.

25. **Dean, D. H., C. L. Fort, and J. A. Hoch.** 1978. Characterization of temperate phages of *Bacillus subtilis*. *Curr. Microbiol.* **1:**213–217.
26. **Dean, D. H., J. C. Orrego, K. W. Hutchison, and H. O. Halvorson.** 1976. New temperate bacteriophage for *Bacillus subtilis*, ρ11. *J. Virol.* **20:**509–519.
27. **Delarue, M., O. Poch, N. Tordo, D. Moras, and P. Argos.** 1990. An attempt to unify the structure of polymerases. *Protein Eng.* **3:**461–467.
28. **Dhaese, P., M.-R. Dobbelaere, and M. Van Montagu.** 1985. The temperate *B. subtilis* phage φ105 genome contains at least two distinct regions encoding superinfection immunity. *Mol. Gen. Genet.* **200:**490–492.
29. **Dhaese, P., C. Hussey, and M. Van Montagu.** 1984. Thermoinducible gene expression in *Bacillus subtilis* using transcriptional regulatory elements from temperate phage φ105. *Gene* **32:**181–194.
30. **Dhaese, P., J. Seurinch, B. De Smet, and M. Van Montagu.** 1985. Nucleotide sequence and mutational analysis of an immunity repressor gene from *Bacillus subtilis* temperate phage φ105. *Nucleic Acids Res.* **13:** 5441–5455.
31. **Dhaese, P., L. Van Kaer, R. De Clercq, and M. Van Montagu.** 1988. The *Bacillus subtilis* phage φ105 repressor-operator interaction: mutational analysis and in vitro binding studies. *J. Cell. Biochem. Suppl.* **12D:**132.
32. **Duncan, C. H., G. A. Wilson, and F. E. Young.** 1977. Transformation of *Bacillus subtilis* and *Escherichia coli* by a hybrid plasmid pCD1. *Gene* **1:**153–167.
33. **Duncan, C. H., G. A. Wilson, and F. E. Young.** 1978. Mechanism of integrating foreign DNA during transformation of *Bacillus subtilis*. *Proc. Natl. Acad. Sci. USA* **75:**3664–3668.
34. **Ehrlich, S. D., I. Bursztyn-Pettegrew, I. Stroynowski, and J. Lederberg.** 1976. Expression of the thymidylate synthetase gene of the *Bacillus subtilis* bacteriophage φ3T in *Escherichia coli*. *Proc. Natl. Acad. Sci. USA* **73:**4145–4149.
35. **Eiserling, F. A.** 1964. Ph.D. thesis. University of California, Los Angeles.
36. **Ellis, D. M., and D. H. Dean.** 1985. Nucleotide sequence of the cohesive single-stranded ends of *Bacillus subtilis* temperate bacteriophage φ105. *J. Virol.* **55:**513–515.
37. **Errington, J., and N. Pughe.** 1987. Upper limit for DNA packaging by *Bacillus subtilis* bacteriophage φ105: isolation of phage deletion mutants by induction of oversized prophages. *Mol. Gen. Genet.* **210:**347–351.
38. **Estrela, A. I., H. de Lencastre, and L. J. Archer.** 1986. Resistance of a *Bacillus subtilis* mutant to a group of temperate bacteriophages. *J. Gen. Microbiol.* **132:**411–415.
38a. **Estrela, A.-I., H. M. Pooley, H. de Lencastre, and D. Karamata.** 1991. Genetic and biochemical characterization of *Bacillus subtilis* 168 mutants specifically blocked in the synthesis of the teichoic acid poly(3-*O*-β-D-glucopyranosyl-*N*-acetylgalactosamine 1-phosphate): *gneA*, a new locus, is associated with UDP-*N*-acetylglucosamine 4-epimerase activity. *J. Gen. Microbiol.* **137:**943–950.
39. **Fernandes, R. M., H. de Lencastre, and L. J. Archer.** 1983. Two newly isolated temperate phages of *Bacillus subtilis*. *Broteria-Genet.* **4(79):**27–33.
40. **Fernandes, R. M., H. de Lencastre, and L. J. Archer.** 1986. Three new temperate phages of *Bacillus subtilis*. *J. Gen. Microbiol.* **132:**661–668.
41. **Fernandes, R. M., H. de Lencastre, and L. J. Archer.** 1989. Specialized transduction in *Bacillus subtilis* by the phages IG1, IG3 and IG4. *Arch. Virol.* **105:**137–140.
41a. **Fernandes, R. M., H. de Lencastre, and L. J. Archer.** 1990. Action of 6-(*p*-hydroxyphenylazo)-uracil on bacteriophage IG1. *Arch. Virol.* **113:**177–182.
42. **Fink, P. S., R. Z. Korman, J. M. Odebralski, and S. A. Zahler.** 1981. *Bacillus subtilis* bacteriophage SPβ*c1* is a deletion mutant of SPβ. *Mol. Gen. Genet.* **182:**514–515.
43. **Fink, P. S., and S. A. Zahler.** 1982. Restriction fragment maps of the genome of *Bacillus subtilis* bacteriophage SPβ. *Gene* **19:**235–238.
44. **Fink, P. S., and S. A. Zahler.** 1982. Specialized transduction of the *ilvD-thyB-ilvA* region mediated by *Bacillus subtilis* bacteriophage SPβ. *J. Bacteriol.* **150:**1274–1279.
45. **Fink, P. S., and S. A. Zahler.** 1983. SPβ*c2pilvA*: plaque-forming bacteriophages that transduce the *Bacillus subtilis ilvA* gene, abstr. H32, p. 111. Abstr. Annu. Meet. Am. Soc. Microbiol. 1983.
46. **Freeman, A. G., K. M. Schweikart, and L. L. Larcom.** 1987. Effect of UV irradiation on the *Bacillus subtilis* phages SPO2, SPP1 and φ129 and their DNA. *Mutat. Res.* **184:**187–196.
47. **Gardner, A., J. Odebralski, S. Zahler, R. Z. Korman, and A. I. Aronson.** 1982. Glutamine synthetase subunit mixing and regulation in *Bacillus subtilis* partial diploids. *J. Bacteriol.* **149:**378–380.
48. **Graham, R. S., F. E. Young, and G. A. Wilson.** 1977. Effect of site-specific endonuclease digestion on the *thy*P3 gene of bacteriophage φ3T and the *thy*P11 gene of bacteriophage ρ11. *Gene* **1:**169–180.
49. **Graham, S., S. Sutton, Y. Yoneda, and F. E. Young.** 1982. Correlation of the genetic map and the endonuclease site map of *Bacillus subtilis* bacteriophage SPO2. *J. Virol.* **42:**131–134.
50. **Graham, S., Y. Yoneda, and F. E. Young.** 1979. Isolation and characterization of viable deletion mutants of *Bacillus subtilis* bacteriophage SPO2. *Gene* **7:**69–77.
51. **Greer, H.** 1975. The *kil* gene of bacteriophage lambda. *Virology* **66:**589–604.
52. **Günthert, U., R. Lauster, and L. Reiners.** 1986. Multispecific DNA methyltransferases from *Bacillus subtilis* phages: properties of wild-type and various mutant enzymes with altered DNA affinity. *Eur. J. Biochem.* **159:**485–492.
53. **Günthert, U., B. Pawlek, J. Stutz, and T. A. Trautner.** 1976. Restriction and modification in *Bacillus subtilis*: inducibility of a DNA methylating activity in nonmodifying cells. *J. Virol.* **20:**188–195.
54. **Günthert, U., L. Reiners, and R. Lauster.** 1986. Cloning and expression of *Bacillus subtilis* phage DNA methyltransferase genes in *Escherichia coli* and *B. subtilis*. *Gene* **41:**261–270.
55. **Gussin, G. N., A. D. Johnson, C. O. Pabo, and R. T. Sauer.** 1983. Repressor and Cro protein: structure, function, and role in lysogenization, p. 93–121. *In* R. W. Hendrix, J. W. Roberts, F. W. Stahl, and R. A. Weisberg (ed.), *Lambda II*. Cold Spring Harbor Laboratory, Cold Spring Harbor, N.Y.
56. **Haas, M., and H. Yoshikawa.** 1969. Defective bacteriophage PBS H in *Bacillus subtilis*. I. Induction, purification, and physical properties of the bacteriophage and its deoxyribonucleic acid. *J. Virol.* **3:**248–260.
57. **Haas, M., and H. Yoshikawa.** 1969. Defective bacteriophage PBSH in *Bacillus subtilis*. II. Intracellular development of the induced prophage. *J. Virol.* **3:**248–260.
58. **Hemphill, H. E., I. Gage, S. A. Zahler, and R. Z. Korman.** 1980. Prophage-mediated production of a bacteriocin-like substance by SPβ lysogens of *Bacillus subtilis*. *Can. J. Microbiol.* **26:**1328–1333.
59. **Hemphill, H. E., and H. R. Whiteley.** 1975. Bacteriophages of *Bacillus subtilis*. *Bacteriol. Rev.* **39:**257–315.
60. **Imae, Y., and T. Fukasawa.** 1970. Regional replication of the bacterial chromosome by derepression of prophage lambda. *J. Mol. Biol.* **54:**585–597.
61. **Inselburg, J. W., T. Eremenko-Volpe, L. Greenwald, W. L. Meadow, and J. Marmur.** 1969. Physical and genetic mapping of the SPO2 prophage on the chromosome of *Bacillus subtilis*. *J. Virol.* **3:**627–628.

62. **Ionesco, H., A. Ryter, and P. Schaeffer.** 1964. Sur un bactériophage hébergé par la souche Marburg de *Bacillus subtilis*. *Ann. Inst. Pasteur* **107:**764–776.
63. **Iwabe, N., K. I. Kuma, and T. Miyata.** 1989. Sequence similarity of bacteriophage SPO2 DNA polymerase with *Escherichia coli* polymerase I. *Nucleic Acids Res.* **17:**8866.
64. **Kawamura, F., H. Saito, Y. Ikeda, and J. Ito.** 1979. Viable deletion mutants of *Bacillus subtilis* phage ρ11. *J. Gen. Appl. Microbiol.* **25:**223–226.
65. **Kenny, E., T. Atkinson, and B. S. Hartley.** 1985. Nucleotide sequence of the thymidylate synthetase gene (*thyP3*) from the *Bacillus subtilis* phage φ3T. *Gene* **34:**335–342.
66. **Lange, C., A. Jugel, J. Walter, M. Noyer-Weidner, and T. A. Trautner.** 1991. 'Pseudo' domains in phage-encoded DNA methyltransferases. *Nature* (London) **352:**645–648.
67. **Lange, C., M. Noyer-Weidner, T. A. Trautner, M. Weiner, and S. A. Zahler.** 1991. M.H2I, a multispecific 5C-DNA methyltransferase encoded by *Bacillus amyloliquefaciens* phage H2. *Gene* **100:**213–218.
68. **Lauster, R.** 1989. Evolution of type II DNA methyltransferases: a gene duplication model. *J. Mol. Biol.* **206:**313–329.
69. **Lauster, R., T. A. Trautner, and M. Noyer-Weidner.** 1989. Cytosine-specific type II DNA methyltransferases: a conserved enzyme core with variable target-recognizing domains. *J. Mol. Biol.* **206:**305–312.
70. **Lipsky, R. H., R. Rosenthal, and S. A. Zahler.** 1981. Defective specialized SPβ transducing bacteriophages of *Bacillus subtilis* that carry the *sup-3* or *sup-44* gene. *J. Bacteriol.* **148:**1012–1015.
71. **Mackey, C. J., and S. A. Zahler.** 1982. Insertion of bacteriophage SPβ into the *citF* gene of *Bacillus subtilis*, and specialized transduction of the *ilvBC-leu* genes. *J. Bacteriol.* **151:**1222–1229.
72. **Marrero, R., F. A. Chiafari, and P. S. Lovett.** 1981. High-frequency elimination of SPO2 prophage from *Bacillus subtilis* by plasmid transformation. *J. Virol.* **39:**318–320.
73. **Marrero, R., and P. S. Lovett.** 1982. Interference of plasmid pCM194 with lysogeny of bacteriophage SPO2 in *Bacillus subtilis*. *J. Bacteriol.* **152:**284–290.
74. **Marrero, R., and R. E. Yasbin.** 1988. Cloning of the *Bacillus subtilis recE*$^+$ gene and functional expression of *recE*$^+$ in *B. subtilis*. *J. Bacteriol.* **170:**335–344.
75. **Mauel, C., and D. Karamata.** 1984. Characterization of proteins induced by mitomycin C treatment of *Bacillus subtilis*. *J. Virol.* **49:**806–812.
76. **McLaughlin, J. R., H. C. Wong, Y. E. Ting, J. N. Van Arsdell, and S. Chang.** 1986. Control of lysogeny and immunity of *Bacillus subtilis* temperate bacteriophage SPβ by its D gene. *J. Bacteriol.* **167:**952–959.
77. **Mele, J.** 1972. Biological characterization and prophage mapping of a lysogenizing bacteriophage for *Bacillus subtilis*. Ph.D. thesis. University of Massachusetts, Amherst.
78. **Neuhard, J., A. R. Price, L. Schack, and E. Thomassen.** 1978. Two thymidylate synthetases in *Bacillus subtilis*. *Proc. Natl. Acad. Sci. USA* **75:**1194–1198.
79. **Noyer-Weidner, M., S. Jentsch, B. Pawlek, U. Günthert, and T. A. Trautner.** 1983. Restriction and modification in *Bacillus subtilis*: DNA methylation potential of the related bacteriophages Z, SPR, SPβ, φ3T, and ρ11. *J. Virol.* **46:**446–453.
80. **Odebralski, J. M., and S. A. Zahler.** 1982. Specialized transduction of the *kauA* and *citK* genes of *Bacillus subtilis* by bacteriophage φ3T, abstr. H 101, p. 130. Abstr. Annu. Meet. Am. Soc. Microbiol. 1982.
81. **Okamoto, K., J. A. Mudd, J. Mangan, W. M. Huang, T. V. Subbaiah, and J. Marmur.** 1968. Properties of the defective phage of *Bacillus subtilis*. *J. Mol. Biol.* **34:**413–428.
82. **Okamoto, K., J. A. Mudd, and J. Marmur.** 1968. Conversion of *Bacillus subtilis* DNA to phage DNA following mitomycin C induction. *J. Mol. Biol.* **34:**429–437.
83. **Osburne, M. S., R. J. Craig, and D. M. Rothstein.** 1985. Thermoinducible transcription system for *Bacillus subtilis* that uses control elements from temperate phage φ105. *J. Bacteriol.* **163:**1101–1108.
84. **Osburne, M. S., and A. L. Sonenshein.** 1980. Inhibition by lipiarmycin of bacteriophage growth in *Bacillus subtilis*. *J. Virol.* **33:**945–953.
85. **Parker, A. P., and D. H. Dean.** 1986. Temperate *Bacillus* bacteriophage SP16 genome is circularly permuted and terminally redundant. *J. Bacteriol.* **167:**719–721.
86. **Peterson, A. M., and L. Rutberg.** 1969. Linked transformation of bacterial and prophage markers in *Bacillus subtilis* 168 lysogenic for bacteriophage φ105. *J. Bacteriol.* **98:**874–877.
87. **Piggot, P. J., M. Amjad, J.-J. Wu, H. Sandoval, and J. Castro.** 1990. Genetic and physical maps of *Bacillus subtilis* 168, p. 493–540. *In* C. R. Harwood and S. M. Cutting (ed.), *Molecular Biology Methods for Bacillus*. John Wiley & Sons, Inc., Chichester, United Kingdom.
88. **Ptashne, M.** 1986. A genetic switch: gene control and phage λ. Blackwell Scientific Publications, Palo Alto, Calif.
89. **Raden, B., and L. Rutberg.** 1984. Nucleotide sequence of the temperate *Bacillus subtilis* bacteriophage SPO2 DNA polymerase gene L. *J. Virol.* **52:**9–15.
90. **Rettenmeier, C. W., B. Gingell, and H. E. Hemphill.** 1979. The role of temperate bacteriophage SPβ in prophage-mediated interference in *Bacillus subtilis*. *Can. J. Microbiol.* **25:**1345–1351.
91. **Romig, W. R.** 1968. Infectivity of *Bacillus subtilis* bacteriophage deoxyribonucleic acids extracted from mature particles and from lysogenic hosts. *Bacteriol. Rev.* **32:**349–357.
92. **Rosenthal, R., P. A. Toye, R. Z. Korman, and S. A. Zahler.** 1979. The prophage of SPβ*c2dcitK*$_1$, a defective specialized transducing phage of *Bacillus subtilis*. *Genetics* **92:**721–739.
93. **Rowe, D. B., T. P. Iismaa, and R. G. Wake.** 1986. Nonrandom cosmid cloning and prophage SPβ homology near the replication terminus of the *Bacillus subtilis* chromosome. *J. Bacteriol.* **167:**379–382.
94. **Rudinski, M. S., and D. H. Dean.** 1978. Evolutionary considerations of related *B. subtilis* temperate phages φ105, ρ14, ρ10 and ρ6 as revealed by heteroduplex analysis. *Virology* **99:**57–65.
95. **Rutberg, L.** 1969. Mapping of a temperate bacteriophage active on *Bacillus subtilis*. *J. Virol.* **3:**38–44.
96. **Rutberg, L.** 1973. Heat induction of prophage φ105 in *Bacillus subtilis*: bacteriophage-induced bidirectional replication of the bacterial chromosome. *J. Virol.* **12:**9–12.
97. **Rutberg, L.** 1982. Temperate bacteriophages of *Bacillus subtilis*, p. 247–268. *In* D. A. Dubnau (ed.), *Molecular Biology of the Bacilli*. Academic Press, Inc., New York.
98. **Rutberg, L., R. W. Armentrout, and J. Jonasson.** 1972. Unrelatedness of temperate *Bacillus subtilis* bacteriophages SPO2 and φ105. *J. Virol.* **9:**732–737.
99. **Rutberg, L., B. Raden, and J.-I. Flock.** 1981. Cloning and expression of bacteriophage SPO2 DNA polymerase gene L in *Bacillus subtilis*, using the *Staphylococcus aureus* plasmid pC194. *J. Virol.* **39:**407–412.
100. **Salas, M.** 1988. Phages with protein attached to the DNA ends, p. 169–191. *In* R. Calendar (ed.), *The Bacteriophages*, vol. 1. Plenum Press, New York.
101. **Sargent, M. G., S. Davies, and M. F. Bennett.** 1985. Potentiation of a nucleolytic activity in *Bacillus subtilis*. *J. Gen. Microbiol.* **131:**2795–2804.

102. **Seaman, E., E. Tarmy, and J. Marmur.** 1964. Inducible phages of *Bacillus subtilis. Biochemistry* **3:**607–613.
103. **Shapiro, J. M., D. H. Dean, and H. O. Halvorson.** 1974. Low-frequency specialized transduction with *Bacillus subtilis* bacteriophage φ105. *Virology* **62:**393–403.
104. **Siegel, E. C., and J. Marmur.** 1969. Temperature-sensitive induction of bacteriophage in *Bacillus subtilis* 168. *J. Virol.* **4:**610–618.
105. **Smith, I., and H. Smith.** 1973. Location of the SPO2 attachment site and the bryamycin resistance marker on the *Bacillus subtilis* chromosome. *J. Bacteriol.* **114:** 1138–1142.
106. **Spancake, G. A., S. D. Daignault, and H. E. Hemphill.** 1987. Genome homology and divergence in the SPβ-related bacteriophages of *Bacillus subtilis. Can. J. Microbiol.* **33:**249–255.
107. **Spancake, G. A., and H. E. Hemphill.** 1985. Deletion mutants of *Bacillus subtilis* bacteriophage SPβ. *J. Virol.* **55:**39–44.
108. **Spancake, G. A., H. E. Hemphill, and P. S. Fink.** 1984. Genome organization of SPβ *c*2 bacteriophage carrying the *thyP3* gene. *J. Bacteriol.* **157:**428–434.
109. **Steensma, H. Y.** 1981. Adsorption of defective phage PBSZ1 to *Bacillus subtilis* 168 Wt. *J. Gen. Virol.* **52:**93–101.
110. **Steensma, H. Y.** 1981. Effect of defective phages on the cell membrane of *Bacillus subtilis* and partial characterization of a phage protein involved in killing. *J. Gen. Virol.* **56:**275–286.
111. **Steensma, H. Y., L. A. Robertson, and J. D. van Elsas.** 1978. The occurrence and taxonomic value of PBS X-like defective phages in the genus *Bacillus. Antonie van Leeuwenhoek* **44:**353–366.
112. **Stewart, C.** 1988. Bacteriophage SPO1, p. 477–515. *In* R. Calendar (ed.), *The Bacteriophages,* vol. 1. Plenum Press, New York.
113. **Stickler, D. J., R. G. Tucker, and D. Day.** 1965. Bacteriophage-like particles released from *Bacillus subtilis* after induction with hydrogen peroxide. *Virology* **26:** 142–145.
114. **Stroynowski, I. T.** 1981. Distribution of bacteriophage φ3T homologous deoxyribonucleic acid sequences in *Bacillus subtilis* 168, related bacteriophages, and other *Bacillus* species. *J. Bacteriol.* **148:**91–100.
115. **Stroynowski, I. T.** 1981. Integration of the bacteriophage φ3T-coded thymidylate synthetase gene into the *Bacillus subtilis* chromosome. *J. Bacteriol.* **148:**101–108.
116. **Terschueren, P. A., M. Noyer-Weidner, and T. A. Trautner.** 1987. Recombinant derivatives of *Bacillus subtilis* phage Z containing the DNA methyltransferase genes of related methylation-proficient phages. *J. Gen. Microbiol.* **133:**945–952.
117. **Thorne, C. B., and J. Mele.** 1974. Prophage interference in *Bacillus subtilis* 168. *Microb. Genet. Bull.* **36:**27–29.
118. **Thurm, P., and A. J. Garro.** 1975. Bacteriophage-specific protein synthesis during induction of the defective *Bacillus subtilis* bacteriophage PBSX. *J. Virol.* **16:**179–183.
119. **Thurm, P., and A. J. Garro.** 1975. Isolation and characterization of prophage mutants of the defective *Bacillus subtilis* bacteriophage PBSX. *J. Virol.* **16:**184–191.
120. **Tran-Betcke, A., B. Behrens, M. Noyer-Weidner, and T. A. Trautner.** 1986. DNA methyltransferase genes of *Bacillus subtilis* phages: comparison of their nucleotide sequences. *Gene* **42:**89–96.
121. **Trautner, T. A., T. S. Balganesh, and B. Pawlek.** 1988. Chimeric multispecific DNA methyltransferases with novel combinations of target recognition. *Nucleic Acids Res.* **16:**6649–6658.
122. **Trautner, T. A., T. Balganesh, K. Wilke, M. Noyer-Weidner, E. Rauhut, R. Lauster, B. Behrens, and B. Pawlek.** 1988. Organization of target-recognizing domains in the multispecific DNA (cytosine-5) methyltransferases of *Bacillus subtilis* phages SPR and φ3T. *Gene* **74:**267.
123. **Trautner, T. A., B. Pawlek, U. Gunthert, U. Canosi, S. Jentsch, and M. Freund.** 1980. Restriction and modification in *Bacillus subtilis*: identification of a gene in the temperate phage SPβ coding for a *Bsu*R specific modification methyltransferase. *Mol. Gen. Genet.* **180:**361–367.
124. **Tucker, R. G.** 1969. Acquisition of thymydylate synthetase activity by a thymine-requiring mutant of *Bacillus subtilis* following infection by the temperate phage φ3. *J. Gen. Virol.* **4:**489–504.
125. **Van Kaer, L., Y. Gansemans, M. Van Montagu, and P. Dhaese.** 1988. Interaction of the *Bacillus subtilis* phage φ105 repressor with operator DNA: a genetic analysis. *EMBO J.* **7:**859–866.
126. **Van Kaer, L., M. Van Montagu, and P. Dhaese.** 1987. Transcriptional control in the *Eco*RI-F immunity region of *Bacillus subtilis* phage φ105: identification and unusual structure of the operator. *J. Mol. Biol.* **197:**55–67.
127. **Van Kaer, L., M. Van Montagu, and P. Dhaese.** 1989. Purification and *in vitro* DNA-binding specificity of the *Bacillus subtilis* phage φ105 repressor. *J. Biol. Chem.* **264:**14784–14791.
128. **Walter, J., M. Noyer-Weidner, and T. A. Trautner.** 1990. The amino acid sequence of the CCGG recognizing DNA methyltransferase M.BsuFI: implications for the analysis of sequence recognition by cytosine DNA methyltransferases. *EMBO J.* **9:**1007–1013.
129. **Warner, F. D., G. A. Kitos, M. P. Romano, and H. E. Hemphill.** 1977. Characterization of SPβ: a temperate bacteriophage from *Bacillus subtilis* 168M. *Can. J. Microbiol.* **23:**45–51.
130. **Weiner, M. P.** 1986. Characterization of phage H2. Ph.D. thesis. Cornell University, Ithaca, N.Y.
131. **Weiner, M. P., and S. A. Zahler.** 1988. Genome homology and host range of some SPβ-related bacteriophages of *Bacillus subtilis* and *Bacillus amyloliquefaciens. J. Gen. Virol.* **69:**1307–1316.
132. **Weiss, A. S., M. T. Smith, T. P. Iismaa, and R. G. Wake.** 1983. Cloning DNA from the replication terminus region of the *Bacillus subtilis* chromosome. *Gene* **24:**83–91.
133. **Wilke, K., E. Rauhut, M. Noyer-Weidner, R. Lauster, B. Pawlek, B. Behrens, and T. A. Trautner.** 1988. Sequential order of target-recognizing domains in multispecific DNA-methyltransferases. *EMBO J.* **7:**2601–2610.
134. **Williams, M. T., and F. E. Young.** 1977. Temperate *Bacillus subtilis* bacteriophage φ3T: chromosomal attachment site and comparison with temperate bacteriophages φ105 and SPO2. *J. Virol.* **21:**522–529.
135. **Wilson, G. A., M. T. Williams, H. W. Baney, and F. E. Young.** 1974. Characterization of temperate bacteriophages of *Bacillus subtilis* by the restriction endonuclease *Eco*RI: evidence for three different temperate bacteriophages. *J. Virol.* **14:**1013–1016.
136. **Wood, H. E., M. T. Dawson, K. M. Devine, and D. J. McConnell.** 1990. Characterization of PBSX, a defective prophage of *Bacillus subtilis. J. Bacteriol.* **172:**2667–2674.
137. **Wood, H. E., K. M. Devine, and D. J. McConnell.** 1990. Characterization of a repressor gene (*xre*) and a temperature-sensitive allele from the *Bacillus subtilis* prophage PBSX. *Gene* **96:**83–88.
138. **Wulff, D. L., and M. Rosenberg.** 1983. Establishment of repressor synthesis, p. 53–73. *In* R. W. Hendrix, J. W. Roberts, F. W. Stahl, and R. A. Weisberg (ed.), *Lambda II.* Cold Spring Harbor Laboratory, Cold Spring Harbor, N.Y.
139. **Yarmolinsky, M. B., and N. Sternberg.** 1988. Bacteriophage P1, p. 291–438. *In* R. Calendar (ed.), *The Bacteriophages,* vol. 1. Plenum Press, New York.

140. **Yasbin, R. E., P. I. Fields, and B. J. Andersen.** 1980. Properties of *Bacillus subtilis* 168 derivatives freed of their natural prophage. *Gene* **12:**155–159.
141. **Yasbin, R. E., V. C. Maino, and F. E. Young.** 1976. Bacteriophage resistance in *Bacillus subtilis* 168, W23 and interstrain transformants. *J. Bacteriol.* **125:**1120–1126.
142. **Yasbin, R. E., G. A. Wilson, and F. E. Young.** 1973. Transformation and transfection in lysogenic strains of *Bacillus subtilis* 168. *J. Bacteriol.* **113:**540–548.
143. **Yasbin, R. E., G. A. Wilson, and F. E. Young.** 1975. Transformation and transfection in lysogenic strains of *Bacillus subtilis*: evidence for selective induction of prophage in competent cells. *J. Bacteriol.* **121:**296–304.
144. **Yasbin, R. E., G. A. Wilson, and F. E. Young.** 1975. Effect of lysogeny on transfection and transfection enhancement in *Bacillus subtilis*. *J. Bacteriol.* **121:**305–312.
145. **Yasbin, R. E., and F. E. Young.** 1972. The influence of temperate bacteriophage ϕ105 on transformation and transfection in *Bacillus subtilis*. *Biochem. Biophys. Res. Commun.* **47:**365–371.
146. **Yasunaka, A., H. Tsukamato, S. Okubo, and T. Horiuchi.** 1970. Isolation and properties of suppressor-sensitive mutants of *Bacillus subtilis* bacteriophage SPO2. *J. Virol.* **5:**819–821.
147. **Yoneda, Y., S. Graham, and F. E. Young.** 1979. Restriction-fragment map of the temperate *Bacillus subtilis* bacteriophage SPO2. *Gene* **7:**51–68.
148. **Zahler, S. A.** 1982. Specialized transduction in *Bacillus subtilis*, p. 269–305. *In* D. A. Dubnau (ed.), *Molecular Biology of the Bacilli*, vol. 1. Academic Press, Inc., New York.
149. **Zahler, S. A.** 1988. Temperate bacteriophages of *Bacillus subtilis*, p. 559–592. *In* R. Calendar (ed.), *The Bacteriophages*, vol. 1. Plenum Press, Inc., New York.
150. **Zahler, S. A., R. Z. Korman, J. M. Odebralski, P. S. Fink, C. J. Mackey, C. G. Poutre, R. H. Lipsky, and P. J. Youngman.** 1982. Genetic manipulations with phage SPβ, p. 41–50. *In* J. A. Hoch, S. Chang, and A. T. Ganesan (ed.), *Molecular Cloning and Gene Regulation in Bacillus*. Academic Press, Inc., New York.
151. **Zahler, S. A., R. Z. Korman, R. Rosenthal, and H. E. Hemphill.** 1977. *Bacillus subtilis* bacteriophage SPβ: localization of the prophage attachment site, and specialized transduction. *J. Bacteriol.* **129:**556–558.
152. **Zahler, S. A., R. Z. Korman, C. Thomas, P. S. Fink, M. P. Weiner, and J. M. Odebralski.** 1987. H2, a temperate bacteriophage isolated from *Bacillus amyloliquefaciens* strain H. *J. Gen. Microbiol.* **133:**2937–2944.
153. **Zahler, S. A., R. Z. Korman, C. Thomas, and J. M. Odebralski.** 1987. Temperate bacteriophages of *Bacillus amyloliquefaciens*. *J. Gen. Microbiol.* **133:**2933–2935.

58. Replication and Transcription of Bacteriophage ϕ29 DNA

MARGARITA SALAS and FERNANDO ROJO

REPLICATION

Characterization of TP at 5′ Ends of ϕ29 DNA

The genome of *Bacillus subtilis* phage ϕ29 consists of a linear, double-stranded DNA 19,285 bp long (40, 104, 115) with a 6-bp-long inverted terminal repeat (AAAGTA; 32, 114) and a terminal protein (TP) covalently linked at the 5′ ends (Fig. 1). The first evidence for the existence of a TP at the ϕ29 DNA ends was the finding of circular structures and concatemers when the DNA was isolated from viral particles and their conversion into unit-length linear DNA after treatment with proteolytic enzymes (74). Moreover, the capacity to transfect competent *B. subtilis* cells was lost by treatment of ϕ29 DNA with proteolytic enzymes, and the DNA isolated from a ϕ29 temperature-sensitive mutant in gene 3 was thermolabile for transfection (112). Later on, it was found that a protein of 31,000 Da, characterized as the product of viral gene 3 (90), is covalently linked at the two 5′ ends of ϕ29 DNA (55, 90, 113). Other *B. subtilis* phages morphologically similar to ϕ29, such as ϕ15, PZA, PZE, Nf, M2, B103, and GA-1, also have linear double-stranded DNA with short inverted terminal repeats and TP covalently linked at the 5′ ends (89).

The linkage between ϕ29 TP (266 amino acids long; 33) and DNA is a phosphoester bond between the OH group of serine residue 232 and 5′-dAMP (48, 49). TP can be released from the DNA by treatment with alkali. Secondary-structure predictions suggest that serine 232 is located in a β turn preceded by an α helix (48).

In Vivo Replication

Replicative intermediates

Analysis by electron microscopy of the replicative intermediates produced in ϕ29-infected *B. subtilis* reveals two main types of replicating molecules, named type I and type II (54, 97; Fig. 2). Type I molecules, accounting for about 20% of the total number of replicating molecules, consist of double-stranded DNA with single-stranded tails; the single-stranded tails are the same length as one of the double-stranded regions, indicating that ϕ29 DNA replication starts at or close to the DNA ends. Type II molecules, which account for about 55% of the total number of replicating molecules, consist of unit-length DNA partially double stranded and partially single stranded. Partial denaturation experiments indicate that replication can start with similar frequency from either ϕ29 DNA end. Type I molecules with a single-stranded branch from each DNA end are also seen; their presence offers support of a model in which type II molecules are formed when two replication forks, one from each DNA end, merge (see below).

Genes required for replication

Viral genes 1, 2, 3, 5, 6, and 17 (Fig. 1A) are required for ϕ29 DNA replication (28, 47, 81, 102). Genes 2 and 3 are involved in initiation, whereas gene 5 is involved in elongation; the results with gene 6 suggest that it is involved in some elongation step, although the possibility that it is also involved in initiation was not ruled out (70). Three genes (G, E, and T) are involved in phage M2 DNA replication. Genes G and E correspond to ϕ29 genes 2 and 3, respectively (65).

In vivo ϕ29 DNA replication does not require bacterial DNA polymerases I or III, since replication occurs in PolA$^-$ mutants and in the presence of 6-(*p*-hydroxyphenylazo)uracil. Development of ϕ29 also takes place normally in most of the *B. subtilis* replication mutants available (reviewed in reference 88).

In vivo replication of both ϕ29 and M2 is inhibited by aphidicolin (51), a known inhibitor of eukaryotic DNA polymerase α (reviewed in reference 53).

In Vitro Replication

Viral proteins essential for initiation of replication

Extracts of ϕ29-infected *B. subtilis* incubated with [α-^{32}P]dATP in the presence of ϕ29 DNA-TP complex as a template give rise to the formation of a ^{32}P-labeled protein with the electrophoretic mobility of the ϕ29 TP. This labeled protein is not found in the presence of anti-TP serum or when extracts from uninfected cells are used (79, 95, 96, 107, 108). Incubation of the ^{32}P-labeled protein with piperidine under conditions that hydrolyze the linkage between the TP and DNA release dAMP, indicating the formation of a TP-dAMP complex (79). When extracts from M2-infected *B. subtilis* are used in the presence of M2 TP-DNA as template, a covalent complex of the M2 TP and dAMP is found (65).

By using extracts from *B. subtilis* infected under restrictive conditions with ϕ29 mutants in replication genes, it was shown that genes 2 and 3 are essential for the formation of the TP-dAMP initiation complex (16, 65). In the case of phage M2, genes G and E were also

Margarita Salas and Fernando Rojo • Centro de Biología Molecular (CSIC-UAM), Universidad Autónoma, Cantoblanco, 28049 Madrid, Spain.

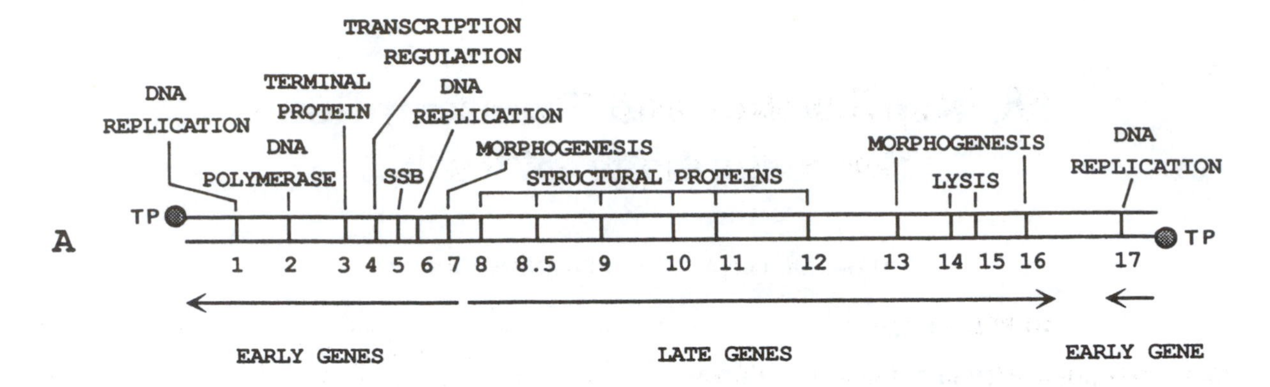

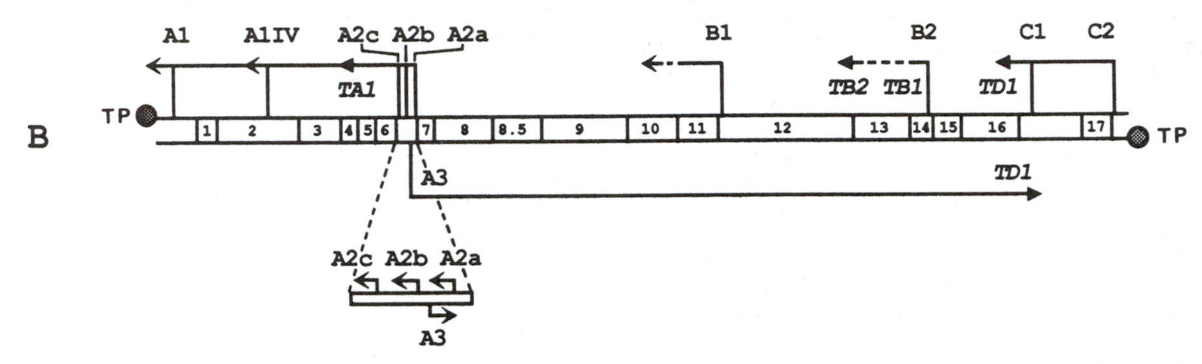

Figure 1. Genetic and transcriptional maps of phage ϕ29 genome. (A) Genetic map (adapted from reference 69). Arrows indicate directions of transcription. The TP is shown attached to the 5′ ends of the genome. (B) Transcription map. Arrowheads indicate directions of transcription from the different promoters (see text). The region containing the main early and late promoters is magnified to show the relative position of each promoter. Transcription terminators (*TA1*, *TB1*, *TB2*, and *TD1*) are represented by thick, filled arrowheads. The position and approximate length of each gene is indicated. SSB, single-stranded-DNA-binding protein.

shown to be essential for in vitro formation of the M2 TP-dAMP complex (65).

The gene 2 product (p2), which is about 66,000 Da, is a DNA polymerase that is able to catalyze both the initiation and the polymerization steps of ϕ29 DNA replication (20, 105). It also has 3′ → 5′ exonuclease activity (21, 106) that is about 10-fold more efficient with single-stranded than double-stranded DNA and is able to excise mispaired nucleotides better than correctly paired ones (37). Therefore, it has properties expected for an enzyme involved in proofreading. In addition, the ϕ29 DNA polymerase has pyrophosphorolytic activity (24), which is the reverse of polymerization.

The in vitro initiation reaction with ϕ29 TP and DNA polymerase is strongly stimulated by NH_4^+ ions (22) because of the formation of a stable heterodimer between the two proteins (19; Fig. 2). More recently, the formation of the TP-dAMP initiation complex was shown to exhibit a strong preference for Mn^{2+} as metal activator. The molecular basis for this preference is mainly the result of a large increase in affinity for dATP (34).

In agreement with the existence of a functional heterodimer between ϕ29 DNA polymerase and TP is the fact that in the absence of template and the presence of [α-^{32}P]dATP and Mn^{2+} as divalent cation, ϕ29 DNA polymerase is able to catalyze the formation of a covalent complex between TP and dAMP in a reaction stimulated by NH_4^+ ions (12). The TP-dAMP linkage is identical to that found between TP and ϕ29 DNA. However, unlike the template-dependent reaction, the DNA-independent deoxynucleotidylation of TP by ϕ29 DNA polymerase does not show dATP specificity, as any of the four TP-deoxynucleoside monophosphate (dNMP) complexes can be obtained with a similar yield. In addition, the DNA-independent reaction is strongly dependent on the deoxynucleoside triphosphate (dNTP) concentration, indicating that the presence of template probably contributes to stabilizing a complementary nucleotide, thereby giving base specificity to the protein-primed initiation reaction.

A purified system consisting of TP (80), DNA polymerase (20), and ϕ29 DNA-TP as template (79) not only generates a TP-dAMP initiation complex but also carries out the synthesis of unit-length ϕ29 DNA (13, 22). These results suggest that ϕ29 DNA polymerase is able to carry out strand displacement. Indeed, when primed M13 single-stranded DNA is used as template, ϕ29 DNA polymerase is able to synthesize DNA chains of more than 70 kb, indicating the ability of this polymerase to carry out strand displacement in the absence of accessory proteins (13). Moreover, ϕ29 DNA polymerase is a very processive enzyme (13).

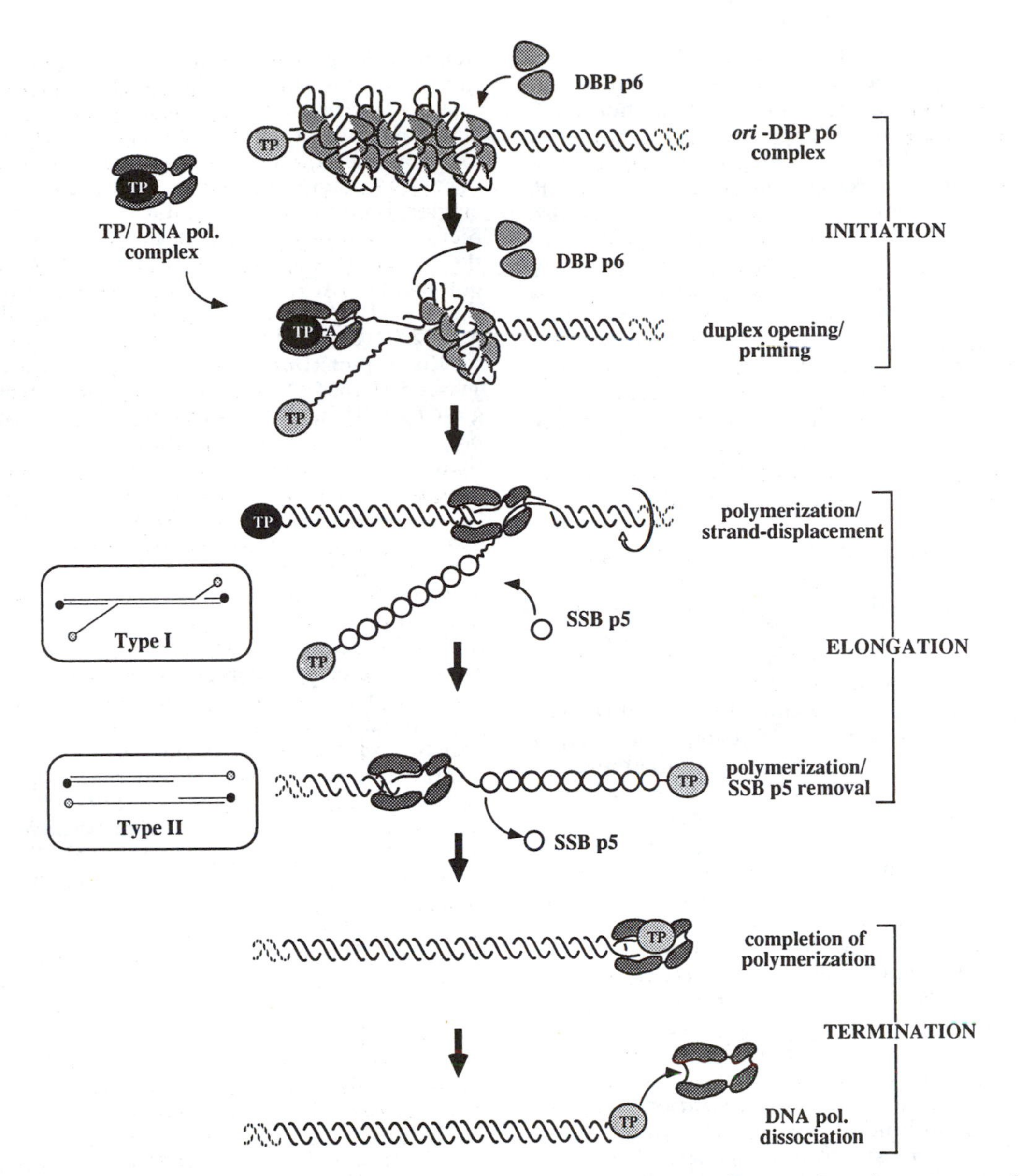

Figure 2. Different stages and viral gene products involved in ϕ29 DNA replication. Based on structural and functional studies (see text for details), a two-domain structure analogous to that of the Klenow fragment of polymerase I is extrapolated for ϕ29 DNA polymerase. ϕ29 primer TP is indicated by black, and parental TP is shaded. DBP, DNA-binding protein; SSB, single-stranded-DNA-binding protein.

Functional domains in ϕ29 DNA polymerase

ϕ29 DNA polymerase is inhibited by drugs that are known inhibitors of eukaryotic DNA polymerase α (53, 56) such as aphidicolin, phosphonoacetic acid, and the nucleotide analogs butyl-anilino dATP (BuAdATP) and butyl-phenyl dGTP (BuPdGTP) (10, 23).

In agreement with these observations, three main regions of amino acid sequence homology found at the carboxyl regions of several eukaryotic α-like DNA polymerases of viral and cellular origin are also found in ϕ29, M2, and other DNA polymerases from TP-containing genomes (10; reviewed in reference 89). These three regions, named 1, 2, and 3, are characterized by the amino acid sequence motifs D--SLYP, K---NS-YG, and YGDTDS, respectively. These regions have been proposed to form the catalytic site involved in dNTP binding (41). More recently, significant amino acid sequence similarity has been found in the C-terminal portions of 27 DNA-dependent DNA polymerases belonging not only to the α-like DNA polymerases described above but also to the *Escherichia coli* DNA polymerase I-like prokaryotic DNA polymerases. The six most conserved regions (1, 2a, 2b, 3, 4, and 5), which together span about 340 amino acids, are located in the same linear arrangement and contain highly conserved motifs and critical residues involved in the polymerization function (11). Regions 1, 2a, and 3 correspond to regions 1, 2, and 3 previously described by Bernad et al. (10).

ϕ29 DNA polymerase mutants in regions 1, 2a, 3,

and 4 have been obtained, overproduced, and purified, and their activities have been determined. Thus, mutant Tyr-254 → Phe (Y254F), at region 1, has normal 3′ → 5′ exonuclease activity but is strongly impaired in the protein-primed initiation step of φ29 DNA replication, showing a decreased affinity for Me^{2+}-dATP, the initiating nucleotide; the mutant protein is also affected in polymerization when Mg^{2+}-dNTPs are used as substrates but not when Mn^{2+}-dNTPs are used. These results indicate that residue Tyr-254 is involved, directly or indirectly, in Me^{2+}-dNTP binding (25b). Mutants Y390F and Y390S, at region 2a, have normal 3′ → 5′ exonuclease and initiation activities, but they are affected in polymerization in the presence of Mg^{2+}-dNTP but not of Mn^{2+}-dNTP. The two mutant proteins are hypersensitive to the nucleotide analogs BuAdATP and BuPdGTP in a template-dependent way (25b). Mutant G391D, also at region 2a, has normal 3′ → 5′ exonuclease activity but undetectable polymerization and φ29 DNA-dependent initiation activities. Interestingly, DNA-independent deoxyadenylation of the TP is higher than that of the wild-type DNA polymerase (25c). The results given above suggest that residues Tyr-390 and Gly-391 are involved in DNA-dependent dNTP binding.

Mutants Y454F, C455G, D456G, T457P, and D458G, at region 3, have normal 3′ → 5′ exonuclease activities. Initiation and elongation activities (both processive and nonprocessive) are undetectable in the T457P and D458G mutant proteins. The remaining mutant proteins, except the C455G one, have greatly reduced processive elongation activity (although the nonprocessive elongation is normal) and initiation activity except in the D456G mutant protein, which has reduced initiation activity due to a decrease in the affinity for dATP (9). On the basis of these results and of homology of these sequences with known or putative metal-binding amino acid sequences, it has been proposed that the YCDTDS motif in φ29 DNA polymerase is involved in synthetic activities of the polymerase (i.e., initiation and polymerization), particularly in the metal binding associated with the dNTP site (9). The pyrophosphorolytic activity of region 3 mutant DNA polymerases except for the C455G mutant protein is very reduced (24).

Mutant K498R, at region 4, has normal 3′ → 5′ exonuclease activity, high initiation activity, essentially normal nonprocessive elongation activity, and reduced processive polymerization; in contrast, mutant K498T has undetectable initiation activity (25a). Residue Lys-498 has been proposed (11) to correspond to residue Lys-758 in *E. coli* DNA polymerase I, which is involved in dNTP binding (77, 87).

In addition to the mutants obtained by site-directed mutagenesis, temperature-sensitive mutants with mutations in φ29 gene 2, ts2(24), and ts2(98) (101) were characterized. Mutation ts2(24) causes an A-to-V substitution at residue 355, which is located between regions 1 and 2a in a conserved region present only in protein-primed DNA polymerases (25); the ts2(24) DNA polymerase has a temperature sensitivity phenotype in both the initiation and the elongation reactions. Mutation ts2(98) causes an A-to-V substitution at residue 492, which is in region 4; the DNA polymerase is insensitive to temperature in vitro, but initiation and elongation activities are stimulated by Mn^{2+} to a greater extent than the wild-type enzyme is (25).

On the other hand, three regions of conserved amino acids, named Exo I, Exo II, and Exo III, have been found at the amino-terminal ends of 33 DNA-dependent DNA polymerases of prokaryotic and eukaryotic origins, both viral and cellular (8, 11, 15, 96a). These regions contain the critical amino acids involved in the 3′ → 5′ exonuclease activity in *E. coli* DNA polymerase I (29, 35, 73). Site-directed mutations in the φ29 DNA polymerase at amino acid residues Asp-12 (D12A) and Glu-14 (E14A) in region Exo I, at residue Asp-66 (D66A) in region Exo II, and at residues Tyr-165 (Y165F, Y165C) and Asp-169 (D169A) in region Exo III have been obtained. The mutant DNA polymerases were overproduced and purified, and their activities were analyzed (8, 96a). None of the mutations affect initiation or polymerization activities; however, they drastically reduce the 3′ → 5′ exonuclease activity to less than 1% that of the wild-type φ29 DNA polymerase except in mutants Y165F and Y165C, in which 3′ → 5′ exonuclease activities are reduced to 8 and 4%, respectively, relative to that of the wild-type protein. These results indicate that the 3′ → 5′ exonuclease domain of the φ29 DNA polymerase is located at the amino end of the protein, involving critical amino acids in conserved regions Exo I, Exo II, and Exo III, as is the case with the Klenow fragment of *E. coli* DNA polymerase I. Interestingly, all the φ29 DNA polymerase mutants with mutations in regions Exo I, Exo II, and Exo III have ~100-fold reduced their capacities to replicate φ29 DNA because of their inabilities to produce strand displacement, which is an intrinsic property of wild-type φ29 DNA polymerase. Taking into account these results, the strand displacement ability of φ29 DNA polymerase seems to reside in the amino-terminal domain, probably overlapping the 3′ → 5′ exonuclease active site (96a; Fig. 2).

The results obtained with φ29 DNA polymerase together with the amino acid sequence homology data make it likely that DNA polymerases have a general three-dimensional structure similar to that of the Klenow fragment of *E. coli* DNA polymerase I (73) and that a "Klenow-like core" containing the DNA polymerase and 3′ → 5′ exonuclease activities has evolved from a common ancestor that gave rise to the present-day prokaryotic and eukaryotic DNA polymerases (11).

Fidelity of φ29 DNA polymerase

The insertion discrimination of φ29 DNA polymerase during DNA polymerization in the presence of Mg^{2+} ranges from 5×10^4 to 2×10^6, depending on the specific nucleotide pair involved. This discrimination is reduced 1- to 100-fold when Mn^{2+} instead of Mg^{2+} is used as metal activator. The efficiency of mismatch elongation in the presence of Mg^{2+} is 10^5- to 10^6-fold lower than the extension efficiency of a properly paired primer terminus. These facts indicate that in the presence of Mg^{2+} as metal activator, DNA polymerization catalyzed by φ29 DNA polymerase is a highly accurate process (34a).

On the other hand, the insertion fidelity of protein-primed initiation is quite low, with an insertion discrimination factor of about 10^2. This factor is unaffected when Mn^{2+} instead of Mg^{2+} is used, because Mn^{2+} increases the affinity for both correct and incorrect nucleotides to a similar extent. Moreover, mismatch elongation discrimination is also rather low: mismatched TP-dNMP complexes are elongated from two- to sixfold more slowly than the correct TP-dNMP complex. In addition, the 3′ → 5′ exonuclease activity of ϕ29 DNA polymerase is unable to act on the TP-dNMP initiation complex, precluding the possibility that a wrong dNMP covalently linked to a TP can be excised and corrected (34a). Therefore, taking into account the three fidelity parameters indicated, protein-primed initiation could be predicted to be an inaccurate reaction. Maintenance of correct sequences at the DNA ends will be discussed below in the context of a new mechanism for protein-primed initiation.

Functional domains in ϕ29 TP

By using deletion mutants, two ϕ29 DNA polymerase-binding regions located at positions 72 to 80 and 241 to 261 were found in the ϕ29 TP (116). In addition, the amino acid regions in the TP at positions 13 to 18, 30 to 51, and 56 to 71 are important for DNA binding (116). A cluster of positively charged amino acids, KKKY, at positions 32 to 35 is flanked by aromatic and hydrophobic residues; this sequence is conserved in the TP of phage Nf (62), is present in several α-like DNA polymerases (103), and has been proposed as being involved in DNA binding (83).

By using site-directed mutagenesis at or near the DNA-linking site of the ϕ29 TP, it has been shown that serine 232, which is involved in covalent linkage to dAMP, cannot be changed to threonine (39). When serine 232 is changed to cysteine, the purified mutant TP has 0.7% of the priming activity of the wild-type TP (38). Change of residue leucine 220, serine 223, or serine 226 into proline gives rise to TP mutants with 3, 140, and 1% of the priming activity of the wild-type protein, respectively (38). All these mutant TPs are able to interact with ϕ29 DNA polymerase and DNA, suggesting that in addition to serine 232, leucine 220 and serine 226 form part of a functional domain involved in the process of initiation of DNA replication. These three amino acids are conserved in the TP of phage Nf (62).

The amino acid sequence RGD, which is found in cell adhesion proteins, is present at positions 256 to 258 in the ϕ29 and M2 TPs (60). The synthetic peptide RGD but not peptide RGE inhibits transfection of ϕ29 and M2 DNAs as well as the in vitro initiation reaction. Recent evidence suggests that the RGD sequence of the TP is responsible for the interaction of the TP-DNA polymerase heterodimer with the parental TP bound to ϕ29 DNA (58).

In addition, the ϕ29 and M2 TPs have the sequence KKIPPDD, which is similar to the sequence KKGCP PDD found in the β subunit of fibronectin receptor protein. A synthetic 20-mer peptide containing the KKIPPDD sequence interferes with the inhibitory effect of the RGD peptide, which suggests that the KKIPPDD sequences of the ϕ29 and M2 parental TPs are the receptor sequence for RGD in the primer TPs (59).

Gene 1 and 17 products

The functions of the gene 1 (ORF6 in the ϕ29 DNA sequence) (115) product, which is 10,000 Da (81), and the gene 17 product, which is 19,000 Da, are unknown (40). Transfection of *B. subtilis* with ϕ29 DNA-TP complex isolated from mutant sus17(112) or with recombinant ϕ29 DNA-TP molecules lacking gene 17 indicated that protein p17 is partially dispensable (31). The possibility that some bacterial protein can substitute for protein p17 remains open. From a comparison of the gene 17 nucleotide sequences of phages ϕ29, ϕ15, and PZA it can be concluded that deletions at the central and carboxy-terminal parts of the gene are tolerated, whereas 83 amino acids from the amino-terminal end of the protein have to be conserved to produce a functional protein (7).

Gene 5 product

Gene 5, which corresponds to ORF10 in the ϕ29 DNA sequence (115), encodes a protein of 13,000 Da (p5) (64). p5 is very abundant in ϕ29-infected cells and binds to single-stranded DNA (63). Electron microscopy of p5 bound to M13 single-stranded DNA shows a twofold reduction in the DNA length upon p5 binding (45). p5 greatly stimulates ϕ29 DNA-TP replication at incubation times when replication in the absence of p5 levels off. The effect of p5 is not on the formation of the TP-dAMP initiation complex or on the rate of elongation (63). p5 binds to single-stranded DNA in the ϕ29 replicative intermediates produced in vitro, which are similar to those observed in vivo. In addition to the stimulation of ϕ29 DNA-TP replication, p5 stimulates primed M13 single-stranded-DNA replication catalyzed by ϕ29 DNA polymerase. Other single-stranded-DNA-binding proteins, such as *E. coli* SSB, T4 gp32, adenovirus DBP, and human replication factor A, can functionally substitute for p5 (45). All these results strongly suggest that protein p5 functions as a single-stranded-DNA-binding protein during ϕ29 DNA replication (Fig. 2).

Gene 6 product

The product of gene 6 is predicted to be 12,000 Da, but when purified, the product behaves as a 23,600-Da species, indicating that the native form of p6 is a dimer (78). p6 stimulates formation of the TP-dAMP initiation complex (78) by a decrease in the K_m for dATP (18). p6 also stimulates ϕ29 DNA-TP replication, the effect being on the transition from the TP-dAMP complex to the first elongation products, TP-$(dAMP)_2$ and TP-$(dAMP)_3$ (14).

Gel retardation experiments showed that p6 binds to double-stranded DNA. DNase I footprinting experiments showed that binding is cooperative and occurs preferentially to fragments containing the left or right terminal sequences of ϕ29 DNA, producing a pattern of hypersensitive bands regularly spaced 24 bp apart that flank protected regions (82). This pattern extends 200 to 300 bp from each ϕ29 DNA end (93). Deletion analysis showed that p6 recognition signals are chiefly located between nucleotides 62 and 125 at the right ϕ29 DNA end and between nucleotides 46 and 68 at the left end. Since there is no sequence similarity in these regions, p6 does not seem to recognize a specific sequence in the DNA but rather a structural feature such as bendability. In fact, the p6 recognition regions were predicted to be bendable every 12 bp (93), which is the periodicity of binding one p6 monomer, as shown by hydroxyl radical footprinting (94). Bendability was shown to be a determinant in p6 recognition by the fact that direct repeats of a 24-bp sequence predicted to be bendable are recognized by p6 in the same positions and with higher affinity than ϕ29 DNA terminal sequences (93a). Footprint studies reveal protections to hydroxyl radical every 12 bp in both strands across the minor groove and to DNase I hypersensitivity every 24 bp between contact sites at alternating strands. This pattern suggests that p6 bends or kinks DNA every 12 bp. Taking into account these results and the fact that p6 restrains positive supercoiling in covalently closed circular DNA (82), a model in which a DNA right-handed superhelix tightly wraps around a multimeric p6 core, with a p6 dimer binding every 24 bp, was proposed (94; Fig. 2).

Plasmids containing different numbers of p6 binding units (24 bp) were used to visualize by electron microscopy the formation of the nucleoprotein complex, taking into account the fact that p6 prevents DNA from psoralen cross-linking. These plasmids were also used to measure the linking number change restrained by a single p6 dimer. In addition, the p6-DNA complex was visualized directly by glutaraldehyde fixation, and DNA compaction was measured. These data, together with the DNA helical repeat in the complex (12 bp), indicate that one superhelical turn is about 63 bp long (93a).

Activation of the initiation of ϕ29 DNA replication by p6 requires not only formation of the complex but also its correct positioning relative to the ϕ29 DNA ends, suggesting that other proteins involved in the initiation of ϕ29 DNA replication (TP and/or DNA polymerase) recognize p6 at a precise position (94). The fact that stimulation by p6 is highest at low temperature (0 to 10°C; 93b) and high salt (over 100 mM NaCl; 14) suggests that p6 favors the opening of the ϕ29 DNA ends.

The DNA-binding region in p6 has been mapped by construction of deletion and site-directed mutants of the protein. The activity of p6 decreases greatly when 5 amino acids are removed from the amino end and is undetectable by deletion of 13 amino acids. The two mutant proteins are unable to interact specifically with the ϕ29 DNA ends, although they can interfere with binding of the wild-type p6, suggesting that the amino end of p6 is involved in DNA binding (75). In contrast, the 14 carboxy-terminal amino acids of p6 are fully dispensable (76).

Secondary-structure predictions indicate the existence at the amino end of p6 of an amphipathic α-helix in which polar and charged amino acids are at one side of the α helix and nonpolar amino acids are at the other side. To test the possibility that the charged residues at the amino-terminal end of p6 are involved in DNA interaction, Lys-2 and Arg-6 were changed into Ala. The initiation activity of the R6A mutant p6 and its ability to interact with DNA were completely abolished, whereas the K2A mutant protein had reduced but detectable DNA binding and initiation activities (35a).

Replication of the Displaced DNA Strand

Recombinant ϕ29 DNA molecules containing parental TP at only one DNA end have been constructed and tested for their abilities to be replicated in vivo and in vitro. No replication is obtained after transfection of *B. subtilis* protoplasts with such recombinant molecules, suggesting that a completely displaced strand cannot be used as a template in vivo (31). In agreement with the in vivo results, when the above recombinant DNA molecules are used as templates in the ϕ29 in vitro replication system, type I but not type II molecules are synthesized as replicative intermediates, and unit-length single-stranded DNA accumulates (46). Moreover, when natural ϕ29 DNA-TP is used as template in the in vitro system, type II replicative intermediates appear well before unit-length ϕ29 DNA molecules are synthesized (46). These results, together with the fact that type I replication molecules with one single-stranded tail coming from each DNA end are observed both in vivo (54) and in vitro (46), support a model in which initiation of replication occurs from both DNA ends of the same molecule nonsimultaneously; type II molecules are therefore produced by separation of the replicating molecules when the two replication forks meet rather than by replication of fully displaced single strands.

Replication Origin

Terminal fragments from left or right ϕ29 DNA ends containing parental TP are active templates for TP-primed initiation and elongation reactions, whereas proteinase K-treated fragments, in which a 10-amino-acid peptide remains attached, are inactive (36). However, the same fragments treated with piperidine, which hydrolyzes the linkage between TP and DNA, are active, although the activity is reduced 5- to 10-fold from that obtained with TP-containing fragments (43). Piperidine-treated terminal fragments from the left and right ϕ29 DNA ends were cloned to study DNA sequence requirements. Fragments released by restriction nuclease digestion and containing ϕ29 DNA terminal sequences at the DNA ends are active templates for TP-primed initiation and elongation. No template activity is obtained when ϕ29 DNA sequences are present in the circular plasmid or are not placed at the DNA end (43). The minimal origins of replication are located within the terminal 12 bp at each ϕ29 DNA end, as indicated from the analysis of deletion derivatives (44). Point site-directed and random mutagenesis showed that a change of the second or third A into a C abolishes TP-primed initiation and elongation, whereas changes at positions 4 to 12 do not greatly affect the template activity (44). These results suggest that sequence requirements at the end-proximal region of the replication origin are stricter than those at the distal region.

A Sliding-Back Mechanism for TP-Primed DNA Replication

In agreement with the idea that p6 favors the opening of the double helix at ϕ29 DNA ends to initiate replication, single-stranded oligonucleotides

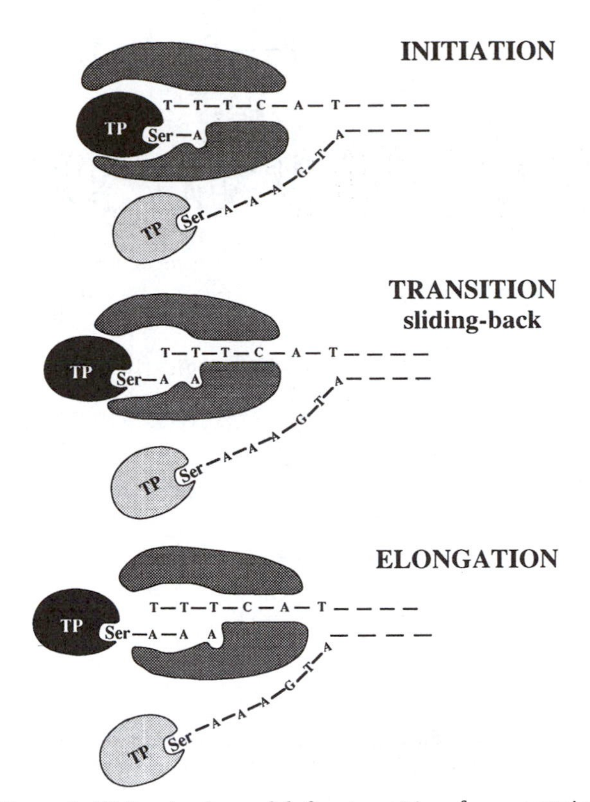

Figure 3. Sliding-back model for transition from protein-primed initiation to DNA elongation. Darkly shaded areas correspond to polymerization domains of ϕ29 DNA polymerase that define a cleft proposed to be used both as DNA- and TP-binding sites. ϕ29 TP, serving as primer for the nascent strand, is indicated by black, whereas parental ϕ29 TP, which is covalently bound to the displaced strand, is indicated by light shading. The sequence indicated corresponds to the 6-bp inverted terminal repeat present at each ϕ29 DNA end. See text for details.

with 3′-terminal sequences corresponding to the left or right ϕ29 DNA ends give rise to the TP-dAMP initiation complex, which is further elongated (70a). By using 12-mer synthetic oligonucleotides, a mutational analysis of the ϕ29 DNA right end was carried out. No strict sequence requirements for TP-primed initiation on single-stranded DNA were found. Interestingly, initiation of replication occurs opposite the second nucleotide at the 3′ end of the template. Nonetheless, the first nucleotide is not lost during replication. When the sequence requirements for elongation were studied, a terminal repetition of at least two nucleotides was shown to be required to efficiently elongate the initiation complex (70a). A sliding-back mechanism for the transition from initiation to elongation that can account for the necessity of a terminal repetition for efficient elongation and for the recovery of the information corresponding to the 3′ end of the template has been proposed. As shown in Fig. 3, the second nucleotide of the template (T) directs ϕ29 DNA polymerase-catalyzed linkage of dAMP to the TP. Then, the TP-dAMP initiation complex slides backwards, pairing dAMP to the first nucleotide of the template (T). After this transition, which probably leads to dissociation of the DNA polymerase-TP heterodimer, the next nucleotide (A) is incorporated, again using the second T of the template as a director.

As indicated above, the fidelity of TP-primed initiation of ϕ29 DNA replication (TP-dAMP complex formation) is rather low; moreover, this step is not subjected to 3′ → 5′ exonucleolytic proofreading. Therefore, it is important to have a mechanism for increasing the fidelity of initiation. Thus, if a mismatched initiation occurs in the transition step, the nucleotide linked to the TP will be mispaired opposite the first nucleotide of the template (T), and the incorrect initiation complex could dissociate from the DNA, providing a second chance for discrimination. Nonetheless, if the incorrect initiation complex is elongated, the second 3′ nucleotide of the template would be used again, restoring the terminal repetition at the end of the molecule. According to the sliding-back model, errors in the first replication event would be either counterselected because of a low elongation after sliding-back or lost if elongation succeeds, because the 3′-terminal nucleotide is not used as template.

It is interesting that all the genomes that contain terminal protein have some sequence repetition at the DNA ends (3′ TTT in *B. subtilis* phages ϕ29, ϕ15, PZA, Nf, M2, B103, and GA-1 and in *Streptococcus pneumoniae* phage Cp-1; 3′ TTTT in linear plasmids S1, S2, pSKL, and pGKL2; 3′ CCCC in *E. coli* phage PRD1; 3′ GGG and 3′ TGTG in linear plasmids pSLA2 and pGKL1, respectively; and 3′ GTAGTA in adenovirus type 2; reviewed in reference 89). This fact suggests that the sliding-back mechanism proposed for ϕ29 DNA replication could be extrapolatable to other systems that use proteins as primers.

TRANSCRIPTION

In Vivo Transcription

Initiation sites

Transcription of the ϕ29 genome takes place in two stages. At the beginning of the infection only the genes involved in DNA replication and transcription regulation are expressed. Genes coding for structural components of the phage particle and for proteins involved in morphogenesis and cell lysis are expressed later on in infection. The arrangement of the different genes reflects this temporal regulation: all late genes are clustered in the central part of the genome and are transcribed from the same DNA strand, while early genes are located at both ends of the genome and are transcribed from the complementary strand (Fig. 1A).

Binding sites for σ^A-RNA polymerase were initially located by electron microscopy (98). They were named, from left to right on the genetic map, A1, A2, A3, B1, B2, C1, and C2. Thereafter, transcription initiation sites were precisely localized by S1 mapping, which led to the identification of 10 promoters (5, 67; Fig. 1B). The weak early promoter A1, which is located at the left end of the genome, gives rise to a small transcript, which is thought to participate in the packaging of DNA into proheads (42, 110). An additional weak early promoter, named A1-IV, was localized within the coding region for gene 2. This promoter is close to a site known to be able to bind in vitro *E. coli* σ^{70}-RNA polymerase but not *B. subtilis*

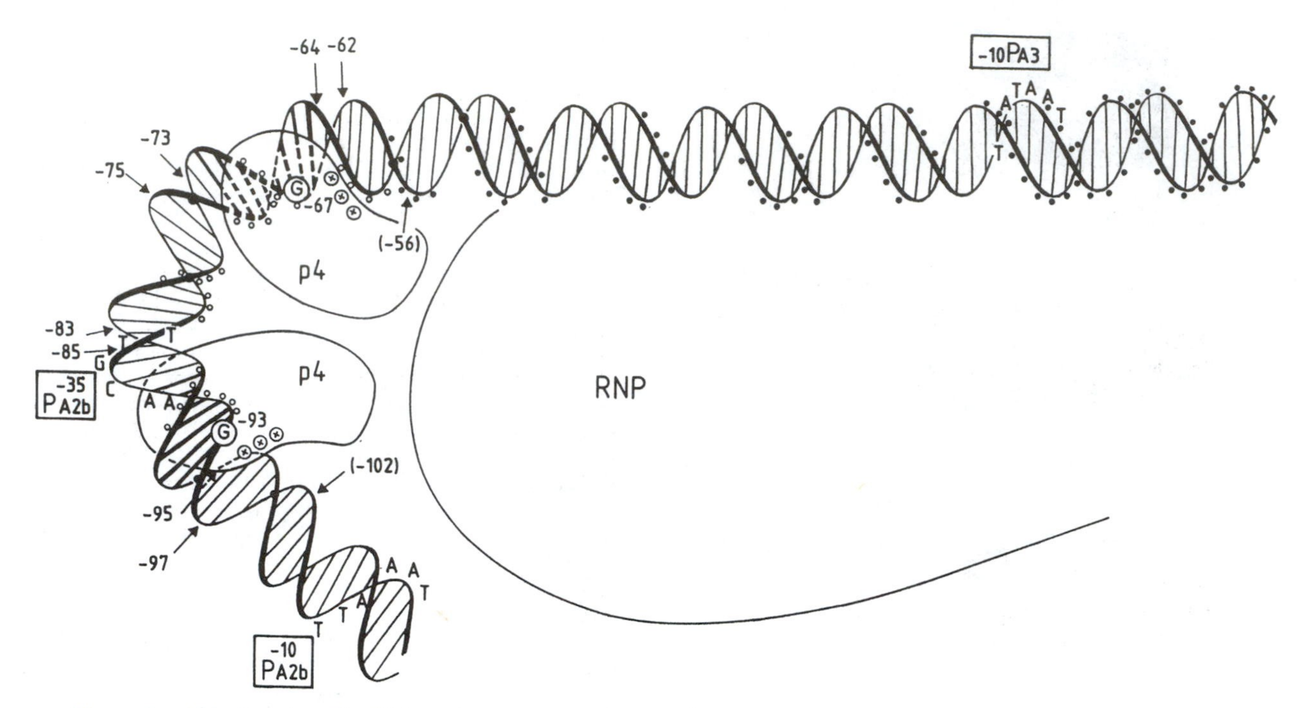

Figure 4. p4-binding site. The DNA region represented spans the area from the early A2b promoter (bottom left) to the divergent late A3 promoter (right), with a p4-binding site between them. RNA polymerase (RNP) is bound to the late A3 promoter. Positions protected from the attack of hydroxyl radicals by p4 or by the RNA polymerase bound at the late A3 promoter are shown by open and filled dots, respectively. p4-binding sequences are represented by thickened lines in the DNA sequence. Guanine residues whose methylation interferes with p4 binding to DNA are also shown (Ⓖ). Arrows indicate positions that become hypersensitive to DNase I cleavage upon p4 binding, most likely as a consequence of the protein-induced DNA curvature. Positions −56 and −102 do not appear when both p4 and RNA polymerase are bound to the late A3 promoter and are shown in parentheses. Note that the p4-binding site partially overlaps the early A2b promoter, thus displacing the RNA polymerase from it and directing the polymerase to the late A3 promoter.

σ^A-RNA polymerase (99; see below) and influences the expression levels of gene 1 (26a). The A2 binding region for σ^A-RNA polymerase contains three early promoters, which were named (from left to right) A2c, A2b, and A2a (67, 68). These promoters would be responsible for the expression of genes 6, 5, 4, 3, 2, and 1. Nevertheless, the activity of the A2a promoter was detected only late during infection (to be discussed below), and the A2c promoter was shown to be much weaker than A2b (2, 68). This indicated that promoter A2b is the main early promoter responsible for in vivo expression of genes 6 to 1. Another strong early promoter, named C2, was mapped at the right end of the genome, from which gene 17 would be expressed. Three additional weak early promoters, named C1, B2, and B1, were also detected. Promoters B2 and B1 would give rise to transcripts that do not contain open reading frames of significant length. These transcripts were proposed to act as antisense RNAs that would down regulate the expression of particular late genes (6). Transcripts starting from B1 would include sequences complementary to the start of gene 12, which codes for the protein forming the appendages of the viral neck (28) and thus could regulate this protein. The transcript originating at the B2 promoter was proposed to act in a similar way, regulating the levels of p14 needed for cell lysis. No function has been assigned to the transcript originating at the C1 promoter, although it includes two open reading frames of unknown function that are highly conserved among ϕ29-related phages (7). The sequences of the −10 and −35 regions of the early promoters show good homology to those recognized by *B. subtilis* σ^A-RNA polymerase.

All late genes are expressed from the late A3 promoter (5, 98), which was mapped between the early A2b and A2a promoters (5, 67). The expression of the late A3 promoter, which lacks a −35 consensus sequence, was shown to require the early p4, since it is inactive after restrictive infection with nonsense mutants in gene 4 (98).

When the activities of all mentioned promoters were compared in vivo and in vitro by S1 mapping assays, it was concluded that the promoters recognized in vitro by purified *B. subtilis* σ^A-RNA polymerase are the same as those used in vivo except for promoter A1-IV, which could be detected in vivo but not in vitro (68).

Termination sites

The main early and late transcription termination sites in vivo were localized by S1 mapping (Fig. 1B). Stem-loop structures followed by uridine-rich sequences are present at the 3′ ends of the early and late RNAs, suggesting that transcription termination in phage ϕ29 occurs through a rho-independent mechanism (6). Four transcriptional terminators were detected, one of them bidirectional. Transcription originating at the main early promoter, A2b, which is responsible for the expression of genes 6 to 1, encounters a terminator named TA1 at the beginning of gene 4. A significant number of transcripts terminate at this point; this would allow high levels of expression of proteins p6 and p5, which are known to be synthesized in large amounts in infected cells (28, 63). Terminators TB2 and TB1, located one after the other, stop transcripts coming from the weak B2 promoter, therefore limiting the effect of the RNA presumably produced to down regulate gene 12. Finally, the bidirectional terminator TD1 would prevent overlapping of the early transcription originating at the C1 and C2 promoters with that coming from the late genes in the opposite direction. The overall distribution of promoters, genes, and terminators allows symmetrical transcription that carefully avoids the appearance of complementary transcripts except in the two cases already mentioned that give rise to antisense RNAs, which are thought to down regulate certain late genes. No precise termination point was detected for transcripts originating at the weak early B1 promoter.

Control of ϕ29 early and late transcription

A ϕ29 derivative with a nonsense mutation in gene 4 is unable to induce the synthesis of late RNA, which suggests that p4 is involved in the control of late transcription (98). The in vivo activities of the early and late promoters at different stages of the infection were studied following production of the corresponding transcripts by nuclease S1 analysis (2). When protein synthesis is inhibited at the beginning of the infection by the addition of chloramphenicol, early genes 6 to 1 are transcribed only from the A2b promoter. The A2c promoter, known to be weak in vivo, does not give a signal strong enough to be significant, and the A2a promoter is not active. Similarly, the late A3 promoter remains inactive when protein synthesis is inhibited. When no chloramphenicol is added, the activity of the early A2b promoter declines slightly late in infection, while the A2a promoter becomes active, giving a signal stronger than that of A2b. Note that the same genes are transcribed from both promoters. Activation of A2a is paralleled by the induction of transcription of late genes from the A3 promoter. Therefore, the activities of the early A2a and late A3 promoters seem to depend on synthesis of an early viral protein, which, as suggested by earlier experiments in the case of the A3 promoter, could be p4. When the infection is performed with phage ϕ29 harboring a nonsense mutation in gene 4, the promoters behave as they do when protein synthesis is inhibited, suggesting that p4 mediates, directly or indirectly, the activation of both the early A2a and the late A3 promoters late in infection. Activation of the A2a promoter could be a strategy for maintaining high levels of expression of early genes 6 to 1 (including gene 4) late in infection.

All other promoters except C2 seem to be expressed constitutively throughout infection. The early C2 promoter was shown to be repressed both in vivo and in vitro by the viral early protein p6 late in infection (3, 109). Protein p6 in vitro forms at both ends of the ϕ29 genome a nucleoprotein complex (94) that could occlude this promoter. The C2 promoter directs the expression of p17, which would be expressed only at the beginning of infection.

In Vitro Transcription

Characterization of p4 as a transcriptional activator of the late A3 promoter

Sequencing of gene 4 revealed that it codes for a small basic protein 125 amino acids long with a molecular weight of 15,133 (33). Since the protein was expressed during infection in rather low amounts, its purification was achieved after cloning and overexpression of gene 4 in *E. coli* (1a). p4 showed a marked tendency to aggregate when overproduced, although it could be purified from the soluble fraction in an active form. It was realized later on that the protein could also be purified in a much simpler way by solubilizing the aggregated fraction with guanidinium chloride, which leads to higher yields of active protein (86).

Purification of p4 and the setup of an in vitro transcription assay (1a) allowed us to characterize this protein as a transcriptional activator of the late A3 promoter. Transcription from A3 required not only p4 but also the σ^A factor of RNA polymerase, ruling out the possibility that p4 is an alternative sigma factor (4).

By DNase I footprinting assays, the binding site for p4 was localized at a region spanning positions −56 to −102 relative to the transcription start site of the A3 promoter; this region was shown to have static curvature. As a consequence of p4 interaction with the DNA, certain positions of the binding site become hypersensitive to DNase I digestion (4). Since these positions are spaced about one helix turn from each other and since DNase I is known to cut DNA preferentially at positions being particularly flexible or already bent (52, 100), it was proposed that the DNase I hypersensitive sites appear as a consequence of an increase in DNA curvature originated by p4 upon its binding to DNA (4). This was later on confirmed by circular permutation assays, which indicated that p4 increases the curvature of its binding site from about 45° to at least 85° without changing the bend center (86). The DNase I hypersensitive sites allowed deduction of which positions should lie on the inner and outer faces of the induced bend (Fig. 4), since the hypersensitive sites should be on the outermost face of the bend. The pattern of hydroxyl radical sensitivity in the absence of p4 also indicated that the orientation of the static curvature is the same as that induced by p4, thereby supporting the possibility that p4 is increasing the preexisting static curvature rather than creating a new one (4). This region of the promoter contains an 8-bp inverted repeat that has been proposed to be the p4 recognition target. The first data supporting this idea came from methylation interference assays, which indicated that the guanine residues whose methylation interfered with p4 binding are those located in the inverted sequences (4). Hydroxyl radical interference assays showed that the nucleosides whose removal was detrimental to p4 binding lie precisely in the above-mentioned inverted repeat, which confirmed their involvement in promoter recognition (72). The symmetry of the binding site, the small size of the protein, and the relatively great distance separating the inverted sequences (2.5 helix turns) suggest that p4 probably binds DNA as a dimer (or a tetramer). Dimerization should be a DNA-dependent process, since p4 is thought to be a monomer in solution (1a).

It is not clear which domain of p4 recognizes the target sequences in the A3 promoter. A systematic analysis of the amino acid sequence using the algorithms of Dodd and Egan (30) or Brennan and Mathews (27) does not allow prediction of any classical Cro-like helix-turn-helix DNA binding motif such as is found in many DNA-binding proteins (91). On the basis of partial homology in what would be the first helix and the turn, a domain spanning amino acids 77 to 96 has been proposed as one of these DNA-binding motifs (71). The score that the algorithms cited above give to this domain is too low for us to consider the domain a likely candidate to mediate DNA binding, particularly considering that there is no genetic or biochemical evidence to support this possibility. Interestingly, deletion of the first five amino acids of the N terminus of p4 gives rise to a stable protein that can no longer bind to the promoter (84). Point mutations in this region of the protein give similar phenotypes (1). Therefore, the N terminus of p4 could play a role in the DNA-binding process. The structure predicted for this region of the protein is similar to that of the N termini of phage P22 Arc and Mnt repressors, which are known to recognize their target sequences through the first amino acids of their N termini, regions having a β-sheet structure (57). It is tempting to speculate that p4 and the Arc and Mnt repressors bind to DNA by using similar motifs. Nevertheless, it should be taken into account that the N terminus of p4 could participate in DNA binding in several ways, for example, by mediating dimerization of the protein when p4 is binding to the promoter.

DNA bending and direct contacts with RNA polymerase in activation of late A3 promoter

Detailed characterization of the p4-DNA complex gave several hints on the molecular mechanism of activation. The first data came from the study of deletion mutations in the carboxyl-end of p4, which is rich in basic amino acids (86). Removal of the last 12 amino acids of the activator, 6 of which are positively charged, or their replacement with negatively charged residues renders the protein inactive as a transcriptional activator. Interestingly, the activities of the mutants depend more on the net positive charges of the carboxyl ends than on the exact sequences or extents of the deletions. Both active and inactive mutants bind to the p4 recognition sequences, but those unable to activate transcription, i.e., those that have lost most of their basic residues from the carboxyl end, produce only partial bending of the DNA. These negative mutants can still give rise to the DNase I hypersensitive positions located between the recognition sequences but not to those located at both sides of the binding site, suggesting that bending at these regions has been lost. Circular permutation assays confirmed that these mutants produce less overall bending than the wild-type p4. This allowed us to propose a model describing how p4 bends the DNA. According to our model, binding of two monomers of p4 to the inverted recognition sequences and a subsequent interaction between them would bend the DNA between the inverted repeats. Bending at both sides of the binding site would arise as a consequence of a nonspecific electrostatic interaction between the positively charged carboxyl end of the protein and the negatively charged DNA backbone (86). The facts that the mutations inducing only a partial DNA bending are not active as transcriptional activators and do not favor the binding of σ^A-RNA polymerase to the promoter suggest that the p4-induced bend can be involved in the activation process. Indeed, there is a significant correlation between the activities of the mutants, their abilities to induce a stable bending of the DNA, and the extent of RNA polymerase binding to the promoter. Of the four mutants analyzed, one is rather active and bends the DNA as the wild-type p4 does, two are inactive and do not bend the DNA at the ends of the binding site, and the fourth has an intermediate behavior: it has lost about 70% of its activity, is hardly able to favor the binding of RNA polymerase to the promoter, and generates DNA bending that is unstable in the sense that it is detected by DNase I footprints (all hypersensitive positions are present) but not by circular permutation assays.

DNA bending could participate in the activation process in two ways. It is apparent from Fig. 4 that a loss of curvature would change the orientation of the activator towards the RNA polymerase, thus impairing direct contacts between both proteins (see below). Considering that the proteins are very close to each other and that the DNA is rather rigid on such short distances, it is unlikely that both proteins can interact correctly once the p4-induced DNA bend is lost. On the other hand, the induced DNA curvature could create a DNA structure that favors transcription initiation events, as has been described in other cases (26, 61, 66, 111, 117).

DNase I and hydroxyl radical footprints and gel retardation assays (85, 92) show that the RNA polymerase can bind to the promoter only in the presence of p4, suggesting that a major role of the activator is to stabilize the RNA polymerase at the promoter. Kinetic analysis of the transcription initiation process from the A3 promoter has confirmed that the role of p4 is to favor binding of the RNA polymerase to the promoter as a closed complex without significantly affecting further steps of the initiation process (72a).

Four lines of evidence indicate that direct contacts between p4 and RNA polymerase very likely exist and are important in the activation process.

(i) The binding of both proteins to the A3 promoter is a cooperative process. Indeed, when part of p4 recognition sequences is deleted, p4 can bind to the promoter only in the presence of RNA polymerase, which stabilizes the activator in its binding site (72a).

(ii) The effect of DNA insertions of different lengths between the binding sites for p4 and RNA polymerase shows that the activator needs to be stereospecifically aligned relative to the RNA polymerase to activate transcription (92). Activation is abolished when 4 bp are inserted between the binding sites but partially restored when the insertion is about one helical turn

(10 bp) long. Activation is still possible with longer insertions (20 and 30 bp) but only if the intervening DNA is bent towards the RNA polymerase, suggesting that the activator needs to contact the RNA polymerase to activate transcription. Most interesting, DNase I footprinting of the complex formed by p4 and RNA polymerase bound to the promoters harboring inserts 20 or 30 bp long shows clear positions hypersensitive to DNase I digestion, spaced about one helical turn from each other, in the DNA intervening between the two proteins. This suggests that activation at a distance by p4 requires the formation of a DNA loop that may help the activator contact the RNA polymerase and that can be formed only if the intervening DNA is bent in the correct direction.

(iii) Hydroxyl radical footprinting of the binding of p4 and RNA polymerase indicates that the two proteins bind to the same face of the DNA helix adjacent to each other (Fig. 4). Indeed, a minor groove is protected by both proteins, the upstream part of it by p4 and the downstream part by RNA polymerase, indicating that the two proteins are in very close proximity (85).

(iv) By using gel retardation assays, it has been possible to detect complexes formed by p4 and RNA polymerase when the DNA supplied contains only the activator-binding site or the RNA polymerase-binding site (72a).

Characterization of p4 as a repressor of the main early A2b promoter

p4 binding sequences overlap the −35 region of the main early promoter, A2b, which suggests that p4 can repress this early promoter while activating the late A3 promoter. Both DNase I footprints and runoff transcription assays show that p4 functions as a switch, excluding σ^A-RNA polymerase from the early A2b promoter and directing it to the adjacent but divergently transcribed late A3 promoter (85). Repression is very efficient in vitro, which contrasts with the fact that transcripts arising from the A2b promoter are still detectable, though at a reduced level, late in infection (2). Therefore, either these transcripts are very stable in the cell or p4 is not binding to all the viral genomes present. The need for activating the A2a promoter late in infection suggests that this promoter is switched on to compensate for the repression of the A2b promoter, since the same genes are transcribed from both of them. It should be recalled that p4 is also a likely candidate to activate the A2a promoter, although how this could be achieved is still unknown. It is intriguing that p4 has to come along in these multiple regulatory events to activate late transcription without impairing expression of the early genes.

Role of DNA bending in transcription repression

The partial overlap of the p4 binding site and the −35 region of the early A2b promoter suggests that transcription repression is exerted by steric hindrance, as is frequently assumed for most repressors. Nevertheless, many repressors bend the DNA at their binding sites (118), which suggests that this distortion of the DNA can participate in the repression process. Four site-directed mutations were introduced into the p4-binding site to increase its static curvature in the same direction as p4 does and to analyze how static DNA curvature resembling that induced by p4 can influence the activity of the early A2b promoter (85). Mutations were introduced sequentially, so that three mutants in which the original static curvature (45°) was increased to about 58, 64, and 72° were obtained. The curvature induced by p4 is about 85°. The activity of the A2b promoter in the mutant derivatives was analyzed by runoff, DNase I footprinting, and kinetic analyses. It was concluded that the increase in curvature produces a decrease both in the affinity of the RNA polymerase for the promoter and in its ability to initiate transcription once it is bound to the promoter. The overall strength of the mutant having the most pronounced curvature is almost 1 order of magnitude lower than that of the wild-type promoter. The decrease in affinity is explained by postulating that the DNA curvature creates a structure that the RNA polymerase cannot recognize properly as a promoter. On the other hand, RNA polymerase has been shown to bend the DNA at certain promoters when open complexes are formed (50). This bending is likely to be extensive to most if not all promoters, and the direction of the curvature would be opposite that induced in the mutant A2b templates. Therefore, the unfavorable rigid curvature introduced into the A2b promoter is likely to disturb the transition from closed to transcriptionally active open complexes, as was shown by kinetic analysis. It was therefore proposed that the curvature induced by p4 in its binding site can play a role in repression of the A2b promoter. The repression efficiency of the steric hindrance presumably exerted by p4 could thus be increased by the induction of DNA curvature that prevents the RNA polymerase from recognizing the promoter. In this way, the competition between p4 and RNA polymerase for their respective overlapping binding sites is efficiently reduced in favor of p4. Since many repressors bend the DNAs in their binding sites, this could be a frequent strategy used by many repressors to increase their repression efficiencies.

Acknowledgments. We thank all colleagues in the laboratory for their contribution and for many helpful discussions. Work was supported by grants from the National Institutes of Health (5 RO1 GM27242-12), Dirección General de Investigacion Científica y Técnica (PB 90-0091), and European Economic Community (BIOT CT 91-0268) and by an institutional grant from Fundación Ramón Areces.

REFERENCES

1. **Alonso, J. C., F. Rojo, and M. Salas.** Unpublished data.

1a. **Barthelemy, I., J. M. Lázaro, E. Méndez, R. P. Mellado, and M. Salas.** 1987. Purification in an active form of the phage ϕ29 protein p4 that controls the viral late transcription. *Nucleic Acids Res.* **15:**7781–7793.

2. **Barthelemy, I., R. P. Mellado, and M. Salas.** 1988. Symmetrical transcription in bacteriophage ϕ29 DNA. *Biochimie* **70:**605–609.

3. **Barthelemy, I., R. P. Mellado, and M. Salas.** 1989. In vitro transcription of bacteriophage ϕ29 DNA: inhibition of early promoters by the viral replication protein p6. *J. Virol.* **63:**460–462.

4. **Barthelemy, I., and M. Salas.** 1989. Characterization of a new prokaryotic transcriptional activator and its DNA recognition site. *J. Mol. Biol.* **208:**225–232.

5. **Barthelemy, I., M. Salas, and R. P. Mellado.** 1986. In vivo transcription of bacteriophage φ29 DNA: transcription initiation sites. *J. Virol.* **60:**874–879.
6. **Barthelemy, I., M. Salas, and R. P. Mellado.** 1987. In vivo transcription of bacteriophage φ29: transcription termination. *J. Virol.* **61:**1751–1755.
7. **Benes, V., L. Arnold, J. Smrt, and V. Paces.** 1989. Nucleotide sequence of the right early region of *Bacillus* phage φ15 and comparison with related phages: reorganization of gene 17 during evolution. *Gene* **75:**341–347.
8. **Bernad, A., L. Blanco, J. M. Lázaro, G. Martín, and M. Salas.** 1989. A conserved 3′ → 5′ exonuclease active site in prokaryotic and eukaryotic DNA polymerases. *Cell* **59:**219–228.
9. **Bernad, A., J. M. Lázaro, M. Salas, and L. Blanco.** 1990. The highly conserved amino acid sequence motif Tyr-Asp-Thr-Asp-Ser in α-like DNA polymerases is required by phage φ29 DNA polymerase for protein-primed initiation and polymerization. *Proc. Natl. Acad. Sci. USA* **87:**4610–4614.
10. **Bernad, A., A. Zaballos, M. Salas, and L. Blanco.** 1987. Structural and functional relationships between prokaryotic and eukaryotic DNA polymerases. *EMBO J.* **6:**4221–4225.
11. **Blanco, L., A. Bernad, M. A. Blasco, and M. Salas.** 1991. A general structure for DNA-dependent DNA polymerases. *Gene* **100:**27–38.
12. **Blanco, L., A. Bernad, J. A. Esteban, and M. Salas.** 1992. DNA-independent deoxynucleotidylation of the φ29 terminal protein by the φ29 DNA polymerase. *J. Biol. Chem.* **267:**1225–1230.
13. **Blanco, L., A. Bernad, J. M. Lázaro, G. Martín, C. Garmendia, and M. Salas.** 1989. Highly efficient DNA synthesis by the phage φ29 DNA polymerase. Symmetrical mode of DNA replication. *J. Biol. Chem.* **264:**8935–8940.
14. **Blanco, L., A. Bernad, and M. Salas.** 1988. Transition from initiation to elongation in protein-primed φ29 DNA replication: salt-dependent stimulation by the viral protein p6. *J. Virol.* **62:**4167–4172.
15. **Blanco, L., A. Bernad, and M. Salas.** 1992. Evidence favouring the hypothesis of a conserved 3′ → 5′ exonuclease active site in DNA-dependent DNA polymerases. *Gene* **112:**139–144.
16. **Blanco, L., J. A. García, M. A. Peñalva, and M. Salas.** 1983. Factors involved in the initiation of phage φ29 DNA replication *in vitro*: requirement of the gene 2 product for the formation of the protein p3-dAMP complex. *Nucleic Acids Res.* **11:**1309–1323.
17. **Blanco, L., J. A. García, and M. Salas.** 1984. Cloning and expression of gene 2, required for the protein-primed initiation of the *Bacillus subtilis* phage φ29 DNA replication. *Gene* **29:**33–40.
18. **Blanco, L., J. Gutiérrez, J. M. Lázaro, A. Bernad, and M. Salas.** 1986. Replication of phage φ29 DNA *in vitro*: role of the viral protein p6 in initiation and elongation. *Nucleic Acids Res.* **14:**4923–4937.
19. **Blanco, L., I. Prieto, J. Gutiérrez, A. Bernad, J. M. Lázaro, J. M. Hermoso, and M. Salas.** 1987. Effect of NH_4^+ ions on φ29 DNA-protein p3 replication: formation of a complex between the terminal protein and the DNA polymerase. *J. Virol.* **61:**3983–3991.
20. **Blanco, L., and M. Salas.** 1984. Characterization and purification of a phage φ29-encoded DNA polymerase required for the initiation of replication. *Proc. Natl. Acad. Sci. USA* **81:**5325–5329.
21. **Blanco, L., and M. Salas.** 1985. Characterization of a 3′ → 5′ exonuclease activity in the phage φ29-encoded DNA polymerase. *Nucleic Acids Res.* **13:**1239–1249.
22. **Blanco, L., and M. Salas.** 1985. Replication of phage φ29 DNA with purified terminal protein and DNA polymerase: synthesis of full-length φ29 DNA. *Proc. Natl. Acad. Sci. USA* **82:**6404–6408.
23. **Blanco, L., and M. Salas.** 1986. Effect of aphidicolin and nucleotide analogs on the phage φ29 DNA polymerase. *Virology* **153:**179–187.
24. **Blasco, M. A., A. Bernad, L. Blanco, and M. Salas.** 1991. Characterization and mapping of the pyrophosphorolytic activity of the phage φ29 DNA polymerase. *J. Biol. Chem.* **266:**7904–7909.
25. **Blasco, M. A., L. Blanco, E. Parés, M. Salas, and A. Bernad.** 1990. Structural and functional analysis of temperature sensitive mutants of the phage φ29 DNA polymerase. *Nucleic Acids Res.* **18:**4763–4770.
25a. **Blasco, M. A., J. A. Esteban, J. Méndez, L. Blanco, and M. Salas.** 1993. Structural and functional studies on φ29 DNA polymerase. *Chromosoma* **102:**32–38.
25b. **Blasco, M. A., J. M. Lázaro, A. Bernad, L. Blanco, and M. Salas.** 1992. φ29 DNA polymerase active site: mutants in conserved residues Tyr^{254} and Tyr^{390} are affected in dNTP binding. *J. Biol. Chem.* **267:**19427–19434.
25c. **Blasco, M. A., J. M. Lázaro, L. Blanco, and M. Salas.** Unpublished data.
26. **Bracco, L., D. Kotlarz, A. Kolb, S. Diekmann, and H. Buc.** 1989. Synthetic curved DNA sequences can act as transcriptional activators in *Escherichia coli*. *EMBO J.* **8:**4289–4296.
26a. **Bravo, A.** Unpublished data.
27. **Brennan, R. G., and B. W. Mathews.** 1989. The helix-turn-helix DNA binding motif. *J. Biol. Chem.* **264:**1903–1906.
28. **Carrascosa, J. L., A. Camacho, F. Moreno, F. Jiménez, R. P. Mellado, E. Viñuela, and M. Salas.** 1976. *Bacillus subtilis* phage φ29: characterization of gene products and functions. *Eur. J. Biochem.* **66:**229–241.
29. **Derbyshire, V., P. S. Freemont, M. R. Sanderson, L. Beese, J. M. Friedman, C. M. Joyce, and T. A. Steitz.** 1988. Genetic and crystallographic studies of the 3′ → 5′ exonucleolytic site of DNA polymerase I. *Science* **240:**199–201.
30. **Dodd, I. B., and J. B. Egan.** 1990. Improved detection of helix-turn-helix DNA binding motives in protein sequences. *Nucleic Acids Res.* **18:**5019–5026.
31. **Escarmís, C., D. Guirao, and M. Salas.** 1989. Replication of recombinant φ29 DNA molecules in *Bacillus subtilis* protoplasts. *Virology* **169:**150–160.
32. **Escarmís, C., and M. Salas.** 1981. Nucleotide sequence at the termini of the DNA of *Bacillus subtilis* phage φ29. *Proc. Natl. Acad. Sci. USA* **78:**1446–1450.
33. **Escarmís, C., and M. Salas.** 1982. Nucleotide sequence of the early genes 3 and 4 of bacteriophage φ29. *Nucleic Acids Res.* **10:**5785–5798.
34. **Esteban, J. A., A. Bernad, M. Salas, and L. Blanco.** 1992. Metal activation of synthetic and degradative activities of φ29 DNA polymerase, a model enzyme for protein-primed replication. *Biochemistry* **31:**350–359.
34a. **Esteban, J. A., M. Salas, and L. Blanco.** 1993. Fidelity of φ29 DNA polymerase: comparison between protein-primed initiation and DNA polymerization. *J. Biol. Chem.* **268**, in press.
35. **Freemont, P. S., J. M. Friedman, L. S. Beese, M. R. Sanderson, and T. A. Steitz.** 1988. Cocrystal structure of an editing complex of Klenow fragment with DNA. *Proc. Natl. Acad. Sci. USA* **85:**8924–8928.
35a. **Freire, R., M. Serrano, M. Salas, and J. M. Hermoso.** Unpublished data.
36. **García, J. A., M. A. Peñalva, L. Blanco, and M. Salas.** 1984. Template requirements for the initiation of phage φ29 DNA replication *in vitro*. *Proc. Natl. Acad. Sci. USA* **81:**80–84.
37. **Garmendia, C., A. Bernad, J. A. Esteban, L. Blanco, and M. Salas.** 1992. The bacteriophage φ29 DNA polymer-

ase, a proofreading enzyme. *J. Biol. Chem.* **267:**2594–2599.

38. **Garmendia, C., J. M. Hermoso, and M. Salas.** 1990. Functional domain for priming activity in the phage ϕ29 terminal protein. *Gene* **88:**73–79.
39. **Garmendia, C., M. Salas, and J. M. Hermoso.** 1988. Site-directed mutagenesis in the DNA linking site of bacteriophage ϕ29 terminal protein: isolation and characterization of $Ser_{232} \rightarrow$ Thr mutant. *Nucleic Acids Res.* **16:**5727–5740.
40. **Garvey, K. J., H. Yoshikawa, and J. Ito.** 1985. The complete sequence of the *Bacillus* phage ϕ29 right early region. *Gene* **40:**301–309.
41. **Gibbs, J. S., H. C. Chiou, J. D. Hall, D. W. Mount, M. J. Retondo, S. K. Weller, and D. M. Coen.** 1985. Sequence and mapping analysis of the herpes simplex virus DNA polymerase gene predict a C-terminal substrate binding domain. *Proc. Natl. Acad. Sci. USA* **82:**7969–7973.
42. **Guo, P., S. Erickson, and D. Anderson.** 1987. A small viral RNA is required for *in vitro* packaging of bacteriophage ϕ29 DNA. *Science* **236:**690–694.
43. **Gutiérrez, J., J. A. García, L. Blanco, and M. Salas.** 1986. Cloning and template activity of the origins of replication of phage ϕ29 DNA. *Gene* **43:**1–11.
44. **Gutiérrez, J., C. Garmendia, and M. Salas.** 1988. Characterization of the origins of replication of bacteriophage ϕ29 DNA. *Nucleic Acids Res.* **16:**5895–5914.
45. **Gutiérrez, C., G. Martín, J. M. Sogo, and M. Salas.** 1991. Mechanism of stimulation of DNA replication by bacteriophage ϕ29 single-stranded DNA-binding protein p5. *J. Biol. Chem.* **266:**2104–2111.
46. **Gutiérrez, C., J. M. Sogo, and M. Salas.** 1991. Analysis of replication intermediates produced during bacteriophage ϕ29 DNA replication *in vitro*. *J. Mol. Biol.* **222:** 983–994.
47. **Hagen, E. W., B. E. Reilly, M. E. Tosi, and D. L. Anderson.** 1976. Analysis of gene function of bacteriophage ϕ29 of *Bacillus subtilis*: identification of cistrons essential for viral assembly. *J. Virol.* **19:**501–517.
48. **Hermoso, J. M., E. Méndez, F. Soriano, and M. Salas.** 1985. Location of the serine residue involved in the linkage between the terminal protein and the DNA of ϕ29. *Nucleic Acids Res.* **13:**7715–7728.
49. **Hermoso, J. M., and M. Salas.** 1980. Protein p3 is linked to the DNA of phage ϕ29 DNA through a phosphoester bond between serine and 5′-dAMP. *Proc. Natl. Acad. Sci. USA* **77:**6425–6428.
50. **Heumann, H., M. Ricchetti, and W. Werel.** 1988. DNA-dependent RNA polymerase of *Escherichia coli* induces bending or an increased flexibility of DNA by specific complex formation. *EMBO J.* **7:**4379–4381.
51. **Hirokawa, H., K. Matsumoto, and M. Ohashi.** 1982. Replication of *Bacillus* small phage DNA, p. 45–46. *In* D. Schlessinger (ed.), *Microbiology—1982*. American Society for Microbiology, Washington, D.C.
52. **Hogan, M. E., M. W. Roberson, and R. Austin.** 1989. DNA flexibility may dominate DNase I cleavage. *Proc. Natl. Acad. Sci. USA* **86:**9273–9277.
53. **Huberman, J. A.** 1981. New views of the biochemistry of eukaryotic DNA replication revealed by aphidicolin, an unusual inhibitor of DNA polymerase. *Cell* **23:**647–648.
54. **Inciarte, M. R., M. Salas, and J. M. Sogo.** 1980. Structure of replicating DNA molecules of *Bacillus subtilis* bacteriophage ϕ29. *J. Virol.* **34:**187–199.
55. **Ito, J.** 1978. Bacteriophage ϕ29 terminal protein: its association with the 5′ termini of the ϕ29 genome. *J. Virol.* **28:**895–904.
56. **Khan, N. W., G. E. Wright, L. W. Dudycz, and N. C. Brown.** 1984. Butylphenyl dGTP: a selective and potent inhibitor of mammalian DNA polymerase alpha. *Nucleic Acids Res.* **12:**3695–3706.
57. **Knight, K. L., and R. T. Sauer.** 1989. DNA binding specificity of the Arc and Mnt repressors is determined by a short region of N-terminal residues. *Proc. Natl. Acad. Sci. USA* **86:**797–801.
58. **Kobayashi, H., K. Kitabayaski, K. Matsumoto, and H. Hirokawa.** 1991. Primer protein of bacteriophage M2 exposes the RGD receptor site upon linking the first deoxynucleotide. *Mol. Gen. Genet.* **226:**65–69.
59. **Kobayashi, H., K. Kitabayashi, K. Matsumoto, and H. Hirokawa.** 1991. Receptor sequence of the terminal protein of bacteriophage M2 that interacts with an RGD (Arg-Gly-Asp) sequence of the primer protein. *Virology* **185:**901–903.
60. **Kobayashi, H., K. Matsumoto, S. Misawa, K. Miura, and H. Hirokawa.** 1989. An inhibitory effect of RGD peptide on protein-priming reaction of bacteriophages ϕ29 and M2. *Mol. Gen. Genet.* **220:**8–11.
61. **Lavigne, M., M. Herbert, A. Kolb, and H. Buc.** 1992. Upstream curved sequences influence the initiation of transcription at the *Escherichia coli* galactose operon. *J. Mol. Biol.* **224:**293–306.
62. **Leavitt, M. C., and J. Ito.** 1987. Nucleotide sequence of *Bacillus* phage Nf terminal protein gene. *Nucleic Acids Res.* **15:**5251–5259.
63. **Martín, G., J. M. Lázaro, E. Méndez, and M. Salas.** 1989. Characterization of phage ϕ29 protein p5 as a single-stranded DNA binding protein. Function in ϕ29 DNA-protein p3 replication. *Nucleic Acids Res.* **17:** 3663–3672.
64. **Martín, G., and M. Salas.** 1988. Characterization and cloning of gene 5 of *Bacillus subtilis* phage ϕ29. *Gene* **67:**193–201.
65. **Matsumoto, K., T. Saito, and H. Kirokawa.** 1983. *In vitro* initiation of bacteriophage ϕ29 and M2 DNA replication: genes required for formation of a complex between the terminal protein and 5′ dAMP. *Mol. Gen. Genet.* **191:**26–30.
66. **McAllister, C. F., and E. C. Achberger.** 1989. Rotational orientation of upstream curved DNA affects promoter function in *Bacillus subtilis*. *J. Biol. Chem.* **264:**10451–10456.
67. **Mellado, R. P., I. Barthelemy, and M. Salas.** 1986. *In vivo* transcription of bacteriophage ϕ29 DNA early and late promoter sequences. *J. Mol. Biol.* **191:**191–197.
68. **Mellado, R. P., I. Barthelemy, and M. Salas.** 1986. *In vitro* transcription of bacteriophage ϕ29 DNA. Correlation between *in vitro* and *in vivo* promoters. *Nucleic Acids Res.* **14:**4731–4741.
69. **Mellado, R. P., F. Moreno, E. Viñuela, M. Salas, B. E. Reilly, and D. L. Anderson.** 1976. Genetic analysis of bacteriophage ϕ29 of *Bacillus subtilis*: integration and mapping of nonsense mutants of two collections. *J. Virol.* **19:**495–500.
70. **Mellado, R. P., M. A. Peñalva, M. R. Inciarte, and M. Salas.** 1980. The protein covalently linked to the 5′ termini of the DNA of *Bacillus subtilis* phage ϕ29 is involved in the initiation of DNA replication. *Virology* **104:**84–96.

70a. **Méndez, J., L. Blanco, J. A. Esteban, A. Bernad, and M. Salas.** 1992. Initiation of ϕ29 DNA replication occurs at the second 3′ nucleotide of the linear template: a sliding-back mechanism for protein-primed DNA replication. *Proc. Natl. Acad. Sci. USA* **89:**9579–9583.

71. **Mizukami, Y., T. Sekiya, and H. Hirokawa.** 1986. Nucleotide sequence of gene F of *Bacillus* phage Nf. *Gene* **42:**231–235.
72. **Nuez, B., F. Rojo, I. Barthelemy, and M. Salas.** 1991. Identification of the sequences recognized by phage ϕ29 transcriptional activator: possible interaction between the activator and the RNA polymerase. *Nucleic Acids Res.* **19:**2337–2342.

72a. **Nuez, B., F. Rojo, and M. Salas.** 1992. Phage ϕ29 regulatory protein p4 stabilizes the binding of the RNA polymerase to the late promoter in a process involving

direct protein-protein contacts. *Proc. Natl. Acad. Sci. USA* **89:**11401–11405.
73. **Ollis, D. L., R. Brick, R. Hamlin, N. G. Xuong, and T. A. Steitz.** 1985. Structure of the large fragment of *Escherichia coli* DNA polymerase I complexed with dTMP. *Nature* (London) **313:**762–766.
74. **Ortín, J., E. Viñuela, M. Salas, and C. Vásquez.** 1971. DNA-protein complex in circular DNA from phage ϕ29. *Nature New Biol.* **234:**275–277.
75. **Otero, M. J., J. M. Lázaro, and M. Salas.** 1990. Deletions at the N terminus of bacteriophage ϕ29 protein p6: DNA binding and activity in ϕ29 DNA replication. *Gene* **95:**25–30.
76. **Otero, M. J., and M. Salas.** 1989. Regions at the carboxyl end of bacteriophage protein p6 required for DNA binding and activity in ϕ29 DNA replication. *Nucleic Acids Res.* **17:**4567–4577.
77. **Pandey, V. N., K. R. Williams, K. L. Stone, and M. J. Modak.** 1987. Photoaffinity labeling of the thymidine triphosphate binding domain in *Escherichia coli* DNA polymerase I: identification of histidine-881 as the site of cross-linking. *Biochemistry* **26:**7744–7748.
78. **Pastrana, R., J. M. Lázaro, L. Blanco, J. A. García, E. Méndez, and M. Salas.** 1985. Overproduction and purification of protein p6 of *Bacillus subtilis* phage ϕ29: role in the initiation of DNA replication. *Nucleic Acids Res.* **13:**3083–3100.
79. **Peñalva, M. A., and M. Salas.** 1982. Initiation of phage ϕ29 DNA replication *in vitro*: formation of a covalent complex between the terminal protein, p3, and 5′-dAMP. *Proc. Natl. Acad. Sci. USA* **79:**5522–5526.
80. **Prieto, I., J. M. Lázaro, J. A. García, J. M. Hermoso, and M. Salas.** 1984. Purification in a functional form of the terminal protein of *Bacillus subtilis* phage ϕ29. *Proc. Natl. Acad. Sci. USA* **81:**1639–1643.
81. **Prieto, I., E. Méndez, and M. Salas.** 1989. Characterization, overproduction and purification of the product of gene 1 of *Bacillus subtilis* phage ϕ29. *Gene* **77:**195–204.
82. **Prieto, I., M. Serrano, J. M. Lázaro, M. Salas, and J. M. Hermoso.** 1988. Interaction of the bacteriophage ϕ29 protein p6 with double-stranded DNA. *Proc. Natl. Acad. Sci. USA* **85:**314–318.
83. **Reha-Krantz, L. J.** 1988. Amino acid changes coded by bacteriophage T4 DNA polymerase mutator mutants. Relating structure to function. *J. Mol. Biol.* **202:**711–724.
84. **Rojo, F., and M. Salas.** 1990. Short N-terminal deletions in the phage ϕ29 transcriptional activator protein p4 impair its DNA binding ability. *Gene* **96:**75–81.
85. **Rojo, F., and M. Salas.** 1991. A DNA curvature can substitute phage ϕ29 regulatory protein p4 when acting as a transcriptional repressor. *EMBO J.* **10:**3429–3438.
86. **Rojo, F., A. Zaballos, and M. Salas.** 1990. Bend induced by phage ϕ29 transcriptional activator in the viral late promoter is required for activation. *J. Mol. Biol.* **211:** 713–725.
87. **Rush, J., and W. H. Konigsberg.** 1990. Photoaffinity labeling of the Klenow fragment with 8-azido-dATP. *J. Biol. Chem.* **265:**4821–4827.
88. **Salas, M.** 1988. Phages with protein attached to the DNA ends, p. 169–191. *In* R. Calendar (ed.), *The Bacteriophages*, vol. 1. Plenum Press, New York.
89. **Salas, M.** 1991. Protein-priming of DNA replication. *Annu. Rev. Biochem.* **60:**39–71.
90. **Salas, M., R. P. Mellado, E. Viñuela, and J. M. Sogo.** 1978. Characterization of a protein covalently linked to the 5′ termini of the DNA of *Bacillus subtilis* phage ϕ29. *J. Mol. Biol.* **119:**269–291.
91. **Schleif, R.** 1988. DNA binding by proteins. *Science* **241:**1182–1187.
92. **Serrano, M., I. Barthelemy, and M. Salas.** 1991. Transcription activation at a distance by phage ϕ29 protein p4. Effect of bent and non-bent intervening DNA sequences. *J. Mol. Biol.* **219:**403–414.
93. **Serrano, M., J. Gutiérrez, I. Prieto, J. M. Hermoso, and M. Salas.** 1989. Signals at the bacteriophage ϕ29 DNA replication origins required for protein p6 binding and activity. *EMBO J.* **8:**1879–1885.
93a. **Serrano, M., C. Gutiérrez, M. Salas, and J. M. Hermoso.** The superhelical path of the DNA in the nucleoprotein complex that activates the initiation of phage ϕ29 DNA replication. *J. Mol. Biol.*, in press.
93b. **Serrano, M., and M. Salas.** Unpublished data.
94. **Serrano, M., M. Salas, and J. M. Hermoso.** 1990. A novel nucleoprotein complex at a replication origin. *Science* **248:**1012–1016.
95. **Shih, M. F., K. Watabe, and J. Ito.** 1982. *In vitro* complex formation between bacteriophage ϕ29 terminal protein and deoxynucleotide. *Biochem. Biophys. Res. Commun.* **105:**1031–1036.
96. **Shih, M. F., K. Watabe, H. Yoshikawa, and J. Ito.** 1984. Antibodies specific for the ϕ29 terminal protein inhibit the initiation of DNA replication *in vitro*. *Virology* **133:** 56–64.
96a. **Soengas, M. S., J. A. Esteban, J. M. Lázaro, A. Bernad, M. A. Blasco, M. Salas, and L. Blanco.** 1992. Site-directed mutagenesis at the EcoIII motif of ϕ29 DNA polymerase: overlapping structural domains for the 3′-5′ exonuclease and strand-displacement activities. *EMBO J.* **11:**4227–4237.
97. **Sogo, J. M., J. A. García, M. A. Peñalva, and M. Salas.** 1982. Structure of protein-containing replicative intermediates of *Bacillus subtilis* phage ϕ29 DNA. *Virology* **116:**1–18.
98. **Sogo, J. M., M. R. Inciarte, J. Corral, E. Viñuela, and M. Salas.** 1979. RNA polymerase binding sites and transcription of the DNA of *Bacillus subtilis* phage ϕ29. *J. Mol. Biol.* **127:**411–436.
99. **Sogo, J. M., M. Lozano, and M. Salas.** 1984. *In vitro* transcription of the *Bacillus subtilis* phage ϕ29 DNA by *Bacillus subtilis* and *Escherichia coli* RNA polymerases. *Nucleic Acids Res.* **12:**1943–1960.
100. **Suck, D., A. Lahm, and C. Oefner.** 1988. Structure refinement to 2 Å of a nicked DNA oligonucleotide complex with DNase I. *Nature* (London) **332:**464–468.
101. **Talavera, A., F. Jiménez, M. Salas, and E. Viñuela.** 1971. Temperature-sensitive mutants of bacteriophage ϕ29. *Virology* **46:**586–595.
102. **Talavera, A., M. Salas, and E. Viñuela.** 1972. Temperature-sensitive mutants affected in DNA synthesis in phage ϕ29 of *Bacillus subtilis*. *Eur. J. Biochem.* **31:**367–371.
103. **Tomalsky, M. D., J. Wu, and L. K. Miller.** 1988. The location, sequence, transcription and regulation of a baculovirus DNA polymerase gene. *Virology* **167:**591–600.
104. **Vlcek, C., and V. Paces.** 1986. Nucleotide sequence of the late region of *Bacillus* phage ϕ29 completes the 19285-bp sequence of ϕ29 genome. Comparison with the homologous sequence of phage PZA. *Gene* **46:**215–225.
105. **Watabe, K., M. Leusch, and J. Ito.** 1984. Replication of bacteriophage ϕ29 DNA *in vitro*: the roles of terminal protein and DNA polymerase. *Proc. Natl. Acad. Sci. USA* **81:**5374–5378.
106. **Watabe, K., M. Leusch, and J. Ito.** 1984. A 3′ to 5′ exonuclease activity is associated with phage ϕ29 DNA polymerase. *Biochem. Biophys. Res. Commun.* **123:** 1019–1026.
107. **Watabe, K., M. F. Shih, and J. Ito.** 1983. Protein-primed initiation of phage ϕ29 DNA replication. *Proc. Natl. Acad. Sci. USA* **80:**4248–4252.
108. **Watabe, K., M. F. Shih, A. Sugino, and J. Ito.** 1982. *In vitro* replication of bacteriophage ϕ29 DNA. *Proc. Natl. Acad. Sci. USA* **79:**5245–5248.

109. **Whiteley, H. R., W. D. Ramey, G. B. Spiegelman, and R. D. Holder.** 1986. Modulation of *in vivo* and *in vitro* transcription of bacteriophage ϕ29 early genes. *Virology* **155:**392–401.
110. **Wichitwechkarn, J., D. Johnson, and D. Anderson.** 1992. Mutant prohead RNAs in the in vitro packaging of ϕ29 DNA-gp3. *J. Mol. Biol.* **223:**991–998.
111. **Wu, H.-W., and D. M. Crothers.** 1984. The locus of sequence-directed and protein-induced DNA bending. *Nature* (London) **308:**509–513.
112. **Yanofsky, S., F. Kawamura, and J. Ito.** 1976. Thermolabile transfecting DNA from temperature-sensitive mutant of phage ϕ29. *Nature* (London) **259:**60–63.
113. **Yehle, C. O.** 1978. Genome-linked protein associated with the 5′ termini of bacteriophage ϕ29 DNA. *J. Virol.* **27:**776–783.
114. **Yoshikawa, H., T. Friedmann, and J. Ito.** 1981. Nucleotide sequences at the termini of ϕ29 DNA. *Proc. Natl. Acad. Sci. USA* **78:**1336–1340.
115. **Yoshikawa, H., and J. Ito.** 1982. Nucleotide sequence of the major early region of bacteriophage ϕ29. *Gene* **17:**323–335.
116. **Zaballos, A., and M. Salas.** 1989. Functional domains in the bacteriophage ϕ29 terminal protein for interaction with the ϕ29 DNA polymerase and with DNA. *Nucleic Acids Res.* **17:**10353–10366.
117. **Zinkel, S. S., and D. M. Crothers.** 1991. Catabolite activator protein-induced DNA bending in transcription initiation. *J. Mol. Biol.* **219:**201–205.
118. **Zwieb, C., J. Kim, and S. Adhya.** 1989. DNA bending by negative regulatory proteins: gal and lac repressors. *Genes Dev.* **3:**602–611.

59. Morphogenesis of Bacteriophage ϕ29

DWIGHT ANDERSON and BERNARD REILLY

Bacteriophages are very useful models for studying the protein-derived principles of form determination in virus assembly. The formation of a precapsid or prohead with the aid of a scaffold protein prior to double-stranded DNA packaging may imply a subtle threshold of complexity that distinguishes this class of assembly from that of the small plant and animal viruses. Here, the structural information of the prohead proteins is expressed sequentially in the context of the protein pools, and the expression is completed only in the context of genome packaging (40). Thus, subassemblies form that in the aggregate produce competent proheads, and the conformation of the prohead changes as the DNA is packaged.

Bacteriophage models cannot be ignored because of the assembly of the virus tail at the prohead portal vertex or connector, the entry point of double-stranded DNA. From our perspective, the assembly events prior to tail construction may be general properties common to all viruses of this level of structural complexity. The competent proheads of prolate double-stranded DNA bacteriophages like T4 and ϕ29 are characterized by a connector with sixfold symmetry that functions first as a platform for scaffold-directed prohead assembly, then serves as a site for the creation of a DNA-packaging machine, and finally contains the foundation for tail assembly or addition. The marvelous DNA-packaging machine is both complex and efficient and may have properties representative of the whole as opposed to its components. For example, the ATPase and putative gyrase functions in ϕ29 DNA-gp3 packaging (see below) seem to be properties of the entire machine, not specific enzymatic components of the machine.

It is typical of bacteriophages that the prohead has the approximate size and shape of the virus but in fact assumes its true form only after DNA packaging and partial destruction of the packaging machine. These considerations suggest that bacteriophage assembly can result in a new perspective on viral assembly and that the relationship of function and form at this level has an intrinsic beauty. It follows that simple systems and models may give us more insight into the mechanistic details when the assembly involves a simple competent prohead and an efficient DNA-packaging system. Bacteriophage ϕ29 of *Bacillus subtilis* (2, 52), illustrated in the family portrait with bacteriophage T2 of *Escherichia coli* (Fig. 1), features simplicity, efficient in vitro assembly, and advanced genetic and biochemical characterizations. For example, the precursor viral capsid or prohead can be composed of only the major capsid protein (gene product 8 [gp8]) and the portal protein or connector (gp10) (25, 47). Practically every purified prohead can package the 19-kb DNA-gp3 complex in vitro with the aid of ATPase gp16. Then, the three proteins of the neck and tail can be assembled to yield the mature virion (8, 30). The genes encoding the structural proteins have been marked and mapped (41, 53). The genes that encode the connector, scaffold, and shell of the prohead and gp16 have been cloned, and the proteins have been overproduced in *E. coli* and purified (21, 29, 30). This chapter emphasizes ϕ29 structure and the mechanisms of prohead assembly and DNA packaging, the major thrusts of current ϕ29 research in Minneapolis.

STRUCTURE OF ϕ29

The ϕ29 prohead is composed of about 220 copies of the major capsid protein (gp8), 180 copies of the scaffold (gp7), and 12 copies of the head-tail connector or portal protein (gp10) (44, 47). The scaffold exits during DNA-gp3 packaging (9), and the subunits of the major capsid protein are conserved in this process though conformationally altered (see below). When present, the nonessential head fibers (gp8.5) radiate from the apical regions of the head (2, 51, 56). Six copies of the lower collar (gp11), 12 appendages (dimers of gp12* cleaved from gp12), and three or four copies of the tail protein (gp9) are assembled onto the DNA-filled head in a single morphogenetic pathway to yield the mature virion (Fig. 2; 15, 34, 42, 57).

Electron microscopy of frozen-hydrated proheads and mature ϕ29 demonstrates that the head is about 38 by 51 nm in size (7). The prolate mature capsid was modeled with trimeric aggregates of gp8 dimers arranged on a T = 1 icosahedral lattice, with the apical regions separated by 10 additional trimers in the equatorial zone; this structure would contain 90 dimers of gp8, or 180 subunits (15, 38, 60). The total number of subunits would be reduced to 170 or 175 if five gp8 dimers or subunits, respectively, were missing at the portal vertex to accommodate the connector. A reevaluation of the data of Carrascosa and colleagues (15) and the true molecular weight of the gp8 derived from the DNA sequence indicate that the number of capsid subunits would be 211. Similarly, a reevaluation of our published biochemical data indicates that the capsid would contain 218 ± 6 subunits of gp8 (34, 47). Finally, recent determinations of particle mass by scanning transmission electron microscopy (STEM) at the Brookhaven National Laboratory suggest that the prohead is composed of about 217 subunits of gp8 (45). There are too many copies of gp8

Dwight Anderson and Bernard Reilly • University of Minnesota, 18-246 Moos Tower, 515 Delaware Street, S.E., Minneapolis, Minnesota 55455.

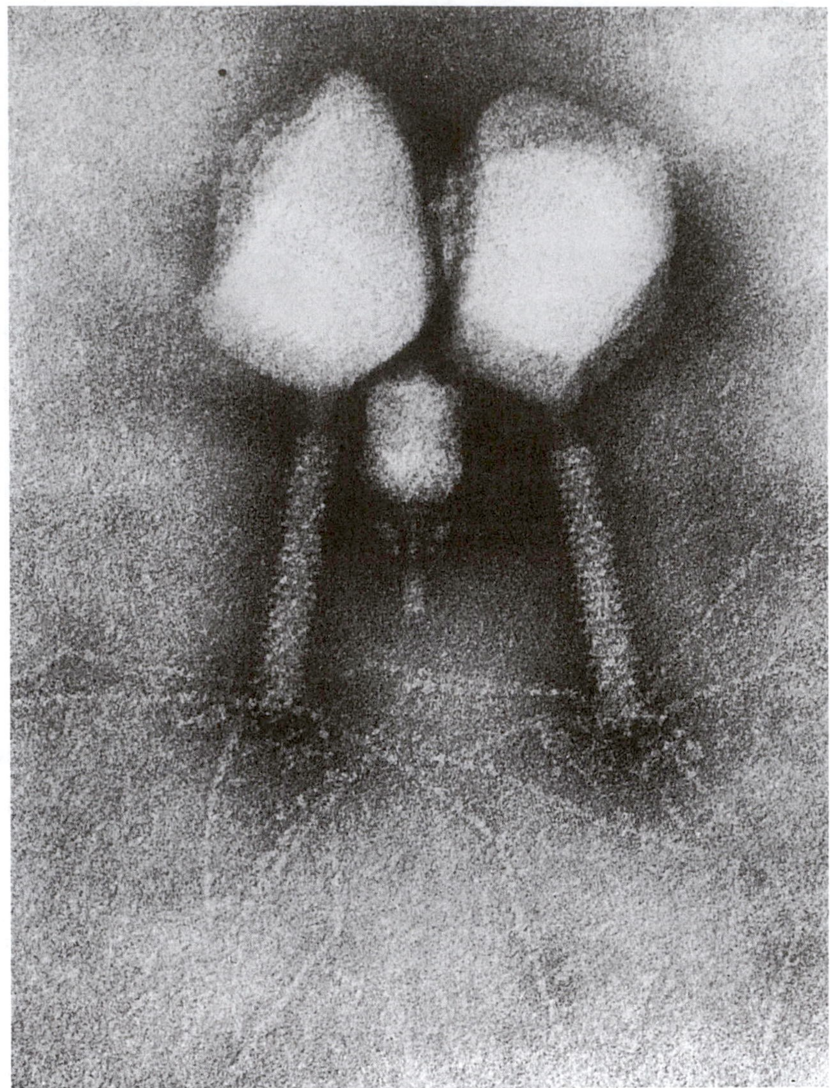

Figure 1. Family portrait of phages ϕ29 and T2. Magnification, ×428,000.

to fit the T = 1 model, and it is more likely that the prohead and ϕ29 are modified T = 3 structures containing about 235 copies of gp8. (The model in Fig. 3 was proposed by S. Casjens and R. Hendrix at the XIIth International Conference on Bacteriophage Assembly, Cable, Wisc., on the basis of biochemical data and mass determinations by STEM.) The gp8 subunits would be arranged as 30 hexamers and 11 pentamers. Moreover, the length/width ratio of this hypothetical T = 3 structure corresponds to the particle dimensions observed in electron microscopy of negatively stained and frozen-hydrated proheads and ϕ29 (2, 43).

The scaffold gp7 is required for prohead assembly in vivo (34, 42, 55) but not for DNA-gp3 packaging in vitro (25, 47). The gp7 contents of ϕ29 proheads are variable, ranging from about 50 to 130 copies (34). By using *B. subtilis* DB431, an organism defective for several proteases (18), as host, proheads with about 180 copies of gp7 have been isolated (5). The gp7 (M_r, 11,267) has only about 14% of the mass of the gp8 (M_r, 49,847) in the prohead.

Three-dimensional reconstructions of the connector demonstrate a spool with an axial hole that tapers from a proximal diameter of about 14 nm with 12-fold axial symmetry to a distal diameter of about 8 nm with 6-fold symmetry; the distal part may bind DNA-gp3 in packaging and subsequently gp11 in neck-tail assembly (12, 13, 39). The connector may occupy distinct environments relative to the major capsid protein in the prohead and the mature phage. In the prohead, the connector, with an axial height of about 8 nm, does not protrude from the base of the particle; therefore, the wide part is inside the prohead. In the phage, the narrow part, with the gp11 attached distally (14, 16), may protrude from the particle so that the wide portion is flush with the shell. Thus, the connector may be able to move up and down in the head axially, and it may do so in DNA-gp3 packaging (see below).

To determine detailed prohead and ϕ29 structures, attempts are being made to reconstruct three-dimensional information from images of frozen-hydrated

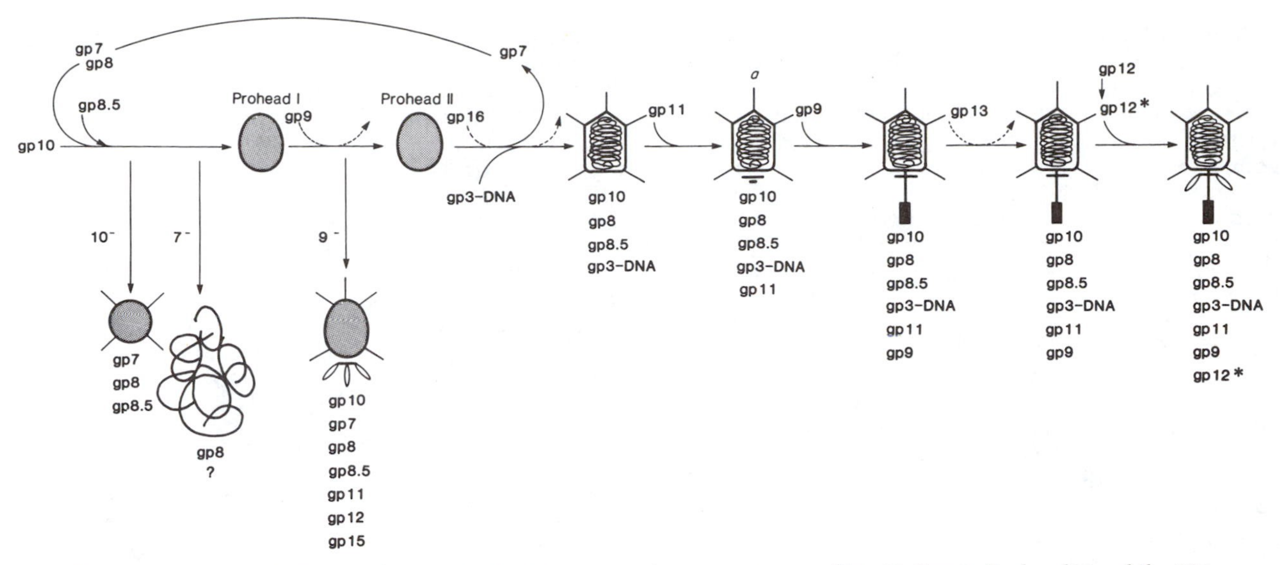

Figure 2. Pathway of ϕ29 morphogenesis. Proteins composing a structure are listed below it. Prohead II and the 11^-, 12^-, and 13^- particles are true intermediates in the pathway. The 10^-, 7^-, and 9^- structures shown below the main pathway are abortive structures. Particle a has been observed only in restrictive *sus*12(716) infection together with the 12^- particle; its composition is inferred from morphology.

specimens (43) and to produce three-dimensional crystals for analysis by X-ray diffraction.

MODEL FOR FORM DETERMINATION IN PROHEAD ASSEMBLY

The basic problem of ϕ29 morphogenesis is the mechanism by which a prolate shell of particular dimensions is assembled from subunits that are also capable of being assembled into incorrect structures. Perhaps the most revealing fact about ϕ29 prohead assembly is that the connector-defective suppressor-sensitive mutants form isometric particles in the non-permissive host, and particles assembled in *E. coli* in the absence of the connector protein are isometric or variable in size and shape (29, 34). Thus, the connector has a central role in determining the prolate shape of the normal prohead. However, the other essential prohead proteins, the shell gp8 and the core scaffold gp7, also have morphopoietic (form-determining) functions in prohead morphogenesis. Isometric particles form in *ts*7(224)-infected cells at the nonpermissive temperature, suggesting a defective gp7-gp10 interaction (55). In *sus*7 mutant-infected nonpermissive cells, the shell protein gp8 exhibits intrinsic curvature in structures assembled in the absence of gp7 (34). Finally, at the nonpermissive temperature, gp10 of *ts*10(256) forms a prolate particle with a distal apex that resembles the proximal apex (i.e., the distal apex does not have the typical flattened appearance) (4), whereas gp8 of *ts*8(211) is assembled into a prolate particle with prohead morphology under restrictive conditions (55). These results are consistent with the in vitro interaction of purified gp7 and gp10 (29) and the idea that if gp7 can interact with a subassembly of defective proteins (either gp8 in a hexamer or gp10 in

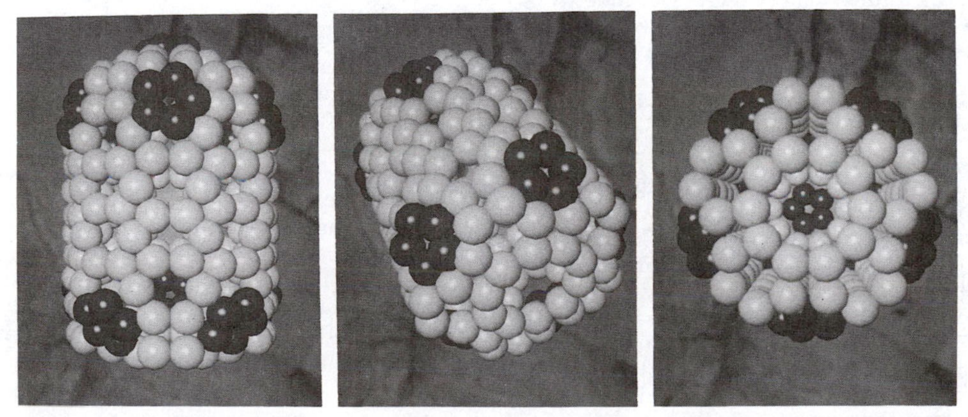

Figure 3. Three views of the modified T = 3 model of the ϕ29 prohead according to S. Casjens and R. Hendrix. The model contains 30 hexamers (light) and 11 pentamers (dark) of the major capsid protein. The panel on the right shows a view of the base, where the connector (not shown) is attached. The computer-generated schematics are by E. Egelman.

a dodecamer), then the mechanics of the interaction will produce a stable particle of the approximate size and form of the mature prohead.

Following is a working model of ϕ29 assembly. The initiator for prohead assembly is the dodecameric connector. Under the influence of host chaperonins, gp7-gp8 heterodimers and gp8 monomers are assembled into hexamers (6-mers) and pentamers (5-mers), respectively. Five 6-mers pack around the periphery of the connector as a result of gp7-gp10 interaction. The interaction might be symmetry independent by virtue of a uniform distribution of positive charge over some surface of the connector. The connector holds the surrounding five 6-mers in a particular angular relationship that forms the peculiar flattened base of the ϕ29 head and is sufficient to direct assembly of 5-mers and 6-mers to fill out the prolate prohead to the proper size and shape. Finally, the 174-base ϕ29-encoded RNA (pRNA) that is essential for DNA-gp3 packaging is attached (see below).

Although there is no direct evidence that gp7 and gp8 form heterodimers, gp7 and gp8 form an isometric particle in the absence of gp10. The provisional T = 3 model of prohead structure requires 30 6-mers of gp8 (180 copies), and the prohead contains 180 subunits of gp7; thus the 6-mers may be constructed of gp7-gp8 heterodimers, and the gp7 pool size may regulate the number of 6-mers. Host chaperonins are evoked because some host mutants, called "vam" (virus assembly mutants), cannot assemble ϕ29 proheads even though protein production is normal (49), and assembly of shell 6-mers and 5-mers from guanidinium chloride-dissociated proheads of the coliphage HK97 requires GroEL, GroES, and ATP (36). Although the gp7 of heterodimers and 6-mers may form a stabilizing network, we suggest that the assembly of five 6-mers around the connector (simply the number that pack around the connector periphery) to form the distal apex is a result of gp7-gp10 interaction. Connectors are purified by phosphocellulose chromatography, and their apparent positive charges may be distributed to effect packing of five 6-mers in symmetry-independent interactions.

The assembly of proheads in *E. coli* from products of cloned ϕ29 genes and reconstitution of these proheads in vitro with pRNA show that pRNA is not needed for prohead assembly and that host factors unique to *B. subtilis* are not required for prohead assembly (29, 33).

B. SUBTILIS MUTANTS DEFECTIVE FOR ϕ29 PROHEAD ASSEMBLY

Mutants of asporogenous *B. subtilis* that cannot assemble ϕ29 (vam) were selected by the use of antibodies that reacted more strongly with the free connectors than with the portal vertex of proheads or phage (48). Phage adsorption and the synthesis of phage proteins and DNA-gp3 were normal in these mutants, but prohead and phage production was greatly reduced. The assembly defect was transferred to *B. subtilis* by transformation and transduction.

Two properties of the vam may provide leads to the nature of the defect(s) (49). First, connectors assembled in vam are more homogeneous and slower in sucrose density gradient centrifugation than are connectors assembled in the wild-type host; thus, the structure and/or composition of the connectors produced in vam may be unusual or different. Second, some vam cannot assemble isometric particles when infected with the mutant *sus*10(302)-*sus*14(1241) that is defective for the connector, even though the scaffold gp7 and the major capsid protein gp8 are produced in quantity; this suggests that the scaffold and shell proteins cannot interact or be assembled into oligomers, perhaps because of the lack of chaperonin (e.g., *groE*) function. By PBS1 transduction, the *vam* locus is closely linked to Tn*917* located at 317° on the *B. subtilis* chromosome, a region that is near the reported *groE* locus (61).

PACKAGING OF ϕ29 DNA-gp3

Bacteriophage DNA-packaging machines seem to have in common two packaging proteins and a connector with sixfold symmetry. One protein binds DNA, and the second is a DNA-dependent ATPase with magnesium- and prohead-binding domains (32). The problem of packaging seems to be quite complex in a physical sense. The viral genomes can vary from 1 to 10 times the length of the cell and have a length more than 200 times the prohead diameter. In the ϕ29 system, about 10 bacterial genome equivalents of viral DNA are packaged with an efficiency of more than 50% and possibly at a rate of about 3,000 bp/s per genome. Demonstration of the roles played by negative repulsion and hydration is almost intractable in cell extracts, and the effectiveness of packaging with respect to orientation and exit during infection remains a mystery. However, useful information on DNA packaging can be obtained with a simple, efficient system.

In vitro DNA-gp3 packaging and assembly in extracts has been highly efficient from the beginning; about one-third of the proheads were converted to ϕ29, with a yield of about 200 phage per cell equivalent of proheads (8). In extracts, the concentration dependence for proheads and DNA-gp3 was first order, while multiple copies of ATPase gp16 were required (10). In the completely defined system, more than 20% of the DNA-gp3 added was packaged into proheads (30). Currently, virtually every prohead is competent to package DNA-gp3, and every genome can be packaged. ATPase gp16, produced in *E. coli* and purified to near homogeneity, is maximally efficient at about six copies per prohead (46).

Thus, in vitro DNA-gp3 packaging and tail assembly in extracts is of the same order of magnitude as that occurring in vivo. A model system for DNA-gp3 packaging in which each component save DNA-gp3 can be produced from cloned genes and highly purified has been developed. All of these components are accessible to experimental manipulation in a system that is a very close analog of its natural progenitor, and results obtained with this system have a good chance of being biologically relevant.

In Vitro Packaging Systems

A typical extract complementation involves mixing and incubating a prohead donor extract with a gp16 donor extract and assaying for production of ϕ29 (8). Purified gp16 or proheads can complement a defective extract, and exogenous [^{3}H]DNA-gp3 is used to mark the position of filled heads and to quantify packaging in sucrose density gradient centrifugation. The defined system consists of purified proheads, DNA-gp3, ATPase gp16, and ATP (30). Unpackaged DNA-gp3 is digested with DNase I, and packaged DNA-gp3 is extracted and quantified by agarose gel electrophoresis (22).

Events and Intermediates

Packaging intermediates can be isolated after DNase I interruption of in vitro complementation in extracts (9). Restriction enzyme digestion of DNA molecules extracted from DNase I-treated proheads shows that DNA-gp3 packaging is oriented with respect to the physical map. Left-end restriction fragments of DNA-gp3 are selectively packaged, and packaging is quantized. Packaging of the left one-third of the genome results in exit of the scaffold gp7 and a transition to particle angularity.

Events in initiation of DNA-gp3 packaging have been studied in the defined system (31, 32). The binding of gp16 to the prohead is followed by attachment of DNA-gp3, binding of ATP, and hydrolysis of 1 ATP per 2 bp of DNA packaged. The selective packaging of left ends reflects recognition of the left-end protein and, possibly, DNA sequence(s) near the inverted terminal repeat sequence AAAGTA. In the defined system, after some rate-limiting initiation event(s), DNA-gp3 packaging is too fast to allow isolation of packaging intermediates (30). Also, both left- and right-end restriction enzyme fragments of the DNA-gp3 can be packaged, albeit the left ends are more stably packaged (22). Nonend restriction fragments are packaged at an efficiency 1 order of magnitude lower. The selective packaging of left-end fragments observed in extracts (9) is restored in the defined system by the addition of glycerol or extracts of either infected or uninfected *B. subtilis* (22). M2Y DNA, which has distinct terminal-protein and base sequences, is packaged into ϕ29 proheads in the defined system but not in the presence of glycerol or bacterial extracts. The results suggest that the stringency for DNA-gp3 recognition is higher in extracts than in the defined system. The terminal protein gp3 forms two bands in sodium dodecyl sulfate-polyacrylamide gel electrophoresis (SDS-PAGE) (3). These proteins are cleaved by seryl proteases to the same number of peptides, but these peptides differ in their mobilities in SDS-PAGE; moreover, the gp3 of the right end of purified DNA-gp3 is more sensitive to proteolysis (22). Thus, the gp3 of the left and right DNA ends differ, presumably because of posttranslational modification, and this difference may affect the orientation and stability of DNA-gp3 packaging.

Role for ϕ29-Encoded RNA

A novel 174-base ϕ29-encoded RNA (pRNA) is essential for DNA-gp3 packaging in extracts and in the defined in vitro system (1, 27, 28). The RNA is a transcript of bases 320 through 147 at the left early region of the genome (27) and is produced from promoter PE1(A1) (17, 54). Six copies of pRNA are found on proheads isolated in the presence of RNase inhibitor (62), and occasionally, the pRNA is cleaved from 174 to 120 bases, presumably by adventitious nucleases, during prohead isolation. The attachment site for pRNA is the connector of the prohead (27). pRNA is not present on isometric particles produced in nonpermissive infections with the mutant *sus*10(302) or *sus*10(940), which both lack the connector, nor on prolate particles produced in *ts*10(256) nonpermissive infection (4, 28). A large amber fragment of gp10 is produced in *sus*10(940) infection, suggesting that a short segment at the C terminus is essential for initiation of prohead assembly. The bulk of the pRNA can be removed from proheads by incubation in Tris-borate buffer (pH 8.3) containing 2.5 mM EDTA or by RNase treatment (27, 28); alternatively, proheads without pRNA can be produced in *E. coli* HMS174(DE3)(pARgp7-8-8.5-10), which expresses the cloned capsid genes (29). Proheads devoid of pRNA can be reconstituted to DNA-gp3 packaging activity by the addition of pRNA either isolated from proheads (27, 28) or purified from *B. subtilis* Spo0A12(pUM102), which expresses the cloned gene (33, 63).

Determination of the secondary structure of pRNA has been facilitated by a phylogenetic approach utilizing pRNA sequences of the ϕ29-like phages M2-Nf, SF5, and GA1 (Fig. 4; 6). The 5′ domain is composed of 113 to 117 residues and contains four helices. A smaller 3′ domain is composed of 40 to 44 residues and consists of two helices. The two domains are separated by 8 to 13 unpaired residues, where cleavage occurs to generate the biologically active 120-base ϕ29 pRNA.

Both the prohead connector and the ATPase gp16 utilize RNA recognition motifs characteristic of a number of RNA-associated proteins, and the binding of gp16 by the prohead shields pRNA from RNase A (24). Nuclease footprint analysis of the binding of pRNA to free connectors or proheads has shown that free connectors protect most of the molecule, probably by forming a distinctive oligomer with the pRNA, while the prohead protects an area of the pRNA principally between residues 30 and 90 (5′ to 3′) (50). The ATPase activity of gp16 is stimulated fourfold by pRNA; the pRNA is needed continuously, suggesting that the gp16 and pRNA are in a complex. A 10-fold stimulation of the ATPase activity of gp16 by proheads depends on the pRNA (24).

Truncated forms of pRNA have been generated in situ on the prohead by limited digestion with RNase A, and the effects on packaging of restriction enzyme fragments of DNA-gp3 in the defined system have been determined (23). Proheads with a pRNA of 174 or 120 bases (5′ residues 1 through 120; Fig. 4) package both the left and right ends of DNA-gp3, particles with pRNA of 95 bases (5′ residues 26 through 120) package only the left end, and surprisingly, proheads with pRNA of 71 bases (5′ residues 50 through 120) package only internal (nonend) fragments, which generally is 1 order of magnitude less efficient. Thus, a domain for recognition of the gp3 of DNA-gp3 may exist between residues 1 and 50. Proheads with 69-base (5′ residues 50 through 118) or 54-base (5′ residues 50 through

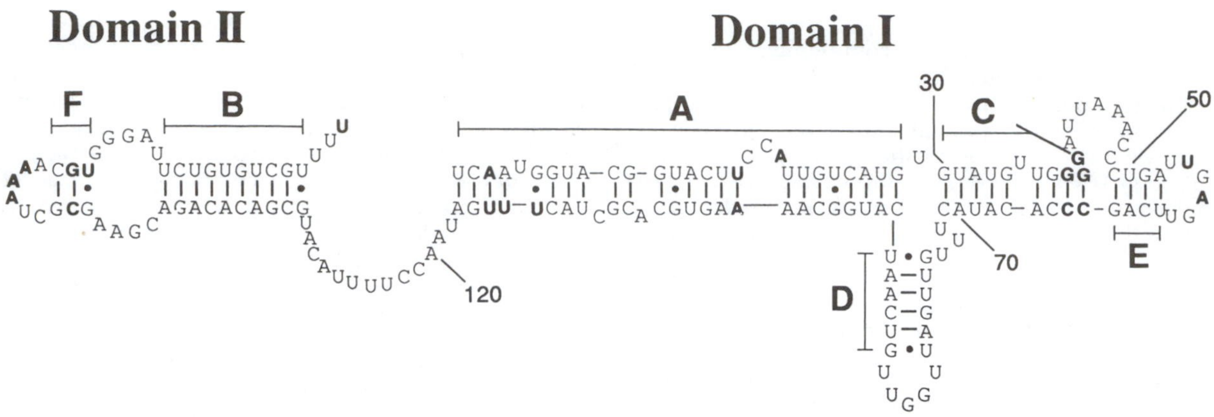

Figure 4. Secondary structure of ϕ29 pRNA derived by phylogenetic analysis (6). A through F, helical stems. Residues conserved in the ϕ29-PZA, M2-Nf, SF5, and GA1 pRNA sequences are designated in boldface letters.

103) pRNA are inactive, perhaps because of defects in gp16 or DNA binding or in ATPase function.

Computer programs for RNA structure analysis have been used to fold hypothetical pRNA mutants and thus to target oligonucleotide mutagenesis for correlation of structure and function (63). Changes of highly conserved bases (Fig. 4) that retain the predicted secondary structure of the pRNA have modest effects on DNA-gp3 packaging in the defined system, while changes predicted to alter the pRNA secondary structure drastically reduce DNA-gp3 packaging. Therefore, changes of conserved bases that have a modest effect in the laboratory probably do not occur in the more stringent natural environment.

In summary, the results suggest that pRNA functions in assembly of the DNA-packaging machine and in recognition of the DNA termini. In addition, pRNA may be a part of the DNA-gp3-translocating ATPase.

A Dynamic Role for the Connector

It has been proposed that the phage connector may rotate relative to the shell, utilizing the head-tail symmetry mismatch and ATP hydrolysis, to drive DNA packaging (37). This idea and our recent demonstration that the ϕ29 connector wraps supercoiled plasmid DNA on the outside form the basis for a new model of DNA packaging in which the connector has a dynamic role. We have proposed that the connector may wrap DNA, rotate relative to the shell with the aid of ATP hydrolysis, and act as a winch to translocate DNA (58). The axial hole in the connector may function only in DNA ejection.

The evidence for the wrapping of DNA around the outside of the connector is provided in part by comparison of the mass and the mean radial-density distribution of the connector and connector-pBR322 DNA complexes by STEM (58). With a radius of binding of the DNA of 70 Å (1 Å = 0.1 nm), the connector wraps about 180 bp of DNA in 1.4 turns. Connectors wrap only supercoiled pBR322 DNA, removing negative supercoils, although they bind linear and open circular DNAs. Topoisomerase I digestion of connector–supercoiled-DNA complexes followed by protein digestion and gel electrophoresis shows that supercoils are restrained by the connector. Moreover, electron microscopy demonstrates a decrease in the contour length of the connector-DNA complexes with an increase in the number of connectors bound, further evidence that the connector wraps about 180 bp of DNA.

A rotating connector cannot be rigidly bound by the shell protein 6-mers. The larger part of the connector is inside the prohead, and the smaller part is flush with the prohead base. The smaller part, reported to bind DNA (39), may drop down, wrap DNA, shift back up, and rotate to deposit the DNA within the shell. The six copies each of ATPase gp16 and pRNA may be spaced on the sixfold symmetric connector to effect the axial and rotational motions in discrete steps. Both the connector and gp16 have RNA recognition motifs (24), and pRNA may be the mediator in the molecular motion, being held and released alternately by the connector and gp16. ATP hydrolysis may drive conformational change of ATPase gp16, which is more active in the presence of pRNA (24).

A gyrase action would be needed to generate supercoiled DNA for packaging. DNA-gp3 packaging in the defined system is sensitive to the gyrase inhibitors novobiocin, coumermycin A1, oxolinic acid, and nalidixic acid (26, 28).

Structural Transitions

In part, the rationale for crystallizing both the ϕ29 prohead and the mature head (see above) is to define the nature and extent of shell protein conformational change during DNA-gp3 packaging. The prohead is a rounded, flexible structure, easily distorted during negative staining for electron microscopy, while the DNA-filled head and the empty head that has lost DNA-gp3 are rigid and angular. The prohead moves faster than the DNA-filled head during particle gel electrophoresis in agarose, suggesting that the particles have different surface charges. Some missense mutants in the shell, detected by SDS-PAGE of viral proteins, cannot assemble the normal head fibers and therefore produce fiberless mature phage; surprisingly, the proheads of some of these mutants have fibers, suggesting that the fiber-binding site of the shell is altered or rearranged in DNA-gp3 packaging. These changes in shell rigidity, charge, and interaction with head fibers must reflect changes in shell subunit conformation and bonding angles that are of intrinsic interest.

Chimeric Proheads

Phage lambda proheads that have a φ29 connector have been assembled in vitro and reported to package both lambda and φ29 DNA-gp3 (19, 20). The connectors of φ29, lambda, and T3 are similar in size, and although the proteins share little or no amino acid homology, they are reported to display similar predicted secondary structures (39). In contrast, tail proteins of unrelated coliphages share significant amino acid homology and may therefore have been transferred horizontally (35). Did the phage connector proteins evolve independently to the same structure and task? Tantamount to connector interchangeability is shell swapping; i.e., if two or more connectors can interact with one shell, then two or more shells may be able to interact and function with one connector. Perhaps a very large shell can be assembled onto the φ29 connector, and accordingly, perhaps a very large DNA can be packaged. Moreover, the initiating complexes in prohead formation that include the connector and shell (e.g., the hypothetical connector-hexamer complex proposed above for φ29) must be remarkably similar in φ29 and lambda.

Summary

The φ29 prohead shell and connector together with the detachable pRNA and ATPase gp16 constitute the relatively simple and dynamic DNA-gp3-packaging machine. The wrapping of supercoiled DNA and the proposed rotation of the connector change the perspective of mechanistic studies, because the DNA-gp3 substrate may be supercoiled in some phase of packaging, and one or more components of the packaging machine must have topoisomerase activity. Further, there is new incentive to rationalize quantized packaging as the focus is shifted away from the axial hole of the connector to the outside domains that wrap DNA and interact with the shell. The challenge is to extend the structural and biochemical basis of knowledge to the dynamic, i.e., the motion of DNA in packaging.

TESTING THE MODELS

The T = 3 model will be verified by the use of protein quantification, mass determination by STEM, three-dimensional reconstruction, and ultimately, X-ray crystallography. The model for prohead assembly can be tested by the demonstration of 6-mers composed of gp7-gp8 heterodimers. The proteins (gp7 and gp8) expressed from the cloned genes and the chaperonins are readily attainable in highly purified form.

Connector symmetry is an important consideration in construction of proheads and the DNA-gp3-packaging machine and in understanding the mechanism that drives the dynamic connector in this model. What is the symmetry of the connector prior to DNA packaging, and is the sixfold connector symmetry a requirement for packaging or the result of DNA packaging?

Why has this group of viruses evolved a pRNA with conserved secondary structure of two domains even though the primary base sequences are dissimilar? Is this pRNA unique to this type of virus? The counterpart to pRNA of φ29 has not been identified in other phages, although the packaging of both lambda and Mu DNA is sensitive to RNase (11, 19).

The role of ATP is defined in part by the positioning of gp16 and pRNA, the symmetry of the connector during packaging, and the topology of the packaging substrate. The six copies each of gp16 and pRNA that are involved in packaging may relate directly to connector symmetry and dynamic action. Spectroscopic molecular probes have been attached to the connector to evaluate its rotational motion relative to the shell (59).

The demonstration of chimeric proheads (19, 20) suggests that both the connectors and the shell contacts among unrelated phages may be quasi-identical. Even though each virus seems to have unique structural and assembly features, the production of φ29 proheads in *E. coli* (29, 33) shows that there are general assembly mechanisms.

The critical consideration for the proposed dynamic role of the connector in packaging is the structural relationship of the connector and the shell. Is it relevant to the DNA packaging dynamic or a requirement for viral stability during the critical interval between the completion of packaging and the subsequent stages of tail assembly? After φ29 DNA-gp3 packaging, pRNA must be removed, the connector must be moved with reference to the shell proteins, and the right end of the DNA-gp3 must be positioned for entry into the cell. During wild-type particle formation, a significant proportion of packaging events results in the production of angular empty heads; with *sus*13 nonpermissive infection, many empty proheads are visible within the infected cell (34); and during in vitro extract packaging, virus production is less efficient than DNA packaging, particularly if there is a significant time interval between packaging and the addition of the DNA-filled heads to a tail-forming extract (8). These observations are consistent with instability of DNA-filled particles, and the stabilization required may obscure the DNA-packaging process.

Acknowledgments. The research was supported by grants DE-03606, DE-08515, and GM-39931 from the National Institutes of Health. We thank Charlene Peterson for preparing the manuscript.

REFERENCES

1. **Anderson, D., and J. W. Bodley.** 1990. Role of RNA in bacteriophage φ29 DNA packaging. *J. Struct. Biol.* **104:** 70–74.
2. **Anderson, D. L., D. D. Hickman, and B. E. Reilly.** 1966. Structure of *Bacillus subtilis* bacteriophage φ29 and the length of φ29 deoxyribonucleic acid. *J. Bacteriol.* **91:** 2081–2089.
3. **Anderson, D. L., and B. E. Reilly.** 1974. Analysis of bacteriophage φ29 gene function: protein synthesis in suppressor-sensitive mutant infection of *Bacillus subtilis*. *J. Virol.* **13:**211–221.

4. **Atz, R., and D. Anderson.** Unpublished data.
5. **Atz, R., B. Rajagopal, and D. Anderson.** Unpublished data.
6. **Bailey, S., J. Wichitwechkarn, D. Johnson, B. E. Reilly, D. L. Anderson, and J. W. Bodley.** 1990. Phylogenetic analysis and secondary structure of the *Bacillus subtilis* bacteriophage RNA required for DNA packaging. *J. Biol. Chem.* **265:**22365–22370.
7. **Baker, T. S. (Purdue University).** 1990. Personal communication.
8. **Bjornsti, M. A., B. E. Reilly, and D. L. Anderson.** 1981. *In vitro* assembly of the *Bacillus subtilis* bacteriophage ϕ29. *Proc. Natl. Acad. Sci. USA* **78:**5861–5865.
9. **Bjornsti, M. A., B. E. Reilly, and D. L. Anderson.** 1983. Morphogenesis of bacteriophage ϕ29 of *Bacillus subtilis*: oriented and quantized in vitro packaging of DNA-gp3. *J. Virol.* **45:**383–396.
10. **Bjornsti, M. A., B. E. Reilly, and D. L. Anderson.** 1984. Bacteriophage ϕ29 proteins required for in vitro DNA-gp3 packaging. *J. Virol.* **50:**766–772.
11. **Burns, C. M., H. L. B. Chan, and M. S. DuBow.** 1990. *In vitro* maturation and encapsidation of the DNA of transposable Mu-like phage D108. *Proc. Natl. Acad. Sci. USA* **87:**6092–6096.
12. **Carazo, J. M., L. E. Donate, L. Herranz, J. P. Secilla, and J. L. Carrascosa.** 1986. Three-dimensional reconstruction of the connector of bacteriophage ϕ29 at 1.8 nm resolution. *J. Mol. Biol.* **192:**853–867.
13. **Carazo, J. M., A. Santisteban, and J. L. Carrascosa.** 1985. Three-dimensional reconstruction of bacteriophage ϕ29 neck particles at 2.2 nm resolution. *J. Mol. Biol.* **183:**79–88.
14. **Carrascosa, J. L., J. M. Carazo, and N. Garcia.** 1983. Structural localization of the proteins of the head to tail connecting region of bacteriophage ϕ29. *Virology* **124:** 133–143.
15. **Carrascosa, J. L., E. Mendez, J. Corral, V. Rubio, G. Ramirez, M. Salas, and E. Vinuela.** 1981. Structural organization of *Bacillus subtilis* phage ϕ29. A model. *Virology* **111:**401–413.
16. **Carrascosa, J. L., E. Vinuela, N. Garcia, and A. Santisteban.** 1982. Structure of the head-tail connector of bacteriophage ϕ29. *J. Mol. Biol.* **154:**311–324.
17. **Davison, B. L., T. Leighton, and J. C. Rabinowitz.** 1979. Purification of *Bacillus subtilis* RNA polymerase with heparin-agarose. *J. Biol. Chem.* **254:**9220–9226.
18. **Doi, R. (University of California, Davis).** 1991. Personal communication.
19. **Donate, L. E., and J. L. Carrascosa.** 1991. Characterization of a versatile *in vitro* DNA-packaging system based on hybrid lambda/ϕ29 proheads. *Virology* **182:**534–544.
20. **Donate, L. E., H. Murialdo, and J. L. Carrascosa.** 1990. Production of lambda-ϕ29 phage chimeras. *Virology* **179:** 936–940.
21. **Garcia, J. A., E. Mendez, and M. Salas.** 1984. Cloning, nucleotide sequence and high level expression of the gene coding for the connector protein of *Bacillus subtilis* phage ϕ29. *Gene* **30:**87–98.
22. **Grimes, S., and D. Anderson.** 1989. *In vitro* packaging of bacteriophage ϕ29 DNA restriction fragments and the role of the terminal protein gp3. *J. Mol. Biol.* **209:**91–100.
23. **Grimes, S., and D. Anderson.** 1989. Cleaving the prohead RNA of bacteriophage ϕ29 alters the *in vitro* packaging of restriction fragments of DNA-gp3. *J. Mol. Biol.* **209:** 101–108.
24. **Grimes, S., and D. Anderson.** 1990. RNA dependence of the bacteriophage ϕ29 DNA packaging ATPase. *J. Mol. Biol.* **215:**559–566.
25. **Grimes, S., P. Guo, C. Peterson, and D. Anderson.** 1986. Protein stoichiometry and biological activity of ϕ29 proheads. Abstr. Xth Biennial Meet. Bacteriophage Assembly 1986.
26. **Guo, P., and D. Anderson.** Unpublished data.
27. **Guo, P., S. Bailey, J. W. Bodley, and D. Anderson.** 1987. Characterization of the small RNA of the bacteriophage ϕ29 DNA packaging machine. *Nucleic Acids Res.* **15:** 7081–7090.
28. **Guo, P., S. Erickson, and D. Anderson.** 1987. A small viral RNA is required for *in vitro* packaging of bacteriophage ϕ29 DNA. *Science* **236:**690–694.
29. **Guo, P., S. Erickson, W. Xu, N. Olson, T. S. Baker, and D. Anderson.** 1991. Regulation of the phage ϕ29 prohead shape and size by the portal vertex. *Virology* **183:**366–373.
30. **Guo, P., S. Grimes, and D. Anderson.** 1986. A defined system for *in vitro* packaging of DNA-gp3 of the *Bacillus subtilis* bacteriophage ϕ29. *Proc. Natl. Acad. Sci. USA* **83:**3505–3509.
31. **Guo, P., C. Peterson, and D. Anderson.** 1987. Initiation events in *in vitro* packaging of bacteriophage ϕ29 DNA-gp3. *J. Mol. Biol.* **197:**219–228.
32. **Guo, P., C. Peterson, and D. Anderson.** 1987. Prohead and DNA-gp3-dependent ATPase activity of the DNA packaging protein gp16 of bacteriophage ϕ29. *J. Mol. Biol.* **197:**229–236.
33. **Guo, P., B. S. Rajagopal, D. Anderson, S. Erickson, and C.-S. Lee.** 1991. sRNA of phage ϕ29 of *Bacillus subtilis* mediates DNA packaging of ϕ29 proheads assembled in *Escherichia coli*. *Virology* **185:**395–400.
34. **Hagen, E. W., B. E. Reilly, M. E. Tosi, and D. L. Anderson.** 1976. Analysis of gene function of bacteriophage ϕ29 of *Bacillus subtilis*: identification of cistrons essential for viral assembly. *J. Virol.* **19:**501–517.
35. **Haggard, E., C. Halling, and R. Calendar.** 1991. DNA sequence of three tail genes of bacteriophage P2: further evidence for horizontal transfer of tail fiber genes among unrelated bacteriophages. Abstr. XIIth Int. Conf. Bacteriophage Assembly 1991.
36. **Hendrix, R., Z. Xie, R. Duda, and K. Martincic.** 1991. Genetics of HK97 head assembly and *in vitro* shell assembly. Abstr. XIIth Int. Conf. Bacteriophage Assembly 1991.
37. **Hendrix, R. W.** 1978. Symmetry mismatch and DNA packaging in large bacteriophages. *Proc. Natl. Acad. Sci. USA* **75:**4779–4783.
38. **Hendrix, R. W.** 1985. Shape determination in virus assembly: the bacteriophage example, p. 170–203. *In* S. Casjens (ed.), *Virus Structure and Assembly*. Jones and Bartlett, Portola Valley, Calif.
39. **Herranz, L., J. Bordas, E. Towns-Andrews, E. Mendez, P. Usobiaga, and J. L. Carrascosa.** 1990. Conformational changes in bacteriophage ϕ29 connector prevents DNA-binding activity. *J. Mol. Biol.* **213:**263–273.
40. **Kellenberger, E.** 1990. Form determination of the heads of bacteriophages. *Eur. J. Biochem.* **190:**233–248.
41. **Mellado, R. P., F. Moreno, E. Vinuela, M. Salas, B. E. Reilly, and D. L. Anderson.** 1976. Genetic analysis of bacteriophage ϕ29 of *Bacillus subtilis*: integration and mapping of reference mutants of two collections. *J. Virol.* **19:**495–500.
42. **Nelson, R. A., B. E. Reilly, and D. L. Anderson.** 1976. Morphogenesis of bacteriophage ϕ29 of *Bacillus subtilis*: preliminary isolation and characterization of intermediate particles of the assembly pathway. *J. Virol.* **19:**518–532.
43. **Olson, N. H., W. Xu, W. D. Grochulski, D. L. Anderson, and T. S. Baker.** 1990. Electron microscopy of negatively stained and frozen-hydrated bacteriophage ϕ29, abstr, p. 270. Abstr. Proc. XIIth Int. Congr. Electron Microscopy 1990.
44. **Peterson, C., R. Atz, and D. Anderson.** Unpublished data.
45. **Peterson, C., R. Atz, M. Simon, E. Egelman, and D. Anderson.** Unpublished data.
46. **Peterson, C., S. Grimes, and D. Anderson.** Unpublished data.
47. **Peterson, C., S. Grimes, R. Atz, E. Egelman, and D. Anderson.** 1991. The mass, composition and biological

activity of ϕ29 proheads. Abstr. XIIth Int. Conf. Bacteriophage Assembly 1991.

48. **Rajagopal, B., B. Reilly, and D. Anderson.** 1991. *Bacillus subtilis* participation in phage ϕ29 assembly. Abstr. XIIth Int. Conf. Bacteriophage Assembly 1991.
49. **Rajagopal, B., B. Reilly, and D. Anderson.** Unpublished data.
50. **Reid, R., J. Wichitwechkarn, and D. Anderson.** 1991. Structure and function of phage ϕ29 prohead RNA in DNA-gp3 packaging. Abstr. XIIth Int. Conf. Bacteriophage Assembly 1991.
51. **Reilly, B. E., R. A. Nelson, and D. L. Anderson.** 1977. Morphogenesis of bacteriophage ϕ29 of *Bacillus subtilis*: mapping and functional analysis of the head fiber gene. *J. Virol.* **24:**363–377.
52. **Reilly, B. E., and J. Spizizen.** 1965. Bacteriophage deoxyribonucleate infection of competent *Bacillus subtilis*. *J. Bacteriol.* **89:**782–790.
53. **Reilly, B. E., M. E. Tosi, and D. L. Anderson.** 1975. Genetic analysis of bacteriophage ϕ29 of *Bacillus subtilis*: mapping of the cistrons coding for structural proteins. *J. Virol.* **16:**1010–1016.
54. **Sogo, J. M., M. R. Inciarte, J. Corral, E. Vinuela, and M. Salas.** 1979. RNA polymerase binding sites and transcription map of the DNA of *Bacillus subtilis* phage ϕ29. *J. Mol. Biol.* **127:**411–436.
55. **Thacker, K.** 1984. Ph.D. thesis. University of Minnesota, Minneapolis.
56. **Tosi, M., and D. L. Anderson.** 1973. Antigenic properties of bacteriophage ϕ29 structural proteins. *J. Virol.* **12:** 1548–1559.
57. **Tosi, M. E., B. E. Reilly, and D. L. Anderson.** 1975. Morphogenesis of bacteriophage ϕ29 of *Bacillus subtilis*: cleavage and assembly of the neck appendage protein. *J. Virol.* **16:**1282–1295.
58. **Turnquist, S., M. Simon, E. Egelman, and D. Anderson.** 1992. Supercoiled DNA wraps around the bacteriophage ϕ29 head-tail connector. *Proc. Natl. Acad. Sci. USA* **89:** 10479–10483.
59. **Turnquist, S., D. Thomas, and D. Anderson.** Unpublished data.
60. **Vinuela, E., A. Camacho, F. Jimenez, J. L. Carrascosa, G. Ramirez, and M. Salas.** 1976. Structure and assembly of phage ϕ29. *Phil. Trans. R. Soc. Lond. Sect. B* **276:**29–35.
61. **Wetzstein, M., U. Volker, J. Dedio, U. Zuber, C. Herget, A. Schmidt, M. Hecker, and W. Schumann.** 1991. Analysis of the heat shock response of *Bacillus subtilis*. Abstr. 6th Int. Conf. Bacilli 1991.
62. **Wichitwechkarn, J., S. Bailey, J. W. Bodley, and D. Anderson.** 1989. Prohead RNA of bacteriophage ϕ29: size, stoichiometry and biological activity. *Nucleic Acids Res.* **17:**3459–3468.
63. **Wichitwechkarn, J., D. Johnson, and D. Anderson.** 1992. Mutant prohead RNAs in the *in vitro* packaging of bacteriophage ϕ29 DNA-gp3. *J. Mol. Biol.* **223:**991–998.

X. PRODUCTION OF COMMERCIAL PRODUCTS

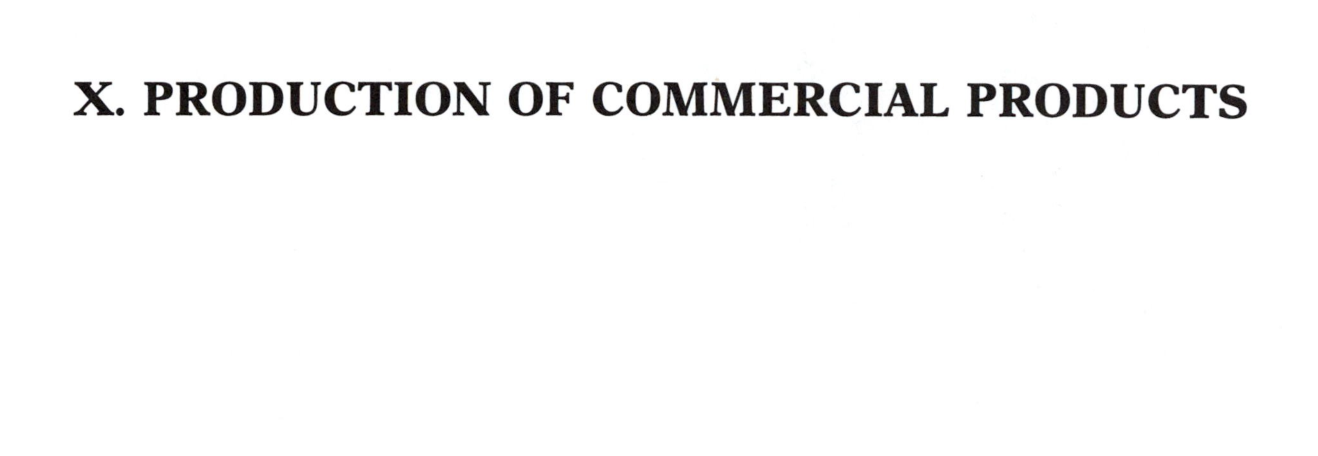

60. Fermentation of *Bacillus*

M. V. ARBIGE, B. A. BULTHUIS, J. SCHULTZ, and D. CRABB

INTRODUCTION

The genus *Bacillus* includes a variety of industrially important species that are commonly used as hosts in the fermentation industry. *Bacillus* strains from different sources can be cultivated under extreme temperature and pH conditions to give rise to products that are in turn stable in a broad variety of harsh environments. The large divergence in physiological types found in *Bacillus* spp. (which in part has been attributed to the genetic diversity of the genus [165]) along with the facts that most members of the genus are nonpathogenic, relatively easy to manipulate by genetics, good secretors of proteins and metabolites, and simple to cultivate make *Bacillus* one of the preferred hosts for production by fermentation. However, *Bacillus* spp. sporulate, and an in-depth understanding of this developmentally regulated process and its impact on product formation is necessary for the development of an optimal fermentation process.

The products in commerce today that are produced by *Bacillus* fermentations include enzymes (Table 1) (which account for over half of the total enzyme sales [$700 to 800 million] annually) (4), antibiotics (82), and insecticides (14). Other products are nucleotides and nucleosides for food flavor enhancement (33) and amino acids (138).

The fastest-growing area of production by *Bacillus* spp. is in the expression of heterologous gene products. A small sampling of compounds that have been expressed through cloning and secretion is given in Table 2. This list is growing rapidly because of the large base of knowledge of both *Bacillus* genetics and large-scale production with this organism.

Fermentation Issues

In developing a fermentation process for any microorganism, a variety of applied issues must be taken into consideration. The first to consider is the development of a suitable production medium that meets the organism's basic needs for sources of carbon, nitrogen, phosphorus, sulfur, potassium, and trace elements in order to maximize growth and product yield. Statistical design methods (104) are important tools in this optimization process. A recent review (138) has extensively explored this area for *Bacillus* spp.

Other critical applied issues include fermentation monitoring and control. The classic parameters that need to be both monitored and controlled for optimal production include temperature, aeration rate, pressure, feed rate, revolutions per minute, foaming, volume, pH, O_2 and CO_2 contents of broth, dissolved oxygen, power input, and draw. Other measured analytical needs include substrate concentration (carbon, nitrogen, phosphate), biomass levels, and product levels. The advent of many new sensors for use during fermentation allows measurement of cell mass, mineral ion levels, nucleic acid pools, sugar and nitrogen concentrations, NAD-NADH, etc. Automated sampling devices now allow for on-line analysis of product concentrations and other compounds that can be analyzed by gas chromatography or high-pressure liquid chromatography. This information can be analyzed by and interfaced with computers, which in turn can change the physicochemical parameters of the reactor in response to the changing environment. We suggest consulting a series of references for the details needed to explore these issues further (142).

This chapter focuses on the basic principles necessary for an understanding of growth and product production in *Bacillus* spp. We show the important link between some very basic observations and how these can be (and need to be) applied in medium development processes, choice of equipment, and elimination of bottlenecks in production.

It will be necessary to burden the reader with some models and equation development, as the reference material is cumbersome to piece together. These equations, in fact, become the tools necessary for tracking the physiological responses of *Bacillus* spp. (or any microorganism) to nutrient and environmental stimulation.

Finally, using protease production as a model, examples show the potential value of using controlled fermentations in identifying some of the limitations of product formation in *Bacillus* spp.

Bacillus Cells as a Desirable Host

Safety and regulatory issues

Bacillus spp. have a history of safe use in both food and industry (137, 138). A number of products from *Bacillus* spp. are generally recognized as safe for specific uses by the U.S. Food and Drug Administration. These products include enzymes for food processing (amylases and proteases) as well as whole foods produced from microorganisms (natto). Proteases from *Bacillus* spp. (like other proteases and, indeed, proteins in general) may elicit allergenic responses in susceptible individuals. As proteases were first being developed for large-scale commercial use in the 1960s, worker exposure led to sensitivity and allergic reactions among the work force. Present-day safety prac-

M. V. Arbige, B. A. Bulthuis, J. Schultz, and D. Crabb • Genencor International Inc., 180 Kimball Way, South San Francisco, California 94080.

Table 1. Examples of *Bacillus* enzymes in commerce

Enzyme	*Bacillus* species
Metalloproteases	*B. amyloliquefaciens*
Serine proteases	*B. licheniformis, B. lentus, B. alcalophilus*
α-Amylases	*B. amyloliquefaciens, B. licheniformis, B. stearothermophilus*
Glucose isomerase	*B. coagulans*
Pullulanase	*B. acidopullulyticus*

tices typically result in lower exposure levels in production plants. However, in no cases were these problems associated with pathogenicity or invasiveness of the host microorganism. Microorganisms used for production purposes must comply with regulations for general large-scale industrial practices. For recombinant organisms, these guidelines specify that the host organism must be well characterized and safe for handling at large-scale production facilities.

Secretion capabilities

Bacillus spp. as a class are known to secrete a large number of extracellular enzymes, including several proteases, amylase, cellulase, and other degradative enzymes. By extensive mutagenesis and selection programs, a number of natural isolates that produce commercial quantities of enzymes of interest have been developed. It is clear that *Bacillus* spp. in general are capable of expressing and secreting individual gene products at very high levels. Because of the extensive genetic knowledge of *Bacillus subtilis*, this organism can be an appropriate model host for developing expression-secretion systems for novel proteins (67). Though details of the secretion pathways for *B. subtilis* are not as well dissected as those in *Escherichia coli*, work described in other chapters in this book is beginning to delineate both the similarities and the differences between the two organisms. As our understanding of the genetics of expression (including induction by two-component regulatory systems and control of stationary-phase promoters) increases, our ability to exploit the genetics of *B. subtilis* makes the organism especially amenable to commercial use.

Table 2. Examples of recombinant proteins synthesized and secreted in *Bacillus* spp.

Compound	Reference
Pertussis toxin subunits	66
β-Lactamase	88
Human proinsulin	118
α-Galactosidase	122
Diphtheria toxin	61
Pneumolysis	174
Human epidermal growth factor	203
Human atrial α-factor	194
Protein A (*Staphylococcus* spp.)	191
Staphylococcal nuclease	102
Human serum albumin	152
Semliki Forest virus	97
Macrophage plasminogen activator	161
Catechol 2,3-dioxygenase	206
Human growth hormone	68
α_2-Interferon	154

MICROBIAL GROWTH KINETICS AND *BACILLUS* PRODUCT FORMATION

Fermentors in Microbial Product Formation

In some sections below, it will be assumed that the reader is familiar with the fundamentals of bacterial growth kinetics and fermentor (bioreactor) types for submerged cultures. For those who are uncertain on the subject, this section provides an introduction. We will briefly describe the basic types of fermentors as background for subsequent discussions of growth kinetics.

Ideal types of fermentors

Fermentors (bioreactors) for submerged microbial fermentations come in a large variety. Which type of fermentor to use is dictated first by the process to be carried out, that is, by the particular type of organism and fermentation: unicellular or mycelial-filamentous microbes, aerobic or anaerobic, (semi)continuous or batchwise, aimed at producing biomass or microbial products, etc. Once the process has been decided on, the methods of gas distribution (mechanical stirring, airlift, pumps, or pressurized injection), substrate dispersion, and gas (oxygen) and heat transfer; presence of hard or fibrous materials in the broth; viscosity of the broth; presence of flammable and explosive substrates or products (methane, alcohols); feasibility of long-term asepsis; in situ or prerun liquid sterilization; air sterilization; foaming control; (dis)continuous product removal and harvest; and power and water consumption must be considered. These are mainly aspects of careful engineering for which introductory reviews can be found in textbooks (27, 84, 101, 107, 155, 156).

Despite the large variety of fermentor designs, most types can be described as one of a small number of archetypes (at least with regard to growth and production kinetics of microbial cultures). In its simplest form, a fermentor can be a closed vessel of constant volume as depicted in Fig. 1A. For aerobic growth, either air already in the vessel or added air provides the necessary oxygen. This is a *batch fermentor*, identical (in its kinetic properties) to the well-known laboratory shake-flask. Upon inoculation of the nutrients (all present in excess of demand), growth occurs until a substrate component is exhausted or an inhibitor accumulates. A plot of biomass versus time results in the well-known batch curve. The several stages in a batch culture have been described elaborately (see, for example, references 162 and 196).

Adding a feedline to the batch system (Fig. 1B) results in a fermentor type referred to as a *fed-batch fermentor* (FB). During fermentation, substrate is added, and the amounts of biomass and product and the volume increase gradually (if the substrate is gaseous, the volume may remain constant). This fermentor type is very commonly used in industrial fermentations (for a review, see reference 204).

Figure 1C shows a *continuous fermentor* (CF), characterized by volumetric constancy. In time, a steady-state situation develops in which all physiological parameters (biomass concentration, growth rate, etc.) are dynamic constants.

Figure 1D shows a combination of fermentor types

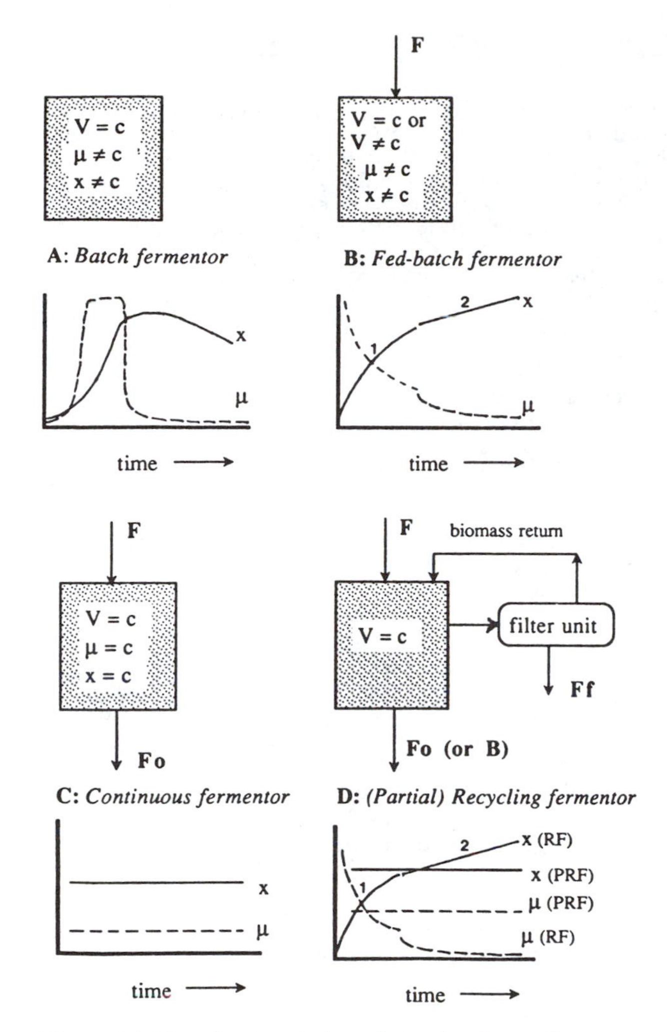

Figure 1. Main characteristics of predominant fermentor types. Many variations on these archetypes are possible (e.g., the feed rate in an FB need not be constant). *V*, volume of culture; c, constant, μ, specific growth rate; x, biomass. (A) Batch fermentor (BcSc). Nutrient is present from the beginning; no feed. No steady state can be reached. (B) FB (BcSc). Nutrient is fed at a constant or variable rate; there is (usually) no flow out of the fermentor. As described in the text, growth in a carbon-limited FB can initially proceed exponentially (domain 1), after which (at least) one domain (domain 2) of linear growth occurs. No steady state can be reached. (C) CF, e.g., chemostat (BoSo). Nutrient is fed at a constant rate. Rate of inflow (F) equals rate of outflow (Fo); dilution rate $(D) = F/V = c$. In steady state, $\mu = D$. (D) Combination of fermentor types. If inflow (*F*) equals outflow (Fo) while the filtrate flow (Ff) equals 0, this type is, again, a chemostat with $V = c$, $\mu = c$, and $x = c$. If Ff > 0 and Fo = 0, then D1 is a 100% RF (BcSo). $V = c$, $\mu \neq c$, $x \neq c$. Nutrient is fed at a constant rate. Rate of inflow (F) equals rate of filtrate-flow (Ff). Culture broth is continuously recycled over a cell separation unit (e.g., filter), biomass is returned to the fermentor, and filtrate is pumped off. No steady state can be reached. If Ff > 0 and Fo > 0, then D2 is a PRF (BoSo). $V = c$, $\mu = c$, $x = c$. As in the 100% RF, culture broth is continuously recycled over a cell separation unit. Not as in the RF, here Ff < F; part of the culture broth is pumped off at rate Fo; so, F = (Ff + Fo). In fact, a PRF is a CF type like the chemostat; here, too, $D = F/V = c$, but $\mu = Fo/V \neq D$ (the Fo of a PRF is usually referred to as "bleed" [*B*], so $\mu = B/V$). Nutrient is fed at a constant rate, and a steady state can be reached.

and how they are interrelated. In a *partial recycling fermentor* (PRF), feed enters at rate *F*, while part of the broth leaves the fermentor at rate *B* (<*F*). Broth is continuously circulated via a filter system (or any cell separation unit). Biomass is returned to the fermentor ("recycled"), and filtrate is pumped off at such a rate that the two outflows (*B* and F_f) equal the inflow (*F*) (so the volume is constant). Both the CF shown in Fig. 1C and the PRF are continuous open systems with steady states.

If, in Fig. 1D, flow rate *B* is 0 (while $F = F_0$), the system is closed with regard to biomass, and no steady state will be reached; as in an FB, biomass will gradually increase and growth rates will drop. This is a *100% recycling fermentor* (RF).

Classification of fermentors

Both the batch fermentor and the FB constitute so-called closed systems. The operational classification of open and closed systems was introduced by Herbert (64). Closed systems are those in which biomass is fully retained (and usually increases); they can be run until a nutrient component is exhausted or an inhibitor accumulates. Consequently, closed systems are characterized by ever-changing conditions in which no time-independent steady-state condition can be reached. This is in contrast to the open systems like many CF and the PRF in which such a steady state can be reached. These latter systems are characterized by volumetric constancy and, importantly, by the fact that a constant fraction of biomass leaves the fermentor at the same fractional rate as it increases (in steady state) in the fermentor.

The terms "continuous" and "steady state" are sometimes mistakenly presumed to be synonymous. For instance, a much-used way to run an FB in fermentation industries is to perform the fermentation in the FB (a closed system), harvest the bulk of the volume, refill the fermentor, and run the FB again (the so-called dump-and-fill process). From an operational standpoint, this fermentation will be more or less continuous, but steady-state conditions are never reached. Neither will an RF culture, even if it is run continuously for several weeks, ever reach a steady state.

The classification closed or open (c or o) can be extended in order to refer to biomass (B) as well as to substrate (S) (185). Thus, four types are distinguished: biomass open, substrate closed (BoSc), exemplified by some dialysis fermentors (resembling the RF); biomass open substrate open (BoSo), exemplified by many CF and PRF; biomass closed, substrate open (BcSo), exemplified by the 100% RF and FB; and finally, biomass closed, substrate closed (BcSc), exemplified by the batch fermentor. Essentially, any type of fermentor can be classified according to this scheme. Only the biomass-open systems can reach steady states and can therefore (in theory, at least) be run continuously for an unlimited time. As will become apparent later on, BoSo systems are in general preferable for biomass production. For fermentation end products, BcSo systems might be preferred, because although growth rates will ultimately be lower than in BoSo systems, the higher biomass densities in BcSo systems allow for higher volumetric output rates.

Table 3. Some examples of adapted Monod equations

Equation	Comment
$\mu = \mu_{max}\, s/[(K_s + s)(1 + s/K_i)]$	Haldane equation (e.g., used for *Nitrobacter* spp. by Boon and Laudelot [16]) for inhibition by high concentrations of s
$\mu = \mu_{max}\, s^n/(K_s + s^n)$	Rearrangement by Panikov and Pirt (125) for substrate-inhibited growth of *Chlorella* spp. in which n is the so-called Hill number (dimensionless), expressing extent of cooperativity between different enzyme systems
$\mu = \mu_{max}\, s/[(K_s + s)(1 - P/P_m)^n]$	Rearrangement by Levenspiel (91) for product (P) inhibition of growth by ethanol; P_m is maximum product concentration; n is fitting constant
$\mu = \mu_{max}\,(1 - x/x_f)$	So-called logistic equation (not a Monod equation) in which x_f is final biomass concentration in batch culture (e.g., see reference 21)

Within BcSo systems, an RF can be preferable to an FB, as the first allows for continuous soluble-product removal via the filtrate outlet.

A number of similar fermentors can be combined in one system (a homogeneous cascade, e.g., coupled CF), or different types of fermentors can be combined into one heterogeneous system. Reasons for doing this can be a temporal division between biomass production and product formation, product formation with concomitant toxin accumulation, or production of mRNA in the exponential phase of a batch culture and translation of the RNA into product in the stationary phase only. Regarding the last possibility, α-amylase production by *B. subtilis* has been carried out in two coupled continuous cultures (44). The first fermentor functioned as a means for biomass growth at a high rate while the second fermentor gave rise to relatively high rates of product formation by the slowly growing biomass. Another example is the establishment of highly sporulating cultures of parasporal crystalline protein-producing *Bacillus thuringiensis* (used as bioinsecticide; 43, 89, 139; also see other chapters in this volume) with a high rate of biomass production in the first continuous culture and a high level (80%) of sporulating cells in the second continuous culture (81). Note that in these multiple-stage continuous fermentations (SoBo), the steady-state character (time independency) of the final fermentors in the system is lost, since they are fed by the first fermentors. The nonhomogeneity of those final fermentor populations with regard to the physiological state of the cells will increase.

Kinetics of Microbial Growth and Production

Basic and Monod-type equations for growth

In a population of cells, a certain fraction consists of cells in the process of division, giving rise to daughter cells and thus to an increase in cell mass of the total population. This increase in cell mass is given by the following equation:

$$dx/dt = \mu x \quad (1)$$

in which dx/dt is the growth rate (grams per hour), x is the initial cell mass (in grams), and the proportionality factor μ is the specific growth rate (grams per gram·hour). Under unlimited conditions (when all nutrients are present in excess of demand), μ equals μ_{max} (the maximum specific growth rate) and will be a constant, determined only by the intrinsic properties of the cells under those particular conditions. Integration of equation 1 yields

$$x_t = x_0 e^{\mu t} \quad (2)$$

(t is elapsed time), from which can be derived the mean mass doubling time (T_D, in hours) during exponential growth:

$$T_D = \ln 2/\mu \quad (3)$$

If all components of the biomass change by the same constant factor per time unit, growth is said to be balanced.

As the substrate is depleted, growth becomes restricted and μ becomes dependent on s (the growth-limiting substrate concentration). This dependency of μ on s is and has been a lively subject. Major pioneering work was carried out by Monod (105), who, by analogy with Michaelis-Menten enzyme kinetics, formulated the μ-s relation as follows:

$$\mu = \mu_{max}\, s/(K_s + s) \quad (4)$$

in which s is the growth-limiting substrate concentration (e.g., moles per liter) and K_s is the phenomenological saturation (or affinity) constant (usually, K_s has a magnitude of milligrams per liter of perhaps up to 100 μM for carbohydrates and a lower magnitude of micrograms per liter of up to 100 nM for, e.g., amino acids).

Although other equations have been derived from enzyme kinetics in order to describe bacterial growth, Monod's are among the most widely used and adapted equations. Adaptations have been made for both very low (s close to K_s) and high concentrations of s and for situations with inhibiting-product formation (a common problem in industrial fermentations). Table 3 shows some examples.

Since the basis of all these equations (except the logistic) comes from enzyme kinetics, it follows that when they are applied to bacterial growth kinetics, they constitute simplified, unstructured models of the growth process. Monod's equation actually simplifies growth to a process in which the total metabolic rate is set by the substrate uptake rate (28, 31). Problems with the Monod-type equations will occur, for instance, when there is switching to different affinity systems for substrate uptake or disturbances of balanced growth, leading to transient responses (205). Another drawback, recognized early by Málek (99) and Herbert (64), is the mistaken assumption that the chemical composition of cells does not change with μ (18, 38, 108). And finally, there are problems with the actual value of s (123) and the presumed constancy of K_s (it was pointed out by Bazin et al. [12] that for open systems [like many CFs] K_s is not constant but is μ dependent, albeit to a negligible extent) (15). An attempt to establish the relation between μ and s under conditions of steady-state growth (explained below) of *Klebsiella pneumoniae* showed that s declined over a period of 50 T_Ds and that the relation between μ and s could be described by a hyperbolic (Monod-type) curve equally as well as by a logarithmic curve (117). This confirms early reports that the history or life span of a culture can be of major importance (8, 109, 135, 136).

All the above-mentioned objections are realistic and form a major problem for those who try to model microbial growth. One should not be led to believe, though, that the Monod-type equations are unimportant. When combined with material balances and equations for substrate consumption, they constitute a so-called unstructured model (i.e., with minimal input from detailed cellular biochemical pathways), and these unstructured models are relatively easy to handle. Thatipamala et al. (177) successfully used Levenspiel's rearrangement (Table 3) to describe changes in μ with increasing ethanol concentrations. Observations on deviating behavior in industrial fermentations can often be met by additions and substitutions to the unstructured model's equations (cf. reference 54) that result in satisfactory curve fits of the data. This at least allows for the all-important possibilities of process control and steering, even without extensive knowledge of the microbial processes involved. As Lee and Ramirez wrote (90), "Our past experience in process optimization . . . demonstrates the importance of developing an applicable model set which, even though it may lack some of the structural meaning, can lead to the solution of an optimal strategy without loss of the generality."

Growth yields, substrate utilization rates, and maintenance energy demand

Two examples of CF are the *turbidostat* (in which biomass is kept constant through control of the optical density) and the *chemostat* (in which flow rate is the primary control) (117). Essentially, these two CFs are the same (64). The chemostat has been used most often, and a wealth of information on growth kinetics and product formation has been obtained from studies at steady-state conditions for μ values ranging from μ_{max} to about 0.10 h^{-1}. Chemostat kinetics were explored by Monod (106), Novick and Szilard (117), and Herbert et al. (65). Elaborate derivations of the kinetic characteristics can be found in the publications of those authors and more recently in those of Pirt (132), Mandelstam et al. (100), and Slater (163).

A chemostat is characterized by a constant volume (V): the inflow of fresh medium (with relevant substrate concentration [S_R] and flow rate [F]) equals the outflow of culture fluid. Growth proceeds exponentially, and equation 1 applies: $r_x = \mu x$, giving rise to biomass increase (from now on, rates will be written as r, so $dx/dt = r_x$ and $ds/dt = r_s$). At the same time, the outflow of culture fluid leads to biomass decrease: $r_x = Dx$ (D is the dilution rate, defined as F/V). The overall rate of biomass change in the culture vessel is therefore given as follows:

$$\text{overall } r_x = (\mu - D)x \tag{8}$$

In a chemostat (or CF), all environmental conditions are kept constant. Usually, growth is limited by some substrate component, leading to anabolic limitation (carbon, nitrogen) or catabolic limitation (energy), while all other necessary nutrients are considered to be present in excess of demand (more than one substrate can be limiting at the same time [10, 37, 113]). Given some time and provided that $D < \mu_{max}$, a steady-state condition will develop in which no net changes in biomass occur (overall $r_x = 0$, while $\mu = D$). The relation with the Monod equation can be seen by applying equation 4 to equation 8:

$$\text{overall } r_x = \{[(\mu_{max}s)/(K_s + s)] - D\}x \tag{9}$$

The rate of substrate utilization [r_s] can be determined as follows. Any increase in biomass (r_x) because of growth is the result of and proportional to a decrease in substrate (r_s); the proportionality is determined by the particular cells' metabolic abilities and by the substrate and environment. This can be expressed as $r_x = Yr_s$, in which the proportionality factor Y is the (observed) growth yield (grams of biomass per moles of substrate). Introduction of another parameter, the specific utilization (or production) rate ($q = \mu/Y$, in moles of substrate per grams of biomass per hour) leads to the following equation:

$$r_s = \mu x/Y \tag{10}$$

In a chemostat, the overall r_s is the sum of substrate addition ($=DS_R$), substrate outflow ($=Ds$), and substrate consumption ($=\mu x/Y$), so that

$$\text{overall } r_s = DS_R - Ds - \mu x/Y \tag{11}$$

(Under substrate limitation, the substrate concentration [s] in the fermentor will be virtually zero.)

As to the growth yield (Y), it was in fact introduced by Monod (105) as the yield constant, representing the growth rate r_x as a constant fraction of the substrate utilization rate r_s while assuming an independence of Y from either μ or s. This concept of constancy turned out to be wrong. Monod (106) did allow for a variable Y, while Herbert (63) introduced the concept of endogenous metabolism to explain the observation of dropping Y values at low growth rates in carbon-limited continuous cultures (deviations from Monod kinetics can also be caused by substrate toxicity, production of storage materials, change in limiting substrate, etc.; 20). Endogenous metabolism can be defined as the consumption of their own biomass by cells deprived of a carbon or energy source. Nowadays, it is common to refer to maintenance metabolism (131), which is defined as metabolic processes that use up carbon and energy without a concomitant net increase in biomass. This concept was formulated in 1965 by Pirt with the following linear equation for substrate utilization:

$$q = \mu/Y^{max} + m \text{ (or } r_s = \mu x/Y^{max} + mx) \quad (12)$$

in which Y^{max} is the maximal growth yield (Y corrected for maintenance processes) and m is the maintenance coefficient, a μ-independent fraction of q that is used for maintenance processes like turnover of macromolecules, maintenance of transmembrane ion gradients, etc. (in a linear plot of q versus μ, Y^{max} is the inverse of the slope, while m is the intercept with the ordinate). Clearly, the observed growth yield (Y) will be strongly influenced by the magnitude of m and the value of μ. Notably at lower growth rates, the relative importance of m increases. Since a large number of industrial processes are carried out in FB, which at later stages are characterized by very low growth rates, maintenance processes are of both scientific and industrial interest (24, 170).

Ever since Pirt's formulation of 1965, the subject of maintenance has been much debated. For a number of cases, equation 12 has had to be adapted; the concept of μ-dependent (both linear and nonlinear) maintenance was introduced, the possible impact of various percentages of viability of cultures was taken into account, and adaptations of Pirt's initial proposal were made for both carbon-limited and carbon-excess situations (41, 112, 113, 133, 134, 178). It has also been shown that an overestimate of m in bacterial cultures can be caused by such factors as fermentor design (or rather feed addition regime [19, 24, 176]) and, notably at low μ values, the increasing costs of bacterial biomass synthesis with decreasing μ (22).

Very low growth rates

Batch culture studies were most important in the description and elucidation of the complex temporal regulation of *Bacillus* sporulation and the distinction between that process and vegetative growth. Genetic and physiological characteristics of sporulating cells are still described with reference to their timing in batch cultures (T1 [1 h after onset of stationary phase], T2, etc.). Batch fermentors are of limited use, though, not only because of their inherently unstable character but also because the only μ values that are sustained for a long time are μ_{max} (exponential phase) and $\mu = 0$ (stationary phase). Studies on growth rate-related events must therefore make use of other fermentor types.

In a chemostat, setting of the pump speed (= nutrient addition rate; Fig. 1C) will determine the value of D (and thus μ) anywhere between 0 and μ_{max}. In practice, however, D values below about 0.10 h^{-1} in chemostats are characterized by an increasingly pulsewise substrate addition and a loss of the continuous and steady-state character; there are reports on decreasing viability of chemostat cultures at low D values (52, 175). This technical problem can be solved by turning to another type of CF, the PRF (Fig. 1D), which has the same characteristics (kinetics) as a chemostat (except that for a given r_s, the steady-state biomass concentration in a PRF will be higher than that in a chemostat). Most important is the fact (outlined above) that μ can be low while D is high ($\mu = B \neq D$; Fig. 1D), thus preventing pulsewise substrate addition at low μ values (<0.10 h^{-1}).

Very low growth rates (<<0.05 h^{-1}, as occurs in many industrial fermentations) can readily be reached by cultivating cells in an FB or an RF. At a constant addition rate (r_s) of limiting substrate, biomass will initially start to increase because of exponential growth. From Pirt's equation 12, it follows that

$$r_s = r_x/Y^{max} + mx \quad (13)$$

and

$$r_x = (r_s - mx)Y^{max} \quad (14)$$

so that biomass at any time would be given by

$$x = r_s/m + (x_0 - r_s/m)e^{-mtY\max} \quad (15)$$

(t = elapsed time). The specific growth rate will fall continuously as less and less substrate is available per unit of biomass; since $\mu = r_x/x$ (equation 1) while $x = x_0 + r_x t$ (a biomass-closed system), it follows that

$$\mu = (Yr_s)/(tYr_s + x_0) \quad (16)$$

Growth in FB and RF, however, does not always proceed according to these equations, as explained in the next paragraph.

Stringent response and growth domains

For growth in FB and RF, continuous equations 14 and 15 can properly describe the initial growth phase (r_x, μ, Y) but cannot describe growth after μ has dropped to very low values (about 0.05 h^{-1} and much smaller). At those very low μ values, biomass was found to increase linearly with time, resulting in constant values for r_x and Y. Exactly how many domains (one or two) of linear growth (prestringent and stringent responses) follow the initial domain of Pirt-type growth (if occurring; 146) is not yet clear; this may be related to the particular species and to the status of its *rel* genes. This behavior of cultures at very low growth rates, initially described by Chesbro et al. (29), has by now been described for a number of organisms (6, 187), including *Bacillus polymyxa* (2). So, in FB and RF, one can distinguish growth domains in time that are characterized by the Pirt equation at higher μ values and by linear growth at lower μ values. Note that the maintenance coefficient, which results from a curve-fitting procedure (188), may be low in the domain of linear growth but that the maintenance energy demand as a fraction of total substrate metabolism can be as high as 50 to 75% (28) and even higher (169). In a production process in which the product of interest is being produced in a domain of linear growth, it can be a waste to have to spend such a large fraction of substrate on maintenance processes. Some possibilities for remedying this problem will be discussed below. (The present discussion centers on submerged cultures with readily soluble and available growth substrate. Solid-surface fermentations and submerged cultures fed with solid or slowly dissolving or desorbing substrates often show linear growth, but in that case, the linear growth is due to the rate-limiting substrate transfer from the solid to the dissolved, readily available substrate and not directly to a physiological response of the cells [192; for a discussion of linear cell growth, see references 85 and 124].)

Linear growth is thought to be caused by the stringent response, since cellular concentrations of the regulatory nucleotide guanosine-5′-diphosphate-3′-diphosphate (ppGpp) were found to coincide with the different growth domains (2, 3, 29, 185, 187; for a review of the stringent response, see reference 26). The occurrence of the stringent response in FB and RF can have a major impact on the fermentation process and products. The situation for *Bacillus* species is especially complicated, as the usual response of these bacteria to deteriorating environmental conditions, a common trigger for the stringent response, is sporulation (119, 120). The abundant use of (partially) sporulation-deficient strains in fermentation industries means that during the production stages, stringent (or relaxed) responses may coincide with the characteristic developments of the early sporulation stages.

Strains incapable of accumulating ppGpp (relaxed strains) will not show a growth domain of stringent response. For a very low-sporulating, relaxed *Bacillus licheniformis* growing in an RF, a first domain of Pirt-type growth (with concomitant exoprotease production) was followed by a second domain characterized by linear growth (188). Upon entering this second domain, the culture showed an upshift in μ (instead of a drop in μ, as in stringent strains), after which μ dropped continuously. Importantly, van Verseveld et al. (188) explained this behavior by the observed shutoff in exoprotease production that allowed the biosynthesis rate (μ) to rise, thus evading growth rates that would have led to a full relaxed response. This illustrates the importance of the genotype of a production strain.

Material balance equations and reduction degree

Important tools in description and analysis of fermentations are the macroscopic material (mass and energy) balances. These balances relate the utilization of carbon (substrate), oxygen, and nitrogen to the production of biomass, product, carbon dioxide, and water. The elementary cell composition of bacteria can be formulated as $CH_pO_nN_g$, and aerobic metabolism can be described as follows:

$$\begin{aligned} CH_mO_k \text{ (substrate)} + aNH_3 + bO_2 \rightarrow\ & Y_cCH_pO_nN_g \text{ (biomass)} + zCH_rO_sN_t \text{ (product)} + \\ & cH_20 + dCO_2 + \text{heat} \end{aligned} \tag{17}$$

(39, 103, 144, 145, 167, 170); respectively, $C_6H_{10.8}O_{3.0}N_{1.2}$ and $C_6H_{9.48}O_{2.04}N_{1.74}$ can be used to describe biomass and [exo]protein [5]). The fractions of substrate turned into biomass, product, and substrate are, respectively, Y_c, z, and d (a, b, and c are moles), so that the carbon balance is given by $Y_c + z + d = 1.0$, and the nitrogen balance is given by $a = gY_c + zt$. With the introduction of the generalized reduction degree, i.e., the amount of available electrons that may be transferred to oxygen, electron balances can be formulated (103, 128, 144, 145). For $\gamma_c = +4$, $\gamma_H = +1$, $\gamma_O = -2$, and $\gamma_N = -3$, the following reduction degrees (γ_s, γ_b, and γ_p for, respectively, substrate, biomass, and product) can be inferred from the notation in equation 17:

$\gamma_s = 4 + m - 2k$ (for glucose, $\gamma_s = 4.0$)

$\gamma_b = 4 + p - 2n - 3g$ (for general biomass [see above], $\gamma_b = 4.2$)

$\gamma_p = 4 + r - 2s - 3t$ (for general exoprotein [see above], $\gamma_p = 4.03$)

From the mass material balance (compare equation 17) and the expression for γ_s, γ_b, and γ_p, the following balances can be constructed. The electron balance is

$$\gamma_s - 4b = \gamma_b Y_c + z\gamma_p \tag{18}$$

From this, the O_2 consumption b (=moles of O_2 per mole of substrate) can be inferred.

$$b = 0.25(\gamma_s - \gamma_b Y_c - z\gamma_p) \tag{19}$$

Another balance is constructed by comparing the average oxidation-reduction quotient (O/R value) for metabolism (equation 17); by definition, the O/R value for a compound with the formula $(CH_2O)_x$ is 0, while each surplus (or deficit) of 2 hydrogen atoms means a value of −1 (or +1) for O/R (32). The (absolute) values for carbon balance, redox balance, and O/R balance should be 1.0; for experimental conditions, the results are considered reliable when the values lie between some arbitrary limits like 0.95 and 1.05.

The material balances are an indispensable tool in any fermentation process. Discrepancies between measured values and values inferred from the balances can indicate unanticipated products or product formation (195). For anaerobic growth of *B. licheniformis*, for instance, for which predominant catabolism through glycolysis was presumed, reliable O/R and electron balances could be obtained only by assuming the simultaneous functioning of glycolysis and the pentose phosphate route (25). Likewise, by applying the so-called metabolic pathway method (essentially a compilation of metabolic stoichiometries during anaerobic fermentation, using mass and energy balances), Papoutsakis and Meyer (126) showed how for a set of *B. subtilis* butanediol fermentations, the balance calculations indicated variable involvement of the pentose phosphate route together with glycolysis.

Growth parameters and efficiencies

Growth yields (Y) can be obtained for essentially all components in the mass material balance (equation 17); for instance, the number of cells produced per mole of oxygen reduced is Y_{O_2}. With regard to utilization and production rates (q), the amount of CO_2 or exoproduct produced per gram of cells per hour is the q_{CO_2} or q_p. The maximal growth yield (Y^{max}), i.e., growth yield corrected for maintenance processes, is obtained by taking the inverse of the slope of a linear plot of q versus μ; in this plot, the ordinate cutoff represents the maintenance coefficient m.

Product-corrected growth yields are used when extracellular products are synthesized (they can also be applied when storage materials like polyhydroxybutyrate [PHB] are synthesized). Since part of the substrate will have been used for the synthesis of those products, that part must be substracted from the total amount of utilized substrate in order to calculate how much biomass has really been synthesized. Those growth yields are denoted by a superscript (Y^{corr}). An example will illustrate their use. Bulthuis et al. (23) described two variants (one producing exoprotease and the other not producing it) of a *B. licheniformis* strain that were characterized for their bioenergetic parameters in chemostat cultures by Frankena et al. (50). The $Yglu^{max}$ values for these variants (respectively, 83.8 and 90.9 g/mol) were significantly different, indicating a difference in growth efficiency. This difference, however, might be expected to be due to extra costs for the producing variant because of exoprotease production. Indeed, when these parameters were calculated with a correction for the amount of glucose that had been used for synthesis of the exoprotease, the values for $Yglu^{max,corr}$ were respectively 100.3 and 97.2 g/mol and were no longer significantly different. Apparently, both variants had the same efficiency of growth, a conclusion that could be important in choosing the right host for production purposes.

Yet another growth yield, the Y_{ATP} (grams of biomass per mole of ATP), indicates the efficiency of growth. This parameter can be calculated if the ATP yield (molecules of ATP produced per mole of glucose used) is known. For aerobic growth, that constitutes a problem, since the ATP yield depends on the variable composition of the electron transfer chain (and the stoichiometry of the ATP synthase), which in turn determines the efficiency of energy conservation reflected by the P/O ratio, i.e., molecules of ATP produced per molecule of O_2 reduced. Since energy status (adenylate energy charge, $[NADH + H^+]/[NAD^+]$ ratio, and Y_{ATP}) of the cells seems to be related to the capacity for and extent of exocellular enzyme synthesis (49), manipulation of the energy status could be a feasible approach to increased production (for reviews, see references 72 and 170). In general, rates of energy supply and turnover have a large impact on growth and production.

Energy spilling and product formation

The impact of a substrate's reduction degree on growth is significant. In general, substrates with a γ_s of 4 or higher (a low carbon/energy ratio) give rise to carbon-limited growth (their catabolism yields more energy than is called for in biosynthesis), while growth on substrates with a γ_s lower than 4 tends to be energy limited. This was shown by Linton and Stephenson (94); a plot of $Y_{substrate}$ versus γ_s initially shows a linear increase of $Y_{substrate}$ with increasing γ_s and a constant value of $Y_{substrate}$ for γ_s values higher than 4 (Fig. 2A). This "critical" value of γ_s has also been taken as 4.2 (= γ_b) and actually depends on the P/O ratio and Y_{ATP}. Since both these parameters (can) depend on μ, there is no fixed value for the critical γ_s; it is even possible that growth proceeds energy limited at low values and C limited at high μ values. This can be a reflection of the cell's efforts to adapt the efficiency of energy conservation (or rather, the ATP supply rate) to the needs of biosynthesis (13, 144, 170, 199).

The potential for energy spilling appears to be innate to most bacteria. Application of the mosaic nonequilibrium thermodynamics theory (MNET) to this subject has led to the conclusion that bacteria seem to have evolved to obtain maximal growth rates at simultaneous optimum efficiency of energy conversion (growth on highly reduced substrates might correspond to optimization of a maximal yield and/or growth rate only [199]). For instance, glucose-limited aerobic growth of *B. licheniformis* in a chemostat above $\mu = 0.3\ h^{-1}$ was characterized by increased production of overflow metabolites (24) (i.e., metabolic intermediates and compounds usually produced as fermentation end products [114]). This was explained by the necessity to obtain higher rates of ATP turnover in order to sustain the higher biosynthesis (growth) rates. Thus, efficiency of energy conservation was sacrificed for a higher growth rate. This is also exemplified by glucose-excess (potassium-limited) aerobic chemostat cultures of *Bacillus stearothermophilus* that showed acetate production of up to 34% of the total carbon consumed. Similarly, in any aerobic batch culture (excess glucose; μ_{max}), a large part of the glucose is not fully oxidized to CO_2 and H_2O but is dumped in the form of overflow metabolites. For an exoprotease-producing *B. licheniformis*, up to 48% of the glucose carbon ended up (in early stationary phase) in overflow products; batch cultures of *B. stearothermophilus* and *B. subtilis* converted, respectively, 18 and 22% of glucose carbon to overflow products (25, 130).

For obvious reasons, the rate of ATP production from catabolism needs to be more or less coupled to the ATP utilization rate in cell synthesis. Mismatches can arise during growth on substrates with a high degree of reduction (γ_s) and under conditions of substrate excess. For instance, Neijssel and Tempest (112, 113) showed the occurrence of an overall higher oxygen uptake rate (q_{O_2}) for glucose-excess than for glucose-limited chemostat cultures of *Klebsiella aerogenes*. The differences were largest at low μ values and very small at high μ values. So, the Y_{O_2} for the glucose-limited cultures was larger than for the glucose-excess cultures. The observed differences were explained by the occurrence of energy spilling (overflow metabolism) during glucose-excess conditions; ATP was apparently synthesized much faster than it could be usefully degraded. Similar results were found for *B. subtilis* (114) and *B. licheniformis* (49).

As a general truth, then, high rates of synthesis (of biomass, product, etc.) can be sustained only if the turnover rates of ATP and NADH are high as well (59). At quantitative imbalances between production and consumption of NADH and ATP, synthesis rates will inevitably drop unless the cells start making use of systems whereby excesses of NADH and ATP can be wasted. With regard to this ability to sustain high product synthesis rates, Linton and Rye (93) and Linton (92) concluded (generalizing) that the lower the $Y_{O_2}{}^{max}$, the higher the rate of metabolite produc-

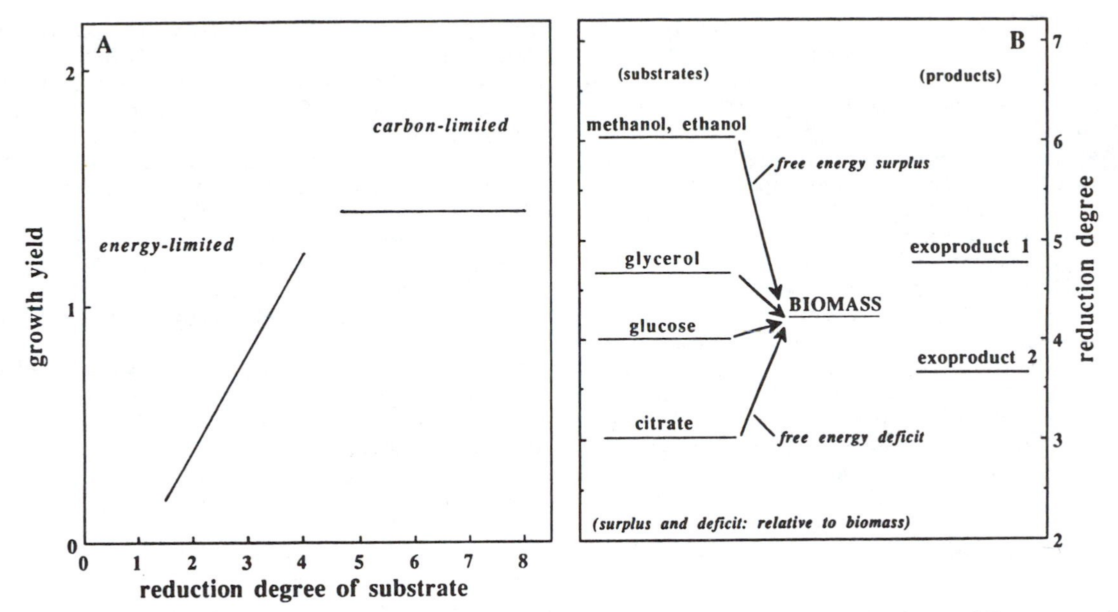

Figure 2. Dependence of growth yield [Y_{xs}] on reduction degree of substrate (γ_s) (A) and relative differences in free energies of substrate, biomasses, and exoproducts (B). (A) Growth on a substrate with $\gamma_s > 4$ tends to be carbon limited and energy sufficient; for $\gamma_s < 4$, the reverse will hold. Initially ($\gamma_s < 4$), the increase in free-energy content of the substrate is reflected in an increase in biomass yield (Y_{xs}). Above a certain γ_s (about 4 to 4.5), ATP yield from the substrate can exceed ATP demands for biomass synthesis, and in order to sustain high metabolic rates, this surplus needs to be turned over by energy-spilling reactions (including energy-requiring exoproduct formation) (see text). (B) The free-energy content of, e.g., citrate is lower than that of biomass; bridging that gap is obtained by combusting a significant portion of the substrate to CO_2 and H_2O to produce energy. Concomitant synthesis of exoproduct 2 is possible; that of exoproduct 1 is very unlikely. At growth on, e.g., ethanol, a surplus of free energy that could be shunted toward synthesis of both exoproducts 1 and 2 will be available. Glucose ($\gamma_s = 4$) holds an intermediate position among these substrates, with a γ very close to γ_b (=4.2). Obviously, simultaneous consumption of two substrates with different γ_s values could also be a means of steering carbon and energy flows toward a more-desirable product formation than is possible with one substrate only. These schemes should be regarded in relation to biochemical routes. Growth on methanol, for instance, has been found to be energy limited despite the high γ_s of methanol. This was due to the occurrence of CO_2 fixation via the costly Calvin cycle (189). For that type of growth, the substrate is actually not methanol but methanol + CO_2, with a lower γ_s than for methanol. (idealized drawing in panel A is based on one in reference 94; reduction degrees are as defined in reference 144).

tion; or, a high growth efficiency will indirectly mean a limited possibility of energy dissipation. They proposed that all obligate aerobes are capable of dissipating energy to an extent that is inversely proportional to $Y_{O_2}{}^{max}$ (under glucose-excess conditions). On the basis of the principle of energy efficiency and NADH-ATP turnover rates, three major groups of (exo)products are distinguished (92, 93).

The first group of exoproducts consists of exopolysaccharides (alginate, succinoglycan, xanthan, etc.). During production of the oxidized constituents of these exopolysaccharides, NADH is generated, while subsequent synthesis of the polymer requires NADH. A high rate of exopolysaccharide synthesis is found predominantly in organisms with relatively low $Y_{O_2}{}^{max}$ values, while the reduction degree (γ_p) of the polysaccharide is inversely related to $Y_{O_2}{}^{max}$. This is easily understood if one considers the amount of reducing equivalents generated during synthesis of the oxidized constituents of the exopolysaccharide. At a high P/O ratio (i.e., molecules of ATP produced per molecules of O_2 reduced), the ATP generated from these reduction equivalents will outnumber the ATP (NADH) needed in the final steps of exopolysaccharide production. This would make exopolysaccharide synthesis useless as a means of energy spilling, as would, also, a too-high value of γ_s. Maximization of the r_p of exopolysaccharides will thus be obtained by a fine-tuned integration of carbon and energy flows (Fig. 2B).

The second group of exoproducts consists of organic acids and some secondary metabolites, whose synthesis is accompanied by a net production of NADH-ATP. By analogy to the exopolysaccharides, maximization of r_p can be obtained by energy dissipation.

The third group of exoproducts consists of biosurfactants and other secondary metabolites. Characterized by moieties of different reduction degrees, often with high ATP demands in synthesis, maximization of the r_p for these compounds can be achieved by growth on more than one carbon-energy source in order to obtain a proper difference between γ_p and γ_s.

While production processes for substances such as amino acids and exopolysaccharides can apparently profit from an induced extent of energy spilling, product formation involving net ATP costs usually will not be enhanced under those conditions. Although the grouping of products just explained is illuminating, it also emphasizes the need for each production process to be examined individually.

This discussion also relates to the efficient use or waste of substrate in non-growth-associated processes. Again, a situation of substrate (energy) spilling can be desirable for product formation of one type and undesirable for another. If biomass is the desired product and if growth proceeds on a substrate with a γ_s of about 4 (many carbohydrates), growth will probably be carbon limited and energy is split; the substrate's potential is not fully used (Fig. 2B). The addition of a second (inexpensive) substrate with a higher γ_s can lead to more satisfactory efficiencies of substrate conversion if that second substrate is used for energy generation; a larger fraction of the first substrate will then be turned into biomass (170, 186). If biomass formation is most undesirable (e.g., in wastewater treatment), then expensive maintenance processes and/or an extensive mismatch between catabolism and anabolism is preferred. By choosing the right substrate(s) and reduction degrees, enforcing certain growth rates and growth domains, and imposing suitable environmental conditions (temperature, osmolarity, O_2 tension, etc.) that will or will not force the cells into a state of energy spilling, a process can in principle be steered towards the product of choice (83, 168).

Costs of exoprotein formation and product formation rates

Costs of exoprotein production can be estimated from material balances, molecular formulae, and growth parameters. Results of these calculations can be a key element in choosing a fermentor type and mode of operation for desired product formation. For the yield Y, subscripts x (denoting biomass) and s (denoting substrate) are added; if applicable, m is added to denote the maximum growth yield (Y^{max}). So, Y_{xsm} is the maximum (m) growth yield for biomass (x) on substrate s; likewise, Y_{psm} is the maximum (m) product (p) yield on substrate s.

If considerable amounts of exoprotein (or any exoproduct) are formed, then equation 13 must be extended:

$$r_s = (r_x/Y_{xsm}) + mx + (r_p/Y_{psm}) \quad (20)$$

($r_p = dP/dt$ is the rate of [exo]product formation.) By analogy to the conversion of equation 4 to equation 8, Monod's equation could be applied and a term for inhibition (K_p) could be incorporated (144)

$$r_s = (r_s^{max}\, xsP)/[(K_s + s)(K_p + P)] \quad (21)$$

There is no way to estimate Y_{xsm}, Y_{psm}, and m independently, so one actually has to assume values for the one and then, knowing r_s, r_x, and r_p, find values for the others. This problem is simplified if one assumes a (nearly) linear dependency of the rate of exoproduct formation (r_p) on the rate of biomass synthesis (r_x) (96, 188):

$$r_p = ax + br_x \quad (22)$$

The assumption of linearity and equations 20 and 22 result in equation 23:

$$r_s = [(1/Y_{xsm}) + (b/Y_{psm})]r_x + [m + (a/Y_{psm})]x \quad (23)$$

This equation is used in a procedure of simultaneous parameter optimization that establishes best-fitting curves for experimental data for biomass, exoproduct, CO_2, and O_2. (Detailed rate equations for r_s, r_x, r_p, r_{CO_2}, and r_{O_2}, including the components of the carbon balance Y_c [z and d] and reduction degrees [γ] of the participating components, are given in reference 188); calculations on continuous culture experiments are more straightforward, since Y_{xsm} and m follow from simple linear regression of q_s versus μ.) Fitted parameter values for Y_{psm} can then be compared with the possible range of Y_{psm} calculated as described in the Appendix.

Applying this method of calculation to RF and chemostat experiments with exoprotease-producing *B. licheniformis* (51) under glucose-limited (aerobic) conditions resulted in the conclusion that (in chemostat) $80 < Y_{psm} < 95$ (in grams of exoprotein per mole of glucose), while accompanying values for the total costs were 1.46 to 1.73 glucose per exoprotein (i.e., $1.03 + A/B$), representing 24 to 29% (i.e., $C_p + d_p$) of the total glucose flux. About 30 to 40% of the total costs of exoprotein production are due to energy supply and export of the exoprotein. In general, carbon costs of exoprotein production outweigh the energy costs (including export) (51). The overall substrate costs to support these energy costs can increase substantially and the Y_{xsm} can decrease substantially if the efficiency of oxidative phosphorylation (P/O ratio) and thus the ATP yield per mole of substrate is low (171). Experiments with RF cultures showed the occurrence of two domains of growth and production in time (51, 188), as mentioned above. Different rates of exoprotein production were observed in the two domains. In both domains, equation 22 could be applied; experimental values for the rates of exoprotein production were $r_p = 0.021x + 0.095r_x$ for the first domain (with nonlinear, continuously decreasing values of r_x) and $r_p = 1.057r_x$ for the second domain (relaxed response, with a low, constant value for r_x). It is interesting to compare these data from the 100% RF with the above mentioned results with the chemostat cultures (CF), since they illustrate the difference in production performance between fermentor types. In Table 4, several parameter values are compared for 40 h run time (after which, in the RF, the second domain with a less favorable exoprotein production started).

The exoprotein yield in the two fermentors (Table 4) is virtually the same. The RF seems to have some definite advantages over the CF: 30% lower glucose costs, a higher carbon conversion rate, and a smaller, cleaner effluent volume (i.e., no biomass). However, the RF culture will have a continuously decreasing μ; continuous (repeated) long-term operation is therefore not possible. Also, the constant composition of the CF effluent (though less clean than that of the RF) is an advantage of the CF over the RF.

So, with only these data (Table 4), a choice between these two fermentor types cannot be made. A somewhat similar dilemma can be found for a two-stage FB culture of β-galactosidase-producing *B. amyloliquefaciens* (69). Compared with growth on glucose alone, growth on first glucose and then lactose gave 34% higher specific activities of the exoenzyme. Final total productivity, however, was higher for the glucose FB (18 U/cm^3) than for the two-stage FB (15 U/cm^3).

Table 4. Comparison of aerobic growth and production of *B. licheniformis* in RF and CF

Parameter or characteristic	RF	CF
Total glucose input (mmol)	60	92
Total exoprotein output (g)	2	2.1
Total biomass output (g)	2.7	7.0
Total effluent output (liters)	6	9.2
% carbon conversion (glucose → exoprotein)[b]	24	17
$r_p = ax + br_x$[c]	$a = 0.021$, $b = 0.095$	$a = 0$, $b = 0.27$
Major organic components of effluent	Exoprotein, no biomass; composition variable in time	Exoprotein and biomass; composition constant in time
Continuity of run	Relatively short run time; after 35 to 40 h, μ must be increased to avoid relaxed response	Theoretically, unlimited run time
Downstream processing	Continuous processing possible, although complicated by variable composition of effluent	Continuous processing possible; no complications as in RF, but effluent is less clean

[a] Minimal glucose nutrient. Data apply to 40-h run time. Calculations were carried out with data taken from reference 51. Working volumes, 1 liter. For RF, r_s = 1.5 mmol of glucose per liter · h), and x_0 (dry weight at zero time) = 0.62 g/liter. For CF, r_s = 2.3 mmol/liter · h) at dilution rate [D] = 0.23 h^{-1}, and x_t (dry weight) = 0.76 g/liter (constant).
[b] Based on C-mol/C-mol.
[c] *a* is expressed as grams of exoprotein per grams of biomass per hour. *b* is expressed as grams of exoprotein per grams of biomass.

Despite that, depending on a.o. the costs of downstream processing, the higher specific activity of the two-stage FB with a final amount of biomass in the fermentor of only 36% of the glucose FB might be advantageous.

In a comparison of several continuous processes and repeated fed batches, Heijnen et al. (56) point out a series of bottlenecks in continuous processes and conclude that these can often be circumvented by the use of CF in the (P)RF mode and repeated fed-batch systems. The best choice will be determined not only by the productivity of the process but also by the full range of production plant factors and economic and market aspects (4, 56).

Screening for desirable product formation rates

In Table 4, the *a* and *b* values for production rate r_p are given. Clearly (see equation 22), a high value of *b* (and a low value of *a*) means that production will be highest in fast-growing cultures, which have high values of r_x. Since in FB the growth rate can drop quite fast to low values, r_p will drop fast, too, and it will not make much sense to continue the fermentation in that fermentor type for long; a CF might be preferable. On the other hand, a high value of *a* (and a low value of *b*) is best exploited by running fermentations at very high cell densities, like those in an FB. For an RF (in the present context, the same as an FB), van Verseveld et al. (188) showed how, at given values of Y_{xsm}, Y_{psm}, *m*, and r_s, the productivity was influenced by different *a* and *b* values (Fig. 3B). A higher value for *b* will have its largest impact in the early stages of the fermentation, when r_x still has an appreciable value; the impact of a high *a* value will become larger and larger as the fermentation (at very low r_x but high cell density) continues. Since so many fermentation industries use FB, the value for *a* is rather important. Therefore, screening methods for production may have to be adapted to select for strains with high *a* values. Quite often, one of the screening stages consists of growing the strain in a batch culture (shake-flask). If final production is then assayed in the early stages of stationary phase, one selects for strains with a high *b* value (188), and production with that organism in an FB might be disappointing. This is illustrated in Fig. 3. If the batch cultures (Fig. 3A) had been sampled after 12 h (after growth), the screening procedure would have resulted in a ranking of strains as 5 > 4 > 3 > 1 > 2. In the subsequent carbon-limited fed-batch culture (Fig. 3B), this ranking does not hold. Straight from the beginning, strains 2 and 3 perform better than strains 4 and 5. Moreover, with increasing fed-batch run time, the differences between strains 2 and 3 and between strains 4 and 5 increase (Fig. 3B), while in the batch cultures, the differences decrease (Fig. 3A). While strain 1 would have been classified as either a poor or a very good producer (after, respectively, 12 and 22 h; Fig. 3A), it does not perform very differently from strains 2 and 3 during 48 h. Two conclusions can be drawn from this example. First, sampling for the amount of product in the screening procedure is insufficient unless it is done frequently enough to allow for an estimate of parameters *a* and *b*. Second, the feed rate in the fed batch must obviously be adapted to the values of *a* and *b*; e.g., strain 5 is clearly not very well suited for a system in which the growth rate is low most of the run time ($a = 0.0$), while strain 1 is quite versatile (substantial value for both *a* and *b*) and, depending on the feed rate/growth rate ratio, can be useful in several fermentor types. Needless to say, one can probably not indiscriminately increase the feed rates to accommodate strains with high *b* values, since repression and other unwanted affects are likely to occur. Those problems will have to be solved for each individual case. Moreover, values for *a* and *b* will differ for different environmental conditions (e.g., carbon or nitrogen limitation or excess, etc.).

For the sake of completeness, it should be emphasized that a linear relation between r_p and r_x is not the rule; usually, that assumption holds for relatively small intervals of growth rates. For chemostat cultures (large range of μ values) of *B. licheniformis*, for instance, nonlinear relations of r_p and r_x were found

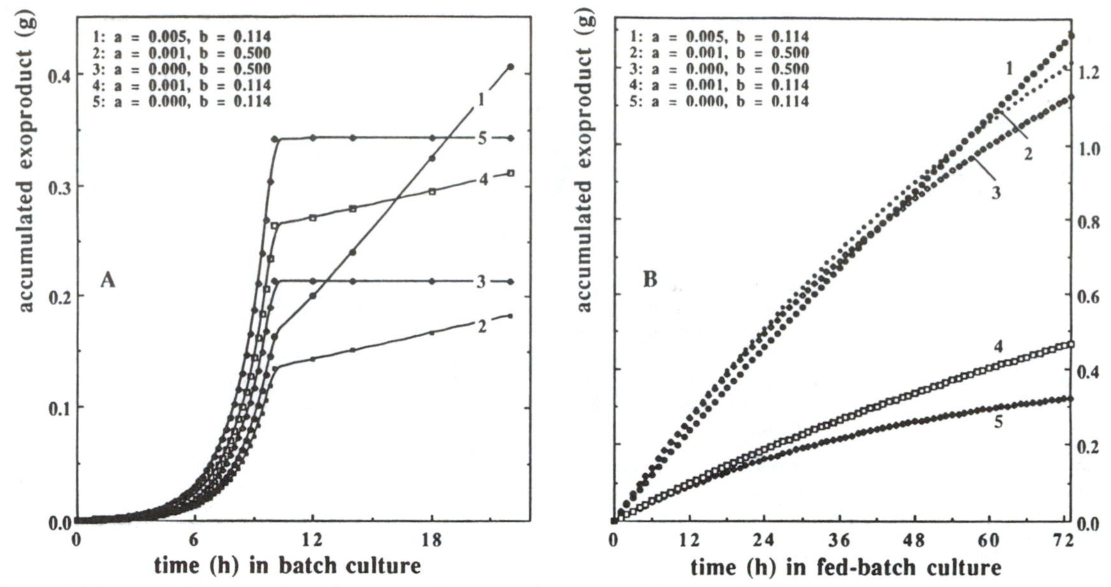

Figure 3. Theoretical exoproduct formation in batch (A) and fed batch (B) cultures. In panel A, five strains are evaluated for exoproduct formation by using batch cultures (i.e., screening stage). The same five strains are run in a carbon-limited fed-batch culture (i.e., production stage) (B). The production profiles show how the amounts of product in batch cultures at any time of sampling (without knowing the *a* and *b* values in equation 22) cannot lead to reliable predictions of exoproduct formation in fed-batch cultures. For further discussion, see text. Assumptions are that in the batch culture, growth of all five strains halted after consumption of 50 mmol of glucose because a compound other than the carbon and nitrogen source became limited. Growth in the fed-batch culture proceeded glucose limited (constant feed rate $[r_s] = 0.001$ mol/$[g \cdot h]$). Exoproduct formation is described by $r_p = axt + brx$ (21). The value of *a* (in grams per gram · hour) is constant; *b* (in grams per gram) was assumed to be described as follows: 1, $b = 0.117 - 0.14\mu$; 2, $b = 0.530 - 0.83\mu$; 3, $b = 0.530 - 0.80\mu$; 4, $b = 0.117 - 0.10\mu$; 5, $b = 0.117 - 0.07\mu$ (relations that can be obtained from, e.g., chemostat experiments). Other parameter values: $V = 1$ liter; $Y_{xsm} = 100$ g/mol; $Y_{psm} = 60$ g/mol; $m_s = 0.00024$ mol/(g · h); $x_0 = 0.01$ g (A) or 0.5 g (B); $P_0 = 0$ g (calculated according to van Verseveld et al. [188]; values for strains 2 through 5 in panel B are from that publication but modified).

because of changing values of *b* with μ ($b = 2 - 4.5\mu$ or $b = 2.2 \times 10^{-1.48\mu}$) (24, 51).

From the equation $r_s = (r_x/Y_{xsm}) + mx + (r_p/Y_{psm})$ (equation 20), it will be appreciated how the maintenance coefficient *m* can influence productivity (and growth). In many industrial fed-batch fermentations, after the initial phase of fast growth, a considerable amount of substrate is spent on maintenance processes in the phase of production-very slow growth. Van Verseveld et al. (188) showed how a fourfold reduction in maintenance costs can result in a more than fourfold increase in productivity.

Finally, apart from changing environmental conditions, there are some possibilities for manipulating *a* and *b* values, such as tampering with the temporal expression of the gene promoter systems for exoenzymes. For instance, the *b* value for exoprotease production by *Bacillus* species in batch cultures will be near zero, with production occurring as a sporulation-associated event in the late exponential and stationary phases only (11, 181). Manipulations that lead to earlier activity of the exoprotease promoter or to putting the exoprotease gene under the control of a constitutive promoter can give rise to a *b* value larger than zero. This would both improve productivity and increase the flexibility of the production organism with regard to which fermentor system can be used, thus making possible even higher product outputs.

Models of microbial growth

The kinetic descriptions of growth by Monod-type equations are examples of so-called unstructured models. These descriptions are phenomenological. Little, if any, information on the detailed biochemical routes by which initial and final situations are connected is incorporated; usually, the dynamics of biomass concentration (or of another, single variable) are deemed sufficient to describe the growing culture (variation of biomass composition is ignored).

A more sophisticated and necessarily much more complicated approach incorporates part of the increasingly available amount of information on biochemical pathways that constitute the cells' metabolism, giving rise to the structured models. In these models, both quality and quantity of the variables are considered; they contain information on the internal metabolic mechanism and the composition of the biomass. Usually, the culture is divided into a small number of components or pools that can be more or less defined (e.g., chemically, as DNA, RNA, precursor metabolites, etc.). An additional classification is made by distinguishing segregated (corpuscular) and nonsegregated (distributive or continuum) models that, respectively, do and do not take the individuality of cells within the culture into account.

Several unstructured and simply structured models were described and compared by Harder and Roels

(55) and Esener et al. (40–42) for batch, fed-batch, and continuous culture; by von Stockar and Auberson (193) for continuous culture; by Shuler (159); and by Groen and Westerhoff (53) for some control models (see below).

Among the models that started out as unstructured but had structure added to them through the years is the description by MNET (201). Among the incentives for development of this model were the need to account for lower-than-predicted values for Y_{ATP}^{max} (grams of biomass synthesized per mole of ATP, maximally) and for energy-spilling phenomena like overflow metabolism (e.g., see references 114 and 168). To what extent do these differences imply that microbial growth is thermodynamically inefficient, and what might be a rationale for energy spilling (60, 182, 201)? Initially, the cells were described as a black-box system (as in any unstructured model) characterized by catabolic input flux and anabolic output flux and by the difference in free energy for the substrates and products (catabolites, anabolites plus biomass) of the two processes. From this formulation, it was concluded that "thermodynamic efficiency of microbial growth is low but optimal for maximal growth rate" (172, 199, 201). Likewise, it provided an explanation for the observation that with increasing reduction of the substrate (γ_s), the thermodynamic efficiency of growth decreases (201). As outlined before (cf. Fig. 2), at high γ_s, the free-energy content of the substrate is higher than that of the biomass, and this energy has to be split, leading to low and negative thermodynamic efficiencies.

To incorporate structure (biochemical mechanisms) into the MNET description, metabolism is divided into three subprocesses: anabolism, catabolism, and leakage (i.e., ATP hydrolysis without coupling to anabolism or catabolism), again described by mutually dependent relations between the process rates or fluxes and the intracellular phosphate potential (ΔG_p) as the driving force. The mere existence of leakage reflects the absence of perfect coupling between catabolism and anabolism (the leakage phenomenon incorporates the maintenance energy demand from other models). Predictions of the model in line with experimental results include substrate consumption at zero growth rate (maintenance) and linear relations between the rates of anabolism and catabolism. The slopes of the lines relating anabolism and catabolism indicate whether a growth limitation has a more catabolic or anabolic character. Furthermore, under certain conditions, it is possible not only to recognize ATP leakage but also to assign it to a specific part of metabolism. Predicted changes in fundamental physiological parameters like maintenance (μ dependent and independent) and growth yield as a consequence of metabolic uncoupling have been substantiated in a number of instances (147, 150, 182). Mulder et al. (108), extending MNET by incorporating chemical structure (i.e., anabolic fluxes subdivided in protein, DNA, RNA, lipid, and polysaccharide), showed how the line relating catabolism to anabolism was influenced by changes in biomass composition (at different growth rates in a chemostat). The changes in slope of the line, however, were very small. Those authors mentioned how the impact of variable biomass composition could be larger if lipids, for example, were produced in excessive amounts (in general, the same would hold for storage compounds like PHB and polysaccharides).

The metabolic control theory (MCT), initiated by studies of Kacser and Burns (75) and Heinrich and Rapoport (58) on flux control through linear, enzymatic pathways, focuses on the optimization of metabolic fluxes (or any variable) relevant to production. This theory seeks to determine which enzymes most determine the flux through a pathway and what manipulations (genetical, physiological, or enzymological) will increase the flux through a pathway (78). As stated by Kacser (74), MCT provides the link between enzymology and physiology. Since any biotechnological production process should optimize the volumetric productivity (i.e., the amount of product per unit time and fermentor) and since a major impact on this productivity will be exerted by the magnitude of the flux through an enzyme pathway (or metabolic route), answers to the questions asked above are essential.

At the core of MCT lie the control coefficients and elasticity coefficients. For example, the flux control coefficients (C^J) consider changes (in flux [J]) between steady-state conditions. By definition, $C^J_e = [(dJ/J)/(de/e)]_{ss} = [(d \ln J/d \ln e)]_{ss}$ (74, 76, 78), where ss is steady state. So, for a small, fractional change in enzyme concentration ([e]), the flux control coefficient gives a dimensionless measure for the importance of enzyme e as indicated by the fractional change in pathway flux [J]. Usually, C^J has a value between 0 (no control) and 1 (total control). Distribution of control over the pathway enzymes is indicated by the ratio of the relevant control coefficients. Importantly, for any given (near-ideal) pathway, the sum of the C^J values equals 1 (summation theorem; total control is shared by all the pathway enzymes), so the average flux control of an enzyme will be close to $1/n$ (for a pathway containing n enzymes). This leads to two important conclusions. First, since the control of flux is spread out, it is more realistic to establish a ranking of relative importance of the pathway enzymes with regard to flux than to ask for the single rate-limiting step. Second, the concept of flux control explains why multiple rounds of mutation are necessary for strain improvement, since upon changing one enzyme, the total pathway control will still be 1 and the average control of any enzyme will still be close to $1/n$. A given mutation would cause control to be reshuffled with no major impact on the flux.

The elasticity coefficient ε is defined as the fractional change in the turnover number of an enzyme divided by the fractional change in the effector considered. Thus, $\varepsilon^e_x = (dv/v)/(dX/X) = (d \ln v)/(d \ln X)$ (here, X is the effector, e.g., the substrate [S] of the enzyme). A low elasticity value, for instance, reflects a small change in turnover for a significant change in the enzyme's substrate concentration. The elasticity coefficients are interrelated with the control coefficients by connectivity theorems that (like the summation theorem) allow extensive quantitative treatment of metabolic control (53, 74, 77–80, 200).

The quantitative character of MCT allows some rational decisions to be made. It becomes possible to decide whether it is more useful to increase the enzyme concentration (by overexpression of its genes) or to engineer the enzyme itself with respect to the K_m, for instance (77–79). Successful experimental applications of MCT to microbial metabolism are given by Heinisch (57), Nimmo and Cohen (116), and Kell et al. (77); these applications include increasing the enzyme concentration, introducing plasmids with extra gene copies for the enzyme, regulating promoter induction, titrating with specific enzyme inhibitors, etc.

Rutgers et al. (150) have expressed control coefficients in MNET parameters. Among other things, their formulations predict how the anabolic system always has a positive flux control on the growth rate, while the catabolic system and leakage process exert variable levels of control at different growth rates from zero to μ_{max} (at constant control by the anabolic system). The same group showed how glucose and ammonium can simultaneously be growth controlling, depending on the ratio of the two substrates in the feed (149). Transition from one form of limitation to the other was shown to proceed smoothly, through intermediate *Y* values. Applying MCT, they showed that both flux control coefficients (for ammonium and for glucose on the growth rate) simultaneously exceeded zero, proof of the existence of multiple-substrate-limited growth (the term control rather than limitation is used, since limitation suggests 100% control, while actual control can be only partial [182]). Since thermodynamic efficiency of growth is significantly dependent on the substrate and product concentrations (148), suggesting high efficiency of growth during multiple substrate controlled growth (when both concentrations are very low), it is essential to establish to what extent the multiple substrates present in the feed exert control. For instance, a large number of publications on *Bacillus* protease production deal with nitrogen-limited growth, but whether control is really predominantly exerted by only one feed component or by more than one is seldom assessed. Since multiple-nutrient-limited growth can occur over a large range of carbon/nitrogen or carbon/magnesium ratios (with strong dependency on the growth rate), an increased insight into the character and extent of control can be advantageous.

An example of a relatively simple structured model is provided by Jöbses et al. (71), who extended the unstructured model described above to a two-compartment model incorporating growth rate-related changes in biomass composition. Their model (with earlier work [42, 55]) describes biomass in a K and a G compartment, comprising RNA, carbohydrates, and monomers of, respectively, protein, DNA, and lipids (with maintenance energy demand taken up as part of this G compartment). Incentives for construction of the model were the needs to incorporate data on biomass composition and to accumulate evidence for variable maintenance energy demands at different growth rates.

Recognition of the individuality of cells in a population has led to segregated models. The extreme complexity of the resulting equations induced the development of single-cell models. It is assumed that since the single cell under discussion will be representative of a subpopulation of the culture, the meaning of the single-cell model can be extrapolated to the population level (7, 159, 160) with or without the addition of an extra, population-related parameter (35, 36). The initial model of Shuler and Domach (160), developed for growth on glucose and ammonium, predicts changes in macromolecular composition, cell size and shape, duration of C and D periods of the cell cycle, and growth rate. With about 35 equations describing the rate of changes in cell components (and over 100 related parameters), it gave good predictions of experimental observations. The model was recently extended by including the kinetics and biochemistry of glutamine metabolism, giving it predictive value for changes in glucose, glutamine, and ammonium concentrations (158); situations of both carbon and nitrogen limitation can be described. According to Shuler and Domach (160), discrepancies between experimental and predicted values can probably be remedied when a more detailed regulatory system for the tricarboxylic cycle is incorporated.

It is hard not to admire the intricacies of such very detailed models and the enormous effort that has to be made before they can be constructed. It is equally hard for the industrial biotechnologist not to break down and cry upon realizing that in large-scale fermentations, it is impossible to get accurate data even on biomass concentration, let alone some insight into the precise kinetics and biochemistry of the production system. Joshi and Palsson (73) describe a still complex but simplified version of the model by Shuler and colleagues, reasoning that "a balance between physiological reality and simplicity calls for reduction in the complexity of the mathematical description to develop a tractable model that is of practical utility." Their approach to simplification starts with the recognition of characteristic time scales for all biochemical events, the meshing of these events into integrated subsets of several time scales, and the practical realization that for industrial applications, one is most interested in those processes of the microbial integrated system that have the same order of time scale as the relevant external processes. Essentially, internal mechanisms can be ignored if they are faster than environmental changes (for the internal process will be in a quasi steady state with respect to the external variable); if the internal process is slower, it can also be ignored from a short-term view, as in industrial process control. Applying these principles, Joshi and Palsson (73) developed a three-pool model from the single-cell model of Shuler and colleagues, eliminating the metabolic transients faster (occurring in seconds to minutes) and slower (occurring in days) than growth. The three pools consist of (i) RNA and nucleotides, (ii) amino acids and protein, and (iii) the cell envelope and its precursors. Predictions from this model compare favorably with experimental data and are very similar to the predictions that follow from the more complex single-cell model from which it was derived.

In the cybernetic models of Ramkrishna et al. (9, 141), a relatively small number of rate equations describing saturation kinetics for growth, maintenance, and critical resource synthesis is formulated. Resource (for cellular enzymes) represents (a part of) the protein synthesizing machinery and is introduced in the model in recognition of its possible limiting effect under several transient conditions (in the absence of this factor, discrepancies between model predictions and experimental data were found under those conditions [179, 180]). A series of cybernetic variables that modify the process rates by control of enzyme activity and enzyme synthesis is then introduced. The final description consists of rate equations (with cybernetic and concentration variables) for growth, maintenance, and resource and for synthesis of the enzymes of these processes (as in virtually any model, "enzyme" is understood as a lumped entity, comprising, for instance, enzymes for the whole of glycolysis and other major subsets of metabolism). Disrupting steady-state growth in continuous cultures leads to transient responses that, unlike in previous model formulations, were predicted well. Especially when low growth rates occur, the authors find better predictive power than with previous models. This is ascribed not only to incorporation of the variable term for the critical resource but also to incorporation of two terms for maintenance energy demand. A basal-maintenance process, growth associated, is distinguished from a low-maintenance process, which operates predominantly at low growth rates (less than about 0.1 h^{-1}). This distinction was introduced to account for maintenance substrate consumption in transient chemostats that outranged the predictions of previous models (179). Reassuringly, the successful recognition of such an increased bacterial maintenance demand at low growth rates in this cybernetic model is completely in line with observations by others on maintenance and stringent regulation at low growth rates (2, 24, 26, 28, 133, 134, 169, 185, 187). Baloo and Ramkrishna (8) have extended their model to a description of transients during feed switching and mixed substrate growth in continuous cultures. The present model (9) reduces to previous models (34, 179, 180) in cases where the effects of the resource level and low maintenance are insignificant, i.e., in batch cultures (exponential phase) and steady-state continuous cultures in which sudden transient conditions do not occur.

Not many structured models exist exclusively for *Bacillus* species. This might not be very surprising, given the extreme complexity of the cell differentiation processes that occur during sporulation (87, 95). Recently, though, an elaborate structured (nonsegregated) model was proposed by Jeong et al. (70). Their model accounts for 35 cellular events and/or components; there is emphasis on the inclusion of purine metabolism for its obvious involvement in sporulation initiation signaling (120, 129, 164). With some 39 coupled differential equations and about 200 parameters and constants (maximal reaction rates and saturation-inhibition constants compiled from published data), the model was used to simulate exponential and postexponential (sporulation initiation) phases in a glucose batch culture. Predictions from the model are compared with data in the literature, and given the complexity of the model, they agree quite reasonably. Starzak and Bajpai (166) developed a structured-nonsegregated model for growth of and sporulation-associated δ-endotoxin production by *B. thuringiensis*. Intracellular pools distinguished are amino acids, three forms of protein (vegetative, spore, and the δ-endotoxin product), nucleotides and reducing equivalents, nucleic acids, and organic phosphate carriers. With some 20 equations for stoichiometry of vegetative-growth reactions (including energy generation from substrate and oxidative phosphorylation) and associated rate equations as well as expressions for the kinetics of spore and crystal protein formation, Starzak and Bajpai found their model predictions to be in qualitative agreement with data published by others on batch cultures growing with glucose and ammonia. The model could be improved by incorporating the consumption of organic acids (produced during exponential phase) and the turnover of storage materials like PHB in the stationary phase.

DEVELOPING STRAINS FOR INDUSTRIAL USES

General Considerations

The functions included under strain development are identification and isolation of an organism exhibiting the trait (or traits) of interest; analysis and optimization of cultural conditions that allow expression of the trait; and mutagenesis, selection, and screening for isolates that hyperproduce the trait. The utility of a strain developed for the production of a commercial product will depend on (i) factors associated with the strain itself, (ii) fermentation conditions, and (iii) economics of the product to be manufactured. Economic considerations include the type of product (pharmaceutical versus commodity), desired recovery and/or purification costs, raw-material costs, fermentation plant capacity, and projected sales price of the product. The combination of these factors drives development of the production organism.

For a given production process, the strain and the fermentation are usually optimized as a unit. In practical terms, this means that the strain development program must, at a minimum, be conscious of any limiting parameters of the proposed fermentation process. Since fermentation costs are generally very sensitive to raw-material costs, strains that can utilize readily available, inexpensive sources of protein and carbohydrate are often developed. However, the type of medium used can have a major impact on the yield of a protein and downstream processes such as cell separation and product recovery. For example, residual protein or carbohydrate in the fermentation broth can drastically affect flux rates during subsequent steps of diafiltration (to remove salts, etc.) or ultrafiltration (to concentrate the broths). Similarly, in high-value-added products, for which purity is a prime concern, medium components can necessitate additional steps such as selective precipitation or column chromatography to achieve product specification targets.

When cells are cultured on a rapidly utilizable carbohydrate source, the production of many enzymes involved in the breakdown of complex nutrients (48, 115) is repressed. This repression is termed catabolite repression and is mediated via a cyclic AMP monitoring system in *Escherichia coli* (for a review, see reference 98). In *Bacillus* spp., catabolite repression is evident, but the actual mechanism is not known (157; see chapter 15, this volume). Regardless of how catabolite repression is controlled, *Bacillus* manufacturing strains must be able to produce products in the presence of high levels of carbohydrates in the fermentor. Alternatively, strains that produce under carbon-limiting conditions may be designed.

Many of the compounds naturally produced by

bacilli are not made during vegetative growth but rather during the transition phase between vegetative growth and sporulation (see chapter 51, this volume). The use of sporulation-deficient strains is commonplace in the industry (137); however, since natural production is often controlled by sporulation factors (181), it is essential that Spo$^-$ strains retain their abilities to produce during the transition phase. Mutations early in the sporulation process (*spo0A*, for example) decrease sporulation but render the cells incapable of protease production (46, 47) and are therefore inappropriate for strain development, while genes that block the sporulation process at stage II or later can prove useful (197). In addition to increasing productivity, sporulation-deficient strains have benefits in lowering the release of viable microorganisms from a production facility, which is important both for regulatory compliance and for maintaining the proprietary nature of production cultures. It must be noted, however, that most strains used for large-scale production were derived from natural isolates of various *Bacillus* spp., and the identification of specific spore markers remains obscure.

Classic Approaches

The development of a novel microbial product occurs in a series of stages. Historically, the first step entails a search for a natural isolate and a screening program to identify organisms that have the requisite activity. During the second step, culture conditions for the isolate are optimized to achieve maximal productivity. The third stage is a series of iterative rounds of mutagenesis, selection, and screening to isolate mutants of the original parent strain that have enhanced abilities to produce the product. This stage is aided by direct selection of overproducers and rapid screens or plate assays, which allow easily observable identification of those individual colonies that overproduce the product of interest. Typical of these approaches are the visualization of clearing zones on starch containing agar by overproducers of bacterial amylases, halo formation on skim milk plates by protease producers, and growth inhibition of indicator organisms by antibiotic producers.

A major disadvantage of the classic approach is that knowledge gained during the strain development and optimization process may be of extremely limited value in any future pursuits. The development of novel related products may need to go through the same labor- and time-intensive process of strain improvement as that outlined above.

Molecular Approaches

Generic host production systems

Molecular approaches to strain development can replace classic steps throughout the entire process. Protein engineering to design the appropriate characteristic into a preexisting enzyme can, for example, replace the search for natural isolates (198); however, identifying a novel activity is of limited commercial interest in the absence of a production mechanism. By designing a host production system with multiple expression capabilities, the development process can be shortened considerably. Using a generic host system for a variety of related products gives the advantage of a single fermentation optimization program that can be used for multiple products (30). The disadvantage, however, is that a molecular approach to strain development requires more detailed knowledge of the control mechanism involved in production, and more basic research is required than would be necessary for a classic approach. The use of stronger promoters or multiple copies of a gene for overexpression allows rapid yield increases (183). In *Bacillus* spp., the *aprE* promoter has been used for expression of a number of individual genes because of our extensive knowledge regarding its regulation (30, 46, 47, 181). Similar systems are being developed to optimize expression from other *Bacillus* promoters, including amylase (*amyE*) (111, 151, 153) and levansucrase (*sacB*) (109, 110, 202).

Pathway engineering

Though the discussion above has focused on the production of single-gene products, similar work for modifying pathways to overproduce specific chemicals is under way. For example, a recent European patent application (129a) describes a *B. subtilis* strain that has been developed through genetic engineering to overproduce the vitamin riboflavin.

Regulatory considerations

Commercial products in the United States are regulated through the Bureau of Alcohol, Tobacco and Firearms, U.S. Department of Agriculture, Food and Drug Administration, or Environmental Protection Agency, depending on their ultimate uses. For products produced by microorganisms, regulatory concerns focus on both the safety of the product and the safety of the organism used for production purposes. By developing host systems based on organisms with a history of safe use for commercial purposes, the regulatory process can be streamlined. Additionally, this opens the opportunity to develop products from hosts that may themselves not be amenable to commercial fermentation for regulatory or physiological reasons. Recent effort has been placed on developing regulatory mechanisms to address the safety of products produced through biotechnology, and a framework has been proposed (1).

PROTEASE PRODUCTION AS A MODEL SYSTEM

Alkaline protease (subtilisin, encoded by the *aprE* gene) is an enzyme that has commercial importance and can be efficiently secreted from *B. subtilis*. The economic importance of alkaline proteases has fueled efforts towards improving the yield, activity, and stability of the enzyme. Yield improvements to date have come primarily from host strain development (*degU* and *scoC* mutations) and mutagenesis of signal sequences. Likewise, protein engineering has produced proteases with increased activities and stabilities (198). Yet, the yields of secreted protein in many *Bacillus* species appear to have definable limits (see below). This has led to an examination of the rate-limiting process of high-yield protein production that focuses on host cell physiology and whether transcrip-

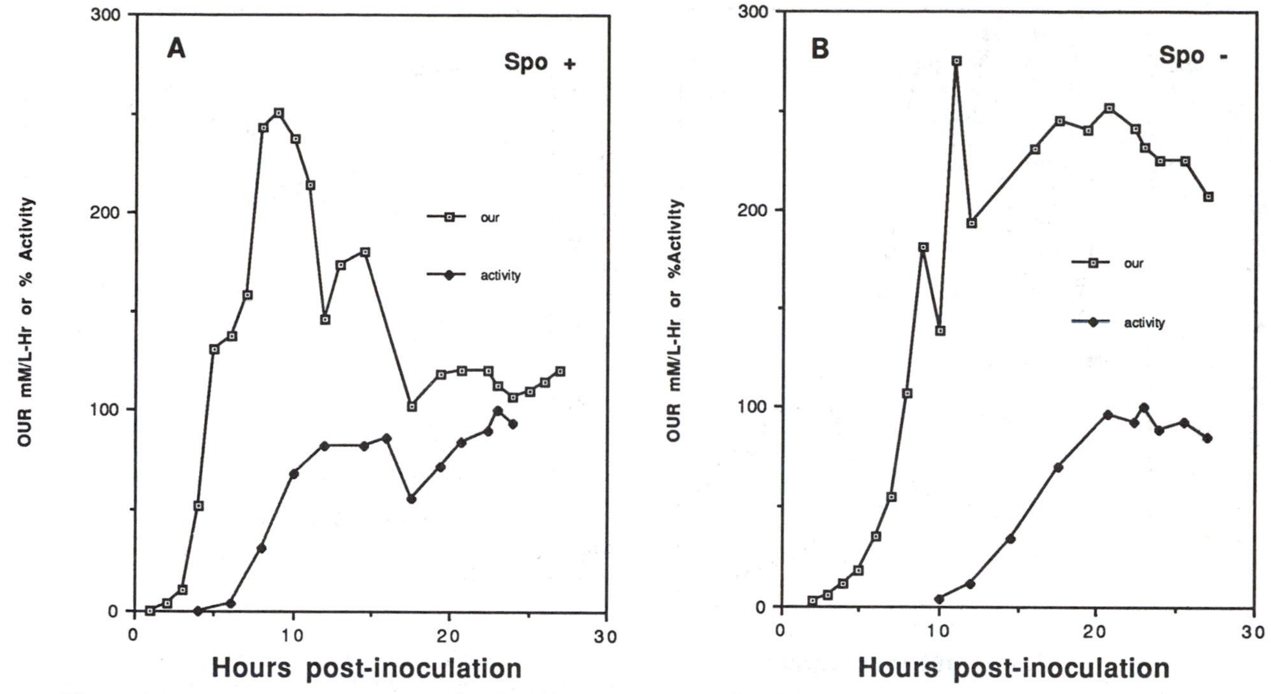

Figure 4. Exoprotease enzyme activity levels and respiration rate in a glucose fed-batch fermentation of *B. subtilis* (45). (A) Wild type (*spo*$^+$); (B) *spo* mutant. OUR, oxygen uptake rate.

tion, translation, or secretion is limiting protease expression. Protease production from *B. subtilis* is a good model system for these studies. It is likely that the identification of a rate-limiting process in the host strain or in production conditions would lead to increases in protein yields and be applicable to other heterologous proteins. Here, we discuss the value of using controlled fermentations for these studies.

Features of the Fermentation Process

Figure 4 shows typical subtilisin production and respiration curves from fed-batch fermentations with Spo$^+$ or Spo$^-$ strains of *B. subtilis* (otherwise isogenic). Subtilisin is expressed and secreted into the culture medium in response to a developmental growth phase-dependent switch. In Spo$^+$ strains, this switch directs cells to sporulate, and the cell population progresses from a respiring, vegetative state to a dormant state. In contrast, Spo$^-$ strains are unable to sporulate, and they continue to respire for extended periods. This, however, has little impact on subtilisin production, which initiates and halts at about the same time in both Spo$^+$ and Spo$^-$ strains (45).

Another feature of the subtilisin fermentation process is that the timing of initiation and cessation of subtilisin synthesis in *B. subtilis* is constant for a given respiration rate. Further, the synthesis rate is proportional to the oxygen uptake rate (Fig. 5), and thus, the rate of protease production is elevated at higher respiration rates and decreased at lower respiration rates. Regardless of the rates of protease accumulation, the final yields remain the same, with the production period varying in length. These observations can be made only by using fed-batch fermentation processes, because it is necessary to manipulate the respiration rate of the culture. Batch (shake-flask) fermentation data do not allow one to unlink respiration and product formation.

The cessation of protease synthesis in these cases appears to be a timed event that is due to both general decrease in the protein synthetic capability and specific inhibition of *aprE* gene expression. In a *spo* mutant strain growing in complex medium, the synthesis rates of total protein and secreted protein decrease simultaneously (Fig. 6). However, the ratio of

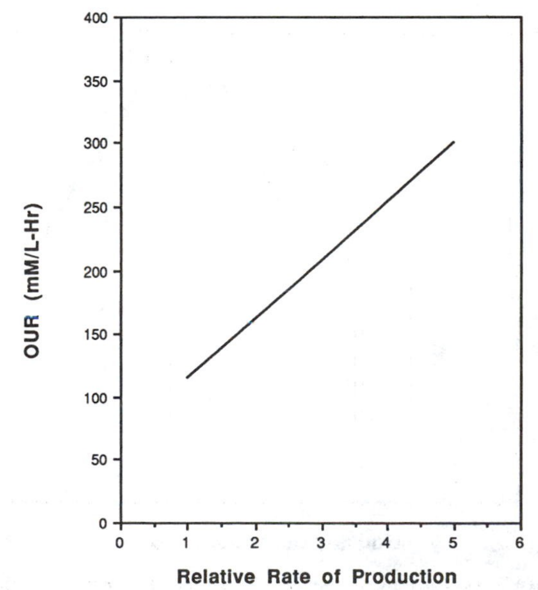

Figure 5. Trend of rates of production versus respiration rates. OUR, oxygen uptake rate.

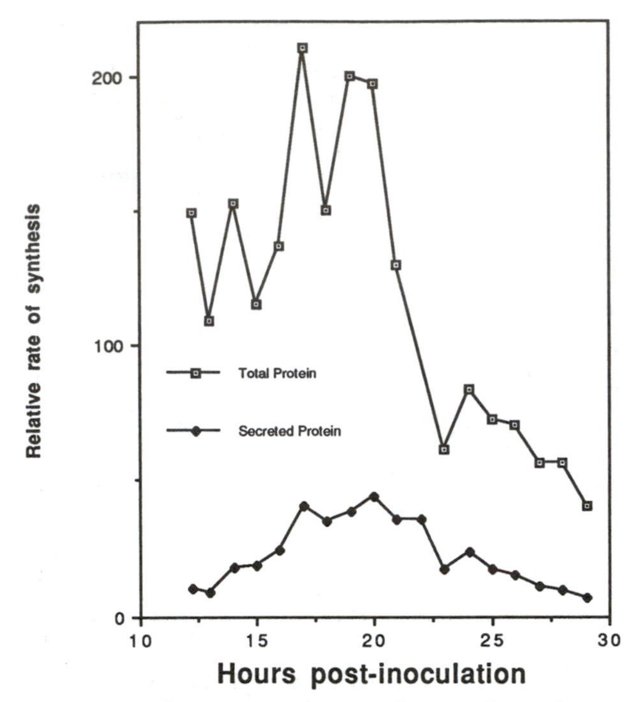

Figure 6. Total protein synthesis and secreted protein synthesis as measured by [^{35}S]methionine counts in a glucose fed-batch fermentation of *B. subtilis.*

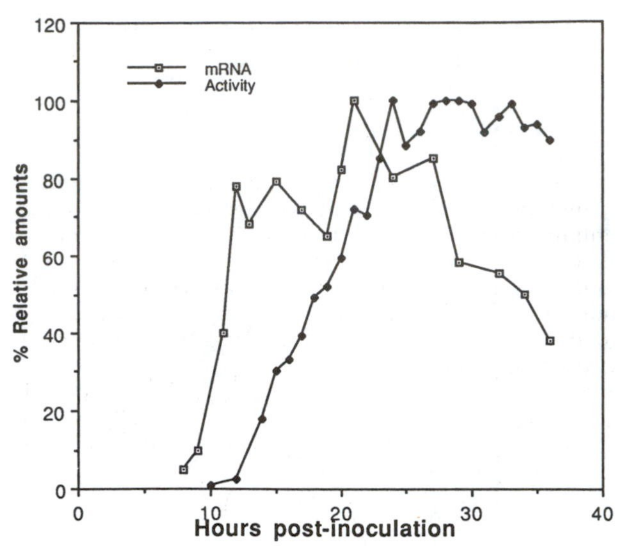

Figure 8. Relative mRNA levels versus exoprotease levels in a glucose fed-batch fermentation of *B. subtilis* (45).

secreted protein synthesis to total protein synthesis declines late in production at an enhanced rate, suggesting *aprE*-specific inhibition (Fig. 7).

These fermentation profiles are apparent whether production occurs in a minimal glucose salts medium or in the presence of complex nitrogen and carbohydrate sources (i.e., grains, hydrolyzed starches). However, the overall protease synthesis rates are higher in complex media than in minimal media, presumably because of the reduced protein synthesis load on the cell.

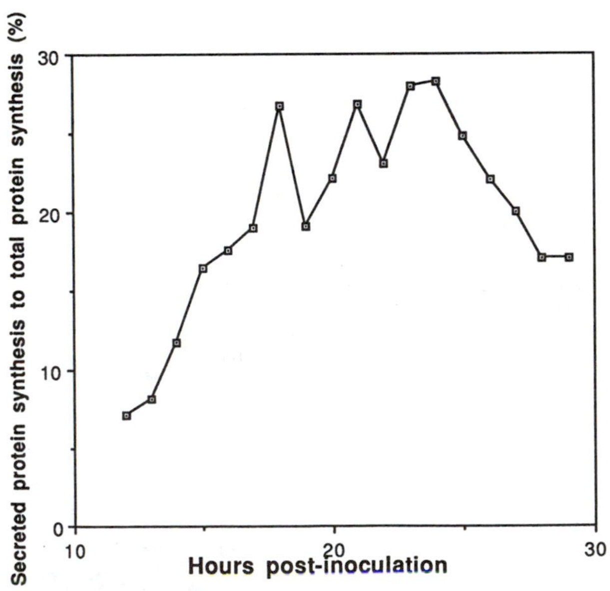

Figure 7. Percentage of secreted protein synthesis as measured by [^{35}S]methionine counts in glucose fed-batch fermentation of *B. subtilis.*

Possible Rate-Limiting Processes

The fact that there appears to be a definable limit to protease yield suggests that increases could result from identifying and overcoming rate-limiting processes in host cell physiology. Initial studies have focused on transcription, since an abundance of functional message is needed for maximal protein production. Steady-state levels of protease mRNA can be increased 5- to 50-fold by mutations in transcriptional activators (*degU*) or repressors (*scoC* [*hpr*]), and this fact is reflected by increases in protease yields (62). The pool of *aprE* message appears to be substantial, accumulates hours before protease is initially secreted, and although declining, is still at reasonable levels even after protease synthesis stops (Fig. 8) (45). Indeed, mRNA pools accumulate sufficiently to support protease synthesis for 80 min in the presence of rifampin (17). O'Connor et al. (121) have shown that mRNA pools have cyclic fluctuations. They hypothesized that during transcription of the protease gene, mRNA is supplied directly for translation and at the same time, a reserve of nontranslatable mRNA is built up. When transcription is reduced, protease synthesis relies on the reserve until it is exhausted, and thereafter, a new cycle of transcription resumes. Conversion of nontranslatable to translatable mRNA is not understood at the molecular level, but it may be a rate-limiting step in protease production.

The *aprE* message is apparently very stable: Resnek et al. (143) found that the *aprE* mRNA of a *B. subtilis* spo^+ strain grown in shake-flasks had a half-life of about 40 min as the cells entered stationary phase. Similarly, little degradation of *aprE* mRNA was observed over a 45-min period at either early or late times in the fermentation (45).

Another possible rate-limiting process in protease

production is translation efficiency, which can be affected by the mRNA secondary structure. An example is the neutral protease, NprT, whose mRNA has eight stacking regions; it is expressed poorly compared with a similar neutral protease, NprM, whose mRNA contains only one stacking region (86). Increasing the energy of the *nprM* stacking region reduced translation efficiency, whereas decreasing the energy enhanced translation. Putative stem-loop structures in the *aprE* mRNA leader have also been identified and may block translation initiation. This can be addressed by site-directed mutagenesis of the mRNA leader, which may weaken hairpins and thus increase translation efficiency.

Secretion is another key area to explore in identifying rate-limiting processes in protease production. Processing of the preproprotease by signal peptidase and release of the mature protein from the membrane are affected not only by mutations in or immediately downstream of the signal sequence but also by the location of the fusion joint between the pre- or preprosequence and a heterologous protein destined for secretion (190). The interaction of the signal peptide with the mature sequence is critical, and different signal peptides are optimum for different proteins (110; see also chapter 49, this volume).

Vectors containing translational fusions to secreted reporter genes have been developed for *Bacillus* spp. One example is *E. coli* alkaline phosphatase (PhoA), which is easily assayed and is active only when secreted (127). This type of vector is useful for screening random or site-directed mutants with an increased capacity to secrete protein.

Although the secretion machinery of *Bacillus* spp. is not yet well characterized, several homologs to *E. coli* *sec* genes have been found in *Bacillus* spp. *secA* and *secY* homologs as well as signal peptidase have been cloned (173, 184), and they can be used to probe the cell cycle during fermentation. In addition, overexpression of these proteins may improve the secretion efficiency of heterologous proteins, as has been demonstrated for signal peptidase (184).

Probing for these roadblocks at the fermentation level may lead to new discoveries, allowing *Bacillus* spp. to be used for a wider range of products when we can extend the organisms' cell cycle and distinguish the periods between product formation and other metabolic functions.

APPENDIX

Calculation of Carbon and Energy Costs of Exoprotein Production (51, 171, 188)

Two criteria are applied to limit the possible range of Y_{psm} values that were obtained in the parameter fit of the experimental data (188). The first (upper) limit to what realistic values Y_{psm} can have is set by the fact that exoprotein formation is more energy expensive than biomass synthesis (so Y_{psm} should not exceed Y_{xsm}). This is based on theoretical calculations (51, 168) that show how ATP requirements for biomass and exoprotein synthesis are, respectively 5.66 and 7.13 mol of ATP/mol (for growth on glucose mineral medium). The higher carbon costs for biomass production (equation 24) than for exoprotein production (equation 25) eliminate this difference, but the 7.13 ATP for exoprotein is an underestimate, as it does not include costs for export. The second (lower) limit to the range of possible Y_{psm} values is obtained from the values for Y_{ATP} (the assumed Y_{psm} values follow from the accompanying values of Y_{xs}^{corr}); obviously, Y_{ATP} should not exceed theoretical Y_{ATP}^{max} (= 26.1 g of biomass/mol of glucose [51]) or experimental values (if available) for Y_{ATP}^{max}.

For assimilation,

$$1.10C_6H_{12}O_6 \text{ (glucose)} + 1.2NH_3 + 0.6NAD^+ \rightarrow \rightarrow 1.00C_6H_{10.8}O_{3.0}N_{1.2} \text{ (biomass)} + 0.6CO_2 + 2.4H_2O + 0.6(NADH + H^+) \quad (24)$$

and for exoprotein formation,

$$1.03C_6H_{12}O_6 \text{ (glucose)} + 1.74NH_3 + 0.27NAD^+ \rightarrow \rightarrow 1.00C_6H_{9.48}O_{2.04}N_{1.74} \text{ (exoprotein)} + 0.18CO_2 + 3.78H_2O + 0.27(NADH + H^+) \quad (25)$$

So, judging from carbon costs only, the maximal product yield (Y_{psm}) would equal the molecular weight of the product, i.e., 138.48 g of exoprotein/mol of glucose; however, since it takes 1.03 glucose to synthesize 1 exoprotein (equation 25), the true or real Y_{psm} (not corrected for energy costs) equals (1/1.03) × 138.48 = 134.45 g of exoprotein/mol of glucose. So, the carbon costs are 1.03 mol of glucose/mol of exoprotein; the energy costs are unknown and need to be estimated.

Suppose that:

1 mol of glucose → (dissimilation) → *B* mol of ATP

1 mol of ATP → (product formation) → 1 mol of exoprotein

The total substrate cost (carbon + energy) for exoprotein formation will be 1.03 (for carbon) plus *A/B* (for energy), and so

$$(1.03 + A/B) \text{ mol of glucose} \rightarrow 1.0 \text{ mol of protein} \quad (26)$$

Apart from this relation (equation 26), fractional carbon flows to biomass and exoprotein are needed. From the carbon balance [$(Y_c + z + d) = 1.0$], the fractions of substrate turned into biomass and exoprotein are Y_c and z (assuming all exoproduct is protein), respectively. Therefore, the carbon flows (C) to biomass and exoprotein are, respectively,

$$C_b = 1.1Y_c \text{ and } C_p = 1.03z \quad (27)$$

Also from the carbon balance, d is the fraction of glucose dissimilated (energy yielding), so the energy flow to biomass and exoprotein (d) equals $d_b + d_p$. The total glucose flow to exoprotein equals

$$d_p + C_p = z(1.03 + A/B) \quad (28)$$

If the carbon costs (C_p) (that follow from the experimental z) are subtracted from these total costs, energy requirements for exoprotein formation (d_p) will be known. Since d ($=d_b + d_p$) also follows from the experiment, while d_p has now been calculated, the energy costs for biomass synthesis will also be known: $d_b = d - d_p$.

To summarize:

$$\text{carbon flows: } C_b = 1.1Y_c \text{ and } C_p = 1.03z$$

$$\text{energy flows: } d_b = d - d_p \text{ and } d_p + C_p = z(1.03 + A/B)$$

The value for Y_{psm} follows from equation 26 and the M_r (=138.48) for exoprotein:

$$Y_{psm} = 138.48/(1.03 + A/B) \quad (29)$$

Once values for Y_{psm} have been obtained, product-corrected (for both carbon and energy costs) growth yields for biomass (Y_{xs}^{corr}) can be calculated from the experimental values for Y_{xs}: $Y_{xs}^{corr} = Y_{xs}/(C_b + d_b)$, which allows calculation of Y_{ATP} ($=Y_{xs}^{corr}$/ATP yield) (some uncertainty on this value stems from the earlier mentioned fact that ATP yields from aerobic metabolism are not precisely known).

REFERENCES

1. **Anonymous.** 1990. Biotechnologies and food: assuring the safety of foods produced by genetic modification. *Regul. Toxicol. Pharmacol.* **12**(Part 2):1–196.
2. **Arbige, M., and W. R. Chesbro.** 1982. Very slow growth of *Bacillus polymyxa*: stringent response and maintenance energy. *Arch. Microbiol.* **132:**338–344.
3. **Arbige, M., and W. R. Chesbro.** 1982. *relA* and related loci are growth rate determinants for *Escherichia coli* in a recycling fermentor. *J. Gen. Microbiol.* **128:**693–703.
4. **Arbige, M. V., and W. H. Pitcher.** 1989. Industrial enzymology: a look towards the future. *Trends Biotechnol.* **7:**330–335.
5. **Babel, W., and R. H. Müller.** 1985. Correlation between cell composition and carbon conversion efficiency in microbial growth: a theoretical study. *Appl. Microbiol. Biotechnol.* **22:**201–207.
6. **Babel, W., and H. W. van Verseveld.** 1987. Theoretical limits of growth yields and an analysis of experimental data, p. 210–219. *In* H. W. van Verseveld and J. A. Duine (ed.), *Microbial Growth on C1 Compounds.* Martinus Nijhoff Publishers, Dordrecht, The Netherlands.
7. **Bailey, J. E.** 1983. Single-cell metabolic model determination by analysis of microbial populations, p. 135–157. *In* H. W. Blanch, E. T. Papoutsakis, and G. Stephanopoulos (ed.), *Foundations of Biochemical Engineering. Kinetics and Thermodynamics in Biological Systems.* ACS symposium series 207. American Chemical Society, Washington, D.C.
8. **Baloo, S., and D. Ramkrishna.** 1991. Metabolic regulation in bacterial continuous cultures. II. *Biotechnol. Bioeng.* **38:**1353–1363.
9. **Baloo, S., and D. Ramkrishna.** 1991. Metabolic regulation in bacterial continuous cultures. I. *Biotechnology* **12:**1337–1352.
10. **Baltzis, B. C., and A. G. Frederickson.** 1988. Limitation of growth rate by two complementary nutrients: some elementary but neglected considerations. *Biotechnol. Bioeng.* **31:**75–86.
11. **Basalp, A., G. Özcengiz, and N. G. Alaeddinoglu.** 1992. Changes in patterns of alkaline serine protease and bacilysin formation caused by common effectors of sporulation in *Bacillus subtilis* 168. *Curr. Microbiol.* **24:**129–135.
12. **Bazin, M., S. Gray, and E. Rashit.** 1990. Stability properties of microbial populations, p. 127–143. *In* R. K. Poole, M. J. Bazin, and W. Keevil (ed.), *Microbial Growth Dynamics.* Society for General Microbiology special publication 28. IRL Press, Oxford.
13. **Behal, V.** 1986. Enzymes of secondary metabolism in microorganisms. *Trends Biochem. Sci.* **11:**88–91.
14. **Bella, L. A., R. M. Faust, R. Andrews, and N. Goodman.** 1985. Insecticidal bacilli, p. 186–210. *In* D. Dubnau (ed.), *Molecular Biology of the Bacilli,* vol. 2. Academic Press, Inc., Orlando, Fla.
15. **Bley, T., and W. Babel.** 1992. Calculating affinity constants of substrate mixtures in a chemostat. *Acta Biotechnol.* **12:**13–15.
16. **Boon, B., and H. Laudelot.** 1962. Kinetics of nitrite oxidation by *Nitrobacter winogradsky. Biochem. J.* **85:** 440–447.
17. **Both, G. W., J. L. McInnes, J. E. Hanlon, B. K. May, and W. H. Elliott.** 1972. Evidence for an accumulation of messenger RNA specific for extracellular protease and its relevance to the mechanism of enzyme secretion in bacteria. *J. Mol. Biol.* **67:**199–217.
18. **Bremer, H., and P. P. Dennis.** 1987. Modulation of chemical composition and other parameters of the cell by growth rate, p. 1527–1542. *In* F. C. Neidhardt, J. L. Ingraham, K. B. Low, B. Magasanik, M. Schaechter, and H. E. Umbarger (ed.), *Escherichia coli and Salmonella typhimurium: Cellular and Molecular Biology,* vol. 2. American Society for Microbiology, Washington, D.C.
19. **Brooks, J. D., and J. L. Meers.** 1973. The effect of discontinuous methanol addition on the growth of a C-limited culture of *Pseudomonas. J. Gen. Microbiol.* **77:**513–519.
20. **Bull, A. T.** 1974. Microbial growth, p. 415–442. *In* A. T. Bull, J. R. Lagnado, J. O. Thomas, and K. F. Tipton (ed.), *Companion to Biochemistry.* Longman, London.
21. **Bull, A. T., M. E. Bushell, T. G. Mason, and J. H. Slater.** 1975. Growth of filamentous fungi in batch culture: a comparison of the Monod and logistic models. *Proc. Soc. Gen. Microbiol.* **3:**62–63.
22. **Bulthuis, B. A.** 1990. Stoichiometry of growth and product-formation by *Bacillus licheniformis.* Ph.D. thesis. Free University, Amsterdam, The Netherlands.
23. **Bulthuis, B. A., J. Frankena, G. M. Koningstein, A. H. Stouthamer, and H. W. van Verseveld.** 1988. Instability of protease production in a rel^+/rel^--pair of *Bacillus licheniformis* and associated morphological and physiological characteristics. *Antonie van Leeuwenhoek Int. J. Gen. Mol. Microbiol.* **54:**95–111.
24. **Bulthuis, B. A., G. M. Koningstein, A. H. Stouthamer, and H. W. van Verseveld.** 1989. A comparison between aerobic growth of *Bacillus licheniformis* in continuous culture and partial-recycling fermentor, with contributions to the discussion on maintenance energy demand. *Arch. Microbiol.* **152:**499–507.
25. **Bulthuis, B. A., C. Rommens, G. M. Koningstein, A. H. Stouthamer, and H. W. van Verseveld.** 1991. Formation of fermentation products and extracellular protease during anaerobic growth of *Bacillus licheniformis* in chemostat and batch-culture. *Antonie van Leeuwenhoek Int. J. Gen. Mol. Microbiol.* **60:**355–371.
26. **Cashel, M., and K. E. Rudd.** 1987. The stringent response, p. 1410–1438. *In* F. C. Neidhardt, J. L. Ingraham, K. B. Low, M. Schaechter, and H. E. Umbarger (ed.), *Escherichia coli and Salmonella typhimurium: Cellular and Molecular Biology,* vol. 2. American Society for Microbiology, Washington, D.C.
27. **Charles, M.** 1985. Fermentor design and scale-up, p. 57–75. *In* C. L. Cooney and A. E. Humphrey (ed.), *Comprehensive Biotechnology; the Principles, Applications and Regulations of Biotechnology in Industry, Agriculture and Medicine,* vol. 2. Pergamon Press, Oxford.
28. **Chesbro, W. R., M. Arbige, and R. Eifert.** 1990. When nutrient limitation places bacteria in the domains of slow growth: metabolic, morphologic and cell cycle behavior. *FEMS Microbiol. Ecol.* **74:**103–120.
29. **Chesbro, W. R., T. Evans, and R. Eifert.** 1979. Very slow growth of *Escherichia coli. J. Bacteriol.* **139:**625–638.
30. **Crabb, W. D.** 1990. Subtilisin: a commercially relevant model for large-scale enzyme production, p. 82–94. *In* G. F. Leatham and M. Himmel (ed.), *Enzymes in Biomass Conversion.* American Chemical Society, Washington, D.C.
31. **Dabes, J. N., R. K. Finn, and C. R. Wilke.** 1973. Equations of substrate limited growth: the case for Blackman kinetics. *Biotechnol. Bioeng.* **15:**1159–1177.
32. **Dawes, E. A., D. J. McGill, and M. Midgley.** 1971. Analysis of fermentation products. *Methods Microbiol.* **6:**53–217.
33. **Demain, A. L.** 1987. Production of nucleotides by microorganisms, p. 178–208. *In* A. H. Rose (ed.), *Economic*

Microbiology, vol. 2. *Primary Products of Metabolism*, Academic Press, London.

34. **Dhurjati, P., D. Ramkrishna, M. C. Flickinger, and G. T. Tsao.** 1985. A cybernetic view of microbial growth: modelling of cells as optimal strategists. *Biotechnol. Bioeng.* **27:**1–9.
35. **Domach, M. M., S. K. Leung, R. E. Cahn, G. G. Cocks, and M. L. Shuler.** 1984. Computer model for glucose-limited growth of a single cell of *Escherichia coli* B/r-A. *Biotechnol. Bioeng.* **26:**203–216.
36. **Domach, M. M., and M. L. Shuler.** 1984. A finite representation model for an asynchronous culture of *E. coli*. *Biotechnol. Bioeng.* **26:**877–884.
37. **Egli, T.** 1991. On multiple-nutrient-limited growth of microorganisms, with special reference to dual limitation by carbon and nitrogen substrates. *Antonie van Leeuwenhoek Int. J. Gen. Mol. Microbiol.* **60:**225–334.
38. **Ellwood, D. C., and D. W. Tempest.** 1972. Effects of environment on bacterial wall content and composition. *Adv. Microb. Physiol.* **7:**83–117.
39. **Erickson, L. E., I. G. Minkevich, and V. K. Eroshin.** 1978. Application of mass and energy balance regularities in fermentation. *Biotechnol. Bioeng.* **20:**1595–1621.
40. **Esener, A. A., J. A. Roels, and N. W. F. Kossen.** 1981. Fed-batch culture: modelling and applications in the study of microbial energetics. *Biotechnol. Bioeng.* **23:** 1851–1871.
41. **Esener, A. A., J. A. Roels, and N. W. F. Kossen.** 1983. Theory and applications of unstructured growth models: kinetic and energetic aspects. *Biotechnol. Bioeng.* **25:**2803–2841.
42. **Esener, A. A., T. Veerman, J. A. Roels, and N. W. F. Kossen.** 1982. Modelling of bacterial growth; formulation and evaluation of a structured model. *Biotechnol. Bioeng.* **24:**1749–1764.
43. **Feitelson, J. S., J. Payne, and L. Kim.** 1992. *Bacillus thuringiensis*: insects and beyond. *Bio/Technology* **10:** 271–275.
44. **Fencl, Z., J. Ricica, and J. Kodesová.** 1972. The use of the multi-stage chemostat for microbial product formation. *J. Appl. Chem. Biotechnol.* **22:**405–416.
45. **Ferrari, E., H. Heinsohn, B. Christensen, J. Schultz, B. A. Bulthuis, D. Crabb, and M. Arbige.** 1991. Biochemical changes during subtilisin production in *Bacillus subtilis*. Abstr. 6th Int. Conf. Bacilli, Stanford, Calif.
46. **Ferrari, E., D. J. Henner, M. Perego, and J. A. Hoch.** 1988. Transcription of *Bacillus subtilis* subtilisin and expression of subtilisin in sporulation mutants. *J. Bacteriol.* **170:**289–295.
47. **Ferrari, E., S. M. H. Howard, and J. A. Hoch.** 1985. Effect of sporulation mutations on subtilisin expression, assayed using a subtilisin–beta-galactosidase gene fusion, p. 180–184. *In* J. A. Hoch and P. Setlow (ed.), *Molecular Biology of Microbial Differentiation*. American Society for Microbiology, Washington, D.C.
48. **Fisher, S. H., and A. L. Sonnenshein.** 1991. Control of carbon and nitrogen metabolism in *Bacillus subtilis*. *Annu. Rev. Microbiol.* **45:**105–135.
49. **Frankena, J., G. M. Koningstein, H. W. van Verseveld, and A. H. Stouthamer.** 1986. Effect of different limitations in chemostat cultures on growth and production of exocellular protease by *Bacillus licheniformis*. *Appl. Microbiol. Biotechnol.* **24:**106–112.
50. **Frankena, J., H. W. van Verseveld, and A. H. Stouthamer.** 1985. A continuous culture study of the bioenergetic aspects of growth and production of exocellular protease in *Bacillus licheniformis*. *Appl. Microbiol. Biotechnol.* **22:**169–176.
51. **Frankena, J., H. W. van Verseveld, and A. H. Stouthamer.** 1988. Substrate and energy costs of the production of exocellular enzymes by *Bacillus licheniformis*. *Biotechnol. Bioeng.* **32:**803–812.
52. **Gottschal, J. C.** 1990. Phenotypic response to environmental changes. *FEMS Microbiol. Ecol.* **74:**93–102.
53. **Groen, A. K., and H. V. Westerhoff.** 1990. Modern control theories: a consumer's test, p. 110–118. *In* A. Cornish-Bowden and M. L. Cárdenas (ed.), *Control of Metabolic Processes*. NATO ASI series A: life sciences, vol. 190. Plenum Press, New York.
54. **Han, K., and O. Levenspiel.** 1988. Extended monod kinetics for substrate, product and cell inhibition. *Biotechnol. Bioeng.* **32:**430–437.
55. **Harder, A., and J. A. Roels.** 1982. Application of simple structured models in bioengineering. *Adv. Biochem. Eng.* **21:**51–107.
56. **Heijnen, J. J., A. H. Terwisscha van Scheltinga, and A. J. Straathof.** 1992. Fundamental bottlenecks in the application of continuous bioprocesses. *J. Biotechnol.* **22:**3–20.
57. **Heinisch, J.** 1986. Isolation and characterization of the two structural genes coding for phosphofructokinase in yeast. *Mol. Gen. Genet.* **202:**75–82.
58. **Heinrich, R., and T. A. Rapoport.** 1974. A linear steady-state treatment of enzymatic chains: general properties, control and effector strength. *Eur. J. Biochem.* **42:**89–95.
59. **Hellingwerf, K. J., and W. N. Konings.** 1985. The energy flow in bacteria: the main free energy intermediates and their regulatory role. *Adv. Microb. Physiol.* **26:**125–154.
60. **Hellingwerf, K. J., J. S. Lolkema, R. Otto, O. M. Neijssel, A. H. Stouthamer, W. Harder, K. van Dam, and H. V. Westerhoff.** 1982. Energetics of microbial growth: an analysis of the relationship between growth and its mechanistic basis by mosaic non-equilibrium thermodynamics. *FEMS. Microbiol. Lett.* **15:**7–17.
61. **Hemila, H., L. M. Glode, and I. Palva.** 1989. Production of diphtheria toxin CRM 228 in *Bacillus subtilis*. *FEMS Microbiol. Lett.* **53:**193–198.
62. **Henner, D. J., E. Ferrari, M. Perego, and J. A. Hoch.** 1988. Location of the targets of the *hpr-97*, *sacU32*(Hy), and *sacQ36*(Hy) mutations in upstream regions of the subtilisin promoter. *J. Bacteriol.* **170:**296–300.
63. **Herbert, D.** 1958. Some principles of continuous cultivation, p. 381–396. *In* G. Tunevall (ed.), *Recent Progress in Microbiology*. Almqvist and Wiksell, Stockholm.
64. **Herbert, D.** 1961. The chemical composition of microorganisms as a function of their environment. *Symp. Soc. Gen. Microbiol.* **11:**391–416.
65. **Herbert, D., R. Elsworth, and R. C. Telling.** 1956. The continuous culture of bacteria; a theoretical and experimental study. *J. Gen. Microbiol.* **14:**601–622.
66. **Himanen, J. P., S. Taira, M. Saruas, and K. Runebery-Nyman.** 1990. Expression of pertussis toxin subunit S4 as an intracytoplasmic protein in *Bacillus subtilis*. *Vaccine* **8:**600–604.
67. **Hoch, J. A.** 1991. Genetic analysis in *Bacillus subtilis*. *Methods Enzymol.* **204:**305–320.
68. **Honjo, M., A. Akaoka, A. Nakayama, and Y. Furutani.** 1986. Secretion of human growth hormone in *Bacillus subtilis* using prepropeptide coding region of *Bacillus amyloliquefaciens* neutral protease gene. *J. Biotechnol.* **4:**63–71.
69. **Iijima, S., K. H. Lin, and T. Kobayashi.** 1991. Increased production of cloned β-galactosidase in two-stage culture of *Bacillus amyloliquefaciens*. *J. Ferment. Bioeng.* **71:**69–71.
70. **Jeong, J. W., J. Snay, and M. M. Ataai.** 1990. A mathematical model for examining growth and sporulation processes of *Bacillus subtilis*. *Biotechnol. Bioeng.* **35:** 160–184.
71. **Jöbses, I. M. L., G. T. C. Egberts, A. van Baalen, and J. A. Roels.** 1985. Mathematical modelling of growth and substrate conversion of *Zymomonas mobilis* at 30 and 35°C. *Biotechnol. Bioeng.* **27:**984–995.

72. **Jones, C. W.** 1988. Membrane-associated energy transduction in Bacteria, p. 1–82. *In* C. Anthony (ed.), *Bacterial Energy Transduction*. Academic Press, Inc., New York.
73. **Joshi, A., and B. O. Palsson.** 1988. *Escherichia coli* growth dynamics: a three-pool biochemically based description. *Biotechnol. Bioeng.* **31:**102–116.
74. **Kacser, H.** 1988. Regulation and control of metabolic pathways, p. 1–23. *In* M. J. Bazin and J. I. Prosser (ed.), *Physiological Models in Microbiology*, vol. 1. CRC Press, Inc., Boca Raton, Fla.
75. **Kacser, H., and J. A. Burns.** 1973. The control of flux. *Symp. Soc. Exp. Biol.* **27:**65–104.
76. **Kacser, H., and J. W. Porteous.** 1987. Control of metabolism: what do we have to measure? *Trends Biotechnol.* **12:**5–15.
77. **Kell, D. B., K. van Dam, and H. V. Westerhoff.** 1989. Control analysis of microbial growth and productivity, p. 61–93. *In* S. Baumberg, I. Hunter, and M. Rhodes (ed.), *Microbial Products: New Approaches. 44th Symposium of the Society of General Microbiology*, Cambridge University Press, Cambridge.
78. **Kell, D. B., and H. V. Westerhoff.** 1986. Metabolic control theory: its role in microbiology and biotechnology. *FEMS Microbiol. Rev.* **39:**305–320.
79. **Kell, D. B., and H. V. Westerhoff.** 1986. Towards a rational approach to the optimization of flux in microbial biotransformations. *Trends Biotechnol.* **20:**137–142.
80. **Kell, D. B., and H. V. Westerhoff.** 1990. Control analysis of organized multienzyme systems, p. 273–289. *In* P. A. Srere, M. E. Jones, and C. K. Mathews (ed.), *Structural and Organizational Aspects of Metabolic Regulation.* Wiley-Liss Inc., New York.
81. **Khovreychev, M. P., A. N. Slobodkin, Z. V. Sakharova, and T. P. Blokhina.** 1990. Growth and development of *Bacillus thuringiensis* in multiple stage continuous cultivation. *Mikrobiologiya* **59:**998–1003.
82. **Kleinkauf, H., and H. von Dohren.** 1983. Non-ribosomal peptide formation on multifunctional proteins. *Trends. Biochem. Sci.* **8:**281–283.
83. **Konings, W. N., B. Poolman, and A. J. M. Driessen.** 1992. Can the excretion of metabolites by bacteria be manipulated? *FEMS Microbiol. Rev.* **88:**93–108.
84. **Kono, T., and T. Asai.** 1969. Kinetics of fermentation processes. *Biotechnol. Bioeng.* **11:**293–321.
85. **Kubitschek, H. E., and S. R. Pai.** 1988. Variation in precursor pool size during the division cycle of *Escherichia coli*: further evidence for linear cell growth. *J. Bacteriol.* **170:**431–435.
86. **Kubo, M., and T. Imanaka.** 1989. mRNA secondary structure in an open reading frame reduces translation efficiency in *Bacillus subtilis. J. Bacteriol.* **171:**4080–4082.
87. **Kunkel, B.** 1991. Compartmentalized gene expression during sporulation in *Bacillus subtilis. Trends Genet.* **7:**167–173.
88. **Lan Wong, S., F. Kawamura, and R. Doi.** 1986. Use of the *Bacillus subtilis* signal peptide for efficient secretion of TEM β-lactamase during growth. *J. Bacteriol.* **168:** 1005–1009.
89. **Lecadet, M. M., J. Chaufaux, J. Ribier, and D. Lereclus.** 1992. Construction of novel *Bacillus thuringiensis* strains with different insecticidal activities by transduction and transformation. *Appl. Environ. Microbiol.* **58:**840–849.
90. **Lee, J., and W. F. Ramirez.** 1992. Mathematic modelling of induced foreign protein production by recombinant bacteria. *Biotechnol. Bioeng.* **39:**635–646.
91. **Levenspiel, O.** 1980. The Monod equation: a revisit and a generalization to product inhibition situations. *Biotechnol. Bioeng.* **22:**1671–1687.
92. **Linton, J. D.** 1991. Metabolite overproduction and growth efficiency. *Antonie van Leeuwenhoek Int. J. Gen. Mol. Microbiol.* **60:**293–311.
93. **Linton, J. D., and A. J. Rye.** 1989. The relationship between the energetic efficiency in different microorganisms and the rate of metabolite overproduced. *J. Ind. Microbiol.* **4:**85–96.
94. **Linton, J. D., and R. J. Stephenson.** 1978. A preliminary study on growth yields in relation to the carbon and energy content of various organic growth substrates. *FEMS Microbiol. Lett.* **3:**95–98.
95. **Losick, R., and P. Stragier.** 1992. Crisscross regulation of cell-type-specific gene expression during development in *B. subtilis. Nature* (London) **355:**601–604.
96. **Luedeking, R., and E. L. Piret.** 1959. A kinetic study of the lactic acid fermentation. *J. Biochem. Microb. Technol. Eng.* **1:**393–412.
97. **Lundstrom, K., I. Palva, L. Kaariainen, H. Garoff, M. Saruas, and R. Pettersson.** 1985. Secretion of Semliki-Forest virus membrane glycoprotein F-1 from *Bacillus subtilis. Virus Res.* **2:**69–83.
98. **Magasanik, B., and F. C. Neidhardt.** 1987. Regulation of carbon and nitrogen utilization, p. 1318–1325. *In* F. C. Neidhardt, J. L. Ingraham, K. B. Low, M. Schaechter, and H. E. Umbarger (ed.), *Escherichia coli and Salmonella typhimurium: Cellular and Molecular Biology*, vol 2. American Society for Microbiology, Washington, D.C.
99. **Málek, I.** 1958. The physiological state of microorganisms during continuous culture, p. 11–28. *In* Continuous Cultivation of Microorganisms: a Symposium. Academia Publishing House of the Czechoslovak Academy of Science, Prague.
100. **Mandelstam, J., K. McQuillen, and I. Davies.** 1982. *Biochemistry of Bacterial Growth*. Blackwell, Oxford.
101. **McDuffie, N. G.** 1991. *Bioreactor Design Fundamentals*. Butterworth-Heinemann, Stoneham, Mass.
102. **Miller, J., S. Kovacevic, and L. Veal.** 1987. Secretion and processing of staphylococcal nuclease by *Bacillus subtilis. J. Bacteriol.* **169:**3508–3514.
103. **Minkevich, I. G., and V. K. Eroshin.** 1973. Productivity and heat generation of fermentation under oxygen limitation. *Folia Microbiol.* **18:**376–386.
104. **Monaghan, R., and L. Koupal.** 1989. Use of the Plackett and Burman technique in a discovery program for new natural products, p. 94–116. *In* A. L. Demain (ed.), *Novel Microbial Products for Medicine and Agriculture* Society for Industrial Microbiology, Arlington, Va.
105. **Monod, J.** 1942. Recherches sur la croissance des cultures bactérienne. Hermann & Cie, Paris.
106. **Monod, J.** 1950. La technique de culture continue. Théorie et applications. *Ann. Inst. Pasteur* **79:**390–410.
107. **Moser, A.** 1985. Imperfectly mixed bioreactor systems, p. 77–98. *In* C. L. Cooney and A. E. Humphrey (ed.), *Comprehensive Biotechnology; the Principles, Applications and Regulations of Biotechnology in Industry, Agriculture and Medicine*, vol. 2. Pergamon Press, Oxford.
108. **Mulder, M. M., H. M. L. van der Gulden, P. W. Postma, and K. van Dam.** 1988. Effect of macromolecular composition of microorganisms on the thermodynamic description of their growth. *Biochim. Biophys. Acta* **936:** 406–412.
109. **Nagarajan, V.** 1990. System for secretion of heterologous proteins in *Bacillus subtilis. Methods Enzymol.* **185:**214–223.
110. **Nagarajan, V., and M. Chen.** 1991. The role of precursor conformation on protein secretion in Bacillus subtilis. Abstract 6th Int. Conf. Bacilli, Stanford, Calif.
111. **Nakazawa, K., H. Sasamoto, Y. Shiraki, S. Harada, K. Yanagi, and K. Yamane.** 1991. Extracellular production of mouse interferon beta by the *Bacillus subtilis* alpha-amylase secretion vectors: antiviral activity and deduced NH_2-terminal amino acid sequences of the secreted proteins. *Intervirology* **32:**216–227.

112. **Neijssel, O. M., and D. W. Tempest.** 1976. The role of energy-spilling reactions in the growth of *Klebsiella aerogenes* NCTC418 in aerobic chemostat culture. *Arch. Microbiol.* **110:**305–311.
113. **Neijssel, O. M., and D. W. Tempest.** 1976. Bioenergetic aspects of aerobic growth of *Klebsiella aerogenes* NCTC 418 in carbon-limited and carbon-sufficient chemostat cultures. *Arch. Microbiol.* **107:**215–221.
114. **Neijssel, O. M., and D. W. Tempest.** 1979. The physiology of metabolite overproduction. *Symp. Soc. Gen. Microbiol.* **29:**53–82.
115. **Nicholson, W. L., and G. H. Chambliss.** 1985. Isolation and characterization of a *cis*-acting mutation conferring catabolite repression resistance to alpha-amylase synthesis in *Bacillus subtilis. J. Bacteriol.* **161:**875–881.
116. **Nimmo, H. G., and P. T. W. Cohen.** 1987. Applications of recombinant DNA technology to studies of metabolic regulation. *Biochem. J.* **247:**1–13.
117. **Novick, A., and L. Szilard.** 1950. Experiments with the chemostat on spontaneous mutations of bacteria. *Proc. Natl. Acad. Sci. USA* **36:**708–719.
118. **Novikov, S., I. Borukhov, and A. Strongin.** 1990. *Bacillus amyloliquefaciens* α-amylase signal sequence fused in frame with human proinsulin is properly processed by *Bacillus subtilis* cells. *Biochem. Biophys. Res. Commun.* **169:**297–301.
119. **Ochi, K., J. C. Kandala, and E. Freese.** 1981. Initiation of *Bacillus subtilis* sporulation by the stringent response to amino acid deprivation. *J. Biol. Chem.* **256:**6866–6875.
120. **Ochi, K., J. Kandala, and E. Freese.** 1982. Evidence that *Bacillus subtilis* sporulation induced by the stringent response is caused by the decrease in GTP or GDP. *J. Bacteriol.* **151:**1062–1065.
121. **O'Connor, R., W. H. Elliott, and B. K. May.** 1978. Modulation of an apparent mRNA pool for extracellular protease in *Bacillus amyloliquefaciens. J. Bacteriol.* **136:**24–34.
122. **Overbeeke, N., H. Geertruida, M. Termorshuizen, M. Giuseppin, D. Underwood, and C. Verrips.** 1989. Secretion of the α-galactosidase from *Cyamopsis tetragonoloba* (guar) by *Bacillus subtilis. Appl. Environ. Microbiol.* **56:**193–198.
123. **Owens, J. D., and J. D. Legan.** 1987. Determination of the Monod substrate saturation constant for microbial growth. *FEMS Microbiol. Rev.* **46:**419–432.
124. **Pagni, M., T. Beffa, C. Isch, and M. Aragno.** 1992. Linear growth and poly(β-hydroxybutyrate) synthesis in response to pulse-wise addition of the growth-limiting substrate to steady-state heterotrophic continuous cultures of *Aquaspirillum autotrophicum. J. Gen. Microbiol.* **138:**429–436.
125. **Panikov, N., and S. J. Pirt.** 1978. The effects of cooperativity and growth yield variation on the kinetics of nitrogen or phosphate-limited growth of *Chlorella* in a chemostat culture. *J. Gen. Microbiol.* **108:**295–303.
126. **Papoutsakis, E. T., and C. L. Meyer.** 1985. Equations and calculations of product yields and preferred pathways for butanediol and mixed-acid fermentations. *Biotechnol. Bioeng.* **27:**50–66.
127. **Payne, M. S., and E. N. Jackson.** 1991. Use of alkaline phosphatase fusions to study protein secretion in *Bacillus subtilis. J. Bacteriol.* **173:**2278–2282.
128. **Payne, W. J.** 1970. Energy yields and growth of heterotrophs. *Annu. Rev. Microbiol.* **24:**17–52.
129. **Perego, M., C. F. Higgins, S. R. Pearce, M. P. Gallagher, and J. A. Hoch.** 1991. The oligopeptide transport system of *Bacillus subtilis* plays a role in the initiation of sporulation. *Mol. Microbiol.* **5:**173–185.
129a. **Perkins, J. B., J. G. Pero, and A. Sloma.** January 1991. Riboflavin overproducing strains of bacteria. European patent application 90111916.4.
130. **Pierce, J. A., C. R. Robertson, and T. J. Leighton.** Physiological and genetic strategies for enhanced subtilisin production by *Bacillus subtilis. Biotechnol. Prog.*, in press.
131. **Pirt, S. J.** 1965. The maintenance energy of bacteria in growing cultures. *Proc. R. Soc. Lond. Sect. B* **163:**224–231.
132. **Pirt, S. J.** 1975. *Principles of Microbe and Cell Cultivation*. Blackwell, Oxford.
133. **Pirt, S. J.** 1982. Maintenance energy: a general model for energy-limited and energy-sufficient growth. *Arch. Microbiol.* **133:**300–302.
134. **Pirt, S. J.** 1987. The energetics of microbes at slow growth rates: maintenance energies and dormant organisms. *J. Ferment. Technol.* **65:**173–177.
135. **Powell, E. O.** 1967. The growth rate of microorganisms as a function of substrate concentration, p. 34–56. *In Microbial Physiology and Continuous Culture. Proceedings of the 3rd International Symposium*. Her Majesty's Stationary Office, London.
136. **Powell, E. O.** 1969. Transient changes in the growth rate of microorganisms, p. 275–284. *In* I. Málek, K. Beran, Z. Fencl, V. Munk, J. Ricica, and H. Smrcková (ed.), *Continuous Cultivation of Microorganisms. Proceedings of the 4th International Symposium*. Acadamia, Prague.
137. **Priest, F.** 1977. Extracellular enzyme synthesis in the genus *Bacillus. Bacteriol. Rev.* **41:**711–753.
138. **Priest, F. G.** 1989. Products from bacilli, p. 293–315. *In* C. F. Harwood (ed.), *Handbooks of Biotechnology*, vol. 2. *Bacillus*. Plenum Press, New York.
139. **Priest, F. G.** 1992. Biological control of mosquitos and other biting flies by *Bacillus sphaericus* and *Bacillus thuringiensis*—a review. *J. Appl. Bacteriol.* **72:**357–369.
140. **Priest, F. G., and R. J. Sharp.** 1989. Fermentation of bacilli, p. 73–132. *In* J. O. Neway (ed.), *Fermentation Process of Development of Industrial Organisms*. Marcel Dekker, Inc., New York.
141. **Ramkrishna, D., D. S. Kompala, and G. T. Tsao.** 1984. Cybernetic modelling of microbial populations: growth on mixed substrates, p. 241–261. *In* L. K. Doraiswamy and R. A. Mashelkar (ed.), *Frontiers in Chemical Reaction Engineering*. Wiley Eastern, New Delhi, India.
142. **Reda, K. D., and D. R. Omstead.** 1990. Automatic fermentor sampling and stream analysis, p. 73–107. *In* D. R. Omstead (ed.), *Computer Control of Fermentation Processes*. CRC Press, Boca Raton, Fla.
143. **Resnek, O., L. Rutberg, and A. von Gabain.** 1990. Changes in the stability of specific mRNA species in response to growth stage in *Bacillus subtilis. Proc. Natl. Acad. Sci. USA* **87:**8355–8359.
144. **Roels, J. A.** 1980. Application of macroscopic principles to microbial metabolism. *Biotechnol. Bioeng.* **22:**2457–2514.
145. **Roels, J. A.** 1981. The application of macroscopic principles to microbial metabolism. *Ann. N.Y. Acad. Sci.* **369:**113–134.
146. **Ross, R., J. D'Elia, R. Mooney, and W. Chesbro.** 1990. Nutrient limitation of two saccharolytic clostridia: secretion, sporulation, and solventogenesis. *FEMS Microbiol. Ecol.* **74:**153–164.
147. **Rutgers, M.** 1990. Control and thermodynamics of microbial growth. Ph.D. thesis. University of Amsterdam, Amsterdam, The Netherlands.
148. **Rutgers, M., P. A. Balk, and K. van Dam.** 1989. Thermodynamic efficiency of bacterial growth calculated from growth yield of *Pseudomonas oxalaticus* OX1 in the chemostat. *Biochim. Biophys. Acta* **973:**302–307.
149. **Rutgers, M., P. A. Balk, and K. van Dam.** 1990. Quantification of multiple substrate controlled growth—simultaneous ammonium and glucose limitation in chemostat cultures of *Klebsiella pneumoniae. Arch. Microbiol.* **153:**478–484.
150. **Rutgers, M., K. van Dam, and H. V. Westerhoff.** 1991. Control and thermodynamics of microbial growth: rational tools for bioengineering. *Crit. Rev. Biotechnol.* **11:**367–395.

151. **Sari, P., S. Taira, U. Airaksinen, A. Palva, M. Sarvas, and K. Runeberg-Nyman.** 1990. Production and secretion of pertussis toxin subunits in *Bacillus subtilis*. *FEMS Microbiol. Lett.* **56:**143–148.
152. **Saunders, C., B. Schmidt, R. Mallonee, and M. Guyer.** 1987. Secretion of human serum albumin from *Bacillus subtilis*. *J. Bacteriol.* **169:**2917–2925.
153. **Saunders, C. W., J. A. Pedroni, and P. M. Monahan.** 1991. Optimization of the signal-sequence cleavage site for secretion from *Bacillus subtilis* of a 34-amino acid fragment of human parathyroid hormone. *Gene* **102:** 277–282.
154. **Schein, C. H., K. Kashiwagi, A. Fujijawa, and C. Weissmann.** 1986. Secretion of mature interferon alpha-2 and accumulation of uncleaved precursor by *Bacillus subtilis* transformed with a hybrid alpha amylase signal sequence interferon alpha-2 gene. *Bio/Technology* **4:**719–725.
155. **Scheller, F., and F. Schubert.** 1992. *Biosensors*. Elsevier Science Publishing, Amsterdam.
156. **Schügerl, K., and W. Sittig.** 1987. *Bioreactors*, p. 179–224. *In* P. Präve, U. Faust, W. Sittig, and D. A. Sukatsch (ed.), *Fundamentals of Biotechnology*. VCH Verlagsgesellschaft, Weinheim, Germany.
157. **Setlow, P.** 1973. Inability to detect cAMP in vegetative or sporulating cells or dormant spores of *Bacillus megaterium*. *Biochem. Biophys. Res. Commun.* **52:**365–372.
158. **Shu, J., and M. L. Shuler.** 1989. A mathematical model for the growth of a single cell of *E. coli* on a glucose/glutamine/ammonium medium. *Biotechnol. Bioeng.* **33:** 1117–1126.
159. **Shuler, M. L.** 1985. Dynamic modelling of fermentation systems, p. 119–131. *In* C. L. Cooney and A. E. Humphrey (ed.), *Comprehensive Biotechnology: the Principles, Applications and Regulations of Biotechnology in Industry, Agriculture and Medicine*, vol. 1. Pergamon Press, Oxford.
160. **Shuler, M. L., and M. M. Domach.** 1983. Mathematical models of the growth of individual cells. Tools for testing biochemical mechanisms, p. 93–133. *In* H. W. Blanch, E. T. Papoutsakis, and G. Stephanopoulos (ed.), *Foundations of Biochemical Engineering. Kinetics and Thermodynamics in Biological Systems*. ACS symposium series 207. American Chemical Society, Washington, D.C.
161. **Shuyler, M., and W. Forman.** 1984. Alvedar macrophage plasminogen activator. *Exp. Lung Res.* **6:**159–169.
162. **Siegele, D. A., and R. Kolter.** 1992. Life after log. *J. Bacteriol.* **174:**345–348.
163. **Slater, J. H.** 1985. Stoichiometry of microbial growth, p. 189–213. *In* C. L. Cooney and A. E. Humphrey (ed.), *Comprehensive Biotechnology: the Principles, Applications and Regulations of Biotechnology in Industry, Agriculture and Medicine*, vol. 1. Pergamon Press, Oxford.
164. **Sonenshein, A. L.** 1989. Metabolic regulation of sporulation and other stationary-phase phenomena, p. 109–130. *In* I. Smith, R. A. Slepecky, and P. Setlow (ed.), *Regulation of Procaryotic Development*. American Society for Microbiology, Washington, D.C.
165. **Stackebrant, F., and C. F. Woese.** 1981. The evolution of prokaryotes. *Symp. Soc. Gen. Microbiol.* **32:**1–32.
166. **Starzak, M., and R. K. Bajpai.** 1991. A structured model for vegetative growth and sporulation in *Bacillus thuringiensis*. *Appl. Biochem. Biotechnol.* **28/29:**699–718.
167. **Stephanopoulos, G.** 1986. Application of macroscopic balances and bioenergetics of growth to the on-line identification of biological reactors. *Ann. N.Y. Acad. Sci.* **469:**332–349.
168. **Stouthamer, A. H.** 1979. The search for correlation between theoretical and experimental growth yields, p. 1–47. *In* J. R. Quayle (ed.), *International Review of Biochemistry*, vol. 21. *Microbial Biochemistry*. University Park Press, Baltimore.
169. **Stouthamer, A. H., B. A. Bulthuis, and H. W. van Verseveld.** 1990. Energetics of growth at low growth rates and its relevance for the maintenance concept, p. 85–102. *In* R. K. Poole, M. J. Bazin, and W. Keevil (ed.), *Microbial Growth Dynamics*. Society for General Microbiology special publication 28. IRL Press, Oxford.
170. **Stouthamer, A. H., and H. W. van Verseveld.** 1985. Stoichiometry of microbial growth, p. 215–238. *In* C. L. Cooney and A. E. Humphrey (ed.), *Comprehensive Biotechnology: the Principles, Applications and Regulations of Biotechnology in Industry, Agriculture and Medicine*, vol. 1. Pergamon Press, Oxford.
171. **Stouthamer, A. H., and H. W. van Verseveld.** 1987. Microbial energetics should be considered in manipulating metabolism for biotechnological purposes. *Trends Biotechnol.* **5:**149–155.
172. **Stücki, J. W.** 1980. The optimal efficiency and the economic degrees of coupling of oxidative phosphorylation. *Eur. J. Biochem.* **109:**269–283.
173. **Suh, J. W., S. A. Boylan, S. M. Thomas, K. M. Dolan, D. B. Oliver, and C. W. Price.** 1990. Isolation of a secY homologue from *Bacillus subtilis*: evidence for a common protein export pathway in eubacteria. *Mol. Microbiol.* **4:**305–314.
174. **Taira, S., E. Julonen, J. Paton, M. Saruas, and K. Runeberg-Nyman.** 1990. Production of pneumolysin, a pneumococcal toxin, in *Bacillus subtilis*. *Gene* **77:**211–218.
175. **Tempest, D. W., D. Herbert, and P. J. Phipps.** 1967. Studies on the growth of *Aerobacter aerogenes* at low dilution rates in a chemostat, p. 240–254. *In* E. O. Powell, C. G. T. Evans, R. E. Strange, and D. W. Tempest (ed.), *Microbial Physiology and Continuous Culture*. Her Majesty's Stationary Office, London.
176. **Tempest, D. W., and O. M. Nijssel.** 1980. Comparative aspects of microbial growth yields with special reference to C_1 utilizers, p. 325–334. *In* H. Dalton (ed.), *Microbial Growth on C_1 Compounds*. Heyden, London.
177. **Thatipamala, R., S. Rohani, and G. A. Hill.** 1992. Effects of high product and substrate inhibitions on the kinetics and biomass and product yields during ethanol batch fermentation. *Biotechnol. Bioeng.* **40:**289–297.
178. **Tsai, S. P., and Y. H. Lee.** 1990. A model for energy-sufficient culture growth. *Biotechnol. Bioeng.* **35:**138–145.
179. **Turner, B. G., and D. Ramkrishna.** 1988. Revised enzyme synthesis rate expression in cybernetic models of bacterial growth. *Biotechnol. Bioeng.* **31:**41–43.
180. **Turner, B. G., D. Ramkrishna, and N. B. Jansen.** 1989. Cybernetic modelling of bacterial cultures at low growth rates: single substrate systems. *Biotechnol. Bioeng.* **34:**252–261.
181. **Valle, F., and E. Ferrari.** 1989. Subtilisin: a redundantly temporally regulated gene?, p. 131–146. *In* I. Smith, R. A. Slepecky, and P. Setlow (ed.), *Regulation of Procaryotic Development*. American Society for Microbiology, Washington, D.C.
182. **van Dam, K., M. M. Mulder, J. Teixera de Mattos, and H. V. Westerhoff.** 1988. A thermodynamic view of bacterial growth, p. 25–48. *In* M. J. Bazin and J. I. Prosser (ed.), *Physiological Models in Microbiology*, vol. 1. CRC Press, Boca Raton, Fla.
183. **van der Laan, J. C., G. Gerriste, L. J. S. M., Mulliners, R. A. C. Van Der Hoek, and W. J. Quax.** 1991. Cloning, characterization, and multiple chromosomal integration of a *Bacillus* alkaline protease gene. *J. Bacteriol.* **57:**901–909.
184. **van Dijl, J. M., A. de Jong, J. Vehmaanpera, G. Venema, and S. Bron.** 1991. *Bacillus subtilis* signal peptidase I. Abstr. 6th Int. Conf. Bacilli, Stanford, Calif.
185. **van Verseveld, H. W., M. Arbige, and W. R. Chesbro.**

1984. Continuous culture of bacteria with biomass retention. *Trends Biotechnol.* **2:**8–12.

186. **van Verseveld, H. W., J. P. Boon, and A. H. Stouthamer.** 1979. Growth yields and efficiency of oxidative phosphorylation of *Paracoccus denitrificans* during two (carbon) substrate-limited growth. *Arch. Microbiol.* **121:** 213–223.
187. **van Verseveld, H. W., W. R. Chesbro, M. Braster, and A. H. Stouthamer.** 1984. Eubacteria have 3 growth modes keyed to nutrient flow. Consequences for the concept of maintenance and maximal growth yield. *Arch. Microbiol.* **137:**176–184.
188. **van Verseveld, H. W., J. A. de Hollander, J. Frankena, M. Braster, F. J. Leeuwerik, and A. H. Stouthamer.** 1986. Modelling of microbial substrate conversion, growth and product formation in a recycling fermentor. *Antonie van Leeuwenhoek J. Microbiol.* **52:**325–342.
189. **van Verseveld, H. W., and A. H. Stouthamer.** 1978. Growth yields and the efficiency of oxidative phosphorylation during autotrophic growth of *Paracoccus denitrificans* on methanol and formate. *Arch. Microbiol.* **118:**21–26.
190. **Vasantha, N., and D. Filpula.** 1989. Expression of bovine pancreatic ribonuclease A coded by a synthetic gene in *Bacillus subtilis. Gene* **76:**53–60.
191. **Vasantha, N., and L. Thompson.** 1986. Fusion of Pro region of subtilisin to staphylococcal protein A and its secretion of *Bacillus subtilis. Gene* **49:**23–28.
192. **Volkering, F., A. M. Breure, A. Sterkenburg, and J. G. van Andel.** 1992. Microbial degradation of polycyclic aromatic hydrocarbons: effect of substrate availability on bacterial growth kinetics. *Appl. Microbiol. Biotechnol.* **36:**548–552.
193. **von Stockar, U., and L. C. M. Auberson.** 1992. Chemostat cultures of yeasts, continuous culture fundamentals and simple unstructured models. *J. Biotechnol.* **22:**69–88.
194. **Wang, L. F., S. L. Wong, S. G. Lee, N. K. Kalyan, P. Hung, S. Hilliker, and R. H. Doi.** 1988. Expression and secretion of human atrial natriuretic α-factor in *Bacillus subtilis* using the subtilisin signal peptide. *Gene* **69:**39–47.
195. **Wang, N. S., and G. Stephanopoulos.** 1983. Application of macroscopic balances to the identification of gross measurement errors. *Biotechnol. Bioeng.* **25:**2177–2208.
196. **Wanner, U., and T. Egli.** 1990. Dynamics of microbial growth and cell composition in batch culture. *FEMS Microbiol. Rev.* **75:**19–44.
197. **Weickert, M. J., L. Larson, W. L. Nicholson, and G. H. Chambliss.** 1990. Negative control of amylase synthesis: mutations which eliminate catabolite repression or temporal turn-off, p. 237–244. *In* M. M. Zukowski, A. T. Ganesan, and J. A. Hoch (ed.), *Genetics and Biotechnology of the Bacilli.* American Society for Microbiology, Washington, D.C.
198. **Wells, J. A., and D. A. Estell.** 1988. Subtilisin—an enzyme designed to be engineered. *Trends Biol. Sci.* **13:**291–297.
199. **Westerhoff, H. V., K. J. Hellingwerf, and K. van Dam.** 1983. Thermodynamic efficiency of microbial growth is low but optimal for maximal growth rate. *Proc. Natl. Acad. Sci. USA* **80:**305–309.
200. **Westerhoff, H. W., and D. B. Kell.** 1987. Matrix method for determining steps most rate-limiting to metabolic fluxes in biotechnological processes. *Biotechnol. Bioeng.* **30:**101–107.
201. **Westerhoff, H. V., J. S. Lolkema, R. Otto, and K. J. Hellingwerf.** 1982. Nonequilibrium thermodynamics of bacterial growth. The phenomenological and the mosaic approach. *Biochim. Biophys. Acta* **683:**181–220.
202. **Wu, X. C., W. Lee, L. Tran, and S. L. Wong.** 1991. Engineering a *Bacillus subtilis* expression-secretion system with a strain deficient in six extracellular proteases. *J. Bacteriol.* **173:**4952–4958.
203. **Yamagata, H., K. Nakahama, Y. Suzuki, A. Kakinuma, N. Tsukayoshi, and S. Udaka.** 1989. Use of *Bacillus*—brews for efficient synthesis and secretion of human epidermal growth factor. *Proc. Natl. Acad. Sci. USA* **86:**3589–3593.
204. **Yamanè, T., and S. Shimizu.** 1984. Fed-batch techniques in microbial processes. *Adv. Biochem. Eng. Biotechnol.* **30:**147–194.
205. **Young, T. B., and H. R. Bungay.** 1973. Dynamic analysis of a microbial process: a systems engineering approach. *Biotechnol. Bioeng.* **15:**377–393.
206. **Zukowski, M., and L. Miller.** 1986. Hyperproduction of an intracellular heterologous protein in a sac U mutant of *Bacillus subtilis. Gene* **46:**247–255.

61. Peptide Antibiotics

PETER ZUBER, MICHIKO M. NAKANO, and MOHAMED A. MARAHIEL

A number of gram-positive bacteria that inhabit complex ecological communities such as those within soil and aquatic environments produce an abundance of special compounds that are believed to enhance their survival capabilities (13, 63, 105, 125). We know these compounds primarily through their use in medicine and industry, in which they are valued for their antimetabolic and pharmacological properties (43, 95). They have been termed secondary metabolites (13) and represent an enormously diverse collection of compounds including aminoglycosides, β-lactams, polyketides, and small polypeptides. The term "secondary metabolites" has been used because these compounds are generally thought of as nonessential for growth and proliferation of the producing organism (13). Although this definition may apply to bacteria grown under laboratory conditions, it can be argued that the laboratory provides a flawed glimpse of life in the natural environment, whether it be soil, aquatic, oral, or intestinal. The production of these secondary metabolites may be crucial to the survival of the producer organism in complex ecological systems. Hence, a more recently employed term that has gained acceptance is "special metabolites" (10).

Although a wealth of information has been gained through the massive effort to isolate and characterize natural products endowed with antimicrobial or specific pharmacological activities, these products represent a fraction of the special metabolites produced by microbes in the environment. Very few if any generalities can be made about special metabolites as a whole with regard to structure or function (148). Those composed in part of short polypeptides are catagorized as peptide antibiotics, an abundant class of special metabolites produced by many microbial species including gram-positive bacteria (63, 70–72, 105). Many special metabolites are produced under conditions of nutritional stress; a common observation is the accumulation of special metabolites in stationary-phase cultures (63, 125). It is believed that their function under these conditions is to help the producing organism compete for limited resources (63, 148). In some cases, they are believed to induce developmental pathways that cause bacteria to differentiate to a cell type impervious to environmental insults (46, 105, 155). One intriguing hypothesis holds that in the ancient world, special metabolites were cofactors involved in fundamental cellular processes such as transcription and translation (18). They now recognize the same targets but exert antagonistic effects on the processes they once facilitated. However, those who have pondered the functions of special metabolites have concluded that no single purpose can be assigned to their production by microorganisms (148). Efforts at characterizing antimicrobial agents, although of great benefit to medicine, industry, and biological research, probably provide us with a myopic view of the primary function of special metabolites.

This review focuses on the peptide antibiotics, as these are the predominant class of *Bacillus* special metabolite that has been characterized biochemically and by the methods of molecular biology and genetics (72, 91, 105). Although the peptide antibiotics are composed of amino acids, they often show little similarity to gene-encoded polypeptides in terms of structure and mechanism of their biosynthesis. Some are gene encoded and synthesized ribosomally (57), but these often undergo posttranslational processing and modifications before being exported out of the producing cell. Those produced nonribosomally are composed of between 2 and 20 amino acids organized in a linear, cyclic, or branched cyclic structure (71, 72). The constituent amino acids of peptide antibiotics often undergo extensive modifications, including N-methylation, acylation, glycosylation, racemization from L to D forms, and covalent linkage to a variety of functional groups including nucleosides. The amino acids can be linked to each other by peptide bonds or through the formation of lactones and esters. In some cases, they contain amino and hydroxy acids linked by alternating peptide and ester bonds, an arrangement found in a class of peptide antibiotics called depsipeptides (70). In all, approximately 300 different constituent compounds can be activated and incorporated into special metabolites (72). This accounts for the enormous diversity of peptide special metabolites produced by microorganisms. Table 1 is a compilation of peptide antibiotics produced by *Bacillus* spp. and other gram-positive bacteria.

RIBOSOMAL SYNTHESIS: THE LANTIBIOTICS

Peptide antibiotics are synthesized by one of two mechanisms, ribosomal and nonribosomal. Many antimicrobial polypeptides, such as the colicins and the microcins produced by gram-negative bacteria, are plasmid encoded and synthesized ribosomally. Of the ribosomally synthesized antibiotics produced by gram-positive bacteria, the lantibiotics (57, 128) have been studied in some detail in terms of structure and the biochemistry of their synthesis. These polypeptides are synthesized as precursors that undergo ex-

Peter Zuber and Michiko M. Nakano • Department of Biochemistry and Molecular Biology, Louisiana State University Medical Center, Shreveport, Louisiana 71130-3932. **Mohamed A. Marahiel** • FB Chemie, Philips-Universität Marburg, Marburg, Germany.

Table 1. Peptide antibiotics produced by gram-positive bacteria, producing organisms, and brief descriptions of structure, function, and uses[a]

Antibiotic	Organism(s)	Structure	Properties
A19009	*Streptomyces collinus*	Acyldipeptide	Antifungal
A21978	*Streptomyces roseosporus*	Acylpeptidolactone	Antibacterial, feed additive
Actinomycin	*Streptomyces antibioticus* *Streptomyces chrysomallus*	Peptidolactone 2(5*n*)-Phenoxaninone	Antibacterial, antitumor Antiviral, clinically used, DNA intercalator
Alboleutin	*Bacillus subtilis*		Antifungal, used in agriculture
Albomycin	*Streptomyces griseus*	Modified, cyclic (5*n*)	Iron binding, membrane active, clinically used
Althiomycin	*Streptomyces althioticus*	Modified tetrapeptide	Antibacterial, thiostrepton analog
Amphomycin	*Streptomyces canus*	Branched acylpeptide (11*n*)	Antibacterial, feed additive, cell wall synthesis inhibitor
Ancovenin	*Streptomyces* spp.	Lantibiotic (19*n*)	Phospholipase inhibitor
Antipain	*Streptomyces michiganensis*, *Streptomyces yokosukanensis*	Modified tetrapeptide	Proteinase inhibitor
Antrimycin	*Streptomyces xanthocidicus*, *Streptomyces cirratus*	Modified peptide (7*n*)	Antibacterial, antimycobacterial
Bacillomycin	*Bacillus subtilis*	Cyclic (8*n*)	Antifungal
Bacilysin	*Bacillus subtilis*	Dipeptide	Antibacterial, antifungal
Bacitracin	*Bacillus licheniformis*	Branched cyclic (12*n*)	Antibacterial, topical antibiotic, metal ion binding, membrane acting, cell wall synthesis inhibitor
BBM-928	*Actinomyces* sp. GU55-101	Acylpeptide (6*n*)	Antibacterial, antitumor, DNA-binding agent
Berninamycin	*Streptomyces berniensis*	Branched cyclic (13*n*)	Antibacterial, translation inhibitor, binds to ribosome
Bestatin	*Streptomyces olivoreticuli*	Dipeptide	Immunomodulator used in cancer therapy, proteinase inhibitor
Bicyclomycin	*Streptomyces sapporoensis*, *Streptomyces trabensis*	Modified cyclic dipeptide	Antibacterial, clinically used, feed additive, effects lipoprotein synthesis
Blasticidin S	*Streptomyces griseochromogenes*, *Streptomyces globifer*, *Streptomyces morookaensis*	Modified dipeptide	Antifungal, antitumor, used in agriculture
Bleomycin	*Streptomyces verticillus*	Modified acylpeptide (8*n*)	Broad-spectrum antibiotic, DNA scission
Botrycidin AJ1316	*Bacillus subtilis*	Polypeptide (62*n*)	Gene encoded, antifungal
Bottromycin	*Streptomyces bottropensis*	Modified branched cyclic (7*n*)	Antibacterial, translation inhibitor
Brevistin	*Bacillus brevis*	Acylcyclic peptidolactone (11*n*)	Antibacterial
Cairomycin	*Streptomyces* sp.	Cyclic dipeptide	Antibacterial
Cerexins	*Bacillus cereus*	Acylpeptide (10*n*)	Antibacterial
Chymostatin	*Streptomyces hygroscopicus*	Modified peptide (4*n*)	Chymotrypsin inhibitor
Chlorotetain	*Bacillus subtilis*	Dipeptide	Antifungal
Cinnamycin	*Streptoverticillium griseoverticillatum*	Lantibiotic (19*n*)	Phospholipase inhibitor
Cycloheptamycin	*Streptomyces* sp.	Modified peptidolactone (7*n*)	Antibacterial, antimycobacterial
Distamycin	*Streptomyces distallicus*	Modified (3*n*)	Broad spectrum, DNA binding, transcription inhibitor, topical antiviral agent
Duramycins	*Streptoverticillium* spp.	Lantibiotic (19*n*)	Phospholipase inhibitor
Edeines	*Bacillus brevis* Vm4	Modified (5*n*)	Antibacterial, antitumor, translation inhibitor, nucleic acid binding
Elastatinal	*Streptomyces griseoruber*	Modified (4*n*)	Elastase inhibitor
EM 49	*Bacillus circulans*	Cyclic acylated (8*n*)	Antibacterial
Enduracidin	*Streptomyces fungicidicus*	Acylated lactone (17*n*)	Antibacterial, antimycobacterial, clinically used, feed additive, cell wall synthesis inhibitor
Epidermin	*Staphylococcus epidermidis*	Antibiotic (22*n*)	Antibacterial
Esperin	*Bacillus mesentericus*	Lactone (8*n*)	Antimycobacterial
Etamycin	*Streptomyces griseus*	Modified acylated lactone (7*n*)	Antibacterial, antimycobacterial, broad spectrum, metal ion binding, membrane active
Fengycin	*Bacillus subtilis* F-29-3	Modified acylpeptide (10*n*)	Antifungal
Ferramidochloromycin	*Streptomyces* sp.		Antibacterial, metal ion binding

(Continued)

Table 1—*Continued*

Antibiotic	Organism(s)	Structure	Properties
Ferrimycin	*Streptomyces griseoflorus, Streptomyces galilaeus, Streptomyces lavendulae*		Antibacterial, metal ion binding
Ficellomycin	*Streptomyces ficellus*	Modified dipeptide	Antibacterial, antifungal, nucleic acid binding
Fk-156	*Streptomyces violaceus*	Acylpeptide (4*n*)	Immunomodulator
Gallidermin	*Staphylococcus gallinarum*	Lantibiotic (34*n*)	Antibacterial
Gramicidin	*Bacillus brevis*	Modified (15*n*)	Antibacterial topical antibiotic, membrane acting, forms ion channels
Gramicidin S	*Bacillus brevis*	Cyclic (10*n*)	Antibacterial, surfactant, nucleotide binding
Griselimycin	*Streptomyces griseus*	Modified acylpeptidolactone	Antibacterial, antimycobacterial
Griseoviridin	*Streptomyces griseoviridus, Streptomyces griseus*	Modified cyclic dipeptide	Antibacterial, antitumor
Ilamycin	*Streptomyces islandicus, Streptomyces insubtus*	Modified cyclic peptide (7*n*)	Antimycobacterial
Iturin	*Bacillus subtilis*	Cyclic lipopeptide (7*n*)	Antifungal, clinically used
Leupeptin	*Streptomyces roseus*	Modified acylpeptide (3*n*)	Proteinase inhibitor
MAPI	*Streptomyces nigrescens* WT-27	Tetrapeptide	Proteinase inhibitor
Micrococcin	*Micrococcus* spp., *Bacillus pumilis*	Modified branched cyclic (12*n*)	Antibacterial, antimycobacterial, ribosome binding
Mycosubtilin	*Bacillus subtilis, Bacillus subtilis* subsp. *niger*	Cyclic (9*n*)	Antifungal
Negamycin	*Streptomyces purpeofuscus*	Modified dipeptide	Antibacterial, clinically used, causes translation misreading
Nisin	*Lactococcus lactis*	Lantibiotic (34*n*)	Gene encoded, antibacterial, antimycobacterial, food preservative
Nosiheptide	*Streptomyces antibioticus*		Antibacterial, feed additive
Octapeptins	*Bacillus circulans* ATCC 31805	Branched cyclic acylpeptide (8*n*)	Antibacterial, antimycobacterial, antifungal, antiprotozoal
Parvulin	*Streptomyces parvulus, Streptomyces pseudogriseolus*		Antibacterial, feed additive
Pep5	*Staphylococcus epidermidis*	Lantibiotic (34*n*)	
Pepstatin	*Streptomyces testaceus*	Modified acylpeptide (5*n*)	Cell wall synthesis inhibitor
Peptidolipin	*Nocardia asteroides*	Cyclic (8*n*)	Antifungal
Phleomycins	*Streptomyces verticillus*	Modified acylpeptide (8*n*)	Broad spectrum, antiviral, DNA binding
Piperazinedione	*Streptomyces griseoluteus*	Modified cyclic dipeptide	Antiviral, clinically used
Polylysine	*Streptomyces albulus*	25–30*n*	Phage inactivation
Polymyxins	*Bacillus polymyxa*	Branched cyclic acylpeptide (10*n*)	Antibacterial, membrane acting
Polyoxins	*Streptomyces cacaoi*	Modified dipeptide	Antifungal, feed additive
Polypeptins	*Bacillus circulans*	Peptidolactone (10*n*)	Antibacterial, broad spectrum, proteinase inhibitor
Quinomycin	*Streptomyces echinatus, Streptomyces lavendulae, Streptomyces griseolus*	Acylated peptidolactone (8*n*)	DNA binding
Rhizocticins	*Bacillus subtilis*	Dipeptide and tripeptide	Antifungal
Ristocetin	*Nocardia lurida*	Modified (7*n*)	Antibacterial, cell wall synthesis inhibitor
Saramycetin	*Streptomyces saraceticus*		Antifungal
Siomycin	*Streptomyces sioyaensis*	Modified acylpeptide (16*n*)	Antibacterial, feed additive, ribosome binding, translation inhibitor
SS-70A	*Streptomyces olivogriseus*	Modified acylpeptide (8*n*)	Antibacterial, antitumor
Stendomycin	*Streptomyces endus*	Acylated peptidolactone (14*n*)	Antibacterial, antifungal
Streptothricin	*Streptomyces lavendulae, Streptomyces griseus, Streptomyces candidus*	Acylpeptide (1–6*n*)	Antibacterial, antifungal
Subtilin	*Bacillus subtilis*	Lantibiotic (32*n*)	Gene encoded, antibacterial, antitumor, used in agriculture
Surfactin	*Bacillus subtilis*	Acylated cyclic (7*n*)	Antimycobacterial, membrane acting
Telomycin	*Streptomyces canus*	Peptidolactone (11*n*)	Antibacterial
Thiopeptin	*Streptomyces tateyamensis*	Acylated modified (16*n*)	Antibacterial, veterinary drug, feed additive
Thiostrepton	*Streptomyces azureus, Streptomyces laurentii*	Acylated modified (16*n*)	Antibacterial, veterinary drug, translation inhibitor

(Continued)

Table 1—*Continued*

Antibiotic	Organism(s)	Structure	Properties
Tridecaptins	*Bacillus polymyxa*	Acylated (13*n*)	Antibacterial
Triostins	*Streptomyces aureus*	Acylated peptidolactone (8*n*)	Antibacterial, DNA binding
Tuberactino-mycin	*Streptomyces griseoverticillatus-tuberacticus*	Modified branched cyclic (6*n*)	Antimycobacterial, clinically used, translation inhibitor
Tyrocidine	*Bacillus brevis* ATCC 8185	Cyclic (10*n*)	Antibacterial, topical antibiotic, hemolytic
Valinomycin	*Streptomyces fuleissimus, Streptomyces tsusimaensis*	Depsipeptide (12*n*)	Antibacterial, antimycobacterial, antifungal, membrane acting
Vancomycin	*Streptomyces orientalis*	Modified (7*n*)	Antibacterial, antimycobacterial, membrane acting, cell wall synthesis inhibitor
Viomycin	*Streptomyces vinaceus, Streptomyces puniceus, Streptomyces olivoreticula*	Modified branched cyclic (6*n*)	Antibacterial, antimycobacterial
Virginiamycin A	*Streptomyces virginiae, Streptomyces pristiniaespiralis, Streptomyces ostreogriseus*	Modified cyclic depsipeptide (2*n*)	Antibacterial, antitumor feed additive, membrane acting, ribosome binding, translation inhibitor
Virginiamycin B	*See virginiamycin A*	Modified acylpeptidolactone (6*n*)	Antibacterial, ribosome binding, translation inhibitor
Zorbamycin	*Streptomyces bikiniensis*	Modified acylated (8*n*)	Antibacterial, antifungal, antimycobacterial, antitumor

[a] List is derived from a previously published list (70) and is supplemented with compounds described in the text. *n* refers to the number of amino acids in the peptide molecule.

tensive processing and modification, resulting in a mature product 19 to 34 amino acids long. The lantibiotics have gained attention as potentially useful topical antibiotics and food additives (57, 128). In fact, nisin, a lantibiotic produced by *Lactococcus lactis*, is used as a food preservative because of its effectiveness against gram-positive bacteria such as *Clostridium* spp. The type A lantibiotics, which are structurally similar to nisin (34 amino acids; Fig. 1), include subtilin (32 amino acids [7]) and subtilosin (32 amino acids [5]) produced by strains of *Bacillus subtilis*, Pep5 (34 amino acids [60]) and epidermin (22 amino acids, [128]) produced by *Staphylococcus epidermidis*, and gallidermin (22 amino acids [65, 127]) produced by *Staphylococcus gallinarum*. The type B lantibiotics, produced by actinomycetes, are the duramycins (19 amino acids [23, 59]), which have been reported to possess pharmacological activity including inhibition of phospholipases C and A_2 as well as immunomodulating and antiviral properties. The duramycins include cinnamycin (Fig. 1 [59]), duramycins B and C (23), and ancovenin (57).

The lantibiotics are so named because of the presence in them of the unusual amino acids lanthionine and methyl lanthionine (57). Lanthionine formation begins posttranslationally and involves the dehydration of serine or threonine, yielding dehydroalanine and dehydrobutyrine, respectively (Fig. 2). These form thioether linkages with cysteine residues, resulting in the formation of the lanthionine (Ala-s-Ala) and methyl lanthionine (ABA-s-Ala, where ABA is amino butaric acid and s is the thioether cross-link) moieties. The thioether cross-links confer upon the lantibiotic a characteristic heterocyclic structure and extreme thermal stability (83). The duramycins, with the exception of ancovenin, contain lysinoalanine cross-links in addition to lanthionine cross-links.

The mechanism of lanthionine-containing antibiotic biosynthesis has long been thought to involve the ribosome. This hypothesis was based on the observation that protein synthesis inhibitors block the synthesis of nisin in cultures of *L. lactis* (58). Antibody raised against the peptide antibiotic subtilin of *B. subtilis* was used to show that subtilin is synthesized as a prepropeptide (108) that undergoes posttranslational processing resulting in formation of the propeptide, a polypeptide that is the length of the mature product but is incompletely modified. Chemical methods of amino acid sequence determination were used in an attempt to elucidate lantibiotic primary structure (66). This proved to be difficult because of the unusual properties of lantibiotics, such as their high pK_i and the presence in them of the modified amino acids that block the Edman degradation reactions (66). A combination of proteolytic digestion and Edman degradation was used to determine the primary structure of Pep5 (66).

The current strategy for determining the primary structure of the preprolantibiotic is isolation and characterization of the gene encoding the lantibiotic. This has been accomplished by designing oligonucleotides that, according to protein sequence data, should encode part of the lantibiotic polypeptide. The oligonucleotide is then used as a hybridization probe to identify plasmid or phage recombinant clones that contain the lantibiotic gene. Thus, the genes encoding nisin (12, 58), Pep5 (60), gallidermin (127), subtilin (7), cinnamycin (59), and epidermin (128) were isolated. With the lantibiotic gene in hand, it is possible to determine its nucleotide sequence, thereby identifying the open reading frame encoding the lantibiotic. This analysis has revealed that the lantibiotics examined to date are synthesized ribosomally as longer prepropeptides. This fact is supported by the observation (151) that Pep5 prepropeptide accumulates when cultures of *S. epidermidis* are treated with phenylmethylsulfonyl fluoride (PMSF), a finding that also suggests that the processing event generating the propeptide requires the activity of a serine protease. The leader peptide of the preprolantibiotic does not resemble the hydrophobic signal sequences normally found in preproteins targeted to the membrane or

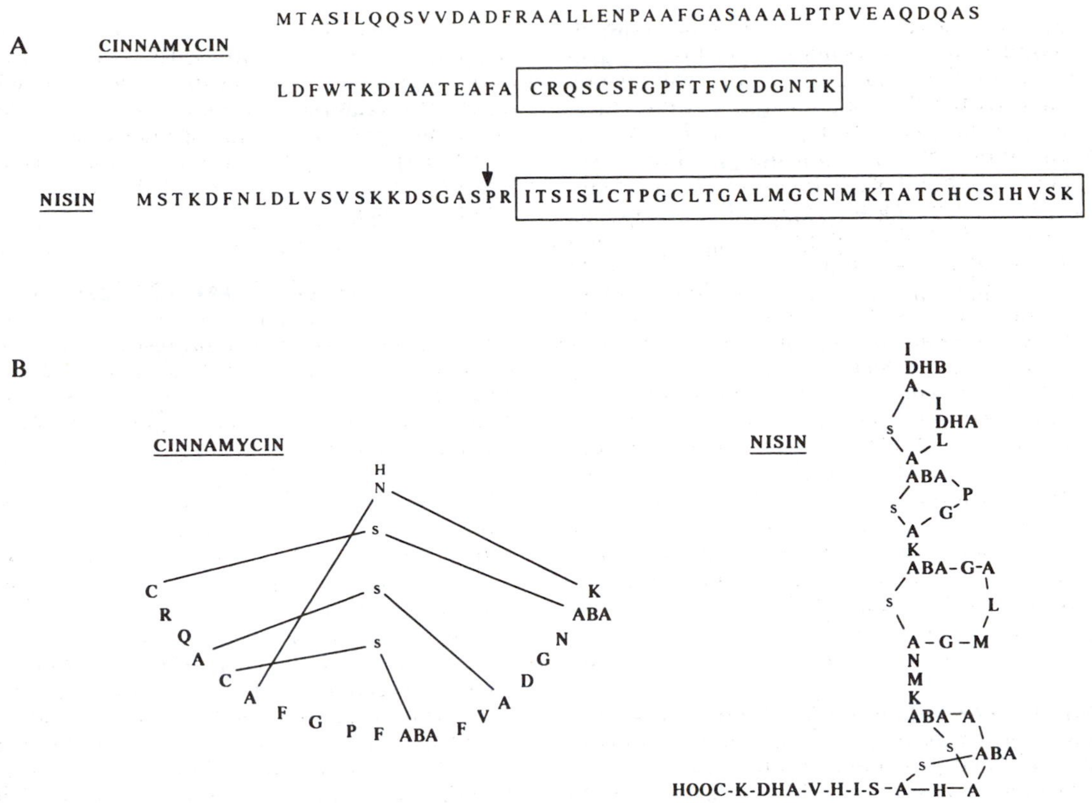

A **CINNAMYCIN** M T A S I L Q Q S V V D A D F R A A L L E N P A A F G A S A A A L P T P V E A Q D Q A S

L D F W T K D I A A T E A F A | C R Q S C S F G P F T F V C D G N T K |

NISIN M S T K D F N L D L V S V S K K D S G A S P R ↓ | I T S I S L C T P G C L T G A L M G C N M K T A T C H C S I H V S K |

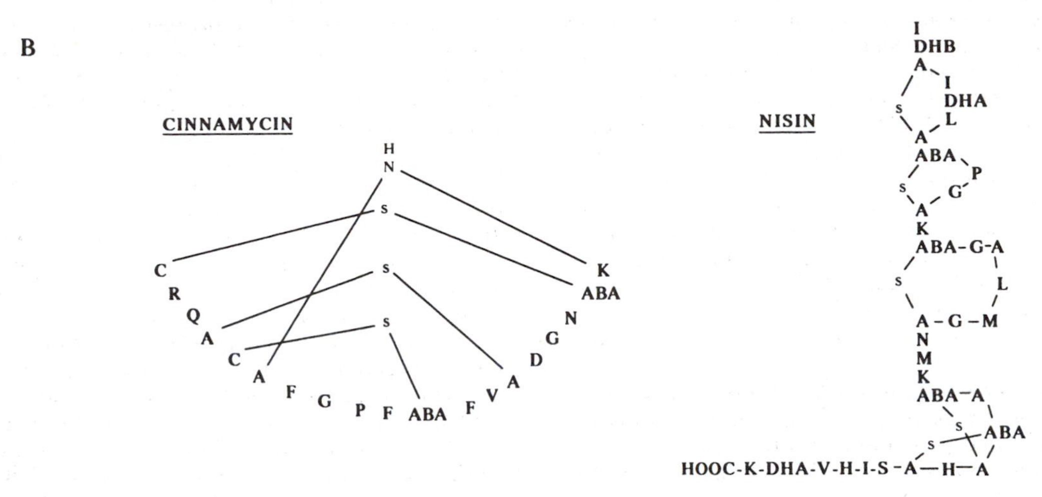

Figure 1. Primary structure of lantibiotics nisin (type A) and cinnamycin (type B). (A) Amino acid sequences of the prepropeptides of cinnamycin and nisin are based on nucleotide sequences of the open reading frames of the nisin and cinnamycin genes. Boxes contain amino acid sequences of mature products. The arrow indicates the conserved proline of the type A lantibiotics. (B) Amino acid sequences of cinnamycin and nisin. s identifies the thioether cross-links in each sequence.

processed for export (Fig. 1). Instead, it contains a number of charged residues and has a net negative charge. Structure prediction analysis identified a propensity to form an α-helix within the leader region of the prepropeptides of nisin, epidermin, and Pep5 (57, 58, 128, 151). The helix is broken by a proline residue

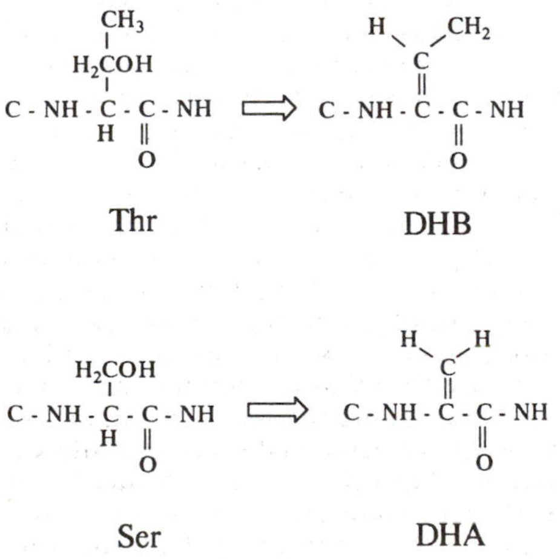

Thr ($C\text{-}NH\text{-}C(H)(H_2COH\text{-}CH_3)\text{-}C(=O)\text{-}NH$) ⟹ DHB ($C\text{-}NH\text{-}C(=C(H)CH_2)\text{-}C(=O)\text{-}NH$)

Ser ($C\text{-}NH\text{-}C(H)(H_2COH)\text{-}C(=O)\text{-}NH$) ⟹ DHA ($C\text{-}NH\text{-}C(=CH_2)\text{-}C(=O)\text{-}NH$)

Figure 2. Dehydration of serine and threonine in lantibiotics which yields dehydroalanine (DHA) and dehydrobutyrine (DHB).

positioned two amino acids from the processing site (Fig. 1), cleavage at which releases the propeptide. This proline is conserved among the lantibiotics except in the case of the duramycin-like lantibiotics (59). Cinnamycin, a 19-amino-acid duramycin-like peptide, is made from a large precursor of 77 amino acids (59). The leader protein in this case is very long (58 amino acids), and there is no proline residue near the processing site, although there are prolines at other positions. As with the prelantibiotics of the A type (such as nisin), there exists within the leader a long stretch of amino acids that can potentially form an α-helix and is followed by a helix-breaking proline. However, this sequence is situated 37 amino acids from the cleavage site, which is preceded by another α-helix. The way the processing mechanism recognizes the prelantibiotic probably differs between the duramycins and the nisin-subtilin group.

A recent report (151) described attempts to determine the stage of expression at which the dehydration of Ser and Thr and the formation of lanthionine take place. The prepropeptide of Pep5 was isolated from phenylmethylsulfonyl fluoride-treated cultures of *S. epidermidis*. Amino acid analysis revealed that there were no lanthionines or methyl lanthionines in the prepropeptide. However, Edman degradation reactions were observed to stop at the first Thr residue of the prolantibiotic sequence, indicating that the Thr had undergone a modification. The primary unmodified translation product was not obtained from phenylmethylsulfonyl fluoride-treated cells. The experiments demonstrated that modification is initiated at the prepropeptide stage, perhaps occurring on the nascent translation product. The absence of lanthionine or methyl lanthionine in the prepropeptide indicates that thioether formation occurs after processing, perhaps when the prolantibiotic is associated with the membrane in preparation for export. The modifications are confined to the Thr and Ser residues of the prolantibiotic sequence within the prepropeptide, although Thr and Ser are present in the leader region. A speculative model to explain the specificity of the dehydration modifications for the C-terminal end of the prepropeptide is that the N-terminal leader is associated with a cellular component that functions in translocation of the lantibiotic and recognizes the characteristic secondary structure of the leader peptide. This association renders the Ser and Thr residues of the leader sequence inaccessible to the mechanism that catalyzes dehydration. The secondary-structure conservation of the leader region of the lantibiotics suggests that the α-helical character may be important in lantibiotic translocation, processing, and export.

Of major interest in lantibiotic research is the complete characterization of the lantibiotic biosynthetic pathway. Although much has been learned recently about lantibiotic structure and biosynthesis, nothing is known about the cellular mechanisms that carry out modifications, formation of the thioether cross-links, and lantibiotic export. With respect to the last, it may be instructive to examine the systems that function in the export of bacteriocins and peptide antibiotics of gram-negative organisms. Colicin V (35) and microcin B17 (26) utilize a transport system whose components are members of the "abc" family of export proteins (48, 144). Thus, *mcbE* and *mcbF* encode components of a "pump" that serves to export microcin B17. McbE is an integral membrane protein, and McbF is an ATP-binding protein that is homologous to a family of prokaryotic and eukaryotic transport proteins (48). Likewise, colicin V export requires the *cvaB* product, which is also a member of the abc family. There is evidence that these dedicated transport systems exist in gram-positive organisms, including *Streptococcus pneumoniae*, which contains the pleiotropic *ami* locus (3). The *spo0K* (109, 121) and *dciA* (92) operons of *B. subtilis* also encode transport proteins of the abc family. It is reasonable to predict that the export of lantibiotics also involves this class of transport proteins.

The precursors of the lantibiotics are encoded by genes located on plasmids and in the chromosome. The *S. epidermidis* gene for preepidermin is located on a 54-kbp plasmid (128), whereas the epidermin homolog, gallidermin, is chromosomally encoded in *S. gallinarum* (127). The ability of *L. lactis* to synthesize the antibiotic nisin was observed to be genetically associated with the ability to utilize sucrose as a sole carbon source (21, 27). There is substantial evidence that the precursors of both nisin and sucrose utilization are encoded by a 70-kbp genetically transmissible transposable element called Tn*5301* (21, 47). Within Tn*5301* and adjacent to the nisin gene is the insertion element IS*904*. However, it is not known whether this element functions in transposition. Curing of the nisin production phenotype results in loss of nisin resistance. By analogy with microcin biosynthesis and immunity (26), one would predict that the nisin export functions may be responsible for resistance and that these functions are encoded by the transmissible genetic element.

NONRIBOSOMAL SYNTHESIS

The biochemistry of multienzyme-catalyzed synthesis of peptide antibiotics has been investigated in detail over the past 30 years (72, 80, 82). From these studies, one model was formulated to describe the mechanism employed by antibiotic-producing microorganisms to carry out peptide synthesis. According to this "multienzyme thiotemplate" model, synthesis is catalyzed by a large multisubunit enzyme complex. This peptide synthetase complex is composed of some of the largest polypeptides produced in living cells, ranging from 100,000 to >600,000 Da. These are known as multienzymes or polyenzymes, since each polypeptide subunit can catalyze a number of biochemical reactions, including activation of the amino acids to the amino acyladenylates, covalent attachment of the amino acid to the subunit through the formation of a carboxyl thiolester, and a number of modification reactions. According to the multienzyme thiotemplate model (Fig. 3), the covalently linked amino acids are organized on the thiotemplate so that they are aligned in the order of their addition to the growing peptide chain. Transpeptidation (peptide bond formation) is facilitated by a 4′-phosphopantetheine cofactor that was thought to be a "swinging arm" that transferred the growing peptide chain from one thiotemplate position to the next. The peptide intermediate is linked to the cofactor by a thiolester that is acted upon presumably by a thioesterase during peptide bond formation. The termination of peptide synthetase-catalyzed synthesis can involve cyclization, dimerization or trimerization of short peptide intermediates, or transfer of the peptide to a functional group such as a phospholipid.

The original impetus to study enzyme-catalyzed peptide synthesis was the notion that this kind of synthesis represented an ancestral form of ribosomal polypeptide synthesis (82). In fact, peptide synthesis of the multienzyme thiotemplate mechanism is unrelated to the ribosomal mechanism and instead has more in common with the mechanism of fatty acid synthesis, since it involves the use of a pantetheine cofactor and possibly one or more thioesterases. The activation of amino acids by adenylation is similar to that carried out by amino acyl tRNA synthetases, but the mechanism is clearly different, as is evident by the fidelity of the adenylation reactions carried out by the two enzymes. Compared with the amino acyl tRNA synthetases, the peptide synthetases exhibit a reduced specificity for substrate amino acids and for ATP (71). The peptide synthetase can carry out the activation of amino acids when 2′-dATP is substituted for ATP, a characteristic that can be exploited during purification of peptide synthetase subunits in order to distinguish them from amino acyl tRNA synthetases (77).

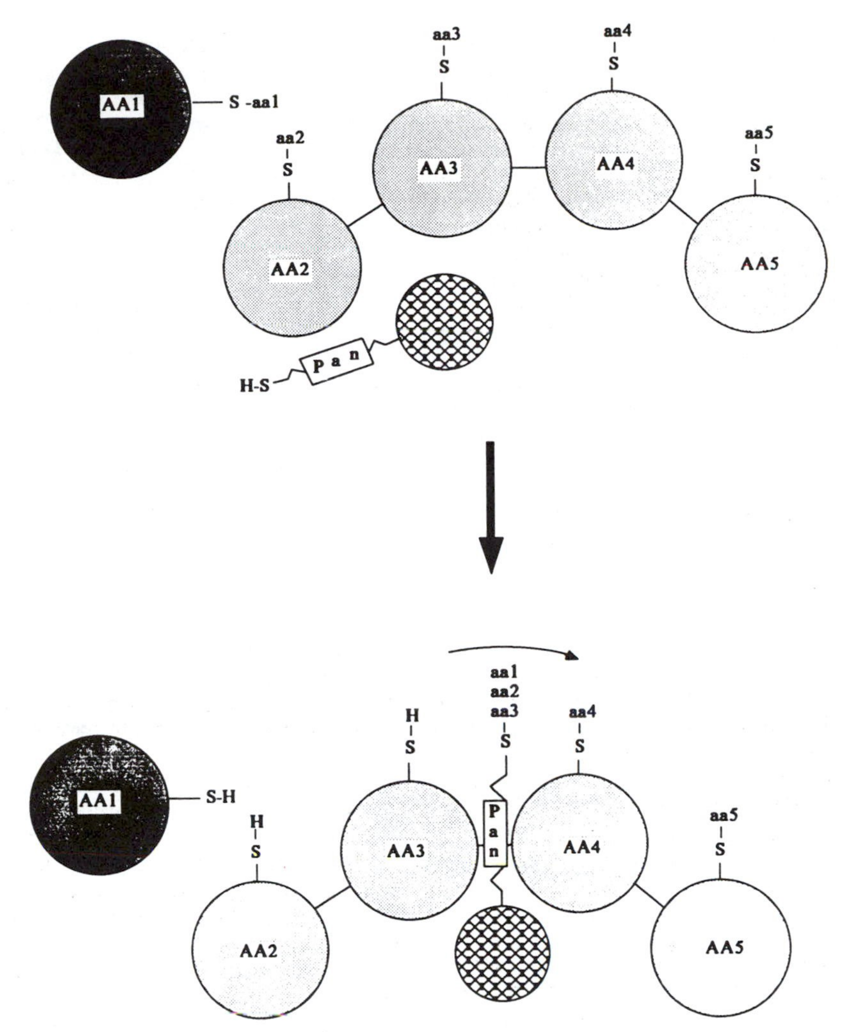

Figure 3. Schematic diagram of the multienzyme thiotemplate mechanism of peptide synthesis. Enzyme domains (AA1, AA2, AA3, AA4, and AA5) that activate and covalently bind to the constituent amino acids (aa1, aa2, aa3, aa4, and aa5) by a thiolester linkage (S) are shown. They are arranged in the order that corresponds to the amino acid sequence of the peptide. Transpeptidation proceeds with the aid of a pantetheine cofactor (pan) attached to an enzyme subunit. The arrow shows the direction in which the cofactor moves in order to transfer the growing peptide chain from one amino acid position to the next.

Indications of reduced substrate specificity and a loosened fidelity of peptide synthesis emerge from structural analysis of the peptide antibiotics. The first phenylalanine of tyrocidine and gramicidin can be replaced with other aromatic amino acids and in some cases by leucine (71). This reduced substrate amino acid specificity exhibited by peptide synthetases can be exploited in an operation called directed biosynthesis (or precursor feed experiment). An excess of precursor analog can be added to cultures of peptide antibiotic-producing organisms, resulting in the desired substitution in 1 to 10% of the purified product (71). However, a much more controlled system of directed biosynthesis has been achieved by using cell extracts of *Bacillus brevis* cells harboring gramicidin S synthetase activity. Amino acid analogs or amino acids different from the constituent amino acid were incorporated at high frequency by supplying the heterologous precursor in millimolar amounts (71). These experiments demonstrate the potential of using a large-scale version of a cell-free synthetase system to manufacture peptides of defined compositions and bioactive properties.

The multienzyme thiotemplate mechanism is not a universal strategy for producing peptide metabolites. For example, the synthesis of glutathione is carried out by the enzymes γ-glutamylcysteine synthetase and glutathione synthetase, which bear little resemblance to the peptide synthetases that catalyze peptide antibiotic synthesis (93). The antifungal agent mycobacillin, produced by *B. subtilis* B3, is also synthesized by a different mechanism. The biosynthesis of both mycobacillin and glutathione involves activation of the constituent amino acids, but an amino acylphosphate is formed instead of an amino acyladenylate. These features of peptide synthesis are detailed below.

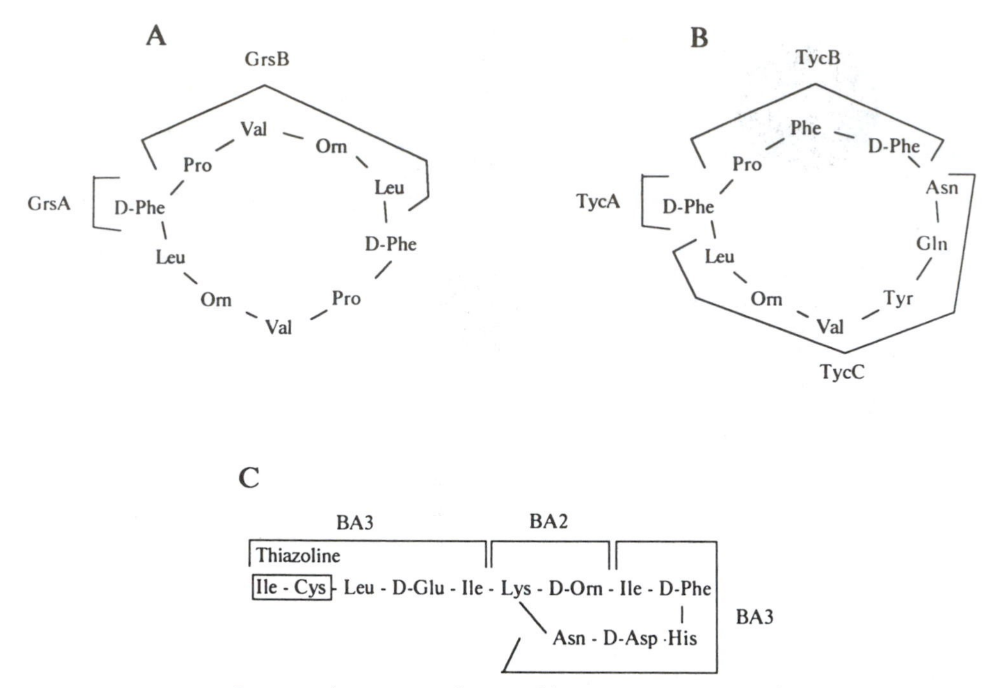

Figure 4. Primary structures of gramicidin S, tyrocidine, and bacitracin. Amino acid sequences and enzymes that catalyze the synthesis of each are shown. The amino acids activated by each enzyme are indicated.

BIOCHEMISTRY AND MOLECULAR BIOLOGY OF PEPTIDE ANTIBIOTIC BIOSYNTHESIS IN *BACILLUS* SPP.

Gramicidin S and Tyrocidine

Much of what is known about peptide antibiotic biosynthesis by the thiotemplate mechanism has come from the study of the cyclic decapeptide gramicidin S (Fig. 4). This decapeptide is composed of two molecules of the pentapeptide D-Phe–Pro–Val–Orn–Leu linked in a head-to-tail fashion. Gramicidin S synthetase is a complex of two peptide synthetase subunits, gramicidin S synthetase 1 (GrsA), which activates and racemizes L-Phe to D-Phe, and gramicidin S synthetase 2 (GrsB), which activates Pro, Val, Orn, and Leu. The genes that encode the two enzymes have been isolated, and their nucleotide sequences have been determined (76, 145), resulting in an accurate estimation of their molecular weights (126,663 Da for GrsA and 510,287 Da for GrsB). Earlier work that focused on the reactions catalyzed in cell extract from *B. brevis* cells established the first steps of peptide synthesis, i.e., activation of the constituent amino acids followed by their covalent attachment to a protein thiotemplate by a thiolester (28, 29, 61, 143). Purification of the gramicidin S synthetase complex is facilitated by an assay of amino acid-charging activity called the ATP-PP_i exchange reaction (82). This reaction relies on the reversibility of the amino acid adenylation reaction when $^{32}PP_i$ is added to the cell-free gramicidin S synthetase system, resulting in the production of [^{32}P]ATP. Covalent attachment of constituent amino acids is demonstrated by size exclusion chromatography of an in vitro peptide synthesis reaction using ^{14}C-amino acids as substrates (29, 61, 69, 139). Quantitative release of the amino acids from the peptide synthetase complex is accomplished with alkali, performic acid, or dithiothreitol, which is consistent with the thiotemplate model, in which a carboxyl thioester bond links the constituent amino acid with the peptide synthetase following activation (69, 82). The involvement of a 4′-phosphopantetheine in the transpeptidation reaction was inferred on the basis of isolation of peptide intermediates bound by a thiolester to the cofactor (34, 82). Some mutants defective in gramicidin synthesis produced gramicidin S synthetase devoid of a pantetheine group (44). Cell-free synthesis experiments showed that the mutant Grs was able to activate the constituent amino acids but could not carry out peptide formation.

A common observation from experiments involving cell-free synthesis of gramicidin S was decomposition of the peptide synthetases into subunits of approximately 70,000 Da that retained amino acid activation activity (82). These subunits were thought to be the domains of the peptide synthetase that recognized and activated individual amino acids. Further support of the domain organization of gramicidin synthetase was provided by the isolation and expression in *Escherichia coli* of a segment of the *grsB* gene that encoded the ornithine-activating domain of gramicidin S synthetase 2 (75). More recent attempts at defining the domains of gramicidin S synthetase involve proteolytic dissection of purified GrsB and purification of amino acid-activating fragments (81, 134, 135). Thus, the amino-terminal, proline-activating domain of GrsB was isolated and was shown to accept the D-phenylalanine from GrsA, resulting in production of the N-terminal dipeptide D-Phe–Pro of gramicidin S (81). Interestingly, formation of the dipeptide in vitro by the combined GrsA and GrsB-Pro fragment occurred in the presence of very little detectable pantetheine cofactor, although it was acknowledged that higher concentrations of the cofactor might accelerate dipeptide synthesis. Proteolysis of GrsB was also used to localize the site of amino acylation of the peptide synthetase following activation (126). This was accomplished by purifying the proteolytic product of GrsB that contained a ^{14}C-labeled substrate amino acid. The sequence of the peptide and a comparison of it with the coding sequence of the *grsB* gene showed that the amino acylation site very closely resembled the 4-phosphopantetheine attachment site of fatty acid synthetase acyl carrier proteins. Peptide sequencing also revealed that the site did not contain a cysteine, which could account for the thioester linkage between peptide synthetase and constituent amino acid. Instead, what was observed was a modified serine within the putative pantetheine-binding sequence. This suggested that a pantetheine thiol was required for attachment of the activated amino acid to the peptide synthetase.

The two genes encoding gramicidin S synthetases 1 and 2 (*grsA* and *grsB*) form an operon with an open reading frame (*grsT*) that encodes a putative product that shows striking similarity to thioesterases of fatty acid synthetase complexes (76, 145). The three genes are arranged in the order *grsT-grsA-grsB*, are transcribed from a single promoter, and are immediately followed by a sequence resembling a factor-independent transcription termination site. Through sequence analysis, the *grsB* product was found to be composed of four homologous domains, each approximately 700 amino acids long. These domains are homologous to 700-amino-acid segment of GrsA. Several experiments lent support to the hypothesis that these domains were responsible for activating the constituent amino acids of gramicidin S. A fragment of the *grsB* gene encoding the amino-terminal homologous region of GrsB was expressed in *E. coli*, resulting in synthesis of a proline-activating product (45). The *grsB* fragment expressed in *E. coli* that resulted in ornithine-activating product was found by sequence analysis to encode the third homologous region of GrsB (145). Analysis of the amino acylated peptides of proteolytically cleaved GrsB and comparison with the deduced amino acid sequence of the *grsB* product confirmed that the Leu-activating domain was located at the carboxy-terminal end of GrsB and that the valine-activating domain was the second homologous region (126, 145). The identification of the amino acid-activating domains of GrsB supports the view that the linear arrangement of the domains of *grsB* (Pro-Val-Orn-Leu) determines the order in which the constituent amino acids of gramicidin S are incorporated into the growing peptide chain, a model that is in keeping with the multienzyme thiotemplate mechanism.

Tyrocidine synthetase complex is composed of three enzymes (Fig. 4; 72, 82): TycA, which activates and racemizes the first Phe of the peptide; TycB, which synthesizes the Pro–Phe–D-Phe tripeptide; and TycC, which is responsible for the synthesis of the Asn-Gln-Tyr-Val-Orn-Leu segment. The complex catalyzes cyclization of the linear decapeptide to yield tyrocidine. The mechanism of synthesis is largely the same as that employed by gramicidin S synthetase, judging from biochemical studies and primary structure analysis (150) of *tycA* and *tycB*, which reveal a domain organization of their products very similar to those of GrsA and GrsB. The *tyc* genes are organized as a large operon, *tycA-tycB-tycC* (94), that is transcriptionally induced when cells enter stationary phase, a time when the antibiotic tyrocidine begins to accumulate (90). Studies of the *tyc* operon promoter in *B. subtilis* (25, 90) showed that *tyc* is under the control of the *spo0A-abrB* system (110, 141, 156). The AbrB protein, which interacts with sequences upstream and downstream of the *tyc* promoter (25, 118), controls stationary-phase expression of *tyc*, as *abrB* mutations result in constitutive *tyc* transcription (90).

There is striking similarity between GrsA and TycA at the primary structure level. They also cross-react with the same antiserum and activate and racemize the same constituent amino acid, L-Phe. However, one cannot be substituted for the other in cell-free synthesis of gramicidin S or tyrocidine (82). That the mechanism of synthesis is essentially the same as that of gramicidin S has been shown through ATP-PP_i exchange reaction and through the isolation of pantetheine-bound peptide intermediates.

Bacitracin

The branched cyclic peptide antibiotic bacitracin A (Fig. 4) is produced by *Bacillus licheniformis* (63). It is a dodecapeptide with four amino acids in the D configuration. The compound contains a cyclic hexapeptide and a thiazoline ring. Like gramicidin S and tyrocidine, its synthesis is catalyzed by a large multienzyme complex composed of three subunits (24, 72). Subunit BA1 is a polypeptide of approximately 335,000 Da that activates the linear portion of the peptide, i.e., the amino acids isoleucine, cysteine, glutamic acid, and an additional isoleucine. Subunit BA2 is a protein of 240,000 that activates lysine and ornithine. Subunit BA3 is 380,000 Da and activates the amino acids of the cyclic portion of bacitracin A, i.e., isoleucine, phenylalanine, histidine, aspartic acid, and asparagine.

The observations that the peptide intermediates are covalently bound to bacitracin synthetase and that 4′-phosphopantetheine is associated with the synthetase subunits indicate that the multienzyme thiotemplate mechanism is utilized in the synthesis of bacitracin (24, 120). There are some unique features of bacitracin synthesis. Bacitracin contains a thiazoline ring composed of the isoleucylcysteine dipeptide in the linear portion of the molecule (52). Chemical analysis of the isoleucylcysteine intermediate isolated from preparations of BA1 showed that the thiazoline ring is formed at the dipeptide stage (52). Another unique property of bacitracin synthetase is its substrate recognition, which differs from that of gramicidin S synthetase. D-Amino acids are able to substitute the L isomers to some extent in cell-free bacitracin synthetase systems, whereas only the L isomers of the gramicidin S constituent amino acids are recognized for activation by gramicidin S synthetase.

The genes for the bacitracin A synthetase subunits are clustered and are likely to be organized in an operon like the genes of tyrocidine and gramicidin S synthetases (51). Mutations in *B. licheniformis* that

Iturin A

$CH_3(CH_2)_2$- $(CH_2)_8$ - CH - CH_2 CO - L-Asn - D-Tyr - D-Asn
variable NH L-Ser - D-Asn - L-Pro - L-Gln

Surfactin

CH_3 CH_3 CH - $(CH_2)_{7-9}$ - CH - CH_2 - CO - L-Glu - L-Leu - D-Leu
O L-Leu - D-Leu - L-Asp L-Val

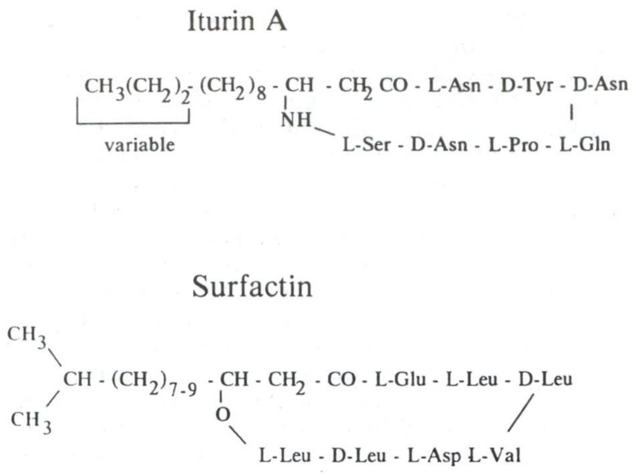

Figure 5. Primary structures of the lipopeptides iturin A and surfactin. The variable region of the iturin β-amino fatty acid can also be $CH_3CH(CH_3)CH_2$-, $CH_3CH_2CH(CH_3)$-, $CH_3CH(CH_3)(CH_2)_2$-, or $CH_3(CH_2)_4$-.

render bacitracin synthesis defective have been isolated (37, 114). One mutation, *bac-54*, causes a defect in covalent attachment of the constituent amino acids to subunits BA2 and BA3, whereas *bac-319* results in an apparent defect in transpeptidation. Another mutation, *bcr-1*, causes sensitivity to bacitracin (115). By analogy to *mcbE* and *mcbF* mutations, which affect microcin B17 resistance in *E. coli*, this mutation could be in genes that function in bacitracin export (26). All mutations reside within a single locus according to SP15 generalized transduction mapping (114, 115).

Linear Gramicidin

The unusual peptide linear gramicidin is produced by the strain of *B. brevis* that produces tyrocidine (9, 72). The membrane-affecting activities of gramicidin are well known, as it forms ion channels in lipid bilayers. The pentadecapeptide is blocked by a formyl group at its amino-terminal end and by an ethanolamine at its carboxy end (9, 78). It is composed of alternating L and D amino acids: fVal–Gly–Ala–D-Val–Val–D-Val–Trp–(D-Leu–Trp)$_3$–ethanolamine. The ion channel-forming capabilities depend in part on the tryptophan residues, according to amino acid substitution studies with synthetic derivatives of gramicidin (17).

Formylation of the amino terminus is necessary for efficient biosynthesis of gramicidin (1). This reaction is carried out by the enzyme subunit that initiates synthesis. The reaction that appends ethanolamine to the carboxy-terminal end occurs at the cytoplasmic membrane and is believed to direct the compound into the membrane (78). The enzymology of linear gramicidin synthesis is not completely characterized, although two of the synthetase subunits have been isolated (72). LG1 is 140,000 to 180,000 Da and catalyzes formylation and formation of the first dipeptide. LG2 is a protein of 350,000 Da that catalyzes elongation of the dipeptide to the heptapeptide. The remaining enzyme(s) has not been purified, but there is evidence of a subunit that activates D-Val, Leu, and Trp. This suggests that there is a single enzyme that catalyzes the synthesis of the repeating (D-Leu–Trp)$_3$ hexapeptide.

Lipopeptides

Several strains of *B. subtilis* produce cyclic lipopeptides. Of those characterized thus far, most are composed of rings of eight members, seven of which are amino acids. The eighth member is a fatty acid moiety of various chain lengths and configurations. The iturin group of peptide antibiotics include iturins A (Fig. 5), C, D, and E; bacillomycins F and L; and mycosubtilin (111–113, 147). All contain a β-amino fatty acid linked by amide bonds to the constituent amino acid residues of the iturin ring. The iturin lipopeptides share a common sequence (β-hydroxy fatty acid–L-Asp–D-Tyr–D-Asn) and show variation at the other four positions.

Surfactin (Fig. 5), which is produced by *B. subtilis* ATCC 21332, contains a β-hydroxy fatty acid (usually a β-hydroxydecanoic acid) linked to a glutamate residue by an amide bond and to the D-Leu residue by an ester linkage (147). Synthesis of the surfactin peptide chain is carried out by the multienzyme thiotemplate mechanism (146, 147). This has been investigated by using a cell-free system from extracts of *B. subtilis* OKB105 (101, 147). The surfactin synthetase complex is thought to be composed of two subunits, one of 280,000 Da that activates Val and one of 600,000 Da that is endowed with Asp-, Glu-, Leu-, and Val-charging activity. What is not known is the mechanism of fatty acid incorporation into the lipopeptide. Studies of the lipopeptide polymyxin provide a clue to the role of the lipid moiety in lipopeptide synthesis (73–75). Polymyxin is a branched cyclic compound containing an isooctanoyl group at its linear end. Initiation of synthesis requires the presence of isooctanoyl coenzyme A (CoA). Indeed, *N*-octanoyl-dehydroaminobutyrine was identified as an intermediate in polymyxin E biosynthesis. By analogy to polymyxin, perhaps the lipid moieties of the iturins and surfactin are incorporated in the very first step of lipopeptide synthesis.

Surfactin is known for its surface-affecting activity and is the most powerful biosurfactant known. It has been described as an antifungal agent and an inhibitor of cyclic-AMP-dependent phosphodiesterase (147). There is evidence that the enzymes that catalyze surfactin synthesis are involved in the regulation of competence development and sporulation in *B. subtilis* (100, 133). Genetic studies utilizing transposon Tn*917* identified genes required for surfactin production (101). Three loci were characterized. *sfp* encodes a product of 224 amino acids with considerable sequence similarity to the putative product of the open reading frame (*orfX*) associated with the *grs* operon of *B. brevis* (99). The function of the *sfp* product or the putative *orfX* product is not known. *srfA* encodes at least some of the enzymes that catalyze surfactin synthesis (100, 104). Nucleotide sequence analysis shows significant homology to the peptide synthetases TycA and GrsA (100, 133). Amino acid sequence analysis of one of the subunits that catalyze surfactin synthesis in vitro showed that it is encoded by the first open reading frame of *srfA*. A mutation called *srfB* (103) is a large deletion that results in loss of the early competence genes *comQ*, *comP*, and *comA* (22, 152–154). *comP* and *comA* are thought to be two-component regulatory partners (2) that activate the competence developmental pathway. *comP* encodes a putative membrane-bound histidine protein kinase, and *comA* encodes a response regulator protein that when activated is believed to be a positive transcriptional regulator. Under conditions that promote competence development, the transcription of *srfA* requires ComQ, ComP, and ComA, where the ComA is thought to act as a positive transcriptional regulator of *srfA* (22, 102, 103, 107). Two regions of dyad symmetry that lie upstream of the *srfA* promoter and are required for *srfA-lacZ* expression have been identified (102, 107). Mutational analysis suggests that these are the binding sites for a ComA-dependent factor, probably the ComA protein itself. A proposed model holds that the ComA protein interacts as a tetramer, with one dimer at each dyad symmetry sequence, in order to activate *srfA* transcription (107).

Several mutations have been identified in *srfA*. Four are Tn*917* mutations: *comL*::Tn*917* (133), *csh-293*::Tn*917* (53), *srfA*::Tn*917* OK120 (101), and *srfA*::Tn*917* OK223 (100). All four mutations block the production of surfactin, but *comL* and *csh-293*, which are located in the 5′ half of *srfA*, also inhibit competence development and efficient sporulation. This fact, together with the findings that *srfA* transcription requires ComA and that *srfA* is required for the expression of late competence genes (22), suggests that *srfA* occupies an intermediate position in the regulatory pathway that controls the development of competence. This possibility was confirmed by experiments in which an inducible promoter was used to control *srfA* transcription, thus rendering *srfA* transcription and competence independent of ComA (39, 106). The *srfA* product(s) functions in the synthesis of a lipopeptide and in the regulation of a process of cell specialization (competence development). It has been proposed that SrfA proteins catalyze the synthesis of a peptide, perhaps a precursor of surfactin, that is a signal that regulates expression of competence genes and, in an unknown way, facilitates the sporulation process (100).

Dipeptides and Tripeptides

A number of small peptides composed of unusual amino acid derivatives are produced by *Bacillus* species. Bacilysin (Fig. 6) was isolated from fluid of *B. subtilis* Marburg 168 cultures as a substance with bactericidal activity against cells of *Staphylococcus aureus* (119). It was later found to possess activity against a wide range of organisms, including *Candida albicans*. Bacilysin is composed of alanine and anticapsin (Fig. 5), an amino acid derived from the aromatic amino acid intermediate prephenate (41), as predicted from the structural studies of Walker and Abraham (149). That prephenate is a precursor of anticapsin was shown by determining the levels of bacilysin in aromatic amino acid auxotrophs. *tyr* and *phe* mutants produced bacilysin, but little was detected in *aroG* mutants that lack chorismate mutase. It is not known how the epoxyhexanone ring of anticapsin is formed from prephenate. Bacilysin synthesis

Chlorotetain

Bacilysin

Rhizocticin A

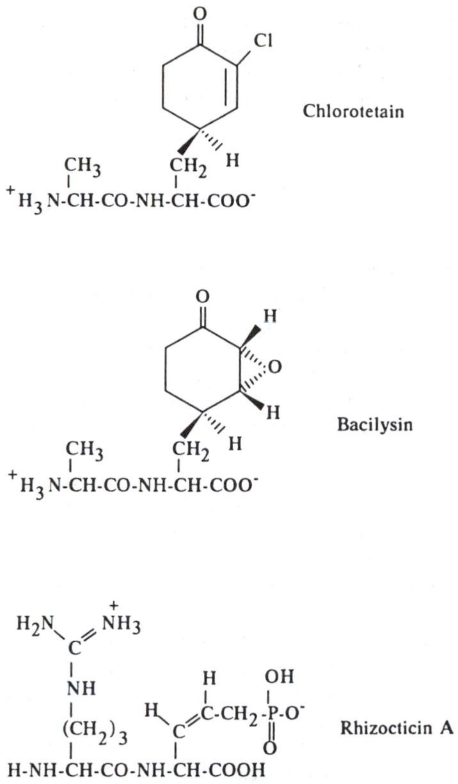

Figure 6. Dipeptide antibiotics produced by *B. subtilis*.

activity has been detected in a cell-free system, but the enzyme(s) that carries out the reaction is uncharacterized (123). A mutation, *bac-1*, located in the vicinity of *ctr* and *sacA* at 11 o'clock on the *B. subtilis* chromosome that renders *B. subtilis* cells unable to synthesize bacilysin has been identified (42). The mutant is not an aromatic amino acid auxotroph, nor does it exhibit pleiotropic behavior like that of a *spo* mutation. The mutation could identify a gene encoding bacilysin synthetase or an enzyme subunit.

The target of its action is glucosamine-6-phosphate synthetase of bacteria and yeast cells; it is thus a potent inhibitor of cell wall synthesis. Anticapsin alone shows activity towards *E. coli*, but the dipeptide is more effective (68). This is believed to be due to transport of the bacilysin through the oligopeptide and dipeptide transport systems, an idea that is supported by the observation that peptides antagonize bacilysin activity.

New antibiotics have been recently isolated from "old" *B. subtilis* strains. These include the rhizocticins, which are di- and tripeptides with activity against several plant pathogenic fungi and *Caenorhabditis elegans* (Fig. 5). Rhizocticins (Fig. 6; 79) are produced by *B. subtilis* ATCC 6633, which is known to produce subtilin. All rhizocticins contain arginine and L-2-amino-5-phosphono-3-*cis*-pentenoic acid (L-APPA). Chlorotetain (Fig. 6; 117) is a recently identified bacilysinlike compound that possesses an unusual chlorine-containing amino acid (3′-chloro-4′-oxo-2′-cyclohexenyl)alanine. Like bacilysin, the rhizocticins and chlorotetain are members of the so-called "missile" and "warhead" antibiotics. The anticapsin of bacilysin and the L-APPA of the rhizocticins are the toxic components of the antibiotics, but they require the α-amino acid portions of the antibiotics for delivery into susceptible organisms via peptide transport systems.

Mycobacillin

Cells of *B. subtilis* B_3 produce a membrane-affecting antifungal agent called mycobacillin (89). It is a cyclic tridecapeptide (L-Pro–D-Asp–D-Glu–L-Tyr–L-Asp–L-Tyr–L-Ser–D-Asp–L-Leu–D-Glu–D-Asp–L-Ala–D-Asp) whose synthesis is catalyzed by a large enzyme complex (31–33, 97). In many ways, the synthesis of mycobacillin is different from that of gramicidin S and other peptide antibiotics that utilize the multienzyme thiotemplate mechanism. Three enzyme fractions that catalyze mycobacillin synthesis have been isolated from *B. subtilis* cells. Each fraction appears to contain a single enzyme polypeptide. Fraction A contains a protein of 252,000 Da that catalyzes polymerization of the first pentapeptide (Pro-Asp-Glu-Tyr-Asp). Fraction B (198,000 Da) binds to the pentapeptide and sequentially adds the next four amino acids, thereby forming a nonapeptide (Pro-Asp-Glu-Tyr-Asp-Tyr-Ser-Asp-Leu). Fraction C (108,000 Da) catalyzes the addition of the last four amino acids, resulting in formation of the finished product. The enzymes in isolation do not activate the mycobacillin constituent amino acids unless they are provided with the cognate mycobacillin precursor (30). Thus, fraction A will activate the first five amino acids only if provided with L-Pro, the first amino acid of the mycobacillin biosynthetic pathway. Fraction B will activate amino acids 6 through 9 if provided with the substrate pentapeptide. Fraction C will activate amino acids 10 through 13 if provided with its nonapeptide substrate. This situation is in contrast to that with gramicidin S synthetase 2, which can activate Pro, Val, Orn, and Leu in the absence of the GrsA–D-Phe complex. Activation of mycobacillin constituent amino acids results in formation of amino acylphosphates, thus resembling the activation of glutamate and cysteine during glutathione biosynthesis (33). Mycobacillin-negative mutants of *B. subtilis* B_3 have been observed to accumulate tripeptide, pentapeptide, or nonapeptide precursors in culture fluid (87, 88). This suggests that these precursors are not covalently linked to an enzyme template in the way the nascent gramicidin S peptide is linked to the Grs complex. In keeping with this observation is the finding that mycobacillin fractions A, B, and C do not contain detectable levels of 4′-phosphopantetheine (97).

PEPTIDE ANTIBIOTICS OF ACTINOMYCETES

Bacteria of the group *Actinomycetes* are known for the abundance of special metabolites they produce. These are gram-positive eubacteria, many members of which grow as mycelia. The production of special metabolites is one of several complex responses to growth limitation, as is the case in *Bacillus* spp. (14). Several kinds of special metabolites, including β-lactams, aminoglycosides, polyketides, and peptides, are produced by actinomycetes. Many of those produced by *Streptomyces* species are peptides (70, 71; Table 1) that have found wide use against infectious diseases and are valued as antibacterial, antifungal, and antiparasitic agents. In agriculture, some special *Streptomyces* metabolites serve as herbicides. Still others have useful antitumor and immunomodulatory effects. Hence, there is a massive effort to exploit *Streptomyces* spp. for identifying new metabolites with useful properties and for overproducing important antibiotics and pharmacological agents (43).

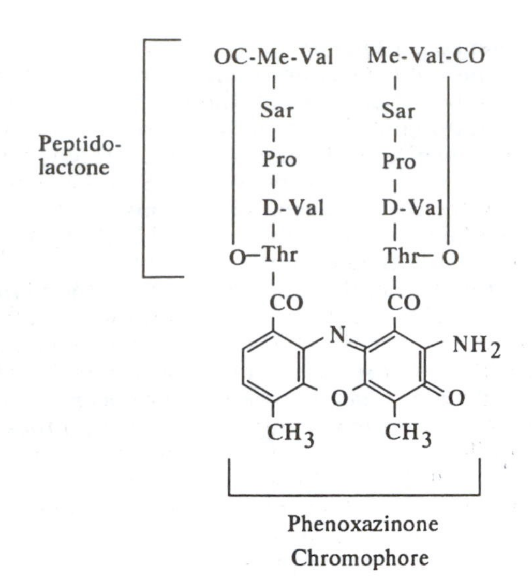

Figure 7. Actinomycin.

Like *Bacillus* spp., actinomycetes such as *Streptomyces* spp. undergo cellular differentiation upon nutritional deprivation (14). This differentiation involves the formation of aerial hyphae and results ultimately in the production of exospores. The onset of aerial-hypha formation is often coupled temporally with the production of special metabolites, which are thought to function as antibiotics that allow the developing hyphae to cannibalize the substrate mycelium in order to scavenge nutrients for further cell differentiation. Other special metabolites are known to act as extracellular regulatory signals for accelerating antibiotic production and sporulation (46, 155), as has been observed in *Streptomyces griseus*, which produces A factor, a butyrolactone that controls streptomycin production and differentiation in the producing organism (46).

A growing list of peptide antibiotics has emerged from efforts aimed at identifying metabolites with antimicrobial activity. A few studies have focused on the molecular biology of peptide antibiotics in *Streptomyces* spp., with the goal of understanding both the regulation and the mechanism of synthesis.

Actinomycin

Several species of *Streptomyces* produce actinomycin (Fig. 7), a highly toxic DNA-binding peptide lactone (62, 70). It is composed of two 4-methyl-3-hydroxyanthranilic acid (4-MHA) pentapeptide lactones that are joined through a six-electron oxidation reaction to form the chromophore phenoxazinone (8). Two multifunctional enzymes, actinomycin synthetases II and III (ASII and ASIII), are believed to be peptide synthetases that catalyze the synthesis of the pentapeptides (64, 72). ASII is 225,000 Da and activates Thr and Val. ASIII is 280,000 Da and activates the remaining three amino acids, Pro, Gly, and Val. Gly and Val are modified by ASIII through N methylation to give sarcosine and *N-methyl*-L-Val, respectively. ASI is a 47,000-Da protein that activates 4-MHA and catalyzes its attachment to the pentapeptide. The last step, phenoxazinone formation, is catalyzed by phenoxazinone synthase (55), which is an 88,000-Da, Cu-containing protein that joins the two 4-MHA pentapeptides. Other enzymes catalyze the formation of 4-MHA from tryptophan; these are thought to include tryptophan pyrrolase, kynurenine formidase, and hydroxykynureninase (62). Other as-yet-unidentified activities, such as a C-methyltransferase for attaching methyl groups to the aromatic moieties of 4-MHA, are believed to be involved.

Attempts at identifying and isolating the genes that encode the actinomycin biosynthesis enzymes have utilized both classic and molecular genetic approaches (38, 56, 86). The genes of the actinomycin C biosynthetic pathway of *S. chrysomallus* were investigated by using low-resolution genetic mapping procedures involving conjugation between actinomycin-negative (*acm*) cells and cells of various auxotrophic mutants (38). The *acm* mutations were located within a large region of the chromosome but appeared not to be a contiguous cluster. Most of the mutations were pleiotropic with respect to actinomycin biosynthesis, conferring defects in several of the biosynthetic activities. These were thought to be regulatory mutations, although it is just as plausible that these were large deletions, as the chromosomes of several *Streptomyces* species are known to exhibit instability (11, 129). Only one class of mutants appeared to contain mutations in a single gene, as these conferred defects in the activity of ASI, which activates 4-MHA.

Isolation of the gene encoding phenoxazinone synthase was carried out by taking advantage of the transformation system developed for use in *Streptomyces lividans* (56). Use was also made of the inability of *S. lividans* to produce actinomycin or detectable phenoxazinone synthase activity. A plasmid library of *Streptomyces antibioticus* DNA was used to transform *S. lividans* protoplasts, and clones were screened for phenoxazinone synthase activity. The gene was isolated as judged by in vitro transcription-translation of one of the phenoxazinone synthase plasmid clones. However, two plasmid clones appeared to activate a cryptic *S. lividans* phenoxazinone synthase gene (55, 86). The mechanism of cryptic phenoxazinone synthase gene activation is not known but could involve the titration of a regulatory factor repressing phenoxazinone synthase gene transcription or overproduction of a positive control factor that leads to derepression. It is well known that several *Streptomyces* species contain silent copies of antibiotic biosynthesis genes that can be activated by the introduction of heterologous DNA (43).

The regulation of production of actinomycin and expression of actinomycin biosynthetic genes was examined in *S. antibioticus* (54). The activity of phenoxazinone synthase and the appearance of actinomycin in culture filtrates were suppressed in medium containing glucose but activated in late-growth-phase cultures containing galactose. Glucose-dependent regulation was exerted at the level of mRNA abundance, although phenoxazinone synthase stability was also enhanced in late-growth-phase cells. The stringent response has also been implicated in regulation of actinomycin biosynthesis enzymes and transcription of the phenoxazinone synthase gene, as enzyme activity and mRNA levels were reduced in a *relC* mutant (67). However, because of the pleiotropy of *relC* alleles (140), care must be taken in interpreting these observations as indicating a role for ppGpp in antibiotic gene regulation.

Bialaphos

The tripeptide bialaphos is a commercially important herbicide produced by cells of *Streptomyces hygroscopicus* and *Streptomyces viridochromogenes* (131). It is a missile-warhead antibiotic composed of two alanine residues and the glutamic acid analog phosphinothricin (PT). Thus, the alanine-alanine portion is required for delivery of the antibiotic into the cell via peptide transport. The PT moiety is a potent inhibitor of glutamine synthetase. Synthesis of the PT moiety begins with nine reaction steps that result in the formation of demethylphosphinothricin (DMPT) from phosphoenolpyruvate (49, 50, 130–132). The product of the *bar* gene, which confers bialaphos resistance, also catalyzes the formation of *N*-acetyl-DMPT from DMPT. The alanine dipeptide is then attached to *N*-acetyl-DMPT to create *N*-acetyl bialaphos, which is finally deacetylated to form the final product. The reaction steps in PT synthesis were elucidated by analysis of nonproducing mutants that accumulated biosynthetic precursors (49, 50, 130). These mutants (designated NP for nonproducing) were used to isolate by complementation the genes of *S. hygroscopicus* that function in bialaphos biosynthesis (the *bap*, or bialaphos production, genes).

All of the genes observed to complement the NP mutations are clustered within a 35-kbp region of the *S. hygroscopicus* chromosome (98, 116). The *bap* genes of *S. viridochromogenes* were isolated by using a plasmid library and protoplast transformation to select for bialaphos resistance in *S. lividans* (40). Resistance is conferred by the *bar* gene, which was correctly presumed to lie within a cluster of *bap* genes. Analysis of the *S. viridochromogenes* DNA containing the *bar* gene revealed the presence of other *bap* genes and showed that their organization was identical to that of *S. hygroscopicus*. Associated with the *bap* gene cluster are two genes that encode putative thioesterases, which may correspond functionally to the *grsT* product of the gramicidin S system (116). A putative product that showed extensive hydrophobicity and is perhaps involved in bialaphos export was also identified. Also within the *bap* cluster was the *brp* (bialaphos regulatory protein) gene, which is required for expression of the *bap* genes (4). Two-dimensional gel analysis revealed that the synthesis of at least 27 proteins was impaired by the *brp* mutation. That the *brp* product plays a regulatory role in the expression of *bap* genes

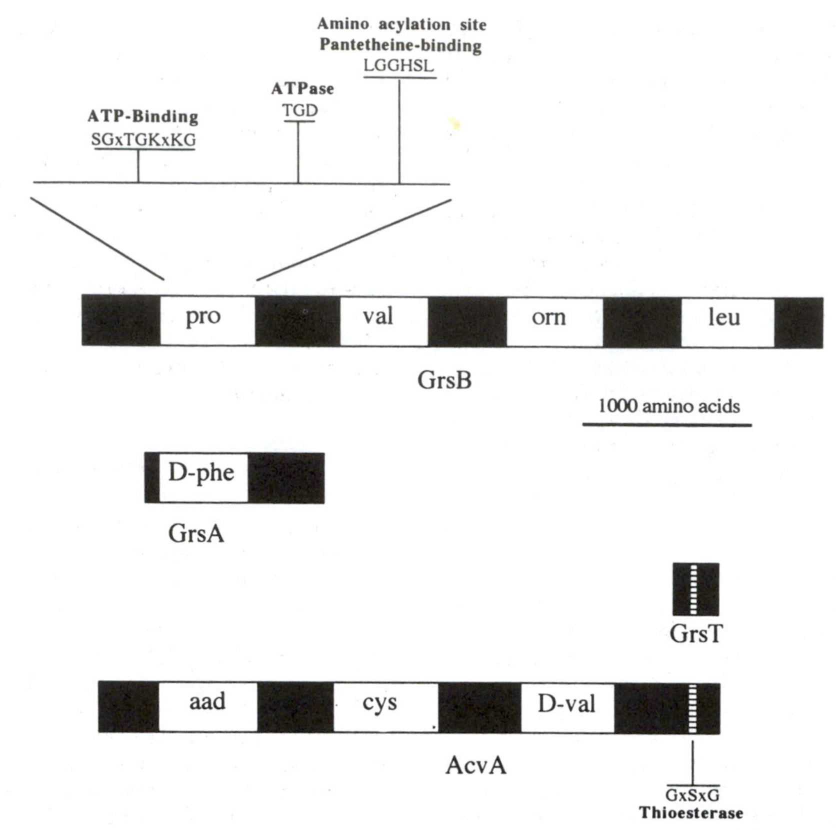

Figure 8. Organization of the homologous amino acid-activating domains of ACV synthetase and gramicidin S synthetase. Shown are locations of the conserved motifs within domain 1 of GrsB that are also found in other peptide synthetase domains. Also shown are the approximate locations of the thioesterase motifs in AcvA and GrsT.

is supported by nucleotide sequence analysis, which showed that Brp shows significant amino acid sequence similarity to a number of proteins of the response regulator class of two-component regulatory proteins (2, 116). Interestingly, the similarity lies within the carboxy end of the protein and not in the amino-terminal end, which is the site of histidine protein kinase-catalyzed phosphorylation of the response regulator protein. Instead, the amino-terminal end of Brp has an abundance of hydrophobic residues, suggesting that the protein may be associated with the membrane. Brp is thought to be a positive regulator of *bap* gene transcription, but how it activates transcription and how Brp activity is regulated are not known.

ACV SYNTHETASE AND THE PEPTIDE SYNTHETASE GENE FAMILY

A structural model of peptide synthetases began to emerge with the isolation and primary-structure determination of a tripeptide-synthesizing enzyme found in β-lactam producers. Several species of eukaryotic and prokaryotic microorganisms produce β-lactam antibiotics such as cephamycin, cephalosporin, and penicillin (71). The gram-positive bacteria of *Flavobacterium* strain SC12,154, *Streptomyces clavuligerus*, and *Nocardia lactamdurans* are among prokaryotic members of the β-lactam producers. Biosynthesis of β-lactams begins with synthesis of the tripeptide δ-(L-α-aminoadipoyl)-L-cysteinyl-D-valine (ACV) followed by the cyclization of ACV to form isopenicillin N. The first step is catalyzed by the enzyme ACV synthetase, which has been isolated from several β-lactam producers. The purified enzyme from *S. clavuligerus* contains 4′-phosphopantetheinate, the cofactor found associated with other peptide synthetases (6). The genes for ACV synthetase and the other enzymes of the β-lactam biosynthetic pathway have been isolated by cloning the DNA that confers β-lactam resistance or by using the β-lactam gene DNA to probe DNA libraries of other producing organisms (16, 20, 36, 85, 91, 137). As was the case with bialaphos and the genes that function in the biosynthesis of other antibiotics, the β-lactam synthesis genes are clustered together, the ACV synthetase gene being in the same transcription unit as the gene encoding isopenicillin N synthetase (136). The nucleotide sequence analysis of *pcbAB* (encoding ACV synthetase) from *N. lactamdurans* revealed a large open reading frame of 3,649 amino acids (16). Its sequence and organization were very similar to those of the ACV synthetases of β-lactam-producing eukaryotic organisms such as *Penicillium chrysogenum*, *Acremonium chrysogenum*, and *Cephalosporium acremonium*, but closer inspection reveals sequence similarity to peptide antibiotic biosynthesis enzymes gramicidin S synthetase 1 and 2 (Fig. 8) and tyrocidine synthetase 1 (137, 145). An interesting difference between ACV synthetase and the *Bacillus* enzymes is the presence of

Table 2. Homology of putative active sites within adenylating and peptide synthetase enzymes

Enzyme[a]	ATP-binding site	ATPase motif	Pantetheine-binding motif
EntE	LSGGTTGTPKL	CSGDLSI	
LcfA	FTSGTTGNPKG	HTGDIGK	
Luc	WSSGSTGLPKG	HAGDIAY	
Acusan	YTSGSTGKPKG	FTGDAGR	
4c1	YSSGTTGLPKG	RTGDIGF	
Cepa1	YTSGTTGFPKG	KTGDLAR	FFRLGGHSI
Cepa2	YTSGTTGMPKG	KTGDLVR	FFSLGGDSL
Cepa3	FTSGTTGKPKG	KTGDLVR	LFARGGDSI
EntF	FTSGSTGRPKG	RTGDVAR	FFALGGHSL
GrsA	FTSGSTGRPKG	KTGDQAR	FYALGGDSI
GrsB1	YTSGTTGKHKR	RTGDLAR	FFSLGGHSL
GrsB2	YTSGSTGKPKG	RTGDLAR	FFELGGHSL
GrsB3	YTSGTTGKPKG	RTGDLAR	FFTIGGHSL
GrsB4	YTSGTTGKPKG	KTGDLAK	FFELGGHSL
Lys2	FTSGSEGIPKG	RTGDLGR	FFKLGGHSI
Noca1	YTSGTTGVPKG	RTGDLAR	FFRLGGQSI
Noca2	YTSGTTGKPKA	KTGDLVR	FFALGGDSI
Noca3	FTSGTTGKPKA	RTGDVVR	FFRCGGDSI
Nida1	YTSGTTGFPKG	RTGDLAR	FFRLGGHSI
Nida2	YTSGTTGKPKG	TRGDLVR	FFTLGGDSL
Nida3	FTSGTSGKPKG	RTGDLFR	LFRLGGDSI
Pena1	YTSGTTGFPKG	KTGDLAR	FFRLGGHSI
Pena2	YTSGTTGRPKG	KTGDLVR	FFSLGGDSL
Pena3	FTSGTSGKPKG	KTGDLVR	LFKLGGDSI
TycA	YTSGTTGKPKG	RTGDLAK	FYSLGGDSI
Consensus	(F/Y) TSGTTGKPKG	(k/r) TGDLxR	FFxLGG (D/H) S (I/L)

[a] EntE, activates 2,3-dihydroxybenzoate with ATP to the acyladenylate (138). LcfA, long-chain acyl-CoA synthetase of rat; activates long-chain fatty acids to acyladenylates (142). Luc, luciferase; activates luciferrin to adenylate (19). Acusan, acetyl-CoA synthetase of *Aspergillus nidulans* and *Neurospora crassa*; activates acetate as acyladenylate followed by ligation with CoA (15). 4c1, 4-coumarate CoA ligase of *Petroselinum crispum* (parsley); activates 4-coumarate to acyladenylate, which is followed by ligation with CoA (84). Cepa1, -2, and -3, ACV synthetase of *C. acremonium* (36). EntF, activates L-serine to acyladenylate in enterobactin biosynthesis in *E. coli* (122). GrsA, gramicidin S synthetase 1 of *B. brevis* ATCC 9999 (76). GrsB1, -2, -3, and -4, the four domains of gramicidin S synthetase 2 of *B. brevis* ATCC 9999 (145). Lys2, activates α-amino adipinate to acyladenylate in lysine biosynthesis in *S. cerevisiae* (96). Noca1, -2, and -3, ACV synthetase of *N. lactamdurans*. nida1, -2, and -3, ACV synthetase of *A. nidulans* (85). Pena1, -2, and -3, ACV synthetase domains of *P. chrysogenum* (137). TycA, tyrocidine synthetase 1 of *B. brevis* ATCC 8185 (150).

a thioesterase motif at the carboxy-terminal end of the former (Fig. 8; 36). This may correspond functionally to GrsT (76) and the putative thioesterases of the bialaphos system, which are encoded by genes that are separate from those encoding peptide and phinothricin biosynthesis. Marahiel and coworkers (145) have undertaken a thorough analysis of peptide synthetase sequence similarities and have proposed the existence of a gene family whose members encode peptide synthetases and several adenylating enzymes such as luciferase (19; Table 2). Shared among its members are an ATP-binding site of the P-loop class (124) and a putative ATPase motif, the two separated by a conserved number of amino acids whose sequence exhibits significant conservation throughout the gene family (145). Those members of the gene family that catalyze the formation of amide or ester bonds possess in addition a putative 4′-phosphopantetheine-binding motif (126, 145). Among the enzymes of this family whose members catalyze amino or hydroxy acid activation and carboxyl thioester formation are the *entF* product of *E. coli*, which activates L-Ser, a component of the depsipeptide enterobactin (122); LYS2 of *Saccharomyces cerevisiae*, which functions in lysine biosynthesis by activating α-amino-adipinate (96); and the product of the *Drosophila melanogaster ebony* gene, β-alanine dopamine synthetase (145). Mutations in *ebony* affect pigmentation as well as vision, courtship behavior, and circadian rhythm. The growing list of peptide synthetases and the realization that they play important roles not only in the synthesis of commercially important compounds but also in the regulation of complex cellular and organismal processes emphasize the need to understand the mechanism of peptide synthetase activity and its physiological role.

CONCLUSION

A wealth of information about the structure and biosynthesis of peptide antibiotics has been gained recently. Those antibiotics produced by gram-positive bacteria are particularly useful, as many can be studied by using facile genetic techniques available for *B. subtilis* and several species of *Streptomyces*. Thus, detailed descriptions of the complex mechanisms of lantibiotic maturation and nonribosomal peptide synthesis are within reach. Of particular interest is the thiotemplate mechanism of peptide synthesis, the model for which will likely be modified to account for the multiple pantetheine-binding sites revealed in the primary-structure analysis of peptide synthetase genes. With the development of powerful molecular genetic techniques, the possibility of manipulating genes encoding the lantibiotics and peptide synthetases may provide a means for producing peptides of desired structures and biological activities. Of growing importance is the role of special metabolites in nature. In addition to their value as antimetabolites, evidence suggests that they may also be involved in coordinating the activities of bacteria within a large population so as to induce a concerted response to environmental stimuli. Certainly, the continued study of special metabolites is of fundamental importance to our understanding of the microbial world.

Acknowledgments. We thank J. Vater, C. Thompson, F. Besson, K. Hori, and G. Jung for providing published and unpublished material for this review. Research conducted in our laboratories and described herein was supported by Public Health Service grant GM34976 (to P.Z.) and by a Deutsche Forschungsgemeinshaft grant (to M.A.M.).

REFERENCES

1. **Akashi, K., K. Kubota, and K. Kurahashi.** 1977. Biosynthesis of enzyme-bound formylvaline and formylvalylglycine: a possible initiation complex for gramicidin A biosynthesis. *J. Biochem.* **81:**269–272.
2. **Albright, L. M., E. Huala, and R. M. Ausubel.** 1989. Prokaryotic signal transduction mediated by sensor and regulator protein pairs. *Annu. Rev. Genet.* **23:**311–336.
3. **Alloing, G., M.-C. Trombe, and J.-P. Claverys.** 1990. The *ami* locus of Gram-positive bacterium *Streptococcus pneumoniae* is similar to binding-protein-dependent transport operons of Gram-negative bacteria. *Mol. Microbiol.* **4:**633–644.
4. **Anzai, H., T. Murakami, S. Imai, A. Satoh, K. Nagaoka, and C. J. Thompson.** 1987. Transcriptional regulation of bialaphos biosynthesis in *Streptomyces hygroscopicus. J. Bacteriol.* **169:**3482–3488.
5. **Babasaki, K., T. Takao, Y. Shimonishi, and K. Kurahashi.** 1985. Subtilosin A, a new antibiotic peptide produced by *Bacillus subtilis* 168: isolation, structural analysis, and biogenesis. *J. Biochem.* **98:**585–603.
6. **Baldwin, J. E., J. W. Bird, R. A. Field, N. M. O'Callaghan, and C. J. Schofield.** 1990. Isolation and partial characterization of ACV synthetase from *Cephalosporium acremonium* and *Streptomyces clavuligerus. J. Antibiot.* **43:**1055–1057.
7. **Banerjee, S., and J. N. Hansen.** 1988. Structure and expression of a gene encoding the precursor of subtilin, a small protein antibiotic. *J. Biol. Chem.* **263:**9508–9514.
8. **Barry, C. E., III, P. G. Nayar, and T. P. Begley.** 1989. Phenoxazinone synthase: mechanism for the formation of the phenoxazinone chromophore of actinomycin. *Biochemistry* **28:**6323–6333.
9. **Bauer, K., R. Roskoski, Jr., H. Kleinkauf, and R. Lipmann.** 1972. Synthesis of a linear gramicidin by a combination of biosynthetic and organic methods. *Biochemistry* **11:**3266–3271.
10. **Bennett, J. W., and R. Bentley.** 1989. What's in a name—microbial secondary metabolites. *Adv. Appl. Microbiol.* **34:**1–28.
11. **Birch, A., A. Hausler, M. Vogtli, W. Krek, and R. Hutter.** 1989. Extremely large chromosomal deletions are intimately involved in genetic instability and genomic rearrangements in *Streptomyces glaucescens. Mol. Gen. Genet.* **217:**447–458.
12. **Buchman, G. W., S. Banerjee, and J. N. Hansen.** 1988. Structure, expression, and evolution of a gene encoding the precursor of nisin, a small protein antibiotic. *J. Biol. Chem.* **263:**16260–16266.
13. **Bu'lock, J. D.** 1961. Intermediary metabolism and antibiotic synthesis. *Adv. Appl. Microbiol.* **3:**293–342.
14. **Chater, K. F., and D. A. Hopwood.** 1990. Antibiotic biosynthesis in *Streptomyces*, p. 129–150. *In* D. A. Hopwood and C. F. Chater (ed.), *Genetics of Bacterial Diversity.* Academic Press, Inc., New York.
15. **Connerton, I. F., J. R. S. Fincham, R. A. Sandeman, and M. J. Hynes.** 1990. Comparison and cross-species expression of the acetyl-CoA synthetase genes of the ascomycete fungi, *Aspergillus nidulans* and *Neurospora crassa. Mol. Microbiol.* **4:**451–460.
16. **Coque, J. J. R., J. F. Martin, J. G. Calzada, and P. Liras.** 1991. The cephamycin biosynthetic genes *pcbAB*, encoding a large multidomain peptide synthetase, and *pcbC* of *Nocardia lactamdurans* are clustered together in an organization different from the same genes in *Acremonium chrysogenum* and *Penicillium chrysogenum. Mol. Microbiol.* **5:**1125–1133.
17. **Daumus, P., F. Heitz, L. Ranjalahy-Rasoloarijao, and R. Lazaro.** 1989. Gramicidin A analogs: influence of the substitution of the tryptophans by naphthylalanines. *Biochimie* **71:**77–81.
18. **Davies, J.** 1990. What are antibiotics? Archaic functions for modern activities. *Mol. Microbiol.* **4:**1227–1232.
19. **De Wet, J. R., K. V. Wood, M. De Luca, D. R. Helsinki, and S. Subrami.** 1987. Firefly luciferase gene: structure and expression in mammalian cells. *Mol. Cell. Biol.* **7:**725–737.
20. **Diez, B., S. Gutierrez, J. L. Barredo, P. Solingen, L. H. M. van der Voort, and J. F. Martin.** 1990. The cluster of penicillin biosynthetic genes. Identification and characterization of the *pcbAB* gene encoding the α-aminoadipyl-cysteine-valine synthetase and linkage to the *pcbC* and *penDE* genes. *J. Biol. Chem.* **265:**16358–16365.
21. **Dodd, H. M., N. Horn, and M. J. Gasson.** 1990. Analysis of the genetic determinant for production of the peptide antibiotic nisin. *J. Gen. Microbiol.* **136:**555–566.
22. **Dubnau, D.** 1991. Genetic competence in *Bacillus subtilis. Microbiol. Rev.* **55:**395–424.
23. **Fredenhagen, A., G. Fendrich, F. Marki, W. Marki, J. Gruner, F. Raschdorf, and H. H. Peter.** 1990. Duramycins B and C, two new lanthionine containing antibiotics as inhibitors of phospholipase A_2: structural revision of duramycin and cinnamycin. *J. Antibiot.* **43:**1403–1412.
24. **Frøyshov, Ø.** 1977. The production of bacitracin synthetase by *Bacillus licheniformis* ATCC 10706. *FEBS Lett.* **81:**315–318.
25. **Furbaß, R., M. Gocht, P. Zuber, and M. A. Marahiel.** 1991. Interaction of AbrB, a transcriptional regulator from *Bacillus subtilis*, with the promoters of the transition state-activated genes *tycA* and *spoVG. Mol. Gen. Genet.* **225:**347–354.
26. **Garrido, M. D. C., M. Herrero, R. Kolter, and F. Moreno.** 1988. The export of the DNA replication inhibitor microcin B17 provides immunity for the host cell. *EMBO J.* **7:**1853–1862.
27. **Gasson, M. J.** 1984. Transfer of sucrose fermenting ability, nisin resistance and nisin production into *Streptococcus lactis* 712. *FEMS Microbiol. Lett.* **21:**7–10.
28. **Gevers, W., H. Kleinkauf, and F. Lipmann.** 1968. The activation of amino acids for biosynthesis of gramicidin S. *Proc. Natl. Acad. Sci. USA* **60:**269–276.
29. **Gevers, W., H. Kleinkauf, and F. Lipmann.** 1969. Peptidyl transfers in gramicidin S biosynthesis from enzyme-bound thioester intermediates. *Proc. Natl. Acad. Sci. USA* **63:**1335–1342.
30. **Ghosh, S. K., S. Majumder, N. K. Mukhopadyay, and**

S. K. Bose. 1986. Role of ATP and enzyme-bound nascent peptides in the control of elongation for mycobacillin synthesis. *Biochem. J.* **240:**265.
31. **Ghosh, S. K., N. K. Mukhopadyay, S. Majumder, and S. K. Bose.** 1983. Fractionation of the mycobacillin synthesizing enzyme system. *Biochem. J.* **215:**539.
32. **Ghosh, S. K., N. K. Mukhopadyay, S. Majumder, and S. K. Bose.** 1985. Functional characterization of constituent enzyme fractions of mycobacillin synthetase. *Biochem. J.* **230:**785.
33. **Ghosh, S. K., N. K. Mukhopadyay, S. Majumder, and S. K. Bose.** 1986. Purification of the constituent enzyme fractions of mycobacillin synthetase. *Biochem. J.* **235:** 81.
34. **Gilhuus-Moe, C. C., T. Kristensen, J. E. Bredesen, R. L. Zimmer, and S. G. Laland.** 1970. The presence and possible role of phosphopantetheinic acid in gramicidin S synthetase. *FEBS Lett.* **7:**287–290.
35. **Gilson, L., H. K. Mahanty, and R. Kolter.** 1990. Genetic analysis of an MDR-like export system: the secretion of colicin V. *EMBO J.* **9:**3875–3884.
36. **Gutierrez, S., B. Diez, E. Montenegro, and J. F. Martin.** 1991. Characterization of the *Cephalosporium acremonium pcbAB* gene encoding α-aminoadipyl-cysteinyl-valine synthetase, a large multidomain peptide synthetase: linkage to the *pcbC* gene as a cluster of early cephalosporin biosynthetic genes and evidence of multiple functional domains. *J. Bacteriol.* **173:**2345–2365.
37. **Haavik, H. I., and S. Thomassen.** 1973. A bacitracin-negative mutant which is able to sporulate. *J. Gen. Microbiol.* **76:**451–454.
38. **Haese, A., and U. Keller.** 1988. Genetics of actinomycin C production in *Streptomyces chrysomallus. J. Bacteriol.* **170:**1360–1368.
39. **Hahn, J., and D. Dubnau.** 1991. Growth stage signal transduction and the requirements for *srfA* induction in development of competence. *J. Bacteriol.* **173:**7275–7282.
40. **Hara, O., T. Murakami, S. Imai, H. Anzai, R. Itoh, Y. Kumada, E. Takano, E. Satoh, A. Satoh, K. Nagaoka, and C. Thompson.** 1991. The bialaphos biosynthetic genes of *Streptomyces viridochromogenes*: cloning, heterospecific expression, and comparison with the genes of *Streptomyces hygroscopicus. J. Gen. Microbiol.* **137:** 351–359.
41. **Hilton, M. D., N. G. Alaeddingoglu, and A. L. Demain.** 1988. Synthesis of bacilysin by *Bacillus subtilis* branches from prephenate of the aromatic amino acid pathway. *J. Bacteriol.* **170:**482–484.
42. **Hilton, M. D., N. G. Alaeddinoglulu, and A. L. Demain.** 1988. *Bacillus subtilis* mutant deficient in the ability to produce the dipeptide antibiotic bacilysin: isolation and mapping of the mutation. *J. Bacteriol.* **170:**1018–1020.
43. **Hopwood, D.** 1989. Antibiotics: opportunities for genetic manipulation. *Phil. Trans. R. Soc. Lond. Sect. B* **194:**549–562.
44. **Hori, K., M. Kanda, T. Kurotsu, S. Miura, Y. Yamada, and Y. Saito.** 1981. Absence of pantotheinic acid in gramicidin S synthetase 2 obtained from some mutants of *Bacillus brevis. J. Biochem.* **90:**439–447.
45. **Hori, K., Y. Yamamoto, K. Tokita, F. Saito, T. Kurotsu, M. Kanda, K. Okamura, J. Furuyama, and Y. Saito.** 1991. The nucleotide sequence for a proline-activating domain of gramicidin S synthetase 2 gene from *Bacillus brevis. J. Biochem.* **110:**111–119.
46. **Horinouchi, S., and T. Beppu.** 1990. Autoregulatory factors of secondary metabolism and morphogenesis in actinomyces. *Crit. Rev. Biotechnol.* **10:**191–204.
47. **Horn, N., S. Swindell, H. Dodd, and M. Gasson.** 1991. Nisin biosynthesis genes are encoded by a novel conjugative transposon. *Mol. Gen. Genet.* **228:**129–135.
48. **Hyde, S. C., P. Emsley, M. J. Hartshorn, M. M. Mimmack, U. Gileadi, S. R. Pearce, M. P. Gallagher, D. R. Gill, R. E. Hubbard, and C. F. Higgins.** 1990. Structural model of ATP-binding proteins associated with cystic fibrosis, multi-drug resistance and bacterial transport. *Nature* (London) **346:**362–365.
49. **Imai, S., H. Seto, T. Sasaki, T. Tsuruoka, H. Ogawa, A. Satoh, S. Inouye, T. Niida, and N. Otake.** 1985. Studies on the biosynthesis of bialaphos (SF-1293). 4. Production of phosphonic acid derivatives 2-hydroxyethylphosphonic acid, hydroxymethylphosphonic acid and phosphonoformic acid by blocked mutants of *Streptomyces hygroscopicus* SF-1293 and their roles in the biosynthesis of bialaphos. *J. Antibiot.* **37:**1505–1508.
50. **Imai, S., H. Seto, T. Sasaki, T. Tsuruoka, H. Ogawa, A. Satoh, S. Inouye, T. Niida, and N. Otake.** 1985. Studies on the biosynthesis of bialaphos (SF-1293). 6. Production of N-acetyldemethylphosphinothricin and N-acetylbialaphos by blocked mutants of *Streptomyces hygroscopicus* SF-1293 and their roles in the biosynthesis of bialaphos. *J. Antibiot.* **38:**687–690.
51. **Ishihara, H., N. Hara, and T. Iwabuchi.** 1989. Molecular cloning and expression in *Escherichia coli* of the *Bacillus licheniformis* bacitracin synthetase 2 gene. *J. Bacteriol.* **171:**1705–1711.
52. **Ishihara, H., and K. Shimura.** 1988. Further evidence for the presence of a thiazoline ring in the isoleucylcysteine dipeptide intermediate in bacitracin biosynthesis. *FEB Lett.* **226:**319–323.
53. **Jaacks, K. J., J. Healy, R. Losick, and A. D. Grossman.** 1989. Identification and characterization of genes controlled by the sporulation regulatory gene *spo0H* in *Bacillus subtilis. J. Bacteriol.* **171:**4121–4129.
54. **Jones, G. H.** 1985. Regulation of phenoxazinone synthase expression in *Streptomyces antibioticus. J. Bacteriol.* **163:**1215–1221.
55. **Jones, G. H., and D. A. Hopwood.** 1984. Activation of phenoxazinone synthase expression in *Streptomyces lividans* by cloned DNA sequences from *Streptomyces antibioticus. J. Biol. Chem.* **259:**14158–14164.
56. **Jones, G. H., and D. A. Hopwood.** 1984. Molecular cloning and expression of the phenoxazinone synthase gene from *Streptomyces antibioticus. J. Biol. Chem.* **259:** 14151–14157.
57. **Jung, G.** 1991. Lantibiotica-ribosomal synthetisierte Polypeptidwirkstoffe mit Sulfidbrucken und α,β-didehydroamino Sauren. *Angew. Chem.* **103:**1067–1084.
58. **Kalletta, C., and K.-D. Entian.** 1989. Nisin, a peptide antibiotic: cloning and sequencing of the *nisA* gene and posttranslational processing of its peptide product. *J. Bacteriol.* **171:**1597–1601.
59. **Kalletta, C., K.-D. Entian, and G. Jung.** 1991. Prepeptide sequence of cinnamycin (Ro 09-0198); the first structural gene of a duramycin-type lantibiotic. *Eur. J. Biochem.* **199:**411–415.
60. **Kaletta, C., K.-D. Entian, F. Kellner, G. Jung, M. Reis, and H. G. Sahl.** 1989. Pep5, a new lantibiotic: structural gene isolation and prepeptide sequence. *Arch. Microbiol.* **152:**16–19.
61. **Kanda, M., K. Hori, T. Kurotsu, S. Miura, Y. Yamada, and Y. Saito.** 1981. Sulfhydryl groups related to the catalytic activity of gramicidin S synthetase 1 of *Bacillus brevis. J. Biochem.* **90:**765–771.
62. **Katz, E.** 1967. Actinomycin, p. 276–341. *In* D. Gottlieb and P. D. Shaw (ed.), *Antibiotics II*. Springer-Verlag, New York.
63. **Katz, E., and A. L. Demain.** 1977. The peptide antibiotics of *Bacillus*: chemistry, biogenesis and possible functions. *Bacteriol. Rev.* **41:**449–474.
64. **Keller, U.** 1987. Actinomycin synthetases: multifunctional enzymes responsible for the synthesis of the peptide chains of actinomycin. *J. Biol. Chem.* **262:**5852–5856.
65. **Kellner, R., G. Jung, T. Horner, H. Zahner, N. Schnell,**

K.-D. Entian, and F. Gotz. 1988. Gallidermin: a new lanthionine-containing polypeptide antibiotic. *Eur. J. Biochem.* **177:**53–59.

66. **Kellner, R., G. Jung, M. Josten, C. Kaletta, K.-D. Entian, and H.-G. Sahl.** 1989. Pep5: structure elucidation of a large lantibiotic. *Angew. Chem. Int. Ed. Engl.* **28:**616–619.
67. **Kelly, K. S., K. Ochi, and G. H. Jones.** 1991. Pleiotropic effects of a *relC* mutation in *Streptomyces antibioticus*. *J. Bacteriol.* **175:**2297–2300.
68. **Kenig, M., and E. P. Abraham.** 1976. Antimicrobial activities and antagonists of bacilysin and anticapsin. *J. Gen. Microbiol.* **94:**37–45.
69. **Kleinkauf, H., W. Gevers, and F. Lipmann.** 1969. Interrelation between activation and polymerization in gramicidin S biosynthesis. *Proc. Natl. Acad. Sci. USA* **62:**226–233.
70. **Kleinkauf, H., and H. von Dohren.** 1984. Peptide antibiotics, p. 283–307. *In* H. Pape and H.-J. Rehm (ed.), *Biotechnology*, vol. 4. *Microbial Products II*. VCH Verlagsgesellschaft mbH, Weinheim, Germany.
71. **Kleinkauf, H., and H. von Dohren.** 1988. Peptide antibiotics, β-lactams, and related compounds. *Crit. Rev. Biotechnol.* **8:**1–32.
72. **Kleinkauf, H., and H. von Dohren.** 1990. Nonribosomal biosynthesis of peptide antibiotics. *Eur. J. Biochem.* **192:**1–15.
73. **Komura, S., and K. Kurahashi.** 1980. Biosynthesis of polymyxin E by a cell-free system. *J. Biochem.* **88:**285–288.
74. **Komura, S., and K. Kurahashi.** 1980. Biosynthesis of polymyxin E. III. Total synthesis by a cell-free enzyme system. *Biochem. Biophys. Res. Commun.* **95:**1145–1151.
75. **Komura, S., and K. Kurahashi.** 1985. Biosynthesis of polymyxin E. IV. Acylation of enzyme-bound L-2,4, deamino butyric acid. *J. Biochem.* **97:**1409–1417.
76. **Kratzschmar, J., M. Krause, and M. A. Marahiel.** 1989. Gramicidin S biosynthesis operon containing the structural genes *grsA* and *grsB* has an open reading frame encoding a protein homologous to fatty acid thioesterases. *J. Bacteriol.* **171:**5422–5429.
77. **Krause, M., M. A. Marahiel, H. von Dohren, and H. Kleinkauf.** 1985. Molecular cloning of an ornithine-activating fragment of the gramicidin S synthetase 2 gene from *Bacillus brevis* and expression in *Escherichia coli*. *J. Bacteriol.* **162:**1120–1125.
78. **Kubota, K.** 1982. Generation of formic acid and ethanolamine from serine in biosynthesis of linear gramicidin by cell-free preparation of *Bacillus brevis* (ATCC 8185). *Biochem. Biophys. Res. Commun.* **105:**688–697.
79. **Kugler, M., W. Loeffler, C. Rapp, A. Kern, and G. Jung.** 1990. Rhizocticin A, an antifungal phosphono-oligopeptide of Bacillus subtilis ATCC6633: biological properties. *Arch. Microbiol.* **153:**276–281.
80. **Kurahashi, K.** 1981. Biosynthesis of peptide antibiotics, p. 325–352. *In* J. W. Corcoran (ed.), *Antibiotics*, vol. IV. *Biosynthesis*. Springer, Berlin.
81. **Kurotsu, R., K. Hori, M. Kanda, and Y. Saito.** 1991. Characterization and location of the L-proline activating fragment from the multifunctional gramicidin S synthetase 2. *Biochem. J.* **109:**763–769.
82. **Lipmann, F.** 1980. Bacterial production of antibiotic polypeptides by thiol-linked synthesis on protein templates. *Adv. Microb. Physiol.* **21:**227–260.
83. **Liu, W., and J. N. Hansen.** 1990. Some chemical and physical properties of nisin, a small-protein antibiotic produced by *Lactococcus lactis*. *Appl. Environ. Microbiol.* **56:**2551–2558.
84. **Loyoza, E., H. Hoffmann, C. Douglas, W. Schulz, D. Scheel, and K. Hahlbrock.** 1988. Primary structure and catalytic properties of isoenzymes encoded by the two 4-courmarate:CoA ligase genes in parsley. *Eur. J. Biochem.* **176:**661–667.
85. **MacCabe, A. P., H. van Liempt, H. Palissa, S. E. Unkles, M. B. R. Riach, E. Pfeifer, H. Von Dohren, and J. R. Kinghorn.** 1991. δ-(L-α-Aminoadipyl)-L-cysteinyl-D-valine synthetase from *Aspergillus nidulans*—molecular characterization of the *acvA* gene encoding the first enzyme of the penicillin biosynthetic pathway. *J. Biol. Chem.* **266:**12646–12654.
86. **Madu, A. C., and G. H. Jones.** 1989. Molecular cloning and in vitro expression of a silent phenoxazinone synthase gene from *Streptomyces lividans*. *Gene* **84:**287–294.
87. **Majumder, S., S. K. Ghosh, N. K. Mukhopadhyay, and S. K. Bose.** 1985. Accumulation of peptides by mycobacillin-negative mutants of *Bacillus subtilis* B_3. *J. Gen. Microbiol.* **131:**119–127.
88. **Majumder, S., N. K. Mudhopadhyay, S. K. Ghosh, and S. K. Bose.** 1988. Genetic analysis of the mycobacillin biosynthetic pathway in *Bacillus subtilis* B_3. *J. Gen. Microbiol.* **134:**1147–1153.
89. **Majumder, S. K., and S. K. Bose.** 1958. Mycobacillin, a new antifungal antibiotic produced by *Bacillus subtilis*. *Nature* (London) **181:**134–135.
90. **Marahiel, M. A., P. Zuber, G. Czekay, and R. Losick.** 1987. Identification of the promoter for a peptide antibiotic gene from *Bacillus brevis* and studies on its regulation in *Bacillus subtilis*. *J. Bacteriol.* **169:**2215–2222.
91. **Martin, J. F., and P. Liras.** 1989. Organization and expression of genes involved in the biosynthesis of antibiotics and other secondary metabolites. *Annu. Rev. Microbiol.* **43:**173–206.
92. **Mathiopoulos, C., J. P. Mueller, F. J. Slack, C. G. Murphy, S. Patankar, G. Bukusoglu, and A. L. Sonenshein.** 1990. A *Bacillus subtilis* dipeptide transport system expressed early during sporulation. *Mol. Microbiol.* **5:**1903–1913.
93. **Meister, A.** 1988. Glutathione metabolism and its selective modification. *J. Biol. Chem.* **263:**17205–17208.
94. **Mittenhuber, G., R. Weckermann, and M. A. Marahiel.** 1989. Gene cluster containing the genes for tyrocidine synthetases 1 and 2 from *Bacillus brevis*: evidence for an operon. *J. Bacteriol.* **171:**4881–4887.
95. **Monaghan, R. L., and J. S. Tkacz.** 1990. Bioactive microbial products: focus upon mechanism of action. *Annu. Rev. Microbiol.* **44:**271–301.
96. **Morris, M. E., and S. Jinks-Robertson.** 1991. Nucleotide sequence of the LYS2 gene of Saccharomyces cerevisiae: homology to Bacillus brevis tyrocidine synthetase 1. *Gene* **98:**141–145.
97. **Mukhopadhyay, N. K., S. Majumder, S. K. Ghosh, and S. K. Bose.** 1986. Characterization of three-fraction mycobacillin synthetase. *Biochem. J.* **235:**639–643.
98. **Murakami, T., H. Anzai, S. Imai, A. Satoh, K. Nagaoka, and C. J. Thompson.** 1986. The bialaphos biosynthetic genes of *Streptomyces hygroscopicus*: molecular cloning and characterization of the gene cluster. *Mol. Gen. Genet.* **205:**42–50.
99. **Nakano, M. M., N. Corbell, J. Besson, and P. Zuber.** Isolation and characterization of *sfp*: a gene required for the production of the lipopeptide biosurfactant, surfactin in *Bacillus subtilis*. *Mol. Gen. Genet.* **232:**313–321.
100. **Nakano, M. M., R. Magnuson, A. Myers, J. Curry, A. D. Grossman, and P. Zuber.** 1991. *srfA* is an operon required for surfactin production, competence development, and efficient sporulation in *Bacillus subtilis*. *J. Bacteriol.* **173:**1770–1778.
101. **Nakano, M. M., M. A. Marahiel, and P. Zuber.** 1988. Identification of a genetic locus required for biosynthesis of the lipopeptide antibiotic surfactin in *Bacillus subtilis*. *J. Bacteriol.* **170:**5662–5668.

102. **Nakano, M. M., L. Xia, and P. Zuber.** 1991. The transcription initiation region of the *srfA* operon which is controlled by the *comP-comA* signal transduction system in *Bacillus subtilis*. *J. Bacteriol.* **173:**5487–5493.
103. **Nakano, M. M., and P. Zuber.** 1989. Cloning and characterization of *srfB*: a regulatory gene involved in surfactin and competence in *Bacillus subtilis*. *J. Bacteriol.* **171:**5347–5353.
104. **Nakano, M. M., and P. Zuber.** 1990. Identification of genes required for the biosynthesis of the lipopeptide antibiotic surfactin in *Bacillus subtilis*, p. 397–405. *In* J. Hoch and A. T. Ganesan (ed.), *Genetics and Biotechnology of Bacilli*, vol. 3. Academic Press, Inc., New York.
105. **Nakano, M. M., and P. Zuber.** 1990. Molecular biology of antibiotic production in *Bacillus*. *Crit. Rev. Biotechnol.* **10:**223–240.
106. **Nakano, M. M., and P. Zuber.** 1991. The primary role of *comA* in the establishment of the competent state in *Bacillus subtilis* is to activate the expression of *srfA*. *J. Bacteriol.* **173:**7269–7274.
107. **Nakano, M. M., and P. Zuber.** 1991. Transcriptional regulation of *srfA* and the involvement of *srfA* in competence development in *Bacillus subtilis*. Int. Conf. Genet. Biotechnol. Bacilli, Stanford, Calif.
108. **Nishio, C., S. Komura, and K. Kurahashi.** 1983. Peptide antibiotic subtilin is synthesized via precursor proteins. *Biochem. Biophys. Res. Commun.* **116:**751–758.
109. **Perego, M., C. F. Higgins, S. R. Pearce, M. P. Gallagher, and J. A. Hoch.** 1991. The oligopeptide transport system of *Bacillus subtilis* plays a role in the initiation of sporulation. *Mol. Microbiol.* **5:**173–185.
110. **Perego, M., G. B. Spiegelman, and J. Hoch.** 1989. Structure of the gene for the transition state regulator, *abrB*, regulator synthesis is controlled by the *spo0A* sporulation gene in Bacillus subtilis. *Mol. Microbiol.* **2:**689–699.
111. **Peypoux, F., D. Marion, R. Maget-Dana, M. Ptak, B. C. Das, and G. Michel.** 1985. Structure of bacillomycin F, a new peptide antibiotic of the iturin group. *Eur. J. Biochem.* **153:**335–340.
112. **Peypoux, F., M.-T. Pommier, B. C. Das, F. Besson, L. Delcambe, and G. Michel.** 1984. Structures of bacillomycin D and bacillomycin L peptidolipid antibiotics from *Bacillus subtilis*. *J. Antibiot.* **37:**1600–1604.
113. **Peypoux, F., M. T. Pommier, D. Marion, M. Ptak, B. C. Das, and G. Michel.** 1986. Revised structure of mycosubtilin, a peptidolipid antibiotic from *Bacillus subtilis*. *J. Antibiot.* **39:**636–641.
114. **Podlesek, Z., and M. Grabnar.** 1987. Genetic mapping of the bacitracin synthetase gene(s) in *Bacillus licheniformis*. *J. Gen. Microbiol.* **133:**3093–3097.
115. **Podlesek, Z., and M. Grabnar.** 1989. Sporulation of a bacitracin-sensitive mutant of *Bacillus licheniformis* is self-inhibited by bacitracin. *J. Gen. Microbiol.* **135:** 2813–2818.
116. **Raibaud, A., M. Zalacain, G. Holt, R. Tizard, and C. J. Thompson.** 1991. Nucleotide sequence analysis reveals linked *N*-acetyl hydrolase, thioesterase, transport, and regulatory genes encoded by the bialaphos biosynthetic gene cluster of *Streptomyces hygroscopicus*. *J. Bacteriol.* **173:**4454–4463.
117. **Rapp, C., G. Jung, W. Katzer, and W. Loeffler.** 1988. Chlorotetain from *Bacillus subtilis*, an antifungal dipeptide with an unusual chlorine-containing amino acid. *Angew. Chem.* **27:**1733–1734.
118. **Robertson, J. R., M. Gocht, M. A. Marahiel, and P. Zuber.** 1989. AbrB, a regulator of gene expression in Bacillus, interacts with the transcription initiation regions of a sporulation and an antibiotic biosynthesis gene. *Proc. Natl. Acad. Sci. USA* **86:**8457–8461.
119. **Rogers, H. J., G. G. F. Newton, and E. P. Abraham.** 1965. Production and purification of bacilysin. *Biochem. J.* **97:**573–586.
120. **Roland, I., Ø. Frøyshov, and S. G. Laland.** 1975. On the presence of pantetheinic acid in the three complementary enzymes of bacitracin synthetase. *FEBS Lett.* **60:** 305–308.
121. **Rudner, D., J. R. LeDeaux, K. Ireton, and A. D. Grossman.** 1991. The *spo0K* locus of *Bacillus subtilis* is homologous to the oligopeptide permease locus and is required for sporulation and competence. *J. Bacteriol.* **173:**1338–1398.
122. **Rusnak, F., M. Sakaitani, D. Drueckhammer, J. Reichert, and C. T. Walsh.** 1991. Biosynthesis of the *Escherichia coli* siderophore enterobactin: sequence of the *entF* gene, expression and purification of EntF, and analysis of covalent phosphopantetheine. *Biochemistry* **30:**2916–2927.
123. **Sakajoh, M., N. A. Solomon, and A. L. Demain.** 1987. Cell-free synthesis of the dipeptide antibiotic bacilysin. *J. Ind. Microbiol.* **2:**201–208.
124. **Saraste, M., P. R. Sibbald, and A. Wittinghofer.** 1990. The P-loop: a common motif in ATP- and GTP-binding proteins. *Trends Biochem. Sci.* **15:**430–434.
125. **Schaeffer, P.** 1969. Sporulation and the production of antibiotics, exoenzymes, and exotoxins. *Bacteriol. Rev.* **33:**48–71.
126. **Schlumbohm, W., T. Stein, C. Ullrich, J. Vater, M. Krause, M. A. Marahiel, V. Kruft, and B. Wittman-Liebold.** 1991. An active serine is involved in covalent substrate amino acid binding at each reaction center of gramicidin S synthetase. *J. Biol. Chem.* **266:**23135–23141.
127. **Schnell, N., K.-D. Entian, F. Gotz, R. Horner, R. Kellner, and G. Jung.** 1989. Structural gene isolation and prepeptide sequence of gallidermin, a new lanthionine containing antibiotic. *FEMS Lett.* **58:**263–268.
128. **Schnell, N., K.-D. Entan, U. Schneider, F. Gotz, H. Zahner, R. Kellner, and G. Jung.** 1988. Prepeptide sequence of epidermin, a ribosomally synthesized antibiotic with four sulphide-rings. *Nature* (London) **333:** 276–278.
129. **Schrempf, H.** 1985. Genetic instability: amplification, deletion and rearrangement within *Streptomyces* DNA, p. 436–439. *In* L. Leive (ed.), *Microbiology—1985*. American Society for Microbiology, Washington, D.C.
130. **Seto, H., S. Imai, T. Tsuruoka, H. Ogawa, A. Satoh, T. Sasaki, and N. Otake.** 1983. Studies on the biosynthesis of bialaphos (SF-1293). 3. Production of phosphinic acid derivatives, MP-103, MP-104, and MP-105, by a blocked mutant of *Streptomyces hygroscopicus* SF-1293 and their roles in the biosynthesis of bialaphos. *J. Antibiot.* **37:**1509–1511.
131. **Seto, H., S. Imai, T. Tsuruoka, A. Satoh, M. Kojima, S. Inouye, T. Sasaki, and N. Otake.** 1982. Studies on the biosynthesis of bialaphos (SF-1293). 1. Incorporation of ^{13}C and ^{3}H labeled precursors into bialaphos. *J. Antibiot.* **35:**1719–1721.
132. **Seto, H., T. Sasaki, S. Imai, T. Tsuruoka, H. Ogawa, A. Satoh, S. Inouye, T. Niida, and N. Otake.** 1983. Studies on the biosynthesis of bialaphos (SF-1293). 2. Isolation of the first natural products with a C-P-H bond and their involvement in the C-P-C bond formation. *J. Antibiot.* **36:**96–98.
133. **Sinderen, D. V., S. Withoff, H. Boels, and G. Venema.** 1990. Isolation and characterization of *comL*, a transcription unit involved in competence development of *Bacillus subtilis*. *Mol. Gen. Genet.* **224:**396–404.
134. **Skarpeid, H.-J., T.-L. Zimmer, B. Shen, and H. von Dohren.** 1990. The proline-activating activity of the multienzyme gramicidin S synthetase 2 can be recovered on a 115-kDa tryptic fragment. *Eur. J. Biochem.* **187:**627–633.
135. **Skarpeid, H.-J., T.-L. Zimmer, and H. von Dohren.** 1990. On the domain construction of the multienzyme

gramicidin S synthetase 2. *Eur. J. Biochem.* **189:**517–522.

136. **Smith, D. J., M. K. R. Burnham, J. H. Bull, J. E. Hodgson, J. M. Ward, P. Browne, J. Brown, B. Barton, A. J. Earl, and G. Turner.** 1990. β-Lactam antibiotic biosynthetic genes have been conserved in clusters in prokaryotes and eukaryotes. *EMBO J.* **9:**741–747.
137. **Smith, D. J., A. F. Earl, and G. Turner.** 1990. The multifunctional peptide synthetase performing the first step of penicillin biosynthesis in *Penicillium chrysogenum* is a 421 073 dalton protein similar to *Bacillus brevis* peptide antibiotic synthetases. *EMBO J.* **9:**2743–2750.
138. **Staab, J. F., M. Elkins, and C. F. Earhart.** 1989. Nucleotide sequence of the *Escherichia coli entE* gene. *FEMS Microbiol. Lett.* **59:**15–20.
139. **Stoll, E., Ø. Frøshov, H. Holm, T. L. Zimmer, and S. G. Laland.** 1976. On the mechanism of gramicidin S formation from intermediate peptides. *FEBS Lett.* **11:**348–352.
140. **Strauch, E., E. Takano, H. A. Baylis, and M. J. Bibb.** 1991. The stringent response in *Streptomyces coelicolor. Mol. Microbiol.* **5:**289–298.
141. **Strauch, M. A., Spiegelman, G. B., M. Perego, W. C. Johnson, D. Burbulys, and J. A. Hoch.** 1989. The transition-state transcription regulator AbrB of *Bacillus subtilis* is a DNA-binding protein. *EMBO J.* **8:**1615–1621.
142. **Suzuki, H., Y. Kawarabayashi, J. Kondo, T. Abe, K. Nishikawa, S. Kimura, T. Hashimoto, and T. Yamamoto.** 1991. Structure and regulation of rat long-chain acyl-CoA synthetase. *J. Biol. Chem.* **265:**8681–8685.
143. **Tomino, S., M. Yamada, H. Itoh, and K. Kurahashi.** 1967. Cell-free synthesis of gramicidin S. *Biochemistry* **6:**2552–2560.
144. **Trowsdale, J., I. Hanson, I. Mockridge, S. Beck, A. Townsend, and A. Kelly.** 1990. Sequences encoded in the class II region of the MHC related to the ABC superfamily of transporters. *Nature* (London) **348:**741–743.
145. **Turgay, K., M. Krause, and M. A. Marahiel.** 1992. Four homologous domains in the primary structure of GrsB are related to domains in a superfamily of adenylate-forming enzymes. *Mol. Microbiol.* **6:**529–546.
146. **Ullich, C., B. Kluge, P. Zbigniew, and J. Vater.** 1991. Cell-free biosynthesis of surfactin, a cyclic lipopeptide produced by *Bacillus subtilis. Biochemistry* **30:**6503–6508.
147. **Vater, J.** 1989. Lipopeptides, an interesting class of microbial secondary metabolites, p. 27–38. *In* U. P. Schlunegger (ed.), *Biologically Active Molecules.* Springer-Verlag, Berlin.
148. **Vining, L. C.** 1990. Functions of secondary metabolites. *Annu. Rev. Microbiol.* **44:**395–427.
149. **Walker, J. E., and E. P. Abraham.** 1970. The structure of bacilysin and other products of *Bacillus subtilis. Biochem. J.* **118:**563–570.
150. **Weckermann, R., R. Furbaß, and M. A. Marahiel.** 1988. Complete nucleotide sequence of *tycA* gene coding the tyrocidine synthetase 1 from *Bacillus brevis. Nucleic Acids Res.* **16:**11841.
151. **Weil, H.-P., A. G. Beck-Sickinger, H. Metzger, S. Stevanovic, G. Jung, M. Josten, and H.-G. Sahl.** 1990. Biosynthesis of the lantibiotic Pep5. *Eur. J. Biochem.* **194:**217–223.
152. **Weinrauch, Y., N. Guillen, and D. A. Dubnau.** 1989. Sequence and transcription mapping of *Bacillus subtilis* competence genes *comB* and *comA*, one of which is related to a family of bacterial regulatory determinants. *J. Bacteriol.* **171:**5362–5375.
153. **Weinrauch, Y., T. Msadek, F. Kunst, G. Rapoport, and D. Dubnau.** 1991. Sequence and properties of *comQ*, a new competence gene of *Bacillus subtilis. J. Bacteriol.* **173:**5685–5693.
154. **Weinrauch, Y., R. Penchev, E. Dubnau, I. Smith, and D. Dubnau.** 1990. A *Bacillus subtilis* regulatory gene product for genetic competence and sporulation resembles sensor protein members of the bacterial two-component signal-transduction systems. *Genes Dev.* **4:**860–872.
155. **Willey, J., R. Santamaria, J. Guijarro, M. Geistlich, and R. Losick.** 1991. Extracellular complementation of a developmental mutation implicates a small sporulation protein in aerial mycelium formation by *S. coelicolor. Cell* **65:**641–650.
156. **Zuber, P., and R. Losick.** 1987. Role of AbrB in the Spo0A- and Spo0B-dependent utilization of a sporulation promoter in *Bacillus subtilis. J. Bacteriol.* **169:**2223–2230.

62. Commercial Production of Extracellular Enzymes

EUGENIO FERRARI, ALISHA S. JARNAGIN, and BRIAN F. SCHMIDT

Although the importance of enzymes as catalysts in industrial processes is widely recognized, no more than a handful have a high market volume for either industrial or household applications (70). The major limitation to using enzymes as catalysts in industrial processes is associated with their production costs, which often makes them noncompetitive with current well-established methods (97). This initially curbed the effort toward exploring and studying new applications. However, the ability to clone and express virtually any desired gene has removed this obstacle and increased the number of enzymes conceivably available for industrial experimentation from a few hundred to several thousand (97). Furthermore, the ability to manipulate genes in vitro combined with recent progress in understanding the structure-function relationship of enzyme molecules allows the generation of novel enzymes that are more suitable for specific applications. For example, protein engineering has led to the isolation of a number of alkaline proteases with properties and specific activities not available in natural isolates, thus generating enzymes that are more stable and efficient in the environment of the target application, e.g., a washing machine (reviewed in references 50 and 70). Additionally, concern over the environmental pollutants often associated with chemical processes will be a driving force behind the effort to replace such approaches with nature-friendly ones. One example in this direction is the recent introduction of cellulases, of both bacterial and fungal origin, in the stone washing process of indigo-dyed denim material. The use of cellulases allows the partial or in some cases complete elimination of pumice from this procedure, with benefit to the environment as well as to the economics of the process. Cellulases with different substrate specificities and pH stability profiles are currently being marketed for this application. The dollar value associated with cellulases sold for this use alone was well above the $20 million mark only 2 years after their introduction to the market. Other possible applications of this group of enzymes, such as in the food and paper industry and for the generation of fuel from renewable sources, for which endoglucanases were highly touted, lag far behind their use for this fashion whim.

Although we survey here the industrial applications of some extracellular enzymes produced by the genus *Bacillus*, a detailed analysis of this subject goes beyond the scope of this chapter and has been dealt with extensively in recent publications (14, 19, 25, 128, 200). Also, throughout this chapter, we refer the reader to some recent reviews dealing with specific enzymes and their current or possible industrial uses.

The major thrust of this chapter is a discussion of production strains and the regulation of extracellular-enzyme expression. We use the subtilisin gene as a model. The chapter also analyzes the properties of amylases and cellulases produced by *Bacillus* spp., with comparisons to similar enzymes made by other microorganisms. Proteases, which currently command a market value of more than $400 million (70), will be dealt with in a separate chapter of this volume (see chapter 63).

PRODUCTION STRAINS

As mentioned above, the use of enzymes in industrial processes has been limited by two major factors: (i) the ability to identify a microorganism synthesizing an enzyme with useful characteristics and (ii) the likelihood of being able to transform this microbe into a high-volume, low-cost enzyme factory. Given the traditional approaches of screening and testing naturally secreted enzymes followed by mutagenesis of the host to improve or modify its productivity, the problem is not easy to overcome. The effort required to achieve these goals is of uncertain outcome, time-consuming, and labor intensive (129) and is therefore necessarily associated with high costs. Recent developments in molecular biology and genetics have simplified the task.

Expression of newly available natural enzymes or novel ones created by protein engineering requires the availability of reliable production hosts that are easy to manipulate. The genus *Bacillus* is one of the best candidates to fulfill this need. The total industrial enzyme market has a 1992 value of about $800 million (70), and about two-thirds of these enzymes are produced by members of the genus *Bacillus*. The typical enzyme yield from a *Bacillus* fermentation process is estimated to be around 20 g/liter of secreted material in a relatively short time with very low-cost carbon and nitrogen sources (21, 200). This demonstrates the abilities of *Bacillus* spp. to produce large quantities of enzymes at competitive costs. It is obvious from other chapters in this book that the tools for genetic manipulation of some members of this genus are highly refined. The use of well-characterized mutants combined with the most advanced molecular biology techniques also allows a rapid and more accurate understanding of the behavioral physiology of *Bacillus* spp. in fermentor conditions. This understanding permits the construction of novel production strains tailored for specific requirements. Furthermore, the fact that these bacteria have been used for decades to

Eugenio Ferrari, Alisha S. Jarnagin, and Brian F. Schmidt • Genencor International Inc., 180 Kimball Way, South San Francisco, California 94080.

produce substances generally recognized as safe makes *Bacillus* spp. an excellent choice for the production of recombinant proteins.

Historically, *Bacillus licheniformis* and *Bacillus amyloliquefaciens* have been the organisms of choice for fermentation production, largely because of the properties of their extracellular enzymes, which are more suitable for certain industrial applications. The major drawback with the industrial strains is that they are very recalcitrant to most genetic manipulation, making it very difficult to modify them to produce proteins coded by rDNA genes of heterologous or homologous origin. There is also a paucity of published material on the physiology and genetics of the strains used for industrial fermentation processes, primarily because of the proprietary nature of large-scale production processes. Also, the diversity of the fermentation equipment as well as the strain employed and the enzyme produced may influence the choice of media and the manufacturing process used (see chapter 60).

Two other major issues are associated with the development of these bacteria as general production hosts. One issue is related to the presence of numerous secreted proteases that are responsible for the degradation of most heterologous secreted gene products (24; see chapter 63). The second issue is associated with the limited secretion of most heterologous gene products. Because the industrial strains are difficult to manipulate genetically, it is troublesome to isolate and characterize mutants deficient in production of proteases or having altered secretion characteristics. This makes *Bacillus subtilis*, with its exquisitely developed genetics, the leading candidate to become the production host of choice. On the other hand, there are no clear reports that *B. subtilis* I-168, the standard transformable strain, has ever been used for industrial fermentations. This raises doubts that laboratory strains of this organism may ever reach production rates similar to those achieved by the classical industrial strains of *B. amyloliquefaciens* and *B. licheniformis*.

The construction of *B. subtilis* I-168 derivatives capable of producing several hundredfold more α-amylase than the parent strain has nonetheless been described elsewhere (58, 193, 196). These strains were constructed by stepwise introduction via transformation of a combination of promoter mutations (*amyR2*), chromosomal amplifications (*tmrB*), and other *trans*-acting mutations (*degU*) that affect the production of this enzyme. Similarly, as shown in Table 1, a *B. subtilis* derivative has been made capable of producing subtilisin at levels 100-fold higher than those obtained in a wild-type strain (31). This was achieved simply by introducing mutations such as *scoC* (23) and *sacU*(Hy) (91), which have synergistic effects on subtilisin production, into a transformable strain derived from I-168.

Table 1. Subtilisins obtained from plasmid-bearing *B. subtilis* strains[a]

Strain	Genotype	Amt of subtilisin produced (mg/liter)
BG3001	Δ*npr* Δ*aprE* pBS7AS	20
BG3002	Δ*npr* Δ*aprE*::pSAI	0.5
BG3004	*sacU32*(Hy) Δ*npr* Δ*aprE*::pSAI	40
BG3008	*scoC* Δ*npr* Δ*aprE*::pSAI	4
BG3009	*sacQ* Δ*npr* Δ*aprE*::pSAI	5
BG3010	*scoC sacU32*(Hy) Δ*npr* Δ*aprE*::pSAI	60
BG3013	*sacQ sacU32*(Hy) Δ*npr* Δ*aprE*::pSAI	30
BG3016	*scoC sacQ* Δ*npr* Δ*aprE*::pSAI	27

[a] Plasmid pSAI is an integrated plasmid, and pBS7AS is a multicopy replicating plasmid. Both plasmids carry the subtilisin gene.

REGULATION OF GENE EXPRESSION

Almost all of the extracellular enzymes in *Bacillus* spp. are under temporal control (possibly because of their scavenging nature). Some of them, such as amylase, are synthesized at the end of vegetative growth, although they do not appear to be controlled by sporulation (70). The expression of other genes, such as those encoding subtilisin and neutral protease, cannot be dissociated from the onset of sporulation (176). Both academic and commercial laboratories have encouraged studies of the elements that regulate expression of these extracellular enzymes. Among these elements, the promoter that regulates subtilisin gene expression stands out for its complexity and has become a paradigm for the study of stage 0 sporulation genes and transitional state regulators (for reviews, see references 152 and 176). We use the term "transition state regulator" as defined by Hoch and collaborators (125, 153, 154) to define those regulatory genes that exert their controlling functions during the transition between growth and sporulation.

Through *aprE-lacZ* fusions and analysis of subtilisin-specific mRNA synthesis, it has been established that several stage 0 sporulation genes effectively control the expression of subtilisin (29, 30). Progressive deletions of the region immediately upstream of the subtilisin promoter, from about −400 to −50, have determined the sites of action of several transcriptional regulators of this gene (52).

Although it was known that the *abrB* gene product was involved in modulating the control of subtilisin expression exerted by the stage 0 genes (59), only with the cloning of the gene has it been possible to elucidate the role of *abrB*. Footprinting and DNAse protection experiments indicate that the AbrB protein is capable of binding to the subtilisin promoter from −60 to +15 (Fig. 1), thus effectively preventing transcription of the gene during vegetative growth (154). The synthesis of AbrB appears to be repressed by Spo0A at the onset of sporulation, and subsequent lowering of AbrB levels allows RNA polymerase to transcribe the *aprE* gene (153, 155). Because of the lack of any sequence similarity among the promoters to which AbrB has been shown to bind in vitro, it has been postulated that recognition occurs at the level of DNA structure (152).

Another controller of subtilisin expression is the *hpr* gene product. The *hpr* gene has been cloned (124), and Hpr protein has been shown to bind at four sites on the subtilisin promoter. Three of these sites are located in the region upstream of the subtilisin promoter, and the fourth one is on the promoter itself (Fig. 1), where it partially overlaps the AbrB binding

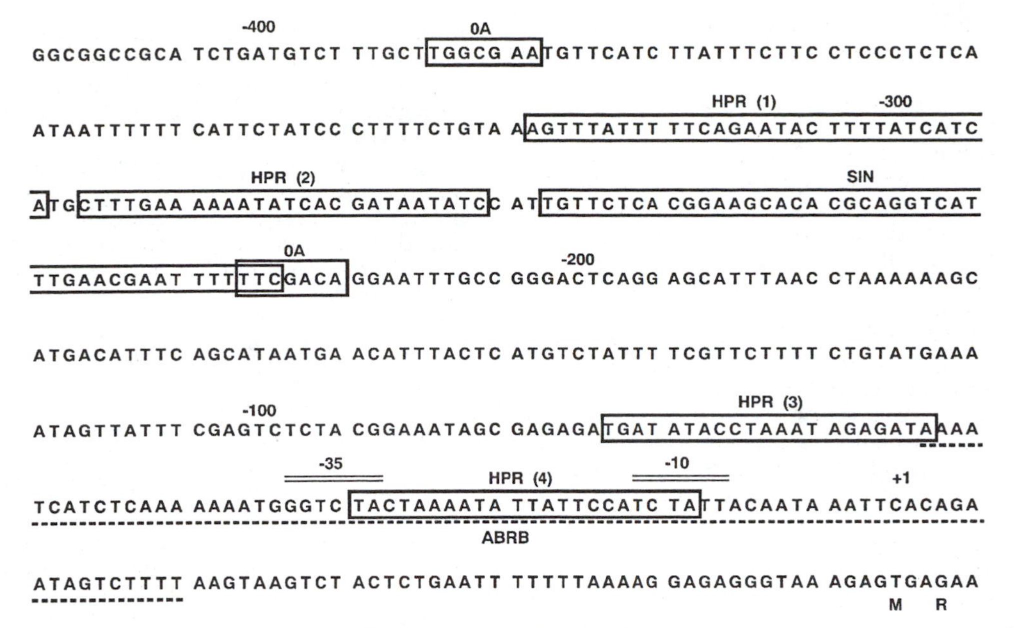

Figure 1. Locations of binding sites of *aprE* regulators on the subtilisin promoter. The regions protected from DNAse digestion by Hpr and Sin are boxed; the dashed line indicates the area of protection due to AbrB. This figure is courtesy of J. A. Hoch.

site (74). Analysis of the Hpr-binding sites suggests that the palindrome AATANTATT is the consensus binding sequence for Hpr.

Another repressor protein, Sin, binds to the region upstream of the *aprE* promoter (44, 74). Like mutations in *hpr*, mutations in the *sin* gene seem to have pleiotropic effects on sporulation (149).

It appears that most of the proteins that bind in vitro to the subtilisin promoter act as repressors. None of the gene products that behave as transcriptional activators bind to the *aprE* promoter in vitro. Possibly the most important regulators of this group are encoded by the *degS-degU* operon (see chapter 50). *degS-degU* mutants show a dramatic effect on the expression of all degradative enzymes tested so far and on a variety of other phenotypes such as motility and competence (30, 53, 91). Mutations in either one of the genes of this operon can dramatically increase (or decrease) the level of subtilisin by modulating the level of *aprE* mRNA (52). The *degS-degU* operon has been cloned and characterized, and the DegS-DegU gene product pair has been shown to belong to the family of two-component signaling systems (54, 87, 105). Although it has been shown that DegU can be phosphorylated in vitro (18, 106) and that "hyper" mutations in the *degU* gene cause a longer half-life of the DegU-phosphate complex (104), the mechanisms of action of these proteins remain unclear. One likely explanation is that DegU controls the expression of molecule X, which in turn acts as a transcriptional activator for the *aprE* gene. The fact that gene X has eluded detection might indicate that X has some essential role in cell growth or sporulation. In this case, any mutation in gene X would either be lethal or have a dramatic effect on the sporulation process as well.

Two other genes, *degQ* and *degR*, are capable of increasing the level of subtilisin-specific mRNA, and they seem to exert their action in the same promoter region that DegU does (52). In fact, these genes are under the transcriptional control of DegS-DegU (53). DegQ and DegR are small proteins, less than 70 amino acids long, with a helix-turn-helix structure typical of DNA-binding proteins, although no DNA-binding properties have been shown so far. Very similar to this pair is the *senS* gene product (187). SenS is also a very short peptide with a helix-turn-helix structure that when present in multicopy increases the level of transcription from the *aprE* promoter (187). Other genes such as *pai* (61) and those involved in the regulation of competence seem to have an effect on the expression of subtilisin.

It is puzzling that a gene such as *aprE* with a nonessential function would require such a complex regulatory scheme on the part of the cell. Perhaps a word of caution is needed. Because of the complexity of the sporulation process and the number of biochemical events occurring in the cell during the transition state, the observed effect of all these gene products on subtilisin expression could be due simply to an imbalance in cell physiology caused by the various mutations rather than to a direct cause-effect relationship. In any case, the subtilisin story is a good example of the complexity involved in developing *Bacillus* production hosts. These complexities are probably similar in magnitude to those encountered in developing other expression systems capable of high-yield, low-cost production. Thus, for rDNA production, it seems advisable to use organisms with well-developed molecular genetics, and *B. subtilis* I-168 has many characteristics necessary for optimal rDNA gene expression.

INDUSTRIAL ENZYMES AND THEIR APPLICATIONS

Starch-Degrading Enzymes

The most important members of the starch-degrading group of enzymes are the amylases, traditionally used in conjunction with other enzymes for the conversion of starch to high-fructose syrup. The 1992 market value for amylases used for this application exceeds $100 million. Because of the conditions at which the liquefaction process is carried out (see below), a major breakthrough for this group would be the isolation or generation of an enzyme with improved performance at high temperature.

New or improved amylases could be used as additives in laundry or dishwashing detergents. The market for enzymes for these applications could easily rival that of the alkaline proteases. There is also a score of other minor applications in a variety of industries such as textile sizing, brewing, and sewage treatment. The abilities of some starch-degrading enzymes to make cyclodextrins has generated high interest. Because of their peculiar shapes and internal hydrophobic regions, these compounds are being studied as vehicles for encapsulating and protecting compounds used in a variety of industrial applications involving food, pharmaceuticals, and cosmetics (51).

Cellulases

Cellulases from bacteria have played a secondary role to their related fungal counterparts. Only recently has a more careful screening of soil bacteria discovered several cellulase producers. The role of bacterial cellulases in commercial applications, although now only in its infancy, will probably increase as the enzymes become better characterized. The bacterial cellulases, several of which are produced by members of the genus *Bacillus*, differ from the fungal cellulases in several ways. With rare exceptions, the fungal enzymes display a higher activity at acidic pH, while the bacterial enzymes seem to perform best at neutral or alkaline pH. In addition, differences in substrate specificity and stability profiles at high temperatures are being investigated (135, 156).

As mentioned in the introduction, the main driving force behind the development of cellulases was the quest for ways to generate alternate sources of energy. Cellulases are a necessary step in the conversion of cellulose to ethanol. While this application is largely experimental, a variety of other interesting uses have appeared. Cellulases are used in fruit juice and olive oil extraction, wine and beer making, peel removal of citrus skin for the fresh-fruit industry, and modification of grain and fibers for baking (16). Additionally, cellulases are widely used in the manufacture of stone-washed denim products, as mentioned in the introduction.

Lipases

Similar to cellulases, most of the lipases commercially sold for large-scale industrial applications are of fungal origin. Only recently has there been an interest in lipases of bacterial origin (48) (particularly those from *Bacillus* spp.) because of their different characteristics, including increased resistance to higher temperatures compared with the fungal enzymes (143).

Lipases are currently being used for a variety of applications, the fastest growing of which is in the detergent industry. Other applications include removal of fats in leather processing, flavor development in dairy products, and processing of meat and cheese products.

Proteases

Because of the important role that proteases have played in the development of molecular genetic studies of *Bacillus* spp., an entire chapter is dedicated to this topic (see chapter 63). We limit ourselves here to brief considerations of a commercial nature. Two proteases, a neutral protease and an alkaline protease (subtilisin), from *Bacillus* spp. are produced for commercial use, and their combined market value is estimated at $400 million in 1992 (70), of which more than $300 million comes from sales of subtilisin for laundry detergent. This makes subtilisin the commercial enzyme with the highest dollar volume of any industrial enzyme. Other applications for these enzymes include baking, brewing, cheese making, fish and leather processing, and use in meat tenderizers and digestive aids (25).

Glucose Isomerase

Although the most commonly used glucose isomerase in corn starch processing is from *Streptomyces* spp., an enzyme from *Bacillus coagulans* is also being used because of its different pH activity profile and temperature stability (200).

STARCH-DEGRADING ENZYMES

Starch is the primary storage polysaccharide found in higher plants. One component of starch, amylose, is a linear polymer composed of $\alpha(1\rightarrow4)$-linked glucose units. The lengths of the amylose polymers can be quite heterogeneous, but they generally average about 1,000 glucose residues. The other component of starch, amylopectin, is a highly branched polymer of glucose with $\alpha(1\rightarrow4)$ linkages along the main backbone and $\alpha(1\rightarrow6)$ linkages forming the branch points (Fig. 2). On average, there are about 30 glucose units between the branch points. Glycogen, the major storage polysaccharide in animal cells, is similar to amylopectin but is more highly branched (every 8 to 12 glucose residues). The actual composition of the starch (length of polymers and branch points, percentage of amylopectin, etc.) depends on the plant source from which it was isolated.

Starch-degrading enzymes have different action patterns on starch (Fig. 2) and have been characterized by a number of different criteria. Exoamylases cleave the $\alpha(1\rightarrow4)$ glucosidic linkages starting from the free nonreducing ends and continuing down the chain. Depending on the enzyme, they may or may not be able to hydrolyze the $\alpha(1\rightarrow6)$ glucosidic bonds at

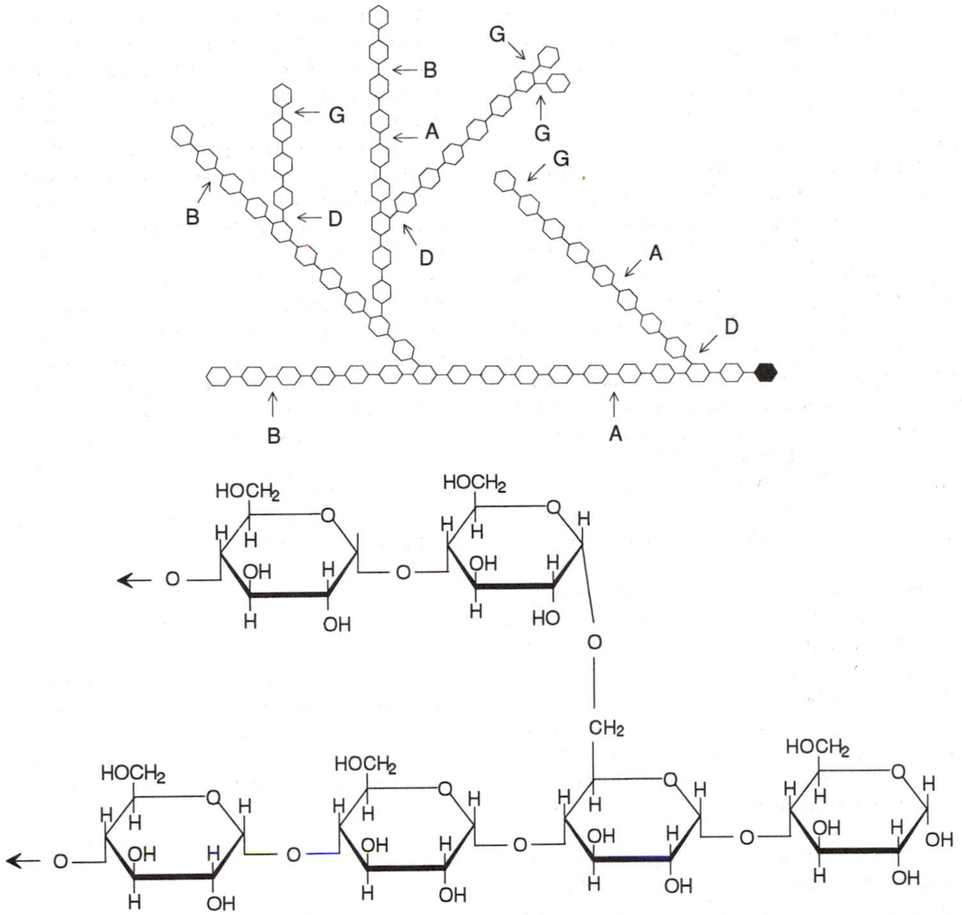

Figure 2. Schematic representation of an amylopectin-type molecule, with an $\alpha(1\rightarrow6)$ branch point shown in more detail below. Potential cleavage sites for α-amylases (A), β-amylases (B), debranching enzymes (D; pullulanases or isoamylases), and glucoamylases (G) are indicated by arrows. The reducing end is indicated by a filled symbol.

the branch points. Endoamylases can cleave internal $\alpha(1\rightarrow4)$ and/or $\alpha(1\rightarrow6)$ linkages. Starch-degrading enzymes can also be characterized by their abilities to hydrolyze different substrates, e.g., amylose, amylopectin, glycogen, and pullulan [a linear polymer of about 480 α-1,4-maltotriose units linked by $\alpha(1\rightarrow6)$ glucosidic bonds]. The types, lengths, and anomeric configurations (α or β) of the products have also been used to help define amylase activity.

Exoamylases

Three different types of exoamylases (α-exoamylases, glucoamylases, and β-amylases) have been described (179). The α-exoamylases cleave the $\alpha(1\rightarrow4)$ glucosidic bonds producing maltose or longer oligosaccharides with α-anomeric configurations. The glucoamylases can hydrolyze both $\alpha(1\rightarrow4)$ and $\alpha(1\rightarrow6)$ terminal linkages [cleavage of $\alpha(1\rightarrow6)$ bonds is much slower], resulting in complete digestion of starch to β-glucose. The nonreducing penultimate $\alpha(1\rightarrow4)$ linkage is cleaved by β-amylases, producing β-anomeric maltose. Complete digestion of amylopectin with β-amylases results in a limit dextrin, since $\alpha(1\rightarrow6)$ linkages cannot be cleaved by this enzyme. In industry, glucoamylases (from fungal sources) are among the enzymes used extensively for the production of high-fructose syrups from starch. The major commercial use of other exoamylases is in the production of high-maltose syrups. Some exoamylases may also be utilized to produce maltotriose, maltotetraose, or longer oligosaccharides of defined lengths.

Only a few α-exoamylases have been characterized in any detail. The gene from *Bacillus stearothermophilus* coding for a maltogenic α-exoamylase has been cloned and expressed in *B. subtilis* (119). An enzyme from *Bacillus circulans* generates maltohexaose, which upon further digestion yields maltotetraose and maltose (167). In general, α-exoamylases are similar to α-amylases with respect to size, pH optimum, thermal stability, and Ca^{2+} requirements. In fact, the determined amino acid sequences of the maltotetraose-forming α-exoamylases from *Pseudomonas stutzeri* (37) and *Pseudomonas saccharophilia* (199) are homologous to α-amylases (see below).

Although glucoamylase is an extremely important commercial enzyme, few glucoamylases have been recovered from bacteria. Glucoamylase activities in *B. stearothermophilus* (151), *Clostridium acetobutylicum* (15), and *Clostridium thermohydrosulfuricum* (66) have been reported, although in some cases, the observed activities could be due to the presence of more than one type of enzyme. Recently, a glucoamylase from *Clostridium thermosaccharolyticum* was purified to homogeneity. This enzyme is ~75 kDa; is thermostable; has a pH optimum of 5.0; and hydrolyzes amylose, amylopectin, and glycogen but has very low activity when pullulan is a substrate (150).

β-Amylase activity has been reported in a number of different *Bacillus* spp., including *Bacillus cereus* (144, 164), *B. circulans* (145), *Bacillus megaterium* sensu stricto (170), *B. megaterium* (56), *Bacillus polymyxa* (98), *B. stearothermophilus* (151), *B. subtilis* (32), and an unidentified *Bacillus* sp. (10). The β-amylase genes from *B. polymyxa* (35, 79, 132, 175), *B. circulans* (145, 146), and *Clostridium thermosulfurogenes* (82) have been cloned and sequenced. The gene cloned from *B. polymyxa* is unusual in that it codes for a ~130-kDa protein that is processed into a β-amylase from the N-terminal portion and a 48-kDa α-amylase from the C-terminal half (175). The two β-amylases of the cloned genes from *B. polymyxa* are 98% identical to each other and are 79 and 48% identical to the *B. circulans* and *C. thermosulfurogenes* enzymes, respectively. The amino acid sequences are also homologous to the β-amylases from barley and soybean (86, 103); a number of regions show a high degree of similarity, and three of these have been suggested as being active-site regions (82, 103, 175).

An acid-base catalysis by starch-degrading enzymes is generally thought to occur either by an oxocarbonium ion mechanism or by a nucleophilic attack (136). Since most starch-degrading enzymes have low pH optima, electrostatic stabilization of the oxocarbonium ion (or nucleophilic attack) is usually considered to be accomplished by a carboxylate, while the proton-donating group may be a carboxyl. Alternative nucleophiles (e.g., cysteine) and proton-donating groups (e.g., histidine) need to be considered, especially for β-amylases, since they generally have pH optima near 7.0 and are sensitive to sulfhydryl reagents. For β-amylases, mechanisms have also been proposed to take into account the inversion of the anomeric configuration (34, 184). In the β-amylase sequences available to date, five conserved aspartates, two glutamates, and three histidines are active-site candidates; two of these aspartates (D89, D328), one glutamate (E163), and one histidine (H81) are in the putative active-site regions (numbering is for the mature *B. polymyxa* protein sequence). One cysteine (C83) is conserved in all the sequenced β-amylases, another (C323) is present in all but the *B. circulans* enzyme, and both are located in putative active-site regions.

Endoamylases

A specific endoamylase may be classified as an α-amylase, isoamylase, pullulanase, isopullulanase, neopullulanase, oligo-1,4-glucosidase, oligo-1,6-glucosidase, or cyclodextrin glucanotransferase (CGTase) depending on the original method used to characterize the enzyme. The debranching enzymes (isoamylases and pullulanases) can effectively cleave the internal α(1→6) glucosidic bonds in starch and/or pullulan and may also hydrolyze internal α(1→4) linkages as an α-amylase depending on the enzyme, substrate, and reaction conditions. By definition, pullulanases can hydrolyze pullulan (isoamylases cannot); in contrast to pullulanase, isopullulanases and neopullulanases cleave the α(1→4) glucosidic linkages of pullulan on the reducing and nonreducing sides of the α(1→6) branch points, respectively. CGTases, like α-amylases, cleave internal α(1→4) bonds in starch but produce a large amount of cyclodextrins as products [α-, β-, or γ-cyclodextrins are α(1→4)-linked rings of six, seven, or eight glucose residues, respectively]. Since cyclodextrins are able to solubilize and/or stabilize a variety of organic and inorganic compounds by forming inclusion complexes, they are attractive for medical and other commercial applications (33, 36, 73). Because of the importance of α-amylases and debranching enzymes in commercial starch hydrolysis, numerous endoamylases from *Bacillus* spp. and other microorganisms (179) have been characterized. For this application, the driving force is for enzymes that quickly produce the desired products at high temperature and low pH.

Homology

Since it is not unusual to find endoamylases with different cross specificities (α-amylases with pullulanase activity, α-amylases with CGTase activity, etc.), it is not surprising that nearly all the endoamylases sequenced to date have homologous regions (94, 95, 110, 111, 137, 161, 162). In addition, a branching enzyme from *Escherichia coli* (2) and an intracellular dextran glucosidase from *Streptococcus mutans* (139) also contain sequences homologous to endoamylases. Numerous sequence comparisons reveal four generally recognized areas of homology among the endoamylases (designated I, II, III, and IV in Fig. 3). There also appear to be three additional small regions of lower similarity (i, ii, and iii in Fig. 3).

From a computer-generated multiple-sequence alignment (28), a dendrogram was generated to represent relationships between various endoamylases from *Bacillus* spp. and other gram-positive bacteria (Fig. 4; additional sequences from other organisms were included for comparative purposes). A large group of related enzymes includes both α-amylases and CGTases from a variety of species (referred to further as the CGTase group). Even the α-amylase from the fungus *Aspergillus oryzae* seems to be related to this CGTase group (Fig. 4). There is a loose linkage of this group to the α-exoamylases from *P. stutzeri* and *P. saccharophilia* (not listed in Fig. 4) and to an alkalophilic α-amylase from a gram-positive bacterium.

In another, smaller group, the thermostable pullulanase from *B. stearothermophilus* is linked to the pullulanase from the gram-negative bacterium *Klebsiella aerogenes* (and somewhat less to the isoamylase from *Pseudomonas amyloderamosa*), but it is not closely related to the oligo-1,6-glucosidase of *B. cereus* or the α-amylase-pullulanase from *C. thermohydrosulfuricum*.

It is apparent that closely related *Bacillus* species do not necessarily have closely related endoamylases. For example, as species, *B. amyloliquefaciens*, *B. licheniformis*, *B. megaterium*, and *B. subtilis* are gener-

```
                 Region i                 Region ii                  Region I

               36         47        62                80        92              108
  A:Bst  <-----LGITALSLP.PAY---------LYDLGEFNQKGTVRTKYGT---------HAAGMQVYADVVFDHKG------
  P:Bst  <-----LGVTHVELL.PFN---------GYNPLHYNA...PEGSYAT---------QSNGIRVIMDVVTNHVY------
  A:Bme  <-----NGIWMMPVN.PSP---------KYDVTDYYN...IDPQYGN---------DKRDVKVIMDLVVNHTS------
 AP:Cth  <-----LGISVIYLN.PIF---------RYDTTDYTK...IDELLGD---------HAKGIKVILDGVFNHTS------
  A:Shy  <-----AGYGYVEVS.PAS---------SYQPVSYKI...AGR.LGD---------HAAGVKVIADAVVNHMA------
  A:Bsu  <-----AGYTAIQTS.PIN---------LYQPTSYQI...GNRYLGT---------EEYGIKVIVDAVINHTT------
C:B17-1  <-----MGVTAIWISQPVE---------GYWARDFKK...TNPAYGI---------HAKNIKVIIDFAPNHTS------
  A:Bpo  <-----MGFTAIWIT.PVT---------GYHTYDFYA...VDGHLGT---------HDKNIAVMVDVVVNHTG------
  A:G+   <-----LGTNAIWISAPWE---------GYYGLDFTA...MDQNMGT---------HSLGIRVVLDIVMNHVG------
                ♦ ◊◊     ♦            ◊   ◊◊          ◊◊         ◊  ◊ ◊◊◊ ♦ ◊ ◊♦◊


                              Region II                                 Region III

              202                                     241               260     268
  A:Bst  ------ADLDMD..HPEVVTELKNWGKWYVNTTNIDGFRLDGLKHIKF----------------FTVGEYWSY------
  P:Bst  ------NDIASE..RKMVRKWIIDSVRFWVEEYHVNGFRFDLMGILDV----------------LVFGEGWDL------
  A:Bme  ------PDLNYD..NPEVRKEMINVGKFWL.KQGVDGFRLDAALHIFK----------------YLTGEVWDQ------
 AP:Cth  ------ADFIIN..NPNAISKYWLNPDGDK.NVGADGWRLDVANEVAH----------------PMVAENWND------
  A:Shy  ------ADLGTG..SDYVRTTIAGYL.GLR.SLGVDGFRIDAAKHISA----------------YGAGEAVRP------
  A:Bsu  ------YDWNTQ..NTQVQSYLKRFLDRAL.NDGADGFRFDAAKHIEL----------------FQYGEILQD------
C:B17-1  ------ADLNHN..NSTVDTYLKDAIKMWL.DLGIDGIRMDAVKHMPF----------------FTFGEWFLG------
  A:Bpo  ------DDLNHE..NPATANELKNWIKWLLNETGIDGLRLDTVKHVPK----------------FTMGEIFHG------
  A:G+   ------AEPYRQDLNIAPKDYLIKWITSWVEEFGIDGFRVDTAKHVEI----------------WMTAEVFGH------
                ◊      ◊  ◊               ◊ ◊◊◊♦ ♦ ◊◊◊                      ◊♦

                              Region IV     Region iii

                             224     332          358        369
                  A:Bst  ------VTFVDNHDT------------EGYPCVFYGDYY------>
                  P:Bst  ------INYVESHDN------------QGIPFLHSGQEF------>
                  A:Bme  ------GIFLTNHDQ------------PGNPYIYYGEEI------>
                 AP:Cth  ------MNLLGSHDT------------PGMPSIYYGDEA------>
                  A:Shy  ------RTFVDNWDT------------YGSPNVYSGYEW------>
                  A:Bsu  ------VTWVESHDT------------GSTPLFFSRPEG------>
                C:B17-1  ------VTFIDNHDM------------RGVPAIYYGTEQ------>
                  A:Bpo  ------GVFIDNHDV------------RGIPIIYQGTEQ------>
                  A:G+   ------LSYVSSHDT------------PGGVQVFYGDET------>
                                 ◊◊ ◊◊♦◊             ◊ ◊  ◊◊◊ ◊
```

Figure 3. Seven regions of homology in amino acid sequence alignment of nine endoamylases chosen from different cluster groups (Fig. 4). Computer alignments (Genetics Computer Group, Inc.) of the 34 amino acid sequences shown in Fig. 4 were determined by pairwise alignments of the most related sequences or sequence clusters (28). Amino acids conserved in at least five of the nine sequences shown (◇) and those conserved in all 34 sequences shown in Fig. 4 (◆) are indicated. Dashed lines indicate gaps in the sequences of various lengths, and a period represents a gap of one amino acid. Sequences are named by whether they were classified as an α-amylase (A:), an α-amylase-pullulanase (AP:), a CGTase (C:), or a pullulanase (P:) and by the organism from which they were isolated: *B. stearothermophilus* (Bst), *B. megaterium* (Bme), *C. thermohydrosulfuricum* (Cth), *S. hygroscopicus* (Shy), *B. subtilis* (Bsu), *Bacillus* sp. strain 17-1 (B17-1), *B. polymyxa* (Bpo), or a gram-positive bacterium (G+). As in the text, numbering is given for the mature α-amylase from *B. stearothermophilus*, and regions with putative catalytic and/or binding residues are denoted by uppercase roman numerals.

ally grouped together (130), yet only the *B. amyloliquefaciens* and *B. licheniformis* α-amylases are highly similar (they form a group with *Bacillus* sp. strain 707 and *B. stearothermophilus* α-amylases). The α-amylase most closely related to the *B. subtilis* enzyme is one produced by *Butyrivibrio fibrisolvens*, and both are loosely linked to the *Streptomyces* spp. (also the α-amylase from *Thermomonospora curvata* [not listed in Fig. 4]) and two gram-negative α-amylases. The *B. megaterium* α-amylase clusters in a completely different group and is most similar to the oligo-1,6-glucosidases from *B. cereus* and *Bacillus thermoglucosidasius* and the dextran glucosidase from *S. mutans*.

Naturally, only after plants began using starch as a storage polysaccharide did endoamylases start evolving into their present forms. Thus, it seems likely that as the ancestral endoamylase gene (or genes) evolved, different gene lines were at various times passed horizontally between potentially highly unrelated species (most likely by transposons or other vectors). Depending on the environmental demands, the ancestral genes could have evolved into the different classes of endoamylases (α-amylases, isoamylases, pullulanases, or CGTases) or in some cases retained activities common to more than one class of enzyme.

Structure

The three-dimensional structures of the α-amylases from *A. oryzae*, *Aspergillus niger*, and pig pancreas and the structure of the CGTase from *B. circulans* have been determined (9, 12, 83, 99). In addition, the structure of the *A. oryzae* α-amylase has been used to model the three-dimensional structure of the *B. stearothermophilus* α-amylase (60).

The basic α-amylase structure has been divided into three domains. The N-terminal catalytic portion con-

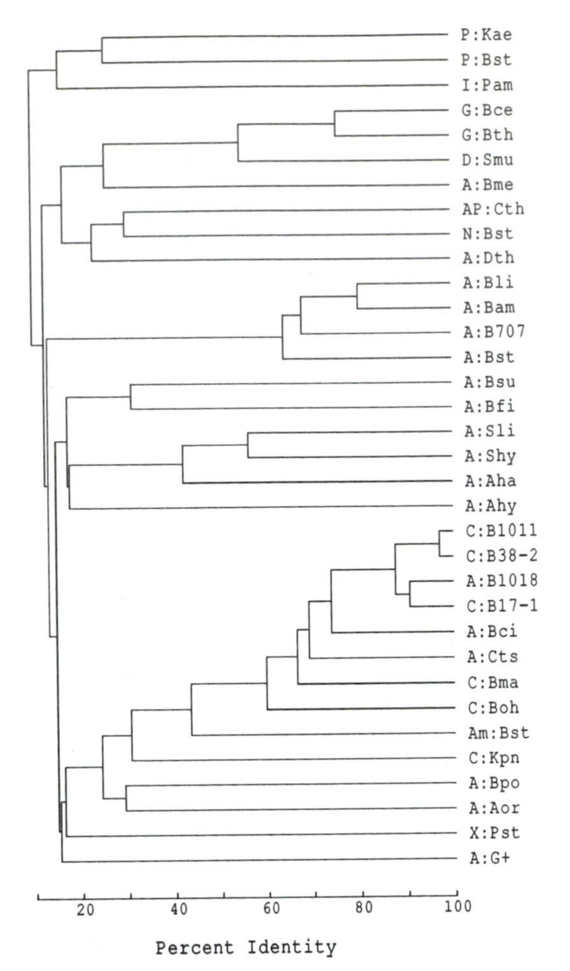

Figure 4. Dendrogram illustrating the relationships among 34 endoamylases aligned as described in the legend to Fig. 3. The nomenclature used is the same as in Fig. 3. Additional classifications are isoamylase (I:), oligo-1,6-glucosidase (G:), dextran glucosidase (D:), neopullulanase (N:), maltogenic α-amylase (Am:), and exoamylase (X:). The sequences in order from top to bottom are: *K. aerogenes* pullulanase (P:Kae; 77), *B. stearothermophilus* pullulanase (P:Bst; 89), *P. amyloderamosa* isoamylase (I:Pam; 1), *B. cereus* oligo-1,6-glucosidase (G:Bce; 189), *B. thermoglucosidasius* oligo-1,6-glucosidase (G:Bth; 188), *S. mutans* dextran glucosidase (D:Smu; 139), *B. megaterium* α-amylase (A:Bme; 102), *C. thermohydrosulfuricum* α-amylase-pullulanase (AP:Cth; 101), *B. stearothermophilus* neopullulanase (N:Bst; 88), *Dictyoglomus thermophilium amyB* amylase (A:Dth; 63), *B. licheniformis* α-amylase (A:Bli; 198), *B. amyloliquefaciens* α-amylase (A:Bam; 166), *Bacillus* sp. strain 707 α-amylase (A:B707; 174), *B. stearothermophilus* α-amylase (A:Bst; 67), *B. subtilis* α-amylase (A:Bsu; 195), *B. fibrisolvens* α-amylase (A:Bfi; 138), *Streptomyces limosus* α-amylase (A:Sli; 93), *S. hygroscopicus* α-amylase (A:Shy; 64), *Alteromonas haloplanctis* α-amylase (A:Aha; 27), *Aeromonas hydrophilia* α-amylase (A:Ahy; 47), *Bacillus* sp. strain 1011 CGTase (C:B1011; 80), *Bacillus* sp. strain 38-2 CGTase (C:B38-2; 75), *Bacillus* sp. strain B1018 α-amylase (A:B1018; 68), *Bacillus* sp. strain 17-1 CGTase (C:B17-1; 76), *B. circulans* α-amylase (A:Bci; 115), *C. thermosulfurogenes* α-amylase (A:Cts; 3), *Bacillus macerans* CGTase (C:Bma; 163), *Bacillus ohbensis* CGTase (C:Boh; 147), *B. stearothermophilus* maltogenic α-amylase (Am:Bst; 22), *K. pneumoniae* CGTase (C:Kpn; 8), *B. polymyxa* α-amylase (A:Bpo; 175), *A. oryzae* α-amylase (A:Aor; 190), *P. stutzeri* exoamylase (X:Pst; 37), and gram-positive bacterium α-amylase (A:G+; 13). Sequences from organisms that are not gram-positive bacteria were included for comparisons. For simplicity, some available sequences of endoamylases were not included, since most are similar to ones already used in the alignments (4, 26, 49, 57, 72, 111, 116, 126, 142, 157, 178, 181, 192, 194, 199). The distance to the branch point connecting two sequences is proportional to the identity between the two sequences, as shown by the scale on this figure (for example, the amino acid sequences of A:Sli and A:Shy are 56% identical). In general, the inclusion of a sequence in a given group should be considered approximate when the sequence is less than 20% identical to the other members of the cluster group.

sists of domains A and B. Domain A forms a typical $(\alpha/\beta)_8$ barrel structure. Domain B, inserted between the third β strand and the third α helix of domain A, consists of two antiparallel β sheets. Domain C, the C-terminal region, is a β barrel of eight antiparallel strands. Depending on the type of enzyme, additional domains may be found on either the N-terminal or the C-terminal side of this basic structure (83, 162, 192).

Recent evidence from nuclear magnetic resonance studies of the porcine pancreatic α-amylase suggests that catalysis occurs by a nucleophilic attack by a carboxylate group (on C-1 of the incipient reducing sugar) that forms a covalent intermediate rather than by a carbonium ion mechanism (168). Both of these mechanisms would also require an electrophile to donate a proton to the sugar-leaving group. On the basis of homology and structure, this electrophile has been proposed to be a carboxyl group (12, 99, 180) but might possibly be another proton-donating group. Three aspartic acids, D101, D234, and D331 (in regions I, II, and IV, respectively; numbering refers to the *B. stearothermophilus* mature α-amylase, since its modeled structure is available [60]), and one glutamic acid, E264 (region III), shown in Fig. 3 are conserved in all the sequences for the enzymes listed in Fig. 4. Another aspartic acid, D203, is conserved in all except the isoamylase from *P. amyloderamosa* (an asparagine is reported at this position) and the α-amylase from an alkalophilic gram-positive bacterium (A:G+ in Fig. 3). D234, E264, and D331 have all been suggested as catalytic residues, and by mutagenesis, each is necessary for enzyme activity but not for substrate binding (12, 60, 99, 107, 113, 165, 180). The most likely other potential proton-donating groups are H106 and H330 (regions I and IV, respectively; H330 is not conserved in the α-amylase from *Streptomyces hygroscopicus*).

In general, calcium ions are necessary for endoamylase stability and activity, and a major Ca^{2+}-binding site has been located in the three-dimensional structure near the active site (9, 12, 83). The side chains of D203 (mentioned above; region II) and N105 (an aspartic acid in *B. stearothermophilus*, a glutamine in the CGTase from *Bacillus macerans*; region I), the main chain carbonyl of H238 (conserved in nearly all sequences), and the main chain carbonyl of an unconserved amino acid some 10 residues upstream of D203 (corresponds to D190 in *B. stearothermophilus*) have been reported to be the Ca^{2+}-binding ligands (9, 12, 60, 83, 180). These residues are conserved even in the oligo-1,6-glucosidases, which reportedly have no calcium requirements (188).

Amino acids implicated in substrate binding have

generally been localized to high-similarity regions (12, 99). These include the highly conserved residues H106 (region I), D229 (region II), H238 (region II, shown by mutagenesis to be important for activity [60, 165, 180]), E264 (region III), and H330 (region IV). In addition, Y63, a conserved amino acid (except in the *C. thermosulfurogenes* α-amylase) in a region of lower similarity (region ii), also seems to be involved in substrate binding. When oligosaccharide substrates were used, the *B. amyloliquefaciens* α-amylase was reported to have nine glucose-binding subsites, with cleavage occurring between the sixth and seventh subsites (numbering from the ultimate nonreducing end) (85, 131, 169). Another conserved amino acid, R232 (region II), is necessary for activity and may be needed to electrostatically hold substrate-binding and/or active-site residues in the proper configuration (180).

As expected, all the extracellular enzymes listed in Fig. 4 except for the neopullulanase from *B. stearothermophilus* have typical signal sequences (182) that are removed during secretion. The *B. stearothermophilus* neopullulanase lacks a normal signal sequence yet is found exclusively in the culture medium except when the cloned gene is expressed in *B. subtilis*, in which only about 20% of the enzyme is produced extracellularly (88).

Stability

As stated above, for commercial starch hydrolysis, endoamylases that perform well at high temperature and low pH are desirable. Thus, many enzymes from acidophiles and thermophiles have been characterized (81, 140, 191). Of course, endoamylases with different activities may be needed for other industrial applications. For the production of high-glucose syrups from starch, the thermostable α-amylase from *B. licheniformis* has been the enzyme of choice for starch liquefaction. Liquefying α-amylases (which initially produce long oligosaccharides) like the one from *B. licheniformis* are preferred over saccharifying enzymes like *B. subtilis* α-amylase. After starch liquefaction, pullulanases from *Bacillus* species have also been added as an adjuvant with the fungal glucoamylases in order to decrease saccharification time (148). Unfortunately, many pullulanases isolated from thermophiles have α(1→4) instead of α(1→6) activity when starch is the substrate (140).

Not surprisingly, a number of studies directed at improving or understanding the thermostability of endoamylases have been undertaken. In several cases, hybrids of thermostable and thermolabile enzymes have been constructed by swapping DNA fragments from related genes (49, 75, 159). From these studies, the thermal instability of the *B. stearothermophilus* α-amylase and *Bacillus* sp. strain 17-1 CGTase mapped to the N-terminal portions of the enzymes, while thermal instability was located in both the N-terminal and the C-terminal ends of the α-amylase from *B. amyloliquefaciens*. It has been suggested that thermal inactivation of the less stable *B. amyloliquefaciens* and *B. stearothermophilus* α-amylases occurs by monomolecular conformational scrambling, while inactivation of the more stable *B. licheniformis* enzyme is due to deamidation of asparagine and glutamine residues (171, 172). In fact, single-point mutations (e.g., A269K) in the *B. amyloliquefaciens* α-amylase can increase the thermal stability nearly to that of *B. licheniformis* (159), possibly by reducing the rate of conformational scrambling. However, a single substitution (H133Y) in the *B. licheniformis* α-amylase increases its thermostability, which seems not to be related to deamidation (20). Other workers have found that more thermostable endoamylases tend to have larger numbers of proline residues, suggesting that prolines may decrease the entropy of unfolding (158, 160). In fact, a thermostable oligo-1,6-glucosidase was found to have 14 more prolines than a more thermolabile enzyme, but otherwise the two had 72% identical residues (188). However, proline composition does not seem to correlate with thermostability in α-amylases, for which hydrophobic interactions are the most important forces that determine irreversible thermoinactivation (6, 11). Thus, a single paradigm for increasing thermostability has not yet been discovered.

CELLULOLYTIC ENZYMES

Cellulases

Cellulose, one of the most prevalent polymeric enzymatic substrates of plant biomass, is a focus of biotechnologists interested in conversion of plant biomass to useful products such as ethanol for fuel (7). Cellulose is an unbranched polymer of anhydro-D-glucose linked in β-1,4 bonds. The polysaccharide polymer contains amorphous regions of random structure and crystalline sections of highly ordered structure. Cellulases have been studied for many years, with the primary focus on cellulases from fungi (17). The classes of cellulase enzymes are defined as endoglucanases (1,4-β-D-glucan glucanohydrolase [EC 3.2.1.4]), exoglucanases (1,4-β-D-glucan cellobiohydrolase [EC 3.2.1.91]), and β-glucosidases (β-D-glucoside glucohydrolase [EC 3.2.1.21]). Endoglucanases cleave the cellulose polymer at random sites in the amorphous structural regions and thereby provide additional start sites for attack by the exoglucanases that cleave the polymer from the nonreducing end, thus releasing cellobiose subunits. Exoglucanases, also called cellobiohydrolases, are necessary for hydrolysis of the crystalline regions of the cellulose polymer. β-Glucosidases cleave cellobiose to glucose and thereby ameliorate product inhibition for the exoglucanases. The degradation of cellulose by fungi is thought to involve all three of these enzymes acting synergistically (17). Most of the fungal cellulose-degrading enzymes have activity optima in the acidic pH range.

Research on bacterial cellulases has been going on for many years, with an increased interest recently in finding cellulases with altered properties (such as higher pH optima) and potential for new industrial applications (e.g., laundry detergents). The spectrum of bacterial cellulases has been extensively reviewed recently (7, 45, 135). In this section, we review the characteristics of cellulases of *Bacillus* species, with some discussion of cellulases of related species.

```
             1                                                    50
Bsu PAP115   MKRSISIFIT CLLITLLTMG GMIASPASAA GTKTPVAKNG QLSIKGTQLV
Bsu BSE616   MKRSISIFIT CLLITLLTMG GMLASPASAA GTKTPVAKNG QLSIKGTQLV
Bsu DLG      MKRSISIFIT CLLIAVLTMG GLLPSPASAA GTKTPVAKNG QLSIKGTQLV
BsuIFO3034   MKRSISIFIT CLLITVLTMG GLQASPASAA GTKTPAAKNG QLSIKGTQLV

             51                                                  100
Bsu PAP115   NRDGKAVQLK GISSHGLQWY GEYVNKDSLK WLRDDWGITV FRAAMYTADG
Bsu BSE616   NRDGKAVQLK GISSHGLQWY GEYVNKDSLK WLRDDWGITV FRAAMYTADG
Bsu DLG      NRDGKAVQLK GISSHGLQWY GDFVNKDSLK WLRDDWGITV FRAAMYTADG
BsuIFO3034   NRDGKAVQLK GISSHGLQWY GDFVNKDSLK WLRDDWGITV FRAAMYTADG

             101                                                 150
Bsu PAP115   GYIDNPSVKN KVKEAVEAAK ELGIYVIIDW HILNDGNPNQ NKEKAKEFFK
Bsu BSE616   GIIDNPSVKN KMKEAVEAAK ELGIYVIIDW HILNDGNPNQ NKEKAKEFFK
Bsu DLG      GYIDNPSVKN KVKEAVEAAK ELGIYVIIDW HILNDGNPNQ NKEKAKEFFK
BsuIFO3034   GYIDNPSVKN KVKEAVEAAK ELGIYVIIDW HILNDGNPNQ HKEKAKDFFK

             151                                                 200
Bsu PAP115   EMSSLYGNTP NVIYEIANEP NGDVNWKRDI KPYAEEVISV IRKNDPDNII
Bsu BSE616   EMSSLYGNTP NVIYEIANEP NGDVNWKRDI KPYAEEVISV IRKNDPDNII
Bsu DLG      EMSSLYGNTP NVIYEIANEP NGDVNWKRDI KPYAEEVISV IRKNDPDNII
BsuIFO3034   EMSSLYGNTP NVIYEIANEP NGDVNWKRDI KPYAEEVISV IRKNDPDNII

             201                                                 250
Bsu PAP115   IVGTGTWSQD VNDAADDQLK DANVMYALHF YAGTHGQFLR DKANYALSKG
Bsu BSE616   IVGTGTWSQD VNDAADDQLK DANVMDALHF YAGTHGQFLR DKANYALSKG
Bsu DLG      IVGTGTWSQD VNDAADDQLK DANVMYALHF YAGTHGQSLR DKANYALSKG
BsuIFO3034   IVGTGTWSQD VNDAADDQLK DANVMYALHF YAGTHGQSLR DKANYALSKG

             251                                                 300
Bsu PAP115   APIFVTEWGT SDASGNGGVF LDQSREWLKY LDSKTISWVN WNLSDKQESS
Bsu BSE616   APIFVTEWGT SDASGNGGVF LDQSREWLKY LDSKTISWVN WNLSDKQESS
Bsu DLG      APIFVTEWGT SDASGNGGVF LDQSREWLNY LDSKNISWVN WNLSDKQESS
BsuIFO3034   APIFVTEWGT SDASGNGGVF LDQSREWLNY LDSKNISWVN WNLSDKQESS

             301                                                 350
Bsu PAP115   SALKPGASKT GGWRLSDLSA SGTFVRENIL GTKDSTKDIP ETPSKDKPTQ
Bsu BSE616   SALKPGASKT GGWRLSDLSA SGTFVRENIL GTKDSTKDIP ETPAKDKPTQ
Bsu DLG      SALKPGASKT GGWPLTDLTA SGTFVRENIR GTKDSTKDVP ETPAQDNPTQ
BsuIFO3034   SALKPGASKT GGWPLTDLTA SGTFVRENIL GNKDSTKERP ETPAQDNPAQ

             351                                                 400
Bsu PAP115   ENGISVQYRA GDGSMNSNQI RPQLQIKNNG NTTVDLKDVT ARYWYKAKNK
Bsu BSE616   ENGISVQYRA GDGSMNSNQI RPQLQIKNNG NTTVDLKDVT ARYWYNAKNK
Bsu DLG      EKGVSVQYKA GDGRVNSNQI RPQLHIKNNG NATVDLKDVT ARYWYNVKNK
BsuIFO3034   ENGISVQYKA GDGGVNSNQI RPQLHIKNNG NATVDLKDVT ARYWYNAKNK

             401                                                 450
Bsu PAP115   GQNFDCDYAQ IGCGNVTHKF VTLHKPKQGA DTYLELGFKN GTLAPGASTG
Bsu BSE616   GQNVDCDYAQ LGCGNVTYKF VTLHKPKQGA DTYLELGFKN GTLAPGASTG
Bsu DLG      GQNFDCDYAQ MGCGNLTHKF VTLHKPKQGA DTYLELGFKT GTLSPGASTG
BsuIFO3034   GQNFDCDYAQ IGCGNLTHKF VTLHKPKQGA DTYLELGFKT GTLSPGASTG

             451                                                 500
Bsu PAP115   NIQLRLHNDD WSNYAQSGDY SFFKSNTFKT TKKITLYDQG KLIWGTEPN.
Bsu BSE616   NIQLRLHNDD WSNYAQSGDY SFFKSNTFKT TKKITLYDQG KLIWGTEPN.
Bsu DLG      NIQLRLHNDD WSNYAQSGDY SFFQSNTFKT TKKITLYHQG KLIWGTEPN.
BsuIFO3034   NIQLRLHNDD WSNYAQSGDY SFFQSNTFKT TKKITLYHQG KLIWGTEPH.
```

Figure 5. Comparison of protein sequences of cellulases from four different strains of *B. subtilis*. Positions marked in boldface indicate differences between the strains. Designations are as follows: Bsu PAP115, *B. subtilis* PAP115 (96); Bsu BSE616, *B. subtilis* BSE616 (122); Bsu DLG, *B. subtilis* DLG (134); BsulFO3034, *B. subtilis* IFO3034 (114).

B. subtilis Cellulases

Cellulases from four different strains of *B. subtilis* have been cloned and sequenced in the past few years: *B. subtilis* PAP115 (96), *B. subtilis* IFO3034 (84, 114), *B. subtilis* BSE616 (122, 123), and *B. subtilis* DLG (133, 134). Figure 5 compares the proteins encoded by these four genes. These four genes show 93% homology in the protein-coding regions, and all code for a protein of 499 amino acids. Although the homologies are not surprising, since these strains should be closely related, the differences in the cellulase sequences are significant enough to warrant designation of the organisms as different variants if not different species. Further homologies have been noted in that the acidic endoglucanase of *B. subtilis* IFO3034 (114) has a partial homology to the alkaline cellulase from the alkalophilic *Bacillus* sp. strain 1139 (40). Other homologies are discussed below.

The activities of the cellulases of these *B. subtilis* strains are fairly similar. The temperature optima for activity for *B. subtilis* IFO3034, BSE616, and DLG are 55, 58, and 60°C, respectively (84, 114, 122, 123, 133, 134). The peak pH optima for these enzymes are more varied, although the ranges of good (>50% of optimum) activity overlap. The pH optimum for *B. subtilis* IFO3034 is 6.25, with a range from pH 5 to 8.5. *B. subtilis* BSE616 cellulase has a pH optimum of 5.5, and for *B. subtilis* DLG, the optimum peak is at pH 4.8, but 90% of peak activity is maintained from pH 4.8 to 6.0. The cellulases of the *B. subtilis* strains were characterized as endoglucanases by their abilities to hydrolyze carboxymethyl cellulose. The *B. subtilis* endoglucanases were not able to degrade crystalline forms of cellulose such as Avicel or filter paper. *Bacillus* strain KSM-330, a relative of *B. subtilis*, expresses an endo-β-1,4-glucanase (sharp pH optimum of 5.2; temperature optimum of 45°C) that hydrolyzes lichenan (a carbohydrate polymer of β-1,3;1,4 linkages) but does not hydrolyze the crystalline forms of cellulose (120).

The sequenced genes for the *B. subtilis* strains all encode a protein of 52 kDa; however, most of the secreted activities were in the size range of 32 to 36 kDa (92, 95, 96, 122, 133, 134). Carboxy-terminal processing of the endoglucanases was indicated, since results of amino-terminal sequencing of secreted products were as predicted from the gene sequence (134). Furthermore, analysis of carboxy-terminal deletion mutants indicated that activity was maintained in mutants missing the C-terminal 164 amino acids, but activity was absent with longer C-terminal deletions (92, 122).

Alkalophilic *Bacillus* Cellulases

The purified endoglucanase of *Bacillus* strain 1139 has a pH 9.0 optimum, molecular weight of 92,000, pI of 3.1, and an associated *trans*-glucosidase activity (38). By analysis of a series of truncated gene products of the endoglucanase gene, *celF*, of *Bacillus* strain 1139, Fukumori et al. (39) determined that the amino-terminal portion of the endoglucanase was important for pH characteristics and catalytic activity. Full activity required at least the N-terminal 404 amino acids (of 800 encoded amino acids); a truncated gene product of 372 amino acids had no activity.

Bacillus sp. strain N-4 was found to produce several carboxymethyl cellulase activities in alkaline medium (62). Two genes, *celA* and *celB*, encoding these cellulase activities have been cloned, sequenced, and expressed in *E. coli* (43, 141). These two cellulases display similar pH optima (a broad range from pH 5.0 to 10.9), thermal stabilities, and immunological properties. The *celA* gene encodes a protein of 409 amino acids (45.6 kDa), while the *celB* gene product is 488 amino acids (54.2 kDa), with a directly repeated sequence of 59 to 60 amino acids in the C-terminal portion of the protein. The repeated sequence units can be deleted without loss of activity (43). The two cellulases of *Bacillus* strain N-4 are highly homologous (89% identical) (43), and the genes are located in tandem in the same orientation on a cloned fragment of *Bacillus* strain N-4 DNA (42). These two cellulases may have resulted from gene duplication via homologous recombination of elements in nearby noncoding regions (42). A portion of a third cellulase gene, *celC*, from *Bacillus* strain N-4 has been cloned and sequenced (41). The endoglucanase expressed by *E. coli* transformed with the *celC* partial gene had an alkaline pH optimum (pH 9.0) and a molecular mass of 100 kDa (822 amino acids). Activity of the enzyme encoded by the partial gene of *celC* was not surprising, since other carboxy-terminal-truncated cellulases retain enzymatic activity (39, 41, 92, 122, 134). The deduced amino acid sequence of the *Bacillus* strain N-4 *celC* gene product showed a strong homology (67%) with the *celF*-encoded protein of *Bacillus* sp. strain 1139 (40, 41). Figure 6 compares the protein sequences of several alkaline cellulases of bacilli, including those of *Bacillus* strains N-4 and 1139.

The neutral cellulase of *B. subtilis* IFO3034 (discussed above) shows significant homology (70% identity within the conserved regions) with the alkaline cellulase encoded by the *celB* gene of *Bacillus* strain N-4 (112). Nakamura et al. (112) examined an extensive set of structural chimeras of these two enzymes to define regions responsible for the different enzymatic properties of the two cellulases. They divided the protein primary structure into five regions and then used gene cassette techniques to replace each region of one of the cellulases with the corresponding region from the other cellulase. From analysis of the enzymatic characteristics of the chimeras, they determined that one region of the *B. subtilis* neutral enzyme would confer the *B. subtilis* pH activity profile in the alkaline region for all of the chimeras tested. However, two regions of the alkaline enzyme of *Bacillus* strain N-4 were required to confer alkaline activity on the chimeric forms. A separate region of the enzymes was responsible for the activity profile of the chimeras in the acidic pH range. On the basis of kinetic analysis of the chimera, it was determined that the effect on activity in the acidic range was due to an effect on catalytic activity and not on substrate binding (112).

The gene encoding an alkaline cellulase of 941 amino acids, which was isolated from alkalophilic *Bacillus* sp. strain KSM-635, has been cloned and sequenced (121). Similar to many of the other *Bacillus* cellulases, this cellulase has specificity for carboxymethyl cellulose but does not hydrolyze crystalline forms of cellulose. The properties of the KSM-635 cellulase were determined with an eye to using the enzyme for improving the efficacy of laundry detergent (69, 197). The optimum pH was pH 9.5, but activity was maintained at 80% even at pH 11. The temperature optimum was 40°C, and the enzyme was stable against a number of chelating agents and surfactants. Homologies of the cellulase of *Bacillus* sp. strain KSM-635 to other alkaline cellulases from *Bacillus* species such as strains 1139 (40) and N-4 (41, 43) and also to the neutral endoglucanases of *B. subtilis* (96, 114, 134) were noted (121) (Fig. 6). Ninety amino acids were conserved in all of these bacterial cellulases, and there was a remarkable conservation of tryptophan residues. Furthermore, a set of 8 residues was conserved only in the alkaline cellulases, and another set of 18 amino acids was conserved in the neutral cellulases.

Cellulases of Other *Bacillus* Species

A series of neutrophilic *Bacillus* strains have been characterized for their abilities to produce alkali-resistant cellulases with potential applications in laundry detergents (78). These neutrophilic strains resemble a variety of species such as *B. cereus*, *B. licheniformis*, and *Bacillus pumilus* but are novel strains. Interestingly, the cellulases produced are capable of hydrolyzing crystalline forms of cellulose such as Avicel and filter paper. The pH optima for the cellulases are in the neutral range, but the enzymes have broad pH activity profiles, with relatively high

```
                1                                                  50
Bac N-4 CelA    .......... .......... .......... .......... ..........
Bac N-4 CelB    .......... .......... .......... .......... ..........
Bac KSM-635     mkikqikqsl slllliitlim slfvpmasan tnesksnafp fsdvkktsws
Bac str.1139    .......... .......... .......... .......... ..........
Bac N-4 CelC    .......... .......... .......... .......... ..........
   Consensus    ---------- ---------- ---------- ---------- ----------

                51                                                100
Bac N-4 CelA    .......... .......... .......... .......... ..........
Bac N-4 CelB    .......... .......... .......... .......... ..........
Bac KSM-635     fpyikdlyeq evitgtsatt fsptdsvtra qftvmltrgl gleasskdyp
Bac str.1139    .......... .......... .......... .......... ..........
Bac N-4 CelC    .......... .......... .......... .......... ..........
   Consensus    ---------- ---------- ---------- ---------- ----------

                101                                               150
Bac N-4 CelA    .......... .......... .......... .......... ..........
Bac N-4 CelB    .......... .......... .......... .......... ..........
Bac KSM-635     fkdrknwayk eiqaayeagi vtgktngefa pnenitreqm aamavrayey
Bac str.1139    .......... .......... .......... .......... ..........
Bac N-4 CelC    .......... .......... .......... .......... ..........
   Consensus    ---------- ---------- ---------- ---------- ----------

                151                                               200
Bac N-4 CelA    .......... .......... .......... .......... ..........
Bac N-4 CelB    .......... .......... .......... .......... ..........
Bac KSM-635     leneLsLpee qReyndSssI stFaqdavqk AyvlElmEgn tDgyFqpk..
Bac str.1139    ....MmLRkK tkqLisSiLI lvLllSLFpt AlaAE..... ..........
Bac N-4 CelC    ......MRnK lRrLlaimMa vlLitSLFap mvsAEegDng dDddLvtpie
   Consensus    ------LR-K -R-L--S--I --L--SLF-- A--AE----- -D--------

                201                                               250
Bac N-4 CelA    .........m KKlTTiFIVF ..TLALLfVg N...stsAnn gsvVEqnGQL
Bac N-4 CelB    .........m KKITTiFVVL lmTLALFiIg N...ttaAdd ysvVEehGQL
Bac KSM-635     .rnsTREqsa KvIsTlLwkv ashdyLYhte aVKsPSEAGA LQLVElnGQL
Bac str.1139    ..gnTREdNf K......... ....hLLGnd NVKRPSEAGA LQLqEVdGQM
Bac N-4 CelC    ieerphEsNy eKypal.... .....LdGgl derRPSEAGA LQLVEVdGQv
   Consensus    ----TRE-N- KKITT-F-V- --TLALLG-- NVKRPSEAGA LQLVEV-GQL

                251                                               300
Bac N-4 CelA    sIqNGqLVnE hGDPVQLkGM SsHGLQWYGq fVNydsiKwL rdDWGitVfR
Bac N-4 CelB    sIsNGeLVnD rGEPVQLkGM SsHGLQWYGq fVNyesmKwL rdDWGitVfR
Bac KSM-635     T.....LagE dGtPVQLRGM STHGLQWFGE IVNENAFvAL sNDWGSNmIR
Bac str.1139    T.....LVdq hGEkIQLRGM STHGLQWFpE IlNDNAYKAL aNDWeSNmIR
Bac N-4 CelC    T.....Ladq dGvPIQLRGM STHGLQWFGE IVNENAFaAL aNDWGSNVIR
   Consensus    TI-NG-LV-E -GEPVQLRGM STHGLQWFGE IVNENAFKAL -NDWGSNVIR

                301                                               350
Bac N-4 CelA    aAMYtssgGY iEdPs.VKEK VkEaVEaAID lgiYVIIDWH IlsdnDPN..
Bac N-4 CelB    aAMYtssgGY iEdPs.VKEK VkEaVEaAID lgiYVIIDWH IlsdnDPN..
Bac KSM-635     LAMYIGENGY atNPE.VKDl VyEGIELAfE hDMYVIVDWH VHAPGDPrAD
Bac str.1139    LAMYVGENGY asNPELIKsr VikGIDLAIE nDMYVIVDWH VHAPGDPrdp
Bac N-4 CelC    LALYIGENaY ryNPDLI.EK VyaGIELAkE nDMYVIIDWH VHAPGDPNAD
   Consensus    LAMYIGENGY -ENPELVKEK V-EGIELAIE -DMYVIIDWH VHAPGDPNAD

                351                                               400
Bac N-4 CelA    IYKE...... ....eAKEFF DEmsaLYGDY PN...VIYEi ANEP.....n
Bac N-4 CelB    IYKE...... ....eAKDFF DEmsELYGDY PN...VIYEi ANEP.....n
Bac KSM-635     VYs....... ....GAyDFF EEIADhYkDh PkNhyIIWEL ANEPSpNNNG
Bac str.1139    VYa....... ....GAeDFF rDIAaLY... PNNPHIIYEL ANEPSSNNNG
Bac N-4 CelC    IYqggvnedg eeylGAKDFF lhIAEkY... PNdPHlIYEL ANEPSSNssG
   Consensus    IYKE------ ----GAKDFF DEIAELYGDY PNNPH-IYEL ANEPSSNNNG

                401                                               450
Bac N-4 CelA    GhnVrwDsh. ...IKpYAEe VIpviR..aN dpnNIVIVGt atWSQdvh.e
Bac N-4 CelB    GsdVTwDnq. ...IKpYAEe VIpviR..nN dpnNIIIVGt gtWSQdvh.h
Bac KSM-635     GPGlTNDEkG WEAVKEYAEP IVEmLREkG. ..DNmIlVGn PNWSQRpDL.
Bac str.1139    GaGIpNnEEG WnAVKEYADP IVEmLRDSGN adDNIIIVGs PNWSQRpDL.
Bac N-4 CelC    GPGITNDEDG WEAVrEYAqP IVDaLRDSGN aeDNIIIVGs PNWSQRmDLa
   Consensus    GPG-TNDE-G WEAVKEYAEP IVE-LRDSGN --DNIIIVG- PNWSQR-DL-

                451                                               500
Bac N-4 CelA    AADNqlDDpN VMYaFHFYaG THGq...... .........q lrnqVdYALs
Bac N-4 CelB    AADNqltDpN VMYaFHFYaG THGq...... .........N lrdqVdYALD
Bac KSM-635     sADNPIDaeN IMYsvHFYTG sHGaShigYP EGTPSSERSN VMANVrYALD
Bac str.1139    AADNPIDDhh tMYtvHFYTG sHaaStESYP peTPnSERgN VMsNtrYALE
Bac N-4 CelC    AADNPIDDhh tMYtLHFYTG THegtnESYP EGisSeDRSN VMANakYALD
   Consensus    AADNPIDD-N VMY-FHFYTG THG-S-ESYP EGTPSSERSN VMANV-YALD

                501                                               550
Bac N-4 CelA    rGaAIFvsEW GTSaAtGDGG vFLDEAqVWI DFMdErNlSW ANWSLThKdE
Bac N-4 CelB    qGaAIFvsEW GTSeAtGDGG vFLDEAqVWI DFMdErNlSW ANWSLThKdE
Bac KSM-635     NGvAVFATEW GTSqAnGDGG PYFDEADVWl nFLNkhNISW ANWSLTNKNE
Bac str.1139    NGvAVFATEW GTSqAnGDGG PYFDEADVWI EFLNEnNISW ANWSLTNKNE
Bac N-4 CelC    kGkAIFATEW GvSeAdGnnG PYLnEADVWl nFLNEnNISW tNWSLTNKNE
   Consensus    NG-AIFATEW GTS-A-GDGG PYLDEADVWI DFLNE-NISW ANWSLTNKNE
```

Figure 6. Comparison of protein sequences of alkaline cellulases from bacilli. Consensus denotes a consensus sequence determined by position similarities in two or more sequences; uppercase amino acid symbols indicate similar or identical residues that generated the consensus sequence. Lowercase symbols designate differences from consensus sequence. Sequence designations are as follows: Bac N-4 CelA, *Bacillus* strain N-4 endoglucanase A (43); Bac N-4 CelB, *Bacillus* strain N-4 endoglucanase B (43); Bac KSM-635, *Bacillus* strain KSM-635 (121); Bac str.1139, *Bacillus* strain 1139 (40); Bac N-4 CelC, *Bacillus* strain N-4 endoglucanase C (41).

```
               551                                                600
Bac N-4 CelA   sSaALmP... ........ga nPtggWTaaE LSpSGaFVRe kIrEsasiPP
Bac N-4 CelB   sSaALmP... ........ga sPtggWTeaE LSpSGtFVRe kIrEsattPP
Bac KSM-635    iSGAFTPFEL GrtDATDLDP GanQVWaPEE LSlSGEYVRa RIk.GiEYtP
Bac str.1139   vSGAFTPFEL GkSnATsLDP GPdQVWvPEE LSlSGEYVRa RIk.GvnYeP
Bac N-4 CelC   tSGAFTPFiL neSDATDLDP GedQVWsmEE LSvSGEYVRs RIl.GeEYqP
   Consensus   -SGAFTPFEL G-SDATDLDP GP-QVWTPEE LS-SGEYVR- RI-EG-EYPP

               601                                                650
Bac N-4 CelA   sDpTPPSDpd pgepDptpps dpGeypawDp nqIytneIvy hNgqLwqakw
Bac N-4 CelB   sDpTPPSDpd pgep...... .......... .......... ..........
Bac KSM-635    IDRT...kFt klvWDFNDGT TQGFqVNGDS PnkEsITlsN nNdALQIeGL
Bac str.1139   IDRT...kYt kvlWDFNDGT kQGFgVNGDS P.VEDVvIEN eagALklsGL
Bac N-4 CelC   IDRTPreEFs eviWDFNDGT TQGFvqNsDS Pl..DVTIEN vNdALQItGL
   Consensus   IDRTPPSDF- ---WDFNDGT TQGF-VNGDS P--EDVTIEN -N-ALQI-GL

               651                                                700
Bac N-4 CelA   wtqNqepGan qyGpWeplgd appsepSDpp ppsepEpdpg EpdpgEPdpg
Bac N-4 CelB   .......... .......... .......... .......... .....EPdpg
Bac KSM-635    nvSNDIS..E ..GNYWdNVR LSADgWSEnV DILGAtELTi DVIVEEPTTV
Bac str.1139   DaSNDVS..E ..GNYWaNaR LSADgWgksV DILGAEkLTM DVIVDEPTTV
Bac N-4 CelC   DeSNaIaGeE ..edYWsNVR iSADeWeEtf DILGAEELsM DVVVDDPTTV
   Consensus   D-SNDISG-E --GNYW-NVR LSAD-WSE-V DILGAEELTM DVIVDEPTTV

               701                                                750
Bac N-4 CelA   epdptPpSdp gEYp.AWdpt qIytnEiVYh ngqLWqAkw. wTqnqePgyp
Bac N-4 CelB   epdptPpSdp gDYp.AWdpn tIytDEiVYh ngqLWqAkw. wTqnqePgdp
Bac KSM-635    sIAAIPQgpa agWaNptraI KVteDDFesf gDg.YKAlvT ITseDsPsLe
Bac str.1139   sIAAIPQgps anWvNpnraI KVeptnFVpl eDk.FKAelT ITsaDsPsLe
Bac N-4 CelC   aIAAIPQSsa hEWaNAsnsV lIteDDFeeq eDgtYKAllT ITgeDaPnLt
   Consensus   -IAAIPQS-- -EW-NAW--I KI--D-FVY- -D-L-KA--T IT--D-P-L-

               751                                                800
Bac N-4 CelA   YgpWEPlN.. .......... .......... .......... ..........
Bac N-4 CelB   YgpWEPlN.. .......... .......... .......... ..........
Bac KSM-635    tIAtsPEdNt MsNIILFVGT EdADVISLDN ITVsG..TEI EIeVIHDeKG
Bac str.1139   aIAmhaENNn iNNIILFVGT EgADVIyLDN IkViG..TEV EIPVVHDpKG
Bac N-4 CelC   nIAeDPEgse LNNIILFVGT EnADVISLDN ITVtGdresV pePVeHDtKG
   Consensus   YIAWEPENN- -NNIILFVGT E-ADVISLDN ITV-G--TEV EIPV-HD-KG

               801                                                850
Bac N-4 CelA   .......... .......... .......... .......... ..........
Bac N-4 CelB   .......... .......... .......... .......... ..........
Bac KSM-635    tAtLPStFED GTRQGWDWht ESGVKTALTI EEANGSNALS WEYAYPEVKP
Bac str.1139   EAvLPSvFED GTRQGWDWag ESGVKTALTI EEANGSNALS WEFgYPEVKP
Bac N-4 CelC   DsaLPSdFED GTRQGWEWds ESaVrTALTI EEANGSNALS WEYAYPEVKP
   Consensus   -A-LPS-FED GTRQGWDW-- ESGVKTALTI EEANGSNALS WEYAYPEVKP

               851                                                900
Bac N-4 CelA   .......... .......... .......... .......... ..........
Bac N-4 CelB   .......... .......... .......... .......... ..........
Bac KSM-635    SDgWATAPRL DFWKDELVRG tsDYIsFDFY IDaV..RAsE GAIsINaVFQ
Bac str.1139   SDnWATAPRL DFWKsDLVRG EnDYVtFDFY lDPV..RATE GAmnINLVFQ
Bac N-4 CelC   SDdWATAPRL tLYKDDLVRG DyEFVaFDFY IDPIedRATE GAIdINLIFQ
   Consensus   SD-WATAPRL DFWKDDLVRG --DYV-FDFY IDPV--RATE GAI-INLVFQ

               901                                                950
Bac N-4 CelA   .......... .......... .......... .......... ..........
Bac N-4 CelB   .......... .......... .......... .......... ..........
Bac KSM-635    PPANGYWqev PtTFEIDLtE LDSATVTsDe LYHYEVKINI RDIEaItDDT
Bac str.1139   PPtNGYWvQA PkTYtInFDE LEepnqv.nG LYHYEVKINV RDItNIqDDT
Bac N-4 CelC   PPAaGYWaQA seTFEIDLEE LDSATVTdDG LYHYEVeINI eDIEN...Di
   Consensus   PPANGYW-QA P-TFEIDL-E LDSATVT-DG LYHYEVKINI RDIENI-DDT

               951                                               1000
Bac N-4 CelA   .......... .......... .......... .......... ..........
Bac N-4 CelB   .......... .......... .......... .......... ..........
Bac KSM-635    ELRNLLLIFA DEDSDFAGRV FVDNVRFE.. .......... ..........
Bac str.1139   lLRNMMiIFA DvESDFAGRV FVDNVRFEga atTepvepEp vdpgeetppV
Bac N-4 CelC   ELRNLMLIFA DDESDFAGRV FlDNVRmDms leTkvevlEr ninelqeqlV
   Consensus   ELRNLMLIFA D-ESDFAGRV FVDNVRFE-- --T-----E- ---------V

               1001      1019
Bac N-4 CelA   .......... .........
Bac N-4 CelB   .......... .........
Bac KSM-635    .......... .........
Bac str.1139   Dekeakteqk eaekeekee
Bac N-4 CelC   Evealmr... .........
   Consensus   ---------- ---------
```

Figure 6. *Continued.*

activity in the alkaline range. One strain, K597, produced a cellulase that maintained 80% of its maximum activity over the pH range of 5 to 10. All of the characterized cellulases were unaffected by detergent components such as surfactants, proteases, and chelating agents (78).

Bacillus lautus PL236, isolated from compost and capable of degrading crystalline cellulose, contains multiple different endo-β-1,4-glucanase genes (71). One of the *B. lautus* PL236 genes, *celB*, was sequenced and found to encode a protein of 566 amino acids with significant homology to endoglucanase E of *Clostridium thermocellum* but only a distant relationship to *Bacillus* cellulases (71).

An endoglucanase of *B. polymyxa* was found through an extensive screen of various *Bacillus* species (5). The gene was cloned and sequenced, and results indicated that the encoded protein (397 amino acids) had homology to *celB* of *C. thermocellum* but no significant homology to other *Bacillus* cellulases (5).

Cellulases of Other Gram-Positive Bacteria

A number of other organisms, particularly those that reside in environments rich in cellulosic materials, have been examined for their cellulase activities. Stutzenberger (156) has provided an extensive review of bacterial cellulases from both gram-positive and gram-negative organisms. Here, we discuss a few of the interesting cases.

The gene for an endoglucanase from an alkalophilic *Streptomyces* strain, KSM-9, with a pH 8.5 optimum and stability to 40°C has been cloned (108). Strain KSM-9 also produced two other endoglucanases with pH 6.0 optima, but none of the three cellulases had much activity on crystalline cellulose (109).

Wachinger et al. (183) surveyed the cellulase activities of 180 *Streptomyces* strains and found 25 that were capable of degrading microcrystalline cellulose (Avicel). *Streptomyces reticuli* cellulase activity consisted of the components Avicelase, two endoglucanases, and β-glucosidase. The purified components had pH optima of 7.0 and temperature optima in the range of 45 to 55°C (183).

Microbispora bispora, another actinomycete, produces a multiple-component cellulase with at least four endoglucanases, two cellobiohydrolases, and a cell-associated β-glucosidase (185, 186). This cellulase complex was able to degrade microcrystalline cellulose, but the isolated component enzymes could not.

Thermomonospora fusca cultures are capable of hydrolyzing microcrystalline cellulose with cellulases that are heat stable and have broad pH optima (156). The DNA sequences for three endoglucanases of *T. fusca* have been obtained, and sequence comparisons indicate that the *T. fusca* cellulase E2 gene is 73% identical to the *celA* gene of *M. bispora* (90). A similarity to CenA of *Cellulomonas fimi* was also identified (39% conserved amino acids). Homologies of *Thermomonospora* cellulase E5 protein sequence with the protein sequences of *Bacillus* sp. strain N-4 *celA* and *celB* and with the cellulase of *B. subtilis* DLG were observed, with particularly strong conservation in two regions (residues 255 to 263 and 292 to 302 of E5) (90).

C. thermocellum produces a cellulase complex that degrades crystalline cellulose and contains 14 to 18 components, many of which are endoglucanases (reviewed in reference 156). Seven genes have been sequenced, and another eight have been cloned (7). Endoglucanase genes from *Ruminococcus albus* and *Ruminococcus flavefaciens* have been cloned and sequenced, and homologies to endoglucanase E of *C. thermocellum* have been detected (100, 117, 127). Ten endoglucanase genes have been observed in *R. albus* (65).

Three cellulase genes of *C. fimi* have been cloned and sequenced (100, 118, 191). The availability of these sequences led to an extensive and detailed analysis of the structural components of these *C. fimi* cellulases that developed into an overall analysis of the domains of all microbial sequences of cellulases. That analysis was presented in an elegant review by Gilkes et al. (45) that will be discussed next.

Classification and Domains of Microbial Cellulases

Through biochemical analyses of the microbial cellulases and comparisons of gene sequence homologies, predicted amino acid sequence homologies, and predicted structural similarities, several investigators have identified common functional domains and classified the cellulases into nine families. This classification scheme was recently reviewed by Gilkes et al. (45) and Béguin (7). There are four types of structural-sequence domains present in fungal and bacterial cellulases and xylanases: carbohydrate-binding domains (CBD), catalytic domains, repeated-sequence domains, and characteristic linker sequences, which connect the other domains. These domains are ordered in a variety of arrays in different cellulases, and not all cellulases contain each kind of domain. The relevance of these domains to *Bacillus* cellulases will be highlighted in the following explanation of the domains and their functions.

The linker regions are identified as short sequences (6 to 59 amino acids) that are rich in proline and/or hydroxyamino acids but not in any specific sequence. Although sequence homologies have been detected within linker regions of cellulases from one organism, no homologies have been detected between species (45). Linker sequences have been identified in endoglucanases (CelA and CelB) of *Bacillus* sp. strain N-4 (43) and also in the endoglucanase of *Bacillus* sp. strain 1139 (40).

There are fewer occurrences of repeated sequences, which are highly conserved sequences (internally) of 20 to 150 amino acids that are repeated at least once within an individual cellulase. *Bacillus* strain N-4 cellulase CelB has a repeat of 61 amino acids at its C terminus (90% identity) with 28 amino acids between the repeats (43). *Bacillus* strain N-4 cellulase CelA has a homologous sequence but only as a single copy at its C-terminal end (43).

For the majority of cellulases, the CBDs have not been characterized in detail, but they have been characterized for the cellulases of *C. fimi* (46) and the cellulases of *Trichoderma reesei* (173, 177). Among the characteristics of the bacterial CBDs are low numbers of charged amino acids, high hydroxyamino acid content, and conserved positions of tryptophan, asparagine, and glycine residues. The position of the CBD can be either N terminal or C terminal in a cellulase, and for the most part, the CBDs do not appear to be required for catalytic activity (45).

The positions of catalytic domains of some cellulases have been identified at both gene and protein levels by using various biochemical techniques (7, 45). Over 60 cellulase and xylanase genes have been sequenced, and the putative catalytic-domain sequences have been compared and grouped. Extension and verification of the groups occurred by the use of hydrophobic cluster analysis, which compares sequences on the basis of similarities in secondary structures (55). Similarities can be detected even when the homologous domains are separated by various intervening sequence lengths (7, 45, 55). Henrissat et al. (55) first identified six cellulase and xylanase families, which have been expanded to nine families as more sequences have become available (45). Most of the *Bacillus* endoglucanases are classified in family A, which contains both bacterial and fungal cellulases. The benefits of this classification system will come when detailed information about the mechanisms of action and the three-dimensional structures for representative enzymes of these families becomes available, with the result that the knowledge may be easily extended to the other members of the families.

CONCLUSION

Bacillus spp. produce the largest market share of enzymes used for industrial applications in terms of both volume and dollar value. Because of their abilities to express and secrete high amounts of proteins, *Bacillus* spp. have the potential of becoming general-purpose expression hosts for rDNA products. Current studies of the molecular genetics of *B. subtilis* will generate a better knowledge of the regulation of gene expression and of the mechanisms involved in translation and secretion. This knowledge will allow use of this bacterium to express and secrete high volumes of homologous modified enzymes and heterologous gene products.

A number of different types of starch-degrading enzymes have been isolated from gram-positive bacteria. The most-utilized of these enzymes for the commercial production of high-fructose corn syrup is the α-amylase isolated from *B. licheniformis*. However, in the near future, the productivity of this industrial process will probably be enhanced by the improvement of existing enzymes by protein engineering techniques and/or the availability of alternative enzymes isolated from novel organisms.

The catalog of *Bacillus* cellulases contains enzymes with multiple activities at various pHs and temperature optima, allowing one to choose an enzyme for a given application on the basis of requirements for that application. Of particular utility is the set of alkaline cellulases, which expands the range of application possibilities beyond the acidic range of optima for the fungal cellulases. One can also choose a *Bacillus* cellulase, either endoglucanase or exoglucanase, on the basis of hydrolytic activity with or without side activities. A final attribute of the variety of *Bacillus* cellulases is the relationship between primary sequence homologies and differences in activities; this relationship provides a fertile milieu for examining structure-function relationships and the effects of various amino acids on mechanisms of action.

Acknowledgments. We thank Randy Berka and Dennis Henner for critically reading the manuscript.

REFERENCES

1. **Amemura, A., R. Chakraborty, M. Fujita, T. Noumi, and M. Futai.** 1988. Cloning and nucleotide sequence of the isoamylase gene from *Pseudomonas amyloderamosa* SB-15. *J. Biol. Chem.* **263:**9271–9275.
2. **Baecker, P. A., E. Greenburg, and J. Preiss.** 1986. Biosynthesis of bacterial glycogen: primary structure of *Escherichia coli* 1,4-α-D-glucan:1,4-α-D-glucan 6-α-D-(1,4-α-D-glucano)-transferase as deduced from the nucleotide sequence of the *glg*B gene. *J. Biol. Chem.* **261:** 8738–8743.
3. **Bahl, H., G. Burchhardt, A. Spreinat, K. Haeckel, A. Wienecke, B. Schmidt, and G. Antranikian.** 1991. α-Amylase of *Clostridium thermosulfurogenes* EM1: nucleotide sequence of the gene, processing of the enzyme, and comparison to other α-amylases. *Appl. Environ. Microbiol.* **57:**1554–1559.
4. **Bahri, S. M., and J. M. Ward.** 1990. Nucleotide sequence of an α-amylase gene isolated from *Streptomyces thermolliolaceus* CUB74. GenBank Genetic Sequence Data Bank, Release 68.0, Locus:Stmamy, Accession: M34957. Unpublished sequence.
5. **Baird, S. K., D. A. Johnson, and V. L. Seligy.** 1990. Molecular cloning, expression, and characterization of endo-β-1,4-glucanase genes from *Bacillus polymyxa* and *Bacillus circulans*. *J. Bacteriol.* **172:**1576–1586.
6. **Bealin-Kelly, F., C. T. Kelly, and W. M. Fogarty.** 1991. Studies on the thermostability of the α-amylase of *Bacillus caldovelox*. *Appl. Microbiol. Biotechnol.* **36:**332–336.
7. **Béguin, P.** 1990. Molecular biology of cellulose degradation. *Annu. Rev. Microbiol.* **44:**219–248.
8. **Binder, F., O. Huber, and A. Böck.** 1986. Cyclodextrin-glycosyltransferase from *Klebsiella pneumoniae* M5a1: cloning, nucleotide sequence and expression. *Gene* **47:** 269–277.
9. **Boel, E., L. Brady, A. M. Brzozowski, Z. Derewenda, G. G. Dodson, V. J. Jensen, S. B. Petersen, H. Swift, L. Thim, and H. F. Woldike.** 1990. Calcium binding in α-amylases: an X-ray diffraction study at 2.1-Å resolution of two enzymes from *Aspergillus*. *Biochemistry* **29:**6244–6249.
10. **Boyer, E. W., and M. B. Ingle.** 1972. Extracellular alkaline amylase from a *Bacillus* species. *J. Bacteriol.* **110:**992–1000.
11. **Brosnan, M. P., C. T. Kelly, and W. M. Fogarty.** 1992. Investigation of the mechanisms of irreversible thermoinactivation of *Bacillus stearothermophilus* α-amylase. *Eur. J. Biochem.* **203:**225–231.
12. **Buisson, G., E. Duée, R. Haser, and F. Payan.** 1987. Three dimensional structure of porcine pancreatic α-amylase at 2.9-Å resolution. Role of calcium in structure and activity. *EMBO J.* **6:**3909–3916.
13. **Candussio, A., G. Schmid, and A. Böck.** 1990. Biochemical and genetic analysis of a maltopentaose-producing amylase from an alkaliphilic Gram-positive bacterium. *Eur. J. Biochem.* **191:**177–185.
14. **Cheetam, P. S. J.** 1985. The applications of enzymes to industry, p. 274–379. *In* A. Wiseman (ed.), *Handbook of Enzyme Biotechnology*, 2nd ed. Ellis Horwood Ltd., Chichester, United Kingdom.
15. **Chojecki, A., and H. P. Blaschek.** 1986. Effect of carbohydrate source on alpha-amylase and glucoamylase formation by *Clostridium acetobutylicum* SA-1. *J. Ind. Microbiol.* **1:**63–67.
16. **Chu, K. (Genencor International).** 1992. Personal communication.
17. **Coughlan, M. P.** 1990. Cellulose degradation by fungi, p. 1–36. *In* W. M. Fogarty and C. T. Kelly (ed.), *Microbial Enzymes and Biotechnology*, 2nd ed. Elsevier Applied Science, New York.
18. **Dahl, M. K., T. Msadek, F. Kunst, and G. Rapoport.** 1991. Mutational analysis of the *Bacillus subtilis* DegU regulator and its phosphorylation by the DegS protein kinase. *J. Bacteriol.* **173:**2539–2547.
19. **Debabov, V. G.** 1982. The industrial use of bacilli, p.

331–370. *In* D. A. Dubnau (ed.), *The Molecular Biology of the Bacilli*. Academic Press, Inc., New York.

20. **Declerck, N., P. Joyet, C. Gaillardin, and J.-M. Masson.** 1990. Use of amber suppressors to investigate the thermostability of *Bacillus licheniformis* α-amylase. *J. Biol. Chem.* **265:**15481–15488.
21. **Demain, A. L.** 1990. Regulation and exploitation of enzyme biosynthesis, p. 331–368. *In* W. M. Fogarty and K. T. Kelly (ed.), *Microbial Enzymes and Biotechnology*, 2nd ed. Elsevier Applied Science, London.
22. **Diderichsen, B., and L. Christiansen.** 1988. Cloning of a maltogenic alpha-amylase from *Bacillus stearothermophilus*. *FEMS Microbiol. Lett.* **56:**53–60.
23. **Dod, B., and G. Balassa.** 1978. Spore control (*sco*) mutations in *Bacillus subtilis*. III. Regulation of extracellular protease synthesis in the spore control mutations *scoC*. *Mol. Gen. Genet.* **163:**57–63.
24. **Doi, R. H.** 1991. Proteolytic activities in *Bacillus*. *Curr. Opin. Biotechnol.* **2:**682–684.
25. **Dordick, J. S.** 1991. An introduction to industrial biocatalysis, p. 3–19. *In* J. S. Dordick (ed.), *Biocatalysts for Industry*. Plenum Press, New York.
26. **Emori, M., and B. Maruo.** 1988. Complete nucleotide sequence of an α-amylase from *Bacillus subtilis* 2633, an amylase extrahyperproducing strain. *Nucleic Acids Res.* **16:**7178.
27. **Feller, G., T. Lonhienne, C. Deroanne, C. Libioulle, J. Von Beeumen, and C. Gerday.** 1992. Purification, characterization, and nucleotide sequence of the thermolabile α-amylase from the Antarctic psychotroph *Alteromonas haloplanctis* A23. *J. Biol. Chem.* **267:**5217–5221.
28. **Feng, D.-F., and R. F. Doolittle.** 1987. Progressive sequence alignments as a prerequisite to correct phylogenetic trees. *J. Mol. Evol.* **35:**351–360.
29. **Ferrari, E., D. J. Henner, M. Perego, and J. A. Hoch.** 1988. Transcription of *Bacillus subtilis* subtilisin and expression of subtilisin in sporulation mutation. *J. Bacteriol.* **170:**289–295.
30. **Ferrari, E., S. M. H. Howard, and J. A. Hoch.** 1986. Effect of stage 0 sporulation mutations on subtilisin expression. *J. Bacteriol.* **166:**173–179.
31. **Ferrari, E., and M. Ruppen.** Unpublished data.
32. **Fogarty, W. M., and E. J. Bourke.** 1983. Production and purification of a maltose-producing amylase from *Bacillus subtilis* IMD 198. *J. Chem. Technol. Biotechnol.* **33B:**145–151.
33. **Folkman, J., P. B. Weisz, M. M. Joullié, W. W. Li, and W. R. Ewing.** 1989. Control of angiogenesis with synthetic heparin substitutes. *Science* **243:**1490–1493.
34. **French, D.** 1975. Chemistry and biochemistry of starch, vol. 5, p. 267–335. *In* W. J. Whelan (ed.), *MTP International Review of Science*. University Park Press, Baltimore.
35. **Friedberg, F., and C. Rhodes.** 1986. Cloning and characterization of the beta-amylase gene from *Bacillus polymyxa*. *J. Bacteriol.* **165:**819–824.
36. **Friedman, R. B.** 1991. Linear and cyclic dextrins, p. 327–347. *In* I. Goldberg and R. Williams (ed.), *Biotechnology and Food Ingredients*. Van Nostrand Reinhold, New York.
37. **Fujita, M., K. Torigoe, T. Nakada, K. Tsusaki, M. Kubota, S. Sakai, and Y. Tsujisaka.** 1989. Cloning and nucleotide sequence of the gene (*amyP*) for maltotetraose-forming amylase from *Pseudomonas stutzeri* MO-19. *J. Bacteriol.* **171:**1333–1339.
38. **Fukumori, F., T. Kudo, and K. Horikoshi.** 1985. Purification and properties of a cellulase from alkalophilic *Bacillus* sp. no. 1139. *J. Gen. Microbiol.* **131:**3339–3345.
39. **Fukumori, F., T. Kudo, and K. Horikoshi.** 1987. Truncation analysis of an alkaline cellulase from an alkalophilic *Bacillus* species. *FEMS Microbiol. Lett.* **40:**311–314.
40. **Fukumori, F., T. Kudo, Y. Narahashi, and K. Horikoshi.** 1986. Molecular cloning and nucleotide sequence of the alkaline cellulase gene from the alkalophilic *Bacillus* sp. strain 1139. *J. Gen. Microbiol.* **132:**2329–2335.
41. **Fukumori, F., T. Kudo, N. Sashihara, Y. Nagata, K. Ito, and K. Horikoshi.** 1989. The third cellulase of alkalophilic *Bacillus* sp. strain N-4: evolutionary relationships within the *cel* gene family. *Gene* **76:**289–298.
42. **Fukumori, F., K. Ohishi, T. Kudo, and K. Horikoshi.** 1987. Tandem location of the cellulase genes on the chromosome of *Bacillus* sp. strain N-4. *FEMS Microbiol. Lett.* **48:**65–68.
43. **Fukumori, F., N. Sashihara, T. Kudo, and K. Horikoshi.** 1986. Nucleotide sequences of two cellulase genes from alkalophilic *Bacillus* sp. strain N-4 and their strong homology. *J. Bacteriol.* **168:**479–485.
44. **Gaur, N. K., J. Oppenheim, and I. Smith.** 1991. The *Bacillus subtilis sin* gene, a regulator of alternate developmental processes, codes for a DNA-binding protein. *J. Bacteriol.* **173:**678–686.
45. **Gilkes, N. R., B. Henrissat, D. G. Kilburn, R. C. Miller, Jr., and R. A. J. Warren.** 1991. Domains in microbial β-1,4-glycanases: sequence conservation, function, and enzyme families. *Microbiol. Rev.* **55:**303–315.
46. **Gilkes, N. R., R. A. J. Warren, R. C. Miller, Jr., and D. G. Kilburn.** 1988. Precise excision of the cellulose binding domains from two *Cellulomonas fimi* cellulases by a homologous protease and the effect on catalysis. *J. Biol. Chem.* **263:**10401–10407.
47. **Gobius, K. S., and J. M. Pemberton.** 1988. Molecular cloning, characterization, and nucleotide sequence of an extracellular amylase gene from *Aeromonas hydrophila*. *J. Bacteriol.* **170:**1325–1332.
48. **Godtfredsen, S. E.** 1990. Microbial lipases, p. 255–274. *In* W. M. Fogarty and K. T. Kelly (ed.), *Microbial Enzymes and Biotechnology*, 2nd ed. Elsevier Applied Science, London.
49. **Gray, G. L., S. E. Mainzer, M. W. Rey, M. H. Lamsa, K. L. Kindle, C. Carmona, and C. Requadt.** 1986. Structural genes encoding the thermophilic α-amylases of *Bacillus stearothermophilus* and *Bacillus licheniformis*. *J. Bacteriol.* **166:**635–643.
50. **Graycar, T. P.** 1991. Protein engineering of subtilisin, p. 257–284. *In* J. S. Dordick (ed.), *Biocatalysts for Industry*. Plenum Press, New York.
51. **Hacking, J. A.** 1991. Biocatalysis in the production of carbohydrates for food uses, p. 63–82. *In* J. S. Dordick (ed.), *Biocatalysts for Industry*. Plenum Press, New York.
52. **Henner, D. J., E. Ferrari, M. Perego, and J. A. Hoch.** 1988. Location of the target of the *hpr-97*, *sacU32*(Hy), and *sacQ36*(Hy) mutations in upstream region of the subtilisin promoter. *J. Bacteriol.* **170:**296–300.
53. **Henner, D. J., M. Yang, L. Band, H. Shimotsu, M. Ruppen, and E. Ferrari.** 1986. Genes of *Bacillus subtilis* which regulate the expression of degradative enzymes, p. 81–90. *In* M. Alacevic, D. Hranueli, and Z. Toman (ed.), *Genetics of Industrial Microorganisms. Proceedings of the Fifth International Symposium on the Genetics of Industrial Microorganisms*. Ognjen Prica Printing Works, Karlovac, Yugoslavia.
54. **Henner, D. J., M. Yang, and E. Ferrari.** 1988. Localization of the *Bacillus subtilis sacU*(Hy) mutations to two linked genes with similarities to the conserved procaryotic family of two-component signalling systems. *J. Bacteriol.* **170:**5102–5109.
55. **Henrissat, B., M. Claeyssens, P. Tomme, L. Lemesle, and J.-P. Mornon.** 1989. Cellulase families revealed by hydrophobic cluster analysis. *Gene* **81:**83–95.
56. **Higashihara, M., and H. Okada.** 1974. Studies on beta-amylase of *Bacillus megaterium* no. 31. *Agric. Biol. Chem.* **38:**1023–1029.
57. **Hill, D. E., R. Aldape, and J. D. Rozzell.** 1990. Nucleotide sequence of a cyclodextrin glucosyltransferase

gene, *cgtA*, from *Bacillus licheniformis*. *Nucleic Acids Res.* **18:**199.

58. **Hitosuyanagi, K., K. Yamane, and B. Maruo.** 1979. Stepwise introduction of regulatory genes stimulating production of α-amylase into *Bacillus subtilis*: construction of an α-amylase extrahyperproducing strain. *Agric. Biol. Chem.* **43:**2343–2349.
59. **Hoch, J. A.** 1976. Genetics of bacterial sporulation. *Adv. Genet.* **18:**69–98.
60. **Holm, L., A. K. Koivula, P. M. Lehtovaara, A. Hemminki, and J. K. C. Knowles.** 1990. Random mutagenesis used to probe the structure and function of *Bacillus stearothermophilus* alpha-amylase. *Protein Eng.* **3:**181–191.
61. **Honjo, M., A. Nakayama, K. Fukazawa, K. Kawamura, K. Ando, M. Hori, and Y. Furutani.** 1990. A novel *Bacillus subtilis* gene involved in negative control of sporulation and degradative-enzyme production. *J. Bacteriol.* **172:**1783–1790.
62. **Horikoshi, K., M. Nakao, Y. Kurono, and N. Sashihara.** 1984. Cellulases of an alkalophilic *Bacillus* strain isolated from soil. *Can. J. Microbiol.* **30:**774–779.
63. **Horinouchi, S., S. Fukusumi, T. Ohshima, and T. Beppu.** 1988. Cloning and expression in *Escherichia coli* of two additional amylase genes of a strictly anaerobic thermophile, *Dictyoglomus thermophilum*, and their nucleotide sequences with extremely low guanine-plus-cytosine contents. *Eur J. Biochem.* **176:**243–253.
64. **Hoshiko, S., O. Makabe, C. Nojiri, K. Katsumata, E. Satoh, and K. Nagaoka.** 1987. Molecular cloning and characterization of the *Streptomyces hygroscopicus* α-amylase gene. *J. Bacteriol.* **169:**1029–1036.
65. **Howard, G. T., and B. A. White.** 1988. Molecular cloning and expression of cellulase genes from *Rumonococcus albus* 8 in *Escherichia coli* bacteriophage λ. *Appl. Environ. Microbiol.* **54:**1752–1755.
66. **Hyun, H. H., and J. G. Zeikus.** 1985. General biochemical characterization of thermostable pullulanase and glucoamylase from *Clostridium thermohydrosulfuricum*. *Appl. Environ. Microbiol.* **49:**1168–1173.
67. **Ihara, H., T. Sasaki, A. Tsuboi, H. Yamagata, N. Tsukagoshi, and S. Udaka.** 1985. Complete nucleotide sequence of a thermophilic α-amylase gene: homology between prokaryotic and eukaryotic α-amylases at the active sites. *J. Biochem.* **98:**95–103.
68. **Itkor, P., N. Tsukagoshi, and S. Udaka.** 1990. Nucleotide sequence of the raw-starch-digesting amylase gene from *Bacillus* sp. B1018 and its strong homology to the cyclodextrin glucanotransferase genes. *Biochem. Biophys. Res. Commun.* **166:**630–636.
69. **Ito, S., S. Shikata, K. Ozaki, S. Kawai, K. Okamoto, S. Inoue, A. Takei, Y. Ohta, and T. Satoh.** 1989. Alkaline cellulase for laundry detergents: production by *Bacillus* sp. KSM-635 and enzymatic properties. *Agric. Biol. Chem.* **53:**1275–1281.
70. **Jarnagin, A. S., and E. Ferrari.** 1992. Extracellular enzymes: gene regulation and structure function relationship studies, p. 191–219. *In* R. Doi and M. McGloughlin (ed.), *Biology of Bacilli: Applications to Industry*. Butterworth-Heinemann, Boston.
71. **Jørgensen, P. L., and C. K. Hansen.** 1990. Multiple endo-β-1,4-glucanase-encoding genes from *Bacillus lautus* PL236 and characterization of the *celB* gene. *Gene* **93:**55–60.
72. **Jørgensen, P. L., G. B. Poulsen, and B. Diderichsen.** 1991. Cloning of a chromosomal α-amylase gene from *Bacillus stearothermophilus*. *FEMS Microbiol. Lett.* **77:** 271–276.
73. **Kainuma, K.** 1984. Starch oligosaccharides: linear, branched, and cyclic, p. 125–152. *In* R. L. Whistler, J. N. Bemiller, and E. F. Paschall (ed.), *Starch: Chemistry and Technology*, 2nd ed. Academic Press, Inc., Orlando, Fla.
74. **Kallio, P. T., J. E. Fagelson, J. A. Hoch, and M. A. Strauch.** 1991. The transition state regulator Hpr of *Bacillus subtilis* is a DNA-binding protein. *J. Biol. Chem.* **266:**13411–13417.
75. **Kaneko, T., T. Hamamoto, and K. Horikoshi.** 1988. Molecular cloning and nucleotide sequence of the cyclomaltodextrin glucanotransferase gene from the alkalophilic *Bacillus* sp. strain no. 38-2. *J. Gen. Microbiol.* **134:**97–105.
76. **Kaneko, T., K. Song, T. Hamamoto, T. Kudo, and K. Horikoshi.** 1989. Construction of a chimeric series of *Bacillus* cyclomaltodextrin glucanotransferases and analysis of the thermal stabilities and pH optima of the enzymes. *J. Gen. Microbiol.* **135:**3447–3457.
77. **Katsuragi, N., N. Takizawa, and Y. Murooka.** 1987. Entire nucleotide sequence of the pullulanase gene of *Klebsiella aerogenes* W70. *J. Bacteriol.* **169:**2301–2306.
78. **Kawai, S., K. Oshino, H. Okoshi, H. Mori, K. Ara, S. Ito, and K. Okamoto.** 1988. Alkali resistant cellulases and microorganisms capable of producing same. European patent application EP 270974 A2.
79. **Kawazu, T., Y. Nakanishi, N. Uozumi, T. Sasaki, H. Yamagata, N. Tsukagoshi, and S. Udaka.** 1987. Cloning and nucleotide sequence of the gene coding for enzymatically active fragments of the *Bacillus polymyxa* β-amylase. *J. Bacteriol.* **169:**1564–1570.
80. **Kimura, K., S. Kataoka, Y. Ishii, T. Takano, and K. Yamane.** 1987. Nucleotide sequence of the β-cyclodextrin glucanotransferase gene of alkalophilic *Bacillus* sp. strain 1011 and similarity of its amino acid sequence to those of α-amylases. *J. Bacteriol.* **169:**4399–4402.
81. **Kindle, K. L.** 1983. Characteristics and production of thermostable α-amylase. *Biochem. Biotechnol.* **8:**153–170.
82. **Kitamoto, N., H. Yamagata, T. Kato, N. Tsukagoshi, and S. Udaka.** 1988. Cloning and sequencing of the gene encoding thermophilic β-amylase of *Clostridium thermosulfurogenes*. *J. Bacteriol.* **170:**5848–5854.
83. **Klein, C., and G. E. Schulz.** 1991. Structure of cyclodextrin glycosyltransferase refined at 2.0 Å resolution. *J. Mol. Biol.* **217:**737–750.
84. **Koide, Y., A. Nakamura, T. Uozumi, and T. Beppu.** 1986. Molecular cloning of a cellulase gene from *Bacillus subtilis* and its expression in *Escherichia coli*. *Agric. Biol. Chem.* **56:**233–237.
85. **Kondo, H., H. Nakatani, R. Matsuno, and K. Hiromi.** 1980. Product distribution in amylase catalyzed hydrolysis of amylose: comparison of experimental results with theoretical predictions. *J. Biochem.* **87:**1053–1070.
86. **Kreis, M., M. Williamson, B. Buxton, J. Pywell, J. Hejgaard, and I. Svendsen.** 1987. Primary structure and differential expression of β-amylase in normal and mutant barleys. *Eur. J. Biochem.* **169:**517–525.
87. **Kunst, F., M. Debarbouillie, T. Msadek, M. Young, C. Mauel, D. Karamata, A. Klier, G. Rapoport, and R. Dedonder.** 1988. Deduced polypeptides encoded by the *Bacillus subtilis sacU* locus share homology with two-component sensor regulatory systems. *J. Bacteriol.* **170:** 5093–5101.
88. **Kuriki, T., and T. Imanaka.** 1989. Nucleotide sequence of the neopullulanase gene from *Bacillus stearothermophilus*. *J. Gen. Microbiol.* **135:**1521–1528.
89. **Kuriki, T., J.-H. Park, and T. Imanaka.** 1990. Characteristics of thermostable pullulanase from *Bacillus stearothermophilus* and the nucleotide sequence of the gene. *J. Ferment. Bioeng.* **69:**204–210.
90. **Lao, G., G. S. Ghangas, E. D. Jung, and D. B. Wilson.** 1991. DNA sequences of three β-1,4-endoglucanase genes from *Thermomonospora fusca*. *J. Bacteriol.* **173:** 3397–3407.
91. **Lepesant, J. A., F. Kunst, J. Lepesant-Kejzlarova, and**

R. Dedonder. 1972. Chromosomal location of the mutations affecting sucrose metabolism in *Bacillus subtilis* Marburg. *Mol. Gen. Genet.* **118:**135–160.

92. **Lo, A. C., R. M. MacKay, V. L. Seligy, and G. E. Willick.** 1988. *Bacillus subtilis* β-1,4-endoglucanase products from intact and truncated genes are secreted into the extracellular medium by *Escherichia coli. Appl. Environ. Microbiol.* **54:**2287–2292.
93. **Long, C. M., M.-J. Virolle, S.-Y. Chang, S. Chang, and M. J. Bibb.** 1987. α-Amylase gene of *Streptomyces limosus*: nucleotide sequence, expression motifs, and amino acid sequence homology to mammalian and invertebrate α-amylases. *J. Bacteriol.* **169:**5745–5754.
94. **MacGregor, E. A., and B. Svensson.** 1989. A supersecondary structure predicted to be common to several α-1,4-D-glucan-cleaving enzymes. *Biochem. J.* **259:**145–152.
95. **MacKay, R. M., S. Baird, M. J. Dove, J. A. Erratt, M. Gines, F. Moranelli, A. Nasim, G. E. Willick, M. Yaguchi, and V. L. Seligy.** 1985. Glucanase gene diversity in prokaryotic and eukaryotic organisms. *BioSystems* **18:** 279–292.
96. **MacKay, R. M., A. Lo, G. Willick, M. Zuker, S. Baird, M. Dove, F. Moranelli, and V. Seligy.** 1986. Structure of a *Bacillus subtilis* endo-β-1,4-glucanase gene. *Nucleic Acids Res.* **14:**9159–9170.
97. **MacQuitty, J. J.** 1988. Impact of biotechnology on the chemical industry, p. 11–29. *In* M. P. Phillips, S. P. Shoemaker, R. D. Middlekauff, and R. M. Ottenbrite (ed.), *The Impact of Chemistry on Biotechnology: Multidisciplinary Discussion.* American Chemical Society, Washington, D.C.
98. **Marshall, J. J.** 1974. Characterization of *Bacillus polymyxa* amylase as an exo-acting(1-4)alpha-D-glucan maltohydrolase. *FEBS Lett.* **46:**1–4.
99. **Matsuura, Y., M. Kusunoki, W. Harada, and M. Kakudo.** 1984. Structure and possible catalytic residues of Taka-amylase A. *J. Biochem.* **95:**697–702.
100. **Meinke, A., N. R. Gilkes, D. G. Kilburn, R. C. Miller, Jr., and R. A. J. Warren.** 1991. Multiple domains in endoglucanase B (*cenB*) from *Cellulomonas fimi*: functions and relatedness to domains in other polypeptides. *J. Bacteriol.* **173:**7126–7135.
101. **Melasniemi, H., M. Paloheimo, and L. Hemiö.** 1990. Nucleotide sequence of the α-amylase-pullulanase gene from *Clostridium thermohydrosulfuricum. J. Gen. Microbiol.* **136:**447–454.
102. **Metz, R. J., L. N. Allen, T. M. Cao, and N. W. Zeman.** 1988. Nucleotide sequence of an amylase gene from *Bacillus megaterium. Nucleic Acids Res.* **16:**5203.
103. **Mikami, B., Y. Morita, and C. Fukazawa.** 1988. Primary structure and function of β-amylase. *Seikagaku* **60:**211–216.
104. **Msadek, T., M. K. Dahl, F. Kunst, and G. Rapoport.** 1992. The DegS/DegU signal transduction pathway in *Bacillus subtilis*, abstr. 17. Program Abstr. 11th Int. Spore Conf.
105. **Msadek, T., F. Kunst, D. J. Henner, A. Klier, G. Rapoport, and R. Dedonder.** 1990. Signal transduction pathway controlling synthesis of class of degradative enzymes in *Bacillus subtilis*: expression of the regulatory genes and analysis of mutations in *degS* and *degU. J. Bacteriol.* **172:**824–834.
106. **Mukai, K., M. Kawata, and T. Tanaka.** 1990. Isolation and phosphorylation of the *Bacillus subtilis degS* and *degU* gene products. *J. Biol. Chem.* **265:**20000–20006.
107. **Nagashima, T., S. Tada, K. Kitamoto, K. Gomi, C. Kumagai, and H. Toda.** 1992. Site-directed mutagenesis of catalytic active-site residues of Taka-amylase A. *Biosci. Biotechnol. Biochem.* **56:**207–210.
108. **Nakai, R., S. Horinouchi, and T. Beppu.** 1988. Cloning and nucleotide sequence of a cellulase gene, *casA*, from an alkalophilic *Streptomyces* strain. *Gene* **65:**229–238.
109. **Nakai, R., S. Horinouchi, T. Uozumi, and T. Beppu.** 1987. Purification and properties of cellulases from an alkalophilic *Streptomyces* strain. *Agric. Biol. Chem.* **51:** 3061–3065.
110. **Nakajima, R., T. Imanaka, and S. Aiba.** 1985. Nucleotide sequence of the *Bacillus stearothermophilus* α-amylase gene. *J. Bacteriol.* **163:**401–406.
111. **Nakajima, R., T. Imanaka, and S. Aiba.** 1986. Comparison of amino acid sequences of eleven different α-amylases. *Appl. Microbiol. Biotechnol.* **23:**355–360.
112. **Nakamura, A., F. Fukumori, S. Horinouchi, H. Masaki, T. Kudo, T. Uozumi, K. Horikoshi, and T. Beppu.** 1991. Construction and characterization of the chimeric enzymes between the *Bacillus subtilis* cellulase and an alkalophilic *Bacillus* cellulase. *J. Biol. Chem.* **266:**1579–1583.
113. **Nakamura, A., K. Haga, S. Ogawa, K. Kuwano, K. Kimura, K. Yamane.** 1992. Functional relationships between cyclodextrin glucanotransferase from an alkalophilic *Bacillus* and α-amylases: site-directed mutagenesis of the conserved two Asp and one Glu residues. *FEBS Lett.* **296:**37–40.
114. **Nakamura, A., T. Uozumi, and T. Beppu.** 1987. Nucleotide sequence of a cellulase gene of *Bacillus subtilis. Eur. J. Biochem.* **164:**317–320.
115. **Nishizawa, M., F. Ozawa, and F. Hishinuma.** 1987. Molecular cloning of an amylase gene of *Bacillus circulans. DNA* **6:**255–265.
116. **Nitschke, L., K. Heeger, H. Bender, and G. E. Schulz.** 1990. Molecular cloning, nucleotide sequence and expression in *Escherichia coli* of the β-cyclodextrin glycosyltransferase gene from *Bacillus circulans* strain no. 8. *Appl. Microbiol. Biotechnol.* **33:**542–546.
117. **Ohmiya, K., T. Kajino, A. Kato, and S. Shimizu.** 1989. Structure of a *Ruminococcus albus* endo-1,4-β-glucanase gene. *J. Bacteriol.* **171:**6771–6775.
118. **O'Neill, G. P., S. H. Goh, R. A. J. Warren, D. G. Kilburn, and R. C. Miller, Jr.** 1986. Structure of the gene encoding the exoglucanase of *Cellulomonas fimi. Gene* **44:**325–330.
119. **Outtrup, H., and B. E. Norman.** 1984. Properties and application of a thermostable maltogenic amylase produced by a strain of *Bacillus* modified by recombinant-DNA techniques. *Starch Stärke* **36:**405–411.
120. **Ozaki, K., and S. Ito.** 1991. Purification and properties of an acid endo-1,4-β-glucanase from *Bacillus* sp. KSM-330. *J. Gen. Microbiol.* **137:**41–48.
121. **Ozaki, K., S. Shikata, S. Kawai, S. Ito, and K. Okamoto.** 1990. Molecular cloning and nucleotide sequence of a gene for alkaline cellulase from *Bacillus* sp. KSM-635. *J. Gen. Microbiol.* **136:**1327–1334.
122. **Park, S. H., H. K. Kim, and M. Y. Pack.** 1991. Characterization and structure of the cellulase gene of *Bacillus subtilis* BSE616. *Agric. Biol. Chem.* **55:**441–448.
123. **Park, S. H., and M. Y. Pack.** 1986. Cloning and expression of a *Bacillus* cellulase gene in *Escherichia coli. Enzyme Microb. Technol.* **8:**725–728.
124. **Perego, M., and J. A. Hoch.** 1988. Sequence analysis and regulation of the *hpr* locus, a regulatory gene for protease production and sporulation in *Bacillus subtilis. J. Bacteriol.* **170:**2560–2567.
125. **Perego, M., G. B. Spiegelman, and J. A. Hoch.** 1988. Structure of the gene for the transition state regulator AbrB: regulator synthesis is controlled by the Spo0A protein. *Mol. Microbiol.* **2:**689–699.
126. **Petříček, M., P. Tichý, and M. Kuncová.** 1992. Characterization of the α-amylase-encoding gene from *Thermomonospora curvata. Gene* **112:**77–83.
127. **Poole, D. M., G. P. Hazlewood, J. I. Laurie, P. J. Barker, and H. J. Gilbert.** 1990. Nucleotide sequence of the

Ruminococcus albus SY3 endoglucanase genes *celA* and *celB*. *Mol. Gen. Genet.* **223**:217–223.

128. **Priest, F. G.** 1977. Extracellular enzyme synthesis in the genus *Bacillus*. *Bacteriol. Rev.* **41**:711–753.
129. **Priest, F. G.** 1984. Extracellular enzymes, p. 54–62. American Society for Microbiology, Washington, D.C.
130. **Priest, F. G., M. Goodfellow, and C. Todd.** 1988. A numerical classification of the genus *Bacillus*. *J. Gen. Microbiol.* **134**:1847–1882.
131. **Reilly, P. J.** 1985. Enzymic degradation of starch, p. 101–142. *In* G. M. A. van Beynum and J. A. Roels (ed.), *Starch Conversion Technology*. Marcel Dekker, Inc., New York.
132. **Rhodes, C., J. Strasser, and F. Friedberg.** 1987. Sequence of an active fragment of *B. polymyxa* beta amylase. *Nucleic Acids Res.* **15**:3934.
133. **Robson, L. M., and G. H. Chambliss.** 1986. Cloning of the *Bacillus subtilis* DLGβ-1,4-glucanase gene and its expression in *Escherichia coli* and *B. subtilis*. *J. Bacteriol.* **165**:612–619.
134. **Robson, L. M., and G. H. Chambliss.** 1987. Endo-β-1,4-glucanase gene of *Bacillus subtilis* DLG. *J. Bacteriol.* **169**:2017–2025.
135. **Robson, L. M., and G. H. Chambliss.** 1989. Cellulases of bacterial origin. *Enzyme Microb. Technol.* **11**:626–644.
136. **Robyt, J. F.** 1984. Enzymes in the hydrolysis and synthesis of starch, p. 87–123. *In* R. L. Whistler, J. N. Bemiller, and E. F. Paschall (ed.), *Starch: Chemistry and Technology*, 2nd ed. Academic Press, Inc., Orlando, Fla.
137. **Rogers, J. C.** 1985. Conserved amino acid sequence domains in alpha-amylases from plants, mammals, and bacteria. *Biochem. Biophys. Res. Commun.* **128**:470–476.
138. **Rumbak, E., D. E. Rawlings, G. G. Lindsey, and D. R. Woods.** 1991. Cloning, nucleotide sequence, and enzymatic characterization of an α-amylase from the ruminal bacterium *Butyrivibrio fibrisolvens* H17c. *J. Bacteriol.* **173**:4203–4211.
139. **Russell, R. R. B., and J. J. Ferretti.** 1990. Nucleotide sequence of the dextran glucosidase (*dexB*) gene of *Streptococcus mutans*. *J. Gen. Microbiol.* **136**:803–810.
140. **Saha, B. C., and J. G. Zeikus.** 1989. Novel highly thermostable pullulanase from thermophiles. *Trends Biotechnol.* **7**:234–239.
141. **Sashihara, N., T. Kudo, and K. Horikoshi.** 1984. Molecular cloning and expression of cellulase genes to alkalophilic *Bacillus* sp. strain N-4 in *Escherichia coli*. *J. Bacteriol.* **158**:503–506.
142. **Schmid, G., A. Englbrecht, and D. Schmid.** 1988. Cloning and nucleotide sequence of a cyclodextrin glycosyltransferase gene from the alkalophilic *Bacillus* 1-1, p. 71–76. *In* O. Huber and J. Szejtli (ed.), *Proceedings of the Fourth International Symposium on Cyclodextrins*. Kluwer Academic Publishers, New York.
143. **Shen, G.-J., K. C. Srivastava, Y. Wang, and H. Y. Wang.** March 1992. U.S. patent 5,093,256.
144. **Shinke, R., H. Nishira, and M. Mugibayashi.** 1974. Isolation of β-amylase producing microorganisms. *Agric. Biol. Chem.* **38**:665–666.
145. **Siggens, K. W.** 1987. Molecular cloning and characterization of the beta-amylase gene from *Bacillus circulans*. *Mol. Microbiol.* **1**:86–91.
146. **Siggens, K. W.** 1987. *Bacillus circulans* gene for β-amylase. GenBank Genetic Sequence Data Bank, Release 68.0, Locus:Bacamyb, Accession: Y00523. Unpublished sequence.
147. **Sin, K., A. Nakamura, K. Kobayashi, H. Masaki, and T. Uozumi.** 1991. Cloning and sequencing of a cyclodextrin glucanotransferase gene from *Bacillus ohbensis* and its expression in *Escherichia coli*. *Appl. Microbiol. Biotechnol.* **35**:600–605.
148. **Slomińska, L., and M. Maczyński.** 1985. Studies on the application of pullulanase in starch saccharification process. *Starch Stärke* **37**:386–390.
149. **Smith, I.** 1989. Initiation of sporulation, p. 185–209. *In* I. Smith, R. Slepecky, and P. Setlow (ed.), *Regulation of Procaryotic Development*. American Society for Microbiology, Washington, D.C.
150. **Specka, U., F. Mayer, and G. Antranikian.** 1991. Purification and properties of a thermoactive glucoamylase from *Clostridium thermosaccharolyticum*. *Appl. Environ. Microbiol.* **57**:2317–2323.
151. **Srivastava, R. A. K.** 1984. Studies on extracellular and intracellular purified amylases from a thermophilic *Bacillus stearothermophilus*. *Enzyme Microb. Technol.* **6**:422–426.
152. **Strauch, M. A., and J. A. Hoch.** 1992. Control of postexponential gene expression by transition state regulators, p. 105–121. *In* R. H. Doi and M. McGloughlin (ed.), *Biology of Bacilli: Applications to Industry*. Butterworth-Heinemann, Boston.
153. **Strauch, M. A., M. Perego, D. Burbulys, and J. A. Hoch.** 1989. The transition state regulator AbrB of *Bacillus subtilis* is autoregulated during vegetative growth. *Mol. Microbiol.* **3**:1203–1209.
154. **Strauch, M. A., G. B. Spiegelman, M. Perego, W. C. Johnson, D. Burbulys, and J. A. Hoch.** 1989. The transition state transcription regulator *abrB* of *Bacillus subtilis* is a DNA binding protein. *EMBO J.* **8**:1615–1621.
155. **Strauch, M. A., V. Webb, G. Spiegelman, and J. A. Hoch.** 1990. The Spo0A protein of *Bacillus subtilis* is a repressor of the *abrB* gene. *Proc. Natl. Acad. Sci. USA* **87**:1801–1805.
156. **Stutzenberger, F.** 1990. Bacterial cellulases, p. 37–70. *In* W. M. Fogarty, and K. T. Kelly (ed.), *Microbial Enzymes and Biotechnology*, 2nd ed. Elsevier Applied Science, London.
157. **Suominen, I., M. Karp, J. Lautano, J. Knowles, and P. Mantsälä.** 1987. Thermostable α-amylase of *Bacillus stearothermophilus*: cloning, expression and secretion by *Escherichia coli*, p. 129–137. *In* J. Chaloupka and V. Krumphanzl (ed.), *Extracellular Enzymes of Microorganisms*. Plenum Press, New York.
158. **Suzuki, Y., K. Hatagaki, and H. Oda.** 1991. A hyperthermostable pullulanase produced by an extreme thermophile, *Bacillus flavocaldarius* KP 1228, and evidence for the proline theory of increasing protein thermostability. *Appl. Microbiol. Biotechnol.* **34**:707–714.
159. **Suzuki, Y., N. Ito, T. Yuuki, H. Yamagata, and S. Udaka.** 1989. Amino acid residues stabilizing a *Bacillus* α-amylase against irreversible thermoinactivation. *J. Biol. Chem.* **264**:18933–18938.
160. **Suzuki, Y., K. Oishi, H. Nakano, and T. Nagayama.** 1987. A strong correlation between the increase in number of proline residues and the rise in thermostability of five *Bacillus* oligo-1,6-glucosidases. *Appl. Microbiol. Biotechnol.* **26**:546–551.
161. **Svensson, B.** 1988. Regional distant sequence homology between amylases, α-glucosidases and transglucanosylases. *FEBS Lett.* **230**:72–76.
162. **Svensson, B., H. Jespersen, M. R. Sierks, and E. A. MacGregor.** 1989. Sequence homology between putative raw-starch binding domains from different starch-degrading enzymes. *Biochem. J.* **264**:309–311.
163. **Takano, T., M. Fukuda, M. Monma, S. Kobayashi, K. Kainuma, and K. Yamane.** 1986. Molecular cloning, DNA nucleotide sequencing, and expression in *Bacillus subtilis* cells of the *Bacillus macerans* cyclodextrin glucanotransferase gene. *J. Bacteriol.* **166**:1118–1122.
164. **Takasaki, Y.** 1976. Production and utilization of β-amylase and pullulanase from *Bacillus cereus* var. *mycoides*. *Agric. Biol. Chem.* **40**:1515–1522.

165. **Takase, K., T. Matsumoto, H. Mizuno, and K. Yamane.** 1992. Site-directed mutagenesis of active site residues in *Bacillus subtilis* α-amylase. *Biochim. Biophys. Acta* **1120:**281–288.
166. **Takkinen, K., R. F. Pettersson, N. Kalkkinen, I. Palva, H. Söderlund, and L. Kääriäinen.** 1983. Amino acid sequence of α-amylase from *Bacillus amyloliquefaciens* deduced from the nucleotide sequence of the cloned gene. *J. Biol. Chem.* **258:**1007–1013.
167. **Taniguchi, H., M. J. Chung, N. Yoshigi, and Y. Maruyama.** 1983. Purification of *Bacillus circulans* F-2 amylase and its general properties. *Agric. Biol. Chem.* **47:** 511–520.
168. **Tao, B. Y., P. J. Reilly, and J. F. Robyt.** 1989. Dectection of a covalent intermediate in the mechanism of action of porcine pancreatic α-amylase by using ^{13}C nuclear magnetic resonance. *Biochim. Biophys. Acta* **995:**214–220.
169. **Thomas, J. A., and J. D. Allen.** 1976. Subsite mapping of enzymes, collecting and processing experimental data: a case study of an amylase malto oligosaccharide system. *Carbohydr. Res.* **48:**105–124.
170. **Thomas, M., F. G. Priest, and J. R. Stark.** 1980. Characterization of an extracellular β-amylase from *Bacillus megaterium sensu stricto. J. Gen. Microbiol.* **118:**67–72.
171. **Tomazic, S. J., and A. M. Klibanov.** 1988. Mechanisms of irreversible thermal inactivation of *Bacillus* α-amylases. *J. Biol. Chem.* **263:**3086–3091.
172. **Tomazic, S. J., and A. M. Klibanov.** 1988. Why is one *Bacillus* α-amylase more resistant against irreversible thermoinactivation than another? *J. Biol. Chem.* **263:** 3092–3096.
173. **Tomme, P., H. Van Tilbeurgh, G. Pettersson, J. Van Damme, J. Vandekerckhove, J. Knowles, T. Teeri, and M. Claeyssens.** 1988. Studies of the cellulolytic system of *Trichoderma reesei* QM 9414. Analysis of domain function in two cellobiohydrolases by limited proteolysis. *Eur. J. Biochem.* **170:**575–581.
174. **Tsukamoto, A., K. Kimura, Y. Ishii, T. Takano, and K. Yamane.** 1988. Nucleotide sequence of the maltohexaose producing amylase gene from an alkalophilic *Bacillus* sp. #707 and structural similarity to liquefying type α-amylases. *Biochem. Biophys. Res. Commun.* **151:** 25–31.
175. **Uozumi, N., K. Sakurai, T. Sasaki, S. Takekawa, H. Yamagata, N. Tsukagoshi, and S. Udaka.** 1989. A single gene directs synthesis of a precursor protein with β- and α-amylase activities in *Bacillus polymyxa. J. Bacteriol.* **171:**375–382.
176. **Valle, F., and E. Ferrari.** 1989. Subtilisin: a redundantly temporally regulated gene?, p. 131–146. *In* I. Smith, R. Slepecky, and P. Setlow (ed.), *Regulation of Procaryotic Development*. American Society for Microbiology, Washington, D.C.
177. **Van Tilbeurgh, H., P. Tomme, M. Claeyssens, R. Bhikhabhai, and G. Pettersson.** 1986. Limited proteolysis of the cellobiohydrolase I from *Trichoderma reesei*. Separation of functional domains. *FEBS Lett.* **204:**223–227.
178. **Vigal, T., J. A. Gil, A. Daza, M. D. García-González, and J. F. Martin.** 1991. Cloning, characterization and expression of an α-amylase gene from *Streptomyces griseus* IMRU3570. *Mol. Gen. Genet.* **225:**278–288.
179. **Vihinen, M., and P. Mäntsälä.** 1989. Microbial amylolytic enzymes. *Crit. Rev. Biochem. Mol. Biol.* **24:** 329–418.
180. **Vihinen, M., P. Ollikka, J. Niskamem, P. Meyer, I. Suominen, M. Karp, L. Holm, J. Knowles, and P. Mäntsälä.** 1990. Site-directed mutagenesis of a thermostable α-amylase from *Bacillus stearothermophilus*: putative role of three conserved residues. *J. Biochem.* **107:**267–272.
181. **Virolle, M.-J., C. M. Long, S. Chang, and M. J. Bibb.** Cloning, characterisation and regulation of an α-amylase gene from *Streptomyces venezuelae. Gene* **74:**321–334.
182. **von Heijne, G., and L. Abrahmsen.** 1989. Species-specific variation in signal peptide design. *FEBS Lett.* **244:**439–446.
183. **Wachinger, G., K. Bronnenmeier, W. L. Staudenbauer, and H. Schrempf.** 1989. Identification of mycelium-associated cellulase from *Streptomyces reticuli. Appl. Environ. Microbiol.* **55:**2653–2657.
184. **Wakim, J., M. Robinson, and J. A. Thoma.** 1969. The active site of porcine pancreatic alpha amylase factors contributing to catalysis. *Carbohydr. Res.* **10:**487–503.
185. **Waldron, C. R., Jr., C. A. Becker-Vallone, and D. E. Eveleigh.** 1986. Isolation and characterization of a cellulolytic actinomycete *Microbispora bispora. Appl. Microbiol. Biotechnol.* **24:**477–486.
186. **Waldron, C. R., Jr., and D. E. Eveleigh.** 1986. Saccharification of cellulosics by *Microbispora bispora. Appl. Microbiol. Biotechnol.* **24:**487–492.
187. **Wang, L.-F., and R. Doi.** 1990. Complex character of *senS*, a novel gene regulating expression of extracellular protein genes of *Bacillus subtilis. J. Bacteriol.* **172:** 1939–1947.
188. **Watanabe, K., K. Chishiro, K. Kitamura, and Y. Suzuki.** 1991. Proline residues responsible for thermostability occur with high frequency in the loop regions of an extremely thermostable oligo-1,6-glucosidase from *Bacillus thermoglucosidasius* KP1006. *J. Biol. Chem.* **266:**24287–24294.
189. **Watanabe, K., K. Kitamura, H. Iha, and Y. Suzuki.** 1990. Primary structure of the oligo-1,6-glucosidase of *Bacillus cereus* ATCC7064 deduced from the nucleotide sequence of the cloned gene. *Eur. J. Biochem.* **192:**609–620.
190. **Wirsel, S., A. Lachmund, G. Wildhardt, and E. Ruttkowski.** 1989. Three α-amylase genes of *Aspergillus oryzae* exhibit identical intron-exon organization. *Mol. Microbiol.* **3:**3–14.
191. **Wong, W. K. R., B. Gerhard, Z. M. Guo, D. G. Kilburn, R. A. J. Warren, and R. C. Miller, Jr.** 1986. Characterization and structure of an endoglucanase gene *cenA* of *Cellulomonas fimi. Gene* **44:**315–324.
192. **Yamane, K., Y. Hirata, T. Furusato, H. Yamazaki, and A. Nakayama.** 1984. Changes in the properties and molecular weights of *Bacillus subtilis* M-type and N-type α-amylases resulting from a spontaneous deletion. *J. Biochem.* **96:**1849–1858.
193. **Yamane, K., and B. Maruo.** 1980. *Bacillus subtilis* α-amylase genes, p. 117–123. *In* K. Sakaguchi and M. Ouishi (ed.), *Molecular Breeding and Genetics of Applied Microorganisms*. Academic Press, Inc., New York.
194. **Yamazaki, H., K. Ohmura, A. Nakayama, Y. Takeichi, K. Otozai, M. Yamasaki, G. Tamura, and K. Yamane.** 1983. α-Amylase genes (*amyR2* and *amyE*$^{+}$) from an α-amylase-hyperproducing *Bacillus subtilis* strain: molecular cloning and nucleotide sequences. *J. Bacteriol.* **156:**327–337.
195. **Yang, M. Y., A. Galizzi, and D. Henner.** 1983. Nucleotide sequence of the amylase gene from *Bacillus subtilis. Nucleic Acids Res.* **11:**237–249.
196. **Yoneda, Y.** 1980. Increased production of extracellular enzymes by the synergistic effect of genes introduced into *Bacillus subtilis* by stepwise transformation. *Appl. Environ. Microbiol.* **39:**274–276.
197. **Yoshimatsu, T., K. Ozaki, S. Shikata, Y. Ohta, K. Koike, S. Kawai, and S. Ito.** 1990. Purification and characterization of alkaline endo-1,4-β-glucanases from alkalophilic *Bacillus* sp. KSM-635. *J. Gen. Microbiol.* **136:**1973–1979.

198. **Yuuki, T., T. Nomura, H. Tezuka, A. Tsuboi, H. Yamagata, N. Tsukagoshi, and S. Udaka.** 1985. Complete nucleotide sequence of a gene coding for heat- and pH-stable α-amylase of *Bacillus licheniformis*: comparison of the amino acid sequences of three bacterial liquefying α-amylases deduced from the DNA sequences. *J. Biochem.* **98:**1147–1156.
199. **Zhou, J., T. Baba, T. Takano, S. Kobayoshi, and Y. Arai.** 1989. Nucleotide sequence of the maltotetraohydralase gene from *Pseudomonas saccharophilia*. *FEBS Lett.* **255:**37–41.
200. **Zukowski, M. M.** 1992. Production of commercially valuable products, p. 311–337. *In* R. H. Doi and M. McGloughlin (ed.), *Biology of Bacilli: Applications to Industry*. Butterworth-Heinemann, Boston.

63. Proteases

JANICE PERO and ALAN SLOMA

Historically, *Bacillus* spp. and other gram-positive bacterial species have been used as sources of industrial enzymes, especially proteases (89). The proteases used industrially are extracellular proteases; hence, much more is known about the gram-positive extracellular proteases than the intracellular proteases. This chapter reviews the molecular biology and genetics of gram-positive endoproteases, focusing on *Bacillus subtilis* proteases. Studies on the industrial production of subtilisin are covered in chapter 60, and a review of the regulatory genes affecting protease gene expression can be found in chapters 50, 51, 52, 54, and 62.

Microbial endoproteases are generally classified into four categories based on their mechanisms of action (reviewed in references 34 and 75). Serine proteases, the most common class in *Bacillus* species, have a serine residue in the active site and are inhibited by hydroxyl-reactive organofluorides such as diisopropylflurophosphate and phenylmethylsulfonyl fluoride (PMSF). Metalloproteases, also a common class in *Bacillus* species, require divalent metal cations (usually zinc) and are inhibited by chelators of these cations such as EDTA and 1,10-phenanthroline. Thiol or sulfhydryl proteases contain a sulfhydryl group in the active site, are stimulated by reducing agents such as dithiothreitol and cysteine, and are inhibited by oxidizing agents and mercurials. Finally, the acid or aspartic proteases have low pH optima and are inhibited by biazoketones. To date, no thiol or acid proteases have been reported in *Bacillus* spp. In addition, both serine and thiol proteases generally have esterase activity and are sometimes referred to as esterases. The metalloproteases and acid proteases usually do not have this esterase activity (75, 138).

B. SUBTILIS: EXTRACELLULAR PROTEASES

The genes for seven different extracellular proteases have been cloned and characterized in *B. subtilis* (10, 105–108, 113, 128, 144, 147, 150). All seven genes have been mapped to scattered regions of the *B. subtilis* chromosome (Fig. 1) and shown to be nonessential for either growth or sporulation. These *B. subtilis* proteases are all synthesized as preproproteins; four are members of the subtilisin family of serine proteases, and three are metalloproteases. Below, we review each of the proteases in detail.

Subtilisin

The two extracellular proteases present in greatest concentration in culture fluids of *B. subtilis* are the neutral, or metalloprotease, and the alkaline serine protease, or subtilisin (72, 88). Subtilisin is one of the most extensively studied of all bacterial proteins (see reviews in references 68 and 104). The study of subtilisin began in the 1950s with the crystallization of an alkaline protease (subtilisin Carlsberg) from a *Bacillus* strain classified at the time as *B. subtilis* but since reclassified as *Bacillus licheniformis* (29). The subtilisin present in "transformable" strains of *B. subtilis* was first characterized in the late 1960s (9, 72, 90). This *B. subtilis* enzyme has a pH optimum of 7 to 9 (9, 72) and a molecular weight of 28,500 (66). Because of the alkaline pH optimum, subtilisin is also known as "alkaline protease." This enzyme is a member of the serine protease family and is inhibited by PMSF but not EDTA (9, 66, 72). Also, this alkaline protease has a high proteolytic activity/esterolytic activity ratio (72). Subtilisin is first produced at the onset of sporulation, and its production is controlled by several sporulation genes such as *spo0A* (40, 86).

Most of the earliest structural studies of *Bacillus* subtilisins were performed on the three industrially important subtilisins, Carlsberg, BPN′, and Novo, which are produced by strains of *B. licheniformis* and *Bacillus amyloliquefaciens*. The amino acid sequences of the Carlsberg and BPN′ subtilisins, which are 274 and 275 residues long, respectively, were determined in 1966 and found to be similar but not identical (67, 109, 110). Subsequent analysis of the amino acid sequence of the Novo subtilisin showed it to be identical to that of BPN′ (82). The tertiary structure of subtilisin BPN′ was first determined in 1969 (19, 145). As a result of the extensive knowledge of *B. amyloliquefaciens* subtilisin, the gene for this enzyme was the first protease gene to be cloned from a *Bacillus* species (142). Analysis of the gene structure indicated that the enzyme was synthesized as a preproprotein with a typical *Bacillus* signal peptide (see chapter 49, this volume) and a proregion of roughly 75 amino acids. At the time, this was particularly interesting, because although proenzymes (trypsinogen, chymotrypsinogen) were common in eukaryotic systems, no example of a propeptide had been reported in *Bacillus* species, and only streptococcal protease had been shown to exist as a proprotease in bacterial species (142).

Following this work, a fragment of the subtilisin gene from *B. amyloliquefaciens* was used to clone the subtilisin gene (*apr*) from *B. subtilis* (113). Nucleotide sequence analysis revealed that the *B. subtilis* gene encoded a mature enzyme of 275 amino acids that was 85% homologous to the protein sequence of subtilisin BPN′ (113). The gene contained a 106-amino-acid leader region that was originally thought to encode a

Janice Pero • OmniGene, Inc., 763D Concord Avenue, P.O. Box 9002, Cambridge, Massachusetts 02139-9002. **Alan Sloma** • Novo-Nordisk Biotech, Davis, California 95616.

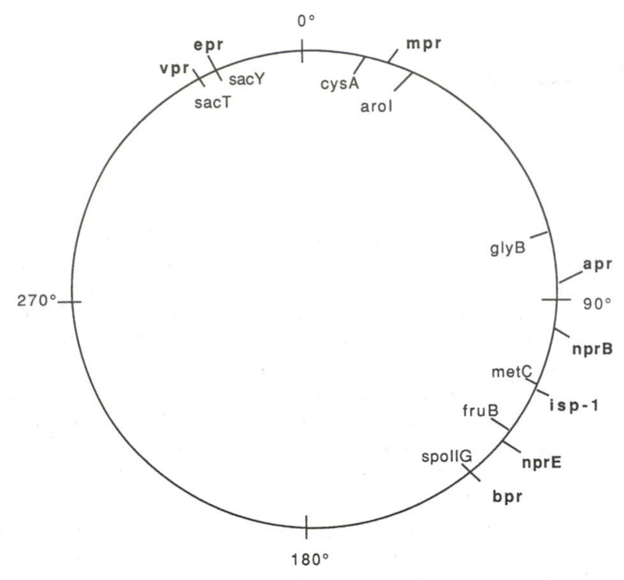

Figure 1. Genetic map of *B. subtilis* protease genes *apr* (113, 144), *bpr* (107, 147), *epr* (10, 105), *isp-1* (54), *mpr* (106), *nprE* (131, 150), *nprB* (128), and *vpr* (108).

signal sequence of 23 amino acids, but other work indicated a 29-amino-acid signal sequence (143, 144). Since the gene encoded approximately 77 amino acids between the end of the signal sequence and the beginning of the determined amino acid sequence of the mature enzyme, it was therefore proposed, as with the enzyme from *B. amyloliquefaciens*, that *B. subtilis* subtilisin was translated as a prepropolypeptide with the proregion removed during protein secretion.

Several studies have focused on the function of the proregion in subtilisin. Power et al. (87) showed that the proregion is removed by autoproteolysis and that this processing is necessary for release of subtilisin from the cell membrane. Studies with fusion proteins containing the subtilisin proregion also support the view that the proregion anchors the enzyme in the membrane (20). Inouye and coworkers (44) further established that the prosequence is required for the production of active subtilisin and argued that it was essential in guiding the folding of subtilisin into an active enzyme. In addition, they showed that exogenously added prosubtilisin mutated in the active site or even a synthetic 77-amino-acid copy of the propeptide could intermolecularly complement the folding of denatured subtilisin to active enzyme (80, 153). On the basis of these results, these workers have argued that the propeptide functions as an intramolecular chaperone (80). Several groups have shown that changing or deleting amino acids in the proregion can prevent processing (20a, 43, 62) and that changing certain areas of the mature enzyme can also block this processing (97). Furthermore, Ohta et al. (80) used the synthetic 77-amino-acid propeptide to establish that the highly charged prosequence binds to the active site of subtilisin and is a specific competitive inhibitor.

The gene for subtilisin (*apr*) was mapped on the *B. subtilis* chromosome by integration of a chloramphenicol resistance gene at the *apr* locus (113, 144). The *apr* gene is 77 to 83% linked to *glyB* and 60% linked to *argC* by PBS1 transduction and is located between *glyB* and *metD* at about 84° on the *B. subtilis* genetic map (Fig. 1; 113, 144).

Since subtilisin or alkaline protease is synthesized at the onset of sporulation and its synthesis is greatly reduced in many *spo0* mutants, it had been argued by some to have an important role in sporulation (5, 40, 65, 86). Cloning of the gene allowed for the construction of an in vitro deletion mutation in *apr* that was then introduced into the bacterial chromosome by gene replacement (50, 113). Strains carrying the *apr* mutation produced only 10% of the wild-type level of serine protease (113) but sporulated normally, with no discernible effect on spore formation (50, 113).

Neutral Protease

The second major extracellular protease in *B. subtilis* is the neutral protease, or metalloprotease (51, 72). Early work established that this metalloprotease is most active at a neutral pH, hence its name. As with subtilisin, the first "*B. subtilis*" neutral protease to be crystallized and characterized was actually from a strain of *B. amyloliquefaciens* (26, 71, 130). This enzyme (30,000 to 40,000 Da) required zinc for activity, was stabilized by calcium ions, and was inhibited by EDTA (71). Millet (72) found a similar enzyme in *B. subtilis* Marburg that appeared at the onset of sporulation.

Isolation of *B. subtilis* Marburg mutants producing a thermosensitive neutral protease allowed the gene for neutral protease (*nprE*) to be mapped (131). More recently, a *cat* gene was integrated at *nprE* and used in mapping the gene (150). The *npr* gene was 36% linked to *cysC* and 63 to 80% linked to *fruB* by transduction, placing it at approximately 127° on the *B. subtilis* chromosome (Fig. 1; 131, 150).

As with subtilisins, the early structural studies of neutral protease from bacilli were not performed with the *B. subtilis* enzyme but instead with a thermostable enzyme isolated from *Bacillus thermoproteolyticus*. Both the primary and the tertiary structures of this neutral protease, called thermolysin, were reported in 1972 (15, 70, 126). Thermolysin, an enzyme of 35,000 Da, has three zinc ligands to bind a zinc atom at the active site and also contains three or four calcium-binding sites (69).

The *npr* genes of both *B. subtilis* (150) and *B. amyloliquefaciens* (134) were cloned and sequenced in 1984. The *nprE* gene of *B. subtilis* encoded a signal sequence of approximately 30 amino acids, with an additional 192 residues rich in charged amino acids preceding the reported amino terminus of the mature enzyme (150). Therefore, like the *apr* gene, the *npr* gene has a prosequence that is removed during secretion. The function of this unusually large prosequence is still under investigation, but presumably it is similar to that for the prosequence of subtilisin, i.e., to maintain the protease in an inactive state and to promote its correct folding (139). As for subtilisin, the presence of a prosequence is essential for the production of active enzyme (120).

Cloning of the *nprE* gene allowed the creation of null mutations in the gene. *B. subtilis* strains carrying a deletion mutation in *nprE* produced only 20% of the level of protease found in wild-type cells (no detectable neutral protease activity) (150). No effects on cell growth, sporulation, or cell morphology were seen in strains carrying an *nprE* deletion mutation or in strains with mutations in both the *apr* and the *nprE* genes (150). These results support the idea that the main role of these extracellular enzymes is as scavengers.

Bacillopeptidase F

In addition to the two major extracellular proteases present in the culture of wild-type *B. subtilis*, several researchers reported the presence of another serine protease (9, 72, 88). This protease, which also appears at the end of exponential growth, has a high esterolytic activity/proteolytic activity ratio and was called an esterase. The enzyme has a pH optimum between 7.2 and 8.1 (64). In addition, the enzyme has a negative charge between pH 7 and 8 and because of this observation was sometimes referred to as the acidic protease (9, 33, 64). The enzyme was further characterized, purified, and named bacillopeptidase F by Roitsch and Hageman (93). They found two active forms, one with a molecular weight of 33,000 and the other with a molecular weight of 50,000. The pIs were 4.4 and 5.4, respectively. Both forms were reported to be glycoproteins (93).

More recent purifications of bacillopeptidase F, prepared for amino-terminal amino acid sequencing, have yielded molecular weights of 48,000 and 50,000 (147) and 47,000 (107). In addition, the 47,000-molecular-weight purified protein did not appear to be a glycoprotein by periodic acid-Schiff staining (107). An oligonucleotide probe based on the amino-terminal sequence of purified bacillopeptidase F was used to clone the gene (*bpr* or *bpf*) encoding this enzyme (107, 147). The gene encoded a primary product of 1,433 amino acids with a preprosequence of approximately 194 amino acids preceding the amino terminus of the mature enzyme (107). The first approximately 30 amino acids of this leader region resembled a typical gram-positive signal sequence, with the remaining approximately 164 amino acids constituting the putative prosequence. Surprisingly, the DNA encoded another 1,239 amino acids after the sequenced amino terminus to the mature protein (107). Since the mature protein is about 50,000 Da, not the 135,000 Da predicted from the DNA sequence of the gene, the protein clearly undergoes C-terminal processing or proteolysis. The function of this long C-terminal region is unclear, since a deletion mutation resulting in a truncated gene product lacking the carboxy-terminal 800 amino acids does not produce a defect in either the secretion or the activity of the enzyme (147). Comparisons of the deduced amino acid sequence of bacillopeptidase F with that of subtilisin showed 30% similarity, with an especially high degree of conservation surrounding the three residues (Asp-227, His-274, and Ser-221) involved in the active site of subtilisin (107, 147).

The *bpr-bpf* gene was mapped near *pyrD* just downstream of the *ftsZ* gene at 135° on the *B. subtilis* chromosome and just upstream of the *spoIIGA* gene (Fig. 1; 107, 115, 147). As was found with the genes for alkaline and neutral protease, deletion mutations in the gene for bacillopeptidase F had no measurable effect on either growth or sporulation (107, 147).

Mpr Protease

Neutral protease, subtilisin, and bacillopeptidase F were the only extracellular proteases originally detectable in the culture fluids of wild-type *B. subtilis* (72, 88). Once the genes corresponding to these enzymes were cloned and deletion mutations were introduced in the chromosome, it became possible to characterize new, heretofore undetectable proteases in the culture fluids. One of the minor proteases isolated from such strains was a novel metalloprotease, Mpr (94). This protease has a molecular weight of 28,000, a basic isoelectric point of 8.7, and a pH optimum of 7.5. An unusual feature of Mpr was its inhibition profile (94). The enzyme was inhibited by dithiothreitol and was insensitive to PMSF. Both mercaptoethanol and EDTA inhibited the enzyme approximately 50%; however, the effects of EDTA and mercaptoethanol were additive: both inhibitors together completely inactivated the enzyme. These results suggested that Mpr is a metalloprotease with disulfide bridges involved in the structural and/or catalytic integrity of the enzyme (94).

The gene for Mpr was cloned by using oligonucleotide probes designed on the basis of a partial amino acid sequence of the purified protein (106). The gene appeared to encode a preproprotein of 313 amino acids consisting of an approximately 34-amino-acid signal sequence, a 58-residue prosequence, and a mature protein of 221 amino acids with a molecular weight of 24,000, a size in apparent agreement with the observed molecular weight of 28,000. The deduced amino acid sequence contained four cysteine residues, suggesting the possible presence of disulfide bonds (106) and confirming the results of the protein studies (94). The presence of these putative disulfide bonds is unique among the extracellular proteases of *B. subtilis* although common in eukaryotic proteases. In fact, Mpr showed significant similarity to human protease E and bovine carboxypeptidase A complex component III in a region surrounding two cysteines that form a disulfide bond in these proteases (106). Surprisingly, Mpr showed no significant similarity to other bacterial proteases (106).

Mapping experiments indicated that the *mpr* gene is located between *cysA* and *aroI* at approximately 20° on the *B. subtilis* chromosome (Fig. 1; 106). As with the other proteases, null mutations in this gene had no effect on growth or spore formation in *B. subtilis* (106).

Epr Protease

Two of the minor extracellular proteases of *B. subtilis* (Epr and NprB) were first discovered by the cloning of their respective genes on multicopy plasmids (10, 105, 128). Only afterward were these proteases characterized. The first protease to be discovered in this manner was Epr. The gene for Epr was discovered by screening a gene bank of chromosomal DNA from *B. subtilis* strains containing deletion mutations in the *apr*, *nprE*, and *isp-1* genes on casein plates for protease overproduction (10, 105). Positive clones were detected and found to be due to the presence of the *epr* gene, which encodes a minor extracellular protease. The *epr* gene encoded a preproprotein with a putative signal sequence of approximately 27 amino acids, a prosequence of approximately 76 amino acids, and an additional 542 amino acids sufficient to yield a mature protein of 58,000 Da (10, 105). However, the mature protease found in the supernatant of cultures with *B. subtilis* strains harboring the *epr* gene on a multicopy plasmid was present as multiple species varying in molecular weight between 34,000 and 40,000 (10, 105). This observation, along with the significant similarity of Epr to subtilisin (40%) throughout the amino-terminal half of Epr, suggested that the Epr protein undergoes incomplete processing at several sites in the C-terminal region. The *epr* gene encodes a C-terminal region of approximately 240 amino acids that has a high number of lysine residues (24%) and a partially homologous sequence of 44 amino acids that is directly repeated five times (105). Deletion analysis indicated that this region is not necessary for enzyme secretion or activity (10, 105). The function of this C-terminal region remains unknown; however, Epr may be a cell wall- or membrane-bound protease with the C-terminal region acting as an anchor (see below).

The Epr protease is completely inhibited by PMSF and EDTA (10, 105). Homology to subtilisin indicates that Epr is a serine protease. Inhibition by EDTA results from degradation of the enzyme, which appears to require Ca^{2+} for stability (10).

The *epr* gene was mapped near *sacA* and found to lie upstream of the *sacY* gene at approximately 335° on the *B. subtilis* chromosome (Fig. 1; 10, 105, 155). As with the other extracellular protease genes, the *epr* gene was found to be nonessential. Null mutations in *epr* have no effect on either growth or sporulation of *B. subtilis* (10, 105).

Neutral Protease B

The second minor extracellular protease to be discovered by cloning its gene on a multicopy plasmid in a *B. subtilis* strain mutated in other protease genes was a novel neutral protease of 37,000 Da named neutral protease B (128). Sequence analysis of the neutral protease B gene (*nprB*) indicated that its primary product is a preproenzyme that has significant sequence homology to other neutral protease genes (128). The mature enzyme was found to have 66% (allowing for conserved amino acids) homology to the thermostable neutral proteases from *B. thermoproteolyticus* and *Bacillus stearothermophilus* and 65, 61, and 56% homology to the thermolabile neutral proteases from *Bacillus cereus*, *B. amyloliquefaciens*, and *B. subtilis*, respectively (128). NprB has a 223-amino-acid preproregion similar to that found in other neutral proteases (128). Like the other neutral proteases, the metalloproteases, NprB has a neutral pH optimum (6.6), is inhibited by the zinc-specific chelator 1,10-phenanthroline and by EDTA, and is stabilized by Ca^{2+} (128).

Genetic mapping experiments indicated that the *nprB* gene is located between *metC* (70% cotransduction) and *glyB* (53% cotransduction) on the *B. subtilis* chromosome and that the gene is not essential for growth or sporulation (Fig. 1; 128).

Vpr Protease

Deletion of the genes encoding alkaline protease, neutral protease, bacillopeptidase F, Epr, and Mpr made it possible to identify and characterize an additional minor protease in culture fluids of *B. subtilis*. This novel protease, Vpr, has a molecular weight of 28,000 but is bound in a complex in the void volume after gel filtration chromatography (108). The enzyme is completely inhibited by PMSF and is insensitive to EDTA, indicating that it is another serine protease.

After the amino-terminal sequence of Vpr had been determined, an oligonucleotide probe was constructed and used to identify and clone the *vpr* gene (108). The gene encoded a primary product of 806 amino acids. Even after allowing approximately 28 amino acids for the putative signal sequence and another 132 amino acids for the putative prosequence, enough amino acids to encode a protein of about 68,000 Da were left. However, the isolated protein had an apparent molecular weight of 28,500, suggesting that Vpr protease, like bacillopeptidase F and Epr, the two other minor extracellular serine proteases in *B. subtilis*, undergoes C-terminal processing or proteolysis.

The Vpr gene maps near *sacA* (108) and is located adjacent to an unknown open reading frame upstream from the *sacT* gene (3) at 332° on the *B. subtilis* chromosome. As with other extracellular protease genes, deletion of the *vpr* gene does not affect the growth or sporulation of *B. subtilis* (108).

B. SUBTILIS: INTRACELLULAR PROTEASES

Intracellular Serine Protease

Two proteases have been isolated from sporulating or stationary-phase cells of *B. subtilis* (52, 92, 112, 117). The first-characterized and most abundant intracellular protease was a serine protease originally called ISP (intracellular serine protease; now called ISP-1). The activity of this enzyme increases dramatically 2 to 3 h after the onset of sporulation (11). ISP was purified and originally found to have a molecular weight of 31,000 (112, 117, 118). More recent studies, however, have indicated that the enzyme is subject to proteolysis during purification and that the form found in vivo is larger, with a molecular weight of 34,000 (100). The shorter form of the protease is completely inhibited by PMSF and EDTA, similar to the Epr protein, and has higher esterolytic activity than subtilisin (117, 118). Inhibition by EDTA apparently results from a requirement for Ca^{2+} for stability and activity (117, 118). Interestingly, however, the larger form of the enzyme does not require calcium and is not inhibited by EDTA (100). Finally, two different groups have reported purifying a protein inhibitor of ISP-1 that disappears from the cells at about the time ISP-1 activity increases (73, 74, 77, 78).

Early amino-terminal protein sequence information

indicated that ISP has significant homology to subtilisins from various *Bacillus* spp. (118). The gene for this major intracellular serine protease (*isp-1*) is the only gene for a nonspecific intracellular protease that has been cloned from *B. subtilis* (54). The DNA sequence of *isp-1* revealed that the gene encodes an additional 17 or 20 amino acids preceding the amino termini of the two shorter forms of purified enzyme (54, 118); however, the size of the gene is in good agreement with the size of the larger form of ISP-1 seen in Western immunoblots, suggesting that the enzyme does not undergo any N-terminal processing (100). As noted previously with a partial amino-terminal amino acid sequence, the deduced amino acid sequence of ISP-1 is similar (45%) to that of the extracellular serine protease subtilisin, suggesting that both genes arose from a common ancestral gene (54). The *isp-1* gene is highly linked to *metC* (99.5%) by transduction, placing it at 118° on the *B. subtilis* chromosome (Fig. 1; 54). Finally, mutant strains carrying a deletion in the *isp-1* gene showed that the gene is not essential for growth or sporulation (6, 54).

Intracellular Serine Protease II

The second intracellular protease, ISP-II, has a molecular weight of 47,000 (112). It is inhibited by PMSF but not EDTA. The enzyme is also inhibited by the chymotrypsin inhibitor tolylsulfonyl phenylalanyl chloromethyl ketone (TPCK), suggesting that ISP-II is a trypsinlike serine protease. A temperature-sensitive mutant of this protease was isolated, and though not effecting growth, it was reported to alter protein turnover rates and the structure of the spore coat (96).

SASP-Specific Protease

In addition to the subtilisin-like and trypsin-like intracellular proteases found in *B. subtilis*, Setlow and coworkers have isolated an endoprotease from spores that specifically degrades a group of small, acid-soluble spore proteins (SASP) at the onset of germination (reviewed in reference 99 and in chapter 55 of this volume). This protease was first purified from *Bacillus megaterium* spores and shown to cleave specifically within a pentapeptide sequence found in all SASP (98). The gene for this protease (termed *gpr*) has been cloned from *B. subtilis* and *B. megaterium* (119). The deduced amino acid sequence of the protease from *B. subtilis* is homologous to that of the corresponding protein from *B. megaterium* (68%) but not to those of other proteases. Since this protease is highly specific, cleaving only a pentapeptide sequence present in the SASP, the lack of homology to other proteases was not surprising. The beginning of the amino terminus of the active form (P_{40}) corresponds to amino acid 18 of the deduced sequence of the protein. It is not clear whether the preceding amino acids are a prosequence (involved in folding or inactivation of the enzyme) or are involved in some other highly specific processing (119).

Synthesis of this highly specific germination protease is regulated much differently than synthesis of the degradative nonspecific endoproteases of *B. subtilis* whose regulation is discussed below. By using a *gpr-lacZ* translational fusion, it was determined that the gene for this protease (*gpr*) is expressed in the forespore compartment of sporulating cells and that its transcription is first controlled by E-σ^F and then by E-σ^G forms of RNA polymerase late in sporulation (119).

Other Specific Proteases

Two other classes of very specific endoproteases have been reported in *B. subtilis*. One class is the signal peptidases responsible for the cleavage of signal peptides from secreted proteins, and the other class is the pro-sigma factor-processing enzymes. A discussion of the signal peptidases can be found in chapter 49 of this volume, and a discussion of the pro-sigma factor-processing enzymes can be found in chapter 45.

COMPARISON OF *BACILLUS* PROTEASES

Serine Proteases

The amino acid sequences of the *B. subtilis* serine proteases Apr (subtilisin), Bpr (bacillopeptidase F), Epr, Vpr, and ISP-1 are compared in Fig. 2. As indicated there, all of the proteases have homology that extends from the beginning of the mature enzyme to the end of subtilisin. The homology between proteases in this region varies between 25 and 40%. Epr and subtilisin (Apr) have some homology at the end of the prosequence (105); however, there is little homology between the other preproregions. There is also no significant homology between the C-terminal regions of bacillopeptidase F, Epr, and Vpr. The most conserved regions, predictably, are those surrounding the Asp, His, and Ser residues in the active site of subtilisin (Fig. 2).

The subtilisin genes from *B. licheniformis* (subtilisin Carlsberg; 45) and *B. amyloliquefaciens* (134, 142) have also been cloned and sequenced. These amino acid sequences are highly homologous (>70%) to each other and to the *B. subtilis* subtilisin. Alkaline protease genes homologous to the subtilisin gene of *B. subtilis* have also been cloned from *Bacillus lentus* (35), *B. subtilis* subsp. *amylosacchariticus* (151), *Bacillus* TA41 (15b), *Bacillus alcalophilus* (133), and the alkalophilic *Bacillus* strain YaB (49). The gene from *Bacillus* strain YaB encodes an alkaline elastase with 82% homology to the alkaline protease from *B. alcalophilus* (133) and 55% homology to subtilisin BPN′ (49). As discussed above, the three-dimensional structures of the *B. amyloliquefaciens* subtilisins BPN′, Novo, and BAS (8, 19, 145) and the *B. licheniformis* subtilisin Carlsberg (7) have been determined. In addition, the crystal structure of the alkaline protease from *B. alcalophilus* has also been resolved (111, 133a). A gene encoding a protein similar (62%) to ISP-1 of *B. subtilis* has also been cloned from *Bacillus polymyxa* (122). A gene analogous to the *lon* gene of *Escherichia coli* has been cloned and characterized from *Bacillus brevis* (44a), and the gene encoding a serine protease inhibitor has also been cloned from *B. brevis* (101a).

```
                                *
Vpr    SAPYIGANDAWDLGYTGKGIKVAIIDTGVEYNHPDLKKNFGQYKGYDFVDNDYDP
Bpr    NVDQIDAPKAWALGYDGTGTVVASIDTGVEWNHPALK--EK-YRGYNPENPNEPE
Epr    NLEPIQVKQAWKAGLTGKNIKIAVIDSGI-SPHDDLS--IA------------GG
Apr    GVSQIKAPALHSQGYTGSNVKVAVIDSGIDSSHPDLK--VA------------GG
ISP-1  GIKVIKAPEMWAKGVKGKNIKVAVLDTGCDTSHPDLKNQII------------GG

                                   *
Vpr    KETPTGDPR-GEATD-------------HGTHVAGTVAANGTIK----GVAPDAT
Bpr    NEMNWYDAVAGEASPY--DDL------AHGTHVTGTMVGSEPDGTNQIGVAPGAK
Epr    YSA-----VSY-TSSY--KDD-----NGHGTHVAGIIGAK-HNGYGIDGIAPEAQ
Apr    ASM-----VPSETNPF--QDN-----NSHGTHVAGTVAAL-NNSIGVLGVAPSAS
ISP-1  --------KNF-SDDDGGKEDAISDYNGHGTHVAGTIAAN-DSNGGIAGVAPEAS

Vpr    LLAYRVLGPG-GSGTTENVIAGVERAV---------QDGA--DVMNLSLGN-SLN
Bpr    WIAVKAFS-EDG-GTDADILEAGEWVLAPKDAEGNPHPEMAPDVVNNSWGGGSGL
Epr    IYAVKALD-QNGSGDLQSLLQGIDWSI---------ANRM--DIVNMSLGTTSDS
Apr    LYAVKVLG-ADGSGQYSWIINGIEWAI---------ANNM--DVINMSLGGPSGS
ISP-1  LLIVKVLGGENGSGQYEWIINGINYAV---------EQKV--DIISMSLGGPSDV

Vpr    NPDWATSTALDWAMSEGVVAVTSNGNSGPNGWTVGSPGTSREAISVGATQLPNEY
Bpr    DEWYRDMVNAWRSA--DIFPEFSAGNTDLFIPGGP----GSIANPA---------
Epr    KI-LHDAVNKAYEQ--GVLLVAASGNDG--------NGKP-VNYPA---------
Apr    AA-LKAAVDKAVAS--GVVVVAAAGNEG----TSG-SSST-VGYPG---------
ISP-1  PE-LEEAVKNAVKN--GVLVVCAAGNEG----DGDERTEEL-SYPA---------

Vpr    AVTFGSYSSAKVMGYNKEDDVKALNNKEVELVEAGIGEAKDFEGKDLTGKVAVVK
Bpr    -------------------------------------------------------
Epr    -------------------------------------------------------
Apr    -------------------------------------------------------
ISP-1  -------------------------------------------------------

Vpr    RGSIAFVDKADNAKKAGAIGMVVYNNLSGEIEANVPGMSVPTIKLSLEDGEKLVS
Bpr    -------------------------------------------------------
Epr    -------------------------------------------------------
Apr    -------------------------------------------------------
ISP-1  -------------------------------------------------------

Vpr    ALKAGETKTTFKLTVSKALGEQVADFSSRGPVMDTWMIKPDISAPGVNIVSTIPT
Bpr    ------NYPESFATGATDINKKLADFSLQGPSPYDEIKPEI-SAPGVNIRSSVPG
Epr    ------AYSSVVAVSATNEKNQLASFSTTG----DEV--EF-SAPGTNITSTYLN
Apr    ------KYPSVIAVGAVDSSNQRASFSSVGPEL-DVM------APGVSIQSTLPG
ISP-1  ------AYNEVIAVGSVSVARELSEFSNANKEI-DLV------APGENILSTLPN

                     *
Vpr    HDPDHPYGYSKQGTSMASPHIAGAVAVIKQAKP     (164-552)
Bpr    QTYEDGWD----GTSMAGPHVSAVAALL---KQ     (202-467)
Epr    QYYATG-S----GTSQATPHAAAMFALL---KQ     (117-341)
Apr    NKY--GAYN---GTSMASPHVAGAAALILS-KH     (114-345)
ISP-1  KKY--GKLT---GTSMAAPHVSGALALI---KS     (25-261)
```

Figure 2. Comparison of the amino acid sequences of five *B. subtilis* serine proteases: Vpr (108), bacillopeptidase F (Bpr) (107, 147), Epr (10, 105), subtilisin (Apr) (113), and ISP-1 (54). Identical residues for all five proteins are enclosed in boxes. Asp, His, and Ser residues in the active site of subtilisin are marked with asterisks. Numbering of the amino acid residues for each protein is in parentheses. Reprinted from reference 108 with permission.

Metalloproteases

In addition to the *nprE* and *nprB* genes of *B. subtilis*, the *npr* genes from *B. stearothermophilus* (24, 59, 79, 121), *B. amyloliquefaciens* (41, 102, 134), *Bacillus caldolyticus* (132), *B. subtilis* subsp. *amylosacchariticus* (152), *B. polymyxa* (122), and *Bacillus brevis* (1) have been cloned and sequenced. The primary structures of the neutral proteases from *B. thermoproteolyticus* (126) and *B. cereus* (103) were also determined, and their crystal structures were solved (70, 84, 113a). The thermostable neutral proteases from *B. thermoproteolyticus* and *B. stearothermophilus* are highly homologous to each other (>85%; 121), whereas the thermostable neutral protease from *B. caldolyticus* is nearly identical to the neutral protease from *B. stearothermophilus*. Although more thermostable (+7 to 8°C) than the *B. stearothermophilus* protease, the *B. caldolyticus* enzyme is different at only three amino acid positions (132). The thermolabile neutral proteases of *B. subtilis* and *B. amyloliquefaciens* are also very homologous (approximately 82%; 121). The two groups are approximately 30 to 50% homologous to each other.

Surprisingly, the Mpr protein from *B. subtilis* showed no significant homology to other metalloproteases or any bacterial proteases. As mentioned above, Mpr showed homology to human protease E and bovine carboxypeptidase A complex component III in a region surrounding two cysteine residues reported to participate in disulfide bond linkages in these other proteases (106). Human protease E is reported to be a serine protease similar to elastase (101). Since Mpr is not a serine protease and does not have elastase activity (94), the evolutionary relationship between Mpr and these eukaryotic proteases is unclear.

REGULATION

Like the onset of sporulation, to which protease production is linked, the regulation of protease production is complex, responding to many environmental stimuli and controlled by many gene products. It has been known for some time that protease synthesis begins at the end of logarithmic growth, that *spo0A* mutations result in deficient synthesis of proteases, and that mutations in the *abrB* gene, a phenotypic suppressor of *spo0* mutations, can restore synthesis (40, 129). It is now apparent that the extracellular proteases are just one class of a group of degradative enzymes whose synthesis is controlled by a large number of regulatory proteins. In fact, at least nine regulatory genes that have pleiotropic effects on the synthesis of degradative enzymes have been described (48; reviewed in reference 116 and chapters 50, 52, and 54). Hoch and coworkers (48, 85, 116) called the proteins encoded by these regulatory genes "transition state regulators because they control functions that are expressed during the transition state between vegetative growth and the onset of stationary phase and sporulation" (48).

In addition to *spo0* genes and *abrB*, the other regulatory genes known to affect extracellular protease expression in *B. subtilis* are *hpr* (39, 85), *degU* (38, 60, 114, 123), *degR* (124, 149), *degQ* (2, 148), *sin* (27), *senS* (141), *pai* (42), and *tenA* and *tenI* (83). A full discussion of all of these regulatory proteins and their effects on extracellular enzyme production can be found in chapters 50, 51, 52, and 54. Here, we limit our discussion to a review of those proteins whose target sites have been mapped on one of the protease genes.

A majority of the more recent work on the regulation of protease production has involved expression of the subtilisin gene from *B. subtilis* and the study of its promoter region. Using an *apr-lacZ* gene fusion, Ferrari et al. (23) measured the effects of various mutations on the initial rate of presubtilisin–β-galactosidase synthesis. They showed that *spo0A* mutations dramatically decrease the rate of β-galactosidase synthesis and that both *hpr* and *sacU* mutations dramatically increase this rate. In addition, *spo0B*, *spo0F*, *spo0E*, and *spo0H* mutations have an intermediate effect in repressing expression of the fused gene (23).

Originally, it was reported that the subtilisin promoter was transcribed in vitro and in vivo by σ^B-containing RNA polymerase and that the start site

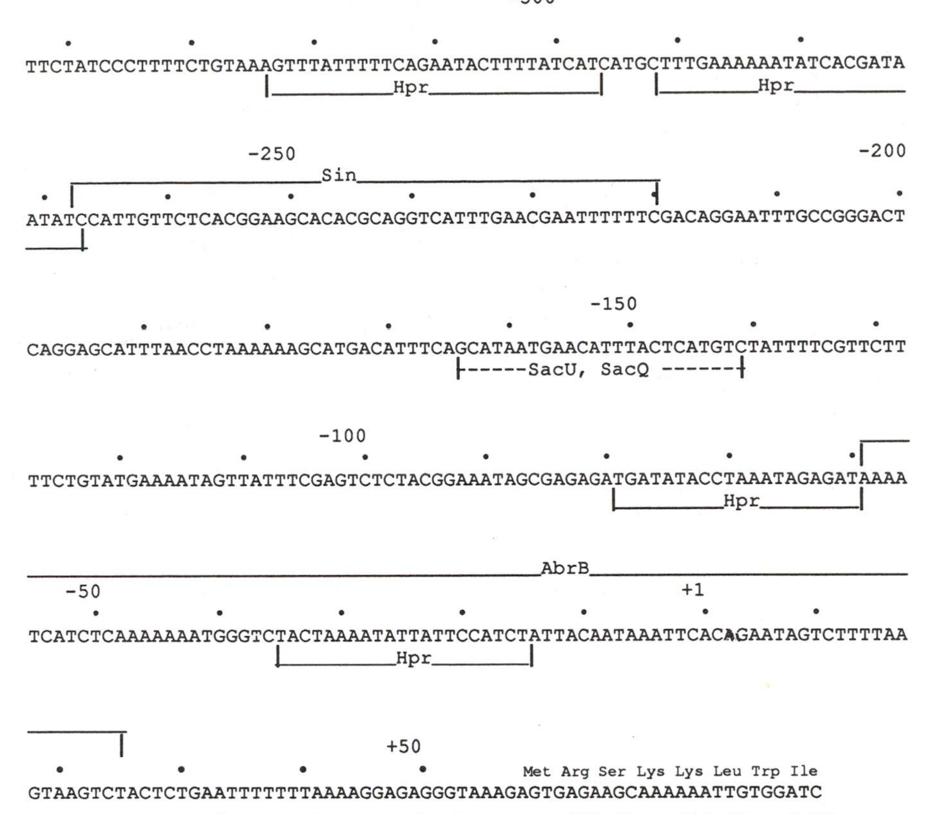

Figure 3. Nucleotide sequence of the subtilisin gene (*apr*) promoter (37). Sites of binding of different transition state regulatory proteins are indicated with brackets: AbrB (116), Hpr (48), and Sin (27). Sites required for activation by SacU and SacQ are indicated by dashed lines (37).

was at +26 and +41 (Fig. 3; 144). However, it was shown later that subtilisin gene expression was not affected in strains carrying a deletion in the σ^B gene (22). Also, the start of transcription was found to be approximately 50 bp upstream of the previously reported start site of transcription (+1 in Fig. 3; 22). Although mutation of the σ^B gene did not effect subtilisin expression, mutations in the *spo0H* gene encoding the σ^H subunit of RNA polymerase decreased expression of an *aprE-lacZ* fusion to 25% of wild-type levels (22). The identification of the sigma factor(s) responsible for subtilisin expression is still unclear.

Henner and coworkers (37) located the targets of the *hpr*, *sacU*, and *sacQ* gene products to a region upstream of the subtilisin promoter. Deletions of the region upstream of the *aprE* promoter greatly reduced the stimulatory effect of mutations in these genes (37). It was determined that a region upstream of −200 was needed for protease stimulation by the *hpr-97* mutation and that the region between −141 and −164 was necessary for stimulation of *aprE* expression by *sacU32*(Hy) and *sacQ36*(Hy) (37).

The *hpr* gene product is a negative regulator of proteases (85) that is capable of binding directly to the promoter region of both the *apr* and *npr* genes (48). In DNase protection studies, four regions of the *apr* promoter region were protected by Hpr protein. Two sites were located upstream of the promoter at positions −324 to −295 and −292 to −267. Two other sites were −79 to −59 and −35 to −14 (Fig. 3; 48). The Hpr protein also protected regions −107 to −82 and −26 to −4 upstream of the *npr* promoter of *B. subtilis* (48). This result agreed with the findings of Toma et al. (127), who found that a deletion from −156 to −90 in the *npr* promoter region caused overproduction of neutral protease.

A similar study of the *aprE* region with purified Sin protein was done by Gaur et al. (27). Elevated levels of Sin repress subtilisin gene expression. Sin binds strongly to the region from −268 to −220 upstream of the subtilisin promoter (27). Furthermore, Sin and Hpr bind independently of each other (48). Analogous binding experiments were also performed with AbrB protein (116). Purified AbrB protein binds specifically to the *aprE* promoter, covering the promoter region from −59 to +25 and acting as a negative regulator of transcription (116). The region bound by AbrB apparently overlaps one of the sites protected by Hpr (Fig. 3).

In addition to altering the expression of subtilisin and neutral protease, some of these transition state regulators may also effect expression of some of the minor extracellular proteases and the major intracellular protease ISP-1 (95). Using an *isp-lacZ* translational fusion, Ruppen et al. (95) showed that *hpr*, *sacU*(Hy) and *sacQ*(Hy) mutations all increased the level of *isp-1* expression while not affecting its temporal expression, similar to the results seen with the extracellular alkaline and neutral proteases.

Apart from the global regulators of protease expression discussed above, one very specific regulatory gene effecting the neutral proteases gene of *B. stearothermophilus* has been reported (79). Just upstream of the thermostable neutral protease gene (*nprS*) of *B. stearothermophilus* TELNE is a regulatory gene, *nprA*, whose product enhances expression of the *nprS* gene about fivefold. The *nprA* gene encodes a protein of 49,097 Da that has consensus regions similar to the DNA recognition domains of DNA-binding proteins (79). The NprA protein has no homology to other regulatory proteins effecting protease production and does not enhance synthesis of the neutral or alkaline proteases of *B. subtilis* (79). Apparently, NprA is specific for the *nprS* promoter. It will be interesting to see whether there are any other such specific regulatory genes that control the expression of other *Bacillus* protease genes.

CONSTRUCTION OF *B. SUBTILIS* STRAINS WITH REDUCED PROTEASE LEVELS

The cloning and sequencing of many protease genes from *B. subtilis* have made it possible to create strains simultaneously containing deletions in multiple protease genes. Strains of *B. subtilis* containing deletion mutations in the *aprE* and *nprE* genes were first reported to produce no extracellular protease activity (150), whereas strains containing a point mutation in the *npr* gene and a deletion in the *apr* gene were reported to have low (1 to 5%) but detectable levels of protease activity (21, 50). These results were complicated by the fact that the *npr* mutation, in at least one case, still produced low levels of neutral protease (140). Other reports of strains containing deletion mutations in the *apr*, *npr*, and *epr* genes indicated that these mutant strains produced approximately 1 to 4% of the total protease activity of wild-type strains (105, 140). All of these strains grew at the same rate as wild-type strains and sporulated with the same efficiency, confirming, as discussed above, that the major proteases are not required for growth or sporulation.

Strains containing deletion mutations in six extracellular protease genes have now been reported (108, 146). Sloma et al. (108) constructed *B. subtilis* strains carrying mutations in *aprE*, *nprE*, *bpr*, *epr*, *mpr*, and *vpr*, whereas Wu et al. (146) made strains carrying mutations in *aprE*, *nprE*, *nprB*, *bpr*, *epr*, and *mpr*. Such strains produce very low levels of extracellular protease activity (approximately 0.5%) compared with wild-type strains. These low levels of protease can allow the production of some proteins that were sensitive to protein degradation in wild-type strains or strains containing deletions in just the major protease genes. In addition, these strains grow and sporulate normally, indicating that the deletion of up to six extracellular protease genes has no adverse effect on the cells (108, 146).

PROTEASES FROM OTHER GRAM-POSITIVE BACTERIA

From the large number of proteases that have been described in various gram-positive bacterial species, we discuss here only those extracellular proteases from other gram-positive bacteria that have been most extensively studied and whose genes have been cloned and characterized.

Lactococcus Proteases

The importance of cell envelope-associated proteolytic activities of *Lactococcus* spp. (*Streptococcus cremoris* and *Streptococcus lactis*) in the production of cheese and other fermented milk products has been well established (125). The activities of these proteases are essential for the degradation of casein and the subsequent rapid growth of *Lactococcus* strains in milk. Different strains of *S. cremoris* have proteases that differ in their abilities to degrade different forms of bovine casein (α_{s1}-, β-, and κ-caseins; 135). The PI-type protease preferentially degrades β-casein, whereas the PIII-type protease degrades not only β-casein but also α_{s1}- and κ-caseins (135). The most active proteolytic activity in all strains is a large, cell wall-associated serine protease(s) (reviewed in reference 55).

Protease genes from three strains of *Lactococcus* have been cloned and sequenced, and in all three cases, the genes were located on a plasmid (53, 57, 58, 136). The protease genes (*prtP*) from *Lactococcus lactis* subsp. *cremoris* Wg2 (57) and *L. lactis* subsp. *lactis* NCDO 763 (53) both encoded a serine protease of 1,902 amino acids, whereas the *prtP* gene from *L. lactis* subsp. *cremoris* SK11 encoded a protein of 1,962 amino acids (136). The protease encoded by the *prtP* gene of strain Wg2 is a PI-type protease, while the protease encoded by the *prtP* gene of strain SK11 is a PIII-type protease. However, the amino acid sequences of all the proteases are highly similar (55).

All the lactococcal proteases are synthesized as preproenzymes and have homology to the subtilisin family of serine proteases (57). The biggest differences between the lactococcal proteases and other serine proteases are their large size and their specificity. The lactococcal proteases are approximately 135 to 140 kDa, compared with about 30 kDa for many other serine proteases (136). In addition, the lactococcal proteases are highly specific, degrading only specific caseins and not other proteins like bovine serum albumin, which is rapidly degraded by subtilisin (28, 57).

The lactococcal proteases also have long C-terminal regions that have C-terminal membrane-anchoring sequences (136). Deletion of 343 amino acids from these C-terminal regions does not affect activity or specificity of the protease or its secretion (57). However the protease is secreted into the culture fluids instead of being located in the cell wall (30, 53, 56, 137).

The regions of the lactococcal proteases that are homologous to subtilisin are separated in the lactococcal proteases by an insert of approximately 200 amino acids of nonhomology between His-64 and Ser-221 of the active site of subtilisin (57). Interestingly, there is partial homology of this 200-amino-acid area to a similar region in the Vpr protease of *B. subtilis* (108). Although there is no homology of the C-terminal regions of the *B. subtilis* proteases Vpr, Bpr, and Epr to the C-terminal regions of the *Lactococcus* proteases, it is interesting to speculate that these minor serine proteases of *B. subtilis* are cell envelope-associated proteases, since they also have long C-terminal regions not needed for activity or secretion.

The plasmids containing the *prtP* gene in the three *Lactococcus* strains all have another gene involved in protease activity, the *prtM* gene, directly upstream and transcribed in the opposite direction from the *prtP* genes (30, 137). This gene encodes a lipoprotein of 32 kDa that is essential for proteolytic activity (30, 31, 137). PrtM affects the autoproteolytic processing of the proregion of the PrtP protease in a manner that is not yet fully understood (31). Release of the lactococcal protease from cells results from C-terminal processing of the *Lactococcus* protease, which also appears to be autocatalytic but independent of *prtM* (32).

Other *Streptococcus* Proteases

A protease gene (*scpA*) from *Streptococcus pyogenes* was also cloned and sequenced (14). Interest in this protease, streptococcal C5a peptidase, results from its ability to specifically cleave the human complement component C5a. This protease has some features similar to those of the *Lactococcus* proteases. The gene for this protease encodes a primary product of 1,167 amino acids that includes a 31-amino-acid signal peptide and putative cell wall-spanning and membrane anchor domains (14). Surprisingly, however, the gene does not appear to encode the typical prosequence seen for all the other extracellular proteases. This protease showed no significant amino acid similarity to any other proteases except that three short regions showed significant homology to the sequences around the active sites of several subtilisins (14).

Streptomyces Proteases

The most studied of the proteases produced by *Streptomyces* spp. is pronase, a commercial mixture of several extracellular proteases secreted by *Streptomyces griseus* (4, 76). Two of the serine proteases present in this mixture, protease A and protease B, have been purified (4, 47), their tertiary structures have been determined (18, 25, 46), and their genes (*sprA* and *sprB*) have been cloned (36). The *sprA* and *sprB* genes show strong similarity, suggesting that they arose by gene duplication (36). As with many other protease genes, both genes encode preproproteins with a presumed signal peptide of 38 amino acids and a propeptide of 76 or 78 amino acids. Interestingly, the mature enzymes begin with the amino terminus, IXGG, which is similar to that seen for many eukaryotic serine proteases (36).

Another component of pronase is a trypsinlike enzyme called *S. griseus* trypsin. Both the primary and tertiary structures of this protease, which has more similarity to mammalian than bacterial proteases, have been determined (81, 91). A protease gene cloned from *Streptomyces glaucesens* ETH 22794 encodes a protease with high sequence similarity (72% at the amino acid level) to this trypsinlike enzyme from *S. griseus* (154). Unlike other protease genes found in *Bacillus* and *Streptomyces* spp., this gene appears not to encode a significant prosequence. Only two to six amino acids appear to be located between the putative signal peptide and the mature enzyme (154).

A number of other protease genes have also been isolated from *Streptomyces* spp. Chang et al. (13) isolated the gene for a neutral metalloprotease (*npr*) from *Streptomyces cacaoi* YM15. This *npr* gene encodes a 60,000-Da preproprotein that is processed to yield a 35,000-Da enzyme. Like other metalloproteases, Npr from *S. cacaoi* is inhibited by EDTA and activated by zinc. Npr contains putative zinc ligand-binding regions similar to those seen in the neutral protease of *B. subtilis* and in thermolysin (13). Unexpectedly, this neutral metalloprotease also has two copies of the Asp-Ser-Gly motif, which is highly conserved in the active sites of aspartic acid and retroviral proteases (13). However, the function of these sequences in this protease is unknown. The presumptive 171-amino-acid prosequence of the enzyme is similar in size and hydrophilic character to that of the neutral proteases of *Bacillus* spp. (13, 134, 150). Surprisingly, however, the Npr protease of *S. cacaoi* has no strong sequence similarity to *Bacillus* proteases or other proteases (13).

Four other laboratories (12, 15a, 61, 63) have described the cloning of the gene for a different novel neutral protease from either *Streptomyces lividans*, *Streptomyces coelicolor* Müller, or *Streptomyces* sp. strain C5. These neutral proteases are small (22,000, 20,000, and 15,500, respectively) and also contain zinc in their active sites (12, 15a, 61, 63). These novel neutral proteases are different from other thermolysinlike neutral protease in having only one zinc-binding domain and not being inhibited by phosphoramidon. Furthermore, although similar to each other (71% identity at the amino acid level), these novel proteases have no significant homology to any other sequenced proteases outside of the conserved zinc ligand site (12, 61).

Genes Regulating Protease Expression in *Streptomyces* spp.

Interestingly, just upstream of the *S. lividans* or the *S. coelicolor* Müller neutral protease genes is a divergently transcribed regulatory gene that encodes a regulator protein that appears to activate protease expression (12, 15a, 63). This protein contains a helix-turn-helix motif at its amino-terminal end and shows discernible similarity to proteins belonging to the LysR family of transcriptional regulators (12, 15a, 63).

Another regulatory gene affecting protease production has been cloned from *S. griseus*. This gene (*saf*) encodes a positive regulatory protein that increases production of proteases and at least four other extracellular enzymes in *Streptomyces* spp. (16, 17). The *saf* gene is reminiscent of the large number of *trans*-acting pleiotropic regulatory genes that control the synthesis of protease and many other extracellular degradative enzymes in *Bacillus* spp.

In summary, gram-positive bacteria have evolved the capacities to produce and secrete a remarkably large number of endoproteases. Most of these proteases are members of either the subtilisin family of serine proteases or the thermolysin family of metalloproteases. Surprisingly, while a few of the proteases discussed play important roles in their cells' physiology, the large number of extracellular proteases produced by *B. subtilis* is seemingly dispensable for growth and sporulation, at least in the laboratory. Presumably, the battery of extracellular hydrolytic enzymes produced by *B. subtilis* serves it well as it scavenges for nutrients in its natural habitat in the soil.

Acknowledgments. We thank W. Strohl, H. Lichtenstein, and M. Butler for providing research results prior to publication; G. A. Rufo, Jr., J. Hoch, J. Perkins, and R. Losick for helpful comments on this review; and D. Potvin for preparation of text and figures.

REFERENCES

1. **Abakov, A. S., A. P. Bolotin, L. G. Kolibaba, A. V. Sorokin, T. M. Shemyakina, M. Paberit, K. Raik, and A. Aaviksaar.** 1990. Cloning and expression of the neutral protease gene of *Bacillus brevis* in *Bacillus subtilis*. *Mol. Biol.* **24:**806–813.
2. **Amory, A., F. Kunst, E. Aubert, A. Klier, and G. Rapoport.** 1987. Characterization of the *sacQ* genes from *Bacillus licheniformis* and *Bacillus subtilis*. *J. Bacteriol.* **169:**324–333.
3. **Arnaud, M., P. Glaser, A. Vertes, A. Danchin, G. Rapoport, and F. Kunst.** Personal communication.
4. **Awad, W. M., Jr., A. R. Soto, S. Siegel, W. E. Skiba, G. G. Bernstrom, and M. S. Ochoa.** 1972. The proteolytic enzymes of the K-1 strain of *Streptomyces griseus* obtained from a commercial preparation (pronase). *J. Biol. Chem.* **247:**4144–4154.
5. **Ballassa, G.** 1969. Biochemical genetics of bacterial sporulation. *Mol. Gen. Genet.* **104:**73–103.
6. **Band, L., D. J. Henner, and M. Ruppen.** 1987. Construction and properties of an intracellular serine protease mutant of *Bacillus subtilis*. *J. Bacteriol.* **169:**444–446.
7. **Bode, W., E. Papamokos, and D. Musil.** 1987. The high-resolution x-ray crystal structure of the complex formed between subtilisin Carlsberg and Eglin C, an elastase inhibitor from the leech *Hirudo-Medicinalis*. *Eur. J. Biochem.* **166:**673–692.
8. **Bott, R., A. K. M. Ultsch, T. Graycar, B. Katz, and S. Power.** 1988. The three-dimensional structure of *Bacillus amyloliquefaciens* subtilisin at 1.8Å and an analysis of the structural consequences of peroxide inactivation. *J. Biol. Chem.* **263:**7895–7906.
9. **Boyer, H. W., and B. C. Carlton.** 1968. Production of two proteolytic enzymes by a tranformable strain of *Bacillus subtilis*. *Arch. Biochem. Biophys.* **128:**442–455.
10. **Brückner, R., O. Shoseyov, and R. H. Doi.** 1990. Multiple active forms of a novel serine protease from *Bacillus subtilis*. *Mol. Gen. Genet.* **221:**486–490.
11. **Burnett, T. J., G. W. Shankweiler, and J. H. Hageman.** 1986. Activation of intracellular serine proteinase in *Bacillus subtilis* cells during sporulation. *J. Bacteriol.* **165:**139–145.
12. **Butler, M. J., C. C. Davey, P. Krygsman, E. Walczyk, and L. T. Malek.** Cloning of genetic loci involved in endoprotease activity in *S. lividans* 66: a novel neutral protease gene with an adjacent divergent putative regulatory gene. *Can. J. Microbiol.*, in press.
13. **Chang, P. C., T.-C. Kuo, A. Tsugita, and Y.-H. W. Lee.** 1990. Extracellular metalloprotease gene of *Streptomyces cacaoi*: structure, nucleotide sequence and characterization of the cloned gene product. *Gene* **88:**87–95.
14. **Chen, C. C., and P. P. Cleary.** 1990. Complete nucleotide sequence of the streptococcal C5a peptidase gene of *Streptococcus pyogenes*. *J. Biol. Chem.* **265:**3161–3167.
15. **Colman, P. M., J. N. Jansonius, and B. W. Matthews.** 1972. The structure of thermolysin: an electron density map at 2.3 Å resolution. *J. Mol. Biol.* **70:**701–724.

15a. **Dammann, T., and W. Wohlleben.** 1992. A metalloprotease gene from *Streptomyces coelicolor* Müller and its transcriptional activator, a member of the LysR family. *Mol. Microbiol.* **6:**2267–2278.

15b. **Davail, S., G. Feller, E. Narinx, and C. Gerdsy.** 1992. Sequence of the subtilisin-encoding gene from an antarctic psychrotroph, *Bacillus* TA41. *Gene* **119:**143–144.

16. **Daza, A., J. A. Gil, T. Vigal, and J. F. Martin.** 1990. Cloning and characterization of a gene of *Streptomyces griseus* that increases production of extracellular enzymes in several species of *Streptomyces*. *Mol. Gen. Genet.* **222:**384–392.
17. **Daza, A., J. F. Martin, T. Vigal, and J. A. Gil.** 1991. Analysis of the promoter region of *saf*, a *Streptomyces griseus* gene that increases production of extracellular enzymes. *Gene* **108:**63–71.
18. **Delbaere, L. T. J., W. L. B. Hutcheon, M. N. G. James, and W. E. Thiessen.** 1975. Tertiary structural differences between microbial serine proteases and pancreatic serine enzymes. *Nature* (London) **257:**758–763.
19. **Drenth, J., W. G. J. Hol, J. N. Jansoius, and R. Koekoek.** 1972. Subtilisin Novo: the three-dimensional structure and its comparison with subtilisin BPN'. *Eur. J. Biochem.* **26:**177–181.
20. **Egnell, P., and J.-I. Flock.** 1991. The subtilisin Carlsberg pro-region is a membrane anchorage for two fusion proteins produced in *Bacillus subtilis*. *Gene* **97:**49–54.

20a. **Egnell, P., and J.-I. Flock.** 1992. The autocatalytic processing of the subtilisin Carlsberg pro-region is independent of the primary structure of the cleavage site. *Mol. Microbiol.* **6:**1115–1119.

21. **Fahnestock, S. R., and K. E. Fisher.** 1987. Protease-deficient *Bacillus subtilis* host strains for production of staphylococcal protein A. *Appl. Environ. Microbiol.* **53:**379–384.
22. **Ferrari, E., D. J. Henner, M. Perego, and J. A. Hoch.** 1988. Transcription of *Bacillus subtilis* subtilisin and expression of subtilisin in sporulation mutants. *J. Bacteriol.* **170:**289–295.
23. **Ferrari, E., S. M. H. Howard, and J. A. Hoch.** 1986. Effect of stage 0 sporulation mutations on subtilisin expression. *J. Bacteriol.* **166:**173–179.
24. **Fujii, M., M. Takagi, T. Imanaka, and S. Aiba.** 1983. Molecular cloning of a thermostable neutral protease gene from *Bacillus stearothermophilus* in a vector plasmid and its expression in *Bacillus stearothermophilus* and *Bacillus subtilis*. *J. Bacteriol.* **154:**831–837.
25. **Fujinaga, M., L. T. J. Delbaere, G. D. Brayer, and M. N. G. James.** 1985. Refined structure of α-lytic protease at 1.7Å resolution; analysis of hydrogen bonding and solvent structure. *J. Mol. Biol.* **183:**479–502.
26. **Fukumoto, J., and J. Negoro.** 1951. Crystallization of bacterial proteinase. *Proc. Jpn. Acad.* **27:**441–444.
27. **Gaur, N. K., J. Oppenheim, and I. Smith.** 1991. The *Bacillus subtilis sin* gene, a regulator of alternate development processes, codes for a DNA-binding protein. *J. Bacteriol.* **173:**678–686.
28. **Geis, A., W. Bockelmann, and M. Teuber.** 1985. Simultaneous extracting and purification of a cell wall-associated peptidase and β-casein specific protease from *Streptococcus cremoris* AC1. *Appl. Microbiol. Biotechnol.* **23:**79–84.
29. **Güntelberg, A. V., and M. Ottesen.** 1952. Preparation of crystals containing the plakalbumin-forming enzyme from *Bacillus subtilis*. *Nature* (London) **170:**802.
30. **Haandrikman, A. J., J. Kok, H. Laan, S. Soemitro, A. M. Ledeboer, W. N. Konings, and G. Venema.** 1989. Identification of a gene required for maturation of an extracellular lactococcal serine proteinase. *J. Bacteriol.* **171:**2789–2794.

31. **Haandrikman, A. J., J. Kok, and G. Venema.** 1991. Lactococcal proteinase maturation protein PrtM is a lipoprotein. *J. Bacteriol.* **173:**4517–4525.
32. **Haandrikman, A. J., R. Meesters, H. Laan, W. N. Konings, J. Kok, and G. Venema.** 1991. Processing of the lactococcal extracellular serine proteinase. *Appl. Environ. Microbiol.* **57:**1899–1904.
33. **Hageman, J. H., and B. C. Carlton.** 1970. An enzymatic and immunological comparison of two proteases from a transformable *Bacillus subtilis* with the "subtilisins." *Arch. Biochem. Biophys.* **139:**67–79.
34. **Hartley, B. S.** 1960. Proteolytic enzymes. *Annu. Rev. Biochem.* **29:**45–72.
35. **Hastrup, S., S. Branner, F. Norris, S. B. Petersen, L. Norskov-Lauridsen, V. J. Jensen, and D. Aaslyng.** 13 July 1989. PCT patent application WO 8906279.
36. **Henderson, G., P. Krygsman, C. Liu, C. C. Davey, and L. T. Malek.** 1987. Characterization and structure of genes for proteases A and B from *Streptomyces griseus. J. Bacteriol.* **169:**3778–3784.
37. **Henner, D. J., E. Ferrari, M. Perego, and J. A. Hoch.** 1988. Location of the targets of the *hpr-97*, *sacU32*(Hy), and *sacQ36*(Hy) mutations in upstream regions of the subtilisin promoter. *J. Bacteriol.* **170:**296–300.
38. **Henner, D. J., M. Yang, and E. Ferrari.** 1988. Localization of *Bacillus subtilis sacU*(Hy) mutations to two linked genes with similarities to the conserved procaryotic family of two-component signalling systems. *J. Bacteriol.* **170:**5102–5109.
39. **Higerd, T. B., J. A. Hoch, and J. Spizizen.** 1972. Hyperprotease-producing mutants of *Bacillus subtilis. J. Bacteriol.* **112:**1026–1028.
40. **Hoch, J. A.** 1976. Genetics of bacterial sporulation. *Adv. Genet.* **18:**69–99.
41. **Honjo, M., K. Manabe, H. Shimade, I. Mita, A. Nakayama, and Y. Furutani.** 1984. Cloning and expression of the gene for neutral protease of *Bacillus amyloliquefaciens* in *Bacillus subtilis. J. Biotechnol.* **1:**265–277.
42. **Honjo, M., A. Nakayama, K. Fukazawa, K. Kawamura, K. Ando, M. Hori, and Y. Furutani.** 1990. A novel *Bacillus subtilis* gene involved in negative control of sporulation and degradative-enzyme production. *J. Bacteriol.* **172:**1783–1790.
43. **Ikemura, H., and M. Inouye.** 1988. *In vitro* processing of pro-subtilisin produced in *Escherichia coli. J. Biol. Chem.* **263:**12959–12963.
44. **Ikemura, H., H. Takagi, and M. Inouye.** 1987. Requirement of pro-sequence for the production of active subtilisin E in *Escherichia coli. J. Biol. Chem.* **262:**7859–7864.

44a. **Ito, K., S. Udaka, and H. Yamagata.** 1992. Cloning, characterization, and inactivation of the *Bacillus brevis lon* gene. *J. Bacteriol.* **174:**2281–2287.

45. **Jacobs, M., M. Eliasson, M. Uhlen, and J.-I. Flock.** 1985. Cloning, sequencing and expression of subtilisin Carlsberg from *Bacillus licheniformis. Nucleic Acids Res.* **13:**8913–8926.
46. **James, M. B. G., A. R. Sielecki, G. D. Brayer, L. T. J. Delbaere, and C. A. Bauer.** 1980. Structures of product and inhibitor complexes of *Streptomyces griseus* protease A at 1.8 Å resolution. *J. Mol. Biol.* **144:**43–88.
47. **Jurásek, L., P. Johnson, R. W. Olafson, and L. B. Smillie.** 1971. An improved fractionation system for pronase on CM-Sephadex. *Can. J. Biochem.* **49:**1195–1201.
48. **Kallio, P. T., J. E. Fagelson, J. A. Hoch, and M. A. Strauch.** 1991. The transition state regulator Hpr of *Bacillus subtilis* is a DNA-binding protein. *J. Biol. Chem.* **266:**13411–13417.
49. **Kaneko, R., N. Koyama, Y.-C. Tsai, R.-Y. Juang, K. Yoda, and M. Yamasaki.** 1989. Molecular cloning of the structural gene for alkaline elastase YaB, a new subtilisin produced by an alkalophilic *Bacillus* strain. *J. Bacteriol.* **171:**5232–5236.
50. **Kawamura, F., and R. H. Doi.** 1984. Construction of a *Bacillus subtilis* double mutant deficient in extracellular alkaline and neutral proteases. *J. Bacteriol.* **160:**442–444.
51. **Keay, L.** 1969. Neutral proteases of the genes *Bacillus. Biochem. Biophys. Res. Commun.* **36:**257–265.
52. **Kerjan, P., E. Keryer, and J. Szulmajster.** 1979. Characterization of a thermosensitive sporulation mutant of *Bacillus subtilis* affected in the structural gene of an intracellular protease. *Eur. J. Biochem.* **98:**353–362.
53. **Kiwaki, M., H. Ikemura, M. Shimizu-Kadota, and A. Hirashima.** 1989. Molecular characterization of a cell wall-associated proteinase gene from *Streptococcus lactis* NCD0763. *Mol. Microbiol.* **3:**359–369.
54. **Koide, Y., A. Nakamura, T. Uozumi, and T. Beppu.** 1986. Cloning and sequencing of the major intracellular serine protease gene of *Bacillus subtilis. J. Bacteriol.* **167:**110–116.
55. **Kok, J.** 1990. Genetics of the proteolytic system of lactic acid bacteria. *FEMS Microbiol. Rev.* **87:**15–42.
56. **Kok, J., D. Hill, A. J. Haandrikman, M. J. B. deReuver, H. Laan, and G. Venema.** 1988. Deletion analysis of the proteinase gene of *Streptococcus cremoris* Wg2. *Appl. Environ. Microbiol.* **54:**239–244.
57. **Kok, J., K. J. Leenhouts, A. J. Haandrikman, A. M. Ledeboer, and G. Venema.** 1988. Nucleotide sequence of the cell wall proteinase gene of *Streptococcus cremoris* Wg2. *Appl. Environ. Microbiol.* **54:**231–238.
58. **Kok, J., J. M. van Dijl, J. M. B. M. van der Vossen, and G. Venema.** 1985. Cloning and expression of a *Streptococcus cremoris* proteinase in *Bacillus subtilis* and *Streptococcus lactis. Appl. Environ. Microbiol.* **50:**94–101.
59. **Kubo, T., and T. Imanaka.** 1988. Cloning and nucleotide sequence of the highly thermostable neutral protease gene from *Bacillus stearothermophilus. J. Gen. Microbiol.* **134:**1883–1892.
60. **Kunst, F., M. Debarbouille, T. Msadek, M. Young, C. Mauel, D. Karamata, A. Klier, G. Rappoport, and R. Dedonder.** 1988. Deduced polypeptides encoded by the *Bacillus subtilis sacU* locus share homology with two-component sensor-regulator systems. *J. Bacteriol.* **170:**5093–5101.
61. **Lampel, J. S., J. S. Aphale, K. A. Lampel, and W. R. Strohl.** 1992. Cloning and sequencing of a gene encoding a novel extracellular neutral proteinase from *Streptomyces* sp. strain C5 and expression of the gene in *Streptomyces lividans* 1326. *J. Bacteriol.* **174:**2797–2808.
62. **Lerner, C. G., T. Kobayashi, and M. Inouye.** 1990. Isolation of subtilisin pro-sequence mutations that affect formation of active protease by localized random polymerase chain reaction mutagenesis. *J. Biol. Chem.* **265:**20085–20086.
63. **Lichenstein, H. S., L. A. Busse, G. A. Smith, L. O. Narbi, M. O. McGinley, M. F. Rohde, J. L. Katzowitz, and M. M. Zukowski.** 1992. Cloning and characterization of a gene encoding extracellular metalloprotease from *Streptomyces lividans. Gene* **111:**125–130.
64. **Mamas, S., and J. Millet.** 1975. Purification et propriétés d'une estérase excrétée pendant la sporulation de *Bacillus subtilis. Biochimie* **57:**9–16.
65. **Mandelstam, J., and W. M. Waites.** 1968. Sporulation in *Bacillus subtilis*: the role of exoprotease. *Biochem. J.* **109:**793–801.
66. **Mantsala, P., and H. Zalkin.** 1980. Extracellular and membrane-bound proteases from *Bacillus subtilis. J. Bacteriol.* **141:**493–501.
67. **Markland, F. S., and E. L. Smith.** 1967. Subtilisin BPN' VII. Isolation of cyanogen bromide peptides and the complete amino acid sequence. *J. Biol. Chem.* **242:**5198–5211.
68. **Markland, F. S., Jr., and E. L. Smith.** 1971. Subtilisins:

primary structure, chemical and physical properties, p. 561–608. *In* P. D. Boyer (ed.), *The Enzymes*. Academic Press, Inc., New York.

69. **Matthews, B. W., P. M. Colman, J. N. Jansonius, K. Titani, K. A. Walsh, and H. Neurath.** 1972. Structure of thermolysin. *Nature* (London) *New Biol.* **238:**41–43.
70. **Matthews, B. W., J. N. Jansonius, P. M. Colman, B. P. Schoenborn, and D. Dupourque.** 1972. Three-dimensional structure of thermolysin. *Nature* (London) *New Biol.* **238:**37–41.
71. **McConn, J. D., D. Tsuru, and K. T. Yasunobu.** 1964. *Bacillus subtilis* neutral proteinase. *J. Biol. Chem.* **239:**3706–3715.
72. **Millet, J.** 1969. Characterization of proteinases excreted by *Bacillus subtilis* Marburg strain during sporulation. *J. Appl. Bacteriol.* **33:**207–219.
73. **Millet, J.** 1977. Characterization of a protein inhibitor of intracellular protease from *Bacillus subtilis*. *FEBS Lett.* **74:**59–61.
74. **Millet, J., and J. Gregoire.** 1979. Characterization of an inhibitor of the intracellular protease from *Bacillus subtilis*. *Biochimie* **61:**385–391.
75. **Morihara, K.** 1974. Comparative specificity of microbial proteinases. *Adv. Enzymol.* **41:**179–243.
76. **Narahashi, Y.** 1972. Pronase. *Methods Enzymol.* **19:**651–664.
77. **Nishino, T., and S. Murao.** 1986. Interaction of proteinaceous protease inhibitor of *Bacillus subtilis* with intracellular protease from the same strain. *Agric. Biol. Chem.* **50:**3065–3070.
78. **Nishino, T., Y. Shimizu, K. Fukahara, and S. Murao.** 1986. Isolation and characterization of a proteinaceous protease inhibitor from *Bacillus subtilis*. *Agric. Biol. Chem.* **50:**3059–3064.
79. **Nishiya, U., and T. Imanaka.** 1990. Cloning and nucleotide sequences of the *Bacillus stearothermophilus* neutral protease gene and its transcriptional activator gene. *J. Bacteriol.* **172:**4861–4869.
80. **Ohta, Y., H. Hojo, S. Aimoto, T. Kobayashi, X. Zhu, F. Jordan, and M. Inouye.** 1991. Pro-peptide as an intermolecular chaperone: renaturation of denatured subtilisin E with a synthetic pro-peptide. *Mol. Microbiol.* **5:**1507–1510.
81. **Olafson, R. W., L. Jurásek, M. R. Carpenter, and L. B. Smillie.** 1975. Amino acid sequence of Streptomyces griseus trypsin. Cyanogen bromide fragments and complete sequence. *Biochemistry* **14:**1168–1177.
82. **Olaitan, S. A., R. J. DeLange, and E. L. Smith.** 1968. The structure of subtilisin Novo. *J. Biol. Chem.* **243:**5296–5301.
83. **Pang, A. S.-H., S. Nathoo, and S.-L. Wong.** 1991. Cloning and characterization of a pair of novel genes that regulate production of extracellular enzymes in *Bacillus subtilis*. *J. Bacteriol.* **173:**46–54.
84. **Pauptit, R. A., R. Karlsson, D. Picot, J. A. Jenkins, A.-S. Niklaus-Reimer, and J. N. Jansonius.** 1988. Crystal structure of neutral protease from *Bacillus cereus* refined at 3.0 Å resolution and comparison with the homologous but more thermostable enzyme thermolysin. *J. Mol. Biol.* **199:**525–537.
85. **Perego, M., and J. A. Hoch.** 1988. Sequence analysis and regulation of the *hpr* locus, a regulatory gene for protease production and sporulation in *Bacillus subtilis*. *J. Bacteriol.* **170:**2560–2567.
86. **Piggot, P. J., and J. G. Coote.** 1976. Genetic aspects of bacterial endospore formation. *Bacteriol. Rev.* **40:**908–962.
87. **Power, S. D., R. M. Adams, and J. A. Wells.** 1986. Secretion and autoproteolytic maturation of subtilisin. *Proc. Natl. Acad. Sci. USA* **83:**3096–3100.
88. **Prestidge, L., V. Gage, and J. Spizizen.** 1971. Protease activities during the course of sporulation in *Bacillus subtilis*. *J. Bacteriol.* **107:**815–823.
89. **Priest, F. G.** 1977. Extracellular enzyme synthesis in the genus *Bacillus*. *Bacteriol. Rev.* **41:**711–753.
90. **Rappaport, H. P., W. S. Riggsby, and D. A. Holden.** 1965. A *Bacillus subtilis* proteinase. I. Production, purification and characterization of a proteinase from a transformable strain of *Bacillus subtilis*. *J. Biol. Chem.* **240:**78–86.
91. **Read, R. J., and M. N. G. James.** 1988. Refined crystal structure of *Streptomyces griseus* trypsin at 1.7 Å resolution. *J. Mol. Biol.* **200:**523–551.
92. **Reysset, G., and J. Millet.** 1972. Characterization of an intracellular protease in *Bacillus subtilis* during sporulation. *Biochem. Biophys. Res. Commun.* **49:**328–334.
93. **Roitsch, C. A., and J. H. Hageman.** 1983. Bacillopeptidase F: two forms of a glycoprotein serine protease from *Bacillus subtilis* 168. *J. Bacteriol.* **155:**145–152.
94. **Rufo, G. A., Jr., B. J. Sullivan, A. Sloma, and J. Pero.** 1990. Isolation and characterization of a novel extracellular metalloprotease from *Bacillus subtilis*. *J. Bacteriol.* **172:**1019–1023.
95. **Ruppen, M. E., G. L. VanAlstine, and L. Band.** 1988. Control of intracellular serine protease expression in *Bacillus subtilis*. *J. Bacteriol.* **170:**136–140.
96. **Sastry, K. J., O. P. Srivastava, J. Millet, P. C. FitzJames, and A. I. Aronson.** 1983. Characterization of *Bacillus subtilis* mutants with a temperature-sensitive intracellular protease. *J. Bacteriol.* **153:**511–519.
97. **Schülein, R., J. Kreft, S. Gonski, and W. Goebel.** 1991. Preprosubtilisin Carlsberg processing and secretion is blocked after deletion of amino acids 97-101 in the mature part of the enzyme. *Mol. Gen. Genet.* **227:**137–143.
98. **Setlow, P.** 1976. Purification and properties of a specific proteolytic enzyme present in spores of *Bacillus megaterium*. *J. Biol. Chem.* **251:**7853–7862.
99. **Setlow, P.** 1988. Small, acid-soluble spore proteins of *Bacillus* species: structure, synthesis, genetics, function, and their degradation. *Annu. Rev. Microbiol.* **42:**319–338.
100. **Sheehan, S. M., and R. L. Switzer.** 1990. Intracellular serine protease 1 of *Bacillus subtilis* is formed in vivo as an unprocessed, active protease in stationary cells. *J. Bacteriol.* **172:**473–476.
101. **Shen, W.-F., T. S. Fletcher, and C. Largman.** 1987. Primary structure of human pancreatic protease E determined by sequence analysis of the cloned mRNA. *Biochemistry* **26:**3447–3452.

101a. **Shiga, Y., K. Hasegawa, A. Tsuboi, H. Yamagata, and S. Udaka.** 1992. Characterization of an extracellular protease inhibitor of *Bacillus brevis* HPD31 and nucleotide sequence of the corresponding gene. *Appl. Environ. Microbiol.* **58:**525–531.

102. **Shimada, H., M. Honjo, I. Mita, A. Nakayama, A. Akaoka, K. Manabe, and Y. Furutani.** 1985. The nucleotide sequence and some properties of the neutral protease gene of *Bacillus amyloliquefaciens*. *J. Biotechnol.* **2:**75–85.
103. **Sidler, W., E. Niederer, F. Suter, and H. Zuber.** 1986. The primary structure of *Bacillus cereus* neutral proteinase and comparison with thermolysin and *Bacillus subtilis* neutral proteinase. *Biol. Chem. Hoppe-Seyler* **367:**643–657.
104. **Siezen, R. J., W. M. deVos, J. A. M. Leunissen, and B. W. Dijkstra.** 1991. Homology modelling and protein engineering strategy of subtiliases, the family of subtilisin-like serine proteinases. *Protein Eng.* **4:**719–737.
105. **Sloma, A., A. Ally, D. Ally, and J. Pero.** 1988. Gene encoding a minor extracellular protease in *Bacillus subtilis*. *J. Bacteriol.* **170:**5557–5563.
106. **Sloma, A., C. F. Rudolph, G. A. Rufo, Jr., B. J. Sullivan, K. A. Theriault, D. Ally, and J. Pero.** 1990. Gene encod-

ing a novel extracellular metalloprotease in *Bacillus subtilis*. *J. Bacteriol.* **172:**1024–1029.

107. **Sloma, A., G. A. Rufo, Jr., C. F. Rudolph, B. J. Sullivan, K. A. Theriault, and J. Pero.** 1990. Bacillopeptidase F of *Bacillus subtilis*: purification of the protein and cloning of the gene. *J. Bacteriol.* **172:**1470–1477. (Erratum, **172:** 5520–5521.)
108. **Sloma, A., G. A. Rufo, Jr., K. A. Theriault, M. Dwyer, S. W. Wilson, and J. Pero.** 1991. Cloning and characterization of the gene for an additional extracellular serine protease of *Bacillus subtilis*. *J. Bacteriol.* **173:**6889–6895.
109. **Smith, E. H., F. S. Markland, C. B. Kasper, R. J. Delange, M. Landon, and W. H. Evans.** 1966. The complete amino acid sequence of two types of subtilisin, BPN' and Carlsberg. *J. Biol. Chem.* **214:**5974–5976.
110. **Smith, E. L., R. J. DeLange, W. H. Evans, M. Landen, and F. S. Markland.** 1968. Subtilisin Carlsberg V. The complete sequence; comparison with subtilisin BPN'; evolutionary relationships. *J. Biol. Chem.* **243:**2184–2191.
111. **Sobek, H., H. J. Hecht, B. Hofmann, W. Aehle, and D. Schomburg.** 1990. Crystal structure of an alkaline protease from *Bacillus alcalophilus* at 2.4 Å resolution. *FEBS Lett.* **274:**57–60.
112. **Srivastava, O. P., and A. I. Aronson.** 1981. Isolation and characterization of a unique protease from sporulating cells of *Bacillus subtilis*. *Arch. Microbiol.* **129:**227–232.
113. **Stahl, M. L., and E. Ferrari.** 1984. Replacement of the *Bacillus subtilis* subtilisin structural gene with an in vitro-derived deletion mutation. *J. Bacteriol.* **158:**411–418.
113a. **Stark, W., R. A. Pauptit, K. S. Wilson, and J. N. Jansonius.** 1992. The structure of neutral protease from *Bacillus cereus* at 0.2 nm resolution. *Eur. J. Biochem.* **207:**781–791.
114. **Steinmetz, M., F. Kurst, and R. Dedonder.** 1976. Mapping of mutations affecting synthesis of exocellular enzymes in *Bacillus subtilis*. *Mol. Gen. Genet.* **148:**281–285.
115. **Stragier, P.** Personal communication.
116. **Strauch, M. A., G. B. Spiegelman, M. Perego, W. C. Johnson, D. Burbulys, and J. A. Hoch.** 1989. The transition state transcription regulator *abrB* of *Bacillus subtilis* is a DNA binding protein. *EMBO J.* **8:**1615–1621.
117. **Strongin, A., D. I. Ya, I. A. Gorodetsky, V. V. Kuznetsova, Z. T. Yanonis, Z. T. Abramov, L. P. Belyanova, L. A. Baratova, and V. M. Stepanov.** 1979. Intracellular serine protease of *Bacillus subtilis* strain Marburg 168. *Biochem. J.* **179:**333–339.
118. **Strongin, A. Y., L. S. Izotaova, Z. T. Abramov, D. I. Gorodetsky, L. M. Ermakova, L. A. Baratova, L. P. Belyanova, and V. M. Stepanov.** 1978. Intracellular serine protease of *Bacillus subtilis*: sequence homology with extracellular subtilisins. *J. Bacteriol.* **133:**1401–1411.
119. **Sussman, M. D., and P. Setlow.** 1991. Cloning, nucleotide sequence, and regulation of the *Bacillus subtilis gpr* gene, which codes for the protease that initiates degradation of small, acid-soluble proteins during spore germination. *J. Bacteriol.* **173:**291–300.
120. **Takagi, M., and T. Imanaka.** 1989. Role of the pre-pro-region of neutral protease in secretion in *Bacillus subtilis*. *J. Ferment. Bioeng.* **67:**71–76.
121. **Takagi, M., T. Imanaka, and S. Aiba.** 1985. Nucleotide sequence and promoter region for the neutral protease gene from *Bacillus stearothermophilus*. *J. Bacteriol.* **163:** 824–831.
122. **Takekawa, S., H. Uozumi, N. Tsukagoshi, and S. Udaka.** 1991. Proteases involved in generation of β- and α-amylases from a large amylase precursor in *Bacillus polymyxa*. *J. Bacteriol.* **173:**6820–6825.
123. **Tanaka, T., and M. Kawata.** 1988. Cloning and characterization of *Bacillus subtilis iep*, which has positive and negative effects on production of extracellular proteases. *J. Bacteriol.* **170:**3593–3600.
124. **Tanaka, T., M. Kawata, Y. Nagami, and H. Uchiyama.** 1987. *prtR* enhances the mRNA level of the *Bacillus subtilis* extracellular proteases. *J. Bacteriol.* **169:**3044–3055.
125. **Thomas, T. D., and G. G. Pritchard.** 1987. Proteolytic enzymes of dairy starter cultures. *FEMS Microbiol. Rev.* **46:**245–268.
126. **Titani, K., M. A. Hermodson, L. H. Ericsson, K. A. Walsh, and H. Neurath.** 1972. Amino-acid sequence of thermolysin. *Nature* (London) *New Biol.* **238:**35–37.
127. **Toma, S., M. D. Bue, A. Pirola, and G. Grandi.** 1986. *nprR1* and *nprR2* regulatory regions for neutral protease expression in *Bacillus subtilis*. *J. Bacteriol.* **167:**740–743.
128. **Tran, L., X.-C. Wu, and S.-L. Wong.** 1991. Cloning and expression of a novel protease gene encoding an extracellular neutral protease from *Bacillus subtilis*. *J. Bacteriol.* **173:**6364–6372.
129. **Trowsdale, J., S. M. H. Chen, and J. A. Hoch.** 1979. Genetic analysis of a class of polymyxin resistant partial revertants of stage 0 sporulation mutants of *Bacillus subtilis*: map of the chromosome region near the origin of replication. *Mol. Gen. Genet.* **173:**61–70.
130. **Tsuru, D., J. D. McConn, and K. Yasunobu.** 1964. *Bacillus subtilis* neutral proteinase. Some physical properties. *J. Biol. Chem.* **240:**2415–2420.
131. **Uehara, H., K. Yamane, and B. Maruo.** 1979. Thermosensitive, extracellular neutral proteases in *Bacillus subtilis*: isolation, characterization, and genetics. *J. Bacteriol.* **139:**583–590.
132. **van den Burg, B., H. G. Enequist, M. E. vanderHaar, V. G. H. Eijsink, B. K. Stulp, and G. Venema.** 1991. A highly thermostable neutral protease from *Bacillus caldolyticus*: cloning and expression of the gene in *Bacillus subtilis* and characterization of the gene product. *J. Bacteriol.* **173:**4107–4115.
133. **van der Laan, J. C., G. Gerritse, L. J. S. M. Mulleners, R. A. C. van der Hoek, and W. J. Quax.** 1991. Cloning, characterization, and multiple chromosomal integration of a *Bacillus* alkaline protease gene. *Appl. Environ. Microbiol.* **57:**901–909.
133a. **van der Laan, J. C., A. V. Teplyakov, H. Kelders, K. H. Kalk, O. Misset, L. J. S. M. Mulleners, and D. W. Dijkstra.** 1992. Crystal structure of the high-alkaline serine protease-PB92 from *Bacillus alcalophilus*. *Protein Eng.* **5:**405–411.
134. **Vasantha, N., L. D. Thompson, C. Rhodes, C. Banner, J. Nagle, and D. Filpula.** 1984. Genes for alkaline protease and neutral protease from *Bacillus amyloliquefaciens* contain a large open reading frame between the regions coding for signal sequence and mature protein. *J. Bacteriol.* **159:**811–819.
135. **Visser, S., F. A. Exterkate, C. J. Slangen, and G. J. C. M. de Veer.** 1986. Comparative study of action of cell wall proteinases from various strains of *Streptococcus cremoris* on bovine α_{s1}-, β-, and κ-casein. *Appl. Environ. Microbiol.* **52:**1162–1166.
136. **Vos, P., G. Simons, R. J. Siezen, and W. M. de Vos.** 1989. Primary structure and organization of the gene for a procaryotic, cell envelope-located serine proteinase. *J. Biol. Chem.* **264:**13579–13585.
137. **Vos, P., M. vanAsseldonk, F. vanJeveren, R. Siezen, G. Simons, and W. M. de Vos.** 1989. A maturation protein is essential for production of active forms of *Lactococcus lactis* SK11 serine proteinase located in or secreted from the cell envelope. *J. Bacteriol.* **171:**2795–2802.
138. **Wagner, F. W.** 1986. Assessment of methodology for the

purification, characterization, and measurement of proteases, p. 17–38. *In* M. J. Dolling (ed.), *Plant Proteolytic Enzymes*, vol. 1. CRC Press, Inc., Boca Raton, Fla.

139. **Wandersman, C.** 1990. Secretion, processing and activation of bacterial extracellular proteases. *Mol. Microbiol.* **3:**1825–1831.
140. **Wang, L.-F., R. Brückner, and R. H. Doi.** 1989. Construction of a *Bacillus subtilis* mutant-deficient in three extracellular proteases. *J. Gen. Appl. Microbiol.* **35:**487–492.
141. **Wang, L.-F., and R. H. Doi.** 1990. Complex character of *senS*, a novel gene regulating expression of extracellular-protein genes of *Bacillus subtilis. J. Bacteriol.* **172:** 1939–1947.
142. **Wells, J. A., E. Ferrari, D. J. Henner, D. A. Estell, and E. Y. Chen.** 1983. Cloning, sequencing, and secretion of *Bacillus amyloliquefaciens* subtilisin in *Bacillus subtilis. Nucleic Acids Res.* **11:**7911–7925.
143. **Wong, S.-L., and R. H. Doi.** 1986. Determination of the signal peptidase cleavage site in the preprosubtilisin of *Bacillus subtilis. J. Biol. Chem.* **261:**10176–10181.
144. **Wong, S.-L., C. W. Price, D. S. Goldfarb, and R. H. Doi.** 1984. The subtilisin E gene of *Bacillus subtilis* is transcribed from a σ^{37} promoter *in vivo. Proc. Natl. Acad. Sci. USA* **81:**1184–1188.
145. **Wright, C. S., R. A. Alden, and J. Krant.** 1969. Structure of subtilisin BPN′ at 2.5 Å resolution. *Nature* (London) **221:**235–242.
146. **Wu, X.-C., W. Lee, L. Tran, and S.-L. Wong.** 1991. Engineering a *Bacillus subtilis* expression-secretion system with a strain deficient in six extracellular proteases. *J. Bacteriol.* **173:**4952–4958.
147. **Wu, X.-C., S. Nathoo, A. S.-H. Pang, T. Carne, and S.-L. Wong.** 1990. Cloning, genetic organization, and characterization of a structural gene encoding bacillopeptidase F from *Bacillus subtilis. J. Biol. Chem.* **265:**6845–6850.
148. **Yang, M., E. Ferrari, E. Chen, and D. J. Henner.** 1986. Identification of the pleiotropic *sacQ* gene of *Bacillus subtilis. J. Bacteriol.* **166:**113–119.
149. **Yang, M., H. Shimotsu, E. Ferrari, and D. J. Henner.** 1987. Characterization and mapping of the *Bacillus subtilis prtR* gene. *J. Bacteriol.* **169:**434–437.
150. **Yang, M. Y., E. Ferrari, and D. J. Henner.** 1984. Cloning of the neutral protease gene of *Bacillus subtilis* and use of the cloned gene to create an in vitro-derived deletion mutation. *J. Bacteriol.* **160:**15–21.
151. **Yoshimoto, T., H. Oyama, T. Honda, H. Tone, T. Takeshita, T. Kamiyama, and D. Tsuru.** 1988. Cloning and expression of subtilisin amylosacchariticus gene. *J. Biochem.* **103:**1060–1065.
152. **Yoshimoto, T., H. Oyama, T. Takeshita, H. Higashi, S. L. Xu, and D. Tsuru.** 1990. Nucleotide sequence of the neutral protease gene from *Bacillus subtilis var. amylosacchariticus. J. Ferment. Bioeng.* **70:**370–375.
153. **Zhu, X., Y. Ohta, F. Jordan, and M. Inouye.** 1989. Pro-sequence of subtilisin can guide the refolding of denatured subtilisin in an intermolecular process. *Nature* (London) **339:**483–484.
154. **Ziegler, R. J., A. H. Ally, D. S. Ally, and G. A. Hintermann.** Unpublished data.
155. **Zukowski, M.** Personal communication.

64. Insecticidal Toxins

ARTHUR I. ARONSON

Several species of *Bacillus* and at least one species of *Clostridium* form parasporal inclusions during sporulation. These inclusions, which are composed of proteinaceous protoxins (called δ-endotoxins in *Bacillus thuringiensis*), come in a variety of shapes and sizes. The ingestion of these inclusions (and usually the spores) by actively feeding larvae with subsequent solubilization and conversion of the protoxins to toxins results in a rapid cessation of feeding and ultimately death. The ability to invade larvae from at least three orders of insects, Lepidoptera, Diptera, and Coleoptera, provides these bacilli with a source of nutrients not readily available to other microorganisms in the soil or phylloplane.

At least 50 protoxins have been well characterized on the basis of gene sequence and toxicity profile and there is an even greater potential for diversity given the hundreds of *B. thuringiensis* isolates, most cataloged only to subspecies, that are present in several stock culture collections: *Bacillus* Genetic Stock Center, Columbus, Ohio; Pasteur Institute collection, C/O Dr. H. deBarjac; the collection of Dr. H. Dulmage at the U.S. Department of Agriculture, Peoria, Ill.; and an undefined number of collections in industry.

B. thuringiensis is unique in that a substantial fraction (10 to 20%) of its total potential genetic information, including most protoxin genes, is present as plasmids (4, 15). Since it is the synthesis of the protoxins and their deposition as inclusions that clearly differentiates these bacilli from a variety of other sporeformers, the properties of the protoxin genes, their regulation, and the mechanism of action of the δ-endotoxins will be the focus of this chapter.

BACILLI PRODUCING INSECTICIDAL TOXINS

The infestation of Japanese beetle grubs by *Bacillus popilliae* is a classic example of bacilli producing insecticidal toxins (13). Infection may require both the spore and some undefined toxic component. Growth and sporulation occur within the infected larvae, leading to a "milky" appearance of the dead larvae. The extensive propagation helps ensure the persistence of this species in the soil. In fact, toxic spores can be obtained only from infected larvae. While cells can be grown in laboratory cultures, sporulation is limited, as is the production of toxic cultures. This problem hampers progress in defining this system.

Bacillus sphaericus produces a binary toxin consisting of two polypeptides which is active against the larvae of certain mosquitoes (8). These proteins are synthesized as protoxins during sporulation and are usually deposited as discernible inclusions. The properties of these toxic proteins and their specificities are virtually identical among a number of isolates. The basis for toxicity and the function of each of the proteins (both processed in the larval midgut to active toxin) are not known.

Recently, a strain of *Clostridium bifermentans* that produces a parasporal inclusion was found to be toxic to mosquito larvae (17). The identification of the toxic component and its relation to other bacterial insect toxins is of great interest.

The most extensive studies have been done with *B. thuringiensis* subspecies that produce proteinaceous inclusions during sporulation. The inclusions are often bipyramidal, but some are cuboidal or multifaceted, and there is a wide variety of other morphologies. Some strains contain more than one type of inclusion in each cell. These inclusions are present within the mother cell adjacent to the spore, but in a few subspecies, they are localized within the exosporium (5). In all cases, inclusions are released, as is the spore, upon cell lysis. Most of the isolates studied to date are active against a variety of lepidopteran larvae, but some protoxins are effective against Diptera, Coleoptera, and perhaps nematodes (25), although the nature of the toxic component has not been defined in the last case. There is also a nonprotein toxin, the β-exotoxin, secreted by some *B. thuringiensis* strains. This toxin, which is assayed on house fly larvae (60), is not as selective as the δ-endotoxins.

Overall, there appears to be a very broad potential for invasion of a wide variety of insect larvae, probably in order to exploit the nutrients present in larval hemolymph. Growth and subsequent sporulation within the larvae, as is most dramatically evident for *B. popilliae*, would provide the opportunity for exchange of genes and/or plasmids. Judged by a number of criteria, the most extensively studied of these bacilli, *B. thuringiensis*, is very closely related to *Bacillus cereus* (54). As discussed below, a major difference between the two species is the presence of a large complement of plasmids, including those encoding protoxins, in *B. thuringiensis*. While little is known about the functions of most of the plasmid genes, the properties of these plasmids are relevant to understanding protoxin synthesis and regulation.

MOST PROTOXIN GENES ARE ON LARGE PLASMIDS

According to an analysis of flagellum antigens, *B. thuringiensis* includes at least 20 serotypes (22), and most isolates contain more than one protoxin gene, with a unique complement in each subspecies. More than 50 protoxin genes have been sequenced, and this information plus some toxicity data has provided a

Arthur I. Aronson • Department of Biological Sciences, Purdue University, West Lafayette, Indiana 47907.

Table 1. Properties of δ-endotoxin genes and their products[a]

Gene designation	Host range[b]	Predicted mass (kDa) of protoxin	Origin of gene (*B. thuringiensis* subspecies)[c]	Unique toxicity properties (reference)[d]
cryIA(a)	L	133.2	*kurstaki* HD1 *sotto*	*Bombyx mori*
cryIA(b)	L(D)	131.0	*kurstaki* HD1 *berliner* *aizawai* *alesti*	Several lepidoptera
cryIA(c)	L	133.3	*kurstaki* HD1 *kurstaki* HD73	*Heliothis virescens*
cryIB	L	138.0	*thuringiensis* HD2	*Pieris brassicae*
cryIC	L	134.0	*aizawai* *entomocidus* *kenyae*	*Spodoptera littoralis*
cryID	L	132.5	*aizawai* HD68	*Manduca sexta*
cryIE[e]	L	133.2	*tolworthi*	*Spodoptera exigua*
cryIF	L	133.6	*aizawai*	*Spodoptera exigua*
cryIG	L	130.0	*galleria*	*Galleria mellonella* (52a)[f]
cryIIA	L/D	70.9	*kurstaki* HD1	Relatively low toxicity for lepidoptera and diptera
cryIIB	L	70.8	*kurstaki* HD1 (cryptic)	*Manduca sexta*
cryIIC	L	69.5	*shanghai*	Good toxicity for *Trichoplusia ni* and *Manduca sexta* (69)
cryIIIA	C	73.1	*san diego* *tenebrionis*	Colorado potato beetle
cryIIIB	C	75	*tolworthi*	Colorado potato beetle
cryIVA	D	134.4	*israelensis* *morrisoni*	Certain diptera (*Aedes aegypti*) and blackflies
cryIVB	D	127.8	*morrisoni*	Certain diptera (*Aedes aegypti*) and blackflies
cryIVC	D	77.8	*morrisoni*	Certain diptera (*Aedes aegypti*) and blackflies
cryIVD	D	72.4	*morrisoni*	Certain diptera (*Aedes aegypti*) and blackflies

[a] Data are from reference 40 unless otherwise specified.
[b] L, lepidoptera; D, diptera; C, coleoptera; L/D, active on both; (D), activity on diptera dependent on protoxin processing (36).
[c] Not exclusive, but published source of gene. HD numbers are per listing in reference 51a.
[d] Special but not exhaustive toxicity properties.
[e] Similar in amino half of toxin to CryIA(b) and in carboxyl half to CryIC (63).
[f] EMBL accession number X58120.

basis for the classification of these genes (40). An overview of the major types of protoxins and some of their unique toxicity spectra are presented in Table 1. As will be discussed in more detail below, gene designations are based on sequence analysis and toxicity profiles (40).

In general, protoxin genes are present on plasmids of ca. 40 to 200 MDa, and in some cases, more than one gene is present on a given plasmid (5, 15). For example, spontaneous loss of a 120-MDa plasmid from a *B. thuringiensis* subsp. *kurstaki* HD1 derivative (strain HD1-7; 15) resulted in the loss of two *cryIA* and two *cryII* genes. *B. thuringiensis* subsp. *israelensis*, the prototype strain active against several dipteran species, contains four genes (designated *cryIVA* through *cryIVD*) plus a gene (*cytA*) encoding a 28-kDa hemolytic factor. Plasmid curing indicated that all of these genes were present on a 72-MDa plasmid (34). The presence of multiple genes on one plasmid raises the possibility of recombination analogous to that found in plasmid constructs introduced into *Escherichia coli* (14; see below) or perhaps in arrangements of genes in operons, as has been found for the *cryIIA* and *cryIIC* genes (67, 71).

Most *B. thuringiensis* isolates contain several different sizes (3 to 200 MDa) of extrachromosomal, double-stranded DNA molecules. At one extreme is *B. thuringiensis* subsp. *kurstaki* HD1, which has 12 size classes, including a linear molecule (15) and one that is 29 MDa (a transducing phage designated TP21) (66). A phage genome has also been found as part of a larger plasmid in another *B. thuringiensis* strain (42). *B. thuringiensis* subsp. *finitimus* has only two plasmids, but they are 77 and 98 MDa (the latter contains a protoxin gene), so the overall plasmid coding capacity is still considerable.

Some of the plasmids contain mobilization functions (56), and others can be mobilized and transferred by cell contact to *B. cereus*, plasmid-cured derivatives of the original *B. thuringiensis* strain, or in a few cases, to other *B. thuringiensis* serotypes (5, 15, 33). Many of the transferable plasmids contain protoxin genes (5, 15). It is also relatively easy to isolate derivatives that lack a protoxin-encoding plasmid (usually of 40 to 50 MDa) by growing them at 42°C and screening the plasmid profiles from random colonies. The screening usually involves 50 to 100 colonies, so the frequency of curing is at least 1% in these cases.

For example, in *B. thuringiensis* subsp. *kurstaki* HD1, which contains three *cryIA* genes (Table 1), about 30% of random colonies lacked a 44-MDa plasmid containing the *cryIA*(b) gene (3). This instability is probably due to plasmid incompatibility, because the 44-MDa plasmid is stable in a derivative of *B. thuringiensis* subsp. *kurstaki* HD1 (strain HD1-7), which has lost several cryptic plasmids (3, 15). In order to sustain the 44-MDa plasmid [and the *cryIA(b)* gene] in the population, the remaining 70% of the cells

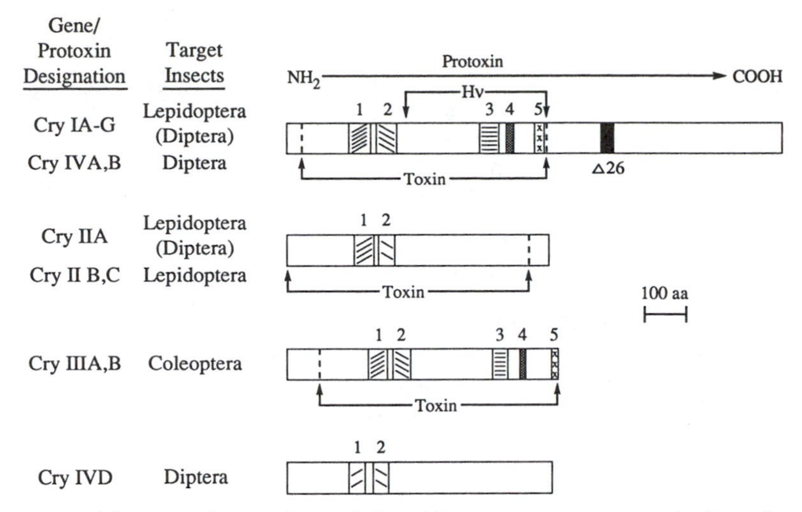

Figure 1. General structural features of protoxins as deduced from gene sequences and other related data. Protoxins designated CryIA through CryIG, CryIVA, and CryIVB contain 1,100 to 1,200 amino acids, and the toxin is processed from within the amino half as shown. The CryII, CryIII, and CryIVD protoxins are smaller, with processing to toxins as indicated (not known for CryIVD). Regions marked 1 through 5 are highly conserved among the CryI, CryIII, CryIVA, and CryIVB toxins and less so (primarily regions 1 and 2) for the CryII and CryIVD toxins. The carboxyl halves of the CryI, CryIVA, and CryIVB protoxins are also extensively conserved. A major difference is the deletion of 26 amino acids (Δ26) in most of the CryIA(b) protoxins. Other portions of the toxins are more or less conserved within a particular class (i.e., those designated CryI or CryII, etc.) but not between these classes.

must be able to transfer this plasmid to cured cells, probably by cell-to-cell contact (33). Such a transfer can be demonstrated by mixing the appropriately marked strains on filters or in a liquid culture.

The overall population of *B. thuringiensis* subsp. *kurstaki* HD1 (a strain used in commercial preparations such as Dipel[a]) is thus composed of cells containing two or three *cryIA* protoxin genes, and so there is a different number of protoxins in the single bipyramidal inclusion produced by each cell. While these three CryIA protoxins are very similar in structure (see below), each has a unique specificity profile for selected lepidoptera (Table 1). The 30% of the cells lacking the CryIA(b) protoxin produce relatively more CryIA(c) protoxin than the remaining 70% (3). In this regard, these cells are similar to *B. thuringiensis* subsp. *kurstaki* HD73, which produces only the CryIA (c) protein and is known to be selective for *Heliothis virescens* (45). Inclusions from *B. thuringiensis* subsp. *kurstaki* HD1 cells lacking the *cryIA*(b) gene are also selective for this insect, especially compared with inclusions from cells producing the three CryIA protoxins (3). Such a dynamic heterogeneity in toxicity profile should be beneficial for adaptation to changes in the population of susceptible insects.

Insertion sequences and in some cases transposons flank these genes. Several insertion sequence elements have been sequenced, and they have been sorted into four groups (23). Their locations close to *cry* genes may be indicative of the origins and perhaps the mobilities of these genes. For example, during a screening of the plasmid profiles of random colonies of *B. thuringiensis* subsp. *israelensis*, changes in the location (and perhaps transcription) of the *cryIVA* through *cryIVD* and *cytA* protoxin genes was found (34). As mentioned above, these genes are all on a 72-MDa plasmid, but in some derivatives, this plasmid was absent, and sequences hybridizing to it were found either in a very large plasmid (or perhaps the chromosome) or on novel plasmids of 49 or 32 MDa. The last two variants were still toxic, so both transposition and plasmid recombination resulted in a change of locale.

In a few cases, insertion sequence-like sequences have been found within protoxin structural genes (46, 58). The function, if any, of these inactive genes is not known. The transposon Tn*4430* is also a substantial fraction of a small cryptic plasmid in *B. thuringiensis* subsp. *thuringiensis* (48). The prevalence of transposon sequences close to or even within protoxin genes may account for gene mobility and perhaps recombination among these genes.

STRUCTURAL FEATURES OF *B. THURINGIENSIS* PROTOXINS

General Properties

All of the protoxins, with the possible exception of CryIVC and CryIVD, have similar-sized toxins and share conserved regions (Fig. 1; 40). A number of protoxins designated CryI as well as the CryIVA and CryIVB protoxins from *B. thuringiensis* subsp. *israelensis* are 130 to 140 kDa. The toxins per se reside between residues 29 and about 610 following removal of the first 29 residues and the carboxyl half by trypsinlike proteases in the larval gut. The toxin region consists of at least two (18–20) and probably three (47) large structural domains. These aspects of the tertiary structure and their functional significance are discussed below.

Judging from the deduced sequences, there are five conserved regions (designated 1 through 5 in Fig. 1) among the CryI and CryIII toxins, and at least regions 1 and 2 are conserved in the CryII and CryIVD toxins. Regions 1 and 2 are hydrophobic and contain potential amphipathic helices, and one or both are important for toxicity (see below). Regions 3 through 5 are in the carboxyl halves of the toxins, and their function(s) are not known. They are present in a so-called hypervariable (HV) region of toxins (Fig. 1), a designation based on a comparison of three related protoxins, CryIA(a), CryIA(b), and CryIA(c), each with a distinct specificity profile for certain lepidoptera (Table 1; 39) and extensive sequence homology among the first 280 residues. The pattern of variation within the carboxyl-proximal 300 to 350 residues (HV region) of these toxins is thus likely to account for the specificity differences among them, a hypothesis supported by the construction of hybrid genes as discussed below. Perhaps of even greater significance for the origin of gene heterogeneity is that the *cryIA(b)* gene appears to have resulted from recombination within the HV region between the *cryIA(a)* and *cryIA(c)* genes (32). These three genes are present in *B. thuringiensis* subsp. *kurstaki* HD1, and the last two are on a 120-MDa plasmid. As discussed below, protoxin genes cloned in tandem in an *E. coli* vector do recombine (14, 68). Perhaps a similar process occurs in *B. thuringiensis* among genes on the same plasmid. Such recombination within HV regions could generate protoxins with novel specificities.

The CryII protoxins produced by *B. thuringiensis* strains that also synthesize CryI types are deposited in separate cuboidal inclusions (73). These toxins are active either against lepidoptera (67, 71) or against lepidoptera and diptera (72). The processing of the 70-kDa protoxin to toxin is less extensive and involves about 20 to 30 residues from the carboxyl end. These toxins contain at least two of the conserved regions found in the CryI toxins. Two novel features of the CryII toxins are the presence of an apparently cryptic gene (*cryIIB*) that is functional when cloned into an *E. coli* promoter-containing plasmid (67) and the existence of an operon containing two open reading frames plus the *cryII* gene (67, 71). The function of the two upstream open reading frames is not known. Neither is present in substantial amounts in the cuboidal inclusions, and one encodes a polypeptide with extensive repeats (i.e., 15 to 21 repeats of 15 or 16 residues). This polypeptide likely has a unique function, and deletion of the gene may prevent inclusion formation (20a).

The CryIII protoxins active against coleoptera are also about 70 kDa, but the processing to toxins occurs primarily at the amino end (16; removal of ca. 50 residues). Hydrophobic stretches analogous to regions 1 and 2 in the CryI toxins and conserved regions 3 through 5 have been identified (Fig. 1), so the overall structures (and modes of action) of the CryI, CryIVA, CryIVB, and CryIII toxins may be the same.

B. thuringiensis subsp. *israelensis* and related subspecies active against certain mosquitoes and blackflies form multifaceted inclusions containing four protoxins (CryIVA through CryIVD) plus a 28-kDa haemolytic or cytolytic factor designated CytA. These inclusions are apparently surrounded by a netlike structure (27), so the entire package may be kept intact following release from the lysing mother cell, a property that may be important in ensuring efficient intake by filter-feeding dipteran larvae. All of these protoxin genes are on a 72-MDa plasmid (34). The CryIVA and CryIVB protoxins are similar in structure to the CryI types (Fig. 1). The CryIVC and CryIVD toxins appear to be novel, judging from sequence comparisons, and CytA is very different in sequence and in its lack of selectivity; i.e., it is generally cytolytic. The packaging of this complex mixture of toxins into one large aggregate suggests related or interdependent functions, but it has been difficult to assess the contribution of each component. Dipteran larvae are filter feeders, so comparisons of the toxicities of purified soluble (or aggregated) fractions with that of the multicomponent inclusion are not readily interpretable. Deletion of the *cytA* gene did not appear to affect inclusion formation or toxicity (24), although there is evidence from mixing of partially purified fractions for synergism between CytA and one or more of the CryIVA through CryIVD toxins (27).

Function of Carboxyl Halves of CryI Protoxins

The carboxyl halves of protoxins that are removed by gut proteases are extensively conserved among the CryI, CryIVA, and CryIVB protoxins. This portion of the molecule is probably important for the deposition of protoxins in inclusions and may also function to protect the toxin. The carboxyl half is unlikely to be critical for toxicity (55), although the extent of oxidation or reduction of the Cys residues may affect processing and thus the specificity profile of a particular toxin (7).

Most of the cysteines in protoxins are present in this region of the molecule, and in inclusions, they are all oxidized probably as intermolecular disulfides (10). Since most *B. thuringiensis* strains contain multiple protoxin genes, it is likely that several CryI-type protoxins are present in a given inclusion, and they may be present as heterodimers or larger disulfide-linked aggregates. If such cross-linking does exist, it could affect the solubility of the protoxins (6) and/or the processing to toxins (see below).

One intriguing difference among the sequences of the carboxyl halves of CryI protoxins is the presence of a 26-amino-acid deletion that is found in most, but not all (46), CryIA(b) protoxins but not in other CryI protoxins (32). This deletion includes four of the cysteines present in the carboxyl half, so it is likely to have structural consequences. These CryIA(b) protoxins are less stable than the related CryIA(a) and CryIA(c) protoxins when each is produced as the only protoxin in cells grown at 30°C (51). When CryIA(b) is present as one of several protoxins, however, it is stable (at 30°C), possibly because of heterologous disulfide cross-linking in the inclusion.

Restriction fragments encoding regions embracing the deleted portions were exchanged between the *cryIA(b)* and *cryIA(c)* genes. The resulting CryIA(c) hybrid protoxin was unstable, whereas the hybrid CryIA(b) protoxin was stable in *B. thuringiensis* transformants (31), confirming the importance of this 26-amino-acid region to stability. The toxicity profiles of these hybrid protoxins for *H. virescens* and *Trichoplusia ni* larvae were the same as those of the parental types. This deletion is present between two direct repeats of 9 or 10 nucleotides (32). Deletions between direct repeats are not uncommon among genes in

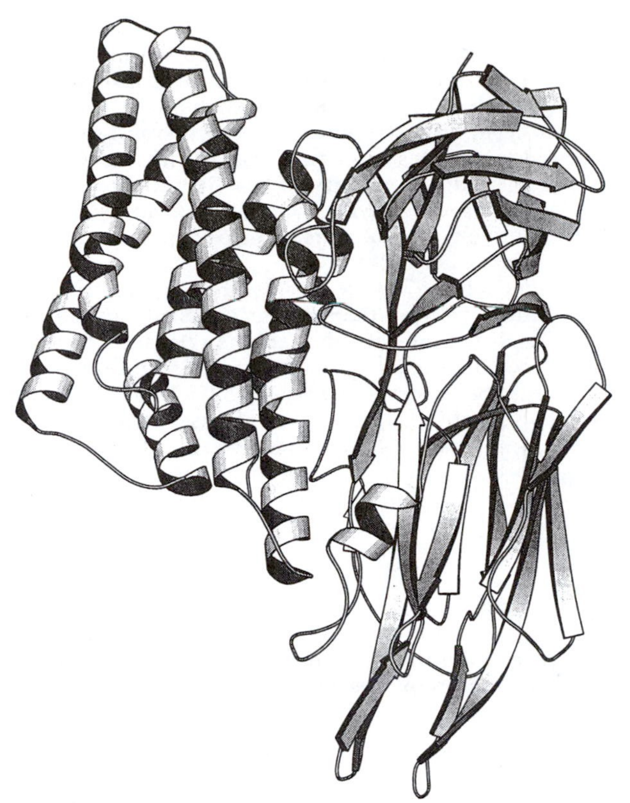

Figure 2. Schematic ribbon representation of the CryIIIA toxin (47). The three domains include a seven-helix bundle (upper left), a three-sheet assembly (bottom right), and a β sandwich (upper right).

general, but this particular deletion has not been found in other CryIA protoxins, all of which contain virtually identical sequences in this region. Perhaps the CryIA(b) protoxin does have a unique function in inclusion formation or structure, as discussed below.

Toxin Domains and Structural Studies

Protease treatment of the CryIA(b) or CryIA(c) toxin resulted in two fragments of about 30 and 36 kDa, which is consistent with a two-domain structure linked at about residue Gly-327 (18–20). Measurements of changes in fluorescence or circular dichroism patterns upon denaturation were also consistent with two domains, i.e., an α-helical amino-proximal portion and a carboxyl part composed primarily of β sheets (19).

The structure of the CryIIIA protoxin has been determined at 2.5-Å (1 Å = 0.1 nm) resolution (47; Fig. 2). Three domains were resolved: a seven-helix bundle near the amino end, a three-sheet domain, and a carboxyl-proximal β sandwich. The seven helices are arranged concentrically, with helix 5 in the center. They all could span the membrane, and they are amphipathic. Helices 5 and 6/7 include two of the conserved toxin regions (1 and 2 in Fig. 1). As discussed below, many of the mutations in a 10-amino-acid stretch within helix 5 resulted in loss of or lower toxicity but had no effect on receptor binding (70). One or more of these helices probably penetrate the membrane and must be involved in toxin action (see below). The five conserved toxin regions are important components of all three of the structural domains, so the general structural features of the CryIII toxin will likely be applicable to other toxins with similar conserved sequences.

MODE OF ACTION

Factors Involved in Toxicity and Specificity

Solubilization of inclusions composed of CryI or CryIV proteins requires alkaline and reducing conditions in vitro, i.e., pH 9.2 to 9.5 plus 0.1 to 0.5% β-mercaptoethanol, conditions apparently similar to those in lepidopteran and dipteran larval midguts. A pH of at least 10 plus the mercaptan is required to solubilize the CryII protoxins in vitro (72), but the significance of the higher pH requirement is not known. The CryIII protoxin active against coleoptera can be solubilized at pH 7 to 8, which is consistent with the less alkaline condition in the midguts of these larvae.

In certain cases, although the soluble protoxins were effective, the inclusions were not (41), implying that the conditions in the guts of some larvae may not be suitable for solubilization of certain inclusion proteins. For example, the solubilities of inclusion proteins from *B. thuringiensis* subsp. *aizawai* and a plasmid-cured derivative differed, as did their toxicities (6). Inclusions from the parental strain contained the CryIA(b), CryIC, and CryID protoxins, whereas those from a strain cured of the 45-MDa plasmid containing the *cryIA(b)* gene produced inclusions containing only the CryIC and CryID protoxins. The CryIC toxin is effective against larvae of the grain storage lepidopteran *Plodia interpunctella* (62), so as anticipated, inclusions or soluble protoxins from the parental strain were active (6). Inclusions containing only the CryIC and CryID protoxins, however, were much less effective than the soluble protoxins extracted from these inclusions. The proteins in these latter inclusions were not as readily solubilized in vitro as those in inclusions from the parental strain, which is consistent with the toxicity differences. Solubility of inclusions containing only the CryIC and CryID protoxins could be restored to that of inclusions from the parental strain, however, by transforming the plasmid-cured strain by electroporation (59) with a clone of the *cryIA(b)* gene (3). These results are consistent with protoxin composition influencing solubility. The unstable CryIA(b) protoxin containing a 26-amino-acid deletion in the carboxyl half may have a special role in this process.

Following solubilization, these protoxins are processed to toxins, probably by trypsinlike enzymes in the larval midgut. Processing can be achieved in vitro with trypsin, and the toxin per se is very resistant to further proteolysis (18, 20). As alluded to earlier, the conformation of the protoxin and probably the extent of disulfide bonding can have an effect on processing (7), as can the array of proteases present in the midguts of particular larvae (35, 36). The selectivity of a CryIA(b) protoxin from a *B. thuringiensis* subsp. *aizawai* strain for either lepidoptera or diptera was dependent on the source of proteases used to convert the protoxin to toxin. If larval gut extract from *Aedes aegypti* was used, the protoxin was processed to a toxin of 53 kDa that was active against diptera. Processing by gut extract from a lepidopteran resulted in a toxin of 55 kDa active only against lepidoptera.

This particular CryIA(b) protoxin differed by only three residues from a "prototype" CryIA(b) protoxin from *B. thuringiensis* subsp. *berliner* according to the sequences of the cloned genes (40). The three differences, Leu537Phe, Ile545Pro, and Ile568Thr (to *B. thuringiensis* subsp. *aizawai* from *B. thuringiensis* subsp. *berliner*), are all close to the carboxyl end of the toxin. The last two are adjacent to Arg residues, which are potential sites of processing by trypsinlike enzymes. A site-directed mutation of Ile545Pro in the *B. thuringiensis* subsp. *aizawai* gene resulted in loss of toxicity for *A. aegypti* larvae or cells in culture, whereas the Ile568Thr change led to loss of toxicity against *Pieris brassicae* larvae or cells (35). These particular residues must be important for the conformation of the protoxin and thus for the sites for processing toxins active against either lepidoptera or diptera.

Following processing, toxins bind to receptors on the surface of brush border membrane cells lining the midgut (38). Subsequently, they insert into the membrane and either form a cation pore or somehow alter a preexisting pore. The ultimate effect is ion flux changes and subsequent osmotic lysis (44).

Receptor Binding

There is saturable, high-affinity binding of labeled toxins to vesicles prepared from the midguts of several species of lepidopteran larvae (38, 61, 62). In general, this binding was observed only for toxins active against the larvae from which the vesicles were prepared. Nonactive toxins did not display saturable binding and did not compete well. The kinetics of association were bimolecular, and the reactions rapidly became nonreversible. In general, there was a correlation of the K_d and/or the number of binding sites with the relative toxicity. There are some exceptions, however, most notably the 10-fold-lower K_d of the CryIA(c) toxin for binding to vesicles from *Lymantria dispar* (gypsy moth) larvae compared with binding of the CryIA(b) toxin (69). The latter larvicide is 400 times more potent than the former. These toxins compete for the same binding sites, so factors other than high binding affinity are probably involved in subsequent steps leading to toxicity.

The most conclusive evidence for specific receptor binding comes from studies with *P. interpunctella* (63). A high level of resistance to a commercial *B. thuringiensis* preparation (Dipel[a]) was obtained by backcrossing and growth under selective conditions (50). Dipel[a] contains *B. thuringiensis* subsp. *kurstaki* HD1 spores plus inclusions, and as previously discussed, the latter contain the closely related CryIA(a), CryIA(b), and CryIA(c) protoxins. The resistant larvae were found, however, to be sensitive to some other species of *B. thuringiensis* containing different protoxins, in particular *B. thuringiensis* subsp. *aizawai*, which contains the CryIC protoxin active against both sensitive and resistant *Plodia* larvae. The affinity of binding (K_d) of CryIA(b) toxin to vesicles prepared from larvae of resistant *P. interpunctella* was 50-fold less than the K_d for vesicles from the sensitive larvae. In contrast, the binding of the CryIC toxin was unchanged, and the number of binding sites was actually increased about threefold in vesicles from the resistant larvae. A similar decrease in the affinity of the CryIA(b) toxin for vesicles prepared from larvae of *Plutella xylostella* (diamond-back moth) resistant to Dipel has been reported elsewhere (28).

Extracts of these vesicles resolved by gel electrophoresis contain polypeptides of various sizes, probably glycoproteins, that can bind toxins (29, 44, 52). The importance of the sugar residues to this binding may vary with the insect species (44, 52). There is some correlation of binding with larval susceptibility, but some toxins that are not very effective do bind, although to different membrane proteins than a more active toxin binds to (52). While binding to specific receptor proteins appears to be necessary, it is obviously not sufficient. The rate of binding and number of sites must be important, but there may also be an induced conformational change of the toxin in order for it to insert into the membrane. Such a change may occur only upon interaction of the toxin with certain specific receptors (or specific sites on these receptors).

Disruption of Cation Flow

Initially, it was demonstrated that toxins could inhibit the potassium-dependent uptake of certain amino acids into vesicles (57). Subsequently, more-general effects on cation flux were found to be consistent with a colloid osmotic-lysis mechanism (44).

Several amphipathic helices present in the amino halves of toxins (47; Fig. 2) may form a pore, or some of these membrane-spanning helices from several toxins may interact to form such a pore (37). There is some evidence for the formation of pores in artificial membranes by the 28-kDa CytA toxin (43). Alternatively, toxins may insert into the membrane and exert their effects on cation flux by altering a preexisting channel. As discussed above, insertion of the CryI through CryIV toxins may require a conformational change in the toxin. Such a structural alteration could be induced by the interaction of the toxin with the appropriate membrane receptor.

DEFINING REGIONS REQUIRED FOR TOXICITY AND SPECIFICITY

Mutations

Some site-directed mutations in a potential amphipathic helical sequence within the first conserved region (helix 5 in Fig. 2) of the *cryIA(c)* gene resulted in lower toxicity (2). In a more detailed study, a number of CryIA(c) toxins with single amino acid changes in this region were isolated by employing a mutagenic oligonucleotide, and about 40% had low or no toxicity for three test larvae (70). Several of the mutated sequences (in conserved region 1 in Fig. 1) were reinserted into the wild-type protoxin gene, and these constructs were electroporated (59) into a plasmid-free *B. thuringiensis* subsp. *kurstaki* HD1 strain. Bioassays were done with the inclusions, and toxins were then generated by trypsin digestion. Following purification, the mutant toxins were found to effectively compete with wild-type toxin for binding to midgut vesicles, indicating that the toxins were stable. However, these toxins had lost the ability to inhibit potassium-dependent amino acid uptake into the vesicles (57). Overall, this particular potential amphipathic region, which is well conserved among all toxins, appears to be very important for toxicity.

Oligonucleotide mutagenesis was also used to generate random mutations in another hydrophobic region closer to the amino end of the toxin, which is probably the sequence at the end of helix 2 where helix 2 joins helix 3 in Fig. 2. In this case, most of the changes within the hydrophobic portion had little or no effect on toxicity for the three test lepidoptera, but mutations of two specific residues, Ala-92 and Arg-93, did result in selective toxicity changes. There was loss of toxicity for *T. ni* and *Manduca sexta* larvae but retention of some toxicity for *H. virescens* larvae. Mutant toxins with loss of a net positive charge at Ala-92–Arg-93 showed this selective loss of toxicity. These mutant toxins did not compete well with wild-type CryIA(c) toxin for binding to vesicles from *M. sexta* larval midguts but did compete for binding to vesicles prepared from *H. virescens* larvae. Arg-93 is probably involved in ionic interactions that directly or indirectly alter the ability of the toxin to bind specifically to some but not all receptors. If this residue is directly involved in specificity, it must interact with amino acids in another region of the toxin, most likely those in the HV portion near the carboxyl end. In that case, a specificity domain would not be composed exclusively of contiguous residues in the HV region but rather would depend on tertiary structural interactions.

Such interactions between residues near the amino and carboxyl ends of a toxin were implied by the retention of toxicity for dipteran larvae by a deleted CryIA(b) toxin from *B. thuringiensis* subsp. *aizawai* that lacked residues 242 to 523 (35). Such an interaction may exist in the native toxin, a possibility consistent with the compact structure of the CryIIIA toxin (Fig. 2).

Some replacements either of single residues (70) or of blocks of residues by exchanges of restriction fragments between related genes resulted in the synthesis of unstable protoxins (31, 64, 70). This limits the mutagenic approach for analyzing the structure of toxins that are apparently very sensitive to small perturbations in their sequences.

Exchange of Restriction Fragments

The HV region (Fig. 1) has already been implicated in specificity on the basis of sequence comparisons among the closely related CryIA(a), CryIA(b), and CryIA(c) toxins (32). Each of these toxins has a unique specificity profile for selected lepidoptera (Table 1); i.e., the CryIA(a) toxin is very active against *Bombyx mori* (silk moth), whereas the CryIA(c) toxin is particularly effective against *H. virescens* (45). Restriction fragments were exchanged between these two genes (between regions 2 and 3 in Fig. 1; 30). A region of the *cryIA(a)* gene encoding residues 332 to 450 conferred toxicity for *B. mori* on the CryIA(c) hybrid toxin. *H. virescens* specificity in a reciprocal hybrid appeared to involve residues 335 to 665, which is almost the entire variable region.

In an analogous study with the *cryIA(b)* and *cryIC* genes, a hybrid consisting of the first 258 residues of the CryIA(b) toxin and the remainder from the CryIC toxin had the specificity of the latter for *Spodoptera littoralis* larvae (64). The reciprocal hybrid behaved as expected, although it was not very stable. The deduced sequence of the CryIE toxin is similar to that of the first hybrid, and this toxin was as effective as the CryIC toxin for *Spodoptera exigua* larvae (65). Specificity domains certainly require residues from the carboxyl halves of toxins, but whether there are interactions with other parts of the toxin such as Arg-93 in the CryIA toxins (70) remains to be determined.

Recombination of Protoxin Genes in *E. coli*

As previously noted, there is extensive sequence homology among the *cryIA* genes and suggestive evidence for the generation of the *cryIA(b)* gene by recombination within the variable regions of the *cryIA (a)* and *cryIA(c)* genes (32). Truncated versions of these latter genes were cloned in tandem in an *E. coli* vector, and plasmids containing recombinant genes were selected (14). The sites of recombination were mapped, and the toxicities of several of the recombinant protoxins appeared to be novel.

A similar study was done with the *cryIIA* and *cryIIB* genes (68). The former is active against certain lepidoptera and diptera, whereas the latter is active only against lepidoptera. Recombinant protoxins were screened for the region of the CryIIA toxin conferring dipteran specificity, and the portion that differed most extensively between the two toxins (residues 300 to 400) appeared to be involved. It should be noted, however, that most of the recombinant toxins had lower activities than the parental toxins against the test larvae. This difference may be due to instability, a property of many toxins with point mutations and of those constructed by oligonucleotide exchange. Alternatively, the lower toxicity may reflect the participation of several regions of the toxin in specificity and/or toxicity.

REGULATION OF PROTOXIN SYNTHESIS

General Considerations

Regulation is complicated by the presence in most strains of multiple protoxin genes on one or more large plasmids. In the best-studied strain, *B. thuringiensis* subsp. *kurstaki* HD1, the CryIA(b) protoxin is more prevalent than the CryIA(a) and CryIA(c) protoxins (49). This difference is reflected in the relative transcription rates of these genes, which are correlated with gene copy number (3). The *cryIA(b)* gene is present on an unstable 44-MDa plasmid in this subspecies (see above), but the gene copy number is still greater than that of the other protoxin genes on a 120-MDa plasmid. As discussed above, this instability results in a dynamic population of at least two cell types, each with a unique specificity profile. The plasmid locale of protoxin genes and the compatibility of these plasmids are thus important factors in the relative amounts and kinds of protoxins produced by a particular subspecies.

Promoters

On the basis of S1 nuclease mapping, two overlapping promoters designated BtI and BtII were found upstream of the *cryIA(a)* gene from *B. thuringiensis* subsp. *kurstaki* HD1 (40). Only the BtI promoter was present in the CryIIA operon. Many of the *cryI*, *cryIII*, and *cryIV* genes have very similar sequences for at least 100 bases upstream of the coding region, so they are all likely to contain the same dual (or perhaps single) promoters.

The sigma subunits of the RNA polymerase required for transcription at the BtI and BtII promoters are 85 and 88% identical to the *B. subtilis* sporulation sigma factors σ^E and σ^K, respectively (1, 12). Clones of these genes can complement *B. subtilis* mutations in either the σ^E or the σ^K structural genes. In addition, RNA polymerases containing these sigma factors have the expected in vitro specificities for transcription at various *B. subtilis* spore gene promoters. In *B. subtilis*, σ^E is utilized from about stages II to IV of sporulation, whereas σ^K functions in the transcription of late sporulation genes (stages IV and V). This combined time interval overlaps that of protoxin synthesis, so the dual promoters could ensure synthesis throughout a prolonged period of sporulation. A single promoter for the CryIIA operon may reflect a more limited time of transcription of these genes.

The remarkable conservation of these sigma subunits between two "distantly related" *Bacillus* species (54) implies that the transcription of certain sporulation genes in *B. thuringiensis* may be competing with protoxin genes for these sigma factors. As previously discussed, the relative transcription of these plasmid-encoded protoxin genes is correlated with copy number, but there is a limitation in the total amount of such transcription in strains containing multiple genes (3). This limitation implies that there may be some mechanisms that ensure a balance between the synthesis of spore components and the synthesis of protoxin. Adequate protoxin synthesis is also aided by the relatively long half-life of the mRNA, i.e., 11 to 13 min versus less than 5 min for most spore mRNAs. The stability of protoxin mRNA is attributable in part to a large, stable stem-and-loop structure at the 3′ end (40), but other factors may be involved.

OTHER PROPERTIES OF *B. THURINGIENSIS* RELATED TO ITS EFFECTIVENESS AS A BIOLOGICAL CONTROL AGENT

Stability and Persistence

Extrachromosomal DNA in *B. thuringiensis* strains, either as a broad size range of plasmids or as a few large plasmids, represents a substantial amount of potential genetic information. Few of these plasmid genes other than those encoding protoxins and perhaps bacteriocins have been identified (26). About 45% of a 9-kb plasmid from *B. thuringiensis* subsp. *thuringiensis* is the transposon Tn*4450* (48), but the function(s) of these small plasmids (8 to 12 kb) is not known. As discussed above, some of the extrachromosomal DNA is bacteriophage (42, 66). Phagelike particles have been noted in thin sections of sporulating cells of a number of *B. thuringiensis* strains, and transducing phage have been isolated from at least three serotypes (4). Conditions for the induction of such phage are not understood, but the phage could influence the stability of these strains.

It has been known for some time that *B. thuringiensis* does not persist in the field, in contrast to *B. popilliae*, which is active against Japanese beetle grubs (13). The lack of persistence may be related to the extreme UV sensitivity of *B. thuringiensis* cells (9). Acrystalliferous, plasmid-cured derivatives are much more UV resistant, indicating that a plasmid or bacteriophage property is involved. In fact, transfer of plasmids (including phage TP21 as a 29-MDa covalently closed circular molecule) from *B. thuringiensis* subsp. *kurstaki* HD1 to *B. cereus* resulted in increased UV sensitivity of the latter. This sensitivity was not readily photoreactivatable, which is consistent with phage induction.

The presence of multiple plasmids has also made it difficult, probably because of incompatibility, to introduce novel protoxin-encoding plasmids either by cell mating or by electroporation of cloned genes (59). In one case, a *B. thuringiensis* recombinant displayed novel toxicity properties not found in either parent, presumably because of the interaction of two or more protoxins (21). This recombinant was unstable, however, and thus of no practical value. In another example, *B. sphaericus* toxin genes were introduced into *B. thuringiensis* subsp. *israelensis* by cell transformation in the hope of broadening the dipteran host specificity (11). Both toxins were produced, but there were no synergistic or additive effects.

Engineering Protoxin Genes into Other Systems

Gene fragments encoding the toxin have been introduced on various vectors into several plants as well as into a bacterium that is a plant epiphyte. Some protection of the plants was obtained, but the results were variable. In order to enhance expression and stability, a *cryIA(b)* gene with altered codon usage and other features designed to improve transcription and translation in plants was synthesized (53). There was substantial expression of this gene in several plants, with no effects detrimental to plant growth, and preliminary field trials have been promising.

Such results are encouraging and should be pursued in order to provide alternative pest management systems. While the practical applications of *B. thuringiensis* and its protoxin genes are very important, the bacterium itself, especially the interactions of plasmid genes, the regulation of inclusion formation, and the properties of these inclusions, warrants further study. Many bacteria have evolved elaborate systems for surviving nutrient limitation, and those employed by *B. thuringiensis* appear to be novel and certainly provide an exciting challenge.

Acknowledgments. J. Li and B. Visser provided prepublication information, and J. Li provided Fig. 2. Chris Baugher made her usual stellar contribution in the preparation of the manuscript.

REFERENCES

1. **Adams, L. F., K. L. Brown, and H. R. Whiteley.** 1991. Molecular cloning and characterization of two genes encoding sigma factors that direct transcription from a *Bacillus thuringiensis* crystal protein gene promoter. *J. Bacteriol.* **173:**3846–3854.
2. **Ahmad, W., and D. J. Ellar.** 1990. Directed mutagenesis of selected regions of a *Bacillus thuringiensis* entomocidal protein. *FEMS Microbiol. Lett.* **68:**97–104.
3. **Aronson, A. I.** Submitted for publication.
4. **Aronson, A. I., W. Beckman, and P. Dunn.** 1986. *Bacillus thuringiensis* and related insect pathogens. *Microbiol. Rev.* **50:**1–24.
5. **Aronson, A. I., and P. C. Fitz-James.** 1976. Structure and morphogenesis of the bacterial spore coat. *Bacteriol. Rev.* **40:**360–402.
6. **Aronson, A. I., E.-S. Han, W. McGaughey, and D. Johnson.** 1991. The solubility of inclusion proteins from *Bacillus thuringiensis* is dependent upon protoxin composition and is a factor in toxicity to insects. *Appl. Environ. Microbiol.* **57:**981–986.
7. **Arvidson, H., P. E. Dunn, S. Strnad, and A. I. Aronson.** 1989. Specificity of *Bacillus thuringiensis* for lepidopteran larvae: factors involved *in vivo* and in the structure of a purified protoxin. *Mol. Microbiol.* **3:**1533–1543.
8. **Baumann, P., M. A. Clark, L. Baumann, and A. H. Broadwell.** 1991. *Bacillus sphaericus* as a mosquito pathogen: properties of the organism and its toxins. *Microbiol. Rev.* **55:**425–436.
9. **Benoit, T. G., G. R. Wilson, D. L. Bull, and A. I. Aronson.** 1990. Plasmid-associated sensitivity of *Bacillus thuringiensis* to UV light. *Appl. Environ. Microbiol.* **56:**2282–2286.
10. **Bietlot, H. P. L., J. Vishmulhatla, P. R. Carey, M. Pozsgay, and H. Kaplan.** 1990. Characterization of the cysteine residues and disulphide linkages in the protein crystal of *Bacillus thuringiensis*. *Biochem. J.* **267:**309–315.
11. **Bourgouin, C., A. Delécluse, F. delaTorre, and J. Szulmajster.** 1990. Transfer of the toxin protein genes of *Bacillus sphaericus* into *Bacillus thuringiensis* subsp. *israelensis* and their expression. *Appl. Environ. Microbiol.* **56:**340–344.
12. **Brown, K. L., and H. R. Whiteley.** 1988. Isolation of a *Bacillus thuringiensis* RNA polymerase capable of transcribing crystal protein genes. *Proc. Natl. Acad. Sci. USA* **85:**4166–4170.
13. **Bulla, L. A., Jr., R. N. Costilow, and E. S. Sharpe.** 1978. Biology of *Bacillus popilliae*. *Adv. Appl. Microbiol.* **23:**1–18.
14. **Caramori, T., A. M. Albertini, and A. Galizzi.** 1991. *In vivo* generation of hybrids between two *Bacillus thuringiensis* insect-toxin-encoding genes. *Gene* **98:**37–44.
15. **Carlton, B. C., and J. M. Gonzalez, Jr.** 1985. The genetics and molecular biology of *Bacillus thuringiensis*, p. 211–249. *In* D. A. Dubnau (ed.), *The Molecular Biology of the Bacilli*, vol. II. Academic Press, Inc., New York.
16. **Carroll, J., J. Li, and D. J. Ellar.** 1989. Proteolytic processing of a coleopteran-specific δ-endotoxin produced by *Bacillus thuringiensis* var. *tenebrionis*. *Biochem. J.* **261:**99–105.
17. **Charles, J.-F., L. Nicolas, M. Sébald, and H. deBarjac.** 1990. *Clostridium bifermentans* serovar *malaysia*: sporulation, biogenesis of inclusion bodies and larvicidal effect on mosquitos. *Res. Microbiol.* **141:**721–733.
18. **Choma, C. T., W. K. Surewicz, P. R. Carey, M. Pozsgay, T. Raynor, and H. Kaplan.** 1990. Unusual proteolysis of the protoxin and toxin from *Bacillus thuringiensis*. *Eur. J. Biochem.* **189:**523–527.
19. **Convents, D., M. Cherlet, J. VanDamme, I. Lasters, and M. Lauwereys.** 1991. Two structural domains as a general fold of the toxic fragment of the *Bacillus thuringiensis* δ-endotoxins. *Eur. J. Biochem.* **195:**631–635.
20. **Convents, D., C. Houssier, I. Lasters, and M. Lauwereys.** 1990. The *Bacillus thuringiensis* δ-endotoxin. Evidence for a two domain structure of the minimal toxic fragment. *J. Biol. Chem.* **265:**1369–1375.

20a. **Crickmore, N., and D. J. Ellar.** 1992. Involvement of a possible chaperonin in the efficient expression of a cloned *CryIIA* δ-endotoxin gene in *Bacillus thuringiensis*. *Mol. Microbiol.* **6:**1533–1537.

21. **Crickmore, N., C. Nicholls, D. J. Eays, T. C. Hodgman, and D. J. Ellar.** 1990. The construction of *Bacillus thuringiensis* strains expressing novel entomocidal δ-endotoxin combinations. *Biochem. J.* **270:**133–136.
22. **deBarjac, H.** 1981. Identification of the H-serotypes of *Bacillus thuringiensis*, p. 35–43. *In* W. H. Burgess (ed.), *Microbial Control of Pest and Plant Diseases, 1970–1980*. Academic Press, Inc., New York.
23. **Delécluse, A., C. Bourgouin, G. Menou, D. Lereclus, A. Klier, and G. Rapaport.** 1990. IS240 associated with the cryIVA gene from *Bacillus thuringiensis israelensis* belongs to a family of gram(+) and gram(−) IS elements, p. 181–190. *In* M. M. Zukowski, A. T. Ganesan, and J. A. Hoch (ed.), *Genetics and Biotechnology of Bacilli*, vol. 3. Academic Press, Inc., San Diego, Calif.
24. **Delécluse, A., J.-F. Charles, A. Klier, and G. Rapaport.** 1991. Deletion by *in vivo* recombination shows that the 28-kilodalton cytolytic polypeptide from *Bacillus thuringiensis* subsp. *israelensis* is not essential for mosquitocidal activity. *J. Bacteriol.* **173:**3374–3381.
25. **Edwards, D. L., J. Payne, and G. G. Soares.** 1990. Novel isolates of *Bacillus thuringiensis* having activity against nematodes. U.S. patent 4,948,734.
26. **Fauret, M. E., and A. A. Yousten.** 1989. Thuricin: the bacteriocin produced by *Bacillus thuringiensis*. *J. Invertebr. Pathol.* **53:**206–216.
27. **Federici, B. A., P. Lüthy, and J. E. Ibarra.** 1990. Parasporal body of *Bacillus thuringiensis israelensis*. Structure, protein composition and toxicity, p. 16–44. *In* H. deBarjac and D. Sutherland (ed.), *Bacterial Control of Mosquitoes and Black Flies*. Rutgers University Press, New Brunswick, N.J.
28. **Ferré, J., M. D. Real, J. VanRie, S. Jansens, and M. Peferoen.** 1991. Resistance to the *Bacillus thuringiensis* bioinsecticide in a field population of *Plutella xylostella* is due to a change in a midgut membrane receptor. *Proc. Natl. Acad. Sci. USA* **88:**5119–5123.
29. **Garczynski, S. F., J. W. Crim, and M. J. Adang.** 1991. Identification of putative insect brush border membrane-binding molecules specific to Bacillus thuringiensis δ-endotoxin by protein blot analysis. *Appl. Environ. Microbiol.* **57:**2816–2820.
30. **Ge, A. Z., N. I. Shivarova, and D. H. Dean.** 1989. Location of the *Bombyx mori* specificity domain on a *Bacillus thuringiensis* δ-endotoxin protein. *Proc. Natl. Acad. Sci. USA* **86:**4037–4041.
31. **Geiser, M. (Ciba-Geigy).** 1990. Personal communication.
32. **Geiser, M., S. Schweitzer, and C. Grimm.** 1986. The hypervariable region in the genes coding for entomopathogenic crystal protein of *Bacillus thuringiensis*: nucleotide sequence of the kurdhl gene of subspecies *kurstaki* HD1. *Gene* **48:**109–118.
33. **Gonzalez, J. M., Jr., B. S. Brown, and B. C. Carlton.** 1982. Transfer of *Bacillus thuringiensis* plasmids coding for δ-endotoxin among strains of *B. thuringiensis* and *B. cereus*. *Proc. Natl. Acad. Sci. USA* **79:**6951–6955.
34. **Gonzalez, J. M., Jr., and B. C. Carlton.** 1984. A large transmissible plasmid is required for crystal toxin pro-

duction in *Bacillus thuringiensis* variety *israelensis*. *Plasmid* **11**:28–38.
35. **Haider, M. Z., and D. J. Ellar.** 1989. Functional mapping of an entomocidal delta-endotoxin: single amino acid changes produced by site-directed mutagenesis influence toxicity and specificity of the protein. *J. Mol. Biol.* **208:** 183–194.
36. **Haider, M. Z., E. S. Ward, and D. J. Ellar.** 1987. Cloning and heterologous expression of an insecticidal delta-endotoxin gene from *Bacillus thuringiensis* subsp. *aizawai* IC1 toxic to both Lepidoptera and Diptera. *Gene* **52:**285–290.
37. **Hodgman, T. C., and D. J. Ellar.** 1990. Models for the structure and function of the *Bacillus thuringiensis* δ-endotoxins determined by compilational analysis. *DNA Sequence* **1:**96–107.
38. **Hofmann, C., H. Vanderbruggen, H. Höfte, J. VanRie, S. Jansens, and H. vanMellaert.** 1988. Specificity of *Bacillus thuringiensis* δ-endotoxins is correlated with the presence of high affinity binding sites in the brush border membrane of target insect midguts. *Proc. Natl. Acad. Sci. USA* **85:**7844–7848.
39. **Höfte, H., J. VanRie, S. Jansens, A. V. Houtven, H. Vanderbruggen, and M. Vaeck.** 1988. Monoclonal antibody analysis and insecticidal spectrum of three types of lepidopteran-specific insecticidal crystal proteins of *Bacillus thuringiensis*. *Appl. Environ. Microbiol.* **54:**2011–2017.
40. **Höfte, H., and H. R. Whiteley.** 1989. Insecticidal crystal proteins of *Bacillus thuringiensis*. *Microbiol. Rev.* **53:**242–255.
41. **Jaquet, F., R. Hütter, and P. Lüthy.** 1987. Specificity of *Bacillus thuringiensis* delta-endotoxin. *Appl. Environ. Microbiol.* **53:**500–504.
42. **Kanda, K., Y. Tan, and K. Aizawa.** 1989. A novel phage genome integrated into a plasmid in *Bacillus thuringiensis* strain AF101. *J. Gen. Microbiol.* **135:**3035–3041.
43. **Knowles, B. H., M. R. Blatt, M. Tester, J. M. Horswell, J. Carroll, G. Menestrina, and D. J. Ellar.** 1989. A cytolytic δ-endotoxin from *Bacillus thuringiensis* var. *israelensis* forms cation-selective channels in plasmid lipid bilayers. *FEBS Lett.* **244:**259–262.
44. **Knowles, B. H., and D. J. Ellar.** 1987. Colloid-osmotic lysis is a general feature of the mechanisms of action of *Bacillus thuringiensis* δ-endotoxins with different insect specificities. *Biochim. Biophys. Acta* **924:**509–518.
45. **Krywienczyk, H., H. T. Dulmage, and P. G. Fast.** 1978. Occurrence of two serologically distinct groups within *Bacillus thuringiensis* serotype 3ab var. *kurstaki*. *J. Invertebr. Pathol.* **31:**372–375.
46. **Lee, C.-S., and A. I. Aronson.** 1991. Cloning and analysis of δ-endotoxin genes from *Bacillus thuringiensis* subsp. *alesti*. *J. Bacteriol.* **173:**6635–6638.
47. **Li, J., J. Carroll, and D. J. Ellar.** 1991. Crystal structure of an insecticidal protein. The δ-endotoxin from *Bacillus thuringiensis* subsp. *tenebrionis* at 2.5 Å resolution. *Nature* (London) **353:**815–821.
48. **Mahillon, J., F. Hespel, A.-M. Pierssens, and J. Delcour.** 1988. Cloning and partial characterization of three small cryptic plasmids from *Bacillus thuringiensis*. *Plasmid* **19:**169–173.
49. **Masson, L., G. Préfontaine, L. Péloquin, P. C. K. Lau, and R. Brousseau.** 1989. Comparative analysis of the individual protoxin components in P1 crystals of *Bacillus thuringiensis* subsp. *kurstaki* NRD-12 and HD-1. *Biochem. J.* **269:**507–512.
50. **McGaughey, W. H.** 1985. Insect resistance to the biological insecticide *Bacillus thuringiensis*. *Science* **229:**193–195.
51. **Minnich, S. A., and A. I. Aronson.** 1984. Regulation of protoxin synthesis in *Bacillus thuringiensis*. *J. Bacteriol.* **158:**447–454.
51a. **Nakamura, L. K., and H. T. Dulmage.** 1988. *Bacillus thuringiensis* cultures available from the U.S. Department of Agriculture. Technical Bulletin no. 1738. U.S. Department of Agriculture, Beltsville, Md.
52. **Oddou, P., H. Hartmann, and M. Geiser.** 1991. Identification and characterization of *Heliothis virescens* midgut membrane proteins binding *Bacillus thuringiensis* δ-endotoxins. *Eur. J. Biochem.* **202:**673–680.
52a. **Osterman, A., et al.** Personal communication.
53. **Perlak, F. J., R. L. Fuchs, D. A. Dean, S. L. McPherson, and D. A. Fischhoff.** 1991. Modification of the coding sequence enhances plant expression of insect control protein genes. *Proc. Natl. Acad. Sci. USA* **88:**3324–3328.
54. **Priest, F. G., M. Goodfellow, and C. Todd.** 1988. A numerical classification of the genus Bacillus. *J. Gen. Microbiol.* **134:**1847–1882.
55. **Raymond, K. C., T. R. John, and L. A. Bulla, Jr.** 1990. Larvicidal activity of chimeric *Bacillus thuringiensis* protoxins. *Mol. Microbiol.* **4:**1967–1973.
56. **Reddy, A., L. Battisti, and C. B. Thorne.** 1987. Identification of self-transmissable plasmids in four *Bacillus thuringiensis* subspecies. *J. Bacteriol.* **169:**5263–5270.
57. **Sacchi, V. G., P. Parent, G. M. Hanozet, B. Giordana, P. Lüthy, and M. G. Wolfersberger.** 1986. *Bacillus thuringiensis* toxin inhibits K^+-gradient-dependent amino acid transport across the brush border membrane of *Pieris brassicae* midgut cells. *FEBS Lett.* **204:**213–218.
58. **Sanchis, V., D. Lereclus, G. Menou, J. Chaufaux, S. Guo, and M.-M. Lecadet.** 1989. Nucleotide sequence and analysis of the N-terminal coding region of the Spodoptera-active δ-endotoxin gene of *Bacillus thuringiensis aizawai* 7.29. *Mol. Microbiol.* **3:**229–238.
59. **Schurter, W., M. Geiser, and D. Mathé.** 1989. Efficient transformation of *Bacillus thuringiensis* and *B. cereus* via electroporation: transformation of acrystalliferous strains with a cloned delta-endotoxin gene. *Mol. Gen. Genet.* **218:**177–181.
60. **Sêbesta, K., J. Farkås, and K. Horská.** 1981. Thuringiensin, the β exotoxin of *Bacillus thuringiensis*, p. 249–281. *In* W. H. Burgess (ed.), *Microbial Control of Pests and Plant Diseases, 1970–1980*. Academic Press, Inc., New York.
61. **VanRie, J., S. Jansens, H. Höfte, D. Degheele, and H. VanMellaert.** 1989. Specificity of *Bacillus thuringiensis* δ-endotoxins. Importance of specific receptors on the brush border membrane of the midgut of target insects. *Eur. J. Biochem.* **186:**239–247.
62. **VanRie, J., S. Jansens, H. Höfte, D. Degheele, and H. VanMellaert.** 1990. Receptors on the brush border membrane of the insect midgut as determinants of the specificity of *Bacillus thuringiensis* delta-endotoxins. *Appl. Environ. Microbiol.* **56:**1378–1385.
63. **VanRie, J., W. H. McGaughey, D. E. Johnson, B. D. Barnett, and H. VanMellaert.** 1990. Mechanism of insect resistance to the microbial insecticide *Bacillus thuringiensis*. *Science* **247:**72–74.
64. **Visser, B., D. Bosch, and G. Honée.** 1991. Domain-function studies of *Bacillus thuringiensis* crystal proteins: a genetic approach. *In* P. Entwistle, H. Cory, D. Bailey, and A. Higgs (ed.), *Bacillus thuringiensis: an Environmental Pesticide*. John Wiley & Sons, Inc., New York, in press.
65. **Visser, B., E. Munsterman, A. Stoker, and W. G. Dirkse.** 1990. A novel *Bacillus thuringiensis* gene encoding a *Spodoptera exigua*-specific crystal protein. *J. Bacteriol.* **172:**6783–6788.
66. **Walter, T. M., and A. I. Aronson.** 1991. Transduction of certain genes by an autonomously replicating *Bacillus thuringiensis* phage. *Appl. Environ. Microbiol.* **57:**1000–1005.
67. **Widner, W. R., and H. R. Whiteley.** 1989. Two highly related insecticidal crystal proteins of *Bacillus thuringiensis* subsp. *kurstaki* possess different host range specificities. *J. Bacteriol.* **171:**965–974.

68. **Widner, W. R., and H. R. Whiteley.** 1990. Location of the dipteran specificity region in a lepidopteran-dipteran crystal protein from *Bacillus thuringiensis*. *J. Bacteriol.* **172:**2826–2832.
69. **Wolfsberger, M.** 1990. The toxicity of two *Bacillus thuringiensis* δ-endotoxins to gypsy moth larvae is inversely related to the affinity of binding sites on midgut brush border membranes for the toxins. *Experientia* **46:**475–477.
70. **Wu, D., and A. I. Aronson.** 1992. Localized mutagenesis defines regions of the *Bacillus thuringiensis* δ-endotoxin involved in toxicity and specificity. *J. Biol. Chem.* **267:**2311–2317.
71. **Wu, D., X. L. Cao, Y. Y. Bai, and A. I. Aronson.** 1991. Sequence of an operon containing a novel δ-endotoxin gene from *Bacillus thuringiensis*. *FEMS Microbiol. Lett.* **81:**31–36.
72. **Yamamoto, T., and R. E. McLaughlin.** 1981. Isolation of a protein from the parasporal crystal of *Bacillus thuringiensis* var. *kurstaki* toxic to the mosquito larva, *Aedes aegyptii*. *Biochem. Biophys. Res. Commun.* **103:**414–419.
73. **Yamamoto, T., and I. Toshihiko.** 1983. Two types of entomocidal toxins in the parasporal crystals of *Bacillus thuringiensis kurstaki*. *Biochim. Biophys. Acta* **227:**233–241.

INDEX